Acknowledgements

Crown Copyright material is reproduced with the permission of the Controller of Her Majesty's Stationery Office.

Extracts from British Standards are reproduced with the kind permission of the British Standards Institution. Complete copies of the documents can be obtained from the British Standards Institution (BSI), 389 Chiswick High Road, London W4 4AL. Telephone: (020) 8996 9000.

Mark Tyler,

MA, LLM, MIOSH

Partner, Shook, Hardy & Bacon International LLP

George Ventris,

formerly secretary of the HSC's Construction Industry Advisory Committee (CONIAC) and a leader of HSE's Construction National Interest Group

Jeffrey Wale,

LLB

Partner, Berrymans Lace Mawer

Ian Wallace,

BSc (HONS), Dip SM, CFIOSH, FLLA

HSE Consultant

Lawrence Waterman,

CFIOSH, MBIOH, ROH

Managing Director, Chairman, Sypol Limited

Angus Withington,

BA Hons

Barrister, Henderson Chambers

John Wintle,

MA (Oxon), MSc, CEng, FIMechE

Fellow of the Institution of Mechanical Engineers, Consultant Engineer

Health and Safety at Work Editorial Board

The following are members of the appointed Health and Safety Editorial Board. Each member brings with them a wealth of experience and knowledge of health and safety matters. The benefits to be gained from introducing this board are to ensure that this publication continues to provide practical, authoritative coverage of health and safety law in all aspects of industrial and office workplaces.

Lawrence Bamber

Lawrence Bamber is currently an Associate Consultant with Norwich Union Risk Services and is also a freelance OSH/ risk management consultant. He has worked in the OSH/ risk management field for over 30 years with General Accident, Stenhouse (Aon), as well as Norwich Union, and is well known for his training, consultancy and presentation skills. Lawrence is a past president of the Institution of Occupational Safety and Health (IOSH) and is the author of many papers and publications on OSH/ risk management.

Mark Tyler

Mark Tyler is a partner at Shook, Hardy & Bacon International LLP. He is a solicitor and a Chartered Safety and Health Practitioner, and is a member of the CBI Health and Safety Panel. Mark is also a co-author of the books *Product Safety*, *Safer by Design* and *Tolley's Workplace Accident Handbook*. He has acted in a number of major cases which include the Southall and Ladbroke Grove Rail Inquiries, the Organophosphate group litigation, and the South Kensington Legionnaires' Disease outbreak.

Contents

List of Illustrations   xxxix

List of Abbreviations   xliii

Introduction   INT-1

Lawrence Waterman, Lawrence Bamber and Mike Bateman

The looseleaf   INT-1

Drivers for health and safety performance   INT-2

The changing world of work   INT-4

Legal framework   INT-5

The structure of UK health and safety law   INT-10

Management of health and safety at work   INT-20

Access, Traffic Routes and Vehicles   A10

Nicola Coote

Introduction   A1001

Statutory duties concerning workplace access and egress – Workplace (Health, Safety and Welfare) Regulations 1992 (SI 1992 No 3004)   A1002

Vehicles   A1008

Common law duty of care   A1021

Confined spaces   A1022

Accident Reporting and Investigation   A30

Mark Tyler

Introduction   A3001

Reporting of Injuries, Diseases and Dangerous Occurrences Regulations 1995 (RIDDOR) – (SI 1995 No 3163)   A3002

Relevant enforcing authority   A3003

Incident Reporting Centre   A3004

Persons responsible for notification and reporting   A3005

What is covered?   A3006

Major injuries or conditions   A3007

Injuries incapacitating for more than three consecutive days   A3008

Specified dangerous occurrences   A3009

Specified diseases   A3010

Duty to notify/report   A3011

Contents

Duty to keep records   **A3014**

Reporting the death of an employee   **A3018**

Road accidents   **A3019**

Duty to report gas incidents   **A3020**

Penalties   **A3022**

Obligations on employees   **A3024**

Objectives of accident reporting   **A3026**

Duty of disclosure of accident data   **A3027**

Accident Investigation   **A3030**

Appendix A – List of Reportable Dangerous Occurrences   **A3035**

Appendix B – List of Reportable Diseases   **A3036**

Appendix C   **A3037**

Appendix D   **A3038**

Appendix E – Health and Safety Executive Regional Offices   **A3039**

Asbestos   **A50**

Lawrence Waterman and Andrea Oates

Introduction   **A5001**

Historical exposures   **A5003**

Asbestos prohibitions   **A5004**

Asbestos removal   **A5005**

Exposure of building maintenance workers   **A5006**

Duties under the Control of Asbestos at Work Regulations 2006   **A5007**

Asbestos building surveys   **A5023**

Asbestos waste   **A5027**

Work With Asbestos   **A5028**

Legislation, Approved Codes of Practice, and Guidance Notes   **A5030**

Business Continuity   **B80**

Lawrence Bamber

What is business continuity planning?   **B8001**

Business continuity planning in context   **B8002**

Why is business continuity planning of importance?   **B8003**

Business continuity drivers   **B8004**

Convincing the board – the business case   **B8010**

The business continuity model   **B8011**

Crisis communication and public relations   **B8038**

Civil Contingencies Act, 2004 **B8043**

Compensation for Work Injuries/Diseases **C60**

David Leckie

Introduction **C6001**

Social security benefits **C6002**

Industrial injuries benefit **C6003**

Accident and personal injury provisions **C6004**

Industrial injuries disablement benefit **C6011**

Constant attendance allowance and exceptionally severe disablement
allowance **C6014**

Other sickness/disability benefits **C6015**

Disability Tax Credit under Working Tax Credit **C6016**

Payment **C6017**

Damages for occupational injuries and diseases **C6021**

Awards of damages and recovery of state benefits **C6044**

Duties of the compensator **C6045**

Complications **C6046**

Appeals against certificates of recoverable benefits **C6050**

Pneumoconiosis **C6059**

Community Health and Safety **C65**

Mike Bateman

Introduction **C6501**

Relevant Legal Requirements **C6502**

Neighbours and Passers-by **C6509**

Trespassers on Work Premises **C6512**

Customers and Users of Facilities and Services **C6515**

Visiting Individuals and Groups **C6518**

Volunteers **C6521**

Major Public Events **C6524**

Use of Contractors **C6527**

Further Information **C6528**

Confined Spaces **C70**

Angus Withington

Introduction **C7001**

Relevant Legal Requirements **C7002**

Contents

Confined Spaces Regulations 1997   **C7003**

Other risks in confined spaces   **C7008**

Safe systems of work   **C7009**

Precautions for work in confined spaces   **C7014**

Emergency arrangements   **C7029**

Construction and Building Operations   **C80**

Gordon Prosser

Introduction   **C8001**

Summary of changes   **C8002**

Typical issues occurring in building and construction projects   **C8018**

Construction, Design and Management   **C85**

Gordon Prosser

Introduction   **C8501**

Information in the Schedules   **C8518**

Information in the Appendices to the Approved Code of Practice L144   **C8519**

Conclusion, the future?   **C8520**

Corporate Manslaughter   **C90**

Andrea Oates

Introduction   **C9001**

The offence   **C9002**

The meaning of "relevant duty of care"   **C9003**

Public policy decisions, exclusively public functions and statutory inspections   **C9004**

Military activities   **C9005**

Policing and law enforcement   **C9006**

Emergencies and the emergency services   **C9007**

Child protection and probation functions   **C9008**

Factors for the jury to consider when deciding if there has been a "gross" breach   **C9009**

Remedial Orders   **C9010**

Publicity Orders   **C9011**

Application to Crown bodies   **C9012**

Application to the armed forces   **C9013**

Application to police forces   **C9014**

Application to partnerships   **C9015**

Procedure, evidence and sentencing   **C9016**

Transfer of functions **C9017**

Director of Public Prosecutions (DPP) consent for proceedings **C9018**

No individual liability for corporate manslaughter or corporate homicide **C9019**

Convictions under the Act and under health and safety legislation **C9020**

Abolition of liability of corporations for manslaughter at common law **C9021**

Extent and territorial application **C9022**

Schedule 1: List of government departments etc **C9023**

Dangerous Goods – Carriage **D04**

John Wintle

Introduction **D0401**

General requirements **D0405**

Exemptions, revocations and radioactive materials **D0439**

Acknowledgement **D0448**

Dangerous Substances and Explosive Atmospheres **D09**

John Wintle

Introduction **D0901**

The Regulations **D0904**

Risk assessment **D0912**

Management of risks **D0915**

Classification of places with explosive atmospheres **D0920**

Equipment and protection systems **D0924**

Specific products and areas **D0930**

Arrangements to deal with incidents and emergencies **D0934**

Provision of information, instruction and training **D0935**

Identification of hazardous contents of pipes and containers **D0936**

Exemption **D0937**

Amendments to modernise petroleum and other legislation **D0938**

Repeals and revocations of existing legislation **D0939**

Enforcement **D0940**

Regulatory impact assessment **D0941**

Approved codes of practice and guidance etc. **D0942**

Other information **D0949**

Case histories **D0951**

Sources and acknowledgement **D0952**

Disaster and Emergency Management Systems (DEMS) **D60**

Contents

Raj Lakha, Christopher Eskell, Mary Spear and Brian Heath

Historical development of disaster and emergency management · · · · · **D6001**

Origin of disaster and emergency management as a modern discipline · · · **D6002**

Definitions · · · · · **D6003**

The need for effective disaster and emergency management · · · · · **D6004**

Disaster and emergency management systems (DEMS) · · · · · **D6034**

External and internal factors · · · · · **D6035**

Establish a disaster and emergency policy · · · · · **D6038**

Organise for disasters and emergencies · · · · · **D6043**

Disaster and emergency planning · · · · · **D6048**

Monitor the disaster and emergency plan(s) · · · · · **D6066**

Audit and review · · · · · **D6067**

Conclusion · · · · · **D6070**

Sources of information · · · · · **D6071**

Display Screen Equipment · · · · · **D82**

Mike Bateman

Introduction · · · · · **D8201**

The Regulations summarised · · · · · **D8202**

DSE workstation assessments · · · · · **D8210**

Special situations · · · · · **D8217**

After the assessment · · · · · **D8221**

Electricity · · · · · **E30**

Chris Buck and Andrea Oates

Introduction · · · · · **E3001**

Understanding electricity · · · · · **E3002**

Legal background and standards · · · · · **E3003**

Dangers of electricity · · · · · **E3004**

Fundamentals of controlling electrical risks · · · · · **E3010**

Competence · · · · · **E3011**

Design and construction of electrical equipment and systems · · · · · **E3012**

Maintenance · · · · · **E3021**

Safe systems of work · · · · · **E3022**

Special situations · · · · · **E3025**

Other relevant legislation · · · · · **E3029**

Emissions into the Atmosphere · · · · · **E50**

Roger Barrowcliffe and Paul Reeve

Introduction   **E5001**

Regulatory position   **E5002**

Regulatory drivers   **E5003**

Key pollutants and issues   **E5017**

The European context   **E5026**

UK Regulations   **E5040**

Employers' Liability Insurance   **E130**

Phil Grace

Introduction   **E13001**

Purpose of compulsory employers' liability insurance   **E13002**

General law relating to insurance contracts   **E13003**

Subrogation   **E13007**

Duty of employer to take out and maintain insurance   **E13008**

Issue, display and retention of certificates of insurance   **E13013**

Penalties   **E13014**

Cover provided by a typical policy   **E13016**

'Prohibition' of certain conditions   **E13020**

Trade endorsements for certain types of work   **E13021**

Insolvent employers' liability insurers   **E13024**

Limitation Act   **E13025**

Long Tail Disease   **E13026**

Vicarious Liability   **E13027**

Making the Market Work   **E13028**

Woolf Reforms   **E13029**

Small Business Performance Indicator   **E13030**

NHS Recoveries   **E13031**

Corporate Manslaughter   **E13032**

Compensation Act   **E13033**

Employment Protection   **E140**

Janet Gaymer

Introduction   **E14001**

Sources of contractual terms   **E14002**

Employee employment protection rights   **E14009**

Enforcement of safety rules by the employer   **E14012**

Contents

Dismissal   E14017

Health and safety duties in relation to women at work   E14023

Enforcement   **E150**

Hannah Wilson

Introduction   E15001

The role of the Health and Safety Commission   E15002

Functions of the new Health and Safety Executive (HSE)   E15005

Enforcing authorities   E15007

'Relevant statutory provisions' covered by the Health and Safety at
Work etc. Act 1974   E15012

Part A: Enforcement Powers of Inspectors   E15013

Part B: Offences and Penalties   E15030

Environmental Management   **E160**

Mark Rutter

Introduction   E16001

Internal drivers   E16002

External drivers   E16005

Environmental management guidelines   E16013

Implementing environmental management   E16017

Benefits of environmental management   E16026

The future   E16027

Equality Act 2010   **E165**

Alexander M S Green

Introduction   E16501

Prohibited conduct   E16502

Health and Safety for Disabled People   E16509

Health and Safety in Relation to Age   E16513

Health and Safety in Relation to Race   E16517

Health and Safety in Relation to Gender   E16520

Remedies and Enforcement in Relation to Unlawful Discrimination   E16522

Ergonomics   **E170**

Margaret Hanson and Jill Cleaver

Introduction   E17001

The ergonomic approach   E17004

Designing for people   E17005

Designing the task   **E17010**

Workstation design   **E17018**

Physical work environment   **E17030**

Job design and work organisation   **E17036**

Musculoskeletal disorders   **E17044**

Ergonomics tools   **E17051**

Relevant legislation and guidance   **E17060**

Facilities Management – An Overview   **F30**

Robert Greenfield

Introduction   **F3001**

The definition of facilities management   **F3002**

Facilities management   **F3003**

Hard and soft facilities management   **F3004**

Managing   **F3005**

Company approach to facilities management   **F3006**

In-house or outsourced facilities management?   **F3008**

Advantages to the company   **F3009**

Responsibilities of facilities management   **F3011**

Four most common health and safety issues for facilities management   **F3024**

How facilities management and building design can help a company   **F3028**

Recognition of facilities management   **F3029**

Innovations within the facilities management industry   **F3030**

The effects of safety, health and environmental legislation on facilities management   **F3031**

Summary of facilities management   **F3035**

Fire Prevention and Control   **F50**

Adair Lewis and Andrea Oates

Elements of fire   **F5001**

Common causes of fires   **F5002**

Fire classification   **F5003**

Electrical fires   **F5004**

Fire extinction   **F5005**

Property risk   **F5006**

Passive and active fire protection   **F5007**

Fire procedures and portable equipment   **F5008**

Contents

Good 'housekeeping'   **F5018**

Pre-planning of fire prevention   **F5019**

Fires on construction sites   **F5024**

Fire and fire precautions   **F5030**

The current position   **F5031**

Fire certification   **F5032**

The Regulatory Reform (Fire Safety) Order 2005   **F5033**

Building Regulations   **F5035**

Appeals against enforcement notices   **F5036**

Fire risk assessment   **F5037**

The Dangerous Substances and Explosive Atmospheres Regulations
(DSEAR) 2002   **F5045**

Fire fighters' switches for luminous signs   **F5046**

Maintenance   **F5047**

Safety assistance   **F5048**

Competent person   **F5049**

Enforcement   **F5056**

Offences and appeals   **F5060**

Application to the Crown   **F5061**

Other dangerous processes   **F5062**

Liability of occupier   **F5073**

Fire insurance   **F5074**

First-Aid   **F70**

Subash Ludhra

Introduction   **F7001**

The employer's duty to make provision for first-aid   **F7002**

The assessment of first-aid needs   **F7003**

Duty of the employer to inform employees of first-aid arrangements   **F7014**

First-aid and the self-employed   **F7015**

Number of first-aiders   **F7016**

Appointed persons   **F7019**

Records and record keeping   **F7020**

First-aid resources   **F7021**

Rooms designated as first-aid areas   **F7025**

Food Safety and Standards   **F90**

Neville Craddock

Introduction   **F9001**

Food Safety Act 1990   **F9006**

Food premises regulations   **F9026**

Food hygiene regulations   **F9035**

Food Safety Manual   **F9070**

Gas Safety   **G10**

Dr R C Slade

Introduction   **G1001**

Gas supply – the Gas Act 1986, as amended   **G1002**

Gas supply management – the Gas Safety (Management) Regulations 1996 (SI 1996 No 551)   **G1003**

Rights of entry – the Gas Safety (Rights of Entry) Regulations 1996 (SI 1996 No 2535)   **G1006**

Pipelines – the Pipelines Safety Regulations 1996 (SI 1996 No 825)   **G1009**

Gas systems and appliances – the Gas Safety (Installation and Use) Regulations 1998 (SI 1998 No 2451)   **G1010**

Interface with other legislation   **G1026**

Summaries of relevant court cases   **G1027**

Gas safety requirements in factories   **G1034**

Pressure fluctuations   **G1035**

Harassment in the Workplace   **H17**

Nick Humphreys

Introduction   **H1701**

The new law of harassment under the Equal Treatment Directive (2002/73/EC)   **H1703**

Definition of harassment   **H1704**

Employers' liability   **H1711**

Legal action and remedies   **H1717**

Preventing and handling claims   **H1735**

Hazardous Substances in the Workplace   **H21**

Mike Bateman and Andrea Oates

Introduction   **H2101**

How hazardous substances harm the body   **H2102**

The COSHH Regulations summarised   **H2104**

What the Regulations require on assessments   **H2106**

Contents

Practical aspects of COSHH assessments   **H2114**

COSHH assessment records   **H2122**

Further requirements of the COSHH Regulations   **H2126**

Other important regulations   **H2136**

Joint Consultation in Safety – Safety Representatives, Safety Committees, Collective Agreements and Works Councils   **J30**
Nick Humphreys

Introduction   **J3001**

Regulatory framework   **J3002**

Consultation obligations for unionised employers   **J3003**

Safety committees   **J3017**

Non-unionised workforce – consultation obligations   **J3020**

Recourse for safety representatives   **J3030**

European developments in health and safety   **J3031**

Lifting Operations   **L30**
Kevin Chicken and Nicola Coote

Introduction   **L3001**

Statutory requirements relating to the manufacture of lifting equipment   **L3002**

Lifts Regulations 1997 (SI 1997 No 831)   **L3003**

Lifting operations and equipment failure   **L3028**

Hoists and lifts   **L3032**

Fork lift trucks   **L3034**

Patient/bath hoist   **L3036**

Vehicle lifting table   **L3037**

Lifting accessories   **L3038**

Once only use accessories   **L3050**

Mobile lifting equipment   **L3051**

Lighting   **L50**
Nicola Coote

Introduction   **L5001**

Statutory lighting requirements   **L5002**

Sources of light   **L5007**

Standards of lighting or illuminance   **L5010**

Qualitative aspects of lighting and lighting design   **L5015**

Machinery Safety   **M10**
Andrea Oates

Contents

Introduction   **M1001**

Legal requirements relating to machinery   **M1002**

Machinery safety   **M1003**

The Provision and Use of Work Equipment Regulations 1998 (PUWER98)   **M1004**

The Supply of Machinery (Safety) Regulations 1992 (as amended)   **M1009**

The Transposed Harmonised European Machinery Safety Standards   **M1019**

Machinery risk assessment   **M1022**

Machinery hazards identification   **M1025**

Options for machinery risk reduction   **M1026**

Types of safeguards and safety devices   **M1027**

Fixed guards   **M1029**

Safety devices   **M1036**

Major accident hazards   **M11**

Roger Bentley

Introduction   **M1101**

Application   **M1104**

Flammable substances and preparations   **M1105**

Explosives   **M1106**

Aggregation   **M1107**

Exclusiions   **M1109**

Notifications   **M1110**

Public register   **M1111**

Duty to report a major accident   **M1112**

Powers to prohibit use   **M1113**

Enforcement   **M1114**

Lower tier duties   **M1116**

Major accident prevention policy (MAPP)   **M1117**

Top tier duties   **M1118**

Purpose of safety reports   **M1119**

Minimum information to be included in safety report   **M1120**

Review and revision of safety reports   **M1121**

On-site emergency plan   **M1122**

Off-site emergency plan   **M1124**

Review, testing and implementing emergency plans   **M1126**

Contents

Local authority charges   **M1127**

Provision of information to the public   **M1128**

Information to be supplied to the public   **M1129**

Statement of policy for managing major incidents   **M1130**

Arrangement for major incident management   **M1131**

Command and control chart of arrangements   **M1132**

Testing and validation   **M1133**

Land-use planning   **M1135**

Legislation   **M1136**

General   **M1137**

Hazardous installations   **M1138**

Mechanisms   **M1139**

Consultation distances and risk contours   **M1140**

Sources of information   **M1141**

Managing Absence   **M15**

Lynda Macdonald

Introduction   **M1501**

Managing long-term sickness absence   **M1502**

Managing short-term absences   **M1518**

Managing Health and Safety   **M20**

Lawrence Bamber and Mike Bateman

Introduction   **M2001**

Legal requirements   **M2002**

The Management Regulations   **M2005**

Construction (Design and Management) Regulations 2007   **M2022**

Health and safety management systems   **M2029**

Policy   **M2031**

Organisation   **M2033**

Planning   **M2038**

Implementation   **M2040**

Monitoring   **M2041**

Audit   **M2042**

Review   **M2043**

Continual improvement   **M2044**

POPIMAR in action   **M2045**

References   **M2046**

Managing Work-related Road Safety   **M21**

Roger Bibbings and Andrea Oates

Overview   **M2101**

The problem   **M2102**

Legal responsibilities   **M2103**

The business case for action   **M2104**

Extending health and safety management systems to cover work-related road safety (WRRS) **M2105**

Assessing risks on the road   **M2111**

Road risk control measures   **M2112**

Monitoring and evaluation   **M2117**

Where is the organisation now?   **M2123**

Useful publications   **M2125**

Useful websites   **M2126**

Manual Handling   **M30**

Mike Bateman and Margaret Hanson

Introduction   **M3001**

What the Regulations require   **M3002**

Risk of injury from manual handling   **M3003**

Avoiding or reducing risks   **M3008**

Planning and preparation   **M3016**

Making the assessment   **M3019**

After the assessment   **M3022**

Noise at Work   **N30**

Roger Tompsett

Introduction   **N3001**

Scale of the problem and latest developments   **N3002**

Overview of the Control of Noise at Work Regulations 2005   **N3002.1**

Hearing damage   **N3003**

The perception of sound   **N3004**

Noise indices   **N3005**

General legal requirements   **N3012**

Specific legal requirements   **N3015**

Compensation for occupational deafness   **N3018**

Contents

Action against employer at common law   **N3022**

General guidance on noise at work   **N3023**

Reducing noise in specific working environments   **N3026**

Workplace noise assessments   **N3029**

Noise reduction   **N3035**

Ear protection   **N3036**

Occupational Health and Diseases – An Overview   **O10**

Leslie Hawkins

What is occupational health?   **O1001**

Why is it important?   **O1002**

Relationships between occupational health and safety   **O1006**

Prevention is better than cure   **O1007**

The occupational health team   **O1010**

Health records   **O1016**

Buying services and priorities   **O1019**

Initiatives and targets   **O1020**

Occupational diseases and disorders   **O1021**

Hand arm vibration syndrome (havs) and whole body vibration   **O1027**

Diseases and disorders of the eye   **O1032**

Diseases of the skin   **O1037**

Occupational respiratory diseases   **O1042**

Musculoskeletal disorders   **O1047**

Psychological disorders (stress, anxiety and depression)   **O1052**

Other occupational health concerns   **O1056**

Occupiers' Liability   **O30**

Alison Newstead

Introduction   **O3001**

Duties owed under the Occupiers' Liability Act 1957   **O3002**

Duty owed to trespassers, at common law and under the Occupiers' Liability Act 1984   **O3008**

Dangers to guard against   **O3012**

Waiver of duty and the Unfair Contract Terms Act 1977 (UCTA)   **O3016**

Risks willingly accepted – 'volenti non fit injuria'   **O3017**

Actions against factory occupiers   **O3018**

Occupier's duties under HSWA 1974   **O3019**

Offshore Operations 070

Ian Wallace

Introduction 07001

Offshore Installations (Safety Case) Regulations 2005 (SI 2005 No 3117) 07010

Offshore Installations and Wells (Design and Construction, etc.) Regulations 1996 (SI 1996 No 913) 07024

Offshore Installations (Prevention of Fire and Explosion, and Emergency Response) Regulations 1995 (SI 1995 No 743) 07028

Offshore Installations and Pipeline Works (Management and Administration) Regulations 1995 (SI 1995 No 738) 07029

Hazardous operations with the potential to cause a major accident 07033

Construction and use of submarine pipelines 07034

Pipelines Safety Regulations 1996 (SI 1996 No 825) 07035

The Diving at Work Regulations 1997 07036

Other offshore specific legislation 07039

Offshore Installations (Logbooks and Registration of Death) Regulations 1972 07040

Offshore Installations (Inspectors and Casualties) Regulations 1973 07041

Offshore Installations (Safety Representatives and Safety Committees) Regulations 1989 07046

First-aid – the Offshore Installations and Pipeline Works (First-Aid) Regulations 1989 07057

The Offshore Electricity and Noise Regulations 1997 07063

Legislation for other hazardous offshore activities 07064

Written procedures and training programmes for dealing with hazards 07069

Hazards in offshore activities 07070

Personal Protective Equipment P30

Nicola Coote

Introduction P3001

The Personal Protective Equipment at Work Regulations 1992 P3002

Work activities/processes requiring personal protective equipment P3003

Statutory requirements in connection with personal protective equipment P3004

Increased importance of uniform European standards P3009

Main types of personal protection P3019

Common law requirements P3026

Pressure Systems P70

Contents

Kevin Chicken

Introduction   **P7001**

Simple pressure vessels   **P7002**

Simple Pressure Vessels (Safety) Regulations (SPVSR) 1991   **P7003**

EC Certificate of Adequacy   **P7004**

EC Type-Examination Certificate   **P7005**

Pressure Equipment Regulations 1999   **P7006**

Essential safety requirements   **P7007**

Conformity assessment procedure   **P7008**

Notified bodies   **P7009**

The Pressure Equipment (Amendment) Regulations 2002   **P7010**

Pressure Systems Safety Regulations 2000   **P7011**

Systems and equipment covered by the regulations   **P7012**

Systems and equipment excepted from some or all the regulations   **P7013**

HSC approved code of practice and guidance   **P7014**

Interpretation   **P7015**

Design, construction and installation   **P7016**

Documentation and marking   **P7017**

Competent persons   **P7018**

Written scheme of examination   **P7019**

Operations, maintenance, modification and repair   **P7026**

Record keeping   **P7030**

Related health and safety legislation   **P7031**

Product Safety   **P90**
Natalie Wood and Alison Newstead

Introduction   **P9001**

General Product Safety Regulations   **P9002**

Industrial products   **P9009**

Regulations made under the Health and Safety at Work etc Act 1974   **P9010**

Criminal liability for breach of statutory duties   **P9011**

Civil liability for unsafe products – historical background   **P9024**

Consumer Protection Act 1987   **P9027**

Contractual liability for substandard products   **P9047**

Unfair Contract Terms Act 1977 (UCTA)   **P9057**

Unfair Terms in Consumer Contracts Regulations 1999 (SI 1999 No

2083)   **P9058**

Radiation   **R10**

Leslie Hawkins and Donald Bruce

Non-ionising and ionising radiation   **R1001**

Rehabilitation   **R20**

Leslie Hawkins

What is Rehabilitation?   **R2001**

Why should we be concerned?   **R2002**

Managing sickness absence   **R2007**

Long term absence and rehabilitation   **R2008**

Using professional advice   **R2009**

REACH: Registration, Evaluation and Authorisation of CHemicals in Europe   **R25**

John Wintle

Introduction   **R2501**

Purpose   **R2502**

Need for REACH   **R2503**

Chemicals covered and exempted   **R2504**

Duty Holders within REACH   **R2505**

Preparation for REACH   **R2506**

The Registration process   **R2507**

Timescale for Registration   **R2508**

Information required for Registration   **R2509**

Safety Data Sheets and downstream use   **R2510**

European Chemicals Agency   **R2511**

Evaluation   **R2512**

Authorisation   **R2513**

Restrictions   **R2514**

UK Competent Authority and Helpdesk   **R2515**

Enforcement   **R2516**

Summary   **R2517**

Further information   **R2518**

Acknowledgement   **R2519**

Risk Assessment   **R30**

Mike Bateman

Introduction   **R3001**

Contents

HSWA 1974 requirements **R3002**

The Management of Health and Safety at Work Regulations 1999 **R3005**

Common regulations requiring risk assessment **R3006**

Specialist regulations requiring risk assessment **R3016**

Related health and safety concepts **R3017**

Management Regulations requirements **R3018**

Children and young persons **R3025**

New or expectant mothers **R3026**

Other vulnerable persons **R3028**

Who should carry out the risk assessment? **R3029**

Assessment units **R3030**

Relevant sources of information **R3031**

Consider who might be at risk **R3032**

Identify the issues to be addressed **R3033**

Variations in work practices **R3034**

Checklist of possible risks **R3035**

Making the risk assessment **R3036**

Assessment records **R3042**

After the assessment **R3044**

Content of assessment records **R3049**

Illustrative assessment records **R3049**

Safe Systems of Work **S30**

Lawrence Bamber

Introduction **S3001**

Components of a safe system of work **S3002**

Which type of safe system of work is appropriate for the level of risk? **S3007**

Development of safe systems of work **S3008**

Permit to work systems **S3011**

Isolation procedures **S3023**

Further reading **S3024**

Statements of Health and Safety Policy **S70**

Mike Bateman

Introduction **S7001**

The legal requirements **S7002**

Content of the policy statement **S7003**

Stress at Work **S110**

Nicola Coote

Introduction   **S11001**

Definition of stress   **S11002**

Understanding the stress response   **S11003**

Why manage stress?   **S11004**

Occupational stress risk factors   **S11010**

The work   **S11016**

Company structure and job organisation   **S11025**

Personal factors   **S11032**

Stress management techniques   **S11036**

Training and Competence in Occupational Safety and Health   **T70**

Hazel Harvey

Introduction   **T7001**

Recognising competence   **T7002**

Legal obligations for training and competence   **T7003**

Setting national standards in health and safety   **T7008**

Education and training towards national standards of competence **T7010**

Maintaining competence   **T7028**

Useful contacts   **T7029**

Ventilation   **V30**

Chris Hartley

Introduction   **V3001**

Sources of contamination   **V3002**

Local exhaust ventilation   **V3003**

Legal requirements   **V3015**

Conclusion   **V3016**

Vibration   **V50**

Roger Tompsett

The scope of this chapter   **V5001**

An introduction to the effects of vibration on people   **V5002**

Vibration measurement   **V5007**

Causes and effects of hand-arm vibration exposure   **V5010**

Vibration-induced white finger as an occupational disease   **V5011**

The Control of Vibration at Work Regulations, 2005 (SI 2005 No 1093)   **V5015**

Contents

Advice and obligations in respect of hand-arm vibration   **V5017**

Personal protection against hand-arm vibration   **V5019**

Advice and obligations in respect of whole-body vibration   **V5023**

HSE guidance on hand-arm and whole-body vibration   **V5029**

Violence in the Workplace   **V80**

Bill Fox and Andrea Oates

Introduction   **V8001**

The nature and extent of the problem   **V8002**

Who is at risk?   **V8003**

Work-related violence – the legislation   **V8004**

Developing and implementing policy   **V8016**

Risk assessment for workplace violence   **V8044**

Reporting, recording and monitoring system   **V8057**

Risk reduction measures   **V8072**

Training   **V8096**

Incident management   **V8116**

Post-incident management   **V8129**

Sources of further information   **V8142**

Vulnerable Persons   **V120**

Mike Bateman

Introduction   **V12001**

Relevant legislation   **V12002**

Children and young persons   **V12006**

New or expectant mothers   **V12017**

Lone workers   **V12027**

Disabled persons   **V12032**

Inexperienced workers   **V12038**

Work at Height   **W90**

George Ventris and Gordon Prosser

Introduction   **W9001**

The Work at Height Regulations 2005 (SI 2005 No 735)   **W9002**

Commencement and scope   **W9003**

Application   **W9004**

Organisation and planning; competence, and avoidance of risks   **W9005**

Work equipment   **W9009**

Fragile surfaces **W9010**

Falling objects and danger areas **W9011**

Inspection **W9012**

Duties of persons at work **W9013**

Protection from falls **W9014**

Conclusion **W9015**

Working Time **W100**

Dorothy Henderson and Adam Rice

Introduction **W10001**

Definitions **W10002**

Maximum weekly working time **W10010**

Rest periods and breaks **W10014**

Night work **W10019**

Annual leave **W10022**

Records **W10025**

Excluded sectors **W10026**

Enforcement **W10038**

Workplaces – Health, Safety and Welfare **W110**

Andrea Oates

Introduction **W11001**

Workplace (Health, Safety and Welfare) Regulations 1992 (SI 1992 No 3004) **W11002**

Safety signs at work – Health and Safety (Safety Signs and Signals) Regulations 1996 (SI 1996 No 341) **W11030**

Disability Discrimination Act 1995 **W11043**

Building Regulations 2000 (SI 2000/2531) **W11044**

Factories Act 1961 **W11045**

Table of Cases **TC-1**

Table of Statutes **TS-1**

Table of Statutory Instruments **TSI-1**

Index **IND-1**

List of illustrations

Plan-Do-Check-Act	Int22
Action to be taken by employers and others when accidents occur at work	A3017
Internal incident notification form	A3026
Elements of the investigation process	A3031
Prescribed form for reporting an injury or dangerous occurrence	A3037
Prescribed form for reporting a case of disease	A3038
Asbestos Management Plan	A5012
Confined spaces – identification of risks	C7010
Permit to work – entry into confined spaces	C7010
Confined spaces – control measures and other precautions	C7010
UN classification system	D0407
ADR Approved List	D0411
Example of UN mark	D0415
Proper Shipping Name (PSN) and UN Number	D0420
Hazard labels	D0420
Truckmarking	D0421
Example of ADR vehicle labelling	D0422
Limited quality package exemptions	D0441
Classification of major incident types	D6003
Disaster and emergency management systems (DEMS)	D6034
Major Incident Matrix	D6050
Decisional Matrix	D6051
Monitoring Matrix	D6066
Distance Scales	E5001
Ergonomic Approach	E17004
The 'S' Shaped Curve of the Spine	E17006
Zone of Convenient Reach	E17022
Viewing Angle	E17026
Body Map	E17057
Detention of food notice	F9014
Withdrawal of detention of food notice	F9014
Food condemnation warning notice	F9014

Improvement notice   F9015

Prohibition order   F9016

Emergency prohibition notice   F9018

Form of application for registration of food premises   F9028

Example of assessment records   H2124

Employee safety representatives   J3007

Relationship between the manufacturers and
the owner/users   L3002

Example of an inspection record sheet   M1007

Example of a Declaration of Conformity   M1015

The Risk Matrix   M1023

Risk assessment framework   M1024

Examples of mechanical hazards   M1025

Design/selection of safeguards   M1028

Example of a fixed guard   M1029

Fixed guard with adjustable element   M1030

Example of an automatic guard   M1031

Main types of electrical interlocking switches   M1032

Trapped key system for power interlocking   M1034

Example of a mechanical interlocking system   M1035

Risk contours and zones around a hazardous installation   M1140

The costs of accidents   M2001

Systematic management cycle   M2008

Continual improvement loop   M2029

ILO model of continual improvement   M2044

Health and safety management
system elements and links   M2105

Suggested risk assessment framework   M2111

Lifting and lowering   M3003

Handling while seated   M3003

Manual handling assessment checklist   M3021

Manual handling issues: Suggested aspects for discussion   M3022

Examples of manual handling assessment records   M3024

Sign for informing that ear protectors must be worn (white
on a circular blue background)   N3013

Normal audiogram with the responses close to 0dB across the

frequency range   O1023

CE mark of conformity for personal protective equipment   P3013

Relationship between Manufacturers' Duties   P7001

Pressure Systems Safety Regulations 2000 – Duties Decision Flow Chart   P7012

Stress Response Curve   S11003

Components of a local exhaust ventilation system   V3003

Duct pressures   V3013

Static pressure measurements in LEV system fault finding   V3013

Model for effective policy development group   V8024

Model for monitoring and evaluating policy   V8042

Risk assessment model   V8050

Likelihood of harm   V8053

Severity of harm   V8053

A model for reporting and monitoring incidents
of workplace violence   V8057

General training and development model   V8097

Identifying the need   V8099

A typical delegate feedback process combining individual
and organisational needs   V8110

Incident management – planning and practice   V8116

Timescale of reactions to workplace violence   V8130

Supporting the victim   V8135

The process of prosecution   V8140

Maybo Risk Management Model   V8142

Examples of prohibitory signs   W11037

Examples of warning signs   W11038

Examples of mandatory signs   W11039

Examples of emergency escape or first-aid signs   W11040

Examples of fire-fighting signs   W11041

Hand signals   W11042

List of abbreviations

Many abbreviations occur only in one section of the work and are set out in full there. The following is a list of abbreviations used more frequently or throughout the work.

Legislation

ACM	Asbestos-containing Material
BRM	Business Risk Management
BS	British Standard
BSE	Bovine Spongiform Encephalepathy
CCTV	Closed Circuit Television
CHAN	Chemical Hazards Alert Notice
CRI	Corporate Responsibility Index
CSR	Corporate Social Responsibility
DSE	Display Screen Equipment
EHO	Environmental Health Officer
ELCI (ELI)	Employer's Liability (Compulsory) Insurance
EMF	Electromagnetic Field
EMS	Environmental Management System
FM	Facilities Management
FMD	Foot and Mouth Disease
FOPS	Falling Object Protective Structure
HAVS	Hand-arm Vibration Syndrome
IR	Infra-red (radiation) or Ionising Radiation
LEV	Local Exhaust Ventilation
EL	Lower Explosive Limit
LOCAE	List of Classified and Authorised Explosives
LPG	Liquefied Petroleum Gas
MDF	Medium Density Fibreboard

MDHS	Methods for the Determination of Hazardous Substances
MEL	Maximum Exposure Limit
MEWP	Mobile Elevating Work Platform
MPE	Maximum Permissible Exposure
MSDS	Material Safety Data Sheet
NOEM	New or Expectant Mother
OEL	Occupational Exposure Limit
OES	Occupational Exposure Standard
OHSAS	Occupational Health and Safety Assessment Series
OHSMS	Occupational Health and Safety Management Systems
PAH	Polycyclic Aromatic Hydrocarbons
PAT	Portable Appliance Testing
PCB	Polychlorinated Biphenyls
PLI	Public Liability Insurance
PPE	Personal Protective Equipment
ppm	parts per million
PTW	Permit to Work
RHS	Revitalising Health and Safety
ROPS	Roll-over Protective Structure
RPE	Respiratory Protective System
RSI	Repetitive Strain Injury
RSP	Registered Safety Practitioner
SHT	Securing Health Together
SWL	Safe Working Load
UEL	Upper Explosive Limit
ULD	Upper Limb Disorder
UV	Ultra-violet (radiation)
VCM	Vinyl Chloride Monomer
VDU	Visual Display Unit
WRULD	Work-related Upper Limb Disorder
WSA	Workers' Safety Advisors
YP	Young Person

Introduction

Lawrence Bamber

The Handbook

In the early days of this publication, health and safety was still a matter for specialists – often professional safety advisers, HR directors and others who were required to 'deal with' health and safety on behalf of their organisations. Now, whether it is a debate on work-related travel, stress or the hazards associated with different forms of electricity generation including nuclear power stations, everyone is engaged in discussing 'risk'. In the boardroom this is often linked to corporate governance and corporate responsibility, on the shop floor it is associated with a radical rejection of a 'compensation culture' in favour of a desire to be able to return home from work healthy and uninjured.

Recent developments of these arguments have focused our attention on the 'health' part of health and safety. This means that absence management, occupational health support and the reduction in the numbers of people on incapacity benefit are uniting management, trades unions and Government in a new vision of the healthy workplace. There are also concerns about overre-action to some health and safety requirements resulting in a variety of 'elf and safety' myths arising, usually emanating from those with little knowledge of the subject. Both the HSE and professionals within health and safety are responding robustly to such myths, and risk aversion generally.

Partially as a result of such a robust response, the coalition Government invited Lord Young to review the current state of health and safety legislation, alongside the compensation culture.

His report "Common Sense: Common Safety" was published by the Prime Minister=s Office on 15th October, 2010. The review makes some thirty-six recommendations. Some, such as the RIDDOR review and the introduction of OSHCR (Occupational Safety and Health Consultant's Register), are either ongoing or completed. A summary of the recommendations is contained on page SP 4 (Stop Press) of Issue 87.

Following Lord Young's departure, the recommendations from his report are to be championed by Employment Minister, Chris Grayling, who announced in early 2011 a new package of changes to Britain's health and safety system: "Good health and safety, good for everyone" (http://www.dwp.gov.uk/policy/health-and-safety).

These changes are intended to support the Government's growth agenda and to ease the regulatory burdens on business.

A full report of the announcement and reactions to it was contained at pages SP 1 and SP 2 (Stop Press) of Issue 88.

One key announcement from Chris Grayling was that the Government have launched a review of all existing health and safety legislation with a view to scrapping measures that are not needed and which put an unnecessary burden on business. The review is being chaired by a leading risk management specialist, Professor Ragnar Lofstedt, who is director of the King's Centre for Risk Management at King's College, London. This evidence-based review, being undertaken by a five-strong advisory panel – chaired by Professor Lofstedt – is due to report back by October 2011, ahead of the EU review of health and safety which is scheduled for 2013.

Running parallel with the Lofstedt review – and to be considered by the review panel – is the Government's Red Tape Challenge (www.lexisurl.com/hsw8376) which had asked members of the public to suggest ways to overhaul UK health and safety laws by, for example, flagging up regulations that should be abolished or amended. The general consensus to date (August 2011) was that osh law is not broken."

This introduction sets the scene for the rest of the publication, identifying current issues and trends, and providing references to individual chapters. The wide range of subjects covered provides considerable detail on legal requirements, practical ways of achieving compliance and good practice generally.

This Introduction is in five main parts

They are:

- drivers for health and safety performance (starting at PARA 3);
- the changing world of work (starting at PARA 6);
- recent trends in UK health and safety law and its enforcement (starting at PARA 7);
- the structure of UK health and safety law (starting at PARA 19);
- the management of health and safety at work (starting at PARA 25).

Drivers for health and safety performance

Why do organisations seek to improve their performance?

This is an issue that has been subject to increasing research in the UK and around the world in recent years. There is also growing evidence of distinct differences between the perception of small businesses and larger companies. It is also worth noting that definitions of smaller businesses vary from less than 10 and less than 100 employees, to the Department for Business, Enterprise and Regulatory Reform (BERR – formerly the Department of Trade and Industry's (DTI)) category of small to medium-sized enterprises (SMEs), at less than 250 employees. In small companies, the attitudes and beliefs of the owners/managers are a dominant force whilst in larger corporations and public bodies the social norms are subject to the broader forces of corporate governance. A combination of the following appear to be of greatest influence, but each overlaps with the other and may be present in different proportions and combinations:

- beliefs and attitudes of owners/managers in very small organisations;
- risk management and reputation risk;
- corporate governance;
- legal compliance.

Beliefs and attitudes of owner/managers in very small organisations

In the UK, across the European Union and in the United States of America, more than half of all workers are employed in organisations of less than 10 people. This trend is expected to continue as large employers in developed countries contract or collapse as a result of the economic downturn. The modern economy increasingly comprises services which are often provided by new organisations and by the self-employed. Identifying what motivates the leaders of such organisations is therefore important to governments and enforcing authorities for health and safety. It appears that the level of awareness of specific health and safety requirements embodied in regulations, is often poor and almost wholly dependent on suppliers, customers and peers. The suppliers of goods and services to small businesses are regarded as a reliable source of information. For example the labels on containers of materials and the accompanying safety data sheets are used by small businesses to establish working practices without, typically, any reference to the requirements of the *Control of Substances Hazardous to Health Regulations 2002 (SI 2002 No 2677 as amended by SI 2003 No 978)*. For some small businesses which supply major clients, those larger organisations increasingly impose operating conditions and provide support which incorporates health and safety, as part of the management of a secure supply chain*. The last significant element of advice and guidance is from the informal contacts with other owners/managers. (* For a study into supply chain management for health and safety, see '*Managing risk – Adding value: How big firms manage contractual relations to reduce risk*' (1998), HSE Books.)

Whatever their source of information as to what they should be doing and how to do it, the owners/managers of small businesses who do act in this area seem to have a clear set of reasons for getting health and safety right. These motivators are:

- the *focus on employees* who are so well known because of daily contact that injury or ill health is treated almost as if a family member were affected (and in many cases family members do work in the small organisation, and children may attend the workplace in the school holidays, etc);
- the *focus on satisfying customers* and responding to guidance from suppliers, a general approach to 'doing the right thing';
- a conviction amongst a minority that looking after worker health and safety and similar matters contributes to a *productive, healthier, happier workforce* and may have benefits such as controlling insurance premiums.

As part of its reducing the burden strategy, the HSE has provided – via its website www.hse.gov.uk – model risk assessments for low hazard workplaces

such as offices, classrooms and shops, thus simplifying the risk assessment p
rocedure for small, low risk organisations.

Why do larger organisations seek to improve their performance?

Most large organisations have been on a learning curve since the post-war
period. At first there was an emphasis on managing the pure risk of getting
health and safety 'wrong', and the impact that this could have on people
(injuries), products (damage) and property (losses such as by fire). By the
1970s this effort had broadened to encompass environmental issues, and also
the concept of business recovery – and as a result events which could lead to
business disruption were being investigated to both prevent their occurrence
but also to respond effectively if they did occur. By the 1980s risk management
had matured into a discipline which also encompassed such matters as brand
management and corporate reputation. This does not mean that large companies have eliminated such risks. The impact on companies involved in rail
crashes or other high-profile health and safety failures is a testament to the
correct evaluation that these are significant matters for corporate value – but
most have implemented systems which are designed to identify and control
such risks. Thus, protecting the reputation of a company as an efficient and
well-managed organisation is one of the most significant drivers to improving
health and safety performance.

From the 1980s up to the present day there have been a series of collapses of
high profile companies. Press reports of 'fat cats' drawing huge salaries and
bonuses for incompetent management have, not surprisingly, resulted in
considerable public criticism and governments have created several committees
to report on remedial steps – Cadbury, Greenbury and Hampel. The latter
called for a new approach to corporate governance which included a component of risk management at the highest level. The Turnbull Report, which
requires organisations to establish a risk management strategy, was adopted by
the London Stock Exchange obliging listed companies (plcs) to publish an
annual risk management report as part of their annual reporting arrangements.
This has prompted many large businesses to adopt formal plan-do-check-act
management systems to achieve the level of assurance required for directors to
sign such reports. The approach to risk management has also dovetailed with
an increasing interest in corporate social responsibility and the pressures on
companies to be seen to look after their staff, the environment and their
neighbouring communities. In the public sector, similar developments have
been encouraged by the Audit Commission, and in the voluntary sector by the
Charities Commission. (See 'Worth the risk: An introduction to risk management and its benefits' (2001), Audit Commission; 'Accounting reporting by
charities: Statement of recommended practice (2000), SORP, which is effectively a mandatory equivalent of the Turnbull requirements.) Across the public
and private sectors, therefore, the development of corporate governance is
linked to and integrated with risk management – all organisations are asked to
identify what could blow them off course, resulting in a diminution of services,
a loss of profitability, the development of an unsustainable approach to their
work. Health and safety is an important part of the matrix of risks identified
in any such exercise, and the management and control of the health and safety

risks has become part of what the board of directors or trustees have to achieve and assure. It was against this background that a guide was developed and published by the Health and Safety Executive (HSE) on the duties of directors (INDG 417 *Leading health and safety at work*).

This guidance– drawn up jointly by the Institute of Directors (IOD) and the Health and Safety Commission (HSC) – sets out an agenda for the effective leadership of health and safety. It is designed for use by all directors, governors, trustees, officers and their equivalents in the private, public and third sectors.It applies to organisations of all sizes.

- Strong and active leadership from the top
    - visible, active commitment from the board
    - establishing effective downward communication systems and management structures
    - integration of good health and safety management with business decisions
- Worker involvement
    - engaging the workforce in the promotion and achievement of safe and healthy conditions
    - effective upward communication
    - providing high quality training
- Assessment and review
    - identifying and managing health and safety risks
    - accessing and following competent advice
    - monitoring, reporting and reviewing performance

The third pressure for health and safety performance is the necessity to achieve legal compliance. Most people, including those who lead organisations, are social beings who accept that the rules they need to live and work by represent the framework for a civilised life. Even if an individual driver occasionally breaks the speed limit, he or she is unlikely to wish for unlimited speeds to be agreed for the street on which they live. For small organisations, there is an often-expressed view that they are very unlikely to be visited by an inspector, and there may be a low awareness of both the statutory requirements and often a misperception of the level of risks associated with their work. It is in the large organisation with a nervous board of directors and a management commitment to legal compliance that represents a significant stimulus. Identifying what regulations need to be complied with, those which are specifically applicable to the mix of work being undertaken, and then assuring that compliance, is a further push towards formal management systems. Greater incentives for effective management of health and safety have been provided by the *Corporate Manslaughter and Corporate Homicide Act 2007* (see the chapter dealing with CORPORATE MANSLAUGHTER), and the increased penalties introduced by the *Health and Safety (Offences) Act 2008*.

These increased penalties are as follows:

Offence	Prior to 16th January, 2009	After 16th January, 2009

Magis- trates' Court	Breach of General Duties under HSWA Act	Fine up to 20,000	Fine up to 20,000 and/or up to 12 months imprisonment
	Breaches of Safety Regula- tions	Fine up to 5,000	Fine up to 20,000 and/or up to 12 months imprisonment
Crown Court	Breaches of Safety Regula- tions	Fine up to 20,000	Fine up to 20,000 and/or up to 2 years impris- onment

The changing world of work

There is a wide range of evidence about the changes in the world of work. In the UK we have many more people working in call centres (about 4 per cent of the workforce) than in mining. This move towards service industries, as traditional manufacturing continues to decline in numbers of people employed, is matched by the increasing dominance of small businesses and the self-employed. Across the western world, it is more common for workers to be employed in companies with less than 10 staff than in companies with more than 250 employees. Flexible work patterns and the temporary nature of much employment, with hardly any young workers anticipating a 'job for life, has also altered the balance of health and safety issues with psychosocial risks giving rise to stress-related sickness absence overtaking musculo-skeletal injuries as the most common work-related illness. In the UK, the great increase in female employment and the pressure on an ageing workforce to stay at work longer is also changing the nature of the 'at risk' group which requires protection whilst at work. This listing of changes being experienced at work suggests that the law, which often moves slowly, the enforcement authorities, employers, employees and professionals safety advisers are each under pressure to adapt and change in turn. Many of these changes are reflected in individual chapters and some are developed further in the remainder of this introduction.

To combat the growing threat of health (as opposed to safety) issues within UK plc, Dame Carol Black, the National Director for Health and Work, published the first ever review into the working age population: "Working for a Healthier Tomorrow" on 17th March, 2008.

To assist with its implementation the HSE has launched a guide which offers practical advice on long term sickness absence and return to work strategies.

This strategy for Workplace Health and Safety outlines six key actions that safety representatives – in partnership with their employers – can take to reduce sickness absence and help sick employees to return to work.

These include:

• helping to identify measures to improve worker health

- suggesting the development of workplace policies for managing sickness absence
- helping to keep workers who are off sick in contact with work
- helping to plan adjustments to enable return to work
- supporting and empowering sick workers to return to work
- helping to promote understanding of ill-health and disability at work

Another useful resource is Workplace Health Connect which is a free advisory service on workplace health for smaller businesses.

This confidential service offers cost-effective and simple solutions to help cut workplace injuries and ill-health. Small businesses in England and Wales can access the service via an advice line: 0845 609 6006. Where necessary, these telephone contacts will be followed up by workplace visits by qualified Workplace Health Connect advisers.

Legal framework

Healthy work, healthy at work and healthy for life

The report *Securing Health Together*, outlined a revised occupational health strategy for UK plc. The occupational health strategy seeks, in addition to reducing the extent to which the workplace is a site for health harm, to improve access to the workplace for those who are ill or disabled by tackling rehabilitation and disability access with real vigour.

The HSE publish annual progress reports on the ambitious statistical targets set in *Revitalising Health and Safety* and these show significant improvements being made in most areas. However, it is debatable whether these improvements result from the effective implementation of the elements of the strategy outlined in PARA 7 above, or from the decreasing numbers employed in higher risk work sectors. There are concerns in many quarters that the reduced funding of the HSE (in real terms) does not reflect a real commitment by the Government to improved standards of health and safety.

Companies and individuals convicted of health and safety offences are now 'named and shamed' on the HSE website – www.hse.gov.uk/enforce/prosecut ions.htm. This was further developed by Dame Carol Black's report "Working for a Healthier Tomorrow" referred to above in section 6.

Risk and Regulation Advisory Council

The Risk and Regulation Advisory Council (RRAC) was established in January 2008 in order to address the key issues related to public risk and the role of policy-makers and regulators. The HSE has a strong interest in working alongside this new body to further promote a commonsense approach to pragmatic risk management.

The role of the European Union

Since the early 1990s the European Union (EU) has provided the principal motor for changes in UK health and safety legislation. Article 118A in the Treaty of Rome 1957 gives health and safety prominence in the objectives of the EU. The Social Charter also contains a declaration on health and safety, although this has no legal force. Whilst the EU can issue its own regulations, it mainly operates through directives requiring Member States to pass their own legislation. The 'Framework Directive' adopted in 1989 as part of the creation of the Single Market contained many broad duties, including the requirement to assess risks and introduce appropriate control measures. Other directives have since been adopted or proposed, including many relating to technical standards and safety requirements for specific products.

EU directives

The UK, along with other Member States, is required to implement EU directives by designated dates. The UK sits about halfway in the league-table for formal implementation of directives by the due date, and there is not at present an effective mechanism for evaluating implementation and enforcement. The pace of new directives has slowed in recent years, providing Member States with greater opportunity to ensure effective implementation.

Implementation and enforcement

Introduction and implementation of health and safety legislation in the UK is mainly overseen by the Health and Safety Executive (HSE). (In 2008 a merger took place between the previous policy-making body, the Health and Safety Commission (HSC), and its implementation arm (the HSE) into a single body under the HSE title.)

For regulatory initiatives the HSE first circulates consultative documents incorporating the proposed regulations together with a related Approved Code of Practice (ACoP) and/or guidance to interested parties (trades unions, employers' organisations, professional bodies, local authorities etc). Following the consultation process, HSC submits a final version to the Secretary of State who then lays the regulations before Parliament. It is at this stage that a date for their coming into force is determined – usually that stipulated by the original EU directive.

Enforcement of health and safety legislation in most high risk work sectors is the province of the HSE although in 2006 the Office of Rail Regulation (ORR) took over enforcement in respect of railways and other guided transport systems. In lower risk work activities enforcement is normally the responsibility of the local authority in whose area the activity is carried out. See the chapter entitled ENFORCEMENT for more details.

Health and Safety Executive – Future Strategy

Early in 2009, HSE completed their consultative process on a new strategy. This proposed strategy sets several general goals for the organisation, which are summarised below:

- investigating work-related accidents and ill health and taking appropriate enforcement action;
- to encourage strong leadership in health and safety and a commonsense approach to the subject;
- to focus on real health and safety issues as distinct from trivia;
- to encourage an increase in competence in health and safety and risk management;
- to reinforce the promotion of worker involvement and consultation;
- to specifically target key health issues;
- to prioritise higher risk activities;
- to customise approaches to help small businesses deal with health and safety;
- to reduce the likelihood of catastrophic incidents; and
- to take account of wider issues that impact on health and safety.

HSE Budgetary Cuts

As a result of the Government's Comprehensive Spending Review (October 2010) the HSE's budget is to be slashed by 35% by 2014 2015, the Department of Work and Pensions (DWP) has announced.

The Government is still committed to a fair and proportionate osh regime but also said that HSE should cut costs in the same way as the rest of the public sector. In seeking to achieve these cost savings, it is the HSE's intention to share more of the cost with those businesses who create risks whilst reducing the burden on low risk organisations.

The All Party Parliamentary Group on Occupational Safety and Health has warned of false economy by stating that any reduction in HSE activities will inevitably lead to increased costs from sickness absence, compensation and benefit payments. The Group thinks that there is scope for more general charging for HSE activities that take place as a result of non-compliance – e.g. where an inspection leads to enforcement action and, as a result, the inspector has to revisit the workplace so as to ensure that the problem has been rectified and legislative compliance achieved.

This "Fee for Fault" approach is supported by HSE Chair, Judith Hackitt, who believes it is a fair and equitable approach that will be welcomed by the vast majority of businesses which are compliant."

HSE Delivery Plan

The HSE Delivery Plan for 1st April, 2011 to 31st March, 2012 outlines the steps that the regulator will be taking to enable innovation that brings economic growth while ensuring that risks are managed properly and proportionately.

The plans are grouped under four main headings:

- transforming our approach
- avoiding catastrophe
- clarifying ownership of risk and improving compliance

- securing justice

Some individual actions already highlighted include:

- Reducing the level of proactive inspections in all but the highest hazard sectors. This means that sectors such as agriculture, quarries, health and social care, transport, electrical generation, textiles and light engineering, will be excluded from proactive HSE inspections
- Inspections will be retained in construction, waste and recycling, and high risk manufacturing.
- Plans for all inspections to fall by one third - approximately 11,000 inspections per annum
- Proposals for the HSE to recover the cost of its inspection and investigation activities to the tune of ,133 per hour as from April 2012
- Closure of the HSE Infoline and Incident Contact Centre from the end of September 2011. Reliance will then be on the HSE website (www.hse.gov.uk) for information and most accident reports
- Creation of a new online resource aimed at small businesses and low risk organisations called "Health and safety made simple". This resource takes SMEs through their basic osh duties, describing what they need to do and how they should do it. It is available at: www.hse.gov.uk/simple-health-safety
- Consultation on RIDDOR
- Creation of OSHCR, the register for competent osh consultants, and its transfer to the professional bodies by the end of 2011, in terms of its control and administration
- Specification of the number of targeted interventions planned for the highest hazard sites
- Definition of the total number of HSE inspections to be made in those sectors deemed to pose significant risk
- Working in partnership with businesses, unions and other stakeholders so as to encourage them to find their own solutions for their own problems, and to take ownership for driving forward improvements in health and safety standards
- Investigation of accidents according to HSE selection criteria, thereby focusing on the most significant failures. Complaints will be investigated based on the level of risk posed.

The full Delivery plan is available at: www.hse.gov.uk/aboutus/strategiesandplans/delivery plans/index.htm.

Penalties and prosecutions

The penalties for breaches of health and safety legislation are increasing steadily with the current record fine for a single offence being the £15 million imposed in 2005 on Transco for a breach of *Section 3(1)* of *HSWA*. This related to a large gas explosion in a residential area of Lanarkshire in 1999 which killed a couple and their two children. Transco were heavily criticised for a lack of effective inspection and maintenance arrangements and inaccurate records of gas pipes. The ductile iron gas main involved in the explosion was badly corroded.

There have been several other fines well in excess of £1 million including those levied on Balfour Beatty and Network Rail (formerly Railtrack) for their parts in the Hatfield derailment in 2000 caused by poor standards of track inspection and maintenance. The Ladbroke Grove rail crash in 1999 also resulted in Network Rail and Thames Trains being fined a total of £6 million. Fines above £100,000 are now commonplace as can be seen on the HSE website.

Until recently custodial sentences could only be imposed under *HSWA* for a limited range of offences – in practice these generally have resulted from working with asbestos-containing materials without the appropriate licence or from continuing to operate in breach of a Prohibition Notice. However, the *Health and Safety (Offences) Act 2008* has considerably expanded the scope for custodial penalties. The same Act also increased the penalties that can be imposed by Magistrates Courts for health and safety offences. (See section 5 above for the specific penalties.)

Custodial sentences have also resulted in the past from manslaughter charges brought in respect of work-related deaths and the *Corporate Manslaughter and Corporate Homicide Act 2007* provides a further opportunity for unlimited fines to be imposed on organisations.

The first company to stand trial under the *Corporate Manslaughter and Corporate Homicide Act, 2007 (CMCHA)* was fined ,385,000 after being found guilty by a jury at Winchester Crown Court on 17th February, 2011.

The fine for Cotswold Geotechnical Holdings Ltd (CGH) related to charges following the death of employee, Alexander Wright, on 5th September, 2008 and will be paid in equal instalments over a ten-year period. Owing to CGH's financial state, the judge did not levy any costs but the fine amounts to 250 percent of CGH's annual turnover.

Mr Wright, twenty-seven, had been working alone in a 3.5 metre deep trench after the Managing Director of CGH, Peter Eaton, had left for the day. A short while later the trench collapsed onto Mr Wright and buried him, On hearing his cry for help, one of the plot owners ran to the trench where he found Mr Wright buried up to his head. He climbed into the trench and removed some of the soil but a further torrent of earth fell into the trench, engulfing Mr Wright who died of asphyxiation.

Peter Eaton, the MD of CGH, had originally been charged with manslaughter by gross negligence, as well as a health and safety offence. These charges were subsequently dropped on the grounds of his ill-health.

In convicting CGH, the jury found that its system for work in digging trial trenches was wholly and unnecessarily dangerous. The court heard that the company had ignored industry guidance which prohibits entry into trenches more than 1.2 metres deep. The trench in question was 3.5 metres deep!

N.B. Further information on Corporate Manslaughter is contained in Chapter C90.

The second case brought under the *CMCHA 2007* involves a company that manufactures storage products in Hyde, Greater Manchester. Lion Steel

Equipment Ltd has been summonsed to appear before Tameside Magis-trates' Court on 2nd August, 2011 following the death of employee Steven Berry, forty-five, from injuries sustained when he fell through a fragile roof panel at Lion's premises in Hyde in May 2008.

As well as being charged with corporate manslaughter, the company is also charged, under *sections 2* and *33* of *HSWA 1974*, for failing to ensure the safety at work of its employees.

Three of the company's directors have each been charged with manslaughter by gross negligence and also, under *section 37* of *HSWA 1974*, for failing to ensure the safety of their employees.

Risk assessments

Much recent UK legislation has included a requirement for some type of risk assessment. The concept was introduced in the early 1980s in regulations applying to asbestos and lead but it came to wider prominence as a core requirement of the *Control of Substances Hazardous to Health Regulations 1988 ('COSHH') (SI 1988 No 1657)*. The group of Regulations which came into force on 1 January 1993 derived from EU directives continued this trend with four sets of regulations requiring an assessment of one kind or another. The most important of these was the general requirement for risk assessment contained in the *Management of Health and Safety at Work Regulations 1999 (SI 1999 No 3242)* ('the *Management Regulations*'), *Reg 3*. Other important regulations requiring more specific types of risk assessment relate to noise, manual handling operations, personal protective equipment, display screen equipment, asbestos and vibration.

In practice a less formal type of risk assessment was already required by health and safety legislation – an assessment of the level of risk, the adequacy of existing precautions and the costs of additional precautions was necessary in order to determine what was 'reasonably practicable'. Risk assessment lies at the heart of the original concept of self-regulation enshrined in the Robens Report 1972 – employers are required to demonstrate that they have identified relevant risks together with appropriate precautions and in most cases must have records available to prove this.

The HSE have provided numerous guides and templates to help SMEs and low risk organisations undertake pragmatic risk assessments with minimal effort.

They may be found on the HSE website at: www.hse.gov.uk/riskassessments.

Supply-chain pressure

Since the introduction of *HSWA 1974*, many court decisions, most notably those involving *Swan Hunter Shipbuilders* ([1984] IRLR 93–116) and *Associated Octel* [1996] 4 All ER 846, have emphasised that employers often have responsibilities for the activities of employees of other organisations. The structure of *HSWA 1974* and much subsidiary legislation is such that responsibilities overlap between employers rather than being neatly appor-tioned between them. Employers must do more than simply not turn a blind eye to the obvious health and safety failings of those with whom they come into contact: they must often take a pro-active interest in the health and safety standards of others.

In recent years a growing number of larger companies, local authorities and other public bodies have had increasingly formalised procedures for checking the health and safety standards of contractors wishing to work for them. This process has been accelerated by the demands of the original *Construction (Design and Management) Regulations 1994 ('CDM') (SI 1994 No 3140)*. The *Construction (Design and Management) Regulations 2007 (SI 2007 No 320)* specifically require clients to satisfy themselves that contractors are *capable of dealing with* the health and safety issues associated with construction work of any type or size. The Regulations also place responsibilities on principal contractors in respect of sub-contractors. Consequently contractors are frequently required to provide details of their health and safety policies and generic risk assessments together with risk assessments and/or method statements for specific projects or activities.

Some companies take an extremely hands-on approach in policing the activities of contractors working on their premises or on projects where they are the principal contractor. Several schemes also exist for vetting the health and safety standards of contractors. There is now a single national scheme – Safety Schemes in Procurement (SSIP) – which is recognised by the majority of those organisations providing contractor vetting schemes. SSIP is an umbrella organisation set up in December 2009 to address the spread of osh pre-qualification schemes. The scheme is voluntary and aims to vastly reduce bureaucracy and to facilitate mutual recognition of core competence criteria within the construction and allied industries. To date (May 2011), thirteen individual schemes have joined SSIP and some 75,000 clients and suppliers have been assessed in accordance with PAS91 – Publicly Available Specification on Construction Pre-qualification – which was launched in October 2010.

Further information is available at: www.ssip.org.uk.

Recent changes in legislation

Changes in health and safety legislation now normally come into force on two dates each year – 6 April and 1 October.

Currently, there are no new major pieces of osh legislation in the pipeline. This may well change following the Lofstedt review referred to elsewhere in this introduction.

The only piece of new legislation that may have an impact on osh management systems is the *Equalities Act, 2010* which came into force on 1st October, 2010. This Act provides a new cross-cutting legislative framework and is aimed at updating, simplifying and strengthening previous equalities legislation and delivering a simple, modern and accessible framework of discrimination law to protect individuals from unfair treatment, thus promoting fairness and equality.

The structure of UK health and safety law

Health and safety in the workplace involves two different branches of the law – criminal law (dealt with below at PARA 20) and civil law (referred to in PARA 24).

Criminal law

Criminal law is the process by which society, through the courts, punishes organisations or individuals for breaches of its rules. These rules, known as 'statutory duties', are comprised in Acts passed by Parliament (eg *HSWA 1974*) or regulations which are made by government ministers using powers given to them by virtue of Acts (eg the *Manual Handling Operations Regulations 1992 (SI 1992 No 2793)* as amended by *SI 2002 No 2174*).

Cases involving breaches of criminal law may be brought before the courts by the enforcement authorities which, in the case of health and safety law, are HSE, the Office of Rail Regulation and local authorities via their environmental health departments. Magistrates' courts hear the vast majority of health and safety prosecutions although more serious cases can be heard by the Crown Courts. As in all criminal prosecutions the case must be proved 'beyond all reasonable doubt'. There is a right of appeal to the Court of Appeal (Criminal Division) and eventually to the House of Lords or even the European courts, although in practice very few health and safety cases go to appeal.

In Magistrates Courts the maximum fine is now £20,000 for most health and safety offences whilst unlimited fines can be imposed by the Crown Court. As a result of the *Health and Safety (Offences) Act 2008* imprisonment is now an option for almost all offences – up to six months in Magistrates Courts and two years in Crown Courts. Deaths involving work activities can result in manslaughter charges which could lead to more severe custodial sentences although such cases are relatively infrequent.

The Health and Safety at Work etc. Act 1974 (HSWA 1974)

HSWA 1974 is the most important Act of Parliament relating to health and safety. It applies to everyone 'at work' – employers, self-employed and employees (with the exception of domestic servants in private households). It also protects the general public who may be affected by work activities. Some of the key sections of the Act are listed below.

Section 2 – Duties of employers

HSWA 1974, s 2(1) is the catch-all provision: 'It shall be the duty of every employer to ensure, so far as is reasonably practicable, the health, safety and welfare at work of all his employees.' See below for further discussion of the term 'reasonably practicable'.

HSWA 1974, s 2(2) goes on to detail more specific requirements relating to:

- the provision and maintenance of plant and systems of work;
- the use, handling, storage and transport of articles and substances;

- the provision of information, instruction, training and supervision;
- places of work and means of access and egress; and
- the working environment, facilities and welfare arrangements.

These are also qualified by the term 'reasonably practicable'.

HSWA 1974, s 2(3) provides that an employer with five or more employees must prepare a written health and safety policy statement, together with the organisation and arrangements for carrying it out, and bring this to the notice of employees (see PARA 27).

Section 3 – Duties to others

HSWA 1974, s 3(1) provides: 'It shall be the duty of every employer to conduct his undertaking in such a way as to ensure, so far as is reasonably practicable, that persons not in his employment who may be affected thereby are not exposed to risks to their health or safety.'

Employers thus have duties to contractors (and their employees), visitors, customers, members of the emergency services, neighbours, passers-by and the public at large. This may extend to include trespassers, particularly if it is 'reasonably foreseeable' that they could be endangered, for example where high-risk workplaces are left unfenced.

Individuals who are self-employed are placed under a similar duty and must also take care of themselves. (If they have employees, they must comply with *section 2.*)

Section 4 – Duties relating to premises

Under *HSWA 1974, s 4* persons in total or partial control of work premises (and plant or substances within them) must take 'reasonable' measures to ensure the health and safety of those who are not their employees. These responsibilities might be held by landlords or managing agents etc, even if they have no presence on the premises.

Section 6 – Duties of manufacturers, suppliers etc.

Those who design, manufacture, import, supply, erect or install any article, plant, machinery, equipment or appliances for use at work, or who manufacture, import or supply any substance for use at work, have duties under *HSWA 1974, s 6*.

Section 7 – Duties of employees

It shall be the duty of every employee while at work:

(a) to take reasonable care for the health and safety of himself and of other persons who may be affected by his acts or omissions at work; and

(b) as regards any duty or requirement imposed on his employer or any other person by or under any of the relevant statutory provisions, to co-operate with him so far as is necessary to enable that duty or requirement to be complied with.

[*HSWA 1974, s 7*]

Consequently employees must not do, or fail to do, anything which could endanger themselves or others. It should be noted that managers and supervisors also hold these duties as employees.

Section 8 – Interference and misuse

No person shall intentionally or recklessly interfere with or misuse anything provided in the interests of health, safety or welfare in pursuance of any of the relevant statutory provisions.

[*HSWA 1974, s 8*]

Section 9 – Duty not to charge

No employer shall levy or permit to be levied on any employee of his any charge in respect of anything done or provided in pursuance of any specific requirement of the relevant statutory provisions.

[*HSWA 1974, s 9*]

Levels of duty

Health and safety law contains different levels of duty:

Absolute

Absolute requirements must be complied with whatever the practicalities of the situation or the economic burden.

Practicable

The term 'practicable' means that measures must be possible in the light of current knowledge and invention.

Reasonably practicable

This term is contained in the main sections of *HSWA 1974* and many important regulations. It requires the risk to be weighed against the costs necessary to avert it (including time and trouble as well as financial cost). If, compared with the costs involved, the risk is small then the precautions need not be taken – it should be noted that such a comparison should be made before any incident has occurred. The burden of proof, however, rests on the person with the duty (usually the employer) – they must prove why something was not reasonably practicable at a particular point in time. The duty holder's ability to meet the cost is not a factor to be taken into account.

In effect, considering what is 'reasonably practicable' requires that a risk assessment be carried out. The existence of a well-documented and carefully considered risk assessment would go a long way towards supporting a case on what was or was not reasonably practicable. Neither risks nor costs remain the same forever and what is practicable or reasonably practicable will change with time – hence the need to keep risk assessments up to date.

Important regulations

Carriage of Dangerous Goods and Use of Transportable Pressure Equipment Regulations 2009 (SI 2009 No 1348)

These regulations apply in relation to the carriage of dangerous goods by road and by rail and to the carriage of dangerous goods by inland waterway but only to the extent relating to the training and examination system for safety advisers and the connected issuing and renewal of vocational training certificates.

See the chapter entitled MANAGING WORK-RELATED ROAD SAFETY.

Construction (Design and Management) Regulations 2007 (CDM) (SI 2007 No 320)

These regulations replaced the 1994 CDM Regulations and the Construction (Health, Safety and Welfare) Regulations 1996, the latter dealing with many practical health and safety aspects of construction work. Duties are placed on clients, designers and contractors in respect of all construction projects. Larger 'notifiable' projects involve additional obligations on these duty-holders, including the appointment by the client of a 'CDM co-ordinator' and a 'principal contractor', whose duties are specified in the regulations. The definition of 'construction work' is extremely wide, bringing many minor repair and maintenance activities within the scope of the regulations.

See the chapters entitled CONSTRUCTION, DESIGN AND MANAGEMENT and CONSTRUCTION AND BUILDING OPERATIONS.

Control of Asbestos Regulations 2006 (SI 2006 No 2739)

The regulations contain many quite detailed requirements relating to work with asbestos-containing materials (ACMs). Some types of asbestos work may only be carried out by organisations holding an appropriate licence but even where this is not the case, strict precautions must be taken. Many employers will be affected by the duty to manage asbestos in non-domestic premises contained in regulation 4. An assessment must be carried out of the possible presence of ACMs in the premises and also their condition. Suitable measures must then be put in place to manage the asbestos risk. This may involve the removal of ACMs but is more likely to require the establishment of an ACM inspection program and arrangements to prevent ACMs being disturbed or damaged.

See the chapter dealing with ASBESTOS.

Control of Major Accident Hazards Regulations 1999 (SI 1999 No 743) as amended by SI 2005 No 1088

These regulations impose requirements with respect to the control of major accident hazards involving dangerous substances. (The expressions 'major accident' and 'dangerous substances' are defined in regulation 2(1).) The Regulations implement Council Directive 96/82/EC on the control of major accident hazards involving dangerous substances, except Article 12 (which relates to land use planning).

The Regulations apply to establishments (as defined in regulation 2(1)) where dangerous substances are present in quantities equal to or exceeding those specified in column 2 of Parts 2 or 3 of Schedule 1, except that regulations 7 to 14 apply to establishments where such substances are present in quantities equal to or exceeding those specified in column 3 of those Parts (regulation 3(1)). The Regulations do not apply in the cases specified in regulation 3(2).

See the chapter entitled MAJOR ACCIDENT HAZARDS.

Control of Noise at Work Regulations 2005 (SI 2005 No 1643)

These Regulations implemented Directive 2003/10/EC on the minimum health and safety requirements regarding the exposure of workers to the risks arising from physical agents. The Regulations impose duties on employers and on self-employed persons to protect both employees who may be exposed to risk from exposure to noise at work and other persons at work who might be affected by that work.

See the chapter NOISE AT WORK.

Control of Substances Hazardous to Health Regulations 2002 (COSHH) (SI 2002 No 2677) as amended by 2004 No 3386

These Regulations require an assessment to be made of all substances hazardous to health in order to identify means of preventing or controlling exposure. There are also requirements for the proper use and maintenance of control measures and for workplace monitoring and health surveillance in certain circumstances.

More detail is available in the chapter HAZARDOUS SUBSTANCES IN THE WORKPLACE.

Control of Vibration at Work Regulations 2005 (SI 2005 No 1093)

Employers are required to carry out assessments of risks to their employees from vibration, whether hand-arm vibration (HAV) or whole-body vibration. If exposure action values are exceeded employers must introduce technical and organisational measures to reduce exposure. The regulations set down exposure limit values which must not be exceeded.

See the chapter dealing with VIBRATION.

Dangerous Substances and Explosive Atmospheres Regulations 2002 (DSEAR) (SI 2002 No 2776)

These regulations apply to substances which have the potential to create risks from fire, explosion and exothermic reactions, including flammable gases and liquids and explosive dusts. Employers are required to carry out risk assessments of activities involving dangerous substances and to implement appropriate control measures.

See the chapter dealing with DANGEROUS SUBSTANCES AND EXPLOSIVE ATMOSPHERES.

Electricity at Work Regulations 1989 (SI 1989 No 635)

These Regulations contain requirements relating to the construction and maintenance of all electrical systems and work activities on or near such systems. They apply to all electrical equipment, from a battery-operated torch to a high-voltage transmission line.

See the chapter entitled ELECTRICITY.

Employers' Liability (Compulsory Insurance) (Amendment) Regulations 2008 (SI 2008 No 1765)

These Regulations amend the Employers' Liability (Compulsory Insurance) Regulations 1998 and are relevant to all employers in Great Britain holding employers' liability compulsory insurance.

See the chapter entitled EMPLOYERS' LIABILITY INSURANCE.

Food Labelling (Amendment) (England) Regulations (SI 2008 No 1188)

These Regulations amend the Food Labelling Regulations to permanently exempt certain food ingredients from labelling as, due to processing, they no longer contain the allergenic component.

See the chapter dealing with FOOD SAFETY.

Health and Safety (Consultation with Employees) Regulations 1996 (SI 1996 No 1513)

These Regulations extended the previous requirements (contained in the *Safety Representatives and Safety Committees Regulations 1977 (SI 1977 No 500)* so that employers must now also consult workers not covered by trade union safety representatives.

See the chapter dealing with JOINT CONSULTATION ON SAFETY.

Health and Safety (Display Screen Equipment) Regulations 1992 (SI 1992 No 2792) as amended by SI 2002 No 2174

Where there is significant use of display screen equipment (DSE), employers must assess DSE workstations and offer 'users' eye and eyesight tests (which may necessitate provision of spectacles for DSE work).

See the chapter entitled DISPLAY SCREEN EQUIPMENT.

Health and Safety (First-Aid) Regulations 1981 (SI 1981 No 917) as amended by SI 2002 No 2174

Basic first-aid equipment controlled by an 'appointed person' must be provided for all workplaces. Higher risk activities or larger numbers of employees may require additional equipment and fully trained first-aiders.

See the chapter entitled FIRST AID.

Health and Safety (Safety Signs and Signals) Regulations 1996 (SI 1996 No 341) as amended by SI 2002 No 2174

These Regulations require safety signs to be provided, where appropriate, for risks which cannot adequately be controlled by other means. Signs must be of the prescribed design and colours.

This subject is dealt with in some detail in the chapter on WORKPLACE HEALTH, SAFETY AND WELFARE.

Health and Safety (Training for Employment) Regulations 1990 (SI 1990 No 1380)

Those receiving 'relevant training' (through training for employment schemes or work experience programmes) are treated as being 'at work' for the purposes of health and safety law. The provider of the 'relevant training' is deemed to be their employer – youth trainees and students on work experience placements therefore have the status of employees and must be protected accordingly. Matters relating to children and young persons are covered in the chapters RISK ASSESSMENT and VULNERABLE PERSONS.

Lifting Operations and Lifting Equipment Regulations 1998 (LOLER) (SI 1998 No 2307)

Lifting equipment is defined in these regulations as 'work equipment for lifting and lowering loads'. As well as cranes, chain-blocks, passenger lifts and mobile elevating work platforms, this definition includes vehicle tail lifts, bath hoists and many other types of equipment. All lifting equipment must be of adequate strength and stability and meet other specific design and installation criteria. It must be subjected to periodic thorough examinations and inspections and all lifting operations must be properly planned, appropriately supervised and carried out in a safe manner.

See the chapter dealing with LIFTING OPERATIONS.

Management of Health and Safety at Work Regulations 1999 (SI 1999 No 3242) as amended by SI 2006 No 438

Employers and the self-employed are required to manage the health and safety aspects of their activities in a systematic and responsible way. The Regulations include requirements for risk assessment, the availability of competent health and safety advice and emergency procedures. Several of these management issues are dealt with later in this Introduction (see PARA 25), but the Regulations are covered in some detail in the chapter entitled MANAGING HEALTH AND SAFETY.

Manual Handling Operations Regulations 1992 (SI 1992 No 2793) as amended by SI 2002 No 2174

Manual handling operations involving risk of injury must either be avoided or be assessed by the employer with steps taken to reduce the risk, so far as is reasonably practicable.

See the chapter entitled MANUAL HANDLING.

Personal Protective Equipment at Work Regulations 1992 (SI 1992 No 2966) as amended by SI 2002 No 2174

Employers must assess the personal protective equipment (PPE) needs created by their work activities, provide the necessary PPE, and take reasonable steps to ensure its use.

See the chapter entitled PERSONAL PROTECTIVE EQUIPMENT.

Provision and Use of Work Equipment Regulations 1998 (PUWER 1998) (SI 1998 No 2306) as amended by SI 2002 No 2174

These Regulations cover equipment safety, including the guarding of machinery. The definition of 'work equipment' also includes hand tools, vehicles, laboratory apparatus, lifting equipment, access equipment etc. The 1998 Regulations introduced additional requirements in respect of mobile work equipment and also replaced previous specific regulations relating to power presses, woodworking machines and abrasive wheels.

See the chapter entitled MACHINERY SAFETY.

Regulatory Reform (Fire Safety) Order 2005 (SI 2005 No 1541)

This Order replaced a vast range of other legal requirements dealing with fire matters (including requirements for Fire Certificates). Enforcement of the Order is normally carried out by the local fire and rescue authority. An assessment must be made of the fire risks and the adequacy of existing fire precautions. The assessment must take into account fire prevention measures, fire detection and alarm equipment, fire escape routes and evacuation procedures, fire fighting equipment and fire-related signs.

See the chapter dealing with FIRE PREVENTION AND CONTROL.

Reporting of Injuries, Diseases and Dangerous Occurrences Regulations 1995 (RIDDOR) (SI 1995 No 3163)

Fatal accidents, major injuries (as defined) and dangerous occurrences (as defined) must be reported immediately to the enforcing authority. Accidents involving four or more days' absence must be reported in writing within seven days.

See the chapter on ACCIDENT REPORTING AND INVESTIGATION.

Safety Representatives and Safety Committees Regulations 1977 (SI 1977 No 500)

Members of recognised trade unions may appoint safety representatives to represent them formally in consultations with their employer in respect of health and safety issues. The functions and rights of safety representatives are detailed in the Regulations. The employer must establish a safety committee if at least two representatives request this in writing.

See the chapter entitled JOINT CONSULTATION IN SAFETY.

Work at Height Regulations 2005 (SI 2005 No 735)

These Regulations impose health and safety requirements with respect to work at height, with certain exceptions relating to recreational climbing and caving.

These Regulations give effect as respects to Directive 2001/45/EC, amending Council Directive 89/655/EEC concerning the minimum safety and health requirements for the use of work equipment by workers at work. They contain additional provisions, including provisions which replace regulations giving effect to certain provisions of Council Directives 89/391/EEC concerning the minimum safety and health requirements for the workplace and 92/57/EEC on the implementation of minimum health and safety requirements at temporary or mobile construction sites.

See the chapter on WORK AT HEIGHT.

Workplace (Health, Safety and Welfare) Regulations 1992 (SI 1992 No 3004) as amended by SI 2002 No 2174 and SI 2005 No 735

Physical working conditions in the workplace together with safe access for pedestrians and vehicles and welfare provisions are covered by these Regulations.

See the chapters entitled WORKPLACES – HEALTH, SAFETY AND WELFARE and ACCESS, TRAFFIC ROUTES AND VEHICLES.

Civil law

A civil action can be initiated by an employee who has suffered injury or damage to health caused by their work. This may be based upon the law of negligence, ie where the employer has been in breach of the duty of care which he owes to the employee. Being part of the common law, the law of negligence has evolved, and continues to evolve, by virtue of decisions in the courts – Parliament has had virtually no role to play in its development.

Civil actions may also be brought on the grounds of breach of statutory duty – it should be noted, however, that *HSWA 1974* and most of the provisions in the *Management of Health and Safety at Work Regulations 1999 (SI 1999 No 3242)* do not confer a right of civil action, although the statutory duties owed by employers to employees under *HSWA 1974* have their equivalent obligations at common law.

Duty of care

Every member of society is under a 'duty of care', ie to take reasonable care to avoid acts or omissions which they can reasonably foresee are likely to injure their neighbour (anyone who ought reasonably to have been kept in mind). What is 'reasonable' will depend upon the circumstances.

Employers owe a duty of care not only to employees but also to such people as contractors, visitors, customers, and people on neighbouring property. In the case of the duty of care owed by employers to employees, it includes the duty to provide:

*   safe premises;
*   a safe system of work;
*   safe plant, equipment and tools; and
*   safe fellow workers.

Occupiers of premises are under statutory duties comprised in the *Occupiers' Liability Acts* of 1957 and 1984 (see OCCUPIERS' LIABILITY), and those suffering injury because of a defect in a product may sue the producer or importer under the *Consumer Protection Act 1987* (see PRODUCT SAFETY).

Vicarious liability

Employers are liable to persons injured by the wrongful acts of their employees, if such acts are committed in the course of their employment. Thus if an employee's careless driving of a forklift truck injures another employee (or a contractor or customer), the employer is likely to be liable. There is no vicarious liability if the act is not committed in the course of employment – thus the employer is not likely to be held liable if one employee assaults another.

Civil procedure

Civil actions must commence within three years from the time of knowledge of the cause of action. In an action for negligence, this will be the date on which the plaintiff knew or should have known that there was a significant injury and that it was caused by the employer's negligence. The plaintiff must be prepared to prove his case in the courts, but in practice most cases are settled out of court following negotiations between the plaintiff's legal representatives and the employer's insurers or their representatives. The *Employers' Liability (Compulsory Insurance) Act 1969* requires employers to be insured against such actions (see EMPLOYERS' LIABILITY INSURANCE), although some public bodies, for example local authorities, are exempt from the provisions of this Act.

As the result of recommendations made by Lord Woolf in his '*Access to Justice*' report of 1996, the *Civil Procedure Rules 1998* introduced widespread changes to civil procedure on 26 April 1999, affecting the progress of civil claims from their commencement to their conclusion. The rules involve a 'pre-action protocol' and govern the conduct of litigation in a way that is intended to limit delay. In most cases a single expert, medical or non-medical, will be instructed rather than each party using separate experts. Even if the case goes to court, the expert's report will usually be in writing, with both parties able to ask written questions of the expert and to see the replies. The new arrangements include a fast track system for lower value personal injury claims.

Damages

Damages are assessed under a number of headings including:

- loss of earnings (prior to trial);
- damage to clothing, property etc;
- pain and suffering (before and after trial);
- future loss of earnings;
- disfigurement;
- medical or nursing expenses; and
- inability to pursue personal or social interests or activities.

Defences

The plaintiff must prove breach of a statutory duty or of the duty of care on a balance of probabilities. However, a number of defences are available to the employer, including:

Contributory negligence

- The employer may claim that the injured person was careless or reckless, for example, that he ignored clear safety rules or disobeyed instructions. Accidental errors are distinguished from a failure to take reasonable care. Damages will be reduced by the percentage of contributory negligence established, which will vary with the facts of each case.

Injuries not reasonably foreseeable
- The employer may claim that the injuries were beyond normal expectation or control (an act of God). In cases of noise-induced hearing damage, mesothelioma (an asbestos-related cancer) or vibration-induced white finger, the courts have established dates after which a reasonable employer should have been aware of the relevant risks and taken precautions.

Voluntary assumption of risk
- If an employee consents to take risks as part of the job, the employer may escape liability. However, this defence (*volenti non fit injuria*) cannot be used for cases involving breach of statutory duty – no one can contract out of their statutory obligations or be deprived of statutory protection.

Other civil actions

Other health and safety related situations may result in civil actions by employees. Employment protection legislation has recently been strengthened in relation to dismissals or redundancies resulting from health and safety activities (including refusal to work in situations of serious and imminent danger). Suspension or dismissal on maternity or medical grounds may also give a right of action. See EMPLOYMENT PROTECTION.

Management of health and safety at work

In 2009 the Institution of Occupational Safety and Health (IOSH) revised its guidance to health and safety management systems, which stated in its introduction that:

> IOSH recognises that work-related accidents and ill-health can be prevented and well-being at work can be improved if organisations manage health and safety competently and apply the same or better standards as they do to other core business activities. We believe that the formal OSHMSs mentioned in this guidance, and others based on similar principles, provide a useful approach to achieving these goals.
> ['*Systems in focus - guidance on occupational safety and health management systems*' (2009) IOSH, Wigston.]

The continuing development and adoption of formal occupational health and safety management systems (OHSMS) is a reflection of this conviction – such systems include those developed by HSE '*Successful health and safety management*' (1997), (HSG65), 2nd edition, by the ILO '*Guidelines on occupational health and safety management systems*', (ILO-OSH 2001), ILO Geneva, BSI '*Guide to occupational health and safety management systems*', (BS8800:2004), BSI London and BS18004: 2008 *Guide to achieving effective occupational health and safety performance* (BSI, London). Each system exhibits differences when compared to the others, but there is a similarity to the core elements of Plan-Do-Check-Act (see Figure 1).

This theme is developed considerably in the chapter entitled MANAGING HEALTH AND SAFETY.

Figure 1: Plan-Do-Check-Act

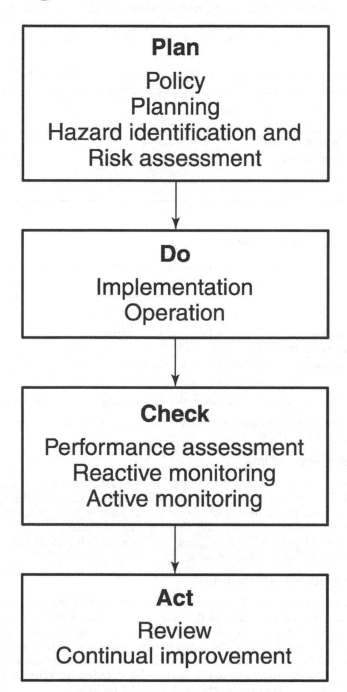

The costs of accidents and ill health

In 2009/2010 the HSE estimated that a total of 28.5 million days were lost in the UK as a result of both work-related ill-health (23.4 million days) and workplace injuries (5.1 million days). The updated cost to society (including employers and individuals of these lost days) was estimated to be between ,26 and ,42 billion.

Using the methods contained in the HSE Accident Cost Ready Reckoner, the specific updated average uninsured accident costs as at June 2011, are as follows:

Lost-time injury accident:	2,901
Non-lost-time injury accident:	47
Damage Accident:	439
Cost/Employee/Year:	439

Health and safety policies

'Successful health and safety management' stated that 'accidents are caused by the absence of adequate management control' and stressed the importance of effective health and safety policies in establishing such control. *Section 2(3)* of the *HSWA 1974* requires employers to prepare in writing:

- a statement of their general policy with respect to the health and safety at work of their employees; and
- the organisation and arrangements for carrying out the policy.

It also requires the statement to be brought to the notice of all employees – employers with fewer than five employees are exempt from this requirement.

Policies are normally divided into three sections, to meet the three separate demands of *HSWA 1974*:

(i) The statement of intent

This involves a general statement of good intent, usually linked to a commitment to comply with relevant legislation. Many employers extend their policies so as to relate also to the health and safety of others affected by their activities. In order to demonstrate clearly that there is commitment at a high level, the statement should preferably be signed by the chairman, chief executive or someone in a similar position of seniority.

(ii) Organisational responsibilities

It is vitally important that the responsibilities for putting the good intentions into practice are clearly identified. In a small organisation this may be relatively simple but larger employers should identify the responsibilities held by those at different levels in the management structure. Whilst reference to employees' responsibilities may be included, it should be emphasised that the law requires the employer's organisation to be detailed in writing. Types of responsibilities to be covered in the policy might include:

- making adequate resources available to implement the policy;
- setting health and safety objectives;
- developing suitable procedures and safe systems;
- delegating specific responsibilities to others;
- monitoring the effectiveness of others in carrying out their responsibilities;
- monitoring standards within the workplace; and
- feeding concerns up through the organisation.

(iii)   Arrangements

The policy need not contain all of the organisation's arrangements relating to health and safety but should contain information as to where they might be found, for example risk assessment records, in a separate health and safety manual or within various procedural documents. Topics which may require detailed arrangements to be specified are:

- operational procedures relating to health and safety;
- training;
- personal protective equipment;
- health and safety inspection programmes;
- accident and incident investigation arrangements;
- fire and other emergency procedures;
- first aid;
- occupational health;
- control of contractors and visitors;
- consultation with employees; and
- audits of health and safety arrangements.

Employees must be aware of the policy and, in particular, must understand the arrangements which affect them and what their own responsibilities might be. They may be given their own copy (for example, within an employee handbook) or the policy might be displayed around the workplace. With regard to some arrangements detailed briefings may be necessary, for example as part of induction training.

Employers must revise their policies as often 'as may be appropriate'. Larger employers are likely to need to arrange for formal review and, where necessary, for revision to take place on a regular basis (eg by way of an ISO 9000 procedure). Dating of the policy document is an important part of this process. Much more detail is provided in the chapter entitled STATEMENTS OF HEALTH AND SAFETY POLICY.

Sources of health and safety advice

Within '*Successful health and safety management*', HSE emphasised the importance of establishing a positive health and safety culture within an organisation as a prerequisite of effective health and safety management. It referred to the 'four Cs' as key components in establishing such a culture: control, competence, communication and co-operation.

While competence in health and safety matters is relevant throughout any workforce, it is particularly important at management levels. The *Management of Health and Safety at Work Regulations 1999 (SI 1999 No 3242)* have taken this concept further by requiring (in *Reg 7*) every employer to appoint one or more competent persons to assist him in complying with the law. The ACoP accompanying the Regulations states that the size and type of resource required will be relative to the size of the organisation and the risks present in its activities. Full-time or part-time specialists may be appointed, or use may be made of external consultants, although the 1999 Management Regulations state a preference for employees. Smaller employers may appoint themselves, provided that they are competent – the ACoP refers to competence as comprising both the possession of theoretical knowledge and the capacity to put it into practice in the work situation. An awareness of the limits of one's own knowledge and capabilities is also important.

Health and safety training is available from many different sources. The following organisations either provide training themselves or oversee training through accredited training centres.

- National Examination Board in Occupational Safety and Health (NEBOSH)
  tel: 0116 263 4700/www.nebosh.org.uk
- Institution of Occupational Safety and Health (IOSH)
  tel: 0116 257 3100/www.iosh.co.uk
- Chartered Institution of Environmental Health (CIEH)
  tel: 0207 928 6006/www.cieh.org.uk
- Royal Society for the Prevention of Accidents (RoSPA)
  tel: 0121 248 2000/www.rospa.co
- British Safety Council (BSC)
  tel: 0208 741 1231/www.britishsafetycouncil.co.uk

As a result of the Lord Young review, the Occupational Safety and Health Consultants Register (OSHCR) was launched at the end of January 2011. This register is live to search – at www.oshcr.org – following an initial sign up by more than 1,600 competent osh practitioners.

Consultants who have joined the freely available register are all Chartered Members or Fellows of IOSH, the CIEH or REHIS, Fellows of IIRSM, or Member of Fellow of BOHS or the IEHF.

In addition, all applicants have declared that they can demonstrate continuing professional development (CPD), have adequate professional indemnity insurance (PII) and will provide sensible and proportionate advice.

Consultants who have joined the register are searchable via industry, topic, location and keyword; many also list the specific services which they are competent to provide.

HSE itself can also be a valuable source of information and advice:

- HSE home page on the internet: www.hse.gov.uk
  An online enquiry service can be accessed from the home page.
- HSE Books
  tel: 01787 881165
  fax: 01787 313995
  website: www.hsebooks.co.uk

HSE's huge range of publications are freely available as pdf downloads from their website. Priced publications are also available from HSE Books, as are a number of free-issue leaflets.

The Essentials of Health and Safety at Work is an excellent starting point for SMEs.

Access, Traffic Routes and Vehicles

Nicola Coote

Introduction

[A1001] With the regular daily flow of labour to and from the workplace and vehicles making deliveries and collecting items, access and egress points constitute potentially hazardous situations. For this reason there is a duty on employers and factory occupiers to 'provide and maintain' safe access to and egress from a place of work both under statute and at common law. As part and parcel of compliance with the *Building Regulations 2010 (SI 2010/2214)*. This includes access facilities for disabled workers and visitors (that is, persons who have difficulty walking or are wheelchair users, or, alternatively, have a hearing problem or impaired vision) (see further **W11043** WORKPLACES – HEALTH, SAFETY AND WELFARE), as well as proper precautions to effect entry or exit to or from confined spaces, where a build-up of gas/combustible substances can be a real though not obvious danger (see **A1022** below). Statutory requirements consist of general duties under the *Health and Safety at Work etc. Act 1974 (HSWA)*, which apply to all employers, and the more specific duties of the *Workplace (Health, Safety and Welfare) Regulations 1992 (SI 1992 No 3004)*.

The *Disability Discrimination Act 1995 (DDA 1995)* also relates to access of buildings and services. The Act, most of which has been replaced by the *Equality Act 2010*, defines a disabled person as someone with:

> a physical or mental impairment which has a substantial and long-term adverse effect on his ability to carry out normal day-to-day activities.

Since 1 October 2004 duties under *DDA 1995* affect all employers with fewer than 15 employees and anyone who provides a service to the public

The provisions, including those that require employers to consider making changes to the physical features of premises that they occupy, have been in force since December 1996. One way in which an employer might unlawfully discriminate against a disabled employee is by not making reasonable adjustments (without justification) and this includes adjusting the working to enable access and safe traffic routes. A Code of Practice – 'Elimination of discrimination in the field of employment against disabled persons or persons who have had a disability' gives general guidance on the main employment provisions of the Act.

The term 'access' is a comprehensive one and refers to just about anything that can reasonably be regarded as means of entrance/exit to a workplace, even if it is not the usual method of access/egress. Unreasonable means of access/egress would not be included, such as a dangerous short-cut, particularly if management has drawn a worker's attention to the danger, though the fact that a worker is a trespasser has not prevented recovery of damages (*Westwood v The Post Office* [1973] 3 All ER 184). Access to a fork lift truck qualified, for the purposes of the *Factories Act 1961, s 29(1)* (repealed as from 1 January 1996), the fork lift truck being a 'place' (*Gunion v Roche Products Ltd*, (1994) Times, 4 November). The proper procedure is for the employer/factory occupier to designate points of access/egress for workers and see that they are safe, well-lit, maintained and (if necessary) manned and de-iced.

In the case of *Fildes v International Computers* (1984), unreported an employee slipped on a patch of ice and injured his back whilst walking from his car in the car park to the factory gates. It was an agreed means of access.

Moreover, the statutory duties apply to access/egress points to any place where any employees have to work, and not merely their normal workplace.

This chapter summarises key statutory and common law duties in connection with:

(a)   access and egress;
(b)   vehicular traffic routes for internal traffic and deliveries;
(c)   work vehicles and delivery vehicles – with particular emphasis on potentially hazardous activities involving such vehicles; and
(d)   work in confined spaces, of necessity involving access and egress points.

In addition, where access routes lead to routes above ground level requirements for safe work at heights may need to be implemented (see further W9001 – WORK AT HEIGHTS)

Statutory duties concerning workplace access and egress – Workplace (Health, Safety and Welfare) Regulations 1992 (SI 1992 No 3004)

[A1002] Statutory duties centre around:

(a)   the organisation of safe workplace transport systems;
(b)   the suitability of traffic routes for vehicles and pedestrians; and
(c)   the need to keep vehicles and pedestrians separate.

Organisation of safe workplace transport systems

[A1003] Every workplace must be so organised that pedestrians and vehicles can circulate in a safe manner. [*Reg 17(1)*].

Suitability of traffic routes

[A1004] Traffic routes in a workplace must be suitable for the persons or vehicles using them, sufficient in number, in suitable positions and of sufficient size. [*Reg 17(2)*].

More particularly:

(a)   pedestrians or vehicles must be able to use traffic routes without endangering those at work;

(b)   there must be sufficient separation of traffic routes from doors, gates and pedestrian traffic routes, in the case of vehicles;

(c)   where vehicles and pedestrians use the same traffic routes, there must be sufficient space between them; and

(d)   where necessary, all traffic routes must be suitably indicated.
 [Reg 17(3), (4)].

Compliance with these statutory duties involves provision of safe access for:

(i)   vehicles, with attention being paid to design and layout of road systems, loading bays and parking spaces for employees and visitors; and

(ii)   pedestrians, so as to avoid their coming into contact with vehicles.

(Traffic routes in existence before 1 January 1993 shall comply with regulation 17(2) and (3) only to the extent that it is reasonably practicable).

Traffic routes for vehicles

[A1005] There should be sufficient traffic routes to allow vehicles to circulate safely and without difficulty. As for internal traffic, lines marked on roads/access routes in and between buildings should clearly indicate where vehicles are to pass e.g. fork lift trucks. Obstructions, such as limited headroom, are acceptable if clearly indicated. Temporary obstacles should be brought to the attention of drivers by warning signs or hazard cones or, alternatively, access prevented or restricted. Both internal and delivery traffic should be subject to sensible speed limits (e.g. 10 mph), which should be clearly displayed. Speed ramps (sleeping policemen), preceded by a warning sign or mark, are necessary, save for fork lift trucks, on workplace approaches. The traffic route should be wide enough to allow vehicles to pass and repass oncoming or parked traffic, and it may be advisable to introduce one way systems or parking restrictions. It may also be necessary to make allowance for pedestrian routes, where these interact with the vehicle traffic route. Traffic signs on roads, for example speed limit signs, must conform with those on public roads, whether or not the road is subject to the *Road Traffic Regulation Act 1984.*

Checklist – safe traffic routes

[A1006] Safe traffic routes should:

(a)   provide the safest route possible between places where vehicles have to call or deliver;

(b)   be wide enough for the safe movement of the largest vehicle, including visiting vehicles (e.g. articulated lorries, ambulances etc.) and should allow vehicles to pass oncoming or parked vehicles safely. One way systems or parking restrictions are desirable;

(c)   avoid vulnerable areas/items, such as fuel or chemical tanks or pipes, open or unprotected edges, and structures likely to collapse;

(d)   incorporate safe areas for loading/unloading;

(e)   avoid sharp or blind bends; if this is not possible, hazards should be indicated (e.g. blind corner);

(f)   ensure that road/rail crossings are kept to a minimum and are clearly signed;

(g)   ensure that entrances/gateways are wide enough; if necessary, to accommodate a second vehicle that may have stopped, without causing obstruction;

(h)   set sensible speed limits, which are clearly signposted. Where necessary, ramps should be used to retard speed, and road humps or bollards to restrict the width of the road. These should be preceded by a warning sign or mark on the road;

(i)   ensure that fork lift trucks should not have to pass over road humps, unless of a type capable of doing so;

(j)   give prominent warning of limited headroom, both in advance and at an obstruction. Overhead electric cables or pipes containing flammable/hazardous chemicals should be shielded, i.e. using goal posts, height gauge posts or barriers;

(k)   ensure that routes on open manoeuvring areas/yards are marked and signposted, and banksmen are employed to supervise the safe movement of vehicles;

(l)   ensure that people at risk from exhaust fumes or material falling from vehicles are screened or protected; and

(m)   restrict vehicle access where high-risk substances are stored (e.g. LPG) and where refuelling takes place.

(n)   consider installation of refuge points (safe havens) where vehicles need to reverse into delivery areas or dead ends, or position barriers to prevent vehicles reversing from colliding into people.

(o)   ensure safety of users from falling if the access route is above ground level (see also section W9001 Work at Height and W9004 Application to Work at Height Regulations).

Traffic routes for pedestrians

Checklist – safe traffic routes

[A1007] In the case of pedestrians, the main object of the traffic route is to prevent their coming into contact with vehicles. Safe traffic routes should:

(a)   provide separate routes/pavements for pedestrians, to keep them away from vehicles;

(b)   where necessary, provide suitable barriers/guard rails at entrances/exits and at the corners of buildings;

(c)   where traffic routes are used by both pedestrians and vehicles, be wide enough to allow vehicles to pass pedestrians safely;

(d)   where pedestrian and vehicle routes cross, provide appropriate crossing points. These should be clearly marked and signposted. If necessary, barriers or rails should be provided to prevent pedestrians crossing at dangerous points and to direct them to designated crossing points;

(e)   where traffic volume is high, traffic lights, bridges or subways should be used to control movement and ensure a smooth, safe flow;

(f)   where crowds use or are likely to use roadways, e.g. at the end of a shift, stop vehicles from using them at such times;

(g)   provide separate vehicle and pedestrian doors in premises, with vision panels on all doors;

(h)   provide high visibility clothing for people permitted in delivery areas (e.g. bright jackets/overalls);

(i)   where the public has access (e.g. at a farm or factory shop), public access points should be as near as possible to shops and separate from work activities.

(See also **W11014** WORKPLACES.)

Vehicles

[A1008] Vehicles account for a high percentage of deaths and injuries at work. Every year, about 70 people are killed and 2000 seriously injured in accidents involving vehicles in and around the workplace. Many of these casualties occur whilst vehicles are reversing, though activities such as loading and unloading, sheeting and unsheeting, as well as cleaning, can similarly lead to injuries, especially where employees are struck by a falling load or a fall from a height on, say, a tanker or HGV. So, too, climbing and descending ladders on tankers during delivery and 'dipping' at petrol forecourts can be hazardous, access onto vehicles and egress being as important as design and construction. Tipping, too, has its dangers, with tipping vehicles, tipping trailers and tankers overturning in considerable numbers. Also, sheeting and unsheeting operations have led to sheeters slipping or losing their grip or falling whilst walking on top of loads or in consequence of ropes breaking; absence of, or inadequate, training being an additional factor in injuries involving work vehicles. This part of the chapter considers the general statutory requirements relating to vehicles at work, precautions in connection with potentially hazardous operations involving vehicles, as well as providing a checklist for vehicle safety. Specific construction and use requirements are not considered. When considering the risk of sheeting operations etc when workers are gaining access to the top of loads employers must also consider risks associated with Work at Heights (see section W9001).

General statutory requirements

[A1009] Both work and private vehicles come within the parameters of health and safety at work. Regarding work vehicles, employers have the direct responsibilities of provision and maintenance generally under *HSWA, s 2*, and, more specifically, under the *Provision and Use of Work Equipment Regulations 1998 (SI 1998 No 2306) (PUWER)*, vehicles qualifying as 'work equipment'. Regarding private vehicles, employers have much less control – at least, as far as design, construction and use are concerned – but should, nevertheless, endeavour to ensure regulated use via:

(a)   restricted routes and access;

(b)   provision of clearly signposted parking areas away from hazardous activities and operations; and

(c)   enforcement of speed limits.

Work vehicles

[A1010] Generally, work vehicles should be as safe, stable, efficient and roadworthy as private vehicles on public roads. As work equipment, they are subject to the controls of *PUWER*, which specifies provision, maintenance, access and safety provisions in the event of rolling or falling over, whilst employers must also ensure that drivers are suitably trained in conformity with the requirements of the *Management of Health and Safety at Work Regulations 1999 (SI 1999 No 3242)*.

Provision

[A1011] All employers must ensure that vehicles:

(a)   are constructed and adapted as to be suitable for its purpose;
(b)   when selected, caters for risks to the health and safety of persons where the vehicles are to be used; and
(c)   are only used for operations specified and under suitable conditions. *[PUWER, Reg 5]*.

Compliance with these requirements on the part of operators of HGVs, fork lift trucks, dump trucks and mobile cranes presupposes conformity with the following checklist, namely:

(i)   a high level of stability;
(ii)   safe means of access and egress to and from the cab;
(iii)   suitable and effective service and parking brakes;
(iv)   windscreens with wipers and external mirrors giving optimum all-round visibility;
(v)   a horn, vehicle lights, reflectors, reversing lights, reversing alarms;
(vi)   suitable painting/markings so as to be conspicuous;
(vii)   provision of a seat and seat belts;
(viii) guards on dangerous parts (e.g. power take-offs);
(ix)   driver protection to prevent injury from overturning, and from falling objects or materials; and
(x)   driver protection from adverse weather.

Maintenance

[A1012] All employers must ensure that work equipment (including vehicles) is maintained in an efficient state, in efficient working order and in good repair. *[PUWER, Reg 6]*. This combines the need for basic daily safety checks by the driver before using the vehicle, as well as preventive inspections and services carried out at regular intervals of time and/or mileage, in accordance with manufacturers recommendations. As regards basic daily safety checks, employers should provide drivers with a log book in which to record visual inspections undertaken and the findings of the following:

–   brakes, tyres, steering, mirrors, windscreen washers and wipers, warning signals and specific safety systems (e.g. control interlocks)

Employers should see that drivers carry out the checks.

Training of drivers

[A1013] All employers must:

(a)   in entrusting tasks to employees, take into account their capabilities as regards health and safety; and

(b)   ensure that employees are provided with adequate health and safety training on recruitment and exposure to new or increased risks.

[Management of Health and Safety at Work Regulations 1999 (SI 1999 No 3242)].

In order to conform with these requirements, employers should ensure that, for general purposes, drivers of work vehicles are over 17 and have passed their driving test or, in the case of drivers of HGVs, that they are over 21 and have passed the HGV test. Moreover, to ensure continued competence, or to accommodate new risks at work or a changing work environment, employers should provide safety updates on an on-going basis as well as refresher training, and require approved drivers to report any conviction for a driving offence, whether or not involving a company vehicle. One method for ensuring this is to require all drivers to submit their licence annually so that a copy can be held in their personnel file.

Contractors and subcontractors

[A1014] Similar assurances (see **A1013** above) should be obtained from drivers of contractors and subcontractors visiting an employer's workplace. If they are not forthcoming, permission to work on site or in-house should be refused until either the contractor's vehicles comply with statutory requirement and/or his drivers are adequately trained. Training of contractor's drivers would normally be undertaken by contractors themselves, though site or in-house hazards, routes to be used etc. should be communicated to contractors by employers or occupiers. Contractors should be left in no doubt of the penalties involved for failure to conform with safe working practices – a useful way of ensuring enforcement on the part of contractors and subcontractors is to issue a licence.

Access to vehicles

[A1015] In addition to *Regulation 5* of *PUWER* (see **A1011** above), employers (and others having control, to any extent, of workplaces (see OCCUPIERS' LIABILITY)), who operate/use vehicles, are subject to:

(a)   the fall prevention requirements of *Regulation 13* of the *Workplace (Health, Safety and Welfare) Regulations 1992 (SI 1992 No 3004)* (see **W9003** WORK AT HEIGHTS). As far as possible, compliance with this regulation would obviate the need for climbing on top of vehicles (by bottom-filling) and also require vehicle operators to ensure that loads are evenly distributed, packaged properly and secured in the interests of drivers going down slopes and up steep hills (see further **A1018** below).

(b)   the co-operation requirements of *Regulation 11* of the *Management of Health and Safety at Work Regulations 1999 (SI 1999 No 3242)*, specifying that where activities of different employers interact, different employers may need to co-operate with each other and co-ordinate preventive and protective measures. This regulation is particularly relevant for example, to tanker deliveries and 'dipping' at petrol forecourts as well as loading and/or unloading operations. For 'dipping' purposes or gaining top access to tankers, access should be by a ladder at the front or rear, such ladders being properly constructed, maintained and securely fixed; ideally, they should incline inwards towards the top. There should be a means of preventing people from falling whilst on top of the tanker. Failing this, employers of tanker drivers and forecourt owners should liaise on potential risks involved in tanker deliveries, e.g. the provision of suitable step-ladders on the part of the latter. As for carriage of goods and loading/unloading operations, consignors should ensure that goods are evenly distributed, properly packaged and secured.

Potentially hazardous operations

[A1016] The following activities and operations are potentially hazardous in connection with vehicles.

Reversing

[A1017] Approximately a quarter of all deaths at work are caused by reversing vehicles; in addition, negligent reversing can result in costly damage to premises, plant and goods. Where possible, workplace design should aspire to obviate the need for reversing by the incorporation of one-way traffic systems. Failing this, reversing areas should be clearly identified and marked, and non-essential personnel excluded from the area. Ideally, banksmen wearing high-visibility clothing should be in attendance to guide drivers through, and keep non-essential personnel and pedestrians away from, the reversing area. Refuge points, also known as safe havens, should be constructed where possible to enable an escape route or safe place for people to stay in the event of a vehicle reversing in an unsafe manner. Vehicles should be fitted with external side-mounted and rear-view mirrors – as, indeed, many now are. Closed-circuit television systems are also advisable for enabling drivers to see round 'blind spots' and corners.

Loading and unloading

[A1018] Because employees can be seriously injured by falling loads or overturning vehicles, loading/unloading should not be carried out:

(a)   near passing traffic, pedestrians and other employees;
(b)   where there is a possibility of contact with overhead electric cables;
(c)   on steep gradients;
(d)   unless the load is spread evenly (racking will assist load stability);
(e)   unless the vehicle has its brakes applied or is stabilised (similarly with trailers); or
(f)   with the driver in the cab.

(See also OFFICES AND SHOPS.)

Tipping

[A1019] Overturning of lorries and trailers is the main hazard associated with tipping. In order to minimise the potential for injuries, tipping operations should only occur:

(a)   after drivers have consulted with site operators and checked that loads are evenly distributed;
(b)   when non-essential personnel are not present;
(c)   on level and stable ground away from power lines and pipework; and
(d)   with the driver in the cab and the cab door closed.

Moreover, after discharge, drivers should ensure that the body of the vehicle is completely empty and should not drive the vehicle in an endeavour to free a stuck load.

Giving unauthorised lifts

[A1020] Giving unauthorised lifts in work vehicles is both a criminal offence and can lead to employers being involved in civil liability. Thus, 'every employer shall ensure that work equipment is used only for operations for which, and under conditions for which, it is suitable'. [*PUWER, Reg 5*].

Where a driver of a work vehicle gives employees and/or others unauthorised lifts, his employer could find himself prosecuted for breach of the above regulation, whilst the driver himself may be similarly prosecuted for breach of *HSWA, s 7*, as endangering co-employees and members of the public. In addition, although acting in an unauthorised manner and contrary to instructions, the employee may well involve his employer in vicarious liability for any subsequent injury to a co-employee and/or member of the public (see further *Rose v Plenty* [1976] 1 All ER 97 at **E11004** EMPLOYERS' DUTIES TO THEIR EMPLOYEES).

Where instructions to employees not to give unauthorised lifts are clearly displayed in a work vehicle, but an employee nevertheless gives an unauthorised lift and a co-employee or member of the public is injured or killed as a result of the employee's negligent driving, it can be argued that the employee has exceeded the scope of his employment and so the employer is absolved from liability. (In *Twine v Bean's Express Ltd* [1946] 1 All ER 202 a driver gave a lift to a third party who was killed in consequence of his negligent driving. There was a notice in the van prohibiting drivers from giving lifts. It was held that the employer was not liable, as the driver was acting outside the parameters of his employment when giving a lift. The injured passenger knew that the driver should not give lifts.)

Conversely, courts have taken the view that such conduct, on the part of drivers, does not circumscribe the scope of employment but rather constitutes performance of work in an unauthorised manner, so leaving the employer liable as the employee is doing what he is employed to do but doing it wrongly (see further *Rose v Plenty and Century Insurance Co v Northern Ireland Road Transport Board* [1942] AC 509 at **E11004** EMPLOYERS' DUTIES TO THEIR EMPLOYEES). Yet again, if the passenger, having seen the prohibition on unauthorised lifts in the vehicle, nevertheless accepted a lift and was injured, arguably, if an adult rather than a minor, he has agreed to run the risk of negligent injury and so will forfeit the right to compensation.

Common law duty of care

[A1021] At common law every employer owes all his employees a duty to provide and maintain safe means of access to and egress from places of work. Moreover, this duty extends to the workforce of another employer/contractor who happens to be working temporarily on the premises. Hence the common law duty covers all workplaces, out of doors as well as indoors, above ground or below, and extends to factories, mines, schools, universities, aircraft, ships, buses and even fire engines and appliances (*Cox v Angus* [1981] ICR 683, where a fireman injured in a cab was entitled to damages at common law).

Confined spaces

[A1022] Accidents and fatalities such as drowning, poisoning by fumes or gassing, have happened as a result of working in confined spaces. See, for instance, the case of *Baker v T E Hopkins & Son Ltd* [1959] 3 All ER 225 where a doctor was overcome by carbon monoxide fumes while going to rescue two workmen down a well – the defence of *volenti non fit injuria* failed). Normal safe practice is a formalised permit to work system or checklist tailored to a particular task and requiring appropriate and sufficient personal protective equipment. Hazards typical of this sort of operation are:

(a)   atmospheric hazards – oxygen deficiency, enrichment (see *R v Swan Hunter Shipbuilders Ltd* [1982] 1 All ER 264 at **C8060** CONSTRUCTION AND BUILDING OPERATIONS), toxic gases (e.g. carbon monoxide), explosive atmospheres (e.g. methane in sewers);

(b)   physical hazards – low entry headroom or low working headroom, protruding pipes, wet surfaces underfoot as well as any electrical or mechanical hazards;

(c)   chemical hazards – concentration of toxic gas can quickly build up, where there is a combination of chemical cleaning substances and restricted air flow or movement.

In order to combat this variety of hazards peculiar to work in confined spaces, use of both gas detection equipment and suitable personal protective equipment are a prerequisite, since entry/exit paths are necessarily restricted.

Prior to entry, gas checks should test for (*a*) oxygen deficiency/enrichment, then (*b*) combustible gas and (*c*) toxic gas, by detection equipment being lowered into the space. This will determine the nature of personal protective equipment necessary. If gas is present in any quantity, the offending space should then be either naturally or mechanically ventilated. Where gas is present, entry should only take place in emergencies, subject to the correct respiratory protective equipment being worn. Assuming gas checks establish that there is no gaseous atmosphere, entry can then be made without use of respiratory equipment. Gas detection equipment should continue to be used whilst people are in the confined space so that any atmospheric change can subsequently be registered on the gas detection equipment. It is essential that an emergency plan is devised when gaining access to confined spaces, which includes effective two way communication between people inside and immediately outside the confined space, contact with emergency services and first aid, and fire prevention personnel on hand.

Statutory requirements

[A1023] It should be noted that the *Confined Spaces Regulations 1997 (SI 1997 No 1713)* came into force on 28 January 1998. These repeal *Factories Act 1961, s 30*, and impose requirements and prohibitions with respect to the health and safety of persons carrying out work in confined spaces.

A 'confined space' is defined in *reg 1(2)* as 'any place, including any chamber, tank, vat, silo, pit, trench, pipe, sewer, flue, well or other similar space in which, by virtue of its enclosed nature, there arises a reasonably foreseeable specified risk'.

A 'specified risk' means a risk of:

(a) serious injury to any person at work arising from a fire or explosion;
(b) without prejudice to paragraph *(a)* –
  (i) the loss of consciousness of any person at work arising from an increase in body temperature;
  (ii) the loss of consciousness or asphyxiation of any person at work arising from gas, fume, vapour or the lack of oxygen;
(c) the drowning of any person at work arising from an increase in the level of a liquid; or
(d) the asphyxiation of any person at work arising from a free flowing solid or the inability to reach a respirable environment due to entrapment by a free flowing solid.

Regulation 4 prohibits a person from entering a confined space to carry out work for any purpose where it is reasonably practicable to carry out the work by other means.

If, however, a person is required to work in a confined space, a risk assessment must be undertaken to comply with the requirements of the *Management of Health and Safety at Work Regulations 1999 (SI 1999 No 3242), Reg 3*. The risk assessment must be undertaken by a competent person and the outcome of the risk assessment process will then provide the basis for the development of a safe system of work (*Confined Spaces Regulations 1997, Reg 3*).

The risk assessment process should make use of all available information such as engineering drawings, working plans, soil or geological information and take into consideration factors such as the general condition of the confined space, work to be undertaken in the space to minimise hazards produced in the area, need for isolation of the space and the requirements for emergency rescue. In particular, information should be collected and assessed on the previous contents of the confined space, residues that still may be present, contamination that may arise from adjacent plant, processes, gas mains, surrounding soil, land or strata; oxygen level and physical dimensions of the space that may limit safe access and/or egress. The work to be undertaken should be assessed to determine if additional risks will be produced as a result of this work and systems developed to control these risks. All information collected should be recorded and a safe system of work developed for safe entry.

The main elements to consider when designing a safe system of work include the following:

(a)   supervision,

(b)   competence levels for personnel working in confined spaces,

(c)   communications,

(d)   testing/monitoring the atmosphere,

(e)   gas purging,

(f)   ventilation,

(g)   removal of residues,

(h)   isolation from gases, liquids and other flowing materials,

(i)   isolation from mechanical and electrical equipment,

(j)   selection and use of suitable equipment,

(k)   personal protective equipment (PPE) and respiratory protective equipment (RPE),

(l)   portable gas cylinders and internal combustion engines,

(m)   gas supplied by pipes and hoses,

(n)   access and egress,

(o)   fire prevention,

(p)   lighting,

(q)   static electricity,

(r)   smoking,

(s)   emergencies and rescue,

(t)   limited working time.

[Confined Spaces Regulations 1997, Reg 4].

Regulation 6 provides for circumstances allowing the Health and Safety Executive to grant exemption certificates.

Accident Reporting and Investigation

Mark Tyler

Introduction

[A3001] Employers and other 'responsible persons' (see **A3003** below) who have control over employees and work premises are required to notify and report to the relevant enforcing authority (see **A3004** below) the following specified events occurring at work:

(a)   accidents causing injuries, fatal and non-fatal, including:
        (i)   acts of non-consensual physical violence committed at work, and
        (ii)   acts of suicide occurring on, or in the course of, the operation of a railway, tramway, trolley or guided transport system
  [*RIDDOR (SI 1995 No 3163), Reg 2(1)*];

(b)   occupational diseases; and
(c)   dangerous occurrences, even where no injury results.

The duty to report applies not only in the case of incidents involving employees, but also to visitors, customers and members of the public killed or injured by work activities [*Reporting of Injuries, Diseases and Dangerous Occurrences Regulations 1995 (RIDDOR) (SI 1995 No 3163), Reg 3(1)*].

Employees also have certain obligations to report accidents to their employers.

It is currently estimated by the HSE that over half of non-fatal injuries to employees (approximately 70,000) are not duly reported, with even higher under-reporting by the self-employed. Nevertheless enforcement over violations has historically been at low levels as well. In the period 2004-2008 only around 40 defendants were convicted of *RIDDOR* offences in proceedings brought by the HSE.

RIDDOR data is often used by organisations and the authorities for monitoring and benchmarking purposes. The HSE annually measures numbers of fatal, major over-three-day injuries (see **A3007–A3008**) per 100,000 employees as well as the actual numbers of such accidents and other reportable matters in order to understand trends nationally. It also analyses the patterns of specific types of injuries, accidents or other occurrences.

As well as providing an invaluable database the *RIDDOR* system also alerts inspectors to events which they may need to investigate for other purposes, in particular to ensure that workplaces have been made safe or to make enquiries with a view to possible enforcement action if the incident is deemed significantly serious. By no means do most matters reported under *RIDDOR* result in any formal action, and often the report is simply logged.

RIDDOR data consists of only very basic information about injuries and other occurrences. It does not comprise a sufficient analysis for the purposes of the responsible employer or other organisation properly investigating accidents. Later in the chapter the rationale for a fuller investigation is explained, and the processes involved are outlined.

Reporting of Injuries, Diseases and Dangerous Occurrences Regulations 1995 (RIDDOR) – (SI 1995 No 3163)

[A3002] The *Reporting of Injuries, Diseases and Dangerous Occurrences Regulations 1995* cover:

(a)   reportable work injuries (see **A3007** below);

(b)   reportable occupational diseases (see Appendix B at **A3036** below) – now 47 in all;

(c)   reportable dangerous occurrences (see **A3009** and Appendix A at **A3035** below) – now 83 in all;

(d)   road accidents involving work (see **A3019** below); and

(e)   gas incidents (see **A3020** below).

Records must be kept by employers and other 'responsible persons' (see **A3005**) of such injuries, diseases and dangerous occurrences for a minimum of three years from the date they were made. In addition, employers must also keep an Accident Book (Form BI510).

Two reporting routes now exist:

•   Injuries and dangerous occurrences can be reported to the relevant enforcing authority on paper – Form F2508 and diseases on Form F2508A (see **A3012** and Appendices C and D at **A3037, A3038** below); or

•   Reports can be made by telephone or electronically to the national Incident Contact Centre (ICC) (see **A3004**).

As for notification of major accident hazards (see **M1112** CONTROL OF MAJOR ACCIDENT HAZARDS), notification and reporting under RIDDOR is sufficient for the purposes of the *Control of Major Accident Hazards Regulations 1999 (SI 1999 No 743), Reg 15(4)*.

Relevant enforcing authority

[A3003] Health and safety law is separately enforced either by the Health and Safety Executive (HSE), or by local authorities, through their environmental health departments, depending on the nature of the business activity in question. Under *RIDDOR*, notifications and reports can be directed to the authority responsible for the premises where the reportable event occurs (or in connection with which the work causing the event is being carried out). Details of HSE offices are listed in Appendix E to this chapter. However, the national Incident Reporting Centre provides a convenient alternative 'one-stop shop' for compliance with most *RIDDOR* obligations.

Incident Reporting Centre

[A3004] Employers and others who are subject to the reporting requirements, can comply with their obligations by contacting the Incident Contact Centre (ICC). This will eliminate the need to identify the local agency or other local enforcing authority. The ICC will then pass the report on to them. Reporting to the local HSE office or local enforcement authority by phone and on the current statutory forms is still an option and this information will be forwarded to the ICC.

Incidents may be reported to the ICC through the following channels:

*   By telephone: Telephone 0845 300 9923 (Monday to Friday, 8.30 a.m. to 5.00 p.m.)
*   By email: riddor@natbrit.com (requires electronic forms downloaded from *RIDDOR* website or scanned hard copy)
*   Online: www.riddor.gov.uk
*   By post: Incident Contact Centre, Caerphilly Business Park, Caerphilly CF83 3GG

In taking advantage of these arrangements, the employer or other reporting person will not have a formal record of a statutory RIDDOR form, so the ICC will send out confirmation copies of reports which should be checked for accuracy and retained.

For fatal accidents and other major incidents which require a more prompt response, the HSE has an out of hours duty officer to contact (0151 922 9235). See www.hse.gov.uk/contact/outofhouse.htm

Persons responsible for notification and reporting

[A3005] The person generally responsible for reporting injury-causing accidents, deaths or diseases is the employer. Failing that, the person having control of the work premises or activity will be the responsible person. [*RIDDOR (SI 1995 No 3163), Reg 2(1)*]. In certain cases these normal rules are displaced and there are specifically designated 'responsible persons', e.g. in the case of mines, quarries, offshore installations, vehicles, diving operations and pipelines (see Table 1 below).

Table 1

Persons generally responsible for reporting accidents

Death, major injury, over-3-day injury or specified occupational disease:	of an employee at work	that person's employer
	of a person receiving training for employment	the person whose undertaking makes immediate provision of the training
	of a self-employed person at work in premises under the control of someone else	the person for the time being having control of the premises in connection with the carrying on by him of any trade, business or undertaking
Specified major injury or condition, or over-3-day injury:	of a self-employed person at work in premises under his control	
Death, or specified major injury or condition:	of a person who is not himself at work (but is affected by the work of someone else), e.g. a member of the public, a shop customer, a resident of a nursing home	the person for the time being having control of the premises in connection with the carrying on by him of any trade, business or undertaking at which, or in connection with the work at which, the accident causing the injury happened

[Reg 2(1)(b), (c)].

Persons responsible for reporting accidents in specific locations

A mine	the mine manager
A quarry	the quarry owner
A closed tip	the owner of the mine or quarry with which the tip is associated
An offshore installation (except in the case of reportable diseases)	the duty holder (see O7029 OFFSHORE OPERATIONS)
A dangerous occurrence at a pipeline	the owner of the pipeline
A dangerous occurrence at a well	the appointed person or, failing that, the concession owner
A diving operation (except in the case of reportable diseases)	the diving contractor
A vehicle	the vehicle operator

[*Reg 2(1)(a)*].

In situations where the responsible person is difficult to identify because of the shared control of a site or operations, arrangements should be made to determine who will deal with *RIDDOR* reporting in line with the duty to co-operate and co-ordinate under the *Management of Health and Safety at Work Regulations 1999 (SI 1999 No 3242), Reg 11.*

What is covered?

[A3006] Covered by these Regulations are events involving:

(a)   employees;
(b)   self-employed persons;
(c)   trainees;
(d)   any person, not an employee or trainee, on premises under the control of another, or who was otherwise involved in an accident (e.g. a visitor, customer, passenger or bystander) (see **A3004** above).

Fatalities, major injuries or conditions

[A3007] Where any person dies or suffers a major injury as a result of, or in connection with, work, such an incident must be notified immediately and details formally reported.

As well as fatal accidents, reportable major injuries and conditions are as follows:

(a)   any fracture (other than to fingers, thumb or toes);
(b)   any amputation;
(c)   dislocation of shoulder, hip, knee or spine;
(d)   loss of sight (whether temporary or permanent);
(e)   a chemical or hot metal burn to the eye or any penetrating eye injury;
(f)   any injury resulting from an electrical shock or electrical burn (including one caused by arcing or arcing products) leading to unconsciousness or requiring resuscitation or admittance to hospital for more than 24 hours;
(g)   any other injury
    (i)   leading to hypothermia, heat-induced illness or unconsciousness,
    (ii)   requiring resuscitation, or
    (iii)   requiring admittance to hospital for more than 24 hours;
(h)   loss of consciousness caused by asphyxia or by exposure to a harmful substance or biological agent;
(i)   either
    (i)   acute illness requiring medical treatment, or
    (ii)   loss of consciousness
resulting from the absorption of any substance by inhalation, ingestion or through the skin;

(j)   acute illness requiring medical treatment where there is reason to
        believe that this resulted from exposure to a biological agent or its
        toxins or infected material.
[*RIDDOR (SI 1995 No 3163), Reg 2(1), Sch 1*].

Injuries incapacitating for more than three consecutive days

[A3008] Where a person at work is incapacitated for more than three
consecutive days from their normal contractual work (excluding the day of the
accident but including any days which would not have been working days)
owing to an injury resulting from an accident at work (other than an injury
reportable as a major injury listed in A3007 above), a report of the accident
must be made direct to the ICC or sent in writing on Form F2508 to the
enforcing authority as soon as is practicable and in any event within ten days
of the accident. [*RIDDOR (SI 1995 No 3163), Reg 3(2)*].

Some problem areas

[A3008.1] Deaths and injuries are reportable when they result from an
accident 'arising out of or in connection with work'. This phrase is only partly
explained by the Regulations as including reference to accidents or dangerous
occurrences 'attributable to the manner of conducting an undertaking, the
plant or substances used for the purposes of an undertaking and the condition
of the premises so used or any part of them' (*RIDDOR, Reg 2(2)(c)*).

The person who dies or suffers injury need not be at work; it is enough if the
death or injury arose from a work activity. For example, reporting requirements would apply to a shopper who fell and was injured on an escalator, so
long as the injury was connected with the escalator (*Woking Borough Council
v British Home Stores* (1994) 93 LGR 396); a member of the public overcome
by fumes on a visit to a factory and who lost consciousness; a patient in a
nursing home who fell over an electrical cable lying across the floor and was
injured; or a pupil or student killed or injured in the course of his curricular
work which was supervised by a lecturer or teacher.

Particular problems can arise with the criteria for reporting acts of violence.
Generally physical harm done to a member of staff would be reportable, but
not violence done to a member of the public if by a staff member.

Advice is available for these and other borderline situations in the HSE guide
to *RIDDOR* (see A3036).

Specified dangerous occurrences

[A3009] 83 types of reportable dangerous occurrences are specified in
RIDDOR (SI 1995 No 3163), Sch 2, ranging from general dangerous

occurrences (e.g. the collapse of a building or structure, the explosion of a pressure vessel, and accidental releases of significant quantities of dangerous substances) to more specific dangerous occurrences in mines, quarries, transport systems and offshore installations (see Appendix A at **A3035** below).

Specified diseases

[A3010] Where a worker suffers from an occupational disease related to a particular activity or process (as specified in *Schedule 3*), a report must be sent to the enforcing authority. [*RIDDOR (SI 1995 No 3163), Reg 5(1)*].

This duty arises only when an employer has received information in writing from a registered medical practitioner diagnosing one of the reportable diseases. (In the case of the self-employed, the duty is triggered regardless of whether or not the information is given to him in writing.) Many of these diseases are those for which disablement benefit is ordinarily prescribed. (In respect of offshore workers, such diseases tend to be communicable.)

Moreover, the *Industrial Diseases (Notification) Act 1981* and the *Registration of Births and Deaths Regulations 1987 (SI 1987 No 2088), Sch 2*, Form 14, require that particulars are to be included on the death certificate as to whether death might have been due to, or contributed to by, the deceased's employment. Such particulars are to be supplied by the doctor who attended the deceased during the last illness. (The full list of specified diseases is contained in Appendix B at **A3036** below.)

Duty to notify/report

Duty to notify

[A3011] 'Responsible persons' must notify enforcing authorities (see **A3003** and **A3004** above) by the quickest means practicable (normally by telephone) of the following:

(a)   death as a result of an accident arising out of, or in connection with, work;

(b)   'major injury' (see **A3007** above) of a person at work as a result of an accident arising out of, or in connection with, work;

(c)   injury suffered by a person not at work (e.g. a visitor, customer, client, passenger or bystander) as a result of an accident arising out of, or in connection with, work, where that person is taken from the accident site to a hospital for treatment;

(d)   major injury suffered by a person not at work, as a result of an accident arising out of, or in connection with, work at a hospital;

(e)   a dangerous occurrence (see **A3009** above)
   [*RIDDOR (SI 1995 No 3163), Reg 3(1)*];

(f)   road injuries or deaths [*RIDDOR (SI 1995 No 3163), Reg 10(2)*] (see **A3019** below); and

(g)   gas incidents [*RIDDOR (SI 1995 No 3163), Reg 6(1)*] (see **A3020** below).

Duty to report

[A3012] Responsible persons must also formally report events causing death or major injury, dangerous occurrences and details associated with workers' occupational diseases. In cases of death, major injury and accidents leading to hospitalisation, the duty to report extends to visitors, bystanders and other non-employees. More particularly, within ten days of the incident, responsible persons must send a written report form to the relevant enforcing authority in relation to:

(a)   all events which require notification (listed in (*a*) to (*g*) at **A3011** above);

(b)   the death of an employee if it occurs within a year following a reportable injury (whether or not reported under (*a*)) [*RIDDOR (SI 1995 No 3163), Reg 4*] (see **A3018** below);

(c)   incapacitation for work of a person at work for more than three consecutive days as a result of an injury caused by an accident at work [*RIDDOR (SI 1995 No 3163), Reg 3(2)*] (see **A3008** above);

(d)   specified occupational diseases relating to persons at work (see **A3009** above and Appendix B Part I at **A3036** below), and also specifically those suffered by workers on offshore installations (see Appendix B Part II at **A3036** below) [*RIDDOR (SI 1995 No 3163), Reg 5(1)*], provided that, in both cases,

    (i)   the responsible person has received a written statement by a doctor diagnosing the specified disease, in the case of an employee, or

    (ii)   a self-employed person has been informed by a doctor that he is suffering from a specified disease.

[*RIDDOR (SI 1995 No 3163), Reg 5(2)*].

Exceptions to notification and reporting

[A3013] There is no requirement to notify or report the injury or death of:

(a)   a patient undergoing treatment in a hospital or a doctor's or dentist's surgery [*RIDDOR (SI 1995 No 3163), Reg 10(1)*]; or

(b)   a member of the armed forces of the Crown [*RIDDOR (SI 1995 No 3163), Reg 10(3)*].

Duty to keep records

[A3014] Records of injury-causing accidents, dangerous occurrences and specified diseases must be kept by responsible persons for at least three years. [*RIDDOR (SI 1995 No 3163), Reg 7(3)*].

Injuries and dangerous occurrences

[A3015] In the case of injuries and dangerous occurrences, such records must contain:

(a)   date and time of the accident or dangerous occurrence;
(b)   if an accident is suffered by a person at work –
    (i)   full name,
    (ii)   occupation, and
    (iii)   nature of the injury;
(c)   in the event of an accident suffered by a person not at work –
    (i)   full name,
    (ii)   status (e.g. passenger, customer, visitor or bystander), and
    (iii)   nature of injury;
(d)   place where the accident or dangerous occurrence happened;
(e)   a brief description of the circumstances;
(f)   the date that the event was first reported to the enforcing authority;
(g)   the method by which the event was reported.
   [*RIDDOR (SI 1995 No 3163), Sch 4, Part I*].

Disease

[A3016] In the case of specified diseases, such records must contain:

(a)   date of diagnosis;
(b)   name of the person affected;
(c)   occupation of the person affected;
(d)   name or nature of the disease;
(e)   the date on which the disease was first reported to the enforcing authority; and
(f)   the method by which disease was reported.
   [*RIDDOR (SI 1995 No 3163), Sch 4, Part II*].

Action to be taken by employers and others when accidents occur at work

[A3017]

Figure 1

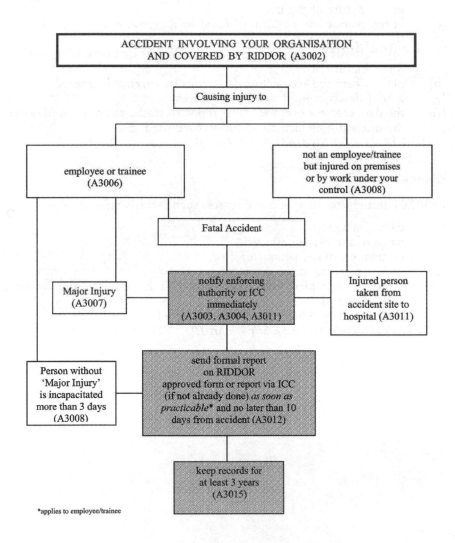

ACCIDENT INVOLVING YOUR ORGANISATION AND COVERED BY RIDDOR (A3002)

Causing injury to

employee or trainee (A3006)

not an employee/trainee but injured on premises or by work under your control (A3008)

Fatal Accident

Major Injury (A3007)

notify enforcing authority or ICC immediately (A3003, A3004, A3011)

Injured person taken from accident site to hospital (A3011)

Person without 'Major Injury' is incapacitated more than 3 days (A3008)

send formal report on RIDDOR approved form or report via ICC (if not already done) *as soon as practicable** and no later than 10 days from accident (A3012)

keep records for at least 3 years (A3015)

*applies to employee/trainee

Reporting the death of an employee

[A3018] Where an employee, as a result of an accident at work, has suffered a reportable injury/condition which is the cause of his death within one year of the date of the accident, the employer must inform the enforcing authority in writing of the death as soon as it comes to his attention, whether or not the accident has been reported under *Regulation 3*. [*RIDDOR (SI 1995 No 3163), Reg 4*].

Road accidents

[A3019] Road accident deaths and injuries are notifiable and reportable if the death or major injury is caused by or connected with:

(a)   exposure to any substance conveyed by road;
(b)   loading or unloading vehicles;
(c)   construction, demolition, alteration or maintenance activities on public roads; or
(d)   an accident involving a train.
   [*RIDDOR (SI 1995 No 3163), Reg 10(2)*].

These provisions should be treated as having potentially wide scope. For example, where a refuse vehicle was involved in a fatal accident when it was driving to empty a bin (ie not yet in the act of unloading or loading) this was reportable (*R (on the application of Aineto) v Brighton and Hove District Coroner* [2003] EWHC 1896 (Admin), [2003] All ER (D) 353 (Jul)).

The person injured, whether fatally or not, may or may not be engaged in the above-mentioned activities. Thus, an employee struck by a passing vehicle or a motorist injured by falling scaffolding, are covered. In addition, certain dangerous occurrences on public highways and private roads are covered (see Appendix A at A3035 below).

Duty to report gas incidents

Gas suppliers

[A3020] Where:

(a)   a conveyor of flammable gas through a fixed pipe distribution system, or
(b)   a filler, importer or supplier (not by way of the retail trade) of a refillable container containing liquefied petroleum gas

receives notification of any death or major injury which has arisen out of, or in connection with, the gas supplied, filled or imported, it must immediately notify the HSE and then send a report of the incident on the approved form within 14 days. [*RIDDOR (SI 1995 No 3163), Reg 6(1)*].

Gas fitters

[A3021] Where an employer or self-employed person who is an approved gas fitter has information that a gas fitting, flue or ventilation is, or is likely to cause, death or major injury, he is required to report that fact to the HSE on a prescribed form within 14 days. [*RIDDOR (SI 1995 No 3163), Reg 6(2)*].

Penalties

[A3022] Contravention of any of the provisions of *RIDDOR* is an offence. The maximum penalty on summary conviction is a fine of £20,000 and/or up to 12 months' imprisonment. Conviction on indictment in the Crown Court carries an unlimited fine and/or up to two years' imprisonment.

Defence in proceedings for breach of Regulations

[A3023] It is a defence under the Regulations for a person to prove that:

(a)   he was not aware of the event requiring him to notify or send a report to the enforcing authority; and

(b)   he had taken all reasonable steps to have all such events brought to his notice.

[*RIDDOR (SI 1995 No 3163), Reg 11*].

Obligations on employees

[A3024] All employees have duties to inform their employers of serious and immediate dangers and recognisable short comings in safety arrangements.

[A3025] There are separate reporting requirements on employees by the *Social Security (Claims and Payments) Regulations 1979 (SI 1979 No 628), Reg 24*. Although not actually bound in with the separate scheme of health and safety legislation, these reporting requirements complement those which employers have under *RIDDOR*, and there is an overlap in the record keeping requirements for both sets of provisions – see further below.

An accident to which the employees' reporting requirements apply is one 'in respect of which benefit may be payable . . . '. [*Social Security (Claims and Payments) Regulations 1979 (SI 1979 No 628), Reg 24(1)*]. Under *Regulation 24*, every 'employed earner' who suffers personal injury by an accident for the purposes of the Regulations is required by to give either *oral or written* notice to the employer in one of number of different ways. These ways include notice given to any supervisor or direct to a person designated by the employer. The method that has probably become the norm through custom and practice is by way of an entry of the appropriate particulars in a 'book kept specifically for these purposes' under the Regulations. These books are available from HSE Books as form BI 510. As well as containing sections for completing details of accidents etc. form BI 150 includes pages of instructions to employees and employers about their responsibilities. It is important only to use the latest (2003) version which has been updated so that it is compliant with the *Data Protection Act 1988*.

The entry in an Accident Book is to be made as soon as practicable after the happening of an accident by the employed earner or by some other person acting on his behalf. [*Social Security (Claims and Payments) Regulations 1979 (SI 1979 No 628), Reg 24(3)*]. Failure to so is an offence punishable by a fine. [*Social Security (Claims and Payments) Regulations 1979 (SI 1979 No 628), Reg 31*].

Under *Reg 25* of the 1979 Regulations, once an accident is reported by the employee, the employer must take reasonable steps to investigate the circumstances and, if there appear to be any discrepancies between the circumstances reported and the findings in these investigations, these should be recorded.

Where the accident in question is one to which *RIDDOR* applies there is, as well as the usual reporting requirements (see **A3014** above), a requirement to keep a record of the accident. [*RIDDOR (SI 1995 No 3163, Reg 7*]. The HSE guide to *RIDDOR* provides (at paragraph 86) that an employer may choose to utilise the form BI 150 Accident Book as this record, and the format of the form includes space for the employer to initial the report as being one reportable under *RIDDOR*.

Objectives of accident reporting

[A3026] There should be an effective accident reporting and investigation system in all organisations. Accident reporting procedures should be clearly established in writing with individual reporting responsibilities specified. Staff should be trained in the system and disciplinary action may have to be taken where there is a failure to comply with it. Moreover, there is a case for all incidents, no matter how trivial they may seem and of the absence of any resulting injury being reported through the internal reporting procedures.

The capture of information about accidents and other incidents can be enhanced by the provision of a simple form used throughout the organisation.

Figure 2: Internal Incident Notification Form

INTERNAL INCIDENT NOTIFICATION FORM	
Time/date:	00.00 hrs dd/mm/yy
Location:	
Description of incident:	
Person(s) injured:	
Nature of injury:	
Other persons involved:	
First aid administered:	
Accident Book completed:	Y/N
Form completed by:	
Date:	dd/mm/yy

Duty of disclosure of accident data

[A3027] Employers are under a duty to disclose accident data to works safety representatives and, in the course of litigation, to legal representatives of persons claiming damages for death or personal injury.

Safety representatives

[A3028] An employer must make available to safety representatives of both unionised and non-unionised workforces the information within the employer's knowledge necessary to enable them to fulfil their functions. [*Safety Representatives and Safety Committees Regulations 1977 (SI 1977 No 500), Reg 7(2)*; *Health and Safety (Consultation with Employees) Regulations 1996 (SI 1996 No 1513), Reg 5(1)*]. The Approved Code of Practice in association with these Regulations states that such information should include information which the employer keeps relating to the occurrence of any accident, dangerous occurrence or notifiable industrial disease and any associated statistical records. (Code of Practice: Safety Representatives and Safety Committees 1976, para 6(c)). See **J3015**.

Legal representatives

[A3029] In the course of litigation (and sometimes before proceedings have actually begun) obligations may arise to give disclosure of relevant documents concerning an accident – even if they are confidential. Legal representatives of an injured party would expect reasonable access to *RIDDOR* information. The decision in *Waugh v British Railways Board* [1979] 2 All ER 1169 established that where an employer seeks to withhold on grounds of privilege a report made following an accident, he can only do so if its dominant purpose is related to actual or potential hostile legal proceedings. In *Waugh* a report was commissioned for two purposes following the death of an employee: (*a*) to recommend improvements in safety measures, and (*b*) to gather material for the employer's defence. It was held that the report was not privileged.

Accident Investigation

Rationale for undertaking investigations

[A3030] Unlike the highly prescriptive legal requirements for notification under *RIDDOR (SI 1995 No 3163)*, there are no specific legal obligations requiring the investigation of accidents or the production of accident reports. The HSC consulted on proposals for a statutory duty to undertake investigations into accidents, diseases and dangerous occurrences in 2001, but the outcome was a decision to develop new investigations guidelines with a view to encouraging the implementation of best practice voluntarily (see **A3035**).

Nevertheless a number of statutory requirements demand some level of investigation is carried out. In particular, one needs to determine whether it is necessary to undertake a review of the adequacy of existing risk assessments for the relevant activity in order to comply to with *Regulation 3(3)* of the *Management of Health and Safety at Work Regulations 1999 (SI 1999 No 3242)* (see **R3022**).

Accident investigation does however does however have wider rationales. The main drivers for it are:

(a)   *learning lessons*: identifying and understanding the immediate and the underlying causes of accidents and identifying ways in which to prevent the recurrence of similar accidents.

(b)   *reassurance and explanation*: particularly the victim of an accident, but also others connected with it often need to be reassured that appropriate action is being taken or that a satisfactory explanation has been given for what has happened.

(c)   *monitoring of performance*: investigations produce data for the reactive monitoring of health and safety performance, measurement of results against an organisation's targets or for benchmarking, recording of essential information in databases ('corporate memory'), and the provision of management reports – ultimately to directors – on the management of operational risks.

(d)   *providing information to other interested parties*: the information is likely to be needed with which to brief insurers, safety representatives, also those who are responsible for dealing with media. Tthere is also an expectation on the part of health and safety inspectors as well that an accident investigation will be produced, and disclosed to them once completed, and an absence of a formal report may be taken as an indicator of management failure.

(e)   *disciplinary procedures*: where there is evidence of deliberate reckless failure to follow procedures an investigation may form part of the disciplinary process and evidence upon which the fairness of action perhaps ultimately resulting in dismissal might be judged.

(f)   *allocation of blame/liability*: this will be part and parcel of any investigation undertaken by enforcement agencies and insurers, any information may need to be obtained and analysed by the organisation involved in the accident in preparation for what are inevitably adversarial proceedings.

Accident investigation literature often tends to be focussed on the risk management and accident prevention benefits that can accrue from the process, while treating disciplinary and liability issues as subsidiary to other objectives. This can result in a number of difficulties and potential conflicts. It is difficult in reality for accidents to be investigated in an entirely blame-free context. Accident investigations can – if there is not careful consideration given to legal and disciplinary implications – prejudice not just the organisation but also the position of individual employees and managers who may be subject to actual or implied criticism or even legal action. It should also be borne in mind that the HSE's internal work instructions for inspectors have been known to steer inspectors towards directing or using an organisation's investigation process in order to achieve enforcement-related goals which include providing *'an early insight into the duty holder's thoughts regarding cause and blame enabling any potential defence or mitigation to any subsequent proceedings to be identified'*.

It is not possible to produce a single system of investigation which ultimately resolves all the potential conflicts between the different rationales and needs for the investigation. However, the processes do need to be operated in a way which takes into account these conflicts and seeks to minimise them by appropriately involving different parts of the organisation, insurers and legal representatives. Particular attention needs to be given to identifying the information which is obtained for advice where legal privilege may apply and where the wider dissemination of that material might cause prejudice in future legal proceedings.

The main elements of investigations

[A3031] The following sections outline the main elements of an accident investigation process, and each stage of this process provides a framework within which more or less details and elaborate examination of relevant issues can be carried out depending on what is considered necessary or proportionate (see FIGURE 3). The process should not however be inflexible, and it may need to be adapted if for example there are parallel investigations taking place by enforcement authorities with a view to future prosecution.

Figure 3: Elements of the investigation process

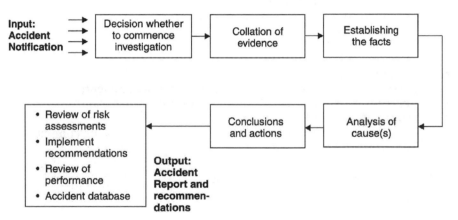

It is possible to devise ad hoc arrangements for each investigation as it arises, but ideally (and essentially in large organisations where there is a steady occurrence of accidents) there should be defined procedures. Responsibilities should be designated, and there should be basic but clear criteria as to the types of occurrence which should be investigated more thoroughly than is necessary merely for the purposes of carrying out RIDDOR notifications. Thought needs to be given to how an investigation team is to be constituted, setting terms of reference, and determining what the reporting lines are during the process and not just for the delivery of the accident report. An agreed format for the contents of accident reports is desirable so that consistent and comprehensive results are produced. Those with responsibility for the investigation process may require training in some areas such as interviewing skills, accident causation principles and report writing.

An HSE investigation or claim for personal injury will raise liability issues and opportunities for internal or external legal advice should be built into the process at each stage.

It is important that the process does not exist in isolation of what is happening elsewhere in the organisation. There may need to be liaison not just with legal advisors, but with others who may be dealing with necessary actions following an accident such as HR managers, insurance managers and of course line managers responsible for the activity in which the accident has occurred who may need immediate guidance on whether (or how) to recommence or continue the activities.

Collating evidence

[A3032] Every situation will be different, but typically the evidence will comprise some or all of the following:

Conditions at the accident scene:
- photographs and videos;
- sketches and plans;
- measurements;
- records of weather/environmental conditions;
- samples of substances;
- equipment or retained parts;
- condition of safety equipment/devices;
- first-aid records;
- information obtained from a reconstruction of the accident;
- lists of eye-witnesses;
- identities of others involved;
- notes of comments made immediately after the accident;
- internal memos and emails dealing with the accident and aftermath;
- statements taken by managers/others;
- other electronic records (e.g. time of emergency calls, or computerised process data).

Safety documentation:
- accident book entries;
- RIDDOR records;
- relevant risk assessments;
- method statements/safe systems of work;
- training materials;
- training records;
- equipment operating instructions;
- equipment logs/maintenance records;
- personnel records of those involved in the accident;
- reports on relevant previous incidents;
- relevant health and safety audits, management reports, consultants' advice;
- relevant emergency procedures.

The most difficult part of this stage of the process is likely to be obtaining statements promptly from witnesses, either because they are distressed or because they are unco-operative (which may be for a variety of reasons including concern that they or their workmates might be criticised). It is nevertheless important to obtain witnesses' evidence as soon as possible after an accident. The accuracy of their information is reduced the more time elapses, and recall can be affected by *post-event enhancement* – the uncon-scious incorporation of ideas and perceptions based on discussing events with others in group setting and by *false confidence* – a hardening of a version of events from recalling and re-telling others or committing evidence to paper.

If witnesses are unco-operative or hostile it may serve no useful purpose to persevere in efforts to interview them. A short period of delay can sometimes assist, in which it may be possible to resolve the underlying problems. Where the witness is an employee, ultimately the failure to co-operate with the employer may be treated as a disciplinary matter.

Another difficulty with interviewing witnesses can be that the HSE, police or other authorities may wish to restrict access to them – so that they can interview them first and take formal statements. There is no legal power to control witnesses in this way, but inspectors sometimes threaten employers with prosecution for obstructing them if they do not agree to postpone their interviews. It is therefore advisable to be open with the authorities about the interview process, and sometimes to offer inspectors or other officers an opportunity to attend interviews with witnesses about whom they have concerns.

Planning the interview

- Identify what information is sought from this interviewee and whether the interviewee is willing or reluctant.
- Draft a list of principal topics and documents to ask about.
- Key interviewees should be interviewed as soon as pos-sible after the event.
- Will the interviewee be accompanied, and if so, by whom?
- Arrange for the venue to be neutral and under the inter-viewer's control if possible.
- Confirm date, time and venue with interviewee.
- Arrange the furniture appropriately, for example no glar-ing spotlights, relatively informal seating arrangements, removal of telephone.
- Availability of refreshments, tissues and props such as plans, models, etc.

Introduce the interview:

- Introductions – interviewer(s), interviewee and accompa-nying observer.
- Summarise purpose of interview, include note-taking, recording, confidentiality.
- Outline structure of interview – chronological, beginning with interviewee's own account, then followed by ques-tions on points of interest.
- Turn off mobile phones, OK to ask for comfort break or pause if needed.

Body of interview:

Planning the interview

- Use broad questions to initiate free account if interviewee is willing.
- Use simple probing questions if the interviewee is reluctant.
- Closed questions used sparingly, avoiding counterproductive questions.
- Use verbal and non-verbal prompts.
- Actively listen to the answers provided, and evaluate and respond appropriately.
- Use a range of questions to obtain more detailed information.
- Keep asking 'why' to find out about critical acts and decisions (human factors).
- Use cognitive interviewing techniques if appropriate.
- Use hunches and normalisation where appropriate.
- Use summary/trailer questions for inconsistencies with other accounts/evidence.
- Pay attention to drop-it cues, returning to issue later if important.
- Deal with interviewee's emotional reactions without becoming involved.
- Ensure your verbal and non-verbal communication are consistent.
- Maintain neutral, open and non-judgemental manner throughout.

Closing the interview:

- Summarise key points and ask for confirmation of account and summary.
- Explain what will happen next (further interviews, testing and analysis, provision of statement and/or transcript).
- Thank interviewee for contribution, mention possibility of follow-up.
- Provide your contact details in case of query or further information.
- Be prepared for any off-the-record statement as interview closes.

(Source: *Tolley's Workplace Accident Handbook* (2nd edition, 2007) ISBN 0 7506 8151 9)

Analysis of causes

[A3033] This stage of the process is concerned with two inter-related questions: 'what happened?' and 'why did it happen?' These apparently simple questions can be subjected to highly sophisticated analytical techniques. (See 'Root causes analysis: Literature review' – HSE Contract Research Report 325/2001 at website: http://www.hse.gov.uk/research/crr_pdf/2001/

crr01325.pdf). The following paragraphs summarise a basic approach, but thos e charged with accident investigations should have a fuller understanding of a ccident causation theory and knowledge of the background to *human factors* (see **E17001**).

Having collated the information about nature and circumstances of the accident the investigation team needs to decide what issues it needs to analyse. A useful starting point is often the organisation's standard template for accident investigation reports (see **A3034** below) but not all the headings will necessarily be relevant in each case, and a degree of flexibility is needed so that the investigation does not become diverted by issues of low or no relevance.

Examples of the questions, which are typically explored, are:

- What was the chain of evens leading up to the accident?
- What planning of the activity had taken place?
- Were the risks known?
- Was the planning deficient?
- Were defective premises, equipment or maintenance a factor?
- Were the correct materials being used?
- Were those involved properly trained?
- Was the training adequate?
- Did people depart from instructions/procedures?
- Was supervision properly undertaken?
- Were the instructions/procedures adequate?
- Did organisational factors of the work play a role?, e.g. excessive hours, rushing to meet a deadline or target.
- Was there a failure to learn lessons from previous incidents or to follow-up previous warnings or complaints?
- Did the emergency procedures and provision of first-aid operate effectively?

It is not unusual for these or other questions to require additional evidence or for further queries to be raised with witnesses, and that the collating analysis stages may have to overlap.

Answers to these questions are used to inform the next stage of analysis which involves reaching an understanding as to the cause or causes of the accident, and determining whether the risk control measures that were in place are in need of improvement.

Causes of accidents can be complex and multi-factorial. It is often the unusual combination of a series of events which leads to injury or damage in what were hitherto routine workplace activities. Each event may be traced back to factors that preceded it, and so it has become common to analysis accidents in terms of three types of 'causes'.

(1) 'Immediate' cause: an unsafe act (or omission) or an unsafe condition which leads directly to the injury or damage. (An example of an unsafe act is improper stacking of a load; an example of an unsafe condition is a machine with a missing guard). Occasionally the immediate cause of an accident may be a 'violation': a deliberate or reckless departure from correct procedures by an individual or group.

(2)   'Underlying' cause: these precede the time of the accident and consist of management control factors (e.g. inadequate risk assessment or the absence of supervision), job factors (e.g. lack of competence or insufficient staffing), or environmental factors (such as working in harsh conditions).

(3)   'Root' causes: a failing from which all the other causes have grown – usually a deficiency in the planning and organisation of management of health and safety.

Caution is needed in using this approach for a number of reasons. First, it may be difficult to separate and distinguish between several different causes as neatly in practice as it is in theory. Second, this approach does not give weight to the *gravity* of causal factors, which may be relevant in the wider employment or liability context. Third, and most importantly, there is a large element of subjectivity in determining what are the relevant underlying, and particular root, causes. The causal analysis therefore needs to be firmly based on the factual analysis, and the reasoning process explained at each stage by reference to the facts as they have found during the first stage of the analysis.

Preparing the report

[A3034] There are no hard and fast rules about the structure or content of accident investigation reports. The essential features however are:

- description of circumstances of the accident;
- consideration of control measures;
- conclusions as to causes;
- recommendations for remedial action.

A basic template which can be used for reports is as follows.

OUTLINE FOR ACCIDENT INVESTIGATION REPORT

Title:	Including details of the time, location and outcome(s) of the accident.
Summary:	Providing background information, which will include factual and historical information. Conclusions and recommendations can be copied and pasted to finish the summary, which will then provide an overview of the investigation.
History of accident:	Including detailed factual and historical information, the sequence of events, and any actions taken in immediate response to mitigate or make the site safe, including emergency response and attendance of first aid or medical personnel.
Investigation and analysis	Including evidence from interviews, engineering and forensic testing and/or analyses, other sources of information.
Conclusions:	Including the most likely or probable immediate and underlying causes, and an indication of the extent to which the conclusions were unanimous. Where there are discrepancies in the evidence, this should be pointed out.
Recommendations:	These should address causes and should have a clear link with the investigation and analysis.

(Source: *Tolley's Workplace Accident Handbook* (2nd edition, 2007) ISBN 0 7506 8151 9)

Before finalising the report it is sometimes desirable to seek comments or corrections from those involved in the accident or connected with the background. Partly this is out of courtesy where individuals might perceive the conclusions as implied criticism, but it also provides a final opportunity to eliminate factual or analytical errors. It is also sensible that proposed recommendations for action are at least canvassed with the relevant managers who may become responsible for implementing them to ensure that they are not impractical or unrealistic. A draft may also be provided to the organisations' legal advisors to consider how it may affect liability issues and whether the conclusions are consistent with the findings of privileged investigations that have been, or are still being, carried out.

When the report is finalised there needs to be a clear and timely management plan of action to implement the recommendations.

The document itself will be disclosable in any subsequent legal proceedings and a copy may be requested by enforcing authorities. Delay or incompleteness in giving effect to the recommendations may expose the organisation to subsequent criticism.

Further guidance

- *Guide to the Reporting of Accidents, Diseases and Dangerous Occurrences Regulations 1995* (HSE Books, L73, ISBN 0 7176 6290 6)
- *Investigating Accidents and Incidents – A Workbook for Employers, Unions, Safety Representatives and Safety Professionals* (HSE Books, HSG245, ISBN 0-7176-2827-2)
- *Reducing Error and Influencing Behaviour* (HSE Books, HSG48, ISBN 0717624528)
- *Tolley's Workplace Accident Handbook*, 2nd Edition (Butterworths-Heinemann, ISBN 978-0-75-068151-3)

Appendix A

List of Reportable Dangerous Occurrences

1995 No 3163

REPORTING OF INJURIES, DISEASES AND DANGEROUS
OCCURRENCES REGULATIONS 1995

SCHEDULE 2

DANGEROUS OCCURRENCES

Regulation 2(1)

PART I

GENERAL

Lifting machinery, etc

1

The collapse of, the overturning of, or the failure of any load-bearing part of any—
- (a) lift or hoist;
- (b) crane or derrick;
- (c) mobile powered access platform;
- (d) access cradle or window-cleaning cradle;
- (e) excavator;
- (f) pile-driving frame or rig having an overall height, when operating, of more than 7 metres; or
- (g) fork lift truck.

Pressure systems

2

The failure of any closed vessel (including a boiler or boiler tube) or of any associated pipework, in which the internal pressure was above or below atmospheric pressure, where the failure has the potential to cause the death of any person.

Freight containers

3

(1) The failure of any freight container in any of its load-bearing parts while it is being raised, lowered or suspended.

(2) In this paragraph, "freight container" means a container as defined in regulation 2(1) of the Freight Containers (Safety Convention) Regulations 1984.

Overhead electric lines

4

Any unintentional incident in which plant or equipment either—

    (a)   comes into contact with an uninsulated overhead electric line in which the voltage exceeds 200 volts; or

    (b)   causes an electrical discharge from such an electric line by coming into close proximity to it.

Electrical short circuit

5

Electrical short circuit or overload attended by fire or explosion which results in the stoppage of the plant involved for more than 24 hours or which has the potential to cause the death of any person.

Explosives

6

(1)   Any of the following incidents involving explosives—

    [(a)   any unintentional fire, explosion or ignition at a site—

        (i)   where explosives are manufactured by a person who holds a licence, or who does not hold a licence but is required to, in respect of that manufacture under the Manufacture and Storage of Explosives Regulations 2005; or

        (ii)   where explosives are stored by a person who holds a licence or is registered, or who is not licensed but is required to be in the absence of any registration, in respect of that storage under those Regulations;

    (aa)   the unintentional explosion or ignition of explosives at a place other than a site described in sub-paragraph (1)(a), not being one—

        (i)   caused by the unintentional discharge of a weapon where, apart from that unintentional discharge, the weapon and explosives functioned as they were designed to do; or

        (ii)   where a fail-safe device or safe system of work functioned so as to prevent any person from being injured in consequence of the explosion or ignition;]

    (b)   a misfire (other than one at a mine or quarry or inside a well or one involving a weapon) except where a fail-safe device or safe system of work functioned so as to prevent any person from being endangered in consequence of the misfire;

    (c)   the failure of the shots in any demolition operation to cause the intended extent of collapse or direction of fall of a building or structure;

    (d)   the projection of material (other than at a quarry) beyond the boundary of the site on which the explosives are being used or beyond the danger zone in circumstances such that any person was or might have been injured thereby;

    (e)   any injury to a person (other than at a mine or quarry or one otherwise reportable under these Regulations) involving first-aid or medical treatment resulting from the explosion or discharge of any explosives or detonator [or from any intentional fire or ignition].

[(2)   In this paragraph—

"danger zone" means the area from which persons have been excluded or forbidden to enter to avoid being endangered by any explosion or ignition of explosives; and

"explosives" has the same meaning as in the Manufacture and Storage of Explosives Regulations 2005.]

Biological agents

7

Any accident or incident which resulted or could have resulted in the release or escape of a biological agent likely to cause severe human infection or illness.

Malfunction of radiation generators, etc

8

(1)   Any incident in which—

    (a)   the malfunction of a radiation generator or its ancillary equipment used in fixed or mobile industrial radiography, the irradiation of food or the processing of products by irradiation, causes it to fail to de-energise at the end of the intended exposure period; or

    (b)   the malfunction of equipment used in fixed or mobile industrial radiography or gamma irradiation causes a radioactive source to fail to return to its safe position by the normal means at the end of the intended exposure period.

[(2)   In this paragraph, "radiation generator" means any electrical equipment emitting ionising radiation and containing components operating at a potential difference of more than 5kV.]

Breathing apparatus

9

(1)   Any incident in which breathing apparatus malfunctions—

    (a)   while in use, or

    (b)   during testing immediately prior to use in such a way that had the malfunction occurred while the apparatus was in use it would have posed a danger to the health or safety of the user.

(2)   This paragraph shall not apply to breathing apparatus while it is being—

    (a)   used in a mine; or

    (b)   maintained or tested as part of a routine maintenance procedure.

Diving operations

10

Any of the following incidents in relation to a [diving project]—

    (a)   the failure or the endangering of—

        (i)   any lifting equipment associated with the diving operation, or

        (ii)   life support equipment, including control panels, hoses and breathing apparatus, which puts a diver at risk;

    (b)   any damage to, or endangering of, the dive platform, or any failure of the dive platform to remain on station,

which puts a diver at risk;

    (c)   the trapping of a diver;

    (d)   any explosion in the vicinity of a diver; or

    (e)   any uncontrolled ascent or any omitted decompression which puts a diver at risk.

Collapse of scaffolding

11

The complete or partial collapse of—

    (a)   any scaffold which is—

        (i)   more than 5 metres in height which results in a substantial part of the scaffold falling or overturning; or

        (ii)   erected over or adjacent to water in circumstances such that there would be a risk of drowning to a person falling from the scaffold into the water; or

(b)   the suspension arrangements (including any outrigger) of any slung or suspended scaffold which causes a working platform or cradle to fall.

Train collisions

12

Any unintended collision of a train with any other train or vehicle, other than one reportable under Part IV of this Schedule, which caused, or might have caused, the death of, or major injury to, any person.

Wells

13

Any of the following incidents in relation to a well (other than a well sunk for the purpose of the abstraction of water)—

(a)   a blow-out (that is to say an uncontrolled flow of well-fluids from a well);

(b)   the coming into operation of a blow-out prevention or diversion system to control a flow from a well where normal control procedures fail;

(c)   the detection of hydrogen sulphide in the course of operations at a well or in samples of well-fluids from a well where the presence of hydrogen sulphide in the reservoir being drawn on by the well was not anticipated by the responsible person before that detection;

(d)   the taking of precautionary measures additional to any contained in the original drilling programme following failure to maintain a planned minimum separation distance between wells drilled from a particular installation; or

(e)   the mechanical failure of any safety critical element of a well (and for this purpose the safety critical element of a well is any part of a well whose failure would cause or contribute to, or whose purpose is to prevent or limit the effect of, the unintentional release of fluids from a well or a reservoir being drawn on by a well).

Pipelines or pipeline works

14

The following incidents in respect of a pipeline or pipeline works—

(a)   the uncontrolled or accidental escape of anything from, or inrush of anything into, a pipeline which has the potential to cause the death of, major injury or damage to the health of any person or which results in the pipeline being shut down for more than 24 hours;

(b)   the unintentional ignition of anything in a pipeline or of anything which, immediately before it was ignited, was in a pipeline;

(c)   any damage to any part of a pipeline which has the potential to cause the death of, major injury or damage to the health of any person or which results in the pipeline being shut down for more than 24 hours;

(d)   any substantial and unintentional change in the position of a pipeline requiring immediate attention to safeguard the integrity or safety of a pipeline;

(e)   any unintentional change in the subsoil or seabed in the vicinity of a pipeline which has the potential to affect the integrity or safety of a pipeline;

(f)   any failure of any pipeline isolation device, equipment or system which has the potential to cause the death of, major injury or damage to the health of any person or which results in the pipeline being shut down for more than 24 hours; or

(g)   any failure of equipment involved with pipeline works which has the potential to cause the death of, major injury or damage to the health of any person.

Fairground equipment

15

The following incidents on fairground equipment in use or under test—

(a)   the failure of any load-bearing part;

(b)   the failure of any part designed to support or restrain passengers; or

(c)   the derailment or the unintended collision of cars or trains.

Carriage of dangerous substances by road

16

(1)   Any incident involving a road tanker or tank container used for the carriage of . . . [dangerous goods] in which—

(a)   the road tanker or vehicle carrying the tank container overturns (including turning onto its side);

(b)   the tank carrying the [dangerous goods] is seriously damaged;

(c)   there is an uncontrolled release or escape of the [dangerous goods] being carried; or

(d)   there is a fire involving the [dangerous goods] being carried.

(2) . . .

17

(1)   Any incident involving a vehicle used for the carriage of . . . [dangerous goods], other than a vehicle to which paragraph 16 applies, where there is—

(a)   an uncontrolled release or escape of the [dangerous goods] being carried in such a quantity as to have the potential to cause the death of, or major injury to, any person; or

(b)   a fire which involves the [dangerous goods] being carried.

(2) . . .

[17A

In paragraphs 16 and 17 above, "carriage" and "dangerous goods" have the same meaning as those terms in regulation 2(1) of [CDG 2007].]

DANGEROUS OCCURRENCES WHICH ARE REPORTABLE EXCEPT IN RELATION TO OFFSHORE WORKPLACES

Collapse of building or structure

18

Any unintended collapse or partial collapse of—

(a)   any building or structure (whether above or below ground) under construction, reconstruction, alteration or demolition which involves a fall of more than 5 tonnes of material;

(b)   any floor or wall of any building (whether above or below ground) used as a place of work; or

(c)   any false-work.

Explosion or fire

19

An explosion or fire occurring in any plant or premises which results in the stoppage of that plant or as the case may be the suspension of normal work in those premises for more than 24 hours, where the explosion or fire was due to the ignition of any material.

Escape of flammable substances

20

(1)   The sudden, uncontrolled release—

    (a)   inside a building—

        (i)   of 100 kilograms or more of a flammable liquid,

        (ii)   of 10 kilograms or more of a flammable liquid at a temperature above its normal boiling point, or

        (iii) of 10 kilograms or more of a flammable gas; or

    (b)   in the open air, of 500 kilograms or more of any of the substances referred to in sub-paragraph (a) above.

(2)   In this paragraph, "flammable liquid" and "flammable gas" mean respectively a liquid and a gas so classified in accordance with regulation 5(2), (3) or (5) of the Chemicals (Hazard Information and Packaging for Supply) Regulations 1994.

Escape of substances

21

The accidental release or escape of any substance in a quantity sufficient to cause the death, major injury or any other damage to the health of any person.

NOTES

Initial Commencement

Specified date

Specified date: 1 April 1996: see reg 1.

Amendment

Para 6: sub-para (1)(a), (aa), substituted, for sub-para (1)(a) as originally enacted, by SI 2005/1082, reg 28(1), Sch 5, Pt 2, para 40(1), (2)(a)(i).
    Date in force: 26 April 2005: see SI 2005/1082, reg 1(1).
Para 6: in sub-para (1)(e) words "or from any intentional fire or ignition" in square brackets inserted by SI 2005/1082, reg 28(1), Sch 5, Pt 2, para 40(1), (2)(a)(ii).
    Date in force: 26 April 2005: see SI 2005/1082, reg 1(1).
Para 6: sub-para (2) substituted by SI 2005/1082, reg 28(1), Sch 5, Pt 2, para 40(1), (2)(a)(iii).
    Date in force: 26 April 2005: see SI 2005/1082, reg 1(1).
Para 8: sub-para (2) substituted by SI 1999/3232, reg 41(1), Sch 9, para 6.
    Date in force: 1 January 2000: see SI 1999/3232, reg 1(a).
Para 10: words "diving project" in square brackets substituted by SI 1997/2776, reg 19, Sch 2, para 5(b).
    Date in force: 1 April 1998: see SI 1997/2776, reg 1.
[Paras 16, 17: word omitted revoked, words in square brackets substituted, and sub-para (2) revoked, by SI 1996/2092, reg 21(7)(a).]

Para 17A (inserted by SI 1996/2092, reg 21(7)(b)): substituted by SI 2004/568, reg 62, Sch 13, para 8(1), (3)(a)(ii).
Date in force: 10 May 2004: see SI 2004/568, reg 1.
Para 17A: words "CDG 2007" in square brackets substituted by SI 2007/1573, reg 94, Sch 8.
Date in force: 1 July 2007: see SI 2007/1573, reg 1.

PART II

DANGEROUS OCCURRENCES WHICH ARE REPORTABLE IN RELATION TO MINES

Fire or ignition of gas

22

The ignition, below ground, of any gas (other than gas in a safety lamp) or of any dust.

23

The accidental ignition of any gas in part of a firedamp drainage system on the surface or in an exhauster house.

24

The outbreak of any fire below ground.

25

An incident where any person in consequence of any smoke or any other indication that a fire may have broken out below ground has been caused to leave any place pursuant to either Regulation 11(1) of the Coal and Other Mines (Fire and Rescue) Regulations 1956 or section 79 of the Mines and Quarries Act 1954.

26

The outbreak of any fire on the surface which endangers the operation of any winding or haulage apparatus installed at a shaft or unwalkable outlet or of any mechanically operated apparatus for producing ventilation below ground.

Escape of gas

27

Any violent outburst of gas together with coal or other solid matter into the mine workings except when such outburst is caused intentionally.

Failure of plant or equipment

28

The breakage of any rope, chain, coupling, balance rope, guide rope, suspension gear or other gear used for or in connection with the carrying of persons through any shaft or staple shaft.

29

The breakage or unintentional uncoupling of any rope, chain, coupling, rope tensioning system or other gear used for or in connection with the transport of persons below ground, or breakage of any belt, rope or other gear used for or in connection with a belt conveyor designated by the mine manager as a man-riding conveyor.

30

An incident where any conveyance being used for the carriage of persons is overwound; or any conveyance not being so used is overwound and becomes detached from its winding rope; or any conveyance operated by means of the friction of a rope on a winding sheave is brought to rest by the apparatus provided in the headframe of the shaft or in the part of the shaft below the lowest landing for the time being in use, being apparatus provided for bringing the conveyance to rest in the event of its being overwound.

31

The stoppage of any ventilating apparatus (other than an auxiliary fan) which causes a substantial reduction in ventilation of the mine lasting for a period exceeding 30 minutes, except when for the purpose of planned maintenance.

32

The collapse of any headframe, winding engine house, fan house or storage bunker.

Breathing apparatus

33

At any mine an incident where—

    (a)   breathing apparatus or a smoke helmet or other apparatus serving the same purpose or a self-rescuer, while being used, fails to function safely or develops a defect likely to affect its safe working; or

    (b)   immediately after using and arising out of the use of breathing apparatus or a smoke helmet or other apparatus serving the same purpose or a self-rescuer, any person receives first-aid or medical treatment by reason of his unfitness or suspected unfitness at the mine.

Injury by explosion of blasting material etc

34

An incident in which any person suffers an injury (not being a major injury or one reportable under regulation 3(2)) which results from an explosion or discharge of any blasting material or device within the meaning of section 69(4) of the Mines and Quarries Act 1954 for which he receives first-aid or medical treatment at the mine.

Use of emergency escape apparatus

35

An incident where any apparatus is used (other than for the purpose of training and practice) which has been provided at the mine in accordance with regulation 4 of the Mines (Safety of Exit) Regulations 1988 or where persons leave the mine when apparatus and equipment normally used by persons to leave the mine is unavailable.

Inrush of gas or water

36

Any inrush of noxious or flammable gas from old workings.

37

Any inrush of water or material which flows when wet from any source.

Insecure tip

38

Any movement of material or any fire or any other event which indicates that a tip to which Part I of the Mines and Quarries (Tips) Act 1969 applies, is or is likely to become insecure.

Locomotives

39

Any incident where an underground locomotive when not used for testing purposes is brought to rest by means other than its safety circuit protective devices or normal service brakes.

Falls of ground

40

Any fall of ground, not being part of the normal operations at a mine, which results from a failure of an underground support system and prevents persons travelling through the area affected by the fall or which otherwise exposes them to danger.

NOTES

Initial Commencement

Specified date

Specified date: 1 April 1996: see reg 1.

PART III

DANGEROUS OCCURRENCES WHICH ARE REPORTABLE IN RELATION TO QUARRIES

Collapse of storage bunkers

41

The collapse of any storage bunker.

Sinking of craft

42

The sinking of any water-borne craft or hovercraft.

Injuries

43

(1)   An incident in which any person suffers an injury (not otherwise reportable under these Regulations) which results from an explosion or from the discharge of any explosives for which he receives first-aid or medical treatment at the quarry.

(2)   In this paragraph, "explosives" has the same meaning [as in regulation 2(1) of the Quarries Regulations 1999].

Projection of substances outside quarry

44

Any incident in which any substance is ascertained to have been projected beyond a quarry boundary as a result of blasting operations in circumstances in which any person was or might have been endangered.

Misfires

45

Any misfire, as defined by [regulation 2(1) of the Quarries Regulations 1999].

Insecure tips

46

Any event (including any movement of material or any fire) which indicates that a tip, [to which the Quarries Regulations 1999 apply], is or is likely to become insecure.

Movement of slopes or faces

47

Any movement or failure of an excavated slope or face which—

(a)   has the potential to cause the death of any person; or

(b)   adversely affects any building, contiguous land, transport system, footpath, public utility or service, watercourse, reservoir or area of public access.

Explosions or fires in vehicles or plant

48

(1)   Any explosion or fire occurring in any large vehicle or mobile plant which results in the stoppage of that vehicle or plant for more than 24 hours and which affects—

(a)   any place where persons normally work; or

(b)   the route of egress from such a place.

(2)   In this paragraph, "large vehicle or mobile plant" means—

(a)   a dump truck having a load capacity of at least 50 tonnes; or

(b)   an excavator having a bucket capacity of at least 5 cubic metres.

NOTES

Initial Commencement

Specified date

Specified date: 1 April 1996: see reg 1.

Amendment

Para 43: in sub-para (2) words "as in regulation 2(1) of the Quarries Regulations 1999" in square brackets substituted by SI 1999/2024, reg 48(2), Sch 5, Pt II.
Date in force: 1 January 2000: see SI 1999/2024, reg 1(1).
Para 45: words "regulation 2(1) of the Quarries Regulations 1999" in square brackets substituted by SI 1999/2024, reg 48(2), Sch 5, Pt II.
Date in force: 1 January 2000: see SI 1999/2024, reg 1(1).
Para 46: words "to which the Quarries Regulations 1999 apply" in square brackets substituted by SI 1999/2024, reg 48(2), Sch 5, Pt II.
Date in force: 1 January 2000: see SI 1999/2024, reg 1(1).

PART IV

DANGEROUS OCCURRENCES WHICH ARE REPORTABLE IN RESPECT OF RELEVANT TRANSPORT SYSTEMS

Accidents to passenger trains

49

Any collision in which a passenger train collides with another train.

50

Any case where a passenger train or any part of such a train unintentionally leaves the rails.

Accidents not involving passenger trains

51

Any collision between trains, other than one between a passenger train and another train, on a running line where any train sustains damage as a result of the collision, and any such collision in a siding which results in a running line being obstructed.

52

Any derailment, of a train other than a passenger train, on a running line, except a derailment which occurs during shunting operations and does not obstruct any other running line.

53

Any derailment, of a train other than a passenger train, in a siding which results in a running line being obstructed.

Accidents involving any kind of train

54

Any case of a train striking a buffer stop, other than in a siding, where damage is caused to the train.

55

Any case of a train striking any cattle or horse, whether or not damage is caused to the train, or striking any other animal if, in consequence, damage (including damage to the windows of the driver's cab but excluding other damage consisting solely in the breakage of glass) is caused to the train necessitating immediate temporary or permanent repair.

56

Any case of a train on a running line striking or being struck by any object which causes damage (including damage to the windows of the driver's cab but excluding other damage consisting solely in the breakage of glass) necessitating immediate temporary or permanent repair or which might have been liable to derail the train.

57

Any case of a train, other than one on a railway, striking or being struck by a road vehicle.

58

Any case of a passenger train, or any other train not fitted with continuous self-applying brakes, becoming unintentionally divided.

59

(1)   Any of the following classes of accident which occurs or is discovered whilst the train is on a running line—
- (a)   the failure of an axle;
- (b)   the failure of a wheel or tyre, including a tyre loose on its wheel;
- (c)   the failure of a rope or the fastenings thereof or of the winding plant or equipment involved in working an incline;
- (d)   any fire, severe electrical arcing or fusing in or on any part of a passenger train or a train carrying dangerous goods;

(e)   in the case of any train other than a passenger train, any severe electrical arcing or fusing, or any fire which was extinguished by a fire-fighting service; or

(f)   any other failure of any part of a train which is likely to cause an accident to that or any other train or to kill or injure any person.

[(2) In this paragraph "dangerous goods" has the meaning assigned to it in regulation 2(1) of [CDG 2007].]

Accidents and incidents at level crossings

60

Any case of a train striking a road vehicle or gate at a level crossing.

61

Any case of a train running onto a level crossing when not authorised to do so.

62

A failure of the equipment at a level crossing which could endanger users of the road or path crossing the railway.

Accidents involving the permanent way and other works on or connected with a relevant transport system

63

The failure of a rail in a running line or of a rack rail, which results in—

(a)   a complete fracture of the rail through its cross-section; or

(b)   in a piece becoming detached from the rail which necessitates an immediate stoppage of traffic or the immediate imposition of a speed restriction lower than that currently in force.

64

A buckle of a running line which necessitates an immediate stoppage of traffic or the immediate imposition of a speed restriction lower than that currently in force.

65

Any case of an aircraft or a vehicle of any kind landing on, running onto or coming to rest foul of the line, or damaging the line, which causes damage which obstructs the line or which damages any railway equipment at a level crossing.

66

The runaway of an escalator, lift or passenger conveyor.

67

Any fire or severe arcing or fusing which seriously affects the functioning of signalling equipment.

68

Any fire affecting the permanent way or works of a relevant transport system which necessitates the suspension of services over any line, or the closure of any part of a station or signal box or other premises, for a period—

(a)   in the case of a fire affecting any part of a relevant transport system below ground, of more than 30 minutes, and

(b)   in any other case, of more than 1 hour.

69

Any other fire which causes damage which has the potential to affect the running of a relevant transport system.

Accidents involving failure of the works on or connected with a relevant transport system

70

(1)   The following classes of accident where they are likely either to cause an accident to a train or to endanger any person—

(a)   the failure of a tunnel, bridge, viaduct, culvert, station, or other structure or any part thereof including the fixed electrical equipment of an electrified relevant transport system;

(b)   any failure in the signalling system which endangers or which has the potential to endanger the safe passage of trains other than a failure of a traffic light controlling the movement of vehicles on a road;

(c)   a slip of a cutting or of an embankment;

(d)   flooding of the permanent way;

(e)   the striking of a bridge by a vessel or by a road vehicle or its load; or

(f)   the failure of any other portion of the permanent way or works not specified above.

Incidents of serious congestion

71

Any case where planned procedures or arrangements have been activated in order to control risks arising from an incident of undue passenger congestion at a station unless that congestion has been relieved within a period of time allowed for by those procedures or arrangements.

Incidents of signals passed without authority

72

(1)   Any case where a train, travelling on a running line or entering a running line from a siding, passes without authority a signal displaying a stop aspect unless—

(a)   the stop aspect was not displayed in sufficient time for the driver to stop safely at the signal; . . .

(b)   . . .

(2)   . . .

NOTES

Initial Commencement

Specified date

Specified date: 1 April 1996: see reg 1.

Amendment

Para 59: sub-para (2) substituted by SI 2004/568, reg 62, Sch 13, para 8(1), (3)(b).
    Date in force: 10 May 2004: see SI 2004/568, reg 1.
Para 59: in sub-para (2) words "CDG 2007" in square brackets substituted by SI 2007/1573, reg 94, Sch 8.
    Date in force: 1 July 2007: see SI 2007/1573, reg 1.
Para 72: sub-para (1)(b) and the word "or" immediately preceding it revoked by SI 1999/2244, reg 7.
    Date in force: 1 April 2000: see SI 1999/2244, reg 1.
Para 72: sub-para (2) revoked by SI 1999/2244, reg 7.
    Date in force: 1 April 2000: see SI 1999/2244, reg 1.

PART V

DANGEROUS OCCURRENCES WHICH ARE REPORTABLE IN RESPECT OF AN OFFSHORE WORKPLACE

Release of petroleum hydrocarbon

73

Any unintentional release of petroleum hydrocarbon on or from an offshore installation which—

    (a)   results in—

        (i)   a fire or explosion; or

        (ii)   the taking of action to prevent or limit the consequences of a potential fire or explosion; or

    (b)   has the potential to cause death or major injury to any person.

Fire or explosion

74

Any fire or explosion at an offshore installation, other than one to which paragraph 73 above applies, which results in the stoppage of plant or the suspension of normal work.

Release or escape of dangerous substances

75

The uncontrolled or unintentional release or escape of any substance (other than petroleum hydrocarbon) on or from an offshore installation which has the potential to cause the death of, major injury to or damage to the health of any person.

Collapses

76

Any unintended collapse of any offshore installation or any unintended collapse of any part thereof or any plant thereon which jeopardises the overall structural integrity of the installation.

Dangerous occurrences

77

Any of the following occurrences having the potential to cause death or major injury—

    (a)   the failure of equipment required to maintain a floating offshore installation on station;

    (b)   the dropping of any object on an offshore installation or on an attendant vessel or into the water adjacent to an installation or vessel; or

    (c)   damage to or on an offshore installation caused by adverse weather conditions.

Collisions

78

Any collision between a vessel or aircraft and an offshore installation which results in damage to the installation, the vessel or the aircraft.

79

Any occurrence with the potential for a collision between a vessel and an offshore installation where, had a collision occurred, it would have been liable to jeopardise the overall structural integrity of the offshore installation.

Subsidence or collapse of seabed

80

Any subsidence or local collapse of the seabed likely to affect the foundations of an offshore installation or the overall structural integrity of an offshore installation.

Loss of stability or buoyancy

81

Any incident involving loss of stability or buoyancy of a floating offshore installation.

Evacuation

82

Any evacuation (other than one arising out of an incident reportable under any other provision of these Regulations) of an offshore installation, in whole or part, in the interests of safety.

Falls into water

83

Any case of a person falling more than 2 metres into water (unless the fall results in death or injury required to be reported under sub-paragraphs (a)–(d) of regulation 3(1)).

NOTES

Initial Commencement

Specified date

     Specified date: 1 April 1996: see reg 1.

Appendix B

List of Reportable Diseases

1995 No 3163

REPORTING OF INJURIES, DISEASES AND DANGEROUS
OCCURRENCES REGULATIONS 1995

SCHEDULE 3

REPORTABLE DISEASES

Regulation 5(1), (2)

PART I

OCCUPATIONAL DISEASES

	Column 1 Diseases	Column 2 Activities
	Conditions due to physical agents and the physical demands of work	
1	Inflammation, ulceration or malignant disease of the skin due to ionising radiation }	
2	Malignant disease of the bones due to ionising radiation }	Work with ionising radiation
3	Blood dyscrasia due to ionising radiation }	
4	Cataract due to electromagnetic radiation	Work involving exposure to electromagnetic radiation (including radiant heat)
5	Decompression illness }	
6	Barotrauma resulting in lung or other organ damage }	Work involving breathing gases at increased pressure (including diving)
7	Dysbaric osteonecrosis }	
8	Cramp of the hand or forearm due to repetitive movements	Work involving prolonged periods of handwriting, typing or other repetitive movements of the fingers, hand or arm
9	Subcutaneous cellulitis of the hand (*beat hand*)	Physically demanding work causing severe or prolonged friction or pressure on the hand

10	Bursitis or subcutaneous cellulitis arising at or about the knee due to severe or prolonged external friction or pressure at or about the knee (*beat knee*)	Physically demanding work causing severe or prolonged friction or pressure at or about the knee
11	Bursitis or subcutaneous cellulitis arising at or about the elbow due to severe or prolonged external friction or pressure at or about the elbow (*beat elbow*)	Physically demanding work causing severe or prolonged friction or pressure at or about the elbow
12	Traumatic inflammation of the tendons of the hand or forearm or of the associated tendon sheaths	Physically demanding work, frequent or repeated movements, constrained postures or extremes of extension or flexion of the hand or wrist
13	Carpal tunnel syndrome	Work involving the use of hand-held vibrating tools
14	Hand-arm vibration syndrome	Work involving:

(a) the use of chain saws, brush cutters or hand-held or hand-fed circular saws in forestry or woodworking;

(b) the use of hand-held rotary tools in grinding material or in sanding or polishing metal;

(c) the holding of material being ground or metal being sanded or polished by rotary tools;

(d) the use of hand-held percussive metal-working tools or the holding of metal being worked upon by percussive tools in connection with riveting, caulking, chipping, hammering, fettling or swaging;

(e) the use of hand-held powered percussive drills or hand-held powered percussive hammers in mining, quarrying or demolition, or on roads or footpaths (including road construction); or

(f) the holding of material being worked upon by pounding machines in shoe manufacture

Infections due to biological agents

15	Anthrax	(a) Work involving handling infected animals, their products or packaging containing infected material; or
		(b) work on infected sites

16	Brucellosis	Work involving contact with: (a) animals or their carcasses (including any parts thereof) infected by brucella or the untreated products of same; or (b) laboratory specimens or vaccines of or containing brucella
17	(a) Avian chlamydiosis	Work involving contact with birds infected with chlamydia psittaci, or the remains or untreated products of such birds
	(b) Ovine chlamydiosis	Work involving contact with sheep infected with chlamydia psittaci or the remains or untreated products of such sheep
18	Hepatitis	Work involving contact with: (a) human blood or human blood products; or (b) any source of viral hepatitis
19	Legionellosis	Work on or near cooling systems which are located in the workplace and use water; or work on hot water service systems located in the workplace which are likely to be a source of contamination
20	Leptospirosis	(a) Work in places which are or are liable to be infested by rats, fieldmice, voles or other small mammals; (b) work at dog kennels or involving the care or handling of dogs; or (c) work involving contact with bovine animals or their meat products or pigs or their meat products
21	Lyme disease	Work involving exposure to ticks (including in particular work by forestry workers, rangers, dairy farmers, game keepers and other persons engaged in countryside management)
22	Q fever	Work involving contact with animals, their remains or their untreated products
23	Rabies	Work involving handling or contact with infected animals

24	Streptococcus suis	Work involving contact with pigs infected with streptococcus suis, or with the carcasses, products or residues of pigs so affected
25	Tetanus	Work involving contact with soil likely to be contaminated by animals
26	Tuberculosis	Work with persons, animals, human or animal remains or any other material which might be a source of infection
27	Any infection reliably attributable to the performance of the work specified in the entry opposite hereto	Work with micro-organisms; work with live or dead human beings in the course of providing any treatment or service or in conducting any investigation involving exposure to blood or body fluids; work with animals or any potentially infected material derived from any of the above

Conditions due to substances

28	Poisonings by any of the following: (a) acrylamide monomer; (b) arsenic or one of its compounds; (c) benzene or a homologue of benzene; (d) beryllium or one of its compounds; (e) cadmium or one of its compounds; (f) carbon disulphide; (g) diethylene dioxide (dioxan); (h) ethylene oxide; (i) lead or one of its compounds; (j) manganese or one of its compounds; (k) mercury or one of its compounds; (l) methyl bromide; (m) nitrochlorobenzene, or a nitro- or amino- or chloro-derivative of benzene or of a homologue of benzene; (n) oxides of nitrogen; (o) phosphorus or one of its compounds	Any activity

29	Cancer of a bronchus or lung	(a) Work in or about a building where nickel is produced by decomposition of a gaseous nickel compound or where any industrial process which is ancillary or incidental to that process is carried on; or
		(b) work involving exposure to bis(chloromethyl) ether or any electrolytic chromium processes (excluding passivation) which involve hexavalent chromium compounds, chromate production or zinc chromate pigment manufacture
30	Primary carcinoma of the lung where there is accompanying evidence of silicosis	Any occupation in:
		(a) glass manufacture;
		(b) sandstone tunnelling or quarrying;
		(c) the pottery industry;
		(d) metal ore mining;
		(e) slate quarrying or slate production;
		(f) clay mining;
		(g) the use of siliceous materials as abrasives;
		(h) foundry work;
		(i) granite tunnelling or quarrying; or
		(j) stone cutting or masonry
31	Cancer of the urinary tract	1 Work involving exposure to any of the following substances:
		(a) beta-naphthylamine or methylene-bis-orthochloroaniline;
		(b) diphenyl substituted by at least one nitro or primary amino group or by at least one nitro and primary amino group (including benzidine);
		(c) any of the substances mentioned in sub-paragraph (b) above if further ring substituted by halogeno, methyl or methoxy groups, but not by other groups; or
		(d) the salts of any of the substances mentioned in sub-paragraphs (a) to (c) above

		2 The manufacture of auramine or magenta
32	Bladder cancer	Work involving exposure to aluminium smelting using the Soderberg process
33	Angiosarcoma of the liver	(a) Work in or about machinery or apparatus used for the polymerisation of vinyl chloride monomer, a process which, for the purposes of this sub-paragraph, comprises all operations up to and including the drying of the slurry produced by the polymerisation and the packaging of the dried product; or
		(b) work in a building or structure in which any part of the process referred to in the foregoing sub-paragraph takes place
34	Peripheral neuropathy	Work involving the use or handling of or exposure to the fumes of or vapour containing n-hexane or methyl n-butyl ketone
35	Chrome ulceration of:	Work involving exposure to chromic acid or to any other chromium compound
	(a) the nose or throat; or	
	(b) the skin of the hands or forearm	
36	Folliculitis }	
37	Acne }	Work involving exposure to mineral oil, tar, pitch or arsenic
38	Skin cancer }	
39	Pneumoconiosis (excluding asbestosis)	1 (a) The mining, quarrying or working of silica rock or the working of dried quartzose sand, any dry deposit or residue of silica or any dry admixture containing such materials (including any activity in which any of the aforesaid operations are carried out incidentally to the mining or quarrying of other minerals or to the manufacture of articles containing crushed or ground silica rock); or

(b) the handling of any of the materials specified in the foregoing sub-paragraph in or incidentally to any of the operations mentioned therein or substantial exposure to the dust arising from such operations

2 The breaking, crushing or grinding of flint, the working or handling of broken, crushed or ground flint or materials containing such flint or substantial exposure to the dust arising from any of such operations

3 Sand blasting by means of compressed air with the use of guartzose sand or crushed silica rock or flint or substantial exposure to the dust arising from such sand blasting

4 Work in a foundry or the performance of, or substantial exposure to the dust arising from, any of the following operations:

(a) the freeing of steel castings from adherent siliceous substance; or

(b) the freeing of metal castings from adherent siliceous substance:

(i) by blasting with an abrasive propelled by compressed air, steam or a wheel, or

(ii) by the use of power-driven tools.

5 The manufacture of china or earthenware (including sanitary earthenware, electrical earthenware and earthenware tiles) and any activity involving substantial exposure to the dust arising therefrom

6 The grinding of mineral graphite or substantial exposure to the dust arising from such grinding

7 The dressing of granite or any igneous rock by masons, the crushing of such materials or substantial exposure to the dust arising from such operations

8 The use or preparation for use of an abrasive wheel or substantial exposure to the dust arising therefrom

9 (a) Work underground in any mine in which one of the objects of the mining operations is the getting of any material;

(b) the working or handling above ground at any coal or tin mine of any materials extracted therefrom or any operation incidental thereto;

(c) the trimming of coal in any ship, barge, lighter, dock or harbour or at any wharf or quay; or

(d) the sawing, splitting or dressing of slate or any operation incidental thereto

10 The manufacture or work incidental to the manufacture of carbon electrodes by an industrial undertaking for use in the electrolytic extraction of aluminium from aluminium oxide and any activity involving substantial exposure to the dust therefrom.

11 Boiler scaling or substantial exposure to the dust arising therefrom

40	Byssinosis	The spinning or manipulation of raw or waste cotton or flax or the weaving of cotton or flax, carried out in each case in a room in a factory, together with any other work carried out in such a room
41	Mesothelioma }	(a) The working or handling of asbestos or any admixture of asbestos;
42	Lung cancer }	(b) the manufacture or repair of asbestos textiles or other articles containing or composed of asbestos;
43	Asbestosis }	(c) the cleaning of any machinery or plant used in any of the foregoing operations and of any chambers, fixtures and appliances for the collection of asbestos dust; or
		(d) substantial exposure to the dust arising from any of the foregoing operations

44	Cancer of the nasal cavity or associated air sinuses	1 (a) Work in or about a building where wooden furniture is manufactured; (b) work in a building used for the manufacture of footwear or components of footwear made wholly or partly of leather or fibre board; or (c) work at a place used wholly or mainly for the repair of footwear made wholly or partly of leather or fibre board 2 Work in or about a factory building where nickel is produced by decomposition of a gaseous nickel compound or in any process which is ancillary or incidental thereto
45	Occupational dermatitis	Work involving exposure to any of the following agents: (a) epoxy resin systems; (b) formaldehyde and its resins; (c) metalworking fluids; (d) chromate (hexavalent and derived from trivalent chromium); (e) cement, plaster or concrete; (f) acrylates and methacrylates; (g) colophony (rosin) and its modified products; (h) glutaraldehyde; (i) mercaptobenzothiazole, thiurams, substituted paraphenylene-diamines and related rubber processing chemicals; (j) biocides, anti-bacterials, preservatives or disinfectants; (k) organic solvents; (l) antibiotics and other pharmaceuticals and therapeutic agents; (m) strong acids, strong alkalis, strong solutions (eg brine) and oxidising agents including domestic bleach or reducing agents; (n) hairdressing products including in particular dyes, shampoos, bleaches and permanent waving solutions;

(o) soaps and detergents;

(p) plants and plant-derived material including in particular the daffodil, tulip and chrysanthemum families, the parsley family (carrots, parsnips, parsley and celery), garlic and onion, hardwoods and the pine family;

(q) fish, shell-fish or meat;

(r) sugar or flour; or

(s) any other known irritant or sensitising agent including in particular any chemical bearing the warning "may cause sensitisation by skin contact" or "irritating to the skin"

| 46 | Extrinsic alveolitis (including farmer's lung) | Exposure to moulds, fungal spores or heterologous proteins during work in:

(a) agriculture, horticulture, forestry, cultivation of edible fungi or malt-working;

(b) loading, unloading or handling mouldy vegetable matter or edible fungi whilst same is being stored;

(c) caring for or handling birds; or

(d) handling bagasse |
| 47 | Occupational asthma | Work involving exposure to any of the following agents:

(a) isocyanates;

(b) platinum salts;

(c) fumes or dust arising from the manufacture, transport or use of hardening agents (including epoxy resin curing agents) based on phthalic anhydride, tetrachlorophthalic anhydride, trimellitic anhydride or triethylene-tetramine;

(d) fumes arising from the use of rosin as a soldering flux;

(e) proteolytic enzymes;

(f) animals including insects and other arthropods used for the purposes of research or education or in laboratories; |

(g) dusts arising from the sowing, cultivation, harvesting, drying, handling, milling, transport or storage of barley, oats, rye, wheat or maize or the handling, milling, transport or storage of meal or flour made therefrom;

(h) antibiotics;

(i) cimetidine;

(j) wood dust;

(k) ispaghula;

(l) castor bean dust;

(m) ipecacuanha;

(n) azodicarbonamide;

(o) animals including insects and other arthropods (whether in their larval forms or not) used for the purposes of pest control or fruit cultivation or the larval forms of animals used for the purposes of research or education or in laboratories;

(p) glutaraldehyde;

(q) persulphate salts or henna;

(r) crustaceans or fish or products arising from these in the food processing industry;

(s) reactive dyes;

(t) soya bean;

(u) tea dust;

(v) green coffee bean dust;

(w) fumes from stainless steel welding;

(x) any other sensitising agent, including in particular any chemical bearing the warning "may cause sensitisation by inhalation"

NOTES

Initial Commencement

SPECIFIED DATE

Specified date: 1 April 1996: see reg 1.

PART II

DISEASES ADDITIONALLY REPORTABLE IN RESPECT OF OFFSHORE WORKPLACES

48

Chickenpox.

49

Cholera.

50

Diphtheria.

51

Dysentery (amoebic or bacillary).

52

Acute encephalitis.

53

Erysipelas.

54

Food poisoning.

55

Legionellosis.

56

Malaria.

57

Measles.

58

Meningitis.

59

Meningococcal septicaemia (without meningitis).

60

Mumps.

61

Paratyphoid fever.

62

Plague.

63

Acute poliomyelitis.

64

Rabies.

65

Rubella.

66

Scarlet fever.

67

Tetanus.

68

Tuberculosis.

69

Typhoid fever.

70

Typhus.

71

Viral haemorrhagic fevers.

72

Viral hepatitis.

NOTES

Initial Commencement

Specified date

     Specified date: 1 April 1996: see reg 1.

Appendix C

Prescribed Form for Reporting an Injury or Dangerous Occurrence

See overleaf

[A3037]

Prescribed Form for Reporting an Injury or Dangerous Occurrence

Health and Safety at Work etc Act 1974
The Reporting of Injuries, Diseases and Dangerous Occurrences Regulations 1995

HSE
Health & Safety
Executive

Report of an injury or dangerous occurrence

Filling in this form
This form must be filled in by an employer or other responsible person.

Part A

About you
1 What is your full name?

2 What is your job title?

3 What is your telephone number?

About your organisation
4 What is the name of your organisation?

5 What is its address and postcode?

6 What type of work does the organisation do?

Part B

About the incident
1 On what date did the incident happen?

/ /

2 At what time did the incident happen?
(Please use the 24-hour clock eg 0600)

3 Did the incident happen at the above address?

Yes ☐ Go to question 4

No ☐ Where did the incident happen?
☐ elsewhere in your organisation – give the name, address and postcode
☐ at someone else's premises – give the name, address and postcode
☐ in a public place – give details of where it happened

If you do not know the postcode, what is the name of the local authority?

4 In which department, or where on the premises, did the incident happen?

F2508 (01/96)

Part C

About the injured person
If you are reporting a dangerous occurrence, go to Part F.
If more than one person was injured in the same incident, please attach the details asked for in Part C and Part D for each injured person.

1 What is their full name?

2 What is their home address and postcode?

3 What is their home phone number?

4 How old are they?

5 Are they
☐ male?
☐ female?

6 What is their job title?

7 Was the injured person (tick only one box)
☐ one of your employees?
☐ on a training scheme? Give details:

☐ on work experience?
☐ employed by someone else? Give details of the employer:

☐ self-employed and at work?
☐ a member of the public?

Part D

About the injury
1 What was the injury? (eg fracture, laceration)

2 What part of the body was injured?

Continued overleaf

3 Was the injury (tick the one box that applies)

☐ a fatality?

☐ a major injury or condition? (see accompanying notes)

☐ an injury to an employee or self-employed person which prevented them doing their normal work for more than 3 days?

☐ an injury to a member of the public which meant they had to be taken from the scene of the accident to a hospital for treatment?

4 Did the injured person (tick all the boxes that apply)

☐ become unconscious?

☐ need resuscitation?

☐ remain in hospital for more than 24 hours?

☐ none of the above.

Part E

About the kind of accident

Please tick the one box that best describes what happened, then go to Part G.

☐ Contact with moving machinery or material being machined

☐ Hit by a moving, flying or falling object

☐ Hit by a moving vehicle

☐ Hit something fixed or stationary

☐ Injured while handling, lifting or carrying

☐ Slipped, tripped or fell on the same level

☐ Fell from a height

How high was the fall?

	metres

☐ Trapped by something collapsing

☐ Drowned or asphyxiated

☐ Exposed to, or in contact with, a harmful substance

☐ Exposed to fire

☐ Exposed to an explosion

☐ Contact with electricity or an electrical discharge

☐ Injured by an animal

☐ Physically assaulted by a person

☐ Another kind of accident (describe it in Part G)

Part F

Dangerous occurrences

Enter the number of the dangerous occurrence you are reporting. (The numbers are given in the Regulations and in the notes which accompany this form)

Part G

Describing what happened

Give as much detail as you can. For instance

- the name of any substance involved
- the name and type of any machine involved
- the events that led to the incident
- the part played by any people.

If it was a personal injury, give details of what the person was doing. Describe any action that has since been taken to prevent a similar incident. Use a separate piece of paper if you need to.

Part H

Your signature

Signature

Date

/	/

Where to send the form

Please send it to the Enforcing Authority for the place where it happened. If you do not know the Enforcing Authority, send it to the nearest HSE office.

For official use

Client number	Location number	Event number	
			☐ INV REP ☐ Y ☐ N

Appendix D

Prescribed Form for Reporting a Case of Disease

See overleaf

[A3038]

Prescribed Form for Reporting a Case of Disease

Health and Safety at Work etc Act 1974
The Reporting of Injuries, Diseases and Dangerous Occurrences Regulations 1995

Report of a case of disease

Filling in this form
This form must be filled in by an employer or other responsible person.

Part A

About you

1 What is your full name?

2 What is your job title?

3 What is your telephone number?

About your organisation

4 What is the name of your organisation?

5 What is its address and postcode?

6 Does the affected person usually work at this address?

Yes ☐ Go to question 7

No ☐ Where do they normally work?

7 What type of work does the organisation do?

Part B

About the affected person

1 What is their full name?

2 What is their date of birth?

/   /

3 What is their job title?

4 Are they
☐ male?
☐ female?

5 Is the affected person (tick one box)
☐ one of your employees?
☐ on a training scheme? Give details:

☐ on work experience?
☐ employed by someone else? Give details:

☐ other? Give details:

F2508A (01/96)

Continued overleaf

Part C

The disease you are reporting

1 Please give:

- the name of the disease, and the type of work it is associated with; or

- the name and number of the disease *(from Schedule 3 of the Regulations – see the accompanying notes).*

2 What is the date of the statement of the doctor who first diagnosed or confirmed the disease?

/ /

3 What is the name and address of the doctor?

Part D

Describing the work that led to the disease

Please describe any work done by the affected person which might have led to them getting the disease.

If the disease is thought to have been caused by exposure to an agent at work *(eg a specific chemical)* please say what that agent is.

Give any other information which is relevant.

Give your description here

Continue your description here

Part E

Your signature

Signature

Date

/ /

Where to send the form

Please send it to the Enforcing Authority for the place where the affected person works. If you do not know the Enforcing Authority, send it to the nearest HSE office.

For official use	
Client number	Location number
Event number	

☐ INV REP ☐ Y ☐ N

Appendix E

Health and Safety Executive Regional Offices

General HSE telephone number: 0845 345 0055

London

[A3039]

Rose Court
2 Southwark Bridge
London SE1 9HS
Fax: 020 7556 2102

East and South-east

AW House
6-8 Stuart Street
Luton LU1 2SJ
Fax: 01582 444320

Wren House
Hedgerows Business Park
Colchester Road
Springfield
Chelmsford CM2 5PF
Fax: 01245 706222

Lakeside 500
Old Chapel Way
Broadland Business Park
Norwich NR7 0WQ
Fax: 01603 828055

Priestley House
Priestley Road
Basingstoke RG24 9NW
Fax: 01256 404 100

Phoenix House
23-25 Cantelupe Road
East Grinstead RH19 3BE
Fax: 01342 334222

International House
Dover Place
Ashford TN23 1HU
Fax: 01233 634827

Midlands

1 Hagley Road
Birmingham B16 8HS
Fax: 0121 607 6349

900 Pavilion Drive
Northampton Business Park
Northampton NN4 7RG
Fax: 01604 738333

City Gate West
Level 6 (First Floor)
Toll House Hill
Nottingham NG1 5AT
Fax: 0115 971 2802

Lyme Vale Court
Lyme Drive
Parklands Business Park
Newcastle Road
Trent Vale
Stoke on Trent ST4 6NW
Fax: 01782 602400

Haswell House
St Nicholas Street
Worcester WR1 1UW
Fax: 01905 723045

Yorkshire and North East Division

Marshalls Mill
Marshall Street
Leeds LS11 9YJ
Fax: 0113 283 4382 (general
enquiries)
Fax: 0113 283 4296 (completed
F10 forms)

Edgar Allen House
241 Glossop Road
Sheffield S10 2GW
Fax: 0114 291 2379

Arden House

Regent Centre
Regent Farm Road
Gosforth
Newcastle upon Tyne NE3 3JN
Fax: 0191 202 6300

PSD – An agency of the Health
and Safety Executive
Mallard House
King's Pool, 3 Peasholme Green
York YO1 7PX
Telephone: 01904 455775
Fax: 01904 455733

North West Division

Knowledge Centre
Health and Safety Executive
(1G) Redgrave Court
Merton Road
Bootle
Merseyside L20 7HS

Marshall House
Ringway
Preston PR1 2HS
Fax: 01772 836 222

Grove House
Skerton Road
Manchester M16 0RB
Fax: 0161 952 8222

2 Victoria Place
Carlisle CA1 1ER
Fax: 01228 548482

Scotland

Belford House
59 Belford Road
Edinburgh EH4 3UE
Fax: 0131 247 2121

1st floor
Mercantile Chambers
53 Bothwell Street
Glasgow G2 6TS
Fax: 0141 275 3100

Lord Cullen House
Fraser Place
Aberdeen AB25 3UB
Fax: 01224 252 662
Offshore Safety Division:
Tel: 01224 252500
Fax: 01224 252662

Longman House
28 Longman Road
Longman Industrial Estate
Inverness IV1 1S
Fax: 01463 713459

Asbestos

Lawrence Waterman and Andrea Oates

Introduction

[A5001] Asbestos is the generic name for a group of naturally occurring fibrous minerals, metallic silicates, which have a wide range of industrial applications. They are divided into two sub-groups: serpentine (chrysotile (white asbestos)),which is the most commonly used type of asbestos and amphiboles, which includes crocidolite (blue asbestos), amosite (brown asbestos), tremolite, actinolite and anthophyllite, of which crocidolite was the most commonly used in the past.

Mined in Canada, South Africa, Russia and elsewhere, the material has been manufactured into products which use its characteristics of heat and chemical resistance. Despite its contribution to fire protection, asbestos is now regarded as one of the most significant causes of occupational disease over the past 125 years: the Health and Safety Executive (HSE) says it is the single greatest cause of work-related deaths in the UK.

The use of asbestos has been extensive, but as a result of prohibitions in the UK on new uses of asbestos-containing materials, products such as brake linings and other friction products are now unlikely to be found in most workplaces. However, its widespread application in materials used in construction has led to many buildings containing asbestos, and the current focus of asbestos controls is largely on the management of this legacy.

Asbestos diseases

[A5002] The diseases associated with asbestos only arise when asbestos fibres are inhaled and penetrate deep into the lung. Extremely small fibres are capable of bypassing the body's defence mechanisms to achieve this. These are therefore called respirable fibres and they are fibres with a diameter less than three micrometres. To illustrate how small these fibres are a human hair is approximately 50 micrometres in diameter.

Asbestosis is a thickening of the wall of the alveoli, the tiny lung sacs where oxygen passes into the blood. Since the lung has a huge surface area one or two areas of such thickening would have little effect. However when millions of fibres are inhaled and reach the deep lung the reduction in gas exchange capacity caused by thickening of the alveoli walls becomes significant. The disease of asbestosis is recognised when the patient has a measurable decrease in lung capacity and even notices a shortness of breath. Signs of the disease can also be detected by x-ray. If exposure is extensive and prolonged the damage to the lungs can be severe and lead eventually to death. The progress of the disease is proportional to exposure, which in modern terms must be massive. It has occurred in industries where raw asbestos was handled in bulk, when raw asbestos was delivered in bales which were manually cut open and fed into hoppers for the manufacturing process. Workers often went home white with asbestos fibres adhering to their body and clothes. A condition associated with asbestosis is pleural plaques: areas of calcification or stiffening on the membrane on the outer surface of the lung. Asbestosis is a very similar disease to coal workers pneumoconiosis and since the level of exposure needed to develop these diseases no longer occurs in Britain today there are virtually no new cases appearing but workers exposed in the past are still alive, suffering from asbestosis and dying of it.

Mesothelioma is a cancer occurring on the outer membrane of the lung. It is normally a very rare disease and so its occurrence amongst asbestos workers was quickly recognised. Typically it is not detected in the early stages and therefore once diagnosed is advanced and usually progresses to death in months rather than years. However the latency period from initial exposure to onset of the disease appears to be long, in the region of decades. The risk of contracting mesothelioma is based on exposure. The greater the concentration of respirable asbestos fibres inhaled the greater is the risk of contracting the disease. In addition, the concentration of respirable asbestos fibres required to cause the disease is far lower than that associated with asbestosis. In the UK mesothelioma was at first associated with crocidolite or blue asbestos and the control limit (the maximum level to which workers could legally be exposed) for blue asbestos was tightened. Historically it was recognised that all amphibole asbestos types were implicated, so amosite, brown asbestos, and crocidolite and the other amphibole asbestos types were assigned tighter control limits. Asbestos products are no longer manufactured in the UK and exposure is now to materials that contain mixtures of fibres rather than to a single asbestos type. Recent legislative updates have recognised this and all asbestos fibre types are now assigned a single lower control limit.

Lung cancer is now recognised as a disease associated with asbestos. The specific diagnosis of lung cancer being an asbestos-related disease is however occupational exposure. Since lung cancer occurring spontaneously, caused by smoking or caused by asbestos is indistinguishable it is the history of the patient which is used in the specific diagnosis. It is firmly established from epidemiology that there is an increased risk of lung cancer for smokers and that there is an increased risk of lung cancer for those exposed to asbestos. Indeed for smokers who are also exposed to asbestos it is estimated that the risk of lung cancer is multiplied rather than the two risks being additive.

When asbestos fibres are inhaled, they have a very long half life in the lungs. Their shape – long and thin – and robustness means that it is difficult for the usual lung clearance mechanisms to remove them from the lungs (in the manner in which dust particles are cleared), and they do not readily dissolve in lung fluid. Whilst the exact mechanisms which cause cancer remain obscure, the toughness and longevity of the fibres helps to explain why they are able to cause ill health many years after first exposure – although the abrasive and chemical effect of the fibres in very high lung concentrations can cause asbestosis on a shorter timescale this is largely an historical disease whilst the cancers continue to take their toll. If workers are exposed to airborne fibres, for example as a result of an incident in which asbestos has been inadvertently disturbed, there is no 'treatment' available to reduce the risk of future disease (except, of course to avoid inhaling cigarette smoke in future). However, it is worth noting that isolated exposures such as this represent a very (immeasurably) low level of risk. The significance of asbestos is that thousands of workers have been exposed on a daily basis to high levels, and that is why the last century saw an 'epidemic' of asbestos disease. This is not to argue that even single exposures to airborne asbestos should not be prevented, rather to encourage a recognition that the emphasis is on routine and continued exposure as this is the root of the problem.

Historical exposures – manufacturing

[A5003] In the 1950s and 1960s the predominant concern arising from exposure to asbestos was for manufacturing workers contracting asbestosis, following exposure to massive concentrations of respirable fibres. These workers were involved in manufacturing asbestos insulating boards and panels, asbestos cement products and friction products such as clutch plates, break shoes and gaskets. Anecdotally, such workers would refer to working in 'snow storms' of the material, which because the fibres are soft and silky did not create any immediate discomfort, as would be the case with glass fibre which is immediately irritating to the eyes, throat and skin. In order to control this risk, and against the background of a well-funded lobbying campaign by the asbestos mining and manufacturing interests (a campaign which still functions, as the Canadian government's challenge through the World Trade Organisation to the EU prohibitions on asbestos importation indicates) Asbestos Regulations were introduced imposing control limits on the maximum concentration of respirable asbestos fibres that workers could be exposed to. Where the concentrations could not be reduced below the control limits it was mandatory to provide suitable respirators. As the risk of lung cancer and mesothelioma was recognised and epidemiological data and analysis became available, the control limits for working with asbestos were gradually reduced. Originally, 'action levels' were adopted and defined in terms of concentrations of respirable airborne fibres measured over a twelve-week period. If the mean concentration of airborne fibres exceeded the action level then the area in the factory had to be designated an 'asbestos zone' and non-essential personnel excluded. If the airborne fibre concentration was likely to exceed the control limit then issuing and wearing respirators became mandatory. The original control regime included a lower control limit and action level for the amphibole asbestos fibre types than that of chrysotile asbestos. The enactment of the *Control of Asbestos Regulations 2006 (SI 2006 No 2739)* have now enforced a single control limit that has been set at 0.1 f/ml over a 4 hour TWA (Time Weighted Average) covering all asbestos fibre types and the action levels have been revoked.

Asbestos prohibitions

[A5004] Historically, the *Asbestos (Prohibition) Regulations 1992 (SI 1992 No 3067)*, as amended, prohibited all imported materials to which asbestos had been added as part of a deliberate manufacturing process, but excluded naturally-occurring minerals which may contain traces of asbestos. Many producers and importers of mineral products carry out comprehensive testing to prevent materials containing significant quantities of asbestos getting into the supply chain. Where asbestos is found the material should not be sold unless the amount is trivially small. Even in those cases suppliers must inform their customers that trace quantities of asbestos may occasionally be found in their products. This will allow 'high-energy' users (for example traces of asbestos in blast-cleaning materials would release airborne fibres) to take appropriate precautions or consider alternative materials. In addition to prohibitions on the sale of asbestos products, where they may be discovered in store rooms or warehouses, they may not be used and must be disposed of safely. Whilst there is no prohibition on asbestos which is already in place, new uses are effectively banned.

Asbestos removal

[A5005] As the manufacture and use of asbestos declined, accelerated by the *Asbestos (Prohibition) Regulations 1992 (SI 1992 No 3067)*, the risk to the health of manufacturing workers also decreased. However, it was recognised that there was a new exposure group and attention turned to those removing asbestos. The *Asbestos (Licensing) Regulations 1983 (SI 1983 No 1649)* were introduced to provide the enforcing authorities with information on which companies were involved in removing asbestos from buildings and where and when the work was being carried out. Approved Codes of Practice (ACOPs) aimed directly at the removal industry provided information and instruction to those in charge of asbestos removal operations on meeting the requirements of these Regulations.

Following the introduction of the *Control of Asbestos Regulations 2006 (SI 2002 No 2739)*, new editions of the Approved Codes of Practice have been published, including *Asbestos: The Licensed Contractor's Guide* (HSG 247) and *Asbestos: The analysts' guide for sampling, analysis and clearance procedures* (HSG 248) which recommends that after asbestos has been removed from an area it should be visibly free from dust and debris and that the airborne fibre concentration measured in the area should be below 0.01 fibres/ml. This is the detection limit of the optical microscopy techniques used to measure airborne fibres and is referred to as the 'clearance indicator'. The clearance indicator is simply a measure of cleanliness of an area and is not a risk based limit such as the control limit. A full list of asbestos legislation, Approved Codes of Practice and Guidance Notes are appended at the end of this chapter (see A5030).

Exposure of building maintenance workers

[A5006] As the *Asbestos (Licensing) Regulations 1983 (SI 1983 No 1649)* were enforced, and the asbestos removal industry became more thoroughly regulated, the risk to the health of the workers in that industry began to be more effectively controlled. However, in the late 1990s it was recognised that there was still a very large group of workers whose exposure to asbestos was not adequately controlled by the existing asbestos control regime. Maintenance workers and those engaged in building refurbishment were still exposed to asbestos because many building owners did not know that asbestos was present and many employers of maintenance workers did not consider that they would be working with asbestos and so took no precautions. For these reasons, the *Control of Asbestos at Work Regulations 2002 (SI 2002 No 2675)*, Regulation 4 introduced a duty to manage asbestos within non-domestic premises which became enforceable in 2004 and continues to be in force within the requirements of the *Control of Asbestos Regulations 2006 (SI 2006 No 2739)*. The provisions require building owners and occupiers to identify where asbestos is present in every building and have a documented plan to control the risks to health presented by that asbestos. The aim is to reduce over time the number of asbestos-related deaths in the UK. Past exposure to asbestos was estimated to result in more than 3,000 deaths per year, and this figure is likely to continue to rise further for another 10 years or so. When we look at current exposure levels the HSE believes that 7,800 individuals will die of asbestos related disease over the next 100 years as a result of exposure over the next 50 years. Of these 4,500 individuals will be from occupational exposure, 2000 from indirect or work related exposure and 1,300 from miscellaneous exposure sources.

Duties under the Control of Asbestos Regulations 2006

[A5007] The following legislation previously governing asbestos within the UK was repealed by the enactment of the Control of Asbestos Regulations 2006 on the 13 November 2006:

— the *Asbestos (Licensing) Regulations 1983 (SI 1983 No 1649)*
— the *Asbestos (Prohibitions) Regulations 1992 (SI 1992 No 3067)*
— the *Control of Asbestos at Work Regulations 2002 (SI 2002 No 2675)*

The *Control of Asbestos Regulations 2006 (SI 2006 No 2739)* ('CAR') amalgamated the requirements of the above regulations along with some key changes to legislative requirements. These include:

— the removal of Decorative Textured Coatings from the licensing regime
— the introduction of a single, lower control limit for all asbestos fibre types
— the removal of the complex 12 week action levels and the introduction of the "Work of Sporadic or Minor Intensity" Derogation
— Stringent emphasis on training requirements for asbestos workers and those sporadically exposed to asbestos.

The *Control of Asbestos Regulations 2006 (SI 2006 No 2739)* ('CAR') place duties on every employer in respect of his employees, and make it clear that these duties, so far as is reasonably practicable, are owed also to any other person, whether at work or not, who may be affected by the work activity undertaken by the employer. The exceptions are in connection with *Regulation 10* of the 2006 Regulations where the duty to inform, instruct and train does not apply to non-employees unless they are in the premises where the work is being carried out, and *Regulation 22* where the duty to carry out medical surveillance and maintain health records for exposed workers only applies to the employer's own employees. The *CAR* apply to the self-employed, applying the duties of both employer and employee (*Regulation 3(1)*).

Under CAR there is a duty which applies to all companies or parties which own, occupy, manage or have any contractual responsibilities for maintenance or repair of non-domestic premises. The duties apply to all business, commercial, industrial, educational, healthcare, military, administrative, retail and leisure properties including hotels and guest houses. Even in the absence of agreement, all parties which control access or maintenance of a building could still be liable should incidental exposure occur where it can be shown that asbestos has not been managed. Duty holders are under a duty either to:

—   manage the risk from asbestos-containing materials ('ACMs'); or
—   co-operate with whoever manages that risk.

Affected premises

[A5008] The duty applies not only to all non-domestic premises, but also applies to the common parts of commercial buildings, housing developments and leasehold flats. Given the widespread use of asbestos, particularly after the 1940s, many buildings still contain asbestos. An estimated 500,000 premises are affected and a number of different parties with any measure of control need to take steps to comply.

Identifying the duty holder

[A5009] Persons with a primary responsibility for a building, such as the owner, sole occupier or part occupier, have duties under the *CAR (SI 2006 No 2739)*. Those with any responsibility for the maintenance of premises by virtue of a contract or tenancy also have a legal duty to manage the risks from asbestos and comply with the *CAR*. If no such formal agreement exists, the owner, sub-lessor or managing agent in charge of a workplace premise, or any employer or self-employed person who occupies commercial premises in which people work, will be subject to the same duty.

Risk assessment

[A5010] The duty holder is required to make a suitable and sufficient risk assessment as to whether asbestos is, or is liable to be present in the premises. It should be noted that wherever a duty such as this applies, compliance requires that the duty holder ensures that it has been carried out. In this case, this means ensuring that a suitable and sufficient assessment has been conducted, documented and is available and used to inform decision-making, advise incoming occupiers, maintenance workers and others. Commissioning others to carry out such work is acceptable practice, so long as the persons employed are competent and allocated sufficient resources to carry out the work properly.

This includes taking into account the age of the premises, building plans and any other relevant information. In carrying out inspections it must be presumed that materials contain asbestos unless there is strong evidence to the contrary. At this stage it is necessary simply to identify whether asbestos may be present and, therefore, determine whether a more thorough investigation is required. It is not, however, sufficient to conclude that a building is free of asbestos simply because it was built after 1990. That conclusion can only be reached if there is documentary evidence from a designer or architect who specified that no asbestos was to be used in the construction of the building. The most likely asbestos materials to be found are:

- Sprayed asbestos and asbestos loose packing – typically used as fire breaks in ceiling voids and to protect steelwork, fire protection in ducts, firebreaks, etc.
- Moulded or preformed lagging – generally used in thermal insulation of pipes and boilers.
- Insulating board – fire protection, thermal insulation, panelling, partitioning, ducts.
- Ceiling tiles.
- Millboard and paper products – insulating electrical products and facing wood fibreboard.
- Woven products – boiler seals and old fire blankets.
- Cement products in flat or corrugated sheets.
- Textured coatings.
- Bitumen roofing materials.
- Vinyl and thermoplastic floor tiles.

Records

[A5011] If asbestos is present, an up-to-date record of the location and condition of asbestos-containing materials ('ACMs') or presumed ACMs in the premises must be made. This is often referred to as an 'asbestos register'. The asbestos register should record where asbestos is present in the building the form in which it is present, e.g. asbestos insulating board, pipe lagging, ceiling tiles, asbestos cement sheets, and the condition of the material. The health risk depends on the inhalation of respirable fibres, so that the condition of the asbestos, as recorded in the asbestos register, should reflect the extent to which fibres may be released from the ACM during normal occupancy and during any maintenance work. Since maintenance work will be required at some stage, and it is necessary to determine the type of asbestos present before such work commences, it is a requirement to identify the type of asbestos present in the ACM and record it in the asbestos register. This requires samples of suspected ACM to be taken for laboratory analysis in accordance with HSG248 "The Analyst's Guide" by a suitably accredited laboratory in accordance with CAR regulation 21. The overall purpose of the asbestos register is to provide information regarding the locations of ACMs to assist in the planning of work to prevent incidental exposure to asbestos fibres. Selecting the correct format of an asbestos register is paramount in ensuring effective asbestos management – for example, electronic asbestos registers are easily updated and controlled and can provide easy access across multiple sites; However hard copy information may be more appropriate for single sites or where computer access is restricted.

Management plan

[A5012] The compilation of an asbestos register is only one part of the duty to manage asbestos. There must also be a documented plan to manage the health risks presented by the presence of ACMs. Where the health risk during normal occupancy is low, the plan may be to maintain the ACM in place and label it. There should then be written procedures for planned maintenance work and procedures for emergency work, and staff familiar with their duties in implementation. These procedures would normally include calling in licensed asbestos contractors to remove the ACM under controlled conditions before the maintenance work proceeded. If the presence of asbestos in the building represents a risk to health under normal occupancy conditions or it is envisaged that emergency situations would involve the removal of ACM, then the asbestos management plan may be to immediately encapsulate or re-move ACMs for particular areas. This work requires the involvement of contractors licensed under the *Control of Asbestos Regulations (SI 2006 No 2739) Regulation 8* working in accordance with the Approved Code of Practice (HSG 247) 'The Licensed Contractors Guide". Where asbestos removal operations are undertaken, it is a requirement under Regulation 20 for air testing to be undertaken by a UKAS accredited laboratory under BS EN ISO/IEC 17025:2005. It is considered best practice to retain the services of a United Kingdom Accreditation Service (UKAS) accredited laboratory to assist in both the monitoring and management of asbestos removal work. The asbestos management plan should also contain the procedures for dealing with suspect ACM. Typically, such procedures would include the requirement to halt the work and call in a UKAS accredited laboratory to take samples and determine whether asbestos is present. The asbestos management plan should be reviewed regularly and the procedures and arrangements monitored to ensure that they are effective. It is important to test the procedures to ensure that they work. For instance, when routine maintenance is undertaken, checks that the tradespeople concerned were furnished with the appropriate informa-tion and that they followed the correct procedures should be made and recorded. In many commercial premises the effect of the *CAR's* duty to manage may create more than one duty holder. Here it is necessary to examine the nature and extent of the repair and maintenance obligation owed by each party, or the level of control in determining their relative contributions to comply.

Asbestos

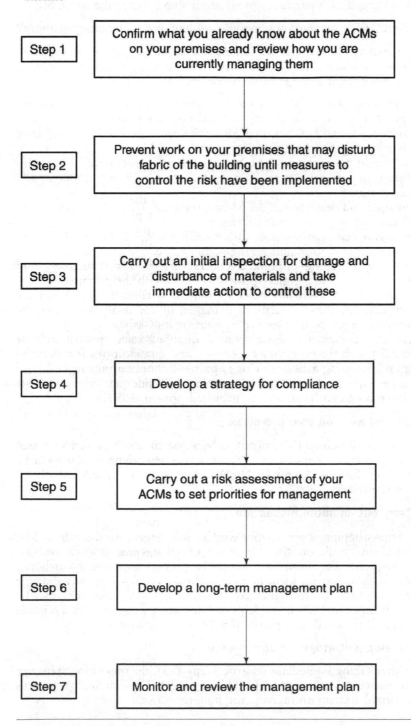

Step 1 — Confirm what you already know about the ACMs on your premises and review how you are currently managing them

Step 2 — Prevent work on your premises that may disturb fabric of the building until measures to control the risk have been implemented

Step 3 — Carry out an initial inspection for damage and disturbance of materials and take immediate action to control these

Step 4 — Develop a strategy for compliance

Step 5 — Carry out a risk assessment of your ACMs to set priorities for management

Step 6 — Develop a long-term management plan

Step 7 — Monitor and review the management plan

To ensure the effective compliance with the duty to manage asbestos in non-domestic premises, there is a simple seven step programme to follow.

Step 1: What ACMs do you know about and how are you managing them?

[A5013] Are there likely to be ACMs on your premises? If there are none, no further action is required. If ACMs are present, or there is uncertainty, a review of current management arrangements is required. Check that:

- Maintenance and building work is effectively controlled, with a system to prevent disturbance of asbestos and unknown material.
- Maintenance and building staff know the rules about ACMs, and how to work safely on or adjacent to them.
- The condition of known ACMs or materials presumed to contain asbestos is known, there is a system to check for damage and deterioration.
- Damaged and deteriorated ACMs are repaired, removed or isolated to prevent exposure to released fibres.
- There is a management plan for recording findings on ACMs, monitoring their condition and taking action as necessary.
- Everyone, including appointed contractors, knows their roles and responsibilities for the management of ACMs on the premises.
- A responsible person has been appointed, is competent and has the resources (premises or facilities manager, or for small businesses the owner/manager) to oversee the management of asbestos.
- Access to competent advice from a qualified safety practitioner or hygienist with the necessary experience and qualifications for identifying and assessing asbestos. This is particularly important in buildings that are known or suspected to contain significant quantities of ACMs or have been constructed or refurbished prior to 1999.

Step 2: Control work on your premises

[A5014] In every building, there needs to be a system which prevents anyone doing building or maintenance work, even running new telephone cables or IT systems, without first checking whether ACMs may be disturbed (usually by reference to the Asbestos Register).

Step 3: Carry out an initial inspection

[A5015] After stopping new activities which could damage or disturb ACMs, it is essential to check on the condition of all existing known or suspected ACMs, and take immediate action to prevent exposure to airborne fibres where such damage is identified. This check is looking for *any* damage to building materials. If any damaged materials are identified, they can be tested to see if they are ACMs, and kept isolated until the results are available and the material removed or repaired if asbestos is confirmed.

Step 4: Develop a strategy for compliance

[A5016] After taking immediate control (steps 1–3), the remaining steps are designed to manage occupancy and building works. The first such step is to develop a formal strategy to manage the long-term risks:

- Commission a survey to establish the extent of ACMs.
- Document the management controls to minimise risk of damage to ACMs, and to identify and act on damage and deterioration effectively.

Step 5: Risk assessment and priorities

[A5017] The HSE have developed a quantitative method for assessing the risks arising from asbestos materials. An asbestos surveyor will normally undertake an assessment of the materials condition in accordance with *MDHS100 "Surveying, Sampling and Assessment of Asbestos-Containing Materials"* this will define the materials potential to release fibres. The second stage of the assessment deals with the likelihood of human exposure and is the responsibility of the duty holder in accordance with CAR Regulation 4. This risk assessment procedure is detailed within HSE Guidance document HSG 227 "A Comprehensive Guide to Managing Asbestos in Premises".

Step 6: Develop a long-term management plan

[A5018] The plan should contain:

- How the location and condition of known or presumed ACMs is recorded.
- Priority assessments and an action plan.
- Decisions about management options and their rationale.
- Action timetable.
- Monitoring arrangements.
- Employees and their responsibilities.
- Contractor procedures including selection, appointment, information and monitoring.
- Training arrangements.
- Procedures and plan for implementation, including controlling work on the fabric of the building.
- Method for passing on information about ACMs to those who need it.
- Responsibility for the management plan and its updating.
- Procedure and timetable for review of the management plan.

Step 7: Monitor and review the management plan

[A5019] Implementation of the management plan should be monitored and checks routinely carried out on the ACMs, and records should be kept up-to-date. Incidents and accidents require careful investigation and lessons learned should be used to improve the plan and its implementation. For most premises it is recommended that the plan is subject to at least a 12 monthly thorough review. Materials should be re inspected according to the level of risk they present. For example an ACM in a locked, disued area is less likely to become damaged than an ACM in a thoroughfare. Therefore the material most likely to be disturbed should be monitored more routinely.

Duties of third parties

[A5020] The *CAR (SI 2006 No 2739)* places any person under a duty to assist in the management of asbestos. Third parties will be required to co-operate with the duty holder so far as is necessary to enable the duty holder to comply with the *CAR* by:

- making all information in relation to the premises available for inspection;
- providing them with information on the locations of asbestos, if known;
- allowing duty holders to identify asbestos or presumed asbestos for all parts of the premises repaired or maintained by the third party;
- assisting in the preparation and implementation of the asbestos management plan to control the health risk presented by ACMs;
- ensuring that everyone potentially at risk from ACMs receives information on the location and condition of the material so far as it is within the third party's control.

CAR contractual and tenancy agreements

[A5021] Organisations need to review the extent of their maintenance and repair obligations under any leases and be clear as to where the responsibility lies for asbestos control. Responsibility for dealing with any asbestos health risks, and therefore each party's contribution, should be considered in pre-contract lease negotiations and specific provisions should be made in the tenancy agreement. This will be especially important in determining who is responsible for common parts of the building where, for example, control over maintenance and repair activities in those parts could be shared.

Minimum requirements

[A5022] It will be a defence to show that an employer has done everything in his power to prevent exposure to asbestos. He will need to be able to produce an up-to-date asbestos register in a suitable format and a documented asbestos management plan containing risk assessments, procedures and arrangements to manage the risk from exposure to asbestos during normal occupancy, planned maintenance and emergency situations. Evidence will also be required to demonstrate that the asbestos management plan has been implemented and that it is effective. The *CAR (SI 2006 No 2739)* is accompanied by an ACoP *'The management of asbestos in non-domestic premises'* (L127), which gives advice about ways to identify asbestos, the associated health risks and how to develop an asbestos management plan to manage those risks.

Asbestos building surveys

[A5023] There are three types of asbestos surveys described in the Health and Safety Executive (HSE) publication MDHS 100 *'Surveying, sampling and assessment of asbestos-containing materials'*. The output from an asbestos survey is a detailed report on the location of all ACMs (whether presumed to contain asbestos or sampled) within a building, in a form which can be readily understood and used by those who may be exposed. This usually means that the information is graphically displayed on a plan of the building. Photographs of the materials identified are also normally included to assist in the identification of the ACM position. The type of survey needed depends on the use of the building, its age and the need for refurbishment.

Type 1 survey: Identification & Condition Assessment

[A5024] The Type I survey involves study of the building history, structure and plans to consider whether asbestos may be present. This is followed by a detailed inspection of the building to assess whether there are materials in place which have the appearance of ACMs. The outcome of a Type I survey may be the conclusion that the building is of a type which should not contain asbestos and that no suspect materials were seen, or that the building may contain asbestos and there were suspect materials which should be investigated, in which case, materials that are suspected of contain asbestos should be assessed and reported in accordance with *MDHS100 Surveying, Sampling and Assessment of Asbestos-containing Materials* The condition of the ACM will be assessed in relation to the propensity for the release of respirable airborne fibres and so the consequent risk to health. This risk assessment will depend not only on the nature and condition of the ACM but also on its location and whether it is likely to be disturbed by normal occupancy or only during maintenance and refurbishment. This information is used to compile the asbestos register required in the *CAR (SI 2006 No 2739)*. At this stage the building owner or occupier may rely on the surveyors presumptions made in the survey report that all suspect materials noted contain asbestos and draw up an asbestos management plan or commission sampling of the materials to prove or refute the surveyor's judgement. Although this process can be undertaken in-house, only a qualified and trained asbestos surveyor will have the necessary knowledge and experience of the different types of ACM to complete a thorough and effective Type 1 asbestos survey.

Type 2 survey: Identification, Sampling & Condition Assessment

[A5025] A Type 2 survey involves the same survey requirements as a Type 1 survey, however samples of suspect materials are taken at the time of inspection for laboratory analysis. Type 2 surveys should only be undertaken by organisations which hold UKAS accreditation for asbestos sampling with accreditation to ISO 17025 and comply with the provisions of 'General Criteria for the Operation of Various Types of Bodies Performing Inspection' BS EN ISO/IEC 17020:2004. Along with all the requirements for a Type 1 Survey report the Type 2 survey report will also include the laboratory analysis certificates of the type of asbestos present, e.g. amosite, crocidolite, chrysotile, "No Asbestos Detected" etc.

Type 3 surveys: Quantitative for Refurbishment & Demolition

[A5026] The outcome of a Type 3 survey is used as part of the specifications given to licensed asbestos removal contractors to tender for asbestos removal contracts prior to refurbishment or building demolition. The Type 3 survey report contains details of the location and type of all ACMs and the type of asbestos present. However, the report also specifies the volumes of each ACM present. This allows contractors to decide how the ACM should be removed and the extent of the operation and, therefore, provide a method statement for removal and an estimate of costs. A Type 3 survey is also required to eliminate asbestos risks when undertaking refurbishment works by intrusive means prior to work commencing.

Asbestos waste

[A5027] Most asbestos waste arising from asbestos removal operations, including asbestos cement is classified as a 'special waste' under the *Hazardous Waste (England and Wales) Regulations 2005 (SI 2005 No 894)*. Consequently, the waste must be transported in suitable containers and be adequately described in a consignment note in accordance with the *Environmental Protection (Duty of Care) Regulations 1991 (SI 1991 No 2839)*.

Work With Asbestos

[A5028] The "Sporadic and Low Intensity Derogation" ultimately replaced the old action levels which placed specific requirements on the employers of those undertaking work with asbestos.

The sporadic and low intensity derogation effectively exempts the duty holder from the detailed record keeping, health surveillance, personal dosimetry etc that is required of those working with licensed materials, should that work be defined as sporadic and low intensity.

Should work with asbestos be defined as sporadic and low intensity all of the other requirements of the regulations, such as risk assessment, training, emergency planning, and plans of work, still apply. The risk assessment process may still in many cases mean that the only people suitably placed to undertake the work without risk to health are licensed asbestos specialists.

In order to clarify and define work that may be of sporadic and low intensity, a fibre count test of 0.6 f/ml over 10 minutes TWA has been set. Work which can be controlled to levels below this figure would be considered as sporadic and low intensity.

Examples of work that fit the sporadic and low intensity derogation are summarised in CAR. For example

—   Work that involves short non-continuous maintenance activities (such as removing 1 gasket)

— Removal of materials in which the asbestos fibres are firmly linked in a matrix (i.e. decorative textured coatings)

— Encapsulation or sealing asbestos materials that are in good condition

All other work with asbestos is legally required to be undertaken by a licensed asbestos contractor.

Licensing of asbestos works

[A5029] Asbestos contractors can be granted a license by the HSE to perform work with ACMs. The licenses are active for a period of up to three years and may contain specific conditions.

The provision within the repealed licensing regulations which stated that if an employer wishes to undertake work with asbestos on their own premises using their own staff they could do so without a license has been removed – this employer would now need a licence.

Notification to the HSE is now required for all work not considered to be of a minor nature, or if it is a condition of a contractor's license that they must notify.

If licensable work is undertaken on an ACM – to repair and encapsulate it, or to remove it, then there is a simple sequence which is required:

(1) Identification of the type of asbestos or assumption that it is asbestos.

(2) Plan of work – detailing the nature and probably duration of the work, the location, the work methods and controls (based on the risk assessment) and the equipment to be used to protect and decontaminate the workers and others nearby. The plan should also detail emergency procedures.

(3) Notification – if a licensed contractor is employed, he may have this requirement as a condition of license.

(4) Information, instruction and training as required for those who may be affected by the works and/or may be required to supervise etc.

(5) Once works are underway, the licensed operator is required to comply with a range of technical duties for the containment of asbestos, the labelling of the containment area, the protection of the workers engaged in ACM work and their decontamination, and the handling of waste. During the works the efficacy of the containment and other controls should be checked by air testing carried out by a body independent of the contractor, should be accredited by UKAS.

(6) On completion of ACM work within a contained area, the premises must be assessed to determine that they are clean and fit for normal occupation. Such testing, to be carried out by a body independent of the contractor, should be accredited by UKAS. This procedure is detailed in HSG248 "The Analysts Guide".

Legislation, Approved Codes of Practice, and Guidance Notes

[A5030] Legislation, Approved Codes of Practice and Guidance Notes dealing with asbestos for England and Wales include, but not exclusively, those listed below. Equivalent documentation exists for Scotland and Northern Ireland respectively.

All legislation, Approved Codes of Practice and Guidance Notes listed together with any subsequent amendments or revisions and any new relevant requirements should be considered before undertaking any work with asbestos or ACMs.

The following legislation, Approved Codes of Practice and Guidance Notes deal primarily with asbestos. Other legislation dealing with health and safety matters has not been listed here, although such legislation still applies to work with asbestos and should be considered at all times.

The following list was last revised in July 2009.

Legislation

- *Health and Safety at Work etc. Act 1974;*
- *Environmental Protection Act 1990;*
- *Environment Act 1995;*
- *Consumer Protection Act 1987 (Commencement No 1) Order 1987 (SI 1987 No 1680);*
- *Control of Asbestos Regulations 2006 (SI 2002 No 2739);*
- *Control of Asbestos in the Air Regulations 1990 (SI 1990 No 556);*
- *Reporting of Injuries, Diseases and Dangerous Occurrence Regulations 1995 (SI 1995 No 3163);*
- *Management of Health and Safety at Work Regulations 1999 (SI 1999 No 3242);*
- *Workplace (Health, Safety and Welfare) Regulations 1992 (SI 1992 No 3004);*
- *Personal Protective Equipment at Work Regulations 1992 (SI 1992 No 2966);*
- *Provision and Use of Work Equipment Regulations 1998 (SI 1998 No 2306);*
- *Construction (Design and Management) Regulations 2007 (SI 2007 No 320);*
- *Carriage of Dangerous Goods and Use of Transportable Pressure Equipment Regulations 2009 (SI 2009 No 1348)*
- *Carriage of Dangerous Goods by Rail Regulations 1996 (SI 1996 No 2089) (revoked in part by SI 2004 No 568) (as amended);*
- *Health and Safety (Fees) Regulations 2009 (SI 2009 No 515);*
- *Environmental Protection (Duty of Care) Regulations 1991 (SI 1991 No 2839 (as amended));*

- *Hazardous Waste (England and Wales) Regulations 2005 (SI 2005 No 894)* (as amended);
- *Waste Management Licensing Regulations 1994 (SI 1994 No 1056) (as amended)*;
- *Controlled Waste Regulations 1992 (SI 1992 No 588)* (as amended);
- *Controlled Waste (Registration of Carriers and Seizure of Vehicles) Regulations 1998 (SI 1998 No 605)*;
- *Environmental Permitting (England and Wales) Regulations 2007 (SI 2007 No 3538)* (as amended).

Approved Codes of Practice

- The Management of Asbestos in Non-Domestic Premises, Regulation 4 of the *Control of Asbestos Regulations 2006* (L127)
- Work with Materials Containing Asbestos *Control of Asbestos Regulations 2006* Approved Code of Practice L143 (2006)
- 'Respiratory Protective Equipment, legislative requirements and lists of HSE approved standards and type approved equipment' (Fourth edition, 1995).

Health and Safety Executive Guidance Notes

Methods for the Determination of Hazardous Substances Series

- MDHS 100 – 'Surveying, sampling and assessment of asbestos-containing materials' (July 2001).
- MDHS 39/4 'Asbestos Fibres in Air' (1995).

Health and Safety Guidance Series

- HSG247 — Asbestos: The Licensed Contractors Guide
- HSG248 — Asbestos: The Analyst's Guide for Sampling, Analysis and Clearance Procedures
- HSG189/2 – Working with Asbestos Cement
- HSG 53 Respiratory Protective Equipment at Work – A Practical Guide (2005)
- HSG 210 – Asbestos Essentials – a task manual for building, maintenance and allied trades on non-licensed asbestos work (2008)
- HSG 213 – Introduction to asbestos essentials: Comprehensive guidance on working with asbestos in the building maintenance and allied trades (2003);
- HSG 227 – A comprehensive guide to managing asbestos in premises (2002);
- HSG 223 – A short guide to managing asbestos in premises (2002).

Business Continuity

Lawrence Bamber

What is business continuity planning?

[B8001] Business continuity planning ('BCP') is not just about disaster recovery, crisis management, risk management or IT. It is a business risk management issue. It presents an opportunity to review the way an organisation performs its business processes, to improve procedures and practices, and increase resilience to interruption and loss.

To quote the Business Continuity Institute, the professional body for BCP:

> BCP is the act of anticipating incidents which will affect critical functions and processes for the organisation, and ensuring that it responds to any incident in a planned and rehearsed manner.

The Turnbull Committee 'Guidance for Directors on Internal Controls' sets out an overall framework of best practice for business, based upon an assessment and control of its significant risks. For many companies, BCP will address some of these key risks and help them to achieve compliance.

Hence BCP as an activity has two primary objectives:

- minimise the risk of a disaster befalling the organisation; and
- maximise the ability of the organisation to recover from a disaster.

It differs from disaster recovery by recognising that approximately 80 per cent of disasters which organisations suffer are generated internally. As a result, BCP places more emphasis on managing out potential causes of disasters, than on recovering from those incidents which may not be avoided.

The BCP is therefore a combination of disaster/risk avoidance and the ability to recover.

Business continuity planning in context

[B8002] Modern businesses cannot avoid all forms of corporate risk or potential damage. A realistic objective is to ensure the survival of an organisation by establishing a culture that will identify, assess and manage those risks that could cause it to suffer:

- inability to maintain customer services;
- damage to image, reputation or brand;
- failure to protect company assets;
- business control failure;

- failure to meet legal or regulatory requirements.

BCP therefore provides the strategic framework to achieve these objectives.

Why is business continuity planning of importance?

[B8003] Very simply the answer to the above question is that disasters kill businesses.

In the UK, 80 per cent of companies who suffered a major disaster and did not have some form of BCP capability went into liquidation within eighteen months. A further 10 per cent suffered the same fate within five years.

Without BCP companies have only a one in ten chance of survival. Also, simply because:

- Customers expect continuity of supply in all circumstances.
- Shareholders expect directors/managers to be fully in control, and to be seen to be in control of any crisis.
- Employees and suppliers expect the business to protect their livelihoods.
- The company's reputation and brand is at risk without BCP.
- It is implicit in good corporate governance and demonstrates best practice in business management.

All the above needs to be taken into account against a business background in which:

- The pressures on business generally are changing and increasing.
- Technology is transforming and underpinning the business environment.
- Consolidation, restructuring, re-engineering, take-overs, acquisitions and mergers etc are now a fact of life.
- New risks and exposures are continually being created.

Business continuity drivers

[B8004] The drivers or arguments that need to be understood by all involved in business continuity may be listed and grouped as follows:

- Commercial.
- Regulatory.
- Standards – Best Practice
- Financial.
- Imported disaster.
- Public relations.

Commercial drivers

[B8005] An organisation may operate in an intensively competitive market where any disruption to normal operation will be exploited by competitors.

For example, when Perrier were forced into a total product withdrawal following a benzene contamination at source, the gaps on supermarket shelves were promptly filled by 30–40 competing brands.

The organisation may depend on a high degree of public confidence to generate/maintain sales, e.g. air transport industry. The issue for many organisations is not merely one of lost revenue, but also of lost customers, who may prove very difficult to get back. On occasions, it has been observed that companies who have suffered a disaster have spent three times their normal PR budget in an attempt to reassure customers that they are still around and worth doing business with.

Conversely, a visible and demonstrable BCP capability can be used to enhance and differentiate products and services.

Quality and regulatory drivers

[B8006] The main regulatory driver within the UK is the *Civil Contingencies Act 2004* (see B8044 below) which predominantly affects the role of local authorities (LAs) in the provision of business continuity support and advice.

The Act came into force in May 2006 and contains emergency powers that place duties and responsibilities on how organisations should respond to disasters.

Standards – Best practice

[B8007] The British Standards Institution (BSI) has published a two-part Standard on Business Continuity Management (BCM).

BS25999-1: 2006
• Business continuity management – Part 1: Code of practice

BS25999-2: 2007
• Business continuity management – Part 2: Specification

Part 1 essentially outlines the content of the Standard and Part 2 gives a steer on how the BCM process should be developed, implemented and maintained.

Part 1: Code of practice

This document presents an overview of the BCM process in section 1, 2 and 3 which are essentially narrative. The key sections are:

• Section 4: BCM policy
• Section 5: Programme management
• Section 6: Understanding the (ie your) organisation
• Section 7: Determining BCM strategy
• Section 8: Developing and implementing a BCM response
• Section 9: Exercising, maintaining and revising the BCM arrangements
• Section 10: Embedding BCM into the organisation's culture

Part 2: Specification for BCM

This document contains six sections. Sections 1 and 2 are descriptive but section 3, 4, 5 and 6 are auditable and follow the 'Plan, Do, Check, Act'

(PDCA) and 'Policy, Organisation, Plan, Implement, Monitor, Audit, Review' (POPIMAR) approaches to systems management. The key sections are:

Section 3: BCM planning
- General requirements;
- establishing, managing and maintaining the BCM system;
- embedding BCM into the organisational culture;
- BCM documentation and records.

Section 4: Implementation and operation
- Understanding the organisation;
- identification of critical activities/critical business processes;
- developing the BCM strategy;
- developing and implementing the BCM response;
- exercising, maintaining and reviewing the BCM arrangements.

Section 5: Monitoring and review
- internal audit: monitor and review of BCM system
- management review of BCM system

Section 6: Maintenance and improvement
- Preventive and corrective actions
- continual improvement.

Financial drivers

[B8008] Traditional financial measures of mitigating or cushioning against risk, such as insurance, are an important protective element within the overall organisational business risk management framework. Insurance on its own however has limitations which do not protect the complete business cycle.

Despite the fact that business interruption insurance is a fairly common component in most commercial insurance packages, cover is normally limited to two years from the event, and certain consequential losses, which may contain significant unquantifiable costs, will be excluded and hence non-recoverable.

For example, BI (business interruption) insurance does not cover the risk of lost customers, lost productivity or the attrition costs of key staff following a disaster. In some cases, this could be as much as 40 per cent of the total loss.

In some organisations, the disaster itself may be caused by, or made worse because of, loss of control over finances and cash flow, e.g. Barings, AIB.

The imported disaster

[B8009] BCP is an approach which recognises that organisations do not operate in isolation of each other and that they rely on suppliers to continue to operate. Hence disasters elsewhere may have a knock-on effect within an organisation, and disasters within an organisation could adversely affect its customers.

With the concepts of single source supply and just-in-time delivery, there is an increasing need to undertake supplier audits to assess their BCP capability.

Also the risk of denial of access should be assessed, as in the case of the Manchester bomb explosion in 1996. Unaffected businesses were denied access by the emergency services for months after the explosion.

Natural disasters can also have the following knock-on effect:

Following the earthquake in Kobe, Japan, the world suffered a severe shortage of silicon chips (microprocessors). The price on the black market rose so dramatically, that it set off a spate of thefts worldwide. This led to any large office block thought to contain large numbers of PCs being hit.

The public relations disaster

[B8010] Increasingly, companies strive to raise their profile in the market place and increase their visibility by promoting themselves as aggressively and publicly as possible. The public awareness they seek, and need, in order to sell their products and services is a window through which to view their operations.

Just as members of the public will be more aware of the products and services the company offer, so they will be more aware of any mistakes, accidents, disasters, poor decisions, poor performance etc. for which the company is ultimately responsible. The ill feeling that can be generated against the company by the general public can be so damaging that confidence in the company is totally undermined for a long period of time.

For example, Gerald Ratner jokingly referred to his products as 'cr*p' and later compounded the felony by suggesting that there was more value in a prawn sandwich than in some of his products. The ultimate consequence of his action was loss of sales and the subsequent break-up of his jewellery business. He lost control of his company and a number of employees lost their jobs through the resulting re-organisation.

Hoover's handling of the massive over-subscription to its 'Free Flights' promotion became a national scandal which caused the company to be pilloried almost weekly on prime time TV consumer programmes. Not only did several directors lose their jobs, but the Hoover crisis severely damaged the Hoover brand name so much that the company even considered changing its name away from one of the most powerful and successful brand names/images in history, hence the 'free flights fiasco!'

More recently (January 2007) following the allegedly racist abuse of a Bollywood actress in the *Celebrity Big Brother* house by Jade Goody (and others), Jade herself effectively caused untold reputation damage to Jade Goody plc, which has an estimated value of £8 million. Indeed PR guru Max Clifford is quoted as saying: "in future when people think of PR blunders they will think of two people, Gerald Ratner and Jade Goody!"

Thus, the ability of a company to avoid, or recover from, a disaster effectively can have a tremendous and positive PR impact, thus vastly improving stakeholders' (investors, public, customers, suppliers, employees etc.) confidence.

This area is covered in more detail in the section on crisis communication (**B8039** below).

Convincing the board – the business case

[B8011] A business case for BCP can be made by taking the following ten key points into account:

(1)   Find out how much time you have for your presentation and make sure you rehearse to make full use of what may be a limited time allocation.

(2)   Check the audience:
  •   who will be there?
  •   have non-executive directors previous experience of disaster and/or BCP?

(3)   Determine, in advance, if there are any regulatory influences that the board may be particularly sensitive to.

(4)   Achieve board buy-in/commitment to BCP by ensuring that they:
  •   understand the negative impact of business interruption;
  •   provide the necessary resources to reduce the risk of disaster;
  •   appreciate the need to prepare and plan.
  •   appoint a corporate champion, at board level, to take overall responsibility for the development and implementation of BCP.

(5)   Stress the benefits of BCP:
  •   peace of mind/sleep at night;
  •   risk reduction.

(6)   Be prepared for the 'It won't happen here' syndrome. Have some statistics (not too many) and some local or industry examples that the board may be aware of.

(7)   Spell out that the board is ultimately responsible for business risk management. N.B. Turnbull/Corporate governance.

(8)   Avoid being drawn into a detailed discussion about how the BCP capability is to be developed.

(9)   Avoid being drawn into a discussion on actual or potential disasters, unless they are specifically relevant.

(10)   Summarise and reinforce the need for BCP by highlighting the unique selling points from the above list.

The business continuity model

[B8012] The model is a phased and active process consisting of five main steps:

Stage 1: Understanding your business
  •   Business impact analysis ('BIA') and risk evaluation/assessment tools are used to: identify the critical business processes within the business; evaluate recovery priorities; and assess risks which could lead to business interruption and/or damage to the organisation's reputation.

Stage 2: Continuity strategies
- This involves determining the selection of alternative strategies available to mitigate loss, assessing the relative merits of these against the business environment; and deciding which are likely to be the most effective in protecting and maintaining the critical business processes.

Stage 3: Plan development
- The development of a proactive BCP plan is designed to improve the risk profile by upgrading optional procedures and practices; introducing alternative business strategies; and using risk financing measures; including insurance.

Stage 4: Plan implementation
- The introduction of BCP capability is achieved via education and awareness of all stakeholders, including employees, customers, suppliers and shareholders. The end result is to establish a BCP culture throughout the organisation. The BCPs should be widely communicated and published to all concerned so that everyone is aware of rules, responsibilities, accountabilities, timescales and KPIs (key performance indicator).

Stage 5: Plan testing and maintenance
- The plan is not the capability. The plan is the means by which the capability is realised.
  Failure to plan is planning to fail!
  BCP is therefore not a 'one-off' exercise. There is a need for ongoing plan testing, maintenance, audit and change in the management of the BCP and its processes. Regular communication and feedback on the results of testing and maintenance are imperative so as to ensure continual improvement of the BCP capability.

The process of developing a BCP capability is a cyclic and evolutionary one in order to take account of the ever-changing organisation.

Understanding your business (stage 1)

Risk evaluation

[B8013] Risk evaluation is the process by which an organisation determines the pattern of risk that is unique to its operation.

Risk is used to describe the relationship between the impact and likelihood for a particular event or scenario:

Risk = Impact × Likelihood

Within the BCP framework, risk evaluation is the process which identifies those aspects of the company's operations, which are essential to the survival of the business, and assesses how well they are protected against serious disruption and/or potential disaster.

As in the field of occupational safety and health ('OSH'), risk evaluation is a combination of identification and assessment. Within BCP, there are three commonly used approaches to risk evaluation:

- scenario driven;
- component;
- critical business processes.

All approaches assume that all concerned fully understand what the company is in business for and what is vital for its survival.

Scenario driven approach

[B8014] This approach involves the listing of all possible risk scenarios that can be thought of. Once the list has been compiled each scenario is considered in isolation with a view to planning specific counter-measures to tackle each risk scenario on an individual (and isolated) basis.

This approach has some serious limitations. It presupposes that risks exist in isolation and that they do not have a dynamic relationship with each other. This is clearly not the case. For example fire creates smoke, fumes and heat. The presence of smoke and fumes increases the risk of inhalation injury and may hinder evacuation or searching activities. If we then throw in the fact that the fumes are toxic or corrosive and also that they could adversely affect neighbouring sites, the single incidence of 'fire' has an exceptional number of associated risks.

Such an approach will have a high cost and also lead to complex planning implications.

Component approach

[B8015] This approach examines each department or unit within the organisation in order to develop their individual risk profiles. Whilst this approach does eventually cover the whole organisation, it fails to prioritise amongst departments and hence does not determine those departments which are important/critical.

Also it sometimes misses linkages and dependencies between departments, e.g. the internal customer. The approach as a whole suffers from a lack of organisation-wide perspective, and hence it is difficult to establish which departments should be reinstated first.

Critical business approach

[B8016] Business can be viewed as a set of linked processes. Some of these business processes, the critical ones, have a greater bearing on the core business or primary activity of the business. These are known as the critical business processes ('CBPs').

The fundamental question to be asked and understood is:

What does my company actually do?

Developing a model may help to clarify thinking and understanding in this regard. Reference to the company's vision, and mission statement and goals will also demonstrate where the organisation is focused. It is on mission critical activities – the critical business processes – where BCP should be focused.

There are four basic questions to be asked:

(1)   What is the business about?
(2)   When are we to achieve our goals?
(3)   Who is involved both internally and externally?
(4)   How are the goals to be achieved?

An organisation has many dependencies, both internally and externally, that support the mission critical business processes. These may include suppliers, customers, shareholders, key employees, IT systems, and manufacturing processes. It is important to identify these at an early stage and to involve people from the CBPs in the development of the BCP.

External influences on the CBPs may include government departments, regulators, competitors, trade bodies, and pressure groups. These must also be taken into account with the BCP framework.

In summary, the critical business approach:

- Needs a common understanding of the company's primary business objective – the focus or mission of the company. This is ultimately what the capability is being designed to protect and recover.
- Some BCPs are doomed to failure because the participants cannot agree to what the company's primary business objective actually is!
- Determines the CBPs by asking how important are they in achieving the business objective. Again agreement amongst the participants regarding the CBPs is vital in developing a relevant BCP.
- Develops a BCP capability to protect all identified CBPs whilst keeping focused on the primary business objective will prevent participants getting sidetracked or bogged down in detail.

Business impact analysis

[B8017] Business impact analysis ('BIA') is the technique used to determine what the business stands to lose if its CBPs are disrupted.

BIA alerts the organisation to the likely cost of a business disruption and assists in the cost benefit analysis of risk reduction measures within the overall BCP capability. Indeed the approval and continuing support of your board for BCP may well hinge on the fact that the costs are justified in the light of potentially catastrophic business losses.

Quantification of business impact

[B8018] It is important that all concerned in the BCP process understand that disasters are characterised by time and not event. What causes the disruption is less important to the survival of the organisation than how long the disruption lasts. BIA therefore should consider the impact of a disruption in any of the CBPs spread over time. Traditionally, the impact of a disruptive event was derived merely by determining how much revenue would be lost over the period during which the company was not operational:

Loss of revenue – *Out for 1 day* –> **Impact** = Annual Revenue/ No of working days/year

Loss of revenue – *Out for a year* –> **Impact** = Annual Revenue

On its own, however, loss of revenue does not give the full picture of the overall impact on the business. Other factors, which need to be taken into account, include:

Erosion of customer base
- The organisation will lose customers if it cannot continue to service their requirements. Some will go simply because they perceive the service is unreliable and some will be forced to go for the sake of their own business continuity. Whatever the reason, it is generally estimated that it costs three times more to sell goods/services to a new customer than to an existing one. In the event of a prolonged disruption, the company could therefore have to treble its marketing, advertising, publicity and sales cost in order to restore its customer base.

Loss of key staff
- Following a disaster, some companies have reported that up to 40 per cent of their workforce have left as a result. The loss of intellectual capital and the cost of recruitment can be considerable.

Loss of reputation
- Most companies depend on their reputation as a reliable service provider to generate further sales. Also a great deal of resources, time and money, will have been spent over the years in cultivating a brand image linked to their reputation. Hence, any event which detracts from that image brings the company's reputation into question and not only jeopardises the previous investment but also seriously affects current and future sales. Such lack of confidence will also have a negative effect on the share price of publicly quoted companies.

Regulatory and contracted impact
- The company may be subject to penalty clauses/payments if it cannot meet contracted obligations. It may also be barred from operating by an inability to comply with regulatory (e.g. prohibition notice requirements), reporting or licensing obligations.

Loss of credit-worthiness
- The organisation may face increased costs of borrowing following a disruption to business, on the basis that it represents a greater investment risk to lenders as a result of the disaster.

Business impact Analysis: outputs

[B8019] BIA enables the organisation to focus on the CBPs and their negative impact on the business, rather than to conduct a global, risk-specific analysis. The process also takes time sensitivity into account, thus enabling the recovery objectives to be agreed.

It is essential to rate the impact of the loss of CBPs on the business. Risk rating may be qualitative: high, medium, low or quantitative, i.e. 1–9 rating scale. Where possible, financial values should be placed on the business impact.

It is also important to involve key personnel who work in the CBPs in the BIA process, especially as the CBPs are cross-functional/cross-departmental, and agreement must be reached on the relative ratings. Agreement from the board should then be obtained to ensure that the output from the BIA, i.e. the plan is put into place.

In summary therefore:

- Ensure involvement of appropriate functions/departments.
- Determine the impact on the business of the loss of CBPs.
- Apply risk ratings, including time dependencies.
- Obtain board approval for BIA outputs, i.e. the plan.

Continuity strategies (stage 2)

[B8020] Having identified those areas where the organisation is most at risk, a decision has to be made as to what approach is to be taken to protect the operation. With the introduction of the Turnbull Guidance on internal control (corporate governance) this decision must be taken at board level.

Many possibilities exist and it is likely that any strategy adopted will comprise a number of these approaches. Whichever are chosen, there are certain considerations to bear in mind, as indicated below:

- Do nothing (risk retention) – in some instances, the board may consider the risk to be commercially acceptable.
- Change or end the process (risk avoidance) – deciding to alter existing procedures have to be done, bearing in mind the organisation's key focus/mission.
- Insurance (risk transfer) – provides financial recompense/support following a loss but does not provide protection for brand, image and reputation.
- Loss mitigation (risk reduction) – tangible loss control programme to eliminate or reduce risk.
- Business continuity planning (risk reduction) – an approach that seeks to improve organisational resilience to interruption, allowing for the recovery of key business systems and processes within the recovery time frame objective whilst maintaining the organisation's critical business processes.

Any strategy must recognise the internal and external dependencies of the organisation and must be accepted by all members of management involved in the CBPs. Hence, in summary, strategy selection is as follows:

- Identify possible BCP strategies.
- Assess suitability of alternative strategies against the outputs of the BIA.
- Propose cost/benefit analyses of the various strategies.
- Present recommendations to the board for decision and approval.

Developing a business continuity plan (stage 3)

[B8021] As previously stated:

- Failing to plan is planning to fail.
- The plan is not the capability.

The BCP is the means by which the capability is realised and can be viewed as the route which takes the organisation from its present state (crisis/disaster) to its desired state (normal operation).

Too often organisations derive a false sense of security from the mere presence of a disaster recovery or BCP and lose sight of the fact that the plan is passive. It has no inherent capability to assist the organisation in any way until it is used. Its value to the organisation can only be assessed by how well it enabled the business to either avoid or recover from a disaster-threatening scenario, i.e. after the event.

The planning dilemma

[B8022] *Question*: How does one include enough detail to provide meaningful instruction and guidance without including so much data that individuals or departments find the plan difficult to use?

There is no simple answer to this question since requirements will vary from business to business. However, via the outputs from the BIA, from the experience of individuals involved in the BCP process, and from regular testing and maintenance of the plan, a balance that suits the organisation will be reached through plan refinement.

Plan components

[B8023] The objective of creating a BCP capability is firstly to design/engineer out as much of the operational risk from the identified CPBs as possible, i.e. risk avoidance.

Thereafter the BCP should provide a structural means of recovering those lost processes within a given timeframe.

Thus the plan will have two main components:

- *Proactive* – range of cost-justified risk reduction measures, procedural, physical, behavioural, processes, i.e. engineering.
- *Reactive* – disaster recovery plan.

Disaster recovery plan

[B8024] The vast majority of disaster recovery plans will comprise three stages:

Emergency response – Immediate
(1) This stage details the actions the company has predefined in immediate response to an emergency situation. Typically, it should include details of evacuation procedures, liaison with emergency services, how the BCP kicks in, notification procedures, damage assessment/control/mitigation, deployment of first aid and other emergency resources.

Fallback procedures – Short term
(2) This stage details the process by which the company starts to cope with the aftermath of the disaster, i.e. the fallback position. This may involve a physical move to alternative premises, alternative working methods,

or the commencement of contractual agreements with third parties to undertake one or more of the CBPs on behalf of the company. A list of potential suppliers of such services, together with contract details is a 'must' in this regard.

Business resumption – Medium term
(3)   This stage details the process which determines when and how the organisation gets on the road back to normal operation. Typically it will contain logistical, commercial, and political criteria which will be used by the company to decide where the business will be resumed, which CBPs will be reinstated first and where resources will be allocated to get the operations back to normality as quickly as possible.

Hierarchical planning structure

[B8025] In order to co-ordinate the recovery process, particularly when some of the decision-makers are not located at the site of the disaster, many companies make use of hierarchical management planning structure:

The command and control team
•   This team is usually staffed by senior managers/decision-makers, supported by a business continuity professional. Their objective is to analyse the information being fed to them by the damage assessment team and to co-ordinate the overall response of the company to the disaster. It is their job to ensure that the recovery is on track; that they respond quickly and effectively to any developments in order to achieve restoration; to allocate emergency response resources; and to handle crisis communications both, internal and external.

The damage assessment team
•   The damage assessment team is generally a multi-disciplinary group able to convey accurate information from the disaster site to the command and control team. They are also typically responsible for on-site OSH issues, site security, access control, liaison with emergency services, providing the command and control team with regular updates on restoration progress, and general tactical decision-making.

Departmental team(s)
•   At some point within the BCP, those departments involved in CBPs will be instructed by the command and control team to incorporate their individual recovery plans. This may involve moving to an alternate location or developing an alternative way of working. Regular progress reports will be provided to the command and control team to enable them to keep track of the overall recovery. In this way, slippage against expected recovery timescales might be quickly identified, thus allowing additional resources to be allocated to that department, if available.

Plan development process

[B8026] The process for developing the BCP follows the criteria for project management methodology, increasingly employed by many organisations. It is essential that the BCP capability is developed in accordance with the following:

- Scope the development in terms of time and resources.
- Identify what skills are required to complete the recovery.
- Set milestones to demonstrate progress.
- Define reporting lines, roles, responsibilities, accountabilities.
- Establish a project tracking method.
- Set sign-off criteria.
- Move into maintenance phase.

Plan implementation (stage 4)

[B8027] The implementation process for the BCP requires the achievement of certain milestones in the following chronological order.

Securing buy-in

[B8028] Hopefully by this time, buy-in by the board will have been secured, using the techniques discussed above. It is vital, however, that this approval manifests itself in some way that allows it to be readily communicated throughout the organisation. A visible commitment to the BCP is required.

Policy

[B8029] Having a clearly stated BCP policy is one good way of visibly demonstrating board level commitment. The policy can be issued under the signature of the MD/CEO as an endorsement of the BCP process. This then provides a mandate to those involved to get the BCP implemented throughout the organisation.

Project authorisation

[B8030] The approval of the board should also include authorisation for the BCP project as a whole. This would generally be based on the information presented via the BIA, risk identification and risk evaluation exercises.

Individual responsibilities

[B8031] All key individuals should, as part of their job, have clearly defined and agreed responsibilities, accountabilities, timescales, KPIs and action plans, so as to ensure smooth plan implementation. In this way individual personal performance is geared to successful BCP implementation and an ongoing BCP capability.

Phased roll-out

[B8032] We have already determined that in order to provide the organisation with the most cost-beneficial BCP capability, there is a need to focus on the CBPs. In both the development and implementation stages, it is essential that the organisation stratifies its efforts to cover and protect those functions most directly involved in the core business (primary business objective) and the CBPs before moving on to less critical functions/areas.

This stratification will normally manifest itself as a phased roll-out of the BCP to the CBPs first, intermediate functions next, and non-business critical functions last, if at all.

Specialist services

[B8033] Quite apart from the emergency services, not all the resources required to implement the plan will be available in-house or on-site.

The following list is by no means exhaustive but serves to illustrate further possible components of the BCP. It is imperative that contact details for all such support services are kept readily available (in at least two known locations) and up to date:

- Data recovery/IT back-up.
- Emergency telecommunications.
- Salvage/decontamination.
- Cleaning/restoration.
- Buildings/facilities management.
- Security.
- Counselling (post-traumatic stress disorder (PTSD)).

Plan testing and maintenance (stage 5)

[B8034] Maintenance and testing are both vital components in ensuring that the BCP continues to support the disaster avoidance and recovery capability required by the organisation. Maintenance is essential to avoid the plan becoming unusable because it no longer reflects the organisation's structure, vision, mission or business priorities.

The objective of testing is to provide a high degree of assurance that the plan will work when it is needed, and also to highlight and rectify any shortcomings in the plan so that the BCP process is as streamlined as possible.

Maintenance should be an ongoing process throughout the lifetime of the BCP. Its primary focus is to change the plan to continuously reflect changes within the organisation. New business lines or processes, new reporting lines, reorganisations, changes of personnel, location, telephone numbers, user IDs etc all need to be reflected in the plan.

It is strongly advisable to allocate responsibilities for plan maintenance and testing to a group of individuals and make it part of their job descriptions, CYOs (current year objectives) and KPIs.

Testing strategies

The desk check/audit

[B8035] Many plans fall into disrepair due to a lack of accuracy and, consequently, integrity in the detail of the plan itself. This situation seriously undermines the effectiveness of the plan and its ability to protect the organisation in a disaster situation. The desk check/audit is a type of testing which seeks to verify that the factual detail contained in the plan is both accurate and current. As well as confirming the factual correctness of the plan, it also provides an excellent indication of how well the maintenance function is being performed.

The walkthrough test

[B8036] This test involves synthesising an incident or fictional set of circumstances, thereafter allowing participants to use the plan to resolve the described situation. Similar in concept to a war game, the rules of engagement are predefined and an external facilitator co-ordinates the introduction of extra information. The logic of the plan is tested conceptionally for robustness against a scenario designed to involve all sections of the plan.

Component testing

[B8037] Component testing normally involves single functions or departments testing in isolation their own parts of the plan by enacting their emergency, fallback and resumption arrangements. The objective is to test the practicality of the plan and to derive an estimate of how effective the plan is in terms of logistics and recovery times. It has the merit of eventually covering the whole organisation but it does not adequately test the linkages between functions and departments, which may more fully represent the flow of activity within the organisation.

Full simulation

[B8038] The organisation deliberately shuts down or denies access to CBPs, sometimes without warning, in order to get the most realistic feel for how groups and individuals react and perform in the event of a disaster.

Whilst this type of testing provides the highest possible degree of assurance that the plan works and the organisation can cope, it needs to be planned extremely carefully, otherwise it may well trigger off an alternative, real-life, disaster!

Normally a test of this type should not be undertaken until the organisation has successfully carried out the other types of test described with consistently favourable results.

Crisis communication and public relations

[B8039] In a crisis, many decisions have to be taken within a very short timescale. Some of those decisions may determine whether the organisation fails or survives! This fact represents one of the most powerful arguments for BCP and hence should ensure that the organisation has a proactive, up-to-date plan to communicate with its internal and external audiences throughout the crisis.

Employees, suppliers, customers, shareholders, the public and the media all expect and need to be kept informed in order for the organisation to retain its reputation and to maintain confidence in the business. An inability to communicate with such audiences before, during and after a crisis can easily cause a public relations disaster, thereby compounding the original incident.

Examples of companies who did not, or chose not to, communicate include:

- Hoffman-La Roche: Seveso (July 1976).

- Eli Lilly: Opren Withdrawal (August 1982).
- Union Carbide: Bhopal (December 1984).
- Delta Airlines: Plane Crash (August 1985).
- Sandoz: Rhine Pollution (November 1986).

In all cases, the share price fell within a short time of the disaster and did not get back up to the market average for at least a year after the event.

The converse of the situation faced by these companies is one where the organisation does communicate and manages to secure excellent PR from a difficult situation for example:

- British Midland: Kegworth Air Crash.
- Commercial Union: City IRA Bomb:

   We don't make a drama out of a crisis!

In order to minimise the risk of negative PR and the damage it can do, many companies include a communications plan within their BCP. This also enables the company to be portrayed in a positive and responsible light.

The communication plan

[B8040] The communications plan should ideally comprise three stages:

Stage 1:   Pre-crisis – to minimise the risk of poor communications making the disaster worse

Stage 2:   During crisis – to manage the volume and complexity of enquiries and information requirements

Stage 3:   After crisis – to let shareholders know that we are back in business

Pre-crisis measures

[B8041]

- Understanding which media are influential in your market and your location and developing a working relationship with them prior to any incident will tend to make them more sympathetic to your version of events in the reporting of a crisis.
- Press statements are an excellent way of ensuring that there is a single source of authorised information being disseminated to the media. They provide a consistent means of updating stakeholders. Having them prepared in advance, for most possible scenarios, will save the company valuable time in a crisis.
- In the event of a disaster, which generally excites the interest of the media, it is likely that directors/senior managers will be approached for comment in an interview format. The physical environment and the circumstances of the interview often conspire to generate a highly charged atmosphere, where the untrained individual may fail to portray the organisation in a good light, hence missing the opportunity to use

this very powerful medium of communication to good effect. Training and rehearsal can dramatically improve individual performance on camera and enhance the organisation's ability to harness the power of the media.

- It is therefore advisable for the company to identify and train not just directors and senior managers, but any individuals who have some detailed knowledge or experience of the organisation's critical business processes. They then become the company spokespersons if problems occur within their area of expertise.

- In the event of a disaster, organisations experience a phenomenon known as call deluge. Organisations can expect to receive up to ten times as many calls and enquiries following a disaster than on a normal day.

Under such a weight of demand, the risk of providing inaccurate or misleading information is vastly increased.

One way of managing call deluge is to set up a dedicated communications office to which are channelled all disaster-related calls. In some cases a separate crisis hotline telephone number should be given out and all received calls be dealt with by trained operators.

From this centre the enquiry can be satisfactorily answered and passed on to one of the company spokespersons, if considered necessary. This approach enables the company to maintain integrity and consistency in the information being released concerning the disaster and its ongoing effects and mitigation.

During crisis

[B8042]

- The communications office swings into action. It is advisable to include the setting up of a dedicated communications office within the testing and maintenance section of the BCP.

- Spokespersons needed for comment and/or interview should be readily available throughout the duration of the crisis. Ideally, they should be on site, or near to the site, working in tandem with the command and control team.

- Press/media statements should be released at regular and frequent intervals to update them on damage mitigation and recovery progress.

- Key stakeholders (e.g. customers) should be updated at regular and frequent intervals with regard to the recovery process and the estimated time for the reinstatement of their service.

- As the situation develops, all staff need to be kept informed, particularly if they are required to participate in the recovery process. In any event, all staff should be aware of the likely timescales for the business getting back into full operation.

- As the recovery from the disaster takes place, a crisis event log should be kept to enable significant events, actions or milestones to be recorded with the purpose of learning from them in the aftermath of the disaster, and possibly reviewing and amending the BCP in the light of unique experiences.

After the crisis

[B8043]

- Use the crisis events log to review and analyse how well the organisation performed towards its agreed recovery targets. Does the BCP need amending in the light of unique experiences during actual recovery?
- Identify those aspects which worked well and those where there is room for improvement to the BCP, in order to make things work more smoothly if ever there is a next time!
- Publicise a summary of performance to all stakeholders in order to demonstrate that the BCP has actually worked in practice. This will breed confidence in the organisation and its ability to rise above and proactively manage any future potential disaster scenarios.
- And finally consider a dedicated marketing campaign to make capital of how the crisis was managed in a positive manner, and to share the lessons learnt with others operating in similar environments. This will again demonstrate the organisation's capability to cope before, during, and after any type of disaster or crisis.

Civil Contingencies Act, 2004

[B8044] This Act contains sweeping emergency powers that place duties and responsibilities on how organisations respond to disasters.

The act is accompanied by two documents:

Emergency Preparedness

This is statutory guidance under Part I of the Act, which sets out exactly what the legalisation requires and also gives best practise, advise to assist respondees in complying with their legal requirements.

Emergency Response and Recovery

This is non-statutory guidance that deals with the response and recovery phases of emergencies.

The Act in detail

[B8045] The CCA came into effect in May, 2006 and was primarily intended to rectify short comings in the UK civil defence and emergency powers legislation, and was therefore enacted to counter civil disasters such as foot and mouth diseases; widespread storm/flood damage/disruption; terrorist attacks/bombings etc.

Part 1 of the Civil Contingencies Act, 2004 covers local arrangements for civil protections by the emergency services — police, fire and rescue services, paramedics — and local authorities.

Part 2 covers emergency powers for wide scale, regional emergencies.

Part 3 contains technical details.

What is an emergency under the CCA?

[B8046] Part 1 of the CCA defines an emergency as:

- an event or situation which threatens serious damage to human welfare in a place in the UK;
- an event or situation which threatens serious damage to the environment of a pace in the UK
- war or terrorism which threatens serious damage to the security of the UK

Local Authorities: duty to provide business continuity advice

[B8047] The CCA places a duty on local authorities (LAs) to provide advice and assistance to businesses on their patch on aspects of business continuity, including emergency preparedness and disaster recovery. In other words, LAs have a duty to promote business continuity planning and management.

In order to comply with this duty, LAs have to demonstrate that they have taken reasonable steps to promote business continuity advice in their areas. This means that they have to identify what businesses need to know when they develop, implement or review their business continuity arrangements.

This implementation includes:

- the kinds of disruption that may occur and the impacts these may have on the business;
- what arrangements the LAs and emergency services have in place to respond to and recover from emergencies, such as evacuation or recovery plans;
- an awareness of sources or warnings, information and advice; and
- the steps that individual businesses can take to prepare for or mitigate the effects of an emergency, i.e. business continuity management procedures.

Government Business Continuity Tool Kit

[B8048] The Cabinet Office has now developed a national business continuity tool kit which is designed to help businesses prepare for potential disasters.

This BCM tool kit will be available on the *'Business Link'* website and also the *'Preparing for Emergencies'* website, www.preparingforemergencies.gov.uk.

The tool kit includes a workbook which sets out the steps that businesses should go through to ensure they have sufficient BCM arrangements in place. It also features a BCM checklist, template and exercise scenarios.

The Civil Contingencies Secretariat, part of the Cabinet Office, has published a promotional leaflet for use by local authorities to encourage businesses to put BCM plans into practice.

Other Sources of Help

[B8049]

Business continuity bench-marking tool

This free bench-marking service is available on line from business advisory firm Deloitte.

ROBUST Business continuity toolkit

This free toolkit is available from RISC Authority, administered by the Fire Protection Association (FPA) at https://robust.riscauthority.co.uk/

Business Continuity Institute (BCI) Good Practice Guidelines

These are available at www.thebci.org/gpgdownloadpage.htm

Compensation for Work Injuries/Diseases

David Leckie

Introduction

[C6001] Compensation for work injuries, diseases and death is payable under the social security system and in the form of damages for civil wrongs (torts). This chapter examines the two systems and the interaction between them.

The social security system is a form of public insurance, funded by employers/employees and taxpayers, and benefit is payable irrespective of liability on the part of an employer, i.e. 'no fault'. However, connection with employment must be established. The social security system was created by legislation, such as the various Social Security Acts, the *Social Security Administration Act 1992*, the *Social Security Contributions and Benefits Act 1992* ('*SSCBA 1992*'), the *Statutory Sick Pay Act 1994*, the *Social Security (Incapacity for Work) Act 1994*, the *Social Security (Recoupment of Benefits) Act 1997*, and other legislation.

Liability in tort depends on proof of negligence or breach of statutory duty against an employer. Employers must be insured for such liability (see EMPLOYERS' LIABILITY INSURANCE). The current law relating to work injuries and diseases is to be found in a variety of Acts, such as the *Law Reform (Personal Injuries) Act 1948*, the *Employers' Liability (Compulsory Insurance) Act 1969*, the *Employers' Liability (Defective Equipment) Act 1969*, the *Damages Act 1996*, various health and safety regulations and a wide body of case law.

Some useful guides worth referring to are:

- *Guidelines for the assessment of general damages in personal injury cases* (2002), 6th edition (ISBN 0 19 925795 7), issued by the Judicial Studies Board, which is a publication which aims to unify judicial approaches to awards of damages. These guidelines are not a legal document and a full examination of the law applicable to each case is required.
- *Industrial injuries handbook for adjudicating medical officers*, was published in 1997. It sets out medical examination procedures and offers guidance on legislation, including discussion of case law.

Social security benefits

[C6002] Under the *Social Security (Incapacity for Work) Act 1994*, and the Regulations made under that Act, the higher rate of incapacity benefit is not payable until after 364 days. The effect that this has had on industrial injuries benefits, including disablement benefit, is that industrial benefits continue to be payable after 90 days, but the benefits which are payable before and at the same time, that is, any national insurance benefits, are less for the relevant period. The higher adult dependant's allowance is delayed for 364 days when the long-term incapacity benefit increases begin. Claimants whose industrial injuries benefit is lower than the rate of incapacity for work benefit and who are thus entitled to claim this benefit are also affected. (See **C6015–C6017** below).

Entitlement to industrial injuries benefits requires compliance with national insurance contribution conditions for incapacity benefit. The incapacity for work tests under the *Social Security (Incapacity for Work) Act 1994* do not apply to industrial disablement benefits. The medical adjudication procedures for these benefits continue with some modifications as before. The incapacity for work tests will apply to sufferers from industrial disablement if they apply for incapacity benefit before the 90 qualifying days for industrial disablement benefits begin, or if they do not qualify for industrial disablement benefit under the percentage or other rules, or if incapacity benefit is payable at a higher rate than industrial disablement pension rates and incapacity benefit is chosen by the claimant.

Under the rules for claimants for incapacity benefit, the employee has to supply information and evidence of sickness and be prepared to submit to a medical examination to decide whether he is fit to work. If he has worked for more than eight weeks in the twenty-one weeks immediately preceding the first day of sickness, a test relevant to his incapacity to do work which he could reasonably be expected to do in the course of his occupation applies. This test criterion continues to the 197th day of incapacity. After that date, and for all claimants who do not qualify for the required period, an 'all work' test arises. Regulations provide for prescribed activities and a person's incapacity, by reason of a specific disease, or bodily or mental disablement, to perform these activities. These tests are detailed and are outside the scope of this publication. [*SSCBA 1992, ss 171A–171G* as inserted by *Social Security (Incapacity for Work) Act 1994, ss 5, 6*]. The *Social Security (Incapacity for Work) (General) Regulations 1995 (SI 1995 No 311)* (as amended by the *Social Security (Incapacity for Work) (General) Regulations 1996 (SI 1996 No 484)*), provide that certain people are deemed to be incapable of work, including those suffering from a severe condition as defined in the Regulations or those receiving certain regular treatment (such as chronic renal failure, hospital in-patients and those suffering from an infectious or contagious disease). 'Welfare to work beneficiaries' will also be treated as incapable of work in the circumstances prescribed by *Regulation 13A* of the 1995 Regulations.

A pregnant woman may also be deemed incapable of work if there is a serious risk to her health, or that of her unborn child, if she does not refrain from work, in the case of the 'own occupation' test; or in the case of the 'all work' test, if she does not refrain from work in any occupation. If she has no entitlement to maternity allowance or statutory maternity pay and the actual date of confinement has been certified, she is deemed to be incapable of work beginning with the first day of the sixth week before the expected week of confinement until the fourteenth day after the actual date of confinement and during this period she will therefore be entitled to claim incapacity benefit.

Whilst most actual work by a claimant disqualifies him from receiving this benefit on any day of such work, certain work does not stop benefit. [*Social Security (Incapacity for Work) (General) Regulations 1995 (SI 1995 No 311), Reg 17*]. Earnings from such work must not exceed £67.50 per week and in most cases must be for less than 16 hours per week. The exempt work is work done on the advice of a doctor which:

- helps to improve, or to prevent or delay deterioration in the disease or bodily or mental disablement which causes that person's incapacity for work; or
- is part of the treatment programme undertaken as a hospital in-patient or out-patient under medical supervision; or
- is done when the claimant is attending a sheltered workshop for people with disabilities.

Voluntary work and duties as a member of a disability appeal tribunal or the Disability Living Allowance Advisory Board are also exempt work.

Industrial injuries benefit

[C6003] The benefits which are available to new claimants for industrial injuries benefits are:

- industrial injuries disablement pension;
- constant attendance allowance; and
- exceptionally severe disablement allowance.

Disabled person's tax credit may also be claimed in some circumstances (see C6016 below). Some older benefits continue to be payable to recipients who were receiving them when they were otherwise abolished, or whose entitlement arose before the relevant dates. (For details of these older benefits and more detailed social security benefits information, see *Tolley's Social Security and State Benefits Handbook*.)

Accident and personal injury provisions

[C6004] The employed earner must have suffered personal injury caused after 4 July 1948 by an accident arising out of and in the course of his employment, being employed earner's employment. [*SSCBA 1992, s 94*]. The *Reporting of Injuries, Diseases and Dangerous Occurrences Regulations 1995 (SI 1995 No 3163)* contain detailed provisions which require reports to be made on prescribed forms to the Health and Safety Executive after the occurrence of an accident or on receipt of a report from a registered medical practitioner of his diagnosis of a prescribed disease. A self-employed person may arrange for this report to be sent by someone else. Records must be kept for three years containing prescribed details of accidents, or the date of diagnosis of the disease, the occupation of the person affected and the nature of the disease. For more information on these Regulations see ACCIDENT REPORTING.

Industrial accident and disease records may be kept for three years on:

- a B510 Accident Book;
- photocopies of completed form F2508; or
- computerised records.

Personal injury caused by accident

[C6005] 'Personal injury' includes physical and mental impairment, a hurt to body or mind, which includes nervous disorders or shocks (R(I) 22/52; R(I) 22/59). The injury must have been caused by an accident. Although this is usually an unintended and unexpected occurrence, such as a fall, if a victim is injured by someone else, may also be considered to be an accident (*Trim Joint District School Board v Kelly* [1914] AC 667). A relevant accident may still have occurred where a series of accidents without separate definite times cause personal injury. An office worker was held to have suffered a series of accidents on each occasion she had been obliged to inhale her colleagues' tobacco smoke and this was held to have caused personal injury.

'Accident' must be distinguished from 'process', that is, bodily or mental derangement not ascribable to a particular event. Injuries to health caused by processes are not industrial injuries, unless they lead to prescribed industrial diseases (see **C6010** below). 'There must come a time when the indefinite number of so-called accidents and the length of time over which they occur, take away the name of accident and substitute that of process' (*Roberts v Dorothea Slate Quarries Co Ltd (No 1)* [1948] 2 All ER 201). In *Chief Adjudication Officer v Faulds* [2000] 2 All ER 961, the House of Lords, in disallowing a claim for damages for psychological injury suffered by a fireman, held there must be at least one identifiable accident that caused the injury. The fact that an employee might develop stress from a stressful occupation would not satisfy the definition of 'accident'.

Accident arising out of and in the course of employment

[C6006] There is no need to show a cause for the relevant accident provided that the employee was working in the employer's premises at the time of the accident. An accident arising in the course of employment is presumed to have arisen out of that employment, in the absence of any evidence to the contrary.

What runs through all the case law is the common requirement giving rise to industrial injuries rights that the employee was doing something reasonably incidental to and within the scope of his employment, including extra-mural activities which the employee has agreed to do (R(I) 39/56). A male nurse who was injured in a football match watched by patients in the hospital grounds succeeded in a claim for benefit as this was reasonably incidental to and within the scope of his employment (R(I) 3/57), but a policeman who was injured whilst playing football for his force was unable to recover benefit despite the fact that his employers had encouraged him to play in the game (*R v National Insurance Commissioner, ex p Michael* [1977] 2 All ER 420). Also, in *Faulkner v Chief Adjudication Officer* [1994] PIQR P 244, a police officer who was injured whilst playing for a police football team was not entitled to industrial injuries benefit despite the benefit to the community resulting from his participation. He was not on duty at the time. Further guidance can be found in *Chief Adjudication Officer v Rhodes* [1999] ICR 178 where it was held that the two main questions to be asked are:

* what are the employee's duties; and
* was he discharging them at the time of the accident.

Supplementary rules

[C6007] Five statutory provisions establish rules under which the employee is deemed to be acting in the course of his employment duties. If the occurrence falls within these rules, the employee will be covered by industrial injuries benefit. These provisions are as follows:

* Illegal employment – if the employee was not lawfully employed, or his employment was actually void because of some contravention of employment legislation. [*SSCBA 1992, s 97*].
* Acting in breach of regulations, or orders of the employer – if the employee is not acting outside his authority under his employment duties, and an accident occurs while the employee is doing something for the purposes of, and in connection with, the employer's business [*SSCBA 1992, s 98*], he will be covered. For instance, a kitchen porter hung up his apron to dry in a recess near to the ovens where he was forbidden to go. He was injured when he fell into a shallow pit. The hanging up of the apron was for the purposes of his employment, so the accident was deemed to have arisen out of, and in the course of, his employment duties (R(I) 6/55).
* Where a dock labourer was employed on loading a ship by the method of two slings, but he instead used a truck which he had not been authorised to use for this purpose, he was held to be acting in the course of his employment. (R(I) 1/70B). This contrasts with the earlier case of *R v D'Albuquerque, ex parte Bresnahan* [1966] 1 Lloyd's Rep 69, where a dock labourer was killed in an accident whilst driving a forklift truck, which he had no authority or permission to use to remove an obstruction. His widow was unable to recover industrial injuries benefit as her husband was held not to have been acting in the course of his employment.
* Travelling in an employer's transport – travel to and from work is not covered except where the employee is travelling in transport provided by the employer with his express or implied permission, whether or not the employee was bound to travel in this transport. [*SSCBA 1992, s 99*]. Outside this express provision, the employee will, in most cases, both be required to be on the employer's premises doing what he was authorised to do, unless his work takes him off the premises. A postman was able to recover benefit when he was bitten by a dog on the street, as his job required him to walk along streets (R(I) 10/57). If the journey is preparatory to the start of timed itinerant duties, for example, as a home help, there will be no entitlement to benefit in respect of injury

sustained on the way to the first home, though if the employee has more discretion about his movements, he may be entitled to benefit on the way to his first call.

• An injury incurred while an employee is trying to prevent a danger to other people, or serious damage to property during an emergency. [*SSCBA 1992, s 100*].

• Accidents caused by another's misconduct, boisterousness or negligence (provided that the claimant did not directly induce or contribute to the accident by his own conduct), the behaviour of animals (including birds, fish and insects). If these cause an accident, or if a person is struck by lightning or by any object, respectively confer entitlement to industrial injuries benefit. [*SSCBA 1992, s 101*].

Relevant employment

[C6008] 'Employed earner's employment' includes all persons who are gainfully employed in Great Britain under a contract of service, or as an office holder, and who are subject to income tax under Schedule E. Self-employed people and private contractors are thus not entitled to industrial injuries benefits.

Certain classes of person are expressly included for the purposes of industrial injuries benefits. They include unpaid apprentices, members of fire brigades (or other rescue brigades), first-aid, salvage or air raid precautions parties, inspectors of mines, special constables, certain off-shore oil and gas workers and certain mariners and air crew. Most trainees on Government training schemes are excluded from the scheme. [*SSCBA 1992, ss 2, 95*].

If an industrial accident occurs outside Great Britain, industrial injuries benefit has since 1 October 1986 been payable when the employee returns to Great Britain, provided the employer is paying UK national insurance contributions, or if the claimant is a voluntary worker overseas and is himself paying UK contributions. Accidents which occur, or prescribed diseases which develop in other EU countries, are covered by common rules. An employee who is entitled to make a claim for industrial benefit in another EU member state should make his claim for benefit in that state, regardless of the country which will actually pay the benefit.

Certain types of employment are specifically excluded from cover. [*SSCBA 1992, s 95; Employed Earners' Employments for Industrial Injuries Purposes Regulations 1975 (SI 1975 No 467), Regs 2–7 and Sch 1, 2*].

Persons treated as employers

[C6009] The *Employed Earners' Employments for Industrial Injuries Purposes Regulations 1975 (SI 1975 No 467), Sch 3* also provide for cases where certain people who may not have a contract with the employee, or who may be an agency employer, to be the relevant employer for the person who has suffered the industrial injury, or who has developed a prescribed disease. An agency that supplies an office cleaner, or a typist, will be the relevant employer. For casual employees of clubs, the club will be the relevant employer.

Benefits for prescribed industrial diseases

[C6010] The rules for certain prescribed diseases (namely deafness, asthma and asbestos related diseases), differ in some respects from the provisions affecting prescribed diseases outlined below. (See C6055–C6059 below.) The different rules applicable to diseases resulting from exposure to asbestos at work are dealt with for industrial injuries purposes by a special medical board and there is a separate state scheme for statutory compensation where one of these diseases develops and there is no remedy against an employer, or entitlement to state benefits (see C6061 below). (See leaflets NI 12, NI 207 and NI 272 for help with making a claim.) To obtain the right to industrial injuries benefits for all other prescribed diseases, the claimant must show:

- that he is suffering from a prescribed disease;
- that the disease is prescribed for his particular occupation (where a disease is prescribed for a general activity, for example, contact with certain substances, he must clearly show that this was more than to a minimal extent); and
- that he contracted the disease through engaging in the particular occupation (there is a presumption that if the disease is prescribed for a particular occupation, the disease was caused by it, in the absence of evidence to the contrary).

(See also OCCUPATIONAL HEALTH AND DISEASES.)

Claims are made on Form BI 100B and there is a right to, and it is advisable to, claim immediately after the disease starts. The 90-day waiting period for receipt of benefit applies as for accidents, as do the percentage disabilities and aggregated assessment rules, except in the case of loss of faculty resulting from diffuse mesothelioma when entitlement begins on the first day of the claim.

Assessments are made by two doctors who will decide on the percentage disability and how long the disability will last. Benefit will then be payable for the period stated in the assessment, but if the doctors are not sure of the period, benefit will be paid for a while with a further review. If the disease recurs during that period, there will be no need to make a further claim, but if the condition has worsened, the assessment may be reviewed. If there is a further attack after the period of the assessment, a fresh claim will have to be made, which will be subject to a further 90-day waiting period.

Industrial injuries disablement benefit

Entitlement and assessment

[C6011] A person is entitled to an industrial injuries disablement pension if:

- he suffers from a prescribed industrial disease (see C6010 above);
- he suffers as a result of the relevant accident, a loss of physical or mental faculty such that the assessed extent of the resulting disablement amounts to not less than 14 per cent [*SSCBA 1992, S 103, Sch 6*]. See leaflet NI6 (July 1999) (update);

- 90 days (excluding Sundays) have elapsed since the date of the accident or onset of the prescribed disease or injury.

An assessment of the percentage disablement up to 100 per cent will be made by an adjudicating medical practitioner who, in the case of accidents, looks at the claimant's physical and mental condition, comparing him in those respects with those of a normal person of the same age and sex. No other factors are relevant. The assessment may cover a fixed period/or the life of the claimant. The degrees of disablement are laid down in a scale so that, for example, loss of one hand is normally 60 per cent and loss of both hands 100 per cent. Disfigurement is included even if this causes no bodily handicap.

Where the claimant suffered from a pre-existing disability before the happening of the industrial accident at the onset of the industrial disease, benefit will only be payable in respect of the industrial accident or disease itself, and the medical adjudicators will compare the original disability with the industrial disability for this purpose. Where one or more disabilities result from industrial accidents or diseases, the level of resulting disability may be aggregated, but not so as to exceed the 100 per cent disability and its corresponding rate of benefit. [*Social Security (General Benefit) Regulations 1982 (SI 1982 No 1408), Reg 11*].

A list of initial application forms is supplied in NI 6 (July 1999) Industrial Injuries Disablement Benefit. Once a claim for industrial injuries benefit has been made, the Secretary of State will send Form B1/76 to the claimant. This form asks for full details of the accident or disease. The claimant must then submit to examination by at least two adjudicating medical practitioners and must agree to follow any appropriate medical treatment. If he fails to comply with these requirements, he will be disqualified from receiving this benefit for six months. A decision in writing will be sent to the claimant. It is possible to appeal a refusal of industrial benefit; details of how to appeal will be sent with the decision if it is negative.

This benefit can only be paid 90 days after the date of the accident excluding Sundays. [*SSCBA 1992, s 103*].

Claims in accident cases

[C6012] Claims should be made on Form BI 100A obtainable from the post office, job centre, or from the Benefits Agency, by claimants disabled by the accident for nine weeks. Claims should be made within six months of the accident to avoid loss of benefit, as the usual period for which benefit will be backdated is three months. [*Social Security (Claims and Payments) Regulations 1987 (SI 1987 No 1968), Reg 19, Sch 4, para 3*]. It is possible to apply to the DSS for a declaration from an adjudicating officer that an accident is covered by the scheme if a claimant suspects that an accident at work may have lasting effects which may not become apparent until some time later. If a positive finding is made, this will bind the DSS on any later claim for benefit. [*Social Security Act 1998, s 29(4)*]. Form BI 95 should be used for making this application. (Claims in respect of prescribed diseases are mentioned at **C6010** above; appeals at **C6018** below.)

Rate of industrial injuries disablement benefit

[C6013] Benefit is paid at one of two rates with the higher being paid to claimants aged over 18.

Constant attendance allowance and exceptionally severe disablement allowance

[C6014] Constant attendance allowance is available if:

- the claimant is receiving industrial injuries disablement benefit based on 100 per cent disablement, or aggregated disablements that total 100 per cent or more;
- the claimant needs constant care and attention as a result of the effects of an industrial accident or disease.

Furthermore, if the carer of the claimant spends at least 35 hours a week looking after him and is of working age and is not earning more than £75 (before tax but after deduction of national insurance and other reasonable expenses) per week, invalid care allowance of £43.15 per week with allowances for the carer's own dependants (if any) may be granted to that person. Claims for industrial injuries benefit are made on Form BI 104. It is granted for a fixed period and may be renewed from time to time. There are four rates of payment:

- part-time (where full-time care is unnecessary);
- normal maximum rate (if the above conditions are fulfilled and full-time care is required);
- intermediate (if the claimant is exceptionally disabled and the degree of attendance required is greater, and the care necessary is greater than under the 'normal' classification, the benefit is limited to one and a half times the normal rate); and
- an exceptional rate (if the claimant is so exceptionally disabled as to be entirely dependent on full-time attendance for the necessities of life).

If the intermediate or exceptional rates are payable, an additional allowance (known as exceptionally severe disablement allowance) will be due if the condition is likely to be permanent. Constant attendance allowance may continue to be paid for up to four weeks if the claimant goes into hospital for free medical treatment.

Other sickness/disability benefits

[C6015] However, most applicants for industrial injuries benefit will, as their incapacity results from accidents or diseases contracted during their employment, be able to claim statutory sick pay or incapacity benefit up to the date of commencement of industrial injuries benefit as 90 days is less than 28 weeks. The rate of statutory sick pay is higher than that for incapacity benefit for 28 weeks when they are then paid at the same rate. Incapacity benefit rates in the first 28 weeks apply to the self-employed and to the unemployed, or those with too few contributions. For the purposes of claiming incapacity benefit before being entitled to claim a disablement benefit, it should be noted that the presumption that a claimant for benefit for industrial or prescribed diseases has satisfied the contribution requirements for incapacity benefit (as was the case with sickness benefit), has been removed. An age addition is added to incapacity benefit where the claimant is under certain age limits. Claims for incapacity benefit for the first time claimant should be made on Form BI 202. Where entitlement to industrial injuries benefit is below the incapacity benefit rates, then that benefit may be claimed to top up industrial injuries benefit to the current rate of incapacity benefit.

It is possible to claim extra for a spouse or person looking after children if the spouse is over 60, or, if younger, child benefit is being paid, and the claimant was maintaining the family to at least the extent of the dependency benefit being claimed. None of these restrictive rules apply to industrial injuries benefit claims.

Income support and other state benefits such as housing benefit and council tax benefit may be available if the claimant with or without dependants does not have sufficient to live on.

Disability Tax Credit under Working Tax Credit

[C6016] This benefit is for people aged 16 years or over who wish to work, but have a physical or mental disability which puts them at a disadvantage in securing a job under criteria set out in the regulations. The applicant must work for at least 16 hours a week to qualify and he receives a credit if he works for 30 hours or more a week. Self-employed people may qualify For a full analysis of the disability tax credit see: www.ircc.inrev.gov/taxcredits.

Payment

[C6017] For the financial year 2003/2004 the weekly benefit rates payable for Statutory sick pay, Incapacity benefit, Industrial injuries disablement benefit, Constant attendance allowance and Exceptionally severe disablement are set by the *Social Security Benefits Up-rating Order 2003 (SI 2003 No 526)*.

Appeals

[C6018]–[C6019] Appeals relating to all industrial injuries are made to an appeal tribunal and should be made within one month from the date that the decision maker sends the decision to the applicant.

Change of circumstances and financial effects of receipt of benefit

* *Hospital*

If a claimant enters hospital, industrial injuries disablement pension continues to be payable, as does exceptionally severe disablement allowance. Constant attendance allowance will stop after four weeks. Statutory sick pay will be reduced after six weeks.

- *Taxation*
Industrial injuries disablement benefit, constant attendance allowance, exceptionally severe disablement allowance and disabled person's tax credit are not taxable. Incapacity benefit is taxable for all claimants who have claimed after 13 April 1995. Those who were in receipt of invalidity benefit before that date are not liable to tax on it. Statutory sick pay is taxable under the PAYE system.

Benefit overlaps

[C6020] Industrial injuries benefits may be taken at the same time as incapacity for work benefit with no reduction. Disability living allowance and attendance allowance may not be received at the same time as constant attendance allowance, but if the claimant should receive a higher rate of either of those benefits, constant attendance allowance will be topped-up to bring it up to the higher benefit rate. (For recoupment of benefit after awards of damages see **C6044** below.)

Damages for occupational injuries and diseases

[C6021] When a person is injured or killed at work in circumstances indicating negligence or breach of duty on the part of an employer, he may be entitled to an award of damages. Damages are assessed by judges in accordance with previously decided cases; very exceptionally they may be assessed by a jury. Damages are also categorised as general and special damages, according to whether they reflect pre-trial or post-trial losses. Calculation of damages is often made by reference to Kemp and Kemp, *The quantum of damages*, and to the Judicial Studies Board's document, *Guidelines for the assessment of general damages in personal injury cases* (2002) (6th edition).

Damages normally take the form of a lump sum; however, 'structured settlements', whereby accident victims are paid a variable sum for the rest of their lives, are now a viable alternative [*Damages Act 1996, s 2*] (see **C6025** below). This may well involve greater reliance on actuarial evidence and a rate of return of interest provided by index-linked government securities.

Legal aid is currently the subject of extensive Government reform. The Legal Services Commission (LSC) replaced the Legal Aid Board in April 2000. Legal aid is no longer available in negligence personal injury cases. It is now only available in non-negligence personal injury claims, i.e. where the injury has been intentionally as opposed to negligently caused. There is one further exception, legal aid may be available where the personal injury claim forms a small part of another claim for which legal aid is available. Legal aid is still available for clinical negligence. The Access to Justice Act 1999 introduced the conditional fee system whereby solicitors represent clients on a 'no win no fee' basis. In the absence of legal aid, such a system will be the way in which most personal injury actions are brought.

Accident Line, which is part of a scheme run by solicitors who are members of a specialist panel of personal injuries lawyers, provides a free half-hour consultation for claimants who have suffered personal injuries, including industrial accidents. The telephone number is 0500 192 939. In addition, the Association of Personal Injuries Lawyers has a helpful website at: www.apil. com.

Basis of claim for damages for personal injury at work

[C6022] The basis of an award of damages is that an injured employee should be entitled to recoup the loss which he has suffered in consequence of the injury/disease at work. 'The broad general principle which should govern the assessment in cases such as this is that the court should award the injured party such a sum of money as will put him in the same position as he would have been in if he had not sustained the injuries' (per Earl Jowitt in *British Transport Commission v Gourley* [1956] AC 185). The *Damages Act 1996*, provides that periodical payments may be made in some cases.

It is not necessary that a particular injury be foreseeable, although it normally would be (see further *Smith v Leech Brain & Co Ltd* [1962] 2 QB 405). Moreover, if the original injury has made the claimant susceptible to further injury (which would not otherwise have happened), damages will be awarded in respect of such further injuries, unless the injuries were due to the negligence of the claimant himself (*Wieland v Cyril Lord Carpets Ltd* [1969] 3 All ER 1006). In this case, a woman, who had earlier injured her neck, was fitted with a surgical collar. She later fell on some stairs, injuring herself, because her bifocal glasses had been dislodged slightly by the surgical collar. It was held that damages were payable in respect of this later injury by the perpetrator of the original act of negligence. Conversely, where, in spite of having suffered an injury owing to an employer's negligence, an employee contracts a disease which has no causal connection with the earlier injury, and the subsequent illness prevents the worker from working, any damages awarded in respect of the injury will stop at that point, since the supervening illness would have prevented (and, indeed, has prevented) the worker from going on working.

[C6023]–[C6024] Listed below are the various losses for which the employee can expect to be compensated. Losses are classified as non-pecuniary and pecuniary.

* *Non-pecuniary losses*
  The principal non-pecuniary losses are:
  (i)   pain and suffering prior to the trial;
  (ii)   disability and loss of amenity (i.e. faculty) before the trial;
  (iii)   pain and suffering in the future, whether permanent or temporary;
  (iv)   disability and loss of amenity in the future, whether permanent or temporary;
  (v)   bereavement.

(Damages for loss of expectation of life were abolished by the *Administration of Justice Act 1982, s 1(1)* (see further 'loss of amenity' at **C6034** below).)

There are four main compensatable types of injury, namely:

(i)   maximum severity injuries (or hopeless cases), e.g. irreversible brain damage, quadraplegia;

(ii) very serious injuries but not hopeless cases, e.g. severe head injuries/loss of sight in both eyes/injury to respiratory and/or excretory systems;

(iii) serious injuries, e.g. loss of arm, hand, leg;

(iv) less serious injuries, e.g. loss of a finger, thumb, toe etc.

There is a scale of rates applicable to the range of disabilities accompanying injury to workers but it is nowhere as precise as the scale for social security disablement benefit. Damages for maximum severity cases can vary from several hundred thousand pounds to millions of pounds. In *Biesheuval v Birrell* [1999] PIQR Q40, the High Court awarded total damages of £9,200,000 (see **C6024** below) to a student who was almost completely paralysed in all four limbs after a car crash. In *Dashiell v Luttitt* [2000] 3 QR 4 a settlement of £5,000,000 was reached for brain damage sustained by a child aged 14 at the time of a school minibus crash. In *Cappoci v Bloomsbury Health Authority* (21 January 2000, unreported) the High Court awarded £2,275,000 damages to a 13-year-old boy who had been asphyxiated at birth and as a result suffered cerebral palsy and other severe physical and mental handicaps. Less serious injuries attract lower damages – for example, in *Williams v Gloucestershire County Council* (10 September 1999, unreported) an out of court settlement of £3,639 was reached in a case where a 6-year-old lost the top of her little finger in an accident.

- *Pecuniary losses*
  These consist chiefly of:
  (i)   loss of earnings prior to trial (i.e. special damages);
  (ii) expenses prior to the trial, e.g. medical expenses;
  (iii) loss of future earnings (see below);
  (iv) loss of earning capacity, i.e. the handicap on the open labour market following disability.

In actions for pecuniary losses, employees can be required to disclose the general medical records of the whole of their medical history to the employer's medical advisers (*Dunn v British Coal Corporation* [1993] ICR 591).

General and special damages

- *General damages*
  General damages are awarded for loss of future earnings, earning capacity and loss of amenity. They are, therefore, awarded in respect of both pecuniary and non-pecuniary losses. An award of general damages normally consists of:
  (i)   damages for loss of future earnings;
  (ii) pain and suffering (before and after the trial); and

(iii)   loss of amenity (including disfigurement).

- *Special damages*

  Special damages are awarded for itemised expenses and loss of earnings incurred prior to the trial. Unlike general damages, this amount is normally agreed between the parties' solicitors. When making an award, judges normally specify separately awards for general and special damages. A statement of special damage must be served with the statement of claim which should suffice to give the defendant a fair idea of the case he has to answer. More detailed information must be supplied after the exchange of medical and expert reports.

EXAMPLE

An example of the way in which damages awards are assessed and broken down is the case of *Biesheuval v Birrell* (referred to in **C6023** above) where the High Court awarded £9,200,000 damages consisting of:

Pain and suffering and loss of amenity	£137,000
Interest on general damages	£6,617
Past loss of earnings	£80,700
Interest on past loss of earnings	£14,929
Other special damages	£360,113
Interest on special damages	£54,516
Tax on interest	£41,215
Loss of future earnings	£3,700,000
Loss of pension rights	£67,491
Initial capital expenditure	£551,803
Recurring costs and future care	£4,267,000

Structured settlements

[C6025] A structured settlement is an agreement for settling a claim or action for damages on terms that the award is made wholly or partly in the form of periodic payments. Such settlements are expected to become more usual in the case of larger awards of damages. Under the *Damages Act 1996*, these periodic payments must be payable in the form of an annuity for life, or for a specified period, and may be held on trust for the claimant if that should be necessary. Provision may be added to the settlements for increases, percentages or adjustments where the court or claimant's advisers secure these variations in his interest. Structured settlements in favour of claimants may be made for the duration of their life. Knowledge of the claimant's special needs is thus vital for the structure to be successful – it should be recognised that structured settlements will not be suitable for all cases. If the claimant dies while in receipt of periodic payments, they pass under his estate. Structured settlements may also be made in awards of damages in respect of fatal accidents. Tax-free annuities are payable directly to the claimant by the Life Office.

Structured settlement awards made by the Criminal Injuries Compensation Authority are also tax free.

When agreeing a settlement (whether structured or not), it is better to agree whether payments are net of repayable benefits. If a settlement offer is silent as to repayable benefits, then a deduction will have to be made in respect of them, possibly with unplanned results for the claimant.

Structured settlements are advisable where brain damage makes the injured person at risk and suggestible to pressure from relatives. They are also useful where the injured person has little experience or interest in investment, or dislikes the possibility of becoming dependent on the State or relatives should funds run out. Disadvantages are that annuities only last for the lifetime of the injured person. There is also a loss of flexibility to deal with changed circumstances and of the better return gained with the skilful investment of a lump sum.

Assessment of pecuniary losses

[C6026] Assessing pecuniary losses, i.e. loss of future earnings, can be a difficult process. As was authoritatively said, 'If (the claimant) had not been injured, he would have had the prospect of earning a continuing income, it may be, for many years, but there can be no certainty as to what would have happened. In many cases the amount of that income may be doubtful, even if he had remained in good health, and there is always the possibility that he might have died or suffered from some incapacity at any time. The loss which he has suffered between the date of the accident and the date of the trial [i.e. special damages (see above)] may be certain, but his prospective loss is not. Yet damages must be assessed as a lump sum once and for all [see 'Provisional awards' at C6040 below], not only in respect of loss accrued before the trial but also in respect of a prospective loss' (per Lord Reid in *British Transport Commission v Gourley* [1956] AC 185). Moreover, if, at the time of injury, a worker earns at a particular rate, it is presumed that this will remain the same. If, therefore, he wishes to claim more, he must show that his earnings were going to rise, for example, in line with a likely increase in productivity – a probable rise in national productivity is not enough.

When the court assesses loss of earnings, the claimant has to mitigate his loss by taking work if he can. In *Larby v Thurgood* [1993] ICR 66, the defendant applied to the court to dismiss an action brought by a fireman who was severely injured in a road traffic accident and who had taken employment as a driver earning £6,000 per annum, unless he agreed to be interviewed by an employment consultant who would then give expert evidence on whether the claimant could have obtained better paid employment. This application was refused. Evidence of whether the claimant could have obtained better paid employment depended partly on medical evidence of his capabilities and the present and future state of the job market where he lived, which could be established by an employment consultant. His general suitability for employment, his willingness and motivation, were matters of fact for the judge; thus expert opinion was not required for that purpose.

A claimant may be earning practically as much as he was before his accident, but may be more at risk of losing his present job, of not achieving expected promotion, or of disadvantages in the labour market. Damages may be claimed for such prospective losses and are known as *Smith v Manchester* damages (after *Smith v Manchester City Council* (1974) 1118 Sol Jo 597).

Capitalisation of future losses

[C6027] Loss of future earnings, often spanning many years ahead, is awarded normally as a once-and-for-all capital sum for the maintenance of the injured victim. The House of Lords considered the way in which lump sums should be calculated in the leading case of *Wells v Wells; Thomas v Brighton Health Authority* [1999] 1 AC 345. This also applies in Scotland (*McNulty v Marshall's Food Group Ltd* 1999 SC 195).

The award of damages is calculated on the basis of the present value of future losses – a sum less than the aggregate of prospective earnings because the final amount has to be discounted (or reduced) to give the present value of the future losses. Inflation is ignored when assessing future losses in the majority of cases (e.g. pension rights), since this was best left to prudent investment (*Lim Poh Choo v Camden and Islington Area Health Authority* [1980] AC 174). And where injury shortens the life of a worker, he can recover losses for the whole period for which he would have been working (net of income tax and social security contributions, which he would have had to pay), if his life had not been shortened by the accident. The present value of future losses can be gauged from actuarial or annuity tables. The net annual loss (based on rate of earnings at the time of trial) (the multiplicand) has to be multiplied by a suitable number of years (multiplier) which takes account of factors such as the claimant's life expectancy and the number of years that the disability or loss of earnings is expected to last. The multiplier normally ranges between 6 and 18, and is set out in actuarial tables. Both the multiplier and the multiplicand can vary for different periods; for example, where medical evidence shows that the need for care could increase or decrease over time – *Wells v Wells; Thomas v Brighton Health Authority* [1998] 3 All ER 481. In *McIlgrew v Devon County Council* [1995] PIQR Q66, the maximum multiplier of 18 was applied for permanent general losses, and the multiplier of 12 was applied to loss of earnings.

The multiplier is calculated on the assumption that the claimant will invest the lump sum prudently. Under *section 1(1)* of the *Damages Act 1996*, the Lord Chancellor may by order prescribe a rate of return which the courts must have regard to. Under this provision, the Lord Chancellor has set the rate of return at 2.5 per cent.

EXAMPLE

In the case of a male worker, aged 30 at the date of trial, and earning £15,000 per year net, on a 2.5 per cent interest yield, the multiplier will be 22.80 (using Table 25 of the Ogden Tables). Hence, general damages will be about £342,000, assuming incapacity to work up to age 65.

In the case of a male worker, aged 50 at date of trial, earning £25,000 net, on a 2.5 per cent interest yield, the multiplier will be 12.06 (using Table 25 of the Ogden Tables). Hence, general damages will be about £301,500 assuming incapacity to work until age 65.

Deductions from awards under this head are made for the actual earnings of the injured claimant. Where the claimant takes a lighter, less well paid job, and thus suffers a loss of earnings, the courts have held that the fact that he gains more leisure through working shorter hours is not to be taken into account to reduce the amount of damages awarded for loss of earnings (*Potter v Arafa* [1995] IRLR 316).

Institutional care and home care

[C6028] Expenses of medical treatment may be claimed, except where the victim is maintained at public expense in a hospital or a local authority financed nursing home, in which case any saving of income during his stay will be set off against the claim for pecuniary loss. Alterations to a home, purchase of a bungalow accessible to a wheelchair, adaptations to a car and equipment (such as lifting equipment), which are necessary for care, may be claimed. Nursing and care requiring constant or less attendance may be claimed whether or not the carer is a professional or voluntary carer. It was confirmed by the House of Lords in *Hunt v Severs* [1994] 2 AC 350, that where an injured claimant is cared for by a voluntary carer, such as a member of his family, that damages could be recovered for this care, but the claimant should hold them in trust for the voluntary carer.

Where voluntary care is undertaken by relatives, compensation for the cost of this care is assessed as a percentage of the Crossroads rate agreed from time to time for community care by most local authorities; the actual percentage awarded being about two-thirds of that rate. A higher percentage of that rate will be awarded where the care being given is beyond the level of care normally provided by home helps (*Fairhurst v St Helens and Knowsley Health Authority* [1995] PIQR Q 1).

Damages for lost years

[C6029] Damages may be payable up to retirement age for lost earnings resulting from the shortening of the claimant's life expectancy by reason of the injury or disease. Estimated costs of living expenses are deductible from these damages. Damages under this head may also be awarded to dependants if the victim has died.

Non-pecuniary losses (loss of amenity)

[C6030] It is generally accepted by the courts that quantification of non-pecuniary losses is considerably more difficult than computing pecuniary losses. This becomes even more difficult where loss of sense of taste and smell are involved, or loss of reproductive or excretory organs. Unlike pecuniary losses, loss of amenity generally consists of two awards: an award for (i) actual loss of amenity and (ii) the impairment of the quality of life suffered in consequence (i.e the psychic loss).

Victims are generally conscious of their predicament, but in very serious cases they may not be, a distinction underlined in the leading case of *H West & Son Ltd v Shephard* [1964] AC 326. If a victim's injuries are of the maximum severity kind (e.g. tetraplegia) and he is conscious of his predicament, damages will be greater. However, where, as is often the case, the injuries shorten the life of an accident victim, damages for non-pecuniary losses will be reduced to take into account the fact of shortened life.

Types of non-pecuniary losses recoverable

[C6031] The following are the non-pecuniary losses which are recoverable by way of damages:

- pain and suffering;
- loss of amenity;
- bereavement.

Pain and suffering

[C6032] Pain and suffering refers principally to actual pain and suffering at the time of the injury and later. Since modern drugs can easily remove acute distress, actual pain and suffering is not likely to be great and so damages awarded will be relatively small.

Additionally, 'pain and suffering' includes 'mental distress' and related psychic conditions; more specifically (i) nervous shock, (ii) concomitant pain or illness following post-accident surgery and embarrassment or humiliation following disfigurement. Claustrophobia and fear are within the normal human emotional experience but even are not compensatable, unless amounting to a recognised psychiatric condition, such as post-traumatic stress syndrome (*Reilly v Merseyside Regional Health Authority* [1995] 6 Med LR 246).

In *Heil v Rankin and Another and joined appeals* [2001] QB 272, the Court of Appeal, reviewed the current awards for pain, suffering and loss of amenity. The court held that whilst payments under £10,000 should not increase, there would be a tapered increase for awards above that up to a maximum of 33 per cent for the highest level of damages.

Nervous shock

[C6033] Nervous shock refers to actual and quantifiable damage to the nervous system, affecting nerves, glands and blood; and, although normally consequent upon earlier negligent physical injury, an action is nevertheless maintainable, even if shock is caused by property damage (*Attia v British Gas plc* [1988] QB 304 where a house caught fire following a gas explosion. The claimant, who suffered nervous shock, was held entitled to damages).

Claimants fall into two categories, namely, (a) primary and (b) secondary victims, the former being directly involved in an accident and the latter are essentially spectators, bystanders or rescuers.

- *Primary victims*
  Primary victims can sue for damages for nervous shock/psychiatric injury, even if they have not suffered earlier physical injury (see *Page v Smith* [1996] AC 155 in which the appellant, who was physically uninjured in a collision between his car and that of the respondent, had developed myalgic encephalomyelitis and chronic fatigue syndrome, which became permanent. It was held by the House of Lords that the respondent was liable for this condition in consequence of his negligent

driving). Foreseeability of physical injury is sufficient to enable a claimant directly involved in an accident to recover damages for nervous shock. Thus, in the case of *Bourhill v Young* [1943] AC 92 a pregnant woman, whilst getting off a tram, heard an accident some fifteen yards away between a motor cyclist and a car, in which the motor cyclist, driving negligently, was killed. In consequence, the claimant gave birth to a stillborn child. It was held on the facts of the case that the unknown motor cyclist could not have foreseen injury to the claimant who was unknown to him.

• *Secondary victims*
As for secondary victims, defendants are taken to foresee the likelihood of nervous shock to rescuers attending to an injured person and to their close relatives, though the precise extent of nervous shock need not have been foreseen (*Brice v Brown* [1984] 1 All ER 997). Persons who witness distressing personal injuries, who are not related in either of these ways to the victim, are not only considered not to have been foreseen by the defendant, but are expected to be possessed of sufficient fortitude to be able to withstand the calamities of modern life. Only persons with a close tie with the victim or their rescuers who are within sight or sound of the accident or its immediate aftermath will be awarded damages for nervous shock as the law stands at the present. Husbands and wives will be presumed to have a sufficiently proximate tie whilst other relationships are considered on the evidence of the proximity of the relationship.

In *Chadwick v British Railways Board* [1967] 1 WLR 912, following a serious railway accident, for which the defendant was held to be liable, a volunteer rescue worker suffered nervous shock and became psycho-neurotic. As administratrix of the rescuer's estate, the claimant sued for nervous shock. It was held that (i) damages were recoverable for nervous shock, even though shock was not caused by fear for one's own safety or for that of one's children, (ii) the shock was foreseeable, and (iii) the defendant should have foreseen that volunteers might well offer to rescue and so owed them a duty of care.

The class of persons who can sue for damages for nervous shock is limited, depending on proximity of the claimant's relationship with the deceased or injured person (*McLoughlin v O'Brian* [1983] 1 AC 410 where the claimant's husband and three children were involved in a serious road accident, owing to the defendant's negligence. One child was killed and the husband and other two children were badly injured. At the time of the accident, the claimant was two miles away at home, being told of the accident by a neighbour and taken to the hospital, where she saw the injured members of her family and heard that her daughter had been killed. In consequence of hearing and seeing the results of the road accident, the claimant suffered severe and recurrent shock. It was held that she was entitled to damages for nervous shock as she was present in the immediate aftermath). This approach was confirmed by the House of Lords in *Alcock v Chief Constable of South Yorkshire Police* [1992] 1 AC 310, where it was stated that the class of persons to whom this duty of care was owed as being sufficiently proximate, was not limited to particular relationships such as husband

and wife or parent and child, but was based on ties of love and affection, the closeness of which would need to be proved in each case, except that of spouse or parent, when such closeness would be assumed. Similarly, in *Hinz v Berry* [1970] 2 QB 40 the appellant left her husband and children in a lay-by while she crossed over the road to pick bluebells. The respondent negligently drove his car into the rear of the car of the appellant. The appellant heard the crash and later saw her husband and children lying severely injured, the former fatally. She became ill from nervous shock and successfully sued the respondent for damages. On this basis, an employee who suffers nervous shock as a result of witnessing the death of a co-employee at work, will be unlikely to be able to claim damages against his employer for nervous shock, as being a 'bystander who happens to be an employee', as distinct from an active participant in rescue (*Robertson v Forth Bridge Joint Board* [1995] IRLR 251 where an employee was blown off the Forth Bridge in a high gale and fell to his death; a co-employee who watched this was unable to sue for damages for nervous shock).

In *Hunter v British Coal Corporation* [1999] QB 140, a claimant who was 30 metres away from the scene of a fatal accident and was told of the victim's death 15 minutes later was unable to recover damages for the psychiatric injury he suffered because he felt responsible for the accident. It was held that he was neither physically nor temporarily close enough to the accident to be a primary victim.

In any event, reasonable fortitude, on the part of the claimant, will be assumed, thereby disqualifying claims on the part of hypersensitive persons (see *McFarlane v EE Caledonia* [1994] 2 All ER 1 where the owner of a rig did not owe a duty of reasonable care to avoid causing psychiatric injury to a crew member of a rescue vessel who witnessed horrific scenes at the Piper Alpha disaster).

In *White v Chief Constable of South Yorkshire Police* [1999] 1 All ER 1, the House of Lords considered the question of psychiatric injury to police officers on duty during the Hillsborough football stadium disaster when 95 spectators were crushed to death. It was held that police officers were not entitled to recover damages against the Chief Constable for psychiatric injury suffered as a result of assisting with the aftermath of a disaster, either as employees or as rescuers. An employee who suffered psychiatric injury in the course of his employment had to prove liability under the general rules of negligence, including the rules restricting the recovery of damages for psychiatric injury.

Loss of amenity

[C6034] Loss of amenity is a loss, permanent or temporary, of a bodily or mental function, coupled with gradual deterioration in health, e.g. loss of finger, eye, hand etc. Traditionally, there are three kinds of loss of amenity, ranging from maximum severity injury (quadraplegias and irreversible brain damage), multiple injuries (very severe injuries but not hopeless cases) to less severe injuries (i.e. loss of sight, hearing etc.). Damages reflect the actual amenity loss rather than the concomitant psychic loss, at least in hopeless cases. Though, if a claimant is aware that his life has been shortened, he will be compensated for this loss. The *Administration of Justice Act 1982, s 1(1)* states:

(i) damages are not recoverable in respect of loss of expectation of life caused to the injured person by the injuries; but

(ii) if the injured person's expectation of life has been reduced by injuries, there shall be taken into account any pain and suffering caused or likely to be caused by awareness that his expectation of life has been reduced.

Bereavement

[C6035] A statutory sum of £10,000 is awardable for bereavement by the *Fatal Accidents Act 1976, s 1A* (as amended by the *Administration of Justice Act 1982, s 3(1)* and the *Damages for Bereavement (Variation of Sum) Order 1990 (SI 1990 No 2575)*). This sum is awardable at the suit of husband or wife, or of parents provided the deceased was under eighteen at the date of death if the deceased was legitimate; or of the deceased's mother, if the deceased was illegitimate. (In *Doleman v Deakin* (1990) Times, 30 January it was held that where an injury was sustained before the deceased's eighteenth birthday, but the deceased actually died after his eighteenth birthday, bereavement damages were not recoverable by his parents.)

Fatal injuries

[C6036] Death at work can give rise to two types of action for damages:

* damages in respect of death itself, payable under the *Fatal Accidents Act 1976* (as amended); and

* damages in respect of liability which an employer would have incurred had the employee lived; here the action is said to 'survive' for the benefit of the deceased worker's estate, payable under the *Law Reform (Miscellaneous Provisions) Act 1934*.

Actions of both kinds are, in practice, brought by the deceased's dependants, though actions of the second kind technically survive for the benefit of the deceased's estate. Moreover, previously paid state benefits are not deductible from damages for fatal injuries. [*Social Security Act 1989, s 22(4)(c)*]. Nor are insurance moneys payable on death deductible, e.g. life assurance moneys. [*Fatal Accident Act 1976, s 4*].

Damages under the Fatal Accidents Act 1976

[C6037] 'If death is caused by any wrongful act, neglect or default which is such as would (if death had not ensued) have entitled the person injured to maintain an action and recover damages, the person who would have been liable if death had not ensued, shall be liable . . . for damages . . . '. [*Administration of Justice Act 1982, s 3(1)*].

Only dependants, which normally means the deceased's widow (or widower) and children and grandchildren can claim – the claim generally being brought by the bereaved spouse on behalf of him/herself and children. [*Administration of Justice Act 1982, s 1(2)*]. The basis of a successful claim is dependency, i.e. the claimant must show that he was, prior to the fatality, being maintained out of the income of the deceased. If, therefore, a widow had lived on her own private moneys prior to her husband's death, the claim will fail, as there is no dependency. The fact that both the deceased and his partner were at the time of the accident on state benefits is irrelevant to the question of loss in assessing damages – *Cox v Hockenhull* [1999] 3 All ER 577. Cointributory negligence on the part of the deceased will result in damages on the part of the dependants being reduced. [*Fatal Accidents Act 1976, s 5*].

Survival of actions

[C6038] Actions for injury at work which the deceased worker might have had, had he lived, survive for the benefit of his estate, normally for the benefit of his widow. This is provided for in the *Law Reform (Miscellaneous Provisions) Act 1934*. Any damages paid or payable under one Act are 'set off' when damages are awarded under the other Act, as in practice actions in respect of deceased workers are brought simultaneously under both Acts.

Assessment of damages in fatal injuries cases

[C6039] Damages in respect of a fatal injury are calculated by multiplying the net annual loss (i.e. earnings minus tax, social security contributions and deductions necessary for personal living (i.e. dependency)) by a suitable number of years' purchase. There is no deduction for things used jointly, such as a house or car. However, where a widow also works, this will reduce the dependency and she cannot claim a greater dependency in future on the ground that she and her deceased husband intended to have children.

It is possible to agree that fatal injury damages should be paid in the form of a structured settlement (see **C6025** above).

Where both parents are dead as a result of negligence or a mother dies, dependency is assessed on the cost of supplying a nanny (*Watson v Willmott* [1991] 1 QB 140; *Cresswell v Eaton* [1991] 1 WLR 1113).

Provisional awards

[C6040] Because medical prognosis can only estimate the chance of a victim's recovery, whether partial or total, or alternatively, deterioration or death, it is accepted that there is too much chance and uncertainty in the system of lump sum damages paid on a once-and-for-all basis. Serious deterioration denotes clear risk of deterioration beyond the norm that could be expected, ruling out pure speculation (*Willson v Ministry of Defence* [1991] 1 All ER 638). Similarly in the case of dependency awards under the Fatal Accidents Act 1976 it can never be known what the deceased's future would have been, yet courts are expected and called upon to make forecasts as to future income. To meet this problem it is provided that provisional awards may be made, e.g 'This section applies to an action for damages for personal injuries in which there is proved or admitted to be a chance that at some definite or indefinite time in the future the injured person will, as a result of the act or omission, which gave rise to the cause of action, develop some serious disease or suffer some serious deterioration in his physical or mental condition'. [*Administration of Justice Act 1982, s 6(1)*]. Moreover, 'Provision may be made by rules of the court for enabling the court to award the injured person:

- damages assessed on the assumption that the injured person will not develop the disease or suffer the deterioration in his condition; and
- further damages at a future date if he develops the disease or suffers the deterioration'.

[*Administration of Justice Act 1982, s 6(2)*].

A claim in respect of provisional damages must be included in the statement of claim to entitle the claimant to such damages. The disease or type of deterioration in respect of which any future applications may be made must be stated (the *Civil Procedure Rules 1998 (SI 1998 No 3132), Part 41 rule 2(2)*). The defendant may make a written offer if the statement of claim includes a claim for provisional damages, offering a specified sum on the basis that the claimant's condition will not deteriorate and agreeing to make an award of provisional damages in that sum.

Interim awards

[C6041] In certain limited circumstances a claimant can apply to the court for an interim payment. This enables a claimant to recover part of the compensation to which he is entitled before the trial rather than waiting till the result of the trial is known – which may be some time away. This procedure is provided for in the *Civil Procedure Rules 1998 (SI 1998 No 3132), Part 25*, but it only applies where the defendant is either:

- insured,
- a public authority, or
- a person whose resources are such as to enable him to make the interim payment.

Prior to making an interim payment a judge is under an obligation to take into account any effect the payment may have on whether there is a 'level playing field' for the hearing – see *Campbell v Mylchreest* [1999] PIQR Q17.

Compensation recovery applies to interim payments as well as to payments into court. Care needs to be taken when applying for interim payments to avoid putting the claimant at a disadvantage. Capital of over £8,000, which could include an interim award, removes entitlement to means tested benefits, particularly income support, and there are also reductions on a sliding scale in such benefits for any capital above £3,000. In the recent judgement of *Beattie v Department of Social Security* [2001] EWCA Civ 498, [2001] 1 WLR 1404 the Court of Appeal held that payments from structured annuity funds constitutes 'income' for the purposes of the *Income Support (General) Regulations 1987 (SI 1987 No 1967)*.

Payments into court

[C6042] A payment into court ('Part 36 payment') may be made in satisfaction of a claim even where liability is disputed. From the defendant's point of view, costs from the date of the Part 36 payment may be saved if the court does not order a higher payment of damages than that paid into court. The claimant may accept a Part 36 payment or Part 36 offer not less than 21 days before the start of the trial without needing the court's permission if he gives the defendant written notice of the acceptance not later than 21 days after the offer or payment was made. If the defendant's Part 36 offer or Part 36 payment is made less than 21 days before the start of the trial, or the claimant does not accept it within the specified period, then the court's permission is only required if liability for costs is not agreed between the parties. [*Civil Procedure Rules 1998 (SI 1998 No 3132), Part 36 rule 11*].

The defendant may, instead of making a Part 36 payment, make a Part 36 offer (formerly known as a Calderbank letter) in which he sets out his terms for settling the action

Interest on damages

[C6043] Damages constitute a judgment debt; such debt carries interest at 8 per cent (currently) up to date of payment. Courts have a discretion to award interest on any damages, total or partial, prior to date of payment (and this irrespective of whether part payment has already been made [*Administration of Justice Act 1982, s 15*]), though this does not apply in the case of damages for loss of earnings, since they are not yet due. Moreover, a claimant is entitled to interest at 2 per cent on damages relating to non-pecuniary losses (except bereavement), even though the actual damages themselves take into account inflation (*Wright v British Railways Board* [1983] 2 AC 773). Under *section 17* of the *Judgments Act 1838* (and *section 35A* of the *Supreme Court Act 1981* and *Schedule 1* to the *Administration of Justice Act 1982*), interest runs from the date of the damages judgment. Thus, where, as sometimes happens, there is a split trial, interest is payable from the date that the damages are quantified or recorded, rather than from the date (earlier) that liability is determined (*Thomas v Bunn, Wilson v Graham, Lea v British Aerospace plc* [1991] 1 AC 362). Moreover, interest at the recommended rate (of 8 per cent) is recoverable only after damages have been assessed, and not (earlier) when liability has been established (*Lindop v Goodwin Steel Castings Ltd*, (1990) Times, 19 June).

Awards of damages and recovery of state benefits

[C6044] The *Social Security (Recovery of Benefits) Act 1997* and accompanying regulations made important changes to the rules for recoupment of benefit from compensation payments. One key change is that recoupment will not be taken from general damages for pain and suffering and for loss of amenity which it has been accepted should be paid in full. With respect to the other heads of compensation, namely loss of earnings, cost of care and compensation for loss of mobility, they are only to be subject to recoupment from specified benefits relevant to each of these heads of compensation. [*Social Security (Recovery of Benefits) Act 1997, s 8*].

Duties of the compensator

[C6045] Before the compensator makes a compensation payment, he must apply to the Secretary of State for a certificate of recoverable benefits. [*Social Security (Recovery of Benefits) Act 1997, s 4*]. The Secretary of State must send a written acknowledgement of receipt of the application and must supply the certificate within four weeks of receipt of the application. [*Social Security (Recovery of Benefits) Act 1997, s 4*].

He must supply the following information with his application:

- full name and address of the injured person;
- his date of birth or national insurance number, if known;
- date of accident or injury when liability arose (or is alleged to have arisen);
- nature of the accident or disease;
- his payroll number (where known), if the injured person is employed under a contract of service and the period of five years during which benefits can be recouped includes a period prior to 1994 [*Social Security (Recovery of Benefits) Regulations 1997 (SI 1997 No 2205), Reg 5*]; and
- the amount of statutory sick pay paid to the injured person for five years since the date when liability first arose, as well as any statutory sick pay before 1994, if the compensator is also the injured person's employer. The causes of his incapacity for work must also be stated.

The certificate of recoverable benefits will show the benefits which have been paid.

The compensator must pay the sum certified within 14 days of the date following the date of issue of the certificate of recoverable benefits. [*Social Security (Recovery of Benefits) Act 1997, s 6*]. If the compensator makes a compensation payment without having applied for a certificate, or fails to pay within the prescribed fourteen days, the Secretary of State may issue a demand for payment immediately. A county court execution may be issued against the compensator to recover the sum as though under a court order. It is wise for the compensator to check the benefits required to be set off against the heads of compensation payment so that he is sure that the correct reduced compensation is paid to the injured person. Adjustments of recoupable benefits and the issue of fresh certificates are possible.

Where the compensator makes a reduced compensation payment to the injured person, he must inform him that the payment has been reduced. Statements that compensation has been reduced to nil must be made in a specific form. Once the compensator has paid the Secretary of State the correct compensation recovery amount and has made the statement as required, he is treated as having discharged his liability. [*Social Security (Recovery of Benefits) Act 1997, s 9*].

Complications

Contributory negligence

[C6046] Where damages have been reduced as the result of the claimant's contributory negligence, the reduction of compensation is ignored and recovery of benefits is set-off against the full compensation sum.

Structured settlements

[C6047] The original sum agreed or awarded is subject to compensation recovery and for this purpose, the terms of the structured settlement are ignored and this original sum is treated as a single compensation payment. [*Social Security (Recovery of Benefits) Regulations 1997 (SI 1997 No 2205), Reg 10*].

Complex cases

[C6048] Where a lump sum payment has been made followed by a later lump sum, both payments are subject to recoupment of benefits where those benefits were recoupable. If the compensator has overpaid the Benefits Agency, he can seek a partial refund. [*Social Security (Recovery of Benefits) Regulations 1997 (SI 1997 No 2205), Reg 9*].

Information provisions

[C6049] Under *section 23* of the *Social Security (Recovery of Benefits) Act 1997*, anyone who is liable in respect of any accident, injury or disease must supply the Secretary of State with the following information within 14 days of the receipt of the claim against him:

- full name and address of the injured person;
- his date of birth or national insurance number, if known;
- date of accident or injury when liability arose (or is alleged to have arisen); and
- nature of the accident, or disease.

Where the injured person is employed under a contract of service, his employer should also supply the injured person's payroll number (if known) if the period of five years during which benefits may be recouped includes a period prior to 1994 and this is requested by the Secretary of State. [*Social Security (Recovery of Benefits) Regulations 1997 (SI 1997 No 2205), Regs 3, 5 and 6*].

If the Secretary of State requests prescribed information from the injured person, it must be supplied within 14 days of the date of the request. This information includes details of the name and address of the person accused of the default which led to the accident, injury or disease, the name and address of the maker of any compensation claim and a list of the benefits received from the date of the claim. If statutory sick pay was received by the injured person, the name and address of the employer who has paid statutory sick pay during the five-year period from the date of the claim or before 6 April 1994.

Appeals against certificates of recoverable benefits

[C6050] Appeals against the certificate of recoverable benefits must be in writing and made not less than three months after the compensation payment was made. The appeal should be made to the Compensation Recovery Unit for a hearing before a tribunal. Leave to appeal to a Commissioner against the decision of the tribunal should be made not later than three months after notice of the tribunal's decision.

Treatment of deductible and non-deductible payments from awards of damages

[C6051] There are three well established exceptions to deductibility of financial gains:

- recovery under an insurance policy to which the claimant has contributed all or part of the premiums paid on the policy;
- retirement pensions;
- charitable or ex-gratia payments prompted by sympathy for the claimant's misfortune.

Deductible financial gains

[C6052] The courts, applying the principles outlined above, have ruled that the following financial gains are deductible:

- tax rebates where the employee has been absent from work as a result of his injuries (*Hartley v Sandholme Iron Co* [1975] QB 600);
- domestic cost of living expenses (estimated) must be set off against the cost of care (*Lim Poh Choo v Camden and Islington Area Health Authority* [1980] AC 174);
- estimated living expenses must be set off against loss of earnings (*Lim Poh Choo*, above);
- payment from a job release scheme must be set off against loss of earnings (*Crawley v Mercer*, (1984) Times, 9 March);
- statutory sick pay must be set off against loss of earnings (*Palfrey v GLC* [1985] ICR 437);
- sick pay provided under an insurance policy must be set off against loss of earnings not paid as a lump sum (*Hussain v New Taplow Paper Mills* [1988] AC 514);
- health insurance payment under an occupational pension plan paid before retirement must be set off against loss of earnings where no separate premium had been paid by the employee who had paid contributions to the pension scheme (*Page v Sheerness Steel plc* [1996] PIQR Q 26);
- reduced earnings allowance (not a disability benefit) must be set off against loss of earnings (*Flanagan v Watts Blake Bearne & Co plc* [1992] PIQR P 144);
- payments under *section 5* of the *Administration of Justice Act 1982* of any saving to the person who has sustained personal injuries through maintenance at the public expense must be set off against loss of earnings.

Non-deductible financial gains against loss of wages

[C6053]–[C6054] The following financial gains are non-deductible:

- accident insurance payments under a personal insurance policy taken out by the employee (*Bradburn v Great Western Railway Co* (1874) LR 10 Exch 1) (contributory);
- lump-sum wage-related accident insurance payment under a personal accident group policy payable regardless of the fault of the employee (*McCamley v Cammell Laird Shipbuilders* [1990] 1 WLR 963) (benevolent);
- incapacity pension (from contributory insurance scheme) both before and after retirement age (*Longden v British Coal Corporation* [1998] AC 653). It should be noted that even though the incapacity pension was triggered by the accident, benefit flows from the prior contributions paid by the injured party (contributory);
- private retirement pensions (*Parry v Cleaver* [1970] AC 1; *Hewson v Downs* [1970] 1 QB 73; *Smoker v London Fire and Civil Defence Authority* [1991] 2 AC 502) (contributory or benevolent);
- redundancy payment unconnected with the accident or disease (*Mills v Hassal* [1983] ICR 330). Where the claimant was made redundant because he was unfit to take up employment in the same trade, however, the redundancy payment was deductible from damages for lost earnings (*Wilson v National Coal Board* 1981 SLT 67);
- ex-gratia payment by an employer (*Cunningham v Harrison* [1973] QB 942; *Bews v Scottish Hydro-Electric plc* 1992 SLT 749) (benevolent);
- ill health award and higher pension benefits provided by the employer (*Smoker v London Fire and Civil Defence Authority* [1991] 2 AC 502) (benevolent);
- moneys from a benevolent fund, paid through trustees (not directly to the injured person or dependant) in respect of injuries;
- charitable donations (but not where the tortfeasor is the donor).

Non-deductible financial gains against cost of care

- Loss of board and lodging expenses awarded despite their being provided voluntarily by parents where the claimant had formerly paid such expenses herself (*Liffen v Watson* [1940] 1 KB 556).

Social security and damages for prescribed diseases

[C6055] The claims for social security benefits are similar to those for personal injury. Claims for damages are also similar, with more latitude for late court applications because of recurrence of, or worsening of, industrial diseases. Lists of prescribed diseases are contained in social security leaflet NI 2. It is advisable always to refer to the latest edition of the leaflet, as the list of diseases is subject to amendment from time to time. There are special provisions applicable to asthma and deafness resulting from industrial processes, and also for pneumoconiosis and similar diseases.

Occupational deafness (see DSS leaflet 207)

[C6056] If the claimant's deafness was caused by an accident at work, then to qualify for disablement benefit his average hearing loss must have been at least 50 decibels in both ears due to damage of the inner ear. The claimant's disablement must be at least 20 per cent or more for him to qualify for disablement benefit (total deafness being 100 per cent). The claimant must have worked for at least the five years immediately before making his claim, or for at least ten years, in one of the listed jobs set out in the current edition of the leaflet. A claimant who has been refused benefit because the rules were not complied with will have to wait for three more years when he may qualify if he has worked in one of the listed occupations for five years. Leaflet NI 196 'Social Security Benefits Rates' shows the rates payable, relevant to each percentage of loss, for a weekly disablement pension.

The claim will be decided by a Medical Board, whose members will also decide on reviews if the claimant believes that his deafness has worsened. They will inform the claimant in detail of their decision. Awards will be made for a period of five years, and there will be a review at the end of that period with another hearing test to determine whether benefit should continue, or be reduced.

Claims must be made on Form BI 100 (OD) which may be obtained free from the Department of the Environment HS ORI Level 4, Caxton House, London SW1H 9NA (tel: 020 7273 5248), or from a post office and some borough libraries.

Asthma because of a job (see DSS leaflet NI 237)

[C6057] As is well known, there has been an increase in the number of sufferers from asthma. To be able to claim disablement benefit because of asthma, a claimant must have worked for an employer for at least ten years and have been in contact with prescribed substances, and his disability must amount to at least 14 per cent. In the list of substances, item 23 includes 'any other sensitising agent encountered at work', so claims may be possible even if the offending sensitising substance is not yet on the list.

Claims must be made on Form BI 100 (OA) which may be obtained from the Department of the Environment (see address and telephone number above), or from a post office and some borough libraries.

Once a claim has been made, the claimant will be seen by a Medical Board whose members will decide whether he has asthma because of his job and, if so, how disabled he is by the asthma. They will also determine how long his disability is likely to last and the extent of his disability, and will then communicate all of this information to the claimant. If they consider that no change is likely, they may award benefit for life. If they expect that there will be changes, there will be a review at the end of the period for which benefit was awarded.

Repetitive strain injuries

[C6058] Repetitive strain injuries are included in the *Social Security (Industrial Injuries) (Prescribed Diseases) Regulations 1985 (SI 1985 No 967), Sch 1,* as item A4. (See OCCUPATIONAL HEALTH AND DISEASES.) The difficulties in establishing causation can be seen from the case of *Pickford v Imperial Chemical Industries* [1998] 3 All ER 462, where the House of Lords reversed the decision of the Court of Appeal and upheld the trial judge's decision that a secretary was not entitled to damages for repetitive strain injury as she had failed to establish the cause of the injury and that the condition was not reasonably foreseeable. However, in many other cases damages for repetitive strain injury have been awarded – for example, *Ping v Esselte-Letraset Ltd* [1992] PIQR P74 where nine claimants who worked at a printing factory were awarded damages ranging from £3,000 to £8,000. In *Fish v British Tissues* (Sheffield County Court) (29 November 1994, unreported), damages of £57,482 were awarded to a factory packer for repetitive strain injuries to arms, hands and elbows.

Pneumoconiosis

[C6059] Pneumoconiosis is compensatable under different heads; first, as a ground for disablement benefit, under the *Social Security Contributions and Benefits Act 1992* and, secondly, by way of claim made under the *Pneumoconiosis etc. (Workers' Compensation) Act 1979* – the latter being in addition to any disablement benefit previously paid. See the Department of Environment guidance on the *Pneumoconiosis etc. (Workers' Compensation) Act 1979* on the DETR website: www.environment.detr.gov.uk//pneumo/index.htm.

Pneumoconiosis/tuberculosis-pneumoconiosis/emphysema – disablement benefit

[C6060] Where a person is suffering from pneumoconiosis accompanied by tuberculosis, then, for benefit purposes, tuberculosis is to be treated as pneumoconiosis. [*Social Security Contributions and Benefits Act 1992, s 110*]. This applies also to pneumoconiosis accompanied by emphysema or chronic bronchitis, provided that disablement from pneumoconiosis, or pneumoconiosis and tuberculosis, is assessed at, at least, 50 per cent. [*Social Security Contributions and Benefits Act 1992, s 110*]. However, a person suffering from byssinosis is not entitled to disablement, unless he is suffering from loss of faculty which is likely to be permanent. [*Social Security Contributions and Benefits Act 1992, s 110*].

Benefit may be payable if the claimant's disablement is assessed at 1 per cent or more, instead of the usual minimum of 14 per cent. If the assessment of disablement is 10 per cent or less (but at least 1 per cent), then the industrial injuries disablement pension will be payable at one-tenth of the 100 per cent rate.

Pneumoconiosis – payment of compensation under the Pneumoconiosis etc. (Workers' Compensation) Act 1979

[C6061] Claims made under the *Pneumoconiosis etc. (Workers' Compensation) Act 1979* are in addition to any disablement benefit paid or payable under the *Social Security (Industrial Injuries) (Prescribed Diseases) Regulations 1985 (SI 1985 No 967)*. Indeed, whereas the latter consists of periodical payments, the former resemble damages awarded against an employer at common law and/or for breach of statutory duty, except that under the above Act fault (or negligence) need not be proved. This statutory compensation is available only if the employee is unable to claim compensation from any of his former employers (for example, if the employer has become insolvent). There are no requirements that the employee must have worked in a particular industry to be able to claim compensation under the scheme. Claimants are entitled to benefits from the date of claim. There is no ninety-day waiting period under the scheme; it is sufficient if the claimant suffers from the prescribed disease. Coal miners are excluded from the scheme because they have their own statutory scheme. Claims made under the *Pneumoconiosis etc. (Workers' Compensation) Act 1979* must be made in the manner set out in the *Pneumoconiosis etc. (Workers' Compensation) (Determination of Claims) Regulations 1985 (SI 1985 No 1645)*. The claim must be made (except in the case of a 'specified disease' – see **C6062** below) within 12 months from the date on which disablement benefit (under the Social Security Regulations) first become payable; or if the claim is by a dependant, within 12 months from the date of the deceased's death. [*Pneumoconiosis etc. (Workers' Compensation) (Determination of Claims) Regulations 1985 (SI 1985 No 1645), Reg 4(1), (2)*]. Awards for pneumoconiosis/byssinosis have slowly declined recently.

Claims for specified diseases

[C6062] Claims for a 'specified disease', i.e:

• pneumoconiosis, including (i) silicosis, (ii) asbestosis and kaolinosis;
• byssinosis (caused by cotton or flax dust);
• diffuse mesothelioma;
• primary carcinoma of the lung coupled with evidence of (i) asbestosis and/or (ii) bilateral diffuse pleural thickening;

[Pneumoconiosis etc. (Workers' Compensation) (Specified Diseases) Order 1985 (SI 1985 No 2034)],

must be made within 12 months from the date when disablement benefit first became payable, or, in the case of a dependant, within 12 months from the date of the deceased's death. [*Pneumoconiosis etc. (Workers' Compensation) (Determination of Claims) Regulations 1985 (SI 1985 No 1645), Reg 4(3), (4)*]. These time periods can be extended at the discretion of the Secretary of State. Moreover, where a person has already made a claim and has been refused payment, he can apply for a reconsideration of determination, on the ground that there has been a material change of circumstances since determination was made, or that determination was made in ignorance of, or based on, a mistake as to material fact.

Community Health and Safety

Mike Bateman

Introduction

[C6501] Since the *Health and Safety at Work etc. Act 1974* came into operation, every employer has had duties in respect of 'persons not in his employment', in addition to duties to his employees. These duties have come into closer focus through the risk assessment requirements contained in the Management of Health and Safety at Work Regulations 1999 and other sets of regulations.

The term 'community' embraces a variety of different categories, such as:

- Neighbours and passers-by
- Trespassers on work premises
- Customers and users of facilities and services
- Visiting individuals and groups
- Volunteers carrying out work

This chapter considers each of these categories in some detail – who is at risk, what types of risk they might be exposed to and the types of precautions likely to be necessary to protect them. In addition, it deals with major events that can involve several community categories and may be subject to the scrutiny of Safety Advisory Groups. The potential impact on the community from contractors used by businesses or hired in by event organisers is also considered.

Relevant Legal Requirements

Health and Safety at Work etc. Act 1974

[C6502] Section 3(1) of the Act places duties on employers in respect of members of the public and other non-employees:

> It shall be the duty of every employer to conduct his undertaking in such a way as to ensure, so far as is reasonably practicable, that persons not in his employment who may be affected thereby are not exposed to risks to their health or safety.

As well as applying to various categories of members of the community (as summarised in C6501), these requirements also apply to others at work, such as contractors and their employees, agency staff, delivery personnel and members of the emergency services. It must also be noted that the duties are qualified by the phrase 'so far as is reasonably practicable' – this means that the level of risk must be considered together with the costs and practicalities of implementing related control measures.

The examples at **C6503** below contain a variety of examples of successful prosecutions under Section 3 of HSWA in respect of risk or actual injury to members of the community.

Substantial fines

[C6503] A record fine under health and safety legislation of £15 million was set in 2005 when Transco was convicted of a breach of Section 3(1) of the Health and Safety at Work etc. This resulted from a gas explosion in 1999 in which a family of four were killed in their bungalow in Lanarkshire. The ductile iron gas main involved was badly corroded, arrangements for its inspection and maintenance were inadequate and records of pipes in the area were inaccurate.

The fatal rail accidents at Hatfield and Ladbroke Grove also resulted in significant fines for train operators, rail network operators and maintenance companies involved. A fine of £10 million (reduced on appeal to £7.5 million) was imposed on the company responsible for maintenance of the track at Hatfield. And in May 2011, the rail infrastructure company Network Rail was fined £3 million over safety failings that lead to the 2002 Potters Bar train derailment which killed seven people and injured more than 70.

Other examples of prosecutions under section 3 of HSWA

[C6504] There have been many other examples of prosecutions under section 3 of the Health and Safety at Work etc. Act 1974. Recent examples include the following:

- Two North East companies were fined a total of £13,000 with costs totalling more than £6,000 after a member of the public was seriously injured when scaffolding collapsed during high winds;
- A waste company was fined a total of £250,000 and ordered to pay costs of £50,000 after a bin lorry ran over and killed a member of the public in Brighton;
- A construction company and its traffic management subcontractor were fined a total of £202,000 after a worker employed by another subcontractor was killed while working on the M25 motorway;
- A care home company was fined £70,000, with £21,818.56 in costs, after an elderly resident suffocated as a result of becoming trapped between her mattress and bed rails intended to stop her falling;
- A construction firm was fined a total of £8,000 and ordered to pay costs of £14,760 after poorly secured panelling fell from a building under refurbishment and injured two passers-by;
- A fairground operator was fined £10,000 and ordered to pay £2,708 costs after an 11-year-old girl suffered a serious head wound when she was thrown from a ride at a Birmingham park;
- A company was fined £20,000 with £15,000 in costs following the death of a driver in a high-speed crash at a tractor-pulling competition in Lancashire;

- A film and television school was fined £17,500 and ordered to pay costs of £4,787 after a volunteer was left permanently paralysed after falling 2.25 metres from a mock staircase on set; and
- A local authority was prosecuted following an incident in which a seven-year old girl was trapped by a water outlet at a swimming pool and was fined £18,000 and ordered to pay costs of £7,500;

The Management Regulations

[C6505] Regulation 3(1) of the Management of Health and Safety at Work Regulations 1999 (MHSWR) (SI 1999 No 3242) states that:

> Every employer shall make a suitable and sufficient assessment of . . . the risks to the health and safety of persons not in his employment arising out of or in connection with the conduct by him of his undertaking.

Consequently risk assessments must take account of relevant members of the community, the risks that they might be exposed to and what precautions are necessary to control those risks to the extent required by the law – including the general requirements imposed the Health and Safety at Work etc. Act 1974.

SI 1999 No 3242, Reg 5 also requires employers to take effective steps to implement the precautions identified through the risk assessment as being necessary. It states:

> Every employer shall make and give effect to such arrangements as are appropriate, having regard to the nature of his activities and the size of his undertaking, for the effective planning, organising, control, monitoring and review of the preventive and protective measures.

Appropriate and effective management techniques must be used to ensure the protection of members of the community, as well as employees and others who are at work. (Detailed guidance is provided in the chapter entitled MANAGING HEALTH AND SAFETY.)

Employers with five or more employees must record both the significant findings of their risk assessments and their arrangements for implementing the related preventive and protective measures.

Other Specific Requirements of Regulations

[C6506] Several sets of regulations contain requirements relating to the protection of members of the wider community and other non-employees. These include:

- *Control of Substances Hazardous to Health Regulations 2002 (COSHH) (SI 2002 No 2677)*
  Reg 3(1) states that:

  > Where a duty is placed by these Regulations on an employer in respect of his employees, he shall, so far as is reasonably practicable, be under a like duty in respect of any other person, whether at work or not, who may be affected by the work carried out by the employer . . .

Whilst there are some exceptions to this requirement (e.g. in relation to health surveillance and training for non-employees), the main requirements of the *COSHH Regulations* apply in respect of members of the community. *COSHH* assessments must take account of risks to the community, particularly the ways in which substances are used or stored.

- *Control of Major Accident Hazards Regulations 1999 (COMAH) (SI 1999 No 743)*
  COMAH applies to a limited range of workplaces where specified quantities of high-risk substances are stored or used, or may be generated e.g. through a chemical process. In developing their Major Accident Prevention Policy (MAPP), employers must consider safety, health and environmental risks both inside and outside their own premises. This includes the development of off-site emergency plans and the provision of information to the public.

- *Construction (Design and Management) Regulations 2007 (CDM) (SI 2007 No 320)*
  Key components of the CDM Regulations are the duty to manage the health and safety aspects of construction projects effectively (*Reg 9*) and, in the case of 'notifiable' projects, the requirement for 'a construction phase plan'. Both are general management arrangements and must take account of the health and safety of the wider community, as well as those carrying out the work. Occupants of and visitors to premises where construction work is taking place must be considered, as also must occupants of neighbouring property and passers-by.

- *Work at Height Regulations 2005 (SI 2005 No 735)*
  The regulations place duties on employers to prevent injuries to any person from falling objects. Members of the community must be protected, as well as those at work. These duties include preventing materials and objects from falling (e.g. through using toe boards, brick guards, tool belts and securing stored items), measures to prevent persons being struck by falling items (e.g. screens, nets etc.) and ensuring that materials or objects are not thrown or tipped from height in circumstances liable to cause injury.

In addition, the Reporting of Injuries, Diseases and Dangerous Occurrences Regulations 1995 (RIDDOR) require that persons having control of premises (generally the employer) report certain accidents and dangerous events that occur there. Accidents which result in members of the public being taken to hospital should be notified to appropriate health and safety enforcement authority, usually the HSE but sometimes the local authority) immediately and followed up by a report on form F2508.

The Regulatory Reform (Fire Safety) Order 2005 (SI 2005 No 1541)

[C6507] Since 1 October 2006, fire safety within workplaces and many other types of premises has been covered by the *Regulatory Reform (Fire Safety) Order 2005*. The Order places duties on the 'responsible person' as defined within the Order. For most workplaces this will be the employer whilst in other types of premises it will be the person who has control of the premises, or the owner.

Under Article 8 of the Order the responsible person must:

(a)   take such general precautions as will ensure, so far as is reasonably practicable, the safety of any of his employees; and

(b)   in relation to relevant persons who are not his employees, take such general fire precautions as may reasonably be required in the circumstances of the case to ensure that the premises are safe.

'Relevant persons' includes any person who is or may be lawfully on the premises and any person in the immediate vicinity who is at risk from a fire — this would include most of the members of the community referred to in C6501.

The requirements of the Order (which is enforced by the fire and rescue authority for the area, not the HSE) are explained in more detail in the chapter entitled FIRE PREVENTION AND CONTROL. A series of Fire Safety Guides are available providing more on the requirements of the Order. A number of these deal with locations where fire safety of members of the community is a major issue e.g. premises providing sleeping accommodation, residential care, places of assembly, theatres and cinemas, educational premises, healthcare premises, transport premises and facilities, and open air events. (See C6528 for further information.) A supplementary guide covers means of escape for disabled people.

Occupiers Liability Acts 1957 and 1984

[C6508] The *Occupiers Liability Act 1957* established the liability of occupiers under civil law to exercise a duty of care in respect of lawful visitors to their property. The occupier's duty of care was extended by the *Occupiers Liability Act 1984* to also include trespassers, in respect of any injury suffered on the premises, either because of any danger due to the state of the premises or things done, or omitted to be done, in the following circumstances:

(1)   If he (the occupier) is aware of the danger or had reasonable grounds to believe that danger exists;

(2)   If he knows or has reasonable grounds to believe that a trespasser is in the vicinity of the danger concerned or that he may come into the vicinity of the danger; and

(3)   If the risk is one against which, in all the circumstances of the case, he may reasonably be expected to offer some protection.

The Act also states that 'no duty is owed . . . to any person in respect of risks willingly accepted by that person'.

Any case brought under this Act would be decided on its merits, but there are two important factors to consider:

•   The Standard of Care Required
     A greater duty of care exists in relation to major risks (e.g. falls from significant height, presence of hazardous substances or deep water) than minor risks (e.g. uneven land or shallow water). The practicality of

taking and maintaining precautions would also have to be taken into account. It may not be reasonable to expect to fence off a large relatively low-risk workplace or site in its entirety, especially in an area where vandalism was prevalent. However, it may be reasonable to fence off individual high-risk areas within the site.

- The Type of Trespasser
  Case law has already established that a greater duty of care exists in respect of child trespassers, although parents are expected to assume the prime responsibility for controlling the whereabouts of very young children. Occupiers must also be aware of the presence of items on sites that may attract children e.g. derelict machinery and opportunities to climb.
  The duty of care owed to adults will also vary. It will be much greater to those simply walking through a site (especially if straying accidentally from a public footpath) than to those indulging in illicit activities such as theft, vandalism or fly-tipping. However, even these latter categories cannot be completely ignored, particularly if the occupier is already aware that significant risk exists.

Activity Centres (Young Persons' Safety) Act 1995 and the *Adventure Activities Licensing Regulations 2004*

The Health and Safety Executive (HSE) has recently carried out a consultation on proposals to replace the Adventure Activities Licensing Authority (AALA), which regulated adventure activities for young people, with a code of practice. The consultation followed a recommendation to abolish AALA and the associated regulations in the Government-commissioned Lord Young report, *Common Sense, Common Safety*.

The current arrangements for regulating adventure activities require that providers of activities for under-18s in four areas — caving, climbing, trekking and water sports — have a licence issued by AALA before they can operate.

The current licensing regime operates under the *Activity Centres (Young Persons' Safety) Act 1995* and the *Adventure Activities Licensing Regulations 2004*. Instead, the HSE proposed a new code of practice to help providers understand how to comply with legal requirements under HSWA. Subject to the outcome of the consultation, it could take the form of a statement of general principles applicable to all existing and emerging activities, or be more specifically aimed at a narrower range of activities.

The code will sit above existing non-statutory schemes such as those run by some national governing bodies (NGBs). The consultation includes consideration of the degree to which participation in an NGB scheme should be taken as sufficient reassurance of safety standards.

The consultation document is available at www.hse.gov.uk/consult/condocs/cd236.htm The consultation period ended on 21 September 2011.

Neighbours and Passers-by

Who Must Be Taken Into Account?

[C6509] There are a number of categories of neighbours and passers-by who must be taken into account when carrying out risk assessments. Some of these will be at work, whilst others will be members of the wider community. They include:

- Persons within the workplace or site
- Those visiting or working within other parts of the premises, such as:
  — other businesses;
  — public buildings;
  — community facilities; and
  — voluntary groups.
- External Neighbours
  Occupants of property close to the workplace and those visiting them must be taken into account. In addition to the types of locations referred to above, this is also likely to include residential property.
- Passers-by
  Pedestrians, drivers and users of recreational areas must be considered – whether or not the roads, footpaths or areas that they are using are designated for public access.

Risks to be considered

[C6510] Neighbours and passers-by may be affected by risks even if they remain outside the workplace but more so if they are able to stray into work areas. The types of risks most likely to be of relevance are:

- Vehicle traffic
  Vehicles are likely to be delivering materials to or collecting them from most workplaces. Heavy items of mobile plant may also be in circulation, particularly on construction sites.
- Lifting operations
  Lifting operations may involve objects being lifted over occupied premises or over areas where pedestrians or vehicles circulate.
- Work at height
  Work at height will inevitably involve a potential risk to neighbours and passers-by from falling objects. There is also a risk of unauthorised persons using access equipment to reach high levels from which they might fall.
- Hazardous and dangerous substances
  Substances may create risks to neighbours and passers-by through 'routine' emissions or from more substantial leakage's. Fires and explosions within workplaces can also affect external areas. (The explosion and subsequent fire at the Buncefield fuel storage depot in 2005 illustrated the extent of the impact of such an event.) Waste substances or materials constitute a risk, particularly if left in unsecured

areas. Many outbreaks of legionnaires' disease within communities have resulted from inadequate cleaning and disinfectant procedures for water systems within workplaces. (Two examples are provided in C6504.)

- Work equipment
  Neighbours and passers-by may be at risk from contact with dangerous parts of equipment being used in areas to which they have access. They may also be hit by particles etc. thrown off from materials being worked on. Any tool, powered equipment or vehicle may be dangerous if it falls into the hands of unauthorised persons, particularly children. The PUWER Regulations contain a specific requirement that 'self-propelled work equipment' must have 'facilities for preventing it being started by an unauthorised person'.

- Dangerous activities
  There are a variety of activities, which might present risks to neighbours or passers-by, even those who are some distance away, such as pressure testing, the use of radioactive substances and the use of explosives. Potentially dangerous quarrying, agricultural and forestry activities are often carried out on land accessible to members of the public.

- Electrical equipment
  Uninsulated electrical cables and other unprotected electrical equipment are obvious risks to any unauthorised person who is able to gain access to them.

Likely Precautions

[C6511] What precautions are necessary will be dependent upon the types of risks identified by the risk assessment, as described in C6510 above. For construction activities this will often be within the context of a CDM 'construction phase plan'. A 2009 Health and Safety Executive (HSE) publication 'Protecting the public – your next move' (HSG 151) (available to download free from the HSE website at www.hse.gov.uk/pubns/priced/hs g151.pdf) provides practical guidance on precautions in respect of construction projects.

- Security of the workplace
  The level of risk may justify the whole workplace or parts of it being made secure against access by neighbours or passers-by. This may involve the use of fencing or other barriers, with locked access gates or doors. Periodic checks will be necessary to ensure that these measures are still in good condition and proving effective. In some cases the presence of those carrying out the work (or a designated sentry) will be adequate to secure the work area against incursion from neighbours or passers-by.

- Traffic management
  Controlling the routes by which traffic enters and leaves work premises may be important in reducing the risks to neighbours and passers-by. It may be possible to avoid high-risk locations, e.g. shopping areas, school entrances. Direction of approach may have to be specified in order to leave an adequate turning circle for larger vehicles. The timing of vehicle

traffic may have to be controlled to avoid times at which there will be large numbers of pedestrians or other vehicles in the vicinity. (There are detailed requirements governing work being carried out on the public highway itself.)

- Other traffic precautions
  Other precautions may be necessary to ensure the safe circulation of tra ffic where work activities are taking place, both on public and private roads. In dark conditions vehicles must possess appropriate lights, and adequate lighting must be provided for activities taking place on roads, e.g. loading or unloading of vehicles. Arrangements to clean roads of mud and other work-related debris may be necessary.

- Planning of lifting operations
  Timing may also be important in minimising the possible presence of p edestrians and vehicles in areas where lifting operations must be carried out. Loads will often need to be lifted over buildings at times when they are unoccupied or when the removal of occupants causes minimum inconvenience. Measures may still be necessary to prevent any pedestrian or vehicle traffic, which may be present from getting into risk areas.

- Storage locations and facilities
  Materials and equipment should be stored in locations where the risks they present to neighbours and passers-by are minimised. Where non-workers have regular access to workplaces, the security of storage facilities will be important. Facilities storing hazardous or dangerous s ubstances must meet relevant standards for the protection both of workers and members of the community.

- Planning of dangerous activities
  As with traffic movements and lifting operations, the timing of other p otentially dangerous work activities should be planned to minimise the risks to others. Precautions in place to protect those carrying out the work should also protect neighbours and passers-by, but measures may still be necessary to keep them away from such activities.

- Emergency procedures
  Such procedures must take account of neighbours and passers-by:
  — how they will be notified of the emergency e.g. telephone messa ges, alarm sounds, loudspeaker announcements;
  — how they should respond e.g. close all windows, evacuate to des ignated locations.

- Worker awareness
  It is essential that workers are made fully aware of how their workplace or work activities may create risks for others. In particular they must be made aware of precautions involving them, e.g. designated traffic routes, permitted times for access or other activities, the importance of securing work areas, equipment or materials and their role in emergency p rocedures.

- Signs and notices
  Signs and notices can play an important role in making workers aware of the need for precautions and also in reinforcing their awareness. Neighbours and passers-by can also be made aware of both risks and p recautions, particularly the need to keep clear of areas where potentially dangerous activities are taking place.

In its guidance on construction activities, 'Protecting the public – your next move' (HSG 151), the HSE points out that vulnerable groups such as the elderly, children and people with certain disabilities may need special attention.

Trespassers on Work Premises

Who must be taken into account?

[C6512] Employers have both statutory duties towards trespassers through the *Health and Safety at Work etc. Act 1974, s 3* and also a civil 'duty of care' (see C6508). These duties are much greater where the trespass is foreseeable, for example:

— pedestrians straying from a right of way;
— persons recovering lost property (such as footballs or golf balls);
— those recovering straying animals or children.

Potential trespass by children is a particular concern because of their lack of awareness of risks and their inability to read or understand warning notices. Children are also generally more inquisitive and adventurous than adults.

Deliberate trespass by adults, e.g. for theft, vandalism or other illicit purposes, presents particular difficulties. However, the actions of employers to protect their assets from such persons will often also protect this type of trespasser from any risks present.

Risks to be considered

[C6513] A number of the risks to neighbours and passers-by mentioned in C6510 may also affect trespassers. However, risks of particular relevance in considering the need to protect trespassers include:

- Unsafe access
  - derelict buildings or structures (unsafe floors, staircases etc.);
  - fragile surfaces;
  - unstable ground or materials (e.g. waste tips); and
  - falls from significant height (shafts, cliffs, scaffolds, towers, gantries, etc.).
- Water
  - deep water (including waste lagoons); and
  - fast-flowing water (both natural watercourses and artificial ones e.g. hydroelectric schemes).
- Railway lines
  - moving trains; and
  - power lines and live rails.
- Dangerous equipment or materials
  - abandoned equipment;
  - insecure items and those capable of toppling over; and
  - equipment capable of being used by unauthorised persons.

- Hazardous or dangerous substances
  - substances stored insecurely;
  - abandoned substances;
  - accessible waste products; and
  - toxic or asphyxiant atmospheres e.g. in sewers, pipes, conduits or other confined spaces.

Likely precautions

[C6514] Many of the precautions described in C6511 to protect neighbours and passers-by may also be appropriate to protect trespassers.

- Fencing and other security measures
  The standard of fencing, locks etc. necessary to prevent access by determined trespassers will generally need to be higher than to prevent more casual entry into work premises. Standards must also relate to the level of risk present within. Fencing and other enclosures may need to be reinforced by periodic security patrols, and in any case they must be inspected regularly to ensure they remain in a satisfactory condition.
- Signs and notices
  The use of signs and notices can be particularly important in making potential trespassers aware of risks within workplaces, particularly those which may not be readily apparent e.g. fragile surfaces, deep water, high voltage electrical equipment. The condition of signs and notices should also be subject to regular inspection.
- Secure storage
  Potentially dangerous materials and equipment should be stored so that they are not accessible to trespassers. The exact nature of the storage will depend upon the items stored, the storage location and the anticipated level of approach by trespassers. Storage facilities on an unattended site accessible to the public will need to be much more robust and secure than in an occupied workplace where any trespass is likely to be less frequent and less prolonged.
- Emergency equipment
  Lifelines or lifebelts may be necessary to safeguard trespassers (as well as workers and any legitimate passers-by) against risks from water. These, too, should be subject to periodic inspection.

Customers and Users of Facilities and Services

Who must be taken into account?

[C6515] The need to protect other persons who may be present within workplaces was considered earlier in the chapter (see C6509). However, many workplaces involve direct contact by members of the community with work premises, work activities and people who are at work. Such members of the community can be described both as customers and as users of facilities and services.

- Customers
  Many businesses have customers who are members of the community:
  — shops, restaurants and other commercial premises;
  — indoor and outdoor markets;
  — transport undertakings (railways, buses, airlines, etc.);
  — suppliers of utilities (gas, electricity, water, sewage, telecommunications)
  — the entertainment industry (including spectator sports and funfairs); and
  — those providing domestic installation and repair services (gas equipment, electrical work, construction work, gardening and similar activities).
- Facility and service users
  This includes members of the community who use facilities or services such as:
  — hotels and similar accommodation;
  — hospitals and care homes;
  — sheltered housing;
  — domiciliary care services;
  — gyms, sports centres and other sports facilities;
  — interactive attractions (e.g. at museums);
  — educational establishments (schools, colleges, universities); and
  — technical training services.
- Many other examples could be added to both of the above categories. (For brevity, all such persons will be referred to as 'customers' in the remainder of this chapter.)

Risks to be considered

[C6516] Many risks to 'customers' are very similar to those to neighbours and passers-by – sections **C6510** and **C6511** deal with such risks and the related precautions. However, some risks relate particularly to the 'customers' themselves – who they are, why they are present in the workplace and what they are doing there. Consideration must be given to the following:

- Access within the premises
  Ways in which 'customers' will access the premises must be taken into account:
  — how many people will be present?
  — what routes will they follow?
  — are those access routes safe?
  — what other risks might be present on access routes?
- Equipment
  What equipment will 'customers' be expected (or have the opportunity) to use and what other equipment (used by workpeople) might they have close contact with? Some situations may involve the use of temporary electrical installations and electrical generators which could represent a risk to 'customers'.
- Dangerous substances

Many portable electrical generators will run on petrol and will need additional supplies to be kept nearby. Temporary food outlets (e.g. in markets or at entertainment locations) will frequently use LPG cylinders for food preparation.

* Activities
  What activities will 'customers' be involved in directly or come into close contact with? Some sporting and recreational activities (e.g. skiing, climbing or pole vaulting) involve an inevitable degree of risk, as do some types of technical training. Activities may also involve risks from hazardous or dangerous substances.

* Accidents and emergencies
  In some workplaces (such as the healthcare sector) there are specific obligations placed on employers to make first-aid provision for their 'customers', and in others there is often a general expectation that they will do so. Fire and other emergencies can affect 'customers' as well as those at work.

* Risks from products
  Risks to 'customers' from food products is a specialist subject in its own right. There may be risks to 'customers' from other materials, substances or equipment supplied to them by a business or simply present on display. This will be of particular importance in plant hire businesses and other establishments offering potentially high-risk products. Gas and electricity present the greatest risks in the utilities sector, with the gas supply industry being closely regulated.

* Capabilities of the 'customer'
  The capabilities of the 'customer' are important in considering the risks associated with all of the above. These may relate to their:
  — physical capabilities ('customers' may be elderly, disabled or have limited physical strength);
  — intellectual capabilities (these may be limited due to the ages of 'customers' or their mental condition);
  — technical capabilities (some activities may require a particular level of skill, knowledge or experience, as well as strength or intellectual ability); and
  — physical size (children in particular may be able to gain access to dangerous equipment or live electrical equipment through small apertures, or may fall underneath guard rails).

Likely precautions

[C6517] The types of precautions necessary will need to relate to the types of 'customers' present and the risks it is foreseeable that they will be exposed to in the workplace. Precautions are likely to include:

* Maintenance of safe access
  Access routes must be maintained in a safe condition for the benefit of staff as well as 'customers'. However, particular consideration must be given to the numbers and types of 'customers' likely to be present. Special arrangements may be necessary where the elderly or those with disabilities (e.g. the partially sighted) are present on the premises.

- Selection and maintenance of equipment

  Where equipment is expected to be used by 'customers', it must be suitable for the purpose. It may need to be simpler to use and more robust than equipment intended for the use of workers. Arrangements must be in place for the maintenance of equipment, and some types of equipment will require formal training of intended users (see below). Where equipment or utilities are supplied into customers' own premises these must be of a suitable construction and design, and arrangements are likely to be necessary for periodic inspection, testing or maintenance.

- Location and storage

  Portable generators and LPG powered equipment (as used at markets or entertainment venues) must be situated in locations where there is adequate ventilation and where they are secure from interference from 'customers' and others. Additional petrol must be kept in suitable containers in safe locations, again secure from interference. Similarly, spare LPG cylinders must be stored safely, and separately from petrol supplies.

- Screening of 'customers'

  In some situations it will be necessary to screen 'customers' in order to establish their physical, intellectual or technical capabilities. This is particularly important where they are likely to be involved in potentially risky activities (e.g. sporting or other outdoor pursuits) or using dangerous equipment. Such screening should establish the level of need for supervision, training etc. and any special individual arrangements. In some cases 'customers' may eventually be judged unsuitable to participate in the activity in question.

- Information and training

  'Customers' may need to be made aware of risks and related precautions through the use of signs and notices or via customer information packs, e.g. in hotels and similar accommodation. Some situations may justify the use of formal briefings of 'customers', particularly where they are likely to be involved in higher risk activities or where there are significant risks present in the workplace. Certain activities or types of equipment will necessitate 'customers' receiving formal training to ensure an adequate level of safety.

- Supervision and control

  An appropriate number of properly trained and informed staff must be available to ensure that 'customers' are adequately supervised. Whilst in some situations (e.g. the retail sector) this supervision will largely be passive, sporting and other outdoor activities will require more proactive forms of supervision, as will entertainment and other public events, through the presence of stewards. Those carrying out supervisory or stewarding duties must monitor the behaviour of their 'customers' and, in particular, restrict them from entering into higher risk areas.

- Monitoring

  As well as monitoring 'customer' behaviour, there is a need to monitor the ongoing effectiveness of all arrangements for ensuring customer safety. In particular areas accessed by 'customers' and equipment used by them should be subjected to periodic formal inspections to ensure that they are still in satisfactory condition. Those in control of locations such as markets, where various small traders are present, will need to ensure that those traders continue to maintain adequate standards of health and safety.

- Emergency arrangements
  In formulating emergency arrangements, account must be taken of 'customers'. Fire evacuation is an obvious case, but other foreseeable emergencies (e.g. chemical leaks or public disorder) may also need to be considered. 'Customers' must be informed about relevant emergency arrangements (see above regarding information and training). Staff also must be fully briefed on their role in respect of 'customers', which is particularly important where large numbers are present (e.g. spectator events, the educational and health care sectors and large retail outlets) and where 'customers' with disabilities may be present. It may be necessary to identify 'customers' with disabilities so that appropriate action can be taken to protect them in an emergency — some disabilities (e.g. hearing defects) may not be readily apparent.

Visiting Individuals and Groups

Who must be taken into account?

[C6518] Most of the immediately preceding sections (**C6515–C6517**) dealing with 'customers' and users of facilities and services is also of relevance to visitors. However, whilst dealing with 'customers' is an integral part of many work activities, other workplaces are less accessible to the community and may only have to deal with visitors now and again. Amongst such visitors are likely to be:

- Schoolchildren and other students
  Individuals may visit as part of work experience and similar programmes. By virtue of the Health and Safety (Training for Employment) Regulations 1990, schoolchildren on formal work experience programmes have the status of employees in respect of health and safety legislation.
- Student and community groups
  Some workplaces are willing to accept groups of visitors from schools and other educational establishments. Similarly, special interest groups or members of the local community may make visits.
- Informal individual visits
  Individual friends and relations of workers may make visits. Whilst employers will be aware of many such visits, others can pose more cause for concern, e.g. where a child is sent with a message for someone working outside normal working hours. During school holidays it is not uncommon for children to accompany parents driving vehicles in making deliveries to workplaces – sometimes even security staff may be unaware that anyone else is in the cab.

Risks to be considered

[C6519] Many of the risks referred to in C6516 in respect of 'customers' are also of relevance to visiting individuals and groups. Particular attention must be given to:

- Areas to be visited
  In order to gain benefit from their visit, it is likely that formal visitors will need to visit some areas of the workplace where risks may be present. A balance must be struck between the intended benefits of the visit and not exposing visitors to an unacceptable degree of risk.
  If informal visitors are not allowed to get beyond a reception area, a canteen or a security lodge, there will be little risk. However, granting informal visitors free access to working areas can introduce considerable risks to safety and also give rise to security issues.
- Activities to be carried out
  Work experience students and similar longer-term visitors are likely to be involved in carrying out actual work or projects related to work activities. A risk assessment is likely to be necessary in each case to identify the risks involved in the work or project and the precautions which must be taken, taking into account the capabilities of the individual (see below).
- Capabilities
  The physical, intellectual and technical capabilities of potential visitors are an important consideration. A visiting group of engineers or safety specialists can be expected to be much more risk aware than a group of schoolchildren. The same applies to individual visitors, whether visiting formally or informally. This is where the potential for children to be making unauthorised visits to workplaces should be of particular concern. Any relevant physical disabilities e.g. mobility, vision, hearing should be identified in advance.

Likely precautions

[C6520] Inevitably the precautions necessary to protect visitors must relate to the areas they are visiting and activities they will be involved in.

- Restricted areas and routes
  Consideration must be given to which areas individual visitors are allowed to visit, particularly those to which they are to be granted unaccompanied access. In planning routes for visiting groups, risks on or close to the proposed route must be taken into account, but so also must the composition of the group – their physical capabilities, awareness of risk, degree of responsibility, the size of the group and the abilities of those in charge to supervise it effectively. In some situations it may be necessary to provide additional signage or barriers to ensure the group keeps to its intended route.
- Restricted activities

Where individuals are making extended visits it must be made quite clear both to the visitor and those providing immediate supervision to them what the visitor is and is not permitted to do. These restrictions will relate to the capabilities of the individual and may be progressively removed as the individual acquires more knowledge and experience.

- Supervision

  The level of supervision necessary for a group will need to relate to its size and type, as well as the areas it will visit. Support may be necessary from teachers or others connected with the visitors in order to ensure effective supervision. Someone must be made responsible for the overall supervision of long-term individual visitors, who may also need to be accompanied when visiting high-risk areas or carrying out high-risk activities. Short-term visitors are likely to need to be accompanied at all times, except when visiting low risk areas – this will particularly be the case for any children. An area may need to be set aside for children to wait (under supervision) whilst their parents are making deliveries to the site.

- Information and instruction

  Visiting groups (and individuals) may need to be provided with information prior to their visit about such matters as:
  — the level of supervision they are expected to provide (for groups);
  — clothing and footwear to be worn; and
  — any special health and safety issues for individuals with specific disabilities e.g. lack of mobility, deafness, visual impairment, asthma.

  At the start of the visit further instruction and information will be necessary on topics such as:
  — risks the visitors may be exposed to;
  — behaviour during the visit;
  — restricted areas, routes and activities;
  — obeying instructions from visit guides/supervisors;
  — emergency arrangements;
  — use of Personal Protective Equipment (PPE) (see below); and
  — what to do if they have questions or concerns.

  During visits further emphasis may need to be given to the risks present or precautions necessary in individual areas. This might involve additional briefings from visit guides and/or use of signs and notices.

- Personal Protective Equipment (PPE)

  Some types of visitors are likely to be able to bring any PPE required for the visit, providing they are advised about this in advance. However, in other cases PPE will need to be made available by the organisation hosting the visit. This is relatively easy for individual visitors – long-term visitors will often, in effect, be treated as employees.

  Provision of PPE for visiting groups can pose practical problems, such as hygiene, availability of suitable sizes, need for adjustment, training in use, etc. Some types of PPE, e.g. safety helmets, high visibility clothing or disposable earplugs, are fairly simple to deal with. However, the issue of safety footwear or respiratory protection to visitors is much more problematical, and areas necessitating their use may need to be avoided. In some situations it may be acceptable for visitors to be instructed to wear sturdy leather footwear for their visit, prohibiting the wearing of trainers, fashion shoes or sandals.

Volunteers

Who must be taken into account?

[C6521] Volunteers carry out many types of work activities that might be carried out by paid staff in other locations. They are exposed to similar risks to workers but their capabilities may not be the same. Volunteers may have considerable knowledge and experience from previous work that they have carried out but this may not always be relevant to their voluntary work. Older volunteers may also have diminishing physical (and even mental) capabilities. There is often a fine line to be followed between ensuring the health and safety of the volunteer and putting them off carrying out the voluntary work completely.

There are many activities in which volunteers are involved:

- Regular volunteers
  Volunteers might work regularly in sectors such as:
  — charity shops and other retail outlets;
  — care work, including home visits;
  — providing transport e.g. for clients or meals on wheels;
  — drop-in centres;
  — coaching or officiating for sporting activities;
  — acting as guides e.g. at museums, historical properties;
  — maintaining equipment e.g. vintage vehicles, steam locomotives; and
  — carrying out environmental work e.g. in country parks or on trails.
- 'One-off' volunteers
  Many people also volunteer to work on a 'one-off' basis at major events, particularly those attracting large numbers, such as:
  — sporting or musical events;
  — carnivals; and
  — fund-raising events.
  Their duties on such occasions often put them in the front line in dealing with other members of the community in roles such as:
  — traffic control;
  — car parking;
  — entry control;
  — staffing registration or enquiry desks; and
  — crowd supervision;
  — providing first aid cover.

Risks to be considered

[C6522] Given the wide range of activities that volunteers can become involved in, they are likely to be exposed to the same variety of risks as those who are at work – as described elsewhere in this publication. This section identifies some of the more common risks that must be taken into account during the risk assessment, including some which may tend to be overlooked.

- Travel
  Some volunteers need to travel about to carry out their work. Apart from the normal risks associated with road traffic, there may be risks to them in the areas they are travelling through or visiting. Such risks will be greater during hours of darkness, particularly for staff who must travel late at night.
- Entry into clients' homes
  Many people providing voluntary care services must enter clients' homes. There may be risks to their personal safety from the clients themselves, their relatives and friends and also from pets. In some cases risks may be due to difficult access (e.g. due to household clutter), defective equipment or dangerous electrical or gas installations.
- Other personal safety risks
  Volunteers will also be prone to personal attack where they are expected to restrict entry to events, handle cash or deal with problems involving the public. Risks will be much higher at certain types of event, with the potential level of alcohol consumption by members of the public often being a major factor.
- The work environment
  In some places where volunteers work it is possible that no one will have been given specific responsibility for cleaning, tidying up or maintaining the premises. Consequently buildings themselves or electrical or gas installations may have deteriorated, possibly to a dangerous condition. Safe storage of equipment and the maintenance of safe access routes within premises will also merit particular attention.
- Equipment
  Voluntary organisations often receive donations of equipment that is no longer required elsewhere. Arrangements must be made to screen out any potentially dangerous equipment and suitable systems put in place to ensure all equipment is properly maintained and regularly inspected. Some types of equipment (e.g. powered machinery) may be dangerous if not used correctly, whilst other types (e.g. storage or display racking, trestle tables) may pose risks if incorrectly assembled. Equipment may also be dangerous if it is left accessible for use by unauthorised persons.
- Manual handling
  Some volunteers may be required to carry out significant amounts of manual handling in respect of goods, materials and equipment. The normal criteria for carrying out manual handling assessments for workers should be applied, namely account must be taken of the task, the load, the working environment and the capabilities of the individual (of particular relevance to many volunteers). For further information see the chapter entitled MANUAL HANDLING.
- Fire and other emergencies

Volunteers must be capable of dealing with foreseeable emergencies both in respect of their own health and safety and also that of clients and other members of the community that they are working with. Volunteers may be responsible for assisting disabled clients in leaving buildings or shepherding members of the public away in the event of fires and similar emergencies.

Likely precautions

[C6523] Some related precautions have already been identified in relation to the risks referred to in C6522 above, whilst others are detailed in other sections of this publication. Precautions of particular relevance to volunteers are identified below:

- Screening and assessment
  Volunteers should be screened in order to identify their existing knowledge and experience in relevant areas. Their physical and intellectual capabilities should also be assessed.
- Information and training
  It is likely that volunteers will need to be briefed on health and safety aspects of their intended work and to be trained in the use of certain equipment or the carrying out of some activities. This needs to be done with sensitivity – whilst volunteers will accept that they might require a certain level of information and training, they will react badly to spending time on subjects they regard as unnecessary or irrelevant. Sessions should be made as enjoyable as possible for the participants, with those who already possess relevant knowledge and experience being encouraged to share this with others.
- Restrictions
  Some activities or use of certain types of equipment may need to be restricted to volunteers who have undergone relevant training. In some cases volunteers may need to be accompanied by a competent paid member of staff.
- Communication
  Effective means of communication need to be established for volunteers working within the community (often remote from colleagues) or at major events. Use of radios, mobile phones and attack alarms may be necessary, depending on the situation. Peripatetic volunteers (e.g. those making home visits) should leave an intended itinerary back at base. There should be arrangements for them to log off on completing their duties and for following up any volunteers who fail to log off or do not keep in touch.
- Supervision and support
  As with paid workers, even very capable volunteers need to have someone that they can communicate with, particularly when problems arise. Support should be readily available to volunteers who need it, especially in emergencies. This may be activated by a radio, telephone message or by the sounding of an alarm.
- Monitoring

The working practices of volunteers should be monitored, as should also be the case for paid workers. However, a more diplomatic approach will need to be taken where corrective actions are necessary to modify the way that volunteers behave. Regular inspections of equipment and premises used by volunteers should take place and appropriate arrangements made for repair and maintenance work.

Major Public Events

[C6524] Major events can create potential health and safety risks for many of the community categories referred to already in the chapter:

—   participants in the event ('customers');
—   volunteer workers (stewards and performers); and
—   neighbours and passers-by.

Not only must event organisers take these members of the community into account, but they must also consider contractors involved in the staging of the event and their potential impact on the community and on each other.

Many major events are now subject to Safety Advisory Groups (SAGs), at which the event is subject to scrutiny by relevant local authorities and the various emergency services. SAGs frequently ask organisers to prepare a properly documented risk assessment for their event so that a full event safety management plan can be developed.

'The Event Safety Guide'

[C6525] Public events can include music events (often requiring public entertainment licences), sporting events, carnivals, fun fairs, agricultural and county shows, firework displays, air shows, car rallies, religious gatherings and many other types of function.

The second edition of 'The Event Safety Guide' (HSG 195) was published by the HSE in 1999 and is primarily aimed at major musical events, although it contains much of value to the organisers of other types of events. The booklet contains general information on the planning and management of events, together with the health and safety responsibilities of the various parties who may be involved. It is available to download free from the HSE website at www.hse.gov.uk/pubns/priced/hsg195.pdf

Risk Assessment

[C6526] Whilst some parts of the HSE publication 'The Event Safety Guide' deal solely with musical events, many of its other headings provide a checklist to use in carrying out a risk assessment for all types of public events. Topics include:

—   venue and site design;
—   fire safety;

—   major incident planning;
—   communication;
—   crowd management;
—   transport management;
—   structures (e.g. stands, gantries, marquees, stages);
—   barriers;
—   electrical installations and lighting;
—   food, drink and water;
—   merchandising;
—   amusements, attractions and promotional displays;
—   sanitary facilities;
—   waste management;
—   special effects, fireworks and pyrotechnics;
—   camping;
—   facilities for people with special needs;
—   medical, ambulance and first-aid management;
—   information and welfare; and
—   children

Of course, some of these topics will not be applicable to all events, whilst some events will necessitate risk topics being broken down in greater detail. For example, a running or cycling race taking place on public roads will need to take account of the direction of arrival and parking of competitors' vehicles and also the interface between competitors and other road users during the event itself. This in turn will require consideration of the race route, barriers, signage, possible road closures, race marshals and the possible need for direct police assistance.

Use of Contractors

[C6527] Event organisers and others engaging contractors must consider the potential risks for members of the community posed by the contractors' equipment and activities. Contractors themselves have many duties under the *Health and Safety at Work etc. Act 1974* and associated legislation, but it has been established for many years that those engaging contractors also have legal responsibilities for risks that the contractors may create, both for those at work and for others who may be affected by their activities. Even where contractors already have good health and safety standards, they may not be fully aware of the risks associated with the environment that they are working in and the members of the community who may be present there.

Advance checks on contractors' health and safety standards should include topics such as:

—   insurance (particularly public liability);
—   health and safety policy;
—   risk assessments and method statements;
—   competence to provide the services required;
—   the competence and training of staff; and
—   availability of relevant licences, certificates, inspection records.

Once contractors have been engaged, they must be informed about precautions that they will be expected to take in their work. Precautions of particular relevance in ensuring that members of the community are adequately protected include:

—　access routes and permitted vehicle access times
—　arrangements for parking;
—　maintenance of safe pedestrian access;
—　segregation of work areas;
—　adequate control of work equipment;
—　safe storage and use of hazardous and dangerous substances; and
—　timing and control of high risk activities.

Their attention must be drawn to relevant characteristics of members of the community who may be present in or around the workplace. This will be of particular importance in relation to customers and other users of buildings and facilities. Special precautions will be necessary in many cases to ensure the safety of children, the elderly and those with disabilities.

In respect of major public events, contractors must be informed about relevant sections of the event safety management plan. In addition to the matters referred to above, they will need to be told about:

—　the layout of the event site;
—　availability of key services and facilities;
—　communications, e.g. with those managing the event and stewards; and
—　emergency arrangements.

The performance of contractors should be monitored both at public events and in other work they carry out. Attention should be paid to their compliance with the health and safety standards set out in their own risk assessments and method statements, as well as to matters of importance in relation to the event or workplace concerned.

Further Information

[C6528] Both of these HSE publications contains details of further references – within other HSE publications and elsewhere:

•　'Protecting the public – your next move' (HSG 151) (2009); and
•　'The Event Safety Guide' (HSG 195) (1999)

The HSE website (www.hse.gov.uk) is another potential source of information. It has sections dealing with school trips, societal risk and various other topics of relevance to community health and safety.

The Fire Safety Guides referred to in **C6507** can be accessed via http://www. communities.gov.uk/fire/firesafety/firesafetylaw/aboutguides/. This web page a lso gives details of how to order hard copies of the guides.

Confined Spaces

Angus Withington

Introduction

[C7001] Several people are killed or seriously injured each year as a result of work in confined spaces. These occur across a range of industries and commonly involve asphyxiation. Recent examples include:

- an employee who was fatally overcome by carbon monoxide and hydrogen sulphide whilst attempting the rescue of another employee similarly overcome in a hopper at a meat rendering plant (*R v John Pointon & Sons* (2008) 2 Cr App R (S) 82).
- two workers who were asphyxiated in a pit surrounding a hot isostatic press, which had filled with argon gas as a result of a leak (*R v Bodycote HIP* [2010] EWCA Crim 802).
- two workers who suffocated in the hold of a barge at a fish farm, where it was believed that rusting metal in a confined space had removed oxygen from the air.

The primary legislative control regulating the risks associated with such work is the *Confined Spaces Regulations 1997 (SI 1997 No 1713)* ("the Regulations"). The particular hazards which the Regulations seek to control arise as a result of the combination of substances or conditions with the confined nature of the place of work, which give rise to particular risks to health and safety. The most common hazards, which must be controlled are:

- flammable substances and oxygen enrichment
- toxic gases, fumes or vapours
- oxygen deficiency
- the ingress or presence of liquids
- free flowing solids or materials (eg. grain, flour, coal dust etc.)
- excessive heat (particularly if personal protective equipment is also required to be worn).

However, other risks (eg the difficulty of access, both generally and in the event of emergencies) also need to be taken into account in relation to confined space work – see **C7007–C2008** below.

A definition of the term 'confined space' is provided within the Regulations (see **C7003**). Confined spaces are enclosures with limited openings such as:

- storage tanks and silos
- reaction vessels
- sewers and enclosed drains
- pipes and ductwork

- rooms lacking in ventilation
- parts of structures under construction.

Relevant Legal Requirements

[C7002] The *Regulations* (see **C7003**) contain specific legal requirements relating to work in confined spaces. They provide a definition of 'confined space' and also define several 'specified risks', which must be addressed in the context of the regulations.

The *Management of Health and Safety at Work Regulations 1999 (SI 1999 No 3242)* require risk assessments to be carried out and effective risk management arrangements to be implemented in respect of all types of work activity. In the case of work in confined spaces, risk assessments must take account not just of 'specified risks' as defined in the regulations but also of other risks which may be present. Common examples of such risks are set out in **C7008** below, which also refers to several specific sets of regulations likely to be relevant to confined space work.

Given the broad definition of 'construction work' contained in the *Construction (Design and Management) Regulations 2007 (SI 2007 No 320)*, these regulations are also of considerable significance in relation to work in confined spaces. The CDM regulations place particular importance on ensuring that persons appointed or engaged in respect of construction work are competent for the purpose. Where work will involve entry into confined spaces, those who plan and manage such work must be competent as well as those actually carrying out the work. For longer projects which are 'notifiable' under the CDM Regulations the 'construction phase plan' must take account of work taking place within confined spaces.

Confined Spaces Regulations 1997

[C7003] *Regulation 1* of the *Confined Spaces Regulations 1997 (SI 1997 No 1713)* contains several definitions of which the most important are as follows:

'confined space'	means any place, including any chamber, tank, vat, silo, pit, trench, pipe, sewer, flue, well or other similar space in which, by virtue of its enclosed nature, there arises a reasonably foreseeable specified risk;
'free flowing solid'	means any substance consisting of solid particles and which is of, or is capable of being in, a flowing or running consistency, and includes flour, grain, sugar, sand or other similar material;
'specified risk'	means a risk of
	(a) serious injury to any person at work arising from a fire or explosion;
	(b) without prejudice to paragraph (a)–

      (i)   the loss of consciousness of any person at work arising from an increase in body temperature;

      (ii) the loss of consciousness or asphyxiation of any person at work arising from gas, fume, vapour or the lack of oxygen;

   (*c*) the drowning of any person at work arising from an increase in the level of liquid; or

   (*d*) the asphyxiation of any person at work arising from a free flowing solid or the inability to reach a respirable environment due to entrapment by a free flowing solid;

'system of work'   includes the provision of suitable equipment which is in good working order.

The definition of 'confined space' contains a list of examples of what might constitute a confined space.

The key words in the definition are in the latter part:

   . . . in which, by virtue of its enclosed nature, there arises a reasonably foreseeable specified risk;.

It should be noted therefore that it is not necessary for the space to be entirely enclosed. The Regulations can therefore apply to open-topped tanks and inadequately ventilated silos, for example.

A roof space in a building would not come within the definition in the normal course of events. However, should significant quantities of a volatile and highly flammable adhesive or paint be used in the roof space, it would become a confined space under the Regulations. Similarly the Regulations would be likely to apply if work was to be carried out there in extremely hot conditions.

The list of 'specified risks' can be paraphrased as meaning a risk of:

* serious injury from fire or explosion;
* loss of consciousness from increase in body temperature;
* loss of consciousness or asphyxiation from gas, fume, vapour or lack of oxygen;
* drowning from an increase in liquid level; or
* asphyxiation or entrapment by a free flowing solid.

Persons with duties under the Regulations

[C7004] *Regulation 3* of the Confined Spaces Regulations 1997 (SI 1997 No 1713) places duties on both employers and the self-employed.

Employers have duties to ensure compliance with the Regulations in respect of any work carried out by:
* their employees; and
* other persons (so far as is reasonably practicable), in relation to matters within their (the employers') control.

Self-Employed Persons have duties in respect of any work by:

- themselves; and
- other persons (so far as is reasonably practicable), in relation to matters within their (the self-employed persons') control.

This partially duplicates the position established by the *Health and Safety at Work etc. Act 1974 (HSWA 1974)*. In particular it places duties on employers who may bring in specialist contractors to work in confined spaces. Contractors and the self-employed also have duties towards each other.

Preventing the need for entry

[C7005] *Regulation 4 (1)* of the *Confined Spaces Regulations 1997 (SI 1997 No 1713)* imposes the primary obligation, namely a preference for avoiding the risk entirely. It states:

> No person at work shall enter a confined space to carry out work for any purpose unless it is not reasonably practicable to achieve that purpose without any such entry

There are many ways in which entry into confined spaces can be avoided. These should be identified as part of a risk assessment which should determine what is reasonably practicable. Examples include:

- testing atmospheres within confined spaces using long probes;
- cleaning or removing residues from outside by water jetting, steam or use of chemicals;
- integrating in-situ cleaning systems into the design of equipment (this also has hygiene and product quality advantages);
- clearing blockages using long tools or remotely-operated flails, vibrating devices or blasts of compressed air;
- providing sightglasses, openings etc. to see what is happening within vessels (internal glasses may need to be fitted with washers and/or wipers); and
- the use of CCTV cameras (additional internal lighting may be necessary).

It may also be reasonably practicable to remove sections of structures or vessels so that they no longer constitute a confined space for the duration of the work.

Designers, manufacturers, importers and suppliers of articles for use at work have duties under *section 6* of the *Health and Safety at Work etc. Act 1974 (HSWA 1974)* to ensure, so far as is reasonably practicable, that the article will be safe when being used, cleaned or maintained. This will include an obligation to ensure that plant and equipment containing confined spaces are properly designed, made or used to eliminate or minimise the need to enter such spaces. Designers also have similar duties under the *regulation 11* of the *Construction (Design and Management) Regulations 2007 (CDM) (SI 2007 No 320)*. Even where it is not reasonably practicable to eliminate the need for entry, the design must still facilitate the establishment of safe systems of work and the provision of emergency arrangements.

Work in confined spaces

[C7006] Where it is not reasonably practicable to prevent entry, *Regulation 4(2)* of the *Confined Spaces Regulations 1997 (SI 1997 No 1713)* states that:

> Without prejudice to paragraph (1) above, so far as is reasonably practicable, no person at work shall enter or carry out work in or (other than as a result of an emergency) leave a confined space otherwise than in accordance with a system of work which, in relation to any relevant specified risks, renders that work safe and without risks to health

Such safe systems of work are, of course, developed as a result of carrying out a risk assessment. Types of risks to be considered and the precautions likely to be necessary to control those risks are dealt with later in the chapter.

Emergency arrangements

[C7007] *Regulation 5* of the *Confined Spaces Regulations 1997 (SI 1997 No 1713)* sets down requirements for emergency arrangements:

(1) Without prejudice to regulation 4 of these Regulations, no person at work shall enter or carry out work in a confined space unless there have been prepared in respect of that confined space suitable and sufficient arrangements for the rescue of persons in the event of an emergency, whether or not arising out of a specified risk.

(2) Without prejudice to the generality of paragraph (1) above, the arrangements referred to in that paragraph shall not be suitable and sufficient unless–

    (a) they reduce, so far as is reasonably practicable, the risks to the health and safety of any person required to put the arrangements for rescue into operation; and

    (b) they require, where the need for resuscitation of any person is a likely consequence of a relevant specified risk, the provision and maintenance of such equipment as is necessary to enable resuscitation procedures to be carried out.

(3) Whenever there arises any circumstance to which the arrangements referred to in paragraph (1) above relate, these arrangements, or the relevant part or parts of those arrangements, shall immediately be put into operation.

The importance of this regulation cannot be under-stated as many fatalities unfortunately occur in the course of attempted rescue. It is estimated that approximately 60% of fatalities occurring in confined spaces are rescuers reacting to an emergency situation.

Several parts of this regulation require further emphasis:

* emergency arrangements must be in place for *any* type of emergency, not just those arising from specified risks – eg. a serious fall within the confined space;
* risks to rescuers must be reduced, so far as is reasonably practicable;
* resuscitation equipment may need to be provided and maintained (a worker who has been asphyxiated is likely to need to be resuscitated quickly (within a few minutes and therefore well before the emergency services can get there); and
* emergency arrangements must be immediately put into operation should an emergency arise.

Practical guidance on emergency arrangements is provided later in the chapter at **C7028** below.

Other risks in confined spaces

[C7008] Many other risks may be present when working in confined spaces. Some of these risks may be greater because of the enclosed nature of the space, whilst rescue of accident victims is likely to be much more difficult. Such risks include:

— restricted openings for entry into and exit from the space;
— restricted access within the space;
— potential for falls from height or into water;
— slips, trips and falls at the same level;
— falling objects;
— lack of visibility within the space;
— mechanical or electrical equipment within the space;
— chemical, biological or dust hazards;
— excessive noise (damaging hearing or interfering with communication);
— excessive cold; and
— the remote location of the confined space.

Even though these other risks are not covered by the *Confined Spaces Regulations 1997 (SI 1997 No 1713)*, other specific regulations will apply eg the *Workplace (Health, Safety and Welfare) Regulations 1992 (SI 1992 No 3004)*, the *Work at Height Regulations 2005 (SI 2005 No 735)*, the *Electricity at Work Regulations 1989 (SI 1989 No 635)*, the *Control of Noise at Work Regulations 2005 (SI 2005 No 1643)*, the *Provision and Use of Work Equipment Regulations 1998 (PUWER) (SI 1998 No 2306)* and the *Control of Substances Hazardous to Health Regulations 2002 (COSHH) (SI 2002 No 2677)*. The *Personal Protective Equipment at Work Regulations 1992 (SI 1992 No 2966*, as amended) will also be relevant in relation to PPE used in confined space work.

The general requirements of *Health and Safety at Work etc. Act 1974* will also apply (*HSWA 1974, s2 (2)(a)* contains a similar requirement for a safe system of work) and the *Management of Health and Safety at Work Regulations 1999 (SI 1999 No 3242)* still require a risk assessment to be carried out.

Figure 1 at **C7010** below provides a checklist to aid in the identification of both 'specified risks' and other risks.

Safe systems of work

[C7009] Both the specific requirements of the *Confined Spaces Regulations 1997 (SI 1997 No 1713)* and the general requirements of *Health and Safety at Work etc. Act 1974* mean that work in confined spaces must be carried out in accordance with a safe system of work. The Approved Code of Practice (ACOP) accompanying the regulations (contained in HSE booklet L101 – 'Safe work in Confined Spaces') makes a clear statement on how safe systems of work should be specified. It states (at paragraph 74):

> To be effective a safe system of work needs to be in writing. A safe system of work sets out the work to be done and the precautions to be taken. When written down it is a formal record that all foreseeable hazards and risks have been considered in advance. The safe procedure consists of all appropriate precautions taken in the correct sequence. In practice a safe system of work will only ever be as good as its implementation.

Dependent upon circumstances it may be more appropriate to specify the safe system of work through a written model procedure (eg where a similar task is carried out on a regular basis) or through a permit to work (more appropriate for variable or infrequent tasks).

Permits to work

[C7010] The ACOP recommends the use of a 'permit to work' procedure as an extension to the general duty of providing a safe system of work where there is a reasonably foreseeable risk of serious injury for an employee entering or working in a confined space. Such a procedure supports the implementation of a safe system by providing the means of recording the results of the risk assessment process and the key elements of the safe system of work, as well as collating information which may be needed during an emergency. A correctly applied permit to work system should ensure that:

- people at work in the confined space are aware of the hazards involved;
- workers know the nature and extent of the work to be carried out;
- appropriate elements of a safe system of work are identified;
- there is a formal check that a safe system of work is in place before people enter or work in the confined space;
- suitable action is taken about other activities which may affect work in the confined space;
- any time limits on entry are specified;
- appropriate means of communication are established; and
- necessary respiratory protective equipment (RPE) and personal protective equipment (PPE) standards are identified.

Figure 2 (below) provides an example of a permit to work form for use in controlling entry into confined spaces. It identifies both the 'specified risks' defined in the Regulations and a range of other risks. However, those intending to use this form should ensure that it addresses fully the risks likely to be present in the confined spaces that they are responsible for – adaptation of the form may well be necessary to deal adequately with local circumstances.

Figure 3 (below) is intended as an aide-memoire to accompany the permit to work form provided in Figure 2. It provides the permit issuer with a list of the precautions which may be necessary to control the risks involved in the work.

Figure 1: Confined Spaces – Identification of Risks

DEPARTMENT

TASK/LOCATION

SPECIFIED RISKS (Confined Spaces Regulations 1997)   significant risk ? possible risk

	or ?	Notes
Serious injury from fire or explosion		
Loss of consciousness – increase in body temperature		
Loss of consciousness or asphyxiation due to:		
— gas, fume or vapour		
— lack of oxygen		
Drowning – increase in level of liquid		
Asphyxiation or entrapment – free flowing solid		
OTHER SIGNIFICANT RISKS		
Restricted entry/exit		
Restricted access within		
Falls from height/into water		
Slips, trips, falls – same level		
Falling objects		
Lack of visibility		
Mechanical equipment		
Electrical		
Chemical		
Biological		
Dust		
Noise		
Excessive cold		
Remote location		
Other (state)		

Should access be controlled by a permit to work?   **YES/NO**

Is a written task procedure required?   **YES/NO**

Signature _____ Name _____ Date _____

Figure 2: Permit to work – Entry into Confined Spaces

Confined   _____   Purpose of   _____
Space   Entry
Date   _____   Can the work be carried   **YES/NO**
                                 out from outside?

RISKS IDENTIFIED	✓ or ✗	CONTROLS REQUIRED (inc. **Tests**) (If necessary, provide details on accompanying documents)
Fire or Explosion		
Increase in Body Temperature		
Gas, Fume, Vapour		
Lack of Oxygen		
Drowning		
Free-flowing Solids		
Restricted Entry/Exit/Access		
Falls. Slips, trips		
Falling Objects		
Lack of visibility		
Mechanical/Electrical		
Chemical/Biological/Dust		
Noise		
Excessive Cold		
Remote Location		
Other (provide details)		

	DETAILS
Communication Methods	
Emergency Equipment & Arrangements	
Person(s) covered Name(s) by this PTW	
Signature(s)	

I confirm that the precautions detailed above have been/will be taken.
I authorise the persons named above to enter this confined space for the purpose stated.
They have been instructed on their roles and responsibilities.

PERMIT VALID	FROM		PERMIT CANCELLED AT	
(Times)	TO		(Time)	(Date)
Signature		(Permit Issuer)	Signature	(Authorised Person)

FIGURE 3: CONFINED SPACES – CONTROL MEASURES AND OTHER PRECAUTIONS

RISKS	POSSIBLE CONTROLS
Fire or Explosion	Cleaning of residues. Atmospheric testing. On-going monitoring.
	Purging. Ventilation. Protected electrical equipment.
	Hot work restriction. No smoking. Fire fighting equipment.
Increase in Body Temperature	Delay before entry. Working time restrictions. Cooling fans.
Gas, Fume, Vapour	} Cleaning of residues. Atmospheric testing. Ongoing monitoring.
Lack of Oxygen (inc. presence of heavy gases nearby)	} Purging. Ventilation (natural or forced). Isolate incoming feeds.
	} Respiratory protection (inc. air-fed respirators or self-contained BA).
Drowning	} Isolate incoming feeds. Drain contents.
Free-flowing Solids	} Harness attached to secure lifeline.
Restricted Entry/Exit	Staff selection. Harness attached to secure lifeline. Instructions to staff.
Falls from height/into water	Harness attached to secure lifeline and inertia reel.
	Temporary scaffolding/platform.
Slips, trips, falls – same level	Clean up surfaces. Suitable footwear.
Falling Objects	Safety helmets. Protective screens.
Lack of visibility	Permanent or temporary lighting. Torches. BACK-UP ESSENTIAL.
Mechanical/Electrical	Secure isolation.
Chemical/Biological/Dust	Drain contents. Clean up. PPE eg gloves, body cover, eye protection.
	RPE – air-fed, self-contained BA, cartridge respirator, dust mask.
	Decontamination.
Noise	Hearing protection. Suitable communications (see below).
Excessive Cold	Suitable clothing and footwear.
Remote Location	Suitable communications (see below). Monitoring from base.
KEY PRECAUTIONS	**OPTIONS AVAILABLE TO CONSIDER**
Communications	Attendant outside. Word of mouth. Telephone. Portable phone.
	Radio. Air horn. Whistle.
	(Ensure that the equipment works in the work location.)

RISKS	POSSIBLE CONTROLS
Emergency Arrangements (For all types of emergencies including falls, heart attacks etc.)	Attendant outside. Harness attached to secure lifeline.
	Recovery equipment eg winch (in position or available nearby).
	RPE/PPE worn by or readily available for rescuers.
	Resuscitation equipment. Stretcher. Other first aid equipment.
	Fire fighting equipment.
	Contact with emergency services/internal assistance.
	Medical staff on standby (high-risk situations only).
	Register of persons inside (large groups only).
	(Ensure relevant staff are trained in use of emergency equipment.)

An essential ingredient of any permit to work system is that those issuing and cancelling the permit are competent for the purpose. Competence and training are dealt with at **C7012** below.

A permit to work system is unlikely to be needed where:

• assessed risks are low and can be controlled easily;
• the system of work is very simple; and
• other work activities cannot affect safe working in the confined space.

However, that decision should also be made by a competent person, taking account of the findings of the risk assessment and the overriding obligation to ensure that a safe system of work is adopted.

Task procedures

[C7011] An alternative way of ensuring that a safe system of work is specified in writing may be to incorporate it into a formal 'task procedure' for carrying out the work. (The terms Job Safe Procedure' (JSP) or Standard Practice Instruction (SPI) are used by some to describe similar documents.) Such formal procedures are usually suitable where certain tasks are to be performed regularly within confined spaces. As with permits to work, they should result from a risk assessment of the task in question, and identify the risks which may be present and the various precautions which must be taken to control those risks to an adequate degree.

However, task procedures do not impose the same degree of formal checking (ie that the safe system of work is actually being implemented) as that provided by the permit to work system. They will usually be more suitable for providing a point of reference for a well-trained in-house work team who are already familiar with the environment, the work involved and the risks it entails, as well as the task procedure itself. A permit to work is more likely to be necessary for controlling outsiders (such as contractors), infrequently performed tasks and/or more hazardous work.

Competence and training

[C7012] Competence for work in confined spaces will be acquired through a combination of adequate training and relevant experience. Consideration should also be given to workers' physical and psychological suitability for work in confined spaces – they may need to be fit enough to wear breathing apparatus and conditions such as claustrophobia may be an issue. Training should include:

- an awareness of the Regulations and their requirements;
- an appreciation of the risks involved in work in confined spaces;
- the systems of work appropriate for different types of work;
- how to use relevant types of equipment eg for communications, access, testing, PPE; and
- emergency and rescue arrangements and equipment

An increasing number of organisations can provide general training on work in confined spaces. However, this will usually need to be augmented by additional training and/or experience to ensure that workers are:

- familiar with the confined spaces they will work in and the risks involved;
- aware of the precautions necessary in their types of work;
- capable of using the equipment required to ensure a safe system of work; and
- capable of using emergency equipment.

Where members of staff are required to issue and cancel permits to work for confined spaces they will need further training, including an appreciation of the purpose of permits to work. They must be familiar both with the range of risks which may be present and the wide variety of precautions which may be necessary to ensure a safe system of work eg isolations, ventilation, gas testing, RPE, PPE etc. There should be a formal system in place for their training, testing and authorisation (as should be the case for all permit to work systems).

Supervision

[C7013] All workers must be supervised effectively and this is particularly important for those working in confined spaces. Supervision will involve:

- identifying when work is to be carried out in confined spaces;
- arranging for necessary permits to work;
- ensuring that workers are aware of the risks involved and the precautions required;
- ensuring that they are capable of implementing the precautions and have all the equipment required;
- monitoring working practices (including the correct operation of the permit to work); and
- dealing with problems which may arise during the work.

For higher risk tasks, the supervisor is likely to remain present whilst the work is carried out. In other cases it may be acceptable to visit periodically, although the supervisor should still be readily available if required.

Precautions for work in confined spaces

[C7014] Figure 3 (see C7010 above) shows the range of precautions that may be necessary when working in confined spaces. Further detail is provided below on the types of precautions required to control 'specified risks'. Additional guidance is available in HSE booklet L101 'Safe work in Confined Spaces'. Other risks which may be present in confined spaces can be controlled using conventional techniques as set out in Figure 3.

Communications

[C7015] Effective communications must be provided between those inside a confined space, between those inside and those outside, and also in order to summon help in an emergency. A variety of means can be utilised, for example, speech, rope tugs, telephone, radio, so as to ensure that rapid and unambiguous communication can be achieved. Consideration should be given to risks of using communication equipment in flammable or explosive atmospheres, to the difficulties of communicating when wearing respiratory protection equipment (RPE) and to the practical difficulties of using telephones or radios in some situations. Air horns or whistles may provide a practical alternative for some environments. Communication equipment and methods should always be checked and tested at the work location before work starts.

Testing and monitoring the atmosphere

[C7016] Testing may be necessary in respect of hazardous gases, fumes or vapours (these may be hazardous to health, or flammable or explosive). The concentration of oxygen present in the atmosphere may also need to be checked. These test results will indicate possible needs for other precautions eg purging, ventilation, RPE. In some cases tests may need to be carried out periodically or prior to each further entry. Constant monitoring of levels may also be necessary – many monitoring devices can give alarms if concentrations of hazardous substances rise above a pre-determined level (or if oxygen levels fall). Those carrying out testing must be competent for the task and test results should be retained.

Gas purging

[C7017] Flammable or toxic gases or vapours may need to be removed by purging with an inert gas (such as nitrogen) for flammable substances, or by an inert gas and/or air for toxic contaminants. Where inert gases are used, further purging with air is likely to be necessary before entry can take place without suitable RPE. Where purging has taken place, there is likely to be a need for further testing or monitoring of the atmosphere and the use of RPE may still be necessary. Consideration must also be given to the safe venting and disposal of the purged gases.

Ventilation

[C7018] Mechanical ventilation (or ventilation openings) may be necessary to remove contaminants already present within the confined space and/or to provide fresh breathing air for those inside the confined space and/or to remove gas, fume or vapour produced by the work. Consideration should be given to the layout of the confined space and the location of any contaminants in order to ensure effective fresh air circulation. Suitable positions should be chosen for air inlets and outlets to avoid possible dead spots and also to prevent contaminated extract air from re-entering. Oxygen must never be used to 'sweeten' any atmosphere, particularly in confined spaces, as oxygen enrichment (above the normal concentration in air) increases the combustibility of many materials and can also have a toxic effect, if inhaled.

Removal of contents and residues

[C7019] Work may only be able to be undertaken safely once the contents of the confined space, together with any remaining residues, have been effectively removed. Methods for removing residues include mechanical means (eg scraping), steam, high-pressure water, or use of chemicals. In some cases the removal work may require entry to be made into the confined space, necessitating its own safe system of work. Testing or inspection after the work may be needed, in order to ensure residues have been removed effectively.

Isolations

[C7020] Frequently the system of work will require the effective prevention of the ingress of substances into the confined space (gases, liquids and other free flowing materials), or the isolation of mechanical and electrical equipment within the space. Such isolations are likely to achieved by precautions such as the removal of sections of supply pipe, the insertion of blanks or the use of a locking valve. Consideration may have to be given to the need to maintain essential services eg lighting, power or water for the work to be carried out whilst preventing undesirable risks eg re-energisation of process equipment or possible flooding. Regular checks should obviously be made to ensure that the method of isolation adopted remains effective.

Electrical equipment

[C7021] Electrical equipment should be chosen in relation to risks within the confined space and the work to be carried out. It may need to be:

- suitable for use in a flammable or explosive atmosphere;
- suitable for use in damp conditions (or chemical exposure);
- low voltage because of electrocution risks (or at least protected by RCD's); or

- protected from mechanical damage or dropping.

Workers may need to be prevented from taking personal electrical equipment such as mobile phones or audio equipment into confined spaces where flammable or explosive atmospheres may be present.

Gas equipment and internal combustion engines

[C7022] Use of gas cylinders within confined spaces should be avoided if possible. If cylinders must be used then mechanical ventilation will usually be necessary. All gas equipment and pipelines should be checked for leaks before being taken into the confined space and removed to a well ventilated area at the end of every work period. This is to avoid possible contamination of the confined space by a slow leak from, for example, a gas cylinder. Where pipes and hoses cannot be removed they must be disconnected from the supply outside the confined space and their contents safely vented.

Petrol engines must not be used within confined spaces and use of diesel engines should be avoided whenever possible. Where the use of diesel engines or portable gas cylinders in confined spaces is unavoidable, adequate mechanical ventilation must be provided to ensure complete combustion takes place. Diesel engine exhausts must be vented well away from the confined space and downwind of any air intakes.

Personal protective equipment (PPE) and respiratory protective equipment (RPE)

[C7023] Although the law expresses a preference for other types of control measures the use of PPE and RPE in confined space work will often be unavoidable. Wearing of harnesses with safety lines is particularly likely to be necessary, as is the use of RPE providing higher levels of protection (self-contained breathing apparatus or air-fed respirators). Wearing PPE and RPE within a confined space may be an additional source of heat stress for workers.

Access and egress

[C7024] Safe routes into and out of the confined space must be provided and maintained. These must also be suitable for emergency access or escape. Sizes of openings must allow for PPE and/or RPE that may need to be worn. The minimum size of an opening to allow access (including in an emergency where self-contained breathing apparatus is used) is 575mm diameter. Some existing plant may have smaller openings (eg road tankers where the standard size is only 410mm). In such circumstances, further precautions should be taken, such as the availability of airline breathing apparatus. Larger workers may find it physically difficult to get through small openings or to move around within the confined space. Signs prohibiting unauthorised entry may also be required. It is recommended that practice drills should be undertaken in order to check that access and egress can be achieved, especially in an emergency situation.

Fire precautions

[C7025] Precautions such as removal of residues, gas purging, ventilation or the control of ignition sources may be necessary as described earlier. Where flammable substances must be present in a confined space (eg paints, cleaning materials) they must be kept to a minimum. Provision of suitable fire fighting equipment will be necessary in such cases. In larger structures consideration may need to be given to providing alternative means of escape in case of fire.

Lighting

[C7026] Adequate and suitable lighting must be provided as for all work-places. Suitable back-up lighting (eg torches) must be available in the event of lighting failure. See C7021 above for relevant criteria for the selection of electrical equipment.

Static electricity, hot work, smoking and sparks

[C7027] Risks from static electricity or other sources of ignition, such as hot work, may need to be considered. Some types of equipment, clothing or activities (eg. steam or water jetting) are prone to static build-up or discharge. Smoking must be prohibited in all confined spaces. Paragraph 71 of the ACOP indicates it may be necessary to exclude smoking for 5 to 10 metres beyond the confined space. Non-sparking tools will be required where flammable or potentially explosive atmospheres may be present.

Limited working time and entry control

[C7028] Working within confined spaces may be physically demanding due to extremes of temperature or humidity, the difficulties of wearing RPE or PPE or cramped working conditions. Use of fans may be necessary to achieve a satisfactory thermal environment. Time limits for individuals to stay in the space and rest periods in between may need to be imposed. These may be enforced by a logging or tally system. Such systems could also be necessary in large confined spaces or for large work groups in order to provide a check as to who is inside at any point in time.

Emergency arrangements

[C7029] Suitable arrangements must be prepared for emergency rescue from a confined space. This may be due to a 'specified risk' or 'other risk' associated with the confined space or for other types of emergency, eg a worker inside suffering a heart attack or other medical condition. These arrangements must reduce risks to rescuers, so far as is reasonably practicable, and include the provision and maintenance of resuscitation equipment, where relevant due to 'specified risks'.

Emergency arrangements must be immediately put into operation should such a situation arise – a fall inside a confined space may allow time for medical and other assistance to be summoned, but a collapse due to gas, fume, vapour or lack of oxygen is likely to require immediate intervention by those on the spot eg. a rescue using a harness, lifeline entry into the space by rescuers using suitable respiratory protective equipment (RPE) and prompt use of resuscitation equipment. Consideration must also be given to lower levels of availability of outside assistance at certain times (eg nights or weekends) and to work taking place in more remote locations.

Examples of emergency equipment and arrangements which may be necessary are:

- an attendant immediately outside the confined space;
- a harness attached to a secure lifeline worn by those inside;
- recovery equipment eg a tripod and winch (either already in position or available nearby, if required);
- suitable RPE or PPE for rescuers (either already worn by them or readily available);
- appropriate resuscitation equipment readily available;
- presence of someone with first aid training;
- stretcher(s) and other appropriate first aid equipment near at hand;
- fire fighting equipment;
- suitable communication equipment (see **C7015**);
- means of contacting and liaising with the emergency services or other sources of assistance (some high risk work may justify informing the emergency services in advance);
- medical staff on standby nearby (only likely to be necessary for high risk situations);
- use of a register of persons inside the confined space (only necessary where large groups or large, complex confined spaces are involved); and
- arrangements to shut down nearby plant (which may be a source of noise or could increase the risks).

Relevant staff must be fully trained in the use of emergency equipment and physically capable of using it. This includes those working within confined spaces, acting as back-up attendants or intended to provide emergency assistance in other ways.

The emergency arrangements required should be identified during the risk assessment process and should normally be recorded – either on the permit to work form or within the task procedure.

Further Reading

[C7030] L101 Safe work in Confined Spaces, Confined Spaces Regulations 1997, 'Approved Code of Practice', Regulations and Guidance (2009; 2nd edition) (HSE Books)

INDG 258 Safe work in confined spaces (free HSE leaflet)

AIS 26 Managing confined spaces on farms (free HSE leaflet)

HSG250 Guidance on permit-to-work systems: A guide for the petroleum, chemical and allied industries (2005) (HSE Books)

Construction and Building Operations

Gordon Prosser

Introduction

[C8001] Since this chapter was last revised, the formal consultation has taken place that has lead to the issue of the *Construction (Design and Management) Regulations 2007*, incorporating the former *Construction (Health, Safety and Welfare) Regulations 1996*.

Tragically 60 workers lost their lives in construction in 2005–06, but this was the lowest figure on record. That there is no room for complacency is illustrated by the fatal injury statistics for the construction industry released by Health and Safety Executive for the following year, 2006–07. These show a 28% increase in the number of fatal accidents to workers.

Despite these figures, the review process identified that work during the construction phase, under the control of Contractors, did not require significant legislative change (see the Construction, Design and Management chapter). That chapter identifies the more significant changes in the ways in which projects are to be planned, designed and managed that are now incorporated into the *CDM 2007 Regulations*.

Summary of changes

[C8002]–[C8017] The following is a summary of the principal changes contained in *Part 4* of *CDM 2007*, with reference to the former *Construction (Health, Safety and Welfare) Regulations 1996*—

Person on whom duties are imposed

* In *Regulation 25*, the two primary duty holders are "contractors carrying out construction work", and "every person who controls the way in which any construction work is carried out". Reference to "employers" and "employees" has been replaced; organisations and the individuals in those organisations or working for themselves are identified.

Safe places of work

— In *Regulation* 26, the phrase "without risks to health has been omitted. "Every place of work shall, so far as is reasonably practicable, shall be made and kept safe for any person at work there"; the presumption is that "safe" includes protection against the risk of both chronic injury (often considered "health" issues) and acute injury.

— Egress as well as access is now covered.

— The relaxation for those engaged in work "for the purpose of making any place safe" has been removed; a simple example would be those erecting or dismantling a scaffold access platform. Those working on an incomplete scaffold are expected to work in accordance with the NASC code of practice developed with the HSE, SG4:05 "*Preventing Falls in Scaffolding and Falsework*".

— Ergonomic considerations, sufficient working space suitable for the person must now take into account any necessary work equipment present.

Good Order

• *Regulation 27* now links site security to good order.

• A site perimeter may be identified by suitable signs alone, but only if the extent of the site is readily identifiable and the level of risk posed is such, (ie *so low*) that fencing is not considered a reasonably practicable precaution in the interests of health and safety. However the presumption is still that fencing and signs will be used. It must be remembered that provision of fencing and "hoards" to protect the public has been a statutory requirement under earlier legislation since the nineteenth century.

• See also the requirements for barriered exclusion zones in the work at height Regulations.

• The requirement not to allow projecting nails to remain in place in timber or other material now includes the term "other similar objects" (eg screws, rivets, coach bolts and the like).

Stability of structures

• Specific reference to the term "including any excavation work" in *Regulation 28* has been omitted as unnecessary. Every cause related to carrying out construction work (which per se includes excavation) that may affect the stability a structure (*or part of a structure, eg during construction, erection, alteration or dismantling*) has to be considered.

- The word "accidentally" has also be omitted; thus where necessary to prevent danger, steps must be taken to prevent collapse, whether accidental or not. An example of non-accidental event would be arson where combustible material had been stored or allowed to accumulate in an unprotected zone near a structure.

- "Any buttress, temporary support or temporary structure must be of such design and so installed and maintained as to withstand any foreseeable loads, which may be imposed on it, and must only be used for the purpose for which it is designed, installed and maintained". How this is to be achieved is not defined in the regulation, but *Regulation 4* in *Part 2*, which applies to all construction sites, now covers the requirement for individual workers and designers to be competent. The previous wording only required that erection and dismantling had to be "under the supervision of a competent person".

- Scaffolding is frequently used to buttress or provide temporary support, and the work at height regulations require that the use of scaffolding must be designed. The new European limit state standard, *EN 12811* has recently replaced the established *British permissible stress Standard BS 5973*. The National Access and Scaffolding Confederation [NASC] are working with industry and the HSE to ensure that the transition is adequately covered and that issues such as wind loading are not ignored.

Demolition or dismantling

- The wording in *Regulation 29* has been simplified. "The demolition or dismantling of a structure, or part of a structure, shall be carried out in such a manner as to prevent danger. Where it is not practicable to prevent danger, danger shall be reduced to as low a level as is reasonably practicable".

- If the extent of demolition or dismantling is such that in itself, or as part of the project to which such work is linked that the 30 working days or 500 man days are not met, then notification of the demolition work is not required.

- The requirement in *Regulation 29*, whether or not the work is notifiable, is that "the arrangements for carrying out such demolition or dismantling shall be recorded in writing before the demolition or dismantling work begins".

Explosives

- Without reference to any other statutory provision (eg *section 25* of the *Explosives Act 1875*, or The *Control of Explosives Regulations 1991*) or guidance relating to explosives, (eg *Health and Safety Executive circular to Chief Officers of Police* [no 1/2005] or the NaCTSO guidance leaflet) *Regulation 30* states that "So far as is reasonably practicable, explosives shall be stored, transported and used safely and security. There was no specific reference to explosives in the 1996 Regulations.

- The HSE document L10, "*A guide to the Control of Explosives Regulations 1991*" (ISBN 0 11 885670 7) provides further information.

Excavations

- The form of wording has been simplified. *Regulation 31* states "All practicable steps shall be taken, where necessary, to prevent danger to any person, including, where necessary, the provision of supports or battering to ensure that: (a) any excavation or part of an excavation does not collapse accidentally; (b) no material from a side or roof, or adjacent, to any excavation is dislodged or falls; and (c) no person is buried or trapped in an excavation by material which is dislodged or falls."

- It is of concern that the requirement in the former *Regulation 12(4)* that "suitable and sufficient equipment for supporting an excavation shall be provided" has been omitted. This requirement discouraged unscrupulous contractors starting work on excavations where there was foreseeable risk without ensuring that the necessary equipment was already on site.

- The requirement for the installation, alteration or dismantling of any support for an excavation to be carried out only under the *supervision* of a competent person has also been omitted. It is recommended that principal contractors or those in control of site-works cover these two omissions with specific site rules in situations where projects involve excavations, and they feel that the requirement of *Regulation 31(4)* stipulating only the more limited requirement of *inspection* by a competent person, and the wording and the requirement for competence in *Regulation 4* being a duty only "as imposed on him by or under any of the relevant statutory provisions", are not sufficient to ensure safe working.

- The wording of *Regulation 31(4)* is as follows: "Construction work shall not be carried out in an excavation where any supports or battering have been provided unless—
  (a) the excavation and any work equipment and materials which affect its safety, have been inspected by a competent person—
     (i) at the start of the shift in which the work is to be carried out;
     (ii) after any event likely to have affected the strength or stability of the excavation; and
     (iii) after any material unintentionally falls or is dislodged; and

(b)   the person who carried out the inspection is satisfied that the work can be carried out there safely"

It is recognised that the wording of this regulation will cover *further* work carried out in a trench or equivalent, but not necessarily the temporary situations during installation and dismantling of support and the like. The risk assessed method statements covering the installation and removal of temporary supports will need careful checking by the person in control of the site. Systems of work that involve the use trench boxes or drag boxes need to be checked to ensure that persons working in the trench are adequately protected from danger in a manner equivalent or better to that specified by this regulation.

- Instability of an excavation caused by surcharge from work equipment (eg mobile plant) or materials (including material excavated or stockpiles for unrelated work) is covered by the simple form of words in *Regulation 31(3)*: "suitable and sufficient steps shall be taken, where necessary, to prevent any part of an excavation or ground adjacent to it from being overloaded by work equipment or material". The person responsible for making such judgement needs to be identified as part of the safe system of work in such circumstances.

Cofferdams and caissons

- This specialist area of work is covered by *Regulation 32*.
- Provision has to be made for shelter or escape "if water or materials" enter.
- The person carrying out the inspection of any cofferdam and caisson has the power to stop work if he is not satisfied as to the safety of the installation.

Reports of inspections

- The requirements for reports of inspection are largely unchanged.
- Particulars to be contained in a report are listed in *Schedule 3*, relating back to *Regulation 33(1)(b)*.
- The *minimum* particulars in a report of inspection must include—
  (i)   name and address of the person on whose behalf the inspection was carried out;
  (ii)   location of the place of work inspected;
  (iii)   description of the place of work or part of that place inspected (including any work equipment and materials);
  (iv)   date and time of the inspection;
  (v)   details of any matter identified that could give rise to a risk to the health or safety of any person;

(vi)   details of any action taken as a result of any matter identified; (this may include what actions were taken to alert workers on site, fence the area off, exclude access and the like);

(vii)   details of any further action considered necessary; and

(viii)   Name and position of the person making report.

Energy distribution systems

- The requirements in relation to services, buried or overhead are covered by Regulation 34. the form of words "suitably located, checked and clearly indicated" replace the former requirement to "identify" contained in the former *Regulation 12*, excavations.
- The focus of the regulation is on electric power cables. Those in control of work where services may be damaged need also to be aware of the significant safety risks associated with water mains (volume of flow and pressure), gas mains (Transco fined £15 million over Larkhall gas explosion, 26 August 2005) and the cost of repairs and business interruption claims for damage to fibre optic and equivalent communication cables.

Prevention of drowning

- There are no substantial change in the requirements of *Regulation 35*: "suitable and sufficient steps shall be taken (a) to prevent, so far as is reasonably practicable, such person from so falling; (b) to minimise the risk of drowning in the event of such a fall; and, (c) to ensure that suitable rescue equipment is provided, maintained and when necessary, used so that such person may be promptly rescued in the event of such a fall.
- The former requirements for "any vessel used to convey any person by water to or from a place of work: (a) shall be of suitable construction; and (b) shall be properly maintained; and (c) shall be under the control of a competent person;" although omitted from the *CDM 2007 Regulations*, are now covered by *PUWER 1998*.

Traffic routes

- Detailed requirements are contained in Regulation 36.
- Segregation of people (*pedestrians*) from vehicles and mobile plant is a high priority area for enforcement due to the history of serious and fatal injuries; planning of construction sites must give consideration to design of loading bays, separation, segregation, inter-visibility (*of pedestrians and vehicles*), choice of crossing points, siting of doors and gates, and the avoidance of obstructions that may cause persons to be trapped or crushed; a bricklayer stepping out from behind a stockpile of bricks en-route back to the canteen was fatally injured by a tele-

handler. The driver of the tele-handler was unaware that the incident had occurred and his link to the incident was only traced back to that item of plant when the accident investigation team located a small quantity of the victim's blood on the vehicle.

• Every traffic route shall be indicated by suitable signs. There is now an additional requirement for traffic routes to checked and properly maintained.

• The *Health and Safety (Safety Signs and Signals) Regulations 1996* require that the "suitable signs" match those used on the Highway, as covered by the Traffic Signs Regulations and General Directions.

Vehicles

• The requirements of Regulation 37 need to be read in conjunction with the *Provision and Use of Work Equipment Regulations 1998, (SI 1998 No 2306)*, regarding the selection, maintenance, inspection and use of mobile work equipment.

• As before, unauthorised use and "unintended movement" of unattended vehicles must be prevented.

• The driver (*"person in control of the vehicle"*) must be able to warn others, but site rules should specify when and where strobe lights and audible warnings and the like are to be used;

• Safe loading, operating, towing as well as diving must be considered;

• Specific reference to rail vehicles (those running on fixed tracks) has now been omitted; Should they be used (for example in tunnelling work) provision to deal with any rail vehicle that becomes derailed will still be required;

• Measures now need to be taken to prevent any vehicle falling into an excavation (not just those involved in the excavation, material handling or tipping process). This is a particular issue when a number of different contractors are working on or passing an area of a site.

Prevention of risk from fire, emergency routes and exits, fire detection and fire fighting

• The statutory duties in *Regulation 38*, and *Regulation 41* are only to protect persons from the risk of injury. Those in control of a site need to be aware of and take measures to protect against the commercial risk of damage to property and materials.

• For example it is recommended that site rules need to cover the issuing of "hot work permits" and the like, and site security to protect against the risk of arson must be a high priority;

- *Regulation 40(4)* requires that the planning and provision of emergency exits and routes need to take account of the type of work being carried out, the work equipment being used on site, and the layout, "characteristics" and size of the site;
- Where the works are being carried out in or adjacent to an existing work place, fire risk assessment and management of emergency procedures need to take account of the requirements of the *Regulatory Reform (Fire Safety) Order 2005 in England and Wales*, or the *Fire (Scotland) Act 2005* in Scotland.
- The emergency procedures for any construction site are a key "site specific" component of the site induction for all person's entering that site.

Facilities for rest

- The two substantial changes between the requirements of the current *Schedule 2* and the former *Schedule 6* (relating to the former *Regulation 21*) are that—
  - (i) "tables and adequate seating *with backs* be provided for the number of persons at work likely to use them at any one time; (ie benches without backs are no longer adequate); and
  - (ii) rooms, huts, cabins or equivalent must be maintained at an appropriate temperature.
- The provision to segregate non-smokers from discomfort (*and potential harm*) caused by tobacco smoke have been overtaken by the *Health Act 2006 Ch 28*, and the subsequent five sets of U.K. "Smokefree" Regulations.
- There is no legal requirement to provide smoking shelters. Although it is common for employers to quote "health" reasons for not spending money creating places for smokers to congregate, the risks of fire and consequent costs due to uncontrolled "clandestine" smoking on site need to be considered. If an outside smoking shelter or area is designated, the consequent fire risks will need to be evaluated and the provider of insurance cover for the site informed. There will be a need to ensure that any shelter is not "enclosed" or "substantially enclosed" as defined by the new law. Currently guidance on this issue varies from local authority to local authority.
- The significance of *Schedule 2* is that it is now the duty of clients (*Regulation 9*), contractors (*Regulation 13*) and principal contractors on notifiable projects (*Regulation 22*) to ensure that welfare facilities, as a minimum, meet these requirements, and that they are in place from the start to the end of the construction phase.
- They may be summarised as follows—

Sanitary conveniences—
— Suitable and sufficient sanitary conveniences shall be provided or made available at readily accessible places. So far as is reasonably practicable, rooms containing sanitary conveniences shall be adequately ventilated and lit.
— So far as is reasonably practicable, sanitary conveniences and the rooms containing them shall be kept in a clean and orderly condition.
— Separate rooms containing sanitary conveniences shall be provided for men and women, except where and so far as each convenience is in a separate room the door of which is capable of being secured from the inside.

Washing facilities—
— Suitable and sufficient washing facilities, including showers if required by the nature of the work or for health reasons, shall so far as is reasonably practicable be provided or made available at readily accessible places.
— Washing facilities shall be provided in the immediate vicinity of every sanitary convenience, whether or not provided elsewhere.
— Washing facilities shall be in the vicinity of any changing rooms, whether or not provided elsewhere.
— Washing facilities shall include—
  • a supply of clean hot and cold, or warm, water (which shall be running water so far as is reasonably practicable);
  • and soap or other suitable means of cleaning; and
  • towels or other suitable means of drying.
— Rooms containing washing facilities shall be sufficiently ventilated and lit.
— Washing facilities and the rooms containing them shall be kept in a clean and orderly condition.
— Separate washing facilities shall be provided for men and women, except where and so far as they are provided in a room the door of which is capable of being secured from inside and the facilities in each such room are intended to be used by only one person at a time.
— Privacy is not required for facilities which are provided for washing hands, forearms and faces only, eg a hand wash basin in food preparation or canteen area.

Drinking water—
— An adequate supply of wholesome drinking water shall be provided or made available at readily accessible and suitable places.
— Every supply of drinking water shall be conspicuously marked by an appropriate sign where necessary for reasons of health and safety.

—   Where a supply of drinking water is provided, there shall also be provided a sufficient number of suitable cups or other drinking vessels unless the supply of drinking water is in a jet from which persons can drink easily.

Changing rooms and lockers—
—   Suitable and sufficient changing rooms shall be provided or made available at readily accessible places if a worker has to wear special clothing for the purposes of his work; and he cannot, for reasons of health or propriety, be expected to change elsewhere. This provision is of particular importance where there is for example, significant contamination on site.
—   For reasons of propriety, separate rooms, or "secure" separate use of the same room must be available for men and women on sites where persons of both genders are working.
—   Changing rooms shall be provided with seating and shall include, where necessary, facilities to enable a person to dry any such special clothing and his own clothing and personal effects.
—   Suitable and sufficient facilities shall, where necessary, be provided or made available at readily accessible places to enable persons to lock away any such special clothing which are not taken home, their own clothing which are not worn during working hours, and their personal effects.

Facilities for rest
—   Suitable and sufficient rest rooms or rest areas shall be provided or made available at readily accessible places.
—   Rest rooms and rest areas shall include suitable arrangements to protect non-smokers from discomfort caused by tobacco smoke. (*Note*: this requirement now has to been considered in the context of the *Health Act 2006* and subsequent "Smokefree" Regulations);
—   Rest rooms shall be equipped with an adequate number of tables and adequate seating with backs (note: not just benches) for the number of persons at work likely to use them at any one time.
—   Where necessary, rest rooms should include suitable facilities for any person at work who is a pregnant woman or nursing mother to rest lying down.
—   Rest rooms should include suitable arrangements to ensure that meals can be prepared and eaten. (*Note*: food hygiene issues therefore need to be managed in terms of the maintenance and cleaning of these areas)
—   Rest rooms should include the means for boiling water and be maintained at an appropriate temperature.

Fresh air—
—   The provision of a safe place of work includes the environment of that work place.

— *Regulation 42* requires that "suitable and sufficient steps shall be taken to ensure, so far as is reasonably practicable, that every place of work or approach thereto has sufficient fresh or purified air to ensure that the place or approach is safe and without risks to health. (Note: this supports the duties in the *Control of Substances Hazardous to Health (COSHH) Regulations 2002 (SI 2002 No 2677)* for example to suppress dust at source and avoid exposure to solvents and the like).

— Equipment providing fresh air must be failsafe or include an effective device to give visible and audible warnings of any failure.

Lighting—

— The requirement in *Regulation 44* for suitable and sufficient lighting, which shall be, so far as is reasonably practicable, by natural light is substantially unchanged. The design of security shutters for site accommodation, cabins, huts and the like needs to take this requirement for day light into account.

— As before, the colour of any artificial lighting provided shall not adversely affect or change the perception of any sign or signal provided for the purposes of health and safety;

— Lighting where IT and display screen equipment is to be used will need to be evaluated, as in any workplace.

— Suitable and sufficient secondary lighting shall be provided in any place where there would be a risk in the event of a failure of the primary source of (artificial) lighting, ie for emergency evacuation.

Typical issues occurring in building and construction projects

[C8018] Traditionally, the main focus for action in the construction and building industry has been—

(a)   falls, including from ladders and from temporary places of works such as scaffolds, in excavations, and through fragile materials such as roofing materials, and not just roof lights; and

(b)   falling material and collapses.

These areas of work are now covered in detail by the *Work at Height Regulations 2005 (SI 2005 No 735)*.

Incidents in the workplace involving transport, moving vehicles and mobile plant, people being struck by excavators, lift trucks, dumpers and other work equipment, have become one of the highest causes of death and serious injury in the workplace. The HSC has published comprehensive statistics in support of the revitalising health and safety programmes relating to work place transport. At their meeting in April 2007, the Health and Safety Commission approved the initiative *"Managing transport hazards—a route map for workplace transport and work related road risks"* which is due to be launched in September 2007. Existing specific guidance is contained in HSG 136, *Workplace transport*, and INDG 148, *Reversing Vehicles*. The HSE enforcement database lists hundreds of notices issued for breach of the former *Regulation 15* of the *Construction (Health, Safety and Welfare) Regulations 1996 (SI 1996 No 1592)*, the equivalent of the new *Regulation 36* in the *CDM Regulations*. Such notices are very effective as the costs to a project of a prohibition notice stopping all work on site often significantly exceeds any fine that a court would impose.

The cause of the highest number of over three day injuries is manual handling (www.hse.gov.uk/statistics/causinj). The *Manual Handling Operations Regulat ions 1992 (SI 1992 No 2793)*, and the updated guidance outline the measures the need to be taken to minimise the costs to industry an individuals associated with this type of injury. It is not just about the weight of an object. Ongoing projects include the review of methods used for laying paving flags, following the successful supply chain project to draw up standards for laying pre-cast concrete kerbs.

To support the *Electricity at Work Regulations 1989 (SI 1989 No 635)*, information is contained in HSE guidance HSG 141. Appropriate competence and experience proportionate to the task is a key factor in reducing or eliminating incidents when working with medium and high voltage electricity.

The *Control of Substances Hazardous to Health Regulations 2002 (SI 2002 No 2677)*, and the *Control of Substances Hazardous to Health (Amendment) Regulations 2004 (SI 2004 No 3386)*, as updated, outline the principals of eliminating or reducing exposure to substances, including respirable dusts such silica from cutting with abrasive wheels, and other hazardous substances; *Asbestos (Control of Asbestos at Work Regulations 2002 (SI 2002 No 2675)* and *Lead (Control of Lead at Work Regulations 2002 (SI 2002 No 2676)* are covered by their own specific.

The *Control of Noise at Work Regulations 2005 (SI 2005 No 1643)*, outline the measures that need to be taken to reduce exposure to damaging levels of noise. On the back of proven research and the requirements of the European directive, the new Regulations reduce the noise levels at which action is required significantly (typically by 5dB at each action level). Health surveillance is covered by *SI 2005 No 1643, Regulation 9(1)*. which outlines when "the employer shall ensure employees are placed under suitable health surveillance, which shall include testing of their hearing".

The *Control of Vibration at Work Regulations 2005, (SI 2005 No 1093)* specify the exposure limits for hand/arm vibration and whole body vibration. However, as with manual handling, it is important to note the requirement of *SI 2005 No 1093, Regulation 6(1)*—"The employer shall ensure that risk from the exposure of his employees to vibration is either eliminated at source or, where this is not reasonably practicable, reduced to *as low a level as is reasonably practicable*".

SI 2005 No 1093, Regulation 7 identifies when health surveillance, intended to prevent or diagnose any health effect linked with exposure to vibration, is required.

The biggest single cause of major injuries is identified (www.hse.gov.uk/statist ics/causinj) as slips and trips. The recent HSE campaign "Watch your step" had the simple catch phrase "See it—Sort it". Commonly referred to as housekeep ing, the requirement for good order. *Regulation 27* of *CDM 2007* states that "every part of a construction site shall, so far as is reasonably practicable, be kept in good order and every part of a construction site which is used as a place of work shall be kept in a reasonable state of cleanliness".

The *Construction (Head Protection) Regulations 1989 (SI 1989 No 2209)* can be looked on as an anomaly, predating the *Management of Health and Safety at Work Regulations 1992 (SI 1992 No 2051)* and the other risk assessment based Regulations that follow. It is important to read their requirements with the requirements of the *Personal Protective Equipment at Work (PPE) Regulations* as updated which emphasise that PPE is a last resort. The *Work at Height Regulations 2005 (SI 2005 No 735)* provide for measures being taken to prevent objects and materials falling, for the creation of signed and fenced exclusion zones. The *Lifting Operations and Lifting Equipment Regulations 1998 (SI 1998 No 2307)*, requires that lifting operations be planned and supervised to ensure that lifting operations are not carried out above or over people.

It is the duty of the person in control of the site to establish and identify the zones or locations where head protection has to be worn, and ensure that signing of the zones is clear and there is adequate supervision in place.

The individual's risk assessment based method statement should identify the most suitable type of head protection to be worn; it may not be the ubiquitous site hard hat. The guidance to the PPE Regulations L25 identifies four different types of widely used head protection, each designed to be most suitable for protecting against a different sort of injury. A non-construction specific example of the development of task specific head protection is the helmet produced for users of quad bikes. Research with HSE involvement has shown that conventional roll over protection, (as per *Provision of Use of Work Equipment Regulations (PUWER) 1998 (SI 1998 No 2306), Regulation 26*) is not appropriate for this type of light mobile work equipment. In addition to the necessary level of head protection to minimise risk to workers, the developed helmet provides ventilation, allows communication, and is lighter and more comfortable to wear for long periods than a motorcycle helmet designed for high speed impacts not associated with working quad bikes.

It is accepted practice to allow Sikhs wearing turbans exemption from the requirement to wear additional head protection on construction sites in accordance with the *Employment Act 1989, sections 11, 12*. The individu- al's risk assessment based method statement will need to record the decision and the reasons for reaching that conclusion and identify any additional precautions required. Reference may be made to a case, *Dhanjal v Brit- ish Steel plc* (Case 50740/91) (1994), unreported.

Some higher risk work, such entry into confined spaces, see the *Confined Spaces Regulations 1997 (SI 1997 No 1713)*, or work in compressed air, see the *Work in Compressed Air Regulations 1996 (SI 1996 No 1656)*, is covered by separate Regulations. Formal safe systems of work, permits to work and effective arrangements for rescue and resuscitation are required.

As with manual handling, the primary duty in the *Confined Spaces Regulations 1997 (SI 1997 No 1713)* is to avoid entry if the necessary work can be carried out without person entry. CCTV survey of sewers and drains is a simple example. It is important to be able to correctly identify potential confined spaces, and the level of risk associated. Sewer access that requires horizontal traverse requires a different level of escape and rescue provision than a vertical entry into a manhole. A confined space can include for example an open trench or inspection chamber, as the multiple fatalities at a construction project on a green-field site in the Peak District tragically illustrated.

Similarly, the health risks of work in compressed air are well known and well documented – see below. The designer and CDM co-ordinator under the *Construction Design and Management) Regulations 2007* would need to justify why one of the many alternative safer methods of work are not being adopted. The requirements of the *Work in Compressed Air Regulations* are onerous but proportionate to the hazards and the known health risks.

Replacing *SI 1958 No 61*, the *1996 Regulations (SI 1996 No 1656)* came into effect on 16 September 1996. They reflect more modern decompression criteria, being more concerned with the long-term effects of rapid return to atmospheric pressure than the short-term effects which were addressed by earlier Regulations. The Regulations require principal contractors to appoint competent compressed air contractors, and the compressed air contractors to appoint contract medical advisers. In addition, greater provision is required in connection with fire prevention and protection measures (including, in particular, emergency means of escape and rescue).

Site inductions and site security

[C8019] Whether or not a construction project is "notifiable" under the CDM Regulations, use of site specific induction's have been shown to be effective in reducing the high rate of incidents that occur in the initial period when people start work on a new site. The induction can help to set the culture of the site, explaining what is expected in relation to health and safety on site. Key personnel on site can be identified. Practical details, where the welfare facilities are located, parking arrangements, access and egress routes, need to be covered. Emergency arrangements will include first aid, routes to the nearest accident and emergency department, as well as fire risk minimisation, sounding of the alarm and the evacuation procedures. The induction gives an opportunity for the site manager to question those coming onto site to see if they understand their own risk assessments and method statements, and assess whether their work will conflict with other parties working on site.

The HSE issued guidance, HS(G)151, *"Protecting the public—your next move"*, in 1997. The minimum standards that document contained have now largely been overtaken by better practice, with contractors assessing their commercial risk, theft, civil claims and the like, and providing fencing barriers and signage proportionate to the risk. Measures for sites in close proximity to schools, housing estates, and work carried out over the summer holidays are all typical factors that have to be considered. On site security, monitored and recorded CCTV and close links with the local police crime prevention officer are all usual precautions adopted by the organisation in control of the site. Initiatives working with local schools, for example sponsoring competitions which involve children in gaining an understanding of the risks of trespass onto construction sites, all help to reduce the incidence of children gaining access into dangerous areas of site.

Under the *Occupiers' Liability Act 1957* and the *Occupiers' Liability Act 1984* a duty of care is imposed on occupiers of existing premises regarding visitors. The duty of care extends to children – it should be noted that a child is regarded as being at greater risk than an adult (see section on occupiers' liability). The 1984 Act further extends the duty of an occupier to people other than lawful visitors, such as trespassers, to ensure that they are not injured whilst on the premises. Reference should be made to the *Health and Safety (Safety Signs and Signals) Regulations 1996 (SI 1996 No 341)*.

Signs are a useful tool to help reduce risk, but are not a substitute for other methods of controlling or eliminating risk. Signboards should use the stipulated shape, colour and pictogram or symbol, with supplementary text added only if necessary. Pictogram signs are particularly important where speakers of other languages may be working. Their induction should ensure that they understand the meaning of signs used.

Signs to regulate traffic off highway, but on site or similar workplaces are required to be as prescribed in the *Road Traffic Regulation Act 1984*, and illustrated in the Highway Code.

In addition, reference should be made to the following related statutory provisions, have aspects relating to public safety during construction or building operations—

(a)   *Highways Act 1980*—
     (i)   Section 168 (building operations affecting public safety);
     (ii)   Section 169 (the control of scaffolding on highways); and
     (iii)   Section 174 (the erection of barriers, signs and lighting etc).
(b)   *New Roads and Street Works Act 1991*—
     (i)   Section 50 (lays down particular safety requirements for work in the street and, specifically, the measures to be taken to minimise inconvenience to the disabled). This Act is supported by a Code of Practice, with additional specialist guidance for high speed and high volume traffic routes.
(c)   *Environmental Protection Act 1990*—
     (i)   Section 79 as amended by the *Noise and Statutory Nuisance Act 1993* (noise or vibration emitted from buildings and from or caused by a vehicle, machinery or equipment in a street).

Waste management

[C8020] Waste produced on construction and demolition sites is classed as controlled waste and as such must be controlled to comply with UK legislation based on EU-based Directives. This includes—

(a)   *Environmental Protection Act 1990*—
     (i)   *Sections 33–46* (waste management and licensing control);

(ii)   *Controlled Waste (Registration of Carriers and Seizure of Vehicles) Regulations 1991 (SI 1991 No 1624)* (as amended) (carriage of controlled waste by registered carriers only);

(iii)   *Waste Management Licensing Regulations 1994 (SI 1994 No 1056)* (as amended) (registers, applications and waste Regulation authorities for the recovery and disposal of waste); and

(iv)   *Special Waste Regulations 1996 (SI 1996 No 972)* (as amended) (hazardous properties of waste).

The major changes relating to waste management and affecting all significant construction projects are the proposals by DEFRA for *mandatory* site waste management plans (SWMPs). Details of the proposed Regulations are available on the DEFRA web site www.defra.gov.uk.

• Construction and demolition waste is the third most significant type of waste that is fly-tipped. In this context, proposals for SWMPs have been developed over the last few years as a way of helping to deal with issues in this area.

• Following consultation, the Government took powers in *section 54* of the *Clean Neighbourhoods and Environment Act 2005* to make Regulat ions introducing SWMPs as a statutory requirement.

• SWMPs are one of a number of measures that will help the construction industry to improve its performance on waste and encourage a more efficient use of materials. By minimising waste wherever possible and t hen considering it as an opportunity and not an inevitable cost, it is s uggested that the construction sector will benefit financially and the impa ct on the environment will be reduced.

• In addition to planning, the SWMP will record actual quantities of waste arising, for example, by recording information included on waste trans fer notes against expected waste quantities.

• SWMPs will be subject to inspection by regulatory bodies, with a system of fixed penalty fines proposed if information is not to hand on site; Provision for further penalties through the courts, in line with other sa fety and environmental legislation will be included in the proposals.

• It is intended that savings made and lessons learned from the SWMP p rocess will help to improve the resource efficiency of construction compa nies when planning future projects.

Construction, Design and Management

Gordon Prosser

Introduction

[C8501] At the time of the last re-drafting this chapter, the legislation relating to construction work was under active review.

The outcome of that review came into effect when the Health and Safety Commission with the consent of the Secretary of State for Work and Pensions, and by virtue of the powers of *Section 16(1)* of the *Health and Safety at Work etc. Act 1974*, approved the revised Code of Practice entitled *"Managing Health and Safety in Construction"* L144. The Code of Practice gives practical guidance with respect to the revised regulations, the *Construction (Design and Management) Regulations 2007*, which came into effect on 6 April 2007. The Code of Practice under the former *Construction (Design and Management) Regulations 1994*, which came into force on 4 September 2001, was replaced with effect from that date, 6 April 2007.

The new regulations were debated by the third delegated Legislation Parliamentary Committee on Thursday 10 May 2007. The debate concluded with all party support and the statement that—*"CDM 2007 is a great leap forward in the way in which we are codifying our regulations and the messages that we are giving out to the industry"*.

The Regulations apply to all construction work in Great Britain and, by virtue of the *Health and Safety at Work etc Act 1974* (Application outside Great Britain) Order 2001, its territorial sea. The term "construction work" is defined in *Regulation 2*, and means the carrying out of any building, civil engineering or engineering construction work and includes—

(a) the construction, alteration, conversion, fitting out, commissioning, renovation, repair, upkeep, redecoration or other maintenance (including cleaning which involves the use of water or an abrasive at high pressure or the use of corrosive or toxic substances), decommissioning, demolition or dismantling of a structure;

(b) the preparation for an intended structure, including site clearance, exploration, investigation (but not site survey) and excavation, and the clearance or preparation of the site or structure for use or occupation at its conclusion;

(c) the assembly on site of prefabricated elements to form a structure or the disassembly on site of prefabricated elements which, immediately before such disassembly, formed a structure;

(d)   the removal of a structure or of any product or waste resulting from demolition or dismantling of a structure or from disassembly of prefabricated elements which immediately before such disassembly formed such a structure; and

(e)   the installation, commissioning, maintenance, repair or removal of mechanical, electrical, gas, compressed air, hydraulic, telecommunications, computer or similar services which are normally fixed within or to a structure.

The regulations apply to both employers and to the self-employed without distinction.

The regulations are divided into five parts

[C8502] *Part 1* of the Regulations deals with matters of interpretation and application. For example defining "construction work", as above, duty holders, including the new role of CDM co-ordinator, documents, including "pre-construction information" and the "health and safety file", and "writing" which includes being kept in electronic form but which can be printed.

Part 2 covers the general management duties that apply to all construction projects, (notifiable and non-notifiable, see Part 3 below).

Part 3 sets out the additional management duties which apply only to those projects of such size or complexity that they are expected to be above a threshold of lasting more than 30 days; or involving more than 500 person days of construction work—for example 50 people working for more than 10 days. This is known as the notification threshold. In addition to notification of the project to the HSE, the additional management duties require particular appointments to be made, a CDM co-ordinator, and a principal contractor. Particular documents including "site rules", "pre-construction information" and the "health and safety file" are required which will assist with the management of health and safety from concept to completion, and during the working life and eventual, demolition, alteration or removal of the constructed project.

Work that involves demolition or dismantling does not per se require notification. However *Regulation 29* in *Part 4* states that "The arrangements for carrying out such demolition or dismantling shall be recorded in writing before the demolition or dismantling work begins".

Part 4 of the Regulations are the subject of a separate chapter. *Part 4* applies to all construction work carried out on construction sites. It also covers the practical and physical safeguards which need to be provided to prevent danger that were previously contained in the *Construction (Health Safety and Welfare) Regulations 1996, (SI 1996 No 1592)*, and that have not been incorporated into the *Work at Height Regulations 2005 (SI 2005 No 735)*.

The duty holders under the regulations in *Part 4* are those who carry out the work, or control the way in which the work is done, irrespective of whether they are contractors or a Principal contractor and irrespective of whether they are employers, employees or are self-employed.

Part 5 tidies up the necessary but secondary administrative issues related to the new regulations.

Civil Liability, enforcement in respect of fire, the transitional provisions, and revocations (listed in schedule 4) and amendments (listed in Schedule 5) are covered in this part.

Aims and Objectives

[C8503] The stated aim of *CDM 2007* is to integrate health and safety into the management of the project and to encourage everyone involved to work together. Specific objectives are that—

(a)   planning and management at the earliest stages of projects is to be improved;

(b)   hazards are to be identified early on, so they can be eliminated or reduced at the design or planning stage; remaining significant risks are to be identified so that they can be properly managed;

(c)   effort is to be targeted where it can do the most good in terms of health and safety; and,

(d)   unnecessary bureaucracy is to be actively discouraged. This sadly may be the hardest objective to achieve, as people are interpreting the requirements for competence of named parties in the regulations as a reason for a paper chase. Appendix 4 in the ACOP states *"Remember that assessments should focus on the needs of the particular job and should be proportionate to the risks arising from the work. Unnecessary bureaucracy associated with competency assessment obscures the real issues and diverts effort away from them."*

Information relating to whole life costs that was included in the consultation document are repeated in the published ACOP as follows—

Typical costs of building are in the ratio—

1   for construction costs;
5   for maintenance costs; and
200   for operating costs.
        from the report of the Royal Academy of Engineering on the long term costs of owning and using buildings (1998)

Understanding the costs and benefits of any action, considering the resources required, is important for those seeking to provide added value to projects by following the requirements of these regulations. *"For which of you intending to build a tower does not first sit down and estimate the cost, to see whether he has enough to complete it? Otherwise when he has laid a foundation and is not able to finish, all who see it will begin to mock".* Luke 14.28.

Some of the revisions to the regulations focus on the client. In *Regulation 9*, the client is identified as the party who must ensure "sufficient resources" are made available to develop and manage the project, and to ensure the safety of the end users in accordance with the *Workplace (Health, Safety and Welfare) Regulations 1992.*

The CDM co-ordinator is a new role replacing and extending the duties formerly carried out by the planning supervisor. There has been debate regarding this role, some due to the misunderstanding that it has to be fulfilled by a single person. For the majority of smaller projects, a single person given sufficient authority and time to fulfill the role will be able to ensure that the statutory requirements are fulfilled and the client's interests are served as a routine part of the way the project is managed. For many client organisations procuring new work there will be the opportunity for the role to be carried out "in-house". For larger notifiable projects, it will be appropriate for this role to be fulfilled by an organisation or team that can call upon specific skills, knowledge or experience where the technicalities of the project require this. For example, if the project were an expansion of an existing heath care facility, the CDM co-ordinator function would involve—

- understanding the needs of the exiting users and management, and to how patient care may be affected during the works;
- the requirements of the future users and management as to how they will operate and maintain the new and expanded facilities;
- the input from persons with experience of the design of the building's structure and also of the specialist equipment and services to be installed; and,
- an understanding of managing and programming work involving a wide range of funding, construction and installation skills.

Management duties applying to Construction Projects, the specific requirements contained in the individual regulations

[C8504] *Regulations 4 to 13, in Part 2* apply to all construction projects.

Regulation 14 to 24, in part 3 contain the additional duties that relate only to those projects that are of such size or duration that they are notifiable.

Competence

[C8505] *Regulation 4, Competence*, is the first management duty applying to *all* construction projects, and requires that all persons who are to carry out or manage design or construction work are to be competent to perform any requirement and avoid contravening any prohibition imposed by or under any of the relevant statutory provisions". To assist with this requirement industry guidance is being developed and made available through the CITB Construction skills web site. The concept of this guidance is that it will be reviewed and updated continuously on the basis of experiences across all industry sectors.

Co-operation

[C8506] *Regulation 5* relates to Co-operation, both a proactive requirement to seek co-operation from others with information, and a passive requirement to co-operate with any party that seeks information. Anything of which a person is aware that is likely to endanger the health or safety of any person must be reported to the person "in control".

Co-ordination

[C8507] *Regulation* 6 relates to the third "C", Co-ordination. It states that "All persons concerned in a project on whom a duty is placed by these Regulations shall co-ordinate their activities with one another in a manner which ensures, so far as is reasonably practicable, the health and safety of persons carrying out the construction work; and affected by the construction work".

The simplicity in the manner in which these statutory duties are drafted gives the opportunity for maximum flexibility in how the information to achieve co-operation and co-ordination is communicated. There is, for example, no necessity to have meetings for the sake of meetings provided the relevant information is passed to those who need it when they need it. Information overload is actively discouraged. *Paragraph 134* of the ACOP states that "information should be brief, clear, precise, and in a form suitable for the users".

Principles of prevention to be followed

[C8508] *Regulation* 7 creates the formal link to the General Principles of Prevention incorporated in the Management of *Health and Safety at Work Regulations 1999* from the associated European Directive. These principles direct the approach to be followed to identify and implement "precautions which are necessary to control risks associated with a project". The word avoid is used in place of eliminate with regard to risk, and only those risks that cannot be avoided are to be evaluated and controlled, at source where possible, managed in the context of coherent overall policy, with collective measures given priority over individual measures, backed up with information and instruction, as required by the *1974 Heath and Safety at Work etc. Act*.

Clients responsibilities

[C8509] A client is defined as "a person (*or organisation*) who in the course or furtherance of a business seeks or accepts the services of another which may be used in the carrying out of a project for him; or carries out a project himself (*itself*)." As with the principal established in the 1974 Act, the domestic home owner having work done on his own private residence is protected from statutory duties under these regulations.

Regulation 9 puts the Client very much in the driving seat in relation to how health and safety will be managed during the life of a project. Clients have an obligation to "take reasonable steps to ensure that the arrangements (*for health and safety*) are maintained and reviewed throughout the project".

The regulations contain a schedule of basic requirements for welfare facilities for those working during the construction phase, *Schedule 2*. It is a duty on the Client to ensure that these provisions for welfare are in place throughout. The client may make arrangements for existing facilities on or adjacent to the site to be made available, or leave it up to others, typically a principal contractor or main contractor, to organise, provide and maintain such facilities.

The client may retain control of parts of the site or work areas adjacent, and as such must ensure that others working or having access to these areas are adequately protected. They may be the employer of end users of the development, and thus have extra incentive to ensure that long-term work related safety issues are addressed during the design stage, and that maintenance work can be carried out safely and without undue costs. Consideration of the ongoing and day to day safety of the end users of a development is a new duty incorporated into CDM 2007. However it reflects the principles of laws drafted over three thousand years ago, "*When you build a new house, you shall make a parapet for the roof; otherwise you might have blood guilt on your house, if anyone should fall from it*, Deuteronomy 22.8.

If the client is a speculative developer, they too have an interest in ensuring the end project is attractive to the end user. This will include ensuring that it can be demonstrated that consideration has been given during the design process to safe operation and low long term operating costs of the development.

Particulars to be notified to the HSE listed in *Schedule 1* for those projects meeting the criteria (over 30 days or 500 man days), the so called form "F10", now include a statement of the time allowed by the client to the principal contractor for planning and preparation for construction work, in addition to the date for the planned start of the construction phase. Time as well as finance is a resource that the client is now to be answerable for.

Regulation 10 is the final duty for clients for projects that are not notifiable. They have to provide or make arrangements for the provision of all pertinent background information relating to the site. Buried services, cables and pipes, previous modifications to buildings, the presence of hazardous materials, asbestos, or contamination due to industrial waste are all examples of the type of detail that information needs to be assembled and passed on to those designing for and managing construction projects.

Designers' responsibilities

[C8510] A designer is defined in the regulations as any person or organisation that makes decisions that affects how a project will be built, including aspects of the structure, (eg its siting and layout), the materials, products, and mechanical and electrical services to be installed, as well carrying out calculations, and preparing drawings, schedules and bills of quantities. A client, a contractor or subcontractor, or a supplier may all make or influence such decisions

Regulation 11 outlines duties for those involved in the design of any project. As with the *1994 Regulations*, there is a duty to inform and ensure that any client using their services is aware of the duties and responsibilities they have commissioning the work. There is a duty on the designer to ensure that, so far as is reasonably practicable and taking due account of other relevant design considerations, foreseeable risks are avoided for those carrying out the work, those who may be affected by the work, those who have to clean or maintain or use and work in the finished project. The hierarchy of eliminating, then reducing remaining risk, and seeking to incorporate collective measures that protect everyone not just individuals (for example, providing railings as opposed to a fall arrest anchor) is repeated in this regulation for emphasis. The Designer must also provide "sufficient information" for other parties, the client (and thus end users and those carrying out future maintenance), other designers (to ensure compatibility) and contractors (those who will have to plan and build the design during the construction phase).

Regulation 12 identifies that wherever in the world the design is prepared, the person commissioning that design is responsible for the requirements of *Regulation 11*, above, being met so far as they affect any construction work carried out in Great Britain. (Northern Ireland has its own set of regulations that match those for the rest of the United Kingdom).

Contractors' responsibilities

[C8511] *Regulation 13* is the first regulation that addresses those carrying out the construction work directly, contractors. Note that a Principal contractor has only to be appointed for those projects that are notifiable (see above); their additional duties are contained in Part 3 of the regulations. The duties of contractors for *all* projects may be summarised as follows—

(i) As with a designer, to check that the client is aware of their duties under the regulations before they start work;

(ii) To plan, manage and monitor, not just turn up and "do";

(iii) To notify and give sufficient time for any sub contractor or other contractor they appoint to plan and prepare before they start work;

(iv) To ensure site induction appropriate to the work and site specific risks is provided. The development of method statements based on site specific risk assessment is key part of this process;

(v) To ensure that emergency procedures, evacuation, first aid, fire safety and the like are in place and the all personnel know the arrangements and who is in control.

(vi) To protect the public and to protect the client and others by ensuring adequate security arrangements are in place. This may include fencing, access control, CCTV and other security arrangements. The likely hood of trespass by children (for example due to proximity of a school or play area) is information that will need to be taken into account.

(vii) To ensure that adequate welfare facilities for the numbers of people working on site and the type of work in progress, as laid out in Schedule 2, are in place and are maintained. For example preliminary site work, when overall numbers on site are low may involve higher exposure to risk to health due to carrying out work in live sewers, removal of asbestos and the like.

Additional duties where projects are notifiable

[C8512] To prevent those managing small projects having to complete paperwork that would not significantly improve the management of safety on site, the additional duties contained in the following regulations only relate to larger projects.

Part 3 of the Regulations, regulations 14 to 24, contains these additional requirements for larger projects, those that are likely to exceed the threshold of 30 working days or 500 man days, referred to as "notifiable".

CDM co-ordinators responsibilities

[C8513] Seemingly out of sequence, the general duties of the new role of CDM co-ordinator are described in *Regulation 20.* In brief, they have to—

(i)   give suitable and sufficient advice and assistance to the client;

(ii)   ensure that suitable arrangements are made and implemented for the co-ordination of health and safety measures during planning and preparation for the construction phase;

(iii)   liaise with the principal contractor regarding the information which the principal contractor needs to prepare the construction phase plan, any design development which may affect planning and management of the construction work; and the contents of the health and safety file; Agreeing the format of the file, and how information will be stored and retrieved in advance will help the gathering of information throughout the design and construction phases;

(iv)   provide **relevant** information to designers and contractors promptly, (*ie when they need it*);

(v)   facilitate the work of designers and the co-operation between designers and contractors;

(vi)   prepare or update the health and safety file, and pass it to the client. Note the change of emphasis, unlike the Planning Supervisor, whose duty under the former *Regulation 14(d)* was to ensure that a health and safety file "*is prepared*", it is the duty of the appointed CDM co-ordinator to "prepare or update" the file,

CDM co-ordinators don't have a statutory duty to "supervise" the principal contractor, but the client should stipulate how monitoring of the health and safety performance of the principal contractor and those working for the principal contractor is achieved, as required by *Regulation 9(2)*. This information should be included in *Part 2* of the pre-construction information, "*Clients considerations and management requirements*", see *Appendix 2* of the regulations.

Regulation 21 covers the procedure for the CDM co-ordinator to notify the HSE of the project, submitting such details specified in *Schedule 1* "as are available".

The requirement is that the notice should be "signed by or *on behalf of* the client", in a manner "approved" by the client. The CDM co-ordinator will need to be able to demonstrate how he is satisfied that the client understands this responsibility.

Additional responsibilities for Clients for notifiable projects

[C8514] *Regulation 15* introduces the relationship between the Client and the CDM co-ordinator. The CDM co-ordinator is a new role defined by this update (2007) of the *CDM regulations*, and replaces the role of Planning Supervisor in the earlier regulations (1994). In simple terms, the CDM co-ordinator has the role of assisting the client fulfill their duties under the regulations, with the opportunity to add value to the project by ensuring that the planning and management of the project delivers not only safety, but consequential lower whole life costs.

This first regulation relates to the Client transferring to the co-ordinator all pertinent information. A summary of the information required to be made available or ascertained is listed in *Appendix 2*, Pre-construction information. This will form the basis of the development of the construction phase plan by the Principal contractor, see *Appendix 3*.

Regulation 16 outlines the requirement for a construction phase health and safety plan to be developed by the appointed principal contractor, and the statutory duty for the Client to prevent the construction phase starting until adequate health and safety arrangements, and welfare facilities as outlined in *Schedule 2*, are in place. This is an area where the client would look for advice or confirmation from the CDM co-ordinator.

Regulation 17 describes the function of the Health and Safety file, to provide information to help the planning process for the proposed works, should one already exist due to development since 1994, and to provide information for the future users, owners and those carrying out work on the development after completion of the current construction project.

Additional responsibilities for Designers for notifiable projects

[C8515] *Regulation 18* reinforces the duty on designers in *Regulation 11* to ensure that clients are aware of their duties, but in particular the requirement for the client to appoint a CDM co-ordinator for notifiable projects. It also reinforces the duties contained in *Regulations 5* and *6* to co-operate and co-ordinate their work with the CDM co-ordinator.

Regulation 19 establishes the link between contractors and the CDM co-ordinator, and covers the development of the construction phase (health and safety) plan by the Principal Contractor, and the notification of the project to the Health and Safety the Executive, (or as the case may be the Office of Rail Regulation), under *Regulation 21*. Details to be included in the construction phase (health and safety) plan are itemised in *Appendix 3*. The following words are in bold in the original: "The level of detail should be proportionate to the risks involved in the project". In addition to a description of the project with details of the existing site specific information available, the management structure, responsibilities, goals and aims, arrangements including those for monitoring, and the identification of and means of controlling significant project specific site risks are to be laid out.

Additional responsibilities for Contractors for notifiable projects

[C8516] Contractors must "promptly" notify further contractors or subcontractors they appoint to the principal contractor, together with, as a minimum, all reportable incidents under *Reporting of Injuries, Diseases and Dangerous Occurrences Regulations 1995* that occur under their control. In practice, the principal contractor's site rules will define the level of incident reporting they require as part of the project's management of health and safety performance.

In addition every contractor must inform the principal contractor if they are aware that the construction phase health and safety plan needs to be altered or added to, or cannot be complied with.

Part 3 of the Regulations is concluded with the duties of the Principal Contractor, *Regulations 22, 23* and *24*.

Regulation 22 sets out the principal contractors statutory duties to—

(i)   plan, manage and monitor the construction phase;
(ii)   facilitate co-operation and co-ordination;
(iii)   liase with the CDM co-ordinator in relation to ongoing design or changes in design;
(iv)   ensure that adequate welfare facilities as specified in schedule 2 are provided throughout the construction phase;
(v)   draw up "appropriate" site rules;
(vi)   identify to each contractor the information relating to that contractor's activity which is likely to be required by the CDM co-ordinator for inclusion in the health and safety file and ensure that such information is promptly provided to the CDM co-ordinator;
(vii) ensure that all persons working on site have a "suitable" site induction, and have received the necessary information instruction and training for them to carry out their tasks without undue risk to themselves sore others. Reviewing the task risk assessment based method statements is a typical part of this process.

Regulation 23 relates to the principal contractor's duties in relation to the construction phase (health and safety) plan. The information that is to be contained in the plan is laid out in *Schedule 3*. The regulation stipulates that the plan must be "updated, reviewed, revised and refined", that the plan must be "implemented" and that the construction phase must be "planned, managed and monitored" in accordance with the plan. *Paragraph 175* of the ACOP points out that principal contractors don't have a statutory duty under *Regulations 22* and *23* to undertake detailed supervision of (*other*) contractors' work. This allows flexibility as to how necessary supervision will be achieved, particularly if the work is of a technical nature. To protect the client's interests and the safety of others on site, the method by which supervision is undertaken, and by whom, needs to be defined in the construction phase plan.

The final regulation in part three relating to notifiable projects is *Regulation 24*. This regulation imposes the duty on principal contractors to "make and maintain" arrangements that enable the principal contractor to consult and co-operate with workers or their representatives on matters relating to health safety and welfare. This mirrors the duties under the *1974 Act* and the Management regulations to utilise the knowledge and information held by all workers on a construction project. This requirement is in place, as research for the HSE as shown that where such arrangements are in place there is a measurable improvement in safety culture and performance. Such arrangements can be used to tackle one of the issues of concern raised in the Parliamentary debate namely that of migrant workers. Their understanding of English as a second language, and their knowledge of the safety culture that the U.K. construction industry is seeking to achieve, may require to be addressed.

General matters contained in Part 5

[C8517] *Part 5, Regulations 45–48* contains general matters concerning—

- civil liability (in relation to persons who are not employees of the duty holder);
- enforcement in respect to fire (in relation to sites which are contained in, or form part of other places of work;
- transitional provisions (which only apply in a limited manner to projects which began before 6 April 2007); and
- revocations and amendments.

Information in the Schedules:

[C8518] For those projects that are notifiable [i.e. projects meeting the criteria, over 30 days or 500 man days, see above and Regulation 21), the information to be forwarded to the Executive (or Office of Rail Regulation if appropriate), is listed in Schedule 1.

The importance of time as a resource is shown by items 1, 8, 9, and 10. The information required is as follows:

(1)   Date of forwarding.
(2)   Exact address of the construction site.
(3)   The name of the local authority where the site is located.
(4)   A brief description of the project and the construction work which it includes.
(5)   Contact details of the client (name, address, telephone number and any Email address).
(6)   Contact details of the CDM co-ordinator (name, address, telephone number and any Email address).
(7)   Contact details of the principal contractor (name, address, telephone number and any Email address).
(8)   Date planned for the start of the construction phase.
(9)   The time allowed by the client to the principal contractor referred to in regulation 15(b) for planning and preparation for construction work.
(10)   Planned duration of the construction phase.
(11)   Estimated maximum number of people at work on the construction site.
(12)   Planned number of contractors on the construction site.
(13)   Name and address of any contractor already appointed.
(14)   Name and address of any designer already engaged.
(15)   A declaration signed by or on behalf of the client that he is aware of his duties under these Regulations.

Schedule 2 contains details of the requirements for welfare facilities. See C080 and regulations 9, 13 and 22.

Schedule 3 contains the minimum particulars that must be contained in a report of inspection. See C080 and regulation 33.

Schedule 4 lists the revocations of statutory instruments, in whole or in part, identifying the extent. See regulation 48.

Schedule 5 lists amendment to Acts (including the Factories Act 1961, and the Fire (Scotland) Act 2005) and regulations, recording the extent of the amendments.

Information in the Appendices to the Approved Code of Practice L144:

[C8519] In addition to the full text of the regulations contained as Appendix 1, the ACOP contains the following information:

Appendix 2 lists the topics to be considered when drawing up the pre-construction information for notifiable projects. It emphasizes that "the level of detail should be proportionate to the risks involved in the project".

The topics to be considered in construction phase plan are listed in Appendix 3. Again the same caveat is included, "the level of detail should be proportionate to the risks involved in the project".

Guidance on the core criteria for demonstrating competence are contained in Appendix 4. This clarifies that for the named duty holders, eg contractor, designer, or CDM co-ordinator, that competence relates to the function. The function can be carried out by companies, by organizations or by a number of individuals.

Because the role of CDM co-ordinator is new, specific guidance relating to this role "for larger or more complex projects, or those with high or unusual risks" is contained in appendix 5.

Appendix 6 illustrates a sample timeline for a worker in construction to develop skills and competencies. It is not the intention of the regulations to hinder persons coming into and working in the construction industry while they develop, through training, working under supervision and the like, the necessary competencies.

The final appendix, Appendix 7 repeats the general principles of prevention set out in Article 6(2) of Council Directive 89/391/EEC and Schedule 1 of the Management of Health and Safety at Work Regulations 1999. It is stated that "duty holders should use these principles to direct their approach to identifying and implementing precautions which are necessary to control risks associated with a project". In addition to the primary requirements to avoid (or eliminate) risks, and evaluate those which remain, clients and designers need to be aware of the need to adapt to technical progress, to replace the dangerous by the non-dangerous or less dangerous, and to have a coherent overall policy for improving safety.

Conclusion, the future?

[C8520] As was confirmed during the Parliamentary debate in May 2007, the efficacy of the revised regulations will be monitored closely during the first years of their implementation by both the Health and Safety Executive and the Health and Safety Commission.

It was stated that work has already begun with the Department for Communities and Local Government (DCLG) in relation to the potential for integrating the Building Control, Planning and CDM regimes, with early Local Authority involvement promised.

The Department for Environment, Food and Rural Affairs consulted earlier this year, (2007), on proposals to make site waste management plans (SWMP) a legal requirement for construction and demolition projects in England and Wales. The proposed regulations would require appropriate persons to prepare and implement a SWMP for works involving construction, demolition or excavation above a specified value. In England and Wales, the construction sector uses some 400 million tones of materials each year and generates an estimated 109 million tonnes of waste. It is estimated that on average between 10% and 15% of all materials delivered to site go into skips without ever being used. The potential for greater resource efficiency is therefore considerable.

Construction and demolition waste currently accounts for about one-fifth of all reported fly-tipping incidents and about a third of the more serious incidents dealt with by the Environment Agency.

The feed back to the consultation already received indicates that such environmental management requirements should be more closely linked to the existing CDM 2007 Regulations.

Finally, *Corporate Manslaughter and Corporate Homicide Act 2007, Chapter 19*, has now received Royal Consent. In a prosecution under this Act, under *Section 8*, jurors may "consider the extent to which the evidence shows that there were attitudes, policies, systems or accepted practices within the organisation that were likely to have encouraged" any outcome. These are matters specifically covered by the *CDM 2007* Regulations. Jurors may also "have regard to any health and safety guidance that relates to the alleged breach". "Health and safety guidance" means any code, guidance, manual or similar publication that is concerned with health and safety matters and is made or issued (under a statutory provision or otherwise) by an authority responsible for the enforcement of any health and safety legislation., "*Managing Health and Safety in Construction*", L144, the ACOP linked to the *CDM 2007* regulations is such a code containing particular detailed guidance.

Corporate Manslaughter

Andrea Oates

Introduction

[C9001] After many years in the making, the *Corporate Manslaughter and Corporate Homicide Act 2007* (*CMCHA 2007*) finally came into force on 6 April 2008, after receiving Royal Assent on 26 July 2007. Under the Act, companies, organisations, and Government bodies face an unlimited fine, as well as remedial and publicity orders, if they are found to have caused death due to their gross corporate health and safety failures.

At the time of its enactment, the Ministry of Justice MoJ said: "The Corporate Manslaughter Act is a landmark in law and the culmination of ten years of campaigning by unions and other groups. Well-run businesses that already have effective systems in place for managing health and safety have nothing to fear from the new legislation. But employees of companies, consumers and other individuals will be offered greater protection against the worst cases of corporate negligence."

Prior to the new legislation, a company could be prosecuted for the common law offence of gross negligence manslaughter – generally referred to as "corporate manslaughter". In order to be guilty of the offence, a company had to be in gross breach of a duty of care owed to the victim who was killed.

However it proved virtually impossible to successfully prosecute a large company for corporate manslaughter under the common law, even where management failures lead to a death, and instead, prosecutions were nearly always taken under health and safety legislation. This lead to calls, particularly from the trades unions and safety campaigners, for the law to be reformed to allow companies whose gross negligence lead to a death to be convicted of the more serious offence of corporate manslaughter, with corresponding harsher penalties.

The key problem with the law as it stood was known as the "identification principle". Before a company could be prosecuted for corporate manslaughter, a single individual at the top of the company – "a directing mind" who could be said to embody the company in his or her actions and decisions – had to be shown to be personally guilty of manslaughter. If this could not be shown, the company escaped liability.

Although 3425 workers were killed between 1992 and 2005, there were only seven successful work-related corporate manslaughter prosecutions over the same period and all these were against small companies or sole traders.

In larger companies it proved extremely difficult to identify an individual who embodied the company and was culpable, as there are often complex management structures and lines of control, and overall responsibilities for safety matters can be unclear in large organisations.

Corporate manslaughter prosecutions were taken against a number of large companies following disasters in which many people died, but all were unsuccessful.

For example on 6 March 1987, the roll-on roll-off P&O European ferry the Herald of Free Enterprise, sank off the coast of Zeebrugge, Belgium, as a result of sailing with its bow doors open, killing 150 passengers and 38 crew. In the prosecution case that followed against seven individuals and the company, the court found that there was insufficient evidence to convict any of the seven personal defendants with gross negligence manslaughter. The case against the company therefore also failed. The company was automatically acquitted at the same time as the directors.

The judge in the case ruled that there was insufficient evidence to show that the risk of the ferry leaving port with its bow doors open was "obvious and serious". He also ruled that there was insufficient evidence of wrongdoing by a director or senior manager. The case failed because the various acts of negligence could not be aggregated and attributed to any individual who was a directing mind of the company.

No other criminal prosecution was taken. When the ferry sank, the *Health and Safety at Work etc Act 1974* did not apply to deaths at sea, and no appropriate charges under merchant shipping legislation were available. A new offence was created in the *Merchant Shipping Act 1988* in response to this situation.

The P&O case was, however, a landmark decision in that it acknowledged that a company could properly be charged with corporate manslaughter, and the collapse of the case also first triggered demands for a reform of the law in this area. These demands were strengthened by other unsuccessful prosecutions against large organisations.

For example, a corporate manslaughter prosecution taken against Great Western Trains after one of its high speed passenger trains, travelling from Swansea to London on the 19 September 1997, went through a red signal and collided with an empty freight train at Southall, killing seven people and injuring 151. The Automatic Warning System that would normally have alerted the driver if he passed a signal at danger was malfunctioning in one of the train's two power units.

The judge in the case ruled in a pre-trial hearing that a company could not be prosecuted for manslaughter by gross negligence unless a named person deemed to be a "controlling mind" of the company was prosecuted, and no such person had been charged.

But he also said that even having a named director would have made no difference in this case, because no director was personally responsible for ordering the running of the train that crashed.

In July 1999 Great Western Trains (GWT) pleaded guilty to breaches of *section 3(1)* of the *Health and Safety at Work etc Act 1974*, for failing to ensure that members of the public were not exposed to risks to their health and safety. A record fine of £1.5 million was imposed for "a serious fault of senior management", but relatives of those killed called this "derisory" in view of the company's £300 million turnover. The Crown Prosecution Service (CPS) also dropped manslaughter charges against the driver of the train.

However, this case was also significant as the Attorney-General referred the legal issues to the Court of Appeal – the first time that an appeal court had considered the law relating to corporate manslaughter.

Manslaughter charges were also brought against both Network Rail (formerly Railtrack PLC) and Balfour Beatty, together with six managers from both companies following the derailment, on 17 October 2000, of a London to Leeds train travelling at 115 mph. The derailment was caused by a defective rail near Hatfield, which had been identified as suffering from a form of metal fatigue, commonly found where track curves, 21 months earlier. The derailment led to the loss of four lives, and 70 people were injured.

Corporate manslaughter charges against Network Rail were subsequently dropped in September 2004, and in July 2005 the judge in the case against Balfour Beatty and five rail executives dismissed the manslaughter charges, five months into the trial.

Mr Justice Mackay, the trial judge commented "This case continues to underline a long and pressing need for the long-delayed reform of the law in this area of unlawful killing".

According to the head of the CPS, the Director of Public Prosecutions (DPP): "The ruling turned on the level of negligence that the jury was being invited to consider, and the judge took the view that it was not sufficient to invite the jury properly to conclude that it was grossly negligent. He therefore directed the jury to find the defendants not guilty."

Network Rail and Balfour Beatty were found guilty of breaching health and safety law. Network Rail was fined a record £3.5 million, and Balfour Beatty was fined £10 million (although this was reduced on appeal to £7.5 million). They were also ordered to each pay £300,000 in costs. The five individuals facing health and safety charges were found not guilty.

The 2008 Act replaced the identification principle with a new test, focusing on how the overall picture of how an organisation's activities are managed by its senior management, rather than focusing on the actions of one individual.

Instead of having to demonstrate that one or more individuals are guilty, liability for the new offence of corporate manslaughter (called corporate homicide in Scotland) depends on a finding of gross negligence in the way in which the activities of the organisation are run. As the explanatory notes to the Act set out—

> In summary, the offence is committed where, in particular circumstances, an organisation owes a duty to take reasonable care for a person's safety and the way in which activities of the organisation have been managed or organised amounts to a gross breach of that duty and causes the person's death. How the activities were managed or organised by senior management must be a substantial element of the gross breach.

The Ministry of Justice set out that the *Corporate Manslaughter and Corporate Homicide Act*—

— Makes it easier to prosecute companies and other large organisations when gross failures in the management of health and safety lead to death by delivering a new, more effective basis for corporate liability;
— Has reformed the law so that a key obstacle to successful prosecutions has now been removed. Until now, a company could only be convicted of manslaughter if a 'directing mind' (such as a director) at the top of the company was also personally liable;
— Means that both small and large companies can be held liable for manslaughter where gross failures in the management of health and safety cause death, not just health and safety violations;
— Does not apply to individual directors, senior managers or other individuals: it is concerned with the corporate liability of the organisation itself. However, where there is sufficient evidence, individuals can already be prosecuted for gross negligence manslaughter and for health and safety offences. The Act does not change this position; and
— Lifts Crown immunity to prosecution (see below). The Act meant that for the first time, crown bodies, such as government departments, were now liable to prosecution. The Act applies to companies and other corporate bodies, in the public and private sector, government departments, police forces and certain unincorporated bodies, such as partnerships, where these are employers.

A number of Government bodies, and quasi Government bodies, such as Government departments, had been able to claim immunity from prosecution because they were said to be acting as a servant or agent of the Crown – know as Crown immunity. In addition, many Crown bodies did not have a separate legal identity for the purposes of a prosecution and the 2008 Act deals with these issues.

The Act will also apply to deaths in custody. This was a key demand of the House of Lords as the Bill progressed through parliament. After the Bill had gone through all its stages in the House of Commons and House of Lords, it then passed between the two houses until they could reach agreement on the issue as to whether deaths in custody should be included. Only once agreement was reached could the Bill receive Royal Assent and become an Act.

Eventually agreement was reached when the Labour government tabled an amendment meaning that on the face of the bill, the new offence would apply to deaths in custody. However, before this part of the Act comes into force, a further resolution must be agreed by both houses of parliament, known as the affirmative resolution procedure. The previous government was committed to passing the resolution, although the then Minister for Justice, Jack Straw, set out a timetable of around three to five years before this will happen.

Although it had been widely assumed since reform of the law in this area was first discussed by the Law Commission back in 1994 that there would be separate legislation for Scotland, the Act applies to the whole of the UK.

In October 2007, the MoJ published detailed guidance on the implementation of the Act, *Reforming corporate liability for work-related death: a guide to the Corporate Manslaughter and Corporate Homicide Act 2007*, together with a leaflet intended to provide a general introduction to the Act, *Understanding the Corporate Manslaughter and Corporate Homicide Act 2007*.

Following a consultation exercise, the Sentencing Guidelines Council (SGC) set out guidelines for the courts on sentencing for corporate manslaughter and for health and safety offences that cause death at work. These guidelines set out that fines for organisations found guilty of corporate manslaughter should be measured in millions of pounds and should not generally be below £500,000. For health and safety offences that cause death at work, fines in the order of hundreds of thousands of pounds and generally not less than £100,000 are appropriate.

This chapter sets out the main requirements of the *Corporate Manslaughter and Corporate Homicide Act 2007*, with reference to the MoJ guidance, and summarises the SGC guidelines on sentencing.

The main requirements set out under the *Corporate Manslaughter and Corporate Homicide Act 2007* are summarised below:

A SUMMARY OF THE CORPORATE MANSLAUGHTER AND CORPORATE HOMICIDE ACT 2007

—   The offence is called corporate manslaughter in England, Wales and Northern Ireland, and corporate homicide in Scotland;

—   An organisation is guilty of the offence if the way in which it organises or manages its activities causes a death, and this amounts to a gross breach of a relevant duty of care it owed to the victim;

—   The offence applies where the organisation owed a duty of care to the victim under the law of negligence;

—   It will apply to deaths in detention and custody, but this part of the Act was not be enacted on 6 April 2008. It is due instead to be enacted some time after April 2011;

—   The offence applies to corporations, Government departments (and other bodies listed in a schedule to the Act), police forces, partnerships, trade unions and employer associations (where these are employers); but does not apply to corporations sole;

—   Crown bodies are not immune from prosecution under the Act;

—   Public policy decisions, exclusively public functions and statutory inspections are excluded from the scope of the offence;

—   The Act includes a number of exemptions: for certain military activities and for police and other law enforcement bodies in particular situations (including terrorism and civil unrest); in relation to responding to emergencies; and for statutory functions relating to child protection and probation functions;

—   The Act sets out factors for the jury to consider when deciding if there has been a gross breach of a relevant duty of care, including whether health and safety legislation was complied with; whether the culture of the organisation tolerated breaches; and any relevant health and safety guidance.

—   Consent of the Director of Public Prosecutions (DPP) is needed for proceedings for the offence to be instituted;

—   There is no secondary liability for the offence. Individuals will not be liable for aiding, abetting, counselling or procuring the commission of, or in Scotland, being art and part of, the offence. Individuals can still be prosecuted for gross negligence manslaughter under the common law (see below);

—   The Act abolishes the common law offence of gross negligence manslaughter in relation to corporations; and

—   An organisation found guilty of corporate manslaughter (or corporate homicide) face an unlimited fine, and the courts have the power to make remedial orders and publicity orders.

The offence

[C9002] The offence is called corporate manslaughter in England, Wales and Northern Ireland, and corporate homicide in Scotland (*CMCHA 2007 s 1(5)*).

An organisation is guilty of the offence of corporate manslaughter (or corporate homicide) if the way in which it manages or organises its activities causes a death and this amounts to a gross breach of a relevant duty of care it owed to the victim (*CMCHA 2007 s 1(1)*). (Relevant duties of care are set out in *Section 2* of the Act – see below).

The MoJ guidance explains that: "The offence is concerned with the way in which an organisation's activities were managed or organised. Under this test, courts will look at management systems and practices across the organisation, and whether an adequate standard of care was applied to the fatal activity."

Although this test is not linked to a particular level of management – instead it considers how an activity was managed within the organisation as a whole – in order to be convicted of the offence, the way in which its activities are managed or organised by its senior management must be a substantial element in the breach (*CMCHA 2007 s1(3)*).

Senior management is defined as the people who play a significant role in decision-making in how all, or a substantial part, of the activities are managed or organised; or actually managing or organising all, or a substantial part, of the activities (*CMCHA 2007 s 1(4)*). Those in the direct chain of management as well as those in strategic or regulatory compliance roles are included in the definition.

The MoJ guidance provides further advice on this:

> These are the people who make significant decisions about the organisation, or substantial parts of it. This includes both those carrying out headquarters functions (for example, central financial or strategic roles or with central responsibility for, for example, health and safety) as well as those in senior operational management roles.

> Exactly who is a member of an organisation's senior management will depend on the nature and scale of an organisation's activities. Apart from directors and similar senior management positions, roles likely to be under consideration include regional managers in national organisations and the managers of different operational divisions.

The offence does not require individual failings by senior managers to be identified. It is sufficient to show that senior management collectively were not taking adequate care and that this was a substantial part of the failure.

The MoJ guidance also makes clear that the offence cannot be avoided by senior management delegating responsibility for health and safety, and inappropriately delegating health and safety matters will leave organisations vulnerable to a charge of corporate manslaughter or homicide. The courts will consider how responsibility was discharged at different levels of the organisation.

It advises: "This does not mean that responsibility for managing health and safety cannot be made a matter across the management chain. However, senior management will need to ensure that they have adequate processes for health and safety and risk management in place and are implementing these."

Here the guidance makes reference to the guidance on directors' health and safety responsibilities produced by the Health and Safety Executive (HSE) and the Institute of Directors (IoD) (see **C9009** below)

In order to amount to a "gross" breach of duty of care, the conduct of the organisation must have fallen far below what can be reasonably expected in the circumstances (*CMCHA 2007 s 1(4)*), reflecting the threshold for the common law offence of gross negligence manslaughter.

And the way in which the organisation's activities were managed or organised must have caused the death. The usual principles of causation in criminal law apply – the management failure need not have been the sole cause of death; it need only be a cause. However, it must have made more than a minimal contribution to the death and an intervening act must not have broken the chain of events linking the management failure to the death. An intervening act will only break the chain of causation if it is extraordinary.

The MoJ guidance provides further guidance on causation. It advises that although it will not be necessary for the management failure to have been the sole cause of death, ""but for" the management failure (including the substantial element attributable to senior management), the death would not have occurred".

It also sets out that the law does not recognise very remote causes, and in some circumstances, an intervening Act may mean that the management failure is not considered to have caused the death.

Section 1(2) of the Act sets out that the offence applies to—

*   Corporations—defined as any body corporate, whether incorporated in the United Kingdom or elsewhere. Companies incorporated under companies legislation, and bodies incorporated under statute (as is the case with many non-Departmental Public Bodies and other bodies in the public sector) or by Royal Charter are all covered. But corporations sole* which cover a number of individual offices in England and Wales and Northern Ireland are specifically excluded (*CMCHA 2007 s 25*);
*   Government departments and other bodies listed in *Schedule 1*. These are Crown bodies which do not have a separate legal personality – *Section 11* (see below) sets out that there is no Crown Immunity under the Act;
*   Police forces; and
*   Partnerships, trade union and employers' associations where these are employers – Partnerships within the *Partnership Act 1890* and limited partnerships registered under the Limited *Partnerships Act 1907* and similar firms and entities come within the scope of the Act (*CMCHA 2007 s 25*);

The list of organisations to which the offence applies can be further extended by the Secretary of State, for example to further types of unincorporated association. This is subject to approval in both Houses of Parliament before it would come into effect (known as the affirmative resolution procedure) (*CMCHA 2007 s 21*).

The MoJ guidance provides more guidance on how the offence applies, or does not apply, to organisations in a number of circumstances. It sets out that:

- a parent company cannot be convicted because of failures within a subsidiary: "Companies within a group structure are all separate legal entities and therefore subject to the offence separately. In practice, the relevant duties of care that underpin the offence are more likely to be owed by a subsidiary than a parent";
- the offence applies to foreign companies: "the new offence applies to all companies and other corporate bodies operating in the UK, whether incorporated in the UK or abroad";
- where a company incorporated abroad is operating through a locally registered subsidiary, it is the subsidiary that is likely to be investigated and prosecuted;
- where sub-contractors are involved, whether a particular contractor could be liable for the offence will firstly depend on whether they owed a relevant duty of care to the victim. The offence applies in respect of existing obligations on the main contractor and sub-contractors for the safety of worksites, employees and other workers they supervise;
- the offence applies to charities and voluntary organisations where these have been incorporated, as a company or as a charitable incorporated organisation under the Charities Act 2006, for example. It will also apply where a charity or voluntary organisation operates as any other form of organisation to which the offence applies, such as a partnership with employees;

* The 2000 report, *Reforming the Law on Involuntary Manslaughter: The Government's Proposals*, provides the following definition of a corporations sole: "a corporation constituted in a single person in right of some office or function, which grants that person a special legal capacity to act in certain ways. Examples include many Ministers of the Crown and government officers, for example the Secretary of State for Defence and the Public Trustee and a bishop (but not a Roman Catholic bishop), a vicar, archdeacon, and canon."

The meaning of "relevant duty of care"

[C9003] The offence only applies where an organisation owed a duty of care, under the law of negligence, to the person who died, reflecting the common law offence of gross negligence manslaughter. It lists the following as a "relevant duty of care":

- a duty owed to its employees or to other persons working for the organisation or performing services for it;
- a duty owed as occupier of premises;
- a duty owed in connection with—
  - the supply by the organisation of goods or services (whether for payment or not),
  - the carrying out by the organisation of any construction or maintenance operations,

- the carrying out by the organisation of any other activity on a commercial basis, or
- the use or keeping by the organisation of any plant, vehicle or other thing.

In addition, a relevant duty of care arises where the organisation is responsible for the safety of person in custody or detention (*CMCHA 2007 s 2(1)* and *s 2(2)*). Various forms of custody and detention are covered, including prisons, the custody area of courts and police stations, immigration detention facilities, prison escort arrangements, secure accommodation for children and young people; and detention under mental health legislation.

There is also provision for the categories of people to whom a duty of care is owed because they are in custody or detention to be extended by the Secretary of State. However, extension of the offence to deaths in custody did not come into effect on the 6 April 2008, but is due to come into force at a later date. The previous government said that it would keep implementation under review, but the timescale it set out for this was between three and five years.

The following are "relevant" duties under the Act—

- The duty to provide a safe system of work for employees. The breach of a duty owed to other people whose work the organisation controls or directs, but who are not formally employed, such as contractors, volunteers and secondees can also trigger the offence;
- Duties to ensure that buildings the organisation occupies are kept in a safe condition;
- Duties owed by organisations to their customers, such as those owed by transport providers to their passengers and by retailers for the safety of their products. It also covers the supply of services by the public sector, such as NHS bodies providing medical treatment;
- The duty of care owed by public sector bodies to ensure that adequate safety precautions are taken, when repairing a road for example, even where duties do not arise because they are not supplying a service or operating commercially; and
- The duty of care owed by organisations, such as farming and mining companies, for example, which are carrying out activities on a commercial basis, though not supplying goods and services.

The MoJ guidance explains that statutory duties owed under health and safety law are not "relevant" duties for the new offence – only a duty of care owed in the law of negligence. It goes on to explain:

> In practice, there is a significant overlap between these types of duty. For example, employers have a responsibility for the safety of their employees under the law of negligence and under health and safety law (see for example section 2 of the Health and Safety at Work etc Act 1974 and article 4 of the Health and Safety at Work (Northern Ireland) Order 1978). Similarly, both statutory duties and common law duties will be owed to members of the public affected by the conduct of an organisation's activities.

> The common law offence of gross negligence manslaughter in England and Wales and Northern Ireland is based on the duty of care in the law of negligence, and this has been carried forward to the new offence. In Scotland, the concepts of negligence and duty of care are familiar from the civil law.

The offence applies where the duty of care owed under the common law of negligence has been superseded by statutory provision (including where this imposes strict liability). For example the duty of care owed by an occupier, which is now owed under the *Occupiers' Liability Acts 1957* and *1984* and the *Defective Premises Act 1972 (CMCHA 2007 s 2(4))* is included.

The MoJ guidance sets out that in some cases where a person cannot be sued under the civil law of negligence, the offence may still apply. For example, this would be the case where a "no fault" scheme for damages has been introduced.

And the offence is not affected by common law rules precluding liability in the law of negligence where people are jointly engaged in a criminal enterprise ("ex turpi causa non oritur actio") or because a person has accepted a risk of harm "("volenti non fit injuria") (*CMCHA 2007 s 2(6)*).

Whether a duty of care exists in a particular case is a matter of law for the judge to decide (*CMCHA 2007 s 2(5)*). Normally in criminal proceedings, questions of law are decided by the judge, whilst questions of fact, and the application of the law to the facts of the case, are for the jury, directed by the judge. However because of the heavily legal nature of the tests relating to the existence of a duty of care in the law of negligence, the Act sets out that the judge will need to determine some facts, rather than the jury – for example whether the person killed was an employee of the organisation.

Public policy decisions, exclusively public functions and statutory inspections

[C9004] Public policy decisions, exclusively public functions and statutory functions are excluded from the scope of the offence – any duty of care owed by a public authority in respect of these is not a "relevant duty of care" (*CMCHA 2007 s 3(1)–(3)*).

Deaths alleged to have been caused by decisions of public authorities – the definition of which includes Government departments, local councils and other public bodies, are therefore outside the scope of the offence. This would include, for example, decisions made by Health Care Trusts about the funding of particular treatments.

The MoJ guidance sets out that strategic funding decisions and other matters involving competing public interests are exempt, but decisions about how resources were managed are not.

An organisation will not be liable for a breach of any duty of care owed in respect of things done in the exercise of "exclusively public functions", unless the organisation owes the duty in its capacity as an employer or as an occupier of premises.

Any organisation, not just Crown or other public bodies, performing that particular type of function is excluded, although this does not affect individual liability. These functions will continue to be subject to other forms of accountability such as independent investigations, public inquiries and the accountability of Ministers through Parliament.

Organisations with a duty of care owed in connection with the carrying out of statutory inspections, such as those carried out by health and safety enforcement authorities, will not be liable unless they owe duties as an employer or occupier of premises (*CMCHA 2007 s 3(3)*).

"Exclusively public functions" are those falling within the prerogative of the Crown, such as providing services in a civil emergency, and activities that require a statutory or prerogative basis and cannot be independently performed by private bodies, such as licensing drugs (*CMCHA 2007 s 3(4)*).

The MoJ guidance sets out that:

> This does not exempt an activity simply because statute provides an organisation with the power to carry it out (as is the case, for example, with legislation relating to NHS bodies and local authorities). Nor does it exempt an activity because it requires a licence (such as selling alcohol). Rather, the activity must be of a sort that cannot be independently performed by a private body. The type of activity involved must intrinsically require statutory or prerogative authority, such as licensing drugs or conducting international diplomacy.

It also explains that private companies that carry out public functions are broadly in the same position as public bodies: "Overall the Act is intended to ensure a broadly level playing field under the new offence for public and private sector bodies when they are in a comparable situation."

Military activities

[C9005] Certain military activities are exempt in respect of all categories of relevant duty of care (*CMCHA 2007 s 4*).

These include peacekeeping operations, operations for dealing with terrorism, civil unrest or serious public disorder, in the course of which members of the armed forces come under attack or face the threat of attack or violent resistance; as well as activities preparing for or supporting these operations. The exemption extends to training involving hazardous activities, and to the activities carried out by members of the special forces.

Policing and law enforcement

[C9006] Reflecting the existing law of negligence, the police and other law enforcement bodies are exempt in respect of all categories of relevant duty of care, including those as an employer or occupier, with respect to the following circumstances:

- operations dealing with terrorism, civil unrest or serious disorder in which an authority's officers or employees come under attack or the threat of attack; or
- where the authority in question is preparing for or supporting such operations; or
- where it is carrying on training with respect to such operations (*CMCHA 2007 s 5(1), (2)*).

A wider range of policing and law enforcement activities are excluded from the offence where the pursuit of law enforcement activities has resulted in a fatality to a member of the public, but not in respect of the duty of care owed as an employer or occupier (*CMCHA 2007 s 5(3)*).

Decisions about and responses to emergency calls, the manner in which particular police operations are conducted, the way in which law enforcement and other coercive powers are exercised, measures taken to protect witnesses, and the arrest and detention of suspects are not, for example, covered by the Act.

The exemption is not confined to police forces and extends to other bodies operating similar functions and to other law enforcement activity, such as action by the immigration authorities to arrest, detain, or deport an immigration offender. (Note however, that the Act does not have any bearing on the question of individual liability.)

Emergencies and the emergency services

[C9007] The offence does not apply to the emergency services when responding to emergencies (*CMCHA 2007 s 6(1)*). However, the emergency services still have duties as employers and occupiers, so must still provide a safe system of work for their employees and secure the safety of their premises.

Organisations to which the exemption applies are—

- English and Welsh fire and rescue authorities;
- a fire and rescue authority or joint fire and rescue board in Scotland;
- the Northern Ireland Fire and Rescue Service Board;
- any other organisation responding to an emergency either for one of the organisations listed above, or if not, otherwise than operating commercially;
- NHS bodies;
- an organisation providing ambulance services for a relevant NHS body or the Secretary of State or Welsh Ministers;
- an organisation providing services for the transport of organs, blood, equipment or personnel in pursuance of arrangements of the kind mentioned in the paragraph above;
- an organisation providing a rescue service (such as the Coastguard and the Royal National Lifeboat Institution (RNLI);
- the armed forces.

Emergency circumstances are defined as being life-threatening, or causing, or threatening to cause, serious injury or illness or serious harm to the environment or buildings or other property (*CMCHA 2007 s 6(7)*). Circumstances believed to be emergency circumstances are also covered (*CMCHA 2007 s 6(8)*).

Medical treatment, and decisions relating to this, other than those to establish the priority for treating patients, is not included in the exemption (*CMCHA 2007 s 6(3), (4)*). Nor does the exemption apply to duties that do not relate to the way in which a body responds to an emergency – so duties to maintain vehicles in a safe condition, for example, remain.

The MoJ guidance sets out that with regard to NHS trusts (including ambulance trusts) duties of care relating to medical treatment in an emergency, other than triage decisions (which determines the order in which injured people are treated), are not exempt.

Child protection and probation functions

[C9008] The offence does not apply in relation to carrying out (or failing to carry out) statutory functions relating to child protection and probation (*CMCHA 2007 s 7(1)*). However, local authorities and probation services are covered by the offence with respect to their responsibilities to employees, and as occupiers of premises.

This applies with regard to:

* Parts 4 and 5 of the Children Act 1989;
* Part 2 of the Children (Scotland) Act 1995; or
* Parts 5 and 6 of the Children (Northern Ireland) Order 1995

And it applies to duties owed under:

* Chapter 1 of Part 1 of the Criminal Justice and Court Services Act 2000;
* section 27 of the Social Work (Scotland) Act 1968; or
* Article 4 of the Probation Board (Northern Ireland) Order 1982.

However, local authorities and probation services are covered by the offence with respect to their responsibilities to employees, and as occupiers of premises.

Factors for the jury to consider when deciding if there has been a "gross" breach

[C9009] Where it has been established that an organisation owed a relevant duty of care to the victim and it falls to the jury to decide whether there was a gross breach of the duty, the jury must consider whether the evidence shows that the organisation failed to comply with health and safety legislation relating to the breach. If that is the case, it must look at how serious the failure was and how much of a risk of death it posed (*CMCHA 2007 s 8 (1), (2)*).

The jury may also consider the wider context, including cultural issues within the organisation, such as attitudes and accepted practices that tolerated breaches, and it may consider any relevant health and safety guidance.

Health and safety guidance does not provide an authoritative statement of required standards and the jury is therefore not required to consider the extent to which this is not complied with. However, where breaches of relevant health and safety duties are established, health and safety guidance may assist a jury in considering how serious this was.

These factors are not exhaustive, and the jury can also have regard to any other matters they consider relevant (*CMCHA 2007 s8 (4)*).

The MoJ guidance sets out that relevant health and safety guidance includes statutory Approved Codes of Practice and other guidance published by regulatory authorities that enforce health and safety legislation.

It advises: "Employers do not have to follow guidance and are free to take other action. But guidance from regulatory authorities may be helpful to a jury when considering the extent of any failures to comply with health and safety legislation and whether the organisation's conduct has fallen far below what could reasonably have been expected.

And it advises companies that in addition to guidance from the Health and Safety Executive (HSE) (and in Northern Ireland from the Health and Safety Executive Northern Ireland (HSENI)), and local authorities: "There are specific regulatory bodies, and in some cases separate legislation too, for certain sectors of industry (for example, in the various transport sectors: rail, marine, air and roads) and for dealing with particular safety issues (such as food and environmental safety). Further information about the standards that apply in these circumstances should be obtained from the relevant regulatory authority."

It goes on to say: "Factors that might be considered will range from questions about the systems of work used by employees, their level of training and adequacy of equipment, to issues of immediate supervision and middle management, to questions about the organisation's strategic approach to health and safety and its arrangements for risk assessing, monitoring and auditing its processes. In doing so, the offence is concerned not just with formal systems for managing an activity within an organisation, but how in practice this was carried out."

The Health and Safety Executive (HSE) and the Institute of Directors (IoD) guidance on directors' responsibilities on health and safety *Leading health and safety at work – leadership actions for directors and board members* is addressed to directors (and their equivalents) of corporate bodies and organisations in the public and voluntary sectors – including governors, trustees and officers.

Although the guidance is not obligatory, the HSE and IoD advise that this guidance could be a relevant consideration for a jury depending on the circumstances of the particular case brought under the CMCH Act.

The guidance sets out that directors can be personally liable for breaches of health and safety law, and that members of the board have both collective and individual responsibility for health and safety

Penalties and Sentencing

[C9010] Companies convicted of the offence of corporate manslaughter will face very large fines.

Guidelines for judges when passing sentence in court produced by the Sentencing Guidelines Council (SGC) set out that fines for organisations found guilty of corporate manslaughter should be measured in millions of pounds and should not generally be below £500,000. (They also say that for health and safety offences that cause death at work, fines in the order of hundreds of thousands of pounds and generally not less than £100,000 are appropriate.)

The final definitive guideline, *Corporate Manslaughter & Health and Safety Offences Causing Death*, differed from recommendations originally set out by the Sentencing Advisory Panel (SAP) in 2007. The SAP provided advice to the Council and had recommended a series of starting points and ranges of fines based on turnover for convictions under the *Corporate Manslaughter and Corporate Homicide Act 2007* (CMCHA 2007) and *Health and Safety at Work etc. Act 1974* (HSWA 1974), where the offence caused death.

Fines

[C9011] An organisation found guilty of corporate manslaughter is liable on conviction on indictment to an unlimited fine (*CMCHA 2007 s 1(6)*) and the court may also make a remedial order and publicity orders. The offence is triable in the Crown Court in England and Wales and the High Court of Justiciary in Scotland (*CMCHA 2007 s 1(7)*), both of which involve trials by jury.

The view of the SAP was that a fine for a conviction under the CMCHA 2007 should be set at a significantly higher level than for an offence under the Health and Safety at Work etc Act 1974 (HSWA) and proposed that for an offence of corporate manslaughter, committed by a first time offender pleading not guilty, the starting point should be 5% of annual turnover averaged over the previous three years. Where there were significant mitigating or aggravating factors, it said that this could be reduced to 2.5% or increased to 10%, or more for the worst cases (see box below).

(With regard to an offence under HSWA 1974 involving death, it proposed a starting point of 2.5% of average annual turnover, with a range of between 1 and 7.5% depending on whether there were aggravating or mitigating factors.)

The consultation paper also sought views as to whether it would be appropriate to set a minimum fine both for corporate manslaughter and offences under HSWA 1974 involving death, to ensure that the harm involved in such offences is properly reflected in the sentence, even where the offending company has a very low annual turnover.

The SGC decided that a fixed correlation between the amount of the fine and a company's turnover or profit was not appropriate because this:

• could inadvertently risk an unfair outcome;
• was particularly difficult to apply to public and third sector bodies; and
• was likely to create a perverse incentive to adjust corporate structures to avoid the proper consequences of offending.

The Council therefore concluded that it did not provide the most effective way of assessing the level of fine. Instead the Council has proposed a level below which a fine should not normally fall, supported by a general indication concerning the extent to which a fine should be above that level. As set out above the guideline sets out that fines for organisations found guilty of corporate manslaughter should be measured in millions of pounds and should not generally be below £500,000.

Assessing the seriousness of the offence

The SGC guideline advises that:

Seriousness should ordinarily be assessed first by asking:

- How foreseeable was serious injury?
- How far short of the applicable standard did the defendant fall?
- How common is this kind of breach in this organisation? and
- How far up the organisation does the breach go?

It also sets out aggravating and mitigating factors. Other factors that are likely to aggravate the offence include:

- more than one death, or very grave personal injury in addition to death;
- failure to heed warnings or advice, whether from officials such as the Inspectorate, or by employees (especially health and safety representatives) or other persons, or to respond appropriately to 'near misses' arising in similar circumstances;
- cost-cutting at the expense of safety;
- deliberate failure to obtain or comply with relevant licences, at least where the process of licensing involves some degree of control, assessment or observation by independent authorities with a health and safety responsibility; and
- injury to vulnerable people.

Mitigating factors include:

- a prompt acceptance of responsibility;
- a high level of cooperation with the investigation, beyond that which will always be expected;
- genuine efforts to remedy the defect;
- a good health and safety record; and
- a responsible attitude to health and safety, such as the commissioning of expert advice or the consultation of employees or others affected by the organisation's activities.

Remedial Orders

[C9012] In addition to the power to impose an unlimited fine, an organisation convicted of corporate manslaughter or corporate homicide may also be issued with a remedial order by the court, requiring it to take specific steps to remedy the breach; any matter the court believes to have resulted from the breach and caused the death; and any deficiencies in the organisation's health and safety policies, systems or practices of which the breach appear to be an indication (*CMCHA 2007 s 9(1)*).

Remedial orders can only be made on an application by the prosecution specifying the terms of the proposed order and after consultation with the health and safety enforcement authority, and there is the opportunity for the convicted organisation to make representations in court (*CMCHA 2007 s 9(2), (3)*).

The order must specify how long the organisation has to carry out the specified steps, and may require that evidence that the order has been complied is given to the enforcement authority (*CMCHA 2007 s 9(4)*. Failure to comply with the order is an offence and liable on conviction to an unlimited fine (*CMCHA 2007 s 9(5)*.

The MoJ guidance sets out that remedial orders will be used in relatively rare circumstances. The sanction is already available under *HSWA 1974 s 42* for health and safety offences, but were not previously possible in relation to a manslaughter conviction. However, remedial orders for health and safety offences have only been used in a handful of cases.

The SGC guideline says that a defendant ought, by the time of sentencing, to have remedied any specific failings involved in the offence and if it has not will be deprived of significant mitigation. It sets out that if it has not, a remedial order should be considered if it can be made sufficiently specific to be enforceable. It also says that the cost of compliance with a remedial order should not ordinarily be taken into account in fixing the fine since the order requires only what should already have been done.

Publicity Orders

[C9013] The court, after consulting the enforcing authorities and listening to representations from the prosecution can also require that an organisation convicted of corporate manslaughter or corporate homicide publicises the fact that it has been convicted of the offence; and specify the particulars of the offence, the amount of any fine, and the terms of any remedial order (*CMCHA 2007 s 10(1)*.

Again the order must specify how long the organisation has to comply, and may require that evidence that the order has been complied is given to the enforcement authority (*CMCHA 2007 s 9(4)*. Failure to comply with the order is an offence and liable on conviction to an unlimited fine (*CMCHA 2007 s 9(5)*.

The SGC guideline says that a publicity order should normally be imposed in the case of corporate manslaughter and that the object of the order is both deterrence and punishment.

It sets out that the order should normally:

- specify the place where public announcement is to be made, and consideration should be given to indicating the size of any notice or advertisement required; and

- ensure that the conviction becomes known to shareholders in the case of companies and local people in the case of public bodies and that consideration should be given to requiring a statement on the defendant's website.

While it says that a newspaper announcement may not be necessary if the proceedings are certain to receive news coverage in any event, but if an order does require publication in a newspaper it should specify the paper, the form of announcement to be made and the number of insertions required.

It also advises that consideration should be given to a stipulation that any comment placed by the defendant alongside the required announcement should be separated from it and clearly identified as such.

Finally it sets out that a publicity order is part of the penalty. Any exceptional cost of compliance should be considered in fixing the fine; but it is not necessary to fix the fine first and then deduct the cost of compliance.

The guideline is available online at: www.sentencingcouncil.org.uk/docs/web__guideline_on_corporate_manslaughter_accessible.pdf

Application to Crown bodies

[C9014] Crown bodies are not immune from prosecution under the Act (*CMCHA 2007 s 11(1)*). Crown bodies that are either bodies corporate or are listed in Schedule 1 to the Act are subject to the offence. A Crown body is to be treated as owing the duties it would owe if it were a corporation that was not a servant or agent of the Crown (*CMCHA 2007 s 11(2)*). *Sections 11(3)* and *(4)* deal with the technicality which means that civil servants in government departments are employed by the Crown rather than the government department they work in, ensuring that the activities and functions of government departments and other Crown bodies can be attributed to the relevant body. Similarly *section 11(5)* ensures that the provisions apply to Northern Ireland department in the same way as they apply to bodies listed in *Schedule 1*.

Application to the armed forces

[C9015] A duty of care is owed to personnel in the armed forces – including the Royal Navy, Army and Air Force – by the Ministry of Defence for the purposes of the offence (*CMCHA 2007 s 12*).

Application to police forces

[C9016] Unlike police authorities, which are bodies corporate, police forces are not incorporated bodies. However under the Act, a police force is treated as owing the duties of care it would owe if it were a body corporate (*CMCHA 2007 s 13*). Therefore police officers, as well as police cadets and police trainees in Northern Ireland, and police officers seconded to the Serious Organised Crime Agency or the National Policing Improvement Agency are treated as the employees of the police force, or other organisation, for which they work and therefore owed the employer's duty of care.

Application to partnerships

[C9017] Partnerships which are not limited liability partnerships – the latter are corporate bodies and so covered by the offence – are not corporations and therefore do not have a distinct legal personality for the purpose of owing a duty of care in the law of negligence. However in the Act, partnerships are treated as though they owed the same duties of care as a corporate body for the purposes of the offence *(CMCHA 2007 s 14(1))*. Any proceedings will be brought in the name of the partnership rather than any of its members *(CMCHA 2007 s 14(2))* and any fine imposed on conviction must be paid out of the funds of the partnership *(CMCHA 2007 s 14(2))*.

The MoJ guidance explains further that this approach reflects that taken under other legislation, such as the Companies Act 2006 and means that partnerships will be dealt with in a similar manner to companies and other incorporated defendants.

It sets out that it took a cautious approach in extending the offence to unincorporated associations, since it represents a new extension of the criminal law to these organisations: "Extending the offence to partnerships will ensure that an important range of employing organisations, already subject to health and safety law, is within the offence and that large firms are not excluded because they have chosen not to incorporate.

In addition, the Act makes provision for the range of organisations covered by the offence to be extended by secondary legislation (section 21).

Procedure, evidence and sentencing

[C9018] Any statutory provision that applies in relation to criminal proceedings against a corporation also applies (unless an order made by the Secretary of State has prescribed any adaptations or modifications) to the organisations listed in Schedule 1 of the Act (see below); police forces, partnerships, trade unions and employers' associations. This includes provisions in relation to procedure, evidence and sentencing *(CMCHA 2007 s 15)*.

An order made under this section is subject to the negative resolution procedure, which means that it would be laid before Parliament and become law, unless disapproved by Parliament *(CMCHA 2007 s 15(4))*.

Transfer of functions

[C9019] Where a death has occurred in connection with functions carried out by a Government department, one of the bodies listed in Schedule 1, incorporated Crown bodies or police forces, and there has been a subsequent transfer of those functions within the public sector, prosecutions will be taken against the body with current responsibility for the relevant function. Where the function has been transferred out of the public sector (in the case of privatisation, for example) proceedings will be taken against the public organisation which last carried out the function (*CMCHA 2007 s 16(1)–(3)*).

However, in some circumstances it may be appropriate for liability to lie with a different body, for example where a function transfers between Government departments but there is no transfer of personnel. In this case, there is provision for the Secretary of State to make an order specifying that liability rests with a different body. Again an order made under this section is subject to the negative resolution procedure (see above).

Director of Public Prosecutions (DPP) consent for proceedings

[C9020] Consent of the Director of Public Prosecutions (DPP) is needed for proceedings for the offence of corporate manslaughter to be instituted. This applies in England, Wales and Northern Ireland. (In Scotland all proceedings on indictment must already be instigated by the Lord Advocate.) (*CMCHA 2007 s 17*).

No individual liability for corporate manslaughter or corporate homicide

[C9021] There is no secondary liability for the offence. An individual cannot be guilty of aiding, abetting, counselling or procuring the commission of, or in Scotland of being art and part of, the offence (*CMCHA 2007 s 18*). (However, an individual can still be directly liable for the offence of gross negligence manslaughter, culpable homicide, or health and safety offences.)

Under the common law offence of gross negligence manslaughter, individual officers of a company (directors or business owners) can be prosecuted for the offence if their own grossly negligent behaviour causes death. The offence is punishable by a maximum of life imprisonment (Offences against the Persons Act 1861 s 5).

The leading case on gross negligence manslaughter is *R v Adomako* [1995] 1 AC 171 which involved the death of a patient who was undergoing an eye operation. In this case, the House of Lords set down a four stage test for the offence of gross negligence manslaughter:

- Did the defendant owe a duty of care towards the victim who has died?
- If so, has the defendant breached that duty of care?
- Has such breach caused the victim's death? and
- If so, was that breach of duty so bad as to amount, when viewed objectively, to gross negligence warranting a criminal conviction?

According to the Home Affairs and Work and Pensions Committees joint report on the *Draft Corporate Manslaughter Bill*, 15 directors or business owners were personally convicted of manslaughter by gross negligence between April 1999 and September 2005 and there have been further convictions since 2005.

In addition, the MoJ guidance makes clear that although prosecutions under the CMCH Act 2007 will be brought against organisations and not specific individuals, individual directors, managers and employees may be called as witnesses and stand in the dock.

A number of individuals have been prosecuted for gross negligence manslaughter under the common law in relation to work-related deaths. For example, builder Colin Holtom and contractor Darren Fowler were jailed in 2009 (for three years and one year respectively) after being found guilty of the manslaughter of 15-year old labourer Adam Gosling. Adam was crushed to death by an unstable wall he had been sent to demolish. The court was told that Adam and his 18-year-old brother were working with minimal supervision, inadequate personal protective clothing, no risk assessments, no training and no welfare facilities.

Also in 2009, a factory owner and his son were found guilty of gross negligence manslaughter after two fire crew died in an explosion while attempting to put out a fire at their fireworks company, Alpha Fireworks Ltd. Another twenty people were injured.

The Crown Prosecution Service (CPS) said the pair were experts in handling and storing everyday fireworks, they knew that fireworks should be treated as explosives and they knew they had a duty to take reasonable care to protect the public.

The jury in the case decided that both defendants breached that duty of care. They were fully aware of the legislation and the different hazard classifications given to fireworks and they knew that storing certain fireworks with others in a metal container posed a high risk of mass explosion – which is what happened.

Martin Winter, owner of Alpha Fireworks Ltd (formerly Festival Fireworks UK Ltd), was jailed for seven years for gross negligence manslaughter. His son and employee, Nathan Winter, was also jailed for five years for the gross negligence manslaughter of the two firefighters. Alpha Fireworks Ltd was fined a total of £30,000 for breaches of the *Manufacture and Storage of Explosives Regulations 2005*.

Convictions under the Act and under health and safety legislation

[C9022] A conviction for corporate manslaughter does not preclude an organisation being convicted for health and safety offences on the same facts (*CMCHA 2007 s 19*) where this is required in the interests of justice. (An individual can be convicted on a secondary basis for an offence under *section 37* of the *Health and Safety at Work etc Act 1974*).

Abolition of liability of corporations for manslaughter at common law

[C9023] The application of the common law offence of gross negligence manslaughter is abolished in relation to corporations, and any application it has to other organisations to which section 1 applies ie unincorporated associations to which the offence applies (*CMCHA 2007 s 20*).

However in Scotland, where the law on culpable homicide differs in certain respects from the law on gross negligence manslaughter, the common law will continue to be in force. The Procurator Fiscal will determine the appropriate charge in light of the circumstances of each individual case.

Where cases occur before 6 April 2008, the MoJ guidance sets out that section 27(4) allows for cases that occur wholly or partly before the new offence comes into force, and prosecutions in those cases will continue to be possible on the basis of existing common law.

Extent and territorial application

[C9024]–[C9025] The Act applies to England and Wales, Scotland and Northern Ireland, and also applies to other locations where criminal jurisdiction currently applies. For example, applies where a death occurs as a result of an incident involving a British ship (or aircraft or hovercraft) but the victim was not on board because they were shipwrecked and drowned.

The MoJ guidance sets out the key areas where the offence applies:

"The Act applies across the UK.

—   The new offence can be prosecuted if the harm resulting in death occurs:
  • in the UK
  • in the UK's territorial waters (for example, in an incident involving commercial shipping or leisure craft)
  • on a British ship, aircraft or hovercraft
  • on an oil rig or other offshore installation already covered by UK criminal law."

And it provides further guidance on jurisdiction: "Harm resulting in death" will typically be physical injury that is fatal, and in most cases the injury and death will occur at the same time and in the same location. However, in some cases death may occur some time after the injury or harm takes place. The courts have jurisdiction in cases where the relevant harm was sustained in the UK, even if the death occurs abroad.

In the case of fatalities related to ships, aircraft and hovercraft, the offence will apply where the death does not occur on board, as long as it relates to an on-board incident.

The Act does not apply to British companies responsible for deaths abroad. The harm leading to death must occur within the UK or one of the places described above.

The guidance points out that where a death occurs abroad, there are acute practical issues for investigators, since they will not have control over the crime scene or the gathering of evidence relating to the death, while the evidence will be a crucial part of the investigation.

Schedule 1: List of government departments etc

- Attorney General's Office
- Cabinet Office
- Central Office of Information
- Crown Office and Procurator Fiscal Service
- Crown Prosecution Service
- Department for Business, Innovation and Skills
- Department for Communities and Local Government
- Department for Culture, Media and Sport
- Department for Education
- Department for Environment, Food and Rural Affairs
- Department for International Development
- Department for Transport
- Department for Work and Pensions
- Department of Energy and Climate Change
- Department of Health
- Export Credits Guarantee Department
- Foreign and Commonwealth Office
- Forestry Commission
- General Register Office for Scotland
- Government Actuary's Department
- Her Majesty's Land Registry
- Her Majesty's Revenue and Customs
- Her Majesty's Treasury
- Home Office
- Ministry of Defence
- Ministry of Justice (including the Scotland Office and the Wales Office)
- National Archives
- National Archives of Scotland
- National Audit Office
- National Savings and Investments
- National School of Government
- Northern Ireland Audit Office
- Northern Ireland Court Service
- Northern Ireland Office
- Office for National Statistics
- Office of Her Majesty's Chief Inspector of Education and Training in Wales
- Ordnance Survey
- Public Prosecution Service for Northern Ireland
- Registers of Scotland Executive Agency
- Revenue and Customs Prosecutions Office

- Royal Mint
- Scottish Executive
- Serious Fraud Office
- Treasury Solicitor's Department
- UK Trade and Investment
- Welsh Assembly Government

In addition, the MoJ guidance sets out that the offence also applies to Crown bodies that are incorporated, such as the Northern Ireland departments, Charity Commission, Office of Fair Trading and Postal Services Commission. And it explains that fatalities caused by Executive Agencies come within the scope of the offence. Executive Agencies come under the responsibility of a parent department, which are all covered by the offence. The Schedule can be amended by an order made by the Secretary of State (*CMCHA 2007 s 22*).

Investigation and prosecution

The Act does not change the current responsibilities of the police to investigate, and the Crown Prosecution Service (CPS) in England and Wales, the Public Prosecution Service in Northern Ireland and the Procurator Fiscal in Scotland to prosecute, corporate manslaughter. The Health and Safety Executive (HSE) and other health and safety authorities will also continue to use their expertise in investigations in order to look at whether there is liability under more specific legislation, and to provide advice and assistance to the police in investigating corporate manslaughter. The joint approach of the HSE and other enforcement authorities, the police and the CPS is set out in protocols on investigating work-related deaths:

- In England and Wales, the Work-Related Deaths Protocol – the National Liaison Committee on the Work-Related Deaths Protocol has also published an "Investigators' guide" to improve consistency in its application.
- Liaison in Northern Ireland is covered by a separate and broadly equivalent document, Investigation of Work-related Deaths: Northern Ireland agreement for liaison; and
- In Scotland, a Protocol on Work-Related Deaths was published in October 2006.

In addition, the Act does not change the role and powers of the independent accident investigation branches which investigate air, marine and rail accidents in order to establish the cause, independently of any criminal investigation.

Ministry of Justice Guidance

The Ministry of Justice (MoJ) had produced two guidance documents providing further information about the Corporate Manslaughter and Corporate Homicide Act 2007 and its implementation.

A general introduction to the Act is provided in a leaflet, *Understanding the Corporate Manslaughter and Corporate Homicide Act 2007*, explaining how the new offence of corporate manslaughter/homicide works, and where it will apply. This is intended to provide fundamental information about the new legislation and is aimed at employers, senior managers and others looking for an overview of the law.

Detailed guidance on the implementation of the Act is provided in *Reforming corporate liability for work-related death: a guide to the Corporate Manslaughter and Corporate Homicide Act 2007*. This explains which organisations are covered, the sort of incident to which it applies, those that are exempt, and the test that will be applied by the court. It is aimed at those who need to know how the new Act will work in some detail, including health and safety managers and professionals.

This more detailed guidance is summarised and referred to throughout this chapter.

Both guides can be downloaded from the MoJ website but hard copies of the guidance will not be issued.

Health and Safety Executive (HSE) Guidance

The Health and Safety Executive (HSE) has guidance on the Act on its website at www.hse.gov.uk/corporatemanslaughter/faqs.htm in the form of Frequently Asked Questions. It also has linked to relevant HSE and HSC publications.

The first corporate manslaughter trial under the Act has been delayed due to the ill-health of the defendant. It relates to the death of junior geologist, Alexander Wright, who was killed in September 2008 when taking soil samples inside a pit which had been excavated as part of a site survey. He was crushed when the sides of the pit collapsed. The Crown Prosecution Service authorised a charge of corporate manslaughter against Cotswold Geotechnical Holdings Ltd, which has also been charged with breaching *HSWA 1974*. Peter Eaton, a director of the company was charged with gross negligence manslaughter and also faces a health and safety charge.

Dangerous Goods – Carriage

John Wintle

Introduction

Scope

[D0401] This chapter presents the UK legislation covering the carriage (transportation) of dangerous goods. This is the *Carriage of Dangerous Goods and Use of Transportable Pressure Equipment Regulations 2007 (2007 No 1573)*, known as *CDG 2007*. The Regulations cover all carriage of dangerous goods by road, rail and inland waterway and now include radioactive substances.

As the carriage of dangerous goods can cross national boundaries, the UK legislation refers extensively to European Agreements and Directives, namely the *ADR* (Accord European relatif au transport international des marchandises dangereuses par route) for road transportation and *RID* (Reglement concernant le transport international ferroviare des marchandises dangereuses) for rail transportation. These are maintained from international standards recommended by the United Nations Committee of Experts through the UN Economic Commission for Europe. In general, these international standards apply to journeys wholly within the UK, but the UK legislation has certain national derogations (deviations) agreed with the European Commission for this type of journey.

The scope of this chapter primarily refers to journeys by road; similar principles apply to journeys by rail and inland waterway. Journeys made by sea or air are covered by other legislation in line with international agreements. This chapter does not currently cover the transportable pressure equipment aspects in *CDG 2007 Part 4*.

The chapter will be of use to those businesses involved with the carriage of dangerous goods. It will be relevant to businesses producing dangerous goods for carriage, haulage companies and specialist conveyors, and to businesses receiving dangerous goods. All aspects of the supply chain are covered: classification and identification, packaging and containment, consignment procedures (including marking, labelling, placarding, plating and documentation), and carriage (including loading, unloading and handling), the transport vehicle and equipment, and the training and duties of the crew.

The definition of dangerous goods for carriage is dealt with in some detail in the section on classification and identification. It covers goods whose release would cause a hazard to the general public, passengers, transport crews, equipment, property or the environment. This covers a wide range of goods, including those which are flammable, corrosive, toxic, compressed gases and radioactive.

The aim of CDG 2007 is to protect against the hazards and danger that release of these goods might cause within the transport system. Protection from substances within the workplace is covered by other regulations, namely the *Control of Substances Hazardous to Health Regulations 1999 (COSHH)*, the *Dangerous Substances and Explosive Atmosphere Regulations 2006 (DSEAR)* and the *Pressure Systems Safety Regulations 2000 (PSSR)*. All these regulations are made under the general provisions of the *Health and Safety at Work Act 1975 (HSWA 1975)*.

The carriage of dangerous goods legislation is prescriptive and complex. It is not the aim of this chapter to provide a comprehensive guide or all the necessary detail; for these the reader is directed to CDG 2007 and ADR, and to various publications issued by the Health and Safety Executive. This chapter will, however, enable businesses to gain an overview of the nature of their responsibilities and direct them to where more detail can be found.

Development of legislation in the UK

[D0402] Legislative control of dangerous substances began with the *Petroleum Act 1879* and the *Petroleum (Consolidation Act) 1928*. The latter remained with (subsidiary regulations) the main legislative control on the transport of all dangerous goods until the 1980's. In the 1980's, influenced by the 1978 tanker disaster in Spain, separate regulations were introduced regulating classification, labelling and packaging (1984), the transport by road of dangerous goods in tankers and tank containers (1981) and in packages (1986). The latter two were subsequently replaced in 1992 with the addition of regulations for the training of drivers of vehicles carrying dangerous goods.

In 1992 the classification, labelling and packaging regulations were replaced by a set covering these aspects for transport by road and rail. These regulations were subsequently replaced in 1996, and aspects concerning the use of transportable pressure receptacles were added. In the same year the separate regulations for tankers and tank containers and for packages were combined into the *Carriage of Dangerous Goods by Road Regulations*, and the driver training regulations were updated. In 1999 various amendments were made, the most significant being the requirement by many duty holders to appoint safety advisers under the *Transport of Dangerous Goods (Safety Advisers) Regulations*.

In order to harmonise national and international regulations for the carriage of dangerous goods, the UK was a signatory to European agreements with Directives concerning the international carriage of dangerous goods by road (ADR) and by rail (RID). This necessitated aligning UK legislation with these Directives. As ADR and RID do not have provisions relating to national enforcement, this aspect had to be dealt with at the national level.

On 10 May 2004 a consolidating set of regulations came into force, the *Carriage of Dangerous Goods and Use of Transportable Pressure Equipment*

Regulations 2004 (SI 2004 No 568). These were substantially restructured in 2007 as the *Carriage of Dangerous Goods and Use of Transportable Pressure Equipment Regulations 2007 (SI 2007 No 1573)*, known as *CDG 2007*.

CDG 2007 cover all road and rail carriage of dangerous goods, and now include radioactive substances that were previously treated separately. The regulations directly reference *ADR* and *RID*, which contain the detailed prescriptive requirements. Therefore it is necessary to consult *ADR* and *RID* to know what to do. Fortunately these documents are available in English and can be easily downloaded.

There are some differences in *CDG 2007* from *ADR* and *RID* for domestic transport. These are mainly for the carriage of explosives, but also to retain the UK system of marking road tankers. There are also some exemptions from the requirements of *ADR* and *RID* for purely domestic transport, but for international journeys the full requirements of *ADR* and *RID* apply.

ADR and *RID* do not contain provisions and arrangements for enforcement in any country. This comes under the national legislation *CDG 2007*. The HSE is one of the enforcement agencies for many aspects of *CDG 2007*, but the Department for Transport is the Competent Authority for most purposes. Suitably trained police and the Vehicles and Operators Standards Agency have responsibilities for enforcement on the road in the UK.

Structure and content of ADR

[D0403] *ADR* is a two volume document. *Volume I* comprises an introduction, which is a brief summary of the duties, the legal text of the *ADR Agreement*, and *Parts 1–3* of *Annex A*. These Parts cover general provisions concerning dangerous goods and an outline of the duties of the consignor of the goods. It identifies hazard classes, allocates dangerous materials into these classes according to UN Number, and provides the information to be used by the consignor to classify dangerous goods in preparation for transportation. *Volume II* contains the duties to be followed by the carrier of the goods such as specific packing instructions, labelling requirements, package testing information, placarding and vehicle suitability/driver training arrangements.

Part 1 of *Annex A* details the scope and availability of *ADR*, along with basic definitions, general obligations and transitional measures. *Part 2* then details the general provisions and principles of hazard classification along with specific class provisions and approved test methods for classification and allocation of dangerous goods to a particular class and packing group. *Part 3* of *Annex A* within *Volume I* is a comprehensive dangerous goods list which details hazard class, packing group and other pertinent information for items falling into each specific UN number category. Its scope is the same as the ACL for UK regulations, but in addition includes details for explosives and radioactive materials (Classes 1 and 7).

Volume II begins with the conclusion of *Part 3*, detailing special provisions and exemptions for certain materials where appropriate. *Part 4* describes packing and tank provisions, which covers general provisions along with special provisions for portable tanks, intermediate bulk containers ('IBCs') and other large vessels. Special provisions for explosives, gases, organic peroxides, self reactive substances and radioactive materials are also laid out.

Part 5 describes consignment procedures including general provisions, marking and labelling requirements for packages, tanks and vehicles along with documentation requirements. *Part 6* describes detailed requirements for the construction and testing of packages, IBC's, gas cylinders and tanks, etc and the resultant UN mark for approved packaging / receptacles. *Part 7* details provisions concerning conditions of carriage, loading, unloading and handling of dangerous goods such as general provisions, mixed loading prohibitions, handling and stowage, cleaning vehicles after unloading and any additional provisions that may apply to specific items or goods.

Annex B outlines provisions concerning transport equipment and transport operations, and details requirements for vehicle crews, equipment, operation and documentation. Matters such as the provision of fire fighting equipment and spillage kits are covered, along with driver training requirements and other miscellaneous requirements to be complied with by the vehicle crew.

The *ADR* parts are broken down into chapters, sections and subsections in a logical manner that allows easy reference to information throughout both volumes of the document.

International and multi-modal context

[D0404] In order to maintain an international uniformity of approach to the transport of dangerous goods legislation, the United Nations has established guidelines on the regulation. These guidelines are published as the '*Recommendations on the Transport of Dangerous Goods: Model Regulations*' which is commonly referred to as the '*Orange Book*'. The *Orange Book* recommends example procedural systems for the classification of substances by types of danger, the identification of dangerous goods, containment system performance and use, documentation and training.

The UN Recommendations form the basis of a series of codes for the transport of dangerous goods by road, rail, sea and air. They have no legal force, but are adopted worldwide as the basis of national legislation, and are therefore of key importance to legislators in Europe and the UK. In Europe the transport of dangerous goods by road is covered by the *ADR* agreement, which covers not only EU and other parts of Europe, but stretches to North Africa and the Middle East.

In the UK, the HSE is the organisation most concerned with the regulation and enforcement of transport of dangerous goods by road and rail. Sea transport is subject to the international IMDG code which is enacted into UK law through the *Merchant Shipping (Dangerous Goods and Marine Pollution) Regulations 1990*, which are enforced by the Department of Transport. Harbour areas are regulated by the *Dangerous Substances in Harbour Areas Regulations 1987*, enforced by HSE.

For air transport the international ICAO "technical instructions" set the relevant standards, with UK enforcement by the Civil Aviation Authority.

Airlines generally work to IATA rules which are based on ICAO technical instructions. Prohibited items are much are common for air transport than transport by road or rail due to the severe consequences of any loss of containment, and certain types of goods are only regarded as dangerous for specific modes of transport. For example, magnetised materials are dangerous goods for air transport but are not regulated for road or rail.

While different modal regulations are all based upon UN Recommendations, separate specific requirements for road, rail, sea and air transport can cause the potential for conflict between various modes of transport. Differences between UK national requirements and the *ADR* requirements of international journeys can also lead to confusion about which requirements to follow. Luckily, the UK regulations generally follow the requirements of *ADR* and are flexible enough to accommodate such concerns.

UK regulations allow compliance with *ADR* for goods being transported within Britain as part of an international journey. A specific implication of international journeys ending or originating within the UK is that almost all such journeys will have to involve either travel by sea, air or channel tunnel. The requirements for sea or air transport, governed by the IMDG Code or IACO Technical Instructions, are tougher than road transport requirements. To prevent having to label goods twice, compliance with UK national road regulations is not required as long as the goods have been correctly classified, packaged and labelled in accordance with applicable air and sea regulations. In the UK, the Department for Transport is the lead department and competent authority on the transport of dangerous goods in whatever mode.

General requirements

Dangerous goods safety adviser

[D0405] Subject to some exemptions discussed below, *CDC 2007 reg 43(3)* requires organisations involved in the carriage of dangerous goods to appoint a Dangerous Goods Safety Adviser (DGSA) in line with *ADR section 1.8.3*. Unless exempted, it applies to carriers, fillers and loaders and prescribes the training and certification regime required for the DGSA and the appointment and duties. When requested by the competent or enforcing authority, the person within the organisation required to appoint a safety adviser shall provide a copy of the annual report prepared in accordance with *ADR subsection 1.8.3.3*, the identity of the DGSA in accordance with *ADR subsection 1.8.3.5*, and a copy of any accident report in compliance with *ADR subsection 1.8.3.6*.

The first exemption is where the main or secondary activity is not the carriage or the related loading of dangerous goods and where any engagement in such activity is only carried out within Great Britain occasionally and that carriage poses little or danger or risk of pollution. These terms pose practical difficulties of interpretation and if in doubt further advice should be sought. This exemption only applies to journeys in Great Britain and not to international carriage. The second exemption applies where the quantity of dangerous goods in a transport unit, wagon or large container or vessel is limited and less than that specified in *ADR subsections 1.1.3.6, 1.8.3.1* or *chapters 3.3* and *3.4*. Limited quantities of dangerous goods in small receptacles or small loads may be exempt on this basis (see [D0441] and [D0442]).

Classification and identification

CDG 2007 requirements

[D0406] *CDG 2007 reg 47* provides the basis for implementing the classification and identification aspects of *ADR* in the UK. This is the duty of the consignor, though he may have to rely on others for information and data. Basically, the consignor shall not consign dangerous goods, or goods that could potentially be dangerous, for carriage unless the hazards of the goods have been identified and the goods have been classified in accordance with the general, specific or test requirements of *ADR*. The classification and identification aspects of *ADR* are based on the United Nations (UN) system.

The United Nations classification/identification system

[D0407] The purpose of the classification is twofold; firstly, to determine which goods are dangerous during transport; secondly, to show what kind(s) of danger are to be found in a particular article or substance.

The UN Committee of Experts on the Transport of Dangerous Goods developed the system for classification. The UN guidelines, published as the '*Recommendations on the Transport of Dangerous Goods: Model Regulations*', commonly referred to as the '*Orange Book*', has a classification system that splits dangerous goods into nine hazard classes according to the kind of danger inherent to the item. Some of these classes are further broken down into smaller groupings, which are known as divisions.

Various regional and national bodies have then modified this to produce the most appropriate controls for different modes of transport. All base their controls upon the UN system of nine classes, and follow the UN system of testing for the various classes. The nine classes and their divisions within *ADR* are as shown—

Class	Goods	Division	Description	Typical technical reference
1	Explosives			Pyrotechnic; non-detonating self-sustaining exothermic chemical reaction
2	Gases	2.1	Flammable gas	Vapour pressure above 300kPa at 50°C; gaseous at 20°C at 101.3kPa.
		2.2	Non-flammable Non toxic	
		2.3	Toxic gas	
3	Flammable liquid			Initial boiling point; flashpoint.

4	Flammable solids	4.1	Flammable substance	Burning time in seconds per 100mm; rate of detonation or deflagration; self-reactive.
		4.2	Spontaneously combustible	Pyrophoric substance; self-heating substance.
		4.3	Water reactive	Water reactive; rate of evolution of flammable gas; spontaneous ignition.

Class	*Goods*	*Division*	*Description*	*Typical technical reference*
5		5.1	Oxidising substance	Increased burning rate.
		5.2	Organic peroxide	Assigned only by competent authority.
6		6.1	Toxic	LD_{50} oral; LD_{50} dermal; LC_{50} inhalation.
		6.2	Infectious	Genetically modified; biological products; diagnostic specimens.
7	Radioactive materials			Specific activity more than 70kBq/kg.
8	Corrosive			Full thickness destruction of intact skin tissue.
9	Miscellaneous			Allocated by regulatory authority.

In addition to indicating the kind of danger presented by a substance, the classification system also gives qualitative guidance as to the degree of danger of that substance in relation to other types of dangerous goods within the same class. The packing group indicates the degree of danger presented by a substance, of which there are three levels as defined below.

UN Packaging Group	*Meaning*
PG1	HIGH danger
PG2	MEDIUM danger
PG3	LOW danger

The packing group plays a wide role – from the determination of acceptable packaging and the definition of threshold limits, to stowage and segregation requirements. The packing group system does not apply to Classes 1, 2, 5.2, 6.2 or 7, as different specification criteria and special packaging requirements apply.

[D0408] In an emergency the emergency services, vehicle drivers, other transport staff and the general public need to know what dangers they face from certain substances. This creates a need for a product reference that can be quickly interpreted for an effective response.

The UN has devised 2 such references—

- Proper Shipping Names ('PSNs') – These are recommended names for a wide range of commonly moved substances, listed in the UN Orange Book. However, they are often complex chemical names and the need for the identification to be achieved on a worldwide basis led to the development of a second system.
- UN Numbers – These are four digit identification numbers directly linked to the PSN.

ADR classification procedures

[D0409] Dangerous substances (including articles) are very widely defined and classified according to the hazard they present. The rules for classification are in *ADR Part 2* with descriptions and criteria given in some detail. These largely conform to the UN system, including the nine hazard classes (see above) and the use of packing groups. *ADR* allocates a substance that does not meet the criteria of any other hazard class, but which is regarded as an 'aquatic pollutant', to Class 9 as an 'environmentally hazardous substance'. Solutions and mixtures with a concentration of 25 per cent or more of such a substance, are to be classified as such.

The Approved List

[D0410] Consignors have a duty to identify the hazards of the goods they intend to transport. They must also assign a proper shipping name and UN number to the substances being transported. The proper shipping name (the proper chemical name, not a trade-name) may also be qualified by the addition of descriptive terms such as Solution, Liquid, Molten or Solid (see *ADR subsections 3.1.2.3–3.1.2.7*).

[D0411] UK regulations and ADR are in line with the UN Recommendations and require the use of the UN Proper Shipping Name (PSN) and UN Numbers. For consignment and transport, the PSN and corresponding UN number can be found in the ADR Approved List. The Approved List is given in *Table A* of *ADR Part 3 chapter 3.2* ordered by UN Number and may be cross referenced using the PSN, or, alternatively, in the corresponding alphabetical Proper Shipping Name list in *Table B* of *chapter 3.2*.

The ADR Approved List contains a large range of substances and generic groups (eg paints) that have been previously classified and assigned packaging groups. The 'Approved List' contains lists of all categories of dangerous goods authorised for road transport. If a substance is listed by name, or its hazardous properties are described, then it is subject to control. The classification and packing group of a substance can be determined by looking up the UN number in the Approved List.

A solution or mixture that contains a dangerous substance and one or more non-dangerous substances is identified using the PSN for the listed one, with the addition of the word 'solution' or 'mixture' eg methanol solution.

This does not apply where—

- The solution or mixture is identified in the Approved List.
- The entry in the Approved List only applies to the pure substance.
- The class, physical state or packing group is different to that of the pure substance.
- There is a significant change in the measures to be taken in an emergency.

If this occurs, then the solution or mixture is identified under the appropriate generic or not otherwise specified (n.o.s.) entry in the Approved List. The index contains a range of generic names that cover substances belonging to particular families (eg Alcohols n.o.s. and Ethers n.o.s.) and also involves more descriptive names that identify their danger (eg Flammable n.o.s. and Corrosive n.o.s.).

Each entry in the Approved List provides information to be used by the consignor to fulfil classification, identification, package selection, labelling and documentation completion duties. For example—

UNNO	Proper Shipping Name	Class	CLS Code	Packing Group	Labels	Limited Quantities	Packaging		
							Packaging instructions	Special packing provisions	Special provisions
1106	Amylamine	3	FC	II	3 + 8	LQ4	P001 IBC02		MP19

UN Portable Tank Instructions		ADR Tank		Vehicle for tank carriage	Transport category	Special provisions for carriage			HIN
Instructions	Special provisions	Tank code	Special provision			Packages	Bulk	Loading, unloading and handling	Operation
T7	TP1	L4BH	TE1	FL	2			S20	S20	38

UNNO = United Nations Number
CLASS = Hazard Class
CLS Code = Hazard Classification Code
HIN = Hazard Identification Number

It is essential to check the implication of any classification codes in the table above as these can significantly affect consignor duties. These codes are described in *chapter 3.2 of ADR*. The carrier of the dangerous goods can use the Approved List to confirm the accuracy of the consignor's transport documentation, to determine the competence required by the driver carrying the load and to decide what vehicle markings are appropriate for the journey.

There is a hierarchy of classification (*ADR subsection 2.1.3.5.3*) and there are rules about choosing the most appropriate entry, and hence UN number, (*ADR section 3.1.2*).

Many preparations will not be found in *Table A of ADR*. In these cases the *ADR* rules for classification from first principles need to be followed. Sometimes items may show danger characteristics of more than one class. In this case, the material is assigned to the hazard class with the greatest risk, and the lesser risks are assigned as subsidiary hazard classes.

Packing groups

[D0412] Some substances with the same name will have different degrees of danger (for example flash point). This is reflected in the packing group which is found in column 4 of *Table A*. The relevant part of *ADR* dealing with packing groups is *subsection 2.1.1.3*. Where a substance is not listed and its classification has been determined from first principles, its packing group is determined by its properties. *ADR subsection 2.2.3.1.3* shows, for example, how unlisted flammable liquids are assigned a packing group.

Some substances (for examples gases and explosives) are not assigned a packing group. These substances do have a transport category which is relevant with regard to limited load exemptions. Where it is not practicable to fully classify a substance before transport, *ADR Section 2.1.4* allows conservative over-classification on the basis of information already available, with a more stringent packing group than would otherwise be necessary.

Controls of radioactive materials follow a significantly different pattern and refer to further sections of *ADR*.

Low quantities of dangerous goods in small receptacles or small loads may be exempt from some or all of these regulations under limited quantities concessions (see [D0441] and [D0442]).

Packaging and containment

CDG 2007 requirements

[D0413] *CDG 2007* requirements for implementing the packaging parts of ADR and, in consequence, UN specification, are given in *CDG 2007 reg 51*. The duty, as far as reasonably practicable, lies with the consignor and the packer, who may be the same person. *Regulation 93* allows a "due diligence" defence if it can be shown that the consignor and packer acted in good faith.

ADR requirements

[D0414] Within the *ADR* agreement, the *General Packaging Conditions* and the detail of UN specification are in *Part 4*. *Chapter 4.1* covers the use of packaging, intermediate bulk containers and large packaging, while *chapters 4.2–4.5* cover various sorts of tanks. Class specific provisions are detailed in *Part 2*. The particular packaging requirements of each package type are identified in specific subchapters within *Part 4*.

The Approved List allows all details of permitted packaging and containment to be accessed. Each packing instruction is identified in the Approved List and describes a range of packing options for substances depending upon their Packing Groups. These must be carefully examined in order to determine any volume or mass limits that may apply to a particular package type.

Requirements for packages

[D0415] All packages and containment used for the transport of dangerous goods have to meet three criteria. In particular they must—

- Meet the general requirements established by the regulation.
- Be authorised to UN specification standards.
- Meet the detailed directions within the applicable regulation for the packaging of the product.

All packages that meet these criteria are available for use to carry the product in the particular transport system. This means that, for most substances and modes of transport, a wide range of package types and materials are available. However, in the worst-case scenario only one system of containment will be authorised for transport use.

UN specification sets the standards for packages used in the transportation of dangerous goods and these are translated in *ADR Part 6* into detailed requirements for the construction and testing of packaging and tanks. The UN specification requires packaging design and materials to have proven their competence to a national competent authority by passing practical tests. Firstly the packs are weathered to their most vulnerable condition if this is likely to be a factor in their performance. They are then subjected to a number of physical tests such as being dropped, held in a stack and placed under pressure demands to simulate transport situations.

The basic set of standards in line with the UN specifications that must be met by every package is wide ranging. However, there are some additional requirements in *ADR* and *CDG 2007*, such as the compatibility of the product and packaging material, the provision of ullage and the inspection of all packages before transport.

Approved packages are given markings that are clear, legible and readily visible. Such packages are marked in particular ways, prefixed by the UN logo and followed by codes, as detailed in *ADR Part 6*, which details the significance of the packaging code with examples (*section 6.1.3*). As an example consider the packaging code shown below—

UN 1A1/X/550/095/GB/0103

UN = UN packaging symbol

1A1 = Type of packaging and material

X = Packing Group competence and maximum density of liquids

550 = Hydraulic pressure test result (kPa)

95 = Year of manufacture

GB = Authorising country

0103 = Packaging ID number

An example of a marking as it may appear on a fibreboard (cardboard) package is shown below.

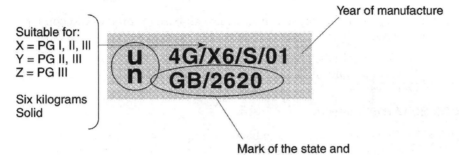

Year of manufacture

Suitable for:
X = PG I, II, III
Y = PG II, III
Z = PG III

Six kilograms
Solid

Mark of the state and manufacturer

Requirements for Intermediate bulk container and tank systems

[D0416] Intermediate bulk containers ('IBCs') are designed for mechanical handling and have a general maximum capacity of 3m³ (3,000 litres). The selection procedures are similar to those for packages. IBCs that are acceptable packaging options in *ADR* are listed in *Volume II Ch 4.1*.

There are many types of tank used in the transport of dangerous goods, but they all are all built to tightly controlled criteria. A range of descriptions are used to identify the different types, but unfortunately, terminology is not consistent across the various transport modes. In each modal control, definitions will be found indicating what kind of equipment is being referred to by any particular name.

The various controls offer two avenues of control of tank traffic—

* Identification of how tanks must be built and equipped to meet the demands of the regulation
* Description of the type tank and ancillary equipment that must be used for the transport of a particular substance

ADR Ch 4.2–4.5 contain constructional and equipment details for the various types of *ADR* approved tanks, which detail requirements for portable tanks, fixed tanks, demountable tanks, tank containers, tank swap bodies, reinforced plastic tanks, welded tanks and vacuum waste tanks.

Consignment procedures

Basic principles

[D0417] To ensure that everyone involved in handling or moving dangerous goods are aware of the hazards, information warning of the dangers are required to be clearly displayed on the outside of packages, tanks, vehicles, etc. This is also vital for rapid and effective action from the emergency response teams faced with dangerous goods incidents.

Terminology is not yet fully harmonised and a distinction is drawn between packages, tanks and transport units (vehicles)—

- Package: drums, jerricans, boxes, gas cylinders, IBCs, etc.
- Transport unit: road freight or road tanker vehicles, rail wagons or rail tanks, multi modal freight containers and tank containers.

Requirements for all modes of transport are all based on the UN Recommendations, the Orange Book, which identify the need for—

- Marking and labelling on packages.
- Placarding and plating of tanks, containers, and transport units (vehicles)

CDG 2007 requirements

[D0418] *CDG 2007 reg 57* is the basis for implementing the requirements of *ADR* in respect of matters immediately prior to the transportation taking place. These include the marking and labelling of packages, and the placarding and marking (plating) of tanks, containers and vehicles. It also covers the documentation to be provided with the goods, matters connected with the use of overpacks and radioactive materials.

There are specific extra requirements for require GB registered vehicles carrying tanks or tank containers on GB domestic journeys *CDG 2007 reg 91* and the associated Schedule. These are discussed under paragraph D0422 below. They are intended to retain elements of good custom and practice established by industry in GB.

Depending on circumstances, the duties under *reg 57* can fall on the packer, loader, consignor or the carrier, which may be separate individuals or organisations. Marking and labelling of packages is typically the duty of packers and the consignor. Placarding and marking of containers, tanks and vehicles is typically the duty of the loader, consignor and carrier. Documentation is the duty of the consignor and the carrier. The consignor should exercise overall control over these aspects, while the carrier must ensure that the required documentation is carried and that the vehicle carries the necessary placards.

ADR requirements

[D0419] The detailed requirements under *ADR* are given in *Part 5*, with some linked requirements in *Part 8*. The *ADR* requirements for marking and labelling also apply to rail (*RID*).

Marking and labelling of packages

[D0420] Marking and labelling both apply to packages and are covered in *ADR Ch 5.2*. Every package must be marked and labelled as follows. The marks and labels required for certain products can be established through the relevant substance entries in the first six columns of the Approved List. Special packing provisions apply variations to the standard marking and labelling. Marking and labelling of packages mean something different as follows.

Marking on packages (*ADR section 5.2.1*) is the application of the UN Number and complete Proper Shipping Name ('PSN') in a readily visible and legible location on the outside of all packages. Other information is also required to be marked for packages containing explosives, gases and radioactive materials (hazard classes 1, 2, and 7). In addition, the UN package certification details are also to be marked. An example of package marking is shown in the figure below—

S28555/   2

Shipping Label for Fisher Scientific

DICHLOROMETHANE

UN 1593

Technical Name:

Fisher Item: D/1856/17
2.5L DICHLOROMETHANE HPLC

Supplied by:
Fisher Scientific UK Ltd., Bishop Meadow Road
Loughborough, Leics, LE11 5RG

Chemical Class   –   6.1   Poisonous
Sub Class 1   –
Sub Class 2   –

Labelling, described in *ADR section 5.2.2*, is the application of diamond shaped warning labels representing both the class of primary danger and any subsidiary risks. Class labels are distinguished by the appropriate class number in the bottom corner, with the upper half containing a pictorial symbol representing the danger. Text indicating the nature of the hazard eg 'Corrosive' may also be shown in the lower half – this is mandatory for Class 7, radioactive materials. Division numbers must also be shown on Class 1 and Class 5 labels.

The standard size for such labels is 100mm x 100mm, with a border line of 5mm. Symbols, text and numbers must be in black, except those containing the Class 8, corrosive, number and any associated text, which must be in white. Each class is identified by means of a diamond label that indicates the nature of the hazard on packaging and vehicles carrying dangerous goods. These are reproduced in the figure below—

Class 1
Explosives

Class 3
Flammable
Liquids

Class 3.2
Organic Peroxides

Class 4.1
Flammable
Solids

Class 4.2
Spontaneously
Combustible

Class 4.3
Water Reactive

Class 5.1
Oxidising
Substances

Class 6.1
Toxic Substances

Class 6.2
Infectious
Substances

Class 7
Radioactive
Material

Class 8
Corrosive
Substances

Class 9
Miscellaneous
Items

Limited quantity concessions of dangerous goods are not subject to the standard marking and labelling duties. The regulations for limited quantities are not yet aligned for different modes of transport and therefore the application varies depending on the type of journey.

Placarding and marking (plating) of vehicles, tanks or containers etc

[D4021] Road vehicles, tanks and containers and other types of transport units must bear the appropriate placards, plates and markings to warn of the dangers of the goods being transported (*ADR Ch 5.3*).

Placards, described in *ADR section 5.3.1*, are larger versions of the diamond hazard warning labels used for packages, and are placed on the tank or container carrying the goods. Generally placards must be at least 250 x 2,500mm with a border line of 12.5mm (*ADR 5.3.1.7*). Smaller placards can be used for small tanks and containers (*ADR subsection 5.3.1.7.3*).

Placards also have to be displayed on the vehicle according to the type of load (*ADR subsections 5.3.1.2–5.3.1.6*). For example, vehicles carrying packages containing explosive (Class 1) or radioactive (Class 7) substances must display placards on both sides and the rear of the vehicle. Where such packages are carried in freight containers, the container should display the relevant placards on all four sides. There is no requirement in the UK to placard vehicles carrying packages containing other Classes of substance.

Marking (otherwise plating) is described in *ADR section 5.3.2* refers to the plain orange plates carried at the front of vehicles (and on the back of vehicles carrying hazardous packages). Sometimes the plain orange plate can be divided by a horizontal black line and shows the UN and hazard identification number or emergency action code (see *ADR subsection 5.3.2.2.1* or *CDG 2007 reg 91*). Marking also applies to tanks and containers etc, and to other marks on the sides and backs of vehicles.

Examples of marking of UK vehicles carrying packages and a container are shown below—

Truck marking
Package load

No danger label on the truck

Container

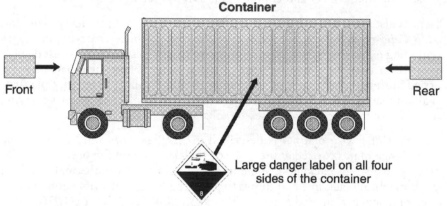

Large danger label on all four
sides of the container

For certain journeys specific markings are required, the type, number and location of these vary depending on the transport mode and the nature of the journey – national or international. For example, the relevant UN number must be displayed on the outside of tank transport units. An elevated temperature mark is required for those units transporting liquids at temperatures of 100°C or more, or solids at 250°C or more. An emergency action code ('EAC'), which provides guidance for the emergency services in the event of an incident involving the load, along with other markings may also be needed for tank or bulk load transport, depending on the type of journey.

GB domestic and international journeys

[D0422] *Regulation 91* and the associated Schedule of *CDG 2007* require GB registered vehicles carrying tanks or tank containers on GB domestic journeys to be marked with "Emergency Action Codes (EAC)" (sometimes called "Hazchem codes"), and to include a telephone number for advice in the event of emergency. This is usually in the form of an integrated hazard warning panel, displaying the UN number, the EAC and telephone number, and the hazard placard, placed on both the sides and rear of the vehicle. This is in addition to the plain orange marking plate at the front of the vehicle.

Vehicles carrying tanks or tank containers on international journeys (defined at *ADR subsection 1.1.2.4*) must carry a Hazard Identification Number (HIN - sometimes called the Kernier code) as shown at *ADR subsection 5.3.2.2.3*. The orange marking plate now has a black border and is divided by a horizontal black line, above which is the HIN and below which is the UN number of the substance. These plates are placed at the rear and on both sides of vehicles, with the plain orange plate at the front, and are in addition to the diamond hazard placards. There is no requirement to display an emergency telephone number, although this would always be good practice.

An example of *ADR* road tanker vehicle marking and placarding is given below—

Road tanker

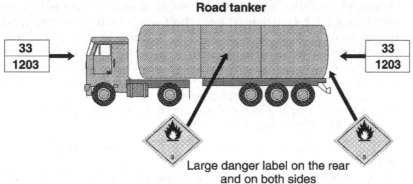

| 33 |
| 1203 |

| 33 |
| 1203 |

Large danger label on the rear
and on both sides

Documentation requirements

General

[D0423] Documentation relating to a dangerous goods shipment describes important aspects of the cargo and its containment as well as indicating its point of origin and eventual destination. This information is required to ensure that all individuals involved in the transport process are fully aware of the dangers of the material that they are handling and employ suitable equipment and procedures to minimise any risks involved. The documentation also provides vital information in a concise form to emergency personnel. Description of packages etc should be as clear as possible to enable rapid identification, and substance descriptions must conform exactly to the requirements of the regulations in order to minimise misunderstanding and errors during an emergency.

Transport documentation

[D0424] With the exception of goods carried under limited quantities exemptions, the consignor must under *CDG 2007*, provide each shipment with documentation according to *ADR subsection 5.4.1.1* detailing—

(1)   The UN Number, preceded by the letters 'UN'
(2)   The proper shipping name, (the designation of the goods along with the designation 'waste' if applicable)
(3)   The hazard class of each item of goods preceded by the word 'Class' (with subsidiary hazard, if any, in brackets)

(4)   The packing group (where assigned)

(5)   The number and description of packages

(6)   The total quantity (mass/volume) of each item of different UN number

(7)   The consignor name and address

(8)   The consignee names and addresses (where there are multiple consign-ees not known at the start of the journey the words "Delivery Sale" may be used

(9)   Any other information necessary for the vehicle operator to comply with their transport documentation requirements.

(10) The emergency action code

(11) The prescribed temperature (where applicable)

(12) A container packing certificate

On the documentation, the first four items must appear in the order given (*ADR subsection 5.1.1.1.1*), but apart from this there is no requirement for all information to be on one document. There are special rules about documentation for wastes, salvage packaging, empty uncleaned packaging (*ADR subsections 5.4.1.1.3–5.4.1.1.6*) and for tanks and bulk (*ADR subsection 5.4.1.1.6*). There are other rules given under *ADR subsection 5.4.1.2* for class 1 (explosives), class 2 (gases and particularly gas mixtures), class 4.1 (flammable solids), class 5.2 (organic peroxides), class 6.2 infectious substances and class 7 (radioactive). Other special rules cover—

- Loads in a transport chain that includes air or sea (*ADR subsection 5.4.1.1.7*)
- Carriage in date expired IBCs (*ADR subsections 5.4.1.1.11, 4.1.2.2*)
- Multi compartment tanks or transport units with more than one tank (*ADR subsection 5.4.1.1.13*)
- Elevated temperature substances (*ADR subsection 5.4.1.1.14*)
- Substances stabilised by temperature control (*ADR subsection 5.4.1.1.15*)

Transport documentation for explosives must detail the total number of packages carried, the total net mass of explosives carried and the name and address of the carrier.

The language of the documentation should be that of the forwarding country and one of English, French or German if not already on the document.

The requirement to carry emergency information and instructions in writing is separate from the goods documentation and is covered under Crew and Vehicle below.

ADR documentation

[D0425] The ADR Agreement imposes specific duties relating to the provision and carriage of documentation when goods are moved by roads on international journeys. Similar to UK requirements the consignor, vehicle operator and driver all have duties under the provisions of ADR.

The consignor has a duty to provide a transport document (information on the goods consigned), a declaration of compliance with ADR, written emergency instructions (actions to be implemented in the event of an accident) and a container packing certificate (only needed if the goods are loaded onto a freight container exceeding 3m^3 capacity) to the vehicle operator.

It is the vehicle operator's responsibility to ensure that the appropriate documents are carried. These could include—

- The transport document and declaration of compliance;
- A container packing certificate, a vehicle approval certificate;
- The driver training certificate;
- The written emergency instructions; and
- A journey authorisation permit.

It is the driver's duty to ensure that the relevant documents are kept readily available and that any irrelevant documents are kept separate and clearly identified to avoid confusion.

Carriage, loading, unloading and handling

CDG 2007 requirements

[D0426] *CDG 2007 reg 62* implements the requirements of *ADR Part 7*, which covers all operations involving handling, loading, carriage and unloading. *ADR Part 7* makes particular provisions for packages, bulk product, tanks and tankers. More general duties in *ADR Part 1* to safeguard any immediate risk to the general public and to stop the journey if there is an infringement that could threaten safety are implemented through *CDG 2007 reg 39*. All these duties fall on the carrier or filler or loader depending on circumstances.

ADR requirements

[D0427] The relevant sections are—

- *ADR 7.1* – general provisions
- *ADR 7.2* – special packaging provisions
- *ADR 7.3* – special bulk carriage provisions
- *ADR 7.4* – special carriage in tanks provisions
- *ADR 7.5* – covers mixed loading, foodstuffs, limited quantities (mainly explosives and organic peroxides), handling and stowage, cleaning after packages have leaked, prohibition of smoking, precautions against electro-static discharge (when carrying substances with a flash point lower than 61°C), and special provisions "CV" specific to particular substances or groups of substances.

The special provisions for packages, bulk carriage and tanks are contained in *ADR chapters 7.1, 7.2* and *7.3* and require reference to the relevant column of *ADR Table A*.

Filling and loading

[D0428] All persons have a duty to ensure that the manner in which goods are loaded, filled or unloaded is not liable to create a significant risk or significantly increase any existing risk to the health and safety of any person. ADR makes provisions for particular hazard classes of dangerous goods. For example, where flammable goods are being transported by tank, the chassis must be properly earthed and flow rates controlled to prevent static discharges. It also provides more general safety requirements, such as prohibiting smoking in the vicinity of vehicles whilst loading and unloading (*ADR section 7.5.9*), and also cleaning tanks and freight containers after use (*ADR section 7.5.8*).

Stowage

[D0429] There are particular stowage requirements for certain dangerous goods. These requirements, detailed in *ADR section 7.5.7*, provide information on the various classes of dangerous goods. For example, flammable solids must be loaded in a fashion that ensures free air circulation and the maintenance of a uniform temperature. There is also general duty to ensure that all dangerous goods are suitably stowed and secured on vehicles to prevent load shifting during transit. This sets a high standard and supplements more general laws on the stowage of loads on goods vehicles.

Segregation

[D0430] From time to time a road traffic accident may occur, a tank valve may fail, or a fire may break out. These incidents could result in two substances reacting dangerously if they came into direct contact with each other. Segregation ensures that acceptable levels of safety are maintained in the transport system, in case one of these events does occur.

Dangerous goods must not be packed together in the same outer packaging with dangerous or other goods if there is a possibility that they will react dangerously together to cause: combustion and/or the evolution of considerable heat; the evolution of flammable, toxic or asphyxiant gases; or the formation of corrosive or unstable substances. Toxic or infectious substances must not be carried on the same vehicle as food, unless they can be 'effectively separated'.

It is the responsibility of the carrier to assess whether the loading arrangements are satisfactory. The circumstances under which the mixed packing of substances in a particular class are permitted, either with other items from the same class, or with products from other classes, are laid out in *ADR chapter 5.2* and *section 7.5.2*.

Crew and vehicle

CDG 2007

[D0431] This part covers aspects that are generally the responsibility of the carrier and crew and includes driver training, equipment to be carried on board, requirements related to the transport unit and documents to be carried, including emergency instructions. *CDG 2007 regs 63, 64* are the basis for implementing *ADR Part 8*. *Regulation 63* deals with the carrier's duty to limit the transport to one trailer and carry fire extinguishers and other equipment as needed, and with the carrier's and crew's duty to carry the transport documents, while *regulation 64* deals with driver training. *Regulation 65* is concerned with the construction and approval of vehicles for carriage by road and implements *ADR Part 9*.

UK and ADR vehicle requirements

[D0432] The vehicle and its ancillary equipment is an integral part of the containment system, which must be effective in order to allow the goods to be carried in a safe manner.

[D0433] A range of requirements relating to the vehicle and its load carrying competence are laid down in the ADR Agreement. The information is presented at two levels: a set of duties that underpin all decision making on vehicle provision, equipment and use; and another set of directions related to each danger class and in some cases to specific item numbers within a class.

ADR Part 9 contains a number of general requirements regarding the physical competence of the vehicle and deals with any extra provisions or variations particular to a class, or a certain substance within a class. The degree of supplementary information varies between classes and covers a range of issues. For example, Class 1 has the greatest level of detail, ranging from vehicle construction requirements, to bodywork and exhaust systems, where as Class 4 provisions cover aspects such as temperature control and the use of closed or sheeted vehicles. *Part 9* should therefore always be checked to confirm any additional requirements.

As a general requirement, a check is to be carried out on the vehicle, its equipment and the driver's documentation before loading the vehicle, to ensure that the journey can be undertaken in a compliant manner.

Operational procedures

[D0434] The vehicle crew (driver, assistant(s), and attendant(s)) is the primary channel of safety during road journeys involving the carriage of dangerous goods. The working procedures, duties, training and instruction of the crew are both a safeguard against the likelihood of an incident occurring involving their hazardous load and a primary means of minimising the consequences of any such incident.

Driver training

[D0435] When carrying dangerous goods, special training and certification requirements apply to many drivers. These requirements are covered in *ADR chapter 8.2*. The term vocational training certificate ('VTC') is no longer used and has been replaced with *ADR Training Certificate* consistent with the wording of *ADR subsection 8.2.2.8* and the model certificate shown at *ADR subsection 8.2.2.8.3*. The *ADR* training requirements, which applied from 1 January 2007, are summarised in the table below.

Vehicle/load	Driver training	ADR reference
All vehicles carrying dangerous goods except those carrying packages under the small load limit	General training plus an ADR training certificate for the class of goods and mode of carriage	8.2.1
Any vehicle carrying packaged dangerous goods under the small load limit	General training	8.2.3, 1.3
Vehicle with a small tank (up to 1m^3)	General training	8.2.1.3, 8.2.3

The general requirements for driver training are detailed in *ADR chapter 1.3*. Carriers must keep a record of training of their drivers as required by *ADR section 1.3.3.* Under *CDG 2007* drivers are required to carry their training certificates (see *reg 24(6)*). These can be obtained by attending a government approved training course, (information may be found on the Department for Transport website) and passing the examinations.

HSE inspectors and the police are entitled to inspect a driver's training certificate under *CDG 2007 reg 63(7)*. They will ensure that the certificate is in the same name as the drivers licence, the expiry date has not passed, the certificate is valid for the class of goods being carried and for the mode of carriage (eg tanks/other than tanks) and, where possible, that the certificate number matches that on the drivers licence.

Duties of the crew

[D0436] There are various duties which must be carried out by the vehicle crew. In particular, the driver must—

- Ensure that any relevant transport documentation is kept readily available.
- Ensure that any relevant placards and markings required for the journey are displayed and removed when not relevant.
- If the load exceeds a specified limit then the special supervision and parking requirements must be satisfied.
- Ensure that any segregation, stowage or load safety instructions are followed.
- Ensure that the crew complies with any emergency information relating to the goods in the event of an accident and also ensure that the appropriate emergency services are contacted.
- Ensure that steps are taken to minimise the risk of fire or explosion.
- Ensure that the engine is shut off during loading and unloading.

Other duties laid down for the vehicle crew are contained in *ADR chapter 7.5* 'Special Provisions Concerning Loading, Unloading and Handling'. These duties state that—

- If the load exceeds a specified limit then the vehicle must be supervised at all times unless it is in a secure factory or depot site.
- Crew members are the only personnel permitted to be carried on dangerous goods journeys.
- Crew members must know how to operate their fire extinguishers.

Vehicle equipment requirements

[D0437] *ADR section 8.1.5* details the vehicle equipment that must be carried on journeys involving dangerous goods. Vehicle equipment requirements on dangerous goods journeys include—

- At least one wheel scotch (information about the suitability of wheel chocks can be found on the VOSA website).
- Two self standing warning signs
- A suitable warning vest or warning clothing for each crew member
- One intrinsically safe handlamp for each crew member (Note the handlamp may have to be suitable for use in a flammable atmosphere in certain circumstances.)

- A general or load-specific safety kit including PPE as listed on the written 'emergency instructions and information', eg where toxic gases are being carried, the crew must be provided with suitable respiratory equipment

New specific requirements for the provision of fire extinguishers have been in force since 2008. The table below sets out the current minimum requirements.

Vehicle	Minimum dry power fire extinguisher provision
Up to 3.5 tonne	2kg for cab plus another 2kg elsewhere
Over 3.5 tonne	2kg for cab plus another 6kg elsewhere
Any vehicle carrying dangerous goods under the small load limit or carrying infectious substances	2kg for cab

Emergency information and other documents

[D0438] The Dangerous Goods Safety Advisor ('DGSA') must ensure the provision of proper emergency procedures and emergency information, and monitor their implementation in the event of any accident or incident that may affect the safety during the transport of dangerous goods.

CDG 2007 through *ADR* requires two types of documentation to be carried. The Transport Documents concerning the load are covered in paragraph [D0424] above. In addition, "Instructions in Writing containing emergency information must also be carried.

Emergency information, usually in the form of a Tremcard, comprises details of the measures to be taken by the driver in the event of an accident or an emergency and other safety information regarding the goods being carried. So long as the information that is set out in *ADR section 5.4.3* is carried and is valid for the substances being transported, then the requirement is satisfied. This information covers areas such as inherent dangers, measures to be taken if someone comes in to contact with the substance, measures to deal with a spillage and measures to be taken in the event of a fire. Additional requirements for specific types of material are given in *ADR 7.5.11*.

Under the provisions of the *CDG 2007*, the driver of the vehicle must ensure that any documentation relevant to the load being carried is kept readily available on the vehicle, and must produce the documentation on request to any authorised person. The driver must also ensure that any documentation relating to dangerous goods not being carried is removed from the vehicle or is secured in a closed container and clearly marked to show that such information inside does not relate to the load currently on board. Once a journey is complete, the vehicle operator must maintain copies of transport documentation (but not emergency information) for three months.

Exemptions, revocations and radioactive materials

Types of exemptions

[D0439] Exemptions from some or all of the *CDG 2007* and *ADR* regulations can arise in three ways: either within *ADR* itself, or through provisions in *CDG 2007* or by Authorisation of the Secretary of State for Transport, which are applicable only to Great Britain.

ADR exemptions

[D0440] The exemptions in *ADR* are given in *section 1.1.3*. The main ones are summarised in the following list. For full detail, *ADR section 1.1.3* should be consulted.

(a)   Private use of vehicles
(b)   Carriage of machinery which happens to contain dangerous goods
(c)   Carriage that is ancillary to the main activity - this is not easy to define
(d)   Dangerous goods for use with some machine or process operated on arrival - resupply of the machine or process is not exempt under this exemption
(e)   Carriage under the control of the emergency services
(f)   Emergency transport intended to save life or protect the environment
(g)   Uncleaned empty 'static' storage vessels that have contained certain gases, flammable liquids or class 9 (miscellaneous) substances. In this context static means not designed for the transport of dangerous goods
(h)   Some carriage of gases (*ADR subsection 1.1.3.2*) and liquid fuels (*ADR 1.1.3.3*)
(i)   Empty uncleaned packaging (*ADR subsection 1.1.3.5*), but not everything is exempt
(j)   Limited quantity packages containing small receptacles
(k)   Small load exemptions

The last two are probably the most frequently encountered and are explained as follows.

Limited quantity packages exemptions

[D0441] Often dangerous goods are transported in such small receptacles within a package that there is no significant risk from them in the event of a serious incident. This exemption provides that, depending on the nature of the goods, small receptacles packed within a box or on a shrink wrapped tray are subject to either simplified dangerous goods controls, or, in certain circumstances, are exempt from the scope of the regulations altogether. This means that even if the substance is dangerous, the quantity in an individual receptacle is so small that the package of receptacles presents no particular danger for transport and is not considered to be a regulated package. The principle is illustrated below—

Limited quantity – concessionary package. Full scope of Regulations does not apply

Not limited quantity – regulated load. Full scope of Regulations applies

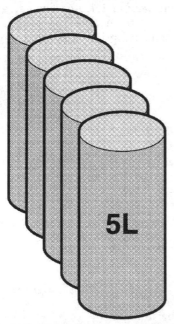

The definition of what is deemed to be a small receptacle varies greatly from class to class and packing group to packing group. Careful attention to detail is required to enable consignors to take advantage of limited quantity exemptions. Limited quantity concessions are available for shipments placed in combination packs, and there is usually a maximum weight specified for the package as well as a maximum size for the receptacles within the package.

All persons have a duty to ensure that the manner in which goods are loaded, filled or unloaded is not liable to create a significant risk or significantly increase any existing risk to the health and safety of any person. ADR makes provisions for particular hazard classes of dangerous goods. For example, where flammable goods are being transported by tank, the chassis must be properly earthed and flow rates controlled to prevent static discharges. It also provides more general safety requirements, such as prohibiting smoking in the vicinity of vehicles whilst loading and unloading (*ADR section 7.5.9*), and also cleaning tanks and freight containers after use (*ADR section 7.5.8*).

The symbol for limited quantities of dangerous goods is a diamond (at least 100mm x 100mm) with the UN number within it where there is just one substance in a package, or, where there is more than one substance, all the UN numbers or the letters LQ within it. This must appear on the package and sometimes on the vehicle. Precise details are given at *ADR section 3.4.4(c)*.

Small load exemptions

[D0442] Small load exemptions relate to the total quantity of dangerous goods carried in packages by the vehicle (or trailer). The transport category of the dangerous goods determines the load limits for small load exemption. The transport categories are allocated on the basis of packing groups and class specific allocation in accordance with *ADR subsection 1.1.3.6*. The category for each specific material is listed in column 15 of Table A, (the Approved List), and are summarised below: Note that small load exemption does not apply to tankers or bulk carriage.

Transport category		Maximum mass/ volume threshold per transport unit
0	Specific materials listed in *ADR Table 1.1.3.6.3* and in the Approved List	0
1	Packing Group I goods (unless otherwise specified) Specific materials listed in *ADR Table 1.1.3.6.3* and in the Approved List	20 litre/kilogram
2	Packing Group II goods (unless otherwise specified) Specific materials listed in *ADR Table 1.1.3.6.3* and in the Approved List	333 litre/kilogram
3	Packing Group III goods (unless otherwise specified) Specific materials listed in *ADR Table 1.1.3.6.3* and in the Approved List and any other dangerous goods not listed in other categories.	1000 litre/kilogram
4	Specific materials listed in *ADR Table 1.1.3.6.3* and in the Approved List. These materials are effectively exempt from *ADR* controls	Unlimited

If a vehicle is carrying goods under the small load threshold many of the requirements of *ADR* are not applicable. However, some are still applicable, depending on circumstances, and it is necessary to consult the relevant part of *ADR* for precise details. The relevant parts of *ADR* are *chapter 5.3* (placarding and marking), *section 5.4.3* (instructions in writing - emergency information), *chapter 7.2* (details of package requirements), *section 7.5.11* (prohibition of loading/unloading in a public place), *Part 8* (vehicle crews, equipment, documentation, operation and driver training), and *Part 9* (construction and approval of vehicles). In most cases the obligatory remaining requirements are—

- General driver training (*ADR section 1.3.2*) and training record (*ADR section 1.3.3*)
- Carrying one 2kg powder fire extinguisher or equivalent (*ADR subsection 8.1.4.2*)
- Proper stowage of the goods (*ADR section 7.5.7*)

It should be noted that any packages that meet the limited quantity criteria are not counted for the purposes of assessing small load exemption. Packages counting towards small load exception have to comply with the relevant standards. The use of small load exemption is optional and carriers may still wish to display the orange warning plates, provided the vehicle is actually carrying some dangerous goods. Orange plates must be removed when no dangerous goods are being carried.

CDG 2007 exemptions

Specific exemptions

[D0443] *CDG 2007* contains a number of further exemptions from *ADR* which are only applicable within Great Britain. Those most frequently encountered relate to the topics in the list below. For details and limitations the regulations themselves should be consulted.

- Type of vehicle (*reg 12(4)*) Includes mobile machinery (not defined but may cover special road construction vehicles such as used for white lining), and agricultural or forestry tractors.
- Movements within premises (*reg 13*).
- Movements between nearby premises and between premises and vehicles (*reg 14*) Except for explosive this provides exemption for movements between private premises and vehicles in the 'immediate vicinity' for which HSE has made an interpretation pending a court decision.
- Explosives – domestic transport (*reg 19*). These effectively maintain existing arrangements and are for domestic transport only.
- Retail distribution (*reg 26*) Allows dangerous goods which have been packaged as limited quantities to be removed from their outer packaging and carried from the distribution depot to the retail outlet (and back) without the packaging having the be marked with UN certification or the hazard symbols. The Department for Transport has issued *Guidance Note 7* which explains this in more detail.
- Documentation (*reg 30*) For small loads there is no need to carry the documentation (except for explosives and radioactive materials).
- Appointment of dangerous goods safety adviser (*reg 43(1)*) This has been discussed above [D0405].

Exemption by "authorisation"

[D0444] *CDG 2007 reg 9* empowers the Health and Safety Executive (in respect of explosives) or the Secretary of State for Transport (in respect of other classes of goods) to authorise the carriage of dangerous goods contrary to the prohibitions or requirements of the regulations for the carriage of dangerous goods. (The power does not apply to transportable pressure equipment.) Authorisations are added or deleted as the need arises or recedes, and they are all time limited. Authorisations granted under the 2004 regulations continue to apply.

A complete list of current authorisations is available on the Department for Transport website as pdf files. Some cover very special situations. Those of most general importance are No 1 (bowsers used for diesel) and No 24 (the provision of just one wheel chock for GB registered vehicles on domestic journeys).

Revocations

[D0445] The following Regulations were completely revoked by the *CDGTPE Regulations 2004 (SI 2004 No 568)*—

- the *Gas Cylinders (Pattern Approval) Regulations 1987 (SI 1987 No 116)*;
- the *Pressure Vessels (Verification) Regulations 1988 (SI 1988 No 896)*;
- the *Packaging of Explosives for Carriage Regulations 1991 (SI 1991 No 2097)*;
- the *Carriage of Dangerous Goods (Classification, Packaging and Labelling) and Use of Transportable Pressure Receptacles Regulations 1996 (SI 1996 No 2092)*;
- the *Carriage of Explosives by Road Regulations 1996 (SI 1996 No 2093)*;
- the *Carriage of Dangerous Goods by Road (Driver Training) Regulations 1996 (SI 1996 No 2094)*;
- the *Carriage of Dangerous Goods (Amendment) Regulations 1998 (SI 1998 No 2885)*;
- the *Carriage of Dangerous Goods (Amendment) Regulations 1999 (SI 1999 No 303)*;
- the *Transport of Dangerous Goods (Safety Advisers) Regulations 1999 (SI 1999 No 257)*;
- the *Transportable Pressure Vessels Regulations 2001 (SI 2001 No 1426)*;
- the *Carriage of Dangerous Goods and Transportable Pressure Vessels (Amendment) Regulations 2003 (SI 2003 No 1431)*

The following regulations were completely revoked by CDG 2007—

- the *Radioactive Material (Road Transport) Regulations 2002 (SI 2002 No 1093)*
- the *Radioactive Material (Road Transport) (Amendment) Regulations 2003 (SI 2003 No 1867)*
- the *Carriage of Dangerous Goods and Use of Transportable Pressure Equipment Regulations 2004 (SI 2004 No 568)*

Radioactive materials

CDG 2007 requirements

[D0446] As noted above, CDG 2007 includes radioactive substances (hazard Class 7) and revokes the *Radioactive Material (Road Transport) Regulations 2002 and Amendment 2003*. In CDG 2007 the particular regulations for radioactive materials are *regulation 31* (exemption from the requirements relating to fire fighting equipment), *regulation 42* (requirements relating to carriage) and *regulation 57* (construction, testing and approval of special form radioactive material, low dispersible radioactive material packages and packaging). There are, however, specific requirements relating to radioactive materials within the general requirements.

CDG 2007 refers to the detailed requirements of *ADR* relating to radioactive materials. In particular, it refers to the need for a radiation protection programme (*ADR section 1.7.2*) and a quality assurance programme (*ADR section 1.7.3*), and also the provisions for special arrangements (*ADR 1.7.4*), subsidiary risk (*ADR section 1.7.5*), and non compliance with applicable radiation or contamination levels (*ADR section 1.7.6*). *Regulation 57* implements the requirements of *ADR chapter 6.4* relating to the design, construction, inspection and marking of packages containing radioactive materials. The enforcement of the legislation dealing with the carriage of radioactive materials is the responsibility of the Department of Transport's radioactive materials team.

ADR requirements

[D0447] The *ADR* requirements derive from requirements set by the International Atomic Energy Authority (IAEA). General consignor duties are outlined in *ADR Part 7*, which provides the general requirements for transporting radioactive materials. These focus upon the need for the containment of radioactive contents, control of external radiation levels, prevention of criticality and heat damage. *ADR chapter 6.4* covers the requirements for the construction, testing and approval of packages and materials for containing radioactive materials.

In addition to radioactive and fissile properties, any subsidiary hazards of dangerous goods items such as flammability, corrosivity, etc, must be taken into account in the documentation, packaging, labelling, marking, placarding, stowage, segregation and carriage requirements.

Acknowledgement

[D0448] The information presented in this chapter has been derived from *CDG 2007* and supporting information issued by the Health and Safety Executive, in particular the *Carriage of Dangerous Goods Manual* (October 2007 as amended April 2008), to which acknowledgement is made. It is a compilation and summary of complex regulations and is therefore not complete in all detail. For full detail and clarification, the reader is strongly advised to consult the regulations themselves (*CDG 2007* and *ADR*), the Health and Safety Executive or the Department for Transport and their publications.

Dangerous Substances and Explosive Atmospheres

John Wintle

Introduction

Background

[D0901] Until the *Dangerous Substances and Explosive Atmospheres Regulations 2002 (SI 2002 No 2776) ('DSEAR')* came into force in December 2002, the regulation of substances that could give rise to the risks of fire and explosion had built up incrementally for different types of substances in different industries. There were separate regulations for highly flammable liquids and liquefied petroleum gases, celluloid in factories and workshops, dry cleaning, cinematograph film manufacture and petroleum. However, other substances and activities giving rise to these risks, such as flour dust and certain grinding operations, were not covered by health and safety regulations specific to the risks they created, and it was becoming increasingly impracticable to legislate on a substance or activity specific basis as technology advanced. There was a need for new legislation that would apply generically to all substances with these potential risks.

The new legislation that would become DSEAR would remove the need to have to separate regulations for particular substances or groups of substances, and would cover substances for which specific regulation did not previously exist. As a result, around 20 pieces of old health and safety legislation could be repealed making a considerable simplification for duty holders and regulators alike. Significantly, the *DSEAR* replaced the *Highly Flammable Liquids and Liquefied Petroleum Gases Regulations 1972 (SI 1972 No 917)*.

European Directives

[D0902] The opportunity to modernise UK legislation, which led to the enactment of the *DSEAR,* arose from the need for the UK government to implement two European Union Directives into UK law: the Chemical Agents Directive and the ATEX 137 (1992/92/EC) Directive. (The term ATEX is derived from the French 'Atmospheres Explosive').The Chemical Agents Directive ('CAD') requires employers to protect workers from fire, explosion and health risks arising from chemical agents present in the workplace. The health requirements of CAD have been implemented by changes to existing health legislation, mainly the *Control of Substances Hazardous to Health Regulations 2002 (SI 2002 No 2677) ('COSHH')*, with some changes to legislation on asbestos and lead.

The ATEX 137 (Explosive Atmospheres) Directive is concerned with workers health and safety in those workplaces where potentially explosive atmospheres may be present and requires employers to protect workers from the risks of fire and explosion caused by explosive atmospheres. It was decided that the safety (ie fire and explosion) aspects of the Chemical Agents Directive and the ATEX 137 Directive dealing with explosive atmospheres would be implemented together in one set of completely new safety regulations. Thus, *DSEAR* covers all substances that could give rise to fires, explosive atmospheres, explosions and similar energetic events and the management of places where explosive atmospheres may arise.

Another European Directive ATEX 95 (94/9/EC) is concerned with the supply of equipment, protective systems, components etc where these are for use in potentially explosive atmospheres. This is a trade Directive whose aim is to harmonise the manufacture and supply of equipment intended for use in potentially explosive atmospheres to common standards of safety. ATEX 95 was implemented in UK law under the Equipment and Protective Systems for Use in Potentially Explosive Atmospheres Regulations 1996 (EPS), by the Department for Trade and Industry, and now comes under the Department for Business Enterprise and Regulatory Reform (BERR), where further information can be obtained.

The DSEAR and EPS Regulations are complementary. DSEAR is concerned with the safe use of dangerous substances and requires employers to zone workplaces according to the hazard and to select equipment and protective systems appropriate for that zone that meets EPS requirements. The EPS puts a duty on the manufacturer and supplier to manufacture or supply equipment that is suitable and marked for use in different types of zone.

Guide to the chapter

[D0903] *Dangerous Substances and Explosive Atmospheres Regulations 2002 (SI 2002 No 2776) (DSEAR)* follow the trend of modern goal setting regulation based on the concept of risk assessment and management by the employer or the person responsible for the risk. This allows the Regulations to deal generically with the risks of fire and explosion, where prescriptive requirements relate directly to the risks, and not to the characteristics of any particular substance or activity. The amount of prescriptive requirement is thereby minimised.

The onus in law is placed squarely on those responsible for the risks to assess the risks within their situation and to manage them safety. It requires duty holders to think for themselves about safety and to take the necessary action. In support of the Regulations and to assist duty holders, the Health and Safety Commission has published a series of Approved Codes of Practice giving practical advice on how to comply with specific aspects of the Regulations and additional Guidance.

This chapter covers the requirements of the *DSEAR* in a form that should assist duty holders to understand easily what is needed to comply with the law.

The first part of the chapter is concerned with:

- aspects relating to the application of the Regulations;
- typical activities;
- who the duty holders are;
- what workplaces are covered;
- what dangerous substances and explosive atmospheres are; and

The rest of the chapter then deals with the requirements under the Regulations, and more specifically:

- the requirement for and considerations of a risk assessment;
- the principles of risk management: elimination, reduction, control and mitigation;
- the classification of places with explosive atmospheres into zones;
- equipment and protection systems;
- arrangements to deal with incidents and emergencies; and
- provision of information, instruction and training and marking, and identification of containers.

Some examples and guidance are given on how to comply with the Regulations, but these should not be taken out of context.

The chapter will be useful to all employers and the self-employed where there is a possibility of dangerous substances or explosive atmospheres being present in their workplace. In practice, this will include a very wide range of businesses and public services, including manufacturing industries, service companies, some retailers, transport and shipping, garages, schools, universities and experimental research institutes, and local authorities. While all have a duty to comply, the manner of compliance can be commensurate with the risk. What is commensurate should be consistent with existing good practice and common sense, but there will no doubt be case law to refer to in time.

The Regulations

Scope

[D0904] The *Dangerous Substances and Explosive Atmospheres Regulations 2002 (SI 2002 No 2776) (DSEAR)* set minimum requirements for the protection of persons from the risks of fire and explosion arising from dangerous substances and explosive atmospheres in the workplace. The Regulations assist duty holders to comply with the general requirement to manage risks from substances having these potential hazards under the *Management of Health and Safety at Work Regulations 1999 (SI 1999 No 3242) (MHSWR)*. While the *DSEAR* protect against substances with the potential to cause fire and explosion, they stand alongside the *Control of Substances Hazardous to Health Regulations (SI 2002 No 2677) (COSHH)* which protect against and control substances hazardous to health.

Many substances are classified as dangerous under the criteria given in the *DSEAR*. These include petrol, liquefied petroleum gas, welding gases, paints, varnishes and certain types of combustible and explosive dusts produced in, for example, machining and sanding operations. The Regulations are therefore relevant to the many businesses in Great Britain at which these substances may be present.

Typical activities

[D0905] The *Dangerous Substances and Explosive Atmospheres Regulations 2002 (SI 2002 No 2776) (DSEAR)* apply to many types of activities commonly found at work in businesses and public services, of which the following are typical:

- Storage and dispensing petrol as a fuel for cars, motor boats, horticultural machinery.
- Handling, storage and use of flammable gases, such as acetylene for welding.
- Handling and storage of waste dusts in a range of manufacturing industries.
- Handling and storage of flammable wastes such as fuel oils.
- Hot work on tanks or drums that have contained flammable material.
- Work activities that could release naturally occurring methane.
- Mining or processing of coal producing dusts.
- Use of flammable solvents such as ether in pathology and school laboratories.
- Storage/display of flammable goods, such as paints, in the retail sector.
- Filling, storage and handling of aerosols with flammable propellants such as LPG.
- Transport of flammable liquids in containers around the workplace.
- Deliveries from road tankers, such as petrol or bulk powders.
- Chemical manufacture, processing and warehousing.
- Extraction and processing of petrochemicals – onshore and offshore.

The examples given in the list above are only representative of the very wide range of relevant activities. All employers and the self-employed must consider if the *DSEAR* 2002 could apply to them.

General conditions for application

[D0906] The *Dangerous Substances and Explosive Atmospheres Regulations 2002 (SI 2002 No 2776) (DSEAR)* apply whenever:

(a) There is work being carried out by an employer or self-employed person.

(b) A dangerous substance is present or liable to be present at the workplace.

(c) The dangerous substance presents a risk to the safety of persons (as opposed to a risk to health).

Duty holders

[D0907] The responsibility for complying with the *Dangerous Substances and Explosive Atmospheres Regulations 2002 (SI 2002 No 2776) (DSEAR)* falls on the employer, or the self-employed person, whichever is appropriate, who is responsible for the workplace where dangerous substances or explosive atmospheres may be present [*SI 2002 No 2776, Reg 4*]. This is normally the owner, lessee or occupier of the workplace. (Note: references to the 'employer' throughout this chapter also apply to self-employed persons where appropriate.)

The duties extend not only to protecting employees but also to any other person, whether at work or not, who may be put at risk by dangerous substances and explosive atmospheres in the workplace. This includes:

*   employees working for other employers;
*   visitors to the site; and
*   members of the public.

However, when making arrangements for dealing with accidents, incidents and emergencies and the provision of information, instruction and training, employers only have duties to persons who are normally at their workplace *SI 2002 No 2776, Reg 4(1)(b)*]. Further, their duties in law in respect of provision of suitable personal protective equipment and appropriate work clothing do not extend to persons who are not employees [*SI 2002 No 2776, Reg 4(1)(a)*]. Responsible employers will, however, ensure that the number of persons who need to be in hazardous places is minimised and that all those that are in hazardous workplaces are suitably equipped.

Sometimes two or more employers will share the same workplace (whether permanently or temporarily) where an explosive atmosphere may occur. Here the employer responsible for the workplace (normally the owner or lessee) has the duty to co-ordinate the implementation of the measures to protect employees from the risk of an explosive atmosphere. The duty for implementing the measures relating to dangerous substances would be expected to lie with the employer responsible for those substances [*SI 2002 No 2776, Reg 11*].

Workplaces

[D0908] The *Dangerous Substances and Explosive Atmospheres Regulations 2002 (SI 2002 No 2776) (DSEAR)* define 'workplace' widely as being any premises or part of premises used for or in connection with work [*SI 2002 No 2776, Reg 2*]. It includes any place to which an employee has access to while at work, and covers any room, lobby, corridor, staircase or road or other place used for access or where facilities are provided. Public roads are not workplaces within the definition of the Regulations. Premises include:

*   all industrial and commercial sites;
*   land based installations;
*   vehicles and vessels;
*   shared buildings, private roads and paths on industrial estates and business parks;

- houses and other domestic premises where there is a work activity.

A few types of workplaces and activities are specifically exempt from some or all of the requirements of the *DSEAR* because there is other legislation fulfilling these requirements, for example, at offshore installations [*SI 2002 No 2776, Reg 3*]. The Regulations do not in general apply to activities on board ships, except for construction, reconstruction, conversion and dry dock repairs carried out in Great Britain. The requirements in the *DSEAR* concerning zoning and co-ordination of safety in shared workplaces do not apply at workplaces in respect of the following substances, activities or types of equipment, where other controlling regulations exist:

(a)   areas used directly for medical treatment;
(b)   the use of gas appliances and fittings for cooking, heating, lighting etc;
(c)   the manufacture, handling, use, storage and transport of explosives or chemically unstable substances;
(d)   any activity at a mine quarry or borehole; and
(e)   land transport regulated by international agreements.

The *DSEAR* apply within Great Britain, and apply outside Great Britain to the extent defined by virtue of the *Health and Safety at Work etc. Act 1974 (Application outside Great Britain) Order 2001 (SI 2001 No 2127)* [*SI 2002 No 2776, Reg 12*].

Dangerous substances

[D0909] The *Dangerous Substances and Explosive Atmospheres Regulations 2002 (SI 2002 No 2776) (DSEAR)* apply to any substance or preparation (or mixture of substances) with the potential to create a risk to persons from energetic events such as fire, explosions, or thermal runaway from exothermic reactions etc. The definition of 'dangerous' includes all substances that meet the criteria for being explosive, oxidising, extremely flammable, highly flammable or flammable according to the *Chemicals (Hazard Information and Packaging for Supply) Regulations (SI 2002 No 1689)* ('CHIP'). Substances classified as being explosive, oxidising, extremely flammable, highly flammable or flammable under the *CHIP Regulations* are therefore dangerous substances for the purposes of the *DSEAR* [*SI 2002 No 2776, Reg 2*].

The definition of 'dangerous' also includes other substances that are not classified under the *CHIP Regulations*, but meet the criteria by virtue of the way they are used or are present at the workplace. For example, the *DSEAR* apply to substances that decompose or react exothermically when mixed with other substances. Peroxides, wood, flour and many other dusts involved with grinding or machining can, when mixed with air, ignite and explode. While diesel (or other high flash point oils) at ambient temperature are not classified as flammable under the *CHIP Regulations*, if they are heated and used at sufficiently high temperatures in a process that creates a fire risk, they may become a dangerous substance for the purposes of the *DSEAR* [*SI 2002 No 2776, Reg 2*].

Thus, in order to determine if a substance present is dangerous, two steps are necessary:

(a)   a check to find out if the substance has been classified under the *CHIP Regulations*; and

(b)   an assessment of whether the physical and chemical properties of the substance and the circumstances of its use in the workplace can create a safety risk to persons from a fire or explosion.

The *DSEAR* are only intended to protect persons from energetic events and the harmful physical effects from thermal radiation (burns), over-pressure (blast injuries) and oxygen depletion (asphyxiation) arising from fires and explosions. They implement the European Explosive Atmospheres Directive (99/92/EC) (ATEX 137) into UK law.

Many of the substances will, however, also create health risks and it should be noted that the *DSEAR* does not address these. Many solvents are toxic as well as being flammable. Health risks are dealt with by the *Control of Substances Hazardous to Health Regulations 2002 (SI 2002 No 2677) (COSHH)*, which have been amended to implement the health side of the European Chemical Agents Directive (98/24/EC).

The *Notification of Installations Handling Hazardous Substances Regulations 1982 (SI 1982 No 1357)* (NIHHS) require that anyone who handles or stores (or intends to handle or store) hazardous substances in sufficient quantities to notify the Health and Safety Executive. The Regulations coverage includes substances that are flammable and/or explosive, and which would be classified as dangerous under *SI 2002 No 2776*. The notification for ammonium nitrate, as used for fertilisers and other purposes, was changed in 2002 under *Notification of Installations Handling Hazardous Substances Regulations (Amendment) Regulations 2002 (SI 2002 No 2979)*.

In respect of ammonium nitrate notifications, these should be sent to the Health and Safety Executive at the address given below. *SI 1982 No 1357* requires that anyone who handles or stores (or intends to handle or store) 150 tonnes or more of ammonium nitrate, or of mixtures containing ammonium nitrate where the nitrogen content exceeds 15.75% of the mixture by weight, should notify the HSE.

Notifications should be submitted to HSE either:

•   electronically to ANNIHHS.Notifications@hse.gsi.gov.uk;
•   in writing to HSE, Agriculture and Food Sector, National Agricultural Centre, Stoneleigh, Kenilworth, Warwickshire CV8 2LG;
•   by telephone on 02476 698350; or
•   by fax on 02476 696542.

Notifications should contain the information set out in *SI 1982 No 1357, Sch 2*. A form is available on the HSE website to assist in recording the necessary information, at www.hse.gov.uk/agriculture/nihhs.htm

HSE has produced a *Self-Help Checklist for the Storage and Handling of Ammonium Nitrate Fertiliser*, to help duty holders determine whether they have taken the necessary precautions to ensure its safe storage and handling. It is available at http://www.hse.gov.uk/explosives/ammonium/chklist.pdf

HSE's free leaflet *Storing and Handling Ammonium Nitrate* gives more detailed guidance and is available at http://www.hse.gov.uk/pubns/indg230.pdf

SI 1982 No 1357 was amended by the *Notification of Installations Handling Hazardous Substances Regulations (Amendment) Regulations 2002, (SI 2002 No 2979)* available at http://www.opsi.gov.uk/si/si2002/20022979.htm

For further information call HSE's InfoLine, tel: 0845 345 0055, visit http://www.hse.gov.uk/contact, or write to: HSE InfoLine, Caerphilly Business Park, Caerphilly CF83 3GG.

Explosive atmospheres and potentially explosive atmospheres

[D0910] An explosive atmosphere is a mixture, under atmospheric conditions, of air and one or more dangerous substances in the form of gases, vapours, mists or dusts in which, after ignition has occurred, combustion spreads to the entire unburned mixture. A potentially explosive atmosphere is one that could become explosive due to local and operational conditions. These would include maintenance activities and fault conditions such as leaks of volatile substances and gases. An atmosphere that is not explosive during the course of normal operations may be potentially explosive if dangerous substances are present.

Commencement

[D0911] The *Dangerous Substances and Explosive Atmospheres Regulations 2002 (SI 2002 No 2776) (DSEAR)* came fully into force on 30 June 2003, although certain transitional arrangements extended to 30 June 2006 [*SI 2002 No 2776, Regs 1* and *17*]. Since 30 June 2006 all workplaces must comply with the relevant requirements of DSEAR.

Risk assessment

Requirement and considerations

[D0912] When a dangerous substance is, or is liable to be present at the workplace, the employer is required to carry out a risk assessment before commencing any new work activity involving that substance [*Dangerous Substances and Explosive Atmospheres Regulations 2002 (SI 2002 No 2776) (DSEAR), Reg 5(1)*]. The risk assessment required by the *DSEAR* is an identification and careful examination of the dangerous substance(s) that may be present, the work activities associated with the substance(s), and how the work activities might give rise to a risk of fire or explosion. The *DSEAR* require that the risk assessment considers certain aspects as follows [*SI 2002 No 2776, Reg 5(2)*]:

(a)   the hazardous properties of the substance;

(b)   information on safety or any safety data sheets that the supplier may provide;

(c) the circumstances of the work including:

    (i) the work processes and substances used and their possible interactions;

    (ii) the amount of substance involved;

    (iii) the risk when two or more dangerous substances are used in combination;

    (iv) the arrangements for the safe handling, storage and transport of dangerous substances and waste containing dangerous substances;

(d) activities such as maintenance where there is the potential for a high level of risk;

(e) the effect of measures taken or to be taken to comply with the Regulations;

(f) the likelihood that an explosive atmosphere will occur and persist;

(g) the likelihood that ignition sources, including electrostatic discharges, will be present and become active and effective;

(h) the scale of the anticipated effects of a fire or an explosion;

(i) any places which are or can be connected via openings to places in which explosive atmospheres may occur;

(j) such additional safety information as the employer may need to complete the risk assessment.

The requirements for a risk assessment and the management of risk are similar in principle to those required under the *Control of Substances Hazardous to Health Regulations 2002 (SI 2002 No 2677) (COSHH)*. A risk assessment may have already been carried out on dangerous substances under the *COSHH* Regulations if they were hazardous to health. However, because of the different nature of the hazards of fire or explosion, the risk assessment under the *DSEAR* has different considerations and must be treated separately. The management of dangerous substances will share aspects in common with the management of substances hazardous to health, but there will be individual elements of management required in each case.

The degree of detail in the risk assessment needs to be commensurate with the degree of risk. For a small workshop, a table showing the substances stored and the hazards that could be created (eg spillages) and the precautions in place (eg open windows when handling), with any recommended improvements (eg provision of an extra fire extinguisher) would probably suffice. For a large factory with many dangerous substances, a more sophisticated risk assessment would be expected that would analyse the threats, detail the protection systems and emergency procedures and the remedial actions in the event of an emergency.

A practical example of a fire risk assessment, and information about the risk assessment requirements contained in the *DSEAR* is given in *Tolley's Practical Risk Assessment Handbook*, 4th edition (November 2003).

Recording findings

[D0913] The purpose of the risk assessment is to enable employers to decide what measures are needed to eliminate, or reduce as far as reasonably practicable, the safety risks from dangerous substances. An employer with five or more employees is required to record in writing the significant findings of the risk assessment as soon as possible after the assessment is made [*Dangerous Substances and Explosive Atmospheres Regulations 2002 (SI 2002 No 2776) (DSEAR), Reg 5(4)*], including:

- the measures (technical and organisational) that have been taken or will be taken to eliminate and/or reduce the risk;
- sufficient information to show that the workplace and work equipment will be safe during operation and maintenance, including details of hazardous zones, co-ordination of safety in shared workplaces, arrangements to deal with accidents, incidents and emergencies and measures to inform, instruct and train employees.

After completing the risk assessment, the *DSEAR* require that the measures identified by the risk assessment are implemented before any new work commences [*SI 2002 No 2776, Reg 5(5)*].

Review

[D0914] The risk assessment must be kept up to date and the employer is required to review it regularly [*Dangerous Substances and Explosive Atmospheres Regulations 2002 (SI 2002 No 2776) (DSEAR), Reg 5(3)*], particularly if:

(a)   there is reason to suspect that it is no longer valid, or
(b)   there has been a significant change in the workplace, work processes or organisation of work.

When changes to the risk assessment are required as a result of the review, the risk assessment must be updated and any resulting safety measures implemented before work commences again [*SI 2002 No 2776, Reg 5(5)*].

Management of risks

Safety principles

[D0915] Employers are required to ensure that the risks to the safety of employees and others from dangerous substances are either eliminated or reduced as far as reasonably practicable [*Dangerous Substances and Explosive Atmospheres Regulations 2002 (SI 2002 No 2776) (DSEAR), Reg 6(1)*]. Where it is not reasonably practicable to eliminate or reduce the risks, employers are required to take, so far as is reasonably practicable, measures to control the risks and measures to mitigate the detrimental consequences should a fire, explosion or similar event occur. The *DSEAR* therefore advocate the established safety management principles of elimination, reduction, control and mitigation of risk.

Elimination and reduction of risk

[D0916] The avoidance of risk by replacement of the dangerous substance with another substance or process that eliminates or reduces the risk is preferable if possible [*Dangerous Substances and Explosive Atmospheres Regulations 2002 (SI 2002 No 2776) (DSEAR)*]. When it is not reasonably practicable to eliminate the risk, it may be possible to reduce the risk by replacing the dangerous substance with another that is less dangerous (eg by replacing a low flashpoint solvent with a high flashpoint one). Alternatively, it may be possible to redesign the process so as to reduce the quantities of dangerous substances involved.

In support of these measures, there are guides published by the Health and Safety Executive on 'Steps to successful substitution of hazardous substances' HSG 110) which can be purchased from HSE Books) and by the Department of Trade and Industry on 'Process intensification' (which can be obtained from the DTI Publications Unit). When substituting other substances or redesigning the process, it is important to take care that the changes do not create new or increased safety or health risks from other sources.

Control measures

[D0917] When dangerous substances are present, the *Dangerous Substances and Explosive Atmospheres Regulations 2002 (SI 2002 No 2776) (DSEAR)* require employers to apply control measures, as far as reasonably practicable, consistent with the risk assessment and appropriate to the nature of the activity or operation [*SI 2002 No 2776, Regs 6(3), (4)*], including the following in priority order:

(a) reducing the quantity of dangerous substance to a minimum (eg limit volume stored);
(b) avoiding or minimising releases (eg keep in sealed containers);
(c) controlling the release at source (eg ensure that valves and stoppers are used);
(d) preventing the formation of an explosive atmosphere (eg apply appropriate ventilation);
(e) collecting, containing and removing any releases to a safe place (eg fume extraction);
(f) avoiding ignition sources (eg sparks, naked flames etc.);
(g) avoiding adverse conditions that could lead to danger (eg overheating, overpressure);
(h) keeping incompatible substances apart (eg by segregation).

Mitigation

[D0918] When dangerous substances are present, the *Dangerous Substances and Explosive Atmospheres Regulations 2002 (SI 2002 No 2776) (DSEAR)* require employers to apply mitigation measures, as far as reasonably practicable, consistent with the risk assessment and appropriate to the nature of the activity or operation [*SI 2002 No 2776, Regs 6(3) and 6(5)*], including the following:

(a) reducing the numbers of persons exposed to the risk (eg restricting access);

(b) providing plant that is explosion resistant (eg robust firmly mounted equipment);

(c) providing explosion suppression or relief equipment (eg pressure vents);

(d) taking measures to control or minimise the spread of fires or explosions (eg reducing combustible material, fire doors etc.);

(e) providing suitable equipment for the protection of personnel (eg helmets, visors etc.).

General safety measures

[D0919] In addition to the specific measures relating to dangerous substances and their immediate vicinity, the *Dangerous Substances and Explosive Atmospheres Regulations 2002 (SI 2002 No 2776) (DSEAR)* also specify a number of general safety measures that employers must take, as far as reasonably practicable [*SI 2002 No 2776, Reg 6(8)* and *Sch 1*]. These measures relate to the workplace and work processes as a whole and include:

(a) Ensuring the workplace is designed, constructed and maintained so as to minimise risk (eg use of fire resistant materials, installation of appropriate blast walls etc.).

(b) Providing work processes that are suitably designed, constructed, assembled, installed so as to reduce risk and ensuring that they are used properly and maintained in an efficient state, working order and good repair.

(c) Ensuring that equipment and protective systems meet the requirements for use in hazardous places (see below), and:

 (i) can be maintained in a safe state of operation independently of the rest of the plant in the event of a power failure;

 (ii) can be manually overridden when incorporated within automatic processes which deviate from the intended operating conditions;

 (iii) can dissipate accumulated energy quickly and safely on operation of emergency shutdown;

 (iv) contain measures to prevent confusion between connecting devices.

(d) Applying appropriate systems of work including written instructions, permits to work and other procedural systems of organising and controlling work.

(e) Identifying the hazardous contents of containers and pipes [*SI 2002 No 2776, Reg 10*]. Many will already be marked or labelled under existing legislation. For those that are not, 'identification' may require labelling, marking or warning signs. It could include training, information or verbal instruction.

(f) Arranging for the safe handling, storage and transport of dangerous substances and waste containing dangerous substances [*SI 2002 No 2776, Reg 6(6)*].

Classification of places with explosive atmospheres

Classification scheme

[D0920] Gases, vapours, mists and dusts can all form explosive atmospheres with air. The source of the gases, vapours, mists and dusts may be a consequence of the natural state of the substance (eg hydrogen gas), or it may be a result of the means of storage or handling (as in the case of a flammable liquid), or as a result of a work activity (eg grinding or powder processing). Where there is any prospect of an explosive atmosphere, a hazardous area classification should be carried out as an integral part of the risk assessment, and for those areas classified as hazardous, *special precautions* over the sources of ignition are needed to prevent fires and explosions.

In workplaces where an explosive atmosphere may occur, employers must first classify the workplace into hazardous and non-hazardous places [*Dangerous Substances and Explosive Atmospheres Regulations 2002 (SI 2002 No 2776) (DSEAR)*]. The criteria for deciding whether a place is hazardous is whether an explosive atmosphere could occur in such quantities as to require *special precautions* to protect the health and safety of workers concerned within the meaning of the *DSEAR*. Similarly, a non-hazardous place is one where an explosive atmosphere is not expected to occur in such quantities as to require special precautions to protect the health and safety of workers concerned [*SI 2002 No 2776, Sch 2 para 1*].

Thus, in the first instance, employers must assess the quantity of explosive atmosphere that could be present in a particular place. The term 'not expected to occur in such quantities' is deliberate and means that employers should also consider the likelihood or possibility of a release of an explosive atmosphere as well as the potential quantities of such a release. Places where it is credible that an explosive atmosphere could exist, even when this is not within normal operating limits should be classified as hazardous. If, on the other hand, a release is extremely unlikely to occur and/or if the quantities released are small, it may not be necessary to classify the area as hazardous.

As an example, a dangerous substance is being carried though a seamless pipe that has been properly installed, maintained and inspected. It is unlikely that the substance will be released. An explosive atmosphere would not be expected to occur from this source and the area surrounding the pipe would be non-hazardous. However, if the pipe contained flanged joints or fittings where there was the possibility of a leak or its condition was uncertain, a release would be a credible event and the area should be classified as hazardous.

With regard to quantity, a spillage from a small bottle of solvent used in a laboratory would release so little flammable vapour that no special precautions are needed other than general control of ignition sources (no smoking, for example) and cleaning and disposing of the spillage. The area would not be classified as hazardous. If, however, the vapour released also had a strong anaesthetic effect, and spillages were known to occur from time-to-time, then the area might be classified as hazardous. When deciding whether it is necessary to classify as 'hazardous' areas with small quantities of dangerous substances, the actual circumstances of use and any specific product guidance should be taken into account.

Where dangerous substances in small pre-packaged containers are stored or are on display for sale in retail premises, for example solvents or aerosols, the area would not normally need to be classified as hazardous. An exception to this might, however, be with storage in poorly ventilated basements. When such substances are held in large quantities, as in a warehouse, with the opportunity for a build up of vapours or a larger spillage, a hazardous classification would be expected. Procedures to control ignition sources and to clean up and dispose of any spillage/release would be needed.

In identifying hazardous and non-hazardous areas and assigning zones, the following matters should be considered:

- The hazardous properties of the dangerous substances involved.
- The amount of dangerous substances involved. The size of any potentially explosive atmospheres is, in part, related to the amount of dangerous substances present. Guidance is given in industry specific codes on the quantities of dangerous substances that may be safety stored. See, for example, the 'Code of Practice on the storage of full and empty LPG cylinders and cartridges' produced by the LPG Association.
- The work processes, and their interaction, including any cleaning, repair or maintenance activities that will be carried out.
- The temperatures and pressure of the dangerous substances. This will affect the nature and extent of any release. Some substances do not form explosive atmospheres unless they are heated (eg diesel oil), while others if released under pressure will form a fine mist that can explode, even if there is insufficient vapour.
- The containment system and controls to prevent liquids, gases, vapours or dusts escaping into the general atmosphere (eg trays, seals).
- The possibility of an explosive atmosphere forming in an enclosed plant or storage vessel. A release of a dangerous substance into an enclosed space always increases the risk of an explosive atmosphere forming.
- The ventilation (natural or a fan extract system) or any other measures designed to dilute sources of release and ensure that any explosive atmosphere does not persist for an extended time. Well-designed ventilation may prevent the need for any zoned area or reduce it so it has a negligible extent.

When considering the potential for dangerous atmospheres, it is important to consider all substances that may be present in the workplace. Waste products, residues, cleaning agents, materials used for maintenance and fuel could all be potentially dangerous. The possibility that combinations of substances could react to create an ignition source or an explosive atmosphere should not be overlooked.

The refuelling of cars or loading and unloading of petrol from tankers intended for use on public roads involve the introduction of potential sources of ignition in an area where a spill is possible. This would normally meet the criteria for being a hazardous area. In these circumstances, safety can be achieved by isolating power sources (turning off engines) where a transfer of fuel is taking place and making suitable checks before and after transfer and before and after vehicles are moved into or out of the refuelling area. Garage owners have a particular responsibility to monitor the forecourt and to ensure that these practices are implemented.

Zoning

[D0921] For places classified as hazardous according to D0920, employers are required to further classify these places into zones on the basis of the frequency and duration of the explosive atmosphere occurring [*Dangerous Substances and Explosive Atmospheres Regulations 2002 (SI 2002 No 2776) (DSEAR), Reg 7(2)* and *Sch 2 para 2*]. The *DSEAR* give the following classification scheme for mixtures with air of dangerous substances (Zones 0, 1 and 2) and combustible dusts (Zones 20, 21, and 22). In decreasing order of risk:

- *Zone 0* – A place in which an explosive atmosphere consisting of a mixture with air of dangerous substances in the form of a gas, vapour or mist is present continuously or for long periods or frequently.
- *Zone 1* – A place in which an explosive atmosphere consisting of a mixture with air of dangerous substances in the form of a gas, vapour or mist is likely to occur in normal operation occasionally.
- *Zone 2* – A place in which an explosive atmosphere consisting of a mixture with air of dangerous substances in the form of a gas, vapour or mist is not likely to occur in normal operation, but if it does occur, will persist for a short period only.
- *Zone 20* – A place where an explosive atmosphere in the form of a combustible dust in air is present continuously or for long periods or frequently.
- *Zone 21* – A place where an explosive atmosphere in the form of a combustible dust in air is likely to occur in normal operation occasionally.
- *Zone 22* – A place where an explosive atmosphere in the form of a combustible dust in air is not likely to occur in normal operation, but if it does occur, will persist for a short period only.

Within this scheme, terms such as 'occasionally' or 'frequently' are not defined but are assumed to be that which a reasonable person would use. Normal operation is any situation when installations are used within their design parameters. Combustible dusts in the form of layers, deposits or heaps must be considered as a source of an explosive atmosphere and classified accordingly.

The basic principles of area classification are explained within the European standards, BS EN 60079/10 for gases and vapours and BS EN 61241/3 for dusts. These standards form a suitable basis for assessing the extent and type of the zone and can be used as a guide to complying with the requirements of the *DSEAR*. However, the guidance must be applied to the site-specific factors in order to determine the extent and type of zone in any particular case. It should be remembered that an explosive atmosphere may spread into areas away from the source of the hazard, for example through ducts, and these areas must also be included in the classification scheme.

Various organisations have published industry specific codes (eg Energy Institute). Providing they are applied appropriately, they are valuable in encouraging a consistent interpretation of the requirements.

Area classification

In areas where dangerous quantities and concentrations of flammable gas or vapour may arise, protective measures by zoning are required to reduce the risk of explosions. Essential criteria against which ignition hazards can be assessed and zone areas determined are given in Part 10 of International Standard ;IEC 60079-10-1:2008, which has been adopted as a European Standard and a British Standard, BS EN 60079-10-1:2009. This gives advice on design and control parameters which can be used to quantify the hazard in order to bound and limit such areas.

BS EN 60079-10-1:2009 gives guidance on the procedure for classifying areas where there may be an explosive gas atmosphere. The procedures are quite complex and should only be carried out by those who understand the relevance and significance of the properties of flammable materials and those who are familiar with the process and the equipment and the layout of the areas in which the equipment is located. Where the necessary electrical, mechanical and other qualified engineering personnel are not available, it is necessary to use consultants specialising in hazardous area classification.

This section is intended to provide an overview of the procedure but is in no way a substitute for reference to BS EN 60079-10-1:2009. It may assist users in determining whether they have the necessary capability and expertise to undertake the procedure themselves. Where consultants are used it will enable users to be better prepared to provide the necessary information.

For plant in design and construction, area classification should be carried out when the initial process and instrumentation line diagrams and layout plans are available so that the design and layout may be optimised to reduce zoned areas to a minimum. Area classification should be confirmed before start-up and reviewed during the life of the plant. For plants in service, area classification needs to take account of local conditions (eg ventilation and other equipment) and as these may change during service, classification needs to be reassessed as part of the management of change procedure.

The procedure comprises firstly of identifying sources, rate and grade of release and then considers factors that determine the type and extent of the zone including ventilation. The detail is provided in four Annexes. Annexes A and B provide information and examples on sources of release and release rate and the assessment of ventilation, while Annexes C and D give examples of areas classification and the treatment of flammable mists.

Sources of explosive atmospheres

Each item of process equipment containing a flammable gas or vapour (and flammable liquids and solids which may on release give rise to them) should be considered as a potential source of flammable material. Where process plant or containment (e.g. storage tanks) is not totally enclosed, or where air can enter the system, it is necessary to decide whether a flammable mixture with air can exist inside the equipment. For all items it is necessary to consider whether a release of flammable materials can create a flammable atmosphere outside the equipment. It is important to realise that explosive atmospheres can be generated by mists of leaking flammable liquid, even though the liquid may be below its flash point.

If the item cannot contain flammable materials under reasonably foreseeable circumstances it will not give rise to a hazardous area around it. The same applies if the item contains a flammable material but cannot release it to the atmosphere because the system is completely sealed. For example a continuous welded pipeline is not considered to be a source of release, however, valves, seals and flanges, which may potentially leak, are considered to be a potential source. If the total quantity of flammable material available for release is small (eg in some laboratory equipment), the area classification procedure may not be appropriate if other risk control measures are applied (eg high ventilation).

Having identified potential sources, these are then graded as continuous, primary or secondary according to the frequency and duration of the release. These normally give rise to zones 0, 1 and 2 as defined above. All times when a release may take place, including commissioning, normal operation, opening equipment, batch filling, cleaning, purging, start-up and shutdown cycles and when the equipment is mothballed and dormant, should be considered.

Extent of zone

The extent of zone is defined by the calculated or estimated distance over which a potentially explosive atmosphere from sources exists before it disperses to a concentration in air below its lower explosive limit with an appropriate safety factor. (The lower explosive limit is the concentration of flammable gas, vapour or mist in air below which an explosive gas atmosphere will not be formed.) The assessment of the spread of gas or vapour and its dispersion and dilution to below its lower explosive limit requires judgment and the advice of experts should be sought.

The key factors determine the extent of a zone are the release rate of gas or vapour, its lower explosive limit, ventilation, and whether the gas is heavier or lighter than air (relative density). Physical barriers (walls, sealed doors etc), and atmospheric barriers (maintaining an external over pressure and air purging) also need to be taken into account. Other parameters that should be considered include the climatic conditions (wind, air temperature) and topography (eg water courses where a flammable liquid may reside on the surface over a large area or trenches, pits and drains where an explosive atmosphere may not easily disperse and dilute).

Determining the rate of release of a gas or vapour can be a complex operation. In general, it depends on the geometry (area) and type of the source of release (eg open surface, leaking flange), the release velocity (related to the pressure differential, source geometry and the physical properties of the flammable material), and the concentration of flammable vapour or gas in the release (some liquids may comprise flammable and non flammable mixtures). After release of flammable liquid, the rate of gas or vapour formation depends on the volatility of the flammable liquid (determined by the vapour pressure as a function of liquid temperature and the enthalpy of vaporisation, to which the boiling point and flash point are roughly related), and the liquid temperature after release (which would need to take account of the ambient temperature and any hot/cold surfaces in the vicinity etc).

Application

Annex C of BS EN 60079-10-1:2009 gives ten worked examples that illustrate the principles of hazardous area classification. Examples range from mechanical pumps at ground level either indoors or outdoors pumping flammable liquid, a fixed process mixing vessel being opened regularly for operational reasons, a mixing room in a paint factory with multiple mixing vessels and pumps. Other examples include a flammable liquid storage tank situated outdoors with a fixed roof and no internal floating roof, a single tanker filling installation for gasoline, top filling with no vapour recovery and an oil water separator in a petroleum refinery.

The discussion on ventilation gives seven example calculations to ascertain the degree of ventilation under different circumstances. The calculations are based on determining a hypothetical volume over which the mean concentration of flammable gas or vapour will be typically between 0.25 to 0.5 times the lower explosive limit, depending on the factor of safety appropriate to the conditions. (The hypothetical volume is related to but not the same as the hazardous area dimensions.) The examples cover a range of flammable materials (eg toluene, propane gas, ammonia, methane) stored under different indoor and outdoor ventilation conditions.

Recent work by the Health and Safety Laboratory (HSE Research Report 630, 2008) has sought a more soundly based methodology as it applies to secondary releases from low pressure natural gas systems with the possibility of removing a significant amount of conservatism from the method given in BS EN 60079-10-1:2009. An extensive programme of theoretical and experimental work has lead to an alternative approach to measuring ventilation effectiveness in enclosures. This has the potential to assist the gas industry and businesses relying on supplies of natural gas for heating and other purposes.

Documentation

Area classification normally takes the form of drawings or plans identifying the hazardous areas and zones. Information about the dangerous substances that will be present, the work activities that have been considered, and other assumptions made will be given. These drawings or plans form part of the risk assessment record that the *DSEAR* require, and should be available for examination by those affected and regulators. They become particularly important when there is a change in the work activity and/or new equipment to be introduced into a zoned area.

Classification drawings or plans may need to be reviewed when maintenance is carried out. If the dangerous substances normally present have been removed it may be possible to treat the area as non-hazardous. Alternatively, if the maintenance creates a larger than normal risk of a release of a dangerous substance, for example if it is necessary to open a normally sealed pipe system or enter a tank where vapours could exist – a larger area may need to be treated as hazardous or the area may need to be to temporarily reclassified to a higher zone. It is not, however, normally necessary to create new classification drawings or plans for the duration of maintenance work, but the process needs to be carefully controlled.

Warning signs on entry

[D0922] Employers are required to ensure that places classified as hazardous are marked at their points of entry with warning signs [*Dangerous Substances and Explosive Atmospheres Regulations 2002 (SI 2002 No 2776) (DSEAR), Reg 7(3)* and *Sch 4*]. The warning sign at entry into places where explosive atmosphere may occur is specified in Annex III of the European Council Directive 99/92/EC and is:

(a)   triangular in shape; and
(b)   contains black letters 'EX' on a yellow background with black edging (the yellow part taking up at least 50 per cent of the area of the sign).

Verification of safety prior to first use

[D0923] Employers are required to ensure that before workplaces containing places classified as hazardous are used for the first time, they are confirmed as being safe (verified) by a person (independent or employed by an organisation) competent in the field of explosion protection [*Dangerous Substances and Explosive Atmospheres Regulations 2002 (SI 2002 No 2776) (DSEAR), Reg 7(4)*]. The person carrying out the verification must be competent to consider the particular risks at the workplace, the adequacy of the risk assessment, and the control and other measures put in place. As a result the person must have suitable experience, professional training or both.

Equipment and protection systems

Compliance with EPS Regulations

[D0924] Employers are required to take *special precautions* to ensure that equipment and protective systems in places classified as hazardous are selected so as to avoid sources of ignition from, for instance, sparks or flames [*Dangerous Substances and Explosive Atmospheres Regulations 2002 (SI 2002 No 2776) (DSEAR), Reg 7(2)* and *Sch 3*]. Specifically, equipment and protective systems must be selected on the basis of the requirements set out in the *Equipment and Protective Systems Intended for Use in Potentially Explosive Atmospheres Regulations 1996 (SI 1996 No 192)* (EPS) unless the risk assessment indicates otherwise.

EPS defines two Equipment Groups. Equipment Group 1 is equipment for use in underground parts of mines, and to those parts of surface installations of mines liable to be endangered by firedamp and/or combustible dust. Equipment Group II is intended for use in non-mining places liable to be endangered by explosive atmospheres. Within each Equipment Group different Categories of equipment are designated.

In a non-mining environment DSEAR specifies that the following categories of equipment (as defined within the EPS Regulations) must be used in the zones indicated, providing the category is suitable:

- in Zone 0 or Zone 20 – Group II Category 1 equipment;
- in Zone 1 or Zone 21 – Group II Category 1 or 2 equipment;
- in Zone 2 or Zone 22 – Group II Category 1, 2 or 3 equipment.

Equipment already in use before 1 July 2003 can continue to be used indefinitely providing the risk assessment shows that it is safe to do so.

The *DSEAR* provide definitions of the terms 'equipment', 'protective systems', 'devices', 'components'. A 'potentially explosive atmosphere' is an atmosphere that could become explosive due to local and operational conditions. These would include maintenance activities and fault conditions. This part of the *DSEAR* complements the *Equipment and Protective Systems Intended for Use in Potentially Explosive Atmospheres Regulations 1996*. For further information see the DTI website at: www.dti.gov.uk/strd/atex.htm

Manufacturers, suppliers and others supplying equipment for use in potentially explosive atmospheres have a duty under the EPS Regulations to mark their equipment to show which Group and Category it falls under, and therefore which zone it may be suitable for. These markings are in addition to the CE sand Ex markings required. Users have the duty to zone their workplaces and to use equipment suitable for that zone.

Equipment built to the requirements of the EPS Regulations can be identified as it will carry the explosion protection symbol'Ex' in a hexagon, the equipment category (1, 2,or 3) and the letter 'g' or 'd' depending on whether it is intended for use in gas or dust atmospheres. In many cases a temperature rating, expressed as a 'T' marking, will also be included. These indicate the limitations to safe use.

For most electrical equipment designed for use in explosive atmospheres there will be little change required, except in the details of the marking on the equipment. However, the 1996 Regulations also apply to mechanical equipment as a potential ignition source. A harmonised European standard for category 3 mechanical equipment is now available as BS EN 13463 Part 1.

Equipment generating a potential explosive atmosphere

[D0925] Where the use of equipment may generate an explosive atmosphere, such as dust, then before installation and use the user has to make a risk assessment under Reg 5 of DSEAR. The assessment should consider the likelihood of explosive concentrations being generated under normal use and fault conditions. In addition, it should identify any additional precautions needed to prevent such concentrations and to mitigate the effects of any explosion to a safe level, such as local ventilation and filtration systems.

A milling machine is an example of equipment where an explosive dust may be generated. For the machine to be ATEX certified it has to be intended for use in a potentially explosive atmosphere and have its own ignition source. If the machine while containing an explosive dust is not operated or exposed to that dust (ie the dust is contained), it may not as a whole need to be ATEX certified.

However, the explosion risks from the processing of materials do have to be assessed and controlled by the machine's manufacturer to meet the essential health and safety requirements of the Machinery Directive (MD) (98/37/EC). Any equipment or protection systems installed as part of the machine to meet these requirements have, in turn, to meet the requirements of ATEX 95 (94/9/EC). For examples, if motors, switches etc could possibly be exposed to an explosive atmosphere of the dust generated, they would, individually, have to be ATEX certified. Explosion relief panels where these are fitted are protections systems and therefore also come within the scope of ATEX and must be certified.

Risks from existing mechanical equipment

[D0926] Equipment already in use before 1 July 2003 can continue to be used indefinitely providing the risk assessment shows that it is safe to do so. For most individual pieces of mechanical equipment there is frequently no documentation suggesting that anyone has previously considered the ignition risk at the time it was manufactured. The risk assessment therefore needs to demonstrate that the equipment is acceptably safe.

In general very little mechanical equipment is an ignition risk in normal operation, but may become a risk when a fault arises. The risk assessment would be a thorough assessment of the equipment under all credible conditions. The users own experience of operating the equipment is important, particularly if faults were known to occur and whether these faults created actual ignition or the obvious potential for ignition.

Contact with the manufacturer may still be possible. If so the manufacturer may be able to provide information about the risks, records of any ignitions and the types and frequency of faults that have been known to occur. The wider experience of the manufacturer may have identified faults that create an ignition risk that the user was not aware of.

The amount and adequacy of maintenance of the equipment are important considerations. Records of maintenance and the original maintenance instructions form an input to the risk assessment. There should be a reasonable basis for saying that the equipment remains as safe as when it was new.

Where new equipment of the same type has a relevant standard, a comparison of the existing equipment with this standard may identify risks that were not fully understood at the time when the old equipment was constructed. While there is no obligation to bring the existing equipment fully up to modern standards, simple changes that would improve safety where reasonably practicable should be adopted. This is the current HSE policy towards to continued use of old equipment.

Storage silos and bins

[D0927] Storage silo and bins that may be used for storing explosive products such a fine dusts are not ATEX 137/EPS equipment, unless they include, as an integral part, equipment or fittings that create an explosion risk. Where silos and bins have explosive vent panels for which there is no information about the standard to which they have been built and tested, nor any details about the operating pressure under normal and explosive conditions, the panels cannot be easily shown to be fit-for-purpose. A risk assessment under DSEAR would be likely to conclude that an assessment of the pressures was needed and new correctly rated panels carrying the CE and Ex mark should be fitted. Where the inside of such storage silos and bins is zoned under DSEAR, any equipment used in the zone would need to be ATEX/EPS rated.

Second hand equipment

[D0928] Both DSEAR and PUWER (Provision and Use of Work Equipment Regulations) are relevant to bringing second hand equipment into service. The EPS Regulations relate to the supply of new equipment. In general, the design, operation and condition of any new or second hand equipment brought into service needs to be reviewed and verified as fit for purpose. PUWER Regulation 10 makes clear that equipment brought into service at a new site where there is potential for an explosive atmosphere needs to comply with the APEX Directives or other relevant single market Directives that applied when it was first supplied or brought into service.

Work clothing

[D0929] Employers are required to ensure that appropriate work clothing that does not give rise to electrostatic discharges is provided to employees entering places classified as hazardous [*Dangerous Substances and Explosive Atmospheres Regulations 2002 (SI 2002 No 2776) (DSEAR), Reg 7(5)*]. Sparks from clothing could create a risk of igniting an explosive atmosphere. Advice should be sought from an industrial clothing supplier if the employer is in any doubt as to the suitability of any item.

Specific products and areas

Flammable substances in public and laboratory areas

[D0930] Where flammable substances are used in areas where the public is present, such as in nail treatments, it is difficult to maintain constantly an effective control over all sources of ignition, such as smoking or portable electrical equipment. Consequently hazardous area classification is not appropriate. It is better to limit the amount of the substance present to the amount used in half a day's work, to use containers that minimise the risk of spills and that will break if knocked over, and to ensure adequate ventilation.

A similar principle applies to areas that handle sacks of substances in the form of dusts, such as supermarkets with in-store bakeries that handle 25 kg sacks of flour sugar and custard powder that are listed as explosible substances in bulk storage areas. In shops there is usually no means to generate a large dust cloud unless a sack fails or tears during handling, and even then only a small proportion of the escaping product will be raised into a dust cloud. Unless there is a some way that dust will be kept as a cloud for a longer period, the largest release foreseeable from a single failure is 25kg of dust, hazardous area classification is not normally appropriate.

Laboratories that handle flammable products but in small volumes. An HSE document 'ATEX, A Guide for Laboratories' addresses this situation.

Liquified petroleum gas (LPG) issues

[D0931] The DSEAR applies wherever a dangerous substance or explosive atmosphere is present or is liable to be present in a workplace, and it therefore applies wherever LPG is stored. Places for bulk LPG storage, LPG cylinder filling and cylinder storage areas therefore fall within the Regulations. The appropriate regulations apply, including the need for a risk assessment and a possible need for zoning.

The requirement to zone areas where LPG cylinders are stored outdoors in a ventilated area has been discussed with the LP Gas Association. Whilst an explosive atmosphere is not likely in normal operation, damage to cylinders has been known to occur during mechanical handling operations resulting in a leak and a short term build-up of gas. The LP Gas Association has published a Code of Practice 7 – The Storage of Full and Empty LPG Cylinders and Cartridges. Provided the outdoor storage area is designed, constructed and maintained fully in accordance with the Code of Practice, the HSE has accepted that it is not essential to designate the LPG cylinder outdoor storage area as Zone 2 as would otherwise be the case. Thus, ATEX compliant equipment and vehicles are not needed in such storage areas. Where the outdoor storage area does not comply with the conditions given in Code of Practice 7, a Zone 2 area may exist.

With regard to the storage of LPG cylinders indoors, Code of Practice 7 requires a full site specific risk assessment to be carried out. The zoning level would depend on a number of factors including ventilation provision. Where an area has been zoned, then suitable ATEX equipment (see below), including vehicles, would be required for work in that area.

Storage of aerosols

[D0932] Warehouses and other areas used for storing aerosols in bulk should be considered for zoning. Most aerosols use liquified flammable gases as the propellant and may also contain flammable liquids as part of the product, but it is best to check with suppliers for details of the amounts and types in the products stored. There are examples where aerosols have been damaged during mechanical handling, leaks their gases and caused major fires. It is good practice to separate aerosols from other less hazardous products in a fire separated part of the warehouse.

The issue has been discussed with the British Aerosol Manufacturers Association, and they have published a guide to safe storage. Provided the conditions set out in the guide are followed closely, HSRE has accepted that it is not necessary to designate warehouse storing aerosols as zone 2, and ATEX compliant vehicles are not needed (see below). Where the conditions in the guide cannot be met, the warehouse should normally be classified as hazardous, zone 2.

Wood dust

[D0933] Wood dust can explode if dispersed in a cloud, and there may be a need to assign hazardous areas to a wood-working machine room if dense clouds can form in the case of some equipment fault or operator error. For example, a dense cloud can form if the local exhaust ventilation fails or machines continue to operate after the ventilation becomes ineffective. A dense cloud is also possible when dust deposits on horizontal surfaces are disturbed, for instance by a sudden blast of air.

IN many cases such as these there will be a need to assign Zone 22 areas, but only a very limited need for zone 21. Where this is the case electrical equipment will need to be ATEX rated (see below). In zone 21 the machine or activity that generates the dust cloud must be stopped before any attempts are made to disperse the cloud or clean up the area.

Arrangements to deal with incidents and emergencies

[D0934] The *Dangerous Substances and Explosive Atmospheres Regulations 2002 (SI 2002 No 2776) (DSEAR)* require employers to make arrangements to protect employees (and other persons who are at the workplace) in the event of incidents and accidents involving dangerous substances and explosive atmospheres [*SI 2002 No 2776, Reg 8(1)*]. The provisions clarify what already needs to be done in relation to the safety management of dangerous substances and will not require any duties in addition to those already present in existing legislation. Specifically they build on existing requirements in *Regulation 8* of the *Management of Health and Safety at Work Regulations 1999 (SI 1999 No 3242)*.

The arrangements required are as follows:

(a)   Giving suitable warnings (including audible and visual alarms) where necessary to alert employees (and other persons who are at the workplace) immediately of a release of a dangerous substance or the presence an explosive atmosphere before explosive conditions are reached, so they may be withdrawn to a safe place.

(b)   Escape facilities (eg designated exit routes, stairways etc.) identified as necessary by the risk assessment to ensure that employees (and other persons who are at the workplace) can, in the event of danger, leave the endangered place promptly and safety.

(c)   Warning and communication systems to enable an appropriate response to be made immediately when such an event occurs, including remedial actions and rescue operations to mitigate the effects and restore the situation to normal [*SI 2002 No 2776, Reg 8(3)(a)*].

(d)   Emergency procedures to be followed in the event of an emergency, including defining responsibilities and actions to be taken.

(e)   Providing equipment and clothing for essential personnel to deal with the incident [*SI 2002 No 2776, Reg 8(3)*].

(f)   Organising relevant safety drills at regular intervals designed to test the emergency procedures.

(g)   The provision of appropriate first-aid facilities in a safe place.

(h)   Making information available to employees (eg through a display at the workplace) on the warnings, escape facilities, emergency procedures, relevant work hazards and hazard identification (eg zoning and marking), and on specific hazards likely to arise at the time of an incident [*SI 2002 No 2776, Reg 8(2)(b)*].

(i)   Contacting the relevant accident and emergency services to let them know that information on emergency procedures is available (or providing them with any information that they consider necessary) [*SI 2002 No 2776, Reg 8(2)(a)*].

The scale and nature of these arrangements should be proportional to the risks, taking account of the frequency of such events and the possible consequences. When the results of the risk assessment show that because of the quantity of each dangerous substance at the workplace there is only a slight risk to employees and risk has been eliminated or reduced as far as reasonably practicable, specific emergency arrangements are not necessary [*SI 2002 No 2776, Reg 8(4)*]. In this case, the general provisions for safety at the workplace will apply.

Provision of information, instruction and training

[D0935] Employers are required to provide employees and other people at the workplace who might be at risk from dangerous substances with suitable information, instruction and training on precautions and actions they need to take to safeguard themselves and others so as to reduce the possibility of an incident taking place [*Dangerous Substances and Explosive Atmospheres Regulations 2002 (SI 2002 No 2776) (DSEAR), Reg 9*]. This should include:

•   names of the substances in use and the risks they present;
•   access to any relevant data sheets;
•   any specific legislative provisions concerning the hazardous properties of the substance;
•   the significant findings of the risk assessment.

Other aspects may include procedures for correct handling, protective clothing and equipment to be used, and the use of safe work processes. Information, instruction and training should be adapted if significant changes in the type or methods of work occur. Existing health and safety legislation already covers much of this.

Information, instruction and training need only be provided to non-employees where it is required to ensure their safety. Where it is provided, it should be in proportion to the level of risk. In general, non-employees should not be exposed to such risks.

Identification of hazardous contents of pipes and containers

[D0936] Employers are required to identify pipes, vessels, tanks and other containers of dangerous substances in accordance with the legislation listed below where this is applicable [*Dangerous Substances and Explosive Atmospheres Regulations 2002 (SI 2002 No 2776) (DSEAR), Reg 10 and Sch 5*].

- the *Chemicals (Hazard Information and Packaging for Supply) Regulations 2002 (SI 2002 No 1689)*;
- the *Health and Safety (Safety Signs and Signals) Regulations 1996 (SI 1996 No 341)*;
- the *Good Laboratory Practice Regulations 1997 (SI 1997 No 654)*;
- the *Classification and Labelling of Explosives Regulations 1983 (SI 1983 No 1140)*;
- the *Carriage of Dangerous Goods by Rail Regulations 1996 (SI 1996 No 2089)*;
- the *Packaging, Labelling and Carriage of Radioactive Material by Rail Regulations 1996 (SI 1996 No 2090)*;
- the *Carriage of Dangerous Goods (Classification, Packaging and Labelling) and Use of Transportable Pressure Receptacles Regulations 1996 (SI 1996 No 2092)*;
- the *Carriage of Dangerous Goods by Road Regulations 1996 (SI 1996 No 2095)*;
- the *Carriage of Explosives by Road Regulations 1996 (SI 1996 No 2093)*;
- the *Radioactive Material (Road Transport) (Great Britain) Regulations 1996 (SI 1996 No 1350)*.

Where containers of dangerous substances are not marked in accordance with this legislation, for reasons of applicability, employers must ensure that the contents of those containers are clearly identifiable together with the nature of the associated hazards.

Exemption

[D0937] The Health and Safety Executive (HSE) has the power to exempt any person or class of persons or any dangerous substance or class of substances from all or any of the requirements or prohibitions imposed by or under the *DSEAR* [*Dangerous Substances and Explosive Atmospheres Regulations 2002 (SI 2002 No 2776) (DSEAR), Reg 13*]. It is expected that any such exemption would only be used in exceptional circumstances. Such exemption is granted by the issue of a certificate in writing and may be subject to conditions and a limit of time and may be revoked at any time.

In exercising this power, the HSE must, having due regard to the circumstances of the case, be satisfied that the health and safety of persons likely to be affected will not be prejudiced. In addition, it must be satisfied that exemption will be compatible with the requirements of European Council Directives (98/24 and 99/92) relating to the protection of workers from risks from chemical agents and explosives, which take precedence.

The Secretary of State for Defence may, in the interests of national security, by a certificate in writing exempt from all or any of the requirements or prohibitions imposed by or under the *DSEAR* the following [*SI 2002 No 2776, Reg 14*]:

- any of Her Majesty's Armed Forces;
- any visiting force and members attached to a headquarters;
- any person engaged in work involving dangerous substances, if that person is under the direct supervision or control of a representative of the Secretary of State for Defence.

Where any such exemption is granted, suitable arrangements must be made for the assessment of risk to safety created by the work and for adequately controlling the risk to the persons to whom the exemption relates. This power of exemption could apply at defence establishments and centres of defence related research.

Amendments to modernise petroleum and other legislation

[D0938] Petroleum legislation is modernised as part of the *Dangerous Substances and Explosive Atmospheres Regulations 2002 (SI 2002 No 2776) (DSEAR)* by means of a number of amendments. Previously, the keeping of petrol in significant quantities was controlled by licences with conditions issued under the *Petroleum (Consolidation) Act 1928*. However, as petrol is a dangerous substance, the *DSEAR* apply to it and duplicate these controls. The *DSEAR* remove licensing requirements for holding petrol, except for petrol that is being kept for dispensing into vehicles (retail and non-retail) at, for example, garages. Accordingly the following Acts, Regulations and Orders are amended [*DSEAR (SI 2002 No 2776), Reg 15* and *Sch 6 Pts 1* and *2*]:

- the *Petroleum (Consolidation) Act 1928*;
- the *Petroleum Spirit (Motor Vehicles etc) Regulations 1929 (SI 1929 No 952)*;
- the *Petroleum (Liquid Methane) Order 1957 (SI 1957 No 859)*;
- the *Petroleum (Consolidation) Act 1928 (Enforcement) Regulations 1979 (SI 1979 No 427)*;
- the *Petroleum Spirit (Plastic Containers) Regulations 1982 (SI 1982 No 630)*.

In addition, the *DSEAR* amend a number of other pieces of legislation:

- the *Celluloid and Cinematograph Film Act 1922*;
- the *Dangerous Substances in Harbour Areas Regulations 1987 (SI 1987 No 37)*;

- the *Fire Precautions (Workplace) Regulations 1997 (SI 1997 No 1840)*;
- the *Fire Certificates (Special Premises) Regulations 1976 (SI 1976 No 2003)*;
- the *Carriage of Dangerous Good by Road Regulations 1996 (SI 1996 No 2095)*.

It should be noted that the amendment of the *Fire Precautions (Workplace) Regulations 1997* changes the definition of a highly flammable liquid within *Schedule 1* of the 1997 Regulations (premises for which a fire certificate is required). The amendment to the *Carriage of Dangerous Goods by Road Regulations 1996* prohibits the filling of fuel tanks for cars or other containers direct from a road tanker.

Repeals and revocations of existing legislation

[D0939] The *Dangerous Substances and Explosive Atmospheres Regulations 2002 (SI 2002 No 2776) (DSEAR)* repeal and revokes a large number of older Regulations and Orders relating to substances that now come under the general provisions of the 2002 Regulations. In this way the *DSEAR* are an important piece of modernising and reforming legislation. The following Acts and statutory instruments are completely repealed and revoked [*SI 2002 No 2776), Reg 16 and Sch 7 Pts 1* and *2*]:

- the *Celluloid etc. Factories and Workshops Regulations 1921 (SI 1921 No 1825)*;
- the *Manufacture of Cinematograph Film Regulations 1928 (SI 1928 No 82)*;
- the *Cinematograph Film Stripping Regulations 1939 (SI 1939 No 571)*;
- the *Petroleum (Carbide of Calcium) Order 1929 (SI 1929 No 992)*;
- the *Petroleum (Carbide of Calcium) Order 1947 (SI 1947 No 1442)*;
- the *Petroleum (Compressed Gases) Order 1930 (SI 1930 No 34)*;
- the *Magnesium (Grinding of Castings and other Articles) Special Regulations 1946 (SI 1946 No 2017)*;
- the *Dry Cleaning Special Regulations 1949 (SI 1949 No 2224)*;
- the *Dry Cleaning (Metrication) Regulations 1983 (SI 1983 No 977)*;
- the *Factories (Testing of Aircraft Engines and Accessories) Special Regulations 1952 (SI 1952 No 1689)*;
- the *Factories (Testing of Aircraft Engines and Accessories) (Metrication) Regulations 1983 (SI 1983 No 979)*;
- the *Highly Flammable Liquids and Liquefied Petroleum Gases Regulations 1972 (SI 1972 No 917)*;
- the *Abstract of Special Regulations (Highly Flammable Liquids and Liquefied Petroleum Gases) Order 1974 (SI 1974 No 1587)*.

All the Regulations (except for the final two relate) to particular industries or activities of a specialised kind. The repeal and revocation of Regulations relating to highly flammable liquids and liquefied petroleum gases has wider effect. The more specific requirements of these Regulations are covered by the general goal setting provisions of the *DSEAR* but remain good practice. Extensive guidance on the safe management of flammable liquids and liquefied petroleum gases is available from the LPG Association.

For completeness, the *DSEAR* also repeal and revoke particular parts of the following legislation:

- the *Petroleum (Consolidation) Act 1928*;
- the *Factories Act 1961*;
- the *Shipbuilding and Ship Repairing Regulations 1960 (SI 1960 No 1932)*;
- the *Dangerous Substances in Harbour Areas Regulations 1987 (SI 1987 No 37)*;
- the *Carriage of Dangerous Goods by Road Regulations 1996 (SI 1996 No 2095)*;
- the *Carriage of Dangerous Goods (Classification, Packaging and Labelling) and Use of Transportable Pressure Receptacles Regulations 1996 (SI 1996 No 2092)*;
- the *Workplace (Health, Safety and Welfare) Regulations 1992 (SI 1992 No 3004)*.

Enforcement

[D0940] The application of the *Dangerous Substances and Explosive Atmospheres Regulations 2002 (SI 2002 No 2776) (DSEAR)* is enforced through existing powers by:

- the HSE or local authorities depending on the allocation of premises under the *Health and Safety (Enforcing Authority) Regulations 1998 (SI 1998 No 494)*. In the main, the HSE will enforce at industrial premises and local authorities (environmental health officers) elsewhere, eg in retail premises.
- fire brigades at most premises subject to the *DSEAR* in relation to general fire precautions such as means of escape.
- petroleum licensing authorities at retail petrol filling stations in relation to the storage and dispensing of petrol, LPG and other fuel subject to the 2002 Regulations.

Regulatory impact assessment

[D0941] The Health and Safety Commission prepared a regulatory impact assessment (RIA) on the *Dangerous Substances and Explosive Atmospheres Regulations 2002 (SI 2002 No 2776) (DSEAR)*. RIA showed that businesses complying with current health and safety legislation would expect little or no additional cost as a result of the *DSEAR*. Businesses with work processes involving potentially explosive atmospheres that had not formally recorded zones where these could occur in their risk assessments would need to zone and mark zoned areas with a sign as a specific requirements of *DSEAR*.

Approved codes of practice and guidance etc.

General

[D0942] In support of the *Dangerous Substances and Explosive Atmospheres Regulations 2002 (SI 2002 No 2776) (DSEAR)* and to assist Duty Holders in the interpretation and meeting the requirements of the Regulations, the Health and Safety Commission, with the consent of the Secretary of State, approved the following Approved Codes of Practice (ACoPs) and accompanying Guidance in October 2003:

- DSEAR – Dangerous substances and explosive atmospheres (L138);
- DSEAR – Design of plant equipment and workplaces (L134);
- DSEAR – Storage of dangerous substances (L135);
- DSEAR – Control and mitigation measures (L136);
- DSEAR – Safe maintenance, repair and cleaning procedures (L137);
- DSEAR – Unloading petrol from road tankers (L133);

These ACoPs are published for the Health and Safety Commission by HSE Books. They give practical advice on how to comply with the Regulations and the law. Duty Holders are strongly recommended to have a copy and to have studied them and taken their advice.

Approved Codes of Practice have special legal status. By following the advice, the Duty Holder will be doing sufficient to comply with the Regulations on those matters to which the advice relates. If Duty Holders are prosecuted for an alleged breach of health and safety law, and it is shown that they did not follow the advice given in the ACoP, then the Duty Holder will need to demonstrate that they complied with the law in some other way. Otherwise, the court will find the Duty Holder at fault.

The Regulations and AcoP are accompanied by Guidance that is more wide ranging than the ACoP. This does not form part of the AcoP, and does not have the same legal status, but if Duty Holders follow the Guidance, then they will normally be doing enough to comply with the law. It is not compulsory to follow the Guidance, but Health and Safety Inspectors may refer to this Guidance as illustrating good practice as they ascertain compliance with the Regulations.

Aspects of these ACoPs and Guidance have been incorporated into this text, but this is by no means comprehensive or complete. Duty Holders are strongly advised to rely on the ACoPs and Guidance themselves. In some respects the scope of one ACoP overlaps with another, but the following gives a summary of the scope of each one.

Dangerous substances and explosive atmospheres (L138)

[D0943] This ACoP and Guidance provides an overview on how employers can meet their duties under the *Dangerous Substances and Explosive Atmospheres Regulations 2002 (SI 2002 No 2776) (DSEAR)*. It details all the DSEAR Regulations including the Schedules. A summary of the Regulations is provided and the relationship of DSEAR to other relevant and related health and safety legislation is set out. There is a comprehensive list of references to other Regulations and standards documents.

Guidance is given on the interpretation of terms used in the Regulations, such as dangerous substances, explosive atmospheres, physico-chemical or chemical property, combustion, fires and explosions, hazard, risk, and workplace and work processes. There is Guidance on the application of the Regulation in respect of activities including maritime activities, the medical treatment of patients, gas safety and fittings, the manufacture of explosives and mineral extracting industries, and use of means of transport. Advice is given on the duties of employers and self-employed under the Regulations.

There is substantial Guidance on *SI 2002 No 2776, Reg 5*, which is the requirement for employers to carry out a risk assessment and the nature of the risk assessment. It covers the large number of factors to consider when assessing risks from dangerous substances. The Guidance draws attention to a range of measures for assessing the overall risk presented by dangerous substances, both from the combination of factors as well as assessing each factor individually, and from more global risk management systems that may be in place.

The elimination or reduction of risks from dangerous substances covered within *SI 2002 No 2776, Reg 6* is dealt with in more detail in the other ACoPs and Guidance. Within this document, elimination and reduction of risks are discussed within the context of the overall principle of 'so far as is reasonably practicable'. The elements of risk elimination or reduction by substitution and/or by control and mitigation measures are described, together with some specific issues.

The Guidance relating to *SI 2002 No 2776, Reg 7* regarding places where explosive atmosphere may occur deals with the classification of areas containing explosive atmospheres, including zoning, the selection of equipment with an 'EX' mark and anti-static clothing for use in hazardous areas, and the marking and verification of areas containing explosive atmospheres.

SI 2002 No 2776, Reg 8 deals with accidents, incidents and emergencies, and the Guidance sets out the overall approach. Further Guidance covers the approach for assessing the detection of accidents, incidents and emergencies, assessing the requirements for emergency arrangements, including first aid and safety drills. There is Approved Practice and Guidance on warning and communication systems, escape facilities and mitigation measures and restoring the situation to normal. Additional Guidance is given on making information available to employees and emergency services, the need to review arrangements and the relationship with other legislation including fire safety.

There is Approved Practice on the information that employers must provide their employees to meet the requirements of *SI 2002 No 2776, Reg 9*. This covers the identity of dangerous substances that could be present and where they are used, the type and extent of the risks and the control and mitigation measures adopted and several other aspects. The Guidance draws attention to the need for information, training and instruction, particularly for new employees, and the level and means for delivering it so as to maximise effectiveness. Guidance is given on the need for training and information for non-employees who may be present on site, including members of the public. The Guidance reminds employers to update information, instruction and training whenever changes to the type of work or work methods are made.

The Guidance in support of *SI 2002 No 2776, Reg 10* deals with the identification of the hazardous contents of containers and pipes. It considers means of identification such as labelling or colour coding etc. The duty of co-ordination where two or more employers share the same workplace is covered under *SI 2002 No 2776, Reg 11*, and the Guidance distinguishes between separate individual and shared workplaces.

Further Approved Practice and Guidance is provided covering the Transitional Provisions (*SI 2002 No 2776, Reg 17*), and General Safety Measures (Schedule 1). There is Guidance on the Amendments (Schedule 6), including amendments to regulations covering the handling of the petroleum and other substances, and Repeals and Revocations (Schedule 7).

Design of plant equipment and workplaces (L134)

[D0944] This ACoP and Guidance gives practical advice on what employers need to do to assess, and eliminate or reduce the risks from dangerous substances through the design of plant equipment and workplaces. It helps them to meet the requirements of the *Dangerous Substances and Explosive Atmospheres Regulations 2002 (SI 2002 No 2776) (DSEAR), Regs 5, 6*. It gives advice on the design and use of plant, equipment and workplaces which handle or process dangerous substances, including measures to make redundant plant and equipment safe.

The ACoP covers the selection and supply of equipment for use where dangerous substances or potentially explosive atmospheres may be present. It highlights the need to comply with the European Directive for equipment intended for use in potentially explosive atmospheres, which must carry the EX mark. Control and mitigation measures that can be a result of the design and conduct of the work activity are discussed. These include reducing to a minimum the volume of substance held, measures to avoid or minimise the release of a dangerous substance, and preventing the formation of an explosive atmosphere, and/or adverse conditions that could allow a dangerous substance to react or decompose.

The avoidance of ignition sources is dealt with in summary as more comprehensive treatment is given in ACoP L136 Control and mitigation. Attention is drawn to reducing to a minimum the number of people exposed to the risk and measures to avoid the propagation of fires or explosions from plant and equipment and mitigate the effects of an explosion. Finally the ACoP covers the rendering safe of redundant plant and equipment using for storing, handling or treating dangerous substances, and the safe disposal of such plant and equipment, including LPG and petrol tanks.

Storage of dangerous substances (L135)

[D0945] This ACoP and Guidance gives practical advice for employers on places where dangerous substances may be stored, including the safe disposal of waste materials. This supports meeting the requirements of the *Dangerous Substances and Explosive Atmospheres Regulations 2002 (SI 2002 No 2776) (DSEAR), Regs 5, 6*. It gives advice on aspects such as containment (tanks etc.) in storage and during material transfer, separation (from site boundaries, people, ignition sources etc.), and segregation (separating two or more incompatible dangerous substances). There is also advice on the siting of storage containers (eg LPG tanks) to ensure adequate ventilation, the identi-fication of storerooms and containers, and the maintenance and security of such facilities.

The ACoP highlights the requirements relating to storage, treatment and disposal of waste materials and by-products that are dangerous substances. It considers the dangers that can arise from mixed waste streams where there could be interactions between substances that may not on their own be dangerous. It draws attention to other legislation covering the disposal of waste, environmental protection and the avoidance of pollution.

Control and mitigation measures (L136)

[D0946] This ACoP and Guidance gives practical advice for employers on the hazards (eg fire and explosion) arising from dangerous substances and the measures that can be taken to control and mitigate the risks. Like the ACoPs discussed, it is also to help employers meet the requirements of the *Dangerous Substances and Explosive Atmospheres Regulations 2002 (SI 2002 No 2776) (DSEAR), Regs 5, 6*. ACoP L136 deals particularly with the requirements for adequate ventilation, the avoidance of ignition, and the separation of sub-stances from potential sources of ignition, and routes for the escape of people and access for emergency services.

Regarding ventilation, the ACoP discusses the requirements for adequate ventilation and the means for achieving this through natural (eg vents, windows) or mechanical (eg ducts, fans) methods. Advice is given on the need for the monitoring of ventilation and ventilation equipment, particularly in confined spaces, which may have to be entered for the purposes of mainte-nance, repair or cleaning. Further advice on safe maintenance, repair and cleaning is given in ACoP L137.

There is a detailed list of potential ignition sources that can arise from heat, electrical energy, mechanical interactions and chemical reactions. You should be aware of these in order to ensure adequate separation from and control of ignition sources. Detailed advice is given on separating dangerous substances from other features that could pose a threat or be threatened by the substances though the provision of physical barriers, fire walls and other and fire and explosion resisting facilities.

Safe maintenance, repair and cleaning procedures (L137)

[D0947] This ACoP and Guidance provides practical advice on what employ-ers need to do at places where maintenance and cleaning and other similar non routine activities are carried out. There is advice on identifying hazards and implementing appropriate control procedures and systems of work. Permit to work and other procedures for hot work in places where dangerous substances or explosive atmospheres are potentially or actually present are dealt with in detail, since these activities carry high risk.

High risk activities are those where the foreseeable consequences of an error or omission could result in immediate and serious injuries to people, from for example, an explosion, or a fire or entrapment by fire. It includes hot work on plant containing or that may have contained a dangerous substance or in the vicinity of such plant or in places where an explosive atmosphere could be present. It also includes all work in a confined space containing or that may have contained a dangerous substance, and where plant and equipment containing or that may have contained a dangerous substance is opened.

The ACoP requires Duty Holders to implement a Permit-to-Work system for high risk activities. The Guidance describes what a Permit-to-Work system entails in terms of authorising certain people to carry out specific work in a specific way in a specified time frame under specific conditions, and other important requirements. There is Approved Practice for employers on who can issue a Permit to Work and how it should be operated in terms of those people affected, and where contractors, subcontractors and self employed control their activities through their own Permit to Work system.

Hot work is any activity requiring or generating heat, sparks or flame, and includes activities such as welding, flame cutting, soldering, braising grinding or using disc cutters etc. The ACoP emphasises the need to eliminate hot work wherever reasonably practicable, and there is Guidance on how to conduct hot work where it cannot be eliminated. It specifically discusses preparation and procedures for hot work, cleaning and gas-freeing plant including the inspection and monitoring of atmospheres before hot work is carried out. Where residual levels of dangerous substances may remain, there is Approved Practice and Guidance on the use of inerting techniques, using a substance such as water, nitrogen, or carbon dioxide, to reduce the oxygen level. Finally, there is Approved Practice and Guidance on working on live (on-line) plant and on using gas welding and cutting equipment.

Unloading petrol from road tankers (L133)

[D0948] This ACoP and Guidance gives practical advice and details the necessary measures with regard to the safe unloading of petrol tankers at petrol filling stations, depots and other places. It is not intended to cover the unloading of liquefied petroleum gas (LPG), compressed natural gas (CNG) or liquid natural gas (LNG), but can be applied to the unloading of diesel. Nor is it intended to cover the unloading of petrol at major hazard sites or at sites other than filling stations where it is intended to store more than 100,000 litres.

The *Dangerous Substances and Explosive Atmospheres Regulations 2002 (SI 2002 No 2776) (DSEAR)* revokes, repeals or modifies the licensing controls under the *Petroleum Consolidation Act 1928*. The good practices of the old legislation are being maintained through the DSEAR and the AcoPs. DSEAR removes licensing requirements for petrol, except for petrol that is being kept for dispensing into vehicles. All parts of workplace premises, including non-workplace premises (such as boat or motor clubs etc) where dispensing of petrol takes place remain subject to licensing.

The ACoP and Guidance are concerned with *SI 2002 No 2776, Reg 6* on how the risk to safety from a dangerous substance, such as petrol, can be removed or eliminated. In this respect, they deal with preventing a fire through preventing the overfilling of a storage tank, controlling sources of ignition, during unloading and dealing with any spillages that may occur. However, in addition, the ACOP and Guidance also provide advice on preventing falls from petrol tankers unloading petrol at filling stations etc, in pursuance of the *Workplace (Health, Safety and Welfare) Regulations 1992, Regs 13(1)–(3)*.

The ACoP and Guidance highlights the general duty of everyone involved in the unloading of petrol from a road tanker under the *Health and Safety at Work etc. Act 1974*. It references duties on employers and the self-employed under the *Management of Health and Safety at Work Regulations 1999*, and duties for employers under the *Safety Representatives and Safety Committees Regulations 1977* and the *Health and Safety (Consultation with Employees) Regulations 1996*. All employers and the self-employed are required to carry out a risk assessment of activities that create a fire and explosion risk, such the unloading of petrol, under *SI 2002 No 2776, Reg 5*.

There is Approved Practice and Guidance detailing the responsibilities in respect of the unloading of petrol of the following parties:

(a)   road tanker operators;
(b)   site operators; and
(c)   tanker drivers.

Additional requirements are given for pumped deliveries. There is a section on the prevention of falls from road tankers through the elimination of the need for routine access, or if this cannot be achieved, a safe means of access and a safe place to work.

Other information

General

[D0949] In addition to the ACoPS and Guidance, the Health and Safety Executive publish a free leaflet *Fire and explosion: How safe is your workplace?*, which provides a short guide to the *Dangerous Substances and Explosive Atmospheres Regulations 2002 (SI 2002 No 2776) (DSEAR)* and is aimed primarily and small and medium-sized businesses. Further information on DSEAR can be obtained on HSE's website: www.hse.gov.uk, which is updat ed regularly.

Safe handling of combustible dusts

[D0950] HSE has published fully revised and updated guidance aimed at industries dealing with combustible dusts, informing them of the safest way of handling them.

Safe handling of combustible dusts seeks to lower the risk of a dust explosion, present in industries such as food production (sugar, flour, custard powder), animal feed production and places handling sawdust, many organic chemicals, plastics, metal powders and coal.

The booklet is targeted at those who operate plants handling dusts which can explode. It describes in simple language the tests used on dusts to assess their explosive properties, the precautions used to control the risks and an outline of the health and safety law that applies.

Dust explosions occur when fine materials are disbursed to a certain concentration and an effective ignition source is present. If dust deposits around premises form a cloud an initial small explosion is often followed by a much larger one. Much can be done at the design stage to prevent such explosions, but some risks usually remain. There are many ongoing precautions that need to be followed by those who work in the plant. These are outlined in the booklet.

The booklet takes account of European Directive 99/92/EC on the Protection of Workers Potentially at Risk from Explosive Atmospheres (The ATEX Directive) which was implemented in the UK by the *Dangerous Substances and Explosive Atmospheres Regulations 2002 (SI 2002 No 2776)*. In particular, the guidance describes how the requirement for hazardous area classification, brought in by the Regulations, applies to dust handling plants, and explains which types of new equipment need to be 'ATEX compliant' i.e. be properly marked after undergoing specified tests and checks for their use in hazardous areas.

Copies of *Safe handling of combustible dusts: Precautions against explosions* (HSG103), ISBN 0 7176 2726 8, price £10.95 are available from HSE Books, PO Box 1999, Sudbury, Suffolk, CO10 2WA (tel: 01787 881 165; fax: 01787 313 995).

Case histories

[D0951] The explosion and fire at the Buncefield Oil Depot on 11 December 2005 illustrated the power and devastating effect of dangerous substances and explosive atmospheres. The Final Report of the Major Incident Investigation Board was published by HSE on 11 December 2008.. The cause of the incident was attributed to overfilling a large petrol storage tank. Among its 78 recommendations are lessons for users of dangerous substances and owners and operators of workplaces with the potential for an explosive atmosphere. The effects of the incident in causing over a billion pounds of damage draw attention to the importance of the *Dangerous Substances and Explosive Atmospheres Regulations 2002 (SI 2002 No 2776)* (DSEAR) and the need for good management of such substances and places where explosive atmospheres may be present.

The Health and Safety Executive (HSE) and Sussex Police have warned motor vehicle repair garages about the importance of having a safe system in place for handling and storing petrol. The warning follows the conclusion of proceedings brought against the owner of a Sussex garage. Howard Hawkins, the owner, was found guilty of breaching *section 2(1) of the Health and Safety at Work etc. Act 1974 ('HSWA')*. He was sentenced at Lewes Crown Court receiving a fine of £10,000 with costs of £15,000.

The prosecution followed the death of an apprentice mechanic, Lewis Murphy (18), who died four days after becoming engulfed in flames in an explosion at the Anchor garage, Peacehaven, Sussex on 19 February 2004.

Passing sentence, Judge Richard Hayward, said of Howard Hawkins:

> To say that you were complacent about health and safety is an understatement. You regard health and safety as a tiresome intrusion into your business and a matter of common sense that you could leave to the experience of your mechanics. Being a dinosaur can sometimes be endearing but not on health and safety matters.

While the prosecution was made under *HSWA 1974*, DSEAR [*SI 2002 No 2776*] and the Approved Practice and Guidance would be have been relevant to the case.

Sources and acknowledgement

[D0952] In preparing this chapter, extensive use has been made of the *Dangerous Substances and Explosive Atmospheres Regulations 2002 (SI 2002 No 2776) (DSEAR)*, and of the commentary provided by the Safety Policy Directorate of the HSE, and the Approved Codes of Practice and Guidance. It should again be emphasised that this chapter is not a substitute for having and implementing these documents. It is primarily designed to increase awareness of the responsibilities under DSEAR and to draw attention to the detailed Approved Practice and Guidance that is available.

Disaster and Emergency Management Systems (DEMS)

Raj Lakha, Christopher Eskell and Mary Spear,
revised by Brian Heath and Roger Bentley

Historical development of disaster and emergency management

[D6001] The terrorist attacks with airliners on the twin towers in New York on 11 September 2001 have marked a paradigm shift in the thinking, planning and perception of 'man-made disasters'. Throughout the 1990s the Federal Emergency Management Agency (FEMA) of the USA (a competent authority (CA) dealing with natural and man-made disasters) was being criticised by Congress and the media for being too resource consuming and inefficient. After '9/11' it was being hailed for its expertise in emergency response management. Suddenly 'corporate America' and the democratic world began to take more seriously the importance of 'disaster and emergency management', 'civil defence management', 'civil protection management' and 'business continuity'. In the United Kingdom since June 2001 the 'civil contingencies' function has been transferred from the Home Office to the Cabinet Office, giving the Prime Minister's Office a greater strategic oversight of national 'major incidents'. In 2004 the *Civil Contingencies Act 2004* came into force, followed in 2005 by the *Civil Contingencies Act 2004 (Contingency Planning) Regulations 2005 (SI 2005 No 2042)*. The provisions were put in place to provide a clear framework for civil protection in the United Kingdom. The terrorist attacks which occurred on public transport in London in July 2005 have further served to evidence the need for a co-ordinated, cross-organisational response to emergencies.

The UK experience broadly indicates that disaster and emergency management (DEM) has evolved in three phases:

Phase 1: pre-Control of Industrial Major Accident Hazards Regulations 1984 (CIMAH) (SI 1984 No 1902)
* DEM was largely confined to local government, the emergency services, large organisations and civil protection agencies (environmental, military etc.). This phase is marked by large scale macro plans and detailed, sequential procedures to be followed in the event of a 'disaster' or 'emergency'. It is exemplified in the guidance prevalent on 'incidents' involving radioactivity, available from central government, the United Nations, Red Cross and others.

Phase 2: liberalisation phase
*   With the advent of decentralisation of industry and the growth of privatisation, organisations (large or otherwise) began to focus on 'procedures' and 'business continuity planning'. British Telecom exemplified this with the establishment of its Disaster Recovery Unit and the offering of its expertise to customers at large. 'Major incidents' such as Chernobyl, *Piper Alpha*, Kings Cross, *The Marchioness* and *Challenger* to name a few, which hit the world headlines in the 1980s also focused planners' minds on effective DEM not just isolated procedures. Simultaneously the introduction of the *CIMAH* Regulations in 1984 required that industrial sites with major accident potential prepared on-site plans and co-operated with local authorities and other agencies in developing off-site plans for the area at risk around the installation.

Phase 3: holistic phase
*   This has two dimensions, first the role of Europe. The impact of the *Framework and Daughter Directives*, as transposed into the '6 Pack Regulations' of 1992, introduced for the first time explicit and strict duties on organisations in general to plan for 'serious and imminent dangers'. The European Commission also began to take the risks of trans-national 'disasters' more seriously, as highlighted by the greater role given to the Civil Protection Unit, in DG XI of the European Commission. Secondly, this phase sees domestic UK legislation becoming more cognisant of disaster and emergency issues (*Environment Act 1995* makes provisions for environmental emergencies; *Control of Major Accident Hazards Regulations 1999 (COMAH) (SI 1999 No 743)* as amended by the *Control of Major Accident Hazards Regulations (Amendment) Regulations 2005 (COMAH) (SI 2005 No 1088)* replaces the *Control of Industrial Major Accident Hazards Regulations 1984 (CIMAH) (SI 1984 No 1902)*; also the need for effective planning when carrying dangerous goods via road legislation, *Carriage of Dangerous Goods by Road Regulations 1996 (SI 1996 No 2095)*).

The most noticeable index of how DEM has changed is the availability of information and templates on 'disaster planning' and 'emergency planning' to organisations of all sizes. There is also a greater media focus on such issues, with the BBC, for instance, having dedicated web resources on 'disasters'.

Origin of disaster and emergency management as a modern discipline

[D6002] DEM is largely a fusion of four branches of knowledge:

Occupational safety and health (OSH)
*   This concerns 'internal' or work-based causes of systems failures, with major impact on life and property. It has been exemplified by Heinrich in his publication *Unsafe Acts and Unsafe Conditions* in the early part of the 1900s, Bird and Loftus with their 'management failing' expla-

nations of accidents and incidents and Turner with his 'incubation' explanation of man-made disasters in the 1970s. OSH academics have led the forefront in terms of analysing the branch and root causes of major industrial disasters.

Security management
- In the 1970s, security threats in the UK such as bomb explosions, terrorism and electronic surveillance failures have given insights to causation and the motivation behind man-made emergencies. Also, guidance from the Home Office as well as the emergency services has enabled a practical understanding of how to cope with emergency situations.

Business management
- The late 1980s saw a shift in the academic paradigm in economics and business management from 'static' or closed business planning – where businesses were told to make the assumption of *cetaris paribus,* that is assume all things are constant with the business acting as if it was the only one in the market place – to 'dynamic' or open planning. The latter sees uncertainty and risk being factored into decision making models. This influenced the development of 'business continuity management', developing strategies when the business faces major corporate uncertainty and crises as well as 'contingency planning'.

Insurance
- The fourth significant influence comes from insurance and loss control. The occurrence of disasters and accidents has involved loss adjusters and actuarial personnel. The former have developed methods of analysing the basic and underlying causes of an event, whilst the latter have been developing statistical methods for calculating the chance of failure and the risk premiums needed to indemnify that failure.
  DEM uses both quantitative (statistics, quantified risk assessments, hazard analysis techniques, questionnaires, computer simulation etc.) and qualitative methods (inspections, audits, case studies etc.).

Definitions

[D6003] Stan Kaplan (*The Words of Risk Analysis,* 1997, Risk Analysis, Vol 17, No 4) stated that 50 per cent of the problems with communication are due to individuals using the same words with different meanings. The remaining 50 per cent are due to individuals using different words with the same meaning. In DEM this is a basic problem. Some authors have argued distinguishing definitions has no practical value and can be ' . . . highly undesirable to try to control how others use them . . . ' (*Managing the Global Consequences of a Disaster,* Richard Read, Paper at the 2nd International Disaster and Emergency Readiness Conference, The Hague, October 1999, page 130 of IDER Papers). Nevertheless there is much confusion over the meaning given to core terms, so that basic definitions can assist in avoiding the Kaplan dilemma. Legislation or approved codes have not provided definitions of 'disaster' or 'emergency' for instance. The *New Oxford Dictionary of English*, OUP, (1999) provides the following primary meanings:

(a)   Catastrophe: 'an event causing great and often sudden damage or suffering: a disaster', page 287.

(b)   Crisis: 'a time of intense difficulty or danger', page 435.

(c)   Disaster: 'a sudden event, such as an accident or a natural catastrophe, that causes great damage or loss of life', page 524.

A catastrophe and a crisis are types of disaster, namely more severe.

(d)   Emergency: 'a serious, unexpected and often dangerous situation requiring immediate action', page 603.

Whilst both disasters and emergencies can be sudden, the former has a macro, large scale impact whilst the latter requires an immediate response.

The definitions provided in Parts 1 and 2 of the *Civil Contingencies Act 2004* are key to understanding of the nature and consequences of emergencies and to ensuring the establishment of appropriate safeguards and responses.

(e)   Accident: '1. an unfortunate incident that happens unexpectedly and unintentionally, typically resulting in damage or injury, . . . 2. an event that happens by chance or that is without apparent or deliberate cause', page 10.

It should be noted that the *Control of Major Accident Hazards Regulations 1999 (COMAH) (SI 1999 No 743)* have introduced the term 'major accident'; this is an event due to:

(i)   unexpected, sudden, unplanned developments in the course of the operation;

(ii)   leads to serious danger to people and the environment both on site at the place of work and off site; and

(iii)   the event involves at least one dangerous substance as defined by *COMAH*.

Given the potential for both human and property loss, and given the *COMAH* requirements for emergency on-site and off-site plans, there seems to be little practical difference between a major accident and an emergency. Both are *response based* concepts. However, if the legislators wished a major accident to be different from an emergency then they would have either said so or implied so. One can regard a major accident as a type of emergency situation.

(f)   Incident: ' an event or occurrence', page 923.

The *Reporting of Injuries, Diseases and Dangerous Occurrences Regulations (RIDDOR) (SI 1995 No 3163)* do not define an accident or incident, but classify the types.

(g)   Major incident

In *Dealing with Disaster* (Home Office, revised third edition 2003), a 'major incident' is the only term explicitly defined by the Home Office: A 'major incident' is any emergency that requires the implementation of special arrangements by one or more of the emergency services, the NHS or the local authority for:

(i)   the initial treatment, rescue and transport of a large number of casualties;

(ii)   the involvement either directly or indirectly of large numbers of people;

(iii)   the handling of a large number of enquiries likely to be generated both from the public and the news media, usually to the police;

(iv)   the need for large scale combined resources of two or more emergency services;

(v)   the mobilisation and organisation of the emergency services and supporting organisations, e.g. local authority, to cater for the threat of death, serious injury or homelessness to a large number of people, page 43.

This definition is accepted by the police, fire service, local government and broadly the NHS (they also have a specific definition of major incident).

Thus the term 'major incident' is a broad phrase encompassing an array of events. 'Accident' has not been included as a type of major incident given the definition of the latter requiring 'major' mobilisation of human and physical resources which in most accidents is not necessarily the case. As an example:

(1)   two trains missing each other would be an 'incident' (near miss);

(2)   an employee or member of the public being injured on a train – this would be an 'accident', irrespective of injury type or fatality (following *RIDDOR*);

(3)   an event at a *COMAH* site where dangerous substances ignite causing damage to the plant, injury to personnel and emissions into the local community, would be a 'major accident' under *COMAH*;

(4)   the immediate event after a train collision and the response needed to the chaos – this would be an 'emergency'.

(5)   if two trains collide causing multiple fatalities and immediate property and environmental damage, that would be referred to as a 'disaster'. For example Ladbroke Grove rail crash in 1999;

(6)   if the collision, with the multiple fatalities and property damage is difficult to access, manage and control, this would be a 'crisis'. Ladbroke Grove fell short from being a 'crisis' in contrast to Clapham Junction in 1988 where a triple train crash caused major access and logistical problems;

(7)   if the event generated major environmental, public and social harm that has 'longer term' implications, over and above the immediate human and property loss, this would be a 'catastrophe' For example, Kings Cross underground fire (1987), Chernobyl (1986), and *Piper Alpha* (1988) to name a few that had wider consequences over and above the immediate impact.

Events (1)–(7) would be major incidents.

Figure 1 overleaf summarises the essential differences between the above events.

FIGURE 1: CLASSIFICATION OF INCIDENTS, ACCIDENTS AND MAJOR INCIDENT TYPES

Characteristic:	Event:	Incident	Accident	Major Accident	'MAJOR INCIDENTS' Emergency	Disaster	Crisis	Catastrophe
				Types of emergency		Types of disaster		
1. MPL	Low							Very High
2. RISK:								
Severity	Near miss etc							Multiple Fatalities
Consequence	Minor							Major
3. NUMBERS	1-5							100 +
4. SOCIO-LEGAL IMPACT	No change likely							New Laws or Guidance
5. TIME	Short							Longer Impact
6. COST:								
Individual	Short Term							Irreparable
Social	None Usually							Irreparable
Environment	None Usually							Irreparable

Key:

MPL=	Maximum Potential Loss (economic and property loss)
Risk=	Severity x Consequence (Severity refers to the quantum of harm generated by the event whilst Consequence measures the scale of impact)
Numbers=	The number of individuals affected
Socio-Legal Impact=	The impact on social attitudes and legislation/guidance as a result of the event
Time=	The length of the event
Cost=	The loss suffered by the individual, society or nature

Such classifications are important, firstly from a philosophical perspective one needs to know how they differ, and secondly from a planning perspective as the resource allocation will accordingly differ. Thirdly, from a response perspective, the response to an incident differs from a disaster, with the organisation needing to define and clarify when an event is an incident and not a disaster.

The need for effective disaster and emergency management

[D6004] There are several reasons why DEM is needed: legal reasons (see D6005–D6029); insurance reasons (see D6030); corporate reasons (see D6031); humanitarian and societal reasons (see D6032); and environmental reasons (see D6033).

Legal reasons

[D6005] In 2004, a fundamental change in the way in which the United Kingdom approaches the preparation for and management of emergencies took place. The *Civil Contingencies Act 2004* came into force and had the effect of establishing a framework for civil protection capable of meeting the variety of challenges likely to be met and as defined in the Act. The Act also had the effect of repealing or revoking those Acts or sections of Acts which dealt with issues of emergency planning or response. In particular, the *Civil Defence Act 1948*; the *Civil Defence (Grant) Regulations 1953 (SI 1953 No 1777)*; the *Local Government Act 1972, s 138*; the *Civil Protection in Peacetime Act 1986*; and the *Emergency Powers Act 1920* have all ceased to have effect.

The Civil Contingencies Act 2004

[D6006] The structure of the *Civil Contingencies Act 2004 (CCA 2004)* separates it into two substantive parts: CCA *2004, Pt 1 (ss 1–18)*, focuses on local arrangements for civil protection and imposes a series of duties on local bodies in the United Kingdom which the Act classes as category I responders. Part 1 also establishes category 2 responders who are required to co-operate with and provide information to category 1 responders in pursuance of their civil protection duties. *CCA 2004, Pt 2 (ss 19–31)*, deals with emergency powers and in repealing the *Emergency Powers Act 1920* allows for a Minister of Sate, in certain defined situations, the power to make regulations if an 'emergency' has occurred or is about to occur. The definition of what constitutes an emergency is key to understanding the purpose and operation of CCA 2004 and by extension the duties and responsibilities of the category 1 and category 2 responders.

The meaning of 'emergency' is defined in both substantive sections of *CCA 2004*. *CCA 2004, s 1(1)* defines 'emergency' for the purposes of *CCA 2004, Pt 1* as an event or situation which threatens damage to human welfare in the United Kingdom, serious damage to the environment of the UK or any act of war or terrorism which threatens serious damage to the security of the United Kingdom. The meaning of event or situation is contained in *CCA 2004, s 1(2)–(3)* and includes loss of human life, human illness and injury, damage to property, disruption to transport and communications, disruption to essential supplies, disruption to health services, contamination of the environment and disruption or destruction of plant and animal life.

The meaning of 'emergency' given in *CCA 2004, s 19(1)–(6)* for the purposes of *CCA 2004, Pt 2* are substantially the same as that given in *CCA 2004, Pt 1* with one exception, that of scale. In order to satisfy the definition given in *CCA 2004 Pt 2*, the threat must be serious and must apply to the United Kingdom, a part or a region. *CCA 2004, s 19(5)* provides the Secretary of State with the power to amend the list of events or situations to allow appropriate response should a system or service become so necessary that disruption of it fall within the given definition of 'emergency'.

Category 1 responders are listed in *CCA 2004, Sch 1 Pt 1* and are at the heart of emergency response: they include local authorities, the emergency services, health services, which include NHS trusts, primary care trusts, local health boards, the Health Protection Agency and port health authorities constituted under the *Public Health (Control of Disease) Act 1984, s 2(4)*. Also included are the Environment Agency and the Secretary of State in respect of his obligations to respond to maritime and coastal emergencies. The list of responders can be amended by a Minister of the Crown under *CCA 2004, s 13*.

Category 2 responders are listed in *CCA 2004, Sch 1, Pt 3* and include public utilities, transport agencies, airport operators, harbour authorities, the Secretary of State in respect of matters for which he is responsible by virtue of the *Highways Act 1980, s 1* (c66) (highway authorities) and the Health and Safety Executive (HSE). The role of category 2 responders is to be co-operating bodies. Although involved in the management of major hazard accidents within their own areas of influence they will be less directly involved in contingency planning for emergencies.

CCA 2004, s 2 establishes the primary duties of category 1 responders to assess, plan and advise in the context of contingency planning. In principle the category 1 responders are required to assess the risk of emergencies occurring and use this information to inform contingency planning, put in place and maintain as necessary emergency plans, put in place business contingency management plans, put in place arrangements to warn, inform, advise the public in the event of an emergency, share information with other local responders to enhance co-ordination and co-operate with other local responders to enhance co-operation. They are required to act in a manner which will prevent the occurrence of an emergency or will reduce, control or mitigate its effects and efficiency. The duties are given further form under the *Civil Contingencies Act 2004 (Contingency Planning) Regulations 2005 (SI 2005 No 2042)* (see **D6007**). In addition, local authorities are required to provide advice about business continuity management to local businesses and voluntary organisations. *CCA 2004, s 2* is supported by *CCA 2004, s 4* which imposes a duty on responders to provide advice and assistance to the public.

Throughout *CCA 2004, Pt 1* the emphasis is on responsiveness with power given to Ministers through general measures to amend definitions and make regulations as required to give effect to the regulatory framework.

CCA 2004, Pt 2 continues this emphasis on responsiveness and provides under *CCA 2004, s 20* the power to make emergency regulations to meet the requirements of the situations defined in *CCA 2004, s 19* and to apply these regulations to specific areas or regions through the appointment of regional and emergency co-ordinators (*CCA 2004, s 24*). The emergency regulations are made by Her Majesty by Order in Council or, in the event of urgency, by a 'senior Minister of the Crown'. A 'senior Minister of the Crown' is defined by *CCA 2004, s 24(3)* to mean the First Lord of the Treasury (the Prime Minister), any of Her Majesty's Principal Secretaries of State and the Commissioners of Her Majesty's Treasury. Limitations are placed on this power by the definition of 'emergency' contained in *CCA 2004, s 19* and by the number of provisions in the remainder in the remainder of the section which establish limitations in respect of emergency regulations (*CCA 2004, s 23*), limitations on duration (*CCA 2004, s 26*) and Parliamentary scrutiny (*CCA 2004, ss 17 and 28*).

CCA 2004 seeks to be responsive to dynamic situations through the making of specific, temporary legislation and to balance this power with specific safeguards as indicated above in that an emergency that threatens serious damage to human welfare, the environment or national security has occurred or is about to occur, it is demonstrably necessary to bring legislation urgently because existing legislation is insufficient to deal with the situation and normal routes are to slow and that the legislation provided is proportional to the emergency faced. The emergency regulations cannot be used to counter strike or other industrial action or to make substantive changes to existing legislation. They must be compatible with European legislation and the *Human Rights Act 1998 (HRA)* and they must be open to challenge in the courts.

CCA 2004 provides for the making of regulations and have given rise to the *Civil Contingencies Act 2004 (Contingency Planning) Regulations 2005 (SI 2005 No 2042)*.

Civil Contingencies Act 2004 (Contingency Planning) Regulations 2005 (SI 2005 No 2042)

[D6007] The Regulations are made under the *Civil Contingencies Act 2004 (CCA), s 2(3)* and relate to the duties imposed on category 1 responders under *CCA 2004, ss 2 and 4* to assess and plan for emergencies and to provide advice and assistance to the public. The Regulations are split into ten parts and comprise some 58 Regulations and one schedule under *SI 2005 No 2042, Reg 3(1)*.

Part 2 of the Regulations, comprising *SI 2005 No 2042, Regs 4–12* inclusive, makes general provisions relating to the extent and the performance of the duties established in *CCA 2004*. *SI 2005 No 2042, Reg 4* relates to co-operation and local resilience forums and requires that category 1 responders co-operate with each other within a single forum, the local resilience forum, meeting at least once every six months and inform and invite category 2 responders as appropriate. *SI 2005 No 2042, Reg 5* makes the same provision for Scotland but referring to the forum for co-operation as the strategic co-ordinating group. *SI 2005 No 2042, Reg 6* provides the same requirement for co-operation in Northern Ireland. The co-operation required may be facilitated by the setting up of protocols (*SI 2005 No 2042, Reg 7*) or by establishing areas of joint responsibility (*SI 2005 No 2042, Reg 8*). *SI 2005 No 2042, Regs 9–11* relate to the establishment of a lead category 1 responder; *SI 2005 No 2042, Reg 9* allows for the identification of a lead category 1 responder in respect of a relevant civil protection duty under *CCA 2004, s 2* whilst *SI 2005 No 2042, Regs 10, 11* relate to the duty to inform and co-operate falling upon the lead category 1 responder and the category 1 responders who do not a have a lead responsibility respectively. *SI 2005 No 2042, Reg 12* excludes responsibility for emergency planning duties covered under existing legislation: this includes major accident hazards under the *Control of Major Accident Hazard Regulations 1999 (SI 1999 No 743)* and the *Control of Major Accident Hazard (Amendment) Regulations 2005 (SI 2005 No 1088)*.

Part 3 of the Regulations, comprising *SI 2005 No 2042, Regs 13–18* inclusive, relates to the duties of category 1 respondents under *CCA 2004, s 2(1)(a)* and *(b)* to assess the risk of an emergency occurring. *SI 2005 No 2042, Reg 13* limits that duty to emergencies in areas where the category 1 responder functions. This duty extends under *SI 2005 No 2042, Reg 15* to collaborating with other category 1 responders to maintain a 'community risk register' of the risk assessments made by the local resilience forum. *SI 2005 No 2042, Regs 16–18* relate this duty to the sharing of information with other responders throughout England, Scotland and Wales. *SI 2005 No 2042, Reg 14* enables Ministers of State in England, Scotland and Wales to issue guidance on emergencies, the likelihood of emergencies occurring or the likely impact of the same.

Part 4 of the Regulations, comprising *SI 2005 No 2042, Regs 19–26* inclusive, relates to the duty of category 1 responders under *CCA 2004, s 2(1)(c)* or *(d)* to maintain plans necessary to respond to an emergency made in respect of any risk assessment carried out under *CCA 2004, s 2(1)(a)* or *(b)*.and, under *CCA 2004, s 2(1)(g)* with regard to the provision to maintain advice and information to the public *(SI 2005 No 2042, Reg 21)*. *SI 2005 No 2042, Reg 23* requires that in considering these plans that category 1 responders take into account the activities of voluntary organisations. In maintaining plans, category 1 responders should consider whether those plans can be generic *(SI 2005 No 2042, Reg 21)* or specific *(SI 2005 No 2042, Reg 22)* in respect of the emergency under consideration. *SI 2005 No 2042, Reg 24* requires that a procedure be put in place to determine whether or not an emergency has occurred. *SI 2005 No 2042, Reg 25* requires that the effectiveness of plans be determined through appropriate arrangements being made for exercises and training. *SI 2005 No 2042, Reg 26* requires the consideration of plans and necessary revisions in light of any guidance issued by ministers of state in the UK.

Part 5 of the Regulations *(SI 2005 No 2042, Reg 27)* requires that in publishing any plans due care should be taken not to unnecessarily alarm the public.

Part 6 of the Regulations, comprising *SI 2005 No 2042, Regs 28–35* inclusive, relates to the duty of category 1 responders to maintain arrangements to warn, inform and advise members of the public if an emergency occurs or is likely to occur in respect of plans made under *CCA 2004, s 2(1)(g)*. As with *Part 4* of the Regulations the arrangement made may be either generic or specific as appropriate *(SI 2005 No 2042, Reg 29)*, must be supported by training and exercises *(SI 2005 No 2042, Reg 31)* and must not unduly alarm the public *(SI 2005 No 2042, Reg 30)*. *SI 2005 No 2042, Regs 32–34* require that arrangement be put in place to identify a lead category 1 responder to have lead responsibility to inform the public. *SI 2005 No 2042, Reg 35* requires that responders have due regard to information provided to the public by other bodies including the Meteorological Office and the Food Standards Agency.

Part 7 of the Regulations, comprising *SI 2005 No 2042, Regs 36–44* inclusive, relates to the duty owed by local authorities, as defined by *CCA 2004, Sch 1, para 1* referred to in *SI 2005 No 2042, Reg 36(a)* as a 'relevant responder', under *CCA 2004, s 4(1)* to provide advice and assistance to the public for the continuance of commercial activities *(SI 2005 No 2042, Reg 39)* or voluntary activities *(SI 2005 No 2042, Reg 40)* against which provision a charge may be levied *(SI 2005 No 2042, Reg 44)*. In common with other sections provision is made for co-operation between relevant responders and the establishment of a lead relevant responder *(SI 2005 No 2042, Regs 41–43)*, with *Regulation 42* relating to cross border communications with Scotland.

Part 8 of the Regulations, comprising *SI 2005 No 2042, Regs 45–54* inclusive, relates to information and the control of the same. *SI 2005 No 2042, Regs 45–46* define 'sensitive information' and provide for control of information where the release of such information may be prejudicial to national security or public safety, may be commercially sensitive or may be personal as defined by the *Data Protection Act 1998 (DPA), s 1(1)(a)–(d)*. *SI 2005 No 2042, Regs 47–50* allow category 1 and 2 responders and Scottish category 1 and 2 responders to seek information from other category 1 and 2 responders in connection with their duties under *CCA 2004*. It is for responders to determine whether the information requested is in fact sensitive. Where sensitive information has been received *SI 2005 No 2042, Reg 51* prohibits its disclosure except in exceptional circumstances. *SI 2005 No 2042, Reg 52* in the same context limits the use of such information to the specific function for which the user has requested it. *SI 2005 No 2042, Reg 53* requires that information once requested and received be kept secure. *SI 2005 No 2042, Reg 54* applies specifically to the HSE and amends the *Health and Safety at Work etc. Act 1974 (HSWA), ss 28(2)* and *28(7)* to allow disclosure of information under the requirements of *CCA 2004*.

Part 9 of the Regulations, comprising *SI 2005 No 2042, Regs 55, 56,* applies the function of the regulations to London and provides for the London Fire Brigade and the emergency planning authority to take the lead role in all functions under the Regulations.

Part 10 of the Regulations, comprising *SI 2005 No 2042, Regs 57, 58,* relates to the performance of duties under *CCA 2004* in Northern Ireland.

The Civil Defence (General Local Authority Functions) Regulations 1993 (SI 1993 No 1812)

[D6008] These Regulations became law in August 1993. Local government is no longer under a duty to provide 'protected emergency centres' for war time preparation and planning. It was a requirement that county and regional councils have a main and standby centre, whilst, district councils provide one emergency centre.

However, county and regional councils are required to:

* create, assess, review, update and exercise plans;
* ensure county and district council staff/other relevant persons are trained;
* action the plans created (not just a 'paper exercise'); and
* to work with and consult other neighbouring authorities.

District councils are required to:

* ensure that relevant information is given to the county council;
* provide the necessary support in the making/revising of plans;
* assist where necessary in preparations;
* assist in implementation of new plans; and
* to ensure staff are adequately trained.

Unitary authorities are also covered and have similar functions to county and regional councils.

It should be noted that Scotland, Wales and Northern Ireland have additional and specific legislation e.g. the *Civil Defence (Scotland) Regulations 2001 (SI 2001 No 139)*.

The Health and Safety at Work etc. Act 1974

[D6009] The intentions of the *Health and Safety at Work etc. Act 1974* are captured by *HSWA 1974, s 1(1)(a)–(d)* which states that the Act is concerned with 'securing the health, safety and welfare of persons at work', protecting the public, effective control over dangerous substances/explosives and the effective control of emission of noxious substances into the atmosphere. Implicit at least is the intention that this Act will influence actions that contribute to disaster and emergency situations, whether industrial, chemical or environmental. The Act does give expressed powers to the Health and Safety Commission to investigate and to hold inquiries in relation to: ' . . . any accident, occurrence, situation . . . '. [*HSWA 1974, s 14(1)*].

The Management of Health and Safety at Work Regulations 1999 (MHSWR) (SI 1999 No 3242)

[D6010] These Regulations provide the main detail applicable to all organisations, for providing 'procedures for serious and imminent danger and for danger areas' [*Management of Health and Safety at Work Regulations 1999 (SI 1999 No 3242), Reg 8*] and 'contacts with external services'. [*SI 1999 No 3242, Reg 9*]. *Regulation 1* (Citation, commencement and interpretation) does not define 'procedure for serious and imminent danger', although the ACOP L21 to *SI 1999 No 3242* exemplifies such situations, 'e.g. a fire, or for the police and emergency services an outbreak of public disorder' (page 20).

SI 1999 No 3242, Reg 8(1) says that a strict duty exists on all employers to:

(a)   'establish and where necessary give effect to appropriate procedures to be followed in the event of serious and imminent danger to persons at work in his undertaking'.

(b)   'nominate a sufficient number of competent persons to implement those procedures in so far as they relate to the evacuation from premises of persons at work in his undertaking'.

(c)   'ensure that none of his employees has access to any area occupied by him to which it is necessary to restrict access on grounds of health and safety unless the employee concerned has received adequate health and safety instruction'.

SI 1999 No 3242, Reg 8(2) says 'Without prejudice to the generality of *paragraph (1)(a)*, the procedures referred to in that sub-paragraph shall:

(a)   'so far as is practicable, require any persons at work who are exposed to serious and imminent danger to be informed of the nature of the hazard and of the steps taken or to be taken to protect them from it;

(b)   enable the persons concerned (if necessary by taking appropriate steps in the absence of guidance or instruction and in the light of their knowledge and the technical means at their disposal) to stop work and immediately proceed to a place of safety in the event of their being exposed to serious, imminent and unavoidable danger; and

(c)   save in exceptional cases for reasons duly substantiated (which cases and reasons shall be specified in those procedures), require the persons concerned to be prevented from resuming work in any situation where there is still a serious and imminent danger'.

SI 1999 No 3242, Reg 8(3) says: 'A person shall be regarded as competent for the purposes of paragraph *(1)(b)* where he has sufficient training and experience or knowledge and other qualities to enable him properly to implement the evacuation procedures referred to in that sub-paragraph'.

SI 1999 No 3242, Reg 8 can be summarised:

- Procedures need to be sequential, logical, documented and clear (clarity of procedures).
- Procedures need to be justified and authorised (legitimise procedures).
- A 'hierarchy of procedural control' seems to be advocated by the Regulation i.e:
    — give information to those potentially affected by serious and imminent dangers;

— take actions or steps to protect such people;
— stop work activity if necessary to reduce danger; and
— prevent the resuming of work if necessary until the danger has been reduced or eliminated.

- Appoint competent persons preferably from within the organisation who will assist in any evacuations. It is implied that the role and responsibility of such persons needs to be clearly demarcated.
- Generally prohibit access to dangerous areas (site management). If access to a 'danger area' is required i.e. a place which has an unacceptable level of risk but must be accessed by the employee, then appropriate measures must be taken as specified by other legislation (see below).

In addition, the Regulations imply:

- Risk assessments will identify foreseeable events that may need to be covered by procedures. Such assessments may also identify 'additional risks' that need additional procedures. Thus the risk assessment, as shall be discussed below, is a vital tool to keep procedures in tune with current generic and specific risks.
- Procedures need to be dynamic – reflecting the fact that events can occur suddenly.
- There may be a need to co-ordinate procedures where workplaces are shared.
- Procedures should also reflect other legislative requirements (see below).

SI 1999 No 3242, Reg 9 is a new addition to the *MHSWR*. *Regulation 9* reads: 'Every employer shall ensure that any necessary contacts with external services are arranged, particularly as regards first-aid, emergency medical care and rescue work'. It can be inferred that 'necessary contacts with external services' does not only relate to the emergency services but the organisation needs to identify both private sector and voluntary organisations that can be called on for assistance. The organisation needs to identify contact names, addresses and contact numbers of such bodies and develop relations with them.

The Control of Major Accident Hazards Regulations 1999 (SI 1999 No 743) and the Control of Major Accident Hazards (Amendment) Regulations 2005 (SI 2005 No 1088)

[D6011] The *Control of Major Accident Hazards Regulations 1999 (SI 1999 No 743)* implement the requirements of the 'Seveso II' Directive (96/82/EC) on the control of major accident hazards involving dangerous substances.

The HSE and the Environment Agency (in England and Wales) or the Scottish Protection Agency are jointly responsible as the CA for *COMAH*. The CA is the author of a report (available on its website www.hse.gov.uk/comah) listing recent major accidents at industrial premises subject to *COMAH*.

On-site emergency plan

[D6012] *Control of Major Accident Hazards Regulations 1999 (SI 1999 No 743), Reg 9(1)*, requires operators of top-tier establishments to prepare an on-site emergency plan. The plan must be adequate to secure the objectives specified in *SI 1999 No 743, Sch 5, Pt 1,* namely:

— containing and controlling incidents so as to minimise the effects, and to limit damage to persons, the environment and property;
— implementing the measures necessary to protect persons and the environment from the effects of major accidents;
— communicating the necessary information to the public and to the emergency services and authorities concerned in the area; and
— providing for the restoration and clean-up of the environment following a major accident.

The plan must contain the following information:

— names or positions of persons authorised to set emergency procedures in motion and the person in charge of and co-ordinating the on-site mitigatory action;
— name or position of the person with responsibility for liaison with the local authority responsible for preparing the off-site emergency plan (see **D6013**);
— for foreseeable conditions or events which could be significant in bringing about a major accident, a description of the action which should be taken to control the conditions or events and to limit their consequences, including a description of the safety equipment and the resources available;
— arrangements for limiting the risks to persons on site including how warnings are to be given and the actions persons are expected to take on receipt of a warning;
— arrangements for providing early warning of the incident to the local authority responsible for setting the off-site emergency plan in motion, the type of information which should be contained in an initial warning and the arrangements for the provision of more detailed information as it becomes available;
— arrangements for training staff in the duties they will be expected to perform, and where necessary co-ordinating this with the emergency services; and
— arrangements for providing assistance with off-site mitigatory action.
[SI 1999 No 743, Sch 5, Pt 2.]

New establishments must prepare such a plan before start-up.

When preparing an on-site emergency plan, the operator must consult:

— employees, persons employed in that the establishment; [*SI 2005 No 1088, Reg 9(3)(a)*]. This amendment is to ensure that subcontractors and the like are consulted as well as the operators own employees.
— the Environment Agency (or the Scottish Environment Protection Agency);
— the emergency services;
— the health authority for the area where the establishment is situated; and
— the local authority, unless it has been exempted from the requirement to prepare an off-site emergency plan.

Off-site emergency plan

[D6013] The local authority for the area where a top-tier establishment is located must prepare an adequate emergency plan for dealing with off-site consequences of possible major accidents. As with the on-site plan, it should be in writing.

The objectives set out in *SI 1999 No 743, Sch 5, Pt 1* (see **D6012** above) also apply to off-site emergency plans.

The plan must contain the following information:

— names or positions of persons authorised to set emergency procedures in motion and of persons authorised to take charge of and co-ordinate off-site action;
— arrangements for receiving early warning of incidents, and alert and call-out procedures;
— arrangements for co-ordinating resources necessary to implement the off-site emergency plan;
— arrangements for providing assistance with on-site mitigatory action;
— arrangements for off-site mitigatory action;
— arrangements for providing the public with specific information relating to the accident and the behaviour which it should adopt; and
— arrangements for the provision of information to the emergency services of other member states in the event of a major accident with possible transboundary consequences.

An operator must supply the local authority with the information necessary for the authority's purposes, plus any additional information reasonably requested in writing by the local authority.

In preparing the off-site emergency plan, the local authority must consult:

— the operator;
— the emergency services;
— the CA;
— the Agency (*SI 2005 No 1088, Reg 11* requires that the Environment Agency and SEPA is consulted on off-site plans. This provides the Agency with the status of consultee and enables it to comment on plans from an emergency response perspective);
— each health authority for the area in the vicinity of the establishment; and
— such members of the public as it deems appropriate.
  [SI 1999 No 743, Reg 10(6) as amended.]

In the light of the safety report, the CA may exempt a local authority from the requirement to prepare an off-site emergency plan in respect of an establishment [*SI 1999 No 743, Reg 10(7)*].

Reviewing, testing and implementing emergency plans

[D6014] *Control of Major Accident Hazards Regulations 1999 (SI 1999 No 743), Reg 11*, requires that emergency plans are reviewed and, where necessary, revised, at least every three years. Reviewing is a key process for addressing the adequacy and effectiveness of the components of the emergency plan – it should take into account:

—   changes occurring in the establishment to which the plan relates;
—   any changes in the emergency services relevant to the operation of the plan;
—   dvances in technical knowledge;
—   knowledge gained as a result of major accidents either on-site or elsewhere; and
—   lessons learned during the testing of emergency plans.

There is a requirement to test emergency plans at least every three years. Such tests will assist in the assessment of the accuracy, completeness and practicability of the plan: if the test reveals any deficiencies, the relevant plan must be revised. Agreement should be reached beforehand between the operator, the emergency services and the local authority on the scale and nature of the emergency plan testing to be carried out. Under *SI 2005 No 1088, Reg 12* this alters *SI 1999 No 743, Reg 11(1)* and requires local authorities to consult with such members of the public as it deems appropriate in it review of off site emergency plans.

Where there have been any modifications or significant changes to the establishment, operators should not wait for the three-year review before reviewing the adequacy and accuracy of the emergency planning arrangements.

When a major accident occurs, the operator and local authority are under a duty to implement the on-site and off-site emergency plans [*SI 1999 No 743, Reg 12*].

Guidance is to be found in the HSE publications *A guide to the Control of Major Accident Hazards Regulations 1999 (as amended)*, HSG111, and *Emergency planning for major accidents*, HSG 191.

The Control of Substances Hazardous to Health Regulations 2002 (SI 2002 No 2677) as amended

[D6015] Where emergency procedures drawn up to comply with the *Management of Health and Safety at Work Regulations 1999 (SI 1999 No 3242)* (see D6010) are insufficient to contain and control any risk to health that hazardous substances might pose during an emergency, then the *Control of Substances Hazardous to Health Regulations 2002, Reg 13* requires employers to extend their emergency procedures to ensure they are capable of:

(a)   mitigating the effects of an incident;
(b)   restoring the situation to normal as soon as possible; and
(c)   limiting the risks to health of employees and anyone else likely to be affected.

The HSE have published an *Approved Code of Practice and guidance*, L5, which provides considerable detail on the requirements. To deal with situations which could present significantly greater risks on account of substances hazardous to health, employers should extend their emergency procedures to include:

(a)   the identity of the relevant substances present at the workplace, where they are stored, used, processed or produced, and an estimate of the amount in the workplace on an average day;

(b)   the foreseeable types of accidents, incidents or emergencies which might occur involving those substances, and the hazards they could present, e.g. failure of controls, spills, uncontrolled releases of vapours, dusts or fumes into the workplace, accidents with machinery transporting substances in the workplace, leaks, or fire. Consider where such incidents might occur; what effect they might have; the other areas that might be affected by the incident spreading and any possible repercussions that might be caused;

(c)   the special arrangements to deal with an emergency situation not covered by the general procedures; and the steps to be taken to mitigate the effects;

(d)   the safety equipment and personal protective equipment to be used in the event of an accident, incident or emergency, where it is stored, and who is authorised to use it. Judgements about the type of safety equipment and personal protective equipment (including respiratory protective equipment) to be used should be made with regard to the level and type of risk, and a worst case estimate of the likely concentration of a hazardous substance in the air in the workplace;

(e)   first-aid facilities sufficient to deal with an incident until the emergency services arrive; where the facilities are located and stored; the likely effects on the workforce of the accident, incident or emergency, e.g. burns, scalds, shock, the effects of smoke inhalation etc.;

(f)   the role, responsibilities and authority of the people nominated to manage the accident, incident or emergency and the individuals with specific duties in the event of an incident e.g. the people responsible for; checking that specific areas have been evacuated; shutting down plant that might otherwise compound the danger; contacting and liaison with the emergency services on their arrival and making sure that they are aware of the hazardous substance(s) that are the cause of or are affected by the emergency;

(g)   procedures for clearing up and safely disposing of any substance hazardous to health damaged or 'contaminated' during the incident;

(h)   If an incident results in the uncontrolled release of a carcinogen or mutagen into the workplace, the equipment the employer provides in accordance with *Regulation 13(3)(b)* must always include suitable respiratory protective equipment, which is capable of providing adequate control of exposure to the carcinogenic or mutagenic substance concerned.

The Control of Asbestos Regulations 2006 (SI 2006 No 2739)

[D6016] Asbestos is one substance that is not included in the *Control of Substances Hazardous to Health Regulations 2002* but has its own specific legislation (lead is the other). The *Control of Asbestos Regulations 2006, Reg 15* makes provisions for dealing with accidents, incidents and emergencies. The Approved Code of Practice published by the HSC, *Work with materials containing asbestos* (L143) gives advice on dealing with uncontrolled releases of asbestos.

The Dangerous Substances and Explosive Atmospheres Regulations 2002 (SI 2002 No 2776)

[D6017] (See also D0934.)

Regulation 8 requires arrangements to deal with accidents, incidents and emergencies that closely mirror those required by the *Control of Substances Hazardous to Health Regulations 2002.*

Radiation (Emergency Preparedness and Public Information) Regulations 2001 (SI 2001 No 2975)

[D6018] The *Radiation (Emergency Preparedness and Public Information) Regulations 2001 (SI 2001 No 2975)* revoke the *Public Information for Radiation Emergencies Regulations 1992 (SI 1992 No 2997)* (subject to savings). The 2001 Regulations came into force on 20 September 2001.

The Regulations implement the emergency planning aspects of Council Directive 96/29/Euratom, which lays down basic safety standards for the protection of health of workers and the general public against the dangers arising from ionising radiation. The Regulations impose requirements on operators of premises where radioactive substances are present (in quantities exceeding specified thresholds). They also impose requirements on carriers transporting radioactive substances (in quantities exceeding specified thresholds) by rail or conveying them through public places, with the exception of carriers conveying radioactive substances by rail, road, inland waterway, sea or air or by means of a pipeline or similar means.

Essential guidance is contained in the HSE publication 'Guide to the Radiation (Emergency Preparedness and Public Information) Regulations 2001' (L126).

The Regulations:

(a)   impose a duty on the operator and the carrier to make an assessment as to hazard identification and risk evaluation and, where the assessment reveals a radiation risk, to take all reasonably practicable steps to prevent a radiation accident of limit the consequences should such an accident occur [*SI 2001 No 2975, Reg 4*];

(b)   impose a duty on the operator and carrier to send the HSE a report of an assessment containing specified matters at specified times and empower the HSE to require a detailed assessment of such further particulars as may reasonably require [*SI 2001 No 2975, Reg 6 and Sch 5 and 6*];

(c)   impose a duty on the operator and the carrier to make a further assessment following a major change to the work with the ionising radiation or within three years of the date of the last assessment, unless there has been no change of circumstances which would affect the last report of the assessment, and send the HSE a report of that further assessment [*SI 2001 No 2975, Regs 5 and 6*];

(d)   where an assessment reveals a reasonably foreseeable radiation emergency arising, impose a duty on the operator or the carrier (as the case may be) and, in the case of an operator, the local authority in whose area the premises in question are situated, to prepare emergency plans [*SI 2001 No 2975, Regs 7, 8 and 9 and Sch 7 and 8*];

(e)   require operators, carriers and local authorities to review, revise and test emergency plans at suitable intervals not exceeding three years [*SI 2001 No 2975, Reg 10*];

(f)   make provision as to consultation and co-operation by operators, carriers, employers and local authorities [*SI 2001 No 2975, Reg 11*];

(g)   make provision as to charging by local authorities for performing their functions under the Regulations in relation to emergency plans [*SI 2001 No 2975, Reg 12*];

(h)   in the event of the occurrence of a radiation emergency or of an event which could reasonably be expected to lead to such an emergency, make provision as to the implementation of emergency plans, and in the event of the occurrence of a radiation emergency, the making of both provisional and final assessments as to the circumstances of the emergency [*SI 2001 No 2975, Reg 13*];

(i)   where an emergency plan provides for the possibility of an employee receiving an emergency exposure, impose a duty on the employer to undertake specified arrangements for employees who may be subject to exposures, such as dose assessments, medical surveillance and the determination of appropriate dose levels and impose further duties on employers in the event that an emergency plan is implemented [*SI 2001 No 2975, Reg 14*]; and

(j)   impose requirements on operators and carriers, where an operator or carrier carries out work with ionising radiation which could give rise to a reasonably foreseeable radiation emergency, and on local authorities, where there has been a radiation emergency in their area, to supply specified information to the public [*SI 2001 No 2975, Regs 16 and 17 and Sch 9 and 10*].

Public Information for Radiation Emergencies Regulations 1992 (SI 1992 No 2997)

[D6019] The *Public Information for Radiation Emergencies Regulations 1992 (SI 1992 No 2997)* were revoked by the *Radiation (Emergency Preparedness and Public Information) Regulations 2001 (SI 2001 No 2975)*. However, *SI 1992 No 2997, Reg 3* continues in force to the extent that it applies in relation to the transport of radioactive substances by road, inland waterway, sea or air. Any other provisions of the 1992 Regulations continue in force so far as is necessary to give effect to *Reg 3*.

An employer (or self-employed person) must supply prior information where a radiation emergency is reasonably foreseeable from his undertaking in relation to the transport of radioactive substances by road, inland waterway, sea or air. He must:

(a)   supply information to members of the public likely to be in the area in which they are liable to be affected by a radiation emergency arising from the undertaking with without their having to request it, concerning:

(i)   basic facts about radioactivity and its effects on persons/environment,

(ii)   the various types of radiation emergency covered and their consequences for people/environment,

      (iii)   emergency measures to alert, protect and assist people in the event of a radiation emergency;

      (iv)   action to be taken by people in the event of an emergency,

      (v)   authority/authorities responsible for implementing emergency measures,

(b)   make the information publicly available; and

(c)   update the information at least every three years.

[Public Information for Radiation Emergencies Regulations 1992 (SI 1992 No 2997), Reg 3 and Sch 2.]

The Reporting of Injuries, Diseases and Dangerous Occurrences Regulations 1995 (SI 1995 No 3163)

[D6020] The *Reporting of Injuries, Diseases and Dangerous Occurrences Regulations (RIDDOR) 1995 (SI 1995 No 3163)* will apply to 'Accidents' through to 'Major Incidents'. *RIDDOR* is a reporting requirement that must be followed by the employer or 'responsible person' in notifying the enforcing authority by the quickest means practicable, when there is an event resulting in a reportable injury, reportable dangerous occurrence, gas incidents, and road incidents involving work if fatal or fulfilling certain other criteria. In addition, work injuries lasting for three days or more and reportable occupational diseases, are reported to the enforcing authority (see **A3001** INTRODUCTION for further details).

The types of incident are defined or listed in Schedules.

The Health and Safety (First-Aid) Regulations 1981 (SI 1981 No 917)

[D6021] These Regulations do not explicitly mention first-aid arrangements necessary in the event of an emergency or disaster. However, they will apply in all types of accidents and major incidents as defined above. The Regulations are a general statement of best practice, of ensuring the existence of adequate medical equipment, competent and trained first-aiders or appointed persons and information on first-aid facilities and equipment to staff/others. The HSE have published an *Approved Code of Practice and guidance* (L74).

The Construction (Design and Management) Regulations 2007 (SI 2007 No 320)

[D6022] *Regulation 39* requires that suitable and sufficient arrangements shall be prepared and implemented for dealing with any foreseeable emergency. HSE have published an *Approved Code of Practice and guidance* (L144).

The Construction (Health, Safety and Welfare) Regulations 1996 (SI 1996 No 1592)

[D6023] The *Construction (Health, Safety and Welfare) Regulations 1996 (SI 1996 No 1592), Reg 20* required employers to make emergency arrangements to cope with foreseeable emergency situations. They have been superseded by the *Construction (Design and Management) Regulations 2007 (SI 2007 No 320)*.

The Confined Spaces Regulations 1997 (SI 1997 No 1713)

[D6024] (See also C7029.)

Confined Spaces Regulations 1997 (SI 1997 No 1713), Reg 5 imposes duties on employers and others to make arrangements for emergencies in confined spaces. It should be noted that 'confined spaces' does not refer to merely a small area or congested place. The Regulations give a very specific definition based on a place where there arises a reasonably foreseeable specified risk listed as:

(a)   serious injury to any person at work arising from a fire or explosion;
(b)   without prejudice to paragraph (*a*):
    (i)   the loss of consciousness of any person at work arising from an increase in body temperature;
    (ii)   the loss of consciousness or asphyxiation of any person at work arising from gas, fume, vapour or the lack of oxygen;
(c)   the drowning of any person at work arising from an increase in the level of a liquid; or
(d)   the asphyxiation of any person at work arising from a free-flowing solid or the inability to reach a respirable environment due to entrapment by a free-flowing solid;

In summary:

- 'suitable and sufficient arrangements' need to be made for rescue of persons working in confined spaces, with no person accessing a confined space until such arrangements have been developed [*SI 1997 No 1713, Reg 5(1)*];
- so far as is reasonably practicable, the risks to persons required to put the arrangements into operation must also be considered as should resuscitation equipment in the event of it being required [*SI 1997 No 1713, Reg 5(2)*]; and
- there is also a duty to act, to implement the arrangements when an emergency results [*SI 1997 No 1713, Reg 5(3)*].

'Suitable and sufficient arrangements' for rescue and resuscitation should include appropriate equipment, rescue procedures, warning systems that an emergency exists, fire safety systems, first-aid, control of access and egress, liaison with the emergency services, training and competence of rescuers, etc.

The Carriage of Dangerous Goods and Use of Transportable Pressure Equipment Regulations 2007 (CDG 2007) (SI 2007 No 1573)

[D6025] Carrying goods by road or rail involves the risk of traffic accidents. If the goods carried are dangerous, there is also the risk of an incident, such as spillage of the goods, leading to hazards such as fire, explosion, chemical burn or environmental damage. Most goods are not considered sufficiently dangerous to require special precautions during carriage. Some goods, however, have properties which mean they are potentially dangerous if carried.

Dangerous goods are liquid or solid substances, and articles containing them, that have been tested and assessed against internationally agreed criteria – a process called classification – and found to be potentially dangerous (hazardous) when carried.

Carriage of dangerous goods by road or rail is regulated internationally by agreements and European Directives, with biennial updates of the Directives take account of technological advances. New safety requirements are implemented by Member States via domestic regulations.

The *Carriage of Dangerous Goods and Use of Transportable Pressure Equipment Regulations 2007* (*CDG 2007*) came into force on 1 July 2007. They succeed the 2004 Regulations and now include the carriage of radioactive materials. *CDG 2007* implements ADR (with a number of exceptions). They refer directly to ADR and there are some additional or alternative requirements to ADR.

Regulations place duties upon everyone involved in the carriage of dangerous goods to ensure that they know what they have to do to minimise the risk of incidents and guarantee an effective response to protect everyone either directly involved (such as consignors or carriers), or who might become involved (such as members of the emergency services and public).

[D6026] A duty is imposed on the operator of a 'container', 'vehicle' or 'tank' to provide information to other operators engaged/contracted, regarding the handling of emergencies or accident situations. Such information includes:

- hazardness of the goods being carried and controls needed to make safe such hazards;
- actions required if a person is exposed/makes contact with the goods being carried;
- actions required to avert a fire and the equipment that should or should not be used to fight the fire;
- the handling of a breakage/spillage; and
- any additional information that should be given that can assist the operators or others.

The Carriage of Explosives by Road Regulations 1996 (SI 1996 No 2093)

[D6027] These were superseded by the *Carriage of Dangerous Goods and Use of Transportable Pressure Equipment Regulations 2004*, which have since been revised as the *Carriage of Dangerous Goods and Use of Transportable Pressure Equipment Regulations 2007*.

The Environment Act 1995

[D6028] Whilst the *Environment Act 1995* (*EA 1995*) does not explicitly deal with DEM issues, its main objective is the mitigation or prevention of such events. For example, *EA 1995, ss 14–18* creates the flood defence committees to co-ordinate the prediction, consequence and response needed in the event of floods.

The Corporate Manslaughter and Corporate Homicide Act 2007

[D6029] Disasters such as *Piper Alpha*, Kings Cross, Clapham Junction, *Herald of Free Enterprise* or Ladbroke Grove, for example, have seen multiple fatalities, allegations of managerial failure and negligence contributing to the disasters. However, health and safety law did not permit the successful prosecution and imprisonment of directors and senior managers because of the difficulty in establishing that the individuals in charge were 'the embodiment of the company'. All of the major disasters have involved large and complex organisations with many management layers, so who is the 'embodiment of the company' or the 'directing mind'? Neither was there any possibility in law of prosecuting a company for manslaughter. This has now been rectified through the *Corporate Manslaughter and Corporate Homicide Act* 2007.

Insurance reasons

[D6030]–[D6031] There are also insurance based reasons for effective DEM:

- given the positive correlation between risk and premium, the existence of DEM indicates hazard and risk control, consequently the premium ought to be less; and
- if insurers are not satisfied or dissatisfied with the DEM system in place then they may not insure the operation, in turn increasing corporate risk as well as reducing corporate credibility. It may also prohibit the operation from tendering for contracts.

Corporate reasons

- The corporate experience of companies like P&O European Ferries (Dover) Ltd, indicates that proactive DEM systems would have saved the company considerable money, publicity and reputation. In March 1987, the *Herald of Free Enterprise*, the roll-on roll-off car ferry, left the Belgian port of Zeebrugge for Dover whereupon it sunk losing 187 lives. The case against the company and the five senior corporate officers collapsed for reasons cited above ('embodiment of the company' and 'controlling mind', see **D6042** above). With the subsequent introduction of the *Corporate Manslaughter and Corporate Homicide Act 2007*, there are now better prospects for a prosecution of a company to succeed.
- Failures to have sound DEM systems can also have dire financial consequences as disasters such as *Piper Alpha* indicates. Occidental Petroleum were generating 10 per cent of all the UK's North Sea output from *Piper Alpha*. Lord Cullen, who led the inquiry said: 'The safety policy and procedures were in place: the practice was deficient'. *Piper Alpha* showed that the company had ineffective emergency response procedures resulting in persons being trapped, dying of smoke inhalation or jumping into the cold North Sea (something the procedures prohibited but a significant number of those that did jump into the sea survived). Such a Disaster had consequences for Occidental's financial reputation, share value, ability to attract investment and growth potential.

Humanitarian and societal reasons

[D6032] The human cost for not planning for worst case outcomes is the most significant. It is bad enough that incidents such as explosions claim victims, but when the death toll increases because the emergency response is inadequate, that may be seen as a worse crime. Society expects its organisations to plan and prepare for the worst case scenarios and when this does not happen, society seeks the closure, forefeiture or expulsion of the organisation from society, e.g. the legal and political costs to Union Carbide in India with its mis-management of its plant in Bhopal. The company has been banned from operating in India.

Environmental reasons

[D6033] Failure to prepare and plan for disasters and emergencies will also have environmental costs, no better illustrated than the Sandoz warehouse fire in Basel, Switzerland in 1986, with its impact on the river Rhein not only in Switzerland, but also affecting Germany on the opposite bank, then France and Holland further downstream. Retention of the fire-fighting run-off water could have prevented much environmental damage.

Disaster and emergency management systems (DEMS)

[D6034] A DEMS is outlined in Figure 2 overleaf:

Figure 2: Disaster and Emergency Management Systems (DEMS)

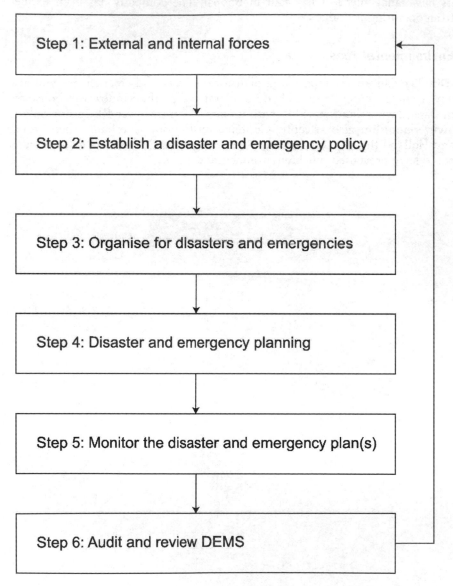

External and internal factors

[D6035]–[D6036] The starting point with DEMS is understanding the variables that can influence or affect DEMS. These are both internal to the organisation and wider societal variables. It would be an erroneous assumption for the organisation to make, if it believed that disaster and emergency planning can be in isolation of wider societal variables; these variables need to be understood and factored into any DEMS. The larger the organisation and the more hazardous its operation, the more it needs to provide specific detail.

External factors influencing DEMS

* The natural environment
  The organisation needs to identify physical variables that can influence its reaction to a disaster or emergency. The organisation needs to identify:
  (i) Seasonal weather conditions – including weather type(s) and temperature range. Effective rescue can be hampered by failing to know, note and record such variables. A simple chart, identifying in user-friendly terms on a monthly basis the weather conditions can assist the reader of any DEM plans.
  (ii) Geography around the site – a description of the physical environment such as terrain, rivers or sea, soil type, geology as well as longitude and latitude positioning of the site. These should be brief descriptions unless the site is remote and could have difficulty being accessed in an emergency.
  (iii) Time and distance – of the site from major emergency and accident centres, fire brigades, police etc. Both long and short routes need to be determined. Longitude and latitude co-ordinates should be identified.
* Societal Factors
  These include:
  (i) Demographics – the organisation needs to identify the age range, gender type, socio-economic structure of the immediate vicinity around the site. A major incident with impact on the local community requires the organisation to calculate risks to the community. *COMAH* off-site emergency plans indicates why planning is not just an on-site 'in these four walls' activity.
  (ii) Social attitudes – the organisation needs to investigate briefly and be aware of attitudes (reactions and responses) of the local town or city and the country to disasters and emergencies. Union Carbide made the assumption that the population and authorities in Madhya Pradesh, where Bhopal is located, would react like the communities in the USA. The response rates, awareness levels and information available to different communities differs.

(iii)   Perception of risk – how does the local community view the site or operation? Is the risk 'tolerable' and 'acceptable' to them for having the operation in their community? Both Chernobyl and Bhopal highlight that economic necessity can alter the perception of risk in contrast to the statistical level of risk.

(iv)   History of community response – in brief the organisation needs to determine how many major incidents there have been in the past and how effectively have the community assisted (emergency services and volunteers). Such support will be critical in a disaster.

(v)   The built environment – a brief description of the urban environment, namely the street layout, urban concentration level, population density, access/egress to railways, motorways are useful. These issues could be covered by the inclusion of a map of the area.

• Government and political factors

Relevant factors are:

(i)   The policy of national government to disasters and emergencies. Are they proactive in advising organisations, what information and guidance do they give? The organisation needs to identify its local or regional office of the HSE, the Environment Agency or a similar body and make contact with officials and seek early input into any planning.

(ii)   Local government – their plans, information and guidance. DEM's need to be aware of any local government restrictions on managing major incidents. In the case of *COMAH* sites, they will need to involve the local authority in preparing off-site emergency plans.

(iii)   Committees and agencies – there are many legal and quasi-legal bodies created by legislation which have guidance, information and templates on response. For example the flooding committees mentioned above under the *Environment Act 1995*.

(iv)   Emergency and medical services – making contact, obtaining addresses/contact numbers and knowing the efficiency of the emergency services/accident and emergency are critical actions.

• Legal factors

The organisation needs to know the legal constraints it has to operate within, not only to comply with the law but also to follow the best advice contained in legislation as a means of preventing disaster and emergency.

• Sources of information issues

These include:

(i)   The local and national media – will not only report any incidents but can be critical in relaying messages. Any planning requires the identification of local and national newspapers (their addresses/contact numbers), local/national radio details and any public sector media (through local government).

(ii)   Local library – they will in turn have considerable contact details and networks with other libraries so can be an efficient medium to relay urgent information. Making contact with the local librarian and identifying such contact details in any planing will be required.

(iii)   Business associations – chambers of commerce, business links, training and enterprise councils need to be identified for the same reasons cited for local libraries. In the event of a major incident they can convey and provide information on a local rescue or occupational health organisation.

(iv)   Voluntary organisations – such as the British Red Cross and St. John Ambulance should be identified and noted.

The organisation needs to build information networks locally and nationally. The experience of major disasters such as Chernobyl or Kings Cross indicate that letting others know of the incident is not a shameful or embarrassing matter – rather it can warn others and halt others from attending the site.

- Technological factors
  These concern what technology and equipment exists in the local community; what does not; lifting equipment, rescue, computers, monitoring equipment measuring equipment etc.

- Commercial factors
  These include:

  (i)   Insurance – liaise with insurers from the beginning and ensure that all plans are drawn to their attention and if possible approved by them. This can avoid difficulties of any claim that may arise due to a disaster or emergency as well as ensure that best advice from the insurer has been factored into the plans.

  (ii)   Customer and supplier response – larger organisations need to identify the responses from customers and suppliers, their willingness to work with the organisation in the event of a worst case scenario and possible alternatives. Suppliers need to be identified. Issues of customer convenience and loyalty also need to be addressed.

  (iii)   Attitudes of the bank – liquidity issues will arise when a major incident strikes. Most organisations do not and cannot afford to make financial provisions for such eventualities. Therefore, if the operation is highly hazardous, establishing financial facilities before hand with the bank is necessary to ensure availability of liquid cash.

(iv)   Strength of the economic sector and economy – larger organisations that are significant players in a sector or the economy need to be cognisant of the 'multiplier effect' that damage to their operation can do to the community and suppliers, employees and others.

Internal factors affecting DEMS

[D6037] DEMS also need to be constructed after accounting for various internal or organisational variables.

*   Resource factors
    Cash flow and budgetary planning whether annual or a longer period needs to account for resource availability for major incidents. This includes physical, human and financial resources. This is wider than banking facilities and encompasses equipment, trained personnel and contingency funds.
*   Design and architecture
    Issues include – is the workplace physically/structurally capable of withholding a major incident? What are the main design and architectural risks? Also layout, access/egress, emergency routes, adequacy of space for vehicles etc.
*   Corporate culture and practice
    The organisation needs to identify its own collective behaviour, attitude, strengths and weaknesses to cope with a major incident. Being a 'large organisation' does not mean it is a 'coping organisation'; there is also the delay in response that is associated with hierarchical structures. In this case, the organisation needs to consider small 'matrix' or project team cell in the hierarchy dedicated to incident response.
*   Individual's perceived behaviour
    (a)   The *Hale and Hale Model* is an attempt to explain how individuals internalise and digest perceived information of danger; make decisions/choices according to the cost/benefit associated with each decision/choice; and the actual actions they take as well as any reactions that result from their actions. In short, these five variables need to be understood in the organisation setting and a picture built up of the behavioural response of an individual.
    (b)   The *Glendon and Hale Model* is a macro model of how the organisation (behaving like a system), being dynamic and fluid, with objectives and indeed limitations (systems boundary) can shape behaviour and in turn influence human error. Following Rasmussen, human error can be skill based (failing to perform an action correctly), rule based (have not learned the sequences to avoid harm) or knowledge based (breaching rules or best practice). If all three error types are committed then the danger level in the system is also greater. The model is a focus on how wider systems can contribute to human error and how that error

can permeate into the organisation. The inference is the organisation needs to clearly define and communicate its intention and objectives and continuously monitor individual response.

(c)   'Reducing Error and Influencing Behaviour', HSG 48, Health and Safety Executive, 1999 identifies the role of human error and human factors in major incident causation. The HSE says 'a human error is an action or decision which was *not intended*, which involved a deviation from an accepted standard, and which led to an undesirable outcome' (page 13). Following Rasmussen they classify four types:

(i)   *slips* (unintended action);

(ii)   *lapses* (short term memory failure) with slips and lapses being skill based;

(iii)   *mistakes* (incorrect decision) which are rule based; and

(iv)   *violations* (deliberate breach of rules) which are knowledge based.

Different types of human errors contribute in different ways to major incidents the HSE say and exemplify. Organisations need to identify from reported incidents the main types of human errors and why they are resulting and the negative harm generated. (Note that in the previous version of HSG 48 called 'Human Factors in Industrial Safety', five types of human errors were defined:

(A)   *misperception* (tunnel vision, excluding wider factors from one's senses e.g. the belief that smoking is safe on the underground or that a 'smouldering' is not a significant fire as in the case of Kings Cross in 1987);

(B)   *mistaken action* (doing something under the false belief it is correct e.g. the pilot switching off the good engine under the belief he was switching off the one with the fire, so both are off, hence the crash landing in the Kegworth disaster);

(C)   *mistaken priority* (a clash of objectives, such as safety and finance as implied in the *Herald of Free Enterprise* disaster in 1987);

(D)   *lapse of attention* (short term memory failure, not concentrating on a task e.g. turning on a valve under repair as in *Piper Alpha* in 1988); and

(E)   *wilfulness* (intentionally breaching rules e.g Lyme Bay disaster when the principal director received a two year imprisonment for the death of teenagers at a leisure centre under his control).

The HSE says that understanding human factors is a means of reducing human error potential. Human factors is a combination of understanding the person's behaviour, the job they do (ergonomics) and the wider organisational system. Major incidents can result if the organisation does not analyse and understand these three variables.

●   Information systems at work

The types of information systems, their effectiveness, and accuracy need to be identified. Thus telephone, fax, email, cellular phone, telex etc. need to be assessed for performance and efficacy during a worst case scenario.

Establish a disaster and emergency policy

[D6038] The DEM policy is a concise document that highlights the corporate intent to cope with and manage a major incident. It will have three parts: statement of policy for managing major incidents (see **D6039**); arrangements for major incident management (see **D6040**); and command and control chart of arrangements (see **D6041**).

Statement of policy for managing major incidents

[D6039]–[D6040] This should be a short (maximum 1 page) missionary and visionary statement covering the following:

- Senior management commitment to be responsible for the co-ordination of major incident response.
- To comply with the law, namely:
  - the protection of the health, safety and welfare of employees, visitors, the public and contractors;
  - to comply with the duties under the *Management of Health and Safety at Work Regulations 1999 (SI 1999 No 3242), Regs 8* and *9* (see **D6010**);
  - to comply with any other legislation that may be applicable to the organisation.
- To make suitable arrangements to cope with a major incident and to be proactive and efficient in the implementation process.
- That this statement applies to all levels of the organisation and all relevant sites in the country of jurisdiction.
- The commitment of human, physical and financial resources to prevent and manage major incidents.
- To consult with affected parties (employees, the local authority and others if needed).
- To review the statement.
- To communicate the statement.
- Signed and dated by the most senior corporate officer.

Arrangements for major incident management

- This refers to what the organisation has done, is doing and will do in the event of a major incident and *how* it will react in those circumstances. The arrangements are a legal requirement under the *Management of Health and Safety at Work Regulations 1999 (SI 1999 No 3242), Regs 5, 8* and *9*.

- Arrangements should be realistic and achievable. They should focus on major actions to be taken and issues to be addressed rather than being a 'shopping list'. The arrangements will have to be verified (in particular for *COMAH* sites).
- Arrangements could be under the following headings with explanations under each. To repeat, the larger the organisation and the more complex the hazard facing it, the more detailed the arrangements need to be. For example:

  (i)   Medical assistance – including first-aid availability, first-aiders, links with accident and emergency at the medical centre, other specialists that could be called upon, rules on treating injured persons, specialised medical equipment and its availability etc.

  (ii)   Facilities management – the location, site plans and accessibility to the main facilities (gas, electricity, water, substances etc.), rendering safe such facilities, availability of water supply in-house and within the perimeter of the site etc.

  (iii)   Equipment to cope – identification of safety equipment available and/or accessible, location of such equipment, types (personal protective equipment, lifting, moving, working at height equipment etc.).

  (iv)   Monitoring equipment – measuring, monitoring and recording devices needed, including basic items such as measuring tapes, paper, pens, tape recorders, intercom and loud-speakers.

  (v)   Safe systems – procedures to access site, working safely by employees and contractors under major incident conditions (what can and cannot be done), hazard/risk assessments of dangers being confronted etc., risks to certain groups and procedures needed for rescue (disabled, young persons, children, pregnant women, elderly persons).

  (vi)   Public safety – ensuring non access to major incident site by the public (in particular children, trespassers, the media and those with criminal intent), warning systems to the public etc.

  (vii)   Contractor safety – guidance and information to contractors at the major incident on working safely.

  (viii)   Information arrangements – the supply of information to staff, the media and others (insurers, enforcers) to inform them of the events. Where will the information be supplied from, when will it be done and updates?

  (ix)   The media – managing the media, confining them to an area, handling pressure from them, what to say and what not to say etc.

  (x)   Insurers/loss adjusters – notifying them and working with them at the earliest opportunity.

(xi)   Enforcers – notifying them of the major incident, working with them including several types e.g. HSE, Environment Agency (or Scottish equivalent) as well as local authority (environmental health, planning, building control for instance).

(xii) Evidence and reporting arrangement – to cover strict rules on removal or evidence by employees or others, role and power of enforcers, incident reporting e.g under the *Reporting of Injuries, Diseases and Dangerous Occurrences Regulations 1995 (SI 1995 No 3163)* etc.

(xiii) The emergency services – working with the police, fire, ambulance/NHS, and other specialists (British Red Cross, search and rescue), rules of engagement, issues of information supply and communication with these services.

(xiv) Specialist arrangements for specific major incidents such as bomb threats and explosions – issues of contacting the police, ordnance disposal, access and egress, rescue and search, economic and human impact to name a few issues were most evident during the London Docklands and Manchester bombings in the 1990s.

(xv)   Human aspects – removing, storing and naming dead bodies or seriously injured persons during the incident. Informing the next-of-kin, issues of religious and cultural respect. Issues of counselling support and person-to-person support during the incident.

This is not an exhaustive list. The arrangements should not repeat those in the safety policy, rather the latter can be abbreviated and attached as a schedule to the above, so that the reader can have access succinctly to specific OSH arrangements such as fire safety, occupational health, safe systems at work, dangerous substances for instance.

Command and control chart of arrangements

[D6041]–[D6042] This highlights who is responsible for the effective management of the major incident.

• It should be a graphical representation preferably in a hierarchical format, clearly delineating the division of labour between personnel in the organisation and the emergency services/others.

• The chart should display three broad levels of command and control, namely strategic, operational and tactical. The first relates to the person(s) in overall charge of the major incident. Will this be the person who signed the statement of policy for major incidents or will it be another (disaster and emergency advisor, safety officer, others)? This person will make major decisions. The second relates to co-ordinators of teams. Operational level personnel need to have the above arrangements assigned to them in clear terms. The third refers to those at the front end of the major incident, for instance first-aiders.

It is most important to note that internal command and control of arrangements does not mean *overall* command and control of the major incident. This can (will be) vested with the appropriate emergency service, normally the police or the fire authority in the UK. In the event of any conflict of decisions, the external body such as the police will have the final veto. Therefore, the chart and the arrangements must reflect this variable.

- The chart should list on a separate page names/addresses/emergency phone, fax, email, cellular numbers of those identified on the chart. It should also list the numbers for the emergency services as well as others (British Red Cross, specialist search and rescue, loss adjusters, enforcing body).
- The chart should also clearly ratify a principle of command and control, as to who would be 'In-charge 1', 'In-charge 2', if the original person became unavailable.
- The chart and the list of numbers should also be accompanied by a set of 'rules of engagement' in short 'bullet points' to remind personnel of the importance of command and control e.g. safety, obedience, communication, accuracy, humanity for instance.

Summary

- The three parts of the DEM policy need to be in one document. Any detailed procedures can be separately documented ('disaster and emergency procedures') and indeed could be an extensive source of information. However, unlike the safety policy and any accompanying safety manual, the same volume of information cannot apply to the DEM policy. For obvious reasons it must be concise, clearly written, very practical and without complex cross-referencing.
- The DEM policy must 'fit' with the safety policy, there can be no conflict so the safety officer and the DEM officer need to cross check and liaise. The DEM policy must also fit with the broader corporate/business policy of the organisation.
- The DEM policy must be proactive and reactive. The former concerned with preventing/mitigating loss and the latter concerned with managing the major incident when it does arise in a swift and least harmful manner.
- The DEM policy needs to be reviewed 'regularly'. This could be when there is 'significant change' to the organisation, or as a part of an audit (semi-annual or annual).
- It must be remembered that the DEM policy is a 'live' document so that it must be accessible and up-to-date.

- Although accessibility is important, the policy should also have controlled circulation to core personnel only (for example those identified in the chart, the legal department). If the policy was to be accessed by individuals wishing to harm the organisations, this will enable such persons to pre-empt and reduce the efficacy of the policy.
- Finally and most crucially, the core contents of the policy need to be communicated to all staff and others (contractors, temporary employees, possibly local authority).

Organise for disasters and emergencies

[D6043] Once the establishment has accounted for external and internal factors and has produced a DEM policy taking account of such factors, it is then necessary to ensure personnel and others are aware of the issues raised in the DEM policy. The '4' c's approach of HSG 65 provides a logical framework to generate this ('Successful Health and Safety Management', HSE): communication (see **D6044**); co-operation (see **D6045**); competence (see **D6046**); and control (see **D6047**).

Establish effective communication

[D6044] Communication is a process of transmission, reception and feedback of information, whether that information is verbal, written, pictorial or intimated. Effective communication of the DEM policy therefore is not a matter of circulating copies but requires the following:

(1)   *Transmission*
- The whole policy should not be circulated as it will mean little to employees and others. Rather an abridged version, possibly in booklet format or as an addition to any OSH documentation supplied will make more sense. Such copies must be clear, user friendly, non-technical as possible, be aware of the end user's capabilities to digest the information, be logical/sequential in explanation and use pictorial representation as much as possible.
- Being aware of the audience is central to the effective communication of the DEM policy. The audience is not one entity but will consist of:
    —   direct employees;
    —   temporary employees;
    —   contractors;
    —   the media;
    —   the enforcers (*COMAH* sites);
    —   the local authority (*COMAH* sites);
    —   insurers;
    —   emergency services (*COMAH* sites); and
    —   the public (*COMAH* sites).

This does not necessarily mean separate copies for each of these entities but the abridged copy will need to satisfy the needs of all such groups.

- Transmission should start from the board, through to departmental heads, and disseminated downwards and across.

(2)   *Reception*
- What format will the end-user receive the abridged copy in (hard copy, electronic on disc, via email, etc.)?
- When will the copy be circulated – upon induction, upon training, ad hoc?

(3)   *Feedback*
- Will the end-user have the opportunity to raise questions, make suggestions, be critical if they spot inconsistencies in the DEM policy?
- There should also be 'tool box talks' and other general awareness programmes to inform individuals of the DEM policy. This could be combined with general OSH programmes or wider personnel programmes, so that a holistic approach is presented.
- The importance of feedback is that the policy becomes owned by all individuals, which in turn is the single most important factor in successful pre-planning to prevent major incidents.

In general it may be useful to retain copies of the abridged and the full policy with other safety documentation in an 'in-house' company library. For smaller organisations, this could be one or two folders on a shelf through to a dedicated room for larger organisations. Second, the abridged copy could be pasted onto an intranet site.

Co-operation

[D6045] Co-operation is concerned with collaborating, working together to achieve the shared goal and objectives:

- Firstly, co-operation between strategic, operational and tactical level management is critical. This reflects the chart in the DEM policy, as discussed above. This could be consolidated as a part of a broader corporate meeting or preferably dedicated time to cover OSH and fatal incident issues. This could be a semi-annual event, with a dedicated day allotted for all grades of management to interface. This is not the same as a board level discussion or a management discussion.
- To give responsibility to either the safety committee or the safety group to co-ordinate review, debate and assessment of the DEM policy or to fuse this function within a broader business/corporate review committee. The former has advantages as it is safety dedicated whilst the latter would integrate DEM policy issues into the wider business debate.
- Involvement of Safety Representatives/Representatives of Employee Safety (ROES), is both a legal requirement as well as inclusive safety management. These persons can be central 'nodes' in linking 'management' and the 'workers' together. In the UK, there has been an increasing realisation that trained safety representatives are a knowl-

edgeable resource with many being trained to IOSH/NEBOSH standard (as with the Amalgamated Engineering and Electrical Union (AEEU) (electrical engineers) trade union, with their National Academy of Safety and Health, NASH).

- Co-operation also needs to extend to contractors. The person responsible for interfacing with contractors needs to up-date them and make them aware of the DEM policy and seek their support and suggestions. It may also be valuable to invite contractors to OSH/fatal incident awareness days or the general committee meetings as observers.

- Co-operation is also needed between the organisation and external agencies such as the local authority, insurers, enforcers, media etc. This can be achieved through providing abridged copies of minutes or a 'newsletter' (1 to 2 sides of A4) distributed semi-annually informing such bodies of the DEM policy and any changes as well as other OSH issues. *COMAH* sites will have to demonstrate as a legal requirement that plans and policies are up-to-date and effective.

Competence

[D6046] Competence is a process of acquiring knowledge, skill and experience to enhance both individual and corporate response. Thus, competence is about enhancing and achieving standards (set by the organisation or others).

- Competence is important, so that certain key persons are trained to understand the DEMS process. The above issues and their link to OSH in particular require personnel that can assimilate, digest and convey the above issues. Training does not necessarily mean formal or academic training but can be vocational or in-house. Neither does it mean the organisation expending vast sums but can be a part of a wider in-house OSH awareness programme (e.g. one day per quarter of a year).

- For larger organisations, they may be able to recruit 'competent persons' to advice on OSH and fatal incident matters.

- In short, all employees need to be brought up to a minimal standard. *Piper Alpha* showed that whilst Occidental Petroleum had detailed procedures, the employees generally did not fully understand them nor had they the minimal understanding of major incident evacuation. The organisation had failed to impart knowledge, skill and experience sufficient to cope with fires and explosions on off-shore sites.

Control

[D6047] Control refers to establishing parameters, constraints and limits on the behaviour and action. This ensures that on the one hand an effective DEM policy exists and on the other, personnel will act and react in a co-ordinated and responsible manner. Controls can be achieved via for example:

- Contractual means – as a term of a contract of employment that instruction and direction on OSH and fatal incident matters must be followed by individuals.

- Corporate means – the organisation continuously makes individuals aware of following rules and best practice.
- Behavioural means – by establishing clear rules, training, supply of information, leading-through-example, showing top-level management commitment etc.
- Supervisory means – ensuring that supervisors monitor employee safety attitudes and risk perceptions.

The cumulative effect of the 4 C's should be a positive and proactive culture in which not only OSH issues but the DEM policy issues are understood. Factors that can mitigate or prevent a positive and proactive culture developing include lack of management commitment, lack of awareness of requirements, poor attitudes to work and life, misperception of the risk facing the organisation, lack of resources or the unwillingness to commit resources, fatalistic beliefs etc.

Disaster and emergency planning

[D6048] 'Planning' is a process of identifying a clear goal and objectives and pursuing the best means to achieve that goal/objectives. Although 'disaster' and 'emergency' are two separate but related terms, in the case of planning the two need to be viewed together. This is because in practice one cannot divorce the serious event (the disaster) from the response to that serious event (emergency).

Disaster and emergency planning can be viewed in three broad stages:

- stage 1: before the event (see **D6049**);
- stage 2: factors to consider during the event (see **D6056**); and
- stage 3: after the event (see **D6057**).

Stage 1: before the event

[D6049] Once the DEM policy has been established and a culture created where the policy has been understood and positively received, next one needs to be establish a 'state-of-preparedness', that is addressing issues, speculating on scenarios and developing support services when the major incident does strike.

The risk assessment of major incident potential and consequent contingency planning

[D6050] What is the probability of the major incident resulting? What would be the severity? What type of major incident would it be? Figure 3 depicts a 'major incident matrix':

Figure 3: major incident matrix

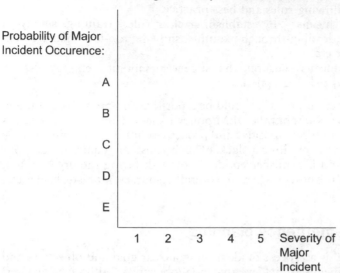

Where:

A=   Certainty of Occurrence (Probability = 1)
B=   Highly Probable
C=   0.5 Probability of Occurrence
D=   Low Probability
E=   Most Unlikely
And:
1=   Serious Injuries
2=   Multiple Fatalities
3=   Multiple Fatalities and Serious Economic/Property Damage
4=   Fatal Environmental, Bio-Sphere and Social Harm
5=   Socio-Physical Catastrophe

The planning responses therefore will differ according to the cell one finds oneself in. For example, cell 1E requires the least complex state of preparedness in terms of resources, detail of planning and urgency. Whilst 5A is a major societal event that requires macro and holistic responses, in which case the organisation's sole efforts at planning are futile. The organisation needs to assess in which cell the major incidents it will confront, will mainly fall in and accordingly develop plans. A conglomerate or an 'exposed operation' such as oil/gas refinery work may require planning at all 25 levels. This is 'contingency planning' – an analysis of the alternative outcomes and providing adequate responses to those outcomes.

During the concerns over the millennium bug, HSBC Bank provided some general advice on contingency planning (*Tackling the Millennium Bug: A final check on preparations including Contingency Planning*, HSBC, 1999). They suggested a 10-point strategy. 'A Contingency Plan is a carefully considered set of alternatives to your usual business processes' (page 18). This advice is generic and can be used more broadly for everyday planning, as follows:

(1)   *Identify* – assume things will go wrong – focus on where the probability and severity would be of failure.

(2)   *Prioritise* the processes into three categories:
    (a)   essential for business continuity;
    (b)   important for business continuity; and
    (c)   non-essential for business continuity.

(3)   *Analyse* e.g.:
    (a)   identify the effects of risks on operation(s);
    (b)   assess 'domino effects'; and
    (c)   grade risk using a suitable risk assessment approach.

(4)   *Develop* solutions e.g.:
    (a)   for each operation develop a solution/control measure to the risk;
    (b)   encourage participation – involve suitable and critical stakeholders; and
    (c)   ensure new risks are not created.

(5)   *Costs* – carry out a 'cost benefit analysis'.

(6)   *Formalise* documents e.g.:
    (a)   write and keep any assessments and inspections;
    (b)   top management involvement and 'joined up business';
    (c)   develop an approach to implementation;
    (d)   regular training to be conducted to ensure improved competence;
    (e)   identify and deploy resources needed (if available);
    (f)   communications to be clear and simple – before, during and after; and
    (g)   contractual issues to be dealt with so that contractors, customers and insurers are kept informed.

(7)   *Co-ordinate* – managers and team to ensure implementation of contingency plan.

(8)   *Test* e.g. controlled testing of the contingency plan(s) (see below).

(9)   *Update* e.g.:
    (a)   update the plan;
    (b)   quality assurance and control of the plan; and
    (c)   accuracy – ensure that the organisation's lawyer and insurer has sight of the plan.

(10)   *Communicate* e.g.:
    (a)   plan signed off by top management; and
    (b)   communicate salient features to employees.
Appendices can be included containing any relevant information.

Decisional planning

[D6051] Planning before the event also involves co-ordinating different levels of decision making. Figure 4 depicts a decisional matrix:

FIGURE 4: DECISIONAL MATRIX

	Production Level	1. INPUT	2. PROCESS	3. OUTPUT
Management Level				
A. STRATEGIC				
B. OPERATIONAL				
C. TACTICAL				

Where:

A=   Strategic or Board/Shareholder/Controller Level (Gold Level)

B=   Operational or Departmental Level (Silver Level)

C=   Tactical or Factory/Office Level (Bronze Level)

And:

1=   Inputs or those resources that make production possible, thus raw materials, people, information, machinery and financial resources.

2=   Process or the manner in which the inputs are arranged or combined to enable production (method of production)

3=   Output or the product or service that is generated.

Decisional planning requires major incident risks to be identified for each cell. Essentially this is assessing where in the production process and at which management level can problems arise that can lead to major incidents. This matrix is not 'closed' but is 'open' and 'dynamic', which means external factors, which were discussed above as well as internal factors need to be accounted for. For example, in a commercial operation at board level, they need to consider:

•   INPUTS: purchasing policy of any raw materials used (safety etc.), recruitment of stable staff, adequacy of resources to enable safe production, known and foreseeable risks of inputs used, reliability of supply, commitment to environmental safety and protection etc.

•   PROCESS: external factors and their impact on production, production safety commitment, commitment to researching/investigating in the market for safer production processes etc.

- OUTPUT: boardroom commitment to quality assurance, environmental safety of the service and/or product produced, customer care and social responsibility etc.

The end result should be, in each cell major risks to production and their impact on a major incident needs to be identified. This need not be a major, time consuming exercise, rather it can be determined through a 'brainstorming' session or each management level fills out the matrix during their regular meeting and then it is jointly co-ordinated in a short (2–5 page) document.

Testing

[D6052] This is considered to be a crucial aspect of pre-planning by all the emergency services in the UK. Testing is an objective rehearsal to examine the state of preparedness and to determine if the policy and the plan will perform as expected.

The Home Office guidance, *Dealing with Disaster* (1998, third edition) identifies training and exercising as types of testing. Training is more personnel focused, aiming to assess how much the human resource knows about the policy and plan and means of enhancing the knowledge, skill and experience of that resource. Exercising is a broader approach, examining all aspects of the policy and plan, not just the human response but also physical and organisational capability to deal with the major incident.

Training

[D6053] Carrying out a training needs analysis (TNA) to determine the quantity (in terms of time) and quality of training needed is essential. It may be that personnel and others adequately understand the policy and plan, therefore the training response should be proportionate. Excessive training is a motivational threat to the interest and enthusiasm of the person as is under-training. The HSE in HSG 65, 'Successful Health and Safety Management' suggest that the TNA consists of issues such as:

- Is the training 'necessary'? In other words, is there an alternative way of ensuring competence and awareness of the policy and plan, rather than just training? Could circulating the information be better? What about regular 'tool box talks'? It is important to weigh the costs and benefits of each option.
- Is it 'needed'? Will the training meet personal, job or organisational needs thereby enhancing awareness and appreciation of the policy and plan?
- What will be the intended learning objectives of the training? Will it be 'this is a course on the content of the DEM policy and plan'?
- What type of training will be offered? Class-room, on-site or both? What are the costs and benefits associated with each?
- Management of the training – where will the training be conducted? Who will deliver it? When will it be delivered – day or evening? What training aids will be provided – if any?
- How does one measure the effectiveness of such training?

Exercising

[D6054] The Home Office cite three types of exercising:

- Seminar exercises
  - a broad, brain-storming session, assessing and analysing the efficacy of the policy and plan,
  - seminars need to be 'inclusive', that is bring staff and others together in order to co-ordinate strategic, operational and tactical issues.
- Table-top exercises
  This is an attempt to identify visually using a model of the production or office site and surrounding areas, what types of problems could arise (access/egress, crowd control, logistics etc.).
- Live exercises
  This is a rehearsal of the major incident, actively and pro-actively testing the responses of the individual, organisation and possibly the community (emergency services, the media etc.). London Underground, the railway sector commonly, and the civil aviation sector, tend to annually assume that a major incident has occurred and the consequential conditions are created.
- Synthetic simulation
  Whilst this is not mentioned in Home Office or other emergency service guidance, this fourth approach is becoming a most popular option. It applies the techniques used in flight simulation and military simulation, to major incidents. Thus, using computers one can model the site and operation and in 3-d format move around the screen. This then enables various eye-points around the site, achieved by moving the mouse. Some advanced systems enable 'computer generated entities' to be included into the data-set. For example, a collision can be simulated outside the production plant or the rate of noxious substance release modelled and calculated across the local community.
  There are costs and benefits associated with each of the four approaches. The need is for developing exercise budgets which enable all four to be used in proportion and to compliment each other.
  Training and exercising are both necessary to ensure that human resources understand the policy and plan, as well as to ensuring that problems can be identified and competent responses developed.
  It should be noted that the guidance to the *Control of Major Accident Hazards Regulations 1999 (SI 1999 No 743)* does not distinguish between 'training' or 'exercising'. It identifies drills, which test a specific aspect of the emergency plan (e.g. fire drills), seminar exercises, walk through exercises, which involves visiting the site, table-top exercises, control post exercises, which assess the physical and geographical posting of the personnel and emergency services during a major incident and live exercises.

Business continuity

[D6055]–[D6056] 'Business continuity' is a planning exercise to ensure critical facilities, processes and functions are operational and available during and immediately after the major incident, thereby enabling the organisation to function commercially and socially.

There are many approaches to business continuity. Two are highlighted below – first, a 'generic approach', and secondly the advice from the Business Continuity Institute (BCI). In addition, there are approaches from the Department of Trade and Industry, Disaster Recovery Institute of the USA, Australian National Audit Office as well as guidance contained in BS/ISO/ICE 17799:2000 (which is an information technology standard).

(1)   *Generic approach*
       Business continuity is a forecasting process of recovery, assessment and ensuring the adequacy of resources for the organisation to continue its operation.
       (a)   Recovery
              Recovery is a state of regaining or salvaging assets that otherwise would have been permanently lost. In the event of a major incident, recovery phase needs to focus on:
              (i)   human resource recovery:
                     —   ensuring personnel are safely evacuated;
                     —   others such as lawful visitors and trespassers are evacuated; and
                     —   all persons in general are removed in the quickest and most practicable means;
              (ii)   information resource recovery:
                     —   essential documents and software; and
                     —   private and confidential documents;
              (iii) physical resource recovery:
                     —   primary electrical and mechanical facilities such as power supply, water and gas services; and
                     —   essential work equipment, if possible and moveable;
              (iv)   financial resource recovery; and
              (v)   valuable assets recovery (if practicable).
              The organisation also needs to rate the recovery potential – how possible is recovery? This could be rated from 'certain' (1) through to 'not possible' (5). This calculation requires different recovery responses; when the major incident strikes, those on the scene need to decide if the rating is 1 or 5. Accordingly, the response will vary – if 5, then there is little purpose risking life and resources to salvage assets.
       (b)   Assessment
              Upon the immediate recovery, the organisation must ask, how much damage (deterioration, infliction of harm or erosion of value) has been inflicted to the operation by the major incident? Damage assessment is a three-stage activity:
              (i)   identify the type of damage:
                     —   human resource damage: physical and/or behavioural;
                     —   informational resource damage: primary, secondary and tertiary documentation;

      — physical resource damage: work equipment, property, facilities, environment; and

      — financial resource damage: cash, art, etc.;

(ii)   identify severity of damage:

this can be a qualitative scale which says 'low' or 'high' to the age or a quantitative scale – the damage is rated on some scale e.g. from 1 to 5. The severity needs to be forecasted for human, informational, physical and financial resources:

      — human resources could be scaled from serious injury (1) through to fatality and multiple fatality (5);

      — informational: from minor harm (1) through to destroyed (5);

      — physical: from reparable (1) to unsalvageable (5); and

      — financial: from no impact on cash flow (1) to financial ruin (5).

(iii)   consequence of damage:

how much does the damage affect the chance of the operation being resumed immediately or in the next few days? The longer it will take to resume, the worse has been the consequence. Consequence could be rated in quantitative terms also, such as a 1 for 'resume immediate' so the harm has been minimal through to say 5 for 'resumption will take months'. The consequence needs to be examined in each case:

      — human: how long will it take people to get back to work?

      — informational: how long will it take for manual and electronic systems to be operational?

      — physical: how long will it take for necessary facilities to be operational?

      — financial: how long will it take for cash flow to become positive or to access financial facilities?

severity multiplied by consequence will give an index of forecasted potential loss, which can then be assessed as a spreadsheet over time. Loss adjustors use various detailed statistical models based upon such generic principles.

(c)   Adequacy of resources

Once the organisation has forecast the potential for recovery and hypothesised about damage assessment, next it needs to ensure that the organisation will have adequate resources to carry on operating in light of the resource losses identified by the assessment. Adequacy needs to consider:

(i)   human resources:

      — adequacy of competent and trained personnel at strategic, operational and tactical levels of the organisation;

         — availability of key advisory support services (lawyer, accountant, etc.); and

         — if the major incident is classified as a 'crisis', then there is a chance some of the key personnel are not available. Pre-planning therefore requires liaison with recruitment and selection specialists;

    (ii) physical resources:

         — telephone, fax, email, cellular connectivity and reliability;

         — furniture and fittings;

         — stationery;

         — work equipment (including computers/typewriters, filing cabinets);

         — working stock;

         — working space;

         — vehicles;

         — safety equipment (personal protective equipment);

         — etc.;

    (iii) informational resources:

         — legal documents (organisation's certificates of incorporation, insurance liability certificates etc.);

         — personnel documents (PAYE, NIC, personnel records);

         — financial documents (availability of bank books); and

         — sales/marketing documents – this is at the heart of the operation and there will be a need to develop databases, contact potential customers and re-establish commercial functionality;

    (iv) financial resources:

         — adequacy of working capital.

(2) *Business Continuity Institute's approach*

In *Business Continuity Management: A Strategy for Business Survival* by BCI, they say: 'BCM is not just about disaster recovery, crisis management, risk management or about IT. It is a business issue. It presents you with an opportunity to review the way your organisation performs its processes, to improve procedures and practices and increase resilience to interruption and loss. Business Continuity Management is the act of anticipating incidents which will affect critical functions and processes for the organisation and ensuring that it responds to any incident in a planned and rehearsed manner.'

Business continuity management (BCM) is the process of proactively identifying, anticipating, resourcing and planning to ensure that business operations can continue after the disaster. In contrast, business continuity planning (BCP) is a stage in the BCM process. It is the tactical and operational document that will be used when disaster strikes. The BCI defines BCM as 'an approach that seeks to improve organisational resilience to interruption, allowing for the recovery of key business and systems processes within the recovery timeframe objective, whilst maintaining the organisation's critical functions.'

The BCI present a five-step approach to BCM.

- STEP 1: *'understand your business'*
Ensure that the main strengths, weaknesses, opportunities and threats are identified. Also ensure that key 'threats' and 'risks' are identified. Further, the support of the board/strategic management team (SMT) is necessary for the necessary resource allocations.
- STEP 2: *'consider continuity strategies'* when the worst case scenario strikes, e.g.:
    - (a) *Do nothing* – it maybe that no resource deployment or salvage strategy is required as the risks of significant loss are quite low.
    - (b) *Changing or ending the process* – the board/SMT may consider that production or service processes/procedures need to be modified.
    - (c) *Insurance* – should it be used? What are the costs?
    - (d) *Loss mitigation* – what actual/tangible actions can be taken to reduce risks?
    - (e) *Business continuity planning* – is a more detailed plan needed given the severity and long term nature of the risk/threat?
- STEP 3: *'develop the response'*
If BCP is needed then develop the plan. The BCI suggest the following sections to a BCP:
    - — Section 1: general introduction and overview:
        - (i) objectives;
        - (ii) responsibilities;
        - (iii) exercising; and
        - (iv) maintenance.
    - — Section 2: plan invocation:
        - (i) disaster declaration;
        - (ii) damage assessment;
        - (iii) continuity actions and procedures;
        - (iv) team organisation and responsibilities; and
        - (v) emergency (crisis) operations centre.
    - — Section 3: communications:
        - (i) who should be informed;
        - (ii) contacts; and
        - (iii) key messages.
    - — Section 4: suppliers:
        - (i) list of recovery suppliers; and
        - (ii) details of contract provision.
- STEP 4: *'establish a continuity culture'* – through education, training, information and awareness.
- STEP 5: *'exercise, plan maintenance and audit'*:
    - (a) Exercise the plan frequently with participation of key stakeholders.
    - (b) Plan maintenance accounting for internal and external changes.
    - (c) Use auditing techniques for 'gap analysis' purposes.

Stage 2: during the event

- Activate disaster and emergency policy and pre-planning
  Major incident response and management ought to be clear and effective if one has followed the previous stages in the DEMS; accounted for external and internal factors and uncertainties, developed a chain of command as well as arrangements to cope, Organised the operation to cope with a major incident and finally pre-planned if the event actually arises. In the main, activating the arrangements made is the central action (refer above).
- Co-ordination
  This is the central action to ensuring efforts are in unison. This includes both liasing with external bodies such as the emergency services, media and internally at strategic, operational and tactical levels (as discussed above). The police will normally have overall command and control, so their guidance must be followed.
- Code of conduct
  A code of conduct is a set of 'golden rules' that staff and contractors need to follow if the major incident arises. This includes:
  (i)   following instruction from those with command and control responsibilities;
  (ii)   using self initiative;
  (iii)   facilities management (gas, electricity, water) – making safe, when and when not to switch on or off;
  (iv)   medical assistance – liaison with health service, first-aiders, when not to administer and when to;
  (v)   rules of evacuation;
  (vi)   site access and egress;
  (vii)   site security;
  (viii)   work and rescue equipment – its safe use and logistics of use;
  (ix)   record keeping of the major incident (audio-visual, verbal and written);
  (x)   resources needed for effective major incident management;
  (xi)   media and public relations management; and
  (xii)   issues of care and compassion of injured persons.
  The code of conduct is a reflection of the disaster and emergency policy and the pre-planning. It should be reinforced verbally and in writing (1 page of A4) and reiterated to the major incident team before and during the major incident. Whilst this seems bureaucratic, it must be noted that if the core team (internal and external) fail to follow best practice and safe-guard themselves, this increases the risk factor of the major incident and could even lead to a double tragedy. A five minute or so reiteration is a minor time cost.

Stage 3: after the event

[D6057] Major incident planning does not cease as soon as the major incident is physically over. There are continuous issues over time that need to be addressed (see **D6062–D6065**)).

I.   Immediate term (immediately after the major incident and within a few days)

[D6058] The following paragraphs explain what steps should be taken.

Statutory investigation

[D6059] This will involve the HSE, local authority, Environment Agency/SEPA for instance. Under statute, these bodies have powers and a duty to investigate a major incident.

- The organisation investigated must fully co-operate with these bodies and afford any assistance they require.
- There must be no removal or tampering with anything at the major incident site. There may be forensic or other data gathering required by these bodies.
- There must be provision for all documentation and information to be provided to these bodies.

Business continuity

[D6060] Implement plans (see **6065** above).

Additional matters to consider

[D6061]–[D6062] Also under the 'immediate term', in addition to the above, one needs to consider:

- Insurers/loss adjusters
  Assuming that the insurance contract covers direct and consequential damage from a major incident (and not all will), the organisation needs to notify the insurer and ensure all paperwork is completed promptly. Most insurance contracts stipulate a time limit by which the paperwork has to be lodged with the insurer; the major incident will divert attention to other issues, so ensuring that a person is appointed to activate this insurance task is vital (this should be the organisation's lawyer).
  The insurer in turn will notify their loss adjusters to investigate the basic and underlying causes of the event. Again, full disclosure and co-operation are implied insurance contractual requirements. Although, the organisation must check all documents that the loss adjusters completes, to ensure they are accurate and cover all aspects of the event.
- The police
  In the suspicion that criminal neglect played a role in the event, then the police will need to interview all core board members, senior management and others.
- Building contractors

The organisation needs to plan for building contractors to visit the site and make it safe and secure. This may have to be done forthwith after the event, even if insurance issues have not been resolved. Therefore, adequacy of resources, as discussed above becomes vital.

• Visitors

Major incidents also lead to the public and media making visits. The arrangements made to handle such groups must extend to after the event.

• Counselling support

Counselling support to affected employees and possibly contractors is not just a personnel management requirement which shows 'caring management' but increasingly a legal duty of the organisation to provide such support. Issues of 'post traumatic stress', 'nervous shock' and 'bereavement' means that the organisation has common law obligations to offer medical and psychological support. This should involve a medical practitioner and an occupational nurse. The insurance policy can be extended before the event to cover the cost of such services.

II Short term (a week onwards after the event)

• Investigation and inquiry

This can be both a statutory inquiry (although most are called within days) and/or the in-house investigation of the event and lessons to be learnt. Issues to consider include:

— Basic causes: was it a fire, bomb, an explosion or a natural peril?
— Underlying causes: what led up to such a peril occurring? Examine the managerial, personnel, legal, technical, organisational and natural factors that could have caused the event.
— Costs and losses involved.
— Lessons for the future.
— Did the disaster and emergency plan operate as expected? Were there any failings? What improvements are required? There needs to be complete de-briefing and examination of the entire process involving internal personnel and external agencies.

The organisation should weigh-up the possibility of external persons carrying out this exercise or whether in-house staff are objective and dispassionate enough to assess what went wrong.

• Visit by enforcers

The organisation also needs to be prepared for further visits from the enforcers and the possibility of statutory enforcement notices being served to either regulate or prohibit the activity. Multiple notices are possible, from health and safety officers, fire authority, Environment Agency/SEPA, building control or planning officers. This will affect the operation, production process and have economic implications. This ought not to occur if the organisation has taken due care to pre-plan for the major incident and had continuous safety monitoring of the operation. Notices will be served if there is a failure to make safe the site.

• Coping with speculation

The public, the media and employees will speculate about causation and there is the risk of adverse publicity. Public relations is a central activity. For legal and moral reasons, it is best practice to disclose all known facts unless the statutory investigation prohibits otherwise.

III   Medium term (a month plus)

[D6063]–[D6064] The 'normalisation process' will begin. The organisation needs to carry on the operation, be prepared for further visits from enforcers, loss adjusters and re-assess its corporate/financial health.

IV.   Long term (six months onwards)

- Systems review
    - A review of the impact of the major incident and whether the organisation is recovering from it.
    - Impact on reputation.
    - Legal threats.
    - Any positive outcomes – learning from mistakes, improving technical know-how, wider industrial benefits from knowing the chain reaction of events etc.

V.   Longer term (one year onwards)

[D6065] The organisation's memory and experience needs to be included in:

- in-house training programmes;
- factored into systems and procedures;
- review of entire *espirit de corps* and corporate philosophy.

Disaster and emergency planning is an extensive exercise, being dynamic and accounting for a diverse array of phenomena as industrial, man-made, environmental, socio-technical, radiological and natural events.

Monitor the disaster and emergency plan(s)

[D6066] Monitoring is a process assessing and evaluating the value, efficiency and robustness of the disaster and emergency plan(s). This involves looking at all three stages as discussed above (before, during and after the event stages) and not just the core document, 'the plan'. It is comprehensive and holistic in its questioning of the entire planning process.

Monitoring can be classified as proactive or reactive. The former attempts to identify problems with the plan(s) before the advent of a major incident. It is a case of continuously comparing the plan(s) with even minor incidents and loop-holes identified in any training/exercising sessions. Reactive monitoring occurs after a major incident or occurrences that could have led up-to a major incident, thereby reflecting back and assessing if the plan(s) need improvement. Both types are important.

FIGURE 5: MONITORING MATRIX

Stages in Planning	1. Before the Major Incident	2. During the Major Incident	3. After the Major Incident
Monitoring Types			
A. PROACTIVE	• Risk Assessments	• Inspections	• Inquiries
	• Testing via Training or Exercises	• Live Interviews	• Systems Review
	• Major Incident Assessment	• Feedback	• Counselling Reports
	• Facilities Inspections		
B. REACTIVE	• Incident Statistics	• Critical Assessments	• Brainstorming
	• Incident Reports	• Incident Levels	• Loss Assessments
	• Warnings	• Audio-Visual Assessment	• Enforcement
	• Notices		

The above cells are not strictly mutually exclusive; many techniques are both proactive and reactive. A third dimension is added in *COMAH* cases, that of, on-site and off-site emergency plans (types of planning).

- Cell A1 (proactive before the major incident)
  - Risk Assessments and hazard analysis techniques will identify significant hazards and their risk level. Monitoring such risk is an index of danger, which in turn is a variable in the type and potential of the major incident, outlined in Figure 1. The HSE's *Five Steps approach* or their *Quantified Risk Assessment* methodology provides outlines of assessing risk. Hazard techniques include hazard and operability studies, HAZANS, fault and event tree analysis etc.
  - Testing will identify any problems or concerns with the disaster and emergency plan. For example, a live exercise or a synthetic simulation could identify factors the plan has not considered or which may not be practicable if the major incident was to arise. Thus, enabling questioning, critical appraisal and comments should not be perceived as a threat or being awkward with the plan(s) but can provide vital information.
  - Major incident assessment, which is a periodical overview of the plan(s), every quarter or semi-annual by both internal and external persons. This could identify areas of concern. This assessment compares the plan(s) with the potential threat – can the former cope with the threat? Threats change as technology and know-how changes, so such assessments become another vital source of information.
  - Facilities inspections of gas, electricity, water, building structure, equipment available/not available etc. can highlight issues of physical resourcing and adequacy of such resourcing.
- Cell B1 (reactive after the major incident)

- Incident statistics will show the type of incidences, the type of injury, when and where it occurred. This enables the organisation to hypothesise/build a picture of the potential and severity of a bigger incident. One cannot divorce occupational health and safety data from 'disaster and emergency management'.

- Analysing incident reports should enable issues of causation to be assessed. What type of occurrence could trigger a major incident? Identifying and developing a pattern of causes will enable one to assess if the plan(s) account for such causes.

- Warnings from employees, contractors, enforcers, public and others of possible and serious problems are to be treated with seriousness. All such warnings are to be analysed and a common pattern and trend spotted.

- Any notices served by enforcers will identify failings in the operation and the remedial actions required. These can be factored into the plan(s).

- Cell A2 (proactive during the major incident)
  - Inspections will be made even as the event occurs. Inspections can range from the stability of the structure through to how personnel behaved and coped. As these inspections are made, the command and control team needs to evaluate if any aspect of the plan(s), which is a 'live document' need immediate changes.
  - Live interviews with internal personnel and emergency services' personnel will enable a continuous appraisal of any difficulty with procedures, arrangements and instructions that emanate from the plan(s). Again, these can lead to immediate changes to the plan(s).
  - Feedback is a proactive technique of requesting regular, interval information on and off-site. This enables a picture to be constructed of what could happen next; trying to anticipate the next sequence and if the plan(s) can cope with it.

- Cell B2 (reactive during the major incident)
  - Critical assessments are carried out after some unexpected occurrence, which causes uncertainty and may even threaten the efficacy of the plan(s). The critical assessment is by the command and control team as a whole. Why did this happen? Why did we not account for it in the plan(s)?
  - Incident levels – in particular if serious injury or fatalities are increasing, then at a moral or philosophical level one needs to ask if the plan(s) have been overwhelmed by reality. All forms of planning, including the statutory *COMAH* planning must not be viewed with rigidity. If the plan(s) are failing, it is better to re-appraise and re-plan. The *Control of Major Accident Hazards Regulations 1999 (SI 1999 No 743)* does not overtly allow for this, although it stresses flexibility and continuous appraisal of the event. In such a case, there has to be quick and clear decision-making, with consequential command and control, as well as immediate communication of this 'alternative plan'. Training and exercising sessions need to factor in this dimension and equip people with decisional techniques.

—   Audio-visual assessments can be a dramatic means of understanding the actual event. This can be video or photographic footage shot by the incident personnel or from the emergency services. This enables monitoring of the extent, potential and actual threat from the major incident.

- A3 (proactive after the major incident)
  —   Inquiries are proactive, even though the event has happened, the inquiry (whether internal or external) will identify strengths and weaknesses in the plan(s), which can lead to future improvements in planning.
  —   Systems review, is an overhaul of the entire reaction and holistic experience of the organisation to the trauma of a major incident. This involves developing future coping strategies for personnel and issues of how well did the organisation respond? Were there adequate resources in place to cope?
  —   Counselling reports will identify the experiences, perceptual and cognitive issues that affected personnel and others. This can provide probably the most significant information on behavioural response of the command and control team and those that were injured. Which in turn can be factored into training and exercise programmes, which itself will lead to personnel skill improvements.
- Cell B3 (reactive after the major incident)
  —   Brainstorming is an open-ended, participative and indeed critical analysis of what went wrong and what was right with the plan(s). Brainstorming should also be inclusive, involving emergency services and possibly enforcers as well, so that their guidance is factored in.
  —   The loss adjusters report will be a vital document as to the chain reaction that lead to the event and the consequences that followed. For legal reasons, their findings may have to be applied before insurance cover is available.
  —   Enforcement notices and enforcers reports will contain recommendations, which need to be viewed as lessons for the future.

Audit and review

[D6067]–[D6069] The final stage of the DEMS process is audit and review. An audit is a comprehensive and holistic examination of the entire DEMS process (Figure 2, see **D6034**). An audit will identify stages in this process that need improvement. A Review is an act of 'zooming in' into that particular stage and carrying out those improvements.

Major incident auditing

- Major incident auditing can be qualitative or quantitative. The former adopts a 'yes' or 'no' response format to questions. The latter asks the auditor to rate the issue being examined from say 0–5.

- The audit must be comprehensive, assessing every aspect of the DEMS process. This means that the Audit will take time to be completed. The audit is not some 'inspection' which is more random focusing upon a hazard rather than the complete system.
- Audits essentially benchmark (compare and contrast) performance. This can be against the DEMS process identified above or against legislation (e.g. *Control of Major Accident Hazards Regulations 1999 (SI 1999 No 743)*) as amended. The benchmarking could also be against another site or wider industry standards.
- Should audits be carried out in-house or rely on external consultants? There are costs and benefits associated with both, with no definitive answer. The *Management of Health and Safety at Work Regulations 1999 (SI 1999 No 3242)* in the UK, emphasises the need to develop and use in-house expertise in relation to general OSH issues, with a reliance on external specialists as a last resort. This may be interpreted as best practice for DEM.
- Audits can be annual or semi-annual. The more complex the operation and risk it poses, the greater the need for semi-annual audits.
- Audits should be proactive, that is learning from the weaknesses in the DEMS process and reducing or eliminating such weaknesses for the future.
- Finally, the results of the audit need to be fed-back into the DEMS process and all affected persons informed of any changes and risk management issues arising.
- Audits are holistic (assess anything associated with major incidents), systemic (assess the entire DEMS process, i.e. the 'system') and systematic (that is logical and sequential in analysis).

Review

(1) The review of any specific problems needs to be actioned by the organisation. The consultant will identify the areas of concern and make recommendations but the final discussion and implementation lies with the organisation. This needs to be led by senior officials in the organisation.

(2) Reviews are by definition 'diagnostic', meaning the organisation needs to look at causation and cure of the failure in any part of the DEMS process.

(3) Budgeting both in time and resource terms is critical in the review, as it will require management and external agency involvement.

(4) A review can be carried out at the same time as an audit. A review can also be a legal requirement, as with *Control of Major Accident Hazards Regulations 1999 (SI 1999 No 743), Reg 11*, which requires a review and where necessary a revision of the on-site and off-site emergency plans for top-tier establishments.

Conclusion

[D6070] All organisations and societies need to prepare for worst case scenarios. DEMS provide a logical framework to understand the main stages in effective preparation. The larger the organisation, the more detailed and analytical the preparation needs to be. DEMS is a live and open system requiring continuous monitoring.

Sources of information

[D6071] These include:

(1)   Legislation
- *Control of Major Accident Hazards Regulations 1999 (SI 1999 No 743).*
- *Control of Major Accident Hazards (Amendment) Regulations 2005 (SI 2005 No 1088).*
- *The Control of Substances Hazardous to Health Regulations 2002 (SI 2002 No 2677)* as amended.
- *The Radiation (Emergency Preparedness and Public Information) Regulations 2001 (SI 2001 No 2975).*
- *Public Information for Radiation Emergencies Regulations 1992 (SI 1992 No 2997).*
- *The Health and Safety (First-Aid) Regulations 1981 (SI 1981 No 917).*
- *Reporting of Injuries, Diseases and Dangerous Occurrences Regulations 1995 (SI 1995 No 3163).*
- *The Confined Spaces Regulations 1997 (SI 1997 No 1713).*

(2)   Guidance notes
- *Emergency Planning for Major Accidents: Control of Major Accident Hazards Regulations 1999* (HSG 191) is joint Guidance from the HSE, Environment Agency and the Scottish equivalent, SEPA. This interprets *COMAH* in a user-friendly manner and recommends implementation approaches.
- Guide to the Radiation (Emergency Preparedness and Public Information) Regulations 2001 (HSE L126).
- First aid at work. The Health and Safety (First-Aid) Regulations 1981. Approved Code of Practice and Guidance (L74) HSE Books, 1997.
- *Dealing with Disaster* by the Home Office (third edition) is a broad but useful outline of the managerial issues involved in planning. It is aimed at the Police, fire service and voluntary bodies.
- *NHS Emergency Planning Guidance 2005*, published by the Department of Health. This is available only online at http://www.dh.gov.uk/en/Publicationsandstatistics/Publications/PublicationsPolicyAndGuidance/DH_4121072.
- There are many other useful guidance documents that can be searched for at the following websites:
    — http://www.hse.gov.uk

This enables one to search for specific guidance, case st
udies and documents.

— http://www.epcollege.gov.uk

The Cabinet Office Emergency Planning College can be
contacted at this website where further assistance can be
obtained.

— http://www.environment-agency.gov.uk

The Environment Agency will have details of specific nat
ural environment or flood related guidance.

(3)   The European Commission
DG XI is the department responsible for environment, nuclear safety and
civil protection. It can be accessed via the EU website: http://www.
europa.eu.int

(4)   International agencies
The United Nations website is http://www.un.org; search for 'humanita
rian affairs'.

(5)   Professional bodies such as:
- International Institute of Risk and Safety Management: www.iirs
m.org
- Institution of Occupational Safety and Health: www.iosh.co.uk
- Business Continuity Institute: www.thebci.org
- Fire Protection Association: www.thefpa.co.uk
- SIESO (originally the Society of Industrial and Emergency
Services Officers): www.sieso.org.uk; telephone 01642 816281.

Display Screen Equipment

Mike Bateman and Andrea Oates

Introduction

[D8201] The *Health and Safety (Display Screen Equipment) Regulations 1992 (SI 1992 No 2792)* ('the *DSE Regulations*'), together with the other 'six pack' Regulations, came into operation on 1 January 1993. They originated from the European Union (EU) so-called framework and daughter directives. The aim of the DSE Regulations is to combat musculo-skeletal disorders (MSDs – also known as repetitive strain injuries or RSI) in the upper limbs, eye and eyesight effects, together with general fatigue and stress associated with work at display screen equipment ('DSE').

There has also been concern about whether electromagnetic radiation emissions from DSE are harmful, particularly in the early stages of pregnancy. The Health and Safety Executive (HSE) states in the guidance booklet to the Regulations (*Work with display screen equipment*(L26)) that they do not consider there are any radiation risks from working with DSE or special problems for pregnant women.

The HSE says that, taken as a whole, research has not shown any link between miscarriages or birth defects and work on DSE. But it advises that while pregnant women do not need to stop work with DSE, to avoid problems caused by stress and anxiety, women who are pregnant or planning children and are worried about working with DSE should be given the opportunity to discuss their concerns with someone adequately informed of the current authoritative scientific information and advice.

The Regulations require assessments to be carried out to identify any health and safety risks at workstations used by 'users' or 'operators' (as defined in the Regulations). 'Users' have the right to an eye and eyesight test and any spectacles (or contact lenses) found to be necessary for their DSE work must be provided by their employer. Several minor amendments were made to the Regulations by the *Health and Safety (Miscellaneous Amendments) Regulations 2002 (SI 2002 2174)*. The latest HSE guidance booklet (L26) incorporates these amendments.

The Regulations summarised

Definitions

[D8202] *Regulation 1* of the *DSE Regulations (SI 1992 No 2792)* contains a number of important definitions which are summarised below:

- Display screen equipment
  - Any alphanumeric or graphic display screen, regardless of the display process involved.
  - A European Court of Justice (ECJ) case involving a film cutter, *Dietrich v Westdeutscher Runddfunk*, C-11/99 [2000] ECR I-5589 ECJ, ruled that the term 'graphic display screen' had to be interpreted to include screens displaying film recordings.
- Workstation
  An assembly comprising:
  - display screen equipment (whether provided with software determining the interface between the equipment and its operator or user, a keyboard or any other input device)
  - any optional accessories to the display screen equipment
  - any disk drive, telephone, modem, printer, document holder, work chair, work desk, work surface or other item peripheral to the display screen equipment, and
  - the immediate work environment around the display screen equipment.
- User
  - An employee who habitually uses DSE as a significant part of his normal work.
- Operator
  - A self-employed person who habitually uses DSE as a significant part of his normal work.

The HSE guidance states that the main factors to consider in determining whether a person is a 'user' or 'operator' are:

- use of DSE normally for continuous or near-continuous spells of an hour or more;
- use of DSE more or less daily;
- the need to transfer information quickly to or from the DSE;

Other factors are:

- high levels of attention and concentration are required;
- high dependence on DSE with little choice about using it;
- special training or skills are needed to use the DSE.

Detailed examples are given in the HSE guidance booklet providing pen-portraits of definite users, possible users and those who are definitely not users. Examples of the type of employees likely to be classed as users include word-processing operators, secretaries and typists, data input operators and journalists. Those who are possibly users include scientists and technical advisers, client managers in large management accountancy consultancies, building society customer support officers, airline check-in clerks, community care workers and receptionists. Those who would not be users include, for example, senior managers who use DSE only for occasional monitoring.

Homeworkers, teleworkers and agency workers all come within the scope of the Regulations, if they fulfil the definition of user or operator – see **D8213** below.

Exclusions

[D8203] The *DSE Regulations (SI 1992 No 2792)* do not apply to:

- drivers' cabs or control cabs for vehicles or machinery;
- DSE on board a means of transport;
- DSE mainly intended for public operation;
- portable systems not in prolonged use;
- calculators, cash registers or other equipment with small displays;
- window typewriters.

However, the *Health and Safety at Work etc. Act 1974* and regulations made under it, such as the *Workplace (Health, Safety and Welfare) Regulations 1992 (SI 1992 No 3004)*, still apply to the use of such equipment. The HSE advises that particular attention should be paid to ergonomics in this context – that is the science of making sure that work tasks, equipment, information and the working environment are suitable for every worker, so that work can be done safely and productively.

Assessment of workstations and reduction of risk

[D8204] *Regulation 2* of the *DSE Regulations (SI 1992 No 2792)* requires employers to perform a 'suitable and sufficient' analysis of all workstations which are:

- used for their purposes by 'users';
- provided by them and used for their purpose by 'operators';

to assess the health and safety risks in consequence of that use. They must then reduce the risks identified to the lowest level reasonably practicable. As for other types of assessments, an assessment must be reviewed if there is reason to suspect it is no longer valid or there have been significant changes.

Further guidance on assessment, including an assessment checklist (at **D8217**), is provided later in the chapter.

Requirements for workstations

[D8205] *Regulation 3* of the *DSE Regulations (SI 1992 No 2792)* states that all workstations must meet the requirements laid down in the *Schedule* to the Regulations. Many of the requirements of the Schedule have been incorporated into the assessment checklist (at **D8217**). The full Schedule is contained in the HSE guidance booklet (L26). However, the requirement must relate to a component present in the workstation concerned and be relevant in relation to the health, safety and welfare of workers. For example there is no need to provide a document holder (referred to in the *Schedule*) if there is little or no inputting from documents and the HSE advises that some individuals with back complaints may benefit from a fixed back rest or a special chair without a back rest. The HSE guidance booklet provides further examples of where the requirements of the *Schedule* may not be appropriate.

Daily work routine of users

[D8206] Employers are required by *Regulation 4* of the *DSE Regulations (SI 1992 No 2792)* to plan the activities of 'users' so that their DSE work is periodically interrupted by breaks or changes of activity. Breaks should be taken before the onset of fatigue and preferably away from the screen. Short frequent breaks are better than occasional longer breaks. It is best if users are given some discretion in planning their work and are able to arrange breaks informally rather than having formal breaks at regular intervals. Some software tools provide a means of ensuring that users take regular breaks but HSE guidance draws attention to the limitations of such software.

In terms of the frequency of breaks, the general advice from trade unions, for example, is a break from intensive DSE work every 30 minutes.

Eyes and eyesight

[D8207] *Regulation 5* of the *DSE Regulations (SI 1992 No 2792)* gives 'users' the right (at their employer's expense) to an appropriate eye and eyesight test by a competent person before becoming a 'user' and at regular intervals thereafter.

The HSE guidance to the regulations says that employers should be guided by the clinical judgement of the optometrist or doctor on the regularity of testing and trade union advice is generally for repeat testing at specified intervals, such as two years, or on the advice of the optometrist.

'Users' are also entitled to tests on experiencing visual difficulties which may reasonably be considered to be caused by DSE work.

Tests are normally carried out by opticians and involve a test of vision and an examination of the eye. Where companies have vision screening facilities, 'users' may opt for a screening test to see if a full eye test is needed. If the eye test shows a need for spectacles (other than the 'user's' normal spectacles) the basic cost of these must be met by the employer, but employees must pay for extras, e.g. designer frames or tinted lenses.

The HSE makes clear that vision screening tests are not an 'eye and eyesight test' and hence do not satisfy the DSE Regulations, but it says that some employers may wish to offer them as an extra.

Provision of training

[D8208] 'Users' are required by *Regulation 6* of the *DSE Regulations (SI 1992 No 2792)* to be provided with adequate health and safety training in the use of any workstation upon which they may be required to work. Training may also be required where workstations are substantially modified. The training should include:

- the causes of DSE-related problems, e.g. poor posture, screen reflections;
- the user's role in detecting and recognising risks;

- the importance of comfortable posture and postural change;
- equipment adjustment mechanisms, e.g. chairs, contrast, brightness;
- use and arrangement of workstation components;
- the need for regular screen cleaning;
- the need to take breaks and for changes of activity;
- arrangements for reporting problems with workstations, or ill health symptoms;
- information about the Regulations (especially eyesight tests and breaks); and
- the user's role in assessments.

The HSE has published a leaflet (*Working with VDUs* (INDG36)) which provides a useful reference for training purposes.

Provision of information

[D8209] *Regulation 7* of the *DSE Regulations (SI 1992 No 2792)* requires employers to ensure that operators and users at work within their undertaking are provided with adequate information. The table below shows the responsibility of the 'host' employer in this respect (and also gives a good guide to his responsibilities generally under the Regulations).

In 2008, a Court of Appeal judgment found in favour of a worker whose repetitive strain injury (RSI) was made worse as a result of keyboard use. She was not required to make an excessive number of keystrokes and had opportunities during the working day to do other work away from the keyboard. However the judgment accepted that her work had aggravated the RSI. The Court found that the company involved had not complied with the DSE regulations, because having been informed of her condition, it provided no information or training or further reductions in her keyboard use (*Goodwin v Bennetts UK Ltd*[2008] EWCA Civ 1374, [2008] All ER (D) 220 (Dec)).

The 'host' employer must provide information about:	*Regulation:*	*Own users:*	*Other users (i.e. agency staff):*	*Operators (self-employed):*
DSE and workstation risks		YES	YES	YES
Risk assessment and reduction measures	2 and 3	YES	YES	YES
Breaks and activity changes	4	YES	YES	NO
Eye and eyesight tests	5	YES	NO	NO
Initial training	6(1)	YES	NO	NO
Training when workstation substantially modified	6(2)	YES	YES	NO

DSE workstation assessments

Decide who will carry out the assessments

[D8210] As with other types of risk assessments, DSE workstation assessments within an organisation may be carried out by an individual or by members of an assessment team. Those responsible for making assessments should have received appropriate training so that they are familiar with the requirements of the *DSE Regulations (SI 1992 No 2792)* and they should have the ability to:

- identify hazards (including less obvious ones) and assess risks from the workstation and the kind of DSE work being done;
- use additional sources of information or expertise as appropriate (recognising their own limitations);
- draw valid and reliable conclusions;
- make a clear record and communicate the findings to those who need to take action;
- recognise their own limitations so that further expertise can be utilised where necessary.

They may be health and safety specialists, or other in-house staff who have received appropriate training or who have the appropriate abilities to carry out assessments (see above). Safety representatives should also be involved in the risk assessment process. The HSE advises that safety reps should also be encouraged to report any problems in DSE work that come to their attention.

Identify the 'users'

[D8211] An important first step is to identify the users' (together with any 'operators') of DSE within the organisation. The definitions in *Regulation 1* of the *DSE Regulations (SI 1992 No 2792)* refer to habitual use of DSE. The HSE guidance booklet (L26) provides considerable advice on the factors which must be taken into account, of which time spent using DSE is the most significant. Some organisations have adopted a rule of thumb that anyone spending more than 50 per cent of their time in DSE work is a DSE 'user'. However, the HSE guidance indicates that a less simplistic approach should be taken.

The assessment checklist provided at **D8217** includes reference to the other factors which the HSE states should be taken into account in deciding whether an individual is a 'user' or an 'operator'. It should be noted that employers have duties to assess workstations used for the purposes of their undertaking by *all* 'users' or 'operators'. This includes 'users' employed by others (e.g. agency-employed staff), 'operators' (e.g. self-employed draughtsmen or journalists), peripatetic staff (e.g. journalists, sales staff, careers advisors) and homeworkers or teleworkers (see **D8213** below).

In practice, most employers do not find it too difficult to decide who their 'users' and 'operators' are. Many have taken the approach of assessing the workstations of those individuals where doubt exists and, if necessary, making a final decision then. More time and expense can often be wasted debating a few borderline cases than would be involved in including them in the definition.

Decide on the assessment approach

[D8212] The HSE guidance to the regulations sets out the principal risks that may arise from DSE work relate to musculoskeletal problems, visual fatigue and mental stress. As in other types of work, it says that ill health can result from poor equipment or furniture, work organisation, working environment, job design and posture, and from inappropriate working methods.

And it says that the known health problems associated with DSE work can be prevented in the majority of cases by good ergonomic design of the equipment, workplace and job, and by worker training and consultation.

It sets out that employers must assess the extent to which any of these risks arise for DSE workers using their workstations who are:

- users employed by them;
- users employed by others (for example agency employed 'temps'); or
- operators – self-employed contractors who would be classified as users if they were employees (for example self-employed agency 'temps', self-employed journalists).

It makes clear that individual workstations used by any of these workers must be analysed and the risks assessed. If employers require their employees to use workstations at home, these too will need to be assessed (see **D8213**below).

In some situations, especially smaller workplaces, it is practicable to carry out a personal assessment of each individual 'user's' or 'operator's' workstation, using a simple checklist (a completed example of such a checklist is provided at **D8217**). However, in larger organisations the number of DSE workstations may be so great that this approach becomes impracticable (particularly where there are regular office rearrangements taking place) or would require an unnecessarily large input of resources. In such cases the issuing of a self-assessment checklist to identified 'users' will often be more appropriate. A sample checklist, together with guidance to 'users' on its completion, is provided at **D8217**. The HSE guidance booklet (L26) provides a checklist covering similar topics. Providing that workers are given the necessary training and guidance in how to use such checklists, this approach is perfectly acceptable. However, it must be supported by other actions such as:

- an inspection of areas where DSE workstations are situated (evaluating general issues such as lighting, blinds, housekeeping, desk space and the standards of chairs and DSE hardware);
- providing employees with the option of an assessment of their individual workstation by a specialist;
- review of completed self-assessment checklists by a competent person;
- responding promptly to problems identified in the completed self-assessment checklists.

Risk assessments should be reviewed if there is reason to suspect that it is no longer valid; or if there have been significant changes and the employer should implement any changes the review shows to be necessary (see **D8223**).

Homeworkers and teleworkers

[D8213] Homeworkers and teleworkers (working away from their employer's premises) are subject to the *DSE Regulations (SI 1992 No 2792)*, whether or not their workstation is provided by their employer. They will be subject to the normal risks from DSE work, some of which may be increased because of their social isolation, the absence of supervision and the practical difficulties of carrying out risk assessments. Risks can be reduced by:

- training such staff to carry out workstation self-assessments (see above);
- requiring them to carry out a self-assessment of their main workstation;
- encouraging them to carry out ad hoc assessments of temporary workstations, e.g. hotel rooms;
- emphasising the importance of ensuring good posture and taking adequate breaks;
- providing clear communication routes for reporting equipment defects and possible health problems;
- responding promptly and effectively to reports of defects or problems.

A free HSE guidance leaflet (*Homeworking: Guidance for employers and employees on health and safety* (INDG226)) provides general guidance on risk assessment in the home environment and the employer's responsibilities for home electrical systems and equipment.

Observations at the workstation

[D8214] Where workstations are to be assessed on an individual basis the assessment checklist questions are intended to identify the principal factors that the assessor(s) need to look out for (as illustrated in the completed example).

Some of the more common problems identified are related to:

- Posture:
  - height of screen;
  - height of seat or position of backrest;
  - position of the keyboard or mouse;
  - keyboard and mouse technique;
  - need for footrest or document holder.
- Vision:
  - angle of screen;
  - position of lights or need for diffusers;
  - the availability, condition and effectiveness of blinds;
  - adjustment of brightness or contrast controls.

Discussions with 'users' and 'operators'

[D8215] An individual assessment provides an opportunity for a two-way dialogue between the assessor(s) and the 'user' and allows the assessor to evaluate whether any problems reported are related to deficiencies in the workstation.

In addition, the HSE advises employers to encouraging users to report any ill health that may be due to their DSE work. It says that this is a useful check that risk assessment and reduction measures are working properly and that reports of ill health may indicate that reassessment is required. DSE training should include the need to report and the organisational arrangements for making a report.

Common causes of problems are:
* Back, shoulders, neck:
  — the height or position of the screen;
  — positioning of the seat, including the backrest;
  — the need for a footrest or document holder.
* Hands, wrists, arms:
  — position of the keyboard or mouse;
  — keyboard or mouse technique.
* Tired eyes or headaches:
  — lack of regular breaks or activity changes;
  — reflected light (artificial or sunlight).
* Discussions with 'users' can also reveal other important pieces of information, e,g:
  — there are problems with sunlight at certain times of day or periods of the year;
  — the 'user' does not know how to adjust their chair or brightness/contrast controls;
  — the 'user's' chair is broken and incapable of being adjusted;
  — the 'user' has never been offered an eye test.
  In addition, problems can arise from excessive workloads and high key stroke rates.

Reducing the risks

[D8216] The HSE advises that postural problems may be dealt with through simple adjustments to the workstation, such as repositioning equipment or adjusting the chair. They may also indicate that training may need to be reinforced with information about correct hand position, posture, or how to adjust equipment, for example. New equipment, such as a footrest or document holder, may be necessary.

It says that visual problems can often be tackled by straightforward methods such as repositioning the screen or using blinds to avoid glare; ensuring that the screen is at a comfortable viewing distance, or by ensuring the screen is kept clean. In some cases, in appropriate lighting may be causing the problem.

And it says that fatigue and stress may also be alleviated (in addition to the measures above) by ensuring that:

- software is appropriate to the task;
- the task is well designed;
- users have a degree of personal control over the pace and nature of their tasks; and
- proper provision is made for training and information, not only on health and safety risks but also on the use of software.

HSE guidance to the regulations says

'It is important to take a systematic approach to risk reduction and recognise the limitations of the basic assessment. Observed problems may reflect the interaction of several factors or may have causes that are not obvious. For example backache may turn out to have been caused by the worker sitting in an abnormal position in order to minimise the effects of reflections on the screen. If the factors underlying a problem appear to be complex, or if simple remedial measures do not have the desired effect, it will generally be necessary to obtain expert advice on corrective action'.

Assessment records

[D8217] As for other types of assessment, there is no standard format for DSE workstation assessment records. The completed sample assessment checklist and the self-assessment checklist (both below), together with the guidance on its completion are offered as examples of record formats that have been found successful in practice. A longer assessment checklist is provided in HSE booklet L26 and also in HSG90 'The law on VDUs. An easy guide' and on the HSE website (www.hse.gov.uk)

The HSE guidance booklet states that records may be stored in electronic as well as paper form. Self-assessment checklists would particularly lend themselves to being completed and submitted electronically. No guidance is provided on how long records should be kept. Prudent employers may prefer to retain them indefinitely, bearing in mind that civil claims for DSE-related conditions may be submitted many years after the condition first arose.

Display Screen Equipment Workstation Assessment	
USER'S NAME: *Christine Jones*	LOCATION: *Sales*
FACTOR	COMMENT
1 WORK PATTERNS	
1.1 Most time spent per day at DSE	*5 to 6 hours*
1.2 Average time per day at DSE	*4 hours*
1.3 Number of days per week at DSE	*5*
1.4 Longest spell without break	*2 hours*
1.5 Can breaks be taken?	*Yes – at Christine's discretion*
1.6 Concentration important?	*Accuracy is important*

1.7	Speed of operation important?	*Sometimes*
	'USER' STATUS CONFIRMED	*Yes*
2	**PROBLEMS EXPERIENCED** Has the user significant experience of problems with:	
2.1	Back, shoulders or neck	*Regular pains in shoulder and neck*
2.2	Hands wrists or arms	*Occasional aching wrists*
2.3	Tired eyes or headaches	*Sometimes*
2.4	Suitability of software	*Suitable for all tasks*
2.5	Other problems	None
3	**LIGHTING/ENVIRONMENT**	
3.1	Artificial lighting:	
	— adequate to see documents	*Yes*
	— any reflection or glare problems	*None – fittings recessed and diffused*
3.2	Sunlight:	
	— any reflection or glare problems	*On winter mornings from window behind*
	— suitable blinds available (if necessary)	*Effective vertical blinds provided*
3.3	Noise:	
	— hindering to communication	*No*
	— distracting or stressful	*Occasionally distracting*
3.4	Temperature and ventilation:	
	— satisfactory in summer and winter	*Office hot and stuffy in summer*
4	**SCREEN**	
4.1	Set at suitable height	*Screen height too low*
4.2	Stable image with clear characters	*Good. Able to vary colours*
4.3	Brightness and contrast adjustable	*Both have adjustable controls*
4.4	Swivels and tilts easily	*Yes*
4.5	Cleaning materials available	*In stationery cupboard*

5	**KEYBOARD**	
5.1	Separate and tiltable	*Yes*
5.2	Sufficient space in front	*Too near edge of desk (wrists bent)*
5.3	Keys clearly visible	*Yes*
6	**DESK AND CHAIR**	
6.1	Desk size adequate	*Satisfactory*
6.2	Sufficient leg room	*Materials being stored in desk well*
6.3	Desk surface low reflectance	*Yes – wooden surface*

6.4	Suitable document holder (if required)	*Not provided*
6.5	Chair comfortable and stable	*Yes*
6.6	Chair height adjustable	*Yes*
6.7	Back adjustable (height and tilt)	*Christine did not know how to adjust*
6.8	Footrest available (if required)	*Not required*

OTHER COMMENTS

Christine was advised to take short breaks more regularly.
Some lengthy jobs without breaks seem to be the cause of her tired eyes and headaches.
Repositioning of the screen and provision of a document holder should overcome the shoulder and neck problems.
Christine was shown how to adjust her chair back.

No.	Actions required	Responsibility
3.2	Close blinds when sunlight bright.	*C Jones*
3.4	Free up office windows (seized up by paint).	*Facilities Manager*
4.1	Provide screen stand and keep top of screen at eye level.	*Office Manager/C Jones*
5.2	Keep space in front of keyboard – wrists horizontal.	*C Jones*
6.2	Remove items from desk well.	*C Jones*
6.4	Provide document holder.	*Office Manager*
6.7	Ensure induction includes chair adjustment mechanisms.	*Training Department*

Assessor: *B WRIGHT*   Signature: *B Wright*   Date: *11/12/10*

PROGRESS WITH ACTIONS

All recommendations acted upon although Christine still needs to remember to take regular breaks on lengthy jobs. Pains in shoulder and neck have ceased and Christine only very occasionally experiences headaches and tired eyes.

B. Wright 6/2/11

Planned date for assessment review: February 2012

DISPLAY SCREEN EQUIPMENT WORKSTATION SELF ASSESSMENT CHECKLIST	
USER'S NAME:	LOCATION:
1 LIGHTING AND WORK ENVIRONMENT	COMMENTS
1.1 Is artificial lighting adequate?	
1.2 Does it cause any reflection or glare problems?	
1.3 Any reflection or glare problems from sunlight?	

1.4	Are suitable blinds available (if necessary)?	
1.5	Are temperature and ventilation satisfactory in summer and winter?	
2	**SCREEN AND KEYBOARD**	
2.1	Is your screen set at a suitable height?	
2.2	Stable image with clear characters?	
2.3	Brightness and contrast adjustable?	
2.4	Screen swivels and tilts easily?	
2.5	Cleaning materials available?	
2.6	Is your keyboard tiltable?	
2.7	Have you sufficient space in front of it?	
3	**DESK AND CHAIR**	
3.1	Is your desk size adequate?	
3.2	Is there sufficient leg room under it?	
3.3	Do you have a suitable document holder (if required)?	
3.4	Is your chair comfortable and stable?	
3.5	Can you adjust your seat height?	
3.6	Can you adjust the height and tilt of your chair back?	
3.7	Do you have a footrest (if required)?	
4	**HAVE YOU HAD SIGNIFICANT EXPERIENCE OF PROBLEMS WITH:**	
4.1	Your back, shoulders or neck?	
4.2	Your hands, wrists or arms?	
4.3	Tired eyes or headaches?	
4.4	The suitability of the software you use?	
4.5	Other problems?	

ANY OTHER PROBLEMS OR COMMENTS?

Would you like a further assessment of your workstation? Yes/No

Signature: Date:

COMPLETED FORMS MUST BE SENT TO THE HUMAN RESOURCES DEPARTMENT

Guidance on completing the DSE workstation self-assessment

Lighting and work environment

1.1 Artificial lighting should be adequate to see all the documents you work with.

1.2 Recessed lights with diffusers shouldn't cause problems. Lights suspended from ceilings might.

1.3 There may be problems in the early morning or afternoon, especially in winter when the sun is low.

1.4 Blinds provided should be effective in eliminating glare from the sun.

1.5 Strong sunlight may create significant thermal gain at times.

Screen and keyboard

2.1 The top of your screen should normally be level with your eyes when you are sitting in a comfortable position.

2.2 There should be little or no flicker on your screen.

2.3 You should know where the brightness and contrast controls are.

2.4 The screen should swivel and tilt so that you can avoid reflections.

2.5 You should know where to get cleaning items for your screen (and keyboard, if necessary).

2.6 Small legs at the back of your keyboard should allow you to adjust its angle.

2.7 Space in front the keyboard allows you to keep your wrists horizontal and to rest your hands and wrists when not keying in.

Desk and chair

3.1 Your desk should have sufficient space to allow you to have your screen and keyboard in a comfortable position and accommodate documents, document holder, phone etc.

3.2 There should be enough space under the desk to allow you to move your legs freely.

3.3 If you are inputting from documents, using a document holder helps avoid frequent neck movements.

3.4 Chairs with castors must have at least five (four is very unstable).

3.5 You should be able to adjust your seat height to work in a comfortable position (arms approximately horizontal and eyes level with the top of the screen).

3.6 The angle and height of your back support should be adjustable so that it provides a comfortable working position.

3.7 DSE users who are shorter may need a footrest to help them keep comfortable when sitting at the right height for their keyboard and screen (see 3.5).

Possible problems

4.1 Problems with back, shoulders or neck might indicate your screen is at the wrong height, an incorrectly adjusted chair or need for a document holder.

4.2 Problems with hands, wrists or arms might indicate incorrect positioning of the keyboard or a poor keying technique. Hands should not be bent up at the wrist and a soft touch should be used on the keyboard, not overstretching the fingers.

4.3 Tired eyes or headaches could indicate problems with lighting, glare or reflections.

They may also indicate the need to take regular breaks away from the screen. Persistent problems might need an eye test – contact Human Resources to request one.

4.4 The software should be suitable for the work you have to do.

IF YOU HAVE A PROBLEM DISCUSS IT WITH YOUR LINE MANAGER. REFER UNRESOLVED PROBLEMS OR REQUESTS FOR FURTHER ASSESSMENTS TO THE HUMAN RESOURCES DEPARTMENT.

Special situations

[D8218] The HSE guidance booklet (L26) provides further guidance on several aspects of DSE work which have developed in recent years as information technology has changed.

Shared workstations

[D8219] 'Hot desking' arrangements mean that many workstations are shared, sometimes by workers of widely differing sizes. Assessments of such workstations should take into account aspects such as:

- whether chair adjustments can accommodate all the workers involved;
- the availability of footrests;
- adequacy of leg room for taller workers;
- arrangements for adjusting the heights of screens (e.g. adjustable mountings or stands).

In some situations the use of desks with adjustable heights may be appropriate.

Work with portable DSE

[D8220] Work with laptop computers as well as mobile phones or personal digital assistants (PDAs) which can be used to compose, edit or view text is becoming increasingly common. Where mobile phones or PDAs are used in this way for prolonged periods they are subject to the *DSE Regulations (SI 1992 No 2792)*, as is all work with laptops.

Mobile phones or PDAs will not generally be used for sufficiently long periods to require a workstation assessment but work involving laptops may be of much longer duration. The revised HSE booklet (L26) contains detailed guidance on the selection and use of laptop computers.

The design of laptops is such that postural problems are much more likely to result and there will be a far greater need to ensure that sufficient breaks are taken. Where the laptop is used in an office or at home for extended periods, the risks can be reduced considerably by the use of a docking station or the provision of separate monitors, keyboards etc. which can be connected to the laptop. The HSE booklet L26 provides further guidance on this. Other risks associated with the use of portable equipment (eg manual handling issues and possible theft involving assault) should also be considered.

Pointing devices

[D8221] Much work at DSE workstations involves the use of a mouse, trackball or similar device to move the cursor around the screen and carry out operations. The HSE booklet (L26) provides guidance on:

Choice of pointing devices
*   Suitability of the device for:
    — the environment (space, position, dust, vibration);
    — the individual (right or left handed, physical limitations, existing upper limb disorder);
    — the task (some devices are better than others in respect of speed or accuracy).

Use of pointing devices
*   Issues to be considered include:
    — positioning (close to the midline of the user's body, not out to one side);
    — technique (wrist straight, arm not stretched, forearm supported, do not grip too tightly);
    — work surface (particularly its height and degree of support for the arm);
    — mousemats (smooth, large enough, without sharp edges);
    — software settings (suitable for the individual user);
    — task organisation (point device use mixed with other activities);
    — training (how to set up and use devices);
    — cleaning and maintenance.

The booklet also contains specific guidance on touch screens and speech interfaces.

After the assessment

Review and implementation of recommendations

[D8222] The general guidance provided in the chapter on **RISK ASSESS-MENT** is equally applicable to recommendations made as a result of DSE workstation assessments. Some issues of general importance will need to be reviewed with others, particularly if there are significant cost implications or practical problems. Responsibility for implementing recommendations should be allocated clearly and all recommendations followed up. It should be noted that the sample assessment checklist includes a space which can be used at a follow-up of an assessment to describe 'Progress with actions'.

Assessment review/re-assessments

[D8223] Assessments must be kept up to date. Changes to the layout of office accommodation often take place with bewildering rapidity. Some of these changes have significant implications for DSE workstations, others do not. Both management and DSE 'users' should be aware that a review or reassessment should be requested from the assessor or assessment team whenever significant changes take place. The use of self-assessments is particularly relevant in such situations. This can be supported by observations by the assessor(s) of changes which have occurred or are in progress.

Because of the frequency of changes a periodic review of DSE workstation assessments would be beneficial. A 2008 HSE evaluation of the DSE regulations found that three quarters of businesses conducted risk assessments every 12 months. Significant changes in DSE equipment, software, furniture, lighting etc. would obviously justify a review of relevant workstation assessments.

Further information

- HSE Publications
  - — L26 Work with display screen equipment
  - — HSG 90 The law on VDUs. An easy guide
  - — INDG 36 Working with VDUs
  - — INDG 226 Homeworking: guidance for employers and employees on health and safety
- HSE Website (www.hse.gov.uk)
  Its content includes:
  - — the DSE Regulations and guidance on their interpretation;
  - — downloadable copies of the HSE publications listed above;
  - — frequently asked questions;
  - — DSE case studies;
  - — links to further sources of information.

Electricity

Chris Buck and Andrea Oates

Introduction

[E3001] Electricity has become fundamental to our way of life – both at work and in the home. It provides an energy source to meet an ever-increasing range of needs and, used with care, is a good servant. However, treated with disrespect, it is a poor master and can result in serious injury or even death.

Some 1000 electrical accidents at work are reported to the Health and Safety Executive (HSE) each year and about 25 people die of their injuries.

The main causes of these deaths and injuries are:

*   the use of poorly maintained electrical equipment;
*   work near overhead power lines;
*   contact with underground power cables during excavation work;
*   mains electricity supplies (230 volt);
*   the use of unsuitable electrical equipment in explosive areas such as car paint spraying booths; and
*   fires started by poor electrical installations and faulty electrical appliances.

Understanding electricity

[E3002] A basic knowledge of electrical principles is important to an understanding of the dangers of electricity and, more importantly, the control measures that need to be put in place to avoid these dangers and thereby prevent injury. The content of this section has been restricted to aiding understanding of the remainder of this chapter and a good textbook on electrical theory should be consulted for more extensive knowledge.

Because electricity is an unseen energy form its nature is often difficult to appreciate. A simple analogy is to compare it with the flow of water in a pipe, where the water flow is determined by the water pressure and resistance to the flow arising from the size of pipe. Fundamentally, the flow of electricity (electric current) arises from the movement of electrons in an electrical circuit. The amount of current (I) flowing in a circuit is determined by the electrical pressure or voltage (V) and the resistance (R) in the circuit. These three basic parameters are related by an equation known as Ohm's Law:

$I = V/R$

Thus varying the electrical pressure will alter the current in direct proportion whilst varying the resistance will vary the current in inverse proportion.

Electric current is measured in amperes or 'amps' for short (A), voltage in 'volts' (V) and resistance in ohms (Ω). The range of values encountered for these basic parameters is very great, often from millions to millionths of the basic unit. Standard prefixes are used to denoted multiples and sub-multiples of the basic unit. For example, 1 mA is one thousandth of an amp whilst 1 mΩ is one thousandth of an ohm. Similarly, 1 kA is one thousand amps, 1 kV is one thousand volts and 1 MΩ is one million ohms.

In any electrical circuit, resistance is encountered in two forms, as continuity or conductor resistance and as insulation resistance. Continuity resistance is the conductor resistance measured end-to-end. This should be as low as possible in order not to unduly restrict the flow of current. Typical values of continuity resistance will usually be less than an ohm, and are often quoted in milliohms (mΩ). Insulation resistance is the resistance between conductors and is therefore a measure of the quality of the electrical insulation surrounding the conductors. Because we do not want electricity to leak across or short-circuit from one conductor to another, or out of the circuit, the resistance value should be as high as possible. The insulation resistance of a circuit should therefore be a very large number of ohms, typically several million ohms (MΩ).

There are two forms of current, alternating (a.c.) and direct (d.c.) and both are likely to be encountered in the workplace. Alternating current is the most common because this is the form of 'mains' electricity. Direct current is the form provided by batteries and is therefore encountered in workplaces where battery-powered works vehicles are used. Certain specialist work processes also use d.c. obtained by conversion (rectification) of the a.c. mains supply. The process of rectification is not perfect and some oscillation generally appears on the d.c. output. This is referred to as ripple and has implications relating to electric shock. In the case of electrical circuits operating from the 'mains' the circuit resistance is more complicated because the oscillating waveform of the current brings into play additional characteristics, namely the magnetising and charging effects of the current. Suffice to say, the resistance of an a.c. circuit is referred to as its impedance (Z) which represents a fairly complex mathematical relationship involving the inductance and capacitance displayed by the circuit, in addition to its resistance. As with resistance, impedance is also measured in ohms.

The flow of an electric current represents power or energy. Electric power is the product of the current and voltage, measured in watts (W) or (kW), whilst electrical energy is the product of electric power and time, measured in watt seconds (Ws) or kilowatt hours (kWh).

Legal background and standards

[E3003] The fundamental piece of legislation regulating electrical safety at work is the *Electricity at Work Regulations 1989 (SI 1989 No 635) ('EWR')*. This chapter is mainly devoted to explaining the requirements of these Regulations but from E3029 onwards some information is included concerning other electrical legislation of relevance in specific circumstances.

Reference is also made to electrical standards and approved codes of practice (ACOPs). A full list of commonly used electrical standards and ACOPs is set out on the HSE website at www.hse.gov.uk/electricity/standards.htm.

The *EWR* comprise 33 individual regulations grouped into four parts:

* *Part I*: Introductory matters (*Regulations 1–3*);
* *Part II*: General technical provisions (*Regulations 4–16*);
* *Part III*: Provisions relating specifically to mines (*Regulations 17–28*);
* *Part IV*: Miscellaneous and general provisions (*Regulations 29–33*).

As these Regulations have been made under the *Health and Safety at Work etc. Act 1974* ('*HSWA*') they apply in the same circumstances as the parent Act, ie they are work-related rather than applying to specific categories of premises. Thus they apply to all types of work establishment and, like the parent Act, place responsibilities on employers, the self-employed and employees, in this case to avoid danger and prevent injury from electricity. The responsibilities of these duty holders are defined in *Regulation 3* and require action to be taken where the matter is under the control of the duty holder. An example of matters under the control of an employer would be the preparation of an appropriate policy relating to electrical safety and the organisation and arrangements for putting it into effect, such as for the maintenance of electrical equipment. An example of corresponding matters under the control of an employee would be to comply with those electrical maintenance procedures and use any tools and equipment, provided in connection with the work, in the correct manner. Some of the duties imposed by the Regulations are qualified, ie 'so far as is reasonably practicable', as in the parent Act; the others are absolute. In the case of a prosecution for a breach of an absolute duty, it is available to the employer to utilise the defence provided by *Regulation 29*. This states that it shall be a defence for any person to prove that he took all reasonable steps and exercised all due diligence to avoid the commission of that offence. Continuing with the example of electrical equipment maintenance, 'taking all reasonable steps' would involve establishing a maintenance policy and putting into place the organisation and arrangements for its implementation, whilst 'exercising all due diligence' would require on-going monitoring to ensure that the policy remained effective and that the maintenance was undertaken to the required standard.

The *EWR* apply to all electrical systems which, by definition (*Regulation 2*) means electrical equipment connected to a common source of energy. Electrical systems range from power stations and the national grid at one end of the spectrum to electrical installations within buildings and electrical equipment at the other. They require action to be taken only where there is danger to be avoided or injury prevented (see E3004).

When originally enacted, the *EWR* did not apply to offshore installations, ie outside territorial waters. However, one of the outcomes from the enquiry into the Piper-Alpha oil rig fire was to extend the application of the *EWR* to such installations through the introduction of the *Offshore Electricity and Noise Regulations 1997 (SI 1997 No 1993)*. Separate but virtually identical Regulations are in force in Northern Ireland, as the *Electricity at Work Regulations (Northern Ireland) 1991*.

Modern health and safety legislation tends to be goal-setting and therefore non-prescriptive. The *EWR* are no exception to this principle! They are well-founded, laying down the fundamental principles for achieving good standards of electrical safety but do not prescribe the specific means for compliance. There is a wide range of guidance published by the HSE, as well as British, European and International Standards, to assist employers in determining how compliance is to be achieved. A key HSE publication is the *Memorandum of Guidance on the Electricity at Work Regulations 1989* (HSE booklet HSR25 (ISBN 9780717662289)). This provides guidance of general application to workplaces. Because of the higher level of risk relating to the use of electricity in mines, guidance in the form of a specific Approved Code of Practice (ACoP) has been published: *The use of electricity in mines* (HSE booklet L128) (ISBN 0 7176 2074 3). An ACoP on the use of electricity in quarries has been replaced by a page on the HSE website: www.hse.gov.uk/qua rries/electricity.htm. This gives practical guidance on electrical safety in higher risk areas in quarries. It does not cover offices or other low risk parts of quarries, and is intended for quarry management, rather than electrical experts. It repla ces the Approved Code of Practice (COP 35), which has been withdrawn.

In the introduction to the HSE Memorandum of Guidance reference is made to one very important British Standard, BS 7671 – *Requirements for electrical installations* (ISBN 978 0 86341 844 0) otherwise known as the non-statutory Institution of Electrical Engineers (IEE) Wiring Regulations. This standard specifies requirements concerning the design, construction, maintenance, inspection and testing of low voltage electrical installations. The HSE states that compliance with BS 7671 is likely to achieve compliance with the relevant aspects of the *EWR*. Furthermore, that an electrical installation installed to an earlier edition of the IEE Wiring Regulations, but current at that time, would not in itself mean that it did not comply with the *EWR*. The seventeenth edition was published in January 2008 and came into effect on 1 July 2008.

In should be borne in mind that other health and safety legislation of a more general application may also be of relevance to electrical safety. For example:

*   the *Management of Health and Safety at Work Regulations 1999 (SI 1999 No 3242)* which, among other things, impose a need for the assessment of electrical risks and the provision of training relevant to electrical work activities;
*   the *Provision and Use of Work Equipment Regulations 1998 (PUWER) (SI 1998 No 2306)*, the requirements of which apply equally to electrical equipment, eg concerning the need for isolation from all energy sources prior to commencing work;
*   the *Supply of Machinery (Safety) Regulations 1992 (SI 1992 No 3073)*; and
*   the *Reporting of Injuries, Diseases and Dangerous Occurrences Regulations 1995 (SI 1995 No 3163)*.

Dangers of electricity

General

[E3004] *Regulation 2* of the *EWR (SI 1989 No 635)* gives the definitions of words used with special meaning in the Regulations. These include 'danger' and 'injury'. Danger is defined as risk of injury whilst injury means:

> . . . death or personal injury from electric shock, electric burn, electrical explosion or arcing, or from fire or explosion initiated by electrical energy, where any such death or injury is associated with the generation, provision, transmission, transformation, rectification, conversion, conduction, distribution, control, storage, measurement or use of electrical energy.

The 1989 Regulations require action to be taken to avoid danger or, where appropriate, injury. The reason for distinguishing between danger and injury is that in some situations, eg live working, the electrical danger is not eliminated because of the presence of live parts and, therefore, precautions must be put in place to prevent injury from that danger. Ideally, injury is prevented by completely removing or eliminating the danger, eg as in the case of equipment isolation prior to undertaking maintenance work.

It is important to appreciate that injury does not always require physical contact to be made with a live conductor. At high voltages electricity has the capability of jumping an air gap (flashover), the flashover distance increasing as the voltage increases. Environmental conditions are also relevant. Damp or wet conditions will reduce the effectiveness of air as an insulator. For this reason the scope of the Regulations covers the prevention of injury from electrical danger arising from work near an electrical system or equipment as well as on it.

The Regulations do not, per se, make a distinction between high and low voltage systems because system voltage is not necessarily the only factor in determining danger. A 1.5 V torch battery would not be expected to present a shock or short-circuit hazard in a normal work environment but might provide sufficient energy to cause ignition in a potentially explosive atmosphere.

The definition of injury effectively encompasses the dangers of electricity which may be summarised as electric shock, electric burn, fire, arcing and explosion. These now will be briefly explained.

Electric shock

[E3005] Electric shock may arise either through direct or indirect contact with electricity. The former involves direct contact with a live part, eg an exposed live terminal or conductor, whilst the latter is a shock received from an exposed-conductive-part, eg the metal casing of electrical equipment, made live under a fault condition. An exposed-conductive-part is any conductive part of equipment which can be touched and which is not a live part but which may become live under fault conditions. The *EWR (SI 1989 No 635)* specify control measures to deal with each of these circumstances (*Regulations 7* and *8* respectively – see **E3014** and **E3015**).

The severity of an electric shock depends on a number of factors, in particular the magnitude and duration of the shock current passing through the body. The current magnitude will depend on the contact voltage and resistance or, to be more strictly correct in the case of a mains supply, which is an alternating current (a.c.), the impedance (Z) of the shock path. Thus, in the case of contact with mains electricity the shock current is:

I = V/Z (Ohms law)

Therefore, the outcome from an electric shock will be very dependent on the circumstances in which it is received, in particular the environment, since this will determine the shock path resistance or impedance. Indoors, standing on a floor covering acting as a good insulator, a hand-to-feet shock at mains voltage (230V nominal) may result in no more than a tingling sensation. Outdoors, however, with little insulation from earth, a shock at the same voltage would prove more severe, even with the possibility of being fatal. A hand-to-hand shock, where one hand makes contact with a live part whilst the other is simultaneously in contact with earthed metal, such as the earthed case of electrical equipment, represents an even higher risk because the shock path impedance is only that of the body. At mains voltage such a situation is quite likely to result in a fatality.

Electric current is measured in amperes or 'amps' for short (A), voltage in 'volts' (V) and resistance in ohms (Ω). The International Electrotechnical Commission (IEC) published information on the physiological effects of current flow through the human body (IEC report 479-1). This information was replicated in British Standard BS PD 6519: *Effects of current passing through the body (Parts 1 and 2)*. Part 1 of this standard has been superceded by DD IEC/TS 60479-1:2005 *Effects of current on human beings and livestock* although BS PD 6519-2:1998 *Guide to effects of current on human beings and livestock, special aspects relating to human beings* was current at May 2009.

In summary the physiological effects of alternating current are:

Around 0.5 mA	Threshold of perception
Up to 10 mA	Unpleasant but usually no harmful physiological effects
Above 10 mA	Increasing likelihood of muscular contractions and, at higher currents, possibility of respiratory system failure
30 mA	Typical current trip rating for a residual current device (RCD)
Around 50 mA	Risk of ventricular fibrillation and cardiac arrest (increasing risk with increasing magnitude and duration of current flow)
250 mA	Current drawn by a 60 W light bulb

The above values serve only as a guide since the effect will vary from one person to another. With higher shock currents the effect on the body, in particular the probability of death, becomes increasingly dependent on how long the current flows for. Currents as low as 50mA (0.05A) can prove fatal, particularly if flowing through the body for several seconds. Higher currents will have a similar effect in a much shorter time.

Whilst a.c, per se, represents a high shock risk, contact with direct current (d.c.) is also hazardous and can disturb the normal heart rhythm. A direct current of approximately 3.75 times the a.c. value will have the same probability of inducing ventricular fibrillation. However, if ripple exists at more than 10 per cent of the nominal d.c. value due to imperfections in the process of rectification then it must be considered as hazardous as a.c.

The HSE advises that a voltage as low as 50 volts applied between two parts of the human body causes a current to flow that can block the electrical signals between the brain and the muscles. It says that this could:

- stop the heart beating properly;
- prevent the person from breathing; and
- cause muscle spasms.

It says: "The exact effect is dependent upon a large number of things including the size of the voltage, which parts of the body are involved, how damp the person is, and the length of time the current flows. Electric shocks from static electricity such as those experienced when getting out of a car or walking across a man-made carpet can be at more than 10,000 volts, but the current flows for such a short time that there is no dangerous effect on a person. However, static electricity can cause a fire or explosion where there is an explosive atmosphere (such as in a paint spray booth)".

Electric burn

[E3006] Body tissue burns may result from the passage of electric current. The nature of such burns is that they tend to be deep seated and difficult to heal. Contact with high voltage is often characterised by body burn marks at the current entry and exit points, eg hand palm and soles of the feet in the case of a hand-to-feet shock path. Electric burns may also arise through exposure to high power sources of electro-magnetic radiation such as from radio transmission antennae or induction heating processes.

Fire

[E3007] Fire may be initiated as a result of heat generation due to the overheating of cables or electrical equipment, which may occur for a number of reasons. Examples are:

- overloading due to insufficient current carrying capacity for the connected load;
- cables bunched or covered with thermal insulation, significantly reducing current carrying capacity;
- lack of, or incorrectly rated, excess current protection (eg fuses or circuit breakers) for circuits and equipment;
- blocked ventilation apertures in equipment enclosures, eg motors;
- leakage currents due to deterioration of insulation surrounding live parts.

Arcing arising from short-circuit flashover or sparks, generated within electrical equipment such as motors or switching devices, may provide an ignition source for adjacent flammable materials (solids, gases, vapours or dusts).

Arcing

[E3008] Arcing or flashover will occur when short-circuiting or bridging between live parts at different voltages, eg between circuit conductors (phase to neutral or across phases in the case of a three-phase system) or from a live part to earth. As well as the risk of serious injury there is also the likelihood of damage to the equipment concerned. The severity of such a flashover will depend on the fault level at the point of short-circuit, ie the magnitude of the short-circuit current, which in turn will be determined by the upstream impedance in the electrical system back to the source of supply. Since this will be very low the prospective short-circuit current will be very high (Ohm's law again). In a factory situation prospective short-circuit currents as high as 10kA may well be encountered.

Short-circuit flashover may also result as a consequence of equipment failure, eg failure of a circuit protective device to disconnect the supply safely following a fault. Where failure has arisen because of a malfunction during a switching operation the operator is at risk of sustaining arc eye and severe burns.

Explosion

[E3009] In the case of a severe short-circuit flashover there is often little to distinguish between the effects of arcing and an explosion. Where the short-circuit occurs in a restricted space, such as within a busbar chamber or inside switchgear, the explosive forces are likely to result in considerable damage to the equipment involved, eg rupture of the external housing or covers blown off. If the switchgear is of the oil-filled type this may lead to the ejection of burning oil, a further hazard for anyone with the misfortune to be close by.

Electricity presents special problems in potentially explosive atmospheres (PEAs), where an electric spark may act as an ignition source. For this reason particular attention must be given to the design and construction of electrical equipment intended for such environments – see E3027 below.

The assessment of electrical risks first requires the likelihood of electrical danger to be evaluated, following which the severity of injury can be predicted. The likelihood of danger will centre on the presence of hazards that may give rise to electric shock, burns, fire, etc. This will include situations where there may be exposed live parts, missing earth connections, incorrect fuse ratings, etc. The severity of the outcome will depend on factors such as the supply voltage, shock path and work environment in the case of electric shock and the fault level and bridging path in the case of short-circuit flashover. With electricity there is a very thin dividing line between the accident being survivable and death. It is therefore best to work on the basis of endeavouring to reduce the likelihood of an electrical accident to as near zero as possible, which is the objective of the *EWR (SI 1989 No 635)*.

Fundamentals of controlling electrical risks

[E3010] The control measures aimed at preventing injury from electrical danger in general situations are prescribed in *Part II* of the *EWR (SI 1989 No 635)*, ie in *Regulations 4–16*. These cover:

Regulation	Topic
4	Systems, work activities and protective equipment
5	Strength and capability of electrical equipment
6	Adverse or hazardous environments
7	Insulation, protection and placing of conductors
8	Earthing or other suitable precautions
9	Integrity of referenced conductors
10	Connections
11	Means for protecting from excess of current
12	Means for cutting off the supply and for isolation
13	Precautions for work on equipment made dead
14	Work on or near live conductors
15	Working space, access and lighting
16	Persons to be competent to prevent danger and injury

Regulation 4 of the *EWR (SI 1989 No 635)* is effectively an 'umbrella' requirement, covering everything relating to the life of an electrical system, which includes electrical equipment as well as fixed electrical installations. The Regulation is in four parts, requiring:

(1) the construction of electrical systems such as to prevent danger;
(2) the maintenance of electrical systems so as to prevent danger;
(3) all work activities to be carried out in such a manner as not to give rise to danger;
(4) any equipment provided under the Regulations for the purpose of protecting persons at work on or near electrical equipment to be suitable for use, maintained in a suitable condition and properly used.

The first three parts of the Regulation are qualified duties, ie 'so far as is reasonably practicable'.

The subsequent regulations in *Part II* cover three important aspects relating to electrical safety:

• the need for competence, which underlies the effective implementation of the *EWR* duties (*Regulation 16*);
• the design, construction and specification of the hardware, eg cables, switchgear circuit protection, etc (*Regulations 5–12* and *15*);
• safe systems of work (*Regulations 13* and *14*).

These aspects will now be considered in turn in more detail.

Competence

[E3011] *Regulation 16* of the *EWR (SI 1989 No 635)* specifically requires that any person engaged in any work activity where technical knowledge or experience is necessary to prevent danger, or where appropriate injury, is competent to prevent danger and injury. Competence means the possession of such knowledge or experience as may be appropriate having regard to the nature of the work. If this is not the case, that person must be under an appropriate degree of supervision, again having regard to the nature of the work. This requirement for competence extends to non-electrical work activities in any situation where electrical danger may be present (see *EWR (SI 1989 No 635), Reg 4(3)*). The HSE has provided guidance concerning the scope of the phrase 'technical knowledge or experience'. This embraces:

*   adequate knowledge of electricity;
*   adequate experience of electrical work;
*   adequate understanding of the system to be worked on and practical experience of that class of system;
*   understanding of the hazards which may arise during the work and the precautions required to be taken;
*   ability to recognise at all times whether it is safe for work to continue.

The above effectively provides a framework training specification. This can be used as a starting point for preparing a more detailed specification to suit a particular training need, eg to inspect and test portable electrical equipment. Thus, competence requires an understanding of the tasks to be performed, as well as of the equipment on which the work is to be undertaken, coupled with the necessary underpinning technical knowledge relating to electricity. As an example, an operative required to inspect and test portable electrical equipment would need a thorough knowledge of the different types and constructions of equipment likely to be encountered, in order to know what to look for when carrying out an inspection and what tests would be appropriate. An understanding of the test instruments to be used would also be required, in particular their functions and methods of use. The underpinning technical knowledge would enable meaningful measurements to be taken, properly recorded and the results correctly interpreted.

A higher level of competence is required for live work since the dangers of electricity remain present. It is necessary for those involved to understand fully the shock and short-circuit hazards in relation to the equipment being worked on and the work being undertaken, in particular possible shock and bridging paths that might arise during the course of the work. The fact that such hazards are always present whilst the work is in progress also must be fully appreciated.

Qualifications based on City & Guilds syllabuses exist for certain electrical activities, eg installation work, inspection and testing of fixed wiring and maintenance of electrical equipment, for which courses are offered by colleges and training organisations. Other routes are also available to acquire the necessary competence, such as on-the-job training linked to NVQs based on engineering occupational standards. Many electricians belong to one or more trade organisations such as the Electrical Contractors Association (ECA) for

England and Wales or its Scottish equivalent (SELECT). The National Inspection Council for Electrical Installation Contracting (NICEIC) also exists, essentially as a consumer safety body, to maintain the standards of the electricians on its roll. However, membership of such organisations does not confer a particular level of electrical competence.

HSE advice on the training required to demonstrate competence is as follows: "A person can demonstrate competence to perform electrical work if he or she has successfully completed an assessed training course that has included the type of work being considered, run by an accredited training organisation, and has been able to demonstrate an ability to understand electrical theory and put this into practice. A successfully completed electrical apprenticeship, with some post apprenticeship experience is a good way of demonstrating competence for general electrical work. More specialised work such as maintenance of high voltage switchgear or control system modification is almost certainly likely to require additional training and experience."

Design and construction of electrical equipment and systems

Strength and capability

[E3012] *Regulation 5* of the *EWR (SI 1989 No 635)* requires that no electrical equipment shall be put into use where its strength and capability may be exceeded in such a way as may give rise to danger. This includes mechanical as well as electrical capability. Thus an overhead cable must have the mechanical strength not to break as well as the electrical capability to carry the required load current. Furthermore, electrical capability includes the capability to handle safely any prospective short-circuit currents that may flow under fault conditions.

Adverse or hazardous environments

[E3013] *Regulation 6* of the *EWR (SI 1989 No 635)* requires that electrical equipment shall be constructed or as necessary protected to prevent danger, so far as is reasonably practicable, from reasonably foreseeable adverse or hazardous environments, namely:

- mechanical damage;
- the effects of the weather, natural hazardous, temperature or pressure;
- the effects of wet, dirty, dusty or corrosive conditions;
- any flammable or explosive substance, including dusts, vapours or gases.

Whilst exposure to earthquakes in the UK might not be reasonably foreseeable, the siting of electrical equipment outdoors means that it is highly likely to be exposed to wet and possibly dirty, dusty or even corrosive conditions, such as salt spray in coastal locations. An appropriate construction specification would be needed, therefore, to accommodate this type of environment.

Protection from electric shock by direct contact

[E3014] *Regulation 7* of the *EWR (SI 1989 No 635)* deals with the prevention of electric shock from direct contact with live conductors. It requires that all conductors in a system, which may give rise to danger, shall either:

(a)   be suitably covered with insulating material and as necessary protected so as to prevent, so far as is reasonably practicable, danger; or

(b)   have such precautions taken in respect of them (including, where appropriate, their being suitably placed) as will prevent, so far as is reasonably practicable, danger.

The normal means for preventing direct contact with live conductors is:

- electrical insulation (eg cable insulation);
- enclosure (eg busbar chamber);
- use of barrier fencing or screening (eg high voltage test bay).

In some circumstances the use of an extra-low voltage system (eg 12 V) may provide an adequate safeguard because at such a low voltage any accidental contact could not give rise to a serious electric shock. Nevertheless a short-circuit hazard may still exist, eg as with a 12 V high ampere-hour capacity battery.

The thickness of electrical insulation will depend on the conductor operating voltage. In some situations additional protection may be required to prevent damage to the electrical insulation. In industrial premises, steel wire armoured (swa) or mineral insulated copper sheathed (mics) cables are often used. Alternatively, cables may be placed in steel conduit or trunking, or placed out of reach on cable trays, to provide additional protection. Where enclosures are used it is important to ensure adequate protection against the ingress of foreign objects or fingers, in accordance with a British Standard known as the IP Code (BS EN 60529: Specification for degrees of protection provided by enclosures).

Barrier fencing is often used in test bays where there is a need to undertake live testing. Access gates must be interlocked with the test supply (eg key type) to ensure that entry cannot be gained whilst the supply is energised. In some circumstances it might be impracticable to enclose completely all live parts. Examples are:

- for functional reasons, eg the supply to electric overhead travelling cranes,
- railway traction supplies and overhead power distribution systems;
- in connection with operational requirements, eg fault finding and diagnostic live testing.

In such circumstances it is then necessary for other appropriate precautions to be implemented to minimise the risk of contact. Placing out of reach, coupled with the posting of warning notices, is a commonly adopted safeguard. However, it is important to recognise that occasions may arise where it becomes necessary to encroach within the prescribed safety clearances, in which case it is then necessary to implement a safe system of work based on isolation of the supply to remove the electrical danger at source.

Protection from electric shock by indirect contact

[E3015] *Regulation 8* of the *EWR (SI 1989 No 635)* requires that precautions are taken, either by earthing or other suitable means, to prevent danger arising from any conductor (other than a circuit conductor) which may reasonably foreseeably become charged as a result of either the use of the system or a fault in the system. For example, an internal equipment fault may cause the metal case to be made live at the supply voltage, creating a shock risk. A range of techniques can be employed, either singly or in combination:

- earthing;
- equipotential bonding;
- double insulation;
- earth-free non-conducting environments;
- current/energy limitation;
- use of safe voltages;
- separated or isolated systems;
- connection to a common reference point on the system.

Earthing (for class I equipment) and double insulation (for class II equipment) are the techniques most commonly adopted. To achieve effective installation earthing it is most important that the earth fault loop impedance (the conductor continuity impedance/resistance from the supply source to the point of fault on the installation and back to source through the earth connection) is kept as low as possible in order that sufficient fault current will flow to enable the circuit protective device to operate to disconnect the fault current quickly. In buildings, connections (referred to as main equipotential bonding) are also made from the main earthing terminal to other services, as well as to the metal structure of the building if applicable. Further connections (supplementary bonding) are made between conductive parts in high shock risk situations (bathrooms, shower rooms, kitchens, etc.). This bonding serves a different role to earthing, namely to minimise the risk of voltage differences arising between exposed-conductive-parts during the time that the fault exists and before the supply is automatically disconnected.

In high risk situations a residual current device (RCD) may be used to supplement other protective measures, through very fast disconnection of the supply in the event of something going wrong (typically 20–40 ms). An RCD is a very sensitive device, capable of responding to the comparatively small currents typical of a shock situation. A common trip current rating is 30mA. Whilst an RCD will not prevent shock – a shock current must flow to create an out-of-balance situation, which is the basis of its operation – the shock should be survivable.

Referenced conductors

[E3016] *Regulation 9* of the *EWR (SI 1989 No 635)* requires that nothing shall be placed in the neutral conductor of an electrical system to cause it to become open-circuited whilst the phase conductor(s) remain live.

This is for two main reasons:

(i)   on modern distribution supply networks the neutral conductor is likely to be used also for earthing purposes, referred to as a protective multiple earthing (PME) system. An open-circuit in the neutral conductor therefore would result in loss of earthing;

(ii)   the secondary (output) windings of three-phase LV distribution supply transformers are generally 'star' connected, with the star point earthed and serving as the system neutral. This provides a reference point for the phase voltages, to help maintain them within the prescribed statutory limits for supply. Loss of the neutral connection back to the supply transformer may result in undue voltage fluctuation depending on the loading on each phase.

Examples of prohibited devices are fuses and thyristors which cannot be relied upon not to give rise to danger by either becoming open-circuited or giving rise to a high conductor resistance. In the case of a three-phase four-pole switch the switch must break the neutral conductor last and make it first when operated. This situation does not arise with a triple-pole and neutral (TP&N) switch, where only the phase conductors are switched and the neutral remains connected.

Connections

[E3017] *Regulation 10* of the *EWR (SI 1989 No 635)* requires that all joints and connections should be mechanically and electrically suitable for use. This applies to both temporary and permanent connections and to connections in protective (earth) conductors as well as circuit conductors (phase and neutral). A broken phase connection could result in a short-circuit to a neutral conductor or to earth via an earthed metal equipment case. Failure of a joint or connection in a protective conductor could result in a loss of earthing and therefore loss of effective protection in the event of electric shock arising from indirect contact.

Excess current protection

[E3018] *Regulation 11* of the *EWR (SI 1989 No 635)* requires efficient means, suitably located, to be provided to protect every part of a system from excess current, as may be necessary to prevent danger. The abnormal conditions likely to give rise to excess current flow are:

- an overload, which can result in the overheating of cables or equipment and the possibility of fire;
- a short circuit, which can result in arcing or explosion, possibly accompanied by fire;
- an earth fault, giving rise to a shock risk through indirect contact.

In the case of electric shock protection it is important that the protective device (eg a fuse or circuit breaker) disconnects the supply as quickly as possible. BS 7671 specifies maximum disconnection times for different system types at different voltages.

For final circuits not exceeding 32A this is not exceeding 0.4 seconds for a TN system. For a TT system this is reduced to 0.2 seconds, although: 'Where, in a TT system, disconnection is achieved by an overcurrent protective device and protective equipotential bonding is connected to all the extraneous conductive parts with the installation in accordance with Regulations 411.3.1.2, the maximum disconnection times applicable to a TN system may be used'.

Distribution circuits have a disconnection time of 5 seconds for TN systems and 1 second for TT Systems.

Isolation

[E3019] *Regulation 12* of the *EWR (SI 1989 No 635)* interfaces with *Regulation 19* of *PUWER (SI 1998 No 2306)*. It requires both a means for cutting off the supply as well as for isolation. In practice, switchgear often performs both functions. However, the fact that these are separately stated in this Regulation implies a difference. The distinction is made clear in the second part of the regulation, which explains the meaning of isolation in the electrical context. This amounts to a legal definition. The requirement for electrical isolation may be summed up as secure disconnection and separation from all supply sources. Key characteristics for compliance would be:

• adequate labelling, ie unambiguous and legible;
• adequate contact separation;
• means for securing the point of isolation, preferably by locking off.

Many accidents occur due to a failure to isolate plant correctly. Compliance with the requirements of *Regulation 12* is an important precursor to satisfying the requirements of the *Regulation 13* of the *EWR (SI 1989 No 635)*, which is concerned with work on de-energised systems (see **E3023** below).

Working space, access and lighting

[E3020] *Regulation 15* of the *EWR (SI 1989 No 635)* requires adequate working space, access and lighting. This is relevant to safe systems of work, particularly in the case of live work. For example, live testing within control panels for fault finding purposes requires accessibility, freedom of hand/arm movement, minimal risk of accidentally being pushed onto exposed live terminals and a lighting level sufficient to enable the live connections to be clearly observed, so that the testing can be undertaken safely. In terms of guidance for determining suitable accessibility, see Appendix 3 in the HSE Memorandum of Guidance, which reproduces a regulation from the former *Electricity (Factories Act) Special Regulations 1944 (SI 1944 No 739)*. The 1944 Regulations extended the provisions of the original 1908 Regulations, which were revoked by the *EWR*. As a guide, a clear distance of the order of one metre should be allowed in front of fixed equipment where access is required for live diagnostic testing and other such work.

Maintenance

[E3021] *Regulation 4(2)* of the *EWR (SI 1989 No 635)* states:

As may be necessary to prevent danger, all systems shall be maintained so as to prevent, so far as is reasonably practicable, such danger.

This legal requirement sets an objective, to ensure that sufficient maintenance is carried out on all electrical systems to prevent danger. The duty is a qualified one and therefore the extent and frequency of the maintenance required should be proportionate to the level of risk. By virtue of the definitions contained in *Regulation 2* of the *EWR (SI 1989 No 635)*, the term 'system' embraces electrical equipment of all forms. In the case of equipment, examples of risk factors would include:

- age;
- operating voltage;
- environment of use (eg indoor or outdoor);
- type of equipment (hand held, portable, fixed, etc);
- class of construction for protection from electric shock (class I, II or III) (BS 2754: *Memorandum. Construction of electrical equipment for protection against electric shock*);
- nature of use (or even abuse).

The high end of the risk spectrum would be a hand-held class I mains powered tool used on outdoor construction work at different sites whilst the low end could be stationary class II IT equipment used in an office environment.

As a minimum, maintenance would include inspection to check for deterioration or damage and to verify safe condition for use by any user. However, some faults cannot be detected by inspection alone and therefore testing may also be required, eg to verify the effective earthing of exposed conductive parts such as a metallic equipment case.

The HSE has published guidance (*Maintaining portable and transportable electrical equipment* (HSG107) ISBN 0 7176 0715 1) to assist duty holders in complying with the requirements of this Regulation. The Institute of Engineering and Technology (IET – formerly the IEE) has published a code of practice which provides detailed guidance for both managers, with responsibilities for establishing policy and putting into place arrangements for electrical equipment maintenance, as well as staff undertaking inspection and maintenance tasks. It includes information relating to the different types and constructions of equipment as well as inspection criteria and test procedures (IEE Code of Practice for in-service inspection and testing of electrical equipment 3rd edition (ISBN 978 0 86341 833 4).

Safe systems of work

General

[E3022] *Regulation 4(3)* of the *EWR (SI 1989 No 635)* imposes a general duty that all work activities, including work near an electrical system, shall be carried out in such a manner as not to give rise, so far as is reasonably practicable, to danger. The *EWR* provide for two basic approaches to achieving a safe system of work, covered respectively by *Regulations 13* and *14*:

(i)   removal of any electrical danger by disconnection from all sources of supply and the implementation of a system of work aimed at verifying that the danger from electricity has been removed and remains removed for the duration of the work;

(ii)   live working, recognising that the danger of electricity remains so that it is then necessary to implement a system of work aimed at preventing injury arising from the danger.

However, the second option is not an automatic choice since the intention to work live first must be properly justified, which forms the starting point for an assessment of electrical risks.

The HSE has published guidance specific to safe working practices involving electricity (*Electricity at work – safe working practices* (HSG85) (ISBN 0 7176 2164 2).

Work on isolated equipment

[E3023] A model safe system of work would require consideration of the following:

*   isolation from all points of supply;
*   earthing to discharge any residual electrical energy or to prevent the build up of an induced charge;
*   proving dead at the point of work;
*   demarcation of the safe zone of work;
*   safety documentation.

PUWER (SI 1998 No 2306), Reg 19 places duties on employers in respect of isolation from all forms of energy, which therefore includes electrical energy. These duties require that:

*   where appropriate, the work equipment shall be provided with suitable means to isolate it from all sources of energy;
*   the means of isolation shall be clearly identifiable and readily accessible;
*   appropriate measures shall be taken to ensure that re-connection of any energy source to work equipment does not expose any person using the equipment to any risk to his health or safety.

Electrical isolation is defined in *EWR (SI 1989 No 635), Reg 12* (also discussed at E3019 above) and means the secure disconnection and separation from all sources of electrical energy. The preferred means of securing points of isolation is by locking off, using special safety locks operated by unique keys.

Other safeguards include, where appropriate, the removal of fuses and the posting of warning notices. Where a number of different individuals or separate parties are working on the plant simultaneously a special locking device can be used to enable each party to apply its own safety lock, thereby maintaining personal control of the point of isolation for each person involved in the work.

Many accidents have arisen through incorrect plant isolation, coupled with a failure to prove dead. For example, this may be due to ambiguous, incorrect or missing circuit identification labelling or a failure to identify properly the isolation requirements appropriate to the intended work, particularly in the case of multiple supply sources. Confusion often arises concerning what has been made dead by the isolation of power and control circuits associated with factory production line equipment. For this reason, proving dead at the actual point of work is an essential element to achieving a safe system of work. Proprietary testers are available for this purpose and, because of the importance of using the right type of tester, the HSE has published a guidance note on the subject (*Electrical test equipment for use by electricians* (GS38) (ISBN 0 7176 0845 X)). It must be recognised also that certain electrical equipment has the capability of retaining energy, in the form of an electrical charge, for a period of time following its isolation. This applies to plant containing capacitors, eg power factor correction equipment, or systems exhibiting capacitive effects, such as a long run of high voltage cable. Similarly, in the case of high voltage systems, a charge may be induced into an isolated circuit from an adjoining live circuit. Earthing is used to safeguard from static or induced charge.

Demarcation of the safe zone of work may be required to minimise the risk of the worker inadvertently seeking to gain access to adjacent live plant. In complex situations it is often useful to confirm the isolation, the equipment to be worked on and the work to be undertaken through the issue of a permit-to-work (PTW). The PTW also can serve to demarcate the safe zone of work. A PTW does not constitute the safe system of work but serves to strengthen communication and act as a control document to hold plant out of service until signed off by the person in charge of the work. The HSE recommends that a PTW should be used for all high voltage work (ie above 3kV) as well as for low voltage work involving multiple points of isolation. Where PTWs are adopted it is important that a management procedure exists to specify and control their use.

Live work

[E3024] *Regulation 14 of the EWR (SI 1989 No 635)* states that no person shall be engaged in any work activity on or so near any live conductor (other than one suitably covered with insulating material so as to prevent danger) that danger may arise unless:

(a)   it is unreasonable in all the circumstances for it to be dead; and
(b)   it is reasonable in all the circumstances for him to be at work on or near it whilst it is live.

Thus, justification centres around two issues. *First*, there must be good reason for the equipment to remain live whilst the work is carried out, eg the nature of certain electrical testing may require this to be undertaken live. *Second*, the equipment must be safe to work on whilst it remains live. This relates to the equipment design and construction, eg working space, extent of exposed live parts, shock and bridging paths (particularly involving any earthed metal case). *Regulation 14* applies to all work activities, including testing, ie there are no exemptions! The criterion is whether electrical danger is present or could arise during the work activity. It is also relevant to non-electrical activities, eg work in the vicinity of underground cables or overhead lines (discussed at E3028 below). The only exception to compliance with *Regulation 14* is where the live parts are suitably covered with insulating material and as necessary protected so as to prevent danger, thus instead satisfying the requirements of *Regulation 7* of the *EWR (SI 1989 No 635)* for the prevention of electric shock by direct contact.

Provided that live working can be justified in accordance with the first two parts of *Regulation 14*, the remaining part of that Regulation requires that suitable precautions shall be taken to prevent injury. This is an absolute duty. Thus, any accident occurring during the course of live work is prima facie evidence that the precautions taken were not suitable, ie in breach of *Regulation 14*.

Therefore, a safe system of work for live working must centre on the prevention of injury which, in turn, needs to recognise the types of injury and circumstances in which they might arise. The principal dangers are electric shock and short-circuit flashover. An understanding of shock and bridging paths, and how these might arise in the particular work activity, is therefore very important. Safeguards need to be devised and effectively implemented to prevent electric shock and short-circuit flashover.

A model safe system of work would require consideration of the following:

- capability of those undertaking the work;
- provision of appropriate information concerning the proposed work;
- use of suitable tools and other protective equipment;
- use of suitable instruments and test probes;
- consideration of the need for accompaniment;
- control of the work area.

There are important considerations concerning the competence of personnel intending to undertake live work, mentioned previously (see E3011 above). Additionally, adequate information needs to be provided concerning the work so that those directly involved can be satisfied that it is within their capability and experience.

Protective equipment must satisfy the requirements of *Regulation 4(4)* of the *EWR (SI 1989 No 635)*, ie it must be:

- suitable for the use for which it is provided;
- maintained in a condition suitable for that use;
- properly used.

Suitability for use infers that management action has been taken to determine what protective equipment is needed and the circumstances in which it is to be used in contributing to the overall safe system of work. It is important that such equipment is of a suitable specification for the intended use, for which a number of British Standards exist:

- BS EN 60900:2004 *Live working. Hand tools for use up to 1000 V a.c. and 1500 V d.c.*

- BS EN 60903: 2003 *Live working. Gloves of insulating material.*
- BS 921: 1976 *Specification. Rubber mats for electrical purposes.*

These same standards include information to assist in the implementation of appropriate in-service care regimes for the protective equipment. Correct use infers the provision to users of relevant information and training, particularly concerning the limitations of such equipment.

Whilst applied insulation may minimise the risk of electric shock, large areas of exposed metal, such as at tool working heads or test probe tips, may create a risk of short-circuit when used in a confined space with limited clearance between exposed live terminals or conductors.

The *EWR* requirements concerning protective equipment are supplemented by the *Personal Protective Equipment at Work Regulations 1992 (PPE) (SI 1992 No 2966)*. One important issue concerns compatibility between different items, eg insulated tools and gloves.

There is no automatic requirement for accompaniment in the case of live work. If it is considered necessary to have a second person present to render first aid in the event of an accident this implies that there could be doubt concerning the proposed system of work or its effective implementation. However, a second person may be necessary to enable safe working, eg by directly assisting with the job as 'another pair of hands' or to control the work area.

Special situations

Lightning protection systems

[E3025] Although lightning is a natural phenomenon Regulation 6 of the *EWR (SI 1989 No 635)* requires electrical equipment which may reasonably foreseeably be exposed to the effects of the weather, such as lightning, to be of such construction or as necessary protected as to prevent, so far as is reasonably practicable, danger arising from such exposure. In reality, it is not just electrical equipment that needs protection but buildings and the contents therein. A lightning strike onto a building may find its way to earth via the electrical installation. This risk can be minimised by careful attention to the design and installation of a lightning protection system for the building as a whole. Lightning protection is outside the scope of BS 7671 but is covered by another British Standard (BS EN 62305: *Protection against lightning*).

The standard comprises four parts:

- BS EN 62305-1:2006 *Protection against lightning. General principles*;
- BS EN 62305-2:2006 *Protection against lightning. Risk management*;
- BS EN 62305-3:2006 *Protection against lightning. Physical damage to structures and life hazard*; and
- BS EN 62305-4:2006 *Protection against lightning. Electrical and electronic systems within structures.*

A practical guidebook has also been published by the British Standards Institute (BSI): *Protection against lightning. A UK guide to the practical application of BS EN 62305* (ISBN 978 0 580 50899 8.

Static electricity

[E3026] This is a common phenomenon. Most people have experienced the effects of static electricity as a short sharp shock when the body or another object appears to have become charged with electricity and discharge occurs as body contact, usually a hand, is made with the object. Frequently occurring examples are alighting from a vehicle and opening a metal filing cabinet in an office. The mechanism surrounding the build up of static charge is not fully understood but involves interaction between different materials aided by particular atmospheric conditions. Certain work processes, involving the generation and movement of fine dusts or the flow of liquids through pipes and orifices, may give rise to the generation of static in work processes. The *EWR (SI 1989 No 635)*, in relating to electrical systems and associated equipment are not intended to apply to such situations. However, other legislation serves to ensure that appropriate safeguards are put in place. In particular, *PUWER (SI 1998 No 2306), Reg 12* requires that measures shall be taken to prevent or, where that is not reasonably practicable, to adequately control specified hazards associated with work equipment, such as the premature explosion of any article or substance produced, used or stored in it. Preventive measures might include redesigning the process or equipment or using different construction materials to prevent the build up of static. A common practice is to electrically bond together all conductive parts, to reduce the possibility of charge build up between different surfaces.

Some work processes depend on electrostatic, eg electrostatic paint spraying and chimney flue gas filters. In these cases *Regulation 6* of the *EWR (SI 1989 No 635)* is applicable.

The Energy Institute Model Code of Safe Practice, Part 21 (IP21), gives guidelines for the control of hazards arising from static electricity.

Electricity in potentially explosive atmospheres

[E3027] The use of electricity in an area with a potentially explosive atmosphere (PEA) poses particular problems. An explosive atmosphere could be present in paint spray booths, near fuel tanks, in sumps, or places where aerosols, vapours, mists, gases, or dusts exist. As the use of electricity can generate hot surfaces or sparks, it can ignite an explosive atmosphere.

Ideally, the risk should be minimised either by siting any electrical equipment outside the area or the work location designed and constructed so that natural ventilation will ensure that a PEA will never arise within it. In practice, the latter option may be difficult to confirm and attention will then need to be given to the design and construction of the equipment, taking account of the likelihood of a PEA arising, to achieve the required level of safety. A British Standard provides a classification for potentially explosive atmospheres based upon the likelihood and duration of a PEA arising (BS EN 60079). BS EN 60079-10-1 sets out the essential criteria against which the ignition hazards can be assessed, and gives guidance on the design and control parameters which can be used in order to reduce such a hazard.

The HSE advises that electrical and non-electrical equipment and installations in PEAs must be specially designed and constructed to eliminate or reduce the risks of ignition. This can be done by sealing electrical equipment so that the explosive atmosphere cannot come into contact with electrical components, reducing the power of electrical equipment, and de-energising electrical equipment where a fault or an explosive atmosphere is detected.

Recently installed equipment should be marked with an 'Ex' to show that it is suitable for use in a PEA. All new equipment must comply with the *Equipment and Protective Systems Intended for Use in Potentially Explosive Atmospheres Regulations 1996* that implemented the European ATEX Directive.

This requires it to be assessed as suitable for a particular explosive atmosphere type and for this to be marked on the equipment along with CE and ATEX markings. Most new equipment being sold in the UK for use in potentially explosive atmospheres must have an ATEX certificate.

The HSE says that equipment for use in explosive atmospheres should be regularly inspected and maintained to ensure it does not pose an increased risk of causing a fire or explosion, and maintenance of the equipment should only be carried out by people who are competent to do so.

A further British Standard provides guidance in respect of electrical equipment for use in the presence of flammable dusts (BS EN 61241-10:2004 Electrical apparatus for use in the presence of combustible dust. Classification of areas where combustible dusts are or may be present).

In addition, in areas where it is possible that an explosive atmosphere may exist, the requirements of the *Dangerous Substances and Explosive Atmospheres Regulations 2002 (SI 2002 No 2776) (DSEAR)* apply. The regulations require that such areas be risk assessed before any new work is carried out in them and that measures be taken to control the risks. The HSE has produced guidance on *DSEAR* that explains how the Regulations can be complied with. This can be accessed via the HSE website at www.hse.gov.uk/fireandexplosion/dsear.htm. It advises: "Care should be taken to prevent static discharges in pot entially explosive atmospheres. Measures such as earth bonding and the select ion of antistatic work clothing and footwear can help to reduce the risk of static discharges."

The *Dangerous Substances and Explosive Atmospheres Regulations 2002 (SI 2002 No 2776)* control fire and explosion risks in workplaces and are examined in more detail in chapter **D09 DANGEROUS SUBSTANCES AND EXPLOSIVE ATMOSPHERES**).

A common example of where a PEA may arise is in battery rooms. The charging of lead-acid batteries, even those that are described as maintenance free, generates hydrogen gas. Lack of ventilation may allow the build up of a PEA since the lower explosive limit (LEL) is only 4 per cent. As hydrogen is lighter than air the PEA is likely to occur first at ceiling level where a light fitting could produce a spark to cause ignition. For this reason care must be given to the ventilation requirements for battery rooms, with vents at both high and low level to achieve a good level of natural ventilation. Additionally, light fittings of an appropriate type, should be wall-mounted. Because of the high number of accidents involving electric storage batteries the HSE has produced a leaflet concerning safe charging and use *Using electric storage batteries safely* (INDG139 rev1).

Work near underground cables and overhead lines

[E3028] By means of *Regulation 4(3)* the *EWR (SI 1989 No 635)* regulate work carried out in close proximity to live conductors, whether overhead lines or buried cables, in situations where there is electrical danger to be avoided. Many accidents involving inadvertent contact with live overhead lines arise during construction or agricultural activities. In the case of the former, examples are the movement or operation of cranes, dumpers, excavators and access equipment too close to a line and for the latter, the use of crop spraying or other irrigation equipment often poses particular problems. With high voltage overhead lines it is important to recognise that the electricity can flash across a gap and therefore actual contact does not have to be made for injury to occur. A safe system of work depends on establishing a safety clearance distance from the line and height clearances where plant is required to pass underneath. These measures are described in more detail in HSE guidance (*Avoidance of danger from overhead electrical lines* (GS6) (ISBN 0 7176 1348 8)) and, where there is any doubt, advice should be sought from the line owner. Excavation work accounts for many cable damages, some of which result in serious injury or even death. Again, it is essential to implement a safe system of work founded on:

- establishing the presence of any buried cables by reference to plans (electricity distribution companies are required to make such information available);
- verifying the position of cables by the use of a cable locating device;
- employing safe excavation techniques involving the digging of trail holes.

The HSE has published guidance on avoiding danger from all types of underground service, which includes electricity cables (*Avoiding danger from underground services* (HSG47) (ISBN 0 7176 1744 0)).

Other relevant legislation

The Plugs and Sockets etc (Safety) Regulations 1994 (SI 1994 No 1768)

[E3029]–[E3030] These Regulations were made under the *Consumer Safety Act 1978* in response to concern that consumer safety was being compromised by the substantial quantity of counterfeit and unsafe electrical plugs and sockets being placed on the UK market and by the provision of electrical equipment without an appropriate means to connect it to the mains supply in the consumer's home. The Regulations are complementary to the *Electrical Equipment (Safety) Regulations 1994* and are administered by the Department for Business, Enterprise and Regulatory Reform (BERR). They contain requirements for the safety of plugs, sockets, adaptors and fuse links designed for use at a voltage of not less than 200 V. For example, under these Regulations, the circuit conductor pins (L and N) of 13 A plugs are required to be partially insulated.

The Electrical Equipment (Safety) Regulations 1994 (SI 1994 No 3260)

[E3031] These Regulations consolidate, with amendments, the *Low Voltage Electrical Equipment (Safety) Regulations 1989 (SI 1989 No 728)* which, in turn, implement the requirements of the EU Low Voltage Directive concerning the harmonisation of requirements across the EU relating to the design and construction of low voltage (ie up to 1000 V) electrical equipment. They are administered by BERR and also include provisions relating to 'CE' marking. It is important to understand the limitations of such marking which does not signify product approval. The 'CE' mark is applied by the manufacturer to show that electrical equipment meets the relevant product standard(s). A disadvantage is that the scheme operates on the basis of self-certification by the manufacturer rather than using the services of an independent test house. A further problem is that counterfeit marks have been known to be applied to domestic type electrical products, such as plugs, originating from countries outside the EU. However, under the Regulations, the manufacturer is required to make available a written declaration of product conformity comprising:

- manufacturer's name and address (or details of authorised representative);
- a description of the electrical equipment;
- reference to the harmonised standards;
- where appropriate, references to the specifications with which conformity is declared;
- identification of the signatory who has been empowered to enter into commitments on behalf of the manufacturer or authorised representative;
- the last two digits of the year in which the CE mark was affixed.

The Electricity Safety, Quality and Continuity Regulations 2002 (SI 2002 No 2665)

[E3032] These Regulations are administered by BERR and cover electricity supply systems and the provision of electricity supplies to consumers. Their prime purpose is to ensure public safety. Whilst mainly directed at those with responsibilities for public supply networks they will also be of interest to organisations operating private distribution systems and highway power supply networks, particularly if these encroach on public land. Duties are placed mainly on generators, distributors, suppliers and meter operators. Some companies may be undertaking more than one of these roles. Some responsibilities are placed also on consumers. The Regulations prescribe the nominal supply voltage and frequency and the statutory limits of variation for these. For standard low voltage supplies this is 400/230 V +10 per cent/–6 per cent, 50 Hz ±1 per cent. The Regulations specify important safety requirements concerning the design and construction of electricity networks, eg minimum ground clearances of overhead line conductors, the safeguarding of buried electricity cables and the protection of distribution equipment in substations, etc. Some of the requirements mirror those contained in the *EWR (SI 1989 No 635)*. The Regulations also specify fundamental requirements for safety concerning consumers' installations, prior to connection. These link in with the fundamental principles for the safety of electrical installations contained in chapter 13 of BS 7671. This is not only to ensure the safety of those using an installation but also to minimise the risk of the supply to other consumers being adversely affected, eg in terms of loss of supply or its quality, in consequence of how it is being used.

The regulations were amended by the Electricity Safety, Quality and Continuity (Amendment) Regulations 2006. The amendments:

- updated references to British Standard documents;
- more clearly exclude trams, trolleybuses and other forms of guided transport from the requirements of the Regulations on 'distributors';
- improved the resilience and reliability of overhead electricity distribution networks; and
- extended the scope of the Regulations in order to apply to offshore generating installations.

The Building Regulations 2000 (SI 2000 No 2531) (as amended)

[E3033] The Building Regulations for England and Wales were amended to introduce a new *Part P*, which came into effect from 1 January 2005. The requirements of this part apply only to low voltage electrical installations in dwellings or the common parts of a building serving one or more dwellings, including outbuildings and gardens. Although not related to workplaces these requirements will be of interest to those organisations managing residential property, such as local authorities and housing associations. The Regulations cover the design, installation, inspection and testing of electrical installations and require:

- reasonable provision to be made in the design, installation, inspection and testing of electrical installations in order to protect persons from fire or injury; and
- sufficient information to be provided so that persons wishing to operate, maintain or alter an electrical installation can do so with reasonable safety.

The Regulations are supported by an approved document.

The legislation has the effect of defining a legal standard for electrical installations in dwellings which, as with the *EWR (SI 1989 No 635)* in the case of work premises, can be met by fulfilling the requirements of BS 7671. With certain defined limited exceptions all work must be undertaken by a suitably competent person. Electricians must be registered under a *Part P* self-certification scheme.

Emissions into the Atmosphere

Roger Barrowcliffe and Paul Reeve

Introduction

[E5001] Pollution of the atmosphere is a subject that encompasses the whole range of distance scales (from personal to global) and can be important over minutes or many lifetimes. Regulation of emissions, therefore, tries to account for all of these effects and, as a consequence, it has to take a variety of forms. The concept of the different distance scales is illustrated in Figure 1 below:

Figure 1: Distance Scales

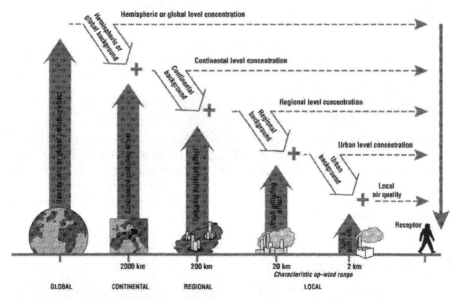

A 'receptor' (a human being, a population, an ecosystem) experiences air pollution as a sum of the contribution from various sources. The extent to which the sources contribute might depend on the pollutant of concern. A pollutant such as sulphur dioxide is relatively short lived in the atmosphere and has the capacity to affect the lungs of people within tens of kilometres of the point of emission. At distances of hundreds of kilometres, however, the sulphur dioxide will have been transformed into sulphate, which may be absorbed into cloud water and be deposited on the ecosystems of another country. This is an example of trans-boundary pollution and popularly known as 'acid rain'.

Pollutants that are chemically more stable than sulphur dioxide have the potential to be dispersed throughout the globe. Some organic pollutants are an excellent example of this. Poly-chlorinated biphenyls were used widely in the middle of the twentieth century as a component part of electrical transformers. They are highly toxic and accumulate in body fats, so that animals (and humans) at the top of the food chain are most exposed. Hence, polar bears living in the Arctic Circle are far removed in distance and time from the original emission but still experience the impact.

Regulation of emissions from industry must recognise all these effects – from the large coal-fired power station with sulphur emissions to the small manufacturer using solvents. Where the potential exists for an air pollution impact, regulators will be seeking to restrict emissions. This chapter examines the means by which this goal is currently being achieved.

Regulatory position

[E5002] For air pollution, there are three main sources of legislation and regulation to which industry has to respond and which this chapter seeks to cover. These are the following:

- European Union directives on air quality and emissions;
- Environment Agency (and local authority) implementation of Integrated Pollution Prevention and Control (see E5027) and the environmental permitting regime; and
- Government Air Quality Strategy for England, Wales, Scotland and Northern Ireland.

In addition, emissions to atmosphere also encompass gases with the potential to contribute to global warming (climate change) and stratospheric ozone depletion and these can be subject to international protocols and other agreements. While health effects occur in the 'troposphere' (the part of the atmosphere where we live), some of the global effects (such as ozone depletion) occur in the 'stratosphere'.

Regulatory drivers

Human health and toxicology

[E5003]–[E5006] The effects of air pollution on human health have long been recognised in the UK. The often quoted example is the prohibition of coal burning in London in 1273, on the grounds that it was 'prejudicial to health'.

Coal burning has been the main cause of air pollution problems since then and has provoked most of the legislation aimed at improving air quality and human health. The infamous four day 'smog' of December 1952 was shown to have caused nearly 4,000 additional deaths through bronchitis and other diseases of the respiratory system. There was also an increase in heart disease.

Today, health effects include irritation and loss of organ functions (e.g. reduced lung capacity). The Committee on the Medical Effects of Air Pollutants estimated in 1998 that 14,000–24,000 admissions a year are associated with short term air pollution, and that it causes between 12,000–24,000 premature deaths per year.

The government claims that improvements made over the period from 1990 to 2001 have helped to avoid an estimated 4,200 premature deaths and 3,500 hospital admissions each year. Nevertheless, its 2007 Air Quality Strategy for England, Scotland, Wales and Northern Ireland sets out that air pollution is estimated to reduce the life expectancy of every person in the UK by an average of 7-8 months and that the health costs of this are estimated at up to £20 billion each year.

It is not only the 'classical' air pollutants that provoke concern with regard to their impact on human health through impairment of the respiratory and cardiovascular systems. Fears also exist about releases of substances with the potential for damage to other parts of the body through long-term exposure at low levels. In particular, 'persistent organic pollutants' (POPs) and heavy metals have the potential to cause human health effects through inhalation or entry into the food chain.

The medical effects of air pollutants

Particulate matter (PM$_{10}$)

PM$_{10}$ are particles in the air that have an aerodynamic diameter of less than 10µm. Particulate air pollution is associated with a range of health effects including effects to the respiratory and cardiovascular systems. PM$_{10}$ is responsible for causing premature deaths to those with pre-existing heart and lung disease. The smaller the particle, the greater the likelihood that it can reach the furthest part of the lung where the most delicate tissues are.

Oxides of nitrogen (NO$_x$)

Nitrogen dioxide (NO$_2$) can cause the inflammation of the airways when in high concentrations, and long-term exposure can affect lung function. NO$_2$ exposure can also enhance the response that many people have to allergens and is particularly harmful to asthmatics.

Ozone

This gas can cause slight irritation to the eyes and nose and may exacerbate symptoms of those with respiratory conditions (e.g. asthma, pneumonia). Very high levels can damage the airways and lead to inflammatory reactions. Its tropospheric health impact is very different to its beneficial impact in the stratosphere, where it helps to protect the Earth against most of the ultraviolet radiation emitted by the Sun.

Sulphur dioxide (SO$_2$)

This gas constricts the airways of the lung by stimulating the nerves in the lining of the respiratory tissues. Asthmatics and people with chronic lung disease are particularly vulnerable.

Polycyclic aromatic hydrocarbons ('PAHs')

Occupational exposure to PAHs has been shown to increase the incidence of tumours in the lung, skin, possibly the bladder, and other sites. While the carcinogenicity of some PAHs is unknown or uncertain, others have been classified as probably or possibly carcinogenic to humans.

Benzene

This gas is a genotoxic human carcinogen, and no safe airborne level has been defined. There is a risk of leukaemia when people are exposed to high concentrations.

1-3 butadiene

This gas is also a genotoxic human carcinogen that can cause the induction of cancers of the lymphoid system, blood forming tissue, lymphomas and leukaemias.

Carbon monoxide

[E5007]–[E5013] This gas causes the formation of carboxyhaemoglobin. This reduces the capacity of the blood to carry oxygen round the body. Those with an existing disease connected to the heart or brain such as angina are particularly vulnerable.

Lead

Lead causes problems in the synthesis of haemoglobin, it has effects on the kidneys, the gastrointestinal tract, joints and reproductive system, and the nervous system. The greatest cause for concern is the effect of lead on brain development in young children. Food and water are the two main sources of ingestion for most people. Lead in air contributes to the lead levels in food through the deposition of dust and rain on crops and the soil. The advent of unleaded petrol has done much to reduce air emissions both from everyday transport and from storage accidents.

Ammonia

Ammonia contributes to the ill-health effects caused by PM10 and PM 2.5 since it is a precursor to secon-dary PM.

Environmental impacts

Acid deposition and damage to ecosystems

[E5014] When atmospheric pollutants such as sulphur dioxide (SO_2), nitrogen oxide (NO_2), and ammonia (NH_3) are deposited at the surface, this is referred to as acidification. This can occur through direct 'dry' deposition of the gases or through 'wet' deposition after transformation into sulphates and nitrates and absorption into rain and cloud water. This has been recognised as a threat to ecosystems, resulting in major international research projects and negotiations to reduce emissions. It is usually referred to by the more popular term of 'acid rain'.

Pollutants can be transported over considerable distance, affecting the air quality and ecosystems in adjacent or distant countries. Such effects can extend over several thousands of kilometres. Deposition of excess acid can result in changes to the chemical composition of soil and surface water, in turn interfering with ecosystems.

Consequences of acidification

[E5015] Acidification has been a major problem in Europe with the highest deposition rates found in the highly populated and industrialised zone extending between Poland and the Netherlands. Within the next ten years, acid deposition in Europe is expected to decrease following the agreed reductions in European emissions of SO^2 and NO^x (while ammonia emissions are not likely to change significantly).

Organisms in water are affected when the concentration of hydrogen ions increases (decreasing the pH of the water). As a result lake beds are covered in mosses, and there is an increase in the death of fish and other animals. Plants are affected by the changing chemistry and biology of the soil, altered by an increased amount of hydrogen ions. There has been a decline in the health of forests in central Europe because of acidification.

Modern forestry and agriculture can contribute to the pollution as well as suffer from it, through the use of fertilisers which release nitrogen and hydrogen ions into the air in the form of ammonia.

Deposition of excess nitrogen can lead to the eutrophication of fresh and marine waters.

Acidification in urban and industrial areas increases the deterioration of many buildings and construction materials. In sheltered positions, a black crust surface layer forms on calcareous materials, and the growth of lichen, mosses and algae is common. Historic monuments and buildings can be especially susceptible to damage.

Acidifying substances also contribute to climate change. Both NO2 and SO2 are radiatively active gases with a warming potential, although sulphate and nitrate aerosols result in cooling through reflection of sunlight in the upper part of the troposphere. Oxides of nitrogen are implicated strongly in the atmospheric chemistry of ozone formation, causing an excess of tropospheric ozone, and the depletion of stratospheric ozone.

Nuisance

[E5016] Air pollution can cause harm to the senses as well as harm to human health and the natural environment. Whilst not as damaging, odour and dust nuisance can result in costly disputes and generally harm stakeholder relations.

Various emissions from small-scale processes (e.g the outputs of local exhaust ventilation) are covered by statutory nuisance legislation (see E5062) and, depending on a range of factors, may be regarded as creating a local nuisance.

Key pollutants and issues

Introduction

[E5017] The issues outlined above at E5003–E5016 are some of the main areas of concern that drive regulation aimed at reducing air pollution. The section below from E5018–E5024 provides background material on the main pollutants and their origin, as well as further material on the regional and global issues that result in industry being asked to emit less substances of a damaging nature to atmosphere.

The UK priority pollutants

[E5018] Until quite recently, national air pollution control legislation was focussed on reducing air pollution from particular sectors of industry. It is only since the late 1980s that this has been added to by 'effects based' legislation. This approach takes as its starting point an air quality standard or measure of harm in the environment and then devises policies to enable progress against this criterion.

This philosophy reached a logical conclusion with the publication of the 2000 Air Quality Strategy (see below at E5067) which has been superceded by the 2007 Air Quality Strategy. The current document targets priority pollutants and explains how the government will achieve air quality objectives relating to airborne concentrations of all of these pollutants.

The priority pollutants

[E5019]

Pollutant	Origin	Comments
Particles (PM_{10})	PM_{10} are particles with an aerodynamic diameter of less than 10mm. This is emitted from road transport, stationary combustion sources, and other industrial processes, quarrying, construction and non-road mobile sources.	Road transport produces primary particles from engine emissions, tyre and break wear and other non-exhaust emissions. Secondary PM is formed from emissions of ammonia, sulphur dioxide and nitrogen oxides, and from emissions of organic compounds from combustion sources and vegetation.

Pollutant	Origin	Comments
Ozone	Ozone is not directly emitted from human-made sources in significant quantities. Instead it arises from chemical reactions in the atmosphere caused by strong sunlight. NOx and volatile organic compounds ('VOCs') are the most important precursors of elevated levels of ozone. Its production can also be stimulated by CO, methane, and other VOCs that come from natural sources.	
Sulphur dioxide (SO_2)	SO_2 dissolves in water to give an acidic solution (sulphuric acid). In the UK the main source of SO_2 is the combustion of fossil fuels containing sulphur, and the combustion of heavy oils.	At the beginning of the 20th century, SO_2 was emitted from the domestic sector, commercial and industrial premises, and power stations by the combustion of coal. The smogs of the 1950s prompted the Clean Air Act 1956. Cleaner fuels have since replaced coal in all sectors. Currently, the main sources are power stations, and in Northern Ireland particularly, coals for domestic use is still a significant source.
Oxides of nitrogen (NO_x)	All combustion processes in air produce oxides of nitrogen. NO2 and nitric oxide are both oxides of nitrogen and together they are referred to as NOx. In 2004, UK emissions were over 1600 tonnes.	Road transport is the main source, followed by the electricity supply industry and other industrial and commercial sources.

Pollutant	Origin	Comments
Polycyclic aromatic hydrocar-bons (PAHs)	There are many different PAHs arising from a variety of sources. The Air Quality Strategy uses benzo[a]pyrene (B[a]P) as a marker for the most hazardous PAHs. The main sources of this PAH are domestic coal and wood burning, fires (including bonfires and accidental fires), and industrial processes such as coek production.	Road transport is the largest source for total PAHs, although these are thought to be mainly less hazardous forms.
Benzene	This gas is a VOC. In the UK the main sources are domestic and industrial combustion and road transport.	
1,3-Butadiene	There are trace amounts of this gas in the atmosphere from the combustion of petrol and other materials. It is used in industry for the production of synthetic rubber for tyres, however motor vehicles continue to be the dominant source.	
Carbon monoxide (CO)	This gas is formed by the incomplete combustion of carbon containing fuels. The main source is road transport. .Residential and industrial combustion are also significant contributors.	
Lead	Lead is a widely used non-ferrous metal. The largest use is for the manufacture of batteries.	The sale of leaded petrol was banned from 1 January 2000 in the UK. This has led to significant decreases in emissions in (particularly) urban areas.

Pollutant	Origin	Comments
Ammonia	Mainly derived from agricultural sources including manure, slurry and fertilisers. Transport and waste disposal are responsible for a small proportion.	

Other important pollutants

[E5020] The principal reason why the UK Air Quality Strategies have focussed on the priority pollutants is to protect the health of the population. It could be argued that there are other pollutants that have the potential to harm health at sufficiently high concentrations. This might be especially true within the industrial sector where the possibility exists of localised 'hotspots' of concentrations of substances arising out of a particular industrial process. Examples are metals and individual organic compounds. For occupational hygiene and the indoor environment, the Health and Safety Executive (HSE) publishes limits for the airborne concentrations of several hundred substances. For the outdoor environment the same principles apply, but target concentrations are much lower than those set for occupational environments.

The following is an indicative list of other pollutants that require regulation from industrial processes:

• cadmium	•	polycyclic aromatic hydrocarbons ('PAHs')
• Mercury	•	polychlorinated bipheyls
• Nickel	•	polychlorinated dibenzo-p-dioxins
• Arsenic	•	polychlorinated dibenzofurans
• vanadium	•	hydrogen fluoride
• hydrogen chloride		

Trans-boundary pollution

[E5021] In the 1970s and 1980s much of the air pollution control policy was aimed at reducing the long-range transport of acidic pollution across national boundaries. The reported decline in the health of trees and the loss of life in lakes from Scandinavia to eastern Europe had provoked a strong response from the public and politicians. This is still an issue, although much has been done to reduce national emissions of sulphur and nitrogen, especially from large power plants burning coal and oil. The UK was once perceived as being the 'dirty man' of Europe in the context of sulphur dioxide emissions, but can now reasonably claim to have reduced national emissions substantially. The United Nations Economic Commission for Europe ('UNECE') Convention on long-range trans-boundary air pollution ('CLRTAP') is an important framework for environmental assessment and policy in Europe.

Climate change

[E5022] Gradually, this has become widely regarded as the most pressing airborne environmental impact. The natural presence of 'radiatively-active' gases in the atmosphere is essential for life. These gases trap heat in the lower atmosphere, creating a 'greenhouse effect'. The six greenhouse gases are:

- carbon dioxide;
- methane;
- nitrous oxides;
- chlorofluorocarbons ('CFCs');
- PFCs; and
- ground level ozone.

Increasing concentrations of these greenhouse gases enables more infa-red radiation to be absorbed in the lower atmosphere, upsetting the earth's radiation balance. This increase allows the troposphere to be warmed more significantly, and the upper layers to be cooled. Without any greenhouse effect at all, the earth would be uninhabitable for humans. In geological history, the earth has variously been warmer and colder than at present. However, a fundamental environmental issue today is the rapid rate at which the climate appears to be changing.

Much of the emission of greenhouse gases comes from anthropogenic (human) activity, especially combustion of fuel and anthropogenic emissions expected to rise, doubling by 2030. As a result, the atmosphere is being gradually heated; the trend of global mean surface temperatures indicates an increase of 0.45C over the past hundred years. UK Climate Projections issued in 2009 (medium emissions scenario) shows that by the 2080s there could be:

- An increase in average summer temperatures of between 2 and 6 degrees C in the South East with a central estimate of 4 degrees;
- A 22% decrease in average summer rainfall in the South East and an increase of 16% in average winter rainfall in the North West; and
- sea level rise of 36cm.

Emissions of carbon dioxide due to the combustion of fossil fuels are a key problem, notably because of the huge amount generated and because emission abatement options are few. This means emissions must be reduced by changes in the amount (and the type) of fossil fuel burnt.

Consequences

[E5023] The consequences of rapid climate change are potentially massive and they are not all predictable: regional sea currents may change dramatically due to polar melting. Vegetation will be in greater competition as optimum conditions become scarce. This may cause shifts of plant and animal life. Extra warmth can trigger insect plagues and plant diseases. Such consequences will affect agriculture, which is also directly affected by changes in rainfall. Changes to the land and coastal areas can lead to population and wildlife displacement, and land degradation.

The depletion of stratospheric ozone

[E5024] The depletion of stratospheric ozone during the last few decades has been a global problem. Stratospheric ozone protects the Earth's surface from harmful UV radiation. The loss is greatest nearer the poles and is sometimes referred to as the 'ozone hole'. The cause of this problem was the increase of industrially produced CFCs, halogens and other halogenated substances in the upper atmosphere, and although most emissions have stopped due to international bans on these substances, the ozone layer will take many years to recover. In addition, it may not recover to previous levels.

Consequences

[E5025] Any depletion of ozone means that certain wavelengths of ultraviolet solar radiation are able to penetrate through to the earth's surface, damaging human health and leading to an increase in skin cancer, as well as damaging ecosystems. This can also lead to a change in global circulation and climate, as the absorption of the radiation by ozone can lead to heat formation in the atmosphere.

The European context

The policy makers

[E5026] Various agreements have been drawn up at international level. Key agreements include the Montreal Protocol on the Control of Ozone-depleting Substances (1988) (ozone-depleting substances), United Nations Framework Convention on Climate Change (1992) (carbon dioxide, unofficially known as the 'Climate Change Treaty'), Sulphur Protocol (1985) (sulphur), Nitrogen Oxides Protocol (1991), the VOC (Volatile Organic Compounds) Protocol (1994) and the Gothenburg Protocol (1999). Many of the protocols have then been implemented by European action.

There are two main European bodies that drive legislation and protocols, namely the European Union and UNECE (see **E5021** above). The latter body brings together 56 countries located in the European Union, non-EU Western and Eastern Europe, South-East Europe and Commonwealth of Independent States (CIS) and North America. Its chief role is to examine the consequences of trans-boundary pollution through CLRTAP (see **E5021** above). The Convention states that all countries should;

> endeavour to limit and, as far as possible, gradually reduce and prevent air pollution, including long range trans-boundary pollution.

In all, the Convention has been responsible for eight protocols, including:

- *The 1985 Sulphur Protocol* required signatories to reduce national sulphur emissions by 30%, based on 1980 levels, on or before 1993. The Protocol was subsequently revised and signed in Oslo in June 1994 and came into force in August 1998. The UK ratified the Protocol (December 1996) and agreed to reduce SO2 emissions by 50% by 2000,

70% by 2005 and 80% by 2010, (again based on 1980 levels). This required 'fuel switching' and advances in abatement technology. The government has drawn up a national plan to keep national emissions within these targets. In the main, these targets have most impact on the power generation and oil refining industries.

- The 1988 Nitrogen Oxides Protocol was adopted in Sophia in 1991. It froze emissions of nitrogen oxides by 1994 using 1987 as a baseline. The *Gothenburg Protocol to Abate Acidification, Eutrophication and Ground Level Ozone (1999)* also concerns nitrogen oxides as well as VOCs and ammonia.

- *The 1991 Volatile Organic Compounds Protocol* was signed in Geneva and came into force in September 1997. The UK ratified the Protocol in June 1994. VOCs are defined as *'all organic compounds of anthropogenic nature, other than methane, that are capable of producing photochemical oxidants by reactions with nitrogen oxides in the presence of sunlight.'* The Protocol required most parties to reduce overall VOC emissions by 30% on or before 1999, using a base year of 1988. Additionally, the protocol obliged signatories to control emissions from industries through new emission limits and to introduce abatement technologies as well as reduce solvent use and reduce VOC emissions through petrol distribution and refuelling.

- *The 1998 Heavy Metals Protocol* was signed in June 1998 at Aarhus. It requires emissions of cadmium, lead and mercury to be reduced below 1990 levels. Its aim is to reduce emissions through the use of stricter emissions limit values for industry and the use of best available technology (see **E5027** below). In 2002, the Department for Environment, Food and Rural Affairs (DEFRA) published a draft strategy for UK ratification of this Protocol.

- *The Persistent Organic Pollutants Protocol* was also signed in June 1998 at Aarhus. It aims to phase out the production of and use of a defined list of substances, as well as imposing requirements to eliminate discharges, emissions and losses and to ensure safe disposal methods. The list covers 16 substances in three categories, (pesticides, industrial chemicals and by products or contaminants, e.g. dioxins).

- *The Acidification, Eutrophication and Ground-level Ozone Protocol* – also referred to as the Multi-pollutant, Multi-effect Protocol, the 2nd NOx Protocol or the Gothenburg Protocol – was signed in December 1999. In Europe, it has a similar scope to EU directives in terms of pollutants, but applies to all European countries, not just EU Member States. When the Gothenburg Protocol is fully implemented, Europe's sulphur emissions should be cut by at least 63%, its NOx emissions by 41%, its VOC emissions by 40% and its ammonia emissions by 17% compared to 1990.

  This Protocol also sets limit values for specific emission sources (e.g. combustion plant, electricity production, transport) and requires 'best available techniques' to be used. VOC emissions from products as paints or aerosols will also have to be reduced.

The effects of these Protocols on industry are not usually seen immediately or directly. Normally they take effect through the subsequent actions of national governments and regulatory agencies who propose action and legislation designed to achieve the aims of the Protocols. In this sense the European Commission is strongly interlinked with the CLRTAP Protocols. Not only is the Community a signatory in itself, but the Commission will frame directives so as to meet the aims of the Protocols.

The most active part of the Commission with regard to air quality legislation is Directorate General Environment ('DG Environment'). The units within this Directorate have direct responsibility for drafting new directives that set air quality standards and implement emissions reductions in key sectors.

The DG Environment also has responsibility for some legislation relevant to industry and air pollution control, such as the implementation of Integrated Pollution Prevention and Control and the Waste Incineration Directive.

The work of the DG Environment in air pollution is characterised by a tension between the contribution from transport and from industry.

Other international protocols

[E5026.1] The Montreal Protocol was drawn up in 1988 and signed by 50 countries. Since then, over 170 countries have signed up. The original protocol set up targets for controlling CFCs, and halons. Carbon tetrachloride, 1,1,1 – trichloroethane, hydrochlorofluorocarbons (HCFCs) and methyl bromide and halons 1211, 1301 and 2402 were added to amendments to the protocol.

The Montreal Protocol affects many industries, e.g. refrigeration and air conditioning, aerosol manufacture and foam blowing. It is directly binding on the UK through EU Regulation EC/2037/2000. This Regulation, which came into force on 1 October 2000, affects the users, producers, suppliers of ozone depleting substances, and others such as those who dispose of them. The *Environmental Protection (Controls on Ozone-Depleting Substances) Regulations 2002 (SI 2002 No 528)* implement the Regulation in the UK.

Hydrochlorofluorocarbons were prohibited from June 1995 except for use as solvents, refrigerants, insulating foams, carrier gas for various sterilising systems and feedstock for chemicals manufacture. Under the Beijing amendment (1999), production by developed countries was frozen in 2004 at 1989 levels. Exports to countries who have not signed up to the Montreal Protocol were banned from 2004.

From January 2000 the use of HCFCs was prohibited in new equipment in cold stores and warehouses, and for equipment of 150 kW and over.

The Climate Change Treaty

[E5027] The Climate Change Treaty was agreed in Rio de Janeiro in 1992 and latterly in Kyoto in 1997 (the *Kyoto Protocol to the 1992 United Nations Framework Convention on Climate Change (1997)*). At the Kyoto conference, cuts in the emission of six greenhouse gases-carbon dioxide, methane, nitrous oxide, hydrofluorocarbons, perfluorocarbons and sulphurhexafluoride were agreed. All developed countries should reduce emissions by specified amounts, to reduce the potential for global warming from sources such as road traffic. Ratification of the Protocol means that the UK is legally required to reduce emissions of these six greenhouse gases by 12.5% below 1990 levels in 2008-2012. Negotiations over a successor to the Kyoto Protocol were due to take place in Copenhagen at the end of 2009.

Integrated pollution prevention and control ('IPPC') and the Environmental Permitting Regime IPPC Directive (Directive on Integrated Pollution Prevention Control 96/61/EC)

The IPPC directive (Directive on Integrated Pollution Prevention Control 96/61/EC) was designed to prevent, reduce, and eliminate pollution at the source, through the careful use of natural resources. The directive was aimed at helping industries move towards greater environmental sustainability, and applied to the installation of the following activities:

*   energy industries (power, oil and gas);
*   the production and processing of metals (ferrous and non-ferrous);
*   mineral industries (cement and glassworks);
*   chemical industries (organic, inorganic, and pharmaceuticals);
*   waste management (landfill sites and incinerators); and
*   other (slaughter houses, food/milk processing, paper, tanneries, animal carcass disposal, pig/poultry units and organic solvent users).

IPPC covers emissions to air, land and water, as well as heat, noise, vibration, energy efficiency, environmental accidents, site protection, and many more processes.

The IPPC directive required that applications must show that installations are run in a way that prevented emissions, using the following principles:

*   must apply the best available techniques ('BAT') to control emissions. Account must be taken of the relative costs and advantages of the available techniques;
*   waste is to be minimised and recycled where possible;
*   energy is to be conserved;
*   accidents are to be prevented, and their environmental consequences to be limited; and
*   the site is to be returned to a satisfactory state after operations cease.

The overall objective for the directive was a high level of protection for the environment as a whole, and a system of permit was set up to achieve this relating to:

*   plant operating conditions;
*   emission limits for certain substances to air, land and water; and
*   annual reporting of pollutant releases.

The *Pollution Prevention and Control Act 1999* paved the way for Regulations to be made which implemented the IPPC Directive (*Pollution Prevention and Control (England and Wales) Regulations 2000* (SI 2000 No 1973). These have now been superceded by the *Environmental Permitting (England and Wales) Regulations 2007* (see below) as a result of the consolidating and more stringent IPPC Directive 2008/1/EC.

IPPC Directive 2008/1/EC

The current IPPC Directive (Directive 2008/1/EC) has been implemented in the UK by the *Environmental Permitting (England and Wales) Regulations 2007 (SI 2007 No 3538)*, with separate regulations applying the IPPC Directive in Scotland, Northern Ireland and to the offshore oil and gas industries (see 5040 below). The 2008 Regulations replaced the previous pollution prevention control regime as well as the waste licensing system derived from the Environmental Protection Act 1990 Part 2. They also replaced more than 40 sets of regulations.

The directive applies an integrated environmental approach to the regulation of particular industrial activities. Emissions to air, water and land, as well as a range of other environmental effects must be considered t-gether. IPPC aims to prevent emissions (and waste production) and where that is not practicable, reduce them to acceptable levels. It also extends this integrated approach beyond permitting to the restoration of sites when industrial activities cease. Regulators must set permit conditions to achieve a high level of environmental protection.

Best Available Techniques (BAT)

These conditions are based on the use of "Best Available Techniques" which takes into account the costs to the operator and the benefits to the environment. It involves identifying options, assessing environmental effects and considering the economic factors.

Determining BAT involves comparing the techniques available to prevent or control emissions and then identifying those that will have the lowest overall impact on the environment. Once options have been identified, the environmental effects (particularly those which are significant) should be assessed. The environmental assessment should focus on identifying and quantifying the effects of releases. Carbon dioxide emissions must be considered (although this is not included in a list of the main polluting substances in Annex III of the directive.

The option which minimises the environmental impact of the installation should be used unless economic factors rule this out.

BREF Notes

[E5028] All permits for IPPC must be based on the concept of BAT, which is defined in Article 2 of the Directive. Article 12 states that Member States should exchange information on BAT. Consequently the European Commission is required to organise the exchange of information between industry, environmental organisations, and experts from the EU member states. This work is co-ordinated by the European IPPC Bureau. The Commission then publishes the results of the information exchange every three years in the form of BAT reference documents, known as BREF Notes. These are prepared for around 30 industrial sectors .

The BREF Notes form the basis of guidance issued in the UK. The BREFs are not directly-applicable guidance notes and so they need to be supplemented by domestic guidance. This is produced by the Environment Agency, drawing on the information contained within BREFs. Where there is no domestic guidance available, operators and regulators should refer directly to the relevant BREFs.

The IPPC Bureau was set up in 1996 in Seville to control the information management system for the BREF Notes. There are four main bodies involved in the information exchange. These are the EC Directorate responsible for the environment (the DG Environment) that has overall control, the IPPC bureau, the information exchange forum (which is a committee of representatives from member states, industry and non-government organisations), and the technical working groups ('TWGs') which are made up of experts. The TWGs are where the information exchange really takes place. Information, BAT studies, and details on specific sectors and emission levels are provided by the experts, who meet at least twice for each industrial sector. The table below at **E5029** shows an alphabetical list of the full series of reference documents the IPPC works on (together with the date the current document was adopted and where a review is taking place, the current draft document).

The IPPC website provides more information about the latest documents and where draft guidance is being prepared at http://eippcb.jrc.es/reference/.

BREF Notes

[E5029]–[E5032]

Cement, Lime and Magnesium Oxide Manufacturing Industries 12.2001 (Current draft 5.2009)
Ceramic Manufacturing Industry 8.2007
Chlor-Alkali Manufacturing Industry 12.2001
Common Waste Water and Waste Gas Treatment/Management Systems in the Chemical Sector 2.2003 (Current draft 10.2009)
Economics and Cross-Media Effects 7.2006
Emissions from Storage 7.2006
Energy Efficiency 2.2009
Ferrous Metals Processing Industry 12.2001
Food, Drink and Milk Industries 8.2006
General Principles of Monitoring 7.2003
Glass Manufacturing Industry 12.2001 (Current draft 7.2009)
Industrial Cooling Systems 12.2001
Intensive Rearing of Poultry and Pigs 7.2003
Large Combustion Plants 7.2006
Large Volume Inorganic Chemicals – Ammonia, Acids and Fertiliser Industries 8.2007
Large Volume Inorganic Chemicals – Solids and Others Industry 8.2007
Large Volume Organic Chemical Industry 2.2003
Management of Tailings and Waste-Rock in Mining Activities 1.2009
Manufacture of Organic Fine Chemicals 8.2006
Mineral Oil and Gas Refineries 2.2003
Non-ferrous Metals Industries 12.2001 (Current draft 7.2009)

Production of Iron and Steel 12.2001 Current draft 7.2009
Production of Polymers 8.2007
Production of Speciality Inorganic Chemicals 8.2007
Pulp and Paper Industry 12.2001
Slaughterhouses and Animals By-products Industries 5.2005
Smitheries and Foundries Industry 5.2005
Surface Treatment of Metals and Plastics 8.2006
Surface Treatment Using Organic Solvents 8.2007
Tanning of Hides and Skins 2.2003 (Current draft 2.2009)
Textiles Industry 7.2003
Waste Incineration 8.2006
Waste Treatments Industries 8.2006

Framework directive on air quality assessment and management

[E5033] Directive 96/62/EC on ambient air quality assessment, the Air Quality framework directive, provides a framework for the EU to use to set limit values for pollutants. This directive identifies twelve pollutants for which limit or target values will be set in daughter directives. These are:

- sulphur dioxide (SO2);
- nitrogen dioxide (NO2);
- particulate matter (PM10);
- lead;
- carbon monoxide;
- benzene;
- ozone;
- polycyclic aromatic hydrocarbons;
- cadium;
- arsenic;
- nickel; and
- mercury.

This directive is particularly important, as it prepared the ground for a number of significant daughter directives on air quality standards and control measures, as set out below.

Daughter directives

[E5034] This directive provides a framework for 'daughter directives'. These have been prepared for sulphur dioxide, nitrogen dioxide, oxides of nitrogen, particulate matter, lead, carbon monoxide and benzene, ozone and arsenic, cadmium, mercury, nickel and polycyclic aromatic hydrocarbons.

Daughter directive for limit values of SO2, oxides of nitrogen, particulate matter, and lead in ambient air

[E5034.1] The first daughter directive in 1998 established legally binding limit values for SO2, NO2 and PM10, to be achieved by 1 January 2005 and

2010. This directive was adopted in April 1999 and member states are required to implement it by July 2001.

Daughter directive to reduce the sulphur content of liquid fuels

[E5035] This daughter directive aims to reduce further the emissions of SO2 resulting from the combustion of liquid fuels. For heavy fuel oil member states have to ensure that the sulphur content is not over 1% by mass. For gas oil, limits are 0.2% by mass and 0.1% by mass from 1 January 2008. Certain classes of fuel used by ships are excluded.

Daughter directive on limit values for benzene and carbon monoxide in ambient air

[E5036] Daughter Directive 2000/69/EC sets limit values for benzene and carbon monoxide. It has been adopted and is being implemented in the UK.

Proposal for national emissions ceilings directive and ozone daughter directive

[E5037]–[E5038] In June 1999 proposals were put forward for two further daughter directives. One was to set target values for ozone and the other emission ceilings for various pollutants. The latter is referred to as the *National Emissions Ceilings Directive* ('NECD'). This directive sets ceilings for national emissions of SO2, NO2, NH3 and VOCs to be achieved by 2010. The NECD is the main instrument for attaining air quality targets for ambient ozone, which are also included in the proposal.

As a result of these proposals, the following daughter directives were adopted and implemented in the UK:

- Directive 99/30/EC relating to limit values for sulphur dioxide, nitrogen dioxide and oxides of nitrogen, particulate matter and lead in ambient air;
- Directive 2000/69/EC relating to limit values for benzene and carbon monoxide in ambient air;
- Directive 2002/3/EC relating to limit values for ozone in ambient air; and
- Directive 2004/107/EC relating to arsenic, cadmium, mercury, nickel and polycyclic aromatic hydrocarbons in ambient air (known as the Fourth Daughter directive).

The *Air Quality Standards Regulations 2007 (SI 2007 No 64)* apply to England and set air quality standards to protect human health and the environment. Similar Regulations apply to Scotland, Wales and Northern Ireland.

[E5039] The Ambient Air Quality and Cleaner Air for Europe Directive came into force in 2008 consolidating the Framework Directive and daughter directives into one piece of legislation and introducing new provisions for controlling PM 2.5, discounting natural sources. It also introduced options for time extensions to meet the PM10 and NO2 deadlines.

A Defra consultation on new regulations to transpose the new directive into UK law ended in January 2010. These would introduced a new framework for PM 2.5 which is a combination of a new exposure reduction approach and a "backstop" limit value.

A national exposure reduction target will be established by the Secretary of State, defined as a percentage reduction in average concentration in urban background. This would be achieved by 2020 and set once data is available for 2009-2011, early in 2012. The target is likely to be 10% and this compares with the 6% reduction likely to be achieved by current measures. The average exposure indicator for 2015 must not exceed 20ug/m3 and the UK is on track to meet this obligation, according to current projections.

Limit and target values will apply across England. The target value is 25ug/m3 by January 2010 and the limit value is 25ug/m3 to be met by January 2015.

UK Regulations

Introduction

[E5040] Air pollution control for industry is dominated by the introduction of the IPPC Directive into UK law. This replaced the Integrated Pollution Control and Local Air Pollution Control ('LAPC') Regulations that existed since 1991. The last 15 years or so have seen the removal of a system of controls and concepts that can be traced back to 1842. In 2008, the *Environmental Permitting (England and Wales) Regulations 2007 (SI 2007 No 3538)* (and regulations for Scotland, Northern Ireland and the offshore oil and gas industries) came into effect implementing the requirements of the 2008 IPPC Directive.

Definition of useful acronyms and terms

[E5041]–[E5042]

IPC	Integrated Pollution Control
IPPC	Integrated Pollution Prevention Control
	IPPC is a regulatory system that employs an integrated approach to control the environmental impacts of certain industrial activities. It involves determining the appropriate controls for industry to protect the environment through a single permitting process.
LAPC	Local Air Pollution Control
BAT	Best Available Technique (IPPC)

	The Best Available Techniques are those that prevent or minimise pollution, and can be implemented effectively and are economically and technically viable while meeting the overall aims of the Directive.
BPEO	Best Practical Environmental Option
Part A1 installations	The largest and most complex processes, regulated for releases to air, water and land by the Environment Agency in England and Wales.
Part A2 Installations	Processes with releases to air, water and land regulated by Local Authorities in England and Wales (but by SEPA in Scotland).
Part B installations	The smaller processes, with releases to air only regulated by Local Authorities in England and Wales (but by SEPA in Scotland).

Useful websites

- *Environment Agency:* www.environment-agency.gov.uk
- *Department for Environment, Food and Rural Affairs (Defra):* www.defra.gov.uk/
- *European Commission:* http://ec.europa.eu
- *The UK National Air Quality Information Archive:* www.airquality.co.uk
- *European Integrated Pollution Prevention and Control Bureau:* http://eippcb.jrc.es/
  N.B: BREF notes can be downloaded from the site above.
- *Scottish Executive:* www.scotland.gov.uk/environment/airquality

The Environmental Permitting Regime requires operators to obtain permits for particular facilities (and registration of exemptions for others) and aims to protect the environment while effectively and efficiently delivering permitting and compliance. It also encourage regulators to promote best practice in the operation of regulated facilities.

The new regime covers facilities previously regulated under the 2000 Pollution Prevention and Control Regulations, waste management licensing and exemptions and mining waste operations.

The Regulations set out:

- The facilities that must have environmental permits or be registered as exempt;
- The process for registering exemptions;
- Permit applications and determinations;
- Changing and surrendering permits;
- The standard rules (a simplified permitting system);
- Compliance obligations, enforcement and penalties – the principle offences are operating without a permit (Regulation 12) or in contravention of permit conditions and failing to comply with enforcement notices;

- Public participation;
- Powers and functions of regulators;
- Transition arrangements; and
- Appeals against permitting decisions.

Schedules to the Regulations set out the requirements of various EU directives which must be delivered through the permitting regime. Each directive has a specific schedule and where a facility falls under more than one directive, all the requirements must be met:

- Part A installations and Part A mobile plant are covered by the Integrated Pollution Prevention and Control Directive (Schedule 7);
- Domestic Part B installations and Part B mobile plant (Schedule 8);
- The Waste Framework Directive (Schedule 9);
- Waste motor vehicles (the End of Life Vehicle Directive) (Schedule 11);
- The Waste Electronic Electrical and Electronic Equipment (or WEEE) directive (Schedule 12);
- The Waste Incineration Directive (Schedule 13);
- The Solvents Emission Directive (Schedule 14);
- The Large Combustion Plants Directive (Schedule 15);
- The Asbestos Directive (Schedule 16);
- The Titanium Dioxide Directive (Schedule 17);
- The Petrol Vapour Recovery Directive (Schedule 18);
- The Batteries Directive (Schedule 18A); and
- The Mining Waste Directive (Schedule 18B).

Regulated facilities

The regulations set out which "regulated facilities" require an environmental permit and provide for the exemption of some waste operations. These are:

- an installation which carries out the activities listed in Schedule 1 to the Regulations. These include installations involved in energy activities, metal production and processing, the mineral industries, the chemical industry, waste management and activities including paper, pulp and board manufacture and the manufacture of dyestuffs, printing and textile treatments;
- a waste operation not carried out at an installation or by means of mobile plant;
- a mobile plant carrying out a Schedule 1 activity or waste operation; and
- a mining waste operation.

An Environmental Permit can cover more than one regulated facility where both the regulator and operator are the same for each facility and (with exceptions) all the facilities are on the same site (Regulation 17).

Applications

Regulation 7 defines an operator as the person who has control over a regulated facility, and only an operator can obtain or hold an Environmental Permit. In most cases, a single operator will need to obtain a single environmental permit, but there are exceptions to this. Applications should normally be made at the design stage. The Regulations contain transitional arrangements for existing PPC permits and waste management licences so new applications for environmental permits are not required.

The application procedure is set out in Schedule 5 to the Regulations. The application must:

- Be made by the operator/s (or their agent);
- By the current operator and future operator in the case of a transfer application;
- Be made on the form provided by the regulator;
- Include information required by the application form; and
- Include the relevant fee.

The regulator must decide whether to grant or refuse the proposal in an application, and where applicable, what permit conditions to impose. Various periods for determination are set out in the regulations.

The application to the regulator will include an assessment of environmental risk (under normal and abnormal operating conditions) and the regulator must be satisfied that this assessment is robust.

When permitting facilities, the regulator must take into account the requirements of European directives which are set out in Schedules 7 to 18B (see above).

Applications can be refused where:

- The operator is not competent to run the regulated facility;
- The environmental impact would be unacceptable;
- The information provided is not adequate for the regulator to determine the permit conditions; or
- The requirements of the European Directive cannot be met.

The regulations set out provisions to vary, transfer and surrender as well as apply for permits.

Standard rules

The regulations contain provisions for the Secretary of State, Welsh Ministers and the Environment Agency to make Standard Rules consisting of requirements common to a particular class of facilities. These can be used instead of site-specific permit conditions. Standard rules do not require a site-specific assessment of risk for a standard facility, but they cannot be appealed against. It is the operators decision as to whether they wish to operate under these rules.

Operator competence

In making the decision as to whether an operator is competent, the regulator will consider management systems, technical competence, including where relevant whether an approved scheme is being complied with, compliance with previous regulatory requirements and financial competence.

Consultation and public participation

The regulations require public consultation on environmental permit applications, although the requirement does not apply to applications for mobile plant, standard permits other than for Part A installations, certain small part B installation or a mining operation not involving a mining waste facility to which Article 7 of the Mining Waste Directive applies. A public consultee is anyone who could be affected by the application or has an interest in the application.

Compliance Assessment, Enforcement and Review

The regulations set out that risk-based compliance assessment should focus on those activities that pose the greatest environmental risk to the environment or human health, have poorer standards of operation, are failing to comply with the terms and conditions of the permit, or are having a greater adverse impact. Regulators must undertake periodic inspections of regulated facilities (regulation 34(2)).

Regulators can serve an enforcement notice if an operator has contravened, is contravening or is likely to contravene any permit conditions (regulation 36). These specify the steps necessary to remedy the problem and the timescale in which they must be taken.

Suspension notices (regulation 37) can be served where there is a serious risk of pollution, whether or not an operator has breached a permit condition. This must describe the nature of the risk of pollution. the action necessary to remove the risk, and it must specify a deadline.

When a suspension notice is served, the permit ceases to authorise the operation of the facility or specified activities, although activities can continue unless their cessation is necessary to address the risk of pollution.

In the case of a prosecution, conviction in a magistrates court carries a fine of up to £50,000 and up to 12 months imprisonment, and conviction in the Crown court can lead to an unlimited fine and up to five years imprisonment.

Operators have an emergency defence which provides a defence where operators can show that they have acted in an emergency in order to avoid danger to human health, they have taken all reasonable steps to minimise pollution and the regulator is informed promptly (regulation 40).

The details of any conviction or formal caution must be placed by the regulator on the public register.

The regulator can revoke a permit by serving a revocation order (regulation 22) where appropriate, for example where other enforcement tools have failed to protect the environment. Where a revocation notice is appealed, it does not come into effect until the appeal is determined (or withdrawn). It must specify any additional steps the operator must take to avoid any risk of pollution to or return the site to a satisfactory state.

Where a regulated facility causes the risk of serious pollution, the risk can be removed by the regulator under regulation 57. This gives the regulator the power to arrange for steps to be taken to remedy the pollution at the expense of the operator if an offence that causes pollution is committed. Site protection must be addressed throughout the life of a permit rather that allowing an operator to contaminate a site during operation and decontaminating it when the facility closes.

The Crown is bound by the Regulations. While the Crown is not criminally liable if it contravenes the regulations, and the regulator cannot take legal proceedings if it does not comply with an enforcement or suspension notice, it may apply to the High Court to declare the contravention unlawful. Schedule 4 of the Regulations sets out the provisons relating to the Crown.

The conditions of the permit can be varied at any time by the regulator (regulation 20). Variations could be required where the findings of a permit review show this to be necessary, or because there is a new environmental quality standard (EQS). Where the conditions of a permit are to be varied, the regulator serves a variation order and the operator may be required to pay a fee. In addition, consultation will take place.

Permits must be reviewed periodically (regulation 34) to check that the conditions remain adequate in the light of experience or new knowledge and that they continue to reflect current standards.

Charging

There are two sets of charging arrangements for facilities regulated by the Environment Agency and by local authorities and different charges apply depending on the installation, sector, and the stage of regulation. Charges are incurred for applications, and subsistence charges are levied to contribute to the regulator's ongoing costs, for checking monitoring data or carrying out compliance assessments for example.

Appeals

An operator can appeal where an application for a permit (or for a variation, transfer or surrender of a permit) has been refused (or deemed to have been refused), The operator can also appeal where they disagree with the conditions set out in the permit, the regulator has deemed an application to be withdrawn. There are also provisions relating to appeals concerning closures, revocation, enforcement, and suspension notices and relating to landfill and mining waste. Decisions about the information that must be included on the public register can also be appealed. The time limits for making appeals varies according to the basis of the appeal. Appeals are made to the Secretary of State or Welsh Ministers.

Public registers and Information

Regulators must maintain public registers for environmental permits containing information on all the regulated facilities for which they are responsible. Local authority registers must also contain information on facilities in their area regulated by the Environment Agency (other than mobile plant). Members of the public must be allowed to inspect the registers, free of charge and at all reasonable times and copies of any entry must be made available for a reasonable charge. Where information is not available due to confidentiality, the register must state that this information exists.

Registers can be internet or computer based. They must contain the information set out in paragraph 1 of Schedule 19 to the Regulations, including copies of permits, applications enforcement notices and monitoring information.

There is an excemption for information which the Secretary of State or Welsh Ministers deem would be contrary to the interests of National Security; and the regulator may judge information to be industrially or commercially confidential and withhold it from the public register.

Information is also available under the Freedom of Information Act and the Environmental Information Regulations 2004 (SI 2004 No 3391) (see www. informationcommissioner.gov.uk for further information).

There are provisions in the Regulations for waste operations that pose a very low risk to be exempt from the need to hold a licence.

Detailed guidance on Environmental Permitting and the IPPC Directive is available on the Department for Environment Food and Rural Affairs (DEFRA) at:

www.defra.gov.uk/environment/policy/permits/documents/ep-core-guida nce.pdf and
www.defra.gov.uk/environment/policy/permits/documents/ippc-parta-guida nce.pdf

The regulators

The environment agencies

[E5043] In England and Wales the Environment Agency's pollution control responsibilities are:

* Authorisations, licences and consents for emissions to air, water and land. Monitoring compliance and enforcement, including prosecution permitting of installations and enforcement of Regulations under IPPC.
* Advice and guidance to industry and others on best environmental practice.

Scottish Environment Protection Agency

[E5044] With regard to the Scottish Environment Protection Agency ('SEPA'), Scotland is divided into the west, east and north for operational purposes. The roles for SEPA are similar for the Environment Agency for England and Wales. In addition it regulates Part B processes for air pollution under the *Environmental Protection Act 1990, Part I*. In other words in England and Wales this function has not been transferred to the Agency and remains with local authorities. SEPA also controls the installations regulated for IPPC.

Northern Ireland

[E5045] The Department of Environment ('DoE') (Northern Ireland) has overall responsibility for air pollution control in Northern Ireland. It has introduced a system similar to the IPPC and air pollution control regimes operating in the rest of the UK.

Part A processes are regulated for IPPC by the Industrial Pollution and Radiochemical Inspectorate within the DoE (Northern Ireland). Part B processes are also regulated by the same Inspectorate.

Local authorities

[E5046]–[E5060] Local authorities have a wide range of responsibilities covering the whole spectrum of pollution control and environmental protection. The local authority associations and the Environment Agency have signed a Memorandum of Understanding (1997) covering those areas of environmental protection for which they have responsibility.

Local authorities regulate so-called 'medium-polluting' installations under the Local Authority Pollution Prevention and Control regime – often referred to as Part B installations. The installations are classified into 82 sectors, and a separate guidance note on air pollution standards is issued for each sector.

The Clean Air Act 1993

[E5061] The IPPC legislation, and its predecessors IPC and LAPC, probably account for some 20,000 individual industrial processes. For those industrial processes not considered to be sufficiently polluting to warrant inclusion, it may be that other Regulations still apply to their activities. This is particularly true of processes that require the burning of fuel or waste. Historically, the *Clean Air Acts* of *1956* and *1968* were concerned very much with the control and reduction of smoke, largely through the burning of coal. These Acts, and other clean air legislation such as the *Control of Pollution Act 1974* and the *Control of Pollution (Amendment) Act 1989* were consolidated in the *Clean Air Act 1993*.

There are three parts to this Act that are relevant:

(1)   *Part I: dark smoke*
      Prohibition of dark smoke from chimneys and from industrial or trade premises.
(2)   *Part II: smoke, grit, dust and fumes*
      The Regulations control the emissions of particles emitted from furnaces and boilers. Direction is also given on the height of chimneys.
(3)   *Part V: Information about air pollution*
      Regulations grant Local Authorities the power to provide information on emissions from specified premises, or to enter premises and make measurements.

Statutory nuisance

[E5062] Air pollution control legislation has been rooted historically in the concept of nuisance to humans, rather than harm to the environment. The causes of such nuisance have traditionally been the burning of fuel or waste materials, giving rise to 'fumes' and 'smoke'. Odour is another form of nuisance.

The main legislation relating to nuisance is now in the *Environmental Protection Act 1990, Part III* as amended by the *Noise and Statutory Nuisance Act 1993*. It enables Local Authorities and individuals to secure the abatement of a statutory nuisance.

A nuisance is defined by the legislation, although only parts of the definition relate to air pollution. Others relate to noise and the state of premises that might be 'prejudicial to health'. In essence a nuisance can arise because of smoke or gases emitted from either private or business premises that are prejudicial to health or are a nuisance. A distinction is made between private dwellings and industrial premises in terms of the emissions. Private dwellings are exempt from causing a nuisance through emissions of 'dust, steam, smell or other effluvia'.

Unless the Secretary of State has granted consent, a local authority cannot bring a prosecution against a nuisance where a prosecution under the environmental permitting regime could be brought. However, activities not covered by environmental permitting, even if on the site of a regulated facility, may be regulated under the statutory nuisance provisions. In addition, private prosecutions can be brought.

Air quality standards and guidelines

Background

[E5063] For much of the twentieth century, UK air pollution control was exercised purely by reference to emissions, especially for industrial air pollution. Only in the 1990s did any concept of effects-based policies begin to emerge. This came about partly because European Community directives on air quality for pollutants such as SO_2 and NO_2 had been adopted, but also because the then Department of the Environment promoted such policies. In the acid rain debate, the effects were very much the start point for framing policies to reduce damage and the concept of a 'critical load' was used to define the extent of environmental damage. The trans-boundary issues were the dominant issue in air pollution policy debates in the 1980s, but subsequently interest in urban air quality and human health was re-awakened. No assessment of impacts in this topic can be made without reference to threshold concentrations at which health effects occur.

Air quality standards are one of the primary regulatory influences on industry, transport and other emitters. For the UK, the main originators of air quality guidelines are the World Health Organisation and the DETR's Expert Panel on Air Quality Standards. These bodies are responsible for recommending guidelines on air quality that represent concentrations at which the most sensitive members of the population will be protected from the damaging effects of air pollutants. For most pollutants, it is possible to decide upon a threshold concentration below which no effects are discernible. For some pollutants, e.g. carcinogens, it is theoretically possible to see an effect right down to near zero concentrations.

A most important point to recognise is that air quality guidelines have no legal standing. They do provide the basis, however, for governments and the European Commission to develop standards. Usually, such standards reflect a degree of 'economic dilution' and the fact that it is rarely possible to achieve a target concentration for 100% of the time. It is more usual to permit a certain number of occasions on which a threshold value may be exceeded. For example, the UK Government in its 1997 version of the air quality strategy proposed that the 24 hour average value of inhalable particulate matter could be exceeded on several days in a year specifically to allow for national festivities, i.e. 'Bonfire Night'.

Different pollutants require different periods of time over which they should apply, depending on the health effect. Sulphur dioxide, for example, provokes a reaction in the human respiratory system over 10–15 minutes. Therefore, it is the short term peak fluctuations of SO2 that need to be regulated. Lead, on the other hand, has an effect over a long period of time, as it accumulates in the human body. In this case, it is the annual average concentration of airborne lead that is the appropriate measure.

The multiplicity of ways in which air quality standards are expressed makes them rather confusing for those unfamiliar with the subject. In essence, however, they fall into two broad categories; those with averaging periods over a year and those with averaging periods over a short period of time, a day or less. The latter often have some expression of the number of occasions for which the threshold concentration may be exceeded. This is frequently described in terms of a 'percentile concentration'. For example, the most recent European Union directive for NO2 ambient concentrations specifies that the hourly concentration of 200 µg m should not be exceeded more than 18 times per year. A total of 18 hours represents 0.2% of the year. Hence, the standard can be expressed in an alternative manner as 200 µg m, as the 99.8 percentile concentration.

World Health Organisation

[E5064] The World Health Organisation ('WHO') formally produced a set of air quality guidelines for Europe in 1987. This was a much referenced document, despite having no legal status, and it was updated in 2005. The most recent WHO recommended air quality guidelines are used as a basis for setting standards in EU directives, and were also taken into account by the UK Expert Panel on Air Quality Standard when making recommendations for UK air quality standards. The table below at **E5066** shows WHO air quality guidelines. The table below at **E5068** shows the standards and guidelines that the UK has adopted to meet with the directive.

WHO air quality guidelines (including 2005 update)

[E5065]

Substances	Time-weighted average	Averaging time
nitrogen dioxide	200 µg m^{-3}	1 hour
	40 µg m^{-3}	annual
ozone	100 µg m^{-3}	8 hours
sulphur dioxide	500 µg m^{-3}	10 mins
	20 µg m^{-3}	24 hour
	50 µg m^{-3}	annual
carbon monoxide	100 mg m^{-3}	15 mins
	60 mg m^{-3}	30 mins
	30 mg m^{-3}	1 hour
	10 mg m^{-3}	8 hours
lead	0.5 µg m^{-3}	annual
cadium	5×10^{-3} µg m^{-3}	annual
mercury	1.0 µg m^{-3}	annual
PM $_{10}$	20 µg m^{-3}	annual
	50 µg m^{-3}	24 hours

In addition to the guidelines expressed as a threshold concentration, guidelines are also given as risk factors for exposure to carcinogens such as benzene and benzo(a)pyrene.

Air quality standards in the UK

[E5066] The most recent version of the Air Quality Strategy (see E5069), and the Air Quality Standards Regulations 2007 (SI 2007 No 64) (and similar regulations covering Scotland, Wales and Northern Ireland), incorporate the EU directives on ambient air quality standards for the relevant pollutants.

These are summarised below at E5068.

Summary of the objectives in the 2007 Air Quality Strategy

[E5067]

Pollutant	Concentration	Measured as	Date to be achieved by
Benzene	16.25 µg m^{-3}	running annual mean	Dec 31, 2003
1,3-Butadiene	2.25 µg m^{-3}	running annual mean	Dec 31, 2003
Carbon monoxide	10 mg m^{-3}	maximum daily running 8 hour mean (in Scotland as running 8 hour mean)	Dec 31, 2003
Lead	0.5 µg m^{-3}	annual mean	Dec 31, 2004
	0.25 µg m^{-3}	annual mean	Dec 31, 2008
Nitrogen Dioxide	200 µg m^{-3}	1 hour mean not to be exceeded more than 18 times a year.	Dec 31, 2005
	40 µg m^{-3}	annual mean	Dec 31, 2005
Particles (PM10) (UK)	50 µg m^{-3}	24 hour mean not to be exceeded more than 35 times a year.	Dec 31, 2004
	40 µg m^{-3}	annual mean	Dec 31, 2004
Scotland	50 µg m^{-3}	24 hour mean not to be exceeded more than 7 times a year.	Dec 31 2010
Scotland	18 µg m^{-3}	Annual mean	Dec 31 2010
Particles (PM2.5) Expo-sure Reduc-tion	25 µg m^{-3} (UK except Scotland)	Annual mean	2020
	12 µg m^{-3} (Scotland)	Annual mean	2020
	40 µg m^{-3} (UK rban areas)	Annual mean	2020
Sulphur dioxide	350 µg m^{-3}	1 hour mean not to be exceeded more than 24 times a year.	Dec 31, 2004
	125 µg m^{-3}	24 hour mean not to be exceeded more than 3 times a year.	Dec 31, 2004

Pollutant	Concentration	Measured as	Date to be achieved by
	266 µg m^{-3}	15 minute mean not to be exceeded more than 35 times a year.	Dec 31, 2005
Ozone	100 µg m^{-3}	8 hour mean not to be exceeded more than 10 times a year	Dec 31, 2005
Polycyclic aromatic hy-drocarbons	0.25 µg m^{-3} B[a]P	As annual average	Dec 31, 2010

Note that the UK standards have been expressed as 'objectives', with a target date by which these objectives are attained.

The Air Quality Strategy ('AQS')

[E5068] The AQS for England, Wales, Scotland and Northern Ireland describes the policies by which air quality will be improved. As noted above, it gives objectives for priority pollutants, in terms of achieving compliance with air quality standards. These are now based on the implementation of the EU directives. The strategy describes the current and future concentrations of the pollutants, providing a framework to help everyone to identify what they can do to help reduce emissions.

The AQS fulfils a requirement under the *Environment Act 1995*, setting out policies for managing ambient air quality. The AQS identifies the need to work at international, national and local level to achieve the objectives. This strategy is essential to anybody interested in achieving the objectives of the directives, not just those who are responsible for them. The strategy has five guiding principles:

- the best practicable protection to human health and the environment;
- the expert panel on AQS recommendations should be the basis for objectives except where an objective derives from an air quality daughter directive limit value based on WHO guidelines;
- it should allow the UK to comply with EU daughter directives, but allow for stricter national objectives for some pollutants;
- objectives should reflect the practicability of the measures needed to reduce pollutants, their costs and benefits and other social and economic factors; and
- objectives should take into account EU legislation and scientific advances.

Local air quality management

[E5069] Local Authorities have a major role to play in delivering cleaner air. They are responsible for land-use planning and traffic management, and for controlling industrial pollution sources. Local Authorities have a long history of local environmental control. They have responsibilities in controlling:

- industrial pollution;
- local pollution hotspots under Local Air Quality Management ('LAQM'); and
- pollution from domestic sources.

In England and Wales local authorities control certain industrial processes, in Scotland SEPA, and in Northern Ireland the Industrial Pollution and Radio-chemical Inspectorate control the equivalent. Air quality has been given increased consideration more recently when they fulfil their strategic planning and transport roles. This has been a natural progression for Local Authorities.

Local authorities are also required to carry out regular reviews and assess-ments of air quality in their areas against standards and objectives prescribed in The *Air Quality Limit Values Regulations 2003 (SI 2003 No 2121)*. Air quality management areas must be designated where these are not being met. Once a local authority has designated an air quality management area, it must prepare an action plan (within 12 months) setting out how it will use its powers to achieve the air quality objectives and a timetable for action.

The LAQM action plans are the major tool for tackling pollution hotspots often caused by road and transport. The LAQM regime is currently being reviewed.

Climate change

The EU Emissions Trading Directive (2003/87/EC) established a scheme to allow greenhouse gas emissions trading. The EU Emissions Trading System is a carbon cap and trade scheme aimed at reducing direct emissions from large installations in the power generation sector and heavy industry such as steel production.

In the UK, the directive was transposed by the *Greenhouse Gas Emissions Trading Scheme Regulations 2005 (SI 2005 No 925)*. The directive applies to a limited range of industrial activities involved in (for example) processing energy, producing and processing ferrous metals (eg steel production), glass manufacture-and other activities including the production of pulp from timber, paper and board.

Under the regulations, an installation carrying out an activity listed in the regulations must hold an emissions permit, which are issued by the Environ-ment Agency. A permit contains a number of allowances (to emit a tonne of carbon dioxide equivalent during a specified period). If less is emitted, the remaining allowances can be sold, but if more is emitted, the installation must buy more allowances.

The Climate Change Act 2008 came into force in November 2008 and sets out provisions including:

- A legally binding target of at least an 80 percent cut in greenhouse gas emissions by 2050 and a reduction in emissions of at least 34 percent by 2020 (against a 1990 baseline);
- A carbon budgeting system which caps emissions over five-year periods, with three budgets set at a time, to help the UK stay on track for the 2050 target. The first three carbon budgets run from 2008-12, 2013-17 and 2018-22;

- The creation of the Committee on Climate Change - a new independent, expert body to advise the government on the level of carbon budgets and on where cost-effective savings can be made;
- The inclusion of international aviation and shipping emissions in the Act (or an explanation to Parliament why not) by 31 December 2012;
- Limits on International credits;
- Further measures to reduce emissions, including powers to introduce domestic emissions trading schemes;
- A requirement for the government to report at least every five years on the risks to the UK of climate change, and to publish a programme setting out how these will be addressed. (The Act also introduces powers for the government to require public bodies and statutory undertakers to carry out their own risk assessment and make plans to address those risks); and
- Requirements for the government to issue guidance by 1 October 2009 on the way companies should report their greenhouse gas emissions, to review the contribution reporting could make to emissions reductions by 1 December 2010; and to use powers under the Companies Act to make reporting mandatory (or explain to Parliament why it has not done so by 6 April 2012).

Defra published the guidance for UK businesses and organisations on how to measure and report their greenhouse gas (GHG) emissions on 30 September 2009 (www.defra.gov.uk/environment/business/reporting/pdf/ghg-guidance.pdf).

Employers' Liability Insurance

Phil Grace

Introduction

[E13001] Most employers carrying on business in Great Britain are under a statutory duty to take out insurance against claims for injuries/diseases brought against them by employees. When such a compulsory insurance policy is taken out the insurance company issues the employer with a certificate of insurance, and the employer must keep a copy of this displayed in a prominent position at his workplace, or supply copies in electronic form so that employees can see it. It is a criminal offence to fail to take out such insurance and/or to fail to display a certificate (see E13013–E13015); however, such a failure does not give rise to any civil liability on the part of a company director in England (see E13008). The above duties are contained in the *Employers' Liability (Compulsory Insurance) Act 1969* (referred to hereafter as the '1969 Act') and in the *Employers' Liability (Compulsory Insurance) Regulations 1998 (SI 1998 No 2573)* (referred to hereafter as the '1998 Regulations'). In addition, the requirements of the *1969 Act* extend to offshore installations but do not extend to injuries suffered by employees when carried on or in a vehicle, or entering or getting onto or alighting from a vehicle, where such injury is caused by, or arises out of use, by the employer, of a vehicle on the road. [*SI 1998 No 2573, Reg 9 and Sch 2, para 14*]. Such employees would normally be covered under the *Road Traffic Act 1988, s 145* as amended by the *Motor Vehicles (Compulsory Insurance) Regulations 1992 (SI 1992 No 3036)*. As from 1 July 1994, liability for injury to an employee whilst in a motor vehicle has been that of the employer's motor insurers.

Employees suffering from occupational diseases with long development periods such as those caused by exposure to asbestos face particular difficulties in making claims, especially with regard to tracking down the insurer of their old employer, who may, by now have ceased trading (see E13026).

In November 2005 the Government published its Compensation Bill. It forms part of a wider Government programme aimed at tackling the 'compensation culture' and also improving the system for obtaining compensation for those with a valid claim. The Lord Chancellor has said that the Bill will 'discourage false expectations that compensation is available for any untoward incident'. The first part of the Act came into effect in November 2006 (see E13033).

A key element of the Bill is the move to regulate the 'claims farmers' that is those firms who offer 'no win no fee' type arrangements. Such firms are thought to be fuelling the growing compensation culture in the UK.

But aside from this and one or two other specific changes the Bill has been described as not adding much to existing legislation.

After many years and numerous delays the Government introduced a Bill dealing with corporate killing in 2005. This was in response to public concerns that the majority of manslaughter cases against firms failed eg following various rail crashes. In contrast, prosecutions of small firms were often successful with fines and even custodial sentences being made against directors. The bill became law in 2008 with the passing of the Corporate Manslaughter and Corporate Homicide Act 2007. (see **E13032**).

Purpose of compulsory employers' liability insurance

[E13002] The purpose of compulsory employers' liability insurance is to ensure that employers are covered for any legal liability to pay damages to employees who suffer bodily injury and/or disease during the course of and as a result of employment. It is the liability of the employer towards his employees which has to be covered; there is no question of compulsory insurance extending to employees, since employers are under no statutory or common law duty to insure employees against risk of injury, or even to advise on the desirability of insurance; it is the employer's potential legal liability to employees which must be insured against (see **E13008**). Such liability is normally based on negligence or breach of statutory duty though not necessarily personal negligence on the part of the employer. Moreover, case law suggests that employers' liability is becoming stricter and there is a belief, in some quarters, that employers' liability claims are incapable of being defended, that the burden of proof has been reversed. While that is not strictly true, the task of defending claims is not made easier in those circumstances where the employer is unable to substantiate their contention that they have complied with legislation, supplied safe plant and equipment, trained employees etc. Good documentation is vital for a sound defence.

An employers' liability policy is a legal liability policy. Hence, if there is no legal liability on the part of an employer, the policy will not respond and no money will be paid out by the insurer. Moreover, if the employee's action against the employer cannot succeed, an action for damages cannot be brought against an employer's insurer (see **E13024**).

The policy indemnifies an employer against claims made by employees; an employee as such is not covered since he normally incurs no liability. Where an act of an employee causes injury to another employee and a claim results the employer is held to be liable for the acts of the employee through what is known as vicarious liability (see **E13027**). Employers' liability policies are unusual in that any compensation agreed or awarded by the court is paid direct to the employee rather than to the employer.

Although offering wide cover an employers' liability policy does not provide cover in respect of claims from third parties (eg independent contractors and members of the public). Such liability is covered by a public liability policy which, though advisable, is not compulsory.

This section examines:

—   the general law relating to contracts of insurance (see E13003–E13006);

— the insurer's right of recovery (ie subrogation) (see **E13007**);
— the duty to take out employers' liability insurance (see **E13008–E13012**);
— issue and display of certificates of insurance (see **E13013**);
— penalties (see **E13014, E13015**);
— scope and cover of policy (see **E13016–E13019**);
— 'prohibition' of certain conditions (see **E13020**);
— trade endorsements for certain types of work (see **E13021,**);
— measures of risk and assessment of premium (**E13022**)
— extensions to cover (see **E13023**)
— insolvent employers' liability insurers (see **E13024**);
— Limitation Act (see **E13025**);
— long tail disease (see **E13026**);
— vicarious liability (see **E13027**).
— Making the Market Work (see **E3028**)
— Woolf Reforms (see **E13029**)
— Small Business Performance Indicator (see **E13030**)
— NHS Recoveries (**E13031**)
— Corporate Manslaughter (**E13032**)
— Compensation Act (**E13033**)

General law relating to insurance contracts

[E13003] Insurance is a contract. When a person wishes to insure, for example, himself, his house, his liability towards his employees, valuable personal property or even loss of profits, he (the proposer) fills in a proposal form for insurance, at the same time making certain facts known to the insurer about what is to be insured. On the basis of the information disclosed in the proposal form, the insurer will decide whether to accept the risk and if so at what rate to fix the premium. If the insurer elects to accept the risk, a contract of insurance is then drawn up in the form of an insurance policy. The vast majority of commercial insurance is arranged through a broker. The broker advises the proposer on the nature and type of covers required in addition to advising on the levels of cover required, the Sum Insured for property insurance or Limit of Indemnity for liability insurance. Brokers are regulated by the Financial Services Authority and are required to act in a professional manner. Brokers are responsible for the advice they give and in the event that the advice proves wrong or incorrect they can be sued and thus brokers are required to take out Professional Indemnity insurance.

There is a growth of what is known as direct sales where companies, generally small firms, purchase insurance 'direct' using the telephone or internet.

Incidentally, it seems to matter little whether the negotiations leading up to contract took place between the insured (proposer) and the insurance company or between the insured and a broker, since the broker is often regarded as the agent of one or the other, generally of the proposer (*Newsholme Brothers v Road Transport & General Insurance Co Ltd* [1929] 2 KB 356). However, a lot depends on the facts. If the broker is authorised to complete blank proposal forms, he may well be the agent of the insurer but this is rare. It is common practice for brokers to complete proposal forms and present them to the proposer for signing before return to the insurer.

Extent of duty of disclosure

[E13004] Most commercial contracts are governed by the doctrine of 'caveat emptor', which translates as 'let the buyer beware'. But the law relating to insurance contracts is based on the legal principle of 'uberrima fides'. This translates as 'utmost good faith' and places a higher duty on both parties but especially the proposer. Whilst the proposer can examine the insurance policy the insurer is unable to examine all aspects of the risk being presented. A proposer must disclose to the insurer all material facts within his actual knowledge. This does not extend to disclosure of facts which he could not reasonably be expected to know. 'The duty is a duty to disclose, and you cannot disclose what you do not know. The obligation to disclose, therefore, necessarily depends on the knowledge you possess. This, however, must not be misunderstood. The proposer's opinion of the materiality of that knowledge is of no moment. If a reasonable man would have recognised that the knowledge in question was material to disclose, it is no excuse that you did not recognise it. But the question always is – Was the knowledge you possessed such that you ought to have disclosed it?' (*Joel v Law Union and Crown Insurance* Co [1908] 2 KB 863 per Fletcher Moulton LJ).

The knowledge of those who represent the directing mind and will of a company and who control what it does, eg directors and officers, is likely to be identified as the company's knowledge whether or not those individuals are responsible for arranging the insurance cover in question (*PCW Syndicates v PCW Reinsurers* [1996] 1 All ER 774).

An element of consumer protection, in favour of insureds, was introduced into insurance contracts by the Statement of General Insurance Practice 1986, a form of self-regulation applicable to many but not to all insurers. This has consequences for the duty of disclosure, proposal forms (**E13005**), renewals and claims. In particular, with regard to the last element (claims), an insurer should not refuse to indemnify on the grounds of:

(a)   non-disclosure of a material fact which a policyholder could not reasonably be expected to have disclosed; or

(b)   misrepresentation (unless it is a deliberate non-disclosure of, or negligence regarding a material fact). Innocent misrepresentation is not a ground for avoidance of payment.

The trend towards greater consumer protection in (inter alia) insurance contracts is reflected in the *Unfair Terms in Consumer Contracts Regulations 1999 (SI 1999 No 2083)*. These Regulations apply in the case of 'standard form' (or non-individually negotiated) contracts [*SI 1999 No 2083, Reg 5*]. A contractual term in such a contract shall:

(i)   be regarded as 'unfair' if contrary to the requirement of 'good faith' it 'causes a significant imbalance in the parties' rights and obligations arising under the contract, to the insured's detriment' [*SI 1999 No 2083, Reg 5(1), Sch 2*];

(ii)   always be regarded as having been individually negotiated where it has been drafted in advance and the consumer has not been able to influence the substance of the term [*SI 1999 No 2083, Reg 5(2)*];

(iii)   require a written contract term to be expressed in plain, intelligible language. If there is doubt about the meaning of terminology, a construction in favour of the insured will prevail [*SI 1999 No 2083, Reg 7*]; and

(iv)   where the insurer claims that a term was individually negotiated, he must prove it [*SI 1999 No 2083, Reg 5(4)*].

Complaints (other than ones which are frivolous or vexatious) or which are to be handled by a qualified body [*SI 1999 No 2083, Sch 1*] relating to 'unfair terms' in standard form contracts, are considered by the Director General of Fair Trading, who may prevent their continued use [*SI 1999 No 2083, Reg 12*]. See also **P9056** PRODUCT SAFETY.

Filling in proposal form

[E13005] Generally only failure to make disclosure of relevant facts will allow an insurer subsequently to void the policy and refuse to compensate for the loss. The test of whether a fact was or was not relevant is whether its omission would have influenced a prudent insurer in deciding whether to accept the risk, or at what rate to fix the premium.

The effect or impact of 'uberrima fides' (ie utmost good faith) is significant. If, when filling in a proposal form, a statement made by the proposer is at that time true, but is false in relation to other facts which are not stated, or becomes false before issue of the insurance policy, this entitles the insurer to refuse to indemnify. In *Condogianis v Guardian Assurance Co Ltd* [1921] 2 AC 125 a proposal form for fire cover contained the following question: 'Has proponent ever been a claimant on a fire insurance company in respect of the property now proposed, or any other property? If so, state when and name of company'. The proposer answered 'Yes', '1917', 'Ocean'. This answer was literally true, since he had claimed against the Ocean Insurance Co in respect of a burning car. However, he had failed to say that in 1912 he had made another claim against another insurance company in respect of another burning car. It was held that the answer was not a true one and the policy was, therefore, invalidated.

Loss mitigation

[E13006] There is an implied term in most insurance contracts that the insured will take all reasonable steps to mitigate loss caused by one or more of the insured perils. Thus, in the case of burglary cover of commercial premises, this could extend to provision of security patrols, the fitting of burglar alarm devices and guard dogs. Again, in the case of fire cover, steps to mitigate the extent of the loss on the part of the insured might well extend to the installation of a sprinkler system, regular visits by the local fire and rescue service and/or securing competent advice on storage of products and materials. Such requirements may be made mandatory by means of policy conditions, especially for high risk trades where large losses are expected. Failure to observe these conditions may result in the insurer deciding not to pay the claim. However, such mandatory requirements can not be placed on Employers' Liability policies (see **E13020**).

In the case of employers' liability, reasonable steps would be regarded as compliance with health and safety legislation, the management of risk and the securing of competent advice through the appointment of an accredited safety officer and/or occupational hygienist, either permanently or temporarily. This last aspect is a specific requirement of the *Management of Health and Safety at Work Regulations 1999 (SI 1999 No 3242)*.

Subrogation

[E13007] Subrogation is a legal principle that provides the right for one person to stand in place of another and avail themselves of all the rights and remedies of that other person. Subrogation enables an insurer to make certain that the insured recovers no more than exact replacement of loss (ie indemnity). 'It [the doctrine of subrogation] was introduced in favour of the underwriters, in order to prevent their having to pay more than a full indemnity, not on the ground that the underwriters were sureties, for they are not so always, although their rights are sometimes similar to those of sureties, but in order to prevent the assured recovering more than a full indemnity' (*Castellain v Preston* (1883) 11 QBD 380 per Brett LJ). Subrogation does not extend to personal accident insurance, policies whereby the insured (normally self-employed but who could in reality be any person) is promised a fixed sum in the event of injury or illness (*Bradburn v Great Western Railway Co* (1874) LR 10 Exch 1 where the appellant was injured whilst travelling on a train, owing to the negligence of the respondent. He had earlier bought personal accident insurance to cover him for the possibility of injury on the train. It was held that he was entitled to both damages for negligence *and* insurance moneys payable under the policy (see further COMPENSATION FOR WORK INJURIES/DISEASES C6001)

The right of subrogation does not arise until the insurer has paid the insured in respect of his loss, however, it should be noted that in respect of employer's liability cases payments are made direct to the injured party rather than the insured, the employer. Subrogation has been invoked infrequently in employers' liability cases. In some circumstances an employer may be entitled to an indemnity from employees for example where the act of an employee has resulted in an injury to another employee and the employer has been held to be vicariously liable. However, it is convention that such recoveries are not sought or pursued. (see **E13027**).

Duty of employer to take out and maintain insurance

[E13008] 'Every employer carrying on business in Great Britain shall insure, and maintain insurance against liability for bodily injury or disease sustained by his employees, and arising out of and in the course of their employment in Great Britain in that business' [*Employers' Liability (Compulsory Insurance) Act 1969, s 1(1)*].

Such insurance must be provided by an insurer authorised to transact Employers' Liability insurance under one or more 'approved policies'. An 'approved policy' is a policy of insurance not subject to any conditions or exceptions prohibited by regulations (see **E13020**) [*Employers' Liability (Compulsory Insurance) Act 1969, s 1(3)*]. This now includes insurance with an approved EU insurer [*Insurance Companies (Amendment) Regulations 1992 (SI 1992 No 2890)*].

There is no duty under the *1969 Act* to warn or insure the employee against risks of employment outside Great Britain (*Reid v Rush Tompkins Group plc* [1989] 3 All ER 228) although the *1998 Regulations* require the employer to insure employees employed on or from offshore installations or associated structures – see **E13009**.

The failure to take out Employers' Liability insurance does not give rise to any civil liability on the part of a company director in England. In *Richardson v Pitt-Stanley* [1995] 1 All ER 460 the plaintiff suffered a serious injury to his hand in an accident at work, and obtained judgment against his employer, a limited liability company, for breach of the *Factories Act 1961, s 14(1)* (failure to fence dangerous parts of machinery). Before damages were assessed, the company went into liquidation and there were no assets remaining to satisfy the plaintiff's judgment. The company had also failed to insure against liability for injury sustained by employees in the course of their employment, as required by the *Employers' Liability (Compulsory Insurance) Act 1969, s 1*. *Section 5* of that Act makes failure to insure a criminal offence. The plaintiff then sued the directors and company secretary who, he alleged, had committed an offence under *s 5*, claiming as damages a sum equal to the sum which he would have recovered against the company, had it been properly insured. His action failed. It was held that the *Employers' Liability (Compulsory Insurance) Act 1969* did not create a civil as well as criminal liability. In Scotland however the Sheriff Principal in *Quinn v McGinty* 1999 SLT 27, Sh Ct, on similar facts did not follow the Court of Appeal in *Richardson v Pitt-Stanley*.

Employees covered by the Act

[E13009] Cover is required in respect of liability to employees who:

(a)   are ordinarily resident in Great Britain; or

(b)   though not ordinarily resident in Great Britain, are present in Great Britain in the course of employment here for a continuous period of not less than 14 days; or

(c)   though not ordinarily resident in the United Kingdom, have been employed on or from an offshore installation or associated structure for a continuous period of not less than seven days.

[*Employers' Liability (Compulsory Insurance) Regulations 1998 (SI 1998 No 2573), Reg 1(2)*].

Employees not covered by the Act

[E13010] An employer is not required to insure against liability to family members, that is employees who are:

(a)   a spouse;
(b)   father;
(c)   mother;
(d)   son;
(e)   daughter;
(f)   other close relative.

[*Employers' Liability (Compulsory Insurance) Act 1969, s 2(2)(a)*].

Those who are not ordinarily resident in the UK are not covered by the Act except as above. Nor are employees working abroad covered. Such employees can sue under English law in limited circumstances (*Johnson v Coventry Churchill International Ltd* [1992] 3 All ER 14 where an employee, working in Germany for an English manpower leasing company, was injured when he fell through a rotten plank. He was unable to sue his employer under German law; although he was working in Germany. A claim was thus made in England against the English firm on the basis that they were, in essence, the employer. Legal precedents stated that a claim of this type could not be made unless the case was of a type or nature that was covered by both overseas and English law. However, it was held this was an exception to the general principle, that England was the country with the most significant relationship with the claim because the injured person had made the contract in England; the firm had paid him, making deductions for National Insurance and held control over where and when he worked. The injured person was thus entitled to expect the firm to compensate him through their insurers for any personal injury sustained in Germany.

Degree of cover necessary

[E13011] The amount for which an employer is required to insure and maintain insurance is £5 million in respect of claims relating to any one or more of his employees, arising out of any one occurrence [*Employers' Liability (Compulsory Insurance) Regulations 1998 (SI 1998 No 2573), Reg 3(1)*].

Between 1 January 1972 (when the *Employer's Liability (Compulsory Insurance) Act 1969* came into force) and 1994, insurers provided unlimited cover under employers' liability policies. As from 1 January 1995, as a result of payments made in respect of claims exceeding the amount of premiums received during the period 1989–1993 across the entire insurance market, unlimited liability was withdrawn, but most insurers continued to offer a minimum of £10 million indemnity for onshore work. A consultative document issued by the Department of the Environment, Transport and the Regions entitled The Draft Employers' Liability (Compulsory Insurance) General Regulations [C4857 September 1997] (hereafter referred to as 'the 1997 consultative document') which preceded the *1998 Regulations* assumed that this practice would continue.

Where an employing company has subsidiaries, there will be sufficient compliance if they insure/maintain insurance for themselves *and* on behalf of their subsidiaries for £5 million in respect of claims affecting any one or more of its own employees and any one or more employees of its subsidiaries arising out of any one occurrence [*Employers' Liability (Compulsory Insurance) Regulations 1998 (SI 1998 No 2573), Reg 3*].

Insurers and the courts have interpreted the legislation to mean that all injuries resulting from one incident (eg an explosion) are treated as one occurrence, and each individual case of gradually occurring injury or disease is treated as an individual occurrence – the only exception being a situation where a sudden and immediate outbreak of a disease amongst the workforce is clearly attributable to an identifiable incident (eg the escape of a biological agent). The introduction to the 1997 consultative document suggested that this interpretation might be challenged and set out a possible alternative regulation to be used instead of what is now *Reg 3* of the *1998 Regulations* if clarification was felt necessary. This was not adopted and no new regulations are currently proposed and therefore it would appear that the Government are now satisfied that the position is clear.

Exempted employers

[E13012] The following employers are exempt from the duty to take out and maintain insurance:

(a)   nationalised industries;

(b)   any body holding a Government department certificate that any claim which it cannot pay itself will be paid out of moneys provided by Parliament;

(c)   any Passenger Transport Executive and its subsidiaries, London Regional Transport and its subsidiaries;

(d)   statutory water undertakers and certain water boards;

(e)   the Commission for the New Towns;

(f)   health service bodies, National Health Service Trusts;

(g)   probation and after-care committees, magistrates' court committees, and any voluntary management committee of an approved bail or approved probation hostel;

(h)   governments of foreign states or commonwealth countries and some other specialised employers;

(i)   Network Rail and its subsidiaries (the exemption ceasing when it is no longer owned by the Crown);

(j)   the Qualifications & Curriculum Authority.

There are other types of employer specified in the Regulations, but these are the main exceptions [*Employers' Liability Compulsory Insurance Act 1969, s 3; Employers' Liability (Compulsory Insurance) Regulations 1998 (SI 1998 No 2573), Sch 2*].

The above situation was changed by the enactment of amending legislation in 2004 [*Employers' Liability (Compulsory Insurance) (Amendment) Regulations 2004 (SI 2004 No 2882)*]. With effect from 1 February 2005 certain small firms, known as sole-employee incorporated companies (SEICs), are exempt from the requirement to take out Employers' Liability insurance. An SEIC is defined as a company with one employee with that employee owning

51% or more of the issued share capital. This step has been taken to reduce the burden of unnecessary costs on small businesses. Such firms are not barred from taking out insurance if they wish to and if they were to take on extra employees, for example temporary staff to enable the company to undertake larger contracts, then insurance would be required.

Issue, display and retention of certificates of insurance

[E13013] The insurer must issue the employer with a certificate of insurance, which must be issued not later than 30 days after the date on which insurance was commenced or renewed [*Employers' Liability (Compulsory Insurance) Act 1969, s 4(1); Employers' Liability (Compulsory Insurance) Regulations 1998 (SI 1998 No 2573), Reg 4*]. Where there are one or more contracts of insurance which jointly provide insurance cover of not less than £5 million, the certificate issued by any individual insurer must specify both the amount in excess of which insurance cover is provided by the individual policy, and the maximum amount of that cover [*Employers' Liability (Compulsory Insurance) Regulations 1998 (SI 1998 No 2573), Reg 4(3)*].

A copy or copies of the certificate must be displayed at each place of business where there are any employees entitled to be covered by the insurance policy and the copy certificate(s) must be placed where employees can easily see and read it and be reasonably protected from being defaced or damaged [*Employers' Liability (Compulsory Insurance) Regulations 1998 (SI 1998 No 2573), Reg 5*]. The exception is where an employee is employed on or from an offshore installation or associated structure, when the employer must produce, at the request of that employee and within ten days from such request, a copy of the certificate [*Employers' Liability (Compulsory Insurance) Regulations 1998 (SI 1998 No 2573), Reg 5(4)*].

The situation changed in 2008 with the passing of *Employers' Liability (Compulsory Insurance) (Amendment) Regulations 2008 (SI 2008 No 1765)*. These Regulations made minor changes to the duties of employers. It is now possible for employers to make copies of their Certificate available to employees in an electronic format (amendment of Regulation 5). Whilst this may well prove of interest to those employers whose workforce is desk based or has ready access to computers, company intranets etc it is most likely that the majority of employees will find that posting a copy on notice boards remains the most economic and practicable approach.

An employer must, if a notice has been served on him by the Health and Safety Executive, produce a copy of the policy to the officers specified in the notice and he must permit inspection of the policy by an inspector authorised by the Secretary of State to inspect the policy [*Employers' Liability (Compulsory Insurance) Regulations 1998 (SI 1998 No 2573), Regs 7, 8*].

A change introduced by the *1998 Regulations* is that employers are now required by law to retain any certificate of employers' liability insurance (or a copy) for a period of 40 years beginning on the date on which the insurance to which it relates commences or is renewed [*Employers' Liability (Compulsory Insurance) Regulations 1998 (SI 1998 No 2573), Reg 4(4)*]. Companies may retain the copy in any eye-readable form in any one of the ways authorised by the *Companies Act 1985, ss 722 and 723 [Employers' Liability (Compulsory Insurance) Regulations 1998 (SI 1998 No 2573), Reg 4(5)]*.

This requirement has been removed from the Regulations (deletion of paragraphs (4) and (5) from Regulation 4) as part of a reduction in the administrative burden on business. However, it is strongly recommended that employers continue to retain their EL insurance certificates despite the removal of a specific legislative requirement to do so.

Penalties

Failure to insure or maintain insurance

[E13014] Failure by an employer to effect and maintain insurance for any day on which it is required is a criminal offence, carrying a maximum penalty on conviction of £2,500 [*Criminal Justice Act 1982, s 37(2)*].

It should be noted that prosecution for failure to insure is rare – just one case in 2006 and another in 2007, attracting fines of £1,000 and £500 respectively. The small number of prosecutions may be a result of the fact that the lack of insurance only becomes known during investigation into other, more serious matters, for example following an accident and where there are breaches of other regulations. In addition, breach of the legislation relating to compulsory insurance is a summary offence and a charge must laid within a period of 6 months. This is difficult to achieve since an employer must be given time to produce a Certificate and this can delay verification that there is no insurance in place.

Failure to display a certificate of insurance

[E13015] Failure on the part of an employer to display a certificate of insurance in a prominent position in the workplace is a criminal offence, carrying a maximum penalty on conviction of £1,000 [*Criminal Justice Act 1982, s 37(2)*].

In the 1997 consultative document, the Government suggests that penalties should be increased to become the same as those under the *Health and Safety at Work etc Act 1974 (HSWA 1974)*, ie fines of £20,000 in a magistrates' court and unlimited in the Crown Court. To implement this change will require primary legislation and currently it would appear that the Government has no plans to introduce such legislation.

Cover provided by a typical policy

Persons

[E13016] Cover is limited to protection of employees. Independent contractors are not covered; liability to them should be covered by a public liability policy. Directors who are employed under a contract of employment are covered, but directors paid by fees who do not work full-time in the business are generally not regarded as 'employees'. Liability to them would normally be covered by a public liability policy. Similarly, since the judicial tendency is to construe 'labour-only' subcontractors in the construction industry as 'employees' (see CONSTRUCTION AND BUILDING OPERATIONS), employers' liability policies often contain an endorsement that will state: 'An employee shall also mean any labour master, and persons supplied by him, any person employed by labour-only subcontractors, any self-employed person, or any person hired from any public authority, company, firm or individual, while working for the insured in connection with the business'. The public liability policy should then be amended to exclude the insured's liability to 'employees' so designated.

It is commonplace, especially for smaller firms, for the same insurer to provide both the Employers' and Public Liability insurance and in such cases the wordings will be aligned. Current market practice is that few insurers provide Employers' Liability cover without also holding the corresponding Public Liability insurance. However, whenever different insurers are used to provide the two covers it will be necessary for the policyholder or their broker to check that the policy wordings are suitably aligned.

It is commonplace, especially for smaller firms, for the same insurer to provide both the Employers' and Public Liability insurance and in such cases the wordings will be aligned. However, whenever different insurers are used to provide the two covers it will be necessary for the policyholder or their broker to check that the policy wordings are suitably aligned.

In the 1997 consultative document, the Government points out that the issue as to what constitutes 'an employee' cannot be completely resolved without primary legislation, but proposes to issue guidance on interpretation, although it was proposed that guidance on interpretation would be issued. Such guidance has not been issued and it must be assumed that the Government remains satisfied that the position is clear.

The definition of employee will usually be extended to include children of school age on Work Experience and young people employed under a Youth Training Scheme or similar. Persons who have been borrowed or hired by the company are also included. If the firm employs outworkers or homeworkers they are usually counted as employees.

The Health and Safety Executive have issued guidance to the effect that voluntary helpers or workers, eg persons who work in charity shops or on heritage railway lines, should be treated as employees. Most Employers' Liability policies include voluntary workers within the definition of employee.

The policy provides cover for the business, ie the declared trade of business of the company, but will in addition cover a number of other activities:

—   repair and maintenance of buildings occupied by the insured;
—   repair and maintenance of vehicles owned by the insured;
—   provision of canteen and welfare services including sports and social clubs;
—   private work for directors undertaken by employees with the consent of the insured.

A recent case has considered the situation regarding agency workers. There has been much debate in the recent past about the status of agency workers and in particular the question "Who is their employer?" since it may be either the agency or the employer on whose premises they are working that might be liable. There have been arguments presented for regarding both 'employers' as being responsible for the agency worker and this was considered in the case of *Brook Street Bureau [UK] Ltd v Dacas* [2004] EWCA Civ 217, [2004] ICR 1437. This case concerned a cleaner who had worked for a local authority for over a year, being placed with them by the employment agency Brook Street Bureau. Leaving aside the somewhat complex details of the case – it concerned employment rights rather than personal injury – the Court of Appeal decided that in view of the fact that she had been employed by the local authority for more than one year she was, in essence, their employee. This was contrary to previous case law and might change the view of some employers with respect to agency labour.

The more recent case of *James v London Borough of Greenwich Council* [2008] EWCA Civ 35, [2008] ICR 545concerned an allegation of unfair dismissal. It was heard at an Employment Tribunal and an Employment Appeal Tribunal before going to the Court of Appeal. The finding was that there was no contract, explicit or implied. This finding contradicts previous case law which had suggested that the duration of a placement, exclusivity of services and control and supervision of agency workers were all factors suggesting that the end user (the firm or business using the agency worker) may be an implied employer, even in the absence of a contract. This case clarifies that an agency-supplied and contracted worker is **not** deemed to be the employee of the end-user, and a contract of employment could **not** be implied.

In brief, the claimant was employed by the Council as an asylum support worker. She began working again for them through an agency, from Sept 2001, and in 2003 moved to another agency which paid better. The claimant went off sick in 2004 and the agency provided another worker in her absence. When she returned, she was told she was no longer required as the agency had replaced her. The claimant tried to claim unfair dismissal from the council.

There was no express contract between the claimant and the council (ie end-user) but there was a contract between an agency and the worker and a separate contract between the end-user and the agency. However there was NO contract between the worker and the end-user and a contract was NOT implied to exist between the worker and the end-user.

Cases such as these will be judged on their own facts and concern the law of employment rather than personal injury and the law of negligence. The decision in this case is unlikely to be relevant in personal injury claims. There are well established tests that are used to establish a "master and servant" relationship which do not necessarily rely on the existence of a contract.

Scope of cover

[E13017] The policy provides for payment of:

(a) compensation for injury or disease, whether agreed or negotiated or awarded by a court

(b) costs and expenses of litigation, incurred with the insurer's consent, in defence of a claim against the insured (ie civil liability);

(c) claimants costs and expenses for which the insured is legally liable;

(d) solicitor's fees, and other legal costs incurred with the insurer's consent, for representation of the insured at proceedings in any court of summary jurisdiction (ie magistrates' court), coroner's inquest, or a fatal accident inquiry (ie criminal proceedings), arising out of an accident resulting in injury to an employee; it does *not* cover payment of a fine imposed by a criminal court;

(e) solicitor's fees, and other legal costs incurred with the insurer's consent, for representation of the insured at proceedings resulting from a prosecution for a breach of health and safety and related legislation, that is a breach of statutory duty; this is subject to certain provisos for example the breach must not have resulted from a deliberate act or omission by the insured and no indemnity will be paid if there is a legal expenses policy in force;

(f) compensation for any director, partner or employee attending court as a witness in a case for which the insurer is providing an indemnity.

The situation is changing with many insurers amending the wordings following the introduction of the Corporate Manslaughter legislation (see **E13032.**)

The policy may contain an excess negotiated between the insurer and employer, ie a provision that the employer pay the first £x of any claim. For the purposes of the *1969 Act*, any condition in a contract of insurance which requires a relevant employee to pay, or an insured employer to pay the relevant employee, the first amount of any claim or any aggregation of claims, is prohibited. Agreements which provide that the insurer will pay the claim in full and may then seek some reimbursement from the employer are permitted [*Employers' Liability (Compulsory Insurance) Regulations 1998 (SI 1998 No 2573), Reg 2*].

A contribution towards a claim is never sought from an employee; however, the amount of damages agreed or awarded may be reduced to reflect a degree of contributory negligence. If it is found that the employee was in some way to blame for the accident, for example they deliberately removed a guard, or failed to follow their training the damages will almost certainly be reduced.

The policy conditions require that as soon as a claim is received the insured, that is the employer, must notify their insurer as soon as possible. The insured must NOT make any admission of liability or offer or promise payment without the 'prior written consent' of the insurer. The insurer will take over the handling of the claim, communicating with the claimant's solicitors, investigating the circumstances and deciding on liability. The insured must supply the insurer with such particulars and information as may be required or requested. Any subsequent correspondence received by the insured must be forwarded to the insurer without delay.

The handling of claims has been dramatically influenced by the Woolf Reforms. These introduced changes that have considerably changed the way in which personal injury claims are handled by insurers (see **E13029**).

If the circumstances are clear cut an offer will be made and settlement negotiated. If a settlement can not be agreed, if there is a dispute on facts or there is no legal precedent to follow then the claim may proceed to court. A judge will then be asked to decide on liability, the amount of damages that should be awarded or both. A very small percentage of claims go to court, in general less than 10%.

Geographical limits

[**E13018**]–[**E13019**] Cover is normally limited to Great Britain, Northern Ireland, the Channel Islands and the Isle of Man, in respect of employees normally resident in any of the above, who sustain injury whilst working in those areas. Cover is also provided for such employees who are injured whilst temporarily working abroad, but many policies limit this cover to periods not exceeding six months in any one year. This cover is usually granted on the basis that the action for damages is brought in a court of law of Great Britain, Northern Ireland, the Channel Islands or the Isle of Man – though even this proviso is omitted from some policies.

Employers must also have employers' liability insurance in respect of employees who, though not ordinarily resident in the United Kingdom, have been employed on or from an offshore installation or associated structure for a continuous period of not less than seven days; or who, though not ordinarily resident in Great Britain, are present in Great Britain in the course of employment for not less than fourteen days [*Employers' Liability (Compulsory Insurance) Regulations 1998 (SI 1998 No 2573), Reg 1(2)*].

Conditions which must be satisfied

(a)   Cover only relates to bodily injury or disease; it does not extend to employee's property. This latter cover is provided by an employers' public liability policy. The only exception to this situation is where an employee's property is damaged in an event that gives rise to a claim under the policy. For example an employee could claim for clothing damaged in an incident that resulted in injury and if the employer's negligence was proven the employee would be entitled to compensation both for any injuries received and any damage to clothing or property. Such situations are rare and the property damage costs are generally far outweighed by any compensation for injury received.

(b)   Injury must arise out of and during the course of employment (see **E11004** EMPLOYERS' DUTIES TO THEIR EMPLOYEES). If injury does not so arise, cover is normally provided by a public liability policy.

(c)   Bodily injury must be caused during the period of insurance. Normally with injury-causing accidents there is no problem, since injury follows on from the accident almost immediately. Certain occupational diseases take many years to develop and may not manifest themselves until much later, eg asbestosis, mesothelioma, pneumoconiosis or deafness. Legal liability arises when the disease was caused ie when the exposure to the harmful substance or noise took place. Moreover, at least as far as occupational deafness is concerned, liability between employers can be apportioned, giving rise to contribution between insurers. This can be illustrated as follows:

—   20 years of employment spread across four employers A, B, C and D

—   15 years of negligent noise exposure involving employers A, B and C

—   the compensation would be awarded against these three employers in proportion to the length of time the claimant was employed by each

—   if there were five insurers who had variously insured the three employers A, B and C over the 15 years of negligent exposure they would contribute to the compensation in proportion to the length of time they had been on cover.

(see further NOISE AT WORK and VIBRATION) (see E13024)

(d)   Claims must be notified by the insured to the insurer as soon as possible, or as stipulated by the policy (see E13017).

'Prohibition' of certain conditions

[E13020] All liability policies contain conditions with which the insured must comply if the insurer is to 'progress' his claim, eg notification of claims. Failure to comply with such condition(s) could jeopardise cover under the policy: the insured would be legally liable but without insurance protection. In the case of an employers' liability policy, an insurer is not entitled to avoid liability under the policy if the condition requiring the insured to take reasonable care to prevent injuries to employees was not complied with.

The insurer is not entitled to insist on conditions requiring compliance with the provisions of any relevant statutes/statutory instruments (eg *HSWA 1974; Ionising Radiations Regulations 1999 (SI 1999 No 3232)*), or to keep records, although the keeping of records is vital to enable the insurer to mount a reasonable defence to a claim.

The object of the 1969 Act was to ensure that an employer who had a claim brought against him would be able to pay the employee any damages awarded. Regulations made under the Act, therefore, seek to prevent insurers from avoiding their liability by relying on breach of a policy condition, by way of 'prohibiting' certain conditions in policies taken out under the Act. More particularly, insurers cannot avoid liability in the following circumstances.

(a)   Some specified thing being done or being omitted to be done after the happening of the event giving rise to a claim (eg omission to notify the insurer of a claim within a stipulated time) [*Employers' Liability (Compulsory Insurance) Regulations 1998 (SI 1998 No 2573), Reg 2(1)(a)*].

(b)   Failure on the part of the policy-holder to take reasonable care to protect his employees against the risk of bodily injury or disease in the course of employment [*Employers' Liability (Compulsory Insurance) Regulations 1998 (SI 1998 No 2573, Reg 2(1)(b)*]. As to the meaning of 'reasonable care' or 'reasonable precaution' here, 'It is eminently reasonable for employers to entrust . . . tasks to a skilled and trusted foreman on whose competence they have every reason to rely' (*Woolfall and Rimmer Ltd v Moyle and Another* [1941] 3 All ER 304). The prohibition is therefore not broken by a negligent act on the part of a competent foreman selected by the employer. Where, however, an employer acted wilfully (in causing injury) and not merely negligently (though this would be rare), the insurer could presumably refuse to pay (*Hartley v Provincial Insurance Co Ltd* [1957] Lloyd's Rep 121, where the insured employer had not taken steps to ensure that a stockbar was securely fenced for the purposes of the *Factories Act 1937, s 14(3)* in spite of repeated warnings from the factory inspector, with the result that an employee was scalped whilst working at a lathe. It was held that the insurer was justified in refusing to indemnify the employer who was in breach of statutory duty and so liable for damages). This was confirmed in *Aluminium Wire and Cable Co Ltd v Allstate Insurance Co Ltd* [1985] 2 Lloyd's Rep 280.

(c)   Failure on the part of the policy-holder to comply with statutory requirements for the protection of employees against the risk of injury [*Employers' Liability (Compulsory Insurance) Regulations 1998 (SI 1998 No 2573, Reg 2(1)(c)*] – the reasoning in *Hartley v Provincial Insurance Co Ltd* (see head (*b*) above), that wilful breach may not be covered, probably applies here too.

(d)   Failure on the part of the policy-holder to keep specified records and make such information available to the insurer [*Employers' Liability (Compulsory Insurance) Regulations 1998 (SI 1998 No 2573, Reg 2(1)(d)*] (eg accident book or accounts relating to employees' wages and salaries (see ACCIDENT REPORTING)).

(e)   By means of the use of an excess in policies [*Employers' Liability (Compulsory Insurance) Regulations 1998 (SI 1998 No 2573, Reg 2(2)*] – see **E13017**.

Trade endorsements for certain types of work

[E13021] There are no policy exceptions to the standard employers' liability cover. The *Employers' Liability (Compulsory Insurance) Regulations 1998 (SI 1998 No 2573), Reg 2* does not prevent insurers, ie underwriters, from applying certain conditions in connection with intrinsically hazardous work; for instance, exclusion of liability for accidents arising out of demolition work, or in connection with the use of explosives. Insurers are also permitted to make decisions about which types of risks, that is trades, they wish to insure. There is a legal requirement for employers to take out insurance but no corresponding legal requirement for insurers to offer the cover. Thus it is possible for an insurer to underwrite, ie offer to provide insurance, for retail risks but not construction risks or for builders but not demolition firms. Trade endorsements are used frequently in underwriting employers' liability risks, and there is nothing in the 1969 Act to prevent insurers from applying their normal underwriting principles and applying trade endorsements where they consider it necessary, ie they will amend their standard policy form to exclude certain risks. Thus, there may be specific exclusions of liability arising out of types of work, such as demolition on a policy for a general builder, or the use of mechanically driven woodworking machinery, or work above certain heights for a window cleaner. This does not mean that there are circumstances where an employee will not obtain compensation from his employer since they are required to obtain employer's liability insurance cover for such activities by means of another, additional policy taken out with another insurer. In practice such conditions are usually used to prevent or restrict an employer from developing their business into such areas without informing the insurer and paying the appropriate premium.

Measure of risk and assessment of premium

[E13022] Certain trades or businesses are known to be more dangerous than others. For most trades or businesses insurers have their own rate for the risk, expressed as a rate per cent on wages (other than for clerical, managerial or non-manual employees for whom a very low rate applies). This is known as the book rate and is derived from the insurer's experience ie the balance of premium against claims for all risks of that type. For small firms this rate is used as a guide and is altered upwards or downwards depending on:

(a)   previous history of claims and cost of settlement;
(b)   size of wage roll;
(c)   whether certain risks are not to be covered, eg the premium will be lower if the insured elects to exclude from the policy certain risks, such as the use of power driven woodworking machinery;
(d)   the insured's attitude towards safety.

For very small firms a minimum premium may be set.

For large firms that may have several claims each year it is possible to calculate the coming year's premium by basing it directly on the cost of claims in previous years, generally the preceding five years. The premium will be adjusted to take into account the firm's efforts to reduce risk, train staff, complete risk assessments etc. This is known as experience rating.

Many insurers survey premises with the object of improving the risk and minimising the incidence of accidents and diseases. This is an essential part of their service, and the insurer's risk surveyors generally work in conjunction with the insured's own management and/or safety staff. However, it should be noted that this practice is restricted to the largest of risks and/or those with the greatest risk potential since insurers employ limited numbers of risk surveyors.

Related covers

[E13023] In addition to employers' liability insurance, it is becoming increasingly common for companies to buy insurance in respect of directors' personal liability. Indeed, in the United States, some directors refuse to take up appointments in the absence of such insurance being forthcoming.

There is a growing trend to purchase legal expenses cover. This is primarily intended to cover the expenses incurred in defending employment tribunal claims such as arise from employment disputes eg relating to dismissal or to discrimination on the grounds of sex, gender or disability. (see **E13032**, **C9001**)

It should be noted that almost all liability insurance policies, both Employers' and Public, include an element of expenses cover. The costs and expenses of defending civil claims for compensation are always covered. If the claimant is successful, the policy will cover both the damages and the claimant's legal costs. However, there will usually be cover for the costs and expenses involved in the defence of criminal prosecutions arising from alleged breaches of health and safety and related legislation. Such prosecutions almost always take place before civil claims for compensation are settled. Thus insurers are interested in arranging for the defence of such claims since it provides early access to the facts of the accident, as determined by the enforcing authorities and enables an early view of potential liabilities to be formed.

Insolvent employers' liability insurers

[E13024] If the employee's action against the employer cannot succeed, the action for damages cannot be brought against an employer's insurer (*Bradley v Eagle Star Insurance Co Ltd* [1989] 1 All ER 961 where the employer company had been wound up and dissolved before the employer's liability to the injured employee had been established) (see below for the transfer of an employer's indemnity policy to an employee). The effect of this decision has been reversed by the *Companies Act 1989, s 141*, amending the *Companies Act 1985 (CA 1985), s 651* which allows the revival of a dissolved company within two years of its dissolution for the purpose of legal claims and, in personal injuries cases, the revival can take place at any time subject to the existing limitation of action rules contained in the *Limitation Act 1980*. For example, in the case of *Re Workvale Ltd (No 2)* [1992] 2 All ER 627, the court exercised its discretion under *section 33* of the *Limitation Act 1980* to allow a personal injuries claim to proceed after the three-year limitation period had expired. This meant that the company could also be revived under the provisions of *CA 1985, s 651(5)* and *(6)* (as amended). Thus, proceedings under the *Third Parties (Rights Against Insurers) Act 1930, s 1(1)(b)* may be brought in this manner.

If the employer becomes bankrupt or if a company becomes insolvent, the employer's right to an indemnity from his insurers is transferred to the employee who may then keep the sums recovered with priority to his employer's creditors. This is only so if the employer has made his claim to this indemnity by trial, arbitration or agreement before he is made bankrupt or insolvent (*Bradley v Eagle Star Insurance Co Ltd* [1989] 1 All ER 961). The employee must also claim within the statutory limitation period from the date of his injury (see **E13007** for subrogation rights generally).

The problems that arise following the insolvency of an employer's liability insurer gained particular prominence following the much publicised liquidation of Chester Street Holdings Limited ('Chester Street') on 9 January 2001. It was initially feared that individuals whose claims pre-dated 1972 (the date when employers' liability insurance became compulsory) would be at risk of not receiving compensation. The situation has, however, largely been clarified following the establishment of the Financial Services Compensation Scheme ('the FSCS') which was created under the *Financial Services and Markets Act 2000* and acts as a 'safety net' for customers of finance sector companies who are unable to play claims against authorised companies. The FSCS came into effect on 1 December 2001. Under the FSCS policy holders (save for any period of cover when the policy holder was a nationalised industry which is specifically excluded from the provisions of the scheme) are eligible for protection if they are insured by an authorised insurance company under a contract of insurance issued in the UK, Channel Islands or Isle of Man. The FSCS scheme pays 100 per cent compensation for post-1972 compulsory employers' liability insurance claims. If the employer still exists and is solvent then the employer initially pays the compensation to the claimant although is able to recoup 100 per cent of the compensation outlay from the FSCS. The FSCS pays 90 per cent compensation for pre-1972 claims providing that the employer is not still in existence and solvent. If however the employer is still in existence – the employer must meet its liability to the claimant and is not compensated by the FSCS.

Special provisions apply to Chester Street for pre-1972 claims where:

- liability was established and quantified before insolvency on 9 January 2001, 90 per cent compensation is payable. Such claims are administered under the *Financial Services and Markets Act 2000* using the guidelines established by the Policyholders Protection Board;
- liability was agreed and quantified between 9 January and 30 November 2001. Such claims are subject to the requirements of a scheme set up by the Association of British Insurers who pay 90 per cent of the award.

Responsibility for claims handled by the Policyholder Protection Board (PPB) was transferred to the FSCS on 1 December 2001. PPB rules continue to apply to any claim arising from the insolvency of an insurance company where it occurred before 1 December 2001.

Further information can be obtained from FSCS, 7th Floor, Lloyds Chambers, 1 Portsoken Street, London E1 8BN, tel: 020 7892 7301.

Limitation Act

[E13025] Potential claimants have always been subject to a limitation on the time period within which they must bring a claim. Limitation is a practical approach to avoid the courts having to deal with a claim long after the event, when memories have faded, details have become uncertain and witnesses unreliable. Limitation is a defence against claims made after a lengthy delay and will be raised on behalf of the defendant, by the insurer acting on behalf of the employer. The law on limitation is contained in the *Limitation Act 1980*. There are many different periods dependent on the nature of the claim, a claim for personal injury through negligence, a claim under contract law etc but the main principle for Employers' Liability insurance is as follows.

An injured person has three years in which to make a claim for damages. This period runs from what is known as the 'date of accrual', usually the date on which the injury occurred. The date of accrual may also be the 'date of knowledge' which is taken to be the date on which the injured person realised they had a right to claim.

Where a specific injury has resulted, such as a broken bone, amputation of part of the body, a burn or serious flesh injury leading to scarring, there is no dispute. The three year period runs from the date of the accident. When a person is found to be suffering from a disease that might have an occupational cause the date of knowledge is usually regarded as being the date when a medical authority such as their doctor or specialist explains that their condition might have been caused by exposure to harmful substances encountered in the workplace.

There is a special case where the potential claimant has died. This may occur with mesothelioma which can result in death within a short period of time following diagnosis. The limitation period is three years from the date when the deceased might have been regarded as having knowledge. However, if the death occurs within the three year period the dependents have three years from the date of death or from the date of their knowledge.

Claims are usually initiated within a reasonable period after the accident but are sometime delayed and only reach insurers towards the very end of the three year period.

Long Tail Disease

[E13026] Employers' Liability policies provide cover in the event that an employee alleges that their disease was caused by exposure to harmful substances or situations in the working environment. Typically claims will arise from exposures to:

— asbestos giving rise to asbestosis, lung cancer or mesothelioma;
— harmful substances resulting in cancer, asthma etc;
— noise giving rise to noise induced deafness and/or tinnitus;
— vibration giving rise to hand arm vibration syndrome (vibration white finger).

Some diseases arise quickly after exposure and can be regarded as injuries, for example dermatitis, however some develop gradually over time eg deafness. There are others that have very long development periods or exhibit latency, not appearing for decades after exposure, mesothelioma which can take anything from 15 to 35 years to manifest itself.

Since the policy cover relates to the time when the harmful exposure took place the employee or ex-employee should make their claim against the employer they think is responsible and that employer will then have to identify the insurer for the relevant period. However, it is common for the claim to be made against the current employer (or the employer from which the employee retired) and that employer's insurer will undertake the task of determining which insurer is responsible. The task of determining which insurers are responsible is made especially difficult for diseases such as noise which develop over a period of time as the employer may have changed insurer and the employee may have worked for several employers and been exposed to noise at some or all of them. Once liability is established and all responsible, ie liable, employers and insurers have been identified the cost of the claim will be apportioned between them according to the degree of liability agreed.

Employees suffering from occupational diseases with long development periods such as mesothelioma, have been helped to track down their employer's insurers thanks to a Code of Practice issued in November 1999. The Code was drawn up by the Department of the Environment, Transport and the Regions, the Association of British Insurers, and the Non-Marine Association at Lloyds, and sets out the procedures and standards of service required of insurers. The Code does not have statutory authority and is a voluntary code that commits member insurers to make a thorough search when requested of all records which exist. The Code obliges insurers to keep records of all policies issued for 60 years. There is no charge levied for conducting a search.

Copies of the Code and enquiry forms can be obtained from the Association of British Insurers, Employers Liability Enquiry Unit, 51 Gresham Street, London EC2V 7HQ, tel: 020 7216 7656; fax: 020 7696 8999.

Discussions are taking place between the Government and the insurance industry regarding the setting up of an Employers' Liability Insurance "database". There exists within the motor insurance market a database that holds details of all motor insurance policies, the details of the vehicle and policyholder. This performs several functions, including allowing the police to establish whether a vehicle is insured and also assists in the tracing of the insured party after accidents in which a vehicle has not stopped.

It is proposed that the creation of such a database for employers' liability would assist in the tracing of policyholders when a claimant seeks compensation for an occupational disease. Whilst this view has some merits there are considerable hurdles to be overcome before the full value of such a database could be realised.

In the first instance the creation of a database from "now" would only have real value or benefit when those who are currently exposed to potentially harmful substance seek compensation at some time in the future, in perhaps 25 or 30 years. It would be necessary to "back load" the database with historic data in order for those currently seeking compensation who are finding it difficult to trace their employer's insurer to realise any value from the existence of such records. This could prove a mammoth task for insurers even if they had complete records dating back over many decades. It is already established that insurer's records are far from perfect eg there are incomplete records of subsidiary and associated companies, an essential aspect in proving which insurer held the cover and should pay the claim.

A further comparison is being made with the Motor Insurers Bureau (MIB). The MIB is funded by contributions from all insurers who provide motor insurance to the public and its purpose is to provide compensation where there has been an accident resulting from the actions of an uninsured driver. A Private Members bill has been placed before Parliament suggesting that a corresponding body should be set up to compensate those who are unable to secure compensation for occupational disease.

The All Party Parliamentary Group on Occupational Safety and Health has reported that of the 1,047 enquiries passed to the ABI over the period April 2007 to December 2007 only 35% were successful in tracing insurance cover. The Group believes that between 1 in 10 and 1 in 20 of those with mesothelioma are unable to claim compensation. The Group further states that the setting up of a Motor Insurance Database and the creation of the Motor Insurers Bureau has been successful for the c26 million vehicles UK so that it should not be too difficult to set up corresponding arrangements for the 1.2 million UK businesses. However, this view oversimplifies the difficulties involved.

Vicarious Liability

[E13027] Vicarious liability is where one person assumes liability for the acts and omission of another. It can arise in several ways but is most commonly encountered in Employers' Liability where the employer accepts liability for the acts of his employees. This is in accordance with the common law principle that sets out that an employee can expect their employer to provide competent fellow employees. There are some exceptions, eg if an employee was doing something expressly forbidden or carrying out a task in an unacceptable way, but generally the employer will be held liable. Also if one employee is responsible for injuring another then as a general rule the employer will be liable.

This principle can give rise to circumstances where an employer may be entitled to an indemnity from an employee. This right is set out in the Civil Liability (Contribution) Act 1978. This right was recognised in the case of *Lister v Romford Ice and Cold Storage* [1957] AC 555. In this case a father and son worked for the same firm and while driving a lorry the son knocked down and injured his father. The father's claim was successful and the employer's insurers then sought to recover from the son by use of the doctrine of subrogation. The insurance industry, through its trade body the British Insurance Association (now the Association of British Insurers), reached a market agreement that they would not sue an employee of an insured employer in respect of injury caused to a co-employee, unless there was either (*a*) collusion and/or (*b*) wilful misconduct on the part of the employee.

In *Morris v Ford Motor Co Ltd* [1973] 2 All ER 1084 Ford had subcontracted cleaning at one of its plants to X, for which Morris worked. While engaged on this work at the plant, Morris was injured owing to the negligence of an employee of Ford. Morris claimed damages and his claim was settled by Ford's insurers. However, the insurers sought to recover from Morris's employer on the basis of an indemnity clause in their contract with Ford. The clause bound X to indemnify Ford for all losses and/or claims arising out of the cleaning operations. However, X was not insured and sought to recover from the Ford employee who had injured Morris. Although accepting that it was bound by the terms of the indemnity, X argued that its liability should be subrogated against the employee who caused the accident, on the ground that the employee had carried out his work negligently. By a majority decision the Court of Appeal held that the claim for subrogation must fail. X had to borrow Ford's name in order to bring the claim and it was stated in the judgement that this was inequitable. And since subrogation was an equitable right the court had the discretion to refuse it.

A recent decision has changed the thinking about vicarious liability. It has been accepted for many years that only one person or organisation can be held to be vicariously liable. In other words the person who has acted negligently can only work for one employer. However, the outsourcing of activity by industrial firms and the growth of self employed contractors, agency workers and similar makes that view less clear.

Although the legal profession maintained the view that two employers could not both be found to be vicariously liable, other legal experts were coming to the opinion that this presumption should be tested in court. *Viasystems (Tyneside) Ltd v Thermal Transfer (Northern) Ltd and ors* [2005] EWCA Civ 1151, [2005] 4 All ER 1181 provided that opportunity. Viasystems (V) engaged Thermal Transfer (TT) to install air conditioning in its factory. TT subcontracted the work to SPD who in turn contracted with CMS to supply fitters and fitters' mates for the actual installation of the equipment. A self-employed fitter, H, contracted to SPD to act as supervisor, was supervising another fitter, M, and his mate, S, both employed by CMS. All three individuals were working in a roofspace using crawling boards laid across the purlins. S was sent to get some fittings but in doing so he disturbed some of the already installed ducting which in turn caused damage to the sprinkler piping which leaked and flooded the factory. It was agreed that V could recover from TT in contract. In the first instance the Judge was asked to consider who was liable for the actions of S and decided that it was CMS but an appeal was lodged. The Court of Appeal had to consider whether both CMS and SPD could be jointly liable for the actions of the fitters mate. While it was obvious that the fitters mate had been negligent there was a question as to whether SPD or CMS were vicariously liable. The Court of Appeal considered who was entitled to give orders about how the work should be done. It was decided that H and M were both entitled and obliged to stop S's actions and thus both their employers were vicariously liable. After analysing the long standing assumption that two separate employers could not be vicariously liable for the actions of an employee the Court held that this could be possible in a modern context. Since both SPD and CMS were found to be vicariously liable there existed the question of contribution. The Court found that as both firms were liable they must be equally liable, and that each should thus contribute 50%.

Making the Market Work

[E13028] In an effort to facilitate the working of the Employers' Liability market and improve the flow of information between industry and insurers, the insurance industry has set up the Making the Market Work Scheme. This attempts to take account of the efforts of trade associations to improve health and safety standards amongst their members.

The Association of British Insurers has set out which aspects of good health and safety management are of interest to insurers. Trade associations are invited to submit details of what actions they have taken and activities they offer to their members. The ABI has set up a standing committee to review the information received and provide constructive feedback to the trade associations. The details of what the trade associations are doing to improve health and safety is passed to insurers. This reduces the work that brokers must do when presenting firms to insurers since insurers will already be aware of the benefits of trade association membership.

Details of the scheme can be obtained from the Association of British Insurers, 51 Gresham Street, London EC2V 7HQ, tel: 020 7600 3333; fax: 020 7696 8999.

Woolf Reforms – the Pre-Action Protocol

[E13029] The Woolf Reforms, which were enacted as the *Civil Procedure Rules 1998 (SI 1998 No 3132)*, came into force on 26 April 1999 and apply to England and Wales. The aim of the new rules, also known as the CPR, was to speed up the processes by which people obtained compensation through the civil courts and make justice more accessible via management of costs.

Pre-action protocols are an important element of the new rules, which should ensure that more claims are settled without litigation. They are aimed at personal injury cases involving less than £15,000 in compensation. Should settlement not be reached through these protocols, the litigation process then goes through the courts. The reforms affect any class of insurance business where there is a third party injury, principally Motor, Public Liability and Employers Liability and occasionally Household.

The reforms were an attempt to resolve the dispute *before* court proceedings are even commenced; thus minimising both cost and court time. If the case is not settled via the pre-action protocols, it may be taken to court. The personal injury pre action protocol sets out the following stages:

— letter of claim;
— reply/acknowledgement;
— investigation;
— disclosure of documents;

— appointment of expert witnesses;
— negotiation and settlement;
— litigation in the courts.

Under the new rules all cases will be allocated to one of three routes:

— Small Claim: (previously the arbitration or Small Claims Court) property claims with a value of up to £5,000 or up to £1,000 for personal injury claims;
— Fast Track: claims with values between £5,000 and £15,000 (Property) or £1,000 and £15,000 (personal injury);
— Multi-track: all claims in excess of £15,000 and any complex claim of lower value.

Fast track cases will generally proceed quickly to a hearing, which is usually limited to one day, as most evidence has been provided in writing and agreed in advance.

The new rules affect policyholders in a number of ways. Any delay in notifying the claim may affect mounting a suitable defence and increase the cost of the claim. Policyholders must understand the reasoning behind decisions on liability; if there is no hard evidence insurers are unable to run a defence. And policyholders need to co-operate with claims handlers and solicitors by supplying information and documents. In particular policyholders must:

— promptly report all incidents likely to give rise to a claim to your insurer;
— carry out an investigation;
— retain all evidence and relevant documents (eg Accident Book or report form, accident investigation report, photos, plans and security videos, copies of risk assessments);
— forward any letter of claim to the insurer as soon as it is received but should not acknowledge its receipt.

Since 1998 the CPR have been regularly reviewed and amended/updated. The most recent update, the 49th came into effect in April 2009, amends some procedures and takes account of changes in legislation eg the Serious Crimes Act 2007. Details of the amendments and updates can be found on the Ministry of Justice website www.justicegov.uk/civil/procrules_fin/index.htm.

Scotland

As stated above the Woolf Reforms only apply to England and Wales. However, following lengthy debate a voluntary Pre-Action Protocol has been agreed between insurers and the Law Society of Scotland and came into effect for all claims intimated after 1 January 2006.

This is a major step forward in Scottish pre-litigation procedure and is the culmination of much work between insurers and the Forum of Scottish Claims Managers. The main points to note about the Protocol are as follows:

— It is a voluntary protocol and does not have legal effect (although parties may seek to draw breaches in the protocol to the attention of the Court on the issue of costs).

—   It applies only to claims intimated on or after 01/01/2006 and thereafter handled within the Protocol.
—   It is primarily aimed at Motor claims with a value below £10,000 although can be agreed on **all** claims at all levels.
—   Any existing claim notified prior to 01/01/2006 will be dealt with as per existing arrangements.
—   The Protocol does not apply to property damage only claims — but will apply where a Property damage claim has a Personal Injury element.
—   The Protocol does not apply to disease claims.

Admissions of Liability are binding where the value of the claim is below £10,000. Solicitors will send a Letter of Claim in similar format to the format of the letter of claim used in England / Wales and insurers will have 21 days to respond from date of receipt. Insurers then have 3 months to investigate liability and respond with a decision.

Medical evidence should be instructed by the Pursuer's (NB Scottish term for Claimant) solicitor within 5 weeks of a liability decision and medical experts will be instructed on an agreed basis (not joint instruction). Medical evidence should be disclosed by the Pursuer's solicitor within 5 weeks of receipt. Questions can be put to the expert by either side by mutual consent.

Pursuer's solicitors will serve a Statement of Valuation and insurers have 5 weeks in which to respond.

Where damages are agreed insurers have 5 weeks to issue cheques or else interest will be payable. The 2003 Fee Scale still applies but this now includes payment of the Investigation Fee

Small Business Performance Indicator

[E13030] The Health and Safety Executive (HSE), with the support of the DTI's Small Business Service, has launched a new web-based tool to assist SME's track and assess how well they are managing their own health and safety performance. The Health and Safety Performance Indicator ('the Indicator') can be accessed at www.hspi.info-exchange.com.

The Indicator was developed to help SME's regularly assess their health and safety performance eg from one year to the next. It is also intended to help companies tell their insurers how well they are managing health and safety so they can more accurately calculate insurance premiums based on individual performance.

Development of the Indicator involved HSE working closely with key stake-holders including the Department for Work and Pensions, the Small Business Service, the Association of British Insurers, the British Insurance Brokers Association and the Federation of Small Businesses.

The development of the Indicator arose in part from the Government's review of Employer's Liability Compulsory Insurance in 2003. In the report of the review Government called on HSE to develop a tool, for SME's in particular, which would enable insurers to determine levels of insurance premiums that better reflected how well those employers were managing risks to health and safety and show HSE's continued commitment in helping businesses to improve their health and safety performance.

The Indicator is an internet based tool. It is free to use and works by asking a series of questions on key hazards that the developers found most SME's encounter and incident frequency. A score out of ten is calculated for each – ten the best and zero the poorest. The results can be used in two ways. The questionnaire can be completed on a regular basis and the results used to monitor improvement over time. Alternatively, a user can benchmark their performance against their peers, for example firms in the same business or trade sector or firms of a similar size. As of May 2009 there was a total of nearly 10,000 records.

SME's can also decide if they wish to share their results with anyone else, for example to let their insurers and brokers know how they are performing at health and safety, so this can be taken into account when insurance terms are set. For the main features of the indicator, go to www.hspi.info-exchange.com/ or alternatively access it via: www.businesslink.gov.uk/bdotg/action/layer?r.l2=1074298750&r.l1=1073858799&r.s=tl&topicId=1074299774.

There is a similar index designed for larger firms with greater than 250 employees. The Corporate Health and Safety Index (CHaSPI) focuses more on management systems than hazards and can be found at www.chaspi.info-exchange.com/default.asp.

NHS Recoveries

[E13031] The passing of the Road Traffic (NHS Charges) Act 1999 enabled hospitals to recover the costs of treatment of those injured in road traffic accidents. The collection of costs was centralised under the auspices of the Compensation Recovery Unit (CRU) that also recovers, from insurers of negligent employers, the benefits paid to personal injury claimants. It has been estimated that the recoveries channelled to the NHS totalled around £105m (2006 prices).

In 2002 the Department of Health investigated the possibility of extending the scheme to include all cases of personal injury where the injured person receives compensation. This would include any claimants seeking compensation from their employer or third parties such as visitors and customers seeking compensation from a shop owner, property owner or similar. This expansion of the scheme was included in the Health and Social Care (Community Health and Standards) Act 2003 and following further consultation the proposals were revised and amended and formed part of the Health Act 2006, coming into effect in January 2007.

The revised scheme is known as the NHS Injury Costs Recovery Scheme (ICR) and provides for the recovery of costs associated with the treatment of people who have suffered injuries as a result of:

—   road traffic accidents (as already provided)

—   workplace accidents
—   other accidents eg those involving members of the public

and who subsequent make a successful claim for compensation. The recovery of costs is triggered when the compensation is paid to the injured person. The costs are recovered from the insurer who pays the compensation on behalf of the negligent party, the employer in the case of workplace accidents or the property owner, firm or other organisation where the injury involved a member of the public, customer or similar. The insurer will forward the appropriate amounts to the CRU who will then forward them to the appropriate NHS hospital, ambulance trust, primary care trust or similar.

A schedule of costs has been drawn up:

—   Flat fee for admission to hospital—£505
—   Daily rate for treatment as an inpatient—£620
—   Ambulance charges (per journey)—£159

with a ceiling of £37,100 per incident. The scale of charges will be reviewed each April and increased in line with hospital cost inflation. The inflationary increase in April 2006 was 4.5% and this serves as an indication of likely future increases.

The ICR scheme will take account of any contributory negligence on the part of the claimant. Thus if an employee was held to be 30% to blame and their compensation reduced the costs recoverable would be reduced by the same amount.

The ICR scheme also recognises that some, mainly larger firms, may have excesses on their policy. Under such arrangements the employer, for example agrees to pay a proportion of each settled claim, perhaps the first £5,000 or £10,000. The usual arrangement is that the insurer will pay the claimant the full amount of damages agreed or awarded by the courts and the employer will then reimburse the insurer. The ICR requires that the insurer pay the full amount of any NHS costs owed irrespective of whether there is an excess in place. The insurer will then recover the insured's proportion of the NHS costs.

Corporate Manslaughter

[E13032] (Further information can be found at C9001)

Pressure groups and campaigning bodies have long held the view that the existing law on corporate manslaughter was inadequate and failed to deliver a suitable punishment following cases of death resulting from work activities, involving either employees or members of the public.

Whilst it is possible to bring a charge of corporate manslaughter against a company or corporate body there are difficulties in securing a conviction. It is currently an agreed legal principle that a corporate body does not possess a "mind" and is thus unable to commit a violent crime. Thus it is necessary to charge both the company and an individual employee with manslaughter. And unless the individual is successfully convicted the company can not be found guilty. This situation has resulted in few cases going to court and low conviction rates.

In the 2007/08 year there were 229 deaths of workers/employees and 358 work related deaths to members of the public, including 263 due to acts of suicide or trespass on the railways. Deaths of workers are fairly static and have dropped slowly over the last 10 years whilst deaths of members of the public are more volatile and do not display such a regular trend. In the period 1992 to 2006, a total of 160 cases of workplace death were referred to the Crown Prosecution Service, 45 of these proceeded to trial and a total of 10 convictions were achieved. The successful prosecutions also tend to involve smaller firms where it has proved easier to establish a link between the actions of senior management and the circumstances surrounding the accident that resulted in the death. Opponents of the existing arrangements thus maintain that the current approach is biased against smaller firms and allows larger firms that have caused deaths to "escape" justice.

In 1996 the Law Commission published a report on the subject of "involuntary manslaughter". This report was a response to growing public pressure for changes in the law. In the period following the Law Commission report the Government carried out various consultation exercises and published a draft bill in 2005. The final form of the Bill was published on late 2006 and after challenges in the House of Lords the Royal assent and into effect in April 2008.

The key elements of the Bill are as follows:

— A new offence of Corporate Manslaughter has been created (Corporate Homicide in Scotland).
— For a charge of manslaughter to be made an organisation must have owed a duty of care to the victim and have breached that duty as the result of way in which its activities were managed or organised by senior manager(s).
— The duty of care can result from the employment of persons, ownership or occupation of premises, the supply of good or services, construction or maintenance activities or the keeping/use of vehicles.
  Note: These duties match exactly the existing duties that are protected by liability policies.
— The breach of duty, termed a senior management failure, must have been gross and caused the death. In determining guilt consideration must be given to the risk of death presented by the breach, the seriousness of the breach and any breach of health and safety legislation.
— The gross breach must involve conduct falling far below what would be reasonably expected in the circumstances.
— Senior managers are described as being those who make decisions about the whole or a substantial part of the organisation's activities.
— All incorporated bodies are covered by the proposed legislation although certain classes of business are exempt eg partnerships, as are certain Government bodies.
— The offence will be tried in the Crown Court and the penalties available are unlimited fines or remedial measures.
— There is specific guidance that the jury should consider the firm's corporate safety culture, policies and procedures and any relevant safety information or guidance.

It should be noted that it is no longer necessary to secure a conviction of senior manager and thus it is thought that it will be easier to secure conviction of a firm.

The existing protocol for joint investigation of work related deaths by the Police and the HSE will remain in force.

The possibility of joint charges for breaches of health and safety legislation and corporate manslaughter remain – there is no change resulting from the proposed legislation. Thus there is the distinct possibility that a firm could be charged with breaches of health and safety legislation eg the Health and Safety at Work Act or specific legislation, and corporate manslaughter and senior managers or directors could be charged with a breach of s 37 of Health and Safety at Work Act.

Guidance on sentencing and the level of fines was published for consultation in November 2007 – the consultation closed in February 2008 (www.sent encing-guidelines.gov.uk/docs/SAP%20(07)K3%20-%20Corporate%20mans laughter%202007-10-31-v%203.10.AR.pdf)

The main thrust has been that fines for Corporate Manslaughter should be between 2.5% and 10% of average annual turnover.

In addition to paying damages and compensation Employers' Liability insurance also covers a range of related costs and expenses (see E13017). Prior to 2008 policy wordings generally restricted the payment of defence costs to prosecutions made under health and safety and related legislation. However, it was common practice for insurers to agree to fund the costs of defending the few manslaughter prosecutions that did occur.

Following the introduction of the Corporate Manslaughter legislation the majority of insurers have modified their wordings. In general the effect has been to explicitly state that the policy will cover the costs of defending any Corporate Manslaughter charge and pay any prosecution costs that may be awarded against the defendant employer. In addition the indemnity will extend, if the policyholder so requests, to any director, partner or employee. This recognises that such a person may be prosecuted for breaches of health and safety legislation at the same time as the company is prosecuted for Corporate Manslaughter.

In April 2009 the Crown Prosecution Service issued a press release stating that the first charge of Corporate Manslaughter had been made. An employee of Cotswold Geothechnical Holdings limited had died as a result of a trench collapse whilst working on a site in September 2008. The firm has been charged with breaches of the Health and Safety at Work etc Act 1974 in addition to the charge of Corporate Manslaughter. A director of the firm has been charged with Gross Negligence Manslaughter in addition to breaches of the 1974 Act.

Compensation Act 2006

[E13033] The Compensation Act is part of the Government's programme to tackle the 'compensation culture' and in the case of those suffering from mesothelioma improve the system for obtaining compensation where there is a valid claim. The Act received the Royal Assent in November 2006 and further parts are expected in 2007. The Act contains a number of provisions and will enable the drafting of specific legislation to reform particular aspects and areas of the law relating to the claiming of compensation. The main elements of the Act are described as follows:

Part 1: Standard of Care
- Deterrent effect of potential liability
- Apologies, offers of treatment etc
- Mesothelioma

Part 2: Claims Management Services
Part 3: General

Potential Effect of Potential Liability (s 1): This section of the Act attempts to overcome the situation where, it is alleged, activities are not undertaken for fear of litigation in the event that something unforeseen occurs and legal action follows. The usual example is that of school trips being cancelled because of fear that teachers will be sued if a pupil is injured in an accident. The Act states that when a court is considering a claim in negligence or breach of statutory duty it should consider whether the taking of reasonable steps to meet a standard of care might prevent a desirable activity from being undertaken or discourage persons from undertaking functions in connection with a desirable activity. The effect of this text is dependant upon more detailed guidance to the courts about the exact meaning of words such as "desirable activity" and "standard of care".

Apologies (s 2): This part of the Act seeks to clarify that an apology or an offer of treatment or "other redress" shall not amount to an admission of negligence or breach of statutory. It is thought that in some cases injured parties are less concerned about claiming compensation and value an apology but that potential defendants are concerned that making an apology and offering treatment eg physiotherapy will be seen as an admission of guilt. This section of the Act seeks to overcome these perceptions and hopefully reduce claims and speed the delivery of rehabilitation.

Mesothelioma (s 3) this section of the Act will enable the making of Regulations to improve the delivery of compensation to those suffering form mesothelioma. It should be noted that this is the only part of the Act that applies in Scotland.

Claims Management Services (Part 2): This part of the Act provides for the introduction of regulations to control the activities of claims management services, often known as "claims farmers". It is thought that the activities of such firms, for example "door stepping" people in shopping malls and at railway stations only serves to increase claims numbers. The Government has already drafted regulations under this part of the Act.

Employment Protection

Kitty Debenham and Janet Gaymer

Introduction

[E14001] The nature of employment protection is continually changing. At a domestic (national) level, recent years have seen the extension of certain employment protection rights to workers as well as employees. For example, both the *Public Interest Disclosure Act 1998* and the *Working Time Regulations 1998 (SI 1998 No 1833)* apply to 'workers' which is a wider category than persons categorised as employees under traditional English employment law analysis.

A number of new individual rights have been also introduced including the right to be accompanied at a disciplinary or grievance hearing and the right to request flexible working. Collective rights have also been enhanced with trade unions being given the right to statutory recognition in certain circumstances.

Europe has also continued to have a significant impact on the regulation of employment relations. Directives issued by Europe have required the introduction of legislation prohibiting discrimination on the grounds of sexual orientation, religion or belief and age, and requiring employers with more than 50 employees to set up national information and consultation procedures if requested by employees.

The relationship between an employer and an employee is a contractual one and as such must have all the elements of a legally binding contract to render it enforceable. In strict contractual terms an offer is made by the employer which is then accepted by the employee. As in the case of the offer, the acceptance may be oral, in writing, or by conduct, for example by the employee turning up for work. The consideration on the employer's part is the promise to pay wages and on the employee's part to provide his services for the employer. Once the employer's offer has been accepted, the contract comes into existence and both parties are bound by any terms contained within it (*Taylor v Furness, Withy & Co Ltd* (1969) 6 KIR 488).

The contractual analysis of the employment relationship is not entirely satisfactory in explaining the relationship between employee and employer. To fit the contract model, various elements comprising the reality of the employment relationship become part of the contract by implication.

An employment contract is unlike many other contracts, because many of the terms will not have been individually negotiated by the parties. The contract will contain the express terms that the parties have agreed – most commonly hours, pay, job description – and there will be a variety of other terms which will be implied into the contract from other sources and which the parties have not agreed. Many of these are relevant to health and safety. If any of the express or implied terms in the contract are breached, the innocent party will have certain remedies. The fact that various employee rights, particularly in relation to health and safety, are implied into the contractual terms and conditions is important for this reason.

In addition to terms implied into the contract by the common law, statute (such as the *Employment Rights Act 1996*) has created additional employment protection rights for employees, including some specific rights in relation to health and safety. These are in addition to detailed rights and duties arising from health and safety legislation and regulations which are discussed elsewhere. An employer will often lay down health and safety rules and procedures. While the law allows an employer the ultimate sanction of dismissal as a method of ensuring that safety rules are observed, such dismissals should be lawful, that is generally with notice, and should be fair. Furthermore, statute has created specific protection from victimisation for employees who are protecting themselves or others against perceived health and safety risks. All of these provisions are the subject of this section.

As a specific health and safety protection measure, the *Health and Safety at Work etc Act 1974, s 2 (HSWA 1974)* lays a general duty on all employers to ensure, so far as is reasonably practicable (for the meaning of this expression, see E15039 ENFORCEMENT), the health, safety and welfare of all their employees. An employer is also under a duty to consult about health and safety matters. In addition, a number of codes of practice have been issued under the *HSWA 1974* by the Health and Safety Executive (HSE), for example relating to safety representatives and allowing them time off to train.

Sources of contractual terms

Express terms

[E14002] These are the terms agreed by the parties themselves and may be oral or in writing. Normally the courts will uphold the express terms in the contract because these are what the parties have agreed. However, if the term is ambiguous the court may be called upon to interpret the ambiguity, for example what the parties meant by 'reasonable overtime'.

Generally the express terms cause no legal problems and the parties can insert such terms into the contract as they wish. There are, however, a number of restrictions which include:

(a)   An employer cannot restrict his liability for the death or personal injury of his employees caused by his negligence. Further, he can only restrict liability for damage to his employee's property if such a restriction is reasonable (*Unfair Contract Terms Act 1977, s 2*).

(b)   The terms in the contract cannot infringe discrimination legislation including:

- the *Equal Pay Act 1970* and the *Sex Discrimination Acts 1975 and 1986*;
- the *Race Relations Act 1976*;
- the *Disability Discrimination Act 1995*;
- the *Employment Equality (Sexual Orientation) Regulations 2003 (SI 2003 No 1661)*;
- the *Employment Equality (Religion or Belief) Regulations 2003 (SI 2003 No 1660)*;
- the *Employment Equality (Age) Regulations 2006 (SI 2006 No 1031)*.

(c) The terms in the contract cannot infringe the *Part-time Workers (Prevention of Less Favourable Treatment) Regulations 2000 (SI 2000 No 1551)*. The Regulations provide that unless justified on objective grounds, a part-time worker has the right not to be treated less favourably than a comparable full-time worker on the ground that the worker is a part-timer in relation to the terms of the contract.

(d) The terms in the contract cannot infringe the *Fixed-term Workers (Prevention of Less Favourable Treatment) Regulations 2002 (SI 2002 No 2034)*. The Regulations provide that unless justified on objective grounds, a fixed-term worker has the right not to be treated less favourably than a comparable full-time worker on the ground that the worker is a fixed-term worker.

(e) The employer cannot have a notice provision which gives the employee less than the statutory minimum notice guaranteed by the *Employment Rights Act 1996, s 86*.

(f) Some judges have suggested that any express terms regarding hours are subject to the employer's duty to ensure his employee's safety and must be read subject to this, so that a term requiring an employee to work 100 hours a week will not be enforceable (see for example Stuart-Smith LJ in *Johnstone v Bloomsbury Health Authority* [1991] IRLR 118). More specifically, the provisions of the *Working Time Regulations 1998 (SI 1998 No 1833)* affect the contractual term in relation to working hours. The 1998 Regulations, save in the case of specified exemptions, set a maximum working week of 48 hours averaged over a 17 week reference period. In addition, the Regulations provide for an obligatory daily rest period, weekly rest, rest breaks, limits on night work and minimum annual leave and otherwise regulate working time. In *Barber v RJB Mining UK Ltd* [1999] IRLR 308, the court decided that the maximum imposed on weekly working time by the Regulations was part of the employees' contract. This decision gives some protection to employees who refuse to work beyond the statutorily stated maximum. It is possible under the 1998 Regulations for a worker to agree with his or her employer in writing to opt-out of the 48 hour working week, subject to complying with certain requirements.

(g) The terms in the contract cannot infringe the *Maternity and Parental Leave etc Regulations 1999 (SI 1999 No 3312)*, which as a result of amendments which came into effect in 2007. These regulations now entitle women who are still in work in the 15th week before the baby is due to a full year of maternity leave, regardless of their length of service.

(h)   Confidentiality provisions should be subject to an employee's right to make a protected disclosure in accordance with the provisions set out in the *Employment Rights Act 1996*.

Whilst the *Sex Discrimination Acts 1975* and *1986* removed some of the restrictions on women and their employment generally and health and safety specifically, some restrictions/prohibitions on certain types of employment by women, in the interests of health and safety at work, still remain.

Under the *Control of Lead at Work Regulations 2002 (SI 2002 No 2676)* an employer is prohibited from employing women of reproductive capacity (or young people) in particular activities relating to lead processes as follows:

(a)   In the lead smelting and refining process:
    (i)   handling, treating, sintering, smelting or refining any material containing 5 per cent or more of lead; or
    (ii)   cleaning where any of the above activities have taken place.
(b)   In the lead acid manufacturing process:
    (i)   manipulating lead oxides;
    (ii)   mixing or pasting;
    (iii)   melting or casting;
    (iv)   trimming, abrading or cutting of pasted plates; or
    (v)   cleaning where any of the above activities have taken place.

Common law implied duties – all contracts

[E14003] Both the employer and employee owe duties towards each other. These are duties implied into every contract of employment and should be distinguished from the implied terms discussed below which are implied into a particular individual contract. Although there are a number of different duties, three are of major importance in relation to health and safety:

(a)   the duties on the part of the employee to obey lawful reasonable orders and to perform his work with reasonable care and skill; and
(b)   the duty on the part of the employer to ensure his employee's safety.

The duty to obey lawful reasonable orders ensures that the employer's safety rules can be enforced and, as it is a contractual duty, breach will allow the employer to invoke certain sanctions against the employee, the ultimate of which may be dismissal.

The same is true of the duty to perform his work with reasonable care and skill. Should the employee be in breach of this duty and place his or others' safety at risk, the employer may impose sanctions against him including dismissal.

The imposition of the employer's duty is to complement the statutory provisions. Statutes such as the *HSWA 1974* provide sanctions against the employer should he fail to comply with the legislation or any regulations made thereunder. The common law duty provides the employee with a remedy should the duty be broken, either in the form of compensation if he is injured, or, potentially, with a claim of unfair dismissal. The employer's duty to ensure his employees' safety is one of the most important aspects of the employment relationship. At least one judge has argued that it is so important that any express term must be read subject to it (see *Johnstone v Bloomsbury Health Authority* [1991] IRLR 118 at **E14002** above). Breach of this duty can lead to the employee resigning and claiming constructive dismissal (see below). In *Walton & Morse v Dorrington* [1997] IRLR 488, an employee claimed that she had been constructively dismissed (unfairly) because her employer had breached the implied term of her contract of employment that it would provide, so far as reasonably practicable, a suitable working environment. The employee had been forced to work in a smoke-filled environment for a prolonged period of time and her employer did not take appropriate steps to redress the problem when she raised the issue. The Employment Appeal Tribunal agreed that she had been constructively dismissed because the employer had breached its duty to her. Further, the employer's duty to ensure his employee's safety has been held to cover stressful environments resulting in injury to the employee. In *Walker v Northumberland County Council* [1995] IRLR 35, the High Court held that an employer was liable for damages on the basis that they owed their employee a duty not to cause him psychiatric damage by the volume and/or character of work that he was required to undertake. In *Fraser v The State Hospitals Board for Scotland* 2001 SLT 1051, the Court of Session held that there was no reason to qualify an employer's duty to take reasonable care for the safety of employees so as to restrict the nature of the injury suffered to a physical one in circumstances where the employee claimed damages for psychological damage as a result of disciplinary measures. The claim failed on the basis of lack of foreseeability.

Common law implied terms – individual contracts

[E14004] The court may imply terms into the contract when a situation arises which was not anticipated by the parties at the time they negotiated the express terms. As such, the court is 'filling in the gaps' left by the parties' own negotiations. The courts use two tests to see if a term should be implied, (i) the 'business efficacy' test (*The Moorcock* (1889) 14 PD 64) or (ii) the 'officious bystander' or 'oh of course' test (*Shirlaw v Southern Foundries Ltd* [1939] 2 KB 206). This describes a term so obvious that it goes without saying that the parties must have intended it. Once the court has decided, by virtue of one of these tests, that a term should be implied, it will use the concept of reasonableness to decide the content of the term. Often this will involve looking at how the parties have worked the contract in the past. For example, if the contract does not contain a mobility clause, but the employee has always worked on different sites, the court will normally imply a mobility clause into the contract (*Courtaulds Northern Spinning Ltd v Sibson* [1988] IRLR 305). (See also *Aparau v Iceland Frozen Foods plc* [1996] IRLR 119, where the Employment Appeal Tribunal (EAT) refused to imply a mobility clause on the basis that there were other ways of achieving the necessary flexibility.) In relation to dismissal, a term has been implied that, except in the case of summary dismissal, the employer will not terminate the employment contract while the employee is incapacitated where the effect would be to deprive the employee of permanent health insurance benefits (*Aspden v Webbs Poultry and Meat Group (Holdings) Ltd* [1996] IRLR 521). Terms can also be implied by the conduct of the parties or by custom and practice in a particular industry or area. The test for this is relatively difficult to fulfil – the term must be notorious and certain and, in effect, everyone in the industry/enterprise must know that it is part of the contract. Arguments based on custom and practice come into play in relation to issues such as statutory holidays and redundancy policies.

Collective agreements

[E14005] Collective agreements are negotiated between an employer or employer's association and a trade union or unions. This means that they are not contracts between an employer and his individual employees because the employee was not one of the negotiating parties. Some terms of the collective agreement will be procedural and will govern the relationship between the employer and the union; some, on the other hand, will impact on the relationship between the employer and each individual employee, for example a collectively bargained pay increase. As the employee is not a party to the collective agreement, the only way he can enforce a term which is relevant to him is if the particular term from the collective agreement has become a term of his individual employment contract. Procedural provisions, policy and more general aspirations are not suitable for incorporation into an individual contract of employment. Incorporation is important because the collective agreement is not a legally binding contract between the employer and the union (*Trade Union and Labour Relations (Consolidation) Act 1992, s 179(1)*) and thus needs to be a term of an employment contract to make it legally enforceable.

The two main ways that a term from a collective agreement becomes a term of an employment contract is by express or implied incorporation. Until recently, implied incorporation was the most common and was complex. It generally required the employee to be a member of the union which negotiated the agreement, to have knowledge of the agreement and of the existence of the term, and to have conducted himself in such a way as to indicate that he accepted the term from the collective agreement as a term of his contract. A decision of the Employment Appeal Tribunal case, *Healy v Corporation of London* (24 June 1999, unreported), illustrates that habitual acceptance of the benefits of a collective agreement does not, in itself, lead to the conclusion that the terms of that collective agreement have become contractually binding on an individual employee. There can be many reasons for an individual to accept the benefits of collective bargaining which do not amount to an acceptance that the underlying agreement forms part of his or her contract.

Express incorporation meant that the employee had expressly agreed (normally in his contract) that any term collectively agreed would become part of his contract. This used to be unusual, but with the change made to the statutory statement which must be given to all employees (see below) employees must be told of collective agreements which apply to them, and this has been held as expressly incorporating those agreements into the contract.

Statutory statement of terms and conditions

[E14006] By the *Employment Rights Act 1996, s 1* every employee no later than two months after starting employment, must receive a statement of his basic terms and conditions. The statement must contain:

(a)   the names of the employer and employee;
(b)   the date the employment began;
(c)   the date the employee's continuous employment began;
(d)   the scale or rate of remuneration and how it is calculated;
(e)   the intervals when remuneration is paid;
(f)   terms and conditions relating to hours;
(g)   terms and conditions relating to holidays;
(h)   terms relating to sickness or injury, including provision for sick pay (if any);
(i)   terms and conditions relating to pensions;
(j)   notice requirements;
(k)   job description;
(l)   title of the job;
(m)   if the job is not permanent, the period of employment;
(n)   place of work, or if various the address of the employer;
(o)   any collective agreements which affect terms and conditions and, if the employer is not a party to the agreements, the persons with whom they were made;
(p)   if the employee is required to work outside the UK for more than one month, the period he will be required to work, the currency in which he will be paid, any additional benefits paid to him and any terms and conditions relating to his return to the UK; and
(q)   details of the employer's disciplinary and grievance procedures.

The terms in (*a*), (*b*), (*c*), (*d*), (*e*), (*f*), (*g*), *(k)*, (*l*) and (*n*) must all be contained in a single document. In relation to pensions and sickness absence and sick pay the employer may refer the employee to a reasonably accessible document, and in respect of notice the employer can refer the employee to a reasonably accessible collective agreement or to the *Employment Rights Act 1996, s 86* which contains provisions relating to minimum notice periods.

There is no duty on an employer to give details of any disciplinary or grievance procedures relating to health and safety. Given the employer's duties under the *HSWA 1974, s 2*, however, and given the law relating to unfair dismissal, it is good industrial relations practice to ensure that all employees know of all the disciplinary procedures which could be invoked against them. The written statement must also provide (either by instalments or in one single document) details of whether the employment is contracted out of the State Second Pension.

Works rules

[E14007] Works rules may or may not be part of the contract. If they are part of the contract and thus contractual terms, they can be altered only by mutual agreement, that is, the employee must agree to any change. It is unusual, however, for such rules to be contractual. To be so, there would have to be some reference to them within the contract and an intention that they are terms of the contract. The more usual position with regard to the employer's rules was stated in *Secretary of State for Employment v ASLEF (No 2)* [1972] 2 QB 455 where Lord Denning said that they were merely instructions from an employer to an employee. This means that they are non-contractual and the employer can alter the rules without the consent of the employees. The fact that they are not contractual does not mean that they cannot be enforced against an employee. All employees have a duty to obey lawful, reasonable orders (see **E14003** above) and thus failing to comply with the rules will be a breach of this duty and therefore a breach of contract. The only requirement that the law stipulates is that the order must be lawful and reasonable and it is unlikely that an order to comply with any health and safety rules would infringe these requirements.

Disciplinary and grievance procedures

[E14008] It has already been noted that the employer must give details of grievance and disciplinary procedures to all employees. Failing to do so could lead to the employee resigning and claiming constructive dismissal (*W A Goold (Pearmak) Ltd v McConnell* [1995] IRLR 516).

The *Employment Act 2002 (Dispute Resolution) Regulations 2004 (SI 2004 No 752)* introduced statutory procedures which applied from October 2004. These were intended to make it easier to resolve disciplinary and grievance issues within the workplace, without the need to involve external parties. However, a review of the employment dispute resolution procedures led by Michael Gibbons recommended in 2007 that they should be repealed and replaced with "clear, simple, non-prescriptive guidelines on how to resolve a dispute".

As a result, a revised Advisory, Concilliation and Arbitration Service (ACAS) *Code of practice on disciplinary and grievance procedures* came into effect on 6 April 2009 (although transitional provisions apply) and re-placed the statutory dispute resolution procedure. It was issued under section 199 of the *Trade Union and Labour Relations (Consolidation) Act 1992*. (The Code does not apply to dismissals due to redundancy or the non-renewal of fixed term contracts on their expiry.)

It sets out what the features of a disciplinary and grievance procedure should be, but is not legally binding and failure to follow the Code will not make a dismissal arising out of a disciplinary issue automatically unfair. But where an employer or employee has unreasonably failed to follow the Code, an employment tribunal can increase or reduce any compensation award (by up to 25%).

While many potential disciplinary or grievance issues can be resolved informally, it important that where they are pursued formally, they must be pursued fairly. The Code sets out the basic principles of fairness and recommends that:

- Clear and specific rules and procedures for handling disciplinaries and grievances are set down in writing. Disciplinary rules should give examples of what the employer regards as acts of gross misconduct, such as theft or fraud, physical violence, gross negligence or serious insubordination;
- Employees and, where appropriate, their representatives are involved in the development of the rules and procedures;
- Employees and managers are helped to understand what the rules and procedures are, where they can be found and how they are to be used;
- Employers and employees raise and deal with issues promptly and do not unreasonably delay meetings, decisions or confirmation of those decisions;
- Both employers and employees act consistently;
- Employers carry out any necessary investigations to establish the facts of the case;
- Employers inform employees of the problem and give them the opportunity to put forward their case in response before any decisions are made;
- Employers allow employees to be accompanied at any formal disciplinary or grievance meeting; and
- Employers allow an employee to appeal against any formal decision made.

Employees have the right to be accompanied at a disciplinary or grievance hearing under section 10 of the *Employment Relations Act 1999*. There is a statutory right to be accompanied by a companion where the disciplinary meeting could result in: a formal warning being issued; the taking of some other disciplinary action; or the confirmation of a warning or some other disciplinary action (appeal hearings). The companion can be a fellow employee, a trade union official, or a trade union representative who has been certified by their union as being competent for this purpose. This right is enforceable in the employment tribunal and compensation is payable for any failure. It is worth noting that fellow employees are under no duty to perform the role of accompanying individual.

The Code recommends that employers keep written records of any disciplinary or grievance cases they deal with, and that they should consider dealing with issues involving bullying, harassment and whistleblowing under a separate procedure.

Subject to the above, employers can establish their own disciplinary procedures. Such procedures may become part of the contract. If, for example, the employer gives the employee a copy of the procedures with the contract, and the contract refers to the procedures and the employee signs for receipt of the contract and the procedures, it is likely that they will be contractual. Employers may prefer their disciplinary procedures not to be contractual. If the procedures are contractual, any employee will be able to claim that his or her contract has been breached if they are not followed. This possibility also applies to those employees who have been employed for less than the one year qualifying period required to bring a claim for unfair dismissal. In response to a claim of breach of contract, a court may award damages against the employer. These damages are based on an assessment of the time for which, if the procedure had been followed, the employee's employment would have continued.

Employee employment protection rights

Right not to suffer a detriment in health and safety cases

[E14009] By the *Employment Rights Act 1996, s 44* every employee has the right not to be subjected to a detriment, by any act or any failure to act, by his employer on the grounds that:

(a)   having been designated by the employer to carry out activities in connection with preventing or reducing risks to health and safety at work, the employee carried out (or proposed to carry out) any such activities;

(b)   being a representative of workers on matters of health and safety at work or a member of a safety committee –

    (i)   in accordance with arrangements established under or by virtue of any enactment; or

    (ii)   by reason of being acknowledged as such by the employer;
            the employee performed (or proposed to perform) any functions as such a representative or a member of such committee;

(ba)   the employee took part (or proposed to take part) in consultation with the employer pursuant to the *Health and Safety (Consultation with Employees) Regulations 1996 (SI 1996 No 1513)* or in an election of representatives of employee safety within the meaning of those Regulations (whether as a candidate or otherwise);

(c)   being an employee at a place where –

    (i)   there was no such representative or safety committee; or

    (ii)   there was such a representative or safety committee but it was not reasonably practicable for the employee to raise the matter by those means;
            he brought to his employer's attention, by reasonable means, circumstances connected with his work which he reasonably believed were harmful or potentially harmful to health or safety;

(d)   in circumstances of danger which the employee reasonably believed to be serious and imminent and which he could not reasonably have been expected to avert, he left (or proposed to leave) or (while the danger persisted) refused to return to his place of work or any dangerous part of his place of work; or

(e)   in circumstances of danger which the employee reasonably believed to be serious and imminent, he took (or proposed to take) appropriate steps to protect himself or other persons from the danger.

In considering whether the steps the employee took or proposed to take under (e) were reasonable, the court must have regard to all the circumstances including the employee's knowledge and the facilities and advice available to him (*Employment Rights Act 1996, s 44(2)*). In *Kerr v Nathan's Wastesavers Ltd* (1995) IDS Brief 548, however, the Employment Appeal Tribunal stressed that tribunals should not place too onerous a duty on the employee to make enquiries to determine if his belief is reasonable. Danger under *(d)* does not necessarily have to arise from the circumstances of the workplace, but can include the risk of attack by a fellow employee (*Harvest Press Ltd v McCaffrey* [1999] IRLR 778). Danger under *(e)* can include danger to others as well as the employee himself (*Masiak v City Restaurants (UK) Ltd* [1999] IRLR 780). Various actions by the employer can constitute a detriment to the employee (such as disciplining the employee). Likewise, a failure to act on the part of the employer can also constitute a detriment (for example not sending the employee on a training course). Furthermore, the section is not restricted to the health and safety of the employee or his colleagues. In *Barton v Wandsworth Council* (1995) IDS Brief 549 a tribunal ruled that an employee had been unlawfully disciplined when he voiced concerns over the safety of patients due to what he considered to be the lack of ability of newly introduced escorts. This shows that the legal protection is triggered in relation to any health and safety issue and includes cases where the employee voices concerns, and is not limited to only those circumstances where the employee commits more positive action. The *Employment Rights Act 1996, s 44(3),* however, provides that an employee is not to be regarded as subjected to a detriment if the employer can show that the steps the employee took or proposed to take were so negligent that any reasonable employer would have treated him in the same manner. Furthermore, if the detriment suffered by the employee is dismissal, there is special protection under *s 100* (see below).

If the employee should suffer a detriment within the terms of *s 44* he may present a complaint to an employment tribunal (*Employment Rights Act 1996, s 48*). The complaint must be presented within three months of the act (or failure to act) complained of, or, if there is a series of acts, within three months of the date of the last act. The tribunal can waive this time limit if it was not reasonably practicable for the employee to present his complaint in time. If the tribunal finds the complaint well founded, it must make a declaration to that effect and may make an award of compensation to the employee, the amount of compensation being what the tribunal regards as just and equitable in all the circumstances (*Employment Rights Act 1996, s 49(2)*). The amount of compensation shall take into account any expenses reasonably incurred by the employee in consequence of the employer's action and any loss of benefit caused by the employer's action. Compensation can be reduced because of the employee's contributory conduct.

The protection from being dismissed or subjected to a detriment on the health and safety grounds specified in *s 44* and *s 100* of the *Employment Rights Act 1996* has been reinforced by more general protection for whistleblowers.

The Public Interest Disclosure Act 1998 provides protection to workers who make disqualifying disclosures about health and safety matters, as well as about criminal acts, failure to comply with legal obligations, miscarriages of justice, damage to the environment and deliberate concealment of any of these matters. Under the Act, which inserts new sections into the *Employment Rights Act 1996*, a worker who makes a 'qualifying disclosure' which he reasonably believes shows one of these matters, may be protected against dismissal or being subjected to a detriment.

Disclosures are only protected if they are made to appropriate persons - which often means that the employer must to be approached in the first instance.

There are other possibilities available under the Act: disclosure to a legal adviser, disclosure to a prescribed person (e.g. the Financial Services Authority or the Commissioners of the Inland Revenue), and a more general category for disclosures made provided that certain specified conditions are met.

In terms of health and safety risks, protection for qualifying disclosures is not limited to cases of imminent or serious danger – it can apply where the health and safety of any individual has been, is being or is likely to be, endangered. In all cases the worker must have a reasonable belief and make the disclosure in good faith (except in the case of disclosure to a legal adviser).

If a disclosure is protected, and an employee is subjected to any detriment or dismissed as a result, it is unlawful. A dismissal in these circumstances is deemed to be automatically unfair and the tribunal is not required to consider whether or not the employer's actions were reasonable. There is no minimum qualifying period for entitlement to make an unfair dismissal claim for this reason and there is no limit on the compensation available to whistleblowers who are unfairly dismissed because they have made a protected disclosure.

Dismissal on health and safety grounds

[E14010] In addition to the normal protection against dismissal (see below), where an employee is dismissed (or selected for redundancy) and the reason or principal reason for the dismissal is one of the grounds listed in the *Employment Rights Act 1996, s 44*, the dismissal will be automatically unfair. The only defence available to an employer applies to a dismissal taken by the employee to protect himself or others from danger that the employee reasonably believed was serious and imminent (*Employment Rights Act 1996, ss 44(1)(e) and 100(1)(e)*). The employer can escape a finding of unfair dismissal if he can show that the actions taken or proposed by the employee were so negligent that any reasonable employer would have dismissed the individual concerned. In respect of a dismissal falling within *s 100*, the normal qualifying period of employment does not apply (*Employment Rights Act 1996, ss 108(3)(c)*). Thus an employee who has only been employed for a few weeks can claim unfair dismissal for a breach of *s 100*. A dismissal which is not automatically unfair under *s 100* may nevertheless be unfair under the general reasonableness test under *s 98*.

There is no limit on the amount of compensation that may be awarded for an unfair dismissal on health and safety grounds.

Dismissal for assertion of a statutory right

[E14011] By the *Employment Rights Act 1996, s 104*, an employee is deemed to be unfairly dismissed where the reason or principal reason for that dismissal was that the employee –

(a)   brought proceedings against an employer to enforce a right of his which is a relevant statutory right, or
(b)   alleged that the employer had infringed a right of his which is a relevant statutory right.

It is immaterial whether or not the employee has the right or whether or not the right has been infringed, as long as the employee made it clear to the employer what the right claimed to have been infringed was and the employee's claim is made in good faith. A statutory right for the purposes of the section is any right under the *Employment Rights Act 1996* in respect of which remedy for infringement is by way of complaint to an employment tribunal, a right under *s 86* of the 1996 Act (minimum notice requirements), rights in relation to trade union activities under the *Trade Union and Labour Relations (Consolidation) Act 1992*, rights conferred by the *Working Time Regulations 1998 (SI 1998 No 1833)*(and other regulations relating to working time in particular industries and rights conferred by the *Transfer of Undertakings (Protection of Employment) Regulations 2006 (SI 2006 No 246)*.

This is an important right for employees. If, for example, after the employee has successfully claimed compensation from his employer for a breach of s 44 he is dismissed, the dismissal will be automatically unfair under s 104. Again, if the employer unlawfully demotes or suspends without pay as a disciplinary sanction for breach of health and safety rules, and after proceedings against him for an unlawful deduction from wages the employer dismisses the employee, this will be unfair under s 104. As with dismissal in health and safety cases under s 100, the normal qualifying period of employment does not apply.

Enforcement of safety rules by the employer

The rules

[E14012] Given the statutory duty on the employer, under the *HSWA 1974, s 2*, to have a written statement of health and safety policy, and the common law duty on the employer to ensure his employees' safety, the employer should lay down contractual health and safety rules, breach of which will lead to disciplinary action against the employee. These rules must be communicated to the employee and be clear and unambiguous so that the employee knows exactly what he can and cannot do.

The employer's disciplinary rules will often classify misconduct, e.g. as minor misconduct, serious misconduct and gross misconduct. It is unlikely that a tribunal would uphold as fair a dismissal for minor misconduct. It will underline the importance of health and safety rules if their breach is deemed to be serious or gross misconduct. The tribunal will, however, look at all the circumstances of the case – it does not automatically follow, therefore, if an employer has stated that a breach of a particular rule will be gross misconduct, that a tribunal will find a resultant dismissal fair.

West London Mental Health NHS Trust v Sarkar [2009] IRLR 512 involved an unfair dismissal claim brought by a consultant psychiatrist who was dismissed following complaints that he bullied and harassed staff. In dismissing him, the employer used two different procedures to dismiss him. It had a conflict resolution procedure which could only result in a formal written warning and a formal disciplinary procedure. The employer initially attempted to use the conflict resolution procedure to deal with the complaints, but when this failed, it conducted a disciplinary hearing and Mr Sarker was found guilty of gross misconduct and dismissed.

Mr Sarker went to an employment tribunal which found in his favour. It said that because the Trust had used it's conflict resolution procedure, this implied that it considered the misconduct to be of a relatively minor nature, and therefore dismissal was not a reasonable response. The employer appealed and the Employment Appeal Tribunal (EAT) ruled in its favour. It said that the use of the conflict resolution procedure did not preclude using the disciplinary procedure if the matter could not be resolved under the first.

The procedures

Disciplinary issues

[E14013] Once an employer has laid down his rules, he must ensure that he has adequate procedures (which should be non-contractual) to deal with a breach. The procedures used by an employer are scrutinised by a tribunal in any unfair dismissal claim and past cases indicate that many employers have lost such claims due to inadequate procedures. As discussed below, an employer in an unfair dismissal claim must show the tribunal that he acted reasonably. This concentrates on the fairness of the employer's actions and not on the fairness to the individual employee (*Polkey v A E Dayton Services Ltd* [1987] IRLR 503). This means that an employer cannot argue that a breach of procedures has made no difference to the final outcome and that he would have dismissed the employee even if he had adhered to his procedure. Breach of procedures themselves by an employer is likely to render a dismissal unfair regardless of which rule was broken.

Essentially any disciplinary procedure should contain three elements: an investigation, a hearing (or meeting) and an appeal; and should observe the principles of natural justice.

(a) Investigation

The law requires that the employer has a genuine belief in the employee's 'guilt', and that the belief is based on reasonable grounds after a reasonable investigation (*British Home Stores v Burchell* [1978] IRLR 379). If the employer suspends the employee during the investigation, this suspension should be with pay and in accordance with the disciplinary procedure. The ACAS Code of practice recommends that where a period of suspension with pay is considered necessary, it should be as brief as possible, be kept under review, and made clear that the suspension is not considered a disciplinary action.

An investigation is important because it may reveal defects in the training of the employee, or reveal that the employee was not told of the rules, or that another employee was responsible for the breach. In all of these cases, disciplinary action against the suspended employee will be unfair. Any investigation should be as thorough as possible and documented. It should also take place as soon as possible since memories fade quickly and this is particularly important if other employees are to be questioned as witnesses. Likewise, taking too long to start an investigation may lead the employee to think that no action will be taken and to then discipline them may itself be unfair.

The Code recommends that it may be necessary, in some cases, to hold an investigatory meeting with the employee before proceeding to any disciplinary hearing in order to establish the facts of the case without reasonably delay. If an investigatory meeting is held, this should not alone result in any disciplinary action.

In other cases, it says that the investigatory stage will be the collation of evidence by the employer for use at any disciplinary hearing. It also recommends that where practicable, different people should carry out the investigation and disciplinary hearing in misconduct cases.

No disciplinary action should be taken until a careful investigation has been concluded.

(b)   Hearing (or meeting)

Once the employer has investigated, he must conduct a hearing (the ACAS Code makes reference to a meeting) to make a decision as to the sanction, if any, he will impose. To act fairly, the employer must comply with the rules of a fair hearing. These are as follows.

(i)   The employee must know the case against him to enable him to answer the complaint. This also means that the employee should be given sufficient time before the hearing with copies of relevant documents to enable him to prepare his case.

(ii)   The employee should have an opportunity to put his side of the case, i.e. the employer should listen to the employee's side of the story and allow the employee to put forward any mitigating circumstances.

(iii)   The employee must be allowed to be accompanied at the hearing by a fellow employee or a trade union representative of his choice (*Employment Relations Act 1999*).

(iv)   The hearing should be unbiased, i.e. the person chairing the hearing should come to it with an open mind and not have prejudged the issue.

(v)   The employee should be provided with an explanation as to why any sanctions are imposed.

(vi)   The employee should be informed of his right to appeal (and the way in which he should go about it) to a higher level of management which has not been involved in the first hearing. If the employee fails to exercise his right of appeal, however, he will not have failed to mitigate his loss, if ultimately a tribunal finds that he has been unfairly dismissed and thus his compensation will not be reduced (*William Muir (Bond 9) Ltd v Lamb* [1985] IRLR 95). Failing to allow an employee to exercise a right of appeal will almost certainly render any dismissal unfair (*West Midlands Co-operative Society Ltd v Tipton* [1986] IRLR 112).

The ACAS Code says that where it is decided that there is a disciplinary case to answer, the employee should receive written notification of this. They should receive sufficient information about the alleged misconduct or poor performance and the possible consequences in order to prepare to answer the case at a disciplinary meeting. Although the meeting should be held without unreasonable delay, the employee should be given sufficient time to prepare their case.

At this point, copies of any written evidence, such as witness statements, should be provided to the employee. The employer should also inform the employee of the time and venue for the meeting, and advise them that they have the right to be accompanied by a companion (see **E14008** above).

With regard to the conduct of the meeting, the Code sets out that:

"Employers and employees (and their companions) should make every effort to attend the meeting. At the meet-ing the employer should explain the complaint against the employee and go through the evidence that has been gathered. The employee should be allowed to set out their case and answer any allegations that have been made. The employee should also be given a reasonable opportunity to ask questions, present evidence and call relevant witnesses. They should also be given an opportunity to raise points about any information provided by witnesses. Where an employer or employee intends to call relevant witnesses they should give advance notice that they intend to do this."

Following the meeting, the employer should decide whether or not disciplinary or any other action is justified and inform the employee about any action to be taken in writing.

(c) Appeal

In an unfair dismissal case a tribunal is required to consider the reasonableness of the employer's action taking into account the resources of the employer and the size of the employer's undertaking. This means that the tribunal will expect the employer to have provided an appeal for the employee. If an employer does not allow an appeal, they will not be complying with the ACAS Code of practice on disciplinary and grievance procedures, which sets out that employees should have the opportunity to appeal if they feel that the disciplinary action taken against them is wrong or unjust. All the rules of a fair hearing equally apply to an appeal. Only an appeal which is a complete rehearing of the case (rather than merely a review of the written notes of the disciplinary hearing) can rectify procedural flaws committed earlier on in the procedure (*Jones v Sainsbury's Supermarkets Ltd* (2000), unreported). An appeal, however, cannot endorse the sanction imposed by the earlier hearing for a different reason, unless the employee has had notice of the new reason and has been given an opportunity to put forward his argument in respect of it. Workers have a statutory right to be accompanied at appeal hearings.

Sanctions other than dismissal

[E14014] There are a variety of sanctions apart from dismissal that an employer may impose. It is important however that the 'punishment fits the crime'. The imposition of too harsh a sanction may entitle the employee to resign and claim constructive dismissal (see below).

(a) Warnings

The ACAS Code states that the usual first step in a formal procedure (where misconduct or unsatisfactory performance has been confirmed) is to give the employee a written warning. A further act of misconduct (or failure to improve performance) within a set period would then normally result in a final written warning. However, if the conduct or under performance is sufficiently serious, it may be appropriate to move directly to a final written warning.

In health and safety cases, a minor breach of a rule may justify a final written warning given the potential seriousness and consequences of breaches of such rules.

Dismissal is the final stage in the disciplinary process, and should not be a penalty for a first offence, except in the case of gross misconduct. The Code sets out that some acts (termed gross misconduct) are so serious, or have such serious consequences that they may call for dismissal without notice for a first offence. Breaches of health and safety rules have been held to be gross misconduct. Even so, a fair disciplinary process should still be followed before dismissing for gross misconduct and any decision to dismiss should only be taken by a manager who has the authority to do so.

(b) Fines or deductions

The employer must have contractual authority or the written permission of the employee before he can make a deduction from the employee's wages as a disciplinary sanction. Deducting without such authority is a breach of the *Employment Rights Act 1996, s 13* and gives the employee the right to sue for recovery in the employment tribunal. It will also lead to a potential constructive dismissal claim.

(c) Suspension without pay

Any suspension without pay will have the same consequences as a fine or deduction if there is no contractual authority or written authorisation from the employee to impose such a sanction.

(d) Demotion

Most demotions will involve a reduction in pay, and thus without written or contractual authority to demote the employer will be in breach of the *Employment Rights Act 1996, s 13* and liable to a constructive dismissal claim.

Suspension from work on maternity grounds – Suitable alternative work

[E14015] Where an employee is suspended from work on maternity grounds, the employer must offer available suitable alternative work. Alternative work will only be suitable if:

(a)   the work is of a kind which is both suitable in relation to the employee and appropriate for the employee to do in the circumstances; and

(b)   the terms and conditions applicable for performing the work are not substantially less favourable than corresponding terms and conditions applicable for performing the employee's usual work (*Employment Rights Act 1996, s 67*).

If an employer fails to offer suitable alternative work, the employee may bring a claim before an employment tribunal which can award 'just and equitable' compensation. Such complaint must normally be lodged within three months of the first day of the suspension (*Employment Rights Act 1996, s 70(4)*).

Remuneration on suspension from work on medical or maternity grounds

[E14016] An employee who is suspended on medical grounds is entitled to normal remuneration for up to 26 weeks. An employee who is suspended on maternity grounds, if no suitable alternative work is available, is entitled to normal remuneration for the duration of the suspension. However, in either case, if the employee unreasonably refuses an offer of suitable alternative work, no remuneration is payable for the period during which the offer applies. An employee may bring a complaint to an employment tribunal if an employer fails to pay the whole or any part of the remuneration to which the employee is entitled (*Employment Rights Act 1996, ss 64, 68, 70(1)*). (See *British Airways Ltd v Moore* [2000] IRLR 296, in which the Employment Appeal Tribunal upheld a purser's claim to a flying allowance on the basis that suitable work must be on terms and conditions not substantially less favourable.)

Dismissal

[E14017] Dismissal is the ultimate sanction that an employer can impose for breach of health and safety rules. All employees are protected against wrongful dismissal at common law, but, in addition, some employees have protection against unfair dismissal. The protection against unfair dismissal comes from statute (the *Employment Rights Act 1996*) and therefore the employee must satisfy any qualifying criteria laid down by the statute before he/she can claim. Given that the protection against wrongful and unfair dismissal rest alongside each other, an employee may claim for both, although he will not be compensated twice. Wrongful dismissal is based on a breach of contract by the employer and compensation will be in the form of damages for that breach – that is, the damage the employee has suffered because the employer did not comply with the contract. Unfair dismissal, on the other hand, is statute based and is not dependent on a breach of contract by the employer. Compensation for such dismissal is based on a formula within the statute (see **E14022** below).

Wrongful dismissal

[E14018] A dismissal at common law is where the employer unilaterally terminates the employment relationship with or without notice. A wrongful dismissal is where the employer terminates the contract in breach, for example, by giving no notice or shorter notice than is required by the employee's contract and the employee's conduct does not justify this. An employer is entitled to dismiss without notice only if the employee has committed gross misconduct. In all other circumstances the employer must give contractual notice to end the relationship, or pay wages in lieu of notice. However, this does require qualification. Firstly, the law decides what is gross misconduct and not the employer. Just because the employer has stated that certain actions are gross misconduct does not mean that the law will regard it as such. Only very serious misconduct is regarded by the law as gross, such as refusing to obey lawful and reasonable orders, gross neglect, and theft. Secondly, contractual notice periods are subject to the statutory minimum notice provisions contained in the *Employment Rights Act 1996, s 86*. Any attempt by the contract to give less than the statutory minimum notice is void. These periods apply to all employees who have been employed for one month or more and are:

(a)   not less than one week if the employee has been employed for less than two years;

(b)   after the employee has been employed for two years, one week for each year of service, up to a statutory maximum of 12 weeks.

Where an employer terminates the contract and pays the employee in lieu of notice, in the absence of an express right to do so, this will be a technical breach of contract. The employee can waive his right to notice or accept wages in lieu of notice. If the contract gives notice periods which are longer than the statutory minimum, the contractual notice prevails. Therefore, if the employer has an employee who has been employed for six years, and the employer sacks him with four weeks' notice, the employee can sue for a further two weeks' wages in the employment tribunal. Finally, if the employer fundamentally alters the terms of the employee's contract, without his consent, in reality the employer is terminating (repudiating) the original contract and substituting a new one. The employee should therefore be given the correct notice before the change comes into effect. A unilateral change by the employer to a fundamental term of the contract (followed by resignation by the employee) will amount to constructive dismissal (i.e. repudiation) and compensation in the form of damages for breach of contract. The employee must take all reasonable steps to mitigate his loss by seeking other employment.

Unfair dismissal

Dismissal

[E14019] While all employees are protected against wrongful dismissal, generally employees must be employed for one year or more before they gain protection against unfair dismissal. In certain circumstances, however, an employee is protected immediately and does not need a year of employment. One of these is dismissal on certain health and safety grounds discussed above.

Once an employee is protected against unfair dismissal the *Employment Rights Act 1996, s 95* recognises three situations which the law regards as dismissal. These are:

(a)   the employer terminating the contract (with or without notice);
(b)   a fixed term contract which expires and is not renewed;
(c)   the employee resigning in circumstances in which he is entitled to do so without notice because of the employer's conduct – a constructive dismissal.

In the first situation the employer is unilaterally ending the relationship. Even if the employer gives the correct amount of notice so that the dismissal is lawful, it does not necessarily follow that the dismissal will be fair.

The second situation needs no explanation. If a fixed term contract has come to an end and is not renewed, this is, in effect, the employer deciding to end the relationship. As from 25 October 1999, it is no longer possible for employees to validly waive unfair dismissal rights in fixed term contracts (*Employment Relations Act 1999*).

The third situation, constructive dismissal, is much more complex. On the face of it the employee has resigned. However if the reason for his resignation is the employer's conduct, then the law treats the resignation as an employer termination. The action on the part of the employer which entitles the employee to resign and claim constructive dismissal is a repudiatory breach of contract. In other words, the employer has committed a breach which goes to the root of the contract and has, therefore, repudiated it. This means that not all breaches by the employer are constructive dismissals but that serious breaches may be. It is also important to recognise that, as discussed above, the terms of the contract may include those which have not been expressly agreed by the parties and therefore rules, disciplinary procedures, terms collectively bargained, and all the implied duties discussed in **E14003** above, may all be contractual terms. Breach of the health and safety duties owed to all employees is likely to give rise to a constructive dismissal claim (*Day v T Pickles Farms Ltd* [1999] IRLR 217). In addition, the law requires as an implied term of the contract that both the employer and employee treat each other with mutual respect and do nothing to destroy the trust and confidence each has in the other. Breach of this duty may give rise to a constructive dismissal claim. In one case, a demotion imposed as a disciplinary sanction was held to be excessive by the Employment Appeal Tribunal. Its very excessiveness was a breach of the duty of mutual respect which entitled the employee to resign and claim constructive dismissal. It has also been held that an employer's disclosure in a reference on behalf of an employee of complaints against the employee, before first giving the individual an opportunity to explain, was a fundamental breach of the implied term of mutual trust and confidence which amounted to constructive (since the employee resigned) and unfair dismissal (*TSB Bank v Harris* [2000] IRLR 157). In *(1) Reed (2) Bull Information Systems Ltd v Stedman* [1999] IRLR 299, the Employment Appeal Tribunal held that in a case where the employer was aware of an employee's deteriorating health, and the employee concerned had complained to colleagues at work about harassment, it was encumbent on the employer to investigate and their failure to do so was enough to justify a finding of breach of trust and confidence and thus constructive dismissal.

Where an employee with one or more year's continuous service (unless one of the specified reasons apply which render a dismissal automatically unfair) is constructively dismissed, he or she will also have the right to claim unfair dismissal. Obviously, the employee must resign before he can make a claim for unfair dismissal. In the majority of cases the repudiatory breach by the employer is a fundamental alteration of the contractual terms (for example hours). In this situation, the employer still wishes to continue the relationship, albeit on different terms. The employee has two choices: he can resign or he can continue to work under the new terms. If the employee continues to work and accepts the changed terms, the contract is mutually varied and no action will lie, provided that the employee was given the correct notice before the change was implemented. If the employee resigns, however, he will have been dismissed. In *Walton & Morse v Dorrington* [1997] IRLR 488, the employee waited to find alternative employment before she resigned. The Employment Appeal Tribunal decided that, in her circumstances, this was a reasonable thing to have done and agreed that she had not accepted her employer's breach of its duty to her and had been constructively dismissed (see **E14003** above).

Reasons for dismissal

[E14020] *Section 98* of the *Employment Rights Act 1996* gives the following potentially fair reasons for dismissal. These are:

(a)   capability or qualifications;
(b)   conduct;
(c)   retirement;
(d)   redundancy;
(e)   contravention of statute;
(f)   some other substantial reason.

Dismissal on health and safety grounds could potentially fall within most of these reasons. It should, however, be remembered that, where an employee is dismissed in circumstances where continued employment involves a risk to the employee's health and safety, the employer may nevertheless face claims of unfair dismissal. Before terminating employment, an employer should consider all the circumstances of the case and assess the risk involved and take measures which are reasonably necessary to eliminate the risk.

Illness may make it unsafe to employ the employee; breach of health and safety rules will normally fall under misconduct; to continue to employ the employee may contravene health and safety legislation or it may be that the employer has had to reorganise his business on health and safety grounds and the employee is refusing to accept the change. This latter situation could be potentially fair under 'some other substantial reason'. (However, an employer may be liable under the *Disability Discrimination Act 1995*.)

Reasonableness

[E14021] Merely having a fair reason to dismiss does not mean that the dismissal is fair. *Section 98(4)* of the *Employment Rights Act 1996* requires the tribunal in any unfair dismissal case to consider whether the employer acted reasonably in all the circumstances (including the size and administrative resources of the employer's undertaking). This means that the tribunal will look at two things – (i) was the treatment of the employee procedurally fair, and (ii) was dismissal a reasonable sanction in relation to the employee's actions and all the circumstances of the case.

Procedures have already been discussed at **E14013** above. If the employer has complied with his procedures, he will not be found to have acted procedurally unfairly unless the procedures themselves are unfair. This is unlikely if the employer is following the ACAS Code.

In respect of the fairness of the decision, the tribunal should consider whether the employer's decision to dismiss fell within the band of reasonable responses to the employee's conduct which a reasonable employer could adopt. The tribunal will look at three things – (i) has the employer acted consistently, (ii) has he taken the employee's past work record into account, and (iii) has the employer looked for alternative employment. The latter aspect is of major importance in relation to redundancy, incapability due to illness or dismissal because of a contravention of legislation, but will not usually be relevant in dismissals for misconduct. It is important to note in addition that special obligations apply to an employer in the case of a disabled employee within the meaning of the *Disability Discrimination Act 1995*.

When looking at consistency, the tribunal will look for evidence that the employer has treated the same misconduct the same way in the past. If the employer has treated past breaches of health and safety rules leniently it will be unfair to suddenly dismiss for the same breach, unless he has made it clear to the employees that his attitude has changed and breaches will be dealt with more severely in the future. Employees have to know the potential disciplinary consequences for breaches of the rules, and if the employer has never dismissed in the past he is misleading employees unless he tells them that things have changed. The law, however, only requires an employer to be consistent between cases which are the same. This is where a consideration of the employee's past work record is important. It is not inconsistent to give a long-standing employee with a clean record a final warning for a breach of health and safety rules and to dismiss another shorter-serving employee with a series of warnings behind him, as long as both employees know that the penalty for breach of the rules could be dismissal. The cases are not the same. It would, however, be unfair if both the employees had the same type of work record and length of service and only one was dismissed, and dismissal had never been imposed as a sanction for that type of breach in the past. In order for a misconduct dismissal to be fair, the employer must have had a reasonable belief in the guilt of the employee of the misconduct in question on the basis of a reasonable investigation.

Remedies for unfair dismissal

[E14022] The remedies for unfair dismissal are:

(a) Reinstatement

The first remedy that the tribunal is required to consider is reinstatement of the employee. When doing so the tribunal must take into account whether the employee wishes to be reinstated, whether it is practicable for the employer to reinstate him and, if the employee's conduct contributed to or caused his dismissal, whether it is just to reinstate him. In order to resist an order for reinstatement, an employer must provide evidence to show that it is not practicable because the implied term of mutual trust and confidence between employer and employee has broken down (*IPC Magazines Ltd v Clements*

(EAT/456/99); *Gentle v Perkins Engines Peterborough Ltd (formerly Perkins Group Ltd)* [2001] All ER (D) 360 (Jul), EAT). Reinstatement means that the employee must return to his old job with no loss of benefits. If reinstatement is ordered and the employer refuses to comply with the order, or only partially complies, compensation will be increased. Reinstatement, however, is rarely ordered by tribunals.

(b) Re-engagement

If the tribunal does not consider that reinstatement is practicable, it must consider whether to order the employer to re-engage the employee. In making its decision the tribunal looks at the same factors as when it considers reinstatement. Re-engagement is an order requiring the employer to re-employ the employee on terms which are as favourable as those he enjoyed before his dismissal, but it does not require the employer to give the employee the same job back. Failure on the part of the employer to comply with an order of re-engagement will lead to increased compensation, although tribunals rarely make re-engagement orders.

(c) Compensation

Compensation falls under a variety of different heads. In an unfair dismissal case the employee will receive:

Basic award: This is based on his age, years of service and salary –

(i)   one and a half weeks' pay for each year of service over the age of 41;
(ii)   one week's pay for each year of service between 41 and 22;
(iii)   half a week's pay for each year of service below the age of 22.

This is subject to a statutory maximum of £350 a week (2009/2010 figures), and a maximum of twenty years' service. The maximum basic award is therefore £10,500 (2009/10). Compensation is reduced by one-twelfth for each month the employee works during his 64th year. Where the employee is unfairly dismissed for health and safety reasons under the *Employment Rights Act 1996*, s 100, the minimum basic award is £4,700 (2009/10 figures).

Compensatory award: This is payable in addition to the basic award to compensate the employee for loss of future earnings, benefits etc, which are in excess of the basic award. As with the basic award the compensatory award can be reduced for contributory conduct. The present maximum compensatory award that can be made in most cases is £66,200 (2009/10 figures).

Additional award: If the employer fails to comply with a reinstatement or re-engagement order, the tribunal may make an additional award. This will be between 26 and 52 weeks' pay (subject to a maximum of £350 per week (2009/10 figures)).

Note: There is no limit on:

•   the compensatory award for employees who are dismissed for health and safety reasons or in whistleblowing cases (see **E14010**);
•   awards of compensation in sex discrimination and equal pay cases (and interest can be included in such awards);
•   the amount of the compensation that a tribunal may award in a case of discrimination under the *Disability Discrimination Act 1995*.

In cases of sex and disability discrimination, compensation may include compensation for injury to feelings.

Health and safety duties in relation to women at work

Sex discrimination

[E14023] Since health and safety issues may give rise to sex discrimination claims under the *Sex Discrimination Act 1975 (SDA 1975)*, it is appropriate to examine what particular considerations an employer needs to bear in mind in its relations with female employees.

The steps necessary to be taken by an employer, in order to comply with his duties under *HSWA 1974, s 2*, may differ for women.

New or expectant mothers are particularly vulnerable to adverse working conditions. Indeed, most employers have probably taken measures to guard against risks to new and expectant mothers, in accordance with their general duties under *HSWA 1974, s 2*, and the *Management of Health and Safety at Work Regulations 1999 (SI 1999 No 3242)*. In addition, employers are required to protect new and expectant mothers in their employment from certain specified risks if it is reasonable to do so, and to carry out a risk assessment of such hazards. If the employer cannot avoid the risk(s), he must alter the working conditions of the employee concerned or the hours of work, offer suitable alternative work and, if no suitable alternative work is available, suspend the employee on full pay (see **E14024** below).

Although *SDA 1975* prohibits discrimination on grounds of sex, *s 51(1)* provides that any action taken to comply with certain existing health and safety legislation (e.g. *HSWA 1974*) will *not* amount to unlawful discrimination. In *Page v Freight Hire (Tank Haulage) Ltd* [1981] IRLR 13, the complainant was an HGV driver. The employer, acting on the instructions of the manufacturer of the chemical dimethylformamide (DMF), refused to allow her to transport the chemical which was potentially harmful to women of child bearing age. She brought a complaint of sex discrimination. It was held that the fact that the discriminatory action was taken in the interests of safety did not of itself provide a defence to a complaint of unlawful discrimination. However, the employer was protected by *s 51(1)* of *SDA 1975* because the action taken was necessary to comply with the employer's duty under *HSWA 1974*.

Pregnant workers, new and breastfeeding mothers

[E14024] *The Management of Health and Safety at Work Regulations 1999 (SI 1999 No 3242)* impose a duty on employers to protect new or expectant mothers from any process or working conditions or certain physical, chemical and biological risks at work (see **E14025** below). The phrase 'new or expectant mother' is defined as a worker who is pregnant, who has given birth within the previous six months, or who is breastfeeding. 'Given birth' is defined as having delivered a living child or, after 24 weeks of pregnancy, a stillborn child.

Risk assessment

[E14025] The 1999 Regulations require employers to carry out an assessment of the specific risks posed to the health and safety of pregnant women and new mothers in the workplace and then to take steps to ensure that those risks are avoided. Risks include those to the unborn child or child of a woman who is still breastfeeding – not just risks to the mother.

An interesting development in relation to this requirement is the case of *Day v T Pickles Farms Ltd* [1999] IRLR 217, where the employee suffered nausea when pregnant as a result of the smell of food at her workplace. As a result of the nausea, she was unable to work and was eventually dismissed after a prolonged absence.

The Employment Appeal Tribunal found that she had not been constructively dismissed. However, the Employment Appeal Tribunal held that the obligation to carry out a risk assessment which considers possible risks to the health and safety of a pregnant female employee is relevant from the moment an employer employs a woman of childbearing age. The question of whether the applicant had been subjected to a detriment was remitted to the employment tribunal.

If an employer suspends a new or expectant mother to avoid health and safety risks, they must provide evidence of the risk and show that it could not otherwise have been avoided. In *New Southern Railway Ltd v Quinn* [2006] IRLR 266, a pregnant employee was removed from her duties as a duty station manager because of the risk of physical assault. Her salary was also reduced to reflect her change of duties and she brought a claim for sex discrimination. The EAT upheld her claim, finding that her employer had suspended her because of a "paternalistic and patronising attitude" rather than for any real health and safety reasons.

The HSE has published a booklet entitled '*New and expectant mothers at work – A guide for employers*'. The booklet provides guidance on what employers need to do to comply with the legislation. The booklet includes a list of the known risks to new and expectant mothers and suggests methods of avoidance.

The main risks to be avoided are as follows:

- Physical agents:
    - shocks/vibrations/movement (including travelling and other physical burdens),
    - handling of loads entailing risks,
    - noise,
    - non-ionising radiation,
    - extremes of heat and cold.
- Biological agents:
    - such as listeria, rubella and chicken pox virus, toxoplasma, cytomegalovirus, hepatitis B and HIV.
- Chemical agents:
    - such as mercury, antimiotic drugs, carbon monoxide, chemical agents of known and percutaneous absorption and chemicals listed under various Directives.

- Working conditions:
  - such as mining work and work with display screen equipment (VDUs).

Where a risk has been identified following the assessment, affected employees or their representatives should be informed of the risk and the preventive measures to be adopted. The assessment should be kept under review.

In particular, employers must consider removing the hazard or seek to prevent exposure to it. If a risk remains after preventive action has been taken, the employer must take the following course of action:

(i)   temporarily adjust her working conditions or hours of work (*Management of Health and Safety at Work Regulations 1999 (SI 1999 No 3242), Reg 16(2)*).

If it is not reasonable to do so or would not avoid the risk:

(ii)   offer suitable alternative work (*Employment Rights Act 1996, s 67*).

If neither of the above options is viable:

(iii)   suspend her on full pay for as long as necessary to protect her health and safety or that of her child (*Management of Health and Safety at Work Regulations 1999 (SI 1999 No 3242), Regs 16(2)* and *16(3); Employment Rights Act 1996, s 67*).

Appendix 1 of the booklet lists aspects of pregnancy such as morning sickness, varicose veins, increasing size etc. that may affect work and which employers may take into account in considering working arrangements for pregnant and breastfeeding workers. These are merely suggestions and not requirements of the law.

Night work by new or expectant mother

[E14026] Where a new or expectant mother works at night and has been issued with a certificate from a registered doctor or midwife stating that night work would affect her health and safety, the employer must first offer her suitable alternative daytime work, and suspend her as detailed at **E14015** above if no suitable alternative employment can be found (*Management of Health and Safety at Work Regulations 1999 (SI 1999 No 3242), Reg 17*).

Notification

[E14027] An employer is not required to alter a woman's working conditions or hours of work or suspend her from work under *Management of Health and Safety at Work Regulations 1999 (SI 1999 No 3242), Reg 16(2)* or *(3)* until she notifies him in writing that she is pregnant, has given birth within the previous six months or is breastfeeding. Additionally, the suspension or amended working conditions do not have to be maintained if the employee fails to produce a medical certificate confirming her pregnancy in writing within a reasonable time if the employer requests her to do so. The same applies once the employer knows that the employee is no longer a new or expectant mother or if the employer cannot establish whether she remains so (*Management of Health and Safety at Work Regulations 1999 (SI 1999 No 3242), Reg 18*).

However, an employer has a general duty under *HSWA 1974* and the *Management of Health and Safety at Work Regulations 1999* to take steps to protect the health and safety of a new or expectant mother, even if she has not given written notification of her condition.

Maternity leave

[E14028] All pregnant workers have a right to 26 weeks' ordinary maternity leave (OML) plus a further 26 weeks' additional maternity leave (ADL), regardless of length of service and number of hours worked (although to get OML she must still be in work into the 15th week before the baby is due). The ordinary maternity leave period starts from:

(a)   the notified date of commencement; or
(b)   the first day of absence because of pregnancy or childbirth after the beginning of the fourth week before the expected week of confinement; or
(c)   the date of childbirth,

whichever is the earlier, and continues for 26 weeks or until the end of the compulsory leave period if later. Additional maternity leave continues until the end of the period of 26 weeks from the day after the last day of ordinary maternity leave.

If an employee is prohibited from working for a specified period after childbirth by virtue of a legislative requirement (e.g. under the *Public Health Act 1936, s 205*), her maternity leave period must continue until the expiry of that later period. The *Employment Rights Act 1996, s 72(1),* provide that an employee entitled to maternity leave should not work or be permitted to work by her employer during the period of two weeks beginning with the date of childbirth.

Enforcement

Hannah Wilson

Introduction

[E15001] Prior to the introduction of the *Health and Safety At Work etc Act 1974 (HSWA 1974)*, the principal sanction against health and safety offences was prosecution. *HSWA 1974* provides enforcing authorities with a wider range of enforcement powers, which do not necessarily rely on prosecution for their efficacy. Most notable is the power to serve improvement and prohibition notices.

The HSE aims to achieve effective enforcement that is transparent, fair and proportionate to the risk. The current approach to enforcement is shaped by the '*Strategy for Workplace Health and Safety in Great Britain to 2010 and beyond*' launched in February 2004 by the HSC. The HSE's Enforcement Guides for England and Wales and for Scotland are available online at www.hse.gov.uk/enforce/index.htm. Whilst the guidance contained in these documents is informative on how HSE uses its discretion in making decisions on enforcement it does not override the requirement for HSE to consider each case in accordance with the latest HSE Enforcement Policy Statement dated February 2009, the Code for Crown Prosecutors in the context of prospective prosecutions, and the law in general. The Code for Crown Prosecutors and the HSE Enforcement Policy Statement are also available to download from the HSE's website.

The stated aims of the current Enforcement Policy Statement include the continuing protection of people through the provision of information and advice, the promotion of a goal-setting system of regulation, and the enforcement of the law where appropriate. The HSE also states its intention to work to encourage the proper recognition of and respect for high standards of health and safety within business, and in particular to show that conforming to these standards can have economic rewards.

The Enforcement Policy states that the purpose of enforcement is to—

- Ensure that duty-holders take action to deal immediately with immediate risks;
- Promote and achieve sustained compliance with the law;
- Ensure that duty-holders who breach health and safety requirements, and directors or managers, who fail in their responsibilities, may be held to account, which may include bringing alleged offenders before the courts in England and Wales, or recommending prosecution in Scotland.

There have been recent fundamental developments in the context of enforcement, notably the introduction of two new acts – the *Corporate Manslaughter and Corporate Homicide Act 2007* (which came into force on 6 April 2008) and the *Health and Safety (Offences) Act 2008* (which came into force on 16 January 2009) – and the publication of the Sentencing Guidelines Council's guideline on sentencing convictions for corporate manslaughter and health and safety offences causing death (which became effective on 15 February 2010).

Policy in England and Wales is that all deaths related to business activities are investigated for possible manslaughter charges. Corporate manslaughter was first formally recognised as an offence in the prosecution that followed the Herald of Free Enterprise disaster in 1987. Since then a number of prosecutions for corporate manslaughter have followed, although in a large proportion the common law offence was found to be wanting. The central difficulty was that in order to secure a conviction it was incumbent on the prosecution to show that one of the company's "directing" or "controlling minds" had been guilty of gross negligence manslaughter. This meant that successful prosecution for corporate manslaughter of large companies was all but impossible. The 2007 Act has overhauled the offence: it now falls on the prosecution to show that the death resulted from serious management failures giving rise to a gross breach of a duty of care.

The *Health and Safety (Offences) Act 2008* revises the mode of trial and maximum penalties applicable to certain health and safety offences. In particular, the 2008 Act has increased the maximum fines that can be imposed by magistrates from £5,000 to £20,000 in respect of many health and safety offences, including breaches of *sections 2 to 6*. The range of offences punishable by a term of imprisonment has also been broadened, and the magistrates' court is now entitled to send a larger number of offences to the Crown Court in order for more stringent penalties to be imposed.

The sentencing guideline for corporate manslaughter and health and safety offences causing death states that companies should be fined minimums of £500,000 if convicted of corporate manslaughter and £100,000 if convicted of breaches of the *HSWA 1974* and/or associated legislation which caused death.

Currently, prosecutions and other enforcement procedures are the responsibility of the appropriate 'enforcing authority', either the HSE or a local authority (see **E15007–E15011** below).

This section deals with the following aspects of enforcement—

- The role of the Health and Safety Executive (see **E15002–E15006**).
- The 'enforcing authorities' (see **E15007–E15011** below).
- The 'relevant statutory provisions', which can be enforced under *HSWA 1974* (see **E15012** below).

Part A: Enforcement Powers of Inspectors'

- Improvement and prohibition notices (see **E15013–E15018** below).
- Appeals against improvement and prohibition notices (see **E15019, E15020** below).
- Grounds for appeal against a notice (see **E15021–E15025** below).
- Inspectors' investigation powers (see **E15026** below).
- Inspectors' powers of search and seizure (see **E15027** below).

- Indemnification by enforcing authority (see **E15028** below).
- Public register of notices (see **E15029** below).

Part B: Offences and Penalties

- Prosecution for contravention of the relevant statutory provisions (see **E15030, E15031** below).
- Main offences and penalties (see **E15032–E15038** below).
- Offences committed by particular types of persons, including the Crown (see **E15039–E15045** below).
- Sentencing guidelines (see **E15043** below).

The role of the Health and Safety Executive

Merger of the HSE and HSC

[E15002]–[E15004] Readers will be aware that the HSC and the HSE were originally created as two separate, non-departmental bodies with working practices and delegated powers designed to distance the Commission from the everyday workings of the Executive. The HSC, which consisted of a chairman and six to nine members, promoted the general aims of *HSWA 1974*, and was instrumental in the development of law, codes of practice and arrangements for research and training. The HSE was tasked with the practical enforcement of existing law.

On 1 April 2008 the HSE and the HSC merged to form a single national regulatory body, called the HSE, formally tasked with the general responsibility of promoting improved health and safety in the workplace. The decision to merge the HSC and the HSE followed consultation with interested parties, and was arrived at pursuant to the process set out in the *Legislative and Regulatory Reform Act 2006*.

A press release emanating from the Department of Work and Pensions set out the practical ramifications of the merger—

- The creation of a single national regulatory body responsible for promoting the cause of better health and safety at work
- The current Chair of the HSC will become the Chair of the Board of the new Executive
- Existing Commissioners will be appointed as non-executive directors of the new Executive for the remainder of their terms of office with the relevant responsibilities of the new roles
- The potential size of the Board of the new Executive will be no more than eleven members, plus the Chair and members will continue to be appointed by the Secretary of State
- All fundamental contents of *HSWA 1974* will remain
- None of the statutory functions of the previous Commission and Executive will be removed
- There will be no change in the health and safety requirements, how they are enforced or how stakeholders relate to the health and safety regulator – no health and safety protections will be removed

Functions of the new Health and Safety Executive (HSE)

[E15005] The new HSE now has a dual function having assumed the responsibilities of the HSC. *HSWA 1974 s 11* has been amended, and sets out the HSE's duties in respect of the "general purposes" (broadly, the securing of the health, safety and welfare of persons at work as defined in *HSWA 1974 s 1*) as follows—

In connection with the general purposes of this Part, the Executive shall—

(a)   assist and encourage persons concerned with matters relevant to those purposes to further those purposes;

(b)   make such arrangements as it considers appropriate for the carrying out of research and the publication of the results of research and the provision of training and information, and encourage research and the provision of training and information by others;

(c)   make such arrangements as it considers appropriate to secure that the following persons are provided with an information and advisory service on matters relevant to those purposes and are kept informed of and are adequately advised on such matters—

(i)   government departments,

(ii)   local authorities,

(iii)   employers,

(iv)   employees,

(v)   organisations representing employers or employees, and

(vi)   other persons concerned with matters relevant to the general purposes of this Part.

The HSE will also draw up proposals for regulations, replacing/updating existing statute and statutory instruments, and will prepare approved codes of practice. Under *HSWA 1974 s 14* the HSE has the power, previously the sole domain of the HSC, to conduct and direct investigations and inquiries with a view to formulating new regulations. It will also adopt one of the most important functions of the old HSC, which was the preparation of draft regulations for approval by the Secretary of State.

As with the old HSE, the new HSE is tasked with the enforcement of health and safety legislation (unless responsibility lies with another enforcing body, such as a local authority). [*HSWA 1974 s 18(1)*].

The HSE's enforcement functions are principally performed by its inspectors and include the carrying out of routine inspections of premises, and the investigation of workplace accidents, dangerous occurrences or cases of ill health. Inspectors also provide advice to companies and individuals on the legal requirements under health and safety legislation and in respect of compliance with notices. The HSE publishes guidance documents (see E15006 below), provides an information service, carries out research and licenses or approves certain hazardous operations, such as nuclear site licensing.

The *Health and Safety (Fees) Regulations 2009 (SI 2009 No 515)* also provide that the HSE may, in certain limited circumstances, charge for their services. In particular, fees are payable for (inter alia)—

(a)   an original approval or the amendment or renewal of an existing approval under any of the mines and quarries provisions (*reg 2*);

(b)   an application for approval of plant and equipment under the *Agriculture (Tractor Cabs) Regulations 1974 (SI 1974 No 2034) (reg 3)*;

(c)   an application for approval of a scheme or programme for examination of freight containers under the *Freight Containers (Safety Convention) Regulations 1984 (SI 1984 No 1890) (reg 4)*;

(d)   an application for a licence under the *Control of Asbestos Regulations 2006 (SI 2006 No 2739) (reg 5)*;

(e)   various applications in connection with the *Ionising Radiations Regulations 1999 (SI 1999 No3232)* and the *Radiation (Emergency Preparedness and Public Information) Regulations 2001 (SI 2001 No 2975) (reg 8)*;

(f)   various applications under the *Manufacture and Storage of Explosives Regulations 2005 (SI 2005 No 1082)* and certain other provisions concerning explosives, including acetylene, and under the *Petroleum (Consolidation) Act 1928* and the *Petroleum (Transfer of Licences) Act 1936 (reg 9)*;

(g)   an application for or changes to an explosives licence under *Part IX* of the *Dangerous Substances in Harbour Areas Regulations 1987 (SI 1987 No 37) (reg 11)*;

(h)   notifications and applications under the *Genetically Modified Organisms (Contained Use) Regulations, SI 2000/2832 (reg 13)*; and

(i)   enforcement provisions in relation to offshore installations (*reg 14*).

HSE publications

[E15006] The HSE, through HSE Books (http://books.hse.gov.uk/hse/public/home.jsf), also publishes guidance notes and advisory literature for employers, local authorities, trade unions etc. on most aspects of health and safety of concern to industry.

Publications include books on wide-ranging subjects, such as *"Essentials of Health and Safety at Work"*, and leaflets on more specialised subjects, such as gas safety in the context of the catering and hospitality industries.

Many of the publications are free and can be downloaded in pdf or Word format.

Enforcing authorities

[E15007] Whilst the HSE is the central body entrusted with the enforcement of health and safety legislation, in any given case enforcement powers rest with the body which is expressed by statute to be the 'enforcing authority'. Here the general rule is that the 'enforcing authority', in the case of industrial premises, is the HSE and, in the case of commercial premises within its area, the local authority (the enforcing authority in over a million premises), except that the HSE cannot enforce provisions in respect of its own premises, and similarly, local authorities' premises are inspected by the HSE. Each 'enforcing authority' is empowered to appoint suitably qualified persons as inspectors for the purpose of carrying into effect the 'relevant statutory provisions' within the authority's field of responsibility. [*HSWA 1974 s 19(1)*]. Inspectors so appointed can exercise any of the enforcement powers conferred by the *HSWA 1974* (see **E15013–E15018** below) and bring prosecutions (see **E15030, E15031** below).

The appropriate 'enforcing authority'

[E15008] The general rule is that the HSE is the enforcing authority, except to the extent that—

(a) regulations specify that the local authority is the enforcing authority instead; the regulations that so specify are the *Health and Safety (Enforcing Authority) Regulations 1998 (SI 1998 No 494)* or;

(b) one of the 'relevant statutory provisions' specifies that some other body is responsible for the enforcement of a particular requirement.

[*HSWA 1974 s 18(1), (7)(a)*].

Activities for which the HSE is the enforcing authority

[E15009] The HSE is specifically the enforcing authority in respect of the following activities (even though the main activity on the premises is one for which a local authority is usually the enforcing authority in accordance with the *Health and Safety (Enforcing Authority) Regulations 1998 (SI 1998 No 494) Sch 2* (see **E15010** below)—

(1) Any activity in a mine or quarry;

(2) Any activity in a fairground;

(3) Any activity in premises occupied by a radio, television or film undertaking in which the activity of broadcasting, recording or filming is carried on;

(4) The following work carried out by independent contractors—

    (a) certain construction work including where the project includes the work is notifiable within the meaning of *reg 2(3)* of the *Construction (Design and Management) Regulations 2007 (SI 2007 No 320)*;

    (b) installation, maintenance or repair of gas systems or work in connection with a gas fitting;

    (c) installation, maintenance or repair of electricity systems;

    (d) most work with ionising radiations;

(5) Use of ionising radiations for medical exposure;

(6) Any activity in premises occupied by a radiography undertaking where work with ionising radiations is carried out;

(7) Agricultural activities, including agricultural shows where livestock or agricultural equipment is handled;

(8) Any activity on board a sea-going ship;

(9) Ski slope, ski lift, ski tow or cable car activities;

(10) Fish, maggot and game breeding (but not in a zoo);

(11) Any activity in relation to a pipeline within the meaning of *reg 3* of the *Pipelines Safety Regulations 1996*;

(12) The operation of a guided bus system or any other system of guided transport, other than a railway, that employs vehicles which for some or all of the time when they are in operation travel along roads; and

(13)   The operation of a trolley vehicle system.

[*Health and Safety (Enforcing Authority) Regulations 1998 (SI 1998 No 494) reg 4(4)(b)* and *Sch 2*].

The HSE is the enforcing authority against the following, and for any premises they occupy, including parts of the premises occupied by others providing services for them. (This is so even though the main activity is listed in *Sch 1* of the *1998 Regulations*, see **E15010** below)—

(1)   a county council;
(2)   any other local authority (as defined in *reg 2*);
(3)   a parish council in England or a community council in Wales or Scotland;
(4)   a police authority or the Receiver for the Metropolitan Police District;
(5)   a relevant authority as defined in *s 6* of the *Fire (Scotland) Act 2005*;
(6)   a fire and rescue authority under the *Fire and Rescue Services Act 2004*;
(7)   a headquarters or an organisation designated for the purposes of the *International Headquarters and Defence Organisation Act 1964*; or a service authority of a visiting force within the meaning of *s 12* of the *Visiting Forces Act 1952*;
(8)   the United Kingdom Atomic Energy Authority;
(9)   the Crown, although not where the premises are occupied by the HSE itself

[*Health and Safety (Enforcing Authority) Regulations 1998 (SI 1998 No 494) reg 3(3)*].

The HSE is also the enforcing authority for the premises set out below, even if occupied by more than one occupier—

(1)   the Channel Tunnel;
(2)   an offshore installation
(3)   a building or construction site;
(4)   a campus or other premises occupied by an educational establishment;
(5)   a hospital.

[*Health and Safety at Work (Enforcing Authority) Regulations 1998 (SI 1998 No 494) reg 3(5)*].

Finally, the HSE is the enforcing authority for the following—

(1)   Common parts of domestic premises (*reg 3(1)*);
(2)   Certain areas within an airport (*reg 3(4)(b)*);

Activities for which local authorities are the enforcing authorities

[E15010] Where the main activity carried on in non-domestic premises is one of the following, the local authority is the enforcing authority (ie the relevant county, district or borough council)—

(1)   Sale or storage of goods for retail/wholesale distribution (including sale and fitting of motor car tyres, exhausts, windscreens or sunroofs), except—
       (a)   at container depots where the main activity is the storage of goods which are of transit to or from dock premises, an airport or railway;

(b) where the main activity is the sale or storage for wholesale distribution of dangerous substances;

(c) where the main activity is the sale or storage of water or sewage or their by-products or natural or town gas.

(2) Display or demonstration of goods at an exhibition, being offered or advertised for sale.

(3) Office activities.

(4) Catering services.

(5) Provision of permanent or temporary residential accommodation, including sites for caravans or campers.

(6) Consumer services provided in a shop, except—

(a) dry cleaning;

(b) radio/television repairs.

(7) Cleaning (wet or dry) in coin-operated units in laundrettes etc.

(8) Baths, saunas, solariums, massage parlours, premises for hair transplant, skin piercing, manicuring or other cosmetic services and therapeutic treatments, except where supervised by a doctor, dentist, physiotherapist, osteopath or chiropractor.

(9) Practice or presentation of arts, sports, games, entertainment or other cultural/recreational activities, save where the main activity is the exhibition of a cave to the public.

(10) Hiring out of pleasure craft for use on inland waters.

(11) Care, treatment, accommodation or exhibition of animals, birds or other creatures, except where the main activity is—

(a) horse breeding/horse training at stables;

(b) agricultural activity;

(c) veterinary surgery.

(12) Undertaking, but not embalming or coffin making.

(13) Church worship/religious meetings.

(14) Provision of car parking facilities within an airport.

(15) Childcare, playgroup or nursery facilities.

[*Health and Safety (Enforcing Authority) Regulations 1998 (SI 1998 No 494) reg 3(1) and Sch 1*].

Transfer of responsibility between the HSE and local authorities

[E15011] Enforcement can be transferred (though not in the case of Crown premises), by prior agreement, from the HSE to the local authority and vice versa. Parties who are affected by such transfer must be notified. [*Health and Safety (Enforcing Authority) Regulations 1998 (SI 1998 No 494) reg 5*]. Transfer is effective even though the above procedure is not followed (ie the authority changes when the main activity changes) (*Hadley v Hancox* (1987) 85 LGR 402, decided under the previous Regulations).

Where there is uncertainty, these Regulations also allow responsibility to be assigned by the HSE and the local authority jointly to either body. [*Health and Safety (Enforcing Authority) Regulations 1998 (SI 1998 No 494) reg 6(1)*].

'Relevant statutory provisions' covered by HSWA 1974

[E15012] The enforcement powers conferred by *HSWA 1974* extend to any of the 'relevant statutory provisions'. These comprise—

(a)   the provisions of *HSWA 1974 Part I* (ie *ss 1–53*);

(b)   any health and safety regulations passed under *HSWA 1974*, eg the *Ionising Radiations Regulations 1999 (SI 1999 No 3232)*, the *Management of Health and Safety at Work Regulations 1999 (SI 1999 No 3242)* as amended and the *Workplace (Health, Safety and Welfare) Regulations 1992 (SI 1992 No 3004)* as amended; and

(c)   the 'existing statutory provisions', ie all enactments specified in *HSWA 1974 Sch 1*, including any regulations etc. made under them, so long as they continue to have effect; that is, the *Explosives Acts 1875–1923*, the *Mines and Quarries Act 1954*, the *Factories Act 1961*, the *Public Health Act 1961*, the *Offices, Shops and Railway Premises Act 1963* and (by dint of the *Offshore Safety Act 1992*) the *Mineral Workings (Offshore Installations) Act 1971* [*HSWA 1974 s 53(1)*].

Part A: Enforcement Powers of Inspectors

Improvement and prohibition notices

[E15013] It was recommended by the Robens Committee that 'inspectors' should have the power, without reference to the courts, to issue a formal improvement notice to an employer requiring him to remedy particular faults or to institute a specified programme of work within a stated time limit.' (*Cmnd 5034, para 269*). 'The improvement notice would be the inspector's main sanction. In addition, an alternative and stronger power should be available to the inspector for use where he considers the case for remedial action to be particularly serious. In such cases he should be able to issue a prohibition notice.' (*Cmnd 5034, para 276*). *HSWA 1974* put these recommendations into effect.

Crown bodies are exempt from the provisions relating to improvement and prohibition notices. Instead, Crown notices may be issued against a relevant body. The notice is not legally binding, and the Crown cannot be prosecuted for breach of the notice [see below].

Improvement notices

[E15014] An inspector may serve an improvement notice if he is of the opinion that a person—

(a)   is contravening one or more of the 'relevant statutory provisions' (see E15012 above); or

(b)   has contravened one or more of those provisions in circumstances that make it likely that the contravention will continue or be repeated.

[*HSWA 1974 s 21*].

In the improvement notice the inspector must—

(i)   state that he is of the opinion in (*a*) and (*b*) above; and
(ii)   specify the provision(s) in his opinion contravened; and
(iii) give particulars of the reasons for his opinion; and
(iv)   specify a period of time within which the person is required to remedy the contravention (or the matters occasioning such contravention).

[*HSWA 1974 s 21*].

The period specified in the notice within which the requirement must be carried out (see (iv) above) must be at least 21 days – this being the period within which an appeal may be lodged with an Employment Tribunal (see **E15019** below). [*HSWA 1974 s 21*]. In order to be validly served on a company, an improvement notice relating to the company's actions as an employer must be served at the registered office of the company, not elsewhere. Service at premises occupied by the company will only be valid if the notice relates to a contravention by the company in the capacity of occupier (*HSE v George Tancocks Garage (Exeter)* [1993] Crim LR 605 and *HSWA 1974 s 46*).

There is no requirement that inspectors give duty-holders written notice of an intention to serve an improvement notice. However, the HSE's website invites inspectors to discuss with the duty-holder the contents of the notice and the date for compliance so that any difficulties can be identified and resolved. In particular, inspectors are advised to make potential recipients of an improvement notice aware of the fact that the time for compliance can be extended. It is also recommended that inspectors explain the consequences of non-compliance.

Failure to comply with an improvement notice can have serious penal consequences (see **E15036** below).

Under *s 23(5)* an improvement notice can be withdrawn before the time for compliance has expired. This will be appropriate where the situation changed after the improvement notice was issued, or where further information has come to light which has made the notice inappropriate. If a new improvement notice is necessary the old one should be formally withdrawn.

Prohibition notices

[E15015] If an inspector is of the opinion that, with regard to any activities to which *s 22(1)* applies (see below), the activities involve or will involve a risk of serious personal injury, he may serve on the person carrying out the activities a notice (a prohibition notice). [*HSWA 1974 s 22(2)*]. Personal injury is defined in *HSWA 1974 s 53*.

A prohibition notice must—

(a)   state that the inspector is of the opinion that the activities create a risk of serious personal injury;
(b)   specify the matters which, in the inspector's opinion, create or will create the risk in question;

(c)   where there is actual or anticipatory breach of provisions and regulations, state that the inspector is of the opinion that this is so and give reasons;

(d)   direct that the activities referred to in the notice must not be carried out on, by or under the control of the person on whom the notice is served, unless the matters referred to in (b) above have been remedied.

[*HSWA 1974 s 22(3)*].

Failure to comply with a prohibition notice can have serious penal consequences (see **E15028** below).

It is incumbent on an inspector to show, on a balance of probabilities, that there is a risk to health and safety (*Readmans Ltd and Another v Leeds* CC [1993] COD 419 where an environmental health officer served a prohibition notice on the appellant regarding shopping trolleys with child seats on them, following an accident involving an eleven-month-old baby. The appellant alleged that the Employment Tribunal had wrongly placed the burden of proof on it by requiring it to show that the trolleys were not dangerous. It was held by the High Court (Roch J allowing the appeal), that it was for the inspector to prove that there was a health and/or safety risk).

The risk of personal injury does not need to be imminent (*Tesco v Kippax* (COIT No 7605, HSIB 180 p 8)

It should be noted however, that a company's temporary cessation of an activity does not preclude an inspector from issuing a prohibition notice. In the case of *Railtrack v Smallwood* [2001] EWHC Admin 78, [2001] ICR 714, Mr Justice Sullivan considered an appeal by Railtrack following the decision of an Employment Tribunal that a prohibition notice issued by an HSE inspector, Mr Smallwood, following the Ladbroke Grove train crash, was valid notwithstanding that activities on the part of the track to which the notice applied, had ceased. Mr Justice Sullivan stated that—

> 'In my judgment, it would be very surprising, and a significant lacuna in the Act if an Inspector, in the aftermath of such a serious accident, when operations have been suspended precisely because of the gravity of the accident, was unable to issue a prohibition notice upon the basis of a present risk of serious personal injury.'

He went on to consider the intention behind *s 22* and concluded that—

> 'Looking at the words of *s 22(1)* in this context, and bearing in mind in particular that *s 22* must have been intended to confer powers upon Inspectors, not merely prior to, but also in the aftermath of, the most serious accidents, I am satisfied that "activities" are (still) being carried on for the purposes of *s 22* if they have been temporarily interrupted or suspended as a result of a major accident. I do not consider that such an interpretation does any violence to the ordinary meaning of the words in *s 22*, provided they are considered in a realistic context.'

In his judgment Mr Justice Sullivan stated that all the surrounding circumstances must be considered, including the reason for the temporary inactivity—

> 'There may well be a distinction to be drawn on the facts between a factory that is inactive because the employer has gone bankrupt and dismissed all of his employees, and a factory that is inactive because it has closed for the summer holidays. If,

however, the factory has just closed, and the workers had been sent home, because of a tragic accident, I do not consider that activities will have ceased for the purposes of s 22. The position may be different if, not as a result of an accident, but pursuant to a pre-planned closure programme, for example, to replace outdated and unsafe machinery, the factory is at a standstill for some weeks or months.

. . . I am satisfied that . . . even though there are temporary interruptions when nothing is actually taking place on the ground, so "activities" will continue for the purposes of s 22 even though they have been interrupted as a result of a serious accident.'

Railtrack's appeal was subsequently dismissed.

In *Chilcott v Thermal Transfer Ltd* [2009] EWHC 2086 (Admin), [2009] All ER (D) 94 (Aug) analogous circumstances produced a different outcome. Charles J considered *Railtrack v Smallwood* and affirmed the approach that is to be taken by a tribunal hearing an appeal from an improvement or prohibition notice. In this case, the appellant company acted as a main contractor on a construction site, and had engaged sub-contractors to undertake platform steelwork to enable the installation of cooling towers. Risk assessments and method statements had been prepared which identified the potential risk to operatives of falls from height. Handrails were to be installed on the platforms to protect against this risk; the installation was to be done by operatives working on and from a mobile platform ("mewp"). On 29 July 2008 one of the sub-contractor's employees, a Mr Campbell, was seen to be standing on the platform. He was not stopped because the appellant's representative was not aware that the method statement stated that all work was to be done from the mewp without operatives standing on the platform. Later that day, Mr Campbell fell from the platform and broke both of his ankles. Mr. Chilcott, the HSE's inspector, served a prohibition notice forbidding further "work at height on the access platform for the new cooling towers" until contraventions of s 3 of the *HSWA 1974* and *Work at Height Regulations 2005* had been remedied. On 30 July 2008 Mr Chilcott indicated that he would be happy to "lift" the notice, even though once made, a prohibition notice of immediate effect cannot latterly be withdrawn.

The company appealed to an Employment Tribunal, which cancelled the notice on the basis that, with hindsight, there was no risk that the sub-contractor would go on to the platform that night, and that in those circumstances the inspector could simply have obtained an assurance that no work on the handrail would be done in the following 24 hours, and that all work thereafter would be carried out in accordance with the method statement. The tribunal deemed the service of the notice premature in the circumstances. The HSE appealed.

Charles J held that the Employment Tribunal had taken the wrong approach to the appeal that was before it and had effectively erred in law (see paragraphs 21 and 22 of the judgment). Charles J quoted with approval, at paragraph 5 of his judgment, the following passage from paragraph 44 of *Railtrack v Smallwood* [2001] EWHC Admin 78, [2001] ICR 714 in which Sullivan J stated:

In light of those factors [as contained in s 24 HSWA 1974], and of the authorities cited in *De Smith Woolf & Jowell's Judicial Review of Administrative Law* (1999), pp 251 – 252, paragraph 6-010, I expressed the provisional view during the course of argument that a Tribunal hearing an appeal under *section 24* of the *1974 Act* was not limited to reviewing the genuineness and/or reasonableness of the Inspector's opinions. It was required to form its own view, paying due regard to the Inspector's expertise, see in particular *Sagnata Investments Ltd v Norwich Corporation* [1971] 2 QB 614.

Charles J went on to say that it was open to the Employment Tribunal to reach its own decision rather than simply to apply public law principles to the issue of whether the notice should have been served. In reaching its own decision, it was incumbent on the tribunal to focus its attention on the situation "on the ground" at the time that the notice was served, and specifically, to consider whether there was a risk flowing from an activity being carried out at that time. He said, at paragraph 19:

> The Employment Tribunal, on the basis of the evidence it has on the approach I have indicated, can and should itself, having due regard to the view of the Inspector and the Inspector's expertise, and indeed the expertise of the assessors of the Tribunal, assess the risk as at the relevant date. The assessment of risk is a multi-faceted exercise; some factors will be more important than others, and it seems to me inappropriate for me to seek to set guidelines in that context.

The notice in this case was cancelled by Charles J on the basis that the appellant had in place at the time that the notice was served a plan to protect against falls from height. Therefore, the risks from falls, as identified by the Inspector, had actually been considered and plans put in place by the appellant at the time of the notice. The notice consequently was not warranted.

Differences between improvement and prohibition notices

[E15016] Prohibition notices differ from improvement notices in two important ways—

(a)   with prohibition notices, it is not necessary that an inspector believes that a provision of *HSWA 1974* or any other statutory provision is being or has been contravened. Instead, the inspector must consider that there are activities which are or may be capable of causing serious personal injury;

(b)   prohibition notices are therefore served in anticipation of danger.

It is irrelevant that the hazard or danger perceived by the inspector is not mentioned in *HSWA 1974*; it can exist by virtue of other legislation, or even in the absence of any relevant statutory duty. In this way notices are used to enforce the later statutory requirements of *HSWA 1974* and the earlier requirements of the *Factories Act 1961* and other protective occupational legislation.

Unlike an improvement notice, where time is allowed in which to correct a defect or offending state of affairs, a prohibition notice can take effect immediately.

A direction contained in a prohibition notice shall take effect—

(a)   at the end of the period specified in the notice; or

(b)   if the notice so declares, immediately.

[*HSWA 1974 s 22(4)* as substituted by *Consumer Protection Act 1987 Sch 3*].

An improvement notice gives a person upon whom it is served time to correct the defect or offending situation. A prohibition notice, which is a direction to stop the work activity in question rather than put it right, can take effect immediately on issue; alternatively, it may allow time for certain modifications to take place (ie deferred prohibition notice). Both types of notice will generally contain a schedule of work which the inspector will require to be carried out. If the nature of the work to be carried out is vague, the validity of the notice is not affected. If there is an appeal, an Employment Tribunal may, within its powers to modify a notice, rephrase the schedule in more specific terms (*Chrysler (UK) Ltd v McCarthy* [1978] ICR 939).

Effect of non-compliance with notice

[E15017] If, after expiry of the period specified in the notice, or in the event of an appeal, expiry of any additional time allowed for compliance by the Employment Tribunal, an applicant does not comply with the notice or modified notice, he can be prosecuted. It is the HSE's policy, as stated in its Enforcement Guide (which provides legal guidance to its inspectors and members of staff), that prosecution should "normally follow" a failure to comply with a notice. If convicted of contravening a prohibition notice, he may be imprisoned. [*HSWA 1974 s 33(1)(g), (2A)(b)*]. In *R v Kerr; R v Barker* (1996, unreported) the directors of a company were each jailed for four months after allowing a machine which was subject to a prohibition notice – following an accident in which an employee lost an arm – to continue to be operated.

Prosecution for a failure to comply with a notice may fail if—

•   the notice was not properly served
•   is incapable of being complied with because the wording is vague, confusing or unclear; or
•   it was, in the opinion of the court, complied with.

If the court considers that there has been a failure to comply with an improvement or prohibition notice, it may order that the recipient of the notice to take steps to remedy the matters within his/her control within a specified time.

Service of notice coupled with prosecution

[E15018] Where an inspector serves a notice, he may at the same time decide to prosecute for the substantive offence specified in the notice. The fact that a notice has been served is not relevant to the prosecution. Nevertheless, an inspector will not normally commence proceedings until after the expiry of 21 days, i.e. until he is satisfied that there is to be no appeal against the notice or until the Employment Tribunal has heard the appeal and affirmed the notice, since it would be inconsistent if conviction by the magistrates were followed by cancellation of the notice by the Employment Tribunal. The fact that an

Employment Tribunal has upheld a notice is not binding on a magistrates' court hearing a prosecution under the statutory provision of which the notice alleged a contravention; it is necessary for the prosecution to prove all the elements in the offence (see **E15031** below).

Employment Tribunals are mainly concerned with hearing unfair dismissal claims by employees; only a tiny proportion of cases heard by them relate specifically to health and safety. Moreover, they are not empowered to determine breaches of criminal legislation.

Appeals against improvement and prohibition notices

[E15019] A person on whom either type of notice is served may appeal to an Employment Tribunal within 21 days from the date of service of the notice. The Employment Tribunal may extend this time where it is satisfied, on application made in writing (either before or after expiry of the 21-day period), that it was not reasonably practicable for the appeal to be brought within the 21-day period. On appeal the Employment Tribunal may either affirm or cancel the notice and, if it affirms it, may do so with modifications in the form of additions, omissions or amendments. [*HSWA 1974 ss 24(2), 82(1) (c)*; *Employment Tribunals (Constitution and Rules of Procedure) Regulations 2004 (SI 2004 No 1861) Sch 4*].

The HSE's Enforcement Guide sets out the test the Employment Tribunal will apply, following the decision in Chilcott:

> Where the Inspector's decision is appealed the Tribunal is not limited to reviewing the genuineness and/or reasonableness of the Inspector's opinions. It was required to form its own view paying due weight to the Inspector's expertise. The test was not the judicial review test as to whether the decision was reasonable. The court should focus on the point at which the notice was served rather than look at the situation with the benefit of hindsight. Their task was to decide what they would have done at that point in time.

Effect of appeal

[E15020] Where an appeal is brought against a notice, the lodging of an appeal automatically suspends operation of an improvement notice, but a prohibition notice will continue to apply unless there is a direction to the contrary from the Employment Tribunal. Thus—

(a)   in the case of an improvement notice, the appeal has the effect of suspending the operation of the notice [*HSWA 1974, s 24(3)(a)*];
(b)   in the case of a prohibition notice, the appeal only suspends the operation of the notice if the Employment Tribunal so directs, on the application of the appellant. The suspension is then effective from the time when the Employment Tribunal so directs.

[*HSWA 1974, s 24(3)(b)*].

Grounds for appeal

[E15021]–[E15023] The most common grounds for appeal involve assertions that—

(a)   The inspector incorrectly interpreted the law or acted in excess of his powers;

(b)   The notice is fundamentally flawed;

(c)   The inspector did not genuinely hold the necessary opinion, or the opinion was based on unreasonable grounds;

(d)   The inspector's opinion, although honestly held, is not one that should be endorsed by the tribunal;

(e)   No contravention of the relevant statutory provision has occurred/will occur, or there is no risk of serious personal injury;

(f)   Breach of law is admitted but the proposed solution is not "practicable" or not "reasonably practicable", or there are no means which would reduce the risk;

(g)   Breach of the law is admitted but the breach is so insignificant that the notice should be cancelled.

Prior to the hearing of an appeal it will be open to the inspector to accept criticisms made of the notice by withdrawing it and serving a new notice in amended terms. Alternatively, the tribunal may affirm the notice and modify it to reflect the criticism.

Inspector's incorrect interpretation of the law

It is doubtful whether many cases have, or indeed would, succeed on this ground. Where regulations impose a strict duty (for example the duty under *reg 6(1)* of the *Provision and Use of Work Equipment Regulations 1992 (SI 1992 No 2932)* to provide work equipment in good repair) there is no scope for argument by the employer. However, where the statute provides a defence, for example, it requires the duty to be carried out 'so far as reasonably practicable' or it provides for a due diligence defence, there is some scope to argue that the inspectors' interpretation of the law is incorrect. For example, in *Canterbury City Council v Howletts and Port Lympne Estates Limited* (1996) 95 LGR 798, a prohibition notice was served on Howletts Zoo following the death of a keeper while he was cleaning the tigers' enclosure. It was Howletts' policy to allow their animals to roam freely; the local authority argued that the zoo's keepers could have carried out their tasks in the tigers' enclosure with the animals secured. The High Court affirmed the Employment Tribunal's decision to set aside the notice, holding that *s 2* of *HSWA 1974* was not intended to render illegal certain working practices simply because they were dangerous.

It can happen that an inspector exceeds his powers under statute by reason of misinterpretation of the statute or regulation (*Deeley v Effer* (COIT No 1/72, Case No 25354/77)). This case involved the requirement that 'all floors, steps, stairs, passages and gangways must, so far as is reasonably practicable, be kept free from obstruction and from any substance likely to cause persons to slip'. [*OSRPA 1963 s 16(1)*] (now replaced by equivalent duties under the *Workplace (Health, Safety and Welfare) Regulations 1992 (SI 1992 No 3004)*. The inspector considered that employees were endangered by baskets of wares in the shop entrance. The Employment tribunal ruled that the notice had to be cancelled, since the only persons endangered were members of the public, and *OSRPA 1963* was concerned with dangers to employees.

Notice is fundamentally flawed

A clear instance of fundamental flaw will be where a notice fails to tell a recipient what is wrong, why it is wrong and what can be done to reduce or remove the risks posed by the activity (see Upjohn J in *Miller Mead v Minister of Housing and Local Government* [1963] 2 QB 196).

In *BT Fleet Ltd v McKenna* [2005] EWHC 387 (Admin), [2005] All ER (D) 284 (Mar) the need for the remedial instructions accompanying a notice to be clear was considered. The case concerned a perceived breach of the manual handling regulations. The inspector serving an improvement notice specified the need for lifting machinery to be installed, and the notice was expressed such that this was the only way that the breach could be remedied. Under the relevant legislation, employers are permitted to take alternative steps to minimise manual handling. Evans-Lombe J, in cancelling the notice, held that where an inspector had elected to include the instructions in the notice he was obliged to ensure that they were clearly worded. He said, at paragraph [18]:

> If the provisions of the relevant statute provide an option to proscribe how the recipient can comply with the notice, and that option is taken, then the specification of how compliance can be effected form part of the notice and, if confusing, may operate to make it an invalid notice.

Inspector did not hold the necessary opinion

Whilst the decision in *Chilcott* has emphasised the need for the Employment Tribunal to make its own decision on the validity and need for a notice as at the date on which it was served, the wording of the judgment makes clear that there will still be circumstances where the Tribunal is required to make an assessment of the inspector's opinion at that time. In order to make this assessment, the Tribunal will need to address:

(a)   Whether the inspector genuinely held the view stated in the notice; and

(b)   If so, whether, at the date of the notice the activities complained of involved or would involve a contravention of the statutory provisions or a risk of serious personal injury.

The inspector will be required to satisfy the Tribunal that he did hold that view, and that it was based on reasonable grounds.

Inspector's opinion not to be endorsed

An example of the inspector holding an honest opinion which was ultimately deemed to be incapable of endorsement is to be found in *Chilcott v Thermal Transfer Limited* (see **E15015** above). In that case, the Inspector genuinely perceived there to be a risk of employees and subcontractors suffering serious personal injury such that the service of a prohibition notice was warranted. The Court of Appeal held that the notice was not, in fact, required because at the forefront of the Inspector's mind was the need for work at height to be planned to avoid or reduce the risks of falls from height, In actuality, a safe system of work and a risk assessment had been devised in that regard. The prohibition notice was therefore cancelled.

No contravention or risk

Employers should note that contesting a prohibition notice on the ground that there has been no legal contravention is pointless, since valid service of a prohibition notice does not depend on legal contravention (*Roberts v Day* (COIT No 1/133, Case No 3053/77)). All that the Inspector must believe is that the activities pose a risk of serious personal injury.

It is suggested that appealing a notice on the basis that the activities do not contravene statutory provisions or pose a risk of injury will require deployment of arguments similar to those outlined above in relation to asserting that an inspector has incorrectly interpreted the law. It is also suggested that it is likely to be extremely difficult to persuade a Tribunal that there has not been or will not be any contravention or any risk in circumstances where an Inspector has deemed it necessary to serve a notice (see *Railtrack v Smallwood*).

Proposed solution not practicable

[E15024] The position in a case where, although breach of the law is admitted, the proposed solution is not considered practicable, depends upon the nature of the obligation. The duty may be strict, or have to be carried out so far as practicable or, alternatively, so far as reasonably practicable. In the first two situations the cost of compliance is irrelevant; in the latter case, where a requirement has to be carried out 'so far as reasonably practicable', cost effectiveness is an important factor but has to be weighed against the risks to health and safety involved in failing to implement the remedial measures identified in the enforcement notice. Where there is a real danger of serious injury the cost of complying with the notice is unlikely to be decisive. Thus in a leading case, the appellant was served with an improvement notice requiring secure fencing on transmission machinery. An appeal was lodged on the ground that the proposed modifications were too costly (£1,900). It was argued that because of the intelligence and integrity of the operators a safety screen costing £200 would be adequate. The Employment tribunal dismissed the appeal: the risk justified the cost (*Belhaven Brewery Co Ltd v McLean* [1975] IRLR 370).

The tribunal in *Associated Dairies v Hartley* [1979] IRLR 171 distilled the following principles from cases, including *Belhaven Brewery*, in respect of treatment of costs considerations—

(a)   The questions for the Tribunal appear to be—
      (i)   Is the requirement of the inspector practicable?
      (ii)   In the whole circumstances is the requirement reasonable?
(b)   The question whether in the whole circumstances the requirement is reasonable, should not be determined having regard solely to the proportion which the risk to be apprehended bears to the sacrifice in money, time or trouble involved in meeting the risk, but it is proper to consider whether the time, trouble and expense of the inspector's requirement are disproportionate to the risk involved to the employees if the Tribunal cancel the improvement notice or affirm it with modifications.

The cost of complying with a notice is likely to carry less weight in the case of a prohibition notice than an improvement notice, as there must be 'a risk of serious personal injury' for a prohibition notice to be served (*Nico Manufacturing Co Ltd v Hendry* [1975] IRLR 225, where the company argued that a prohibition notice in respect of the worn state of their power presses should be cancelled on the ground that it would result in a 'serious loss of production' and endanger the jobs of several employees'. The Employment tribunal dismissed this argument, having decided that using the machinery in its worn condition could cause serious danger to operators). Similarly, an undertaking by a company to take additional safety precautions against the risk of injury from unsafe plant until new equipment was installed was not sufficient (*Grovehurst Energy Ltd v Strawson (HM Inspector)* (COIT No 5035/90).

Where cost is a factor this is not to be confused with the current financial position of the company. A company's financial position is irrelevant to the question whether an Employment tribunal should affirm an enforcement notice. Thus in *Harrison (Newcastle-under-Lyme) Ltd v Ramsay (HM Inspector)* [1976] IRLR 135, a notice requiring cleaning, preparation and painting of walls had to be complied with even though the company was on an economy drive.

Employment tribunals have power under *HSWA 1974 s 24* to alter or extend time limits attaching to improvement and prohibition notices (*D J M and AJ Campion v Hughes (HM Inspector of Factories)* [1975] IRLR 291, where even though there was an imminent risk of serious personal injury, a further four months were allowed for the erection of fire escapes as it was not practicable to carry out the remedial works within the time limits set). Extensions of time in which to comply with notices are most commonly granted where the costs of the improvements and modifications required by the notice are significant.

Breach of law is insignificant

[E15025] Employment tribunals will rarely cancel a notice which concerns breach of an absolute duty where the breach is admitted but the appellant argues that the breach is trivial: *South Surbiton Co-operative Society Ltd v Wilcox* [1975] IRLR 292, where a notice had been issued in respect of a cracked wash-hand basin, being a breach of an absolute duty under the *Offices, Shops and Railway Premises Act 1963*. It was argued by the appellant that, in view of their excellent record of cleanliness, there was no need for officials to visit the premises. The appeal was dismissed.

Inspectors' investigation powers

[E15026] Inspectors have wide ranging powers under *HSWA 1974* to investigate suspected health and safety offences. These include powers to—

(a)   enter and search premises;

(b)   direct that the premises or anything on them be left undisturbed for so long as is reasonably necessary for the purpose of the investigation;

(c)   take measurements, photographs and recordings;

(d)   take samples of articles or substances found in the premises and of the atmosphere in or in the vicinity of the premises;

(e)   dismantle or test any article which appears to have caused or be likely to cause danger;
(f)   detain items for testing or for use as evidence;
(g)   interview any person;
(h)   require the production and inspection of any documents and to take copies; and
(i)   require the provision of facilities and assistance for the purpose of carrying out the investigation.

[*HSWA 1974 s 20*]

Under these powers inspectors may require interviewees to answer such questions as they think fit and sign a declaration that those answers are true. [*HSWA 1974 s 20(1)(j)*]. However, evidence given in this way is inadmissible in any proceedings subsequently taken against the person giving the statement or his or her spouse. [*HSWA 1974 s 20(7)*]. Where prosecution of an individual is contemplated the inspector will, therefore, usually exercise his evidence-gathering powers under the *Police and Criminal Evidence Act 1984* (*PACE 1984*). Evidence given in this way is admissible against that person in later proceedings. Interviews conducted under *PACE 1984* are subject to strict legal controls, for example, interviewees must be cautioned before the interview takes place, they have certain 'rights to silence' (although these were qualified by the *Criminal Justice and Public Order Act 1994*, in particular *s 34*), and there are rules relating to the recording of the interview.

Inspectors' powers of search and seizure in case of imminent danger

[E15027] Where an inspector has reasonable cause to believe that there are on premises 'articles or substances ('substance' includes solids, liquids and gases – *HSWA 1974 s 53(1)*) which give rise to imminent risk of serious personal injury', he can—

(a)   seize them; and
(b)   cause them to be rendered harmless (by destruction or otherwise).

[*HSWA 1974 s 25(1)*].

Enforcing authorities are given similar powers regarding environmental pollution, under the *Environment Act 1995 ss 108, 109*.

Before an article forming 'part of a batch of similar articles', or a substance is rendered harmless, an inspector must, if practicable, take a sample and give to a responsible person, at the premises where the article or substance was found, a portion which has been marked in such a way as to be identifiable. [*HSWA 1974 s 25(2)*]. After the article or substance has been rendered harmless, the inspector must sign a prepared report and give a copy of the report to—

(i)   a responsible person (eg safety officer); and
(ii)   the owner of the premises, unless he happens to be the 'responsible person'. (See **A3005** ACCIDENT REPORTING for the meaning of this term.)

[*HSWA 1974 s 25(3)*].

Indemnification by enforcing authorities

[E15028] Where an inspector has an action brought against him in respect of an act done in the execution or purported execution of any of the 'relevant statutory provisions' (see E15012 above), and is ordered to pay damages and costs (or expenses) in circumstances where he is not legally entitled to require the enforcing authority which appointed him to indemnify him, he may be able to take advantage of the *HSWA 1974 s 26*. By virtue of that provision, the authority nevertheless has the power to indemnify the inspector against all or part of such damages where the authority is satisfied that the inspector honestly believed—

(a)   that the act complained of was within his powers; and
(b)   that his duty as an inspector required or entitled him to do it.

In practice there will be very few circumstances where an inspector is held liable for advice given or enforcement action taken as part of his statutory duties. The Court of Appeal has ruled (in *Harris v Evans* [1998] 3 All ER 522, recently applied in the context of the *Registered Homes Act 1984* (now no longer in force) in *Trent Strategic Health Authority v Jain* [2009] UKHL 4, [2009] 1 AC 853) that an enforcing authority giving advice which leads to the issue of enforcement notices does not owe a duty of care to the owner of the premises affected by the notice and a claim for economic loss arising from such allegedly negligent advice cannot therefore succeed. The court said that if enforcing authorities were to be exposed to liability in negligence at the suit of owners whose businesses are adversely affected by their decisions it would have a detrimental effect on the performance by inspectors of their statutory duties. *HSWA 1974* contains its own statutory remedies against errors by inspectors and the court was not prepared to add to those measures. The court did, however, suggest that a possible exception might arise if a requirement imposed by the inspector introduced a new risk or danger which resulted in physical damage or economic loss.

Public register of improvement and prohibition notices

[E15029] Improvement and prohibition notices relating to public safety matters have to be entered in a public register as follows—

(a)   within 14 days following the date on which notice is served in cases where there is no right of appeal;
(b)   within 14 days following the day on which the time limit expired, in cases where there is a right of appeal but no appeal has been lodged within the statutory 21 days.
(c)   within 14 days following the day when the appeal is disposed of, in cases where an appeal is brought.

[*Environment and Safety Information Act 1988 s 3*].

Notices which impose requirements or prohibitions solely for the protection of persons at work are not included in the register. [*Environment and Safety Information Act 1988 s 2(3)*].

In addition, registers must be kept of notices served by—

(i)   fire authorities, under the *Schedule* to the *Environment and Safety Information Act 1988* for the purpose of *Articles 29, 30 and 31* of the *Regulatory Reform (Fire Safety) Order 2005 (SI 2005 No 1541)*;

(ii)   local authorities, under the Schedule to the *Environment and Safety Information Act 1988* for the purpose of *s 10* of the *Safety of Sports Grounds Act 1975*;

(iii) responsible authorities (as defined by the *Environment and Safety Information Act 1988 s 2(2)*) and the Minister of Agriculture, Fisheries and Food under *s 2* of the *Environment and Safety Information Act 1988* for the purpose of *s 19* of the *Food and Environment Protection Act 1985*;

(iv)   enforcing authorities, under *s 20* of the *Environmental Protection Act 1990*; and

(v)   enforcing authorities, under the *Radioactive Substances Act 1993* (see D1068 DANGEROUS SUBSTANCES I).

These registers are open to inspection by the public free of charge at reasonable hours and, on request and payment of a reasonable fee, copies can be obtained from the relevant authority. [*Environment and Safety Information Act 1988 s 1*]. Such records can also be kept on computer.

Part B: Offences and Penalties

Prosecution for breach of the 'relevant statutory provisions'

[E15030] Prosecutions can follow non-compliance with an improvement or prohibition notice, but equally inspectors will sometimes prosecute without serving a notice. Service of notices remains the most usual method of enforcement. In 2007–08 approximately 7,740 enforcement notices were issued by the HSE compared with 1,028 concluded prosecutions for health and safety offences which were heard in the same period. Prosecutions are most commonly brought after a workplace accident or dangerous incident (such as a fire or explosion). Investigations by the enforcing authorities may take many months to complete and, in complex cases, it is not unusual for prosecutions to be commenced up to a year after the original incident. Prosecutions normally take place before the magistrates; however, the *Health and Safety Offices Act 2008* has enabled certain offences that were previously only prosecuted in the magistrates' court capable of being prosecuted in the Crown Court. Whilst the decision whether to prosecute, issue an enforcement notice, or simply to give advice lies within the discretion of the inspector concerned, the enforcing authorities aim to pursue a policy which is open, consistent and proportionate to the risks in deciding what enforcement action to take (see the HSE's *Enforcement Policy Statement of February 2009*).

In practice, the enforcing authorities usually only investigate the most serious workplace accidents; it continues to be the HSE's policy to investigate all work-related deaths which are reported to it. The most dangerous industry sector remains construction, with 311 convictions secured out of 417 prosecutions brought in 2007–08. This is followed closely by manufacturing with 310 convictions secured out of 502 offences prosecuted. The average penalty across all industries was £12,896.

Burden of proof

[E15031] Throughout criminal law, the burden of proof of guilt is on the prosecution to show that the accused committed the particular offence (*Woolmington v DPP* [1935] AC 462). The burden is a great deal heavier than in civil law, requiring proof of guilt beyond a reasonable doubt as distinct from on a balance of probabilities. While not eliminating the need for the prosecution to establish general proof of guilt *HSWA 1974 s 40* makes the task of the prosecution easier by transferring the onus of proof to the accused for one element of certain offences. *Section 40* states that in any proceedings for an offence consisting of a failure to comply with a duty or requirement to do something so far as is practicable, or so far as reasonably practicable, or to use the best practicable means to do something, the onus is on the accused to prove (as the case may be) that it was not practicable, or not reasonably practicable to do more than was in fact done to satisfy the duty or requirement, or that there was no better practicable means than was in fact used to satisfy the duty or requirement. However, *s 40* does not apply to an offence created by *HSWA 1974 s 33(1)(g)* – failing to comply with an improvement notice.

Generally speaking, statute which requires a defendant to prove a fact is a prima facie affront to the presumption of innocence encapsulated in *Article 6(2)* of the *ECHR*. However, it has been held a reverse burden of proof may not necessarily be incompatible with a defendant's human rights. Such a view was articulated by Lord Bingham in *Sheldrake v DPP* [2004] UKHL 43, [2005] 1 AC 264, who stated that provided a reverse burden is reasonable and proportionate, it will not offend against the principle that it is for the prosecution to prove its case against the defendant.

The burden of proof and its application in prosecutions under *HSWA 1974* was recently considered by the House of Lords in *R v Chargot Ltd (t/a Contract Services)* [2007] EWCA Crim 3032, [2008] 2 All ER 1077; affd [2008] UKHL 73, [2009] 2 All ER 645. The case required the Court of Appeal and then the House of Lords to consider, amongst other things, the impact of *HSWA 1974 s 40*.

The first and second defendants, both companies, were members of the Ruttle Group, of which the third defendant was the managing director. Ruttle owned a farm at which a car park was being built. The deceased was driving a dumper truck at the farm in relation to the construction of the car park. He was not the usual dumper truck driver, but had been asked to drive it that day by the foreman on the site. As the deceased drove down a ramp, the dumper truck fell onto its side and he was buried under the soil and was killed. There were no witnesses to the accident, and its precise cause was never established.

The first defendant was charged with contravening *HSWA 1974 s 2(1)*. The second defendant was charged with contravening *HSWA 1974 s 3(1)*, and the third defendant, as an individual, was charged pursuant to *HSWA 1974 s 37* (it was alleged that, through his connivance, consent or neglect, he had caused the second defendant to breach *HSWA 1974 s 3(1)*). The defendants accepted that whilst there had been no risk assessments in relation to the use of the dumper truck, nor any training provided to the deceased, they had nonetheless done everything which was reasonably practicable to ensure the safety of the deceased and the other workers.

The prosecution based its case against the first and second defendants on the proposition that it was sufficient for it to identify and prove a risk of injury arising from the state of affairs which had prevailed at the site, rather than from a series of specified acts and/or omissions. The case against the third defendant was that he had given the specific directions in respect of how works at the site were to be carried out. The prosecution submitted Routes to Verdict in questionnaire form to the jury in relation to the first and second defendants' charges. The defendants were duly convicted on this basis on 10 November 2006. They appealed against conviction and sentence.

The appeal centred on the effect of the burden placed on the defendants under *s 40* in relation to what the prosecution had to prove to establish the prime facie breach of duty which would trigger the need for the defence to prove that it was not reasonably practicable to do more than had in fact been done to satisfy the duty.

The Court of Appeal dismissed the defendants' appeal against conviction, holding that *HSWA 1974 ss 2, 3* imposed a duty to ensure a state of affairs, so far as was reasonably practicable. *HSWA 1974 s 40* imposed an obligation on the defence to establish that they had done everything reasonably practicable to ensure the existence of that state of affairs. The policy behind *HSWA 1974* was clearly to impose a positive burden on employers rather than simply disciplining them for breaches of specific obligations. That being so, the prosecution was entitled to point to a state of affairs as amounting to a breach of the statutory duty. The prosecution had clearly established that a real risk of injury had been caused by driving the dumper truck (as evidenced by the fact that the accident had occurred). This was enough to justify the requirement that the first and second defendant should have had the burden of proving that they had done all that was reasonably practicable to protect against that risk. It might have been different if the state of affairs which had been alleged to amount to a breach of duty by the employer or undertaker could not be shown to have any causal link to the employment or the undertaking in question, but that was not the case here.

Appeal to the House of Lords was refused by the Court of Appeal. However, the defendants sought permission to appeal from the House of Lords which was granted. Their Lordships delivered their decision on 10 December 2008.

The issues which fell to be determined were—

(a)   What the prosecution had to prove in order to show that, subject to reasonable practicability, an offence had been committed;

(b)   What the prosecution had to prove in order to demonstrate that the officer had committed an offence under *s 37*;

(c)   What further particulars, if any, the prosecution had to produce in the interests of fairness when a prosecution was mounted.

The defendants argued that the prosecution had to identify and then prove the acts and omissions which it asserted had led to the breach of duty, a generalised state of affairs being insufficient to lead to a finding of culpability.

The House of Lords, in affirming the Court of Appeal's decision, held as follows (Lord Hope delivering the leading opinion)—

(a)   That the duties contained in *HSWA 1974 ss 2, 3* were expressed in broad and general terms. The matters set out in *s 2(2)* were those to which employers had to have regard and could be taken as describing a result which employers had to achieve. If that result was not achieved an employer would have to show that it had not been reasonably practicable for it to achieve it. The prosecution had to show that the results set out in *ss 2(1), 3(1)* had not been achieved or prevent in order to establish a case. Once the breach had been established the onus was on the defendant to show that certain measures had not been reasonably practicable within the meaning of *s 40*.

(b)   The allegations contained within the prosecution's case summary were not ingredients in the offence and so did not have to be proved. *HSWA 1974 ss 2(1), 3(1)*set out the various results that employers had to achieve, not the means by which they were to be achieved. It was the results with which the jury had to concern themselves.

(c)   That reversing the burden of proof by placing it on the defendant pursuant to *HSWA 1974 s 40* was not disproportionate or incompatible with the presumption of innocence. The risks contemplated by*HSWA 1974 ss 2, 3* were risks which included serious injury or even death. Once a prima facie contravention of *ss 2, 3* has been established, it is entirely proportionate that the onus should be on the defence to establish that certain preventative measures were not reasonably practicable.

(d)   That fixed rules as to what the prosecution had to identify and then prove in relation to the charge under *s 37* (consent, connivance or neglect) were not capable of being laid down because of the extent to which each case differed on its facts. Where it was shown that a company had failed to achieve/prevent the results contained in *ss 2, 3* it would be a relatively short step for an inference to be drawn that there had been some connivance or neglect on the part of an officer who was controlling the workplace or circumstances under which the relevant work activity was taking place.

(e)   That in the present case, the particulars submitted by the prosecution had been sufficient, and it had not been necessary for it to specify the respects in which risk had been associated with the activity or to identify the cause of the accident. The fact that the accident had occurred at all was sufficient to show that the risk was real; Lord Hope commented that "where a person sustains injury at work, the facts will speak for themselves". The test of how much detail the particulars of the offence needed to go into was one of "fair notice". However, the particulars were not the ingredients of the offence and therefore did not need to be proved by the prosecution.

The effect of the clarification given by the House of Lords in R v Chargot was considered and applied by the Court of Appeal in *R v EGS Ltd* [2009] EWCA

Crim 1942, an appeal by the prosecution against the trial judge's finding that there was no case to answer in respect of certain charges on the indictment. After having set out at length passages from Lord Hope's decision, the Court of Appeal in EGS held that the trial judge had erred when he found that prosecution was unable to establish a connection between the defendant's conduct and the accident, or that the risk was foreseeable. In allowing the appeal, Dyson LJ held that the judge had confused himself on the role that foreseeability and causation played in the prosecution's formulation of its case, and said (emphasis added):

> 23.... Thus, in the present case, the burden is on the prosecution to prove to the criminal standard that EGS failed to conduct its undertaking in such a way as to ensure that persons not in its employment who night affected thereby were not exposed to risks to their health and safety. As Lord Hope made clear at [27] of his speech in Chargot, the risks must be "material" risks to health or safety. The judge may have misunderstood what Lord Hope meant by a "material" risk. In our judgment, Lord Hope was referring back to the earlier part of [27], where he said that the legislation is only concerned with risks that are not trivial or fanciful. A risk that is trivial or fanciful is not material.

> [...]

> 27.... In any event, it is strictly inapt to speak of a risk being foreseeable. **A risk is a present potential danger the existence of which may or may not be appreciated:** see per Steyne in *R v Board of Trustees of the Science Museum* [1993] 1 WLR 1171, 1177F, approved in Chargot at [20]. **If the risk eventuates and an accident occurs, then a question may arise in the context of a section 40 defence as to whether the accident was foreseeable or unforeseeable:** see *R v HTM Ltd* [2006] EWCA Crim 1156. **But it is not relevant to the issue of whether the prosecution has proved the existence of a material risk.** It may be that the judge used the word "foreseeable" inaccurately and that he used it interchangeably with "would have been appreciated".

> 28.... For the same reasons, the judge was also wrong to say at para 39 that "the prosecution cannot establish...that the risk was foreseeable.. or that it should have been foreseeable. It was too remote." Further, we regret that we cannot agree with the judge's statement that "the prosecution cannot...establish a connection between [EGS's] conduct and the accident." Causation is not an essential ingredient of the offence. The prosecution did not have to establish that EGS caused the accident, although in the present case, as in most, they did in fact rely on a causal connection between the EGS's acts and omissions as going to establish risk. They merely had to prove that EGS exposed persons not in its employment to risks to their health or safety. Nor do we understand the reference to remoteness. It would appear that the judge was treating the prosecution as if it were a civil claim for damages for breach of statutory duty. But it is clear from Chargot that this is the wrong approach.

Main offences and penalties

[E15032]–[E15034] Health and safety offences are either (a) triable summarily (ie without jury before the magistrates), or (b) triable summarily and on indictment (ie triable either way), or (c) triable only on indictment. Most health and safety offences, however, fall into categories (a) and (b). The main offences falling into these two categories are set out below. The *Health and*

Safety Offences Act 2008 has inserted a new *HSWA 1974 Sch 3A*, setting out the mode of trial for various offences under *s 33*.

Summary only offences

(a)   Contravening any requirement imposed by or under regulations under *HSWA 1974 s 14* or intentionally obstructing any person in the exercise of his powers under that section;

(b)   Intentionally obstructing an inspector in the exercise or performance of his powers or duties or obstructing a customs officer in the exercise of his powers under *HSWA 1974 s 25A*;

(c)   Falsely pretending to be an inspector.

'Either way' offences

(a)   Failure to carry out one or more of the general duties of *HSWA 1974 ss 2–7*;

(b)   Contravening either—

    (i)   *HSWA 1974 s 8* – intentionally or recklessly interfering with anything provided for safety;

    (ii)   *HSWA 1974 s 9* – levying payment for anything that an employer must by law provide in the interests of health and safety (eg personal protective clothing);

(c)   Contravening any health and safety regulations;

(d)   Contravening a requirement imposed by an inspector under *HSWA 1974 s 25*(power to seize and destroy articles and substances);

(e)   Contravening a requirement of a prohibition or improvement notice (including any notice which is modified on appeal);

(f)   Intentionally or recklessly making false statements, where the statement is made—

    (i)   to comply with a requirement to furnish information; or

    (ii)   to obtain the issue of a document;

(g)   Intentionally making a false entry in a register book, notice etc. which is required to be kept;

(h)   Using a document, with intent to deceive, which has been issue or authorised under any of the relevant statutory provisions or is required for any purpose thereunder, or making and possessing a document which resembles such a document so as to be calculated to deceive;

(i)   failing to comply with a remedial court order made under *HSWA 1974 s 42*.

[*HSWA 1974 s 33(1)*].

In England and Wales there is no time limit for bringing prosecutions for indictable offences [*Interpretation Act 1978 Sch 1*]. Where an offence can only be tried summarily in the magistrates' courts the time limit is 6 months from the date of the commission of the offence or from the time when the matters complained of became apparent [*Magistrates' Courts Act 1980 s 127(1)*]. In Scotland the 6-month time limit extends to 'either way' offences tried summarily. The period may be extended in the case of special reports, coroners' court hearings or in cases of death generally. [*HSWA 1974 s 34(1)*].

Summary trial or trial on indictment

[E15035] Most offences triable either way are tried summarily. However an increasing number of serious offences are being referred to the Crown Court, which has increased sentencing powers, for trial on indictment. The Court of Appeal has advised magistrates to exercise caution in accepting jurisdiction in health and safety cases where the offence may require a penalty greater than they can impose or where death or serious injury has resulted from the offence (see *R v F Howe & Son (Engineering) Ltd* at E15043 below). The defendant may also refuse to consent to summary trial and opt for trial on indictment.

Penalties for health and safety offences

[E15036] The *Health and Safety (Offences) Act 2008* has set new penalties for certain offences and has also increased the level of fines and terms of imprisonment that magistrates can impose.

The following penalties will now apply to offences committed after the 16 January 2009 where a summary conviction is secured (the former provisions will apply to offences committed before that date—

(a)   For failing to discharge a duty pursuant to *HSWA 1974 ss 2–6* – imprisonment for a term not exceeding 12 months or a fine not exceeding £20,000 or both;

(b)   For failing to discharge a duty pursuant to *HSWA 1974 s 7* – imprisonment for a term not exceeding 12 months or a fine not exceeding £20,000 or both;

(c)   For contravention of *HSWA 1974 s 8* – imprisonment for a term not exceeding 12 months, or a fine not exceeding £20,000 or both;

(d)   For contravention of *HSWA 1974 s 9* – a fine not exceeding £20,000;

(e)   For an offence under *HSWA 1974 s 33(1)(d)* (includes contravention of regulations) – a fine not exceeding level 5 on the standard scale (currently £5,000);

(f)   For offences under *s 33(1)(e)–(g)* (includes contravention of a requirement contained in an improvement or prohibition notice - imprisonment for a term not exceeding 12 months, or a fine not exceeding £20,000 or both;

(g)   For an offence under *s 33(1)(h)* (intentional obstruction of an inspector) – imprisonment for a term not exceeding 51 weeks in England and Wales or 12 months in Scotland, or a fine not exceeding level 5 on the standard scale (currently £5,000), or both;

(h)   For an offence under *s 33(1)(i)* – a fine not exceeding the statutory maximum;

(i)   For an offence under *s 33(1)(j)* – imprisonment for a term not exceeding 12 months or a fine not exceeding the statutory maximum, or both;

(j)   For an offence under *s 33(1)(k)* – imprisonment for a term not exceeding 12 months, or a fine not exceeding £20,000, or both;

(k)   For an offence under *s 33(1)(n)* (falsely pretending to be an inspector) – a fine not exceeding level 5 on the standard scale (currently £5,000);

(l)   For an offence under *s 33(1)(o)* (failure to comply with a remedial order pursuant to *HSWA 1974 s 42*) – imprisonment for a term not exceeding 12 months or a fine not exceeding £20,000, or both.

Where no penalty has been specified for an offence under existing statutory provisions and summary conviction is secured, the penalty will be imprisonment for a term not exceeding 12 months, or a fine not exceeding £20,000, or both.

Except in respect of offences under *s 33(1)(b)* and *(i)*, and where offences can only be tried summarily, the Crown Court can impose a penalty of imprisonment for term not exceeding two years, or a fine, or both. Those penalties also apply where no penalty has been specified for an offence under existing statutory provisions.

In the recent case of *R v Jarvis Facilities Ltd* [2005] EWCA Crim 1409, [2006] 1 Cr App Rep (S) 247, the Court of Appeal decided that where there is a significant public element, the court is entitled to take a more severe view of breaches of health and safety at work provisions. In this case the defendant, Jarvis Facilities Ltd had been required to carry out repair works to a set of points on a railway. Trained and qualified workers carried out the work but no specific method statement had been prepared. However, certain necessary processes had been omitted which resulted in the derailment of a freight train. The train had been travelling at 25 mph and remained upright and consequently there was no significant injury or damage. The defendant pleaded guilty to an offence contrary to *HSWA 1974 s 3(1)* but appealed against the £400,000 fine imposed by the court of first instance. The Court of Appeal concluded that public services cases would be likely to be treated more seriously than those where a breach is confined to a private sector, even where there is comparability between gravity of breach and economic strength of defendant. The Appeal court however considered the fine imposed as being excessive and accordingly reduced the original fine of £400,000 to a sum of £275,000 plus costs.

R v Jarvis was quoted by Mackay J in the remarks he made when sentencing Balfour Beatty Rail Infrastructure following the Hatfield rail disaster which resulted in the deaths of four passengers. The Court of Appeal endorsed his summary of the applicable sentencing principles (*R v Balfour Beatty Rail Infrastructure Services Ltd* [2006] EWCA Crim 1586, [2007] 1 Cr App Rep (S) 370), although it reduced the fine from £10 million to £7.5 million.

The overall effect of these decisions is expected to result in more cases being referred to the Crown Courts for sentence and correspondingly higher fines for corporate defendants in particular where the offence involves a public element.

Defences

[E15037] Although no general defences are specified in the *Health and Safety at Work etc Act 1974*, some regulations passed under the Act carry the defence of 'due diligence' (for example, *Regulation 21* of the *Control of Substances Hazardous to Health Regulations 2002 (SI 2002 No 2677)* as amended by *SI 2004 No 3386*. Currently, where a defence of 'due diligence' is available it is for the defendant to prove, on the balance of probabilities, that all appropriate steps to avoid the offence had been taken. Of course, in respect of *HSWA 1974 ss 2, 3*, it will be open to a defendant to assert that steps which would have enhanced health and safety were not reasonably practicable [see below].

However, this is not a defence as such, rather the words "so far as reasonably practicable" qualify the duties contained in *HSWA 1974 ss 2, 3*.

Manslaughter

[E15038] The *Corporate Manslaughter and Corporate Homicide Act 2007* came into force on 1 April 2008. It has re-formulated the test that must be satisfied by the prosecution before a conviction for corporate manslaughter can be secured. There have been no convictions under this act at the time of writing.

Formerly and under the common law, it was necessary to establish gross negligence on the part of one or more of a company's directors or other 'controlling mind' of the company in order for a conviction for corporate manslaughter to follow. In large companies where there were several tiers of management below board level it was almost impossible to show gross negligence on the part of one or more senior executives who could be said to be the 'controlling mind' and who were directly responsible for the commission of the offence. Successful prosecutions were therefore only really possible in respect of small companies.

The new offence

[E15038.1] The *Corporate Manslaughter and Corporate Homicide Act 2007* creates an offence whereby an organisation is guilty of corporate manslaughter (in England and Wales) or corporate homicide (in Scotland) if the way in which it is managed or organised by senior management causes a person's death and amounts to a gross breach of a relevant duty of care. As can be seen, this removes the need for a directing or controlling mind to be found guilty of gross negligence manslaughter. The offence applies to corporations and partnerships, government departments, police forces, and trade unions and employers' associations.

A duty of care is defined as any of the following duties owed under the law of negligence—

(a) a duty owed to employees or other persons working for a company or performing services for it (this is clearly designed to encapsulate relationships between a company and a deceased which fall beyond the usual employer-employee relationship);
(b) a duty owed as an occupier;
(c) a duty owed in connection with—
    (i) the supply by an organisation of goods and services;
    (ii) the carrying on by an organisation of any construction or maintenance operations;
    (iii) (iii) the carrying on by an organisation of any other activity on a commercial basis, or
    (iv) (iv) the use or keeping by an organisation of any plant, vehicle or other thing.
(d) a duty owed to a prisoner or a detained patient, for whose safety an organisation is responsible.

[*Corporate Manslaughter and Corporate Homicide Act 2007 s 2*]

The Act makes clear that reference to the "law of negligence" means a duty that would be owed in common law but for the existence of a statutory duty which is imposed. It is also clear in respect of the questions that fall to the jury: after the existence of a duty of care has been established, they will have to consider whether there has been a gross breach. The jury will also need to reach a decision on whether there has been a breach of health and safety legislation, how serious any breach was, and how great the risk of death the breach posed [s 8 of the *Corporate Manslaughter and Corporate Homicide Act 2007*].

Special provisions apply in respect of the Ministry of Defence [s 4 of the *Corporate Manslaughter and Corporate Homicide Act 2007*], police authorities [s 5 of the *Corporate Manslaughter and Corporate Homicide Act 2007*] and the emergency services [s 6 of the *Corporate Manslaughter and Corporate Homicide Act 2007*].

The new Act creates a new penalty imposable by the Court following a conviction – the publicity order [s 10 of the *Corporate Manslaughter and Corporate Homicide Act 2007*]. Application of this provision will compel an organisation to make public the fact of its conviction, specified particulars of the offence, the amount of any fine that has been imposed, and the terms of any remedial order.

Offences committed by particular types of persons

Corporate offences – delegation of duties to junior staff

[E15039] Companies cannot avoid liability for breach of general duties under *HSWA 1974 ss 2–6* by arguing that the senior management and/or the 'directing mind' of the company had taken all reasonable precautions, and that responsibility for the offence lay with a more junior employee or an agent who was at fault. *HSWA 1974* generally imposes strict criminal liabilities on employers and others (subject to the employer being able to establish that all reasonably practicable precautions had been taken) and it is not open to corporate employers to seek to avoid liability by arguing that their general duties have been delegated to someone lower down the corporate tree. In *R v British Steel plc* [1995] IRLR 310, British Steel was prosecuted under *HSWA 1974 s 3* after the death of a subcontractor who was carrying out construction work under the supervision of a British Steel engineer. British Steel argued that it was not responsible under s 3 for the actions of the supervising engineer as the engineer was not part of the 'directing mind' (following *Tesco Supermarkets Ltd v Nattrass* [1972] AC 153) of the company and all reasonable precautions to ensure the safety of the work had been taken by senior management. The Court of Appeal dismissed this argument; Steyn LJ commented that even passing negligence of an employee could give rise to liability under *HSWA 1974*. In *R v Gateway Foodmarkets Ltd* [1997] IRLR 189, the defendant was charged with breach of s 2 when an employee fell through a trap door in the floor of a lift control room. The accident occurred while the store manager was manually attempting to rectify an electrical fault in the lift in accordance with a local practice which was not authorised by Gateway's head office. The Court of Appeal held that the failure at store manager level was attributable to the employer.

Regulation 21 of the *Management of Health and Safety at Work Regulations 1999 (SI 1999 No 3242)* make clear that an employer cannot avoid liability in criminal proceedings by blaming an employee or "appointed person".

However, it does not follow that an employer will automatically be held criminally responsible for an isolated act of negligence by an employee performing work on its behalf. This is because it may still be possible for the employer to establish that it has done everything reasonably practicable in the conduct of its undertaking to ensure that employees and third parties are not exposed to risks to their health and safety by virtue of the way it has conducted its business (see *R v Nelson Group Services (Maintenance) Limited* [1999] IRLR 646, CA).

The definition of 'reasonably practicable' was authoritatively laid down in *Edwards v National Coal Board* [1949] 1 All ER 743 where it was said that: 'Reasonably practicable' is a narrower term than 'physically possible', and seems to imply that a computation must be made by the owner in which the quantum of risk is placed on one scale and the sacrifice involved in the measures necessary for averting the risk (whether in money, time or trouble) is placed in the other, and that, if it be known that there is a gross disproportion between them – the risk being insignificant in relation to the sacrifice – the defendants discharge the onus on them.'

The availability of a defence that a certain measure was not "reasonably practicable" was central *R v HTM Ltd* [2006] EWCA Crim 1156, [2007] 2 All ER 665, in which the Court of Appeal considered the previous decisions in *R v Gateway Foodmarkets* and *R v Nelson Group*. Two of the defendant company's employees had been fatally injured; it was consequently charged with breach of *HSWA 1974 s 2*. At trial, the company sought to adduce evidence that it had done all that was reasonably practicable to do, and that the accidents had resulted from the actions of the two employees who had perished. The company also asserted that it could not have been foreseen that the employees would behave as they did. The prosecution argued that foreseeabiliy was not relevant and that *reg 21* of the *Management of Health and Safety at Work Regulations 1999* [see above] precluded the company from relying on the act or default of its employees. The Court of Appeal held, in accordance with the judge at first instance, that the likelihood that the accident would happen was material in that it would assist in determining whether the risk would eventuate, and consequently what measures would be reasonable to put in place to prevent that. The Court of Appeal also held that the phrase "so far as is reasonably practicable" did not operate as a defence; instead it qualified the duty incumbent on employers. *Regulation 21* did not apply in the circumstances. It was therefore open to the defendant to attempt to demonstrate that the accident had been caused by the employees' actions.

Offences of directors or other officers of a company

[E15040] Where an offence is committed by a body corporate, senior persons in the hierarchy of the company may also be individually liable. Thus, where the offence was committed with the consent or connivance of, or was attributable to any neglect on the part of a director or officer of the company, that person is himself guilty of an offence and liable to be punished accordingly. Those who may be so liable are—

(a)   any functional director;
(b)   a manager (which does not include an employee in charge of a shop while the manager is away on a week's holiday (*R v Boal* [1992] QB 591, concerning *s 23* of the *Fire Precautions Act 1971* – identical terminology to *HSWA 1974 s 37*));
(c)   a company secretary;
(d)   another similar officer of the company;
(e)   anyone purporting to act as any of the above.

[*HSWA 1974 s 37(1)*].

It is not sufficient that the company through its 'directing mind' (its board of directors) has committed an offence – there must be some degree of personal culpability in the form of proof of consent, connivance or neglect by the individual concerned. Evidence of this sort can be difficult to obtain and prosecutions under *s 37(1)* have, in the past, been rarely compared with prosecutions of companies (although they are increasing in number).

Directors, managers and company secretaries can be personally liable for ensuring that corporate safety duties are performed throughout the company (for example, a failure to maintain a safe system of work can give rise to personal liability). Liability may also arise as a result of a failure to perform an obligation placed on individuals by their employment contracts and job descriptions – for example, obligations imposed under a safety policy – not just in relation to duties imposed by law. In the case of *Armour v Skeen (Procurator Fiscal, Glasgow)* [1977] IRLR 310, an employee fell to his death whilst repairing a road bridge over the River Clyde. The appellant, who was the Director of Roads, was held to be under a duty to supervise the safety of council workmen. He had not prepared a written safety policy for roadwork, despite a written request that he do so, and was found to have breached *HSWA 1974 s 37(1)*.

Similar duties exist under the *Environmental Protection Act 1990 s 157* and the *Environment Act 1995, s 95(2)–(4)*. (See **E5023** emissions into the atmosphere.) Directors convicted of a breach of *HSWA 1974 s 37* may also be disqualified, for up to two years, from being a director of a company, under the provisions of the *Company Directors Disqualification Act 1986 s 2(1)* as having committed an indictable offence connected with (inter alia) the management of a company. In *R v Chapman* (1992), unreported, a director of a quarrying company was disqualified and fined £5,000 for contravening a prohibition notice on an unsafe quarry where there had been several fatalities and major injuries.

The HSE and the Institute of Directors have produced a publication entitled "Leading Health and Safety at Work". This guidance sets out the duties and obligations incumbent upon directors, governors, trustees and equivalent officers in public, private and third sector organisations. Its primary aim is to ensure effective "top-down" leadership on health and safety. The HSE's inspectors will deploy the guidance in the context of routine inspections and in investigations.

Directors' insurance

[E15041] The *Companies Act 2006 s 232* prevents companies from 'immunising' their directors against liability by providing them with indemnities and similar arrangements. *Sub-section (1)* of the section states—'Any provision that purports to exempt a director of a company (to any extent) from any liability that would otherwise attach to him in connection with any negligence, default, breach of duty or breach of trust in relation to the company is void.'

The *Companies Act 2006 s 232* does allow companies to purchase and maintain for a director insurance against any of the liabilities mentioned in *s 232*. In practice, companies have since 1990 (under the terms of *Companies Act 1985 s 310*, now repealed) been able to buy insurance for their directors (usually known as Directors' and Officers', or D & O, insurance) which protects them in the event that they are named in civil or criminal proceedings.

Such insurance may (subject to the terms of the policy), protect directors against claims made during the policy period for breach of contract, negligence, misrepresentation and negligent misstatement caused by the person in his capacity as a director. Some policies also cover breach of statutory duty, employment claims, misfeasance, disqualification proceedings, proceedings by regulatory authorities (such as the HSE or Financial Services Authority) and corporate manslaughter claims. Policies typically allow directors to recover their legal costs and expenses incurred in defending or settling such claims, but fines and other penalties are usually excluded.

Offences due to the act of another person

[E15042] *HSWA 1974 s 36(1)* makes clear that although provision is separately made for the prosecution of less senior corporate staff (e.g. safety officers and works managers) this does not prevent a further prosecution against the company itself. The section states that where an offence under *HSWA 1974* is due to the act or default of some other person, then—

(a)   that other person is guilty of an offence; and
(b)   a second person can be charged and convicted, whether or not proceedings are taken against the first-mentioned person.

Where the enforcing authorities rely on *HSWA 1974 s 36*, this must be made clear to the defendant. In *West Cumberland By Products Ltd v DPP* [1988] RTR 391, the conviction of a company operating a road haulage business for breach of regulations relating to the transport of dangerous substances was set aside as the offence charged related to the obligations of the driver of the vehicle and, in prosecuting the operating company, reliance was not placed on *HSWA 1974 s 36*.

Sentencing Guidelines

[E15043] The level of fine imposed for health and safety offences will ultimately depend on the facts and circumstances of the case, including the gravity of the offence, whether the breach resulted in death or serious injury, and any mitigating evidence the defendant is able to put forward (including details of its means and ability to pay any fine imposed). In the past there have been wide variations in the sentences handed down by different courts for similar breaches of the legislation and there has also been concern at the general low level of fines imposed for health and safety offences. In the face of this mounting concern the Sentencing Guidelines Council has issued guidance, effective from 15 February 2010, to sentencing courts determining fines for convictions for corporate manslaughter and for breaches of the *HSWA 1974* which cause death.

R v F Howe & Son (Engineering) Limited [1999] 2 All ER 249 remains the leading guideline case on appropriate levels of sentencing in health and safety cases which have not resulted in death. It applies where the defendant is either an individual or a corporate entity. In it, the Court of Appeal laid down guidelines to assist magistrates and judges in sentencing health and safety offences. The case concerned an appeal by the company against fines totalling £48,000 and an order for costs of £7,500 imposed in respect of four health and safety offences arising from a fatal accident to one of the company's employees who was electrocuted while using an electric vacuum machine to clean a floor at the company's premises. In reducing the level of fine imposed to reflect the company's limited financial resources, the Court of Appeal laid down the following general principles —

Sentencing Guidelines

General Principles

The level of fine should reflect—

- the gravity of the offence and the standard of the defendant's conduct;
- the degree of risk and extent of danger;
- the extent of the breach and how far short the defendant fell from doing what was reasonably practicable – an isolated incident may attract a lower fine than a continuing unsafe state of affairs;
- the defendant's resources and the effect of the fine on its business.

Aggravating Factors

- failure to heed warnings;
- if the defendant deliberately flouts safety legislation for financial reasons;
- if the offence results in a fatality.

Mitigating factors

- prompt admission of liability and guilty plea;
- steps taken to remedy deficiencies;
- a good safety record.

In applying these factors, the Court of Appeal made clear that every case needs to be considered on its own facts. It declined to lay down a tariff for particular offences or to link the level of fine directly to the defendant's turnover or net profit. However, it emphasised the importance of the defendant's means, as well as the gravity of the offence, in determining the appropriate level of fine, stating that this should be large enough to impress on both the management and shareholders of the defendant company the importance of providing a safe working environment. Although there might be cases where the offences are so serious that the defendant company ought not to be in business, in general the fine 'should not be so large as to imperil the earnings of employees or create a risk of bankruptcy'. In essence, the courts must answer two questions in determining the appropriate level of fine—

(1)   What financial penalty does the offence merit?
(2)   What penalty can the defendant reasonably be ordered to pay?

The Court of Appeal specifically made clear that the size and resources of the defendant company and its ability to provide safety measures or to employ in-house safety advisers are not a mitigating factor: the legislation imposes the same standard of care irrespective of the size of the organisation.

In *R v Rollco Screw & Rivet Co Ltd* [1999] IRLR 439, the Court of Appeal approved the principles laid down in *Howe* and indicated that the fine imposed should make clear that there is a personal responsibility on directors for their company's health and safety arrangements. It did, however, acknowledge that caution was necessary in the case of smaller companies, where the directors were also shareholders, to avoid imposing a fine which amounted to double punishment of the individuals concerned.

The court went on to suggest that, in appropriate circumstances, corporate defendants may be ordered to pay fines and costs by instalments over many years. In reducing *Rollco's* total payment period to 5 years and 7 months, the court held that there was no maximum period for payment of fines and costs. Although there might be good reason to limit the period of payment of fines and costs by a personal defendant, who may suffer anxiety because of his continuing financial obligations, the same considerations do not apply to a corporate defendant. The Court of Appeal indicated that, in proper circumstances, it might be appropriate to order payment of fines and costs by a corporate defendant over a 'substantially longer period' than would be appropriate in the case of an individual.

Penalties for Corporate Manslaughter and Health and Safety Offences Causing Death

In February 2010 the Sentencing Guidelines Council published definitive guidance on sentencing convictions for corporate manslaughter and of corporate bodies for health and safety offences which cause, or are one of the causes of, death. The guideline applies to all such sentencing (it does not apply where the death simply occurred, and was not causally related to the offence) which takes place on or after 15 February 2010. It is applicable in respect of public bodies, as well has corporate entities.

The guideline sets out the principles which are to govern the approach taken by sentencing judges. At the outset, the distinction between corporate manslaughter, as defined in the *Corporate Manslaughter and Corporate Homicide Act 2007*, and offences under the *HSWA 1974* is set out. The former will involve both a gross breach of duty of care and senior management failings, and will therefore often involve systemic failures. Convictions under the *HSWA 1974* result where a defendant is unable to show that it was not reasonably practicable to avoid a risk of injury. Those failings can be operational rather than systemic, with convictions often being attributable to an unauthorised act of an employee. The implication is therefore that less culpability attaches.

The guideline states that the seriousness of the offence should be gauged by addressing the following questions, which are clearly based on the guidance that was contained in *Howe*:

(a) How foreseeable was serious injury? The more foreseeable, the more serious the offence.

(b) How far short of the applicable standard did the defendant fall?

(c) How common is this kind of breach in this organisation? Indications of systemic failures are likely to be deemed more serious.

(d) How far up the organisation does the breach go?

The following factors, if present, are likely to be considered aggravating features of the case (the list is not exhaustive and is again based on the factors set out in *Howe*).

(a) More than one death or very grave personal injury has been caused in addition to the death.

(b) There has been a failure to heed warnings or advice, or there has been a failure to respond appropriately to "near misses" arising in comparable circumstances.

(c) There has been cost-cutting at the expense of safety.

(d) There has been a deliberate failure to obtain or comply with relevant licences.

(e) Injury has been caused to vulnerable persons.

Mitigating features can include, non-exhaustively:

(a) A prompt acceptance of responsibility. An early guilty plea will be recognised by the appropriate reduction.

(b) A high level of co-operation with the investigating authorities, beyond that which will always be expected.

(c) Genuine efforts having been made to remedy the defects.

(d) A company's good health and safety record.

(e) A responsible attitude to health and safety, such as the commissioning of expert advice or the consultation of employees or others affected by the organisation's activities.

It is recommended that the prosecution set out in writing the aggravating and mitigating features which it considers apply (pursuant to *R v Friskies Petcare (UK) Ltd* [2000] EWCA Civ 95, [2000] 2 Cr App Rep (S) 401. Those acting for the defendant can then be required to set out in writing any points on which they differ.

Section C of the guideline states that the sentencing court should require information relating to the defendant's financial circumstances. Usually, information for a three-year period, including the year of the offence, will be necessary. It is then advised to "look carefully" at turnover, profit and assets to gauge the corporate defendant's means in order that any fine imposed is one that the defendant is capable of paying, over an appropriate period if necessary. It should be noted that failure to provide the required information may result in a court making an adverse assumption as to the defendant's means and ability to pay. The guideline states that the effect of a fine on company shareholders and directors will not be relevant, nor will the possible inflation of prices charged by the defendant. Whether the fine will put the defendant out of business will be relevant. In particularly egregious cases it will be an acceptable consequence of the offence and conviction.

The most newsworthy section of the guideline is that which deals with the level of fine that is to be imposed. Where a conviction for corporate manslaughter is secured, "[t]he appropriate fine will seldom be less that £500,000 and may be measured in **millions of pounds**". Health and safety offences which have resulted in death should result in a fine which will "seldom be **less than** £100,000 and may be measured in **hundreds of thousands of pounds or more**" (emphasis original).

Publication of convictions for health and safety offences and enforcement notices

[E15044] In an attempt to improve compliance with health and safety legislation, the HSE are also pursuing an active policy of naming companies that breach the legislation. The HSE 'name and shame' website publicises all prosecution cases initiated by the HSE which resulted in a conviction as well as a register of improvement and prohibition notices issued by the HSE. This information can be accessed at: www.hse.gov.uk/enforce/prosecutions.htm. Generally, information relating to prosecutions against and to notices served on individuals will remain on the public register for five years before being removed. Information relating to enforcement measures taken against compa nies will remain on the public register for longer.

Position of the Crown

[E15045] The general duties of *HSWA 1974* bind the Crown. [*HSWA 1974 s 48(1)*]. (For the position under the *Factories Act 1961*, see **W11031** WORKPLACES – HEALTH, SAFETY AND WELFARE.) However, improvement and prohibition notices cannot be served on the Crown, nor can the Crown be prosecuted [*HSWA 1974 s 48(1)*], although Crown employees can be prosecuted for breaches of *HSWA 1974* [*HSWA 1974 s 48(2)*]. Non-statutory procedures are in place for the issue of Crown improvement and prohibition notices, and for the censure of Crown bodies in circumstances in which a prosecution would otherwise have been brought. Crown censures are available to view at www.hse.gov.uk/enforce/prosecutions.htm.

Crown immunity is no longer enjoyed by health authorities, nor premises used by health authorities (defined as Crown premises) including hospitals (whether NHS hospitals or NHS trusts or private hospitals). [*National Health Service and Community Care Act 1990 s 60*]. Health authorities are also subject to the *Food Safety Act 1990*. Most Crown premises can be inspected by authorised officers in the same way as privately run concerns, though prosecution against the Crown is not possible. [*Food Safety Act 1990 s 54(2)*].

Environmental Management

Mark Rutter

Introduction

[E16001] The management of environmental performance is a critical issue for organisations in both the public and private sectors.

Broadly speaking, environmental management refers to the controls implemented by an organisation to minimise the adverse environmental impacts of its operations. It includes the conservation of natural resources, the protection of habitats and the control of hazards. Historically, environmental management tended to be driven by a complex and interacting array of external pressures, to which organisations somewhat reluctantly responded. Now it is widely recognised that a positive and proactive approach towards environment issues can benefit business performance.

Most large, as well as many medium and smaller-sized organisations have responded by integrating policies that encourage good environmental management into their management systems and routines. External pressures are still important influences, but increasingly they tend to shape the nature and scope of environmental management practices, rather than triggering them in the first place. The emergence of co-operative and constructive stakeholder dialogue as an element of environmental management is an encouraging indication of the growing recognition of the value of effective and proactive environmental management. Furthermore, the ongoing development of new and increasingly innovative approaches to environmental management reflects the fact that there is commercial value to be gained from continuous improvement.

This positive approach to environmental considerations will be even more important with the move towards a low carbon economy to combat climate change. The most successful companies are likely to be those that show an early awareness of the risks and opportunities arising, and that face up to the changing economic values of businesses and assets under a carbon constrained society. There may well be a tendency to believe that it won't happen or to put off addressing it until it does happen. But numerous direct and indirect policies, laws and economic and fiscal instruments are coming into force around the world. Parallels could be drawn with the recent banking and financial crisis, with severe correction for those businesses that have failed to respond to the impacts of a low carbon world.

Sources of pressure on business to adopt more sustainable management practices include:

- corporate social responsibility (CSR) expectations;

- management information needs;
- employees;
- legislation;
- market mechanisms;
- the financial community;
- the supply chain;
- community and environmental groups;
- environmental crises;
- the business community;
- customers; and
- competitor initiatives.

Internal drivers

Corporate social responsibility (CSR)

[E16002] The concept of CSR has gradually crept into mainstream business practice, reflecting the changed conditions in which business operates. Environmental management is an integral part of effective CSR. The removal of trade barriers, the subsequent growth and political influence of trans-national companies, and the opening of previously restricted marketshave contributed to radical changes in the way business operates, including increasing the extent to which it controls its own performance. Consequently, the notion of CSR has also changed, with its increasing recognition by the public and its rise up the boardroom agenda. One reason for this has been a tendency for protest groups to target companies' wider impacts on global society as well as over environmental impact issues. Society is looking less to government to control the social and environmental impacts of business activity, and instead is seeing business as being accountable for those impacts. Good corporate governance is no longer merely a reflection of responsible fiscal performance. As a result, the mandate of corporate directors and managers is expanding as they recognise the need to operate in a more transparent and inclusive manner. Proactive environmental management is a key aspect of that.

Governments have also taken an interest in CSR. In the UK, the Government set out its interpretation of CSR in 2004, describing it as essentially the business contribution to sustainable development. It defined CSR as the action private and public sector organisations take voluntarily over and above the minimum legal requirements for social and environmental performance. This includes environmental protection, as well as equal opportunities, employment terms and conditions and health and safety. In order to help organisations develop their CSR skills and integrate these into day-to-day business practice, the Government then launched a CSR Academy. It was transferred to Business in the Community (BITC) in March 2007. BITC is a business-led charity with a membership of around 850 companies, which includes large multinational household names, small local businesses and public sector organisations. It re-launched the new CR (Corporate Responsibility) Academy in 2008 as a 'one stop shop' offering training, support and advice on CSR for organisations of any size and sector to help incorporate CSR measures.

Although some might predict a waning of interest in CSR during an economic downturn, the recent financial crisis has clearly shown that the business case is stronger than ever. As well as restoring public confidence in corporate activity, a recent study demonstrated a high correlation between responsible business practice and improved financial performance. The research, carried out by Ipsos MORI, showed that FTSE companies that actively managed their environmental and social impacts outperformed the FTSE 350 on total shareholder return by between 3.3% and 7.7% over the period 2002–07.

The UK is also leading the way in the international arena by publishing an International Strategic Framework for CSR in March 2005. This framework sets out the objectives, priorities and overall approach of the UK Government to CSR at a global level. It is designed to encourage relevant international and government institutions to collaborate on best practice and innovation to address sustainable development. Recognising that successful globalisation and sustainable business requires human rights, labour standards and the environment to be respected, the Foreign and Commonwealth Office (FCO) published its own CSR strategy in 2007. This strategy sets out CSR priorities aimed at supporting the future prosperity and global presence of UK business by encouraging governments, businesses, civil society and labour organisations to work together.

At an EU level, the European Commission's Green Paper on CSR, published in July 2001, defined CSR in terms of both company internal management and its impact on society and argued for a European framework for CSR. The Green Paper was followed by two Communications, published in 2002 and 2006 respectively, on new strategies for CSR. The latter Communication launched the European Alliance for CSR. CSR Europe consists of an informal umbrella network of 75 multinational corporations and 25 national partner organisations, including BITC. Its function is to encourage companies to share best practice on CSR and to support its members in integrating CSR into their every business activities. The European Commission made a new commitment in March 2010 to renew the EU strategy for CSR. It issued a call for proposals aimed at encouraging the investment community to better use environmental, social and governance information, thereby increasing the incentives for incorporating sustainable and responsible into business practices.

A new international standard providing guidelines for social responsibility, was issued by the International Organisation for Standardisation (ISO) in November 2010. ISO 26000 is a voluntary standard suitable for public and private sectors in developed and developing countries. As it contains guidance, rather than strict requirements, it is not suitable for use as a management system standard or appropriate for certification purposes or regulatory use. The guidance is meant to assist organisations with contributing to sustainable development, and to encourage them to go beyond legal compliance. It is intended to complement other instruments and initiatives for social responsibility, rather than replace them.

Management information needs

[E16003] Effective business management relies on timely and reliable information on the multitude of factors that influence it. As managers' understanding of the relationship between environmental performance and business performance increases, so too does their requirement for information pertaining to environmental performance. Such information helps to increase their control over those factors.

This reflects the acceptance of the sustainability or sustainable development concept, which recognises that long-term business success requires environmental, social and economic factors to be balanced. While there is no clear guidance or agreement on how such a balance should be achieved, it is clear that it must be based on appropriate information on all three primary elements. Thus the recognition of the relevance of environmental performance to business performance, the value of controlling it and the need for expanded management information is an increasingly important driver of environmental management practices. Structured environmental management systems not only provide a means of controlling environmental performance *per se*, but also allow more informed strategic and operational decisions to be made by management.

Employees

[E16004] Employees have a potentially strong influence over the environmental management practices of an organisation. The desire to minimise staff turnover means that companies and other organisations have to be more responsive to employee enquiries and suggestions regarding environmental performance. Conversely, employers wishing to attract high calibre recruits must take on board the importance of maintaining a strong and positive corporate image, which is often dependent on environmental performance, amongst a number of other things.

External drivers

Legislation

[E16005] UK companies are influenced by a range of international treaties, conventions and protocols; European regulations and directives; and domestic legislation. The latter may be a tool for implementing European directives, or they may have been enacted independently of any requirement of the European Union (EU).

Enforcement of the various legal instruments within England and Wales is primarily the responsibility of the Environment Agency, although local authorities also play a role. In Scotland, responsibility lies with the Scottish Environment Protection Agency.

Good corporate environmental management requires a thorough understanding of the legal requirements imposed on a company, in addition to evidence that the company has made reasonable attempts to ensure ongoing compliance with them, either through technological, procedural and/or administrative mechanisms. Many published corporate environmental policies now commit to going "beyond compliance", so that, legislative compliance is seen as the minimum standard.

This not only applies to large companies. It has often been assumed that due to their small size and lack of resources, few small and medium-sized enterprises (SMEs) would go beyond the bare minimum required to comply with environmental legislation. Indeed, a study published by the Institute of Environmental Management and Assessment (IEMA) in 2011 found low levels of knowledge, understanding and compliance with environmental legislation among SMEs. This was despite clear evidence that the impact of environmental legislation on SMEs is over-estimated and the high risk that they pose collectively to the environment as result of non-compliance.

A recent Spanish study published in the *Journal of Environmental Management* however found that a significant proportion of SMEs have proactive environmental policies. Their small size means that they have shorter lines of communication, close personal links, less bureaucracy and the ability to initiate change quickly. Moreover, the study went on to conclude that those with the most proactive environmental policies also had the best financial performance.

Other initiatives have forced companies that have not voluntarily responded, to give more systematic consideration to environmental management and performance in making strategic and operational business decisions.

The Company Law Review, launched in March 1998 began the most fundamental review of company law in 150 years.

Proposals in a White Paper, issued in 2002, and in a series of consultation papers, which followed, were intended to obligate around 1000 of the largest companies to publish an Operating and Financial Review (OFR), outlining the environmental performance and future direction of the business. Although legislation requiring quoted companies to prepare a statutory OFR for the financial year beginning on or after 1 April 2005 was published, the Government gave in to pressure from business, and revoked the OFR Regulations in November 2005.

But, for some companies, there is an obligation to prepare a Business Review as part of an 'enhanced directors' report' under the *EU Accounts Modernisation Directive (2003/51/EC)*. From financial years beginning on or after 1 April 2005, this Directive requires directors of quoted and large private companies to report on business practice to the "extent necessary" for an understanding of a company's development, performance or position, in the Business Review. This requirement includes the disclosure of significant non-financial matters through the use of key performance indicators such as the impact of the company on the environment and the interests its employees. While similar to the OFR in its disclosure obligations, the Directive does not introduce a duty to discuss and assess future impacts. Medium private companies must also produce a Business Review, although they are not bound to produce performance indicators relating to non-financial information.

The *Companies Act 2006* contains a wide range of provisions applying to all sizes of companies in the UK. One of these provisions introduced significant changes to corporate responsibility and non-financial disclosure requirements. From the financial year starting on or after 1 October 2008 company directors must act in a way that they consider most likely to promote the success of a company for the benefit of its shareholders. This includes having regard to the impact of the company's operations on the community, employees and the environment, in recognition of the fact that it is in a company's long-term interests to take account of these factors.

The *Companies Act 2006* contains a wide range of provisions applying to all sizes of companies in the UK extends the scope of the Business Review for quoted companies to bring it closer to the OFR. The Business Review must inform shareholders, and help them assess how the directors have performed their duty to promote the success of the company, including the company's environmental impacts and the risks. Though it must be a forward-looking narrative, it is intended to be less of an administrative burden than the OFR. Some environmental groups were disappointed with the absence of any statutory standard in the Act for environmental reporting.

The Department for Business, Innovation and Skills (BIS) recently carried out a study of 1001 businesses in order to evaluate awareness of and compliance with, or the adoption of, key measures implemented through the *Companies Act 2006*. Awareness and adoption of the Business Review (81% and 64% respectively) was high compared with other measures in the Act. Despite some early concerns, the majority of companies said that there was no significant additional burden associated with providing the required information in the Business Review, and that it could be beneficial to shareholders. However, it concluded that more guidance was needed on the process of producing the Review in order to improve the quality of information provided.

The *UK Corporate Governance Code* and the associated guidance replaced the An amended version of the 2008 *Combined Code on Corporate Governance* in May 2010, and apply to financial years beginning on or after 29 June 2010. The new Code contains broad principles and sets out standards of good practice for FTSE 350 companies, as well as their effectiveness, remuneration, accountability and relations with shareholders, It also includes specific requirements for disclosure which must be provided in order to comply and obligates companies to comment on how they have applied the Code in their annual report and accounts. Listed companies have to make a disclosure statement in two parts in relation to the Code. In the first part of the statement, the company has to report on how it applies the main principles in the Code. In the second part of the statement the company has either to confirm that it complies with the Code's provisions or, where it does not, to provide an explanation.

The nature of a company's internal controls depends on a proper identification and evaluation of the risks facing it, which could include environmental issues. Principle C.2 of the 2008 Code states that the company should maintain a sound system of internal controls to safeguard shareholders' investment and the company's assets. Such controls could include environmental issues. Under the provisions, the directors are required, at least annually, to conduct a review of the effectiveness of internal controls and to report to shareholders that they have done so. If the company does not have an internal audit function it should periodically review the need for one. In the 2010 Code, Principle C.2 has been re-written to cover the board's responsibility for determining the nature and extent of the significant risks they were willing to take in achieving their strategic objectives. In addition, Supporting Principle E.1 on dialogue with shareholders now makes it the responsibility of the chairman for ensuring that all directors are made aware of shareholders' concerns.

Guidance on assessing how the company has applied Principle C.2 is provided in an annexe to the Code. Originally published by the Institute of Chartered Accountants in England and Wales' Internal Control Working Party in 1999, this guidance has become known as the "Turnbull Guidance", after the Working Party's chairman. Revised Turnbull Guidance on Internal Control was published in October 2005. It set out some questions that the board should consider and discuss with Management when reviewing reports on internal control. One of these questions asked whether the company communicates clearly to its employees what is expected of them and the scope of their freedom to act in a number of areas, including environmental protection. The Financial Reporting Council (FRC) announced in December 2010 that it was deferring its planned review of the 2005 Turnbull Guidance until it was able to assess how companies were responding to the new Principle C. It will be holding meetings and consulting with companies, investors and advisers from early 2011 to discuss the issue before deciding on any further action.

Market mechanisms

[E16006] While the command and control approach embodied in environmental legislation and regulations represents a significant pressure on business, policy makers have introduced new tools to encourage better management of environmental performance.

A range of measures is beginning to emerge designed to influence the economics of polluting activities. These so-called market-based or economic instruments impose costs on pollution-causing activities and provide incentives for companies to look for ways of minimising environmental damage. Further recent concerns over environmental damage, particularly in relation to greenhouse gas emissions, have been reflected in the Government's increasing focus on using fiscal instruments to tackle the problem.

Such instruments include:

(a)   Climate Change Levy
       The climate change levy ('CCL'), which came into effect in April 2001, is a tax on the business use of fossil fuels. It is designed to encourage energy conservation and a switch to cleaner fuels and renewable energy sources to help the UK meet its legally binding commitment to reduce greenhouse gas emissions. All revenues are recycled back to business through a 0.3 per cent cut in employers' National Insurance contributions and additional support for energy-efficiency measures and energy-saving technologies. The levy is applied as a specific rate per nominal unit of energy, with different rates for each type of energy.
(b)   Landfill Tax

The Landfill Tax was introduced in October 1996 to encourage companies to reduce their volume of waste produced. From 1 April 2011, the standard rate for landfill tax for biodegradable waste will be £56 per tonne. The 2010 Budget announced that this rate tax would increase by £8 per tonne each year until at least 2014. The rate applying to inactive or inert waste increased from £2 to £2.50 per tonne from 1 April 2008. The tax is intended to be revenue-neutral to business, with tax receipts going to fund the Landfill Communities Fund (formerly the Landfill Tax Credit Scheme). This enables landfill site operators to claim tax credit for contributions they make to approved environmental bodies for spending on projects that benefit the environment.

(c) Aggregates Levy

An aggregates levy on extracting virgin aggregates, mainly sand, gravel and rock, was launched in 2002. All the revenue raised is returned to business and the local communities affected by quarrying, through a 0.1 per cent cut in employers' National Insurance contributions and a new sustainability fund. Between April 2002 and March 2008, the rate was £1.60 per tonne. However, the Aggregates Levy was increased to £1.95 per tonne from 1 April 2008, and to £2 per tonne from 1 April 2009. The rate was maintained at this level during 2011-2-12. It is intended that the levy will help to ensure that the environmental impact of aggregate extraction is reflected in the price. This is aimed at encouraging more efficient use of aggregates and the development of alternatives including waste glass, tyres and recycled construction and demolition waste.

(d) Vehicle and Fuel Duty

Lower levels of duty on cleaner fuels and changes to company car taxation and vehicle excise duty (VED) are designed to reduce pollution, in particular carbon dioxide, from vehicles. VED rates for cars registered on or after 1 March 2001 are split into 13 bands depending on CO_2 emissions. While there is no VED for vehicles with low CO_2 emissions (up to 100 g/km), the rate for the most polluting cars is being increased significantly. For example the rate of VED for cars emitting over 255 g CO_2 /km rose from £435 in 2010–11 to £460 in 2011-12. In addition, differential First-Year Rates of VED were introduced to all new cars from 1 April 2010. With effect from April 2011, the nine highest bands were liable for additional VED, ranging from an extra £10 to an extra £550 for vehicles with emissions greater than 135 g CO_2/km and 255 g CO_2 /km, respectively. There are also differential rates of VED for buses, lorries and other large commercial buses, designed to encourage the use of more fuel-efficient transport.

(e) Emissions Trading

Companies that reduce their emissions below a quota can sell the unused part of their quota to other firms, thus providing an incentive to improve emissions performance. Such a scheme is included in the Kyoto Protocol, which was adopted in 1997 and came into force in February 2005. It sets targets for cuts in greenhouse gas emissions by developed countries. Article 17 of the Protocol allows developed countries that reduce their emissions by more than their assigned target to gain credits, which can be sold to other developed countries. EU wide emissions trading between companies started in January 2005 under a European Directive. The scheme has since been modified, with a second phase beginning January 2008, and lasting until December 2012. At present, it only covers carbon dioxide and is mandatory for power generation companies and heavy industrial users of electricity. Phase III of the EU Emissions Trading System (previously known as the Emissions Trading Scheme) will run from 1 January 2013 to 31 December 2020. It will introduce major changes including, additional greenhouse gases and emission sources, such as those from aviation, and will have a more ambitious EU-wide cap on emissions.

(f)   Air passenger duty (APD)

In response to increasing concerns over the adverse impact of aviation on climate change, the rates of APD have increased significantly from 1 February 2007. From 1 November 2009, APD was structured to incorporate two rates according to the class of travel, and four distance bands, set at intervals of 2,000 miles from London, so that those flying further pay more. The rates payable from 1 November 2010 range from £12 to £85 for the lowest class of travel, and from £24 to £170 for other classes.

The financial community

[E16007] The emerging realisation of the link between environmental performance and business performance has encouraged investors, shareholders and insurers to develop a direct interest in the environmental performance of companies. Investors are increasingly concerned that companies that fail to manage their social and environmental exposure will suffer. The financial community is therefore now seeking information on how environmental issues will potentially impact on the long-term viability of the companies in which they have a commercial interest. In particular, they are concerned about the extent to which environmental risks are being controlled. Brand reputation accounts for an increasing proportion of stock market valuations. Therefore, they want reassurance that companies are not in breach of legal requirements with the consequent threats of fines, damage to reputation and the need for unanticipated expenditure. They need to be sure that assets, in the form of plant, equipment, property and brand value, against which they have lent money, are correctly valued. Raw materials, by-products and end products may need to be replaced, modified and/or discontinued, which may require provisions or contingent liabilities to be included in the corporate accounts. Such a situation may arise as a result of substances being phased out (for example, legal controls on global warming chemicals), or it may reflect changing market attitudes such that the demand for environmentally damaging products begins to decline. Additional research and development costs are likely to be associated with such changes.

Consideration of environmental risks and performance is now an established component of acquisitions, mergers, flotations, buyouts or divestments. Management of companies involved in any of these deals must be able to demonstrate that environmental liabilities do not constitute an unacceptable risk for investors or insurers. Companies that emphasise their environmental performance are viewed by the City as forward-looking and actively managing their reputation. An increasing number of companies are also paying greater attention to their indirect impacts, such as the environmental implications of investment decisions.

Arguably the most striking example of progress is the rapid growth in socially responsible investment (SRI). The Dow Jones sustainability indexes and FTSE4Good are two examples of indexes designed to track stocks for the purpose of SRI. They include only those companies that meet globally recognised corporate responsibility and environmental performance standards.

Another initiative, which is managed by BITC, is the business-led CR Index. This has become one of the UK's leading benchmarks of responsible business. It is intended to help organisations to take a systematic approach to managing, measuring and reporting potential impacts arising from business operations, products and services. Such impacts may affect the local community, marketplace and workplace. The CR Index consists of an online survey that provides invited large organisations (i.e. those with revenues of over £250 million) with a framework for managing CR. This framework is designed to integrate and improve CR throughout the organisation by providing a systematic approach to managing, measuring and reporting on business impacts in society and on the environment. Participants are grouped into performance bands (Platinum, Gold, Silver and Bronze) according to the extent to which responsible practices are embedded within an organisation's corporate strategy and operations, The CR Index annual ranking is published on the BITC website. Companies not eligible to participate in the public CR Index, or not yet prepared to publicly reveal their performance, can still use the management tool and benchmark themselves again others but they will not be included in the annual CR Index public rankings.

Results from the BITC CR index have shown consistently that companies can gain competitive advantage by moving ahead of legislation in managing their environmental impacts. Furthermore, progressive organisations are now thinking strategically on environmental matters, such as supply chain, climate change and water consumption.

Hermes, the independent fund manager has published 'The Hermes Principles', setting out ten investment principles to address what owners should expect from UK public companies and what these companies expect from their owners. The Hermes Principles are designed to encourage companies to communicate clearly the plans they are pursuing and the likely financial and wider consequences of those plans. Principles 9 and 10 deal with social, ethical and environmental issues. They call for companies to manage effectively relationships with their employees, suppliers and customers and say that they should behave ethically and have regard for the environment and society as a whole. In addition, they require that companies should support voluntary and statutory measures designed to minimise the externalisation of costs to the detriment of society.

The financial consequences of climate change are rapidly becoming a priority for many companies and investors. The risks, as well as the opportunities, posed by climate change are widespread and varied. They include physical risks such as asset damage and project delays resulting from changing weather patterns. Regulatory risks resulting from tighter legislation on greenhouse gas emissions are greater than ever. In addition, there are possible competition risks due to a decline in consumer demand for energy-intensive products and a rise in costs for energy intensive processes. Risks to brand and reputation can arise through a perceived lack of action to implement measures to help counter climate change.

The Carbon Disclosure Project (CDP) works with the world's largest collaboration of institutional investors to annually survey 500 of the world's largest publicly listed companies (FTSE Global Equity Index Series) on the implications of climate change to their business. It collates and makes publicly available greenhouse gas emissions data obtained from these companies, as well as advising investors on the risks and opportunities presented to them by climate change. The CDP is now established as the gold standard for carbon disclosure methodology.

Since its earliest survey in 2003, the CDP has found a significant increase in the amount of climate change-related information communicated to investors. The CDP says that the majority of the Global 500 companies that reported cited regulation as a key risk factor. It has also stated that it believes corporations are making significant progress in understanding and disclosing the risks and opportunities associated with climate change, and that there is a growing determination to act on this. The CDP survey is also attracting increasing support from institutional investors, with three-quarters factoring climate change information into their investment decisions.

In April 2011, the CDP announced a new initiative of requesting detailed climate change action plans from the largest Global 500 companies. The Carbon Action scheme is aimed at getting organisations to deliver emissions cuts, and to identify and implement initiatives that give a positive return on investment. It is being supported by 34 investors with around $7.6 trillion in assets under management, The CDP wants companies to go beyond disclosure by providing information on how they are attempting to reduce their greenhouse gas emissions, as well as getting them to set a public emissions reduction target.

A partnership of leading global institutional investors, environmental organisations and other public interest groups released a 'Global Framework' for climate risk disclosure in October 2006, in response to concerns of institutional investors over the risks of actual and predicted climate change. It contained specific guidance for companies on the information they should supply to investors, including new and emerging regulations, impacts from changing weather and demand for new technology. Investors can use the framework to persuade companies to use existing reporting mechanisms, such as the Global Reporting Initiative or the Carbon Disclosure Project, when submitting financial returns to regulators in order to get comprehensive and consistent climate-related disclosure.

The oil company Exxon Mobil, has in the past been subjected to a boycott campaign by environmental pressure groups because of it opposition to the

Kyoto Protocol. It has come under pressure from investors and some main-stream financial analysts concerned that the companies' environmental stance threatens shareholder value. A number of resolutions have been lodged at annual shareholder meetings urging the company to improve its environment disclosure and performance.

A set of principles, known as the Equator Principles (EPs), has been developed for use by financial institutions when providing loans for large projects. These Principles are intended to ensure that projects being financed are developed in a socially responsible manner that reflects sound environmental management practices. They consist of a voluntary commitment based on the International Finance Corporation (IFC) performance standards on social and environmental sustainability and on the World Bank Group's environmental, health and safety general guidelines and provide the framework for identifying the issues that banks and borrowers need to be aware of. Most UK, and many other international banks have adopted the EPs, realising their use in helping them to document and manage risk exposure. The Equator Principles Association launched a strategic review of the EPs in October 2010. This is the first step in a longer-term process to determine the future of the EPs that would include an EP III Update Process beginning in the latter half of 2011.

The United Nations launched its Principles for Responsible Investment (PRI), backed by world's largest institutional investors, in April 2006. This contains six overarching Principles relating to 35 possible actions that institutional investors can take to integrate environmental, social and corporate governance considerations into their investment decisions. Within these voluntary Principles, there is a recognition that while the global economy is driven by financial demands, better long-term investment returns and more sustainable markets are achieved through taking account of environmental and social considerations. By April 2011, over 850 investment institutions, with approximately US$ 25 trillion of assets under management, had signed up to the PRI. The PRI Initiative was created after the launch of the Principles to help investors to implement the Principles. The Initiative supports investors by sharing best practice, facilitating collaboration and managing a variety of activities.

Supply chain

[E16008] Focus is shifting increasingly from organisations' own emissions to their supply chains. Most are both purchasers and suppliers of a range of goods and services. Introducing environmental criteria into procurement decision-making processes emphasises the importance of issues other than price and quality in purchasing goods or services. Examples of environmental criteria are selecting materials, components or products that were manufactured using relatively less energy than alternatives, or that require relatively less energy in operations, or the substitution of chemical substances of concern with safer alternatives.

The Carbon Disclosure Project (CDP) estimate that over 50% of an average corporation's carbon emissions come from the supply chain. As a result, many companies are taking active steps to control their environmental performance throughout the supply chain. In its survey of 178 companies, which included FTSE100 and FTSE250 companies and sector leaders from the Dow Jones Sustainability Index, Business in the Community found that just over 25% of participants gave priority to ensuring that their procurement follows good environmental practice.

A report issued by the Carbon Trust at the end of 2006 entitled 'Carbon Footprints in the supply chain: the next step for business' is intended for companies looking to assess the greenhouse gas emissions arising from their supply chain. The report outlines the steps that businesses can take to investigate their supply chains, how to identify the carbon footprint of their products and services at every stage of the lifecycle, and how to recognise opportunities to reduce emissions.

One example of the successful use of this scheme to make savings and improvement across the whole supply chain is the strategy employed PepsiCo UK & Ireland. The company implemented a five year plan to reduce by 50% the carbon and water impacts of the oats, apples and potatoes from the UK that make their Quaker Oats, Copella Apple Juice and Walkers Crisps. It worked with farmers to identify their carbon 'hotspots' where the most efficient emissions reductions could be achieved. Another initiative was to use only 100% British potatoes to reduce food miles. In just two years, the carbon footprint of Walkers crisps was improved by 7%. Moreover, the measures resulted in savings of around £400,000.

The CDP Supply Chain Leadership Collaboration (SCLC) was launched in October 2007, in partnership with Wal-Mart. It is aimed at encouraging organisations to use a common methodology to measure and manage their supply chain emissions. Data is collected by incorporating an additional questionnaire on carbon emissions and climate strategy of the supply chain into its annual survey of major corporations (see E16007). One of the main objectives of the SCLC was to better understand how supply chain companies were considering climate change and how they were working to reduce their greenhouse gas emissions.

Many suppliers now show a willingness to disclose information. In 2011, the CDP published a report of its work with 57 major global corporations in the so-called Supply Chain program. This was aimed at implementing supplier engagement strategies and risk management relating to greenhouse gas emissions and climate change. It involved 1,000 participating suppliers across industries all over the world. The results showed that only one third of responding suppliers have a target for carbon reduction, and of those in place, most are not sufficient. However, there was evidence of improved reporting, increased board level responsibility, and a greater realisation that carbon management presents a wider cost and revenue opportunity rather than being a pure risk mitigation activity. It was also found that CDP Members are increasingly using their influence to drive change, with all members, and half of suppliers, now having a formal strategy for climate change.

Pressure from government is also influencing the use of environmental considerations in procurement. Given that public authorities in EU countries spend around an average of 17% of GDP on procurement, the EU see public purchasing as a useful policy tool for tackling environmentally damaging products and services. The process began in 2004 with the launch of the European Commission's handbook for greener public purchasing in response to a revision of EU public procurement rules. The EU launched a help desk for green public procurement (GPP) in January 2010 to promote and disseminate information about GPP and to respond to queries to stakeholders' enquiries.

A recent a study carried out on behalf of the European Commission found that environment credentials are only taken into account in a meaningful way during public procurement in seven out of the 25 EU countries. The study report called for target setting, as well as for more environmental management systems to be implemented, in order to encourage the uptake of GPP. In response to this, the Commission adopted a Communication in July 2008 proposing a target of 50% of all tendering procedures in EU countries to be green by 2010. EU member countries were encouraged to draft national action plans (NAPs) on green public procurement. Most EU countries have now adopted NAPs with voluntary or mandatory targets and specific measures to promote and implement GPP. So far, the EU has developed GPP criteria for 18 product and service groups, such as food, transport and energy, with more planned in the near future. A recent Green Paper on the modernisation of public procurement policy, available for consultation until April 2011, re-iterated the EU's commitment to encourage and support GPP. It asked whether EU public procurement rules should be amended to take more account of environmental considerations, and if obligations to buy only products con-forming to specific environmental conditions should be introduced.

The UK Government published its NAP to reform public sector purchasing of goods and services in 2006 entitled 'Procuring the Future', it provided an analysis of the barriers to sustainable procurement and made six key recom-mendations, alongside details of the actions that should be taken, with clear target dates for the future. It concluded that the UK would gain significantly from being a leader in sustainable procurement. Among the benefits it said, would be better stewardship of taxpayers' money, environmental and social improvements and more support for environment-friendly technologies. With an annual budget in the region of £150 billion, public sector goods and services in the UK has the potential to transform markets, encouraging much greater participation by householders and the private sector.

GPP is making some in-roads into the private sector. Research carried out by the Environment Agency in 2009 found that 95% of the UK's large construc-tion firms give preference to subcontractors who can prove their environmen-tal credentials. A large majority of these firms stated that they have more confidence in subcontractors with proven green policies and procedures in place, as they believed there was a reduced risk of prosecution. Many also believed that green policies would save subcontractors money.

An increase in green procurement is leading to new opportunities for those with suitable green products and services. Conversely, it also poses a threat to traditional products and services that do not take into account environmental impacts.

It is quite common for purchasing companies to require their suppliers to demonstrate ongoing compliance with formal environmental management standards such as ISO 14001 or EMAS.

The EU Eco-label voluntary award scheme has been in operation since 1993, when the first product groups were established, and was comprehensively revised in 2000. An Eco-label (the "Flower symbol") is awarded to products possessing characteristics that contribute to improvements in environment protection. Ecolabel criteria are based on the impact of the product or service on the environment throughout its life-cycle, from raw material extraction through to production, distribution and disposal. The main objective of the scheme is to encourage business to market greener products by providing information to allow consumers to make informed environmental choices when purchasing. Although there has been a steady increase in both the number of applications from manufacturers and the number of eco-labelled products marketed since the inception of the scheme, the European Commission has continually strived to improve it.

In December 2006, it consulted on a range of issues including how the scheme could be improved to increase its uptake and what other product groups could be included. The consultation was also intended to collect views on how the scheme could be used to improve "green" procurement and to support other environmental measures operated by the Commission and EU countries, such a the Eco-design Directive setting requirements for energy related products. The outcome of this was the replacement of *Regulation (EC)* No 1980/2000 by *Regulation (EC)* No 66/2010. The new Regulation should increase the scope of product groups included in the scheme. It was also intended to speed up the development process by simplifying the assessment procedure and criteria documents and to provide better guidance for GPP.

In the UK, the Carbon Trust has devised a label to provide information on a product's carbon footprint across its life cycle. As well as providing information to purchasers, it allows producers to demonstrate commitment to managing and reducing the carbon emissions of their products and services. The carbon reduction label is supported by a number of documents, including PAS 2050 – the standard method for the measurement of the lifecycle greenhouse gas emissions of goods and services. This standard, which was developed by the BSI, is applicable to a wide range of sectors and product categories. It is available to download free from the BSI website. Other documents available to support the labelling initiative are: a Code of Good Practice setting out requirements for organisations making claims for their products' emissions reductions; a Guide to PAS 2050; and the Business Case for carbon footprinting, setting out the potential benefits to companies for carrying out product carbon footprinting.

A new standard for sustainable procurement, BS 8903: Principles and Framework for Procuring Sustainably, was launched in 2010. This standard provides guidance to all sizes and types of organisation on good practice for adopting and integrating sustainable procurement principles and practices. It covers all stages of the procurement process and is applicable across industry, public, private and third sector organisations. Implementation of BS 8903 will enable a uniform approach to reporting and measuring sustainable procurement throughout the supply chains. It will also allow more accurate benchmarking of performance with others, therefore providing an opportunity for profile-raising and increasing competitive advantage.

Community and environment group pressure

[E16009] The majority of public pressure on companies to improve their environmental performance is initiated by local communities that experience the direct effects of pollution. Most companies recognise the importance and value of working co-operatively with local communities. In fact many have established community liaison panels, which comprise representatives of the local community. Such panels interact with the company on a regular basis and provide input on environmental and other community issues.

Environmental pressure groups, often supported by a high level of legal and technical expertise, are also influential. Previously the relationship between environmental pressure groups and business was characterised by mutual mistrust and reactive criticism. The emergence of the concept of "stakeholder engagement", whereby companies take a more inclusive approach to business management, means that environmental pressure groups can be expected to have more direct access to companies and management in the future. In fact their opinions are already actively sought by many organisations, which recognise the importance of constructive dialogue. While companies will not necessarily implement all suggestions made by external pressure groups, the trend towards more timely and constructive dialogue is likely to continue.

An example of the potential for corporate reputation damage that could occur when a company fails to engage with environmental groups is Green-peace's Stop Esso campaign. This campaign targeted Esso's parent company ExxonMobil worldwide, for its negative stance on climate change and lack of investment in renewable energy sources, by calling for a boycott of the company's products. Greenpeace claimed that its boycott campaign had resulted in around one million motorists, a quarter of the number of regular buyers of Esso petrol, boycotting its filling stations. Due largely to this campaign, Deutsche bank warned that Exxon Mobil was being labelled as "environmental enemy number one" and that this posed a significant risk to its business.

Similarly, BP has in the past come under pressure from the Worldwide Fund for Nature (WWF), which sold its shares in the company in protest at its environment and employment policies in Alaska. Although the value of the shares was insignificant in financial terms, it is likely to impact on the environmental reputation of BP. This was followed by an announcement by one of the UK's leading ethical investment funds, Henderson Global Investors, that it intended to sell several million pounds worth of BP shares.

Another tactic of pressure groups is to target customers of companies supplying goods or services that damage the environment. This is often seen as more effective when companies are proving difficult to engage with, particularly when the supplier is a private company, and therefore not accountable to shareholders. Greenpeace's lobbying of McDonald's and other key customers led to the signing a moratorium on Amazonian soya, which in turn influenced the suppliers such as Cargill to address the issue of illegal deforestation in the Amazon.

The evolution of the internet and other sophisticated communication media has significantly increased public awareness of and access to information about corporate environmental performance, as well as increasing the speed with which such information can be transferred and responded to. Under the EU Directive implementing the first objective of the Aarhus Convention concerning public access to information and the *Freedom of Information Act 2000*, the public have gained stronger rights to access information on companies' impacts on the environment. Although this legislation only provides a right of access to recorded information held by public authorities, many industrial projects that could potentially affect the public would be included, thereby extending the rules to cover much more environmental and health and safety information. However, legally classified and commercially confidential information can still be excluded. The result is that pressure on companies to improve environmental performance has increased and this has spread into the international arena. Nowadays, companies are finding it necessary to develop and adopt consistent environmental and social performance standards in all markets in which they operate, since their performance is increasingly subject to international scrutiny.

The not-for-profit organisation, AccountAbility, whose members include leading companies, civil society organisations and service providers, has developed the AA1000 series of standards designed to help organisations become more accountable, responsible and sustainable. These include a Stakeholder Engagement Standard (AA1000SES), which can be used by companies to improve and assure the quality of its communication with stakeholders. The AA1000SES is valid for a range of engagements, whatever their size, including customer care and human rights. It allows stakeholders to assess and comment on the quality of an engagement using set principles. Originally published in 2008, the engagement standard is currently being updated. An 'exposure draft' released in early 2011 revealed that AA1000SES (2011) would consist of four parts. These include a description of the standard's purpose and scope; how to integrate a commitment to stakeholder engagement in strategy and operations; how to define the purpose, scope, and stakeholders of the engagement; and what a quality stakeholder engagement process looks like. The new standard is part of a suite of stakeholder engagement tools that includes the AA1000SE Manual and an AA1000SE WikiHub – an open-access collaborative database managed by stakeholders, containing information on stakeholder engagement.

Environmental crises

[E16010] In many cases, high profile environmental crises are a catalyst for improved environmental management. There are two main types:

(a)   crises that are generated as a result of the actions of an individual company, the effects of which are generally experienced at a local scale;

(b)   crises that are generated by collective action or by natural forces, the effects of which are often experienced at the national or international scale.

Individual companies that have been associated with environmentally damaging events such as oil spills generally find themselves exposed to intense pressure to improve their environmental management practices in the immediate future. The need to correct the damage to reputation caused by environmental crises is a further incentive to respond quickly. Obviously the costs of responding to pressures arising from catastrophic environmental incidents can be extremely high, and most companies seek to avoid those by incorporating environmental issues into their corporate risk management programmes. However, the impacts, in terms of clean-up costs, compensation, reputation damage and share price of such events can go far beyond those foreseen. This was the case with the BP Deepwater Horizon oil rig explosion in 2010 and the devastating effects of the subsequent pollution. The situation was made worse for the company by its initial failings to manage its communication with outside agencies and the public, and was seen by many as a PR catastrophe,

Global or national environmental crises tend to emerge more gradually, and with considerably more debate about accountability and appropriate responses. Nevertheless, there are a number of examples of global environmental issues that have facilitated more systematic and intensive environmental management practices than may otherwise have occurred in the same time period. The most obvious examples are the depletion of the ozone layer and climate change. At a national level, issues such as water shortages, soil erosion and regional air pollution have triggered the adoption of improved environmental management practices on an extensive scale.

Business community

[E16011] Trade associations and business groups such as the Confederation of British Industry (CBI) and the International Chamber of Commerce (ICC) have played a leading and effective role in encouraging businesses to adopt environmental management practices. Increasingly the importance of doing this within a sustainable development framework is being accepted. Sustainable development is generally recognised as the inter-relationship and interdependence of the three core elements of economic, environmental and social consideration, although it is a concept that is open to wide interpretation. However, it is generally understood to mean achieving a better quality of life with effective environmental protection. The most widely used definition is contained in the 1987 United Nations Brundtland report to the World Commission on Environment and Development. This definition states that it is "development which meets the needs of the present without compromising the ability of future generations to meet their own needs."

There is a desire among employers and employees to improve the environment impacts of the workplace. A website promoting partnership action in the workplace to help achieve sustainable development has been initiated by the Advisory Committee for Business and the Environment and the Trade Union Sustainable Development Advisory Committee. Sustainable Workplace (www.s ustainableworkplace.co.uk) provides case studies of employer and union initiat ives that have successfully improved organisations' environmental performa nce. It also contains links to other websites and organisations where business support, information on education for sustainable development, environmental management and reporting, as well as the principles of sustainable development , can be found.

Thousands of companies worldwide have been helped by the ICC voluntary business initiative "Business Charter for Sustainable Development". Launched in 1991 as a tool to help companies tackle the challenges and opportunities of the environmental issues that emerged in the 1980s and early 1990s, the Charter has established a set of 16 principles to guide company strategies and operations towards sustainable development. Its principles for environmental management have provided a global alignment of business to common objectives and have helped thousands of companies worldwide establish the foundation on which to build their own integrated environmental management systems. The Charter highlights such areas as employee and customer education, research facilities and operations, contractors and suppliers and emergency preparedness. It requires organisations to support the transfer of technology, be open to concerns expressed by the public and employees and carry out regular environmental reviews and report progress.

In 1995 the World Industry Council on the Environment (WICE) and the Business Council for Sustainable Development (BCSD) merged to become the World Business Council for Sustainable Development (WBCSD). WBCSD consists of some 200 companies dealing exclusively with business and sustainable development. These were drawn from more than 30 countries and 20 major industrial sectors. The organisation also has a global network of about 60 national and regional business councils and regional partners. By focussing four key areas, energy and climate, development, the business role, and ecosystems, its aim is to promote the effective implementation of sustainable development principles. This is carried out through a combination of advocacy, research, education, knowledge-sharing and policy development. Members of the WBCSD include Proctor & Gamble, Vodafone Group, Sony, BP, Du Pont and Unilever.

The Responsible Care Programme is an example of an international initiative from a specific industry sector. The Programme has been adopted by over 50 national chemical industry associations, including the Chemical Industries Association (CIA) in the UK. All of those associations have made acceptance of Responsible Care requirements compulsory for individual member companies. Responsible Care is designed to promote continuous improvement, not only in environmental management, but also health and safety. The companies must also adopt a policy of openness by releasing information about their activities. Adherence to the principles and objectives of Responsible Care is a condition of membership of CIA.

Individual companies have also contributed to the development of improved environmental performance standards. As the relevance of good environmental management to overall business performance has become more apparent, progressive companies have voluntarily adopted a number of innovative and unique approaches to corporate environmental management. This has had the effect of constantly moving the frontiers of "acceptable" environmental management practices and created substantial peer pressure which in turn has encouraged other companies to adopt similar or even more effective practices.

Customers

[E16012] In the late 1980s and early 1990s some of the major retailers started to market "green" products, including phosphate-free washing powders and biodegradable cleaning products, in response to consumer concern about high-risk chemicals.

It was claimed in a Government report published in 2002 that around half of consumers have identified CSR as an important factor when choosing to buy, and that 20% will boycott or select products on these grounds. More recently, in its 2008 "Framework for Pro-Environmental Behaviours", the Government reported the results of research carried out on public understanding of and behaviour in relation to environmental matters. It concluded that the most likely public actions are to reduce food waste and water usage, and the least likely to stop flying. The research also showed that there is potential for increasing the purchasing of low environment impact goods, particularly in relation to energy efficient products. Nearly half the people surveyed said they would be prepared to pay more in general for environmentally friendly products, with two thirds prepared to do so for appliances with high energy efficiency ratings. In addition, it was found that most people were aware of certification or assurance schemes such as Fair Trade and schemes for ensuring that timber originated from sustainable sources.

Retailers are coming under increasing scrutiny as a result of benchmarking by consumer groups. Consumer Focus, the statutory consumer organisation for the UK, found in its 2009 study that although some UK supermarkets were making improvements in helping customers shop for green products, others were lagging well behind. Sainsbury's and Marks and Spencer were commended for making the biggest advances by achieving the first ever overall 'A' (excellent) score.

Greater public awareness of environment issues will inevitably feed through into consumer demand, which will drive companies to produce "greener" products and services. There is some evidence of the role of customers in the uptake by companies of a sound stance on environmental and other ethical issues. The Co-operative Bank is a well-established proponent of active ethical and environmental policies under which some customers are denied banking. It attributed one fifth of its 2001–2002 pre-tax profits directly to these policies. Surveys for the bank showed that 14 per cent of account holders regarded ethical issues as the most important of a list of factors when choosing it, whilst 26 per cent viewed them as an important factor. There is also some evidence from academic studies that some consumers choose green products because they believe it enhances their social status. This is most evident when this is displayed in public, The recent trend of high-profile celebrities choosing to drive hybrid cars is likely to enhance this behaviour.

Environmental management guidelines

Standards for environmental management

[E16013] Various guidelines exist for responding to those many pressures to minimise damage to the environment and health. Effective environmental management, like quality management or financial management, requires *inter alia*:

(a)   the setting of objectives and performance measures;
(b)   the definition and allocation of responsibilities for implementing the various components of environmental management;
(c)   the measurement, monitoring and reporting of information on performance;
(d)   a process for ensuring feedback on systems and procedures so that the necessary changes can be actioned.

There are two main instruments that influence current approaches to environmental management. The first is the EU Eco-Management and Audit Scheme (EMAS) Regulation. The second is the International Standards Organisation Series of Environmental Management Standards.

The Eco-Management and Audit Scheme ("EMAS") Regulation

[E16014] This voluntary scheme came into force in July 1993 and has been open for participation by companies in all Member States since April 1995. It was initially established by *European Regulation* 1836/93, which has since been updated twice in order to attract more registrations and make the scheme more competitive. The most recent legislative instrument is *EU Regulation (EC)* No 1221/2009, which came into force in January 2010.

EMAS has three main aims:

(a)   the establishment and implementation of environmental policies, programmes and management systems.
(b)   the systematic, objective and periodic evaluation of these measures.
(c)   the provision of information to the public on environmental performance.

The Regulation sets out a number of elements of systematic environmental management. Sites that can demonstrate ongoing compliance with those requirements to an independent assessor have the right to be registered under EMAS.

Following the adoption of a company policy, an initial environmental review is made to identify the potential impacts of a site's operations, and an internal environmental protection system must be established. The system must include specific objectives for environmental performance and procedures for implementing them. The system, and the results of the initial environmental review, must be described in an initial environmental statement. The statement must be validated by an accredited external organisation before being submitted to nominated national authorities in individual Member States for registration of the site under the scheme.

There are a number of key points here:

(a)   the first stage in developing environmental management is to carry out a thorough review of impacts on the environment;

(b)   setting up an environmental management system is a prerequisite of registration under the scheme;

(c)   external validation of the environmental statement is intended to ensure consistency in environmental management systems;

(d)   the description of the environmental management system within the statement will be on the public record.

Once a site has been registered under the scheme, it will require regular audits to review the effectiveness of the environmental management system as well as giving information on environmental impacts of the site. Here it is sufficient to note that the development of procedures for internal auditing is a crucial part of an environmental management system. In addition to the audit, the preparation and external validation of an environmental statement, submitted to the competent authority for continued registration and made public, are elements of an on-going procedure.

The scheme is open to all sectors of the economy, including financial companies, transport and local and public bodies, although industry still accounts for approximately 70% of all EMAS registrations. In addition, registered organisations can use an official logo to publicise their participation in EMAS as well as gaining regulatory benefits. The scheme also encourages more involvement by employees in implementation and strengthens the role of the environmental statement to improve the transparency of organisations and their stakeholders.

The 2009 Regulation introduced revised audit cycles to make EMAS more applicable for small organisations, and offered the opportunity for a single corporate registration, rather than requiring registration for individual sites as under the old rules. It also sets out 'environmental core indicators' that are to be included in the environmental statement and to describe the environmental performance. All the existing explanatory guidelines are now included in the EMAS Regulation. Some of the provisions are optional for EU countries. For example, the UK has chosen not to make use of the article which provides for organisations outside the EU to apply for EMAS registration in the UK. Neither will it appoint a separate licensing body in the UK, or allow the licensing of natural persons as environmental verifiers for the purposes of EMAS.

In March 2011, there were a total of 7910 EMAS registered sites owned by 4634 organisations in a wide range of economic sectors throughout Europe. In the UK, there were 337 sites and 70 organisations with an EMAS registration.

International standards on environmental management (the ISO 14000 series)

[E16015] The International Standards Organisation (ISO) has developed a series of standards for various aspects of environmental management, which it keeps under review. All are designed to assist organisations with implementing more effective environmental management systems. Table 1 outlines the various standards and guidelines within the series.

The most high profile of these standards is ISO 14001, first published in June 1996, which sets out the characteristics for a certifiable environmental management system (EMS). ISO 14001 was based on the British Standard on Environmental Management Systems (BS 7750), although the latter has been superseded by the international standard.

The format of ISO 14001 reflects the procedures and manuals approach of the ISO 9000 quality management systems series. In practice this means that organisations that operate to the requirements of ISO 9000 can extend their management systems to incorporate the environmental management standard, although the existence of a certified quality management system is not a prerequisite for ISO 14001.

ISO 14001 requires an organisation to develop an environmental policy that provides the foundation for the rest of the system. The standard includes guidance on the development and implementation of other elements of an EMS, which is ultimately designed to allow an organisation to manage those environmental aspects over which it has control, and over which it can be expected to have an influence. ISO 14001 does not itself stipulate environmental performance criteria.

Revised versions of ISO 14001 and ISO 14004 were published in November 2004. ISO 14001:2004 is designed to be more closely aligned with ISO 9001, and to clarify a number of the requirements. ISO 14004:2004 is more consistent and compatible with ISO 14001:2004, to encourage their joint use, and to make it more accessible to SMEs. Although the new 14001 standard is very similar to the previous version in many respects, there are significant differences in the scope, impacts, and periodic evaluation of legal and other requirements.

By the end of 2009, 223,149 ISO 14001 certificates had been issued in 159 countries and economies, a significant increase from 2007. The popularity of ISO 14001 over EMAS, is due to the fact that until recently, EMAS did not extend beyond Europe, and that it is much more exacting than ISO 14001.

Table 1: The ISO 14000 environmental management series	
ISO 14001: 2004 ISO 14001:2004/Cor 1:2009	Environmental management systems — Requirements with guidance for use
ISO 14004: 2004	Environmental management systems — General guidelines on principles, systems and supporting techniques
ISO 14005: 2010	Environmental management systems — Guidelines for a staged implementation of an environmental management system, including the use of environmental performance evaluation
ISO/FDIS 14006	Environmental management systems — Guidelines on eco-design (under development)

ISO 14015: 2001	Environmental assessment — Sites and organisations
ISO 14031: 1999	Environmental management — Environmental performance evaluation – Guidelines
ISO 14020: 2000	Environmental labels and declarations — General principles
ISO 14021:1999	Environmental labels and declarations — Self-declared environmental claims (Type II environmental labelling)
ISO 14024:1999	Environmental labels and declarations — Type I environmental labelling — Principles and procedures
ISO 14025:2006	Environmental labels and declarations — Type III environmental declarations — Principles and procedures
ISO/TR 14032:1999	Environmental management — Examples of environmental performance evaluation (EPE)
ISO 14040: 2006	Environmental management — Life cycle assessment — Principles and framework
ISO 14044: 2006	Environmental management — Life cycle assessment — Requirements and guidelines
ISO/DIS 14045	Eco-efficiency assessment — Principles and requirements (under development)
ISO/TR 14047: 2003	Environmental management — Life cycle impact assessment — Examples of application of ISO 14042
ISO/TS 14048:2002	Environmental management — Life cycle assessment — Data documentation format
ISO/TR 14049: 2000	Environmental management — Life cycle assessment — Examples of application of ISO 14041 to goal and scope definition and inventory analysis
ISO 14050: 2009	Environmental management — Vocabulary
ISO/CD 14051	Environmental management — Material flow cost accounting — General principles and framework (under development)
ISO/TR 14062: 2002	Environmental management — Integrating environmental aspects into product design and development
ISO 14063: 2006	Environmental management — Environmental communication — Guidelines and examples
ISO 14064: 2006 (parts 1, 2 and 3)	Environmental management — Greenhouse gases — accounting and verification
ISO 14065:2007	Environmental management — Greenhouse gases — Requirements for greenhouse gas validation and verification bodies for use in accreditation or other forms of recognition
ISO 14066:2011	Environmental management — Greenhouse gases — Competence requirements for greenhouse gas validation teams and verification teams
ISO 19011:2002	Guidelines for quality and/or environmental management systems auditing

ISO 26000: 2010	Guidance on social responsibility

Differences between ISO 14001 and EMAS

[E16016] The key differences between ISO 14001 and EMAS are:

(a)   ISO 14001 does not currently include a requirement for public reporting of environmental performance information, whereas sites registered under EMAS must produce an independently validated, publicly available environmental statement;

(b)   Unlike EMAS, ISO 14001 does not include active involvement of employees;

(c)   the level of control of contractors and suppliers required in EMAS is not matched in ISO 14001, which stipulates only that required procedures are communicated to them.

(d)   EMAS specifies environmental core indicators to assess performance and to be included in the environmental statement. ISO 14001 does not include core indicators.

(e)   a verified initial environmental review is required for EMAS but is only a recommendation in ISO 14001

(f)   EMAS requires a demonstration that the implications of legal requirements relating to the environment have been identified, provide for legal compliance, and have procedures in place that enable the organisation to meet the requirements. ISO 14001 does not require organisations to demonstrate legal compliance, but to show commitment to comply.

The revised EMAS allows integration of an existing ISO 14001 certification, to allow a smoother transition and avoid duplication when upgrading from ISO 14001 to EMAS.

Implementing environmental management

Practical requirements

[E16017] Both EMAS and ISO 14001 set a pattern for companies wishing to develop environmental management systems. The key steps in implementing such systems involve:

(a)   conducting an initial review, designed to establish the current situation with respect to legislative requirements, potential environmental impacts and existing environmental management controls;

(b)   developing an environmental policy which will provide the basis of environmental management practices as well as informing day to day operational decisions;

(c)   establishing specific objectives and performance improvement targets;

(d)   developing a programme to implement the objectives and establish operational control over environmental performance;

(e)   ensuring information systems are adequate to provide management with complete, reliable and timely information; and

(f)   auditing the system to compare intended performance with actual performance.

Review

[E16018] In order to be able to actively manage its interactions with the environment, an organisation needs to understand the relationship between its business processes and its environmental performance. An environmental review should therefore be conducted, which clarifies how various business activities could potentially affect, or be affected by, the quality of different components of the environment (air, land, water and the use of natural resources). This will allow an organisation to understand which of its activities have the greatest potential impact on the environment. Usually, but not always, these will relate to procurement and/or manufacturing processes. It is also important to consider which elements of environmental management and performance have the greatest potential impact on business performance, for example, high profile environmental prosecutions can have a significant impact on corporate reputation and brand value.

The initial and subsequent review should also consider the current and likely future legislative requirements that the company must comply with. It should consider the overall organisational strategy, any other relevant corporate policies, customer specifications and community expectations that could influence environmental management practices. The review also offers the opportunity to establish a baseline of actual management organisation, systems and procedures, its compliance record and the range of initiatives already in place to improve performance.

Ideally, an environmental review should involve input from a range of stakeholders, both internal and external to the organisation. This promotes a wider perspective on potential environmental impacts, and ensures that the resulting policies and programmes to be developed by the organisation, reflects a comprehensive range of issues and risks. Consequently, the chances of unidentified and therefore uncontrolled risks emerging will be minimised.

Policy

[E16019] The results of the initial review should inform the development of a written environmental policy. The policy directs and underpins the remainder of the EMS, and represents a statement of intent with regard to environmental performance standards and priorities.

The policy should be endorsed by the highest level of management in the company, and should be communicated to all stakeholders.

Objectives

[E16020] It is important that the policy be supported by objectives that are both measurable and achievable. They should be cascaded throughout an organisation, so that at each level there are defined targets for each function to assist in the achievement of the objectives. A key part of ensuring continuous improvement in environmental performance is to review and update objectives in the light of progress and changing regulations and standards.

Objectives should be developed in conjunction with the groups and individuals who will have responsibility for achieving them. They should also be clearly linked to the overall business strategy and as far as possible with operational objectives. This ensures that environmental management is viewed as relevant and integral to business performance, rather than being seen as an isolated initiative.

The process of objective setting also needs to consider the most appropriate performance measures for tracking progress towards the ultimate objective. For example, if an objective is to reduce waste by 20 per cent over five years, a number of parameters could be used to reflect different aspects of the organisation's waste reduction efforts towards that, including volumes of waste recycled, efficiency with which raw materials are converted to product, and proportion of production staff that have received waste management training.

ISO 14031 provides useful guidance on the principles to be applied in the selection of appropriate environmental performance indicators. Performance in whichever parameters are selected should be measured regularly to enable corrective actions to be taken in a timely manner. As far as possible, performance measures and the achievement of quantified targets should be linked to existing appraisal systems for business units or individuals.

Responsibilities

[E16021] Allocation of responsibilities is vital for successful environmental management. Its implementation will typically involve changes in management systems and operations, training and awareness of personnel at all levels and in marketing and public relations.

A wide range of business functions will therefore need to be involved in developing and implementing environmental management systems. The commitment of senior personnel to introduce sound environmental management throughout the organisation, and to communicate it to all staff is vital.

Companies have adopted a range of organisational approaches as part of their environmental management systems. In some cases there is a single specialist function with responsibility for monitoring and auditing the system. An alternative is to have a central environmental function with only an advisory role, which can also undertake verification of the internal audits carried out by other divisions or departments. The approach needs to be adapted to the culture and structure of the organisation, but whatever system is adopted, there are a number of crucial elements:

(a)   access to expertise in assessing environmental impacts and developing solutions;

(b)   a degree of independence in the auditing function;

(c)   a clear accountability for meeting environmental management objectives;

(d)   adequate information systems to help those responsible for evaluating performance against objectives, to identify problem areas and to ensure that action is taken to solve them.

Training and communications

[E16022] Both ISO 14001 and EMAS include training and communications as a key requirement of the EMS. In particular, they focus on ensuring that employees at all levels, in addition to contractors and other business partners, are aware of company policy and objectives; of how their own work activities impact on the environment and the benefits of improved performance; what they need to do in their jobs to help meet the company's environmental objectives; and the risks to the organisation of failing to carry out standard operating procedures.

This can involve a significant investment for companies, but if integrated with existing training modules and reinforced regularly, many hours of essential training can be achieved. An important benefit of training is that it can be a fertile ground for new ideas to minimise adverse environmental impacts.

Communication with, for example, regulators, investors, public bodies and local communities is an important part of good environment management and requires a preparedness to be open, honest and informative. It also requires clear procedures for liaising with external groups.

Operational controls

[E16023] The operational control elements of an EMS define its scope and essentially set out a basis for effective day-to-day management of environmental performance. Typically, such controls would include:

(a)   a register of relevant legislation and corporate policies that must be complied with;

(b)   a plan of action for ensuring that the policy is met, objectives are achieved and environmental management is continuously improved. This sets out the various initiatives to be proactively implemented and milestones to be achieved, and could be considered a "road map" for guiding environmental performance;

(c)   a compilation of operating procedures which define the limits of acceptable and unacceptable practices within a company and which incorporate consideration of the environmental interactions identified in the review and the objectives and targets that were defined subsequently;

(d)   an emergency response plan to be implemented in the event of a sudden, unexpected and potentially catastrophic event which could potentially influence a company's environmental performance in an adverse manner;

(e)   a programme for monitoring, measuring, recording and reviewing environmental performance. This should incorporate a mechanism for regularly reporting back to senior management, since they are the key enablers of the EMS and because they retain ultimate responsibility of business performance.

These operational controls, and indeed all elements of an EMS, should be documented.

Information management and public reporting

[E16024] The critical factor in an EMS is the quality and timeliness of conveying information to internal and external users. The importance of providing performance information to external users is increasing as expectations of greater transparency and CR continue to grow. Such expectations have been supported in the UK by strong encouragement from Government for companies to voluntarily and publicly report on their environmental performance. In a recent survey, the non-financial report tracking agency CorporateRegister revealed that 82% of the UK's top 100 listed firms publish a stand-alone non-financial report. While nearly all such reports published up to 1997 were based on environment issues, it says that the majority are now broader and are focused on CSR.

Regardless of whether they have adopted EMAS, ISO 14001, or neither, many companies have a statutory duty to report on some aspects of their environmental performance to demonstrate legislative compliance. Some of that information will be publicly available. An increasing number of companies are choosing to provide information on their environmental performance, either in their annual report and accounts, or in a stand-alone document. In most cases, the information provided extends well beyond a demonstration of legislative compliance, and leans towards a general overview of all significant aspects of environmental management and performance. Furthermore, many companies, recognising the importance of sustainability, now measure and report on their performance in terms of social impact. A requirement under the EU Accounts Modernisation Directive and the Companies Act 2006 for directors of quoted and large private companies to produce a Business Review has also raised the level of corporate environment and social disclosure for these companies.

As stakeholders and communication media become more sophisticated, and our understanding of environmental interactions increases, so information requirements become more complex. This means that management information systems must incorporate database management, modelling, measuring, monitoring and flexible reporting. This trend also means that environmental reports that have been produced solely as a means of improving public relations and to defuse external pressures have become less acceptable to many stakeholders. The main reasons for this appear to be that they are seldom generated in response to internal management information needs, which in turn means that they tend to focus on statements of management intent, qualitative claims and descriptive anecdotes rather than actual performance. They are therefore less likely to include detailed and verifiable information.

Some organisations do recognise the value of maintaining a constructive and open dialogue with stakeholders, and external reporting is a major part of this.

This has been encouraged by the setting up of award schemes that give recognition to the best reports. CorporateRegister have since 2007 organised CR Reporting Awards — independent global annual awards for corporate responsibility reporting aimed at providing recognition for the best corporate non-financial reports. The Association of Chartered Certified Accountants (ACCA) has also been very active in this area in the past.

Institutional investors needing reassurance that companies are aware of their environmental risks, are also driving companies into providing more environmental information. For example, Morley Fund Management has stated that unless a FTSE 100 company publishes an annual environmental report, it would vote against the adoption of its annual reports and accounts. The Association of British Insurers (ABI) issued investment guidelines in October 2001 after consultation with a wide range of fund managers, corporate executives and NGOs. These guidelines aimed to improve disclosure by companies of their approach to external social, ethical and environmental risks and single out environmental risks as among the most significant CSR-related risks faced by companies. The guidelines set out the business case for CSR and recommend that CSR reporting should be integrated into annual financial reports to allow for independent verification. By setting out what institutional investors expect to see disclosed in the annual reports of companies in which they hold stakes, the guidelines were intended to increase transparency and allow shareholders to engage with companies where they consider significant risks have not been assessed adequately. The introduction in July 2000 of a requirement for occupational pension funds to state the extent to which they consider environmental, social and ethical factors in investment decisions, has also increased the pressure for environmental reporting amongst companies.

While there is no unified standard for environmental reporting, a number of sources of guidance are available. One of the first in the UK was the United Nations Environment Program (UNEP)/SustainAbility benchmark study into corporate environmental reporting, later known as the Engaging Stakeholders programme, then as the Global Reporters Programme. Although the focus in the early 1990s was on environmental reporting and engagement, the Programme now provides assistance to organisations wishing to improve their understanding of their stakeholders' needs for communication and accountability, and to disseminate best practice for reporting.

The Global Reporting Initiative (GRI) is the leading CSR and Sustainability reporting guidelines used currently. Established in 1997 to develop globally applicable guidelines for reporting on the economic, environmental and social performance of a range of organisations. It receives input from corporations, NGOs, accountancy organisations, business associations and other stakeholders from around the world. The GRI's Sustainability Reporting Guidelines, released in June 2000, provided an expanded model for voluntary non-financial reporting and reflected the move towards broader sustainability reporting. They included a core set of indicators, which, for the most part, were applicable to organisations in all sectors of commerce and included a list of performance indicators aimed at placing a greater emphasis on social and economic issues, rather than just concentrating on environmental issues. The guidelines also insisted that statements on a company's vision and strategy regarding sustainable development, as well as specific company information on governance, should be addressed before a report could be described as in accordance with GRI guidelines.

New guidelines (G3 Guidelines) were released in October 2006. Relevant to organisations of any size, sector, or location, they were made more user-friendly by incorporating a flexible outline of the core content of a report. Unlike previous revision cycles, when the entire set Guidelines were subject to revision, the G3 Guidelines are updated incrementally. This involves targeting specific revisions on an annual basis, where only certain portions of the Guidelines are amended. Stakeholders can make suggestions for specific revisions directly to the GRI. 'Sector-specific supplements' are also available to complement the core information in the general Guidelines. These cover a range of sectors including public authorities, financial services, mining and metals, tour operators, logistics and transportation, NGOs and the automotive and telecommunications industries. Over 1,500 companies have used them to produce voluntary standalone reports, including sector leaders such as Barclays Bank, Rabobank, Royal Dutch Shell, Ford, Vodafone and Anglo American Plc. The improved G3 Guidelines have given rise to continued growth in such reporting.

Updated guidance, G3.1 Sustainability Reporting Guidelines, was launched by the GRI in March 2011, alongside guidance to help companies to determine what to measure and report. The 'Technical Protocol – Applying the Report Content Principles' should work for organisations looking to produce more relevant reports. Both versions of the guidance remain valid until the next generation of GRI Guidelines is in place, although the GRI recommends the use of G3.1. This next generation is due to be published in 2013 and will become the only valid version of the Guidelines in 2015.

The first Guidelines intended to aid companies in the UK to report on their interactions and impacts on the environment, were issued by the Department for Environment, Food and Rural Affairs (DEFRA) in 2001. New voluntary guidelines were published in January 2006, partly in response to the need for preparing a Business Review under the Modernisation Directive and *Companies Act 2006*. They recommend using Key Performance Indicators (KPIs) as a tool for measuring environmental impacts. In theory this should allow costs to be reduced through the use of standard business data that may have been collected for other purposes. The DEFRA guidelines also contain information on how environmental impacts arising from the supply chain and from the use of products can be taken into account. They recommend that quantitative data should be provided where necessary, and provide a standard calculation method to help ensure consistent reporting. Businesses responded favourably to a consultation on the guidelines, commenting that they were relatively simple to follow. The government estimated that 80% of UK businesses would only have five or fewer of the 22 KPIs to report their performance against.

DEFRA, in partnership with the Department for Energy and Climate Change (DECC), now provides guidance for businesses and organisations on how to measure and report their greenhouse gas emissions. This guidance, which is aimed at all sizes of business as well as public and third sector organisations, also sets targets for emissions reduction. It also contains methodology for calculating transport emissions from freight and work-related travel, as well as both direct and indirect emissions from the supply chain. DEFRA says that emissions reporting can benefit Companies in terms of reputation and brand value.

The European Commission adopted a Recommendation in 2001 providing guidelines on the information relating to environmental expenditures, liabilities and risks that companies should publish in their annual accounts and reports. It advocated that where relevant, this should contain quantitative measures in areas such as emissions and consumption of water, energy and materials. Unlike a Directive, however, a Recommendation is not binding. In fact, no EU Member State has fully implemented the Recommendation. Instead the emphasis has been on adhering to the Modernisation Directive on corporate financial accounts, passed in 2003, as a legal framework.

In an attempt to increase the credibility of published reports and identify opportunities for improving management information systems, a number of companies are seeking independent third party assurance on the reliability, completeness and likely accuracy of information contained in their reports. One initiative is the global Assurance Standard (AA1000AS) for corporate public reporting on social, environmental and economic performance, launched by the British organisation AccountAbility (AA) in March 2003. Developed in conjunction with the investment community, NGOs and business, and openly accessible on a non-commercial basis it has been used by a number of leading companies in the UK. The Standard places new demands on external auditors and verifiers of CSR and environmental reports. They are required to demonstrate their independence and impartiality by publicly disclosing commercial relationships with their clients, as well as proving their competency and commenting where a report has omitted information that could be important to stakeholders. Over 200 companies have used it to assure their sustainability and corporate responsibility reports, and it is endorsed by the GRI. All the sustainability reports assured against the AA1000AS are contained in an online directory administered by Corporate Register.

The AA1000AS 2003 was replaced by 3 separate documents in 2008. AA1000AS 2008 requires the assurance provider to evaluate the extent of adherence to a set of principles rather than simply assessing the reliability of the data. In addition, the assurance provider must look at the underlying management approaches, systems and processes and how stakeholders have participated. It encourages integration of sustainability into organisations' routine operations, so that it is aligned with traditional financial accounting, reporting and auditing. The Assurance Standard is accompanied by Account-Ability Principles Standard 2008 (AA1000APS 2008), which provides a framework for an organisation to better identify, understand, prioritise and respond to its sustainability challenges. Guidance for the use of AA1000AS 2008 is also now available. The new standards and guidance documents are suitable for any organisation, from multinational businesses, to SMEs, governments and civil society organisations, although a license fee is now required for each commercial use of AA1000AS.

With recent advances in developing assurance standards for environmental and sustainability reports, external assurance is heading towards becoming the norm. KPMG reported in 2008 that formal third party assurance increased from 30% to 40% of the Global 250 reports in the previous three years. Furthermore, 27% of reports contained other types of third party commentary, such as stakeholder panels or subject matter expert statements.

Auditing and review

[**E16025**] Auditing of environmental management systems, whether conducted by internal or external parties, is a means of identifying potential risk areas and can assist in identifying actions and system improvements required to facilitate ongoing system and performance improvement.

Benefits of environmental management

Effective environmental management

[**E16026**] Effective environmental management will involve changes across all business functions. It requires commitment from senior management and is likely to need additional human and financial resources initially. However, it can also offer significant benefits to businesses. These include:

(a) avoidance of liability and risk. Good environmental management allows businesses to choose when and how to invest in better environmental performance, rather than reacting at the last minute to new legislation or consumer pressures. Unforeseen problems will be minimised, prosecution and litigation avoided;

(b) gaining competitive advantage. A business with sound environmental management is more likely to make a good impact on its customers. The business will be better placed to identify and respond rapidly to opportunities for new products and services, to take advantage of "green" markets and also respond to the increasing demand for information on supplier environmental performance;

(c) achievement of a better profile with investors, employees and the public. Increasingly, investors and their advisers are avoiding companies with a poor environmental record. The environmental performance of businesses and other organisations is an increasing concern for existing staff and potential recruits. Some are finding that a good environmental record helps to boost their public image;

(d) cost savings from better management of resources and reduction of wastes through attention to recovering, reusing and recycling; and reduced bills from more careful use of energy;

(e) an improved basis for corporate decision-making. Effective environmental management can provide valuable information to corporate decision-makers by expanding the basis of such decisions beyond financial considerations. Organisations that understand the interactions between their business activities and environmental performance are in a strong position for integrating environmental management into their business, thereby incorporating key elements of the principles of sustainable development.

Effective environmental management can turn environmental issues from an area of threat and cost to one of profit and opportunity. As standards for environmental management systems are adopted and are applied widely, the question will increasingly become, as with quality management, can a company or large organisation afford not to adopt environmental management? The external pressures to improve environmental performance are unlikely to abate. Environmental management systems can help to respond to the pressures in a timely and cost-effective way.

It has often been suggested that companies that implement an environmental management system should be rewarded with lesser regulatory control. This policy has been implemented in the UK as a result of Government guidance to the Environment Agency that it should take account of robust environmental management systems, in particular EMAS and ISO 14001, in its regulatory approach to companies. The Agency has integrated the possession of a certified management system into its risk-based Operator and Pollution Risk Appraisal (OPRA) regulation scheme. Firms with a recognised EMS are awarded additional points in the OPRA scoring system, allowing the amount of administration required to be reduced, as it is perceived by the Agency as easier to regulate.

The future

[E16027] Environmental management is now well recognised by many as an essential component of effective business management. Environmental management systems have been widely adopted and in most cases these have facilitated demonstrable and ongoing improvements in environmental performance. The involvement of a range of stakeholders in corporate environmental management is no longer the exception; the constructive contribution that they make is actively sought. Improvement in the quality and in the numbers of environmental reports published continues to increase, and is becoming relatively common among leading companies; most include a mechanism for obtaining feedback from external stakeholders, and an increasing number are independently assured in a similar way to annual financial reports and accounts.

Companies' annual reports have grown in length over the recent decades, largely due to the increased amount of narrative reporting, as well as greater legislative demands. This is most likely the result of increased public scrutiny of business activities and the need for clearer explanations of company activity. In addition to stand-alone social and environmental reports, stakeholders now routinely scrutinise company websites for information. Technological developments, globalisation, and the growth of the knowledge economy have radically increased expectations for comprehensive validated data on companies' environmental impacts to be made available. With the proliferation in social networking and blogging sites, and the speed at which information now travels around the world, it is more important than ever that the accuracy and quality of environmental reports is maintained, even in the current economic downturn. Another source of pressure on business to report on environmental issues are initiatives from a number of organisations, including WBCSD, UNEP and the GRI consortium, seeking to implement their guidelines. An effective environment management system will allow business to respond rapidly to these pressures and demands.

Business must also be prepared for new and emerging environmental concerns. Although these may pose a threat to some companies, they will provide opportunities for some sectors, such as environmental technology companies. Water scarcity and other water-related issues have been identified as potential problem areas that could impact on business in the near future. Robust environmental management will help organisations to respond quickly to such emerging threats.

The trend towards sustainability management, whereby companies are attempting to systematically balance economic, environmental and social considerations in business strategies and operations continues to gather pace. The strong inter-relationships and inter-dependence between these three elements of sustainable development make it imperative for companies to move towards a more integrated approach to managing them.

Continued development of this approach will be greatly influenced by shareholders, market analysts and financial institutions recognising the value of non-financial performance information as a basis for evaluating management competence and predicting business performance. Consequently they can increasingly be expected to insist on changes to the information presented to them by companies, thereby generating a radical shift in the traditional business paradigm.

Therefore, while environmental management will remain a critical issue for business to address, it is increasingly becoming integrated with other aspects of business management, reflecting a growing acceptance of the importance of sustainability.

The Equality Act 2010

Alexander M S Green

Introduction

[E16501] Over the last forty years, a significant body of anti discrimination legislation has evolved covering individuals with a variety of 'protected characteristics' (e.g. race, age, disability, sex and sexual orientation etc). On 1 October 2010, The *Equality Act 2010* ('*EA 2010*') came into force thereby codifying the anti-discrimination legislation into one statute. Much of the case law on the earlier legislation and which pre-dates the EA will, however, remains relevant as the *EA 2010* draws heavily on the earlier legislation to which those cases refer.

Under the *EA 2010*, discrimination is unlawful if it is because of one or more of the 'protected characteristics' as defined by the statute. The 'protected characteristics' are:

- age;
- disability;
- gender reassignment;
- marriage and civil partnership;
- race;
- religion or belief;
- sex;
- sexual orientation;
- pregnancy and maternity.

The effect of the *EA 2010* is that employees and other workers are protected against direct discrimination, indirect discrimination, combined discrimination, harassment, victimisation and discrimination arising from a disability and from a failure to make reasonable adjustments for disabled people. Employers and principals are generally liable for the discriminatory acts of their employees and agents.

The purpose of this chapter is to focus on the specific health & safety issues arising from individuals who have certain 'protected characteristics' under the *EA 2010* namely: disability, age, race and gender. Should readers require a general understanding of anti-discrimination law and employment rights they should consult the standard texts on the subject such as *Tolley's Employment Handbook* (20th Ed).

Prohibited Conduct

[E16502] The *EA 2010* prohibits certain types of conduct. The majority of the cases that are heard in the employment tribunal relate to direct discrimination but it is also important to note that prohibited conduct extends to cover: indirect discrimination, victimisation, harassment and, in the case of disability, discrimination arising from a disability.

Direct discrimination

[E16503] *EA 2010, s 13* defines direct discrimination as differential treatment because of a protected characteristic. A twofold test is applied:

*   was a person treated less favourably than an actual or hypothetical comparator was or would have been treated in circumstances that were the same or not materially different?; and
*   if so, was that less favourable treatment because of a protected characteristic?

Indirect discrimination

[E16504] Indirect discrimination arises where everyone is treated in the same way, but the consequences of that treatment are that people holding a protected characteristic are impacted more disparately than those who do not. *EA 2010, s 19* outlaws indirect discrimination. Indirect discrimination applies to all persons holding protected characteristics except pregnancy and maternity.

Combined discrimination

[E16505] *EA 2010, s 14* makes it unlawful to treat a person less favourably because of a combination of the following protected characteristics:

*   age;
*   disability;
*   gender reassignment;
*   race;
*   religion or belief;
*   sex;
*   sex;

An example of such combined discrimination given by the Government Equalities Office is:

> An older woman applies for a job as a driving instructor. She is unsuccessful in her application and when she asks for feedback she is told that she was not appointed to the job because it is not considered a suitable job for an older woman. The driving school advises her that they don't think she would have the strength and agility needed to grab the steering wheel or be able to brake quickly. She is told that she would have been appointed had she been an older man or a younger woman.

In the case above the woman could not succeed on a sex or age discrimination claim alone, as the reason for her treatment was not her sex or age, but the combination of the two (http://www.lge.gov.uk/lge/core/page.do?pageId=1154559#contents-8).

EA 2010, s 14 has not yet come into force. It will come into effect on a date to be appointed.

Discrimination arising from a disability

[E16506] *EA 2010, s 15* provides that disabled people are protected from unfavourable treatment arising in consequence of the disability where the same cannot be shown to be a proportionate means of achieving a legitimate aim. An example of this is given by the Equality and Human Rights Commission:

> A woman is disciplined for losing her temper at work. However, this behaviour was out of character and is a result of severe pain caused by cancer, of which her employer is aware. The disciplinary action is unfavourable treatment. This treatment is because of something which arises in consequence of the worker's disability, namely her loss of temper. There is a connection between the 'something' (that is, the loss of temper) that led to the treatment and her disability. It will be discrimination arising from disability if the employer cannot objectively justify the decision to discipline the worker.

(http://www.equalityhumanrights.com/uploaded_files/EqualityAct/employercode.pdf.)

If a disabled person is treated less favourably than he or she would otherwise be treated by reason of something arising in consequence of the disability, an act of discrimination will have occurred (subject to the defence of justification). *EA 2010, s 15(1)(b)* sets out the test of justification. The treatment must be a proportionate means of achieving a legitimate aim.

An employer will also have a defence against this type of claim in circumstances where he did not know or could not reasonably have been expected to have known that the disabled person had a disability (*EA 2010, s 15(2)*).

Harassment

[E16507] *EA 2010, s 26* outlaws harassment. There are three separate species of harassment:

*   harassment which is related to a protected characteristic;
*   sexual harassment;
*   less favourable treatment arising out of harassment.

Victimisation

[E16508] Victimisation is prohibited by *EA 2010, s 27*. This outlaws retaliation by employers where a person has done a protected act or where it is considered has done or may do a protected act.

Protected acts consist of:

- bringing proceedings under *EA 2010*;
- giving evidence or information in connection with proceedings under *EA 2010*;
- doing any other thing for the purposes of or in connection with proceedings under the EA 2010;
- making an allegation that a person has contravened *EA 2010*.

If the allegation, evidence or information is false and given or made in bad faith, it will not be a protected act (*EA 2010, s 27(3)*).

Health And Safety for Disabled People

Introduction

[E16509] The *Health and Safety at Work etc Act 1974, s 2* (*HSWA 1974*) imposes a general duty on all employers to ensure, so far as is reasonably practicable (for the meaning of this expression, see **E15039** Enforcement), the health, safety and welfare of all their employees. This includes disabled employees. The HSE estimates that 2% of the UK working age population becomes disabled every year. Furthermore, there are approximately 10 million disabled people in Great Britain who are covered by the *EA 2010*. This represents around 18% of the population (http://www.hse.gov.uk/disability/index.htm).

Employers should avoid discriminating against their disabled workers and carry out appropriate risk assessments to take account of an employee's disability which may entail making reasonable adjustments to accommodate the disability.

The meaning of 'disability'

[E16510] The starting point for understanding the health and safety implications of disability is to consider the meaning of 'disability'. Disability is defined by *EA 2010, s 6* as being where 'a person has a physical or mental impairment and that impairment has a substantial and long term adverse effect on the person's ability to carry out normal day to day activities'. The Office for Disability Issues of HM Government has issued guidance on matters to be taken into consideration in determining questions relating to the definition of disability which can be found at:

http://www.equalityhumanrights.com/uploaded_files/EqualityAct/odi_equalit
y_act_guidance_may.pdf.

These guidance notes provide a useful set of examples of the type of factual questions which are likely to be determined by employment tribunals in determining whether a person is or is not disabled. The Employment Appeal Tribunal ('EAT') has also provided guidance to employment tribunals as to the correct approach to establish whether a person is disabled or not. In *Goodwin v The Patent Office* [1999] IRLR 4, the EAT held that there were four questions which an employment tribunal should consider in determining whether a person was disabled or not:

- does the individual have a mental or physical impairment?
- does the impairment have an adverse effect of the ability of the individual to carry out normal day to day activities?
- is the adverse effect substantial?
- if the adverse effect is substantial, does it have a long term effect?

Whilst the concept of impairment is not defined by *EA 2010* the case law on the topic considers it to be a wide ranging term. A person may suffer from a physical impairment even where there is no underlying fault or defect (e.g. where the person suffers from the effects of an illness but it is not possible to identify the root cause of the illness) (*College of Ripon and York St John v Hobbs* [2002] IRLR 185, [2001] All ER (D) 259 (Nov), EAT). There is no longer a requirement for mental impairment to be linked to a clinically well recognised condition. Consequently, a person can be suffering from a mental impairment even though it does not amount to mental illness (*Dunham v Ashford Windows* [2005] ICR 1584).

Whether an impairment has an adverse effect on the ability of an individual to carry out normal day to day activities involves a consideration of what the individual is able to do and the way in which the individual is able to do it as against what would be the case if the individual did not have the impairment. In particular, the enquiry should focus on what the individual cannot do as opposed to what the individual can do (*Leonard v Southern Derbyshire Chamber of Commerce* [2001] IRLR 19). If a person can come to work and is able to do his job this does not mean that he is not necessarily disabled (*Law Hospitals NHS Trust v Rush* [2001] IRLR 611). It may also be necessary to consider what normal day to day activities an individual is able to carry out in the workplace if the working environment exacerbates the individual's impairment (*Cruickshank v VAW Motorcast Limited* [2002] IRLR 24). Any measures taken to treat or to correct an impairment are disregarded (e.g. medication) unless those measures mean that the individual is restored to being able to carry out normal day to day activities.

To be long term, the effects of the impairment must:

- have lasted for at least 12 months; or
- be likely to last for at least 12 months; or
- be likely to last for the rest of the life of the person.

If the impairment is likely to be recurrent it will be treated as though it is continuing.

EA 2010, Sch 1, para 3 provides that certain specified impairments are considered to be disabilities without more. These are:

- severe disfigurement;
- cancer;
- HIV;
- Multiple Sclerosis.

If a person suffers from a progressive condition which is likely to result in that person having impairment this will be deemed to be a disability.

The duty to make reasonable adjustments

[E16511] *EA 2010, ss 20, 21, 22* and *Schedule 8* impose the duty to make reasonable adjustments in respect of disabled people. *Schedule 8* specifically deals with the duty to make reasonable adjustments in the workplace.If an employer fails to comply with his duty to make reasonable adjustments he is guilty of unlawful discrimination.

The duty to make reasonable adjustments consists of three requirements which are set out in *EA 2010, s 20(3), (4) & (5)*:

(3)   The first requirement is a requirement, where a provision, criterion or practice of A's puts a disabled person at a substantial disadvantage in relation to a relevant matter in comparison with persons who are not disabled, to take such steps as it is reasonable to have to take to avoid the disadvantage.

(4)   The second requirement is a requirement, where a physical feature puts a disabled person at a substantial disadvantage in relation to a relevant matter in comparison with persons who are not disabled, to take such steps as it is reasonable to have to take to avoid the disadvantage.

(5)   The third requirement is a requirement, where a disabled person would, but for the provision of an auxiliary aid, be put at a substantial disadvantage in relation to a relevant matter in comparison with persons who are not disabled, to take such steps as it is reasonable to have to take to provide the auxiliary aid.

The first requirement addresses changing the way that things are done (e.g. changing a practice). The second requirement deals with making changes to the built environment (e.g. providing access to a building). The third requirement covers auxiliary and services (e.g. providing special computer software). In relation to the second requirement *EA 2010, s 20(9)* provides that a reference to avoiding a substantial disadvantage includes a reference to:

•   removing the physical feature in question;
•   altering it; or
•   providing a reasonable means of avoiding it.

With regard to the third requirement, *EA 2010, s 20(11)* provides that reference to an auxiliary aid includes a reference to an auxiliary service.

EA 2010, Schedule 8 applies to the duty to make reasonable adjustments in the workplace. In this context, the duty is owed to an 'interested disabled person'. For the purposes of *Schedule 8*, an 'interested disabled person' is defined in relation to employers (referred to as 'employer A') as:

•   an employee of employer A;
•   an applicant for employment with employer A;
•   a person who is or who has notified employer A that the person may be an applicant for employment.

The description of a disabled person is broad in that it also covers:

•   principals in contract work;
•   partnerships;
•   limited liability partnerships;
•   barristers/ advocates and their clerks;
•   those making appointments to offices;

- qualification bodies;
- employment service providers;
- trade organisations;
- local authorities; and
- occupational pensions.

The Equality and Human Rights Commission have produced a Statutory Code of Practice which should be read in conjunction with the *EA 2010* (http://www. equalityhumanrights.com/publications/guidance-and-good-practice-publicat ions/codes-of-practice/ (the 'Code of Practice'). The Code of Practice sets out, in chapter 6, the principles and the application of the duty to make reasonable adjustments for disabled people who are in employment. The duty is referred to as a 'cornerstone' of the Act requiring employers to take positive steps to ens ure that disabled people can access employment and progress in employment. The duty is not simply one of avoiding treating disabled workers, job applicants and potential job applicants unfavourably. The duty is broader and encompass es taking additional steps to which non-disabled workers and applicants are not entitled (*Archibald v Fife Council* [2004] IRLR 651).

The duty to make adjustments is qualified by the word 'reasonable'. The *Disability Discrimination Act 1995, s 18B* provides a useful list of what would be considered to constitute reasonable adjustments. These provisions have not been re-enacted by *EA 2010* but they do, nonetheless, provide a useful checklist for employers and the HSE have replicated these in their own guidance on the subject and have provided some examples of reasonable adjustments which include:

- adjustments to the workplace to improve access or layout;
- giving some of the disabled person's duties to another person, e.g. employing a temp;
- transferring the disabled person to fill a vacancy;
- changing the working hours, e.g. flexi-time, job-share, starting later or finishing earlier;
- time off, e.g. for treatment, assessment, rehabilitation;
- training for disabled workers and their colleagues;
- getting new or adapting existing equipment, e.g. chairs, desks, comput-ers, vehicles;
- modifying instructions or procedures, e.g. by providing written mat-erial in bigger text or in Braille;
- improving communication, e.g. providing a reader or interpreter, having visual as well as audible alarms;
- providing alternative work (this should usually be a last resort).

(http://www.hse.gov.uk/disability/law.htm#adjustments)

The duty to make adjustments is qualified by the word 'reasonable'. Examples of adjustments to working arrangements include:

- allowing a phased return to work;
- changing individual's working hours;
- providing help with transport to and from work;
- arranging home working, providing a safe environment can be main-tained;

- allowing an employee to be absent from work for rehabilitation treatment.

Examples of adjustments to premises include:

- moving tasks to more accessible areas;
- making alterations to premises.

Examples of adjustments to a job include:

- providing new or modifying existing equipment and tools;
- modifying work furniture;
- providing additional training;
- modifying instructions or reference manuals;
- modifying work patterns and management systems;
- arranging telephone conferences to reduce travel;
- providing a buddy or mentor;
- providing supervision;
- reallocating work within the employee's team;
- providing alternative work.

(http://www.hse.gov.uk/sicknessabsence/reasonableadjustments.htm)

Further guidance on what would be constituted as reasonable adjustments can also be found in Chapter 6 of the Code of Practice. An adjustment would be unlikely to be reasonable to an employer if would involve little benefit to the disabled person. The test of reasonableness is essentially an objective one. It may not be necessarily met by an employer who shows that he personally believed that the making of the adjustment would be too disruptive or costly (*Smith v Churchills Stairlifts plc* [2006] IRLR 41).

The employer's knowledge of a person's disability is relevant to the question of the duty to make reasonable adjustments. *EA 2010, Schedule 8, Part 3, para 20* states:

A is not subject to a duty to make reasonable adjustments if A does not know, and could not reasonably be expected to know—(a) in the case of an applicant or potential applicant, that an interested disabled person is or may be an applicant for the work in question; (b) in any other case referred to in *Part 2* of this *Schedule*, that an interested disabled person has a disability and is likely to be placed at a substantial disadvantage referred to in the first, second or third requirement.

In essence this means:

- did the employer know both that the employee was disabled and that his disability was liable to affect him in the manner set out in the definition of disability? If the answer to that question is 'no', then there is a second question, namely;
- ought the employer to have known both that the employee was disabled and this disability was liable to affect him in the manner set out in the definition of disability?

If the answer to the second question is 'no' then there is no duty to make reasonable adjustments.

Risk assessment and disability

[E16512] Risk assessment lies at the heart of health and safety management and consists of a careful examination of what could harm people and how likely this is to happen, so that employers can weigh up whether or not the steps they have taken are sufficient to comply with health and safety law. Clearly in relation to disabled people, the duty to make reasonable adjustments under the *EA 2010* interacts with an employer's duty to conduct risk assessments under general health and safety law and this fact should always be taken into consideration.

The quality of the risk assessment in relation to disabled people is critical. This point is emphasised by the Equality and Human Rights Commission who state that:

> Your employer may need to use specialist staff and the person doing the assessment must:
>
> - focus on you as an individual, not people with your condition in general;
> - consider the facts;
> - not make assumptions;
> - get individual specific medical advice; and
> - talk to you about how reasonable adjustments can be made.

The risk assessment should also consider the essential elements of the job; the length of time and frequency of any hazardous situations; and any reasonable adjustments that can be made to reduce the risk.

If there is still an unacceptable risk, even with adjustments, then the employer could lawfully dismiss or not employ the employee. The question is what is 'unacceptable': this can only be tested in the courts. Increasing case law precedent is being set, which gives further guidance to employers.

The employer is the person who must take the decision about whether to employ or retain the employee in a job. If they seek expert advice from medical services or health and safety specialists, these are agents of the employer, and the employer has a duty to ensure that the specialists have considered all the facts.

(http://www.equalityhumanrights.com/advice-and-guidance/your-rights/disability/disability-in-employment/health-and-safety-in-the-workplace)

Risk assessments can be carried out in five steps:

- identify the hazards (what, in the work, could cause harm to people);
- decide who might be harmed and how;
- evaluate the risks and decide on precautions;
- record the findings and act on them;
- review the assessment and update if necessary.

(http://www.hse.gov.uk/disability/started.htm)

The HSE identify Step 2 as being about 'who might be harmed?'. To answer this question, employers must think about their whole workforce including disabled employees. There may be specific risks that arise in relation to a disabled employee, and in addressing these the employer should consult and involve the employee in the risk assessment. This helps avoid the following:

- people making assumptions about disabled people which can lead to poor practice or discrimination;
- people hiding an impairment that might have health and safety implications for fear they won't get or keep a job.

Examples of assumptions include:

- thinking a driver who loses an arm can't drive anymore. Steering wheels can be modified or replaced, e.g. with a joystick;
- believing a deaf person can't be warned of fire. Flashing lights can supplement sirens and bells.

Employers should:

- make sure they manage work risks for everyone;
- take account of disability, avoiding assumptions;
- involve disabled workers in doing risk assessments and making 'reasonable adjustments';
- consult others with appropriate expertise where necessary;
- review the situation if necessary, working with the disabled person and/or their representative.

Specific risks to disabled people arise in relation to fires in the workplace. The Equality and Human Rights Commission's guidance on the topic is as follows:

One thing that employers often worry about when thinking about employing a disabled person is what would happen in the event of a fire. This is often based on lack of information, with employers wrongly believing that wheelchair users should not be employed because they would not be able to escape from a building on fire where lifts were out of use. Deaf, hearing impaired blind people may also be discriminated against because people wrongly believe they won't know there is a fire alarm or may 'get in the way' of colleagues trying to escape.

If you have a disability that may present a difficulty during a fire at your workplace, you need to have a discussion and draw up an agreed plan with your line manager and/or premises manager about this. There may be very simple changes that can be made to stop this from ever being a problem:

- providing flashing lights as alarms, as well as things that make a noise;
- making sure that colleagues working with deaf people have a basic awareness of sign language and deaf issues;
- the establishment of a 'buddy' system to ensure wheelchair users are helped in an emergency;
- providing a visually impaired person with named guides in case of fire;
- providing temporary places of refuge for wheelchair users protected by fire resistant doors and from which there is a safe route to a final exit (the refuge may also have a means of communication to a central control point); or
- training for staff in evacuation plans.

These plans should be made known to all staff concerned and tested at regular intervals. A well thought through evacuation plan should take into account the needs of every individual in the building, including, for example, women in the later stages of pregnancy. Very often, this can help to improve procedures and safety overall. You may wish to discuss your needs in confidence, in which case only certain key individuals need know about your unique plan.

(http://www.equalityhumanrights.com/advice-and-guidance/your-rights/disa
bility/disability-in-employment/health-and-safety-in-the-workplace/)

Health and Safety in Relation to Age

Introduction

[E16513] Age is a protected characteristic (*EA 2010, s 5*). The concept of age includes a reference to a 'person of a particular age'. The protected characteristic encompasses the concept of an age group which is referred to in *EA 2010, s 5(2)* as a group of persons defined by reference to age whether by reference to a particular age or to a range of ages.

It is unlawful to discriminate against young workers as well as older workers. All workers and people who apply for work and vocational training are protected from: direct discrimination, indirect discrimination, harassment and victimisation. Discrimination because of age can be justified as a proportionate means of achieving a legitimate aim (*EA 2010, s 13(2)*).

Employers should not use 'health and safety' as an excuse for not continuing to employ older workers or for not receiving training.

Ageing workers

[E16514] It has been estimated that by 2020, approximately 1/3 of the UK's workforce will be over the age of 50. Some industries such as the UK offshore oil and gas industry have an ageing workforce. In 2008, the average age of people working offshore was 45.4 years (source: UK Oil and Gas http://www.oilandgasuk.co.uk/cmsfiles/modules/publications/pdfs/EM007.pdf)

In other industries such as manufacturing and construction older workers tend to be leaving at faster rates. Older workers are also likely to be self-employed. The HSE maintain that age is not an equivalent to personal capacity to work although it is accepted that certain cognitive functions such as memory abilities deteriorate with age.

Certain industries such as the agriculture and construction sectors have demonstrated persistent problems with workers over 65. In those sectors the fatality statistics show that the rate of fatal injuries is higher in workers aged 45 plus compared with younger workers (http://www.hse.gov.uk/diversity/age.htm).

With the abolition of the UK's statutory default retirement age of 65 in April 2011 combined with the general economic downturn and the 'pensions' crisis, it is inevitable that the proportion of older people continuing to work beyond the age of 65 will increase. Furthermore, people are living longer and if they continue to work into their old age their employers will need to assess the health and safety risks which are associated with an ageing workforce.

Risk assessment

[E16515] As people grow older their ability to do their job may change as well as their perception of how they do their job. As part of their general duty to conduct risk assessments, employers should take account of their ageing employees to ensure that they can continue to perform their work safely. There are specific risks which need to be assessed in relation to older workers although each case should always to be assessed on its own particular circumstances. In its guidance on the matter the HSE states that employers should:

- carry out risk assessments routinely and not simply when an employee reaches a certain age;
- assess the activities involved in jobs and modify workplace design if necessary;
- make adjustments on the basis of individual and business needs, not age;
- consider modifying tasks to help people stay in work longer – with appropriate training;
- allow staff to change work hours and job content;
- not assume that certain jobs are too demanding for older workers. Decisions should be based on capability and objective risk and not age;
- encourage or provide regular health checks for staff, regardless of age;
- persuade staff to take an interest in their health and fitness;
- consider legislative duties such as the *EA 2010* or flexible working legislation. These could require businesses to make adjustments to help an employee with a health issue or consider a request to work flexibly.

Young workers

[E16516] Young workers (i.e. above school leaving age and under 18) are protected to at least the same level as adult workers. Furthermore, young people, especially those new to the workplace, will encounter unfamiliar risks from the jobs they will be doing and from the working environment. The Key risks for young people when starting work may arise because of their lack of experience or maturity and not having the confidence to ask for or knowing where they can get help. Employers should take account of this in their risk assessments. (For a more detailed discussion see CHILDREN AND YOUNG PERSONS V12001–V120026)

The HSE provides guidance on the key risks which need to be assessed in relation to young workers (http://www.hse.gov.uk/youngpeople/risks/index.ht m#ypkey). In summary these are as follows:

Prior to employing a young person, the health and safety risk assessment must take the following into account:

- the fitting-out and layout of the workplace and the particular site where they will work;
- the nature of any physical, biological and chemical agents they will be exposed to, for how long and to what extent;
- what types of work equipment will be used and how this will be handled;

- how the work and processes involved are organised;
- the need to assess and provide health and safety training; and
- risks from the particular agents, processes and work.

The Management of Health and Safety at Work Regulations require that young people are protected at work from risks to their health and safety which are a consequence of the following factors:

- physical or psychological capacity;
- pace of work;
- temperature extremes, noise or vibration;
- radiation;
- compressed air and diving;
- hazardous substances;
- lack of training and experience.

There are also risks to young people associated with specific industries or processes:

- agriculture;
- carriage of dangerous explosives and goods;
- shipbuilding and Ship-repairing Regulations;
- provision and use of work equipment;
- power presses;
- woodworking machines;
- mechanical lifting operations (including lift trucks).

The HSE state that there is no need for employers to carry out a new risk assessment each time they employ a young person, as long as their current risk assessment takes account of the characteristics of young people and activities which present significant risks to their health and safety.

The HSE recommend that employers may wish to consider developing generic risk assessments for young people as these could be useful when they are likely to be doing temporary or transient work, and when the risk assessments could be modified to deal with particular work situations and any unacceptable risks.

In all cases, employers will need to review the risk assessment if the nature of the work changes or when they have reason to believe that it is no longer valid.

In carrying out the risk assessment employers should identify the measures they need to take to control or eliminate health and safety risks.

Except in special circumstances, employers should not employ young people to do work which:

- is beyond their physical or psychological capacity;
- exposes them to substances chronically harmful to human health, e.g. toxic or carcinogenic substances, or effects likely to be passed on genetically or likely to harm the unborn child;
- exposes them to radiation;
- involves a risk of accidents which they are unlikely to recognise because of e.g. their lack of experience, training or attention to safety;
- involves a risk to their health from extreme heat, noise or vibration.

These restrictions will not apply in 'special circumstances' where young people over the minimum school leaving age are doing work necessary for their training, under proper supervision by a competent person, and providing the risks are reduced to the lowest level, so far as is reasonably practicable. Under no circumstances can children of compulsory school age do work involving these risks, whether they are employed or under training such as work experience.

Training includes Government-funded training schemes for school leavers, modern apprenticeships, in-house training arrangements and work qualifying for assessment for National/Scottish Vocational Qualifications, e.g. craft skills.

(http://www.hse.gov.uk/youngpeople/risks/index.htm#ypkey)

Health and Safety in Relation to Race

Introduction

[E16517] Race is a protected characteristic. It is defined by *EA 2010, s 9(1)* to include:

- colour;
- nationality;
- ethnic or national origins.

EA 2010, s 9(2)(b) makes reference to persons who share a colour, nationality and/or ethnic origins as being persons of the same racial group. A racial group is a group of persons defined by reference to race (*EA 2010, s 9(3)*).

Examples of the concepts of colour, nationality, ethnic or national origins and a racial group are:

- colour includes being black or white;
- nationality includes being a British, Australian or Swiss;
- ethnic or national origin includes being from a Roma background or of Chinese heritage;
- a racial group could be 'Black Britons' which includes those people who are black and who are British citizens.

Jews and Sikhs have been held be members of a racial group. English and Scottish can be national origins even where the individual can describe his nationality as British.

Employers should not use 'health and safety' as an excuse not to employ people from different races or to continue to employ or to provide them with appropriate training.

Migrant workers

[E16518] The HSE have indentified that race can be a relevant factor in protecting the health and safety of workers. Race is a relevant factor to consider because of:

- differences in vulnerability;
- networks and communication channels for promoting health and safety; and
- the challenges of working with a different language.

These points are starkly illustrated in relation to the employment of migrant workers who are commonly from different racial backgrounds (e.g. Chinese).

In 2006, the HSE published its report 'Migrant Workers in England & Wales' (http://www.hse.gov.uk/research/rrpdf/rr502.pdf).

The Report was based on interviews with 200 migrant workers based in five regions of England & Wales and highlights the risks that migrant workers face and makes recommendations on how the health and safety of this group of workers can be protected. The Report found that in general, migrant workers are often over-qualified for the work that they do and they work long and anti social hours. The Report states that without migrant workers, in some areas of the country, hospitals and care homes would be without staff, construction would not boom, hotels and restaurants would be unable to service customers and farmers and food processing firms would not be able to distribute their produce to the shops.

The Report revealed that the vast majority of migrant workers were working with other migrant workers. In some cases a particular nationality might be dominant but in others the workforce could consist of workers from many different countries speaking different languages and with different skills and experience and knowledge of health and safety systems.

Few checks were made on migrant workers' skills and qualifications for undertaking the work that they were doing. There were cases where migrant workers were performing skilled and potentially dangerous work such as scaffolding who had no previous experience in the task. In food production and in catering most workers were not tested for their knowledge of food hygiene and only a minority were offered training in food hygiene.

Whilst many migrant workers were recruited directly by the employer, a significant number were supplied through recruitment agencies. Many of the migrant workers had bad experiences of dealing with recruitment agencies which included:

- being paid less;
- suffering unexplained deductions from wages;
- irregular work patterns;
- lack of clarity where health and safety responsibilities lay.

One third of the migrant workers interviewed had received no training in health and safety. The remaining two thirds had received cursory training, normally limited to a short session at induction. The problem was exacerbated where there was no common language amongst the work force which complicated effective communication of health and safety training.

Whilst personal protective equipment ('PPE') was generally provided, many migrant workers did not know how to use it properly.

25% of the migrant workers interviewed had either experienced an accident at work or had witnessed one involving a migrant co-worker. This suggested a higher level of accidents than would be expected by UK workers. Many of the accidents were associated with fatigue brought about by working excessively long or anti-social hours. Many migrant workers did not report accidents as they feared dismissal and deportation. Some of the migrant workers under estimated the risks that they faced in doing their job. The HSE attributed this to circumstances where the worker was undertaking work which they perceived to be below their qualifications and skills and they consequently tended to be less conscious of the risks associated with the jobs that they were doing. They took fewer measures to reduce health and safety risks. Risk assessment strategies should, therefore, take account of whether the migrant worker is engaged in work in which he or she has no previous experience.

Many of the workers suffered from discrimination at work because of their race or nationality. Many of the workers interviewed believed that they were allocated to the worst shifts and had less favourable terms and conditions. They frequently referred to name calling and harassment by supervisors and co-workers. The HSE hypothesised that that such discrimination might have an impact on worker health and safety in circumstances where discriminatory treatment combined not simply to contribute to stress at work but an inability to raise concerns about health and safety at work. Discrimination impacts on a worker's ability to challenge unfair and unsafe practices.

Lack of English was also a problem. For example, some of the migrant workers interviewed admitted that they were unable to follow the health and safety training that they were offered. Lack of English also impacted on issues such as supervision of work.

The Migrant workers also expressed a low level of knowledge of their health and safety rights and of how to enforce them. There was a generally held view that responsibility for health and safety lay with the individual worker and that accidents and incidents at work were the fault of individual workers. Consequently, they did not assume that their employers (or recruitment agencies or labour providers) had responsibility for their health and safety.

Risk assessment

[E16519] The HSE's Report on migrant workers highlights specific issues which should be risk assessed. The HSE have made the following recommendations:

- whilst there are no requirements in health and safety legislation for employers to ensure their staff are fluent in English the HSE recommends steps should be taken to ensure understanding of health and safety issues;
- the law requires that employers provide workers with comprehensible and relevant information about risks and about the procedures they need to follow to ensure they can work safely and without risk to health. This does not have to be in English;
- the employer may make special arrangements, which could include translation, using interpreters or replacing written notices with clearly understood symbols or diagrams;

- any health and safety training provided must take into account the worker's capabilities, including language skills;
- workers who do not speak English may need to understand key words and commands relating to danger, e.g. 'Fire' and 'Stop'. Employers will need to ensure that this is communicated clearly and simply, and check understanding afterwards;
- employers may wish to consider suitably tailored ESOL (English for Speakers of Other Languages) provision for longer-term workers.
- there will be few situations where health and safety considerations alone justify not employing workers with poor or no English.

(http://www.hse.gov.uk/diversity/race.htm).

Health and Safety in Relation to Gender

Introduction

[E16520] Gender or sex is a protected characteristic. *EA 2010* defines references to sex as references to a man or to a woman. *EA 2010, s 13(6)* includes within the definition of sex (or more specifically less favourable treatment) the fact that a woman is breast feeding but excludes from the definition of sex the special treatment of women connected with pregnancy or childbirth.

Risk Assessment

[E16521] The HSE have estimated that women make up 42% of the employed population in the EU. Men and women are not the same and the jobs they do, their working conditions and how they are treated by society are not the same. These factors can affect the hazards they face at work and the approach that needs to be taken to assess and control them. The HSE states that factors to take into account include:

- women and men are concentrated in certain jobs, and therefore face hazards particular to those jobs;
- women and men face different risk to their reproductive health.

The impact of gender on both men's and women's occupational health and safety is generally under-researched and poorly understood. However, discrimination against new and expectant mothers is well known and the HSE has been working closely with other government departments to tackle this. There are specific provisions which address risk assessment in relation to pregnant workers which are dealt with elsewhere in this work [E14025].

The European Agency for Safety and Health at Work ('EASHW') have produced research into gender issues in relation to risk assessment which is summarised in its Fact Sheet 43 'Including gender in risk assessments'

(http://osha.europa.eu/en/publications/factsheets/43/view)

It is argued that taking a gender neutral approach to risk assessment and prevention can result in risks to female workers being underestimated or even being ignored altogether. It is further suggested that when we consider hazards at work there is a tendency to think about men working in high accident risk areas such as building site or a fishing vessel rather than women working in health and social care or call centres. It is argued that a careful evaluation of real work circumstances shows that men and women can face significant risks at work. Gender risks should be built into workplace risk assessments. The table below shows examples of hazards and risks which are found in female dominated work areas.

Work Area	Risk Factors and health problems include:			
Healthcare	Infectious diseases, e.g. blood borne, respiratory, etc.	Manual handling and strenuous postures; ionising radiation.	Manual handling and strenuous postures; ionising radiation.	'Emotionally demanding work'; shift and night work; violence from clients and the public.
Nursery workers	Infectious diseases, e.g. particularly respiratory.	Manual handling, strenuous postures.		'Emotional work'
Cleaning	Infectious diseases; dermatitis.	Manual handling, strenuous postures; slips and falls; wet hands.	Cleaning agents.	Unsocial hours; violence, e.g. if working in isolation or late.
Food production	Infectious diseases, e.g. animal borne and from mould, spores, organic dusts.	Repetitive movements, e.g. in packing jobs or slaughterhouses; knife wounds; cold temperatures; noise.	Pesticide residues; sterilising agents; sensitising spices and additives.	Stress associated with repetitive assembly line work.

Catering and restaurant work	Dermatitis.	Manual handling; repetitive chopping; cuts from knives and burns; slips and falls; heat; cleaning agents.	Passive smoking; cleaning agents.	Stress from hectic work, dealing with the public, violence and harassment.
Textiles and clothing	Organic dusts.	Noise; repetitive movements and awkward postures; needle injuries.	Dyes and other chemicals, including formaldehyde in permanent presses and stain removal solvents; dust.	Stress associated with repetitive assembly line work.
Laundries	Infected linen, e.g. in hospitals.	Manual handling and strenuous postures; heat.	Dry cleaning solvents.	Stress associated with repetitive and fast pace work.
Ceramics sector		Repetitive movements; manual handling.	Glazes, lead, silica dust.	Stress associated with repetitive assembly line work.
'Light' manufacturing		Repetitive movements, e.g. in assembly work; awkward postures; manual handling.	Chemicals in microelectronics.	Stress associated with repetitive assembly line work.
Call centres		Voice problems associated with talking; awkward postures; excessive sitting.	Poor indoor air quality.	Stress associated with dealing with clients, pace of work and repetitive work.

Education	Infectious diseases, e.g. respiratory, measles	Prolonged standing; voice problems.	Poor indoor air quality.	'Emotionally demanding work', violence.
Hairdressing		Strenuous postures, repetitive movements, prolonged standing; wet hands; cuts.	Chemical sprays, dyes, etc.	Stress associated with dealing with clients; fast paced work.
Clerical work		Repetitive movements, awkward postures, backpain from sitting.	Poor indoor air quality; photocopier fumes.	Stress, e.g. associated with lack of control over work, frequent interruptions, monotonous work.
Agriculture	Infectious diseases, e.g. animal borne and from mould, spores, organic dusts.	Manual handling, strenuous postures; unsuitable work equipment and protective clothing; hot, cold, wet conditions.	Pesticides.	

The EASHW recommend a model for making risk assessment more gender sensitive to take account of gender issues, differences and inequalities. They recommend a holistic approach to risk assessment to take account of the broad context of gender issues such as sexual harassment, discrimination, involvement in decision making in the workplace and conflicts between work and home. A further aim of a gender orientated risk assessment is to identify less obvious hazards and health problems that are more common with female workers.

A gender sensitive risk assessment should:

*   have a positive commitment to gender issues;
*   look at the real working situation;
*   involve all workers, women and men, at all stages;
*   avoid making assumptions about the hazards are and who is at risk.

It is suggested that hazard identification could include gender by:

*   examining hazards prevalent in both male and female dominated jobs;
*   looking for health hazards in addition to safety hazards;

- asking both female and male workers what problems they have in their work, in a structured way;
- avoiding making initial assumptions about what may be 'trivial';
- considering the entire workforce, e.g. cleaners, receptionists;
- not forgetting part-time, temporary or agency workers, and those on sick leave at the time of the assessment;
- encouraging women to report issues that they think may affect their safety and health at work, as well as health problems that may be related to work;
- looking at and asking about wider work and health issues.

Risk assessment could include gender:

- looking at the real jobs being done and the real work context;
- not making assumptions about exposure based purely on job description or title;
- being careful about gender bias in prioritising risks according to high, medium and low;
- involving female workers in risk assessment. Consider using health circles and risk mapping methods. Participative ergonomics and stress interventions can offer some methods;
- making sure those doing the assessments have sufficient information and training about gender issues in occupational safety and health (OSH);
- making sure instruments and tools used for assessment include issues relevant to both male and female workers. If they do not, adapt them;
- informing any external assessors that they should take a gender-sensitive approach, and checking that they are able to do this;
- paying attention to gender issues when the OSH implications of any changes planned in the workplace are looked at.

Once the risk assessment has been completed, it should be implemented to include gender by:

- aiming to eliminate risks at source, to provide a safe and healthy workplace for all workers. This includes risks to reproductive health;
- paying attention to diverse populations and adapting work and preventive measures to workers. For example, selection of protective equipment according to individual needs, suitable for women and 'non-average' men;
- involving female workers in the decision-making and implementation of solutions;
- making sure female workers as well as men are provided with OSH information and training relevant to the jobs they do and their working conditions and health effects. Ensure part-time, temporary and agency workers are included.

The risk assessment should be monitored and reviewed to include gender by involving female workers. Employers should be aware of new information about gender related occupational health issues.

Remedies and Enforcement in Relation to Unlawful Discrimination

[E16522] The employment tribunals have exclusive jurisdiction to hear claims of unlawful discrimination. Strict time limits apply for presentation of claims. *EA 2010, s 120* provides that for claims relating to discrimination in employment the application to the employment tribunal must be made:

- 3 months from the date when the discriminatory act was done (6 months for complaints within the armed forces);
- If a discriminatory act extends over a period as opposed to being a one off act, it is treated as done at the end of the period.

There is no qualifying period of service required for a claimant to make a complaint.

If an employment tribunal finds that a complaint of unlawful discrimination is well founded it must, if it considers it to be just and equitable, make one of the following orders:

- a declaration;
- an award of compensation; and/or
- a recommendation.

Compensation is assessed in the same manner as any other claim in tort or delict (Scotland) for breach of statutory duty. The award may include compensation for injury to feelings and, if appropriate injury to health and patrimonial loss flowing from the breach.

The award for injury to feelings ranges from a minimum band of between £750 and £7,000, to a middle band ranging from £7,000 to £18,000 to the maximum band of £18,000 to £30,000. The scale of compensation for injury to feelings depends on the gravity of the employer's unlawful act and the employee's reaction to it. These bands will be uprated to allow for inflation. If the discriminator acted in a high handed manner, malicious, insulting or oppressive manner the employment tribunal can award aggravated damages. The award must be compensatory and not punitive. Occasionally the employment tribunal will award 'exemplary damages' in circumstances where there has been oppressive, arbitrary or unconstitutional action by the state (or state employers). The employment tribunal may increase or decrease the award by up to 25% if it considers this just and equitable to do so if the claimant or the respondent have failed to comply with the ACAS Code of Practice.

In indirect discrimination cases, the employment tribunal may award compensation but only if the employment tribunal is satisfied that the 'provision, criterion or practice' was not applied with the intention of discriminating against the claimant.

Compensation may also be awarded in respect of financial loss flowing from the discriminatory act. Loss covers pecuniary loss, loss of benefits and expenses and are normally quantified according the same principles as in an unfair dismissal case.

The employment tribunal may also make a recommendation that the respondent takes action within a specified time for the purpose of obviating or reducing the adverse effect of any matter to which the proceedings relate both in relation to the claimant and to any other person. Failure to comply with a recommendation may lead to an increase in the award of compensation.

Unlike the position with unfair dismissal, there is no statutory limit to the amount of compensation which the employment tribunal may award.

Ergonomics

Margaret Hanson, Jill Cleaver and Andrea Oates

Introduction

[E17001] Ergonomics is concerned with the fit between people and the things they use. People vary considerably in a number of attributes and abilities, height, strength, visual ability, capacity to handle information and so on. Ergonomics applies scientific information about human abilities, attributes and limitations to ensure that the tools or equipment used, tasks undertaken, workstation, environment and work organisation are designed to suit those using them. Ergonomics adopts a people-centred approach, aiming to fit the work to the person rather than forcing the person to adapt to poorly designed equipment, furniture, environments or work systems. Correct application of ergonomics will produce a work system that optimises human performance and minimises the risk to workers' health and safety. Ergonomics can be applied to any environment or system (work, leisure, travel, home, etc.) with which people interact.

The word 'ergonomics' comes from the Greek words 'ergos', meaning 'work,' and 'nomos', meaning 'natural law'. Historically the term 'ergonomics' has been used in the UK and Europe, while 'human factors' has been used in North America, although this distinction is now blurring. Essentially, the terms 'ergonomics' and 'human factors' are synonymous, although in some quarters a distinction has been drawn between physical workplace design (which is referred to as 'ergonomics') and system design and human behaviour (which is referred to as 'human factors'). This chapter does not make that distinction and instead the term 'ergonomics' is used to cover all aspects of the design of equipment, environment or system in relation to people.

The benefits of applying ergonomics

[E17002] The benefits of ergonomically designed equipment, furniture, workplaces and tasks are as follows:

- *Improved safety*, as people are less likely to make a mistake if equipment is designed to take account of abilities (e.g. if characters on displays are of a suitable size and in a colour that can be read accurately).
- *Reduced ill health and sickness absence*, as equipment and workstations are designed to fit people (e.g. seats are comfortable) and tasks are designed to take account of abilities (e.g. work rates and weights handled are within the person's capabilities).
- *Reduced fatigue and stress*, as tools, equipment and systems are easier to use.

- *Increased efficiency, performance, productivity and quality of product* (e.g. if tools are designed to fit the hands of the users, they will be easier to hold, users will be more comfortable and may work more efficiently and make fewer mistakes).
- *Increased job satisfaction*, as tasks are easier to perform.

Although people are highly adaptable and resourceful and are often able to use poorly designed equipment, systems etc., their use can lead to stress, errors, fatigue and injury. The consequences of not applying ergonomics can be enormous, potentially leading to human suffering through physical discomfort and disability, stress and accidents. It may also lead to reduced efficiency and productivity.

Ill health resulting from poorly designed furniture, tasks, lifting and handling, etc. is also extremely costly.

Work-related musculoskeletal disorders (MSDs), are the most common type of occupational ill health in the UK. According to Labour Force Survey (LFS) statistics, in 2007/8 around 539,000 people in Great Britain, who had worked in the last year, believed they were suffering from a musculoskeletal disorder (MSD) that was caused or made worse by their current or past work. The HSE estimates that 11.6 million working days a year are lost to work-related MSDs.

Manual handling injuries account for a significant proportion of these. More than a third of all reported injuries which result in someone being off work for more than three days are caused by manual handling.

Not applying ergonomics can also be one of the factors that contribute to stress. Stress arises as a result of a mismatch between the demands of the job or situation and the abilities of the individual and results in a significant amount of ill health. In 2007/08 an estimated 442,000 people in Britain who worked in the last year believed that they were experiencing work-related stress at a level that was making them ill, according to the LFS. In the same year, it is estimated that work-related stress, depression or anxiety accounted for around 13.5 million lost working days in Britain.

Health effects of stress include headaches, indigestion, disturbed sleep and fatigue, changes in appetite, increased alcohol consumption, smoking or drug taking, loss of concentration, irritability and loss of self esteem. Work-related stress may be caused by poor communication, lack of appropriate training, high workload, lack of control over work, inadequate feedback about work, or repetitive or boring work – namely if tasks, jobs and systems are not designed with full consideration for the users. Individuals vary in their response to these factors but the employer has a responsibility to reduce the risks by designing work appropriately.

As well as increasing the potential for musculoskeletal injuries and increased stress, poor design that results in increased fatigue and error can contribute to accidents and system failures. Inadequate attention to human factor issues has been cited as a contributing element in many catastrophic failures (e.g. Ladbrook Grove rail disaster, Three Mile Island disaster, Kegworth air disaster – see DISASTER AND EMERGENCY MANAGEMENT SYSTEMS (DEMS)). These usually arise as a consequence of a series of errors resulting from inappropriate design and management. The application of ergonomics can help to reduce these problems. More information on human error is contained in the Health and Safety Executive (HSE) publication *Reducing error and influencing behaviour* (HS(G)48) (see **E17069** below).

Core disciplines

[E17003] Ergonomics adopts a multi-disciplinary approach to an issue and thus draws on a number of other key disciplines, using this integrated knowledge to obtain a holistic view of the work system. The core disciplines include anatomy (the structure of the body), physiology (the function and capabilities of the body), biomechanics (the effect of movement and forces on the body), psychology (the performance of the mind and mental capability, including perception, memory, reasoning, concentration, etc.) and anthropometry (the physical dimensions of the body).

The following text outlines:

• an overview of the ergonomic approach (E17004);
• design guidelines and principles relating to different aspects of the system (E17005–E17043);
• musculoskeletal disorders (E17044–E17050);
• tools that can be used to assess work and assist with design (E17051–E17059);
• outline of legislation that promotes ergonomic design (E17060–E17069).

The ergonomic approach

[E17004] The ergonomic approach considers all aspects of a person's inter-action with his or her task, workstation, environment and work system, so that these can be designed to fit the abilities of the users and maximise ease of use. The model opposite can be used to illustrate the framework of this approach.

Figure 1: Ergonomic Approach

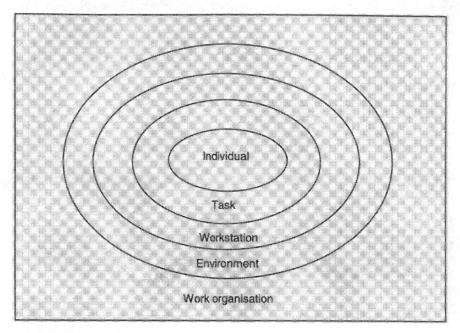

Ergonomics takes account of the physical and mental capabilities and needs of the potential users/workers/population (e.g. the size, strength and ability to handle information), before considering the task and the context in which it will be undertaken. By establishing the physical and mental capabilities of the user group, certain criteria or constraints can be designed for the task. For example, most people can accurately remember up to seven numbers in short-term memory. A picking task requiring order numbers to be remembered should therefore be designed so that the order number is seven digits or less.

Likewise, the workstation should be designed in relation to the physical requirements of the person and the operational requirements of the task. For example, the users' height and reach will be relevant to specifying the dimensions of the workstation; the sequence of use of tools will be relevant to their arrangement on the workstation.

The work environment needs to be designed in relation to the person, task and workstation requirements. For example, the lighting and noise should be within acceptable limits for human work but also appropriate to the requirements of the task (see also LIGHTING and NOISE AND VIBRATION).

The work system is the organisation within which these elements sit and which ultimately affect the whole work. This includes the way in which the person, task and workstation are organised, as well as the requirements and constraints that affect the way work is conducted. It should complement and not constrain the effectiveness of the person, task, workstation and environment interactions.

Ergonomic criteria for these elements are discussed in more detail in paragraphs E17005–E17043 below.

Designing for people

[E17005] When adopting an ergonomic approach, the person is the central focus of the design of any system. Work should be designed to optimise the person's ability to complete the work, so that it is within their mental and physical capabilities and so that they can work in a comfortable and effective manner – the equipment, furniture, environment and task activities should all support this. It is therefore important to understand the abilities and characteristics of the potential users.

It is clear that there are wide differences between people in their physical and psychological capabilities. Understanding the normal range of human abilities and potential variations within this is important in being able to design appropriately. In most work, transport and domestic situations it will not be possible or appropriate to select who uses equipment or systems, and therefore these should be designed to be appropriate for all potential users.

Physical characteristics of interest will include:

- the size and shape of people, so that equipment and furniture can be designed to 'fit' the users;
- strength, so that loads and tasks are designed to be within users' capabilities without leading to fatigue or injury;
- visual abilities, in terms of the items to be viewed, taking account of differences between people in terms of colour blindness, long and short sightedness and variations in this with age;
- hearing abilities, such that instructions and warnings can be clearly heard.

Psychological characteristics of interest will include:

- short- and long-term memory;
- uptake and processing of information;
- reaction time;
- motivation;
- concentration;
- perception.

These physical and psychological characteristics will vary between people and will depend on individual factors such as age, gender, training and skills, health and fitness, and previous injury or disability. For any individual these may also vary over time (e.g. visual ability tends to deteriorate with age).

Posture

[E17006] In order to work comfortably and reduce the risk of musculoskeletal injury or discomfort, people should be able to adopt 'neutral' postures. A neutral posture is one in which the joints of the body are at about the midpoint of their comfortable range of movement and with the muscles relaxed (e.g. arms hanging by the side of the body). Deviations from this place a strain on the muscles and soft tissue.

The trunk is in a neutral posture when the natural 'S' curve of the spine is maintained – as when standing. The lower part of the spine (lumbar region) flattens when in a sitting posture due to rotation of the pelvis; this increases pressure in the inter-vertebral discs. Suitable back support (with adequate lumbar support) should be provided when sitting. It should be recognised that the amount of curvature in the small of the back (lumbar area) varies from person to person and therefore that one shape of backrest will not suit everyone.

The neck is in a neutral posture when the head is held upright and facing forward.

The shoulder is in a neutral posture when the upper arm is relaxed by the side of the body; the wrist is in a neutral posture when the muscles are relaxed and the hand is in line with the forearm.

Figure 2: The 'S' Shaped Curve of the Spine (S Pheasant *Bodyspace: Anthropometry, ergonomics and the design of work* **(1996) p 69 fig. 4.1, reproduced with permission of Taylor and Francis)**

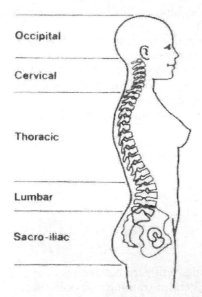

Occipital

Cervical

Thoracic

Lumbar

Sacro-iliac

Depending on the degree of deviation and the duration of maintaining the posture, awkward postures (when the posture deviates significantly from the neutral) and static postures (postures in which there is no movement for a period of time) can both lead to fatigue, discomfort, strain and possibly injury. Awkward postures such as bending, twisting and stretching, reaching the arms behind the shoulders, and reaching above shoulder height or below knee height should be avoided as far as possible; this can be achieved through careful positioning and adjustment of the equipment and furniture used. Introducing movement and changes in posture will help alleviate the discomfort that can arise from static postures.

Anthropometry

[E17007] Anthropometry (the measurement of body dimensions) can be used to ensure that people will be able to reach, use and operate tools, equipment and workstations, by structuring and positioning items appropriately (e.g. so they are within easy reach, adjust sufficiently, allow or prevent access, etc.).

The size of the body varies with age, gender and ethnicity. Tables of anthropometric data are available which give data for many different body dimensions, representing different population groups (see for example Stephen Pheasant and Christine M. Haslegrave *Bodyspace: Anthropometry, Ergonomics And The Design Of Work* CRC Press (2003)). Variations in body size usually follow a normal Gaussian distribution, with most people falling within a central value for a given body dimension, with fewer people tailing off at either side. It is common when designing to take account of the middle 90 per cent of the population, for example to design for those who are taller than the smallest 5 per cent of the population, and smaller than the tallest 5 per cent of the population (namely to design within the fifth to ninety-fifth percentile data). However, in some situations it is necessary to design for the extremes of the population. For example, the height of a door should take account of the tallest potential user; guarding on a machine should take account of the finger width of the smallest potential user.

Population stereotypes

[E17008] Population stereotypes are cues and expectations which provide information regarding the meaning, state or operation of items. We have certain expectations concerning colours, directions of movement, shapes, sounds and abbreviations. For example, a red tap indicates that it supplies hot water; turning a control dial clockwise turns the power up. Designs that take account of population stereotypes will result in increased accuracy and faster reaction times. Population stereotypes can vary by culture (e.g. pushing a light switch up turns it on in the US) and often by industry.

Examples of these stereotypes are shown below.

Population stereotypes in the UK
Colour
• Red = hot, danger, warning, stop
• Green = safe, go
• Yellow = caution
• Blue = cold, information
Shape of signs
• Round = enforceable
• Triangle = caution/ attention

- Square = advisory
  Direction
- Clockwise = on, tightens (screws), increases
- Anti-clockwise = off, loosens, decreases
- Upwards = switches off, increases
- Downwards = switches on, decreases

Allocating tasks to people or machines

[E17009] In most systems people interact with machines. When designing a system, tasks can be allocated to either the person or the machine. Decisions on whether a task should be undertaken by a person or by a machine will depend on an understanding of their relative capabilities. Some general guidelines on the relative capabilities of people and machines are shown below:

FIGURE 3: GUIDANCE FOR ALLOCATING FUNCTIONS (ADAPTED FROM SANDERS AND MCCORMICK, HUMAN FACTORS IN ENGINEERING AND DESIGN (1987) MCGRAW HILL)

Person's Capabilities	Machine's Capabilities
• Sensing stimuli – both at low levels and over background interference	• Sensing stimuli outside range of human sensitivity
• Recognising patterns of stimuli	• Monitoring for prescribed events, especially infrequent ones
• Sensing unexpected occurrences	• Storing coded information quickly and in large quantities
• Decision making	• Retrieving coded information quickly and accurately
• Information recall	• Calculating
• Creating alternatives	• Exerting considerable force
• Interpreting subjective data	• High speed operations
• Adapting responses	• Repetitive tasks
	• Sustained accuracy and reliability
	• Counting or measuring
	• Simultaneous operations

In allocating functions between people and machines, economic or social factors may also need to be considered beside performance. In some situations the 'best' performance may not be required, and contact with people may be preferable to dealing with machines. Also, in order to optimise job design, retaining some 'interesting' tasks for people may be more important than mechanising.

Designing the task

[E17010] The task is the collection and sequence of activities and events that allow the work to be completed. The task should be designed to be within the capabilities of the person, but must take account of the constraints of the work process, environment or system. Mismatches between task requirements and individuals' capabilities increase the potential for human error.

Task design will influence the activities of the person, for example the postures adopted, amount of force applied, repetition and duration of the activities. These may have an influence on performance outcome or on health (e.g. musculoskeletal disorders). Issues to consider include:

- *Duration of the task* – long durations of work without a break may lead to reduced performance and increased discomfort. Decline in concentration usually becomes evident after 30 minutes; physical discomfort increases if static postures are maintained or movements are repeated without a break.

- *Breaks and changes in activity* – infrequent breaks can lead to reduced performance (e.g. on a vigilance task such as inspection) and increased discomfort. Regular changes in posture and movement help reduce physical discomfort and injury. Prolonged periods viewing a display or screen without a break can also lead to visual discomfort. Varying the tasks that are undertaken will also help to reduce boredom.

- *Amount of repetition of movements required* – highly repetitive work which involves the same muscle groups and movements can lead to physical discomfort and boredom.

- *Frequency of the task* – frequently performed tasks may lead to boredom and increased physical discomfort (as the same movements are made). Conversely, there may be training and information needs for infrequently performed tasks.

- *Pacing* (if the person is required to work at a speed set by a machine or process) – this will dictate the speed of work and amount of recovery the person has between tasks or operations. This can have an impact on discomfort and performance. A comfortable work rate should be established.

- *Work rate* – excessively busy periods or quiet periods can lead to stress and discomfort. There should not be benefits for completing the task early, as this can lead to people rushing the task.

- *Complexity of the task* – a number of sub-tasks are usually required to complete the operation, and more than one operator may be required to assist in the operation. Tasks can be divided so that operators each undertake one sub-routine or a more complex combination of tasks. Jobs should be designed to facilitate development of skills, interest, variety and commitment. Several sub-tasks may be undertaken by one operator, but to reduce boredom and to prevent overload and the risk of musculoskeletal injury these should vary in the demands they place on the physical and psychological capabilities of the workers.

- *Awkward or static postures required* – these may be dictated by task requirements or poor tool or workstation design. These can contribute to musculoskeletal disorders and should be avoided.

- *Force required to undertake the task* – application of frequent or excessive force, particularly in relation to the capabilities of the body part applying the force, can lead to discomfort. It may be appropriate to mechanise the task or use a tool or equipment to assist in generating force if required.
- *Sequence of use of tools/equipment* – appropriate layout of equipment and tools will facilitate ease of use, so that, for example, those used most frequently are positioned closest to the user.
- *Levels of concentration and attention required* – the level of arousal (the state of consciousness) helps to maintain attention and affects performance. Low levels of arousal are found in undemanding jobs, where activities are either very mundane or very infrequent, e.g. production line packing, night security work. High levels of arousal can be found in stressful and demanding jobs, e.g. air traffic control and ticket inspection. Both high and low levels of arousal can lead to poor performance. Tasks should be designed to maintain the vigilance and arousal necessary for the task. Methods of facilitating concentration and vigilance include reducing the task time and introducing frequent breaks; providing music or background noise for routine tasks; and exaggerating the size or colour of the stimulus.
- *Information and training required for the task* – some tasks will require personnel to be trained in how to undertake them. For other tasks simple information or instructions may be adequate. The need for training and information will depend on the complexity of the task and the experience and abilities of potential users.

Task design will be influenced by the equipment and workstation design, the organisation of the work and the abilities of the individuals. Application of anthropometric and biomechanical knowledge will help ensure the physical elements of the task are within the physical capabilities of the person. Understanding the psychological abilities of the person, such as their memory and concentration, will help to design a task within limits acceptable to the person (or group of people). The provision of any information or instruction should also account for the way in which people process information.

In general it is those who are undertaking the task who have the most knowledge about the needs of the work system. Involving workers in evaluation and re-design is an important component in ensuring the design meets their needs and those of the system, and that any changes made are accepted by workers.

Design of equipment

[E17011] Equipment should be designed and selected for those who will use it, taking account of the task to be completed and the conditions under which it is used.

Hand tool design

[**E17012**] Tools enable the use of the hand to be extended, for example to extend its ability to apply force, manipulate items, make precise movements and so on. Appropriate design of hand tools can facilitate the task and reduce the risk of musculoskeletal injury.

Hand tools should be designed to allow neutral arm, wrist and finger postures in operation. The hand is in a neutral posture when the hand is in line with the wrist, the thumb facing upwards and the palm facing inwards. This can be seen when the shoulder is relaxed and the arm is allowed to hang relaxed by the side of the body. The maximum grip strength can be applied when the wrist is in a neutral posture; deviations from this (ulnar or radial, namely side to side, flexion or extension – up or down) will reduce the amount of force that can be applied, with the hand able to apply the least force in a flexed posture. Working with the hand in a neutral posture also minimises the strain placed on joints and other soft tissue, which can be compressed or experience friction in awkward postures.

Angling either the work piece or the tool handle can help reduce the amount of wrist deviation required when using a hand tool; however, the potential for variation in orientation of the tool on the work piece may limit the benefit of this.

Key points in hand tool design are:

(1)   Handles should be long enough to fit the whole hand – at least 100 mm, although a length of 120 mm is preferable.

(2)   In general, larger handles decrease the amount of muscular activity required when using the tool. A handle thickness of approximately 40 mm is generally recommended. Slightly thicker handles may be beneficial when applying torque (e.g. for screwdrivers).

(3)   Tools with two handles which require a hand span (e.g. pliers, scissors) should have a span of approximately 60 mm. If the tool is used repetitively, an automatic spring opener will reduce the strain on the weaker finger extensor muscles (used to open the hand).

(4)   Ideally the handle surface should be compressible, textured and non-conductive. It should be free from sharp edges; avoid ridges or finger contouring on the handle, as this can place pressure on the soft tissue of the hand. Avoid cold surfaces (e.g. metal), particularly if the tool is powered by compressed air.

(5)   An excessively smooth handle surface requires the user to grip the handle more tightly, increasing the amount of force required, and this should be avoided. If the hands are sweaty or if the user is wearing gloves, they will also have to grip the handle more tightly. Tools used in these circumstances should therefore be made easier to grip.

(6)   It should be possible to use the tool in either hand; if this is not possible, specific tools for left-handed workers should be provided.

(7)   In terms of power-assistance, power assisted tools can greatly increase the speed of task completion, remove a large degree of force exertion from the operator and reduce some awkward postures, such as rotating the wrist. However, the user may experience vibration from the tool.

Pneumatically powered tools can blow cold air exhaust over the hands, increasing the risk of discomfort and musculoskeletal problems such as Hand Arm Vibration Syndrome (see OCCUPATIONAL HEALTH AND DISEASES). If power tools are used, ensure they are low-vibration and well maintained to reduce vibration transmitted to the hand and arm. Mounting power tools in a jig can reduce vibration transmission.

(8)   If force has to be applied through the tool, where possible use power tools rather than tools requiring manual application of force. However, power tools often weigh more than manually operated tools and handling them can increase the risk of discomfort.

(9)   In terms of weight, frequently used tools should weigh as little as possible; a maximum of 0.5 kg is recommended. The distribution of weight in the tool should be even, with the centre of gravity as close to the hand as possible. Counterbalances may be required to support heavier tools or those with an off-set centre of gravity.

Design of controls

[E17013] Displays and controls are the mechanisms by which people and machines interact. The person gives instructions to the machine through the controls (e.g. knobs, switches, levers, buttons). Controls can be discrete, having a set number of conditions (e.g. on/off/standby), or continuous, where the condition varies along a scale (e.g. volume). Controls should be designed to facilitate the changes that they allow and the amount of effort required to operate them. The following factors should be considered:

• The amount of force required to operate the control (and the amount of resistance to prevent accidental operation). Fingers and hands should be used for quick, precise movements; arms and feet for operations requiring force.

• The size of the control should be appropriate for the force required and body part used to activate it.

• The location of controls should facilitate their use. Hand-operated controls should be easily reached and grasped between elbow and shoulder height. Controls should be sufficiently far apart to allow space for the fingertips. Sequence and frequency of use, and importance may also dictate the location of controls.

• Appropriate feedback should be provided to indicate activation of the control.

• Controls should be coded by colour, shape, texture or size (to allow identification by touch).

Design of displays

[E17014] The state of the machine is relayed to the person through a display. Displays should be designed to enable users to easily and accurately assess information from the machine. The following principles should be followed:

(1) Qualitative displays showing a small number of conditions should be used for discrete information (e.g. on/off). Quantitative displays should be used to present numerical or continuous information, e.g. temperature, speed etc. Qualitative displays may be lights or words etc. Quantitative displays may be scales, counters etc.

(2) Display scales should be clear, unobscured and concise. Scale intervals should increase left to right, bottom to top, preferably increasing in units of 10s, 100s, 1000s, etc. as appropriate.

(3) Displays should conform to population stereotypes in terms of colours used (e.g. red = danger).

(4) Ensure labelling, if used, is clear.

(5) Limit the number of warning lights to avoid confusion and aid identification.

(6) On dials with pointers, avoid parallax (the difference in scale reading depending on the viewing angle) by keeping the dial and pointer close together.

(7) Text should be of a suitable size and font to allow easy reading. If this is not appropriate, B and 8, O and 0 can be confused, and if resolution is not adequate, F and P can also be confused.

Grouping of controls and displays

[E17015] The result of activating a control is often indicated in a display. The relationship between displays and controls should be clear; this can be achieved through location, arrangement, text, shapes, colours and responsiveness. Grouping of controls and displays can improve their association with their function and facilitate their ease of use. Controls and displays can be grouped according to sequence of use (e.g. start, run, finish), by function (e.g. keeping all controls concerned with lights together) or by frequency of use, such that those most frequently used are within convenient reach and within the immediate viewing arc. Locating controls by importance of use can also aid their operation (e.g. emergency controls should be placed within convenient reach).

Text

[E17016] Text used as information instruction or labelling should be clear and concise. Bold, italics, large fonts, underlining and colour can all be used to draw attention to or highlight information but should be used sparingly to preserve the meaning when they are used. Lower case should be used for phrases or sentences, as UPPERCASE DECREASES THE DIFFERENTIATION BETWEEN LETTERS AND IS SLOWER TO READ.

Positive instructions should be used, not negative or double negatives. For example: 'When the alarm sounds turn the machine off' – not 'When the alarm sounds do not leave the machine on'.

Software design

[E17017] The field of Human Computer Interaction (HCI) is a specialist area; in this context it is suffice to say that software should be designed to make the use of a system intuitive and accessible (e.g. through the arrangement of icons, use of colours and text, structure of menus and commands).

Workstation design

[E17018] The workstation (the area where the task is conducted) should be designed so that it is appropriate for the person using it. The workstation may be a desk or work surface, but may equally be at a conveyor belt or a driver's cab. Attention needs to be given to the physical dimensions of the workstation and the arrangement of any necessary equipment/tools/components in it to allow good posture and acceptable movements.

The issues discussed below cover the main points that should be considered in the design of workstations.

Worksurface height

[E17019] Many tasks will involve workers using a work surface. The height of the worksurface should permit a relaxed, upright posture and facilitate the use of equipment at the workstation. The surface height (whether designed for sitting or standing tasks) should be such as to avoid users stooping or reaching to its surface. In most situations the height of the worksurface is fixed. For seated tasks a height adjustable chair should be provided so that the working height can be set appropriately. A footrest may be required to support the feet when sitting at a comfortable height.

An appropriate worksurface height is dependent on the work that is done at that worksurface. As a general rule, the top of the worksurface should be level with the user's elbow height (the user may be sitting or standing, as required at the workstation). The height of the worksurface may need to be reduced to take account of the thickness of the item being worked on, so that this is at an appropriate height.

There may be a range of users utilising the same workstation, and unless easy height adjustment is provided, it will not be possible to obtain the optimum height for all potential users. For a fixed height workstation some compromise will have to be made, based on the task requirements and the likely users.

The following working heights are recommended for different tasks in relation to the user's height (S Pheasant *Ergonomics, Work and Health* (1991) Taylor and Francis).

Task	Height
Manipulative, requiring force and precision	50 mm–100 mm BELOW elbow height

Task	Height
Delicate (including writing)	50 mm–100 mm ABOVE elbow height
Heavy – requiring downward pressure	100 mm–250 mm BELOW elbow height
Lifting and handling	Hands between knuckle and elbow height
Two-handed pushing/pulling	Hands just below elbow height
Hand-operated controls	Located between elbow and shoulder height

The variation in recommended working heights for different tasks arises from the aim to minimise the effort required to perform the tasks, by optimising the posture and muscle groups used.

Using this data, as an example tasks such as component assembly where pneumatic screwdrivers are used should be conducted at between 100 mm and 250 mm below elbow height, so that downward pressure can be applied. If the unit being assembled stands 50 mm above the worksurface, the worksurface will need to be 150 mm–300 mm below elbow height. This assumes that all work at the workstation is of this nature. Workstations that are also used for writing tasks, for example, should be split-level, providing a writing surface at or just above elbow level. If a fixed-height workstation and height-adjustable chair are provided, users should adjust the height of the chair depending on the task. For example, VDU users who also write at their workstation will find it more comfortable to sit higher when they are keying and lower when they are writing.

Height adjustable workstations can be beneficial, particularly for standing tasks and where a range of different users may work at the same workstation. Where height adjustable equipment is provided, users should be trained in how to adjust it and how to identify an appropriate working height.

Workstation characteristics

[E17020] Other characteristics recommended for the workstation include:

- The workstation should be a suitable size such that all equipment can be located and arranged conveniently for the person to complete the task.
- The workstation surface properties should not present a risk to the user – there should be no sharp edges or unprotected hot/cold surfaces. A front edge of 90° can cause discomfort if the arms are rested or pivoted on it.
- Surrounding workstations or other items in the environment, e.g. columns or posts, should not constrain the user's posture and ability to get into and out of the workstation easily.
- There should be sufficient legroom underneath the worksurface to allow the user to be able to sit comfortably, sufficiently close to the workstation, without their posture being constrained.

Workstation layout

[E17021] To facilitate a good working posture, the equipment and items used to complete the tasks should be within convenient reach of the user, so they do not have to stretch or lean. The zone of convenient reach and the normal working area can be used to establish a suitable layout of the workstation.

Zone of convenient reach

[E17022] The zone of convenient reach is defined by the area from the shoulder to the fingertips with the arm outstretched (upward, downward and to the sides), which can be reached without any undue exertion. Items required for the task (e.g. control buttons, handles, work surfaces, tools) should be placed within this zone. The extent of this area obviously depends on the length of the arm and its arc. The fifth percentile arm length (95 per cent of the population will be able to reach items if placed within this area) is 735 mm for British men and 655 mm for British women when reaching forwards (S Pheasant *Peoplesize (Open Ergonomics); Anthropometrics: An introduction for schools and colleges (1984) (BSI PP 7310)*. If designing for both male and female users, the shorter dimension (655 mm) should be used.

Figure 4: Zone of Convenient Reach (E Grandjean Fitting the Task to the Man (1988) p 51 fig. 42 reproduced with the permission of Taylor and Francis)

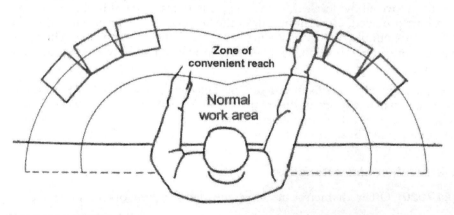

Normal working area

[E17023] The normal working area is defined by the comfortable sweep of the forearm with the elbow bent at 90° and the upper arm in line with the trunk. This provides an area that can be reached without extension of the arm at the shoulder and requiring no trunk movement. The extent of the area depends on the length of the arm from the elbow to the fingertips. The fifth percentile forearm length is 443 mm for men and 402 mm for women (S Pheasant *Peoplesize (Open Ergonomics)*). Again, if designing for both male and female users, the shorter dimension (402 mm) should be used.

In a case highlighted by the retail union USDAW, a shopworker developed a repetitive strain injury (RSI) after the checkouts at the store she worked in were redesigned. Credit card readers and touch screens were installed, but cashiers had to stretch out of their chairs to reach these. The chip and pin machines were 535mm from the edge of the till and the touch screens 430mm away, making the overall workstation outside the comfortable reach of 95% of the women workers at the store.

One woman was particularly badly affected, because she was only four feet nine inches tall. The chip and pin machine was 23cm too far from her reach and within a month she developed tenosynovitis, a form of RSI

Arrangement of items on the workstation

[E17024] Frequently used items and critical items (e.g. emergency controls) should be placed in the normal working area. Occasionally used items can be sited within the zone of convenient reach. Items can be grouped by function or sequence of use, or importance.

Sufficient space should be provided at the workstation to allow a flexible arrangement of all the equipment that is required for the task. Items should be arranged so that operators do not have to reach across their body, for example the phone should be placed on the left hand side if it is answered with the left hand.

Visual considerations

[E17025] The following factors should be taken into consideration.

Up/Down Viewing

[E17026] Items can be viewed both by movement of the eyes and by movements of the head, neck and back.

Awkward neck and back postures may be required if the viewed item is low, high or to one side, and this may cause discomfort. Items to be viewed should be positioned such that no twisting or bending of the neck, head or back is required to see them, and so that the eye muscles are in a neutral position.

The neutral position of the eye is generally taken as about 15° below the horizontal line of sight. The eyes can comfortably move about 30–45° below the horizontal line of sight before the head must be inclined; an upward gaze of approximately 15° is achievable before the head must be tilted backwards to view further. An upward gaze is fatiguing when sustained for any length of time, as the muscles controlling upward movement of the eyes are not as strong as the muscles controlling downwards movement. The most comfortable eye position is an arc of 30° from the horizontal position of the eyes downward, and this is the acceptable viewing angle.

Figure 5: Viewing Angle (S Pheasant *Ergonomics, Work and Health* (1991) reproduced with permission of Palgrave Macmillan)

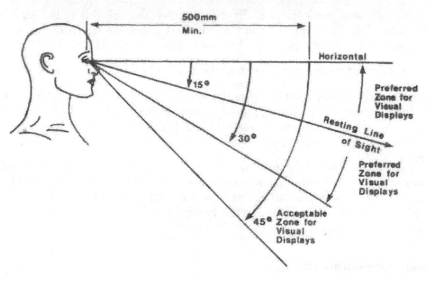

Side/Side Viewing

[E17027] Items outside about a 60° arc in front of the person will involve neck twist to view. Items to be viewed regularly or for long periods should not be positioned outside this area.

Viewing distance

[E17028] The required viewing distance generally depends on the person's visual ability, the size of the item being viewed and the lighting levels. A viewing distance of between 500 mm and 750 mm is likely to be suitable for most items, but a shorter viewing distance may be required for fine work.

Sitting versus standing

[E17029] There are biomechanical benefits to both sitting and standing. Sitting can help to reduce the development of fatigue in the lower limbs, as long as the chair provides adequate support and the feet are also supported. Tasks that require generally static postures (with limited or no back movement) will benefit from being undertaken while sitting. However, inappropriate seating and prolonged sitting can lead to discomfort and fatigue, and may prevent good posture (e.g. armrests may prevent the chair being brought sufficiently close to the workstation) (see WORKPLACES – HEALTH, SAFETY AND WELFARE W11008).

Standing is beneficial if the task requires a range of body movement or handling of heavy items. However, prolonged standing, particularly with little walking, can result in workers experiencing tired or swollen legs and lower back discomfort. The *Workplace (Health, Safety and Welfare) Regulations 1992 (SI 1992 No 3004)* require a suitable seat be provided for each person at work whose task (or a substantial part of it) can or must be done seated. Whether sitting or standing, workers should be able to vary their posture and should be able to take short, frequent breaks away from the workstation.

Sit/stand stools can be used to allow work to be completed at standing height while removing some of the strain in the legs and back associated with standing.

Anti-fatigue matting can be used at standing workstations to cushion the feet, and this is especially beneficial on concrete floors, as these are particularly fatiguing.

Physical work environment

[E17030] The work environment (lighting, noise, temperature, humidity and general layout of the place of work) should also be designed to suit the person and the work. In general, the effects of the environment can impact on performance, subjective comfort, perception, attitudes, safety and health.

Lighting

[E17031] Appropriate levels of lighting are necessary to maintain acceptable levels of quality and performance, and to help avoid visual discomfort. Poor lighting can lead to visual fatigue (symptoms include red or sore eyes, blurred vision, headaches). Inappropriate lighting can also force users to adopt poor postures in order to view items more easily, and this can also lead to discomfort.

Issues to consider in designing visual environments are the brightness of the light sources, their position, the evenness of the light distribution, the contrast between the different work items and the potential for glare/reflections from other items in the work area.

The amount of illuminance required should be determined by the task demands, for example visually demanding tasks such as inspection or fine and precise work require a higher level of light than that required in walkways. Insufficient lighting can result in eye strain and discomfort, while too much light or a bright light source within the visual field will cause glare and result in visual discomfort and reduced visual performance. Where light levels are already high, it may be appropriate to improve the legibility of the source document or item being viewed (e.g. increasing size and contrast) rather than further increasing the lighting levels.

The location of lights will also affect visual performance and comfort – light sources within the visual field may cause glare (which may be visually disabling or cause discomfort). Excessive differences in lighting levels within the work areas should be avoided. Surfaces of a highly reflective material should be covered or replaced to reduce glare. Directional lighting may be appropriate for some inspection tasks but diffused light is generally more satisfactory, as it reduces the amount of glare and shadows produced.

Light may come from a natural source (from windows) or an artificial source (electric lights). It is important to remember that lighting levels may vary by the time of day or year and therefore appropriate adjustments may need to be made, for example provision of local lighting (lamps) or blinds.

The furniture and equipment in the room should be arranged so that the sources of light do not cause light to fall directly onto reflecting surfaces or into the eyes (e.g. sit at right angles to a window rather than facing or with the back to the window).

Colours should be chosen to provide contrast to aid detection but not to exaggerate glare or reflections. Colours with good contrast should be used for displays (e.g. black and white, blue and yellow, etc.). Where lighting levels are high, dark characters on a light background are generally easier to read.

In general, it is appropriate to allow individuals to have control over the lighting levels in their area, as individuals vary in what they find comfortable. In addition, having control over the work environment can help reduce psychosocial stress.

Increased lighting levels may be required for those with poorer eyesight (e.g. older people), although this should be done with care, as older people are generally more susceptible to glare. For further details see LIGHTING. Additional guidance is provided in the HSE Guidance Note *Lighting at Work* (HS(G) 38).

Noise

[E17032] Noise can have an effect both on hearing ability (hearing loss) and on performance. Too much or too little noise can adversely affect concentration. Where work is monotonous, noise can be introduced to provide a stimulus; but where noise is excessive it should be reduced to facilitate concentration. Noise levels should not exceed approximately 55 dB(A) for tasks requiring concentration.

Noise can be used effectively as a warning or alarm, although excessive use of noise for warnings can be detrimental to their effectiveness. Hearing generally deteriorates with age; this and the differences in hearing ability between people should be taken into account when designing critical alarms.

There are several harmful effects of noise, which include:

(1)   Physical damage to the eardrum and ossicles induced by excessively high noises e.g. explosives.
(2)   Hearing damage due to exposure to high levels of noise. Damage can be divided into temporary threshold shift (reversible) or noise-induced hearing loss (irreversible).
(3)   Annoyance and stress – which can lead to reduced concentration. Noises that are annoying are generally unexpected, infrequent, high frequency and those over which the listener has no control.
(4)   Hindering communication – which can lead to an increased rate of accidents as well as stress.

Under the *Control of Noise at Work Regulations 2005 (SI 2005 No 1643)*, where the noise level reaches 80 decibels employers must assess the risk to workers' health and provide them with information and training. If the level reaches 85 decibels (daily or weekly average exposure), they must provide hearing protection and hearing protection zones. There is also an exposure limit value of 87 decibels, taking account of any reduction in exposure provided by hearing protection, above which workers must not be exposed.

For further information see Noise and Vibration.

Vibration

[E17033] Exposure to vibration is associated with musculoskeletal disorders, other health problems and reduced performance. There are two main categories of vibration transmission to the body: whole body and hand/arm (see N3027).

Whole body vibration usually occurs via the feet or seat when the person is working on/in a vibrating environment (e.g. transport systems). Low frequencies of vibration (such as those from motor vehicles) can lead to chest pains, difficulties in breathing, back pain and impaired vision.

Hand-arm vibration affects only the hands and arms, and can be experienced when using power tools such as pneumatic screwdrivers and chainsaws. Exposure of the hands and arms to vibration is a known risk factor in the development of Hand Arm Vibration Syndrome (HAVS), a condition which affects the nerves and blood supply to the hand, resulting in tingling or numbness, impairment or loss of function and blanching of the fingers, notably on exposure to the cold. It is particularly higher frequencies of vibration (such as those associated with hand-held tools) that have been associated with muscle, joint and bone disorders affecting the hand and arm. Transmission of vibration has been found to increase as the gripping force increases and where tight gloves are worn.

Vibration, as with noise, should be tackled at its source. Larger machines can be isolated or the vibrations insulated, for example by introducing better mountings on the machine. Regular maintenance of machines can also reduce vibration. Relocation of the machine and insulation of the source should be considered. Floors, seats and handgrips can all be fitted with damping material. Hand tools should be designed to reduce the vibration created, the weight and the amount of force required, so that tools do not have to be gripped tightly. Where this cannot be achieved, exposure time should be limited and the hands kept warm.

The *Control of Vibration at Work Regulations 2005 (SI 2005 No 1093)* require employers to take action to prevent their employ-ees from developing diseases caused by exposure to vibration at work from equipment, vehicles and ma-chines.

For further information see Noise and Vibration.

Thermal environment

[E17034] The internal body temperature needs to be maintained within a relatively narrow range for health, and the ambient conditions that facilitate this are also relatively narrow. Deviations from comfortable temperatures can lead to reduced performance, fatigue, discomfort and ill health (hypo and hyperthermia as the body temperature rises or drops below those necessary for good health); in extreme circumstances this can be fatal.

Body temperature, and therefore comfort, is dependent on the ambient conditions (air temperature, humidity, any radiant heat, and air velocity), the activity being undertaken (and therefore the heat produced within the body) and the clothing worn. Individuals' response to the thermal environment varies and some people are more affected by high or low temperature than others; it is most satisfactory if employees are able to control the environment to suit themselves (e.g. opening windows etc.), although this can be difficult in large, open plan offices.

As far as possible, environments should be designed and controlled to allow a comfortable working temperature (e.g. through heating, air conditioning, etc.). Where this is not possible (e.g. due to a work process or outdoor working), appropriate clothing can be used as a control measure or in extreme situations it may be necessary to limit exposure time. There is currently no legal upper temperature beyond which workers should not be exposed, although the TUC have called for a 30°C upper limit for indoor work areas. In terms of a lower limit, the temperature in workrooms should normally be at least 16°C. Further guidance is provided in *Workplace (Health, Safety and Welfare) Regulations 1992 (SI 1992 No 3004)*.

Low temperatures reduce manual dexterity, and where this is required the hands should be kept warm (e.g. through appropriate gloves). Manual handling injuries are more likely in extreme thermal environments and account should be taken of this when planning tasks.

Guidelines for control of employees' work in hot/humid/cold environments are shown below.

(1) Maintain temperatures between 20–24°C for sedentary work and 13–20°C for physical work.
(2) Maintain humidity at between 40 and 70 per cent.
(3) Reduce air velocity (draughts) to less than 0.1 ms^{-1} in moderate and cold environments. Increasing air velocity can be an effective control measure in hot environments, provided the air temperature is not close to or above body temperature (37°C), when increased air velocity will increase the thermal load on the body.
(4) Decrease physical workload in hot environments.
(5) Schedule rest pauses or rotate personnel around more strenuous tasks in hot conditions.
(6) In hot environments allow rest periods to be taken in cooler environments.
(7) Schedule outdoor work to avoid periods of very high or low temperature.
(8) Permit gradual acclimatisation to hot or cold environments (7–10 days).
(9) Maintain hydration by consuming sufficient drinking water, particularly in hot environments.

(10) Avoid long exposure periods in cold environments.

(11) Avoid rest for long periods in cold environments.

(12) Provide protection from wind and foul weather when working outside.

Physical hazards

[E17035] Other physical hazards may be present in the workplace (e.g. slipping, falling, tripping hazards). Ideally these should be eliminated at source through appropriate design (e.g. removing the need for an operator to enter a potentially dangerous area). Where hazards are still present, guarding, safety barriers, good housekeeping, appropriate lighting and good workplace layout will reduce the possibility of all types of accident. Raising awareness through education and training can be effective, although it should not be relied on as a first line of control. Personal protective equipment (PPE) should always be seen as the last resort control measure, due to issues related to fit, compatibility, impact on performance, loading on the body, etc., which may mean that it is not worn correctly or possibly at all.

Job design and work organisation

[E17036] The design and organisation of work can have a significant impact on health. The demands of the job should be matched to the skills and abilities of the individuals. Excessive demands or, conversely, under utilisation of skills can both lead to a decrease in performance. The design of the job or work task is important for obtaining the maximum performance from the person for the minimum effort and facilitating a level of job satisfaction. Factors to consider include job rotation, job enlargement and job enrichment, as well as job scheduling, job demands and shift patterns.

Job rotation

[E17037] Job rotation, whereby people are moved from one task to another, allows changes in posture, movement and mental demands. This may lead to the acquisition of additional skills and may help the person identify more with the completed product or service. Job rotation can also help to overcome boredom, particularly in repetitive jobs, and help to prevent fatigue, loss of concentration and deterioration in performance. To be effective as a control measure for musculoskeletal discomfort, people should rotate to tasks that are significantly different in terms of postures, movements and forces required.

Job enlargement

[E17038] Job enlargement, whereby the number and range of tasks undertaken by the individual is increased, allows workers to become more flexible, develop other skills and have greater variety in their tasks.

Job enrichment

[E17039] Jobs can be enriched by incorporating motivating or growth factors, such as increased responsibility and involvement, opportunities for advancement and a greater sense of achievement. This should provide the workers with greater autonomy over the planning, execution and control of work.

Work scheduling

[E17040] Ideally workers should have control over how the work is scheduled. As far as is possible, work should be self-paced rather than machine-paced. Machine pacing may be too fast or too slow for an individual, both of which can lead to problems (lack of recovery time for muscles and soft tissue; frustration, etc.). Machine pacing has been implicated as being a contributory risk factor in upper limb disorders, as well as stress and stress-related illnesses.

Rest breaks help alleviate both mental and physical fatigue. Allowing the worker to select when and how often they take breaks is preferable to fixed breaks, although there is some evidence that people work until fatigue occurs rather than stopping before this point. Therefore, under some circumstances (e.g. tasks that require high concentration or those requiring physical effort) it may be preferable to give fixed rest breaks before the person is likely to become fatigued.

Breaks should be taken away from the workstation, not only because this allows for physical movement and a change of scene, but also because it may facilitate reducing exposure to hazards such as noise. Short, frequent breaks are better than long, less frequent breaks.

Shift-work

[E17041] It is well recognised that shift-work affects health and safety at work. In particular, where shifts involve people changing their eating and sleeping patterns, symptoms such as irritability, depression, tiredness and gastrointestinal problems (e.g. indigestion, loss of appetite and constipation) can occur. These health problems are largely due to the disturbance of the body's physiological cycles. Although many shift-workers adapt to the disruption, it is widely recognised that productivity levels are lower on night shifts than day shifts.

Night shifts should therefore be avoided if at all possible. Younger and older workers (those below the age of 25 and above 50) may be more vulnerable to ill health effects from shift work. Those with digestive disorders, sleep problems or other health problems may find shift work aggravates these. Shift rotations should be short and followed by at least 24 hours' rest.

Job support

[E17042] Appropriate design of jobs, the structure of the organisation and the way it operates are important design criteria. The following points are recommended for providing adequate support in a job:

- Provide opportunity for learning and problem solving within the individual's competence.
- Provide opportunity for development and flexibility in ways that are relevant to the individual.
- Enable people to contribute to decisions affecting their jobs and their objectives.
- Ensure that the goals and other people's expectations are clear and provide a degree of challenge.
- Provide adequate resources (e.g. training, information, equipment, materials).
- Provide adequate supervision, support and assistance from contact with others.
- Provide feedback on performance and communication of objectives.

Organisational support

[E17043] The following factors are considered to be necessary to provide an acceptable background against which people can work:

- Employment relations policies and procedures should be agreed and understood, and issues handled in accordance with these arrangements.
- Payment systems should be seen as fair and reflect the full contribution of individuals and groups.
- Other personnel policies and practices should be fair and adequate.
- Physical surroundings and the health and safety provisions should be satisfactory.

Musculoskeletal disorders

[E17044] The term 'musculoskeletal disorders' (MSDs) refers to problems affecting the muscles, tendons, ligaments, nerves, joints or other soft tissues of the body. These disorders most commonly affect the back, neck and upper limbs. Musculoskeletal disorders are the most common work-related condition affecting the general population in Britain. They account for almost half of all self-reported occupational ill health. Of these, 42 per cent affected the back, 40 per cent the upper limbs or neck, and 17 per cent the lower limbs. The HSE estimates that 11.6 million working days a year are lost to work-related MSDs.

Musculoskeletal disorders give rise to clinical effects in the individual, namely symptoms (e.g. pain, numbness, tingling) and signs (e.g. changes in the appearance of the limb which may be identified on medical examination). They usually also result in some functional changes (e.g. reduced ability to use the part of the body affected; restrictions to movement or strength). They can also have a general effect, such that there may be a resultant reduction in general health or quality of life (e.g. through restriction of activities).

MSDs are frequently work related, but not necessarily caused by work. MSDs have been the subject of much civil litigation in recent years and a significant number of personal injury cases have been successfully brought against employers (see Occupational health and diseases). The employer's duty of care to their employees with respect to upper limb disorders is now well established in the civil courts. This civil law duty runs parallel to the employer's statutory responsibility under health and safety legislation.

There is currently no specific legislation relating to musculoskeletal disorders. However, the *Manual Handling Operation Regulations 1992 (as amended) (SI 1992 No 2793)* and the *Health and Safety (Display Screen Equipment) Regulations 1992 (as amended) (SI 1992 No 2792)* were introduced with the aim of reducing musculoskeletal disorders associated with these activities. There are also general duties under the *Health and Safety at Work etc. Act 1974* and the *Management of Health and Safety at Work Regulations 1999 (SI 1999 No 3242)*. (For more details on legislation see **E17060–E17069**.)

Activities outside of work may also present similar types of risk of MSDs to those experienced in work activities. However, non-work activities are usually not as repetitive, forceful or prolonged as work tasks, and the individual is likely to have more control over whether the activities are undertaken and how long for.

Back disorders

[E17045] Back discomfort is extremely common in adults. Over 80 per cent of adults will experience an episode of back pain during their life, but the vast majority of these (90 per cent) will recover within two weeks, although recurrence of back pain is common. Back pain is rarely caused by a single injury but is generally cumulative in nature. However, a minor incident may be seen as the final straw that triggers discomfort.

Causes of back discomfort include manual handling activities; awkward postures, such as twisting, leaning, bending, stretching and prolonged static postures; poor seat design (including car seats); low physical fitness; and exposure to whole body vibration. The significance of psychosocial factors in experience of back pain is now well recognised.

Useful self-help guidance on how to cope with back pain is given in The Back Book (The Stationery Office – ISBN 9780117029491).

Upper limb disorders

[E17046] Upper limb disorders (ULDs) are a sub-category of MSDs and the term refers to injuries occurring in the upper limbs (fingers, hands, wrists, arms, shoulders and neck). It covers strains, sprains, injury and discomfort, including specific conditions such as tenosynovitis, tennis elbow and carpal tunnel syndrome, as well as non-specific disorders. These non-specific disorders have often been called 'RSI' (Repetitive Strain Injury) but could more accurately be referred to as Non-Specific Arm Pain. The cause of some of these disorders can be related specifically to the work undertaken and become known as work-related upper limb disorders (WRULDs). These disorders are usually cumulative in nature, arising from a series of micro-traumas, which lead to discomfort if the body does not have sufficient rest time to recover.

The main physical risk factors in the development of upper limb disorders are application of force, repetitive movements and awkward or static postures. It is usually the interaction of at least two of these risk factors which leads to disorders. Duration of exposure to these factors is obviously also significant, with long durations of exposure increasing the risk of injury.

(1) *Force* – application of excessive force in relation to the capabilities of the upper limb muscle group will place a strain on the muscles involved and may lead to injury. Different muscle groups are able to apply different levels of force (e.g. shoulder muscles can apply more force than finger muscles).

(2) *Repetition* – frequent, repetitive movements require the same muscles to contract and relax over and over again. If this activity is very rapid or continues for a long period of time the muscle can become fatigued and this can lead to discomfort.

(3) *Posture* – poor postures (both awkward postures and static postures) can lead to injury and discomfort by placing the muscles and joints under unnecessary strain.

(4) *Vibration* – exposure to hand/arm vibration is a recognised risk factor in the development of Hand Arm Vibration Syndrome (HAVS).

WRULDs are not confined to particular activities or industries but are widespread throughout the workforce. Some jobs have particularly been associated with WRULDs, and these involve the recognised risk factors. Those who are such as assembly line workers, construction workers, garment machinists, meat and poultry processors, and display screen equipment workers (particularly intensive data entry tasks) suffer a risk of WRULDs, but this list of jobs is not exhaustive and is provided for illustration only.

Common ULDs are listed in **O1037**. Legislation and guidance relating to ULDs is detailed in **E17063** and **E17067**.

Lower limb discomfort

[E17047] Lower limb disorders are less common and often less debilitating than other musculoskeletal disorders. Although they often do not lead to significant limitations to activity, they cause both local and general body fatigue. Lower limb discomfort can be caused by prolonged standing, particularly on concrete floors; inappropriate (particularly hard-soled), heavy or ill-fitting footwear; operation of foot controls; and a lack of adequate foot support if sitting. Tasks that involve repetitive or prolonged kneeling (e.g. shelf stacking on low shelves) can lead to knee disorders.

Psychosocial risk factors

[E17048] As well as the physical factors that can lead to musculoskeletal disorders, there is strong evidence of the role of work-related psychosocial factors (the worker's psychological response to work and workplace conditions) in the development of these disorders. Relevant factors include the design, organisation and management of work, the context of the work (overall social environment) and the content of the work (the specific impact of job factors).

Specifically, repetitive, monotonous tasks, excessive or undemanding work-loads, lack of control over the task or organisation of the workplace, working in isolation, poor communication and lack of involvement in decision making have been implicated as psychosocial risk factors.

It is thought that many of the effects of these psychosocial factors occur via stress-related processes, which result in biochemical and physiological changes that can result in discomfort. Some can also have a direct impact on working practices and behaviours, for example work pressure may mean workers do not adjust the workstation to suit themselves at the start of the shift or may forego rest breaks.

Individual differences

[E17049] For biological reasons some people may be more likely to develop a musculoskeletal injury. These include new employees who may need time to develop the necessary skills/rate of work; those returning from holiday or sickness absence; older/younger workers; new/expectant mothers; and those with particular health conditions. Account should also be taken of differences in body size which may require awkward postures/reaches.

Prevention of WRMSDs

[E17050] The greatest risk reduction benefits will be achieved by tackling both physical and psychosocial risk factors in the workplace. These factors are best identified and tackled through consultation with the workforce. Appropriate design of tasks, workstations and tools, as outlined in E17005–E17043, can help to prevent injury to the musculoskeletal system by ensuring work is designed to be within the physical capabilities of people. Tasks or equipment may need to be modified for those returning to work following an absence related to a work-related musculoskeletal disorder.

Ergonomics tools

[E17051] The following techniques can be used to facilitate thorough and comprehensive investigation of all parts of the work so that these can be designed appropriately.

Risk assessments

[E17052] Ergonomic risk assessments can identify the risk of injury, accident or reduced performance due to ergonomic deficiencies in work or tasks. Any assessment should be systematic and comprehensive, accounting for all elements of the work, including irregular activities. Checklists can be used to ensure all elements within the task are covered, namely the person, the work task, the workstation, the work environment and work system.

All hazards should be identified and recorded and an assessment of the risk posed by the hazard should be made. All hazards must be addressed, eliminated where possible or reduced as far as practicable.

The HSE has produced guidance risk assessment checklists for the following:

- Manual handling activities – A detailed and comprehensive risk assessment checklist is provided in the *Manual Handling Operations Regulations 1992 (as amended) (SI 1992 No 2793)*; the HSE's *Manual Handling Assessment Charts (MAC)* (INDG383) were produced as a filter for identifying high risk manual handling activities which should then be more fully assessed.
- Display Screen Equipment workstations – The *Health and Safety (Display Screen Equipment) Regulations 1992 (as amended) (SI 1992 No 2792)* include a risk assessment checklist for assessing display screen equipment workstations.
- Upper limb disorder risks – The HSE guidance *Upper limb disorders in the workplace (HS(G) 60 rev)* contains a thorough risk assessment checklist for considering the risks associated with tasks involving upper limb movements (excluding display screen equipment activities).

Several other checklists and assessment tools exist to facilitate in the assessment of working postures and risk, and include:

- the Rapid Upper Limb Assessment tool (RULA).
- See L McAtamney and E N Corlett 'RULA: A survey method for the investigation of work-related upper limb disorders' (1993) *Applied Ergonomics* 24(2) pp 1991–1999;
- the Quick Exposure Checklist (QEC).
  See G Li and P Buckle 'Evaluating change in exposure to risk for musculoskeletal disorders: A practical tool' (1999) HSE Contract Research Report 251/1999;
- the Rapid Entire Body Assessment tool (REBA).
  See S Hignett and L McAtamney 'Rapid Entire Body Assessment (REBA)' (2000) *Applied Ergonomics* 31, pp 201–205.

See also RISK ASSESSMENT.

Task analysis

[E17053] A task analysis provides information about the sequence and inter-relationship of activities undertaken in a task. It can be used to help in the evaluation of existing tasks, or the design of new ones, by identifying the demands on the individual, connections between activities and unplanned or unusual activities. From this tasks can be allocated to different individuals (or the decision may be taken to mechanise), difficulties with equipment, tasks, workstations can be identified and equipment and training needs can be identified.

Workflow analysis

[E17054] In a workflow analysis the movements of an individual at the workstation (e.g. to the equipment that is used) or within a work area are assessed. This helps to identify the pattern of activities and the routes within the workplace to ensure they are optimised. Equipment can be arranged to ensure that items used together or in sequence are placed appropriately.

Frequency analysis

[E17055] Information on the frequency with which tools and equipment are used can be important in positioning items or controls. The frequency of use of tools or adoption of postures can be recorded regularly using a frequency count.

User trials

[E17056] Mock ups of prototype equipment or workstations and trials of new furniture can be useful ways of assessing the suitability of new equipment. A representative sample of the user population should be used to evaluate the design. Rating scales can be used to allow people to score their opinion of a new item in terms of comfort, ease of use (clarity, adjustment etc.) and so on. Rating scales should have a mid-point (neutral); either 5 or 7 point rating scales are generally adequately discriminating.

Data collection

[E17057] Questionnaires and (formal or informal) interviews can also be useful means of collecting information and obtaining users' views of equipment, tasks or workplaces. Open (free response) or closed (series of options) questions can be used depending on the information that is required. Anonymity may be required in some situations (e.g. asking about health issues).

Discomfort surveys can be conducted to gain an understanding of the extent of any discomfort or disorders that are experienced by the workforce. In these, individuals report any discomfort they experience at the end of their shift according to the part of the body affected, and rate its severity (e.g. on a scale of 1–5). The responses of individuals from a work area can be collated and this combined data can be used to identify in which parts of the body operators are experiencing discomfort; it can also identify any trends in relation to the work tasks or work areas. This information can similarly be used as a baseline against which the benefit of any intervention can be evaluated. It is a useful tool to use in relation to risk assessment, as it can assist in the interpretation of risks and prioritisation of risk-reduction measures. An example of a body map is shown in figure 6.

Figure 6: Body Map

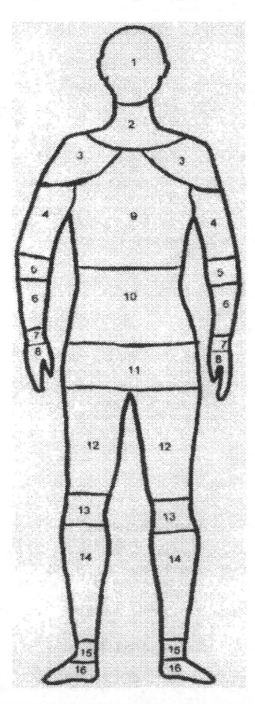

Alternative ways of using the body map are for groups of employees to apply stickers to a body map to indicate any discomfort experience, thus building up a pattern of where discomfort occurs.

Accurate assessment of tasks requires good observational skills and sufficient time to assess the task thoroughly. Observers should ensure that a representative sample of the workforce or tasks undertaken are observed. Irregularly undertaken tasks (e.g. maintenance) should not be overlooked in assessments. In some tasks movements may be rapid or only last for a short period. It can be useful, therefore, to collect video material of tasks to allow subsequent analysis.

Using existing data

[E17058] In some cases data will be available that may indicate ergonomic issues. Productivity, accident and ill health data may all be used to identify any particular issues and the extent of the problem. An accident book may also provide useful information on problems experienced. Data can be analysed by work area or type of injury in order to help prioritise areas for risk reduction measures.

Regular communication (e.g. through safety meetings) can also provide a useful route for identifying problem tasks or equipment.

Involving employees

[E17059] A number of people can contribute to the design of the workplace or task. These may include planners, procurers, human resources, as well as those who undertake the task itself. Involvement of the relevant people in any proposed change is important to ensure that it is acceptable. In particular, involvement of those who undertake the task in risk assessment, redesign and evaluation is a key element in ensuring that the task is thoroughly understood, any redesign is suitable and that any new equipment or changes are acceptable to the users. This can be achieved, for example, through questionnaires, discussions and trials of new equipment.

Relevant legislation and guidance

[E17060] Ergonomics is mentioned in several pieces of legislation, although there is no single, specific piece of legislation concerning this subject. Requirements for ergonomic design (explicit or implicit) in legislation are summarised below. Specific pieces of legislation are discussed in more detail elsewhere in this publication. Selected relevant HSE guidance is also summarised.

Health and Safety at Work etc Act 1974

[E17061] Under this Act employers have a general duty to ensure (as far as is reasonably practicable) the health, safety and welfare at work of their employees. This includes the provision of machinery and equipment that is without risks to health, the duty to keep workplaces in a safe condition and without risks to health, and to provide the information, instruction, training and supervision necessary to ensure employees' health and safety at work.

Management of Health and Safety at Work Regulations 1999 (SI 1999 No 3242)

[E17062] These Regulations require that any risks to health from work have to be assessed; employers must take appropriate preventative and protective measures, and employees must be informed of the risks and preventative measures taken.

Further information on these Regulations is contained in RISK ASSESSMENT.

Manual Handling Operations Regulations 1992 (as amended 2002) (SI 1992 No 2793)

[E17063] These Regulations take an ergonomic approach to handling tasks. They apply to all manual handling tasks and place a duty on employers to avoid hazardous manual handling activities as far as is reasonably practicable. Where this is not practical, employers should assess the risk of injury and then take appropriate steps to reduce the risk of injury as far as possible through appropriate design of the task, load and environment, taking account of the capabilities of the individual.

Further information on these Regulations is contained in MANUAL HANDLING.

Health and Safety (Display Screen Equipment) Regulations 1992 (as amended) (SI 1992 No 2792)

[E17064] Particular ergonomic issues relate to work at display screen equipment (DSE), namely VDUs. The *Health and Safety (Display Screen Equipment) Regulations 1992* set out particular duties on the employer to assess the health and safety risks associated with DSE use and to reduce the risks identified. Minor amendments were made to these regulations under the *Health and Safety (Miscellaneous Amendments) Regulations 2002 (SI 2002 No 2174)*.

The health risks particularly associated with DSE use include musculoskeletal disorders, eyestrain and stress. Specific requirements for the display screen equipment, workstation, software and task design are set out in the Schedule to these Regulations. These requirements ensure that it is possible to make certain adjustments to the equipment and furniture. It is advisable as part of the assessment to ensure that equipment is positioned in such a way for the user as to prevent awkward postures (e.g. twist to view screen) and that the adjustments are suitable for the user (e.g. seat height adjustment is adequate). The guidance to the Regulations includes a DSE risk assessment checklist. The full regulations and guidance are contained in the HSE publication *Work with display screen equipment*, L26, ISBN 978 0 7176 2582 6; This is aimed at large firms and health and safety professionals. A less technical summary (including the as-sessment checklist) is provided in *The law on VDUs: An easy guide*, HSG90, ISBN 978 0 7176 2602 1. This is aimed at small businesses.

The guidance was revised to include advice on the selection and use of the mouse/trackball and other pointing devices, speech interfaces and the use of laptops.

Laptops in prolonged use are subject to the provisions of the DSE Regulations.

Laptop computers pose a particular problem, as it is not possible to independently adjust the distance between the screen and the keyboard, and this may result in visual and musculoskeletal discomfort. Laptops may be used in environments that do not allow adjustment to the workstation (e.g. trains, hotel rooms). There are also manual handling issues related to the transport of laptops, and there can be concern over security and theft. These risks should be considered in a risk assessment. Good practice when issuing laptops includes:

- Provide a docking station for the laptop when office based. Alternatively, provide a separate keyboard and mouse, and view the laptop screen (preferable); or provide a separate screen and use the laptop keyboard. If viewed, the height of the laptop screen may need to be raised so it is at a suitable height.
- Provide information and training on adapting a work area to facilitate a comfortable posture (e.g. sitting in the passenger seat when working in a vehicle, using pillows to raise the height of the chair in a hotel).
- Encourage good working practices (e.g. regular breaks are particularly important because of the potential postural problems arising from laptop use).
- Provide a rucksack or suitable carrying case (e.g. wheeled), for easy transport of laptops. Rucksacks enable the weight to be carried close to the spine, reducing the amount of asymmetric loading and resultant stress on the spine and muscles that can occur if the bag is carried on one shoulder.
- Provide additional power supplies at regularly used locations, so these do not have to be carried.

If users are required to work at home, suitable furniture and equipment should be provided and a risk assessment should be undertaken to ensure the workstation is appropriate and that the user can achieve a comfortable posture.

Further guidance on laptop use is given in *Work with display screen equipment, Appendix 3*. Further information on the DSE Regulations is contained in Offices and shops O5005.

Provision and Use of Work Equipment Regulations 1998 (SI 1998 No 2932)

[E17065] The Regulations require employers to ensure that work equipment is suitable for the purpose and safe to use for the work so it does not pose any health and safety risk. General duties include:

When selecting work equipment, employers should take account of ergonomic risks . . . Operation of the equipment should not place undue strain on the user. Operators should not be expected to exert undue force or stretch or reach beyond their normal strength or physical reach limitation to carry out a task. This is particularly important for highly repetitive work.

Further information on these Regulations is contained in MACHINERY SAFETY M1014.

Personal Protective Equipment at Work Regulations 1992 (SI 1992 No 2966)

[E17066] These Regulations place a duty on employers to ensure that suitable personal protective equipment (PPE) is provided to employees who may be exposed to a risk to their health and safety while at work, in circumstances where such risks cannot be adequately controlled by other means. PPE should take into account the ergonomic requirements of the wearer and be capable of fitting them correctly.

PPE is designed to protect the wearer from a hazard; however, it may introduce other risks, such as reduced vision and hearing, which may restrict ability to detect warnings. PPE may also restrict movement or the person's ability to perform a task (e.g. through reduced dexterity when wearing gloves). Some chemical protective clothing (particularly water vapour impermeable garments) can contribute to heat strain, as the wearer has limited potential to evaporate sweat from the body. In selecting PPE the compatibility of different forms of PPE should be considered: it is often difficult to wear a hard hat with hearing defenders, or a chemical protective suit with a hard hat, although some integrated forms of PPE are available.

Reporting of Injuries, Diseases and Dangerous Occurrences Regulations 1995 (RIDDOR) (SI 1995 No 3163)

[E17067] Under the RIDDOR Regulations certain work-related accidents, diseases and dangerous occurrences have to be reported to the enforcing authorities. In terms of musculoskeletal disorders (which may arise due to poor ergonomics and inappropriate design), RIDDOR specifies that 'cramps in the hands or forearm due to repetitive movements' must be reported if they arise from 'work involving prolonged periods of handwriting, typing or other repetitive movements of the fingers, hands or arms'.

Further information on these Regulations is contained in ACCIDENT REPORTING A3002.

Upper limb disorders in the workplace (HS(G) 60 (rev))

[E17068] This revised guidance on upper limb disorders in the workplace provides a very useful and comprehensive approach to tackling these issues. It models a management approach, outlining seven stages in addressing these issues. These seven stages are:

(1)   Understand the issues and commit to action on ULDs.

(2)   Create the right organisational environment.
(3)   Assess the risks of ULDs in the workplace.
(4)   Reduce the risk of ULDs.
(5)   Educate and inform the workforce of ULDs.
(6)   Manage any episodes of ULDs.
(7)   Carry out regular checks on programme effectiveness.

These stages are not necessarily sequential and different stages may well interact. For instance, educating the workforce concerning ULDs may be part of creating the right organisational environment where these issues are taken seriously and tackled supportively.

The guidance also includes a two-stage risk assessment checklist. The first stage is a screening tool to help identify tasks where there may be a risk of injury and to assist in prioritising assessments. It contains five sections, asking a small number of questions concerning any signs and symptoms of ULDs, repetition, working postures, application of force and exposure to vibration. More detailed risk assessment worksheets are also included. These are divided into eight sections, concerning repetition, posture of the fingers, hands and wrist, posture of the arms and shoulders, posture of the head and neck, force, working environment, psychosocial factors, and individual differences. The risk assessment form encourages the identification of appropriate risk reduction measures for the risks identified.

The guidance also contains useful suggestions for reducing the risk of injury, medical aspects of ULDs and case studies illustrating how organisations have successfully tackled these issues.

Aching arms (or RSI) in small businesses (INDG171) is a free leaflet available from the HSE, aimed at reducing ULDs due to work activities other than those caused by using display screen equipment (DSE). It offers advice for identifying risk factors and gives practical ideas and tips for preventing ULDs.

Reducing error and influencing behaviour (HS(G) 48)

[E17069] This useful guidance promotes the consideration of human elements as a key factor in effective health and safety management. It provides practical advice on identifying, assessing and controlling risks arising from humans' interaction with the working environment.

The Disability Discrimination Act 1995

[E17070] Under the Act the employer has a duty not to discriminate against an individual either in the selection process or the workplace, and to make reasonable adjustments to enable a disabled employee to carry out their duties. These adjustments may include alterations to physical features within the workplace, assigning some duties to other employers and the provision of auxiliary aids and services. Taking account of the individual's abilities and needs when selecting equipment, furniture and designing the task and environment requires an ergonomic approach to ensure these suit the individual. The Act has been significantly amended since it came into force, including by the Disability Discrimination Act 2005.

Facilities Management – An Overview

Robert Greenfield

Introduction

[F3001] *Facilities management? – that's the department that moved us to our new premises – the building manager – the guy that negotiated the contract with the cleaning company – it's a helpdesk number that you ring if anything goes wrong.*

These are just a few examples of the perception that people have of facilities management and the services that may be provided in this relatively young but rapidly emerging industry. Facilities Management is now recognised by government, commerce and researchers, as providing a business edge within today's highly competitive world and is a highly developed strategic discipline which provides a well managed and utilised built environment at best value for the company. Apart from the day to day operational running of activities Facilities Management is an essential element of the high level corporate end of a business in the delivery an effective strategy. In the current economic climate Facilities Management can make or break an organisation and will no doubt move forwards in the future to gain even more recognition as the Facilities Management industry rises to the challenges posed in this challenging world.

The definition of facilities management

[F3002] The British Institute of Facilities Management has adopted an official definition of Facilities Management developed by Central European Normalisation (CEN) the European Committee for Standardisation and which has also been ratified by the British Standards Institute (BSI), as follows:

> "Facilities management is the integration of processes within an organisation to maintain and develop the agreed services which support and improve the effectiveness of its primary activities".

An important element of this definition is the integrated management approach. Even fifteen years ago there is no doubt that the sites and departments in an average company would most likely have been doing their own thing and managing themselves independently of each other with little cross pollination of best practice. Tasks and activities would have been duplicated due to the missing element of an integrated approach and this is what makes facilities management such an important element of an organisations overall strategy.

Facilities management

[F3003] The modern company has to support a diverse range of activities from its core business through to the provision of activities such as maintenance and cleaning, which enables a company to concentrate on its core business. However, the company may not have the expertise or the desire to manage such non-core functions and may not be supported by the most ideal workplace environment. Facilities management is able to effectively manage the non-core activities and use them to support the core business as well as adding value.

The built environment and all associated assets are a major expense for a company and is something which must suit the needs of the core business but also be flexible enough to accommodate change, especially in a dynamic work environment. Being such an expense it is important that all steps are taken to protect the assets of the organisation for the future and as a result facilities management may cover the management of a very broad spectrum of activities that could include the following:

- maintenance of plant equipment;
- property management;
- energy and environmental management;
- health and safety management;
- human resources management;
- sustainability;
- financial management;
- change management;
- domestic services e.g. cleaning, catering, security, postroom, and reprographics;
- utility procurement; and
- lifecycle costing.

An illustration of good facilities management can be seen in the following example.

We all use buildings on a daily basis and to some extent take them for granted, but have you ever stopped and thought about just what goes on behind the scene to provide the lighting, heating, cooling, ventilation, catering, postroom, reprographics, grounds maintenance and all the other functions that go to make your time within a building, as comfortable as possible? I thought not, but then who would and yet when you consider it we use buildings all the time. When we are comfortable and are able to undertake our intended work activity without problem then we do not give the environment within the building a second thought. It is not until the building or something within the building goes wrong that we can become uncomfortable and that's when we as humans experience some discomfort, and start to give our environment some thought.

Imagine the discomfort of not being able to open your windows properly at home in the middle of summer on an extremely hot day. Your whole family will suffer from the heat during the day, which will cause fatigue and possible heat stroke if you are not careful and they will not be able to sleep properly at night, which will lead to tiredness and a lack of concentration the following day and so on. Imagine equating this up to a company employing a few hundred people in a building of modern design with no opening windows that relies upon mechanical ventilation and air-conditioning equipment that continually enables cooled air to be fed into the building. What if the ventilation or air-conditioning equipment behind the scenes, broke down? The staff would soon find themselves in a very uncomfortable hot and stuffy environment, which from the health and safety aspect, would rapidly become untenable and would inevitably lead to closure and loss of production. Imagine the cost to the company.

Hard and soft facilities management

[F3004] The activities associated with facilities management and the support of the core business is normally split into the following two discipline categories.

Hard FM – which includes the provision of all activities involved in the provision of maintenance services, refurbishment, minor and/or additional works of the actual structure of the built environment. In addition the provision of utilities could also fall into this category.

Soft FM – includes provision of all the remaining activities such as human resources, change management, health & safety, security, cleaning, catering, help desk, postroom, reprographics and reception duties.

Managing

[F3005] Changes in society and within the workplace have seen the previous ten years experience major upheaval. The technology available to all companies is now staggering and affects staff in a variety of ways. Modern buildings are designed and operated in a very high-tech manner which makes full use of a vast amount of computer soft and hard ware. The same applies for manufacturing processes, which tend to be very automated and which rely on a massive amount of unseen infrastructure to support this process. The same is true for people management, due to the extensive range of well tested management techniques. During this period of change, which has seen technology and management techniques grow, the concept of facilities management has also grown. Gone are the days of the caretaker running the old school boilers, ordering paper towels and a few cleaning products whilst running his team of cleaners. Today's companies demand an integrated approach to the provision of buildings, workplaces and their management, and this responsibility would fall to a specialist facilities management team, depending on the size of the task. Failure to embrace an effective facilities management strategy for some organisations could lead to failure especially in the current challenging global financial climate.

The person, or in the larger company, the team, responsible for the ventilation and air-conditioning equipment in the example above would be a facilities manager or team who could have been responsible for the design concept of the building, the selection and procurement of the ventilation and air-conditioning equipment, the maintenance and repair of the equipment and even the procurement of the energy actually powering the equipment. As can be seen facilities managers have a very difficult and responsible role, as the impact upon the business of a company, if they get it wrong, can be considerable. Most companies increasingly seem to outsource everything to a facilities management company, with the exception of their core business, which means that the facility manager is being expected to manage an even more diverse range of activities.

If a company is to place facilities management at the forefront of its strategy, then the lead facilities manager will be required to be highly skilled and, ideally, have a seat on the board of directors. If a board placement is not possible then there should be a communications line directly up to a strategic level. This is important as the company that has a facilities manager or a facilities management team will probably have embarked upon a high level facilities management strategy, which would have input to the management decisions regarding all aspects concerning the built environment.

Company approach to facilities management

[F3006] There are a number of ways in which a company could approach the introduction of facilities management and this could depend upon it size, location, types of buildings, core business and objectives, just as an example. It is important that any approach must be tailored to the individual company's needs.

It is important that the company is able to effectively distinguish between its core activities, e.g. production of engineering components or provision of a service and its non-core activities, e.g. security, cleaning or maintenance. If the decision is made within the facilities management strategy to outsource, then it is the non-core activities that should be managed in this way, allowing the company to concentrate their efforts on their core business.

In order to manage the facilities of a company effectively a facilities strategy needs to be developed. The company must decide where ideally it would like to position itself from the facilities management perspective and then decide how it is going to get there. Generally it is good practice to set a clear mission statement together with objectives and targets to enable the development of the facilities management strategy. Scenario planning and business simulation may be useful in formulating a strategy. In this technique a series of 'what if?' questions are applied to produce a table of consequences enabling the organisation to form clearer decisions.

When determining the facilities management strategy it is essential to build in flexibility and to have alternatives, should problems be encountered along the way as survival could depend upon the ability of a company to react to change quickly.

However the strategy is developed, the key is to take a systematic approach.

Facilities strategy

[F3007] *Analysis* of the company's present and future needs. To look closely at objectives, policies, resources, processes, and procedures. It is important to know where the company stands in the market place and, if available, some idea as to its customer's perception. What sort of image is the company generating now and is this in alignment with its desired objectives? What needs to be changed? Where could improvements be made? A full understanding of how the company's resources and assets are being utilised needs to be made. It is important to understand all the business processes surrounding the company's core business to enable a facilities management strategist to establish where improvements can be made, so as to achieve cost savings, add value and increase efficiency by eliminating duplication.

Development of Solutions to make up the facilities management strategy can then be made by careful interpretation of the data and resulting information gained within the analysis. It is essential that solutions add flexibility, efficiency and are innovative e.g. by use of modern technology for communication of information. It can be useful to assemble a team who are capable of looking out side the 'box'. New ideas are crucial and whatever decisions are made they must be able to convince the senior management or board of the company that they will work, and ideally should be financially justified within any resulting report. Consider the use of a risk register to record the various scenarios and be sure to consider positives as well as negatives in this process. Any changes must ultimately be viewed as being cost effective and adding efficiency and value instead of being a cost burden.

Implementation of the solutions has to be carefully planned as without the use of effective 'change management' techniques will result in resistance to any changes being introduced and will probably result in staff problems taking up management time, thereby reducing efficiency. It is important to communicate with staff by way of group meetings and on a one-to-one basis to provide information and to allay any fears that they may have to proposed changes. Also listen to feedback from staff that may have some interesting ideas of their own and it is worthwhile engaging the actual staff undertaking the activities as they will be the ones who know the working practices. It is essential that the implementation plan is developed and includes time lines for each task with key milestones, performance measurement, proper ownership and account-ability . It is important to ensure that the implementation plan includes all activities and elements of the company that will be affected by the change.

A typical facilities management strategy should be capable of the effective delivery of the following:

(a)   in-house/outsourced services – catering, cleaning, maintenance, security etc.;
(b)   specifications for required accommodation standards (RAS);
(c)   procurement strategy;
(d)   IT strategy;
(e)   human resources strategy;
(f)   provision of change management;
(g)   service level agreements (SLA);
(h)   performance measurement;
(i)   financial planning;
(j)   accommodation/real estate strategy; and
(k)   business systems and procedures.

In-house or outsourced facilities management?

[F3008] The decision taken is dependent upon the facilities management strategy of each individual company and may even be a combination of the following two options.

In-house – this is the retention and integration of key staff within the company itself and with the expertise and management skills to enable the provision of effective facilities management services from within the organisation. This however may detract from the core business activities but may suit an individual organisations needs.

Outsourced – most companies consider that facilities management is such a specialist skill that they would prefer to outsource, as all the problems and issues of running their building are left with the total facilities management provider or a contractor, and allows the company to concentrate on its core business. The facilities company, who will in effect be a contractor, will have to be managed, but the added advantage is that facilities will be provided at a guaranteed price for the contract period which makes for easier budgeting. Consideration must also be given to the cost associated with the management of the facilities management contract to ensure that the agreed service is being delivered and that the service provider is complying with their health and safety duties.

The outsourced activities would be placed with a service provider, which would be one of the following:

- managing agent;
- maintenance company; or
- total FM provider.

Advantages to the company

[F3009] The advantages of using effective integrated facilities management within an organisation are vast and are essential to, in some cases, its very survival. Gone are the days where the various departments virtually managed themselves by doing their own thing using practices which had been handed down over the years. In such situations it was often found that tasks were carried out throughout the company with little or no communication, and certainly, no best practice. Today the facilities management department is responsible for advising a company's board and will most probably have an essential role in setting the overall business strategy for the future. With facilities management encompassing all of the Hard and Soft elements of facilities management of a large organisation such a transition provides a totally joined-up management approach towards all elements of the business.

The facilities management integrated approach can add tremendous value to a company by identifying needs in all areas and looking at ways to satisfy such needs at a strategic level by innovation. For example, a department may need to bring in a temporary worker for short-term data entry. However, when viewed at strategic level it may be the case that other departments have administrative pressures at certain times and are also bringing in temporary staff. A view could be taken to either increase the head count of staff and use the new member of staff as a float to be used filling administrative requirements, or if this were not possible, the facilities management team could agree on more time for the temporary worker to act as the float in various departments and could therefore negotiate a better price.

By giving consideration to the needs of the entire company for the supply of an item or a service and grouping the procurement of goods and services together will result in added purchasing power to obtain the best deal from the suppliers and manufacturers. This concept could be taken further by establishing when, for example, a manufacturer is at their quietist. If a company's production run can coincide with a manufacturing quiet period, a more competitive rate could be negotiated with the added advantage to the manufacturer of not having to close the plant temporarily or be forced to lay off staff, thereby potentially losing skilled labour. If consideration is given to the service industry then a client could decide that for example the maintenance of an asset could be undertaken in say a three month period by the contracted specialist and therefore the contractor could maintain the asset when they happen to be in the area thus enabling an even more competitive price to be achieved.

Main advantages

[F3010]

- **Cost certainty** – by inclusion within the facilities management contract of a detailed set of specifications and service level agreements, the desired services can be delivered at an agreed price aiding budgeting within the company. By careful identification of the needs of the company, specifications may be tailored and delivered in a more efficient way adding value.
- **Reduced cost** – by the integrated management of all services, more efficient procurement, utilisation of company assets and staff management, huge savings can be realised which will lead to increased profits and will ensure competitiveness and the long-term survival of the company. The claim that the facilities management department is an overhead, which perhaps costs a company more than it could ever save, can be refuted. It has been proven that good strategic facilities management can improve the efficiency of a company's systems, thereby reducing running costs and assisting in the delivery of the company core objectives. The use of Six Sigma and other management techniques can identify areas of duplication of activities within the organisation thus allowing even greater cost savings to be achieved.
- **Service delivery** – by careful facilities management contract negation and management of company expectations, services can be delivered to an agreed level which, if subject to the contract, may be measured periodically to ensure long-term delivery to the required standard. A system of continual improvement could form part of the initial contract and this could be effected by both parties taking a proactive approach towards developing innovations in their working practices.
- **Improved communications** – will ensue following the integrated management of all services through one department. Issues can be quickly resolved or brought to senior management for a decision. Having all information collated in one department will ensure wide-ranging, accurate and meaningful information can be utilised in the decision making process.

- **Utilisation of support services** – by outsourcing to a total FM provider the client organisation can take advantage of a range of support services already in situ within the FM provider for example call centre, Computer Aided Facilities Management (CAFM) software, health and safety and administrative support to name but a few.

Responsibilities of facilities management

[F3011] As demonstrated earlier, the skills expected of the facilities manager are diverse and are defined by the British Institute of Facilities Management (BIFM) competencies as 20 broad categories of know-how which have been broken down into numerous individual elements. An overview of a range of competencies is shown as follows.

Development of a maintenance policy

[F3012] Maintenance of any building or process plant is crucial as without this equipment the organisations's environment will cease to function efficiently. One aspect of facilities management is to protect the assets of a company and to ensure they are utilised to maximum effect. The facilities management department would be responsible for the development and implementation of the maintenance policy for the company. This will depend on the culture of the company, the type of work carried out by the company, the age of the building, and type of plant equipment. For example, it would not be advisable to adopt a breakdown maintenance policy if the company was reliant on one piece of very complicated plant equipment which was business critical. Therefore, the development and implementation of the maintenance policy for the company is important in the provision of services to enable the company to function, but could also save money in the long term through the by selection of the most suitable policy.

Before any decision is made on the selection of a maintenance policy, each asset must be entered in an asset register and a condition survey completed. This will provide the facilities manager with information, which can be used when considering the policy to adopt. The type of company and business processes conducted will be taken into account during this decision making process, as will the available budget.

Companies always try to cut costs and maintenance is often seen as an area costing too much that a few more per cent taken off will hardly make any difference. It is for the facilities manager to demonstrate that cost cutting in this area could be counterproductive, leading to unreliability of plant equipment and general dilapidation.

It is important to make the correct choice of maintenance policy with current and future trends being considered, together with the long-term effects that a particular maintenance policy could have upon the company's assets and future capital expenditure. By adopting a reduced maintenance policy now, of say, reduced frequency, or perhaps even a breakdown policy, which will no doubt have a detrimental effect upon the assets in the future and will perhaps reduce the life expectancy of the asset. This will either result in expensive maintenance in the future or complete replacement and therefore capital outlay before the expected life of the asset.

There are various maintenance policies available and a combination of all could be implemented, as follows.

Planned preventative maintenance (PPM)

[F3013] In this policy, a maintenance schedule is developed for every item on the asset register and may include periodic inspection of the condition of assets, items of plant equipment being adjusted, and filters replaced or cleaned as appropriate. This is initially the most expensive form of maintenance, but will prevent unexpected breakdowns, will ensure that equipment runs efficiently and will prolong equipment life. This policy would normally be used to maintain any 'business critical' assets within the company.

Planned corrective maintenance

[F3014] This is where the assets are inspected periodically but are not touched in anyway unless they are running very inefficiently or having an impact upon the running of other assets, or perhaps where there is a high risk of the asset failing in the near future. This maintenance policy is cheaper than planned preventative maintenance but will make it difficult to predict breakdown and an asset could run so far outside its design tolerance that long-term damage is inflicted, thus reducing the life expectancy of the asset. This is very much a hands off approach.

Breakdown maintenance

[F3015] Assets are neglected and only touched when they fail. This is therefore purely a reactive system and by far the least expensive policy in the short term, but it's use makes it difficult to predict breakdowns. This policy does not protect assets for the future; dilapidation of equipment could prove costly.

Before deciding on the maintenance policy to be adopted the main aims and objectives of maintenance within the company need to be considered, some of which could include the following:

(1)   upkeep of the company's image;
(2)   to enable the built environment to fulfil its function;
(3)   compliance with legislation to provide a safe and healthy environment;
(4)   ensure plant runs efficiently to provide a comfortable environment;
(5)   minimum cost rather than reliability;
(6)   reliability of plant equipment; and
(7)   creation of a forward maintenance register to allow funds to be amortised for lifecycle replacement of equipment.

Reliability Centered maintenance (RCM)

[F3015A] This is where assets are grouped into categories based upon the impact on the business if a particular asset were to fail. All assets that are operational or business critical or required for statutory compliance are maintained to manufacturer's specification or higher whilst other non essential assets are run to failure. The advantage is that available budget is focused towards all of the critical assets to achieve 100% up-time and greater resilience whilst achieving overall cost savings.

Life cycle costing

[F3016] It is important to consider the building environment, it's condition and likely maintenance requirements for the future. The facilities management team can assist the company by using life cycle costing, also known as whole life costing. Life cycle costing provides the facilities manager with a cost profile which can be broken down into three basic elements:

• acquisition cost;
• operational cost; and
• disposal cost.

This technique may be applied to services as well as actual assets.

All components, whether they are the outer protective layer of a building, a heating pump, or even a light bulb, will have been designed with a life expectancy which the manufacturers will quote. Therefore, a whole building or section may be designed to provide a certain number of years' service before requiring total replacement, subject to the manufacturer's ongoing maintenance being completed throughout its expected life. The cost element will also be considered as a cheaper product may have to be included in the design which will require replacement earlier. However, the ongoing maintenance costs may be far higher than a more expensive product and this will also have to be taken into consideration. This highlights the fine balance a designer will have make to achieve the client's requirements and deliver within a prescribed budget. The compromise is that the client has to accept that although their initial capital expenditure is reduced at the construction phase, at some point in the future there will be an ongoing maintenance cost and further expenditure to replace the component.

In the case of some items of equipment such as a boiler its life expectancy could be 25 years however after say 15 years its efficiency may have reduced significantly and maintenance costs increased due to breakdowns. In this scenario there will come a time when a business case could be developed by the facilities management team to replace the boiler before the manufacturers stated life expectancy. Although there will be an initial capital expenditure over a period of time the cost savings gained through greater efficiency and reduced maintenance costs would far outweigh this expenditure.

There are a number of computer software programs available to assist a facilities management team to create an asset register for the building. Such programs may include everything within the building from the very fabric of the construction to the fixed items of plant equipment, general office furniture and carpets. By inputting full details and the condition of each asset a lifecycle capital expenditure programme can be produced to provide information indicating when assets should be replaced and guidance costs to assist in budgeting.

Customer service

[F3017] One of the difficulties encountered when services are outsourced is the management of customer expectations which comes from the desire to have everything work all the time, or for items to be delivered exactly on time. A key driver in the decision making process to outsource, may have been to save costs and it may be that the customer (e.g. the client company staff) could be expecting a gold service when in fact the company is only now paying for a silver service. It is essential to manager customer expectations and to avoid confusion it is advisable to ensure that staff within the client company are provided with clear information as to the level of service that they should expect.

The facilities manager will be able to measure customer service by use of metrics which should be agreed within the initial contract. Such measures are normally referred to as Key Performance Indicators (KPI's) and will provide an ongoing indication of performance.

The KPI's which could be utilised for a typical facilities management help desk may be defined under the following three headings.

Timeliness – needs to be defined in a very precise way, but most probably that the help desk telephone is answered within five rings or perhaps within 15 seconds.

Appropriateness – when a help desk operator answers a call, was the customer's request understood or was the most appropriate action taken? Was the customer's expectation completely met or did the operator fail to establish a solution to the problem?

Accuracy – most probably the only acceptable standard in this category is 100%. In the example above, did the help desk operator enter the correct information into the system or were there inaccuracies that affected the resolution of the customer's problem?

The measuring and collating of this information is important to the facilities manager because it confirms whether or not the appropriate service is being delivered. With continuous performance measurement in this area it can be ensured that the service is delivered at the agreed level. This may be taken a stage further by publishing the service measures, or KPI's, to demonstrate how the help desk is performing and to encourage staff to provide suggestions on how service delivery may be improved. The performance may also be compared with those of help desks in other companies, provided the KPI measures are the same.

Service delivery is a key driver within the facilities management industry and is taken very seriously. In the example above, it would be expected that a department would be responsible for the completion of customer feedback surveys the results of which is then collated and measured.

Environmental and energy management

[F3018] Much emphasis is being placed on environmental protection and everybody has a duty to ensure that they use energy as efficiently as possible by carrying out basic tasks such as switching off lighting in areas that are not in use or lowering the temperature on a thermostat by a degree or so. Simple things that can be easily performed in everyday life, whether at home or in the workplace, and yet they can make such a difference. Expand this to all of the staff and processes contained within an average working environment and it can then be demonstrated that a company can make a huge difference. Energy consumption within a company burns fossil fuels, which adds to the greenhouse gas effect, which in turn creates global warming by retaining heat within the earth's atmosphere. Organic waste dumped in landfill sites also produces greenhouse gasses, which again adds to the problem.

Facilities management can apply their skills not only to comply with legislation concerning environmental issues, but also to save the company costs on energy bills. A number of companies have also developed a corporate and social responsibility strategy which would probably be part of the facilities management function to ensure that there is compliance with this as well.

The facilities management department would be able to assist energy management within a company at both a strategic and operational level. Typically, part of the environmental strategy would be for a company to agree upon its energy management strategy, which would normally consist of the following.

Strategic

- policy statement;
- objectives;
- roles and responsibilities;
- process of review;
- arrangements for the following:
    - heating – space temperatures, heating periods, and
    - cooling – air conditioning policy – natural/mechanical ventilation.

It is important that all staff are made aware of the environmental and energy policy, together with full details of the practical ways in which staff can really make a difference. For the policy to be effective a clear message must be provided from the top down in the company, with directors and senior managers setting a good example and being supported by the facilities management team whose overall responsibility would be to manage energy issues on a day-to-day basis, as follows.

Operational

- housekeeping issues;

Operational

- switch off lighting,
- doors and windows closed in winter,
- switch off equipment not being used,
- use water wisely,
- wear sensible clothing for weather conditions,
- engineering issues;
    - install or replace existing lamps with energy efficient equipment,
    - ensure external lighting controlled by photo cell or time clock,
    - installation of waterless urinals in gents' toilets,
    - installation of internal lighting control by motion detectors (PIR),
    - use of improved ventilation rather than air-conditioning, and
    - installation of window film and or blinds to control solar gain.

The facilities management team would have a direct effect on the environmental impact of the company and would be able to reduce any adverse impact by careful management of the following service functions:

(a)   products are procured from suppliers with environmental policies;
(b)   all wood used is from sustainably managed sources;
(c)   where possible, water is recycled in production or other processes;
(d)   only use paints with a low volume of volatile organic compounds (VOC's);
(e)   procure products made from recycled materials;
(f)   recycle paper, cardboard, toner and printer cartridges, and mobile phones;
(g)   manage ordering of marketing publications and leaflets to reduce waste;
(h)   ensure that waste from refurbishment projects is recycled; and
(i)   procure energy from renewable energy sources.

These are some of the activities that could be undertaken by a skilled facilities management team, all of which will ensure compliance with current legislation and above all will realise considerable savings in the company's energy consumption and resulting expenditure on energy bills.

Sustainability

[F3018A] Sustainability within an organisation is a commitment to make ethical business decisions that benefit the staff, environment and society in general and will encourage innovative working. The facilities manager will need to be aware of how to develop and implement sustainable strategy within the organisation as there are more and more reasons for organisations to go down this path. Case studies have proven time and again that true sustainable development can not only reduce wastage and increase profit but will also contribute towards compliance with impending legislation such as the Carbon Reduction Commitment. Although there are many definitions of sustainability and sustainable development most probably the simplest was developed by the World commission on Environment and development which states the following:

'Sustainable development meets the needs of the present without compromising the ability of future generations to meet their own needs.'

Some typical keys drivers for sustainability are as follows:

- Legislation
- Customers
- Economic value of the business is enhanced
- Staff motivation
- Shareholders demands and expectations
- Longevity of the organisation
- Financial stability

There are a number of business models that may assist the organisation towards a strategy for sustainable development such as The Five Capitals model, as follows:

(1) Natural – Purchasing of green/renewable energy or reduction in carbon emissions
(2) Human – Development of staff or recruitment of apprentices/trainees
(3) Social – Staff projects in the local community
(4) Manufactured – Reduction of waste to landfill
(5) Financial – Low risk financial decisions or sales and profit growth

An organisation can truly claim to be a sustainable organisation when it is living off the interest and not just the capital of the world.

Acquisition and disposal of buildings

[F3019] There are very few companies with buildings in their estate who could say that their buildings are completely suitable for their operation. Compromises are often made as companies continually change and even if a building was designed specifically for a company, and even if there was a certain amount of flexibility built into the design to cope with future needs, there is no doubt that by the time the building was finally ready for occupation, the company would have changed.

The facilities management team can use their skills in advising a company on the acquisition and disposal of buildings by taking an integrated approach, in which they would consider the company's objectives and current and future needs. The team may also advise on the suitability of a building the value of which may be severly affected by issues such as poor energy efficiency or the presence of asbestos.

If the organisation for example required more space in which to conduct its operations, then areas for consideration could be made, as follows:

- Is this temporary or permanent?
- What has created the need?
- Can some staff be allowed to work from home with a laptop?
- Is it possible to move a department into a satellite office?
- Is it financially viable to take short-term space nearby?
- Should the entire company relocate?
- What is the current lease situation on the present building?
- Is there a lease break or could the building be sub-let?
- What location would suit the company?
- Cost of any necessary refurbishment to new location?
- Maintenance and running costs?

There are so many areas to be considered that usually three options would be selected, each providing a range of advantages and disadvantages, although ultimately, senior management would have to compromise.

Similar skills would have to be applied to the disposal of a building. It could be that only part of a building is no longer required, so consideration could be given to establish whether that particular area could be sub-let, or if there is a lack of a lease break, then it may be feasible to sub-let the entire building. If the building is no longer required but has to be retained due to the lease, then consideration has to be given to reducing building services to an absolute minimum whilst preserving the assets for future final disposal.

Space planning

[F3020] Further to the acquisition and disposal of buildings (at **F3019** above) it can also be demonstrated that, due to the constantly changing company, space within a building will not be utilised to its full potential and facilities management can be called upon to find an appropriate solution.

A building occupied by an organisation is its largest overhead apart from the cost of the staff and thus if the facilities manager is able to reduce overheads then profits will be increased, thus ensuring survival in the marketplace. If space planning is undertaken effectively then the company will further benefit from improved production created by better flow of work and processes through the building. Staff will be working more efficiently as they will have a better environment in which to work and with better management control, defining space requirements and reducing clutter by good housekeeping. A well-planned and managed workplace will enhance the company's image, which will provide a good impression to potential clients visiting the premises.

Project management

[F3021] A company will often have many projects running concurrently, such as the development and introduction of a new information technology system, to the construction of an extension of a building. In most companies a facilities manager would normally manage small to medium sized projects, which would place further demands on the skills base required and workload. Often, it is unclear how to define what a project actually is. Each project will have its own unique set of co-ordinated activities being undertaken by an individual or a team brought together for this specific purpose. There will be defined start and finish points with objectives, time constraints and pre-set cost and performance parameters.

It is most likely that any construction type of project undertaken by the facilities management team will be governed by The Construction (Design & Management) Regulations 2007 known as CDM 2007. From a health and safety aspect when undertaking a project then the facilities manager is advised to adhere to the specific duties of CDM 2007 placed upon clients, designers and contractors to rethink their approach to health and safety. It should be taken into account and then co-ordinated and managed effectively throughout all stages of the construction project and is applicable to all elements from conception, design and planning through to the execution of works on site and subsequent maintenance and repair, and even to final demolition and removal.

The key aim of CDM 2007 is to integrate health and safety into the management of the project and to encourage everyone involved to work together to:

- Identify hazards and risks as early as possible
- Focus efforts where it can do most good from the safety perspective
- Improve planning and management
- Discourage unnecessary bureaucracy and paperwork

A facilities manager would be aware, from the media, of the many larger projects that have failed elsewhere, so facilities managers must use their experience to prevent any similar failure from occurring by ensuring that the following criteria are managed at the outset of the project:

(1) agreeing a detailed project specification;
(2) timescales set are realistic;
(3) selection of appropriate staff in the team;
(4) management of customer expectations; and
(5) management of change when a project is implemented.

Successful projects are those that have the intended project sanctioned at a senior level within a company, containing a detailed project brief outlining the desired objective. It is important that time is spent defining a project brief, as this can save much effort and frustration at a later stage.

A typical project brief could include the following:

(a) goals and objectives – how will the project be deemed as being successful;
(b) scope of the project – remember to include areas out of scope;
(c) risks that will affect the project; and
(d) assumptions made.

Usually, a facilities manager will have to prepare a business case to put to senior management in order to obtain the desired budget. Part of the document would have to define cost, time limitations and benefits as well as operational risks of undertaking, or not undertaking the project.

During the project management of, say, a new IT system, the risks to a company in the changeover maybe considered too high until the new system has been proven. Both old and new systems could be run in parallel with the facilities management team analysing the performance of the new system until such a time as data is transferred completely and the old system discontinued.

Contract management

[F3022] This is a very complex skill, but one in which it is essential that the facilities management team succeed, as a poorly defined contract will lead to poor performance and eventually a complete breakdown in communications between the two parties.

The contract management process itself is one that enables both parties to agree a contract that enables them to deliver the objectives required. It also involves building a good relationship between the parties in order to be able to work proactively. It is unhelpful having a relationship built around an adversarial approach, as the aim is not only to achieve delivery of the agreed services or products, but also to achieve value for money by looking at innovations in delivery and in some cases aiming for continuous improvement.

There are two types of contract that a facilities management team would normally be expected to deal with, which are: service delivery and construction contracts.

Service delivery contracts are generally divided into three areas of:

(1)   service delivery management – to ensure that this is being delivered at the agreed level of performance;
(2)   relationship management – to ensure that the relationship between the parties is maintained; and
(3)   contract administration – to handle changes to the contract and to manage the more formal aspects.

The construction contract is very complex and the contract strategy will depend upon what suits the individual projects itself, and may be influenced by pricing and risk transfer. In this instance, the facilities manager will require much experience of the construction industry.

Change management

[F3023] Everything covered so far will involve an element of change. Humans tend to be creatures of habit and will often resist any changes, partly due to a fear or uncertainty of the unknown, created by a lack of knowledge of the intentions of the company and the reasoning behind the change.

It is important for the facilities management team to manage changes within a company, which is best carried out by effectively communicating with the staff affected. Such communication should be done at the earliest opportunity to reduce the risk of rumours escalating across the workforce.

The first stage in the process is to analyse what effect the intended changes will have on staff, so that a chart can be prepared to detail the likely resistance to change. Management can have in place a series of methods to alleviate any areas of resistance, so ensuring the effective transition.

The second stage will be to hold regular staff meetings to keep the workforce informed of the changes and the drivers behind such changes, e.g. the introduction of a new management information system for improved efficiency in reporting. It is most likely that the staff most affected will need to be met on a one-to-one basis.

The final stage before the go live date will be for the company to agree and produce new procedures that can be communicated to staff through training.

Four most common health and safety issues for facilities management

Asbestos

[F3024] Asbestos is likely to be a major project for a facilities manager to deal with, especially in older buildings of pre-1981 construction, which potentially could contain large quantities of asbestos containing materials (ACM's). Buildings of pre-1999 construction can contain limited amounts of ACM's, which have also been found in recent new constructions where old gaskets have been fitted in pipe work.

Although the *Control of Asbestos Regulations 2006* have replaced the *Control of Asbestos at Work Regulations 2002 (SI 2002 No 2675)*, they are still referred to as the duty to manage regulations, impose a number of requirements upon the 'duty holder' who could be for example the facilities manager, landlord or managing agent but has been defined as the party who has 'control' of the building or part of a building.

The duties imposed on the duty holder may be summarised as follows.

- It is essential that, unless already completed, a competent asbestos surveyor must take reasonable steps to locate Asbestos Containing Materials (ACM's) within a building and to assess their condition.
- There is a further requirement to presume that materials contain asbestos unless there is strong evidence to prove that they do not. Instances where it could be presumed that there could be ACM's present is in areas where access was not gained, such as in lift shafts (fire compartmentalisation/suppression), locked areas, electrical switch gear (electrical flash protection) or in areas where samples are unable to be taken such as in listed buildings, where such damage would be prohibited.
- The findings of the asbestos survey must be maintained in a written record indicating the location and condition of known and presumed ACM's. This would normally be in the form of an asbestos register which should indicate the areas where known ACM's were located, together with non-asbestos samples and areas of no-access. The findings, ideally, would also be supported by the use of photographs and annotated drawings indicating locations where samples were taken.

- The final requirement is to produce an active asbestos management plan which must be kept up to date at all times. This plan should indicate how known asbestos is to be managed and how you as the facilities manager is going to prevent exposure to the ACM e.g. removal, repair and encapsulation and periodic monitoring as well as persons responsible for management, emergency procedures and a record of all activities undertaken on the ACM's.

It is essential that ACM's are managed safely for the protection of all persons likely to come into contact with them and there is no doubt that this task would fall to the facilities management team.

Work at height

[F3025] Facilities managers will have to pay considerable attention to working at height. Working at height may appear to be relatively simple, from changing a lamp using a small pair of stepladders, to a major refurbishment project being undertaken from scaffolding. However, in each of these examples the inherent risk of falling from height may be high unless control measures are in place to mitigate such risk. It is important that the facilities manager is aware of the risks and is able to ensure that persons working at height are working in a safe manner and that they not only protect themselves, but also others who may be nearby.

The new *Work at Height Regulations 2005 (SI 2005 No 735)* came into force on 6 April 2005 and apply to all work at height where there is a risk of a fall liable to cause personal injury. The old two-metre rule has been replaced by the new regulations which apply to all heights where there is a risk of personal harm. As in the *Control of Asbestos Regulations 2006* there are duties placed on dutyholders who are defined as employers, the self-employed and any person who controls the work of others e.g. a facilities manager.

Duty holders are required to ensure the following.

(a) Avoid work at height if at all possible.
(b) All work at height is planned and organised in a systematic way. Consider how long the work will take and if it can be successfully carried out with one hand from a ladder. Is there some other activity in the area that may affect the work?
(c) All persons working at height are competent. Ensure that if contractors are carrying out the work that their health and safety has been evaluated and that the operatives have been suitably trained.
(d) Risks from working at height have been assessed and that suitable control measures and work equipment are in place before work commences.
(e) Risks from working on fragile surfaces have been assessed and are properly controlled. Consider fitting moving bridges and gantries to span fragile surfaces requiring some form of maintenance.
(f) All equipment used for working at height is marked with a serial number or colour coding and is properly maintained and inspected. Operatives should also undertake their own inspection prior to using the equipment.

The facilities manager must ensure that there is an effective safe system of work in place such as a permit to work system for evaluating the competency of operatives, checking their risk assessments and method statements, checking that equipment to be used is suitable and has been maintained. Once work has commenced the facilities manager should supervise the work to ensure that the operatives are carrying out the work in the manner described in the risk assessment/method statement and that there are no other unforeseen hazards.

Water quality

[F3026] In the UK between 30 and 50 deaths are reported per annum, as being as a result of legionnaires' disease and these have been widely reported in the media over recent years, in spite of the introduction of the Code of Practice 'The Prevention or Control of Legionellosis' (including legionnaires' disease) L8 rev and its associated guidance document HS(G) 70. These deaths have resulted from poor maintenance of cooling towers and evaporative condensers and hot and cold water systems in buildings.

The legionella bacteria requires a number of components to proliferate such as Iron and L cysteine which is present in water, as well as the right temperature. Legionella grows at between 20 and 50 degrees centigrade and its optimum temperature seems to be 38 degrees centigrade. Therefore if the facilities manager can maintain the temperature of the hot and cold water within the building outside of these temperatures then this will help prevent to prevent the growth of the bacteria. In addition legionella will also build up within water outlets such as showers that are seldom used as well as dead legs in water pipework that are still connected to the system.

An essential component of the above L8 code of practice is to ensure that a water risk assessment (WRA) is undertaken for the company's building and to make recommendation to reduce the risk to the lowest possible. Within this document there will be an organisational chart indicating persons responsible for ensuring that water services are properly maintained. In a small to medium sized company the facilities manager will have responsibility for ensuring that all recommendation in the WRA are completed and that maintenance of the water system is carried out in accordance with the maintenance scheme shown in the WRA.

The following should be considered when dealing with the water services of a building:

(1)   ensure that a WRA has been carried out by a competent person;
(2)   ensure that all recommendations made in the WRA are completed and records maintained;
(3)   ensure that the maintenance regime shown in the WRA is undertaken at the prescribed intervals by a competent person;
(4)   ensure that all water samples taken are tested by a UKAS accredited laboratory;
(5)   ensure that records are maintained for all maintenance and testing; and
(6)   ensure that the WRA is reviewed on a regular basis or after any alterations have been made to the water system.

If the facilities manager adheres in essence to the above guidance and has maintained good, accurate and auditable records then the incidence of legionalla should be avoided.

Management of contractors

[F3027] Any contractors working on site pose one of the most notable health and safety risks. Some senior managers share the mistaken belief that they are able to transfer any Health and Safety risk by contracting work out. In fact any such risk is shared because the company on whose premises work is undertaken also has a duty to others under Health and Safety at Work etc Act 1974 s 3(1) to ensure that persons not in their employ are working safely. This duty extends to contractors and their employees, visitors, customers, emergency services, neighbours, passers by and the general public at large. It is highly probable that the facilities manager will be the person responsible for employing such contractors and will have to manage them whilst they are on site.

Before employing contractors, facilities management should ensure that the contractors have been thoroughly vetted by way of a health and safety evaluation questionnaire. Such a questionnaire can be designed to verify their areas of competency with points such as accident record, insurance, procedures and risk assessment systems being a minimum. Some companies keep an 'approved contractors' list which shows all contractors who have been subjected to such an evaluation process. A word of warning, however, facilities management should check to ensure that the contractor has been approved for the type of skill required for the work activity to be undertaken. It could be that the contractor has been added to the approved list for some other activity.

Once facilities management is satisfied that the contractor is competent to undertake the allotted work activity, facilities management can allow them to conduct a site survey before holding a meeting to agree how the work should proceed. At this stage the contractor should be preparing their method statements and risk assessments while the facilities manager should inform the contractor of any known hazards connected with the site and should also provide rules or contractor's guidance notes. The next stage would be for the facilities manager to review the method statements and risk assessments to ensure that the work is to be undertaken safely and will not impact upon any other activity on site.

When the contractor arrives on site the facilities manager must be satisfied that the contractor has all the necessary control measures in place to protect not only the contractors but also company staff and any one else who may be affected by the work activity, such as members of the public or visitors. It is important that work is undertaken as agreed and is recorded in the method statements and risk assessments with the facilities manager providing as much supervision as necessary. Provided there are good communications between both parties and with good preparation, any work should proceed smoothly. However, unforeseen circumstances can prevent work progress, in which case it is important that any changes in work procedure must be agreed and the facilities manager must ensure that any such changes will still result in safe working.

How facilities management and building design can help a company

[F3028] The way in which facilities are managed within the work environment is important in creating the desired corporate image, whilst concurrently achieving its corporate objectives. It is pointless creating the desired corporate image if the efficiency of the company is undermined by bad building design or its location.

It is not uncommon to see many fine buildings, which are difficult to maintain with lights or air conditioning cassette units located high up in large atriums of buildings. In these instances the changing light bulbs or cleaning and replacing filters suddenly becomes a major part of the maintenance procedure as safe access becomes a more significant priority with, for example, the constructing scaffolding or mobile towers in order to undertake such a relatively simple task. It would be ideal if an experienced and skilled facilities manager were involved at the design stage, as they could consider the maintenance and operation of the building in a practical way. This situation has now improved as designers have duties, under the *Construction (Design and Management) Regulations 2007 (CDM 2007)*, where they have to consider the construction, maintenance and demolition of the building. However, it is still a problem in some areas.

A facilities manager is expected to be multi-skilled in a wide range of subjects, so that they are best able to advise a company how to create a safe and efficient workplace which provides a comfortable and positive atmosphere for all its staff.

Most visitors will quickly form a first impression of a company by noticing the building location, its design and its general condition. How a company's logo is displayed is all part of a carefully constructed corporate package. A skilled facilities manager with clever use of colours, furniture design and layout will display the company logo in such a way as to create the desired synergy and corporate image in order to help achieve the company's objectives.

Recognition of facilities management

[F3029] Facilities management of built environments is gaining greater recognition and credibility from a wide range of government departments, such as the Health and Safety Executive (HSE), the Office of the Deputy Prime Minister and the Sector Skills Council due, in part, to facility managers working in all types of companies who use a wide variety of skills. Consequently, facilities management is being recognised as an industry in its own right and facility managers are being used to provide feedback for government consultation documents, speak at conferences and attend innovation workshops. In fact, the facilities management industry has progressed to such an extent that it now employs more people than those working in the construction industry.

As mentioned in the introduction at the beginning of this chapter both the Central European Normalisation (CEN) and the British Standards Institute

(BSI) recognise facilities management as an important industry and in fact this has established to such an extent that the British Institute of Facilities Management (BIFM) in conjunction with the Institute of Leadership and Management has developed a new qualification at award level 3 in Facilities Management which is also recognised by the Qualifications and Curriculum Authority (QCA).

BIFM is working closely with the Office of the Deputy Prime Minister (ODPM) on a range of issues concerning energy and environmental management of the built environment and has produced a paper on Approved Document L2 of the *Building Regulations 2000 (SI 2000 No 2531)*. In addition, BIFM also has its own accredited qualification based around 20 core competencies, which further demonstrates the very diverse nature of facilities management.

Asset Skills, the Sector Skills Council for the places in which we live and work, has now created a facilities management sector board which is comprised of representatives from the facilities management industry and BIFM. The intention of the board is to:

- raise the standards of learning and professionalism;
- develop a National Vocational Qualification (NVQ) for facilities management;
- develop a research programme for facilities management (current research is being undertaken by Asset Skills Council, local councils and university estates management departments); and
- work closely with the facilities management industry and large companies to develop facilities management skills and best practice.

The Royal Institution of Chartered Surveyors (RICS) now has a facilities management faculty and a qualification of Chartered Facilities Management Surveyor. The institute is now actively undertaking research into best practice and is actively publishing facilities management articles on their website.

HSE recognise that the facilities management industry has considerable experience, retains broad skills and as a result holds regular update meetings with BIFM to discuss current trends and innovations. HSE is also contributing to a work at height guide designed for use by facilities managers. BIFM is also involved in the provision of feedback concerning government consultation documents by the collation of responses from specialists within its membership.

Industry is also recognising facilities management as an essential element of their strategy to ensure that they remain competitive in the market place. Increasingly, more companies are joining BIFM as corporate members and are actively involved in providing input into the institute and also to putting forward representation on committees across the United Kingdom. Some of the larger blue-chip companies run their own facilities management training and, through their considerable experience, are laying down the foundations of best practice, which in some cases are being adopted throughout the facilities management industry.

Universities are offering degrees in facilities management with a number of the more traditional degrees e.g. architecture and surveying including a number of modules consisting of facilities management disciplines. The same applies to health and safety qualifications where a number of course providers are delivering the requisite course content but are focusing this towards facilities management using practical examples to demonstrate the operational implications within the facilities environment.

Innovations within the facilities management industry

[F3030] New products and methods of working are being developed constantly as manufacturing companies have seen the opportunities to be realised within facilities management, particularly within:

- information technology software manufacture;
- new construction products and materials;
- cleaning;
- security;
- environmental; and
- energy management.

The development of telecommunications and software operating on the world-wide-web has provided an incredible choice of management systems to assist the facilities management industry. The large impact of such choice has led to the development of the 'virtual office'. A company has the opportunity to have staff based at home working from a laptop and phone line. Such working practices has advantages for both parties and can save the company costs in office space and staff equipment. Basically if an employee has a laptop, mobile telephone and a 3G card to link to the internet then within reason they are able to work from anywhere.

Travel can also be reduced these days through the use of Video and Teleconferencing as staff no longer have to travel vast distances to attend meetings. The same may be said of online interactive training courses that are widely available on the internet as a large volume of training is achievable in a short space of time at low cost and once again with zero travel requirements.

Within the cleaning industry, there have been considerable advances in cleaning products which are safe and easy to use and which have such a tolerance that they can be used by an unskilled operator and still provide a good finish. In other areas of cleaning, soap dispensers in toilets are being designed with large reservoirs to reduce the number of times they need to be refilled, helping reduce costs whilst dispensing soap over a longer period.

Floor cleaning equipment has progressed beyond large floor polishers with trailing leads and their inherent problems of trip, electrical and manual handling hazards. Modern machines tend to be smaller, lighter and are powered by maintenance-free batteries, thus eliminating the need for trailing cables. New flooring products manufactured from revolutionary materials are currently entering the marketplace and tend to be virtually maintenance-free. Flooring, especially in reception areas, can require much maintenance yet still look worn very quickly. Worn flooring may create a poor corporate image and as a result modern floor coverings are generally very durable and require little maintenance other than periodic vacuuming.

Technological advances have helped so-called 'intelligent buildings' realise much reduced running costs which have helped save energy and lower the environmental impact caused by the company. An intelligent building can detect when there is a lack of movement i.e. staff in an area, and can automatically switch off lighting and reduce the amount of heating and ventilation generated. Facilities and maintenance managers are being assisted by new hardware which can collate information from the central building management system (BMS) and transmit this information remotely to a handheld wireless computer such as a Blackberry, enabling facilities management real-time information. Recent developments in such building management software enable facilities management to pre-set parameters within the building and as soon as any readings extend outside those parameters a warning is sent to the Blackberry, warning the facilities manager, regardless of his location. This allows a facility manager to be proactive and take corrective action, hopefully before a problem arises.

The security industry has also seen much innovation with the introduction of new camera and software technology that can transmit pictures from remote, unmanned sites. The equipment runs in a hibernating mode until movement is detected, whereupon pictures and sound are transmitted back to a centrally manned site where they are then recorded. A mobile security patrol or the police may be called to deal with the situation. This equipment eliminates the need for a security guard to be exposed to lone working in a building and reduces security costs for the company.

Facilities management help desks have benefited from help desk software systems and the introduction of operative vehicle tracking modules, which not only show the location of the nearest operative, but their skills and competency. The helpdesk operative can locate and send the closest operative with the requisite skills for the work activity. Using wireless Blackberry hand-held computers jobs can be sent to operatives electronically and once completed the operative is then able to close the job down in the system remotely which means that this data within the help desk system is live.

The effects of safety, health and environmental legislation on facilities management

[**F3031**] A strong interrelationship exists between facilities management, safety, health and environmental (SHE) due to a large amount of compliance and legislation that is applicable to the built environment. Even just keeping abreast with ever changing legislation places a facilities manager under considerable pressure by not only having to acquaint themselves with new legislation, but by knowing also how to apply it to the practical aspect of how and when it will impact upon a company and the budget required for any remedial works, training and implementation. HSE has recognised that frequent changes in legislation has caused problems and are now due to release legislative changes twice-a-year in April and October.

Facilities management is becoming even more demanding in the field of SHE, as most small to medium sized companies will be unable to justify employing a SHE specialist or buying-in such advice from a consultant. A facilities manager is now expected to add health and safety to the ever-increasing range of facilities management skills. BIFM continually reviews its full range of facilities management competencies and the section concerning SHE legislation is forever expanding to include the working environment, environmental protection, energy efficiency, development of health and safety policy and risk management to name but a few subjects applicable.

As previously mentioned, building design is critical and if facilities management have some input at the design stage of a new build or refurbishment, then hopefully they can ensure that the building can be operated and maintained safely. A good, experienced facilities manager can make compromises when required and will take a proactive and flexible approach towards practical solutions to enable the safe operation of the building.

Every aspect of the facilities management role in the workplace, whether it is people management, or managing equipment, will at some stage be subject to an aspect of health and safety legislation. Staff within a company have to be managed to ensure that they have a safe and healthy place in which to work and that the actual way in which they work is also safe and will not endanger others. The facilities management role can influence the culture of a company in an effort to ensure that health and safety is viewed as a positive element of its business activities with benefits, rather than one that uses up valuable resources and impedes operations. It has often been noticed that a health and safety management system can generally identify operational issues before they become a problem, so helping to create a more efficient workplace.

Conversely, a building and all it's associated assets, will be subject to legislation throughout it's entire lifecycle, from design, through to its final demolition or disposal, thus making it essential that a facilities manager is kept fully aware of such legislation. This maybe demonstrated by consideration of the legislation and best practice compliance concerning for example a passenger lift within a building at F3032–F3034 below. However this three stage lifecycle model could be applied to any asset within the built environment.

Design

[F3032] *Construction (Design and Management) Regulations 2007 (CDM 2007) (SI 2007 No 320)* – place duties on the designer of the lift to ensure that the equipment is safe to construct, use, maintain and finally de-commission.

Provision and Use of Work Equipment Regulations 1998 (PUWER) (SI 1998 No 2306) – place duties on the manufacturers and suppliers to ensure that the equipment is suitable for the purpose for which is has been designed and manufactured, and is safe for use and maintenance.

Construction

[F3033] *Construction (Design and Management) Regulations 2007 (CDM 2007)* – place duties on the client and principal contractor to ensure that the construction work is undertaken in a safe manner and is a safe system of work.

There is also a requirement to keep a health and safety file, which is a record of all materials and equipment used and the methods of construction. This file must be kept in the building throughout its entire life as this contains valuable information.

Operation

[F3034] *Health and Safety at Work etc Act 1974* – places a duty on the owner/operator to ensure that the lift is safe.

Lifting Operations and Lifting Equipment Regulations 1998 (SI 1998 No 2307) – place duties upon owners and operators of lifts to ensure that they are subjected to preventative maintenance and also a thorough examination at regular intervals.

Construction (Design and Management) Regulations 2007 (CDM 2007) – ensure that the lift is demolished in a safe manner and information from the health and safety file would be used to assist in this process.

There are also environmental issues for a facilities manager to consider and increasingly more companies are developing and implementing environmental management systems (EMS) based upon the ISO 14001 model. Consequently, the suppliers to such companies are also expected to have an EMS.

Approved Document L2 of the *Building Regulations 2000 (SI 2000 No 2531)* deals with the conservation of fuel and power and therefore as a result has consequences for facilities managers as this requires a building logbook be maintained for all new or refurbished buildings. The logbook is designed to provide details of the plant equipment installed, it's controls and maintenance in order that energy consumption can be monitored and controlled. The object of such logbooks is to improve the efficiency of buildings by reducing energy costs and the emission of CO_2 gas into the atmosphere.

The Energy Performance of Buildings Directive is part of European legislation that is now being implemented by the use of Energy Performance Certificates and Display Energy Certificates. The construction of a building and the way in which it is insulated, heated and ventilated and the type of fuel used, all contributes to its energy consumption and carbon emissions and the facilities manager will no doubt be responsible in the day to day running of their portfolio, for the maintenance of this legislation.

The Energy Performance Certificate is one measure introduced to help improve the energy efficiency of our buildings. Other changes include requiring larger public buildings to display certificates showing the energy efficiency of the building and requiring inspections for air conditioning systems.

Please see below a summary of the implementation in England and Wales. Scotland and Northern Ireland have introduced their own regulations.

Energy Performance Certificates

From 1 October 2008 when a building is sold, built or rented an Energy Performance Certificate (EPC) is required.

The certificate provides energy efficiency A-G ratings and recommendations for improvement. The ratings - similar to those found on products such as fridges - are standard so the energy efficiency of one building can easily be compared with another building of a similar type.

Acting on an EPC is important to cut energy consumption, save money on bills and help to safeguard the environment.

EPCs were first introduced for the marketed sale of domestic homes, as part of the Home Information Pack. If you are buying or selling a home it is now law to have a certificate but from April 2008 this was extended to newly built homes and large commercial properties. Since 1 October 2008 when buildings are built, sold or rented, an EPC is required. This includes homes on the market before the phased introduction of EPCs for domestic properties in 2007.

It is essential for the facilities manager to ensure that EPCS are produced by accredited energy assessors.

Display Energy Certificates for public buildings

Since October 2008 Display Energy Certificates (DECs) are required for larger public buildings enabling everyone to see how energy efficient our public buildings are.

The DEC should be displayed at all times in a prominent place clearly visible to the public - and they are accompanied by an Advisory Report that lists cost effective measures to improve the energy rating of a building.

A DEC is valid for one year and the Advisory Report is valid for seven years.

Summary of facilities management

[F3035] Facilities management is gaining considerable recognition as an essential industry and efforts are being made to advertise facilities management within universities and colleges as a worthwhile and challenging career. It will be sometime before careers advisers are recommending facilities management to school leavers, but this should change in due course.

There is collaboration between BIFM in the UK and various facilities management institutes world-wide including Mainland Europe, the United States and Australia. These institutes have an agreement to share best practice and research, which will provide a powerful resource and will be invaluable to companies.

Future innovation and resultant advancements in facilities management will lead to improved efficiency, improved working conditions for staff, more added value and cost efficiencies for companies, reduced energy consumption and will reduce the impact to the environment.

Fire Prevention and Control

Adair Lewis and Andrea Oates

Elements of fire

[F5001] Fire is a chemical reaction resulting in heat and light. For a fire to occur the following need to be combined in the correct proportions:

- oxygen (supplied by the air around us which contains about 21 per cent oxygen);
- fuel (combustible or flammable substances either solids, liquids or gases);
- source of heat energy (ignition sources such as open flame, hot surfaces, overheated electrical components).

For a fire to burn the fuel must be in a gaseous or vapour form. This means that solids and some liquids such as oils will need to be heated up until sufficient vapour is given off to burn. The form that the fuel takes will also influence the ease with which it ignites and the speed with which it burns. Generally, finely divided materials are easier to ignite and burn more quickly than fuels in a solid form.

Once initiated, a fire will continue to burn as a result of the reinvestment of energy during the process provided sufficient fuel vapour, heat and oxygen are available and the process is not interrupted.

The 'triangle' of fire symbolises the fire process. Each side of the triangle represents fuel, heat or oxygen. Take any one side away and the fire will be extinguished or more importantly, stop all three sides from coming together and a fire can be prevented.

The first part of this chapter (F5001–F5029) considers generally the practical aspects of fire prevention and control. It should be borne in mind that *special risks* involving flammable or toxic liquids, metal fires or other hazards should be separately assessed for loss prevention and control techniques. For precautions against fire hazards from flammable liquids, see the Health and Safety Executive (HSE) Guidance documents HSG 51, HSG 140, HSG 176 and HSG 178. Control of these hazards is among the areas dealt with under the *Dangerous Substances and Explosives Atmospheres Regulations 2002 (SI 2002 No 2776)*.

It is the responsibility of management to consider how safe is safe: that is, to balance the costs of improvement against the financial consequences of fire. Considerable improvement can often be made immediately at little or no cost. Other recommendations which may require a financial appraisal must be related to loss effect values. In certain cases, however, due to high loss effect, special protection may be needed almost regardless of cost.

Modern developments in fire prevention and protection can now provide a solution to most risk management problems within economic acceptability. It must be pointed out, however, that it is a waste of time and money installing protective equipment unless it is designed to be functional, the purpose of such equipment is understood and accepted by all personnel and the equipment is adequately inspected and maintained. The reasons for providing such equipment should, therefore, be fully covered in any fire-safety training course. Fire routines should also be amended as necessary to ensure that full advantage is taken of any new measures implemented.

Common causes of fires

[F5002] The following, in no particular order of significance, are the most common causes of fire in the workplace:

- wilful fire raising and arson;
- careless disposal of cigarettes and matches;
- combustible material left near to sources of heat;
- accumulation of easily ignitable rubbish or paper;
- carelessness on the part of contractors or maintenance workers, usually involving hot work;
- electrical equipment left on inadvertently when not in use;
- misuse of portable heaters;
- obstructing ventilation of heaters, machinery or office equipment;
- faulty electrical appliances; and
- inadequate supervision of cooking activities.

Fire classification

[F5003] There are six classes of fire which are related to the fuel involved and the method of extinction, as follows.

- Class A
  This relates to fires generally involving solid organic materials, such as coal, wood, paper and natural fibres, in which the combustion takes place with the formation of glowing embers. Extinction is achieved through the application of water in jet or spray form. Water extinguishes the fire by removing or limiting the heat by cooling.

- Class B
  This relates to fires involving:
  (i)   liquids, which can be separated into those liquids which mix with water, e.g. acetone, acetic acid and methanol; and those which do not mix with water, e.g. waxes, fats, petrol and hydrocarbon solvents; and
  (ii) liquefiable solids, e.g. animal fats, solid waxes, certain plastics.

Foam, carbon dioxide (CO_2) and dry powder can be used on all these types of fire. However, some types of foam break down on contact with water-miscible liquids and thus special alcohol-resistant foam is needed for large volumes of such liquids. Water must not be used on fats, petrol, etc. With foam, carbon dioxide and dry powder, extinction is principally achieved by limiting or removing the oxygen by smothering.

- **Class C**
  This relates to fires involving flammable gases and should only be extinguished by shutting off the gas supply if it is safe to do so. Burning gas should not be extinguished as to do so without isolating the supply may result in a build-up of unburnt gas which might explode.

- **Class D**
  This relates to fires involving combustible metals, such as aluminium or magnesium. These fires burn with very high temperatures, and their extinction can only be achieved by the use of special powders. These powders form a crust over the surface isolating the burning metal from the surroundings. It is extremely dangerous to use water on burning metals. Special training is required where the use of extinguishers on combustible metals is concerned.

- **Class E**
  This relates to electrical fires.

- **Class F**
  This relates to fires involving cooking oils and fats. Extinction is achieved primarily by limiting or removing the oxygen by smothering and in some cases by a degree of cooling. It is extremely dangerous to use water on fires involving cooking oils and fats.BS EN 3–7:2004+A1:2007 specifies the characteristics, performance requirements and test methods for portable fire extinguishers. It is considered hazardous for powder and carbon dioxide fire extinguishers to be used on Class F fires. For this reason powder and carbon dioxide fire extinguishers are excluded for conformance with regard to Class F in BS EN 3–7.

The table below classifies fires which can be controlled by portable fire appliances (see also BS EN 2:1992 Classification of fires and BS EN 3–9:2006 Portable fire extinguishers. Additional requirements to EN 3–7 for pressure resistance of CO_2 extinguishers).

SUITABILITY OF EXTINGUISHING TYPES

Class of fire	Description	Appropriate extinguisher
A	Solid materials, usually organic, with glowing embers	Water (foam, dry powder or CO2 will work but may be less effective than water due to a lack of cooling properties).
B	Liquids and liquefiable solids:	Foam, CO2, dry powder
	• miscible with water e.g. ac-etone, methanol	Alcohol-resistant foam, CO2, dry powder and the skilled use of water spray

Class of fire	Description	Appropriate extinguisher
	• immiscible with water e.g. petrol, benzene, fats, waxes	Foam, dry powder, CO2
F	• cooking oils and fats	Type F, special wet chemical extinguisher

Electrical fires

[F5004] Fires involving electrical apparatus must always be tackled by first isolating the electricity supply and then by the use of carbon dioxide or dry powder, both of which are non-conducting extinguishing mediums. Dry powder and carbon dioxide extinguish the fire by limiting or removing the oxygen by smothering.

EU Regulations, which apply in all EU member countries including the UK, prohibited the use of halon for extinguishing fires in most applications after 31 December 2003. Portable extinguishers and fixed systems using halon should no longer be available for use.

Fire extinction – active fire protection measures

[F5005] Extinction of a fire is achieved by one or more of the following:

*   *Starvation* – this is achieved through a reduction in the concentration of the fuel. It can be effected by:
    (i)   removing the fuel from the fire;
    (ii)   isolating the fire from the fuel source; and
    (iii)   reducing the bulk or quantity of fuel present.
*   *Smothering* – this brings about a reduction in the concentration of oxygen available to support combustion. It is achieved by preventing the inward flow of more oxygen to the fire, or by adding an inert gas to the burning mixture.
*   *Cooling* – this is the most common means of fire-fighting, using water. The addition of water to a fire results in vaporisation of some of the water to steam, which means that a substantial proportion of the heat is not being returned to the fuel to maintain combustion. Eventually, insufficient heat is added to the fuel and continuous ignition ceases. Water in spray or mist form is more efficient for this purpose as the droplets absorb heat more rapidly than water in the form of a jet.

Property risk

[F5006] Fire safety precautions needed for the protection of life are the subject of legislation, as detailed in **F5030–F5071**.

In order to ensure the survival of a business in the event of fire, property protection must also be considered and a business fire risk assessment carried out. This can be best undertaken as part of the fire risk assessment undertaken in compliance with the *Regulatory Reform (Fire Safety) Order 2005 (SI 2005 No 1541)*. The involvement of a company's insurer is essential as insurers have considerable experience in this field. The business fire risk assessment follows similar stages to the life safety risk assessment (see **F5037**) but in addition to the life safety elements it assesses the importance of each area to the function of the business and how vulnerable it is to fire.

For example, serious damage to an IT suite or area where essential records or documents are stored is likely to have a serious effect on the continued smooth running of business operations should a fire occur. Essential equipment, plant or stock which, if destroyed or severely damaged by fire, might be difficult to replace or have a serious effect on production would require special consideration, and often high fire protection requirements to minimise such an effect.

These risks should be determined by management; they need to be identified, considered and evaluated. A report should be produced by each departmental head, outlining areas which may require special consideration. Such a report should also include protection of electronic databases, essential drawings and other essential documents.

A typical area of high loss effect would be the communication equipment room. The loss of this equipment could have a serious and immediate effect upon communications generally. Fire separation (to keep a fire out) is therefore considered essential, and automatic fire suppression by a self-contained extinguishing system should be strongly recommended.

Essential data should be duplicated and stored with a copy of the contingency plans in a safe area off-site. Paper records can be converted to electronic format and similarly stored.

Passive and active fire protection

[F5007] Passive fire protection utilises inherently fire resistant elements of the structure to divide a building into fire-resisting compartments. These serve two functions. One is to limit the spread of fire and can be a property protection measure as well as life safety. The other is to protect escape routes by making the escape route a fire-resisting compartment. All internal fire escape stairways are also fire-resisting compartments.

In day-to-day work the integrity of these compartments should be maintained where there are doorways through the compartment walls. The doors in these openings should therefore be fire resisting door sets fitted with self closing doors. It is essential that these doors are not obstructed and are allowed to self-close freely at all times. Any glazing in such doors must also be fire resistant and if damaged must be repaired to the appropriate standard. (Guidance on the most common types of fire door is given in BS 8214:2008 *Code of practice for fire door assemblies*.)

The breaching of fire resistant walls with services such as pipes, ducts and cables should be avoided. Where there is no alternative, it is important that all the openings are suitably protected to the same rating as the fire wall. Ducts should be fitted with fire dampers while openings around services should be suitably sealed or fire stopped. (Guidance on the fire safety aspects of the design and construction of air handling duct work is given in BS 5588-9: 1999 *Fire precautions in the design, construction and use of buildings: Code of practice for ventilation and air conditioning ductwork.*)

Active fire protection involves systems that are activated when a fire occurs, for example automatic fire detection or automatic sprinkler systems. As such systems are only required in an emergency it is essential that they are designed, installed, tested, inspected and maintained in accordance with acceptable standards and good practice.

Fire procedures and portable equipment

Fire procedures

[F5008] The need for effective and easily understood fire procedures cannot be over-emphasised and form a key element of the fire risk assessment for the premises. It may be necessary to provide a fire procedure manual, so arranged that it can be used for overall fire defence arrangements, and sectioned for use in individual departments or for special risks.

It is essential that three separate procedures are considered:

- procedure during normal working hours;
- procedure during restricted manning on shifts; and
- procedure when only security staff are on the premises.

All procedures should take into consideration absence of personnel due to sickness, leave, etc. The fire brigade should be called immediately if any fire occurs, irrespective of the size of the fire. It should be remembered that any delay in calling the fire brigade must be added to the delay before the fire brigade's actual arrival, which will be related to the traffic conditions or the local appliances already attending another fire.

In a large organisation it may be necessary to identify the actions that should be taken by certain members of staff (such as security staff and till operators) when the alarm is raised. A person should be given the responsibility for ensuring that pre-planned action is carried out and that everyone in the premises can be accounted for when a fire occurs.

Large fires often result from a delayed call, which may be due not to delayed discovery but to wrong action being taken in the early stages following discovery of a fire. A pre-planned fire routine and suitable training to ensure that all staff understand their responsibilities in the event of a fire is essential for fire safety. The fire brigade, when called, should be met on arrival by a designated person available to guide them directly to the area of the fire. It is essential that all fire routines, when finalised, be made known to the fire brigade.

Fire equipment

[F5009] The need to provide fire fighting equipment is set out in the *Regulatory Reform (Fire Safety) Order 2005 (SI 2005 No 1541)*. All areas of the workplace should be provided with a suitable number of appropriate fire extinguishers and staff should be trained in their use. (See *BS 5306–8:2000 Fire extinguishing installations and equipment on premises. Selection and installation of portable fire extinguishers.* Code of practice.) Non-automatic fire fighting equipment should be easily accessible, simple to use and indicated by signs.

There have been a number of cases where a person using an extinguisher has been seriously injured and investigations have shown that either the wrong type of extinguisher was being used or the operator had no training in the correct use of the appliance. The need for training staff cannot be over-emphasised, particularly in areas of special risk, and where deep fat fryers, furnaces, highly flammable liquids or gas cylinders are in use.

The following recommendations are given in order to allow an evaluation of an existing problem and may need to be related to process risks:

- It is essential that persons be trained in the use of extinguishers, especially in areas where special risks require a specific type of extinguisher to be provided.
- Any person employed to work, who is requested to deal with a fire, should be clearly instructed that at no time should that person jeopardise their own safety or the safety of others.
- Persons who may be wearing overalls contaminated with oil, grease, paint or solvents should not be instructed to attack a fire. Such contaminated materials may vaporise due to heat from the fire, and ignite.

Types of fire extinguisher

[F5010] The type of extinguisher provided should be suitable for the hazard involved, adequately maintained and appropriate records kept of all inspections, tests etc. All fire extinguishers should be fitted on wall brackets or located in purpose built floor stands. It has been found that if this is not done, extinguishers are removed, and hence may be missing when required, or be knocked over and damaged. Extinguishers should be sited near exits or on the route to an exit.

Water extinguishers

[F5011] This type of extinguisher is suitable for ordinary fires involving combustible materials such as wood and paper, but is not suitable for flammable liquid fires. Spray type water extinguishers are recommended. Water extinguishers should be labelled 'not to be used on fires involving live electricity'.

Foam extinguishers

[F5012] These are suitable for small liquid spill fires or small oil tank fires where it is possible for the foam to form a blanket over the surface of the

Means of escape

[F5020] Requirements for means of escape are outlined in the *Regulatory Reform (Fire Safety) Order 2005 (SI 2005 No 1541)*. This indicates that in order to safeguard the safety of staff, visitors and others on the premises, the responsible person must ensure that routes to emergency exits from premises and the exits themselves are kept clear at all times.

The following requirements must be complied with in respect of premises where necessary (whether due to the features of the premises, the activity carried on there, any hazard present or any other relevant circumstances) in order to safeguard the safety of relevant persons—

(a)   emergency routes and exits must lead as directly as possible to a place of safety;

(b)   in the event of danger, it must be possible for persons to evacuate the premises as quickly and as safely as possible;

(c)   the number, distribution and dimensions of emergency routes and exits must be adequate having regard to the use, equipment and dimensions of the premises and the maximum number of persons who may be present there at any one time;

(d)   emergency doors must open in the direction of escape;

(e)   sliding or revolving doors must not be used for exits specifically intended as emergency exits;

(f)   emergency doors must not be so locked or fastened that they cannot be easily and immediately opened by any person who may require to use them in an emergency;

(g)   emergency routes and exits must be indicated by signs; and

(h)   emergency routes and exits requiring illumination must be provided with emergency lighting of adequate intensity in the case of failure of their normal lighting.

The means of escape provided should be identified in the fire risk assessment for the workplace.

Further guidance with regard to means of escape is set out in the guides to the *Regulatory Reform (Fire Safety) Order (SI 2005 No 1541)*, these are published by the Department of Communities and Local Government (DCLG) (www. communities.gov.uk).

Any dimensions set out in guidelines or other forms of good practice should always be reviewed during the fire risk assessment for the workplace. In some instances it may be that the fire hazards present dictate that the travel distance (or travel time) be minimised. In other cases a longer travel distance (or time) may be acceptable, especially if there are appropriate compensatory features present.

Care must always be taken when the arrangement of the work areas incorporates 'rooms within rooms'. In these cases the occupants of the inner rooms must be made aware of any threat to the integrity of their escape route at an earliest time as possible. Thus:

•   the access (outer) room should have automatic fire detection installed;

- there should be a suitably sized vision panel between the inner and outer rooms; or
- the wall or partition between the two areas should not extend to within 0.5m from the ceiling.

The following are essential:

- all doors affording means of escape in case of fire should be maintained easily and readily available for use at all times that persons are on the premises;
- doors should be hung so as to open in the direction of travel;
- all doors not in continuous use, affording a means of escape in case of fire, should be clearly indicated;
- sliding doors should not be used on escape routes;
- doors on escape routes should not be locked or fastened in such a way that they cannot be easily and immediately opened without the use of a key; and
- all gangways and escape routes must be kept clear at all times.

Unsatisfactory means of escape

[F5021] The following are unsatisfactory means of escape and should not be used in the event of fire:

- lifts (unless specially designed for use in the event of fire);
- portable ladders;
- spiral staircases; and
- lowering lines.

Fire drills

[F5022] The *Regulatory Reform (Fire Safety) Order 2005 (SI 2005 No 1541)* states that:

> The responsible person must establish, and where necessary, give effect to appropriate procedures, including safety drills, to be followed in the event of serious and imminent danger to relevant persons.

The responsible person must also 'nominate a sufficient number of competent persons to implement these procedures in so far as they relate to the evacuation of relevant persons from the premises'.

Employers should acquaint the workforce with the actions they should take in the event of fire. This consists of putting up a notice in a prominent place (normally next to each fire alarm call point) stating the action employees should take on:

- hearing the alarm, or
- discovering the fire.

In addition, employees should receive regular fire drills, at least once, and preferably twice a year, even though normal working is interrupted.

Employees should be designated and trained as fire wardens to ensure that everyone leaves the building safely and to assist members of the public or colleagues with a disability, as necessary. In addition, selected employees should be trained in the proper use of fire extinguishers.

Periodical visits by the local fire authority should be encouraged by employers, since this provides a valuable source of practical information on fire fighting, fire protection and training and enables the fire brigade to become familiar with the layout of the premises and the hazards that may be present.

Typical fire action notice

[F5023] When the fire alarm sounds:

(1)   Switch off electrical equipment and leave room, closing doors behind you.
(2)   Walk quickly along the nearest available route to open air.
(3)   Do not use lifts.
(4)   Report to fire warden at assembly point.
(5)   Do not re-enter building.

When you discover a fire:

(1)   Raise alarm (normally by operating a break glass call point).
(2)   Leave the room, closing doors behind you.
(3)   Leave the building by the nearest available route.
(4)   Report to fire warden at assembly point.
(5)   Do not re-enter building.

Fires on construction sites

[F5024] Each year there are numerous fires on construction sites and in buildings undergoing refurbishment. For that reason *Fire Prevention on Construction Sites the Joint Code of Practice on the Protection from Fire of Construction Sites and Buildings Undergoing Renovation'* (published jointly by the Fire Protection Association and the Construction Confederation) indicates that the main contractor should appoint a *site fire safety co-ordinator*, responsible for assessing the degree of fire risk and for formulating and regularly updating the *site fire safety plan*; and he should liaise with the co-ordinator for the design phase (see **F5025**). The site fire safety plan should detail:

- organisation of and responsibilities for fire safety;
- general site precautions, fire detection and warning alarms;
- requirements for a hot work permit system;
- site accommodation;
- fire escape and communications system (including evacuation plan and procedures for calling the fire brigade);
- fire brigade access, facilities and co-ordination;
- fire drill and training;
- effective security measures to minimise the risk of arson; and

- materials storage and waste control system.

Role of site fire safety co-ordinator

[F5025] The site fire safety co-ordinator must:

- ensure that all procedures, precautionary measures and safety standards (as specified in the site fire safety plan) are clearly understood and complied with by all those on the project site;
- ensure establishment of hot work permit systems;
- carry out weekly checks of fire fighting equipment and test all alarm and detection devices;
- conduct weekly inspections of escape routes, fire brigade access, fire fighting facilities and work areas;
- liaise with local fire brigade for site inspections;
- liaise with security personnel;
- keep a written record of all checks, inspections, maintenance of fire protection equipment, tests and fire drill procedures;
- monitor arrangements/procedures for calling the fire brigade;
- during the alarm, oversee safe evacuation of site, ensuring that all staff/visitors report to assembly points; and
- promote a safe working environment.

Emergency procedures

[F5026] The following emergency procedures should be implemented, where necessary:

- establish a means of warning of fire, e.g. handbells, whistles etc.
- display written emergency procedures in prominent locations and give copies to all employees;
- maintain clear access to site and buildings;
- alert security personnel to unlock gates/doors in the event of an alarm; and
- install clear signs in prominent positions, indicating locations of fire access routes, escape routes and positions of dry riser inlets and fire extinguishers.

Designing out fire

[F5027] Construction works should be designed and sequenced to achieve the early installation and operation of:

- permanent fire escape stairs, including compartment walls;
- fire compartments in buildings under construction, including installation of fire doors;
- fire protective materials to structural steelwork;
- planned fire fighting shafts duly commissioned and maintained;
- lightning conductors;
- automatic fire detection systems; and
- automatic sprinkler and other fixed fire fighting installations.

Moreover, adequate water supplies should be available and hydrants suitably marked and kept clear of obstruction.

Other fire precautions on site

[F5028] Portable fire extinguishers can represent the difference between a conflagration and a fire kept under control. Therefore, personnel should be trained in the use of portable fire fighting equipment and adequate numbers of suitable types of portable extinguishers should be available. They should be located in conspicuous positions near exits on each floor. In the open, they should be 500 mm above ground bearing the sign 'Fire Point' and be protected from both work activities and adverse weather conditions. In addition, all mechanically-propelled site plant should carry an appropriate fire extinguisher, and extinguishers, hydrants and fire protection equipment should be maintained and regularly inspected by the site fire safety co-ordinator.

Plant on construction sites also constitutes a potential danger. All internal combustion engines of powered equipment, therefore, should be positioned in the open air or in a well-ventilated non-combustible enclosure. They should be separated from working areas and sited so that exhaust pipes/gases are kept clear of combustible materials. Moreover, fuel tanks should not be filled whilst engines are running and compressors should be housed singly away from other plant in separate enclosures.

Consequences of failure to comply with code

[F5029] Non-compliance with the provisions of this code carries the threat of insurance ceasing to be available or being withdrawn, thereby constituting a breach of a Standard Form contract (see **C8141** CONSTRUCTION AND BUILDING OPERATIONS). Where fire damage is caused to property by the negligence of employees of a subcontractor, then, in accordance with Clause 6.2 of the JCT Contract, to the effect that the contractor is liable for 'injury or damage to property', the contractor is liable and not the employer (or building owner); even though, under Clause 6.3B, the employer is required to insure against loss or damage to existing structures, to the work in progress, and to all unfixed materials and goods intended for, delivered to or placed on the works (*National Trust for Places of Historic Interest or Natural Beauty v Haden Young Ltd*, (1994) 72 BLR 1). The practical implementation of fire precautions on construction sites forms part of the overall construction health and safety plan (see **C8033** CONSTRUCTION AND BUILDING OPERATIONS).

Fire and fire precautions – legislation

[F5030] Fire safety legislation in the UK is concerned principally with life safety issues, including the safety of firefighters and others who may need to enter a building on fire as part of their professional duties. There are two main sets of regulations that control this area: (*a*) the imposition of controls on the design and construction of buildings and (*b*) requirements leading to the safe management of buildings that are in use.

The provision of fire safety in new buildings relates to the physical provisions at the time of construction and is controlled through the Building Regulations in England and Wales, by Building Standards (Scotland) Regulations in Scotland and Building Regulations (Northern Ireland).

Buildings in use are currently controlled by a range of statutes relating to workplace fire risk assessment and licensing of premises for consumption of alcohol, entertainment and similar licensed activities. These statutes generally relate to physical provisions and management requirements.

This section deals, almost exclusively, with the safe management of the workplace. The principle legislation relating to fire safety in the workplace is the *Regulatory Reform (Fire Safety) Order 2005 (SI 2005 No 1541)*. This legislation is concerned with the fire safety provisions and management of non-domestic premises. The Order replaced fire certification under the *Fire Precautions Act 1971* with a general duty to ensure, as far as reasonably practical, the safety of employees, a general duty, in relation to non-employees to take such fire precautions as may reasonably be required in the circumstances to ensure that premises are safe and a duty to carry out a fire risk assessment (see section **F5053**).

The current position – overview

[F5031] Until the introduction of the *Regulatory Reform (Fire Safety) Order 2005 (SI 2005 No 1541)* the statute law relating to fire and fire precautions was extensive and complex. The legislation included:

- *Fire Precautions Act 1971* ('FPA 1971') (and regulations and orders made thereunder);
- *Fire Safety and Safety of Places of Sport Act 1987*;
- *Fire Precautions (Workplace) Regulations 1997 (SI 1997 No 1840)* as amended by the *Fire Precautions (Workplace) (Amendment) Regulations 1999 (SI 1999 No 1877)*;
- *Health and Safety at Work etc Act 1974* ('HSWA 1974') (in the form of the *Fire Certificates (Special Premises) Regulations 1976 (SI 1976 No 2003)*);
- Petroleum Acts (and regulations made thereunder);
- *Public Health Acts 1936–1961*;
- *Building Act 1984* and the Building Regulations made thereunder; and
- *Fire Services Act 1947* and the *Fires Prevention (Metropolis) Act 1774*;

as well as certain regulations, made under the *Factories Act 1961*.

The *Regulatory Reform (Fire Safety) Order 2005 (SI 2005 No 1541)*, introduced in October 2006, has greatly simplified fire safety legislation, with many Acts being repealed and Regulations being revoked. Other pieces of legislation have been amended to streamline the legislative basis of fire safety in non-domestic premises. However, the Fire Safety Order sits comfortably alongside the *Dangerous Substances and Explosive Atmospheres Regulations 2002 (SI 2002 No 2776)*, which remain in force.

Fire certificates are no longer issued by fire and rescue services but fire inspectors or duly authorised officers of the fire brigade have powers to visit and enter premises in order to carry out their duties (see **F5032** and **F5056**).

Fire certification

[F5032] Until the introduction of the *Regulatory Reform (Fire Safety) Order 2005 (SI 2005 No 1541)* in October 2006, one of the prime tools in respect of fire safety in the workplace was the *Fire Precautions Act 1971*. In the *Fire Precautions Act 1971* there was the requirement for 'designated' premises to be 'fire-certificated' – normally by the local fire authority, though in the case of exceptionally hazardous industrial premises (i.e. special premises) by the Health and Safety Executive (HSE). Two designation orders were made affecting factories, offices, shops and railway premises *and* hotels and boarding houses.

The fire certificate had to be kept on the premises to which it referred and thus this important document will still be present to form a vital element when undertaking or reviewing the fire risk assessment and determining the fire safety policies and procedures for the building. Although fire certificates do not now have the legislative authority that they previously enjoyed they should be kept safely together with the fire safety risk assessment for the premises.

Fire certificates contained valuable information. For example they specified:

- use/uses of premises;
- means of escape in case of fire;
- how means of escape can be safely and effectively used;
- alarms and fire warning systems; and
- fire-fighting apparatus to be provided in the building.

Moreover, at its discretion, the fire authority could, additionally, impose requirements relating to:

- maintenance of means of escape and fire-fighting equipment;
- staff training; and
- restrictions on the number of people within the building.

(It should be remembered that statutory restrictions on the occupancy levels of certain areas of the premises may still be in force as a result of licenses issued by the local authority. Such restriction may apply, for example, to bars and other places of assembly.)

The Regulatory Reform (Fire Safety) Order 2005

[F5033] The *Regulatory Reform (Fire Safety) Order 2005 (SI 2005 No 1541)* replaces fire certification under the *Fire Precautions Act 1971* with a general duty to ensure, so far as is reasonably practicable, the safety of employees, a general duty in relation to non-employees to take such fire precautions as may reasonably be required in the circumstances to ensure that premises are safe, and a duty to carry out a risk assessment. The Order imposes a number of specific duties in relation to the fire precautions to be taken and also provides for the enforcement of the Order. It amends or repeals other primary legislation concerning fire safety to take account of the new system and provides for minor and other consequential amendments, repeals and revocations. For example safety certificates continue to be issued under the *Safety of Sports Grounds Act 1975* or the *Fire Safety and Safety of Places of Sport Act 1987* but are not allowed to require anyone to do anything that would cause them to contravene *SI 2005 No 1541*.

The Order applies to all non-domestic premises other than a restricted number of exceptions (see **F5034**).

The Order applies only in England and Wales; in Scotland similar legislation is introduced by the *Fire Safety (Scotland) Regulations 2006 (SSI 2006 No 456)* made under the *Fire (Scotland) Act 2005*. The *Fire and Rescue Services (Northern Ireland) Order 2006 (SI 2006 No 1254)* introduces the requirements, which are based on EU Directives.

The Regulatory Reform Order therefore continues with the requirement that a fire risk assessment be undertaken for the premises, a requirement introduced by the *Fire Precautions (Workplace) Regulations 1997 (SI 1997 No 1840)* (as amended). One new concept introduced is that the main duty holder in relation to the premises is the 'responsible person' who should be identified in the assessment. The responsible person means the employer, if the workplace is under his control; otherwise the duties of the responsible person are extended to any person who has control of the premises.

There are a number of duties imposed on the responsible person, who must:

- take such general fire precautions as will ensure, so far as is practicable, the safety of his employees;
- in relation to relevant persons who are not his employees, take such general fire precautions as may reasonably be required in the circumstances to ensure that the premises are safe; and
- make a suitable and sufficient assessment of the risks to which relevant persons are exposed for the purpose of identifying the general fire precautions he needs to take.

Application of the Regulatory Reform (Fire Safety) Order 2005

[F5034] The Order applies to all non-domestic premises other than a restricted number of exceptions. The Order does **not** apply in relation to:

- single private dwellings;
- an offshore installation within the meaning of the *Offshore Installation and Pipeline Works (Management and Administration) Regulations 1995 (SI 1995 No 738), Reg 3*;
- a ship, in respect of the normal ship-board activities of a ship's crew which are carried out solely by the crew under the direction of the master;

- fields, woods or other land forming part of an agricultural or forestry undertaking but which is not inside a building and is situated away from the undertaking's main buildings;
- an aircraft, locomotive or rolling stock, trailer or semi-trailer used as a means of transport or a vehicle for which a licence is in force under the *Vehicle Excise and Registration Act 1994* or a vehicle exempted from duty under that Act;
- a mine within the meaning of the *Mines and Quarries Act 1954, s 180*, other than any building on the surface at a mine; and
- a borehole site to which the *Borehole Sites and Operations Regulations 1995 (SI 1995 No 2038)* apply.

Building Regulations

[F5035] Where the *Building Regulations 2000 (SI 2000 No 2531)* as amended, apply to premises, with respect to requirements as to means of escape in case of fire, the fire authority cannot serve an improvement notice requiring structural or other alterations. However, a notice can be served when the building is occupied if the fire authority is satisfied that the means of escape in case of fire are inadequate, by reason of matters/circumstances of which particulars were not required by the Building Regulations. Full plans should be deposited with the local authority where building work is intended to be carried out in relation to a building to which the Regulatory Reform (Fire Safety) Order 2005 applies, or will apply on completion of the work.

The *Building Regulations 2000* (as amended) include requirements to make 'appropriate' provisions for the early warning of fire and 'reasonable facilities to assist fire fighters in the protection of life'.

The Department for Communities and Local Government (DCLG) publishes guidance on meeting the requirements in what are known as '*Approved Documents*'. The current document is *Approved Document B - Volume 2 - Buildings other than dwellinghouses (2006 Edition)*. Volume 1 covers dwelling houses.

The following requirements are specified in respect of new premises:

- The building must be designed/constructed so that there are *means of escape* in case of fire, to a place of safety outside.

Internal fire spread

- To inhibit internal fire spread, internal linings must:
  (i)   adequately resist flame spread over surfaces; and
  (ii)   if ignited, have a reasonable rate of heat release.
- The building must be designed and constructed so that, in the event of fire, its stability will be maintained for a reasonable period; and a common wall should be able to resist fire spread between the buildings.
- The building must be designed and constructed so that unseen fire/smoke spread within concealed spaces in its fabric and structure, is inhibited.

External fire spread

- External walls shall adequately resist fire spread over walls and from one building to another.
- A roof should be able to resist fire spread over the roof and from one building to another.
- The building must be designed and constructed to provide facilities to firefighters and enable fire appliances to gain access.

Means of warning and escape

- The building shall be designed and constructed so that there are appropriate provisions for the early warning of fire, and appropriate means of escape in case of fire from the building to a place of safety outside the building capable of being safely and effectively used at all material times.

Access and facilities for the fire service

- The building shall be designed and constructed so as to provide reasonable facilities to assist firefighters in the protection of life and reasonable provision made within the site of the building to enable fire appliances to gain access.

A number of changes came into force on 6 April 2007 affecting building work in England and Wales. For non-domestic buildings, the key changes included the introduction of a maximum unsprinklered compartment size for single storey warehouses, new guidance on residential care homes (including on the use of sprinklers) and a new requirement to ensure occupiers are made aware of their building's fire protection measures so as to assist with the preparation of fire risk assessments under the new *Regulatory Reform (Fire Safety) Order* regime.

The technical guidance can be downloaded from www.planningportal.gov.uk/england/professionals/en/1115314683691.html.

The Building Regulations are enforced by measures that fall outside the scope of this chapter.

Appeals against enforcement notices

[F5036] An appeal must be lodged against an alterations, enforcement or prohibition notice made under the *Regulatory Reform (Fire Safety) Order 2005 (SI 2005 No 1541)* within 21 days from the date of service. When an appeal is brought against an alterations or enforcement notice the effect of an appeal is to suspend operation of the notice until the appeal is finally disposed of or withdrawn. When an appeal is made against a prohibition notice the brining of the appeal does not have the effect of suspending the operation of the notice unless the (magistrates) court so directs.

The court may either cancel or affirm the notice and may affirm it either in its original form or with such modifications as the court may see fit. Failure to do so carries with it similar penalties as explained in **F5060** below.

Fire risk assessment

[F5037] The *Regulatory Reform (Fire Safety) Order 2005 (SI 2005 No 1541)* requires the responsible person for the premises to assess the risk a fire could pose to the employees and other persons who may be present in the workplace.

Although there is no prescriptive way in which the fire risk assessment should be carried out, it may involve six stages:

Stage 1 – identifying the fire hazards;
Stage 2 – identifying the people at risk;
Stage 3 – removing or reducing the hazards;
Stage 4 – assigning a risk category;
Stage 5 – deciding if existing arrangements are satisfactory or need improvement; and
Stage 6 – recording the findings – a statutory requirement if more than five people are employed.
The assessment should be reviewed periodically and when any significant changes occur to, for example, layout, management or staffing.

Carrying out the fire risk assessment

[F5038] The responsible person must make a suitable and sufficient assessment of the fire risk to which the staff and others on the premises are exposed, this should:

- identify any group of persons identified by the assessment as being specially at risk;
- record the significant findings, including the measures which have, or will be taken by the responsible person;
- identify the general fire precautions needed to be taken; and
- include consideration of the presence, use and handling of any hazardous materials that are present.

The fire risk assessment should be recorded where there are five or more employees, there is a licence in force in relation to the premises and where an alterations notice requiring this is in force in relation to the premises (see F5057).

No new work involving a dangerous substance may commence unless a risk assessment has been made and the measures identified during that exercise have been implemented.

The principles of prevention to be applied when undertaking the assessment (*Regulatory Reform (Fire Safety) Order 2005 (SI 2005 No 1541), art 10*) are:

- avoiding risks;
- evaluating the risks which cannot be avoided;

- combating the risks at source;
- adapting to technical progress;
- replacing the dangerous by the non-dangerous or less dangerous;
- developing a coherent overall prevention policy which covers technology, organisation of work and the influence of factors relating to the working environment;
- giving collective protective measures priority over individual protective measures; and
- giving appropriate instructions to employees.

Controlling the residual risk

[F5039] Having eliminated and reduced fire hazards as far as practicable, there are a number of measures that should be taken to control the residual risks. These include the provision of:

- emergency routes;
- fire safety policies and procedures;
- provisions for fire fighting and fire detection; and
- additional measures in respect of dangerous substances.

Emergency routes

[F5040]–[F5041] In order to safeguard the safety of employees and other persons on the premises in case of fire:

- routes to emergency exits must be kept clear at all times;
- emergency routes and exits must lead as directly as possible to a place of safety;
- in the event of danger it must be possible for persons to evacuate the premises as quickly and as safely as possible;
- the number, distribution and dimensions of emergency routes must be adequate regarding their use, equipment and dimensions of the premises and the maximum number of persons who may be present there at any one time;
- emergency doors must open in the direction of travel;
- sliding or revolving doors must not be used for exits specifically intended as emergency exits;
- emergency doors must not be so locked or fastened that they cannot be easily and immediately opened in an emergency;
- emergency routes and exits must be indicated by signs; and
- emergency routes requiring illumination must be provided with emergency lighting of adequate intensity in the case of failure of the normal lighting.

Fire safety policy and procedures

- the responsible person must establish appropriate procedures, including safety drills, to be followed in the event of serious and imminent danger to persons;

- a sufficient number of competent persons must be nominated to implement those procedures in so far as they relate to the evacuation of the premises; and
- the responsible person must ensure that no person has access to any area to which it is necessary to restrict access on grounds of safety unless the person concerned has received adequate safety instruction.

Provisions for fire fighting and fire detection

[F5042] The responsible person must:

- ensure that the premises is, to the extent that is appropriate, equipped with appropriate fire fighting equipment and with fire detectors and alarms;
- ensure that any non-automatic fire fighting equipment so provided is easily accessible, simple to use and indicated by signs;
- take measures for fire fighting in the premises, adapted to the activities carried on there and the size of the undertaking and the premises concerned;
- nominate competent persons to implement these fire fighting measures – the number of these persons, their training and the equipment available to them must be adequate, taking into account the size of the premises and the specific hazards there; and
- establish any necessary contacts with external emergency services, particularly as regards fire fighting, rescue work, first aid and emergency medical care.

Additional measures in respect of dangerous substances

[F5043] In order to safeguard staff and others in respect of an incident arising from the presence of a dangerous substance, the responsible person must ensure (*Regulatory Reform (Fire Safety) Order 2005 (SI 2005 No 1541), art 16*) that:

- information is available regarding hazard identification, work hazards and specific hazards likely to arise at the time of an accident or incident – this information should be displayed in the premises and also be made available to the emergency services to allow them to prepare their own response procedures;
- suitable warning and communication systems are established to enable an appropriate response to be made when an incident occurs;
- where necessary visual or audible warnings are given before explosive conditions are reached and relevant persons are withdrawn from the area; and
- where the risk assessment indicates it to be necessary, escape facilities are provided and maintained to enable persons can leave the endangered areas promptly and safely.

In the event of a fire relating to the presence of a dangerous substance the responsible person must ensure that:

- immediate steps are taken to mitigate the effect of the fire, restore the situation to normal and inform the relevant person who may be affected; and
- only those persons essential for carrying out repairs and other necessary work are allowed into the affected area and that they are provided with appropriate personal protective clothing and equipment.

The above points do not apply, however, where the results of the risk assessment show that the quantity of the dangerous substances on the premises are such that there is only slight risk to relevant persons.

Measures to be taken in respect of dangerous substances

[F5044] In applying measures to control risks the responsible person must, (*Regulatory Reform (Fire Safety) Order 2005 (SI 2005 No 1541), art 12*) in order or priority:

- reduce the quantity of dangerous substances to a minimum;
- avoid or minimise the release of a dangerous substance;
- control the release of a dangerous substance at source;
- prevent the formation of an explosive atmosphere, including the application of appropriate ventilation;
- ensure that any release of dangerous substance which may give rise to risk is suitably collected, safely contained, removed to a safe place or otherwise rendered safe;
- avoid ignition sources, including electrostatic discharges and other adverse conditions that could result in harmful effects from a dangerous substance; and
- segregate incompatible dangerous substances.

The responsible person must ensure that mitigation measures are applied which include:

- reducing the number of persons exposed to a minimum;
- measures to avoid the propagation of fires and explosions;
- providing explosion relief arrangements;
- providing plant that is constructed to withstand the pressure likely to be produced by an explosion; and
- providing suitable personal protective equipment.

Included in a series of additional requirements, the responsible person must also:

- ensure that premises are designed, constructed and maintained so as to reduce risk;
- ensure that mitigating measures are designed, constructed, assembled, installed, provided and used so as to reduce risk;
- ensure that special technical and organisational measures are maintained in an efficient working order and good repair; and
- ensure that appropriate systems of work are issued in writing and a suitable system of permits to work is instituted and maintained prior to work being carried out.

The Dangerous Substances and Explosive Atmospheres Regulations (DSEAR) 2002

[F5045] The *Dangerous Substances and Explosive Atmospheres Regulations (DSEAR) 2002 (SI 2002 No 2776)* sit comfortably alongside the *Regulatory Reform (Fire Safety) Order 2005 (SI 2005 No 1541)* in that DSEAR requires employers and the self employed to:

• carry out a risk assessment of any work activities involving dangerous substances;
• provide technical and organisational measures to eliminate or reduce, as far as is reasonably practical, the identified risks;
• provide equipment and procedures to deal with accidents and emergencies;
• provide information and training to employees; and
• classify places where explosive atmospheres may occur into zones and mark the zones where necessary.

Fire fighters' switches for luminous signs

[F5046] In addition to the fire safety duties set out in *Regulatory Reform (Fire Safety) Order 2005 (SI 2005 No 1541), Part 2*, there are additional miscellaneous requirements in *Part 5*; foremost among these are measures to protect fire fighters in the event of a fire.

SI 2005 No 1541, art 37 is concerned with the provision of switches for fire fighters' use for isolating luminous signs operating at a voltage in excess of the normal supply voltage. Where such apparatus is installed it must be provided with a cut-off switch so placed and marked as to be readily recognisable by and accessible to fire fighters.

Even though a switch may comply with the requirements of BS 7671 (the requirements for electrical installations) the responsible person must give notice to the fire and rescue authority at least 42 days before work is to commence to install the apparatus. The proposal should be deemed to satisfy the requirements of the fire authority unless they serve a counter notice indicating that they are not satisfied with the proposals within 21 days from the date of the service of the notice.

This article does not apply to premises licensed under the *Licensing Act 2003* for the exhibition of a film or to premises where apparatus has already been installed in compliance with the *Local Government (Miscellaneous Provisions) Act 1982*.

In addition to the specific requirement for cut-off switches, *SI 2005 No 1541, art 38* requires the responsible person to safeguard the safety of fire fighters by ensuring that the premises and any facilities, equipment and devices provided for the use or protection of fire fighters under the Order be subject to a suitable maintenance regime and be maintained in efficient working order and good repair.

Maintenance

[F5047] Where necessary in order to safeguard the safety of relevant persons the responsible person must ensure that the premises and any facilities, equipment and devices provided in compliance with the *Regulatory Reform (Fire Safety) Order 2005 (SI 2005 No 1541)* are subject to a suitable system of maintenance and are maintained in an efficient state, in efficient working order and in good repair (*SI 2005 No 1541, art 17*).

Where the premises form part of a building then the responsible person may make arrangements with the occupiers of other parts of the building to ensure that the requirements outlined in the paragraph above are met. The occupiers of other parts of the building have a duty to co-operate in this respect.

Safety assistance

[F5048] *Regulatory Reform (Fire Safety) Order 2005 (SI 2005 No 1541), art 18* requires that the responsible person appoint one or more competent persons to assist him in undertaking the preventive and protective measures. There should be arrangements for ensuring adequate co-operation between the responsible person and the safety assistant(s).

Sufficient time must be made available to allow the safety assistant(s) to properly fulfil their functions.

Where there is a competent person in the responsible person's employment, that person must be appointed as safety assistant in preference to an outsider. Where, however, the safety assistant appointed is not in the responsible person's employment, he must be provided with the information necessary for him to undertake his task. This will include access to relevant safety data relating to the dangerous substances present and the risk that they present.

Competent person

[F5049] A competent person is defined in *Regulatory Reform (Fire Safety) Order 2005 (SI 2005 No 1541), art 18(5)* as:

> A person is regarded as being competent where he has sufficient training and experience or knowledge and other qualities to enable him properly to assist in undertaking the preventive and protective measures.

Provision of information to employees

[F5050] The responsible person must provide his employees with comprehensive and relevant information (*Regulatory Reform (Fire Safety) Order 2005 (SI 2005 No 1541), art 19*) on:

- the risks to them identified in the risk assessment;
- the preventive and protective measures;
- the identities of the staff nominated to assist with fire fighting and the evacuation of relevant persons; and

- the risks in other parts of the building that have been notified to him.

With regard to dangerous substances on the premises, the responsible person must also provide his employees with:

- the details of such substances, including the name and the risk that it presents, access to relevant data sheets and information relating to any legislative provisions applying to those substances; and
- the significant findings of the risk assessment.

The information should be adapted to take into account the methods of work in use and be provided in a manner appropriate to the risk identified in the risk assessment.

Provision of information to other persons

[F5051] In addition to providing employees with information, the responsible person also has a duty (*Regulatory Reform (Fire Safety) Order 2005 (SI 2005 No 1541), art 20*) to inform non-employees in certain circumstances:

- the employer of employees working in or on the premises must be provided with comprehensible and relevant information on the risks to their employees and the preventive and protective measures that are in place; and
- employees of other undertakings who are working on the premises must be provided with appropriate instructions and comprehensible and relevant information regarding any risk to those persons.

Employment of a child

[F5052] Before employing a child the responsible person must (*Regulatory Reform (Fire Safety) Order 2005 (SI 2005 No 1541), art 19(2)*) provide a parent of the child with information on:

- the risks to that child identified by the fire risk assessment;
- the preventive and protective measures; and
- the risk in other parts of the building that have been notified to him.

Training

[F5053] The responsible person must (*Regulatory Reform (Fire Safety) Order 2005 (SI 2005 No 1541), art 21*) ensure that his employees are provided with adequate fire safety training:

- at the time when they are first employed; and
- on their being exposed to new or increased risks (such as being given a change of responsibilities or the introduction of new equipment, new work processes or new technology).

The training should:

- be repeated periodically where appropriate;
- be adapted to take account of new or changed risks to the safety of the employees;

- be provided in a manner appropriate to the risk identified by the risk assessment; and
- take place during working hours.

Co-operation and co-ordination

[F5054] Where two or more responsible persons share duties in respect of premises each must (*Regulatory Reform (Fire Safety) Order 2005 (SI 2005 No 1541), art 22*):

- co-operate with the other responsible persons(s) to enable them to comply with the requirements of the Order; and
- inform the other responsible persons(s) of the risks to relevant persons in connection with their undertaking.

Where two or more responsible persons share premises where an explosive atmosphere may arise the responsible person with overall responsibility for the premises must co-ordinate the implementation of the measures to protect relevant persons from any risk from the explosive atmosphere.

Duties of employees at work

[F5055] In addition to the duties of the responsible person, every employee, while at work, must (*Regulatory Reform (Fire Safety) Order 2005 (SI 2005 No 1541), art 23*):

- take reasonable care for the safety of himself and other relevant persons who may be affected by his actions or omissions;
- co-operate with their employer to enable the employer to comply with his duties and requirements under the Order; and
- inform his employer or safety representative of any situation considered to be a serious and immediate threat to safety, or any matter which is considered to be a shortcoming in the employer's protection arrangements.

Enforcement

[F5056] *Regulatory Reform (Fire Safety) Order 2005 (SI 2005 No 1541), art 25* indicates that the enforcing authority for the Regulatory Reform Order is the fire and rescue authority for the area in which the premises are situated, apart from the points below.

(a) The following, for which the Health and Safety Executive (HSE) undertake the role:
  (i) any premises for which a licence is required is required in accordance with the *Nuclear Installations Act 1965 (NuIA 1965), s 1* (or for which a permit is required under *NuIA 1965, s 2*);
  (ii) any Crown premises which would otherwise be required to have a licence or permit in accordance with the provisions referred to in sub-paragraph (i) above;

        (iii)   a ship, including a naval vessel, that is in the course of construction, reconstruction., conversion or repair by persons other than the master and crew of the ship; and

        (iv)   any workplace which is or is on a construction site within the meaning of regulation 2(1) of the *Construction (Design and Management) Regulations 2007 (SI 2007 No 320)*, (other than construction sites referred to in regulation 46 of those Regulations);

(b)   The following, for which the Defence Fire Service are the enforcing authority:

        (i)   premises (other than premises specified in paragraph (b)(iii), occupied solely for the purposes of the armed forces;

        (ii)   premises occupied solely by any visiting force or international headquarters or defence organisation designated for the purposes of the *International Headquarters and Defence Organisations Act 1964*; and

        (iii)   premises (other than premises specified in paragraph (b)(iii), which are situated within premises occupied solely for the purposes of the Crown but are which are not themselves so occupied.

(c)   The local authority in relation to:

        (i)   sports grounds requiring a safety certificate under the *Safety of Sports Grounds Act 1975*; and

        (ii)   a regulated stand with respect to the *Fire Safety and Safety of Places of Sport Act 1987*.

(d)   a fire inspector, or any person authorised by the Secretary of State for the purposes of the Order in relation to:

        (i)   premises owned or occupied by the Crown (with some specified exceptions); and

        (ii)   premises in relation to which the United Kingdom Atomic Energy Authority is the responsible person (with some specified exceptions).

Every enforcing authority must (*SI 2005 No 1541, art 26*) enforce the provisions of the *Regulatory Reform (Fire Safety) Order 2005* and any regulations made under it. They may enforce the Order by the serving of one of three types of notices on the responsible person for the premises:

- alterations notices;
- enforcement notices; and
- prohibition notices.

Alterations notice

[F5057] An alterations notice (*Regulatory Reform (Fire Safety) Order 2005 (SI 2005 No 1541), art 29*) does not require the responsible person, or anyone else, to make alterations to the premises. It is served when the enforcing authority believes that the premises constitute a serious risk to relevant persons or there may be such a risk if a change is made to the premises or the use to which they are put.

The notice must state the matters which in the opinion of the enforcing authority constitute such a risk if the changes to the premises or their use is made.

Where such a notice has been served the responsible person must notify the enforcement authority of the proposed changes, and enclose a copy of the fire risk assessment and a summary of the changes he proposes to make to the existing general fire precautions if the alterations go ahead.

Enforcement notice

[F5058] If the enforcing authority is of the opinion that the responsible person has failed to comply with any provision of the Order or of any regulations made under it, the authority may serve on him an enforcement notice (*Regulatory Reform (Fire Safety) Order 2005 (SI 2005 No 1541), art 30*).

The enforcement notice must specify the provisions that have not been complied with and require the responsible person to take steps to remedy the failure within a stated period of time (but not less than 28 days).

An enforcement notice may include directions as to the measure that are considered necessary and give a choice between different ways of remedying the situation.

Prohibition notice

[F5059] If the enforcing authority is of the opinion that the use of the premises involves or will involve a risk to relevant persons so serious that use of the premises ought to be prohibited or restricted the authority may serve a prohibition notice (*Regulatory Reform (Fire Safety) Order 2005 (SI 2005 No 1541), art 31*) on the responsible person.

A prohibition notice must specify the matters which give rise to the risk and direct that the use to which the prohibition notice relates is prohibited or restricted to such extent as may be specified in the notice until the specified matters have been remedied.

A prohibition notice may include directions as to the measure which have to be taken to remedy the matters specified in the notice and any such measures should be framed so as to afford a choice between different ways of remedying the matters.

A prohibition notice takes effect immediately it is served if the enforcing authority is of the opinion that the risk of serious personal injury is sufficiently serious, otherwise it takes effect at the end of the time specified in the notice.

Offences and appeals

[F5060] It is an offence for a responsible person to fail to comply:

- with requirements, thereby placing one or more persons at risk of death or serious injury in case of fire;

- with any requirements of alteration notices, enforcement notices or prohibition notices; and
- in relation to the provision of luminous tube signs (*Regulatory Reform (Fire Safety) Order 2005 (SI 2005 No 1541), art 37*).

Any person guilty of an offence under points referred to above is liable:

- on summary conviction to a fine not exceeding the statutory maximum; or
- on conviction on indictment to a fine or to imprisonment for a term not exceeding two years, or to both.

It is an offence for any person to:

- fail to comply with the general duties of employees at work (*SI 2005 No 1541, art 23*) where that failure places one or more persons at risk of death or serious injury in case of fire;
- knowingly make a false entry into any register, book, notice or other document required to be kept under the Order;
- knowingly or recklessly give false information in response to any enquiry made under the Order;
- intentionally obstruct an inspector in the performance of his duties under the Order;
- fail without reasonable excuse to comply with any requirements imposed by an inspector under the Order;
- pretend, with intent to deceive, to be a inspector under the Order;
- fail to comply with the prohibition with regard to the charging of employees for things done or provided in connection with the Order (*SI 2005 No 1541, art 40*); and
- fail to comply with any prohibition or restriction imposed by a prohibition notice.

Any person guilty of other offences listed above is also liable on conviction to a fine, apart from failure to comply with a prohibition notice, which attracts a more serious punishment.

Where an offence under this Order has been committed by a corporate body is proved to have been committed with the consent or connivance of any director, manager, secretary or similar officer of the body corporate he, as well as the corporate body is guilty of that offence and is liable to be proceeded against and punished accordingly.

Application to the Crown

[F5061] As with other health and safety duties and regulations, generally speaking, statutory fire duties and fire regulations apply to the Crown and this continues with the introduction of the *Regulatory Reform (Fire Safety) Order 2005*, art 49. However owing to Crown immunity in law, proceedings cannot be enforced against the Crown (see further E15045 Enforcement). This has the effect that the responsible person in Crown premises, that is, government buildings such as the Treasury and the Foreign Office as well as royal palaces, is required to comply with the requirements of the Order, including the preparation of a fire risk assessment. The articles relating to alterations and enforcement notices and consequent offences and appeals do not, however, apply.

Further guidance on specific types of premises

[F5061A] A suite of guidance documents has been published by the Department for Communities and Local Government. These have the general title 'Fire safety risk assessment' and address the following occupancies:

- offices and shops;
- factories and warehouses;
- sleeping accommodation;
- educational premises;
- small and medium places of assembly;
- large places of assembly;
- theatres, cinemas and similar premises;
- open air events and venues;
- healthcare premises;
- transport premises and facilities;
- means of escape for disabled people (supplementary guide); and
- animal premises and stables.

These can be downloaded free of charge at: www.communities.gov.uk/publicat ions/fire/firesafetyrisk5. *Regulatory Reform (Fire Safety) Order 2005* – a short guide to making your premises safe from fire has also been published by the DCLG which can also be downloaded from the DCLG website at: www. communities.gov.uk.

In addition, detailed guidance is available in *Fire risk management in the workplace – a guide for employers* is published by the Fire Protection Association.

Other dangerous processes

[F5062]–[F5072] Until the *Dangerous Substances and Explosive Atmospheres Regulations (DSEAR) 2002 (SI 2002 No 2776)* came into force, fire prevention measures for certain particularly dangerous processes were controlled by specific regulations. Most of these have been repealed or revoked and hazards are now controlled by the requirements of the *DSEAR*. Some of the regulations replaced by *DSEAR* are:

(a)   the *Celluloid (Manufacture, etc.) Regulations 1921 (SR & O 1921 No 1825)* – applying to the manufacture, manipulation and storage of celluloid and the disposal of celluloid waste;

(b)   the *Manufacture of Cinematograph Film Regulations 1928 (SR & O 1928 No 82)* – applicable to the manufacture, repair, manipulation or use of cinematograph film;

(c)   the *Cinematograph Film Stripping Regulations 1939 (SR & O 1939 No 571)* – applicable to the stripping, drying or storing of cinematograph film;

(d)   the *Highly Flammable Liquids and Liquefied Petroleum Gases Regulations 1972 (SI 1972 No 917)* – applicable to premises containing highly flammable liquids and liquefied petroleum gases;

(e)   the *Magnesium (Grinding of Castings and Other Articles) Special Regulations 1946 (SR & O 1946 No 2017)* – prohibition on smoking, open lights and fires;

(f)   the *Factories (Testing of Aircraft Engines & Accessories) Special Regulations 1952 (SI 1952 No 1689)* – applicable to the leakage or escape of petroleum spirit;

Some regulations with specific fire safety requirements that are still on the statute book are given below.

* the *Offshore Installations (Prevention of Fire and Explosion, and Emergency Response) Regulations 1995 (SI 1995 No 743)* – to prevent and minimise the effects of fire and explosion on offshore installation;

* the *Construction (Design and Management) Regulations 2007 (SI 2007 No 320)*;

* the *Work in Compressed Air Regulations 1996 (SI 1996 No 1656), Reg 14*;

* the *Electricity at Work Regulations 1989 (SI 1989 No 635), Reg 6(d)* – electrical equipment which may reasonably foreseeably be exposed to any flammable or explosive substance, must be constructed or protected so as to prevent danger from exposure;

* the *Dangerous Substances in Harbour Areas Regulations 1987 (SI 1987 No 37)* – applicable to risks of fire and explosion in harbour areas. A fire certificate is not required for these premises.

Liability of occupier

[F5073] An occupier of premises where fire breaks out can be liable to:

* lawful visitors to the premises injured by the fire or falling debris (and is also liable to unlawful visitors, i.e. trespassers, as the principle of 'common humanity', enunciated in *Herrington v British Railways Board* [1972] 1 All ER 749 applies as does the *Occupiers' Liability Act 1984*, see OCCUPIERS' LIABILITY);
* firemen injured during fire-fighting operations; and
* adjoining occupiers.

In order to ensure therefore that occupiers and others involved may minimise their liability, insurance cover, though not compulsory, is highly desirable. The basic principles relating to fire cover are considered below (see **F5080**).

Fire insurance

[F5074] Insurance is intended to provide the insured with cover in the event of an accident that is not predictable. This applies to fire insurance and is reflected in the terms of insurance policies. Thus if a fire occurs in a premises where it is thought that the insured might be involved in deliberately starting the fire the insurance company may refuse to pay out under the policy. However if the insured is innocent of any involvement, even if the fire was started deliberately, the insurer will pay out as arson is not predictable in the context of this issue. Where a high fire risk has been identified and adequate measures are not taken to manage the risk there may be a case for non-payment on the grounds of negligence. In practice insurance plays a very important role in the management of many risks in industry, including fire, and adequate appropriate fire cover is considered an essential part of modern business risk management.

First-Aid

Subash Ludhra

Introduction

[F7001] The *Health and Safety (First-Aid) Regulations 1981 (SI 1981 No 917)*, which came into operation on 1 July 1982, require employers to have arrangements for the provision of first-aid in their place of work. The current regulations were formally reviewed by the HSE in 2004/05, with agreement that no significant change was required at that time. However following further extensive consultation with employer duty holders, employees, first-aiders and first-aid training providers it was agreed that changes would be made to training requirements and a new approved code of practice and guidance was issued in 2009 with and the new training requirement commenced on 1 October 2009.

The main drivers for the change were to help employer duty holders access competent first-aid, provide a more proportionate response to their first-aid needs and to minimise the burden on business.

Employers have to decide whether their first-aiders should be trained in first-aid at work (FAW) or emergency first-aid at work (EFAW). If the assessment indicates that first-aiders should be trained to FAW standard, it will not be acceptable to only provide first-aiders that possess an EFAW certificate.

People can and do suffer injury or fall ill at work. This may or may not be as a result of work-related activity. However, it is important that they receive immediate attention.

Medical treatment should be provided at the scene promptly, efficiently and effectively before the arrival of any medical teams that may have been called. First-aid can save lives and can prevent minor injuries from becoming major ones. Employers are responsible for making arrangements for the immediate management of any illness or injury suffered by a person at work. First-aid at work covers the management of first-aid in the workplace – it does not include treating ill or injured people at work with medicines.

However, the Regulations do not prevent specially trained staff taking action beyond the initial management of the injured or ill at work.

Interpretation

First-aid means:

(1)   In cases where a person will need help from a medical practitioner or nurse, treatment for the purpose of preserving life and minimising the consequences of injury and illness until such help is obtained and

(2)   treatment of minor injuries which would otherwise receive no treatment or which do not need treatment by a medical practititioner or nurse.

(3)   First-aid at work covers the arrangements that need to be made to manage injuries or illness suffered at work. The Regulations do not prevent staff, who are specially trained, from taking action beyond the initial management stage.

The employer's duty to make provision for first-aid

[F7002] Employers must provide or ensure that there is equipment and facilities provided which are adequate and appropriate for rendering first-aid if any of their employees are injured or become ill at work. Employers must also ensure that there are an adequate and appropriate number of suitable persons who:

(a)   are trained and have such qualifications as the Health and Safety Executive may approve for the time being; and

(b)   have additional training, if any, as may be appropriate in the circumstances.

Where such a suitable person is absent in temporary and exceptional circumstances, an employer can appoint a person or ensure that a person is appointed:

(a)   to take responsibility for first-aid in situations relating to an injured or ill employee who needs help from a medical practitioner or nurse;

(b)   to ensure that equipment and facilities are adequate and appropriate in the circumstances.

With regard to any period of absence of the first-aider, consideration must be given to:

(a)   the nature of the undertaking;
(b)   the number of employees at work; and
(c)   the location of the establishment.

The assessment of first-aid needs

[F7003] It is necessary for the employer to assess his first-aid needs and requirements as appropriate to the circumstances of his workplace. The employer's principal aim must be the reduction of the effects of injury and illness at the place of work. Adequate and appropriate first-aid personnel and facilities should be available for rendering assistance to persons with common injuries or illnesses and those likely to arise from specific hazards at work. Similarly, there must be adequate facilities for summoning an ambulance or other professional assistance.

The extent of first-aid provision in a particular workplace that an employer must make depends upon the circumstances of that workplace. There are no fixed levels of first-aid – the employer must assess what personnel and facilities are appropriate and adequate (see appendix B — assessment of first-aid needs). Employers with access to advice from occupational health services (internal or external to their organisation) may wish to use these resources for the purposes of conducting such an assessment and then take the advice given as to what first-aid provision would be deemed appropriate.

In workplaces that employ:

- qualified medical doctors registered with the General Medical Council;
- nurses whose names are registered with the Nursing and Mid-wifery Council; or,
- paramedics registered with the Health Professions Council.

the employer may consider that there is no need to appoint first-aiders (however this judgement must be made as part of the risk assessment exercise).

There is no legal requirement for the results of the first-aid risk assessment to be recorded in writing. However, it may be a useful exercise for the employer – for he may subsequently be asked to demonstrate how he came to the conclusion that the first-aid provision available is adequate and appropriate for the workplace (see appendix C — record of first-aid provision).

When assessing first-aid needs, employers must consider the following:

- the nature of the work and workplace hazards and risks;
- the size of the organisation;
- the history of accidents in the organisation;
- the nature and distribution of the workforce;
- the distance from the workplace to emergency medical services;
- work patterns;
- travelling, distant and lone workers' needs and requirements;
- employees working on shared or multi-occupied sites;
- annual leave and other absences of first-aiders and appointed persons;
- first-aid provision for non employees.

The nature of the work and workplace hazards and risks

[F7004] *The Management of Health and Safety at Work Regulations 1999 (SI 1999 No 3242)* require employers to make a suitable and sufficient assessment of the risks to health and safety at work of their employees. The assessment must be designed to identify the measures required for controlling or preventing any risks to the workforce: highlighting what types of accidents or injuries are most likely to occur will help employers address such key questions as the appropriate nature, quantity and location of first-aid personnel and facilities.

Where the risk assessment conducted by an employer identifies a low risk to health and safety for example in a shop or an office, employers may only need to provide (i) a first-aid box clearly identified and suitably stocked, and (ii) an appointed person to look after first-aid arrangements and resources, and to take control in emergencies. However even in low risk environments it is still possible for an accident or sudden illness to occur and it is recommended that employers consider having a qualified first-aider available.

Where risks to health and safety are greater, employers may need to consider the following:

- the provision of an adequate number of trained first-aiders so that first-aid can be given immediately;
- the additional training of first-aiders in specialist skills to deal with specific risks or hazards;
- informing the local emergency services in writing of the risks and hazards on the site where hazardous substances or processes are in use;
- the requirement for additional first-aid equipment;
- the provision of one or more first-aid room(s).

Employers will need to consider the different risks within each part of their company or organisation. Where an organisation occupies a large premises with different processes being performed in different parts of the premises or within different buildings, each area's or buildings risks must be assessed separately. It would not be appropriate to conduct a generic assessment of needs to cover a variety of activities – the parts of the building with higher risks will need greater first-aid provision than those with lower risk.

A list of common hazards found in workplaces can be found in appendix A.

The size of the company

[F7005] In general, the level of first-aid provision that is required will increase according to the number of employees present. Employers should be aware, however, that in some organisations there may be few employees but the risks to their health and safety might be high – and, as a result, their first-aid needs will also be high.

The history of incidents and accidents

[F7006] When assessing first-aid needs, employers might find it useful to collate data on accidents and near misses that have occurred in the past and then analyse them, for example, the numbers and types of accidents or near misses, their frequency and consequences. Organisations with large premises should refer to such information when determining the first-aid equipment, facilities and personnel that are required to cover specific areas.

The nature and distribution of the workforce

[F7007] The particular needs of young workers, trainees, temporary workers, students on work experience, pregnant workers, employees with disabilities or other special health problems should be addressed. Consideration must also be given to the gender of the employees and any ethnic or cultural needs that may need to be addressed including language barriers.

The employer should bear in mind that the size of the premises can affect the time it might take a first-aider to reach an incident. If there are a number of buildings on the site, or the building in question comprises several storeys, the most suitable arrangement might be for each building or floor to be provided with its own first-aiders.

The distance from the workplace to emergency medical services

[F7008] Where a workplace is far from emergency medical services, it may be necessary to make special arrangements for ensuring that appropriate transport can be provided for taking an injured person to the emergency medical services or providing additional facilities at the workplace to treat the injured person until the emergency medical services arrive.

In every case where the place of work is remote, the very least that an employer should do is to give written details to the local emergency services of the layout of the workplace, and any other relevant information, such as information on specific hazards.

Work patterns

Where employees work shifts or work out of normal working hours it is important to ensure that sufficient provision is always available when they are at work. It may be necessary to have separate arrangements for each shift.

The needs and requirements of travelling, distant and lone workers

[F7009] Employers are responsible for meeting the first-aid needs of their employees whilst they are working away from their main company premises.

When assessing the needs of staff who travel long distances or who are constantly mobile, consideration should be given to what they are required to do, the hazards and risks they are typically exposed to and their medical condition to help decide whether they ought to be provided with a personal first-aid kit.

Organisations with staff working in remote areas or working alone must make special arrangements for those employees in respect of communications (this may include providing mobile phones), special training and arranging emergency transport.

Personnel on sites that are shared or multi-occupied

[F7010] Employers with personnel working on shared or multi-occupied sites can agree to have one employer on the site who is solely responsible for providing first-aid cover for all of the workers. It is strongly recommended that this agreement is written to avoid confusion and misunderstandings between the employers. It will highlight the risks and hazards of each company on the site and will ensure that the shared provision is suitable and sufficient. After the employers have agreed the arrangement, the personnel must be informed accordingly.

When employees are contracted out to other companies, their employer must ensure they have access to first-aid facilities and equipment. The host employer bears the responsibility for providing such facilities and equipment and making the employees aware of it.

First-aiders on annual leave or absent from the workplace

[F7011] Adequate provision of first-aid must be available at all times. Employers should therefore ensure that the arrangements they make for the provision of first-aid at the workplace are adequate to cover for any annual leave of their first-aiders or appointed persons. Such arrangements must also be able to cover for any unplanned or unusual absences from the workplace of first-aiders or appointed persons.

First-aid provisions for non employees

[F7012] Employers are not obliged by these regulations to provide first-aid cover or facilities for members of the public, as the regulations are aimed at employees. However, many organisations like health authorities, schools and colleges, places of entertainment, fairgrounds and shops do make first-aid provision for persons other than their employees, and this practice is strongly recommended by the HSE, in fact they further state that there may be an opportunity for employers to harness the skills and devotion of their first-aiders to further promote health and safety in the workplace. In addition to its general guidance on first-aid at work the HSE has also produced more specific guidance, which incorporates first-aid, in relation to specific activities/ sectors where there might be a large public presence, these include the event safety guide – a guide to health, safety and welfare at music and similar events. HSG195 1999 HSE Books ISBN 0 7176 2453 6, Managing health and safety in swimming pools HSG179 2003 HSE Books ISBN 0 7176 2686 5, Fairgrounds and amusement parks – guidance on safe practice. Practical guidance on the management of health and safety for those involved in the fairgrounds industry. HSG175 2006 HSE Books ISBN 0 7176 1174 4 and Health and safety in care homes. HSG220 2001 HSE Books ISBN 0 7176 2082 4. The Department for children schools and families have produced guidance on first-aid provisions in schools and the *Road Traffic Act 1988* regulates first-aid provision on buses and coaches.

Where employers extend their first-aid provision to cover more than merely their employees, the provision for the employees must not be diminished and should not fall below the standard required by these regulations.

The compulsory element of employers' liability insurance does not cover litigation resulting from first-aid given to non-employees. It is advised that employers check their public liability insurance policy on this issue.

Reassessing first-aid needs

[F7013] In order to ensure that first-aid provision continues to be adequate and appropriate, employers should from time to time review the first-aid provision in the workplace, particularly when changes have been made to working practices. It is advisable to record details of any reviews made and their findings.

Duty of the employer to inform employees of first-aid arrangements

[F7014] Employers are under a duty to inform employees of the arrangements that have been made for first-aid in their workplace. This may be achieved by:

- Distributing guidance to all employees which highlights the key issues in the first-aid arrangements, such as listing the names of all first-aiders and describing where first-aid resources are located.
- Nominating key employees to ensure that the guidance is kept up to date and is distributed to all staff, and to act as information officers for first-aid in the workplace.
- Internal memos can be used as a method of keeping the personnel informed of any changes in the first-aid arrangements.
- Displaying announcements up on notice boards informing employees of the first-aid arrangements and of any changes to those arrangements.
- Providing new, or transferring, employees with the information as part of their induction training.

Any person with a reading or language problem must be given the information in a way that they can understand.

First-aid and the self-employed

[F7015] The self-employed should provide, or ensure that there is provided, such equipment, if any, as is appropriate in the circumstances to enable them to render first-aid to themselves whilst at work. The self-employed who work in low-risk areas, for example at home, are required merely to make first-aid provision appropriate to a domestic environment.

When self-employed people work together on the same site, they are each responsible for their own first-aid arrangements. If they wish to collaborate on first-aid provision, they may agree a joint arrangement to cover all personnel on that particular site.

Number of first-aiders

[F7016] Sufficient numbers of first-aiders should be located strategically on the premises to allow for the administration of first-aid quickly when the occasion arises. The assessment of first-aid needs may have helped to highlight the extent to which there is a need for first-aiders. Appendix E gives suggested numbers of first-aiders (FAW/ EFAW) or appointed persons who should be available at the workplace. The suggested numbers are not a legal requirement – they are merely for guidance.

There are no hard and fast rules on numbers – employers will have to make a judgement, taking into account all of the relevant circumstances of their organisation. If the company is a long way from a medical facility or there are shift workers on site or the premises cover a large area, the numbers of first-aid personnel set out below may not be sufficient – the employer may have to make provision for a greater number of first-aiders to be on site.

Selection of first-aiders

[F7017] First-aiders must be reliable and of a good disposition. Not only should they have good communication skills, but they must also possess the ability and aptitude for acquiring new knowledge and skills, and must be able to handle physical and stressful emergency incidents and procedures. Their position in the company should be such that they are able to leave their place of work immediately to respond to an emergency.

Employers should note that trained first-aiders are also more likely to promote health and safety awareness in the workplace and can have a positive impact in improving the overall safety culture within an organisation.

The training and qualifications of first-aid personnel

[F7018] Before taking up their first-aid duties, a first-aider should hold a valid certificate of competence in either;

- first-aid at work (FAW) issued by a training organisation approved by the HSE; or
- emergency first-aid at work (EFAW) issued by a training organisation approved by the HSE or a recognised awarding body of Ofqual/Scottish Qualifications Authority.

Appendix D details the contents of the FAW and EFAW courses.

Where possible organisations that are contracted to train first-aid personnel should be notified of any particular risks or hazards in a workplace so that the first-aid course provided can be tailored to include the risks specific to that workplace.

Additional special training may be undertaken to deal with unusual risks and hazards (the content of these additional training courses is not specified or approved by the HSE. It may be undertaken as an extension to the FAW/ EFAW training or as a standalone course and a certificate should be issued separately from the FAW/ EFAW certificate). This will enable the first-aider to be competent in dealing with such risks.

It is important for employers to understand that FAW / EFAW certificates are valid for three years. Employers should make every effort to ensure that first-aiders with a current 3 year FAW certificate attend a FAW re-qualification course within the 3 month period prior to the certificate expiry date. However, where this has not been possible, the HSE will allow extension of the certificate for 28 days beyond the expiry date, within which a re-qualification course should be completed. There is no need to contact the HSE to request a certificate extension in such circumstances. During the extension period, the HSE will continue to recognise the FAW qualification and the first-aider will continue to be a suitable person that the employer can use for the purpose of providing first-aid to employees.

Any first-aider who is not able to complete a re-qualification course up to a maximum of 28 days after the expiry date of their 3 year certificate, will be required to retake a full FAW course. Anyone re-qualifying within a period of certificate extension will have their new certificate dated from the expiry date of the previous one.

Employees attending the EFAW course will also need to undertake a requalifiacation course within 3 months prior to their certificate expiry date (every 3 years) to become re-qualified and the same 28 day grace period will apply.

In addition to re-qualification the HSE strongly recommended that employers provide their employees with annual refresher training for FAW /EFAW courses.

It is advisable for employers to keep a record of first-aiders in the company, together with their certification dates, in order to assist them in organising refresher/re-qualification training. Employers should develop a programme of knowledge and skills training for their first-aiders to enable them to be updated on new skills and to make them aware of suitable sources of first-aid information, such as occupational health services and training organisations qualified by the HSE to conduct first-aid at work training.

Note

The EFAW course will be delivered over a minimum of 6 contact hours and the FAW course will be delivered over a minimum of 18 contact hours. After 3 years, first-aiders will need to complete another course (either a 6 contact hour EFAW or 12 contact hour FAW re-qualification course, as appropriate) to obtain a new certificate (Contact hours' refer to teaching and practical time and do not include lunch and breaks etc).

Appointed persons

[F7019] An appointed person is an individual who takes charge of first-aid arrangements for the company, including looking after the facilities and equipment and calling the emergency services when required. The appointed person is allocated these duties when it is found through the first-aid needs assessment that a first-aider is not necessary. The appointed person is the minimum requirement an employer can have in the workplace. Clearly, even if the company is considered a low health and safety risk and, in the opinion of the employer, a first-aider is unnecessary, an accident or illness still may occur, therefore somebody should be nominated to call the emergency services, if required.

Appointed persons are **not** first-aiders – and therefore they should not be called upon to administer first-aid if they have not received the relevant training. However, employers may consider it prudent to send their appointed persons on first-aid appointed persons training courses. Such a course normally last for four hours and includes the following topics:

• action necessary in an emergency;

- cardio-pulmonary resuscitation;
- first-aid treatment for the unconscious casualty;
- first-aid treatment for the bleeding or the wounded.

HSE approval is not required for this training.

The only time an appointed person can replace a first-aider is when the first-aider is absent, due to circumstances that are temporary, unforeseen and exceptional. Appointed persons cannot replace first-aiders who are on annual leave. If the first-aid assessment has identified a requirement for first-aiders, they should be available whenever there is a need for them in the place of work.

Records and record keeping

[F7020] It is considered good practice to keep records of incidents that required the attendance of a first-aider and treatment of an injured person. It is advisable for smaller companies to have just one record book, but for larger organisations this may not be practicable and more than one may be needed, in these circumstances each book should be numbered to aid identification.

The data entered in the record book should include the following:

- the date, time and place of the incident;
- the injured person's name and job title;
- a description of the injury or illness and of the first-aid treatment administered;
- details of where the injured person went after the incident, namely hospital, home or back to work;
- the name and signature of the first-aider or person who dealt with the incident.

This information may be collated to help the employer improve the environment with regard to health and safety in the workplace. It could be used to help determine future first-aid needs assessment and will be helpful for insurance and investigative purposes. The statutory accident book is not the same as the first-aid record book but they may be combined (provided the requirement of each are still met).

Remember employers, the self employed and controllers of premises have a duty to report some accidents and incidents at work under the Reporting of Injuries, Diseases and Dangerous Occurrences Regulations 1995 (RIDDOR) and any data or records held must be held in accordance with the *Data Protection Act 1998*.

First-aid resources

[F7021] Having completed the first-aid needs requirements assessment, the employer must provide the resources; that is the equipment, facilities, materials and time for first-aiders to carry out their duties, which will be needed to ensure that an appropriate level of cover is available to the employee's at all relevant times. First-aid equipment, suitably marked and obtainable, must be made available at specific places where working conditions require it.

First-aid containers

[F7022] First-aid equipment must be suitably stocked and contained in a properly identifiable container (white cross on a green background). At least one first-aid container with a sufficient quantity of first-aid materials must be made available for each worksite – this is the minimum level of first-aid equipment. Larger premises, for example, will require the provision of more than one container.

First-aid containers should be easily accessible and, where possible, near hand-washing facilities. The containers should be used only for first-aid equipment. Tablets and medications (potions, lotions, creams or sprays) should not be kept in them. The first-aid materials within the containers should be protected from damp and dust. It may be practical to provide first-aiders with their own individual containers and then make them responsible for their own specific containers.

Having completed the first-aid needs assessment, the employer will have a good idea as to what first-aid materials should be stocked in the first-aid containers. If there is no specific risk in the workplace, a minimum stock of first-aid materials would normally comprise the following (there is no mandatory list):

- a leaflet giving guidance on first-aid (e.g. HSE leaflet *Basic advice on first-aid at work*);
- 20 individually wrapped sterile plasters (assorted sizes), appropriate to the type of work being carried out and the needs of the employees);
- two sterile eye pads;
- four individually wrapped triangular bandages (preferably sterile);
- six safety pins;
- six medium-sized individually wrapped sterile unmedicated wound dressings – approximately 12 cm × 12 cm;
- two large sterile individually wrapped unmedicated wound dressings – approximately 18 cm × 18 cm;
- one pair of disposable gloves.

This list is a suggestion only – other equivalent materials will be deemed acceptable.

An examination of the first-aid kits should be conducted frequently (i.e. a recorded monthly inspection). Stocks should be replenished as soon as possible after use and ample back-up supplies should be kept on the company premises. Any first-aid materials found to be out of date should be carefully discarded.

Additional first-aid resources

[F7023] If the results of the assessment suggest a need for additional resources such as scissors, adhesive tape, disposable aprons or individually wrapped moist wipes, they can be kept in the first-aid container if space allows. Otherwise they may be kept in a different container, as long as they are ready for use if required.

If the assessment highlights the need for such items as protective equipment, they must be securely stored next to first-aid containers or in first-aid rooms or in the hazard area itself. Only persons who have been trained to use these items may be allowed to use them.

If there is a need for eye irrigation and mains tap water is unavailable, at least a litre of sterile water or sterile normal saline solution (0.9%) in sealed, disposable containers should be provided. If the seal is broken, the containers should be disposed of and not reused. Such containers should also be disposed of when their expiry date has been passed.

Some employees may have their own medication that has been prescribed by their doctor. If an individual needs to take their own prescribed medication, the first-aiders role will be limited to helping them to do so and then contacting the emergency services as appropriate.

If the employer decides to provide a defibrillator in the workplace, it is important to ensure that only those people who are appropriately trained and authorised to use it do so. Employers will need to ensure that staff are appropriately trained to use the equipment provided (the HSE does not currently specify the contents of a defibrillator training course).

First-aid kits for travelling

[F7024] First-aid kits for travelling may contain the following items:

- a leaflet giving general guidance on first-aid (eg HSE leaflet *Basic advice on first-aid at work*);
- six individually wrapped sterile plasters;
- one large sterile unmedicated dressing – approximately 18 cm x 18 cm;
- two triangular bandages;
- two safety pins;
- individually wrapped moist cleansing wipes;
- one pair of disposable gloves.

This list is a suggestion only – it is not mandatory. However, if a kit is provided it must then be regularly inspected and topped up from a back-up store at the home site.

Rooms designated as first-aid areas

[F7025] A suitable room, or rooms, should be made available for first-aid purposes where the first-aid needs assessment found such a room, or rooms, to be necessary. Such room(s) should have sufficient first-aid resources, be easily accessible to stretchers and be easily identifiable, and where possible should be used only for administering first-aid.

First-aid rooms are normally necessary in organisations operating within high-risk industries. Therefore they would be deemed to be necessary on chemical, ship building and large construction sites or on large sites remote from medical services. A person should be made responsible for the first-aid room.

On the door of the first-aid room a list of the names and telephone extensions of all of the first-aiders should be displayed, together with details of how and where they may be contacted on site.

First-aid rooms should:

- have enough space to hold a couch with space in the room for people to work, a desk, a chair and any other resources found necessary;
- where possible, be near an access point in the event that a person needs to be taken to hospital;
- have heating, lighting and ventilation;
- have surfaces that can be easily washed;
- be kept clean and tidy; and
- be available and ready for use whenever employees are in the workplace.

The following is a list of resources that may be found in a first-aid room:

- a record book for logging incidents where first-aid has been administered;
- a telephone;
- a storage area for storing first-aid materials;
- a bed/couch with waterproof protection and clean pillows and blankets;
- a chair;
- a foot-operated refuse bin with disposable yellow clinical waste bags or some receptacle suitable for the safe disposal of clinical waste;
- a sink that has hot and cold running water – also drinking water and disposable cups;
- soap and some form of disposable paper towel.

Where first-aid rooms are provided, employers must also make provision for regular cleaning, emptying of bins (including sharp objects and bodily fluids) and for the laundering of any bed linen or blankets used. If the designated first-aid room has to be shared with the working processes of the company, the employer must consider the implications of the room being needed in an emergency and whether the working processes in that room could be stopped immediately. Can the equipment in the room be removed in an emergency so as not to interfere with any administration of first-aid? Can the first-aid resources and equipment be stored in such a place as to be available quickly when necessary? Lastly, the room must be appropriately identified and, where necessary, be signposted by white lettering or symbols on a green background.

Appendix A

Common hazards found in the workplace

Hazard	Causes of accidents	Examples of injury requiring first-aid
Chemicals	Exposure during handling, spillages, splashing, leaks	Poisoning, loss of consciousness, burns, eye injuries
Electricity	Failure to securely isolate electrical systems and equipment during work on them, poorly maintained electrical equipment, contact with either overhead power lines, underground power cables or mains electricity supplies, using unsuitable electrical equipment in explosive atmospheres	Electric shock, burns
Machinery	Loose hair or clothing becoming tangled in machinery, being hit by moving parts or material thrown from machinery, contact with sharp edges.	Crush injuries, amputations, fractures, lacerations, eye injuries
Manual handling	Repetitive and/ or heavy lifting, bending and twisting, exerting too much force, handling bulky or unstable loads, handling in uncomfortable working positions.	Fractures, lacerations, sprains and strains
Slip and trip hazards	Uneven floors, trailing cables, obstructions, slippery surfaces due to spillages, worn carpets and mats.	Fractures, sprains and strains, lacerations
Work at height	Over-reaching or over-balancing when using ladders, falling off or through a roof.	Head injury, loss of consciousness, spinal injury, fractures, sprains and strains
Workplace transport	Hit by, hit against or falling from a vehicle, being hit by part of a load falling from a vehicle, being injured as a result of a vehicle collapse or overturn.	Crush injuries, fractures, sprains and strains

Appendix B

Assessment of first-aid needs – checklist

The minimum first-aid provision for each worksite is:

- a suitably stocked first-aid container;
- an appointed person to take charge of first-aid arrangements;
- information for employees on first-aid arrangements.

This checklist below will help you assess what additional first-aid provision you need to make for your workplaces.

Factors to Consider

Impact on first-aid provision

Hazards – use the findings of your risk assessment and take account of any parts of your workplace that have different work activities/ hazards which may require different levels of first-aid provision

Does your workplace have low hazards such as those that might be found in offices and shops?

The minimum provision is:
(*a*) an appointed person to take charge of first-aid arrangements;
(*b*) a suitably stocked first-aid box.

Does your workplace have higher hazards such as chemicals or dangerous machinery (see Table 1)?
Do your work activities involve special hazards such as hydrofluoric acid or confined spaces?

You should consider:
(*a*) providing first-aiders;
(*b*) additional training for first-aiders to deal with injuries resulting from special hazards;
(*c*) additional first-aid equipment
(*d*) precise siting of first-aid equipment;
(*e*) providing a first-aid room;
(*f*) informing the emergency services.

Employees

How many people are employed on site?

Where there are small numbers of employees, the minimum provision is:
(*a*) an appointed person to take charge of first-aid arrangements;
(*b*) a suitably stocked first-aid box.
Even in workplaces with a small number of employees, there is still the possibility of an accident or sudden illness so you should consider providing a qualified first-aider.
Where there are large numbers of employees you should consider providing:
(*a*) first-aiders;
(*b*) additional first-aid equipment;
(*c*) a first-aid room.

Are there inexperienced workers on site, or employees with disabilities or special health problems?

You should consider:
(*a*) additional training for first-aiders;
(*b*) additional first-aid equipment;
(*c*) local siting of first-aid equipment.
Your first-aid provision should cover any work experience trainees.

Record of accidents and ill health

What is your record of accidents and ill health? What injuries and illness have occurred and where did they happen?

Ensure your first-aid provision will cater for the type of injuries and illness that might occur in your workplace. Monitor accidents and ill health and review your first-aid provision as appropriate.

Working arrangements

Do you have employees who travel a lot, work remotely or work alone?

You should consider:
(*a*) issuing personal first-aid kits;
(*b*) issuing personal communicators to remote workers;
(*c*) issuing mobile phones to lone workers.

Do any of your employees work shifts or work out of hours?

You should ensure there is adequate first-aid provision at all times people are at work.

Are the premises spread out, for example are there several buildings on the site or multi-floor buildings?

You should consider provision in each building or on each floor.

Is your workplace remote from emergency medical services?

You should:
(*a*) consider special arrangements with the emergency services;
(*b*) inform the emergency services of your location.

Do any of your employees work at sites occupied by other employers?

You should make arrangements with other site occupiers to ensure adequate provision of first-aid. A written agreement between employers is strongly recommended.

Do you have sufficient provision to cover absences of first-aiders or appointed persons?

You should consider:
(*a*) consider special arrangements with the emergency services;
(*b*) what cover is needed for unplanned and exceptional absences.

Non-employees

Do members of the public visit your premises?

Under the Regulations, you have no legal obligation to provide first-aid for non-employees but HSE strongly recommends that you include them in your first-aid provision. This is particularly relevant in workplaces that provide a service for others such as schools, places of entertainment, fairgrounds and shops.

Do not forget to allow for leave or absences of first-aiders and appointed persons. First-aid personnel must be available at all times when people are at work.

Appendix C

Record of First-aid Provision

First-aid personnel	Required yes / no	Number needed
First-aider with a first-aid at work certificate		
First-aider with an emergency first-aid at work certificate		
First-aider with additional training		
Appointed person		
First-aid equipment and facilities	**Required yes / no**	**Number needed**
First-aid container		
Additional equipment (specify)		
Travelling first-aid kit		
First-aid room		

Appendix D

Content of a first-aid at work course

On completion of training, successful candidates should be competent in:

- emergency first-aid at work (see below);
- recognising the presence of major illness and applying general first-aid principles in its management.

In addition, candidates should be able to demonstrate the correct first-aid management of:

- soft tissue injuries;
- injuries to bones including suspected spinal injuries;
- chest injuries;
- burns and scalds;
- eye injuries including how to irrigate an eye;
- sudden poisoning and anaphylactic shock.

Content of an emergency first-aid at work course

On completion of training, successful candidates should be able to:

- understand the role of the first-aider including reference to the use of available equipment, the need for recording incidents and actions and the importance of avoiding cross contamination;
- understand the importance of basic hygiene in first-aid procedures;
- assess the situation and circumstances in order to act safely, promptly and effectively in an emergency;
- administer first-aid to a casualty who is unconscious and/or in seizure;
- administer cardiopulmonary resuscitation;
- administer first-aid to a casualty who is wounded or bleeding and/or in shock;
- administer first-aid to a casualty who is choking;
- provide appropriate first-aid for minor injuries.

Appendix E

Guide to the category and number of first-aid personnel to be available at all times people are at work

First-aid personnel

Category of risk	Numbers employed at any location	Suggested number of first-aid personnel
Lower hazard – eg shops, offices, libraries	less than 25	at least one appointed person**
	25–50	at least one first-aider trained in EFAW
	more than 50	at least one first-aider (FAW) per 100 employees or part thereof
Higher Hazard – eg light engineering and assembly work, food processing, warehousing, extensive work with dangerous machinery or sharp instruments, construction, slaughterhouse, chemical manufacturer, work involving special hazards* such as hydrofluoric acid or confined spaces.	less than 5	at least one appointed person**
	5–50	at least one first-aider (EFAW or FAW)***
	more than 50	at least one first-aider (FAW) per 50 employees or part thereof

* additional training may be needed for first-aiders to deal with injuries resulting from special hazards.

** where first-aiders are shown to be unnecessary, there is still a possibility of an accident or sudden illness, so employers should consider providing qualified first-aiders.

*** the type of injuries that might arise in working with those hazards identified, will influence whether the first-aider should be trained in FAW or EFAW.

Food Safety and Standards

Neville Craddock

Introduction

[F9001] Legislation has governed the sale of food for centuries. In Europe we can trace it back at least to the Middle Ages, and the ancient Hebrew food laws, found notably in the Book of Deuteronomy, show that it goes back even further. In the UK, the trade guilds were instrumental in the introduction of legislation aimed at stopping the adulteration of a wide variety of foods and the adulteration of tea, coffee and bread was prohibited by legislation as long ago as the early-mid 18th century.

Two themes have existed from the start:

*   *Food Safety* – the protection of the health and well being of anyone eating food; and
*   *Food Standards* – the control of composition and adulteration of food for the prevention of fraud.

UK legislation is structured principally around the *Food Safety Act 1990* supported by a whole raft of more detailed regulations. Increasingly over the past 35 years Directives and Regulations from the European Union have determined UK legislation – and the last eight years in particular have seen an almost total revision of food safety and consumer protection rules. Devolution within the UK has seen the responsibility for food policy shifted to the Scottish parliament and regional assemblies, with the resultant introduction of a number of local initiatives in areas such as nutrition but little impact, to date, in respect of traditional food safety.

Food safety

[F9002] The principal legislation in the UK is the *Food Safety Act 1990*, to which significant amendments have been introduced since January 2005 in order to align the technical details with the EU General Food Law Regulation 178/2002/EC. In addition, fundamental changes to the UK food hygiene legislation have been introduced with effect from January 2006.

EU Regulation 178/2002/EC introduced new food safety requirements, new traceability requirements, and also measures to ensure effective product recall/withdrawals and notification to competent authorities. Although as an EU Regulation it was directly applicable in Member States, it was still necessary to amend the basic UK Act to introduce the new provisions and to ensure conformity with the EU law.

The *Food Safety Act 1990 (Amendment) Regulations 2004 (SI 2004 No 2990)* align the definition of 'food' in the *Food Safety Act 1990* with the definition in the General Food Law Regulation (EC) 178/2002/EC, and also make some minor amendments to the *Food Safety Act 1990* in respect of public consultation requirements, to remove duplication with directly applicable provisions in Regulation 178/2002/EC.

The *General Food Regulations 2004 (SI 2004 No 3279)* introduce consequential amendments relating to definitions of safety, in particular to *ss 7* and *8* which provide the necessary enforcement powers in respect of the new food safety requirements.

Since the inception of the European Economic Community (EEC), now the European Union (EU), European 'hygiene' regulations had developed on the basis of specific measures for specific food sectors, the so-called 'vertical' regulations. A full complement of these was already in place by the 1990s covering the processing of most foods of animal origin such as meat, fish, eggs, milk and related products, and even honey. These 'vertical' regulations did not in general cover businesses selling food direct to the ultimate consumer, for example catering or retail businesses. The 'horizontal' Food Hygiene Directive 93/43/EEC had previously established general principles for food hygiene and set hygiene standards for those businesses not covered by 'vertical' legislation, such as food factories processing non-animal products and also retail or catering outlets.

However, in 2004 the EU completed a radical overhaul of all food hygiene legislation, consolidating and recasting both the general requirements and those relating to animal-derived products into two principal Regulations (*Regulation 852/2004 on the Hygiene of Foodstuffs and Regulation 853/2004 laying down Specific Hygiene Rules for Products of Animal Origin*). These have been transposed into UK legislation as the *Food Hygiene Regulations 2006 (SI 2006 No 14, amended by SI 2007 No 56)*. Each of the devolved regions has introduced their own set of hygiene legislation but, for the purposes of this summary, they will be grouped under the generic description *Food Hygiene Regulations 2006 (as amended)*. With the exception of the temperature control requirements in Scotland, the provisions are technically virtually identical (the reader is however advised to consult the original text of the regulations appropriate to his region. These Regulations are wide-ranging and have replaced numerous previous provisions.

The following account will be largely restricted to the provisions of the *Food Safety Act 1990 (as amended)* and the *Food Hygiene Regulations 2006 (SI 2006 No 14), as amended.*

Food standards

[F9003] Food standards are governed by a complex range of regulations. Despite its name, the *Food Safety Act 1990* is also the primary legislation covering most aspects of food standards and a number of regulations have been made under it. Over recent years, there has been a significant change in emphasis. Whereas previously there was a tendency to prescribe compositional standards for foods, many of these have disappeared and relatively little of this legacy remains. The approach proved insufficiently flexible in a climate of rapid innovation and product development, and particularly when exposed to the range of competing products traded within the expanding EU Internal Market. The emphasis therefore shifted to allow more diversity of composition, but supported by informative labelling. The *Food Labelling Regulations 1996 (SI 1996 No 1499, plus several amendments)* play an important role in providing a substantial amount of information to consumers, including nutritional information in many cases, about the composition and properties of individual foods and how they should be safely stored and prepared. The *Food Labelling (Amendment) Regulations 1998 (SI 1998 No 1398)* took this even further with a specific requirement that pre-packed foods must carry a 'quantitative ingredient declaration' or 'QUID'. This dictates that the label must carry a declaration of the percentage quantity of characterising or emphasised ingredients of any food.

In 2004, these Regulations were further amended to require 12 specified allergens to be declared when they or their derivatives are deliberately added to pre-packed food, including alcoholic drinks, *(Food Labelling (Amendment) (England) (No. 2) Regulations SI 2004 No 2824)*. Lupins and molluscs were added to the list by the *Food Labelling (Declaration of Allergens) (England) Regulations SI 2007 No 3256*. However, not all highly-refined derivatives of these specified allergens are allergenic. The *Food Labelling (Amendment) (England) (No 2) Regulations 2005 (SI 2005 No 2057)*, amended by *SI 2005 No 2969*, therefore set out temporary exemptions from the allergen labelling rules. These temporary exemptions expired in 2007 but Directive 2007/68/EC established a list of permanent exemptions, which has now been implemented into UK legislation *(Food Labelling (Amendment) (England) Regulations 2008 (SI 2008 No 1188))*. Again, separate, but parallel legislation on allergen declarations applies in Scotland, Wales and Northern Ireland.

Food standards and labelling have also been enforced, historically, through the more general Trades Descriptions Act 1968. However, many of the provisions of this Act have been repealed and superseded by new Consumer Protection from *Unfair Trading Regulations (SI 2008 No 1277)* and the *Business Protection from Misleading Marketing Regulations 2008 (SI 2008 No 1276)*, which implement Directive 2005/29/EC on Unfair Commercial Practices and Directive 2006/114/EC on Misleading and Comparative Advertising, respectively.

Administrative regulations

[F9004] Finally, there are more general regulations made under the *Food Safety Act 1990*. These may regulate the law enforcement process, setting criteria for inspectors, food analysts or examiners. A new regime of Official Control legislation has been introduced from January 2006 to implement EU Regulation 854/2004, covering animal-derived products, and Regulation 882/2004, covering all aspects of other food and feed controls including imported products. Previous UK legislation on the registration of premises and temperature control is now covered under the *Food Hygiene Regulations 2006 (SI 2006 No 14), amended by SI 2007 No 56*. In addition, many other pieces of consumer protection legislation are applicable to food; notably the *Weights and Measures Act 1985*, and secondary legislation under it, which controls the way in which food is sold by weight or other measures, the *Consumer*

Protection Act 1987, which establishes a civil law right of redress for death, or injury, caused by using defective consumer goods (the so-called 'product liability' provisions), and the *Prices Act 1974* and the various Price Marking Orders laid under it. Detailed guidelines on the application of these pieces of legislation can be obtained from the Department for Business, Enterprise and Regulatory Reform (BERR), formerly DTI.

Policy and enforcement

[F9005] Since 2000, UK policy on food safety and standards has derived principally from the Food Standards Agency (FSA) although much of the underlying policy and legislation is initiated from Europe. The FSA is a non-ministerial government department that was created and empowered by the *Food Standards Act 1999* in the wake of BSE and other food 'crises' to combine the food policy functions of the Department of Health (DH) and former Ministry of Agriculture, Fisheries and Food (MAFF) into an organisation which operates at arm's length from the government with a primary role of protecting consumers' interests. Complementary FSA structures exist in Scotland, Wales and Northern Ireland; these have limited, regional food policy responsibilities such that minor differences exist between the devolved regions but, generally, food safety legislation is almost identical across the UK.

Complex arrangements are in place for the enforcement of food legislation. Traditionally, local authorities have had the responsibility although central government officials have particular responsibilities in 'upstream' parts of the food chain. This is particularly true of 'meat hygiene' in abattoirs and cutting plants. This role was taken away from local authorities to a central Meat Hygiene Service (MHS) with the MHS itself becoming a branch of the FSA from April 2000. The FSA has also developed a role of local authority monitoring and, increasingly, plays a direct role in enforcement issues that cross local authority borders such as food fraud and contamination.

In April 2009, the FSA established a new Food Fraud Advisory Unit, specifically to support local authorities in their work to tackle food fraud in relation to any illegal activity relating to food or feed. This could cover, for example, the sale of food that is unfit and potentially harmful, or the deliberate misdescription of food in a way that, while not necessarily unsafe, deceives the consumer as to the nature of the product.

Most food enforcement at the production and retail level falls to two groups of enforcement officers employed by local authorities or Port Health Authorities ('authorised officers'):

- Environmental Health Practitioners (EHPs) have responsibility for food safety legislation. They are generally employed at district council (town hall) level.
- Trading Standards Officers (TSOs) take control of food standards issues, including weights and measures. Generally they are employed at county council level.

The distinction between district and county councils has become increasingly blurred with much local government organised on a unitary basis. This includes most English metropolitan areas and all authorities in Scotland. In these cases TSO and EHP functions are usually combined in the same department.

Veterinary inspectors may also be involved in controlling and inspecting imports and exports of foods containing animal-based products. Specialist regimes also operate for wines and spirits, and for organic foods. Following numerous cases of excessive contamination of certain imported foods of non-animal origin, such as nuts, these are increasingly subject to enforcement inspection at officially-designated Border Inspection Posts.

Food Safety Act 1990

[F9006] The *Food Safety Act 1990,* as amended, defines the offences that may be committed if food legislation is contravened and specifies defences that may be available if charges are instigated. It sets up the mechanisms to enforce the laws and fixes penalties that may be levied. The Act also enables the government to make regulations that include more detailed food safety and consumer protection measures.

The Act is divided into four principle sections:

(i)   Part I: Definitions and responsibilities for enforcement;

(ii)   Part II: Main provisions;

(iii)   Part III: Administrative and enforcement issues such as powers of entry; and

(iv)   Part IV: Miscellaneous arrangements, notably the power to issue codes of practice.

EU Regulation 178/2002/EC (the '*General Food Law Regulation*') introduced new food safety requirements and new measures to ensure effective product traceability, recall/withdrawals and notification to competent authorities. Although, as an EU Regulation, 178/2002/EC is directly applicable in Member States, consequential changes to domestic legislation were necessary to introduce new enforcement provisions, and to ensure conformity with this EU law.

The *Food Safety Act 1990 (Amendment) Regulations 2004 (SI 2004 No 2990)* made minor changes to the definition of 'food' in the *Food Safety Act 1990* to align it with the new EU definition and also made some minor amendments to the Act itself in respect of public consultation requirements, and to remove duplication with the directly applicable provisions in Regulation 178/2002/EC. The *General Food Regulations 2004 (SI 2004 No 3279)* provide enforcement powers in respect of the food safety and traceability requirements introduced under Regulation 178/2002/EC, designate competent authorities, specify enforcement authorities, make provision for offences and penalties, and introduce some consequential amendments to the *Food Safety Act 1990*, in particular *sections* 7 and 8.

The following is a brief summary of the main provisions contained within the Act.

Section 1: definitions

[F9007] Under the *Food Safety Act 1990*, as amended, 'food' has a wide meaning and includes drink, chewing gum and any substance, including water, intentionally incorporated into the food during its manufacture, preparation or treatment. It includes water after the point of compliance as defined in Article 6 of Directive 98/83/EC (Water Quality for Human Consumption) and without prejudice to the requirements of Directives 80/778/EEC (Natural Mineral Waters) and 98/83/EC.

'Food' does not include live animals, birds or fish (although shellfish that are eaten raw and alive such as oysters and similar are classed as 'food'). Neither does 'food' include animal feed, residues and contaminants, nor controlled drugs that might be taken orally, medicines or cosmetics that are covered under separate Community legislation.

One key difference is that unlicensed medicinal products are excluded from the definition of 'food' if they are medicinal products within the meaning of the Medicines Directive 2001/83/EC, as amended. However, certain 'borderline' medicinal products (which do not fall within the definitions of Directive 2001/83/EC) will now be included in the new definition of 'food'. The Medicines and Healthcare Products Regulatory Agency (MHRA) will determine whether a given product is a 'medicinal product' or a food on a case by case basis, having regard to the overall presentation and function of the product. Directive 2004/24 defines a range of so-called traditional herbal remedies and brings these into the medicines legislation. This will, ultimately, impact many products containing exotic plant extracts currently considered as food supplements.

The revised definition of 'food' automatically applies to other legislation in related areas that uses the Food Safety Act definition, for example the *Food Standards Act 1999* and the *Food and Environment Protection Act 1985*, as well as to Regulations and Orders made under these Acts.

The scope of the *Food Safety Act 1990* is very wide and covers all commercial businesses, ranging from farmers, manufacturers, wholesalers, distributors, retailers to businesses such as canteens, clubs, schools, hospitals, care homes and so on, whether or not they are run for profit. Government establishments that once had 'crown immunity' are treated no differently from other food businesses. Charity events that sell food are also subject to the safety provisions of the Act but may be exempt from certain administrative requirements.

The Act thus covers any premises from an abattoir to a retail superstore, from a street vendor to a five star hotel. It includes any place (including premises used only occasionally for a food business such as a village hall), any vehicle, mobile stalls and temporary structures.

The *Food Safety Act 1990, s 2* provides an extended, very wide meaning to the term 'sale' of food, which now encompasses the European concept of 'placing on the market'. The Act does not apply only to sales where money changes hands and the business is run for profit. It also covers food given as a prize or as a reward by way of business promotion or entertainment. Entertainment includes social gatherings, exhibitions, games, sport and so on. Thus, if food, which turns out to be unfit, is given as a prize for a competition, it could be subject to an action under the 1990 Act. Similarly, a food business cannot avoid its obligations under the Act by giving away food with other non-food items for which payment is accepted.

The *Food Safety Act 1990, s 3* establishes a presumption that food is intended for human consumption. If a food business has any food or substance capable of being used in the preparation of food in its possession, the Act presumes that the intention is to sell it, unless it can be proved otherwise. Where food raw materials, ingredients, additives or finished products are on trade premises but are either not ready for consumption or have been rejected, the presumption is that they are intended for manufacture or sale and it is prudent to keep them in separate rooms, areas or batches clearly identified as being 'not for human consumption'. For example, food past its 'use by' date should be segregated from food that is for sale and marked clearly, otherwise an enforcement officer could presume that it was for sale.

The *Food Safety Act 1990, ss 4–6*, defines the roles and responsibilities of local and central government and the officials authorised to act on their behalf. Previous formal Codes of Practice made under s 40 of the Act, to offer guidance to non-metropolitan county and district councils, on these roles have been replaced by a comprehensive Food Law Code of Practice (June 2008) and associated Food Law Practice Guidance (March 2006), which encompass enforcement of all aspects of food hygiene legislation and the requirements of Regulation 882/2004 on Official Food and Feed Controls. Parallel documents are in place for the devolved administrations.

Food Authorities must also have regard to the *Framework Agreement on Local Authority Food Law Enforcement* (the Framework Agreement), which reflects the requirements of the Food Law Code. The Framework Agreement is also consistent with the principles of the Enforcement Concordat published by the Cabinet Office, Better Regulation Unit.

Additionally, from April 2008, Food Authorities are also required to give due regard to the Regulators' Compliance Code in relation to their general approach to enforcement (*D-BERR Statutory Code of Practice for Regulators, Dec. 2007*).

Part II: main provisions of the Act

[F9008] The main provisions of the *Food Safety Act 1990* fall within *ss 7–22*. The offences are divided into two types: food safety and consumer protection, which each break down into two separate offences:

(i) *Food Safety*
 • *Section 7*: Rendering food injurious to health.
 • *Section 8*: Selling food not complying with food safety requirements.
(ii) *Consumer Protection*
 • *Section 14*: Food not of the nature, substance or quality demanded.

- *Section 15*: Falsely presenting or describing food.

Section 7: rendering food injurious to health

[F9009] It is an offence to do anything intentional that would make food harmful to anyone that eats it. Even if one did not know that it would have that effect, an offence would still have been committed. For example, subjecting food to poor temperature control that allows bacteria to multiply could render food injurious to health. An example of a more deliberate offence would be the addition to, or removal from, food of substances or components so as to render the food injurious to health.

Section 7 has been amended by the *General Food Regulations 2004 (SI 2004 No 3279)* to require regard to be had not only to the probable immediate and/or short-term and/or long-term effects of that food on the health of a person consuming it, but also on subsequent generations. It also requires the probable cumulative toxic effects and particular health sensitivities of a specific category of consumers, where the food is intended for that category of consumers, to be taken into account. Food would be injurious to health if it were likely to cause immediate harm to anyone that ate it, for example foods contaminants with pathogenic micro-organisms such as listeria, salmonella, etc. It would also be injurious to health if the harm were cumulative over a long period, for example fungal toxins in nuts or cereal products. There is no Court precedence to say whether an adverse allergic reaction to a food (eg anaphylactic shock in response to undeclared peanut) would fall under this section of the Act or remain under the more general provisions of the *Consumer Protection Act 1987*.

Section 8: selling food not complying with food safety requirements

[F9010] Whilst *s 7* deals with the person who renders food injurious to health, the *Food Safety Act 1990, s 8* made it an offence for him or anyone else in the food supply chain to sell food which 'fails to comply with food safety requirements' as defined in the EU Regulation. This offence has now been incorporated into the *General Food Regulations 2004 (SI 2004 No 3279)* which link the offence to contravention of Article 14(1) of EC Regulation 178/2002.

Food would be deemed to be unsafe if it was considered to be either 'injurious to health' or 'unfit for human consumption'. The concept of 'injurious to health' relates to safety, but the concept of 'unfit' relates to unacceptability. Unfit food is not necessarily unsafe in the ordinary sense of the word (eg sour milk and certain mouldy foods).

In determining whether any food is 'unsafe', consideration would be given to the normal use of the food and its handling at each stage of the chain, and to any information provided to the consumer, such as labelling, to help him avoid specific adverse health effects.

In respect of whether a food is 'injurious to health', regard would be had not only to the probable immediate and/or short-term and/or long-term effects of the food on the health of someone who had eaten it, but also on subsequent generations and to any probable cumulative toxic effects. The particular health sensitivities of a specific category of consumers would also be relevant if the food is specifically intended for them. However, where a food is not specifically intended for such a group, the fact that it may harm someone in that group does not automatically mean that it is 'unsafe' for general consumption.

It should be noted that food may be 'injurious to health' under Article 14 in circumstances that were not previously covered by *section 8(2)(a)* of the *Food Safety Act*, which only referred to food which had been rendered injurious to health by means of certain operations. Thus, for example, the requirements would now cover jelly cup sweets intended for consumption by young children, where there is a risk of choking.

As with *s 7* (above), there is no court precedence to indicate whether an adverse allergic reaction to a food would fall under this section of the Act or remain under the more general provisions of the *Consumer Protection Act 1987*. However, the view of the Food Standards Agency is that foods containing undeclared allergens are to be considered as 'unsafe', resulting in frequent and numerous recalls and withdrawals of such products from the market.

Food would be 'unfit' (ie unacceptable) for consumption if it was putrid or contained serious foreign material, for example a dead mouse. This section gives wide scope to prosecute someone for selling food that fails to meet a customer's expectations. For example, if the food is mouldy or contains a rusty nail or there are excessive antibiotic residues in meat.

(The *Food Safety Act 1990, ss 9–13* provide enforcement procedures for food safety offences. These sections are dealt with from **F9014–F9019**.)

Section 14: selling food not of the nature, substance or quality demanded

[F9011] It is an offence to supply to the prejudice of the purchaser food which is not of the nature, substance or quality demanded. The purchaser in this case does not have to be a customer in a retail store or catering outlet. One company, large or small may purchase from another and an offence is committed if the food is inferior in nature or substance or quality to that which they demanded. Once again this allows the law to be applied at any point of the food chain.

The purchaser does not have to buy the food for his or her own use to be prejudiced. They may intend to give it to someone else, for example members of their family.

The three offences created by this section are separate and a charge will be brought under one of the three according to the circumstances. These offences have provided a flexible and far reaching means of preventing the adulteration, contamination and mis-description of food and have been the foundation of food law in the United Kingdom for over 130 years. The wide application of this section has been the subject of many cases in law, and a detailed consideration of them is outside the scope of this brief overview.

If a purchaser asks for cod and gets coley, or beef mince contains a mixture of lamb or chicken, it would be not of the 'nature' demanded.

If a purchaser asked for diet cola and was served regular cola, or expected a sheep's milk cheese and it was made from cow's milk, it would be not of the 'quality' demanded.

Food that was not of the quality or substance demanded often formed the basis for complaints of mouldy food or foreign material. Such complaints are likely to be taken under the *Food Safety Act 1990, s 8* (ie that the food is so contaminated that it would be unreasonable to expect it to be eaten).

The above provision has also been used when a susceptible customer has requested information about peanuts or similar ingredients to which they may have a severe allergic reaction. If, notwithstanding such a request, the consumer is sold food that triggers a reaction, cases have been brought on the grounds that the food was not of the 'substance' demanded.

Section 15: falsely presenting or describing food

[F9012] *The Food Safety Act 1990, s 15* provides the principal protection for consumers in respect of food composition and description. It prohibits false and misleading descriptions of food by way of labelling and advertising and is backed up with very detailed requirements in the *Food Labelling Regulations 1996 (SI 1996 No 1499)* and other 'vertical' regulations that impose specific, additional labelling and compositional requirements on particular foods. It is also an offence if material that is technically accurate is presented in such a way as to mislead the consumer. The scope of this section includes misleading pictorial representation. For some offences, especially labelling by caterers on menus or chalkboards, TSOs have also historically invoked the more general requirements under the, now revoked, *Trades Descriptions Act 1968, s 1(1)*.

Article 16 of Regulation 178/2002 (relating to "Presentation") stipulates that the labelling, advertising and presentation of food shall not mislead consumers. This provision is not limited to 'food business operators'. It applies additionally to *Sections 14* and *15* of the *Food Safety Act 1990*. Unlike *sections 14* and *15*, Article 16 applies in the case of one-off supply free of charge, and also covers cases where a consumer is misled, whether or not it relates to the nature, substance or quality of the food.

The Trade Descriptions Act 1968 has been repealed and superseded from mid-2008 by new *Consumer Protection from Unfair Trading Regulations 2008 (SI 2008 No 1277)*, which implement the Unfair Commercial Practices Directive 2005/29/EC.

The new Regulations impose a 'general prohibition' on the use of unfair commercial practices. A commercial practice will be unfair if it contravenes the requirements of professional diligence and it materially distorts, or is likely to materially distort, the economic behaviour of the typical consumer.

A commercial practice is a misleading action if it contains false information or in any way deceives or is likely to deceive the typical consumer in relation to a large number of matters specified in the legislation, and this causes or is likely to cause the consumer to take a transactional decision he would not otherwise have taken. If a commercial practice misleads in relation to a matter not directly specified in the regulations, then its unfairness will have to be assessed against the 'general prohibition'. As with the previous *Trades Descriptions Act 1968*, the omission of relevant particulars may also be considered a misleading action.

Notwithstanding the requirements of food law and the procedures in place for their formal enforcement, food manufacturers, retailers and traders should also be aware of the role played by the Advertising Standards Authority (ASA) in matters related to misleading broadcast and non-broadcast advertising under the Advertising Standards Code for Broadcast Committee of Advertising Practice (BCAP Code) and the British Code of Advertising, Sales Promotion and Direct Marketing (CAP Code).

ASA is an independent, non-statutory body responsible for administering these Codes so that all broadcast and non-broadcast advertisements are perceived to be legal, decent, honest and truthful. The ASA receives and investigates complaints from the public and industry, and decisions on compliance are taken by the ASA Council; if a complaint is upheld, the advertisement must be withdrawn or amended. ASA adjudications are published and made available to the media.

BCAP is the regulatory body responsible for maintaining a number of standards Codes under a contracting-out agreement with the Office of Communications (Ofcom), the latter having statutory responsibility, under the Communications Act 2003, for maintaining standards in TV and radio advertisements.

CAP is the self-regulatory body that creates, revises and enforces the CAP Code, which covers UK-originated, non-broadcast marketing communications, such as advertisements placed in traditional and new media, sales promotions and direct marketing communications. Parties that do not comply with the CAP Code are liable to sanctions including the denial of media space and adverse publicity.

Enforcement

[F9013] The *Food Safety Act 1990* gives enforcement officers strong powers, namely:

- To enter food premises to investigate possible offences.
- To inspect food.
- To detain or seize suspect food.
- To take action that requires a business to put things right.
- If all else fails:
  - to prosecute,
  - to prohibit the use of premises or equipment,
  - to bar an individual from working in a food business.

Under the *Food Safety Act 1990*, ss 32 and 33, enforcement officers have rights of access at any reasonable time and an offence will be committed if they are obstructed. They must be given reasonable information and assistance. They can inspect premises, processes and records. If records are kept on a computer, they are entitled to have access to them. They can take samples, photographs and even videos, and (under specific circumstances, eg where there is reason to believe an offence has been committed) copy records. If they are refused access to any of these, they can apply to a magistrate for a warrant. They must give 24 hours notice to enter private houses used in connection with a food business.

Of course, none of these requirements to co-operate with officers and give them access to information cancel the basic right to avoid self-incrimination. In the extreme situation, one has the right to remain silent and to consult a solicitor.

Officers themselves commit an offence if they reveal any trade secrets learned in the course of official duties.

The *Food Safety Act 1990, ss 29–31*, allows officers to procure samples and control the sampling and testing procedures. Section 6 of the FSA Food Law Practice Guidelines covers 'Sampling for Analysis'. Samples that are taken carelessly may give unreliable results and may be not be acceptable as evidence in court.

The *Food Safety Act 1990*, ss 9–13, provides enforcement officers with a series of enforcement measures to control food or food businesses that present a health risk.

Section 9: detention and seizure of food

[F9014] *Section 9* provides that, if an officer *suspects* that food is unfit or fails to meet food safety requirements, he can issue a detention notice. This requires the person in charge of the food to keep it in a specified place and not to use it for human consumption pending investigation. Chapter 3.4 of the Food Law Code of Practice describes the circumstances when the use of detention and seizure powers under the *Food Safety Act 1990, s 9*, as amended, is appropriate, including after food has been certified in accordance with the *Food Hygiene (England) Regulations 2006 (SI 2006 No 14), Reg 27*. It also covers the procedures for serving and withdrawal of notices; voluntary surrender; and the destruction or disposal of food. The detention notice must be written on a prescribed form, *Detention of Food (Prescribed Forms) Regulations 1990 (SI 1990 No 2614)*. It is an offence to ignore or contravene a detention notice even if the business may believe that it is unjustified. The officer has 21 days to complete his investigations but must clearly act quickly where the food is highly perishable. If he concludes that the food was actually safe, he must withdraw the notice (using another prescribed form) and restore the food to the person in charge. All relevant forms can be found in Annex 7 of the FSA Food Law Code of Practice.

Authority

Food Safety Act 1990 - Section 9

DETENTION OF FOOD NOTICE

Reference Number:

1. To: ---
    Of: ---

2. Food to which this notice applies:
    Description: ---
    Quantity: ---
    Identification marks: ---

3. *THIS FOOD IS NOT TO BE USED FOR HUMAN CONSUMPTION.*
    In my opinion, the food does not comply with food safety requirements because: ------------------
    ----
    ----

4. The food must not be removed from:

    ----
    *unless it is moved to:

    (*Officer to delete if not applicable)

5. Within 21 days, either this notice will be withdrawn and the food released, or the food will be
    seized to be dealt with by a justice of the peace, or in Scotland a sheriff or magistrate, who may
    condemn it.

    Signed: --- Authorised Officer
    Name in capitals: --
    Date: --
    Address: --

    Tel: ------------------------- Fax:---

> *Please read the notes overleaf carefully. If you are not sure of your rights or the implications of this notice, you may want to seek legal advice.*

NOTES

1. The food described in this notice has been detained pending official investigation.

2. The food must not be used for human consumption until it is released by the officer.

3. The food must remain where it is. If it is moved, it may only be moved to the place stated in paragraph 4 of the notice.

4. If for some reason you need to move the food after receiving this notice, you should contact the officer.

5. Within 21 days the officer must tell you if the notice is being withdrawn or if he is seizing the food for it to be dealt with by a justice of the peace, or in Scotland a sheriff or magistrate, who may condemn it.

6. *COMPENSATION:* If this notice is withdrawn and the food released for human consumption, then you may be entitled to compensation from the authority. This compensation will be payable for any loss in value of the food resulting from the effect of the notice.

> *WARNING*

FAILURE TO COMPLY KNOWINGLY WITH THIS NOTICE IS AN OFFENCE

Offenders will be liable:

···*on summary conviction, to a fine of up to £2000 and/or 6 months in prison,*

or

—*on conviction on indictment, to an unlimited fine and/or up to 2 years in prison.*

Authority: _____   FORM 2

Food Safety Act 1990—Section 9

WITHDRAWAL OF DETENTION OF FOOD NOTICE

1. To: ....
    Of: ....
    ....

2. Detention Notice Number .... dated .... and served on you on
    .... (date) is now withdrawn. The food described in paragraph 3 below can
    now be used for human consumption.

3. Food released for human consumption:
    Description: ....
    Quantity: ....
    Identification marks: ....

    Signed: .... Authorised Officer
    Name in capitals: ....
    Date: ....
    Address: ....
                    ....
                    ....
    Tel: .... Fax ....

> *Please read the notes overleaf carefully. If
> you are not sure of your rights or the
> implications of this notice, you may want to
> seek legal advice.*

NOTES

1. The food described in this notice has now been released for human consumption.

2. If this notice does not relate to all of the food originally detained, then the rest has been seized
    under section 9(3)(b) of the Food Safety Act 1990.

3. *COMPENSATION:* If you can show that any of the food now released for human consumption
    has lost value, you may be entitled to compensation from the authority. Compensation will be
    payable for any loss in value resulting from the effect of the notice.

NOTES

1. You are being warned that the Authority will be applying to a justice of the peace, or in Scotland a sheriff or magistrate, for the food that has already been seized to be condemned.

2. The justice of the peace, or in Scotland the sheriff or magistrate, will listen to the authority's case that the food fails to comply with food safety requirements and should be condemned. You may say why it should not be condemned.

3. You may bring your own evidence and witnesses to challenge the views of the authority and you may be represented by a lawyer.

4. You are not being charged with an offence. The hearing is only to decide whether the food complies with food safety requirements. But the court may order the food to be condemned. However you may be prosecuted for offences under the Food Safety Act 1990.

5. *EXPENSES:* If the justice of the peace, or in Scotland the sheriff or magistrate, orders the food to be condemned, then the owner of the food will have to pay reasonable expenses for it to be destroyed or disposed of.

6. *COMPENSATION:* If the justice of the peace, or in Scotland the sheriff or magistrate, does not condemn the food, the owner of the food may be entitled to compensation from the authority for any loss in its value as a result of the action taken by the authority.

The hearing before the JP may not only decide the fate of the batch of food but may also be a forerunner to other legal proceedings. The accused is allowed to make representations at this hearing and to call witnesses. It is probably advisable to take specific legal advice from a solicitor when the first detention notice is issued. It is wise to have legal representation at a hearing following seizure.

If the JP decides that the food is unsafe, he can order it to be destroyed and he may order the offender to cover the costs of its disposal. If the seized food is only part of a larger batch, the court must legally presume that the whole batch fails unless evidence is provided to the contrary. Disposal must follow the requirements of the relevant waste disposal legislation, using licensed waste disposal contractors, and must also be in accordance with the relevant provisions of Regulation (EC) No 1774/2002 on 'health rules concerning animal by-products not intended for human consumption', implemented by the *Animal By-Products Regulations 2005 (SI 2005 No 2347)*. Further details of the requirements for the disposal of animal by-products are available from DEFRA.

If it turns out that the authorised officer was wrong, the offender may be entitled to compensation. If the food has deteriorated as a result, they may be entitled to compensation equal to the loss in value. If a reasonable sum cannot be agreed with the authority there is an arbitration procedure.

Section 10: Improvement Notices

[F9015] *Section 10* provides authorised officers with a range of powers to deal with unsatisfactory premises from informal advice through to prosecution and emergency prohibition. Chapter 3.2 of the Food Law Code of Practice describes the circumstances when it is appropriate to use Improvement Notices under the *Food Safety Act 1990, s 10* and the use of Hygiene Improvement Notices under the *Food Hygiene (England) Regulations 2006, Reg 6*. If an authorised officer has reasonable grounds for believing that a business does not comply with hygiene or processing regulations, he can issue a formal improvement notice demanding that it puts matters right. This provides a quicker and cheaper remedy than taking the business to court. An improvement notice should not be ignored. It is an offence to fail to comply with a notice and either matters must be put right as indicated on the notice or an appeal lodged against it.

An improvement notice must contain four elements:

(i)   the reasons for believing that the requirements are not being complied with;
(ii)   the ways in which regulations are being breached;
(iii)   the measure which should be taken to put things right;
(iv)   the time allowed for putting things right (this cannot be less than 14 days).

Authority: ....   FORM 1

Food Safety Act 1990—Section 10

IMPROVEMENT NOTICE

Reference Number:

1. To: .... (Proprietor of the food business)

   At: ....

   ....

   .... (Address of proprietor)

2. In my opinion the:

   ....

   ....

   [Office to insert matters which do not comply with the Regulations]

   in connection with your food business ....

   .... (Name of business)

   at ....

   .... (Address of business)

   do/does* not meet the requirements of ....

   of the .... Regulations

   because:

   ....

   ....

   [* Officer to delete as appropriate]

3. In my opinion, the following measures are needed for you to comply with these Regulations: ....

   ....

   ....

4. These measures or measures that will achieve the same effect must be taken by: .... (date)

5. *It is an offence not to comply with this improvement notice by the date stated.*

   Signed: .... Authorised Officer

   Name in capitals: ....

   Date: ....

   Address: ....

   ....

   Tel: .... Fax: ....

> *Please read the notes overleaf carefully.*
> *If you are not sure of your rights or the*
> *implications of this notice, you may want to*
> *seek legal advice.*

NOTES

1. In the opinion of the officer you are not complying with the Regulations under Part II of the Food Safety Act 1990 described in paragraph 2 of the notice. The work needed in the officer's opinion to put matters right is described and it must be finished by the date set.

2. You are responsible for ensuring that the work is carried out within the period specified, which must be at least 14 days.

3. *You have a right to carry out work that will achieve the same effect as that described in the notice.* If you think that there is another equally effective way of complying with the law, you should first discuss it with the officer.

YOUR RIGHT OF APPEAL

4. If you disagree with all or part of this notice, you can appeal to the magistrates' court, or in Scotland to the sheriff. You must appeal within one calendar month of the date of the notice or the period ending with the date stated in paragraph 4 of the notice, whichever ends earlier.

5. If you decide to appeal, the time set out in the notice is suspended and you do not have to carry out the work described until the appeal is heard. *However, if you are not complying with the Regulations mentioned in the notice, you may still be prosecuted for failure to comply with those Regulations.*

6. When the appeal is heard, the magistrates' court, or in Scotland the sheriff, may confirm, cancel or vary the notice.

The business can choose to put things right in a different way to that suggested by the authorised officer provided that the authorised officer is satisfied that it will have the required effect.

If a party does not agree with an improvement notice, the *Food Safety Act 1990, s 37* gives the right to appeal to a magistrate within one month of its issue or within the time specified on the notice if shorter than one month. The authorised officer must inform the party about their rights to appeal and provide the name and address of the court when he serves the notice.

The notice may contain a number of points but the whole notice does not have to be appealed. The proprietor may appeal just one point or simply ask for a time extension. The notice is effectively suspended during the appeal, but it is best to let the authority, as well as the court, know about the appeal. The *Food Safety Act 1990, s 39* allows the magistrate to modify or cancel any part of the notice.

If an authorised officer of an enforcement authority has reasonable grounds for believing that a food business operator is failing to comply with the *Food Hygiene Regulations*, he may serve a Hygiene Improvement Notice. The provisions for Hygiene Improvement Notices fall under the *Food Hygiene Regulations 2006 (SI 2006 No 14), Reg 6.*

Food Authorities must continue to use the prescribed forms set out in the *Food Safety (Improvement and Prohibition - Prescribed Forms) Regulations 1991 (SI 1991 No 100)* when using powers under the *Food Safety Act 1990 s 10.* A model form for use in connection with the *Food Hygiene (England) Regulations 2006* can be found at Annex 7 of the FSA Food Law Code of Practice.

Section 11: Prohibition Orders

[F9016] Chapter 3.3 of the Food Law Code of Practice describes the circumstances when it is appropriate to use Prohibition Orders under the *Food Safety Act 1990, s 10* and the use of Hygiene Prohibition Orders under the *Food Hygiene Regulations 2006, Reg 7*.

If a proprietor is convicted of food safety or processing offences, including failing to comply with an Improvement Notice, the Food Authority can also apply for a prohibition order. If the court is satisfied that any processing, or the construction or state of the premises, or use of equipment, poses a risk to public health, it must issue a prohibition order to the proprietor or any manager of the business. The order can deal with any of one three issues depending upon the actual risk:

- to prohibit a process or treatment,
- to deal with the *structure* of the premises or the use of equipment,
- to deal with a problem that relates to the *condition* of the premises or equipment.

In either of the last two cases, the order may prohibit the use of the equipment or the premises.

Any prohibition order must be served on the proprietor of the business. If it relates to equipment, it will also be fixed to the equipment and if it relates to the premises, it will be fixed prominently to the premises. It is automatically an offence to knowingly breach a prohibition order.

PROHIBITION ORDER

(Food Safety Act 1990, s. 11)

Magistrates' Court

(Code)

Date   :

Accused   :

of   :

WHEREAS

being the proprietor of a food business carried on at premises at

has today been convicted by this court of an offence under regulations to which section 11 of the Food Safety Act 1990 applies [by virtue of section 10(3)(b) of the said Act], namely:-

Decision   :

AND WHEREAS the Court is satisfied that there exists a risk of injury to health by reason of

[the use for the purposes of the said food business of a certain [process] [treatment], namely

]

[the [construction] [state or condition] of the premises at

used for the purposes of the said food business]

[the [use for the purposes of the said food business] the [state or condition] of certain equipment [used for the purposes of the said food business], namely

]

and it is ORDERED that

Order   :

[the use of the said [process][treatment] for the purposes of the business] [the use of the said [premises] [equipment] for the purposes of [the business] [any other food business of the same class or description, namely

]

[any food business]]
[and that the participation by the accused in the management of any food business [of the said class or description]
is prohibited.

[By Order of the Court]

[Justice of the Peace] [*Justices' Clerk*]

Delete any words within square brackets which do not apply.

FS 46   Copyright forms are reproduced by permission of the publishers, Shaw & Sons Ltd., Shaway House, Crayford, Kent, DA1 4BZ, from whom copies may be purchased.

As soon as a party believes that they have put matters right and that the health risk no longer exists, they can apply to the Food Authority to have the prohibition lifted. The Food Authority must respond by reaching a decision within a fortnight and, if it believes that things are satisfactory, issue a certificate within another three days. The certificate will state that enough has been done to ensure that there is no longer an unacceptable risk to public health. If the Food Authority disagrees and refuses to issue a certificate, they must give their reasons and the business can appeal to a magistrate to have the order lifted. The Food Authority must give information on the right to appeal, and who to contact as well as the time scale (the appeal must be made within one month).

If a food business operator is convicted of an offence under the *Food Hygiene Regulations* and the court is satisfied that the health risk condition is fulfilled with respect to the food business concerned, the court must impose the appropriate prohibition. The provisions for Hygiene Prohibition Orders fall under the *Food Hygiene Regulations 2006 (SI 2006 No 14), Reg 7*.

Personal Prohibition under section 11

[F9017] Under the *Food Safety Act, s 11*, the court can also prohibit a person from running or managing a food business. An application to the court to lift a personal prohibition cannot be made within six months of its imposition and if an appeal is unsuccessful, it cannot be appealed again for a further three months.

In addition, if a business operator is convicted of an offence under the *Food Hygiene Regulations* and the court thinks it appropriate, taking into account all the circumstances of the case, it may impose a prohibition order on the food business operator, banning him from participating in the management of any food business, or any specified class of food business.

Section 12: Emergency Prohibition Notices and Orders

[F9018] Whereas *s 11* prohibitions are imposed only by a court following conviction, *s 12* permits, in an emergency, an authorised officer to act on his or her own authority, and with immediate effect. Where the Food Authority is satisfied that a business presents an imminent risk to health, it can serve an Emergency Prohibition Notice without prior reference to a court. Emergency Prohibition can apply to the whole premises or a specific part. The notice will be fixed in a prominent place and anyone removing it or deliberately ignoring it is guilty of an offence.

Authority: .... FORM 2

Food Safety Act 1990–Section 12

EMERGENCY PROHIBITION NOTICE

Reference Number:

1. To: .... (Proprietor of the food business)

    At: ....

    ....

    .... (Address of proprietor)

2* I am satisfied that: ....

    ....

    ....

    at ....

    .... (Address of business)

POSES AN IMMINENT RISK OF INJURY TO HEALTH because:

    ....

    ....

    ....

(*See Note 1 overleaf)

3. *YOU MUST NOT USE IT FOR THE PURPOSE OF THIS/ANY/THIS OR ANY SIMILAR* FOOD BUSINESS.*

[* Officer to delete as appropriate]

    Signed: .... Authorised Officer

    Name in capitals: ....

    Date: ....

    Address: ....

    ....

    Tel: .... Fax: ....

> *Please read the notes overleaf carefully.*
> *If you are not sure of your rights or the*
> *implications of this notice, you may want to*
> *seek legal advice.*

NOTES

1. *When you receive this notice you must IMMEDIATELY stop using the premises, process, treatment or equipment described by the officer in paragraph 2 of the notice and located at the address stated.*

2. Within 3 days of service of this notice, the authority must apply to a magistrates' court, or in Scotland to a sheriff, for an order confirming the prohibition. You will be told the date of the hearing which you are entitled to attend and at which you may call witnesses if you wish.

3. If you believe that you have acted to remove the imminent risk of injury to health, you should apply in writing to the authority for a certificate which would allow you to use the premises, process, treatment or equipment again. You can do this even if the court hearing has not taken place.

4. You are not allowed to use the premises, process, treatment or equipment for the purpose specified in paragraph 3 of the notice (see section 11(3) of the Food Safety Act 1990) until (a) a court decides you may do so; (b) the authority issues you with a certificate as in paragraph 3 above; (c) 3 days have passed since the service of the notice and the authority has not applied to the court as in paragraph 2 above; or (d) the authority abandons the application.

5. A copy of this notice must, by law, be fixed on the premises or equipment which is not to be used. It is an offence (under section 1 of the Criminal Damage Act 1971 or, in Scotland, section 78 of the Criminal Justice (Scotland) Act 1980) to deface it.

6. *COMPENSATION:* If the authority does not apply to the magistrates' court, or in Scotland to the sheriff, for an order confirming its action within 3 days of the date of service of this notice, you will be entitled to compensation for any losses you have suffered because you could not use the premises, process, treatment or equipment because you were complying with this notice. You will also be entitled to such compensation if the magistrates' court, or in Scotland the sheriff, decide at the hearing that the authority's action was wrong.

The authority must then take the matter before a magistrates' court within three days of serving the Emergency Prohibition Notice. At least one day before it goes to court, the authority must serve a notice on the proprietor of the business telling him of the court hearing. If an order is not issued by the court or if no application for one is made within the three day period, the Emergency Prohibition Notice lapses. Once again, if a business finds itself faced with action of this kind, it is advisable to seek specific legal advice.

If the court agrees that the prohibition actions were justified, it will make an Emergency Prohibition Order which replaces the Emergency Prohibition Notice. The arrangements for lifting the Order are the same as for a s 11 Order, above. If the court does not uphold an Emergency Prohibition Notice, the proprietor may seek compensation for loss or damages. An Emergency Prohibition Order cannot be made against an individual. Prohibition of a person can only be made under s 11 following a conviction.

Similarly, and subject to the same procedures, if an authorised officer of an enforcement authority is satisfied that the health risk condition is fulfilled with respect to any food business, he may served a "Hygiene Emergency Prohibition Notice" on the relevant food business operator. The provisions for Hygiene Emergency Prohibition Notices and Orders fall under the *Food Hygiene Regulations 2006 (SI 2006 No 14), Reg 8*.

Section 13: Emergency Control Orders

[F9019] *Section 13* empowers the Secretary of State or the Food Standards Agency to issue Emergency Control Orders where it appears that commercial food-related operations involve, or may involve, imminent risk of injury to health. In such circumstances either of the above may prohibit such operations. It is an offence to knowingly contravene an Emergency Control Order. If an Emergency Control Order is invoked, there is no right of appeal and no compensation arrangements.

It may be that an unsafe batch of food has already been distributed around the country. Emergency prohibition of the manufacturing site would stop further production but it would not control the risk from food already in the system. The *Food Safety Act 1990, s 13* allows an Emergency Control Order to require all steps to be taken that will remove the threat. This is a wide-ranging power that will only be used in exceptional circumstances. In most cases, any business told that it has received unsafe food from a supplier would stop selling it simply to protect their own reputation.

Voluntary arrangements are in place to co-ordinate food hazards and issue warnings and organise recalls when food is distributed more widely than the local area in which it is produced. Following a greater commitment to openness and transparency by all parties, warnings are circulated by electronic mail to Food Authorities around the country. Where possible, trade organisations and the media are also informed and food hazard warnings and recalls are now better publicised, for example through the FSA website at: www.foodstandards.gov.uk or by direct emailings to interested parties, including the public.

Remedial Action Notices

A new type of enforcement notice was introduced under the *Food Hygiene Regulations 2005* (now Regulation 9 of the 2006 Regulations) whereby, in the case of an establishment subject to approval under Article 4(2) of Regulation 853/2004 (animal products hygiene legislation), an authorised officer may serve a Remedial Action Notice or Detention Notice where any of the requirements of the Hygiene Regulations is being breached or inspection under the Hygiene Regulations is being hampered. This may:

- prohibit the use of any equipment or any part of the establishment (and must specify the reason and the action needed to remedy it);
- impose conditions upon or prohibit the carrying out of any process;
- require the rate of operation to be reduced to such extent as to ensure hygienic operation, or to be stopped completely;
- require the detention of any animal or food of animal origin for sampling and examination.

A Remedial Action Notice must be served as soon as practicable and state why it is being served. The notice must be withdrawn by a further written notice as soon as the authorised officer is satisfied that the necessary action has been taken.

Failure to comply with a Remedial Action Notice is an offence, although the right to appeal to a magistrate within one month is granted.

The use of Remedial Action or Detention Notices must be proportionate to the risk to public health and where immediate action is required to ensure food safety. A Remedial Action Notice may be used if a continuing offence requires urgent action owing to a risk to food safety or when corrective measures have been ignored by the food business operator and there is a risk to public health. As soon as the authorised officer is satisfied that the action specified in a Remedial Action Notice has been taken, the notice must be withdrawn by means a further notice in writing. Similarly, in respect of a Food Detention Notice, if the authorised officer is satisfied that the food need no longer be detained, the relevant notice must also be withdrawn by means a further notice in writing.

The use of Remedial Action Notices and Detention Notices falls under the *Food Hygiene Regulations (SI 2006 No 14), Reg 9*. Models of a Remedial Action Notice, Detention Notice and Notice of Withdrawal of a Remedial Action Notice/Detention Notice are provided in Annex 7 of the FSA Food Law Code of Practice.

Sections 16 to 18: power to make regulations

[F9020] The *Food Safety Act 1990, ss 16–18* and *26* give the Secretary of State very wide-ranging powers to make regulations and orders for general food safety and consumer protection; for special provisions for particular foods; for the registration and licensing of food premises; for food hygiene training; and for facilities for sampling and the giving of information. For example:

- To prohibit specified substances in foods.

- To prohibit a certain food process.
- To require hygienic conditions in commercial food premises.
- To dictate microbiological standards for foods.
- To make standards on food composition.
- To specify requirements for food labelling.
- To restrict and control claims made in the advertising of food.

They also provide the power to ban from sale food originating from potentially diseased sources; such powers were used to prohibit the sale of certain parts of beef cattle for human food on the basis that they may have been suffering from BSE.

Much of the detail of food law is in the regulations made under *s 16, which also contains a requirement for consultation prior to the introduction of individual Statutory Instruments.*

Section 17 allows the government to bring EU provisions into UK law. *Section 18* regulations are intended to cover very particular circumstances, such as genetically modified and other novel foods, although in practice these are now harmonised at EU level and the UK has only very limited room in which to act independently. For example, food safety approval, labelling and traceability of foods and food ingredients derived from GM crops are controlled under EU Regulations 1829 and 1830/2003, and implemented by the *Genetically Modified Food (England) Regulations 2004 (SI 2004 No 2335)* and *Genetically Modified Organisms (Traceability and Labelling) (England) Regulations 2004 (SI 2004 No 2412)* respectively. Environmental releases of GM crops for food/feed use may fall under these Regulations or under Directive 2001/18, implemented in the UK by the *Genetically Modified Organisms (Deliberate Release) Regulations 2002 (SI 2002 No 2443),* as amended.

The *Food Standards Act 1999, s 17,* provides limited powers to the FSA to make emergency orders. Under the *Food and Environment Protection Act 1985, ss 1 and 2* and the *Food Safety Act 1990, s 13* the Secretary of State may make emergency orders in response to circumstances or incidents which pose a threat to public health in relation to food. The Secretary of State retains these powers but the FSA may be empowered by him to make Emergency Orders itself on his behalf. This power does not give the FSA the ability to make legislation itself in other areas and, in practice, the FSA will only make orders in emergency situations where the Secretary of State is not available. The Secretary of State is ultimately answerable for emergency legislation made by the FSA on his behalf and anything done by the FSA is, in law, done by the Secretary of State.

Section 19: registration and licensing

[F9021] The *Food Safety Act 1990, s 19* permits the government to bring in regulations demanding licensing or registration of certain food businesses.

Registration and licensing confer significantly different powers on enforcement authorities. Registration is simply an administrative exercise, designed primarily to inform enforcement agencies what food businesses are operating in their area. It also provides some basic information on the size of the business and the type of food that they produce. No conditions are attached to registration, the

Food Authority cannot refuse to register a business, there is no charge, and it does not have to be renewed periodically.

Licensing (or 'prior approval') is quite different. It is a control measure. Precise licensing criteria will be specified and there would normally be prior inspection and approval before a new business is allowed to open. There is often a charge for a licence, which must be renewed periodically. The enforcement authority will usually have the right to withdraw a licence if standards fall and the business would have to close.

Currently, licensing only applies to those businesses subject to the requirements of the EU 'vertical', animal product legislation, such as slaughter houses, cutting plants and dairy operations. Most other types of food businesses are subject only to registration. Article 6(2) of EC Regulation 852/2004 requires food business operators to register their establishments (ie each separate unit of their food businesses) with the appropriate Food Authority, unless they are subject to approval under EC Regulation 853/2004 (animal products hygiene) or fall outside the scope of the EC Regulation 852/2004 (in practice, very few food businesses fall outside the scope). EC Regulation 882/2004, Article 31(1)(a) requires the Food Authority to establish procedures for food business operators to follow when applying for the registration of their establishments. Food business operators must provide the authority with full details of the activities to be undertaken at least 28 days before food operations commence. A model registration form is available in Annex 8 of the FSA Food Law Code of Practice.

EC Regulation 882/2004, Article 31(1)(b) requires the authority to draw up and maintain a list of food establishments that have been registered. Any changes to details previously supplied, eg a change of business operator, a change to the activities carried out, the closure of an establishment etc. must be notified by the current operator. Notification of a change to the operator of a food establishment should be made by the new operator.

Although the *Food Premises (Registration) Regulations 1991 (SI 1991 No 2825)* (as amended by *SI 1993 No 2022* and *SI 1997 No 723*) were technically revoked with effect from January 2006, the principles of registration in accordance with the terms of the EC Regulation 852/2004 are maintained through the provisions of the *Food Hygiene Regulations 2006 (SI 2006 No 14)* – see **F9026–F9033**.

Defences against charges under the Food Safety Act 1990

[F9022] The following paragraphs explain the defences available under the *Food Safety Act 1990*.

Section 21: defence of due diligence

[F9023] The so called 'due diligence' defence first appeared within food safety legislation when the *Food Safety Act 1990* was passed. Previously, the provision had been tried and tested within a number of other pieces of legislation after it first appeared just over 100 years earlier in the *Merchandise Marks Act 1887*. Indeed, legislation that included a due diligence defence, such as the *Trades Descriptions Act 1968*, already overlapped with food legislation.

So, although the defence was new to food law, there were numerous precedents to signal the implications of the due diligence provision.

The *Food Safety Act 1990* creates offences of strict liability. The prosecution is not required to show that the offence was committed intentionally. However the due diligence defence is intended to balance the protection of the consumer against the right of traders not to be convicted for an offence that they have taken all reasonable care to avoid committing. The intention of the due diligence defence is to encourage traders to take proper responsibility for their products and processes.

The *Food Safety Act 1990, s 21* states:

> in any proceedings for an offence under any of the preceding paragraphs of this Part . . . it shall be . . . a defence for the person charged to prove that he took all reasonable precautions and exercised all due diligence to avoid the commission of the offence by him or any person under his control.

The onus is on the business to establish its defence on the balance of probabilities. Part of the due diligence defence may be to establish that someone else was at fault. If that is what is intended, the prosecutor must be informed at least seven days before the hearing or, if a business has been in court already, within one month of that appearance. Under the 1990 Act, reliance upon a supplier's warranty alone is not possible, although a warranty may be one element of a due diligence defence.

It is impossible to give precise advice on what any business would need to do to have a due diligence defence. Cases that have been decided by the courts serve only to demonstrate that each case is decided on its own facts and may also be determined by the size and resources otherwise available to a business. What is certain is that doing nothing will never provide a due diligence defence. Positive steps must be taken if a defence is to succeed.

The defence has two distinct parts:

- Taking *reasonable precautions* involves setting up a system of control. The business operator must consider the risks that threaten the operation and take *all reasonable precautions* that may be expected of a business of the size and type.
- Exercising *due diligence* means that those precautions are taken on a continuous basis.

All reasonable precautions and *all* due diligence are needed. The courts have decided consistently that if a precaution could reasonably have been taken, but was not taken, then the defence would not succeed.

The test is 'what is reasonable?' A business will be expected to take greater precautions with high risk, ready-to-eat foods than are needed for boiled sweets or biscuits. What is reasonable for a large-scale food business may not be reasonable for a smaller enterprise.

The business must ensure that the scope of 'due diligence' covers all aspects of food law requirements. It is not adequate to have excellent control of food safety hazards if there are no precautions in place to ensure that the composition and labelling of food complies with the regulations.

Documentation is important. Unless the precautions are written down, it will be difficult to persuade the authorised officer or the court that a proper system is in place. The same is true of records of any checks made to demonstrate that the system is followed diligently. The burden of proof lies with the business to establish that the defence is satisfied.

A good 'HACCP' plan (see **F9036**), as now required by Article 5 of EC Regulation 852/2004, enforced in the UK by the *Food Hygiene Regulations 2006 (SI 2006 No 14)* may be helpful in showing that a systematic approach to identifying the food safety precautions needed in the operation has been adopted. Cross-reference to industry guidelines or codes of practice may show that the system has a sound basis. Documentation from the hazard analysis will add to the evidence that precautions have been taken. Records such as specifications, cleaning schedules, training programmes and correspondence with suppliers or customers may all play a part in establishing the defence. An organisation chart and job descriptions that identify roles and responsibilities will also be useful evidence of the precautions taken.

Not all food businesses have the same defence. Manufacturers and importers must satisfy the full defence. Some traders within the retail and catering sectors may use a simpler defence of 'deemed due diligence' deriving from the *Food Safety Act 1990, s 21(3)* and *(4)*. Even then, 'deemed due diligence' is not available against the *s 7* offence of rendering food injurious to health. Larger businesses may be expected to take more precautions than smaller businesses. Those selling 'own label' packs may have greater responsibility for upstream activity than those selling manufacturers' 'branded' products. However, the basic responsibilities of retailers to ensure the accuracy of the labelling of all products sold by them was reinforced by the ECJ judgment in *Lidl Italia Srl v Comune di Arcole (VR):* C-315/05 [2006] ECR I-11181, in which it was held that a retailer may be held liable for food labelling infringements by the producer, even if he simply markets a product as delivered by that producer.

Section 35: penalties under the Food Safety Act 1990

[F9024] The courts decide penalties on the merits of individual cases but they are limited by maximum figures included in the *Food Safety Act 1990, s 35*. For most offences, the crown court can send offenders to prison for up to two years and/or impose an unlimited fine. The magistrates' court, where most cases are heard, can set fines of up to £5,000 per offence and up to six months imprisonment, or both. For the two food safety offences in *ss 7* and *8* and the first of the consumer protection offences (*s 14*), the fine may be up to £20,000. In Scotland the sheriff may impose equivalent penalties.

Each set of regulations made under the Act will have its own level of penalties that will not exceed the levels mentioned here.

Section 40: codes of practice

[F9025] The *Food Safety Act 1990, s 40* allows for the development of codes of practice. *Section 40* codes of practice are primarily intended to instruct and inform enforcement officers and related bodies or personnel, for example food analysts or examiners. In carrying out their duties in the execution and enforcement of the Act or Regulations laid under it, enforcement authorities and their authorised officers must 'have regard to' these *s 40* codes of practice. This means, in effect, that Food Authorities must follow and implement the provisions of the Code that apply to them. Any failure to follow a code requirement might result in their decisions or actions being successfully challenged, and evidence gathered during a criminal investigation being ruled inadmissible by a court, thus seriously jeopardising their ability to bring a successful prosecution. Twenty individual codes were originally made under *s 40*, some of which underwent several revisions. The subjects of these original codes were general matters, such as the demarcation of responsibility for enforcing different parts of the Act or general inspection procedures, and more specific information on the enforcement of particular hygiene-related regulations. The most significant was Code of Practice Number 9 on food hygiene inspections, which introduced detailed advice to enforcement officers on how to conduct such inspections and included a 'risk rating' system whereby inspectors could prioritise different food businesses and determine the necessary inspection frequency.

The 20 codes were subsequently consolidated by the FSA into a single Food Law Code of Practice but, to ensure a clear division between what Food Authorities must do in enforcing food law, which must be included in the statutory code, other advice was published as Practice Guidelines accompanying the Code. These major documents had to be revised and updated to take account of EU Food Hygiene and Official Feed and Food Controls Regulations and the UK national implementing regulations, all of which were applicable from 1 January 2006, and were further reviewed by the FSA, which introduced the current Food Law Code of Practice in June 2008.

The latest changes replace the previous enforcement policy focused primarily on inspections, with a new policy for a suite of interventions that will move all food establishments into a risk-based inspection regime and allow authorities to choose the most appropriate action to be taken to improve levels of food law compliance by food establishments. This takes account of the recommendations in 'Reducing Administrative Burdens: effective inspection and enforcement' report by Philip Hampton, published March 2005.

Section 40 codes should not be confused with 'industry guides' that are promoted by the EU General Food Hygiene Directive and the General Food Hygiene Regulations. These 'guides' provide more detailed practical explanation of the implications of the particular regulations in a food business sector. There are also numerous other guides and codes of practice published by industry, government, and other organisations.

Food premises regulations

[F9026] Previous specific UK legislation for the registration of food premises (*Food Premises Registration Regulations 1991 (SI 1991 No 2825)*) has been revoked but the principles of registration in accordance with the terms of the EU Regulation 852/2004, Article 6(2) are maintained through the provisions of the *Food Hygiene Regulations 2006 (SI 2006 No 14)*.

Under EU food law, 'food business' is very widely defined as meaning any undertaking, whether for profit or not and whether public or private, carrying out any of the activities related to any stage of production, processing and distribution of food. Thus, all such businesses must be registered unless they are specifically exempted from the requirement. Food business operators must register their establishments (ie each separate unit of their food businesses) with the appropriate Food Authority, unless they are subject to approval under EC Regulation 853/2004 (animal products hygiene) or fall outside the scope of the EC Regulation 852/2004 (in practice, very few food businesses fall outside the scope).

EC Regulation 882/2004, Article 31(1)(a) requires the authority to establish procedures that business operators must follow when applying for the registration of their establishments. Food business operators must provide the authority with full details of the activities to be undertaken at least 28 days before food operations commence. Any changes to details previously supplied, eg a change of business operator, a change to the activities carried out, the closure of an establishment, etc. must be notified by the current operator. Notification of a change to the operator of a food establishment should be made by the new operator. A model registration form is available in Annex 8 of the FSA Food Law Code of Practice.

The authority must, by law, draw up and maintain a list of food establishments that have been registered.

One change introduced by the EU requirements relates to wholesale markets, where the registration procedure will enable action against a single, infringing unit rather than seeking the closure of an entire market operation.

The following paragraphs cover the principal requirements.

[F9027] Food businesses must notify the local authority of all food premises under their control that carry out any production, processing and distribution of food (including retail and catering premises), unless they are already required by other legislation to be licensed (ie approved under 'vertical' legislation relating to handling products of animal origin). The legislation requires the notification of individual food *premises* and not food businesses. Thus, premises that may be used only occasionally by food businesses, even by different businesses, (eg a church hall, village hall or scout hut used from time to time for the sale of food) may have to be registered by the owner of the premises, depending on the frequency of use. The criterion is that they are used by commercial food businesses for five or more days (which do not have to be consecutive) in any period of five weeks.

Similarly, wholesale markets or other premises used by more than one food business must also be registered. The market operator or the person in charge of the premises is responsible. If mobile food premises are used they may have to be registered. These can range from a handcart in a market to a forty-foot trailer used for hospitality. Privately-owned, moveable premises used in a market must be registered even though the market is registered by the operator. (If stalls provided by the market are used, registration is not necessary). If mobile food premises are operated, the premises at which they are normally kept must be registered with the food authority local to the base.

Staff restaurants must be registered. Contractors and clients should have a clear agreement as to who will action this.

New premises must notify the authorities at least 28 days before opening. This is designed to allow the Food Authority the opportunity to inspect a business before it opens. However, if the authority does not take that opportunity, it is not necessary for the business to wait for a visit or formal permission before opening, provided that the necessary registration form has been sent. To register, the operator must supply a range of information on a registration form obtained from the authority, which is normally contacted via the Town Hall. In case of doubt, it is best to keep a note of when the form was sent and, if possible check that it has been received. Sending the form by fax may be helpful in establishing the transmission date.

FORM OF APPLICATION FOR REGISTRATION OF FOOD PREMISES

1. Address of premises ....
    (or address at which movable   Post code ....
    premises are kept)

2. Name of food business..   Telephone no: ....
    (trading name)

3. Type of premises Please tick ALL the boxes that apply

    Farm/smallholding ☐   Staff restaurant/canteen/kitchen ☐
    Food/manufacturing/processing ☐   Catering ☐
    Slaughterer ☐   Hospital/residential home/school ☐
    Packer ☐   Hotel/pub/guest house ☐
    Importer ☐   Private house used for a food business ☐
    wholesale/cash and carry ☐   Premises used by a number of
                                                    businesses ☐
    Distribution/warehousing ☐   Moveable premises ☐
    Retailer ☐
    Market ☐   Other: please give details ....
    Restaurant/cafe/snack bar ☐   ....

4. Does your business handle or involve any of the following? Please tick ALL the boxes that apply

    Chilled foods ☐   Alcoholic drinks ☐
    Frozen foods ☐   Canning ☐
    Fruit and vegetables ☐   Vacuum packing ☐
    Fish/fish products ☐   Bottling and other packing ☐
    Fresh/frozen meat ☐   Table meals/snacks ☐
    Fresh/frozen poultry ☐   Takeaway food ☐
    Meat products or delicatessen ☐   Accommodation ☐
    Dairy products ☐   Delivery service ☐
    Eggs ☐   Chilled food storage ☐
    Bakery ☐   Bulk storage ☐
    Sandwiches ☐   Use of private water supply ☐
    Confectionery ☐   Other: please give details ....
    Ice cream ☐   ....

5. Are vehicles or ships   Are vehicles, stalls or ships   Number of
    used for transporting   used for preparing or selling   vehicles/stalls/ships kept at or
    food kept at or used   food, kept at or used from   used from the premises, and
    from the premises?   the premises?   Yes/No   used for preparing, selling or
                  Yes/No   transporting food.
                                                                    5 or less ☐   6–10 ☐
                                                                    11-50 ☐   51 plus ☐

6. Name(s) of proprietor(s) of food business ....
    Address of business head office or registered office ....
    if different from address of premises
       ....   Post code ....

7. Name of manager if different from proprietor ....

8. If this is a new business........................   9. If this is a seasonal business ....
    Date you intend to open   Period during which you intend to be open
                                                                 each year

10. Number of people engaged in food business   0–10 11–50 51 plus (Please tick one box)
    Count part-timer(s) (25 hrs per week or less)   ☐   ☐   ☐
    as one-half

                                                         It is an offence to give false or incomplete
                                                         information

    The completed form should be sent to:

    [   ]   Signature ....
                                                         Date ....
                                                         Name ....
    [   ]   (BLOCK CAPITALS)
                                                         Position in company/business............................

Food business operators must also ensure that the authority always has up-to-date information on premises, including any significant change in food-related activities (eg change of product handled) and any closure of an existing establishment. The authority must be notified of the change within 28 days. Many of the other pieces of information on the application form may also change, for example the name of the manager, the phone number, the number of employees and so on but these changes do not need to be notified to the authority. Failing to register, giving false or incomplete information or failing to notify any of the changes mentioned above, can trigger prosecution proceedings.

Some food premises must by law be licensed, in which case additional notification/registration is not required. These premises are generally those (other than retail and catering) that handle raw/unprocessed products of animal origin, which fall into a number of categories under EU Regulation (EC) No 853/2004, and are approved ('licensed') by the competent authority, following at least one on-site visit.

Premises licensed or registered under other measures

Establishments handling raw products of animal origin

[F9028] The Meat Hygiene Service is responsible for the approval of meat-handling establishments where Regulation 853/2004 (animal products hygiene) requires control by an official veterinarian. These will include slaughterhouses, game handling establishments, cutting plants *and* any of these in which minced meat, meat preparations, mechanically separated meat, or meat products are *also* produced.

Local Authorities ('Food Authorities') are responsible for the approval of establishments where control does *not* fall to an official veterinarian. Such 'product-specific establishments' will be producing any, or any combination, of the following: minced meat, meat preparations, mechanically separated meat, meat products, live bivalve molluscs, fishery products, dairy products, egg products, gelatine and collagen, and will include cold stores and certain wholesale markets.

Establishments producing 'composite products'

Regulation 853/2004 does not apply to food comprising mixtures of plant origin and *processed* products of animal origin. However, establishments producing such 'composite products' which are also engaged in activities which would otherwise require *approval* will need to be approved in relation to those activities. Otherwise, Regulation 852/2004 will apply to the part of the establishment where the composite products are assembled. Only the notification process is required in respect of establishments engaged solely in assembling products originating from approved establishments to create 'composite products'.

Premises used only occasionally

[F9029] The EU hygiene legislation states that the requirements should apply only to undertakings, the concept of which implies a certain continuity of activities and a certain degree of organisation, but does not specify precisely how this concept is to be applied. In practice, this has historically been taken to include premises used for less than five days (not necessarily consecutive) in five consecutive weeks. However, the operators of all such premises are advised to seek advice on their specific circumstances from their local Food Authority.

Premises and areas exempt unless used for retail sales

[F9030] The following are examples of 'premises', areas and locations which are not generally subject to registration/licensing under the food hygiene legislation (although farms, smallholdings etc may be subject to measures under other legislation) but which may need to be so registered if the sale of products from them is other than the direct supply, by the producer, of small quantities of primary products to the final consumer or to local retail establishments directly supplying the final consumer. In general, however, the premises and the products obtained from these become subject to food hygiene controls as soon as they are harvested, slaughtered, caught etc if they are destined for sale to third parties.

- Where game is killed in sport (grouse moors).
- Where fish is taken for food (but not processed).
- Where crops are harvested, cleaned, stored or packed. However, if the crops are put into the final consumer pack the premises must be registered.
- Where honey is harvested.
- Where eggs are produced or packed.
- Livestock farms and markets.
- Shellfish harvesting areas (these are, however, subject to specific classi-fication and related controls under Regulations 853/2004 and 854/2004).
- Places where no food is kept, for example an administrative office or head quarters of a food business.

Some domestic premises

[F9031] EU Regulation 852/2004 does not apply to primary production for private domestic use; or to the domestic preparation, handling or storage of food for private domestic consumption; or the direct supply, by the producer, of small quantities of primary products to the final consumer or to local retail establishments directly supplying the final consumer. In addition, the legis-lation states that the requirements should apply only to undertakings, the concept of which implies a certain continuity of activities and a certain degree of organisation but does not specify precisely how this concept is to be applied.

Thus, the use of domestic premises for a number of situations may, technically, bring them within the scope of "food businesses" and subject to registration under the legislation.

Some vehicles

[F9032] -

- Private cars.
- Aircraft.
- Ships unless permanently moored, eg as restaurants, or used for pleasure excursions in inland or coastal waters.
- Food vehicles normally based outside the UK.
- Vehicles or stalls kept at premises that are themselves registered or exempt.
- Tents, marquees and awnings.

Other exemptions

[F9033] -

- Places where the main activity has nothing to do with food but where light refreshments such as biscuits and drinks are served to customers without charge (for example hairdressers' salons).
- Premises where food is sold only through vending machines. Vending machines themselves are subject to the relevant provisions of EU Regulation 852/2004 but, in practice, it is recognised that there is little practical value in the registration of individual vending machines or the premises on which they are sited if the only food-related activity on those premises relates solely to vending machines. However, all ancillary activities related to the vending machines, such as distribution centres where food for stocking them is stored and/or despatched must be registered. The delivery vehicles are subject to the general hygiene requirements for transport.
- Places run by voluntary or charitable organisations, and used only by those organisations, provided that no food is stored on the premises except tea, coffee, dry biscuits etc. For example, some village or church halls, but not if they are used more than five days in five weeks by commercial caterers.
- Crown premises exempted for security reasons.
- Places supplying food and drink in religious ceremonies.
- Stores of food kept for an emergency or national disaster.

Food hygiene regulations

[F9034] In 2004, the EU completed an extensive and fundamental overhaul of all food hygiene legislation, consolidating and recasting both the general requirements and those relating to animal-derived products into two principal Regulations (Regulation 852/2004 on the *Hygiene of Foodstuffs and Regulation 853/2004* laying down Specific Hygiene Rules for Products of Animal Origin). The requirements of these have been transposed into UK legislation as the *Food Hygiene Regulations (England) 2006 (SI 2006 No 14)*. Similar, parallel provisions have been introduced by the devolved administrations. These UK Regulations are wide-ranging and subsume/revoke numerous previous provisions, including the *Food Safety (General Food Hygiene) Regulations 1995 (SI 1995 No 1763)* ('*the General Food Hygiene Regulations*'), the *Food Safety (Temperature Control) Regulations 1995 (SI 1995 No 2200)* ('the Temperature Control Regulations') and the *Food Premises (Registration) Regulations 1991 (SI 1991 No 2825)*.

Two further EU regulations now define the requirements and responsibilities for official controls by the authorities: Regulation 854/2004 laying down specific rules for the organisation of official controls on products of animal origin intended for human consumption and Regulation 882/2004 on official controls performed to ensure the verification of compliance with feed and food law, animal health and animal welfare rules.

The following account will concentrate on the *Food Hygiene Regulations 2006 (SI 2006 No 14)* that stem from the Regulations 852/2004 and 853/2004. It will also cover the associated temperature control requirements that are implemented by the *Food Hygiene Regulations 2006 (SI 2006 No 14), Reg 30 and Sch 4*, which largely re-enact the *Food Safety (Temperature Control) Regulations 1995 (SI 1995 No 2200)* ('the Temperature Control Regulations'). These requirements apply to all food businesses except those specifically covered by EU Regulation 853/2004 or to food operations carried out on a ship or aircraft.

A detailed account of the hygiene requirements within the vertical Regulation 853/2004 is beyond the scope of this publication.

Food Hygiene Regulations 2006

[F9035] The requirements of the *Food Hygiene Regulations 2006 (SI 2006 No 14)* are based on the HACCP principles established by Codex Alimentarius, the joint FAO/WHO Food Standards programme. The significant difference is that the hazard analysis requirement within the 2006 Regulations is less formally structured than Codex and there is no overt requirement for documentation or records. The legislation also recognizes that the application of HACCP principles to primary production is not generally feasible but, nevertheless encourages their use as far as possible.

For the purposes of the legislation, 'hygiene' is defined as the measures necessary to ensure fitness for human consumption of a foodstuff taking into account its intended use. The concepts of 'fitness' and food safety are now defined in the general food legislation, described above. There is no reference to 'wholesomeness', as was the case under previous legislation.

Since the UK legislation is implementing directly-applicable EU Regulations, it is necessary to refer to these in order to establish the precise requirements. Regulation 852/2004 (Article 4, Part A of Annex I and the 12 chapters of Annex II) sets out the general principles to be followed.

Article 5 requires food business operators to put in place, implement and maintain a permanent procedure or procedures based on the HACCP principles and establishes the seven principles to be applied. The Regulation recognises the need to provide sufficient flexibility to avoid undue burdens for very small businesses by requiring that documentation and records need only be commensurate with the nature and size of the business to demonstrate the effective application of their HACCP measures.

Hazard analysis and primary production

[F9036] The requirement to establish procedures based on HACCP principles does not currently apply to primary production but the feasibility of doing so is under on-going review. Regulation 852/2004 specifies seven elements that must be included in the process of identifying steps in the activities of the business that are 'critical' to ensuring that food is safe by identifying and/or establishing:

- hazards that must be prevented, eliminated or reduced to acceptable levels;
- critical control points ('CCPs') at the step(s) at which control is essential to prevent, eliminate or reduce a hazard to acceptable levels;
- critical limits at CCPs to separate acceptability from unacceptability for the prevention, elimination or reduction of identified hazards;
- effective monitoring procedures at CCPs;
- corrective actions that need to be taken when monitoring indicates that a CCP is not under control;
- procedures, to be carried out regularly, to verify that the measures are working effectively;
- documents and records (appropriate to the nature and size of the food business) to demonstrate the effective application of the measures.

HACCP is a powerful and sophisticated management tool that can be used by businesses to develop a food control plan. Full and formal HACCP requires detailed expert knowledge and training. A significant amount of literature has built up over the years that is useful to those larger businesses that employ staff with technical training but smaller businesses tend to find HACCP literature to be too full of jargon and inaccessible.

Clear guidance for some specific food sectors has been developed by national governments and international agencies around the world. Critical control points (CCP) for microbiological food poisoning hazards will tend to focus on a small number of areas.

COOKING: In most processes, there is likely to be a risk of pathogens in raw ingredients, so cooking is frequently a CCP. Adequate cooking temperatures will be the control measure, and target temperatures should be established. These can be monitored either by checking the food temperature directly or by monitoring the process, for example by using a cooking time and temperature that you know will achieve a satisfactory product centre temperature.

CONTAMINATION: Process steps at which ready-to-eat food may become contaminated are also likely to be CCPs. Contamination could come from contact with raw food ('cross contamination'), or from contact with contaminated equipment or personnel. Separation of raw and ready-to-eat foods, and cleaning and disinfection routines will be appropriate control measures. In a good hazard analysis system, these procedures should be specified objectively to allow them to be monitored. Written cleaning schedules will help. Nowadays, there are also rapid test systems that will assess effectiveness of cleaning.

TIME AND TEMPERATURE CONTROLS: Cooling of food after cooking is often a CCP, as are storage times and temperatures of ready-to-eat foods. In most cases, controls will be expressed as a combination of time and temperature. The critical limits may be just a few hours if the food is kept at warm room temperature, or many days if it is kept under good refrigeration. In this area, technology is providing increasingly sophisticated monitoring options including automatic alarms and real-time data monitoring. The temperature control requirements of the *Food Hygiene Regulations 2006 (SI 2006 No 14)* also intervene by prescribing certain targets.

Documentation and records

[F9037] A full, Codex-defined HACCP system would be fully documented. Any monitoring or verification procedures would be recorded and records kept. There is no explicit requirement for either of these in Regulation 852/2004, *Article 5* of which states that Food Business Operators must put in place, implement and maintain a permanent procedure or procedures *based on* the HACCP principles. The introductory *Recital 15* explains that, whilst the HACCP requirements introduced by the Regulation should take account of the principles contained in the *Codex Alimentarius*, they should provide sufficient flexibility to be applicable in all situations, including in small businesses. In particular, *Recital 15* recognises that, in certain food businesses, it is not possible to identify critical control points and, in some cases, good hygienic practices can replace the monitoring of critical control points. Similarly, the requirement to establish 'critical limits' is not to be taken to imply that it is necessary to fix a numerical limit in every case. In addition, the requirement of retaining documents needs to be flexible in order to avoid undue burdens for very small businesses.

Most businesses would, nevertheless, be advised to keep some concise and succinct documentation. Without it, it may not be easy to demonstrate compliance with the legislation and it would be difficult to muster a convincing due diligence defence if that should become necessary. However, the legislation recognises the need to avoid undue burdens for very small businesses by explicitly stating that documentation and records need only be commensurate with the nature and size of the business to demonstrate the effective application of their HACCP measures.

Hygiene 'pre-requisites'

[F9038] Codex Alimentarius, in its definitive *Guidelines for the Application of the HACCP System* (2003), says that any HACCP system must be supported by basic hygiene controls. These have become known as the pre-requisites to HACCP. EU Regulation 852/2004 takes a similar approach by establishing certain requirements for structures and services, covering the following subjects within twelve chapters:

Chapter I: General requirements for food premises.
Chapter II: Specific requirements for rooms where food is prepared.
Chapter III: Movable or temporary premises, etc.
Chapter IV: Transport.
Chapter V: Equipment.
Chapter VI: Food waste.

Chapter VII: Water Supply.
Chapter VIII: Personal hygiene.
Chapter IX: Protection of food from contamination.
Chapter X: Wrapping and packaging
Chapter XI: Heat treatment
Chapter XII: Training

Chapter I: General requirements for food premises, equipment and facilities

[F9039] Food premises must be kept clean, and in good repair and condition. The layout, design, construction, and size must permit good hygiene practice and be easy to clean and/or disinfect and should protect food against external sources of contamination such as pests, dirt, condensation etc. Cleaning agents and disinfectants are not to be stored in areas where food is handled.

Where necessary, suitable temperature-controlled handling and storage conditions should be available, with sufficient capacity for maintaining foodstuffs at appropriate temperatures and designed to allow those temperatures to be monitored and, where necessary, recorded.

Adequate sanitary and hand-washing facilities must be available and lavatories must not lead directly into food handling rooms. Sanitary conveniences are to have adequate natural or mechanical ventilation. Washbasins must have hot and cold (or better still mixed) running water and materials for cleaning and drying hands. Where necessary there must be separate facilities for washing food and hands. Drainage must be suitable and there must be adequate changing facilities.

Premises must have suitable natural or mechanical ventilation, and adequate natural and/or artificial lighting. Ventilation systems must be accessible for cleaning, for example to give easy access to filters.

Chapter II: Specific requirements in food rooms

[F9040] Food rooms should have surface finishes that are in good repair, easy to clean and, where necessary, disinfect. This would, for instance, apply to wall, floor and equipment finishes. The rooms should also have adequate facilities for the storage and removal of food waste. Of course, every food premises must be kept clean. How they are cleaned and how often, will vary. For example, it will be different for a manufacturer of ready-to-eat meals than for a bakery selling bread.

There must be facilities, including hot and cold water, for cleaning and where necessary disinfecting tools and equipment. There must also be facilities for washing food wherever this is needed, using potable water.

Chapter III: Temporary and occasional food businesses and vending machines

[F9041] Premises and vending machines must, as far as reasonably practicable, be sited, designed, constructed and kept clean and maintained in good repair so as to avoid the risk of contamination, in particular by animals and pests.

Most of the requirements outlined in **F9040** and **F9041** apply equally to food businesses trading from temporary or occasional locations like marquees or stalls although for reasons of practicability, some requirements are slightly modified.

Chapter IV: Transport of food

[F9042] The design of containers and vehicles must allow cleaning and disinfection. Businesses must keep them clean and in good order to prevent contamination and place food in them so as to minimise risk of contamination. Precautions must be taken if containers or vehicles are used for different foods or, in particular, for both food and non-food products. Different products should be separated to protect against the risk of contamination, and vehicles or containers should be cleaned effectively between loads. In general, bulk foodstuffs in liquid, granular or powder form should be transported in receptacles and/or containers/tankers reserved for the transport of foodstuffs.

Chapter V: Equipment

[F9043] All articles, fittings and equipment with which food comes into contact must be designed, constructed and maintained in a way that permits them to be effectively cleaned and, where necessary, disinfected at a frequency sufficient to avoid any risk of contamination. One interpretation of this requirement is that wooden cutting boards are not deemed suitable for use with ready-to-eat foods. Equipment should be installed so to allow adequate cleaning of the equipment and the surrounding area.

Where necessary to ensure the safety of the food product, equipment must be fitted with appropriate control devices, eg temperature recorders. If chemical additives are used to prevent corrosion of equipment and containers, they should be used in accordance with good practice.

Chapter VI: Food waste

[F9044] Food and other waste must not accumulate in food rooms any more than is necessary for the proper functioning of the food business. Containers for waste must be kept in good condition and be easy to clean and disinfect. Waste storage containers should be lidded to keep out pests. The waste storage area must be designed so that it can be easily cleaned and prevent pests gaining access. There should be arrangements for the frequent removal of refuse and the area should be kept clean.

All waste must be disposed in accordance with relevant legislation, using licensed waste disposal contractors and, in the case of products containing animal-derived ingredients, must also be in accordance with the relevant provisions of Regulation (EC) No 1774/2002 (as amended) and numerous supplementary and implementing Regulations and Decisions laid under it on 'health rules concerning animal by-products not intended for human consumption', implemented by the *Animal By-Products Regulations 2005 (SI 2005 No 2347)*. This requires animal by-products and waste containing them to be processed or disposed of in approved premises, and specifies the standards to which those premises must operate; it also specifies how animal by-products the products must be transported and identified. The Regulation introduces new disposal methods such as biogas, composting and co-incineration and allows for the approval of further alternative methods, based on scientific evidence.

Chapter VII: Water supply

[F9045] There must be an adequate supply of potable (drinking) water. Ice must be made from potable water (there are special rules for ice and water used for cooling/washing fish). Recycled water used in processing or as an ingredient must be of potable standard, unless it can be shown not to affect the wholesomeness of the food.

Water used for cooling cans etc after thermal treatment must not be a source of contamination. Non-potable water may be used for fire control, steam production, refrigeration and other similar purposes but must be in a separate, duly-identified system and not connect with potable water systems.

Steam used directly in contact with food must not contain any substance that presents a hazard to health or is likely to contaminate the food.

Chapter VIII: Personal hygiene

[F9046] Anyone who works in a food handling area must maintain a high degree of personal cleanliness. The way in which they work must also be clean and hygienic. Food handlers must wear clean and, where appropriate, protective clothes. Anyone whose work involves handling food should—

*   follow good personal hygiene practices;
*   wash their hands routinely when handling food;
*   never smoke or eat in food handling areas;
*   be excluded from food handling if carrying an infection that may be transmitted through food if there is a risk of contamination of the food.

Nobody with infected wounds, skin infections, sores or diarrhoea, or suffering from, or being a carrier of a disease that is likely to be transmitted through food, is to be permitted to handle food or enter any food-handling area in any capacity if there is any likelihood of direct or indirect contamination. This obligation on the business operator is supplemented by an obligation on every employee contained within the *Food Hygiene Regulations 2006 (SI 2006 No 14)*. Personnel working in food handling areas are obliged to report any illness (like diarrhoea or vomiting, infected wounds, skin infections) and if possible their causes, immediately to the proprietor of the business. If a manager receives such notification, he or she may have to exclude them from food handling areas. Such action should be taken urgently. If there is any doubt about the need to exclude, it is best to seek urgent medical advice or consult the local authority.

These personal hygiene provisions are wide ranging. They apply not only to food handlers, but anyone working in food handling areas. There have been well-documented cases of cleaners contaminating working surfaces or equipment with pathogens that are later transmitted to food.

Longstanding guidelines on *Food Handlers Fitness to Work* issued by the Department of Health have recently been updated and re-issued in draft form by the Food Standards Agency.

Chapter IX: Preventing food contamination

[F9047] A business operator must not accept ingredients or other material for use in food products, if they are known or might be expected to be, contaminated with parasites, pathogenic microorganisms or toxic, decomposed or foreign substances to the extent that, even after sorting, preparation and processing, the final product would be unfit for consumption.

Food and ingredients must be stored in conditions that prevent harmful deterioration and protected against contamination that may make them unfit for human consumption or a health hazard. For example, raw poultry must not be allowed to contaminate ready-to-eat foods.

Products likely to support the growth of pathogenic micro-organisms or toxin formation must be kept at appropriate temperatures (more details about temperature controls and management are given below at **F9050**). Frozen materials must be thawed in a way that similarly minimises the risk of micro-organisms growth or toxin formation. Any run-off liquid from the thawing process that may present a health risk must be adequately drained.

Adequate procedures to control pests and to prevent access by domestic animals must be in place. Hazardous and/or inedible substances, including animal feed, must be adequately labelled and stored in separate, secure containers.

Chapter X: Wrapping and packaging

[F9047.1] Materials used for wrapping and packaging must be stored and used in a way which ensures that they are not to be a source of contamination. Where appropriate and, in particular, in the case of cans and glass jars, the integrity of the container's construction and its cleanliness must be checked prior to use.

Wrapping and packaging material re-used for foodstuffs must be easy to clean and, where necessary, disinfect.

Chapter XI: Heat processing

[F9047.2] Specific requirements apply to foods processed and sold in hermetically-sealed containers to ensure safe processing and to avoid defective containers or re-contamination.

Chapter XII: Training and supervising food handlers

[F9048] Whereas eleven of the chapters deal with structural and physical pre-requisites, the final chapter deals with the higher-level issue of competence of personnel to do their job. This provision is more subtle than it is often given credit for. The provision, carried forward from previous legislation, is still subject to occasional criticism that it is too lenient on training requirements but, in fact, it requires three complementary elements, instruction, training and supervision.

Food business operators must ensure that food handlers are supervised and instructed and/or trained in food hygiene matters commensurate with their work activity and that those responsible for the development and maintenance of the HACCP-based procedure or for the operation of relevant guides have received adequate training in HACCP principles. In practice, therefore, food handlers must be instructed on how to do their particular jobs properly and supervised to make sure that they follow instructions. The requirements will differ according to the nature of the business and the job of the individual staff member.

Businesses must also comply with any national legal requirements concerning training programmes for persons working in certain food sectors, eg the meat industry.

Industry guides to good hygiene practice

[F9049] EU Regulation 852/2004 retains the concept of voluntary industry guides to good hygiene practice. These provide more detailed guidance on complying with the Regulations as they relate to specific industry sectors. They are usually produced by trade associations and recognised by the Food Standards Agency. Importantly, enforcement officers are obliged to have regard for them when examining how businesses are operating and industry-demonstrated compliance is likely to contribute towards a potential defence of 'due diligence' should the need arise. Numerous guides have now been published in the UK. Sectors covered include:

* Vending.
* Catering.
* Retail.
* Baking.
* Wholesale.
* Markets and Fairs.
* Fresh Produce.Sandwich bars.
* Specialist cheesemakers.
* Flour Milling.
* Bottled Water.
* Soft Drinks.
* Crustaceans and Finfish.

Temperature Control Requirements *[Food Hygiene Regulations 2006 (SI 2006 No 14) Schedule 4]*

[F9050] Growth of pathogenic bacteria in food will significantly increase the risk of food poisoning. Indeed, growth of any micro-organisms in food will compromise its wholesomeness. Temperature controls at certain food holding or processing steps will be CCPs in most HACCP plans. The importance of good control of food temperatures is emphasised by the fact that some controls are prescribed in regulations.

The *Food Hygiene Regulations 2006 (SI 2006 No 14) Schedule 4* effectively re-enact the principal provisions of the revoked *Food Safety (Temperature Control) Regulations 1995 (SI 1995 No 2200)* by requiring food business operators to observe certain temperatures during the holding of food if this is necessary to prevent a risk to health. (In Scotland, the requirements for boiling and holding temperatures for gelatine, prior to use, have been dropped but the requirement to reheat food to 82°C is retained.) The temperature requirements relate only to matters of food safety. For example, hard cheese that may go mouldy if kept at room temperature is not covered because it would *not* support the growth of *pathogens*.

The temperature control requirements in the *Food Hygiene Regulations 2006 (SI 2006 No 14) Schedule 4* do not apply to foods and establishments that fall under EU Regulation 853/2004 which sets more specific rules for most foods of animal origin during processing and handling prior to the retail or catering outlet.

The legislation makes a fairly simple issue extraordinarily complex. Furthermore, Scotland does not have the same regulations as the rest of the UK. (EU Regulation 852/2004 does not set specific temperature controls and, thus, detailed decisions are made nationally.)

The following table contains a very brief summary of the requirements. The Regulations prescribe temperatures for some but not all steps. The Regulations in Scotland also have similar gaps but not always at the same step. The industry guides, particularly the catering guide, fill in the gaps with recommendations of good practice.

	Rest of the UK	*Scotland*	*Recommended good practice targets*
Chill store	8°C maximum	Not specified	5°C
Cook	Not specified	Not specified	70°C for 2 minutes or 75°C minimum
Hot hold	63°C	63°C	63°C min
Cool	Not specified	Not specified	Below 10°C in 4 hours
Reheat	Not specified	82°C	70°C for 2 minutes or 75°C minimum

Frozen storage is not covered by the Food Hygiene Regulations and will not be a food safety CCP. Other legislation and best practice indicate a frozen storage temperature of −18°C for most foods but somewhat lower (−23°C for ice cream).

Relationship between time and temperature

[F9051] Almost invariably the control of micro-organisms in food depends upon a combination of time and temperature. Temperature alone does not have an absolute effect. For example, destruction of micro-organisms can be effected in a very short time (seconds) at temperatures above 100°C or in a couple of minutes at around 70°C. A similar thermal destruction can be achieved even at temperatures as low as 60°C but exposure for around 45 minutes will be necessary. Similarly in chilled storage, food can remain safe and wholesome for many days if the temperature can be kept close to freezing point at around −1°C. However, it will have a much shorter life at higher storage temperatures around 10°C.

The above remarks are generalisations and not all micro-organisms will react in the same way. Some organisms show greater resistance to heating; some are able to grow at storage temperatures as low as 3°C whereas others will show little growth at temperatures cooler than 15°C. But the general point that control is a function of time and temperature remains true.

Broadly speaking, this relationship is recognised in the Regulations. In some instances, time is actually prescribed in the temperature control regulations. An example is the so called 'four-hour rule' for display of food at a temperature warmer than 8°C. In other cases, times are implicit in the temperature control regulations, such as the necessity to cool food rapidly after heating. Actual parameters for compliance are outlined in the industry guides. For other situations there is an inter-relationship with other regulations. For example the storage life of pre-packed, microbiologically perishable food must be indicated on-pack under the requirements of Directive 2000/13 (the *Food Labelling Regulations 1996 (SI 1996 No 1499)*). A food must be labelled with an indication of its 'minimum durability' that takes the form of a 'use by' date and, in the case where the life of the product in the unopened pack has been extended, for example by the use of a modified atmosphere, the safe life after opening must also be displayed. There is an onus on the producer to ascertain a safe shelf life having regard for the nature of the food and its likely storage conditions, including allowances for potential temperature abuse during the product's life, and to use this as the basis of the date mark.

Temperature controls – England, Wales and N. Ireland

(N.B. Requirements for Scotland are described in detail below, see F9063)

[F9052] The regulations governing temperature controls were largely new provisions when introduced and refined in the early 1990s, and are continued, largely unchanged, under the new hygiene legislation. The structure of the Regulations in England, Wales and N. Ireland has the main requirement that chilled storage should be at 8°C or cooler, followed by a series of exemptions for particular circumstances. The general requirement for temperature control under *Reg 30, Sch 4* of the *Food Hygiene Regulations 2006 (SI 2006 No 14)* comes from the EU Regulation 852/2004, Article 4.3 and it is generally assumed that compliance with the other more specific regulations will deliver compliance. The requirements are outlined in the table below.

Provision within the Food Hygiene Regulations	*Requirement*	*Comment*
Reg 30, Sch 4	A general requirement for all food to be kept under temperature control if that is needed to keep it safe.	Applies to raw materials and foods in preparation. Limited periods outside temperature control are permitted for certain practicalities. No temperature is specified. A combination of time and temperature will be important.
Sch 4, para 2(1)	Chilled food must be kept at 8°C or cooler.	Applies only to foods that would become unsafe.

Provision within the Food Hygiene Regulations	Requirement	Comment
Sch 4, para 3	Various cold foods are exempt: shelf stable, canned foods, raw materials, cheeses during ripening, and others where there is no risk to health.	Soft cheeses once ripe, and perishable food from opened cans must be kept below 8°C.
Sch 4, para 4	Manufacturers may recommend higher storage temperature/shorter storage life (provided that safety is verified).	Caterers must use the food within the 'use by' date indicated.
Sch 4, para 5(1)	Cold food on display or for service can be warmer than 8°C. A maximum of 4 hours is allowed.	Any item of food can be displayed outside temperature control only once. The burden of proof is on the caterer.
Sch 4, para 5(2)	Tolerances outside of temperature control are also allowed during transfer or preparation of food, and defrost or breakdown of equipment.	No time/temperature limits specified. Both should be minimized consistently with food safety.
Sch 4, para 6, 7	Hot food should be kept at 63°C or hotter.	Food may be kept at a temperature cooler than 63°C for maximum 2 hours if it is for service or on display.
	Food must be cooled quickly after heating or preparation.	No limits are specified – must be consistent with food safety.

Note that all temperatures specified are food temperatures not the air temperature of refrigerators, vehicles or hot cabinets.

Chilled storage and foods included in the scope of the regulations

[F9053] The principal rule for chilled storage is contained within the *Food Hygiene Regulations 2006 (SI 2006 No 14), Sch 4 para 2*:

"Subject to sub-paragraph (2) and paragraph 3, any person who keeps any food—

(a)   which is likely to support the growth of pathogenic micro-organisms or the formation of toxins; and

(b)   with respect to which any commercial operation is being carried out, at or in food premises at a temperature above 8 degrees C shall be guilty of an offence."

Schedule 4 para 2 only applies to food that will support the growth of pathogenic micro-organisms. Such foods must be kept at 8°C or cooler. The types of food that will be subject to temperature control are indicated in the following table.

It is often good practice to keep foods at temperatures cooler than 8°C either to preserve quality or to allow longer-term storage and to allow a margin of error below the legal standard. Industry guides suggest a target food temperature of 5°C. This is especially important in cabinet fridges where there can be significant temperature rises during frequent door opening.

Food type	Comment
Cooked meats and fish, meat and fish products.	Includes prepared meals, meat pies, pates, potted meats, quiches and similar dishes based on fish.
Cooked meats in cans that have been pasteurised rather than fully sterilised.	Typically large catering packs of ham or cured shoulder.
Cooked vegetable dishes.	Includes cereals, rice and pulses. Some cooked vegetables or dessert recipes may have sufficiently high sugar content* (possibly combined with other factors like acidity) to prevent the growth of pathogenic bacteria. These will not be subject to mandatory temperature control.
Any cooked dish containing egg or cheese.	Includes flans, pastries etc.
Prepared salads and dressings.	Includes mayonnaise and prepared salads with mayonnaise or any other style of dressing. Some salads or dressings may have a formulation (especially the level of acidity**) that is adequate to prevent growth of pathogens.
Soft cheeses/mould ripened cheeses (after ripening).	Cheeses will include Camembert, Brie, Stilton, Roquefort, Danish Blue and any similar style of cheese.
Smoked or cured fish, and raw scombroid fish.	For example smoked salmon, smoked trout, smoked mackerel etc. Also raw tuna, mackerel and other scombroid fish.
Any sandwiches whose fillings include any of the foods listed in this Table.	
Low acid** desserts and cream products.	Includes dairy desserts, fromage frais and cream cakes. Some artificial cream may be 'ambient stable' due to low water activity and/or high sugar. Any product that does not support the growth of pathogenic micro-organisms does not have to be kept below 8°C. It may be necessary to get clarification from suppliers.
Fresh pasta and uncooked or partly cooked pasta and dough products.	Includes unbaked pies and sausage rolls, unbaked pizzas and fresh pasta.
Smoked or cured meats which are not ambient stable.	Salami, parma hams and other fermented meats will not be subject to temperature controls if they are ambient shelf stable.

* Technically Aw (water activity) is the key criterion.

** Technically, pH 4.5 (or more acid) is the critical limit.

Mail order foods

[F9054] There is a controversial exemption from the specific 8°C requirement if the food is being conveyed by post or by a private or common carrier to the ultimate consumer. The sender still has a responsibility for the safety of the food but the specific 8°C temperature does not apply. This exemption for mail order foods cannot apply to supplies to caterers or retailers neither of whom meet the definition of 'ultimate consumers'. (N.B. there are no specific provisions relating to mail order in Scotland.)

Short 'shelf life' products

[F9055] Some perishable foods are allowed to be kept at ambient temperatures for the duration of their shelf life with no risk to health. This may be because they are intended to be kept for only very short periods (eg certain types of sandwiches). Bakery products like fresh pies, pasties, custard tarts can also be kept without refrigeration for limited periods. However, good practice is to keep all such products that contain animal-derived products such as meats and dairy ingredients under temperature controlled conditions.

Canned foods and similar

[F9056] Sterilised cans or similar packs do not have to be stored under temperature control until the hermetically sealed pack is opened. After that, perishable food must be kept chilled (for example corned beef, beans, canned fish, and dairy products).

However, some canned meats are not fully sterilised (eg large catering packs of ham or pork shoulder) and must be kept chilled even before the can is opened.

High acid canned or preserved foods (some fruit, tomatoes, etc) do not have to be kept chilled after opening for safety reasons. However, it is advisable to remove these from the can for chilled storage after opening. It will inhibit mould growth and avoid any chemical reaction with the metal can body.

Raw food for further processing

[F9057] Raw food that is intended for further processing does not have to be kept at 8°C or cooler, providing the further processing will render the food fit for consumption. Thus it is not against the Regulations to keep raw meat, poultry and fish out of the fridge. Of course, for quality and safety reasons it is usually best kept in the fridge; exceptions would include game and similar meats being "hung" to develop their characteristic maturity. Processed foods must be kept at 8C or cooler, even if they are to be heated again.

Raw meat or fish that is intended to be eaten without further processing (for example beef for steak tartare, or fish for sushi) will not be exempted by this provision. It must be kept chilled. Raw scombroid fish (tuna, mackerel, etc) will also not be exempted by this provision. The 'scombrotoxin' is heat stable and processing will not render contaminated food fit for consumption. Scombroid fish must be kept at 8°C or cooler.

Variations from 8 °C storage

[F9058] The *Food Hygiene Regulations 2006 (SI 2006 No 14) Reg 30, Sch 4 para 4* allow the producer of a food to recommend that it can be kept safely at a temperature above 8°C. He must label the food clearly with the recommended storage temperature and the safe shelf life at that temperature, or convey the information by other written form. He must also have good scientific evidence that the food is safe at that temperature. In practice, little use has been made of this provision.

The four-hour rule

[F9059] The 'four-hour rule' allows cold food to be kept above 8°C when it is on display. This is crucial for many catering outlets, and some retailers, which depend upon this exemption for at least some of their operation. Without it, food could not be served on a buffet without refrigeration. The time that the food is on display must be controlled. The maximum time allowed is four hours.

Only one such period of display is allowed, no matter how short. For example, if a dish of food is placed on display for one hour at the end of a service period, it cannot be displayed for a further three hours above 8C at the next service. Food uneaten at the end of a display period does not have to be discarded provided that it is still fit for consumption. The food must be cooled to 8°C or cooler and kept at that temperature until it can be used safely.

The burden of proof is on the business, which must be able to demonstrate that the time limit is observed. Good management of food displays is therefore important. The amount of food on display must be kept to a minimum consistent with the pattern of trade.

There needs to be systems to help keep to the time limits and to demonstrate that they are being adhered to. These may include the labelling of dishes to indicate when they went onto display. Avoid topping-up of bulk displays of food. Food at the bottom of the dish will remain on display for much longer than four hours if topping-up is allowed.

Exemptions for other contingencies

[F9060] The Regulations allow for the fact that food may rise above 8°C for limited periods of time in unavoidable circumstances such as:

- transfers to or from vehicles;
- during handling or preparation;
- during defrost of equipment;
- during temporary breakdown of equipment.

The Regulations do not put specific figures on the length of time allowed or how warm the food might become. This tolerance is allowed as a defence and the burden of proof will be on the business to show that:

- the food was unavoidably above 8°C for one of the reasons allowed;
- that it was above 8°C for only a limited period;
- that the break in temperature control was consistent with food safety.

The acceptable limits will obviously depend upon the combination of time and temperature. Under normal circumstances, a single period of up to two hours is unlikely to be questioned.

All transfers of food must be organised so that exposure to warm ambient temperatures is reduced and rises in food temperature are kept to a minimum. For example, deliveries should be put away quickly and chilled food moved first, then frozen, and finally ambient grocery.

If food is being collected from a cash and carry warehouse, insulated bags or boxes should be used for any chilled foods to which the Regulations apply. The chilled food should be taken directly to the outlet and quickly put into chilled storage.

Risks of equipment breakdown should be minimised by ensuring planned and regular maintenance.

Hot food

[F9061] Hot food must be kept at 63°C or hotter if this is necessary for safety. This will apply to the same types of food described in the earlier table (see **F9053** above). For example, the rule does not apply to hot bread or doughnuts.

Food must be kept at 63°C or hotter, whether:

- it is in the kitchen or bakery awaiting service or dispatch;
- or in transit to a serving point, no matter how near or far;
- or actually on display in the serving area.

Some tolerance or limited exemptions are allowed for practical handling reasons. As a parallel to the 'four-hour' rule for chilled food, hot food can be kept for service or display at less than 63°C for *one period* of up to *two hours*.

Again the operator must be able to show that:

- the food was for service or on display for sale;
- it had not been kept for more than two hours;
- it had only had one such period.

There is a further general exemption if a well-founded scientific assessment can show that storage and safety at below 63°C does not present any risk to health.

Generally there should be fewer problems with meeting the 63°C target in hot display equipment than encountered in achieving 8°C in 'chilled' display units. However, operators must be aware that hot display units are not designed to heat or cook foods from cold. They should be used for holding only, not heating.

Cooling

[F9062] Previous legal requirements for food that becomes warmer than 8°C during processing to be cooled again to 8°C 'as quickly as possible' have not been re-enacted and the necessary practices will therefore fall under 'good practice' and the general requirements to supply safe food. Over the years, a wide variety of cooling times have been recommended by different sources. Cooling times as short as one-and-a-half hours were specified in UK Department of Health guidelines on cooking chilled foods and published in the early 1980s. Many authorised officers have regarded this figure as definitive. The UK industry guides recommend that cooling between 60°C and 10°C should be accomplished in a maximum of four hours, although research suggests that this may be conservative.

Note that rapid cooling is not only necessary for food that has been heated. Food that has become warm during processing should also be returned to chilled storage below 8°C as quickly as possible after the process is completed.

Temperature controls – Scotland

[F9063] The Scottish Regulations have remained largely unchanged for many years and differ in significant respects from those in the rest of the UK. No chilled storage temperature is specified, yet the 82°C requirement for re-heating pre-cooked food appears unnecessarily high and unsupported by the science (although there is an exemption from this temperature if it can be shown to be impossible to achieve without a deterioration in the quality in the product). However, the long-standing requirements continue largely unchanged under the 2006 regulations, the only change being the deletion of requirements to boil gelatine or to hold it above 71°C for 30 minutes prior to use.

Provision within the Food Hygiene Regulations	*Requirement*	*Comment*
Schedule 4	A general requirement for all food to be kept under temperature control if that is needed to keep it safe. Additionally it includes the need for rapid cooling.	Applies to raw materials and foods in preparation. Limited periods outside temperature control are permitted for certain practicalities. No temperature is specified. A combination of time and temperature will be important.
Sch 4, para 2.1(a)	Cold food must be kept in a cool place or refrigerator.	No temperature is specified.
Sch 4, para 2.1(b)	Hot food must be kept at 63°C or hotter.	There are exemptions for food on display, during preparation, etc.
Sch 4, para 3.1	Food that is reheated must reach 82°C.	Exemption in place if 82°C is detrimental to food quality.
		Unused glaze must be chilled quickly and kept in a refrigerator.

Again, note that all temperatures specified are food temperatures not the air temperature of refrigerators, vehicles or hot cabinets.

Cool storage

[F9064] The *Food Hygiene (Scotland) Regulations 2006, (SI 2006 No 3) Reg 30, Sch 4* apply to any food that may support the growth of food poisoning organisms within the 'shelf life' for which the food will be kept.

The types of food that will be subject to the provision are the same as those that fall under *Reg 30, Sch 4* of the Regulations (applicable to England, Wales and N. Ireland, described above, see **F9052–F9062**).

If cold, this food must be kept either in a refrigerator or a cool well-ventilated place. The Scottish requirements do not specify a temperature but in practice, it is advisable to follow the same rules and recommendations that apply in the rest of the UK. In any case, the HACCP (hazard analysis) requirement of the General Food Hygiene Regulations applies in Scotland, and thus appropriate controls at CCPs will need to be established.

Hot food

[F9065] The target for hot food is the same as England, Wales and N. Ireland, namely 63°C or hotter (but see **F9068** for re-heating temperature rules).

Cooling

[F9066] The Scottish requirements (*Sch 4, para 2(c)*) demand food to be cooled under hygienic conditions as quickly as possible to a temperature which would not result in a risk to health. This requirement is considered to have the same effect as *Sch 4, para 4* in the legislation applicable to England, Wales and Northern Ireland and, in both cases, to fall under the general requirement for safe processing and handling of food.

Exemptions during practical handling

[F9067] In certain circumstances food does not have to be kept cold or above 63°C:

- if it is undergoing preparation for sale;
- if it is exposed for sale or it has already been sold;
- if it is being cooled;
- if it is available to consumers for sale;
- if it is shelf stable.

Again these Regulations have remained unchanged for many years and are less detailed than those that apply in the rest of the UK.

Reheating of food

[F9068] In Scotland only, food that has been heated and is being reheated must be raised to a temperature of 82°C or hotter. Fortunately, this is qualified by a clause stating that this temperature is not necessary if it is detrimental to the quality of the food, which therefore provides the opportunity for businesses to reheat food to the perfectly adequate and scientifically verified temperature of around 70–72°C.

Gelatine

[F9069] The previous rules requiring gelatine to be boiled or held at 71°C or hotter for at least 30 minutes have not been retained.

Food Safety Manual – food safety policy and procedures

[F9070] Both the *Food Safety Act 1990* and the Regulations made under it provide an obligation on proprietors of food businesses to operate within the law. In addition, the *Food Safety Act 1990, s 21* provides the defence of 'due diligence' (see **F9023**). One way for a business to demonstrate its commitment to operating safely and within the law is to document its management system, including the relevant policies and procedures. This documentation may form a significant part of a due diligence defence. The policy and procedures will illustrate the precautions that should be taken by the business and its staff in their day-to-day operations. The procedures will normally require the keeping of various records, and these will in turn help to demonstrate that the *'reasonable precautions'* are being followed *'with all due diligence'*.

It is often convenient to keep these policies and procedures together in a 'Food Safety Manual'. The following provides a brief outline of the contents of a Food Safety Manual for a typical food business.

Policy and procedures

[F9071] The 'policy' will be a statement of intent by the business. A Food Safety Manual may begin with a fairly broad statement of policy. Later, there may be more specific policy statements linked to particular outcomes, for example a policy to take all reasonable precautions to minimise the contamination of food with foreign material.

Procedures will detail the ways in which the business intends to put the policies into effect. A typical layout of any 'Food Safety Manual' is to begin each section or subject heading with a statement of policy, and to follow that with the procedures intended to deliver it. Procedures are often referred to by the acronym SOPs – 'Standard Operating Procedures'.

General policy and organisation

[F9072] A Food Safety Manual would normally begin with a top-level statement of the business's policy with regard to food safety. Recognising the wider scope of 'food safety' within the meaning of the 1990 Act and the EU Regulation 178/2002, it is advisable to add a statement about commitment to true and accurate labelling of food and adherence to compositional requirements.

The statement should be signed by the proprietor of the business, who is the person with ultimate legal responsibility. In a large group this may be the group chief executive.

The early part of the manual should also illustrate the organisational structures and management hierarchy that has responsibility for implementing the policy and procedures. There may be technical support services, (eg pest control, testing laboratories) either from inside or outside the business. Their roles and responsibilities can also be identified within the organisational structures in this part of the manual.

Hazard Analysis (HACCP)

[F9073] The law requires that food safety policy must be centred on a hazard analysis approach and the Food Safety Manual should include a policy statement to that effect. This will be followed by procedures to describe how the business establishes the necessary HACCP procedure and record its outcomes and implementation.

Pre-requisites to Hazard Analysis

[F9074] The modern approach to HACCP focuses on the small number of process steps that *must* be controlled (and monitored) to ensure that the food is safe. But this control must operate against a background of good hygiene practice. The so-called 'pre-requisites' to HACCP must all be in place and the Food Safety Manual should include policies and procedures for all of the following:

- Design – structures and layout
  The Food Safety Manual should include a policy that the design, structure and layout will promote hygienic operation. The procedures should detail specific arrangements to achieve that. This section should also cover points such as the services, especially water and ventilation; provision for washing of hands, equipment and food; and removal of waste.

- Equipment specification
  The policy should be that the business will only acquire and use equipment that is fit for its purpose. The focus should extend beyond just cleanability alone. In many cases, the ability of equipment to do the job can be even more important. For example, can the refrigeration equipment keep food at a specified temperature in operational conditions?

- Maintenance
  A policy for maintenance should address both the premises in general and the equipment. Procedures should detail the arrangements for routine and emergency maintenance. There should be a record of any maintenance whether routine or emergency.

- Cleaning and disinfection
  Detailed cleaning schedules for sections of the building or particular pieces of equipment should be available. Cleaning should be recorded.

- Pest Control
  Employing a pest control contractor is only one small part of the pest control procedures. The prevention of pest access, the integrity of pest proofing, cleaning and the need for vigilance should also be included in the procedures. Records would normally include any sightings of or evidence of pests, positions of bait stations, visits by the contractor together with observations and action taken, and even details such as the changing of UV tubes on electronic flying insect killers.

- Personal hygiene
  The business should have a very clear policy on personal hygiene. The arrangements will include a very clear statement of the dress code and personal hygiene requirements for all staff and contractors working in or passing through food areas; appropriate rules should also apply to visitors to such areas. There must also be very particular policies and procedures to deal with staff suffering illness that may potentially be food borne.

- Food purchasing
  The policy should state that the business will purchase only from reputable suppliers. However, it may be more difficult to deliver such a policy, especially for smaller businesses that do not have the resources to inspect their suppliers. The Manual should, therefore, detail how this policy will be implemented. For assessments, it may be possible to rely

on third party inspection systems such as those that have been developed during and since the late 1990s; especially those operated by accredited inspection or certification bodies.

Supply specifications should also be drafted. Such specifications will usually cover food safety and 'quality' in its broadest sense. In addition to specific safety parameters such as chemical and microbiological criteria, they are the ideal route through which to state food safety conditions such as delivery temperature, shelf life remaining after delivery, packaging and so on.

• Storage

Most businesses will have documented procedures for the storage of various categories of food as well as for stock checks to ensure proper storage temperature and stock rotation.

• Food handling and preparation

The general policy should be to manage all steps in food handling and preparation in order to minimise contamination with any hazardous material or pathogenic micro-organisms. There should also be precautions to minimise the opportunities for growth of any micro-organisms that might contaminate the food. The procedures will be unique to the individual business, its particular range of foods and preparation methods.

• Foreign material

The policy should require that all precautions are taken to minimise the risk of foreign material contamination, including chemical hazards such as cleaning materials. In general terms, procedures will cover three points:

(a) Excluding or controlling potential sources of contamination;
(b) Protecting food from contamination;
(c) Taking any complaints seriously, tracking down what caused them and stopping them from happening again. Complaint trend analysis is important.

Allergy (Anaphylaxis)

[F9075] A responsible business will have a policy to protect susceptible customers from exposure to foods to which they may be allergic. In general terms procedures will rely upon:

(a) Awareness amongst staff of the types of food that may cause reaction.
(b) Labelling of the foods that include such ingredients.
(c) Careful handling, storage and use of these ingredients at all stages to avoid 'cross-contamination' to other products.
(d) An ability to supply reliable information about all ingredients on request from consumers.

Legislation now requires comprehensive labelling of the known presence of 12 specified allergens to be declared when they or their derivatives are deliberately added to pre-packed food, including alcoholic drinks (see F9003).

The detailed requirements of this legislation are outside the scope of this overview of food safety legislation.

Vegetarians and other dietary preferences

[F9076] Most food businesses today recognise the need to cater for customers with particular dietary preferences, notably vegetarians but also vegan, kosher and Halal. There should be a policy to be able to supply vegetarian and / or other specified dishes properly formulated and accurately labelled. The FSA has issued Guidance on the use of the terms 'vegetarian' and 'vegan' on food labels. Guidance on Halal or kosher requirements may be obtained from, for example, the Islamic Cultural Centre (London) or the London Beth Din Kashrut Division, respectively.

Consumer concerns

[F9077] From time to time particular issues will come to the forefront of consumer consciousness. For example, genetically modified ingredients, sustainability, food miles and carbon footprints or animal welfare issues, such as veal production, foie gras and intensive rearing of poultry. The Food Safety Manual could include a statement of the policy on whichever of these issues that the business considers relevant.

Procedures will be needed to deliver them. In the case of food and food ingredients derived from genetically modified organisms, comprehensive legislation requires the traceability and labelling of all such materials throughout the food supply chain, based on EU Regulations 1929/2003 and 1830/2003. Numerous "private" schemes have been developed in recent years to cover areas such as "ethical trading", carbon footprint labelling etc but, as yet, no formal legislation specifically controls these (other than the general rules on misleading labelling and advertising).

Labelling and composition of foods and 'fair-trading'

[F9078] In view of the fact that the *Food Safety Act 1990* deals not only with health and hygiene, and in the light of growing interest in these areas, it may be considered desirable to include policies and procedures on topics such as food composition and nutrition labelling. Ethical and fair-trade issues might also be relevant to the company's product range and intended market; schemes and controls on these are as described under F9077.

Product recall

[F9079] Traceability and the requirement to initiate recall/withdrawal procedures if a company has reason to believe one of its products may be 'unsafe' are now legal requirements. Businesses involved in central production and distribution of food that will have several days of 'shelf life' (if not more) should have contingency plans for product recall by ensuring that products are labelled in such a way that individual batches can be identified and records kept to show where/when it might have been delivered.

A formal recall procedure should be an integral part of its food safety management policy.

Complaints

[F9080] Complaints are an important source of information. A business should have a policy and procedures to deal with them. These will normally include keeping a record of every complaint and making every attempt to track down its cause. Complaint trends should be analysed regularly as a means to identify developing problems in the manufacturing process or supply chain.

Training, instruction and supervision

[**F9081**] Cutting across all of the other policies and procedures will be a policy on staff training, instruction and supervision. Typically a business will keep a personal training record for all members of staff.

Summary

[**F9082**] Modern businesses recognise the value of a properly documented management system. A Food Safety Manual can provide the focus of such a system. The design of a Food Safety Manual must be a fine balance between conflicting objectives. On the one hand, it must be reasonably comprehensive if it is to stand up as part of a 'due diligence' defence. On the other hand, it must be sufficiently concise to be usable as an effective working tool within the business. It must form an integral part of the business management's working tools – it must not be seen as a "nice-to-have" which, once developed languishes on an office shelf, gathering dust.

Copyright forms are reproduced by permission of the publishers, Shaw & Sons Ltd, Shaway House, Crayford, Kent DA1 4BZ, from whom copies may be purchased.

Gas Safety

Dr R C Slade

Introduction

[G1001] Not surprisingly, explosions caused by escaping gas have been responsible for large-scale damage to property and considerable personal injury, including death. Although escape of metered gas or gas in bulk holders might possibly have attracted civil liability under the rule in *Rylands v Fletcher* (1868) LR 3 HL 330 (see further **F5054** FIRE AND FIRE PRECAUTIONS), statutory authority to perform a public utility (i.e. supply gas) used to constitute a defence in such proceedings, coupled with the fact that negligence had to be proved in order to establish liability for personal injury (*Read v J Lyons & Co Ltd* [1947] AC 156). This situation has now changed in favour of imposition of strict product liability for injury/damage caused by escape of gas and incomplete or inefficient combustion causing carbon monoxide poisoning. Transco PLC was fined a total of £15m at the High Court of Justiciary in Edinburgh, under the *Health and Safety at Work etc. Act 1974 (HSWA 1974), s 3*. The jury returned the guilty verdict following a six-month trial concerning an explosion which took place on 22 December 1999, destroying a house and killing its four occupants.

An investigation by officials from the Health and Safety Executive showed that there were holes in the 250mm medium pressure ductile iron pipe that ran through the front garden of the house. Gas leaking from the main found its way into the under-floor void and subsequently the kitchen of the property where it ignited. The precise source of ignition could not be determined but it could have been a gas hob or other source in the kitchen. Commenting outside the court, John Sumner, Head of the Health and Safety Executive's Chemicals Unit in Scotland said:

> This was a very detailed investigation. I would like to pay tribute to the excellent work carried out by all the investigators involved – HSE's own staff, Strathclyde police and the Crown Office and Procurator Fiscal service . . .

> With regard to pipeline safety, it should be noted that HSE served an improvement notice in September 2000 to accelerate Transco's mains replacement programme. This resulted in the replacement of all known sections of the sort of main that ruptured at Larkhall in the three years after the accident. In other words, some 2,500 km of ductile iron medium pressure gas mains were replaced by plastic pipeline.

> The conviction sends a message not just to Transco but to all operators of hazardous plant of the need to keep accurate records, operate effective management systems and properly maintain pipelines and equipment

On 20 October 2005 Transco PLC, now trading as National Grid, was fined £1m and ordered to pay costs of £134,000 after a gas explosion in November 2001 killed a resident at the five storey Cavendish Mill flats, Ashton-under-Lyne, Greater Manchester.

The explosion was caused by gas leaking from a fractured mains supply in front of the converted mill. Although the main did not supply gas to the building, the leaking gas entered the building through voids below ground floor rooms. The gas escape was noticed on the morning of 14 November 2001 and Transco evacuated the flats, ventilated the building and checked individual flats for residual gas. However, they failed to identify and test the building voids. Just before 11.00 pm on that day residents were allowed back into the building. One, a Mr Ian Brady entered his flat, lit a cigarette and ignited residual gas causing the explosion. He died in hospital four days later. The subsequent investigation revealed that Mr Brady's flat and two voids below it contained significant quantities of gas.

Transco admitted a breach of *HSWA 1974, s 3(1)* in that it failed to ensure that residents were not exposed to risks to their health and safety arising from its management of the gas escape. They were fined £1M, Judge Anthony Hammond saying that the 'only penalty is a financial one' and that the degree of negligence 'pales into insignificance' when compared to Larkhall.

GylesHyder, the HSE's investigating inspector said:

> Transco identified some of the voids where gas could accumulate; but their identification and the fact that unusual void spaces are common in converted buildings did not prompt Transco to carry out a more thorough assessment of the building for similar spaces.

After the hearing Tranco said:

> The tradegy at Cavendish Mill was a unique set of circumstances, but we accept that we made mistakes in managing the incident, for which we apologise. Safety is and always will be our number one priority.

According to research undertaken by the Consumer's Association, over half of the appliances surveyed in dwellings had not been serviced in accordance with the requirements of the Gas Safety Regulations. As some well-publicised court cases have demonstrated, landlords and those they employ to deal with gas installations face heavy fines and possibly imprisonment for breach of the regulations (see **G1026** below).

Changes have also been made to the gas industry itself, stemming from the Government's policy of creating competition in the gas supply field. These changes have resulted in the introduction of new and revised legislation, namely:

- *Gas Act 1986*
  This has been considerably modified and extended by the *Gas Act 1995*.
- *Gas Safety (Installation and Use) Regulations 1998 (SI 1998 No 2451)*
  These Regulations were made under the authority of the *Health and Safety at Work etc Act 1974*.
- *Gas Safety (Management) Regulations 1996 (SI 1996 No 551)*

These Regulations were made under the authority of the *Health and Safety at Work etc Act 1974.*
- *Pipelines Safety Regulations 1996 (SI 1996 No 825)*
These Regulations were made under the authority of the *Health and Safety at Work etc Act 1974.*
- *Gas Safety (Rights of Entry) Regulations 1996 (SI 1996 No 2535)*
These Regulations were made under the authority of the *Gas Act 1986* (as amended).

In practice, many of the legislative changes were introduced for the purposes of increasing responsibilities for safety of gas processors, gas transporters and gas suppliers and are therefore outside the scope of this handbook. This chapter concentrates on the legislation which deals with the knowledge required by landlords, managing agents, health and safety managers, employers and other responsible persons who must ensure that gas appliances and fittings are installed safely and checked by a competent person every twelve months.

It should be noted that although the legislation discussed in this chapter is current at the date of publication, the Health and Safety Executive (HSE) is at present considering changes that could affect the *Gas Act 1986* (as amended) and some of the regulations referred to in this section. The proposed changes are outlined in a HSE document entitled *'Fundamental Review of Gas Safety Regime Proposals for Change'*, which is the result of consultation with interested parties throughout the gas industry.

Gas is defined in the *Gas Act 1986* (as amended) and the *Gas Safety (Installation and Use) Regulations 1998 (SI 1998 No 2451)* to include methane, ethane, propane, butane, hydrogen and carbon monoxide mixtures of any two or more of these gases together with inert gases or other non-flammable gases; or combustible mixtures of one or more of these gases and air. However, the definition does not include gas consisting wholly or mainly of hydrogen when used in non-domestic premises. The use and storage of gas, as defined, in any workplace is subject also to the *Dangerous Substances and Explosives Atmospheres Regulations 2002 (SI 2002, No 2776)*. In practical terms, the application of the *SI 2002, No 2776* has the effect of bringing the standards of *SI 1998 No 2451* in the workplace.

Gas supply – the Gas Act 1986, as amended

[G1002] The *Gas Act 1986* (as amended) requires the Secretary of State for the Environment to establish a Gas Consumers' Council and appoint an officer – the Director General of Gas Supply – to perform the functions relating to the supply of gas as set out in *Part I* of the Act.

As far as safety in the supply of gas is concerned, the Secretary of State and the Director each have a principal duty to exercise the functions assigned to them under Part I for the protection of the public from dangers arising from the conveyance or from the use of gas conveyed through pipes.

The following persons are defined in the *Gas Act 1986* as having duties and responsibilities in respect of the safe supply of gas:

(a) Domestic customer

A domestic customer is a person who is supplied by a gas supplier with gas conveyed to particular premises at a rate which is reasonably expected not to exceed 2,500 therms a year.

(b) Owner

In relation to any premises or other property, 'owner' includes a lessee, and cognate expressions are construed accordingly.

—   in relation to dangers arising from the conveyance of gas by a public gas transporter, or from the use of gas conveyed by such a transporter, 'relevant authority' means that transporter; and

—   in relation to dangers arising from the conveyance of gas by a person other than a public gas transporter, or from the use of gas conveyed by such a person, 'relevant authority' means the Secretary of State for the Environment.

With regard to safety, the *Gas Act 1986* (as amended) provides for:

—   licensing and general duties [*Gas Act 1986, ss 4A, 7–10* and *23*];
—   security [*Gas Act 1986, s 11*];
—   safety regulations regarding rights of entry for inspection of connected systems and equipment and the making safe and investigation of gas escapes [*Gas Act 1986, ss 18* and *18A*];
—   pipeline capacity [*Gas Act 1986, ss 21, 22, 22A*];
—   standards of performance [*Gas Act 1986, s 33* as amended by the *Competition and Service (Utilities) Act 1992*].

Gas supply management – the Gas Safety (Management) Regulations 1996 (SI 1996 No 551)

[G1003] Principally, the *Gas Safety (Management) Regulations 1996 (SI 1996 No 551)* require an appointed person – the 'network emergency co-ordinator' – to prepare a safety case for submission to and acceptance by HSE before a gas supplier can convey gas in a specified gas network [*SI 1996 No 551, Reg 3*].

The safety case

[G1004] The safety case document must contain:

—   the name and address of the person preparing the safety case (the duty holder);
—   a description of the operation intended to be undertaken by the duty holder;
—   a general description of the plant, premises and interconnecting pipes;
—   technical specifications;
—   operation and maintenance procedures;

— a statement of the significant findings of the risk assessment carried out pursuant to the *Management of Health and Safety at Work Regulations 1999 (SI 1999 No 3242), Reg 3*;

— particulars to demonstrate the adequacy of the duty holder's management system to ensure the health and safety of his employees and of others (in respect of matters within his control);

— particulars to demonstrate adequacy in the dissemination of safety information;

— particulars to demonstrate the adequacy of audit and reporting arrangements;

— particulars to demonstrate the adequacy of arrangements for compliance with the duty of co-operation [*Gas Safety (Management) Regulations 1996 (SI 1996 No 551), Reg 6*];

— particulars to demonstrate the adequacy of arrangements for dealing with gas escapes and investigation [*Gas Safety (Management) Regulations 1996 (SI 1996 No 551), Reg 7*];

— particulars to demonstrate compliance with the content and characteristics of the gas to be conveyed in the network [*Gas Safety (Management) Regulations 1996 (SI 1996 No 551), Reg 8*];

— particulars to demonstrate the adequacy of the arrangements to minimise the risk of supply emergency;

— particulars of the emergency procedures [*Gas Safety (Management) Regulations 1996 (SI 1996 No 551), Reg 3(1) and Sch 1*].

Duties of compliance and co-operation

[G1005] The duty holder must ensure that the procedures and arrangements described in the safety case and any revision of it are followed. In addition, a duty of co-operation is placed upon specified persons including:

— a person conveying gas in the network;
— an emergency service provider;
— the network emergency co-ordinator in relation to a person conveying gas;
— a person conveying gas in pipes which are not part of a network;
— the holder of a licence issued under the *Gas Act 1986, s 7*;
— the person in control of a gas production or processing facility [*Gas Safety (Management) Regulations 1996 (SI 1996 No 551), Regs 5 and 6*].

The *Gas Safety (Management) Regulations 1996 (SI 1996 No 551), Reg 7* deals with the duties and responsibilities of gas suppliers, gas transporters and 'responsible persons' regarding actions to be taken in the event of gas escape incidents. All incidents must be investigated including those resulting in an accumulation of carbon monoxide gas from incomplete combustion of a gas fitting.

Anyone discovering or suspecting a gas leak must notify British Gas plc immediately by telephone. British Gas are obliged to provide a continuously manned telephone service in Great Britain. The person appointed 'responsible person' for the premises must take all reasonable steps to shut off the gas supply.

The reporting of gas incidents to the HSE is laid down in the *Reporting of Injuries, Diseases and Dangerous Occurrences Regulations 1995 (SI 1995 No 3163) (RIDDOR)* in conjunction with HSE Form F2508G.

For further information regarding compliance with these Regulations see the HSE publication L80: '*A guide to the Gas Safety (Management) Regulations 1996*' (ISBN 0 7176 1159 0).

Rights of entry – the Gas Safety (Rights of Entry) Regulations 1996 (SI 1996 No 2535)

[G1006] Where an officer authorised by a public gas transporter has reasonable cause to suspect an escape of gas into or from premises supplied with gas, he is empowered to enter the premises and to take any steps necessary to avert danger to life or property. [*Gas Safety (Rights of Entry) Regulations (SI 1996 No 2535), Reg 4*].

Inspection, testing and disconnection

[G1007] On production of an authenticated document, an authorised person must be allowed to enter premises in which there is a service pipe connected to a gas main for the purpose of inspecting any gas fitting, flue or means of ventilation. Fittings and any part of a gas system may be disconnected and sealed off when it is necessary to do so for the purpose of averting danger to life or property.

It is incumbent on the authorised officer carrying out any disconnection or sealing off activities to give a written statement to the consumer within five days. The notice must contain:

— the nature of the defect;
— the nature of the danger in question;
— the grounds and the manner regarding the appeal procedure available to the consumer.

Prominent and conspicuous notices must also be affixed at appropriate points of the gas system regarding the consequences of any unauthorised reconnection to the gas supply. [*Gas Safety (Rights of Entry) Regulations (SI 1996 No 2535), Regs 5–8*].

Prohibition of reconnection

[G1008] It is an offence for any person, except with the consent of the relevant authority, to reconnect any fitting or part of a gas system. This provision is qualified by the term 'knows or has reason to believe that it has been so disconnected'. [*Gas Safety (Rights of Entry) Regulations (SI 1996 No 2535, Reg 9*].

Pipelines – the Pipelines Safety Regulations 1996 (SI 1996 No 825)

[G1009] The *Pipelines Safety Regulations 1996 (SI 1996 No 825)* apply to all pipelines in Great Britain, both on and offshore, with the following exceptions:

— pipelines wholly within premises;
— pipelines contained wholly within caravan sites;
— pipelines used as part of a railway infrastructure;
— pipelines which convey water.

For the purposes of the Regulations, a pipeline for supplying gas to premises is deemed not to include anything downstream of an emergency control valve, that is, a valve for shutting off the supply of gas in an emergency, being a valve intended for use by a consumer of gas. [*SI 1996 No 825, Reg 3(4)*].

The Regulations complement the *Gas Safety (Management) Regulations 1996 (SI 1996 No 551)* and include:

— the definition of a pipeline [*SI 1996 No 825, Reg 3*];
— the general duties for all pipelines [*SI 1996 No 825, Regs 5–14*];
— the need for co-operation among pipeline operators [*SI 1996 No 825, Reg 17*];
— arrangements to prevent damage to pipelines [*SI 1996 No 825, Reg 16*];
— the description of a dangerous fluid [*SI 1996 No 825, Reg 18*];
— notification requirements [*SI 1996 No 825, Regs 20–22*];
— the major accident prevention document [*SI 1996 No 825, Reg 23*];
— the arrangements for emergency plans and procedures [*SI 1996 No 825, Regs 24–26*].

Generally, the Regulations place emphasis on 'major accidents' and the preparation of emergency plans by local authorities. As with the *Gas Safety (Management) Regulations 1996 (SI 1996 No 551)*, the detail relates to the specification and characteristics of gas and the design of the pipes to convey it. The design of gas service pipelines is specifically addressed in the HSE Approved Code of Practice, L81: '*Design, construction and installation of gas service pipes*' (1996) (ISBN 0 7176 1172 8). For further information on and guidance to the *Pipelines Safety Regulations 1996*, see the HSE Publication L82: '*A guide to the Pipelines Safety Regulations 1996*' (ISBN 0 7176 1182 5).

The *Pipelines Safety (Amendment) Regulations 2003 (SI 2003, No 2563)* amend the *Pipelines Safety Regulations 1996 (SI 1996, No 825)* by inserting a new *Regulation 13A* (iron pipelines). The new Regulation requires that the HSE should approve any programme for the decommissioning of any description of iron (not including steel) pipe used in a pipeline.

Gas systems and appliances – the Gas Safety (Installation and Use) Regulations 1998 (SI 1998 No 2451)

[G1010] The *Gas Safety (Installation and Use) Regulations 1998 (SI 1998 No 2451)* are supported by the HSE Approved Code of Practice and Guide, L56: *'The Gas Safety (Installation and Use) Regulations 1998'*.

The Regulations aim at protecting the gas-consuming public and cover natural gas, liquefied petroleum gas (LPG), landfill gas, coke, oven gas, and methane from coal mines when these products are 'used' by means of a gas appliance. Such appliances must be designed for use by a gas-consumer for heating, lighting and cooking. However, a gas appliance does not include a portable or mobile appliance supplied with gas from a cylinder except when such an appliance is under the control of an employer or self-employed person at a place of work. Similarly, the Regulations do not cover gas appliances in domestic premises where a tenant is entitled to remove such appliances from the premises.

In addition to the general provisions governing the safe installation of gas appliances and associated equipment by HSE approved and competent persons, the 1998 Regulations lay down specific duties for landlords.

Except in the case of 'escape of gas' *(SI 1998 No 2451, Reg 37)*, and certain types of valves to control pressure fluctuations, the Regulations do not apply to the supply of gas when used in connection with:

— bunsen burners in an educational establishment [*SI 1998 No 2451, Reg 2*];
— mines or quarries [see *Mines and Quarries Act 1954*];
— factories [see *Factories Act 1961 (FA61), s 175*] or electrical stations [*FA61, s 123*], institutions [*FA61, s 124*], docks [*FA61, s 125*] or ships [*FA61, s 126*];
— agricultural premises;
— temporary installations used in connection with any construction work within the meaning of the *Construction (Design and Management) Regulations 1994 (SI 1994 No 3140)*;
— premises used for the testing of gas fittings; or
— premises used for the treatment of sewage.

Note: The Regulations would apply in relation to the above premises (or parts of those premises) if used for domestic or residential purposes or as sleeping accommodation, but not to hired touring caravans. [*SI 1998 No 2451, Reg 2*].

Generally, gas must be used in order to come within the scope of the Regulations. Therefore, the venting of waste gas from coal mines and landfill sites is excepted unless the gas is collected and intended for use. Gas used as motive power or from grain drying is also not covered by the Regulations. Service pipes (that is, the pipes for distributing gas to premises from a distribution main and the outlet of the first emergency control downstream of the main) remain the property of the gas supplier or gas transporter. At least 24 hours' notice should be given to the relevant organisation before any work can be performed on a service pipe. [*L56, paras 2–11*].

'Work' in relation to a gas fitting is defined in the Regulations as including any of the following activities carried out by any person, whether an employee or not, that is to say:

(a)   installing or reconnecting the fitting;
(b)   maintaining, servicing, permanently adjusting, disconnecting, repairing, altering or renewing the fitting or purging it of air or gas;
(c)   where the fitting is not readily movable, changing its position; and
(d)   removing the fitting.

The Approved Code of Practice, L56: '*The Gas Safety (Installation and Use) Regulations 1998*', provides further guidance on the meaning of work and states that 'work' for the purposes of these Regulations also includes do-it-yourself activities, work undertaken for friends and work for which there is no expectation of reward or gain, such as charitable work. [*L56, para 12*].

Duties and responsibilities

[G1011] Duties are placed on a wide range of persons associated with domestic premises and on persons concerned with the supply of gas to those premises and to industrial premises which have accommodation facilities.

Most of the legal requirements are 'absolute', i.e. they are not qualified by 'as far as is practicable' or 'so far as is reasonably practicable'; in all cases the duty must be satisfied to avoid committing an offence. Persons affected by the Regulations include:

—   *individuals*. These include householders and other members of the general public;
—   *responsible person*. This is the occupier of the premises or, where there is no occupier or the occupier is away, the owner of the premises or any person with authority for the time being to take appropriate action in relation to any gas fittings;
—   *landlord*. In England and Wales the landlord is defined as follows:
    (i)   where the relevant premises are occupied under a lease, the person for the time being entitled to the reversion expectant on that lease or who, apart from any statutory tenancy, would be entitled to possession of the premises; and
    (ii)   where the relevant premises are occupied under a licence, the licensor, save where the licensor is himself a tenant in respect of those premises.
    In Scotland, the landlord is the person for the time being entitled to the landlord's interest under a lease;
—   *tenant*. In England and Wales the tenant is defined as follows:
    (i)   where the relevant premises are so occupied under a lease, the person for the time being entitled to the term of that lease; and
    (ii)   where the relevant premises are so occupied under a licence, the licensee.
    In Scotland, the tenant is the person for the time being entitled to the tenant's interest under a lease.

Duties are imposed on employers and the self-employed to ensure that all persons carrying out work in relation to gas fittings are competent to do so and are members of HSE approved organisations. [*Gas Safety (Installation and Use) Regulations 1998 (SI 1998 No 2451), Reg 3*].

Persons connected with work activities include:

— *supplier.* In relation to gas, a 'supplier' means:
  (i) a person who supplies gas to any premises through a primary meter; or
  (ii) a person who provides a supply of gas to a consumer by means of the filling or refilling of a storage container designed to be filled with gas at the place where it is connected for use, whether or not such container is or remains the property of the supplier; or
  (iii) a person who provides gas in refillable cylinders for use by a consumer whether or not such cylinders are filled, or refilled, directly by that person and whether or not such cylinders are or remain the property of that person.
  Note: A retailer is not a supplier when he sells a brand of gas other than his own.

— *transporter.* This is defined as meaning a person, other than a supplier, who conveys gas through a distribution main.

— *gas installer.* This is any person who installs, services, maintains, removes or repairs gas fittings whether he is an employee, self-employed or working on his own behalf (for example, a do-it-yourself activity).

Escape of gas

[G1012] It is the duty of the responsible person for the premises to take immediate action to shut off the gas supply to the affected appliance or fitting and notify the gas supplier (or the nominated gas emergency call-out office if different from the supplier). All reasonable steps must be taken to prevent further escapes of gas, and the gas supplier must stop the leak within twelve hours from the time of notification. This may entail cutting off the gas supply to the premises. [*Gas Safety (Installation and Use) Regulations 1998 (SI 1998 No 2451), Reg 37(1)–(3)*].

An escape of gas also includes an emission of carbon monoxide resulting from incomplete combustion in a gas fitting. However, the legal duties regarding the action to be taken by the gas supplier are limited to making safe and advising of the need for immediate action by a competent person to examine, and if necessary carry out repairs, to the faulty fitting or appliance. [*Gas Safety (Installation and Use) Regulations 1998 (SI 1998 No 2451), Reg 37(8)*].

Competent persons and quality control

[G1013] No work is permitted to be carried out on a gas fitting or a gas storage vessel except by a competent person. [*Gas Safety (Installation and Use) Regulations 1998 (SI 1998 No 2451), Reg 3(1)*].

Where work to a gas fitting is to any extent under their control – or is to be carried out at any place of work under their control – employers (and self-employed persons) must ensure that the person undertaking such work is registered with an HSE-approved body such as the Council of Registered Gas Installers (CORGI). [*Gas Safety (Installation and Use) Regulations 1998 (SI 1998 No 2451), Reg 4*].

A gas fitting must not be installed unless every part of it is of good construction and sound material, of adequate strength and size to secure safety and of a type appropriate for the gas with which it is to be used. [*Gas Safety (Installation and Use) Regulations 1998 (SI 1998 No 2451), Reg 5(1)*].

Competent persons – qualifications and supervision

[G1014] To achieve competence in safe installation, a person's training must include knowledge of purging, commissioning, testing, servicing, maintenance, repair, disconnection, modification and dismantling of gas systems, fittings and appliances. In addition, a sound knowledge of combustion and its technology is essential, including:

— properties of fuel gases,
— combustion,
— flame characteristics,
— control and measurement of fuel gases,
— gas pressure and flow,
— construction and operation of burners, and
— operation of flues and ventilation.

To reach the approval standard required by *Gas Safety (Installation and Use) Regulations 1998 (SI 1998 No 2451), Reg 3*, gas installers and gas fitters should know:

(a)   where/how gas pipes/fittings (including valves, meters, governors and gas appliances) should be safely installed;
(b)   how to site/install a gas system safely, with reference to safe ventilation and flues;
(c)   associated electrical work (e.g. appropriate electrical power supply circuits, that is, overcurrent and shock protection from electrical circuits, earthing and bonding);
(d)   electrical controls appropriate to the system being installed/maintained/repaired;
(e)   when/how to check the whole system adequately before it is commissioned;
(f)   how to commission the system, leaving it safe for use.

In addition, they should know how to recognise and test for conditions that might cause danger and what remedial action to take, as well as being able to show consumers how to use any equipment they have installed or modified, including how to shut off the gas supply in an emergency. They should also alert customers to the significance of inadequate ventilation and gas leaks and the need for regular maintenance/servicing.

To meet the criteria for HSE approval, individual gas fitting operatives must be assessed (or reassessed) by a certification body accredited by the United Kingdom Accreditation Service. The scheme requires every registered gas fitter or gas installer to possess a certificate, which consumers can ask to see. The HSE has announced changes to the Nationally Accredited Certification Scheme (ACS) for Individual Gas Fitting Operatives. The changes mean that a prepared and experienced operative will be able to complete a 'tailored' domestic natural gas initial assessment in less than three days and a 'tailored' re-assessment in one day. This follows a recommendation of the Health and Safety Commission's 'Fundamental Review of Gas Safety'.

For those entering the ACS regime for the first time the 'tailored' assessment will cover basic gas safety issues plus appliance specific safety issues associated with the installation, maintenance and repair of four types of appliance-cookers, gas fires, water heaters and central heating boilers. The 'tailored' re-assessment for domestic work will cover changes in legislation, standards and technology demonstrated across the range of four appliances. The re-assessment will also ensure that operatives have retained their gas safety competence with regard to carrying out gas work safely and commissioning appliances, e.g. gas tightness, ventilation and flueing, and the ability to recognise and take action to remove dangerous and risk situations.

For those operatives wishing to be assessed on fewer or more appliances covered by the 'tailored' assessments, certification bodies, through their assessment centres, can offer packaged assessments which cover the required variation without unnecessary duplication of tasks successfully completed.

Assessment centres are already offering domestic natural gas and LPG in permanent dwellings 'tailored' and packaged initial assessments. Further work is being undertaken to extend the 'tailored' approach to the remainder of LPG work and to commercial catering and other commercial work.

'Tailored' re-assessments for all gas work have been available from 1 August 2003.

Materials and workmanship

[G1015] It is incumbent on gas installers to acquaint themselves with the appropriate standards about gas fittings and to ensure that the fittings they use meet those standards. Most gas appliances are subject to the *Gas Appliances (Safety) Regulations 1992 (SI 1992 No 711)* and therefore should carry the CE mark or an appropriate European/British Standard.

It is an offence to carry out any work in relation to a gas fitting or gas storage vessel other than in accordance with the appropriate standards and in such a way as to prevent danger to any person. [*Gas Safety (Installation and Use) Regulations 1998 (SI 1998 No 2451), Reg 5(3)*]. Gas pipes and pipe fittings installed in a building must be metallic or of a type constructed in an encased metallic sheath and installed, so far as is reasonably practicable, to prevent the escape of gas into the building if the pipe should fail. Pipes or pipe fittings made from lead or lead alloy must not be used. [*Gas Safety (Installation and Use) Regulations 1998 (SI 1998 No 2451), Reg 5(2)*].

General safety precautions

[G1016] A general duty is imposed on all persons in connection with work associated with gas fittings to prevent a release of gas unless steps are taken which ensure the safety of any person [*SI 1998 No 2451, Reg 6(1)*].

The *Gas Safety (Installation and Use) Regulations 1998 (SI 1998 No 2451), Regs 6(2)–(6)* provide that the following precautions must be observed when carrying out work activities:

— a gas fitting must not be left unattended unless every complete gasway has been sealed with an appropriate fitting;
— a disconnected gas fitting must be sealed at every outlet;
— smoking or the use of any source of ignition is prohibited near exposed gasways;
— it is prohibited to use any source of ignition when searching for escapes of gas;
— any work in relation to a gas fitting which might affect the tightness of the installation must be tested immediately for gas tightness.

With regard to gas storage vessels it is an offence for any person intentionally or recklessly to interfere with a vessel, or otherwise do anything which might affect it so that the subsequent use of that vessel might cause a danger to any person. [*SI 1998 No 2451), Reg 6(9)*]. In addition:

— gas storage vessels must only be installed where they can be used, filled or refilled without causing danger to any person [*SI 1998 No 2451, Reg 6(7)*];
— gas storage vessels, or appliances fuelled by LPG which have an automatic ignition device or a pilot light, must not be installed in cellars or basements [*SI 1998 No 2451, Reg 6(8)*]; methane gas must not be stored in domestic premises [*SI 1998 No 2451, Reg 6(10)*];
— storage of dangerous substances such as LPG and natural gas should comply with the general safety measures of the *Dangerous Substances and Explosives Atmospheres Regulations 2002 (SI 2002, No 2776), Reg 6(8)*. The Approved Code of Practice and Guidance, L135 gives further details.

Gas appliances

[G1017] Precautions to be observed regarding gas appliances apply to everyone, not just gas installers. Generally, it is an offence for the occupier, owner or other responsible person to permit a gas appliance to be used if at any time he knows, or has reason to suspect, that the appliance is unsafe or that it cannot be used without constituting a danger to any person. [*SI 1998 No 2451, Reg 34*]. It is also an offence for anyone to carry out work in relation to a gas appliance which indicates that it no longer complies with approved safety standards. [*SI 1998 No 2451, Reg 26(7), (8)*]. The *Gas Safety (Installation and Use) Regulations 1998 (SI 1998 No 2451), Reg 27* imposes similar safety requirements on persons who install or connect gas appliances to flues (see **G1019** below).

All gas appliances must be installed in a manner which permits ready access for operation, inspection and maintenance. [*SI 1998 No 2451, Reg 28*]. Manufacturer's instructions regarding the appliance must be left with the owner or occupier of the premises after its installation. [*SI 1998 No 2451, Reg 29*]. If the appliance is of the type where it is designed to operate in a suspended position, the installation pipework and other associated fittings must be constructed and installed as to be able to support the weight of the appliance safely. [*SI 1998 No 2451, Reg 31*].

During installation it is essential to ensure that all gas appliances are:

—   connected (in the case of a flued domestic gas appliance) to a gas supply system by a permanently fixed rigid pipe [*SI 1998 No 2451, Reg 26(2)*];
—   installed with a means of shutting off the supply of gas to the appliance unless it is not reasonably practicable to do so [*SI 1998 No 2451, Reg 26(6)*].

With the exception of the direct disconnection of the gas supply from a gas appliance, or the purging of gas or air which does not adversely affect the safety of that appliance, all commissioning operations and any other work performed on a gas appliance must be immediately examined to ensure the effectiveness of flues, and checks must be made with regard to:

—   the supply of combustion air;
—   the operating pressure or heat input; and
—   the operation of the appliance to ensure its safe functioning.
  [*SI 1998 No 2451, Reg 26(9)*].

Any defects must be rectified and reported to the appropriate responsible person as soon as is practicable; the defect must be reported to the appropriate gas supply organisation if a responsible person or the owner is not available. [*SI 1998 No 2451, Reg 26(9)*].

A series of leaflets has been published by the HSE providing guidance for landlords and important safety advice for gas consumers in 12 different languages.

The leaflet '*Gas appliances: Get them checked, keep them safe*', provides new information on the need to obtain urgent medical advice if a person suspects that they or their family have been exposed to carbon monoxide (CO) poisoning. The doctor would need to take a blood or breathe sample but if delayed for more than four hours after exposure has ceased the test results may be inaccurate as CO rapidly leaves the blood.

Symptoms of CO poisoning can mimic many common ailments and may be confused with flu or simple tiredness. If in doubt consult your doctor.

The leaflet also advises on how to ensure that a gas installer is registered with the Council for Registered Gas Installers (CORGI) and provides new easy-to-understand diagrams showing the difference between safe and dangerous gas appliances. It also mentions the potential contribution of CO alarms as a useful back-up precaution, but emphasises that they must not be regarded as a substitute for proper installation and maintenance by a CORGI registered installer.

'Landlord–a guide to landlord's duties: Gas Safety (Installation and Use) Regulations 1998', also available in a variety of languages, explains landlord's duties to ensure safety of gas appliances and flues, including annual safety checks. This leaflet also tells landlords how to ensure their gas installer is registered with CORGI.

HSE's Gas Safety website, at: www.hse.gov.uk/gas/domestic/index.htm, offers printable versions of these information leaflets, as well as useful advice and information for members of the public, including a range of videos for cons umers to watch.

The HSE Catering Information Sheet No 23 Gas safety in catering and hospitality advises the installation of a suitable device or system to interlock the gas supply with the mechanical ventilation system in commercial kitchens so that in the event of failure of the flue gas extract system the gas supply is automatically shut down. See: BS 6173: 2001 Specification for the installation of gas-fired catering appliances for use in catering establishments.

Room-sealed appliances

[G1018] A room-sealed appliance is an appliance whose combustion system is sealed from the room in which the appliance is located and which obtains air for combustion from a ventilated uninhabited space within the premises or directly from the open air outside the premises. The products of combustion must be vented safely to open air outside the premises.

A room-sealed appliance must be used in the following rooms:

— a bathroom or a shower room [*SI 1998 No 2451, Reg 30(1)*];
— in respect of a gas fire, gas space heater or gas water heater (including instantaneous water heaters) of more than 14 kilowatt gross heat input, in a room used or intended to be used as sleeping accommodation [*SI 1998 No 2451, Reg 30(2)*].

Gas heating appliances which have a gross heat input rating of less than 14 kilowatt or an instantaneous water heater may incorporate an alternative arrangement, i.e. an approved safety control designed to shut down the appliance before a build-up of dangerous combustion products can occur [*Reg 30(3)*].

For the purposes of the *Gas Safety (Installation and Use) Regulations 1998 (SI 1998 No 2451), Reg 30(1)–(3)*, a room also includes:

— a cupboard or compartment within such a room; or
— a cupboard, compartment or space adjacent to such a room if there is an air vent from the cupboard, compartment or space into such a room [*SI 1998 No 2451, Reg 30(4)*].

Flues and dampers

[G1019] A flue is a passage for conveying the products of combustion from a gas appliance to the external atmosphere and includes the internal ducts of the appliance. Generally, it is prohibited to install a flue other than in a safe position, and, in the case of a power-operated flue, it must prevent the operation of the appliance should the draught fail [*Gas Safety (Installation and Use) Regulations 1998 (SI 1998 No 2451), Reg 27(4), (5)*].

A flue must be suitable and in a proper condition for the operation of the appliance to which it is fitted. It is prohibited to install a flue pipe so that it enters a brick or masonry chimney in such a manner that the seal cannot be inspected. Similarly, an appliance must not be connected to a flue surrounded by an enclosure, unless the enclosure is sealed to prevent spillage into any room or internal space other than where the appliance is installed. [*Gas Safety (Installation and Use) Regulations 1998 (SI 1998 No 2451), SI 1998 No 2451, Reg 27(1)–(3)*].

Manually operated flue dampers must not be fitted to serve a domestic gas appliance, and, similarly, it is prohibited to install a domestic gas appliance to a flue which incorporates a manually operated damper unless the damper is permanently fixed in the open position.

In the case of automatic dampers, the damper must be interlocked with the gas supply so that the appliance cannot be operated unless the damper is open. A check must be carried out immediately after installation to verify that the appliance and the damper can be operated safely together and without danger to any person.

Where a gas appliance is connected to a flue, the installer must ensure that the flue and the means of connection are suitable and effective and complies with applicable building regulations. The *Building Regulations 2000 (SI 2000, No 2531)*, as amended *(SI 2001, No 3335)*, and the *Building Standards (Scotland) Regulations 1990 (SI 1990, No 2179)*, as amended *(SI 1999, No 173)*, impose comprehensive requirements for the safe installation of heat-producing appliances in buildings. Gas burning appliances with a rated input to 60 kW (Type 2 appliances, including cookers, balanced and open-flued boilers, water and convestor heaters and decorative log and solid fuel-effect fires, are covered in Approved document J: 2002 Edition: 'Guidance and supplementary information on the UK implementation of European standards for chimneys and flues'.

Gas fittings

[G1020]–[G1021] Gas fittings are those parts of apparatus and appliances designed for domestic consumers of gas for heating, lighting, cooking or other approved purposes for which gas can be used (but not for the purpose of an industrial process occurring on industrial premises), namely:

— pipework;
— valves; and
— regulators, meters and associated fittings.

[*Gas Safety (Installation and Use) Regulations 1998 (SI 1998 No 2451), SI 1998 No 2451, Reg 2*].

An emergency control is a valve for use by a consumer of gas for shutting off the supply of gas in an emergency. Emergency controls must be appropriately positioned with adequate access, and there must be a prominent notice or other indicator showing whether the control is open or shut. A notice must also be posted either next or near the emergency control indicating the procedure to be followed in the event of an escape of gas. [*Gas Safety (Installation and Use) Regulations 1998 (SI 1998 No 2451), SI 1998 No 2451, Reg 9*].

Meters and regulators

[G1022] The *Gas Safety (Installation and Use) Regulations 1998 (SI 1998 No 2451), Reg 12* provides that gas meters must be installed where they are readily accessible for inspection and maintenance. A meter must not be so placed as to adversely affect a means of escape or where there is a risk of damage to it from electrical apparatus. After installation, meters and other other associated fittings should be tested for gastightness and then purged so as to remove safely all air and gas other than the gas to be supplied. [*SI 1998 No 2451, Reg 12(6)*].

Meters must be of sound construction so that, in the event of fire, gas cannot escape from them and, where a meter is housed in an outdoor meter box, the box must be so designed as to prevent gas entering the premises or any cavity wall, i.e. any escaping gas must disperse to the air outside. Combustible materials must not be kept inside meter boxes. Where a meter is housed in a box or compound that includes a lock, a suitably labelled key must be provided to the consumer. [*SI 1998 No 2451, Reg 13*].

The *Gas Safety (Installation and Use) Regulations 1998 (SI 1998 No 2451), Regs 14–17* prescribe detailed precautions to be observed during the installation of the various types of meters. Most of these are for compliance by gas installers and should be included in work procedures. In essence, a meter must not be installed in service pipework unless:

(a)   there is a regulator to control the gas pressure [*SI 1998 No 2451, Reg 14(1)(a),(b)*];

(b)   a relief valve or seal is fitted which is capable of venting safely [*SI 1998 No 2451, Reg 14(1)(c)*]; and

(c)   the meter contains a prominent notice (in permanent form) specifying the procedure in the event of an escape of gas [*SI 1998 No 2451, Reg 15*].

Similar requirements are imposed regarding gas supplies from storage vessels and re-fillable cylinders. [*SI 1998 No 2451, Reg 14(2)–(4)*].

Where gas is supplied from a primary meter to a secondary meter, a line diagram must be provided and prominently displayed showing the configuration of all meters, installation pipework and emergency controls. It is incumbent on any person who changes the configuration to amend the diagram accordingly. [*SI 1998 No 2451, Reg 17*].

Installation pipework

[G1023] For the purposes of this chapter, 'installation pipework' covers any pipework for conveying gas to the premises from a distribution main, including pipework which connects meters or emergency control valves to a gas appliance, and any shut-off devices at the inlet to the appliance. The term 'service pipe' is used to describe any pipe connecting the distribution main with the outlet of the first emergency control downstream from the distribution main. Similarly, 'service pipework' are those pipes which supply gas from a gas storage vessel. All service pipes normally remain the property of the gas supplier or transporter and their permission must be obtained before any work associated with such pipes can be performed.

Installation pipework must not be installed:

— where it cannot safely be used, having regard to other pipes, pipe supports, drains, sewers, cables, conduits and electrical apparatus;
— in or through any floor or wall, or under any building, unless adequate protection is provided against failure caused by the movement of these structures;
— in any shaft, duct or void, unless adequately ventilated;
— in a way which would impair the structure of a building or impair the fire resistance of any part of its structure;
— where deposit of liquid or solid matter is likely to occur, unless a suitable vessel for the reception and removal of the deposit is provided. (It should be noted that such clogging precautions do not normally need to be taken in respect of natural gas or LPG as these are 'dry' gases.)

[*Gas Safety (Installation and Use) Regulations 1998 (SI 1998 No 2451), Regs 18–21*].

Following work on installation pipework, the pipes must immediately be tested for gastightness and, if necessary, a protective coating applied; followed, where gas is being supplied, by satisfactory purging to remove all unnecessary air and gas. Except in the case of domestic premises, the parts of installation pipework which are accessible to inspection must be permanently marked as being gas pipes. [*Gas Safety (Installation and Use) Regulations 1998 (SI 1998 No 2451), Regs 22, 23*].

Testing and maintenance requirements

[G1024] According to the *Gas Safety (Installation and Use) Regulations 1998 (SI 1998 No 2451), Reg 33(1)*, a gas appliance must be tested when gas is supplied to premises to verify that it is gastight and to ensure that:

(a) the appliance has been installed in accordance with the requirements of the 1998 Regulations;
(b) the operating pressure is as recommended by the manufacturer;
(c) the appliance has been installed with due regard to any manufacturer's instructions accompanying the appliance; and
(d) all gas safety controls are in proper working order.

If adjustments are necessary in order to comply with (*a*) to (*d*) above, but cannot be carried out, the appliance must be disconnected or sealed off with an appropriate fitting. [*SI 1998 No 2451, Reg 33(2)*].

A general duty is imposed on employers and self-employed persons to ensure that any gas appliance, installation pipework or flue in places of work under their control is maintained in a safe condition so as to prevent risk of injury to any person. [*SI 1998 No 2451, Reg 35*].

A similar duty is imposed on landlords of premises occupied under a lease or a licence to ensure that relevant gas fittings and associated flues are maintained in a safe condition. [*SI 1998 No 2451, Reg 36(2)*]. A relevant gas fitting includes any gas appliance (other than an appliance which the tenant is entitled to remove from the premises) or installation (not service) pipework installed in such premises.

It is the responsibility of landlords to ensure that each gas appliance and flue is checked by a competent person (HSE approved person) at least every 12 months. [*SI 1998 No 2451, Reg 36(3)*]. Such safety checks shall include, but not be limited to, the following:

(a)   the effectiveness of any flue;
(b)   the supply of combustion air;
(c)   the operating pressure and/or heat input of the appliance;
(d)   the operation of the appliance to ensure its safe functioning.

Nothing done or agreed to be done by a tenant can be considered as discharging the duties of a landlord in respect of maintenance except in so far as it relates to access to the appliance or flue for the purposes of carrying out such maintenance or checking activities. [*SI 1998 No 2451, Reg 36(10)*].

In addition, it is the duty of landlords to ensure that appliances and relevant fittings are not installed in any room occupied as sleeping accommodation such as to cause a contravention of *Reg 30(2)* or *(3)* (see **G1018** above).

Inspection and maintenance records

[G1025] The *Gas Safety (Installation and Use) Regulations 1998 (SI 1998 No 2451)* mark a significant increase in the obligations now placed upon landlords.

A landlord is under a duty to retain a record of each inspection of any gas appliance or flue for a period of two years from the date of the inspection. The record must include:

—   the date on which the appliance or flue is checked;
—   the address of the premises at which the appliance or flue is installed;
—   the name and address of the landlord of the premises at which the appliance or flue is installed;
—   a description of, and the location of, each appliance and flue that has been checked together with details of defects and any remedial action taken.

The person who carries out the safety check must enter on the record his name and his particulars of registration. Then he should sign the record confirming that he has examined the following:

—   the effectiveness of any flue;
—   the supply of combustion air;
—   the operating pressure and/or heat input of the appliance;
—   the operation of the appliance to ensure its safe functioning.
[*SI 1998 No 2451, Reg 36(3)*].

Within 28 days of the date of the check, the landlord must ensure that a copy of the record is given to each tenant of the premises to which the record relates. [*SI 1998 No 2451, Reg 36(6)(a)*]. In addition, before any new tenant moves in, the landlord must provide any such tenant with a copy of the record – however, where the tenant's right to occupation is for a period not exceeding 28 days, a copy of the record may instead be prominently displayed within the premises. [*SI 1998 No 2451, Reg 36(6)(b)*]. A copy of the inspection record given to a tenant under *Reg 36(6)(b)* need not include a copy of the signature of the person who carried out the inspection, provided that it includes a statement that the tenant is entitled to have another copy, containing a copy of such signature, on request to the landlord at an address specified in the statement. Where the tenant makes such a request, the landlord must provide the tenant with such a copy of the record as soon as is practicable. [*SI 1998 No 2451, Reg 36(8)*].

Under the *Housing Act 2004 (HA 2004)*, which received Royal Assent on 18 November 2004, owners of residential property will be required to provide documents relating to the condition of property when it is sold. Regulations made under *HA 2004, s 163* prescribe what documents must be included in the 'Home Information Pack' It is anticipated that gas safety certificates (Declaration of Safety) providing proof that an appliance has been professionally installed will be among those required as part of the pack.

The *Gas Safety (Installation and Use) Regulations 1998 (SI 1998 No 2451), Reg 36(7)* provides that where any room occupied, or about to be occupied, by a tenant does not contain any gas appliance, the landlord may – instead of giving the tenant a copy of the inspection record in accordance with *Reg 36(6)* (see above) – prominently display a copy of the record within the premises, together with a statement endorsed upon it that the tenant is entitled to have his own copy of the record on request to the landlord at an address specified in the statement. Where the tenant makes such a request, the landlord must provide the tenant with a copy of the record as soon as is practicable.

Interface with other legislation

[G1026] Although wide-ranging in their scope the *Gas Safety (Installation and Use) Regulations 1998 (SI 1998 No 2451)* are not concerned directly with product safety. Certain premises, gas fittings and uses are also subject to exceptions. However, in many of these instances where exceptions are made similar requirements for gas safety are to be found in the *Health and Safety at Work etc Act 1974* and supporting legislation. The most important of these are:

- *Gas Appliances (Safety) Regulations 1995 (SI 1995 No 1629)*
  These Regulations were made under the authority of the *Consumer Protection Act 1987*. They implement European directives that require gas appliances and fittings to conform with essential safety requirements and to be safe when used normally.
- *Management of Health and Safety at Work Regulations 1999 (SI 1999 No 3242)* (see also RISK ASSESSMENT)

These Regulations were made under the *Health and Safety at Work etc Act 1974*. They require employers and the self-employed to undertake a suitable and sufficient assessment of the risks to the health and safety of their employees and others who may be affected by the work that they do. These Regulations further require employers (including those who install or supply gas or gas appliances or who undertake gas maintenance work) to ensure controls are in place and kept under review.

- *Provision and Use of Work Equipment Regulations 1998 (SI 1998 No 2306)* as amended *(SI 2002, No 2174)* (see also MACHINERY SAFETY)
  These Regulations were made under the authority of the *Health and Safety at Work etc Act 1974*. They impose health and safety requirements on work equipment including any gas appliance, apparatus, fitting or tool used or provided for use at work. Equipment must be suitable for its intended use, be maintained and inspected and users provided with appropriate health and safety information and where necessary written instructions.
  The HSE has recently investigated several complaints relating to flame failure devices (FFD's) on gas forges used primarily by farriers. The key issues under the *Provision and Use of Work Equipment Regulations 1998 (SI 1998 No 2306)* are the purchase of suitable equipment and safe operation of the forge. A build-up of unburned gas can lead to explosions, so gas appliances must be constructed so that during ignition and flame extinction a build-up of gas is avoided. This means that a FFD should be fitted that shuts off the gas supply if it is not burning.
  Tony Mitchell, of the HSE's Agriculture Safety Section, said:

  > HSE cannot condone in any way the deliberate overidding of FFD's and certainly expects that whenever these types of forges are used in a workroom the FFD must be operating correctly. Should an incident occur, as well as risking serious injury, the forge operator could face legal and civil action which might jeopardise the business.

  > When choosing a gas furnace, users should check for evidence that the equipment meets basic safety standards. If work in the open air is envisaged and the FFD is likely to cause problems, the use of cowls and wind breaks should be considered. Manufacturers of mobile forges can also help by looking at their designs to minimise the problem.

- *Pressure Systems Safety Regulations 2000 (SI 2000 No 128)*
  These Regulations were made under the authority of the *Health and Safety at Work etc Act 1974*. Their objective is the prevention of serious injury from the failure of a pressure system or its components, including those that contain gas in excess of 0.5 bar above atmospheric pressure and pipelines used to convey gas at 2 bar above atmospheric pressure. Under these Regulations duties are placed on those who design, manufacture, use and maintain gas systems overlapping with similar duties under the *Gas Safety (Installation and Use) Regulations 1998 (SI 1998 No 2451)*
- *Building Regulations 1991 (SI 1991 No 2768)* (see also WORKPLACES – HEALTH, SAFETY AND WELFARE)

These Regulations (as amended in 1994) were made under the authority of the *Building Act 1984*. They are supported by non-mandatory approved documents that provide practical guidance on for example the safe installation of gas appliances used for heating – Approved Document J is concerned with air supply to gas heaters and discharge of the products of combustion to the open air.

• *Dangerous Substances and Explosive Atmospheres Regulations 2002 (SI 2002 No. 2776)* apply to all workplaces where dangerous substances are used or generated irrespective of the quantity involved. Such dangerous substances include gases and liquefied gases such as methane, propane and butane (LPG) which are extremely flammable and may form explosive atmospheres if mixed in air within the explosive limits. Employers are required to eliminate or reduce the risk to safety from fire and explosion arising from the use or occurrence of such substances and to have in place a combination of control and mitigation measures to ensure the safety of employees and others. *SI 2002 No. 2776, Reg 7* (classification into hazardous and non-hazardous areas) and *SI 2002 No. 2776, Reg 11* (co-ordination between employers sharing a workplace) do not however apply to workplace where gaseous fuels are used for cooking, heating and hot water production.

Summaries of relevant court cases

[G1027] The importance of the provisions relating to the correct installation and regular maintenance aspects of gas appliances, even in the earlier Gas Safety Regulations, is demonstrated in the following summaries of recent court cases.

Unsafe gas fittings

[G1028] A Nottingham landlord narrowly escaped a charge of manslaughter when he was tried and convicted of carrying out unsafe work practices which resulted in the deaths of two people.

The court heard of makeshift repairs to a gas boiler which then leaked carbon monoxide into an adjacent flat. The gas supplier (British Gas) gave evidence of having previously made the boiler inoperable because of its unsafe condition but it was reconnected by the landlord without any authority to do so, in December 1992. The landlord was charged under the *Gas Safety (Installation and Use) Regulations 1984* and the *Health and Safety at Work etc Act 1974, s 3(2)*. He was accused of failing in his duty as a self-employed person to conduct his undertaking so as not to put himself or others at risk and of failing to carry out work in relation to a gas fitting in a proper and workmanlike manner.

The judge made it clear that if the legislation had carried a penalty of imprisonment, he would have imposed it. The seriousness of the case could only be reflected by a heavy fine and he sentenced the landlord with a fine of £32,000 with £24,800 costs.

[Leicester Crown Court, June 1996].

Note: The *Gas Safety (Installation and Use) Regulations 1994* introduced a wider scope of control and increased the penalties for contravention to an unlimited fine and a maximum of two years' imprisonment.

Local authority (as landlord) guilty of bad gas safety management

[G1029] A local authority found itself in court for nine breaches of the *Gas Safety (Installation and Use) Regulations 1994* (failing to carry out safety checks and to keep records) and for failing to ensure the safety of tenants, contrary to the *Health and Safety at Work etc Act 1974, s 3(1)*. The case was brought after the discovery of more than 150 flues which were left disconnected after refurbishment work on a London council estate; and on other estates, annual safety checks had not been carried out.

The magistrate commented that there had been serious risk to hundreds of householders and imposed fines on the council of £27,000 for the breaches of the *Gas Safety (Installation and Use) Regulations 1994* and £17,000 for the offences which contravened the *Health and Safety at Work etc Act 1974*. [*Clerkenwell Magistrates' Court, January 1997*].

Note: The HSE pointed out that local authorities were the largest landlords in the country and that failure to manage in this case had put many tenants at risk of carbon monoxide poisoning and fire and explosion.

The London Borough of Hammersmith and Fulham was fined £350,000 at Blackfriars Crown Court in December 2001. The council pleaded guilty to one breach of the *Management of Health and Safety at Work Regulations 1992 (SI 1992 No 2051)* and another of the *Health and Safety at Work etc. Act 1974* with £30,750 awarded in costs to the HSE which brought the case.

The charges related to the deaths of two persons in Sharnbrook House on the West Kensington Estate, London in January 1998. They were poisoned by carbon monoxide given off from a gas boiler between 18 and 19 January 1998 in the flat they let from the council. The HSE investigation showed that the boiler had not been properly maintained and its annual gas safety check was overdue. Other gas boilers on the estate were also found to be in use beyond their gas safety check dates.

Gas explosion results in three separate prosecutions

[G1030] A major gas explosion at a twenty two-storey tower block resulted in the prosecution of a borough council, an installation company and a self employed electrician. The explosion blew off the roof of a boiler house, only minor injuries were sustained but there was considerable property damage from flying debris. There had been serious risk to hundreds of householders.

An investigation revealed that new gas burners and gas pressure booster pumps had been inadequately installed and not commissioned correctly.

The borough council was fined £75,000.00 for failing to manage the project.

The installation company was fined £10,200.00 for responsibility for the inadequate installation and management of the project.

The electrician was fined £3,000.00 for failing to install and commission the boilers correctly, not being competent, and not registered with CORGI. [*Middlesex Guildhall Crown Court, September 1999*].

Following a gas explosion on 1 May 2004 at 55 Temple View Terrace, Leeds, a rented domestic premise, Mr Glyn Robinson, the Managing Agent was prosecuted under the *Gas Safety Regulations 1972* and fined £200 by Leeds Magistrates Court. He had failed to carry out the duty delegated to him to carry out annual gas safety checks.

Student death – landlord and gas fitter imprisoned

[G1031] A landlord and an unregistered gas fitter were charged and convicted of manslaughter following the death of a student who inhaled carbon monoxide fumes from an inappropriate type of gas boiler which was not correctly ventilated.

The investigation into the incident by the HSE originated with the *Gas Safety (Installation and Use) Regulations 1994* which required gas fitters to be registered with the Council of Registered Gas Installers (CORGI). The charge of manslaughter was brought because of the seriousness of the offence: it had resulted in the death of one student and the hospitalisation of four other tenants who had been affected by poisonous fumes.

Both the landlord and the gas fitter pleaded guilty to manslaughter: the landlord was judged to be most responsible and was imprisoned for two years. The gas fitter was imprisoned for fifteen months. [*Stafford Crown Court, December 1997*].

Note: Broadly, the law of manslaughter is part of the law of homicide. It is less serious than murder with the sentence at the discretion of the court. For health and safety purposes, manslaughter may be tentatively defined as killing without the intention to kill or to cause serious bodily harm. This case was the second health and safety manslaughter charge to result in imprisonment.

Gas fitter jailed – chimney flue capped with concrete caused death

[G1032] This case involved an unregistered gas fitter who failed to notice that the flue into which he had connected a gas fire was capped with a concrete slab.

The fitter, the landlord and the landlord's son were charged with manslaughter following the death of a tenant in a bedsit. The prosecuting counsel told the court that the landlord of the bedsit, and his son who was looking after the property on behalf of his father, were also responsible for the death. He gave evidence of negligence in 20 other properties where failure to maintain gas appliances properly could have had similar tragic consequences.

In the case in question the judge said that the landlord and his son had been lulled into a false sense of security by a gas safety certificate provided by the fitter. He imposed the following sentences:

Gas fitter: 12 months' imprisonment, six of which were suspended;

Landlord and his son: 9 months' imprisonment suspended for 18 months – the landlord's son was also ordered to pay £16,000 costs.
[*Norwich Crown Court, April 1998*].

Unsafe gas appliance – landlord guilty

[G1033] A landlord was charged with:

—   failing to maintain gas appliances in a safe condition, contrary to the *Gas Safety (Installation and Use) Regulations 1994, Reg 35A*; and with
—   providing false information in connection with his failure to comply with an improvement notice, contrary to the *Health and Safety at Work etc Act 1974, ss 20* and *21*.

Two tenants who suffered dizziness and headaches informed the gas emergency company which then inspected the gas boiler and declared it unsafe. The boiler was switched off and a label was attached, warning that it should not be used.

The landlord instructed an unregistered gas fitter to repair the boiler; he failed to check for carbon monoxide leaks after switching the boiler back on. Two days later the two tenants collapsed after inhaling poisonous fumes and were hospitalised.

The HSE found that both the boiler and a gas cooker in the same property were leaking dangerously high levels of carbon monoxide gas. The landlord gave false information about the identity of the fitter who had carried out the work and switched the boiler back on.

The Crown Court judge said that the landlord was very lucky not to be facing manslaughter charges and imposed fines and costs totalling £24,000.
[*Cardiff Crown Court, May 1998*].

A landlord was imprisoned following two fatalities from carbon monoxide poisoning from an open flued gas fire in Flat 1, 29 Kitchener Road, Great Yarmouth. The landlord and his helper had installed the fire into a flue set designed for a room-sealed appliance in breach of the *Gas Safety (Installation and Use) Regulations 1998, Regs 3(3), 26(1) and 27(1)* and the *Health and Safety at Work etc. Act 1974, s 3(2)* following a joint investigation by the HSE and Norfolk CID. The defendant was also charged with two counts of manslaughter.
[*Bury St Edmunds Crown Court 2004*]

Gas safety requirements in factories

[G1034] Where part of plant contains explosive/flammable gas under pressure greater than atmospheric, that part must not be opened unless:

(a)   before the fastening of any joint of any pipe, connected with that part of the plant, any flow of gas has been stopped by a stop-valve;
(b)   before such fastening is removed, all practicable steps have been taken to reduce the gas pressure in the pipe or part of the plant, to atmospheric pressure; and

(c)   if such fastening is loosened or removed, inflammable gas is prevented from entering the pipe/part of the plant, until the fastening has been secured or securely replaced.

No hot work must be permitted on any plant or vessel which has contained gas until all practicable steps have been taken to remove the gas and any fumes arising from it, or to render it non-explosive or non-flammable. Similarly, and if any plant or vessel has been heated, no gas shall be allowed to enter it until the metal has cooled sufficiently to prevent any risk of ignition.
  [*Factories Act 1961, s 31(3), (4)*].

Pressure fluctuations

[G1035] The *Gas Safety (Installation and Use) Regulations 1998 (SI 1998 No 2451), Reg 38,* provides that where gas is used in plant which is liable to produce pressure fluctuations in the gas supply, such as may cause danger to other consumers, the person responsible for the plant must ensure that any directions given to him by the gas transporter to prevent such danger are complied with.

If it is intended to use compressed air or any gaseous substance in connection with the consumption of gas, at least 14 days' written notice must be given to the gas transporter.

Any device fitted to prevent pressure fluctuations or to prevent the admission of a gaseous substance into the gas supply must be adequately maintained.

Harassment in the Workplace

Nick Humphreys

Introduction

[H1701] The law relating to harassment in the workplace was substantially amended in the period following the passing of the Equal Treatment Directive (2002/73/EC). It was responsible for, inter alia, extending the categories of unlawful discrimination which are actionable under EU law, and providing a free standing claim of harassment under various aspects of national law (these being the *Sex Discrimination Act 1975 (SDA 1975)*, the *Race Relations Act 1976 (RRA 1976)*, the *Disability Discrimination Act 1995 (DDA 1995)*, the *Employment Equality (Religion and Belief) Regulations 2003 (SI 2003 No 1660) (Religion and Belief Regulations 2003)*, the *Employment Equality (Sexual Orientation) Regulations 2003 (SI 2003 No 1661) (Sexual Orientation Regulations 2003)* and the *Employment Equality (Age) Regulations 2006 (SI 2006 No 1031) (Age Regulations 2006)*(collectively "the Old Discrimination Legislation")).

Prior to the changes introduced into the Old Discrimination Legislation by the Equal Treatment Directive (2002/73/EC), the law relating to harassment was but a specific example of the general law relating to discrimination in the workplace under the various strands of the Old Discrimination Legislation, albeit an area of such significant importance that in the context of sex discrimination, the European Commission was moved to publish a Recommendation (O.J.1992, L49/1) and annexed Code of Practice on the Protection of the dignity of men and women at work (92/131/EEC).

The law has now moved on with the implementation of the *Equality Act 2010 (EA 2010)* in October 2010.

The core architecture of the *EA 2010* in the field of employment law is that *EA 2010, s 4* sets out the protected characteristics to which the *EA 2010* applies (which will include all those covered by the Old Discrimination Legislation), and as regards harassment, this is now dealt with under *EA 2010, s 26* as a freestanding code.

Under the *EA 2010, s 26* there are three types of harassment, as follows.

The first type of harassment will apply to all the protected characteristics (apart from pregnancy and maternity, and marriage and civil partnership), and involves unwanted conduct which is related to a relevant circumstance (which includes employment) and has the purpose or effect of creating an intimidating, hostile, degrading humiliating or offensive environment for the claimant or violating the claimant's dignity.

The second type is sexual harassment which is unwanted conduct of a sexual nature where this has the same purpose or effect as the first type of harassment.

The third type is treating someone less favourably because they have either submitted to or rejected sexual harassment, or harassment related to sex or gender reassignment.

History

[H1702] The law has also moved on in the way that harassment is considered as a concept.

Historically, harassment was considered as a form of standard discrimination law. The first significant case on the issue was under the *Sex Discrimination Act 1975*, *Porcelli v Strathclyde Regional Council* [1986] IRLR 134. Here, the claimant was a female laboratory assistant working at a school alongside two male colleagues. The male assistants did not like the employee and they subjected her to a campaign of harassment in the form of brushing against her in the workplace and making sexually suggestive remarks to her and about her in order to make her leave her job. The campaign was a 'success' in that it did indeed force the claimant to apply for a transfer to another school. Having done so, the claimant then brought a claim against her employer claiming that the actions of her former colleagues amounted to direct sex discrimination for the purposes of the *SDA 1975, s 1(1)(a)*, in that she had been subjected to a detriment. The employer tried to defend the claim by calling in evidence the male employees responsible for the course of harassment. During their evidence, they ventured to suggest that the conduct that had been meted out to the claimant would also have been displayed towards a hypothetical male comparator employee that they did not like and, consequently, the employer could not have committed sex discrimination against the claimant since the treatment was gender neutral.

The Scottish Court of Session refused to accept the defence. Lord Emslie stated of sexual harassment that it was:

> . . . a particularly degrading and unacceptable form of treatment which it must be taken to have been the intention of Parliament to restrain.

The court graphically likened the use of sexual harassment as a 'sexual sword' which had been 'unsheathed and used because the victim was a woman'. It added that, although an equally disliked male employee may well have been subjected to a campaign of harassment, the harassment would not have had as its cutting edge the gender of the victim. The court then went on to give general guidance as to what amounted to harassment. It stated that sexual harassment amounted to:

> . . . [u]nwelcome acts which involve physical contact which, if proved, would also amount to offences at common law, such as assault or indecent assault; and also conduct falling short of such physical acts which can be fairly described as sexual harassment.

The concept of what amounted to harassment was then refined following the *Porcelli* decision, most notably in the re-statement by the Employment Appeal Tribunal (EAT) in the case of *(1) Reed (2) Bull Information Systems Ltd v Stedman* [1999] IRLR 299 and *Macdonald v Advocate General for Scotland; Pearce v Governing Body of Mayfield Secondary School* [2003] UKHL 34,

[2003] IRLR 512. The *MacDonald* and *Pearce* cases showed that whilst the judicial construct of harassment was effective, it was also limited in its scope since comparison under the relevant constituent of the Old Discrimination Legislation was a necessary component of a claim for harassment.

Following the passing of the Race Directive (2000/43/EC), the General Framework Directive (2000/78/EC), and the Equal Treatment Directive (2002/73/EC), the law of harassment has been amended to where it now stands under the *EA 2010, s 26* covering the following:

- harassment on grounds of sex, the law having been amended so as to comply effectively with the definition of harassment under the Equal Treatment Directive (76/207/EC) (this occurring initially from 6 April 2008 following changes made by the *Sex Discrimination Act 1975 (Amendment) Regulations 2008 (SI 2008 No 656)* in the light of *Equal Opportunities Commission v DTI* [2007] EWHC 483 (Admin), [2007] IRLR 327, which held that the definition of 'harassment' under the *Sex Discrimination Act 1975* should be recast so as to provide that it arises from 'association' with sex, not causation by it);
- harassment on grounds of sex and gender reassignment (with effect from 1 October 2005 following changes made by the *Employment Equality (Sex Discrimination) Regulations 2005 (SI 2005 No 2467)* to the *Sex Discrimination Act 1975*, by introducing to it *s 4A*;
- harassment on grounds of race, this being from 19 July 2003 following changes made by the *Race Relations Act 1976 (Amendment) Regulations 2003 (SI 2003 No 1626)* to the *Race Relations Act 1976*;
- harassment on grounds of religion or religious belief, under the *Employment Equality (Religion or Belief) Regulations 2003 (SI 2003 No 1660), Reg 5*;
- harassment on grounds of sexual orientation, under the *Employment Equality (Sexual Orientation) Regulations 2003 (SI 2003 No 1661), Reg 5*; and
- harassment on grounds of age, under the *Employment Equality (Age) Regulations 2006 (SI 2006 No 1031), Reg 6*.

All provisions of the Old Discrimination Legislation insofar as they relate to harassment were repealed by the *EA 2010* with effect from 1 October 2010. However, the provisions of the Old Discrimination Legislation remain good law for acts of harassment which occurred before 1 October 2010 and which were not continuing act cases (albeit if the cases have not been commenced in the Employment Tribunal there was likely to be a limitation issue); cases of harassment which are continuing act cases will now be governed by the principles set out in the *EA 2010* under the transitional provisions set out in the *Equality Act 2010 (Commencement No. 4, Savings, Consequential, Transitional, Transitory and Incidental Provisions and Revocation) Order 2010, Art 7*.

The effect of these changes (under both the Old Discrimination Legislation and also under the *EA 2010, s 26*) was considered by the EAT in *Richmond Pharmacology v Dhaliwal* [2009] IRLR 336. In the *Richmond Pharmacology* case, Underhill P confirmed the codifying nature of the changes to the law of harassment and held that harassment claims should be treated in a broadly consistent fashion, regardless of the particular protected characteristic of discrimination in a given case. Accordingly, 'harassment' will now, for cases both before and after the commencement of the *EA 2010*, focus on three elements, these being:

(1)   unwanted conduct;
(2)   where the same has the purpose *or* effect of either:
    (a)   violating the claimant's dignity; or
    (b)   creating an adverse environment for the claimant; and
(3)   in case before 1 October 2010 on, or in cases after 1 October 2010, related to, the protected statutory grounds.

The EAT recommended that Employment Tribunals in harassment cases should consider each of these points (stating that to do so would be a 'healthy discipline'), and should ensure that factual findings are made on each of the points.

The EAT then went on to add four general heads of guidance for Employment Tribunals in harassment cases, as follows:

(1)   Case law decided before the amendments made by various European Directives on discrimination is 'unlikely to be helpful', and it is more likely to hinder. Further, still less assistance can be sourced from the 'entirely separate provisions' of the Protection from Harassment Act 1997.
(2)   Element (2) above of Underhill P's analysis shows that there are two separate heads of liability, purpose or effect (albeit the EAT acknowledged that Employment Tribunals will commonly focus on cases where harassment is the effect rather than the purpose).
(3)   It is not enough that harassment within the statutory test is alleged to have been suffered by a claimant; it must also be reasonable that the conduct complained of should have that effect. This therefore addresses the situation of the overly sensitive claimant who, although it is stated by the claimant that the claimant has suffered harm is acting unreasonably. The EAT held that whether a claimant is found to be unreasonably prone to take offence is 'quintessentially a matter for the factual assessment of the tribunal'.
(4)   Finally, element (3) of Underhill P's test above, ('on the grounds that' or 'by reason that', ie "the reason why" the alleged harasser acted) is a separate element and has to be determined by the Employment Tribunal.

Accordingly, it is necessary now to ensure when pleading a claim for harassment that it is expressly pleaded as a separate statutory head of claim setting out the matters in the preceding paragraphs and which type of harassment is being alleged.

Elements of a harassment claim

[H1703] The EA 2010, s 26 contains a definition of harassment in the following terms:

Harassment

(1)   A person (A) harasses another (B) if—
    (a)   A engages in unwanted conduct related to a relevant protected characteristic, and
    (b)   the conduct has the purpose or effect of—
        (i)   the conduct has the purpose or effect of—
        (ii)   creating an intimidating, hostile, degrading, humiliating or offensive environment for B.

(2)   A also harasses B if—
    (a)   A engages in unwanted conduct of a sexual nature, and
    (b)   the conduct has the purpose or effect referred to in subsection (1)(b).

(3)   A also harasses B if—
    (a)   A or another person engages in unwanted conduct of a sexual nature or that is related to gender reassignment or sex,
    (b)   the conduct has the purpose or effect referred to in subsection (1)(b), and
    (c)   because of B's rejection of or submission to the conduct, A treats B less favourably than A would treat B if B had not rejected or submitted to the conduct.

(4)   In deciding whether conduct has the effect referred to in subsection (1)(b), each of the following must be taken into account—
    (a)   the perception of B;
    (b)   the other circumstances of the case;
    (c)   whether it is reasonable for the conduct to have that effect.

(5)   The relevant protected characteristics are—
    age;
    disability;
    gender reassignment;
    race;
    religion or belief;
    sex.

The consolidated definition of harassment under the *EA 2010 s 26* now provides a broad view of harassment, with the conduct in relation to which a complaint is made being conduct "related to" a relevant protected characteristic (for the purposes of the *EA 2010, s 26(5)*) rather than the characteristic being a personal one of the claimant.

The definition is divided into three types of harassment being:

•   a general head which applies to all "relevant protected characteristics" (defined in *EA 2010 s 26(5)* so as to exclude marriage and civil partnership, and pregnancy and maternity). This head relates to "conduct" which is unwanted by the claimant, the same being related to a relevant characteristic which has the purpose or effect of creating an intimidating, hostile, degrading, humiliating or offensive atmosphere for the claimant, or which violates the claimant's dignity (an "unlawful purpose or effect") (*EA 2010, s 26(1)*);

•   unwanted conduct of a sexual nature which has an unlawful purpose or effect (*EA 2010, s 26(2)*); or

- treatment of a person less favourably because that person has either submitted to or rejected sexual harassment or harassment related to sex or gender reassignment (*EA 2010, s 26(3)*).

Further, although marriage (including civil partnership) and pregnancy and maternity are not covered as relevant protected characteristics for the purposes of a claim of harassment under the *EA 2010, s 26*, where conduct is meted out to a claimant in respect of any of marriage, civil partnership, pregnancy or maternity, there is a possibility that a general discrimination claim might be raised, For example, in *Nixon v Ross Coates Solicitors and another* UKEAT/0108/10ZT, the EAT held that workplace rumour concerning an employee's pregnancy was capable of being both pregnancy discrimination and sexual harassment.

Furthermore, the *EA 2010 s 40* provides that the protection against third-party harassment applies to relevant protected characteristics. Accordingly, an employer could be liable for the harassment of an employee by, eg, a customer, in the event of such harassment occurring on at least two prior occasions in circumstances where the employer has failed to take reasonable steps to prevent the harassment.

Yet further, the *EA 2010* has made some changes to the Old Discrimination Legislation in that there is now protection against colour and nationality discrimination (EA 2010, s 9), and protection for all of the relevant protected characteristics under the *EA 2010, s 26(5)* in respect of perceived and associative discrimination thereby extending and harmonising the spread of case law (see, for example: *Saini v All Saints Haque Centre* [2009] IRLR 74 in relation to religion/belief discrimination; *EBR Attridge Law LLP v Coleman (No 2)* [2010] IRLR 10 in relation to disability discrimination; and, *Showboat Entertainment Centre Limited v Owens* [1984] IRLR 7, *Weathersfield Limited v Sargent* [1999] IRLR 94 and *Redfearn v Serco Limited* [2006] IRLR 623 in relation to race discrimination).

Commission of harassment

[H1704] It follows from the foregoing that a claim for harassment can arise in a variety of different formats. However, whichever format is argued as the claim, all of the elements of the claim must be present, as follows.

(i) The unwanted nature of the conduct giving rise to a claim

[H1705] The conduct giving rise to a claim must be unwanted by the victim of the harassment. Accordingly, if the victim participates willingly in the conduct, the conduct cannot be unwanted.

To this end, in the case of *Thomas Sanderson Blinds Limited v English* UKEAT/0316/10/JOJ, a case involving allegations of perceived sexual orientation discrimination, the EAT had to determine whether an Employment Tribunal had directed itself correctly in looking at the claimant's own perceptions and feelings in order to decide whether the alleged unwanted conduct had the effect of violating his dignity or creating an intimidating, hostile, degrading, humiliating or offensive environment for him in a situation where the Claimant willingly participated in banter about his alleged sexual orientation, and, indeed, responded in kind, writing articles for a work newsletter which the Tribunal found to "riddled with sexist and ageist innuendo" (the Tribunal also recorded an occasion when the Claimant had been obliged to apologise to a woman for an offensive remark about her breasts). The Tribunal concluded that the claimant had not been discriminated against.

Each case turns on its own facts and it is up to the Tribunal to determine whether the conduct alleged is lawful or unlawful: one act of conduct may be found not to be discrimination, even where the conduct is unwanted, but repeated conduct may be found to overstep the mark; conversely, conduct on one occasion may be extreme and unwelcome, and therefore harassment. However, the intention of the party engaged in the conduct is irrelevant in determining whether the conduct amounts to harassment.

Distinguishing between whether conduct is lawful or not can cause real problems especially where the parties might view the conduct in different ways. Further, drawing the line can be difficult; a one off act, such as asking a colleague out for a drink, may not be harassment, but could become so if the rejected party then repeats the invitation. This type of situation was commented upon by HH Judge McMullen QC in *Fearon v The Chief Constable of Derbyshire* UKEAT/0445/02/RN, 16 January 2004.

It is no defence to a claim for harassment that the protected characteristic is shared by both perpetrator and victim, or that neither of them has the protected characteristic: it is entirely possible for an employee of the same gender to treat another employee of that gender in a way that is sexually harassing fashion notwithstanding that the harasser would not have treated a different gender employee similarly to the harasser's actual victim (see, for example, *Walker v BHS Ltd* [2005] All ER (D) 146 (May), EAT under the old law, and note the *EA 2010, s 24* which provides that the characteristics of the alleged discriminator will be irrelevant to the question of liability under the new law).

(ii) The nature of harassing conduct

[H1706] There is no definition of what amounts to conduct under the *EA 2010*. Accordingly, it falls to an Employment Tribunal to determine what conduct is on a case by case basis; it can be anything, whether written, oral or physical.

The older case law provides various examples of conduct related cases in three broad categories, these being course of conduct cases, single incident cases and failure to investigate cases.

Course of conduct cases

As the heading suggests, the hallmark of this genus of cases is that an employee is subjected to repeated harassment in the course of her employment. The following are reported examples of such cases, albeit decided under the Old Discrimination Legislation and therefore needing to be treated with some caution.

(a) *Porcelli v Strathclyde Regional Council* [1986] IRLR 134. The claimant was a female laboratory assistant and was the victim of a campaign of harassment by fellow male employees to make her leave her job. The male employees would frequently brush past the claimant in a sexually suggestive manner and direct comments of a sexual nature at her. The Court of Session held that the harassment the claimant had suffered amounted to sex discrimination by way of a detriment.

(b) *Tower Boot Company Limited v Jones* [1995] IRLR 529. The claimant was a black employee who was subjected to such acts as being whipped with a leather belt, 'branded' with a red-hot screwdriver and subjected to racist name calling. The Court of Appeal held that, although employees were not employed to harass other employees (the employer's defence being that it could not be vicariously liable on that basis), the acts had been committed during the course of the employees' employment and therefore the employer was liable for the discrimination by way of detriment through the harassment.

(c) *Reed and Bull Information Systems v Stedman* [1999] IRLR 299. The claimant was a junior secretary responsible to the marketing manager of the employer. The claimant catalogued fifteen separate incidents ranging from the telling of dirty jokes by the manager to other colleagues in her presence to the pretence of looking up her skirt in the office smoking room. The EAT upheld the findings of the Employment Tribunal that, although the incidents by themselves would have been insufficient to amount to a detriment, collectively they amounted to a course of conduct of innuendo and general sexist behaviour amounting to harassment.

(d) *Ministry of Defence v Fletcher* [2010] IRLR 25. This case concerned the bullying of a lesbian soldier by a male superior with the bullying comprising conduct and comments made to the victim to the effect that: the victim's partner was "ugly"; and, that the victim should have sex with the superior. Further, the superior sent text messages of a sexual nature to the victim. The victim presented a grievance and was then subsequently discharged from the army for being temperamentally unsuitable, notwithstanding that her grievances were upheld. An Employment Tribunal found that the complaint of sex discrimination under the *SDA 1975* was proven (albeit the EAT held on appeal that the Tribunal had erred in relation to the amount of compensation ordered by the Tribunal).

Single incident cases

There is no requirement that conduct should be a course of conduct and it is perfectly possible that conduct could arise from a single incident. Examples of such single incidents cases are as follows.

(a) *Bracebridge Engineering Ltd v Darby* [1990] IRLR 3. The claimant was physically manhandled into the office of the works manager by both a chargehand and the works manager, and then indecently assaulted. She was warned not to complain because no one would believe her. The claimant subsequently complained but the complaint was then not properly investigated by the employer (in respect of which, see below). It was held by the EAT that the assault was sufficiently serious to amount to sexual harassment.

(b) *Insitu Cleaning Co Ltd v Heads* [1995] IRLR 4. The claimant was a cleaning supervisor who attended a company meeting. Also present was the son of one of the directors of the company who greeted her by saying 'Hiya, big tits'. The claimant found the remarks particularly distressing as the director's son was approximately half her age. The EAT held that the remark was capable of subjecting the claimant to discrimination and rejected the employer's contention that the remark was of the same effect as a statement made to a bald male employee about his head.

(c) *Chief Constable of the Lincolnshire Police v Stubbs* [1999] IRLR 81. The claimant was a Detective Constable in the Lincolnshire Constabulary. She attended an off-duty drink with some colleagues in a public house one evening after work. During the course of the evening a male colleague pulled up a stool next to her, flicked her hair and re-arranged her collar so as to give the impression that there was a relationship between himself and the claimant. The claimant found this attention to be distressing and moved away from him. On another occasion at a colleague's leaving party that she had attended with her boyfriend, she was accosted on her way to the toilet by the same officer who stated to her 'Fucking hell, you look worth one. Maybe I shouldn't say that it would be worth some money'. The claimant found the remark to be humiliating and brought proceedings for sex discrimination. The EAT upheld the decision of the Employment Tribunal that the statement amounted to harassment and a detriment.

(d) *Dhaliwal v Richmond Pharmacology* [2009] IRLR 336. The comment made "We will probably bump into each other in the future unless you are married off in India", was found not to be harassment due to the unreasonable perception of the victim of the allegedly harassing nature of the comment.

Failures to investigate or protect

Where an employee raises an allegation of harassment by another employee, their employer has a duty to carry out a proper investigation into the complaint and to take such further steps as are necessary in the light of the findings of the investigation. A failure to carry out such a proper investigation can amount to a further acts of harassment in its own right. The following cases are examples of alleged failures to investigate or protect employees.

(a) *Bracebridge Engineering Ltd v Darby* [1990] IRLR 3. The claimant was sexually assaulted by male staff at the factory where she worked. The employer failed properly to investigate the claimant's complaint (the personnel manager simply accepted that it was the word of two employees against one and that there was insufficient evidence to warrant further investigation), whereupon she resigned and claimed in respect both of constructive dismissal and unlawful sex discrimination. The EAT held that where serious allegations of sexual harassment are made, employers have a duty to investigate them properly.

(b)   *Burton v De Vere Hotels* [1996] IRLR 596. The case took place under
        the provisions of the Old Discrimination Legislation, specifically the
        SDA 1975 and the *RRA 1976*. The case was brought by a waitress at
        a hotel and concerned the detriment of suffering sexual and racial
        harassment by a third party not connected to the employer but whom
        over the employer was able to exercise control (at a function at which
        the claimant was working, she was subjected to sexist and racial abuse,
        masquerading as comedy, from the third party, a well known
        comedian). Although the employer apologised to the claimant for the
        actions of the comedian, the claimant presented proceedings for sexual
        and racial harassment and succeeded in the claim. The EAT held that
        where an employer was in a position to exercise control over a third
        party so that discrimination would either not occur, or the extent of the
        harassment would be reduced, the employer would be subjecting the
        claimant to a detriment by failing to take the steps necessary to control
        the third party.

The reasoning in *Burton* was overturned by the House of Lords in the joined
appeal in *MacDonald v Advocate General for Scotland and Pearce v Governing Body of Mayfield Secondary School* [2003] UKHL 34, [2003] IRLR 512,
where it was held that Burton was wrongly decided, and had gone too far.
Their Lordships held that the failure to take reasonable steps to prevent an
employee from racial or sexual abuse was discrimination only where the
reason for that failure to act amounted to race or sex discrimination. Their
Lordships held that the Burton decision was vulnerable because it treated an
employer's inadvertent failure to take such steps to protect their employees
from racial or sexual abuse as discrimination, even though the failure had
nothing to do with the sex or race of the employees.

Following *MacDonald*, the Old Discrimination Legislation was amended in
consequence of the Equal Treatment Directive (2002/73/EC), (the amendment
carrying over into the *EA 2010*). One of the first recorded decisions under this
amended legislation was *May & Baker Limited v Okerago* [2010] IRLR 394,
a case where there was a statement by an agency worker that a full time
employee should "go back to [her] own fucking country". It was held that an
employer, without more, would not be responsible for the racially harassing
actions of such third parties.

In the subsequent case of *Sheffield City Council v Norouzi* [2011]
UKEAT/0497/10/RN, the EAT held that the employer (a local authority) was
liable for acts of racial harassment carried out by a child in a care home against
one of the employer's employees. This decision was arrived at following a
consideration of the decision of Burton J in *R (Equal
Opportunities Commission) v Secretary of State for Trade and Industry* [2007]
ICR 1234, on the basis that an employer can be liable for the conduct of a third
party in circumstances where there is a continuing course of offensive conduct
about which the employer is aware but does nothing to safeguard against.

Further, under the *EA 2010, s 40*, employers are liable for harassment arising
due to sex, race, disability, age, gender reassignment, religion or belief, or
sexual orientation arising as a result of a third party's conduct. Such liability
will arise where the third party subjects an employee to harassment in the

course of the employee's employment and the employee's employer has failed to take such steps as would have been reasonably practicable to prevent the third party from doing so. The scope of this head of liability is limited by the *EA 2010, s 40(3)*, which provides that the employer will only be liable where the employer knows that the woman has been subject to harassment in the course of the employee's employment on at least two other occasions by a third party. The *EA 2010, s 40* extends substantially the scope of such third party liability, since, prior to its implementation, liability existed only for sex discrimination under this head (under the (now repealed) *SDA 1975, s 6(2)(b)* (as amended by the *Sex Discrimination Act 1975 (Amendment) Regulations 2008 (SI 2008 No 656)).*

This head of liability will be particularly relevant in the context of service industries where behaviour of an employer's customers may be impaired by alcohol (such as pubs, restaurants, hotels, night clubs, and the like).

Finally, conduct does not even have to be conduct which is specific to the claimant; it can comprise conduct of which a claimant becomes aware (for example, widespread downloading of pornography within the workplace on screens to which the Claimant has access *Moonsar v Fiveways Express Transport Ltd* [2005] IRLR 9, EAT).

(iii) Purpose or effect

[H1707] Under the *EA 2010, s 26(1)(b)*, it is necessary for the conduct of which the claimant has raised a complaint to have either the purpose or the effect of violating the claimant's dignity or creating an intimidating, hostile, degrading, humiliating or offensive environment.

Given the distinction purpose based claims and effect based claims, it is necessary to consider what Employment Tribunals will look at to prove such claims.

In purpose based claims, Employment Tribunals will examine the motive or intention of the person who is alleged to have carried out the act(s) of harassment.

Such an examination can involve an Employment Tribunal having to draw inferences as to what that true motive or intent actually was, and the Tribunal will also be required to take into account the transferring burden of proof under the *EA 2010, s 136*.

Where the harassment is purpose based harassment, it does not matter whether the conduct was undertaken against the victim due to the victim's protected status under the *EA 2010*: all that is necessary is that, objectively in all the circumstances, the wrongdoer has committed the act.

Conversely, in order to establish a claim of harassment based on the effect of conduct, the Employment Tribunal must consider the perception by the victim of what was done. Although decided before the changes to the Old Discrimination Legislation by the Equal Treatment Directive (2002/73/EC), the guidance provided by Morison P in *Reed and Bull Information Systems v Stedman* [1999] IRLR 299, is still useful for Employment Tribunals when determining whether harassment may have been committed. In that case, his Lordship stated of harassment (on grounds of sex) that—

(i)   A characteristic of harassment is that it undermines the victim's dignity at work and constitutes a detriment . . . lack of intent is not a defence.

(ii)   The words or conduct must be unwelcome to the victim and it is for her to decide what is acceptable or offensive. The question is not what (objectively) the Tribunal would or would not find offensive.

(iii)   The Tribunal should not carve up a course of conduct into individual incidents and measure the detriment from each; once unwelcome interest has been displayed, the victim may be bothered by further incidents which, in a different context, would appear unobjectionable.

(iv)   In deciding whether something is unwelcome, there can be difficult factual questions for a Tribunal; some conduct . . . may be so clearly unwanted that the [victim] does not have to object to it expressly in advance. At the other end of the scale is conduct which normally a person would be unduly sensitive to object to, but because it is for the individual to set the parameters, the question becomes whether that individual has made it clear that she finds that conduct unacceptable. Provided that that objection would be clear to a reasonable person, any repetition will generally constitute harassment.

Ultimately though, since it is a question of objective fact for an Employment Tribunal to determine whether there may have been an act of harassment committed, it is still possible that a victim may be perceived as what has been termed 'hypersensitive' (*Driskel v Peninsula Business Services Limited* [2000] IRLR 151 at 155); it is not enough under the statutory definition of harassment under the *EA 2010, s 26* that a claimant personally suffers a sense of grievance arising from what he or she felt to be unlawful conduct if that is not objectively borne out by the facts.

Further, harassment may exist without proof of the elements needed for direct discrimination. In particular, no requirement exists now to see how a person in what would otherwise be a comparator group is or would be treated. Under the old case law (i.e., *Porcelli v Strathclyde Regional Council* [1986] IRLR 134), it was necessary to show that less favourable treatment had been meted out to a victim on the grounds of his or her protected status.

Now, harassment must arise within the scope of the statutory definition under the relevant head of the *EA 2010, 26*: a claim may not arise where a person carries out generalised indiscriminate workplace bullying (sometimes referred to as the 'bastard defence', i.e., 'I am a bastard to everyone', although see in this regard the possible claims which may arise under the *Protection from Harassment Act 1997*; there is also the fact that such generalised indiscriminate bullying will be likely to amount to a breach of the implied term not without reasonable cause to undermine the relationship of trust and confidence: *Horkulak v Cantor Fitzgerald* [2003] EWHC 1918 (QB), [2003] IRLR 716.

To succeed, the claim must be founded on one of the protected forms of harassment under the EA 2010, s 26.

In this regard, the decision in *Brumfitt v Ministry of Defence* [2005] IRLR 4 (which pre-dates the amendments under the Equal Treatment Directive

(2002/73/EC) to the *SDA 1975*), which was brought by a woman in connection with abusive language made to a mixed-sex group of which she was a member, would in all likelihood still stand today.

Further, in the case of *English v Thomas Sanderson Blinds Limited* [2008] IRLR 342, a case brought under the *Employment Equality (Sexual Orientation) Regulations 2003*, the importance of a claim being within the ambit of the relevant aspect of the discrimination legislation was brought home. The claimant, a heterosexual male, had been made the subject of workplace bullying based upon him having certain personality traits which were apparently stereotypical of homosexual males in the minds of those bullying him. His claim for sexual orientation-based discrimination failed, with the EAT holding that the treatment was not meted out on grounds of the claimant's actual sexual orientation.

However, on appeal, the Court of Appeal held that the perceived discrimination of the claimant's possession of homosexual characteristics could give rise to a claim for harassment ([2008] EWCA Civ 1421, [2009] IRLR 206, CA).

Furthermore, the *EA 2010, s 26(1)* makes it clear that the definition of harassment applies to both of associative and perceived discrimination in that covers conduct "related to a relevant protected characteristic".

(iv) Violation of claimant's dignity or the creation of an intimidating, hostile, degrading, humiliating or offensive work environment

[H1708] The final hurdle to be crossed in order to bring a successful claim is to show that a claimant has suffered, what under the previous case law on the point was, a detriment. Now, following the changes made by the Equal Treatment Directive (2002/73/EC), although a detriment does not need to be shown, there still must be some form of adverse consequence which the claimant suffers, either in the form of a violation of the claimant's dignity, or by the creation of an intimidating, hostile, degrading, humiliating or offensive work environment.

Unsurprisingly, an unjustified sense of grievance will not give rise to a claim (*Barclays Bank plc v Kapur (No 2)* [1995] IRLR 87).

However, beyond this, there are many situations in which a claim for harassment can be triggered. In *Reed and Bull Information Systems v Stedman* [1999] IRLR 299, the claimant had resigned claiming that the behaviour of her manager amounted to sexual harassment. The employee cited a list of fifteen alleged incidents that she considered amounted to a course of harassment. The Employment Tribunal concentrated on four of the allegations, these being such matters as telling the employee that sex was a beneficial form of exercise, telling dirty jokes to colleagues in her presence, making a pretence of looking up the employee's skirt and then laughing when she angrily left the room, and stating to her when she was listening to a trial presentation that 'You're going to love me so much for my presentation that when I finish you will be screaming out for more and you will want to rip my clothes off'. The Employment Tribunal stated that, of themselves, each of the allegations was insufficient to amount to sexual harassment. However, if the claimant was able to prove the incidents, they would amount to a course of conduct that would itself be sufficient to amount to harassment. The Employment Tribunal went on to find that the acts had occurred and that the employer and the manager had therefore discriminated against the claimant. The employer and manager appealed to the EAT. The EAT dismissed the appeal agreeing that the actions by the manager, although not meant to be sexually harassing, were indeed discriminatory.

That said, not every utterance or act in the workplace which prima facie appears to be harassing on prohibited grounds will give rise to a claim of harassment. For example, the EAT had to consider in *Dhaliwal v Richmond Pharmacology* [2009] IRLR 336 whether a comment made ("We will probably bump into each other in the future unless you are married off in India"), by an employee's line manager to an Indian employee who had tendered her resignation, to emphasise that they both worked in a small business world, was harassing to the extent necessary to give rise to a claim in law. The EAT held that an employer will only be liable for harassment where a claimant feels or perceives that his or her dignity has been violated and that it is reasonable for that perception to arise. To that end, it is necessary to consider all the relevant circumstances including the purpose for which an act was undertaken or comment made. On the facts of the case, it was not reasonable for the employee to have felt that she was being harassed given the reason for the making of the statement.

Limitations to the definition of harassment

[H1709] The *EA 2010* does have its limitations in relation to what amounts to harassment.

First, as stated above, it is possible that an employer may argue the 'bastard defence' and attempt to show that the treatment meted out to a victim is indiscriminate and that the employer would always behave in that fashion, regardless of any protected status of a claimant (albeit that such a defence has its limitations).

Second, it is entirely possible that although a victim may have been subjected to what he or she believes to be unpleasant conduct, the conduct is found not to be unlawful under the *EA 2010*.

For example, in *Balgobin v London Borough of Tower Hamlets* [1987] IRLR 401, complaints were made by women employees that they had been subjected to the detriment of being required to work with their alleged harasser after an investigation into their allegations of harassment were found to be inconclusive. The EAT refused to accept that the claimants had been treated less favourably on grounds of sex since it was possible that if a male employee had been harassed, the employer would have treated the male in a similar way. This decision would probably be decided in the same way under the *EA 2010*.

Third, under the EA 2010, there are three heads of claim which are not allowed to present a free-standing claim for harassment, these being claims arising from a victim's:

(i)   marital status;

(ii)   pregnancy; or

(iii)   maternity leave.

In respect of these types of claim, it would be necessary to attempt pursue claims in the form of a claim of direct discrimination under the *EA 2010 Part II, Chapter II.*

Employers' liability

Vicarious liability

[H1710] A problem that occurs in connection with harassment claims is that, often, it is not the employer who directly subjects a victim to harassment; rather it is a fellow employee of the victim/claimant who is responsible. The employer may then be held responsible for the employee's actions on the grounds of vicarious liability.

The problem here is that employers do not employ staff to harass other members of the workforce. Therefore, it is necessary to determine what amounts to 'the course of employment' in order to see when an employer will be liable for the harassment of its employees by other members of staff.

The course of employment

[H1711] The first case to take as an employer's line of defence the fact that a harasser was not acting in the course of his employment when committing extreme acts of harassment was *Tower Boot Co Ltd v Jones* [1997] IRLR 168. The case concerned proceedings for racial harassment brought under the provisions of the now repealed *RRA 1976* prior to its amendment by the Equal Treatment Directive (2002/73/EC). The applicant had been subjected to a particularly vicious campaign of name-calling and physical assaults. An Employment Tribunal found that this constituted racial harassment and that the employer was vicariously liable for the acts of the employees engaged in the course of conduct. The employer appealed to the EAT which found for the employer on the ground that the common law test for vicarious liability had not been satisfied. The EAT took the view that for this test to be met the aggressor employees had to be engaged in doing the business of the employer, albeit in an improper manner, i.e. the act had to be a different mode of doing what the aggressor employees were employed to do. The EAT then held that on the facts, the actions of the aggressor employees could not 'by any stretch of the imagination' be found to be an improper method of doing what the aggressors were properly employed to do.

The EAT decision was widely condemned on the grounds that the more serious that harassment was, the less likely that an employer would be found to be vicariously liable for it. Indeed, the logical conclusion to the EAT's decision was that, since people were not employed to harass other members of staff, employers would never be liable, a surprising result indeed. Unsurprisingly, the decision was appealed to the Court of Appeal which readily found for the employee. In doing so, the court held that the words, 'in the course of his employment' (the test under the now repealed *RRA 1976, s 32(1)* and its replacement the *EA 2010, s 109(1)*), were not subject to the old common law rules on vicarious liability which would artificially restrict the natural everyday sense that every layman would understand of the wording in the *EA 2010, s 109(1)*.

Consequently, harassment by an aggressor employee did not have to be connected with acts authorised to be done as part of the aggressor's work. In so deciding, the Court of Appeal attacked the reasoning of the EAT, stating that its interpretation of the section creating liability:

> . . . cut across the whole legislative scheme and underlying policy which was to deter racial and sexual harassment through a widening of the net of responsibility beyond the guilty employees themselves by making all employers additionally liable for such harassment.

Requirement of causal link with employment

[H1712] It is not every act of harassment that is committed by an employee that will create liability on the part of the employer: there must still be a causal link with employment. A good example of this point is the case of *Waters v Commissioner of Police of the Metropolis* [1997] IRLR 589. In this case, the complaint was one of sexual harassment in the form of an alleged sexual assault by a male police officer on a female police officer in her section house room after they had returned from a late night walk together. Given that the attack took place outside the normal working hours of the claimant, the Employment Tribunal refused to follow the decision in the *Tower Boot* case and held that the claimant was in the same position as if the alleged assailant had been a total stranger to the employer. The alleged assailant did not live in the section house, both officers were off duty and there was nothing linking the alleged assailant to the employer other than the fact that he was a police officer. The Court of Appeal ultimately upheld the decision of the Tribunal on appeal.

This said, the decision in *Waters* does not provide an absolute defence to harassment undertaken after employees leave work for the day. Whether there is a causal link between the harassment and the employment, thereby attaching liability to the employer, will be a question of fact in each particular situation. In the subsequent proceedings brought by Ms Waters in respect of bullying and failure to investigate her complaint, the House of Lords held that an employer owes a duty of care to investigate allegations of rape, bullying and harassment (*Waters v Commissioner of Police for the Metropolis (No 2)* [2000] 4 All ER 934). An example of this point is *Chief Constable of the Lincolnshire Police v Stubbs* [1999] ICR 547, [1999] IRLR 81. The claimant had been the subject of distressing personal comments of a sexual nature from a fellow police officer made at a colleague's leaving party. The Employment Tribunal at first instance held that since the party was an organised leaving party, the incidents were 'connected' to work and the workplace. It stated that the incident:

> . . . would not have happened but for [WPC Stubbs'] work. Work-related social functions are an extension of employment and we can see no reason to restrict the course of employment to purely what goes on in the workplace.

The EAT emphasised, however, that it would have been different 'had the discriminatory acts occurred during a chance meeting between the employee and the aggressor outside the workplace.

The decisions give a wide degree of discretion to Employment Tribunals on the facts of a particular case to decide whether harassment has been committed in the course of employment. If a Tribunal finds that it is not committed in such circumstances, the effect is that, subject to *Waters (No 2)*, whilst an employee may be able to sue the harasser in tort, the remedy that is provided may well be empty if the harasser does not have the means to pay for the wrongdoing, let alone meet the costs of the claimant's action.

Employer's defence

[H1713] A defence is provided to employers in relation to allegations that they are vicariously liable under the *EA 2010, s 109(4)*. The defence provides that:

> In proceedings against A's employer (B) in respect of anything alleged to have been done by A in the course of A's employment it is a defence for B to show that B took all reasonable steps to prevent A—
>
> (a)   from doing that thing, or
> (b)   from doing anything of that description.

The problem for employers with the defence is that, whilst it provides a defence where an employer has a full procedure dealing with harassment claims (including such matters as monitoring, complaints handling and disciplinary proceedings), the defence requires that the employer must pay more than mere lip service to the existence of such procedures and must also be able to show that its employees were aware of the procedures and, where relevant, that the employer had fully implemented the procedures in practice. A good example of a case where the defence succeeded was *Balgobin v London Borough of Tower Hamlets* [1987] IRLR 401. Here, the alleged harasser was suspended following the initial allegations, his conduct was fully investigated and it was found that there was insufficient evidence against the cook to justify the claims. The EAT was moved to comment that it was difficult to see what further steps the employer could have taken to avoid the alleged discrimination.

Ultimately, if an employer has gone to the expense of having policies and procedures drafted then they should not be kept in the office safe. Employees should be left under no illusions as to what will happen in cases of harassment.

Liability of employees

[H1714] Although an employer may be vicariously liable for acts of harassment, primary responsibility for such action will remain with the employee who has committed the discrimination (*EA 2010, s 110*).

Further, the *EA 2010, s 112* provides that a person who knowingly aids another person to do an act made unlawful by the relevant element of the Discrimination Legislation is to be treated for the purposes as himself doing an unlawful act of the like description.

The scope of these provisions was given a liberal interpretation in *AM v WC & SPV* [1999] IRLR 410. Here, the EAT held that an Employment Tribunal had erred in not allowing a police officer to bring a complaint of harassment against another individual officer. The case arose under the now repealed *SDA 1975* and the EAT considered that the construction of the *SDA 1975, s 41(1), (3)* (*EA 2010, ss 109(1) and (4)*), and *SDA, s 42* (*EA 2010, s 112*), was to make an employer vicariously liable for acts of discrimination committed by employees *EA 2010, s 109(1)*), to provide a conscientious employer with a defence for the actions of its employees where the actions were effectively a 'bolt from the blue' (now *EA 2010, 109(4)*) and to allow an employee who has committed acts of discrimination against another to be joined as a 'partner' to proceedings commenced against an employer (now *EA 2010, s 112*). The EAT added that even where a defence did exist for an employer under what is now *EA 2010, s 109(4)*, there is nothing in the *EA 2010* that prevented the victim of discrimination from launching proceedings against a discriminator employee alone.

That said, in order for an employer to be responsible, there must be a genuine employer or principal relationship; if there is no such relationship, there can be no liability (see *May & Baker Limited v Okerago* [2010] IRLR 394, in this regard).

The employer's defence has also been considered by the House of Lords in *Anyanwu v South Bank Student Union* [2001] IRLR 305, where it was held that the words 'knowingly aids' means that the person providing aid must have given some kind of assistance to the person committing the discriminatory act which helps the discriminator to do the act. The amount or value of the help is of no importance. All that is needed is an act of some kind, done knowingly, which helps the discriminator to do the unlawful act.

Narrowing the extent of employers' liability

[H1715] Where an employee is able to establish that he or she has been the victim of unlawful harassment, there can be circumstances where the damages that would otherwise be awarded to a victim of such discrimination will be dramatically reduced. In the context of sexual harassment, this will be where the victim's conduct shows that she would not have suffered a serious detriment as a consequence of the harassment. For example, in *Snowball v Gardner Merchant Ltd* [1987] IRLR 397, the claimant complained that she had been sexually harassed by her manager. The allegations were denied and the claimant was subjected to questioning at the Tribunal hearing as to her attitude to matters of sexual behaviour so as to show that even if she had been harassed, her feelings had not been injured. In particular, evidence was called as to the fact that she referred to her bed as her 'play pen' and that she had stated that she slept between black satin sheets. On appeal, the EAT held that the evidence had been correctly admitted so as to show that the claimant was not likely to be offended by a degree of familiarity having a sexual connotation.

Likewise, in *Wileman v Minilec Engineering Ltd* [1988] IRLR 144, the EAT refused to overturn an award of compensation in the sum of £50 that had been made for injury to feelings. In this case, the claimant had made a complaint of sexual harassment against one of the directors of her employer. The award had been set at this figure on account of the fact that the victim had worn revealing clothes and it made it inevitable that comment would be passed. The extent of the compensation awarded in the *Wileman* case would now have to be reconsidered in the light of the guidelines given in *Vento v Chief Constable of West Yorkshire Police (No 2)* [2002] EWCA Civ 1871, [2003] IRLR 102 (updated in *Da'Bell v National Society for the Prevention of Cruelty to Children* [2009], UKEAT/0227/09) as to the appropriate amount of compensation to award in a given case. The minimum award of compensation for injury to feelings would now be in the order of £500.

With respect to the EAT, these cases, whilst reflecting that harassment can still occur in cases where a claimant has a high degree of confidence in sexual banter, unfortunately appear to send out the wrong message in that they can be construed as being a 'harasser's charter' and do not seem to recognise that a person may be particularly upset by the unwelcome attentions of a specific person.

Legal action and remedies

[H1716] Where a person unlawfully harasses another, the consequences of such action can be extremely far-reaching and potentially spread across the full range of actions in both civil and criminal law.

Unlawful discrimination

Elements of the tort

[H1717] The most obvious form of action that a victim of harassment may consider is a claim for harassment under the relevant provision of the old Discrimination Legislation.

Remedies

[H1718] The remedies that are open to a victim of unlawful harassment are:

(a)   a declaration;
(b)   a recommendation; or
(c)   compensation.

A declaration is a formal acknowledgement from an Employment Tribunal that discrimination in the form of harassment has occurred. It will be granted in every successful case brought before a Tribunal.

Recommendations are used where an claimant to an Employment Tribunal remains in the employment of her employer. The purpose of a recommendation is for a Tribunal to give guidance to an employer relating to its business for the purposes of eliminating discrimination.

However, the main remedy that victims of harassment will seek is compensation for the loss that has been suffered as a result of the harassment including, particularly, injury to feelings.

Unfair dismissal

[H1719] Unfair dismissal can arise as a claim where an employer fails to control unlawful harassment in the workplace or fails to investigate properly complaints of unlawful harassment. Such failures may (and probably will) also act to undermine the employee's trust and confidence in the employer. The result of such failures is that an employee may be able to claim that the employer has constructively dismissed the employee, thereby amounting to a dismissal for the purposes of the *Employment Rights Act 1996, s 95(1)(c)*. Indeed, where a unlawful harassment claim has been brought on the back of an employee's departure from his or her employment, most cases will also include a claim for unfair dismissal as well.

Proving constructive dismissal

[H1720] The first hurdle that an employee has to overcome is to prove that he or she has been dismissed for the purposes of the *Employment Rights Act 1996 (ERA 1996), s 95(1)(c)*. It is trite law that for an employee to show that the employee has been constructively dismissed the acts of his or her employer must show that the employer, by its conduct, evinced an intention not to be bound by the terms and conditions of the contract of employment (see, to this effect, *Western Excavating (ECC) Limited v Sharp* [1978] IRLR 27). The problem that an employee may face in this regard is that if an employer has in place a full harassment procedure that is properly implemented, the employee may fail to establish that the employer has acted in breach. For an example of this occurring in practice see *Balgobin v London Borough of Tower Hamlets* [1987] IRLR 401.

Likewise, an employee may also fail to show that an employer has breached the implied term of trust and confidence where the act of harassment is a serious one-off act of harassment that the employer was unable to predict would happen and where the employer properly deals with the matter immediately upon a complaint being made. To the extent that the employer does not investigate properly a complaint in such circumstances it is almost inevitable that the employer will breach the contract of employment (see *Bracebridge Engineering Ltd v Darby* [1990] IRLR 3). Further, the High Court confirmed in the case of *Horkulak v Cantor Fitzgerald* [2003] EWHC 1918 (QB), [2003] IRLR 756 that workplace bullying (bullying being a form of harassment) of an employee may amount to grounds for claiming that the employee has been constructively dismissed.

Remedies

[H1721] The remedies that can be provided for unfair dismissal are those set out in *ERA 1996, ss 113–118*, namely:

(a)   an order for reinstatement;
(b)   an order for re-engagement; or
(c)   an order for compensation.

The practical consequence of an employee claiming that he or she has been constructively dismissed as a result of an employer's breach of the implied duty of trust and confidence is that an Employment Tribunal is likely to be unwilling to make an order of reinstatement or re-engagement and therefore the reality is that most cases will provide for compensation to be awarded. The downside to claiming unfair dismissal as opposed to bringing a claim for unlawful discrimination is that, whereas the compensation for discrimination is both unlimited as regards the amount that can be awarded and can contain an award for injury to feelings, the compensation that is available for unfair dismissal is subject to a cap (currently) of £68,400 and there is no right to an award of compensation for injury to feelings.

Actions in tort

[H1722] Broadly speaking, torts are the category of actionable civil wrongs not including claims for breach of contract or those arising in equity. Acts of unlawful harassment, depending upon the particular characteristics of the harassment concerned, are capable of amounting to a variety of different torts.

Assault and battery

[H1723] Assault and battery are forms of trespass to the person. They are separate torts and occur where a person fears that either he or she will receive or has been the subject of unlawful physical contact. As was stated of the torts in *Collins v Wilcock* [1984] 3 All ER 374:

> An assault is an act which causes another person to apprehend the infliction of immediate, unlawful, force on his person; a battery is the actual infliction of unlawful force on another person.

For the tort of assault to be committed it is irrelevant that the victim is not in fear of harm. Consequently, it is possible for an assault to occur where a victim of harassment is threatened with the possibility that he or she will receive unwanted physical contact. All that must exist for the tort to be committed is that the victim must have a reasonable apprehension that she will imminently receive unwanted contact.

Further, for either tort, the use of the word 'force' in the formulation of the tort does not mean physical violence. It is not necessary that the victim is subjected to a full-scale attack (as occurred, for example, in *Bracebridge Engineering Ltd v Darby* [1990] IRLR 3 and *Tower Boot Company Limited v Jones* [1995] IRLR 529). It is possible for the torts to be committed where a person brushes past another suggestively (as in *Porcelli v Strathclyde Regional Council* [1986] IRLR 134) or even kisses another without consent. The torts therefore cover the full range of possible forms of contact related harassment from physical injury to unwanted molestation and are capable of providing a cause of action even though no actual physical harm has been occasioned to the victim of the particular tort.

Negligence

[H1724] While negligence may imply that an act committed against a person has been committed carelessly, it is perfectly possible for the tort to be committed intentionally by a person. The classic phrasing of the tort is derived from the case of *Donoghue v Stevenson* [1932] AC 562, where Lord Atkin stated of the elements of the tort:

> You must take reasonable care to avoid acts or omissions which you can reasonably foresee would be likely to injure your neighbour . . . [who are] persons who are so closely and directly affected by my act that I ought reasonably to have them in contemplation as being so affected when I am directing my mind to the acts or omissions which are called in question.

The test thus framed ensures that liability will be created where there is reasonable foresight of harm being inflicted upon persons foreseeably affected by an actor's conduct. Consequently, the conduct can be intentional or reckless as well as careless.

Where it is the case that an employee harasses a fellow employee, the harasser and victim are likely to be within the relationship of proximity required by the law (it being trite law that fellow employees owe a duty of care to each other).

Once it has been established that a duty is owed, two other factors become relevant these being whether harm has been sustained by a victim and whether the magnitude of the harm suffered by the victim is far greater than the actor intended.

(a)   *Requirement for loss*. For the tort of negligence to be committed, the victim must suffer loss or damage flowing from the negligent act, an obvious form of which is stress related illnesses that can occur from prolonged harassment (for example, nervous shock). It is well recognised in law that where a person suffers from medically treatable stress related illnesses that are caused by the negligence of an employer, it is possible to recover damages for the injuries so inflicted (see to this end the non-harassment case of *Walker v Northumberland County Council* [1995] ICR 702).

(b)   *'Eggshell-skulls'*. The extent of the injury suffered, provided that a victim could foreseeably have suffered a particular form of harm (e.g. distress) from the commission of the unlawful act, is irrelevant. This principle is known as the 'eggshell-skull' rule. The law recognises that it is possible for 'eggshell-personalities' to exist (see, for example, *Malcolm v Broadhurst* [1970] 3 All ER 508). It is irrelevant that another person possessed of the same characteristics of the claimant would not have suffered injury as a consequence of harassment being meted out provided that the hypothetical person would have been distressed by it.

Given that it is possible to recover damages for injury to feelings occasioned as a consequence of unlawful discrimination, it may well be the case that an employee will not want to pursue a claim for negligence. This said, it is not possible to obtain an injunction in respect of unlawful harassment under the Discrimination Legislation.

Remedies in tort

[H1725] Where an employee commits an actionable tort two remedies will commonly be available to the victim. These will be:

(a)   damages (which is available as of right for a tort); and
(b)   an injunction (where damages are not adequate either in whole or in part).

In tort, the correct measure of damages is that the victim should be given such a sum as would put the victim in the position as if the tort had not been committed (*Livingstone v Rawyards Coal Co* (1880) 5 App Cas 25).

As to how much this will be is a question of fact in each case and, potentially, in the case of a course of harassment that induces, say, nervous shock preventing an employee from working again, loss of career damages could run into hundreds of thousands of pounds.

However, the problem incurred with the common law solution of awarding damages to the victim of a tort in the context of unlawful harassment is that the victim may not be properly compensated for on-going acts. In the circumstances, damages may not be adequate in such a situation and the better remedy may be to sue for an injunction as well.

Further, the use of an injunction to stop on-going harassment may be of more practical benefit to a currently distressed employee than would a future award of damages. This is because injunctions can be applied for and obtained relatively speedily as an interim remedy pending the full trial of an action.

The problem with injunctions is that, assuming that an interim injunction is applied for whilst both the harasser and the victim are still in the employment of their employer, the victim's trust and confidence in his or her employer may well have become exhausted by the employer's failure to take preventative action in relation to the aggressor employee. Consequently, in many cases, the better course of action may well be to leave employment on the grounds of constructive dismissal and sue for compensation under the relevant element of the Discrimination Legislation.

Protection from Harassment Act 1997

[H1726] An offence and tort of harassment was created by the *Protection From Harassment Act 1997* ('*PHA 1997*'). The *PHA 1997* was passed to protect victims of 'stalking' but this has not proved a barrier to it being used in the workplace as a means of seeking to prevent workplace bullying (see for example the case of *Majrowski v. Guy's and St Thomas's NHS Trust* [2005] EWCA Civ 251, [2005] IRLR 340). Not all cases of alleged harassment will be successful, and although a standard of criminal conduct had been held necessary to establish liability for harassment claims in order to establish liability under the *PHA 1997* (according to the Court of Appeal in *Conn v Sunderland City Council* [2007] EWCA Civ 1492, [2008] IRLR 324), this was reviewed in *Veakins v Kier Islington Ltd* [2009] EWCA Civ 1288, [2010] IRLR 132, where the Court of Appeal developed this proposition in holding that the court must focus on whether the conduct complained of is oppressive and unacceptable, as opposed to merely unattractive, unreasonable or regrettable, albeit the court must keep in mind that the conduct must be of an order that would sustain criminal liability. Further, in *Marinello v City of Edinburgh Council* [2010] CSOH 17, the Scottish Court of Session, following a full review of the authorities held that what is required for conduct to become criminal is that the conduct amounts to harassment.

Harassment is not specifically defined by the *PHA 1997*, although *PHA 1997, s 7* provides that harassment must amount to a course of conduct causing alarm or distress to a person.

The phrase 'a course of conduct' is defined to be conduct on two or more occasions and 'conduct' includes words (*PHA 1997, s 7(4)*). It has been held in the case of *Lau v DPP* [2000] All ER (D) 224 that where there are few incidents separated by a long period of time, it will be difficult to establish a course of conduct.

It is possible to obtain an injunction against a harasser under the *PHA 1997 s 3* and also to claim damages.

Breach of contract

[H1727] Where an employee harasses a fellow employee, the harasser will almost certainly breach one or more of the terms of his or her employment contract. In the first instance, most disciplinary procedures provide that harassment of fellow employees on grounds of sex, race or disability amounts to gross misconduct. Given that many contracts of employment provide that a disciplinary procedure is contractual, harassment *per se* may provide grounds for summary dismissal. In any event, gross misconduct would act to undermine the implied term not without lawful and reasonable cause to undermine the relationship of trust and confidence between employer and employee.

Even if an employer fails to provide a contractual disciplinary procedure, an employee may well be in breach of several of the implied terms under his contract of employment if he or she is engaged in unlawfully harassing a fellow employee. This is because the employee will owe a duty of care to the employer that requires the employee to take reasonable care for fellow employees. Deliberate unlawful harassment will in all likelihood breach this term and fix the employer with potential vicarious liability for the harassing employee's wrongdoing.

Remedies

[H1728] As with remedies provided for torts, the remedies for breach of contract will be either damages or an injunction.

The law places a different emphasis on the method of computing damages for breach of contract to that provided in relation to torts. Practically though, in the context of a contract of employment, the final outcome will usually be the same. In the law of contract, the victim is to be placed so far as possible in the position as if the contract had been properly performed by the employer (*Robinson v Harman* (1848) 1 Exch 850). The amount of damages payable will vary according to the loss that has been suffered by a victim of a breach of contract. Ordinarily, if the employee leaves his/her employment claiming that he or she has been constructively dismissed, the employee will be able to claim damages for loss of contractual benefits during her notice period (subject to the employee's duty to mitigate). However, it seems that in spite of the decision of the House of Lords in *Malik v BCCI* [1997] IRLR 462, following the subsequent decision of the House of Lords in *Johnson v Unisys Ltd* [2001]

UKHL 13, [2001] IRLR 279, it is not possible to recover damages for injury to feelings occasioned as a result of the humiliating manner of a dismissal.

This said, if the claim is framed by an employee as one going to the employer's duty of care to the employee during employment, it is well established that an employee will be able to claim damages for the illness so caused (see, to this effect, *Walker v Northumberland County Council* [1995] ICR 702).

Criminal action

[H1729] The conduct of a person who harasses another may also amount to one or more of the following crimes:

(a) battery;
(b) harassment under the *Protection from Harassment Act 1997*;
(c) intentional harassment under the *Public Order Act 1986*; or
(d) racially aggravated harassment under the *Crime and Disorder Act 1998*.

Battery

[H1730] Battery is a common law offence. The elements of the offences are that the harasser intentionally and unlawfully inflicts personal violence upon the victim. In *Collins v Wilcock* [1984] 3 All ER 374 the Court of Appeal held that there was a general exception to the crime which embraced all forms of physical contact generally acceptable. However, it is clear from this that if the victim makes known to his/her harasser the fact that the victim does not welcome physical contact from him, the harasser will commit the offence simply by touching the victim.

The offence can only be tried summarily.

Harassment under the Protection from Harassment Act 1997

[H1731] The *PHA 1997* can act to criminalise a course of conduct pursued by a person who 'knows or ought to know' that the conduct amounts to harassment. The *PHA 1997, s 1(2)* provides that a person will be deemed to have knowledge where a reasonable person in possession of the same information as the harasser would think that the course of conduct amounted to harassment.

The offence is one that is triable either way. If the offence is tried on indictment, the maximum penalty is five years imprisonment or a fine or both. If tried summarily, the maximum penalty is a term of imprisonment not exceeding six months, a fine subject to the statutory maximum or both.

Harassment under the Public Order Act 1986

[H1732] The offence exists under the *Public Order Act 1986 (POA 1986), s 4A* and criminalises threatening, abusive or insulting words or behaviour thereby causing harassment, alarm or distress. The offence can be committed in a public or private place (other than a dwelling). Consequently, sexual harassment in the workplace is capable of being covered by the *POA 1986*.

The offence can only be tried summarily and a fine on level 3 of the standard scale can be levied upon conviction.

Racially aggravated crimes under the Crime and Disorder Act 1998

[H1733] The offence of racially aggravated harassment arises under the *Crime and Disorder Act 1998 (CDA 1998), s 32*. The offence is committed where a person commits an offence against another and the offence is racially aggravated (as defined by *CDA 1998, s 28*).

The offence in either case is triable either way. In the event that a person is convicted on indictment of racially aggravated harassment under *CDA 1998, s 32(1)(a)*, the person can be sentenced to imprisonment for a term of up to two years, a fine or to both, or for putting people in fear of violence under *CDA 1998, s 32(1)(b)*, the person can be sentenced to imprisonment for a term of up to seven years, a fine or to both.

Where a person is summarily convicted of racially aggravated harassment under *CDA 1998, s 32(1)(a)*, the person can be sentenced to imprisonment for a term of up to six months, or a fine not exceeding the statutory maximum or to both, or for putting people in fear of violence under *CDA 1998, s 32(1)(b)*, the person can be sentenced to imprisonment for a term of up to six months or a fine not exceeding the statutory maximum or to both.

Preventing and handling claims

Harassment policy

[H1734] Codes of practice covering unlawful harassment under the headings of gender, race and disability in respect of the Discrimination Legislation have been issued by the Equality and Human Rights Commission.

In order for an employer's policy to be effective and properly used, it must have the full support of senior managers within the business and must be used fairly and properly on each occasion that its provisions become relevant. So as to achieve this, it is important that the policy is clear and made known to all employees. A harassment policy is an essential feature of a proper equal opportunities policy and should exhibit the following characteristics:

(a)   a complaints procedure;
(b)   confidentiality;
(c)   a method for informal action to be taken; and
(d)   a method for formal action to be taken.

Complaints procedure

[H1735] The hallmark of an effective policy is that provision is made for a complaints procedure that both encourages justifiable complaints of harassment to be made and protects the maker of the complaint. By way of guidance, in the context of sexual harassment, the European Commission ('EC') Code of Practice states in this regard (at section B paragraph (iii)) that a formal complaints procedure is needed so as to:

. . . give employees confidence that the organisation will take allegations of sexual harassment seriously.

The EC Code also states that such a procedure should be used where employees feel that informal resolution of a dispute is inappropriate or has not worked.

In order for the complaints procedure to be effective, the EC Code recommends that employees should know clearly to whom to make a complaint and should also be provided with an alternative source of person to whom to complain in the event that the primary source is inappropriate (e.g. because that person is the alleged harasser).

The EC Code adds that it is good practice to allow complaints to be made to a person of the same sex as the victim.

Confidentiality

[H1736] In order for complaints to be made effectively, it is necessary that some measure of confidentiality be maintained in the complaint making procedure. The problem that arises here is that there is an inevitable trade off between the right of the victim of harassment to be protected against recrimination and the right of an alleged harasser to know the case that he/she is meeting for disciplinary purposes. Indeed, to the extent that complaints are made against a person who is not told where, when and whom he/she is alleged to have harassed and the alleged harasser is then disciplined as a result of the complaint, there is a chance that the employer will be acting in breach of the duty to maintain the alleged harasser's trust and confidence.

To this end, the requirement for confidentiality exists to the extent that although a harasser may have the right to know the full nature of the complaint that he/she has to meet, the victim should be protected from recrimination in the workplace by the employer ensuring that all details surrounding the complaint are suppressed from other employees in the workplace save to the extent that it is necessary to involve them to either prove or disprove the allegations of harassment.

Further, the requirement of confidentiality also extends to ensuring that the allegations, to the extent that they are made in good faith, are not subsequently used against the claimant for any other purposes connected with the claimant's employment (thereby tying in with the duty of the employer to ensure that there is no victimisation occasioned to the claimant — see *Nagarajan v London Regional Transport* [1999] IRLR 572 dealing with the issue of victimisation).

Informal action

[H1737] Most victims of harassment simply want the unwanted attention or conduct towards them to stop. Whilst the employer's policy should leave employees in no doubt as to the fact that deliberate harassment will amount to a case of gross misconduct being brought against an employee, it may be the case that the form of the harassment is innocent in nature and can be remedied by informal action.

In the context of sexual harassment, it may be the case that the victim of harassment can simply explain to the harasser that the conduct that he/she is receiving is unwelcome and makes the victim feel uncomfortable in the workplace. Where the victim finds it embarrassing or uncomfortable to address the harasser, it may be possible for a fellow employee (in the case of a larger organisation possibly a confidential counsellor) to make the requests to the harasser. Equal opportunities policies should reflect this.

This said, in some cases it simply will not be possible or appropriate for the matter to be dealt with informally and where this is the case, the only possible resolution will be through formal channels.

Formal action

[H1738] In the event that formal action is required to end harassment, both harassers and their victims should be made fully aware as to the procedures that an employer will adopt in relation to a claim of harassment.

Equal opportunities policies should include details as to:

(a)   how investigations are to be carried out;
(b)   the time frame for investigations (which should be as short as possible for the benefit of both the victim of actual harassment and the alleged harasser);
(c)   whether there is to be a period of suspension whilst a claim is investigated; and
(d)   the penalty for an employee who is found to have engaged in harassment.

It may be necessary to require employees to be transferred to different duties or sites (in some cases even where a case is not proven against an alleged harasser so as to prevent the employees being forced to work with each other against their will). Indeed, where they are required to do so in the face of an unproved allegation, unless the matter has been thoroughly investigated by the employer, it may be a recipe for litigation as the case of *Balgobin v London Borough of Tower Hamlets* [1987] IRLR 401 illustrates.

Training

[H1739] In the context of unlawful discrimination based harassment, there should training of the workforce is essential in order to ensure that harassment does not occur in the first place. Specifically, there should be training for supervisory staff and managers. The recommendations for training include:

(a)   identifying the factors contributing to a working environment free of sexual harassment;
(b)   familiarisation of employees with the employer's harassment policy;
(c)   specialist training of those employees required to implement the harassment policy; and
(d)   training at induction days for new employees as to the employer's policy on harassment.

Monitoring and review

[H1740] The acid test of an employer's equal opportunity policy is not how it looks on paper but whether and how it is implemented and monitored in practice. As to monitoring, much will depend upon the facts relating to particular employers.

In the case of smaller employers, it may simply require the employer to be aware of any recommendations that are proposed by bodies such as Equality and Human Rights Commission and/or the Chartered Institute of Personnel and Development. However, where employers are larger, the monitoring function may require a full comparative analysis of complaints made in order to determine what matters need to be implemented with regard to training and whether any changes need to be made to the employer's policy for the purposes of such matters as investigating complaints and tightening up the policy generally.

In short, once the policy is adopted, it must be kept up to date with regard to what is acceptable within the workplace.

Hazardous Substances in the Workplace

Mike Bateman and Andrea Oates

Introduction

[H2101] For most employers, the main requirements relating to hazardous substances in the workplace are those contained in the *Control of Substances Hazardous to Health Regulations SI 2002 (SI 2002 No 2677)* ('*COSHH Regulations*'). This chapter is therefore mainly concerned with the requirements under those Regulations and the assessments they require.

The *COSHH Regulations* were first introduced in 1988 but have been subject to many changes and amendments since. Currently the *Control of Substances Hazardous to Health Regulations 2002* are in force, but these have been amended several times since.

However, there are many other legal requirements affecting what would generally be regarded as hazardous substances, even though some of these do not fall within the definition of 'substance hazardous to health' contained in the *COSHH Regulations*. Relevant regulations include:

- *Chemicals (Hazard Information and Packaging for Supply) Regulations 2009 (SI 2009 No 716)* (commonly known as 'CHIP'). These regulations are being progressively replaced by a European Regulation on the Classification, Labelling and Packaging of Substances and Mixtures (the CLP Regulation) which adopts in Europe the Globally Harmonised System for the Classification and Labelling of Chemicals (GHS) and the REACH regime (see **H2137**);
- *Control of Asbestos at Work Regulations 2006 (SI 2006 No 2739)*;
- *Control of Lead at Work Regulations 2002 (SI 2002 No 2676)*;
- *Control of Major Accident Hazards Regulations 1999 (SI 1999 No 743)* (known as 'COMAH');
- *Dangerous Substances and Explosive Atmospheres Regulations 2002 (SI 2002 No 2776)* (known as 'DSEAR')
- *Dangerous Substances (Notification and Marking of Sites) Regulations 1990 (SI 1990 No 304)*;
- *Ionising Radiations Regulations 1999 (SI 1999 No 3232)*.

The requirements of these regulations are summarised at the end of this chapter and some of them are dealt with more fully in other chapters.

How hazardous substances harm the body

Routes of entry

[H2102] Hazardous substances can be present in the workplace in a variety of forms – solids, dusts, fumes, smoke, liquids, vapour, mists, aerosols and gases. In deciding what risks are posed by these substances, three potential entry routes into the body must be considered:

- *Inhalation* – the hazardous substance may cause damage directly to the respiratory tract or the lungs. Inhalation also allows substances to enter the bloodstream via the lungs and thus affect other parts of the body.
- *Ingestion* – accidental ingestion of hazardous substances is always possible, particularly if containers are not correctly labelled. Eating, drinking and smoking in the workplace introduce the risk of inadvertently ingesting small quantities of hazardous substances.
- *The skin* – substances can damage the skin (or eyes) through their corrosive or irritant effects. Some solvents can enter the body by absorption through the skin. Damage is more likely when there are cracks or cuts in the skin.

Occupational ill health

[H2103] The adverse effects of hazardous substances are many and varied and are a field of study in their own right. Many types of occupational diseases must be reported to the enforcing authorities under the requirements of the *Reporting of Injuries, Diseases and Dangerous Occurrences Regulations 1995 (SI 1995 No 3163)* ('RIDDOR') and some are eligible for social security benefits under government schemes. Employees suffering from occupational ill health can of course bring civil actions for damages against their employers, former employers or others they consider responsible for their condition.

Some examples of occupational ill health caused by hazardous substances are:

- Respiratory problems
  — *Pneumoconiosis* – e.g. asbestosis, silicosis, byssinosis, siderosis.
  — *Respiratory irritation* – caused by inhaling acid or alkali gases or mists.
  — *Asthma* – sensitisation of the respiratory system which may be caused by many substances, e.g. isocyanates, flour, grain, hay, animal fur, wood dusts.
  — *Respiratory cancers* – e.g. those caused by certain chemicals, asbestos, pitch or tar.
  — *Metal fume fever* – a flu-like condition caused by inhaling zinc fumes.
- Poisoning
  Acute (short-term) or chronic (long-term) poisoning may be caused by:
  — *Metals and their compounds* – particularly lead, manganese, mercury, beryllium, cadmium.
  — *Organic chemicals* – which may affect the nervous system, the liver, the kidneys or the gastro-intestinal system.

- — *Inorganic chemicals* – as above plus possible asphyxiant effects such as those caused by carbon monoxide.
- Skin conditions
  - — *Dermatitis*:
    - (i) caused by primary irritants where skin tissue is damaged by acids or alkalis or natural oils are removed by solvents, detergents etc.
    - (ii) due to the effects of sensitising agents, e.g. isocyanates, solvents, some foodstuffs.
  - — *Skin cancer* – caused by contact with pitch, tar, soot, mineral oils etc.
- Biological problems
  From contact with animals, birds, fish (including their carcasses and products) or with micro-organisms from other sources, e.g:
  - — *Livestock diseases* – e.g. anthrax or brucellosis.
  - — *Allergic Alveolitis* – caused by inhaling mould or fungal spores present in grain etc.
  - — *Viral Hepatitis* – usually due to contact with blood or blood products.
  - — *Legionnaires' Disease* – caused by inhaling airborne water droplets containing the legionella bacteria (which can occur extensively in water at certain temperatures).
  - — *Leptospirosis* – Weil's disease, caused by contact with urine from small mammals, particularly rats.

Information on the harmful effects associated with individual products may be available on packaging but certainly should be contained in a material safety data sheet provided by the manufacturer or supplier.

The COSHH Regulations summarised

Substances hazardous to health

[H2104] The definition of 'substance hazardous to health' is contained in the *COSHH Regulations 2002 (SI 2002 No 2677), Reg 2(1)* and includes:

- substances designated as very toxic, toxic, corrosive, harmful or irritant under product labelling legislation (the *CHIP Regulations*) – these substances should be clearly identified by hazard symbols. The orange and black symbols required under the *CHIP Regulations* are being replaced under the CLP regulation (see **H2137**);
- substances for which the Health and Safety Commission has approved a Workplace Exposure Limit (WEL);
- biological agents (micro-organisms, cell cultures or human endoparasites);
- other dust of any kind, when present at a concentration in air equal or greater to 10 mg/m^3 (inhalable dust) or 4 mg/m^3 (respirable dust) as a time weighted average over an 8-hour period; and

• any other substance creating a risk to health because of its chemical or toxicological properties and the way it is used or is present at the workplace.

HSE advice is that "COSHH covers chemicals, products containing chemicals, fumes, dusts, vapours, mists and gases, and biological agents (germs). If the packaging has any of the hazard symbols then it is classed as a hazardous substance. COSHH also covers asphyxiating gases. COSHH covers germs that cause diseases such as leptospirosis or legionnaires' disease: and germs used in laboratories."

Given this widely drawn definition, in any cases of doubt it is always advisable to treat the Regulations as applying, although they do not cover lead, asbestos or radioactive substances because these are all covered by their own specific regulations.

Duties under the Regulations

[H2105] The *COSHH Regulations* place duties on employers in relation to their employees and also on employees themselves. [*SI 2002 No 2677, Reg 8*]. The self-employed have duties as if they were both employer and employee. *Regulation 3(1)* extends the employer's duties 'so far as is reasonably practicable' to 'any other person, whether at work or not, who may be affected by the work carried on'. Consideration must be given to:

• Others at work in the workplace, e.g. contractors, visitors or co-tenants.
• Visitors.
• Members of the public, particularly in public places and buildings.
• Customers and service users.
• Members of the emergency services.

Some of these (e.g. children) cannot be expected to behave in the same way as employees or contractors. An early prosecution under the *COSHH Regulations* was of a doctor's surgery when a young child gained access to an insecure cleaner's cupboard and drank the contents of a bottle of carbolic acid, suffering serious ill effects.

The main requirements of the Regulations are summarised below. Some will be explained in more detail later. Full details of the Regulations together with associated Approved Codes of Practice ('ACoPs') and guidance are contained in a single Health and Safety Executive ('HSE') booklet – *Control of substances hazardous to health* (L5).

• *Prohibitions relating to certain substances* [*SI 2002 No 2677, Reg 4*] – the manufacture, use, importation and supply of certain specified substances is prohibited (full details are contained in *Schedule 2* to the Regulations).
• *Application of Regulations 6 to 13* [*SI 2002 No 2677, Reg 5*] – exceptions are made from the application of these Regulations where other more specific regulations are in place. These relate to respirable

dust in coal mines, lead, asbestos, the radioactive, explosive or flammable properties of substances, substances at high or low temperatures or high pressure and substances administered in the course of medical treatment.

- *Assessment [SI 2002 No 2677, Reg 6]* – an assessment of health risks must be made to identify steps necessary to comply with the Regulations and these steps must be implemented.

- *Prevention or control of exposure [SI 2002 No 2677, Reg 7]* – this must be achieved preferably through elimination or substitution or alternatively the provision of adequate controls, e.g. enclosure, LEV, ventilation, systems of work, personal protective equipment ('PPE'). This regulation requires a 'hierarchical' approach to be taken – this is explained in **H2107** below.

- *Use of control measures etc. [SI 2002 No 2677, Reg 8]* – employees are required to make 'full and proper use of control measures, PPE etc.', and employers must 'take all reasonable steps' to ensure they do.

- *Maintenance, examination and testing of control measures [SI 2002 No 2677, Reg 9]* – control measures, including PPE, must be maintained in an efficient state. Controls must also be examined and tested periodically.

- *Monitoring exposure at the workplace [SI 2002 No 2677, Reg 10]* – monitoring (e.g. through occupational hygiene surveys) may be necessary in order to ensure adequate control or to protect employees' health.

- *Health surveillance [SI 2002 No 2677, Reg 11]* – surveillance of employees' health may be required in some circumstances.

- *Information, instruction and training [SI 2002 No 2677, Reg 12]* – must be provided for persons who may be exposed to hazardous substances (and also for those carrying out COSHH assessments).

- *Arrangements to deal with accidents, incidents and emergencies [SI 2002 No 2677, Reg 13]* – emergency procedures must be established where significant risks exist, and information on emergency arrangements provided.

What the Regulations require on assessments

Assessing the risk to health

[H2106] Sub-paragraph (2) of *Regulation 6* of the *COSHH Regulations 2002 (SI 2002 No 2677)* details many specific factors which must be taken into account during the assessment. The Regulation states:

(1)   An employer shall not carry on any work which is liable to expose any employees to any substance hazardous to health unless he has—
    (a)   made a suitable and sufficient assessment of the risk created by that work to the health of those employees and of the steps that need to be taken to meet the requirements of these Regulations; and
    (b)   implemented the steps referred to in sub-paragraph (*a*).
(2)   The risk assessment shall include consideration of—

      (a)   the hazardous properties of the substance;

      (b)   information on health effects provided by the supplier, including information contained in any relevant safety data sheet;

      (c)   the level, type and duration of exposure;

      (d)   the circumstances of the work, including the amount of the substance involved;

      (e)   activities, such as maintenance, where there is the potential for a high level of exposure;

      (f)   any relevant occupational workplace exposure limit or similar occupational exposure limit;

      (g)   the effect of preventive and control measures which have been or will be taken in accordance with regulation 7;

      (h)   the results of relevant health surveillance;

      (i)   the results of monitoring of exposure in accordance with regulation 10;

      (j)   in circumstances where the work will involve exposure to more than one substance hazardous to health, the risk presented by exposure to such substances in combination;

      (k)   the approved classification of any biological agent; and

      (l)   such additional information as the employer may need in order to complete the risk assessment.

(3)   The risk assessment shall be reviewed regularly and forthwith if—

      (a)   there is reason to suspect that the risk assessment is no longer valid;

      (b)   there has been a significant change in the work to which the risk assessment relates; or

      (c)   the results of any monitoring carried out in accordance with regulation 10 show it to be necessary,

and where, as a result of the review, changes to the risk assessment are required, those changes shall be made.

(4)   Where the employer employs 5 or more employees, he shall record—

      (a)   the significant findings of the risk assessment as soon as practicable after the risk assessment is made; and

      (b)   the steps which he has taken to meet the requirements of regulation 7.'

In carrying out assessments, it should be noted that *Regulation 3(1)* also requires employees to take account, so far as is reasonably practicable, of others who may be affected by the work, e.g. visitors, contractors, customers, passers-by and members of the emergency services.

Prevention or control of exposure

[H2107] The *COSHH Regulations 2002 (SI 2002 No 2677), Reg 6(1)(a)* state that the assessment must take into account the steps needed to meet the requirements of the Regulations. *Regulation 7 of the COSHH Regulations 2002* sets out a clear hierarchy of measures which must be taken to prevent or control exposure to hazardous substances. It states:

(1)   Every employer shall ensure that the exposure of his employees to substances hazardous to health is either prevented or, where this is not reasonably practicable, adequately controlled.

(2)   In complying with his duty of prevention under paragraph (1), substitution shall by preference be undertaken, whereby the employer shall avoid, so far as is reasonably practicable, the use of a substance hazardous to health at the

workplace by replacing it with a substance or process which, under the conditions of its use, either eliminates or reduces the risk to the health of the employees.

(3)   Where it is not reasonably practicable to prevent exposure to a substance hazardous to health, the employer shall comply with his duty of control under paragraph (1) by applying protection measures appropriate to the activity and consistent with the risk assessment, including in order of priority—

(a)   the design and use of appropriate work processes, systems and engineering controls and the provision and use of suitable work equipment and materials;

(b)   the control of exposure at source, including adequate ventilation systems and appropriate organisational measures; and

(c)   where adequate control of exposure cannot be achieved by other means, the provision of suitable personal protective equipment in addition to the measures required by sub-paragraphs (a) and (b).

Thus a clear order of preference is established:

(1)   Prevention of exposure (by elimination or substitution).

(2)   Appropriate work processes, systems and engineering controls (e.g. enclosure) and suitable work equipment and materials.

(3)   Adequate control at source (by ventilation or organisational measures).

(4)   Adequate control through PPE.

(*Sub-paragraph (4)* of *Regulation 7* contains further details of what the measures referred to in *paragraph (3)* must include – these are referred to in H2109.)

Prevention of exposure

[H2108] The first preference should always be to prevent exposure to hazardous substances. This could be achieved by:

•   Changing work methods, e.g. adopting ultrasonic techniques or high-pressure water jets for cleaning purposes rather than using solvents.

•   Using non-hazardous (or less hazardous) alternatives, e.g. replacing solvent-based paints or inks by water-based ones (or ones utilising less hazardous solvents).

•   Using the same substances but in less-hazardous forms, e.g. substituting powdered materials by granules or pellets, using more dilute concentrations of hazardous substances.

•   Modifying processes, e.g. eliminating the production of hazardous by-products, emissions or waste by changing process parameters such as temperature or pressure.

Exposure must be prevented 'so far as is reasonably practicable' – a definition of this qualifying phrase is provided in the chapter on RISK ASSESSMENT. The level of risk will be determined by the type and severity of the hazards associated with a substance and its circumstances of use.

Considerations weighing against the prevention of exposure being reasonably practicable might be the costs associated with using alternative methods or materials, the detrimental effect of alternatives on the product or additional risks which may be introduced by the alternatives.

For example:

- ultrasonic cleaning may not achieve the required standards of cleanliness;
- use of high pressure water can involve additional risks;
- water-based paints may result in increased problems from corrosion; and
- alternative solvents may be more flammable.

Control measures, other than PPE

[H2109] The *COSHH Regulations 2002 (SI 2002 No 2677), Reg 7(4)* states that the control measures required by *paragraph (3)* shall include:

(a)   arrangements for the safe handling, storage and transport of substances hazardous to health, and of waste containing such substances, at the workplace;

(b)   the adoption of suitable maintenance procedures;

(c)   reducing, to the minimum required for the work concerned—
(i)   the number of employees subject to exposure,
(ii)   the level and duration of exposure, and
(iii)   the quantity of substances hazardous to health present at the workplace;

(d)   the control of the working environment, including appropriate general ventilation; and

(e)   appropriate hygiene measures including adequate washing facilities.

Examples of such control measures include:

- Enclosed or covered process vessels.
- Enclosed transfer systems, e.g. use of pumps or conveyors, rather than open transfer.
- Use of closed and clearly labelled containers.
- Plant, processes and systems of work which minimise the generation of, or suppress and contain, spills, leaks, dust, fumes and vapours. (This often involves a combination of partial enclosure and the use of local exhaust ventilation ('LEV').)
- Adequate levels of general ventilation.
- Appropriate maintenance procedures.
- Minimising the quantities of hazardous substances stored and used in workplaces.
- Restricting the numbers of persons in areas where hazardous substances are used.
- Prohibiting eating, drinking and smoking in areas of potential contamination. Legislation banning smoking in enclosed workplaces and public places in England came into effect on 1 July 2007.
- Providing showers in addition to normal washing facilities.
- Regular cleaning of walls and other surfaces.
- Using signs to indicate areas of potential contamination.
- Safe storage, handling and disposal of hazardous substances (including hazardous waste).

Paragraphs (5) and *(6)* of *Regulation 7* set out measures required to control exposure to carcinogens and biological agents – these are dealt with later in H2113.

Control using PPE

[H2110] Control of exposure using PPE should be the last option, where prevention or adequate control by other means are not reasonably practicable. This might be the case where:

- the extent of use of hazardous substances is very small;
- adequate control by other means is impracticable from a technical viewpoint;
- control by other means is excessively expensive or difficult in relation to the level of risk;
- PPE is used as a temporary control measure pending the implementation of adequate control by other means;
- employees are dealing with emergency situations;
- infrequent maintenance activities are being carried out.

PPE necessary to control risks from hazardous substances might be:

- respiratory protective equipment (RPE);
- protective clothing;
- hand or arm protection (usually gloves or gauntlets);
- eye protection;
- protective footwear.

The *COSHH Regulations 2002 (SI 2002 No 2677), Reg 7(9)* state that:

> Personal protective equipment provided by an employer in accordance with this regulation shall be suitable for the purpose and shall—
>
> (a) comply with any provision in the Personal Protective Equipment Regulations 2002 which is applicable to that item of personal protective equipment; or
>
> (b) in the case of respiratory protective equipment, where no provision referred to in sub-paragraph *(a)* applies, be of a type approved or shall conform to a standard approved, in either case, by the Executive.

HSE publishes detailed guidance on PPE generally and on RPE in particular.

Adequate control

[H2111] The *COSHH Regulations 2002 (SI 2002 No 2677), Reg 7(7))* (as amended) and *(11)* define what is meant by the term 'adequate control' in respect of risks from hazardous substances:

> (7) Without prejudice to the generality of paragraph (1), where there is exposure to a substance for which a maximum exposure limit has been approved, control of exposure shall, so far as the inhalation of that substance is concerned, only be treated as being adequate if the level of exposure is reduced so far as is reasonably practicable and in any case below the maximum exposure limit.

Regulation 7(11) goes on to say:

(11) In this regulation, "adequate" means adequate having regard only to the nature of the substance and the nature and degree of exposure to substances hazardous to health and "adequately" shall be construed accordingly.

Workplace exposure limits

[H2112] Workplace exposure limits (WELs) replace the terms previously used in the regulations – maximum exposure limits and occupational exposure standards. Listings of WELs are published regularly by the HSE and are now available on their website (see http://www.hse.gov.uk/coshh/table1.pdf).

In some cases it will be relatively easy to make a judgement as to whether exposure is controlled to levels well within the WEL but in other situations it may be necessary to carry out an occupational hygiene survey or even arrange for routine or periodic monitoring of exposure.

Carcinogens, mutagens and biological agents

[H2113] The *COSHH Regulations 2002 (SI 2002 No 2677), Reg 7(5)* and *(6)* set out a number of specific measures which must be applied where it is not reasonably practicable to prevent exposure to carcinogens, mutagens and biological agents. (These are in addition to the measures required by *paragraph (3)* of the regulation.)

Measures specified for controlling carcinogens and mutagens are:

- totally enclosing process and handling systems (unless not reasonably practicable);
- prohibition of eating, drinking and smoking in areas that might be contaminated. Legislation banning smoking in enclosed workplaces and public places in England came into effect on 1 July 2007;
- cleaning of floors, walls and other surfaces regularly and whenever necessary;
- designating areas of potential contamination by warning signs;
- storing, handling and disposing of carcinogens or mutagens safely, using closed and clearly labelled containers.

Appendix 1 to the HSE ACoP booklet provides further detail on the control of carcinogenic and mutagenic substances whilst Annex 1 of the booklet contains a background note on occupational cancer.

Measures for controlling biological agents are:

- use of warning signs (a biohazard sign is shown in *Schedule 3* of the Regulations);
- specifying appropriate decontamination and disinfection procedures;
- safe collection, storage and disposal of contaminated waste;
- testing for biological agents outside primary confinement areas;
- specifying work and transportation procedures;
- where appropriate, making available effective vaccines;
- instituting appropriate hygiene measures;
- control and containment measures where human patients or animals are known (or suspected to be) infected with a Group 3 or 4 biological agent.

The *COSHH Regulations 2002 (SI 2002 No 2677), Reg 7(10)* require work with biological agents to be in accordance with *Schedule 3* to the Regulations which is contained in the HSE COSHH ACoP booklet (L5). This sets out several detailed requirements, including standards for containment measures.

Practical aspects of COSHH assessments

The requirements summarised

[H2114] COSHH assessments must involve the following:

- identification of risks to the health of employees or others;
- consideration of whether it is reasonably practicable to prevent exposure to those hazardous substances which create risks (by elimination or substitution);
- if prevention is not reasonably practicable, identification of the measures necessary to achieve adequate control of exposure, as required by the *COSHH Regulations 2002 (SI 2002 No 2677), Reg 7* (these control measures are described at **H2109** and **H2110**);
- identification of other measures necessary to comply with the *COSHH Regulations 2002 (SI 2002 No 2677), Regs 8–13*, e.g:
  — use of control measures;
  — maintenance, examination and test of control measures etc;
  — monitoring of exposure at the workplace;
  — health surveillance;
  — provision of information, instruction and training;
  — arrangements for accidents, incidents and emergencies.
  (These are described more fully at **H2127–H2135** below).

The assessment must be 'suitable and sufficient', and its complexity will vary considerably dependent on the circumstances. In some cases it may only be necessary to study relevant safety data sheets and decide whether existing practices are sufficient to achieve adequate control. However, for more complex processes and variable activities involving greater numbers of hazardous substances, the assessment must be much more detailed. The planning, preparation and implementation of a COSHH assessment will be very similar to other types of risk assessment and much of the guidance given in the chapter on RISK ASSESSMENT will be valid here.

Who should carry out COSHH assessments?

[H2115] *Regulation 7* of the *Management of Health and Safety at Work Regulations 1999 (SI 1999 No 3242)* requires employers to appoint a competent person or persons to assist them in complying with their duties, which will include carrying out COSHH assessments. *Regulation 12(4)* of the *COSHH Regulations 2002 (SI 2002 No 2677)* also states:

> Every employer shall ensure that any person (whether or not his employee) who carries out any work in connection with the employer's duties under these Regulations has suitable and sufficient information, instruction and training.

The degree of knowledge and experience required will depend upon the circumstances. An assessment in a workplace containing a few simple uses of hazardous substances may be carried out by someone without any specialist qualifications or experience but with the capability to understand and apply the principles contained in this chapter. (However, such a person may need access to someone with greater health and safety knowledge and a more detailed understanding of the *COSHH Regulations*.) Some qualified and experienced individuals may be capable of carrying out detailed assessments of relatively high-risk situations themselves, although they are likely to need to consult others during the assessment process.

Higher risk situations, with much more complex and significant use of hazardous substances are more likely to require multi-disciplinary teams, some of whom have specialist knowledge and experience. Possible candidates for such a team might be:

- health and safety officers or managers;
- occupational hygienists;
- occupational health nurses;
- physicians with occupational health experience;
- chemical or process engineers;
- maintenance or ventilation engineers.

How should the assessments be organised?

[H2116] In larger workplaces it is often best to divide them into manageable assessment units which could be based on:

- Departments or sections.
- Buildings or rooms.
- Process lines.
- Activities or services.

Where an assessment team is used, those individuals most suited for a particular situation can be selected to assess that unit.

Gathering information together

[H2117] Prior to the assessment, information should be gathered together, particularly about which hazardous substances are present and what hazards are associated with them. Substances to be taken into account include:

- Raw materials used in processes.
- The contents of stores and cupboards.
- Substances produced by processes:
  — intermediate compounds;
  — products;
  — waste and by-products;
  — emissions from the process.
- Substances used in maintenance and cleaning work.
- Buildings and the work environment:
  — surface treatments;

—   pollution or contaminants;
—   bird droppings, animal faeces etc;
—   possible legionella sources.
- Substances involved in the activities of others, e.g. contractors.

Potential sources of information about the hazards associated with these substances include:

- Manufacturers' or suppliers' safety data sheets.
- Information provided on containers or packaging (including the symbols required by the *CHIP Regulations (SI 2009 No 716)*.
- HSE publications and the HSE website.
- Other reference books and technical literature.
- Direct enquiries to specialists or suppliers.

This information will primarily concern the hazards associated with the substance but is also likely to contain recommendations on control methods. However, the substance's actual circumstances of use may constitute a greater or lesser degree of risk than the manufacturer or supplier anticipated – it is for the assessor(s) to determine what control measures are necessary in each situation.

Other useful information may be available from previous assessments or surveys. Even if previous COSHH assessments were inadequate or are now out of date, they may still contain information of value in the assessment process. The results of previous occupational hygiene surveys (e.g. dust, gas or vapour concentrations) are also likely to be useful as will the collective results of any health surveillance work which has been carried out.

Details of existing control measures may be contained in operating procedures, health and safety rule books or listings of PPE requirements or may be referred to within training programmes. Specification information may be available for some control measures, e.g. design flow rates for ventilation equipment or performance standards for PPE such as respiratory protection or gloves. (Some of this information may also be sought during the assessment itself or prior to the preparation of the assessment record.)

Observations in the workplace

[H2118] Time spent in making COSHH assessments can be reduced and made more productive by good planning and preparation. Nevertheless, observations of workplaces and work activities must still be an integral part of the COSHH assessment process.

Sufficient observations must be made to arrive at a conclusion on the level of risk and the adequacy of control measures for all the hazardous substances in the workplace – be they raw materials, products, by-products, waste, emissions etc. This should include the manner and extent to which employees (and others) are exposed and an evaluation of the equipment and working practices intended to achieve their control.

Aspects to be considered include:

- Are specified procedures being followed?

- Does local exhaust ventilation or general ventilation appear effective?
- Is PPE being used correctly?
- Is there evidence of leakage, spillage or dust accumulation?
- What equipment is available for cleaning?
- Are dusts, fumes or strong smells evident in the atmosphere?
- Are there any restrictions relating to eating, drinking or smoking? (Legislation banning smoking in enclosed workplaces and public places in England came into effect on 1 July 2007.)
- What arrangements are there for storage, cleaning or maintenance of PPE?
- Are there suitable arrangements for washing, showering or changing clothing?
- Are special arrangements for laundering clothing required?
- What arrangements are in place for storage and disposal of waste?
- Is there potential for major spillages or leaks?
- What emergency containment equipment or PPE is available?

Discussions with staff

[H2119] It is also important to talk to those working with hazardous substances, their safety representatives and those managing or supervising their work. Questions might relate to working practices, awareness of risks or precautions, the effectiveness of precautions or experience of problems. These might include:

- Why tasks are done in particular ways?
- Are the present work practices typical?
- How effective is the LEV/general ventilation?
- What happens if it ever breaks down?
- What types of PPE are required for the work?
- Are there any problems with the PPE?
- Have workers experienced any health problems?
- How often is the workplace cleaned?
- What methods are used for cleaning?
- How is waste disposed of?
- Is there any possibility of leaks occurring or emergencies arising?
- What would happen in such situations?
- What are the arrangements for washing/cleaning/maintaining PPE?

Further assistance

[H2120] Guidance on COSHH assessments is available in the HSE booklet *A step by step guide to COSHH assessment*. The HSE has also developed a range of *COSHH essentials* publications and there is a *COSHH essentials* web site at www.coshh-essentials.org.uk.

Further investigations

[H2121] Assessment in the workplace may reveal the need for further tests on the effectiveness of control measures before the assessment can be concluded.

The tests may be in the form of occupational hygiene surveys – usually to measure the airborne concentrations of dust, fume, vapour or gas and compare them with the relevant WEL. Alternatively there may be a need to measure the effectiveness of local exhaust ventilation or general ventilation.

Further investigations may also be necessary into matters such as:

- hazards associated with substances which were not identified during the preparatory phase of the assessment;
- whether the specifications for PPE found in use are adequate for the exposure involved;
- the feasibility of using alternative work methods (e.g. preventing exposure) or alternative methods of control to those currently in place.

COSHH assessment records

Preparation of assessment records

[H2122] Much of the guidance on note taking and the preparation of assessment records contained in the chapter on RISK ASSESSMENT is relevant to COSHH assessments. Illustrative examples of completed assessment records are provided at **H2124** below, although it should be stressed that no single record format will automatically cater for all types of workplaces or work activities. The essential components which must be included in COSHH assessment records are:

- the work activities involving risks from hazardous substances;
- information as to the hazardous substances involved (and their form, e.g. liquid, powder, dust etc.);
- control measures which are (or should be) in place;
- improvements identified as being necessary.

The COSHH ACoP states that employers with less than five employees are exempt from the requirement to record assessments, although it strongly advises keeping records as good practice. In all other cases the significant findings must be recorded and kept readily accessible to those who may need to see them. Employees or their representatives should be informed of the results of COSHH assessments.

The illustrative COSHH assessment records demonstrate the contrasting levels of detail appropriate for different situations. The amount of information should be proportionate to the risks posed by the work. In many cases it will be appropriate to state that in the circumstances of use there is little or no risk and that no more detailed assessments are necessary.

Review and implementation of recommendations

[H2123] Once the assessment has been completed and the record has been prepared it is important that recommendations are reviewed, implemented and followed up. The review process is likely to involve people who were not involved in the original assessments – senior managers, or staff with relevant technical expertise. Whilst the assessment team may be prepared to make changes to their assessment findings or recommendations, they should not allow themselves to be pressurised into making changes they find unacceptable. If senior management choose not to implement COSHH assessment recommendations then they must accept the responsibility for their actions.

Responsibilities for implementing each recommendation within a designated timescale must be clearly allocated to individuals, possibly as part of an overall action plan. Each of these recommendations must then be followed up to ensure that they have actually been carried out and also to check that additional risks have not been created inadvertently. The assessment record should then be annotated to indicate that the recommendation has been completed (alternatively a new record could be prepared reflecting the improved situation).

Example assessment records

[H2124]

ACORN ESTATE AGENTS, NEWTOWN

Risk Assessment

Risk topic/issue: Hazardous substances

Sheet 1 of 1

Reference number: 6	Precautions in place	Recommended improvements
Risks identified The substances listed below could present risks to both staff and clients: **Office materials**	Office supplies are kept in a cupboard in a part of the office not normally accessible to clients.	
With warning symbols Spray adhesive (harmful)	If the spray adhesive is used for more than a couple of minutes, a nearby window is opened which provides adequate ventilation.	
Without warning symbols Photocopier and printer toners	The photocopier is used in well-ventilated areas and there are no noticeable ozone smells, even on long copying runs.	Provide disposable gloves for cleaning significant spillages of photocopier or printer toner.
Cleaning materials **With warning symbols** Thick bleach (Irritant) Polish stripper (Irritant)	The cleaner's cupboard is kept locked except when substances are being removed. Suitable gloves are provided (and worn) for handling the irritant and corrosive substances in concentrated form.	Investigate replacing the thick bleach by a more dilute solution.
Acid descaler (Corrosive)	The cleaner is aware of the risks of mixing bleach with other substances e.g. the acid descaler.	Ensure the relief cleaner is also made aware of these risks.
Without warning symbols Furniture polish Floor polish Window and glass cleaner		

Risk Assessment

Reference number: 6	Risk topic/issue: Hazardous substances	Sheet 1 of 1
Signature(s) K Stephenson, R Lewis	Name(s) K Stephenson, R Lewis	Date 9/1/010
Dates for	Recommendation follow up: February 2010	Next routine review: January 2014

RAINBOW PRODUCTS

COSHH Assessment	
	Mixing Hall
Assessment unit	
Activities	Manufacture of solvent-based paints and other surface treatments. Solvents are pumped into the mixing vessels from an external tank farm. Solid constituents and some liquid components are charged into the mixing vessels from platforms above. After mixing, the products are pumped directly to filling stations in a neighbouring building.
Substances used (Data sheet file references)	4, 7, 14, 23, 49, 50, 51, 60, 73, 80, 92, 106, 117. Detailed formulations for each product are available from the Quality Control Department.
Main risks	Dust from solid constituents during charging. Solvent vapours: – escaping from the charging hatches; – from minor leaks at valves and pipe connections; – from liquid components during charging. Entry into mixing vessels for cleaning or maintenance purposes.
Those at risk	Mixing Hall production employees. Maintenance staff when working in the area. Contractors involved in vessel cleaning or maintenance work
Controls and other precautions in place	Good general ventilation (specification 20 air changes per hour). Annual surveys of solvent vapours show levels well below all WELs. Hoods with LEV over each vessel charging position. These are examined and tested annually by the Maintenance Department. Exposure Monitoring Surveys (Sept 2002, Nov 2005) show production employees as well within WELs for all dusts and solvents. Disposable dust masks are available for vessel charging, although their use is not compulsory. A portable vacuum cleaning unit with suitable filter is available when required. Maintenance carry out an annual physical inspection of all mixer vessels and pipelines. A weigh station is provided in a fume cupboard for weighing out smaller quantities of solid constituents (this also is examined and tested annually). Entry into vessels is controlled by the permit to work system. All employees are subject to the company's annual health screening programme.

COSHH Assessment

Mixing Hall

Assessment unit	
Observations	Discarded empty paper sacks were strewn on several loading platforms. Liquid component transfer containers had been left open on platforms 1 and 4 (still containing residual materials). The ventilation at charging stations 2 and 4 appeared inadequate. There were dust accumulations on all loading platforms. A brush appears to have been used to sweep up dust on some platforms. A pipe flange below mixer 3 had developed a small leak.
Recommendations	1. Improve the sack disposal containers on all loading platforms.
	2. Remind staff of the importance of disposing of sacks correctly.
	3. Remind staff that liquid transfer containers should have their lids replaced after use.
	4. Rectify the ventilation at charging stations 2 and 4.
	5. Introduce simply weekly checks on the ventilation at all charging stations.
	6. Introduce weekly cleaning for all loading platforms.
	7. Remind staff of the importance of using vacuum methods for cleaning.
	8. Investigate obtaining vacuum cleaning units which can be lifted onto the platforms more easily.
	9. Repair the leaking flange below mixer 3.
	10. Increase the frequency of pipeline inspections to six monthly.
	11. Develop a standard procedure for entering vessels for cleaning or maintenance purposes (linked to the permit to work system).
	12. Investigate alternative cleaning methods avoiding the need for entry, e.g. immersion in solvent over weekend periods.
Signature(s) A Storm, P Gold	Date 7 February 2010
Recommendation follow up: May 2010	
Assessment Review: February 2012	

Ongoing reviews of assessments

[H2125] The *COSHH Regulations 2002 (SI 2002 No 2677), Reg 6(3)* require COSHH assessments to:

be reviewed regularly and forthwith if—

(a)   there is reason to suspect the risk assessment is no longer valid;
(b)   there has been a significant change in the work to which the risk assessment relates; or
(c)   the results of any monitoring carried out in accordance with regulation 10 show it to be necessary.

- *Assessments no longer valid*
  Assessments might be shown to be no longer valid because of:
  — new information received about health risks, e.g. information from suppliers, revised HSE guidance, changes in the WEL;
  — results from inspections or through examinations or tests (*Regulation 9*), e.g. indicating fundamental flaws in engineering controls;
  — regular reports or complaints about defects in control arrangements;
  — the results from workplace exposure monitoring (*Regulation 10*), e.g. showing the WEL is regularly being exceeded (or approached);
  — results from health surveillance (*Regulation 11*), e.g. demonstrating an unsatisfactory or deteriorating position;
  — a confirmed case of an occupational disease.
- *Significant changes*
  Changes necessitating a review of an assessment might involve:
  — the types of substances used, their form or their source;
  — equipment used in the process or activity (including control measures);
  — altered methods of work or operational procedures;
  — variations in volume, rate or type of production;
  — staffing levels and related practical difficulties or pressures.
- *Regular review*
  The periods elapsing between reviews should relate to the degree of risk involved and the nature of the work itself. The COSHH ACoP previously stated that assessments should be reviewed at least every five years, but no longer contains such a recommendation. However, this would seem to be a reasonable maximum period to use for practical purposes.
  A review would not necessarily require a revision of the assessment – it may conclude that existing controls are still adequate despite changed circumstances. However, where changes are shown to be required, *paragraph (3)* of *Regulation 6* requires that these be implemented.

Further requirements of the COSHH Regulations

[H2126] Some recommendations from COSHH assessments could relate to the ongoing maintenance of effective control measures and many of these aspects are covered by further specific requirements of the *COSHH Regulations.*

Use of control measures

[H2127] The *COSHH Regulations 2002 (SI 2002 No 2677), Reg 8* place duties on both employers and employees in respect of the proper use of control measures, including PPE.

(1)   Every employer who provides any control measure, other thing or facility in accordance with these Regulations shall take all reasonable steps to ensure that it is properly used or applied as the case may be.

(2)   Every employee shall make full and proper use of any control measure, other thing or facility provided in accordance with these Regulations and where relevant shall—

    (a)   take all reasonable steps to ensure it is returned after use to any accommodation provided for it; and,

    (b)   if he discovers any defect therein, report it forthwith to his employer.

Employers have duties under the *Management of Health and Safety at Work Regulations 1999 (SI 1999 No 3242)* to monitor all of their health and safety arrangements. Areas where monitoring of COSHH control measures are likely to be necessary are:

- correct use of LEV equipment;
- compliance with specified systems of work;
- compliance with PPE requirements;
- storage and maintenance of PPE;
- compliance with requirements relating to eating, drinking or smoking. Legislation banning smoking in enclosed workplaces and public places in England came into effect on 1 July 2007;
- condition of washing and showering facilities and personal hygiene standards;
- whether defects are being reported by employees.

Where employees are unwilling to use control measures properly, employers should consider the use of disciplinary action, particularly in relation to persistent offenders.

Maintenance, examination and testing of control measures

[H2128] The *COSHH Regulations 2002 (SI 2002 No 2677), Reg 9* contain both general and specific requirements for the maintenance of control measures:

(1)   Every employer who provides any control measure to meet the requirements of regulation 7 shall ensure that, where relevant, it is maintained in an efficient state, in efficient working order, in good repair and in a clean condition.

(2)   Where engineering controls are provided to meet the requirements of regulation 7, the employer shall ensure that thorough examination and testing of those controls is carried out—

(a)   in the case of local exhaust ventilation plant, at least once every 14 months, or for local exhaust ventilation plant used in conjunction with a process specified in Column 1 of Schedule 4, at not more that the interval specified in the corresponding entry in Column 2 of that Schedule; or

(b)   in any other case, at suitable intervals.

(3)   Where respiratory protective equipment (other than disposable respiratory protective equipment) is provided to meet the requirements of regulation 7, the employer shall ensure that thorough examination and, where appropriate, testing of that equipment is carried out at suitable intervals.

(4)   Every employer shall keep a suitable record of the examinations and tests carried out in pursuance of paragraphs (2) and (3) and of repairs carried out as a result of those examinations and tests, and that record or a suitable summary thereof shall be kept available for at least 5 years from the date on which it was made.

Paragraphs (5), (6) and *(7)* contain requirements specifically relating to PPE – these are examined later.

General maintenance of controls

[H2129] The COSHH ACoP states that, where possible, all engineering control measures should receive a visual check at appropriate intervals, and in the case of LEV and work enclosures, at least once every week. Such checks may simply confirm that there are no apparent leaks from vessels or pipes and that LEV or cleaning equipment appear to be in working order. No records of such checks need to be kept, although it is good practice (and prudent) to do so.

The *COSHH Regulations 2002 (SI 2002 No 2677), Reg 9(2)* require thorough examinations and tests of engineering controls. Requirements relating to LEV are reviewed below at **H2130** but for other engineering controls such examinations and tests must be 'at suitable intervals', and suitable records must be kept for at least five years. The nature of examinations and tests will depend upon the engineering control involved and the potential consequences of its deterioration or failure. Examples might involve:

- detailed visual inspections of tanks and pipelines;
- inspections and non-destructive testing of critical process vessels;
- testing of detectors and alarm systems;
- planned maintenance of general ventilation equipment;
- checks on filters in vacuum cleaning equipment.

Persons carrying out maintenance, examinations and testing must be competent for the purpose in accordance with the *COSHH Regulations 2002 (SI 2002 No 2677), Reg 12(4)*.

Local exhaust ventilation (LEV) plant

[H2130] Most LEV systems must be thoroughly examined and tested at least once every 14 months, although *Schedule 4* of the *COSHH Regulations 2002 (SI 2002 No 2677)*, requires increased frequencies for LEV used in conjunction with a handful of specified processes. Dependent upon the design and purpose of the LEV concerned, the examination and test might involve visual inspection, air flow or static pressure measurements or visual checks of efficiency using smoke generators or dust lamps. In some cases air sampling to confirm efficiency levels, filter integrity tests or checks on air flow sensors, may be appropriate.

Personal Protective Equipment (PPE)

[H2131] All types of PPE are subject to the general maintenance requirements contained in the *COSHH Regulations 2002 (SI 2002 No 2677), Reg 9(1)* whilst *Regulation 9(3)* contains specific requirements relating to non-disposable respiratory protective equipment ('RPE').

Further requirements in respect of PPE are contained in *Regulation 9(5), (6)* and *(7)* of the 2002 Regulations that state:

(5)   Every employer shall ensure that personal protective equipment, including protective clothing, is:

(a)   properly stored in a well-defined place;
(b)   checked at suitable intervals; and
(c)   when discovered to be defective, repaired or replaced before further use.

(6)   Personal protective equipment which may be contaminated by a substance hazardous to health shall be removed on leaving the working area and kept apart from uncontaminated clothing and equipment.

(7)   The employer shall ensure that the equipment referred to in paragraph (6) is subsequently decontaminated and cleaned or, if necessary, destroyed.

Some types of PPE can easily be seen to be defective by the user whilst in other cases (e.g. for gloves or clothing providing protection against strongly corrosive chemicals) it may be appropriate to introduce more formalised inspection systems.

For non-disposable RPE the COSHH ACoP states that thorough examinations and, where appropriate, tests should be made *at least* once every month, although it suggests that for RPE used less frequently, periods up to three months are acceptable. Alternatively, examination and testing prior to next use may be more appropriate. An HSE booklet *Respiratory protective equipment at work* provides detailed guidance on the subject.

Monitoring exposure at the workplace

[H2132] Reference was made earlier in the chapter at **H2111** and **H2121** to the possible need to carry out air testing in the workplace as part of the COSHH assessment process in order to determine the adequacy of control measures. Such testing may also be necessary in order to ensure that adequate control continues to be maintained and a requirement for this is contained in the *COSHH Regulations 2002 (SI 2002 No 2677), Reg 10*:

(1)   Where the risk assessment indicates that—
(a)   it is requisite for ensuring the maintenance of adequate control of the exposure of employees to substances hazardous to health; or
(b)   it is otherwise requisite for protecting the health of employees,

the employer shall ensure that the exposure of employees to substances hazardous to health is monitored in accordance with a suitable procedure.

(2)   Paragraph (1) shall not apply where the employer is able to demonstrate by another method of evaluation that the requirements of regulation 7(1) have been complied with.

(3)   The monitoring referred to in paragraph (1) shall take place—
    (a)   at regular intervals; and
    (b)   when any change occurs which may affect that exposure.

(4)   Where a substance or process is specified in Column 1 of Schedule 5, monitoring shall be carried out at least at the frequency specified in the corresponding entry in Column 2 of that Schedule.

(5)   The employer shall ensure that a suitable record of any monitoring carried out for the purpose of this regulation is made and maintained and that record or a suitable summary thereof is kept available—
    (a)   where the record is representative of the personal exposures of identifiable employees, for at least 40 years; or
    (b)   in any other case, for at least 5 years,
    (c)   from the date of the last entry made in it.

(6)   Where an employee is required by regulation 11 to be under health surveillance, an individual record of any monitoring carried out in accordance with this regulation shall be made, maintained and kept in respect of that employee.

(7)   The employer shall—
    (a)   on reasonable notice being given, allow an employee access to his personal monitoring record;
    (b)   provide the Executive with copies of such monitoring records as the Executive may require; and
    (c)   if he ceases to trade, notify the Executive forthwith in writing and make available to the Executive all monitoring records kept by him.

Schedule 5 to the *COSHH Regulations 2002 (SI 2002 No 2677)* automatically requires monitoring to be carried out in processes involving vinyl chloride monomer and electrolytic chromium plating.

A number of techniques are available for monitoring air quality in the workplace. Most of these involve the use of chemical indicator tubes, direct reading instruments, or sampling pumps and filter heads. An HSE booklet *Monitoring strategies for toxic substances* provides further guidance on the subject.

Health surveillance

[H2133] The *COSHH Regulations 2002 (SI 2002 No 2677), Reg 11(1)* and *(2)* contain the main requirements relating to the need for health surveillance.

(1)   Where it is appropriate for the protection of the health of his employees who are, or are liable to be, exposed to a substance hazardous to health, the employer shall ensure that such employees are under suitable health surveillance.

(2)   Health surveillance shall be treated as being appropriate where—
    (a)   the employee is exposed to one of the substances specified in Column 1 of Schedule 6 and is engaged in a process specified in Column 2 of that Schedule, and there is a reasonable likelihood that an identifiable disease or adverse health effect will result from that exposure; or

(b)   the exposure of the employee to a substance hazardous to health is such that—

    (i)   an identifiable disease or adverse health effect may be related to the exposure,

    (ii)   there is a reasonable likelihood that the disease or effect may occur under the particular conditions of his work, and

    (iii)   there are valid techniques for detecting indications of the disease or the effect,

and the technique of investigation is of low risk to the employee.

The remaining parts of *Regulation 11 (paragraphs (3)–(11))* relate to the manner in which health surveillance is conducted and used, together with the maintenance of and access to surveillance records.

The decision as to whether health surveillance is appropriate to protect the health of employees is one that would normally be taken at the time of a COSHH assessment or during its subsequent review. However, for those processes and substances specified in *Schedule 6* to the Regulations, surveillance must be carried out.

Normally health surveillance programmes would be initiated and carried out under the overall supervision of a registered medical practitioner, and preferably one with relevant occupational health experience. However, the surveillance itself may be carried out by an occupational health nurse, a technician or a responsible member of staff, providing that individual was competent for the purpose.

There are many different procedures available for health surveillance including:

- Biological monitoring, e.g. tests of blood, urine or exhaled air.
- Biological effect monitoring, e.g. through lung function testing.
- Medical surveillance, e.g. through physical examinations.
- Interviews about possible symptoms, e.g. skin abnormalities or shortness of breath.

For health surveillance to be 'appropriate' there must firstly be a significant enough risk to justify it and there must also be a valid technique for detecting indications of related occupational diseases or ill-health effects. HSE provide specialist guidance in their booklets *Health surveillance at work* and *Biological monitoring in the workplace*. Records of health surveillance must contain information specified in the COSHH ACoP and must be retained for at least 40 years.

Information, instruction and training

[H2134] The *COSHH Regulations 2002 (SI 2002 No 2677), Reg 12* contain requirements relating to information, instruction and training for persons who may be exposed to substances hazardous to health.

(1)   Every employer who undertakes work which is liable to expose an employee to a substances hazardous to health shall provide that employee with suitable and sufficient information, instruction and training.

(2)   Without prejudice to the generality of paragraph (1), the information, instruction and training provided under that paragraph shall include—

    (a)   details of the substances hazardous to health to which the employee is liable to be exposed including—
        (i)   the names of those substances and the risk which they present to health,
        (ii)   any relevant workplace exposure limit or similar occupational exposure limit,
        (iii)   access to any relevant safety data sheet, and
        (iv)   other legislative provisions which concern the hazardous properties of those substances;
    (b)   the significant findings of the risk assessment;
    (c)   the appropriate precautions and actions to be taken by the employee in order to safeguard himself and other employees at the workplace;
    (d)   the results of any monitoring of exposure in accordance with regulation 10 and, in particular, in the case of any substance hazardous to health for which a workplace exposure limit has been approved, the employee or his representatives shall be informed forthwith, if the results of such monitoring show that the workplace exposure limit is exceeded;
    (e)   the collective results of any health surveillance undertaken in accordance with regulation 11 in a form calculated to prevent those results from being identified as relating to a particular person; and
    (f)   where employees are working with a Group 4 biological agent or material that may contain such an agent, the provision of written instructions and, if appropriate, the display of notices which outline the procedures for handling such an agent or material.

(3)   The information, instruction and training required by paragraph (1) shall be—
    (a)   adapted to take account of significant changes in the type of work carried out or methods of work used by the employer; and
    (b)   provided in a manner appropriate to the level, type and duration of exposure identified by the risk assessment.

(4)   Every employer shall ensure that any person (whether or not his employee) who carries out work in connection with the employer's duties under these Regulations has suitable and sufficient information, instruction and training.

(5)   Where containers and pipes for substances hazardous to health used at work are not marked in accordance with any relevant legislation listed in Schedule 7, the employer shall, without prejudice to any derogations provided for in that legislation, ensure that the contents of those containers and pipes, together with the nature of those contents and any associated hazards, are clearly identifiable.

The general requirements of *paragraph (1)* match those found in various other regulations but the contents of *paragraph (2)* are much more prescriptive than the requirements of previous versions of the *COSHH Regulations*. Whilst it is important that workers are aware of the substances they are exposed to and the risks that they present, only a minority are likely to comprehend fully the significance of WELs, or understand all of the contents of a typical safety data sheet. However, a good awareness of the risks will mean employees are more likely to take the appropriate precautions. Instruction and training will be particularly relevant in relation to:

- awareness of safe systems of work;
- correct use of LEV equipment;
- use, adjustment and maintenance of PPE (especially RPE);

- the importance of good personal hygiene standards;
- requirements relating to eating, drinking and smoking;
- emergency procedures;
- arrangements for cleaning and the disposal of waste;
- contents of containers and pipes (see *Regulation 12(5)* above).

Employers must also inform employees or their representatives of the results of workplace exposure monitoring forthwith if the WEL is shown to have been exceeded. Employees must also be informed of the *collective* results of health surveillance, e.g. the average results within a department or on a particular shift from biological monitoring, or the numbers of employees referred for further investigation following a skin inspection.

The type of information, instruction and training provided must be appropriate for the level, type and exposure involved and adapted to take account of significant changes.

Any person carrying out work on the employer's behalf (whether or not an employee) is required by *paragraph (4)* to have the necessary information, instruction and training. Thus a consultant carrying out a COSHH assessment or an occupational hygienist conducting exposure monitoring must be verified by the employer as being competent for the purpose and be provided with the information necessary to carry out their work effectively.

Accidents, incidents and emergencies

[H2135] *Regulation 13* was introduced in the *COSHH Regulations 2002 (SI 2002 No 2677)* and requires employers to ensure that arrangements are in place to deal with accidents, incidents and emergencies related to hazardous substances. (This is in addition to the general duty to have procedures 'in the event of serious and imminent danger' contained in *Regulation 8* of the *Management of Health and Safety at Work Regulations 1999 (SI 1999 No 3242.)* Based on HSE guidance in the COSHH ACoP booklet (L5) such emergencies might include:

- a serious process fire posing a serious risk to health;
- a serious spillage or flood of a corrosive agent;
- a failure to contain biological, carcinogenic or mutagenic agents;
- a failure that could lead (or has led) to a sudden release of chemicals;
- a threatened significant exposure over an OEL, e.g. due to a failure of LEV or other controls.

The regulation is rather lengthy and prescriptive in its requirements which are summarised below.

Arrangements should include:

- emergency procedures, such as:
    — appropriate first aid facilities;
    — relevant safety drills (tested at regular intervals);
- providing information on emergency arrangements:
    — including details of work hazards and emergencies likely to arise;

— made available to relevant accident and emergency services;
— displayed at the workplace, if appropriate;
• suitable warning and other communications:
— to enable an appropriate response, including remedial actions and rescue operations.

In the event of an accident, incident or emergency, the employer must:

• take immediate steps to:
— mitigate its effects;
— restore the situation to normal;
— inform employees who may be affected;
• ensure only essential persons are permitted in the affected area and that they are provided with:
— appropriate PPE;
— any necessary safety equipment and plant;
• in the case of a serious biological incident, inform employees or their representatives, as soon as practicable of:
— the causes of the incident or accident;
— the measures taken or being taken to rectify the situation.

Regulation 13(4) states that such emergency arrangements are not required where the risk assessment shows that because of the quantities of hazardous substances the risks are slight and control measures are sufficient to control that risk. (However, this 'exception' does not apply in the case of carcinogens or biological agents.)

Regulation 13(5) requires employees to report possible releases of a biological agent which could cause severe human disease, forthwith to their employer (or another employee with specific responsibility).

HSE guidance in the ACoP booklet states that whether or not arrangements are required under *Regulation 13* is a matter for judgement, based on the potential size and severity of accidents and emergencies which may occur. Many incidents will be capable of being dealt with by the control measures required by the *COSHH Regulations 2002 (SI 2002 No 2677), Reg 7*. Even if an emergency does occur, the response should be proportionate – a small leak would not necessarily justify an evacuation of the workplace. The ACoP booklet provides further details of what emergency procedures might need to include, such as:

• the identity, location and quantities of hazardous substances present;
• foreseeable types of accidents, incidents or emergencies;
• special arrangements for emergencies not covered by general procedures;
• emergency equipment and PPE, and who is authorised to use it;
• first-aid facilities;
• emergency management responsibilities e.g. emergency controllers;
• how employees should respond to incidents;
• clear up and disposal arrangements;
• arrangements for regular drills or practice;
• dealing with special needs of disabled employees.

Other important regulations

[H2136] The introduction to this chapter referred to several other legal requirements relating to hazardous substances in the workplace and these are summarised below.

Chemicals (Hazard Information and Packaging for Supply) Regulations 2009

[H2137] The *Chemicals (Hazard Information and Packaging for Supply) Regulations 2009 (SI 2009 No 716)* (commonly known as 'CHIP') impose duties on suppliers (and importers into the EU) of substances classified as dangerous for supply.

The CHIP regulations are being gradually be replaced by the European Classification, Labelling and Packaging (CLP) Regulation. There is a transitional period before requirements of the Regulation for classification, labelling and packaging become mandatory until 1 December 2010 for substances and 1 June 2015 for mixtures.

Safety data sheets are no longer covered by the CHIP regulations and the requirement to provide them has been transferred from regulation 5 of the CHIP Regulations to the European REACH Regulation – REACH stands for Registration, Evaluation, Authorisation and restriction of CHemicals.

The HSE leaflet *REACH and Safety Data Sheets* explains to manufacturers, importers, downstream users and distributors supplying substances or mixtures when a safety data sheet has to be provided (http://www.hse.gov.uk/reach/resources/reachsds.pdf).

This is where substances or mixtures have been classified as dangerous, persistent, bioaccumulative and toxic (or very persistent and very bioaccumulative) or are included in the European Chemicals Agency's Candidate List of substances of very high concern. The HSE advises that safety data sheets also have to be provided by suppliers where the customer requests one for certain other mixtures.

The HSE sets out that the safety data sheet must be dated and contain information under the following headings:

(1)   Identification of the substance/mixture and of the company/undertaking;
(2)   Identification of the hazards;
(3)   Composition or information about the ingredients;
(4)   First-aid measures;
(5)   Fire-fighting measures;
(6)   Accidental release measures;
(7)   Handling and storage;
(8)   Exposure controls or personal protection;
(9)   Physical and chemical properties;
(10) Stability and reactivity;
(11) Toxicological information;

(12)   Ecological information;
(13)   Disposal considerations;
(14)   Transport information;
(15)   Regulatory information; and
(16)   Other information.

Control of Asbestos at Work Regulations 2006 (SI 2006 No 2739)

[H2138] Earlier regulations on asbestos predated the original *COSHH Regulations* and thus provided a framework on which the *COSHH Regulations* were based. They continue to follow a similar pattern, involving an assessment of work which exposes employees to asbestos and the preparation of a suitable written plan of work which must prevent or reduce exposure to asbestos. Many other requirements mirror COSHH requirements, e.g. information, instruction and training; use of control measures; maintenance of control measures; requirements for air monitoring (plus standards for air testing and analysis); and health records and medical surveillance.

There are also a number of other specific requirements including those relating to the notification of work with asbestos (to HSE), to the storage, distribution and labelling of raw asbestos and asbestos waste and to the supply of products containing asbestos (which in both cases must carry specified labels).

The Control of Asbestos Regulations 2006 simplified the regulatory regime for asbestos, reduced exposure limits and introduced mandatory training for work with asbestos. The Regulations include:

• a duty to manage asbestos;
• a single control limit of 0.1 fibres per cm^3 of air for work with all types of asbestos;
• specific mandatory training requirements for anyone liable to be exposed to asbestos;
• a requirement to analyse the concentration of asbestos in the air with measurements in accordance with the 1997 World Health Organisation (WHO) recommended method; and
• practical guidelines for the determination of 'sporadic and low intensity exposure' as required by the EU Directive.

Most work with asbestos needs to be undertaken by a licensed contractor, but any decision on whether particular work is licensable is determined by the risk. Most work with textured decorative coatings containing asbestos (TCs) is removed from the licensing regime. However, all work with asbestos-containing materials, including TCs, must be undertaken by trained workers following a risk assessment and in accordance with appropriate controls to prevent exposure to asbestos fibres.

No exposure to asbestos can be considered to be sporadic and of low intensity if the concentration of asbestos in the atmosphere is liable to exceed 0.6 fibers per cm^3 of air measured over 10 minutes in any working day.

More details of what work is licensable, what training is necessary and how to undertake work with asbestos-containing materials can be found in the Approved Code of Practice (ACoP), *Work with materials containing asbestos.*

Further guidance on the duty to manage asbestos in premises can be found in the ACoP, *The management of asbestos in non-domestic premises.*

Further details on the subject, including the important 'duty to manage asbestos in non-domestic premises' are contained in the chapter dealing with ASBESTOS.

Control of Major Accident Hazard Regulations 1999 (COMAH)

[H2139] The *Control of Major Accident Hazard Regulations 1999 (SI 1999 No 743)* as amended in 2005 only apply to sites containing specified quantities of dangerous substances. They require the preparation of both on-site and off-site emergency plans with the objectives of:

- containing and controlling incidents so as to minimise their effects, and to limit damage to persons, the environment and property;
- implementing the measures necessary to protect persons and the environment from the effects of major accidents;
- communicating the necessary information to the public and to the emergency services and authorities concerned in the area;
- providing for the restoration and clean-up of the environment following a major accident.

Such plans must be prepared utilising risk assessment techniques. HSE have a number of booklets providing guidance on the methodology to be followed and the parameters to be taken into account. Further details are provided in the chapter CONTROL OF MAJOR ACCIDENT HAZARDS.

Control of Lead at Work Regulations 2002

[H2140] Like asbestos, previous regulations on lead preceded the original *COSHH Regulations* and the *Control of Lead at Work Regulations 2002 (SI 2002 No 2676)* continue to follow similar principles. The Regulations together with an ACoP and related guidance are contained in an HSE booklet *Control of Lead at Work Regulations 2002. Approved code of practice and guidance* which is supported by other HSE guidance material.

Dangerous Substances and Explosive Atmospheres Regulations 2002 (DSEAR)

[H2141] The *DSEAR Regulations (SI 2002 No 2776)* are concerned with protecting against risks from fire and explosion. They replaced several other regulations including the *Highly Flammable Liquids and Liquified Petroleum Gases Regulations 1972 (SI 1972 No 917)* and (apart from the dispensing of petrol from pumps) the requirement for licensing under the *Petroleum (Consolidation) Act 1928.*

Dangerous substances

[H2142] The *DSEAR Regulations (SI 2002 No 2776)* contain a detailed definition of 'dangerous substances' including:

- explosive, oxidising, extremely flammable, highly flammable or flammable substances or preparations (whether or not classified under *CHIP Regulations*), e.g. petrol, solvents, paints, LPG;
- other substances or preparations creating risks due to their physico-chemical or chemical properties;
- potentially explosive dusts e.g. flour, sugar, custard, pitch, wood.

Explosive atmosphere

[H2143] This is defined as 'a mixture, under atmospheric conditions of air and one or more dangerous substances in the form of gases, vapours, mists or dusts, in which, after ignition has occurred, combustion spreads to the entire unburned mixture'.

The *DSEAR Regulations (SI 2002 No 2776)* require employers to:

- Carry out a risk assessment of work activities involving dangerous substances [*SI 2002 No 2776, Reg 5*].
- Eliminate or reduce risks as far as is reasonably practicable [*SI 2002 No 2776, Reg 6*].
- Classify places where explosive atmospheres may occur in zones, and ensure that equipment and protective systems in these places meet appropriate standards, marking zones with signs, where necessary [*SI 2002 No 2776, Reg 7*] (this duty is being phased in).
- Provide equipment and procedures to deal with accidents and emergencies [*SI 2002 No 2776, Reg 8*].
- Provide employees with suitable and sufficient information, instruction and training [*SI 2002 No 2776, Reg 9*].

Regulation 6 requires control measures to be implemented according to a specified order of priority which can be summarised as:

- Avoid the presence or use of a dangerous substance at the workplace by replacing it with a substance or process which either eliminates or reduces the risk.
- Reduce the quantity of dangerous substances to a minimum.
- Avoid or minimise releases of dangerous substances.
- Control releases at source.
- Prevent the formation of an explosive atmosphere (including providing appropriate ventilation).
- Collect, contain and remove any releases to a safe place or otherwise render them safe.
- Avoid the presence of ignition sources (including electrostatic discharges).
- Avoid adverse conditions that could lead to danger (e.g. by temperature or other controls).
- Segregation of incompatible dangerous substances.

The regulation also requires measures to mitigate the detrimental effects of a fire or explosion (or other harmful physical effects). These include minimising the number of employees exposed and the provision of explosion pressure relief arrangements or explosion suppression equipment. Further information on DSEAR is provided in the chapter DANGEROUS SUBSTANCES AND EXPLOSIVE ATMOSPHERES and detailed guidance on the subject is available from the HSE.

Dangerous Substances (Notification and Marking of Sites) Regulations 1990

[H2144] The *Dangerous Substances (Notification and Marking of Sites) Regulations 1990 (SI 1990 No 304)* are primarily for the benefit of the fire service and other emergency services. Both the fire authority and HSE must be provided with notification of dates when it is anticipated that a total quantity of 25 tonnes of 'dangerous substances' will be present on the site. (The definition of 'dangerous substances' is contained in the Regulations.) Further notifications must be made where significant changes occur in substances present, including cessation or reductions. Sites must also be marked with standard signs as specified in *Schedule 3* to the Regulations. A booklet providing detailed guidance on the regulations is available from HSE.

Ionising Radiation Regulations 1999

[H2145] *Regulation 7* of the *Ionising Radiation Regulations 1999 (SI 1999 No 3232)* require employers to carry out a risk assessment before commencing any new activity involving work with ionising radiation. The assessment must be:

. . . sufficient to demonstrate that—

(a)   all hazards with the potential to cause a radiation accident have been identified; and
(b)   the nature and magnitude of the risks to employers and other persons arising from those hazards have been evaluated.

Where such radiation risks are identified, all reasonably practicable steps must be taken to:

*   prevent any such accident;
*   limit its consequences should such an accident occur;
*   provide employees with necessary information, instruction, training and equipment necessary to restrict their exposure.

The Regulations contain many detailed requirements on the prevention of radiation accidents and related control measures. An HSE booklet contains the Regulations and an ACoP and guidance on the regulations. Much more guidance is available from HSE on specific aspects of radiation safety. Further details are provided in the chapter dealing with RADIATION.

Joint Consultation in Safety – Safety Representatives, Safety Committees, Collective Agreements and Works Councils

Nick Humphreys

Introduction

[J3001] The provision of 'information, consultation and participation' and of 'health protection and safety at the workplace' are outlined as two of the 'fundamental social rights of workers' in accordance with the Community Charter of Fundamental Rights of 1989. Combined with Article 138 of the Consolidated Version of the Treaty Establishing the European Community (i.e. Treaty of Rome 1957), the Health and Safety Directive 1989 (resulting in 'the six pack' collection of regulations) and the United Kingdom's signing of the Social Chapter in 1998, the EU has provided great stimuli for the introduction of extensive obligations to consult on health and safety issues.

At national level, there has already been change to the obligations owed by employers to their workers. The *Working Time Regulations 1998 (SI 1998 No 1833) ('Working Time Regulations')* were introduced in October 1998 and have subsequently been amended.

Furthermore, the obligations owed under the European Works Councils Directive 1994 have been transposed into national law under the *Transnational Information and Consultation of Employees Regulations 1999 (SI 1999 No 3323) ('Works Councils Regulations')*.

Most recently, there has been the Information and Consultation Directive (Directive 2002/19/EC) (transposed into national law under the *Information and Consultation of Employees Regulations 2004 (SI 2004 No 3426)*), which is being introduced on a phased basis into all undertakings with 50 or more employees (currently it applies to those undertakings employing 150 or more employees). Consultation with the workforce is likely to become a central issue for industrial relations and health and safety over the next few years.

The Health and Safety Commission ('HSC') seeks to promote change in health and safety at work. In this regard, it has published a number of consultative documents. The period of consultation for the most recent of such consultative documents, *'Improving worker involvement – Improving health and safety (2006)'*, expires on 8 September 2006.

Under *'Improving worker involvement – Improving health and safety'*, the HSC states that there are three pillars to a strategy for improving worker involvement in health and safety, these being:

- legislation, which acts to set standards on which to build health and safety consultation;
- guidance on good practice, which helps employers and workers to reach at least minimum standards of workplace health and safety consultation; and
- encouragement to strive continually towards best practice in workplace health and safety consultation.

The HSC sets out in *'Improving worker involvement – Improving health and safety'* the current position in relation to each of the three pillars and suggests options for strengthening each. The current options are to provide and promote better good practice guidance, the development of a framework of voluntary standards of best practice and to amend the existing legislation on worker health and safety consultation.

At a further non-statutory level, so as to increase worker involvement and consultation, the Health and Safety Executive ('HSE') runs a Workers' Safety Adviser Challenge Fund, which was established in 2004 and is funded by the Department for Work and Pensions which is providing £3 million over a three-year period. The focus of the Challenge Fund is small and medium-sized enterprises lacking arrangements for involving their workforce in the management of their own and others health and safety. The purpose of the Challenge Fund is to encourage projects based on organisations working in partnerships to generate and improve collaboration between employers and workers in small firms in occupational health and safety. The Challenge Fund was set up due to the estimated loss in Great Britain each year of over 39 million workdays through work-related accidents and ill health.

Successful projects are awarded substantial sums (in 2006, the third round winners of the Challenge Fund received awards of between £52,777 and £100,000).

Another non-statutory measure is the HSE research report on the HSC's Worker Safety Advisor ('WSA') Pilot.

The pilot, which was up and running by August 2001, introduced specially trained WSAs into small businesses that did not have safety representatives in the retail, hospitality, voluntary, construction and automotive fabrication sectors. Their function was to provide support, advice and training to get the workforce more involved in health and safety, and create channels for improving dialogue between employers and employees.

The purpose of the pilot was to test the effectiveness of WSAs in:

- promoting greater consultation on health and safety;
- raising health and safety standards; and
- broadening and increasing employers' and workers' knowledge of health and safety.

Among the findings published in *The Worker Safety Advisors (WSA) Pilot Report* are that:

- over 75 per cent of employers have made changes to their approach to health and safety as a result of the pilot; these included revising or introducing new policies and procedures, regular health and safety discussions with staff, and joint working on risk assessments and staff training; and
- involvement of WSAs led to improvements in the approach of small non-unionised workplaces to health and safety.

The report concludes that the development of the scheme should continue with the core WSA model, this being a voluntary approach.

The HSE is presently working on developing the pilot and is commissioning an action plan setting down the steps required to identify the right organisation(s) to run it and recruit the WSAs and employers, the steps required to ensure that the WSA approach would be sustainable, and options for taking it forward in various sectors and geographical regions.

The Worker Safety Advisors (WSA) Pilot Report (RR144), is available from HSE Books, PO Box 1999, Sudbury, Suffolk CO10 6FS (tel: 01787 881165; fax: 01787 313995), price £20.00

The main thrust of this chapter though is the existing domestic legislation that places an employer under an obligation to consult with its workforce and this will be examined in the following sections.

Regulatory framework

[J3002] The legal requirements for consultation are more convoluted than might be expected, principally because there are two groups of workers who are treated as distinct for consultation purposes. Until 1996 only those employers who recognised a trade union for collective bargaining purpose were obliged to consult the workforce through safety representatives. The *Safety Representatives and Safety Committees Regulations 1977 (SI 1977 No 500) (as amended) ('Safety Representatives Regulations')*, which were made under the provisions of the *Health and Safety at Work etc. Act 1974 ('HSWA 1974')*, *s 2(6)*, came into effect in 1978 and introduced the right for recognised trade unions to appoint safety representatives. The *Safety Representatives Regulations* were amended in 1993 – by the *Management of Health and Safety at Work Regulations 1992 (SI 1992 No 2051)* (now superceded by the *Management of Health and Safety at Work Regulations 1999 (SI 1999 No 3242))* – to extend employers' duties to consult and provide facilities for safety representatives.

The duty to consult was extended on 1 October 1996 by virtue of the *Health and Safety (Consultation with Employees) Regulations 1996 (SI 1996 No 1513) ('Consultation with Employees Regulations')*, which have subsequently been amended. The *Consultation with Employees Regulations* were intro-duced as a 'top up' to the *Safety Representatives Regulations*, extending the obligation upon employers to consult all of their employees about health and safety measures, including those who are not represented by a trade union. The *Consultation with Employees Regulations* expanded the obligation beyond just trade union appointed representatives. When introduced, the *Consultation with Employees Regulations* addressed the general reduction of union recognition over the preceding five years. However, the more recent trend in workforces, driven by the twin forces of Europe and the current Government, is for greater worker participation in employers' undertakings and the current framework facilitates this whether or not a union is the conduit for worker consultation.

The obligation of consultation, its enforcement and the details of the role and functions of safety representatives and representatives of employee safety, whether under the *Safety Representatives Regulations* or the *Consultation with Employees Regulations*, are almost identical but, for the sake of clarity, are dealt with separately below. The primary distinction is that different obligations apply depending on whether the affected workers are unionised or not. The respective Regulations also cover persons working in host employers' undertakings.

There are specific supplemental provisions which apply to offshore installations, which are governed by the *Offshore Installations (Safety Representatives and Safety Committees) Regulations 1989 (SI 1989 No 971)*. Likewise, in the case of the education sector, the HSC has published the 'Safety Representatives' Charter' ('the Charter'). The Charter has been developed to:

* Promote and emphasise the rights, roles and functions of safety representatives and to actively promote the involvement of safety representatives in the education sector's efforts to improve health and safety.
* Motivate employers, safety representatives and employees to work in partnership to develop a positive safety culture throughout the education sector.
* Raise awareness amongst employers of the important contribution of safety representatives towards the development of such a culture.
* Encourage employers to demonstrate their full commitment towards consulting and involving safety representatives in matters of health, safety and welfare.
* Increase the participation of education sector employees and their safety representatives in health and safety activities.
* Contribute towards an improved health and safety performance which aims to reduce accidents and ill health in the education sector.

In particular, the Charter sets out the legal obligations that employers in the education sector owe under the *Safety Representatives Regulations* and the *Consultation with Employees Regulations*.

While it is not legislation itself, the Charter is evidence of best practice within the education sector.

Although the *Consultation with Employees Regulations* and the *Safety Representatives Regulations* are based upon good industrial relations practice, there is still the possibility of a prosecution by HSE inspectors of employers who fail to consult their workforce on health and safety issues. There is substantial overlap, however, with employment protection legislation and the obligations in respect of consultation – as can be seen from the protective rights which are conferred upon safety representatives, such as the right not to be victimised or subjected to detriment for health and safety activities, together with the consequential right to present a complaint to an employment tribunal if they are dismissed or suffer a detriment as a result of carrying out their duties.

The *Consultation with Employees Regulations* also extend these rights to the armed forces. However, armed forces representatives are to be appointed rather than elected, and no paid time off is available. The *Consultation with Employees Regulations* do not apply to sea-going ships.

The Working Time Directive 1994 has resulted in the *Working Time Regulations 1998 (SI 1998 No 1833)* (subsequently amended) under national law. The *Working Time Regulations* allow collective modification of the night working requirements in *Reg 6*, the daily rest provisions in *Reg 10*, the weekly rest provisions in *Reg 11*, the rest break provisions in *Reg 12* and also allow modification of the averaging period for calculating weekly working time under *Reg 4* where the same is either for objective, technical or organisational reasons. The method of modification is either by collective agreement in the case of a unionised workforce or by workforce agreement (as defined in the *Working Time Regulations 1998 (SI 1998 No 1833), Sch 1*) in the case of non-unionised employees.

In relation to pan-European employers, obligations are owed to workers under the European Works Councils Directive 1994. The United Kingdom adopted the Directive on 15 December 1997 and provisions were enacted into national law by the *Works Councils Regulations (SI 1999 No 3323)* and came into force on 15 January 2000. While the *Works Councils Regulations* do not specifically encompass health and safety obligations, these issues are within the remit of a European Works Council.

National level consultation requirements have also been introduced under the *Information and Consultation of Employees Regulations 2004 (SI 2004 No 3426)*, derived from the Information and Consultation Directive (Directive 2002/19/EC). As with European Works Councils, health and safety is not specifically included in the scope of the information and consultation requirements under those Regulations, but may be included under the general remit of an employee information and consultation procedure.

Consultation obligations for unionised employers

[J3003] Under the *Safety Representatives Regulations (SI 1977 No 500), Regs 2* and *3*, an independent trade union recognised by an employer has the right to appoint an individual to represent the workforce in consultations with the employer on all matters concerning health and safety at work, and to carry out periodic inspections of the workplace for hazards. Every employer has a duty to consult such union-appointed safety representatives on health and safety arrangements (and, if they so request him, to establish a safety committee to review the arrangements – see J3017 below) [*HSWA 1974, s 21*].

General duty

[J3004] The general duty of the employer under the *Safety Representatives Regulations (SI 1977 No 500)* is to 'consult with safety representatives with regard to both the making and maintaining of arrangements that will enable the employer and its workforce to co-operate in promoting and developing health and safety at work, and monitoring its effectiveness'.

Employers are also required under *Safety Representatives Regulations (SI 1977 No 500), Reg 4A(2)* to provide such facilities and assistance as safety representatives may require for the purposes of carrying out their functions under the *Heath and Safety at Work etc. Act 1974, s 2(4)*.

General guidance has been issued by HSC in the form of the Codes of Practice *'Safety Representatives and Safety Committees'* and *'Time Off for the Training of Safety Representatives'* to which regard should be had generally.

Appointment of safety representatives

[J3005] The right of appointment of safety representatives was, until 1 October 1996, restricted to independent trade unions who are recognised by employers for collective bargaining purposes. The *Consultation with Employees Regulations (SI 1996 No 1513)* extended this to duly elected representatives of employee safety, as detailed above. In a case where there are no Safety Representatives under the *Safety Representatives Regulations (SI 1977 No 500)* and where the employees do not elect representatives under the *Consultation with Employees Regulations (SI 1996 No 1513)*, the obligation will fall on employers to consult with the entire workforce as to the matters within the scope of the *Consultation with Employees Regulations (SI 1996 No 1513)* (per *Reg 3*).

The terms 'independent' and 'recognised' are defined in the *Safety Representatives Regulations (SI 1977 No 500)* and follow the definitions laid down in the *Trade Union and Labour Relations (Consolidation) Act 1992, ss 5* and *178(3)* respectively. The *Safety Representatives Regulations* make no provision for dealing with disputes which may arise over questions of independence or recognition (this is dealt with in the *Trade Union and Labour Relations (Consolidation) Act 1992, ss 6* and *8* – the amendments that have been made to the *Trade Union and Labour Relations (Consolidation) Act 1992* by the *Employment Relations Act 1999* in relation to recognition of unions do not help as *Sch A1* is confined to recognition disputes concerning pay, hours and holiday).

Safety representatives must be representatives of recognised independent trade unions, and it is up to each union to decide on its arrangements for the appointment or election of its safety representatives [*Safety Representatives Regulations (SI 1977 No 500), Reg 3*]. Employers are not involved in this matter, except that they must be informed in writing of the names of the safety representatives appointed and of the group(s) of employees they represent [*Safety Representatives Regulations (SI 1977 No 500), Reg 3(2)*].

The *Safety Representatives Regulations (SI 1977 No 500), Reg 8*, state that safety representatives must be employees except in the cases of members of the

Musicians' Union and actors' Equity. In addition, where reasonably practicable, safety representatives should have at least two years' employment with their present employer or two years' experience in similar employment. The HSC guidance notes advise that it is not reasonably practicable for safety representatives to have two years' experience, or employment elsewhere, where:

- the employer is newly established;
- the workplace is newly established;
- the work is of short duration; or
- there is high labour turnover.

The same general guidance is followed for employee safety representatives under the *Consultation with Employees Regulations (SI 1996 No 1513)*.

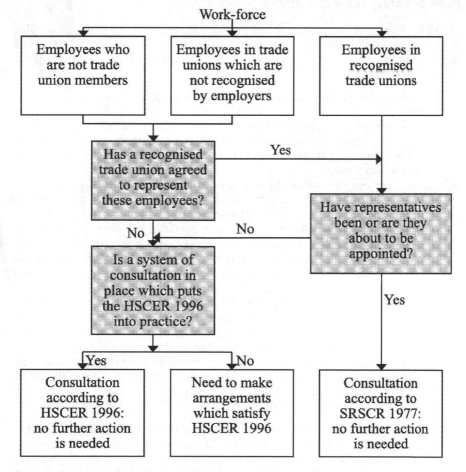

Number of representatives for workforce

[J3006] The *Safety Representatives Regulations 1977 (SI 1977 No 500)* do not lay down the number of safety representatives that unions are permitted to appoint for each workplace. This is a matter for unions themselves to decide, having regard to the number of workers involved and the hazards to which they are exposed. The HSE's view is that each safety representative should be regarded as responsible for the interests of a defined group of workers. This approach has not been found to conflict with existing workplace trade union organisation based on defined groups of workers. The size of these groups varies from union to union and from workplace to workplace. While normally each workplace area or constituency would need only one safety representative, additional safety representatives are sometimes required where workers are exposed to numerous or particularly severe hazards; where workers are distributed over a wide geographical area or over a variety of workplace locations; and where workers are employed on shiftwork.

Role of safety representatives

[J3007] The *Safety Representatives Regulations 1977 (SI 1977 No 500), Reg 4(1)* (as amended by the *Management of Health and Safety at Work Regulations 1999 (SI 1999 No 3242)*) lists a number of detailed functions for safety representatives:

(a)   to investigate potential hazards and causes of accidents at the workplace;

(b)   to investigate employee complaints concerning health etc. at work;

(c)   to make representations to the employer on matters arising out of (*a*) and (*b*) and on general matters affecting the health etc. of the employees at the workplace;

(d)   to carry out the following inspections (and see **J3008** below):

 • of the workplace (after giving reasonable written notice to the employer – see *Safety Representatives Regulations, Reg 5*),

 • of the relevant area after a reportable accident or dangerous occurrence (see accident reporting) or if a reportable disease is contracted, if it is safe to do so and in the interests of the employees represented (see *Safety Representatives Regulations 1977 (SI 1977 No 500), Reg 6*),

 • of documents relevant to the workplace or the employees represented which the employer is required to keep (see *Safety Representatives Regulations 1977 (SI 1977 No 500), Reg 7*) – reasonable notice must be given to the employer; and

(e)   to represent the employees they were appointed to represent in consultations with HSE inspectors, and to receive information from them (see **J3016** below); and

(f)   to attend meetings of safety committees.

These functions are interrelated and are to be implemented proactively rather than reactively. Safety representatives should not just represent their members' interests when accidents or near-misses occur or at the time of periodic inspections, but should carry out their obligations on a continuing day-to-day basis. The *Safety Representatives Regulations* make this obligation clear by stating that safety representatives have the functions of investigating potential hazards and members' complaints *before* accidents, as well as investigating dangerous occurrences and the causes of accidents *after* they have occurred.

These functions are only assumed when the employer has been notified in writing by the trade union or workforce of the identity of the representative.

Thus safety representatives may possibly be closely involved not only in the technical aspects of health, safety and welfare matters at work, but also in those areas which could be described as quasi-legal. In other words, they may become involved in the interpretation and clarification of terminology in the *Safety Representatives Regulations*, as well as in discussion and negotiation with employers as to how and when the regulations may be applied. This would often happen in committee meetings.

Under the *Safety Representatives Regulations 1977 (SI 1977 No 500), Reg 4A(1)* (introduced by the *Management of Health and Safety at Work Regulations 1992*), the subjects on which consultation 'in good time' between employers and safety representatives should take place are:

- the introduction of any new measure at a workplace which may substantially affect health and safety;
- arrangements for appointing competent persons to assist the employer with health and safety and implementing procedures for serious and imminent risk;
- any health and safety information the employer is required to provide; and
- the planning and organisation of health and safety training and health and safety implications of the introduction (or planning) of new technology.

The safety representative's terms of reference are, therefore, broad, and exceed the traditional 'accident prevention' area. For example, the *Safety Representatives Regulations 1977 (SI 1977 No 500), Reg 4(1)* empowers safety representatives to investigate 'potential hazards' and to take up issues which affect standards of health, safety and welfare at work. In practice it is becoming clear that four broad areas are now engaging the attention of safety representatives and safety committees:

- health;
- safety;
- environment; and
- welfare.

These four broad areas effectively mean that safety representatives can, and indeed often do, examine standards relating, for example, to noise, dust, heating, lighting, cleanliness, lifting and carrying, machine guarding, toxic substances, radiation, cloakrooms, toilets and canteens. The protective standards that are operating in the workplace, or the lack of them, are now coming under much closer scrutiny than hitherto.

Workplace inspections

[J3008] Under the *Safety Representatives Regulations 1977 (SI 1977 No 500), Regs. 5* and *6*, safety representatives are given a power to undertake health and safety inspections. Arrangements for three-monthly and other more frequent inspections and reinspections should be by joint arrangement.

The TUC advises that the issues to be discussed with the employer can include:

- more frequent inspections of high risk or rapidly changing areas of work activity;
- the precise timing and notice to be given for formal inspections by safety representatives;
- the number of safety representatives taking part in any one formal inspection;
- the breaking-up of plant-wide formal inspections into smaller, more manageable inspections;
- provision for different groups of safety representatives to carry out inspections of different parts of the workplace;
- the kind of inspections to be carried out, e.g. safety tours, safety sampling or safety surveys; or
- the calling in of independent technical advisers by the safety representatives.

Although formal inspections are not intended to be a substitute for day-to-day observation, they have on a number of occasions provided an opportunity to carry out a full-scale examination of all or part of the workplace and for discussion with employers' representatives about remedial action. They can also provide an opportunity to inspect documents required under health and safety legislation, e.g. certificates concerning the testing of equipment. It should be emphasised that, during inspections following reportable accidents or dangerous occurrences, employers are not required to be present when the safety representative talks with its members. In workplaces where more than one union is recognised, agreements with employers about inspections should involve all the unions concerned. It is generally agreed that safety representatives are also allowed under the *Safety Representatives Regulations 1977 (SI 1977 No 500)* to investigate the following:

- potential hazards;
- dangerous occurrences;
- the causes of accidents; and
- complaints from their members.

This means that imminent risks, or hazards which may affect their members, can be investigated right away by safety representatives without waiting for formal joint inspections. Following an investigation of a serious mishap, safety representatives are advised to complete a hazard report form, one copy being sent to the employer and one copy retained by the safety representative.

Rights and duties of safety representatives

Legal immunity

[J3009] Ever since the *Trade Disputes Act 1906*, trade unions have enjoyed immunity from liability in tort for industrial action, taken or threatened, in contemplation or furtherance of a trade dispute (although such freedom of action was subsequently curtailed by the *Trade Union and Labour Relations (Consolidation) Act 1992, s 20*). Not surprisingly, perhaps, this immunity extends to their representatives acting in a lawful capacity. Thus, the *Safety*

Representatives Regulations 1977 (SI 1977 No 500) state that none of the functions of a safety representative confers legal duties or responsibilities [*Safety Representatives Regulations 1977 (SI 1977 No 500), Reg 4(1)*]. As safety representatives are not legally responsible for health, safety or welfare at work, they cannot be liable under either the criminal or civil law for anything they may do, or fail to do, as a safety representative under the *Safety Representatives Regulations*. This protection against criminal or civil liability does not, however, remove a safety representative's legal responsibility as an employee. Safety representatives must carry out their responsibilities under the *HSWA 1974, s 7* if they are not to be liable for criminal prosecution by an HSE inspector. These duties as an employee are to take reasonable care for the health and safety of one's self and others, and to co-operate with one's employer as far as is necessary to enable him to carry out his statutory duties on health and safety.

Time off with pay

General right

[J3010] Under the *Safety Representatives Regulations 1977 (SI 1977 No 500)*, safety representatives are entitled to take such paid time off during working hours as is necessary to perform their statutory functions, and reasonable time to undergo training in accordance with a Code of Practice approved by HSC.

Definition of 'time off'

[J3011] The *Safety Representatives Regulations 1977 (SI 1977 No 500), Reg 4(2)* provides that the employer must provide the safety representative with such time off with pay during the employee's working hours as shall be necessary for the purposes of:

- performing his statutory functions; and
- undergoing such training in aspects of those functions as may be reasonable in all the circumstances.

Further details of these requirements are outlined in the Code of Practice attached to the *Safety Representatives Regulations* and the HSC Approved Code of Practice on time off for training. The Code of Practice is for guidance purposes only – its contents are recommendations rather than requirements. However, it is guidance that an employment tribunal can and will take into account if a complaint is lodged in relation to an employer's unreasonable failure to allow time off.

The combined effect of the ACAS Code No 3: *'Time Off for Trade Union Duties and Activities'* (2010) and the HSC Approved Code of Practice on time off is that shop stewards who have also been appointed as safety representatives are to be given time off by their employer to carry out both their industrial relations duties and their safety functions, and also paid leave to attend separate training courses on industrial relations and on health and safety at work – this includes a TUC course on *COSHH* (*Gallagher v The Drum Engineering Co Ltd, COIT 1330/89*).

An employee is not entitled to be paid for time taken off in lieu of the time he had spent on a course. This was held in *Hairsine v Hull City Council* [1992]

IRLR 211 when a shift worker, whose shift ran from 3 pm to 11 pm, attended a trade union course from 9 am to 4 pm and then carried out his duties until 7 pm. He was paid from 3 pm to 7 pm and he could claim no more.

A similar decision was arrived at in *Calder v Secretary of State for Work and Pensions*(UKEAT/0512/08/LA) [2009] All ER (D) 106 (Aug)in relation to an employee who worked part-time Tuesdays, Wednesdays and Thursdays. The employee was a health and safety secretary for a major trade union in the public sector. In December 2006, the employee applied to the employer to attend the stage 3 health and safety representatives training course provided by the Trade Union Congress. The course was held on Fridays for a period of 36 weeks. The employer refused permission for the employee to attend on a paid basis since because of the nature of the employee's duties and the nature of the course curriculum, it was not reasonable for her to go on the course nor necessary for her duties. In the employee's application to the Employment Tribunal for her health and safety representatives paid time off claim, and the employee's appeal to the EAT, the employee lost; it was held that unless there was a refusal to permit time off from working hours (which Fridays were not for the employee), the *Safety Representatives Regulations 1977 (SI 1977 No 500)*did not come into effect.

However, where more safety representatives have been appointed than there are sections of the workforce for which safety representatives could be responsible, it is not unreasonable for an employer to deny some safety representatives time off for fulfilling safety functions (*Howard and Peet v Volex plc (HSIB 181)*).

A decision of the EAT seems to favour jointly sponsored in-house courses, except as regards the representational aspects of the functions of safety representatives, where the training is to be provided exclusively by the union (*White v Pressed Steel Fisher* [1980] IRLR 176). Moreover, one course per union per year is too rigid an approach (*Waugh v London Borough of Sutton* (1983) HSIB 86).

Definition of 'pay'

[J3012] The amount of pay to which the safety representative is entitled is contained in the *Safety Representatives Regulations (SI 1977 No 500), Sch 2*.

Recourse for the safety representative

[J3013] Where the employer's refusal to allow paid time off is unreasonable, he must reimburse the employee for the time taken to attend [*Safety Representatives Regulations 1977 (SI 1977 No 500), Reg 4(2), Sch 2*]. In the case of *Scarth v East Herts DC (HSIB 181)*: the test of reasonableness is to be judged at the time of the decision to refuse training.

Safety representatives who are refused time off to perform their functions or who are not paid for such time off are able to make a complaint to an employment tribunal [*Safety Representatives Regulations 1977 (SI 1977 No 500), Reg 11*].

Facilities to be provided by employer

[J3014] The type and number of facilities that employers are obliged to provide for safety representatives are not spelled out in the Regulations, Code of Practice or guidance notes, other than a general requirement in the *Safety Representatives Regulations 1977 (SI 1977 No 500), Reg 5(3)* which states, *inter alia*, that 'the employer shall provide such facilities and assistance as the safety representatives shall require for the purposes of carrying out their functions'. Formerly, the requirement to provide facilities and assistance related only to inspections.

Trade unions consider that the phrase 'facilities and assistance' includes the right to request the presence of an independent technical adviser or trade union official during an inspection, and for safety representatives to take samples of substances used at work for analysis outside the workplace. The TUC has recommended that the following facilities be made available to safety representatives:

- a room and desk at the workplace;
- facilities for storing correspondence;
- inspection reports and other papers;
- ready access to internal and external telephones;
- access to typing and duplicating facilities;
- provision of notice boards;
- use of a suitable room for reporting back to and consulting with members; and
- other facilities should include copies of all relevant statutes, regulations, Approved Codes of Practice and HSC guidance notes; and copies of all legal or international standards which are relevant to the workplace.

Disclosure of information

[J3015] Employers are required by the *Safety Representatives Regulations* to disclose information to safety representatives which is necessary for them to carry out their functions [*Safety Representatives Regulations 1977 (SI 1977 No 500), Reg 7(1)*]. A parallel provision exists under the *Management of Health and Safety at Work Regulations 1999, Reg 10(2)* in relation to the information that is to be provided to the parents of a child to be employed by an employer.

Regulation 7 is consolidated by paragraph 6 of the Code of Practice which details the health and safety information 'within the employer's knowledge' that should be made available to safety representatives. This should include:

- plans and performance and any changes proposed which may affect health and safety;
- technical information about hazards and precautions necessary, including information provided by manufacturers, suppliers and so on;
- information and statistical records on accidents, dangerous occurrences and notifiable industrial diseases; and
- other information such as measures to check the effectiveness of health and safety arrangements and information on articles and substances issued to homeworkers.

The exceptions to this requirement are where disclosure of such information would be 'against the interests of national security'; where it would contravene a prohibition imposed by law; any information relating to an individual (unless consent has been given); information that would damage the employer's undertaking; and information obtained for the sole purpose of bringing, prosecuting or defending legal proceedings [*Safety Representatives Regulations 1977 (SI 1977 No 500), Reg 7(2)*].

However, the decision in *Waugh v British Railways Board* [1979] 2 All ER 1169 established that where an employer seeks, on grounds of privilege, to withhold a report made following an accident, he can only do so if its dominant purpose is related to actual or potential hostile legal proceedings. In this particular case, a report was commissioned for two purposes following the death of an employee:

- to recommend improvements in safety measures; and
- to gather material for the employer's defence.

It was held that the report was not privileged.

This was followed in *Lask v Gloucester Health Authority* (1995) Times, 13 December where a circular '*Reporting Accidents in Hospitals*' had to be discovered by order after an injury to an employee while he was walking along a path.

Where differences of opinion arise as to the evaluation or interpretation of technical aspects of safety information or health data, unions are advised to contact the local offices of HSE, because of HSE expertise and access to research.

Technical information

[J3016] HSE inspectors are also obliged under the *HSWA 1974, s 28(8)*, to supply safety representatives with technical information – factual information obtained during their visits (i.e. any measurements, testing and results of sampling and monitoring), notices of prosecution, copies of correspondence and copies of any improvement or prohibition notices issued to their employer. The latter places an absolute duty on an inspector to disclose specific kinds of information to workers or their representatives concerning health, safety and welfare at work. This can also involve personal discussions between the HSE inspector and the safety representative. The inspector must also tell the representative what action he proposes to take as a result of his visit. Where local authority health inspectors are acting under powers granted by the *HSWA 1974* (see ENFORCEMENT), they are also required to provide appropriate information to safety representatives.

Safety committees

[J3017] There is a duty on every employer, in cases where it is prescribed (see below), to establish a safety committee if requested to do so by safety representatives. The committee's purpose is to monitor health and safety measures at work [*HSWA 1974, s 2(7)*]. Such cases are prescribed by the *Safety Representatives Regulations 1977 (SI 1977 No 500)* and limit the duty to appoint a committee to requests made by trade union safety representatives.

Establishment of a safety committee

[J3018] If requested by at least two safety representatives in writing, the employer must establish a safety committee [*Safety Representatives Regulations 1977 (SI 1977 No 500), Reg 9(1)*].

When setting up a safety committee, the employer must:

*   consult with both:
    —   the safety representatives who make the request,
    —   the representatives of recognised trade unions whose members work in any workplace where it is proposed that the committee will function; and
*   post a notice, stating the composition of the committee and the workplace(s) to be covered by it, in a place where it can easily be read by employees; and
*   establish the committee within three months after the request for it was made.

[*Safety Representatives Regulations 1977 (SI 1977 No 500), Reg 9(2)*]

Function of safety committees

[J3019] In practical terms, trade union appointed safety representatives are now using the medium of safety committees to examine the implications of hazard report forms arising from inspections, and the results of investigations into accidents and dangerous occurrences, together with the remedial action required. A similar procedure exists with respect to representatives for tests and measurements of noise, toxic substances or other harmful effects on the working environment.

Trade unions regard the function of safety committees as a forum for the discussion and resolution of problems that have failed to be solved initially through the intervention of the safety representative in discussion with line management. There is, therefore, from the trade unions' viewpoint, a large measure of negotiation with its consequent effect on collective bargaining agreements.

If safety representatives are unable to resolve a problem with management through the safety committee, or with HSE, they can approach their own union for assistance – a number of unions have their own health and safety officers who can, and do, provide an extensive range of information on occupational health and safety matters. The unions, in turn, can refer to the TUC for further advice.

The 'Brown Book', which contains the *Safety Representatives Regulations*, Code of Practice and guidance, was revised in 1996 to include the amendments made in 1993 by the *Management of Health and Safety at Work Regulations 1992* (now superceded by the *Management of Health and Safety at Work Regulations 1999 (SI 1999 No 3242))* (see above) and the *Consultation with Employees Regulations* (see **J3021**).

Non-unionised workforce – consultation obligations

[J3020] A representative of employee safety is an elected representative of a non-unionised workforce who is assigned with broadly the same rights and obligations as a safety representative in a unionised workforce.

General duty

[J3021] The *Consultation with Employees Regulations 1996 (SI 1996 No 1513)* introduced a duty to consult any employees who are not members of a group covered by safety representatives appointed under the *Safety Representatives Regulations*. Under the *Consultation with Employees Regulations*, employers must consult either with elected employee representatives or in the absence of such representatives, directly with the whole workforce (the *Consultation with Employees Regulations, Reg 3*). The obligation imposed on such employers is to consult those employees in good time on matters relating to their health and safety at work.

Number of representatives for workforce

[J3022] Guidance notes on the *Consultation with Employees Regulations 1996 (SI 1996 No 1513)* state that the number of safety representatives who can be appointed depends on the size of the workforce and workplace, whether there are different sites, the variety of different occupations, the operation of shift systems and the type and risks of work activity. A DTI Workplace Survey has concluded that in non-unionised workplaces which have appointed worker representatives, it is usual for there to be several representatives, with the median number of such representatives being three. The survey estimated that there are approximately 218,000 representatives across all British workplaces with 25 or more employees.

Role of safety representatives

[J3023] The functions of representatives of employee safety are:

* to make representations to the employer of potential hazards and dangerous occurrences at the workplace which affect or could affect the group of employees the representative represents;
* to make representations to the employer on general matters affecting the health and safety at work of the group of employees the representative represents, and in particular on such matters as the representative has been consulted about by the employer under the *Consultation with Employees Regulations*; and
* to represent that group of employees in consultations at the workplace with inspectors appointed under the *HSWA 1974*.

Rights and duties of representatives of employee safety

Time off with pay

[J3024] Under the *Consultation with Employees Regulations 1996 (SI 1996 No 1513), Reg 7(1)(b)*, the right that a representative of employee safety has to take time off with pay is generally the same as that for safety representatives.

Definition of 'time off'

[J3025] An employer is under an obligation to permit a representative of employee safety to take such time off with pay during working hours as shall be necessary for:

- performing his functions; and
- undergoing such training as is reasonable in all the circumstances.

A candidate standing for election as a representative of employee safety is also allowed reasonable time off with pay during working hours in order to perform his functions as a candidate [*SI 1996 No 1513, Reg 7(2)*].

Definition of 'pay'

[J3026] The *Consultation with Employees Regulations (SI 1996 No 1513), Sch 1* deal with the definition of pay, and generally the definition is the same as that for union safety representatives.

Provision of information

[J3027] The employer must provide such information as is necessary to enable the employees or representatives of employee safety to participate fully and effectively in the consultation. In the case of representatives of employee safety, the information must also be sufficient to enable them to carry out their functions under the *Consultation with Employees Regulations 1996 (SI 1996 No 1513)*.

Information provided to representatives must also include information which is contained in any record which the employer is required to keep under *RIDDOR 1995 (SI 1995 No 3163)* and which relates to the workplace or the group of employees represented by the representatives. Note that there are exceptions to the requirement to disclose information under the *Consultation with Employees Regulations, Reg 5(3)*, these being similar to those under the *Safety Representatives Regulations 1977 (SI 1977 No 500), Reg 7* (see **J3015** above).

Relevant training

[J3028] Under the *Consultation with Employees Regulations 1996 (SI 1996 No 1513), Reg 7(1)*, representatives of employee safety must be provided with reasonable training in respect of their functions under the *Consultation with Employees Regulations*, for which the employer must pay.

Remedies for failure to provide time off or pay for time off

[J3029] A representative of employee safety, or candidate standing for election as such, who is denied time off or who fails to receive payment for time off, may make an application to an employment tribunal for a declaration and/or compensation. As in the case of safety representatives, the remedies obtainable (set out in the *Consultation with Employees Regulations 1996 (SI 1996 No 1513), Sch 2* are similar to those granted to complainants under the *Trade Union and Labour Relations (Consolidation) Act 1992, s 168* (time off for union duties).

Recourse for safety representatives

[J3030] Safety representatives (whether they are appointed under the *Safety Representatives Regulations 1977 (SI 1977 No 500)* or the *Consultation with Employees Regulations 1996 (SI 1996 No 1513)* are provided with statutory protection for the proper execution of their duties.

An employee who is:

• designated by his employer to carry out a health and safety related function;
• a representative of employee safety; or
• a candidate standing for election as such,

has the right not to be subjected to any detriment or unfairly dismissed on the grounds that:

— having been designated by the employer to carry out a health and safety related function, he carried out, or proposed to carry out, the function;
— he undertook, or proposed to undertake, any function(s) consistent with being a safety representative or member of a safety committee;
— he took part in, or proposed to take part in, consultation with the employer; or
— he took part in an election of representatives of employee safety.

In relation to a detriment claim, where the employer infringes any of these rights, the employee has the right to make a complaint to an employment tribunal under the *Employment Rights Act 1996, s 44(1)* (as amended by the *Consultation with Employees Regulations 1996 (SI 1996 No 1513), Reg 8*. An employment tribunal can make a declaration and also award compensation [*Employment Rights Act 1996, ss 48–49*]. There is neither a minimum qualifying period of service nor an upper age limit for bringing such a claim.

If the employer unfairly dismisses such an employee for one of the above reasons, or where it is the principal reason for the dismissal, that dismissal shall be deemed automatically unfair [*Employment Rights Act 1996, s 100, as amended by the Consultation with Employees Regulations 1996 (SI 1996 No 1513), Reg 8*]. Conversely, if the dismissal is not connected with the health and safety issues which have arisen, the dismissal will not be automatically unfair [*Dunn v Ovalcode Limited* [2003] ALL ER (D) 241 (Apr), EAT].

It will also be an automatically unfair dismissal to select a representative or candidate for redundancy for such a reason [*Employment Rights Act 1996, s 105*]. However, there is no presumption of a right to positive discrimination for such representatives where the employer is undertaking a redundancy programme [*Shipham & Co Limited v Skinner* [2001] All ER (D) 201 (Dec), EAT].

The normal minimum qualifying period of service, the normal upper age limit and the cap on the compensatory award for unfair dismissal claims do not apply [*Employment Rights Act 1996, ss 108, 109 and 124(1A)* as amended by the *Employment Relations Act 1999, s 37(1)*)].

Under the *Employment Rights Act 1996, s 103A* a right to automatically claim unfair dismissal would exist for a worker who is dismissed for making a protected disclosure. A right also exists under the *Employment Rights Act 1996, s 47B* for workers to claim that they have been subjected to a detriment for making a protected disclosure. 'Protected disclosures' in this regard are capable of covering the situation where the health and safety of an individual has been or is likely to be endangered [*Employment Rights Act 1996, s 43B(1)(d)*].

European developments in health and safety

Introduction

[J3031] Health and safety law will continue to be subject to change in the future with the implementation of further European directives and HSC programme of modifying and simplifying health and safety law. At a European level the most important legislation is the directives made under Art 138 of the Treaty of Rome (as amended). The Framework Directive, from which the *Consultation with Employees Regulations 1996 (SI 1996 No 1513)* were derived, will continue to drive forward developments in UK health and safety law.

Further developments have occurred under European law which are having an impact at national level these being in the form of the Working Time Directive (93/104) and the European Works Councils Directive (94/45). These Directives have been implemented into national law under the *Working Time Regulations 1998 (SI 1998 No 1833)* (as amended) and the *Works Councils Regulations 1999 (SI 1999 No 3323)* respectively.

The Working Time Regulations 1998 (as amended)

[J3032] A full discussion of the impact of the *Working Time Regulations 1998 (SI 1998 No 1833)* is beyond the scope of this chapter (see WORKING TIME).

That said, the *Working Time Regulations* have introduced a joint consultation function into the operation of these Regulations by use of collective, workforce and relevant agreements. The *Working Time Regulations* provide that it is possible to vary the extent to which the *Working Time Regulations* must be strictly complied with through the use of these devices. The various types of agreements can be described as follows:

- collective agreements – these are defined by *s 178 of the Trade Union and Labour Relations (Consolidation) Act 1992* as being agreements between independent trade unions and employers;
- workforce agreements – these were created by the *Working Time Regulations* and are defined in *the Working Time Regulations 1998 (SI 1998 No 1833), Reg 2* and *Sch 1*. They amount to agreements between an employer and either duly elected worker representatives of the employer or, in the case of an employer employing less than 20 workers, a majority of the individual workers themselves, where the agreement concluded:
  - — is in writing;
  - — has effect for a specified period not exceeding five years;
  - — applies to either:
    - (i) all of the relevant members of the workforce; or
    - (ii) all of the relevant members of the workforce who belong to a particular sub-group;
  - — is signed:
    - (i) by the worker representatives or by the particular group of workers; or
    - (ii) in the case of an employer having less than 20 employees on the date on which the agreement is first concluded, either by appropriate representatives or by a majority of the workers working for the employer; and
  - — before being made available for signature, was provided in copy form to all of the workers to whom the agreement was intended to apply, together with such guidance as the workers might reasonably require in order to understand the draft agreement;
- relevant agreements – these are workforce agreements that cover a worker, any provision of a collective agreement that is individually incorporated into the contract of employment of a worker, or any other agreement in writing between a worker and his employer that is legally enforceable (e.g. a staff handbook).

By the *Working Time Regulations 1998 (SI 1998 No 1833), Reg 23(a)* it is possible to modify the provisions relating to:

- the length of night work (see *Working Time Regulations 1998 (SI 1998 No 1833), Reg 6*); and
- the minimum daily and weekly rest periods and rest breaks (see *Working Time Regulations 1998 (SI 1998 No 1833), Regs 10–12* respectively).

Modification must be by way of a collective or workforce agreement and such an agreement must make provision for a compensatory rest period of equivalent length [*Working Time Regulations 1998 (SI 1998 No 1833), Reg 24*].

Further, by the *Working Time Regulations 1998 (SI 1998 No 1833), Reg 23(b)* it is possible for an employer and its workers to agree by collective or workforce agreement to vary the reference period for calculating the maximum working week from the usual 17 weeks to a 52-week period if there are objective or technical reasons relating to the organisation which justify such a change.

The variation provisions allow a degree of flexibility where it is necessary for the interests of an employer's business to effect such change for operational reasons whilst still ensuring the protection of the health and safety of the workforce. The provisions also ensure that any change that is to be made must survive collective scrutiny of the employer's workforce.

European Works Councils and information and consultation procedures

[J3033] The provisions relating to European Works Councils are at present confined to large pan-European entities.

As stated at **J3001**, the Information and Consultation Directive 2002/14/EC will impose information and consultation obligations on any employer employing more than 50 workers. Transitional provisions will apply initially limiting the impact of Directive 2002/14/EC to employers employing more than 150 employees. All employers in the United Kingdom employing 50 or more employees will have to comply with the Directive by 23 March 2007. National legislation to implement Directive 2002/14/EC was introduced in the form of the *Information and Consultation of Employees Regulations 2004 (SI 2004 No 3426)*. Currently, the requirement to inform and consult under the *Information and Consultation of Employees Regulations 2004* only arises in the case of employers employing at least 150 employees in their undertaking. The *Information and Consultation of Employees Regulations 2004* is phased so that on 6 April 2007, employers employing 100 employees, and on 6 April 2008, employers employing 50 or more employees, will be caught by its provisions. The *Information and Consultation of Employees Regulations 2004* do not apply directly to health and safety consultation, although such consultation could be voluntarily covered under an information and consultation procedure under those Regulations.

The European Work Council Directive (94/45) was incorporated into national law by the *Works Councils Regulations 1999 (SI 1999 No 3323)*. A full discussion of the operation of the *Works Councils Regulations* is beyond the scope of this chapter, which instead focuses upon the health and safety aspect of the Regulations.

The *Works Councils Regulations* govern employers employing a total of 1,000 or more workers where at least 150 workers are so employed in each of two or more member states.

Their main purpose is procedural. *Part IV* of the *Works Council Regulations 1999 (SI 1999 No 3323)* creates machinery between workers and their employer for the purpose of establishing either a European Works Council ('EWC') or an Information and Consultation Procedure ('ICP'). *Regulation 17(1)* of the *Works Councils Regulations 1999 (SI 1999 No 3323)* provides that the central management of the employer and a special negotiating body (defined in *Part III* of the Regulations) are bound to:

. . . negotiate in a spirit of co-operation with a view to reaching a written agreement on the detailed arrangements for the information and consultation of employees in a Community-scale undertaking or Community-scale group of undertakings.

Regulation 17(3) of the *Works Councils Regulations 1999 (SI 1999 No 3323)* leaves the choice of whether to proceed with an EWC or an ICP to the parties.

The parties are free to include in the agreement reference to whatever matters are likely to affect the workers of the employer at a trans-national level. Health and safety is clearly such an issue.

If:

- the parties fail to agree the content of the agreement; or
- within six months of a valid request being made to the central management of an employer the employer fails to negotiate so as to create either an EWC or ICP; or
- after three years of negotiation to produce an agreement for an EWC or ICP the parties cannot agree as to the constitution,

default machinery is provided by the *Schedule* to the *Works Councils Regulations* [*Works Councils Regulations 1999 (SI 1999 No 3323), Reg 18(1)*]. *Paragraph 6* of the *Schedule* provides in relation to EWCs that:

> The competence of the European Works Council shall be limited to information and consultation on the matters which concern the Community-scale undertaking or Community-scale group of undertakings as a whole or at least two of its establishments or group undertakings situated in different Member States.

In relation to ICPs, *para 7(3)* of the *Schedule* to the *Works Councils Regulations 1999 (SI 1999 No 3323)* provides that meetings of ICPs:

> . . . shall relate in particular to the structure, economic and financial situation, the probable development of the business and of production and sales, the situation and probable trend of employment, investments, and substantial changes concerning organisation, [and] introduction of new working methods or production processes . . .

Although not expressly providing for discussion of health and safety issues, the provisions relating to both EWCs and ICPs will, by implication, include debate of health and safety matters.

Lifting Operations

Kevin Chicken and Nicola Coote

Introduction

[L3001] General requirements for the provision and maintenance of safe plant and equipment set out in the *Health and Safety at Work etc. Act 1974 section 2(2)(a)* and *(b)* apply.

More specific requirements applying to all workplaces came into force on 5 December 1998 in the *Lifting Operations and Lifting Equipment Regulations 1998 (LOLER) (SI 1998 No 2307)* which encompasses all lifting equipment for use at work.

LOLER demands that four main principals are complied with:

- the initial integrity of the equipment is suitable for its intended use;
- the lifting operation is correctly planned and assessed;
- the operation is undertaken in a safe manner; and
- thereafter, that the equipments safe integrity is maintained.

There is a duty on the employer to select suitable lifting equipment based on a fitness for purpose criteria in both *Regulation 4* of the *Provision and Use of Work Equipment Regulations 1998 (SI 1992 No 2932)* (see **M1004**) and its sister legislation, LOLER.

LOLER has replaced some earlier terminology, namely:

LOLER	Previous Terminology
Accessories	Lifting tackle
Carrier	Lift, personnel carrier, working platform, etc.
Safety Co-efficient	Factor of safety
Rated capacity indicator	Automatic safe load indicator

Accessories for lifting are also lifting equipment, this encompasses and adds to the formerly termed 'Lifting tackle' and includes: chains, ropes, slings, hooks, jigs, eyebolts and other components kept for attaching loads to lifting machinery..

Statutory requirements relating to the manufacture of lifting equipment

[L3002] The statutes that are applied to the manufacture of lifting equipment are:

- *Lifts Regulations 1997 (SI 1997 No 831)*; and
- the Machinery Directive which is implemented into UK law by the *Supply of Machinery (Safety) Regulations 1992 (SI 1992 No 3073)*, as amended by *SI 1994 No 2063)*, Machinery Directive 2006/42/EC becomes applicable on 29 December 2009, superseding the existing Machinery Directive 98/37/EC. While the two are broadly similar, there are significant differences that will affect machine builders, those performing final assembly and CE marking of machinery, and those placing imported machinery on the market in the European Economic Area.

The relationship between the manufacturers and the owner/users duties are shown in Figure 1 opposite:

Figure 1 Relationship between the manufacturers and the owner/users

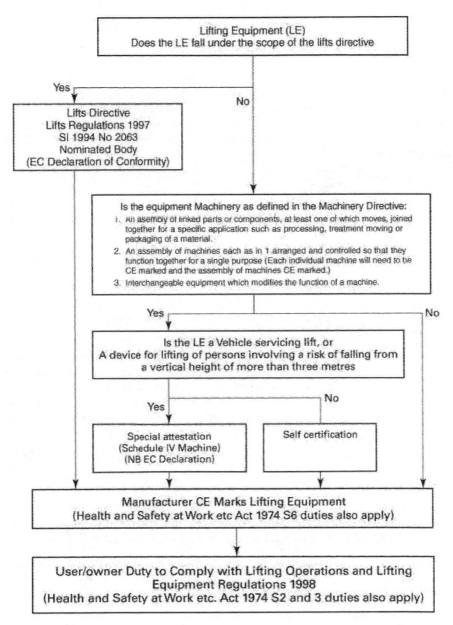

In addition to the specific requirements above, there is also a necessity for lifting equipment incorporating electrical equipment to comply with:

- *Electrical Equipment (Safety) Regulations 1994 (SI 1994 No 3260)* implementing the Low Voltage Directive; and
- *Electromagnetic Compatibility Regulations 1992 (SI 1992 No 2372)*, as amended by *SI 1994 No 3080.*

Lifts Regulations 1997 (SI 1997 No 831)

[L3003] 'Lift' – this means an appliance serving specific levels, having a car moving along guides which are rigid – or along a fixed course even where it does not move along guides which are rigid (for example, a scissor lift) – and inclined at an angle of more than 15 degrees to the horizontal and intended for the transport of (there is difference in terminology when LOLER refers to Lifts as Carriers.):

- persons;
- persons and goods; or
- goods alone if the car is accessible, that is to say, a person may enter it without difficulty, and fitted with controls situated inside the car or within reach of a person inside.

A lift operating company was fined £400,000 at Southampton Crown Court today, Friday 7 April 2006. The prosecution was brought by the HSE following a breach of *HSWA 1974.* Otis Investments (formerly Otis Plc) and Otis Limited (formerly Express Lifts Company Limited) – sentenced as a single company, Otis – were fined £400,000 following an incident that claimed the lives of two young men.

Michael Dawson, 27 and Daniel Digby, 25, were killed when they fell down a lift shaft of Shirley Towers, a high–rise block of flats in Southampton. The incident took place in the early hours of 4 February 2001 when the pair fell against the lift doors, which swung open in the style of a cat flap due to inadequate fixings on the lower rail of the opening. Both men plunged about 30 metres and are believed to have died instantly.

Otis Investments and Otis Limited pleaded guilty to a single breach of HSWA 1974 s 3(1) *(HSWA 1974)* at an earlier hearing. The judge also awarded costs of £145,000 against the defence.

The HSE provided the following guidance after this incident through their Operational Circular OC232/29 isued for guidance to Health and Safety Executive / Local Authorities Enforcement Liaison Committee (HELA)

Owners/users of lifts used in public areas and public buildings should ensure that landing entrances and doors are designed and constructed in order to withstand the anticipated risk of vandalism and physical abuse.

Effective and appropriate arrangements should be in place to ensure that lift landing doors are maintained in a safe operating condition including regular checks by the lift maintenance engineer of the door retaining system. Checks will include the condition of the guide channel(s), guide shoes (sometimes referred to as guide blocks or gibs), fixings, rollers etc to ensure there are no defects which could affect the effectiveness of the door retaining system. Such checks are expected to form part of a preventive maintenance programme for the lift.

Competent persons, as part of their responsibilities for the thorough examination, should examine among other things the condition of the landing door retaining system. The risks which could arise from the failure of the door retaining system will determine how thorough the examination needs to be.

The Regulations

[L3004]–[L3005] These Regulations apply to lifts permanently serving buildings or constructions; and safety components for use in such lifts. [*Lifts Regulations 1997 (SI 1997 No 831), Reg 3*].

These Regulations do not apply to:

(a)   the following lifts – and safety components for such lifts:
—   cableways, including funicular railways, for the public or private transportation of persons,
—   lifts specially designed and constructed for military or police purposes,
—   mine winding gear,
—   theatre elevators,
—   lifts fitted in means of transport,
—   lifts connected to machinery and intended exclusively for access to the workplace,
—   rack and pinion trains,
—   construction-site hoists intended for lifting persons or persons and goods;
[*SI 1997 No 831, Reg 4, Sch 14*].

(b)   any lift or safety component which is placed on the market (ie when the installer first makes the lift available to the user, but see **L3009** below) and put into service before 1 July 1997 [*SI 1997 No 831, Reg 5*];

(c)   any lift or safety component placed on the market and put into service on or before 30 June 1999 which complies with any health and safety provisions with which it would have been required to comply if it was to have been placed on the market and put into service in the United Kingdom on 29 June 1995.
This exclusion does not apply in the case of a lift or a safety component which;
—   unless required to bear the CE marking pursuant to any other Community obligation, bears the CE marking or an inscription liable to be confused with it; or
—   bears or is accompanied by any other indication, howsoever expressed, that it complies with the Lifts Directive.
[*SI 1997 No 831, Reg 6*]; and

(d)   any lift insofar as and to the extent that the relevant essential health and safety requirements relate to risks wholly or partly covered by other Community directives applicable to that lift [*SI 1997 No 831, Reg 7*].

General requirements

General duty relating to the placing on the market and putting into service of lifts

(i)   Subject to *Lifts Regulations 1997 (SI 1997 No 831), Reg 12* (see **L3010** below), no person who is a responsible person shall place on the market and put into service any lift unless the requirements of paragraph (ii) below have been complied with in relation to it [*SI 1997 No 831, Reg 8(1)*].

(ii)   *SI 1997 No 831, Reg 8(2)* provides that the requirements in respect of any lift are that:
— it satisfies the relevant essential health and safety requirements and, for the purpose of satisfying those requirements:
— where a transposed harmonised standard covers one or more of the relevant essential health and safety requirements, any lift constructed in accordance with that transposed harmonised standard shall be presumed to comply with that (or those) essential health and safety requirement(s);
(EC type-examination is the procedure whereby a notified body, ascertains and certifies that a model lift, or a lift for which there is no provision for an extension or variant, satisfies the requirements of the Directive. This procedure is conducted by examination of the technical dossier and a representative model of the lift and to perform or have performed the appropriate checks and tests necessary to check that the solutions adopted by the installer of the lift meet the requirements of the Directive and allow the lift to comply with them).
— by calculation, or on the basis of design plans, it is permitted to demonstrate the similarity of a range of equipment to satisfy the essential safety requirements;
— the appropriate conformity assessment procedure in respect of the lift has been carried out in accordance with *SI 1997 No 831, Reg 13(1)* (see **L3011** below);
— the CE marking has been affixed to it by the installer of the lift in accordance with *SI 1997 No 831, Sch 3*;
— a declaration of conformity has been drawn up in respect of it by the installer of the lift; and
— it is in fact safe.
Note: SI 1997 No 831, Reg 18(1) provides that a lift which bears the CE marking, and is accompanied by an EC declaration of conformity in accordance with *SI 1997 No 831, Reg 8(2)*, is taken to conform with all the requirements of the *Lifts Regulations 1997 (SI 1997 No 831)*, including *SI 1997 No 831, Reg 13* (see **L3011** below), unless reasonable grounds exist for suspecting that it does not so conform.

(iii)   Any technical documentation or other information in relation to a lift
          required to be retained under the conformity assessment procedure used
          must be retained by the person specified in that respect in that
          conformity assessment procedure for any period specified in that
          procedure [*SI 1997 No 831, Reg 8(3)*].

In these Regulations, 'responsible person' means:

—   in the case of a lift, the installer of the lift;
—   in the case of a safety component, the manufacturer of the component
      or his authorised representative established in the Community; or
—   where neither the installer of the lift nor the manufacturer of the safety
      component nor the latter's authorised representative established in
      the Community, as the case may be, have fulfilled the requirements of
      SI 1997 No 831, Reg 8(2) (see above) or *SI 1997 No 831, Reg 9(2)* (see
      L3006 below), the person who places the lift or safety component on
      the market.

*General duty relating to the placing on the market and putting into service
of safety components*

[L3006] Safety components include devices for locking landing doors; devices
to prevent falls; overspeed limitation devices; and electric safety switches.

(i)   Subject to *Lifts Regulations 1997 (SI 1997 No 831), Reg 12* (see **L3010**
         below), no person who is a responsible person shall place on the market
         and put into service any safety component unless the requirements of
         paragraph (ii) below have been complied with in relation to it [*SI 1997
         No 831, Reg 9(1)*].

(ii)   *SI 1997 No 831, Reg 9(2)* provides that the requirements in respect of
         any safety component are that:
         —   it satisfies the relevant essential health and safety requirements
               and for the purpose of satisfying those requirements where a
               transposed harmonised standard covers one or more of the
               relevant essential health and safety requirements, any safety
               component constructed in accordance with that transposed
               harmonised standard shall be presumed to be suitable to enable
               a lift on which it is correctly installed to comply with that (or
               those) essential health and safety requirement(s);
         —   the appropriate conformity assessment procedure in respect of
               the safety component has been carried out in accordance with *SI
               1997 No 831, Reg 13(1)* (see **L3011** below);
         —   the CE marking has been affixed to it, or on a label inseparably
               attached to the safety component, by the manufacturer of that
               safety component, or his authorised representative established in
               the Community, in accordance with *SI 1997 No 831, Sch 3*
               (which specifies the requirements for CE marking);
         —   a declaration of conformity has been drawn up in respect of it by
               the manufacturer of the safety component or his authorised
               representative established in the Community; and
         —   it is in fact safe.

Note: SI 1997 No 831, Reg 18(1) provides that a safety component – or its label – which bears the CE marking, and is accompanied by an EC declaration of conformity in accordance with *SI 1997 No 831, Reg 9(2)*, is taken to conform with all the requirements of the *Lifts Regulations 1997*, including *SI 1997 No 831, Reg 13* (see **L3011** below), unless reasonable grounds exist for suspecting that it does not so conform.

(iii)   Any technical documentation or other information in relation to a safety component required to be retained under the conformity assessment procedure used must be retained by the person specified in that respect in that conformity assessment procedure for any period specified in that procedure [*SI 1997 No 831, Reg 9(3)*].

General duty relating to the supply of a lift or safety component

[L3007] Subject to *Lifts Regulations 1997 (SI 1997 No 831), Reg 12* (see **L3010** below) any person who supplies any lift or safety component but who is not a person to whom *SI 1997 No 831, Reg 8* or *9* applies (see **L3005** and **L3006** above) must ensure that that lift or safety component is safe [*SI 1997 No 831, Reg 10*].

Penalties for breach of Regs 8, 9 or 10

[L3008]–[L3009] A person who is convicted of an offence under *Lifts Regulations 1997 (SI 1997 No 831), Reg 8, 9* or *10* above is liable to imprisonment (not exceeding three months) or to a fine not exceeding level 5 on the standard scale, or both. *SI 1997 No 831, Reg 22*, however, provides a defence of due diligence: ie the defendant must show that he took all reasonable steps and exercised all due diligence to avoid committing the offence.

Specific duties relating to the supply of information, freedom from obstruction of lift shafts and retention of documents

(i)   The person responsible for work on the building or construction where a lift is to be installed and the installer of the lift must keep each other informed of the facts necessary for, and take the appropriate steps to ensure, the proper operation and safe use of the lift. Shafts intended for lifts must not contain any piping or wiring or fittings other than that which is necessary for the operation and safety of that lift.

(ii)   Where, in the case of a lift, for the purposes of *Lifts Regulations 1997 (SI 1997 No 831), Reg 8(2)* (see **L3005** above) the appropriate conformity assessment procedure is one of the procedures set out in *SI 1997 No 831, Reg 13(2)(a), (b)* or *(c)* (see **L3011** below), the person responsible for the design of the lift must supply to the person responsible for the construction, installation and testing all necessary documents and information for the latter person to be able to operate in absolute security.

A person who is convicted of an offence under (i) or (ii) above is liable to imprisonment (not exceeding three months) or to a fine not exceeding level 5 on the standard scale, or both. *SI 1997 No 831, Reg*

22, however, provides a defence of due diligence: ie the defendant must show that he took all reasonable steps and exercised all due diligence to avoid committing the offence.

(iii)   A copy of the declaration of conformity mentioned in *SI 1997 No 831, Reg 8(2)* or *9(2)* must:

—   in the case of a lift, be supplied to the EC Commission, the member states and any other notified bodies, on request, by the installer of the lift together with a copy of the reports of the tests involved in the final inspection to be carried out as part of the appropriate conformity assessment procedure referred to in *SI 1997 No 831, Reg 8(2)*; and

—   be retained, by the person who draws up that declaration, for a period of ten years – in the case of a lift, from the date on which the lift was placed on the market; and in the case of a safety component, from the date on which safety components of that type were last manufactured by that person.

A person who fails to supply or keep a copy of the declaration of conformity, as required above, is liable on summary conviction to a fine not exceeding level 5 on the standard scale. *SI 1997 No 831, Reg 22*, however, provides a defence of due diligence: ie the defendant must show that he took all reasonable steps and exercised all due diligence to avoid committing the offence.

[SI 1997 No 831, Reg 11].

Exceptions to placing on the market or supply in respect of certain lifts and safety components

[L3010]   For the purposes of *Lifts Regulations 1997 (SI 1997 No 831), Reg 8, 9* or *10*, a lift or a safety component is not regarded as being placed on the market or supplied:

(i)   where that lift or safety component will be put into service in a country outside the Community; or is imported into the Community for re-export to a country outside the Community – but this paragraph does not apply if the CE marking, or any inscription liable to be confused with such a marking, is affixed to the lift or safety component or, in the case of a safety component, to its label; or

(ii)   by the exhibition at trade fairs and exhibitions of that lift or safety component, in respect of which the provisions of these Regulations are not satisfied, if:

—   a notice is displayed in relation to the lift or safety component in question to the effect that it does not satisfy those provisions; and that it may not be placed on the market or supplied until those provisions are satisfied; and

—   adequate safety measures are taken to ensure the safety of persons.

[SI 1997 No 831, Reg 12].

Conformity assessment procedures

[L3011] For the purposes of *Lifts Regulations 1997 (SI 1997 No 831), Reg 8(2)* or *9(2)* (see **L3005** and **L3006** above), the appropriate conformity assessment procedure is as follows:

For lifts – one of the following procedures:

— if the lift was designed in accordance with a lift having undergone an EC type-examination as referred to in *SI 1997 No 831, Sch 5*, it must be constructed, installed and tested by implementing:
  — the final inspection referred to in *SI 1997 No 831, Sch 6*; or
  — the quality assurance system referred to in *SI 1997 No 831, Sch 11* or *Sch 13*; and
  the procedures for the design and construction stages, on the one hand, and the installation and testing stages, on the other, may be carried out on the same lift;

— if the lift was designed in accordance with a model lift having undergone an EC type-examination as referred to in *SI 1997 No 831, Sch 5*, it must be constructed, installed and tested by implementing:
  — the final inspection referred to in *SI 1997 No 831, Sch 6*; or
  — one of the quality assurance systems referred to in *SI 1997 No 831, Sch 11* or *Sch 13*; and
  all permitted variations between a model lift and the lifts forming part of the lifts derived from that model lift must be clearly specified (with maximum and minimum values) in the technical dossier required as part of the appropriate conformity assessment procedure;

— if the lift was designed in accordance with a lift for which a quality assurance system pursuant to *SI 1997 No 831, Sch 12* was implemented, supplemented by an examination of the design if the latter is not wholly in accordance with the harmonised standards, it must be installed and constructed and tested by implementing, in addition, the final inspection referred to in *SI 1997 No 831, Sch 6* or one of the quality assurance systems referred to in *SI 1997 No 831, Sch 11* or *Sch 13*;

— the unit verification procedure, referred to in *SI 1997 No 831, Sch 9*, by a notified body (see **L3013** below); or

— the quality assurance system in accordance with *SI 1997 No 831, Sch 12*, supplemented by an examination of the design if the latter is not wholly in accordance with the transposed harmonised standards.

For safety components – one of the following procedures:

— submit the model of the safety component for EC type-examination in accordance with *SI 1997 No 831, Sch 5* and for production checks by a notified body (see **L3013** below) in accordance with *SI 1997 No 831, Sch 10*;

— submit the model of the safety component for EC type-examination in accordance with *SI 1997 No 831, Sch 5* and operate a quality assurance system in accordance with *SI 1997 No 831, Sch 7* for checking production; or

— operate a full quality assurance system in accordance with *SI 1997 No 831, Sch 8*.

[SI 1997 No 831, Reg 13].

Requirements fulfilled by the person who places a lift or safety component on the market

[L3012] Where in the case of a lift or a safety component, any of the requirements of *Lifts Regulations 1997 (SI 1997 No 831), Regs 8, 9, 11* and *13* to be fulfilled by the installer of the lift or the manufacturer of the safety component or, in the case of the latter, his authorised representative established in the Community, have not been so fulfilled such requirements may be fulfilled by the person who places that lift or safety component on the market [*SI 1997 No 831, Reg 14*].

This provision, however, does not affect the power of an enforcement authority to take action in respect of the installer of the lift, the manufacturer of the safety component or, in the case of the latter, his authorised representative established in the Community in respect of a contravention of or a failure to comply with any of those requirements.

Notified bodies

[L3013] For the purposes of these Regulations, a notified body is a body which has been appointed to carry out one or more of the conformity assessment procedures referred to in *Lifts Regulations 1997 (SI 1997 No 831), Reg 13* which has been appointed as a notified body by the Secretary of State in the United Kingdom or by a member State.

[*SI 1997 No 831, Reg 15*].

Lifting Operations and Lifting Equipment Regulations 1998 (LOLER) (SI 1998 No 2307)

[L3014] The *Lifting Operations and Lifting Equipment Regulations 1998 (LOLER) (SI 1998 No 2307)* uses the term employer rather than duty-holder and duties specifically assigned to the employer can be assumed to apply to the duty holder, if they have any control over lifting operations.

Where a company provides personnel to undertake work which will involve the use of lifting equipment then that company is regarded as an employer and they have a duty under LOLER to provide persons competent to undertake the work.

Duties under LOLER apply to anyone with control, to any extent, of:

— lifting equipment;
— a person at work who uses or supervises or manages the use of lifting equipment; or
— the way in which lifting equipment is used,

and to the extent of his control.

The following general definitions apply to LOLER:

- 'lifting equipment' – work equipment for lifting or lowering loads and includes its attachments for anchoring, fixing or supporting it;
- 'accessory for lifting' – work equipment for attaching loads to machinery for lifting (pendant, sling, shackle, etc.);
- 'load' – includes material or people lifted by the lifting equipment;
- 'examination scheme' – suitable scheme drawn up by a competent person for such thorough examination of lifting equipment at such intervals as may be appropriate for the purpose described in *SI 1998 No 2307, Reg 9*; and
- 'thorough examination' – means a thorough examination by a competent person including such testing as is appropriate for the purpose.

The lifting equipment that could fall under the range of LOLER is all encompassing and the following list of equipment is covered by the regulations is by no means exhaustive:

— pedestal cranes;
— mobile cranes;
— overhead gantry cranes;
— lifting accessories – slings, shackles, eyebolts, jigs, etc;
— lifts for the lifting of persons or goods;
— abseiling equipment;
— patient-hoists;
— vehicle lifting tables;
— vacuum hoists; and
— vehicle tail-lifts.

Some of these specific items of lifting equipment are looked at in more detail in **L3024** below onwards.

General requirements

Strength and stability

[L3015] Every employer must ensure that lifting equipment is of adequate strength and stability for each load, having regard in particular to the stress induced at its mounting or fixing point – and that every part of a load and anything attached to it and used in lifting it is of adequate strength [*Lifting Operations and Lifting Equipment Regulations 1998 (LOLER) (SI 1998 No 2307), Reg 4*].

To ensure adequate strength and stability the employer must:

- take account of the combination of forces which the equipment may be subjected to;
- assess the implication of the weight of any accessories;
- ensure that the equipment is not susceptible to in-service failure modes (fracture, wear or fatigue);
- have an appropriate factor of safety against foreseeable failure modes;
- take account of any combination of destabilising forces; and
- provide resistance to overturning.

Although the load does not fall within LOLER it is incumbent upon the employer to ensure that any lifting points on the load are of adequate strength.

The HSE investigation into the collapse of a Wolff 320 BF tower crane at Canada Square, London, E14 on 21 May 2000 in which three workers were killed has been produced in a public report, available since 27 June 2005, which explains that despite a very technical and complex investigation involving HSE inspectors, the Metropolitan Police, the Health and Safety Laboratory, lifting specialists and external experts, there is no conclusive explanation for the incident.

Following a detailed independent review of evidence, HSE has decided that there is insufficient evidence to support any enforcement action in relation to this tragic incident and it is not in the public interest to bring a prosecution in relation to other matters. HSE appreciates that this decision may disappoint the families of those who died.

Michael Whittard, an erection supervisor from Leeds, Martin Burgess an erector from Castleford, and Peter Clark, a crane driver from Southwark, south London, were part of a team who were in the process of using a large 'climbing frame' incorporating a hydraulic lifting device to raise the height of the tower crane when the incident occurred. They fell more than 120 meters when the top of the crane and the climbing frame overturned as they were approaching the end of a weekend of climbing operations on this and a sister crane on the site. Two other members of the crew escaped into the tower of the crane and survived.

Commenting on the announcement, Rosi Edwards, acting Chief Inspector of Construction said: 'In 2003 HSE published a discussion paper that explained concerns about crane climbing systems. Work now being undertaken by the British Standards Institute will lead to British Standard BS 7121: Part 5 Tower Cranes being revised and an improved approach to climbing in the future. Work is also underway by the Construction Industry Research and Information Association on guidance about tower crane stability and the relevant CEN and ISO committees are considering how the lessons learnt might be reflected in standards. I urge crane companies to consider the issues discussed in this report and ensure such operations are effectively managed. Meanwhile, my thoughts are with the families of Mr. Whittard, Mr. Burgess and Mr. Clark.'

According to HSE this incident was only the second collapse of a tower crane during climbing. The first occurred in San Francisco on 28 November 1989.

The summary of the report can be viewed at www.hse.gov.uk/construction/crane/overview.pdf. The full report can be viewed at www.hse.gov.uk/construction/crane. The Discussion paper on the safe use of 'external' frames on tower cranes can be viewed at www.hse.gov.uk/construction/crane/discussion.pdf.

Lifting equipment for lifting persons

[L3016] *Lifting Operations and Lifting Equipment Regulations 1998 (LOLER) (SI 1998 No 2307), Reg 5(1)(b)* does contain the exception in the applied duty under LOLER, in this case the standard is qualified **by so far as is reasonably practicable (SFAIRP)**, in that the employer shall SFAIRP prevent a person being injured whilst carrying out activities from a carrier eg mobile elevated work platform, suspended cradle etc. All other parts of this regulation demand an **absolute duty**.

Every employer must ensure that lifting equipment for lifting persons:

(i)   is such as to prevent a person using it being crushed, trapped or struck or falling from the carrier;

(ii)   is such as to prevent so far as is reasonably practicable a person using it, while carrying out activities from the carrier, being crushed, trapped or struck or falling from the carrier;

(iii)   has suitable devices to prevent the risk of a carrier falling, and, if the risk cannot be prevented for reasons inherent in the site and height differences, the employer must ensure that the carrier has an enhanced safety coefficient suspension rope or chain which is inspected by a competent person every working day; and

(iv)   is such that a person trapped in any carrier is not thereby exposed to danger and can be freed.

[*SI 1998 No 2307, Reg 5*].

Common examples of equipment NOT designed for lifting persons but which are used are forklift trucks and telescopic handlers. It should be emphasising that this type of lifting equipment is not primarily designed for lifting persons and should only be used in exceptional circumstances.

HSE has alerted owners of Mobile Elevating Work Platforms ('MEWP') to an incident in which a work basket became detached from its supporting boom. No one was injured during this incident, but it has raised concerns about the safety of other MEWPs of the same construction and make. The MEWP was made by Access Machines Ltd. It was a Micro 95 trailer mount, class A1, made in 1998, with a Safe Working Load of 120 Kg.

The basket of the MEWP that failed was bolted to a round plate by four bolts and in turn the plate was welded to a pin or stub axle on the main boom. In this instance the pin to plate weld failed allowing the basket to fall off the end of the boom. There was some evidence of fatigue in the weld but HSE is concerned that poor quality welding may have contributed to the failure.

This is possibly a one off incident but nevertheless HSE strongly recommends that owners/operators of these and similar machines have this pin/plate connection thoroughly examined by a competent person at the earliest possible opportunity.

Positioning and installation

[L3017] Employers must ensure that lifting equipment is positioned or installed in such a way as to reduce to as low as is reasonably practicable the risk of the lifting equipment or a load striking a person, or the risk from a load:

—   — drifting;
—   — falling freely; or
—   — being released unintentionally,

and that otherwise it is safe. Further consideration should be given to:

•   the need to lift loads over people, and were this unavoidable minimise the risk;

•   crushing is prevented at extreme operating positions;

- loads moving along a fixed path are suitably protected to minimise the risk of the load or equipment striking a person; and
- trapping points are prevented or access limited on travelling or slewing equipment.

Employers must also ensure that there are suitable devices for preventing anyone from falling down a shaft or hoistway and that lifting equipment cannot be unintentionally released during a loss of power to the lifting equipment or through the collision of equipment or their loads.

The use of hooks with safety catches, motion limiting devices and safe systems of work are possible means of minimising these risks. [*Lifting Operations and Lifting Equipment Regulations 1998 (LOLER) (SI 1998 No 2307), Reg 6*].

The HSE has warned of the need for an effective risk assessment for all persons involved in the operation of cranes after the prosecution of the Port of Felixstowe.

The warning followed the death of 51-year-old Dennis Burman from Brantham, Essex. Mr Burman was crushed between railings as he moved between a fixed and moveable walkway on a crane's platform during a dock familiarisation course on 17 June 2003. He then fell approximately 120 feet to the ground.

The Felixstowe Dock and Railway Company Ltd were fined £250,000 with £27,288 costs at Ipswich Crown Court on Wednesday 8 September 2004 after pleading guilty at an earlier hearing. The firm admitted failing to ensure that Mr Burman and other workers were not exposed to risks to their safety and thereby breaching *section 2 (1) of the Health and Safety at Work etc Act (HSWA) 1974.*

HSE Inspector, David Gregory, who investigated the death, said:

> In this instance the crane driver intentionally moved the cab. However his view of the cross-over point between the moving and fixed access walkway was obscured. In any event he was not able to observe the cross-over point whilst at the same time watching where he was driving the crane.

A suitable and sufficient risk assessment would have identified the potential for a fatal or serious injury at the cross-over point. Preventing access to all personnel excluding the driver whilst the crane was in operation would have prevented this incident. Alternatively, interlocking the access gates on the walkways to the movement of the cab would have achieved the same result. I hope this prosecution will remind companies who operate this type of machinery that safety must be the foremost consideration.

Marking of lifting equipment

[L3018] Employers must ensure that:

(i) machinery and accessories for lifting loads are clearly marked to indicate their safe working loads;

(ii) where the safe working load of machinery for lifting loads depends on its configuration, either the machinery is clearly marked to indicate its safe working load for each configuration, or information which clearly indicates its safe working load for each configuration is kept with the machinery;

(iii)   accessories for lifting are also marked in such a way that it is possible to identify the characteristics necessary for their safe use;

(iv)   lifting equipment which is designed for lifting persons is appropriately and clearly marked to this effect; and

(v)   lifting equipment which is not designed for lifting persons but which might be mistakenly so used is appropriately and clearly marked to the effect that it is not designed for lifting persons.

[Lifting Operations and Lifting Equipment Regulations 1998 (LOLER) (SI 1998 No 2307), Reg 7].

Organisation of lifting operations

[L3019] Every employer must ensure that any lifting or lowering of a load which involves lifting equipment is properly planned by a competent person, appropriately supervised and carried out in a safe manner.

The *Lifting Operations and Lifting Equipment Regulations 1998 (LOLER) (SI 1998 No 2307)* demands as a minimum standard that all lifting operations are carried out safely under adequate supervision and following a lifting plan. The competent person planning the operation should have adequate practical and theoretical knowledge and experience of planning lifting operations.

The plan will need to address the risks identified during a risk assessment and should identify all resources, procedures and responsibilities necessary for safe operation. The degree of planning will vary considerably depending on the type of lifting equipment and complexity of the lifting operation and degree of risk involved. This regulation does allow for routinely repeated lifting operations to be assessed in the first instance and thereafter as appropriate.

There are two elements to the plan: the suitability of the lifting equipment as per *Regulation 4* of the *Provision and Use of Work Equipment Regulations 1998 (PUWER) (SI 1998 No 2306)* and the individual lifting operation to be performed.

SI 1998 No 2307, Reg 8 requires that the employer or controller of lifting operations ensures that suitable persons are appointed for planning and supervising of such operations.

For any lifting operation it is necessary to:

(a)   carry out a risk assessment under the *Management of Health and Safety at Work Regulations 1999 (SI 1999 No 3242)*;

(b)   select suitable equipment for the range of tasks; and

(c)   plan the individual lifting operation.

The term 'Competent Person' is not prescriptively described in LOLER and is used to identify a number of different roles under the regulations. In practical terms the competency of a person may be confirmed by formal, vocational qualification or through first hand knowledge of planning or supervising the lifting operations.

The guidance to the Regulations/ACoP does state that the competent person carrying out the planning task is unlikely to be the same person as the competent person undertaking the thorough examination and testing. This

planning must take into account the location, the load to be lifted, the duration and the specific operation to be carried out.

Reference should be made to the following publications for more explicit guidance on the safe use of particular items of lifting equipment.

Regulation 4 of *(PUWER) (SI 1998 No 2306)* requires suitable work equipment to be provided for the task. There is therefore a close link between this regulation and the requirement for planning. Factors to be considered when selecting lifting equipment so that it is suitable for the proposed task should include:

•   the load to be lifted;
•   its weight, shape and centre of gravity;
•   positioning of the load before and after lifting;
•   how often the lifting equipment will be used to carry out the task;
•   where the lifting equipment will be used; and
•   the personnel available and their knowledge, training and experience (ie competence).

John Doyle Construction Limited of Welwyn Garden City, Herts and Exterior International Plc of London, EC2 were fined a total of £350,000 following the investigation into a fatal incident on a building site on 6 August 2002, at Albion Riverside Development, Hester Road, Battersea, London SW11.

Jack Tangney, a 29 year old from New Zealand died when a large timber panel called a 'shutter', which fell to the ground whilst it was being lifted from the ground to the 9th floor, from the north west corner of the site, struck him.

John Doyle Construction Limited subcontractors on the site and Exterior International Plc, principal contractor's on site, previously pleaded guilty to the above charges on 11 April 2006 at City Magistrates' Court, 1 Queen Victoria Street, London EC4N 4XY

Planning of individual operations

[L3020] Planning for routine operations this will usually be a matter for the persons undertaking the task, such as a trained and certified a forklift operator. The competent person carrying out this exercise will have to have appropriate knowledge and experience. The main elements that would need to be considered in any lifting plan could include:

—   assessment of the weight of the load, including any lifting accessories;
—   establish that the crane and accessories are in date and covered by a current thorough examination;
—   choice of the correct slings and other accessories dependant upon the local environment and conditions;
—   check the anticipated travel path;
—   prepare a place to set down the load;
—   establish the centre of gravity and that this is compensated for if off centre;
—   make a trial lift to check the lifting equipments safety devices (as appropriate);
—   use tag lines if necessary to prevent the load swinging;

— lift the load and move to the setting down position;
— release the load; and
— clear the work area.

For routine operations the initial plan will only be required for the first lifting operation. This will however need to be reviewed to ensure that nothing has changed and that the plan has remained valid.

Examples where generic lifting plans could be provided for routine lifting operations may include:

(i)   a passenger lift in an office block;
(ii)   fork lift trucks used in a warehouse operation;
(iii) teagle hoist for unloading HGVs;
(iv)   construction site hoist;
(v)   mobile elevated work platforms (MEWP's) used for general mainte-
       nance; and
(vi)   suspended cradle used for window cleaning.

[*Lifting Operations and Lifting Equipment Regulations 1998 (LOLER) (SI 1998 No 2307, Reg 8*].

Thorough examination and inspection

[L3021] An employer is under a duty to ensure:

(i)   before lifting equipment is put into service for the first time by him, that
       it is thoroughly examined for any defect unless:
       — the lifting equipment has not been used before; and
       — in the case of lifting equipment for which an EC declaration of
         conformity could or (in the case of a declaration under the *Lifts
         Regulations 1997 (SI 1997 No 831)* should have been drawn up,
         the employer has received such declaration made not more than
         twelve months before the lifting equipment is put into service; or
       — if obtained from the undertaking of another person, it is
         accompanied by physical evidence referred to in (iv) below;
       [*Lifting Operations and Lifting Equipment Regulations 1998
       (LOLER) (SI 1998 No 2307), Reg 9(1)*].
(ii)   where the safety of lifting equipment depends on the installation
       conditions, that it is thoroughly examined – after installation and
       before being put into service for the first time and after assembly and
       before being put into service at a new site or in a new location, to ensure
       that it has been correctly installed and is safe to operate [*SI 1998 No
       2307, Reg 9(2)*];
(iii) that lifting equipment which is exposed to conditions causing deterio-
       ration which is liable to result in dangerous situations is:
       thoroughly examined;
       — at least every six months, in the case of lifting equipment for
         lifting persons or an accessory for lifting;
       — at least every twelve months, in the case of other lifting
         equipment; or
       — in either case, in accordance with an examination scheme; and
       — whenever exceptional circumstances which are liable to jeop-
         ardise the safety of the lifting equipment have occurred, and

if appropriate for the purpose, is inspected by a competent person at suitable intervals between thorough examinations,

to ensure that health and safety conditions are maintained and that any deterioration can be detected and remedied in good time;
[*SI 1998 No 2307, Reg 9(3)*].

(iv) that no lifting equipment leaves his undertaking; or, if obtained from the undertaking of another person, is used in his undertaking, unless it is accompanied by physical evidence that the last thorough examination required to be carried out under this regulation has been carried out. [*SI 1998 No 2307, Reg 9(4)*].

Competent person

[L3022] The competent person responsible for undertaking the thorough examination must remain objective in their approach to this activity and be totally independent of any commercial or other conflicts of interest in the item of lifting equipment that they are to examine.

External competent persons eg Engineering Insurance Companies are expected to be accredited to the relevant BS EN 45004 standard.

A Prohibition Notice was served by the HSE on Falcon Crane Hire Ltd of Shipdam, Norfolk which required them, to take out of service all tower cranes in their fleet that had not been subject to a thorough examination by an independent competent person.

HSE took this action following the collapse of two of the company's tower cranes in less than four months at sites in Battersea (London) and Liverpool. Both incidents are the subject of on–going investigations and it is therefore too early for us to be able to identify the exact causes of either failure. Nevertheless, HSE decided to adopt a precautionary approach and required the company to demonstrate those cranes which have been thoroughly examined by competent persons employed by them, are safe to continue in operation. Any lessons learnt from the investigations will be shared with the industry as soon as possible.

The Notice affected up to 180 tower cranes which are erected on construction sites throughout Great Britain. Cranes that had already been examined by an independent competent person were not affected by the Notice and continued in service.

HSE issued a safety alert on 17 October 2006 following the incident at Battersea which can be seen at: www.hse.gov.uk/construction/pdf/towercranes.pdf. HSE re– issued the same alert on 16 January 2007 following the incident in Liverpool.

Thorough examination

[L3023]–[L3025] The employer (owner of the equipment) must determine the level of thorough examination based on manufacturer's information and other factors. The employer may need to seek specialist advice to comply with this requirement. Factors that must be taken into account by the employer should include the work being carried out, any specific risks at the location that may affect the condition of the equipment, and the intensity of use of the equipment.

A thorough examination may include visual examination, a strip down of the equipment and functional tests. Advice should be sought from manufacturer's instructions, and a competent person for guidance on what an inspection should include for each piece of equipment.

A thorough examination must be carried out under the following circumstances.

• When the equipment is put into service for the first time. If it is new equipment that has not been used before there should be a declaration of conformity, which confirms that the equipment has undergone a thorough examination.

• Where safety depends upon the conditions after installation and before being used for the first time.

If the equipment is obtained from another undertaking, as in the case of hired equipment, then a copy of the previous certificate of thorough examination must accompany the equipment.

Installed equipment applies to lifting equipment erected or built on site, such as tower cranes, hoists or gantry cranes ie lifting equipment which is intended to be there for a period of time. It would not apply to equipment such as a mobile crane as it is not 'installed'. For equipment such as a mobile crane, there must be a copy of the previous thorough examination certificate.

Thorough examination periods

(i) Lifting equipment for lifting persons must be thoroughly examined at least every six months.

(ii) Lifting accessories must be thoroughly examined at least every six months.

(iii) All other lifting equipment must be thoroughly examined at least every twelve months

(iv) Each time that exceptional circumstances that are liable to jeopardise the safety of the lifting equipment have occurred, a thorough examination is required. These may include:

— any accident involving lifting equipment, eg collapse of a crane boom;

— after extended periods without proper servicing or maintenance; and

— after long periods out of use after substantial modifications and repairs.

In general **below the hook** items are examined at six monthly intervals with **above the hook** items examined at twelve monthly intervals unless the equipment is used for lifting persons in which case the lifting equipment defaults to a six monthly interval.

The employer may decide to examine specific items of lifting equipment at different intervals in accordance with an examination scheme approved by a competent person eg an occasionally used overhead crane could be examined depending on the number of lifting operations, stipulated in the scheme of examination which be may greater than the previous frequency.

Reports and defects

(i)   A person making a thorough examination for an employer under *Lifting Operations and Lifting Equipment Regulations 1998 (LOLER) (SI 1998 No 2307), Reg 9* must:
—   immediately notify the employer of any defect in the lifting equipment which in his opinion is or could become a danger to anyone;
—   write a report of the thorough examination (see **L3022** below) to the employer and any person from whom the lifting equipment has been hired or leased;
—   where there is in his opinion a defect in the lifting equipment involving an existing or imminent risk of serious personal injury send a copy of the report to the relevant enforcing authority ('relevant enforcing authority' means, where the defective lifting equipment has been hired or leased by the employer, the Health and Safety Executive – and otherwise, the enforcing authority for the premises in which the defective lifting equipment was thoroughly examined).

(ii)   A person making an inspection for an employer under *SI 1998 No 2307, Reg 9* must immediately notify the employer of any defect in the lifting equipment which in his opinion is or could become a danger to anyone, and make a written record of the inspection.

(iii)   Every employer who has been notified under (i) above must ensure that the lifting equipment is not used before the defect is rectified; or, in the case of a defect which is not yet but could become a danger to persons, after it could become such a danger.
  [SI 1998 No 2307, Reg 10].

Prescribed information

[L3026] *Lifting Operations and Lifting Equipment Regulations 1998 (LOLER) (SI 1998 No 2307), Sch 1* specifies the information that must be contained in a report of a thorough examination, made under *SI 1998 No 2307, Reg 10* (see **L3021** above). This information can be stored electronically in a secure manner eg read only with write access protected, and available only to the competent person.

The information within the report of thorough examination must include the following.

(i)   The name and address of the employer for whom the thorough examination was made.
(ii)   The address of the premises at which the thorough examination was made.
(iii)   Particulars sufficient to identify the lifting equipment including its date of manufacture, if known.
(iv)   The date of the last thorough examination.
(v)   The safe working load of the lifting equipment or, where its safe working load depends on the configuration of the lifting equipment, its safe working load for the last configuration in which it was thoroughly examined.

(vi)   In respect of the first thorough examination of lifting equipment after installation or after assembly at a new site or in a new location:
　　— 　 that it is such thorough examination; and
　　— 　 if in fact this is so, that it has been installed correctly and would be safe to operate.

(vii) In respect of all thorough examinations of lifting equipment which do not fall within (vi) above:
　　(a) 　 whether it is a thorough examination under *SI 1998 No 2307, Reg 9(3)*:
　　　　— 　 within an interval of six months;
　　　　— 　 within an interval of twelve months;
　　　　— 　 in accordance with an examination scheme; or
　　　　— 　 after the occurrence of exceptional circumstances;
　　(b) 　 if in fact this is so, that the lifting equipment would be safe to operate.

(viii) In respect of every thorough examination of lifting equipment:
　　(a) 　 identification of any part found to have a defect which is or could become a danger to anyone, and a description of the defect;
　　(b) 　 particulars of any repair, renewal or alteration required to remedy a defect found to be a danger to anyone;
　　(c) 　 in the case of a defect which is not yet but could become a danger to anyone:
　　　　— 　 the time by which it could become such a danger;
　　　　— 　 particulars of any repair, renewal or alteration required to remedy the defect;
　　(d) 　 the latest date by which the next thorough examination must be carried out;
　　(e) 　 particulars of any test, if applicable; and
　　(f) 　 the date of the thorough examination.

(ix)   The name, address and qualifications of the person making the report; that he is self-employed or, if employed, the name and address of his employer.

(x)   The name and address of a person signing or authenticating the report on behalf of its author.

(xi)   The date of the report.

Alternative 'multiple exception report' forms are utilised by most competent persons for lifting accessories and simple items of lifting equipment. This method of reporting satisfies the requirements above, but is restricted to multiple items examined during the one visit.

The report clearly identifies the equipment being examined and lists the repairs if necessary as either an A or B defect.

A being the most serious type with an immediate repair required, whereas the B defect is usually time stipulated.

Keeping of information

[L3027] An employer who obtains lifting equipment to which the 1998 Regulations apply, and who receives an EC declaration of conformity relating to it, must keep the declaration for so long as he operates the lifting equipment [*Lifting Operations and Lifting Equipment Regulations 1998 (LOLER) (SI 1998 No 2307), Reg 11(1)*].

SI 1998 No 2307, Reg 11(2)(a) provides that the employer must ensure that the information contained in every report made to him under *SI 1998 No 2307, Reg 10(1)* (see **L3025** above) is kept available for inspection:

—   in the case of a thorough examination under *SI 1998 No 2307, Reg 9(1)* (see **L3021** above) of lifting equipment other than an accessory for lifting, until he stops using the lifting equipment;

—   in the case of a thorough examination under *SI 1998 No 2307, Reg 9(1)* (see **L3021** above) of an accessory for lifting, for two years after the report is made;

—   in the case of a thorough examination under *SI 1998 No 2307, Reg 9(2)* (see **L3021** above), until he stops using the lifting equipment at the place it was installed or assembled;

—   in the case of a thorough examination under *SI 1998 No 2307, Reg 9(3)* (see **L3021** above), until the next report is made under that paragraph or the expiration of two years, whichever is later.

The employer must ensure that every record made under *SI 1998 No 2307, Reg 10(2)* (see **L3025** above) is kept available until the next such record is made [*SI 1998 No 2307, Reg 11(2)(b)*].

Lifting operations and equipment failure

General

[L3028] Lifting equipment and accessories are not subjected to frequent failure but should failure occur it is often catastrophic. Lifting equipment is subject to statutory inspection to ensure that a fitness for purpose examination, this examination must be current throughout the equipments working life. *Lifting Operations and Lifting Equipment Regulations 1998 (LOLER) (SI 1998 No 2307)* has placed further demands than those that previously existed under repealed lifting legislation in that *SI 1998 No 2307, Reg 9(3)(b)* requires that if appropriate, that a competent person undertakes an inspection between thorough examinations (See **L3021** above). The specific requirements of this intermediate inspection are based on risk exposure and subsequent determination of the equipment.

Some failures do however still occur and they are often attributed to a combination of procedural and hardware faults.

Examples of specific causes of failure are examined in the following chapters. These examples look at possible causations, but they are by no means exhaustive and human error will almost certainly be a factor in most failures.

Cranes

[L3029] Cranes are widely used in lifting/lowering operations in construction, dock and shipbuilding works. The main hazard, generally associated with overloading or incorrect slewing, is collapse or overturning. Contact with overhead power lines is also a danger. In such cases, the operator should normally remain inside the cab and not allow anyone to touch the crane or load; the superintending engineer should immediately be informed. (For reportable dangerous occurrences in connection with cranes etc. see ACCIDENT REPORTING at A3027.)

There are a number of different types of cranes available to perform varying tasks and while there are variants they generally fall into the following categories.

(i)   Fixed crane.
(ii)   Tower crane.
(iii)   Mobile crane.
(iv)   Overhead travelling crane.

There are in addition, on construction sites, rough terrain cranes as well as crawler and wheeled cranes. Other variants include ship to shore cranes (portainers) in use at container terminals.

The HSE is to run a statutory tower crane registration scheme to improve tower crane safety and public confidence in their safety. The move follows a spate of tower crane collapses in the last two or three years. The new register has been agreed alongside a package of measures to improve tower crane safety and is likely to be in place by April 2010. The register is part of a package of proposals to improve tower crane safety.

The proposals for a statutory tower crane registration scheme stem from the April 2008 report, by the Work and Pensions Select Committee, into the role of the HSC and HSE. The report recommended that HSE should bring forward such proposals, to include information such as ownership, age, design type and other relevant factors.

Safe lifting operations by mobile cranes

[L3030] A mobile crane is a crane which is capable of travelling under its own power.

Safe lifting operations – as per BS 7121 'Safe use of cranes' – depend on co-operation between supervisor (or appointed person), slinger (and/or signaller) and crane driver.

Appointed person
(a)   Overall control of lifting operations rests with an 'appointed person', who can, where necessary, stop the operation. Failing this, control of operations will be in the hands of the supervisor (who, in some cases, may be the slinger). The appointed (and competent) person must ensure that:

      (i)   lifting operations are carefully planned and executed;

(ii)   weights and heights are accurate;

(iii)   suitable cranes are provided;

(iv)   the ground is suitable;

(v)   suitable precautions are taken, if necessary, regarding gas, water and electricity either above or below ground;

(vi)   personnel involved in lifting/lowering are trained and competent; and

(vii)   access within the vicinity at where the lifting operation is undertaking is minimised.

Supervisor

(b)   Supervisors must:

(i)   direct the crane driver where to position the crane;

(ii)   provide sufficient personnel to carry out the operation;

(iii)   check the site conditions;

(iv)   report back to the appointed person in the event of problems;

(v)   supervise and direct the slinger, signaller and crane driver; and

(vi)   stop the operation if there is a safety risk.

Crane driver

(c)   Crane drivers must:

(i)   erect/dismantle and operate the crane as per manufacturer's instructions;

(ii)   set the crane level before lifting and ensure that it remains level;

(iii)   decide which signalling system is to apply;

(iv)   inform the supervisor in the event of problems; and

(v)   carry out inspections/weekly maintenance relating to:

—   defects in crane structure, fittings, jibs, ropes, hooks, shackles,

—   correct functioning of automatic safe load indicator, over hoist and derrick limit switches.

Drivers should always carry out operations as per speeds, weights, heights and wind speeds specified by the manufacturer, mindful that the weight of slings/lifting gear is part of the load. Such information should be clearly displayed in the cab and not obscured or removed. Windows and windscreens should be kept clear and free from stickers containing operational data. Any handrails, stops, machinery guards fitted to the crane for safe access should always be replaced following removal for maintenance; and tools, jib sections and lifting tackle properly secured when not in use.

After a load has been attached to the crane hook by the lifting hook, tension should be taken up slowly, as per the slinger's instructions, the latter being in continuous communication with the driver. In the case of unbalanced loads, drivers/slingers should be familiar with a load's centre of gravity – particularly if the load is irregularly shaped. Once in operation, crane hooks should be positioned directly over the load, the latter not remaining suspended for longer than necessary. Moreover, suspended loads should not be directed over people or occupied buildings.

Slingers
(d)   Slingers must:
        (i)   attach/detach a load to/from the crane;
        (ii)   use correct lifting appliances;
        (iii)   direct movement of a load by correct signals. Any part of a load
                likely to shift during lifting/lowering must be adequately secured
                by the slinger beforehand. Spillage or discharge of loose loads
                (eg scaffolding) can be a problem, and such loads must be
                properly secured/fastened. Nets are useful for covering palletised
                loads (eg bricks).

Signallers
(e)   Quite frequently, signallers are responsible for signalling in lieu of
        slingers. Failing this, their remit is to transmit instructions from slinger
        to crane driver, when the former cannot see the load.

Crane failure

[L3031] Crane failure may be attributed to:

(a)   failure to lift vertically, eg dragging a load sideways along the ground
        before lifting;
(b)   'snatching' loads, ie not lifting slowly and smoothly;
(c)   exceeding the maximum permitted moment, ie the product of the load
        and the radius of operation;
(d)   excessive wind loading, resulting in crane instability;
(e)   defects in the fabrication of the crane, eg badly welded joints;
(f)   incorrect crane assembly in the case of tower cranes;
(g)   brake failure (rail-mounted cranes); or
(h)   in the case of mobile cranes:
        (i)   failure to use outriggers,
        (ii)   lifting on soft or uneven ground, and
        (iii)   incorrect tyre pressures.

The HSE Construction Division have issued a technical alert to the construction iIndustry on 25 January 2007 regarding High tensile bolt connections on
tower cranes – the full text can be viewed at www.hse.gov.uk/construction/pdf/t
owercranes.pdf.

This technical alert is issued as reminder and supplementary guidance to that
contained in BS 7121 Part 2:2003 "Code of practice for safe use of cranes –
Inspection, testing and examination" and the HSE have issued it with the
following qualification "This information is issued without prejudice to any
ongoing investigation."

Those who own, operate and hire tower cranes are instructed that high tensile
bolt connections, including those on masts, jibs and slew rings of their tower
cranes, need to be correctly installed and pre-loaded (tensioned). Failure to do
so could lead to the bolt connection failing, with catastrophic consequences.

High tensile bolt connections consist of a bolt, nut, and hardened washer, and
on occasions a spacing sleeve. All of the components should be manufactured

from high strength materials. Bolts and nuts are typically manufactured to BS EN ISO 898-1 and BS EN 20898-2, and are marked with a manufacturer's logo or name, strength grade and a batch number. Only components of the same strength grade should be used together.

High tensile bolt connections are subject to repeated cyclic loading as the crane lifts and lowers loads and slews. When correctly installed and preloaded high tensile bolt connections can transmit very large loads.

Hoists and lifts

[L3032] The safe use and maintenance of hoists and lifts is governed primarily by the *Lifting Operations and Lifting Equipment Regulations 1998 (SI 1998 No 2307)*. Although these regulations would not apply where the lift is not in a workplace and is predominantly used by members of the public eg within a shopping precinct.

Examples include: goods lifts; man hoists, scissors lifts and passenger lifts.

Powered working platforms are commonly used for fast and safe access to overhead machinery/plant, stored products, lighting equipment and electrical installations as well as for enabling maintenance operations to be carried out on high-rise buildings. Their height, reach and mobility give them distinct advantages over scaffolding, boatswain's chairs and platforms attached to fork lift trucks. Typical operations are characterised by self-propelled hydraulic booms, semi-mechanised articulated booms and self-propelled scissors lifts.

Platforms should always be sited on firm level working surfaces and their presence indicated by traffic cones and barriers. Location should be away from overhead power lines – but, if this is not practicable the safe system of work in operation needs to control this additional risk. A key danger arises from overturning as a consequence of overloading the platform. Maximum lifting capacity should, therefore, be clearly indicated on the platform as well as in the manufacturer's instructions, and it is inadvisable to use working platforms in high winds (ie above Force 4 or 16 mph). Powered working platforms should be regularly maintained and only operated by trained personnel.

Hoist and lift failure

[L3033] Hoist and lift failure can be attributed to:

(a)   excessive wear in wire ropes;
(b)   excessive broken wires in ropes;
(c)   mechanical failure (eg worm and gearing);
(d)   failure of the overload protection device;
(e)   failure of the overrun device;
(f)   failure of the speed governor;
(g)   safety gear failure; or
(h)   failure of the landing door interlocks.

Fork lift trucks

[L3034] Fork lift trucks are the most widely used item of mobile mechanical handling equipment. There are several varieties which are as follows:

(1)   *Pedestrian-operated stackers – manually-operated and power-operated*
Manually-operated stackers are usually limited in operation, for example, for moving post pallets, and cannot pick up directly from the floor. Whereas power-operated stackers are pedestrian-operated or rider-controlled, operate vertically and horizontally and can lift pallets directly from the floor.

(2)   *Reach trucks*
Reach trucks enable loads to be retracted within their wheel base. There are two kinds, namely, (*a*) moving mast reach trucks, and (*b*) pantograph reach trucks. Moving mast reach trucks are rider-operated, with forward-mounted load wheels enabling carriage to move within the wheel base – mast, forks and load moving together. Pantograph reach trucks are also rider-operated, reach movement being by pantograph mechanism, with forks and load moving away from static mast.

(3)   *Counterbalance trucks*
Counterbalance trucks carry loads in front counterbalanced to the weight of the vehicle over the rear wheels. Such trucks are lightweight pedestrian-controlled, lightweight rider-controlled or heavyweight rider-controlled.

(4)   *Narrow aisle trucks*
With narrow aisle trucks the base of the truck does not turn within the aisle in order to deposit/retrieve load. There are two types, namely, side loaders for use on long runs down narrow aisles, and counterbalance rotating load turret trucks, having a rigid mast with telescopic sections, which can move sideways in order to collect/deposit loads.

(5)   *Order pickers*
Order pickers have a protected working platform attached to the lift forks, enabling the driver to deposit/retrieve objects in or from a racking system. Conventional or purpose-designed, they are commonly used in racked storage areas and operate well in narrow aisles.

Forklift trucks are involved in far too many incidents although manufacturers are increasing the number of safety devices an attempt to limit human error, these include:

- speed limiters with the forks elevated;
- pressure sensitive seats, that cut off the fuel supply when the driver disembarks;
- TV monitors when the truck has obscured vision;
- intelligent start up systems, swipe cards or similar;
- lap straps; and
- falling object protection.

Forklift trucks are mobile work equipment and with the exception of rough terrain vehicles all are protected from rolling over beyond 90 degrees should an overturn occur by their mast's. Lap straps should be used when fitted although the requirement for their use is risk based.

Forklift truck failure

[L3035] Forklift truck failure can be attributed to:

(a)   uneven floors, steeply inclined ramps or gradients, ie in excess of 1:10 gradient;
(b)   inadequate room to manoeuvre;
(c)   inadequate or poor maintenance of lifting gear;
(d)   the practice of driving forwards down a gradient with the load preceding the truck;
(e)   load movement in transit;
(f)   sudden or fast braking;
(g)   poor stacking of goods being moved;
(h)   speeding;
(i)   turning corners too sharply;
(j)   not securing load sufficiently eg pallets stacked poorly;
(k)   hidden obstructions in the path of the truck;
(l)   use of the forward tilt mechanism with a raised load; or
(m)   generally bad driving, including driving too fast, taking corners too fast, striking overhead obstructions, particularly when reversing and excessive use of the brakes.

Patient/bath hoist

[L3036] Patient/bath hoists are used for the lifting of persons and require a six-month thorough examination.

Patient hoist failure:

(a)   mechanical damage to pulleys sheaves and drums;
(b)   anchor point failure;
(c)   breaking mechanism failure;
(d)   boom arm distortion or corrosion;
(e)   electrical faults (ere fitted); and
(f)   hook catch failure.

BUPA Care Homes Ltd has been fined £90,000 and ordered to pay £19,247 costs to HSE in a prosecution by the HSE, following the death of 95–year–old woman, Mrs Charlotte Wood from Mottingham.

Mrs Wood became a resident at the Abbotsleigh Mews Residential and Nursing Home in Sidcup run by BUPA Care Homes Ltd in 2001. She was unable to walk or move independently. On 28 November 2003, Mrs Wood slipped from a Sarita hoist, which was being used to get her out of a bath, and fractured her shoulder. While waiting for surgery on her shoulder, Mrs Wood contracted pneumonia and subsequently died on 2 December 2003.

Following the sentencing hearing at Southwark Crown Court on 7 September 2006, HSE Inspector Hazel McCallum said: 'All too often we hear of cases where vulnerable, elderly people are not afforded the standard of care they deserve because the systems in place are not properly followed. The tragedy is that Mrs Wood's death was entirely avoidable.'

The care assistant who was attending Mrs Wood had been employed by BUPA for about six weeks but had not received training and had not used this type of hoist before. Risk assessments and procedures for manual handling and safe bathing were not brought to the attention of care assistants and the supervision of staff carrying out lifting operations was inadequate.

Mrs Wood's family say she had other minor falls from hoists while at Abbottsleigh Mews. Her son, Geoff Wood said: 'My Mother was a much loved and central part of our family and her loss was a great blow to us all. Although she was frail she was in good health and we had fully expected her to get a telegram from the Queen.'

BUPA were found guilty at the hearing at Southwark Crown Court of an offence as provided by *HSWA 1974 s 33(1)(a)*.

BUPA Care Homes Ltd pleaded guilty in November 2004 to two other charges under the Lifting Operations and Lifting Equipment Regulations 1998. In this case a 90–year–old woman died after a brain haemorrhage resulting from a fall from a sling when she was being hoisted out of a bath. BUPA was fined £2,500 and the HSE awarded full costs.

Vehicle lifting table

[L3037] The primary function is as a vehicle lift but personnel are at risk when the vehicle is descending, by definition a twelve monthly thorough examination would normally satisfy the inspection frequency laid down in the *Lifting Operations and Lifting Equipment Regulations 1998 (SI 1998 No 2307)* most of industry have adopted a six monthly examination frequency. Lifting tables are normally either wire suspension type or hydraulic actuated. BS EN 1433:1999 refers.

Vehicle lifting table failure:

(a) mechanical damage to pulleys sheaves and wires;
(b) anchor point failure;
(c) breaking mechanism failure;
(d) electrical faults;
(e) hydraulic failure;
(f) foundation bolt failure;
(g) catching device failure causing vehicles to fall from table; and
(h) carrying arm failure.

Lifting accessories

[L3038] 'Accessory for lifting' – work equipment for attaching loads to machinery for lifting (pendant, sling, shackle, etc.).

It has become common practice within industry to use a simple colour coded cable tie to identify that the accessory is within its inspection cycle. This colour tie will normally be displayed at the lifting equipment store for ease of recognition.

Lifting accessories refer to:

(a)   ropes;
(b)   chains;
(c)   slings;
(d)   shackles etc.

Ropes

[L3039] Ropes are used quite widely throughout industry in lifting/lowering operations, the main hazard being breakage through overloading and/or natural wear and tear. Ropes are either of natural (eg cotton, hemp) or man-made fibre (eg nylon, terylene). Natural fibre ropes, if they become wet or damp, should be allowed to dry naturally and kept in a well-ventilated room. They do not require certification prior to service, but six months' examination thereafter is compulsory (all lifting equipment accessories require a six monthly examination to meet the requirements under the *Lifting Operations and Lifting Equipment Regulations 1998*). If the specified safe working load (SWL) cannot be maintained by a rope, it should forthwith be withdrawn from service. Of the two, man-made fibre ropes have greater tensile strength, are not subject so much to risk from wear and tear, are more acid/corrosion-resistant and can absorb shock loading better.

Fibre rope failure

[L3040] Fibre rope failure can be attributed to:

(a)   bad storage in wet or damp conditions resulting in rot and mildew;
(b)   inadequate protection of the rope when lifting loads with sharp edges;
(c)   exposure to direct heat to dry, as opposed to gradual drying in air;
(d)   chemical reaction; or
(e)   Ultra Violet degradation.

Wire (steel) ropes

[L3041] Wire ropes are much in use in cranes, lifts, hoists, elevators etc. and as such become part of the lifting equipment, they are also commonly utilised as accessories and they need to be lubricant-impregnated to minimise corrosion and reduce wear and tear unless they are manufactured from corrosive resistive materials for specialist applications. BS EN 12385 refers.

Wire rope failure(s)

[L3042] Wire rope failure can be attributed to:

There have been instances where ropes have had to be changed due to visible damage to the wires. Ropes showing no external visible sign of damage have failed during normal operations with catastrophic consequences as a result of fatigue failure often brought about by reverse cycle leading which significantly reduces the ropes operational life.

(a)   excessive broken wires;
(b)   failure to lubricate regularly;
(c)   *fatigue failure of the rope*

(d)   frequent knotting or kinking of the rope; or

(e)   bad storage in wet or damp conditions which promotes rust.

Chain

[L3043] A chain is a classic example of a lifting accessory, in spite of an increase in use of wire ropes. There are several varieties, including mild, high-tensile, and alloy steel chains. The principal risk is breakage, usually occurring in consequence of a production defect in a link, or through application of excessive loads. A crucial part of the thorough examination is to measure the chain wear for which a simple gauge is used which allows measurement of elongation and link/pin wear.

Chain failure

[L3044] Chain failure can be attributed to:

(a)   mechanical defects in individual links;

(b)   application of a static in excess of the breaking load;

(c)   snatch loading; and

(d)   hydrogen embrittlement (manufacturing welding defects refer HSE publication PM 39, Hydrogen cracking of grade T(8) chain and components)).

Slings

[L3045] Slings are manufactured in natural or man-made fibre or chain. The safe working load of any sling varies according to the angle formed between the legs of the sling.

The relationship between sling angle and the distance between the legs of the sling is also important. (See Table 1 below.)

Table 1 Safe working load for slings	
Sling Angle	*Distance between legs*
30°	$^1/_2$ leg length
60°	1 leg length
90°	$1^1/_3$ leg length
120°	$1^2/_3$ leg length

For a one tonne load, the tension in the leg increases as shown in Table 2 below.

Table 2 Safe working load for slings – increased tension	
Sling leg angle	*Tension in leg (tonnes)*
90°	0.7
120°	1.0
151°	2.0

Table 2
Safe working load for slings – increased tension

Sling leg angle	*Tension in leg (tonnes)*
171°	6.0

Other causes of failure in slings are:

(a)   cuts, excessive wear, kinking and general distortion of the sling legs;

(b)   failure to lubricate wire slings;

(c)   failure to pack sharp corners of a load, resulting in sharp bends in the sling and the possibility of cuts or damage to it; and

(d)   unequal distribution of the load between the legs of a multi-leg sling.

Eyebolts

[L3046] Eyebolts are mainly for lifting heavy concentrated loads; there being several types, namely, dynamo, collar and eyebolt incorporating link.

Eyebolt failure

[L3047] Eyebolt failure can be attributed to:

(a)   mechanical defects;

(b)   incorrect thread (metric and imperial threads, cross threaded);

(c)   dynamo eyebolt used where collar eyebolt was required;

(d)   eyebolt not hardened down to the shoulder; or

(e)   oil remaining in the tapped hole receiving the eyebolt, thus destroying the receiving housing under hydraulic pressure.

Shackles

[L3048] Shackles come in two main types 'Dee' shackle and 'Bow' so named because of there shapes. The bow shackle has the capability to accept a larger number of ropes, slings etc. The shackle and pin are matched pairs and should not be separated and mixed up.

Shackle failure

[L3049] Shackle failure can be attributed to:

(a)   mechanical defects;

(b)   interchange of pin and shackle;

(c)   nut and bolt used instead of the shackle pin; or

(d)   wear on the shackle.

Once only use accessories

[L3050] Also included are 'one off' accessories which are disposed of after there use and examples of these are:

- industrial bulk carriers (IBC's) bags that are commonly used in the building and chemical industry where it would not be possible certify integrity following there first use;
- health care, under body sling used with patient hoists, the problems associated with sterilising such equipment through autoclaving when it would deteriorate and the cost of examination by a competent person have made it both a practical solution and an economic proposition to discard such equipment after use; and
- dynamo eyebolt, are designed for a single lift only.

Mobile lifting equipment

[L3051] Mobile work equipment is any equipment which carries out work whilst travelling. It may be self-propelled, towed or remotely controlled and it may incorporate attachments. An example of mobile lifting equipment is an excavator involved in digging tasks.

The risk of mobile lifting equipment overturning is another cause for concern, and the problem was addressed by the additional requirements introduced in the *Provision and Use of Work Equipment Regulations 1998 (SI 1998 No 2306)*. These implement into UK legislation the amending directive to the *Use of Work Equipment Directive (AUWED)*. Under this legislation, workers must be protected from falling out of the equipment and from unexpected movement eg overturning.

To prevent such equipment overturning, the following action should be taken:

— increase stability when reasonably practicable (this may be done by reducing the equipment's speed)
— fit stabilisers (eg outriggers or counterbalance weights);
— ensure that roll over protection structure is installed which ensures that it does no more than fall on its side;
— ensure the structure gives sufficient clearance to anyone being carried if it overturns more than on its side;
— provision of harnesses or a suitable restraining system to protect against being crushed;
— use only on firm ground; and
— avoid excessive gradients.

Lighting

Nicola Coote

Introduction

[L5001] Increasingly, over the last decade or so, employers have come to appreciate that indifferent lighting is both bad economics and bad ergonomics, not to mention potentially bad industrial relations. Conversely, good lighting uses energy efficiently and contributes to general workforce morale and profitability – that is, operating costs fall whilst productivity and quality improve. Alternatively, poor lighting reduces efficiency, thereby increasing the risk of stress, denting workforce morale, promoting absenteeism and leading to accidents, injuries and even deaths at work.

Ideally, good lighting should 'guarantee' (*a*) employee safety, (*b*) acceptable job performance and (*c*) good workplace atmosphere, comfort and appearance. This is not just a matter of maintenance of correct lighting levels. Ergonomically relevant are:

(a)   horizontal illuminance;
(b)   uniformity of illuminance over the job area;
(c)   colour appearance;
(d)   colour rendering;
(e)   glare and discomfort;
(f)   ceiling, wall, floor reflectances;
(g)   job/environment illuminance ratios;
(h)   job and environment reflectances;
(i)   vertical illuminance.

Increasingly there are growing financial pressures to review lighting more from an energy conservation point of view. This can have many financial and environmental benefits, provided the energy efficient system being used does not detriment the lighting levels. The Chartered Institute of Building Service Engineers (CIBSE www.cibse.org). Their CIBSE *Guide F: Energy efficiency in buildings* (chapter 9) contains further information about energy efficiency lighting.

Statutory lighting requirements

General

[L5002] Adequate standards of lighting in all workplaces can be enforced under the general duties of the *Health and Safety at Work etc. Act 1974*, which require provision by an employer of a safe and healthy working environment (see EMPLOYERS' DUTIES TO THEIR EMPLOYEES). *Regulation 8* of the *Workplace (Health, Safety and Welfare) Regulations 1992 (SI 1992 No 3004)* requires that all workplaces have suitable and sufficient lighting and that, so far as is reasonably practicable, the lighting should be natural light.

More particularly, however, people should be able to work and move about without suffering eye strain (visual fatigue) and having to avoid shadows. Local lighting may be necessary at individual workstations and places of particular risk. Outdoor traffic routes used by pedestrians should be adequately lit after dark. Lights and light fittings should avoid dazzle and glare, and be so positioned that they do not cause hazards, whether fire, radiation or electrical. Switches should be easily accessible and lights should be replaced, repaired or cleaned before lighting becomes insufficient. Moreover, where persons are particularly exposed to danger in the event of failure of artificial lighting, emergency lighting must be provided (*Workplace (Health, Safety and Welfare) Regulations 1992 (SI 1992 No 3004), Reg 8(3)* (and ACOP)).

To accommodate these 'requirements', refresh rates (flickering), a combination of general and localised lighting (not to be confused with 'local' lighting, e.g. desk light) is often necessary. Moreover, state of the art visual display units have thrown up some occupational health problems addressed by the *Health and Safety (Display Screen Equipment) Regulations 1992 (SI 1992 No 2792) as amended by the Health and Safety (Miscellaneous) Regulations 2002* (see O5005 *et seq.* OFFICES AND SHOPS).

The Chartered Institute of Building Services Engineers (CIBSE) produces a wide range of technical and industry-specific guidance on design of lighting in various environments, commissioning and installation, etc. As an example, *Lighting Guide 7: office lighting* covers all aspects of the lighting of offices, from board room through general offices to the post room. This organisation provides guidance on considerations to be taken when designing energy reduction systems for lighting in various work areas as well as recommended luminescence for different types of work environment. There are several lighting guides for a wide range of working environments and further industry specific guidance can be obtained via the CIBSE website.

The Society of Light and Lighting acts as the professional body for lighting in the UK. It operates worldwide and offers a range of publications and advice on specific areas of lighting.

Specific processes

Lighting and VDUs

[L5003] Annex A to BS EN 9241-6-1999 provides guidance on illuminance levels, specifically in relation to use with display screen equipment (BS EN 9241 – Ergonomic requirements for office work with visual display terminals (VDTs); Part 6 – Guidance on work environment). BS EN 12464-1–2002 (Light and lighting. Lighting of work places, Indoor work places) also provides advice and guidance in this respect.

Machinery and work equipment – Provision and Use of Work Equipment Regulations 1998 (SI 1998 No 2306)

[L5004] General lighting requirements for work with machinery comes under the *Provision and Use of Work Equipment Regulations 1998 (SI 1998 No 2306), Reg 21*. This Regulation is very broad and merely requires that suitable and sufficient lighting is provided, taking into account the operations to be carried out. To ascertain what is required to meet this duty, the HSE guidance has been produced. This refers to occasions when additional or localised lighting may be needed such as:

- when the task requires a high perception of detail;
- when there is a dangerous process;
- to reduce visual fatigue;
- during maintenance operations.

Additional guidance is provided in the HSE Guidance Note HS(G) 38, 'Lighting at Work.

Lighting requirements on construction sites – *Construction (Design & Management) Regulations 2007 (SI 2007 No 320)*

[L5005] Lighting requirements for construction sites can be found in Regulation 44. These regulations state that every place of work, including traffic routes around the construction site and the approaches to the workplace must be provided with suitable and sufficient lighting, which shall be, so far as is reasonably practicable, by natural light. The colour of any artificial lighting provided must not adversely affect or change the perception of any sign or signal provided for the purposes of health and safety.

Suitable and sufficient secondary lighting shall be provided in any place where there would be a risk to the health or safety of any person in the event of failure of primary artificial lighting.

Lighting requirements for electrical equipment

[L5006] So as to prevent injury, adequate lighting must be provided at all electrical equipment on which or near which work is being done in circumstances that may give rise to danger (*Electricity at Work Regulations 1989 (SI 1989 No 635), Reg 15*).

Sources of light

Natural lighting

[L5007] Daylight is the natural, and cheapest, form of lighting, but it has only limited application to places of work where production is required beyond the hours of daylight, at all seasons, and where daytime visibility is restricted by climatic conditions. But however good the outside daylight, windows can rarely provide adequate lighting alone for the interior of large floor areas. Single storey buildings can, of course, make use of insulated opaque roofing materials, but the most common provision of daylight is by side windows. There is also the fact that the larger the glazing area of the building, the more other factors such as noise, heat loss in winter and unsatisfactory thermal conditions in summer must be considered.

Modern conditions, where the creation of pleasant building interior environment requires the balanced integration of lighting, heating, air conditioning, acoustic treatment, etc., are such that lighting cannot be considered in isolation. At the very least, natural lighting will have to be supplemented for most of the time with artificial lighting, the most common source for which is electric lighting.

Solar lighting

[L5008] This technology is still fairly new but is beginning to expand rapidly, and government grants are now frequently available to assist with installation. However, they still provide quite limited results in a number of workplace environments. However, there are some environments when use of solar lighting is useful and effective. Examples include isolated satellite work premises that are occupied for short periods, some outdoor areas for additional evening lighting and areas where low levels of luminance are sufficient. The CIBSE guide J: *Weather, Solar and Illuminance Data* can help provide further information on this subject.

Electric lighting

[L5009] Capital costs, running costs and replacement costs of various types of electric lighting have a direct bearing on the selection of the sources of electric lighting for particular application. Such costs are as important a consideration as the size, heat and colour effects required of the lighting. The efficiency of any type of lamp used for lighting is measured as light output, in lumens, per watt of electricity. Typical values for various types of lamp are as follows (the term 'lumen' is explained in the discussion of standards of illuminance in L5010 below).

Type of lamp	Lumens per watt
Incandescent lamps	10 to 18
Tungsten halogen	22
High pressure mercury	25 to 55
Tubular fluorescent	30 to 80 (depending on colour)
Mercury halide	60 to 80
High pressure sodium	100

In general, the common incandescent lamps (coiled filament lamps, the temperature of which is raised to white heat by the passage of current, thus giving out light) are relatively cheap to install but have relatively expensive running costs. The sale of incandescent bulbs is being phased out in the UK from 2011. A discharge or fluorescent lighting scheme (which works on the principle of electric current passing through certain gases and thereby producing an emission of light) has higher capital costs but higher running efficiency, lower running costs and longer lamp life. In larger places of work the choice is often between discharge and fluorescent lamps. The normal mercury discharge lamp and the low pressure sodium discharge lamp have restricted colour performance, although newly developed high pressure sodium discharge lamps and colour corrected mercury lamps do not suffer from this disadvantage.

Energy saving bulbs are being widely used, and these have the benefit of producing the same luminance with a fraction of the energy required to activate a common incandescent bulb. Their lifespan is also significantly increased. However, they take a lot longer to become fully activated to full output and this can have a detrimental affect on safety in areas where lights are being turned on and off on a regular basis, as visibility will be reduced until they have sufficiently warmed up.

Standards of lighting or illuminance

The technical measurement of illuminance

[L5010] The standard of illuminance (the amount of light) required for a given location or activity depends on a number of variables, including general comfort considerations and the visual efficiency required. The unit of illuminance is the 'lux', which equals one lumen per square metre; this unit has now replaced the 'foot candle' which was the number of lumens per square foot. The term 'lumen' is the unit of luminous flux, describing the quantity of light received by a surface or emitted by a source of light.

Light measuring instruments

[L5011] For accurate measurement of the degree of illuminance at a particular working point, a reliable instrument is required. Such an instrument, suitable for most measurements, is a pocket lightmeter; this incorporates the principle of the photo-electric cell, which generates a tiny electric current in proportion to the light at the point of measurement. This current deflects a pointer on a graduated scale measured in lux. Manufacturers' instructions should, of course, be followed in the care and use of such instruments.

Average illuminance and minimum measured illuminance

[L5012] HSE Guidance Note HS(G) 38 'Lighting at work' (1998) relates illuminance levels to the degree or extent of detail which needs to be seen in a particular task or situation. Recommended illuminances are shown in Table 1 at L5013 below.

This guidance note makes recommendations both for average illuminance for the work area as a whole and for minimum measured illuminance at any position within it. As the illuminance produced by any lighting installation is rarely uniform, the use of the average illuminance figure alone could result in the presence of a few positions with much lower illuminance, which pose a threat to health and safety. The minimum measured illuminance is therefore the lowest illuminance permitted in the work area taking health and safety requirements into account.

The planes on which the illuminances should be provided depend on the layout of the task. If predominantly on one plane, for example horizontal, as with an office desk, or vertical, as in a warehouse, the recommended illuminances are recommended for that plane. Where there is either no well-defined plane or more than one, the recommended illuminances should be provided on the horizontal plane and care taken to ensure that the reflectances of surfaces in working areas are high.

Illuminance ratios

[L5013] The relationship between the lighting of the work area and adjacent areas is significant. Large differences in illuminance between these areas may cause visual discomfort or even affect safety levels where there is frequent movement, for example forklift trucks. This problem arises most often where local or localised lighting in an interior exposes a person to a range of illuminance for a long period, or where there is movement between interior and exterior working areas exposing a person to a sudden change of illuminance. To reduce hazards and possible discomfort, specific recommendations shown in Table 1 below should be followed.

Where there is conflict between the recommended average illuminances shown in Table 1 and maximum illuminance ratios shown in Table 2, the higher value should be taken.

Table 1 Average illuminances and minimum measured illuminances for different types of work			
General activity	Typical locations/types of work	Average illuminance (Lx)	Minimum measured illuminance (Lx)
Movement of people, machines and vehicles*	Lorry parks, corridors, circulation routes	20	5
Movement of people, machines and vehicles in hazardous areas; rough work not requiring any perception of detail	Construction site clearance, excavation and soil work, docks, loading bays, bottling and canning plants	50	20

Work requiring limited perception of detail**	Kitchens, factories, assembling large components, potteries	100	50
Work requiring perception of detail	Offices, sheet metal work, bookbinding	200	100
Work requiring perception of fine detail	Drawing offices, factories assembling electronic components, textile production	500	200

Notes

Only safety has been considered, because no perception of detail is needed and visual fatigue is unlikely. However, where it is necessary to see detail to recognise a hazard or where error in performing the task could put someone else at risk, for safety purposes as well as to avoid visual fatigue the figure should be increased to that for work requiring the perception of detail.

** The purpose is to avoid visual fatigue: the illuminances will be adequate for safety purposes.

Table 2

Maximum ratios of illuminance for adjacent areas

Situations to which recommendation applies	Typical location	Maximum ratio of illuminances		
		Working area		Adjacent area
Where each task is individually lit and the area around the task is lit to a lower illuminance	Local lighting in an office	5	:	1
Where two working areas are adjacent but one is lit to a lower illuminance than the other	Localised lighting in a works store	5	:	1
Where two working areas are lit to different illuminances and are separated by a barrier but there is frequent movement between them	A storage area inside a factory and a loading bay outside	10	:	1

Maintenance of light fitments

[L5014] The lighting output of a given lamp will reduce gradually in the course of its life but an improvement can be obtained by regular cleaning and maintenance, not only of the lamp itself, but also of the reflectors, diffusers and other parts of the luminaire. A sensible and economic lamp replacement policy is called for (e.g. it may be more economical, in labour cost terms, to change a batch of lamps than deal with them singly as they wear out).

Qualitative aspects of lighting and lighting design

[L5015] While the quantity of lighting afforded to a particular location or task in terms of standard service illuminance is an important feature of lighting design, it is also necessary to consider the qualitative aspects of lighting, which have both direct and indirect effects on the way people perceive their work activities and dangers that may be present. The quality of lighting is affected by the presence or absence of glare, the distribution of the light, brightness, diffusion and colour rendition.

Glare

[L5016] This is the effect of light which causes impaired vision or discomfort experienced when parts of the visual field are excessively bright compared with the general surroundings. It may be experienced in three different forms:

(a)   disability glare – the visually disabling effect caused by bright bare lamps directly in the line of vision;

(b)   discomfort glare – caused by too much contrast of brightness between an object and its background, and frequently associated with poor lighting design. It can cause discomfort without necessarily impairing the ability to see detail. Over a period it can cause visual fatigue, headaches and general fatigue;

(c)   reflected glare – the reflection of bright light sources on shiny or wet work surfaces, such as plated metal or glass, which can almost entirely conceal the detail in or behind the object that is glinting.

N.B. The Illuminating Engineering Society (IES) publishes a Limiting Glare Index for each of the effects in (*a*) and (*b*) above. This is an index representing the degree of discomfort glare which will be just tolerable in the process or location under consideration. If exceeded, occupants may suffer eye strain or headaches or both.

Distribution

[L5017] Distribution is concerned with the way light is spread. The British Zonal Method classifies luminaires (light-fittings) according to the way they distribute light from BZ1 (all light downwards in a narrow column) to BZ10 (light in all directions). However, a fitting with a low BZ number does not necessarily imply less glare. Its positioning, the shape of the room and the reflective surfaces present are also significant.

The actual spacing of luminaires is also important when considering good lighting distribution. To ensure evenness of illuminance at operating positions, the ratio between the height of the luminaire and the spacing of it must be considered. The IES spacing: height ratio provides a basic guide to such arrangements. Under normal circumstances, for example offices, workshops and stores, this ratio should be between $1\frac{1}{2}:1$ and $1:1$ according to the type of luminaire.

Brightness

[**L5018**] Brightness or 'luminosity' is very much a subjective sensation and, therefore, cannot be measured. However, it is possible to consider a brightness ratio, which is the ratio of apparent luminosity between a task object and its surroundings. To ensure the correct brightness ratio, the reflectance (the ability of a surface to reflect light) of all surfaces in the working area should be well maintained and consideration given to reflectance values in the design of interiors. Given a task illuminance factor (the recommended illuminance level for a particular task) of 1, the effective reflectance values should be 0.6 for ceilings, 0.3 to 0.8 for walls and 0.2 to 0.3 for floors.

Diffusion

[**L5019**] This is the projection of light in all directions with no predominant direction. The directional flow of light can often determine the density of shadows, which may prejudice safety standards or reduce lighting efficiency. Diffused lighting will reduce the amount of glare experienced from bare luminaires.

Colour rendition

[**L5020**] Colour rendition refers to the appearance of an object under a specific light source compared to its colour under a reference illuminant, for example natural light. Good standards of colour rendition allow the colour appearance of an object to be properly perceived. Generally, the colour rendering properties of luminaires should not clash with those of natural light, and should be just as effective at night when there is no daylight contribution to the total illumination of the working area.

Stroboscopic effect

[**L5021**] One aspect of lighting quality that formerly gave trouble was the stroboscopic effect of fluorescent tubes, which gave the illusion of motion or even the illusion that a rotating part of machinery was stationary. With modern designs of fluorescent tubes, this effect has largely been eliminated.

Machinery Safety

Hani Raafat and Andrea Oates

Introduction

[M1001] The introduction of the harmonised European Union ('EU') regulations and standards in health and safety has changed the approach to the prevention of accidents and ill health at work. In the place of a prescriptive set of standards and regulations, the harmonised regulations and standards represent a remarkable breakthrough in the identification, assessment and control of machinery and work equipment risks.

As far as machinery safety is concerned, the introduction of the EU concept for the proactive 'goal-setting' approach has fundamentally changed the approach to the prevention of accidents and ill health at work. In the place of prescriptive legislation and standards, the new risk-based approach to machinery and work equipment safety has fundamentally changed the way for the identification, assessment and control of machinery hazards. The new approach is essentially based on structured and systematic assessment of risks.

This chapter aims to review this approach, which is based on the EU Machinery and Use of Work Equipment Directives as well as the Transposed Harmonised EU Machinery Safety Standards. It aims to give guidance and examples on the link between the structured approach for hazard identification and analysis as well as risk evaluation. This helps in developing action (health and safety) plans for the implementation of a package of measures for machinery and work equipment accident/ill health prevention in a practical way.

Currently, statutory duties are dually (but separately) laid on both manufacturers of machinery for use at work and employers and users of such machinery and equipment, prior compliance, on the part of manufacturers leading to compliance on the part of employers/users. General and specific duties are imposed on manufacturers by the *Supply of Machinery (Safety) Regulations 1992 (SI 1992 No 3073)* as amended by the *Supply of Machinery (Safety) (Amendment) Regulations 1994 (SI 1994 No 2063)*.

Duties of employers are incorporated into the *Provision and Use of Work Equipment Regulations 1998 (SI 1998 No 2306)*; as well as the *Lifting Operations and Lifting Equipment Regulations 1998 (SI 1998 No 2307)*.

Legal requirements relating to machinery

[M1002] The introduction of the *Health and Safety at Work etc. Act ('HSWA 1974')* in 1974, in addition to placing general duties on employers under *section 2* of the Act, placed the responsibility of machinery safety on designers, manufacturers and suppliers. This is still applicable only in the UK, in addition to the EU requirements.

The original version of the text in *section 6(1)* of *HSWA 1974* placed the following duties on designers, manufacturers and suppliers:

> It shall be the duty of any person who designs/supplies . . . any article for use at work to ensure, so far as is reasonably practicable, that the article is so designed . . . as to be safe and without risks to health . . . when properly used.

The original *section 6* concentrated on protection against physical injuries during normal use of machinery and placed obligations on the user to follow the manufacturer instructions. The term 'so far as is reasonably practicable' has been interpreted by the courts as to indicate that in deciding an adequate level of safety, both the risk and cost of dealing with must be taken into account. This is a difficult concept for the majority of machine designers who understandably are cost driven.

The introduction of the European Directive on product liability, which was implemented in the UK under the *Consumer Protection Act* in 1987, in addition to the deficiencies shown by several court cases, resulted in the revision of *section 6* in 1988.

The current *section 6(1)* of the 1974 Act requires that:

> It shall be the duty of any person who designs/ supplies . . . any article for use at work to ensure, so far as is reasonably practicable, that the article is so designed . . . that it will be safe without risk to health at all times: when being set, used, cleaned or maintained by a person at work.

It should be noted that the term 'when properly used' had disappeared, as it is expected that designers should take account of normal behaviour of operators and maintenance personnel and the machine design should accommodate foreseeable misuse.

Machinery safety – the risk based approach

[M1003] In May 1985 the EU Ministers agreed to a New Approach to Technical Harmonisation and Standards to overcome the problem of trade between European partners. The Machinery Safety Directive (89/392/EEC), subsequently amended, is one of the Product Safety Directives and sets out the essential health and safety requirements ('EH&SR's') for machinery which must be met before machinery is placed on the market anywhere within the EU.

EH&SRs are expressed in general terms and it is intended that the European Harmonised Standards should fill in the detail so that machinery designers and suppliers have clear guidance on how to achieve conformity with the Directive.

The Use of Work Equipment Directive (89/655/EEC) ('UWED') was also introduced to outline the responsibilities of employers for the protection of workers from machinery and work equipment. In the UK this became the

Provision and Use of Work Equipment Regulations 1992 (SI 1992 No 2932) ('*PUWER*'92). These regulations were replaced, as a result of the amending EU Directive on the Use of Work Equipment Directive (95/63/EC) by the *Provision and Use of Work Equipment Regulations 1998 (SI 1998 No 2306)* ('*PUWER '98'*).

PUWER '98 have since been amended by the following legislation:

(a)   regulation 4(4) has been amended by the *Police (Health and Safety) Regulations 1999 (SI 1999 No 860)*;

(b)   regulation 6(5) has been amended by the *Work at Height Regulations 2005 (SI 2005 No 735)* and the *Construction (Design and Management) Regulations 2007 (SI 2007 No 320)*;

(c)   regulations 10(1) and (2), 11(2) and 18(1) have been amended by the *Health and Safety (Miscellaneous Amendments) Regulations 2002 (SI 2002 No 2174)*.

Tho Provision and Use of Work Equipment Regulations 1998 (PUWER'98)

[M1004] The first EU Directive aimed at the use of work equipment was the Use of Work Equipment by Workers at Work Directive (89/655/EEC) ('UWED'). This Directive required all member states of the EC to have the same minimum requirements for the selection and use of work equipment, and was implemented in the UK by *PUWER'92 (SI 1992 No 2932)*. The Directive was subsequently amended and the non-lifting aspects of the amendments by Directive 95/63/EC ('AUWED') were implemented in the UK by *PUWER'98 (SI 1998 No 2306)*. These Regulations came into force on 5 December 1998 and on the same date *PUWER'92* was revoked. The lifting aspects of AUWED have been implemented by the *Lifting Operations and Lifting Equipment Regulations 1998 (SI 1998 No 2307)* ('LOLER').

The 1998 Regulations are aimed at employers, and are designed to implement EU Directive 89/655/EEC, made under Article 118A of the *Single European Act*. This is aimed specifically at protecting the health and safety of workers in the Community. PUWER is also part of the so-called '6 pack Regulations'. They came into force on 1 January 1993 and applied immediately to equipment supplied after that date, with a transitional period to 1997 for some requirements to apply to existing equipment.

PUWER'98, was made under the *HSWA 1974*. The Health and Safety Executive (HSE) states that the primary objective of *PUWER'98* is to ensure that work equipment should not result in health and safety risks, regardless of its age, condition or origin.

Structure of PUWER'98

[M1005] The Regulations are structured in five main parts. These are set out as follows:

Part I: Regulations 1–3

This includes interpretation, definitions and application of *PUWER'98*.

Part II: Regulations 4–10 (General – software)

This sets out the 'management' duties of *PUWER'98* covering: selection of suitable equipment, maintenance, inspection, specific risks, information, instruction and training. *Regulation 10* covers the conformity of work equipment with the requirements of EU Directives on product safety.

Part II: Regulations 11–24 (General – hardware)

This deals with the hardware aspects of *PUWER'98*. It covers the guarding of dangerous parts of work equipment, the provision of appropriate stop and emergency stop controls, stability, lighting and suitable warning or devices.

Part III: Regulations 25–30 (Mobile work equipment)

This deals with specific risks associated with the use of self-propelled, towed and remote controlled mobile work equipment.

Part IV: Regulations 31–35 (Power presses)

This deals with the management requirements for the safe use and thorough examination of power presses. (Readers are advised to refer to a HSE publication 'Power presses: Maintenance and thorough examination' (HSG236), which provides practical advice on what, when and how to maintain presses.)

Part V: Regulations 36–39

Covers miscellaneous issues including transitional provisions, repeal of Acts and revocation of instruments.

What is work equipment?

The definition of 'work equipment' covers machinery, appliances, apparatus, tools and any assembly of components which function as a whole. Although not in relation to these Regulations, in *Knowles v Liverpool City Council* [1993] IRLR 588, it was held that a flagstone used for repairing a pavement was 'equipment'.

More recently, *Smith v Northamptonshire County Council* [2008] EWCA Civ 181, [2008] 3 All ER 1054, involved a carer/driver who was injured while pushing a wheelchair down a wooden ramp outside the home of the wheelchair user. The court found that the ramp, which had given way, was not work equipment. But in the House of Lords case, *Spencer-Franks v Kellogg Brown and Root Ltd* [2008] UKHL 46, [2009] 1 All ER 269, a door closer on the door of the central control room of an offshore oil platform was held to be work equipment.

The term 'use' covers starting, stopping, modifying, programming, setting, transporting, maintaining, servicing and cleaning.

Risk assessment and PUWER'98

[M1006] There is no requirement in *PUWER'98 (SI 1998 No 2306)* to carry out a risk assessment. The general risk assessment requirement is made under *Regulation 3(1)* of the *Management of Health and Safety at Work Regulations 1999 (SI 1999 No 3242)*, which is explained in general terms in the associated HSE publications.

The HSE publication *Safe use of work equipment: Provision and Use of Work Equipment Regulations 1998* (L22), contains more specific guidance on risk assessment in relation to work equipment. This publication indicates that the factors to be considered in a risk assessment to meet the requirements of *PUWER'98* should include: type of work equipment, substances and electrical or mechanical hazards to which people may be exposed. Action to eliminate/control any risk might include, for example, during maintenance: disconnection of power supply, supporting parts of the work equipment which could fall, securing mobile equipment so that it cannot move and depressurising pressurised systems. Reference is also made to the use of the HSE publication '5 steps to risk assessment' (INDG163) (rev 2), revised 06/06.

In dealing with risk assessment L22 refers to 'significant risk' which is defined by in that document as 'one which could foreseeably result in a major injury or worse'. This definition implies that if the consequences of exposure to a hazard could result in a minor injury, there is no need to comply with many of *PUWER'98* requirements.

'Risk' is defined as the combination of the probability (or chance) of a harm being realised coupled with the consequences (severity) as a result of exposure to the harm. Depending on which criteria is used to evaluate risks, if the chance of a minor injury is likely, then this would constitute a significant risk. The lack of guidance on agreed criteria for the evaluation of health and safety risks makes the task of deciding whether risks are tolerable somewhat subjective and arbitrary. The HSE's '5 steps to risk assessment' may be too generic to result in adequate hazards identification and analysis or in the evaluation of risks, particularly in relation to machinery.

Summary of main requirements of PUWER'98

Part II: General (Regulations 4–24)

[M1007] *Part II of PUWER'98 (SI 1998 No 2306)* applies to all work equipment, including machinery; mobile plant and work equipment.

Regulation 4 –   Suitability

Equipment must be suitable, by design, construction or adaptation, for the actual work it is provided to do. In selecting the work equipment, the employer must take account of the environment in which the equipment is to be used and must ensure that the equipment is used only for the operations, and under the conditions, for which it is suitable.

Regulation 4 was amended by the *Police (Health and Safety) Regulations 1999 (SI 1999 No 860)* with regard to offensive weapons provided for use as self-defence or as deterrent equipment and work equipment provided for use for arrest or restraint by police officers and cadets.

In the ACOP to the regulations, the HSE advises that the risk assessment carried out under regulation 3 (1) of the Management Regulations will help in selecting work equipment and assessing its suitability. It says that most dutyholders will be capable of making the risk assessment themselves using expertise within the organisation. But where there are complex hazards or equipment, it may need to be done with the help of external health and safety advisors appointed under regulation 7 of the Management Regulations.

Regulation 5 – Maintenance

Equipment must be maintained in an efficient state, in efficient working order and in good repair. (The term 'efficient' relates to how the condition of the equipment might affect health and safety, not productivity). There is no requirement to keep a maintenance log, but where one exists (and this is good management practice) it must be kept up to date.

Regulation 6 – Inspection

Regulation 6 of *PUWER'98 (SI 1998 No 2306)* includes a requirement for the inspection of work equipment. The purpose of an inspection is to identify whether the work equipment can be operated, adjusted and maintained safely and that any deterioration, eg defects, damage or wear can be detected and repaired before it results in unacceptable risks.

The guidance to *PUWER'98* states that inspection is only necessary where there is a significant risk resulting from incorrect installation or re-installation, deterioration or as a result of exceptional circumstances which could affect the safe operation of the work equipment. Examples given in the guidance of work equipment unlikely to need an inspection include hand tools, non-powered machinery and powered machinery such as a reciprocating fixed blade metal cutting saw.

The extent of the inspection required will depend on the level of risk associated with the work equipment, and should include, where appropriate visual and functional checks to the more comprehensive inspection which may require some dismantling and/or testing. The frequency of inspections made under *Regulation 6* is to be decided by a 'competent person'.

A fundamental requirement under *Regulation 6* is that every employer must ensure that the result of an inspection is recorded and kept until the next inspection is recorded. Although records do not have to be kept in a particular form, the example shown in FIGURE 1 below identifies the main items to be recorded on an inspection record sheet.

Components, which must receive particular attention as part of an inspection, are the safety-related and safety-critical parts, for example interlocking devices, safeguards, protection devices, controls and control systems.

Regulation 6 also states that no work equipment should leave an undertaking, or be obtained from an undertaking of another person (for example by hiring), unless it is accompanied by physical evidence that the last inspection has been carried out.

Regulation 6(5) was amended by the *Work at Height Regulations 2005 (SI 2005 No 735)* to exclude the inspection of work equipment where regulation 12 of those regulations, which also deals with inspection, applies from the regulations; and by the *Construction (Design and Management) Regulations 2007 (SI 2007 No 320)* to exclude work equipment required to be inspected by under those regulations.

Figure 1: Example of an inspection record sheet

INSPECTION RECORD

Type and model of equipment: Press Brake-Komato ....
Serial number/identification mark: KP/012 ....
Normal location of equipment: Press Shop ....
Date inspection carried out: 08/09/2009 ....

Type of Inspection:
 Visual check:
 Functional check:
 Dismantling/testing: ✓

Who carried out inspection: John Smith ....

Items and safety-related parts inspected

1. Guard-Interlocking Device ....
2. Photo-electric Trip Device ....
3. Fixed and Interlocking Safeguards ....
4. Pneumatic Clutch and Brakes Systems ....
5. Machine Control and Electrical Systems ....

Faults and defects found:
1. Small air leak near clutch ....
2. Interlocking device wiring loose ....
....

Action/corrective measures taken:
1. Seals replaced for pneumatic system ....
2. Re-wiring of interlocking system ....
....

To whom faults have been reported: Kevin Taylor ....

Date repairs/actions were carried out: 12/09/2009 ....

Regulation 7 – Specific risks

Regulation 7 of *PUWER'98 (SI 1998 No 2306)* requires employers to ensure that, where the use of work equipment is likely to involve a specific risk, the use, repairs, modifications, maintenance or servicing of such equipment is restricted to those persons who have been specifically designated to perform operations of that description. It is further required that employers must provide adequate training to perform such tasks safely.

The Approved Code of Practice ('ACoP') and Guidance to *PUWER'98* does not define what is meant by 'specific risks', or what makes such risks 'specific'. The examples given in the Guidance L22 on machinery which poses specific risks, eg a platen printing machine or a drop-forging machine does not make reference to which specific risks are associated with these machines.

The hazards associated with the use, maintenance, repair etc. of the above example of machines should be identified if a suitable and sufficient risk assessment is carried out. For example, hazards associated with a drop forge might include: noise, vibration, crushing, shearing, heat, splashes of molten metal, toxic fumes, etc. It may not be impractical to provide adequate protection against all these hazards by means of physical safeguards. As a result, significant 'residual' risks remain which may require the use of adequate personal protective equipment and safe systems of work. Are these meant to be 'specific' risks?

The ACoP relating to *Regulation 7* requires employers to ensure that risks are always controlled by (in the order given):

(a)   eliminating the risks, or if that is not possible;
(b)   taking 'hardware' (physical) measures to control risks such as the provision of guards; but if risks cannot be adequately controlled;
(c)   taking appropriate 'software' measures to deal with the residual risks, such as following safe systems of work and the provision of information, instruction and training.

Regulation 8 – Information and instructions

Managers, supervisors and users of work equipment must be provided with adequate health and safety information and, where appropriate, written instructions relating to work equipment. This must include the conditions in which, and the methods by which, the equipment may be used, the limitations on its use together with any foreseeable difficulties that might arise and the required actions.

Regulation 9 – Training

Managers, supervisors and all users of work equipment must receive adequate health and safety training in the methods, which may be adopted when using equipment, any risks, which may arise from such use and precautions to be taken. The level of training will depend on the circumstances of use and the competence of the employee. Other regulations may contain specific training requirements. Particular attention must be paid to the training and supervision of young persons. Self-propelled work equipment, including any attachments or towed equipment must only be driven by workers who have received the appropriate training in the safe driving of the equipment, and the ACOP provides specific guidance regarding the training, certificates of competence in relation to the use of chainsaws.

The case of *Allison v London Underground Ltd* [2008] EWCA Civ 71, [2008] IRLR 440 provided important guidance on the duty to ensure that all persons who use work equipment receive adequate health and safety training. It establishes that the duty to provide training is mandatory, and that the test for adequacy is what training is needed in the light of what the employer ought to have known about the risks arising from the activities of the business.

Lady Justice Smith said that 'the test for the adequacy of training for the purposes of health and safety is what training was needed in the light of what the employer ought to have known about the risks arising from the activities of his business.'

Regulation 10 – Conformity with Community requirements

Regulation 10 of *PUWER'98 (SI 1998 No 2306)* (amended by the *Health and Safety (Miscellaneous Amendments) Regulations 2002 (SI 2002 No 2174)*), applies to items of work equipment provided for use in the premises or undertaking of the employer for the first time after 31 December 1992. *Regulation 10(1)* requires employers to ensure that work equipment complies at all times with any EH&SR's that applied to that kind of work equipment at the time of its *first supply* or when it is *first put into service*. It places a duty on the employer that complements those on manufacturers and suppliers in other jurisdictions regarding the initial integrity of equipment.

Regulation 10(2) defines the term 'essential requirements' – which are the essential health and safety requirements listed in the Regulations implementing the relevant directives in the UK (in *Schedule 1* of the 1998 Regulations).

The employer can demonstrate compliance with *Regulation 10* if the following conditions are satisfied:

- that the employer where appropriate checks to see that the equipment bears a CE marking and asks for a copy of the EU Declaration of Conformity;
- the employer also needs to check that adequate operating instructions have been provided and that there is information about residual hazards such as noise and vibration;
- more importantly, the employer should check the equipment for obvious faults.

The HSE booklet 'Buying new machinery' (INDG271) is written in easy to understand language and contains two checklists: Checklist A – What should I talk to a supplier (or manufacturer) about? and Checklist B – What do I do when I have bought new machinery? In both checklists, the employer relies on guidance given by the machine designers/suppliers. This guidance is aimed at small/medium-size enterprises and the checklists are somewhat superficial and over simplified. Basic requirements such as the format of Declarations of Conformity/Incorporation, or the structure/relevance of the Transposed Harmonised EU Standards are not discussed.

Regulation 11 – Dangerous parts of machinery

Regulation 11(1) of *PUWER'98 (SI 1998 No 2306)* requires employers to ensure that access to dangerous parts of the machine is prevented, or if access is needed to ensure that the machine is stopped before any part of the body reaches the danger zone. *Regulation 11(2)* sets out the following hierarchy of preventive measures (see also **M1027** onwards):

(a)   fixed enclosing guards;
(b)   other guards or protective devices;
(c)   protection appliances, eg jigs, holders, push sticks, etc.
(d)   the provision of information, instruction, training and supervision.

This hierarchy follows the traditional preference of fixed guards and is open to criticism. If an individual needs frequent access into the machine danger zone, a fixed guard could be the worst option, as the machine can operate without it.

The European Commission did express concern that the original *Regulation 11(2)(d)* – the provision of information, instruction, training and supervision, could be used as the sole solution to working with the unguarded parts of dangerous machinery.

To address this and other concerns, *Regulation 11* was amended so that the provision of information, instruction, training and supervision is now seen as an additional requirement in each case and not as part of the hierarchy of measures.

Regulation 11(3) describes the need for the provision of safeguards/safety devices which are robust, difficult to defeat, well maintained and meet the requirements of relevant EN standards. Illustrated examples of the options given in *Regulation 11* are shown in the guidance to *PUWER'98*.

Regulation 12 – Protection against specified hazards

Appropriate measures must be taken to prevent or adequately control exposure to 'specified hazards' arising from the use of work equipment. These measures should be other than the provision of personal protective equipment or of information, instruction, training and supervision, so far as is reasonably practicable. The 'specified hazards' are: falling or ejected articles or substances; rupture or disintegration of parts of equipment; work equipment catching fire or overheating; and unintended or premature discharges or explosions.

Regulation 13 – High or very low temperature

Work equipment, parts of work equipment and any article or substance produced, used or stored in work equipment must be protected so as to prevent burns, scalds or sears.

Regulations 14 to 18 – Controls and control systems

Work equipment must be provided with:

• One or more controls for starting the equipment or for controlling the operating conditions (speed, pressure etc.), where the risk after the change is greater than or of a different nature from the risks beforehand.

- One or more readily accessible stop controls that will bring the equipment to a safe condition in a safe manner and which operate in priority to any control that starts or changes the operating conditions of the equipment.
- One or more readily accessible emergency stop controls – unless by the nature of the hazard, or by the adequacy of the normal stop controls (above), this is unnecessary. Emergency stop controls must operate in priority to other stop controls.
- Controls should be clearly visible and identifiable (including appropriate marking). The position of controls should be such that the operator can establish that no person is at risk as a result of the operation of the controls. Where this is not possible, safe systems of work must be established to ensure that no person is in danger as a result of equipment starting or where this is not possible, there should be an audible, visible or other suitable warning whenever work equipment is about to start. Sufficient time and means shall be given to a person to avoid any risks due to the starting or stopping of equipment.
- Control systems should not create any increased risk, which ensure that any faults or damage in the control systems or losses of energy supply do not result in additional or increased risk, and which do not impede the operation of any stop or emergency stop control.

Regulation 18 – Control systems

Regulation 18 of *PUWER'98 (SI 1998 No 2306)* relates to work equipment control systems. This requires that:

(a)   Every employer shall—
    (a)   ensure, so far as is reasonably practicable, that all control systems of work equipment are safe; and
    (b)   are chosen making due allowance for failures, faults and constraints to be expected in the planned circumstances of use.

A control system is defined as: a system which responds to input signals and generates an output signal which causes the equipment to perform controlled functions.

The input signals may be made by the operator (manual control), or automatically controlled by the equipment through sensors or protection devices, eg guard interlocking devices, photoelectric device, emergency stop or speed limiters.

The objective of *Regulation 18* is to prevent or reduce the likelihood that a single failure in the interlocking device, wiring, contactors, hardware, software could cause injury or ill health.

The subject of control systems generally and the safety related functions of control systems in particular are complex, even for engineers.

The ACOP refers to national, European and international standards, both current and in preparation (BS EN 60204-1, BS EN ISO-1: 2006) which provide guidance on the design of control systems so as to achieve high levels of performance related to safety. Aimed at new machinery, the HSE advises that they may be used as guidance for existing work equipment.

Regulation 19 – Isolation from sources of energy

Equipment must be provided, where appropriate, with identifiable and readily accessible means of isolating it from all its sources of energy. Reconnection of any source of energy must not expose anyone to any health and safety risk.

Regulation 20 – Stability

All work equipment must be stabilised where necessary – by clamping, tying or fastening.

Regulation 21 – Lighting

Suitable and sufficient lighting, taking account of the operations carried out, must be provided. This may entail additional lighting where the ambient lighting is insufficient for the required tasks.

Regulation 22 – Maintenance operations

Work equipment must be constructed or adapted in such a way that maintenance operations, which involve a risk to health or safety, can be carried out while the equipment is shut down. If this is not possible, the work should be carried out in such a way as to prevent exposure to risk and appropriate measures taken to protect those carrying out the maintenance work. Such measures might include, for instance, functions designed into the equipment that limit the power, speed or range of movement of dangerous parts during maintenance.

Regulation 23 – Markings

Equipment must be marked for the purposes of ensuring health and safety with clearly visible markings.

Regulation 24 – Warnings

Equipment must incorporate appropriate warnings or warning devices, which must be unambiguous, easily perceived and easily understood.

A warning is normally in the form of a notice and can be positive instructions, eg "Hard hats must be worn", prohibitions, eg "Not to be operated by people under 18 years", or restrictions eg "Do not heat above 60°C". A warning device is an active unit giving a signal, typically audible or visible.

Second-hand machinery

Second-hand machinery supplied within the EU is not required to be 'CE' marked, nor issued with a Declaration of Conformity/Incorporation. These however should subject to a suitable and sufficient risk assessment and should comply with *Regulations 11 to 24 of PUWER'98 (SI 1998 No 2306)*.

Second-hand machinery supplied from outside the EU should be treated as new machinery and will be subject to CE marking requirements.

It is the management responsibility to ensure that second-hand machinery and modification work do comply with the requirements of the *PUWER'98*.

Part III: Mobile work equipment

[M1008] *Part III of PUWER'98 (SI 1998 No 2306)* added new requirements for the management of mobile equipment.

Mobile equipment is defined as any work equipment, which carries out work while it is travelling or which travels between different locations where it is used to carry out work. Mobile work equipment may be self-propelled, towed or remote controlled and may incorporate attachments. *Regulations 25 to 30*, in addition to other requirements of *PUWER'98* (eg training, guarding and inspection), apply to all work equipment including mobile work equipment. Since December 1998, all new mobile equipment taken into use should comply with all requirements of *PUWER'98*.

Regulation 25 – Employees carried on mobile work equipment

Regulation 25 of *PUWER'98 (SI 1998 No 2306)* contains a general requirement relating to the risks to people (drivers, operators and passengers) carried by mobile equipment, when it is travelling. This includes risks of people falling from the equipment or from unexpected movement while it is in motion or stopping.

This Regulation requires every employer to ensure that no employee is carried by mobile work equipment unless:

(a)   it is suitable for carrying persons; and
(b)   it incorporates features for reducing to as low as is reasonably practicable risks to their safety, including risks from wheels or tracks.

Mobile work equipment can be made suitable for carrying persons by means of operator stations, seats or work platforms to provide a secure place.

Features for reducing risks may include the following:

- Falling object protective structures ('FOPS'). This may be achieved by a strong safety cab or protective cage.
- Restraining systems – this can be full-body seat belts, lap belts or purpose-designed restraining system. This should provide adequate protection against injury through contact with or being flung from the mobile equipment if it comes to a sudden stop moves unexpectedly or rolls over.
- Speed adjustment – when carrying people, mobile equipment should be driven within safe speed limits for stability.
- Guards and barriers fitted to mobile work equipment to prevent contact with wheels and tracks.

Regulation 26 – Rolling over of mobile work equipment

In addition to the main general requirements of *Regulation 25*, *Regulation 26* of *PUWER'98 (SI 1998 No 2306)* covers the measures necessary to protect employees where there are risks from roll-over while travelling. If this risk is significant, then the following measures should be considered:

- Stabilisation – this might include fitting appropriate counterbalance weights, wider wheels and locking devices.
- Structures to prevent rolling over by more than 90?, eg boom of a hydraulic excavator, when positioned in its recommended travel position.
- Roll-over protective structures ('ROPS') – normally fitted on mobile equipment to withstand the force for roll over through 180?.

Regulation 27 – Overturning of forklift trucks

Regulation 27 of *PUWER'98 (SI 1998 No 2306)* applies to FLT's fitted with vertical masts, which effectively protect seated operators from being crushed between the FLT and the ground in the event of roll-over. FLT's capable of rolling over 180° should be fitted with ROPS.

FLT's with a seated ride-on operator can roll-over in use. There is also a history of accidents on counterbalanced, centre control, high lift trucks. Restraining systems will normally be required on these trucks.

Regulation 28 – Self-propelled work equipment

- Preventing unauthorised start-up – this could be achieved by a starter key or device, which is issued only to authorised personnel.
- Minimise consequences of a collision of rail-mounted work equipment – this can be achieved by introducing safe systems of work, buffers or automatic means to prevent contact.
- Devices for stopping and braking – all self-propelled work equipment should have brakes to enable it to slow down and stop in a safe distance and park safely.
- Emergency braking and stopping facilities – a secondary braking system may be required if risk is significant.
- Driver's field of vision – where the driver's direct field of vision is impaired then mirrors or more sophisticated visual or sensing facilities may be necessary.
- Lighting for use in the dark – to be fitted to equipment if outside lighting is inadequate.
- Carriage of appropriate fire-fighting equipment – where escape from self propelled work equipment in the case of fire is not easy.

Regulation 29 – Remote-controlled, self-propelled work equipment

- This is self-propelled work equipment that is operated by controls which have no physical link with it, for example radio control.
- Alarms or flashing lights – to be considered for the prevention of collision with others.
- Sensing or contact devices – to prevent or reduce risk of injury following contact.
- Hold-to-run control devices – so that any hazardous movement can stop when the controls are released.
- Control range – once equipment leaves its control range should stop and remain in a safe state.

Regulation 30 – Drive shafts

- A 'drive-shaft' is a device, which conveys the power from the mobile work equipment to any work equipment connected to it. In agriculture these devices are known as power take-off shafts.
- Shaft seizure – if seizure can result in significant risk, for example the ejection of parts due to equipment break-up, guards should be fitted in accordance with *Regulation 12* of *PUWER'98 (SI 1998 No 2306)*.
- Isolation from source of energy – to reduce the risk associated with drive shaft stalling.

- Support on cradle – when drive shaft and its guards are not in use, the drive shaft should be supported on cradle or equivalent for protection against damage.

The Supply of Machinery (Safety) Regulations 1992 (as amended)

[M1009] General and specific duties relating to machinery safety are laid on designers and manufacturers of machinery for use at work by the *Supply of Machinery (Safety) Regulations 1992 (SI 1992 No 3073)*, as amended by the *Supply of Machinery (Safety) (Amendment) Regulations 1994 (SI 1994 No 2063)*. These regulations implemented the EU Machinery Directive 89/392/EEC, as amended by Directives 91/368/EEC, 93/44/EEC and 93/68/EEC.

Supply of safe machinery (including safety components, roll-over protective structures and industrial trucks) is governed by these Regulations. The 1994 Regulations extend to machinery for lifting people (eg elevating work platforms). [*SI 1994 No 2063, Reg 4, Sch 2, para 10*]. The combined effect of the two sets of Regulations requires that, when machinery or components are properly installed, maintained and used for their intended purposes, there is no risk (except a minimal one) of their being a cause or occasion of death or injury to persons, or damage to property. [*SI 1992 No 3073, Reg 2(2)*, as amended by *SI 1994 No 2063, Reg 4, Sch 2, para 5*].

The EU Machinery Directive (89/392/EEC) as amended is aimed at machinery designers and suppliers. This is a so-called 'Product Directive' made under Article 100A of the *Single European Act* to ensure fair competition by eliminating barriers to trade. The relevant products have to meet certain essential health and safety requirements ('EH&SRs') before they can be placed on the market within the Community.

The manufacturer affixing a CE-Mark to the machinery claims compliance with all relevant EU Directives.

EU Machinery Directive

[M1010] The EU Machinery Directive requires:

- The clear allocation of responsibilities to machinery suppliers.
- Type approval: 'attestation' via a technical construction file and a Declaration of Conformity before the CE mark is affixed to the machine.
- Intended use of machinery has to be clearly defined.
- Principles of Safety Integration. Manufacturers must demonstrate that the following order of options for machinery safety are followed:
  (i) elimination or reduction of risks by seeking to design inherently safe machinery;
  (ii) necessary protection measures taken (eg safeguarding or safety devices) to deal with the risks which cannot be eliminated;

(iii)   users of residual risks are informed.

The majority of the requirements of the Directive are couched as 'objects to be achieved'.

The Directive and EN standards show that while traditional mechanical hazards continue to merit attention, manufacturers must also consider: discomfort, fatigue and psychological stress; electricity, including static; sources of energy supply other than electricity; errors of fitting; temperature extremes; fire and explosion; noise and vibration; radiation; laser equipment; and emissions of gases, etc.

Definition of a machine

[M1011] A 'machine' is defined by the EU Machinery Directive as 'an assembly of linked parts or components, at least one of which moves, with the appropriate actuators, control and power circuits, joined together for a specific application, in particular for the processing, treatment, moving or packaging of a material'. The definition includes an assembly of machines which functions as an integral whole as well as interchangeable equipment (not being a spare part or tool) which can be assembled with a machine by the operator – for instance, an item of agricultural equipment attached to a tractor.

The main elements in the framework for demonstrating compliance are summarised as follows.

Essential health and safety requirements ('EH&SR')

[M1012] The route to demonstrating compliance starts with the identification of relevant EH&SRs relating to the design and construction of machinery. These are set out in Annex I of the EU Machinery Directive, and include the following:

- Principles of safety integration.
- Materials and products health and safety.
- Lighting.
- Design of the machine to facilitate its handling.
- Safety and reliability of the control system.
- Control devices.
- Starting, change-over and stopping devices.
- Emergency stops.
- Mode selection.
- Power supply failure.
- Failure of the control system, including software.
- Stability.
- Risk of break-up during operation.
- Risks due to falling or ejected objects.
- Sharp surfaces, edges or angles.
- Hazards associated with moving parts.
- Required characteristics of guards and protection devices.
- Fire, explosion, noise, vibration, radiation, laser equipment, etc.

Suppliers should identify relevant EH&SRs applicable to the machine and show how these have been considered.

Conformity assessment procedure carried out

[M1013] The responsibility for demonstrating that the machinery satisfies EH&SR's rests with the manufacturer, supplier or the importer into the EU ('the responsible person').

The EU Machinery Directive sets out various ways in which a supplier can demonstrate compliance with EH&SR's. This is known as 'conformity assessment procedure'. There are three different conformity assessment procedures, detailed in *Regulations 13, 14* and *15* of the *Supply of Machinery (Safety) Regulations 1992 (SI 1992 No 3073)*. These are:

- for most machinery other than that listed in *Schedule 4* of *SI 1992 No 3073* (Annex IV of the Directive);
- Schedule 4 (Annex IV) machinery manufactured in accordance with the EU Transposed Harmonised Standards; or
- Schedule 4 machinery not manufactured in accordance with the EU Harmonised Standards or where they do not exist.

Schedule 4 machinery lists categories of machines that pose special risks. These include mechanical presses, press brakes, injection and compression plastics moulding machines.

For machinery listed in *Schedule 4* of the *Supply of Machinery (Safety) Regulations 1992 (SI 1992 No 3073)* that is manufactured in conformity with Harmonised Transposed Standards, the machinery technical file is submitted to an approved body for EC type-examination. The approved body may simply verify that the Transposed Harmonised Standards have been correctly applied and draw up a certificate of adequacy.

The technical file

[M1014] The conformity procedure requires the supplier to assemble records, which technically describe the rationale behind the approach to ensuring that the machine satisfies EH&SR's. These records are referred to as the 'technical file'. The technical file should include a risk assessment (if applicable) and other technical information which includes: technical drawings, test results, a list of the EH&SR's, Harmonised Standards, other standards and technical specifications that were used in the design of the machinery. It should also include a description of the methods adopted to eliminate hazards, other relevant technical information and a copy of the instructions for the machinery.

For machinery posing special hazards (listed in Annex IV of the EU Machinery Directive), further action is necessary which usually involves verification by an approved body that either recognised standards have been correctly applied or an EU type-examination of the machine shows that it satisfies the relevant provisions of the Directive.

The supplier must retain the technical file and all documentation relating to particular machinery for a minimum of 10 years after the production of the last unit of that machinery.

EU Declaration of Conformity/Incorporation

[M1015] An EU Declaration of Conformity must be drawn up which states that the relevant machinery complies with all the EH&SR's that apply to it.

Such a declaration must include a description of the machinery (with make, type and serial number), indicate all relevant provisions with which the machinery complies, give details of any EU-type examination and specify any standards and technical specifications which have been used. Reference should also be made to all other relevant EU Product Directives, eg Low Voltage, EMC and Simple Vessel Directives.

The machine supplier prior to the commissioning or operation of the machine should issue the Declaration of Conformity. A typical example of a Declaration of Conformity is shown in FIGURE 2 below.

Figure 2: Example of a Declaration of Conformity

EU DECLARATION OF CONFORMITY

In accordance with:

**The Supply of Machinery
(Safety) Regulations 1992**

**WE DECLARE THAT THIS MACHINE CONFORMS WITH THE
ESSENTIAL HEALTH AND SAFETY REQUIREMENTS**

Description of the machine, type and serial number:

Business name and full address of manufacturer:

All relevant product directives complied with:
**Machinery Directive (89/392/EEC) and all amendments thereto.
Low Voltage, CE marking and EMC Directives, ...**

The machine complies with the following transposed harmonised standards:
**EN 292-1, EN 292-2, EN 294, EN 60204-1, EN 954-1, EN 1050,
EN 1088**

Other standards and technical specifications used:
IEC 1508

IDENTIFICATION of the person
empowered to sign on behalf of   Usually Managing Director
the Company:

I certify that on 14th of January 1996, the above machine conformed
with the EH&SR's of all the relevant Product Directives.

Signature: ....   Date: ....

The three most relevant sections on this form are:

(1)   Relevant EU Directives applicable to the product;
(2)   The Transposed Harmonised Standards used; and
(3)   Other standards and technical specifications relevant to the machinery.

The declaration must identify European Transposed Harmonised Standards applicable to the machine design and construction. Other national and international standards relevant to the machine, which are not covered by relevant European standards should also be identified on the certificate. For example IEC 61508, which is an international electro-technical standard for the design and selection of programmable electronic systems with safety related applications. The Declaration of Conformity should also make reference to all relevant EU product directives, eg Low Voltage and Electromagnetic Compatibility Directives.

A Declaration of Incorporation is required where the machinery is intended for incorporation into another machinery or assembly with other machinery to constitute machinery covered by the Directive. The Declaration of Incorporation is similar to the declaration of conformity where relevant Harmonised Standards are listed. The only difference is that a 'CE' mark is not required, and a condition included in the declaration, which states that: 'the equipment must not be put into service until the machinery into which it is incorporated has been declared in conformity with the provision of the Machinery Directive'. See Annex IIB of the Directive for more detail.

'CE' marking

[M1016] A 'CE' marking (meaning European Conformity) on a product or system indicates a legal declaration by the supplier that the product complies with EH&SRs of the EU Machinery Directive and all other EU directives relevant to it.

The CE mark is regarded as properly affixed only if: (a) an EU declaration of conformity has been issued, (b) it is affixed in a distinct, visible, legible and indelible manner, and (c) the machinery complies with the requirements of other relevant directives. It is an offence to affix the CE mark to machinery, which does not satisfy the EH&SRs, or to machinery that is not safe. The CE mark is described in the Machinery Directive and consists of a symbol 'CE' and the last two figures of the year in which the mark is affixed.

Requirements of other EU directives

[M1017] Machinery suppliers need to identify, apart from the Machinery Directive, all other EU Directives relevant to their products. It is then necessary to identify specific legislation, which implement them. Usually more than one EU directive may apply to particular machinery. For example, in addition to the Machinery Directive, an item of electrically driven machine would be subject to the following Directives:

- Electromagnetic Compatibility Directive (89/336/EEC) as amended. This was implemented in the UK as the *Electromagnetic Compatibility Regulations 1992 (SI 1992 No 2372)* and 2006 regulations now apply. The *Electromagnetic Compatibility Regulations 2006 (SI 2006 No 3418)* apply to electrical and electronic equipment liable to cause electromagnetic disturbance or the performance of which is liable to be affected by such disturbance.
  The purpose of the Regulations is to ensure that the electromagnetic disturbance generated by electrical or electronic equipment does not exceed a level above which radio and telecommunications equipment and other equipment cannot operate as intended, and that the equipment itself has an adequate level of immunity to electromagnetic disturbance. These Regulations do not deal with safety-related matters.

- Low Voltage Directive (72/23/EEC) implemented in the UK under the *Electrical Equipment (Safety) Regulations 1994 (SI 1994 No 3260)*. (see **E30 ELECTRICITY**).
- Other EU directives, which may be relevant to some machinery include Simple Pressure Vessel Directive (87/404/EEC) as amended, Potential Explosive Atmosphere Directive (94/9/EEC) and Telecom Terminal Equipment Directive (91/263/EEC).

The machine is in fact 'safe'

[M1018] The key objective of the EU Machinery Directive is that the onus is placed on suppliers to demonstrate that a machine complies with relevant EH&SRs before it is placed for sale within the market place of the EU. The requirement for the machine to be in fact safe is an UK rather than EU requirement.

'Safe' is defined by the *Supply of Machinery (Safety) Regulations 1992 (SI 1992 No 3073)* as, 'when the machinery is properly installed and maintained and used for the purpose for which it is intended, there is no risk (apart from one reduced to a minimum) of it being the cause or occasion of death or injury to persons or, where appropriate, to domestic animals or damage to property. The practicability at the time of manufacture, of reducing the risk may be taken into account when considering whether or not the risk has been reduced to a minimum'.

Failure to comply with relevant EU requirements will mean that the machinery cannot legally be supplied in the UK and could result in prosecution and penalties on conviction of a fine up to £5,000 or in some cases, of imprisonment for up to three months, or of both.

The same rules apply everywhere in the EU, so machinery complying with the EU regime may be supplied in any Member State.

The Transposed Harmonised European Machinery Safety Standards

[M1019] Application of the European Transposed Harmonised Machinery Safety Standards is fundamental to the risk-based approach. These standards have now replaced corresponding British Standards relating to machinery safety, as well as other EU member countries national standards.

A process for incorporating the European Machinery Safety Standards into international standards has been adopted by ISO/IEC since 1990. The American National Standards Institute (ANSI) is also considering the EU risk-based approach to machinery safety.

Status and scope of the Harmonised Standards

[M1020] The role and status of the European Harmonised Standards has been frequently misunderstood, mainly due to the previously prescriptive nature of the health and safety legislation and standards. The 'new approach' is based on the legal status of the EH&SRs as an objective, but the Harmonised Standards represent ways for fulfilling the details of the individual EH&SRs. The Harmonised Standards are not of the same legal status as the EH&SRs, but do reflect recognised practice and should be the first consideration.

EH&SRs are expressed in general terms and it is intended that the European Harmonised Standards should fill in the detail so that designers have clear guidance on how to achieve conformity with the directive".

The Machinery Directive states 'objectives to be achieved' where the ES&HRs constitute the legal objective requiring to be complied with and the Harmonised Standards are the current recognised methods, and therefore do not constitute the legal duty as to the method of compliance.

Structure of the Harmonised Standards

[M1021] The structure of safety standards for machinery is as follows:

- 'A' type Standards provide general requirements applicable to all machines.

- The key standards relating to the EU Machinery Directive are BS EN ISO 12100 and BS EN ISO 14121.
  BS EN ISO 12100 is introduced in two parts:
  - BS EN ISO 12100–1:2003 Safety of machinery. Basic concepts, general principles for design. Basic terminology, methodology
  - BS EN ISO 12100–2:2003 Safety of machinery. Basic concepts, general principles for design. Technical principles

- BS EN ISO 14121–1:2007 is entitled: Safety of machinery. Risk assessment. Principles and provides a framework for risk assessment and the range of hazards to be considered in the assessment

- 'B' type Standards are standards on techniques or components, which could be applicable to a large number of machines. 'B1' Standards provide generic guidance and information, eg BS EN ISO 13857:2008 Safety of machinery. Safety distances to prevent hazard zones being reached by upper and lower limbs; whereas 'B2' Standards outline generic safety standards for 'hardware', eg electro-sensitive safety systems, two-hand controls, control systems). The B type or horizontal group safety standards therefore deal with only one safety aspect or one safety-related device at the time.

'C' type Standards are drawn up to cover a particular type or class of machines. Examples of C Standards include BS EN 289:2004+A1:2008 Plastics and rubber machines. Presses. Safety requirements and EN 692 for mechanical presses. A large number of 'C' Standards are yet to be produced or introduced in provisional form. These may be classed as vertical standards. In following a particular 'C' Standard, cross-reference is extensively made to the relevant sections of 'A' and 'B' type Standards.

It is important to stress that the use of European Harmonised Standards is not mandatory. However, compliance with relevant EN standards is one of the most effective means for demonstrating that relevant EH&SRs are met. The other alternative would be the application of a robust risk assessment, which considers all relevant hazards and hazardous situations listed in BS EN ISO 12100and BS EN ISO 14121.

Machinery risk assessment

[M1022] The risk-based approach involves a structured and systematic risk assessment. Risk assessment is essentially a proactive means to foresee how accidents can happen and to decide what corrective and preventive actions are needed in advance to prevent such accidents.

Risk assessment is defined as ' . . . A structured and systematic procedure for *identifying hazards* and *evaluating risks* in order to prioritise decisions to reduce risks to a *tolerable* level'.

This section will provide guidance on the following key elements of a structured and systematic risk assessment:

- techniques for hazard identification and analysis;
- criteria for the estimation and evaluation of risks;
- the concept of 'ALARP' and risk tolerability, and
- the preferred hierarchy for risk control options.

In addition to demonstrating compliance with relevant legislation/standards, risk assessment is now crucial as it aids in the consistency of decision making and the cost effectiveness in allocation of resources for health, safety and environmental issues.

Risk assessment is vital for two reasons: the level of risk determines priority which should be accorded to and the selection of appropriate health, safety and environmental control measures to deal with different hazards, and the standard of integrity of corrective/preventive measures will depend on the level of risk.

Risk assessment techniques are complementary to the more pragmatic ways of problem identification and assessment. They highlight systematically how hazards can occur and provide a clearer understanding of their nature and possible consequences, thereby improving the decision-making process for the most effective way to prevent injury and damage to health/environment.

The techniques range from relatively simple qualitative methods of hazard identification and analysis to the advanced quantitative methods for risk assessment in which numerical values of risk frequency or probability are derived.

Hazard and risk – definitions

[M1023] A *hazard* is defined as 'a potential to cause harm'.

Harm is defined as:

- injury or damage to health;

- damage to the environment;
- economic losses (interruption to production/asset damage).

Risk is defined as:

- 'the chance (probability) of the harm being realised, combined with its consequences; or
- Risk = Chance of exposure to the hazard x Consequences (severity).

FIGURE 3 below illustrates the concept of risk. If for example it was estimated that the chance of an accident involving the collision between a fork-lift truck and an operator as 'remote', and the resulting injury severity as fatal 'catastrophic', then the Risk Matrix shows this as risk level 'A' or high, which should warrant some urgent attention. If on the other hand it was estimated that the chance of slipping on a wet surface in part of the workplace as 'occasional', which would result in 'minor' injury. The risk matrix shows this to be risk level 'C' or low, which is broadly acceptable.

Figure 3: The Risk Matrix

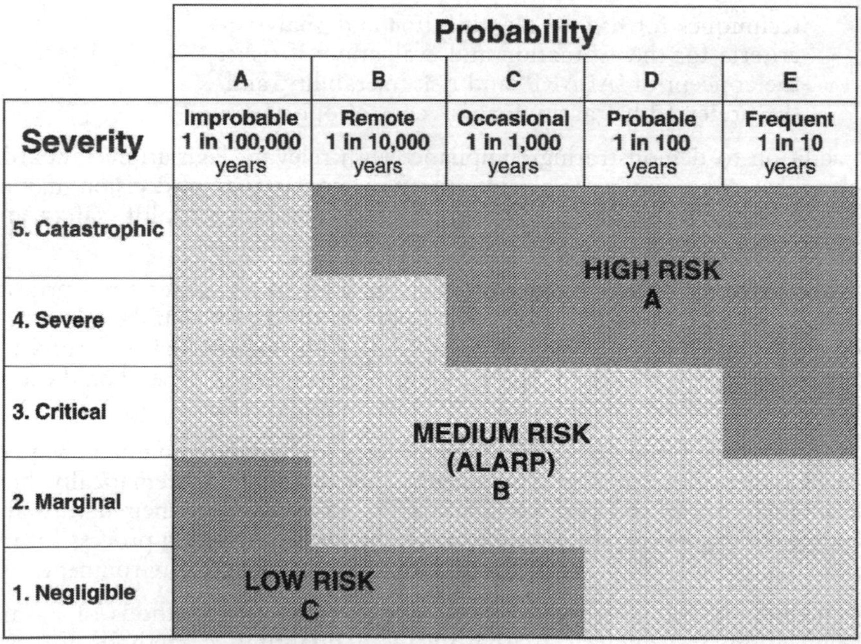

Severity	Probability				
	A	B	C	D	E
	Improbable 1 in 100,000 years	Remote 1 in 10,000 years	Occasional 1 in 1,000 years	Probable 1 in 100 years	Frequent 1 in 10 years
5. Catastrophic				HIGH RISK A	
4. Severe					
3. Critical			MEDIUM RISK (ALARP) B		
2. Marginal					
1. Negligible	LOW RISK C				

Risk assessment framework

[M1024] Hazard analysis is an essential part of the overall risk assessment framework. The flow-chart shown in FIGURE 4 outlines elements of the overall framework. The effectiveness of hazard analysis and risk evaluation processes involves a number of practical considerations.

Figure 4: Risk assessment framework

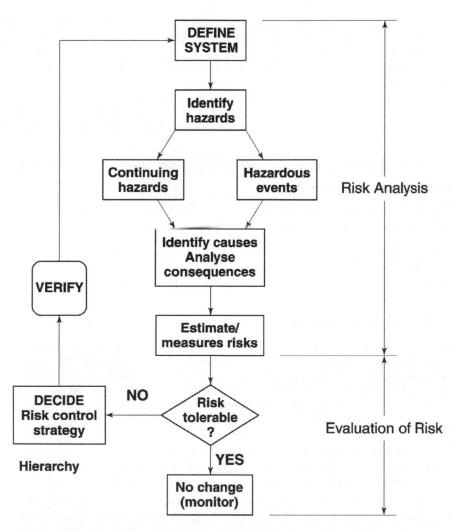

Before commencing the study, its objectives and scope must be defined explicitly. The purpose of the study and the fundamental assumptions made must be clearly stated. In order to limit the analysis, the human/system boundaries must also be defined, as well as its intended design, use, operation and layout.

The risk assessment procedure contains two essential elements:

(1)   *Risk analysis*. Where hazards and hazardous situations are systematically identified and their consequences are analysed. The level of risk associated with each hazardous situation is estimated or measured.

(2)   *Risk evaluation*. In which a judgement is made as to whether the level of risk is acceptable/tolerable or whether some corrective/preventive actions are needed.

Following any modifications in the plant/process design or operation/maintenance procedures, there is a need to verify that the original hazards/hazardous situations are controlled, and that new hazards are not introduced as a result of these modifications. It is crucial that the risk assessment is updated as a result of design/procedural evolution. The safety management system developed by the employer as a result of risk assessment is also a living system, capable of learning from experience as the people who run it, and should reflect what is actually taking place, and not what should happen.

The main elements of risk assessment, ideally, should include the following:

Define the system/activity

The first step in any risk assessment is to provide terms of reference to the detailed description of the process/activity being studied. This clear description is needed at this stage in order to avoid confusion at a later stage and to define the boundary and interfaces with other processes/activities. This step also involves the identification of the design objectives, material being handled/processed and exposure patterns for individuals involved in operation and maintenance.

Define the study objective

Because risk assessment is used as means to demonstrating compliance with a number of codes/regulations, it is important to limit the study to the range of hazards and potential consequences in relation to specific requirements, eg COSHH ie control of substances hazardous to health', noise at work, the environment, machinery safety or economic risks.

Identify the hazards

Having defined the scope and objectives, the identification of hazards is the third, but the most important step in any risk assessment study because any hazard omitted at this stage will result in the associated risk not to be assessed.

It is important in this respect to distinguish between continuing hazards (those inherent in the work activity/equipment/machinery/substances under normal conditions), eg noise, toxic substances, mechanical hazards, and hazards which can result from failures/error, eg hardware/software failures as well as foreseeable human error.

The checklist shown in TABLE I at **M1025** may be used to identify continuing hazards.

The range of continuing hazards outlined in the checklist is based on the EU-Machinery Directive and is intended to be used as an aid-memoire. This approach to hazard identification will focus attention on parts of the work activity/equipment/machinery to identify the sources of each relevant hazard, who may be exposed to them, eg the operator, maintenance, and the task involved, eg loading, removing, tool setting, cleaning, adjusting.

Hazards resulting from equipment failure and human error require an open-ended type of analysis, based on brain storming. This approach normally involves the following:

(1)   A detailed hazard identification to be carried out for all stages of the process life-cycle. This ideally should include the identification of contributory causes for each hazards.
(2)   Analysis of systems of work and established procedures to identify, who might be exposed to a hazard and when, eg operator under normal conditions, maintenance/tool fitter/cleaner, etc.

Structured and systematic techniques which could assist in the identification of hazardous events include the following:

•   *Hazard and Operability Study* ('HAZOP') – a qualitative technique to identify hazards resulting from hardware failures and human error. This involves potential causes and consequences
•   *Failure Modes and Effects Analysis* ('FM&EA') – an inductive technique to identify and analyse hardware failures. This technique can be used to quantify risks.
•   *Task Analysis* – an inductive technique to identify the opportunity for human error. An approach, based on task-based hazards identification and analysis is introduced in this document.

Machinery hazards identification

[M1025] A person may be injured at machinery as a result of one or more of the following:

(a)   contact or entanglement with the machinery;
(b)   crushing between a moving part of the machine and fixed structures;
(c)   being struck by ejected parts of the machinery;
(d)   being struck by material ejected from the machinery.

These are regarded as mechanical hazards.

Figure 5: Examples of mechanical hazards

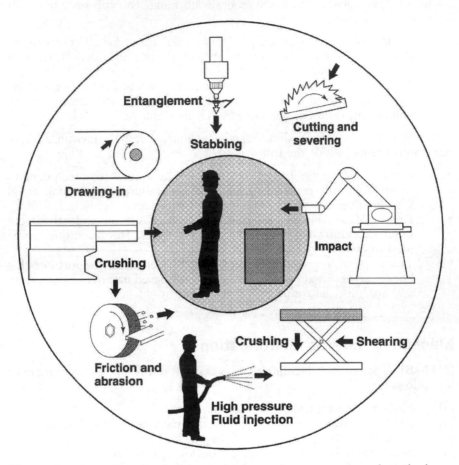

The main types of mechanical hazards shown in FIGURE 5 are described as:

- **Crushing:** Occurs when part of the body is caught between a moving part of a machine against a fixed object, eg underneath scissor lift or between the tools of a press.
- **Shearing:** Parts of the body may be sheared by scissor action caused by parts of the machine, eg mechanism of scissor lift or oscillating pendulum.
- **Cutting and severing:** Cutting hazards include contact with circular saws, guillotine knife, rotary knives or moving sheet metal.
- **Entanglement:** Occurs as a result of clothing or hair contact with rotating objects or catching on projections or in gaps, eg drills, rotating workpiece or belt fasteners.
- **Drawing-in or trapping:** Occurs when part of the body is caught between two counter-rotating parts, eg gears, mixing mills or between belt and pulley or chain and chain wheel.

- **Impact:** Impact hazards are caused by a moving object striking the body without penetrating it. Examples include: being hit by a robot arm or by moving traffic.
- **Stabbing and puncture:** flying objects, swarf or rapid moving parts, eg sewing machine.
- **Friction and abrasion:** contact with moving rough or abrasive surfaces, eg abrasive wheels.
- **High pressure fluid injection:** sudden release of fluid under pressure can cause tissue damage similar to crushing. Examples include: water jetting, compressed air jets and high pressure hydraulic systems.

Since a hazard is defined as 'a potential to cause injury or damage to health', it is therefore important to include health hazards in addition to the mechanical hazards which are aimed at physical injury. Other hazards include the following:

- Electrical hazards – to include direct and indirect contact with live electricity and electro-static phenomena.
- Radiation hazards – to include ionising and non-ionising sources as well as electro-magnetic effect.
- Work environment hazards – to include hot/cold ambient conditions, noise, vibration, humidity and poor lighting.
- Hazardous materials and substances – to include toxic, flammable and explosive substances.
- Work activity hazards – to include highly repetitive actions, stressful posture, lifting/handling heavy items, visual fatigue and poor workplace design.
- The neglect of ergonomic principles.

TABLE 1: HAZARDS IDENTIFICATION CHECKLIST

Type of hazard	Source	Task involved (Who is exposed and when?)
1. Mechanical hazards		
1.1 Crushing 1.2 Shearing 1.3 Cutting/severing 1.4 Entanglement 1.5 Drawing-in/trapping 1.6 Impact 1.7 Stabbing/puncture 1.8 Friction/abrasion 1.9 High pressure fluid injection 1.10 Slips/trips/falls 1.11 Falling objects 1.12 Other mechanical hazards		
2. Electrical hazards		
2.1 Direct contact 2.2 Indirect contact 2.3 Electrostatic phenomena 2.4 Short circuit/overload 2.5 Source of ignition 2.6 Other electrical hazards		
3. Radiation hazards		

Type of hazard	Source	Task involved (Who is exposed and when?)
3.1 Lasers 3.2 Electro-magnetic effects 3.3 Ionising/non-ion radiation 3.4 Other radiation hazards		
4. Hazardous substances		
4.1 Toxic fluids 4.2 Toxic gas/mist/fumes/dust 4.3 Flammable fluids 4.4 Flammable gas/mist/fumes/dust 4.5 Explosive substances 4.6 Biological substances 4.7 Other hazardous substances		
5. Work activities hazards		
5.1 Highly repetitive actions 5.2 Stressful posture 5.3 Lifting/handling heavy items 5.4 Mental overload/stress 5.5 Visual fatigue 5.6 Poor workplace design 5.7 Other workplace hazards		
6. Work environment hazards		
6.1 Localised hot surfaces 6.2 Localised cold surfaces 6.3 Significant noise 6.4 Significant vibration 6.5 Poor lighting 6.6 Hot/cold ambient temperature 6.7 Other work environment hazards		

Options for machinery risk reduction

[M1026] The most effective risk control measures are those implemented at the machine/work equipment design stage. It is a legal requirement in the EU that all machinery supplied after January 1995 should carry a 'CE' mark. This demonstrates that the machine complies with all relevant EU machinery directives, which include meeting the ES&HRs. Fundamental to this approach, manufacturers and suppliers need to carry out a risk assessment and demonstrate that all risks are adequately controlled by design, rather than by procedures. The following risk control options reflect the preferred order of priority:

Technical measures to be taken at the design stage

(1)   To make the machinery inherently safer (design out hazards) as a priority:
—   to eliminate hazards at the design stage;
—   to substitute hazardous substances used by the machine or process by less hazardous ones.

(2)   To provide protection from the hazards by the machine design, egto avoid the need for access to hazardous parts of the machine.

(3)   To improve machinery reliability by reducing the chance for fail-to-danger of critical components/systems.

(4)   To provide protection from the hazards by adequate safeguarding.

(5)   To reduce ease of access to danger zones (eg safeguards or safety devices which allow safe access, but prevent access at other times).

(6)   To reduce the chance of potential human error:
  - by improving the ergonomics of machine control layout to reduce the chance for unintended errors;
  - by reducing the chance for violations by making maintenance, cleaning, adjustments, etc. tasks convenient, easy and safe.

Procedural measures (make work tasks safer)

These may be achieved by the following measures:

- Planned preventive maintenance and regular inspection of machines and safeguards/safety devices.
- Safe systems of work (which would minimise the needs for access into the danger zone).
- Permit-to-work procedures (to formalise precautions in the face of a hazard).
- Adequate personal protection equipment ('PPE') and planning for emergencies.

Behavioural measures (develop safer people)

This involves the following:

- Selection and certification of personnel against specific work tasks.
- Training: basic skills; systems and procedures as well as knowledge of hazards.
- Adequate instructions, warnings and supervision.
- Improve safety culture within the business.
- Ask: why should safeguards/safety devices be violated/defeated? *
  * Note that violations are prevented if the perceived inconvenience of defeating a safeguard/safety device exceeds the benefit of so doing.

Types of safeguards and safety devices

[M1027] The main types of safeguards and safety devices can be classified as follow:

(1)   Fixed guards.
(2)   Fixed guards with adjustable element.
(3)   Automatic guards.
(4)   Interlocked guards.
(5)   Safety devices.
       These are described as:

5.1 Trip devices, eg photoelectric light curtains, pressure sensitive devices.

5.2 Two-hand control devices.

Machinery safeguards – general considerations

Figure 6: Design/selection of safeguards

[M1028]

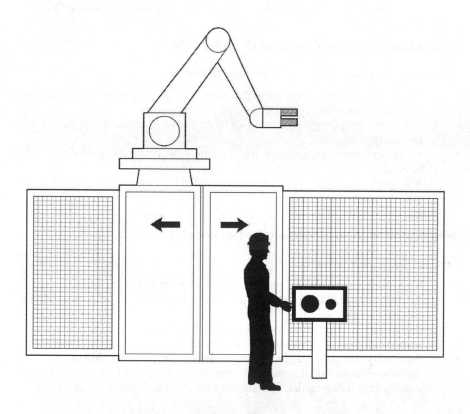

It is a general requirement that all machinery safeguards and safety devices must have the following characteristics outlined below.

Safeguards and safety devices should:

- Be of robust construction: strength, stiffness and durability to prevent ejected parts of the machine/components or material penetrating the guard.
- Not give rise to additional hazards.
- Not be easy to bypass or render non-operational.
- Be located at an adequate distance from the danger zone.
- Cause minimum obstruction of view for machine operators.

- Enable essential work (eg maintenance, cleaning) to be done without safeguard removal.

The two most applicable EU Standards relevant to safeguards are BS EN 953:1997+A1:2009 *Safety of machinery. Guards. General requirements for the design and construction of fixed and movable guards* and BS EN ISO 13857:2008 *Safety of machinery. Safety distances to prevent hazard zones being reached by upper and lower limbs*. The two most applicable EU Standards relevant to safeguards are BS EN 953: 1994 (design and construction), and BS EN 294: 1994 (safety distances).

Fixed guards

[M1029] Appendix 2 of the ACOP sets out that fixed guards have no moving parts and are fastened in a constant position in relation to the danger zone. They are kept in place either permanently, by welding for example, or by means of fasteners, screws and nuts for example. The ACOP sets out that. "If by themselves, or in conjunction with the structure of the equipment, they enclose the dangerous parts, fixed guards meet the requirements of the first level of the hierarchy". It also points out that fixed enclosing guards, and any other types of guard can have openings as long as they comply with appropriate safe reach distances (set out in BS EN ISO 13857: 2008).

A fixed guard should when fitted be incapable of being displaced casually, so the method of fixing is important. The ideal method of fixing the guard would be of a captive type, which would make unofficial access difficult.

It is clear from experience that if the need for access into hazardous parts of a machine fitted with a fixed guard, there will be a tendency not to replace the guard back onto the machine. Therefore an interlocked guard is more suited in this situation.

Figure 7: Example of a fixed guard

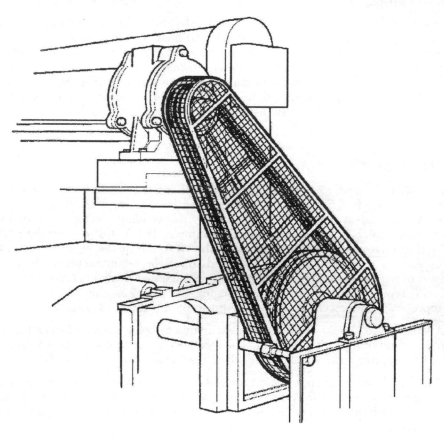

Adjustable guards

[M1030] Adjustable guards comprise a fixed guard with adjustable elements that the setter or operator has to position to suit the job being worked on. They are widely used for woodworking and tool-room machines. Where adjustable guards are used, the operators should be familiar and trained in how to adjust them so that full protection can be obtained.

FIGURE 8 shows the application of a telescopic guard to a heavy duty-drilling machine. The idea here is that as the drill descends into the workpiece, the fixed guard will always enclose the drill. The approach to selecting the most appropriate safeguard should start by hazard identification. Hazards associated with the use of the drilling machine include the following:

- entanglement with the drill;
- stabbing/puncture by the swarf;
- ejection of broken drill bit;
- stabbing by the drill bit; and
- possible ejection of the work-piece.

Figure 8: Fixed guard with adjustable element

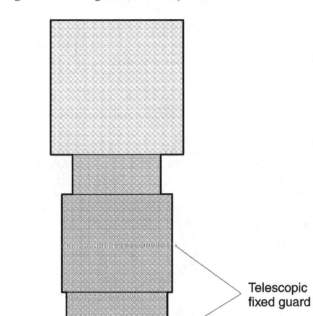

It is noted that the telescopic guard should provide protection against most of the above hazards apart from stabbing/puncture by the drill and ejection of the work-piece.

Self-adjusting guards are also classified as a fixed guard with adjustable element. This guard prevents access to the hazard except when the guard is forced open by the passage of the work. They usually incorporate a spring-loaded pivoted element. These types of guards are mainly used on woodworking machinery.

Automatic guards

[M1031] An automatic guard is a guard, which is moved into position automatically by the machine, thereby removing any part of a person from the hazardous area of the machine. This is sometimes known as 'sweep away' guard and is used on some paper guillotines and large mechanical or hydraulic presses.

There are however many factors which could limit the effectiveness of these types of guards. The example of the automatic guard shown in FIGURE 9 is used to illustrate how this type of guard can cause injury.

Figure 9: Example of automatic guard

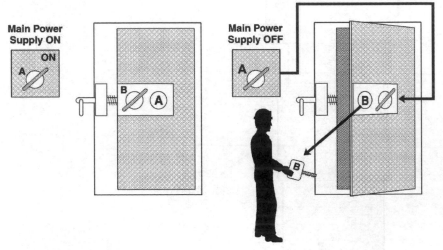

Some possible hazards associated with this automatic guard include the following:

• shearing hazard between the moving and fixed parts of the guard;
• impact with the moving part of the guard;
• crushing hazard against fixed structures nearby as well as the ergonomics of the guard, eg height of guard in relation to operators and the size of guard gaps.

Interlocking guards

[M1032] Interlocking guards are usually movable, eg they could be hinged, sliding or removable. These guards are used, where frequent access to hazardous parts of a machine may be needed, and are connected to the machine controls by means of 'position sensors'. These interlocking elements could be electrical, mechanical, magnetic, hydraulic or pneumatic. The position sensors interlock the guard with the power source of the hazard. When the guard is open, the power is isolated thus allowing safe access into the relevant part of the machine.

Figure 10: Main types of electrical interlocking switches

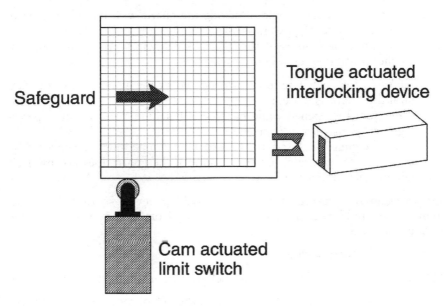

Safeguard

Tongue actuated
interlocking device

Cam actuated
limit switch

BS EN 1088:1995+A2:2008 Safety of machinery. Interlocking devices associated with guards. Principles for design and selection provides detailed guidance on the design and selection of the appropriate type of devices. Risk assessment plays a vital role in the selection of type and level of integrity on interlocking devices.

The integrity of the interlocking mechanism must ensure that the device is reliable, capable of resisting interference, difficult to defeat and the system should not, as far as possible, fail-to-danger. The term 'fail-safe' is no longer used, as it can give the impression that if a component fails in a certain mode this will always result in the machine stopping even when the safeguard is closed. Fail-to-danger of an interlocking device means that the machine can be operated when the guard is open. Therefore to minimise the probability of failure-to-danger is a preferred option.

Interlocking guards are convenient, as they give ready access into the danger zone, while ensuring the safety of the operator. However, some of these guards may be defeated to carry out necessary tasks, eg maintenance, setting, fault finding etc. Therefore, additional measures such as locking off or isolation may need to be introduced under a safe system of work. It is however a requirement that the machine designer should consider how these tasks can be safely performed at the design stage.

The choice of interlocking method must be compatible with the level of risk and should take into account the probability of failure of the interlocking element(s), proportion of time a person is exposed to the hazard and the potential consequences (eg injury severity and/or ill health).

There are two broad categories of interlocking guards:

•   Guard is interlocked with the source of power.

- Guard is interlocked with respect to the motion itself.

There are clear advantages for interlocking the guard with the moving parts rather than the source of power, as the power might fail-to-danger due to wiring/earth faults or failures of the interlocking switch. The main requirements for the movable interlocking guards are:

(a)   the hazardous machine functions covered by the guard cannot operate until the guard is closed;

(b)   if the guard is opened while hazardous machine functions are operating, a stop instruction is given;

(c)   when the guard is closed, the hazardous machine functions covered by the guard can operate, but the closure of the guard does not by itself initiate their operation.

Interlocking guards may be fitted with a locking device so that the guard remains closed and locked until any risk of injury from the hazardous machine functions has passed.

Types of interlocking devices

[M1033] Some common types of interlocking elements include the following:

- *Cam-operated limit switch interlocks*: are one of most popular types of interlocks used generally as they are versatile, effective and easy to install and use. Another type of mechanically actuated device is the 'tongue' actuator, which opens or closes the contacts of a switch. Both types of interlocks are shown in FIGURE 13.

- *Trapped key interlocks*: a master key, which controls the power supply through a switch at the master key box, has to be turned OFF before the keys for individual guards can be released. The master switch cannot be turned to ON until all the individual keys are replaced in the master box. This type of interlock is commonly used in electrical isolation.

- *Captive-key interlocking*: involves a combination of an electrical switch and a mechanical lock in a single assembly.

- *Direct manual switch or valve interlocks*: where a switch or valve controlling power source cannot be operated until a guard is closed, and the guard cannot be opened at any time the switch is in run position (position sensors).

- *Mechanical interlocks*: provide a direct linkage from guard to power or transmission control. In other words the switch cannot be reached when the guard is open.

- *Magnetic switches* with the actuating 'coded' magnet attached to the guard: have the disadvantage that they can be defeated easily. However, types using a shaped magnet have been used with removal guards on screw conveyors. Normal reed switches are not acceptable unless they incorporate special current limiting features. Reed switches are often encapsulated and have application in flammable atmospheres. A similar type of switch relies on inductive circuits.

- *Electro-mechanical device*: can provide a time delay where the first movement of the bolt trips the machine, but the bolt has to be unscrewed a considerable distance before the guard is released. Time delay arrangements are necessary when the machine being guarded has a long run down time. A solenoid-operated bolt can also be used in conjunction with a time delay circuit.

- *Mechanical restraints (scotches)*: are used as a back-up to other forms of interlocks on certain types of presses to protect the operator against crushing injury between platens.

Trapped key interlocking (Power isolation)

[M1034] The electrical interlocking switches interrupt the power source of the hazard by switching of a circuit, which controls the power-switching device. This is known as control interlocking.

Power interlocking on the other hand involves the direct switching of the power supply to the hazard, eg isolation of power. The most practical means of power interlocking is a trapped key system.

An example of such a system is illustrated in FIGURE 11. The power supply isolation switch is operated by key 'A' which is trapped in position while the isolation switch is in the ON position. When the key is turned and released, the isolation switch contacts are locked open thus isolating the power supply. The machine safeguard door in the meantime is locked closed. To open the guard, the isolation key 'A' is inserted into the guard-locking unit and turned to release the door. The key is then trapped in position and cannot be removed until the guard is closed and locked again.

One of the drawbacks of this system is that unknown to others, a person may still be inside the machine while another person decides to start-up the machine by closing and locking the guard. To overcome this foreseeable problem, the use of a personal key 'B' which is also trapped in the guard locking unit and is released when the guard is open. Key 'A' cannot be released from the guard unless key 'B' is inserted and turned into the guard locking unit. Key 'B' can also be used for other tasks, eg inch mode for presses or programming mode for robotics.

Figure 11: Trapped key system for power interlocking

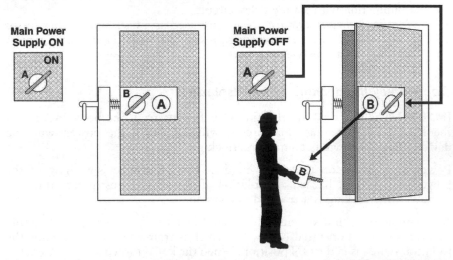

This type of system is reliable and does not require electrical wiring to the guard. The main disadvantage is that because if requires the exchange of the keys every time, it may not be suitable if access is frequently required. Strict control of the keys is essential.

Mechanical interlocking

Figure 12: Example of mechanical interlocking system

[M1035]

The system shown in FIGURE 12 is known as guard inhibited interlocking. This is achieved in the above example by a sliding guard, when open the guard retains the lever of a switch or a control valve in the power OFF position. When the guard has been closed, the lever or switch can be moved to the ON position, thus inhibiting the guard from being open.

This system however is easy to defeat by removing one of the door-stops, or even by removing the guard.

Other forms of mechanical interlocks include mechanical restraints (scotches) to prevent gravity fall of vertical machinery, eg scissors lifts and presses.

Safety devices

[M1036] A safety device is a protective appliance, other than a guard, which eliminates or reduces risk, alone or associated with a guard. There are many forms of safety device available.

(a) Trip device

A trip device is one which causes a machine or machine elements to stop (or ensures an otherwise safe condition) when a person or a part of his body goes beyond a safe limit. Trip devices take a number of forms – for example, mechanical, electro-sensitive safety systems and pressure-sensitive mat systems.

Trip devices may be:

(i)   mechanically actuated – eg trip wires, telescopic probes, pressure-sensitive devices; or

(ii)   non-mechanically actuated – eg photo-electric devices, or devices using capacitive or ultrasonic means to achieve detection.

(b) Enabling (control) device

This is an additional manually operated control device used in conjunction with a start control and which, when continuously actuated, allows a machine to function.

(c) Hold-to-run control device

This device initiates and maintains the operation of machine elements for only as long as the manual control (actuator) is actuated. The manual control returns automatically to the stop position when released.

(d) Two-hand control device

A hold-to-run control device which requires at least simultaneous actuation by the use of both hands in order to initiate and to maintain, whilst hazardous condition exists, any operation of a machine thus affording a measure of protection only for the person who actuates it. BS EN 574:1996+A1:2008 Safety of machinery. Two-hand control devices. Functional aspects. Principles for design specifies the safety requirements relating to two-hand control devices.

(e) Limiting device

Such a device prevents a machine or machine elements from exceeding a designed limit (eg space limit, pressure limit).

(f) Limited movement control device

The actuation of this sort of device permits only a limited amount of travel of a machine element, thus minimising risk as much as possible; further movement is precluded until there is a subsequent and separate actuation of the control.

(g) Deterring/impeding device

This comprises any physical obstacle which, without totally preventing access to a danger zone, reduces the probability of access to this zone by preventing free access.

Major Accident Hazards

Roger Bentley

The Control of Major Accident Hazards Regulations 1999 (SI 1999 No 743) and the Control of Major Accident Hazards (Amendment) Regulations 2005 (SI 2005 No 1088)

History

[M1101] European Directive 82/501/EC (the 'Seveso' Directive), which was implemented into UK legislation by the *Control of Industrial Major Accident Hazards (CIMAH) Regulations 1984 (SI 1984 No 1902)*, applied to fixcd installations with major accident hazard potential, and imposed duties on operators of sites with inventories of hazardous substances in excess of given thresholds. The thresholds distinguished two tiers of requirements, with differences for storage as opposed to usage that created some confusion. Amendments followed in 1988, 1990 and 1994. Although the regulations referred to limiting the consequences of major accidents not only to persons but also to the environment, there were actually few specific requirements for the latter. Neither did the *CIMAH* regulations address land-use planning.

[M1102] The *Control of Major Accident Hazards (COMAH) Regulations 1999 (SI 1999 No 743)* implement the requirements of the 'Seveso II' Directive (96/82/EC) on the control of major accident hazards involving dangerous substances. *COMAH* came into force on 1 April 1999. The provisions of Article 12 of Seveso II concerning land-use planning were implemented by the *Planning (Control of Major Accident Hazards) Regulations 1999 (SI 1999 No 981)*.

SI 1999 No 743 has now been amended by the *Control of Major Accident Hazards (Amendment) Regulations 2005 (SI 2005 No 1088)* implementing Directive 2003/105/EC on the control of major accident hazards involving dangerous substances. The main effect of the amendment is to broaden the scope of *COMAH* through changes to the schedule of named substances and generic categories, also revisions to the qualifying quantities.

The Health and Safety Executive (HSE) together with either the Environment Agency (EA) in England and Wales or the Scottish Environment Protection Agency (SEPA) are jointly responsible as the Competent Authority (CA) for enforcing *COMAH*. Northern Ireland makes separate legislative and administrative arrangements, but has implemented the Seveso Directive with corresponding regulations.

[M1103] The *Control of Major Accident Hazards Regulations*, as amended, impose requirements where dangerous substances are present in quantities equal to or exceeding those specified in *Schedule 1* (reproduced in Tables 1 and 2 below), irrespective of whether they are raw materials, products, by-products, residues or intermediates. It includes those which it is reasonable to believe may be generated during the loss of control of an industrial chemical process. The emphasis is on controlling risks to both people and the environment through demonstrable safety management systems, which are integrated into the routine of business rather than dealt with as an add-on.

An occurrence is regarded as a major accident if:

—   it results from uncontrolled developments (i.e. they are sudden, unexpected or unplanned) in the course of the operation of an establishment to which the Regulations apply;
—   it leads to serious danger to people or to the environment, on- or off-site; and
—   it involves one or more dangerous substances defined in the Regulations.

Major emissions, fires and explosions are the most typical major accidents.

The duties placed on operators by *COMAH* fall into two categories, defined by the amounts of dangerous substances: these are now generally known as 'lower-tier' and 'top-tier' though these are not terms used in the Regulations. Lower-tier duties fall upon all operators with inventories exceeding the lower threshold. Where inventories exceed the higher threshold, operators are subject to additional top-tier duties in *Regs 7–14*, and these are far more onerous.

An 'operator' is defined as a person (including a company or partnership) who is in control of the operation of an establishment or installation. Where the establishment or installation is yet to be constructed or operated, the 'operator' is the person who proposes to control its operation. Where that person is not known, the operator is the person who has commissioned its design and construction.

Unlike *CIMAH*, *COMAH* make no distinction between processing and storage of chemicals.

Once it has been established that the Regulations apply to an establishment, the duties for the appropriate tier then apply to all dangerous substances present, not just those that are in excess of the threshold.

A charging regime has been introduced and the fees are payable by the operator for work carried out by the HSE and the Environment Agency. Details of the current charging regime may be found on the HSE website at: www.hse.gov.uk/comah/charging/index.htm.

The main HSE guidance to the Regulations is contained in 'A guide to the Control of Major Accident Hazards Regulations 1999 (as amended)' (L111, second edition, 2006).

The consolidated 1999 Regulations:

(a)   impose a duty on the operator of an establishment to take all measures necessary to prevent major accidents and limit their consequences for persons and the environment (*Reg 4*);

(b)   impose a duty on the operator to prepare within a time limit a major accident prevention policy document containing specified information and to revise it in specified circumstances (*Reg 5 and Sch 2*);

(c)   require the operator to notify the competent authority within a time limit of specified matters at specified times (*Reg 6 and Sch 3*);

(d)   require the operator to send at specified times a safety report to the competent authority containing specified information, to revise that report in specified circumstances, and not to start the construction or operation of the establishment until he has received from the competent authority the conclusions of its examination of the report (*Reg 7 and Sch 4*);

(e)   require the operator to review and revise the safety report in specified circumstances and to notify the CA even if the review does not result in a revision of the report (*Reg 8*);

(f)   require the operator to prepare within a time limit an on-site emergency plan for specified purposes and containing specified information (*Reg 9 and Sch 5*);

(g)   require, subject to any exemption that may be granted by the competent authority, the local authority to prepare an off-site emergency plan for specified purposes and containing specified information and require the operator to supply to the local authority the information necessary to enable the plan to be prepared (*Reg 10 and Sch 5*);

(h)   require on-site and off-site emergency plans to be reviewed, tested and implemented (*Regs 11 and 12*), and empower the local authority to charge the operator a fee for preparing, reviewing and testing the off-site emergency plan (*Reg 13*);

(j)   require the operator to provide, after consulting the local authority, specified information to specified persons at specified times and to make that information publicly available (*Reg 14 and Sch 6*);

(k)   require the operator to demonstrate to the competent authority that he has taken all measures necessary to comply with the Regulations (*Reg 15(1)*);

(l)   require the operator to provide, when requested to do so by the competent authority, information to the authority for specified purposes (*Reg 15(2)*);

(m)   require the operator to notify major accidents to the competent authority (*Reg 15(3) and (4)*);

(n)   require the operator to provide information to, and to co-operate with, other establishments in a group designated by the competent authority (*Reg 16*);

(o)   impose functions on the competent authority with respect to:

  (i)   its consideration of the safety report sent by the operator (*Reg 17*);
  (ii)   prohibiting the operation of an establishment (*Reg 18*);
  (iii)   inspections and investigations (*Reg 19*);
  (iv)   enforcement (*Reg 20*); and
  (v)   the provision of information (*Reg 21 and Schs 7 and 8*);

(p)   provide for fees to be payable by the operator to the HSE in relation to the performance of specified functions of the Executive or competent authority (*Reg 22*);

(q)   amend the *Petroleum Consolidation Act 1928*, the *Petroleum-Spirit (Motor Vehicles etc.) Regulations 1929* and the *Petroleum-Spirit (Plastic Containers) Regulations 1982 (SI 1982 No 630)* so as to disapply those Regulations to establishments subject to these Regulations and to sites subject to the *Notification of Installations Handling Hazardous Substances Regulations 1982 (SI 1982 No 1357) (Reg 23)*.

Application

[M1104] The *Control of Major Accident Hazards Regulations* (as amended) apply to an *establishment* where:

(a)   dangerous substances are present;
(b)   their presence is anticipated; or
(c)   it is reasonable to believe that they may be generated during the loss of control of an industrial chemical process.

'Loss of control' excludes expected, planned or permitted discharges.

'Industrial chemical process' means that premises without such chemical process do not fall within the scope of the Regulations solely because of dangerous substances generated during an accident.

An 'establishment' refers to the whole area under the control of the same person where dangerous substances are present in one or more installations. Two or more areas under the control of the same person and separated only by a road, railway or inland waterway are to be treated as one whole area.

'Dangerous substances' are defined as those which:

(a)   are named at the appropriate threshold (see Table 1 below); or
(b)   fall within a generic category at the appropriate threshold (see Table 2 below).

There are two threshold levels: lower tier (Column 2 of the Tables) and top tier (Column 3).

Table 1		
Column 1	*Column 2*	*Column 3*
Dangerous substances	*Quantity in tonnes*	
Ammonium nitrate (see note 1 below)	5,000	10,000
Ammonium nitrate (see note 2 below)	1,250	5,000
Ammonium nitrate (see note 3 below)	350	2,500
Ammonium nitrate (see note 4 below)	10	50
Potassium nitrate (see note 5 below)	5,000	10,000
Potassium nitrate (see note 6 below)	1,250	5,000
Arsenic pentoxide, arsenic (V) acid and/or salts	1	2
Arsenic trioxide, arsenious (III) acid and/or salts	0.1	0.1
Bromine	20	100
Chlorine	10	25
Nickel compounds in inhalable powder form (nickel monoxide, nickel dioxide, nickel sulphide, trinickel disulphide, dinickel trioxide)	1	1
Ethyleneimine	10	20
Fluorine	10	20

Table 1		
Column 1	*Column 2*	*Column 3*
Dangerous substances	*Quantity in tonnes*	
Formaldehyde (concentration=>90%)	5	50
Hydrogen	5	50
Hydrogen chloride (liquefied gas)	25	250
Lead alkyls	5	50
Liquefied extremely flammable gases (including LPG) and natural gas (whether liquefied or not)	50	200
Acetylene	5	50
Ethylene oxide	5	50
Propylene oxide	5	50
Methanol	500	5,000
4, 4-Methylenebis (2-chloraniline) and/or salts, in powder form	0.01	0.01
Methylisocyanate	0.15	0.15
Oxygen	200	2,000
Toluene diisocyanate	10	100
Carbonyl dichloride (phosgene)	0.3	0.75
Arsenic trihydride (arsine)	0.2	1
Phosphorus trihydride (phosphine)	0.2	1
Sulphur dichloride	1	1
Sulphur trioxide	15	75
Polychlorodibenzofurans and polychlorodibenzodioxins (including TCDD), calculated in TCDD equivalent	0.001	0.001
The following CARCINOGENS at concentrations above 5% by weight (see note 7 below):	0.5	2
4-Aminobiphenyl and/or its salts, Benzotrichloride, Benzidine and/or salts,		
Bis(chloromethyl) ether, Chloromethyl methyl ether, 1,2-Dibromoethane, Diethyl sulphate, Dimethyl sulphate,		
Dimethylcarbamoyl chloride, 1,2-Dibromo-3-chloropropane, 1,2-Dimethylhydrazine, Dimethylnitrosomine,		
Hexamethylphosphoric triamide, Hydrazine, 2-Naphthylamine and/or salts,		
1, 3 Propanesultone and 4-nitrodiphenyl		
Automotive petrol and other petroleum spirit Petroleum Products (see note 8 below)	2,500	25,000
(*a*) gasolines and naphthas		
(*b*) kerosenes (including jet fuels)		

Table 1		
Column 1	Column 2	Column 3
Dangerous substances	Quantity in tonnes	
(c) gas oils (including diesel fuels, home heatingoils and gas oil blending streams)		

[SI 1999 No 743, Sch 1, Pt 2]

Ammonium nitrate
The entries for ammonium nitrate were changed in the 2005 amending regulations in response to an explosion at a fertiliser plant in Toulouse, France. The two existing entries for fertiliser grade and technical grade were maintained, but two new entries were added to cover fertilisers capable of self-sustaining decomposition and 'off-spec' material.

1.
The 5,000/10,000 quantities apply to ammonium nitrate and ammonium nitrate compound/composite fertilisers (compound or composite fertilisers containing ammonium nitrate with phosphate and/or potash) in which the nitrogen content as a result of the ammonium nitrate is:

(a)   between 15.75 and 24.5% by weight and either with not more than 0.4% total combustible or organic materials or which satisfy the detonation resistance test describes in the *Ammonium Nitrate Materials (High Nitrogen Content) Safety Regulations 2003 (SI 2003 No 1082), Sch 2*; or

(b)   15.75% or less by weight and unrestricted combustible materials,

and which are capable of self-sustaining decomposition according to the UN Trough Test specified in United Nations Recommendations on the Transport of Dangerous Goods: Manual of Tests and Criteria (3rd revised edition) Part III Subsection 38.2.
15.75% nitrogen content by weight as a result of ammonium nitrate corresponds to 45% ammonium nitrate.
24.5% nitrogen content by weight as a result of ammonium nitrate corresponds to 70% ammonium nitrate.

2.
The 1250/5000 quantities apply to straight ammonium-nitrate based and to ammonium nitrate based compound/composite fertilisers in which the nitrogen content as a result of ammonium nitrate is:

(a)   more than 24.5% by weight, except for mixtures of ammonium nitrate with dolomite, limestone and/or calcium carbonate with a purity of at least 90%;

(b)   more than 15.75% by weight for mixtures of ammonium nitrate with ammonium sulphate; or

(c)   more than 28% by weight for mixtures of ammonium nitrate with dolomite, limestone and/or calcium carbonate with a purity of at least 90%, and which satisfy the detonation resistance test.

28% nitrogen content by weight as a result of ammonium nitrate corresponds to 80% ammonium nitrate.

3.

The 350/2500 quantities apply to ammonium nitrate and preparations of ammonium nitrate in which the nitrogen content as a result of the ammonium nitrate is:

(a) between 24.5 and 28% by weight, and which contain not more than 0.4% combustible substances; or

(b) more than 28% by weight, and which contain not more than 0.2% combustible substances,

and aqueous ammonium nitrate solutions in which the concentration of ammonium nitrate is more than 80% by weight.

4.

The 10/50 quantities apply to 'off spec' material and fertilisers not satisfying the detonation resistance test. This applies to:

(a) material rejected during the manufacturing process and to ammonium nitrate and preparations of ammonium nitrate, straight ammonium nitrate-based fertilisers and ammonium nitrate based compound/composite fertilisers referred to in notes 2 and 3, that are being or have been returned from the final user to a manufacturer, temporary storage or reprocessing plant for reworking, recycling or treatment for safe use, because they no longer comply with the specifications of Notes 2 and 3; or

(b) fertilisers which do not fall within Notes 1(a) and 2 because they do not satisfy the detonaton resistance test, or other fertilisers which:

(i) at the time of delivery to a final user satisfied the detonation resistance test but:

(ii) later became degraded or contaminated; and

(iii) are temporarily present in the establishment of the final user prior to their return for reworking, recycling or treatment for safe use or to their being applied as fertiliser.

Potassium nitrate

Because of its similarities with ammonium nitrate, potassium nitrate has been included as a named substance by the 2005 amendment with two entries distinguishing between potassium nitrate in a granular form and potassium nitrate in crystalline form. Potassium nitrate is less combustible than ammonium nitrate and this is reflected in the threshold quantities.

5.

The 5,000/10,000 quantities apply to composite potassium nitrate-based fertilisers composed of potassium nitrate in prilled/granular form.

6.

The 1,250/5,000 quantities apply to composite potassium nitrate-based fertilisers composed of potassium nitrate in crystalline form.

Named carcinogens

7.

The dilution cut-off was raised by the 2005 amendment from 0.1% to 5% by weight, and the qualifying quantities have been raised. These quantities are one order of magnitude lower than those for substances classified as 'very toxic' and are intended to reflect public concern about carcinogens and the application of the precautionary principle in the absence of full scientific data.

Petroleum products

8.

As a result of rising concerns relating to petroleum products as substances dangerous to the environment, the previous entry in the 1999 Regulations was replaced in the 2005 amendment by a petroleum products category containing three classes of products. The category provides a definitive listing. Only those products listed in this category will be considered to be petroleum products. Other substances, not so defined, such as pentane or paraffin, would fall under the generic categories contained in Table 2 below unless specifically named in Table 1

Table 2
Substances and preparations shall be classified according to Reg. 4 of the *Chemicals (Hazard Information and Packaging for Supply) Regulations 2002 (SI 2002 No 1689)* as amended.

Column 1	Column 2	Column 3
Categories of dangerous substances	*Quantity in tonnes*	
1. VERY TOXIC	5	20
2. TOXIC	50	200
3. OXIDISING	50	200
4. EXPLOSIVE (see D0706) where the substance, preparation or article is an explosive within UN/ADR Division 1.4	50	200
5. EXPLOSIVE (see D0706) where the substance, preparation or article is an explosive within UN/ADR Division 1.1, 1.2, 1.3, 1.5 or 1,6 or risk phrase R2 or R3	10	50
6. FLAMMABLE (see D0705)	5,000	50,000
7a. HIGHLY FLAMMABLE (see D0705)	50	200
7b. HIGHLY FLAMMABLE liquids (see D0705)	5,000	50,000
8. EXTREMELY FLAMMABLE (see D0705)	10	50
9. DANGEROUS FOR THE ENVIRONMENT in combination with risk phrases:		
R50: 'Very toxic to aquatic organisms' (including R50/R53)	100	200
R51/R53: 'Toxic to aquatic organisms'; and R53: 'May cause long-term adverse effects in the aquatic environment'	200	500

Table 2 Substances and preparations shall be classified according to Reg. 4 of the *Chemicals (Hazard Information and Packaging for Supply) Regulations 2002 (SI 2002 No 1689) as amended.*		
Column 1	*Column 2*	*Column 3*
Categories of dangerous substances	*Quantity in tonnes*	
10. ANY CLASSIFICATION not covered by those given above in combination with risk phrases:		
R14: 'Reacts violently with water' (including R14/15)	100	500
R29: 'in contact with water, liberates toxic gas'	50	200

[SI 2005 No 1088, Sch 1, Pt 3]

Changes were made by the 2005 amending Regulations in respect of substances dangerous to the environment which significantly reduce the qualifying quantities, recognising that even relatively small quantities of hazardous materials can cause severe environmental damage.

Where a substance or group of substances named in Table 1 also falls within a category in Table 2, the qualifying quantities set out in Table 1 must be used.

Isolated storage of dangerous substances in quantities not exceeding 2% of the relevant qualifying quantity are ignored for the purposes of calculating the total quantity present, provided that their location is such that they cannot initiate a major accident elsewhere on site. Guidance on the interpretation of this is to be found in Appendix 2 of the HSE's *A guide to the Control of Major Accident Hazards Regulations 1999 (as amended)*.

Flammable substances and preparations

[M1105] Categories 6, 7a, 7b and 8 in Table 2 above are defined in *SI 2005 No 1088, Sch 1, Pt 3, note 3*:

Category 6: flammable liquids – substances and preparations having a flash point equal to or greater than 21°C and less than or equal to 55°C (risk phrase R10), supporting combustion.

Category 7: highly flammable liquids:

(a)   *Category 7a*: substances and preparations which may become hot and finally catch fire in contact with air at ambient temperature without any input of energy (risk phrase R17); substances and preparations which have a flash point lower than 55°C and which remain liquid under pressure, where particular processing conditions, such as high pressure or high temperature, may create major-accident hazards.

(b)   *Category 7b*: substances and preparations having a flash point lower than 21°C and which are not extremely flammable (risk phrase R11, second indent).

Category 8: extremely flammable gases and liquids:

(i)   liquid substances and preparations which have a flash point lower than 0°C and the boiling point (or, in the case of a boiling range, the initial boiling point) of which at normal pressure is less than or equal to 35°C (risk phrase R12, first indent);

(ii)   gases which are flammable in contact with air at ambient temperature and pressure (risk phrase R12, second indent), which are in a gaseous or supercritical state; and

(iii)   flammable and highly flammable liquid substances and preparations maintained at a temperature above their boiling point.

A gas is defined as any substance that has an absolute vapour pressure equal to or greater than 101.3 kPa at a temperature of 20°C. A liquid is any substance that is not a gas and is not in the solid state at a temperature of 20°C and at a standard pressure of 101.3 kPa.

Explosives

[M1106] Also under the *Control of Major Accident Hazards (Amendment) Regulations 2005 (SI 2005 No 1088)*, the definitions of the two categories of explosives were revised to align with the UN/ADR classification to distinguish between explosives on the basis of the hazard they represent, instead of the previous reference to risk phrases contained in EC Directives on classification, packaging and labeling of dangerous substances. The revised definitions are intended to more accurately describe the hazards of different types of explosives, particularly consumer (domestic) fireworks. The amended entries are contained in the table above, with the following explanatory notes provided in the Regulations.

An 'explosive' means—

(a)   a substance or preparation which creates the risk of an explosion by shock, friction, fire or other sources of ignition (risk phrase R2);

(b)   a substance or preparation which creates extreme risks of explosion by shock, friction, fire or other sources of ignition (risk phrase R3); or

(c)   a substance, preparation or article covered by Class 1 of the European Agreement concerning the International Carriage of Dangerous Goods by Road (UN/ADR), concluded on 30 September 1957, as amended, as transposed by Council Directive 94/55/EC of 21 November 1994 on the approximation of the laws of the Member States with regard to the transport of dangerous goods by road.

Included in this definition are pyrotechnics, which for the purposes of these Regulations mean substances (or mixtures of substances) designated to produce heat, light, sound, gas or smoke or a combination of such effects through self-sustained exothermic chemical reactions.

Where a substance or preparation is classified by both UN/ADR and risk phrase R2 or R3, the UN/ADR classification shall take precedence over assignment of risk phrases.

Substances and articles of Class 1 are classified in Divisions 1.1 to 1.6 in accordance with the UN/ADR classification scheme. The Divisions concerned are—

(a)   Division 1.1: substances and articles which have a mass explosion hazard (a mass explosion is an explosion which affects almost the entire load virtually instantaneously).

(b)   Division 1.2: substances and articles which have a projection hazard but not a mass explosion hazard.

(c)   Division 1.3: substances and articles which have a fire hazard and either a minor blast hazard or a minor projection hazard or both, but not a mass explosion hazard—
   (i)   combustion of which gives rise to considerable radiant heat; or
   (ii)   which burn one after another, producing minor blast or projection effects or both.

(d)   Division 1.4: substances and articles which present only a slight risk in the event of ignition or initiation during carriage. The effects are largely confined to the package and no projection of fragments of appreciable size or range is to be expected. An external fire shall not cause virtually instantaneous explosion of virtually the entire contents of the package.

(e)   Division 1.5: very insensitive substances having a mass explosion hazard which are so insensitive that there is very little probability of initiation or of transition from burning to detonation under normal conditions of carriage. As a minimum requirement they shall not explode in the external fire test.

(f)   Division 1.6: extremely insensitive articles which do not have a mass explosion hazard. The articles contain only extremely insensitive detonating substances and demonstrate a negligible probability of accidental initiation or propagation. The risk is limited to the explosion of a single article.

Included in this definition are also explosive or pyrotechnic substances or preparations contained in articles. In the case of articles containing explosive or pyrotechnic substances or preparations, if the quantity of the substance or preparation contained is known, that quantity shall be considered for the purposes of these Regulations. If the quantity is not known, then, for the purposes of these Regulations, the whole article shall be treated as explosive.

Aggregation

[M1107] Where an establishment fails to reach a threshold in respect of any single named substance in Table 1 above, or the threshold for any category in Table 2 above, but has more than one of these, then it is necessary to consider whether those items should be aggregated.

To take a simple illustration, if an establishment has four tonnes of a substance classified as very toxic, this is below the five-tonne lower-tier threshold. Likewise, if the site has 30 tonnes of materials classified as toxic, this is below the 50-tonne lower-tier threshold. But if these inventories are both present, then the proportion of threshold of both (0.80 on very toxic and 0.60 on toxic) must be added together (1.40) and, as this is greater than one, then the threshold is deemed to be exceeded and lower tier applies.

Similarly oxidising, explosive, flammable, highly flammable and extremely flammable categories are aggregated.

The same principle also applies to the two subdivisions of materials dangerous for the environment.

Furthermore, named substances in Table 1 must be included in the appropriate categories, but using the threshold specific to that named substance.

[M1108] The precise rule for aggregation of dangerous substances is set out in Note 4 to *Schedule 1 Part 3* of the *Control of Major Accident Hazards Regulations (SI 1999 No 743)*, as amended.

In the case of an establishment where no individual substance or preparation is present in a quantity above or equal to the relevant qualifying quantities, the following rules shall be applied to determine the application of these Regulations to the establishment.

If the sum—

$q_1/Q_{U1} + q_2/Q_{U2} + q_3/Q_{U3} + q_4/Q_{U4} + q_5/Q_{U5} + \ldots$ is greater than or equal to 1, where—

(a) qx = the quantity of dangerous substance x (or category of dangerous substances) falling within Part 2 or 3 of this Schedule; and

(b)   QUX = the relevant qualifying quantity for substance or category x from column 3 of Part 2 or 3,

then these Regulations apply.

If the sum—

$q_1/Q_{L1} + q_2/Q_{L2} + q_3/Q_{L3} + q_4/Q_{L4} + q_5/Q_{L5} + \ldots$ is greater than or equal to 1, where—

(a) qx = the quantity of dangerous substance x (or category of dangerous substances) falling within Part 2 or 3 of this Schedule; and

(b)   QLX = the relevant qualifying quantity for substance or category x from column 2 of Part 2 or 3,

then these Regulations, save Regulations 7 to 14, apply.

These rules shall be used to assess the overall hazards associated with toxicity, flammability and eco-toxicity. They must therefore be applied three times—

(a)   for the addition of substances and preparations named in Part 2 and classified as toxic or very toxic, together with substances and preparations falling into category 1 or 2;

(b)   for the addition of substances and preparations named in Part 2 and classified as oxidising, explosive, flammable, highly flammable or extremely flammable, together with substances and preparations falling into category 3, 4, 5, 6, 7a, 7b or 8; and

(c)   for the addition of substances and preparations named in Part 2 and classified as dangerous for the environment (R50 (including R50/53) or R51/53), together with substances and preparations falling into category 9(a) or 9(b),

and the relevant provisions of these Regulations shall apply if any of the sums thereby obtained is greater than or equal to 1.

Definitive guidance on aggregation can be found in HSE's *A guide to the Control of Major Accident Hazards Regulations 1999 (as amended)*'.

Exclusions

[M1109] The *Control of Major Accident Hazards Regulations (SI 1999 No 743)* does not apply to:

— Ministry of Defence establishments;
— extractive industries exploring for, or exploiting, materials in mines and quarries;
— waste land-fill sites;
— transport related activities; and
— substances at nuclear licensed sites which create a hazard from ionising radiation.

However *COMAH does* apply to explosives and chemicals at nuclear installations.

Notifications

[M1110] Notification requirements are set out in *Control of Major Accident Hazards Regulations 1999 (SI 1999 No 743), Reg 6* as amended by the *Control of Major Accident Hazards (Amendment) Regulations 2005 (SI 2005 No 1088)*.

'Notify' under amendment *SI 2005 No 1088, Reg 3* means notify in writing, including in an email or by such other means as the recipient will allow. Postal notifications should be sent either to the local HSE office or to the HSE Hazardous Installations Directorate, Chemical Industries Division – C14a 4N.2, Redgrave Court, Merton Road, Bootle, Merseyside L20 7HS. Alternatively it may be sent electronically to comah.notifications@hse.gov.uk.

Within a reasonable period of time before the start of construction of an establishment and before the start of the operation of an establishment, the operator must send to the CA a notification containing the information specified in *SI 1999 No 743, Sch 3*.

(a) the name and address of the operator;
(b) the address of the establishment concerned;
(c) the name or position of the person in charge of the establishment;
(d) information sufficient to identify the dangerous substances or category of dangerous substances present;
(e) the quantity and physical form of the dangerous substances present;
(f) a description of the activity or proposed activity of the installation concerned; and
(g) details of the elements of the immediate environment liable to cause a major accident or to aggravate the consequences thereof.

The HSE's website has forms that may be used for the notification.

Guidance on the amount of information required under each heading is given in the HSE's publication *A guide to the COMAH Regulations 1999 (as amended)*.

There is no need for the notification sent before start-up to contain any information which has already been included in the notification sent before the start of construction, if that information is still valid.

Amendment Regulation *SI 2005 No 1088, Reg 6(3A)* has been added requiring that where changes to legislation make an establishment subject to *COMAH*, the operator of the establishment shall send to the competent authority a notification containing the information specified in *Sch 3* within three months after the establishment becomes subject to this Regulation

Notification is a continuing duty. *SI 1999 No 743, Reg 6(4)* provides that an operator must further notify the CA forthwith in the event of:

— any significant increase in the quantity of dangerous substances previously notified;
— any significant change in the nature or physical form of the dangerous substances previously notified, the processes employing them or any other information notified to the CA in respect of the establishment;
— *SI 1999 No 743, Reg 7* ceasing to apply to the establishment as a result of a change in the quantity of dangerous substances present there; or
— permanent closure of an installation in the establishment.

Amendment *SI 2005 No 1088, Reg 6(4)(ba)* has been added which requires the operator to notify the CA of any modifications to establishment or an installation which could have significant repercussions with respect to the prevention of major accidents.

Information that has been included in a safety report does not need to be notified.

In addition, there are special notification arrangements for petroleum products under *Sch 3, para 5 and Reg 6* to specify the quantity falling within each class. This will provide the CA with information about the types of substances present and the potential risk.

Public register

[M1111] Under *Control of Major Accident Hazards Regulations 1999 (SI 1999 No 743), Sch 8*, the CA must maintain a public register which will include:

— the information included in the notifications submitted by the operators under *SI 1999 No 743, Reg 6*;
— top-tier operators' safety reports; and
— the CA's conclusions of its examination of safety reports.

An application can be made to withhold from the register any of the information that is personally or commercially, and the Secretary of State can require information to be excluded in the interests of national security. The latter power has been extensively exercised in recent years, and very little information is available to the public.

Duty to report a major accident

[M1112] Where a major accident has occurred at an establishment, the operator must immediately inform the CA of the accident (*Control of Major Accident Hazards Regulations 1999 (SI 1999 No 743), Reg 15(3)*).

This duty will be satisfied when the operator notifies a major accident to the HSE in accordance with the requirements of the *Reporting of Injuries, Diseases and Dangerous Occurrences Regulations 1995 (SI 1995 No 3163)*.

The CA must then conduct a thorough investigation into the accident (*SI 1999 No 743, Reg 19*).

Powers to prohibit use

[M1113] The CA is required to prohibit the operation or bringing into operation of any establishment or installation or any part of it where the measures taken by the operator for the prevention and mitigation of major accidents are seriously deficient. [*Control of Major Accident Hazards Regulations 1999 (SI 1999 No 743), Reg 18(1)*].

The CA may prohibit the operation or bringing into operation of any establishment or installation or any part of it if the operator has failed to submit any notification, safety report or other information required under the Regulations within the required time. Where the CA proposes to exercise its prohibitory powers, it must serve on the operator a notice giving reasons for the prohibition and specifying the date when it is to take effect. A notice may specify measures to be taken. The CA may, in writing, withdraw any notice.

Enforcement

[M1114] The Regulations are treated as if they are health and safety regulations for the purpose of the *Health and Safety at Work etc. Act 1974*. The provisions as to offences of *HSWA 1974, ss 33–42*, apply. A failure by the CA to discharge a duty under the Regulations is not an offence. [*Control of Major Accident Hazards Regulations 1999 (SI 1999 No 743), Reg 20(2)*], although the remedy of judicial review is available.

[M1115] The CA is the author of an annual report to the European Commission listing recent major accidents at industrial premises subject to *COMAH*. The reports from 1999–2005 are available at www.hse.gov.uk/coma h/accidents.htm.

Lower-tier duties

General duty

[M1116] Under the *Control of Major Accident Hazards Regulations (SI 1999 No 743), Reg 4*, every operator, both of lower-tier and top-tier sites, is under a general duty to take *all measures necessary* to prevent major accidents and limit their consequences to persons and the environment.

This is the general duty on all operators and underpins the whole of the Regulations. It is a high standard which applies to all establishments within the scope of the Regulations, by requiring measures both for prevention and mitigation. The wording of the duty recognises that risk cannot be completely eliminated. There must be some proportionality between the risk and the measures taken to control the risk. The phrase 'all measures necessary' will be interpreted to include this principle. Where hazards are high, high standards will be expected by the enforcement agencies to ensure that risks are acceptably low. The ideal should always be, wherever possible, to avoid a hazard altogether. This is known as inherent safety. Where reliance is placed on people as part of the necessary measures, human factor issues (including human reliability) should be addressed with the same rigour as technical and engineering measures.

Operators should look at how activities can be made safer by reducing hazards, for example by reducing the inventory. The required standard is that risks have been reduced to a level as low as is reasonably practicable (ALARP) and this is generally done by adopting good practice. (Note: the CA view of ALARP is explained briefly in Appendix 4 of the HSE guidance *Preparing Safety Reports*, HSG 190. Much more detailed treatises appear at www.hse.gov.uk/comah/circular/perm09.htm and www.hse.gov.uk/comah/circular/perm12.htm.)

Good practice represents a consensus on what constitutes proportionate action to control a given hazard. Among other things it takes account of what is technically feasible and the balance between the costs and benefits of the measures taken. Sources of good practice include Approved Codes of Practice and standards produced by organisations such as the British Standards Institution (BSI) and Comite Europeen de Normalisation (CEN).

In most cases, good practice will mean defence in depth by adopting sound engineering design principles, along with good operating and maintenance practices.

Operators may employ a risk management approach to prevention and mitigation based on first principles as an alternative to compliance with established good practice, but the competent authority will require this to be thoroughly justified. In cases where no suitable standard for good practice exists, this may be the only possible course of action.

Risk management systems typically include the following elements:

(a)   identifying the hazards and risks;
(b)   examining the control options available and their merits, including the human factor aspects;
(c)   adopting decisions for action informed by the findings of (a) and (b) above;
(d)   implementing the decisions; and
(e)   evaluating the effectiveness of the actions taken and revising where necessary.

Based on this system, operators must be able to demonstrate that the control measures adopted are adequate for the risks identified.

It will not usually be necessary to prepare any special documents to comply with this regulation. For top-tier establishments the safety report, emergency plans, hazardous substances consent and planning permissions should provide sufficient evidence. For lower-tier establishments the major accident prevention policy MAPP, hazardous substance consent and planning permissions will normally be enough.

'All measures necessary' includes measures for mitigating the effects of major accidents. This includes land-use planning which helps to mitigate the effects of major accidents by ensuring adequate separation of people and the environment from their consequences. The hazardous substances authority and planning authorities deal with land-use planning but information from the operator in the hazardous substances consent application is essential for them to perform this function (see **D0735**).

Major accident prevention policy (MAPP)

[M1117] The *Control of Major Accident Hazards Regulations, Reg 5(1)*, as amended, requires that every operator without delay, but at all events within three months after the establishment becomes subject to this regulation, prepares, and thereafter keeps, a document setting out a policy on the prevention of major accidents (in the Regulations referred to as a 'major accident prevention policy document').

The MAPP document must be in writing and must include sufficient particulars to demonstrate that the operator has established an appropriate safety management system. *SI 1999 No 743, Sch 2*, sets out the principles to be taken into account when preparing a MAPP document and at *para 4* lists the following specific issues to be addressed by the safety management system:

(a) *organisation and personnel* – the roles and responsibilities of personnel involved in the management of major hazards at all levels in the organisation. The identification of training needs of such personnel and the provision of the training so identified. The involvement of employees and, where appropriate, sub-contractors;

(b) *identification and evaluation of major hazards* – adoption and implementation of procedures for systematically identifying major hazards arising from normal and abnormal operation and the assessment of their likelihood and severity;

(c) *operational control* – adoption and implementation of procedures and instructions for safe operation, including maintenance of plant, processes, equipment and temporary stoppages;

(d) *management of change* – adoption and implementation of procedures for planning modifications to, or the design of, new installations, processes or storage facilities;

(e) *planning for emergencies* – adoption and implementation of procedures to identify foreseeable emergencies by systematic analysis and to prepare, test and review emergency plans to respond to such emergencies; provide specific training to all people working on the establishment on the procedures to be followed in the event of emergencies;

(f) *monitoring performance* – adoption and implementation of procedures for the on-going assessment of compliance with the objectives set by the operator's major accident prevention policy and safety management system, and the mechanisms for investigation and taking corrective action in the case of non-compliance. The procedures should cover the operator's system for reporting major accidents or near misses, particularly those involving failure of protective measures, and their investigation and follow-up on the basis of lessons learnt; and

(g)   *audit and review* – adoption and implementation of procedures for periodic systematic assessment of the major accident prevention policy and the effectiveness and suitability of the safety management system; the documented review of performance of the policy and safety management system and its updating by senior management.

The MAPP is a concise but key document for operators which sets out the framework within which adequate identification, prevention/control and mitigation of major accident hazards is achieved. Its purpose is to compel operators to provide a statement of commitment to achieving high standards of major hazard control, together with an indication that there is a management system covering all the issues set out in (a)–(g) above. The guidance to the Regulations suggests that the essential questions which operators must ask themselves are:

—   does the MAPP meet the requirements of the Regulations?
—   will it deliver a high level of protection for people and the environment?
—   are there management systems in place which achieve the objectives set out in the policy?
—   are the policy, management systems, risk control systems and workplace precautions kept under review to ensure that they are implemented and that they are relevant?

Not only must the MAPP be kept up-to-date (*SI 1999 No 743, Reg 5(4)*), but also the safety management system described in it must be put into operation (*SI 1999 No 743, Reg 5(5)*).

For top-tier sites the MAPP can be part of the safety report rather than a standalone document.

Top-tier duties

Safety report

[M1118] Operators of top-tier sites are required to produce a safety report. The key requirement is that operators must show that they have taken all necessary measures for the prevention of major accidents and for limiting the consequences to people and the environment of any that do occur.

Safety reports are required before construction as well as before start-up. The HSE publication '*Preparing safety reports*' (HSG 190), gives practical and comprehensive guidance to site operators in the preparation of a *COMAH* safety report. Furthermore on the HSE website is a Safety Report Assessment Manual (SRAM) along with guides for several specific examples such as the storage of LPG.

Control of Major Accident Hazards Regulations 1999 (SI 1999 No 743), Reg 7(1) requires that within a reasonable period of time before the start of construction of an establishment the operator must send to the CA a report:

—   containing information which is sufficient for the purpose specified in *SI 1999 No 743, Sch 4, Pt 1, para 3(a)*; and
—   comprising at least such of the information specified in *SI 1999 No 743, Sch 4, Pt 2* as is relevant for that purpose.

Within a reasonable period of time before the start of the operation of an establishment, the operator must send to the CA a report containing information which is sufficient for the purposes specified in *SI 1999 No 743, Sch 4, Pt 1* and comprising at least the information specified in *SI 1999 No 743, Sch 4, Pt 2 [SI 1999 No 743), Reg 7(5)]*.

The report sent before start-up is not required to contain information already contained in the report sent before the start of construction.

An operator must ensure that neither construction of the establishment, nor its operation, is started until he has received the CA's conclusions on the report. The CA must communicate its conclusions within a reasonable period of time of receiving a safety report [*SI 1999 No 743, Reg 17(1)(a)*].

An existing establishment that is not subject to *COMAH* requirements, or is only designated lower-tier, may become a top-tier site due to changes in classification or qualifying quantities. In these circumstances *Control of Major Accident Hazards (Amendment) Regulations 2005 (SI 2005 No 1088), Reg 7* requires that the operator shall without delay, but at all events within one year after the establishment becomes subject to this regulation, send to the competent authority a report which is sufficient for the purpose specified in *SI 1999 No 743, Sch 4, Pt 1* and comprising at least the information specified in *SI 1999 No 743, Pt 2*.

Purpose of safety reports

[M1119] -

(1)   D-emonstrating that a major accident prevention policy and a safety management system for implementing it have been put into effect in accordance with the information set out in *Control of Major Accident Hazards Regulations 1999 (SI 1999 No 743), Sch 2*.

(2)   Demonstrating that major accident hazards have been identified and that the necessary measures have been taken to prevent such accidents and to limit their consequences to persons and the environment.

(3)   Demonstrating that adequate safety and reliability have been incorporated into the:
(a)   design and construction; and
(b)   operation and maintenance;
of any installation and equipment and infrastructure connected with its operation, and that they are linked to major accident hazards within the establishment.

(4)   Demonstrating that on-site emergency plans have been drawn up and supplying information to enable the off-site plan to be drawn up in order to take the necessary measures in the event of a major accident.

(5)   Providing sufficient information to the competent authority to enable decisions to be made in terms of the siting of new activities or developments around establishments.

[Reproduced from Control of Major Accident Hazards Regulations 1999 (SI 1999 No 743), Sch 4, Pt 1.]

Minimum information to be included in safety report

[M1120] -

(1)   Information on the management system and on the organisation of the establishment with a view to major accident prevention.
This information must contain the elements set out in *Control of Major Accident Hazards Regulations 1999 (SI 1999 No 743), Sch 2*.

(2)   Presentation of the environment of the establishment:

(a)   a description of the site and its environment including the geographical location, meteorological, geological, hydrographic conditions and, if necessary, its history;

(b)   identification of installations and other activities of the establishment which could present a major accident hazard; and

(c)   a description of areas where a major accident may occur.

(3)   Description of installation:

(a)   a description of the main activities and products of the parts of the establishment which are important from the point of view of safety, sources of major accident risks and conditions under which such a major accident could happen, together with a description of proposed preventive measures;

(b)   description of processes, in particular the operating methods;

(c)   description of dangerous substances:

(i)   inventory of dangerous substances including–

—   the identification of dangerous substances: chemical name, the number allocated to the substance by the Chemicals Abstract Service, name according to International Union of Pure and Applied Chemistry nomenclature;

—   the maximum quantity of dangerous substances present;

(ii)   physical, chemical, toxicological characteristics and indication of the hazards, both immediate and delayed, for people and the environment;

(iii)   physical and chemical behaviour under normal conditions of use or under foreseeable accidental conditions.

(4)   Identification and accidental risks analysis and prevention methods:

(a)   detailed description of the possible major accident scenarios and their probability or the conditions under which they occur including a summary of the events which may play a role in triggering each of these scenarios, the causes being internal or external to the installation;

(b)   assessment of the extent and severity of the consequences of identified major accidents including maps, images or, as appropriate, equivalent descriptions, showing areas which are liable to be affected by such accidents arising from the establishment;

(c)   description of technical parameters and equipment used for the safety of the installations.

(5)   Measures of protection and intervention to limit the consequences of an accident:

(a)   description of the equipment installed in the plant to limit the consequences of major accidents;

(b)   organisation of alert and intervention;

(c)   description of mobilisable resources, internal or external; and

(d)   summary of elements described in sub-paragraphs (*a*), (*b*) and (*c*) necessary for drawing up the on-site emergency plan.

(6)   The names of relevant organisations involved in the drawing up of the report.

[Reproduced from Control of Major Accident Hazards Regulations 1999 (SI 1999 No 743), Sch 4, Pt 2.]

All or part of the information required to be included in a safety report can be so included by reference to information contained in another report or notification furnished by virtue of other statutory requirements. This should be done only where the information in the other document is up to date, and adequate in terms of scope and level of detail.

If an operator can demonstrate that particular dangerous substances are in a state incapable of creating a major accident hazard, the CA can limit the information required to be included in the safety report [*SI 1999 No 743), Reg 7(12)*].

An operator must provide the CA with such further information as it may reasonably request in writing following its examination of the safety report [*SI 1999 No 743, Reg 7(13)*].

Review and revision of safety reports

[M1121] Where a safety report has been sent to the CA, the operator must review it:

—   whenever the operator makes a change to the safety management system (referred to in the *Control of Major Accident Hazards Regulations 1999 (SI 1999 No 743), Sch 4, Pt 1, para 1*) which could have significant repercussions with respect to the prevention of major accidents or the limitation of consequences of major accidents to persons and the environment];

—   whenever such a review is necessary because of new facts or to take account of new technical knowledge about safety matters; and

—   fully at least every five years,

and where in consequence of that review it is necessary to revise the report, the operator shall do so forthwith and notify the competent authority of the details of such revision. [*Control of Major Accident Hazards (Amendment) Regulations 2005 SI 2005 No 1088, Reg 8(1)*].

A change will have significant repercussions with respect to the prevention or control of major accidents if it changes the nature of the major accident risks, so requiring changes in the measures taken to ensure that those risks remain as low as reasonably practicable (ALARP). Whether a change has significant repercussions will depend on the degree to which it:

—   introduces a new major accident hazard;

—   changes the risk from an existing hazard;

—   affects control or mitigation measures (including off-site emergency plans).

Changes that have a positive impact on the risk profile are also important.

Table 3 provides some examples of changes that could have significant repercussions. Table 4 gives examples of what those repercussions might be.

Table 3	
Changes that could have significant repercussions	
	Reorganisation of the management structure.
	Contractorisation, delayering, demanning, or multi-skilling in relation to the operation or maintenance of the establishment.
SMS changes:	Changes in health and safety policy, procedures, standards, aims, objectives or priorities, including changes to the MAPP or SMS.
	A change in the quantity of a dangerous substance.
	A change in the phase of a dangerous substance; for example, a change from liquid to gaseous chlorine.
	The introduction of new dangerous substances or removal of existing dangerous substances.
	New processes.
	Changes to storage facilities.
	Changes to the control systems of safety-critical plant.
	Changes to the mode of delivery or transport of dangerous substances.
	Changes to the design or location of control rooms and/or the number of people present within them.
	Changes to the location of occupied buildings and/or the number of people present within them.
	Changes to the original design parameters such as process operating conditions or practices, changed throughput, design life extensions or removal of safety-critical plant.
	Construction of a new installation on an existing site.
	A small modification which could have large consequences, such as a change to the valve type used in a particular line.
	Introduction of temporary equipment.
	Repairs to structures or any plant and equipment.
Modifications:	Decommissioning of plant and installations.

Table 3	
New facts/knowledge:	A substance which is present on-site, but not previously classified as a dangerous substance, is reclassified as dangerous (or the reverse).
	A change in the risk phrases assigned to a dangerous substance.
	Incidents which reveal potentially hazardous reactions or loss of control scenarios not previously considered.
	Recommendations made following a public inquiry or major incident.
	Lessons from worldwide incidents.
	Advances in technology that might render parts of the safety report out of date very rapidly (though in general, steady advances in technical knowledge can be accommodated at the five-year review stage).
	New scientific or technical research, or other advances (such as reduction in the cost of safety measures) that may affect the decisions previously made about which measures are necessary.
	Population changes on- and off-site.
	Changes in the land-use of surrounding areas.
	Changes in the conservation designation of surrounding land.
	Classification or reclassification of surrounding land as environmentally sensitive areas.

Table 4	
Examples of significant repercussions	
Change in	**Examples**
Risk profile	New major accident scenario/loss of existing major accident scenario[*].
Frequency	Increase/decrease in frequency of scenarios by a factor of ten. Movement between qualitative risk bands in an existing risk matrix.
Severity	Increase/decrease in severity by a factor of two. Movement between qualitative consequence bands in an existing risk matrix.
Control measures	Need for additional or different risk control measures (including SMS).
Emergency procedures	Changes (other than admin arrangements) required to on-site plans – changes to off-site plans foreseeable.

[*] *Includes 'additional repeat' major accidents from increases/changes in inventory, e.g. if there is a major accident involving flammables, and either new flammables are stored or increase in storage volumes of existing flammables, then there are 'significant repercussions'.*

Key to the effective review of a safety report is the existence of well-developed change-management procedures. If implemented properly and in a timely manner, such procedures will enable identification of changes that could have significant repercussions with respect to the prevention of major accidents or the limitation of their consequences.

An operator must notify the CA when he has reviewed the safety report even if it has not been necessary to revise it [*Control of Major Accident Hazards (Amendment) Regulations 2005 (SI 2005 No 1088), Reg 8(2)*]. The HSE has indicated it is helpful to include:

—   a description of how you conducted your review, with justification for your conclusions; and
—   a description of your change-management system, how you applied it and what the outcome was.

Such a notification must be made available in the register available to the public. [*Sch 8, Pt 1(ba)*].

The amendment at *SI 2005 No 1088, Reg 8(1)* changes the order of the requirements on the operator to focus attention on the need to review. This is reinforced throughout amended *SI 2005 No 1088, Reg 8* by the change in the requirement to inform the CA which is now a requirement to notify.

When modifications which could have significant repercussions are proposed (to the establishment or an installation in it, to the process carried on there or the nature or quantity of dangerous substances present there), the operator must review the safety report and where necessary revise it, in advance of any such modification. Details of any revision must be notified to the CA. [*Reg 8(4)*].

The CA should be consulted as early as possible before the changes are made, as enforcement action could be taken if changes are implemented which the CA later judges to be to an unsatisfactory standard. A new installation at an existing establishment will require at least six months' notice, but for most modifications a much shorter period will be appropriate.

The *COMAH* Regulations do not specify any particular format for a safety report and the options for a revision include:

—   resubmission of the entire report;
—   submission of a supplementary document/appendix to the safety report;
—   submission of individual pages where the original report was in looseleaf format and the changes are not too extensive.

Copies of revisions to a safety report are to be sent to the CA together with a covering letter containing the following information:

—   reasons for reviewing the report;
—   an outline of the nature of the revisions;
—   reference to any supplementary information that has been consolidated at the five-year review;
—   where the entire report has been reviewed at an interim change review this should be clearly stated, so the period to the next five-year review is reset.

A minimum of three copies of the revisions are required: one for the HSE, one for the Environment Agency and one for the public register. If more detailed assessment involving HSE specialists is needed the CA may require additional copies. If you would prefer to know exactly how many copies will be required in advance of submission, to enable you to submit them all together, you should discuss the nature and extent of the revisions with your site inspector and seek advice. Where the revision contains information previously excluded from the public register on the grounds of national security and/or personal or commercial confidentiality, or if you consider that any new information in the revision should be similarly excluded, you should follow the procedure for exclusion of information from public registers on the HSE website.

On-site emergency plan

[M1122]–[M1123] *Control of Major Accident Hazards Regulations 1999 (SI 1999 No 743), Reg 9(1)*, requires operators of top-tier establishments to prepare an on-site emergency plan. *Reg 10(3)* also requires them to provide information to the local authority which is required to prepare an off-site emergency plan in respect of that establishment (see **D0724**). Guidance is given in the HSE's publications *A guide to the Control of Major Accident Hazards 1999 (as amended)* and *Emergency planning for major accidents*, HSG191.

The on-site emergency plan must be adequate to secure the objectives specified in *SI 1999 No 743, Sch 5, Pt 1*, namely:

— containing and controlling incidents so as to minimise the effects, and to limit damage to persons, the environment and property;
— implementing the measures necessary to protect persons and the environment from the effects of major accidents;
— communicating the necessary information to the public and to the emergency services and authorities concerned in the area; and
— providing for the restoration and clean-up of the environment following a major accident.

The plan must contain at least the following information:

— names or positions of persons authorised to set emergency procedures in motion and the person in charge of and co-ordinating the on-site mitigatory action;
— name or position of the person with responsibility for liaison with the local authority responsible for preparing the off-site emergency plan (see **D0724**);
— for foreseeable conditions or events which could be significant in bringing about a major accident, a description of the action which should be taken to control the conditions or events and to limit their consequences, including a description of the safety equipment and the resources available;
— arrangements for limiting the risks to persons on site including how warnings are to be given and the actions persons are expected to take on receipt of a warning;

—   arrangements for providing early warning of the incident to the local authority responsible for setting the off-site emergency plan in motion, the type of information which should be contained in an initial warning and the arrangements for the provision of more detailed information as it becomes available;

—   arrangements for training staff in the duties they will be expected to perform, and where necessary co-ordinating this with the emergency services; and

—   arrangements for providing assistance with off-site mitigatory action, such as special equipment that can be made available.

[SI 1999 No 743, Sch 5, Pt 2.]

New establishments must prepare such a plan before start-up. An existing site that becomes subject to the requirement because of changes to the regulations or classification must comply within one year.

When preparing an on-site emergency plan, the operator must consult:

—   employees, and persons employed in that the establishment *[SI 2005 No 1088, Reg 9(3)(a)]*. This requires that contractors are consulted as well as the operators own employees;

—   the Environment Agency (or the Scottish Environment Protection Agency);

—   the emergency services;

—   the health authority for the area where the establishment is situated; and

—   the local authority, unless it has been exempted from the requirement to prepare an off-site emergency plan.

(a)   A 'site main controller' needs to be appointed. This is a strategic level person, who oversees the company's response to the major incident from the emergency control centre. This person must:

    —   oversee and take control over the event;

    —   control the activities of operational and tactical level co-ordinators;

    —   must contact or confirm that the emergency services have been contacted;

    —   must confirm that on-site plan has been initiated;

    —   mobilise key personnel, identified in the policy and pre-planning;

    —   continuously review and assess the major incident;

    —   authorise the evacuation process;

    —   authorise closing down of plant and work equipment;

    —   ensure casualties are being cared for and notify their relatives, as well as listing missing persons;

    —   monitor weather conditions;

    —   liaise with the emergency services, HSE, local authority, and the Environment Agency;

    —   account for all personnel and contractors;

    —   ensure traffic management, site access/egress etc. controlled;

    —   record keeping of all decisions made, for later assessment and inquiry;

    —   provide for welfare facilities (food, clothing etc.) of personnel;

-     through the media and in liaison with the emergency services, issue information;
-     ensure compliance with the law – non-removing of evidence etc; and
-     control affected areas after the major incident.

(b)   On-site emergency control centre (ECC):
-     this is a room or area near the major incident site, where the command and control decisions will be made;
-     the ECC personnel co-ordinate liaison with the media and the public; and
-     the ECC should contain: telecommunication equipment, cellular communication equipment, detailed site plans and maps, technical drawings of the operation and its shut-off points, list of critical/hazardous substances and waste on-site, location of all safety equipment, location of fire safety equipment and access/egress points etc.

Off-site emergency plan

[**M1124**] The local authority for the area where a top-tier establishment is located must prepare an adequate emergency plan for dealing with off-site consequences of possible major accidents. As with the on-site plan, it should be in writing.

The objectives set out in *SI 1999 No 743, Sch 5, Pt 1* (see **D0722** above) also apply to off-site emergency plans.

The plan must contain the following information:

-     names or positions of persons authorised to set emergency procedures in motion and of persons authorised to take charge of and co-ordinate off-site action. This should be senior, strategic level management that will co-ordinate links with the local authority and the public;
-     arrangements for receiving early warning of incidents. This can be via on-site team, technology etc. In addition, there must be arrangements for alerting others of the event and procedures for call-out of the emergency services, specialist assistance, other search and rescue bodies;
-     arrangements for co-ordinating resources necessary to implement the off-site emergency plan. This should identify those organisations that can assist in the off-site emergency, how each of these organisations will be alerted, how personnel of the organisation affected will recognise and identify the emergency services and specialist services and vice-versa, channels of communication between personnel and those other emergency services, specialist organisations, meeting place, off-site needs to be identified, for direct communication between personnel and others, access to the operation to use equipment;
-     arrangements for providing on-site assistance, such as from the fire brigade;
-     arrangements for off-site occurrences and the mitigating actions needed to protect the public and the environment;

—   arrangements for providing the public with specific information relating to the accident and the behaviour which it should adopt; In addition, details of how the media could be utilised to disseminate information to the community and emergency organisations; and

—   arrangements for the provision of information to the emergency services of other member states in the event of a major accident with possible transboundary consequences.

An operator must supply the local authority with the information necessary for the authority's purposes, plus any additional information reasonably requested in writing by the local authority.

In preparing the off-site emergency plan, the local authority must consult:

—   the operator;
—   the emergency services;
—   the CA;
—   the Agency (*SI 2005 No 1088, Reg 11* requires that the Environment Agency and SEPA is consulted on off-site plans. This provides the Agency with the status of consultee and enables it to comment on plans from an emergency response perspective);
—   each health authority for the area in the vicinity of the establishment; and
—   such members of the public as it deems appropriate.

[SI 1999 No 743, Reg 10(6) as amended.]

In the light of the safety report, the CA may exempt a local authority from the requirement to prepare an off-site emergency plan in respect of an establishment [*SI 1999 No 743, Reg 10(7)*]. Derogations are explained in Appendix 1 of the HSE publication *A guide to the Control of Major Accident Hazard Regulations 1999* (as amended).

[M1125] Off-site emergency control centre (ECC):

—   this parallels the on-site ECC and is at a safe distance from the site. It is a command and control centre for all off-site liaison; and
—   it must also link up with the on-site ECC to ensure that plans are in unison.

Thus, under *COMAH*, the on-site and off-site emergency plans are required to reflect, account for and provide details of the arrangements and command and control variables identified in stage 2 of the DEMS process (see **D6056**).

Reviewing, testing and implementing emergency plans

[M1126] *Control of Major Accident Hazards Regulations 1999 (SI 1999 No 743), Reg 11,* requires that emergency plans are reviewed and, where necessary, revised, at least every three years. Reviewing is a key process for addressing the adequacy and effectiveness of the components of the emergency plan – it should take into account:

—   changes occurring in the establishment to which the plan relates;
—   any changes in the emergency services relevant to the operation of the plan;

— advances in technical knowledge;
— knowledge gained as a result of major accidents either on-site or elsewhere; and
— lessons learned during the testing of emergency plans.

There is a requirement to test emergency plans at least every three years. Such tests will assist in the assessment of the accuracy, completeness and practicability of the plan: if the test reveals any deficiencies, the relevant plan must be revised. Agreement should be reached beforehand between the operator, the emergency services and the local authority on the scale and nature of the emergency plan testing to be carried out. *SI 2005 No 1088, Reg 12* alters *SI 1999 No 743, Reg 11(1)* and requires local authorities to consult with such members of the public as it deems appropriate in its review of off-site emergency plans.

Where there have been any modifications or significant changes to the establishment, operators should not wait for the three-year review before reviewing the adequacy and accuracy of the emergency planning arrangements.

When a major accident occurs, the operator and local authority are under a duty to implement the on-site and off-site emergency plans. [*SI 1999 No 743, Reg 12*].

Local authority charges

[M1127] A local authority may charge the operator a fee for performing its functions under *Control of Major Accident Hazards Regulations 1999 (SI 1999 No 743), Regs 10* and *11*, i.e. for preparing, reviewing and testing off-site emergency plans.

The charges can only cover costs that have been reasonably incurred. If a local authority has contracted out some of the work to another organisation, the authority may recover the costs of the contract from the operator, provided that they are reasonable.

In presenting a fee to an operator, the local authority should provide an itemised, detailed statement of work done and costs incurred.

Provision of information to the public

[M1128] The operator must supply information on safety measures to people within an area without their having to request it. The area known as the public information zone (PIZ) is notified to the operator by the CA as being one in which people are liable to be affected by a major accident occurring at the establishment. The minimum information to be supplied to the public is specified in *Control of Major Accident Hazards Regulations 1999 (SI 1999 No 743), Sch 6*.

Under the *Control of Major Accident Hazards (Amendment) Regulations 2005 (SI 2005 No 1088), Reg 13* alters *SI 1999 No 743, Reg 14(1)* and requires operators to provide information of safety measures to every school, hospital or other establishment serving the public which is situated in such area, and to make all such information permanently available to the public.

Information to be supplied to the public

[M1129] -

— name of operator and address of the establishment;
— identification, by position held, of the person giving the information;
— confirmation that the establishment is subject to these Regulations and that the notification referred to in the *Control of Major Accident Hazards Regulations 1999 (SI 1999 No 743), Reg 6* or the safety report has been submitted to the competent authority;
— an explanation in simple terms of the activity or activities undertaken at the establishment;
— the common names or, in the case of dangerous substances covered by *SI 1999 No 743, Sch 1, Pt 3*, the generic names or the general danger classification of the substances and preparations involved at the establishment which could give rise to a major accident, with an indication of their principal dangerous characteristics;
— general information relating to the nature of the major accident hazards, including their potential effects on the population and the environment;
— adequate information on how the population concerned will be warned and kept informed in the event of a major accident;
— adequate information on the actions the population concerned should take, and on the behaviour they should adopt, in the event of a major accident;
— confirmation that the operator is required to make adequate arrangements on site, in particular liaison with the emergency services, to deal with major accidents and to minimise their effects;
— a reference to the off-site emergency plan for the establishment. This should include advice to co-operate with any instructions or requests from the emergency services at the time of an accident; and
— details of where further relevant information can be obtained, unless making that information available would be contrary to the interests of national security or personal confidentiality or would prejudice to an unreasonable degree the commercial interests of any person.

[*COMAH, Sch 6*].

Statement of policy for managing major incidents

[M1130] This should be a short (maximum 1 page) missionary and visionary statement covering the following:

• Senior management commitment to be responsible for the co-ordination of major incident response.
• To comply with the law, namely:
    — the protection of the health, safety and welfare of employees, visitors, the public and contractors;
    — to comply with the duties under the *Management of Health and Safety at Work Regulations 1999 (SI 1999 No 3242), Regs 8* and *9* (see **D6010**);

— to comply with any other legislation that may be applicable to the organisation.

- To make suitable arrangements to cope with a major incident and to be proactive and efficient in the implementation process.
- That this statement applies to all levels of the organisation and all relevant sites in the country of jurisdiction.
- The commitment of human, physical and financial resources to prevent and manage major incidents.
- To consult with affected parties (employees, the local authority and others if needed).
- To review the statement.
- To communicate the statement.
- Signed and dated by the most senior corporate officer.

Arrangements for major incident management

[M1131]

- This refers to what the organisation has done, is doing and will do in the event of a major incident and *how* it will react in those circumstances. The arrangements are a legal requirement under the *Management of Health and Safety at Work Regulations 1999 (SI 1999 No 3242), Regs 5, 8* and *9*.
- Arrangements should be realistic and achievable. They should focus on major actions to be taken and issues to be addressed rather than being a 'shopping list'. The arrangements will have to be verified (in particular for *COMAH* sites).
- Arrangements could be under the following headings with explanations under each. To repeat, the larger the organisation and the more complex the hazard facing it, the more detailed the arrangements need to be. For example:
  - (i) Medical assistance – including first-aid availability, first-aiders, links with accident and emergency at the medical centre, other specialists that could be called upon, rules on treating injured persons, specialised medical equipment and its availability etc.
  - (ii) Facilities management – the location, site plans and accessibility to the main facilities (gas, electricity, water, substances etc.), rendering safe such facilities, availability of water supply in-house and within the perimeter of the site etc.
  - (iii) Equipment to cope – identification of safety equipment available and/or accessible, location of such equipment, types (personal protective equipment, lifting, moving, working at height equipment etc.).
  - (iv) Monitoring equipment – measuring, monitoring and recording devices needed, including basic items such as measuring tapes, paper, pens, tape recorders, intercom and loud-speakers.
  - (v) Safe systems – procedures to access site, working safely by employees and contractors under major incident conditions (what can and cannot be done), hazard/risk assessments of

dangers being confronted etc., risks to certain groups and procedures needed for rescue (disabled, young persons, children, pregnant women, elderly persons).

(vi)   Public safety – ensuring non access to major incident site by the public (in particular children, trespassers, the media and those with criminal intent), warning systems to the public etc.

(vii)   Contractor safety – guidance and information to contractors at the major incident on working safely.

(viii)   Information arrangements – the supply of information to staff, the media and others (insurers, enforcers) to inform them of the events. Where will the information be supplied from, when will it be done and updates?

(ix)   The media – managing the media, confining them to an area, handling pressure from them, what to say and what not to say etc.

(x)   Insurers/loss adjusters – notifying them and working with them at the earliest opportunity.

(xi)   Enforcers – notifying them of the major incident, working with them including several types e.g. HSE, Environment Agency (or Scottish equivalent) as well as local authority (environmental health, planning, building control for instance).

(xii)   Evidence and reporting arrangement – to cover strict rules on removal or evidence by employees or others, role and power of enforcers, incident reporting e.g under the *Reporting of Injuries, Diseases and Dangerous Occurrences Regulations 1995 (SI 1995 No 3163)* etc.

(xiii)   The emergency services – working with the police, fire, ambulance/NHS, and other specialists (British Red Cross, search and rescue), rules of engagement, issues of information supply and communication with these services.

(xiv)   Specialist arrangements for specific major incidents such as bomb explosions – issues of contacting the police, ordnance disposal, access and egress, rescue and search, economic and human impact to name a few issues were most evident during the London Docklands and Manchester bombings in the 1990s.

(xv)   Human aspects – removing, storing and naming dead bodies or seriously injured persons during the incident. Informing the next-of-kin, issues of religious and cultural respect. Issues of counselling support and person-to-person support during the incident.

This is not an exhaustive list.

Command and control chart of arrangements

[M1132] This highlights who is responsible for the effective management of the major incident.

•   It should be a graphical representation preferably in a hierarchical format, clearly delineating the division of labour between personnel in the organisation and the emergency services/others.

- The chart should display three broad levels of command and control, namely strategic, operational and tactical. The first relates to the person(s) in overall charge of the major incident. Will this be the person who signed the statement of policy for major incidents or will it be another (disaster and emergency advisor, safety officer, others)? This person will make major decisions. The second relates to co-ordinators of teams. Operational level personnel need to have the above arrangements assigned to them in clear terms. The third refers to those at the front end of the major incident, for instance first-aiders.

  It is most important to note that internal command and control of arrangements does not mean *overall* command and control of the major incident. This can (will be) vested with the appropriate emergency service, normally the police or the fire authority in the UK. In the event of any conflict of decisions, the external body such as the police will have the final veto. Therefore, the chart and the arrangements must reflect this variable.

- The chart should list on a separate page names/addresses/emergency phone, fax, email, cellular numbers of those identified on the chart. It should also list the numbers for the emergency services as well as others (British Red Cross, specialist search and rescue, loss adjusters, enforcing body).

- The chart should also clearly ratify a principle of command and control, as to who would be 'In-charge 1', 'In-charge 2', if the original person became unavailable.

- The chart and the list of numbers should also be accompanied by a set of 'rules of engagement' in short 'bullet points' to remind personnel of the importance of command and control e.g. safety, obedience, communication, accuracy, humanity for instance.

Testing and validation

[M1133] This is considered to be a crucial aspect of pre-planning by all the emergency services in the UK. Testing is an objective rehearsal to examine the state of preparedness and to determine if the policy and the plan will perform as expected.

The Home Office guidance, *Dealing with Disaster* (1998, third edition) identifies training and exercising as types of testing. Training is more personnel focused, aiming to assess how much the human resource knows about the policy and plan and means of enhancing the knowledge, skill and experience of that resource. Exercising is a broader approach, examining all aspects of the policy and plan, not just the human response but also physical and organisational capability to deal with the major incident.

Training and exercising are both necessary to ensure that human resources understand the policy and plan, as well as to ensuring that problems can be identified and competent responses developed.

It should be noted that the guidance to the *Control of Major Accident Hazards Regulations 1999 (SI 1999 No 743)* does not distinguish between 'training' or 'exercising'. It identifies drills, which test a specific aspect of the emergency plan (e.g. fire drills), seminar exercises, walk-through exercises, which involves visiting the site, table-top exercises, control-post exercises, which assess the physical and geographical posting of the personnel and emergency services during a major incident and live exercises.

[M1134] The Home Office publication cites three types of exercising:

- Seminar exercises
    - a broad, brain-storming session, assessing and analysing the efficacy of the policy and plan;
    - seminars need to be 'inclusive', that is bring staff and others together in order to co-ordinate strategic, operational and tactical issues.
- Table-top exercises
  This is an attempt to identify visually using a model of the production site and surrounding areas, what types of problems could arise (access/egress, control, logistics, etc.). It involves a limited number of people, an advantage in terms of time commitment, but a disadvantage in that others do not have the opportunity to learn from it. It is perhaps most valuable in validation of new plans.
- Live exercises
  This is a rehearsal of the major incident, actively and proactively testing the responses of the individual, organisation and possibly the community (emergency services, the media, etc.) to a simulated event.

Though not mentioned in Home Office or other emergency service guidance, a fourth approach – synthetic simulation – is becoming a most popular option. It applies the techniques used in flight simulation and military simulation, to major incidents. Thus, using computers one can model the site and operation and in 3-d format move around the screen. This then enables various eye-points around the site, achieved by moving the mouse. Some advanced systems enable 'computer generated entities' to be included into the data-set. For example, a collision can be simulated outside the production plant or the rate of noxious substance release modelled and calculated across the local community.

There are costs and benefits associated with each of the four approaches. The need is for developing exercise budgets which enable all four to be used in proportion and to compliment each other, though synthetic simulation may well be out of reach for smaller organisations.

Land-use planning

[M1135] Since the early 1970s, arrangements have existed for local planning authorities (PAs) to obtain advice from HSE about risks from major hazard sites and the potential effect on populations nearby. The Advisory Committee on Major Hazards (ACMH), set up in the aftermath of the Flixborough disaster in 1974, laid down a framework of controls which included a strategy of mitigating the consequences of major accidents by controlling land-use developments around major hazard installations.

Legislation

[M1136] Land-use planning is covered by planning legislation in Great Britain, in particular, the *Planning (Hazardous Substances) Act 1990* and the *Planning (Hazardous Substances) Regulations 1992 (SI 1992 No 656)* as amended by the *Planning (Control of Major Accident Hazards) Regulations 1999 (SI 1999 No 981)*. These latter regulations implemented the land-use planning requirements of the Seveso II Directive (Council Directive 96/82/EC of 9 December 1996).

Enforcement of the *Planning (Hazardous Substances) Act 1990* and the Regulations are the responsibility of the appropriate hazardous substances authority (usually the local planning authority).

General

[M1137] The flammable, oxidising, explosive or toxic properties of some of the substances used and produced by chemical industry are such that accidents involving them may have the potential to cause harm to people or property, or to the environment. Where the substances are present in the scale associated with the major chemical industries, the consequences of accidents may cause loss of life, injury and damage to property over a wide area. Communities where such accidents have occurred have paid a high price for the hazard on their doorstep, for example Seveso, Bhopal, Mexico City and, in the UK, Flixborough.

Much has already been achieved through controls under *COMAH* to reduce the likelihood of major accidents and to mitigate the consequences of those which do occur, by emergency planning and information to people in the surrounding area. But there is a further and very important way in which the consequences of major accidents can be mitigated: land-use planning. Land-use planning allows decisions to be made about the siting of major hazard installations and about the development of land around existing installations.

In an ideal world, industries using large quantities of hazardous substances would be located far away from centres of housing and other developments that could be affected in case of accident. In reality, the situation is very different. Factories, housing, schools and shops have developed close to each other; indeed, in many cases these industries provide the economic heart of the local community.

Greater control is possible when planning the location of new hazardous activities, but even here the options may be limited. There are few locations where new hazardous installations can be 'shoehorned' into place without creating some risk to an existing community. The UK is a small, densely populated island and such undeveloped areas as do exist are often so remote or of such environmental value as to be unsuitable for industrial use.

Hazardous installations

[M1138] These have a number of definitions: the first is that in the *Control of Major Accident Hazard Regulations 1999* and applies where a dangerous substance listed in *Sch 1* of the Regulations is present in a quantity equal to or exceeding the threshold for that named substance or a category. Additionally quantities of similar substances must be aggregated (see **D0707**.

The second is a derivative of the first and is taken from *The Planning (Hazardous Substances) Regulations 1992* as amended by *The Planning (Control of Major Accident Hazards) Regulations 1999*. Again, a hazardous installation is defined by the presence of a substance at or above the threshold quantity for that substance if named, or otherwise for categories, including aggregation. However in this case there is a far longer list of named substances. The thresholds in many cases coincide with the lower tier of *COMAH*, but there are exceptions. The quantities are listed in *Sch 1* of the Regulations.

Table 5		
Named Substances		
Column 1	*Column 2*	*Column 3*
Hazardous substances	*Controlled quantity (Q) in tonnes*	*Quantity for purposes of note 4 to the notes to Parts A and B (Q *)*
1. Ammonium nitrate to which Note 1 of the notes to Part A applies	350.00	–
2. Ammonium nitrate to which Note 2 of the notes to Part A applies	1000.00	1250.00
3. Arsenic pentoxide, arsenic (V) acid and/or salts	1.00	–
4. Arsenic trioxide, arsenious (III) acid and/or salts	0.10	–
5. Bromine	20.00	–
6. Chlorine	10.00	–
7. Nickel compounds in inhalable powder form (nickel monoxide, nickel dioxide, nickel sulphide, trinickel disulphide, dinickel trioxide)	1.00	–
8. Ethyleneimine	10.00	–
9. Fluorine	10.00	–
10. Formaldehyde (**concentration 90%**)	5.00	–
11. Hydrogen	2.00	5.00
12. Hydrogen chloride (liquefied gas)	25.00	–
13. Lead alkyls	5.00	–
14. Liquefied petroleum gas, including commercial propane and commercial butane, and any mixture thereof, when held at a pressure greater than 1.4 bar absolute	25.00	50.00
15. Liquefied extremely flammable gases excluding pressurised LPG (entry no. 14)	50.00	–

Table 5		
Named Substances		
Column 1	Column 2	Column 3
Hazardous substances	Controlled quantity (Q) in tonnes	Quantity for purposes of note 4 to the notes to Parts A and B (Q *)
16. Natural gas	15.00	50.00
17. Acetylene	5.00	–
18. Ethylene oxide	5.00	–
19. Propylene oxide	5.00	–
20. Methanol	500.00	–
21. 4, 4-Methylenebis (2-chloraniline) and/or salts, in powder form	0.01	–
22. Methylisocyanate	0.15	–
23. Oxygen	200.00'	–
24. Toluene diisocyanate	10.00	–
25. Carbonyl dichloride (phosgene)	0.30	–
26. Arsenic trihydride (arsine)	0.20	–
27. Phosphorus trihydride (phosphine)	0.20	–
28. Sulphur dichloride	1.00	–
29. Sulphur trioxide (including sulphur trioxide dissolved in sulphuric acid to form Oleum)	15.00	–
30. Polychlorodibenzofurans and polychlorodibenzodioxins (including TCDD), calculated in TCDD equivalent (see Note 3 to the notes to Part A)	0.001 0.001	–
31. The following carcinogens: 4-Aminobiphenyl and/or its salts; Benzidine and/or salts; Bis(chloromethyl)ether; Chloromethyl methyl ether; Dimethylcarbamoyl chloride; Dimethylnitrosomine; Hexamethylphosphoric triamide; 2-Naphthylamine and/or salts; 1, 3-Propanesultone; 4-Nitrodiphenyl		
32. Automotive petrol and other petroleum spirits	5000.00	–
33. Acrylonitrile	20.00	50.00
34. Carbon disulphide	20.00	50.00
35. Hydrogen selenide	1.00	50.00
36. Nickel tetracarbonyl	1.00	5.00
37. Oxygen difluoride	1.00	5.00
38. Pentaborane	1.00	5.00
39. Selenium hexafluoride	1.00	50.00

Table 5		
Named Substances		
Column 1	*Column 2*	*Column 3*
Hazardous substances	*Controlled quantity (Q) in tonnes*	*Quantity for purposes of note 4 to the notes to Parts A and B (Q *)*
40. Stibine (Antimony hydride)	1.00	5.00
41. Sulphur dioxide	20.00	50.00
42. Tellurium hexafluoride	1.00	5.00
43. 2,2-Bis(tert-butylperoxy) butane (>70%)	5.00	50.00
44. 1,1–Bis(tert-butylperoxy) cyclohexane (>80%)	5.00	50.00
45. tent-Butyl peroxyacetate (>70%)	5.00	50.00
46. tent-Butyl peroxyisobutyrate (>80%)	5.00	50.00
47. tent-Butyl peroxyisopropylcarbonate (>80%)	5.00	50.00
48. tent-Butyl peroxymaleate (>80%)	15.00	150.00
49. tert-Butyl peroxypivalate (>77%)	5.00	50.00
50. Cellulose Nitrate other than: (1) cellulose nitrate to which the Explosives Act 1875(a) applies; or, (2) cellulose nitrate where the nitrogen content of the cellulose nitrate Does not exceed 12.3% by weight and contains not more than 55 parts of Cellulose nitrate per 100 parts by weight of solution	50.00	
51. Dibenzyl peroxydicarbonate (>90%)	5.00	50.00
52. Diethyl peroxydicarbonate (>30%)	5.00	50.00
53. 2,2-Dihydroperoxyprbpane (>30%)	5.00	50.00
54. Di-isobutyryl peroxide (>50%)	5.00	50.00
55. Di-n-propyl peroxydicarbonate (>80%)	5.00	50.00
56. Di-sec-butyl peroxydicarbonate (>80%)	5.00	50.00
57. 3,3,6,6,9,9-Hexamethyl-1,2,4,5-tetroxacyclononane (>75%)	5.00	50.00
58. Methyl ethyl ketone peroxide (>60%)	5.00	50.00
59. Methyl isobutyl ketone peroxide (>60%)	5.00	50.00
60. Peracetic acid (>60%)	5.00	50.00
61 Sodium chlorate	25.00	50.00
62. Gas or any mixture of gases (not covered by entry 16) which is flammable in air, when held as a gas	15.00	–

Table 5 Named Substances		
Column 1	Column 2	Column 3
Hazardous substances	Controlled quantity (Q) in tonnes	Quantity for purposes of note 4 to the notes to Parts A and B (Q *)
63. A substance or any mixture of substances which is flammable in air When held above its boiling point (measured at 1 bar absolute) as a liquid or as a mixture of liquid and gas at a pressure of more than 1.4 bar absloute (see Note 4 to the Notes to Part A)	25.00	

Table 6	
Categories of dangerous substances	
Column 1	Column 2
1. Very toxic	5
2. Toxic	50
3. Oxidising	50
4. Explosive, with risk phrase R2	50
5. Explosive, with risk phrase R3	10
6. Flammable	5,000
7. Highly flammable	50
8. Highly flammable	5,000
9. Extremely flammable	10
10. Dangerous for the environment in combination with risk phrases:	
R50: 'Very toxic to aquatic organisms'	200
R51: 'Toxic to aquatic organisms'; and R53: 'May cause long-term adverse effects in the aquatic environment'	500
11. Any classification not covered by those given above in combination with risk phrases:	
R14: 'Reacts violently with water' (including R14/15)	100
R29: 'in contact with water, liberates toxic gas'	50

The Department of the Environment, Transport and the Regions (as it was known at that time) published in 2000 *Hazardous substances consent – a guide for industry*. This also contains these listings together with explanatory notes, as Annex 1.

Mechanisms

[M1139] There are a number of mechanisms for controlling the risks from hazardous installations.

The first is on the siting of new hazardous installations and the primary control is under the *Health and Safety at Work etc. Act 1974* via the *Control of Major Accident Hazard Regulations 1999*. These require operators to submit a pre-construction safety report to HSE before construction can begin. Operators must also apply for a hazardous substances consent from the hazardous substance authority (HSA: usually the planning authority). On application, the HSA is required to consult HSE as to the advisability or otherwise of the location of the installation. HSE will then advise on the residual risk that still remains when all reasonably practicable steps have been taken to ensure safety. HSE's role is purely advisory: it is for the HSA to take into account other economic or social factors that should be considered. If the consent is granted, HSE notifies to the planning authority (PA), a consultation zone around the installation within which it must be consulted on any further developments such as housing, shops, schools, hospitals and the like.

This brings us to the second mechanism for controlling risks around hazardous installations, the siting of other developments. All applications for planning permission beyond a certain size are required to be notified to HSE by the PA for advice as to their suitability in the vicinity of a major hazard site. This is required by the *Town and Country Planning (General Development Procedure) Order 1995 (SI 1995 No 419), Art 10(d)* and *(zb)*. As above, HSE advises on the basis of the residual risk that remains after all reasonably practicable steps have been taken to ensure safety in compliance with HSWA. Again, HSE's role is purely advisory: it is for the PA to take into account other economic or social factors that should be considered.

It also remains the case that, to be economically viable, industries need to be sited where they are accessible to main transport routes and to sources of labour. This inevitably means that in making planning decisions about hazardous installations, safety, however important, is only one of the elements to be considered. A balance has to be struck between the needs of industry, the needs of the community and the interests of safety. The HSE's role is to provide advice which will allow an informed decision to be made from a proper understanding of the risks involved. HSE may specify conditions that should be imposed by the HSA, over and above compliance with statutory health and safety requirements, to limit risks to the public (e.g. limiting which substances can be stored on site, or requiring tanker delivery rather than on-site storage).

LPAs may be minded to grant permission against HSE's advice. In such cases HSE will not pursue the matter further as long as the LPA understands and has considered the reasons for our advice. However HSE has the option, if it believes for example that the risks are sufficiently high, to request the decision is 'called in' for consideration by the Secretary of State, in England and Wales (a very rare situation). In Scotland, if the planning authority is minded to grant permission they have to notify the Scottish Ministers who can decide to call-in the application.

Consultation distances and risk contours

[M1140] The HSE undertakes a detailed assessment of the hazards and risks from the installation and produces a map with three risk contours representing defined levels of risk or harm which any individual at that contour would be subject to. The risk or harm to an individual is greater the closer to the installation. In each case the risk relates to an individual sustaining the so-called 'dangerous dose' (see definition below) or specified level of harm.

Dangerous dose would lead to:

Severe distress to all;

—   a substantial number requiring medical attention;
—   some requiring hospital treatment; and,
—   some (about 1%) fatalities.

The three contours represent levels of individual risk of 10, 1 and 0.3 chances per million per year respectively of receiving a dangerous dose or defined level of harm. The contours form three zones (see Figure 1), with the outer contour defining the consultation distance (CD) around major hazard sites. The PA consults HSE on relevant proposed developments within this CD.

Figure 1: Risk contours and zones around a hazardous installation

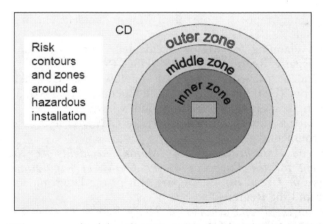

When consulted by the PA, HSE firstly identifies which of the three defined zones the proposed development is in. Secondly, the proposed development is classified into one of four 'Sensitivity Levels'. The main factors that determine these levels are the numbers of persons at the development, their sensitivity (vulnerable populations such as children and old people) and the intensity of the development. With these two factors known, a simple decision matrix is used to give advice to the PA, as shown below:

Level of sensitivity	Development in Inner Zone	Development in Middle Zone	Development in Outer Zone
1. e.g. factories	√	√	√
2. e.g. housing	X	√	√

Level of sensitivity	Development in Inner Zone	Development in Middle Zone	Development in Outer Zone
3. e.g. school	X	X	√
4. e.g. hospital	X	X	X

Sources of information

[M1141] These include:

(1)   Legislation
- *Control of Major Accident Hazards Regulations 1999 (SI 1999 No 743)*;
- *Control of Major Accident Hazards (Amendment) Regulations 2005 (SI 2005 No 1088).*
- *The Planning (Hazardous Substances) Regulations 1992 (SI 1992 No 656).*
- *The Planning (Control of Major-Accident Hazards) Regulations 1999 (SI 1999 No 981).*

(2)   Guidance
- 'A guide to the Control of Major Accident Hazard Regulations 1999' (HSG L111);
- 'Preparing safety reports: Control of Major Accident Hazards Regulations 1999' (HSG 190);
- 'Emergency Planning for Major Accidents: Control of Major Accident Hazards Regulations 1999' (HSG 191);
- 'Major accident prevention policies for lower-tier COMAH establishments' (HSE Information Sheet, Chemical Sheet No. 3);
- HSE website: www.hse.gov.uk/comah/index.htm;
- *Guidance on environmental risk aspects of COMAH safety reports*, Environment Agency/SEPA;
- *Guidance on Interpretation of Major Accident to the Environment for the Purposes of the COMAH Regulations* (known as *The Green Book*), Department of the Environment, Transport and the Regions;
- *Guidance on the Preparation of a Safety Report to meet the Requirements of Council Directive 96/82/EC (Seveso II)* edited by GA Papadakis and A Amendola, European Commission 1997, ISBN 92 828 1451 3;
- *Hazardous substances consent – a guide for industry*, Department of the Environment, Transport and the Regions, 2000. This is now out of print, but can be downloaded from the Communities and Local Government website: www.communities.gov.uk/publications/planningandbuilding/hazardoussubstancesguide.

(3)   Professional bodies such as:
- International Institute of Risk and Safety Management: www.iirsm.org;
- Institution of Occupational Safety and Health (www.iosh.co.uk), particularly the Major Hazards subject group;

- SIESO (originally the Society of Industrial and Emergency Services Officers): www.sieso.org.uk;
  and services:
- visiting individuals and groups;
- volunteers carrying out work.

This chapter considers each of these categories in some detail – who is at risk, what types of risk they might be exposed to and the types of precautions likely to be necessary to protect them. In addition, it deals with major events that can involve several community categories and may be subject to the scrutiny of Safety Advisory Groups. The potential impact on the community from contractors used by businesses or hired in by event organisers is also considered.)

Managing Absence

Lynda Macdonald

Introduction

[M1501] It is advisable for all employers, irrespective of their size or industry sector, to take active steps to manage employee absence, both long and short-term. The causes of absences should be recorded and analysed so that the employer can review whether any patterns exist or whether factors in the workplace are causing or contributing to employee absence. Records should also be kept of all employees' sickness absences when they occur. In this way, employers may be able to minimise the incidence of employee absence, especially short-term absence.

Where, despite good management, employees are nevertheless absent from work, the employer should adopt good practice techniques to manage those absences. This chapter aims to help managers to understand the various employment laws that impact on employee sickness absence and to apply sound management techniques to the management of both long-term and short-term absences.

Research published by the Chartered Institute of Personnel and Development (CIPD) in October 2010 (CIPD Annual Survey Report on Absence Management) shows that employee absence during 2009 was running at an average of 7.7 days per employee per year, representing 3.4 per cent of working time, and at a median cost of £600 per employee per year.

Absence levels in the public sector were highest at an average of 9.6 days per employee. In the private sector, in manufacturing and production, the average level was 6.9 days, whilst in private sector services organisations it was 6.6 days. Non-profit sector organisations experienced an average of 8.3 days absence per employee per year. Larger organisations tend on the whole to have higher average levels of absence than smaller organisations.

CIPD reported that the most common causes of long-term absence were acute medical conditions (for example stroke, heart attack and cancer), musculo-skeletal injuries, stress, mental ill-health and back pain. Amongst manual workers, the main causes of long-term absence were musculoskeletal injuries and back pain, whilst absences due to stress were more common for non-manual workers.

The main cause of short-term absence for both manual and non-manual workers was minor illness such as colds, flu and stomach upsets. For manual employees, the next most common causes of short-term absenteeism were musculoskeletal injuries and back pain, followed by stress, home/family responsibilities and recurring medical conditions such as asthma, angina and allergies. Stress remains the second most common cause of short-term absence among non-manual workers, followed by musculoskeletal injuries, back pain and recurring medical conditions.

Managing sickness absence effectively is likely to result in:

- a reduction in the potential duration of some employees' sickness absence;
- an increase in the likelihood of long-term sick employees returning to work;
- a reduction in the likelihood of repeated spells of absence after employees have returned to work; and
- a decrease in the costs to the business of employee absence.

Managing long-term sickness absence

[M1502] Long-term absence is usually defined as absence of more than four weeks (although there is no legal definition). CIPD reported in their annual survey on Absence Management, published in 2010, that 20% of absence from work relates to periods of absence of four weeks or longer.

There are two stages to managing an employee's long-term sickness absence. The first is to manage the employee's absence from work and the second to manage their return to work.

The management of long-term sickness absence requires a pro-active approach from management in order that:

(a)   the employee's absence can be regularly reviewed;
(b)   appropriate steps can be taken to support the absent employee;
(c)   actions can be taken to facilitate the employee's return to work;
(d)   the employee's return to work can be effectively managed; and
(e)   where the employee does not recover sufficiently to resume working, the employer can fairly dismiss the employee.

Reviewing the employee's absence

[M1503] Employers should adopt a 'case management' approach when dealing with employees on long-term sickness absence. This means regularly reviewing the employee's absence and state of health/fitness to see whether there is any improvement, and whether the employer can do anything to facilitate the employee's recovery and return to work.

Case reviews would normally be held quarterly or even monthly, and these would involve the employee's line manager, health and safety personnel, HR department, medical specialists and of course the employee who is absent.

Keeping in touch with an employee who is absent

[M1504] Where an employee is absent from work due to a long-term condition, it is good practice for the employer to nominate an appropriate person to keep in touch with them. This has a number of advantages:

- the employee who is absent will feel reassured of their employer's ongoing support and is less likely to feel 'abandoned' or gain the impression that their employer does not care about their welfare;
- the employer will be able to keep up-to-date on the employee's state of health and progress and the likelihood of a return to work;
- the employer will be able to organise temporary cover more effectively;
- the employer can keep the employee informed of any developments in the business and the work of their department; and
- the employee is more likely to return to work sooner, if they feel confident that their employer is adopting a supportive attitude towards them.

Naturally, managers should take care to ensure that contact is not viewed as unwelcome or intrusive. Equally, it is important to make sure that no unfair pressure is put on the employee through such contact and that the employee understands that this is not the case. It is, however, a common assumption that contacting an employee who is off sick will be inappropriate, whilst in fact many employees who are absent long-term would welcome regular contact The absent employee may feel very isolated and excluded whilst off sick and regular contact may help to reduce these feelings and make the employee feel that they are still viewed as part of the team even though they are not able to work.

The type of contact and its frequency should be agreed directly with the employee. This can be instigated by writing to the employee explaining that the manager would like to maintain contact and asking the employee whether they would prefer telephone calls, occasional visits at home (perhaps by a workmate), e-mail communication or a combination of these. The letter should make clear the manager's interest and concern about the employee's welfare and progress and offer any support that is reasonable and practicable. The employee's views on how contact should be maintained should always be sought and respected.

The employee's line manager should take responsibility for arranging and maintaining such contact, but he or she will not always be the best person to maintain the contact. It may be that a close colleague of the absent employee can be nominated instead, in particular where visits to the employee's home are agreed.

Medical statements

From 6 April 2010, the traditional GP's medical certificate, or 'sick note', was replaced by a new 'statement of fitness for work', also known as a 'fit note'. The new statement allows doctors to state either that the employee is 'not fit for work' or that he/she 'may be fit for work' taking account of advice specified by the doctor. There is no 'fit for work' option as it was considered that GP's do not have sufficient knowledge about individual employees' roles or the risks inherent in them to assess this properly.

The purpose of the fit note scheme is to facilitate return to work in circumstances where adjustments by the employer would help the employee resume working sooner than might otherwise be the case.

The fit note system gives doctors the opportunity to highlight one (or more) of four options to help facilitate the employee's return to work. These are:

- a phased return;
- amended job duties;
- altered hours of work;
- workplace adaptations.

The GP may also write in any other option that they consider may be appropriate in the circumstances.

There is no legal obligation on an employer to comply with any recommendation made on a GP's fit note and what is written on the fit note may therefore be regarded as advisory rather than obligatory. Equally, any changes to employees' hours or job duties, whether temporary or permanent, should be made only with the agreement of the employee. Nevertheless, managers should take what the employee's doctor has written seriously and give fair consideration – in consultation with the employee – as to whether any of the changes recommended by the doctor can be accommodated.

If the employer is unable to facilitate the change(s) recommended by the doctor, this will be evidence that the employee remains unfit to carry out their normal job.

Employment laws that impact on long-term sickness absence

[M1505] There are a number of laws that impact on long-term sickness absence:

(1) *Equality Act 2010*;
(2) *Employment Equality (Repeal of Retirement Age Provisions) Regulations 2011*;
(3) *Access to Medical Reports Act 1988*;
(4) *Employment Rights Act 1996, s 1* (the duty on employers to provide employees with a written statement of employment particulars);
(5) *Social Security Contributions and Benefits Act 1992, Pt XI* (the duty to pay statutory sick pay);
(6) *Working Time Regulations 1998*;
(7) *Social Security (Medical Evidence) and Statutory Sick Pay (Medical Evidence) (Amendment) Regulations 2010*.

When an employee's condition may amount to a disability

[M1506] The Disability Rights Commission (which has now been integrated into the Equality and Human Rights Commission) estimated in 2004 that around 9.8 million people in Britain would qualify as disabled in a legal sense (the relevant provisions are contained in the *Equality Act 2010*). The definition of disability in the *Equality Act* is very wide and includes both physical and mental impairments. An individual who has a serious disfigurement may qualify as disabled under the Act, depending on the severity and positioning of the disfigurement.

Definition of disability

[M1507] Specifically, a person is disabled if he or she has a physical or mental impairment that has a substantial and adverse long-term effect on his or her ability to carry out normal day-to-day activities. 'Long-term' in this context means twelve months or more. 'Substantial' is defined as 'not minor or trivial' (which is a low hurdle). There is no need for an employee to be 'registered' as disabled to qualify for protection under the *Equality Act 2010*.

'Normal day-to-day activities' stands to be interpreted according to the lay-person's understanding of the phrase, in other words any activity that people do on a reasonably regular basis in their every-day lives will be a normal day-to-day activity for the purpose of the Act.

Scope of the Act

[M1508] All workers are protected by the disability discrimination provisions of the *Equality Act 2010* and all employers, with the exception of the armed forces, are under a duty to comply with the Act's provisions.

In line with the European Court of Justice (ECJ) ruling in *Coleman v Attridge Law*: [2008] IRLR 722 the *Equality Act 2010*provides that workers are protected against 'discrimination by association'. This means that individuals who, for example, care for a disabled child, parent or partner are protected against discriminatory treatment (including harassment) on any grounds related to the fact that they have an association with or responsibility towards the disabled person. For this type of claim to succeed, however, the claimant would have to show that the alleged discrimination against him/her occurred specifically because the associated person had a disability.

There is no qualifying period of employment for an employee to bring a claim of disability discrimination to an employment tribunal, and no ceiling on the amount of compensation that can be awarded if a claim is successful.

The main duties in respect of disability under the Act are:

(a)   the duty not to discriminate directly, ie the duty not to treat a disabled worker (or job applicant) unfavourably directly because of a disability;

(b)   the duty not to discriminate indirectly, ie not to apply any provision, criterion or practice that puts, or would put, people who share a particular type of disability at a particular disadvantage, unless the provision, criterion or practice is objectively justified;

(c)   the duty not to discriminate for a reason 'arising in consequence of' a disability (unless the particular treatment of the person is objectively justified), for example discrimination against someone because of sickness absence that is occasioned by their disability;

(d)   the duty to make reasonable adjustments to any provision, criterion or practice, or to physical features of premises, that may place a disabled worker at a substantial disadvantage.

Each of these is considered separately below.

Conditions that may amount to a disability

[M1509] A wide range of physical and mental conditions and illnesses may amount to disabilities, depending always on whether the effect of the condition on the person is substantial and long-term. Conditions that are intermittent, or that fluctuate in their effects, will entitle the person to protection under the disability discrimination provisions of the *Equality Act 2010* at all times (provided the condition is likely to recur – 'likely' has been defined by the House of Lords as meaning 'could well happen'), even if at a particular point in time the condition is in remission. A further important point is that a condition will amount to a disability even if the person, as a result of medication or other form of support, experiences no adverse effects on a day-to-day basis. The question that determines whether an employee is disabled is how the condition would affect them if they did not take their medication or use the support. The one exception to this is in respect of sight impairments. If a sight impairment is capable of correction by spectacles or contact lenses, then the question of whether the individual is disabled will stand to be judged against the assessment of the person's sight whilst wearing the spectacles or lenses.

Progressive conditions such as Alzheimer's, Parkinson's disease and muscular dystrophy (for example) are covered under the definition of disability as soon as the condition is diagnosed and the person's normal day-to-day activities are affected in any way, whether substantial or not. Since October 2005 (when changes to the legislation were implemented), any employee diagnosed with cancer, multiple sclerosis or HIV is automatically deemed disabled whether or not they are experiencing any symptoms at the time in question and whether or not it is likely that the effects of the condition will, in time, become substantial.

The cause of an illness or other condition is irrelevant to the assessment of whether it amounts to a disability. Equally, there is no need for an employee to prove a specific medical diagnosis in order to claim protection under the Act, nor to prove that some kind of illness underlies their impairment. Instead it will be sufficient for them to show that the adverse effects of the condition are substantial and long-term. There are some exceptions to the Act, ie conditions that are specifically deemed not to be disabilities. These include alcohol addiction, addiction to illegal drugs and hay fever. A small number of mental conditions are also excluded.

The duty not to discriminate

[M1510] Employers are under a duty not to treat a disabled employee unfavourably:

(1)   directly because they have a particular disability;
(2)   indirectly, eg by applying a policy or provision which has a dispropor-tionate adverse impact on people with a particular type of disability and which is not justified;
(3)   for a reason arising from disability, unless justified.

Direct disability discrimination, which can never be justified, is where an employer treats a disabled employee or job applicant unfavourably just because they have a disability. An example could be the automatic rejection of a job applicant just because they had disclosed that they had a history of mental illness.

In contrast, it is possible for an employer to justify indirect disability discrimination if the provision in question can be shown to be appropriate and necessary with a view to achieving a legitimate business aim.

It is also possible to justify discriminating against an employee for a reason arising in consequence of their disability, for example if an existing employee had, following an accident, become completely unable to perform the job for which he or she was employed. In these circumstances it might be justifiable to terminate the employee's employment, provided the employer had first explored whether any reasonable adjustments could be made (seeM1511below), for example whether the employee could reasonably be transferred to different job duties. To bring a claim of this type, the individual is not required to demonstrate that their treatment was less favourable than that afforded to anyone else. Justification will be made out if the employer can show that the treatment of the employee represented a proportionate means of achieving a legitimate business aim.

Unfavourable treatment can take many forms and may include:

- rejection for employment or promotion;
- exclusion from training opportunities;
- being afforded less favourable terms and conditions;
- removal of job duties or transfer to less interesting or less responsible work;
- exclusion from contractual benefits or non-contractual perks;
- harassment, which can be physical, verbal or non-verbal; or
- dismissal or selection for redundancy.
- The list is not exhaustive.

The duty to make reasonable adjustments

[M1511] The *Equality Act 2010*places a duty on employers to make reasonable adjustments to any provision, criterion or practice applied by the employer and to physical features of premises in order to accommodate the needs of disabled workers and job applicants. There is also a duty to provide an 'auxiliary aid' (eg specialised computer software) for a disabled employee where it is reasonable for the employer to do so. The duty to make reasonable adjustments arises whenever any aspect of the employer's working practices or premises, puts a disabled person at a substantial disadvantage or where the disabled person would, but for the provision of an auxiliary aid, be put at a substantial disadvantage. An employer who fails to comply with this duty will be acting in contravention of the *Equality Act 2010* unless they can show that it was not reasonable for them to make adjustments generally, or to make a particular adjustment.

An important point to bear in mind is that the duty to make reasonable adjustments places the employer under a positive duty to take the initiative and consider what adjustments would be possible and practicable for a particular disabled employee (or job applicant). It is advisable for employers to make a full assessment of the employee's condition, its effects on their ability to perform their job duties and the steps that might reasonably be taken to support them.

Although the duty to make reasonable adjustments is on the employer, it is nevertheless advisable for managers to consult the employee concerned about possible adjustments as the employee will, in any event, know more about their condition and its effects than the manager, and will often be able to suggest adjustments that would be helpful for them. Managers should be open-minded and willing to take on board any reasonable suggestions from the employee for adjustments.

Adjustments can include:

(a)   transferring the employee to another job, for example lighter work, provided the employee consents to such a move;

(b)   adjusting the duties of the job, for example exempting an employee with a back condition from doing any heavy physical work;

(c)   changing the method of doing the job, for example allowing an employee who cannot drive on account of a medical condition to use taxis to travel on business;

(d)   adjusting working hours, for example allowing someone with a mental illness a later or flexible start time, or more frequent rest breaks, or exempting an employee with a debilitating condition from a requirement to work overtime;

(e)   changing the place of work, for example moving someone with limited mobility to a ground floor location or allowing home-working for part of the working week;

(f)   adjusting procedural requirements, for example allowing an employee with a recurring illness to take more time off work than would normally be acceptable;

(g)   providing additional or tailored training, coaching or mentoring, for example for someone with learning difficulties;

(h)   providing special equipment, for example voice activated software for someone with a visual impairment;

(i)   modifying instructions or reference manuals, for example providing them in Braille;

(j)   providing a reader or interpreter, for example for someone who is profoundly deaf;

(k)   providing an auxiliary aid, for example specialised computer software;

(l)   providing additional or special supervision; and

(m)   modifying premises, for example widening a doorway for a wheelchair user, or relocating door handles or shelves for someone who has difficulty reaching.

The Employment Appeal Tribunal said, however, (*in O'Hanlon v HM Revenue and Customs* [2006] IRLR 840) that it is unlikely to be a reasonable adjustment for an employer to adjust their sick pay scheme so as to allow a disabled employee to receive full pay indefinitely when absent from work for a disability-related reason, unless there are exceptional circumstances. The Court of Appeal subsequently upheld the EAT's decision in *O'Hanlon v Commissioners of HM Revenue and Customs* [2007] EWCA Civ 283, [2007] IRLR 404.

Unless an employer has done all they reasonably can to make adjustments for a disabled employee, any subsequent unfavourable treatment (for example dismissal) will always be discriminatory and unlawful.

The ECJ stated in the *Coleman* case (see above) that the duty to make reasonable adjustments is not applicable to individuals on the grounds that they are associated with a disabled person.

Insurance benefits

Under the *Employment Equality (Repeal of Retirement Age Provisions) Regulations 2011*, an exemption was introduced (on 6 April 2011) for employers who offer or provide their employees with certain types of insurance benefits on a group risk basis (ie through third party insurers). Examples include life assurance, income protection schemes, sickness/accident insurance and private medical insurance.

The exemption permits employers who offer or provide these benefits to their employees to stop providing them in respect of employees who have reached the age of 65 (or state pension age, if higher, ie when it is raised above age 65 in the future). Such a policy would otherwise be age discriminatory.

Access to medical reports

[M1512] The *Access to Medical Reports Act 1988* places certain restrictions on organisations that wish to obtain medical information about someone from his or her doctor. The Act also gives individuals a variety of rights in relation to any such medical reports.

The Act applies whenever a medical report is 'prepared by a medical practitioner who is or has been responsible for the clinical care of the individual'. The effect of this is that the Act applies to medical reports prepared by an employee's own doctor, specialist or consultant, but would not normally apply to reports prepared by an occupational doctor.

Employees have a number of rights under the Act:

- to be informed in writing if the employer wishes to seek their consent to contact their doctor for a medical report;
- (at the same time) to be informed by the employer of their rights under the Act;
- to decline to give their consent for the employer to apply to their doctor for a medical report;
- (if they have given consent) to ask their doctor for a copy of the report once it has been prepared;
- to ask the doctor to amend the report, if, in their opinion, it contains anything inaccurate or misleading; and
- to refuse to allow the report to be released to the employer.

The employee's right to refuse to give consent for their employer to apply to their doctor for a medical report and their right to refuse to allow a report, once prepared, to be released to the employer, are statutory rights. This means that the employer cannot override these rights by inserting a clause in employment contracts to the effect that employees are required to give such consent. Any contractual term to this effect will be unenforceable.

Since the Act does not normally apply to reports prepared by occupational doctors, it is open to an employer to include clauses in employees' contracts to the effect that they must agree, on request by the employer, to be examined by an occupational doctor, and allow the doctor to provide a report to the employer.

General Medical Council Guidance

Although the *Access to Medical Reports Act 1988* does not normally apply to reports prepared by occupational doctors, Guidance from the General Medical Council, published in October 2009, requires all doctors (including occupational doctors) preparing reports for employment purposes to obtain the employee's written consent before passing any report to the employer. The doctor is also required to be satisfied that the employee concerned is fully informed of the purposes and likely results of disclosing the report to the employer, and offer to show the employee the report before passing it to the employer. This means, amongst other things, that employees may in effect be able to refuse to consent to an occupational doctor's report being disclosed to the employer as it is unlikely a doctor would be prepared to disclose a report in breach of the General Medical Council's Guidance.

Employees' contractual rights

[M1513] Under the *Employment Rights Act 1996, s 1,* employers are under a duty to provide a written statement of key terms and conditions of employment to all employees whose employment is to continue for more than one month. The 'written statement' as it is usually known must include information on a number of defined issues, including any terms and conditions relating to 'incapacity for work due to sickness or injury, including any provision for sick pay'.

There is no duty on an employer to pay an employee 'occupational sick pay', ie their normal wages or salary, if they are absent from work due to sickness or injury. That is a matter for each employer to decide for themselves. Employers must, however, inform each employee in writing (as part of the written statement) what their entitlement to sick pay will be in the event of sickness absence, and of any conditions attached to the payment of sick pay.

Many employers pay occupational sick pay for defined periods of time, whilst others make the payment of sick pay discretionary. Employers should bear in mind, however, that a discretionary policy carries with it the danger that employees who are not paid sick pay will perceive the policy (and management) to be unfair and that this may lead to grievances. Where the employer wishes to retain discretion over the payment of sick pay, it is advisable therefore to devise clear and transparent management guidelines on how discretion should be exercised to ensure consistency and fairness across different departments. In *Clark v Nomura International plc* [2000] IRLR 766, the High Court held that management discretion in relation to pay must not be exercised perversely or irrationally. It would also be normal practice to make the payment of occupational sick pay dependent on certain conditions being fulfilled, for example to require employees to:

(1) telephone a designated person as soon the employee knows that they will be unable to attend work;

(2) provide a self-certificate for the first week of absence and doctors' statements for all periods of absence thereafter;

(3) maintain contact with the employer at regular intervals (for example weekly) with a view to advising the employer how long they expect to be absent;

(4) agree to be examined by an occupational doctor on request and to the doctor providing a report to the employer; and

(5) attend a 'return to work' interview with a designated person (usually the line manager) on return to work.

These conditions allow the employer to manage absence and sick pay effectively.

Statutory sick pay (SSP) is an entirely different matter and is dealt with at M1514 below.

Statutory sick pay

[M1514] Although the employer may decline to make payment of wages or salary during sickness absence, they cannot contract out of their obligation to pay SSP.

There is no length of service required for an employee to qualify for SSP, and so both temporary and permanent employees can, potentially, qualify. However, complex provisions are in place with regard to when payment is due and there is a requirement for certain record keeping.

Both full-time and part-time employees will qualify, but an employee who earns less than the equivalent of the lower limit for the purposes of national insurance contributions (£102.00 per week as from April 2011) are excluded. Where an employee is paid occupational sick pay, SSP must be offset against this, ie SSP cannot be paid over and above normal wages or salary.

To qualify for SSP, an employee must comply with their employer's conditions for notification of absence and provide suitable evidence of incapacity.

SSP is a fixed weekly rate, payable to an employee who is absent from work due to personal sickness for up to 28 weeks during any 'qualifying period'. The first three days of absence during any qualifying period are 'waiting days' and no SSP is payable for these days. The rate in force (as from 6 April 2011 is £81.60 per week (or the appropriate pro-rata daily rate).

When the rules on SSP were first introduced, employers were able to recoup amounts paid out by offsetting payments against their national insurance contributions. These provisions were, however, subsequently abolished, which means that employers now foot the bill for SSP themselves.

Accrual of annual leave during periods of sickness absence

In January 2009, the European Court of Justice (ECJ) published an important judgement (*Stringer v HM Revenue and Customs* [2009] IRLR 214) concerning the question of whether statutory annual holidays continue to accrue whilst an employee is absent from work due to sickness. Technically, the judgment applies only in respect of the first four weeks of statutory annual holiday leave, ie the portion of leave governed by the EC Working Time Directive, although it is not yet known whether the UK courts may rule that it applies to all statutory annual leave (ie 5.6 weeks).

The ECJ ruled as follows:

*   employees who are absent from work due to sickness continue to accrue statutory annual holiday entitlement in the usual way;
*   this is the case irrespective of whether the employee is absent for a whole holiday year or only part of it (and irrespective of whether or not the employee is being paid whilst on sick leave);
*   the employer may agree, if the employee requests it, to convert a period of sickness absence into paid annual leave;
*   if an employee has been unable to take holidays as a result of sickness absence, the employer must permit them to take their full statutory entitlement after they have returned to work;
*   if the employee's absence has spanned more than one holiday year (ie if they have accrued annual leave whilst off sick during one holiday year and return to work only during the next holiday year), they must be permitted to carry forward any untaken holidays to the next holiday year and take them during that year;
*   if an employee is dismissed whilst absent from work, the employer must make a payment in lieu of all accrued statutory annual leave which has not been taken, even in circumstances where the employee was on sick leave for one or more full holiday years.

In a subsequent case, *Pereda v Madrid Movilidad SA* [2009] IRLR 959, the ECJ ruled that where an employee goes off sick just before a period of planned statutory annual leave, the employer must permit them, on request, to reschedule the period of annual leave to another time once they have recovered and returned to work. Employees have the right to choose to take their annual leave at a time other than during a period that overlaps with a period of sickness absence.

Rehabilitation of an employee who has been absent from work

[M1515] Where an employee has been absent from work for a lengthy period of time, it will be essential to manage their return to work carefully. The employee's line manager should take responsibility for discussing the employee's return to work directly with them and for agreeing a detailed plan to support the employee on their return.

Many employees will feel anxious about the prospect of returning after a lengthy period away from the workplace and they may be worried about how they will cope with being back at work, or concerned about their colleagues' reactions.

The first action point will be for the manager to speak to the employee (preferably prior to their return if that is possible, but if not on the employee's first day back) to discuss the details of their return. One way of dealing with this would be to suggest to the employee that they might make a social visit to the workplace shortly before their agreed return date to allow a meeting with their line manager and/or HR department and also allow them to chat informally with their colleagues about any ongoing relevant matters.

The discussion with the line manager should include:

(a)   the employee's opinion about their capabilities, for example whether the employee is confident that they are capable of full job performance or only partial performance;

(b)   whether the employee's return should be to full-time duties or whether a phased return would be beneficial;

(c)   whether there are any recommendations from the employee's GP regarding adaptations that might be made to facilitate their return to work (ie recommendations contained in a doctor's 'fit note');

(d)   whether the employee will still be taking any medication after their return to work that might have side effects, for example tiredness;

(e)   any special arrangements, additional support or adjustments to the employee's duties or environment that would help the employee to reintegrate into the workplace; and

(f)   whether a tailored induction programme is desirable or necessary, for example if a number of organisational or procedural changes have taken place during the employee's absence, a 'mini-induction' may be very helpful.

The employer should also arrange for the employee to be medically examined by an occupational doctor to confirm that they are genuinely capable – both physically and mentally – of returning to work. The doctor should also be asked to comment on whether the employee should be exempted initially from any particular job duties and if any special arrangements are necessary or advisable.

After the employee's return, the line manager should:

•   monitor the situation over the first few weeks to ensure the employee is coping with their work and the day-to-day pressures of working life;

•   make sure the employee is not 'thrown in the deep end', for example by requiring them to deal with a huge backlog of work caused by their absence; and

•   take all reasonable steps to ensure the employee's reintegration into the workplace.

When dismissal on the grounds of long-term absence can be fair

[M1516] Long-term sickness absence can be a fair reason for dismissal for the purposes of the *Employment Rights Act 1996, s 98* under the heading of 'capability'. 'Capability' is defined as having reference to 'skill, aptitude, health or any other physical or mental quality'. For a dismissal to be fair, however, the employer would have to show that the employee's long-term absence was sufficient to justify dismissal and that they had acted reasonably in dismissing the employee for this reason, taking into account the size and administrative resources of the organisation. Employment tribunals, in practice, will assess, based on all the available evidence, whether the dismissal for the stated reason was within what the Courts have termed the 'band of reasonable responses'.

In the event of a claim to tribunal, the burden of proving the reason for the dismissal is on the employer.

All employees are entitled to bring a claim for unfair dismissal to an employment tribunal provided that they:

(1)   are employed under a contract of employment;
(2)   have a minimum of one year's continuous service with the employer.

There are no upper or lower age limits on the right to claim unfair dismissal.

Before contemplating dismissal, the employer should review the circumstances to establish whether grounds for dismissal exist. In doing so, the employer should consider:

(a)   the effect of the employee's past and likely future absences on the organisation, for example to what extent colleagues are being burdened by the employee's absence;
(b)   the degree to which it is possible (if at all) to cover the absent employee's work, for example by using temporary agency workers;
(c)   how similar situations have been handled in the past;
(d)   medical advice;
(e)   the likelihood of the employee being able to return to work in the reasonably near future; and
(f)   the availability of suitable alternative work which the employee would be capable of doing, ie whether there is any alternative to dismissal.

Procedure for ensuring a dismissal is fair

[M1517] Over and above having a potentially fair reason for dismissal (ie ill-health absence), an employer who is challenged in an employment tribunal must also be able to demonstrate that they:

•   acted reasonably towards the employee in an overall sense; and
•   followed their own in-house procedures fully.

Courts and tribunals have, over the course of many years, developed tests of fairness in cases of dismissal. In relation to dismissals on the grounds of long-term ill-health absence, the employer should, as a minimum, ensure that they:

(1)   review the employee's absence record to assess whether it is sufficient to justify dismissal;
(2)   set time limits for review and inform the employee of these;
(3)   consult the employee regularly;
(4)   obtain up-to-date medical advice;
(5)   advise the employee as soon as it is established that termination of employment has become a possibility;

(6)   advise the employee that they have the right, if they attend any formal meetings with the employer to discuss their ongoing employment, to bring a colleague or trade union representative along with them to the meeting if they wish;

(7)   review whether there are any other jobs that the employee could do prior to taking any decision on whether to dismiss; and

(8)   act reasonably throughout.

There is no time limit in law after which it is fair to dismiss an employee who is absent from work. The key question is whether in all the circumstances the employer can reasonably be expected, in light of the requirements of their business, to wait any longer for the employee to recover and return to work. This will depend, amongst other things, on the size and resources of the business and the degree of disruption or difficulty that the employee's long-term absence is causing.

The fairness of a dismissal stands to be judged in line with the House of Lords' ruling in the case of *Polkey v A E Dayton Services Ltd* [1987] IRLR 503. In this case, the House of Lords said that any procedural shortfall will make a dismissal unfair, even in circumstances where the employer had a legitimate reason to dismiss the employee. This means that any failure of the employer to adhere to its own in-house procedures when dismissing a member of staff (for any reason) is likely, in almost all cases, to render the employee's dismissal unfair.

In April 2009, ACAS published a revised Code of Practice on Disciplinary and Grievance Procedures. The Code provides statutory guidance to employers (and employees) on matters of discipline, performance and grievances. The Code does not specifically address termination of employment on the grounds of sickness absence, nevertheless many of its provisions may be broadly relevant to such dismissals as they relate to fair treatment in a general sense. The Code is accompanied by a comprehensive non-statutory Guide, 'Discipline and Grievances at Work', which contains a section titled 'Dealing with Absence' (in appendix 4).

Managing short-term absences

[M1518] Frequent, persistent short-term absences can be extremely disruptive for an employer. Where an employee is absent long-term, the employer can at least make plans, as they will have some idea about how long the employee is likely to be absent. The nature of short-term absenteeism, however, is that it tends to be unpredictable and uncertain.

The CIPD reported in 2010 (in their annual survey on absence management) that 66 per cent of employee absences are for periods of seven days or less and 16% is accounted for by absences of between eight days and four weeks. They reported the most common causes of short-term absences as minor illnesses, musculoskeletal injuries, back pain, stress, recurring medical conditions (such as asthma, angina and allergies) and home and family responsibilities.

The CIPD reported also that approximately one in five employers believe that absences that are not due to genuine ill-health rank among the top five most common causes of short-term absence for both manual (23%) and non-manual (17%) workers.

Employees' rights in relation to short-term absence will be determined according to the terms of their contracts of employment. Contracts or supplementary documentation should state clearly what employees' obligations and entitlements are when they are unable to attend work, for example what they will be paid when they are absent and how they should notify the employer.

To manage short-term absenteeism effectively, employers should:

- require employees who are unable to come to work personally to notify a designated person of their absence as soon as possible;
- require employees to 'self-certify' for all periods of absence of up to one week and to provide doctors' statements for all periods of absence beyond the first week;
- conduct regular 'return-to-work interviews' after every period of absence, however short;
- investigate the possible causes of frequent absenteeism;
- look for patterns in absenteeism to see whether there is, for example, a problem in a particular department;
- implement a policy with defined trigger points which, when reached, activate a formal review of the employee's attendance;
- set targets and time limits for improvement; and
- keep detailed records of all employee absences.

These issues are explored further below.

Sound management techniques for dealing with short-term absenteeism

[M1519] There is much an employer can do to manage short-term absenteeism effectively with a view to reducing the frequency of its occurrence.

Requirement to notify

[M1520] Employers should have in place a set procedure under which employees are required personally to notify a designated person (usually their line manager) by telephone as soon as they know that they will be unable to come to work. The line manager should record the date and time of the call, the reason given by the employee for inability to attend work and the employee's view of how long they expect to be absent. This creates the start of an absence record. A pre-printed questionnaire could be devised for this purpose, thus ensuring consistency in approach. The questionnaire could also be used as a means of instigating payment of statutory sick pay.

If the manager is unavailable when the employee telephones, they should telephone the employee at home soon afterwards to obtain the necessary information.

The information should subsequently be passed to the HR department (if the organisation has one) or a named senior manager to whom responsibility has been allocated for collecting and analysing sickness absence data across the whole organisation.

Certification

[M1521] Employers should routinely require employees who are absent from work due to sickness to complete a self-certification form when they return to work (at the return-to-work interview – see **M1522**below). The employee should be asked to complete and sign the form in front of their manager who should counter-sign the form, and the obligation to do so should apply to every absence of one day or more. The form should require the employee to write down the reason for their absence, the dates on which their sickness began and ended and whether they consulted a doctor, rather than just tick a series of boxes. This makes the process of self-certification a more active process.

Line managers should check that self-certification forms have been properly completed, for example by ensuring that the reason given for the absence is clearly stated. Any vague, woolly reasons for absence such as 'sickness' or 'debility' should be questioned (although the manager should not ask intrusive questions about the employee's state of health).

It is often a sound idea to make the payment of sick pay dependent on the completion of a self-certification form (and on the provision of doctors' statements in the case of longer periods of absence).

Return to work interviews

[M1522] As well as requiring the employee to complete a self-certification form, employers should adopt a policy whereby their managers routinely conduct 'return-to-work interviews' every time an employee has been absent from work due to sickness.

The interview should be:

(1)   informal (but should be more than just a 'corridor chat')
(2)   private and confidential;
(3)   structured and factual;
(4)   carried out in a positive and supportive way; and
(5)   recorded in writing (with a copy of the record being provided to the employee).

The dual purpose of such an interview will be to establish the reason for the employee's absence (without requiring disclosure of any medical details) and to assess whether the employer can do anything to help or support the employee. It is important to note that this type of interview should not be confused with a disciplinary interview and that fact should be made clear to the employee. It may assist the process of communication if the manager explains to the employee that the purpose of return-to work interviewing is to manage and monitor employees' absences so that any problem areas can be identified and support offered where appropriate.

Because the interview is informal, the right for the employee to be accompanied will not apply.

Adopting this procedure will let employees know that management is serious about monitoring absences and may therefore deter casual absences.

CIPD reported in their 2010 survey on absence management that return-to-work interviews are rated by employers as the most effective approach to managing short-term absenteeism (followed by trigger mechanisms for reviewing attendance and the use of disciplinary procedures to deal with unacceptable levels of absence, restricting sick pay and the provision of sickness absence information to line managers).

Considering the possible causes of frequent absenteeism

[M1523] Most employers have one or more employees who are absent from work on a regular basis. If the employee has an underlying medical condition that is causing these absences, then the employer will know what they are dealing with and medical advice can be sought to support any employment decisions that the employer might contemplate making. The employer should also consider whether the employee might qualify as disabled under the *Equality Act 2010* (see **M1507** above) and if so review what reasonable adjustments might be made to the employee's job or working environment.

There are, however, many other possible causes of frequent absenteeism. Apart from ill-health, the employee may:

(a)   have personal or family difficulties, for example childcare problems;
(b)   have another job which is causing a conflict of interest;
(c)   be de-motivated and disinclined to attend work; or
(d)   have a specific problem in the workplace (see **M1524** below) that is causing or contributing to their frequent absences.

It will be essential for the manager to remain open-minded and to refrain from jumping to the conclusion that the employee is taking time off work without good reason. It may be that the cause of the employee's absences is something outside their control. The manager should speak to the employee to establish whether there is a specific underlying cause and whether anything can be done to support the employee. Understanding the root cause of the absence problem is important because, until the cause is correctly pinpointed, it will not be possible to identify an appropriate course of action to remedy it.

Reviewing whether factors in the workplace are causing absence

[M1524] Whilst a certain level of absenteeism is inevitable due to minor illnesses, there may also be occasions when employees take time off work for reasons associated with factors in the workplace.

Managers should be constantly alert to the possibility that an employee's absences might be caused by, or exacerbated by, factors in the workplace. For example, the problem may be related to:

•   the volume of work being too much for the employee to cope with;

- deadline dates or other pressures that the employee feels unable to face;
- unsatisfactory working relationships or outright conflict with colleagues;
- bullying or harassment;
- de-motivation caused by ineffective management or an authoritarian management style;
- inability to cope with change;
- a feeling of inadequacy or loss of control over the job; or
- other factors at work causing dissatisfaction, for example ineffective procedures or equipment, having no clear goals or targets, etc.

Issues such as those identified above may well lead to absences and managers should be alert to signals that the employee may be suffering from stress to an extent that he or she is not coping. Medical certificates that state 'stress', 'depression' or 'anxiety' should put the manager on notice that there may be a workplace problem that needs to be addressed.

Line managers should always be alert to such issues and should make enquiries of any employee who has frequent short-term absences as to whether there is anything in the workplace causing or contributing to their absences. Such enquiries should be made sympathetically, with the manager reassuring the employee that if he or she is experiencing problems at work, it would be their genuine wish to provide support with a view to resolving those problems.

If a workplace problem is identified, the manager should of course take steps to remove or reduce the factor that is causing the problem, if that is at all possible. Once the cause of the employee's absences is removed, the employee's attendance may well improve. A failure to take steps to support an employee who is experiencing health problems as a result of factors in the workplace may have serious consequences as the employer could be held liable in law if the employee subsequently had a mental breakdown as a result of factors in the workplace and the breakdown was reasonably foreseeable.

Recent court decisions demonstrate the dangers of doing nothing when an employee complains of serious workplace stress. In *Intel Corporation (UK) Ltd v Daw* [2007] IRLR 355, the employer was held liable in negligence when Ms Daw had a complete breakdown, which had been caused by overwork. The employee had, at her manager's request, provided a written account of the problems she was experiencing, but despite this, nothing was done to reduce her workload. The Court held that it would have been clear to the employer, following the written account, that Ms Daw was at risk of personal injury. It followed that the injury to her health as a result of an excessive workload had been reasonably foreseeable. Management should have taken immediate action to reduce Ms Daw's workload upon receiving the written account of her problems, as this was the only course of action that would have properly addressed the problem.

Similarly, in *Dickins v O2 plc* [2008] EWCA Civ 1144, [2009] IRLR 58, the employer was held liable for having caused the depressive illness of an employee who had repeatedly complained of overwork and of being 'at the end of her tether'. The Court of Appeal held that management's failure to reduce Ms Dickins' workload amounted to a breach of the duty of care, and that once they became aware of the extent of her difficulties, they should have sent her home pending urgent investigation by occupational health, even though, at that time, she had not been signed off by her GP.

Looking for patterns

[M1525] Employers who are experiencing high levels of short-term absenteeism should use the records created by line managers about individuals' absences to review the absenteeism, and whether any patterns exist, for example:

(1)   whether absenteeism is higher than average in any particular department or section;

(2)   whether the rates of absence of employees performing any particular type of work are disproportionate when compared to other employees; or

(3)   whether there is any particularly common cause of absence (for example back pain).

In each case, the employer should investigate in order to try to establish the root cause of the particular pattern. Following identification of the likely cause, steps can be taken to remedy the situation. For example if absenteeism is unusually high in a particular manager's section, it could possibly be that the manager is the cause. Well-designed training in supervisory skills or inter-personal skills, or individual coaching may improve the manager's people-management skills and subsequently lead to a reduction in levels of absenteeism in his or her department.

On a more local level, line managers should, when reviewing individual employees' absence records, look to see whether there is a pattern amongst the absences. Examples could include the frequent Monday morning absentee or the employee whose absences tend to occur at particular times, for example just before a detailed month-end report is due.

If such a pattern is identified, the manager should speak to the employee about it and ask the employee if they can explain the pattern. Managers should, of course, not make assumptions, nor blame or accuse the employee, but instead should remain open-minded. The main aim in speaking to the employee about a pattern of absences will, once again, be to try to establish the underlying reasons(s) for the frequent absences. It is only when the underlying cause is identified that it will be possible to decide what to do. For example, if absence is due to tiredness and the reason the employee is tired is because they have a second job, then the manager might proceed by explaining to the employee that the situation is unsatisfactory and seek to reach a compromise on how many hours the employee works in the other job.

Even if the employee is unable to put forward any explanation, such a meeting will have the advantage of alerting the employee to the fact that the line manager has noticed the pattern. This, in turn, may deter further casual absences.

Defined trigger points

[M1526] It can be very effective for employers to devise and implement an attendance procedure with defined trigger points that act to instigate a formal review of the employee's level of absence and its causes. Trigger points are usually defined by reference to:

(a)   the total number of days' absence the employee has had in a defined period (for example in any twelve-month period); or

(b)   the number of occasions when the employee has been absent from work during the same period.

A typical example of trigger points could be three separate absences (whatever their length) or absences totalling 15 days in any twelve-month period.

Setting targets and time limits for improvement

[M1527] Where an employee's level of short-term absences has reached a stage where their level of attendance is unsatisfactory, the line manager should seek to agree with the employee reasonable targets and time limits for improvement in attendance. The manager should try to ensure that the employee is committed to achieving the agreed level of improvement. The manager should of course follow up at the agreed time and review the employee's level of attendance again. If it has not improved to a satisfactory level, formal action may have to be instigated according to the employer's procedures.

Keeping records

[M1528] Full records should always be kept of employees' absences and of all discussions held with the employee about absence and attendance. Self-certificates and medical statements should also be retained. Such records should, of course be held confidentially, preferably by the organisation's HR department (if they have one). The terms of the *Data Protection Act 1998* will have to be adhered to in holding such records.

Giving formal warnings for unsatisfactory attendance

[M1529] Although it is appropriate for managers to be supportive in the first instance towards employees who, for genuine reasons, have frequent absences from work, managers also need to ensure the work of their department is done efficiently and effectively. Even where the reason for an employee's absences is undisputedly genuine, that does not change the fact that the employee's work still needs to be done on a regular and reliable basis. If, therefore, informal measures have not led to an improvement in the employee's attendance, it may be that formal procedures need to be instigated. This will be appropriate when the employee's absences have become excessive or where they are beginning to cause serious disruption or dissatisfaction.

Some employers use their standard disciplinary procedure to deal with this type of situation, whilst others have separate capability procedures which follow a similar format. The key action steps of any such procedure will be that the employer would:

- write to the employee, explaining that their level of absence has reached a level that is causing concern, enclosing a note detailing the dates and duration of all absences over the last six or twelve-month period (as appropriate);
- invite the employee to attend a meeting to discuss the matter, informing them that they have the right to bring a colleague or trade union official along, either to represent them or simply to provide moral support;
- hold the meeting, and explain to the employee that their absence has reached a level that is considered unsatisfactory and the reasons why this is the case;
- give the employee a full and fair opportunity once more to explain their absences and put forward any mitigating factors or other representations;
- decide after the meeting whether it is appropriate to issue a formal warning, following the terms of the employer's own procedure;
- make it clear in any warning that if the employee's level of attendance does not improve to a defined standard within a defined time period, further formal action will be taken and that continuing absences could, eventually, lead to dismissal; and
- set down a date for a further review, typically in three or six months time.

This type of procedure may be instigated irrespective of whether or not the employee's absences are for genuine reasons. If some of the absences are not genuine, the action taken would of course fall under the banner of discipline. Even where the employee's absences are for genuine reasons (for example genuine ill-health, family problems, etc.), formal action can be justified under the banner of 'capability' or 'attendance'.

Fair dismissal on the grounds of short-term absence

[M1530] If, following formal action, the employee's level of attendance does not improve to a satisfactory degree within the defined time-scale, the employer would move to the next stage of the disciplinary or capability procedure. It is usual for two written warnings to be given following which the employee may ultimately be dismissed. Dismissal should not, of course, be undertaken lightly and should normally only ever be a last resort after all other possible courses of action have been fully explored.

The reason for dismissal will be either 'lack of capability' (ie ill-health that has led to the employee being unable to perform their job to a satisfactory standard) or 'some other substantial reason' (ie unsatisfactory attendance, whatever the causes). Both of these are stated in the *Employment Rights Act 1996* as potentially fair reasons for dismissal.

For a dismissal to be fair, however, the employer also has to show that the employee's level of absence was sufficient to justify dismissal, that they had acted reasonably in dismissing the employee for this reason and that they followed their own in-house procedures fully.

Conclusion

[M1531] Many employers are wary of tackling employee absence and uncertain as to what action they can lawfully take. The basic underlying principle is that, if an employee is incapable of performing their job, whether due to genuine ill-health or not, the employer may take action up to and including dismissal. Long before the dismissal stage is reached, however, active management intervention can often help to:

(1)   identify the cause(s) of an employee's absences;
(2)   provide support to the employee, if appropriate;
(3)   deter casual absences;
(4)   identify whether there are any problems inherent in the workplace that are contributing to rates of absenteeism, for example bullying, and ensure these are addressed;
(5)   establish whether or not an employee's level of attendance is likely to improve within a reasonable time-frame;
(6)   improve morale and motivation; and
(7)   lead to a reduction in rates of absenteeism within the organisation and an associated reduction in costs.

Managing Health and Safety

Lawrence Bamber and Mike Bateman

Introduction

[M2001] There are many reasons why employers need to manage health and safety well. These are usually grouped under three main headings:

Humanitarian

- In a typical year in the UK over 300 people (workers and members of the public) are killed in work-related accidents, around 30,000 people suffer major injuries (fractured limbs, amputations etc), well over 2,000 people die as a result of occupational diseases and at least 10,000 other deaths are partly attributable to occupational ill health. Employers would obviously not wish to be involved in these accidents and cases of ill health. The psychological effects of individual incidents on employers, employees and others are sometimes quite traumatic.

  In recent years greater attention has been placed on corporate responsibility with the European Agency for Safety and Health at Work publishing a report in 2004 entitled 'Corporate social responsibility for health and safety at work'. This contains many examples of good practice together with related conclusions and recommendations.

Legal

- Society has been increasingly unwilling to accept such a toll of accidents and ill health. In many respects the UK has led the world in health and safety law and now European directives result in a steady flow of health and safety legislation. Many regulations (particularly the *Management of Health and Safety at Work Regulations 1999 (SI 1999 No 3242)* – see **M2005** below), specifically require the application of appropriate management techniques. The *Corporate Manslaughter and Corporate Homicide Act 2007* requires corporate entities to ensure that their activities are safely managed so as to prevent fatalities.

Financial

- Accidents and ill-health can have a direct financial impact on a business through fines and compensation claims. Fines for breaches of health and safety legislation have increased considerably in recent years, with the present record fine of £15 million for a single offence being set in 2005 following the prosecution of Transco after an explosion at its Larkhall plant in Lanarkshire.

  However, there may also be significant indirect costs associated with the unavailability of injured employees or damaged equipment, or due to the impact of legal sanctions such as prohibition notices. Loss of reputation following serious accidents has had a significant adverse impact on a number of companies, particularly in the rail sector, and supply-chain pressure can result in those whose health and safety standards are found lacking, finding it increasingly difficult to gain business.

Insurance will only protect employers against a limited range of direct and indirect costs; research published in 1993 showed that the ratio of 'insured' costs to 'uninsured' costs could be as high as 1:36. The ways in which such costs may be incurred are described in FIGURE 1. The cost to employers of workplace injuries and work-related ill health was estimated at £48.1 billion in 2005–6. These costs comprise expenditure relating to sick pay, recruitment costs, administrative costs, compensation and insurance.

Figure 1: The costs of accidents

INSURED

DIRECT			INDIRECT
• Employers' liability claims*	• Major business interruption*		
• Public liability claims*	• Product liability*		
• Major damage*			
• External legal costs*			
• Fines*	• Minor business interruption		
• Sick Pay*	• Prohibition notice interruption		
• Minor damage	• Customer dissatisfaction (product delay)		
• Lost or damaged product	• First aid and medical attention		
• Clear-up costs	• Investigation time and internal administration		
	• Diversion of attention/spectating time		
	• Loss of key staff		
	• Hiring/training replacement staff		
	• Hiring replacement equipment		
	• Lowered employee morale		
	• Damage to external image		

UNINSURED

* Only these costs can be quantified without considerable effort

(Expanded version of diagram previously published in HSE booklet HSG 96 (no longer available).)

Legal requirements

[M2002] Almost all recent health and safety legislation contains provisions relating to effective management, often requiring the application of risk assessment and control techniques. Set out below are the requirements of *sections 2* and *3* of the *Health and Safety at Work etc. Act 1974* ('*HSWA 1974*'), together with those contained in the *Management of Health and Safety* *at Work Regulations 1999 (SI 1999 No 3242)* (the '*Management Regulations*').

The *Construction (Design and Management) Regulations 2007 (SI 2007 No 320)* increased the requirements for effective health and safety management of all construction work – see **M2022** below

Duties of employers to their employees

[M2003] *Section 2* of *HSWA 1974* sets out a number of general duties that employers hold towards their employees. *Subsection (1)* of *section 2* contains an all-embracing requirement:

> It shall be the duty of every employer to ensure, so far as is reasonably practicable, the health, safety and welfare of all his employees.

Subsection (2) of *section 2* contains a number of more specific requirements (again qualified by the phrase 'so far as is reasonably practicable') relating to:

- provision and maintenance of plant and systems of work;
- use, handling, storage and transport of articles and substances;
- provision of information, instruction, training and supervision;
- places of work and means of access and egress;
- the working environment, facilities and welfare arrangements.

It is inconceivable that an employer could achieve compliance with these extremely wide-ranging duties without the application of effective management principles.

In addition, *subsection (3)* of *section 2* requires employers to prepare and revise a written statement of their policy with respect to the health and safety at work of his employees and the organisation and arrangements for carrying out that policy. This duty is examined in greater detail at **M2025** below and in the chapter entitled STATEMENTS OF HEALTH AND SAFETY POLICY.

Duties of employers to others

[M2004] *Section 3(1)* of *HSWA 1974* places a general duty on employers in respect of persons other than their employees:

> It shall be the duty of every employer to conduct his undertaking in such a way as to ensure, so far as is reasonably practicable, that persons not in his employment who may be affected thereby are not exposed to risks to their health or safety.

Persons protected by this requirement include contractors (and their employees), visitors, customers, members of the emergency services, neighbours, passers-by and other members of the general public – including trespassers, although the 'so far as is reasonably practicable' qualification is of particular relevance to them. (*Subsection (2)* of *section 3* places a similar duty on the self-employed.)

Once again, effective health and safety management provides the only way of complying with this duty.

The Management Regulations

[M2005] The *Management Regulations 1999 (SI 1999 No 3242)* came into operation on 29 December 1999. They modified the 1992 Management Regulations, which in turn had developed many of the principles already established by the *HSWA* in 1974. The Regulations make it quite clear that health and safety must be managed systematically, like any other aspect of an organisation's affairs. Some of the main requirements of the Regulations relate to:

- *Risk assessments* – to identify risks and related precautions (*Regulation 3*).
- *Management systems* – to ensure that precautions are implemented (*Regulation 5*).
- A *competent source of health and safety advice* being appointed (*Regulation 7*).
- *Emergency procedures* being developed (*Regulation 8*).

HSE booklet L21 contains the Regulations in full together with an Approved Code of Practice (ACOP) and related guidance. The 1999 Regulations incorporated previous amendments to the 1992 Regulations, relating to young persons, new or expectant mothers and fire risks.

The requirements of a number of the regulations (eg risk assessment) are dealt with more fully elsewhere in this publication (see RISK ASSESSMENT) and therefore receive only limited attention here.

Risk assessment (Regulation 3)

[M2006] Employers (and the self-employed) must make 'suitable and sufficient' assessments of the risks to their employees and others who may be affected by their work activities, in order to identify the measures necessary to comply with the law. The 'significant findings' of the assessment must be recorded by those with five or more employees.

Given the widely drawn nature of the requirements of *HSWA 1974* and other legislation, all risks must be included in this exercise. There is no need to repeat other 'assessments' required under more specific regulations, provided these are still valid. The *Management Regulations 1999 (SI 1999 No 3242)* require special consideration to be given to risks to young persons (under 18s) (*Reg 3(4)* and *(5)*) and to new and expectant mothers and their babies by virtue of *Regulation 16*.

Principles of prevention to be applied (Regulation 4)

[M2007] Preventive and protective measures must be implemented on the basis of principles specified in *Schedule 1* to the *Management Regulations 1999 (SI 1999 No 3242)*. These include:

- avoiding risks;
- evaluating risks which cannot be avoided;
- combating risks at source;

- replacing the dangerous by the non-dangerous or less dangerous;
- giving collective protective measures priority over individual measures, i.e. following good principles of health and safety management.

Health and safety arrangements (Regulation 5)

[M2008] Employers must make and give effect to appropriate 'arrangements' for the effective planning, organisation, control, monitoring and review of preventive and protective measures. Records should be kept of these arrangements by those with 5 or more employees.

While *Regulation 3* of the *Management Regulations 1999 (SI 1999 No 3242)* requires precautions to be identified, *Regulation 5* requires them to be put into practice effectively, utilising a systematic management cycle (see FIGURE 2).

Figure 2

The risk assessment cannot remain a meaningless piece of paper – *Regulation 5* of the *Management Regulations (SI 1999 No 3242)* requires it to be put into effect. Much of the remainder of this chapter is concerned with the practical application of these management principles.

Health surveillance (Regulation 6)

[M2009] Employers must ensure that employees are provided with appropriate health surveillance where this is identified by the risk assessment as being necessary. In most cases surveillance should already have been identified as being necessary to comply with other specific regulations (eg COSHH, asbestos). However, there may be other types of health risk not adequately covered by other legislation, eg colour blindness for train drivers or electricians; risks of vibration white finger.

Health and safety assistance (Regulation 7)

[M2010] Employers must appoint one or more 'competent' persons to assist them in complying with the law. The numbers of such persons and the time and resources available to them must be adequate for the size of and the risks present in the undertaking.

Competence in relation to this requirement might involve:

- knowledge and understanding of the work activities involved;
- knowledge of the principles of risk assessment and prevention;
- understanding of relevant legislation and current best practice;
- the capability to apply this in a practical environment;
- an awareness of one's own limitations (in experience or knowledge);
- the willingness and ability to learn.

Although formal health and safety qualifications give a good guide to competence, they are not an automatic requirement. Small employers are allowed to appoint themselves under this Regulation, providing they are competent. The *Management Regulations 1999 (SI 1999 No 3242)* state (in *Regulation 7(8)*) that there is a preference for the competent person to be an employee rather than someone from outside, eg a consultant.

Procedures for serious and imminent danger and for danger areas (Regulation 8)

[M2011] Procedures must be established 'to be followed in the event of serious and imminent danger to persons at work'. These procedures must be implemented where appropriate. Access must also be restricted to areas which are known to be dangerous.

Most employers are familiar with the process of preparing a fire evacuation procedure. Similar procedures must be established for dealing with other types of emergency which may have been identified through the process of risk assessment as being foreseeable. Such emergencies might include bomb threats, leaks or discharges of hazardous substances, runaway processes, power failure, violence, animal escape, severe weather etc. Competent persons must be appointed to implement evacuation where necessary.

Areas identified as being dangerous, eg because of the presence of electrical risks, chemical risks, potentially dangerous animals (or humans), must have access to them restricted to those who have received adequate instruction on the risks present and the precautions to be taken.

Contacts with external services (Regulation 9)

[M2012] Necessary contacts must be arranged with emergency services (and others if appropriate) regarding first aid, emergency medical care and rescue work, particularly in respect of the types of emergencies referred to in **M2011** above.

Information for employees (Regulation 10)

[M2013] Employers must provide employees with comprehensible and relevant information on:

- risks to their health and safety identified by risk assessment;
- the related preventive and protective measures;
- emergency procedures (and evacuation co-ordinators);
- risks notified to them by other employers in shared workplaces.

Regulation 10(2) of the *Management Regulations 1999 (SI 1999 No 3242)* requires parents or guardians of children (under the minimum school leaving age) to be provided with information on risks the child may be exposed to and the precautions identified as necessary by the risk assessment. This is of relevance in relation to school work experience programmes and part-time work by schoolchildren.

Co-operation and co-ordination (Regulation 11)

[M2014] Where two or more employers share a workplace (whether temporarily or permanently) each employer must:

- co-operate with the other employers where necessary for them to comply with the law;
- take all reasonable steps to co-ordinate their precautions with others;
- take all reasonable steps to inform the others of risks arising out of their work.

(Self-employed persons are also covered by this requirement.)

This type of co-operation and co-ordination was in effect already required under *sections* 2 and 3 of *HSWA 1974*. In many cases a main employer may already control the worksite with the other employers assisting. However, in some circumstances it may be necessary to identify an individual or organisation to carry out a co-ordinating role. Even those without employees in the workplace (eg landlords) may still be required to co-operate in order to meet their obligations under *section 4* of *HSWA 1974*.

Persons working in host employers' or self-employed persons' undertakings (Regulation 12)

[M2015] Host employers must ensure that employers whose employees are working in their undertaking are provided with comprehensible information on:

- risks arising out of or in connection with the host's activities; and
- measures being taken by the host to comply with the law.

Hosts have an additional duty to ensure that visiting employees are given instructions and information on the risks present and any emergency or evacuation procedures.

(Under *Regulation 12* of the *Management Regulations 1999 (SI 1999 No 3242)* the self-employed are treated as both employers and employees).

Over and above the requirements of *Regulation 11* of the *Management Regulations 1999 (SI 1999 No 3242)* and *HSWA 1974*, hosts are given twin duties of informing visiting employers and of *ensuring* that visiting employees are instructed and informed. Whether they do this themselves or require others to do it on their behalf will depend on local circumstances. *Regulation 12* and *Regulation 11* inter-relate closely with requirements of the *Construction, Design and Management Regulations. 2007 (SI 2007 No 320)* – see **M2023**.

Capabilities and training (Regulation 13)

[M2016] Employers must take into account employees' capabilities as regards health and safety in entrusting tasks to them. The intellectual and physical capabilities of individual employees must be considered as well as the demands of the task and the knowledge and experience required to perform it. Adequate health and safety training must be provided to employees:

- on recruitment into the undertaking;
- on exposure to new or increased risks because of:
  - transfer or a change in responsibilities;
  - new or changed work equipment;
  - new technology;
  - new or changed systems of work.

Training must:

- be repeated periodically where appropriate;
- be adapted to take account of new or changed risks;
- take place during working hours.

The requirements on training develop on the base provided by *section 2* of *HSWA 1974*. Many other sets of regulations also have specific training requirements. Even where employees have been trained in the same way their capabilities may differ and *Regulation 13* of the *Management Regulations 1999 (SI 1999 No 3242)* requires employers to take ability into account. It also identifies the possible need for refresher training.

Employees' duties (Regulation 14)

[M2017] Employees must use machinery, equipment, dangerous substances etc in accordance with the training and instructions they have been provided with. They must also inform their employer (or his representative) of situations representing serious and immediate danger and of any shortcomings in the employer's protection arrangements. To a large extent, this requirement duplicates those already contained in *section 7* of *HSWA 1974* and other legislation. A failure to report a dangerous situation is likely to be considered a serious 'omission' under *section 7*.

Temporary workers (Regulation 15)

[M2018] *Regulation 15* of the *Management Regulations (SI 1999 No 3242)* creates several detailed requirements in relation to the use of fixed-term contract or agency staff. It should always be remembered that temporary workers are still affected by most of the other requirements of these Regulations.

Risk assessment in respect of new or expectant mothers (Regulations 16, 17 and 18)

[M2019] Where there are:

- women workers of childbearing age, and
- the work could involve risk to the mother or baby, eg processes, working conditions, physical, biological or chemical agents (see Annexes I and II of EC Directive 92/85/EEC),

the risk assessments made under *Regulation 3(1)* of the *Management Regulations 1999 (SI 1999 No 3242)* must take account of such risks.

For individual employees the employer must, *if necessary*:

* alter working conditions or hours of work, or *if necessary*,
* suspend the employee from work,

in order to avoid such risks.

Regulation 17 of the *Management Regulations 1999 (SI 1999 No 3242)* allows doctors (or midwives) to require employers to suspend women from night work where necessary for health and safety reasons. *Regulation 18* of the 1999 Regulations states that employers need not take action under *Regulations 16* and *17* in respect of individual employees unless the employee has notified them in writing of her pregnancy or that she has given birth within the previous 6 months or is breast feeding. (The employer may request confirmation of pregnancy through a certificate from a doctor or midwife.) Issues relating to new and expectant mothers are dealt with in more detail in the chapter entitled VULNERABLE PERSONS.

Protection of young persons (Regulation 19)

[M2020] Employers must ensure young persons are protected from risks which are a consequence of their:

* lack of experience;
* absence of awareness of risks;
* lack of maturity,

eg by appropriate training, supervision, restrictions (see also *Regulation 3* of the *Management Regulations 1999 (SI 1999 No 3242)*).

Young persons may not be employed for work:

* beyond their physical or psychological capacity;
* involving harmful exposure to toxic, carcinogenic, genetically damaging agents etc;
* involving harmful exposure to radiation;
* involving risks they cannot recognise or avoid (due to insufficient attention, lack of experience or training);
* in which there are health risks from extreme cold or heat, noise or vibration.

Such work may be carried out if it is necessary for the training of young persons who are not children, providing they will be supervised by a competent person and risks are reduced to the lowest level reasonably practicable.

This topic is dealt with more fully in the chapter VULNERABLE PERSONS.

Other requirements

[M2021] The remaining requirements under the *Management Regulations 1999 (SI 1999 No 3242)* cover the following:

Regulation 20–Exemption certificates
* (For the armed forces in the interests of national security)

Regulation 21–Provisions as to liability
- (Of employers in relation to criminal proceedings)

Regulation 22–Restriction of civil liability for breach of statutory duty
- (A breach of a duty imposed on an employer by these Regulations does not confer a right of action in civil proceedings insofar as that duty applies for the protection of persons not in his employment. The previous exclusion of right of action under the Regulations by employees was removed in 2003.)

Regulation 23–Extension outside Great Britain
- (The Regulations apply to various offshore activities)

Regulations 23–27
- (Contain minor amendments in respect of first aid, offshore installations and pipeline works, mines and construction)

Regulations 28–30
- (Deal with administrative matters, revocations and transitional arrangements)

Construction (Design and Management) Regulations 2007

[M2022] The *Construction (Design and Management) Regulations 2007 (SI 2007 No 320)* (CDM 2007) came into operation on 6 April 2007, replacing the 1994 CDM Regulations, the Construction (Health, Safety and Welfare) Regulations 1996 and the Construction (General Provisions) Regulations 1961. The Regulations are in five parts:

(1) Introduction (dealing with interpretation of terms and application of the regulations)
(2) General Management Duties applying to construction projects
(3) Additional Duties where project is notifiable
(4) Duties relating to health and safety on construction sites (dealing with a variety of practical health and safety topics relating to construction work)
(5) General (covering civil liability, enforcement etc.)

The definition of 'construction work' contained in the regulations is extremely broad and would include such minor work as replacing a broken window or installing a new electrical socket outlet in a building. The management duties in Part 2 of the regulations apply to all such work, as also do the practical requirements set out in Part 4.

Where a project is notifiable to the enforcing authorities (i.e. the construction phase is likely to involve more than 30 days or 500 person days of construction work) the additional management duties of Part 3 apply.

(The requirements of CDM 2007 are summarised below but full details of the regulations, together with an ACOP and guidance are contained in HSE book L144.)

Management Duties applying to all construction projects

[M2023] Part 2 of CDM 2007 contains requirements applying to all construction projects.

Sub-section (1) of Regulation 4 states that:

No person on whom these Regulations place a duty shall—

(a)   appoint or engage a CDM co-ordinator, designer, principal contractor or contractor unless he has taken reasonable steps to ensure that the person appointed or engaged is competent;

(b)   accept such an appointment or engagement unless he is competent;

(c)   arrange for or instruct a worker to carry out or manage design or construction work unless the worker is—

(i)   competent, or

(ii)   under the supervision of a competent person.

(This requirement relates to competence in respect of compliance with health and safety legislation.)

Consequently all parties managing construction activities must ensure that all those they appoint, engage or instruct in respect of construction work are competent — whether they employ them directly or obtain their services from elsewhere.

Regulations 5 and 6 require persons concerned in construction work to co-operate with each other and co-ordinate their activities, largely duplicating requirements in the Management Regulations (see **M2014**). Similarly Regulation 7 requires that the General Principles of Prevention contained in Schedule 1 of the Management Regulations are applied in relation to construction projects (See **M2007**).

Where there is more than one client involved in a project, Regulation 8 allows a written agreement between clients so that only one (or more) client carries out the majority of the duties of the client in respect of the regulations.

Specific duties are placed on clients, designers and contractors in relation to all construction projects, of whatever size.

Clients

[M2024] Clients are required by Regulation 9 to ensure that suitable management arrangements are in place for the project and that these arrangements are maintained and reviewed.

Regulation 10 requires clients to provide designers and contractors with relevant information, particularly about the site and about the proposed use of the structure as a workplace.

Designers

[M2025] Regulation 11 requires that designers do not commence work on a project unless the client is aware of his duties under the regulations. In preparing or modifying designs for construction work designers must, so far as is reasonably practicable, avoid foreseeable risks to the health and safety of any person

- carrying out construction work;
- liable to be affected by such work;
- cleaning windows or transparent or translucent surfaces;
- maintaining permanent fixtures and fittings;
- using a structure as a workplace.

Designers must provide relevant information about their designs to clients, other designers and contractors to assist them in complying with their duties. Where designs are prepared or modified abroad, Regulation 12 places the responsibilities set out in Regulation 11 with the person commissioning the design or the client.

Contractors

[M2026] Under Regulation 13 contractors also must not carry out any construction work unless the client is aware of his duties. They must plan, manage and monitor construction work in a way which ensures that, so far as is reasonably practicable, it is carried out without risks to health and safety.

Contractors must also provide workers under their control with information and training necessary for the work to be carried out safely and without risks to health. They must not begin work on a site unless reasonable steps have been taken to prevent access by unauthorised persons.

Additional duties where the project is notifiable

[M2027] The duties in Part 3 of CDM 2007 come into operation where a project is notifiable (see **M2022** above). Full details of these requirements are contained in the HSE book L144, but because of their importance to the management of larger construction projects, these regulations are summarised below.

Clients
- Regulation 14 requires that for notifiable projects the client appoints in writing a CDM co-ordinator and a principal contractor (see below). If no such appointment is made the client is deemed to hold either or both of these roles.
  Clients are required by Regulation 15 to provide the CDM co-ordinator with relevant information. They must ensure that the construction phase does not start unless the principal contractor has prepared a satisfactory 'construction phase plan' (Regulation 16). Clients must also provide the CDM co-ordinator with information in their possession necessary for the 'health and safety file' for the structure and subsequently keep the 'health and safety file' available for those who may need it and revise it as appropriate (Regulation 17).

Designers
- Under Regulation 18 designers must not commence work on a project (other than initial design work) unless a CDM co-ordinator has been appointed. They must also provide the CDM co-ordinator with sufficient information to assist him in complying with his duties.

Contractors
- Contractors have additional duties under Regulation 19, particularly in respect of providing the principal contractor with information about relevant parts of their risk assessments and about any sub-contractors they may engage, Additionally they must comply with directions from the principal contractor and any site rules, and take reasonable steps to work in accordance with the construction phase plan.

CDM Co-ordinator
- The CDM co-ordinator has slightly enhanced duties compared to the Planning Supervisor under the 1994 CDM Regulations. These are set out in Regulation 20 and include:
  - giving advice and assistance to the client (particularly in respect of whether the 'construction phase plan' is at a satisfactory stage for construction work to start);
  - ensuring suitable arrangements for co-ordinating health and safety during planning and preparation activities;
  - liasing with the principal contractor;
  - identifying and collecting pre-construction information;
  - providing this in a convenient form to designers and contractors engaged by the client;
  - taking reasonable steps to ensure designers comply with their duties and co-operation between designers and the principal contractor;
  - preparing (or reviewing and updating) the 'Health and Safety File';
  - passing the 'Health and Safety File' to the client at the end of the construction phase.

  Regulation 21 requires the CDM Co-ordinator to ensure that the project is notified to the HSE (or the Office of Rail Regulation).

Principal Contractor
- The principal contractor has many specific duties set out in Regulation 22, particularly in planning, managing and monitoring the construction phase to ensure it is carried out without risk to health and safety, so far as is reasonably practicable. He is required by Regulation 23 to prepare a suitable 'construction phase plan' to ensure that work is planned, managed and monitored so that the construction phase can be started. This plan must be updated, reviewed, revised and refined as the construction work progresses and the principal contractor must ensure that it is implemented.

  Regulation 24 places duties on the principal contractor in respect of co-operation and consultation with workers.

(Guidance on the content of the 'Health and Safety File' and 'construction phase plan' is provided in HSE book L144.)

Accident causation

[M2028] The domino model of accident or loss causation has been used by many to demonstrate how accidents result from failures in health and safety management. The model is based on five dominos standing on their edges in a line:

- The *fifth and final domino* in the line represents all the *potential losses* which can result from an accident or incident – personal injury, damage to equipment, materials, premises, disruption of business etc;
- The *fourth domino* in the line represents the specific *accident or incident* which occurred – for example a failure of an item of lifting equipment;
- This failure could have a number of *immediate causes* which are represented by the *third domino* – overloading or misuse of the lifting equipment, an undetected defect etc;
- Behind these immediate causes are a variety of *basic or underlying causes*, represented by the *second domino* – an absence of training, a lack of supervision, failure to conduct regular inspections and thorough examinations of the lifting equipment;
- These in turn are indicative of a *lack of management control* which is represented by the *first domino* – in a well-managed workplace appropriate training is provided, there is effective supervision, and inspection and thorough examination procedures are in place.

If there is effective management control then all the dominoes stay in position. However, if management control breaks down then the other dominoes will topple, eventually causing an accident or incident and the resultant losses.

Health and safety management systems

[M2029] The vast majority of occupational safety and health ('OSH') management systems (and environmental management systems) follow the guidance given in HSE's publication 'Successful health and safety management', (HSG65) and embellished in BS 8800:2004, Occupational Health and Safety Management Systems Guide, which outlines a system based on:

- establishing a policy with targets and goals;
- organising to implement it,
- setting forth practical plans to achieve the targets/goals;
- measuring performance against targets/goals; and
- reviewing performance.

The whole process is overlaid by auditing. This outline may be remembered via the use of the mnemonic: POPIMAR:

Policy
Organising

Planning
Implementing
Measuring/monitoring
Auditing
Reviewing

These key elements are linked in the form of a continual improvement loop (see FIGURE 3).

Figure 3: Continual improvement loop

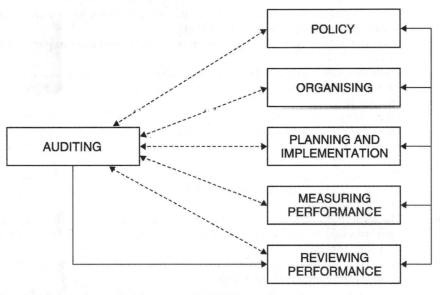

solid links = control   dashed links = information

The more times organisations go round the loop, the better will be their OSH management performance.

Within BS 8800, two alternative approaches are suggested:

*   the HSG65 approach;
*   the ISO 14001 approach (BSI BS EN ISO 14001: 2004, Environmental Management Systems – Requirements with guidance for use).

BS 8800 itself is merely a guide; it is *not* a certifiable standard; whereas ISO 14001 *is* a certifiable standard. To date, there is no equivalent in health and safety to the ISO 9000 standard for quality systems or the ISO 14001 standard for environmental management systems. OHSAS 18001 is not a recognised certifiable ISO standard, but an assessment specification for occupational health and safety management systems.

BS 8800 builds on the HSG65 approach (outlined above) but starts off the loop with an initial status review in order to establish 'Where are we now?' BS 8800 also includes an alternative approach aimed at those organisations wishing to base their OSH management system not on HSG65 but on BS EN ISO 14001, the environmental management systems standards which again incorporates an initial status review.

A review checklist

[M2030] In order to undertake an initial status review, use may be made of the following checklist:

- Does a current OSH policy exist for the organisation/location?
- Is the policy up to date (i.e. not more the three years old)?
- Has the policy been signed (and dated) by a director/senior manager who has (site) responsibility for OSH?
- Does the policy recognise that OSH is an integral and critical part of the business performance?
- Does the policy commit the organisation/location to achieve a high level of OSH performance with legal compliance seen as a minimum standard?
- Does this commitment include the concept of continual improvement?
- Does the policy state that adequate and appropriate resources, i.e. time, money, people, will be provided to ensure effective policy implementation?
- Does the policy allow for the setting and publishing of OSH objectives for the organisation/location and/or for individual directors, managers and supervisors?
- Does the policy clearly place the prime responsibility for the management of OSH on to line management, from the most senior executive to first line supervision?
- Has the policy been effectively brought to the attention of *all* employees/agency staff/temporary employees?
- Are copies of the policy on display throughout the organisation/location?
- Is the policy understood, implemented and maintained at all levels within the organisation/location via the use of suitable arrangements?
- Does the policy ensure that employee involvement and consultation takes place in order to gain their commitment to it and its implementation?
- Does the policy require that it gets periodically reviewed (at least every three years) in order to ensure that management and compliance audit systems are in place?
- Does the policy require that all employees at all levels, including agency staff and temporary employees, receive appropriate training to ensure that they are competent to carry out their duties?
- Does the policy contain a section dealing with the organisational framework – people and their duties – so as to facilitate effective implementation?
- Does the policy contain a section dealing with the need for risk assessments to be undertaken – both general and specific – their significant findings acted upon, and an assessment record-keeping system maintained?

- Does the policy contain a section dealing with the arrangements – systems and procedures – by which the policy will be implemented on a day-to-day basis?
- Does the policy contain a section dealing with the monitoring/measurement of OSH performance?
- Does the policy contain a section dealing with planning for and reviewing the organisation's policy implementation and overall OSH performance? (This would normally be in the arrangements section.)
- The HSE has produced a free leaflet (INDG 343) 'Directors' responsibilities for health and safety' which sets out the critical role directors play in the effective management of health and safety. Further publications are in preparation in anticipation of legislation on 'Corporate Manslaughter' see M2001.

Policy

Requirements under HSWA 1974

[M2031] *Section 2(3)* of *HSWA 1974* states:

> Except in such cases as may be prescribed, it shall be the duty of every employer to prepare and as often as may be appropriate revise a written statement of his general policy with respect to the health and safety at work of his employees and the organisation and arrangements for the time being in force for carrying out that policy, and to bring the statement and any revision of it to the notice of his employees.

Regulation 2 of the *Employers' Health and Safety Policy Statements (Exception) Regulations 1975 (SI 1975 No 1584)* excepts from these provisions any employer who carried on an undertaking in which for the time being he employs less than five employees. In this respect regard is to be given only to employees present on the premises at the same time (*Osborne v Bill Taylor of Huyton Ltd* [1982] IRLR 17).

Contents of policy statements

[M2032] The policy is the first stage in the POPIMAR management system and the statement must reflect that. Reference to its contents was made in the checklist provided in M2030 above. Some contents will be common to all organisations, whatever their size or work activity. Typical of these are the following:

- A commitment to achieving high standards of health and safety (not only in respect of employees but also others who may be affected by the organisation's activities).
- Reference to the importance of people to the organisation and the organisation's moral obligations.
- A statement that compliance with legal obligations must be achieved (although these should be regarded as a minimum standard).

- Effective management of health and safety must be a key management objective (it must continue right through the line of management responsibility).
- Reference to the provision of resources necessary to implement the policy.
- A commitment to consulting employees on health and safety matters.

Where the organisation is fully committed to the type of management system represented by POPIMAR this should be reflected in the policy statement by:

- A reference to achieving continuous improvement in health and safety standards.
- Reference to the setting of health and safety objectives and the allocation of responsibility for their implementation (whether on an annual basis or otherwise).

The statement must be signed by an appropriate director or senior manager to reflect a high level commitment to the policy and must be dated. Review of the policy as a whole, where appropriate, is a legal requirement, as well as a key component of the POPIMAR system. An out of date policy statement signed by someone who has long since left the organisation is not indicative of a real commitment to effective health and safety management.

Organisation

[M2033] The HSE booklet 'Successful health and safety management' (HSG65) refers to the need to establish a 'health and safety culture' which it defines as 'the product of individual and group values, attitudes, perceptions, competencies and patterns of behaviour that determine the commitment to, and the style and proficiency of, an organisation's health and safety management.' Put in other words it is 'the way we do things here'. In order to achieve such a culture, activities are necessary on four fronts (often called the four Cs):

- establishing and maintaining *control* in respect of health and safety matters;
- securing *co-operation* between individuals and groups;
- achieving effective *communication*;
- ensuring the *competence* of individuals to make their contribution;

In effect the four Cs encompass both the organisation and arrangements which are required to implement the health and safety policy statement.

Control

[M2034] Management must take responsibility and provide clear direction in respect of all matters relating to health and safety. Key elements of control include:

Clear definitions of responsibilities
- Responsibilities must be allocated to all levels of line management as well as persons with specialist health and safety roles. These responsibilities should be set out in the health and safety policy (see the chapter STATEMENTS OF HEALTH AND SAFETY POLICY) and also included in individuals' job descriptions.

Nomination of a senior person to control and monitor policy implementation
- This should ensure that health and safety is seen to be important to the organisation. The senior figure must be involved in the development of plans and particularly in monitoring their implementation. They may also be involved in formal means of consultation with the workforce such as health and safety committees.

Establishment of performance standards
- The old adage of 'what gets measured gets done' applies just as much to health and safety as to other aspects of management. Performance standards must be incorporated into health and safety plans and also into health and safety related procedures and arrangements. Standards must specify:
  - what needs to be done;
  - who is responsible for carrying it out;
  - when (or how often) the task must be carried out;
  - what is the expected result,
  - For example, an investigation must be made into all accidents requiring first aid treatment. This must be carried out by the injured employee's supervisor and a report submitted on the company report form within 24 hours.

Adequate levels of supervision
- The precise level of supervision will depend on the risks involved in the work and the competence of those carrying it out. Supervisors are key players in informing, instructing and, to an extent, training the members of their teams. They also act as an important point of reference on health and safety matters, particularly on whether it is safe to continue with a task.

Co-operation

[M2035] Health and safety must become everyone's business – there must be real commitment to health and safety throughout the organisation. Commitment comes where there is a genuine consultation and involvement. This can partially be achieved through formal means of consultation, such as safety representatives and health and safety committees. Hazard report books and suggestion schemes can also play their part. However, much can also be achieved informally by line managers and health and safety specialists by simply 'walking the job' and talking to those at work about the health and safety aspects of their job.

The HSE have taken a number of initiatives to encourage the involvement of workers in health and safety management. Their website contains a 'worker involvement' site which provides tools to aid businesses in increasing the ways in which employers and workers can co-operate in improving health and safety standards.

Communication

[M2036] Good communication is a key element of all aspects of running any organisation. Employees need to be made aware of what is being done to improve health and safety standards (and why), and also of their own role in achieving this improvement. Information needs to move in various directions.

Incoming information
- The organisation must keep abreast of developments in the outside world, eg:
  — potential and actual changes in legal requirements;
  — technical developments in respect of control measures;
  — awareness of new or changed risks;
  — good practices in health and safety management.

Information circulation within the organisation
- Information will need to be passed down from management levels but there must also be means of passing information up the line and across the organisation. Important information to circulate includes:
  — the health and safety policy document itself;
  — senior management's commitment to implementing it;
  — current plans for achieving improvement;
  — progress reports on previous plans;
  — other performance reports, eg accident frequency rates;
  — proposed changes to health and safety procedures and arrangements;
  — employees' views on such proposals;
  — comments and ideas for improvement;
  — details of accidents and incidents, together with lessons learned as a result.

  Information can be communicated in written form (eg policy documents, booklets, newsletters, notices, e-mails etc.) but face to face communication is also important. This can include health and safety committee meetings, health and safety discussions or briefings at other meetings, tool box talks. Once again, 'walking the job' by management can be invaluable in achieving two way communication in an informal setting.

  Increasing use of company 'Intranets' provides a means of ensuring that health and safety related procedures and documents are readily available to those affected by them.

Outgoing information
- Health and safety related information may also need to be communicated outside the organisation, for example:
  — formal reports to enforcing authorities (eg as required by RIDDOR);

— technical information about products supplied or manufactured (eg material safety data sheets, noise and vibration levels);
— information to other organisations involved in the same business (about potential risks or successful control measures);
— emergency planning information (to the emergency services, local authorities and neighbours who may be involved).

Competence

[M2037] All employees need to be capable of carrying out their jobs safely and effectively. What is required to achieve this will vary depending upon the individual's role in the organisation. Important elements include:

Recruitment and appointment arrangements
• Employees must have appropriate levels of intellectual and physical abilities to carry out their jobs. Whilst knowledge and experience can be acquired subsequently, certain levels may be a pre-requisite for some posts.

Training
• Systems should be in place to ensure that staff receive adequate training for their work, particularly following their recruitment, transfer or promotion. Arrangements will also be necessary to train those given particular responsibilities, eg first aiders, fire wardens. Other training needs may also be identified as a result of audits, job appraisals, changes in technology, legislative changes etc. Refresher training may also be necessary for some staff.

Succession and contingency planning
• Account must be taken of the need to have sufficient competent staff available at all times in some safety-critical roles. Staff may be absent for short periods, eg holidays or sickness, for longer periods due to serious injury or illness, or leave permanently, whether for other jobs or on retirement.

CDM 2007 (see **M2023**) place specific responsibilities on duty holders under the regulations to ensure that those they appoint or engage to carry out or manage construction work (and related design work) are competent for the purpose.

The chapter TRAINING AND COMPETENCE IN OCCUPATIONAL SAFETY AND HEALTH provides more information on competence generally.

Planning

[M2038] -

'Failure to plan is planning to fail'.

Effective planning results in an occupational safety and health management system ('OSHMS') which controls risks, reacts to changing demands and sustains a positive OSH culture. Such planning involves designing, developing and installing risk control systems and workplace precautions commensurate to the risk profile of the organisation. It should also involve the operation, maintenance and improvements to the OSHMS to suit changing needs, hazards and risks.

Planning is the first stage of the PDCA cycle: Plan, Do, Check, Act, which is the cornerstone of the vast majority of management systems.

Plan:	Policies, organisation, hazard identification, risk assessments, change management
Do (Implement):	The implementation of the management plans and processes
	Worker consultation is a crucial aspect of planning and doing which helps to create a positive culture
Check (Monitor):	Inspections, audits, exposure monitoring, health surveillance, attendance/absence monitoring, accident reporting and investigation, statistical/causal analyses, testing of emergency arrangements/procedures
Act (Audit/Review):	Take control action based on monitoring activities, keep track of developments/corrective measures, continually improve processes and plans

The planning process

[M2039] The planning process involves:

- Setting objectives.
- Risk control systems.
- Legal and other requirements.
- OSH management arrangements.
- Emergency plans and procedures.

Setting objectives

This relies on having an annual plan for the organisation with clearly outlined individual and collective 'SMART' OSH targets. SMART refers to the need for the targets to be: specific, measurable, achievable, realistic, trackable.

The annual OSH plan should:

- outline individual and collective action plans for the current year;
- target continual improvement in OSH performance;
- fix accountabilities via SMART targets;
- be linked to job descriptions/performance appraisals;
- be adequately resourced – time, money, people.

Risk control systems

These should ideally evolve from an ongoing review of all general and specific risk assessments in place within the organisation. Indeed, as part of the annual plan, all current risk assessments should be reviewed, updated, and recommunicated to all concerned.

Risk should be reduced to tolerable or acceptable levels via an ongoing process of hazard identification and risk assessment within the working environment. Suitable risk control systems should be developed and implemented with the goal of hazard elimination and risk minimisation. As part of the process all significant findings/risk assessments should be recorded in writing.

Legal and other requirements

These need to be incorporated into the annual plan in order to establish and maintain arrangements which ensure that all current and emerging OSH legal and other requirements are complied with and understood throughout the organisation.

Bench-marking within the organisation's sector can ensure that best practice and performance measures are adopted and communicated.

OSH management arrangements

These should reflect both the needs of the business and the overall risk profile and should cover all elements of the POPIMAR model. Specific arrangements should include:

- plans, objectives, personnel, resources;
- operational plans to control risks;
- contingency plans for emergencies;
- planning for organisational arrangements;
- plans covering change management (positive OSH culture);
- plans for interaction with third parties (eg contractors);
- planning for performance measurement, audits and reviews;
- implementing corrective actions.

Emergency plans and procedures

These should be in place and reviewed as part of the annual planning processes. Specific serious and imminent dangers should have been identified through risk assessments and procedures for dealing with them must be developed in accordance with the *Management Regulations (SI 1999 No 3242), Reg 8* (see **M2011**).

Documented emergency procedures must have been communicated effectively to all those affected by them, and necessary emergency equipment must be readily available. Situations requiring such precautions might include fire, flood, structural collapse/subsidence, chemical leak or spillage, transport accident, bomb scare, explosion, etc.

Implementation

[M2040] In order to ensure smooth implementation of the OSH policy and management system, it is imperative to have sufficient health and safety related records and documentation as this is a vital element in organisational communication and continual improvement. However, in order to ensure efficiency and effectiveness of the OSHMS, documentation should be kept to realistic proportions and should be tailored to suit organisational needs. The detail should be in proportion to the level and complexity of the associated hazards and risks.

Specifically, some of the key risk areas requiring implementation of commensurate risk control systems are:

- employee selection and training;
- general/specific risk assessments;
- safe systems of work/permit to work systems;
- management of third parties on site;
- workplace and safe access;
- work equipment;
- fire precautions;
- hazardous and dangerous substances;
- noise and vibration;
- first aid;
- issue and use of personal protective equipment.

These are generally referred to as 'arrangements' in most organisational OSH policies.

As outlined in the section on Planning above at **M2032**, this 'doing' section requires the use of action plans and SMART targets in order to ensure that the written down standards and procedures are actually achieved in practice on a continual basis. The phrase 'Mind the gap(s)' springs to mind!

Most organisations split their arrangements into general and specific. General arrangements require implementation throughout the organisation whereas specific arrangements only operate at certain locations or for certain tasks.

Within the implementation section of the OSHMS, the risk control systems relevant to the organisation should be briefly described, together with who is responsible for their implementation and who should be involved in the implementation process.

In larger organisations the detail may well be contained in a separate OSH manual which is referred to but is not part of the OSH policy. Intranet systems allow documents to be linked directly.

Monitoring

What gets measured gets done!

[M2041] Monitoring of the OSHMS is vital to ensure continual improvement. Monitoring procedures fall into two distinct categories:

- proactive;
- reactive.

Proactive monitoring takes place before an unwanted event takes place (accident, disease, damage etc.) and includes audits, inspections and specific surveys or checks.

Reactive monitoring is always after the event and includes accident and incident reporting and investigation, numerical/causal analysis, compilation of accident and incident statistics, and adherence to the RIDDOR '95 requirements on accident notification.

OSH inspections – proactive monitoring
- This should involve competent individuals going around the workplace on a regular (ideally monthly) basis in order to identify hazards – unsafe acts as well as unsafe conditions – with a view to their rectification.

  Such inspections should preferably be carried out jointly by management and workforce representatives and should always involve talking to the people at risk in the workplace being inspected. This may help to identify unsafe practices, near misses and shortcomings in the established systems and procedures. In most cases use will be made of a specific inspection checklist which ideally should be compiled by the people undertaking the inspection.

  In all cases there needs to be a post-inspection action plan prepared which clearly indicates who is going to take remedial action and by when. There should also be full communication of the findings and recommendations from all such inspections.

Specific surveys and checks – proactive monitoring
- Such surveys and checks may involve: housekeeping inspections; checks on adherence to agreed safe systems of work, permit to work systems, site safety rules; utilisation of personal protective equipment; documentation checks – risk assessments, statutory inspection records, maintenance and test records; training records, etc..

Accident and incident reporting and investigation systems – reactive monitoring
- This should encompass all accidents, damage and near-miss incidents and occupational ill-health in order to ensure RIDDOR '95 compliance and an effective accident database with which to plan further accident reduction strategies. However, even more importantly, all accidents and incidents

  should be promptly investigated in order to identify their causes and to implement commensurate control measures so as to prevent a recurrence. Most organisations will have set up an internal reporting and investigation system to capture accident and incident data. Any investigation culture should be positive, i.e. *not* blame-apportioning or fault-finding.

  The main purposes of accident and incident investigation are therefore to:
  - discover the immediate and underlying causes (note: plural!);
  - prevent recurrences;
  - minimise future legal liability;
  - collect sufficient data on which to base future plans;
  - identify accident/illness/absence trends;
  - ensure RIDDOR '95 compliance;
  - maintain records for organisational purposes;

— continually improve the OSHMS performance.

Audit

[M2042] Formal OSH audits should be undertaken at regular intervals, (possibly annually), by trained and competent auditors, sometimes external to the organisation. The main objectives of any audit are to reduce risks, with avoidance being the best strategy, and to ensure continual improvement of the OSHMS.

All audit findings (successes as well as shortcomings) should be highlighted and communicated to all concerned, including worker representatives. Any recommendations for improvement should be prioritised and allocated to named individuals for action within agreed, finite time-scales. The progress on action completion should also be followed up and communicated.

The audit needs to compare actual OSH performance in the workplace against established standards – mind the gaps! – such as:

- legislative compliance;
- HSE guidance;
- best practice (bench-marking);
- national/international standards.

Essentially the OSH audit, as with financial audits, should be a deep and critical appraisal of all elements (POPIMAR) of the OSHMS and should be seen as being additional to routine monitoring and reviews. Preferably, the auditors need to be independent of the area/activity/location being audited and the audit checklist should be tailored to suit the activities, risks and needs of the organisation.

The audit scope should address key areas such as:

- is the overall OSHMS capable of achieving and maintaining the required standards?
- is the organisation fulfilling all OSH obligations?
- what are the strengths and weaknesses of the OSHMS?
- is the organisation actually doing and achieving what it claims or believes?
- are the audit results acted upon, communicated and followed up?
- are the audit reports used in management reviews?

In 2006 the HSE introduced their Corporate Health and Safety Performance Index (CHaSPI) which through a series of structured questions allows organisations to evaluate their OSHMS. CHaSPI can be accessed via the HSE website.

Review

[M2043] Reviews of the performance of the OSHMS generally fall into two categories:

- periodic status review ('PSR');
- management review.

The main purposes of the PSR are to make judgements on the adequacy of performance and to ensure that the right decisions are made and taken in connection with the nature and timings of the remedial actions.

Specifically, the PSR should consider:

- overall OSHMS performance;
- performance of individual elements of the management system;
- audit findings;
- internal and external factors such as changes in organisational structure, changing business activities and demands, pending legislation, introduction of new technology;
- anticipated future changes;
- information gathered from proactive and reactive monitoring.

The PSR should be a continual process which includes management and supervisory responses to system shortcomings such as non-implementation of agreed risk control systems, substandard performances, or failure to assess the impact of forward plans and objectives on their part of the organisation. Successes should also be highlighted.

The effectiveness of the PSR can be enhanced by clearly establishing responsibilities for implementing the remedial actions identified and by setting deadlines for action completion, using the SMART approach — see **M2039**. The PSR needs to be auditable, trackable, fully communicated and followed up.

The management review should be undertaken annually by the board/senior management. It should ensure that the OSHMS is suitable, adequate, effective and efficient, and that it satisfies and achieves the aims and objectives as stated in the OSH policy.

The review should be based on the audit results and on the periodic status review reports. Specifically it should:

- establish new OSH objectives for continual improvement;
- develop OSH annual plans;
- consider changes to OSHMS arrangements;
- review findings

which should then be documented, communicated to stakeholders via annual reports and followed up to ensure implementation and continual improvement.

Continual improvement

[M2044] Continual improvement is at the heart of the OSHMS. It essentially is a fundamental commitment to manage OSH risks proactively so that accidents and ill-health are continuously reduced (system effectiveness) and the system achieves its desired aims using less resources (system efficiency).

The International Labour Office Guidelines on Occupational Safety and Health Management Systems (ILO, 2001) includes the following model which clearly illustrates the importance of the goal of continual improvement.

Figure 4

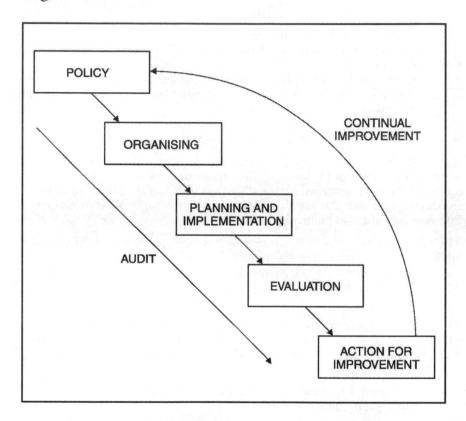

The route to continual improvement has been highlighted throughout the course of this chapter. It involves the use of SMART targets, the setting of OSH Key Performance Indicators ('KPIs'), the use of performance appraisals which include individual and collective OSH performance reviews, and clearly establishing OSH responsibilities at all levels within the organisation.

The goal of continual improvement should manifest itself in a number of ways:

- better year on year results – less injuries, diseases, damage, near misses;
- steadily improving results using less resources;
- better targeted and more efficient and effective OSHMS;
- indications of breakthrough performances;
- a more positive, proactive OSH culture;
- OSHMS elements which are more comprehensive and easier to understand;

In order to achieve continual improvement, the following factors should be taken into account:

- regular OSH audits;
- statistical information;
- bench-marking;
- industry/sector guidelines;
- ownership of health and safety matters:
  — managers, supervisors;
  — workers' representatives;
- results of proactive monitoring in relation to targets;
- effective and efficient operation via workforce involvement;
- generation and evaluation of ideas for improvement(with regular feedback to originators);
- fully resourced, implemented and monitored action plans (with SMART objectives);
- widespread and effective OSHMS training;
- appropriate OSHMS documentation and records.

POPIMAR in action

[M2045] The example below shows how the POPIMAR principles can be followed in practice. It relates to the introduction into an organisation of a permit to work ('PTW') system to control certain high risk activities.

Policy

The PTW system will become part of the 'arrangements' for implementing the health and safety policy statement. The PTW procedure (see below) is likely to contain a general statement about the importance of ensuring that appropriate controls are in place to minimise the risks associated with potentially high risk activities.

Organisation

This will primarily centre around the development of a formal PTW procedure which will involve:

Control
- The procedure must identify:
  — what activities will be controlled by the PTW;
  — who will issue (and cancel) PTWs;
  — the timing sequence for issuing and canceling individual PTWs;
  — the format of the PTW form.

Co-operation
- The draft procedure must be reviewed with those likely to be involved in issuing and working under PTWs. This could involve an ad hoc consultative group in addition to safety representatives and the health and safety committee.

Communication

*   A wider section of the workforce must be informed about the proposed procedure and the reasons for its introduction.

Competence

*   Those who are to issue the PTWs must receive formal training in the procedure and use of the PTW form. There should also be a formal appointment process to ensure that they have successfully assimilated the training and can relate it to their practical work situations. Arrangements for this should be referred to in the procedure.

Planning and implementation

Once the procedure has been finalised the timescales and responsibilities must be identified for achieving the various component parts of its implementation. This will include:

*   delivery of training to PTW issuers;
*   formal appointment of PTW issuers (after an appropriate test process);
*   printing and distributing the PTW forms;
*   obtaining any necessary equipment (eg isolation locks, atmospheric testing equipment);
*   announcement of the formal introduction of the PTW procedure.

Measuring performance

Monitoring of the PTW can be carried out in various ways:

*   a survey soon after the procedure's introduction;
*   ongoing monitoring during formal health and safety inspections;
*   periodic mini-audits of the procedure (eg by health and safety specialists or managers).

Auditing

The operation of the PTW procedure would be included in any general OSH auditing activities.

Reviewing performance

Both monitoring and audit findings must be reviewed by appropriate individuals and groups such as:

*   health and safety specialists;
*   the health and safety committee;
*   senior management;
*   ad hoc review teams.

Where satisfactory standards are not being achieved, the reviewers must consider how the POPIMAR sequence can best be applied to taking remedial action.

References

[M2046] (All HSE publications)

L 21 Management of Health and Safety at work

L 144 Managing health and safety in construction

HSG 65 Successful health and safety management

INDG 343 Directors' responsibilities for health and safety.

Managing Work-related Road Safety

Roger Bibbings and Andrea Oates

Overview

[M2101] According to the Royal Society for the Prevention of Accidents (RoSPA), road accidents and workplace transport accidents are the biggest causes of work-related death in the UK. It says that road accidents that involve someone who is at work at the time are estimated to claim the lives of some 800 to 1,000 people each year, and provisional figures for 2007/08 showed that 65 workers were killed in workplace transport accidents, with thousands more injured.

The human and financial costs to families, businesses and the wider community are massive.

Employers' have clear duties under health and safety legislation to manage occupational road-risk in they same way that they manage other health and safety risks. Essentially this means organisations need to:

- communicate clear messages to their staff about their approach to road safety;
- set up systems and allocate duties to key members of staff (particularly managers);
- carry out 'suitable and sufficient' risk assessments;
- use these to check that they are doing all that is 'reasonably practicable' to avoid risk on the road or to ensure safe driving;
- provide driver training where necessary; and
- monitor and review performance.

The Health and Safety Executive (HSE) has produced general guidance on managing occupational road risk in a leaflet (see HSE/DfT guidance 'Driving at Work' INDG 382) (www.hse.gov.uk/pubns/indg382.pdf) which emphasises that employers' duties under health and safety legislation extend to at work driving. Organisations need to integrate Work-Related Road Safety (WRRS) wit hin their existing policy, organisation and arrangements for managing health a nd safety at work. They need to undertake suitable risk assessments (covering journey, vehicle and driver risk factors) and ensure safe journey planning and other safety measures are in place. Top of the list of possible risk reduction meas ures is 'meeting without moving' (for example, by being video enabled). Next best is going by train or plane.

Vehicles need to be fit-for-purpose and properly maintained with additional safety features where necessary.

Managers need to avoid systems of work which cause people to speed (for example, 'just in time' delivery, payment by number of calls made, 'job and finish', and unrealistic guaranteed call-out or delivery times).

They need to avoid asking staff to drive while tired and at times of day when falling asleep at the wheel is more likely. They also need to address employees' sleep deprivation at home (caused by looking after sick children and dependants for example) and avoid introducing driver distractions like making and receiving mobile phone calls while driving, even with 'hands free'. (No mobile while mobile!) Managers need to be aware of potential health impairments, driver fitness issues, and issues like alcohol and drugs which can affect their people's ability to drive safely.

Drivers and managers need always to plan safest routes, avoiding congestion, crash sites and adverse weather. If a journey at the start of day is over 100 miles, or longer than two hours, staff need to go the night before and stay over. Managers should be alive to the effects of stress and fatigue on road safety arising from poor work/life balance.

Organisations need to assess their drivers' attitudes and driving competence, follow their crash and penalty points histories and analyse and learn from their crashes and 'near-misses'. Driver assessment should be used to target training at those with greatest needs.

Above all, organisations need to train their line managers, consult their safety representatives, require their senior managers to lead by example and recognise, celebrate and reward safe driving achievement. A good approach is to set up a multi-disciplinary team, with driver and safety representative involvement, to ask 'where are we now?' and to develop an action plan with clear targets.

The problem

[M2102] Eight people are killed each day on UK roads, which equates to almost 3,000 deaths every year with nearly 28,000 people suffering serious injury. It is estimated that around a third of all road crashes involve someone who is working at the time. Driving as part of their job is one of the riskiest activities many people will undertake while at work. Road crashes wreck families and no amount of money can compensate for the loss of a loved one or for a severe disabling injury.

Although most drivers tend to believe that they are safe, crashes still happen. Clearly as road users, employee drivers have a direct responsibility to obey road traffic law. Employers' duties under health and safety law are also applicable to at-work driving. They should therefore manage occupational road risk within the framework that they already have in place for managing other workplace health and safety risks. For example, organisations should ensure that their approach to safety at work on the road dovetails with arrangements for safe use of vehicles on their work sites.

The following section describes legal responsibilities.

Legal responsibilities

[M2103] In addition to any existing duties employers may have under specific road traffic law their duties under the *Health and Safety at Work etc. Act 1974* and the *Management of Health and Safety at Work (MHSW) Regulations 1999 (SI 1999 No 3242)* also apply when their employees are engaged in on-the-road work activities. This does not extend however to commuting journeys, except where an employee is travelling from their home to a location that is not their usual place of work.

Under the Act employers have a general duty of care to ensure, so far as is reasonably practicable, the health, safety and welfare of their employees at all times. They also have a similar responsibility to ensure that the safety of others such as contractors' staff and members of the public is not put at risk by their employees' activities.

The driver is responsible for the way a vehicle is driven on the road and should not put themselves or others at risk. They must also co-operate with their employer. But employers can have a significant influence over what the driver does. For example, they could increase the risk of crashes by:

- imposing unrealistic schedules which increase pressure to break speed limits;
- requiring drivers to work excessive hours – particularly when driving is combined with other tasks – which may lead them to become dangerously fatigued;
- failing to provide drivers with adequate or suitable training; and
- failing to provide drivers with the right vehicle for the job and to maintain it in a safe condition.

Under the MHSW Regulations employers have a duty to establish a suitable policy, organisation and arrangements to manage health and safety at work and to have access to competent advice.

These Regulations require employers to:

- carry out an assessment of the risks to health and safety of their employees while at work and others who may be affected by their activities (and record the results where the organisation employs more than five people);
- monitor health and safety performance, review their risk assessments and keep them up-to-date; and
- ensure that employees are competent and receive any necessary training, information and supervision.

They also have a duty to consult with employees over health and safety matters, and where they are appointed, their health and safety representatives.

Failure to comply with these duties could lead to enforcement action being taken. In extreme circumstances this could result in a prosecution leading to a fine and/or imprisonment.

The Department for Transport website on Driving for Work (www.dft.gov.uk/drivingforwork) provides a summary of how the law (including road safety law) applies to employers who have employees who drive road vehicles. In addition to the *Health and Safety at Work etc Act and the Management Regulations*, the following pieces of legislation apply:

- The *Workplace (Health, Safety and Welfare) Regulations 1992* (see W110 WORKPLACES – HEALTH, SAFETY AND WELFARE) contain requirements concerning traffic routes for vehicles within the workplace;
- Road Traffic Acts (supported by the Highway Code) – apply to all road users and include information on signs and markings, road users, the law and driving penalties. The DfT advises: "It is an offence for an organisation to set driver schedules which may cause them to break speed limits, and or have payment reward schemes which incentivise them to do so."
- *EC Drivers' Hours Rules; UK Domestic Drivers' Hours Rules, Tachograph Regulations and the Road Transport (Working Times) Regulations 2005* – under these rules and regulations, both drivers and employers are responsible for ensuring compliance with drivers' hours and tachograph regulations (see W100 WORKING TIME).
- The *Road Vehicles (Construction and Use) Regulations 1986* set out requirements regarding the safety of loads on vehicles; and
- The Corporate Manslaughter and Corporate Homicide Act 2007 introduced a new offence for very serious senior management failures which result in fatality (see C90 CORPORATE MANSLAUGHTER).

The business case for action

[M2104] In addition to meeting their duties under health and safety law, managing work-related road safety (WRRS) should also deliver a wide range of business benefits, many of which are financial. As with other kinds of work-related accidents, the true costs of crashes to organisations are nearly always considerably higher than the resulting repairs and insurance claims. For small businesses and the self-employed, the consequences of a major road crash are likely to be particularly severe. Even relatively minor road crashes can lead to time off work due to whiplash or psychological trauma, and some of the effects can be long lasting.

By improving WWRS and reducing crashes, the benefits to a business can include:

- fewer lost days due to injury;
- reduced risk of work-related ill health;
- reduced stress and improved morale;
- less need for investigation and paperwork;
- less lost time due to work rescheduling;
- fewer vehicles off the road for repair;
- reduced running costs through better driving standards – less wear and tear and lower fuel costs;
- fewer missed orders and business opportunities;
- less risk of damage to the organisation's reputation and image, both to the general public and to potential customers;

- reduced risk of losing the goodwill of customers; and
- preventing key employees from being banned from driving.

In addition to these business benefits, organisations will also have better control over costs, including:

- wear and tear and fuel payments;
- insurance premiums and legal fees; and
- claims from employees and third parties.

Although they should be primarily interested in the safety of their employees while driving for work purposes, from a business perspective organisations also have an interest in promoting the safety of their staff when they are commuting by road or when they are driving for domestic reasons. Whether staff are injured when driving for work purposes or at other times, the resulting costs to an organisation are likely to be very similar. By promoting a safety culture in their business, and encouraging their staff to take improved road safety home, businesses can offer a wider societal benefit by influencing their employees' private driving behaviour and in turn by employees influencing family members and friends.

The following section shows how businesses can extend their existing health and safety management system to ensure that it addresses WRRS issues.

Extending health and safety management systems to cover work-related road safety (WRRS)

[M2105] To manage work-related road safety (WRRS) successfully organisations need to integrate it into their existing system, structure and processes for managing health and safety at work (Figure A). They should already have:

(1)   a **clear policy statement** which sets out their overall safety direction and general objectives, backed by senior management commitment at the highest level;

(2)   **good safety organisation** with clear responsibilities and relationships which promote a positive safety culture and the implementation of safety policies by: securing control; encouraging co-operation and effective communication; and by ensuring competence of staff at all levels;

(3)   a **planned approach** to safety with performance standards for eliminating or controlling risk based on **risk assessment,** backed by clearly prioritised and time based targets for **implementation;**

(4)   adequate means to **measure safety performance** – both **actively** by monitoring compliance with standards and **reactively** by investigating the causes of accidents and incidents; and

(5)   appropriate procedures for **reviewing performance** against targets, **auditing safety management processes** and feeding back information and experience to further develop policies and achieve a cycle of continuous improvement in safety performance.

Organisations need to check that this approach, which is similar to other systems they may have in place for managing quality and environmental protection, is extended so that it addresses WRRS.

Figure A: health and safety management system elements and links

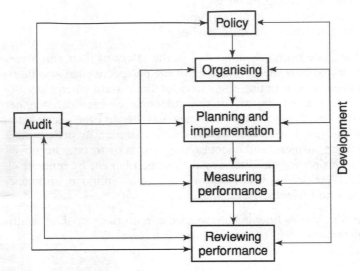

Source: HSE guide: 'Successful Health and Safety Management' HSG65

The development of each element of Figure A is now discussed in more detail.

Policy

[M2106] If an organisation employs five or more people their statutory health and safety policy statement has to be made available to all its staff. Managers must check that their organisation's current document states that their organisation's general policy is to:

- operate vehicles in a safe, efficient and effective manner;
- minimise vehicle related risks to employees and others affected by the organisation's operations; and
- strive for continuous improvement in occupational road safety performance.

Organisations can supplement this with a separate and more detailed policy statement on WRRS.

Key parts of their policy should be:

- explained to relevant staff on induction;
- included in drivers' handbooks and other documentation;
- displayed and referred to in relevant training courses; and
- given to contractors with tender documentation.

Key question

- Has the health and safety policy statement been updated to ensure it addresses risk on the road?

Organising

[M2107] Every organisation needs a structure for meeting its health and safety objectives and this should already be properly documented and widely available. Does the existing structure:

- state the roles, responsibilities and competences for managing WRRS from board level down?
- make WRRS performance a line management responsibility – which is reviewed as part of appraisal?
- make clear the roles of the organisation's health and safety, fleet, insurance and human resource professionals, and trainers?
- Does the organisation's board level 'champion' for health and safety have a leading role on WRRS?
- Are there adequate arrangements for consulting with employees and their safety representatives on WRRS issues to obtain their views and get consensus on the way forward?

Key questions:

- Does the organisation have commitment to WRRS from director level down?
- Do all managers have an understanding of their legal responsibilities for the safety of their staff?
- Have they attended relevant courses and conferences on the subject?
- Does the organisation consult with safety representatives on WRRS?

Planning and implementing

[M2108] Organisations need to develop a planned approach and ensure that WRRS considerations are built into their business planning processes based on the results of risk assessment. (More advice on risk assessment in given at M2117–M2122 below). Planning should also involve prioritising WRRS objectives, preparing an action plan informed by standards, targets and timescales for implementation and selecting indicators with which to measure and evaluate inputs and outcomes.

Some examples of targets include:

- reviewing all risk assessments annually;
- minimising high risk journeys;
- upgrading vehicles;
- specifying reductions in fleet crash rates;
- specifying reductions in crash costs;
- specifying reductions in numbers of road traffic offences; and
- targeted assessment and levels of training achievement for critical employee groups, including drivers and relevant managers.

Safety targets should be SMART (specific, measurable, achievable, realistic, time-based). They should not conflict with other targets set within the organisation, such as those for productivity and if there is conflict, it should be established how this can be resolved.

Key questions:

- Are procedures in place for undertaking routine risk assessments on all aspects of at-work road use, including for at-work pedestrians?
- Are any problems identified dealt with immediately with appropriate action?
- Are achievable targets set for crash and cost reductions?

Measuring performance

[M2109] To know 'where they are now', assess progress towards targets and identify areas for further action, organisations need to be able to monitor and evaluate performance both 'actively' and 'reactively'.

Active monitoring: Procedures need to be put in place to check that road safety standards, such as safe journey planning and expected driver behaviour, are being complied with throughout the organisation.

Reactive monitoring: There need to be procedures for investigating both the immediate and underlying causes of accidents and incidents and to gather, analyse and learn from data following crashes and 'near hits'.

Key questions:

- Does the organisation check to see whether its WRRS policies are adhered to by staff at all levels?
- Are targets monitored to see if safety is improving?
- Are all incidents involving vehicles reported and recorded (including minor bumps and scrapes)?
- Is incident data analysed to determine causes and frequency of incidents?

More advice on monitoring and evaluation is given at **M2117–M2122** below.

Reviewing and auditing performance

[M2110] As with other aspects of health and safety, organisations need periodically to review their WRRS performance. For example, does the organisation's health and safety committee monitor safety to identify problems and the need for further action? Do managers ensure that WRRS is covered when they commission independent sampling audits of the organisation's health and safety management system and procedures?

Key questions:

- Is a date set and are procedures established to review the organisation's performance?
- Are conclusions from monitoring used to help put in place measures to prevent future incidents?
- Do managers keep up-to-date on WRRS issues, for example through updates from industry publications, trade associations, the HSE or safety organisations?

Having reviewed how to extend existing health and safety management systems to WRRS, the following section discusses how to assess risks at work on the road.

Assessing risks on the road

[M2111] Risk assessment has many meanings in a wide range of different contexts. In the context of WRRS risk assessment is an aid to understanding and making judgements about safety. It needs to:

- be appropriate to the circumstances of the organisation;
- not over-complicated and bureaucratic;
- carried out by competent people with a practical knowledge of the work activities being assessed;
- generate 'an inventory of actions'; and
- encourage managers, supervisors, work allocators and drivers to think of themselves as road risk assessors.

A 'hazard' is something with potential to cause harm. Risk is an estimation of the likelihood (probability or chance) that harm will arise from exposure to a hazard. On the road, examples of hazards could include: road junctions where the risk of crashes may be higher; a vehicle having worn tyres, thus increasing the risk of skidding; or a driver failing to wear their glasses and reducing their hazard perception. 'Risk assessment' therefore involves estimating how likely it is that a crash will happen and how serious the consequences might be if it does.

Risks are assessed in this way help an organisation to:

- ensure that potential safety problems are properly understood;
- check on whether existing control and emergency measures are adequate or whether more needs to be done; and to
- prioritise any unacceptable risks for further action.

If they already have a robust health and safety management system in place, organisations will have procedures for carrying out and regularly reviewing risk assessments for their main activities. The same general principles can be adapted to help them assess work-related road risks.

It is important to ensure that a senior manager is in overall charge of the programme and appoints and trains people as risk assessors. A team approach involving line managers and safety representatives can be useful, as can using a suitable *pro-forma* (or software) to organise and record the relevant information.

Organisations also need to determine the level of sophistication required when carrying out risk assessments for their on-road activities. For example, because the potential outcomes of road crashes are nearly always 'severe' or 'very severe' (for other road users such as children, cyclists, and pedestrians even if not for the vehicle occupants), assessment can often be simplified by concentrating on features of the following three areas which are associated with an increased probability of crashes occurring:

(1)   the journey;
(2)   the vehicle; and
(3)   the driver.

By estimating the combined effect of such factors organisations can assign a risk rating (for example, 'high', 'medium', or 'low') to particular driving tasks, prioritise them for attention and then consider to consider what risk reduction measures might be taken.

Figure B: Suggested risk assessment framework

Factors to consider If any of the factors listed below are likely to increase the likelihood of an incident then tick the box and consider the level of risk	Risk Level			Existing control/ Action required
	Low	Medium	High	
The journey – enhancement of risk associated with: • road types ☐ • distances to be covered ☐ • reasonable time allocation ☐ • allowance for suficient breaks ☐ • traffic density (urban or rural) ☐ • areas with a high pedestrian density ☐ • driving at night/darkness ☐ • poor weather conditions ☐				
The vehicle – enhancement of risk associated with: • maintenance to a suitable standard ☐ • performance (power, all-terrain) ☐ • crash resistance ☐ • other safety features (e.g. ABS, traction control, air bags) ☐ • distractions (e.g. mobile phones) ☐ • driver familiarity with vehicle ☐ • loads to be carried ☐				
The driver – enhancement of risk associated with: • age ☐ • experience ☐ • driving competence ☐ • associated skills (loading, checks) ☐ • health and fitness ☐ • stress and fatigue ☐ • attitude ☐ • crash/enforcement history ☐				

Remedial Action. What remedial action should be taken in order of priority?

1 _____

2 _____

3 _____

4 _____

5 _____

This sort of specific risk assessment is likely to be most relevant where journeys follow a regular and predictable pattern, for example planned deliveries or regular sales calls. Line managers, drivers and safety representatives can review risk factors such as routes, traffic conditions, time of day, seasonal conditions, vehicle specifications and fitness and driver profiles to see if there are areas of potential concern and if there is scope for improving safety.

Much can be achieved by applying thorough management practice, but organisations can also use sources of national crash data and analyse their own claim/incident records and investigation findings to give particular attention to specific factors such as start times or time of day, speed, fatigue or specific tasks.

Where driving is a clear job requirement, they can include questions on a candidate's driving record at interview. Line managers can regularly inspect drivers' licences for entitlements and penalty points, track their annual mileages and review their crash involvement. They can also arrange for in-vehicle competence assessments or use computer or questionnaire-based techniques to assess knowledge, attitudes and hazard perception skills.

Risk assessment is not just a matter of assessing drivers. The aim should be to make an overall assessment of all the various factors that can increase the chances of a crash happening.

Where journeys have many factors which are identical or broadly similar (such as same vehicles, same routes and same times) it may be possible to develop 'generic' risk assessments to help identify safety requirements.

Where drivers have considerable autonomy over key features of their journey tasks, they can be trained to apply the risk assessment system themselves ('dynamic assessment'). A simple driver checklist highlighting risk factors and indicating relevant standards or limits to be observed can be a highly effective tool to assist drivers to make safe choices about safest routes, timing, breaks and any other risks.

A very simple approach is to identify those drivers exposed to the highest risks (such as high mileages, night time driving and congested urban areas), with the most penalty points on their licences and the worst crash histories. This enables the organisation to focus on options for reducing risk, for example, by journey task re-design and improving driver competence.

Whatever approach is adopted, the overall aim is to make informed choices about where, when and how to take action to reduce the chances of crashes occurring and to reduce the consequences if they do occur.

The following section discusses some of the options for reducing risks.

Road risk control measures

[M2112] Suitable risk assessment will help to identify where further action needs to be taken and the kinds of risk control measures the organisation needs to consider. They may need to consider what they can do to improve the safety of:

- the journey task;
- the vehicle; and
- the driver.

The following list is not exhaustive and is meant as an introductory overview of risk management options. The control measures which are implemented should be appropriate to the organisation's particular needs and operations.

The journey task

[M2113] Eliminating unnecessary journeys by road

- Can the organisation justify the need to travel by road?
- Rather than face-to-face contact, could the organisation use other means of communication, such as teleconferencing?

Changing mode of travel

- Could alternative, safer methods of transport such as train or plane be used?

Avoiding driving in adverse conditions

- Where possible, are drivers discouraged from driving at night and in bad weather conditions?
- Where possible, are drivers encouraged to stay overnight rather than continue a journey back in bad conditions?

Controlling driver hours

- Does the organisation set in-house limits for driving hours and distances?
- Does this include daily, weekly and monthly driving hours and distances?

Setting safe schedules

- Does the organisation take into account traffic and weather conditions and speed limits when planning journey times and schedules?
- Does the organisation minimise the need for driving during high risk hours for fatigue (between 2am and 6am and 2pm and 4pm) when planning schedules?
- Are the organisation's schedules contributing to driver fatigue and stress?

Encouraging the use of the 'safest' routes

- Does the organisation ensure that drivers take the safest routes (ie: using motorways and dual carriageways and avoiding crash 'black spots')

Avoiding incentives to speed

- Has the organisation eliminated incentives to speed (such as 'job and finish' contracts, payment by number of visits made, or unrealistic customer service promises)?

Discipline

- Are significant and persistent breaches of speed limits dealt with in normal disciplinary procedures?

The vehicle

[M2114] Specifying appropriate vehicles

- Is it the most appropriate vehicle for the job (eg where load carrying is involved)?
- Does it have undesirable features such as 'bull bars' that increase injury severity to vulnerable road users?
- Does it incorporate desirable additional safety features? (for example, primary features such as ABS, high level brake lights, traction control, headway information systems and alcohol ignition interlocks to help prevent crashes happening; and secondary safety features such as crash resistance, side impact bars, air bags and seatbelt interlocks to reduce consequences if crashes do occur).
- Is it ergonomically suited to the driver, particularly to help minimise the risk of musculo-skeletal disorders?

Ensuring effective vehicle maintenance

- Are the organisation's vehicles properly and regularly maintained to ensure they are in a safe condition?
- Are the organisation's vehicles maintained by skilled, knowledgeable and qualified mechanics?
- Where staff provide their own vehicles, does the organisation require them to have them serviced in accordance with the manufacturer's instructions and to provide a current MoT certificate?
- Does the organisation require daily and weekly driver/line manager checks of safety critical features such as tyres, mirrors and lights? If any defects are found, are they reported and repaired immediately?

The driver

[M2115] Ensuring health and fitness

- Does the organisation monitor employee well being through pre-employment and periodic medical and eyesight checks?
- Does the organisation pay particular attention to musculo-skeletal problems such as back pain?

Tackling driver impairment

- Does the organisation have clear policies and procedures on alcohol, drugs and medicines (prescription and non-prescription)?
- Does the organisation address impairment factors such as stress (both work-related and domestic) and poor sleep?
- Does the organisation advise drivers on strategies to manage fatigue?
- Does the organisation prohibit the use of hands held or hands free mobile phones whilst driving? (Both are equally distracting.)

Enhancing driver competence

- Has the organisation put in place a needs/risk-based driver assessment and training programme?
- Does driver assessment and training address specific needs such as safe reversing?
- Does it focus on factors such as attitude, domestic and occupational stress, fatigue and poor preparation for work?

Providing effective information and supervision

- Does the organisation have a comprehensive and frequently updated handbook for drivers?
- Does the handbook explain the organisation's work-related road safety policy and make clear that the organisation requires all employees to comply with key aspects of road traffic law such as compliance with speed limits?
- Does the handbook explain procedures for incident reporting, dealing with emergencies such as crashes or breakdowns and ensuring the personal safety of staff?
- Are the organisation's staff coached and assessed on their understanding and application of the content?
- Does the organisation use internal communications (such as bulletins, the intranet and award schemes) to update drivers on WRRS?
- Is there adequate supervision of drivers by line managers trained in WRRS?

Implementing an organisation's road risk control measures

[M2116] The 'health and safety management system' approach, described at **M2105** above, is the framework that needs to be in place for tackling risks in a planned and methodical way, informed by the results of appropriate risk assessments (at **M2111** above) and control measures such as those suggested above. Management systems by themselves, however, will not deliver continuous improvements in performance unless they are underpinned by proactive management and a positive 'health and safety culture'. Organisations need to build on this to develop a 'corporate road safety culture' in which WRRS becomes a common talking point and is taken seriously by every employee. The following steps can aid this process.

- Avoid focussing solely on drivers and driving, but aim to eliminate or reducing risks at source wherever possible. There is no point in training drivers and then requiring them to use unfit vehicles or to carry out unsafe, unsatisfactory or badly designed driving tasks.
- Consult key players such as managers, safety representatives and drivers (who are likely to have differing points of view) in order to get consensus.
- Use limited resources most appropriately to arrive at the most cost effective combination of measures by trading off the costs and benefits of different options.

- Check to see whether the expected safety gain resulting from a given 'spend' in one direction – say, upgrading to 'safer' vehicles – is justified when compared with the expected safety gain from the same 'spend' on other measures such as awareness raising or setting safer journey standards.

The following section highlights the importance of monitoring and evaluation (introduced at **M2105** above) in the process of improving WRRS. This should be built into programmes at the beginning of the process rather than as an afterthought.

Monitoring and evaluation

[M2117] To manage WRRS successfully organisations need appropriate systems to be able to monitor performance and learn lessons that can help an organisation improve all aspects of its management system.

Traditionally organisations have monitored road safety performance 'reactively' by focusing on 'lagging' indicators such as insurance claims data and costs but it is also important to undertake 'active' monitoring by looking at 'leading' indicators which help managers and safety reps to understand how well the management system is working.

Active monitoring

[M2118] Some of the options for 'active' monitoring include:

- tracking risk assessment records and journey plans to monitor driving patterns and adherence to standards;
- fitting 'black box' technologies to vehicles (after informed consultation with drivers and their representatives) to monitor excessive acceleration/deceleration forces, speeding and compliance with other journey standards such as time and distance limits;
- checking vehicle servicing schedules and records;
- spot checking vehicle conditions and fuel usage;
- monitoring key performance indicators (KPIs) such as near hits, complaints and driving at blackspots;
- quarterly monitoring of driving licences (with their informed consent) to identify any penalty points and details of the offences for which they were incurred;
- regular end of shift or journey debriefs for posts with substantial driving;
- monitoring drivers' attitudes, for example through surveys; and
- checking during staff appraisals to see how well managers and drivers are meeting their WRRS responsibilities.

Reactive monitoring

[M2119]–[M2121] Organisations also need be able to assess the number, type, causes and cost of at-work crashes in order to learn lessons to help it improve prevention and track changes in performance.

Gathering data

- Are procedures for collecting crash data at the scene explained in all driver training programmes, internal communications and driver hand-books?
- Are there clear claim and incident reporting procedures in place, for example, suitable 'bump cards', report and investigation forms?
- Are cameras or miniature tape recorders available to help drivers capture crash/incident information at the scene?
- Is there a similar procedure for reporting and sharing information on 'near hits' or WRRS problems?
- Is there a 'no blame' reporting policy to make it clear that the organisation is more interested in reducing crashes/incidents than in penalising those involved?

Investigation

- Are there criteria and procedures in place to investigate crashes, incidents and near hits in depth to determine immediate and underlying causes?
- Does the organisation adopt a team approach led by a senior manager (and involving managers, driver trainers, senior drivers and safety representatives) to share insights and expertise?
- Is it policy to concentrate most investigation effort on those events that are likely to yield the most significant lessons?
- Are the results of crash/near hit investigations considered as part of periodic performance review?

Evaluation

[M2122] Has the organisation set Key Performance Indicators (KPIs) to help it evaluate the effectiveness of its WRRS programme?

Is it policy, for example, to track:

- the number of crashes and claims?
- the claims rate per vehicle or per 100,000 kilometres per year?
- the number and severity of injuries?
- the type and location of crashes?
- severities (including 'write-offs' and air bag 'go-offs')?

Does the organisation seek to evaluate crash costs (including vehicle, driver, third party and other costs)?

Are there regular discussions with the organisation's insurer, brokers or accident management company about how they can analyse and provide effective feedback from claims?

Have baselines been established from which to measure progress and/or change over time, and to provide a basis for performance benchmarking, for example with other organisations?

Key questions:

- Does the organisation monitor drivers' day-to-day behaviour (for example through incident rates, fuel/tyre/brake pad use, freephone schemes, black box technology, end-of-shift debriefs or other methods)?

- Are the organisation's drivers actively encouraged to report all incidents, and near hits, no matter how small?
- Does the organisation investigate all, but particularly significant, crashes to establish immediate and underlying causes?
- Does the organisation set appropriate KPIs?
- Does the organisation monitor insured and uninsured crash costs?
- Does the organisation feed back results to key managers and other staff (for example, through the staff intranet, email, notice-boards and in-house magazines)?

Monitoring and evaluation is a key element in improving WRRS and helps the organisation to understand 'where it is now', which is discussed next.

Where is the organisation now?

[M2123] If the importance of WRRS is still not fully recognised or understood in an organisation, its full extent needs to be quantified and costed to raise awareness and convince everyone, from board level down, of the need for a comprehensive approach. The most important first step is to undertake an initial status review (ISR) to answer the following questions:

- 'where is the organisation now (and why is WRRS important to it)?'
- 'where does the organisation want to be?' and
- 'how is the organisation going to get there?'.

A useful starting point is the collation of information on:

(1) key fleet parameters such as vehicles, miles and drivers;
(2) safety incidents such as injuries, fatalities, airbag 'go-offs', insurance claims, near hits, wear and tear, maintenance repairs, violations, fuel use and costs; and
(3) policies, people and procedures which may already be in place to manage WRRS.

This should also help the organisation to focus on the nature and scale of its actual and potential WRRS costs and opportunities and its capacity to manage them systematically. In turn, this should help to focus attention on the importance of WRRS and provide a baseline from which progress can be measured.

The results of an ISR should be considered at board level, with the proposal that a WRRS 'action plan' be developed and implemented throughout the organisation. Everyone needs to be convinced that WRRS is an important issue affecting costs, business effectiveness and employee well-being. Simply counting crashes, injuries and costs for the first time and talking about the impact of crashes on individuals, their families and on the business can be extremely motivating.

To develop an 'action plan' a small team can be set up, for example, under the leadership of the board level health and safety 'champion'. It can involve managers and safety representatives and professionals from areas such as health and safety, fleet management, insurance risk management, training and human resources. Others such as insurers, brokers, WRRS service providers, the police and local authority road safety officers can be invited to participate.

Key elements of the plan can include: establishing a clear corporate policy; developing a framework for risk assessment; setting some indicative WRRS standards; improving crash/incident reporting and investigation procedures; and setting a target date for reviewing performance. It is very important that, before investing in driver assessment and training or other safety control measures, appropriate training is provided for all line managers of staff who drive or work on the road.

If an organisation has already developed a modern, proactive risk management approach and has a strong health and safety culture, it will have few difficulties in embracing such an approach to WRRS. Colleagues will also be keen to learn from good practice developed by other businesses by benchmarking through their trade associations and other networks. For those that have yet to find this path, tackling WRRS can often provide the organisational learning experience which will help them establish a cycle of continuous improvement in all areas of health and safety risk management.

The next section describes where to get support.

Where to get help

[M2124] There are many bodies that can provide organisations with help and advice (either free or on a fee-for service basis). Some of these include:

- employer/trade associations;
- professional bodies;
- trades unions;
- the police;
- national road safety organisations;
- national occupational safety organisations;
- Local Authority road safety departments;
- local health and safety or advanced drivers group;
- driver training providers;
- specialist fleet risk management consultants;
- motoring organisations;
- insurers, brokers or accident management companies; and
- vehicle suppliers/hirers.

There is also much useful information available, for example on the HSE's WRRS home page on their website, the Department for Transport's road safety website and on the resources section of the Occupational Road Safety Alliance website. Time spent visiting these sites can pay dividends in helping to identify other sources of guidance and information relevant to an organisation's specific needs.

Useful publications

[M2125] —

(1)   *Management of work-related road safety* – research report 018. Available from HSE Books Ref RR 018

(2)   *The contribution of individual factors to driving behaviour: Implications for managing work-related road safety* – research report 020. Prepared by Entec UK Limited for the Health and Safety Executive and Scottish Executive 2002 (www.hse.gov.uk/research/rrpdf/rr020.pdf).

(3)   *Driving at work (INDG 382)* HSE/DfT free leaflet available from www. hse.gov.uk or copies free by mail order from HSE Books, PO Box 1999, Sudbury, Suffolk, CO10 2WA. Tel: 01787 881165 Fax: 01787 313995 Website: www.hsebooks.co.uk (HSE priced publications are also availa ble from bookshops).

(4)   *Managing Occupational Road Risk the RoSPA guide* RoSPA House Birmingham B5 7ST. Price £20. Visit www.rospa.com.

(5)   *The Highway Code.* Available from The Stationary Office ISBN 0 11 551977 7. Can also be viewed on www.highwaycode.gov.uk

(6)   *Health and Safety at Work etc Act 1974.* Available from The Stationery Office, ISBN 0 10 543774 3.

(7)   *Management of Health and Safety at Work Regulations 1999.* Available from the Stationery Office SI 1999/3242, ISBN 0 11 085625 2

(8)   *Successful Health and Safety Management HSG 65.* Available from HSE Books ISBN 0 7176 1276 7 1997

(9)   *Five steps to risk assessment. INDG 163 (rev1).* Available from HSE Books ISBN 0 7176 1565 0

(10)   *Temporary Traffic Management on High Speed Roads – Good Working Practice.* Available from the Highways Agency website: www.highways. gov.uk

(11)   *Safe Working of Vehicle Breakdown and Recovery Operators Mana gement System Specification – 43 PAS.* Available from British Standards Institution, 389, Chiswick High Road, London W4 4AL. (www.bs i-global.com)

(12)   *Back in Work* CRR 441/2002 www.hse.gov.uk/research

Useful websites:

[M2126] www.dft.gov.uk/drivingforwork – Department for Transport, developed in partnership with the Transport Research Laboratory (TRL), and containing detailed and comprehensive resources for organisations wanting to create or enhance a driving for work policy.

www.rospa.com – Royal Society for the Prevention of Accidents

www.orsa.org.uk – Occupational Road Safety Alliance

www.hse.gov.uk/roadsafety – HSE WRRS home page

www.airso.org.uk – Association of Industrial Road Safety Officers

www.roadsafe.org.uk – Roadsafe – a partnership in road safety

www.pacts.org.uk – Parliamentary Advisory Council on Transport Safety

www.brake.org.uk – Brake – road safety charity

www.roadsafetygb.org – Road Safety GB (formerly LARSOA) – a national road safety organisation that represents local government road safety teams across the UK

Manual Handling

Mike Bateman

Introduction

[M3001] The *Manual Handling Operations Regulations 1992 (SI 1992 No 2793), as amended by SI 2002 No 2174,* came into operation on 1 January 1993. The Health and Safety Executive (HSE) had long been concerned at the large numbers of accidents resulting from manual handling activities — more than a quarter of the accidents reported to the enforcing authorities (the HSE and local authorities) at the time of the introduction of the regulations. Many manual handling injuries have cumulative effects eventually resulting in physical impairment or even permanent disability.

Evaluations of the regulations conducted in 1996 and 2001 found that good progress had been made, particularly in larger organisations. In the 2001 study employers were generally happy with the approach taken by the regulations, with 48% feeling that the benefits of the regulations outweighed increased costs and only 11% thinking that the costs outweighed the benefits. Progress was less marked in medium and smaller organisations with infrequent tasks and tasks carried out away from the employer's base less well controlled. Musculoskeletal Disorders (MSDs) continue to be the most common cause of occupational ill health in Great Britain, affecting some one million people a year and costing society £5.7 billion. They therefore remain an important priority area for the HSE.

What the Regulations require

[M3002] Various terms used in the *Manual Handling Operations Regulations 1992 (SI 1992 No 2973)* are defined in *Regulation 2* which states:

"manual handling operations" means any transporting or supporting of a load (including the lifting, putting down, pushing, pulling, carrying or moving thereof) by hand or by bodily force.

"Injury" does not include injury caused by any toxic or corrosive substance which:

(a)   has leaked or spilled from a load;
(b)   is present on the surface of a load but has not leaked or spilled from it; or
(c)   is a constituent part of a load;

and 'injured' shall be construed accordingly;

(This in effect establishes a demarcation with the *COSHH Regulations (SI 2002 No 2677)* – eg where there is oil on the surface of a load making it difficult to handle that is a matter for the *Manual Handling Operations Regulations*, whereas any risk of dermatitis is covered by the *COSHH Regulations*.)

Load includes any person and any animal.

(This is of great importance to a number of work sectors including health and social care, agriculture and veterinary work.)

HSE guidance accompanying the Regulations states that an implement, tool or machinery being used for its intended purpose is not considered to constitute a load – presumably establishing a demarcation with the requirements of *Provision and use of Work Equipment Regulations 1998 (PUWER 98) (SI 1998 No 2306)*. However, when such equipment is being moved before or after use it will undoubtedly be a load for the purpose of these Regulations.

Regulation 2(2) imposes the duties of an employer to his employees under the Regulations on a self-employed person in respect of himself. *Regulation 3* excludes the normal ship-board activities of a ship's crew under the direction of the master from the application of the Regulations (these are subject to separate merchant shipping legislation).

The duties of employers under the Regulations are all contained in *Regulation 4* which establishes a hierarchy of measures that employers must take. *Regulation 4* states:

(1)   Each employer shall—
    (a)   so far as is reasonably practicable, avoid the need for his employees to undertake any manual handling operations at work which involve a risk of their being injured; or
    (b)   where it is not reasonably practicable to avoid the need for his employees to undertake any manual handling operations at work which involve a risk of their being injured–
      (i)   make a suitable and sufficient assessment of all such manual handling operations to be undertaken by them, having regard to the factors which are specified in column 1 of Schedule 1 to these Regulations and considering the questions which are specified in the corresponding entry in column 2 of that Schedule,
      (ii)   take appropriate steps to reduce the risk of injury to those employees arising out of their undertaking any such manual handling operations to the lowest level reasonably practicable, and
      (iii)   take appropriate steps to provide any of those employees who are undertaking any such manual handling operations with general indications and, where it is reasonably practicable to do so, precise information on–
        (a)   the weight of each load, and
        (b)   the heaviest side of any load whose centre of gravity is not positioned centrally.
(2)   Any assessment such as is referred to in paragraph (1)(*b*)(i) of this Regulation shall be reviewed by the employer who made it if –
    (a)   there is reason to suspect that it is no longer valid; or
    (b)   there has been a significant change in the manual handling operations to which it relates,

and where as a result of any such review changes to an assessment are required, the relevant employer shall make them.

As a result of the *Health and Safety (Miscellaneous Amendments) Regulations 2002 (SI 2002 No 2174)*, a further paragraph was added to *Regulation 4*. This states:

(3)   In determining for the purpose of this regulation whether manual handling operations at work involve a risk of injury and in determining the appropriate steps to reduce that risk regard shall be had in particular to–
  (a)   the physical suitability of the employee to carry out the operations;
  (b)   the clothing, footwear or other personal effects he is wearing;
  (c)   his knowledge and training;
  (d)   the results of any relevant risk assessment carried out pursuant to regulation 3 of the Management of Health and Safety at Work Regulations 1999;
  (e)   whether the employee is within a group of employees identified by that assessment as being especially at risk; and
  (f)   the results of any health surveillance provided pursuant to regulation 6 of the Management of Health and Safety Regulations 1999.

The hierarchy of measures which must be taken by employers where there is a risk of injury can be summarised as follows:

*   Avoid the operation (if reasonably practicable);
*   Assess the remaining operations;
*   Reduce the risk (to the lowest level reasonably practicable);
*   Inform employees about weights.

Reference will be made later in the chapter to a variety of measures which can be taken to avoid or reduce risks. The factors to which regard must be given during the assessment (as specified in *Schedule 1* to the Regulations) are the *tasks, loads, working environment* and *individual capability* together with other factors referred to in *Regulation 4(3)*. These factors and the questions specified by *Schedule 1* are also covered later in the chapter.

Regulation 5 places a duty on employees. It states:

Each employee while at work shall make full and proper use of any system of work provided for his use by his employer in compliance with Regulation 4(1)(*b*)(ii) of the Regulations.

Regulation 6 provides for the Secretary of State for Defence to make exemptions from the requirements of the Regulations in respect of home forces and visiting forces and their headquarters.

Regulation 7 extends the Regulations to apply to offshore activities such as oil and gas installations, including diving and other support vessels. Previous provisions relating to manual handling were replaced or revoked by *Regulation 8*. The Regulations themselves are extremely brief but they are accompanied by considerable HSE guidance, much of which will be referred to during this chapter.

Since the regulations were enacted, a number of cases have provided further interpretation on their requirements.

These include the following:

In *Swain v Denso Martin Ltd*, The Times, 24 April 2000, the Court of Appeal ruled that the fact than an employer had failed to carry out a risk assessment, did not mean that he was not under a duty to take steps to give information about the weight of a load where there was a risk of injury;

In *Sussex Ambulance NHS Trust v King* [2002] EWCA Civ 953, the judges took into account the difficulties an ambulance service has reconciling its duties towards patients with their duties towards their employeesThe case involved an ambulance technician who had injured his back while carrying a patient downstairs from his bedroom;

However, in *Knott v Newham Healthcare NHS Trust* [2002] EWHC 2091, a High Court judge ruled that the employer's arrangements for lifting were inadequate and breached the Regulations by failing to provide sufficient lifting equipment and effectively expecting nurses to lift patients manually; and

In a 2002 Court of Appeal case, *O'Neill v DSG Retail Ltd*, The Times, 9 September 2002, it was held that an employer required to take appropriate steps to reduce a risk of injury to the lowest level reasonably practicable, must in assessing the risk, consider the particular task, the context of where it is performed and the employee required to perform it;

Risk of injury from manual handling

Assessment guidelines

[M3003] Appendix 3 of the guidance accompanying the *Manual Handling Operations Regulations 1992 (SI 1992 No 2973)* (HSE booklet L23 *Manual Handling Operations 1992 (as amended): Guidance on the Regulations*) provides assessment guidelines which can be used to filter out manual handling operations involving little or no risk and help to identify where a detailed risk assessment is necessary. However, it must be stressed that *these are not weight limits* – they must not be regarded as safe weights nor must they be treated as thresholds which must not be exceeded. The HSE guidance states that application of the guidelines will provide a reasonable level of protection to around 95 per cent of working men and women. Some workpeople will require additional protection (eg pregnant women), and this should be considered in relation to the assessment of individual capability.

In the latter part of 2003 the HSE published a free leaflet, Manual Handling Assessment Charts (MAC) (INDG 383), to provide a tool to assist in identifying high risk manual handling operations during assessments. It uses numerical scores and a green, amber, red, purple colour coding system to categorise risks as low, medium, high and very high respectively. Risks from lifting, carrying and team handling are covered, although not from pushing and pulling. The leaflet also contains guidance on assessment technique. This should only be used as a preliminary screening tool to identify workplace manual handling activities which pose a significant risk. The risk associated with these activities should then be assessed as outlined in the Manual Handling Operations Regulations 1992 (as amended). The user of the MAC tool should also be aware that the graphs presenting the risks associated with frequently handled loads are based on one form of experimental data which is recognised to overestimate the individual's handling capacity when loads are handled frequently. The graphs should be interpreted with caution.

The MAC tool is also available online at www.hse.gov.uk/pubns/indg383.pdf.

Lifting and lowering

The guidelines for lifting and lowering operations as set out in the Manual Handling Operations Regulations 1992 (as amended) are shown in the accompanying diagram (see Figure 1 below). They assume that the load can be easily grasped with both hands and that the operation takes place in reasonable working conditions, using a stable body position. Where the hands enter more than one box, the lowest weights apply. A detailed assessment should be made if the weight guidelines are exceeded or if the hands move beyond the box zones.

These guideline weights apply up to approximately 30 operations per hour. They should be reduced for more frequent operations:

- by 30 per cent for one or two operations per minute;
- by 50 per cent for five to eight operations per minute;
- by 80 per cent for more than twelve operations per minute.

Other factors likely to require a more detailed assessment are where:

- workers do not control the pace of work, eg on assembly lines;
- pauses for rest are inadequate;
- there is no change of activity allowing different muscles to be used;
- the load must be supported by the handler for any length of time.

Carrying

Similar guidelines figures can be applied for carrying where the load is held against the body and carried up to 10 metres without resting. For longer distances or loads held below knuckle height or above elbow height a more detailed assessment should be made. The guideline figures can be applied for carrying loads on the shoulder for distances in excess of 10 metres but an assessment may be required for the lifting of the load onto and off the shoulder.

Pushing and pulling

Where loads are slid, rolled or supported on wheels the guideline figures assume that force is applied with the hands between knuckle and shoulder height. The guideline figure for starting or stopping the load is a force of about 20kg (about 200 Newtons) for men and 10kg (about 100 Newtons) for keeping the load in motion (these figures reduce to 15kg and 7kg respectively for women). The guidelines also assume that the distance involved is no more than about 20 metres. In practice it is extremely difficult to make a judgement as to whether forces of these magnitude are being applied. Many employers are likely to make assessments of significant pushing or pulling activities anyway. Factors such as the condition of the floor, space constraints and possible tripping hazards must always be taken into account.

Handling when seated

The accompanying diagram (see Figure 2 below) illustrates the guideline figures for handling whilst seated. Any handling activities outside the specified box zones should be assessed in any case.

Fig. 1: Lifting and lowering

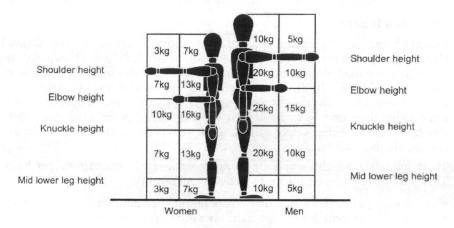

Fig. 2: Handling while seated

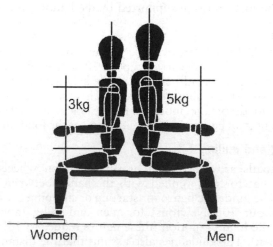

Schedule 1 to the *Manual Handling Operations Regulations 1992 (SI 1992 No 2973)* contains a list of questions which must be considered during assessments of manual handling operations. All of these indicate factors which may increase the risk of injury. (Most of these factors have been included in the checklist for assessments provided later in the chapter.)

Task related factors

Holding or manipulating loads at a distance from the trunk

[**M3004**] Stress on the lower back increases as the load is moved away from the trunk. Holding a load at arms' length imposes five times the stress as the same load when held very close to the trunk.

Unsatisfactory body movement or posture

The risk of injury increases with poor feet or hand placings, eg feet too close together, body weight forward on the toes, heels off the ground.

Twisting the trunk whilst supporting a load increases stress on the lower back – the principle should be to move the feet, not twist the body.

Stooping also increases the stress on the lower back whilst *reaching upwards* places more stress on the arms and back and lessens control on the load.

Reaching upwards places additional stress on the arms and back and makes control of the load more difficult.

Excessive movement of loads

Risks increase the further loads must be *lifted or lowered*, especially when lifting from floor level or from above head height. Movements involving a *change of grip* are particularly risky. Excessive *carrying* of loads increases fatigue – hence the guidance figure of about 10 metres carrying distance before a more detailed assessment is made.

Excessive pushing or pulling

Here the risks relate to the forces which must be exerted, particularly when starting to move the load, and the quality of the grip of the handler's feet on the floor. Risks also increase when pushing or pulling below knuckle height or above shoulder height.

Risk of sudden movement

Freeing a box jammed on a shelf or a jammed machine component can impose unpredictable stresses on the body, especially if the handler's posture is unstable. Unexpected movement from a client in the health care sector or an animal in agricultural work can produce similar effects.

Frequent or prolonged physical effort

Risk of injury increases when the body is allowed to become tired, particularly where the work involves constant repetition or a relatively fixed posture. The periodic inter-changing of tasks within a work group can do much to reduce such risks.

Clothing, footwear etc.

The effects of any PPE or other types of clothing, footwear etc. required for the task may make manual handling more difficult or increase the possibilities of fatigue, dehydration etc.

Insufficient rest and recovery

The opportunity for rest and recovery will also help to reduce risks, especially in relation to tasks which are physically demanding.

Rate of work imposed by a process

In such situations workers are often not able to take even short breaks, or to stretch or exercise different muscle groups. More detailed assessments should be made of assembly line or production line activities. Again periodic role changes within a work team may provide a solution.

Load related factors

Is the load too heavy?

[M3005] The guideline figures provided earlier are relevant here but only form part of the picture. Task-related factors (eg twisting or repetition) may also come into play or the load may be handled by two or more people. Size, shape or rigidity of the load may also be important.

Is the load bulky or unwieldy?

Large loads will hinder close approach for lifting, be more difficult to get a good grip of or restrict vision whilst moving. The HSE guidance suggests that any dimension exceeding 75cm will increase risks and the risks will be even greater if it is exceeded in more than one dimension. The positioning of suitable handholds on the load may somewhat reduce the risks.

Other factors to consider are the effects of wind on a large load, the possibility of the load hitting obstructions or loads with offset centres of gravity (not immediately apparent if they are in sealed packages).

Is the load difficult to grasp?

Where loads are large, rounded, smooth, wet or greasy, inefficient grip positions are likely to be necessary, requiring additional strength of grip. Apart from the possibility of the grip slipping there are likely to be inadvertent changes of grip posture, both of which could result in loss of control of the load.

Is the load unstable?

The handling of people or animals was referred to earlier. Not only do such loads lack rigidity, but they may also be unpredictable and protection of the load as well as the handler is of course an important consideration. Other potentially unstable loads are those where the load itself or its packaging may disintegrate under its own weight or where the contents may shift suddenly, eg a stack of books inside a partially full box.

Is the load sharp, hot etc?

Loads may have sharp edges or rough surfaces or be extremely hot or cold. Direct injury may be prevented by the use of protective gloves or clothing but the possibility of indirect risks (eg where sharp edges encourage an unsuitable grip position or cold objects are held away from the body) must not be overlooked.

Factors relating to the working environment

Space constraints

[M3006] Areas of restricted headroom (often the case in maintenance work) or low work surfaces will force stooping during manual handling and obstructions (eg in front of shelves) will result in other unsatisfactory postures. Restricted working areas or the lack of clear gangways will increase risks in moving loads, especially heavier or bulkier items.

Uneven, slippery or unstable floors

All of the above can increase the risks of slips, trips and falls. The availability of a firm footing is a major factor in good handling technique – such a footing will be a rarity on a muddy construction site. Moving workplaces (eg trains, boats, elevating work platforms) will also introduce unpredictability in footing. (The guideline weights should be reduced in such situations.)

Variation in levels of floors or work surfaces

Risks will increase where loads have to be handled on slopes, steps or ladders, particularly if these are steep. Any slipperiness of their surface, eg due to ice, rain, mud will create further risks. The need to maintain a firm handhold on ladders or steep stairways is another factor to consider. Movement of loads between surfaces or shelving at different levels may also need to be considered, especially if there is considerable height change, eg floor level to above shoulder height.

Extremes of temperature or humidity

High temperatures or humidity will increase the risk of fatigue, with perspiration possibly affecting the grip. Work at low temperatures (eg in cold stores or cold weather) is likely to result in reduced flexibility and dexterity. The need for gloves and bad weather clothing is another factor to consider.

Ventilation problems or gusts of wind

Inadequate ventilation may increase fatigue whilst strong gusts of wind may cause considerable danger when handling larger loads, eg panels being moved on building sites.

Poor lighting

Lack of adequate lighting may cause poor posture eg by encouraging stooping or an increased risk of tripping. It can also prevent workers identifying risks associated with individual loads, eg sharp edges or corners, offset centres of gravity.

Capability of the individual

[M3007] Manual handling capabilities will vary significantly between individual workers. Reference was made earlier in the chapter to the assessment guideline weights providing a reasonable level of protection to around 95 per cent of workers and those guideline figures differed for men and women. *Regulation 4(3)* of the *Manual Handling Operations Regulations 1992 (SI 1992 No 2973)* (as amended) now specifically refers to factors which must be taken into account in respect of individual workers. Individual factors include:

Gender

Whilst there is overlap between the capabilities of men and women, generally the lifting strength of women is less than men.

Age

The bodies of younger workers will not have matured to full strength, whilst with older workers there will be a gradual decline in their strength and stamina, particularly from the mid 40's onwards. Such persons will be the types of groups of employees referred to in *Regulation 4(3)(e)* of the *Manual Handling Operations Regulations 1992 (SI 1992 No 2973)* (as amended).

Experience

More mature workers may be better able to recognise their own capabilities and pace themselves accordingly, and should also have acquired better handling techniques.

Pregnancy

The need to protect new or expectant mothers is emphasised in the chapter RISK ASSESSMENT.

Previous injury or ill-health

Workers with long term musculoskeletal problems, a history of injuries or ill-health or short term injuries (including those from non-work related causes) will all justify additional protection. Such problems should be detected by the health surveillance required by *Regulation 6* of the *Management of the Health and Safety at Work Regulations 1999 (SI 1999 No 3242)*. (See **M3015** later in the chapter.)

Physique and stature

These factors will vary considerably within any workforce. Whilst bigger will often mean stronger, it must not be overlooked that some tasks or work locations may force taller workers to stoop or adopt other unsuitable handling postures.

Generally the objective should be to ensure that all manual handling operations can be performed satisfactorily by most reasonably fit and healthy employees, although a minority of workers may justify restrictions on their handling activities (see later in the chapter).

During the assessment process it is important that special note is taken of any manual handling operations which:

- Require unusual strength, height etc, eg reels of wrapping paper can only be safely positioned in a machine by persons above a certain height.

- Create additional risk to workers who are pregnant, disabled or have health problems.
  Such tasks may need to be reassessed to reduce risks or it may be necessary to restrict some individuals from carrying them out.
- Require special information or training.
  The health care sector provides many examples of situations where both the worker and the client (the load) may be at risk if correct technique is not used. Many other tasks are likely to require specific techniques to be adopted – whether in manual handling itself or the use of handling aids.

Clothing, footwear etc.

Some individuals may be affected more than others by the use of personal protective equipment (PPE) or clothing, eg gloves, protective suits, breathing apparatus. These may significantly affect the user's mobility or dexterity in respect of manual handling. Similarly uniforms or costumes (eg in the entertainment industry) may also be a factor to be taken into account.

Avoiding or reducing risks

[M3008] The *Manual Handling Operations Regulations 1992 (SI 1992 No 2973)* require employers to avoid the need for employees to undertake manual handling operations involving a risk of injury, so far as is reasonably practicable. Measures for avoiding such risks can be categorised as:

- Elimination of handling.
- Automation or mechanisation.
- Reduction of the weight or size of the load.

Elimination of handling

[M3009] It may be possible for employers to eliminate manual handling altogether (or to eliminate handling by their own employees) by means such as:

Redesigning processes or activities

For example, so that activities such as machining or wrapping are carried out in situ, rather than a product being manually handled to a position where the activity take place; a treatment is taken to a patient rather than vice versa (this may have other benefits to the patient).

Using transport better

For example, allowing maintenance staff to drive vehicles carrying tools or equipment up to where they are working, rather than manually handling them into place.

Requiring direct deliveries

For example, requiring suppliers to make deliveries directly into a store rather than leaving items in a reception area (such suppliers may be better equipped in terms of handling aids and their staff may be better trained in respect of manual handling).

Automation or mechanisation

[M3010] The automation or mechanisation of a work activity is best considered at the design stage although there is no reason in principle why such changes cannot be made later, eg as the result of a risk assessment. In making changes it is important to avoid creating additional risks, eg the use of fork lift trucks for a task in an already cramped workplace may not be the best solution. However, as employers have become more aware of manual handling risks (many prompted by the advent of the Regulations) a much greater range of mechanical handling solutions to manual handling problems have become available.

These include:

Mechanical lifting devices

Fork lift trucks, mobile cranes, lorry mounted cranes, tail lifts, mobile elevating work platforms, vacuum devices and other powered handling equipment are commonly used to eliminate or greatly reduce the manual handling of loads.

Manually-operated lifting devices

Pallet and stacker trucks, manually operated chain blocks or lever hoists can all be used to move heavy loads using very limited physical force.

Powered conveyors

As well as fixed conveyors (both belt and driven roller types), mobile conveyors are being used increasingly, particularly for the loading and unloading of vehicles, eg stacking of bundles of newspapers in delivery vans.

Non-powered conveyors, chutes etc.

Free-running roller conveyors, chutes or floor-mounted trolleys allow loads to move under the effects of gravity or to be moved manually with little effort. Such devices may operate between different levels or be set into the workplace floor.

Raw material transfer systems

Raw materials such as powders or liquids can be transferred using gravity feeds or pneumatic transfer equipment, avoiding handling of bags or other containers.

Trolleys and trucks

Trolleys and trucks can be used to greatly reduce the manual handling effort required in transporting loads. Some types also incorporate manual or mechanically-powered lifting mechanisms. Specialist trolleys are designed for carrying drums or other containers whilst an ingenious triple wheel system can be lifted to some trucks to aid them in climbing or descending stairs.

Lifting tools

Special lifting tools are available to reduce the manual handling effort required for lifting or lowering certain loads, eg paving slabs, manhole covers, drums or logs.

All of these and many other types of devices are illustrated in HSE booklet HSG 115 *Manual handling: Solutions you can handle* (which can be downloaded from the HSE website at www.hse.gov.uk/pubns/books/hsg115.htm).

Reduction of the weight or size of the load

[M3011] There is considerable scope to avoid the risk of injury by reducing the sizes of the loads which have to be handled. In some cases the initiative for this has come from suppliers of a particular product whereas in others the impetus has come from customers or from users of equipment, often as a result of their manual handling assessments. Examples include:

- Packaged building materials (eg cement) reduced to 25kg.
- Photocopying paper now supplied in five ream boxes (approximately 12kg) – 10 ream boxes were previously commonplace.
- Newspaper bundle sizes reduced below 20kg (printers reduce the number of copies in a bundle as the number of pages in the paper increases).
- Customers specifying maximum container weights they will accept from suppliers, eg weights of brochures from commercial printers.
- Equipment being separated into component parts and assembled where it is to be used, eg emergency equipment used by fire services.

Where it is not reasonably practicable to avoid a risk of injury from a manual handling operation, the employer must carry out an assessment of such operations with the aim of reducing the risks to the lowest level reasonably practicable. The four main factors to be taken into account during the assessment are *the task, the load, the working environment* and *individual capability*. Possible risk reduction measures related to each of these factors are described below.

(Later in the chapter these are summarised in an assessment checklist – see M3020 below.)

Task-related measures

Mechanical assistance

[M3012] The types of mechanical assistance described earlier can be used to reduce the risk of injury, even if such risks cannot be avoided entirely. The use of roller conveyors, trolleys, trucks or levers can reduce considerably the amount of force required in manual handling tasks.

Task layout and design

The layout of tasks, storage areas etc should be designed so that the body can be used in its more efficient modes, eg:

- heavier items stored and moved between shoulder and mid-lower leg height (preferably around waist height);
- allowing loads to be held close to the body;
- avoiding reaching movements, eg over obstacles or into deep bins;
- avoiding twisting movements or handling in stooped positions, eg by repositioning work surfaces, storage areas or machinery;
- providing resting places to aid grip changes whilst handling;
- reducing lifting and carrying of loads, eg by pushing, pulling, sliding or rolling techniques;
- utilising the powerful leg muscles rather than arms or shoulders;
- avoiding the need for sustaining fixed postures, eg in holding or supporting a load;
- avoiding the need for handling in seated positions.

Work routines

Work routines may need to be altered to reduce the risk of injury using such measures as:

- limiting the frequency of handling loads (especially those that are heavier or bulkier);
- ensuring that workers are able to take rest breaks (either formal breaks or informal rest periods as and when needed);
- introducing job rotation within work teams (allowing muscle groups to recover whilst other muscles are used or lighter tasks undertaken).

Team handling

Tasks which might be unsafe for one person might be successfully carried out by two or more. HSE guidance indicates that the capability of a two-person team is approximately two-thirds of the total of their individual capabilities; and of a three-person team approximately one half of their total. Other factors to take into account in team handling are:

- the availability of suitable handholds for all the team;
- whether steps or slopes have to be negotiated;
- possibilities of the team impeding each other's vision or movement;
- availability of sufficient space for all to operate effectively;
- the relative sizes and capabilities of team members;
- good communication between team members.

Load-related measures

Weight or size reduction

[M3013] The types of measures described earlier in the chapter might be adopted to reduce the risk of injury even if it is not possible to avoid the risk entirely. However, reducing the individual load size will increase the number of movements necessary to handle a large total load. This could result in a different type of fatigue and also the possibility of corners being cut to save time. Sizes of loads may also be reduced to make them easier to hold or bring them closer to the handler's body.

Making the load easier to grasp

Handles, grips or indents can all make loads easier to handle (as demonstrated by the handholds provided in office record storage boxes). It will often be easier handling a load placed in a container with good handholds than handling it alone without any secure grip points. The positioning of handles or handholds on a load can also influence the handling technique used, eg help avoid stooping. Handholds should be wide enough to accommodate the palm and deep enough to accommodate the knuckles (including gloves where these are worn).

The use of slings, carrying harnesses or bags can all assist in gaining a secure grip on loads and carrying them in a more comfortable and efficient position.

Making the load more stable

Loads which lack rigidity themselves or are in insecure packaging materials may need to be stabilised for handling purposes. Use of carrying slings, supporting boards or trays may be appropriate. Partially full containers may need to have their contents wedged into position to prevent them moving during handling.

Reducing other risks

Loads may need to be cleaned of dirt, oil, water or corrosive materials for safe handling to take place. Sharp corners or edges or rough surfaces may have to be removed or covered over. There may also be risks from very hot or very cold loads. In all cases the use of containers (possibly insulated) other handling aids or suitable gloves or other PPE may need to be considered.

Improving the work environment

Providing clear handling space

[M3014] The provision of adequate gangways and sufficient space for handling activities to take place will be linked with general workplace safety issues. Low headroom, narrow doorways and congestion caused by equipment and stored materials are all to be avoided. Good housekeeping standards are essential to safe manual handling.

Floor condition and design

Even in temporary workplaces such as construction sites, every effort should be made to ensure that that firm, even floors are provided where manual handling is to take place. Special measures may be necessary to allow water to drain away or to clear promptly any potentially slippery materials (eg food scraps in kitchens). Equipment is now available to measure the slipperiness of floors. Where risks are greater special slip-resistant surfaces may need to be considered. In outdoor workplaces the routine application of salt or sand to surfaces made slippery by ice or snow may need to be introduced.

Differing work levels

Measures may be necessary to reduce the risks caused by different work levels. Slopes may need to be made more gradual, steps may need to be provided or existing steps or stairways made wider to accommodate handling activities. Work benches 02may need to be modified to provide a uniform and convenient height.

Thermal environment and ventilation

Unsatisfactory temperatures, high humidity or poor ventilation may need to be overcome by improved environmental control measures or transferring work to a more suitable area. For work close to hot processes or equipment, in very cold conditions (eg refrigerated storage areas) or carried on out of doors the use of suitable PPE may be necessary.

Gusts of wind

Where gusts of wind (or powerful ventilation systems) could affect larger loads extra precautions, in addition to those normally taken, may be necessary eg:

- using handling aids (eg trolleys);
- utilising team handling techniques;
- adopting a different work position;
- following an alternative transportation route.

Lighting

Suitable lighting must be provided to permit handlers to see the load and the layout of the workplace and to make accurate judgements about distance, position and the condition of the load.

Individual capability

[M3015] In ensuring that individuals are capable of carrying out manual handling activities in the course of their work, account must be taken both of factors relating to the individual and those relating to the activities they are to perform. Steps which may be appropriate include:

Medical screening

Workers may be screened both prior to being offered employment and periodically thereafter to check whether they are physically capable of carrying out the full range of manual handling tasks in the workplace. In some extreme situations (eg the emergency services) an individual may be deemed unsuitable for employment if they are not sufficiently fit. However, legislation prevents employers discriminating against those with disabilities *without good reason*. In most workplaces the results of such screening may simply result in restrictions on the range of tasks individuals are allowed to perform.

Long-term restrictions

Long-term restrictions may be appropriate where workers are identified through screening programmes as not being capable of carrying out certain manual handling activities without undue risk, or they become incapable due to some long term injury or the effects of the ageing process. It may be necessary to place long term restrictions on young workers until they attain an appropriate age at which their capabilities can be re-assessed.

Short-term restrictions

Relatively short-term restrictions on manual handling activities may be necessary because of:

- pregnancy (or a particular stage of pregnancy);
- recently having given birth;
- injury or illness of short duration.

Fitness programmes

An increasing number of employers actively encourage employers to be physically fit, thus reducing their risk of injury during manual handling activities. A number of Fire Services have particularly pro-active programmes.

General manual handling training

Many employers offer general training in manual handling, either to all employees or to all those employees expected to engage in significant manual handling work. Aspects to cover in such training include:

- recognition of potentially hazardous operations;
- avoiding manual handling hazards;
- dealing with unfamiliar handling operations;
- correct use of handling aids;
- proper use of PPE;
- working environment factors;
- importance of good housekeeping;
- factors affecting individual capability (knowing one's limitations);
- good handling technique.

Task-specific training and instruction

Training and instruction will often need to relate to the safety of specific manual handling tasks, eg:

- how to use specific handling aids;
- specific PPE requirements;
- specific handling techniques.

General principles of handling technique are summarised in HSE leaflet INDG 143 *Getting to grips with manual handling: A short guide for employers* and also in *L23*. Many other HSE publications deal with specific occupational areas and often contain guidance on training (see "HSE Guidance" at **M3019**). *Regulation 4(3)(c)* of the *Manual Handling Operations Regulations 1992 (SI 1992 No 2973)* (as amended) requires employees' level of knowledge and training to be considered during the risk assessment.

Planning and preparation

[M3016] The principles of planning and preparation for manual handling assessments are no different from those for any other types of risk assessments.

Who will carry out the assessments?

[M3017] HSE guidance contained in the main HSE booklet on the *Manual Handling Operations Regulations 1992* (L 23) states that while one individual may be able to carry out an assessment in relatively straightforward cases, in others it may be more appropriate to establish an assessment team. It also refers to employers and managers with a practical understanding of the manual handling tasks to be performed, the loads to be handled, and the working environment being better able to conduct assessments than someone outside the organisation (although it is not stated in the guidance that in-house personnel will also usually be more aware of the capabilities of employees carrying out manual handling tasks).

The guidance refers to the individual or team performing the assessments needing to possess the following knowledge and expertise:

- the requirements of the Regulations;
- the nature of the handling operations;
- a basic understanding of human capabilities;
- awareness of high risk activities;
- practical means of reducing risk.

The HSE also acknowledges that there will be situations where external assistance may be required. These might include:

- providing training of in-house assessors;
- assessing risks which are unusual or complex;
- re-designing equipment or layouts to reduce risks;
- training of staff in handling techniques.

How will the assessments be arranged?

[M3018] As for general risk assessments, the workplace or work activities will often need to be divided into manageable assessment units. Depending on how work is organised, these might be based on:

- Departments or sections.
- Buildings or rooms.
- Parts of processes.
- Product lines.
- Work stations.
- Services provided.
- Job titles.

Members of the assessment team can then be allocated to the assessment units where they are best able to make a contribution.

Gathering information

[M3019] It is important to gather information prior to the assessment in order to identify potentially high risk manual handling activities and the precautions which ought to be in place to reduce the risks. Relevant sources might include:

Accident investigation reports

Dependent upon the requirements of the organisation's accident investigation procedure, these may only include the more serious accidents resulting from manual handling operations.

Ill health records

These may reveal short absences or other incidents due to manual handling which may have escaped the accident reporting system. Enquiries may need to be made in respect of the health surveillance records of individuals (see **M3015** above).

Accident treatment records

Accident books or other treatment records may identify regularly recurring minor accidents which do not necessarily result in any absences from work, eg cuts or abrasions from sharp or rough loads.

Operating procedures etc.

Reference to manual handling activities and precautions which should be adopted may be contained in operating procedures, safety handbooks etc.

Work sector information

Trade associations and similar bodies publish information on manual handling risks and how to overcome them.

HSE Guidance

Much guidance is available from the HSE – some of the more important publications were referred to earlier in the chapter. The table below gives details of additional guidance relating to specific work sectors.

HSE publications on manual handling in specific work sectors include:

HSG 196	Moving food and drink: Manual handling solutions for the food and drink industries (2000)
HSG 225	Handling home care. Achieving safe, efficient and positive outcomes for care workers and clients (2002)
HSG 234	Caring for cleaners (2003)
INDG 269	Checkouts and musculoskeletal disorders (1998)
INDG 318	Manual handling solutions in woodworking (2000)
INDG 332	Manual packing in the brick industry (2000)
INDG 348	Mark a parcel save a back (2002)

INDG 390	Choosing a welding set. Make sure you can handle it (2009)
FIS 33	Roll cages and wheeled racks in the food and drinks industries (2003)
IACL 105	Handling the news: Advice for employers on manual handling of bundles (1999)
IACL 106	Handling the news: Advice for newsagents and employees on safe handling of bundles (2006)
IACL 103	Manual handling in the textile industry (1998)
—	Manual handling in the health services (1998)
—	Manual handling in paper mills (2005)
—	Picking up the pieces: Prevention of musculoskeletal disorders in the ceramics industry (1996)
—	Handling rubber: Reducing manual handling injuries in the rubber industry (1999)

Further general sources of guidance on handling are listed in **M3028**.

Making the assessment

[M3020] As with other types of assessments, visits to the workplace are an essential part of manual handling assessments. Manual handling operations need to be observed (sometimes in some detail) and discussions with those carrying out manual handling work, their safety representatives and their supervisors or managers need to take place.

The assessment checklist provided on the following pages provides guidance on the factors which may need to be taken into account during the observations and discussions and also on possible measures to reduce the risk. The checklist utilises the four main factors specified in the Regulations – task, load, working environment and individual capability. In practice each of these will have greater or lesser importance, depending upon the manual handling activity.

MANUAL HANDLING ASSESSMENT CHECKLIST
THE TASK

ASSESSMENT FACTORS	**REDUCING THE RISK**
DISTANCE OF THE LOAD FROM THE TRUNK	TASK LAYOUT
BODY MOVEMENT / POSTURE	USE THE BODY MORE EFFECTIVELY
• Twisting • Reaching • Stooping • Sitting	eg SLIDING OR ROLLING THE LOAD
DISTANCE OF MOVEMENT	SPECIAL SEATS
• Height (? grip change) • Carrying (? over 10m) • Pushing or pulling	RESTING PLACE / TECHNIQUE USE OF TROLLEYS etc TECHNIQUE / FLOOR SURFACE
RISK OF SUDDEN MOVEMENT	
FREQUENT / PROLONGED EFFORT	AWARENESS / TRAINING IMPROVED WORK ROUTINE
	eg Job rotation
CLOTHING / FOOTWEAR ISSUES	
RATE OF WORK IMPOSED BY A PROCESS	ADEQUATE REST OR RECOVERY PERIODS

THE LOAD

ASSESSMENT FACTORS	**REDUCING THE RISK**
HEAVY	MAKE THE LOAD LIGHTER TEAM LIFTING MECHANICAL AIDS
BULKY (Any dimensions above 75cm)	MAKE THE LOAD SMALLER
UNWIELDY (Offset centre of gravity)	INFORMATION / MARKING
DIFFICULT TO GRASP (Large, rounded, smooth, wet, greasy)	MAKE IT EASIER TO GRASP
	• Handles • Handgrips • Indents • Slings • Carrying devices • Clean the load
UNSTABLE • Contents liable to shift • Lacking rigidity	WELL-FILLED CONTAINERS USE OF PACKING MATERIALS SLINGS / CARRYING AIDS
SHARP EDGES / ROUGH SURFACES	AVOID OR REDUCE THEM GLOVES OR OTHER PPE
HOT OR COLD	CONTAINERS (? insulated)

THE WORKING ENVIRONMENT

ASSESSMENT FACTORS	REDUCING THE RISK
SPACE CONSTRAINTS AFFECTING POSTURE	ADEQUATE GANGWAYS, FLOORSPACE, HEADROOM
UNEVEN, SLIPPERY OR UNSTABLE FLOORS	WELL MAINTAINED SURFACES PROVISION OF DRAINAGE SLIP-RESISTANT SURFACING GOOD HOUSEKEEPING PROMPT SPILLAGE CLEARANCE
VARIATION IN WORK SURFACE LEVEL eg steps, slopes, benches HIGH OR LOW TEMPERATURES HIGH HUMIDITY POOR VENTILATION	STEPS NOT TOO STEEP GENTLE SLOPES UNIFORM BENCH HEIGHT BETTER ENVIRONMENTAL CONTROL RELOCATING THE WORK SUITABLE CLOTHING
STRONG WINDS OR OTHER AIR MOVEMENT	RELOCATING THE WORK ALTERNATIVE ROUTE HANDLING AIDS TEAM HANDLING
LIGHTING	SUFFICIENT WELL-DIRECTED LIGHT
MOVING WORKPLACE • Boat • Train • Vehicle	

INDIVIDUAL CAPABILITY

ASSESSMENT FACTORS	REDUCING THE RISK
INDIVIDUAL FACTORS STRENGTH HEIGHT FLEXIBILITY STAMINA PROBLEMS WITH CLOTHING ETC. PREGNANCY INJURY / HEALTH PROBLEM • Back trouble • Hernia • Temporary injury	BALANCED WORK TEAMS 'SELF SELECTION' MAKE ASSISTANCE AVAILABLE ENCOURAGE FITNESS SPECIAL ASSESSMENTS FORMAL RESTRICTIONS CLEAR INSTRUCTIONS

TASK FACTORS OPERATIONS OR SITUATIONS REQUIRING PARTICULAR	FORMALISED WORK PROCEDURES CLEAR INSTRUCTIONS ADEQUATE TRAINING
• Awareness of risks • Knowledge • Method of approach • Technique	• Recognising potentially hazardous operations • Dealing with unfamiliar operations • Correct use of handling aids • Proper use of PPE • Environmental factors • Importance of housekeeping • Knowing one's own limitations • Good technique

Observations

[M3021] Sufficient time should be spent in the workplace to observe a sufficient range of the manual handling activities taking place and to take account of possible variations in the loads handled (raw materials, finished product, equipment etc). Aspects to particularly be considered include:

• comparison of actual methods used with those specified in operating procedures, industry standards etc;
• the level and manner of use of handling aids and their effectiveness;
• handling techniques adopted;
• physical conditions, eg access, housekeeping, floor surfaces, lighting;
• variations between employees, eg size, strength, technique adopted:
• damage to products, containers or equipment.

Discussions

[M3022] In discussing manual handling issues with workpeople an enquiring approach should be taken, utilising open-ended questions wherever possible. Aspects which might be discussed include:

• whether normal conditions are being observed;
• workers' awareness of procedures, rules etc;
• workers' training in the use of handling aids or handling techniques;
• possible variations in work activities, eg product changes, seasonal variations, differences between day and night shift;
• what happens if handling equipment or handling aids break down or are unavailable;
• whether assistance is available if required;
• which manual handling operations are difficult to perform;
• what manual handling problems have been experienced;
• have there been any injuries with manual handling;
• reasons precautions are not being taken.

Notes

[M3023] Notes should be taken during the assessment of anything which might be of relevance such as:

- further detail about manual handling risks found;
- handling risks which are discounted as insignificant;
- handling aids available;
- other control measures in place and appearing effective;
- problems identified or concerns expressed;
- alternative precautions worthy of consideration;
- related procedures, records or other documents.

These rough notes will eventually form the basis of the assessment.

After the assessment

Assessment records

[M3024] HSE guidance states that the significant findings of the assessment must be recorded unless:

- the assessment could very easily be repeated and explained at any time because it is so simple and obvious; or
- the manual handling operations are of low risk, are going to last only a very short time, and the time taken to record them would be disproportionate.

As should be the case in relation to other types of assessment, it is better to make a simple record if there is any doubt.

There is no standard format for recording an assessment. The main HSE Guidance booklet (L23) includes a sample record format together with a useful worked example. This record format includes a checklist of assessment factors under the four main headings (task, load, working environment, capability) in a similar listing to the assessment checklist we have provided. Alongside each checklist item the assessor has space to:

- identify whether the risk from this factor is low, medium or high;
- provide a more detailed description of the problems;
- identify possible remedial action.

Some will no doubt find this checklist format meets their needs. However, many assessment factors often have little or no relevance to a specific manual handling operation. As a result, the eventual record contains a lot of blank paper, with only occasional comments. Consequently it is often preferable to keep a separate checklist (as with the checklist earlier in the chapter — see M3020), and only include within the record a reference to the risks, precautions and recommended improvements which are relevant to the operation being assessed. Two worked examples of this type of record are included below.

MANUAL HANDLING ASSESSMENT		BROADACRES DEVELOPMENT	DATE: 4 April 2011	ASSESSMENT BY: A Backworth
No.	ACTIVITY	RISKS	EXISTING PRECAUTIONS	RECOMMENDATIONS
1	HANDLING OF STATIONERY AND PRINTED MATERIALS	Heaviest stationery items are 5 ream boxes of paper (approximately 12 kg). Some packs of printed promotional material weigh over 20 kg. Items are moved from main store on the ground floor to smaller stores on other floors.	Suppliers deliver items directly to main store. Trolley and lift used for transporting between main store and smaller stores.	Require printing suppliers to supply materials in packs weighing no more than 12 kg. Ensure paper and printed materials are stored on shelves between knee and shoulder height.
2	HANDLING OF ARCHIVED RECORDS	Records are normally stored in file boxes weighing no more than 20 kg. They are moved into and out of the record store on the third floor. (Older records are removed for storage off-site by a specialist company.)	File boxes are kept below shoulder height on substantial racking. They are moved around using the trolley and lift. Movement to off-site storage is carried out by Facilities staff who have had manual handling training.	Access to the Accounts section of the archive store racking was obstructed by various items which must be removed. Some records were stored in oversize boxes — this practice must cease.
3	MOVEMENT OF EQUIPMENT PC monitors, TV sets, projection equipment etc.	These items (of varying weights) are moved between the equipment store on the first floor and locations where they are required.	The trolley and lift are used where appropriate. IT staff have been trained in handling techniques. Some racking is provided in the equipment store but this is inadequate and many items are stored on the floor.	Training section assistants should also receive handling technique training. Provide dust coats (so that staff can carry dirty items close to their bodies). Install additional racking in the equipment store and keep heavier items on more accessible shelves.
4	MOVEMENT OF FURNITURE Desks, seating etc.	Desks and seating are moved as required for individual workstations. (Major moves are carried out by contractors.)	Facilities section staff have had manual handling training. Team lifting is used for larger items, eg desk tops. The lift is used as appropriate.	Training section assistants require training (see above). Suitable racking must be provided in the equipment store for desk and table tops.

MANUAL HANDLING ASSESS-MENT	BROADACRES DEVELOPMENT	DATE: 4 April 2011	ASSESSMENT BY: A Backworth
No. ACTIVITY	RISKS	EXISTING PRECAUTIONS	RECOMMENDATIONS
	Seating and tables are moved by Facilities section staff when required for major meetings and training events.	Desk and table tops are kept in the equipment store.	

GOLD STAR ENGINEERING MAINTENANCE STORES	MANUAL HANDLING ASSESSMENT Assessment by: *T Wrightson*	Date: *26 February 2011*
ACTIVITIES/ RISKS	EXISTING PRECAUTIONS	RECOMMENDATIONS
Handling activities involve the movement of equipment and materials within the stores building and the yard outside, together with two storage containers and a flammable liquid store situated within the yard. Both equipment and materials must also be moved to locations within the factory.	A fork lift truck is used to handle heavier items within the yard. A hand-operated pallet truck and a lightweight lifting truck are available. Several trucks and trolleys are provided including a stair-climbing truck, gas cylinder and fire extinguisher trolleys and a drum transporter. Skates, bogie units and dollies are also available for transporting heavy items.	
There is considerable variation in the sizes and weights of items handled. Some items are palletised, some are supplied in boxes and some are kept in bins and cages.	Most internal and external floor surfaces are in good condition.	A section of the floor just inside the main stores entrance requires repair.
	There is a gradual ramp into the main stores building but not into the storage containers or the flammable liquid store (the latter having a sill at its entrance).	Suitable permanent or removable ramps should be provided to allow handling aids to be used in the storage containers and flammable liquid store.
Drums (up to 100 litres capacity), gas cylinders and large fire extinguishers must also be moved.	Lighting standards are good in the main stores building, the flammable liquid store and the yard, but there are no lights in the storage containers.	Lighting must be provided within the storage containers.
Racking and cupboards are provided within the building and inside the outside storage containers.	Heavier items are generally stored either on the floor or on accessible levels on racking or in cupboards.	Ensure walkways between racking are kept clear.

GOLD STAR ENGINEERING	MANUAL HANDLING ASSESSMENT	
MAINTENANCE STORES	Assessment by: *T Wrightson*	Date: *26 February 2011*
ACTIVITIES/ RISKS	EXISTING PRECAUTIONS	RECOMMENDATIONS
Some items to be handled have sharp or rough edges.	Latex coated knitted gloves (for grip) and gloves with chrome leather palms are available (as are gloves for chemical risks). Most stores' staff have attended a one-day course in manual handling techniques tailored to their needs.	Develop a formal training programme to ensure that all stores staff receive training in manual handling technique and the correct use of the various handling aids available.

Review and implementation of recommendations

[M3025] As for other types of risk assessments, there will be a need to review recommendations and prepare action plans for their implementation. Recommendations will need to be followed up in order to ensure that they have actually been implemented and also to assess any new risks which may have been introduced. This latter point is particularly relevant to manual handling as changes in working practice may have resulted in different techniques being adopted (with their own attendant risks) or the introduction of handling aids may have created additional training needs.

Assessment review

[M3026] *Regulation 4(2)* of the *Manual Handling Operations Regulations 1992* requires a manual handling assessment to be reviewed if:

(a) there is reason to suspect that it is no longer valid; or
(b) there has been a significant change in the manual handling operations to which it relates.

As is the case for other assessments, a review would not necessarily result in the assessment being revised, although a revision may be deemed to be necessary. A significant manual handling related injury should automatically result in a review of the relevant assessment. It is recommended that assessments are reviewed periodically in any case in order to take account of small changes in work practices which will eventually have a cumulative effect. The periods for such reviews might vary between two and five years – depending upon the degree of risk involved and the potential for gradual changes to take place.

Training

[M3027] The need for specific manual handling technique training may frequently be one of the recommendations made as a result of a manual handling assessment. With manual handling activities featuring in a wide range of work sectors many employers have chosen to provide all or a significant proportion of their workforce with general training in good handling technique. Such training is often provided as part of a more comprehensive health and safety induction programme. Guidance on the content of such general training is contained in paragraphs 197 and 198 of HSE Guidance booklet (L23) and also in other HSE publications dealing with specific work sectors (see the table **M3019**) and also **M3028** below.

Further general sources of guidance

[M3028] In addition to the work sector specific guidance in **M3019**, many other HSE publications provide guidance of relevance to manual handling. This includes:

L 23   Manual Handling. Manual Handling Operations Regulations 1992 (as amended).
Guidance on Regulations. (2004)

HSG 60   Upper limb disorders in the workplace. (2002)

HSG 61   Health surveillance at work. (1999)

HSG   Manual Handling. Solutions you can handle. (1994)
115

INDG   Getting to grips with manual handling. A short guide. (2004)
143

INDG   Manual handling assessment charts. (2008)
383

INDG   Are you making the best use of lifting and handling aids?
398   (2011)

The HSE website (www.hsegov.uk) is also a useful source of guidance on manua
l handling. Sections of particular relevance include:

* Moving goods safely;
* Musculoskeletal disorders.

Noise at Work

Roger Tompsett

Introduction

[N3001] This chapter is aimed at company directors, legal advisors, health and safety officers, staff representatives and others who need a comprehensive, concise reference source on the issue of noise at work.

It will be helpful to anyone who needs to commission a workplace noise assessment or to undertake a noise control project. It will also assist those who wish to devise long-term noise reduction and equipment-buying strategies.

Despite decades of warnings and legislation, noise-induced hearing loss is still a huge but underrated problem, with a slow, cumulative and irreversible effect. A particular problem with noise-induced hearing loss is the long time between exposure and the damage becoming apparent. As a result, legislation has been tightened, so that many more workplaces now fall into the net (**N3002**).

This chapter discusses the scale of problems arising from workplace noise exposure, the legal obligations on employers to reduce the risk of noise-induced hearing damage both in the short term and as a long-term obligation (**N3012**); and the risks in terms of compensation claims (**N3022**). An overview of social security benefits is also provided (**N3018**).

The chapter gives a short introduction to the perception of sound and the various indexes used to measure noise (**N3004**). It provides general guidance on methods of assessing and reducing noise (see **N3029** and **N3035**), and suggests sources of more detailed guidance on these matters (see **N3023**).

Scale of the problem and latest developments

[N3002] According to the 2002 General Household Survey undertaken by the UK Office for National Statistics, 16 per cent of adults have hearing difficulties. In 2002, 19 per cent of men had a hearing difficulty compared with 13 per cent of women, and at the age of 75 or over, 52 per cent of men have a hearing difficulty. Four per cent of men and 3 per cent of women wear a hearing aid, and the likelihood of older people wearing a hearing aid doubled between 1979 and 2002. Nevertheless, 62 per cent of people with a hearing aid found that it did not overcome their hearing problems.

It is very likely that the difference in hearing difficulties between men and women is due to men's greater likelihood of workplace noise exposure. A Medical Research Council survey in 1997–98 estimated that there were

507,000 people in Great Britain suffering from hearing difficulties as a result of exposure to noise at work. This is much higher than HSE's own estimate of 87,000 based on its 2001–02 self-reported work-related Illness (SWI) survey. The large difference is probably due to the survey methods, and suggests that even people who suffer hearing difficulties are often unaware that their noisy workplace has contributed to their problems.

The Department of Social Security (DSS) pays disablement benefit to people who suffer at least 50 decibels (dB) of noise-induced hearing loss in both ears. This level of hearing loss is roughly equivalent to listening to the television or a conversation through a substantial brick wall. Moreover, claimants must have been employed for at least ten years in a specified noisy occupation. Despite these restrictions, the number of new awards appears to have stabilised in the range of 260 to 325 per year, after a significant fall during the 1980s and 1990s from around 1,500 a year in 1984–85 to a low of 226 in 2000. There are currently about 14,000 people – 99 per cent of them men – receiving the benefit from DSS.

The number of claimants with between 35 and 49 decibels of hearing loss (still a severe disability, equivalent to listening through a substantial partition wall in a house) shows an improving trend. According to statistics from audiological examinations, the number of claimants has fallen from 1200 in 1995 to 710 in 1999.

It is likely that some of this improvement is a result of the contraction of traditionally noisy industries such as ship-building and coal mining, with the armed forces now showing some of the highest rates of incidence. However, the music and entertainment industry has been shown to be at significant risk of excessive workplace noise exposure. After being given a short period to adjust to the new controls, it is now required to be fully compliant. It is not just amplified music that is risky: classical orchestras and theatre musicians can also be subject to high levels of noise exposure.

Recent research has confirmed the synergistic risk to hearing caused by ototoxic substances (poisonous to the hearing mechanism), which include some quite commonly used solvents, thereby adding justification to the requirement in the Regulations to consider this effect.

Overview of the Control of Noise at Work Regulations 2005

[N3002A] The *Control of Noise at Work Regulations 2005 (SI 2005 No 1643)*, which came into force in April 2006, give effect to the EU's Physical Agents (Noise) Directive (2003/10/EC). These supersede the *Noise at Work Regulations 1989 (SI 1989 No 1790)* and since 6 April 2008, have also applied to the music and entertainment industry. The regulations will apply to places outside Great Britain where the *Health and Safety at Work, etc Act 1974* applies, such as offshore installations. However, the regulations will not normally apply to the master and crew of a seagoing ship, and where they do, the noise limits will not apply until 6 April 2011.

The Regulations, which are fully described later in this chapter, emphasise a risk-based approach. This could mean that in some circumstances noise measurements may not be needed, whilst in other circumstances more stringent precautions may be needed, including ongoing health surveillance (which includes hearing tests). Where hearing damage is identified, the employer must ensure that a doctor examines the employee to consider whether the damage is likely to have been caused by noise exposure, and if so, a range of actions must be taken to reduce the risk to this and other employees.

Assessments must be done by a person who has 'sufficient information, instruction and training'. This change of emphasis means that it is likely that previous 'competent persons' under the 1989 Regulations would need refresher courses to bring them up to date with the present requirements.

Compared with the 1989 Regulations, the 2005 Regulations reduce the action levels by 5 dB, requiring exposure to be assessed when it exceeds 80 dB(A) averaged over eight hours (in some cases, averaged over five days each of eight hours), but there are also many other subtle changes. Hearing protection must be made available at 80 dB(A) personal daily exposure, as well as information and training. Health surveillance must be made available where exposure reaches 85 dB(A) and hearing protectors must be worn. There is a limit on exposure of 87 dB(A), the exposure to be calculated to include the effect of any hearing protection that is provided. Corresponding changes are made to action levels in relation to peak sound pressures.

Hearing damage

[N3003] We all lose hearing acuity with age, and this loss is accelerated and worsened by exposure to excessive noise. Fortunately, in early life, we have much greater hearing acuity than modern life demands, and a small loss is readily compensated by turning up the volume of the radio or television.

Loud noises cause permanent damage to the nerve cells of the inner ear in such a way that a hearing aid is ineffective. At first, the damage occurs at frequencies above normal speech range, so that the sufferer may have no inkling of the problem, although it could be identified easily by an audiogram, which measures the sensitivity of the ear at a number of frequencies across its normal range. As exposure continues, the region of damage progresses to higher and lower frequencies. Damage starts to extend into the speech range, making it difficult to distinguish consonants, so that words start to sound the same. Eventually, speech becomes a muffled jumble of sounds.

Some people are much more susceptible to hearing damage than others. A temporary dullness of hearing or tinnitus (a ringing or whistling sound in the ears) when emerging from a noisy place are both indicative of damage to the nerve cells of the inner ear. Because these symptoms tend to disappear with continued exposure, sufferers may think that their ears have become 'hardened' to the noise, whereas in fact they have lost their ability to respond to the noise. Even when hearing has apparently returned to normal, a little of the sensitivity is likely to have been lost. Therefore, anyone who experiences dullness of hearing or tinnitus after noise exposure must take special care to avoid further exposure. They may also be advised to seek medical advice and an audiogram.

Very loud impulsive sounds, such as those that may arise from cartridge-operated tools, firearms and explosives, can be particularly damaging to hearing, as they occur too suddenly for the ear's defensive mechanisms to operate. These can cause instantaneous damage to hearing.

Apart from such catastrophic exposures, the risk of hearing loss is closely dependent on the 'noise dose' received, and especially on the cumulative effect over a period of time. British Standard BS5330 provides a procedure for estimating the risk of 'hearing handicap' due to noise exposure. Handicap is there defined as a hearing loss of 30 dB, which is sufficiently severe to impair the understanding of conversational speech or the appreciation of music. This may be compared with the loss of 50 dB required for DSS disability benefit.

The perception of sound

[N3004] Sound is caused by a rapid fluctuation in air pressure. The human ear can hear a sensation of sound when the fluctuations occur between 20 times a second and 20,000 times a second. The rate at which the air pressure fluctuates is called the frequency of the sound and is measured in Hertz (Hz). The loudness of the sound depends on the amount of fluctuation in the air pressure. Typically, the quietest sound that can be heard (the threshold of hearing) is zero decibels (0 dB) and the sound becomes painful at 120 dB. Surprisingly perhaps, zero decibels is not zero sound. The sensitivity and frequency range of the ear vary somewhat from person to person and deteriorate with age and exposure to loud sounds.

The human ear is not equally sensitive to sounds of different frequencies (i.e. pitch): it tends to be more sensitive in the frequency range of the human voice than at higher or lower frequencies (peaking at about 4 kHz). When measuring sound, compensation for these effects can be made by applying a frequency weighting, usually the so-called 'A' weighting, although other weightings are sometimes used for special purposes.

The ear has an approximately logarithmic response to sound: for example, every doubling or halving in sound pressure gives an apparently equal step increase or decrease in loudness. In measuring environmental noise, sound pressure levels are therefore usually quoted in terms of a logarithmic unit known as a decibel (dB). To signify that the 'A' weighting has been applied, the symbol of dB(A) is often used. However, current practice tends to prefer the weighting letter to be included in the name of the measurement index. For example, dB(A) L_{eq} and dB L_{Aeq} both refer to the 'A' weighted equivalent sound level in decibels.

Depending upon the method of presentation of two sounds, the human ear may detect differences as small as 0.5 dB(A). However, for general environmental noise the detectable difference is usually taken to be between 1 and 3 dB(A), depending on how quickly the change occurs. A 10 dB(A) change in sound pressure level corresponds, subjectively, to an approximate doubling or halving in loudness. Similarly, a subjective quadrupling of loudness corresponds to a 20 dB(A) increase in sound pressure level (SPL). When two sounds of the same SPL are added together, the resultant SPL is approximately 3 dB(A) higher than each of the individual sounds. It would require approximately *nine* equal sources to be added to an original source before the subjective loudness is doubled.

Noise indices

[N3005] The sound pressure level of industrial and environmental sound fluctuates continuously. A number of measurement indices have been proposed to describe the human response to these varying sounds. It is possible to measure the physical characteristics of sound with considerable accuracy and to predict the physical human response to characteristics such as loudness, pitch and audibility. However, it is not possible to predict *subjective* characteristics such as annoyance with certainty. This should not be surprising: one would not expect the light meter on a camera to be able to indicate whether one was taking a good or a bad photograph, although one would expect it to get the physical exposure correct. Strictly speaking, therefore, a meter can only measure sound and not noise (which is often defined as sound unwanted by the recipient): nevertheless, in practice, the terms are usually interchangeable.

Equivalent continuous 'A' weighted sound pressure level, L_{Aeq}

[N3006] This unit takes into account fluctuations in sound pressure levels. It can be applied to all types of noise, whether continuous, intermittent or impulsive. L_{Aeq} is defined as the steady, continuous sound pressure level which contains the same energy as the actual, fluctuating sound pressure level. In effect, it is the energy-average of the sound pressure level over a period which must be stated, e.g. $L_{Aeq, \, (8-hour)}$.

This unit is now being put forward as a universal noise index, because it can be used to measure all types of noise, although it has yet to supplant older units in certain cases, particularly for the assessment of road traffic noise where calculation techniques and regulatory criteria have not been updated.

Levels measured in L_{Aeq} can be added using the rules mentioned earlier. There is also a time trade-off: if a sound is made for half the measurement period, followed by silence, the L_{Aeq} over the whole measurement period will be 3 dB less than during the noisy half of the period. If the sound is present for one-tenth of the measurement period, the L_{Aeq} over the whole measurement period will be 10 dB less than during the noisy tenth of the measurement period. This is a cause of some criticism of L_{Aeq}: for discontinuous noise, such as may arise in industry, it does not limit the maximum noise level, so it may be necessary to specify this as well.

Daily personal noise exposure level, $L_{EP,d}$

[N3007] This is used in the *Control of Noise at Work Regulations 2005 (SI 2005 No 1643))* as a measure of the total sound exposure a person receives during the day. It is formally defined in a *Schedule* to the Regulations.

$L_{EP,d}$ is the energy-average sound level (L_{Aeq}) to which a person is exposed over a working day, disregarding the effect of any ear protection which may be worn, adjusted to an eight-hour period. Thus, if a person is exposed to 90 dB L_{Aeq} for a four-hour working day, their $L_{EP,d}$ is 87 dB, but if they work a 12–hour shift, the same sound level would give them an $L_{EP,d}$ of 91.8 dB.

Weekly personal noise exposure level, $L_{EP,\ w}$

This is the daily personal noise exposure level averaged over a 5-day working period. For example, if a person has a daily personal noise exposure of 90 dB $L_{EP,d}$ on 2 ½ days, but had no noise exposure for the rest of the week, their $L_{EP,\ w}$ would be 87 dB. However, if they worked for six days a week, their $L_{EP,\ w}$ would be 90.8 dB.

Maximum 'A' weighted sound pressure level, $L_{Amax,T}$

[N3008] The maximum 'A'-weighted root-mean-square (rms) sound pressure level during the measurement period is designated L_{Amax}. Sound level meters indicate the rms sound pressure level averaged over a finite period of time. Two averaging periods (T) are defined, S (slow) and F (fast), having averaging times of 1 second and 1/8th second. It is necessary to state the averaging period. Maximum SPL should not be confused with peak SPL which is a measure of the instantaneous peak sound pressure.

Peak sound pressure

[N3009] Instantaneous peak sound pressure is used in the assessment of explosive sounds, such as from gunshots, cartridge-operated tools and blasting, and in hearing damage assessments of this type of sound. For the *Control of Noise at Work Regulations 2005 (SI 2005 No 1643)*, it is measured in C-weighted decibels. C-weighting is used because it responds equally to sounds over the wide range of frequencies that can occur in an explosive event. It can only be measured with specialist instruments that have special circuitry designed to respond to, and capture, very short, intense events.

Background noise level, L_{A90}

[N3010] L_{A90} is the level of sound exceeded for 90 per cent of the measurement period. It is therefore a measure of the background noise level, in other words, the sound drops below this level only infrequently. (The term 'background' should not be confused with 'ambient', which refers to *all* the sound present in a given situation at a given time).

Sound power level, L_{WA}

[N3011] The sound output of an item of plant or equipment is frequently specified in terms of its sound power level. This is measured in decibels relative to a reference power of 1 pico-Watt (dB re 10^{-12} W), but it must not be confused with sound pressure level. The sound pressure level at a particular position can be calculated from a knowledge of the sound power level of the source, provided the acoustical characteristics of the surrounding and intervening space are known. As a crude analogy, the power of a lamp bulb gives an indication of its light output, but the illumination of a surface depends on its distance and orientation from the source, and the presence of reflecting and obstructing objects in the surroundings.

To give a rough idea of the relationship between sound power level and sound pressure level, then for a noise source which is emitting sound uniformly in all directions close above a hard surface in an open space, the sound pressure level 10m from the source would be 28 dB below its sound power level.

General legal requirements

[N3012] Statutory requirements relating to noise at work *generally* are contained in the *Control of Noise at Work Regulations 2005 (SI 2005 No 1643)*, which replace the *Noise at Work Regulations 1989 (SI 1989 No 1790)* and the previous Department of Employment (voluntary) Code of Practice on Noise (1972). The Regulations require employers to assess the risk of excessive noise at work affecting the health and safety of their employees and to identify the measures needed to meet the requirements of the Regulations.

There are specific provisions relating to vibration exposure which are covered in the VIBRATION chapter. This is addition to the provision contained in the *Social Security (Industrial Injuries) (Prescribed Diseases) Regulations 1985 (SI 1985 No 967)* and the *Reporting of Injuries, Diseases and Dangerous Occurrences Regulations 1985 (SI 1985 No 2023)*. Occupational deafness and certain forms of vibration-induced conditions, i.e. vibration-induced white finger, are prescribed industrial diseases for which disablement benefit is payable (though the 14 per cent disablement rule will obviously limit the number of successful claimants (see **C6011** COMPENSATION FOR WORK INJURIES/DISEASES). Damages may also be awarded against the employer (see further **N3022** below).

Control of Noise at Work Regulations 2005 (SI 2005 No 1643)

[N3013] The following duties are laid on employers:

(1)	To make (and update where necessary) a suitable and sufficient assessment of the risk to the health and safety of employees who are liable to noise exposure at or above any **Lower Exposure Action Levels** which are:
(a)	80 dB(A) daily or weekly personal noise exposure; and
(b)	a peak sound pressure of 135 dB(C).

(2)   To reduce noise exposure to as low a level as is reasonably practicable, by means of organisational and technical measures other than personal hearing protectors, where any employee is likely to be exposed above any **Upper Exposure Action Levels** which are:
    (a)   85 dB(A) daily or weekly personal noise exposure; and
    (b)   a peak sound pressure of 137 dB(C).

(3)   The employer shall ensure that his employees are not exposed to noise above any **Exposure Limit Values**, which are:
    (a)   87 dB(A) daily or weekly personal noise exposure; and
    (b)   a peak sound pressure of 140 dB(C).

(4)   If an Exposure Limit Value is exceeded, the exposure must be immediately reduced to below the Exposure Limit Value. The reason for exceeding the value must be identified and organisational and technical changes made to prevent it happening again.

(5)   In calculating exposure to an Exposure Limit Value, the effect of any personal hearing protectors provided by the employer may be taken into account. However, the effect of personal hearing protectors cannot be taken into account when assessing exposure to the Lower or Upper Exposure Action Values.

(6)   The risk assessment shall be conducted by assessing the noise exposure of employees and relating this to the Action and Limit values. The assessment shall be made by:
    (a)   observing specific working practices;
    (b)   reference to probable levels of noise from the equipment and particular working conditions;
    (c)   noise measurements, if necessary.

(7)   The risk assessment shall include consideration of the level, type and duration of exposure, including the peak sound pressure, and the effect of exposure on people whose health is at particular risk from noise exposure.

(8)   Consideration must also be given to interactions such as between noise and vibration, noise and the use of substances at work that may be ototoxic (poisonous to the hearing mechanism). Indirect effects on health and safety such as the ability to hear audible warning systems or other sounds must also be considered.

(9)   The risk assessment must consider any information provided by the equipment manufacturers; the availability of quieter equipment or noise-reducing equipment; exposure to noise at the workplace beyond normal working hours, such as in rest-rooms; appropriate information from health surveillance, including published information, and the availability of adequate personal hearing protectors.

(10)   Employees or their representatives should be consulted on the risk assessment.

(11)   A record of the risk assessment is to be kept. The record must include the significant findings of the risk assessment and the measures to be taken in order to eliminate or control the risks. This must include details of the provision of hearing protection and the provision of information and training to employees on the detection and prevention of hearing damage.

The regulations require health surveillance (ie hearing checks) for employees who may have a health risk arising from their noise exposure. HSE guidance indicates that this applies to all employees exposed above an upper action value. It suggests that ideally hearing checks would be made before an employee is exposed to noise, and repeated at two or three-yearly intervals, unless a problem indicates more frequent testing is needed.

Hearing protection is to be made available on request to any employee exposed above the lower exposure action value, and must be provided to any employee exposed at or above the upper exposure action level. Any area where an employee is likely to be exposed at or above the upper action level must be designated as a hearing protection zone, and no employee should enter that area unless they are wearing ear protectors.

(From the *Health and Safety (Safety Signs and Signals) Regulations 1996 (SI 1996 No 341)*)

Legal obligations of designers, manufacturers, importers and suppliers of plant and machinery

[N3014] The *Supply of Machinery (Safety) Regulations 1992 (SI 1992 No 3073* as amended) require manufacturers and suppliers of noisy machinery to design and construct such machinery so that the risks from noise emissions are reduced to the lowest level taking account of technical progress. Information on noise emissions must be provided when specified levels are reached.

If a machine is likely to cause people at work to receive a daily personal noise exposure exceeding the first or peak action levels, adequate information on noise must be provided. If the second or peak action levels are likely to be exceeded, this should include a permanent sign or label, or if the machine may be noisy in certain types of use, an instruction label which could be removed following noise testing.

Specific legal requirements

Agriculture

[N3015] The *Agriculture (Tractor Cabs) Regulations 1974 (SI 1974 No 2034)* (as amended by *SI 1990 No 1075*) provide that noise levels in tractor cabs must not exceed 90 dB(A) or 86 dB depending on which annex is relevant in the certificate under Directive 77/311/EEC. [*SI 1974 No 2034, Reg 3(3)*].

Equipment for construction sites and other outdoor uses

[N3016] An EC Directive relating to the noise emission in the environment by equipment for use outdoors (Directive 2000/14/EC) was issued in 2000 to consolidate and update 23 earlier directives relating to the noise emission of construction plant and certain other equipment. The Directive was implemented by the *Noise Emission in the Environment by Equipment for Use Outdoors Regulations 2001 (SI 2001 No 1701)*. The earlier statutory instruments were revoked from 3 January 2002 and replaced by the requirements of the new statutory instrument.

The 2001 Regulations apply to a very wide range of powered equipment used outdoors. They do not apply to most non-powered equipment, equipment for the transport of goods or persons by road, rail, air or on waterways, or equipment for use by military, police or emergency services.

Each type of equipment is subject to a maximum permissible sound power level, dependent on the power of the equipment's engine or other drive system. The permissible sound power level was reduced in January 2006 by 2 or 3 dB, relative to the level permitted in 2002.

Examples of the type of equipment covered include: cranes, hoists, excavators, dozers, dump trucks, compressors, generators, pumps, power saws, concrete breakers, compactors, glass recycling containers, lawnmowers and other gardening equipment.

The Regulations define the noise testing methods in great detail and also require a standardised marking of conformity indicating the guaranteed sound power level. Interestingly, it has been noted that many manufacturers simply use the maximum permitted level as the guaranteed level, which would seem to indicate that they see no marketing advantage in promoting quieter products.

The Supply of Machinery (Safety) Regulations, 1992 (SI 1992 No 3073 as amended) are also relevant to noise emissions from machinery.

Code of practice for noise and vibration control

[N3017] The code of practice for basic information and procedures for noise and vibration control, BS 5228: Parts 1 to 5, gives detailed guidance on the assessment of noise and vibration from construction sites, open-cast coal extraction, piling operations, and surface mineral extraction. Part 1, revised in

1997, gives detailed noise calculation and assessment procedures. Its predecessor (published in 1984) was an approved code of practice under the *Control of Pollution Act, 1974.* Part 1 of the code of practice is principally concerned with environmental noise and vibration, but also briefly recites the *Noise at Work Regulations 1989 (SI 1989 No 1790)* (which are now superseded).

Compensation for occupational deafness

Social Security

[N3018] The most common condition associated with exposure to noise is occupational deafness. Deafness is prescribed occupational disease A10 (see OCCUPATIONAL HEALTH AND DISEASES). Prescription rules for occupational deafness have been extended three times, in 1980, 1983 and 1994. It is defined as: 'sensorineural hearing loss amounting to at least 50 dB in each ear being the average of hearing losses are 1, 2 and 3 kHz frequencies, and being due, in the case of at least one ear, to occupational noise'. [*Social Security (Industrial Injuries) (Prescribed Diseases) Amendment Regulations 1989 (SI 1989 No 1207), Reg 4(5)*]. Thus, the former requirement for hearing loss to be measured by pure tone audiometry no longer applies. Extensions of benefit criteria relating to occupational deafness are contained in the *Social Security (Industrial Injuries) (Prescribed Diseases) Regulations 1985 (SI 1985 No 967)*, which are amended, as regards assessment of disablement for benefit purposes, by the *Social Security (Industrial Injuries) (Prescribed Diseases) Amendment Regulations 1994 (SI 1994 No 2343)*.

Conditions for which deafness is prescribed

[N3019] Occupational deafness is prescribed for a wide range of occupations involving working with, or in the immediate vicinity of:

- mechanically-powered grinding and percussive tools used on metal; on rock and stone in quarries or in mines, in sinking shafts and in tunnelling;
- sawing masonry blocks, jet channelling of stone and masonry, vibrating metal moulding boxes in the concrete products industry;
- forging, cutting, burning and casting metal;
- plasma spray guns engaged in the deposition of metal;
- air arc gouging;
- high pressure jets of water or a mixture of water and abrasive material in the water jetting industry (including work under water);
- textile manufacturing where the work is undertaken in rooms or sheds in which there are machines engaged in weaving man-made or natural (including mineral) fibres or in the high speed false twisting or fibres; or mechanical bobbin cleaning;

- machines for automatic moulding, pressing or forming of glass hollow ware and continuous glass toughening furnaces; and spinning machines using compressed air to produce glass wool or mineral wool;
- a wide range of woodworking and metalworking machines;
- chain saws in forestry;
- work in ships' engine rooms;
- work on gas turbines in connection with performance testing on a test bed, installation testing of replacement engines in aircraft, and acceptance testing in armed service fixed wing combat planes.

The Regulations are detailed and should be consulted for precise definitions of these occupations. [*Social Security (Industrial Injuries) (Prescribed Diseases) Regulations 1985 (SI 1985 No 967); Social Security (Industrial Injuries) (Prescribed Diseases) Amendment No 2 Regulations 1987 (SI 1987 No 2112); Social Security (Industrial Injuries) (Prescribed Diseases) Amendment Regulations 1994 (SI 1994 No 2343)*].

'Any occupation' covers activities in which an employee is engaged under his contract of employment. The fact that the work-force is designated, classified of graded by reference to function, training or skills (e.g. labourer, hot examiner, salvage and forge examiner) does not of itself justify a conclusion that each separate designation, classification or grading involves a separate occupation (*Decision of the Commissioner No R(I) 3/78*).

'Assistance in the use' of tools qualifies the actual use of tools, not the process in the course of which tools are employed. Thus, a crane driver who positions bogies to enable riveters to do work on them and then goes away, assists in the process of getting bogies repaired, which requires use of pneumatic tools – is not assistance in the actual use of tools, for the purposes of disablement benefit. The position is otherwise when a crane holds a bogie in suspension to enable riveters to work safely on them. Here the crane driver assists in the actual use of pneumatic percussive tools (*Decision of the Commissioner No R(I) 4/82*).

Conditions under which benefit is payable

[N3020] For a claimant to be entitled to disablement benefit for occupational deafness, the following conditions under the *Social Security (Industrial Injuries) (Prescribed Diseases) Regulations 1985 (SI 1985 No 967)* currently apply:

- they must have been employed:
  - (i)   at any time on or after 5 July 1948, and
  - (ii)   for a period or periods amounting (in the aggregate) to at least ten years.
- there must be permanent sensorineural hearing loss, and loss in each ear must be at least 50 dB; and
- at least loss of 50 dB in one ear must be attributable to noise at work [*SI 1985 No 967, Sch 1, Pt 1*].
  (There is a presumption that occupational deafness is due to the nature of employment [*SI 1985 No 967, Reg 4(5)*]);

- the claim must be made within five years of the last date when the claimant worked in an occupation prescribed for deafness [*SI 1985 No 967, Reg 25(2)*];
- any assessment of disablement at less than 20 per cent is final [*SI 1985 No 967,Reg 33*].

A person, whose claim for benefit is turned down because he/she had not worked during the five years before the claim in one of the listed occupations, may claim if he/she continued to work and later met the time conditions. If a claim is turned down as the disability is less than 20 per cent, the claimant must wait three years before re-applying. If, by waiting three years, it would be more than five years since the applicant worked in one of the listed occupations, the three-year limit is waived.

Assessment of disablement benefit for social security purposes

[N3021] The extent of disablement is the percentage calculated by:

(a) determining the average total hearing loss due to all causes for each ear at 1, 2 and 3 kHz frequencies; and

(b) determining the percentage degree of disablement for each ear; and then

(c) determining the average percentage degree of binaural disablement.

[Social Security (Industrial Injuries) (Prescribed Diseases) Amendment Regulations 1989, Reg 4(2)].

The following chart (TABLE 1 below), shows the scale for all claims made on/after 3 September 1979.

Table 1	
Percentage degree of disablement in relation to hearing loss	
Hearing loss	Percentage degree of disablement
50–53 dB	20
54–60 dB	30
61–66 dB	40
67–72 dB	50
73–79 dB	60
80–86 dB	70
87–95 dB	80
96–105 dB	90
106 dB or more	100
[*Social Security (Industrial Injuries) (Prescribed Diseases) Regulations 1985, Reg 34, Sch 3, Pt 2, as amended*]	

Any degree of disablement, due to deafness at work, assessed at less than 20 per cent, must be disregarded for benefit purposes. [*Social Security (Industrial Injuries) (Prescribed Diseases) Amendment Regulations 1990 (SI 1990 No 2269)*].

Action against employer at common law

[N3022] There is no separate action for noise at common law; liability comes under the general heading of negligence. Indeed, it was not until as late as 1972 that employers were made liable for deafness negligently caused to employees (*Berry v Stone Manganese Marine Ltd* [1972] 1 Lloyd's Rep 182). Absence of a previous general statutory requirement on employers regarding exposure of employees to noise sometimes led to the law being strained to meet the facts (*Carragher v Singer Manufacturing Co Ltd* 1974 SLT (Notes) 28 relating to the *Factories Act 1961, s 29*: 'every place of work must, so far as is reasonably practicable, be made and kept safe for any person working there', to the effect that this is wide enough to provide protection against noise). Admittedly, it is proper to regard noise as an aspect of the working environment (*McCafferty v Metropolitan Police District Receiver* [1977] 2 All ER 756 where an employee, the plaintiff, who was a ballistics expert, suffered ringing in the ears as a result of the sounds of ammunition being fired from different guns in the course of his work. When he complained about ringing in the ears – the ballistics room had no sound-absorbent material on the walls and he had not been supplied with ear protectors – he was advised to use cotton wool, which was useless. It was held that his employer was liable since it was highly foreseeable that the employee would suffer hearing injury if no steps were taken to protect his ears, cotton wool being useless). Moreover, although there are specific statutory requirements to minimise exposure to noise (in agriculture and offshore operations and construction operations, see **N3015, N3016** above), these have generated little or no case law.

The main points established at common law are as follows.

(a)   As from 1963, the publication date by the (then) Factory Inspectorate of 'Noise and the Worker', employers have been 'on notice' of the dangers to hearing of their employees arising from over-exposure to noise (*McGuinness v Kirkstall Forge Engineering Ltd* (1979), unreported). Hence, consistent with their common law duty to take reasonable care for the health and safety of their employees, employers should 'provide and maintain' a sufficient stock of ear muffs.
This was confirmed in *Thompson v Smiths*, etc. (see (*d*) below). However, more recently, an employer was held liable for an employee's noise-induced deafness, even though the latter's exposure to noise, working in shipbuilding, had occurred *entirely before* 1963. The grounds were that the employer had done virtually nothing to combat the *known* noise hazard from 1954–1963 (apart from making earplugs available) (*Baxter v Harland & Wolff plc* [1990] IRLR 516, NI CA). This means that, as far as Northern Ireland is concerned, employers are liable at common law for noise-induced deafness as from 1 January 1954 – the earliest actionable date. Limitation statutes preclude employees suing prior to that date (*Arnold v Central Electricity Generating Board* [1988] AC 228).

(b)   Because the true nature of deafness as a disability has not always been appreciated, damages have traditionally not been high (*Berry v Stone Manganese Marine Ltd* [1972] 1 Lloyd's Rep 182 – £2,500 (halved because of time limitation obstacles); *Heslop v Metalock (Great*

Britain) Ltd (1981) Observer, 29 November – £7,750; *O'Shea v Kimberley-Clark Ltd* (1982) Guardian, 8 October – £7,490 (tinnitus); *Tripp v Ministry of Defence* [1982] CLY 1017 – £7,500).

(c)   Damages will be awarded for exposure to noise, even though the resultant deafness is not great, as in tinnitus (*O'Shea v Kimberley-Clark Ltd*, (1982) Guardian, 8 October).

(d)   Originally the last employer of a succession of employers (for whom an employee had worked in noisy occupations) was exclusively liable for damages for deafness, even though damage (i.e. actual hearing loss) occurs in the early years of exposure (for which earlier employers would have been responsible) (*Heslop v Metalock (Great Britain) Ltd* (1981) Observer, 29 November). More recently, however, the tendency is to *apportion* liability between offending employers (*Thompson, Gray, Nicholson v Smiths Ship Repairers (North Shields) Ltd*; *Blacklock, Waggott v Swan Hunter Shipbuilders Ltd*; *Mitchell v Vickers Armstrong Ltd* [1984] IRLR 93). This is patently fairer because some blame is then shared by the original employer(s), whose negligence would have been responsible for the actual hearing loss.

(e)   Because of the current tendency to apportion liability, even in the case of pre-1963 employers (see *McGuinness v Kirkstall Forge Engineering Ltd* above), contribution will take place between earlier and later insurers.

(f)   Although judges are generally reluctant to be swayed by scientific/statistical evidence, the trio of shipbuilding cases (see (*d*) above) demonstrates, at least in the case of occupational deafness, that this trend is being reversed (see Table 2 below, the 'Coles-Worgan classification'); in particular, it is relevant to consider the 'dose response' relationship published by the National Physical Laboratory (NPL), which relates long-term continuous noise exposure to expected resultant hearing loss. This graph always shows a rapid increase in the early years of noise exposure, followed by a trailing off.

(g)   Current judicial wisdom identifies three separate evolutionary aspects of deafness, i.e. (i) hearing loss (measured in decibels at various frequencies); (ii) disability (i.e. difficulty/inability to receive everyday sounds); (iii) social handicap (attending musical concerts/meetings etc.). That social handicap is a genuine basis on which damages can be (*inter alia*) awarded, was reaffirmed in the case of *Bixby, Case, Fry and Elliott v Ford Motor Group* (1990), unreported.

General guidance on noise at work

Guidance on the Control of Noise at Work Regulations 2005

[N3023] HSE provides guidance on the Control of Noise at Work Regulations in the publication *Controlling noise at work: Guidance on the Control of Noise at Work Regulations 2005*. (L108, ISBN 0-7176-6164-4 HSE Books). It covers the following areas:

- the revised approach to the management and control of workplace noise;
- legal duties of employers to prevent damage to hearing;
- guidance on the assessment and management of noise risks;
- practical advice on noise control;
- selection of quieter machinery;
- selection and use of personal ear protection;
- Development of health surveillance procedures;
- duties of designers, manufacturers, importers and suppliers of noisy machinery;
- guidance for technical advisors;

Guidance on specific types of equipment

[N3024] HSE has also issued a number of guidance documents for specific types of machine or activity, including:

- Hazards arising from ultrasonic processes (Local Authority Circular 59/1, 2000).
- Metal cutting circular saws (Noise Reduction (Local Authority Circular 59/2, 2000).
- The reduction of noise from pneumatic breakers, hammers, drills, etc (Local Authority Circular 59/4, 2000).
- Reducing noise from CNC presses (Engineering Information Sheet No. 39, 2002).

Other leaflets cover the Food and Drink Industry, Paper and Board Mills, Foundries, and various types of power presses.

Other guidance

[N3025] Some working environments carry particular difficulties in the assessment and control of noise exposure, mainly because of the variability of the noise levels and the length of exposure. The following specific guidance may be of assistance:

- *A guide to reducing the exposure of construction workers to noise*, R A Waller, Construction Industry Research and Information Association, CIRIA Report 120 (1990).
- *Offshore installations: Guidance on design and construction, Part II, Section 5*, Department of Energy (1977).
- *Guide to health, safety and welfare at pop concerts and similar events*, Health and Safety Commission, Home Office and the Scottish Office, HMSO (1993).
- *Practical solutions to noise problems in agriculture*, HSE Research Report 212 (2004).

Reducing noise in specific working environments

[N3026] Although many noisy industries are in decline, noise is now becoming an issue in some newer industries. In particular, the application of the Regulations to the music and entertainment industries has posed new challenges, not least because operators have two conflicting needs: their clients want loud music, but a safe working environment must be provided for staff. It is, of course, an interesting point as to whether clients will in the future be looking for compensation for noise-induced hearing loss.

Discos and night clubs

[N3027] The music levels on the dance floor are often over 100 dB(A) to provide the buzz that clients expect. A number of different techniques may need to be used in order to achieve adequate control of the noise exposure of staff. Loudspeakers should be kept away from the bar area. The bar itself can be designed to be partially screened from the loudspeakers and with sound absorbent to reduce noise levels in the locality. A high-quality sound system with low distortion levels can produce the bass frequencies that clients want, with less high-frequency distortion. This results in lower overall sound pressure levels than a poorer system would produce. Such a system can have the additional advantage of reducing environmental noise levels, which are often problematic. Some clubs provide a quiet 'chill-out' room to which staff and members can retreat.

Ear plugs can be used by staff patrolling the dance floor and seating areas, and jobs can be rotated between staff so that the average noise exposure is reduced. Most ear plugs will reduce sound at speech frequencies by more than at the bass frequencies usual in disco music and this may reduce the ability of staff to understand speech. It is important to ensure that ear plugs give adequate protection at the important bass frequencies.

Concert halls and theatres

[N3027.1] Musicians in orchestras can be subjected to noise exposure in excess of not only the Exposure Action Levels, but also the Exposure Limit Values. In one study, 40% of musicians in one concert hall had a daily personal noise exposure in excess of the Upper Exposure Action Value, indicating that hearing protection must be used. In a theatre orchestra, all sections had a daily personal noise exposure exceeding the Lower Exposure Action Level, but some exceeded the Exposure Limit Value. The regular performances meant that weekly values were also exceeded. Conductors need to be aware of the exposure of musicians and to consider how this can be limited, for example by using a greater range of dynamics, rehearsing the orchestra in sections, or planning and varying the repertoire. In the case of the theatre orchestra, the use of sound-absorbent screens between sections and a perspex sight-panel at the top can reduce sound to a small extent. Drums can also be very noisy and may need to be separately screened off, with the drummer using headphones to reduce noise exposure whilst allowing him to hear the performance.

Outdoor music venues

[N3027.2] Outdoor music venues require considerable thought and attention, as the correct balance between audience sound levels and environmental noise must be obtained, at the same time ensuring that noise exposure of staff is controlled. This will require the various staff duties to be identified and the noise exposure of each duty to be separately assessed. External staff must be included in the assessment. Noise exposure can be modelled by computer simulations to an extent, or it can be based on measurements made at similar venues. It is likely that certain areas will need to be designated as hearing protection zones. It will be important to ensure that everyone is made aware of the risks and the control measures necessary to manage the risks.

Call centres

[N3028] Call centres can be very noisy simply because of the large number of people talking on the telephone. If poorly designed, each operator will inevitably raise their voice so as to be heard above the noise produced by their neighbours. This becomes self-defeating as others then raise their voices further. They may also need to raise the volume of their earpiece, and this could lead to excessive sound exposure. There have been successful cases of compensation for deafness (sometimes only in one ear) suffered by people using telephone and radio headsets, and employers should ensure that staff do not adjust them to be excessively loud.

A number of techniques must be used to achieve a satisfactory acoustic environment. Firstly the basic acoustics of the call centre must be satisfactory. Working areas should be as sound-absorbent as feasible, with acoustic ceiling tiles and a thick, sound absorbent carpet. Acoustic screens should be used between workstations unless they are widely-spaced. The screens need to have thick absorbent surfaces, and should be high enough to screen operators when they are sitting at their workstations.

Staff training must include discussion of appropriate voice and earpiece levels, and the risks of excessive levels. They need to be encouraged to keep their voices down: it should be explained that it is not necessary for them to raise their voices in order for the customer to hear them clearly.

Workplace noise assessments

Competent person

[N3029] The *Control of Noise at Work Regulations 2005 (SI 2005 No 1643)* require that any person who carries out work in connection with the employer's duties under the Regulations has 'suitable and sufficient information, instruction and training.' The Regulations do not stipulate a particular form of certification for such people, although the Institute of Acoustics and other organisations do provide formal training and certification that is intended to meet these requirements.

Sound level meters

[N3030] It is usually more economical to employ an acoustics consultant for workplace noise assessments than to send a member of staff on a training course and to hire a sound level meter to do the measurements in-house. It may be helpful if the report is clearly independent should an employee make a claim. Suitable sound level meters are expensive. A precision meter is required and these cost some thousands of pounds, depending on their capabilities, and they must be sent away for laboratory calibration at regular intervals. Those meters that can be bought cheaply from electronics suppliers are not suitable: they could be misleading even as a 'screening' tool. This is because cheap meters do not have an accurate response to rapid or impulsive sounds, and usually do not contain an L_{Aeq} function as this requires an on-board computer chip for the complex averaging process.

Assessing the daily personal noise exposure

[N3031] It is important to appreciate that noise assessments must determine the daily personal noise exposure of each individual member of staff – it is not sufficient simply to measure the noise level in various parts of the workplace and to mark the noisy areas as 'hearing protection zones'.

In some workplaces, this is not particularly difficult to do: if the employee has a fixed workstation and fixed working routine, it will only be necessary to measure at a location representative of their head position, for a sufficient period of time to cover a few work cycles – maybe five minutes or so.

However, it is quite common for a worker to use a variety of machines or to move between different parts of the works during a normal day. Some workers, such as crane or lorry drivers, may have a work environment where noise levels vary a great deal and it is not practicable for the noise surveyor to accompany them throughout a representative period.

These considerations mean that a workplace noise assessment needs planning before it is made. The first item is to establish the different types of work that are being done on the site: this will need a meeting with the site manager to determine how many staff there are, the jobs that they do, shift-working patterns, and the hours of work, including when and where meal-breaks are taken. It is also important to consider the sensitivities of the workforce: it is helpful to explain in brief terms what you are doing and to confirm what they do. They will also want to know the results, but the surveyor should explain that this cannot be provided immediately as the daily personal noise exposure needs to be calculated. In some workplaces, it may be necessary to ensure that union representatives are consulted.

A scale plan of the workplace will be needed so that measurement locations, noise sources and noisy areas can be recorded for the report. The surveyor will wish to ensure that all machinery and plant is tested, and will want to note anything that is out of use during the survey.

Dosimeters

[N3032] Where there are people such as lorry drivers, whom the surveyor cannot accompany, then it will probably be best to ask them to wear a dosimeter. This is a small electronic device about the size of a mobile phone that clips into the top pocket, with a microphone that clips to the lapel or some convenient part of the clothing near to the ear. The dosimeter does not record the actual sound, only the sound level. (It is not a 'spy in the cab'.) Dosimeters usually contain some protection to deter tampering or falsification and to prevent the worker obtaining the readout, although the surveyor will need to be alert to the risks.

Where workers move between various workstations during their working day, then it is usually acceptable to measure the noise level at each workstation and to calculate the daily personal noise exposure from an estimate of the proportion of the day that each worker typically spends at each one.

Although it may seem that this is less accurate than using a dosimeter, it is important to recognise that each of the work-cycles will usually produce a slightly different noise exposure, as there are so many things that affect the noise levels. Moreover, it is necessary to obtain actual workstation noise levels in order to identify any ear protection zones that need to be marked out. In some borderline cases, maybe dosimetry might settle the matter, but it would be better to look at ways of reducing noise levels, since the effect on any one individual's hearing is difficult to determine, even though the Regulations define action levels precisely.

Hazardous environments

[N3033] Special care is required in hazardous working environments, especially those with an explosion risk (such as gas and petroleum processing plants). Many sound level meters use a high 'polarisation voltage' for the microphone, which could produce a spark. 'Intrinsically safe' meters can be obtained for such instances, although they can be less convenient for ordinary use.

Requirements for the report

[N3034] A written report of the survey must be produced and kept as a record of the survey at least until the next survey is made. The report should state all the information used in the survey, and should contain a plan showing the areas surveyed and hearing protection zones that were identified. It must identify individual staff or jobs where noise exposure exceeds any of the action levels. The report should also contain a statement of the employer's duties resulting from the findings of the survey. However, it is not usual for the report to include detailed specifications for any noise remediation that might be needed. This is usually a separate study, as the issues cannot be known prior to the survey and there will usually be a number of options which will require detailed discussion with the employer.

Noise reduction

[N3035] It is the duty of every employer to reduce the risk of hearing damage from noise at work as far as reasonably practicable. This is an on-going and long-term obligation. Unfortunately, it is often difficult to reduce the noise of many items of equipment, particularly hand-held tools. Remarkably, it is still the case that quietness does not seem to be a primary selling point for tools and equipment, and it should not be assumed that newer equipment will necessarily be quieter. Moreover, although many items of equipment must be labelled with their sound power, the label often shows only the permitted maximum level rather than the actual level, so this is of little help in choosing the quietest items. Furthermore, it should not be assumed that smaller items of plant will be quieter than larger items. This is because larger items often have more scope for noise control in the form of extra casings, insulation and silencers, as size and weight are less of a limitation.

Under the *Control of Noise At Work Regulations*, buying policy has become a very important aspect of noise reduction, since it is often difficult, inconvenient and expensive to retrospectively fit noise control equipment. A noise target level should be set and then equipment bought to comply with this, taking into account the combined effect of many items operating simultaneously, and the effect of reverberant build-up of sound in the workplace.

When designing noise reduction measures, it is essential to identify the major sources of noise first, since no amount of work in reducing minor sources will have any effect on the overall noise level. Noise is usually created by a moving or vibrating part of the equipment inducing vibrations in the surrounding air. Noise can be reduced by preventing such vibrations from reaching the operator of the equipment: the most effective solution is to block 'air-paths' with a solid, heavy casing. Even tiny air-holes can allow the sound to escape, which is a problem when a flow of cooling air is needed, or where there must be continuous access to feed work-pieces.

A different approach is to fit 'damping material' to vibrating surfaces, to reduce the amplitude of vibration. Damping material is especially effective on thin sheet metal, such as equipment casings, hoppers, containers and ductwork. This is because even a small amount of vibration can cause their surfaces to 'radiate' a large amount of noise energy. The vibrational energy is absorbed in the damping material, rather than being radiated. There are a number of ways of applying damping material: a bitumastic or rubber sheet can be glued to the surface, a second sheet of metal can be glued by an elastic adhesive to the surface, or the item can be made from a laminate composed of two sheets of metal glued with an elastic adhesive.

It should be noted that sound absorptive material (mineral fibre, etc) is not used as a damping material. However, it can be used to provide a vibration break when cladding a vibrating surface. For example, if noise is breaking out from a duct, it can be wrapped in mineral fibre and then clad with a thin metal sheet. The mineral fibre separates the inner and outer cladding, thereby preventing much of the vibration in the wall of the duct from reaching the cladding, which therefore radiates much less noise.

Sometimes, a less 'springy' material than steel can be used. At one time lead was popular for being both heavy and floppy, but is now considered too hazardous for general use. However, plasterboard is a popular acoustic material, as it is reasonably heavy, has good natural damping and is inexpensive. It can often be used to prevent noise breakout, as long as there is a vibration break between it and the radiating surface.

Ventilation and cooling is often a problem, since enclosing a piece of equipment can prevent 'natural' cooling. Moreover, fans used to provide a flow of cooling air can be very noisy in themselves, not counting the noise that escapes along the ductwork. The best solution in such cases is to place the fan on the inside of the casing and to fit an attenuator ('silencer') on the external side of the fan. Attenuators are essentially rectangular or circular ducts with an acoustic absorbent lining.

There are many different types of fan and it is important to select one that is appropriate for the task in hand, taking account of the volume of air that needs to be moved and the pressure (resistance) that the fan must work against. Small, fast-running fans tend to be noisier than larger slower-running fans for a given duty. The amount of noise produced by a fan and the amount of attenuation provided by a silencer are well-characterised and so ventilation systems can be designed with confidence over their performance, although this is a specialist job.

Sometimes, developments intended for another purpose can have the additional benefit of reducing noise. For example, there was a need to develop a circular saw blade that gave better cutting accuracy. It was found that vibration of the saw blade caused irregularities in the cut, so slots were cut in the surface of the blade and these were filled with a resin to maintain the strength of the blade. This extra damping was effective in reducing vibration, thereby improving both the cut and the noise levels.

Despite the technology, it is often difficult to obtain adequate noise reduction. This is because of the huge range of the hearing mechanism. A 10 dB reduction requires the escaping noise energy to be reduced by 90 per cent, a 20 dB reduction requires the escaping noise energy to be reduced by 99 per cent, and a 30 dB reduction requires 99.9 per cent of the energy to be removed. Because of these considerations, it may be advisable to employ a specialist who can design a solution with a minimum of trial and error.

Ear protection

[N3036] There are two main categories of ear protectors:

(i)   circumaural protectors (ear muffs) which fit over and surround the ears, and which seal to the head by cushions filled with soft plastic foam or a viscous liquid; and

(ii)   ear plugs, which fit into the ear canal.

Ear protectors will only be effective if they are in good condition, suit the individual and are worn properly. Ear protectors can be uncomfortable, especially if they press too firmly on the head or ear canal, or cause too much sweating. Some users may be tempted to bend the head-band so that the muffs do not press so tightly, which reduces the effectiveness of the seal. Ear muffs are also less effective for people with thick spectacle frames, long hair or beards that prevent the muff from sealing fully against the head.

The amount of protection differs between different designs, and usually ear protectors give more protection at middle frequencies and less protection at low frequencies. Where the noise has strong tones (often described by words such as rumble, drone, hum, whine, screech) then particular care is needed to ensure that the protectors give adequate attenuation at those frequencies. However, it can be counterproductive to choose ear protectors that give far more attenuation than necessary, since these will be heavier, with greater head-band or insert pressure and so less comfortable to wear. And an ear protector that is not worn does not give any protection at all. Ear plugs should not be used by people with ear infections and certain other ear conditions, and it may be desirable for such people to obtain medical advice.

Because individuals differ greatly in their preference, the employer should select more than one type of suitable protector and offer the user a personal choice where possible. British Standard BS EN 485: 1994 gives guidance and recommendations for the selection, use, care and maintenance of hearing protectors.

Ear protectors need to be maintained in a clean and hygienic condition. Insert types should not be shared between people, and muffs should not usually be shared. With ear muffs, it is necessary to check that the sealing cushion is not damaged, as this will seriously reduce its effectiveness. On some makes it is possible to change a damaged cushion. The ear cups have a foam lining which must not be discarded. If it gets dirty, it can usually be taken out and washed in detergent, then dried and replaced.

If the headband tension becomes weak, the muffs should be discarded.

Some earplugs are designed for one-off use, but others are reusable. These need to be kept in a protective container when not in use, and regularly washed according to the manufacturer's instructions. If the seals become damaged or hardened, the plugs must be replaced.

The employer needs to ensure that staff know how to check the condition of their ear protection and how to obtain replacements as soon as the need occurs.

Acoustic booths

[N3037] In some workplaces, it may be possible for operators to be located in a control booth from which they can monitor the plant, only occasionally entering the noisy area. They may still need put on ear defenders when leaving the acoustic booth, since it may require only a few seconds or minutes to exceed the allowable daily noise dose.

Even where it is not practical to have a separate control booth, it may be possible to use an acoustic refuge – a small cabin with an open side – that will give some respite from the noise. However, this should not be expected to give more than a nominal sound reduction in most cases.

Work rotation

[N3038] Work rotation can sometimes be a method of reducing the daily personal noise exposure. The staff work part of the day in a noisy area and the rest of the day in a quiet area. If a person works for half their day in a place with a noise level of 88 dB L_{Aeq} (four hours) and the other half of their day in a place with a noise level of 70 dB L_{Aeq} (4 hours) they would just meet a noise exposure of 85 dB. However, this is not a very satisfactory method as it can easily go wrong. They may spend longer than expected in the noisy workplace, or there may be a higher than expected noise level in the quiet area, for example. And again, whilst this approach could satisfy the legal requirements, some people may suffer hearing loss from shorter exposure to loud noise.

Noise barriers

[N3039] Noise barriers often give very limited noise reduction within buildings. This is because sound can be reflected over the barrier from the ceiling, pipe-work, and other overhead fittings. It is usually necessary to use noise barriers in conjunction with sound baffles suspended from the ceiling. Suspended sound baffles on their own have very limited application. They can sometimes be helpful in very reverberant work-spaces where the staff are not exposed to the direct sound of noisy machines.

PA and music systems

[N3040] Some workplaces use public address and music systems. One study found that they were a major source of excessive noise in lorry cabs. Care must be taken to ensure that in noisy workplaces these system are not so loud that they create excessive noise in themselves. This also applies to people who wear radio or telephone headsets.

Sound systems for emergency purposes

[N3041] Voice messages can be superior to bells or sirens to convey warnings and instructions in emergencies. However, these must be properly audible in noisy places. British Standard BS 7443: 1991 gives specifications for sound systems for emergency purposes.

Occupational Health and Diseases – An Overview

Leslie Hawkins

What is occupational health?

[O1001] Occupational health is concerned with the prevention, identification and management of health issues relating to employment. It is very broad and covers not only the fact that work and the conditions of work may cause ill-health, but also includes the management of employees who have a health problem which may impact on their work or their employment. One definition of occupational health is that it concerns *the effects of work on health and the effects of health on work*. The primary aim of good occupational health practice should always be to prevent ill health. Occupational health practitioners (see O1010 below) are health professionals who are additionally trained to understand the complex relationships between work and the impact this may have on the health of the individual. Occupational ill-health is caused by exposure to a wide variety of chemical, physical and organisational aspects of the work and it's environment and to be effective an occupational health practitioner must have a good understanding of how the environment at work (in it's broadest context) might affect a person's health.

Why is it important?

Sickness absence

[O1002] Occupational health is an important, but often neglected, aspect of an organisation's management. It is important to the organisation for a number of reasons. Employees are the most important, and often the most costly, asset any organisation has. Sickness absence can be very costly to an organisation. Of *short term* sickness, 40–50% is potentially avoidable and can be put down to a variety of reasons ranging from drug and alcohol abuse to lack of commitment and interest in the job and poor management. 20% of absence is *long term* – employees who are absent for weeks or months with a condition that renders them unable to travel to work and/or to carry out their expected duties. This 20% of long-term sickness absence accounts for 70% of the costs of sickness absence. A good occupational health service should be able to considerably reduce the level of *avoidable* sickness absence and hasten the successful return to work of those who are absent long term, with significant financial savings for the organisation (see R2001 Rehabilitation).

Personnel support

[O1003] An occupational health service can be an invaluable support to the Human Resources/Personnel function in managing the many aspects of employment that relate to an individual's medical condition and their consequent fitness to work. As well as managing sickness absence (see **O1002** above), the occupational health service can advise on fitness to work at the point of recruitment or placement (pre-employment or pre-placement health screening). This will reduce the risks to the organisation of recruiting someone to a position for which an existing medical condition renders them vulnerable to further illness or injury or for which they may be simply unfit to undertake the proposed work. It may be necessary at the point of recruitment to consider the implications of the *Disability Discrimination Act 2005*, and to determine what restrictions may have to be placed on an employee's work or what reasonable alterations may have to be made to the work or work-place in order to avoid discrimination. Other medical aspects of employment on which the occupational health service can advise include retirement on grounds of ill-health, temporary restrictions on work or working times to aid recovery or rehabilitation, keeping at-risk employees under health surveillance, and advice on managing medical conditions which are impacting on the capacity of the employee to undertake expected duties.

Legal duties

[O1004] Another important reason for having access to occupational health advice is to ensure that statutory and civil legal obligations are being met. Several regulations place a legal requirement on the employer to place at-risk employees under health surveillance or to have regular medical examinations (usually by a doctor 'appointed' or 'approved' for the particular requirements of the regulation). The *Ionising Radiations Regulations 1999 (SI 1999 No 3232)*, the *Control of Lead at Work Regulations 2002 (SI 2002 No 2676)* and the *Control of Asbestos at Work Regulations* 2006 (SI 2006 No 2739) are examples of regulations which may require medical surveillance by a relevant doctor. In addition, other areas of work, such as working with respirators and working offshore, may require medical examinations either under statute or under industry standards.

Risk assessment

[O1005] The occupational health department may also need to be involved in risk assessments where these involve health considerations. Under the *Management of Health and Safety at Work Regulations 1999 (SI 1999 No 3242)* there is a general requirement to undertake an assessment of the risk to health and safety of employees. In some circumstances the risk may be to the employee's health (a classic example being the risk assessment for work-related stress (see **O1054** below) and the occupational health department can play an important role in helping to determine those aspects of the job that may be stressful. In many instances the risk assessment will have to take account of an individual's vulnerability (perhaps because of age, pregnancy, or previous history of injury or illness) and the occupational health department is in a unique position to translate a previous medical history (or current condition) into the additional risks that this may pose to the employee. As importantly, occupational health can also advise on what needs to be done to avert the risk either by additional protection or modifications to the job or work routine. Examples of where there may be a need to take account of individual vulnerability, in addition to pregnancy, include a history of upper limb disorder, a history or earlier episode of back pain, earlier periods of stress-related disorder, a history of a skin allergy and existing respiratory disorder (such as asthma). There are of course many other examples, but it is extremely important to take such vulnerabilities fully into account in the risk assessment; failure to do so may increase the chances of successful litigation for damages for failure to exercise an appropriate level of duty of care. Other regulations such as the *Control of Substances Hazardous to Health Regulations 2002 (SI 2002 No 2677)*, the *Health and Safety (Display Screen Equipment) Regulations 1992 (SI 1992 No 2792)*, and the *Manual Handling Operations Regulations 1992 (SI 1992 No 2793)*, may require an occupational health input into the risk assessment and may require employees to be placed under health surveillance.

Relationships between occupational health and safety

[**01006**] Generally, it is only the larger organisations which have an occupational health department, although many smaller employers have access to occasional occupational health or medical advice through buying in a service from a local provider or from the local general practice. Most organisations, though, will have some sort of safety provision, either a dedicated safety department (again mostly in the larger organisations) down to a safety person who fits safety in with their main job. Occupational health and safety are two different, but complementary, means of achieving the same goal – ensuring the well being of both the organisation and its employees.

Prevention is better than cure

Being reactive

[**01007**] Inevitably there are cases of employees becoming ill, either as a result of work or for reasons completely unrelated to work. Either way, the role of the occupational health service is to advise management on the employment issues associated with the individual. If the illness is work related or thought to be so, then the advice relates to what can be done to ensure that there is no exacerbation of the condition when the employee returns to work. If it is unrelated to work then the advice may be concerned with how a person's newly acquired health condition may impact on work, either by affecting the way he or she can perform the expected duties or the way in which the medical condition may now render that person more susceptible to the conditions of work. The employer may require advice on employees who are taking frequent short term sick leave or a period of long term sick leave and seek advice on how to manage the sickness absence or manage the rehabilitation and return to work. They may need advice on medical retirement on grounds of permanent ill health. All of these are examples of occupational health, albeit playing an important role, but being *reactive* – reacting to the needs of the organisation in managing the consequences of staff being unfit for work.

A good occupational health service should also be *proactive* – prevention being better than cure. Being proactive involves the anticipation of the possibility of ill health and putting into place measures, which are aimed at preventing or minimising the risk of this occurring. Prevention starts with the pre-employment or pre-placement health assessment (see O1003 above). This is aimed at preventing employees being recruited to jobs for which they are medically unfit or making sure that where preventative measures need to be taken (e.g. vaccinations) then these are undertaken before commencement of work. During employment, prevention is then exercised through a number of initiatives that a good occupational health service will be engaged in. Health education and health promotion programmes are a good way of improving and maintaining the health of the workforce. This may be strictly confined to those health issues related to the work, (for example how to look after your skin, how to prevent hearing damage, or how to prevent musculoskeletal disorders) or it may be concerned with general health promotion such as smoking cessation, women's or men's health, look after your heart campaigns and exercise, diet and obesity. It is obviously good if the organisation is actively involved in promoting health generally and maintaining a healthy workforce will bring benefits and this fits with the Government's initiative in 'Securing Health Together' (see O1020 below). However, not all organisations will see general health promotion as being their responsibility and may not engage in this on account of cost and lack of resources.

Prevention and risk assessment

[O1008] Prevention is also achieved by involving occupational health in the risk assessment process (see O1005 above). This ensures that relevant health issues are fully included in the risk assessment process. As discussed at O1005 above, the occupational health practitioner can help ensure that the medical aspects of the risk assessment are fully included and that individuals' vulnerabilities, where relevant, are taken into account. In addition the occupational health input can ensure that, where required, appropriate health surveillance is set up and carried out as a means of early detection and hence prevention of work-related disorders.

Health surveillance

[O1009] Health surveillance is another important aspect of prevention in which the occupational health department has an important role to play. *Regulation 6* of the *Management of Health and Safety at Work Regulations (SI 1999 No 3242)* requires that:

every employer shall ensure that his employees are provided with such health surveillance as is appropriate having regard to the risks to their health and safety which are identified by the assessment.

The Guidance and Approved Code of Practice to the Regulations (HSE L21), provides the criteria where health surveillance is considered to be necessary. These are where:

(a) there is an identifiable disease or adverse health condition related to the work concerned;

(b) valid techniques are available to detect indications of the disease or condition;

(c) there is reasonable likelihood that the disease or condition may occur under the particular conditions of work; and

(d) surveillance is likely to further the protection of the employees to be covered.

Health surveillance can be carried out by anyone competent to do so; it does not necessarily have to be an occupational health professional. Simple surveillance can be done by a trained supervisor (checking for early signs of dermatitis for example), or even by self-surveillance if the individuals at risk know what to look for and who to report concerns to. However, where surveillance includes physiological or biochemical measurements, such as hearing function, lung function, or urine or blood tests, then an occupational health doctor or nurse would usually be necessary. As mentioned at **O1004** above, under some regulations it is mandatory for the health surveillance to be carried out by an approved or appointed doctor. What is most important, however the health surveillance is carried out, is that there is a recognised means of referral for expert help if the health surveillance reveals someone with early signs or symptoms of a work-related condition.

In summary, whilst occupational health has an important role to play in investigating and managing illness and the consequences of ill health that occur during the course of employment, the core of occupational health practice should be to maintain a healthy workforce and prevent as many work-related conditions as possible from arising in the first place.

The occupational health team

[01010] The occupational health 'team' varies enormously from one organisation to another depending on the complexity of the working environment and the perceived needs of the organisation. It can vary from no provision at all up to a multi-professional team. The majority of organisations who make provision for occupational health, do so with the core service being provided by an occupational health nurse who may be full or part time depending on the size of the organisation. Occupational health nurses, working on their own, will usually have a network of external support, including a physician. The external support should include other occupational health nurses who can participate in clinical audit.

Occupational health nurses

[01011] *Occupational health nurses* (also called occupational health advisors) are registered nurses whose names appear on part 3 of the specialist community public health nursingregister held by the Nursing and Midwifery Council (NMC). In addition, an occupational health nurse will hold a qualification in occupational health; i.e. OHN Certificate, Post Graduate Diploma or Degree and their names will be recorded on the specialist register of the Nursing and Midwifery Council.

Experienced occupational health nurses can undertake a wide range of duties including determining fitness for work by pre-employment health assessment, contributing health considerations to the risk assessment process, undertaking health surveillance, and advising the organisation on health issues relating to individuals, such as sickness absence management and return to work. Occupational health nurses also usually play a key role in preventing work-related illness and promoting good heath among the workforce. Many occupational health nurses are also competent to undertake environmental assessments such as simple measurements of noise, lighting etc., and workstation assessments.

The competent occupational health nurse will recognise his or her limitations and be prepared to refer an individual for a medical opinion from a doctor or refer a problem that it outside of his or her competence for expert help.

Where other advice or practical help is required this may be from any one of a number of sources:

- occupational health physicians;
- occupational hygienists;
- ergonomists;
- physiotherapists; and
- counsellors.

Occupational physicians

[01012] *Occupational physicians* are doctors who have specialised in occupational medicine. To become a specialist in occupational medicine the doctor must first of all undertake a programme of study and examination to acquire Associateship of the Faculty of Occupational Medicine (AFOM). This qualification is an indication of the doctor's expertise in occupational medicine but they are not regarded as accredited specialists (with consultant status in NHS terms) until they have gained full membership of the faculty (MFOM). Full members are admitted to the accreditation register held by the faculty. Some doctors also hold an MSc in Occupational Medicine or a Diploma in Industrial Health (DIH). These are academic qualifications in occupational medicine but holders of these qualifications may or may not be accredited specialists.

The role of the occupational physician is to undertake medical examinations and to diagnose work-related ill health, to determine the work-related causes of ill-health and to advise on the management of health problems which impact on the person's ability to work. Pension funds also often require a doctor to make the decision regarding permanent ill health retirement or admission to pension schemes, although they do not necessarily have to be specialist occupational physicians.

Occupational hygienists

[01013] Occupational hygienists are employed to measure chemical and physical hazards in the workplace. Information on levels of exposure to chemicals, dusts, fumes, noise etc., is often required in order to determine the risks to health. There is also sometimes a statutory need to determine exposure levels and to keep records of individual exposures. Occupational hygienists are also expert in advising on protective equipment such as the suitability of hearing defenders and dust masks.

Ergonomists

[01014] *Ergonomists* have specialist skills, which enable them to design work systems that optimise the match between the person and their work and work environment. The important skill of the ergonomist is to understand the mental and physical capabilities of people and the wide range of human variability, and to design work systems which impact least on the person's health.

Treatment services

[01015] *Physiotherapists, chiropractors and counsellors.* Occupational health is primarily concerned with the prevention of work-related disorders and the management of health problems that affect employees at work and the organisation employing them. Although it is not primarily the function of occupational health to provide treatment, many organisations consider it a worthwhile investment to provide some types of treatment in order to assist the employee in recovering and returning to work more quickly. Such treatment services typically involve those aimed at recovery from musculoskeletal disorders and psychological illness which together form the two most frequent reasons why employees are off sick. Physiotherapy (and/or chiropractic) and counselling are therefore sometimes provided by the occupational health service, or organised with external providers through the occupational health service.

Health records

Health records and confidentiality

[01016] It is sometimes necessary to obtain the medical records of an employee or a prospective employee (at the time of recruitment), in order to make a judgement about the current state of the person's health and their fitness to work. If this is done then it is important that these are obtained by a heath professional employed by the company (or engaged by the company under contract), who can take responsibility for the confidentiality of the records. Medical records should always be kept confidentially under the guardianship of a registered medical practitioner or registered nurse, and should not form part of the person's personnel file.

Obtaining records and reports

[01017] Requesting medical records or a medical report requires the informed consent of the individual and is subject to the legal requirement of the *Access to Medical Reports Act 1988* (in the case of medical *reports*) and the *Data Protection Act 1998*, (in the case of health and medical *records*). The right of personal access to health records used to be under the *Access to Health Records Act 1990*, but this has now been repealed by the *Data Protection Act 1998*. The only provision now contained in the *Access to Health Records Act 1999* is where access is requested to health records pertaining to the deceased. Under the *Access to Medical Reports Act 1988* an individual has the right of access to any medical report relating to him or her which is to be, or has been, supplied by a medical practitioner for employment purposes or insurance purposes.

Subject access to health records under the *Data Protection Act 1998* is much wider than under the *Access to Medical Reports Act 1988*. It applies to any health record kept by any health professional including doctors, nurses, midwives, dentists, chiropodists, physiotherapists and clinical psychologists, who are, or have been, responsible for the care of the individual. An applicant for access to a health record can be the patient or another person that the patient has authorised in writing to make application on his behalf. In the case of an employer seeking access to an employee's health records, the employer is acting on behalf of the patient (his employee) and therefore must seek written informed consent to make an application. Under this act the individual does not have the automatic right to see the record before it is released to the applicant, but does have the right to subsequently request that any misleading or inaccurate entry in the health records is corrected.

The normal way in which an employer seeks to obtain a medical report or access to health records is to explain to the employee what the purpose is in requesting a medical report or copies of health records and asking the employee to sign a consent form. The consent form must clearly spell out the legal rights of the individual under the *Access to Medical Reports Act 1988* and *Data Protection Act 1998*.

Keeping and making use of the occupational health records.

[01018] Some occupational health records have to be kept for a statutory period of time. Under the *Control of Asbestos at Work Regulations 2006 (SI 2006 No 2739)*, for example, the health records of employees who are exposed to asbestos have to be kept for a minimum of 40 years. Under the *Control of Substances Hazardous to Health Regulations 2002 (SI 2002 No 2677)*, health records pertaining to individuals must be kept for a minimum of 30 years.

Health records are kept for management purposes and these obviously relate to individuals and must be kept confidentially. These records are used firstly to provide a record of any health surveillance, referrals for medical opinion and the opinions received, the results of tests and vaccination history etc. These form an important record of the duty of care afforded by the employer and might be necessary to defend possible claims for damages. Health records are also kept for health management purposes such as recalling employees for routine health checks and tests or making sure that vaccinations are kept up-to-date.

Health records on individuals are confidential and that confidentiality must be preserved. However, if the records are kept in such a way as to allow analysis of collective and anonymised data then they can be a useful means of detecting trends in illness or detecting the emergence of a new illness or condition that has not previously been recognised in the organisation. For example, by analysing the health surveillance records of all the employees working in a particular department (perhaps by producing monthly statistics), it might become apparent that since the introduction a new solvent cleaner that there has been an increase in the incidence of dermatitis. Health records should therefore not simply be a passive record of what was done and was found, but be used positively to promote the health of the individual and employees collectively and to protect the liability of the organisation.

Buying services and priorities

[01019] The priority for any manager contemplating buying in an occupational health service is to determine exactly what the organisation needs and what it wants from its occupational health provider. The needs are determined firstly as a 'must have' list and then a 'nice to have' list. 'Must have' would include any statutory requirement for medical examination or health surveillance and whatever else the organisation determines it must have to manage the business effectively and support the personnel function. This would include pre-employment screening, advice on the management of sickness absence and rehabilitation, and any requirement of the pension fund regarding ill-health retirement (see O1003 above). The nice to have list would build on this, depending on the organisation's resources and intention to promote health and be proactive. This list would include non-statutory health surveillance, occupational health involvement in risk assessment, and health promotion and health education activities.

Organisations intending to buy services from an occupational health provider, can locate one near to them by undertaking an internet search The key words 'occupational health provider' (restricted to a UK search) will identify many such independent providers. The National Health Service offers commercial occupational health provision in many areas, through NHS Plus (http://www.nhsplus.nhs.uk). It is important to ensure that the provider is able to offer competent and properly qualified personnel for the work expected from them. Ask to see CV's of the personnel being put forward and seek confirmation of the professional registration of the medical and nursing staff.

Initiatives and targets

[01020] The Government's 'Securing Health Together' initiative (securing health together – a long-term occupational health strategy for England, Scotland and Wales HSC MISC225) lays out a number of strategies and targets for occupational health. The four main aims are:

- to reduce ill-health in workers and the public caused or made worse by work;
- to help people who have been ill, whether caused by work or not, to return to work;
- to improve work opportunities for people currently not in employment due to ill-health or disability; and
- to use the work environment to help people maintain or improve their health.

The targets are:

- a 20% reduction in the incidence of work-related ill health;
- a 20% reduction in ill-health to members of the public caused by work activity;
- a 30% reduction in the number of days lost due to work-related ill-health; and
- everyone currently in employment but off work due to ill health or disability to be made aware of opportunities for rehabilitation.

The aim is that by 2010 everyone has access to appropriate occupational health support. The HSE has been developing an initiative called Workplace Health Connect, in which small and medium-sized enterprises were able to get limited occupational health advice. Although the pilots for this finished in February 2008, and Workplace Heath Connect no longer offers free advice, there is useful information on the Workplace Health Connect website as to where help and advice can be obtained (http://www.workplacehealthconnect. co.uk).

Occupational diseases and disorders

Noise and occupational hearing loss

What is it?

[01021] Noise is defined as unwanted sound. Sound is created when an object that is vibrating or impacting causes a wave of pressure to move through the air. The magnitude of the air pressure wave is a measure of the intensity of the sound which our ears and brain detect as volume or sound level. The number of times the wave repeats itself in a given time is the frequency of the sound. Again the ear and the brain interpret this as the pitch of the sound. By utilising combinations of intensity and pitch we can interpret the meaning of sounds such as speech or music for example.

Sounds from the environment are an important way for us to communicate and be aware of dangers. However, both the ear and the brain can be affected by sounds that are harmful. Sound can be harmful in two ways:

• when the sound level is sufficiently high to cause damage to the mechanism of the ear; and

• when the level and pitch of the sound is such that it interferes with concentration, communication or the peaceful enjoyment of our surroundings.

Noise is therefore a level of sound sufficient to cause hearing damage or sound of sufficient intensity or frequency to interfere with expected levels of communication or mental well being. The latter effect of noise can have very important consequences at work when, for example, noise in a busy office causes annoyance, distraction, loss of concentration and stress. These psychological effects are important to note but will not be dealt with in further detail here. This section looks in particular at the effects of noise on hearing loss.

How is it caused?

[01022] The ear receives the changes in air pressure, by a vibration of the eardrum (tympanic membrane). The vibrations of the eardrum are passed through a chain of small bones (the ossicles), and transmitted into the inner ear (the cochlea). In the cochlea the vibrations distort microscopic hairs attached to hair cells, and it is this distortion of the hairs that triggers nerve impulses which pass to the brain where frequency and sound level are interpreted as recognisable sounds. The cochlea is able to separate the sounds into individual frequency and sound level is interpreted by the magnitude of the distortion of the hair cells. Hearing loss is caused when the sound levels are so great that the hairs of the hair cells are distorted so much that they fracture. Once damaged in this way the hair cells do not recover. Also, because the cochlea separates frequencies, the part of the cochlea that is damaged will depend on the frequency of the sound that has caused the damage. It is for this reason that hearing loss is sometimes confined to a small range of frequencies (see figure 1 at O1023 below). The young human ear can normally detect frequencies in the range of about 20 to 20,000Hz (1Hz (Hertz)) is one cycle per second), although this ability does decline significantly with age and the upper limit may fall to 16,000Hz or less as we get older. The ear can detect sound pressure from as little as 2×10^{-5} Newtons per square metre ($N.m^{-2}$), (the threshold of hearing), and is most sensitive at around 2 to 4kHz (kilo Hertz) which roughly corresponds with human speech frequencies.

Under most circumstances, hearing loss is gradual with the damage being cumulative over many years. Hearing loss, even in the early stages, may also be accompanied by tinnitus.

How is it detected?

[01023] Sound level is measured using a sound level meter which measures the sound pressure level according to an agreed international standard. Conventionally the sound intensity is measured using a logarithmic scale, the decibel (dB) scale. The International and British Standards use reference levels that define 0dB as the threshold of hearing. On this scale 0dB is the average

threshold of hearing and 130dB is the pain threshold; this roughly defines the range of human hearing ability (although not safe levels of exposure). One other characteristic of sound level meters is the convention to relate the measured sound to the sensitivity of the ear at different frequencies. The A weighted scale is most often used (although others exist), as this is weighted to correlate most closely between measured and perceived sound. It also correlates best with the ability of sound to cause hearing damage. When the A weighted scale is used, the letter A is placed in brackets e.g. 60dB(A).

Hearing loss is measured using an audiometer; the resulting graphical expression of the ability to perceive sounds is the audiogram. The audiometer measures the sound level at which the individual can just perceive the sound, at a range of individual frequencies, usually between 125Hz to 8,000 or 12,000Hz. As this uses the dB(A) scale, a normal audiogram would show a fairly straight line across this frequency range at around 0dB(A) (defined as the threshold of hearing). If there is hearing loss, greater levels of sound will be required to elicit hearing. Hearing loss may occur at any frequency or across the whole range of measured frequencies, but most often starts with a loss at around 3–4,000Hz, the frequency at which the ear is most sensitive.

Figure 1 (a) shows a normal audiogram with the responses close to 0dB across the frequency range. Figure 1 (b) shows the typical drop in hearing ability, in this case of between 50 and 60dB, in the frequencies around 4,000Hz. This is a case of occupational hearing loss.

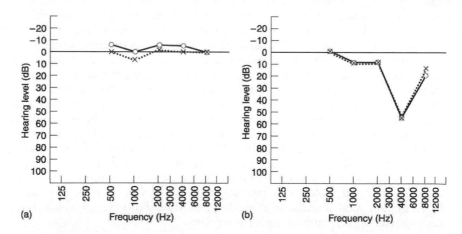

How is it managed?

[O1024] Hearing loss (occupational deafness) is a serious risk in many industries which involve the use of machinery, or physical percussion or impact. To ensure that the risk of hearing loss is minimised (and legal requirements are being met) the following steps should be taken.

- Identify any areas or work tasks, which may exceed the legal threshold for noise at work (see O1025 below).
- Identify individuals who may be exposed to these noisy areas or types of work.

- Obtain a baseline audiogram on any exposed individuals. (Audiometry should be undertaken in an approved soundproof booth to avoid erroneous results due to interference with the test from background sounds).
- Have a noise assessment undertaken on the noisy areas by a competent person and act on the report and advice given (see note under **O1025** below).
- Ensure that every reasonable effort is taken to reduce the noise at source. Protect the hearing of exposed employees by issuing recommended hearing protection.
- Ensure relevant employees are properly trained and informed in the measures to protect themselves from the damaging effects of noise and how to properly wear and maintain the hearing protection provided.
- Have periodic audiometry undertaken and keep the records.

What are the legal requirements?

[01025] The primary legal requirement is contained in the Control of *Noise at Work Regulations 2005 (SI 2005 No 1643)*. The following are the key duties under these regulations.

(a)   Employers must survey the workplace for areas likely to come under the Regulations. Any area likely to be affected must have a formal 'noise assessment' carried out by a competent person (see note below).

(b)   If the sound level is above 80dB(A), ('the lower exposure action level') hearing protection must be provided on request. If it is above 85dB(A), ('the upper exposure action level'), it must not only be provided, but it must be worn (see note below).

(c)   The daily or weekly exposure must not exceed the exposure limit value of 87db(A). The peak sound pressure must not exceed 140dB(C weighted).

(d)   Any area of the workplace that is likely to create exposure at or above the upper exposure value (85bB(A)) must be demarcated and designated as a "Hearing Protection Zone"

(e)   Risk of hearing damage must be reduced to the lowest reasonably practicable level.

(f)   Noise control measures must be fully maintained and properly used.

(g)   Education and training – employees must know the risks from noise and how to properly use the means to protect themselves from these risks.

[01026] Note. Under these regulations it is necessary to assess the cumulative dose of noise during an eight-hour working day. This is done by measurement of the sound levels in the work area (or personal noise dose), and by extrapolation to derive a daily personal exposure over an eight-hour working day (the LEP,d). The values of 85 and 90dB(A) quoted above are LEP,d values. Under the 2005 regulations it is permissible to use weekly exposure, in place of daily exposure, if the noise exposure varies markedly from day to day. It is a legal requirement to have the noise assessment performed by a competent person, which usually means an occupational hygienist with experience of noise assessment work (see **O1013** above).

Regulation 9 of these regulations requires the employer to place an employee who is likely to be exposed to noise under health surveillance. The choice, issue and maintenance of hearing protection is also covered by the *Personal Protective Equipment at Work Regulations 1992 (SI 1992 No 2966)* as amended by the *Health and Safety (Miscellaneous Amendments) Regulations 2002 (SI 2002 No 2174)*.

Hand arm vibration syndrome (havs) and whole body vibration

What is it?

[01027] Hand arm vibration syndrome (HAVS) is a set of symptoms suffered by people whose work exposes them to vibrating hand tools. These include drills, de-scalers, grinders, jackhammers, riveting guns, chain saws, brush cutters etc. Whole body vibration causes a different set of symptoms and is potentially caused by driving or being passenger on a variety of vehicles or work platforms, especially those driven off-road and over rough terrain.

Vibration describes the oscillatory movements of an object. Vibration can be characterised by its magnitude (the amount of displacement measured as the distance the object moves during the course of its motion) or by the speed it moves or the rate of change of speed (the acceleration). Vibration, either of the hand-arm or whole body, is also characterised by the direction of movement. This can be in the fore-aft direction, or the horizontal or vertical plane or a complex combination of these axes of movement. Although all of these characteristics can be measured, the value that most closely correlates with the health risk is the acceleration. The acceleration, measured in the unit of metres per second per second (m.s ^{2}) is the most usual way of describing the magnitude of the vibration in terms of the likely health risk. It is also usual to measure the acceleration in each of the three axes of movement (fore-aft, vertical and horizontal).

The common symptoms of HAVS are:

- painful finger blanching attacks triggered by cold or wet conditions (vibration white finger);
- loss of sense of touch and temperature;
- numbness and tingling;
- loss of grip strength; and
- loss of manual dexterity.

HAVS is a progressive condition, which starts with minor sensory changes to the fingers and progresses either slowly or rapidly to become a serious and debilitating condition. It may cause the sufferer to be unable to continue working.

Whole body vibration causes a number of symptoms, which commonly include:

- back pain, displacement of intervertebral discs, degeneration of the vertebrae, osteoarthritis;
- abdominal pain, digestive disorders;
- increased urinary frequency, prostatitis;
- visual disturbance;
- sleeplessness; and
- headache.

How is it caused?

[01028] The causes of HAVS and symptoms of whole body vibration are not fully understood. HAVS can be due to changes in the blood vessels of the arm and hand, the nerves supplying the blood vessels or in the muscles and bones. The most likely cause of the condition in the majority of cases is damage to the nerves supplying the blood vessels caused by the repeated trauma of the shaking of the upper limb. Vibration white finger is caused by a severe reduction in the blood supply to the fingers but may actually be due to damage of the nerves controlling the blood vessels. Once damaged, the nerves do not recover. HAVS is often triggered, and the condition worsened, by exposure to the cold at the same time as using the vibrating tool or equipment.

The symptoms of whole body vibration is also caused by the trauma of repeatedly being shaken and may involve more generalised symptoms such as fatigue, headache and sleeplessness.

How is it detected?

[01029] HAVS and symptoms of whole body vibration will need to be diagnosed by an occupational physician. Anyone exposed to vibrating equipment should be kept under health surveillance. Health surveillance does not need to be done by an occupational health professional, but the person responsible for undertaking the surveillance (including the exposed individuals if self-surveillance is used) must know what to look for and there must be in place a means of referral for expert occupational health advice for anyone suspected of showing early signs or symptoms of the condition.

Vibration white finger is usually quantified and recorded using the Stockholm Scale.

THE STOCKHOLM SCALE

Stage	Grade	Description
0		No attacks
1	Mild	Occasional attacks only affecting the tips of one or more fingers
2	Moderate	Occasional attacks affecting the distal and middle phalanges of one or more fingers
3	Severe	Frequent attacks affecting all the phalanges of most fingers
4	Very severe	As in stage 3 but with trophic skin changes at the finger tips

How is it managed?

[01030] The health risks from vibration exposure, especially hand arm vibration, are serious and debilitating. The risk of these conditions must be effectively managed and competent advice taken if there is thought to be a need for exposure assessment, diagnosis or management of the health effects.

Jobs, which expose people to the risk of vibration illness, and the individuals who are potentially affected, must be identified. A competent risk assessment must be made and the control measure identified by the risk assessment must be put in place. This may include measures such as job rotation to reduce the risk, improved maintenance of equipment to reduce vibration, replacement of old equipment by new, wearing vibration absorbing gloves, making sure that the hands are protected from the cold, and placing at-risk individuals under health surveillance.

Individuals identified by the health surveillance as having early signs or symptoms of the condition must be referred for expert occupational health advice.

What are the legal requirements?

[01031] The EU Physical Hazards (Vibration) Directive (2002/44/EC) has given rise to the *Control of Vibration at Work Regulations 2005 (SI 2005 No 1093)*. Under these regulations there is a requirement to carry out a health risk assessment and to eliminate or control the risks to health arising from vibration. For hand arm vibration there is a daily exposure *limit value* of $5m.s^{-2}$ and an *action level* of $2.5m.s^{-2}$. For whole body vibration the daily exposure *limit* is set at $1.15m.s^{-2}$ and the daily exposure *action value* is $0.5m.s^{-2}$. These values all refer to the exposure during a working day normalised to an eight-hour day. Whereas the older guidance set limits for vibration of 2.8 m.s^{-2}, in the dominant axis of movement only, the new regulations require the level of vibration to be determined as the sum of each of the three axes of movement (see **O1027** above). The measurement and calculation of daily dose is therefore technically quite complex and will need to be done by a competent person.

The provision of personal protection, which may include gloves to reduce exposure to the cold or to reduce the transmission of vibration, fall under the *Personal Protective Equipment at Work Regulations 1992 (SI 1992 No 2966)*, as amended by the *Health and Safety (Miscellaneous Amendments) Regulation 2002 (SI 2002 No 2174)*.

In addition the *Provision and Use of Work Equipment Regulations 1998 (SI 1998 No 2306)*, apply insofar as equipment must be suitable for the work, and must be properly maintained in good repair. *SI 1998 No 2306, Reg 7* of these regulations requires 'specific risks' in the use of the equipment to be eliminated or adequately controlled. The regulations also require users of equipment to receive adequate health and safety training. Under the *Supply of Machinery (Safety) Regulations 1992 (SI 1992 No 3073)*, as amended by the *Supply of Machinery (Safety) Amendment Regulations 1994 (SI 1994 No 2063)*, manufacturers and suppliers of machinery are obliged to reduce risks to a minimum, to provide data on vibration, and information on risks to health and their control.

Diseases and disorders of the eye

What are they?

[01032] Diseases and disorders of the eye relate to:

* disorders of the eye, including injuries, which are *caused* by the conditions of work; and
* conditions of the eye, such as poor visual acuity or defects in colour vision which affect safety at work or affect the health of the worker.

How are they caused?

[01033] Eye injuries. The eye is prone to injury and to disorders at work. Injuries are caused by a number of hazards such as:

* physical penetration of the eye with sharp objects (metal chips or swarf, wood splinters etc.);
* splashes with acid, alkali or other caustic substances; and
* exposure to high flux of UV light, typically in arc welding operations (arc eye).

Irritation. The surface of the eye (the conjunctiva) can be irritated by exposure to irritant fumes or chemicals (such as formaldehyde, some pesticides, dusts etc., causing conjunctivitis. Conjunctivitis can also be caused by *allergy* to particles in the atmosphere including animal dander (in veterinary work or animal husbandry), plant material including pollens (in horticultural and agricultural workers). Frequently, allergic conjunctivitis is accompanied by irritation of the nasal passages and the respiratory system generally in a condition commonly called hay fever. Employers have a particular responsibility if the employee's work brings them in contact with known allergens, which cause eye or respiratory allergy. However, these are common conditions in the population and many employees will, at certain times of the year, have to work whilst suffering severe irritation of the eye or have to take time off sick if the condition prevents them doing their work or driving to work. In some cases the employer may have to determine if the employee is safe to continue working if, for example, they work at heights, drive on the roads or drive forklift trucks. The medication taken to alleviate these conditions may itself cause a safety issue.

Dryness and low humidity. The surface of the eye is also made sore by low humidity. This causes the surface of the eye to dry and become irritatant. This is often seen in air-conditioned buildings where the air has low humidity or in cases of sick building syndrome.

Photokeratitis and photoconjunctivitis. Exposure to UV light in particular can cause damage to the tissues of the cornea (photokeratitis) and to the conjunctiva (photoconjunctivitis). Photokeratitis (arc eye) is now rare among arc welders but is, nevertheless, a risk if proper precautions are not observed. Other causes are where there is high reflectivity (snow blindness), in the use of sun beds and in laboratories using UV light. The possibility that with increased outdoor UV radiation in recent years, normal outdoor exposure in the summer may now be sufficient to cause photokeratitis and photoconjunctivitis remains uncertain, but there is sufficient evidence that it may be a risk to suggest that outdoor workers should protect their eyes with suitable sun-glasses.

Cataract. Cataracts are caused by the lens of the eye becoming opaque. This may be a natural age-related change or may be caused or hastened by exposures at work. Cataracts may be caused by excessive exposures, or exposures over a period of time, to ultraviolet light, infra-red light or microwave radiation. Sun lamps, sun beds, welding arcs, lasers and furnaces (including glass blowing), are sources of these radiations which may give rise to the risk of cataract. With increasing levels of natural UV in sunlight, there is growing concern that workers who spend long periods of time outdoors may also be at risk. There is sufficient evidence that this may be the case to suggest wearing suitable sunglasses as a precaution.

Visual acuity. The other eye disorders, which have great importance in occupational health, relate to visual acuity. Visual acuity is the ability of the eye to sharply focus on objects from near to far distance. The young eye can accommodate (i.e. change its focal length) from about 10cms up to far distance. However, as we get older the ability to focus on near objects diminishes, so that by the age of 60, the near point of vision has receded to about 83cms. Visual acuity is very important in some jobs where the ability to see accurately may have important implications for safety. Driving, for example, has standards determined by the Driver and Vehicle Licensing Authority. For car drivers the basic standard is the ability to read a car number plate at 20.5 metres (67 feet), although there are also restrictions if the visual field is affected. Group 2 (LGV and PCV drivers) have more stringent standards for visual acuity and for binocular vision.

Visual acuity is important in display screen workers because inability to focus sharply at the average distance of the computer screen is sometimes not corrected by glasses or contact lenses, which correct for near or far sight visual defects. Correction for the intermediate distance of the display screen equipment (DSE) screen may require specific lenses. It is for this reason that there is provision in the *Health and Safety (Display Screen Equipment) Regulations 1992 (SI 1992 No 2792)* for eyesight and vision tests and for the employer to supply suitable corrective lenses (see **O1036** below).

Colour vision. In certain jobs accurate colour perception may be a safety issue. About 8% of males and 0.4% of females are colour blind. The most common form is red-green colour blindness, which makes the distinction between these colours difficult. People working in jobs that require accurate colour perception, or where safety is dependent on colour perception, should have their colour vision tested.

How is it detected?

[O1034] Defects in visual acuity are measured by a number of standard tests. The basic test is the Snellen's Chart which requires a person to read letters of diminishing size from a distance of 6 meters (20 feet). The normal visual acuity is quoted as 6/6 (or 20/20 in imperial) although many people have better acuity than this, for example, 6/4.5 (20/15). The UK driving standard is equivalent to a Snellen's score of 6/10.

Other tests may measure binocular vision, the ability of the two eyes to work together to give proper depth and distance perception. Vision tests should normally be carried out by a registered optician or ophthalmic practitioner. However, screening tests, using a vision-screening instrument, are often carried out by occupational health advisors to detect individuals who may have visual difficulties. Anyone with a suspected visual impairment would be referred to an optician for full examination.

Tests for colour blindness can be carried out by occupational health advisors using a vision screening instrument or standard colour charts (Ishihara charts).

Other eye conditions or suspected eye conditions, such as injuries to the eye, allergic inflammation, cataracts, photokeratits or photoconjunctivitis must be referred to a doctor or medically qualified ophthalmic practitioner for diagnosis and treatment.

How is it managed?

[O1035] The risk of eye injury must always be anticipated in those work environments where there are hazards such as metal swarf, wood dusts or splinters, mineral and cement, brick and concrete dusts or chips, or when handling chemicals. A risk assessment (see **O1036** below) will reveal where such risks may exist and appropriate controls must be put in place. These normally involve the establishment of systems of work (ways of working and adequate supervision), which should prevent such injuries occurring. However good the system of work is, there is always likely to be a residual risk and further safeguards will be necessary. These include the provision and use of eye protection (safety glasses and goggles), and suitable first aid facilities (eye wash bottles) and first aid provision.

Jobs in which visual acuity or colour perception is critical should have eye tests on recruitment and periodically thereafter. The frequency of these will depend on the condition, its likely rate of change and the age of the individual. The occupational health advisor or optician (if referred to one) will normally advise on the frequency of tests. DSE users should have their visual acuity measured at the normal viewing distance of the display screen and, if necessary, be provided with suitable corrective lenses (see **O1036** below).

What are the legal requirements?

[O1036] There is a duty under the *Management of Health and Safety at Work Regulations 1999 (SI 1999 No 3242)*, (as amended by the *Management of*

Health and Safety at Work and Fire Precautions (Workplace) (Amendment) Regulations 2003 (SI 2003 No 2457)), to undertake a risk assessment wherever the risk of eye injury or eye disease may be a possibility. The risk assessment should identify who is at risk (taking into account any existing vulnerability), what the possible eye injury or condition might be and what controls need to be put in place to prevent or minimise the risk.

Where personal eye protection is considered necessary then the *Personal Protective Equipment at Work Regulations 1992 (SI 1992 No 2966)*, as amended by the *Health and Safety (Miscellaneous Amendments) Regulations 2002 (SI 2002 No 2174)*, apply.

The *Health and Safety (Display Screen Equipment) Regulations 1992 (SI 1992 No 2792)*, as amended by *SI 2002 No 2174*, gives employees who are designated as DSE 'users' the right to have an eye and eyesight test carried out by a competent person. Employers have a legal duty to supply corrective lenses, if needed, for the work at the display screen.

Diseases of the skin

What are they?

[01037] About 90–95% of diseases of the skin related to work are *dermatitis*. Because these conditions are caused by contact with substances in the workplace they are referred to as *contact dermatitis*. Contact dermatitis is further classified as:

- irritant contact dermatitis; and
- allergic contact dermatitis.

In addition to contact dermatitis, the other, but much less common diseases of the skin are oil folliculitis, chloracne, leucoderma, ulcerations and skin cancers.

How are they caused?

[01038] *Irritant contact dermatitis.* Contact with a wide variety of agents that are irritant to the skin can cause this disease. The hands are usually affected because they are most likely to be in contact with the irritant agent. These are either chemical substances or physical agents that cause damage to the cells of the outer layer of the skin. Well known highly irritant substances (strong acids and alkalis for example) are normally well protected against and are not the usual causes of contact irritant dermatitis, although accidental splashes may cause skin burns. It is usually the milder substances, which may not cause an immediate reaction, that are the cause. Irritant contact dermatitis often occurs after a lengthy period of contact with agents that do not immediately cause any problem. The onset is therefore insidious and not usually predictable in any particular individual. There is a large individual variation in susceptibility and the only probable risk factor is a history of eczema as a child affecting the hands.

The agents most frequently encountered as causing irritant contact dermatitis are soaps, detergents, mild acids and alkalis, cutting oils, organic solvents, animal and plant products, and oxidising and reducing agents. Physical agents include friction, and low humidity. It is often the case that the dermatitis occurs after exposure to more than one irritant source.

Allergic contact dermatitis. This disease is caused by contact with materials, which can penetrate the skin and provoke an immunological response in the skin cells. The type of response, which characterises dermatitis, is called a type IV allergy. The allergic reaction occurs after a period of sensitisation, which may occur after repeated exposure over a prolonged period (many years in some cases) or as little as a single previous contact. Allergic contact dermatitis is much less easy to control than irritant contact dermatitis because once sensitised the person will react to minute quantities of the substance. Substances known to be capable of causing contact allergy are called contact allergens. Hundreds of contact allergens are now known including chromates, methacrylates, formaldehyde, dyes, epoxy resins, and some woods.

Urticaria. An alternative allergic reaction is called urticaria (nettle rash). This is a typical wheal and flare reaction, which occurs within 20–30 minutes of contact with the allergen. This is not classified as an allergic contact dermatitis because it involves a different immune response (a type I reaction) and may involve a whole body response, including asthma and anaphylaxis (whole body shock). Reactions to substances causing urticaria may therefore be more serious and even life-threatening. The most common substance at work causing urticaria is latex rubber. Other allergens include proteins from animal skin and hair, and foodstuffs. People working in the food industry, with animals and in the health care industry (using latex gloves) may be at risk.

Other conditions. Oil folliculitis (oil acne), chloracne, leucoderma, ulcerations and skin cancers are caused by contact with (or ingestion of) a variety of substances found in the workplace. **Oil folliculitis** is an irritation of the hair follicles caused usually by contact with petroleum oils. **Chloracne** is a serious skin condition caused by ingestion of certain polychlorinated aromatic hydrocarbons. **Leucoderma** is a chemical depigmentation of the skin caused by exposure to a number of chemicals including phenols, catechols and hydroquinones. Such chemicals are commonly found in products such as adhesives. The skin depigmentation may occur at sites remote from the site of skin contact. **Skin ulcers** are usually caused by contact with wet cement or chromium compounds (chrome ulcers). Occupational **skin cancers** are now rare although they were once a common industrial disease from exposure to coal tar and machine oils. Outdoor exposure to sunlight may now be the most likely cause of occupational skin cancers.

How is it detected?

[01039] Skin conditions, of whatever type, are easily detected by visual examination. Workers identified by the risk assessment as being at risk of skin disease (i.e. because they are working with known irritants or allergens), should be placed under health surveillance. It is important that the person undertaking the periodic skin inspection (and the people themselves who are at

risk) should know exactly what to look for and be able to recognise early signs of a skin complaint. Equally important is to have a system in place for rapid referral to a skin specialist or occupational physician. Although abnormal reactions in the skin are easily detected, the diagnosis of the precise nature of the disease, what is causing it and how to manage it will require expert advice from an occupational physician or dermatologist experienced in occupational skin diseases.

How is it managed?

[01040] Occupational skin disease can be a very debilitating condition, which may render the individual unable to continue their work. Allergic contact dermatitis in particular, is difficult to manage once someone has become sensitised because it may be difficult to avoid the very small levels of exposure needed to bring about a reaction. Prevention is therefore always better than cure. The risk assessment should be able to identify those employees who are at risk from exposure to substances and physical agents, which could possibly cause skin conditions. Avoidance of contact with such agents will prevent irritant contact dermatitis and will greatly reduce the possibility of sensitisation and prevent allergic contact dermatitis. Persons with a history of childhood eczema or a past work history of dermatitis should be carefully screened if they are working with possible irritants or contact allergens. If the job is known to expose people to such substances then a pre-placement or pre-employment health screen should identify those at risk.

Protection against exposure should ideally start with attempting to procure engineering solutions. Enclosure of processes, local exhaust ventilation, and remote working which avoids handling irritant substances are possible solutions which should be considered. Personal hygiene is also very important. Staff must be encouraged to wash their hands after handling irritant material and be provided with easily accessible means to do so. Remember though that soaps and detergents may themselves be a cause of dermatitis. Protection may require the use of gloves (but again remembering that latex gloves themselves may be a risk), or barrier creams. Barrier creams may have a place if gloves are difficult to wear but do not afford the same level of protection.

What are the legal requirements?

[01041] Since the major causes of dermatitis and other skin complaints are chemical substances, the primary legal obligation falls under the *Control of Substances Hazardous to Health Regulations (COSHH) (SI 2002 No 2677)*, as amended by the *Control of Substances Hazardous to Health (Amendment) Regulations 2003 (SI 2003 No 978)*, and the *Control of Substances Hazardous to Health (Amendment) (COSHH) Regulations 2004 (SI 2004 No 3386)*. Under these regulations there is a requirement to undertake a risk assessment of the health effects and to devise and implement effective control measures including, if necessary, personal protection. The COSHH regulations also require health surveillance under certain circumstances – these are the same criteria as given by the Management Regulations (see **O1009** above). In the case of skin disease these criteria are met, so health surveillance is mandatory for people at risk. Health records pertaining to individuals have to be kept for 30 years (see **O1016** above). Under the COSHH regulations there is requirement for information, instruction and training.

The use of personal protection falls under the *Personal Protective Equipment at Work Regulations 1992 (SI 1992 No 2966)*, as amended by the *Health and Safety (Miscellaneous Amendments) Regulation 2002 (SI 2002 No 2174)*.

Physical risks, such as friction or low humidity are subject to the general requirements for risk assessment, control, health surveillance, information, instruction and training of the *Management of Health and Safety at Work Regulations 1999 (SI 1999 No 3242)*, as amended by the *Management of Health and Safety at Work and Fire Precautions (Workplace) Regulations 2003 (SI 2003 No 2457)*.

Occupational respiratory diseases

What are they?

[01042] Respiratory diseases are those that affect the respiratory tract from the nose down to, and including, the lung. The majority of occupational diseases of the respiratory system involve the lower respiratory tract within the lung. Historically, respiratory diseases have been a major cause of industrial illness and death. Diseases associated with dust inhalation in the coalmines (coal worker's pneumoconiosis), in the metal and pottery industries (siderosis and silicosis) and in the cotton industry (byssinosis) were scourges of the industrial revolution. Collectively, diseases of the lung associated with dust inhalation are called the pneumoconioses. Some of these diseases are also associated with development of cancer of the lung. Chronic bronchitis and emphysema (chronic obstructive pulmonary disease – COPD) was also very prevalent, although this was related as much to cigarette smoking as to industrial air pollution. Later, diseases associate with asbestos, (asbestosis, mesothelioma, other asbestos related lung cancers and pleural thickening), became the respiratory lung disease of major concern. Today, most of the older diseases associated with industrial pollution are rare, but have not completely gone way. Asbestos-related lung cancers can have a very long lag time between exposure and the onset of disease. About 3,500 people a year still die from asbestos-related lung cancer from past exposure. This figure is predicted to keep rising until about the year 2015. Asbestos was extensively used in building construction, shipbuilding, locomotive construction and in lagging industrial plant, between the 1950's and the1980's. It is therefore still a considerable risk, especially to those who work in asbestos clearance and in demolition. Builders, plumbers, electricians and anyone working in the refurbishment of older buildings may still be at risk.

Notwithstanding the continuing importance of asbestos-related cancers, to-day's most common work-related respiratory diseases are allergies of the upper respiratory tract (allergic rhinitis), and the lung (extrinsic allergic alveolitis and asthma). *Extrinsic allergic alveolitis* is an inflammatory immune response in the lung to the inhalation of a wide variety of both organic and inorganic materials. Examples include bird fancier's lung (from inhaling proteins in avian excreta), farmer's lung (from mould spores in mouldy hay or grain), and mushroom worker's lung (from mushroom spores). Some inorganic substances of industrial importance such as those used in spray painting and polyurethane foam, (diisocyanates and diphenyl methane) can cause the same reaction. Asthma is caused by narrowing of the airways, which is reversible usually over a short period of time. *Occupational asthma* is defined as attacks, which are initiated by inhalation of substances at work. Initiators can include irritants (irritant-induced asthma) or agents which bring about an immune response (hypersensitivity-induced asthma).

Tuberculosis is worth noting also as an occupational respiratory disease of some importance. The prevalence of tuberculosis in the community is rising and although it is rarely caused by work, it has great importance for those working in the health care industry and others, who may come in close contact with vulnerable groups (community social workers, teachers etc.). There is a risk in these cases for infected staff to pass on the infection to those in their care. Other infectious diseases of the lung have occupational importance (see O1043 below).

How are they caused?

[01043] Respiratory diseases are normally caused for one of three main reasons:

- inhalation and retention in the lung of dust and particles;
- inhalation of substances that bring about an allergic reaction; and
- inhalation of infectious organisms.

The resulting pathology is, however, very complex. Inhaled and deposited dusts and particles will have different effects depending on the chemical properties of the substance. Some will be fairly inert and be retained in the lung as patches of dust accumulation. Simple coal worker's pneumoconiosis (the early stages of the disease) is an example of this. Some will trigger the formation of protective fibrous capsules, for example iron dust (siderosis), and some will stimulate wider fibrous changes in the lung leading to serious impairment of lung function and eventually death. Silicosis is an example. Some substances are carcinogenic and their retention in the lung significantly increases the risk of lung cancer. Asbestos has been mentioned as one such substance, others include, arsenic, beryllium and cadmium compounds, silica, tars, soots, diesel-engine exhaust particulates, welding fumes, formaldehyde, vinyl chloride monomer and radioactive gases and particles (including radon). This list is of course not exhaustive. Exposure to inhaled carcinogens not only cause cancer in the lung they may also affect the nose and upper airway.

Allergic responses in the lung occur after a period of sensitisation. The allergic reaction releases substances into the lung which either cause inflammation in the lowest parts of the lung (the bronchioles and alveoli) or may cause constriction of the slightly higher parts of the airway. The inflammatory reaction, which also causes fluid build-up in the lower parts of the lung, is the cause of extrinsic allergic alveolitis. The bronchial constriction results in an asthmatic attack. Asthma may also be triggered by irritant substances which do not involve an allergic reaction.

Work-related *Infectious diseases* of the lung are not common but can have serious occupational implications. Tuberculosis may be caught at work, but the occupational health risk is far more likely to be that an infected employee transmits the infection to someone else. Animal workers are also at risk from catching a variety of animal-borne lung infections (zoonoses). These include brucellosis, psittacosis and Q fever, which are pneumonia-like illnesses. Legionnaire's disease is a serous type of pneumonia caused by the legionella bacterium infecting water systems. To cause infection the water has to be released as fine spray capable of being inhaled. Although it remains an ever-present risk it is now uncommon with elimination of water cooling towers and much better management of building water systems.

How is it detected?

[01044] This depends on the particular disease. It is important for employers to identify individuals who may be at risk from any of the occupational lung diseases. Employees must be made aware of the risks and what symptoms might arise if they become affected. They should be encouraged to report any symptoms such as chest tightness, breathlessness, cough or wheeze and the organisation must have a system in place for immediate referral to an occupational health doctor or nurse (or advise the employee to go to their own doctor with a note of what they are exposed to at work). It may be necessary to place some employees under health surveillance and this typically involves routine lung function testing. Lung function tests are performed using a spirometer, which measures a number of functional capacities. The most useful predictors of early changes that might indicate the onset of lung disease are the forced vital capacity (FVC) and the forced expiratory volume in one second (FEV1). Monitoring the health of someone with asthma is best done with another measure, the peak flow (PF). Modern electronic spirometers calculate the lung function values and indicate abnormal values by comparison with 'predicted' values for the person's age, sex and height. However, this cannot by itself be relied on to show early significant changes in lung function. It is better to compare individuals with themselves over time. Although modern equipment can be used by anyone with minimum training, it is best to have lung function tests done by an occupational health practitioner who can interpret the findings and ask other relevant questions about the individual's respiratory health at the same time.

Other tests for respiratory diseases include chest x-rays, but they are no longer done routinely unless a physician specifically indicates them.

How is it managed?

[01045] Lung disease is usually serious, debilitating and sometimes life-threatening. As with any occupational disease every effort should be put into prevention. This starts with the risk assessment which, by law, must be thorough, systematic and take into account all the relevant factors. Anyone working with any substance or potentially exposed to infectious organisms, capable of causing respiratory illness, must have a risk assessment undertaken on their work. If it is reasonably practicable, then the first action must be to prevent exposure at source. This may mean suitable engineering solutions (such as enclosure or local exhaust extraction), to prevent dust, particulates or fumes from escaping into the work area. It may be necessary, because of the nature of the work, to use respiratory protection, but this should only be used if engineering solutions are not possible or do not provide complete protection. A wide variety of respiratory protection is available, from simple dust masks to respirators. The correct type and specification must be provided for the type of substance it is intended to protect against. It may be necessary to seek the advice of an occupational hygienist.

Individuals at risk of respiratory illness should be placed under health surveillance. The criteria for health surveillance under the *Management of Health and Safety at Work Regulations 1999 (SI 1999 No 3242)* and the *Control of Substances Hazardous to Health Regulations 2002 (SI 2002 No 2677)* are probably met in most cases which means that health surveillance is a legal requirement (see **O1009** above). Health surveillance might include self-surveillance for symptoms, respiratory health questionnaires and lung function tests (see **O1044** above), and should usually be devised and carried out by an occupational health advisor.

What are the legal implications?

[O1046] Occupational lung disease usually results from inhalation of 'substances' at work and so the primary legal duty falls under the *Control of Substances Hazardous to Health Regulations (COSHH) 2002 (SI 2002 No 2677)*, as amended by the *Control of Substances Hazardous to Health (Amendment) Regulations 2003 (SI 2003 No 978)*, and the *Control of Substances Hazardous to Health (Amendment) Regulations 2004 (SI 2004 No 3386)*. Infectious organisms are also defined as 'substances' under these regulations. Under these regulations there may be a requirement to monitor the levels of exposure to ensure that they do not fall over any prescribed limits (these limits are published in EH40 2002 (HSE) with amended guidance 2005).

If the risk assessment under the COSHH regulations determines a need for personal respiratory protection then the choice of this, it's suitability, issue, maintenance and instruction and training all fall within the *Personal Protective Equipment at Work Regulations 1992 (SI 1992 No 2966)*, as amended by the *Health and Safety (Miscellaneous Amendments) Regulations 2002 (SI 2002 No 2174)*.

If the respiratory risk specifically arises from working with asbestos or potentially coming into contact with asbestos, then the *Control of Asbestos at Work Regulations 2006 (SI 2006 No 2739)* applies. It is possible also that the intended work with asbestos may require a licence from the HSE, under the *Asbestos (Licensing) Regulations 1983 (SI 1983 No 1649)*, as amended by the *Asbestos (Licensing) (Amendment) Regulations 1998 (SI 1998 No 3233)*.

Musculoskeletal disorders

What is it?

[01047] The term musculoskeletal disorders (MSDs) refer to a range of conditions which affect the skeletal system and its associated muscles, tendons and ligaments. These conditions are usually separated into those affecting the upper limb (from the shoulder to the tips of the fingers), the lower limb (from the hip to the toes), and the back. The most common conditions are those affecting the upper limb (work-related upper limb disorders or WRULDS) and the back (backache, acute and chronic back pain). Conditions which affect the lower limb are less common but in some industries (such as carpet layers) the knee can be damaged by repeated kneeling, causing a condition called beat knee (bursitis). The other part of the body often affected is the neck, especially in sedentary jobs, (at computer workstations for example) where postural problems result in neck pain.

The common conditions affect the tendons of the arm and hand. These tendons run from the muscles of the forearm through the wrists and hand to insert on the individual bones of the wrist and fingers. In places, as they pass across the wrist and the back and front of the hand, they pass through lubricated sheaths (tendon sheaths). Repetitive and forceful hand-arm movements can damage almost any part of this system.

Among people who have ever worked, 2.7% suffer from some form of musculoskeletal disorder resulting in 10.7 million lost working days (2006/07). On average each affected person takes 16.7 days off in a 12-month period.

The following is a range of musculoskeletal disorders.

Upper limb disorders. These are commonly called repetitive strain injury (RSI). However, RSI is not a medical condition by itself and the courts have rejected claims for damages for RSI on the grounds that it is not a recognised disease. It is usual practice now to refer to these conditions collectively as 'work-related upper disorders' (WRULDS) and then to diagnose one of the recognised conditions that make up this collection of complaints. In practice, however, the precise nature of these complaints is uncertain and drawing a diagnostic distinction between them is difficult.

- *Tenosynovitis*. An inflammatory and painful condition of the tendons as they pass through the tendon sheaths of the wrist and in the hand. It may also involve inflammation and dryness of the tendon sheath itself.
- *Tendinitis*. Inflammation of the tendons but not involving the tendon sheaths. Tendinitis can occur in the wrists and hand and also in the upper arm and shoulder.
- *Peritendinitis*. Inflammation at the point where the tendon joins the muscle in the arm.
- *De Quervain's disease*. Tenosynovitis with pain and swelling at the back of the wrist, particularly involving the tendon to the thumb.

- *Epicondylitis.* Pain at the point where the muscles of the arm join the bone at the elbow (the epicondyles). On the inner side of the elbow this is called medial epicondylitis (golfers elbow). On the outer side this is called lateral epicondylitis (tennis elbow).

- *Carpal tunnel syndrome.* Entrapment or compression of the median nerve as it passes through the front of the wrist (the carpal tunnel). This causes numbness and tingling in the part of the hand supplied by the median nerve – the palm side of the thumb, the second and third fingers and half the fourth (ring) finger.

- *Chronic (non-specific) upper limb pain.* A chronic widespread pain in the upper limb with tenderness to which no certain diagnosis can be made. Linked with repetitive use of the hand and arm at work. Some doctors will diagnose this as RSI, despite it not being a clinical entity, simply because no clearer diagnosis can be made.

- *Beat hand and beat elbow.* Repeated use of picks, shovels and hand tools in labourers, quarrymen and other manual jobs, can cause chronic pain, tenderness, redness and swelling of the hand and the elbow.

- *Cramp.* A condition that appears to affect people in occupations that involve repetitive movements of the hand and arm. There is some dispute about its true nature but it is a prescribed occupational disease. It causes spasm, tremor and pain in the affected muscles when the familiar action is attempted. Typified by writer's cramp.

Back pain. Most back pain is caused by mechanical damage to the spine and associated muscles, ligaments and tendons. *Acute* (sudden onset) pain is usually associated with a torn ligament or muscle around the spine caused by sudden forceful lifting or movement associated with awkward posture. In 90% of cases, such episodes usually recover within six weeks. However, these conditions are cumulative and repeated episodes may take longer to recover and recovery may be progressively less complete. The condition therefore becomes *recurrent.* Back pain can also be caused by rupture (prolapse or herniation) of the intervertebral disc. These soft tissue discs act as 'shock absorbers' between each of the vertebral bones. Severe force on the spine can cause them to rupture and compress the nerve roots as they leave the spine. The most common part of the spine for this to occur is in the lumbar region where the nerves leave to supply the leg – compression of these nerves causes pain to be referred down the leg (sciatica). *Chronic* back pain occurs where there is incomplete recovery from repeated trauma. Disc rupture is often associated with subsequent chronic back pain.

Osteoarthritis. Osteoarthritis is a very common condition in the population as whole, but there is evidence that its prevalence is higher in some occupations. Osteoarthritis is a degeneration and loss of the articular cartilage in some joints particularly of the hand, knee and hip. Articular cartilage forms a smooth lubricated surface between the joint surfaces – its loss means that bone rubs against bone causing severe damage to the joint. Osteoarthritis of the knee is more common in men involved with heavy manual labour and especially if the work involves kneeling. Osteoarthritis of the hip is less clearly associated with occupation although prolonged heavy lifting may be a risk factor.

What causes it?

[01048] The following are risk factors associated with most of the musculosketal disorders.

- **Repetition.** Frequent repeated movements over a relatively long period of time.
- **Force.** The amount of force put into the movement such as pulling, pushing or lifting. The greater the force required, the greater the risk.
- **Duration.** The longer someone is engaged in the same work, the greater is the risk of cumulative damage.
- **Lack of rest.** The risk of musculoskeletal damage is increased if the work is carried out for long periods of time without rest or time for recovery.
- **Posture.** Work carried out in awkward postures greatly increase the risk. This is true of sedentary jobs such as typing or checkout work, or in manual jobs such as lifting.
- **Working in adverse conditions.** Manual work where the hands are cold, the floor uneven or slippery, or where loads are difficult to handle, increase the risk. Working in uncomfortable conditions at computer workstations increases the risk of upper and lower limb disorders.
- **Stress.** Jobs carried out whilst under stress from time pressures etc. increase the risk of musculoskeletal disorder.
- **Individual factors.** Factors such as age and sex, height, weight and strength, a history of such conditions (or a pre-existing condition), and pregnancy will have an influence on the risk of musculoskeletal injury.

How is it detected?

[01049] Musculoskeletal disorders can normally only be detected by self-reporting or by causes of sickness absence. Employees in any at-risk job should be encouraged to report symptoms. Intervention at an early stage can often result in preventing the problem becoming worse. Acute conditions will result in sudden disabling pain and disability, which will of course require medical intervention. The onset of the chronic condition may be insidious and over a period of time. However, during this time there will be early symptoms, which should not be ignored. With upper limb conditions or slowly developing back pain there may be episodes of acute pain or numbness, which quickly resolve. Recurrent episodes or early signs of disorder such as pain, numbness or tingling, should be referred to an occupational physician (or the person's GP) for diagnosis.

How is it managed?

[01050] Almost all jobs carry a risk of musculoskeletal disorder. Heavy manual work may carry a more obvious risk but sedentary jobs are no less likely to be associated with problems. In an office the unplanned task of carrying or moving something heavy may be all that is required to cause an acute back injury. All jobs should be considered as inherently capable of giving rise to a musculoskeletal condition, although of course some jobs, which carry the known risk factors (see **O1048** above), will be much more at risk. All jobs should, therefore, have a risk assessment for MSD that takes into account all of the known and likely factors associated with the job and any individual vulnerabilities. This must be done by a competent person who recognises all the risk factors and has a good understanding of what control measure should be put in place. Having recognised the risks and implemented the necessary controls it is important that these are adhered to – this has to be done partly by effective supervision (e.g. if rest pauses are built into the job make sure they are taken) and partly by providing good information, education and training of the persons involved.

Employees must be made aware of the symptoms to report and to whom they should be reported. A system of referral to an occupational health nurse or doctor (or if that is not available to the person's GP) must be put in place. Referral to the health professional should ask for advice on what needs to done in the workplace or to the job to ensure that the condition does not worsen. Referral to an occupational health professional is much more likely to result in practical advice, which many GPs may not have the experience to offer.

Conditions that have become chronic or disabling to the point where the individual is not able to carry out their expected duties must be referred to an occupational physician with the view that permanent ill-health retirement may be necessary. However, the *Disability Discrimination Act 2005* (see **O1003** above) will almost certainly apply to chronic disabling musculoskeletal disorder.

What are the legal requirements?

[01051] General duties of risk assessment, risk management, information, education and training, competency of those undertaking risk assessment and health surveillance come under the *Management of Health and Safety at Work Regulations 1999 (SI 1999 No 3242)*, as amended by the *Management of Health and Safety at Work and Fire Precautions (Workplace) Regulations 2003 (SI 2003 No 2457)*. These regulations apply entirely where the risk arises from adverse conditions at work such as slipping and tripping hazards, poor environmental conditions (heat, wet and cold), where the primary hazard is poor or awkward posture and where stress may be a contributing factor. Where the risk arises specifically from manual handling (lifting, pushing or pulling loads) the *Manual Handling Operations Regulations 1992 (SI 1992 No 2793)* as amended by the *Health and Safety (Miscellaneous Amendments) Regulations 2002 (SI 2002 No 2174)* applies. Working with display screens (including computer workstations, and video display screens) falls within the *Health and Safety (Display Screen Equipment) Regulations 1992 (SI 1992 No 2792)* as amended by *SI 2002 No 2174*.

Psychological disorders (stress, anxiety and depression)

What is it?

[01052] The HSE defines stress as 'the adverse reaction people have to excessive pressure or other types of demand placed on them'. Pressure is a normal part of work, indeed a normal part of life, but stress occurs when the pressures are perceived by the individual as being greater then their ability to cope. Stress is therefore a very individual phenomenon; people vary greatly both in their perception of the amount of pressure on them and their abilty to cope. Stress often leads to anxiety and depression, but anxiety and depression can occur for many reasons without the underlying cause being stress. Stress also occurs in our everyday lives outside of work and it is important for the employer to understand what contribution, if any, the work has in causing an individual's stress or other psychological problem. An employer can only be responsible for that part of an individual's health complaint that is related to work, but of course most employers will be concerned with an employee who is suffering a psychological illness, whatever the cause, because it is likely to have a major impact on the person's performance at work.

Stress is not by itself an illness, but a set of symptoms which result from the mental and physiological reaction to an environment in which the perceived demands exceed the perceived ability to cope. The *resulting* psychological and physiological illness may be classified as diseases. It is important to recognise this distinction because the *Disability Discrimination Act 1995* only applies to diseases and conditions which are able to be classified as illnesses under the WHO International Classification of Diseases (ICD10). For example, depression is able to be classified under mood (affective) disorders, F30–39 and anxiety states under F40–48. Stress does not have a classiication.

Stress may not only cause psychological illness but may be related to mild to serious physical illness. In 2006/07, 60,000 people reported heart disease related to work, the majority of whom related this to stress Stress may cause high blood pressure, gastrointestinal complaints, disturbed sleep and fatigue and a number of other physical illnesses.

Someone suffering from stress may exhibit any of the following signs, some of which will be obvious to work colleagues and management and should trigger action to discover, and hopefully remedy, the cause.

- Irritability with colleagues.
- Expressed feeling of inability to cope.
- Frequent periods of sickness absence.
- Difficulty concentrating.
- Tearfulness.
- Mood changes.
- Undue tiredness/lethargy.
- Headaches.
- Insomnia.
- Palpitations.
- Appetite change either loss of appetite or over-indulgence.

- Abdominal and chest pains.
- Diarrhoea or constipation.

It is possible too that individuals under stress will drink more alcohol, smoke more cigarettes and indulge in 'comfort' eating. These are attempts at reducing the symptoms of stress but will, particularly in the long term, have serious adverse health effects.

In 2006/07, 530,000 people thought they were experiencing work-related stress that was making them ill. Stress, depression and anxiety accounted for 13.8 million working days lost in 2006/07, with an average of 30.2 working days lost per person affected. Stress is the largest single cause of sickness absence in the UK, accounting for over one-third of the working time lost.

What causes it?

[01053] There are very many factors causing stress at work. Stress may be caused, or the response to it influenced, by any of the following.

- **The work itself and the work environment.** This includes a huge number of possible factors including the workload, time pressures, and complexity (and adequate training for the work expected). It includes also managerial and organisational aspects; how well supported the individual feels, how valued and rewarded they feel, how managers and work colleagues treat them. The physical environment can also be stressful – noise which distracts and interferes with communication and concentration, high or low temperatures which cause physical discomfort, poor lighting etc. Unreliable equipment, frequent system failures, and the pressures brought about by modern technology (including emails) are 'technological' causes of stress in the modern workplace. Bullying, harassment and violence at work are also a cause of stress for some. In some jobs stress arises because of the nature of the job – working in the emergency services, in health care and social work brings emotional stress including traumatic stress. Individual factors that are capable of causing stress are called 'stressors'; *stress* is the adverse reaction people may have to exposure to these stressors. Stress is often caused by an additive effect when several of these stressors are present together.
- **Modifying factors.** The way individuals respond to stressors will depend on a number of modifying factors. These include personality, the support they receive (both at work and outside of work) and their ability to learn successful coping strategies. The effect stressors have will also depend on the person's previous history of stress and their 'stress proneness'. It should usually be taken as true that someone with a history of stress-related disorder is likely to be more vulnerable to continuing or additional levels of stress. Stress frequently arises outside of work and one important modifying factor is the reduced ability to cope with the stressors of work if the person is already suffering stress from non work-related causes.

The HSE has categorised a number of causes of stress which form part of their stress management standards.

- **Demands.** Issues like workload, work pattern and the work environment.
- **Control.** How much say the person has in the way they do their work.
- **Support.** The encouragement and support provided to employees by management and colleagues.
- **Relationships.** Conflict and unacceptable behaviour.
- **Role.** People should understand their role within the organisation and the organisation should avoid creating conflicting roles.
- **Change.** How organisational change is managed and communicated.

How is it managed?

[O1054] As with any occupational health issue the primary aim is to prevent stress happening. However, individuals will become stressed, perhaps mainly for reasons beyond the employer's control (see modifying factors at **O1053** above). The organisation, usually with professional support, will need to carefully manage the individual's recovery and return to productive working and address any work-related factors which may have caused or contributed to the illness.

The key to prevention is to undertake a stress risk assessment. Identify any areas of the work or workplace which might place undue pressures on people. Identify aspects of the organisation which might create stress. The list of factors identified in the HSE management standards (at **O1053** above) serves as a good check list for undertaking the risk assessment although there may be other factors particular to the organisation. The risk assessment will have to be kept under constant review as things within the organisation change. In particular, it is important to reconsider the risk assessment for an individual who has been diagnosed with 'stress' or a stress-related disorder. Failure to take account of an individual's known vulnerability to stress has led to several succesful civil claims for damages.

Consider having a stress management policy. This would detail the way in which stress will be identified in the organisation. It will also detail who has responsibility for what and what support services are available to employees. An important aspect of a stress management policy is that it shows a concern by the management and a committment to address the issues, but only if it is seen to be working.

Even with the best stress management policy there will be individuals who will become stressed and it is important to have a support system in place to deal quickly and effectively with these cases. If the cause is clearly rooted within the organisation (excessive workload, poor working relationships, bullying, harrassment etc.) then it is a management problem that will need to be addressed and resolved. Individuals may need support during this time and the organisation should have available to it a professional support service (occupational health or a counsellor experienced in stress counselling). If the causes are partly or wholly outside the organisation, but it is nevertheless impacting on the work (causing excessive sickness absence for example), then it is also important for the company to provide the necessary support.

What are the legal requirements?

[01055] The legal requirement falls under the *Management of Health and Safety at Work Regulations 1999 (SI 1999 3242)*, as amended by the *Management of Health and Safety at Work and Fire Precautions (Workplace) (Amendment) Regulations 2003 (SI 2003 No 2457)*. Under these regulations there is a need to assess the risks of stress occurring and the source of the stress (the stressors). The risk assessment should identify who is at risk (taking into account any existing vulnerability), what the possible adverse outcome might be and what controls need to be put in place to prevent or minimise the risk. Whilst there is not a legal obligation to do so, the HSE Stress Management Guidelines and the associated Indicator Tool are a useful guide to undertaking a risk assessment in this difficult area (http://www.hse.gov.uk/stress/index.htm).

Other occupational health concerns

[01056] This short review has focussed on the main occupational health issues that confront today's workforce and employers. Inevitably it is not complete and many additional occupational health issues will from time to time arise. Some of these will involve potentially serious health problems and must be addressed pro-actively through the usual practice of risk assessment and risk management. The legal requirement for any of these is much the same – assessment of the risk, devising and implementing suitable controls, education and training, monitoring and review, and keeping at-risk individuals under health surveillance.

Other conditions range for the serious and life threatening to those that affect the well being and comfort of the employee. The more serious (as well as cancers of the lung and skin already mentioned) included cancers of the mouth, the nose, the liver, the bladder and blood cancers. A very large number of chemicals and other agents (such as radiation) used in industry and in manufacturing are known or suspected carcinogens and anyone working with these should be well protected. Other conditions of note are disorders of the nervous system including the brain. Many commonly used substances such as solvents, metals, insecticides, styrene and carbon monoxide are neurotoxic and capable of causing nerve damage. Radiations pose a special risk. Ionising radiations are well known hazards potentially capable of causing cancers and must be well controlled. Protection against the risk of exposure to ionising radiation comes under the *Ionising Radiations Regulations 1999 (SI 1999 No 3232)*. However, the risk from non-ionising radiations, such as extra low frequency (from mains electricity), radio frequency and microwaves is more controversial. High levels of exposure such as that encountered in some occupations do carry a risk of body heating, shock and burn and must be well protected against. However, the possible cancer and other health risks from low levels of non-ionising radiation from such things as hand held transmitters (including mobile phones) are controversial at the present time. If there is a risk it appears to be of very low magnitude. At the less serious end of the spectrum of occupational health concerns are issues such as thermal comfort, low humidity causing eye, nose and throat dryness, postural issues causing aches and pains, and headache and eyestrain caused by poor lighting and glare. Whilst these issues may not cause serious health complaints, they do, nevertheless, impact significantly on the employee's ability to work efficiently and may contribute to stress.

The review is not a complete description of the many and varied diseases and disorders that may be caused or worsened by the conditions of work, nor has it dealt in detail with the various legal requirements. Readers who believe they may have a health issue in their workplace should seek advice on what they may need to do to manage such problems and avoid the inevitable subsequent cost and disruption to work. Occupational health problems rarely go away on their own; many are cumulative and will progress to more serious conditions the longer they are ignored. The key to good occupational health management is to anticipate what might cause ill health and to avert the risk of serious harm to the employee and to the organisation.

Many occupational health conditions will require reporting to the HSE under the *Reporting of Injuries, Diseases and Dangerous Occurrences Regulations 1995 (RIDDOR) (SI 1995 No 3163)*. Under these regulations many of the diseases reviewed here such as dermatitis, oil folliculitis, skin cancer, other types of occupational cancer, occupational hearing loss, most work-related lung diseases, hand arm vibration syndrome, certain musculoskeletal disorders and work-related infections will need to be reported.

The legal duties cited in this review apply mainly to regulations which come under the *Health and Safety at Work etc. Act 1974*. These should be interpreted together with the general provisions of the Act itself.

Occupiers' Liability

Alison Newstead

Introduction

[03001] Inevitably, by far the greater part of this work details the duties of employers (and employees) both at common law, and by virtue of the *Health and Safety at Work etc Act 1974* (*HSWA 1974*) and other kindred legislation. This notwithstanding, duties are additionally laid on persons (including companies and local authorities) who merely occupy premises which other persons either visit or carry out work activities upon, eg repair work or servicing. More particularly, persons in control of premises (see O3004 for meaning of 'control'), that is, occupiers of premises and employers, where others work (although not their employees), have a duty under *HSWA 1974 s 4* to take reasonable care towards such persons working on the premises. Failure to comply with this duty can lead to prosecution and a fine on conviction (see further ENFORCEMENT). This duty applies to premises not exclusively used for private residence, eg lifts/electrical installations in the common parts of a block of flats, and exists for the benefit of workmen repairing/servicing them (*Westminster City Council v Select Managements Ltd* [1985] 1 ALL ER 897). Moreover, a person who is injured while working on or visiting premises, may be able to sue the occupier for damages, even though the injured person is not an employee. Statute law relating to this branch of civil liability (i.e. occupiers' liability) is to be found in the *Occupiers' Liability Act 1957* (*OLA 1957*) and, as far as persons other than visitors (ie trespassers) are concerned, in the *Occupiers' Liability Act 1984* (*OLA 1984*). (In Scotland, the law is to be found in the *Occupiers' Liability (Scotland) Act 1960*.)

Duties owed under the Occupiers' Liability Act 1957

[03002] An occupier of premises owes the same duty, the 'common duty of care', to all his lawful visitors. [*OLA 1957, s 2(1)*]. 'The common duty of care is a duty to take such care as in all the circumstances of the case is reasonable to see that the visitor will be reasonably safe in using the premises for the purposes for which he is invited or permitted by the occupier to be there' [*OLA 1957, s 2(2)*]. Thus, a local authority which failed, in severe winter weather, to see that a path in school grounds was swept free of snow and treated with salt and was not in a slippery condition, was in breach of *s 2*, when a schoolteacher fell at 8.30 a.m. and was injured (*Murphy v Bradford Metropolitan Council* [1992] 2 All ER 908). Similarly, the Court of Appeal held in *Brioland Limited v Searson* [2005] All ER (D) 197 (Jan) that the owner of a hotel had breached *s 2* by not displaying a sign warning guests of the unusual presence of a 2.8cm high upstand in the doorway of the entrance to the hotel over which the claimant tripped whilst looking ahead and sustained severe injuries.

The duty extends only to requiring the occupier to do what is 'reasonable' in all the circumstances to ensure that the visitor will be safe. Some injuries occur where no one is obviously to blame. For example, in *Graney v Liverpool County Council* (1995), unreported an employee who slipped on icy paving slabs could not recover damages against his employer. The Court of Appeal held that many people slip on icy surfaces in cold weather – it did not follow that in this particular case the employer was to blame. In the case of *Clare v Perry* [2005] EWCA Civ 39, [2005] ALL ER (D) 67 (Jan), the claimant deliberately jumped over the edge of a wall running alongside the defendant's hotel car park, to access the road below (even though a safe exit was provided). The claimant sustained serious injuries. The Court of Appeal dismissed the claimant's argument that the defendant had breached s 2 by failing to fence off the highest part of the wall and held that the standard of care imposed on the defendant did not extend to protecting people against their 'foolhardy and unexpected conduct' which was, in this case, unforeseeable.

Nature of the duty

[03003] The *OLA 1957* is concerned only with civil liability. 'The rules so enacted in relation to an occupier of premises and his visitors shall also apply, in like manner and to like extent as the principles applicable at common law to an occupier of premises and his invitees or licensees would apply . . . ' [*OLA 1957, s 1(3)*]. This contrasts with the *HSWA 1974*, in which the obligations are predominantly penal measures enforced by the HSE (or some other 'enforcing authority'). The *OLA 1957* cannot be enforced by a state agency – nor does it give rise to criminal liability. Action under the *OLA 1957* must be brought in a private suit between parties.

The liability of an occupier towards lawful visitors at common law was generally based on negligence; so, too, is the liability under the *OLA 1957*. It is never strict. In *McGivney v Goldersea Ltd* (1997), unreported the defendant owned a block of flats which had been built in 1955 – the glass in the door at the foot of the communal stairs had been installed at the same time. When the claimant, who was visiting a friend on the third floor, came down the stairs, he slipped and his hand went through the glass panel on the front door resulting in personal injury. The claimant claimed that the defendant was in breach of the *OLA 1957 s 2(2)* in that the glass was unsafe and should have been replaced with thicker safety glass. However, at the time the flats were built, the glass satisfied the building regulations then in force. There was therefore no breach of any statutory duty. The Court of Appeal stressed the words in *s 2(2)* "to take such care as in all the circumstances of the case is reasonable", and held that it was impossible to find the defendant in breach of *s 2(2)* of the *OLA 1957* in failing to remove a pane of glass which complied with the regulations when it was installed and to replace it with more up-to-date safety glass. This can be contrasted with the case of *Moloney v Lambeth Borough Council* (1966) 64 LGR 440, in which the child of a tenant of a block of flats tripped on the stairs, fell through the balustrade and was seriously injured. There was no evidence that the child was misbehaving. In this case, the local authority contended that the balustrade conformed with all reasonable standards of safety under the Act since no other accidents had occurred. The Court found that whilst the evidence of absence of other accidents supported the local authority's contention that the balustrade conformed with all reasonable standards of safety under the *OLA 1957*, the local authority should have envisaged that a child would use the stairs unaccompanied and, taking into account the position and gaps in the balustrade, it was a matter of common sense that the staircase did not comply with the occupier's common law duty of case under *OLA 1957*. In *Stevens v Blaenau Gwent BC* [2004] EWCA Civ 715, [2004] All ER (D) 116 (Jun), a case which proceeded on the basis of common law negligence and not under the *OLA 1957*, the Court of Appeal held that the defendant local authority were not liable in negligence for the injuries suffered by a child who had fallen out of the window of a local authority house. The defendant had refused to comply with the claimant's mother's request to install safety locks on the windows of her house to protect her children, on the basis of fire regulations. The Court of Appeal held that the claimant had not demonstrated that the local authority owed him a duty of care. The authority were entitled to apply their policy on locks arrived at having assessed the risks – in the absence of any evidence that it presented exceptional risk to the claimant's family.

The relationship of occupier and visitor is less immediate than that of employer and employee, and is not, as far as occupiers' liability is concerned, under-written by compulsory insurance as would be the case with employers' liability insurance. Such a relationship would be covered by public liability insurance, which is not obligatory but advisable.

Who is an 'occupier'?

[03004] 'Occupation' is not defined in *OLA 1957*, which merely states that the 'rules regulate the nature of the duty imposed by law in consequence of a person's occupation or control of premises . . . but they (shall) not alter the rules of the common law as to the persons on whom a duty is so imposed or to whom it is owed . . . ' [*OLA 1957, s 1(2)*]. The meaning of 'occupation' must, therefore, be gleaned from common law. Where premises, including factory premises, are leased or subleased, control may be shared by lessor and lessee or by sublessor and sublessee.

> Wherever a person has a sufficient degree of control over premises that he ought to realise that any failure on his part to use care may result in injury to a person coming lawfully there, then he is an "occupier" and the person coming lawfully there is his "visitor" and the "occupier" is under a duty to his "visitor" to use reasonable care. In order to be an occupier it is not necessary for a person to have entire control over the premises. He need not have exclusive occupation. Suffice it that he has some degree of control with others.

(Per Lord Denning in *Wheat v E Lacon & Co Ltd* [1965] 2 All ER 700.)

In *Jordan v Achara* (1988) 20 HLR 607, the claimant, who was a meter reader, was injured when he fell down stairs in a basement of a house. The defendant landlord, who was the owner of the house, had divided it into flats. Because of arrears of payment, the electricity supply had been disconnected. The local authority, having arranged for it to be reconnected for the tenants, took over responsibility for the payment of electricity charges and the right to have rents payable to them. However, it was held that the landlord was liable for the injury, under *OLA 1957*, since he was the occupier of the staircase and the passageway (where the injury occurred) and maintained his control of the common parts (and therefore his duties under *OLA 1957*) in spite of the local authority being in receipt of rents.

Liability associated with dangers arising from maintenance and repair of premises will fall upon the person responsible for maintenance and/or repair under the lease or sublease:

> The duty of the defendants here arose not out of contract, but because they, as the requisitioning authority, were in law in possession of the house and were in practice responsible for repairs . . . and this control imposed upon them a duty to every person lawfully on the premises to take reasonable care to prevent damage through want of repair.

(Per Denning LJ in *Greene v Chelsea BC* [1954] 2 All ER 318, concerning a defective ceiling which collapsed, injuring the appellant, a licensee.)

Significantly also, managerial control constitutes 'occupation' (*Wheat v Lacon*, see above, concerning an injury to a lodger at a public house owned by a brewery and managed by a manager – both were held to be 'in control'). The case of *Fischer v CHT Ltd (No 2)* [1966] 1 All ER 88, further demonstrates that two or more people may be occupiers of the same land, each under an obligation to use such care as is reasonable in relation to his degree of control. In this case, the owner of a club and the defendant who ran a restaurant in the club were both found to be occupiers. The more recent case of *Rhind v Astbury Water Park Ltd and another* [2003] EWHC 1029 (QB), [2003] All ER (D) 217 (May) confirms this principle. In this case, the first defendant, *Astbury Water Park Ltd*, obtained a ten year licence from freeholders of a mere and surrounding land. They in turn entered into an agreement with the second defendant, *Maxout Ltd (Astbury Sailing School)*. Although the case was dismissed on the basis that there was no breach of duty on the part of the defendants, it was held that—

> Two or more people may be in occupation of the same premises and are under a duty of care to visitors. Whether a particular occupier is in breach of the common duty of care to a particular visitor will depend on the circumstances [. . .] Thus the circumstances that may give rise to liability will include the extent and purpose of a particular occupier's occupation and the nature of his ability on the premises and his relationship with the visitor.

(Per Morland J)

Moreover, if a landlord leases part of a building but retains other parts, eg roof, common staircase, lifts, he remains liable for that part of the premises (*Moloney v Lambeth BC* see above). The Court of Appeal has now held (*Ribee v Norrie* [2001] PIQR P 8), that the proper question was not what actual control the landlord did or did not exercise but what power of control he had, ie whether or not he had the authority to act and was in a position where he could have taken steps to prevent certain risks.

However, it is not the occupier's duty to make sure that the visitor is *completely safe*, but only reasonably so. For example, in *Berryman v Hounslow London Borough Council* [1997] PIQR P 83, the claimant injured her back when she was forced to carry heavy shopping up several flights of stairs because the lift in the block of flats maintained by the landlord was broken. The landlord's obligation to take care to ensure that the lift he provided was reasonably safe was discharged by employing competent contractors. There was no causal connection between the claimant's injury on the stairs and the absence of a lift service. In other words, the injury to the claimant was not a 'foreseeable consequence' of the broken lift.

Premises

[03005] *OLA 1975* regulates the nature of the duty imposed by law in consequence of a person's occupation of premises [*OLA 1957 s 1(2)*]. This means that the duties are not personal duties but depend on occupation of premises; and extend to a 'person occupying, or having control over, any fixed or movable structure, including any vessel, vehicle or aircraft' [*OLA 1957 s 1(3)(a)*] (eg a car, *Houweling v Wesseler* [1963] 40 DLR(2d) 956-Canada or a sea wall in *Staples v West Dorset District Council* (1995) 93 LGR 536 (see O3006)). In *Bunker v Charles Brand & Son Ltd* [1969] 2 ALL ER 59 the defendant was a contractor, digging a tunnel for the construction of the Victoria Line of London Underground. To this end the contractor used a large digging machine owned by London Transport which moved forward on rollers. The claimant was injured when he slipped on the rollers. The defendant was held to be the occupier of the tunnel, even though it was owned by London Transport, as it retained control of the tunnel and the machine.

To whom is the duty owed?

Visitors

[03006] Visitors to premises entitled to protection under the Act are both (*a*) invitees and (*b*) licensees. This means that protection is afforded to all lawful visitors, whether the visitors enter for the occupier's benefit (clients or customers) or for their own benefit (factory inspectors, policemen), though not to persons exercising a public or private way over premises [*OLA 1957 s 2(6)*] (see O3007 COUNTRYSIDE AND RIGHTS OF WAY ACT 2000). A person entering premises in exercise of rights under this Act is not considered a visitor of the occupier of those premises for the purposes of the *OLA 1957*).

Nevertheless, occupiers are under a duty to erect a notice warning visitors of the immediacy of a danger (*Rae v Mars (UK) Ltd* [1990] 03 EG 80 where a deep pit was situated very close to the entrance of a dark shed, in which there was no artificial lighting, into which a visiting surveyor fell, sustaining injury. It was held that the occupier should have erected a warning). In some instances it may be necessary not only to provide proper warning notices but also to arrange for the supervision of visitors at the occupier's premises (*Farrant v Thanet District Council* (1996), unreported). However, there is no duty on an employer to light premises at night, which are infrequently used by day and not occupied at night. It is sufficient to provide a torch (*Capitano v Leeds Eastern*

Health Authority (Current Law, October 1989) where a security officer was injured when he fell down a flight of stairs at night, while checking the premises following the sounding of a burglar alarm. The steps were formerly part of a fire escape route but were now infrequently used). However, landowners need not warn against obvious dangers (see O3017).

Trespassers

[03007] -

> . . . a trespasser is not necessarily a bad man. A burglar is a trespasser; but so too is a law-abiding citizen who unhindered strolls across an open field. The statement that a trespasser comes upon land at his own risk has been treated as applying to all who trespass, to those who come for nefarious purposes and those who merely bruise the grass, to those who know their presence is resented and those who have no reason to think so.

(*Railways Comr (NSW) v Cardy* (1961) 104 CLR 274, Aus HC.)

Common law defines a trespasser as a person who:

(a)   goes onto premises without invitation or permission; or
(b)   although invited or permitted to be on premises, goes to a part of the premises to which the invitation or permission does not extend; or
(c)   remains on premises after the invitation or permission to be there has expired; or
(d)   deposits goods on premises when not authorised to do so.

The *Countryside and Rights of Way Act 2000* (*CRWA 2000*) now gives the public the right to roam across the open countryside in England and Wales. *CRWA 2000 s 13* amends *OLA 1957 s 1(4)* so that: 'A person entering any premises in exercise of [the right to roam] is not, for the purposes of this Act, a visitor of the occupier of the premises.' Therefore, ramblers under the *CRWA 2000* are also classed as 'trespassers'. Indeed, *CRWA 2000* amends *OLA 1984 s 1* so that no duty of care is owed to a person exercising the right to roam under the Act in respect of: 'a risk resulting from the existence of any natural feature of the landscape, or any river, stream, ditch or pond whether or not a natural feature . . . or a risk of that person suffering injury when passing over, under or through any wall, fence or gate, except by proper use of the gate of or a style.' The 2000 Act therefore explicitly states what the position is in law – whilst 'visitors' falling under the 1957 Act must refer to case law.

Duty owed to trespassers, at common law and under the Occupiers' Liability Act 1984

Common law

[03008] The position under common law used to be that an occupier was not liable for injury caused to a trespasser, unless the injury was either intentional or done with reckless disregard for the trespasser's presence (*R Addie & Sons (Collieries) Ltd v Dumbreck* [1929] AC 358). This is no longer true and in more recent times, the common law has adopted an attitude of humane conscientiousness towards simple (as distinct from aggravated) trespassers. ' . . . the question whether an occupier is liable in respect of an accident to a trespasser on his land would depend on whether a conscientious, humane man with his knowledge, skill and resources could reasonably have been expected to have done, or refrained from doing, before the accident, something which would have avoided it. If he knew before the accident that there was a substantial probability that trespassers would come, I think that most people would regard as culpable failure to give any thought to their safety' (*Herrington v British Railways Board* [1972] 1 All ER 749). The effect of this House of Lords case was that it became possible for a trespasser to bring a successful action in negligence under the common law. However, the precise nature of the occupier's duty to the trespasser remained unclear, leading to Parliamentary intervention culminating in the *OLA 1984*.

Occupiers' Liability Act 1984

[03009] *OLA 1984* introduced a duty on an occupier in respect of trespassers, that is to persons other than visitors, in respect of any risk of them suffering injury on the premises (either because of any danger due to the state of the premises, or things done or omitted to be done on them) [*OLA 1984 s 1(1)(a)*]. Such a duty arises in respect of the occupier if—

(a)   if he was aware of the danger or had reasonable grounds to believe that it exists;

(b)   he knows or has reasonable grounds to believe that a trespasser is in the vicinity of the danger concerned, or that he may come into the vicinity of the danger (whether the trespasser has lawful grounds for being in that vicinity or not); and

(c)   the risk is one against which, in all the circumstances of the case, he may reasonably be expected to offer some protection.

[*OLA s 1(3)*]

Where under *OLA 1984*, the occupier is under a duty of care to the trespasser in respect of the risk, the standard of care is 'to take such care as is reasonable in all the circumstances of the case to see that he does not suffer injury on the premises by reason of the danger concerned' [*OLA 1984, s 1(4)*]. The duty can be discharged by issuing a warning, giving warning of the danger or to discourage persons from incurring the risk. One way to do this would be to post notices warning of hazards. However, these must be explicit and not merely vague. Thus, 'Danger' might not be sufficient whereas 'Highly Flammable Liquid Vapours – No Smoking' would be [*OLA 1984 s 1(5)*] (see O3015). Moreover, under *OLA 1984* there is no duty to persons who willingly accept risks (see O3016). However, the fact that an occupier has taken precautions to prevent persons going on to his land where there is a danger, does not mean that the occupier has reason to believe that someone would be likely to come into the vicinity of the danger, thereby owing a duty to the trespasser under the *OLA 1984 s 1(4)* (*White v St Albans City and District Council*, (1990) Times, 12 March).

In *Revill v Newbery* [1996] 1 All ER 291, the court considered the duty of an occupier to an intruder. It found that the intruder could recover damages despite being engaged in unlawful activities at the time the injury occurred. The defendant was held to have injured the intruder by using greater force than was justified (by shooting the intruder through a locked door) in protecting himself and his property. The doctrine of *ex turpi causa non oritur actio* (ie no cause of action may be founded on an immoral or illegal act) cannot, without more, be relied upon by the defendant to escape liability. A person is entitled to use reasonable and necessary force to protect himself, others and his property (*R v Owino* [1995] Crim LR 743). Violence may be returned with necessary violence, but the force used must not exceed the limits of what is reasonable in all the circumstances. Courts have recognised that in considering what is 'reasonable' a person being attacked may overreact with the measure of his response (*Palmer v R* [1971] 1 All ER 1077). For the extent of the intruder's contributory negligence, see O3011.

It should be noted that dangers arising from the trespasser's own activities rather than the premises themselves will not render the occupier liable (See *OLA 1984 s 1(1), (9)* and *Keown v Coventry Healthcare NHS Trust* [2006] EWCA Civ 39, [2006] 1 WLR 953 in which a twelve year old fell from a fire escape, having climbed up the underside. The judge held that the fire escape itself was not dangerous, and any danger was due to the claimant's own actions on the premises and not due to the state of the premises themselves. See also *Siddorn v Patel* [2007] EWHC 1248 (QB), [2007] All ER (D) 453 (Mar) in which the claimant rented a flat from the defendants. The claimant sustained injuries having fallen through a skylight on a flat roof (which did not form part of the demised premises) and which had been accessed via a lounge window during the course of a party. The Court held that the claimant had failed to establish that the flat roof and skylights were dangerous within the meaning of *OLA 1984 s 1(1)(a)*; the danger had arisen out of the activities of those on the roof rather than the state of the premises.

Children

[03010] 'An occupier must be prepared for children to be less careful than adults' [*OLA 1957 s 2(3)(a)*]. Where an adult would be regarded as a trespasser, a child is likely to qualify as an implied licensee, in spite of the stricture that:

> it is hard to see how infantile temptations can give rights however much they excuse peccadilloes.

(Per Hamilton LJ in *Latham v Johnson & Nephew Ltd* [1913] 1 KB 398).

If there is something or some state of affairs on the premises (eg machinery, a boat, a pond, bright berries, a motor car, forklift truck, scaffolding), this may constitute a 'trap' to a child. If the child is then injured by the 'trap', the occupier will often be liable. (See *Dornan v Department of the Environment for Northern Ireland* [1993] NI 1, in which the concept of what action may constitute a trap was considered). In the case of *Phipps v Rochester Corporation* [1955] 1 All ER 129 the Court noted that the existence of a trench dug for the purposes of laying a sewer did not constitute an allurement for a child, or a danger concealed from adults, but that it was a danger imperceptible to a child of five years of age, because he did not realise that care should be taken to avoid the trench and, in addition, children did not frequently venture into the area unaccompanied. The Court therefore held that the 'licence' allowing the claimant to enter onto the land was a condition that he be accompanied by an adult. As he was not, the claim failed. See also *Southern Portland Cement v Cooper* [1974] AC 623 in which the Court held that the defendants (quarriers of limestone) could have expected the presence of children at the quarry (a sandhill being an allurement to the children) and it would have been easy for them to take steps to present the dangerous situation and the 13 year old claimant's injuries.

Although under *OLA 1984* a duty of care will normally only arise when an occupier knows that a trespasser may come upon a danger on his premises that is latent (see **O3017**, *Titchener v British Railways Board*). The judge held in *Young v Kent County Council* [2005] EWHC 1342 (QB), [2005] All ER (D) 217 (Mar) that an occupier may also owe a duty under the Act to protect children against an obvious risk. In this case, the claimant, who was 12 years old at the time of the accident, had climbed on the roof of a school building and had fallen through a skylight. The judge held that as the claimant was a child and that it was probable on the evidence available that the defendant knew that children were climbing on the roof of the building and it was foreseeable that children might injure themselves. The defendant should therefore have taken the cheap, precautionary measure of fencing off the access to the roof. The notices warning the children against the dangers of going on the roof were not sufficient to discharge this duty. The defendant therefore breached its duty of care under the Act. However, the judge also held that the claimant was 50 per cent contributory negligent as he would have been aware of the warning notices and of the fact that he was misbehaving when climbing on the roof.

The case of *Jolley v Sutton London BC* [2000] 3 All ER 409, concerned an old wooden boat that had been dumped on a grassed area belonging to a block of local authority flats. It became rotten and decayed, but was not covered or fenced off. The council placed a danger sticker on the boat, warning people not to touch it. This did not deter two teenage boys, who decided to repair it so they could take it sailing. Using a car jack and some wood from home, they jacked up the front of the boat so that they could repair the hull. One of the boys was lying on the ground underneath the boat when the prop gave way. The boat fell on him and caused very serious spinal injuries.

The Court of Appeal ruled that although it was foreseeable that children who were attracted by the boat might climb on it and be injured by the rotten planking giving way, it was not reasonably foreseeable that an accident could occur as a result of the boys deciding to work under a propped-up boat. However, the House of Lords decided this was wrong and that the council was liable; the trial judge was right in saying that the wider risk was that children would meddle with the boat at the risk of some physical injury, and what the boys did was not so different from normal play and was reasonably foreseeable.

Contributory negligence

[03011] In a number of cases brought under *OLA 1957*, the question of contributory negligence is often considered and impacts considerably on the amount of damages eventually awarded. This may be because the duty on the occupier is only to take such care as is reasonable to see that the visitor will be reasonably safe in using his premises. Therefore, it is possible to imagine that, even where an occupier is found to have breached his duty of care, the injured party may also, by his own conduct, have contributed to the circumstances giving rise to the injury and may have exacerbated the severity of the injuries sustained. *The Law Reform (Contributory Negligence) Act 1945 s 1(1)* provides that 'where any person suffers damage as the result partly of his own fault and partly of the fault of any other person or persons, a claim in respect of that damage shall not be defeated by reason of the fault of the person suffering the damage, but the damages recoverable in respect thereof shall be reduced to such extent as the court thinks just and equitable having regard to the claimant's share in the responsibility of the damage'. There is nothing to suggest that the above provision does not apply to children or trespassers. However, Lord Denning in *Gough v Thorne* [1966] 3 All ER 398 (a case in which it was held that a 13½ year old girl could not reasonably be expected to stop and check for herself whether it was safe to cross a road, when a lorry driver had beckoned her across) held that—

> a very young child cannot be guilty of contributory negligence. An older child may be. But it depends on the circumstances.

(See *Young v Kent County Council* (at **O3010**).)

In *Revill v Newbery* (see **O3009**), although the defendant's conduct was not reasonable, the plaintiff intruder was found to be two-thirds to blame for the injuries he suffered and his damages were reduced accordingly. In *Betts v Tokley* [2002] All ER (D) 99 (Jan), the Court of Appeal upheld a ruling that the claimant was 60 per cent contributory negligent after she fell on some unlit steps whilst leaving her employer's premises by using the back exit (as the front exit was locked). The rationale behind the decision was that the claimant had pressed on to find the back exit despite the fact that she was fully aware that she had to take great care if she continued in the dark, there was nothing to stop her returning through the lit building to the front door and the steps to the front of the building should have alerted her to the risk of further steps at the back. The claimant's suggestion that the defendant had created a trap was rejected.

In 2002 the Court held that a claimant cannot be guilty of 100 per cent contributory negligence (*Anderson v Newham College of Further Education* [2002] EWCA Civ 505, [2003] ICR 212) if evidence showed that the entire fault was to lie with the claimant then there could be no liability on the defendant. This should be compared to the dicta of the Court of Session in the Scottish case of *McEwan v Lothian Buses Plc* 2006 SCLR 592 in which the court, in stark contrast to the approach in *Anderson*, considered the case of *Jayes v IMI (Knoch) Ltd* [1985] ICR 155 to be much more persuasive in that there was no logical reason why 100 per cent contributory negligence should not apply when the degree of fault on the part of the defendant could be too small to warrant an apportionment.

Assessing a claimant's contributory negligence will often involve a careful examination of the facts. In *Wattleworth v Goodwood Road Racing Co Ltd* [2004] EWHC 140 (QB), [2004] All ER (D) 51 (Feb) the claimant's husband, an experienced driver, died when his car crashed into a tyre wall during an amateur track day. Although the occupier of the premises, as well as the official bodies which had inspected the track prior to the incident, were not held liable for the driver's death, the judge went on to consider whether the driver would have been held contributory negligent had the defendants been liable for his death. Evidence showed that the driver had committed two mistakes prior to the incident. Firstly, he had taken the wrong driving line when approaching the bend where he crashed. Although the mistake was probably due to a lack of concentration, the margin of error was very small and therefore the judge decided that he could not have been held contributory negligent by virtue of this mistake. However, following his first error, he maintained his foot on the throttle and did not attempt to brake despite having 'ample time' (per Davis J at para 178) to do so (he had travelled a distance of over 100 metres before crashing against the tyre wall). The judge decided that had it been necessary, he would have held the driver 20 per cent contributory negligent on the basis of his second mistake.

Dangers to guard against

[03012] The duty owed by an occupier to his lawful visitors is a 'common duty of care', so called since the duty is owed to both invitees and licensees, ie those having an interest in common with the occupier (eg business associates, customers, clients, salesmen) and those permitted by regulation/statute to be on the premises, eg factory inspectors/policemen. That duty requires that the dangers against which the occupier must guard are twofold: (*a*) structural defects in the premises; and (*b*) dangers associated with works/operations carried out for the occupier on the premises.

Structural defects

[03013] As regards structural defects in premises, the occupier will only be liable if either he actually knew of a defect or foreseeably had reason to believe that there was a defect in the premises. Simply put, the occupier would not incur liability for the existence of latent defects causing injury or damage, unless he had special knowledge in that regard (eg a faulty electrical circuit, unless he were an electrician); whereas, he would be liable for patent (ie obvious) structural defects, eg an unlit hole in the road. This duty now extends to 'uninvited entrants', eg trespassers, under the *OLA 1984 s 1*.

Workmen on occupier's premises

[03014] 'An occupier may expect that a person, in the exercise of his calling, will appreciate and guard against any special risks ordinarily incident to it, so far as the occupier leaves him free to do so' [*OLA 1957 s 2(3)(b)*]. This has generally been taken to imply that risks associated with systems or methods of work on customer premises are the exclusive responsibility of the employer (not the customer or occupier) (*General Cleaning Contractors Ltd v Christmas* [1952] 2 All ER 1110, concerning a window cleaner who failed to take proper precautions in respect of a defective sash window: it was held that there was no liability on the part of the occupier, but liability on the part of the employer). In *Hood v Mitie Property Services (Midlands) Limited* [2005] All ER (D) 11 (Jul), the claimant, the employee of an independent contractor hired by the occupier, fell through a skylight whilst working on a roof. The Court of Appeal said that there was no reason why the occupier should not expect a visitor in the position of the claimant to guard against the risks normally incident to this job, especially as the independent contractor had held itself out as able to do roofing work and as having a proper approach to safety. However, where a window was so designed to be cleaned from the inside, and not from the sill, the employer was held to be liable (*King v Smith* [1995] ICR 339).

Further, there is the requirement to ensure that the independent contractor is competent. This is particularly true where an occupier wishes to organise a hazardous event on his land. In *Bottomley v Todmorden Cricket Club* [2003] EWCA Civ 1575, [2003] All ER (D) 102 (Nov) the claimant, who had volunteered to help two independent contractors with a pyrotechnic display held on the defendants' premises, had been filling a mortar tube with gunpowder when the contents of the tube exploded prematurely, causing him severe burns. Although the defendants were occupiers and not employers of the claimant, the Court of Appeal held that the defendants were liable for the claimant's injuries as they had not taken all the necessary precautionary steps to ensure that the independent contractors were competent (in particular, the defendants failed to ensure whether the contractors had public liability insurance and whether they had prepared a written safety plan). The rationale behind this decision was that it was reasonable to impose liability on the Club because it did not do what it ought to have done before it allowed the dangerous event to take place on its land.

The Court of Appeal decisions in the cases of *Gwilliam v West Hertfordshire Hospital NHS Trust* [2002] EWCA Civ 1041, [2003] 3 WLR 1425 (see judgment of Lord Woolf CJ), *Bottomley v Todmorken Cricket Club* [2003] EWCA Civ 1575, [2003] 48 LS Gaz R 18 (see **O3014**) and *Naylor (trading as Mainstreet) v Payling* [2004] EWCA Civ 560, [2004] All ER (D) 83 (May) suggest that when an occupier wishes to organise a hazardous or dangerous activity on its premises, part of an occupier's duty to ensure the competence of an independent contractor will involve ensuring that the contractor holds public liability insurance. When delivering the first instance judgment in *Bottomley*, the judge held that enquiring about a contractor's insurance cover was important, as insurers would wish to reduce the risk of having to meet liability claims and would therefore want to ask questions to clarify the extent of the risk which was to be covered. Further, the judge added that if a contractor was unable to provide evidence of insurance cover, there could be a good reason for this and this could highlight a contractor's incompetence. It must be pointed out that this rule is limited to hazardous activities such as a

'Splat Wall' funfair (*Gwilliam*) or a pyrotechnic display (*Bottomley*) and does not apply to more regular activities such as for example those of a bouncer providing security outside a nightclub (*Naylor*).

'Unusual' Dangers

The occupier will ordinarily only incur liability for a structural defect in premises which the oncoming workman would not normally guard against as part of a safe system of doing his job, ie against 'unusual' dangers.

> And with respect to such a visitor at least, we consider it settled law that he, using reasonable care on his part for his own safety, is entitled to expect that the occupier shall on his part use reasonable care to prevent damage from unusual danger, which he knows or ought to know.

(Per Willes J in *Indermaur v Dames* (1866) LR 1 CP 274, concerning a gasfitter testing gas burners in a sugar refinery who fell into an unfenced shaft and was injured.)

Firemen

The law is not entirely settled in the case of firemen. In *Sibbald v Sher Bros*, (1981) Times, 1 February, the Court examined the issue of liability when a number of firemen were killed following a fire at a warehouse which had an untreated hardboard ceiling which had suddenly burst into flames. It was held that the occupier's duty in relation to firemen was not the same as a duty to other visitors, such as employees. The firemen were, in fact, considered to be the occupier's 'neighbours'.

In *Hartley v British Railways Board* (1981) 125 Sol Jo 169 (a case in which a railway station caught fire and the defendant wrongly indicated the station was staffed, leading to the claimant unnecessarily entering the station and being injured), Waller LJ held—

> It (is) . . . very unlikely that the duty of care owed by the occupier to workers was the same as that owed to "firemen" (per Lord Fraser of Tullybelton), (but) it is arguable that a "fireman" (is) a "neighbour" of the occupier in the sense of Lord Atkin's famous dictum in *Donoghue v Stevenson* [1932] AC 562, so that the occupier owes him some duty of care, as for instance, to warn firemen of an unexpected danger or trap of which he knew or ought to know.

One's neighbour has been defined in law as follows—

> persons who are so closely and directly affected by my act that I ought reasonably to have them in contemplation as being so affected when I am directing my mind to the acts or omissions which are called in question.

(Per Lord Atkin in *Donoghue v Stevenson* [1932] AC 562.)

The case of *Simpson v A I Dairies Farms Limited* [2001] EWCA Civ 13, [2001] All ER (D) 31 (Jan) reinforces this principle, as a farmer who failed to warn a fireman of the presence of an uneven drain in a yard which was concealed by water was deemed liable when the fireman was injured.

The case of *Ogwo v Taylor* [1987] 3 All ER 961 concerned a fire caused by the occupier's negligence when using a blowtorch to remove paint from his house. The fireman was injured by steam created when fighting the fire. The Court considered the question of whether the duty of care to fireman was limited to special or exceptional risks. The Court held that the defendant was responsible for the injuries, regardless of whether the injuries were sustained as a result of exceptional or ordinary risks.

Dangers associated with works being done on premises

[03015] Where work is being done on premises by a contractor, the occupier is not liable if he—

(a)   took care in selecting a competent contractor; and
(b)   satisfied himself that the work was being properly done by the contractor [*OLA 1957 s 2(4)(b)*].

An examination of the apportionment of liability for the negligence of contractors was provided by the Court of Appeal in *Clark v Hosier & Dickinson Ltd* [2003] EWCA Civ 1467, [2003] All ER (D) 225 (Oct). In this case, a theatre company commissioned the claimant to install some gates on its premises. The theatre company also engaged building contractors to carry out ground works where the gates were to be placed. During the course of the ground works the building contractor came across a mains electric cable. He drew this to the attention of the occupier. The occupier left the building company to deal with this matter. The building company provided an inaccurate drawing as to the location and depth of the mains cable to the claimant, who subsequently struck the cable whilst drilling and sustained serious injuries.

The Court of Appeal held that the building company had been in breach of its contractual obligations to carry out its work safely and competently. The Court considered that the building company should have acted on its knowledge, by at least contacting an architect or the Electricity Board. The theatre company was also criticised for failure to seek advice from these parties, but only one third of the liability was attributed to the theatre company occupier, the building contractor being held two thirds responsible on the basis of the inaccurate drawing being the cause of the claimant's injuries.

As regards the checking of works, it may be highly desirable (indeed necessary) for an occupier to delegate the 'duty of satisfaction', especially where complicated building/engineering operations are being carried out, to a specialist, eg an architect or a geotechnical engineer. Not to do so, in the interests of safety of visitors, is probably negligent.

> In the case of the construction of a substantial building, or of a ship, I should have thought that the building owner, if he is to escape subsequent tortious liability for faulty construction, should not only take care to contract with a competent contractor . . . but also cause that work to be supervised by a properly qualified professional . . . such as an architect or surveyor . . . I cannot think that different principles can apply to precautions during the course of construction, if the building owner is going to invite a third party to bring valuable property on to the site during construction.

(Per Mocatta J in *AMF International Ltd v Magnet Bowling Ltd* [1968] 2 All ER 789.)

Waiver of duty and the Unfair Contract Terms Act 1977 (UCTA)

[03016] Where damage was caused to a visitor by a danger of which he had been warned by the occupier (eg by notice), an explicit notice, eg 'Highly Flammable Liquid Vapours – No Smoking' (as distinct from a vague notice such as 'Fire Hazards') used to absolve an occupier from liability. Now, however, such notices are ineffective (except in the case of trespassers, see O3008) and do not exonerate occupiers. Thus: 'a person cannot by reference to any contract term or to a notice given to persons generally or to particular persons exclude or restrict his liability for death or personal injury resulting from negligence' [*UCTA 1977 s 2*]. Such explicit notices are, however, a defence to an occupier when sued for negligent injury by a simple trespasser under *OLA 1984*. As regards negligent damage to property, a person can restrict or exclude his liability by a notice or contract term, but such notice or contract term must be 'reasonable', and it is incumbent on the occupier to prove that it is in fact reasonable [*UCTA 1977 ss 2(2), 11*].

The question of whether the defendant did what was reasonable to bring to the notice of the claimant the existence of a particularly onerous condition regarding his statutory rights, was considered in *Interfoto Picture Library Ltd v Stiletto Visual Programmes Ltd* [1989] QB 433. It was held that for a condition seeking to restrict statutory rights (eg under *OLA 1957*) to be effective it must have been fairly brought to the attention of a party to the contract by way of some clear indication which would lead an ordinary sensible person to realise, at or before the time of making the contract, that such a term relating to personal injury was to be included in the contract. (See also the Court of Appeal case of *O'Brien v MGN Ltd* [2001] EWCA Civ 1279, [2001] All ER (D) 01 (Aug) which considered this issue).

Risks willingly accepted – 'volenti non fit injuria'

[03017] 'The common duty of care does not impose on an occupier any obligation to a visitor in respect of risks willingly accepted as his by the visitor' [*OLA 1957 s 2(5)*]. However, this 'defence' has generally not succeeded in industrial injury claims. This is similarly the case with occupiers' liability claims. See, for example, *Burnett v British Waterways Board* [1973] 2 All ER 631 where a lighterman working on the River Thames was held not to be bound by the terms of a notice erected by the respondent, (stating that the Board accepted no liability for loss, damage or injury arising from the lighterman's use of a lock) even though the lighterman had seen the notice many times and understood it. This decision has since been reinforced by the *Unfair Contract Terms Act 1977*. 'Where a contract term or notice purports to exclude or restrict liability for negligence a person's agreement to or awareness of it is not of itself to be taken as indicating his voluntary acceptance of any risk' [*UCTA 1977 s 2(3)*]. Further, in the case of *Wattleworth v Goodwood Road Racing Co Ltd* (see O3011), it was held that although a driver who died as a result of a crash during an amateur track day, would have accepted the risks inherent in racing, he would only have done so on the basis that those

responsible for the circuit's safety had taken all due steps to see that the circuit was reasonably safe. Therefore, had it been necessary to consider the defendants' defence of *volenti non fit injuria* (the defendants were held not to be liable for the driver's injuries), the judge would have rejected the defence.

The OLA 1957 makes clear that there may be circumstances in which even an explicit warning will not absolve the occupier from liability. Nevertheless, the occupier is entitled to expect a reasonable person to appreciate certain obvious dangers and take appropriate action to ensure his safety. In those circumstances no warning would appear to be required. It was not necessary, for example, to warn an adult of sound mind that it was dangerous to go near the edge of a cliff in *Cotton v Derbyshire Dales District Council* (1994) Times, 20 June. Likewise in *Staples v West Dorset District Council* [1995] PIQR P439 no duty of care was imposed on a local authority to warn a visitor of the obvious danger of slipping on visible algae that had formed on a wall; the visitor being able to evaluate the danger. In *Darby v The National Trust* [2001] EWCA Civ 189, [2001] All ER (D) 216 (Jan) it was held there was no duty on the defendant to warn against the risk of swimming in a pond as the dangers of drowning were no more then an adult should have foreseen. The risk to a competent swimmer was an obvious one. There was therefore no relative causative risk; a sign would have told the deceased no more than he already knew. In *Tomlinson v Congleton Borough Council* [2003] UKHL 47, [2004] 1 AC 46, the claimant sustained severe injuries after diving into the shallow water at the edge of a lake (in a country park owned by the defendant) and striking his neck on the bottom. The House of Lords held that the defendant, who had prohibited swimming and erected notices and distributed leaflets warning against the dangers of swimming in the lake, was under no duty under *OLA 1984* to offer protection to the claimant against a risk which was perfectly obvious (the claimant was held to be a trespasser by virtue of the prohibition on swimming). Further, two Law Lords, Lord Hoffman and Lord Hobhouse, said that there was an important question of freewill at stake and that—

> it is not and should never be the policy of the law to require the protection of the foolhardy or reckless few to deprive, or interfere with, the enjoyment by the remainder of society of the liberties and amenities to which they are rightly entitled.

(Per Lord Hobhouse at para 81.)

Therefore, it would have been unjust for the defendant to be under a legal obligation to block access to the lake. Although this case was concerned with whether a duty arose under *OLA 1984*, the House of Lords made clear that had the defendant owed a duty under *OLA 1957 s 2(2)*, this duty would not have required him to take any steps to warn the claimant against dangers which were 'perfectly obvious' (per Lord Hoffman at para 50).

In the case of *Evans v Kosmar Villa Holidays Plc* [2007] EWCA Civ 1003, [2008] 1 All ER 530, the Court of Appeal considered the issue of whether a tour operator had a duty to protect against obvious risk. In this case, the claimant booked a holiday with a tour operator. Whilst on holiday, he dived into the shallow end of a swimming pool in the early hours of the morning, sustaining serious injuries. The claimant's case was that the positioning and size of the "no diving" signs were inadequate and that the signage relating to the closure of the pool at night should have been explicit and enforced. The claimant accepted that he knew that diving into the shallow end of the pool posed a dangerous risk. The defendant tour operator said that it has no duty to warn of the obvious risk of diving into shallow water or water of an unknown depth. The Court held that the principle that there is no duty to protect against obvious risks, which had been applied in the cases of trespassers and lawful visitors to whom a duty of care was owed under the OLA 1957, also applied to those to whom a duty of care was owed in contract. The defendant's duty of care did not extend to warning about the obvious risk, the claimant being able to make a genuine and informed decision.

OLA 1984 s 1(6) regulates the position with respect to trespassers. OLA 1984 s 1(6) states, 'No duty is owed . . . to any person in respect of risks willingly accepted as his by that person . . . '. This applies even if the trespasser is a child (see **O3010**) (*Titchener v British Railways Board* [1983] 3 All ER 770 where a 15-year-old girl and her boyfriend aged 16 had been struck by a train. The boy was killed and the girl suffered serious injuries. They had squeezed through a gap in a fence to cross the line as a short-cut to a disused brickworks which was regularly used. It was held by the House of Lords that the respondent did not owe a duty to the girl to do more than they had done to maintain the fence. It would have been 'quite unreasonable' for the respondent to maintain an impenetrable and unclimbable fence. However, the Court held that—

   'the duty (to maintain fences) will tend to be higher with a very young or a very old person than with a normally active and intelligent adult or adolescent.

(Per Lord Fraser)

Scott and Swainger v Associated British Ports and British Railways Board [2000] All ER (D) 1937, concerned schoolboys who on separate occasions were both severely injured whilst attempting to 'train surf' ie hitch rides on slow passing freight trains. The case was decided on the narrow issue of causation. The judge found as a question of fact that the provision of a fence would not have deterred them. In this situation, the court seemed to conclude that defendants may not do more than fence in dangers in order to protect children and that by scaling a fence the claimants would have willingly consented to the risk of injury. This position might well be otherwise if anti-trespasser devices were not satisfactorily maintained (*Adams v Southern Electricity Board* (1993) Times, 21 October where the electricity board was held to be liable to a teenage trespasser for injuries sustained when he climbed on to apparatus by means of a defective anti-climbing device). In the curious case of *Arthur v Anker* [1996] 3 All ER 783 a car was parked without permission in a private car park. It was clamped but the driver removed his vehicle with the clamp still attached. The owner of the car park sued the driver for the return of the wheel clamp and payment of the fine for illegal parking. The question of *volenti* was considered in detail. The judge held that the owner parked his car in full knowledge that he was not entitled to do so and that he was therefore consenting to the consequences of his action (payment of the fine); he could not complain after the event. The effect of this consent was to render conduct lawful which would otherwise have been tortious. By voluntarily accepting the risk that his car might be clamped, the driver accepted the risk that the car would remain clamped until he paid the reasonable cost of clamping and de-clamping.

Actions against factory occupiers

[03018] The *Workplace (Health, Safety and Welfare) Regulations 1992 (SI 1992 No 3004)*, as amended apply with respect to the health, safety and welfare of persons in a defined workplace. The application of the Regulations is not confined to factories and offices – they also affect public buildings such as hospitals and schools, but they do not apply to, for example, construction sites and quarries. Employers, persons who have control of the workplace and occupiers are all subject to the Regulations, the provisions of which address the environmental management and condition of buildings, such as the day-to-day considerations of ventilation, temperature control and lighting and potentially dangerous activities such as window cleaning. In many ways the Regulations reflect the obligations imposed by the old *Factories Act 1961*. For example, *Regulation 12(3)* requires floors and the surfaces of traffic routes, 'so far as is reasonably practicable', to be kept free from obstacles or substances which may cause a person to 'slip, trip or fall'. These Regulations also closely mirror the general obligation placed on employers by *HSWA 1974 s 2* to ensure so far as is reasonably practicable, the health, safety and welfare of their employees whilst at work. Breach of these Regulations gives rise to civil liability by virtue of *HSWA 1974 s 47(2)* (for criminal liability, see **O3019** and enforcement). An Approved Code of Practice and guidance to the Regulations is available from the Health and Safety Executive.

The case law on these Regulations is mostly concerned with the interpretation of *Regulations 5* and *12*. *Regulation 5* applies in particular to maintenance of the workplace and workplace equipment failure of which is likely to result in a breach of the Regulations. Under *Regulation 5(1)*, the workplace, equipment, devices and systems to which theRegulation applies, must be maintained in 'an efficient state, in efficient working order and in good repair'. Under *Regulation 12(1)*, every floor in a workplace and the surface of every traffic route in a workplace must be of a construction 'so that the floor or surface of the traffic route is suitable for the purpose for which it is used'. As explained, *Regulation 12(3)* requires that, so far as is reasonably practicable, floors and traffic routes be kept free from obstructions which may cause a person to 'slip, trip or fall'. It must be pointed out that although *Regulation 12* imposes specific obligations, this does not prevent other obligations from arising under *Regulation 5*, as the duties imposed by both provisions are not mutually exclusive (see the first instance decision in *Irvine v Metropolitan Police Commissioner* [2004] EWHC 1536 (QB), [2004] All ER (D) 79 (May)). Further, the Court of Appeal held in *Ricketts v Torbay* [2003] EWCA Civ 613 that the duties imposed by *Regulation 5* were not absolute as this would be difficult to reconcile with the fact that the duties owed under *Regulation 12(3)* are subject to the qualification of reasonable practicability.

In *Lewis v Avidan* [2005] EWCA Civ 670, [2005] All ER (D) 136 (Apr), an employee slipped on a patch of water in a corridor during a night shift. The water had leaked from a concealed pipe that had burst shortly before the incident. The Court of Appeal held that although the pipe could be considered as 'equipment' under *Regulation 5(3)*, the fault in the pipe did not amount to a breach of *Regulation 12(3)*. Such a breach would only have occurred if the employer had failed to take reasonable steps to prevent the floor from being slippery, which the Court considered was not the case here. Further, the Court was of the opinion that the definition of a workplace was limited to areas of work to which employees had access, which would therefore include the floor but not an enclosed pipe. In addition, the Court held that the word 'maintaining' implied the doing of an action to the floor, such as cleaning or repairing it. The court therefore held that an 'entirely unexpected and unpredictable flood' did not mean that the floor was not maintained in an efficient state and the employee's claim under *Regulation 5(1)* was therefore dismissed. In *Jaguar v Coates* [2004] EWCA Civ 337, [2004] All ER (D) 87 (Mar), the Court of Appeal held that the defendant was not liable under *Regulation 5(1)* for the injuries sustained by his employee who fell whilst going up four regular steps which were not equipped with a handrail. The Court held that *Regulation 5(1)* did not apply as this Regulation was concerned with 'maintenance' and the fitting of a handrail on the side of the steps was not a maintenance issue. However, in the Court of Appeal's decision in *Gitsham v CH Pearce & Sons Plc* [1992] PIQR P57, Stocker LJ submitted that a duty to maintain should not necessarily be limited to the state of a structure but could also extend to taking precautions against snow and ice. This case was concerned with *s 29* of the *Factories Act 1961* which imposes on employers a duty, 'so far as reasonably practicable', to provide and maintain 'safe means of access to every place at which any person has at any time to work'. As this duty is similar to those imposed by the Regulations, literature suggests that this decision also applies to the meaning of 'maintaining' under Regulation 5. It must be pointed out that following the decisions of the Court of Appeal in *Ricketts v Torbay* and of the Scottish Inner House (the appeal court of the Scottish Court of Session) in *Donaldson v Hays Distribution Services Ltd* 2005 SLT 733, the duty owed by employers under the Regulations apply only to persons at work in the workplace and not to visitors coming into the premises.

Occupier's duties under HSWA 1974

[03019] In addition to civil liabilities under *OLA 1957* and at common law, occupiers of buildings also have duties, under the *HSWA 1974*. Failure to carry out such duties can lead to criminal liability . . . In accordance with the provisions of *HSWA 1974*, 'each person who has, to any extent, control of premises (ie 'non-domestic' premises) or the means of access thereto or egress therefrom or of any plant or substance in such premises' must do what is reasonably practicable to ensure that the premises, means of access and egress and plant/substances in the premises or provided for use there, are safe and without health risks [*HSWA 1974 s 4(2)*]. This section applies in the case of (*a*) non-employees; and (*b*) non-domestic premises [*HSWA 1974 s 4(1)*]. In other words, to a large extent, it places health and safety duties on persons and companies letting or sub-letting premises for work purposes, even though the persons working in those premises are not employees of the lessor/sublessor. However, it should be noted that the duty also extends to non-working persons, eg children at a play centre (*Moualem v Carlisle City Council* (1994) 158 JP 1110). As with most other forms of leasehold tenure, the person who is responsible for maintenance and repairs of the leased premises is the person who has 'control' (see, by way of analogy, **O3004**) [*HSWA 1974 s 4(3)*]. Included as 'premises' are common parts of a block of flats. These are 'non-domestic' premises (See *Westminster City Council v Select Managements Ltd* [1985] 1 All ER 897 which held that being a 'place' or 'installation on land', such areas are 'premises'; and they are not 'domestic', since they are in common use by the occupants of more than one private dwelling).

The reasonableness of the measures which a person is required to take to ensure the safety of those premises is to be determined in the light of his knowledge of the expected use for which the premises have been made available and of the extent of his control and knowledge, if any, of the use thereafter. More particularly, if premises were not a reasonably foreseeable cause of danger to anyone acting in a way a person might reasonably be expected to act, in circumstances that might reasonably be expected to occur during the carrying out of the work, further measures would not be required against unknown and unexpected events (*Mailer v Austin Rover Group* [1989] 2 All ER 1087 where an employee of a firm of cleaning contractors, whilst cleaning one of the appellant's paint spray booths and the sump underneath it, was killed by escaping fumes. The contractors had been instructed by the appellants not to use paint thinners from a pipe in the booth (which the appellants had turned off but not capped) and only to enter the sump (where the ventilator would have been turned off) with an approved safety lamp and when no one was working above. Contrary to those instructions, an employee used thinners from the pipe, which had then entered the sump below, where the deceased was working with a non-approved lamp, and an explosion occurred. It was held that the appellant was not liable for breach of the *HSWA 1974 s 4(2)*). (See also dicta in *R v HTM Ltd* [2006] EWCA Crim 1156, [2007] 2 All ER 665).

When considering criminal liability of an occupier under *HSWA 1974* in relation to risks to outside contractors, *s 4* is likely to be of less significance as a result of the decision of the House of Lords in *R v Associated Octel* [1996] 4 All ER 846, a decision which in effect enables the *HSWA 1974 s 3* to be used against occupiers. *HSWA 1974 s 3(1)* provides that 'It shall be the duty of every employer to conduct his undertaking in such a way as to ensure, so far as is reasonably practicable, that persons not in his employment who may be affected thereby are not thereby exposed to risks to their health or safety'. The decision establishes that an occupier can still be deemed to be conducting his undertaking by engaging contractors to do work, even where the activities of the contractor are separate and not under the occupier's control. *HSWA 1974 s 3* therefore requires the occupier to take steps to ensure the protection of contractors' employees from risks not merely arising from the physical state of the premises but also from the process of carrying out the work itself.

However, the judgment of *Makepeace v Evans Brothers (Reading) (A firm)* [2000] BLR 737 should be noted in that it was held that an occupier would not usually be liable to an employee of a contractor employed to carry out work at the occupier's premises if the employee was injured as a result of any unsafe system of work used by the employee and the contractor. It was not generally reasonable to expect an occupier of premises, having engaged a contractor whom he had reasonable grounds to regard as competent, to supervise the contractor's activities in order to ensure that the was discharging his duties to ensure a safe systems of work for his employees. Even if the occupier may have reason to suspect that the contractors may have been using an unsafe system of work, this would not, in itself, be enough to impose upon him liability under *HSWA 1974*, or in negligence.

In *Hampstead Heath Winter Swimming Club v Corp of London* [2005] 1 WLR 2930, the defendant refused to grant the members of a swimming club permission to swim unsupervised in a pond which it owned on the basis that it would then be liable to prosecution under *HSWA 1974 s 3*. The judge held that although the defendant had been correct to consider that its power to regulate the admission of swimmers to the pond constituted an undertaking under *HSWA 1974 s 3*, the defendant had made a legal error by refusing admission to the pond to unsupervised swimmers. The judge referred to the concept of freewill which played an important role in the House of Lords' decision in *Tomlinson v Congleton Borough Council* (see **O3017**) and held that the risks incurred by adult swimmers with knowledge of the dangers of swimming unsupervised, could only be attributed to their decision to swim unsupervised and not to the permission to do so given to them by the defendant as part of its undertaking. The defendant could therefore not be liable to conviction under *HSWA 1974 s 3*.

Offshore Operations

Fred Osliff and Ian Wallace

Introduction

[07001] 'Offshore operations' take place, broadly, in three specific stages:

(a) exploration for recoverable reserves of oil and gas;
(b) production of those reservoirs which prove to be economic; and
(c) decommissioning and/or removal of the facilities once the reservoir has been depleted to the economic level.

Activities connected with petroleum exploration and production have a history of at least eighty years when, by an enactment of 1918, the searching for petroleum onshore was made the subject of Government licence. This was intended to be a temporary measure, but the *Petroleum (Production) Act 1934* (now repealed) vested property rights in the Crown, and gave new powers to the Board of Trade to make regulations and impose conditions based on 'model clauses' which were incorporated into exploration/exploitation licences.

Model clauses were drafted in conjunction with the Institute of Petroleum, and matters of safety were covered by the phrase 'licensees must execute all operations in accordance with good oilfield practice' and comply with any instructions which might be given by the Board. Good oilfield practice was not, however, defined. The relevant provisions of earlier legislation governing the issue of licences have been amended by the *Petroleum Act 1998*.

In the 1950s petroleum exploration activities were being undertaken at sea. The rights of coastal countries to exploit the natural resources of their continental shelves were first formally recognised in international law by the Geneva Convention on the Continental Shelf 1958. The Convention was ratified by the United Kingdom in 1964 by the *Continental Shelf Act 1964* (*CSA 1964*) which vested in the Crown all rights exercisable in designated areas outside territorial waters in relation to the sea bed and subsoil in those areas. The *CSA 1964* also provided that the model clauses for offshore petroleum licences should make provision for the health, safety and welfare of employees. Enforcement of the Act and subordinate legislation continued to be the responsibility of the Department of Energy under the Safety Directorate of the Petroleum Engineering Division. However, the only sanction available for the enforcement of the safety procedures was withdrawal of a licence.

Following an accident in 1965 to the exploration rig Sea Gem, which capsized and sank with the loss of 13 lives, the Government accepted recommendations for strengthening statutory provisions and the *Mineral Workings (Offshore Installations) Act 1971* (*MWA 1971*) was enacted specifically to regulate offshore safety matters.

Today, most of the provisions of *MWA 1971* have been repealed and replaced by health and safety regulations made under the *Health and Safety at Work etc. Act 1974*. The provisions of the 1934 Act have been re-enacted with modifications by the *Petroleum Act 1998 (PA 1998)* to cover:

- petroleum (rights vested in Her Majesty, licensing, development fees, mines and workings):
  - — 'petroleum' is defined to include any mineral oil or relative hydrocarbon and natural gas petroleum existing in its natural condition in strata. It does not include coal or bituminous shales or other stratified deposits from which oil can be extracted by destructive distillation;
- offshore activities:
  - — application of criminal and civil law etc.;
  - — prosecutions for offences with respect to carriage of passengers and goods by helicopter (*Civil Aviation Act 1982 – in part*); infringement of safety zones (*Petroleum Act 1987, s 23*); any offence from remaining provisions of the *Mineral Workings (Offshore Installations) Act 1971*;
- submarine pipelines (see also **O7034** below):
  - — construction and use of pipelines; authorisations; compulsory modifications; rights to use pipelines; and
  - — pipeline inspectors and procedures;
- abandonment of offshore installations (see also **O7014** below):
  - — abandonment programmes – duties and default procedures.

In addition, the *Petroleum Act 1998* makes amendments to the following statutes associated with offshore (safety) operations:

- *Continental Shelf Act 1964*;
- *Prevention of Oil Pollution Act 1971*;
- *Offshore Petroleum Development (Scotland) Act 1975*;
- *Sex Discrimination Act 1975*;
- *Fatal Accidents and Sudden Deaths Inquiry (Scotland) Act 1976*;
- *Race Relations Act 1976*;
- *Energy Act 1976*;
- *Telecommunications Act 1984*;
- *Food and Environment Protection Act 1985*;
- *Oil and Pipelines Act 1985*;
- *Gas Act 1986*;
- *Petroleum Act 1987*;
- *Territorial Sea Act 1987*;
- *Food Safety Act 1990*;
- *Aviation and Maritime Security Act 1990*;
- *Social Security Contributions and Benefits Act 1992*;
- *Offshore Safety Act 1992*;
- *Trade Union and Labour Relations (Consolidation) Act 1992*;
- *Merchant Shipping Act 1995*;
- *Employment Rights Act 1996*.

Application of criminal and civil law to offshore installations

[07002] The *Petroleum Act 1998* provides that any act or omission constituting an offence under United Kingdom law which takes place on, under or above an offshore installation in United Kingdom waters or within 500 metres of such an installation (the safety zone) shall be treated as taking place in the United Kingdom. Police powers are granted accordingly. The application of criminal law to offshore installations also extends to corporate bodies, including the provisions relating to ' . . . the consent or connivance of, or to be attributable to any neglect on the part of, any director, manager . . . ' (*HSWA 1974, s 37*).

Matters which may be heard under civil law regarding offshore activities are determined in accordance with the law in force in such part of the United Kingdom as may be specified in an Order in Council. Any jurisdiction conferred on a court under this provision cannot prejudice any jurisdiction exercisable by any other court. The application of civil law also extends to mobile installations in transit.

The Cullen Report

[07003] *MWA 1971* and the regulations made under it remained in force until 1992 when, following the Piper Alpha tragedy in 1988 (167 lives were lost), the fatal accident inquiry (the Cullen Report) made 76 recommendations regarding offshore health and safety. During the inquiry Lord Cullen made reference to the Robens Report (1972) and the Burgoyne Report (1980) and in particular referred to the significance of the following:

— the Robens Report (which led to the enactment of *HSWA 1974*) had implications for safety in onshore petroleum exploitation but it did not explicitly consider offshore safety;

— the decision of the Robens Committee to exclude *MWA 1971* because of the similarities in fact and in relevant (marine) law between offshore installations and ships which are not covered by *HSWA 1974*;

— the subsequent decision of the Government to extend *HSWA 1974* to cover oil and gas workers offshore [*Health and Safety at Work etc. Act 1974 (Application outside Great Britain) Order 1977 (SI 1977 No 1232)* – now revoked] but that its enforcement should remain the responsibility of inspectors appointed by the Department of Energy under the powers provided under *MWA 1971*;

— the ineffectiveness of inspections as compared with those of HSE inspectors regulating similar major hazards onshore, under the *Control of Industrial Major Accident Hazards Regulations 1984 (CIMAH 1984) (SI 1984 No 1902 as amended)*.

Regarding *CIMAH 1984* Lord Cullen, although influenced by the HSC/HSE enforcement policy, recognised that due to the close proximity of working and living areas and the impracticability of rapid evacuation from offshore installations in an emergency, the following additional provisions were required which have no direct counterpart in *CIMAH 1984*:

(a)   that the 'duty holder's' standards for management of health and safety and the control of major hazards must be demonstrated by the submission of a Safety Case and subject to formal acceptance by the regulatory body. Any installation without an accepted Safety Case will not be allowed to operate;

(b)   that measures to protect the workforce must include arrangements for temporary refuge from fire, explosion and associated hazards until evacuation can be effected from an installation;

(c)   that suitable use should be made of quantitative risk assessment (QRA) techniques as an aid to demonstrating the adequacy of the preventive and protective measures;

(d)   that arrangements should be put in place relating to safety management and audit.

Offshore safety statistics for 2004/05 show a reduction in the rate of fatal and major injuries to workers with the number of work-related deaths standing at zero, compared to three in 2003/04.

The provisional statistics, which also reveal that there were 48 major injuries during the reporting period, are contained within the Offshore Safety Statistics Bulletin 2004/05 which was released on 16 August 2005 by the Health and Safety HSE (HSE).

The Bulletin sets out the headline health and safety statistics for the offshore oil and gas industries and provides an important indicator of health and safety performance in the sector. It is prepared specifically for the offshore workforce and their safety representatives and is available to all via the HSE website.

Commenting on the statistics, Taf Powell, Head of HSE's Offshore Division, said:

> The annual statistics bulletin is now, in effect, an "end of year report" on progress towards the industry ambition to make the UK continental shelf the world's safest offshore sector by 2010. The statistics show that we are heading in the right direction, but, in my view, not fast enough. There is a lot of activity to align industry resources behind the 2010 goal but there are a number of critical challenges.

Copies of Offshore Safety Statistics Bulletin 2004/2005 can be viewed on the HSE website at: www.hse.gov.uk/offshore/statistics/stat0405.htm

Repeal and revocation of outdated legislation

[07004] Today, as Lord Cullen recommended, offshore health, safety and welfare is regulated by a specialist division of the HSE – the Offshore Safety Division (HSE-OSD) – and the new regime promotes risk-based and goal-setting legislation made under *HSWA 1974*. The change came about following the enactment of the *Offshore Safety Act 1992 (OSA 1992)* and the *Offshore Safety (Repeals and Modifications) Regulations 1993 (SI 1993 No 1823)*. The principal change was to expressly extend the application of *Part I (General Duties)* of *HSWA 1974* to include:

(a)   securing the safety, health and welfare of persons on offshore installations or engaged on pipe-line works;

(b)   securing the safety of such installations and preventing accidents on or near them;

(c)   securing the proper construction and safe operation of pipe-lines and preventing damage to them; and

(d)   securing the safe dismantling, removal and disposal of offshore installations and pipe-lines. [OSA 1992, s 1(1).]

At the same time the following statutes were expressed to come within the ambit of the enforcement provisions of *HSWA 1974*:

—   the *Pipe-lines Act 1962*;
—   the Mineral Workings (Offshore Installations) Act 1971;
—   the Petroleum and Submarine Pipe-lines Act 1975 (now repealed); and
—   the Petroleum Act 1987.

Subsequent regulations made under *HSWA 1974* have repealed other sections of *MWA 1971* and its subordinate legislation, leaving only a few provisions of the early offshore legislation still extant. These include:

—   *Offshore Installations (Logbooks and Registration of Death) Regulations 1972 (SI 1972 No 1542)*;
—   *Offshore Installations (Inspectors and Casualties) Regulations 1973 (SI 1973 No 1842)*; and
—   *Submarine Pipe-lines (Inspectors etc.) Regulations 1977 (SI 1977 No 835)*.

The *Offshore Installations (Safety Case) Regulations 1992* enacted to implement Lord Cullen's recommendations have been repealed and replaced by the *Offshore Installations (Safety Case) Regulations 2005*. These regulations also amend the following regulations:

—   *Offshore installations (Safety Representatives and Safety Committees) Regulations 1989*;
—   *Offshore Installations and Pipeline Works (Management and Administration) Regulations 1995*;
—   *Offshore Installations (Prevention of Fire and Explosion and Emergency Response) Regulations 1995*;
—   *Reporting of Injuries, Diseases and Dangerous Occurrences Regulations 1995*;
—   *Offshore Installations and Wells (Design and Construction etc) Regulations 1996*;
—   *Diving at Work Regulations 1997*; and
—   *Health and Safety (Fees) Regulations 2005*.

The regulations also implement Article 3(2) of Council Directive 92/91/EEC (OJ No L348 28.11.92, p9) concerning minimum requirements in the mineral-extracting industries through drilling.

What is an offshore installation?

[07005] The *Offshore Installations and Pipeline Works (Management and Administration) Regulations 1995 (SI 1995 No 738 as amended)* specify that for a structure to be classed as an 'installation' means a structure which is, or is to be, or has been used, while standing or stationed in relevant waters, or on the foreshore or other land intermittently covered with water—

(a)   for the exploitation, or exploration with a view to exploitation, of mineral resources by means of a well;

(b)   for the storage of gas in or under the shore or bed of relevant waters or the recovery of gas so stored;

(c)   for the conveyance of things by means of a pipe; or

(d)   mainly for the provision of accommodation for persons who work on or from a structure falling within any of the provisions of this paragraph, and which is not an excepted structure.

Excepted structures are—

(a)   a structure which is connected with dry land by a permanent structure providing access at all times and for all purposes;

(b)   a well;

(c)   a structure or device which does not project above the sea at any state of the tide;

(d)   a structure which has ceased to be used for any of the purposes specified in paragraph (1), and has since been used for a purpose not so specified;

(e)   a mobile structure which has been taken out of use and is not for the time being intended to be used for any of the purposes specified in paragraph (1); and

(f)   any part of a pipeline.

Where two or more structures are, or are to be, connected permanently above the sea at high tide they shall for the purposes of these Regulations be deemed to comprise a single offshore installation. The *Health and Safety at Work etc. Act 1974* (Application outside Great Britain) Order 2001 reconfirmed this definition and extended it to include the 6 towers referred to as NSR M-1, R-1, R-2, R-3, R-4 and R-5 and the related cables between the towers.

Offshore installations are divided into two types; production installation which:

(a)   extracts petroleum from beneath the seabed by means of a well;

(b)   stores gas in or under the shore or bed of relevant waters and recovers gas so stored;

(c)   is used for the conveyance of petroleum by means of a pipe; and

    (i)   includes a:

        —   non production installation converted for use as a production installation for so long as it is so converted,

        —   production installation which has ceased production for so long as it is not converted to a non-production installation; and production installation which has not come into use, and

    (ii)   does not include an installation which, for a period of no more than 90 days, extracts petroleum from beneath the seabed for the purpose of well testing; and

    (iii)   non-production installation which means an installation other than a production installation.

Because of the complexity of defining offshore installations, appropriate advice should be sought in any case of doubt.

Parts of offshore installations

[**07006**] The *Offshore Installations and Pipeline Works (Management and Administration) Regulations 1995, Reg 3(3)* states that the following offshore equipment is deemed to be part of an offshore installation:

— wells, if they are connected by a pipeline or cable;
— Christmas trees or blowout preventers;
— pipelines, apparatus or works, attached to the offshore installation and within the 'safety zone' of that installation.

Safety zones

[**07007**] The threat of collision with an offshore installation was first controlled by the *Continental Shelf Act 1964* and later extended by the *Petroleum Act 1987*. *Section 21* of the *Petroleum Act 1987* provides for the automatic establishment of a 500-metre exclusion zone around every offshore installation 'on station' whilst permitting authorised vessels or helicopters to attend the installation.

The *Offshore Installations (Safety Zones) Order 2004 (SI 2004 No 343)* came into force on 1 March 2004.

The Order establishes safety zones having a radius of 500 metres from the specified point around each installation specified in the Schedule to this Order, and stationed in territorial waters and waters in areas designated under the *Continental Shelf Act 1964*.

Vessels, which include hovercraft, submersible apparatus and installations in transit, are prohibited from entering or remaining in a safety zone except with the consent of the HSE or in accordance with the *Offshore Installations (Safety Zones) Regulations 1987 (SI 1987 No 1331)*.

The prohibition on vessels entering or remaining in a safety zone does not apply to a vessel:

— in connection with pipe laying or inspection activities;
— providing services for transportation of goods or persons or those under the authority of a Government required to inspect an installation;
— performing duties relating to the safety of navigation;
— in connection with the saving or attempted saving of life or property; or
— entering because of distress or adverse weather.

Permission to enter a safety zone is given by the installation manager.

Temporary exclusion zones

[**07008**] In addition to the set safety zones around each offshore installation or sub-sea installation, the Secretary of State for Transport is empowered under marine law to establish temporary exclusion zones around a wrecked ship or structure. In determining whether such a casualty, including those in distress, requires the establishment of an exclusion zone, the Secretary of State will take into account the 'significant harm' factor, that is, in terms of pollution or damage to persons or property. [*Merchant Shipping Act 1995, s 100A* – as introduced by the *Merchant Shipping and Maritime Security Act 1997*].

Which health and safety law applies to offshore installations?

[07009] The key as to whether regulations made under *HSWA 1974* are enforceable on offshore installations is the word 'apply'. The *Health and Safety at Work etc. Act 1974 (Application outside Great Britain) Order 1977 (SI 1977 No 1232)*, now revoked, had the effect of extending the application of most of the provisions of *HSWA 1974*, and certain selected regulations made under *HSWA 1974*, to offshore sites. The following regulations were reapplied to offshore installations in their entirety by virtue of the *Health and Safety at Work etc. Act 1974 (Application outside Great Britain) Order 1989 (SI 1989 No 840)*:

— *Control of Lead at Work Regulations 1980 (SI 1980 No 1248)*;
— *Asbestos (Licensing) Regulations 1983 (SI 1983 No 1649)*;
— *Freight Containers (Safety Convention) Regulations 1984 (SI 1984 No 1890)*;
— *Ionising Radiations Regulations 1985 (SI 1985 No 1333)*;
— *Control of Asbestos at Work Regulations 1987 (SI 1987 No 2115)*; and
— *Reporting of Injuries, Diseases and Dangerous Occurrences Regulations 1995 (SI 1995 No 3163)*.

The *Health and Safety at Work etc. Act 1974 (Application outside Great Britain) Order 1989 (SI 1989 No 840)* was revoked by the *Health and Safety at Work etc. Act 1974 (Application outside Great Britain) Order 1995 (SI 1995 No 263)*, which re-enacted its provisions with modifications. Existing regulations which refer to the 1989 Application Order (*SI 1989 No 840*) (or to a previous Order) should be construed as referring to the appropriate Articles of the current Order.

The *Health and Safety at Work etc. Act 1974 (Application outside Great Britain) Order 1995 (SI 1995 No 263)* was revoked and re-enacted by the *Health and Safety at Work etc. Act 1974 (Application outside Great Britain) Order 2001 (SI 2001 No 2127)*. This order extended the application of the prescribed provision to:

— diving projects involving the survey and preparation of the seabed consequent on the removal of an offshore installation;
— a supplementary unit connected to an offshore installation and all the connections;
— specified activities in relation to an energy structure and the transfer of people or goods to or from any such structure; and
— a mine as defined in the *Mines and Quarries Act 1954* situated within the territorial sea or extending beyond it and any activity in connection with it while it is being worked.

Determining whether or not health and safety regulations apply to offshore operations can be complex. Advice should be sought from company legal representatives or the HSE-OSD in any cases of doubt. The following comments, however, may be useful:

Firstly, regulations made under *HSWA 1974* apply offshore only if they contain a clause in the current Application Order which explicitly states that they do apply. Each set of new regulations issued should detail the extent of their application by referring to the relevant Articles in the Application Order.

Secondly, regulations which contain an 'Application outside Great Britain' clause but which do not refer to any specific Article in the Application Order should be taken as applying to all the categories of offshore descriptions and activities specified in the Application Order.

To understand how a particular set of regulations is applied offshore, it is necessary to take account of several legally defined terms and expressions that may be included in the 'Application outside Great Britain' clause:

— *territorial waters* – i.e. United Kingdom territorial waters adjacent to Great Britain.
  [*The Health and Safety at Work etc. Act 1974 (Application outside Great Britain) Order 1995 (SI 1995 No 263), Reg 2(1)*].
— *within territorial waters* includes on, over or under United Kingdom territorial waters adjacent to Great Britain.
  [*The Health and Safety at Work etc. Act 1974 (Application outside Great Britain) Order 1995 (SI 1995 No 263), Reg 2(1)*].
— *territorial sea* – i.e. the breadth of the territorial sea adjacent to the United Kingdom (the 12-mile limit).
  [*Territorial Sea Act 1987, s 1*].
— *relevant waters* – i.e. (a) tidal waters and parts of the sea in or adjacent to Great Britain up to the seaward limits of 'territorial waters', and (b) any area designated by Order under the Continental Shelf Act 1964, s 1(7).
  [*The Offshore Installations and Pipeline Works (Management and Administration) Regulations 1995 (SI 1995 No 738 as amended), Reg 2(1)*].
— *associated structure* – i.e. in relation to an offshore installation a 'vessel', aircraft or hovercraft attendant on the installation or any floating structure used in connection with the installation.
  [*The Offshore Installations and Pipeline Works (Management and Administration) Regulations 1995 (SI 1995 No 738 as amended), Reg 2(1)*].
— *vessel* – i.e. a hovercraft and any floating structure which is capable of being staffed.
— *jurisdiction* – i.e. (a) the law in force in England and Wales, (b) the law in force in Scotland, (c) the law in force in Northern Ireland,
  and the authority given to the High Court for the determination of any questions of law arising in a designated area made under the *Continental Shelf Act 1964, s 1(7)*, and in particular the application of law as it

would apply in those countries by virtue of the *Civil Jurisdiction (Offshore Activities) Order 1987 (SI 1987 No 2197)* and the *Criminal Jurisdiction (Offshore Activities) Order 1987 (SI 1987 No 2198)*.

— *fixed installation* – means an installation which cannot be moved from place to place without major dismantling or modification, whether or not it has its own motive power;

— *mobile installation* – No longer defined

— *activities within territorial waters* – Article 8 specifies the following activities which may be carried out by non-offshore workers in accordance with the provisions of *HSWA 1974* and its subordinate legislation:

(a) the construction, reconstruction, alteration, repair, maintenance, cleaning, demolition and dismantling (or any preparatory activities) of any building or other structure (other than a vessel);

(b) loading, unloading, fuelling or provisioning of a vessel (which includes a 'stacked' installation, i.e. one which is not in use or intended for use at that time);

(c) the construction, reconstruction, finishing, refitting, repair, maintenance, cleaning or breaking up of a vessel (including a 'stacked' installation) except when carried out by the master or any officer or member of the crew of that vessel;

(d) diving operations;

(e) maintaining a 'stacked' installation on station.

The current Application Order is the *Health and Safety at Work etc. Act 1974 (Application outside Great Britain) Order 2001 (SI 2001 No 2127)*.

Generally, if the most onerous of the regulations were complied with then the requirements of other health and safety legislation may be deemed to be satisfied.

The following regulations apply offshore:

— *Control of Substances Hazardous to Health 2002 (SI 2002 No 2677)*; and

— *Control of Explosives Regulations 1991 (SI 1991 No 1531)*.

Offshore Installations (Safety Case) Regulations 2005 (SI 2005 No 3117)

[07010] As Lord Cullen recommended, new offshore safety regulations, the *Offshore Installations (Safety Case) Regulations 1992* (the *Safety Case Regulations 1992*), were introduced within the *HSWA 1974* framework.

These regulations have been repealed and replaced by the *Offshore Installations (Safety Case) Regulations 2005 (SI 2005 No 3117)* except some regulations relating to combined operations safety cases remain in force until 6 October 2007. The new regulations came into force on the 7 April 2006.

The new regulations require:

- licensees to ensure anyone they appoint as an operator is capable of fulfilling their legal responsibilities for safety and carries out those responsibilities;
- well operators to notify the HSE before commencing well operations;
- duty holders to send an early design notification, instead of a design safety case, to HSE when establishing a new production installation;
- duty holders to send a relocation notification before moving a production installation to a new location (whether from outside relevant waters or not).
- safety cases for production installations and non production installations;
- safety cases be revised and resubmitted before dismantling an offshore installation;
- duty holders to carry out a thorough and fundamental review of their safety cases at least every five years, or as directed by HSE;
- the present requirement to re-submit safety cases every three years has been removed (inspectors will be checking to see that safety cases are being kept up to date through inspection);
- combined operations safety cases have been replaced by notifications, which do not need HSE acceptance;
- the requirement to revise and resubmit a safety case before making a material change to an installation is continued;
- duty holders to ensure compliance with the safety case;
- an appeals procedure has been introduced; and
- the *Offshore Installations (Safety Representatives and Safety Committees) Regulations 1989 (SI 1989 No 971)* have been amended to extend consultation with safety representatives to reviewing and revising a safety case, as well as preparing one.

The primary aim of these regulations is to reduce risks to the health and safety of the workforce employed on offshore installations or in connected activities. The principal matters to be demonstrated in safety cases and submitted to the HSE in compliance with *SI 2005 No 3117*, whilst taking account of the requirements of the above regulations, are that:

(a) the management system is adequate to ensure compliance with all statutory health and safety requirements;

(b) adequate arrangements have been made for audit, and the preparation of audit reports;

(c) all hazards with the potential to cause a 'major accident' have been identified, their risks evaluated, and measures taken to reduce risks to persons to as low as reasonably practicable; and

(d) Adequate arrangements are in place for ensuring the suitability of safety critical elements and emergency escape and evacuation.

The definition of 'major accident' embraces fire, explosion or the release of a dangerous substance involving death or serious personal injury, an event involving major damage to the structure of the installation or plant or loss in the stability of the installation, collision of a helicopter with the installation, the failure of life support systems for diving operations in connection with the installation, the detachment of a diving bell used for such operations or the trapping of a diver in a diving bell or other subsea chamber used for such operations; or any other event arising from a work activity involving death or serious personal injury to five or more persons on the installation or engaged in an activity in connection with it.

For convenience the expression 'duty holder' is used as a shorthand way of referring to the person on whom duties are placed by the Regulations, for a production installation that is the operator and for a non production installation the owner.

The case for safety is required to relate to all activities carried out on, or in connection with, an installation including, for example, relevant activities involving divers, pipelaying or the movement of vessels or helicopters. The various people concerned with these activities are under a duty to co-operate with 'duty holders' in the preparation of installation safety cases.

The new regulations allow the use of electronic communication and storage of information The HSE have issued documents giving the principles that they use when assessing offshore safety cases as follows:

(i)   IND 250 How The HSE Assess Offshore Safety Cases; and
(ii)   Assessment Principles for Offshore Safety Cases.

Both documents can be viewed on the HSE website.

Regulations to underpin the Safety Case Regulations

[07011] In addition to the *Safety Case Regulations 1992 (SI 1992 No 2885)*, now superseded by the 2005 regulations, three further goal-setting regulations have been enacted to implement the recommendations made by Lord Cullen; they complete offshore implementation of the Extractive Industries (Boreholes) Directive and support and complement the objectives of the *Safety Case Regulations 1992*:

—   *Offshore Installations and Pipeline Works (Management and Administration) Regulations 1995 (SI 1995 No 738)*;
—   *Offshore Installations (Prevention of Fire and Explosion, and Emergency Response) Regulations 1995 (SI 1995 No 743)*; and
—   *Offshore Installations and Wells (Design and Construction etc.) Regulations 1996 (SI 1996 No 913)*.

Previously, under *section 5* of the now defunct *Mineral Workings (Offshore Installations) Act 1971*, a general duty was placed on the manager of an offshore installation for all matters affecting safety, health or welfare on his offshore installation. This duty also included the maintenance of order and discipline over all people on or about the installation. This authority and duty was continued by the *Offshore Installation and Pipelines (Management and Administration) Regulations 1995 (SI 1995 No 738)*.

Verification of safety-critical elements of installations

[07012] A major amendment to the *Safety Case Regulations 1992* by the *Offshore Installations and Wells (Design and Construction etc.) Regulations 1996 (SI 1996 No 913)* (the *Design and Construction Regulations 1996*) and continued in the *Offshore Installations (Safety Case) Regulations 2005 (SI 2005 No 3117)*, imposes a duty on the duty holder to set up arrangements for the scrutiny of safety-critical elements of the offshore installation to be verified as suitable by an independent and competent person. This requirement replaces the certification regime established by *the Offshore Installations (Construction and Survey) Regulations 1974 (SI 1974 No 289)* which have been revoked by the *Design and Construction Regulations 1996*.

The purpose of such scrutiny is to obtain assurance of the achievement of the satisfactory condition of the safety critical elements of the installation. The scheme is to be carried out throughout the installation's life cycle (from design through fabrication, construction, hook-up and commissioning, and the whole operating life to decommissioning and dismantling). For the duty holder, the scheme will contribute to the duty holder's ability to demonstrate that the installation is 'fit for purpose'. A summary of the scheme must be included in the installation's Safety Case.

Before commencing operations

[07013] *HSE have the discretion to allow shorter notification period in appropriate cases.*

(a) Production installations to be established or relocated

Reg 6(1) requires the operator of a production installation to submit a Design Notification (see Sch 1) by a date which is early enough to allow them to take account in the design of any matters raised by the HSE within three months of the submission.

Reg 6(2) requires the operator of a production installation which is to be moved to a new location (whether from outside relevant water or not) to submit a relocation notification (see schedule 1) at such time before the submission of a field development programme to the DTI to allow him to take account in the design of any matters raised by the HSE within three months of the submission.

HSE will indicate in writing any changes to the design or any considerations which it would wish to be taken into account at detailed design, construction or commissioning stages, and which would be likely in due course to assist in acceptance of the full operational Safety Case required under *Reg 4(2)*.

No operations may be commenced until six months have elapsed since the operator sent the Safety Case to the HSE and it has been accepted by the HSE. Commencement of an operation includes the first well drilling operation which may require the release of hydrocarbons beneath the sea bed or when hydrocarbons are brought on to the site for the first time.

(b) Production installations

The Operator of a production installation must ensure that it is not operated unless a Safety Case has been sent to the HSE at least six months before commencing the operation of the installation, and the Safety Case has been accepted by the HSE [*Reg 7* and *Sch 2*].

(c)   Non-production installation

The Operator of a non-production installation must ensure that it is not moved in relevant waters with a view to it being operated there unless a safety case has been sent to the HSE at least three months before the movement of the installation and the Safety Case has been accepted by the HSE [*Reg 8* and *Sch 2*].

The Operator of a non-production installation must ensure that it is not converted to enable it to operate as a production installation unless a Design Notification of any particulars not contained in the current Safety Case has been sent to the HSE by a date which is early enough to allow him to take account in the design of any matters raised by the HSE within three months of the submission [*Reg 9* and *Sch 1*].

Decommissioning fixed installations

[07014] Operator of a fixed installation must not decommission the installation unless he has:

(a)   revised the current Safety Case in line with Sch 5;
(b)   sent it to the HSE at least three months before the decommissioning starts; and
(c)   the HSE has accepted the revision to the Safety Case [*Reg 11*].

The submission required by *Reg 11* is intended to ensure that the method and arrangements for abandonment of a fixed installation provide for adequate control of risks to persons on the installation or engaged in connected activities. It is an offence to begin decommissioning before the Safety Case has been accepted.

HSE's acceptance of a Safety Case for decommissioning of an installation is required, and will be given, independently of any other approval such as the obligation under the *Petroleum Act 1998* to submit an abandonment programme.

The abandonment programme

[07015] The process of abandonment of a fixed offshore installation or submarine pipeline is subject to the acceptance of a detailed formal document known as the abandonment programme by the Secretary of State which must:

(a)   contain an estimate of the cost of the measures proposed in the programme;
(b)   specify the periods/times when the programme is to be put into effect, or provide information relevant to determining how the specified times are to be determined;
(c)   include provisions regarding the continuing maintenance requirements of installations or pipelines (including those not fully demolished) which are to be left in position.

Part IV of the *Petroleum Act 1998* provides that when an abandonment programme has been approved by the Secretary of State, it becomes the duty of the persons who submitted the programme to ensure that it is carried out and that any conditions are complied with.

Safety Case specifications

[07016] The particulars to be included in all Safety Cases are required to demonstrate that:

(a)   there is a management system that will make sure that the statutory provisions relating to the installation are complied with;

(b)   adequate arrangements have been made for audit and reporting of the management system;

(c)   all hazards that could cause a major accident have been identified;

(d)   risks have been evaluated and measures will be taken to keep these risks to the lowest level that is reasonably practicable [*Reg 12*].

Matters to be considered include:

(i)   the plant and arrangements for the control of well operations;

(ii)   a description of any pipeline with the potential to cause a major accident and how it is intended to secure safety and how he will ensure compliance with the *Pipelines Safety Regulations 1996 (SI 1996 No 825), Reg 11*;

(iii)   how compliance with the PFEER regulations will be achieved;

(iv)   how personnel will be protected from toxic gas;

(v)   how the duty holder will ensure compliance with the *Offshore Installations and Wells (Design and Construction etc.) Regulations 1996 (SI 1996 No 913)* and the suitability of the safety critical elements; and

(vi)   where appropriate, the arrangement, methods and procedures for dismantling the installation and connected pipelines.

Verification of safety-critical elements

[07017] The duty holder (fixed and mobile installations) must ensure that the installation is not operated unless a suitable verification scheme covering the 'safety-critical elements' of the installation has been put in place. These elements include such parts of an installation and of its plant (including computer programs) the failure of which could cause or contribute substantially to a major accident (*Reg 19* and *Sch 7*).

To be suitable, the written scheme, prepared by a competent person, must demonstrate that the safety-critical elements are in good repair and condition and are verified as such by competent persons carrying out the following actions:

—   examination, including testing where appropriate;

—   examination of any design, specification, certificate, CE marking or other document, marking or standard relating to those elements;

—   examination of work in progress;

—   the taking of appropriate action following reports by such persons.

Competent persons must be independent, that is, their functions will not involve the consideration of elements for which they bear or have borne such responsibility as might compromise objectivity. In addition, it must be shown that they are sufficiently independent of a management system to ensure that they will be objective in discharging their functions.

Review, revision and recording of verification schemes

[07018] The duty holder must ensure that as often as may be appropriate:

— the verification scheme is reviewed and, where necessary, revised or replaced by or in consultation with an independent and competent person;

— a note is made of any reservation expressed by such person in the course of drawing it up;

— verification schemes must be kept (at an address notified to the HSE) for a minimum of six months following cessation of the scheme along with the following documents;

— all examinations and testing carried out;

— the findings;

— remedial actions recommended; and

— remedial actions performed [*Reg 20* and *Sch 7*].

Notification of hazardous activities

[07019] Duty holders must ensure that the following operations are not commenced until the HSE has been notified. However, unlike a Safety Case, notification does not require acceptance by HSE before the activities can be commenced. Should the information provided give cause for concern, HSE may intervene by, for example, requesting further information, inspecting, or by issuing an enforcement notice.

When considering whether intervention is appropriate HSE will take account of the following factors (among others):

— the major accident potential of the operations being undertaken;

— the assessed adequacy of the management and other risk control systems; and

— whether the operations will take place in conditions approaching the operating limits described in the installation Safety Case (for example, extreme or unusual environmental conditions or well operations).

Combined operations

Operators of installations must not engage in a combined operation unless they agree a combined operations notification and one of them sends it to the HSE 21 days before the combined operation is due to start [*Reg 10* and *Sch 4*].

Well operations

[07020]–[07021] Well operations require notification of the HSE at least 21 days before commencing the operation. The notification must contain the information listed in schedule 6. However for well operations on a production installation involving the insertion of a pipe into a well or altering the construction of a well the notice period is reduced to ten days [*Reg 17* and *Sch 6*].

Summary of contents of a production Safety Case

[07022] The production Safety Case which is required to be submitted under *Reg 7* must include a comprehensive account of the following matters:

— a description of the structure, plant, pipelines, wells and any connections to other installations and any limits to safe operation;

— how compliance with the *Offshore Installations and Wells (Design and Construction) Regulations 1996 (SI 1996 No 913)* and the suitability of the safety critical elements will be achieved;

— management of health and safety, control of major hazards and the auditing of these systems;

— compliance with the PFEER regulations and the protection of personnel from toxic gas;

— all foreseeable activities which are intended to be undertaken, or which may need to be undertaken, during the operating lifetime of the installation;

— any occasional activities such as major maintenance projects or diving work and any planned construction or alteration projects;

— activities which may involve other vessels (for example, nearby diving support vessels, supply vessels, floating storage units), aircraft or other installations should be addressed as far as they are foreseeable (*note: where not foreseeable, further details may need to be provided in a revised Safety Case (under *Reg 9(2)*) or a combined operations Safety Case (as required by *Reg 6*);

— well operations so far as they are foreseeable and, in particular, a description of the intended number of wells and the purpose of each (that is, whether for production, gas injection or water injection);

— the drilling facilities and the general drilling programme;

— the expected performance characteristics of the reservoir and individual wells (including well drive mechanisms);

— well control procedures;

— any simultaneous drilling and production operations from the installation and the arrangements for controlling such activities;

— procedures for the suspension or abandonment of any well;

— the safety management system for the operation, control and emergency procedures of the pipelines connected to the installation;

— a summary of the design philosophy for ensuring the safe operation of the installation; and

— management of arrangements with contractors and sub-contractors.

The *Health and Safety (Fees) Regulations 2005 (SI 2005 No 676)* include provisions allowing the HSE to charge for the assessment and acceptance of safety cases. The regulations set the fees payable by an operator or owner for the performance by HSE of the following functions under the *Offshore Installations (Safety Case) Regulations 2005 (SI 2005 No 3117)* are payable as follows:

Function	Person by whom fee is payable
Assessing a safety case (sent to the HSE pursuant to the *Safety Case Regulations 1992 (SI 1992 No 2885), Reg 4(1))* for the purpose of deciding whether to raise matters relating to health and safety and raising such matters	The operator who sent the safety case to the HSE pursuant to that provision
Assessing a safety case or a revision to a safety case (sent to the HSE pursuant to any other provision of the 1992 Regulations) for the purpose of deciding whether to accept that safety case or revision and accepting any such safety case or revision	The operator or owner who sent the safety case or revision to the HSE pursuant to that provision
Providing advice with respect to the preparation of a safety case or a revision to a safety case which is proposed to be sent to the HSE pursuant to any provision of the 1992 Regulations	The operator or owner who has requested that advice
Assessing whether to grant an exemption pursuant to regulation 17 of the 1992 Regulations and granting any such exemption	The operator or owner who has requested the exemption

The fee must not exceed the sum of the costs reasonably incurred by HSE for the performance of the function. It is payable within 30 days from the date of HSE's invoice.

Record keeping

[07023] Copies of the current Safety Case and any summary of any review along with every audit report must be kept at an address in Great Britain and on the installation for at least three years. Documents relating to the verification scheme (see **O7018** above) must be kept for six months.

Reports must be made, and the actions recorded, in respect of audits carried out under *Reg 8(1)(b)*. These requirements are intended to ensure, among other matters, that the reports and action records will be conveniently available for examination by the HSE during the auditing of the duty holder's safety management and audit system.

Documents required to be kept on the installation are normally kept by the installation manager. Arrangements for keeping the documents and revising them as required should be considered as part of the safety management system.

Offshore Installations and Wells (Design and Construction, etc.) Regulations 1996 (SI 1996 No 913)

[07024] The *Offshore Installations and Wells (Design and Construction, etc.) Regulations 1996* aim to ensure (*a*) the structural integrity of offshore installations is maintained throughout their normal life cycle; (*b*) that risks to installation workers from structural failure or loss of stability are minimised to reasonably practicable levels; (*c*) the safe condition of wells at all stages of their life cycle; and (*d*) to implement the relevant aspects of the Extractive Industries (Boreholes) Directive (EID) (92/91/EEC)

The DCR regulations complement the requirements of the *Pipeline Safety Regulations 1996 (SI 1996 No 825)*, PUWER, marine and aviation regulations.

The DCR regulations modified the original 1992 safety case regulations by introducing a requirement for the verification of safety critical elements on an offshore installation, replacing the certification regime specified in the *Offshore Installations (Construction and Survey) Regulations 1974 (Reg 26 and Sch 2)*. Regulation 26 and schedule 2 have been repealed by the *Offshore Installations (Safety Case) Regulations 2005 (SI 2005 No 3117)* and replaced by the requirements of *Regs 19–21* and *Sch 7* of the 2005 regulations.

The definitions in the DCR regulations have been amended by *SI 2005 No 3117* to bring them into line with these regulations.

Integrity of installations – duties of dutyholders

[07025] So far as reasonably practicable, duty holders must ensure that:

(a)   installations are able to withstand such adverse forces as are reasonably foreseeable [*Reg 5(1)(a)*];
(b)   the structural integrity of an installation is not prejudiced by (i) its layout and configuration, and (ii) its fabrication, transportation, construction, operation, modification, maintenance, repair [*Reg 5(1)(b), (c)*];
(c)   in the event of reasonably foreseeable damage, the installation will retain sufficient integrity to enable action to be taken to safeguard the health and safety of personnel on or near the installation [*Reg 5(1)(e)*];
(d)   installations are capable of being decommissioned and dismantled safely [*Regs 5(1)(d), 10*];
(e)   installations are composed of materials sufficiently proof against and/or protected from anything likely to prejudice their integrity [*Reg 5(2)*];
(f)   work to an installation does not impair its integrity [*Reg 6*];
(g)   installations are (i) operated within appropriate limits, and (ii) that environmental conditions in which they may safely operate have been recorded. Such records must be kept on the installation and readily available for inspection by personnel [*Reg 7*];
(h)   installations are subject to periodic maintenance, and arrangements made for (i) periodic assessment, and (ii) execution of any remedial work [*Reg 8*];

(i)   within ten days after the appearance of a significant threat to the integrity of an installation, a report is sent to HSE, identifying the threat and specifying any remedial action [*Reg 9*];

(j)   every helicopter landing area is large enough for landing and departure purposes and is of adequate design and construction [*Reg 11*]; and

(k)   the additional requirements set out in Sch 1, which are largely derived from the HSE Industries (Boreholes) Directive, are complied with.

Integrity of wells – duties of well-operators

[07026] Well-operators must ensure that:

(a)   wells are designed, constructed and operated so that (i) so far as reasonably practicable, there is no unplanned escape of fluids, and (ii) risks to the health and safety of persons are as low as reasonably practicable [*Reg 13*];

(b)   prior to design of a well, assessments have been made of geological strata and formations and fluids therein, including any hazards [*Reg 14*];

(c)   wells are designed and constructed so that, so far as reasonably practicable,
    (i)   they can be suspended or abandoned safely, and
    (ii)   following suspension or abandonment, there is no unplanned escape of fluids [*Reg 15*];

(d)   wells are constructed of suitable materials [*Reg 16*];

(e)   suitable well control equipment is provided [*Reg 17*];

(f)   prior to commencement or adoption of design of a well, written arrangements are made for examination, by independent and competent persons, of a well or parts thereof [*Reg 18*];

(g)   in the case of:
    (i)   escape of fluids,
    (ii)   completion of a well,
    (iii) abandonment operations,
    (iv) drilling operations, and
    (v)   workover operations;
a report is sent to the HSE at such intervals as are agreed (or otherwise at weekly intervals following commencement), containing:
—   identifying number of well,
—   name of installation,
—   summary of activity,
—   diameter and true vertical and measured depths of (*a*) any hole drilled, and (*b*) any casing installed,
—   drilling fluid density,
—   in the case of an existing well, its current operational state [*Reg 19*]; and

(h)   personnel carrying out well operations have received appropriate information, instruction and training and are properly supervised [*Reg 21*].

Duties of other personnel

[07027] Every person concerned (or to be concerned) in whatever capacity, in an operation in relation to a well, must co-operate with the well-operator, so far as is necessary, to enable him to discharge his duties [*Reg 20*].

Offshore Installations (Prevention of Fire and Explosion, and Emergency Response) Regulations 1995 (SI 1995 No 743)

[07028] The regulations have been amended by the *Offshore Installations (Safety Case) Regulations 2005 (SI 2005 No 3117)* to bring the definitions into line with those of the 2005 safety case regulations and to delete the requirements for maintaining the equipment since these are now covered by the verification requirements of the safety case regulations.

These regulations, applicable to both production and non-production installations, require operators, owners and duty holders to:

(a)   protect persons on the installation from fire and explosion, and secure effective emergency response [*Reg 4*];

(b)   carry out, update, where necessary, and keep a record of an assessment of measures (including performance standards) for effective:
   (i)   evacuation,
   (ii)   escape,
   (iii)   recovery, and
   (iv)   rescue [*Reg 5*];

(c)   establish the organisation/arrangements to be implemented in an emergency, and ensure provision of:
   (i)   training/instruction on necessary action, and
   (ii)   written information on use of emergency plant [*Reg 6*];

(d)   ensure availability of suitable equipment in the event of an accident involving a helicopter [*Reg 7*];

(e)   prepare and update the emergency response plan:
   (i)   ensure its availability and that its contents are known, and
   (ii)   ensure that it is tested from time to time [*Reg 8*];

(f)   take effective measures to prevent fire/explosion, e.g.
   (i)   identifying/designating areas where there is a risk of occurrence of flammable/explosive atmospheres,
   (ii)   controlling the carrying on of hazardous activities there,
   (iii)   prohibiting/limiting use of plant (in such areas) unless suitable for use, and
   (iv)   controlling placement (in such areas) of electrical fixtures/sources of ignition [*Reg 9*];

(g)   take steps to detect fire and communicate information to places where control action can be taken [*Reg 10*];

(h)   provide suitable emergency response arrangements:
   (i)   ensure a warning is given of an emergency (see below), and
   (ii)   communicate the purpose of the emergency response.

These arrangements should ideally remain effective during an emergency [*Reg 11(1)*].

Warnings take the form of (A) illuminated signs and (B) acoustic signals.

(A)   Illuminated signs
    These are:
    —   in the case of warning of toxic gas, a red flashing sign, and
    —   in other cases, a yellow flashing sign.

(B)   Acoustic signals
    These are:
    —   in the case of warning of evacuation, a continuous signal of variable frequency,
    —   in the case of warning of toxic gas, a continuous signal of constant frequency, and
    —   in other cases, an intermittent signal of constant frequency [*Reg 11*];

(i)   control/limit extent of emergencies (including fire and explosions), e.g. by provision for remote operation of plant and ensuring that it is capable of remaining effective in an emergency [*Reg 12*];

(j)   protect persons from the effects of fire and explosion [*Reg 13*];

(k)   provide muster areas and:
    (i)   provide safe egress from accommodation and work areas, and
    (ii)   provide safe access to muster areas/evacuation and escape points which must be kept unobstructed, adequately lit with emergency lighting, marked with suitable signs and, so far as is reasonably practicable, remain passable in an emergency [*Reg 14(1), (2)*];

Notes:
    •   doors for use in an emergency must open in the appropriate direction, or, if this is not possible, be sliding doors, and must not be fastened,
    •   accommodation areas must be provided with at least two means of egress situated suitably apart at each level [*Reg 14(3)*], and
    •   each person on an installation must be assigned to a muster area; and muster lists must be kept up-to-date and clearly displayed [*Reg 14(4)*];

(l)   arrange for safe evacuation of persons and their conveyance to a place of safety [*Reg 15*];

(m)   ensure provision of means of escape, where arrangements for systematic evacuation fail [*Reg 16*];

(n)   arrange for:
    (i)   recovery of persons following evacuation/escape,
    (ii)   rescue of persons near the installation, and
    (iii) conveyance of rescued/recovered persons to a place of safety [*Reg 17*];

(o)   provide emergency protective equipment and prepare and operate a written scheme for its examination and testing by a competent person [*Reg 18*] (see further PERSONAL PROTECTIVE EQUIPMENT);

(p)   provide suitable emergency plant and equipment [*Reg 19*]. The requirement to maintain the equipment in good working order etc. is now covered by the verification requirements of *(SI 2005 No 3117)*;

(q)   provide life-saving appliances, e.g. survival craft, life-rafts, life-buoys, life-jackets, which are:

    (i)   of conspicuous colour,

    (ii)   suitably equipped (where necessary), and

    (iii)   kept available in sufficient numbers [*Reg 20*];

(r)   provide information to all persons on the installation, in relation to:

    (i)   areas in which there is a risk of occurrence of flammable/explosive atmospheres,

    (ii)   non-automatic plant for fire-fighting purposes,

    (iii)   plant connected with personal protective equipment and life-saving appliances [*Reg 21*].

In addition to the 'on-board' fire-fighting facilities, the duty holder may elect to utilise the services of a fire brigade from relevant local fire authorities – see *Fire Services Act 1947, s 3(1)(dd)*, as inserted by the *Merchant Shipping and Maritime Security Act 1997*.

The Health and Safety Commission (HSC) has published a revised Approved Code of Practice (ACOP) and guidance* to accompany the *Offshore Installations (Prevention of Fire and Explosion, and Emergency Response) Regulations 1995 (SI 1995 No 743)*. The revised ACOP came into force on 1 June 1997.

The main changes are to paragraphs 167 and 170 of the 1995 ACOP and deal with standby vessels. The revised ACOP sets out the factors to be considered when selecting a standby vessel to provide rescue and recovery services at offshore installations. The factors, such as manoeuvrability, visibility, lighting and communication, are based mainly on Lord Cullen's recommendations following the 1988 Piper Alpha disaster.

---

* The revised ACOP, '*Prevention of fire and explosion, and emergency response on offshore installations: Approved Code of Practice and guidance*' (ref L65) is available from HSE Books.

Offshore Installations and Pipeline Works (Management and Administration) Regulations 1995 (SI 1995 No 738)

[07029] The *Offshore Safety (Miscellaneous Amendments) Regulations 2002 (SI 2002 No 2175)* amended the Offshore Installations and Pipeline Works (Management and Administration) Regulations 1995 to ensure that all parts of an offshore installation are within the scope of appropriate health and safety legislation.

The amendments are relatively minor but the new Regulations were required to ensure that offshore health and safety law is appropriately applied and that the definitions of an offshore installation in *MAR* and in other offshore legislation, such as the *Health and Safety at Work etc. Act 1974 (Application*

Outside Great Britain) Order 2001 (SI 2001 No 2127), are consistent. Because of new developments in technology offshore, the definition of what is an offshore installation needed to be extended to include supplementary units. Such units provide power and other support facilities for offshore installations and can be powered by wind or wave or by conventional means.

Revised guidance to support the amended *Offshore Installations and Pipeline Works (Management and Administration) Regulations 1995* has been published and is intended to help people who may be affected by the Regulations to understand what the Regulations require. It is a simple explanation of the main provisions of the Regulations to assist installation operators, installation owners, employers, managers, safety representatives, safety committee members and others involved with offshore activities. The guidance has been revised to reflect the new definition of what is an offshore installation. The guidance also reflects other changes made since it was first published in 1995 including; defining when a well service vessel (WSV) becomes an offshore installation and clarifying the position of specialist support vessels which provide accommodation (flotels).

The MAR regulations have been further amended by the *Offshore Installations (Safety Case) Regulations 2005 (SI 2005 3117)* to update the definitions where relevant and bring them into line with the new safety case regulations.

These regulations impose managerial and administrative duties on the duty holder who is defined as:

(a)   operators of production installations, and
(b)   owners of non-production installations;

not on licensees or managers (except the duty to co-operate). The installation manager is the delegate of such duties and as such has a limited liability, whilst ultimate residual liability rests with the duty holder (see *McDermid v Nash Dredging and Reclamation Co Ltd* [1987] 2 All ER 878).

Chain of command

Licenee

Duty holder

Installation manager

Duties on operators of fixed installations and owners of mobile installations – duty holders

[07030] Duty holders must:

(1)   notify the HSE in writing of the date of entry/departure of an installation into/out of relevant waters. Where there is a change of duty holder, an installation must not be operated until the HSE has been notified of:
(i)   the date of change, and
(ii)   the name and address of the new duty holder [*Reg 5*];
(2)   appoint a competent installation manager, who must be known or readily ascertainable by everyone on the installation [*Reg 6*];

(3)   keep both an offshore and onshore record of details of all persons on and working on the installation [*Reg 9*];

(4)   where necessary, introduce a permit to work system [*Reg 10*];

(5)   issue written instructions on procedures to be observed, bringing them to the attention of all personnel [*Reg 11*];

(6)   establish a sufficient system for communicating health and safety arrangements between installation and shore, vessels and aircraft [*Reg 12*];

(7)   appoint a competent helicopter landing officer to ensure safe helicopter landings and take-offs [*Reg 13*];

(8)   collect and keep meteorological/oceanographic data and information concerned with the movement of the installation [*Reg 14*];

(9)   ensure that the address and telephone number of the local HSE is known or readily ascertainable by personnel on the installation [*Reg 15*];

(10) provide employees with health surveillance [*Reg 16*];

(11) ensure an adequate supply of clean wholesome drinking water at suitable locations [*Reg 17*];

(12) ensure that all food is fit for human consumption [*Reg 18*];

(13) ensure that the installation is readily identifiable by sea or from the air [*Reg 19*]; and

(14) ensure that the *Employer's Liability (Compulsory Insurance) Act 1969* is fully complied with by all employers of employees who work on any offshore installation under his control.

Certificates of exemption may be issued by the HSE in respect of these duties.

Duty of installation manager

[07031] Where necessary, in the interests of health and safety, an installation manager can restrain a person and put him ashore [*Reg 7*].

Duties of all persons on the installation

[07032] Everyone on the installation must co-operate with the installation manager to enable him to discharge his statutory safety duties and functions; similarly, with the helicopter landing officer [*Reg 8*].

Hazardous operations with the potential to cause a major accident

[07033] Lord Cullen recommended that, in parallel with the move to the Safety Case regime, the existing offshore legislation should be comprehensively reviewed with a view to its progressive replacement by a modernised and rationalised structure of Regulations.

This recommendation took account of experience of similar reforms onshore, developed by the Health and Safety Commission using its powers under *HSWA 1974*. Following enactment of the *Offshore Safety Act 1992* (which brought the existing offshore legislation within the scope of the 1974 Act) the Commission published, in August 1992, its outline plans for the offshore reform programme.

The definition of 'major accident' in the *Offshore Installations (Safety Case) Regulations 1992* embraces the full range of such accidents which can occur offshore and not just those arising from fire and explosion hazards. The Safety Case is required to relate to all activities carried out on, or in connection with, an installation including, for example, relevant activities involving divers, pipelaying or the movement of vessels or helicopters. The Regulations place a duty on the various people concerned with these activities, to co-operate with duty holders in the preparation of installation Safety Cases.

Construction and use of submarine pipelines

[07034] The *Petroleum Act 1998* prohibits the construction of a pipeline (or use of a pipeline constructed after 1975) without the specific written authorisation of the Secretary of State. Once authorised, the pipeline becomes a '*controlled pipeline*' and any work affecting the design, route and boundaries is subject to any terms and conditions specified by the Secretary of State. In particular, it is necessary to ensure that sufficient funds are available regarding liability cover in the event of damage attributable to the release or escape from the pipeline of 'permitted substances', that is, anything or any substance authorised to be transported in that pipeline.

The Secretary of State is granted power to order the compulsory modifications of a pipeline for the purposes of improving flow through that pipeline. A person other than the owner of the pipeline is permitted to make an application to the Secretary of State for a modification order [*PA 1998, s 16*].

Pipelines Safety Regulations 1996 (SI 1996 No 825)

[07035] The *Pipelines Safety Regulations 1996* aim to secure the initial and continuing integrity of pipelines throughout their life cycle. The Operator, in relation to a pipeline means;

(a)   the person who has control over the conveyance of fluid in the pipeline;
(b)   until that person is known, the person who commissions, or is to commission, the design and construction of the pipeline; or
(c)   where the pipeline is no longer in use the person last having control over the conveyance of fluid in it.

To this end, the operator of a pipeline ensure that:

(a)   it is designed so as to be safe within range of all operating conditions [*Reg 5*], provided with appropriate safety systems [*Reg 6*], can be examined and maintained safely [*Reg 7*] and is composed of suitable materials [*Reg 8*];
(b)   it is constructed and installed so that it is fit for the purpose which it was designed [*Reg 9*];

(c)   it is modified and maintained so that its fitness for purpose is not prejudiced [*Reg 10*] and operated within safe operating limits [*Reg 11*];

(d)   adequate arrangements have been made to deal with any accidental loss of fluid, discovery of a defect or other emergency [*Reg 12*];

(e)   the pipeline is maintained in an efficient state, efficient working order and good repair [*Reg 13*];

(f)   abandoned at the end of their life cycle so as not to become a source of danger [*Reg 14*];

(g)   steps are taken to inform persons of its existence and where to ensure that no damage is caused to the pipeline [*Reg 16*]; and

(h)   other persons have a duty not to cause damage to a pipeline as may give rise to danger to persons [*Reg 16*].

Moreover, in the case of pipelines conveying dangerous fluids (e.g. flammable in air, very toxic, oxidising, reacting violently with water):

(i)   the pipeline must be fitted with emergency shut-down valves [*Reg 19*];

(ii)   details relating to construction/fluid to be conveyed must be notified to the HSE at least six months prior to commencement of construction [*Reg 20*];

(iii)   the HSE must be notified at least 14 days before fluid is conveyed in a major accident pipeline [*Reg 21*] or within 14 days of a change in operator [*Reg 22(1)*];

(iv)   the HSE must be notified of specified particulars 3 months before any event listed in schedule 5 takes place [*Reg 22(2)*];

(v)   a major accident prevention policy must be prepared (and revised) prior to completion of design [*Reg 23*];

(vi)   appropriate emergency organisation and arrangements must in place and details of emergency procedures must be finalised [*Reg 24*]; and

(vii)   local authorities, in whose area the pipeline is to pass, must be notified that construction is to take place, and then prepare and update an emergency plan [*Reg 25*], for which it may charge a fee [*Reg 26*].

The HSE has published guidance* on the regulations. The guidance is designed to give pipeline operators and others involved with pipeline activities or who may be affected by the regulations, an understanding of what is required.

The regulations provide for high standards to be maintained for all pipelines onshore and offshore while enabling outdated legislation to be removed; and disentangle legislation dealing with health and safety from the consents and authorisations regime operated by the Department of Trade and Industry. The guidance provides detailed guidance on the requirements of the regulations, for example, the general duties which include design, construction, installation and work on a pipeline; arrangements for incidents and emergencies; maintenance and decommissioning. The guidance also sets out the requirements of the regulations for major accident hazard pipelines, covering notification to the HSE at key stages throughout the life of the pipeline, the preparation of major accident prevention documents and emergency procedures by the pipeline operators, and the requirements for local authorities to prepare pipeline emergency plans.

---

* 'A Guide to the Pipelines Safety Regulations 1996' (Ref: L82) is available from HSE Books.

The Diving at Work Regulations 1997

[07036] The *Diving at Work Regulations 1997 (SI 1997 No 2776)*, which revoked the *Diving Operations at Work Regulations 1981*, impose requirements and prohibitions relating to health and safety upon persons who work as divers. Under the 1997 Regulations, a person 'dives' if he enters water or any other liquid; or a chamber in which he is subject to pressure greater than 100 millibars above atmospheric pressure and in order to survive in such an environment he breathes in air or other gas at a pressure greater than atmospheric pressure.

Environments such as scientific clean rooms or submersible craft subjected to a pressure less than 100 millibars above atmospheric pressure are not covered by the Regulations. The regulations do not cover work carried out in air that is compressed to prevent the ingress of ground water or the treatment of patients not under the control of a diving contractor [Reg 3(1)].

The regulations do apply offshore [Reg 3(2)].

Every person who is engaged in a diving project or is responsible for, or has control over, such a project must take such measures as it is reasonable for a person in his position to take to ensure that the provisions of the 1997 Regulations are complied with [Reg 4].

There could be a number of diving projects taking place on one site. Each of these projects could be separate from each other and may have different 'diving contractors' in charge of them.

The HSE are authorised to approve diving qualifications [Reg 14(1)] and medical examiners of divers [Reg 15(6)].

The HSE is advising owners of diving cylinders to note that from 1 July 2006 the regulations covering the periodic inspection and testing of their cylinders will change. In the future, persons conducting cylinder inspections in the UK will be subject to a third party conformity assessment.

One possible option for people wishing to carry out inspections would be direct assessment by the United Kingdom Accreditation Service (UKAS) to ensure their competency. Those who are successfully assessed will then be formally appointed by HSE. However, the costs of direct conformity assessment and appointment by UKAS could be fairly significant.

The issue of the new requirements placing an increased financial burden on small businesses was recognised some time ago and an alternative option for conformity assessment has been agreed and is being developed. When it is finalised the current HSE guidance will be updated.

The guidelines for appointing diving cylinder conformity assessment bodies, which are required under the *Carriage of Dangerous Goods and Use of*

Transportable Pressure Equipment Regulations 2004 (SI 2004 No 568), have been issued in July 2004. The regulations, which came into force on 10 May 2004, implemented EC Directives 2003/28/EC, 2003/29/EC and completed the implementation of Directive 1999/36/EC. They are designed to radically simplify the regulatory framework covering the carriage of dangerous goods by bringing together, in a single set of regulations, requirements previously set out in 14 separate pieces of legislation.

The current guidance covering the inspection of diving cylinders: Guidelines for the Appointment of Conformity Assessment Bodies Required by *SI 2004 No 568,* can be found on HSE's website at: http://www.hse.gov.uk/cdg/pdf/gfap obod.pdf

Diving personnel responsibilities

The diving contractor

[07037] One person and only one persons shall be appointed to be diving contractor for a diving project [*Reg 5(1)*].

The diving contractor must ensure, so far as is reasonably practicable, that the diving project is planned, managed and conducted in a manner which protects the health and safety of all persons taking part in that project. This requires the preparation of a diving project plan which must be updated as necessary during the continuance of the project.

Other duties placed on the diving contractor include:

— appointing a person to supervise a diving operation and ensuring that that person is competent and suitably qualified to act as supervisor for the operation;
— making a written record of that appointment;
— ensuring that the person appointed is supplied with a copy of any part of the diving project plan which relates to that operation;
— ensuring that there are sufficient people with suitable competence to carry out safely and without risk to health both the diving project and any action (including the giving of first-aid) which may be necessary in the event of a reasonably foreseeable emergency connected with the diving project;
— ensuring that suitable and sufficient plant is available whenever needed to carry out safely and without risk to health both the diving project and any action (including the giving of first-aid) which may be necessary in the event of a reasonably foreseeable emergency connected with the diving project;
— ensuring that all diving equipment is maintained in a safe working condition;
— ensuring, so far as reasonably practicable, that any person taking part in the diving project complies with the requirements and prohibitions imposed on him by or under the relevant statutory provisions and observes the provisions of the diving project plan; and
— ensuring that a record containing the required particulars, given in annex 1, is kept for each diving operation and retained for at least two years after the date of the last entry in it [*Reg 5–9*].

The diving supervisor

[07037.1] Although the main duty for the whole project is placed on the diving contractor, supervisors also have a duty to manage the diving operation safely and give such reasonable directions to any person to enable him to do so. Supervisors should not participate in a diving operation which they consider to be unsafe because insufficient supervisors have been appointed [*Reg 10–11*].

The diver

[07037.2] A diver has the responsibility not to dive unless he has an approved valid qualification [*Reg 12(1)*] and is competent to carry out the activities he may reasonably expect to carry out whilst taking part in the diving project [*Reg 13*]. A diver must maintain a daily record of his diving [*Reg 12(3)* and *Annex 2*] and have a valid medical certificate to dive [*Reg 12(1)*] and must not dive if he knows of anything which makes him unfit to dive [*Reg 13(1)*].

Dive planning and risk assessment

[07038] The diving contractor must ensure that a diving project plan is prepared before commencement of diving and the plan must be updated as necessary [*Reg 5(2)*]. The diving plan must be based on an assessment of the risks to any person taking part in the diving project [*Reg 8*].

Similar duties regarding the safe operation of diving projects are also placed on 'clients' who raise diving contracts. They should be involved in the risk assessment elements of the 'diving project plan' required by *Reg 8*.

The Health and Safety HSE (HSE) has warned divers of the risks of using unsafe electrical equipment underwater during commercial fishing operations. The warning comes after an investigation into illegal diving for razor fish. The investigation revealed that some fishermen were dropping electrified cables, which consist of several un-insulated metal electrodes, into the water that are then dragged by the vessel across the seabed stunning razor fish as they go. A diver who follows the path of the cable then collects the fish. However, if the diver comes into contact or even close proximity to the electrodes there is a real risk of electrocution.

Copies of Code of Practice for the Safe Use of Electricity Under Water, AODC 035, are available from The International Marine Contractors Association, 5 Lower Belgrave Street, London, SW1 0NR (Tel: 020 7824 5521, alternatively email: publications@imca-int.com).

Further Information on regulations covering diving at work safely and guidance to help employers meet their duties under the law can be found on the HSE website at: www.hse.gov.uk/diving

Other offshore specific legislation

[07039] The following Regulations were enacted under the *Mineral Workings (Offshore Installations) Act 1971* and apply only to offshore operations:

— *Offshore Installations (Logbooks and Registration of Death) Regulations 1972 (SI 1972 No 1542)*;
— *Offshore Installations (Inspectors and Casualties) Regulations 1973 (SI 1973 No 1842)*;
— *Offshore Installations (Safety Representatives and Safety Committees) Regulations 1989 (SI 1989 No 971)*;
— *Offshore Installations and Pipeline Works (First-Aid) Regulations 1989 (SI 1989 No 1671)*;
— *Offshore Electricity and Noise Regulations 1997 (SI 1997 No 1993)*.

Offshore Installations (Logbooks and Registration of Death) Regulations 1972

[07040] These Regulations *(SI 1972 No 1542)* have largely been revoked, leaving only those provisions which cover the registration of deaths and of persons lost from offshore installations, and the offences and penalties for failure to comply with the reporting requirements.

Duties are imposed on the installation manager and the owner of an offshore installation to submit a notification form to the Department of Transport in cases where a person dies or is lost overboard from an offshore installation [*Regs 8–11* and the *Schedule*]. The notification to the Department of Transport is additional to the reporting requirements of the *Reporting of Injuries, Diseases and Dangerous Occurrences Regulations 1995 (SI 1995 No 3163)*.

Offshore Installations (Inspectors and Casualties) Regulations 1973

[07041] These Regulations *(SI 1973 No 1842)*, which have largely been revoked, cover duties regarding the inspection of offshore installations. The Regulations should be read in conjunction with the *Health and Safety at Work etc. Act 1974, ss 19–28* and the *Provision and Use of Work Equipment Regulations 1992 (SI 1992 No 2932)*.

The following terms are defined in the Offshore Installations (Inspectors and Casualties) Regulations 1973:

casualty means a casualty or other accident involving loss of life or danger to life suffered by a person:
    (a)   employed on, or working from, an offshore installation; or
    (b)   employed on, or working from, an attendant vessel, in the course of any operation undertaken on or in connection with an offshore installation;
equipment means any plant, machinery, apparatus or system used, formerly used or intended to be used (whether on or from an offshore installation or on or from an attendant vessel) in the assembly, reconstruction, repair, dismantlement, operation, movement or inspection of an offshore installation or the inspection of the sea bed under or near an offshore installation.

Inspection of offshore installations

Functions and powers of inspectors

[07042] In addition to the powers granted by the *Health and Safety at Work etc. Act 1974, s 19*, an inspector may at any time:

— board an offshore installation;
— inspect an offshore installation and any equipment;
— inspect the sea bed and subsoil under or near an offshore installation;
— inspect and take copies from any certificate of insurance or from other records relating to the operation or safety of an offshore installation or of any equipment;
— test any equipment;
— dismantle any equipment or test to destruction or take possession of any equipment, where a casualty has occurred or is apprehended; and
— require the owner or manager or any person on board or near an offshore installation to do or to refrain from doing any act as appears to the inspector to be necessary or expedient for the purpose of averting a casualty (whether the danger is immediate or not) or minimising the consequences of a casualty [*Reg 2*].

Offshore equipment and materials

[07043] Regarding articles, plant and equipment, materials and substances, and any documents relating to the operation of the installation, an inspector also has powers to:

(a) make such requirements of any person (including the owner and manager of the installation) as appear to the inspector to be required for the performance of his functions whether by himself or any other person acting at the direction of the HSE, provided that the inspector consults the owner or manager with a view to maintaining safety and to minimising interference with the operation of the installation;

(b) require any person to produce to him any article to which this Regulation applies and which is in his possession or custody;

(c) make notes, take measurements, make drawings and take photographs of an offshore installation and of any article to which this Regulation applies;

(d) require the owner or manager of the installation to furnish him with any article to which this Regulation applies (other than a document) or, in the case of any article on any vessel, may so require the master, captain or person in charge of the vessel;

(e) require the owner or manager of an offshore installation or any person employed on or in connection with the installation or equipment to carry out or to assist in carrying out any inspection, test or dismantlement of the offshore installation or of any equipment;

(f) require the owner or manager of an offshore installation or the concession owner concerned to assist him in carrying out an inspection of the sea bed or subsoil under or near the installation; and

(g)   require the owner or manager to provide at any reasonable time conveyance to or from the installation of the inspector, any other person acting at the direction of the HSE, any equipment required by the inspector for testing and any article of which he has taken possession pursuant to these Regulations.

Should the inspector require further information, this would normally be provided in writing. If the information is furnished orally, the person providing the information may do so in the presence of another person, including the installation manager if it is practicable to do so and if the person concerned so wishes [*Reg 4*].

Duties to assist an inspector

[07044] The owner of the offshore installation must provide an inspector with reasonable accommodation and afford such facilities and assistance as the inspector may reasonably require to enable him to carry out his functions [*Reg 5*]. The owner is also obliged to provide (at his expense) inspectors with any necessary helicopter transportation.

Disclosure of information

[07045] The Regulations prohibit any person acting at the discretion of the HSE (persons holding Government office excepted) from disclosing any information obtained during an inspection of an offshore installation without the consent of:

—   the owner of the installation;
—   the person who furnished the information; or
—   the Health and Safety HSE.

Offshore Installations (Safety Representatives and Safety Committees) Regulations 1989

[07046] Following the Piper Alpha disaster the present Regulations, the *Offshore Installations (Safety Representatives and Safety Committees) Regulations 1989 (SI 1989 No 971)* were issued by the then enforcing authority, the Department of Energy. The purpose of these Regulations, which were made under the authority of the *Mineral Workings (Offshore Installations) Act 1971* and not the *Health and Safety at Work etc. Act 1974*, is to ensure that the whole workforce (not just individuals who are trade unionists) is formally involved in promoting health and safety through freely elected safety representatives and the safety committee.

Since the publication of the Cullen Report in 1990, much has been written regarding the identification and control of major accident hazards and protection of the workforce from such risks. The Cullen inquiry made it quite clear that high standards of safety and health performance, necessary on an offshore installation, cannot be achieved without the positive and informed commitment of the workforce. Everyone involved with offshore operations must have a safety management system which describes the arrangements for securing this commitment – the provision of information and advanced consultation – whenever it is appropriate to do so.

Information must be given in a form and manner which ensures that it is both comprehensible and relevant to those who are to receive it.

The HSE have published guidance to the Regulations: '*A Guide to the Offshore Installations (Safety Representatives and Safety Committees) Regulations 1989*' (Ref: L110).

The definitions in the safety representatives regulations have been updated to match those in the *Offshore Installations (Safety Case) Regulations 2005 (SI 2005 No 3117)* by those regulations.

Establishment of constituencies

[07047] The first stage in the appointment of safety representatives is the establishment of a system of constituencies by the installation manager in consultation with the safety committee. Thus, unlike onshore safety representatives appointed by recognised trade unions, offshore safety representatives are elected by all the workers in a constituency; in other words, by voluntary decision of the workforce. Their principal role is to act as a conduit on safety from workforce to management. Constituencies will take account of:

(a)   areas of the offshore installation;
(b)   activities undertaken on or from the installation;
(c)   employees of the workforce; and
(d)   other objective criteria which appear to the installation manager to be appropriate to the circumstances of the installation.

There must be, at least, two constituencies and every worker can be assigned to one, but subject to a maximum of forty and a minimum of three. It is the duty of the installation manager to post particulars of a constituency in suitable places on the installation and, if necessary, in appropriate languages. This done, the workers comprising the constituency can elect a safety representative. Any worker can stand for election to represent his constituency as a safety representative, provided that he:

(i)   is a member of that constituency;
(ii)   is willing to stand as a candidate;
(iii)   has been nominated by a second member; and
(iv)   his nomination is seconded by a third member.

A list of duly nominated candidates must be posted within a week after nominations have expired [*Regs 5–10*].

The purpose of dividing an installation into constituencies is to provide for appropriate groupings of the workforce from which safety representatives can be elected. Each installation, no matter how small the permanent workforce, must have at least two constituencies. If an installation manager wishes to create new constituencies, his proposals must be agreed by the safety committee, or both safety committees if two installations are temporarily bridge-linked (for example, production and accommodation platforms). Every new arrival to an installation must be given written notification and details of his constituency. Exceptions to this rule may be permitted where a person is not expected to remain on the installation for longer than 48 hours.

Each installation manager should actively encourage the filling of a constituency vacancy. HSE inspectors may wish to discuss such matters with other safety representatives during visits to the installation [*L110, paras 12–33*].

Functions of offshore safety representatives

[07048] Safety representatives on offshore installations have similar functions to safety representatives generally. More particularly, they can:

(a)   investigate potential hazards/dangerous occurrences and examine causes of accidents;
(b)   investigate *bona fide* health and safety complaints from their members;
(c)   draw matters arising from (*a*) and (*b*) to the attention of the installation manager and any employer;
(d)   approach the installation manager/any employer about general health and safety matters;
(e)   attend meetings of the safety committee;
(f)   represent members in consultation with the Inspectorate;
(g)   consult members on any part of the above matters;
(h)   inspect accident sites/equipment;
(i)   alert the HSE if an imminent risk of serious injury is considered to exist;
(j)   carry out site inspections;
(k)   consult with HSE inspectors;
(l)   receive copies of statutory health and safety documentation.

Exercise of these functions cannot give rise to criminal and/or civil liability on the part of a safety representative [*Reg 16*].

It is important to distinguish between functions and duties – safety representatives cannot be held to account for not carrying out their functions. However, since the functions are the principal basis of the safety representative system, safety representatives should endeavour to carry them out as fully as possible. Although not normally a formal part of the installation's management team, safety representatives should be able to take matters of safety to the installation manager without delay and without the need for a formal written approach. [*L110, paras 52–55*].

Inspection of equipment

[07049] A safety representative can inspect any part of the installation or its equipment, provided that part of the installation or equipment has not been inspected in the previous three months, and that he has given reasonable notice in writing to the manager and, if his employer is not the duty holder, also his employer.

More frequent inspections can take place, by agreement with the manager [*Reg 17*].

Inspection of safety documents and Safety Case

[07050] A safety representative is entitled to see occupational health and safety at work documents which are required by law to be kept on the installation. This entitlement does not extend to a document consisting of or relating to any health record of an identifiable individual.

Regarding the Safety Case for the installation, safety representatives must be supplied with a written summary of the main features of the Safety Case including any particulars concerning remedial work and the time by which it will be done. Appropriate facilities must be made available on the installation for reading the information if it is kept on film or in electronic form [*Regs 18, 18A*].

The *Health and Safety at Work etc. Act 1974, s 28(8)* requires inspectors appointed under that Act to give certain factual and other information concerning their actions to employees or their representatives. Information is normally provided in the form of letters from inspectors, and the Regulations allow for the safety representatives to receive such information. It is recommended that the safety representatives agree that one of their number will act as a contact point with inspectors for this purpose. This representative can then inform his fellow safety representatives, but such arrangements should not preclude any of the representatives establishing working relationships with inspectors and meeting them when they visit.

One of the functions of safety representatives is the carrying out of formal inspections, both on a regular basis and following an incident. When determining the cause of a 'notifiable incident', safety representatives are empowered to inspect relevant parts of the installation and equipment, provided that the installation manager is notified and confirms:

(a)   there has been a notifiable incident;
(b)   it is safe for an inspection; and
(c)   the interests of his members might be involved.

A 'notifiable incident' is any death, injury, disease or dangerous occurrence which is required to be reported under the *Reporting of Injuries, Diseases or Dangerous Occurrences Regulations 1995*. In addition, a safety representative can inspect any part of the installation or its equipment, provided that the relevant part of the installation or equipment has not been inspected in the previous three months, and that he has given reasonable notice in writing to the manager and, if his employer is not the duty holder, also his employer.

More frequent inspections can take place by agreement with the manager [*Reg 17*].

The main purpose of the examination should be to determine the cause so that the possibility of action to prevent a recurrence can be considered. For this reason it is important that the approach to the problem should be a joint one by the installation manager or his representative, relevant employers and the safety representatives. Whilst it may be necessary, following a notifiable incident, for the installation manager to take immediate steps to safeguard against further hazards he should involve the safety representatives at the earliest opportunity. If this is not possible he should notify the safety representatives of the action he has taken and confirm this in writing.

Examinations may include visual inspection, photography (subject to any work permit requirements) and discussions with persons who are likely to be in the possession of relevant information and knowledge regarding the circumstances of the notifiable incident. The examination must not, however, include interference with any evidence or the testing of any machinery, plant, equipment or substance which could disturb or destroy the factual evidence before a HSE inspector has had the opportunity to investigate in accordance with his powers.

The number of safety representatives taking part in a formal inspection should be a matter for agreement with the installation manager. It may be appropriate for an installation safety officer or a specialist to provide advice during an inspection. Where any remedial action has been taken, safety representatives should be given the opportunity to make a re-inspection to satisfy themselves in order to report to the safety committee. [*L110, paras 63–70*].

Risk of serious personal injury

[07051] If two or more safety representatives think that there is an 'imminent risk of serious personal injury', arising from an installation activity, they can:

(a)   make representations to the manager, who must send a written report to an inspector, and

(b)   themselves send a written report to an inspector [*Reg 17(4)*].

Safety representatives are not directly empowered to stop activities which they consider involve imminent risk of serious personal injury. However, where two or more of them consider that there is such imminent risk, they are required by *Reg 17(4)(a)* to make representations on the matter to the installation manager. The manager is then required to report the matter in writing to an inspector as soon as is reasonably practicable.

In addition, the safety representatives may make their own written report on the matter to an inspector, exercising good judgment in invoking this function which should be reserved for genuinely imminent danger. In exercising his powers a safety representative is entitled to seek advice and guidance from persons on the installation or elsewhere. This may usually be informally sought, for example from management, the safety officer, inspectors or from his trade union health and safety adviser. However, if a safety representative considers that there is a need for formal external independent advice, for example a noise survey or a hazard analysis, it is recommended that the matter be raised in the safety committee which may direct the request to the owners [*L110, paras 71–73*].

Safety committees

[07052] Owners of offshore installations, with one or more safety representatives, must establish a safety committee [*Reg 19*].

This must consist of:

(a)   the installation manager (as chairman),

(b)   one other person (to be appointed by the owner or manager),
(c)   all safety representatives, and
(d)   persons who may be co-opted by unanimous vote, e.g. safety officer(s)
       [*Reg 20*].

The person who may be appointed to the committee should preferably be a representative of shore management not directly involved in the day-to-day operation of the installation. Such an appointment would ensure a direct and continuous link with onshore management for the purpose of pursuing any improvements which cannot be made with the resources directly at the disposal of the installation manager. It would also enable solutions to common problems encountered on other installations owned by the same company to be relayed directly to the committee, ensuring consistency of approach.

The Regulations also allow for the safety committee, by unanimous vote, to co-opt on to the committee such additional persons as they wish, but these co-opted individuals are not entitled to vote on the co-option of any further individuals. It is expected that one of them should normally be the installation safety adviser, whilst others might include specialists in particular activities on the installation. Co-option of a minutes secretary may assist the committee [*L110, paras 78–83*].

Safety committee meetings

[07053] A safety committee must be convened by the chairman (normally the installation manager) within six weeks of the date of its establishment, and thereafter at least once every three months. If the committee wishes to meet more frequently, it may establish a policy on the circumstances under which such meetings may be called, for example to consider an urgent unforeseen matter. A safety representative who is unable, for whatever reason, to attend a meeting can nominate another member of his constituency to attend in his stead. The nominee should act on his representative's behalf at the meeting but has no other functions or powers under the Regulations.

Generally, safety committee representatives should cover certain basic items, for example:

—   all incidents involving the installation since the committee last met;
—   any feedback from onshore management including any incidents on other installations which may have implications for their installation;

and any other matter which could possibly affect the health and safety of people working on the installation. Safety committees on bridge-linked installations may decide to co-ordinate their business but it is a legal requirement for each installation to have a safety committee which must meet at least every three months. [*L110, paras 84–89*].

Functions of safety committees

[07054] Safety committees have the following functions:

(i)   to monitor health and safety at the workplace;
(ii)   to keep under review the constituency system;
(iii)   to monitor arrangements for the training of safety representatives;

(iv)   to monitor the frequency of safety committee meetings;

(v)   to consider representations from any member of a safety committee;

(vi)   to consider the causes of accidents, dangerous occurrences and cases of ill- health;

(vii)   to consider any statutory documents relating to occupational health and safety, e.g. registers (but not medical records);

(viii)   to prepare and maintain a record of its business, with a copy being kept for one year from the date of the meeting; and

(ix)   to review safety representative training arrangements [*Reg 22*].

Installation owners/managers must co-operate with safety committees and safety representatives to enable them to perform their functions and they must disclose relevant information to safety representatives [*Regs 23, 24*].

The general philosophy behind the Regulations is that the safety committee will promote co-operation on all matters affecting occupational health and safety between all parties on the installation and will seek to promote and develop measures to ensure the occupational health and safety of the work-force. However, the committee should not attempt to replace the normal day-to-day channels of communication on specific health and safety issues between the individual and his immediate supervisor and ultimately the installation manager, but should serve as a back-up in the event of continuing concern.

The safety committee decides upon the format of its meetings. For example, the installation manager may be required to:

—   report on any feedback from shore management on previous meetings;

—   make a report on all notifiable incidents involving the installation since the committee last met;

—   report any incidents or hazards to health which happened on other installations operated by the same operator which may have repercussions for his own installation (for this he would need briefing from shore);

—   report any major changes in the equipment or method of operation planned for the installations; and

—   report on anything else which could possibly affect the health and safety of persons working on the installation.

The importance of consultation to satisfy the requirements of *Reg 23* cannot be overemphasised. The installation manager should keep the safety committee informed of progress on agreed actions to be taken. [*L110, paras 90–96*].

Training

[07055]–[07056] Employers must permit safety representatives time off from work, without loss of pay, to enable them:

(a)   to perform their functions, and

(b)   to undertake training,

and employers must meet the cost of training, travel and subsistence expenses. [*Regs 26, 27*].

Safety representatives must be permitted time off without loss of pay (including any appropriate offshore allowances) during normal working hours for basic training. This should be as soon as possible after their election. The length of training required is not prescribed but basic training should take into account:

— the functions of safety representatives;
— the training necessary for providing an understanding of the role of the safety representatives and safety committees;
— the legal requirements relating to the health and safety of persons on the installation, particularly the constituency members they directly represent;
— the nature and extent of workplace hazards, and the measures necessary to eliminate or minimise them; and
— the health and safety policy of the employer and the organisations and arrangements for fulfilling that policy.

A major element of safety representative training should be devoted to his representational role and the desirability that the representative should be looked upon as a part-time safety officer. Training should also include inspection and investigation techniques, an appreciation of the appropriate level of response to problems and how to pursue any action with management.

The Offshore Petroleum Industry Training Organisation (OPITO) produces and updates training guidelines for offshore safety representative courses. The guidelines may be adopted by any organisation capable of providing suitable facilities and resources for such training. Although the ultimate choice of courses rests with the employer, it could also be a function of the safety committee.

Further training will be needed where the safety representative has special responsibilities, or in order to meet changes in circumstances or relevant legislation.

Installation owners and managers may make allowances in contractual arrangements in order that contractors' employees may be elected as safety representatives. The responsibility for training then rests with the contractor who must ensure not only that training is provided to any employee who is a safety representative, but also that any such employee is given the necessary time off both for training and for the performance of his role and duties as safety representative.

One of the functions of the safety committee is to keep the training requirements of safety representatives under review. All training providers and courses should be considered in order to ensure that safety representatives have been provided with suitable opportunities to become competent for their role, including the requisite interpersonal skills of communication, presentation and representation. Follow-up training may need to be arranged according to developments affecting the installation, new legislation, changing technologies or differing working practices. [*L110, paras 103–109*].

Offences/defences

(a)   *Offence* – If a duty holder, manager or employer fails to comply with any of his statutory duties, he commits an offence (punishable as with mainline *HSWA 1974* offences – see ENFORCEMENT) [*Reg 28(1)*].

(b)   *Defence* – It is a defence for the accused to prove:
   (i)   that he exercised all due diligence to prevent the commission of the offence, and
   (ii)   that the relevant failure to comply was committed without his consent, connivance or wilful default [*Reg 28(2)*].

Employment protection for offshore safety representatives was provided for by the *Offshore Safety (Protection Against Victimisation) Act 1992*. These regulations have been repealed and replaced by the *Employment Rights Act 1996*. Under the 1996 Act, all employees (both onshore and offshore) irrespective of length of service or hours of work, are entitled to complain to an employment tribunal if they are dismissed, selected for redundancy or subjected to any other detriment by their employer. So far as offshore workers are concerned, this means that they can expect not to be disciplined, dismissed, 'blacklisted' or 'not required back' (NRB), transferred to other duties or to another installation, or disadvantaged in any way, because of any actions they take on health and safety grounds [*L110, App 1*].

First-aid – the Offshore Installations and Pipeline Works (First-Aid) Regulations 1989

Medical personnel

[07057] Installation managers/owners and those in control of pipeline works must make adequate arrangements for first-aid for persons at work. This includes provision of suitably trained personnel and advice or attendance of a registered medical practitioner when needed, and employees must be informed of these arrangements. [*Offshore Installations and Pipeline Works (First-Aid) Regulations 1989 (SI 1989 No 1671), Reg 5(1)* and accompanying Code of Practice (ACOP).]

The Regulations do not apply to diving operations offshore, which are subject to the requirements of the *Diving at Work Regulations 1997*. However, when a diving operation ceases, any members of a diving team who remain present on an installation or barge will be covered by the duty of care which the 1989 Regulations impose upon the person in control.

Numbers and types of 'suitable persons'

[07058] The 'suitable persons' required by *Reg 5(1)(a) and (b)* may be offshore medics or offshore first-aiders. These are defined in the Code of Practice as follows:

offshore medic means a person who holds a current Offshore Medic Certificate issued by a body approved by the HSE to train, examine and certify offshore medics; and

offshore first-aider means a person who holds a current Offshore First-Aid Certificate issued by a body approved by the HSE to train, examine and certify offshore first-aiders.

In most circumstances, the minimum number of offshore first-aiders and offshore medics who should be available at all times where people are at work on an offshore installation or a pipelaying barge or a barge, such as a crane barge used in offshore construction, reconstruction, alteration, repair, maintenance, cleaning, demolition or dismantling activities, or in providing accommodation, should be as follows:

Number of people at work	Number of offshore first-aiders	Number of offshore Medics
Up to 25	2	–
26 to 50	3	1
51 to 100	5	1
101 to 150	7	1
151 to 200	9	1
201 and above	Additional personnel, if required, to ensure reasonable access by all offshore workers	

In determining whether there is a need for a medic, the number of people likely to be regularly at work at one time on the installation should be considered. In certain circumstances, such as relatively low hazard installations or when the installation is within easy reach of onshore medical services, a medic should be available only where the number of people regularly at work exceeds 50.

Manned installations – sick bays

[07059] In particular, manned offshore installations and all pipelaying barges and barges used in offshore construction, repair, maintenance, cleaning, demolition or dismantling activities must have a sick bay. The sick bay should:

(a) be clearly identifiable;
(b) be in the charge of a suitable person (e.g. offshore medic or first-aider) who should always be on call;
(c) be available at all times and not used for other purposes;
(d) be kept locked at all times when not in use (though suitable arrangements should be made for immediate access in cases of emergency);
(e) have attached to the door a notice clearly showing names and locations of:
(i)   offshore medics, or
(ii) offshore first-aiders
(who should be easily identifiable, e.g. by arm bands) and means of contacting them (see further Duty of installation manager, **O7062** below);

(f)   contain suitable furniture, medications and equipment. Medications should be kept in a container and locked, the keys at all times being held by:
(i)   the installation manager, or
(ii)   the master, or
(iii)   the offshore medic;
(g)   have facilities for effective two-way communication with onshore medical services [ACOP, para 6].

First-aid kits in sick bays

[07060] Sick bays should always contain a first-aid kit; but, in addition, offshore first-aiders must be provided with a first-aid kit readily available for use in emergency and located at the first-aider's place of work or around the installation.

First-aid kits in sick bays should contain:

(i)   12 individually wrapped sterile triangular bandages (90 cm × 127 cm);
(ii)   20 packs of sterile gauze pads, each containing five pads (7.5 cm × 7.5 cm); and
(iii)   a suitable device for facilitating mouth to mouth resuscitation [ACOP, paras 3, 4].

Contents of sick bay

[07061] Sick bays on offshore installations should contain the following items:

(a)   medications which an offshore medic can administer without the direction of a doctor, e.g. aluminium hydroxide tablets for indigestion, calamine lotion for irritating rashes;
(b)   standing orders relating to medications which can only be administered under the direction of a doctor, e.g. ampicillin capsules for infections, diazepam tablets for sedation;
(c)   dressings and bandages;
(d)   instruments and appliances, e.g. resuscitation devices, mouth gag, face masks; and
(e)   furnishings and equipment, e.g. alarm bell system, telephone, armchair, accident record book, daily treatment record book [ACOP, paras 31–36, Appendix 2, Parts 1, 2, 3].

Duty of installation manager

[07062] The installation manager must see that all employees (and self-employed persons) are informed of the location of first-aid and medical equipment and how workers can contact the offshore medic or first-aiders rapidly in case of emergency. New workers, in particular, should be told this on first coming aboard [*Reg 5(1)*].

To this end, notices should be put up (if necessary in various languages) in conspicuous positions (including sick bays), giving locations of first-aid and medical equipment and names and, if possible, locations of medic and first-aider. Offshore medics and first-aiders must be regularly supervised by fully registered medical practitioners and, where necessary, given medical advice [*Reg 5(1)(c)*], and written instructions should be drawn up, setting out arrangements for liaison with the medical practitioner. Copies of such instructions should be displayed in:

(a)   the sick bay;
(b)   the radio operator's room; and
(c)   the master or installation manager's office offshore [ACOP, paras 23–30].

The Offshore Electricity and Noise Regulations 1997

[07063] These Regulations *(SI 1997 No 1993)* amend the following:

—   *Electricity at Work Regulations 1989 (SI 1989 No 635)* (see **E3001** ELECTRICITY),
—   *Noise at Work Regulations 1989 (SI 1989 No 1790)* (see **N3001** NOISE AND VIBRATION),

by applying them in their entirety to offshore installations.

The duties imposed by these Regulations do not extend to:

(a)   the master or crew of a sea-going ship or to the employer of such persons, in relation to the normal ship-board activities of a ship's crew under the direction of the master; or
(b)   the crew of any aircraft or hovercraft which is moving under its own power or any other person on board any such aircraft or hovercraft who is at work in connection with its operation.

The Regulations amend the *Offshore Installations and Wells (Design and Construction, etc.) Regulations 1996 (SI 1996 No 913)* (see **O7024** above) by removing references to electrical or noise provisions.

Legislation for other hazardous offshore activities

[07064] Generally, an installation's Safety Case will not be considered for acceptance unless it contains all the particulars specified in the *Safety Case Regulations 1992, Reg 8* and the relevant *Schedule*. In some cases, with respect to hazardous activities carried out on offshore installations, there will be a degree of overlap with the particulars required in accordance with the provisions of associated legislation.

Set out below are brief references to various regulations which relate to hazards that potentially could cause a major accident.

Working in confined spaces

[07065] The *Confined Spaces Regulations 1997 (SI 1997 No 1713)* impose requirements and prohibitions with respect to the health and safety of persons carrying out work in confined spaces.

Although offshore installations are not specifically covered by the regulations, the regulations would apply to certain activities aboard installations which are 'stacked' out of use in territorial waters (see O7009 above) and which are not currently classed as offshore installations, such as the activities of shore-based workers undertaking repair, maintenance or cleaning.

A 'confined space' is defined in *Reg 1(2)* as 'any place, including any chamber, tank, vat, silo, pit, trench, pipe, sewer, flue, well or other similar space in which, by virtue of its enclosed nature, there arises a reasonably foreseeable specified risk'.

Under these regulations a confined space has two defining features. Firstly, it is a place which is substantially (though not always entirely) enclosed and, secondly, there will be a reasonably foreseeable risk of serious injury from hazardous substances or conditions within the space or nearby.

Radiation hazards

[07066] From the site radiography techniques used to assess the integrity of pipeline joints, one can infer the need for an exceptionally high level of radiological protection for operators. There must be total compliance with the *Ionising Radiations Regulations 1985 (SI 1985 No 1333)* in terms of the designation of competent persons, authorised persons and classified workers, medical supervision, personal monitoring procedures following excessive exposure, control of sealed sources of radiation, equipment tests and the maintenance of records (see DANGEROUS SUBSTANCES I).

Oil pollution

[07067] In addition to the environmental aspects, oil and volatile hydrocarbons spilled into the sea in the vicinity of an offshore installation pose a real threat of fire to that installation. In this context the principal legislation in respect of the control of oil pollution is as follows:

— *Merchant Shipping Act 1995 (as amended)*;
— *Merchant Shipping (Prevention of Pollution) Order 1983 (SI 1983 No 1106)* (as amended);
— *Merchant Shipping (Prevention of Pollution) (Law of the Sea Convention) Order 1996 (SI 1996 No 282)*; and
— *Merchant Shipping (Prevention of Pollution) Regulations 1996 (SI 1996 No 2154)*.

In the *Merchant Shipping (Prevention of Pollution) Regulations 1996*, the legal definition of 'discharge' expressly excludes the release of harmful substances directly arising from the exploration, exploitation and associated offshore processing of sea-bed mineral resources [*Reg 1(2)*].

However, there is a prohibition on the discharge into the sea of other harmful substances, or effluents containing such substances, whether such discharge occurred as a result of escaping, disposing, spilling, leaking, pumping and emitting or emptying. In respect of offshore installations, when engaged in the exploration, exploitation and associated offshore processing activities, the

Regulations provide that they must be equipped with suitable oil filtering and oil discharge monitoring and control systems.

Each installation is also required to have a shipboard oil pollution emergency plan and an up-to-date Oil Record Book.

The *Offshore Combustion Installations (Prevention and Control of Pollution) Regulations 2001 (SI 2001 No 1091)* were made under the *Pollution Prevention and Control Act 1999* to implement the requirements of the Integrated Pollution Prevention and Control Directive (Council Directive 96/61 EC) in relation to offshore combustion installations. A combustion installation is any technical apparatus in which fuel is oxidised to generate heat and includes gas turbines and diesel and petrol engines. It does not include apparatus used mainly for the disposal of gas by flaring or incineration. A permit is required to operate a combustion installation. The DTI can issue enforcement and/or prohibition notices. The regulations were modified by the *Offshore Petroleum Activities (Oil Pollution Prevention and Control) Regulations 2005 (SI 2005 No 2055)*.

The *Offshore Chemicals Regulations 2002 (SI 2002 No 1355)* made under the *Pollution Prevention and Control Act 1999* established a regime to implement the obligations under the Convention for the Protection of the Marine Environment of the NE Atlantic (OSPAR) Decision (2000/2). Operators require a permit from the DTI to use and discharge chemicals. *SI 2005 No 2055* amends *SI 2002 No 1355* and authorises the DTI to issue an enforcement notice for contravention of any condition of a permit or a prohibition notice for an imminent risk of serious pollution.

The *Offshore Petroleum Activities (Oil Pollution Prevention and Control) Regulations 2005 (SI 2005 No 2055)* were made under the *Pollution Prevention and Control Act 1999* to phase out the exemptions under the *Prevention of Oil Pollution Act 1971* which permitted certain discharges of oil into the sea and replacement of that system by a permit system. Operators of offshore installations must have a permit to discharge oil into relevant waters. The permits lay down any conditions considered necessary. The regulations also allow for the introduction of a trading scheme for discharge allowances. The DTI have issued Rules of the Dispersed Oil in Produced Water Trading Scheme which have the support of UKOOA.

Carriage of dangerous substances

[07068] In addition to the Regulations relating to the labelling, packaging and transportation of dangerous goods by road or rail (see DANGEROUS SUBSTANCES II), carriage to the installation must comply with the following legislation:

— *Merchant Shipping (Dangerous Goods and Marine Pollutants) Regulations 1997 (SI 1997 No 2367)*; and
— *Air Navigation (Dangerous Goods) Regulations 1994 (SI 1994 No 3187)*.

Reference to these regulations is given for guidance only and must not be regarded as comprising a complete list.

Written procedures and training programmes for dealing with hazards

[07069] Detailed plans, procedures, safe systems of work and written instructions, including those covering the operation of permit to work systems, are a standard feature of offshore safety operations. Such procedures and safe systems of work may include the use of access equipment, e.g. scaffolding, boatswain's chairs, ladders, etc., work over the sea, the safe use of lifting equipment, helicopter operations and general fire and emergency procedures.

High standards of operator training, prior to arrival on a working platform and at frequent intervals during their employment, are crucial to ensure safe operation.

Induction training should incorporate the following aspects:

(a)   *General safety* – dealing with the hazards of offshore work and the personal precautions necessary, in particular the correct use of all forms of personal protective equipment, i.e. foul weather protection, safety helmets, safety harnesses.

(b)   *Safe working procedures* – the use of permit to work systems and safe systems of work covering electricity, welding and ionising radiation in particular; the use of scaffolding, working platforms, the use of plant and equipment, maintenance and inspection procedures and record keeping (see further 43 INTRODUCTION).

(c)   *Health precautions* – including precautions against toxic dusts, fumes, gases and vapours, asbestos, the risk of asphyxiation and anoxia, and physical hazards associated with noise, electricity and radiation.

(d)   *Survival* – instruction in the use of life jackets, survival capsules, life boats and rafts; emergency platform evacuation procedures, particularly in the event of fire or explosion.

(e)   *Helicopter operations* – procedures for safe movement of personnel by helicopter to or from the platform.

(f)   *Fire protection* – the correct use of fire appliances, platform evacuation procedure in the event of fire, platform fire alarm system.

(g)   *First-aid* – elementary first-aid procedures.

Hazards in offshore activities

Cranes and lifting equipment

[07070] The potential for crane and lifting equipment failures, including the use of helicopters in lifting operations during the initial work stage, represents a serious hazard in offshore work. All lifting operations and equipment must comply with the *Lifting Operations and Lifting Equipment Regulations 1998* (see L3023 LIFTING MACHINERY AND EQUIPMENT).

HSE has published technical guidance on the safe use of lifting equipment offshore as part of a key initiative to reduce the incidence of lifting-related injuries and deaths offshore.

The new guidance supplements existing HSE guidance on the safe use of lifting equipment and provides technical information for those who are involved in the operation and control of lifting equipment offshore. It has been written for duty holders, OIMs, managers and people who are directly involved in using lifting equipment offshore, although others, such as safety representatives, manufacturers, suppliers and verification bodies will also find it useful. The guidance covers all frequently used lifting equipment and accessories. It includes a section on personnel carriers.

Copies of *Technical guidance on the safe use of lifting equipment offshore*, (HSG221), ISBN 0 7176 2100 6, are available from HSE Books.

Burns and eye injuries

[07071] These are the two most common forms of injury to operators, in many cases associated with a failure to wear personal protective equipment provided (see PERSONAL PROTECTIVE EQUIPMENT REGULATIONS at P3002).

Access hazards and working at height

[07072] Falls from scaffolds and ladders and whilst working 'over the side' are common emphasising the need for good systems of scaffold erection, inspection and maintenance and adherence to the *Work at Height Regulations 2005 (SI 2005 No 735)* which apply offshore (see WORK AT HEIGHT).

Gassing accidents and incidents

[07073] Precautions against asphyxiation, oxygen enrichment and various fumes, dusts, gases and vapours are now a standard feature of the industry.

Permit to work

[07074] As with many other industries non standard operations are controlled by permit to work systems. Guidance on these systems has been issued by the Offshore Oil Industry Committee set up by the HSC to move towards standardisation of the systems between companies to reduce confusion in the peripatetic contractors, see Oil Industry accounting Committee (OIAC) guidance on permit to work systems in the petroleum industry, HSC 1997. This guidance is in the process of being updated.

Personal Protective Equipment

Nicola Coote

Introduction

[P3001] Conventional wisdom suggests that 'safe place strategies' are more effective in combating health and safety risks than 'safe person strategies'. Safe systems of work, and control/prevention measures serve to protect everyone at work, whilst the advantages of personal protective equipment are limited to the individual(s) concerned. Given, however, the fallibility of any state of the art technology in endeavouring to achieve total protection, some level of personal protective equipment is inevitable in view of the obvious (and not so obvious) risks to head, face, neck, eyes, ears, lungs, skin, arms, hands and feet.

Current statutory requirements for employers to provide and maintain suitable personal protective equipment are contained in the *Personal Protective Equipment at Work Regulations 1992 (SI 1992 No 2966)* as amended by *the Health and Safety (Miscellaneous Amendments) Regulations 2002 (SI 2002 No 2174)*. This legislation requires the employer to carry out an assessment to determine what personal protective equipment is needed and to consider employee needs in the selection process. Adjunct to this, common law insists, not only that employers have requisite safety equipment at hand, or available in an accessible place, but also that management ensure that operators use it (see *Bux v Slough Metals Ltd* [1974] 1 All ER 262). That the hallowed duty to 'provide and maintain' has been getting progressively stricter is evidenced by *Crouch v British Rail Engineering Ltd* [1988] IRLR 404 to the extent that employers could be in breach of either statutory or common law duty (or both) in the case of injury/disease to a member of the employee's immediate family involved, say, in cleaning protective clothing – but the injury/disease must have been foreseeable (*Hewett v Alf Brown's Transport Ltd* [1992] ICR 530).

This section deals with the requirements of the *Personal Protective Equipment at Work Regulations 1992*, the *Personal Protective Equipment (EC Directive) Regulations 1992 (SI 1992 No 3139)* (now revoked and consolidated by *SI 2002 No 1144*), the duties imposed on manufacturers and suppliers of personal protective equipment, the common law duty on employers to provide suitable personal protective equipment, and the main types of personal protective equipment and clothing and some relevant British Standards.

The Personal Protective Equipment at Work Regulations 1992 (SI 1992 No 2966 as amended by SI 2002 No 2174)

[P3002] With the advent of the *Personal Protective Equipment at Work Regulations 1992*, all employers must make a formal assessment of the personal protective equipment needs of employees and provide ergonomically suitable equipment in relation to foreseeable risks at work (see further **P3006** below). However, employers cannot be expected to comply with their new statutory duties, unless manufacturers of personal protective equipment have complied with theirs under the requirements of the *Personal Protective Equipment (EC Directive) Regulations 1992 (SI 1992 No 3139)* and *SI 1994 No 2326* (now revoked and consolidated by *SI 2002 No 1144*) – that is, had their products independently certified for EU accreditation purposes. Although imposing a considerable remit on manufacturers, EU accreditation is an indispensable condition precedent to sale and commercial circulation (see **P3012** below).

It is important that personal protective clothing and equipment should be seen as 'last resort' protection – its use should only be prescribed when engineering and management solutions and other safe systems of work do not effectively protect the worker from the danger, and where risk of injury is still foreseeable after other controls have been implemented (*Health and Safety (Miscellaneous Amendments) Regulations 2002 (SI 2002 No 2174)*). (*Gerrard v Staffordshire Potteries* [1995] PIQR P 169.)

Employees must be made aware of the purpose of personal protective equipment, its limitations and the need for on-going maintenance. Thus, when assessing the need for, say, eye protection, employers should first identify the existence of workplace hazards (e.g. airborne dust, projectiles, liquid splashes, slippery floors, inclement weather in the case of outside work) and then the extent of danger (e.g. frequency/velocity of projectiles, frequency/severity of splashes). Selection can (and, indeed, should) then be made from the variety of CE-marked equipment available, in respect of which manufacturers must ensure that such equipment provides protection, and suppliers ascertain that it meets such requirements/standards [*HSWA s 6*] (see further PRODUCT SAFETY). Typically, most of the risks will have already been logged, located and quantified in a routine risk/safety audit (see RISK ASSESSMENT), and classified according to whether they are physical/chemical/biological in relation to the part(s) of the body affected (e.g. eyes, ears, skin). Good practice dictates that any such assessment should include the input from the users, as simply providing equipment that will meet a technical specification for protection will not meet requirements for a "suitable and sufficient" assessment. To achieve this, a deeper analysis of environment in which it is being used, comfort, fit etc will be needed.

Selection of personal protective equipment is a first stage in an on-going routine, followed by proper use and maintenance of equipment (on the part of both employers and employees) as well as training and supervision in personal protection techniques. Maintenance presupposes a stock of renewable spare parts coupled with regular inspection, testing, examination, repair, cleaning and disinfection schedules as well as keeping appropriate records. Depending on the particular equipment, some will require regular testing and examination (e.g. respiratory equipment), whilst others merely inspection (e.g. gloves, goggles). Generally, manufacturers' maintenance schedules should be followed.

Equipment must be appropriate for the risk or risks involved, the conditions at the place where exposure to the risk may occur, and the period for which it is worn. It must also take into account the ergonomic requirements and the state of health of the person or persons who may wear it, and of the characteristics of the workstation of each such person.

Suitable accommodation must be provided for protective equipment in order to minimise loss or damage and prevent exposure to cold, damp or bright sunlight, e.g. pegs for helmets, pegs and lockers for clothing, spectacle cases for safety glasses.

On-going safety training, often carried out by manufacturers for the benefit of users should combine both theory and practice. Where it is appropriate, and at suitable intervals, the employer should organise demonstrations in the wearing of personal protective equipment. (*Health and Safety (Miscellaneous Amendments) Regulations 2002 (SI 2002 No 2174)*.)

Work activities/processes requiring personal protective equipment

[P3003] Examples abound of processes/activities of which personal protective equipment is a prerequisite, from construction work and mining, through work with ionising radiations, to work with lifting plant, cranes, as well as handling chemicals, tree felling and working from heights. Similarly, blasting operations, work in furnaces and drop forging all require a degree of personal protection.

Statutory requirements in connection with personal protective equipment

[P3004] General statutory requirements relating to personal protective equipment are contained in of the *Health and Safety at Work etc Act 1974, ss 2, 9*, the *Personal Protective Equipment at Work Regulations 1992*, and the *Health and Safety (Miscellaneous Amendments) Regulations 2002 (SI 2002 No 2174)* 'Every employer shall ensure that suitable personal protective equipment is provided to his employees who may be exposed to a risk to their health or safety at work except where, and to the extent that such risk has been adequately controlled by other means which are equally or more effective'. [*Reg 4(1)*]. More specific statutory requirements concerning personal protective equipment exist in the *Personal Protective Equipment at Work Regulations 1992* (in tandem with the *Personal Protective Equipment (EC Directive) Regulations 1992 (SI 1992 No 3139)* and *SI 1994 No 2326* (now revoked and consolidated by *SI 2002 No 1144*)), and sundry other recent regulations applicable to particular industries/processes, e.g. asbestos, noise, construction (see **P3008** below). Where personal protective equipment is a necessary control measure to meet a specific statutory requirement or following a risk assessment, this must be provided free of charge [*HWSA s 9*]. However, where it is optional or non essential, there is nothing to stop the employer from seeking a financial contribution.

General statutory duties – HSWA 1974

[P3005] Employers are under a general duty to ensure, so far as reasonably practicable, the health, safety and welfare at work of their employees – a duty which clearly implies provision/maintenance of personal protective equipment. [*HSWA s 2(1)*].

Specific statutory duties – Personal Protective Equipment at Work Regulations 1992 (SI 1992 No 2966 as amended by SI 2002 No 2174)

Duties of employers

[P3006] Employers – and self-employed persons in cases (*a*), (*b*), (*c*), (*d*) and (*e*) below, must undertake the following:

(a)   Formally assess (and review periodically) provision and suitability of personal protective equipment. [*Reg 6(1), (3)*].
       The assessment should include:
       (i)   risks to health and safety not avoided by other means;
       (ii)   reference to characteristics which personal protective equipment must have in relation to risks identified in (i); and
       (iii)   a comparison of the characteristics of personal protective equipment having the characteristics identified in (ii).
       [*Reg 6(2)*].
       The aim of the assessment is to ensure that an employer knows which personal protective equipment to choose; it constitutes the first stage in a continuing programme, concerned also with proper use and maintenance of personal protective equipment and training and supervision of employees.

(b)   Provide suitable personal protective equipment to his employees, who may be exposed to health and safety risks while at work, except where the risk has either been adequately controlled by other equally or more effective means. [*Reg 4(1)*].
       Personal protective equipment is not suitable unless:
       (i)   it is appropriate for risks involved and conditions at the place of exposure;
       (ii)   it takes account of ergonomic requirements, the state of health of the person who wears it and the duration of its use;
       (iii)   it is capable of fitting the wearer correctly; and

(iv)   so far as reasonably practicable (for meaning, see ENFORCEMENT), it is effective to prevent or adequately control risks involved without increasing the overall risk.

[*Reg 4(3)*].

Where it is necessary to ensure that personal protective equipment is hygienic and otherwise free of risk to health, every employer and every self-employed person shall ensure that personal protective equipment provided is provided to a person for use only by him. [*Reg 4*]

(c)   Provide compatible personal protective equipment – that is, that the use of more than one item of personal protective equipment is compatible with other personal protective equipment. [*Reg 5*].

(d)   Maintain (as well as replace and clean) any personal protective equipment in an efficient state, efficient working order and in good repair. [*Reg 7*].

(e)   Provide suitable accommodation for personal protective equipment when not being used. [*Reg 8*].

(f)   Provide employees with information, instruction and training to enable them to know:

(i)   the risks which personal protective equipment will avoid or minimise;

(ii)   the purpose for which and manner in which personal protective equipment is to be used;

(iii)   any action which the employee might take to ensure that personal protective equipment remains efficient.

[*Reg 9*].

(g)   Ensure, taking all reasonable steps, that personal protective equipment is properly used. [*Reg 10*].

Summary of employer's duties

- duty of assessment;
- duty to provide suitable PPE;
- duty to provide compatible PPE;
- duty to maintain and replace PPE;
- duty to provide suitable accommodation for PPE;
- duty to provide information and training;
- duty to see that PPE is correctly used.

Duties of employees

[P3007] Every employee must:

- use personal protective equipment in accordance with training and instructions [*Reg 10(2)*];
- return all personal protective equipment to the appropriate accommodation provided after use [*Reg 10(4)*]; and
- report forthwith any defect or loss in the equipment to the employer [*Reg 11*].

Specific requirements for particular industries and processes

[P3008] In addition to the general remit of the *Personal Protective Equipment at Work Regulations 1992*, the following regulations impose specific requirements on employers to provide personal protective equipment up to the EU standard.

- *Control of Lead at Work Regulations 2002 (SI 2002 No 2676), Regs 6, 8* – supply of suitable respiratory equipment and where exposure is significant, suitable protective clothing and equipment (see APPENDIX 4);
- *Control of Asbestos Regulations 2006 (SI 2006 No 2739), Reg 14* – provision and cleaning of personal protective equipment (see APPENDIX 4);
- *Control of Substances Hazardous to Health Regulations 2002 (SI 2002 No 2677 as amended by SI 2003 No 978) (COSHH), Regs 7, 8, 9* – supply of suitable protective equipment where employees are foreseeably exposed to substances hazardous to health;
- *Noise at Work Regulations 2005 (SI 2005 No 1643), Reg 7* – personal ear protectors;
- *Construction (Head Protection) Regulations 1989 (SI 1989 No 2209), Reg 3* – suitable head protection (for the position of Sikhs, see **C8115** CONSTRUCTION AND BUILDING OPERATIONS);
- *Ionising Radiations Regulations 1999 (SI 1999 No 3232), Reg 9* – supply of suitable personal protective equipment and respiratory protective equipment;
- *Shipbuilding and Ship-repairing Regulations 1960 (SI 1960 No 1932), Regs 50, 51* – supply of suitable breathing apparatus, belts, eye protectors, gloves and gauntlets (amended in 1994).

Increased importance of uniform European standards

[P3009] To date, manufacturers of products, including personal protective equipment (PPE), have sought endorsement or approval for products, prior to commercial circulation, through reference to British Standards (BS), HSE or European standards, an example of the former being Kitemark and an example of the latter being CEN or CENELEC. Both are examples of *voluntary* national and international schemes that can be entered into by manufacturers and customers, under which both sides are 'advantaged' by conformity with such standards. Conformity is normally achieved following a level of testing appropriate to the level of protection offered by the product. With the introduction of the *Personal Protective Equipment (EC Directive) Regulations 1992* as amended by *SI 1994 No 2326* (now revoked and consolidated by *SI 2002 No 1144*) (see **P3012** below), this voluntary system of approval was replaced by a statutory certification procedure.

Towards certification

[P3010] Certification as a condition of sale is probably some time away. Transition from adherence based on a wide range of voluntary national (and international) standards to a compulsory universal EU standard takes time and specialist expertise. Already, however, British Standards Institution (BSI) has adopted the form BS EN as a British version of a uniform European standard. But, generally speaking, uniformity of approach to design and testing, on the part of manufacturers, represents a considerable remit and so statutory compliance with tougher new European standards is probably not likely to happen immediately. Here, too, however, there is evidence of some progress, e.g. the EN 45000 series of standards on how test houses and certification bodies are to be established and independently accredited. In addition, a range of European standards on PPE have been implemented in BS EN format, for example, on eye protection, fall arrest systems and safety footwear.

Interim measures

[P3011] Most manufacturers have elected to comply with the current CEN standard when seeking product certification. Alternatively, should a manufacturer so wish, the inspection body can verify by way of another route to certification – this latter might well be the case with an innovative product where standards did not exist. Generally, however, compliance with current CEN (European Standardisation Committee) or, alternatively, ISO 9000: Quality Assurance has been the well-trodden route to certification. To date, several UK organisations, including manufacturers and independent bodies, have established PPE test houses, which are independently accredited by the National Measurement Accreditation Service (NAMAS). Such test houses will be open to all comers for verification of performance levels.

Personal Protective Equipment (EC Directive) Regulations 1992 (SI 1992 No 3139 as amended by SI 1994 No 2326)

[P3012] The *Personal Protective Equipment (EC Directive) Regulations 1992 (*as amended by *SI 1994 No 2326)* required most types of PPE (for exceptions, see **P3019** below) to satisfy specified certification procedures, pass EU type-examination (i.e. official inspection by an approved inspection body) and carry a 'CE' mark both on the product itself and its packaging before being put into commercial circulation. Indeed, failure on the part of a manufacturer to obtain affixation of a 'CE' mark on his product before putting it into circulation, is a criminal offence under the *Health and Safety at Work etc Act 1974*, s 6 carrying a maximum fine, on summary conviction, of £20,000 (see ENFORCEMENT). Enforcement is through trading standards officers.

The *Personal Protective Equipment Regulations 2002 (SI 2002 No 1144)* now revoke and consolidate the *Personal Protective Equipment (EC Directive) Regulations 1992*. The 2002 Regulations implement the Product Safety Directive 89/696/EEC, on the approximation of the laws of the Member States relating to personal protective equipment. The new Regulations came into force on 15 May 2002.

The 2002 Regulations place a duty on responsible persons who put personal protective equipment (PPE) on the market to comply with the following requirements:

- the PPE must satisfy the basic health and safety requirements that are applicable to that type or class of PPE;
- the appropriate conformity assessment procedure must be carried out;
- a CE mark must be affixed on the PPE.

The Regulations will be enforced by the Weights and Measures Authorities (District Councils in Northern Ireland).

CE mark of conformity

[P3013] The *Personal Protective Equipment (EC Directive) Regulations 1992* as amended by the *Personal Protective Equipment (EC Directive) (Amendment) Regulations 1994 (SI 1994 No 2326)* (now revoked and consolidated by *SI 2002 No 1144*) specify procedures and criteria with which manufacturers must comply in order to be able to obtain certification. Essentially this involves incorporation into design and production basic health and safety requirements. Hence, by compliance with health and safety criteria, affixation of the 'CE' mark becomes a condition of sale and approval (see *fig. 1* below). Moreover, compliance, on the part of manufacturers, with these regulations, will enable employers to comply with their duties under the *Personal Protective Equipment at Work Regulations 1992*. *Reg 4(3)(e)* of the *Personal Protective Equipment at Work Regulations 1992* requires the PPE provided by the employer to comply with this and any later EU directive requirements.

CE accreditation is not a synonym for compliance with the CEN standard, since a manufacturer could opt for certification via an alternative route. Neither is it an approvals mark. It is rather a quality mark, since all products covered by the regulations are required to meet formal quality control or quality of production criteria in their certification procedures. [*Arts 8.4, 11, 89/686/EEC*].

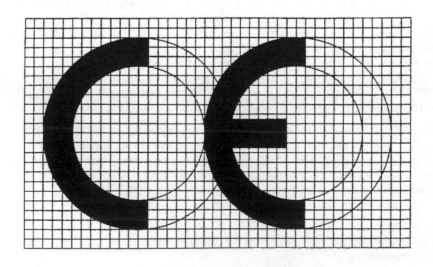

Certification procedures

[P3014] The *Personal Protective Equipment (EC Directive) Regulations 1992* as amended by *SI 1994 No 2326* (now revoked and consolidated by *SI 2002 No 1144)*, identify three categories of PPE:

- PPE of simple design;
- PPE of non-simple design;
- PPE of complex design.

Each of these have different requirements for manufacturers to follow.

In order to obtain certification, manufacturers must submit to an approved body the following documentation (except in the case of PPE of simple design, where risks are minimal).

(1)   **PPE – simple design**
         This includes use of PPE where risks are minimal e.g. gardening gloves, helmets, aprons, thimbles. It is used where the designer assumes that the user can himself assess the level of protection provided against the minimal risks concerned the effects of which, when they are gradual, can be safely identified and managed. This type of equipment must still be subject to a manufacturers declaration of conformity.

(2)   **PPE – non-simple design**
         In order to obtain certification, manufacturers must submit to an approved body the following documentation:
         (a)   Technical file, i.e.
                 (i)   overall and detailed plans, accompanied by calculation notes/results of prototype tests; and
                 (ii)   an exhaustive list of basic health and safety requirements and harmonised standards taken into account in the model's design.
         (b)   Description of control and test facilities used to check compliance with harmonised standards.
         (c)   Copy of information relating to:
                 (i)   storage, use, cleaning, maintenance, servicing and disinfection;
                 (ii)   performance recorded during technical tests to monitor levels of protection;
                 (iii)   suitable PPE accessories;
                 (iv)   classes of protection appropriate to different levels of risk and limits of use;
                 (v)   obsolescence deadline;
                 (vi)   type of packaging suitable for transport;
                 (vii)   significance of markings.
         [*Art 1.4, Annex II; Art 8.1, Annex III*].

This must be provided in the official language of the member state of destination.

(3)   PPE – complex design
This includes use of PPE for protection against mortal/serious dangers where the user cannot identify the risk in sufficient time. It requires compliance with a quality control system. Examples of PPE coming within the complex design category include: firemens helmets, breathing air apparatus for spray painters and fall arrest systems for roof / scaffold workers.

This category of PPE, including:

(a)   filtering, respiratory devices for protection against solid and liquid aerosols/irritant, dangerous, toxic or radiotoxic gases;
(b)   respiratory protection devices providing full insulation from the atmosphere, and for use in diving;
(c)   protection against chemical attack or ionising radiation;
(d)   emergency equipment for use in high temperatures, whether with or without infrared radiation, flames or large amounts of molten metal (100°C or more);
(e)   emergency equipment for use in low temperatures (–50°C or less);
(f)   protection against falls from heights;
(g)   protection against electrical risks; and
(h)   motor cycle helmets and visors;

must satisfy either

(i)
—   an EU quality control system for the final product, or
—   a system for ensuring EU quality of production by means of monitoring;
and in either case,
(ii)   an EU declaration of conformity.
[Art 8.4].

Basic health and safety requirements applicable to all PPE

[P3015]–[P3016] All PPE must:

• be ergonomically suitable, that is, be able to perform a risk-related activity whilst providing the user with the highest possible level of protection;
• preclude risks and inherent nuisance factors, such as roughness, sharp edges and projections and must not cause movements endangering the user;
• be kept in clean and hygienic condition and be of a design that enables good hygienic standards, unless it is disposable;
• provide comfort and efficiency by facilitating correct positioning on the user and remaining in place for the foreseeable period of use; and by being as light as possible without undermining design strength and efficiency; and
• be accompanied by necessary information – that is, the name and address of the manufacturer and technical file (see **P3014** above).
[Annex II].

General purpose PPE – specific to several types of PPE

- PPE incorporating adjustment systems must not become incorrectly adjusted without the user's knowledge;
- PPE enclosing parts of the body must be sufficiently ventilated to limit perspiration or to absorb perspiration;
- PPE for the face, eyes and respiratory tracts must minimise risks to same and, if necessary, contain facilities to prevent moisture formation and be compatible with wearing spectacles or contact lenses;
- PPE subject to ageing must contain date of manufacture, the date of obsolescence and be indelibly inscribed if possible. If the useful life of a product is not known, accompanying information must enable the user to establish a reasonable obsolescence date. In addition, the number of times it can be cleaned before being inspected or discarded must (if possible) be affixed to the product; or, failing this, be indicated in accompanying literature;
- PPE which may be caught up during use by a moving object must have a suitable resistance threshold above which a constituent part will break and eliminate danger;
- PPE for use in explosive atmospheres must not be likely to cause an explosive mixture to ignite;
- PPE for emergency use or rapid installation/removal must minimise the time required for attachment and/or removal;
- PPE for use in very dangerous situations (see **P3015** above) must be accompanied with data for exclusive use of competent trained individuals, and describe procedure to be followed to ensure that it is correctly adjusted and functional when worn;
- PPE incorporating components which are adjustable or removable by user, must facilitate adjustment, attachment and removal without tools;
- PPE for connection to an external complementary device must be mountable only on appropriate equipment;
- PPE incorporating a fluid circulation system must permit adequate fluid renewal in the vicinity of the entire part of the body to be protected;
- PPE bearing one or more identification or recognition marks relating to health and safety must preferably carry harmonised pictograms/ideograms which remain perfectly legible throughout the foreseeable useful life of the product;
- PPE in the form of clothing capable of signalling the user's presence visually must have one or more means of emitting direct or reflected visible radiation;
- multi-risk PPE must satisfy basic requirements specific to each risk.

[Art 2, Annex II].

Additional requirements for specific PPE for particular risks

[P3017]–[P3018] There are additional requirements for PPE designed for certain particular risks as follows.

PPE protection against

- *Mechanical impact risks*
  (i)   impact caused by falling/projecting objects must be sufficiently shock-absorbent to prevent injury from crushing or penetration of the protected part of the body;
  (ii) falls. In the case of falls due to slipping, outsoles for footwear must ensure satisfactory adhesion by grip and friction, given the state and nature of the surface. In the case of falls from a height, PPE must incorporate a body harness and attachment system connectable to a reliable anchorage point. The vertical drop of the user must be minimised to prevent collision with obstacles and the braking force injuring, tearing or causing the operator to fall;
  (iii) mechanical vibration. PPE must be capable of ensuring adequate attenuation of harmful vibration components for the part of the body at risk.
- *(Static) compression of part of the body* – must be able to attenuate its effects so as to prevent serious injury of chronic complaints.
- *Physical injury* – must be able to protect all or part of the body against superficial injury by machinery, e.g. abrasion, perforation, cuts or bites.
- *Prevention of drowning* – (lifejackets, armbands etc.) must be capable of returning to the surface a user who is exhausted or unconscious, without danger to his health. Such PPE can be wholly or partially inherently buoyant or inflatable either by gas or orally. It should be able to withstand impact with liquid, and, if inflatable, able to inflate rapidly and fully.
- *Harmful effects of noise* – must protect against exposure levels of the *Noise at Work Regulations 2005* (i.e. 85 and 90 dB(A)) and must indicate noise attenuation level.
- *Heat and/or fire* – must possess sufficient thermal insulation capacity to retain most of the stored heat until after the user has left the danger area and removed PPE. Moreover, constituent materials which could be splashed by large amounts of hot product must possess sufficient mechanical-impact absorbency. In addition, materials which might accidentally come into contact with flame as well as being used in the manufacture of fire-fighting equipment, must possess a degree of non-flammability proportionate to the risk foreseeably arising during use and must not melt when exposed to flame or contribute to flame spread. When ready for use, PPE must be such that the quantity of heat transmitted by PPE to the user must not cause pain or health impairment. Second, ready-to-use PPE must prevent liquid or steam penetration and not cause burns. If PPE incorporates a breathing device, it must adequately protect the user. Accompanying manufacturers' notes must provide all relevant data for determination of maximum permissible user exposure to heat transmitted by equipment.

- *Cold* – must possess sufficient thermal insulating capacity and retain the necessary flexibility for gestures and postures. In particular, PPE must protect tips of fingers and toes from pain or health impairment and prevent penetration by rain water. Manufacturers' accompanying notes must provide all relevant data concerning maximum permissible user exposure to cold transmitted by equipment.
- *Electric shock* – must be sufficiently insulated against voltages to which the user is likely to be exposed under the most foreseeably adverse conditions. In particular, PPE for use during work with electrical installations (together with packaging), which may be under tension, must carry markings indicating either protection class and/or corresponding operating voltage, serial number and date of manufacture; in addition, date of entry into service must be inscribed as well as of periodic tests or inspections.
- *Radiation*
  (i) Non-ionising radiation – must prevent acute or chronic eye damage from non-ionising radiation and be able to absorb/reflect the majority of energy radiated in harmful wavelengths without unduly affecting transmission of the innocuous part of the visible spectrum, perception of contrasts and distinguishment of colours. Thus, protective glasses must possess a spectral transmission factor so as to minimise radiant-energy illumination density capable of reaching the user's eye through the filter and ensure that it does not exceed permissible exposure value. Accompanying notes must indicate transmission curves, making selection of the most suitable PPE possible. The relevant protection-factor number must be marked on all specimens of filtering glasses.
  (ii) Ionising radiation – must protect against external radioactive contamination. Thus, PPE should prevent penetration of radioactive dust, gases, liquids or mixtures under foreseeable use conditions. Moreover, PPE designed to provide complete user protection against external irradiation must be able to counter only weak electron or weak photon radiation.
- *Dangerous substances and infective agents*
  (a) respiratory protection – must be able to supply the user with breathable air when exposed to polluted atmosphere or an atmosphere with inadequate oxygen concentration. Leaktightness of the facepiece and pressure drop on inspiration (breath intake), as well as in the case of filtering devices purification capacity, must keep the contaminant penetration from the polluted atmosphere sufficiently low to avoid endangering health of the user. Instructions for use must enable a trained user to use the equipment correctly;
  (b) cutaneous and ocular contact (skin and eyes) – must be able to prevent penetration or diffusion of dangerous substances and infective agents. Therefore, PPE must be completely leak-tight but allow prolonged daily use or, failing that, of limited leaktightness restricting the period of wear. In the case of certain dangerous substances/infective agents possessing high penetra-

tive power and which limit duration of protection, such PPE must be tested for classification on the basis of efficiency. PPE conforming with test specifications must carry a mark indicating names or codes of substances used in tests and standard period of protection. Manufacturers' notes must contain an explanation of codes, detailed description of standard tests and refer to the maximum permissible period of wear under different foreseeable conditions of use.

- *Safety devices for diving equipment* – must be able to supply the user with a breathable gaseous mixture, taking into account the maximum depth of immersion. If necessary, equipment must consist of
  (a) a suit to protect the user against pressure;
  (b) an alarm to give the user prompt warning of approaching failure in supply of breathable gaseous mixture;
  (c) life-saving suit to enable the user to return to the surface.
  [Art 3, Annex II].

Excluded PPE

- PPE designed and manufactured for use specifically by the armed forces or in the maintenance of law and order (helmets, shields);
- PPE for self-defence (aerosol canisters, personal deterrent weapons);
- PPE designed and manufactured for private use against:
  (i)   adverse weather (headgear, seasonal clothing, footwear, umbrellas),
  (ii) damp and water (dish-washing gloves),
  (iii) heat (e.g. gloves);
- PPE intended for the protection or rescue of persons on vessels or aircraft, not worn all the time.
  [Annex I].

Main types of personal protection

The main types of personal protection covered by the *Personal Protective Equipment at Work Regulations 1992* are (a) head protection, (b) eye protection, (c) hand/arm protection, (d) foot protection and (e) whole body protection.

In particular, these regulations do not cover ear protectors and most respiratory protective equipment – these areas are covered by other existing regulations and guidance, for example the *Noise at Work Regulations 2005* and HSE's guidance booklet HS(G) 53, 'Respiratory protective equipment: a practical guide for users'.

Head protection

[P3019] This is likely to come under the non-simple or complex categories of PPE and takes the form of:

- crash, cycling, riding and climbing helmets;
- industrial safety helmets – to protect against falling objects;
- scalp protectors (bump caps) – to protect against striking fixed obstacles;
- caps/hairnets – to protect against scalping,

and is particularly suitable for the following activities:

(i)   building work – particularly on scaffolds;
(ii)   civil engineering projects;
(iii)   blasting operations;
(iv)   work in pits and trenches;
(v)   work near hoists/lifting plant;
(vi)   work in blast furnaces;
(vii) work in industrial furnaces;
(viii) ship-repairing;
(ix)   railway shunting;
(x)   slaughterhouses;
(xi)   tree-felling;
(xii) suspended access work, e.g. window cleaning.

Eye protection

[P3020] The majority of these are likely to come under non-simple category but some will come under the complex category. This takes the following forms:

- safety spectacles – these are the same as prescription spectacles but incorporating optional sideshields; lenses are made from tough plastic, such as polycarbonate – provide lateral protection;
- eyeshields – these are heavier than safety spectacles and designed with a frameless one-piece moulded lens; can be worn over prescription spectacles;
- safety goggles – these are heavier than spectacles or eye shields; they are made with flexible plastic frames and one-piece lens and have an elastic headband – they afford total eye protection; and
- faceshields – these are heavier than other eye protectors but comfortable if fitted with an adjustable head harness – faceshields protect the face but not fully the eyes and so are no protection against dusts, mist or gases.

Eye protectors are suitable for working with:

(i)   chemicals;
(ii)   power driven tools;
(iii)   molten metal;
(iv)   welding;
(v)   radiation;
(vi)   gases/vapours under pressure;
(vii) woodworking.

Hand/arm protection

[P3021] Gloves that protect against low risk cleaning fluids etc will come under the simple category but the majority will come under the non-simple or complex categories. They are used to provide protection against:

- cuts and abrasions;
- extremes of temperature;
- skin irritation/dermatitis;
- contact with toxic/corrosive liquids,

and are particularly useful in connection with the following activities/processes:

(i)   manual handling;
(ii)   vibration;
(iii)   construction;
(iv)   hot and cold materials;
(v)   electricity;
(vi)   chemicals;
(vii)   radioactivity.

Foot protection

[P3022] Again, the range of this type of protection will range fro simple to complex. This takes the form of:

- safety shoes/boots;
- foundry boots;
- clogs;
- wellington boots;
- anti-static footwear;
- conductive footwear,

and is particularly useful for the following activities:

(i)   construction;
(ii)   mechanical/manual handling;
(iii)   electrical processes;
(iv)   working in cold conditions (thermal footwear);
(v)   chemical processes;
(vi)   forestry;
(vii)   molten substances.

Respiratory protective equipment

[P3023] The majority of these come under the non-simple or complex categories. This takes the form of:

- half-face respirators,
- full-face respirators,
- air-supply respirators,
- self-contained breathing apparatus,

and is particularly useful for protecting against harmful:

(i)   gases;
(ii)   vapours;
(iii)   dusts;
(iv)   fumes;
(v)   smoke;
(vi)   aerosols.

Whole body protection

[P3024] This takes the form of:

- coveralls, overalls, aprons;
- outfits to protect against cold and heat;
- protection against machinery and chainsaws;
- high visibility clothing;
- life-jackets,

and is particularly useful in connection with the following activities:

(i)   laboratory work;
(ii)   construction;
(iii)   forestry;
(iv)   work in cold-stores;
(v)   highway and road works;
(vi)   food processing;
(vii)   welding;
(viii)   fire-fighting;
(ix)   foundry work;
(x)   spraying pesticides.

Some relevant British Standards for protective clothing and equipment

[P3025] Although not, strictly speaking, a condition of sale or approval, for the purposes of the *Personal Protective Equipment (EC Directive) (Amendment) Regulations 1994 (SI 1994 No 2326)* (now revoked and consolidated by *SI 2002 No 1144*), compliance with British Standards (e.g. ISO 9000: 'Quality Assurance') may well become one of the well-tried verification routes to obtaining a 'CE' mark. For that reason, if for no other, the following non-exhaustive list of British Standards on protective clothing and equipment should be of practical value to both manufacturers and users.

Table 1	
Protective clothing	
BS EN 60903: 2004	Live working. Gloves on insulating materials
BS 3314: 1982	Specification for protective aprons for wet work

Table 1	
Protective clothing	
BS 5426: 1993	Specification for workwear and career wear
BS 5438: 1989	Methods of test for flammability of textile fabrics when subjected to a small igniting flame applied to the face
1989 (1995)	Methods of test for flammability of textile fabrics when subjected to a small igniting flame applied to the face or bottom edge of vertically oriented specimens
BS 6408: (1990)	Specification for clothing made from coated fabrics for protection against wet weather
BS EN 340: 2003	Protective clothing. General Requirements
BS EN 341: (1999)	Personal Protective Equipment against falls from a height. Descender devices
BS EN 348: 1992	Protective clothing. Determination of behaviour of materials on impact of small splashes of molten metal
BS EN 353: 2002	Personal protective equipment against falls from a height. Guided type of fall arresters
BS EN 354: 2002	Personal protective equipment against falls from a height. Lanyards
BS EN 355: 2002	Personal protective equipment against falls from a height. Energy Absorbers
BS EN 360: 2002	Personal protective equipment against falls from a height. Retractable type fall arresters
BS EN 361: 2002	Personal protective equipment against falls from a height. Full body harnesses
BS EN 362: 2004	Personal protective equipment against falls from a height. Connectors
BS EN 363: 2002	Personal protective equipment against falls from a height. Fall arrest systems
BS EN 364: 1993 (1999)	Personal protective equipment against falls from a height. Test methods
BS EN 365: 2004	Personal protective equipment against falls from a height. General Requirements for instructions for use and for marking
BS EN 366: 1993	Protective clothing. Protection against heat and fire
BS EN 367: 1992	Protective clothing. Protection against heat and fires
BS EN 368: 1993	Protective clothing. Protection against liquid chemicals
BS EN 369: 1993	Protective clothing. Protection against liquid chemicals
BS EN 373: 1993	Protective clothing: Assessment of resistance of materials to molten metal splash

Table 1	
Protective clothing	
BS EN 381 Part 1: 2002	Protective clothing for users of hand-held chain saws
BS EN 463: 1995	Protective clothing. Protection against liquid chemicals. Test method
BS EN 464: 1994	Protective clothing. Protection against liquid and gaseous chemicals including liquid aerosols and solid particles. Test method
BS EN 465: 1995	Protective clothing: Protection against liquid chemicals. Performance requirements for chemical protective clothing
BS EN 4661: 1995	Protective clothing: Protection against liquid chemicals. Performance requirements for chemical protective clothing
BS EN 467: 1995	Protective Clothing: protection against liquid chemicals.
BS EN 468: 1995	Protective clothing for use against liquid chemicals. Test method
BS EN 469: 1995	Protective clothing for firefighters
BS EN 470 Part 1: 1995	Protective clothing for use in welding and allied processes General requirements
BS EN 471: 2003	High-visibility warning clothing for professional use
BS EN 510: 1993	Specification for protective clothing for use where there is a risk of entanglement with moving parts
BS EN 530: 1995	Abrasion resistance of protective clothing material
BS EN 531: 1995	Protective clothing for workers exposed to heat
BS EN 532: 1995	Protective clothing. Protection against heat and flame
BS EN 533: 1997	Clothing for protection against heat or flame
Protective footwear	
BS 2723: 1956 (1995)	Specification for fireman's leather boots
BS 4676: 2005	Protective clothing. Footwear and gaiters for use in molten metal foundries.
BS 5145: 1989	Specification for lined industrial vulcanized rubber boots
BS 6159:	Polyvinyl chloride boots
Part 1: 1987	Specification for general and industrial lined or unlined boots
BS EN	Safety, protective and occupational footwear for profes-sional use
BS EN 345: 1993	Safety footwear for professional use

Table 1	
Protective clothing	
BS EN 346: 1997	Protective footwear for professional use
BS EN 347: 1993	Occupational footwear for professional use
BS EN 381: 2002	Protective clothing for users of hand-held chain saws
Head protection	
BS 6658: 1985 (1995)	Specification for protective helmets for vehicle users
BS 3864: 1989	Specification for protective helmets for firefighters
BS 4033: 1978	Specification for industrial scalp protectors (light duty)
BS 4423: 1969	Specification for climbers helmets
BS EN 1384: 1997	Specification for helmets for equestrian activities.
Face and eye protection	
BS 679: 1989	Specification for filters, cover lenses and backing lenses for use during welding and similar industrial operations
BS 1542: 1982 (1995)	Specification for eye, face and neck protection against non-ionising radiation arising during welding and similar operations
BS 2092: 1987	Specification for eye protectors for industrial and non-industrial users
BS 2724: 1987 (1995)	Specification for sunglare eye protectors for general use
BS 4110: 1999	Specification for visors for vehicle users
BS 7028: 1999	Eye protection for industrial and other uses. Guidance on selection, use and maintenance
BS EN 167: 2002	Personal eye protection. Optical test methods
BS EN 168: 2002	Personal eye protection. Non-optical test methods
BS EN 169: 2002	Personal eye-protection. Filters for welding and related techniques.
BS EN 170: 2002	Personal eye-protection. Ultraviolet filters.
BS EN 171: 2002	Personal eye-protection. Infrared filters
BS EN 172: (2000)	Specification for sunglare filters used in personal eye-protectors equipment for industrial use
Respiratory protection	

Table 1	
Protective clothing	
BS 4001	Care and maintenance of underwater breathing apparatus
Part 1 (1998)	Care and maintenance of underwater breathing apparatus. Recommendations for the compressed air open circuit type
Part 2 (1995)	Standard diving equipment
BS 4275: 1997	Guide to implementing an effective respiratory protective device programme
BS 4400: 1969 (1995)	Method for sodium chloride particulate test for respirator filters
BS EN 132: 1999	Respiratory protective devices. Definitions of terms and pictograms
BS EN 133: 2001	Respiratory protective devices. Classification
BS EN 134: 1998	Respiratory protective devices. Nomenclature of compounds
BS EN 135: 1999	Respiratory protective devices. List of equivalent terms
BS EN 136: 1998	Respiratory protective devices. Full face mask requirements for testing and marking
BS EN 137: 1993	Specification for respiratory protective devices self-contained open-circuit compressed air breathing apparatus
BS EN 138: 2000	Respiratory protective devices. Specification for fresh air hose breathing apparatus for use with full face mask, half mask or mouthpiece assembly
BS EN 139: 1995	Respiratory protective devices. Compressed air line breathing apparatus for use with a full face mask, half mask or a mouthpiece assembly. Requirements, testing, marking
BS EN 141: 1991	Specification for gas filters and combined filters used in respiratory protective equipment
BS EN 143: 1991	Specification for particle filters used in respiratory protective equipment
BS EN 144	Respiratory protective devices. Gas cylinder valves
BS EN 145: 1998	Respiratory protective devices. Self-contained closed-circuit compressed oxygen breathing apparatus
BS EN 146: 1992	Respiratory protection devices. Specification for powered particle filtering devices incorporating helmets or hoods
BS EN 147: 1992	Respiratory protection devices. Specification for power assisted particle filtering devices incorporating full face masks, half masks or quarter masks
BS EN 148	Specification for thread connection
BS EN 149: 2001	Respiratory protective devices. Filtering half masks to protect against particles
BS EN 371: 1992	Specification for AX gas filters and combined filters against low boiling organic compounds used in respiratory protective equipment

Table 1	
Protective clothing	
BS EN 372: 1992	Specification for SX gas filters and combined filters against specific named compounds used in respiratory protective equipment
Radiation protection	
BS 1542: 1982 (1995)	Specification for equipment for eye, face and neck protection against non-ionising radiation arising during welding and similar operations
BS 3664: 1963	Specification for film badges for personnel radiation monitoring
BS EN 269: 1995(2000)	Respiratory protective devices. Specification for powered fresh air hose breathing apparatus incorporating a hood.
BS EN 270: 1995	Respiratory protective devices. Compressed air line breathing apparatus incorporating a hood. Requirements, testing, marking
Hearing protection	
BS EN 352:	Hearing protectors. Safety requirements and testing
Part 1: 2002	Ear muffs
Part 2: 2002	Ear plugs
Part 3: 2002	Ear muffs attached to an industrial safety helmet
Hand protection	
BS EN 374:	Protective gloves against chemicals and microorganisms
Part 1: 2003	Terminology and performance requirements
Part 2: 2003	Determination of resistance to penetration
Part 3: 2003	Determination of resistance to permeation by chemicals
BS EN 511: 1994	Specification for protective gloves against cold
BS EN 421: 1994	Protective gloves against ionising radiation and radioactive contamination
BS EN 407: 2004	Protective gloves against thermal risks (heat and/or fire)
BS EN 388: 2003	Protective gloves against mechanical risks

Common law requirements

[P3026] In addition to general and/or specific statutory requirements, the residual combined duty, on employers at common law, to provide and maintain a safe system of work, including appropriate supervision of safety duties, still obtains, extending, where necessary, to protection against foreseeable risk of eye injury (*Bux v Slough Metals Ltd* [1974] 1 All ER 262) and dermatitis/facial eczema (*Pape v Cumbria County Council* [1991] IRLR 463). *Pape* concerned an office cleaner who developed dermatitis and facial eczema after using Vim, Flash and polish. The employer was held liable for damages at common law, since he had not instructed her in the dangers of using chemical materials with unprotected hands and had not made her wear rubber gloves.

Consequences of breach

[P3027] Employers who fail to provide suitable personal protective equipment, commit a criminal offence, under the *Health and Safety at Work etc Act 1974* and the *Personal Protective Equipment at Work Regulations 1992* and sundry other regulations (see **P3006–P3008** above). In addition, if, as a result of failure to provide suitable equipment, an employee suffers foreseeable injury and/or disease, the employer will be liable to the employee for damages for negligence. Conversely, if, after instruction and, where necessary, training, an employee fails or refuses to wear and maintain suitable personal protective equipment, he too can be prosecuted and/or dismissed and, if he is injured or suffers a disease in consequence, will probably lose all or, certainly, part of his damages.

Pressure Systems

Kevin Chicken

Introduction

[P7001] Legislation concerning the safety of pressure systems applied in its infancy to steam generating boilers and is traceable back to the original *Boiler Explosions Act 1882*. At this time statutory requirements were placed on the users and owners of the equipment for the safety of the boiler and little or no consideration was given to the safety of the system incorporating it. Periodic examination of boilers became mandatory following the *Factories and Workshops Act 1901* when requirements were also extended to include other items of pressure equipment such as steam and air receivers.

The legislation governing the construction and use of pressure equipment is divided into two areas where the initial duty falls onto those either manufacturing, installing or supplying pressure equipment and thereafter to users or owners of the equipment.

There are three principal statutes that cover the manufacture, installation and operation of pressure equipment:

- *Simple Pressure Vessel Regulations 1991 (SI 1991 No 2749)* amended by *Simple Pressure Vessels (Safety Amendment) Regulations 1994 (SI 1994 No 3098)*;
- *Pressure Equipment Regulations 1999 (SI 1999 No 2001)* as amended by the *Pressure Equipment (Amendment) Regulations 2002 (SI 2002 No 1267)*;
- *Pressure Systems Safety Regulations 2000 (SI 2000 No 128)*.

These regulations provide a synergistic approach to the management of pressure systems based on a cradle to grave concept.

The Pressure Equipment Directive 87/404/EEC (PED) and the Simple Pressure Vessel Directive (SPVD) (see Figure 1) as implemented by the *Pressure Equipment Regulations 1999 (SI 1999 No 2001)* (PER) as amended by the *Pressure Equipment (Amendment) Regulations 2002 (SI 2002 No 1267)* and the *Simple Pressure Vessel Regulations 1991 (SI 1991 No 2749)* as amended by the *Simple Pressure Vessels (Safety Amendment) Regulations 1994 (SI 1994 No 3098)* (SPVR). This amendment was needed to take account of the CE Marking Directive (93/68/EEC) which required all simple pressure vessels to be CE Marked (see P7004 and 7005).

These Regulations set minimum standards of safety and quality for the manufacture of new pressure systems and equipment in the European Union so that non tariff trade barriers are removed and they present a 'level playing field' for all member states. The Department of Trade and Industry implement these requirements in the UK.

The *Pressure Systems Safety Regulations 2000 (SI 2000 No 128)* are mainly concerned with ensuring that minimum safety standards are satisfied with due regard for in service pressure systems and these are specific to the UK.

Provisions relating to pressurised transportable gas containers (receptacles) have now been removed from the governing legislation (*Pressure Systems Safety Regulations 2000 (SI 2000 No 128)*). These Regulations replaced the *Pressure Systems and Transportable Gas Container Regulations 1989 (SI 1989 No 2169) (PSTGCR)*.

The introduction of the *PSTGCR* in 1989 was a major step forward in the control over pressure systems. They introduced the important concepts of goal setting legislation, the written scheme of examination, and the role of the competent person who was charged within certain limitations of determining the scope and frequency of the thorough examination. Prior to the PSTGCR, pressure systems were covered under the *Factories Act 1961*, supported by prescriptive requirements for their thorough examination at frequencies stipulated within the Act.

Transportable gas containers that were previously considered in the PSTGCR have been divorced and are now covered under *the Carriage of Dangerous Goods and Use of Transportable Pressure Equipment Regulations 2004 (SI 2004 No 568)*, these regulations replaced the earlier *Carriage of Dangerous Goods (Classification, Packaging and Labelling) and Use of Transportable Pressure Receptacles Regulations 1996 (SI 1996 No 2092)* which was in turn replaced by the short lived and now revoked *Transportable Pressure Vessel Regulations 2001 (SI 2001 No 1426)*.

Figure 1. Relationship between Manufacturers' Duties

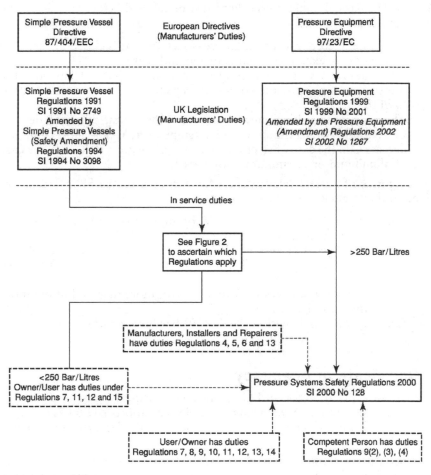

Simple pressure vessels

[P7002] Simple pressure vessels are those that are constructed under the criteria of the *Simple Pressure Vessel Regulations 1991 (SI 1991 No 2749)* amended by the *Simple Pressure Vessels (Safety Amendment) Regulations 1994 (SI 1994 No 3098)*.

A simple pressure vessel is a welded vessel intended to contain air or nitrogen at a gauge pressure greater than 0.5 bars but one that is not intended for exposure to a flame.

Simple Pressure Vessels (Safety) Regulations (SPVSR) 1991

[P7003] The basic design criteria that must be fulfilled before SPVSR can be applied are that:

- The vessel incorporates circular cross section that is closed at each end by either convexed dished ends or flat end plates.
- The maximum working pressure is not more than 30 bar, and the pressure multiplied by the volume in bar litres is not more than 10,000 bar/litres.
- The minimum working temperature is not lower than minus 50 degrees Celcius, and the maximum working temperature is not higher than 300 degrees C in the case of steel vessels, and 100 degrees Celcius in the case of aluminium or aluminium alloy vessels.

The categories of vessels relevant for the purposes of the SPVSR that are defined in the Regulations are—

- *Category A*: Vessels with a PS/V of more than 50 bar-litres, this category being subdivided into:
  - *Category A.1*: Vessels with a PS/V of more than 3,000 bar-litres;
  - *Category A.2*: Vessels with a PS/V of more than 200 but not more than 3,000 bar-litres; and
  - *Category A.3*: Vessels with a PS/V of more than 50 but not more than 200 bar-litres;
- *Category B*: Vessels with a PS/V of 50 bar-litres or less.

EC Certificate of Adequacy

[P7004] An application for an EC certificate of adequacy must be accompanied by the design and manufacturing schedule. The approved bodies to which such an application can be made are available from the Department of Trade and Industry.

The schedule must contain all the required information; and additionally provide evidence that the vessel has been manufactured in accordance with the schedule and that it conforms to a relevant national standard eg BSI EN 13445.

A certificate of adequacy can then be issued by the approved body which allows the simple pressure vessel to be placed on the market as being 'safe'.

If the approved body is not so satisfied and refuses to issue a certificate of adequacy it must inform the applicant in writing the reasons for the refusal.

EC Type-Examination Certificate

[P7005] EC type-examination is the process whereby an approved body ascertains, and certifies by means of an EC type-examination certificate, that a prototype representative of the production envisaged satisfies the requirements of the Directive.

The prototype may be representative of a family of vessels in which case the Approved Body is required to examine the prototype, and perform such tests as it considers are appropriate, and once it is satisfied that the prototype is manufactured in conformity with the schedule, and meets the essential safety requirements specified in Schedule 1, declare that it is safe.

Pressure Equipment Regulations 1999 (SI 1999 No 2001) and the EU Pressure Equipment Directive (97/23/EC)

[P7006] The Directive was taken into UK legislation as the *Pressure Systems Safety Regulations 1999 (SI 1999 No 2001)* and covers pressure equipment and assemblies with a maximum allowable pressure greater than 0.5 bar. Pressure equipment includes vessels, piping, safety accessories and pressure accessories. Assemblies are several pieces of pressure equipment assembled to form an integrated, functional whole.

The intention of legislation is to remove technical barriers to trade by harmonising national laws of legislation regarding the design, manufacture, marking and conformity assessment of pressure equipment. It does not deal with in-use requirements, which may be necessary to ensure the continued safe use of pressure equipment.

The Directive covers a wide range of equipment such as, reaction vessels, pressurised storage containers, heat exchangers, shell and water tube boilers, industrial pipework, safety devices and pressure accessories.

There are some specific exclusions such as: equipment within the scope of other directives (eg simple pressure vessels, aerosols, parts for the functioning of vehicle braking system); and equipment covered by the international conventions for the transport of dangerous goods and equipment for which the pressure hazards are low and are adequately covered by other directives (such as the Machinery Directive).

Essential safety requirements

[P7007] The principle adopted in the *PER Pressure Equipment Regulations 1999 (SI 1999 No 2001)* is similar to other European Directives, and requires the manufacturer to consider hazards associated with pressure equipment and accordingly take these hazards into consideration when considering the design and construction of the equipment.

Compliance with the Essential Safety Requirements (ESRs) provides the necessary safety elements for protecting public interest. Essential safety requirements for design, manufacture, testing, marking, labelling, instructions and materials, are provided in general terms, compliance is mandatory and must be met before products may be placed on the market in the European Community.

The main areas for assessment are:

- General requirements.
- Design requirements:
  - Adequate strength;
  - Experimental design methods;
  - Provisions to ensure safe handling and operation;
  - Means of examination;
  - Means of draining and ventilating;
  - Corrosion;
  - Wear;
  - Assemblies;
  - Provisions for filling and discharge;
  - Protection against exceeding the allowable limits of pressure equipment;
  - Safety accessories;
  - External firing.
- Manufacturing:
  - Manufacturing procedures;
  - Final assessment;
  - Marking and labelling;
  - Operating instructions.
- Materials.
- Fired or Heated Pressure equipment with a Risk of Overheating.
- Piping.
- Specific Quantitative Requirements for Certain Pressure Equipment:
  - Allowable stresses;
  - Joint coefficients;
  - Pressure limiting devices, particularly for pressure vessels;
  - Hydrostatic test pressure;
  - Material characteristics.

The requirements set out above may be met by applying one of the following methods, as appropriate, if necessary as a supplement to or in combination with another method:

- design by formula;
- design by analysis; or
- design by fracture mechanics.

Conformity assessment procedure

[P7008] Schedule IV of the *Pressure Equipment Regulations 1999 (SI 1999 No 2001)*, or annex III of the Pressure Equipment Directive 97/23/EC detail the procedures and responsibilities for the contributing parties. This is a detailed document and it consists of a product classification table and nine classification charts.

A Conformity Assessment must be undertaken by the manufacturer or notified body, depending on the category of the equipment, in order to demonstrate that the essential safety requirements are met.

In order to declare pressure equipment to be safe the manufacturer or their authorised representative must before supplying any equipment satisfy the certification procedure which is a similar but more detailed procedure to that applied to conformity assessment for verification under the Simple Pressure Vessel Directive 87/404/EEC.

The New Approach has introduced a modular approach to conformity assessment, thereby subdividing it into a number of independent activities. Modules differ according to the type of assessment (eg documentary checks, type approval, design approval, quality assurance) and the organisation carrying out the assessment (ie the manufacturer or a third party).

- Draw up a technical document which will allow an assessment to be made of the conformity of the pressure equipment, containing:
  — General description;
  — Conceptual design and manufacturing drawings;
  — Descriptions and explanations necessary to understand the drawings;
  — A list of harmonised standards applied and the solutions adopted to meet the essential safety requirements were harmonised standards have not been applied;
  — Results of design calculations made;
  — Test reports.
- Ensure manufacturing process complies with technical documentation.
- Affix CE Mark.
- Draw up written declaration of conformity.
- Retain declaration of conformity and technical documentation for ten years.

Published Harmonised (European) Standards, list are a specific subset of European Standards (EN, produced by CEN and available from the national Standards Institutes) with particular consideration of the Essential Safety Requirements the reference number of which is published in the Official Journal of the European Commission. The use of a Published Harmonised Standard in the design and manufacture of a product will give the presumption of conformity to those ESRs listed in the particular Harmonized Standard.

There are a number of different conformity assessment modules available for the different categories of equipment; eg the boiler manufacturer's codes:

- BS EN 12592 Water Tube Boilers;
- BS EN 12593 Shell Boilers.

The unfired pressure vessel manufacturer's new code:

- BS EN 13445 Unfired pressure vessels.

To satisfy the requirements of the (*Pressure Equipment Regulations 1999 (SI 1999 No 2001)*). Manufacturers will need to:

- meet essential safety requirements covering design, manufacture and testing;
- satisfy appropriate conformity assessment procedures; and

- carry the CE marking.

By affixing a CE mark the manufacturer declares the completion of conformity assessment and that the equipment or assembly complies with the provisions of the Directive and meets the essential safety requirements.

Pressure equipment and assemblies below the specified pressure / volume thresholds must that needs to conform to "Sound Engineering Practice (SEP)" (SEP applies to equipment that is not subject to conformity assessment but must be designed and manufactured in accordance with the sound engineering practice of a Member State in order to ensure safe use. That the design and manufacture takes into account all relevant factors influencing safety during the intended lifetime. The equipment must be accompanied with adequate instructions for use and must bear the identification of the manufacturer. The responsibility for compliance with the PED lies solely with the manufacturer):

To be safe and meet compliance with SEP the equipment must;

- be designed and manufactured according to sound engineering practice; and
- bear specified markings (but not the CE marking).

Notified bodies

[P7009] The Secretary of State for the Department of Trade and Industry (DTI) is responsible for the appointment of recognised third-party notified bodies, following an assessment against criteria laid down in the directive.

The Notified Body is a semi-official or private technical organisation appointed by Member States, either for approval and monitoring of the manufacturers' quality assurance system or for direct product inspection. A Notified Body may be specialised for certain products/product categories or for certain modules.

The Pressure Equipment (Amendment) Regulations 2002 (SI 2002 No 1267)

[P7010] The *Pressure Equipment (Amendment) Regulations 2002 (SI 2002 No 1267)* exempt an item of pressure equipment or an assembly of exempting pressure equipment that is placed on the market from the requirements of the 1999 Regulations, provided it has not been used in the course of business at all times since its manufacture or import and that prior to its being made so available it has been used otherwise than in the course of business at all times since its manufacture or import.

There have also been some amendments to the penalties available to the courts where a person guilty of an offence under *Regulation 25(a)* shall be liable:

(a) On summary conviction, to a fine not exceeding the statutory maximum or to imprisonment not exceeding three months or to both;

(b) On conviction on indictment, to a fine or to imprisonment for a term not exceeding two years or to both.

Pressure Systems Safety Regulations 2000 (SI 2000 No 128)

Duty holders

[P7011] The *Pressure Systems Safety Regulations 2000* impose statutory duties on individuals and organisations connected with pressure systems.

The *Pressure Systems Safety Regulations 2000* ('the Regulations') came into force on 14th February 2000. [*Pressure Systems Safety Regulations 2000 (SI 2000 No 128), Reg 1.*] They are the principal UK statutory controls over the operation and safety of in-service pressure systems. The aim of the Regulations is to prevent serious injury from the hazard of stored energy as a result of failure of the system or one of its component parts.

- **Users and owners** of pressure systems have primary and wide responsibilities for ensuring the safety of their systems.
- **'Competent persons'** are responsible for drawing up or certifying suitable written schemes of examination, for the examination in accordance with the written scheme, and for action in the case of imminent danger.
- **Designers, manufacturers, importers, suppliers and installers** have responsibilities relating to design, materials and construction, and information and marking.
- Employers of **persons, who install,** modify or repair pressure systems have duties to ensure that their work does not gives rise to danger.
- **The enforcing authority** (Health and Safety Executive or Environmental Health Department of the local authority), has powers under the Regulations to enforce and exempt the application of the Regulations.

In cases where pressure systems are supplied by way of lease, hire or other arrangements, the supplier is allowed to discharge the duties of the user with regard to examinations, operation, maintenance and record keeping. [*Pressure Systems Safety Regulations 2000 (SI 2000 No 128), Reg 3(5) and Sch 2.*]

A duty holder charged with an offence under the Regulations is allowed a defence if the offence was due to an act or default of another person (not being an employee) and that all reasonable precautions and due diligence had been taken to avoid the offence taking place. [*Pressure Systems Safety Regulations 2000 (SI 2000 No 128), Reg 16.*]

Systems and equipment covered by the regulations

[P7012] The Regulations are concerned with systems containing relevant fluids under pressure (See Figure 2 below). A relevant fluid is:

- Steam (above atmospheric pressure);
- A gas or a mixture of gases;
- A liquid which would evaporate as a gas if system failure occurred.

The Regulations therefore cover systems containing compressed air, other compressed gases (eg nitrogen, oxygen or acetylene), hot water contained above its normal boiling point, and a gas dissolved under pressure in a solvent contained in a porous substance.

Except in the case of steam (which is always a relevant fluid when stored above atmospheric pressure), a fluid is relevant if it exerts a pressure or vapour pressure greater than 0.5 bar above atmospheric pressure.

The Regulations cover both installed (fixed) pressure systems and mobile systems that can be readily removed between and used in different locations. Transportable pressure receptacles such as a tanker used for carrying dangerous goods by rail or road, are covered under the *Carriage of Dangerous Goods and Use of Transportable Pressure Equipment Regulations 2004* (if these Regulations cease to apply and a relevant fluid is carried, then the *Pressure Systems Safety Regulations 2000* will apply). Steam locomotives, including model steam engines, are classed as installed systems for the purposes of the Regulations. Within the Regulations, an installed system is the responsibility of the user, while the owner has responsibility for a mobile system.

Within the Regulations, a pressure system comprises of one or more pressure vessels of rigid construction, any associated pipework, and protective devices. Pipework used in connection with transportable pressure receptacles and pipelines with their associated protective devices are also pressure systems. The terms 'pipework', 'pipeline' and 'protective devices' are defined by the Regulations and cover all items of pressure equipment including valves, pumps, hoses, compressors, bursting discs and pressure relief devices. [*Pressure Systems Safety Regulations 2000 (SI 2000 No 128), Reg 2.*]

Users and owners should decide, seeking suitable advice if required, whether equipment for which they are responsible is covered by these Regulations. Examples of equipment for which the Regulations are likely to apply to include:

- Steam boilers and associated pipework and protective devices.
- Compressed air systems with a receiver of a stored capacity exceeding 250 Bars/litres
- Portable hot water steam cleaning units fitted with a pressure vessel.
- Steam sterilising autoclaves and associated pipework and protective devices.
- Steam pressure cookers.
- Gas loaded hydraulic accumulators if the free gas capacity exceeds 250 Bars/litres
- Steam locomotives including model and narrow gauge steam engines.
- Vapour compression refrigeration systems where the power exceeds 25KW.
- Fixed liquid petroleum gas (LPG) storage systems eg supplying fuel for heating in a workplace.

Figure 2. Pressure Systems Safety Regulations 2000 – Duties Decision Flow Chart

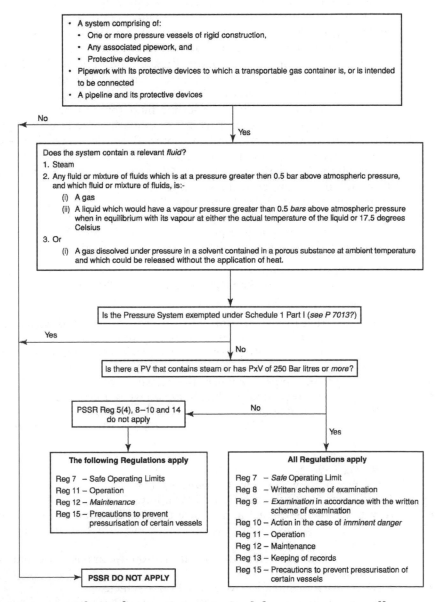

Systems and equipment excepted from some or all the regulations

[**P7013**] Some types of pressure systems are excepted from certain Regulations [*Pressure Systems Safety Regulations 2000 (SI 2000 No 128), Sch 1,*

Part II.] These exceptions are primarily to avoid overlap with other regulations and apply to the following.

(i)   Pressure systems to which the *Medical Devices Regulations 1994 (SI 1994 No 3017)* apply, other than those which contain or are liable to contain steam, are excepted from regulations on design, construction and provisions for information and marking.

(ii)   Pressure equipment subject to the *Pressure Equipment Regulations 1999 (SI 1999 No 2001)* is excepted from regulations on design, construction and provisions for information and marking.

(iii)   Pressure systems containing a relevant fluid (other than steam) are excepted from provisions on marking, examination and record keeping if the product of the pressure and internal volume control in each pressure vessel is less than 250 bar litres. Where the relevant fluid is steam, all the Regulations apply regardless of vessel pressure and size.

(iv)   Tank containers subject to the *Carriage of Dangerous Goods and Use of Transportable Pressure Equipment Regulations 2004 (SI 2004 No 568)* and examined under the provisions of those Regulations are not subject to the corresponding provisions of the *Pressure Systems Safety Regulations 2000* when they have been removed from the vehicle.

Some 25 categories of pressure systems are exempted from all the Regulations [*Pressure Systems and Safety Regulations 2000 (SI 2000 No 128), Sch 1, Part I*] and include:

- Any system that forms part of a ship, hovercraft, aircraft or spacecraft.
- Any system that forms part of a weapon system.
- Any system which is only pressurised by a leak test or unintentionally.
- Some systems which form part of a research experiment.
- Any low pressure pipelines where the pressure does not exceed 2 bars.
- Any equipment regulated by the *Diving at Work Regulations 1997 (SI 1997 No 2776), Reg 6(3)(b)*.
- A chamber, tunnel or airlock which fall under the *Work in Compressed Air Regulations 1996 (SI 1996 No 1656)*.
- A tank to which the *Carriage of Dangerous Goods by Rail Regulations 1996 (SI 1996 No 2089)* apply.
- Any system to which the *Carriage of Dangerous Goods by Road Regulations 1996 (SI 1996 No 2095)* apply.
- Any system being carried in a vehicle engaged international transport.
- A system that forms part of the braking, control or suspension system of a vehicle.
- Any water cooling system on an international combustion engine or compressor.
- Any tyre used or intended to be used on a vehicle.
- Any system fitted to a vehicle within the meaning of the *Road Traffic Act 1988, s 185*.
- Any portable fire extinguisher with a working pressure of less than 25 bars at 60°C.
- Any part of a tool or appliance designed to be held in the hand.

Some of these categories may, however, be subject to other regulations.

Approved code of practice and guidance

[P7014] The Regulations are largely non-prescriptive goal-setting regulations. For example, the Regulations state that pressure systems 'shall be properly designed and properly constructed from suitable material so as to prevent danger'. The goal is to prevent danger; what is 'proper' and 'sufficient' to achieve this goal is left to duty holders and ultimately the courts to decide.

The Health and Safety Commission's (HSC) Approved Code of Practice ('the ACoP') to the *Pressure Systems Safety Regulations 2000 (SI 2000 No 128)* includes guidance and a guide to the Regulations ('*Safety of Pressure Systems, Approved Code of Practice*', L122, ISBN 07176 1767 X, published by HSE Books). This document is essential reading for duty holders. It has special legal status in relation to compliance with the Regulations.

The ACoP gives practical advice on how to comply with the Regulations. Following the advice in the ACoP should be enough to ensure compliance with the law on specific matters, although alternative methods may be used. If the relevant provisions of the code are not followed, then it is necessary, in any prosecution for breach of the law, to show compliance with the law in other ways.

The additional guidance material is not compulsory but illustrates good practice. The guide sets out the law and explains what it means in simple terms.

Interpretation

[P7015] The following words and phrases have a specific defined meaning within the context of the Regulations:

- Competent person.
- Danger.
- Data.
- Examination.
- The Executive.
- Installed system.
- Mobile system.
- Owner.
- Pipeline.
- Pipework.
- Pressure system.
- Protective devices.
- Relevant fluid.
- Safe operating limits.
- Scheme of examination.
- Ship.
- System failure.
- Transportable pressure receptacle.
- User.

[*Pressure Systems Safety Regulations 2000 (SI 2000 No 128), Reg 2.*]

Design, construction and installation

[P7016] The following persons have duties for ensuring that the design and construction of pressure systems and component parts meet the requirements of the Regulations:

(i)   Designers.
(ii)   Manufacturers.
(iii)   Importers.
(iv)   Suppliers.

[Pressure Systems Safety Regulations 2000 (SI 2000 No 128), Reg 4.]

The Regulations require that pressure systems shall be properly designed and properly constructed from suitable material so as to prevent danger. The design and construction must enable necessary examinations to be carried out for preventing danger from defects etc., and that where access to the inside of a pressure system is provided, it can be gained without danger to the inspector. Pressure systems must be provided with suitable protective devices (safety relief valves, bursting discs etc.) to prevent danger, and such devices must be capable of releasing the contents safely.

Pressure systems supplied with a CE marking under the *Pressure Equipment Regulations 1999 (SI 1999 No 2001)* are presumed to have complied with this Regulation (see **P7008**.)

Employers of persons installing a pressure system have a duty to ensure that there is nothing about the way it is installed that could give rise to danger from the system or otherwise impair the operation of protective devices or the facility to examine the equipment. The ACoP draws attention to a number of items that the employer should consider relating to the siting and access to and around the pressure system and the means and manner of installation. Special consideration needs to be given by the manufacturer to any in-service examinations that may require access to constricted areas and the specific requirements of *the Confined Space Regulations 1997 (SI 1997 No 1713)*.

Documentation and marking

[P7017] The following information and documentation is required for each pressure system.

(a)   *Information concerning the design, construction, examination, operation and maintenance*
        The Regulations require that pressure systems (or any component parts) are provided with written information concerning the design, construction, examination, operation and maintenance. Responsibility for providing this information falls on the person who designs or supplies (whether as manufacturer, importer or in some other capacity) the system. When a pressure system is modified or repaired, the responsibility to provide information about the modification or repair falls on the employer of the person who carried out the work or any self employed individual. *[Pressure Systems Safety Regulations 2000 (SI 2000 No 128), Reg 5.]*

The Regulations do not specify the information to be provided, but set the objective that it must be sufficient enough to comply with their provisions. The information must be provided with the design or with the supplied pressure system. When systems are modified or repaired, the information must be provided immediately afterwards. Although it is not possible to give a complete list of all the information which might be needed, the ACoP highlights items that should be considered.

(b)   *A plate on each pressure vessel with the information given in Schedule 3 of the Regulations*
Manufacturers of pressure vessels must ensure that before the vessel is supplied, it is marked or has a plate attached containing the information specified in the *Pressure Systems Safety Regulations 2000 (SI 2000 No 128), Sch 3*. This information includes the manufacturer's name, standard of construction, and the design pressure(s) and temperature(s). Similarly, no one may import a vessel unless it is marked or has a plate containing this information. The Regulations prohibit the removal or falsification of information contained on the plate or on any other mark on the vessel.

(c)   *A statement of the safe operating limits*
The user of an installed pressure system or the owner of a mobile system will have established *safe operating limits* of the system before the system is operated. The nature of the safe operating limits is not specified by the Regulations and will depend on the system. They should, however, be such as to ensure that there is a suitable margin of safety before failure is liable to occur. Typically, operating limits may be set on pressure, temperature, number of cycles, or time in operation, but it is for the duty holder to determine what is appropriate.
In the case of mobile systems, the owner, if not the user, must supply the user with a written statement specifying the safe operating limits or ensure that the system is marked with this information in a legible, durable and visible way.

(d)   *A written scheme of examination* (see **P7019** below);

(e)   *A report on each examination carried out under the written scheme* (see **P7024** below);

(f)   *Instructions for safe operation and action in emergencies* (see **P7026** below).

Under certain circumstances, the following additional documentation is required:

(g)   *An agreement to postpone the date of examination;*

(h)   *A report in case of imminent danger* (see **P7025** below);

(i)   *Information required by the Pressure Equipment Regulations 1999 (SI 1999 No 2001);*

(j)   *A certificate exempting application of the requirements/prohibitions of the Regulations.*

Competent persons

[P7018] Whilst users and owners have responsibility for the management of their pressure systems, the Regulations require and define the role and responsibilities of a 'competent person'. The role of the competent person is primarily in connection with the examination and action in the case of imminent danger. The term 'competent person' is used in connection with two distinct functions:

- drawing up or certifying schemes of examination;
- carrying out examinations and making the examination report.

A single individual may carry out these functions, but equally, separate individuals can also carry them out. The user or owner may, however, also use the competent person in an advisory capacity in conjunction with other matters relating to the Regulations. For example, the competent person may assist users/owners in deciding the scope of the examination.

The level of expertise needed by the competent person depends on the size and complexity of the pressure system in question. The ACoP distinguishes between minor, intermediate and major systems and the attributes required of the competent person and back up specialist services and organisation required in each case. In all but minor systems, the competent person must be a chartered engineer.

For larger and more complex systems, it is unlikely that one individual will have sufficient knowledge and expertise. In this case the competent person should be part of a corporate body with a team of employees which have the necessary expertise and experience. The requirements for 'competence' then applies to the company and not to individual employees.

The following bodies may provide competent person services:

- A user company with its own in-house inspection department.
- An inspection organisation providing such services to clients.
- A self employed person.

The Competent Person should be either an individual or company with sufficient knowledge and experience to undertake the work. As a minimum standard the Competent Person should be able to demonstrate sufficient knowledge and experience to undertake the work required. A national certification scheme has been developed for bodies that provide competent person services known as the RG2, UKAS – Accreditation for In-Service Inspection of Pressure Systems/Equipment who will accredit competent persons to EN 45004:1995.

This standard provides guidance to those requirements in BS EN ISO/IEC 17020:2004 *General criteria for the operation of various types of bodies performing inspection* which need interpretation when applied by Inspection Bodies carrying out in-service inspection of pressure systems / pressure equipment. BS EN ISO/IEC 17020:2004, as applied by UKAS in accordance with IAF/ILAC-A4: 2004.

The following criteria is provided by UKAS as suitable qualifications for competent persons:

Category 1 Chartered Engineer as defined by the Engineering Council or equivalent (eg appropriate degree with relevant experience, NVQ Level V Engineering) including at least 3 years experience within an engineering discipline associated with in-service inspection of pressure systems.

Category 2 Incorporated Engineer as defined by Engineering Council or equivalent (eg appropriate HNC with relevant experience, NVQ Level IV Engineering) including at least 5 years experience within a relevant engineering discipline of which at least one year* shall have been spent working within an engineering discipline associated with in-service inspection of pressure systems.

Category 3 Engineering Technician as defined by Engineering Council or equivalent (eg appropriate ONC with relevant experience, NVQ Level III) having a minimum of 5 years experience within a relevant discipline of which at least one year shall have been spent working within an engineering discipline associated with the in service inspection of pressure systems.

Category 4 Person trained in a relevant engineering discipline with a recognised and documented engineering apprenticeship with a minimum of 5 years experience within a relevant discipline of which at least one year shall have been spent working within an engineering discipline associated with the in-service inspection of pressure systems.

Category 5 Person with less than tradesman's apprenticeship but with a minimum of 5 years spent working with or within the industry associated with pressure systems and has general knowledge of pressure systems and its operating environment. Personnel shall be placed on recognised training courses with appropriate documented tests in in-service inspection of pressure systems. The minimum age for this Category is 21 years.

Persons in Categories 4 and 5 shall pass a qualifying test, established by the Inspection Body, associated with the particular inspection activities relating to pressure systems / equipment and this should cover relevant knowledge of the law, codes of practice and inspection techniques.

* Where a person meets the minimum requirement for a specific discipline and is to be trained in a second discipline, it may not be necessary to have experience of at least one year in the second discipline provided that the required competence can be demonstrated.

Written scheme of examination

Scope

[P7019] At the heart of the Regulations is the requirement for users of installed pressure systems or owners of mobile systems to have a 'suitable' written scheme of examination for every pressure system that they use or own. The user or the owner must ensure that the scheme is drawn up or certified as suitable by a competent person. The written scheme must specify:

- the parts to be examined (the scope);

- the maximum intervals between examinations (the frequency);
- the types of examinations necessary (the nature);
- any special measures necessary to prepare the system for safe examination.

Written schemes of examination may be held as electronic or other forms of data providing that the data is capable of being printed as a written copy and is kept secure from loss or unauthorised interference.

Responsibility for defining the scope of the written scheme of examination rests with the user or owner. The user or owner is likely to be most conversant with the equipment and relevant boundaries between systems. The competent person may be useful in assisting the user or owner to define the scope. The written scheme of examination must cover the relevant parts of the system which are specified by the Regulations as:

(a)   All protective devices.
(b)   Every pressure vessel and every pipeline in which a defect may give rise to danger.
(c)   Those parts of pipework where a defect may give rise to danger.

The user or owner should first identify and establish which parts of the pressure system are pressure vessels, protective devices or pipework as defined by the Regulations. The user or owner will then decide which parts of the system should be included for examination. Where the decision is made to exclude parts from regular examination, users and owners should take advice from a person with appropriate and relevant technical expertise and experience (who need not be the competent person) and be able to justify the decision.

A written scheme of examination must be prepared before the pressure system is operated for the first time. It will specify the examination necessary before the pressure system enters service. The user or owner with the competent person will take account of fabrication and installation quality and inspection reports.

The user or owner must ensure that the scheme is reviewed at appropriate intervals by a competent person to determine whether it is still suitable for the current conditions of use of the system. This normally occurs after every examination. If, as a result of the review, the competent person recommends modifications to the content of the scheme, then the user or owner must ensure that the scheme is modified.

Schemes of examination must be suitable for the purposes of preventing danger arising from the system. Whilst it is for the duty holders to decide how to meet the goal of suitability and whether the scheme achieves it, the ACoP gives detailed advice about the factors that should be considered.

Further explanation regarding the preparation of written schemes of examination can be found in the Safety Assessment Federation (SAFed) Guidelines for the Production of Written Schemes of Examination and the Examination of Pressure Vessels Incorporating Openings to Facilitate Ready Internal Access Ref: PSG4

Frequency of examination

[P7020] The written scheme will specify a suitable frequency of examination ie the maximum intervals between successive examinations. The ACoP and Guidance give advice about the factors that should be taken into account when deciding the frequency between examinations. Separate considerations apply to the initial examination before entry to service, the first examination after entering service, and the regular examinations thereafter.

The Regulations do not prescribe intervals between examinations and therefore allow flexibility and freedom to exercise good engineering judgement in each case. Different parts of the system may be examined at different intervals depending on the degree of risk associated with each part. Much will depend on the extent and degree of confidence in the knowledge that is available about the predicted condition of the equipment, degradation mechanisms and rates, the operating conditions and operating history.

Advice on the frequency of examinations is available. The *Factories Act 1961* sets limits on the intervals between examinations of boilers, steam receivers and air receivers. These limits are still relevant good practice. The Institute of Petroleum Model Code of Practice (Parts 1 and 2) provides advice on examination periods for chemical pressure vessels, heat exchangers and other equipment depending on the results of earlier examinations. The Safety Assessment Federation (SAFed), which represents the major engineering insurers, has published recommendations on the periodicity of examination of pressure equipment (*'Pressure Systems Guidelines on the Periodicity of Examinations'* (PSG1), available from the Safety Assessment Federation, ISBN 1901212106).

Protective device

Means a device that protects control or measuring equipment which is essential to prevent a dangerous situation from arising. Protective devices are often categorised as:

*   Category 1: Devices designed to protect the system against system failure eg safety valve;
*   Category 2: Devices that warn that a system failure could occur eg pressure gauge.

Examples of protective devices commonly found in use include:

For steam systems

*   Safety (or relief) valves;
*   Pressure gauges;
*   Water level gauges;
*   Low water level devices;
*   Pressure limiting devices;
*   High pressure cut out devices;
*   Flame failure protection systems;
*   Pressure reducing vales.

For compressed air systems

- Safety (or relief) valves;
- Fusible plugs;
- Bursting discs;
- Pressure gauges;
- Pressure limiting or unloading devices.

For bulk LPG installations

- Safety (or relief) valves;
- Pressure gauges;
- Bursting discs.

The HSE Guidance to the Regulations (L122) deals with the scheduling of examinations of several identical vessels on the same site. The overriding principle is that each vessel must be examined within the interval specified by the written scheme for that vessel. Within this limit, a form of staged examination may be used.

Nature of examination

[P7021] The written scheme must specify the nature of the examination. This may vary, depending on the system, from a simple visual examination to a requirement for the use of non-destructive testing methods. Pressure vessels within a system will need to be examined from the inside when their accessibility permits this. The Safety Assessment Federation (SAFed), have a number of publications covering specialist examinations that provide the explanation and guidance that the competent person needs to follow for:

- Guidelines for the examination of shell-to-endplate and furnace-to-endplate welded joints Ref: SBG1 (ISBN 1 901212 05);
- Guidelines for the examination of longitudinal seams of shell boilers Ref: SBG2 (ISBN 1 901212 30 0).

Some pressure systems will require an examination both with the system stripped down and out of service and with the system running under normal operating conditions so that the systems integrity and protective devices can be verified as operational to protect the equipment under fault conditions, eg Pressure relief valve lifts under an overpressure condition.

Other guidance documents produced by the Safety Assessment Federation (SAFed) that cover these areas of the thorough examination include:

- Pressure Systems: Guidelines on Periodicity of Examinations Ref: PSG1 (ISBN 1 901212 10 6);
- Guidelines for the Operation of Steam Boilers Ref: PSG2.

Examination

[P7022] The Regulations require users and owners to ensure that their pressure systems are examined by a competent person as specified by the written scheme of examination. [*Pressure Systems Safety Regulations 2000 (SI*

2000 No 128), Reg 9.] The examination must take place within the maximum time interval specified by the scheme. There is a clear duty on users/owners to ensure that the equipment is not operated/supplied beyond the date specified in the last examination report until it has been examined again.

Users and owners must ensure that systems are adequately prepared in time for examination and that appropriate precautions for safety are taken. Adequate advance notice should be given to the competent person.

On 29 June 2005 ConocoPhillips Limited was fined a total of £895,000 and ordered to pay £218,854 costs at Grimsby Crown Court, after pleading guilty to breaching health and safety legislation at an earlier hearing. The case follows an investigation by the HSE into two incidents: a fire and explosion at the Humber Refinery, South Killingholme, North Lincolnshire on 16 April 2001 and a release of liquefied petroleum gas (LPG) at the Immingham Pipeline Centre, Immingham Dock, on 27 September 2001.

At an earlier hearing at Grimsby Crown Court, ConocoPhillips pleaded guilty to seven breaches of *HSWA 1974* and the *Pressure Systems and Transportable Gas Containers Regulations 1989*.

The first incident happened on 16 April 2001 when 170 tonnes of highly flammable LPG was released from ConocoPhillips' (then Conoco Ltd) Saturate Gas Plant at its Humberside oil refinery. The gas cloud ignited causing a massive explosion and fire. As the fire burned it caused failures of other pipework resulting in another explosion and fireball.

The fire burned for approximately two–and–a–half hours. There were no serious injuries but considerable damage was caused to other processing plants and buildings on the refinery and to properties off–site. HSE's investigation found that the initiating event was the failure of a 15 cm diameter pipe at an elbow, due to corrosion and erosion. The most likely source of the ignition was a gas–fired heater in an adjacent processing unit.

During HSE's investigation a second incident occurred at the company's nearby Immingham Pipeline Centre. Over the night of 27 September 2001, approximately 16 tonnes of LPG leaked from a road tanker and the liquid pool and gas cloud dispersed without ignition. As a result, HSE launched a detailed investigation into the cause of the release.

The company pleaded guilty to seven breaches, the first six applying to its Humberside plant:

- Section 2(1) of HSWA 1974 in not ensuring the safety of its employees, for which it was fined £400,000;
- Section 3(1) of the same Act applying to non- employees. It was fined a further £400,000 on this charge;
- Four breaches of reg 9(1)(a) of the Pressure Systems and Transportable Gas Containers Regulations 1989 for not ensuring parts of the pressure system were examined by a competent person and for which it was fined £5000 on each charge; and
- Section 2(1) of HSWA 1974 in not ensuring the safety of employees at the Immingham Pipeline Centre, for which it was fined £75,000.

After the hearing, Kevin Allars, head of HSE's Chemical Industries Division, said:

The incident at the Humber refinery was possibly the most serious chemical incident in Britain since the Flixborough disaster in 1974 and it is fortunate that there were no deaths or very serious injuries. This was mainly because the incident occurred on a Bank Holiday and during a shift change when the limited staff on site were away from the plant. The potential for loss of life was great. However, the extent of the damage to the site and to properties in the nearby village of South Killingholme indicates the violent nature of the explosion. The severity of the events at the Humber Refinery have been reflected in the penalties imposed by the court today.

Allars added: 'Our investigation revealed a systematic failure by the company to inspect the pipework in certain parts of the refinery. It is vital that companies who operate high–hazard sites – such as oil refineries and chemical plants – put rigid and robust systems in place for inspecting pipework to detect corrosion or other defects.'

There is provision to postpone the due date of examination once, by agreement in writing between the user or owner and the competent person. The date of examination may be postponed to a later date provided that the following conditions are satisfied:

(a)   The postponement does not give rise to danger.

(b)   Only one postponement is made for any one examination.

(c)   Such a postponement is notified in writing by the user or owner to the enforcing authority before the date set for the next examination in the previous report.

Where the user or owner is the competent person, such an agreement is unnecessary, but the notification to the enforcing authority must contain a declaration that postponement will not give rise to danger. Any postponement must be agreed by the Competent Person who undertook the previous examination. No periods of deferment are quoted within the Regulations, ACoP and Guidance (L122) but when justified the Competent Persons are unlikely to agree more than 25% of the frequency period.

The competent person must take responsibility for all examinations. Often the competent person will carry out the examination, but a user or owner and the competent person may agree that some examinations are carried out by specialist non-destructive testing technicians who are not classed as competent persons within the meaning of the Regulations. The competent person will accept responsibility for ensuring that the examinations are carried out properly according to the scheme and for the results, interpretation and reporting.

Fitness for service

[P7023]   After obtaining the results of the examination, the competent person will make an assessment of the continued fitness for service of the system. If the competent person is not satisfied about continued fitness for service under the current condition or safe operating limits of the system, then s/he will specify in the report any repairs, modifications, or changes to the safe operating limits deemed necessary.

The competent person will then, having regard to the assessment of continued fitness for service, specify the next date, within the limits set by the written scheme, beyond which the system may not be operated without a further examination. The competent person will also consider whether the written scheme remains suitable or whether it should be modified.

Examination report

[P7024] The Regulations require the competent person to make a written report of the examination and to sign or otherwise authenticate it. [*Pressure Systems Safety Regulations 2000 (SI 2000 No 128), Reg 9.*] The competent person must send the report to the user or owner as soon as is practicable after completing the examination (or for integrated systems, the last examination in the series). In any event, the report must arrive within 28 days of the examination (or the last in the series). The report must also arrive before the date specified by the previous examination report. Where the user or owner is also the competent person, the report must also be made by these times.

The examination report must contain the following information:

(a)   A statement of the parts examined, the condition of those parts and the results of the examination.

(b)   Specification of any repairs, modifications or changes to the operating limits necessary to prevent danger or ensure the continued effective working of protective devices. The date by which any repairs, modifications or changes are to completed must also be specified. The user or owner must ensure that the system is not operated after the date specified for completion of the repairs, modifications or changes until these have been made.

(c)   The next date (within the limits of the scheme of examination) after which the pressure system may not be operated without a further examination under the scheme. The owner of a mobile system must ensure that the date of the next examination is legibly and durably marked on the system and is clearly visible.

(d)   A statement of whether the current written scheme of examination is suitable for preventing danger or whether it should be modified (with reasons).

Imminent danger

[P7025] After the examination of a pressure system and assessment of fitness for service, the competent person must consider whether the condition of the system, or any part thereof, is such that continued operation in its current condition would give rise to imminent danger. If this is the case, then the competent person will immediately make a written report. [*Pressure Systems Safety Regulations 2000 (SI 2000 No 128), Reg 10.*] The report will specify the system and the parts in dangerous condition, and the repairs, modifications or changes to the operating conditions needed in order to make the system safe from imminent danger. A pressure system in a state of imminent danger usually points to wider failings in general management of the user's/owner's organisation.

The competent person will give the report to the user of an installed system or the owner and user of a mobile system. In addition, within 14 days of the examination, the competent person will send a report to the enforcing authority for the place where the pressure system is situated. This will normally be the Health and Safety Executive or the Environmental Health Department of the local authority.

On receipt of this report, the user of an installed system must ensure that the system (or the part effected) is not operated until the repairs, modifications or changes have been carried out. Owners of mobile systems must take all practicable steps to ensure that the system is not operated. Where the competent person is the user or owner, s/he must take responsibility to prepare the report, inform the enforcing authority and ensure that the system is not operated until the repairs, modifications or changes have been carried out.

Operations, maintenance, modification and repair

Safe operation

[P7026] Users and owners have a duty to provide adequate and suitable operating instructions for the safe operation of the system and the action to be taken in an emergency. Users must ensure that their pressure systems are not operated except in accordance with the instructions.

Users of pressure systems must provide adequate training to persons who operate the systems as required by the *Provision and Use of Work Equipment Regulations 1998 (SI 1998 No 2306), Reg 9.*

Precautions to prevent overpressure

[P7027] Users have a duty to take precautions to prevent overpressure to vessels. [*Pressure Systems Safety Regulations 2000 (SI 2000 No 128, Reg 15.*] The Regulation applies to vessels constructed with a permanent outlet to the atmosphere (or to a space where the pressure does not exceed atmospheric pressure) which could become pressurised if the outlet were obstructed. In this case, the user must ensure that the outlet is kept open at all times and is free from obstruction when the vessel is in use.

Maintenance

[P7028] Users and owners must ensure that their systems are properly maintained so as to prevent danger. [*Pressure Systems Safety Regulations 2000 (SI 2000 No 128), Reg 12.*] This should not be confused with the requirement for examinations under the written scheme. Maintenance might include the periodic replacement of parts, flange tightening, and checks on vulnerable areas of pipework systems not included for examination under the written scheme. Advice on maintenance programmes is given in the ACoP.

Systems and equipment covered by the regulations

The HSE has warned of the need for effective operation, inspection and maintenance regimes to ensure the safe use of high-pressure equipment.

Example 1

The warning comes after Giancarlo Coletti, 48, a worker at Clariant Life Science Molecules (UK) Limited's plant in Sandycroft, Flintshire, sustained serious injuries to his right arm in October 2003. The accident happened when the clamping system on a pressure vessel lid he was operating failed, causing the lid to fly off and hit him.

Clariant Life Science Molecules (UK) Limited was fined £100,000, split equally between *Regulations 11* and *12* of the *Pressure Systems Safety Regulations 2000 (SI 2000 No 128)*, with £24,474 costs, at Mold Crown Court on 17 December 2004 after pleading guilty at an earlier hearing. The company admitted failing to provide workers operating the pressure vessel with adequate and suitable instructions for its safe operation and also failing to ensure the vessel was adequately maintained.

HSE Inspector, Dr Stuart Robinson, who investigated the incident, said:

> Our investigation revealed serious deficiencies in Clariant's safe systems of work. Opening and closing of the lid was regarded as a simple process, carried out three times a day, but the hazards had been overlooked. Although the clamping system was designed with a substantial margin of safety, it had been allowed to deteriorate to such an extent that the risk of injury became unacceptably high.

In particular, the HSE's investigation found that Clariant had failed to put in place adequate operating procedures to ensure the system was used correctly. For example, clamps were regularly over-tightened, occasionally causing them to break and the system was allowed to operate with less than its full complement of eight clamps.

In addition, at the time of the accident, one of the clamps was missing and others showed excessive wear and tear, or inadequate repair. Furthermore, it had become common practice for leaks to be nipped up with the system under pressure because the operating procedures failed to state that the system should be depressurised first.

Robinson added:

> The investigation also revealed that although Clariant had arranged for an independent competent person to examine the pressure system periodically, this was insufficient due to the frequent operation of the clamps, and the high level of wear and tear they showed. Instead, Instead, the firm should have introduced a more frequent system of inspection and maintenance.

Example 2

On 12 June 2007, a 44-year-old Loughborough man employed at Authentic World Cuisine Limited, which produced a range of sachet Asian meals for supermarkets, sustained multiple injuries when the door of an autoclave (known as a retort, or pressure cooker) exploded under pressure midway through a cooking cycle.

The detached circular steel door struck him and sent him eight metres across the factory causing multiple fractures. He also sustained burns from the hot contents of the vessel. As a result of the incident, his leg was amputated, he spent more than five months in hospital and is now looked after at home whilst he rehabilitates.

The company which manufactured at Castle Business Park, Pavilion Way, was fined £4,000, and ordered to pay this within 28 days*, at Loughborough Magistrates Court today, after pleading guilty to breaching Regulation 12 of the Pressure Systems Safety Regulations 2000 for failing to ensure that the company's pressure vessel (autoclave) was maintained in a state of good repair.

Roger Amery one of HM Inspectors who deals with Leicestershire said:

> "Major injuries in the food industry arising from the sudden or explosive release of pressure are, I am pleased to say, infrequent events. But this incident shows just how serious the consequences can be when things do go wrong and just how much energy is contained within industrial pressure systems. There might easily have been one or several fatalities in this explosion. The injuries that did result were very severe with long term consequences for the man who had the misfortune of being in the path of this door when the system exploded."

> "I hope this will serve as a reminder to employers and to managers who have specific responsibility for plant and equipment, that as well as ensuring periodic statutory examinations are made, it is equally important to ensure that all maintenance on pressure vessels is undertaken competently and that the protection systems are kept in good working order at all times."

Modification and repair

[P7029] Employers of persons who modify or repair pressure systems have to duty to ensure that the manner in which modifications or repairs are carried out does not give rise to danger, or otherwise impair the operation of any protective devices or inspection facility. [*Pressure Systems Safety Regulations 2000 (SI 2000 No 128), Reg 13).*]

Record keeping

[P7030] The user of an installed system or the owner of a mobile system must keep the originals or copies of the following documents:

(a)   The report of the last examination made by the competent person.

(b)   Reports of previous examinations if they contain information that can assist in assessing whether the system is safe to operate or whether any repairs can be carried out safely.

(c)   Any documents where reports of the last or previous examinations are recorded as data.

(d)   Any agreement, declaration or notification made in relation to the postponement of the date of the next examination.

[*Pressure Systems Safety Regulations 2000 (SI 2000 No 128), Reg 14.*]

For an installed system, these documents must be kept at the premises where the system is installed or at other premises approved by the enforcing authority (normally the Health and Safety Executive). For a mobile system, the documents must be kept at the premises from which the deployment of the system is controlled.

Where the user or owner of a pressure system changes, the previous user or owner must give to the new user or owner all originals or copies of documents relating to the system as soon as practicable.

Related health and safety legislation

[P7031] The Regulations are concerned with the hazard of stored energy. The control of hazards associated with the release of hazardous substances, for example toxic or flammable materials, from containers and systems falls under the *Control of Major Accident Hazards Regulations 1999 (COMAH) (SI 1999 No 743)* as amended by *The Control of Major Accident Hazards (Amendment) Regulations 2005 (SI 2005 No 1088)*. Controls to prevent the release of hazardous materials cover the management of storage systems, containers and tanks.

Other legislation relating to pressure equipment includes:

* *The Carriage of Dangerous Goods and Use of Transportable Pressure Equipment Regulations 2004*;
* *Pipelines Safety Regulations 1996 (SI 1996 No 825)*.

The main legislation covering health and safety at work in general is:

* *Health and Safety at Work etc Act 1974*;
* *Workplace (Health and Safety) Regulations 1992 (SI 1992 No 3004)*;
* *Management of Health and Safety at Work Regulations 1999 (SI 1999 No 3242)*;
* *Provision and Use of Work Equipment Regulations 1998 (SI 1998 No 2306)*.

Product Safety

Alison Newstead and Natalie Wood

Introduction

[P9001] With the expansion of EU directives and other legislation aimed at manufacturers, designers, importers and suppliers, product safety emerges as a fast growth area. Indeed, there has been an identifiable trend towards placing responsibilities on producers and those involved in commercial circulation as well as on employers and occupiers (see MACHINERY SAFETY and PERSONAL PROTECTIVE EQUIPMENT).

Although there is no question of removal of duties from employers and occupiers and users of industrial plant, machinery and products, there is a realisation that the *sine qua non* of compliance on the part of employers and users with their safety duties is compliance with essential health and safety requirements on the part of designers and manufacturers. This trend is likely to continue with more products being brought within the scope of the EU's 'new approach' regime which promotes supply of products throughout the single market provided they meet essential requirements which can be demonstrated by compliance with harmonised European standards. As far as civil liability is concerned, this trend may well result in a long-term shift in the balance of liability for injury away from employers and occupiers towards designers, manufacturers and other suppliers.

It was thalidomide that first focused serious attention on the legal control of product safety, leading to the *Medicinal Products Directive (65/65/EEC)* and the *Medicines Act 1968*, the first comprehensive regulatory system of product testing, licensing and vigilance. Since then a host of regulations has appeared, covering areas as diverse as pencils, aerosols, cosmetics and electrical products. The General Product Safety Directive (92/59/EEC), introduced the general concept that producers of consumer goods must place only safe products on the market and undertake appropriate post-marketing surveillance. The significance of the introduction of these Regulations lies not in their substantive requirements, but rather in the completion of an all-embracing consumer product safety regime across the European Union which covers any gaps in the consumer protection network left by existing regulatory requirements.

The General Product Safety Directive was amended and re-issued as Directive 2001/95/EC and is implemented in the UK by way of the *General Product Safety Regulations 2005 (SI 2005 No 1803)*. The 2005 Regulations came into force in the UK on 1 October 2005 and incorporate even more post-marketing obligations and powers for enforcing authorities in respect of consumer products, even those already covered by sector-specific regulatory requirements.

In contrast, the regulation of products for use in the workplace has developed along separate lines (although there is an increasing overlap by virtue of the 'new approach' directives, such as those for machinery or electromagnetic compatibility, which apply uniform requirements for consumer and workplace products albeit with separate enforcement regimes). Under the *Health and Safety at Work etc. Act 1974 (HSWA 1974)*, s 6 all articles and substances for use at work are required to be as safe as is reasonably practicable, and manufacturers and other suppliers are under obligation to provide information and warnings for safety in use. Various product-specific regulations have also been made under *HSWA 1974*, for example the *Chemicals (Hazard Information and Packaging for Supply) Regulations 2002 (SI 2002 No 1689)*.

As far as commercial considerations are concerned, quality and fitness for purpose of products has been an implied statutory term in contracts for sale of goods in legislation stemming from the late nineteenth century and which is now found in the *Sale of Goods Act 1979*, as amended by the *Sale and Supply of Goods Act 1994*, the *Sale of Goods (Amendment) Act 1995*. The 1994 amendments included a more focused definition of the standard of *satisfactory quality* which a supplier is obliged by contract to provide, the criteria for which now explicitly includes safety, durability, and freedom from minor defects. The *Sale and Supply of Goods to Consumers Regulations 2002 (SI 2002 No 3045)* widened the way in which satisfactory quality is to be determined in consumer contracts, by extending the definition of satisfactory quality to include any public statement on the specific characteristics of the goods, particularly in advertising or labelling. The ability to exclude these statutory terms, or insist that buyers waive the rights they have, is now strictly curtailed by the *Unfair Contract Terms Act 1977* ('UCTA') (banning exclusion of the most significant rights in consumer contracts, and subjecting exclusions in standard form commercial contracts to a statutory test of 'reasonableness'). The *Sale and Supply of Goods to Consumers Regulations 2002 (SI 2002 No 3045)* also amended UCTA such that individuals are treated as consumers even if the goods supplied or hired by them are not of a type ordinarily supplied for private use or consumption. In addition, and based on the Directive on Unfair Terms and Consumer Contracts (93/13/EEC), there are now other restrictions contained in the *Unfair Terms in Consumer Contracts Regulations 1999 (SI 1999 No 2083)* imposing a requirement of fairness on most terms in consumer contracts which are not individually negotiated. These Regulations also lay down a requirement for suppliers to ensure that terms are expressed in plain, intelligible language. The most important restriction of all for present purposes in these two sets of Regulations is the effective ban on notices and contract terms seeking to exclude liability for death or personal injury resulting from negligence, a prohibition which is reinforced in relation to defective products by the *Consumer Protection Act 1987*, s 7.

This section deals with criminal and civil liabilities for unsafe products for consumer and industrial use as well as the duties imposed on sellers and suppliers of consumer products and after-sales personnel as far as contract law is concerned.

Consumer products

General Product Safety Regulations 2005

Duties of producers and distributors

[P9002] The *General Product Safety Regulations 2005 (SI 2005 No 1803)* lay down general requirements concerning the safety of products intended for or likely to be used by consumers. The Regulations implement the provisions of the General Product Safety Directive 2001/95/EC which replace in their entirety the 1992 General Product Safety Directive 92/59/EEC (and therefore the General Product Safety Regulations 1994) and significantly increase the obligations on producers and distributors which were formally in place under the 1994 regime.

The 2005 Regulation defines the scope of a consumer product as:

> . . . a product which is intended for consumers or likely, under reasonably foreseeable conditions, to be used by consumers even if not intended for them, and which is supplied or made available, whether for consideration or not, in the course of a commercial activity, and whether new, used or reconditioned and includes a product which is supplied or made available to consumers for their own use in the context of providing a service [. . .].

The scope of the 2005 Regulations extends the previous definition of 'product' found under the 1994 Regulations in that:

(1) It encompasses products which are made available to consumers, as well as supplied to them. This is likely to include many products provided in beauty centres, theme parks or playgrounds or in connection with transport, sporting events or health treatments.

(2) It includes products supplied or made available in the context of providing a service. This is likely to include repairs and after-sales service, maintenance and cleaning work, hotel and restaurant services, health treatment, provision of gas and electricity etc. or a tyre inflation gauge and compressed air pump at a garage.

Products used solely in the workplace are not covered by the 2005 Regulations, but the fact that a product has a wider commercial application will not prevent it being a consumer product for the purposes of these Regulations, if it is supplied to consumers or it is likely, under reasonably foreseeable conditions to be used by consumers, even if not intended for them.

The Directive makes it clear that, although business products are not included *per se*, the Directive does include products which are designed exclusively for professional use but have subsequently migrated to the consumer market, such as laser pens. This is supported by Guidance produced by the UK Department of Trade and Industry ('Guidance for Businesses, Consumers and Enforcement Authorities' – August 2005). The Guidance states that the 2005 Regulations cover products which were originally designed and intended for professional use, but which subsequently 'migrate' onto the consumer market (e.g. power tools). In such circumstances, where professional products become available to consumers, the 2005 Regulations will apply (to the extent that other specific legislation does not apply). It should be noted that the DTI have indicated that 'migration' does not necessarily mean that a product is unsafe for a consumer, but if it is foreseen that such migration may occur, instructions for consumer use and warnings of non-obvious risks must be provided. On the other hand, the DTI guidance also states that in circumstances in which a professional product could never be safely used by a consumer supply should be strictly controlled, and labeling a product 'strictly for professional use only' is unlikely to be sufficient on its own.

All manner of products are potentially caught by the Regulations, including clothing, primary agricultural products, DIY equipment and motor vehicles.

The Regulations do not apply to:

— second-hand products which are antiques;
— second-hand products supplied for repair or reconditioning before use, provided that the purchaser is clearly informed accordingly;
— products exported direct by the UK manufacturer to a country outside the European Union; or
— products which are used by service providers themselves to supply a service to consumers, in particular equipment on which they ride or travel, which is operated by a service provider.

The Regulations are disapplied in relation to any product where there are specific provisions in European law governing *all aspects* of the safety of a product. Where there are product-specific provisions in European law which do not cover all aspects, the 2005 Regulations apply to the extent that specific provision is not made by European law.
  [General Product Safety Regulations 2005 (SI 2005 No 1803, Reg 3(1).]

The main requirements of the Regulations are that:

(a)   no producer may place a product on the market unless it is safe *[General Product Safety Regulations 2005 (SI 2005 No 1803), Reg 5(1)]*;

(b)   producers provide consumers with relevant information, so as to enable them (i.e. consumers) to assess risks inherent in products throughout the normal or foreseeable period of use, where such risks are not immediately obvious without adequate warnings, and to take precautions against those risks; *[General Product Safety Regulations 2005 (SI 2005 No 1803), Reg 7(1)(a) and (b)]*;

(c)   producers update themselves regarding risks presented by their products and take appropriate action (if necessary recall and withdrawal), for example by:
  (i)   identifying products/batches of products by marking;
  (ii)   sample testing;
  (iii) investigating and, if necessary, keeping a register of complaints concerning the safety of a product;
  (iv)   keeping distributors informed of the results of such monitoring where a product presents a risk or may present a risk;
         [General Product Safety Regulations 2005 (SI 2005 No 1803), Reg 7(3), (4);]

(d) in order to enable producers to comply with their duties, distributors must act with due care and in particular:

    (i) must not supply products to any person which are known or presumed (on the basis of information in their possession and as professionals) to be dangerous products, and

    (ii) must participate in monitoring safety of products placed on the market – in particular, by passing on information on product risks, keeping (and if necessary, producing) the documentation necessary for tracing the origin of the product and co-operating in action taken by a producer or an enforcement authority to avoid risks.

[General Product Safety Regulations 2005 (SI 2005 No 1803), Reg 8(1)(a), (b).]

The 2001 Directive introduced a significant additional obligation on producers and distributors to notify the relevant competent authorities immediately if they know or ought to know, on the basis of information in their possession and as professionals, that a product they have placed on the market poses risks to the consumer that are compatible with the general safety requirement, i.e. that it is dangerous, and of any action taken to prevent such risks [General Product Safety Directive 2001/95/EC, Article 5.3].

Both the European Commission and the DTI in the UK have produced guidance in respect of the new notification obligation. The EU Guidance (*'Guidelines for the Notification of Dangerous Consumer Products to the Competent Authorities of the Member States by Producers and Distributors in accordance with Article 5(3) of Directive 2001/95/EC'*) contains simple and clear criteria on notifications (and criteria as to how to determine whether notification is necessary), together with the content and a standard form for notifications.

The DTI has also produced useful guidance in September 2005 *(The General Product Safety Regulations 2005 – Notification guidance for producers and distributors).*

The notification provisions require producers and distributors to put in place sophisticated procedures to capture and assess safety information. Deciding whether new information means that a product is no longer legally safe can be a major issue in practice and is best undertaken against a background of previous safety data and an assessment that the product's safety is acceptable. Companies' systems may now need to provide for obtaining independent expert advice on these issues from engineers and lawyers. For certain products, at least, it is now necessary to undertake a documented risk analysis before a product is placed on the market, and subsequently to keep it up to date (e.g. with monitoring and sample testing), so as to define the level of anticipated risks with a product's use, to determine that these are acceptable and that the general safety requirement is met, and to provide a background and baseline statement of acceptable risks against which any increase in risk may be assessed.

The obligation to notify applies to both the producer and the distributor, although discussion between the producer and the distributor is encouraged when a potential safety issue arises to avoid duplication of notifications.

The 2005 Regulations stipulate that written notification to the requisite authority should be made 'forthwith' [*General Product Safety Regulations 2005 (SI 2005 No 1803) Reg 9(1)*]. Guidelines produced by the Department of Trade and Industry in September 2005 (*'The General Product Safety Regulations 2005 – Notification guidance for producers and distributors'*) state that producers and/or distributors must make a notification:

(i)   in general, within ten calendar days of concluding, on the basis of a risk assessment, that the product is unsafe;

(ii)   no later than three calendar days of concluding that the product is unsafe and the risk posed is serious.

These guidelines do not have the force of law in the UK and producers and distributors should adhere to the (rather ambiguous) provisions set out in the Regulations which suggest that notification should be made 'without delay'.

Notification in the UK should be made to the home authority (i.e. the local authority for the area where the decision making function of the business is located), and by way of a standard form (as prescribed by the Guidance produced by the European Commission and adopted by the DTI). Notification should not be delayed on the basis that the form is incomplete. Incomplete notifications can be supplemented with information as and when it is received.

Notification is not required in respect of isolated circumstances or products [*General Product Safety Regulations 2005 (SI 2005 No 1803) Reg 9(2)(b)*].

Producers and distributors are now required to co-operate with the competent authorities, at the request of the latter, on action taken to avoid risks posed by products that they supply or have supplied [*General Product Safety Regulations 2005 (SI 2005 No 1803) Reg 9(4)*].

In addition, producers (as part of their duty to update themselves regarding risks and to take appropriate action) are obliged to take appropriate action when necessary including withdrawal, adequately and effectively warning customers as to the risks, or as a last resort recall. (Thus the recall obligation has now been extended beyond the chain of distribution and includes products already supplied or made available to the consumer/user.)
   [*General Product Safety Regulations 2005 (SI 2005 No 1803) Reg 7(1)(b).*]

Recalls are only required as a last resort where other measures would not suffice to prevent the risks involved, where the producer considers it necessary or where they are obliged to do so by a Member State. Practical issues which may arise include: the ability of a system adequately to contact users over safety issues; to measure the number and percentage of products returned in a recall; and decide what level of returns would be acceptable. UK guidance may assist to an extent (see *Consumer Product Recall: A Good Practice Guide* (1999) Department of Trade and Industry and *Product Safety in Europe: A Guide to Corrective Action including Recalls* produced by Prosafe/UNICE/Euro Commerce and BEUC).

Complementary to the obligation to recall products from consumers is a power for enforcement authorities to serve suspension notices on producers (in order to prevent the supply of products whilst safety evaluations are being carried out), to require producers to mark products which could pose risks in certain conditions, to warn consumers of risks that products may post and to order or organise the immediate withdrawal of any dangerous product already on the market and alert consumers to the risks it presents and to order or co-ordinate or, if appropriate, to organise with producers and distributors its recall from consumers and its destruction in suitable conditions [*General Product Safety Regulations 2005 (SI 2005 No 1803) Regs 11–15*].

The revised Directive also includes a provision to prohibit the export from the European Community any products which have been the subject of a European Commission-initiative decision (after consulting the Member States and, whenever it proves necessary also a Community scientific committee) that requires Member States to take measures in relation to that particular product on the grounds that it presents a serious risk (unless the decision specifically provides that the products may still be exported outside the Community).

Presumption of conformity with safety requirements

[P9003] A 'safe' product is a product which, under normal or reasonably foreseeable conditions of use (as well as duration), does not present any risk (or only minimal risks) compatible with the product's use, considered as acceptable and consistent with a high level of protection for the safety and health of consumers, with reference to:

(a)   product characteristics (e.g. composition, packaging, instructions for assembly and, where applicable, installation and maintenance instructions);
(b)   the effect on other products, where use with other products is reasonably foreseeable;
(c)   product presentation (i.e. labelling, any warnings and instructions for use and disposal and any other indication or information concerning the product); and
(d)   categories of consumers at serious risk – in particular, children and the elderly.

[General Product Safety Regulations 2005 (SI 2005 No 1803), Reg 2.]

Where a product conforms with specific UK safety requirements (e.g. the *Plugs and Sockets etc (Safety) Regulations 1994 (SI 1994 No 1768)*) there is a presumption that the product is safe, until the contrary is proved. In addition, where a product conforms to a voluntary national standard in the UK, giving effect to a European Standard (which has been published in the Official Journal of the European Union in accordance with the 2001 Directive) the product is presumed to be safe so far as concerns the risks and categories of risk covered by that national standard [*General Product Safety Regulations 2005 (SI 2005 No 1803) Reg 6(1), (2)*].

In the absence of these two particular circumstances where a presumption of safety exists, the assessment of the safety of a product is to be made by taking into account:

(i)   voluntary national standards implementing European standards other than those listed in the Official Journal with reference to the General Product Safety Directive;
(ii)   standards drawn up in the Member State in which the product is marketed;

(iii)   European Commission recommendations setting guidelines on product safety assessment;

(iv)   product safety codes of good practice in the relevant sector;

(v)   state of the art and technology; and

(vi)   the safety that consumers may reasonably expect.

[General Product Safety Regulations 2005 (SI 2005 No 1803), Reg 6.]

Market surveillance and enforcement

[P9004] The 2001 Directive increased the strength of market surveillance and enforcement provisions and there are now additional obligations on Member States. The European Commission also has an obligation to promote and take part in the operation of a European network of authorities of the Member States.

The obligations and enforcement powers of Member States have been expanded to include the following elements:

(a)   a detailed definition of the tasks, powers, organisation and cooperation arrangements for market surveillance of competent authorities;

(b)   establishment of sectoral surveillance programmes by categories of products or risks and the monitoring of surveillance activities, findings and results;

(c)   the follow-up and updating of scientific and technical knowledge concerning the safety of products;

(d)   periodic review and assessment of the functioning of the control activities and, if necessary, revision of the approach and organisation of surveillance;

(e)   procedures to receive and follow-up complaints from consumers and others on product safety, surveillance or control activities;

(f)   exchange of information between Member States on risk assessment, dangerous products, test methods and test results, recent scientific developments and other aspects relevant for control activities;

(g)   joint surveillance and testing projects between surveillance authorities;

(h)   exchange of expertise and best practices, and cooperation in training activities; and

(i)   improved collaboration at Community level on tracing, withdrawal and recall of dangerous products.

[General Product Safety Directive 2001/95/EC, Articles 6–10]

An extended list of enforcement powers is included in the 2001 Directive and includes the power for a Member State to order or organise the issuance of warnings about, or recall of, dangerous products [General Product Safety Directive 2001/95/EC, Article 8.1].

Other significant changes

[P9005] Other significant changes introduced by the 2001 Directive can be summarised as follows:

(1)   Member States are required to take due account of the precautionary principle when they are taking enforcement measures. The precautionary principle is that Member States should use a precautionary approach to risks where there is scientific uncertainty as to the level of risk *[General Product Safety Directive 2001/95/EC, Article 8.2].*

(2)   Information which is available to the authorities relating to consumer health and safety, in particular on product identification, the nature of any risk and on measures taken, shall in general be available to the public, in accordance with the requirements of transparency. However, information which is obtained by the authorities shall not be disclosed if the information, by its nature, is covered by professional secrecy in duly justified cases.
[General Product Safety Directive 2001/95/EC, Article 16.1].

Offences, penalties and defences

[P9006] Contravention of *Regs 7(1), 7(3), 8(1)(b) or 9(1) of the 2005 Regulations (regarding provision of information to consumers, risk assessment, monitoring and notification)* is an offence, carrying, on summary conviction, a maximum penalty of:

(a)   imprisonment for up to three months, or
(b)   a fine of £5,000, or
(c)   both.

Contravention of *Regs 5 or 8(1)(a) of the 2005 Regulations* (offering or agreeing to supply or supplying an unsafe product) is an offence and liable to conviction, on indictment to:

(a)   imprisonment for a term not exceeding 12 months, or
(b)   a fine not exceeding £20,000; or
(c)   both; or

on summary conviction: (*a*) a term not exceeding 3 months; or (*b*) a fine not exceeding £5000; or (*c*) both.
[General Product Safety Regulations 2005 (SI 2005 No 1803), Regs 20(2)].

Offence by another person

[P9007] As with breaches of *HSWA* and similarly-oriented legislation, where an offence is committed by 'another person' in the course of his commercial activity, that person can be charged, whether or not the principal offender is prosecuted. Similarly, where commission of an offence is consented to or connived at, or attributable to neglect, on the part of a director, manager or secretary, such persons can be charged in addition to or in lieu of the body corporate. [*General Product Safety Regulations 2005 (SI 2005 No 1803), Reg 31.*]

Defence of 'due diligence' – and exceptions

[P9008] It is a defence for a person charged under these Regulations to show that he took all reasonable steps and exercised all due diligence to avoid committing the offence. [*General Product Safety Regulations 2005 (SI 2005 No 1803), Reg 29(1).*] (See further ENFORCEMENT.)

The exceptions to the above are:

(a) Where, allegedly, commission of the offence was due to:
  (i) the act or default of another, or
  (ii) reliance on information given by another,
  a person so charged cannot, without leave of the court, rely on the defence of 'due diligence', unless he has served a notice, within, at least, seven days before the hearing, identifying the person responsible for the commission or default or who gave the information. [*General Product Safety Regulations 2005 (SI 2005 No 1803), Reg 29(2), (3).*]

(b) A person so charged cannot rely on the defence of 'information supplied by another', unless he shows that it was reasonable in all the circumstances for him to have relied on the information. In particular, whether he took reasonable steps to verify the information or whether he had any reason to disbelieve the information. [*General Product Safety Regulations 2005 (SI 2005 No 1803), Reg 29(4).*]

A mere recommendation on the part of an importer that labels should be attached to boxes by retailers does not constitute 'due diligence' for the purposes of *Reg 29(1)*. (In *Coventry County Council v Ackerman Group plc* [1995] Crim LR 140, an egg boiler imported by the defendant failed to contain instructions that eggs should be broken into the container and yolks pricked before being microwaved. The defendant had learned of the problem and had printed instructions which were sent to all retailers, recommending that they be fixed to the boxes.)

Compliance with recognised standards (such as British Standards) will not amount to the defence of due diligence if a product is nevertheless unsafe for the user. In *Whirlpool (UK) Ltd and Magnet Ltd v Gloucestershire County Council* (1993) 159 JP 123 cooker hoods which were intrinsically safe and which met applicable standards for the purposes of the *Low Voltage Electrical Equipment (Safety) Regulations 1989* failed to meet the general safety requirement contained in the then applicable consumer protection legislation because they were liable to result in fires when used in conjunction with certain gas hobs.

Industrial products

[P9009] The *HSWA 1974* was the first Act to place a *general* duty on designers and manufacturers of industrial products to design and produce articles and substances that are safe and without health risks when used at work. Prior to this date legislation had tended to avoid this approach, e.g. the *Factories Act 1961*, on the premise that machinery could not be made design safe (see MACHINERY SAFETY). Statutory requirements had tended to concentrate on the duty to guard and fence machinery, and with the placement of a duty upon the user/employer to inspect and test inward products for safety. There is a general residual duty on the employers/users of industrial products to inspect and test them for safety under *HSWA 1974, s 2*.

These duties notwithstanding, the trend of legislation in recent years has been towards safer design and manufacture of products for industrial and domestic use. Thus, *HSWA 1974, s 6* as updated by the *Consumer Protection Act 1987, Sch 3* imposes general duties on designers, manufacturers, importers and suppliers of products ('articles and substances') for use at work, to their immediate users. Contravention of s 6 carries with it a maximum fine, on summary conviction, of £20,000 or an unlimited fine in the Crown Court (see ENFORCEMENT). In addition, a separate duty is laid on installers of industrial plant and machinery. However, because of their involvement in the key areas of design and manufacture, more onerous duties are placed upon designers and manufacturers of articles and substances for use at work than upon importers and suppliers, who are essentially concerned with distribution and retail of industrial products. However, under *HSWA 1974, s 6(8A)* importers are made liable for the first time for the faults of foreign designers/manufacturers.

There has been a recent development in case law regarding the interaction between *HSWA 1974*, European directives concerning the placing on the market and putting into service of certain product types, and the UK legislation which implements the European directives. In the House of Lords case of *R (on the application of Junttan Oy) v Bristol Magistrates' Court* [2003] UKHL 55, [2004] 2 All ER 555, concerning a judicial review of a decision of the Magistrates' Court involving a prosecution brought by the HSE based on offences under *HSWA 1974, ss 3* and 6 regarding machinery, the House of Lords ruled that in this particular instance (where it was questioned whether the HSE should have brought a prosecution based on *HSWA 1974, s 6* where the equipment's Safety requirements where covered by the *Supply of Machines (Safety) Regulations 1992, Reg 29(a)* (implementing the Machinery Directive 98/37/EC)), there was nothing in the Regulations to prevent the HSE bringing a prosecution under *HSWA 1974, s 6* instead and nothing in the Directive that affected the pre-existing rights of Member States to take action against machinery that is believed to be unsafe. It should be noted that fines are higher under the *HSWA 1974* than that relevant Regulations, and prosecutions under the Regulations are not subject to time limits.

Regulations made under the Health and Safety at Work etc Act 1974

[P9010] Various safety regulations are made under the umbrella of *HSWA 1974* and these often implement European health and safety directives. The *Provision and Use of Work Equipment Regulations 1998 (SI 1998 No 2306)* are such Regulations which implemented European Directive 89/655/EEC.

The 1998 Regulations place obligations on employers and on certain persons having control of work equipment, or of persons who use or supervise or manage the use of work equipment or of the way in which the equipment is used. A key requirement is to ensure that work equipment is maintained in an efficient state, in efficient working order and in good repair. [*Provision and Use of Work Equipment Regulations 1998 (SI 1998 No 2306), Reg 5 (1)*]. It was confirmed in *Stark v The Post Office* [2000] All ER (D) 276, CA that this imposes strict liability on the employer and that, even where a workplace product develops a hidden fault (in this case metal fatigue in a bicycle), the employer will be liable even in situations where through wear and tear the product has become dangerous. The employer's duty to maintain translates into a duty to ensure that there are no latent defects.

Criminal liability for breach of statutory duties

[P9011] This refers to duties laid down in *HSWA 1974* as revised by the *Consumer Protection Act 1987, Sch 3*. The duties exist in relation to articles and substances for use at work and fairground equipment. There are no civil claims rights available to employees under these provisions except for breach of safety regulations. (For the specific requirements now applicable to machinery for use at work, see MACHINERY SAFETY.)

Definition of articles and substances

[P9012] An article for use at work means:

(a)   any plant designed for use or operation (whether exclusively or not) by persons at work; and
(b)   any article designed for use as a component in any such plant.
   [HSWA 1974, s 53(1).]

A substance for use at work means 'any natural or artificial substance (including micro-organisms), whether in solid or liquid form or in the form of a gas or vapour'. [*HSWA 1974, s 53(1).*]

An article upon which first trials/demonstrations are carried out is not an article for use at work, but rather an article which *might* be used at work. The purpose of trial/demonstration was to determine whether the article could safely be later used at work (*McKay v Unwin Pyrotechnics Ltd* [1991] Crim LR 547, DC where a dummy mine exploded, causing the operator injury, when being tested to see if it would explode when hit by a flail attached to a vehicle. It was held that there was no breach of *HSWA 1974, s 6(1)(a)*).

Duties in respect of articles and substances for use at work

[P9013] *HSWA 1974, s 6* (as amended by the *Consumer Protection Act 1987, Sch 3*) places duties upon manufacturers and designers, as well as importers and suppliers, of (*a*) articles and (*b*) substances for use at work, whether used exclusively at work or not (e.g. lawnmower, hair dryer).

Articles for use at work

[P9014] Any person who designs, manufactures, imports or supplies any article for use at work (or any article of fairground equipment) must:

(a)   ensure, so far as is reasonably practicable (for the meaning of this expression see ENFORCEMENT), that the article is so designed and constructed that it will be safe and without risks to health at all times when it is being (i) set, (ii) used, (iii) cleaned or (iv) maintained by a person at work;
(b)   carry out or arrange for the carrying out of such testing and examination as may be necessary for the performance of the above duty;

(c)   take such steps as are necessary to secure that persons supplied by that person with the article are provided with adequate information about the use for which the article is designed or has been tested and about any conditions necessary to ensure that it will be safe and without risks to health at all such times of (i) setting, (ii) using, (iii) cleaning, (iv) maintaining *and* when being (v) dismantled, or (vi) disposed of; and

(d)   take such steps as are necessary to secure, so far as is reasonably practicable, that persons so supplied are provided with all such revisions of information as are necessary by reason of it becoming known that anything gives rise to a serious risk to health or safety.

[HSWA 1974, s 6(1)(a)–(d) as amended by the Consumer Protection Act 1987, Sch 3.]

(See **P9016** and **P9017** below for further duties relevant to articles for use at work.)

In the case of an article for use at work which is likely to cause an employee to be exposed to 80 dB(A) or above, or to peak sound pressure of 135dB(c) or above, adequate information must be provided about noise likely to be generated by that article. [*Control of Noise at Work Regulations 2005 (SI 2005 No 1643), Reg 10*.] (See further NOISE AND VIBRATION).

Substances for use at work

[P9015] Every person who manufactures, imports or supplies any substance must:

(a)   ensure, so far as is reasonably practicable, that the substance will be safe and without risks to health at all times when it is being (i) used, (ii) handled, (iii) processed, (iv) stored, or (v) transported by any person at work or in premises where substances are being installed;

(b)   carry out or arrange for the carrying out of such testing and examination as may be necessary for the performance of the duty in (a);

(c)   take such steps as are necessary to secure that persons supplied by that person with the substance are provided with adequate information about:

    (i)   any risks to health or safety to which the inherent properties of the substance may give rise;

    (ii)   the results of any relevant tests which have been carried out on or in connection with the substance; and

    (iii)   any conditions necessary to ensure that the substance will be safe and without risks to health at all times when it is being (a) used, (b) handled, (c) processed, (d) stored, (e) transported and (f) disposed of; and

(d)   take such steps as are necessary to secure, so far as is reasonably practicable, that persons so supplied are provided with all such revisions of information as are necessary by reason of it becoming known that anything gives rise to a serious risk to health or safety.

[HSWA 1974, s 6(4) as amended by the Consumer Protection Act 1987, Sch 3.]

Additional duty on designers and manufacturers to carry out research

[P9016] Any person who undertakes the design or manufacture of any article for use at work must carry out, or arrange for the carrying out, of any necessary research with a view to the discovery and, so far as is reasonably practicable, the elimination or minimisation of any health or safety risks to which the design or article may give rise. [*HSWA 1974, s 6(2)*.]

Duties on installers of articles for use at work

[P9017] Any person who erects or installs any article for use at work in any premises where the article is to be used by persons at work, must ensure, so far as is reasonably practicable, that nothing about the way in which the article is erected or installed makes it unsafe or a risk to health when it is being (*a*) set, (*b*) used, (*c*) cleaned, or (*d*) maintained by someone at work. [*HSWA 1974, s 6(3)* as amended by the *Consumer Protection Act 1987, Sch 3*.]

Additional duty on manufacturers of substances to carry out research

[P9018] Any person who manufactures any substance must carry out, or arrange for the carrying out, of any necessary research with a view to the discovery and, so far as is reasonably practicable, the elimination or minimisation of any health/safety risks at all times when the substance is being (*a*) used, (*b*) handled, (*c*) processed, (*d*) stored, or (*e*) transported by someone at work. [*HSWA 1974, s 6(5)* as amended by the *Consumer Protection Act 1987, Sch 3*.]

No duty on suppliers of industrial articles and substances to research

[P9019] It is not necessary to repeat any testing, examination or research which has been carried out by designers and manufacturers of industrial products, on the part of importers and suppliers, in so far as it is reasonable to rely on the results [*HSWA 1974, s 6(6)*.]

Custom built articles

[P9020] Where a person designs, manufactures, imports or supplies an article for or to another person on the basis of a written undertaking by that other to ensure that the article will be safe and without health risks when being (*a*) set, (*b*) used, (*c*) cleaned, or (*d*) maintained by a person at work, the undertaking will relieve the designer/manufacturer etc. from the duty specified in *HSWA 1974, s 6(1)(a)* (see **P9013** above), to such extent as is reasonable, having regard to the terms of the undertaking. [*HSWA 1974, s 6(8)* as amended by the *Consumer Protection Act 1987, Sch 3*.]

Importers liable for offences of foreign manufacturers/designers

[P9021] In order to give added protection to industrial users from unsafe imported products, the *Consumer Protection Act 1987, Sch 3* has introduced a new subsection (*HSWA 1974, s 6(8A)*) which, in effect makes importers of unsafe products liable for the acts/omissions of foreign designers and manufacturers. *Section 6(8A)* states that nothing in (*inter alia*) *s 6(8)* is to relieve an importer of an article/substance from any of his duties, as regards anything done (or not done) or within the control of:

(a)   a foreign designer; or

(b)   a foreign manufacturer of an article/substance.
[HSWA 1974, s 6(8A).]

Proper use

[P9022] The original wording of *HSWA 1974, s 6(1)(a), 6(4)(a)* and *6(10)* concerning 'proper use' excluded 'foreseeable user error' as a defence, which had the consequence of favouring the supplier. Thus, if a supplier could demonstrate a degree of operator misuse or error, however reasonably foreseeable, the question of initial product safety was side stepped. Moreover, 'when properly used' implied, as construed, that there could only be a breach of *s* 6 once a product had actually been *used*. This was contrary to the principle that safety should be built into design/production, rather than relying on warnings and disclaimers. The wording in *s* 6 is now amended so that only *unforeseeable* user/operator error will relieve the supplier from liability; he will no longer be able to rely on the strict letter of his operating instructions. [*Consumer Protection Act 1987, Sch 3.*]

Powers to deal with unsafe imported goods

[P9023] *HSWA 1974* does not empower enforcing authorities to stop the supply of unsafe products at source or prevent the sale of products by foreign producers after they have been found to be unsafe, but enforcement officers have the power to act at the point of entry or anywhere else along the distribution chain to stop unsafe articles/substances being imported by serving prohibition notices (see ENFORCEMENT).

Customs officers can seize any imported article/substance, which is considered to be unsafe, and detain it for up to two (working) days. [*HSWA s 25A (incorporated by Schedule 3* to the *Consumer Protection Act 1987).*]

In addition, customs officers can transmit information relating to unsafe imported products to HSE inspectors. [*HSWA 1974, s 27A (incorporated by the *Consumer Protection Act 1987, Sch 3).*]

Civil liability for unsafe products – historical background

[P9024] Originally at common law where defective products caused injury, damage and/or death, redress depended on whether the injured user had a contract with the seller or hirer of the product. This was often not the case, and in consequence many persons, including employees repairing and/or servicing products, were without remedy. This rule, emanating from the decision of *Winterbottom v Wright* (1842) 10 M & W 109, remained unchanged until 1932, when *Donoghue v Stevenson* [1932] AC 562 was decided by the House of Lords. This case was important because it established that manufacturers were liable in negligence (i.e. tort) if they failed to take reasonable care in the manufacturing and marketing of their products, and in consequence a user suffered injury when using the product in a reasonably foreseeable way. More particularly, a 'manufacturer of products, which he sells in such a form as to show that he intends them to reach the ultimate consumer in the form in which they left him with no reasonable possibility of intermediate inspection, and with the knowledge that the absence of reasonable care in the preparation or putting up of the products will result in an injury to the consumer's life or property, owes a duty to the consumer to take reasonable care' (per Lord Atkin). In this way, manufacturers of products which were defective were liable in negligence to users and consumers of their products, including those who as intermediaries repair, maintain and service industrial products, it being irrelevant whether there was a contract between manufacturer and user (which normally there was not).

Donoghue v Stevenson is a case of enormous historical importance in the context of liability of manufacturers, but although the principle has become well established and is still widely applied it did not give a reliable remedy to injured persons, who still had to satisfy the legal burden of proving that the manufacturer had not exercised reasonable care, a serious obstacle to over-come in cases involving technically complex products. In some instances it became possible to avoid this obstacle, by establishing liability on other bases.

Defective equipment supplied to employees

[P9025] At common law an employer obtaining equipment from a reputable supplier was unlikely to be found liable to an employee if the equipment turned out to be defective and injured him (*Davie v New Merton Board Mills Ltd* [1959] AC 604). Although it represented no bar to employees suing manufac-turers direct for negligence, the effect of this decision was reversed by the *Employers' Liability (Defective Equipment) Act 1969* rendering employers strictly liable (irrespective of negligence) for any defects in equipment causing injury. (See **E11004–E11005** EMPLOYERS' DUTIES TO THEIR EMPLOYEES). In such circumstances the employer would be able to claim indemnity for breach of contract by the supplier of the equipment.

Breach of statutory duty

[P9026] Whilst legislation and regulations dealing with domestic and indus-trial safety are principally penal and enforceable by state agencies (Trad-ing Standards officers and Health and Safety Executive inspectors) if injury, death or damage occurs in consequence of a breach of such statutory duty, it may be possible to use this breach as the basis of a civil liability claim. Here there may be strict liability if there are absolute requirements, or the duty may be defined in terms of what it is practicable or reasonably practicable to do (see **E15031** ENFORCEMENT).

As far as workplace products are concerned, *HSWA 1974, s 47* bars a right of action in civil proceedings in respect of any failure to comply with *HSWA 1974, s 6* (or the other general duties under the Act). However a breach of a duty imposed by health and safety regulations will generally be actionable in this way, unless the particular regulations in question contain a proviso to the contrary.

The position is the same in relation to consumer safety regulations made under the *Consumer Protection Act 1987* (*Consumer Protection Act 1987, s 41(1)*). No provision is made in the *General Product Safety Regulations 2005 (SI 2005*

No 1803) for any breach thereof to give rise to civil liability for the benefit of an injured person. Given that the Regulations stem from European law it is unlikely that they would be construed in such a way as to give more extensive rights than those contained in the Product Liability Directive (85/374/EEC) (see **P9031**).

Consumer Protection Act 1987

[P9027] At European level it was deemed necessary to introduce a degree of harmonisation of product liability principles between Member States, and at the same time to reduce the importance of fault and negligence concepts in favour of liability being determined by reference to 'defects' in a product. Thus the nature of the product itself would become the key issue, not the conduct of the manufacturer. After protracted debate the Product Liability Directive (85/374/EEC) was adopted.

Introduction of strict product liability

[P9028] The introduction of strict product liability is enshrined in Britain within the *Consumer Protection Act 1987, s 2(1)*. Thus, where any damage is *caused* wholly or partly by a defect (see **P9031** below) in a product (e.g. goods, electricity, a component product or raw materials), the following may be liable for damages (irrespective of negligence):

(a)   the producer;
(b)   any person who, by putting his name on the product or using a trade mark (or other distinguishing mark) has held himself out as the producer; and
(c)   any person who has imported the product into a Member State from outside the EU, in the course of trade/business.
[Consumer Protection Act 1987, s 2(1), (2).]

Producers

[P9029] Producers are variously defined as:

(a)   the person who manufactured a product;
(b)   in the case of a substance which has not been manufactured, but rather won or abstracted, the person who won or abstracted it; or
(c)   in the case of a product not manufactured, won or abstracted, but whose essential characteristics are attributable to an industrial process or agricultural process, the person who carried out the process.
[Consumer Protection Act 1987, s 1(2)(a)–(c).]

Liability of suppliers

[P9030] Although producers are principally liable, intermediate suppliers can also be liable in certain circumstances. Thus, any person who supplied the product is liable for damages if:

(a)   the injured person requests the supplier to identify one (or more) of the following:
    (i)   the producer,
    (ii)   the person who put his trade mark on the product,
    (iii) the importer of the product into the EU; and

(b)   the request is made within a reasonable time after damage/injury has occurred *and* it is not reasonably practicable for the requestor to identify the above three persons; and

(c)   within a reasonable time after receiving the request, the supplier fails either:
    (i)   to comply with the request, or
    (ii)   identify his supplier.

[Consumer Protection Act 1987, s 2(3).]

Importers of products into the EU and persons applying their name, brand or trade mark will also be directly liable as if they were original manufacturers *[Consumer Protection Act 1987, s 2(2)(b), (c)]*.

Defect – key to liability

[P9031] Liability presupposes that there is a defect in the product, and indeed, existence of a defect is the key to liability. Defect is defined in terms of the absence of safety in the product. More particularly, there is a 'defect in a product . . . if the safety of the product is not such as persons generally are entitled to expect' (including products comprised in that product) [*Consumer Protection Act 1987, s 3(1)*].

Defect can arise in one of three ways and is related to:

(a)   construction, manufacture, sub-manufacture, assembly;
(b)   absence or inadequacy of suitable warnings, or existence of misleading warnings or precautions; and
(c)   design.

The definition of 'defect' implies an entitlement to an expectation of safety on the part of the consumer, judged by reference to *general* consumer expectations not individual subjective ones. (The American case of *Webster v Blue Ship Tea Room Inc* 347 Mass 421, 198 NE 2d 309 (1964) is particularly instructive here. The claimant sued in a product liability action for a bone which had stuck in her throat, as a result of eating a fish chowder in the defendant's restaurant. It was held that there was no liability. Whatever her own expectations may have been, fish chowder would not be fish chowder without some bones and this is a general expectation.)

The court will require a claimant to be specific and be particular about exactly what the alleged defect is (*Paul Sayers and Others v SmithKline Beecham plc and Others (MMR/MR litigation)* [1999] MLC 0117).

Consumer expectation of safety – criteria

[P9032] The general consumer expectation of safety must be judged in relation to:

(a)   the marketing of the product, i.e:

(i)   the manner in which; and

(ii)   the purposes for which the product has been marketed;

(iii)   any instructions/warnings against doing anything with the product; and

(b)   by what might reasonably be expected to be done with or in relation to the product (e.g. the expectation that a sharp knife will be handled with care); and

(c)   the time when the product was supplied (e.g. a product's shelf-life).

A defect cannot arise retrospectively by virtue of the fact that, subsequently, a safer product is made and put into circulation.
[Consumer Protection Act 1987, s 3(2).]

The definition of 'defect' was discussed extensively by the Court in *A v National Blood Authority* [2001] 3 All ER 289. The Claimants in this case had contracted Hepatitis C from infected blood transfusions. The Court found that the blood was defective; the judge concluding that the Claimants were entitled to expect that the blood being supplied to them was free from infection. This case should be read in light of the recent case of *Pollard v Tesco Stores Ltd* [2006] EWCA Civ 393, [2006] All ER (D) 186 (Apr), where the Court gave further consideration as to the definition of defect and the link between the legitimate expectation of the public and national safety standards.

Time of supply

[P9033] Liability attaches to the *supply* of a product (see **P9041** below). More particularly, the producer will be liable for any defects in the product existing at the time of supply (see **P9045**(*d*) below); and where two or more persons collaborate in the manufacture of a product, say by submanufacture, either and both may be liable, that is, severally and jointly (see **P9043** below).

Contributory negligence

[P9034] A person who is careless for his own safety is probably guilty of contributory negligence and, thus, will risk a reduction in damages [*Consumer Protection Act 1987, s 6(4)*]. (See also **E11010** EMPLOYERS' DUTIES TO THEIR EMPLOYEES.) However, in the product liability context, carelessness of the user may mean that there is no liability at all on the part of the producer. If, for example, clear instructions and warnings provided with the product had been disregarded, when compliance would have avoided the accident, it is highly unlikely that the product would be found to be defective for the purposes of the Act. No off-setting of damages for contributory fault of the claimant would arise.

Absence or inadequacy of suitable/misleading warnings

[P9035] The common law required that the vendor of a product should point out any latent dangers in a product which he either knew about or ought to have known about. Misleading terminology/labelling on a product or product container could result in liability for negligence (*Vacwell Engineering Ltd v BDH Chemicals Ltd* [1969] 3 All ER 1681 where ampoules containing boron tribromide, which carried the warning 'Harmful Vapours', exploded on contact with water, killing two scientists. It was held that this consequence was reasonably foreseeable and accordingly the defendants should have researched their product more thoroughly). In a similar product liability action today, the manufacturers would be strictly liable (subject to statutory defences) and, if injury/damage followed the failure to issue a written/pictorial warning, as required by law (e.g. the *Chemicals (Hazard Information and Packaging for Supply) Regulations 2002 (SI 2002 No 1689)*), there would be liability.

The duty, on manufacturers, to research the safety of their products, before putting them into circulation (see *Vacwell Engineering Ltd v BDH Chemicals Ltd*) is even more necessary and compulsory now, given the introduction of strict product liability. This includes safety in connection with directions for use on a product. A warning refers to something that can go wrong with the product; directions for use relate to the best results that can be obtained from products, if the directions are followed. In the absence of case law on the point it is reasonable to assume that in order to avoid actions for product liability, manufacturers should provide both warnings, indicating the worst results and dangers, and directions for use, indicating the best results; the warning, in effect, identifying the worst consequences that could follow if directions for use were not complied with.

In the case of *Worsley v Tambrands Ltd (No 2)* [2000] MCR 0280, it was held that a tampon manufacturer had done what was reasonable in all the circumstances to warn a woman about the risk of toxic shock syndrome from tampon use. They had placed a clear legible warning on the outside of the box directing the user to the leaflet. The leaflet was legible, literate and unambiguous and contained all the material necessary to convey both the warning signs and the action required if any risk were present. The manufacturer could not cater for lost leaflets or for those who chose not to replace them. This gives valuable guidance as to the extent that manufacturers are expected to warn users of the risks associated with their products.

Defect must exist when the product left the producer's possession

[P9036] This situation tends to be spotlighted by alteration of modification to, or interference with, a product on the part of an intermediary, for instance, a dealer or agent. If a product leaves an assembly line in accordance with its intended design, but is subsequently altered, modified or generally interfered with by an intermediary, in a manner outside the product's specification, the manufacturer is probably not liable for any injury so caused. In *Sabloff v Yamaha Motor Co* 113 NJ Super 279, 273 A 2d 606 (1971), the claimant was injured when the wheel of his motor-cycle locked, causing it to skid, then crash. The manufacturer's specification stipulated that the dealer attach the wheel of the motor-cycle to the front fork with a nut and bolt, and this had not been done properly. It was held that the dealer was liable for the motor-cyclist's injury (as well as the assembler, since the latter had delegated the function of tightening the nut to the dealer and it had not been properly carried out).

Role of intermediaries

[P9037] If a defect in a product is foreseeably detectable by a legitimate intermediary (e.g. a retailer in the case of a domestic product or an employer in the case of an industrial product), liability used to rest with the intermediary rather than the manufacturer, when liability was referable to negligence

(*Donoghue v Stevenson* [1932] AC 562). This position does not duplicate under the *Consumer Protection Act 1987*, since the main object of the legislation is to fix producers with strict liability for injury-causing product defects to users. Nevertheless, there are common law and statutory duties on employers to inspect/test inward plant and machinery for use at the workplace, and failure to comply with these duties may make an employer liable. In addition, employers, in such circumstances, can incur liability under the *Employers' Liability (Defective Equipment) Act 1969* (see **E11005** EMPLOYERS' DUTIES TO THEIR EMPLOYEES) and so may seek to exercise contractual indemnity against manufacturers.

Comparison with negligence

[P9038] The similarity between product liability and negligence lies in causation. Defect must be the material *cause* of injury. The main arguments against this are likely to be along the lines of misuse of a product by a user (e.g. knowingly driving a car with defective brakes), or ignoring warnings (a two-pronged defence, since it also denies there was a 'defect'), or – as is common in chemicals and pesticides cases – a defence based on alternative theories of causation of the claimant's injuries.

Product liability differs from negligence in that it is no longer necessary for injured users to prove absence of reasonable care on the part of manufacturers. All that is now necessary is proof of (*a*) defect (see **P9035** above) and (*b*) that the defect caused the injury. It will, therefore, be no good for manufacturers to point to an unblemished safety record and/or excellent quality assurance programmes, or the lack of foreseeability of the accident, since the user is not trying to establish negligence. How or why a defect arose is immaterial; what is important is the fact that it exists.

Liability in negligence still has a role to play in cases where liability under the *Consumer Protection Act 1987* cannot be established because, for example, the defendant is not a 'producer' as defined, or because the statutory defence or time limit would bar a strict liability claim (see **P9044** and **P9048**). There is greater scope under the law of negligence for liability to be established against distributors and retailers who may be held responsible for certain defects, especially where inadequate warnings and instructions have been provided (e.g. *Goodchild v Vaclight*, [1965] CLY 2669).

Joint and several liability

[P9039] If two or more persons/companies are liable for the same damage, the liability is joint and several. This can, for example, refer to the situation where a product (e.g. an aircraft) is made partly in one country (e.g. England) and partly in another (e.g. France). Here both partners are liable (joint liability) but in the event of one party not being able to pay, the other can be made to pay all the compensation (several liability) [*Consumer Protection Act 1987, s 2(5)*].

For injuries arising out of occupational exposures, the traditional approaches to causation and joint and several liability are being revised. The recent asbestos cases of *Fairchild v Glenhaven Funeral Services Ltd* [2002] UKHL 22, [2003] 1 AC 32 and *Barker v Corus (UK) plc* [2006] UKHL 20, [2006] 2 AC

572 were concerned with mesothelioma injuries. The Claimants had been exposed to asbestos by various (negligent) defendants but it was impossible to show which of these exposures had caused the disease. It was held that in such circumstances, the Defendants were nevertheless liable in spite of the lack of proof of causation, but in balancing fairness between the parties, each Defendant was only liable (severally, not joint) for a share of the liability apportionised according to its contribution to the risk (this is likely to be according to the relative times and intensity of exposure).

Parameters of liability

[P9040] For certain types of damage including (a) death, (b) personal injury and (c) loss or damage to property liability is included and relevant to private use, occupation and consumption by consumers [*Consumer Protection Act 1987,s 5(1)*]. However, producers and others will not be liable for:

(a) damage/loss to the defective product itself, or any product supplied with the defective product [*Consumer Protection Act 1987, s 5(2)*];

(b) damage to property not 'ordinarily intended for private use, occupation or consumption' e.g. car/van used for business purposes [*Consumer Protection Act 1987, s 5(3)*]; and

(c) damage amounting to less than £275 (to be determined as early as possible after loss) [*Consumer Protection Act 1987, s 5(4)*].

Defences

[P9041] The following statutory defences are open to producers:

(a) the defect was attributable to compliance with any requirement imposed by law/regulation or a European Union rule/regulation [*Consumer Protection Act 1987, s 4(1)(a)*];

(b) the defendant did not supply the product to another (i.e. did not sell/hire/lend/exchange for money/give goods as a prize etc. (see below 'supply')) [*Consumer Protection Act 1987, s 4(1)(b)*];

(c) the supply to another person was not in the course of that supplier's business [*Consumer Protection Act 1987, s 4(1)(c)*];

(d) the defect did not exist in the product at the relevant time (i.e. it came into existence after the product had left the possession of the defendant). This principally refers to the situation where for example a retailer fails to follow the instructions of the manufacturer for storage or assembly [*Consumer Protection Act 1987, s 4(1)(d)*];

(e) that the state of scientific and technical knowledge at the relevant time was not such that a producer 'might be expected to have discovered the defect if it had existed in his products while they were under his control' (i.e. development risk) [*Consumer Protection Act 1987, s 4(1)(e)*] (see **P9049** below);

(f) that the defect:
   (i) constitutes a defect in a subsequent product (in which the product in question was comprised); and
   (ii) was wholly attributable to:
      (A) the design of the subsequent product; or

(B)   compliance by the producer with the instructions of the producer of the subsequent product.

[*Consumer Protection Act 1987, s 4(1)(f)*.] This is known as the component manufacturer's defence.

The meaning of 'supply'

[P9042] Before strict liability can be established under the *Consumer Protection Act 1987*, a product must have been '*supplied*'. This is defined as follows:

(a)   selling, hiring out or lending goods;
(b)   entering into a hire-purchase agreement to furnish goods;
(c)   performance of any contract for work and materials to furnish goods (e.g. making/repairing teeth);
(d)   providing goods in exchange for any consideration other than money;
(e)   providing goods in or in connection with the performance of any statutory function/duty (e.g. supply of gas/electricity by public utilities); and
(f)   giving the goods as a prize or otherwise making a gift of the goods.
[*Consumer Protection Act 1987, s 46(1)*.]

Moreover, in the case of hire-purchase agreements/credit sales the effective supplier (i.e. the dealer), and not the ostensible supplier (i.e. the finance company),is the 'supplier' for the purposes of strict liability [*Consumer Protection Act 1987, s 46(2)*].

Building work is only to be treated as a supply of goods in so far as it involves provision of any goods to any person by means of their incorporation into the building/structure, e.g. glass for windows [*Consumer Protection Act 1987, s 46(3)*].

No contracting out of strict liability

[P9043] The liability to person who has suffered injury/damage under the *Consumer Protection Act 1987*, cannot be (*a*) limited or (*b*) excluded:

(i)   by any contract term; or
(ii)   by any notice or other provision.
[*Consumer Protection Act 1987, s 7*.]

Time limits for bringing product liability actions

[P9044] No action can be brought under the *Consumer Protection Act 1987, Part I* (i.e. product liability actions) after the expiry of ten years from the time when the product was first put into circulation (i.e. the particular item in question was supplied in the course of business/trade etc.) [*Limitation Act 1980, s 11A(3); Consumer Protection Act 1987, Sch 1*]. In other words, ten years is the cut-off point for liability. An action can still be brought for common law negligence after this time.

However. the 2001 case of *SmithKline Beecham plc and Another v Horne-Roberts* [2001] EWCA CIV 2006, [2002] 1 WLR 1662 has serious implications for this cut-off point. The court allowed a claimant to substitute a new defendant for an existing one in a strict liability claim despite the fact that the substitution was outside the ten year cut-off period.

The recent European Court of Justice case, *O'Byrne v Sanofi Pasteur MSD Ltd*: C-127/04 [2006] ECR I-1313 case further doubt over the finality of the ten-year long-stop. In this case, the Court had to consider whether the passing of a product between two different companies (and therefore legal entities) constituted 'putting the product in circulation'. The European Court concluded that:

> A product must be considered as having been put into circulation [. . .] when it leaves the production process and enters a marketing process in the form in which it is offered to the public in order to be used or consumed.

Switching from the production to the marketing process therefore appears to be key. In addition, the European Court also considered whether, if an individual had sued the wrong defendant, the Product Liability Directive allowed proceedings to continue against the correct defendant outside the ten-year long-stop limit. The Court indicated that this was an issue for national courts to decide. The finality of the ten-year long-stop therefore remains in question.

All actions for personal injury caused by product defects must be initiated within three years of whichever event occurs later, namely:

(a) the date when the cause of action accrued (i.e. injury occurred); or
(b) the date when the injured person had the requisite knowledge of his injury/damage to property.
[Limitation Act 1980, s 11A(4); Consumer Protection Act 1987, Sch 1.]

However, if during that period the injured person died, his personal representative has a further three years from his death to bring the action. (This coincides with actions for personal injuries against employers, except of course in that case there is no overall cut-off period of ten years.) *[Limitation Act 1980, s 11A(5).]*. Of course, the limitation period for a child to bring an action does not commence until that child reaches 18 years of age.

Development risk

[P9045] The development risks defence was initially considered as potentially being the most important defence to product liability actions. However, its use has not been extensive to date. Manufacturers have argued that it would be wrong to hold them responsible for the consequences of defects which they could not reasonably have known about or discovered. The absence of this defence would have the effect of increasing the cost of product liability insurance and stifle the development of new products. On the other hand, consumers maintain that the existence of this defence threatens the whole basis of strict liability and allows manufacturers to escape liability by, in effect, pleading a defence associated with the lack of negligence. For this reason, not all EU states have allowed this defence; Finland and Luxembourg exclude the defence entirely from national law and in France, Germany and Spain, certain products are excluded from the scope of the defence. The states in favour of its retention are the United Kingdom, Germany, Denmark, Italy and the Netherlands. The burden of proving development risk lies on the producer and it seems likely that he will have to show that no producer of a product of that sort could be expected to have discovered the existence of the defect. 'It will not necessarily be enough to show that he (the producer) has done as many tests as his competitor, nor that he did all the tests required of him by a government regulation setting a minimum standard.' (Explanatory memorandum of EC Directive on Product Liability, Department of Trade and Industry, November 1985.)

Additionally, the fact that judgments in product liability cases are 'transportable' could have serious implications for the retention of development risk in the United Kingdom (see **P9050** below).

Transportability of judgments

[P9046] The so-called 'Brussels Regulation' (Regulation 44/2001 of 22 December 2000 on *Jurisdiction and the Recognition and Enforcement of Judgments in Civil and Commercial Matters*) requires judgments given in one of the Member States (except Denmark which will continue to apply the 1968 Brussels Convention) to be enforced in another. The United Kingdom became bound by this Regulation as of 1 March 2002, which, by virtue of the *Civil Jurisdiction and Judgments Order 2001 (SI 2001 No 3929)* (which amends the *Civil Jurisdiction and Judgments Act 1982*), is part of UK law. Where product liability actions are concerned, litigation can be initiated in the state where the defendant is based or where injury occurred and the judgment of that court 'transported' to another Member State. This could pose a threat to retention of development risk in the United Kingdom and other states in favour of it from a state against it, e.g. France, Belgium, Luxembourg.

Contractual liability for sub-standard products

[P9047] Contractual liability is concerned with defective products which are substandard (though not necessarily dangerous) regarding quality, reliability and/or durability. Liability is predominantly determined by contractual terms implied by the *Sale of Goods Act 1979* and the *Sale and Supply of Goods Act 1994*, in the case of goods sold; the *Supply of Goods (Implied Terms) Act 1973*, where goods are the subject of hire purchase and conditional and/or credit sale; and the *Supply of Goods and Services Act 1982*, where goods are supplied but not sold as such, primarily as a supply of goods with services; hire and leasing contracts are subject to the 1982 Act as well.

Exemption or exclusion clauses in such contracts may be invalid by virtue of the *Unfair Contract Terms Act 1977*, which also applies to such transactions. Moreover, 'standard form' contracts with consumers where the terms have not been individually negotiated, which contain 'unfair terms' – that is, terms detrimental to the consumer – will have such terms excised, if necessary, by the Director General of Fair Trading, under the *Unfair Terms in Consumer Contracts Regulations 1999 (SI 1999 No 2083)*. These Regulations extend to most consumer contracts between a seller or supplier and a consumer. [*Unfair Terms in Consumer Contracts Regulations 1999, (SI 1999 No 2083), Reg 4(1).*] (See P9057 below.)

Contractual liability is strict. It is not necessary that negligence be established (*Frost v Aylesbury Dairy Co Ltd* [1905] 1 KB 608 where the defendant supplied typhoid-infected milk to the claimant, who, after its consumption, became ill and required medical treatment. It was held that the defendant was liable, irrespective of the absence of negligence on his part).

Sale of Goods Acts

[P9048] Conditions and warranties as to fitness for purpose, quality and merchantability were originally implied into contracts for the sale of goods at common law. Those terms were then codified in the *Sale of Goods Act 1893*. However, this legislation did not provide a blanket consumer protection measure, since sellers were still allowed to exclude liability by suitably worded exemption clauses in the contract. This practice was finally outlawed, at least as far as consumer contracts were concerned, by the *Supply of Goods (Implied Terms) Act 1973* and later still by the *Unfair Contract Terms Act 1977*, the current statute prohibiting contracting out of contractual liability and negligence. Indeed, consumer protection reached a height with the *Unfair Terms in Consumer Contracts Regulations 1999 (SI 1999 No 2083)*, invalidating 'unfair terms' in most consumer contracts of a standard form nature (see P9062 below),and was further bolstered by the *Sale and Supply of Goods to Consumers Regulations 2002 (SI 2002 No 3045)* (implementing Directive 1999/44/EC on certain aspects of the sale of consumer goods and associated guarantees). These Regulations extended the definition of satisfactory quality, giving a statutory right to repair or replacement, amending the law relating to the passing of risk in consumer contracts, and further amending UCTA to make the position more favourable to consumers.

The initial law relating to sale and supply of goods was updated by the *Sale of Goods Act 1979* and the *Sale and Supply of Goods Act 1994*, the latter replacing the condition of 'merchantable quality' with 'satisfactory quality'. The difference between the 'merchantability' requirement, under the *1979 Act*, and its replacement 'satisfactory quality', under the *1994 Act*, is that, under the former Act, products were 'usable' (or, in the case of food, 'edible'), even if they had defects which ruined their appearance; now they must be free from minor defects as well as being safe and durable. Further, under the previous law, a right of refund disappeared after goods had been kept for a reasonable time; under the 1994 Act, consumers benefited from a right of examination for a reasonable time after buying. Regulation 5 of the 2002 Regulations further amended the *Sale of Goods Act 1979* by allowing consumers who ordered goods which do not conform to contract at the time of deliver, the right to require the seller to repair or replace the product. If a defect is discovered in the product during the first six months after delivery and that defect amounts to non-conformity with the contract of sale, then it will be regarded as having been in existence at the time of delivery, subject to certain minor exceptions.

Sale of Goods Act 1979

[P9049] In 1979 a consolidated *Sale of Goods Act* was passed and current law on quality and fitness of products is contained in that Act. Another equally important development has been the extension of implied terms, relating to quality and fitness of products, to contracts other than those for the sale of goods, that is, to hire purchase contracts by the *Supply of Goods (Implied Terms) Act 1973*, and to straight hire contracts by the *Supply of Goods and Services Act 1982*. In addition, where services are performed under a contract, that is, a contract for work and materials, there is a statutory duty on the contractor to perform them with reasonable care and skill. In other words, in the case of services liability is not strict, but it is strict for the supply of products. This is laid down in the *Supply of Goods and Services Act 1982, s 4*. This applies whether products are simultaneously but separately supplied under any contract, e.g. after-sales service, say, on a car or a contract to repair a window by a carpenter, in which latter case service is rendered irrespective of product supplied.

Products to be of satisfactory quality – sellers/suppliers

[P9050] The *Sale of Goods Act 1979* (as amended) writes two quality conditions into all contracts for the sale of products, the first with regard to satisfactory quality, the second with regard to fitness for purpose.

Where a seller sells goods in the course of business, there is an implied term that the goods supplied under the contract are of satisfactory quality, according to the standards of the reasonable person, by reference to description, price etc. [*Sale of Goods Act 1979, s 14(2)* as substituted by the *Sale and Supply of Goods Act 1994, s 1(2)*.] The 'satisfactory' (or otherwise) quality of goods can be determined from:

(a)   their state and condition;
(b)   their fitness for purpose (see **P9057** below);
(c)   their appearance and finish;
(d)   their freedom from minor defects;
(e)   their safety; and
(f)   their durability.

However, the implied term of 'satisfactory quality' does not apply to situations where:

(i)   the unsatisfactory nature of goods is specifically drawn to the buyer's attention prior to contract; or
(ii)   the buyer examined the goods prior to contract and the matter in question ought to have been revealed by that examination.

[Sale of Goods Act 1979, s 14(2) as substituted by the *Sale and Supply of Goods Act 1994, s 1(2A), (2B) and (2C).]*

The Sale and Supply of Goods to Consumers Regulations 2002 (SI 2002 No 3045, Reg 3) widens the ambit or how satisfactory quality is to be determined, by extending the definition of satisfactory quality to include 'any public statement on the specific characteristics of the goods make about them by the seller, producer or his representative, particularly in advertising or on the labelling'.

Sale by sample

[P9051] In the case of a contract for sale by sample, there is an implied condition that the goods will be free from any defect making their quality unsatisfactory, which would not be apparent on reasonable examination of the sample. [*Sale of Goods Act 1979, s 15(2)* as substituted by the *Sale and Supply of Goods Act 1994, s 1(2)*.]

Conditions implied into sale – sales by a dealer

[P9052] The condition of satisfactory quality only arises in the case of sales by a dealer to a consumer, not in the case of private sales. The *Sale of Goods Act 1979, s 14(2)* also applies to second-hand as well as new products. It is not necessary, as it is with the 'fitness for purpose' condition (see **P9058** below), for the buyer in any way to rely on the skill and judgment of the seller in selecting his stock, in order to invoke *s 14(2)*. However, if the buyer has examined the products, then the seller will not be liable for any defects which the examination should have disclosed. Originally this applied if the buyer had been given opportunity to examine but had not, or only partially, exercised it. In *Thornett & Fehr v Beers & Son* [1919] 1 KB 486, a buyer of glue examined only the outside of some barrels of glue. The glue was defective. It was held that he had examined the glue and so was without redress.

Products to be reasonably fit for purpose

[P9053]

Where the seller sells goods in the course of a business and the buyer, expressly or by implication, makes known:

(a)   to the seller, or
(b)   where the purchase price or part of it is payable by instalments and the goods were previously sold by a credit-broker to the seller, to that credit-broker,

any particular purpose for which the goods are being bought, there is an implied condition that the goods supplied under the contract are reasonably fit for that purpose, whether or not that is a purpose for which such goods are commonly supplied, except where the circumstances show that the buyer does not rely or that it is unreasonable for him to rely, on the skill or judgment of the seller or credit-broker.

[Sale of Goods Act 1979, s 14(3).]

Reliance on the skill/judgment of the seller will generally be inferred from the buyer's conduct. The reliance will seldom be express: it will usually arise by implication from the circumstances; thus to take a case of a purchase from a retailer, the reliance will be in general inferred from the fact that a buyer goes to the shop in the confidence that the tradesman has selected his stock with skill and judgment (*Grant v Australian Knitting Mills Ltd* [1936] AC 85). Moreover, it is enough if the buyer relies partially on the seller's skill and judgment. However, there may be no reliance where the seller can only sell goods of a particular brand. A claimant bought beer in a public house which he knew was a tied house. He later became ill as a result of drinking it. It was held that there was no reliance on the seller's skill and so no liability on the part of the seller (*Wren v Holt* [1903] 1 KB 610).

Even though products can only be used normally for one purpose, they will have to be reasonably fit for that particular purpose. A claimant bought a hot water bottle and was later scalded when using it because of its defective condition. It was held that the seller was liable because the hot water bottle was not suitable for its normal purpose (*Preist v Last* [1903] 2 KB 148). But, on the other hand, the buyer must not be hypersensitive to the effects of the product. A claimant bought a Harris Tweed coat from the defendants. She later contracted dermatitis from wearing it. Evidence showed that she had an exceptionally sensitive skin. It was held that the coat was reasonably fit for the purpose when worn by a person with an average skin (*Griffiths v Peter Conway Ltd* [1939] 1 All ER 685).

Like *s 14(2)*, *s 14(3)* extends beyond the actual products themselves to their containers and labelling. A claimant was injured by a defective bottle containing mineral water, which she had purchased from the defendant, a retailer. The bottle remained the property of the seller because the claimant had paid the seller a deposit on the bottle, which would be returned to her, on return of the empty bottle. It was held that, although the bottle was the property of the seller, the seller was liable for the injury caused to the claimant by the defective container (*Geddling v Marsh* [1920] 1 KB 668).

Like *s 14(2)* (above), *s 14(3)* does not apply to private sales.

Strict liability under the Sale of Goods Act 1979, s 14

[P9054] Liability arising under the *Sale of Goods Act, s 14* is strict and does not depend on proof of negligence by the purchaser against the seller (*Frost v Aylesbury Dairy Co Ltd* (scc **P9051** above)). This fact was stressed as follows in the case of *Kendall v Lillico* [1969] 2 AC 31:

> If the law were always logical one would suppose that a buyer who has obtained a right to rely on the seller's skill and judgment, would only obtain thereby an assurance that proper skill and judgment had been exercised, and would only be entitled to a remedy if a defect in the goods was due to failure to exercise such skill and judgment. But the law has always gone further than that. By getting the seller to undertake his skill and judgment the buyer gets . . . an assurance that the goods will be reasonably fit for his purpose and that covers not only defects which the seller ought to have detected but also defects which are latent in the sense that even the utmost skill and judgment on the part of the seller would not have detected them.

(Per Lord Reid.)

Dangerous products

[P9055] As distinct from applying to merely substandard products, both *s 14(2)* and *(3)* of the *Sale of Goods Act 1979* can be invoked where a product is so defective as to be unsafe, but the injury/damage must be a reasonably foreseeable consequence of breach of the implied condition. If, for instance, therefore, the chain of causation is broken by negligence on the part of the user, in using a product knowing it to be defective, there will be no liability. In *Lambert v Lewis* [1982] AC 225, manufacturers had made a defective towing coupling which was sold by retailers to a farmer. The farmer continued to use the coupling knowing that it was unsafe. As a result, an employee was injured and the farmer had to pay damages. He sought to recover these against the retailer for breach of *s 14(3)*. It was held that he could not do so.

Credit sale and supply of products

[P9056] Broadly similar terms to those under the *Sale of Goods Act 1979, s 14* exist, in the case of hire purchase, credit sale, conditional sale and hire or lease contracts, by virtue of the Acts described above at **P9051** having been modified by the *Sale and Supply of Goods Act 1994*.

Unfair Contract Terms Act 1977 (UCTA)

[P9057] In spite of its name, this Act is not directly concerned with 'unfairness' in contracts, nor is it confined to the regulation of contractual relations. The main provisions are as follows:

(a)   Liability for death or personal injury resulting from negligence cannot be excluded by warning notices or contractual terms. [*UCTA 1977, s 2*]. (This does not necessarily prevent an indemnity of any liability to an injured person being agreed between two other contracting parties: if A hires plant to B on terms that B indemnifies A in respect of any claims by any person for injury, the clause may be enforceable (see *Thompson v T Lohan (Plant Hire) Ltd* [1987] 2 All ER 631).

(b)   In the case of other loss or damage, liability for negligence cannot be excluded by contract terms or warning notices unless these satisfy the requirement of reasonableness.

(c)   Where one contracting party is a consumer, or when one of the contracting parties is using written standard terms of business, exclusions or restrictions of liability for breach of contract will be permissible only in so far as the term satisfies the requirement of reasonableness [*UCTA 1977, s 3*]; as far as contracts with consumers are concerned, it is not possible to exclude or restrict liability for the implied undertakings as to quality and fitness for purpose contained in the *Sale of Goods Act 1979* (as amended) or in the equivalent provisions relating to hire purchase and other forms of supply of goods [*UCTA 1977, s 6(2)*]. Separate provisions make it an offence to include this type of exclusion clause in a consumer contract [*Consumer Transactions (Restrictions on Statements) Order 1976 (SI 1976 No 1813)*].

(d)   In all cases where the reasonableness test applies, the burden of establishing that a clause or other provision is reasonable will lie with the person who is trying to rely on the term in question. The consequence of contravention of the Act is, however, limited to the offending term being treated as being ineffective; the courts will give effect to the remainder of the contract so far as it is possible to do so.

It should be noted that UCTA was amended by the *Sale and Supply of Goods to Consumers Regulations 2002 (SI 2002 No 3045, Reg 14)* in that individuals are treated as dealing as consumers even if the goods which are supplied or hired to them are not of a type ordinarily supplied for private use or consumption. In addition, where an individual buys new goods at auction or by competitive tender, they will be treated as dealing as a consumer i.e. afforded protection against unfair terms in the contract of sale.

Unfair Terms in Consumer Contracts Regulations 1999 (SI 1999 No 2083)

[P9058] Consumers faced with standard form contracts are given more ammunition to combat terminological obscurity, legalese and inequality of bargaining power, by the *Unfair Terms in Consumer Contracts Regulations 1999 (SI 1999 No 2083)*. These Regulations have the effect of invalidating any 'unfair terms' in consumer contracts involving products and services, e.g. sale, supply, servicing agreements, insurance and, as a final resort, empowering the Director General of Fair Trading to scrutinise 'standard form' terms, with a view to recommending, where necessary, their discontinued use. Contracts affected by the Act will remain enforceable minus the unfair terms (which will be struck out), in so far as this result is possible [*Unfair Terms in Consumer Contracts Regulations 1999 (SI 1999 No 2083), Reg 8(2)*].

The contracts affected are only 'standard form' ones, that is, contracts whose terms have not been 'individually negotiated' between seller/supplier and consumer [*Unfair Terms in Consumer Contracts Regulations 1999 (SI 1999 No 2083), Reg 5*]; and a term is taken not to have been 'individually negotiated', where it has been drafted in advance and the consumer has not been able to influence the substance of the term [*Unfair Terms in Consumer Contracts Regulations 1999 (SI 1999 No 2083), Reg 5(2)*]. Significantly, it is incumbent on the seller/supplier, who claims that a term was 'individually negotiated' (i.e. not unfair) to prove that it was so. [*Unfair Terms in Consumer Contracts Regulations 1999 (SI 1999 No 2083), Reg 5(4)*]. However, if the terms of the contract are in plain, intelligible language, fairness, relating to subject matter or price/remuneration, cannot be questioned [*Unfair Terms in Consumer Contracts Regulations 1999 (SI 1999 No 2083), Reg 6(2)*].

The key provisions are:

(a)   an 'unfair term' is one which has not been individually negotiated and causes a significant imbalance in the parties' rights and obligations, detrimentally to the consumer, contrary to the underlying tenet of good faith.
[*Unfair Terms in Consumer Contracts Regulations 1999 (SI 1999 No 2083), Reg 5(1)*]

(b)   an 'unfair term' is not binding on the consumer.
[*Unfair Terms in Consumer Contracts Regulations 1999 (SI 1999 No 2083), Reg 8(1)*]

(c)   terms of standard form contracts must be expressed in plain, intelligible language and, if there is doubt as to the meaning of a term, it must be interpreted in favour of the consumer except in proceedings brought under *Reg 12* (see *(e)* below).
[*Unfair Terms in Consumer Contracts Regulations 1999 (SI 1999 No 2083), Reg 7*]

(d)   complaints about 'unfair terms' in standard forms are to be considered by the Director General of Fair Trading, who may bring proceedings for an injunction to prevent use of the terms in future if the proponent of the unfair terms does not agree to desist from using them.
[*Unfair Terms in Consumer Contracts Regulations 1999 (SI 1999 No 2083), Reg 10*]

(e)   instead of the Director General of Fair Trading, a qualifying body named in *Sch 1* (statutory regulators, trading standards departments and the Consumers' Association) may notify the Director that it agrees to consider the complaint and may bring proceedings for an injunction to prevent the continued use of an unfair contract term.
[*Unfair Terms in Consumer Contracts Regulations 1999 (SI 1999 No 2083), Regs 11 and 12*].

Examples of 'unfair terms'

[P9059] Terms which have the object or effect of:

(a)   excluding or limiting the legal liability of a seller or supplier in the event of the death of a consumer or personal injury to the latter resulting from an act or omission of that seller or supplier;

(b)   inappropriately excluding or limiting the legal rights of the consumer vis-à-vis the seller or supplier or another party in the event of total or partial non-performance or inadequate performance by the seller or supplier of any of the contractual obligations, including the option of offsetting a debt owed to the seller or supplier against any claim which the consumer may have against him;

(c)   making an agreement binding on the consumer whereas provision of services by the seller or supplier is subject to a condition whose realisation depends on his own will alone;

(d)   permitting the seller or supplier to retain sums paid by the consumer where the latter decides not to conclude or perform the contract, without providing for the consumer to receive compensation of an equivalent amount from the seller or supplier where the latter is the party cancelling the contract;

(e)   requiring any consumer who fails to fulfil his obligation to pay a disproportionately high sum in compensation;

(f)   authorising the seller or supplier to dissolve the contract on a discretionary basis where the same facility is not granted to the consumer, or permitting the seller or supplier to retain the sums paid for services not yet supplied by him where it is the seller or supplier himself who dissolves the contract;

(g)   enabling the seller or supplier to terminate a contract of indeterminate duration without reasonable notice except where there are serious grounds for doing so;

(h)   automatically extending a contract of fixed duration where the consumer does not indicate otherwise, when the deadline fixed for the consumer to express this desire not to extend the contract is unreasonably early;

(i)   irrevocably binding the consumer to terms with which he had no real opportunity of becoming acquainted before the conclusion of the contract;

(j)   enabling the seller or supplier to alter the terms of the contract unilaterally without a valid reason which is specified in the contract;

(k)   enabling the seller or supplier to alter unilaterally without a valid reason any characteristics of the product or service to be provided;

(l)   providing for the price of goods to be determined at the time of delivery or allowing a seller of goods or supplier of services to increase their price without in both cases giving the consumer the corresponding right to cancel the contract if the final price is too high in relation to the price agreed when the contract was concluded;

(m)   giving the seller or supplier the right to determine whether the goods or services supplied are in conformity with the contract, or giving him the exclusive right to interpret any term of the contract;

(n)   limiting the seller's or supplier's obligation to respect commitments undertaken by his agents or making his commitments subject to compliance with a particular formality;

(o)   obliging the consumer to fulfil all his obligations where the seller or supplier does not perform his;

(p)   giving the seller or supplier the possibility of transferring his rights and obligations under the contract, where this may serve to reduce the guarantees for the consumer, without the latter's agreement;

(q)   excluding or hindering the consumer's right to take legal action or exercise any other legal remedy, particularly by requiring the consumer to take disputes exclusively to arbitration not covered by legal provisions, unduly restricting the evidence available to him or imposing on him a burden of proof which, according to the applicable law, should lie with another party to the contract.

[Unfair Terms in Consumer Contracts Regulations 1999 (SI 1999 No 2083), Sch 2.]

Radiation

Leslie Hawkins and Donald Bruce

Non-ionising and ionising radiation

Non-Ionising Radiation – What is it?

[R1001] Ionising radiation has sufficient energy to displace tightly bound electrons from atoms, causing the atom to become electrically charged or *ionised*. These ions, although short-lived, can be very reactive, causing tissue damage and changes in DNA, which lead to cancer (see **R2001–R2011**). *Non-ionising radiation* is that part of the electromagnetic spectrum which has insufficient energy to cause ionisation. Its interaction with body tissues, and hence the potential harm it can cause, is quite different, therefore, from ionising radiation.

All radiation can be thought of as waves of energy radiating away from the source that produces it. In the case of non-ionising radiation, this energy can be characterised by both an *electrical* field (the E-field) and a *magnetic* field (the B or H field – see **R1002**). Both of these characteristics can be measured and can be used to quantify the strength of the field. In addition, the energy wave has a *frequency*: the number of times it repeats itself in a given time (a frequency of 1 cycle per second is 1 Hertz, or Hz). The *wavelength* (which is directly related to the frequency) is the distance between the waves measured in the unit of metres (or down to nanometres (nm) for the very short wavelengths in the visible and ultraviolet region). The frequency is an important way to characterise non-ionising radiation, since the effects it can have on the body are determined as much by frequency as by the strength of the field. The frequency of non-ionising radiation extends from just above zero Hz (if it was zero and therefore had no frequency, it would be a *static* electrical or magnetic field) to between 10^{15} and 10^{16} Hz in the ultraviolet region. At this frequency the non-ionising region gives way to the ionising region. The complete range of energies is divided into bands, which are described by the electromagnetic spectrum. (TABLE 1).

TABLE 1. THE ELECTROMAGNETIC SPECTRUM

Region			Wavelength (metres)	Approximate frequency range (Hz)*
		Static	*0*	*0*
Non – Ionising Radiation	Radiofrequencies	ELF	10^8–10^5	0.3 Hz to 3 kHz
		VLF	10^5–10^4	3 kHz to 30 kHz
		LF	10^4–10^3	30 kHz to 300 kHz
		MF	10^3–10^2	300 kHz to 3 MHz
		HF	10^2–10	3 MHz to 30 MHz)
		VHF	10–1	30 MHz to 300 MHz)
	Micro-waves	UHF	1–10^{-1}	300 MHz to 3GHz)
		SHF	10^{-1}–10^{-2}	3 GHz to 30 GHz)
		EHF	10^{-2}–10^{-3}	30 GHz to 300 GHz)
	Infrared	IR	10^{-3}–10^{-6}	10^{11}–10^{14}
	Visible **	Red	740–625 nm	10^{14}–10^{15}
		Orange	625–590 nm	
		Yellow	590–565 nm	
		Green	565–520 nm	
		Cyan	520–500 nm	
		Blue	500–435 nm	
		Violet	435–400 nm	
	Ultraviolet	UVA	400–315 nm	10^{15}–10^{17}
		UVB	315–280 nm	
Ionising Radiation		UVC	280–100 nm	
	X ray	X-ray	10^{-8}–10^{-11}	10^{17}–10^{20}
	Gamma ray	γ- ray	10^{-11}–10^{-14}	10^{20}–10^{22}

*	1 Hz is 1 cycle per second: kHz = one thousand Hertz (kilo Hertz); MHz = one million Hertz (Mega Hertz); GHz = one thousand million Hertz (Giga Hertz)
**	The visible band is a small range of frequencies that the retina of the eye is able to detect and which gives us the sense of vision. The wavelength of the visible part of the spectrum is usually expressed in terms of nanometres (nm; 1 nm is 10^9 of a metre). Visible light extends from 740 nm (perceived as red) to 400 nm (perceived as violet). Above the violet part of the visible spectrum is the ultraviolet region, which is broken down into three bands (A, B and C). The transition from non-ionising radiation to ionising radiation occurs around the wavelength of UVC.

[R1002] Electromagnetic fields can be characterised by their frequency and/or wavelength (see TABLE 1). The frequency is measured in the unit of Hertz (Hz).

One Hz equals one cycle per second. The wavelength is measured in the unit of a metre (or for very high frequency radiations, down to nanometres (nm)). One nm is equal to one thousand-millionth of a metre (10^{-9} m). The strength of the field is measured by either the strength of the electrical component of the field or the strength of the magnetic component. The electrical field (the E-field) is measured as volts per metre (v.m^{-1}). The magnetic field can be measured as the magnetic field strength in units of Amps per metre (A.m^{-1}; the H-field) or the magnetic flux density in units of Tesla (or more usually as micro Tesla – µT; the B field). 1 µT is one-millionth of a Tesla. In addition, electromagnetic fields above 100kHz in the radio frequency part of the spectrum have the ability to heat body tissues, but the risk depends on the amount of radiation received. This is measured as the amount of power falling on a given area of the body surface in the units of Watts per square metre (W.m^{-2}). This measure is the power density. The rate at which energy is absorbed in the body tissues is measured in the units of Watts per kilogram (W.kg^{-1}). This measure, the Specific Absorption Rate (SAR), is a fundamental measure of absorbed dose, which relates to the health and safety risk and is used as the basis for guidelines to ensure safe limits of exposure (see **R1003**).

How Does Non-ionising Radiation Interact with the Body?

[**R1003**] Non-ionising radiation poses a *known* risk and an *uncertain* risk to health. The *known* risks relate to the way in which the radiation is absorbed by the body and causes the heating of body tissues, possible electrical shock and skin burn. The *uncertain* health effects relate to a possible risk of cancers and other diseases that may be linked to exposure to non-ionising radiation. Another way to distinguish these two types of risk is to consider the known effects as being *acute* risks (namely, occurring after a very short period of exposure) and the uncertain health effects as being *chronic* risks (those requiring long and/or intermittent periods of exposure).

The Known Health Risks

[**R1004**] The known health effects are separated into direct effects and indirect effects. Direct effects are those that result from a direct interaction between the body and the electromagnetic field to which the body is exposed. Indirect effects occur as a result of an interaction between the electromagnetic field, an external object (usually a metallic structure) and the human body.

Direct Effects

[**R1005**] The direct effects are usually considered differently for frequencies above and below 100kHz. Below 100kHz the hazard is one of induced electrical current in the body. Induced electrical charge on the surface of the body can be detected by a small proportion of people as a tingling sensation or as a perception of the raising or the vibration of the skin hairs. About 10 per cent of people can detect surface charges in this way, but only when the electrical field to which they are exposed is in the order of 10–15kV.m^{-1} (at ELF frequencies – see TABLE 1; 50Hz is the frequency of the electrical mains distribution and would be the usual source of exposure to fields causing this effect). This is not a health hazard but is annoying. A lot of research has focused on what physiological effects induced currents may have on the body. These have included effects on the nervous control of breathing and the heart, the effects of induced electrical signals in the brain and spinal cord, effects on performance and behaviour, effects on vision and hearing, and effects on the blood and on circadian rhythms (the body's daily cycle). The results of much of this work is equivocal and not reproducible. Conservative analysis of the data that is available has, however, allowed limits of exposure to be set, which should prevent adverse effects (see **R1009**). For frequencies above 100 kHz, the major effects are body heating. Healthy people can tolerate short-term rises in body temperature of about 1°C without significant adverse effects. However, a rise of 1°C will cause increased sweating, increased heart rate and a fall in blood pressure. A body temperature rise of 1°C may also be associated with changes in circulating hormone levels (such as thyroxin and adrenaline), and rises of more than this are considered potentially hazardous. The limits on exposure are set so that the SAR is restricted to a value that will ensure that the rise in body temperature is under 1°C (see **R1009**). There is sufficient margin built into these restrictions on exposure to take account of work rate, environmental temperature and clothing, which in themselves will also influence body temperature.

As well as the hazard of whole body heating, frequencies above 100 kHz can cause localised heating, which may present a risk of damage to particular tissues. The lens of the eye, the testes and the embryo and foetus are the tissues most often considered to be at particular risk. These localised effects are most likely to be a problem with microwave radiations, but in the case of damage to the lens of the eye, infrared and ultraviolet radiations are also a significant risk (see TABLE 1). In the case of the eye, localised heating may cause cataracts, an opacity in the lens causing visual impairment and blindness. The testes may be vulnerable because they are usually maintained at a few degrees below body temperature, so that a rise in testicular temperature can cause reduced sperm counts and transient infertility. The embryo and foetus may be particularly sensitive to heating, which has been linked to birth defects.

Indirect Effects

[R1006] Indirect effects result from an accumulation of charge in the body, which is then subsequently discharged to a conductive (usually metallic) object. The discharge current may be sufficient to cause significant electric shock, and can have serious consequences, such as paralysis of the respiratory muscles and ventricular fibrillation (an erratic beating of the heart, causing collapse and death). At lower levels of current the discharge may cause pain at the point of exit of the discharge (the point where the body is in contact with the metallic object) and sufficient localised heating to result in burns to the skin. Electrical discharges may be sudden or may be continuous. Sudden discharges occur where the body is effectively insulated from the ground, allowing the accumulation of charge, which is then suddenly discharged when contact is made with a metallic object. This is most likely to happen if the body is in an electromagnetic field of less than 100 kHz but is also extremely common with static electrical fields. This is a common occurrence in indoor environments and in cars where static electricity causes an accumulation of charge sufficient to cause a discharge on subsequent contact with a metallic object. The discharge current in such circumstances will not be sufficient to cause serious effects but will result in pain and discomfort, and is particularly distressing if repeated frequently. Alternatively, discharge may occur if the person is in continuous contact with the metallic object (or other grounded conductor) during the time of exposure to the electromagnetic field. The person effectively becomes a conduit for the electrical current to pass to earth. This is most likely to occur when working with fields of below 100 kHz and at the same time in contact with a grounded conductor. The resulting current may be sufficient to cause electric shock if the field strength is sufficient. The limitations on exposure (see **R1009**) are set at a value that should limit the contact current to below 1 milli Amp (1 mA), which has been found to be able to be tolerated by adults without adverse effect.

Implanted Medical Devices

[R1007] Both direct and indirect exposure to electromagnetic field sources may interfere with implanted devices such as cardiac pacemakers, automatic implantable cardioverter defibrillators (AICDs) and cochlea implants. It is impossible to be precise about what levels of exposure are safe for persons fitted with such devices, because each device varies, but there is a risk that induced currents will interfere with their function. In the case of pacemakers and AICDs, the consequence of malfunction could be serious. Generally, the levels of exposure that might give rise to concern will be lower than those recommended in guidelines for general exposure (see **R1009**). Anyone newly fitted with an implantable electronic medical device (or newly employed with one fitted) who is working with sources of electromagnetic radiation, or working in an area where exposure to significant sources may be a problem, must have a competent risk assessment undertaken (see **R1010**). It is also possible that individuals fitted with metallic implants, such as hip or knee replacements, will experience localised heating in certain circumstances. Such problems will again need to be addressed through a risk assessment.

Uncertain Health Effects

[R1008] These effects result from chronic (namely, long-term) exposure to electromagnetic fields, and may be caused by relatively low levels of exposure. However, what the exact risks are, and indeed if there are any risks at all, is most contentious. The risks cited mainly include cancers of various types, including cancers of the breast, brain and nervous system, leukaemia, degenerative brain and spinal cord conditions such as Amyotrophic Lateral Sclerosis (ALS), and behavioural conditions. The cancer risk is the most frequently cited and the most researched. The possibility of a cancer risk from chronic exposure to electromagnetic fields first appeared in 1979 (Wertheimer and Leeper, 1979). This study, from America, demonstrated a two- to three-fold risk in childhood leukaemia, lymphomas and nervous system tumours in houses that had a high level of magnetic fields from the use of mains electricity (in the USA these are 60 Hz magnetic fields). Since then, scores of epidemiological studies have been done to attempt to confirm these findings. Epidemiological studies attempt to look for statistical differences in disease prevalence in people exposed to a particular factor, compared with people who are not (or are less) exposed. In these studies the cancer prevalence rates are compared for people living near, for example, power transmission lines, compared to people who have never lived near a power line. The difficulty encountered by such studies is that the numbers of people exposed to power lines is fairly small and the number of those with relevant cancers is also small. The statistical power of many of the studies is therefore poor. One of biggest studies ever carried out, the UK Childhood Cancer Study, failed to find evidence that exposure to magnetic fields associated with electricity supply and transmission increases the risk of childhood cancers (UKCC, 1999). However, other studies with similar statistical validity have concluded that there is a risk between two to three times higher when the mean exposure exceeds about 0.2–0.3 µT. (see, for example, Savitz et al, 1988; Feychting and Ahlbom, 1993.) The current conclusion of the Health Protection Agency (supported by the independent Advisory Group on Non-Ionising Radiation — AGNIR), are that the low levels of exposure generally in the UK do not provide a population that is large enough to detect any effect on childhood leukaemia incidence at average levels of exposure at or above 0.4 µT. Much of the research over the last 25 years, especially on leukaemia and nervous system tumours, has been concerned with exposure to electrical power fields at 50/60 Hz. However, considerable work has also been done on exposures to radiofrequency fields. To some extent this has been because of concerns about living near to radio or television transmitter masts but probably in greater part because of the concern that the transmitted power from mobile phones may be a health risk. There is also concern that living near to mobile phone base stations poses a health risk.

In the United Kingdom AM radio uses frequencies between about 180 kHz and 1.6 MHz; FM radio between 88 and 108 MHz; and television from 470 to 854 MHz. Mobile phone networks operate between 872 and 960 MHz or 1710 and 1875 MHz. Third generation mobile communication (3G) will utilise frequencies between 1.88 and 2.01 GHz and 2.11 and 2.20 GHz (see TABLE 1). In addition, other frequencies, particularly in the microwave region, are used by the emergency services, for civilian and military radar, for satellite communications and for digital radio and television transmission. There have been a number of epidemiological studies to investigate the health risk of living near to television and radio transmitter sites with almost the same equivocal results as exist for living near to power lines. One UK study found that there was no relationship between living within two kilometres of TV and radio transmitter sites and the prevalence of leukaemia and lymphoma, although the prevalence did decline with increasing distance from the sites. (Dolk et al, 1997a). In addition, the same authors had shown that the incidence of leukaemia and lymphoma around a single transmitter site in the Midlands was increased by a factor of 1.8. (Dolk et al, 1997b). Other studies in Sydney and Hawaii have shown an increased risk of leukaemia to those living close to television and radio masts, (Maskarinec et al, 1994 and Hocking et al, 1996), whilst a study in San Francisco around a single microwave transmitter failed to find any excess incidence of disease (Selvin et al, 1992). These findings are at present too inconclusive to confirm that exposure to radiofrequency radiation is a cancer risk.

One area that has received much attention recently is the suggestion that exposure to electromagnetic radiation may increase the risk of breast cancer. A theoretical link between the two has been suggested for about 20 years. This theory suggest that electromagnetic radiation can influence the production of a hormone, melatonin, and that this in turn increase the risk of breast cancer. The independent Advisory Group on Non-ionising Radiation has concluded that EMFs do not influence the production or action of melatonin and that there is no evidence that EMF exposure is associated with an increased risk of breast cancer. (HPA 2006)

Much the same conclusion can be reached when looking at the research on mobile phone cancer risks. Animal and cellular studies suggest that exposure to the power levels received by the brain during a mobile phone conversation can cause changes in the DNA of cells, which, theoretically, could be associated with cancer (Lai and Singh, 1996). However, there is no evidence so far to suggest that this is a real risk in humans during the normal use of mobile phone technology. A study of the mortality rate and leukaemia and brain tumour prevalence among 250,000 mobile phone users did not show any increased risk. (Rothman et al, 1996). An important study published in 2006 concluded that the use of mobile phones was not associated with an increased risk of glioma (the most common type of brain cancer), (Hepworth et al. 2006). The Independent Expert Group on Mobile Phones (IEGMP) in a wide ranging and detailed review of the evidence linking mobile phones (and other RF sources) to health risks, concluded that RF exposure below the NRPB and ICNIRP guidelines do not cause adverse health effects. (IEGMP 2000). However, this report did recommend that, as a precautionary measure, the ICNIRP guidelines should be adopted in preference to those of the NRPB. The ICNIRP (International Commission on Non-ionising Radiation Protection) guidelines are more stringent than the UK guidelines (See **R1009**). The IEGMP report has subsequently been updated by two NRPB Reports (NRPB 2003, NRPB 2004). The World Health Organisation has similarly concluded that in relation to exposure to mobile phone base station emissions there is to date no scientific evidence that these cause adverse health effects (WHO 2006) Despite earlier concerns that mobile phone technology might be associated with serious adverse health risks, recent evidence suggest that this fear is unfounded. Much research has focused on the risk of leukaemia in children exposed to power frequency fields. With radiofrequency fields, research has concentrated on the risks of living close to transmission sources. In the occupational health context much of this work (especially that concerning childhood disease) is irrelevant and can only be used to determine what might be the general risks of working with power frequency electrical or magnetic fields or sources of radiofrequency transmission. However, several studies have looked specifically at the risks of working with significant sources of electromagnetic energy. Such jobs are found in electronics and electrical engineering, engineering (welding), in the electrical supply industry, on the railways, in radio and television transmission and transmitter maintenance, and in the military. Some of these studies on occupational exposures have concluded that there is a significant risk of brain cancers, leukaemias and breast cancer (in both men and women) in those occupations in which there is significant exposure to electromagnetic radiation (see, for example, Wilkins and Koutras, 1988; Savitz and Loomis,

1995; Guenel et al, 1993; Floderus et al, 1993; Loomis et al, 1994). In most cases the excess risk is between two and four times higher. However, there are also studies in which no association has been found (see, for example, Tynes et al, 1992; Sahl et al, 1993; Rosenbaum et al, 1994). At least six studies have looked at paternal occupational exposure to electromagnetic fields and the risk of brain tumour in their offspring. Three of these showed an increased risk of brain tumour (relative risk between 2 and 3.6) in the children of exposed fathers (Spitz and Johnson, 1985; Wilkins and Koutras, 1988; Johnson and Spitz, 1989), whilst the other three failed to show any relationship (Nasca et al, 1988; Wilkins and Hundley, 1990; Bunin et al, 1990).

In conclusion, health risks from chronic exposure either to community or occupational sources of low level electromagnetic radiation remain uncertain. The vast amount of research in this area has failed to uncover any consistent or reliable evidence that there is a health risk. Many of the studies are too weak to reach definite conclusions on a health risk, but the underlying body of evidence that there might be a health risk cannot be totally ignored. If there is a risk, it appears to be a very small one, perhaps doubling the incidence of diseases that are in any case rare. Brain tumours, leukaemias and lymphomas are the most consistent health effects noted. In the absence of conclusive evidence for a health risk, the suggested method of risk management is to apply the precautionary principle ('prudent avoidance').

Guidelines on Exposure

[R1009] The National Radiological Protection Board (NRPB, now the Radiation Protection Division of the Health Protection Agency) has published guidelines on exposure to electromagnetic fields (NRPB, 1993). These provide guidance on exposure only to prevent the known health risks of acute exposure (electric shock, body heating and heat damage and burn; see **R1004**). They do not provide guidance to avoid the uncertain health risks discussed in **R1008**; if such risks exist, they will occur at much lower levels of exposure. Neither does the guidance ensure that there is no risk of electromagnetic interference with implanted medical devices (**R1007**); again, these risks may be at much lower levels of exposure than the guidance indicates.

In addition to the NRPB (HPA) guidelines, there are guidelines from various other bodies, including, notably, the International Commission on Non Ionising Radiation Protection (ICNIRP). The impending importance of the ICNIRP guidelines is that they will become the basis of the legal standard in the UK, when the EC Physical Agents (Electromagnetic Fields) Directive is adopted (see **R1010**). This is likely to become law in the UK in April 2008. Other countries may use their own guidelines or will have adopted the ICNIRP guidelines. If a company has staff working in countries other than the UK, it will be necessary to check what standards are applied in that country; in the UK it will usually be satisfactory to work to the NRPB (HPA) guidelines. However, with a new legal requirement pending within the next year, it may be prudent to start now to adopt the ICNIRP guidelines (ICNIRP, 1998 and NRPB, 1999).

The principle of the guidelines are firstly that standards for frequencies above 100 kHz are set to limit the rate of absorbed radiation (the SAR; see **R1002**).

Remember that these guidelines relate to thresholds (with a considerable margin for uncertainties) to protect against the well-known direct and indirect effects of acute exposure. In the frequency range from 0 to 100 kHz the basic restriction is set as a magnetic field or magnetic flux density value (see **R1002**).

Above 100 kHz the basic restriction is 0.4 W.kg^{-1} for the whole body, averaged over any 15-minute period. The basic restriction on SAR for the head, foetus, neck and trunk and the limbs alone, have different (higher) values but are averaged over any 6-minute period. Above 10 GHz the basic restriction is expressed as a power density and is 100 W.m^{-2}.

Although the basic restrictions form the basis of the guidelines, they are not very useful in practice, as it is virtually impossible to measure the SAR under field conditions. The NRPB (HPA) have therefore set levels of exposure measured as magnetic field strength, magnetic flux density, electric field strength or power density. A variety of instruments exist for measuring field strength or power density and it is therefore much more practical to ensure compliance with the basic restrictions by measuring these exposure levels rather than by attempting to directly measure the SAR. NRPB (HPA) call these exposure values *investigation levels* (the ICNIRP calls these *reference levels*). Investigation levels are set at values of exposure that should not result in the SAR being exceeded and therefore should not result in the adverse effects of electric shock or body heating. The investigation levels are set with a considerable margin of safety in extrapolating from the SAR to values of field strength, and worse case situations are taken into account. For example, at frequencies above 10 MHz, children may be more at risk because of their body size, and the investigation levels at these frequencies are given for situations where children might be exposed. If it can be certain that no children will be exposed, the investigation levels can be relaxed.

It is beyond the scope of this chapter to reproduce the guidelines contained in the NRPB (HPA) statement or the ICNIRP guidelines. Readers who need to undertake a risk assessment must refer to the guidelines set out in the Board Statement or the ICNIRP guidelines, and be prepared to understand the complexities and nuances of how they should be interpreted. One particular complexity occurs where exposures may be to more than one frequency simultaneously. There may, under these conditions, be additive effects or combined effects, which cannot easily be related to the guideline investigation levels. It may, therefore, be necessary to seek expert advice on how to determine levels of exposure, how to interpret the guidelines and how to devise suitable means of risk control. Although instruments to measure electric and magnetic fields can be hired, it may be necessary to engage someone with experience of undertaking such measurements in order to determine actual levels of exposure.

In summary, restrictions on exposure:

- Are set to ensure that there is no risk from acute exposure to electromagnetic fields (mainly the risks of electric shock and body heating).
- Do not necessarily ensure that there no risk for those wearing implanted electro-medical devices (such as pacemakers). Individual risk assessments must be undertaken for such individuals.

- Do not consider the risk from chronic (long-term) exposure. These risks are very uncertain but cannot categorically be ruled out, so a precautionary approach should be applied in such circumstances.
- Are set to limit the absorbed dose of radiation (the Specific Absorption Rate – SAR). This is the *basic restriction*.
- Are, in practice, applied as restrictions (limitations) on the levels of exposure measured as electric or magnetic field strengths or power density. These are set out in the NRPB (HPA) guidelines as *investigation levels* or in the ICNIRP guidelines as *reference levels*.
- Can be complex to interpret, and expert help may be needed to determine levels of exposure and interpret compliance with the basic restrictions.

Legal Requirements

[R1010] There is no specific legislation dealing with non-ionising radiation. The European Commission has now issued a Directive 'On the Minimum Health and Safety Requirements Regarding Exposure of Workers to the Risks Arising from Physical Agents (Electromagnetic Fields)' (OJ L159 30.4.2004; and a corrigendum in L184/1 24.5.2004). The basic restrictions and guidelines on exposure (called reference levels in the Directive) follow those of the International Commission on Non Ionising Radiation (ICNIRP). The EU has decided to postpone the implementation of the directive in the UK until 30 April 2012. In the meantime the legal requirement is that contained in *s 2* of the *Health and Safety at Work etc. Act 1974*, which places a general duty on the employer 'to ensure, so far as is reasonably practicable, the health, safety and welfare at work of all his employees'. The requirement to undertake a risk assessment for those employees for whom exposure to electromagnetic radiation may be significantly encountered in their work is contained in the *Management of Health and Safety at Work Regulations 1999 (SI 1999 No 3242), Reg 3*. These Regulations also provide a legal requirement to identify and implement the necessary precautions to avert the risk. *Regulation 7* requires that anyone charged with assisting in the management of risk must be competent. It is therefore a legal requirement to ensure that whoever is charged with measuring electromagnetic field strengths, interpreting exposure against accepted guidelines and devising suitable means of controlling the risk must be competent. It is the employer's legal duty to ensure competence in anyone required to undertake such work.

Regulation 5 may require exposed individuals to be placed under health surveillance. The need for this should be determined by the risk assessment, in consultation, if necessary, with an experienced Occupational Health Physician.

Ionising Radiation – What is it?

[R1011] Ionising radiation is radiation with sufficient energy to electrically charge or *ionise* material that the radiation strikes. Ionising radiation arises from a wide range of natural and man-made radioactive sources, and its properties have been harnessed extensively for industrial and medical use. Ionising radiation can, however, adversely influence living tissue and its use must always be assessed and monitored carefully to prevent harmful effects. Types of radiation that cause ionisation include:

- α (alpha) particles;
- β (beta) radiation;
- γ (gamma) radiation;
- x rays;
- neutrons.

Units of Measurement: Radioactivity and Radiation Dose

[R1012] Radiation is emitted from radioactive sources and this radioactivity is measured in units known as *bequerels*. Radiation dose to humans is defined as the amount of energy absorbed by the body and the *gray (Gy)* is the unit of dose. Different types of radiation have, however, different effects on tissue, and in order to take account of these differences, a further unit of dose, the *sievert*, has been developed.

The *sievert (Sv)* is the most commonly used measure of radiation in occupational and environmental exposures. Doses, however, are usually measured in smaller fractions of the sievert, either *millisieverts* or *mSv* (one-thousandth of a sievert), or *microsieverts* or *?Sv* (one-thousandth of a millisievert).

Natural Radiation

[R1013] On earth there are two main sources of natural or background radiation: cosmic radiation from space and natural radiation from the ground. Most cosmic radiation is filtered by the upper layers of the atmosphere, but ionising radiation from space can be significant for those in the space and aviation industry. A typical radiation dose for a flight from the UK to Spain would be about 10 ?Sv (NRPB, 1994).

Certain rock types, such as granite, emit weak ionising radiation but the most significant source of background radiation from the ground in the UK occurs from radon, a naturally occurring gas, which largely accounts for the difference in background radiation across the UK. Typical annual doses from background radiation in the UK range from 2.1 mSv in London, where Radon levels are low, to 7.8 mSv in Cornwall, where they are much higher (NRPB, 1994).

Man-Made Radiation

[R1014] The development of ionising radiation for industrial and medical purposes has produced significant benefits. Major uses include:

- Medicine – Ionising radiation has a number of medical applications including the diagnosis of disease, using techniques such as computed tomography (CT) scanning or X-ray radiography, and in the treatment of cancer.
- Nuclear Power – In UK in 2004, about 20 per cent electricity was generated by nuclear energy.
- Radiography in industry – Radiographic techniques are used in monitoring the integrity of metal structures such as pipelines and bridges and X-ray machines are used widely at ports and airports for baggage security.

- Consumer products – Radioactive materials are used in a number of consumer products including smoke detectors and in luminous dials.

Harmful Effects

[R1015] Human beings, like all organisms on earth, are made up of a structure of *cells*, the 'building blocks' of living tissue. Ionising radiation can damage essential proteins in these cells, and this damage can give rise to both short and long term health effects.

- Short-term effects are generally related to the capacity of ionising radiation to kill cells, and this is proportional to the dose of radiation. Doses above 3 Gy to the skin will produce redness brought about by the death of skin tissue. In humans whole body doses of greater than about 4–5 Gy can be fatal.
- Lower radiation doses can damage but not kill living cells. The long-term effects of ionising radiation arise as a result of imperfect repair of damage to cells leading to abnormal cell growth. In turn, this abnormal cell growth can lead to cancer. These effects are random, but in a working population the possibility of developing cancer is *increased* by about 4 per cent for every sievert of radiation dose received (NRPB, 1995).

Irradiation and Contamination

[R1016] In considering the possible harmful effects of radiation, it is useful to differentiate between *irradiation* and *contamination*. Exposure to penetrating radiation from an external source, *irradiation*, may cause an individual to sustain damage to their body depending upon the dose, but the individual would not remain radioactive and would not become a hazard to others. Once the source of radiation is removed, the individual would not receive a further dose. An example of this would be a chest X-ray.

By contrast, if an individual has radioactive material on them or in their body, they are *contaminated*. As the *contamination* continues to emit radiation, they will continue to receive a radiation dose until the contamination is removed or declines naturally, even if they leave the area where the contamination occurred. They may also transfer the contaminated material to others, who may become at risk.

Protection against Radiation

[R1017] The key to radiation protection lies in understanding the nature of the hazard. Different types of radiation can be blocked or *shielded* by different types of material. For instance, α (alpha) particles can be stopped by a sheet of paper, whereas β (beta) radiation requires a sheet of metal and γ (gamma) radiation a thickness of lead.

Other properties of radiation can be used when considering protection. The effect of radiation, for example, is reduced by distance from the source, and if the distance is doubled, the dose is reduced to a quarter. This is the *inverse square* phenomenon. The duration of exposure is also important. The less time an individual is exposed to a source, the less the received radiation dose will be.

In UK, background information and advice on radiation protection is available from the Radiation Protection Division at the Centre for Radiation, Chemical and Environmental Hazards (formerly known as the National Radiological Protection Board). The Centre for Radiation, Chemical and Environmental Hazards is the responsibility of the Health Protection Agency (HPA) and access to radiation protection information is via the HPA website www.hpa.org.uk.

Legal Requirements

[R1018] In the UK work with ionising radiation is covered by the *Ionising Radiations Regulations 1999 (SI 1999 No 3232 (IRR99))*. Other legislation covering ionising radiation can be accessed via the UK Government Cabinet Office website www.opsi.gov.uk, and includes:

- The *Radioactive Substances Act 1993*, the legislation by which the Environment Agency regulates the use and disposal of radioactive materials.

- The *High-activity Sealed Radioactive Sources and Orphan Sources Regulations 2005 (SI 2005 No 2686)* which implement the EU High-activity Sealed Sources Directive

- The *Radiation (Emergency Preparedness and Public Information) Regulations 2001 (SI 2001 No 2975)*, which regulate emergency planning and information if the public is likely to be involved.

- The *Radioactive Material (Road Transport) Regulations 2002 (SI 2002 No 1093)* and the *Radioactive Material (Road Transport) (Amendment) Regulations 2003 (SI 2003 No 1867)*, which regulate the package and transport of radioactive materials

- The *Justification of Practice Involving Ionising Radiation Regulations 2004 (SI 2004 No 1769)*, by which the Department of Environment, Food and Rural Affairs assesses new processes involving ionising radiation.

Medical exposures to ionising radiation are governed by the Department of Health through:

- The *Ionising Radiation (Medical Exposure) Regulations 2000 (SI 2000 No 1059)*.

- The *Ionising Radiation (Medical Exposures) (Amendment) Regulations 2006 (SI 2006 No 2523)*.

- The *Medicines (Administration of Radioactive Substances) (Amendment) Regulations 1995 (SI 1995 No 2147)*.

In relation to work with ionising radiation, health and safety managers may also have to consider the *Management of Health and Safety at Work Regulations 1999 (SI 1999 No 3242)* (MHSWR) and the *Personal Protective Equipment Regulations 2002 (SI 2002 No 1144)*.

Key Features of the Ionising Radiations Regulations 1999

[R1019] Employers with staff working with ionising radiation, or those who are considering work with ionising radiation, should study carefully the *Ionising Radiations Regulations 1999*, together with the Approved Code of Practice and Guidance (ACOP) (ISBN 0 7176 1746 7). Key sections for the non-specialist include:

- Risk assessment – *Regulation 7* requires employers commencing any new activity with ionising radiation to undertake a 'suitable and sufficient' risk assessment. The ACOP provides useful general advice on risk assessment and lists specific items to be considered.
- Personal protective equipment – *Regulations 8* and *9* make reference to the use of personal protective equipment. Employers must ensure that such equipment is appropriate, properly used and maintained in accordance with the *Personal Protective Equipment Regulations 2002*.
- Management of radiation protection – *Regulation 13* requires employers to consult a suitable Radiation Protection Advisor (RPA) to advise on the Regulations. Employers must ensure that the RPA fulfils the HSE's criteria of competence but will find that an RPA will be particularly useful in providing practical guidance in applying the Regulations. The HSE website, www.hse.gov.uk/radiation/ionising, is helpful for employers new to work with ionising radiation, and the site has lists of organisations with HSE-approved radiation protection advisors.
- Co-operation between employers – *Regulation 15* requires employers sharing the same workplace to co-operate with each other to co-ordinate the measures they take to comply with statutory duties. This reflects the general obligation of the *Management of Health and Safety at Work Regulations 1999 (SI 1999 No 3242, Reg 11* of employers sharing the same workplace to co-operate.
- Designated areas – *Regulations 16 – 19* cover the requirements of employers to designate and monitor areas in which employees may receive a significant radiation dose. These areas may be designated as *controlled* or *supervised*, and employers are obliged to have in place written *local rules* to identify key working instructions to restrict radiation exposure in these areas and the steps to be taken in the event of a radiation accident. Employers are also required to appoint *Radiation Protection Supervisors* to ensure compliance with the Regulations and local rules.
- Designation of classified persons – *Regulation 20* requires employers to designate as *classified persons* (or classified workers) employees who are likely to receive a radiation dose greater than 6 mSv per year or an equivalent dose which exceeds three-tenths of any dose limit. In practice many employees become classified workers because they are required to enter designated controlled areas and, frequently, actual radiation doses may be substantially less than 6 mSv per year. Classified persons must be over 18 years of age and under medical surveillance by an appointed doctor (see below).
- Dose assessment and recording – *Regulation 21* sets out the requirement for employers to appoint an *approved dosimetry service* (ADS) to monitor the radiation dose and to keep the dose records. Dosimetry services have to be approved by the HSE for certain functions, and employers should confirm that the chosen dosimetry service has an appropriate certificate of approval. The HSE website, www.hse.gov.uk/radiation/ionising, lists dosimetry service providers with HSE approval.
- Medical surveillance – *Regulation 24* sets out the requirement for employers who employ classified persons to make arrangements for their medical surveillance. The medical surveillance of classified persons must

be undertaken by a registered medical practitioner who has been appoint ed in writing by the HSE in accordance with the Regulations. Lists of na mes of *appointed doctors* are held by local HSE offices, but employers should note that an HSE Form FOD MS38A (available from local HSE offices) needs to be completed for each organisation for whom an app ointed doctor acts. A *health record* has to be maintained to record the outcome of annual medical surveillance for each classified person, and the requirements of this record are set out in Schedule 7 of the ACOP. Form F2067 is the standard form available from the HSE. Note that this health record is separate from the employee's medical file and should be kept available for scrutiny by a Factory Inspector. Initial health surveilla nce should include a physical examination of the employee by the app ointed doctor. Subsequent annual surveillance may be by review of s ickness absence and radiation dose only ('a paper review') but it is good practice for employees to be examined every three or four years. Certain high-risk groups of employees, such as industrial radiographers, must, however, be examined annually.

- Appeals – In the event that an appointed doctor assesses an employee as unfit, the employee has the right to apply to the HSE for a review of that decision. The appeal must be made in writing or by completing the form IRR (Med) 1, available from the local HSE office.

- Female employees – The IRR make special provision for female emp loyees, and *Regulation 8(5)* sets out the requirement for an assessment of risk for them. The external radiation dose to the abdomen of woman of reproductive capacity should not exceed 13 mSv in any consecutive three-month period (Schedule 4 of IRR), and the employer should con- firm this annually before medical surveillance.

- Pregnant and breast-feeding employees – *Regulation 8* requires emp loyers, on being notified of pregnancy in an employee, to restrict the dose to the foetus to 1 mSv for the remainder of the pregnancy. In employees who are breast-feeding, the employer must ensure that significant ra diological contamination is avoided.

- Outside workers and passbooks – *Outside workers* are employees des ignated as classified persons who carry out services in a controlled area designated by another employer. A key feature of outside workers is their requirement to hold a *radiation passbook (Regulation 21(5))*, which is provided by the ADS.

- Overexposures – *Regulations 25* and *26* detail the action to be taken in the event of an overexposure. The employer must notify the HSE and the appointed doctor as soon as practicable. Care should also be taken to notify other employers in the event that their employees are involved. A special medical examination should be undertaken by the appointed doct or in the event that the effective dose of radiation to an employee is great er than 100 mSv in a year or an equivalent dose of at least twice any relevant annual dose limit. In practice it is recommended that employees are seen by the RPA and the appointed doctor at an early opportunity and provided with as much information and support as possible. Anxiety and the psychological aspects of overexposures are usually much more significant to the employee than the physical factors.

Key Features of Other Legislation relating to work with Ionising Radiation

[R1020] Much other legislation relating to work with ionising radiation is specialised and beyond the scope of this overview. In the event, however, that an employer believes that any of this legislation might apply, the advice of an appropriate RPA should be sought at an early opportunity.

Emergencies and Unforeseen Incidents

[R1021] The *Ionising Radiations Regulations 1999* place an obligation on the employer to prepare a contingency plan if the risk assessment shows that an accident is foreseeable (*Regulation 12*). The *Radiation (Emergency Preparedness and Public Information) Regulations 2001* set out the requirements for emergency planning and the obligation for information to be made available to the public in the event that the public could be affected.

The UK has a scheme, the National Arrangements for Incidents involving Radioactivity (NAIR). NAIR is a set of national arrangements that provide a 'long-stop' to other emergency plans. NAIR has been designed to provide advice and assistance to the police in incidents involving radioactivity where members of the public may require protection. The arrangements have been devised around assistance to the police, since they will normally be among the first informed of any incident in a public place. In an incident the police check whether there are plans to deal with the event. If no suitable plans exist, or if plans fail, NAIR assistance should be sought. NAIR assistance may be obtained by means of a 24-hour national notification telephone number (0800 834 153). This connects to the United Kingdom Atomic Energy Authority Constabulary, Force Communications Centre, who will take details of the incident and contact the nearest respondent (? Health Protection Agency – Radiation Protection Division).

References

[R1022] Bunin G R, Ward E, Kramer S, et al (1990) Neuroblastoma and parental occupation, Am.J. Epidemiol. 1990; 131:776-780.

Dolk H., Elliott P, Shaddick G, Walls P, and Thakrar B (1997a) Cancer incidence near radio and television transmitters in Great Britain. 2. All high powered transmitters, Am.J. Epidemiol. 1997; 145:10.

Dolk H, Shaddick G, Walls P, Grundy C, Thakrar B, Kleinschmidt L, and Elliott P (1997b) Cancer incidence near radio and television transmitters in Great Britain. 1. Sutton Coldfield Transmitter, Am.J. Epidemiol. 1997; 145:1-9.

Feychting M and Ahlbom A (1993) Magnetic fields and cancer in children residing near Swedish high-voltage power lines, Am.J. Epidemiol. 1993; 138:467-81.

Floderus B, Persson T, Stenland C et al (1993) Occupational exposure to electromagnetic fields in relation to leukaemia and brain tumours: A case control study in Sweden. Cancer Causes Control 1993; 4:465-476.

Guenel P, Raskmark P, Andersen J B, and Lynge E (1993) Incidence of cancer in persons with occupational exposure to electromagnetic fields in Denmark, Br.J.Ind.Med 1993; 50:758-764.

Hepworth S.J., Schoemaker M.J., Muir K., Swerdlolw A.J., van Tongeren M.J., and McKinney P.A (2006) Mobile Phone Use and the risk of glioma: case control study. BMJ, doi:10.1136/bmj.38720.687975.55 (20th Jan 2006).

Hocking B, Gordon I R, Grain H L, Hatfield G E (1996) Cancer incidence and mortality and proximity to TV towers, Med J.Aust 1996; 165:601.

HPA (2006) Power Frequency Electromagnetic Fields, Melatonin and the Risk of Breast Cancer. Documents of the Health Protection Agency, Series B: Radiation, Chemical and Environmental Hazards, RCE-1, Feb 2006. ISBN 0 85951 573 3. Available to download from the HPA website at: www.hpa.org. uk/radiation/publications/docs_hpa/index.htm

HPA (2007) Health Protection Agency, www.hpa.org.uk/radiation/faq/emf/ emf28.htm

ICNIRP (1998), Guidelines for limiting exposure to time-varying electric and magnetic fields (up to 300 GHz) Health Physics 1998; 75(4):494-552.

IEGMP (2000), Independent Expert Group on Mobile Phones. Chairman Sir William Stewart. Mobile Phones and Health. (2000), NRPB Chilton, Didcot, Oxon.

Johnson C C and Spitz M R (1989) Childhood nervous system tumours: An assessment of risk associated with paternal occupations involving use, repair or manufacture of electrical and electronic equipment, Int.J. Epidemiol. 1989; 18:756-762.

Lai H and Singh N P (1996) Single and double-strand DNA breaks in rat brain cells after acute exposure to radiofrequency microwave radiation, Int.J.,Radioat. Biol. 1996; 69:513.

Loomis D P, Savitz D A, and Ananth C V (1994) Breast cancer mortality among female electrical workers in the United States, J. Natl Cancer Inst. 1994; 86:921-925.

Maskarinec G, Cooper J and Swygert L (1994) Investigation of increased incidence in childhood leukaemia near radio towers in Hawaii; preliminary observations. (1994), J.Environ. Pathol. Toxicol. Oncol. 1994; 13:33.

Nasca P C, Baptiste M S, MacCubbin P A et al (1988) An epidemiologic case-control study of central nervous system tumours in children and paternal occupational exposures, Am.J. Epidemiol. 1988; 128:1256-1265.

NRPB (1993) Board Statement on Restrictions on Human Exposure to Static and Time Varying Electromagnetic Fields and Radiation, Documents of the NRPB, Volume 4, No.5. 1993.

NRPB (1994) Radiation Doses – Maps and Magnitudes 'At-a-Glance' series of leaflets published by NRPB.

NRPB (1995) Risk of Radiation-induced Cancer at Low Doses and Low Dose Rates for Radiation Protection Purposes, Documents of the NRPB Vol 6 No 1 1995.

NRPB (1999) Board Statement: Advice on the 1998 ICNIRP Guidelines for Time-varying Electric, Magnetic and Electromagnetic Fields (up to 300 GHz), Documents of the NRPB Vol. 10 No. 2 1999.

NRPB (2003). Health Effects from Radiofrequency Electromagnetic fields: Report of an independent Advisory Group on Non-ionising Radiation. Documents of the NRPB Vol 14, No.2 NRPB Chilton, Didcot, Oxon.

NRPB (2004). Mobile Phones and Health 2004: Report by the Board of the NRPB. Documents of the NRPB Vol 15 No 5, NRPB Chilton, Didcot, Oxon.

Rosenbaum P F, Vena J E, Zielezny M A and Michalek A M (1994) Occupational exposures associated with male breast cancer, Am.J. Epidemiol 1994; 139:30-36.

Rothman K J, Loughlin J E., Funch D P, and Dreyer N A (1996) Overall mortality of cellular telephone customers, Epidemiology 1996; 7:303.

Sahl J D, Kelsh M A, and Greenland S (1993) Cohort and nested case-control studies of haematopoietic cancers and brain cancers among electric utility workers, Epidemiology 1993; 4:104-114.

Savitz D A and Loomis D P (1995) Magnetic field exposure in relation to leukaemia and brain cancer mortality among electric utility workers, Am.J. Epidemiol. 1995; 141:123-134.

Savitz D A, Wachtel H A, Barnes F, John E M, Tvrdik J G (1988) Case control study of childhood cancer and exposure to 60 Hertz magnetic fields, Am.J. Epidemiol. 1988; 128:21-38.

Selvin S, Schulman J and Merrill D W (1992) Distance and risk measures for the analysis of spatial data: a study of childhood cancers, Soc.Sci.Med 1992; 34:769.

Spitz M R, and Johnson C C (1985) Neuroblastoma and paternal occupation: A case-control study, Am.J.Epidemiol. 1985; 121:924-929.

Tynes T, Jynge H, and Vistnes A (1994) Leukaemia and brain tumours in Norwegian railway workers: A nested case-control study, Am.J. Epidemiol. 1994; 139:645-653.

UKCC (1999) United Kingdom Childhood Cancer Study. Exposure to power-frequency magnetic fields and the risk of childhood cancer, Lancet 1999; 354: 1925-1931.

Wertheimer N and Leeper E (1979) Electrical wiring configuration and childhood cancer, Am.J. Epidemiol. 1979; 109:273-84.

WHO (2006) Electromagnetic fields and public health. Base Stations and wireless technologies. Fact Sheet No. 304 May 2006. World Health Organisation.

Wilkins J R 3rd and Hundley V.D (1990) Paternal occupational exposure to electromagnetic fields and neuroblastoma in offspring, Am.J. Epidemiol. 1990; 131:995-1008.

Wilkins J R 3rd, and Koutras R.A. (1988) Paternal occupation and brain cancer in offspring, Am.J.Ind. Med. 1988; 14:299-318.

Rehabilitation

Leslie Hawkins

What is rehabilitation?

[R2001] The term vocational rehabilitation is now widely used to describe the process of broadly getting people back to work. However different stakeholders have different views of what is meant by vocational rehabilitation. One view is that it refers to getting people into employment when for some reason they may have been out of work for a long period of time or perhaps have never worked, and indeed this is one important aspect of rehabilitation. The reasons why people may not have worked for a long time, or indeed not have worked at all, are complex and only in some cases may this be due to a disability or long term health impairment. The other view is that vocational rehabilitation relates to the successful return to work of an employee who has been ill or has been injured or who is for some other reason, taking significant periods of absence. The employee may eventually return to his or her original job or may return to a modified job or even a new job within the organisation. Either way this is good for the employer who will reduce the costs of absence and will retain a valued employee. The two views therefore refer in the one instance to getting people into employment who are currently unemployed and in the other instance of getting people who do have a job back into the workplace usually following a period of sickness or injury. It is this latter which will be explored in more detail here since it concerns the relationships between employers and their employees and concerns the successful management of the organisation.

Why should we be concerned?

The business argument

[R2002] All organisations should be concerned to maximise the use of resources and minimise their operational costs. People are usually an organisation's most valuable and most costly asset and it makes good business sense therefore, to ensure as far as possible, that their employees are absent as little as possible. On average in the UK about 2 to 3% of the workforce are absent from work every day because of illness or injury. In some organisations the sickness absence rate can be considerably higher than this.

An annual survey by the Confederation of British Industry (CBI) published in June 2010 calculated that the country's economy lost 190 million working days to absence in 2010, with each employee taking 6.5 days off sick on average. It found that this cost the economy £17 billion, including over £2.7bn from 30.4 million days of non-genuine sickness absence.

An annual survey by the manufacturing employers' organisation EEF, published in July 2010, reported a steady decrease in sickness absence between 2007 and 2010 with the average employee taking five days off sick in 2010, compared with 6.7 days in 2007.

But the Chartered Institute of Personnel and Development (CIPD) annual survey of absence management reported in October 2010 that there had been an overall increase in reported absence, with 9.6 days absence per year per employee in the public sector, compared with 6.6 days in the private sector.

The Chartered Institute of Personnel and Development (CIPD) says that most absence is genuine and clearly a proportion is unavoidable,with much short, self-limiting periods of absence with common causes such as colds, headaches and stomach upsets. About 80% of sickness absence is short termThe remaining 20% of sickness absence is long term – employees who are absent for weeks or months with a condition that renders them unable to travel to work and/or to carry out their expected duties. This 20% of long-term sickness absence accounts for 70% of the costs of sickness absence. From a business point of view therefore:

- reducing the impact of long term sickness absence;
- controlling the non-genuine reasons for absence;
- controlling frequent short-term absences,

are essential in reducing costs. The processes of rehabilitation can play an important role in helping an organisation to reduce the impact of these types of absence.

There is no agreed definition of short term and long term. Often short term refers to absences of 7 days or less which coincides with the period most organisations will accept self-certification. However, long term is not normally taken to be more than 7 days and more usually is defined as three weeks or more. Most organisations will have their own definition of short and long term absence and will have these contained within their absence policy.

In addition to single short term absences, there will be people who take very frequent one or two day absences which over the course of a year add up to a significant amount of lost time. Frequent short term absence has different causes from continuous long term absence but nevertheless the problem needs to be addressed.

Absence costs a great deal of money both in terms of direct salary loss, loss of business or business inefficiency and replacement costs and needs to be managed like any other aspects of an organisation. It is important to focus attention to where efforts to manage absence are most likely to pay dividends. It is pointless devoting energy to unavoidable short term absence but having control over frequent short term absence, avoidable short term absence and long term absence can be very cost effective. Rehabilitation can be seen to be the process by which people can be enabled back to work or to a more reliable work pattern (see **R2007** below).

The legal argument

[R2003] Although it is important to get people back to work as soon as possible and to limit the amount of sickness absence time taken, there is no legal duty to do so. The Health and Safety Executive (HSE) produces information and advice for employers on rehabilitation, but this is guidance and contains no legal obligation. Rehabilitation often involves the intervention of Occupational Health Services but again many organisations will not have access to Occupational Health provision and in the UK there is no legal obligation to do so (see **R2009** below). Rehabilitation therefore is something employers will want to provide because of the benefits to the business rather than any direct legal necessity for it.

Common law

[R2004] Despite there being no legal obligation to provide for rehabilitation, there are nevertheless a number of important legal ramifications. In the case of *Young v The Post Office* [2002] EWCA Civ 661, [2002] IRLR 660, the Court of Appeal ruled that employers were in breach of their duty of care in failing to ensure that arrangements made for the claimant's return to work were adhered to. In this case the claimant had returned after an absence of four months due to a nervous breakdown brought on by stress at work. The arrangements for his return to work were not adhered to with the result that, within seven weeks of returning to work, the claimant suffered a recurrence of his psychiatric illness. An important aspect of this judgement is that the court rejected the suggestion that it was up to the claimant to speak out if he felt that he was under stress. This was a case involving stress, but the principle would equally apply to return to work following other illnesses or injury. The employer is liable for the subsequent injury to the employee if adequate arrangements are not made for the return to work and/or if the arrangements made are not adhered to. It is up to the employer to monitor the employee and not to rely on his or her reporting of any worsening of their condition.

This case and others like it (see *Walker v Northumberland County Council* [1995] 1 All ER 737), have established an important principle. If an employer knows that an employee has been off sick and will be returning with a vulnerability, then the employer has a legal duty to ensure that the employee's residual condition is not gong to be made worse on return to work. However the employer's liability will be subject to the usual principles of law. These are as follows.

Foreseeability   The employee must show that his employer should have been
                    able to reasonably foresee that the conditions of work on his
                    return would have made his condition worse or prevented his
                    recovery. Foreseeability will depend on what an employer
                    knows or ought reasonably to know about the employee's con-
                    dition. For an employee who has been off sick for reasons that
                    are well documented by Medical Statements (now known as fit
                    notes rather than sick notes) then it would be difficult for an
                    employer to argue that he did not know of the employee's con-
                    dition. If the employer has any doubts about the vulnerability
                    the employee may suffer on his return to work and how this
                    vulnerability might be affected by the conditions of work, then
                    it would considered reasonable for the employer to have taken
                    steps to find out. This would normally mean referral to an Oc-
                    cupational Health professional and a request for an opinion on
                    what adjustments might need to be made to ensure any risks
                    inherent in the employee's return to work were minimised (see
                    R2009 below).

Breach of duty   An employer is in breach of his duty of care to his employee if
                    he has failed to take reasonable steps to prevent the employee
                    being harmed. What is reasonable will depend on a number of
                    factors including the size of the undertaking, its resources, and
                    what is feasible. It may eventually be necessary for an employer
                    to argue in court that what he did, or failed to do, was reason-
                    able. It is important to note that breaches of statutory regula-
                    tions can often be used to show breach of common law duty.
                    The most significant change occurred with the *Management of
                    Health and Safety at Work and Fire Precautions (Workplace)
                    (Amendment) Regulations 2003, (SI 2003 No 2457)* which
                    amended regulation 22 of the *Management of Health and
                    Safety at Work Regulations 1999 (SI 1999 No 3242)*. making it
                    possible for employees to bring a civil claim for damages as a
                    result of the failure of an employer to comply with *SI 1999 No
                    3242*. (Employees have been afforded protection against civil
                    claims by third parties in the Management of Health and Safety
                    at Work (Amendment) Regulations 2006 (SI 2006 No 438)). In
                    terms of rehabilitation the most likely implication is that em-
                    ployers will have to ensure that risk assessments and risk con-
                    trols (including the need for health surveillance under
                    regulation 6) required under *SI 1999 No 3242* are competently
                    carried out and amended to include any factors which arise as
                    a result of the employee's altered state of heath on return to
                    work.

Causation   The employee will need to show that the breach of duty has
                    caused, or has materially contributed, to the harm that he has
                    suffered. It would not be sufficient to show that the work by
                    itself caused an exacerbation of the residual illness or injury
                    someone returned to work with. It must be shown that the em-
                    ployer did not do all that was reasonably practicable (and/or
                    did not comply with relevant statutory duties), and that this
                    breach of duty was responsible for the harm caused.

Apportionment and quantification	Where a worsening of a condition on return to work may have more than one cause, then the courts may have to decide on apportioning the damages. The courts may also have to decide on any contribution the claimant may have made to the worsening of his condition (by not complying with the risk controls put in place by the employer for example) and may also have to decide whether the deterioration in a person's condition may have been going to happen anyway (because for example it is a progressive condition). The principle is that the employer should only pay for that proportion of the damage caused by his breach of duty.

The Equality Act 2010

[R2005] Another aspect of the legal argument for ensuring that rehabilitation is firmly on an organisation's agenda, is the requirements of the *Equality Act 2010 (EA 2010)*..

EA 2010 came into effect on 1 October 2010 and replaced a raft of previous discrimination legislation, including the Disability *Discrimination Act 2005 (DDA 2005)*. Disability is a protected characteristic under the Act and it covers direct discrimination, indirect discrimination, discrimination by association, perceptive discrimination and protection against victimisation and harassment. It also introduced a new offence of "discrimination arising from disability".

The Act requires employers to treat people with disabilities no differently from those without that disability and to make reasonable adjustments to the work or the workplace to enable someone with a disability to work. *EA 2010* applies to all aspects of employment including at the point of recruitment, but in the context of rehabilitation its importance is that it may need to be applied when someone returns to work following an illness or injury. However, much depends on the definition of disability and what adjustments might be considered reasonable under the Act. If an employee claims breach of *EA 2010*, redress is generally through an employment tribunal rather than the courts.

Under the Act a disability is regarded as any mental or physical impairment that has a substantial and long-term adverse effect on a person's ability to carry out their normal day-to-day activities. *EA 2010* abolished the list of specified "day-to-day" activities which included mobility, manual dexterity and physical coordination which was widely regarded as unnecessarily restrictive.

If an employee returns to work with a residual condition, following an illness, injury or operation, and the condition is likely only to be temporary before the person returns to normal full capacity, then that would not fall within the requirements of *EA 2010*. However if the condition is expected to be long term (the Act defines this as lasting at least twelve months) then adjustments have to made. Long term may also be defined as recurring, if it is likely that the condition (for example epilepsy) may recur at least once in twelve months.

Examples of adjustments that may be necessary and which would be considered reasonable include making adjustments to premises, re-deploying someone to a different job or different job location, altering the person's work hours, allocating some of the person's job to another person or allowing time off for rehabilitation or treatment. This is not an exhaustive list and many other types of adjustment such as providing new work equipment or modifying existing equipment would need to be considered.

Health and safety and the Equality Act 2010

[R2006] There is a potential conflict between an employer's duties under health and safety law and disability discrimination law. This arises when an employer may decide that it is unsafe to employ a person with a disability or, when considering rehabilitation, that it is unsafe to retain someone in the workplace or allow them back to work. The employer may come to the decision that a person has become a danger to themselves or to other people in the workplace as a result of a disability and uses health and safety law to justify not allowing someone back to work. A Health and Safety Executive (HSE) report on this (HSE Research Report No 167, *The extent of use of health and safety requirements as a false excuse for not employing sick or disabled persons*) (www.hse.gov.uk/research/rrpdf/rr167.pdf), suggest that in 40% of cases heard at employment tribunals this justification was not upheld and the tribunal ruled that adequate adjustments were possible but had not been made. What happens though if an employer decides that it is unsafe for an individual to continue in employment and no reasonable adjustments are possible, but the employee wants to continue working? Case law on this is confused. In the case of *Withers v Perry Chain Co Ltd* [1961] 3 All ER 676 it was concluded that there was no common law duty to dismiss in these cases. Although this precedes both *DDA 1995* and *EA 2010*, a similar judgement was made in the Appeal Court in the case of *Hatton v Sutherland and other appeals* [2002] EWCA Civ 76, [2002] 2 All ER 1. However, in *Coxall v Goodyear Great Britain Ltd* [2002] EWCA Civ 1010 [2002] IRLR 742, Mr Coxall brought a claim for damages against his employer for allowing him to continue to work in an environment which had caused his asthma, even though he had wanted to continue working. The Court of Appeal held that an employer may be under a duty to stop an employee from working if he is put at risk by continuing to do so, even though it was the employee's wish to continue.

When an employee has been off sick with an illness (whether physical or mental) or an injury or develops a chronic illness and when the individual appears ready to return to work, then the implications of discrimination law and health and safety law have to be weighed up very carefully. If there is a residual disability which appears to be both long term and has substantial adverse effects on the person's ability to carry out day-to-day activities, then *EA 2010* applies and the question of reasonable adjustment to avoid discrimination has to be considered. If there appears to be a health and safety reason why the person should not continue in that employment then it may be necessary to cease the person's employment. However, this step should only be made after thorough examination of possible adjustments has been made including additional health and safety precautions.

Managing sickness absence

[R2007] A number of terms are used to describe the systems that should be in place to reduce absence through sickness. These include:

- absence management;
- sickness absence management;
- attendance management.

Absence management addresses the whole problem of non-attendance at work, some of which include:

- genuine absence for sickness and injury,

but may also include:

- absence for other reasons such as routine dental and doctors appointments;
- staying at home with sick children;
- compassionate leave;
- unauthorised absence.

An organisation will normally want to have a policy on such issues and rules by which its employees will be expected to conform as part of their contract of employment. When referred to specifically, sickness absence means absence from work because of mental or physical ill-health, injury or recovery from treatments such as surgery, chemotherapy or radiotherapy. Within the boundaries of sickness absence lies the problematic question of absence for non-genuine sickness. (see **R2002** above).

Attendance management is not dissimilar to absence management but places the emphasis on the more positive aspects of encouraging people to stay at work or return to work rather than the sometimes negative and heavy-handed methods employed to discourage sickness absence. There has been a drive in the public services to reduce the level of short-term sickness absence. This has translated into policies to discourage employees from taking sick leave even though the reasons may be genuine. For example some employers now require a return-to-work interview with the line manager after just one day's sick leave, others require staff who are off sick to report their intended absence on a daily basis to an occupational health nurse, and some have penalised sickness absence by withdrawing privileges such as the right to work overtime.

The reverse of absenteeism is presenteeism. Presenteeism is the trend, encouraged by stigmatising absenteeism, of people being at work who should be off for reasons of ill health. Kivimaki et al (2005) have found that there is twice the risk of coronary heart disease episodes (non-fatal myocardial infarction and fatal coronary heart disease) in men who took no sickness absence despite being ill, compared with those who took moderate levels of sickness absence. This seems to suggest that encouraging people to remain at work, even though they are ill, has damaging health effects.

The conditions that people often endure and continue working include:

- asthma;
- migraine;

- allergies;
- back pain;
- other musculoskeletal disorders;
- depression;
- irritable bowel syndrome.

Consider too that there is a huge cost of having people at work who are not well. There could be safety implications, increased risk of accident and reductions in efficiency and productivity. Two studies by Stewart and his colleagues in America have shown that presenteeism with depression costs US companies $31 billion (£15.8 billion) per year and with common pain conditions $61 billion (£31.2 billion) per year. Pain conditions include headache, back pain, arthritis and other musculoskeletal complaints. The extremely common complaint of headache loses companies on average 3.5 hours per affected employee per week (Stewart et al 2003).

In 2010, the TUC reported that in the month previous to its survey on sickness absence, around a fifth of respondents said that they had gone to work when they were "too ill to do so". A further 36% of respondents reported that, in the last year, they had been to work when they were "too ill to do so". Only 13% of respondents stated that they had never been to work when they had been "too ill to do so".

Sickness absence management therefore has to take a very careful line between exerting too much pressure on genuinely sick people to remain at work and being too lax in allowing those who are not genuinely ill to take time off without question. The key to managing sickness absence is to have a well-developed policy which is written, transparent and made available to everyone The *Employment Rights Act 1996*, requires employers to provide staff with information on 'any terms and conditions relating to incapacity for work due to sickness or injury, including any provision for sick pay'.

The Chartered Institute of Personnel and Development (CIPD) (http://www.cip d.co.uk), suggest that a policy should include:

- details of contractual sick pay terms and its relationship with statutory sick pay;
- the process employees must follow if taking time off sick-covering when and whom employees should notify if they are not able to attend work;
- when (after how many days) employees need a self-certificate form;
- when they require a fit-note from their doctor to certify their absence;
- mention that the organisation reserves the right to require employees to attend an examination by a company doctor and (with the worker's cons ent) to request a report from the employee's doctor;
- provisions for return-to-work interviews as these have been identified as the most effective intervention to manage short-term absence; and
- give guidance on absence during major adverse events, such as snow, pa ndemics; or popular sporting events such as the Olympic Games or World Cup.

A new system of GP medical statements was introduced in April 2010, Fit notes replaced sick notes and these can either indicate that a person is "not fit for work" or that they "may be fit for work taking account of the following advice". The GP can recommend: a phased return to work; altered hours; amended duties; or workplace adaptations. The CIPD recommends that employers should arrange to meet with an employee who is assessed as "may be fit for work" to discuss appropriate ways to manage the return-to-work process.

To control sickness absence it is essential to have an effective system of recording absence (of all types) and to have clear policies of when interventions are required. In larger organisations this will require the full co-operation of line managers and supervisors who will have to ensure that absence is recorded in a central location (usually in HR). Employers must be careful not to breach the *Data Protection Act 1998*, when they store information on employees' health. The Information Commissioner's Office (www.ico.gov.uk) has issued guidance on storing and using data on workers health. (The Employment Pract ices. Data Protection Code. Part 4. Information about Workers Health (www. ico.gov.uk/upload/documents/library/data_protection/practical_application/ coi_html/english/employment_practices_code/part_4-information_about _workers_health_2.html).

Local managers will also have to be able to give return-to-work interviews and might need to be trained to do this sensitively, confidentially and be able to recognise when a case that gives concern (for example when a work-related illness or injury appears to be developing or when there appears to non-genuine causes or persistent short term absences). There should be clear guidance to managers on who to report problem cases to and this is normally a more senior manager or human resources. In addition to this, someone should have responsibility for monitoring the central absence record to pick up particular patterns of absence (e.g. the persistent Monday absence), to identify particular departments or sections of the workforce with high absence rates, and to detect when someone has reached the threshold for long term absence (usually three weeks). Managers should be given regular absence information for their department. Appropriate interventions for long-term absence will then need to be put in place, which might include the need to consider rehabilitation when the individual returns to work.

The CIPD has found that fewer than half of employers monitor the cost of absence or produce data on their absence rates. Sometime this is because sickness absence is not recorded (or not recorded reliably) but sometimes there is just no interest in making use of sickness absence data despite the considerable cost of it to the organisation (see **R2002** above). The most common measure of absence is the lost-time rate. This expresses the percentage of the total time available which has been lost to absence in a given period (usually an annual lost time rate).

It is calculated from:

Total Absence (hours or days) / Possible total hours or days worked × 100.
Thus if someone has 20 days absence in a year and could possibly have worked 220 days then the annual lost time rate is:
20/220×100 = 9.1%

It is very useful as part of an absence policy to have a target for lost time rate and some organisations have set this as low as 2%.

The CIPD suggest a number of useful interventions in managing short term absence these include:

- return-to-work interviews;
- disciplinary procedures for unacceptable absence;
- use of trigger mechanisms to review attendance;
- involving line managers in absence management;
- providing sickness absence information to line managers;
- restricting sick pay;
- managers are trained in absence handling;
- occupational health professionals involvement.

Their suggested interventions for long term absence include:

- occupational health involvement and proactive measures to support staff health and well being;
- line management involvement as part of the absence management programme;
- restricting sick pay;
- changes to work patterns or environment;
- return-to-work interviews;
- rehabilitation programme;
- managers are trained in absence handling.

Long-term absence and rehabilitation

[R2008] Long-term absence may need a carefully thought through plan for return to work. The potential legal pitfalls of not having a comprehensive return-to-work plan, or of not adhering to it, were discussed earlier (see R2003 above). The longer an employee is off sick the more likely that a plan for rehabilitation will be needed. However the need for a return-to-work plan will also depend on the nature of the illness or injury for which the person has been off sick and may be essential even if they have only been absent for a relatively short time. This will be important if the employee has a condition such as stress or a musculoskeletal injury and particularly if the condition may have been caused by the conditions of work. Indeed any condition which may be work-related, such as a developed asthma or dermatitis, (neither of which are likely to entail excessively long-term absence), will need to be carefully managed on the person's return to work.

It is important to keep in touch with the long-term absent employee on a regular basis during their absence. This can be done by, say, a weekly telephone call from HR or the immediate line manager to discuss how they are progressing and, at an appropriate time, to start to discuss when they might return to work and what the return-to-work plan might consist of and what adjustments might need to be made.

Return-to-work options include the following.

- A phased return. A gradual re-introduction to the workplace starting with part-time and building up over a planned number of weeks to full-time. A phased return is an extremely important way of getting someone back to work but will be resisted by some managers in some occupations as being too disruptive and difficult to manage. However, a relatively short period of difficulty will be likely to ensure that the person will successfully return to full productivity; without it they may never return.

- Altering the employee's hours, either temporarily or permanently. Moving the employee from shift work to permanent days. Allowing flexible-working.

- Moving the employee to another job or another department in the organisation. This might be necessary to avoid lifting and carrying following a back injury, to avoid exposure to a chemical that has caused an allergy, for example, or if someone is stressed because of harassment or bullying. It might also be necessary if the job entails driving and if the employee has developed a condition which would now preclude them from holding a licence.

- Considering what adjustments might need to be made to accommodate a new condition that the employee now suffers. This might be, for example, epilepsy following brain surgery, partial paralysis or a speech defect following a stroke, or reduced physical capacity following a heart attack. In these cases there may be no need to redeploy the person; they may be able to carry out their normal work, but with some adjustments made to their workplace, work equipment or working practices. Remember that if the condition falls within the meaning of the *Equality Act 2010*, then there is a legal requirement to make reasonable adjustments so as not to discriminate against the person (see *EA 2010* at **R2005** above).

Whatever the agreed return-to-work plan it is essential that someone, probably the line manager, is made responsible to ensure that it is adhered to, that it is reviewed if necessary and that the employee is kept fully consulted. The legal consequences of not seeing through a plan were highlighted in the case of *Young v The Post Office* [2002] EWCA Civ 661, [2002] IRLR 660, (see **R2004** above).

Using professional advice

[R2009] It was mentioned earlier (at **R2003** above), that there is no legal requirement for an occupational health service, and indeed many organisations in the UK do not have access to occupational health provision. However, a good occupational health practitioner will be invaluable in helping to manage absence generally and ensuring that rehabilitation is successful. Alternatively, and it sometimes becomes the case, an occupational health physician may need to give advice on the necessity for retirement on the grounds of ill-health.

Many different practitioners contribute to the overall practice of occupational health but the two most relevant are occupational health nurses and occupational health physicians. Having said that however, it is possible that in individual circumstances the advice or input of an ergonomist, an occupational hygienist, physiotherapist or safety practitioner may be required. However the occupational health nurse or occupational health physician will be able to advise on the need for additional expert help and one of these two should always be the first point of reference.

Occupational health nurses (also called occupational health advisors), are qualified nurses who have specialised in occupational health. The occupational health nurse is likely to be the key professional advising on the rehabilitation of employees and advising on suitable return-to-work options and any necessary work adjustments. The OH nurse would also play an important role in monitoring the progress on return to work.

Occupational health nurses are registered nurses whose names appear on the statutory register to practice held by the Nursing and Midwifery Council (NMC). In addition, an occupational health nurse will hold a qualification in occupational health, i.e. Occupational Health Nursing Certificate (OHNC), post-graduate diploma or degree and their names may be recorded on the specialist register of the Nursing and Midwifery Council (see www.nmc-uk. org).

Occupational physicians are doctors who have specialised in occupational medicine.

The Health and Safety Executive (HSE) advises that the three levels of qualification in occupational medicine for doctors are:

- Diploma in Occupational Medicine (DOccMed);
- Associateship of the Faculty of Occupational Medicine (AFOM); and
- Membership of the Faculty (MFOM).

It adds: "Doctors who are members of the Faculty and have made a distinguished contribution to the specialty, and who demonstrate a greater depth of experience and expertise in occupational medicine, may also be awarded a Fellowship of the Faculty (FFOM)."

And it advises that: "Doctors without these qualifications who rely solely on experience gained in the workplace may not meet the requirements for competence demanded by some health and safety legislation. It is recommended that the Diploma in Occupational Medicine is used as the minimum standard of qualification. However, occupational health doctors still need to work within the limits of their specific competence and seek more specialist advice when appropriate."

For occupational health doctors, the General Medical Council (GMC) website provides information relating to an individual's registration and fitness to practice (www.gmc-uk.org).

In addition, the Faculty of Occupational Medicine should be able to confirm their occupational health qualification (www.facoccmed.ac.uk).

Organisations who do not have their own occupational health provision can buy into a group practice such as that run by the Robens Centre for

Occupational Health and Safety (http://www.rcohs.com) or can buy services from NHS Plus (http://www.nhsplus.nhs.uk).

Further reading

[R2010] Wardell G., Kim Burton A., Kendall NAS., *Vocational rehabilitation What works for whom, and when?* http://www.dwp.gov.uk/docs/hwwb-vocat ional-rehabilitation.pdf

Tolley's Guide to Employee Rehabilitation. Lexis Nexis Tolley. 2004

Kivimaki M., Head J., Ferrie J.E., Hemingway H., Shipley M.J., Vahtera J., and Marmot M.G. (2005) Working While Ill as a risk factor for Serious Coronary Events: The Whitehall II Study. American Journal of Public Health, 95 (1) 98–102.

Stewart W.F., Ricci J.A., Chee E., Hahn S.R and Morganstein D. (2003) Cost of Lost Productive Work Time Among US Workers with Depression. JAMA 289: 3135–3144

Stewart W.F., Ricci J.A., Chee E., Morganstein D. and Lipton R. (2003). Lost Productive Time and Cost Due to Common Pain Conditions in the US Workforce. JAMA 290: 2443–2454

Health and Safety Executive (HSE) advice on managing sickness absence and return to work is available at: http://www.hse.gov.uk/sicknessabsence/index.ht m

Chartered Institute of Personnel and Development (CIPD) information on absenteeism, the causes of absence, absence monitoring and management, attendance, sickness and long term illness, return to work, fit notes and sick pay can be found at http://www.cipd.co.uk/hr-topics/absence.aspx.

REACH: Registration, Evaluation and Authorisation of CHemicals in Europe

John Wintle

Introduction

[R2501] A new European Community Regulation on chemicals came into force in all the countries of the European Community on 1 June 2007. Known as REACH, the Regulation provides for the Registration, Evaluation and Authorisation of CHemicals manufactured or imported in quantities of one tonne per year or more. It will affect manufacturers, importers, suppliers, distributors, and downstream users of chemicals and substances and articles containing substances, as well as regulatory authorities, all of whom have responsibilities under the Regulation.

Under REACH, all businesses in the community that manufacture or import more than 1 tonne of any given substance each year are responsible for registering information about that substance with the new European Chemicals Agency. Pre-registration and registration of substances will take place from 1 June 2008 onwards. Registrants have a duty to inform the supply chain and downstream users of the hazards and appropriate risk management measures for particular uses of the substance. Downstream users have a duty to apply the risk management measures and a responsibility to report back any new hazards or uses.

REACH will replace several current European Directives and Regulations with a single system, and will help to harmonise the information on substances and their control across the EU. It provides an opportunity to consider the latest data available on the properties of substances so as to classify them accordingly, and to exchange and improve information for users. For substances that pose a particular threat and are of high concern, the Regulation provides an opportunity for authorising or restricting their availability and use.

Purpose

[R2502] The purpose of REACH is to make those businesses and organisations that place substances on the European market responsible for understanding and managing the risks associated with their use. They will be responsible for collecting, collating and registering information about their chemical with the European Chemicals Agency, which is being set-up Helsinki, Finland. The ultimate aim of REACH is to ensure that employees, users, the public and the environment in the EU have a high level of protection from any properties of substances that may be hazardous to human and biological health.

The Regulation is intended to promote the free movement of substances on the EU market by removing any specific national restrictions and introducing common regulation across the EU. It will ensure that substances imported into the EU are registered to the same standards as substances manufactured within the EU. The registration process will give wide-ranging consideration to the properties of substances and will encourage the development of new or alternative methods by which the hazardous properties of substances may be assessed. It will also promote the development of less hazardous substitutes. In these ways, it is hoped that REACH will stimulate the European chemicals industry into more innovation and increased competitiveness in world markets.

Need for REACH

[R2503] There are an estimated 30,000 substances supplied by businesses in quantities of over 1 tonne that are on the market in the EU. In general, very little information is publicly available on the hazards and risks to human health and the environment, and the current system of regulation by authorities has been slow to produce results, particularly for substances existing prior to the requirement for notification. It is estimated that the number of these substances that have been properly assessed over the past 30 years is less than 200, while the number of notifications of new substances is around 1000.

There is increasing public concern over the health risks from chemicals. However, the evidence base of information to address this concern is currently limited and needs to be improved. At a regulatory level, REACH is intended to simplify the current system for the control of substances that is considered confusing for industry to understand and for authorities to administer.

Chemicals covered and exempted

[R2504] Under REACH, 'chemicals', usually referred to as 'substances', are defined as chemical elements or compounds in the natural state or obtained by any manufacturing process. This includes any additive necessary to preserve its stability and any impurity derived from the process used. The term excludes any solvent that may be separated without affecting the stability of the substance or changing its composition. REACH applies to individual substances on their own, in preparations, or in articles. Preparations are a mixture or solution of two or more substances. An article is where a substance is given a special shape, surface or design, which determine its function to a greater degree than its chemical composition.

Generally, REACH applies to substances manufactured or imported into the EU in quantities of one tonne per year or more. The amount of a given substance supplied in excess of one tonne affects the timetable for its registration. A distinction is made for substances supplied at 100 tonnes or more per year and 1000 tonnes or more per year.

Some substances that may not be properly on the market or that are transitory in nature are specifically excluded:

- Substances under customs supervision
- Substances undergoing transport
- Radioactive substances
- Non-isolated intermediates
- Waste
- Some naturally occurring low hazard substances.

Other substances are excluded from REACH by being covered by more specific legislation with tailored provisions and include:

- Substances intended as human and veterinary medicines
- Food and foodstuff additives
- Plant protection products and biocides
- Isolated intermediates
- Substances used for research sand development.

Duty Holders within REACH

[R2505] Many businesses will have responsibilities as duty holders under REACH. There are three main types of REACH duty holder, each with different roles and responsibilities. These types of duty holder are:

- Manufacturers or Importers (Registrants)
- Distributors
- Downstream users

Businesses that manufacture or import more than one tonne of any given substance in the Community each year are duty holders as Registrants under REACH. Because substances in articles also count, it is possible that some producers and importers of articles will also be Registrants, if the substances contained in them exceed the threshold. Where downstream users are also importers of substances, then they will also become Registrants.

Distributors are businesses or persons established within the Community that buy, store, place on the market and sell a substance, on its own, or in a preparation or article, but are not the importer, manufacturer or final user. A retailer who only stores and places on the market substances and articles for sale to third parties is a distributor.

Downstream users include any business or persons established within the Community, other than the manufacturer or importer who uses a substance, either on its own or in a preparation, in the course of his industrial or professional activities. A consumer buying substances for use at home is not a downstream user. A re-importer of substances manufactured or imported in the Community, exported and then re-imported is also regarded as a downstream user.

Manufacturers and importers have two key responsibilities. Firstly they are responsible for registering with the European Chemicals Agency information about the properties and hazards of each substance that they manufacture or import and the uses to which that substance may be put. Secondly, registrants are responsible for providing information directing downstream users to the appropriate risk management measures, and responding to information about use and effects passed back by distributors and downstream users.

Distributors have specific duties to pass information provided by manufacturers and importer down the supply chain to the downstream users. They also have duties to pass information back up the supply chain to their own suppliers when their own clients ask them to do so.

Downstream users have a duty to use substances on their own or in preparations in a safe way, and according to the information on risk management measures passed down the supply chain from manufacturers and importers for that use. Where a downstream user gains new information about the hazardous properties of a substance, the downstream user has a responsibility to pass the information back up the supply chain to the manufacturer or importer and, where appropriate, to the European Chemicals Agency. For any use outside the conditions communicated by the supplier, or for any use that the supplier advises against, the downstream use may have to prepare a chemical safety report including a risk assessment and risk management measures covering this use. If the downstream user wishes to keep its use of the substance confidential, the user shall provide information on the use with the European Chemicals Agency independently of the original manufacturer or importer.

Preparation for REACH

[R2506] Businesses that think they are potential duty holders under REACH should prepare for their responsibilities depending on whether the business is a potential registrant (manufacturer or importer), distributors, or downstream user, or more than one of these. All potential duty holders are advised to compile an inventory of the substances that enter, become part of, or leave the business. This should include feedstock's, intermediates (isolated or otherwise), and substances in preparations, articles and products that are manufactured.

All substances should be identified. For preparations and mixtures, it is necessary to know all the ingredient substances and the proportions by weight of each. A list of the substances can then be drawn up with the estimated tonnage of each substance that enters, is created by, and leaves the business per year.

If the business produces or imports articles, it is necessary to determine if any substance which is intended to be released under normal or reasonably foreseeable conditions of use is present in these articles in quantities which might total more than 1 tonne per year. These substances should to be added to the list.

When this inventory has been established, businesses should consider their supply chain and determine which type(s) of REACH duty holder they are, taking account of alternatives or variations in substances, processes and supply. Potential registrants (manufacturers and importers) should establish the typical uses of their product, while downstream users can be more specific about the uses to which substances are put, and whether their suppliers are likely to be aware of these uses. Users should consider whether they would wish to share information about their use with their suppliers, or where this might be sensitive whether they would prefer to compile a risk assessment themselves.

In some areas, REACH could have commercial impact. Certain substances may in future be subject to authorisation and/or restrictions of use or supply. Businesses should consider the possible impact of any such measures on their business, particularly for substances that are especially hazardous, which should be specifically identified.

Having determined what type(s) of duty holder it is, a business may wish to understand the role and responsibilities for that duty holder, and the roles of responsibilities of other duty holders in its supply chain. It may be helpful for the supply chain to discuss the implications and expectation of REACH for each actor. A focal point for REACH within each business would be useful so as to provide an interface with suppliers and customers on REACH responsibilities. The first step for manufacturers and importers is to prepare for registration by pre-registering substances that already exist.

The Registration process

[R2507] Registration is the requirement of registrants in industry (manufacturers and importers) to collect and collate specific sets of information on the properties of the substances that they manufacture or import. An assessment of the hazards and risks a substance may pose and how those risks can be managed is also required. All registrants must register their substances and the necessary information and assessment with the European Chemicals Agency.

For the existing 30,000 or so substances that are already being manufactured or imported in Europe, registration is being staged over three phases spread over 11 years. In order to benefit from phased registration, manufacturers and importers should pre-register their substances with the European Chemicals Agency between 1 June and 30 November 2008.

It is the intention that for any given substance a single collation of the information and assessment is produced. Where there is more than one manufacturer or importer for a given substance, as will often occur, arrangements are being made to put all like registrants into contact with each other for generating the information and making the assessment jointly.

The European Chemicals Agency will identify and notify all those businesses intending to register the same chemical so that they can come together and form a Substance Information Exchange Forum (SIEF). Through this forum they can collectively share their available data and knowledge, and generate a single collation of the information. All registrants of that substance then share the information and hazard assessment for the purposes of registration.

It is for the members of the Substance Information Exchange Forum to negotiate the sharing of their data and the costs of generating any new data required. Where there is information that is business specific or sensitive, companies can choose to submit information separately to the European Chemicals Agency. It is believed that this process will help industry reduce the costs of registration.

Timescale for Registration

[R2508] Phased registration applies to the 30,000 or so existing substances, known as 'phase-in substances, that meet at least one of the following criteria.

(a)   A substance that is listed on the European Inventory of Existing Commercial Chemical Substances (EINECS), or

(b)   A substance that was manufactured in the Community but not placed on the market at least once in the 15 years prior to REACH entering to force, or

(c)   A substance that was placed on the market by the manufacturer or importer and was considered to be notified in accordance with the first indent of Article 8(1) of Directive 67/548/EEC, but does not meet the definition of a polymer.

Manufacturers or importers should have documentary evidence to support criteria (b) and (c). In order to benefit from phased registration, manufacturers and importers should pre-register their substances between 1 June and 30 November 2008. Phase-in substances will be registered in three phases with the date for registration according to their tonnage and/or hazardous properties.

(a)   Substances that must be registered by 1 December 2010 are:
   —   Substances supplied at 1000 tonnes or more per year, or
   —   Substances classified under CHIP (Chemicals (Hazard Information and Packaging for Supply) Regulations 2002) as very toxic to aquatic organisms that are supplied at 100 tonnes or more per year, or
   —   Substances classified under CHIP as Category 1 or 2 carcinogens, mutagens or reproductive toxicants hat are supplied at 1 tonne or more per year.

(b)   Substances that must be registered by 1 June 2013 are
   —   Substances out-with substances in (a) above that are supplied at 100 tonnes or more per year.

(c)   Substances that must be registered by 1 June 2018 are
   —   Substances out-with substances in (a) and (b) above that are supplied at 1 tonne or more per year.

Non phase-in substances are those that do not meet the criteria (a) to (c) above or are completely new substances or those that have not been pre-registered. They will be subject to registration as soon as reasonably practicable after

1 June 2008. Until then the current regulations for the Notification of New Substances (NONS) continue to apply.

Information required for Registration

[R2509] The information required for registration depends on the amount of the chemical produced. For substances produced in quantities of more than one tonne but less than ten tonnes per year, a Technical Dossier is required. For substances produced in amounts greater than 10 tonnes per year, a more extensive Chemical Safety Report is required.

The Technical Dossier contains information on the properties, uses and classification of the chemical, as well as guidance on its safe use. The reduced information requirements for substances supplied between 1 to 10 tonnes are intended to help the many small and medium sized enterprises (SMEs) manufacturing substances. Two thirds of the substances that will be subject to REACH registration are supplied between 1 to 10 tonnes per year, mainly by small and medium sized enterprises (SMEs).

Under REACH, the classification of substances according to their characteristics follows an established system. For example, established classification characteristics include those that are corrosive, or toxic to fish. Work is underway through REACH to bring the EU in line with the United Nations' Globally Harmonized System for classification and labeling.

In contrast to the Technical Dossier, a Chemical Safety Report details the classification and the hazardous effects of the chemical. It includes an assessment as to whether the chemical is persistent (P), bio-accumulative (B) and toxic (T) or very persistent and very bio-accumulative (vPvB). In addition, the Chemical Safety Report also describes 'Exposure Scenarios' within a Chemical Safety Assessment.

Exposure Scenarios describe the conditions under which the substance is manufactured or is to be used during its life. They specify the controls that the manufacturer or importer has or recommends in order to limit exposure to humans and the environment. These should include the appropriate risk management measures and operational restrictions, limits and controls to ensure that, when these are properly applied, the risk from exposure to the substance is sufficiently low.

The Chemical Safety Assessment should address all the identified and known uses of the substance on its own (but including any major impurities and additives), in a preparation and in an article. All stages of the life-cycle of the substance must be considered, from manufacture to final disposal. A connection is made between the potential adverse effects of the substance with the known or reasonably foreseeable Exposure Scenarios for humans and/or the environment.

Safety Data Sheets and downstream use

[R2510] A key feature of REACH is to ensure good transmission of information up and down the supply chain. Downstream users need to know about the hazards involved in using substances and how to control the risks from the knowledge of manufacturers and importers. However, suppliers need to know from users about new uses of the substances that they supply, and the experience of users so that this information can be passed to other users.

Safety Data Sheets are the means by which manufacturers or importers can inform the supply chain and downstream users of the appropriate safety information on a substance. It is the responsibility of the manufacturer and importer to prepare a Safety Data Sheet and to include the Sheet when supplying the substance. Where a Chemical Safety Report is required for registration, the Safety Data Sheet should include the information given in the Chemical Safety Report and be consistent with information provided for registration.

Downstream users have a duty to follow the information and guidance provided on the Safety Data Sheet. Where a downstream user intends a use with an exposure scenario outside those listed on the Safety Data Sheet, a dialogue with the supplier on the safety of the intended use might be the first step. Where the downstream user does not wish to do this, or is advised against the use by the supplier, the downstream user should normally prepare a Chemical Safety Report unless specifically exempted. This should include a risk assessment and risk management measures for the use, and must be available to the European Chemicals Agency if required.

The Safety Data Sheet is an existing system that will be developed under REACH. In addition to the information that is currently contained, under REACH the Safety Data Sheet will also include information on safe handling and use in the form of exposure scenarios.

European Chemicals Agency

[R2511] A European Chemicals Agency (ECHA) is being established in Helsinki to manage the technical and administrative aspects of REACH across the EU and to ensure consistency at community level. In addition to managing registration, the ECHA will play a role in the evaluation process at a scientific level. The ECHA also has responsibilities with regard to Authorisation and Restriction of substances considered of very high concern.

Evaluation

[R2512] Registration Packages submitted under REACH by industry to the European Chemicals Agency can be evaluated in different ways. At least 5% of Registration Packages will be subject to a full compliance check of the quality and accuracy of the information and assessment submitted. The ECHA will scrutinise all registration packages containing proposals for testing, primarily to ensure that no unnecessary animal testing is carried out, either routinely or as part of an on-going assessment process. For substances registered at higher tonnage levels, a check will be made as to whether any testing plan proposed is appropriate.

Each year Member States and the European Commission will agree on an annual list of prioritised substances to be evaluated in depth for potential regulatory control action because of concerns about their properties or uses. For these substances an in-depth evaluation of all registered information will be made. These in-depth evaluations will be undertaken by Competent Authorities in Member States.

Authorisation

[R2513] As REACH progresses two lists of substances will be created. A 'List of Substances of Very High Concern' and subject to authorisation controls will be declared by the ECHA as Annex XIV of the Regulation. The ECHA will also create another list of 'candidate substances' that are considered to be of concern, but which are not currently subject to formal authorisation controls.

Member States will propose substances for admission to the Annex XIV list. Where substances are admitted to the list they cannot be supplied or used unless an Authorisation for that supply or use has been applied for and granted. Any company wishing to supply or use a substance on the list must apply the ECHA for an authorisation, which may be granted or refused.

Authorisation will be granted where it can be demonstrated that the risks from the substance are under "adequate control". "Adequate control" allows regulatory authorities to prioritise action on those hazardous substances that cannot be so controlled.

Where "adequate control" is not possible, authorisation to supply or use a substance of very high concern may be granted on socio-economic grounds if there is no safer alternative. However, companies will be required to make efforts to find safer substitutes as part of the authorisation process. Any substitute proposed must deliver lower risks and be technically and economically feasible.

Substances of very high concern will include those identified as: carcinogenic, mutagenic or toxic to reproduction (CMRs); persistent, bio-accumulative and toxic (PBTs); very persistent and very bio-accumulative (vPvBs); and substances demonstrated to be of equivalent concern, such as endocrine disruptors.

Restrictions

[R2514] REACH enables restrictions on the supply and/or use of substances to be placed across the EU, where this is shown to be necessary. Either Member States or the European Commission may prepare proposals for restrictions on any particular chemical. When approved, the form of restrictions could vary, for example, from a total ban to not being able to supply it to the general public. Restrictions can apply to any substance that poses a particular threat, including those that do not require registration. This aspect of REACH takes over the provisions of the Marketing and Use Directive.

UK Competent Authority and Helpdesk

[R2515] REACH requires Member States of the European Union to appoint Competent Authorities to manage the obligations of REACH at a national level and liase with the new European Chemicals Agency in Helsinki. For the UK, the Competent Authority (CA) is centred on the Health and Safety Executive (HSE), which working with other government Departments and devolved administrations. DEFRA (the Department for the Environment, Food and Rural Affairs) has overall policy responsibility for REACH and is the lead government Department to ensure the smooth introduction of REACH in the UK.

The HSE has set up a Helpdesk for REACH to provide information and technical advice for UK based industry and business in preparing for REACH. The Helpdesk can be contacted on 0845 408 9575 or via email at ukreachca@hse.gsi.gov.uk. It is already assisting manufacturers, importers and downstream users, and other interested parties understand their respective responsibilities under REACH.

The CA has responsibilities under REACH to help raise awareness and assist the introduction. It will be the primary body for enforcing compliance with registration, but in this regard it is expected, initially, to act reactively on information supplied. It will liase with other enforcing organisations on the implementation of risk management measures by downstream users, the adherence to restrictions, and dealing with the refusal of authorisation.

Where substances are recommended for evaluation, the CA will make assessments and prepare draft decisions for agreement at a European level. It will identify and propose substances of very high concern for authorisation. Where appropriate, the CA will propose restrictions of substances in the national and international interest.

Representing the UK, the CA has nominated members to the ECHA Committees on risk assessment and socio-economic analysis. It will appoint members to the Member State Committee, which will resolve differences of opinion on evaluation matters. In addition, the CA has the responsibility for appointing members to the Forum for Information Exchange. This Forum is being created to discuss enforcement matters across the EU. In support of these roles, the CA will ensure adequate scientific and technical resources are available to the UK members of the committees.

Enforcement

[R2516] National regulations to enforce REACH are required by 1 December 2008. In the UK, DEFRA, who are the policy lead for REACH, have consulted with industry and business on enforcement during 2007. The Statutory Instrument enforcing REACH in UK law is expected to be drafted in early 2008. A further consultation on it will be held in the spring of 2008.

The approach to enforcement that the government is proposing is expected to use best practice to encourage compliance, supported by more formal procedures where necessary. It is currently proposed that, as Competent Authority, the HSE should take lead responsibility for registration compliance relating to UK registrants. It will also enforce supply chain issues of packaging, information availability and transfer. However, a range of authorities (eg local authorities and environmental agencies etc) will enforce REACH at the point of use (eg ensuring use in accordance with the instructions and risk management measures) at the premises where they already enforce other regulations.

Enforcement will ensure compliance with the three broad areas of activity that REACH encompasses. These are:

- Registration,
- The supply chain; and
- End use.

Key areas of REACH where DEFRA recommends that UK enforcement is needed are:

- The manufacture, import, sale or supply of substances without the appropriate registration.
- Using a hazardous substance outside the terms of an authorisation or contrary to a restriction.
- Failure to provide information up and down the supply chain: (this could affect downstream users, as well as manufacturers and distributors).
- Failure to comply with other duties regarding information, for example the right of access to information of workers and consumers.
- Failure of a downstream user to apply risk management measures as contained in Safety Data Sheets.

The full scope of enforcement will not become clear until the relevant Statutory Instrument is published.

Summary

[R2517] The new European REACH Regulation is now in force. Arrangements for its implementation and enforcement at national and international level are in hand. REACH is intended to provide a good balance between improving protection of human health and the environment while maintaining the competitiveness of European industry in an international marketplace. Several major pieces of existing legislation will be replaced with a single EU wide system for substance regulation.

REACH will carry forward the current EU restriction regime and the Safety Data Sheets System. It introduces a new requirement for registration of all substances supplied above 1 tonne per year, and a new requirement for authorisation for Substances of Very High Concern listed in Annex XIV of the Regulation. The responsibility for gathering data and carrying out initial risk assessments of substances is now transferred from authorities to industry. The scope of REACH includes responsibilities for manufacturers, importers, suppliers, and downstream users of substances, preparations and articles containing substances.

Further information

[R2518] The HSE has set up a Helpdesk for REACH to provide information and technical advice for UK based industry and business in preparing for REACH. The Helpdesk can be contacted on 0845 408 9575 or via email at ukreachca@hse.gsi.gov.uk. It will assist manufacturers, importers and downst ream users, and other interested parties understand their respective responsibilit ies under REACH. There is useful information on REACH available on the HSE and DEFRA websites. www.hse.gov.uk/reach, and www.defra.gov.uk.

The ECHA also available to help industry and business in the European Union. It has useful information on its website: www.defra.gov.uk.

The definitive text of the REACH Regulation is contained in the 29 May issue of the Official Journal of the European Union L136/37. This is a corrigenda to the original Regulation passed by he European Parliament on 18 December 2006. The text from OJ L136/37 can be downloaded from the ECHA website and other sources.

REACH is still evolving further developments and guidance can be expected.

Acknowledgement

[R2519] In preparing this chapter, use has been made of official publications and information provided by the Health and Safety Executive and the Department for Food and Rural Affairs (DEFRA).

Risk Assessment

Mike Bateman

Introduction

[R3001] Although the term 'risk assessment' probably first came into common use as a result of the *Control of Substances Hazardous to Health Regulations 1988* (commonly known as the '*COSHH Regulations*' and revised several times since), similar requirements had actually previously been contained in both the *Control of Lead at Work Regulations 1980* and the *Control of Asbestos at Work Regulations 1987*.

In practice a type of risk assessment had already been necessary for some years particularly as a result of the use of the qualifying clause 'so far as is reasonably practicable' in a number of the important sections of the *Health and Safety at Work etc. Act 1974* ('*HSWA 1974*').

Risk assessment is now at the heart of effective health and safety management, with many recent sets of regulations containing specific risk assessment requirements — see **R3006** to **3016** below.

HSWA 1974 requirements

[R3002] *HSWA 1974, s 2* contains the general duties of employers to their employees with the most general contained within *s 2(1)*:

> It shall be the duty of every employer to ensure, so far as is reasonably practicable, the health, safety and welfare at work of all employees.

Other more specific requirements are contained in *s 2(2)* and these are also qualified by the term 'reasonably practicable'.

HSWA 1974, s 3 places general duties on both employers and the self-employed in respect of persons other than their employees. *Section 3(1)* states:

> It shall be the duty of every employer to conduct his undertaking in such a way as to ensure, so far as is reasonably practicable, that persons not in his employment who may be affected thereby are not exposed to risks to their health or safety.

Employers thus have duties to contractors (and their employees), visitors, customers, members of the emergency services, neighbours, passers-by and the public at large. This may to a certain point extend to include trespassers. Self-employed people are put under a similar duty by virtue of *HSWA 1974, s 3(2)* and must also take care of themselves. In each case these duties are subject to the 'reasonably practicable' qualification.

HSWA 1974, s 4 places duties on each person who has, to any extent, control of non-domestic premises used for work purposes in respect of those who are not their employees. *HSWA 1974, s 6* places a number of duties on those who design, manufacture, import or supply articles for use at work, or articles of fairground equipment and those who manufacture, import or supply substances. Many of these obligations also contain the 'reasonably practicable' qualification.

What is reasonably practicable?

[R3003] The phrase 'reasonably practicable' is not just included within the key sections of *HSWA 1974* but is also contained in a wide variety of regulations. Lord Justice Asquith provided a definition in his judgment in the case of *Edwards v National Coal Board* [1949] 1 All ER 743 in which he stated:

> "Reasonably practicable" is a narrower term than "physically possible" and seems to me to imply that a computation must be made by the owner in which the quantum of risk placed on one scale and the sacrifice involved in the measures necessary for averting risk (whether in money, time or trouble) is placed in the other, and that, if it be shown that there is a gross disproportion between them – the risk being insignificant in relation to the sacrifice – the defendants discharge the onus on them. Moreover, this computation falls to be made by the owner at a point in time anterior to the accident.

HSWA 1974, s 40 places the burden of proof in respect of what was or was not 'reasonably practicable' (or 'practicable', see **R3004** below) on the person charged with failure to comply with a duty or requirement. Employers and other duty holders must establish the level of risk involved in their activities and consider the various precautions available in order to determine what is reasonably practicable i.e. they must carry out a form of risk assessment.

The Health and Safety Executive (HSE) website section dealing with Risk Management (www.hse.gov.uk/risk/index.htm) provides guidance on the p rinciples and guidelines which are used by HSE itself in determining what is 'reas onably practicable'. Both 'risk' and 'sacrifice' are considered in some detail. In respect of the latter the HSE states: 'Individual duty-holders' ability to afford a control measure or the financial viability of a particular project is not a legit imate factor in the assessment of its costs. HSE must present duty-holders with a level playing field. Thus HSE cannot take into account the size and financial position of the duty-holder when making judgements on whether risks have been reduced as low as reasonably practicable.'

Further guidance is provided on the influence of 'societal concerns', 'transfer of risks' (e.g. to other employees or members of the public) and 'good practice'. Additional information on material available on the HSE website is contained in **R3053** below.

However, it is important to keep a sense of perspective. The HSE has expressed concern in recent years about what has come to be called 'risk aversion' – where quite excessive precautions have been demanded by some in respect of low levels of risk. This subject is examined further in **R3052** below.

Practicable and absolute requirements

[R3004] Not all health and safety law is qualified by the phrase 'reasonably practicable'. Some requirements must be carried out 'so far as is practicable'. 'Practicable' is a tougher standard to meet than 'reasonably practicable' – the precautions must be possible in the light of current knowledge and invention (*Adsett v K and L Steelfounders & Engineers Ltd* [1953] 2 All ER 320). Once a precaution is practicable it must be taken even if it is inconvenient or expensive. However, it is not practicable to take precautions against a danger which is not yet known to exist (*Edwards v National Coal Board* [1949] 1 All ER 743), although it may be practicable once the danger is recognised.

Many health and safety duties are subject to neither 'practicable' nor 'reasonably practicable' qualifications. These absolute requirements usually state that something 'shall' or 'shall not' be done. However, such duties often contain other words which are subject to a certain amount of interpretation e.g. 'suitable', 'sufficient', 'adequate', 'efficient', 'appropriate' etc.

In order to determine whether requirements have been met 'so far as is practicable' or whether the precise wording of an absolute requirement has been complied with, a proper evaluation of the risks and the effectiveness of the precautions must be made.

The Management of Health and Safety at Work Regulations 1999

[R3005] The Management Regulations were introduced in 1992 and revised by the 1999 Regulations. They were intended to implement the European Framework Directive (89/391) on the introduction of measures to encourage improvements in the safety and health of workers at work. The *Management of Health and Safety at Work Regulations 1999 (SI 1999 No 3242), Reg 3* require employers and the self-employed to make a suitable and sufficient assessment of the risks to both employees and persons not in their employment. The purpose of the assessment is to identify the measures needed 'to comply with the requirements and prohibitions imposed . . . by or under the relevant statutory provisions . . . ' i.e. identifying what is needed to comply with the law.

Given the extremely broad obligations contained in *HSWA 1974, ss 2, 3, 4* and *6*, all risks arising from work activities should be considered as part of the risk assessment process (although some risks may be dismissed as being insignificant). Compliance with more specific requirements of regulations must also be assessed, whether these are absolute obligations or subject to 'practicable' or 'reasonably practicable' qualifications. The requirement for risk assessment introduced in the Management Regulations simply formalised what employers (and others) should have been doing all along i.e. identifying what precautions they needed to take to comply with the law. More detailed requirements of the Management Regulations are covered later in the chapter.

Any such additional precautions must then be implemented (the *Management of Health and Safety at Work Regulations 1999 (SI 1999 No 3242), Reg 5* contains requirements relating to the effective implementation of precautions). Practical guidance is provided in the chapter on MANAGING HEALTH AND SAFETY.

Common regulations requiring risk assessment

[R3006] An increasing number of regulations contain requirements for risk assessments. Several of these regulations are of significance to a wide range of work activities and are dealt with in more detail elsewhere in the looseleaf. The most significant regulations requiring risk assessment include:

Control of Substances Hazardous to Health Regulations 2002 (COSHH)

[R3007] The *Control of Substances Hazardous to Health Regulations 2002 (SI 2002 No 2677 as amended), Reg 6* require employers to make a suitable and sufficient assessment of the risks created by work liable to expose any employees to any substance hazardous to health and of the steps that need to be taken to meet the requirements of the regulations. See HAZARDOUS SUBSTANCES IN THE WORKPLACE, which provides more details of the COSHH requirements.

The Health and Safety Executive ('HSE') booklet L5, contains both the regulations and the associated Approved Code of Practice. There are many other relevant HSE publications including HSG97 '*A step by step guide to COSHH assessment*'.

Control of Noise at Work Regulations 2005

[R3008] *Regulation 5* of the *Control of Noise at Work Regulations (SI 2005 No 1643)* requires employers carrying out work which is liable to expose any employees to noise at or above a lower exposure limit to make an assessment of the risk from that noise. (The lower exposure limit is defined in the regulations. The measures necessary to meet the requirements of the regulations must be identified during the assessment.)

See the chapter dealing with NOISE AT WORK and also the HSE booklet L108 'Controlling Noise at Work' which contains the full regulations and related guidance, both on the interpretation of the regulations and on the assessment process.

Manual Handling Operations Regulations 1992

[R3009] Employers are required by Regulation 4 of the *Manual Handling Operations Regulations 1992 (SI 1992 No 2793 as amended)*, to make a suitable and sufficient assessment of all manual handling operations at work which involve a risk of employees being injured, and to take appropriate steps to reduce the risk to the lowest level reasonably practicable (They must avoid such manual handling operations if it is reasonably practicable to do so).

See MANUAL HANDLING which provides further details of the Regulations and the carrying out of assessments. HSE booklet L23, '*Manual Handling: Manual*

Handling Operations Regulations 1992 (as amended) – *guidance on regulations'* provides detailed guidance on the Regulations and manual handling assessments.

Health and Safety (Display Screen Equipment) Regulations 1992

[R3010] The *Health and Safety (Display Screen Equipment) Regulations 1992 (SI 1992 No 2792 as amended), Reg 2* require employers to perform a suitable and sufficient analysis of display screen equipment (DSE) workstations for the purpose of assessing risks to 'users' or 'operators' as defined in the Regulations. Risks identified in the assessment must be reduced to the lowest extent reasonably practicable.

The HSE booklet L26 '*Work with display screen equipment*' contains guidance on the Regulations and on workstation assessments.

Personal Protective Equipment at Work Regulations 1992

[R3011] Under the *Personal Protective Equipment at Work Regulations 1992 (SI 1992 No 2966* as amended), *Reg 6* employers must ensure that an assessment is made to determine risks which have not been avoided by other means and identify personal protective equipment ('PPE') which will be effective against these risks. The Regulations also contain other requirements relating to the provision of PPE; its maintenance and replacement; information, instruction and training; and the steps which must be taken to ensure its proper use.

See PERSONAL PROTECTIVE EQUIPMENT which provides further details of the requirements under the Regulations. HSE booklet L25, '*Personal protective equipment at work*' provides detailed guidance on the Regulations and the assessment of PPE needs.

Regulatory Reform (Fire Safety) Order 2005

[R3012] From October 2006 the *Regulatory Reform (Fire Safety) Order 2005 (SI 2005 No 1541)* replaced the *Fire Precautions Workplace Regulations 1997 (SI 1997 No 1840)*, many parts of the *Fire Precautions Act 1971* and other pieces of legislation related to workplace fire safety. The Order (which is enforced by the relevant Fire Authority) requires the 'responsible person' (a definition is provided in the Order) to ' . . . make a suitable and sufficient assessment of the risks to which relevant persons are exposed for the purposes of identifying the general fire precautions he needs to take'.

More information is provided in the chapter on FIRE PREVENTION AND CONTROL.

Control of Asbestos Regulations 2006

[R3013] The *Control of Asbestos Regulations 2006 (SI 2006 No 2739)* continue the requirement originating in the 1987 Regulations, for employers to make a risk assessment before carrying out work which is liable to expose employees to asbestos. *Regulation 6* contains several specific requirements on what the risk assessment must involve and other regulations detail many precautions which must be taken in asbestos work. Some types of asbestos work can only be done by companies licensed by the HSE. There are several important HSE reference publications on work with asbestos:

- L143 Work with materials containing asbestos. Control of Asbestos Regulations 2006 — AcoP and Guidance;
- HSG210 asbestos essentials. A task manual for building, maintenance and allied trades on non-licensed asbestos work:
- HSG247 asbestos. The licensed contractors guide.

The 2006 Regulations continue the 'Duty to manage asbestos in non-domestic premises' introduced in the 2002 regulations and contained in *Regulation 4*. Duty holders under this regulation are defined within it and are normally the owner or leaseholder, the employer occupying the premises or a combination of these (depending upon the terms of any lease or contract or who controls the premises).

Duty holders must carry out an assessment of their premises in order to identify any asbestos-containing material ('ACM') and also assess its condition. They must then prepare a written plan for managing the risks from ACM, including such measures as regular inspections of confirmed and presumed ACM, restricting access to or maintenance in ACM areas and, where necessary, protection, sealing or removal of ACM.

Further details on these requirements are contained in the chapter on ASBESTOS.

Important HSE references worth referring to are:

- L127 The management of asbestos in non-domestic premises; and
- HSG227 A comprehensive guide to managing asbestos in premises.

Dangerous Substances and Explosive Atmospheres Regulations 2002

[R3014] These Regulations (known sometimes as 'DSEAR') replaced many older pieces of legislation including the *Highly Flammable Liquids and Liquefied Petroleum Gases Regulations 1972* and many of the licensing requirements of the *Petroleum (Consolidation) Act 1928*. They are concerned with protecting against risks from fire and explosion. The definition of 'dangerous substances' contained within the regulations includes such substances as petrol, many solvents, some types of paint, LPG etc. and also potentially explosive dusts.

Employers are required by the *DSEAR (SI 2002 No 2776), Reg 5* to carry out a risk assessment of work involving dangerous substances. The *DSEAR (SI 2002 No 2776), Reg 6* require a range of specified control measures to be implemented in order to eliminate or reduce risks as far as reasonably practicable. Places where explosive atmospheres may occur must be classified into zones and equipment and protective systems in these zones must meet appropriate standards [*DSEAR (SI 2002 No 2776), Reg 7*]. The *DSEAR (SI 2002 No 2776), Reg 8* require equipment and procedures to deal with accidents and emergencies whilst the *DSEAR (SI 2002 No 2776), Reg 9* states that employees must be provided with information, instruction and training. Further details are contained in HAZARDOUS SUBSTANCES IN THE WORKPLACE.

Work at Height Regulations 2005

[R3015] The *Work at Height Regulations 2005 (SI 2005 No 735)* do not introduce a separate requirement for risk assessment. However, *Reg 6* requires that general risk assessments conducted to comply with the *Management Regulations* take account of a hierarchy of measures set out within the regulation. Many other specific requirements are contained in the regulations.

See the chapter WORK AT HEIGHT. Various HSE publications provide further guidance.

Specialist regulations requiring risk assessment

[R3016] Some regulations requiring risk assessments are of rather more specialist application, although some are still covered in some detail elsewhere in the publication. References to relevant chapters of the looseleaf and to key HSE publications of relevance are included below. These regulations include:

- *Control of Lead at Work Regulations 2002 (SI 2002 No 2676)*
  See HAZARDOUS SUBSTANCES IN THE WORKPLACE.
  L132 Control of lead at work.
- *Supply of Machinery (Safety) Regulations 1992 (SI 1992 No 3073)*
  The Regulations include a variety of procedures which must be followed in assessing conformity of machinery with essential health and safety requirements set out in the Machinery Directive.
  See MACHINERY SAFETY.
  INDG270 Supplying new machinery: Advice to suppliers, free leaflet.
  INDG271 Buying new machinery: A short guide to the law, free leaflet.
- *Control of Major Accident Hazard Regulations 1999 (SI 1999 No 743) (COMAH)*
  The Regulations only apply to sites containing specified quantities of dangerous substances.
  See DISASTER AND EMERGENCY MANAGEMENT SYSTEMS (DEMS).
  L111 A guide to the Control of Major Accident Hazard Regulations 1999 as amended.
  HSG 190 Preparing safety reports: Control of Major Accident Hazard Regulations 1999.
  HSG 191 Emergency planning for major accidents: Control of Major Accident Hazard Regulations.
- *Ionising Radiations Regulations 1999 (SI 1999 No 3232)*
  See RADIATION,
  L121 Work With Ionising Radiation: Ionising Radiations Regulations 1999 – approved code of practice and guidance.
- *Control of Artificial Optical Radiation at Work Regulations 2010 (SI 2010 No 1140)*

HSE guidance: Guidance for Employers on the *Control of Artificial Optical Radiation at Work Regulations (AOR) 2010*

* Control of Vibration at Work Regulations 2005 (SI 2005 No 1093) See VIBRATION.

L140 Hand-arm vibration. Control of Vibration at Work Regulations 2005.

INDG 175 Control the risks from hand-arm vibration. Advice for employers.

L141 Whole -body vibration, Control of Vibration at Work Regulations 2005.

INDG 242 Control back pain risks from whole -body vibration. Advice for employers.

Related health and safety concepts

[R3017] Risk assessment techniques are an essential part of other health and safety management concepts.

* *Safe systems of work*
  Employers are required under *HSWA 1974, s 2(2)(a)* to provide and maintain 'systems of work that are, so far as is reasonably practicable, safe and without risks to health'.
  Both the *Confined Spaces Regulations 1997 (SI 1997 No 1713)* and the *Lifting Operations and Lifting Equipment Regulations 1998 (LOLER) (SI 1998 No 2307)* contain similar requirements for safe systems of work. A safe system of work can only be established through a process of risk assessment.
  See the chapter on SAFE SYSTEMS OF WORK for more information on this concept.
* *'Dynamic risk assessment'*
  The term 'dynamic risk assessment' is often used to describe the day to day judgements that employees are expected to make in respect of health and safety. However, employers must ensure that employees have the necessary knowledge and experience to make such judgements. The employer's 'generic risk assessments' must have identified the types of risks which might be present in the work activities, established a framework of precautions (procedures, equipment etc.) which are likely to be necessary and provided guidance on which precautions are appropriate for which situations.
* *Permits to work*
  A permit to work system is a formalised method for identifying a safe system of work (usually for a high risk activity) and ensuring that this system is followed. The permit issuer is expected to carry out a dynamic risk assessment of the work activity and should be more competent in identifying the risks and the relevant precautions than those carrying out the work.
  Permits to work are dealt with in more detail in the chapter on SAFE SYSTEMS OF WORK.
* CDM *Construction Phase Plans*

The construction phase plan required for 'notifiable projects' under the *Construction (Design and Management) Regulations 2007 (CDM) (SI 2007 No 320)* is in effect a risk assessment in relation to the construction project. The various risks associated with the project must be identified, together with the means for eliminating or controlling those risks. Further detail on the regulations is provided in CONSTRUCTION, DESIGN AND MANAGEMENT and also in the HSE booklet *L144 'Managing health and safety in construction'* in which Appendix 3 sets out topics for inclusion in a construction phase plan.

- *Method statements*
  Method statements usually involve a description of how a particular task or operation is to be carried out and should identify all the components of a safe system of work arrived at through a process of risk assessment.

Management Regulations requirements

[R3018] The general requirement concerning risk assessment is contained in the *Management of Health and Safety at Work Regulations 1999 ('the Management Regulations') (SI 1999 No 3242), Reg 3*. Changes to the original regulations passed in 1992 mean that the risk assessment must now take specific account of risks to both young persons (under 18's) (see **R3025** below) and new and expectant mothers (see **R3026** below). Other regulations require more specific types of risk assessment e.g. of hazardous substances (COSHH), noise, manual handling operations, display screen equipment and personal protective equipment (see **R3006–R3016** above).

Regulation 3(1) of the *Management Regulations* states:

every employer shall make a suitable and sufficient assessment of:

(a) the risks to the health and safety of his employees to which they are exposed whilst they are at work; and

(b) the risks to the health and safety of persons not in his employment arising out of or in connection with the conduct by him of his undertaking,

for the purpose of identifying the measures he needs to take to comply with the requirements or prohibitions imposed upon him by or under the relevant statutory provisions.

Regulation 3(2) of the *Management Regulations* imposes similar requirements on self-employed persons.

Regulation 3(3) requires a risk assessment to be reviewed if:

- there is reason to suspect that it is no longer valid; or
- there has been a significant change in the matters to which it relates.

Regulation 3(4) requires a risk assessment to be made or reviewed before an employer employs a young person, and *Regulation 3(5)* identifies particular issues which must be taken into account in respect of young persons (especially their inexperience, lack of awareness of risks and immaturity). Further requirements in respect of young persons are contained in *Regulation 19*.

Regulation 16 contains specific requirements concerning the factors which must be taken into account in risk assessments in relation to new and expectant mothers. These relate to processes, working conditions and physical, biological or chemical agents. Assessments in respect of young persons and new or expectant mothers are dealt within more detail later in the chapter.

Regulation 3(6) requires employers who employ five or more employees to record:

— the significant findings of their risk assessments; and
— any group of employees identified as being especially at risk.

Methods of recording assessments are described later in this chapter (see **R3042** below).

HSE booklet L21, *'Management of health and safety at work'*, contains the Management Regulations in full, the associated Approved Code of Practice ('ACoP') and Guidance on the Regulations.

Hazards and risks

[R3019] The ACoP to the Management Regulations *(SI 1999 No 3242)* provides definitions of both hazard and risk.

A *hazard* is something with the potential to cause harm.

A *risk* is the likelihood of potential harm from that hazard being realised.

The *extent of the risk* will depend on:

• the likelihood of that harm occurring;
• the potential severity of that harm (resultant injury or adverse health effect);
• the population which might be affected by the hazard i.e. the number of people who might be exposed.

As an illustration, work at height involves a hazard of those below being struck by falling objects. The extent of the risk might depend on factors such as the nature of the work being carried out, the weight of objects which might fall, the distance to the ground below, the numbers of people in the area etc.

Evaluation of precautions

[R3020] The ACoP to the *Management Regulations 1999 (SI 1999 No 3242)* clearly states that risk assessment involves 'identifying the hazards present in any working environment or arising out of commercial activities and work activities, and evaluating the extent of the risks involved, taking into account existing precautions and their effectiveness.'

The evaluation of the effectiveness of precautions is an integral part of the risk assessment process. This is overlooked by some organisations who concentrate on the identification (and often the quantification) of risks without checking whether the intended precautions are actually being taken in the workplace and whether these precautions are proving effective.

Using the illustration in the previous paragraph, the risk assessment must take account of what precautions are taken, such as:

- use of barriers and/or warning signs at ground level;
- use of tool belts by those working at heights;
- provision of edge protection on working platforms; and
- use of head protection by those at ground level.

'Suitable and sufficient'

[R3021] Risk assessments under the *Management Regulations 1999 (SI 1999 No 3242)* (and several other regulations) must be 'suitable and sufficient', but the phrase is not defined in the Regulations themselves. However, the ACoP to the Regulations states that 'The level of risk arising from the work activity should determine the degree of sophistication of the risk assessment.' The ACoP also states that insignificant risks can usually be ignored, as can risks arising from routine activities associated with life in general ('unless the work activity compounds or significantly alters those risks'). Reference was made earlier (see **R3003**) to the concern expressed by the HSE (and others) about so-called 'risk aversion', which can result in precautions being demanded which are out of all proportion to the level of risk existing. (This is dealt with in greater detail in **R3052**.) However, in practice, a risk can only be concluded to be insignificant if some attention is paid to it during the risk assessment process and, if there is any scope for doubt, it is prudent to state in the risk assessment record which risks are considered insignificant.

Winter weather (with its attendant rain, ice, snow or wind) may be considered to pose a routine risk to life. However, driving a fork lift truck in an icy yard or carrying out agricultural or construction work in a remote location may involve a far greater level of risk than normal and require additional precautions to be taken.

The risk assessment must take into account both workers and other persons who might be affected by the undertaking.

For example, a construction company would need to consider risks to (and from) their employees, sub contractors, visitors to their sites, delivery drivers, passers by and even possible trespassers on their sites.

Similarly a residential care home should take into account risks to (and from) their staff, visiting medical specialists, visiting contractors, residents and visitors to residents.

Reviewing risk assessments

[R3022] The *Management of Health and Safety at Work Regulations 1999 (SI 1999 No 3242), Reg 3(3)* require a risk assessment to be reviewed if:

(a) there is reason to suspect it is no longer valid; or
(b) there has been a significant change in the matters to which it relates; and where as a result of any such review changes to an assessment are required, the employer or self-employed person concerned shall make them.

The ACoP to the Management Regulations states that those carrying out risk assessments 'would not be expected to anticipate risks that were not foreseeable.' However, what is foreseeable can be changed by subsequent events. An accident, a non-injury incident or a case of ill-health may highlight the need for a risk assessment to be reviewed because:

- a previously unforeseen possibility has now occurred;
- the risk of something happening (or the extent of its consequences) is greater than previously thought; or
- precautions prove to be less effective than anticipated.

A review of the risk assessment may also be required because of significant changes in the work activity e.g. changes to the equipment or materials used, the environment where the activity takes place, the system of work used or to the numbers or types of people carrying out the activity. The need for a risk assessment review may be identified from information received from elsewhere e.g. from others involved in the same work activity, through trade or specialist health and safety journals, from the suppliers of equipment or materials or from the HSE or other specialist bodies. Routine monitoring activities (inspections, audits etc.) or consultation with employees may also identify the need for an assessment to be reviewed. A review of the risk assessment does not necessarily require a repeat of the whole risk assessment process but it is quite likely to identify the need for increased or changed precautions.

In practice, workplaces and the activities within them are constantly subject to gradual changes and the ACoP states 'it is prudent to plan to review risk assessments at regular intervals'. The frequency of such reviews should depend on the extent and nature of the risks involved and the degree of change likely. It is advisable that all risk assessments should be reviewed at least every five years.

There are many activities where the nature of the work or the workplace itself changes constantly. Examples of such situations are construction work or peripatetic maintenance or repair work. Here it is possible to carry out 'generic' assessments of the types of risks involved and the types of precautions which should be taken. However, some reliance must be placed upon workers themselves to identify what precautions are appropriate for a given set of circumstances or to deal with unexpected situations. Such workers must be well informed and well trained in order for them to be competent to make what are often called 'dynamic' risk assessments (see **R3017** above).

Related requirements of the Management Regulations

[R3023] A number of other requirements within the Management Regulations *(SI 1999 No 3242)* are closely related to the risk assessment process.

- *Regulation 4: Principles of prevention to be applied*
  The *Management of Health and Safety at Work Regulations 1999 (SI 1999 No 3242), Reg 4* and *Sch 1* require preventive and protective measures to be implemented on the basis of a number of specified principles:
- avoiding risks;
- evaluating risks that cannot be avoided;
- combating risks at source;

- adapting the work to the individual, particularly with regard to workplace design, the choice of work equipment and working and production methods, with a view to alleviating monotonous work and work at a predetermined work-rate;
- adapting to technical progress;
- replacing the dangerous with safe or safer alternatives;
- developing a coherent overall prevention policy covering technology, work organisation, working conditions, social relationships, and the working environment;
- giving priority to collective over individual protective measures; and
- giving appropriate instructions to employees.
- *Regulation 5: Health and safety arrangements*
  Under the *Management of Health and Safety at Work Regulations 1999 (SI 1999 No 3242), Reg 5,* employers are required to have appropriate arrangements for the effective planning, organisation, control, monitoring and review of preventive and protective measures. Those employers with five or more employees must record these arrangements. This application of the 'management cycle' to health and safety matters is dealt with in greater detail in **R3024** below.
- *Regulation 6: Health surveillance*
  Under the *Management of Health and Safety at Work Regulations 1999 (SI 1999 No 3242), Reg 6,* where risks to employees are identified through a risk assessment, they must be 'provided with such health surveillance as is appropriate.' (Surveillance is also likely to be necessary to comply with the requirements of more specific regulations e.g. COSHH *(SI 2002 No 2677), Control of Asbestos Regulations 2006 (SI 2006 No 2739), Ionising Radiations Regulations 1999 (SI 1999 No 3232)).* Surveillance may be appropriate to deal with risks such as colour blindness and other vision defects (e.g. in electricians and train or vehicle drivers) or blackouts or epilepsy (e.g. for drivers, operators of machinery and those working at heights).
  HSE guidance is available on this important area (see HSG61 *'Health surveillance at work').*
- *Regulation 8: Procedures for serious and imminent danger and for danger areas*
  The *Management of Health and Safety at Work Regulations 1999 (SI 1999 No 3242), Reg 8* require every employer to 'establish and where necessary give effect to appropriate procedures to be followed in the event of serious and imminent danger to persons at work in his undertaking.' It also refers to the possible need to restrict access to areas 'on grounds of health and safety unless the employee concerned has received adequate health and safety instruction.'
  The need for emergency procedures or restricted areas should of course be identified through the process of risk assessment. Situations 'of serious and imminent danger' might be due to fires, bomb threats, escape or release of hazardous substances, out of control processes, personal attack, escape of animals etc. Areas may justify access to them being restricted because of the presence of hazardous substances, unprotected electrical conductors (particularly high voltage), potentially dangerous animals or people etc.

Application of the management cycle

[R3024] The time spent in carrying out risk assessments will be wasted unless those precautions identified as being necessary are actually implemented. The *Management of Health and Safety at Work Regulations 1999 (SI 1999 No 3242), Reg 5* require the application of a five stage 'management cycle' to this process.

- Planning
  The employer must have a planned approach to health and safety involving such measures as:
  — a health and safety policy statement;
  — annual health and safety plans; and
  — development of performance standards.
- Organisation
  An organisation must be put in place to deliver these good intentions e.g:
  — individuals allocated responsibility for implementing the health and safety policy or achieving elements of the health and safety plan;
  — arrangements for communicating with and consulting employees (using such measures as health and safety committees, newsletters, notice boards); and
  — the provision of competent advice and information.
- Control
  Detailed health and safety control arrangements must be established. The extent of such arrangements will reflect the size of the organisation and the degree of risk involved in its activities but they are likely to involve:
  — formal procedures and systems e.g. for accident and incident investigation, fire and other emergencies, the selection and management of contractors;
  — the provision of health and safety related training e.g. at induction, for the operation of equipment or certain activities, for supervisors and managers; and
  — provision of appropriate levels of supervision.
- Monitoring
  The effectiveness of health and safety arrangements must be monitored using such methods as:
  — health and safety inspections (both formal and informal);
  — health and safety audits; and
  — accident and incident investigations.
- Review
  The findings of monitoring activity must be reviewed and, where necessary, the management cycle applied once again to rectify shortcomings in health and safety arrangements.
  The review process might involve:
  — joint health and safety committees;
  — management health and safety meetings; and
  — activities of health and safety specialists.

The management cycle can also be applied to ensure the effective implementation of individual health and safety management procedures (e.g. for the investigation and reporting of accidents and non-injury incidents) or to specific types of health and safety precautions (e.g. guarding of machinery or precautions for working at heights).

The chapter MANAGING HEALTH AND SAFETY provides more information on management systems.

Children and young persons

[R3025] Previous Acts and regulations identified many types of equipment which children and young persons were not allowed to use or processes or activities that they must not be involved in. Many of these 'prohibitions' were revoked by the *Health and Safety (Young Persons) Regulations 1997 (SI 1997 No 135)*, since incorporated into the *Management of Health and Safety at Work Regulations 1999 (SI 1999 No 3242)*. (A few 'prohibitions' still remain). The emphasis has now changed to restrictions on the work which children and young persons are allowed to do, based upon the employer's risk assessment.

The *Health and Safety (Training for Employment) Regulations 1990 (SI 1990 No 1380)* have the effect of giving students on work experience training programmes and trainees on training for employment programmes the status of 'employees'. The immediate provider of their training is treated as the 'employer'. (There are exceptions for courses at educational establishments, i.e. universities, colleges, schools etc.). Therefore employers have duties in respect of all children and young persons at work in their undertaking: full-time employees, part-time and temporary employees and also students or trainees on work placement with them.

The term 'child' and 'young person' are defined in the *Management of Health and Safety at Work Regulations 1999 (SI 1999 No 3242), Reg 1(2)*.

- *Child* is defined as a person not over compulsory school age in accordance with:
  - the *Education Act 1996, s 8* (for England and Wales); and
  - the *Education (Scotland) Act 1980, s 31* (for Scotland).
  (In practice this is just under or just over the age of sixteen)
- *Young Person* is defined as 'any person who has not attained the age of eighteen'.

Some of the prohibitions remaining from older health and safety regulations use different cut off ages.

See the chapter on VULNERABLE PERSONS which provides much more detail on the requirements of the Management Regulations and the factors to be taken into account when carrying out risk assessments in respect of children and young persons. These factors centre around their:

- lack of experience;
- lack of awareness of existing or potential risks; and
- immaturity (in both the physical and psychological sense).

That chapter also contains a checklist of 'Work presenting increased risks for children and young persons' (see **V12016**).

New or expectant mothers

[R3026] Amendments made in 1994 to the previous Management Regulations implemented the European Directive on Pregnant Workers, requiring employers in their risk assessments to consider risks to new or expectant mothers. These amendments were subsequently incorporated into the 1999 Management Regulations. The *Management of Health and Safety at Work Regulations 1999 (SI 1999 No 3242), Reg 1* contains two relevant definitions:

* '*New or expectant mother*' means an employee who is pregnant; who has given birth within the previous six months; or who is breastfeeding.
* '*Given birth*' means 'delivered a living child or, after twenty-four weeks of pregnancy, a stillborn child.'

The requirements for 'risk assessment in respect of new and expectant mothers' are contained in the *Management of Health and Safety at Work Regulations 1999 (SI 1999 No 3242), Reg 16* which states in *paragraph (1)*:

(1)   Where —
    (a)   the persons working in an undertaking include women of child-bearing age; and
    (b)   the work is of a kind which could involve risk, by reason of her condition, to the health and safety of a new or expectant mothers, or to that of her baby, from any processes or working conditions, or physical, biological or chemical agents, including those specified in Annexes I and II of Council Directive 92/85/EEC on the introduction of measures to encourage improvements in the safety and health at work of pregnant workers and workers who have recently given birth or are breastfeeding,
the assessment required by regulation 3(1) shall also include an assessment of such risk.

Regulation 16(4) states that in relation to risks from infectious or contagious diseases an assessment must only be made if the level of risk is in addition to the level of exposure outside the workplace. (The types of risk which are more likely to affect new or expectant mothers are described in **R3027** below.) *Regulation 16(2)* and *(3)* set out the actions employers are required to take if these risks cannot be avoided. *Paragraph (2)* states:

Where, in the case of an individual employee, the taking of any other action the employer is required to take under the relevant statutory provisions would not avoid the risk referred to in paragraph (1) the employer shall, if it is reasonable to do so, and would avoid such risks, alter her working conditions or hours of work.

Consequently where the risk assessment required under *Reg 16(1)* shows that control measures would not sufficiently avoid the risks to new or expectant mothers or their babies, the employer must make reasonable alterations to their working conditions or hours of work.

In some cases restrictions may still allow the employee to substantially continue with her normal work but in others it may be more appropriate to offer her suitable alternative work.

Any alternative work must be:

—   suitable and appropriate for the employee to do in the circumstances;
—   on terms and conditions which are no less favourable.

Regulation 16(3) states:

> If it is not reasonable to alter the working conditions or hours of work, or if it would not avoid such risk, the employer shall, subject to section 67 of the 1996 Act, suspend the employee from work for so long as is necessary to avoid such risk.

(The 1996 Act referred to is the *Employment Rights Act 1996* which provides that any such suspension from work on the above grounds is on full pay. However, payment might not be made if the employee has unreasonably refused an offer of suitable alternative work.)

The *Management of Health and Safety at Work Regulations 1999 (SI 1999 No 3242), Reg 17* deals specifically with night work by new or expectant mothers and states:

> Where —
>
> (a)   a new or expectant mother works at night; and
> (b)   a certificate from a registered medical practitioner or a registered midwife shows that it is necessary for her health or safety that she should not be at work for any period of such work identified in the certificate,
>   the employer shall, subject to section 67 of the 1996 Act, suspend her from work for so long as is necessary for her health or safety.

Such suspension (on the same basis as described above) is only necessary if there are risks arising from work. The HSE does not consider there are any risks to pregnant or breastfeeding workers or their children working at night per se. They suggest that any claim from an employee that she cannot work nights should be referred to an occupational health specialist. The HSE's own Employment Medical Advisory Service (EMAS) is likely to have a role to play in such cases.

The requirements placed on employers in respect of altered working conditions or hours of work and suspensions from work only take effect when the employee has formally notified the employer of her condition. The *Management of Health and Safety at Work Regulations 1999 (SI 1999 No 3242), Reg 18* states:

> Nothing in paragraph (2) or (3) of regulation 16 shall require the employer to take any action in relation to an employee until she has notified the employer in writing that she is pregnant, has given birth within the previous six months, or is breastfeeding.

Regulation 18(2) states that the employer is not required to maintain action taken in relation to an employee once the employer knows that she is no longer a new or expectant mother; or cannot establish whether this is the case.

The chapter VULNERABLE PERSONS provides more information on these requirements.

Risks to new or expectant mothers

[R3027] The HSE booklet HSG122 *'New and expectant mothers at work: A guide for employers'* provides considerable guidance on those risks which may be of particular relevance to new or expectant mothers, including those listed in the EC Directive on Pregnant Workers (92/85/EEC). These risks might include:

Physical agents
- Manual handling.
- Ionising radiation.
- Work in compressed air.
- Diving work.
- Shock, vibration etc.
- Movement and posture.
- Physical and mental pressure.
- Extreme heat.

Biological agents
Many biological agents in hazard groups 2, 3 and 4 (as categorised by the Advisory Committee on Dangerous Pathogens) can affect the unborn child should the mother be infected during pregnancy.

Chemical agents
- Substances labelled with certain risk phrases.
- Mercury and mercury derivatives.
- Antimitotic (cytotoxic) drugs.
- Agents absorbed through the skin.
- Carbon Monoxide.
- Lead and lead derivatives.

Working conditions
The HSE's frequently stated position in respect of work with display screen equipment (DSE) is that radiation from DSE is well below the levels set out in international recommendations, and that scientific studies taken as a whole do not demonstrate any link between this work and miscarriages or birth defects. There is no need for pregnant women to cease working with DSE. However, the HSE recommends that, to avoid problems from stress or anxiety, women are given the opportunity to discuss any concerns with someone who is well informed on the subject.

Other factors associated with pregnancy may also need to be taken into account e.g. morning sickness, backache, difficulty in standing for extended periods or increasing size. These may necessitate changes to the working environment or work patterns.

Further information on the risks all of the above pose to new or expectant mothers is contained in the chapter dealing with VULNERABLE PERSONS.

Other vulnerable persons

[R3028] As well as children and young persons and new or expectant mothers, there are other vulnerable members of the workforce who merit special consideration during the risk assessment process. These include:

- Lone workers
  Some staff members may work alone continuously or only part of the time. They may be working on the employer's own premises or outside, within the community. The level of risk may relate to the worker's location, the activities they are involved in or the type of people they might encounter.
  Amongst lone workers will be:
  — security staff;
  — cleaners;
  — isolated receptionists or enquiry staff;
  — some retail staff;
  — some maintenance workers;
  — staff working late, at nights or at weekends;
  — keyholders (particularly at risk if called to suspected break-ins);
  — delivery staff;
  — sales and technical representatives; and
  — property surveyors etc.

- Disabled people
  Legislation concerning disability discrimination will affect general access requirements within work premises. The *Health and Safety (Miscellaneous Amendments) Regulations 2002 (SI 2002 No 2174)* introduced an additional requirement into *Regulation 25* of the *Workplace (Health, Safety and Welfare) Regulations 1992 (SI 1992 No 3004)*:

  > Where necessary, those parts of a workplace (including in particular doors, passageways, stairs, showers, washbasins, lavatories and workstations) used or occupied directly by disabled persons at work shall be organised to take account of such persons.

  In considering access in general and also specific risks which may affect disabled persons, account must be taken of all possible types of disability:
  — persons lacking mobility (including wheelchair users);
  — those lacking physical strength (particularly in relation to manual handling);
  — persons with relevant health conditions (especially on COSHH-related matters);
  — blind or visually impaired persons;
  — those who have hearing difficulties; and
  — persons with special educational needs.
  Whilst employers have particular duties towards their own employees, the risk assessment must also consider risks to disabled persons who are:

—   on work placements;
—   visitors;
—   service users; and
—   neighbours or passers-by.

All types of risks must be considered but particular note must be taken of:

—   access posing particular dangers to disabled persons;
—   the presence of hazardous substances;
—   potentially dangerous work equipment;
—   hot items or surfaces; and
—   the need for special emergency arrangements for disabled persons.

•   **Inexperienced workers**
Some adult workers may (like young people) be lacking in experience of some types of work or lack awareness of existing or potential risks of particular work activities. Some may also lack a suitable degree of maturity, particularly those with special needs. This must be taken into account during the risk assessment process, especially in situations such as:

—   training or retraining programmes;
—   work activities with rapid staff turnover; and
—   activities involving people with special needs.

More information on the above is available in the chapter VULNERABLE PERSONS.

Who should carry out the risk assessment?

[R3029] The *Management of Health and Safety at Work Regulations 1999 (SI 1999 No 3242)*, *Reg 7* require employers to appoint competent health and safety assistance. Those appointed must have sufficient training and experience or knowledge together with other qualities, in order to be able to identify risks and evaluate the effectiveness of precautions to control those risks. (*Regulation 7(8)* expresses a preference for such persons to be employees as opposed to others e.g. consultants.)

The HSE guidance on risk assessment suggests that in smaller, low-risk situations the employer or a senior manager may be quite capable of carrying out the risk assessment. However, as size and risks increase, a greater level of health and safety competence will be required. Guidance from the HSE on *Regulation 7* refers to competence as not necessarily depending on the possession of particular skills or qualifications. In simple situations what is required is:

—   an understanding of relevant current best practice;
—   an awareness of the limitations of one's own experience and knowledge; and
—   the willingness and ability to supplement existing experience and knowledge, where necessary by obtaining external help and advice.

The same principles can be applied to selecting persons to carry out risk assessments.

Larger employers may create risk assessment teams who might be drawn from managers, engineers and other specialists, supervisors or team leaders, health and safety specialists, safety representatives and other employees. Some team members may require some additional training or information about the principles of risk assessment and possibly technical aspects of the work activities to be assessed.

Many employers continue to utilise consultants to co-ordinate or carry out their risk assessments. A national register of occupational safety consultants, the Occupational Safety Consultants Register (OSCHR), has been developed by the HSE and a network of professional bodies representing safety consultants. This provides details of consultants who have met the highest qualification standard of recognised professional bodies. To be eligible to join the register, individual consultants must be either Chartered members of the safety bodies the Institution of Occupational Safety and Health (IOSH), Chartered Institute of Environmental Health (CIEH) or Royal Environmental Health Institute of Scotland (REHIS) or a Fellow of the International Institute of Risk and Safety Management (IIRSM). Membership means that they have a commitment to continuous professional development, a degree equivalent qualification, two years' experience, professional indemnity insurance and are bound by a code of conduct to only providing sensible and proportionate advice. The register is freely accessible and searchable for employers at www.os hcr.org.

Whether the assessments are to be carried out by an individual or by a team, others will need to be involved during the process. Managers, supervisors, employees, specialists etc. will all need to be consulted about the risks involved in their work and the precautions that are (or should be) taken.

Assessment units

[R3030] In a small workplace it may be possible to carry out a risk assessment as a single exercise but in larger organisations it will usually be necessary to split the assessment up into manageable units. If done correctly this should mean that assessment of each unit should not take an inordinate amount of time and also allows the selection of the people best able to assess an individual unit. Division of work activities into assessment units might be by departments or sections, buildings or rooms, processes or product lines, or services provided.

As an illustration a garage might be divided into:

- Servicing and repair workshop.
- Body repair shop.
- Parts department.
- Car sales and administration.
- Petrol and retail sales.

Relevant sources of information

[R3031] There is a wide range of documents which might be of value during the risk assessment process including:

- *Previous risk assessments*
  (including risk assessments carried out to comply with specific regulations e.g. COSHH or Noise regulations).
- *Operating procedures*
  (where these include health and safety information).
- *Safety handbooks etc.*
- *Training programmes and records*
- *Accident and Incident records*
- *Health and safety inspection or audit reports*
- *Relevant Regulations and Approved Codes of Practice*
  (Risk assessment involves an evaluation of compliance with legal requirements and therefore an awareness of the legislation applying to the workplace in question is essential.)
- *Relevant HSE publications*
  The HSE publishes a wide range of booklets and leaflets providing guidance on health and safety topics. Some of these relate to the specific requirements of regulations, others deal with specific types of risks whilst some publications deal with sectors of work activity.
  All of these can be of considerable value in identifying which risks the HSE regard as significant and in providing benchmarks against which precautions can be measured. The guidance on specific types of workplaces (which includes engineering workshops, motor vehicle repair, warehousing, kitchens and food preparation, golf courses, horse riding establishments and many others) should form an essential basis for those carrying out assessments in those sectors.
  Details of HSE publications are available via the HSE Books website (http://books.hse.gov.uk/hse/public/home.jsf) which has search and 'a dvanced search' facilities. A detailed catalogue is also published from t ime to time.
  The booklet *'Essentials of Health and Safety at Work'* provides an excellent starting point for those in small businesses needing guidance or carrying out risk assessment. The booklet also contains a useful reference section to other HSE publications which may be of relevance. To obtain these publications contact HSE Books directly at PO Box 1999, Sudbury, Suffolk CO10 2WA (tel: 01787 881165; fax: 01787 313995; website: (http://books.hse.gov.uk/hse/public/home.jsf). There is also a national network of booksellers stocking more important publications.
- *HSE website*
  The HSE website at http://books.hse.gov.uk/hse/public/home.jsf conta ins an ever-growing resource of guidance material. The section dealing with Risk Assessment provides further links to example risk assessments for a wide range of occupational areas and an online risk assessment web tool for office environments which prompts employers to answer a series of questions that generates their risk assessment and action plan (see www.hse.gov.uk/risk/office.htm).

The tool was developed by the HSE in response to a recommendation in the 2010 Young Review of health and safety legislation, 'Common Sense: Common Safety' to simplify the risk assessment procedure for 'low-hazard' workplaces such as offices, classrooms and shops. The HSE has been asked to produce simpler interactive risk assessments and p eriodic checklists for 'low-hazard' workplaces and voluntary sector orga nisations.

Safety organisations including the Chartered Institute of Environmental Health (CIEH) and the Royal Society for the Prevention of Accidents (RoSPA) both point out that while specific sectors may appear to present low hazard workplaces, even seemingly low hazard settings, such as offices, can have significant hazards associated with them.

It is also possible to search for information on particular risk topics or using a word search facility. Much of this information can be downloa ded directly from the website or the HSE Books website. A subscription service (HSE Direct) also allows access to priced HSE publications and these are also available to download free from the website.

- *Trade Association codes of practice and guidance*
- *Information from manufacturers and suppliers*
  (Equipment handbooks or substance data sheets).
- *General health and safety reference books*

Consider who might be at risk

[R3032] The risk assessment process must take account of all those who may be at risk from the work activities i.e. both employees and others. It is important that all of these are identified.

- *Employees*
  Different categories of employees to take into account might include:
  — production workers;
  — maintenance workers;
  — administrative staff;
  — security officers;
  — cleaners;
  — delivery drivers;
  — sales representatives;
  — others working away from the premises;
  — temporary employees.
  Some of these employees may merit special considerations:
  — children and young persons (see **R3025**);
  — women of childbearing age (i.e. potential new or expectant mothers) (see **R3026** and **R3027**); and
  — other vulnerable persons (see **R3028**);
- *Contractors and their staff*
  Contracted services might involve:
  — construction or engineering projects;
  — routine maintenance or repair;
  — hire of specialist plant and its operators;

— support services e.g. catering, cleaning, security, transport;
— professional services e.g. architects, engineers, trainers; and
— supply of temporary staff.
(Contractors also have duties to carry out risk assessments in respect of their own staff.)

- *Others at risk*
The types of people who might be put at risk by the organisation's activities will depend upon the nature and location of those activities. Groups of people to be considered include:
— volunteer workers;
— co-occupants of premises;
— occupants of neighbouring premises;
— drivers making deliveries;
— visitors (both individuals and groups);
— residents e.g. in the care or hospitality sectors;
— passers-by;
— users of neighbouring roads;
— trespassers; and
— customers or service users.

Identify the issues to be addressed

[R3033] The issues which will need to be addressed during the risk assessment process should be identified. This may be done in respect of the assessment overall or separately for each of the assessment units and will be based upon the knowledge and experience of those carrying out the assessment and the information gathered together from the sources described earlier. This will create an initial list of headings and sub-headings for the eventual record of the risk assessment, although in practice this list is likely to be amended along the way. The sample assessment records contained later in the chapter (see **R3050** and **R3051**) will demonstrate how such a list can be built up for typical workplaces.

These headings are likely to consist of the more common types of risk e.g. fire, vehicles, work at heights, together with some specialised types of risk associated with the work activities such as violence, lasers or working in remote locations. A checklist of possible risks to be considered during risk assessments is provided in **R3035** below. This includes some of the risks which have specific regulations associated with them.

In some situations particular notes may also be made to check on the effectiveness of the precautions which should be in place to control the risks e.g. standards of machine guarding, compliance with personal protective equipment (PPE) requirements or the quality and extent of training.

Variations in work practices

[R3034] Consideration should also be given at this stage to possible variations in work activities which may create new risks or increase existing risks. Such variations might involve:

- Fluctuations in production or workload demands.
- Reallocation of staff to meet changing workloads.
- Problems introduced by staff absences.
- Seasonal variations in work activities.
- Abnormal weather conditions.
- Alternative work practices forced by equipment breakdown/unavailability.
- Urgent or 'one-off' repair work.
- Work carried out in unusual locations.
- Difference in work between days, nights or weekends.

Further variations may emerge later on in the assessment process.

Checklist of possible risks

[R3035] This checklist is intended to assist in identifying which issues need to be addressed during risk assessments. In some cases the headings and/or sub-headings might be used in the form shown, in other cases it may be more appropriate to combine them or modify the titles.

Work Equipment
- Process machinery
- Other machines
- Powered tools

- Handtools

- Knives/blades
- Fork-lift trucks
- Cranes
- Lifts
- Hoists
- Lifting equipment
- Vehicles

Access
- Vehicle routes
- Rail traffic
- Pedestrian access

Other factors
- Vibration
- Lasers
- Ultra violet/infra red radiation
- Work related upper limb disorder
- Stress
- Bullying and harassment

Services/power sources
- Electrical installation
- Compressed air
- Steam
- Hydraulics
- Other pressure systems
- Buried services
- Overhead services

Storage
- Shelving and racking

- Glazing

Work at height
- Scaffolding
- Mobile elevating work platforms
- Ladders and stepladders
- Falling objects
- Roof work

Work activity
- Burning or welding
- Entry into confined spaces
- Electrical work
- Excessive fatigue
- Handling cash/valuables
- Use of compressed gases
- Molten metal

External factors
- Violence or aggression
- Robbery
- Large crowds
- Animals
- Clients' activities

- Stacking
- Silos and tanks
- Waste

Fire and explosion prevention
- Flammable liquids *
- Flammable gases *
- Storage of flammables *
- Hot work
* DSEAR requirements

Work locations
- Heat
- Cold
- Severe weather
- Deep water
- Tides
- Remote locations
- Work alone
- Homeworking
- Poor hygiene
- Infestations
- Work abroad
- Work in domestic property
- Clients' premises
- Site security
- Work-related driving

Risks/issues subject to separate assessment requirements	
Hazardous substances (COSHH)	Lead
Noise	Asbestos work
Manual handling	Asbestos in premises
Display screen equipment workstations	Conformity of machinery
PPE needs	Major accident hazards (COMAH)
Fire risks and precautions	Ionising radiation Artificial optical radiation
Dangerous Substances and Explosive Atmospheres (DSEAR)	Vibration

Making the risk assessment

[R3036] Good planning and preparation can reduce the time spent in actually making the risk assessment as well as enabling that time to be used much more productively. However, it is essential that time is spent in work locations, seeing how work is actually carried out (as opposed to how it should be carried out).

Observation

[R3037] Observation of the work location, work equipment and work practices is an essential part of the risk assessment process. Where there are known to be variations in work activities a sufficient range of these should be observed to be able to form a judgement on the extent of the risks and the adequacy of precautions. Evaluations can be made of the effectiveness of fixed guards, the suitability and condition of access equipment, compliance with PPE requirements, the observance of specified operating procedures or working practices and many other aspects. It should also be borne in mind that work practices may change once workers realise they are under observation. Initial or undetected observation of working practices may be the most revealing.

Discussions

[R3038] Discussions with people carrying out work activities, their safety representatives and those responsible for supervising or managing them are also an essential part of risk assessment. Amongst aspects of the work that might be discussed are possible variations in the work activities, problems that workers encounter, workers' views of the effectiveness of the precautions available, the reasons some precautions are not utilised and their suggestions for improving health and safety standards.

Tests

[R3039] In some situations it may be appropriate to test the effectiveness of safety precautions e.g. the efficiency of interlocked guards or trip devices, the audibility of alarms or warning devices or the suitability of access to remote workplaces e.g. crane cabs, roofs. When carrying out such tests care must be taken by those carrying out the assessment not to endanger themselves or others, nor to disrupt normal activities.

Further investigations

[R3040] Frequently further investigations will need to be made before the assessment can be concluded. Such investigations may involve detailed checks on standards or records, or enquiries into how non-routine situations are dealt with. Examples of further checks or enquiries which might be appropriate are:

- Detailed requirements of published standards e.g. design of guards, thickness of glass.

- Contents of operating procedures.
- Maintenance or test records.
- Contents of training programmes or training records.

Once again the potential list is endless although lines of further enquiry should be indicated by the observations and discussions during the initial phase of the assessment.

Notes

[R3041] Rough notes should be made throughout the assessment process. It will be on these notes that the eventual assessment record will be based. It will seldom be possible to complete an assessment record 'on the run' during the assessment itself. The notes should relate to anything likely to be of relevance, such as:

- Risks discounted as insignificant.
- Further detail on risks, or additional risks identified during the assessment.
- Risks which are being controlled effectively.
- Descriptions of precautions which are in place and effective.
- Precautions which do not appear to be effective.
- Alternative precautions which might be considered.
- Problems identified or concerns expressed by others.
- Related procedures, records or other documents.

Assessment records

[R3042] This phase of the risk assessment process concludes with the preparation of the assessment records. It may be preferable to record the findings in draft form initially, with a revised version being produced once the assessment has been reviewed more widely and/or recommended actions have been completed.

- *Decide on the record format*
  Details of the content of assessment records and examples of one assessment record format are provided later in the chapter (see **R3050** and **R3051**) but many alternatives are available. Different types of records may be appropriate for different departments, sections or activities. Use of a 'model' assessment might be relevant for similar workplaces or activities. Some (or all) of the assessment record might be integrated into documented operating procedures.
- *Identify the section headings to be used*
  During the preparatory phase of the assessment a list of risks and other sources to be addressed in each assessment unit was prepared as an aide memoire. This list will now need to be converted into section headings for the assessment records. As a result of the assessment some of the headings may have been sub-divided into different headings whilst others may have been merged.
- *Prepare the assessment records*

The rough notes made during assessment must then be converted into formal assessment records. Preparing the records will normally require at least half of the time that was spent in the workplace carrying out the assessment and sometimes might even take longer. It is wise not to allow too long to elapse between assessing in the workplace and preparation of the assessment record. Notes will seem much more intelligible and memory will often be able to 'colour in' between the notes.

- *Identify the recommendations*
  The assessment will almost inevitably result in recommendations for improvements and these must be identified. As can be seen later, some assessment record formats incorporate sections in which recommendations can be included but in other cases separate lists will need to be prepared. The use of risk rating matrices (see **R3043**) may assist in the prioritisation of recommendations.

Risk rating matrices

[R3043] The ACoP accompanying the *Management of Health and Safety at Work Regulations 1999 (SI 1999 No 3242)* describes how the risk assessment process needs to be more sophisticated in larger and more hazardous sites. Nevertheless it suggests that quantification of risk will only be appropriate in a minority of situations. Some organisations use risk-rating systems to assist in the identification of priorities. Most involve matrices, utilising a simple combination of the likelihood of a hazard having an adverse effect and the severity of the consequences if it did. Some use numbers to produce a risk rating, as in the example below:

Risk Assessment Matrix			*Likelihood of adverse effect*		
			Unlikely	Possible	Frequent
			1	2	3
Severity of consequences	Minor	1	1	2	3
	Moderate	2	2	4	6
	Severe	3	3	6	9

The numbers can be replaced by descriptions of the level of risk as shown in the next example:

Risk Assessment Matrix		*Likelihood of adverse effect*		
		Unlikely	Possible	Frequent
Severity of consequences	Minor	Low	Low	Medium
	Moderate	Low	Medium	High
	Severe	Medium	High	Very high

The author's view is generally against the use of such matrices on the basis that time is often spent considering risk values at the expense of evaluating the effectiveness of the controls which ought to be in place. This view was supported in 2006 by the HSE who, in launching revamped Risk Assessment guidance, urged businesses to spend less time in dotting 'i's and crossing 't's and more time on putting practical actions into effect.

After the assessment

[R3044] Whilst the recording of the risk assessment is an important legal requirement, it is even more important that the recommendations for improvement identified during the assessment are actually implemented. This is likely to involve several stages.

Review and implementation of the recommendations

[R3045] There may be a need to involve others outside (and probably senior to) the risk assessment team. The reasoning behind the recommendations can be explained and various alternative ways of controlling risks can be evaluated. Changes to the assessment findings or recommendations may be made at this stage but the risk assessment team should not allow themselves to be browbeaten into making alterations that they do not feel can be justified. Similarly they should not hold back on making recommendations they consider are necessary just because they believe that senior management will not implement them. The assessment team should carry out their duties to the best of their abilities in identifying what precautions are necessary in order to comply with the law – the responsibility for achieving compliance rests with their employer.

Some recommendations may need to be costed in respect of the capital expenditure or staff time required to implement them. It is unlikely that all recommendations will be able to be implemented immediately; there may be a significant lead time for the delivery of materials or the provision of specialist services from external sources. Once costings and prioritisation have been agreed, the recommendations should be converted into an action plan with individuals clearly allocated responsibility for each element of the plan, within a defined timescale. Some members of the risk assessment team (particularly health and safety specialists) may have responsibility for implementing parts of the plan or providing guidance to others.

Recommendation follow-up

[R3046] Even in well-intentioned organisations, recommendations for improvement that have been fully justified and accepted are often still not implemented. It is essential that the risk assessment process includes a follow-up of the recommendations made. The recording format provided later in the chapter includes reference to this. As well as establishing that the improvements have actually been carried out, consideration should also be given to whether any unexpected risks have inadvertently been created.

Once the follow-up has been carried out, the assessment record should be annotated or revised to take account of the changes made. If recommendations have not been implemented there is a clear need for the situation to be referred back to senior management for them to take action to overcome whatever are the obstacles to progress.

Assessment review

[R3047] The *Management of Health and Safety at Work Regulations 1999 (SI 1999 No 3242)* and the associated ACoP state that assessments must be reviewed in certain circumstances (see **R3022** above). The ACoP also states that 'it is prudent to plan to review risk assessments at regular intervals'. A frequency for reviews should be established which relates to the extent and nature of the risks involved and the likelihood of creeping changes (as opposed to a major change which would automatically justify a review).

The review may, however, conclude that the risks are unchanged, the precautions are still effective and that no revision of the assessments is necessary. The process of conducting regular reviews of assessments and, where appropriate, making revisions may be aided by the application of document control systems of the type used to achieve compliance with ISO 9000 and similar standards.

Content of assessment records

[R3048] The *Management of Health and Safety at Work Regulations 1999 (SI 1999 No 3242), Reg 3(6)* states:

> Where the employer employs five or more employees, he shall record —
>
> (a)   the significant findings of the assessment; and
> (b)   any group of his employees identified by it as being especially at risk.

The accompanying ACoP refers to the record as representing 'an effective statement of hazards and risks which then leads management to take the relevant actions to protect health and safety.' It goes on to state that the record must be retrievable for use by management, safety representatives, other employee representatives or visiting inspectors. The need for linkages between the risk assessment, the record of health and safety arrangements (required by *Reg 5* of the Management Regulations) and the health and safety policy is also identified. The ACoP allows for assessment records to be kept electronically as an alternative to being in written form.

The essential content of any risk assessment record should be:

- Hazards or risks associated with the work activity.
- Any employees identified as especially at risk.
- Precautions which are (or should be) in place to control the risks (with comments on their effectiveness).
- Improvements identified as being necessary to comply with the law.

Other important details to include are:

- Name of the employer.

- Address of the work location or base.
- Names and signatures of those carrying out the assessment.
- Date of the assessment.
- Date for next review of the assessment.

(These might be provided as an introductory sheet.)

Illustrative assessment records

[R3049] In the final pages of this chapter the risk assessment methodology described earlier is used to provide illustrations of how the process can be applied in two different types of workplace. In each case, relevant risks or issues to be addressed are listed, one of those risks or issues is selected, relevant regulations, references and other key assessment points relating to that risk or issue are identified and an illustration of how the completed assessment record might look is provided. The content of the form used is similar to that provided in the HSE's *'Five steps to risk assessment'* leaflet but the layout is felt to be more user-friendly.

The illustrations used are for a supermarket and a newspaper publisher.

In recent years the HSE website has followed a similar path by providing sample assessments for a range of occupational areas. However, the HSE quite rightly stress that such samples are only illustrations of the approach to take. Every business is different and employers must still make sure that they have identified all the hazards and the necessary control measures for their particular workplace.

A supermarket

[R3050] Relevant risk topics are likely to include:

- Food processing machinery – in the delicatessen.
- Knives – delicatessen and butchery.
- Fork-lift trucks – warehousing areas.
- Vehicle traffic – delivery vehicles, customer and staff vehicles.
- Vehicle unloading – in the goods inward area.
- Pedestrian access – external to and inside the store.
- Glazed areas – store windows, display cabinets.
- Electrical installation – the power system within the store.
- Electrical equipment – including tills, cleaning equipment, office equipment.
- Shelving – in the store.
- Racking – in warehouse areas.
- Refrigerators – in the store and warehouse.
- Fire – general risks only, precautions must take account of customers.
- Aggression – e.g. from unhappy customers.
- Possible robbery – the supermarket will hold large quantities of cash.
- Work related upper limb disorders (WRULD) – particularly for check-out operators.

- Hazardous substances – cleaning materials, office supplies.
- Manual handling – e.g. shelf stacking, movement of trolleys.
- Display screen equipment – used at workstations in the office.
- PPE requirements – throughout the supermarket.
- Asbestos – the possible presence of asbestos-containing materials within the premises.
- Use of contractors – for maintenance and repair work.
- Stress, bullying and harassment.

A sample risk assessment record for **vehicle traffic** is provided below.

In conducting this risk assessment:

- The requirements of the *Workplace (Health, Safety and Welfare) Regulations 1992* must be complied with.
- HSE booklet HSG 136 *'Workplace Transport Safety'* and leaflet INDG 199 *'Workplace Transport Safety. An overview'* are likely to be useful.
- Particular attention should be paid to:
  — the condition of roadways and parking areas;
  — signage and road markings;
  — observation of vehicle movements and speeds;
  — conditions during busy periods and hours of darkness.

ABC SUPERMARKETS, NEWTOWN

Risk Assessment

Reference number: 4	Risk topic/issue: Vehicle traffic	Sheet 1 of 1
Cross references: HSE booklet HSG 136 Risk assessments 5 (vehicle unloading) and 6 (pedestrian access)		
Risks identified	Precautions in place	Recommended improvements
Vehicles making deliveries to the 'Goods inward' bay present risks to each other and to any pedestrians on the access road	Prominent 10 mph are in place on the roadway. The speed limit is enforced effectively. Signs prohibit use of the road by pedestrians and by customer or staff vehicles.	
There are also risks to other road users as they leave and rejoin the main road.	There are give way signs and road markings at the junction with the main road. Vehicles are parked in a holding area prior to backing up to the 'Goods Inward' bay. Movement of vehicles is controlled by a designated member of the supermarket staff. The area is well lit by roadside lamps and floodlights on the side of the building.	Reposition the advertising sign which partly blocks visibility on rejoining the main road Provide this designated staff member with a high visibility waterproof jacket.
Customer and staff vehicles circulating in the car park area present risks to each other and to pedestrians in the area.	The access road around the car park is one way and well indicated by signs. The entrance and exit are well separated from each other and the goods access road.	

There are also risks to other road users at the entrance from and access back into the main road.	There are prominent 15 mph signs around the roadways (some vehicles exceed this speed). There are also some give way markings and signs at all roadway junctions. Parking bays are well marked. Pedestrian crossing points are clearly marked and signed. The area is well lit by lighting towers which are protected at the base. Security staff inspect, and where necessary, salt roadways in icy or snowy weather	Provide clearly marked speed ramps at suitable locations. The surface of some bays in the south-west corner of the car park should be repaired. Include the goods access road, Goods inward bay and car park in routine safety inspections.
Signature(s) A Smith, B Jones	**Name(s)** A Smith, B Jones	**Date** 23/4/11
Dates for	**Recommendation follow up** July 2011	Next routine review December 2014

A newspaper publisher

[R3051] A large workplace like this would need to be divided into assessment units (see **R3030** above), which might consist of:

- Common facilities, services etc. – e.g. fire, electrical supply, lifts, vehicle traffic, presence of asbestos.
- Reel handling and stands – e.g. supply of reels of newsprint to the press area.
- Platemaking – e.g. equipment and chemicals used to produce printing plates.
- Printing press – e.g. press machinery, solvents, noise etc.
- Despatch – e.g. inserting equipment, newspaper stacking, strapping and loading.
- Circulation and transport – e.g. distribution and other vehicles, fuel, waste disposal.
- Maintenance – e.g. workshops, garage, maintenance activities.
- Offices – e.g. editorial and administrative areas.

Selecting the **reel handling and stands** unit, risk topics are likely to include:

- Fork-lift trucks – used to unload, transport and stack reels.
- Reel storage – stability of stacks, access issues.
- Reel handling equipment – hoists, conveyors and the reel stands feeding the press.
- Wrappings and waste – removal of wrappings, storage and disposal of waste paper etc.
- Noise and dust – from the operation of the reel stands and nearby press.

A sample assessment record for **reel handling equipment** is provided below.

In conducting this risk assessment:

- The requirements of the *Provision and Use of Work Equipment Regulations 1998 (PUWER)*, the *Lifting Operation and Lifting Equipment Regulations 1998* and the *Manual Handling Operations Regulations 1992* must be complied with.
- Reference to HSE leaflet INDG 290 *'Simple guide to the Lifting Operations and Lifting Equipment Regulations 1998'* may be necessary;
- Reference may need to be made to BS, EN or other standards for conveyors or specialist handling equipment.
- Particular attention should be paid to:
  - guarding standards and the possible presence of unguarded dangerous parts;
  - any need for manual handling of the reels;
  - training issues relating to the above;
  - statutory examination records (for the hoist).

NEWTOWN NEWS

Risk Assessment		
Reference number: B3	Risk topic/issue: **Reel handling and stands – reel handling equipment**	Sheet 1 of 1
Cross references: Risk assessments B1 (Fork-lift trucks), B2 (Reel storage), B5 (Noise and dust)		
Risks identified	**Precautions in place**	**Recommended improvements**
The equipment below presents risks to all staff working in the area.		
Reel hoist – Carries reel down from the reels store to the reel stand basement. It is fed by fork-lift trucks and feeds onto the roller conveyor system.	Slow moving hoist protected by a substantial mesh guard.	
	No need for access within the hoist enclosure.	
	Sign states 'Do not ride on hoist'.	Replace this sign by one complying with the *Safety Signs Regulations*.
	Gap between base of hoist and roller conveyor (no shear trap).	
	Statutory LOLER examinations by Insurance Engineers (kept by Works Engineer).	
Roller conveyor – This is in a T formation and consists of powered and free running rollers with a turntable at the junction. Reels are transferred to holding bays or floor-based trolleys by fork-lift truck.	All drives for the conveyor are fully enclosed and there are no in-running nips. The conveyor is protected by kerbs from fork-lift damage.	
	Signs prohibit climbing on the conveyor system.	Replace these signs as above.

Floor-based trolleys – This system carries reels right up to the transfer carriages which load them onto the reel stands.	Reels can be easily moved by one person using the trolley system. Reels can be safely rolled across the floor onto the trolleys if the correct technique is used.	Include formal training on correct handling techniques in the induction programme for new staff in the area.
Some reels must be transferred manually from the holding bays onto the trolleys.	Two persons are required to move the transfer carriages up to the reel stands.	
Reel stands – The reel stands rotate mechanically ensuring a constant web feed to the press. The rotating drive shaft for the reel stands is approximately 4 metres above the ground level.	There are no accessible in-running nips. Operatives do not need to approach the rotating reels. This shaft is considered to be 'safe by position' for normal operating purposes.	Ensure a safe system of work for maintenance work near the shaft, e.g. by using a permit to work procedure.
Signature(s) J Young, K Old	**Names(s)** J Young, K Old	**Date** 28/3/08
Dates for	**Recommendation follow up** June 2008	**Next routine review** March 2010

Risk aversion

[R3052] 'Health and safety' regularly gets blamed for nonsensical decisions which have nothing to do with the correct application of risk assessment and risk management principles. Often such decisions are made by people with no real knowledge of health and safety, sometimes just as an excuse for a particular course of action or inaction. Both the HSE and the Institution of Occupational Safety and Health (IOSH) have become increasingly concerned about what has come to be described a 'risk aversion'.

In his 2010 review of health and safety legislation and 'the compensation culture', *Common Sense: Common Safety*, Lord Young set out that he believed that sensible health and safety rules that apply to hazardous occupations have been applied across all occupations; and that unregulated health and safety consultants are taking an overzealous approach and attempting to eliminate all risk.

HSE Chair, Judith Hackitt, welcomed the report 'as an important milestone on the road to recovery for the reputation of real health and safety' and said it provided 'a tremendous opportunity to refocus health and safety on what it is really about – managing workplace risks.'

The HSE sets out 'We believe that risk management should be about practical steps to protect people from real harm and suffering – not bureaucratic back covering. If you believe some of the stories you hear, health and safety is all about stopping any activity that might possibly lead to harm. This is not our vision of sensible health and safety we want to save lives, not stop them. Our approach is to seek a balance between the unachievable aim of absolute safety and the kind of poor management of risk that damages lives and the economy.'

The HSE website has a 'Myth of the month' section and the 'Risk assessment' page on the website contains 10 principles of sensible risk management.

Sensible risk management is about:

- Ensuring that workers and the public are properly protected;
- Providing overall benefit to society by balancing benefits and risks, with a focus on reducing real risks — both those which arise more often and those with serious consequences;
- Enabling innovation and learning not stifling them;
- Ensuring that those who create risks manage them responsibly and understand that failure to manage real risks responsibly is likely to lead to robust action;
- Enabling individuals to understand that as well as the right to protection, they also have to exercise responsibility.

Sensible risk management is not about:

- Creating a totally risk free society;
- Generating useless paperwork mountains;
- Scaring people by exaggerating or publicising trivial risks;
- Stopping important recreational and learning activities for individuals where the risks are managed;
- Reducing protection of people from risks that cause real harm and suffering.

In carrying out risk assessments one must ignore the trivia and keep a sense of proportion.

References

[R3053] *HSE publications*

Many specific references are contained earlier in this chapter. The following general references are relevant in risk assessment and/or risk management.

INDG 163	Five steps to risk assessment (2006)
INDG 259	An introduction to health and safety. Health and safety in small businesses (2003)
INDG 417	Leading health and safety at work. Leadership action for directors and board members (2007)
—	Essentials of health and safety at work (2006)
HSG 65	Successful health and safety management (1997)

HSE website

The HSE website (www.hse.gov.uk) contains a section dealing with risk management (accessible via 'Health & safety topics'). This provides access to much more information including:

- Five steps to risk assessment;
- Sensible risk management (the 10 principles set out in R3052 above);
- Example risk assessments (for a range of occupational activities);
- Risk and disability;

HSE publications can be downloaded from the web pages.

Safe Systems of Work

Lawrence Bamber

Introduction

[S3001] This chapter examines the need for safe systems of work ('SSWs') within the framework of an occupational safety and health management system ('OSHMS').

In essence SSWs outline the control measures needed to avoid or reduce risk in the workplace and are usually one of the main results of a suitable and sufficient risk assessment.

This chapter considers:

*   the components of a safe system of work – both legal and others (S3002–S3006);
*   appropriate safe system of work required for level of risk involved (S3007);
*   development of SSWs, including job safety analysis; preparation of job safety instruction and safe operating procedures (S3008–S3010);
*   permit to work systems (S3011–S3022);
*   isolation procedures (S3023).

This particular chapter will be of use to guide those managers and supervisors who have the responsibility to develop and utilise SSWs, including permit to work systems, within their areas of control. It will also be useful to occupational safety and health ('OSH') practitioners who have to advise line management on practical development and implementation of such systems.

Essentially, a safe system of work is written down, and chronologically lists the methods of doing a particular job in such a way so as to avoid or minimise risk within the workplace. The safe system of work is therefore a very important workplace precaution or control measure which should be known by all concerned, agreed with employees via joint consultation, understood via regular communications/toolbox talkes, adhered to by all involved in the job concerned, and rigorously enforced by supervision within the workplace.

All employers should therefore utilise suitable and sufficient safe systems of work commensurate with the degree of risk highlighted in a risk assessment, especially when the legal requirement for SSWs are taken into account (see S3003 and S3004).

The pitfalls of not utilising SSWs within the risk control system include:

*   an increase in the number of accidents/injuries/diseases;
*   the increased risk of prosecution;

- increased employers' liability insurance premiums;
- a more dangerous workplace.

Remember – SAFE SYSTEMS OF WORK MEAN FEWER ACCIDENTS! To illustrate the down side of not utilising SSWs as part of an OSHMs, consider the case of a fatality in a Nottingham paint shop, outlined below:

Nottingham paint shop Leadmaster Ltd was fined £20,000 at Newark Magistrates' Court on 18 April 2005.

The HSE prosecution followed an investigation into the circumstances surrounding the death of Robert Fountain, who had been spray-painting a 300kg fabricated steel grid hanging from a forklift truck in the company's Newark paint shop on 17 June 2004. He was found by his son, Jason Fountain, trapped against a steel table by the grid, which had fallen on him.

Leadmaster Ltd is based on the Colwick Industrial Estate, Nottingham, and has a factory at Beaconside Works, Beacon Hill, Newark, which employs eight people and trades as Parsons (Newark).

The company admitted one charge under *HSWA 1974 s 2(1)*. It was fined £20,000, the maximum at magistrates' court, and ordered to pay costs of £2,922.

HSE Inspector David Appleton said:

> 'This was a tragic incident to a newly recruited employee. It would have been simple and cheap to prevent this death by using a rope or strap to secure the fabrication, or by placing it on trestles for painting. The incident underlines the need for a safe system of work to be devised for all tasks and for everybody involved to understand how to do the job safely.'

More recently, further prosecutions by HSE have highlighted how accidents could and should have been prevented via the use of written SSWS.

9th December 2008: the HSE warned construction companies to ensure they provide written safe systems of work after part of a man's leg was amputated when a tipper truck fell onto him, trapping him against a pile of brick rubble.

M J Curle Ltd of Shifnal were fined £5,000 and ordered to pay costs of £7,500 at Shrewsbury Crown Court after pleading guilty to breaching *section 3(1)* of the *HSW Act 1974*.

Principal Contractor, Anthony Wilson Homes Ltd of Shifnal pleaded guilty to breaching *sections 2(1)* and *3(1)* of *HSW Act 1974* and *Regulation 3(1)(a)* of the *Management of Health and Safety at Work Regulations 1999* (MHSWR).

The court imposed a fine of £25,000 with costs of £10,000 on Anthony Wilson Homes Ltd for its greater responsibility for the lack of site safety, including a lack of SSWS.

14th January 2009: The HSE have called on employers in the removal and haulage business to ensure that proper training and written safe systems of work are in place, even for routine tasks.

This message follows an accident in Louth, Lincolnshire in which a worker was crushed between a seventeen-tonne removals van and a brick wall.

Fox Group (Moving and Storage) Ltd were fined £3,515 and ordered to pay £2,000 costs by Skegness Magistrates Court after pleading guilty to breaching *MHSW 1999* and contravening *Regulation 9(1)* of the *Provision and Use of Work Equipment Regulations 1998* (PUWER) for failing to undertake sufficient risk assessment and training in the resulting SSWS for employees.

29th March, 2011: James Paterson Haulage Ltd appeared at Inverness Sheriff Court and pleaded guilty to breaching *section 2(1)* of the *HSWA 1974* and were fined £13,000. Mackay Steelwork and Cladding Ltd appeared at the same hearing and pleaded guilty to breaching *section 3(1)* of *HSWA 1974* and were fined £40,000.

These prosecutions followed a fatal accident to one of James Paterson's lorry drivers who was killed when two steel gates fell off his vehicle and landed on him during an inadequately planned lifting operation which took place during the delivery of twenty steel gates from Mackay's yard at Belney, Invergordon to a garden centre in Inverness.

When the delivery driver arrived at the garden centre, he removed the securing straps from his load and began to assist a forklift truck (flt) driver in unloading the gates. He directed the flt driver to ensure that the forks were positioned underneath one of the two stacks of gates. It is believed he then walked around to the far side of the lorry to place the straps inside a storage box located next to the vehicle's fuel tank.

As he was bending over to open the box, the flt prongs extended beyond the first stack of gates and struck the second stack, causing four of the gates to fall off the lorry. Two of the gates landed on the lorry driver who died at the scene as a result of fatal neck injuries.

HSE Inspector, Graeme McMinn, explained that both companies involved had failed to adequately liaise with each other or obtain enough information so as to ensure that a written safe system of work was in place, particularly in relation to the role that the lorry driver would play in the unloading of the gates. He went on to say that one method of doing this would have been to create a segregated, clear zone, designed to prevent workers from standing around the lorry as it was being unloaded.

Inspector McMinn said: 'Those involved in arranging and carrying out deliveries should exchange and agree information to ensure lorries can be loaded and unloaded in a safe manner. They must make sure a safe way of working is in place and that workers have clear responsibilities so everyone involved in the lifting operation knows what everyone else is meant to be doing and where they are meant to be.'

Components of a safe system of work

[S3002] The components of a safe system of work comprise of legal elements – common law (S3003) and statute law (S3004) – and other components which explain where SSWs fit into the overall framework of an occupational safety and health management system ('OSHMS') (S3006). Indeed SSWs are considered to be the cornerstone of all successful OSHMSs.

Legal components – common law

[S3003] The need for SSWs can be traced back to 1905 when Lord McLaren in *Bett v Dalmey Oil Co* (1905) 7F (Ct of Sess) 787 judged that:

> the obligation is threefold: the provision of a competent staff of men, adequate material, and a proper system and effective supervision.

In 1938, the common law duty of care was highlighted in the case of *Wilsons and Clyde Coal Co Ltd v English* [1938] AC 57 where Lord Wright said that:

> the whole course of authority consistently recognises a duty which rests on the employer, and which is personal to the employer, to take reasonable care for the safety of his workmen, whether the employer be an individual, a firm or a company, and whether or not the employer takes any share in the conduct of the operations.

This duty of care implies the need for SSWs and also that delegation to an employee does not remove the employer's duty to provide a safe system of work.

The question of what constitutes a safe system of work was further considered by the Court of Appeal in 1943 by Lord Greene, Master of the Rolls, in the case of *Speed v Thomas Swift Co Ltd* [1943] 1 All ER 539, who said:

> I do not venture to suggest a definition of what is meant by a system. But it includes, or may include according to circumstances, such matters as the physical layout of the job, the setting of the stage, the sequence in which the work is to be carried out, the provision of warnings and notices, and the issue of special instructions.

He also said that: 'a system may be adequate for the whole course of the job, or it may have to be modified or improved to meet circumstances which arise: such modifications or improvements appear to me equally to fall under the heading of system.'

He added that: 'the safety of a system must be considered in relation to the particular circumstances of each particular job.'

The conclusion here is that the safe system of work must be tailored to the job to which it applies by co-ordinating the work, planning the arrangements and layout, stating the agreed method of work, providing the appropriate equipment, and giving employees the necessary information, instruction, training and supervision. All the above are designed to prevent harm to persons at work.

Indeed, safe systems of work have aided the improvement in health and safety performance of a number of industries in recent years.

The Health and Safety Commission's, Paper and Board Industry Advisory Committee (PABIAC), welcome the evident improvement in health and safety performance in the paper industry. Statistics provided by the Confederation of Paper Industries (CPI) display a downward trend in the workplace accident rate and significant reductions in major injuries.

On 20 February 2007 PABIAC Chair and HSE Head of Manufacturing Sector, James Barrett, said: 'It is very encouraging to see the positive affect the PABIAC strategy has had on health and safety standards in the paper industry.

By continuing our partnership approach between employer's, trade unions and HSE, we are confident that there will be further reductions in the rate of accidents across the industry.'

Director General of the CPI and PABIAC member Martin Oldman said: 'The papermaking industry has been working diligently over a number of years to significantly improve its health and safety performance and, whilst more can be done, we are heartened by these results and anticipate further progress throughout the industry'.

Since March 2004, members of the CPI Paper and tissue making sector have seen a 33% reduction in accidents in their businesses. The amount of major injuries has been reduced by 32% and the number of slips and trips is down by 30%. There has also been a 17% improvement in the number of manual handling accidents.

CPI Corrugated sector members have successfully reduced their accident rate by 48%. In the key areas of manufacturing and manual handling accidents have been reduced by 53% and 48% respectively.

CPI Recovered Paper Sector, who only recently signed up to the PABIAC strategy have already seen a fall of 25% in major industries between December 2004 and August 2006.

Legal components – statute law

[S3004] The above common law precedents (at S3003) were taken into account during the drafting of the *Health and Safety at Work etc. Act 1974* ('*HSWA 1974*').

Section 2(2)(a) of *HSWA 1974* requires 'the provision and maintenance of plant *and systems of work* that are, so far as is reasonably practicable, safe and without risks to health'.

Furthermore, *section 2(3)* of *HSWA 1974* requires employers having five or more employees to prepare a *written* health and safety policy statement, together with the organisation and *arrangements* for carrying it out.

These 'written arrangements' should include all relevant written safe systems of work.

This general duty has been tightened up and more clearly defined since the enactment of the *Management of Health and Safety at Work Regulations1999 (SI 1999 No 3242)* ('*MHSWR*').

Specifically, *Regulation 3* of MHSWR *(SI 1999 No 3242)* requires 'suitable and sufficient' risk assessments. A safe system of work should be the end result of the vast majority of risk assessments.

Regulation 4 of MHSWR *(SI 1999 No 3242)* requires the 'Principles of Protection' to be applied in connection with the introduction of 'any preventive and protective measures'. These measures are control measures and are 'arrangements' within the written health and safety policy. These principles of protection are listed in *Schedule 1* to the Regulations and should be referred to when developing SSWs.

Further clarification of 'arrangements' is given in *Regulation 5 of MHSWR (SI 1999 No 3242)*: 'arrangements for the effective planning, organisation, control, monitoring and review of the preventive and protective measures'. This infers that as SSWs are part of the preventive and protective (i.e. control) measures, then they should be planned, monitored and reviewed.

An example of a company falling found of the duty to have a known, understood and adhered to safe system of work is outlined below:

A managing director was sentenced on 7 January 2005 to a 16–month custodial sentence following a prosecution brought by the Crown Prosecution Service (CPS). The case, heard at Manchester Crown Court, followed a police led, joint investigation with the HSE into the death of Mr Daryl Arnold on 11 June 2003.

Mr Arnold, aged 27, and several others had been employed by Mr Lee Harper of Cannock, Staffordshire, to remove and replace the roof of a warehouse on the Lynton industrial estate in Salford. No safe system of work had been prepared before the work began and no safety precautions were in place at the time of the incident. Mr Arnold had never worked on a roof before.

Whilst working on the roof, Mr Arnold stepped backwards onto a fragile roof light on an adjoining warehouse, which gave way. Mr Arnold fell approximately 6.75 metres landing on the ground floor directly below. He died as a result of his injuries.

Mr Harper, managing director of Harper Building Contractors Ltd of Cannock Staffordshire pleaded guilty to charges of manslaughter and a breach of *HSWA 1974 s 2*.

Pam Waldron, HSE's Head of Construction for Scotland and the North West, said:

> 'No penalty can make up for the loss of a loved one. However, Lee Harper's sentence properly reflects the seriousness of his failure to ensure that Daryl Arnold was safe and HSE is pleased that the matter has been concluded. There was a fundamental failure to recognise that the roof included fragile roof lights that will not bear a man's weight. Moreover, the equipment to prevent people falling through fragile materials is readily available and relatively cheap. A sensible, straightforward approach to health and safety in managing risks on this job should have prevented this tragic death.'

HSWA 1974 s 2 states: 'It shall be the duty of every employer to ensure, so far as is reasonably practicable, the health, safety and welfare at work of all his employees.'

The breach of *HSWA 1974 s 2* by Mr Harper was brought against him by virtue of *HSWA 1974 s 37*. Mr Harper was the sole Director of Harper Building Contractors Ltd. *HSWA 1974 s 37* states: 'Where a "body corporate" commits a health and safety offence, and the offence was committed with the consent or connivance of, or was attributable to any neglect on the part of, any director, manager, secretary or other similar officer of the body corporate, then that person (as well as the body corporate) is liable to be proceeded against and punished.'

The arrangements for liaison between the police, CPS and HSE are set out in a '*Work-related deaths: A protocol for liaison*', which is available from HSE website at www.hse.gov.uk/pubns/misc491.pdf.

HSE's free leaflet '*Working on Roofs*' includes this advice and gives more information about preventing falls from a variety of types of roof work. '*Health and Safety in Roof Work*', HSG33 gives more detailed information and includes advice about other issues encountered whilst working on roofs.

The construction industry is one where there is a need for specific written safe systems of work to be in place. In the majority of cases, they may well be labelled as Method Statements. However, an absence of such systems increases the risks to construction workers.

"Nearly 1 in 3 construction refurbishment sites inspected put the lives of workers at risk", Stephen Williams, HSE Head of Construction said on 11 September 2007.

This startling figure comes after The HSE carried out over 1500 inspections as part of its rolling inspection programme, resulting in enforcement action on 426 occasions in just two months.

Stephen Williams said: 'We stopped work on site immediately during 244 inspections because we felt there was a real possibility that life would be lost or ruined through serious injury. It is completely unacceptable that so many lives have been put at risk. Our inspectors were appalled at the apparent willingness to ignore basic safety precautions The simple fact is that despite knowing what they should be doing, too many people are prepared to allow bad practices to continue, even though last year 39 people died on refurbishment, repair and maintenance sites. We are determined to tackle this issue head on and will continue to take enforcement action against those rogues who flout safety precautions. Let me be clear to all those who put lives at risk – we will continue to carry out further inspections and will take all action necessary to protect workers, including closing sites and prosecution'.

Work at height remains the biggest concern. Over half of the enforcement action taken during this inspection initiative was against dangerous work at height, which last year led to the death of 23 workers.

Stephen Williams continued: "My advice to those who work in the refurbishment sector is to plan work, use competent workers and if working at height use the right equipment and use it safely".

Welcoming the Secretary of State's decision to hold a construction forum to discuss safety standards in the construction industry, HSE confirmed that inspectors will continue to target falls and trips in the refurbishment sector as part of their ongoing work.

More guidance and advice is available at hse.gov.uk/construction/index.htm.

In order to ensure SSWs are fully implemented within the workplace, it is imperative that they are communicated to all concerned and enforced via adequate supervision.

Companies are reminded of the need to ensure proper training and supervision is given to employees following the death of an untrained demolition worker.

On 22 December 2006 HSE successfully brought criminal charges against five different parties after the death of Mr David Moran. Between them they were fined a total of £87,000 and ordered to pay £57,228 costs at Manchester's Minshull Street Crown Court.

David fell eight metres to his death when he stepped on a fragile roof light at 17 Chesford Grange in Warrington on 20 September 2002. David and another untrained demolition worker, Anthony Harris from Collyhurst, Manchester, were using the roof to access another roof on the site.

HSE construction inspector for Cheshire Nic Rigby who brought the cases to court said: 'This prosecution follows the tragic death of a young man on a site in Warrington. Unfortunately, his death is not unique: on average, one person is killed on a construction site in Great Britain every five or six days, and many more are seriously injured.'

David's employer Elmsgold Haulage Ltd of Clayton House, Piccadilly, Manchester and John McSweeney, of Willow Road, Prestwich, the Managing Director of Elmsgold Haulage Ltd – pleaded guilty to two charges under *HSWA 1974 s 2(1)* in that they failed to provide a safe system of work and failed to ensure that people working on site were properly trained and supervised, and a third charge under *Regulation 9(3) of the Lifting Operations and Lifting Equipment Regulations 1998* in that they failed to ensure that lifting equipment was properly examined and inspected.

Elmsgold Haulage was fined £10,000 for each charge and ordered to pay total costs of £9,756. Mr McSweeney was fined £5,000 for each charge and ordered to pay total costs of £5,000.

Demolition contractor Excavation & Contracting (UK) Ltd of Sandringham Avenue, Denton in Manchester, the principal contractor for the Chesford Grange project, and the company's former Managing Director Bernard O'sullivan, now living in Australia, pleaded guilty to a charge under *HSWA 1974 s 3(1)* in that they each failed to ensure that risks to non employees were adequately controlled.

Excavation & Contracting (UK) Ltd was fined £35,000 and ordered to pay £9,972 costs. Bernard O'sullivan was fined £20,000 and ordered to pay £30,000 costs.

Dennis O'Connor, of St James' Road, Orrell, Wigan, Elmsgold Haulage's site foreman pleaded guilty to a charge under *HSWA 1974 s 7* in that he failed to ensure the safety of other employees. He was fined £2,500 and ordered to pay £2,500 costs.

At an earlier hearing at Warrington Magistrate's Court on 31 January 2006, John Edge of Knight Frank, a property management company acting for the owner of Chesford Grange and planning supervisor for the project, pleaded guilty to two charges under Regulation 15 of the Construction (Design and Management) Regulations 1994 for which Knight Frank was fined a total of £7,000 plus full prosecution costs of £4,500.

Knight Frank operated as a partnership, which does not constitute a legal entity for the purposes of prosecution, and therefore the case was taken against one of the partners, Mr John Edge.

Another recent example of a heavy fine being levied because of a lack of a safe system of work was highlighted in the June 2008 edition of Health and Safety at Work journal.

City of York Council was fined £20,000 plus £20,423 costs after pleading guilty to a breach of section 2(1) of the Health and Safety at Work Act 1974 for failing to ensure employee safety.

This successful prosecution followed the death of a council worker which occurred when his ride-on lawnmower overturned on sloping ground. It was held that the Council had not properly assessed the risk and was also not following the mower manufacturer's safe operating instructions, which should have been referred to in any written SSW.

Frank Smith, aged 54, died on 19 May 2005, whilst using a Hayter ride-on mower to cut long, wet grass on an embankment at Water End in York City Centre. His job was to cut the top of the embankment and then to turn the mower and tidy up an area near a bridge.

There were no witnesses to the accident but the HSE believes that in turning the mower, Mr Smith lost control and he and the mower slid down the sloping bank, hitting a low retaining wall at the bottom. This caused the mower to flip over, throwing him out of the driving seat.

HSE Principal Inspector, Keith King, stated that the Council had no site specific assessment of the risks and has failed to implement a written safe system of work. It has a generic assessment – dated 1997 – which was considered to be 'quite old'. However, although this generic assessment identified the possibility of a mower turning over with fatal consequences, the Council had failed to act. The mowers were not equipped with either seat belts or roll-over protection.

Mr King stated that the main issue was the Council's failure to send anyone competent to assess the degree of the slope of the embankment. At 25 degrees, this was significantly greater than the 19 degree upper limit specified in the manufacturer's instructions for safe use of the mower.

Mr King continued by saying that the Council should have identified all areas of sloping ground so as to ensure that the right equipment was in use and it should also have provided proper supervision and guidance for employees. The only warning provided by the Council was the instruction to take extra care when working on slopes.

Summary of legal components of SSWs

[S3005] From the common and statute law references above (S3004), it may be seen that the key elements of SSWs are:

- adequate plant and equipment:
  — the right tools for the job;
  — plant designed for safe access and isolation.
- competent staff:
  — properly trained and experienced in the work they do;

— instructed clearly in the work to be done;
— provided with the necessary information on substances, safe use of equipment etc.
- proper supervision:
— to monitor the work as it progresses;
— to ensure the safe system of work is adhered to.

Other components

[S3006] The bulk of the SSWs component parts are derived from the common and statute law considerations above.

The safe system of work may be therefore seen as the centre piece of a jigsaw which links together all the other component parts including:

- competent staff;
- premises;
- safe design;
- plant safeguards;
- safe installation;
- safe access/egress;
- tools and equipment;
- instruction;
- supervision;
- training;
- information;
- safety rules;
- planned maintenance;
- monitoring;
- PPE;
- environment.

Hence SSWs need to take account of all of the above components, and they need to be in writing.

The relative importance of each of the components will be task-dependent. Some tasks will be heavily dependent on the intrinsic safety built into the plant at the design stage. Other jobs will rely on the skills and expertise of the competent employee(s).

It may be of use therefore to consider the SSWs components under the sub-headings: hardware and software.

The 'hardware' includes: design, installation, premises/plant, tools and equipment, and working environment – heating, lighting, ventilation, noise control.

The 'software' comprises: planned maintenance, competent employees, adequate supervision, safety rules, training and information, correct use of PPE, and correct use of tools and equipment.

Which type of safe system of work is appropriate for the level of risk?

[S3007] Different jobs will require different systems, depending on the levels of risk highlighted via the risk assessment process.

A very low risk job may require adherence to safety rules or a previously agreed guide, whereas a very high risk job may well require a formal written permit to work system. Both types qualify as safe systems of work.

The following matrix may prove useful:

Risk level	Type of SSW
Very high	Permit to Work
High	Permit (or written SSW)
Medium	Written SSW
Low	Written SSW
Very low	Verbal (with written back-up, such as safety rules)

Once the SSWs are in place, there is a need to ensure that they are reviewed at least annually, so as to ensure:

- continued legislative compliance (N.B. any new legislation);
- continued compliance with most recent risk assessment;
- SSWs still work in practice;
- any plant modifications are incorporated;
- substituted (safer) substances are allowed for;
- new work methods are incorporated;
- advances in technology are exploited;
- control measures are improved in the light of accident experience; and
- continued involvement in, and awareness of the importance of, all relevant SSWs.

It is vital that regular feedback is needed to all concerned following any updates of existing SSWs.

In essence, SSWs are one of the most important foundations of any OSHMS. They need to be:

- known, i.e. brought to the attention of all relevant employees and third parties;
- well understood and agreed with by all concerned;
- adhered to by all concerned; and
- enforced by management and supervision.

Development of safe systems of work

[S3008] The steps in the development of SSWs of all types are as follows:

- Hazard identification.

- Risk assessment.
- Define safety operating procedures/methods ('SOP's).
- Implement agreed system.
- Monitor system performance.
- Review system (at least annually).
- Provide feedback on system updates.

Job safety analysis ('JSA')

[S3009] JSA is also known as job hazard analysis and/or task analysis. It is a technique that has evolved from the work study techniques of method study and work measurement which, in turn, formed part of the scientific management approach in the early twentieth century.

The work study engineer analysed jobs in order to improve methods of production by eliminating unnecessary job steps or changing those steps that were inefficient.

From an occupational safety and health ('OSH') viewpoint, JSA assists in the elimination of hazardous steps, thus eliminating or avoiding risky operations.

In essence, JSA is a useful risk assessment tool as it identifies hazards, describes risks and assists in the development of commensurate control measures.

The original work study approach made use of the 'SREDIM' principle:

Select	=	work to be studied
Record	=	how work is done
Examine	=	the total job
Develop	=	the best method for doing the work
Install	=	the preferred method into the company's way of working
Maintain	=	the agreed method via training, supervision etc.

Work study is used to break the total job down into its component parts and, by measuring (work measurement/time and motion) the quantity of work done in each of the component parts, the total job time could be established. If any of the component parts could be modified to make them more efficient or productive usually by saving time these improvements would be made as part of the overall study.

From measuring a number of similar tasks undertaken by different employees, standard times for each job component – and hence the total job-were derived. These were subsequently used in the development of payment and bonus systems.

JSA uses the SREDIM principle, but measures/assesses the risk content, as opposed to the work content, in each of the component parts of the job. From this detailed breakdown a safe system of work or safe operating procedure can then be compiled.

The basic JSA procedure is as follows:

(1)   Select the job to be analysed (Select).
(2)   Break the job down into its component parts in an orderly and chronological sequence of job steps (Record).
(3)   Critically observe and examine each job step/component part to determine whether there is any risk involved (Examine).
(4)   Develop control measures to eliminate or reduce the risk (Develop).
(5)   Formulate written SSWs and job safety instructions ('JSIs') for the job (Install).
(6)   Review SSWs at regular intervals to ensure continued utilisation (Maintain)

Use may be made of the following three-column layout in order to record the JSA:

Sequence of job steps	Risk factors	Controls advised

It is important to work vertically down the columns in their entirety, rather than to list one or two job steps and then work horizontally across the rows. Experience has shown that the former approach is much less time-consuming!

Once the job to be analysed has been selected, the next stage is to break the job down into its component parts or job steps. On average there should be approximately ten to twenty job steps. If there are more than twenty, too large a job for analysis has probably been chosen and one should therefore split it into two separate jobs. If there are less than ten job steps, then combine two or more jobs together.

The job steps should follow an ordered and chronological approach. Consider the following job – *changing a car wheel*:

The scenario is that you have come out of work and gone to your car where you left it on the works car park. It has a flat tyre. You are not allowed to call for assistance but you have to replace the wheel yourself! The sequence of job steps will be as follows:

(1)   Check handbrake on and check wheels.
(2)   Remove spare tyre from boot, check tyre pressure.
(3)   Remove hub cap/wheel trim.
(4)   Ensure jack is suitable and located on solid ground.
(5)   Ensure jacking point is sound.
(6)   Jack car up part way, but not so wheels leave ground.
(7)   Loosen wheel nuts.
(8)   Jack car up fully.
(9)   Remove nuts, then wheel.
(10)   Fit spare.
(11)   Replace nuts and tighten up.
(12)   Lower car.
(13)   Remove jack and store in boot with replaced wheel.

(14)   Re-tighten wheel nuts.
(15)   Replace trim/hub cap.
(16)   Ensure wheel is secure before driving off.

Once column 1 has been agreed and completed, then commence listing the associated risk factors by completing column 2. Thereafter, assign commensurate control measures in column 3.

The complete JSA should look like this:

Sequence of job steps	Risk factors	Controls advised
1. Check handbrake, check wheels	Strain to wrist/arm	Avoid snatching, rapid movements
2. Remove spare tyre, check tyre pressure	Strain to back	Use kinetic handling techniques
3. Remove hub cap/trim	Strain, abrasion to hand	Ensure correct lever used
4. Ensure jack on solid ground	Vehicle slipping, jack sinking into ground	Check jack and ground
5. Ensure jacking point is sound	Vehicle collapse	Place spare wheel under car floor as secondary support
6. Jack car up part way, wheels still on ground	Strain, bumping hands on car/jack	Avoid snatching, rapid movements. Use gloves
7. Loosen wheel nuts	Hands slipping- bruised knuckles, strain	Ensure spanner/brace in good order; avoid snatching, rapid movements. Use gloves
8. Jack car up fully	Strain, bumping hands on car/jack	Avoid snatching, rapid movements
9. Remove wheel	Strain to back, may drop onto foot	Use kinetic handling techniques. Use gloves to improve grip
10. Fit spare	Strain to back	Use kinetic handling techniques
11.Tighten nuts	Hands slipping- bruised knuckles, strain	Use gloves, avoid snatching, rapid movements
12.Lower car	Strain, bumping hands on jack/car	Avoid snatching, rapid movements
13.Remove wheel and store with jack in boot	Strain to back	Use kinetic handling techniques
14. Re-tighten nuts	Hands slipping- bruised knuckles	Use gloves, avoid snatching rapid movements
15. Replace hub cap	Abrasion to hand	Use gloves
16. Ensure wheel secure	Vehicle collapse, loose wheel	Check wheel and area around car before driving off.

The above example serves to illustrate what a typical JSA should look like. Any job/task within the workplace should be able to be recorded in this way and, ideally, displayed in close proximity to where the job is being undertaken.

The ultimate aim must be to undertake JSA on all jobs within the workplace and to communicate the findings and resultant control measures to all concerned. However, to assist in prioritising which jobs to analyse initially, the following criteria should be taken into account:

- past accident/loss experience;
- worst possible outcome/maximum potential loss ('MPL');
- probability of recurrence;
- specific legal requirements;
- newness of the job;
- number of employees at risk.

JSA may be activity-based or job-based.

Activity-based JSAs may cover:

- all work carried out at height;
- all driving activities:
  — internal (works transport);
  — external (driving on public highways);
- loading and unloading of vehicles.

Job-based JSAs may cover:

- the activities of a maintenance engineer repairing a roof;
- the activities of a chemical process worker obtaining samples for analysis.

The third column 'Controls advised' is in essence the job safety instructions ('JSI') and forms the basis of the written safe system of work.

Preparation of job safety instruction and safe operating procedures

[S3010] The purpose of job safety instructions ('JSIs') and/or safe operating procedures ('SOPs') is to communicate the safe system of work to all concerned – supervisors, employees and contractors.

For each job step (column 1 in table in S3009) there should be a corresponding control action (column 3) designed to reduce or eliminate the risk factor (column 2) associated with the job step. The chronological ordering of the control measures listed in column 3 becomes the JSI for that particular job.

Such JSIs should be utilised in OSH training and communication, both formal (classroom style) and informal (on the job contact sessions).

All managers and supervisors should be fully knowledgeable and aware of all JSIs and SSWs that are operational within their areas of control and they should ensure that they are adhered to at all times.

JSIs may be:

- verbal (for very low risk tasks) – a word in the ear from a supervisor reminding the employee of the do's and don'ts;
- written (for low, medium and some high risk tasks) – as part of an agreed safe system of work;
- written and incorporated into safe operating procedures (usually for medium and some high risk tasks); or
- as part of an all singing, all dancing formal permit to work system (usually for 'high' or 'very high' risk tasks).

From a practical viewpoint, to aid communication and use of JSIs and SSWs, they should be listed on encapsulated cards and posted in the areas where the jobs are to be carried out. They should also be individually issued to all relevant employees, who should receive training in their application and use. All such training and communication – whether formal or informal – should be recorded on individual training records.

The use of the three-column format to communicate JSIs and SSWs is well proven and is an excellent method of revitalising what in most organisations has become a rather moribund set of documents. As stated above, all jobs/tasks should be able to be fitted onto one or two sheets of A4 and use of the technique is a good way to regenerate interest in SSWs via a review.

Most organisations only review SSWs when things have gone wrong, e.g. as part of an accident investigation. This is negative and a form of reactive monitoring.

It is much better to be proactive and to review SSWs at least annually. This can be achieved by setting targets for managers and supervisors to review one safe system of work per month, together with relevant safety representatives, employees and/or contractors. Any findings or changes to the existing SSWs must rapidly be fed back to all concerned, so as to ensure continual improvement in OSH performance.

Permit to work systems

[S3011]

'A permit to work system (PTW) is a formal written system of work used to control certain types of work that are potentially hazardous. A permit to work is a document which specifies the work to be done and the precautions to be taken. Permits to work form an essential part of a safe system of work for many maintenance activities. They allow work to start only after safe procedures have been defined and they provide a clear record that all foreseeable hazards have been considered. A permit is needed when maintenance work can only be carried out if normal safeguards are dropped or when new hazards are introduced by the work. Examples are: entry into vessels, hot work, and pipeline breaking.'

(HSE, 1997)

When may PTW systems be required?

[S3012] A PTW system may be required in the following instances:

- where the risk assessment indicates 'very high' or 'high' risk;

- entry into confined spaces;
- working at heights;
- working over water;
- asbestos removal;
- high voltage electrical work;
- complex maintenance work – usually involving mechanical/electrical/chemical isolation;
- demolition work;
- work in environments which present considerable health hazards, such as:
  — radiation work;
  — thermal stress (work in hot or cold environments);
  — toxic dusts, gases, vapours;
  — oxygen enrichment or deficiency;
  — flammable atmospheres;
- lone working;
- work on or near overhead travelling cranes (7 metre rule);
- work involving contractors/third parties on site.

Essential features of PTW systems

[S3013] HSE Guidance (1997) states that employers should incorporate the following essential features into bespoke PTW systems:

- provision of information;
- selection and training;
- competency-use of competent persons;
- description of work to be undertaken;
- hazards and precautions;
- procedures.

Provision of information in PTW systems

[S3014] The PTW system should clearly state how the system works in practice, the job(s) it is to be used for, the responsibilities and training of those involved, and how to check that it is working effectively.

There must be clear identification in the system of those persons who may authorise particular jobs and also a note as to the limits of their competency.

There should also be a clear identification of who is responsible and competent for specifying the necessary precautions, e.g. isolation, use of RPE/PPE, emergency arrangements etc.

The PTW system paperwork should be clear, unambiguous and not misleading.

The system should be so designed that it can be used in unusual circumstances, especially to cover contractors.

Selection and training for PTW systems

[S3015] Those persons authorised to issue permits must be sufficiently knowledgeable concerning the hazards and precautions associated with the plant/equipment/substances and the proposed work. They should have sufficient imagination and experience to ask enough 'what if' questions to enable them to identify all potential hazards. Lateral thinking is required here, especially if more than one trade is involved; or when own employees are working alongside contractors.

Ideally formal training should be required for all:

- permit issuers/authorisers;
- permit receivers/users;
- contractors' supervision;
- line managers/supervisors.

All staff and contractors should fully understand the importance and assurance of safety associated with PTW systems and should be 'competent persons' within the framework of the system.

What is a 'competent person'?

[S3016] Recent legislation, codes of practice and guidance refer to a competent person in connection with permit to work systems in particular, and the OSHMS in general.

Regulation 3 of the *MHSWR (SI 1999 No 3242)* requires employers to undertake suitable and sufficient risk assessments and *Regulation 7* of the same Regulations requires employers to appoint one or more competent persons to assist them in undertaking such assessments.

MHSWR '99 defines 'competent' as follows: 'a person shall be regarded as competent for the purposes of these Regulations where he has sufficient training and experience or knowledge and other qualities to enable him properly to assist in undertaking the measures referred to in these Regulations.'

This is especially important in connection with the authorisation and receipt of permit to work documentation. Specifically, permit authorisers and receivers need:

- a knowledge and understanding of:
  - the work involved;
  - the principles of risk assessment;
  - the practical operation of commensurate control measures;
  - current OSH best practice.
- the ability to:
  - identify OSH problems;
  - implement prompt solutions;
  - evaluate the PTW system effectiveness;
  - promote and communicate the workings of the PTWS to all involved/concerned.

PTW system: work description

[S3017] The permit documentation should clearly identify the work to be done, together with the associated hazards. Plans and line diagrams may be used to assist in the description, location and limitations of the work to be done.

All plant and equipment should be clearly tagged/identified/numbered, so as to assist permit issuers and users in selecting and/or isolating the correct piece of kit. All isolation switchgear should be similarly numbered or tagged. An asset register listing all discrete numbering should be maintained.

The permit should also include reference to a detailed method statement for the more complicated tasks. This should form part of the overall PTW system.

Hazards and precautions

[S3018] The PTW system should require the removal of hazards – risk avoidance being the best strategy. Where this cannot be achieved then effective control measures designed to minimise the risk should be incorporated.

Any control measures should ensure legislative compliance is achieved as an absolute minimum standard. Specific regulations that may apply include: *COSHH 2002, Confined Spaces Regulations 1997, Control of Asbestos Regulation, 2006*. Permit authorisers and users should be aware of all relevant legislation.

The permit should state those precautions that have already been taken before work commences, as well as those that are needed whilst work is in progress. This may include what physical, chemical, mechanical and electrical isolations are required and how these may be achieved and tested. Also, the need for different types of PPE and/or RPE should be specified.

Reference should also be made to any residual hazards that might remain, together with any hazards that might be introduced during the work – e.g. welding fume, vapour from cleaning solvents.

PTW system procedures

[S3019] The permit should contain clear rules about how the work should be controlled or abandoned in the case of an emergency.

The permit should have a hand-back procedure which incorporates statements that the maintenance work has finished and that the plant has been made safe prior to being returned to production staff.

Time limitations and shift changeovers should be built into the PTW system. Ideally the permit should only remain valid for the time that the authorisation/issuer is present on site.

There should be clear procedures to be followed if the work has to be suspended for any reason.

It is vitally important that there is a method for cross-referencing when two or more jobs (at least one of which is subject to a PTW system) may adversely affect each other.

A copy of the permit must be displayed as close as possible to where the work is being undertaken.

All jobs subject to a PTW system should be checked at regular intervals during the duration of the job, so as to ensure that the system is being followed, is still relevant and is working properly.

General application of PTW systems

[S3020] In order for the PTW system to be applied to risk reduction in the workplace there needs to be a well understood communication system in place that everyone concerned is aware of and becomes involved in.

Most organisations make use of PTW system forms which have been designed to take account of individual site conditions and requirements. In some cases there will be general permits; in other cases special permits will be required for specific jobs, e.g. hot work, entry into confined spaces. This enables sufficient emphasis to be given to the particular hazards present and the precautions advised.

Outline of a PTW system

[S3021] There are many and varied PTW system documents available in the OSH literature. The key headings/subject areas on PTW system documentation should include the following as a minimum requirement:

- permit title;
- permit number;
- date and time of issue;
- reference to other permits;
- job location;
- plant/equipment identification;
- description of work to be done;
- hazard identification;
- precautions necessary – before, during and after job/work;
- protective equipment;
- authorisation;
- acceptance;
- shift handover procedures;
- hand back;
- permit cancellation – work satisfactorily and safety completed.

Operating principles

[S3022] In operating a PTW system the following principles should be observed:

- Title of permit.
- Permit number (including reference to other relevant permits or isolation certificates).
- Precise, detailed and accurate information concerning:
  - job location;
  - plant identification;
- Clear description of work to be done and its limitations.
- Hazard identification – including residual hazards and hazards introduced by the work.
- Specification of plant already made safe, together with an outline of what precautions have been taken, e.g. isolation.

- Specification of what further precautions are necessary prior to the commencing of work-e.g. rpe/ppe
- Specification of what control measures should be in place for the duration of the work.
- Clear statement of when the PTW system comes into effect, and for how long it remains in effect. A re-issue or extension should take place if the work is not completed within the allocated time. Generally, the PTW system should only be valid for the time the authoriser is on site.
- The permit document should be recognised as the master instruction which, unless it is cancelled, overrides all other instructions.
- The authorised/competent person issuing the permit must assure himself/herself – and the persons undertaking the work – that all the precautions/controls specified as necessary to make the job, plant and working environment safe and healthy have in fact been taken. This inevitably should involve an inspection of the job location by the authoriser. The authoriser's signature on the permit document gives this assurance of safety to the people doing the work.
- The person who accepts the permit becomes responsible for ensuring that all specified safety and health precautions continue in force and that only permitted work is undertaken within the location specified on the permit. At this time, confirmation that all permit information has been communicated to all concerned should be obtained. A signature of the accepting person should also be on the permit document.
- A copy of the permit should be clearly displayed in the work area.
- If the permit needs to be extended beyond the initially agreed duration, e.g. because of shift changeover, then signatures of oncoming authorised/competent persons – together with that of the original authoriser – should be added to the appropriate section of the permit to confirm that checks have been made – again involving a visual walk-round inspection to ensure that the plant/equipment remains safe to be worked on. The new acceptor and all involved are then made fully aware of all hazards and associated precautions. A new expiry time is agreed. This should be built into relevant shift hand-over procedures.
- The hand-back procedure should involve the acceptor signing that the work has been satisfactorily completed. The authoriser of the permit also signs off certifying that the work is complete and the plant/equipment is ready for testing and recommissioning.
- The permit is then cancelled, certifying that the work has been tested and the plant/equipment has been satisfactorily completed. The copy permit should also be removed from the work location once the work has been completed.
- It is advisable to keep copies of each PTW system for a period of five years as these constitute maintenance records.

Isolation procedures

[S3023] SSWs and PTW systems frequently require the effective isolation of plant and associated services, including:

- electrical power;
- pneumatic/hydraulic pressure;
- mechanical power;
- steam;
- chemical lines.

The principles of effective isolation require four steps to be taken:

1. Switch off	at the appropriate operating control panels. It is a good idea to post warning signs on the panel to alert other workers that it is switched off.
2. Isolate	by creating a physical break or gap between the source of energy and the part to be worked on. This can be achieved by operating an electrical isolator, removing a section of pipe or blanking off with a plate. N.B. Removing pipe sections is preferable to blanking off.
	Reliance on control valves alone has been a frequent cause of accidents and is not considered to be effective isolation. From a mechanical isolation viewpoint, drive belts and linkages should be physically removed.
3. Secure	the means of isolation by locking off. Multiple hasp padlock devices (e.g. Isolok) can be used when several people are working on the same piece of kit. Each puts his own padlock on the isolator so that the power cannot be reinstated until all personal padlocks are removed. It is therefore imperative to have a suite of different keys/padlocks.
4. Check	the isolation is effective by trying all available operating controls.

Many maintenance procedures will justify either a written safe systems of work or a formal PTW system; both can have the isolation procedure incorporated into the documentation.

Further reading

[S3024] The following publications are available either via the HSE website www.hse.gov.uk or from HSE Books, Sudbury, telephone 01787 881165.

- HS G 250 Guidance on permit to work systems *'A guide for the petroleum, chemical and allied industries'* (2005)
- HS G 253 *'The safe isolation of plant and equipment'* (2006)
- *'Permit to work systems'*, INDG98 (rev) (1997).
- *'Safe work in confined spaces: Confined Spaces Regulations 1997: Approved Code of Practice, Regulations and Guidance (2nd Edition)'*, L101 (2009).
- *'Safe work in confined spaces'*, INDG258 (1999).
- *'Management of health and safety at work: Management of Health and Safety at Work Regulations 1999: Approved Code of Practice and guidance'*, L21 (rev) (2000).

- *'Emergency isolation of process plant in the chemical industry'*, CHIS 2 (1999).
- *'Work at Height Regulations 2005 (as amended)* , IND G 401 (rev) (2007)
- *'Electricity at work. Safe working practices (2nd Edition)* HS G 85 (2003)
- *'Safe working with flammable substances'*, IND G 227 (1996)
- *'Safe maintenance, repair and cleaning procedures, Dangerous Substances and Explosive Atmospheres Regulations 2002'*. Approved Code of Practice and guidance, L137 (2003).
- *'A guide to the Control of Major Accident Hazards, 1999 (as amended)'*. Guidance on Regulations L111 (2006).
- *'Developing process safety indicators: A step by step guide for chemical and major hazard industries'*, HS G 254 (2006)
- *'Safe use of work equipment, Provision and Use of Work Equipment Regulations 1998'*. Approved Code of Practice and guidance (rev), L22 (2008)
- *'Working alone: Health and safety guidance on the risks of working alone'*, IND G 73 (rev 2) (2009)
- *'Work with materials containing asbestos, Control of Asbestos Regulations 2006'* Approved Code of Practice and guidance, L143 (2006)

Statements of Health and Safety Policy

Mike Bateman and Andrea Oates

Introduction

[S7001] Employers with five or more employees have a statutory duty under s 2(3) of the *Health and Safety at Work etc. Act 1974 (HSWA 1974)* to prepare a written statement of their health and safety policy, including the organisation and arrangements for carrying it out, and to bring the statement to the attention of their employees. Employers with less than five employees are excepted from this requirement.

This section sets out the legal requirements relating to health and safety policies, explains the essential ingredients that a policy should contain and provides checklists to assist in the preparation of policies. It also refers to further sources of guidance on the subject.

The legal requirements

[S7002] *Section 2(3)* of the *HSWA 1974* states:

> Except in such cases as may be prescribed, it shall be the duty of every employer to prepare and as often as may be appropriate revise a written statement of his general policy with respect to the health and safety at work of his employees and the organisation and arrangements for the time being in force for carrying out that policy, and to bring the statement and any revision of it to the notice of his employees.

The *Employers' Health and Safety Policy Statements (Exception) Regulations 1975 (SI 1975 No 1584)* except from these provisions any employer who carries on an undertaking in which for the time being he employs less than five employees. In this respect regard is to be given only to employees present on the premises at the same time (*Osborne v Bill Taylor of Huyton Ltd* [1982] IRLR 17).

Directors, managers and company secretaries can be personally liable for a failure to prepare or to implement a health and safety policy. (see further *Armour v Skeen* at **E15040** ENFORCEMENT).

Content of the policy statement

[S7003] The policy statement should have three main parts:

—   *The Statement of Intent*

This should involve a statement of the organisation's overall commitment to good standards of health and safety and usually includes a reference to compliance with relevant legislation. While *s 2(3)* of the *HSWA 1974* only requires the statement to relate to employees, most organisations also make reference to others who may be affected by their activities e.g. contractors, clients, members of the public (This may be of particular relevance if the policy is being reviewed by other parties such as clients). In order to demonstrate that there is commitment at a high level, the statement should preferably be signed by the chairman, chief executive or someone in a similar position of seniority.

—   *Organisation*

It is vitally important that the individuals responsible for putting the good intentions into practice are clearly identified. This may be relatively simple in a small organisation but larger employers are likely to need to identify the responsibilities held by those at different levels in the management structure as well as those for staff in more specialised roles e.g. health and safety officers. In all cases the competent person (or people) who are to assist in complying with health and safety requirements should be identified *[Management of Health and Safety at Work Regulations 1999 (SI 1999 No 3242), Reg 7]*.

—   *Arrangements*

The practical arrangements for implementing the policy (e.g. emergency procedures, consultation mechanisms, health and safety inspections) should be identified in this section. It may not be practicable to detail all of the arrangements in the policy document itself but the policy should identify where they can be found e.g. in a separate health and safety manual or within risk assessment records, organisational procedures or similar documents.

More detailed guidance on what should be included in each of these parts is contained in the following sections, together with guidance on two other important issues:

•   Communication of the policy to employees (see S7008).
•   Revision of the policy (see S7009).

The statement of intent

[S7004] The statement should emphasise the organisation's commitment to health and safety. The detailed content must reflect the philosophical approach within the organisation and also the context of its work activities. Typically it might include:

—   a commitment to achieving high standards of health and safety in respect of its employees;

—   a similar commitment to others involved or affected by the organisation's activities (possibly specifying them e.g. contractors, visitors, clients, tenants, members of the public);

—   a recognition of the organisation's legal obligations under the *HSWA 1974* and related legislation and a commitment to complying with these obligations;

— a reference to the importance of people to the organisation and its moral obligations towards them;
— references to specific obligations e.g. to provide safe and healthy working conditions, equipment and systems of work;
— a commitment to providing the resources necessary to implement the policy;
— a reference to the importance of health and safety as a management objective;
— a commitment to consultation with the organisation's employees;
— a reference to achieving continuous improvement in health and safety standards.

Most employers keep the statement of intent relatively short – less than a single sheet of paper. This allows it to be prominently displayed (e.g. in reception areas or on noticeboards), or to be inserted easily into other documents such as employee handbooks.

EXAMPLE:

The Midshires Housing Association (MHA) is committed to achieving high standards of health and safety not only in respect of its own employees but also in relation to tenants of the Association's property, contractors working on the Association's property or projects, visitors and members of the community who may be affected by the Association's activities. MHA recognises the importance to the Association of its employees and its moral obligations towards ensuring their health and safety.

The Association is also well aware of its obligations under the Health and Safety at Work Act and related legislation and is fully committed to meeting those obligations. MHA will provide the resources necessary to implement this Policy and regards the successful management of health and safety as a key management objective. An annual health and safety plan will be prepared and implemented, with the aim of continuously improving health and safety standards. MHA supports the concept of consultation with its staff on health and safety matters and has established a Health and Safety Committee to provide a forum for such consultation.

The organisation and arrangements for the implementation of this Policy are set out in the full Health and Safety Policy document, which is available on the Association's Intranet. The Policy will be reviewed annually and staff will be informed of any revisions that are made.

A Champion

Chief Executive

April 2011

Organisation

[S7005] The statement of intent is of little value unless the organisation for implementing these good intentions is clearly established. This usually involves identifying the responsibilities of people at different levels in the management organisation and also those with specific functional roles e.g. the health and safety officer. How these responsibilities are allocated will depend upon the structure and size of the organisation. The responsibilities of the managing director of a major company will be quite different from those of a managing director in a small business. The latter will quite often have responsibilities held by a supervisor in a larger concern e.g. ensuring staff comply with personal protective equipment (PPE) requirements.

Further advice on leadership for directors, governors, trustees, officers and their equivalents in the private, public and third sectors is available in the joint Health and Safety (HSE) and Institute of Directors (IOD) publication, *Leading Health and Safety at Work*, and on the HSE website at www.hse.gov.uk/leaders hip.

Most organisations prefer to identify the responsibilities through job titles rather than by name as this avoids the need to revise the document when individuals change jobs or leave. Responsibilities might be allocated to levels in the organisation e.g:

— Managing Director/Chief Executive
— Senior Managers
— Junior Managers
— Supervisory Staff
— Employees

and also to those with specialist roles e.g:

— Health and Safety Manager or Officer
— Occupational Hygienist
— Occupational Health staff
— Personnel / Human Resources Specialists
— Maintenance & Project Engineers
— Training Manager or Officer

Where health and safety advice is provided from outside (e.g. a consultant or a specialist at a different location), this should be stated and the means of contacting this person for advice should be identified.

The following examples are provided as an illustration of how responsibilities might be allocated in a medium sized organisation. The allocation will undoubtedly be different for businesses of greater or lesser size.

EXAMPLES:

• Managing Director
  The Managing Director has overall responsibility for health and safety and in particular for:
  — ensuring that adequate resources are available to implement the health and safety policy;
  — ensuring health and safety performance is regularly reviewed at board level;
  — monitoring the effectiveness of the health and safety policy;
  — reviewing the results of health and safety audits and arranging the implementation of appropriate action;
  — reviewing the policy annually and arranging for any necessary revisions.
• Works Manager
  The Works Manager is primarily responsible for the effective management of health and safety within the factory. In particular this includes:
  — delegating specific health and safety responsibilities to others;
  — monitoring their effectiveness in carrying out those responsibilities;
  — ensuring the company has access to adequate competent health and safety advice;
  — ensuring that safe systems of work are established for activities within the factory;
  — ensuring that premises and equipment are adequately maintained;
  — ensuring that risk assessments are carried out;

- — ensuring that adequate health and safety training is provided;
- — ensuring that appropriate remedial action is taken following accident and incident investigations;
- — ensuring that contractors working within the factory have satisfactory health and safety standards;
- — chairing the health and safety committee;
- — preparing an annual health and safety plan and co-ordinating its implementation.

- Supervisors
  Each Supervisor is responsible for the effective management of health and safety within his or her own area or function. In particular this includes:
  - — ensuring that safe systems of work are implemented;
  - — enforcing PPE requirements and appropriate standards of behaviour;
  - — ensuring that staff are adequately trained for the tasks they perform;
  - — monitoring premises and work equipment, reporting faults where necessary;
  - — monitoring any contractors or visitors who may be present;
  - — identifying and reporting health and safety related problems and issues;
  - — identifying training needs;
  - — investigating and reporting on accidents and incidents;
  - — participating in the risk assessment programme;
  - — setting a good example on health and safety matters.

- All employees
  All employees have a legal obligation to take reasonable care for their own health and safety and for that of others who may be affected by their actions e.g. colleagues, contractors, visitors, delivery staff. Employees are responsible for:
  - — complying with company procedures and health and safety rules;
  - — complying with PPE requirements;
  - — behaving in a responsible manner;
  - — identifying and reporting defects and other health and safety concerns;
  - — reporting accidents and near miss incidents to their supervisor;
  - — suggesting improvements to procedures or systems of work;
  - — co-operating with the company on health and safety matters.

- Health and Safety Officer
  The Health and Safety Officer is responsible for co-ordinating many health and safety activities and for acting as the primary source of health and safety advice within the company. These responsibilities specifically include:
  - — co-ordinating the company's risk assessment programme;
  - — acting as secretary to the Health and Safety Committee;
  - — administering the accident investigation and reporting procedure;
  - — liasing with the HSE, the company's insurers and other external bodies;
  - — submitting reports as required by *Reporting of Injuries, Diseases and Dangerous Occurrences Regulations 1995*(RIDDOR);
  - — co-ordinating the health and safety inspection programme;
  - — arranging annual health and safety audits;
  - — identifying health and safety training needs;
  - — providing or sourcing health and safety training;
  - — providing health and safety induction training to new staff;
  - — identifying the implications of changes in legislation or HSE guidance;
  - — preparing and submitting progress reports on the annual health and safety action plan;
  - — sourcing additional specialist health and safety assistance when necessary;
  - — assisting the Works Manager in the preparation of the annual health and safety plan;
  - — preparing quarterly progress reports on the health and safety plan.

Arrangements

[S7006] The duty to state the arrangements for carrying out the health and safety policy overlaps with the duty imposed by *Regs 3* and *5* of the

Management of Health and Safety at Work Regulations (SI 1999 No 3242) to record the results of risk assessments and of arrangements 'for the effective planning, organisation, control, monitoring and review of the preventive and protective measures'. Rather than providing all of the detail on arrangements within the health and safety policy many employers prefer to provide information on where these details can be found, for example in:

— risk assessment records;
— health and safety manuals, handbooks or other documents;
— formal health and safety procedures (e.g. within an ISO 9000 controlled system. ISO 9000 consists of standards and guidelines relating to quality management systems and related supporting standards).

This has the benefit of keeping the policy document itself relatively short – an aid in the effective communication of the policy to employees. Some key aspects of health and safety arrangements which should be included or referred to in this section of the policy are those for:

— conducting and recording risk assessments;
— specialised risk assessments e.g. Fire, Control of Substances Hazardous to Health (COSHH), Noise, Manual Handling, DSE workstations, Asbestos-containing materials;
— assessing and controlling risks to young workers and new or expectant mothers;
— establishing PPE standards and providing PPE;
— operational procedures (where these relate to health and safety);
— permit to work systems;
— routine training e.g. induction, operator training;
— specialised training e.g. first aiders, fork lift truck drivers;
— health screening e.g. pre-employment medicals, regular monitoring, DSE user eye tests;
— statutory examinations and inspections;
— consultation with employees e.g. health and safety committees, staff meetings, shift briefings;
— health and safety inspections;
— auditing health and safety arrangements;
— investigation and reporting of accidents and incidents;
— selection and management of contractors;
— control of visitors;
— fire prevention, control and evacuation;
— dealing with other emergencies e.g. bomb threats, chemical leaks, aggression or violence;
— first-aid;
— stress, bullying and harassment;
— maintaining and repairing premises and work equipment
— staff working elsewhere e.g. sales representatives, delivery staff, installation engineers, homeworkers, secondees;
— work abroad.

Ensuring arrangements work effectively

[S7007] *Regulation 5* of the *Management of Health and Safety at Work Regulations 1999 (SI 1999 No. 3242)* requires employers to have appropriate arrangements for the effective planning, organisation, control, monitoring and review of preventive and protective measures i.e. to have a suitable 'management cycle' in place.

If a health and safety policy is to be meaningful and to have credibility with employees and others, these arrangements must actually work.

In order to ensure this, the management cycle must be applied to the arrangements themselves. New procedures, systems etc. must be properly thought through and introduced effectively, with particular attention being paid to communicating them to those affected. The monitoring and review of health and safety arrangements can be carried out through health and safety inspections and audits, and the activities of health and safety committees.

The chapter MANAGING HEALTH AND SAFETY goes into some detail on how to ensure that health and safety arrangements work effectively while the chapter on RISK ASSESSMENT provides further guidance on the application of the management cycle.

Communication of the policy

[S7008] Employers must bring the policy (and any revision of it) to the notice of their employees. The most important time for doing this is when new employees join the organisation and communication of the policy should be an integral part of any induction programme. Young people in particular need to be reminded of their personal responsibilities at this stage, as well as any restrictions on what they can and can't do. Even experienced workers may need to be made aware that they are joining an organisation that takes health and safety seriously, as well as knowing what the PPE rules are and where they can obtain the PPE from.

Many employers provide their employees with a personal copy of the policy whilst others include it in a general employee handbook or display it prominently e.g. on noticeboards. In all cases communication will be aided by keeping the policy itself relatively brief and providing the detail (particularly in respect of 'arrangements').

Where organisations have their own Intranet systems these provide an excellent means of not only communicating the health and safety policy itself but also giving links to related procedures, forms, risk assessments and other important health and safety documents. Such systems make it relatively easy to communicate changes to the policy and related documents but can also be used to record the fact that individual employees have been informed about the changes.

Review

[S7009] The policy must be revised 'as often as may be appropriate' but the need for revision can only be identified through a review. The HSE does not stipulate a frequency for review but have in the past suggested carrying one out annually. This should not be too onerous even for small employers, for whom it should be a fairly simple affair.

In larger organisations the policy could be subjected to an annual review by incorporating it within an ISO 9000 document control system (see S7006) or making the review an annual task of the health and safety committee. Alternatively the review could be made the responsibility of an individual (e.g. the Managing Director or the Health and Safety Officer), either formally through specifying this within the organisational responsibilities or informally through a simple diary entry.

The review may well conclude that there is no need for change but it may identify the need for revision such as:

— A new Managing Director should make his personal commitment to the statement of intent by signing and re-issuing it.
— Changes in the management structure necessitate a reallocation of responsibilities for health and safety.
— New 'arrangements' for health and safety have been established (or existing ones have been altered) and the policy needs to be amended to match this.
— Changes in health and safety legislation or HSE guidance.

A policy which is clearly well out of date does not reflect well on the organisation's commitment to health and safety, nor on the effectiveness of its health and safety management.

Contractors' Approval Systems

[S7010] Being able to demonstrate effective standards of health and safety management to others is of increasing importance to employers supplying services to others (see introduction 9 SUPPLY CHAIN PRESSURE). Client organisations e.g. major construction companies, local authorities etc. are increasingly requiring those wishing to work for them to submit details of their health and safety management arrangements as part of an approval system.

Almost all of these increasingly complex and demanding questionnaires require information to be submitted on the contractor's health and safety policy — the three key elements (policy statement, organisation and arrangements), usually accompanied by details of how the policy is communicated to employees and how it is reviewed and revised.

Further Guidance

[S7011] The free HSE booklet *'An introduction to health and safety. Health and safety in small businesses'* (INDG 259) provides valuable guidance on many aspects of health and safety. One section of the booklet contains a template of a health and safety policy into which a small business can insert its own details as appropriate. As well as a statement of intent and a general section dealing with health and safety responsibilities, it provides sections into which details of both responsibilities and arrangements for common topics can be inserted. The document was revised in 2008 and covers:

— responsibilities;
— risk assessments;
— consultation with employees;
— plant and equipment;
— safe handling and use of substances;
— information, instruction and supervision;
— competency and training;
— accidents, first aid and work-related ill-health;
— monitoring (i.e. inspections, audits etc.);
— emergency procedures.

Even where a small business needs to refer to important topics not contained in the above list, it will often be easier to deal with these through accompanying sections, rather than prepare a policy from scratch.

Health & safety in annual reports: Guidance from the Health and Safety Commission (available via the HSE website at http://www.hse.gov.uk/revitalising/annual.htm) will also be of use to some.

Important references in ensuring that adequate arrangements are in place for carrying out the policy are the HSE publication '*Successful health and safety management*' (HSG 65) (which can be downloaded at http://www.hse.gov.uk/pubns/books/hsg65.htm, the principles of which are summarised in the leaflet *Managing health and safety: Five steps to success* (ING 275) (which can be downloaded at http://www.hse.gov.uk/pubns/indg275.pdf). However, these principles are also explained in some details in the chapter MANAGING HEALTH AND SAFETY.

Stress at Work

Nicola Coote

Introduction

[S11001] Stress is becoming an increasing concern for employers and employees alike. More seems to be demanded by organisations in less time. At work, life is becoming more competitive, with more people chasing the same goals.

People are very complex and the way in which they react to events within their lives varies considerably. Sometimes, a single event may cause a crisis in someone's life. How they respond will depend upon a number of factors, including other circumstances that may exist within their life, their own reaction to the event and the length of time it takes to deal with the event. There are people who, when stressed, will make sure that everyone around them knows it. They will become withdrawn, and internalise their stress. Physically retreating from stressful situations, such as by taking frequent smoking breaks, is one way that some people cope. Others will obviously be angry, they care little about anyone except themselves, and their tension is obvious. This may result in behaviour that appears to be aggressive or threatening. Indeed, in some cases the stress may build to an extent that someone does resort to expressing their feelings by using violent behaviour. For others, stress will be a result of them being exposed to violent or threatening situations. Examples where this may occur include: care workers, retailers, police and other emergency support workers, and call centre staff.

Definition of stress

[S11002] Taken in its simplest form, the Oxford English dictionary defines stress as 'demand upon physical or mental energy'. The HSE's definition of stress is given as 'the adverse reaction people have to excessive pressures or other types of demand placed upon them'. This has been seen by some as a rather negative view, as it implies that stress produces only adverse effects.

Stress often occurs when the pressures placed upon a person exceed their capacity to cope. This can be negative or positive. For some people, the pressure of a tight deadline will 'positively stress' them, with the end result being that it enhances their performance. This may be due to their fear of failure or the consequences of not meeting the deadline. This is particularly true of people working in competitive sport and those people who set themselves challenges e.g. to increase sales figures by 50 per cent in the next 12 months. In these cases, the pressures placed upon them have a positive effect in that it produces drive, challenge and excitement.

Understanding the stress response

[S11003] People's ability to respond to difficult or stressful situations has evolved over many million years. The mind recognises a particular need to respond to an external source (both physical and mental). The alert message is sent to the brain which then initiates dramatic changes in the hypothalamus. This is the control centre which integrates our reflex actions and coordinates the different activities within our bodies. When the hypothalamus is activated, the muscles, the brain lungs and the heart are given priority over any other bodily activity. The chemical adrenaline is produced and the message of a threat is passed on to other parts of the body which prepares itself for vigorous physical action – known as the fight or flight instinct. When the threat has abated, the parasympathetic nervous system takes over again and the body returns to a state of equilibrium, (known as homeostasis) where all our functions are in balance. This is demonstrated in the stress response curve.

Figure 1: Stress Response Curve

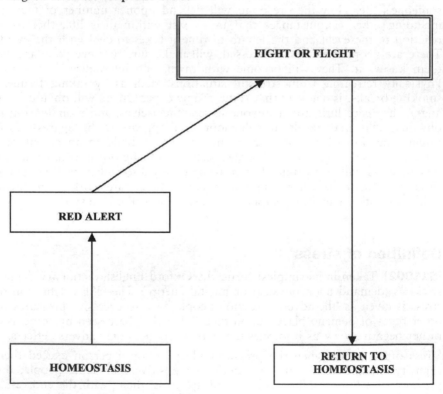

To summarise, given below are the physical responses to the 'fight or flight' mechanism:

- brain goes on red alert and stimulates hormonal changes including the production of adrenalin and noradrenaline (stress hormones);
- muscles tense, ready for action;

- eye pupils dilate to enable the danger to be seen more clearly;
- the heart beat increases to provide more blood to the muscles, therefore blood pressure increases;
- breathing increases to provide extra oxygen for the tensed muscles;
- liver releases glucose to provide extra energy for the muscles;
- digestive system shut down, making the mouth go dry and the sphincters close;
- sweating occurs in anticipation of expending extra energy; and
- immune system slows down.

These reactions are highly complex, but demonstrate the increased demands that are placed upon the body when feelings of stress occur. It is easy to see, therefore, how this response can have a detrimental effect upon well being and health if the stress is prolonged.

Why manage stress?

[S11004] The following paragraphs explain the importance of managing stress.

Legal issues

[S11005] There is no legislation that relates specifically to stress but the following Regulations touch upon the issue.

Health and Safety at Work etc Act 1974

[S11006] Under the *Health and Safety at Work etc Act 1974, s 2(1)* employers have a general duty to ensure, so far as is reasonably practicable, the health, safety and welfare of employees at work. This implies that measures should be in place to ensure that employees do not suffer stress-related illness as a result of their work.

Management of Health and Safety at Work Regulations 1999

[S11007] Under the *Management of Health and Safety at Work Regulations 1999 (SI 1999 No 3242), Reg 3* employers must also take into consideration the risk of stress-related ill health or unsafe working practices that may be caused or exacerbated by stress. Stress needs to be considered when undertaking general risk assessments (e.g. a risk assessment of the work environment, or of a particular task) as people's responses may differ when they are feeling the symptoms of stress, or overload. Some organisations may decide to undertake a specific assessment of occupational stress within their workplace, i.e. factors associated with the work that are known to affect the way people feel and respond.

Working Time Regulations 1998

[S11008] The *Working Time Regulations 1998 (SI 1998 No 1833)* seek to reduce the effects on employees who are subjected to working long hours, or who have insufficient rest periods during and between work shifts (these issues are known to be occupational stressors). The Regulations are detailed, and cover minimum rest times during a shift (20 minutes every six hours), minimum rest time between shifts (11 hours) and maximum number of hours that can be worked within a weekly period (48 hours). The maximum number of hours may be calculated over a 17-week reference period.

There are a number of exclusions that apply to certain categories of personnel under the Regulations, and these are under review in the European Courts at present.

Detailed requirements apply to employers to maintain a record of hours worked for each employee who is covered by this legislation.

For further information see WORKING TIME.

Criminal and civil liability

[S11009] Whilst there is potential for prosecution in criminal law for failure to meet the above duties, the likelihood is that compliance will be driven by civil litigation. To date, there have been no prosecutions under criminal law for failure to manage stress at work. Enforcement of the general duty of care under health and safety legislation has been due to successful claims being made for compensation following allegations by an employee that they have suffered illness or ill-health as a result of being exposed to unacceptable levels of stress at work.

Indeed, there have been some large insurance settlements for employees who have successfully claimed for stress at work. The most notable was *Walker v Northumberland County Council* [1995] IRLR 35. In this case, Mr Walker, a senior social worker, had been employed by the council for 17 years. During the 1980s his workload had increased, and as a result, he suffered a nervous breakdown. Upon his return to work he told his employers that his workload must be reduced. This was not done, and Mr Walker suffered another nervous breakdown which resulted in him being dismissed for permanent ill health. He claimed compensation from his employer, and was successful on the grounds that the employer had been aware that he was under extreme pressure at work. A key question in the case was whether Mr Walker's risk of mental illness was materially higher than that which would normally affect a senior social worker with a heavy workload.

As regards the second breakdown, the employer should have foreseen a risk that Mr Walker's career would end and they should have appreciated that Mr Walker was more vulnerable to damage than he had been before his first breakdown. Because of this, it was deemed reasonable for Mr Walker to have received assistance, but in not providing such assistance, the council was in breach of their common law duty of care.

In February 2002, a landmark decision by the Court of Appeal in *Hatton v Sutherland and Others* [2002] EWCA Civ 76, [2002] 2 All ER 1 overturned three out of four stress claims and quashed the awards for damages initially given to the workers in the High Court. The Appeal Court ruling stated that for a compensation claim to succeed it must first be 'reasonably foreseeable' to the employer that the employee concerned would suffer a psychiatric injury as a result of stress in the workplace. When deciding whether a risk is 'foreseeable' the following needs to be taken into account:

- nature and extent of the work undertaken by the employee (e.g. is the workload more than is normal for that particular job?); and
- signs from the employee themselves of impending harm to their health as a result of work-related stress (e.g. has an employee already suffered harm to their health from work-related stress?).

The Appeal Court ruling further stated that employers are entitled to assume that an employee can withstand the normal pressures of a job, except where they are aware of a particular problem or vulnerability (e.g. the employee has warned the employer that they are suffering from work-related stress). In addition, employers are entitled to take at face value what they have been told by their employees about the effects of work-related stress on their mental health. For example, if an employee has had sickness absence due to stress but does not inform their employer of the cause of the absence, then it would be reasonable for the employer to expect that the worker would be fit to return to their normal work.

The Court of Appeal set out a number of factors that must be taken into account by the courts once it has been established that it was reasonably foreseeable that an employee would suffer harm as a result of work-related stress. These include:

- an employer is only in breach of their duty of care if they fail to take reasonable steps to prevent the employee from suffering as a result of stress;
- indications of impending harm to health must be plain enough for any reasonable employer to realise;
- when considering what actions are reasonable for an employer to take to reduce risk of harm from stress, the size and resources of the employer's operation may be taken into account;
- it is not enough for a claimant to show they are suffering harm from a result of occupational stress – they must demonstrate that their 'injury' was caused by the employer's breach of duty of care by failing to take reasonable steps to protect the worker;
- an employer will not be deemed to be in breach of their duty of care if they allow a willing employee to continue with their job, despite them stating they have been suffering from stress, if the only reasonable alternative is to dismiss the employee; and
- if an employer offers a confidential staff counselling service for those who experience stress at work with referral to treatment services, they will be unlikely to be found in breach of their duty of care.

The court ruling places more responsibility on the employee to let an employer know if they are suffering from symptoms of occupational stress, rather than leaving it totally to the employer to identify a 'stressed' employee. In practice, therefore, an employee is unlikely to have a valid claim unless they can demonstrate that they have told their employer that they are suffering from work-related stress. Note that this does not take away the employer's duty to be proactive and seek to identify stress risk factors in the workplace, and to implement suitable precautions to minimise these risks.

This ruling was used in the case of *Pratley v Surrey County Council* [2003] EWCA Civ 1067, [2004] ICR 159 when a stress compensation claim made by Ms Pratley failed because she had not told her County Council employer that she was feeling under pressure.

Another significant ruling emerged from the case of *Barber v Somerset* [2004] UKHL 13, [2004] 2 All ER 385. Mr Barber was a teacher in East Bridgewater Community School and made a claim for the damages caused to him by suffering a mental breakdown which he attributed to the pressures and stresses of his workload. He was successful in his claim and was awarded £91,000 for loss of earnings plus £10,000 for pain, suffering and loss of amenity. Somerset County Council took the case to the Court of Appeal on the advice on their insurers and were successful in overturning the decision. However, it was referred to the House of Lords who overturned the decision on 1 April 2004 but reduced the damages by almost £30,000. It is important to note that the House of Lords broadly supported the guidance given by the Court of Appeal in dealing with cases concerning stress at work and that the principles established by the Court of Appeal were still valid. The main rulings from this lengthy case were:

- unless the employer knows of some particular problem or vulnerability he is usually entitled to assume that the employee is up to the normal pressures of the job;
- employers are generally entitled to take what they are told by or on behalf of the employee at face value; and
- an employee who returns to work after a period of sickness without making further disclosure or explanation to his employer is usually implying that he believes himself fit to return to the work which he was doing before.

The House of Lords decided on the basis of facts in this particular case that Mr Barber's employer was in breach of its duty of care by failing to take steps to lessen job-related stress that could lead to psychiatric illness, and thus made a compensatory award of £72,500.

The Court of Appeal's judgment gave useful guidance to employers, with practical steps that could be taken in case of complaints relating to psychiatric illness brought about by stress at work.

For an employee to succeed with such a claim, the employer must be able to reasonably foresee whether the employee would suffer from psychiatric harm which could be attributable to stress at work. This of course depends on what the employer knows (or ought to reasonably know) about the individual employee and their illness.

In general, the employer is entitled to take what they are told by the employee at face value, unless they have good reason to think to the contrary. They do not generally have to make searching inquiries of the employee or seek permission to make further enquiries of the employee's medical advisers. However, the employer should always seek further medical advice to clarify the nature of the illness and any recommended actions.

A more recent case tested the defence of reasonable foreseeability. In *Hartman v South Essex Mental Health and Community Care NHS Trust* [2005] EWCA Civ 6, [2005] ICR 782 the Court of Appeal reviewed the general principals that were considered in *Barber v Somerset*. In the *Hartman v South Essex* case, the court did not accept that the employer knew the claimant was vulnerable because she had disclosed the fact that she had suffered a nervous breakdown and was taking medication when completing a confidential questionnaire produced by the occupational health service. The court decided that it was not appropriate to attribute to the employer knowledge of the contents of information submitted by the employee in a confidential medical document to the occupational health department. In addition, the Court of Appeal commented that the fact that an employer offers an occupational health service does not mean that the employer had foreseen the risk of psychiatric injury owing to stress at work to any individual or class of employee.

The case of *Hone v Six Continents Retail Ltd* [2005] EWCA Civ 922, [2006] IRLR 49 is one of the most recently decided cases in this area. The Court of Appeal found unanimously in favour of the Claimant and dismissed his employer's appeal. In this case the claimant was a pub manager who had been working for excessive numbers of hours with little support.

The Court of Appeal held in the case that the test laid down in the *Hatton* case (see above) which further stated that a duty to take steps would be triggered when 'the indications of impending harm to health arising from stress at work must be plain enough for any reasonable employer to realise that he should have done something about it'.

The case therefore supports the notion that a successful claim for occupational stress can be mounted from a breach of the *Working Time Regulations*.

See also OCCUPATIONAL HEALTH AND DISEASES O1052.

Recent research on occupational stress

There have been numerous studies in recent years to identify if workplace stress exists and to what extent it is affecting the health and wellbeing of the workforce. There are too many to include in this chapter but details of some of the reports can be found on the HSE website (www.hse.gov.uk/stress/research/htm). This website includes a report outlining the business benefits attained by Somerset County Council from undertaking a process of stress risk assessment and intervention.

One of the more notable pieces of research is the *Beacons of Excellence* report which describes the authors' work to identify good practice in stress prevention and then to identify organisations within the UK that could be called 'beacons of excellence' in comparison with this model. The length of programmes reviewed varied from a few hours to a ten-year longitudinal study. The report summarises and draws conclusions from all of the substantive academic studies on stress prevention over the last decade and uses this information, as well as advice from a panel of international experts, to develop a comprehensive stress prevention model. This model is then used to describe examples of stress prevention practices within a wide range of UK organisations.

The *Beacons of Excellence* report states that organisational strategies to date have concentrated on employers providing access to specific services, with the intention of assisting employees during stressful periods. These strategies have been referred to as primary, secondary and tertiary levels of stress intervention (Murphy 1988).

Primary interventions seek to eliminate sources of stress by changing the physical and socio-political environment so it meets more with employee needs and provides more control over individuals' work. They also focus on improving communication, involving employees in decision-making processes and redesigning jobs to suit the needs of individuals.

Secondary interventions seek to help employees manage their stress but do not try to eliminate or modify workplace stressors. Such stress management programmes will help workers to identify stress within themselves and to cope it with more effectively.

Tertiary prevention strategies seek to help people who are experiencing on-going problems either from their working life or work environment. The idea behind these strategies is to modify individual lifestyle and behaviour but to make little or no change to organisational practices.

According to Kompier and Cooper (1999) stress intervention practice is currently focusing more on secondary and tertiary strategies rather than working on the primary prevention techniques, and the *Beacons for Excellence* study reiterated this view. Reasons for this are given in detail in the report. It states that the benefits of implementing stress prevention and management strategies are likely to vary according to the management commitment in implementing programmes that are suitably developed, and with the involvement of its employees.

Another study of note is the research project that the British Occupational Health Research Foundation (BOHRF) completed to provide evidence-based answers to some of the key questions related to mental ill health in the workplace. The questions were:

- What is the evidence for preventative programmes at work, and what are the conditions under which they are most effective?
- For employees who are identified as at risk, what interventions most effectively enable them to remain in work?
- For those employees who have experienced bouts of mental health related sickness, what systems most effectively support their rehabilitation and return to work?

The study looked at over 15,000 references in published research worldwide and identified 111 papers that were critically appraised. Based on these it was found there was moderate evidence to suggest that stress prevention strategies can have a beneficial and practical impact. There was stronger evidence to suggest that retention strategies are effective, i.e. properly targeted interventions to individuals can help to prevent long-term illness. The most effective programmes focused on personal support, individual social skills and coping skills training, and using a combination of these suggested that the most long lasting effects were likely. The study also found that rehabilitation of an affected worker can be very effective, and the most effective treatment was Cognitive Behaviour Therapy (CBT). This is a generic term that is used to describe a range of approaches designed to modify the individual's behaviour. The evidence suggested that this treatment was most suitable for those who had a reasonable level of control over their work

In addition to the research a series of discussions were held with over 200 senior managers, HR professionals, health professionals and policy makers from both the private and public sectors. Again, one of the key findings was that good management was seen as being crucial to the effectiveness of a stress management programme. There was strong support for the training of managers and supervisors in how to be pre-emptive in dealing with stress.

Full details of the BOHRF research study can be found at the BOHRF website: www.bohrf.org.uk/content/mhealth.htm

Management standards on occupational stress set by the Health and Safety Executive (HSE)

The HSE's commissioned research has indicated that:

- about half a million people in the UK experience work-related stress at a level they believe is making them ill;
- up to five million people in the UK feel 'very' or 'extremely' stressed by their work; and
- an estimated 9.8 million working days were lost to work related stress in 2009/10.

To tackle the growing problem of absence and ill health caused by occupational stress, the HSE has produced a set of management standards that employers should seek to achieve. Compliance with these is not a legal requirement but will assist the employer in reducing stress at source, i.e. by using primary interventions, and by having systems in place that will support a worker who is experiencing stress problems related to their work.

The Management Standards are based around six key areas; these are:

- demands, which includes issues such as workload, work patterns and the work environment (standard 1);
- control, which concentrates on how much involvement or autonomy a person has in the way they work (standard 2);
- support, which considers the level of encouragement and assistance provided by the organisation and the resources made available to provide support where additional assistance is needed (standard 3);
- relationships, which promotes positive behaviours and cooperation at work, and identifies if there are systems in place to respond to unacceptable behaviour (standard 4);
- role, which considers how clearly job descriptions and organisational structures are defined, whether there are conflicts in workers' roles and whether there is a clear and unambiguous reporting line (standard 5); and
- change, which assesses how organisation change is managed and communicated to the workforce (standard 6).

The standards seek to enable employers to identify the gaps between their current performance and conditions that are set in the standards, and to develop their own solutions where problems are identified. Note that the standards are not numbered by the HSE, but have been numbered for easy reference when reading the next sections in this chapter.

Further information on these standards can be found on the HSE's website: http ://www.hse.gov.uk/stress/standards/index.htm

Occupational stress risk factors

[S11010] The following paragraphs discuss in detail the most common occupational stress risk factors.

The physical environment

[S11011] The physical environment will come under HSE Management Standard No 1 (Demands).

The physical environment can be described as the general conditions in which people have to work. The work environment is often seen as a contributory cause for stress at work, i.e. something that compounds a problem that is being experienced at work. For others the work environment may be the main cause of stress amongst the workforce, especially if the conditions prevent work targets from being achieved, or detract workers from undertaking their work effectively.

Our general work conditions have improved drastically over the past century, but so have our expectations. The environments in which people work will vary considerably, and will be dependant upon the type of work that is being undertaken. For example, a die casting factory is traditionally a hot, dirty, and malodorous environment. Any investments are usually spent upon the equipment to produce the product rather than to enhance the work environment, and managers may be reluctant to spend a heavy outlay on a workplace that is likely to become dusty and dirty within a short time. These types of environment often have less time and effort spent on them in maintaining high standards of cleanliness, good housekeeping or ensuring that lighting remains in good order throughout the premises.

This can, however, have a knock-on effect for staff, both in their mental attitude towards the workplace, in their general conduct and in their general well being. It has been found, following several studies of the workplace, that an environment that is poorly maintained produces a negative attitude amongst the staff – 'if management don't care then why should we' type approach. If the workplace is allowed to become dirty, untidy or cramped, then workers will respond accordingly with the 'why should we tidy up – management don't seem to mind' type attitude. A workplace that is dull, untidy, badly designed or scruffy is likely to make employees feel negative about coming to work, and they are more likely to respond negatively to issues that arise at work.

The condition of the workplace

[S11012] The condition of the workplace can be described as its general standard of maintenance and decoration, both internal and external. House-keeping standards within the workplace could be included under this heading. To understand how a workplace condition can affect how a person feels and responds, consider the following examples.

- *Scenario 1*

  An old hospital building was first constructed during the early 1900s. The hospital management has struggled to keep up with the increase in usage over the years and the structural changes that have been necessary to accommodate more patients and more services. The hospital is crowded and cluttered. Government financial cutbacks have been a common setback and funding has focussed on provision of medical care, rather than on maintenance.

  The walls are grey and flooring is mainly grey concrete. It is cracked in many areas, causing ruts and trip hazards. Lighting is poor and several overhead tubes are not working. Paint on the wall is dirty, marked in places and chipped in others. Some of the windows are cracked.

  Doors are damaged due to the constant thoroughfare of hospital trolleys being pushed against them.

  Sacks of dirty laundry, and black sacks containing rubbish, seem to be constantly accumulating in corners awaiting collection.

- *Scenario 2*

  This scene depicts another hospital that was constructed in the early 1900s. Similar challenges have been faced within the hospital to provide increased services and a higher level of care. Budgets restraints also present a challenge.

  The walls are still grey but have been painted within the last few years. However, different parts of the hospital have received a splash of colour by way of painting the support pillars, door surrounds etc with bright contrasting colours. The doors are less damaged as they have metal plates fitted to reduce damage from the trolleys. Windows are cracked in some places, but have been cleaned, which has resulted in more light coming into the hospital.

  Rubbish bags and dirty laundry bags are out of sight. The floor looks cleaner and there are less pot holes and cracks in the flooring.

- *Scenario 3*

  This scene depicts a hospital that was also built in the early 1900s. However, despite the hospital facing the same challenges as the hospitals in the first two scenarios, the management team have adopted a much more proactive approach to general decoration. Staff have been encouraged to participate in choosing colour schemes when new services have been introduced and the patients' views have been taken into account when planning any refurbishment. This has resulted in local colleges being invited to practise their artistic skills in main areas at little or no cost to the hospital. Waiting rooms have brightly coloured murals along the walls. Corridors are coloured differently for each separate unit (e.g. cardiology, rehabilitation) and collages and other pictures have been displayed along the walls. These have resulted in interesting schemes for people to enjoy as they walk through the corridors.

  Some extra money has been made available each year for ongoing repairs which has resulted in a higher standard of maintenance. A proactive maintenance schedule is in place whereby all lighting tubes are replaced annually, thereby significantly minimising lighting failure. Staff have been encouraged to share their requirements or preferences for lighting in different areas, and for reporting to management when problems are occurring. Windows are kept clean.

  Wards are brightly coloured, and curtains have been fitted at windows which are pleasant to look at. Plants are situated around the hospital and a water feature has been installed in the main reception waiting area.

  Waste collection points have been clearly defined. Laundry collection/delivery areas are clearly marked and waste is not allowed to accumulate.

An employee working in any one of these environments, doing work which is known to be emotionally and physically demanding, would obviously prefer to be working in an environment depicted in scenario 3.

People feel better and therefore respond more positively when their environments are pleasant. This positive factor is increased when staff feel that they have had some level of control over their own environment, even if it is something as basic as deciding on the best areas for storage or waste collection points.

The workplace layout

[S11013] The workplace layout will be considered in HSE Management Standard 1 (Demands) and 2 (Control).

The workplace layout can make a significant impact upon a worker's efficiency, safety and general well being. This is where the study of ergonomics can have a major input into reducing stress levels at work. Ergonomics is a specialist area, which looks at the interaction of the worker, the work equipment and the work environment. The use of an expert may be needed at some stage in workplace layout, but the first step should always be to discuss employee requirements and to see how these can be accommodated within current restraints. See ERGONOMICS E17021.

Simple changes to a work layout can make a drastic improvement in a person's efficiency, safety and general well being, especially when the idea has emanated from the workforce itself. The worker is the expert in the particular task that is being undertaken, and whilst an employer does not want to provide a 'wish list' to the workforce, valuable information can be ascertained merely by finding out if and what causes frustration within a work layout.

An employer might consider asking its staff the following questions.

Question	Yes/No	Comments
Are worksurfaces large enough for the work and equipment that are needed?		
Are worksurfaces of a suitable height (too high or too low) for the worker(s)?		
Is there sufficient space for the worker to get to, from and around their own work area?		
Are storage facilities in close proximity and easy to access?		
Is there sufficient storage?		
Is storage being used to full effect?		
Do workers have somewhere to store their own personal items?		
Are seats or suitable rest areas provided, at or near to the workstations?		
Are walkways wide enough to enable people and equipment to pass through (e.g. trolleys)?		
Are work areas in large areas clearly identifiable, signed and labelled?		

Question	Yes/No	Comments
Is the layout designed to minimise environmental hazards (e.g. glare from windows, noise from nearby machines)?		

Poorly designed layouts can induce other health conditions such as musculoskeletal disorders e.g. back problems from having to stoop, stretch or crouch to undertake a task. Stress creates physical tension which can contribute to these health conditions. Some proposed new EU legislation will require that stress and work pressure be considered as part of risk assessments for back pain and other musculoskeletal disorders.

Staff facilities

[S11014] Staff facilities will need to be considered under HSE's Management Standard 1 (Demands) and perhaps under Support (Standard 3).

Everyone occasionally feels a need to take a break from their work, and this may be more of a need than a desire in some work situations that are known to incur high levels of stress e.g. dealing with the public, working in the care profession. Having nowhere to go to take a break or let off steam may result in stress levels increasing to a point where a worker becomes distressed. This will be compounded when other aspects of the work are problematic, for example, the shift patterns are badly designed or working hours are excessive.

Staff facilities should include basic facilities such as sanitary conveniences and washing facilities, and somewhere from where drinks can be prepared or obtained. Employers should ensure that such facilities:

- accommodate the numbers of people using them;
- take into consideration the need for privacy (especially if workers need to change clothes or shower before or after work);
- are maintained to a good state of hygiene and cleanliness;
- provide basic equipment such as soap, drying equipment, toilet paper and sanitary bins;
- enable workers to obtain both hot and cold drinks, including provision to clean spillages or clean utensils if drinks are being prepared by the workers; and
- provide a ready supply of hot/warm and cold water, and potable water for drinking purposes.

The *Workplace (Health, Safety and Welfare) Regulations 1992 (SI 1992 No 3004)* and supporting guidance contain detailed minimum standards for sanitary conveniences, washing facilities, drinking water facilities for changing clothes.

These Regulations also contain provisions for rest facilities, and require that somewhere that is suitable and sufficient is made available for people to eat meals where meals are regularly eaten in the workplace. Resting and eating facilities must ensure that people who do not smoke are protected from the effects of tobacco smoke.

This legislation also requires the provision of rest facilities for workers who are pregnant, or who are nursing mothers. Some employers have expressed concern at the practicalities of meeting this requirement. However, the female workforce is increasing as more mothers are returning to work earlier and employers have a legal as well as moral duty to ensure that these provisions are available. These types of issue do have an impact upon the stress levels of employees.

Since July 2007 it has been against the law to smoke in virtually all enclosed public places, workplaces and public/work vehicles.

The *Working Time Regulations 1998 (SI 1998 No 1833)* contain requirements for rest periods. During working hours the maximum that someone is allowed to work is six hours, following which they are entitled to a 20 minute break. Rest facilities should be in line with the standards stipulated above.

For further information see WORKPLACES – HEALTH, SAFETY AND WELFARE W11001.

Environmental conditions

[S11015] Environmental conditions will be considered in HSE Management Standard 1 (Demands), but will also need to be considered under Standard 2 (Control), i.e. how much control do they have over their local environment (e.g. temperature, space etc.), and Standard 3 (how much support is available to address concerns that are raised).

When considering stress induced by environmental conditions, an employer should consider issues such as lighting, heating, ventilation, humidity and space.

In some cases, the environmental conditions may be the sole or primary cause of stress, in others it may be a factor that exacerbates other stressful conditions within the workplace. Poor environmental conditions are the cause of other workplace issues such as sick building syndrome.

Several regulations contain specific duties for provision of environmental standards. These include the *Workplace (Health, Safety and Welfare) Regulations 1992 (SI 1992 No 3004)*, the *Schedule* to the *Health and Safety (Display Screen Equipment) Regulations 1992 (SI 1992 No 2792)* and the *Provision and Use of Workplace Regulations 1998 (SI 1998 No 2306)*.

An employer should take the following into account when deciding on the standards that should be achieved in the workplace:

- legal requirements contained in various statutes;
- official guidance issued by the HSE, trade associations or other recognised bodies such as the Chartered Institute of Building Service Engineers;
- the nature of the work (work requiring intricate detail will need a higher level of lighting than one undertaking general tasks);
- the amount of physical effort in a task (a cooler background temperature will be needed where workers exert physical effort);
- the need to combine workers' needs with other stakeholders (e.g. customers, patients);
- the level of humidity in the environment (work in a humid environment will reduce a worker's ability to control their own internal temperature by sweating, thus increasing the risk of heat exhaustion); and

- the amount of space for the work components, the work equipment, and the worker's need for movement.

There are several sources of guidance which will help an employer to interpret conditions that will be deemed as 'reasonable' to meet statutory requirements for a work environment. A good starting point, other than a specific trade association, is the Chartered Institute of Building Service Engineers. This institute can provide detailed guidance on recommended environmental standards, and should be referred to when deciding optimum work conditions for different types of work activity.

Employers should take into account, however, that any such guidance should not be used in isolation. The most effective way of reducing physical and emotional stress from work environments is to involve the staff. It is not possible to please everyone, but by obtaining a general consensus an employer has an opportunity of ascertaining if any problem exists, where the problem might be, and the extent that the environment may be cause or adding to a problem.

Another important key to reducing stress induced by environmental conditions is to provide as much control as possible over the local environment. For example, it is usual in large open workspaces for the lights to be controlled by only a few switches. The lighting circuits could be changed so that a smaller area is controlled by each switch, enabling more control over an employee's immediate working vicinity. The strength and colour of the tubes could also be altered to further meet individual preferences.

The work

[S11016] There are many aspects of work that may induce pressure. Pressure at work is not necessarily bad as it helps to generate enthusiasm and can produce energy. However, prolonged pressure is likely to create a state of nervous tension as the person has no respite from the constant rush of adrenaline that occurs when stress is induced. Unless this is relieved, the end result can lead to health problems.

The trick is to try to achieve a balance between providing a challenge to staff, and overloading them.

The following areas are known to be occupational stressors, and should be reviewed when any stress management strategy is being considered.

Targets, deadlines and standards

[S11017] Targets and deadlines clearly come under HSE's Management Standard 1 (Demands). They may also need to be considered under Standard 2 (Control) and 3 (Role). For example, if a worker is given conflicting roles and job functions this may well have a negative impact on their ability to meet their targets.

Targets are necessary at work. This is because an employee who is not given a target will set their own target which may not be in line with organisational expectations. However, problems arise when the standard that has been set by the employer exceeds the individual's ability to meet it.

When a deadline or target is unachievable, a sense of helplessness may be experienced by the person. Similar emotions may be experienced when each time the target is met the employer increases the next target. Whilst for some employees this will represent a sense of excitement, there will inevitably be a time when an employee begins to feel defeated, undermined or useless. This problem becomes more serious when there is no mechanism for the employee to let the employer know how they are feeling, or when there is no system for monitoring staff behaviour and performance once the target has been set. If the employee feels that their job depends upon meeting the target then the feeling of pressure is increased, as the consequences of failure are more significant to that individual.

There are various reasons why a deadline or target may be unachievable. Managers should consider the following questions, and perhaps agree with the employee what is reasonable when all the restraints have been taken into consideration. This will help the employee to perceive a sense of involvement and control over the amount of pressure being placed upon them.

Assessment of targets			
Question	Yes/No	Comments	Agreed action
Is the employee sufficiently trained/experienced to meet the standard?			
Are there adequate tools/equipment to meet the deadline/target?			
What other demands are being made which may affect the standard/deadline being met?			
Is there any mechanism for the employee to report early if he/she feels they are not likely to meet the standard?			
To what extent can machine failure etc affect the target being met?			
Is the time frame for the amount of work being set reasonable (i.e. is this being achieved reasonably by others)?			
What are the consequences of meeting or failing the target?			
Are there any other factors that the employee feels may detrimentally affect the target/deadline being achieved?			

People who have experienced continual pressure from unrealistic deadlines or targets often explain their feelings with comments such as not being able to see the light at the end of the tunnel and being on a treadmill that they cannot get off. When these feelings become excessive, a common response is for the employee to take time off for sickness absence, or even to leave the company. Early warning signs that an employee may be feeling excessive pressure may include erratic behaviour or mood swings, reduced or erratic work performance, looking worn out or tired or even compulsive behaviour (increased smoking, drinking or eating). Some people, of course, will respond with the opposite effects, i.e. by losing their appetite, becoming very quiet or seeming to become particularly conscientious.

Time demands

[S11018] Time demands clearly fall under HSE Standard No 1 (Demands) and No 5 (Role).

Stress from conflicting demands on time may be a sign of 'role conflict' or 'role ambiguity', i.e. which task is more important. It may be that the time available is insufficient to deal with the extent of work being required. If time demands being made upon the employer are such that everything is expected to be undertaken at the same time, then the worker will suffer.

Different people have different ways of handling their work. Some people prefer to complete one task before giving any attention to another, whilst others appear to thrive on dealing with numerous matters all at the same time. Problems arise when the manager operates to one of these principles whilst the subordinates preferred style is totally different. The end results can be disastrous, with neither party feeling that a job is being completed satisfactorily and both parties feeling that they are being misunderstood. Two-way dialogue between management and employees is needed to ensure that each party understands the other's needs, and to agree parameters for working.

A commonly reported cause of stress is knowing that several other tasks are building up whilst only one task is being addressed. The employee will either rush through all the work, to try to reduce the mountain that is piling, or select one task that they feel is the most important at the expense of others which are perceived to be less important. Neither response is likely to result in staff feeling satisfied that they are coping and doing a good job. A key to managing this type of problem is to:

- review work loads;
- take into account deadlines;
- agree priorities for each task that is building up;
- ensure the most suitable type of person is allocated (some people deal better with a high throughput that requires low levels of accuracy whilst others prefer work that require less output but a higher level of attention); and
- encourage and develop self-organisation for the individual to manage their work load.

Working hours and shift patterns

[S11019] Working hours and shift patterns will be reviewed under a number of the HSE Management Standards, including Demands, Control, Support and maybe even Relationships and Role. This will be particularly relevant for people who are covering more than one job role e.g. during periods of short staffing.

Management of working hours is a key feature in reducing stress levels at work. When considering working hours, the employer should take into consideration:

- the length of the shift (how many hours worked each day);
- the shift pattern, especially if rotating shifts are used;
- rest times between shifts; and
- rest and break times during the shift.

As mentioned previously (see **S11008**) the *Working Time Regulations 1998 (SI 1998 No 1833)* contain requirements on the number of hours that should be worked each week, the minimum rest time between shifts and the minimum rest breaks during shifts. Regardless of this whether an employee is excluded from the Regulations or chooses to 'opt out', the employer has a duty of care to ensure that health, safety and welfare are not detrimentally affected by the work.

Problems with working hours seem to crystalize when the time spent at work impinges on employees' home and social activities. Someone who habitually works long hours often feels too tired to enjoy their private time. An employee will be prepared to 'trade off' a well-designed shift pattern in order meet social demands being made at home. It is clear, therefore, that a balance must be struck between shift patterns that have been deemed to be the most effective, against the social needs of the workers.

The most frequently reported source of stress among shift workers who work night shifts is the interruption in their sleep, or difficulty in being able to sleep during the day. Possible sources of interruptions are numerous and include doorbells, telephones, increased traffic noise, family, pets and neighbours. Workers who operate to rotating shifts are more likely to feel the effects of sleep disturbance.

The need for variety during night shifts in terms of the range of tasks, and the mental and physical output necessary, is more important for night workers. The attention span will reduce, typically between 3.00 a.m. – 4.00 a.m. and this will be exacerbated if the work is monotonous, requiring high levels of concentration or is physically demanding. Fatigue is unavoidable during night work, but will be reduced if the interest of the shift worker is maximised.

Travel during work

[S11020] Time spent on travelling and the modes of transport permitted within the organisation's policy or resources can have a significant impact upon a worker's wellbeing. This is likely to be considered under HSE Management Standards 1 (Demands), 2 (Control), and 3 (Support).

Many employees enjoy travelling as part of their work as it provides a sense of variety, change in environment and a level of autonomy about their work pattern (e.g. by not feeling that they are being watched). Problems arise when the amount of time travelling exceeds the amount of time that can be spent in the workplace, or impinges upon the time available for the remainder of work that needs to be completed. Travelling is also exhausting, both physically and mentally, and can reduce a person's work performance. Other problems associated with travelling include feelings of isolation from colleagues, family and friends, especially when extended stays away from home occur.

Apart from drivers of heavy goods or public services vehicles, there are no maximum times set in legislation that a worker may travel on the road without taking a break. Therefore an employer should undertake a risk assessment, and agree with the workers what travel arrangements are considered reasonable. Factors that need to be taken into account include:

• preferred method of travel;
• manual handling issues of the traveller (cases, files etc);
• the length of the journey (including times for delays);
• how much time the employee has been given to reach their destination (i.e. work time versus private travelling);
• travel routes (especially if using the road);
• the need for regular two-way communication with the worker whilst away from the workplace; and
• suitability of overnight accommodation, including location of accommodation.

Organisations such as ROSPA (Royal Society for the Prevention of Accidents) have produced detailed guidance and literature detailing results of their research into occupational road risk, and this may be a useful source of reference when considering the physical and psychological risks associated with travelling.

Technology and work equipment

[S11021] This section will be considered in HSE's Management Standards on Control, Support and maybe Change (if the technology changes).

There is nothing more frustrating that having a mismatch between the tools necessary to complete a task and the task itself. If the problem is simply being given the wrong tool to complete the task, then this can quickly be rectified (provided the employer is informed). However, stress levels increase when equipment constantly fails to meet its design criteria, breaks down due to overload or poor maintenance, or when there is a skill's shortage amongst the workers.

To overcome the problem the employer should undertake a task analysis to identify what tools and equipment are needed, and to ensure that they are provided. Ongoing maintenance may be needed, including an emergency call-out service where the job is dependant upon the work being completed within a certain time. This needs to be followed by a training needs' analysis, to identify if and what type of training is required to ensure workers can use the equipment effectively and safely.

Most people are instinctively sociable, and respond to both verbal and non-verbal behaviour and communication whenever they meet. Someone making a light-hearted comment can lighten a dull or pressurised meeting, a smile will indicate whether an ambiguous comment is meant as a joke. This type of behaviour is missing from computerised communication i.e. e-mail or even telephone. Messages received without the additional benefit of non-verbal communication may come across as blunt, aggressive or hostile and may make the recipient feel threatened when the opposite was the intention. Stress will result from the perception of aggression or threat, especially if that person feels helpless to respond or defend themselves. This needs to be taken into account when transmitting difficult messages, to avoid inducing unnecessary anxiety. In extreme cases, e-mail communication has been known to cause severe disruption to workplace harmony and cause individuals to become insecure or even sick.

Role conflict or ambiguity

[S11022] Very clearly, this will be predominantly reviewed under HSE's Management Standard 5 (Role), but also under Standard 3 (Support).

People can work much more effectively if they know what they are supposed to do, how it is supposed to be undertaken, why it needs to be undertaken in a particular way and how their role fits into the corporate objective.

When job roles conflict or are ambiguous the end result is confusion and frustration for everyone involved. If this becomes a frequent event then those feelings of frustration may increase into anger and resentment. This may be compounded when the person's level of authority is limited and they have no access or authority to change what is causing the problem. The following scenarios may be indicative of conflicting or ambiguous roles.

- *Scenario 1*

  A secretary is told by her manager to manage the diary and to keep all the clients satisfied when they ring. Several potential clients have telephoned to book meetings which the secretary has entered into the book, but her boss is very displeased when he finds out and tells her to cancel them. Nearly all other bookings made into the diary by the secretary get changed or altered, causing the secretary additional work and upset. When challenged by the manager about why all these bookings have been made, the secretary defends herself by stating that she is doing what she was told to do. The boss responds by saying that nothing should be booked without checking with him first.

- *Scenario 2*

  A departmental manager has responsibility for reducing expenditure in his area of authority. However, he has also been given responsibility for managing health and safety. The health and safety standards in the department are poor. The manager identifies a need for health and safety training which will improve standards but this will take him over the agreed spending threshold which will directly affect profitability and his requirement to reduce expenditure.

Clerical/administration support

[S11023] Many employees now undertake their own basic administration tasks, using basic computer software packages. This is acceptable provided they have the basic skills to use the equipment, and the time allocated within their work patterns to complete the administration. Common indicators that completion of paperwork is becoming problematic may include:

- workers regularly taking work home in the evenings/weekends;
- employees not using their holiday entitlement;
- increase in errors within written correspondence (typographical, grammar or content);
- tasks being half completed;
- paperwork piling up on workstations;
- messages not being answered; and
- comments from workers about bureaucracy.

Employers should ensure that time is made available to enable workers to complete any necessary paperwork, and that the equipment is accessible. This is particularly relevant where the worker is based in a non-office environment, and would not automatically have access to a desk, paper, filing system etc. readily to hand.

Most people do not like excessive paper-work, and some do not like any paper-work. Employers should therefore seek to streamline their systems so that administration is minimised, and simplified.

Cover during staff absence

[S11024] This is an area which will cross through a number of the Management Standards, but is likely to be predominantly considered under Standard 4 (Relationships) and Standard 5 (Role).

Administration often becomes a stress issue when staff are constantly required to cover for colleagues. The additional pressures this causes are unlikely to be problematic if it is for only short duration, e.g. sickness or holiday. However, staff who feel that they are frequently undertaking additional work to cover for shortages will begin to feel the effects of the burden. Early symptoms to look for may include reduced efficiency, increased error rates, lower patience or attention thresh-hold, or increased sickness absence.

Managers should seek to avoid relying on employee goodwill as the sole method for dealing with the workload when a colleague is absent. Use of temporary staff should only be used as a stopgap. Often their effectiveness is limited as they are unfamiliar with the work and work systems. It also takes time to train a temporary worker, and if this becomes a regular exercise then the use of temporary workers will be seen as a hindrance rather than a help.

Employers should therefore adopt a more proactive stance, and consider what alternatives would be suitable to deal with staff shortages. Examples may include use of multi-skilled workers who could be moved around, or 'floating' staff who are employed specifically to cover during other staff absences.

Company structure and job organisation

[S11025] The following paragraphs discuss the relevance of company structure and job organisation on stress levels.

Hierarchy of control

[S11026] The following areas will focus on HSE's Management Standard 2 (Control) and 5 (Role) but might also impact on Standard 4 (Relationships). Where changes in company organisation occur then Standard 6 (Change) will be predominant.

Ambiguity over job roles and the level of control over a task is a major stressor at work.

Problems occur when employees are given responsibility for a task but have no authority to implement necessary systems, training etc to effectively deal with the task. Colleagues and fellow team members will judge both the corporate reaction and the reaction of the particular manager to any decisions that are made within an organisation, and will build their own attitudes as a result when showing commitment back to the organisation. It is therefore very important that everyone knows what their level of responsibility is, and the authority that they have to deal with a given situation. When a challenge is presented at work the employee's physiological state will alter to deal with the event, (increased blood pressure, etc.). Any nervousness or heightened state of alertness to deal with the situation will be better controlled by the individual if they know exactly how they are supposed to respond, and the parameters in which they can deal with the given situation.

The interface between two overlapping parts of an organisation can also be a cause of stress – each will have their own priorities and their own way of working. Good communication and clearly defined lines of authority are vital to identify the essential patterns of overlap and to enhance the organisation's efficiency.

To overcome such problems the employer should review where any overlap in processes or responsibilities may occur, and where any gaps or areas that have not been clearly defined may be present. These should be cross-checked against the recipients (i.e. employee's) perceptions and interpretations of the responsibilities to ensure that misunderstandings do not occur.

Organisational changes and arrangements

[S11027] Any change within the organisation will need to be considered under HSE Standard 6 (Change). Change is stressful to many people but symptoms can be alleviated by good communication and a feeling that they are involved.

People generally do not like change. It upsets the person's comfort factor and invariably leads to a period of uncertainty whilst they are adapting to the changes. Some people adapt better to changes in their lives than others. Having an awareness that any change is likely to cause stress is the first step to managing the problem.

It is necessary to communicate as clearly as possible the nature of any changes, and the consequences of these to the workers who are affected. Even bad news can be dealt with positively if it is communicated honestly, openly and with sensitively.

Responsibility

[S11028] Responsibility links with the level of control an individual has (HSE Standard 2). How clearly this responsibility is defined will need to be reviewed under HSE Standard 4 (Role).

Once a responsibility has been allocated to an individual, there should be sufficient scope allocated to enable the person to carry out the task. This does not mean that the employee is then left to deal with a given responsibility without any mechanism for support or assistance if the responsibility becomes more extensive or complex than first anticipated. Being given a responsibility without any access to support when needed will inevitably lead to stress.

The feeling of a lack of appreciation, through non-realisation of an individual's potential, often arises from being given insufficient responsibility. Many people, especially those in jobs which do not enjoy high status in their organisation, feel frustration through lack of challenge, or feel frustrated when they can see things going on around them when they do not have an opportunity to be involved in making a contribution. Giving people autonomy, i.e. increasing the degree of self-control over their work, will help to counteract this.

Developing a system for employee recognition and staff development will help to identify where an employee needs more or less responsibility within their work. This will also provide a mechanism for a manager to explain why more responsibility has not been given to an individual who feels they want or need it, and for managers and employees to work together to reach a level which is mutually agreeable.

Individual perception of company

[S11029] How people feel about their organisation will reflect in their work attitude and performance, so this area may be considered under Standard 4 (Relationships), 5 (Role) and perhaps also 2 (Control).

If staff have a negative view of the company, they will respond accordingly. Employees who feel that the company is going nowhere, or does not care about its image, will not care about the company's future or its image. This will give them no sense of achievement or pride in their employment. The end result will be a higher staff turnover and the associated problems that go with this. Output will not be as effective as a motivated team of workers, and standards of work are likely to be lower.

Involving staff in the company's vision, or business plan will help employees to see what the company is hoping to achieve and the goals that have been set. Even if the employees do not fully understand or agree with the vision they will at least have an understanding of what is happening when changes begin to occur.

Perception of one's self-esteem and role within the company

[S11030] Everyone needs to feel that they have a valid role to play, in their life and in their work; everyone needs to feel valued and needed. When this does not occur in the workplace, the result is an employee who feels worthless, undervalued and misunderstood. The problem will become compounded if the individual concerned is someone who is ambitious or is trying to achieve promotion or a more fulfilling role.

Previous experience

[S11031] Someone who has experience of the environment will be more aware of the penalties and rewards associated with the work. They are likely to have been exposed to any associated work stressors before and may be more adept at managing them. However, this can sometimes have the opposite effect, in that if someone has been exposed to high levels of stress in a similar type job previously then this may well have reduced their tolerance to pressures.

Personal factors

[S11032] The way in which someone responds to their work can be reviewed under most of the Management Standards, but is perhaps more prevalent under Standard 1 (Demands). The compatibility of the demands being made upon the person with their own personality traits needs close review and monitoring to ensure the job and the person have the right fit.

Everyone responds differently to events that occur in their lives. This very much depends upon their basic personality traits, their upbringing and their experiences in life.

Various studies have been undertaken in behaviour patterns. The research undertaken by Friedman and Rosenman *Type 'A' behaviour and your heart* (1975) (ISBN 0704501589) gave some insight into why some people are more prone to stress-related disease. In these studies, two main risk behaviours were identified, which were called Type A for high risk, and Type B for low risk. In general, Type B behaviours are the opposite of Type A behaviours. These categories should only be considered as a general yardstick as there are no absolutes, and most people will fall somewhere between the two extremes identified.

Type A people were said to be:

- autonomous, dominant and self-confident;
- very self-driven, busy and self-disciplined;
- aggressive, hostile and impatient;
- ambitious and striving for upward social mobility;
- preoccupied with competitive activities, enjoying the challenge of responsibility; and
- believing that their behaviour pattern is responsible for their own success.

Type A people were said to work longer hours and spend less time in relaxation and recreational activities. They received less sleep, communicated less with their partners and tended to derive less pleasure from socialising.

In contrast, type B people tend to:

- find it easier and are more effective at delegating;
- work well in a team;
- allow subordinates some flexibility in how a task is undertaken;
- avoid, or do not set, unrealistic targets;
- keep a sense of balance in the events that occur in their lives; and
- be more accepting of a given situation.

Type B people are often more secure within themselves and more accepting of their strengths and limitations. A very simplistic summary of this type of character could be that they are simply more 'laid back'.

The level of responsibility and control over work

[S11033] Clearly the following sections are most relevant to Standards 2 and 5 (Control and Role).

These are major factors as to whether someone is likely to suffer the effects of stress. Karasek and Theorell *Healthy Work* (1996), a comprehensive study of job demand and autonomy, showed that stress is a function of the combination of psychological demand and 'decision latitude', or autonomy.

The term 'executive stress' has been widely used, insinuating that perhaps people with high levels of responsibility are more likely to suffer from occupational stress. Research has shown, in fact, that this is not true. Whilst the level of responsibility increases with 'white collar' roles, so does the level of autonomy about how the work is to be carried out. Material rewards are usually higher, which helps to enhance self-esteem within the worker.

Karasek and Theorell's studies showed that the greatest risks are among those occupations with least autonomy and of a lower status. Examples include bus drivers, production line workers and care staff. These are classic examples of the conclusion reached by Karasek and Theorell that it 'is the bossed, and not the bosses' who are most likely to suffer from stress. They often have little or no control over how the work is to be undertaken but have to deal with the problems and difficulties that arise when a job is badly designed or managed. They also have less material reward for the work undertaken.

Qualifications and competencies

[S11034] There is a rather cynical saying that people are promoted to their level of incompetence. Whilst this may be an exaggerated view there is some truth in the concept that people who are newly promoted may not immediately be competent. The same could apply to almost any role at work, if there is a mismatch between the task that is required and the skills that are possessed by the people undertaking the work. People need to grow and develop into their roles at work, to a point that they feel comfortable and competent in what they do without becoming bored or complacent. An employee who works beyond their competence is likely to cause problems for the organisation in that work standards may reduce and production may be affected. From the workers' point of view they are not likely to feel confident or able that they can achieve what is required, and this will have a knock-on effect upon their self-esteem. The end result is likely to be that such an employee would feel frustrated and dissatisfied, and maybe very unhappy. The manager is likely to feel equally frustrated and dissatisfied because the work is not being completed satisfactorily.

Personal issues

[S11035] Separating personal problems from work issues is difficult for some people, and impossible for others. Both work and personal issues have an indirect effect upon each other and if an employee is experiencing a stressful situation in their home life then this may impact upon work performance.

An employer cannot be responsible for personal issues in an employee's life, but an awareness that a problem exists for the individual is useful knowledge for the employer. Systems should be developed that enable a manager to offer support and sympathy whilst the difficulty exists. For example, if a family member is in hospital perhaps working hours could be adapted slightly to fit in with the hospital's visiting times. The reaction to an individual who is going through a crisis, in terms of the level of support and flexibility shown by the manager, can stay with that person for a long time. Genuine help being offered by the manager can be an investment which promotes commitment from that person at a later stage.

A pre-agreed workload level may be acceptable in normal circumstances, but intolerable when a long term problem is occurring in someone's personal life. Therefore, the events going on in the employees' lives outside work will, at some stage, have an effect on their ability to cope with work. If these events are positive then the ability to cope will improve, with the opposite effect happening when the personal events are negative.

Managers need to take a balanced approach when responding to an individual's personal issues. An employee who is considered to have a good 'credit' balance in terms of their commitment and enthusiasm to the organisation may be treated more sympathetically by the organisation if a crisis develops in their lives. However, this could in itself have a significant knock-on effect to other employees who feel that they have not been treated so considerately, and this can induce feelings of being undervalued, misinterpreted or simply overlooked by those colleagues.

Stress management techniques

[S11036] The following paragraphs provide detailed guidance and advice on managing stress.

Stress auditing/surveys

[S11037] Stress audits or surveys are a good starting point, as this will give the employer an indication of whether a stress issue exists in the workplace, and the areas where this may be most problematic.

There are several approaches to audits and surveys, ranging from a simple checklist that employees are asked to complete, to detailed analyses with scoring systems that are undertaken by specialists. The type of audit that is used will depend upon the organisational culture and the extent to which an employer feels that a stress problem exists.

A survey in its simplest form may merely be a short questionnaire that asks employees to list their three most positive and three most negative aspects about their work. Guided headings may be included, perhaps with a tick box and a space for the employee to write comments explaining why it is seen as a positive/negative aspect of work. Headings that are used could include areas that the employer suspects are causing pressures upon the employees. This type of approach is often used as a pre-audit review, to enable employers to gain a feel as to any patterns which might be obvious to the workers.

Stress assessments are more detailed, and require planning, thought and sensitivity. They usually take the form of a guided list of questions or areas that the assessor wishes to cover about the work. They can include all members of staff, or selected members of staff from certain job titles to act as a representation. The interview needs to be undertaken by someone who is sufficiently competent to complete such a task. Competencies necessary to achieve this may include:

- counselling skills.
- polished interview techniques.
- knowledge of the organisation and its management structure.
- knowledge of procedures and practices.
- understanding of health and safety issues, including stressors at work.
- understanding knowledge of human behaviour (i.e. human factors at work).
- the ability to gain the confidence of the interviewee.
- the ability to listen.
- understanding of the HSE's Management Standards and what to cover in each topic.

Points to remember:

- Reassure staff that all comments will be confidential.
- Respect the confidentiality of your staff.
- Tell your staff what you are planning to do, what you hope to achieve, why you are doing it, and the benefits that may be in it for them.
- Provide a mechanism by which anyone can discuss any queries or concerns about the audit process beforehand.
- Involve all levels of employment, from senior board members to junior staff.
- Consider issuing a pre-interview questionnaire which you can review before your interview (e.g. three best and three worst aspects of their job).
- Design a simple framework around which you can develop your audit/survey format.
- Ask simple, open questions.
- Give people time to consider questions asked, or points that have been raised.
- Allow plenty of time for the interview, and sufficient time afterwards to write up the key findings.
- Tell your staff what you plan to do with any information collected.
- Involve them as much as possible in any further decisions or developments arising from the survey.

After the audit:

- Inform staff and/or representatives about the key findings and what the company plans to do in response.
- Provide (where possible) an indication of time frames as to when and how these changes can expect to occur.
- Involve staff and/or representatives.
- Explain why any suggestions may not be taken forward by the company, don't just refuse.
- Follow up any changes you make to ensure they are having the effect you intend.
- Keep staff and representatives involved at all times during the design and implementation of any systems or procedures resulting from the audit.
- Lead by example – managers can communicate important signals about the importance of managing stress.

Stress management policy

[S11038] A proactive management strategy for occupational stress should include development of a policy. This should highlight the company's philosophy and recognition of the issue, and clearly state the procedures in place to eliminate or manage stress associated with work.

The policy should include:

- the company's definition of occupational stress;
- recognition that stress is a work issue that will be taken seriously and sympathetically by management;
- acknowledgement that stress is not an illness or a weakness;
- explanation of the procedures the company will implement to identify occupational stressors and implement risk reduction control systems;
- details of personnel who have been allocated with responsibilities for implementing and monitoring risk reduction measures for stress;
- explanation of employees' responsibilities and procedures for reporting to management when they feel they are suffering from stress; and
- training and support packages, including any counselling services that may be available.

On-going monitoring

[S11039] Like any other health and safety system it is necessary to monitor systems and procedures to ensure they are effective. Monitoring will have the added benefit of helping to identify a stressed employee who is unwilling to report they have a problem. Monitoring can be undertaken in many ways including:

- monitoring during team meetings, by encouraging a general discussion about positive and negative aspects of work that have occurred;
- monitoring during appraisals and other one-to-one meetings with staff;
- implementing an anonymous stress report box in which people who are finding a work situation to be continuously problematic can report this without any fear of recrimination;

- circulating staff attitude or opinion surveys which will encourage feedback from workers;
- observation of workers' behaviour patterns and relationships with colleagues (looking for any obvious differences from the norm, or spotting relationships/atmospheres that appear to be tense); and
- monitoring accident and incident statistics, and sickness absence records.

Information and training

[S11040] Provision of information and training is a general duty under the *Health and Safety at Work etc Act 1974, s 2(2)(c)*. This will include information and training in occupational stress. Information should be given to employees to enable them to understand the organisation's views on occupational stress, its policies and its procedures for dealing with stress and the part that employees can play in managing occupational stress for themselves.

Training of managers and supervisors in how to recognise symptoms of stress, and how to be pre-emptive when addressing any problems, was identified as a key benefit in the research project undertaken by the British Occupational Hygiene Research Foundation.

Training can be used to underpin any findings from surveys or assessments that may have been undertaken, and to help the policy on stress to be understood and effectively implemented. Training should be provided to managers and to staff in general. Managers will need to know:

- what stress is and it contrasts with or effects occupational stress;
- symptoms of stress and how to recognise someone who may be suffering from stress;
- health risks associated with stress;
- occupational factors which cause or contribute towards stress;
- identifying, separating and managing personal and management stressors;
- outline of company policy and procedures on stress;
- identifying measures to identify stress; and
- understanding of measures to manage/reduce occupational stress.

Employees will need to understand for themselves exactly what stress is, and how to recognise for themselves when they or a colleague may be experiencing symptoms. The training will enable the organisation to demonstrate that it is a caring employer.

The Chartered Institute of Environmental Health (CIEH) has developed a one day accredited course called 'Stress Awareness'. This provides a detailed programme that covers the key issues associated with occupational stress. The fact that is accredited by a recognised body will help to increase its profile to employees who may be more cynical about the subject itself, or the organisation's commitment to employee welfare.

Other stress management techniques

As part of the above measures, employers should seek to implement stress management techniques that focus on primary interventions, but also have secondary and tertiary interventions to underpin the management system and deal with workers at all stages of a stressful situation. To recap, components of these stages are as follows:

Primary interventions

These seek to eliminate occupational stress symptoms at source, rather than support someone who is experiencing a stressful situation at work. Development of a cohesive stress prevention and management policy is a key to such a pro-active approach, supported by training and other mechanisms to implement the policy objectives.

Specific pro-active interventions that employers could take include:

- developing, or further improving, communication channels with employees so that they feel involved and have a better understanding of workplace issues and changes;
- including aspects of stress assessment and review during job appraisals and personal feedback sessions;
- redesigning jobs to improve efficiency, comfort and productivity;
- involving staff in any changes to their working pattern or equipment, and in the work process;
- keeping up to date with technology and equipment available, and ensuring that tools and equipment do not become too outdated or cumbersome;
- monitor and review working hours and environmental conditions to ensure the workplace is as pleasant and conducive to work as it can reasonably be;
- encourage early reporting of stress by workers, with a cohesive policy of supporting actions following the report being made; and
- more attention to the person/job fit during the recruitment process.

Secondary interventions tend to help people to manage stress without seeking to eliminate or modify workplace stressors. Examples may include:

- training of employees in how to recognise stress in themselves;
- promoting self-help or coping mechanisms when someone is reporting stress;
- introduction of mentoring or 'buddy' systems amongst the workforce when someone is feeling under stress;
- provision of 'quiet rooms' or other such places where someone can go if they are feeling under excess pressure; and
- grievance procedures for staff who feel that they are not being treated fairly.

Tertiary interventions seek to assist workers who are experiencing on-going problems from their working life, without making any changes to the source of the problem:

- counselling service for someone who needs support and assistance in helping themselves to cope;

- access to various specialist treatments such as Cognitive Behavioural Therapy (CBT), Psychoanalysis; or
- yoga, meditation or other relaxation classes offered during lunch breaks.

In reality an effective stress management system should include elements of all the interventions mentioned above. The most effective system will focus predominantly on primary interventions, and will have the full commitment of senior management which is visible to all the workforce.

Training and Competence in Occupational Safety and Health

Hazel Harvey

Introduction

[T7001] Although the term 'incompetent' is widely recognised by people in the world at large and can quite easily be defined as not qualified or able to undertake a defined action or task, when looking at the definition of competent or competence, the boundaries are less easy to define. Someone may be qualified or able to undertake a task to a certain level or in a specified area, so they could be considered competent either generally at a lower level or in part. This particular dilemma is one that has faced most professions and occupations during their development stages and has led to the formation of professional bodies and trade associations, who as one of their major reasons for existence have set standards of practice for their own particular disciplines. This is perhaps where competent performance related to health and safety is very different to other professions and trades in so much that everyone, whether at work or not, needs to have knowledge of how to remain safc and healthy and from many different angles and perspectives.

This becomes particularly relevant in workplaces where both employers and employees have moral and legal responsibilities both for themselves and others in terms of maintaining a safe working environment. Everyone needs to have a level of competence in terms of occupational health and safety, from employees who need to ensure that their actions do not endanger either themselves or their colleagues, to directors of the company who are responsible for the health and safety of the workforce and need to know what this entails. Additionally specialist advisers in health and safety will need to have much higher and broader levels of competence in occupational health and safety issues, so that the advice they give to both the management and the workforce in organisations is clear and of a standard expected from a professional practitioner.

The Health and Safety Commission ('HSC'), made a clear statement regarding health and safety competence in their 'Strategy on Health and Safety Training' which has the vision: 'Everyone at work should be competent to fulfil their roles in controlling risk'. It also makes a statement about the training required to do this:

> Health and safety training covers all training and developmental activities aimed at providing worker, including safety representatives, and managers with:
>
> (a)   greater awareness of health and safety issues;
> (b)   specific skills in risk assessment and risk management;

(c)   skills relating to the hazards of particular tasks and occupations; and
(d)   a range of other skills, including those relating to job specification and design, contract management, ergonomics, occupational health etc.

The specific aims of the strategy are to:

- raise awareness of the importance of health and safety training;
- bring about a substantial improvement in the quality and quantity of health and safety training;
- promote an awareness of the importance of competence in controlling risk;
- influence providers of the education system to provide the necessary framework of basic knowledge and skills.

With the objectives of the strategy being to ensure competence by:

- encouraging employers and trade unions to recognise the need to provide health and safety training of good quality;
- engaging HSE and Local Authority (LA) inspectors to assess the competence of workers and managers and the adequacy of training provided by employers. This forms an important part of inspection, investigation and enforcement activities;
- continuing to seek partnerships elsewhere to provide training of appropriate quality and quantity;
- influencing other government departments to reflect the need for health and safety training provision in their responsibility; and
- ensuring that all parts of the education system provide a foundation of knowledge upon which health and safety training can be built.

This chapter focuses on:

- the components of competence and the acquisition of this;
- how it may be attained and assessed for all the sectors of the workforce;
- the legal and managerial requirements for competence; and
- the various organisations that fulfil roles in the process.

In all cases, the chapter recognises that any level of competence is time bounded. Skills and knowledge become out dated very quickly in the light of new technologies and practices, and is a particular aspect that needs to be considered in maintaining the health and safety of the workforce.

Recognising competence

[T7002] It is worth having a more in depth look at some of the more common definitions of competence, before exploring the ways and means of attaining specific and relevant levels of health and safety competence. These definitions, although not in actual conflict with each other do have slightly different slants on the concept, which very much depends on the source of the definition. Clearly people from outside the health and safety world may well view competence in a slightly different way to those more familiar with the legal concepts. This may well lead to the position where employers believe they have an appropriate level of competent advice (as described in the next section at T7003 below), but may in fact be leaving themselves exposed to both criminal and civil actions in the event of an accident, where it can be proved that the appropriate level of advice was not available. When either selecting individuals or training programmes for employees, employers should be advised to seek appropriate advice to ensure that they are legally and morally covered in terms of health and safety; good advice can also have a considerable effect on a company's profit margins!

Looking at the concept of competence from a plain English perspective *The New Oxford Dictionary of English* defines competence as the ability to do something successfully or efficiently and competent as having the necessary ability, knowledge, or skill to do something successfully.

The National Vocational Standards define competence as 'the ability to perform to the standards required in employment across a range of circumstances and to meet changing demands'.

In case law a competent person is viewed as 'one who is a practical and reasonable man who knows what to look for and how to recognise it when he sees it' (*Gibson v Skibs A/S Marina and Orkla Grobe A/B and Smith and Coggins Ltd* [1966] 2 All ER 476). It could also be added that they should also know what to do with the acquired knowledge once they have found it.

All of these definitions basically amount to the same thing, but with a slightly different slant, that competence is a combination of knowledge, experience and skills. It should be recognised that competence itself is based on the outcomes of this combination of requirements and it is a demonstration that an individual can perform to specified standards rather than simply a record of a person's qualifications and experience, with which it is often confused. It is this simple definition of knowledge, skills and experience that is used throughout this chapter, recognising that the actual determination of an individual's level of competence does require a further stage of assessment by an employer. Only the training and education leading to the development of competence can be identified here.

Legal obligations for training and competence

Health and Safety at Work etc Act 1974

[T7003] The *Health and Safety at Work etc Act 1974* ('*HSWA 1974*') is the principal enabling legislation for the UK. It imposes duties on employers, employees, bodies corporate, manufactures and others, as well as establishing the HSC which recommends policy regarding health and safety, and the Health and Safety Executive ('HSE') which enforces this policy and to serve a range of legal notices to employers.

Training is specifically mentioned within *section 2* of the *HSWA 1974* which prescribes the general duties of employers. *Section 2(2)* of the 1974 Act states that in addition to other requirements it is an employer's duty to ensure 'the provision of such information, instruction, training and supervision as is necessary to ensure, so far as is reasonably practicable, the health and safety at work of his employees'. This in fact does not place an absolute duty on an employer to provide training but recognises that having carried out generic and specific risk assessments of the workplace an employer may well believe that training would be an effective measure of control. Many employers take this approach in addition to other risk control methods.

Competence is not specifically identified within the *HSWA 1974* other than an allusion in *section 19(1)* which refers to the appointment of inspectors. Here it relates to 'such persons having suitable qualifications as it thinks necessary for carrying into effect the relevant statutory provisions within its field of responsibility'.

Perhaps the most important part of the *HSWA 1974* in relation to training and competence is in *section 15* which relates to the power to make health and safety regulations. The first set of these regulations as relevant to training and competence are the *Safety Representatives and Safety Committees Regulations 1977 (SI 1977 No 500)* which require an employer to allow a trade union appointed safety representative to take time off 'to undergo such training as may be reasonable in the circumstances'.

Management Regulations

[T7004] Major changes to the requirement for competence on a general basis came about as a result of the UK's enactment of a series of European directives on health and safety via a series of regulations. The *Management of Health and Safety at Work Regulations 1992 (SI 1992 No 2051)* (now revoked) was a set of generic regulations relating to the overall management of health and safety and worked in conjunction with already existing regulations such as the Control of Substances Hazardous to Health Regulations ('COSHH') and the Noise at Work Regulations. The 1992 Management Regulations actually introduced the concept of 'competent person'. The Regulations were amended in 1999 with *Regulation 6* becoming *Regulation 7* in the *Management of Health and Safety at Work Regulations 1999 (SI 1999 No 3242)* ('MHSWR 1999'). These Regulations are the main focus when considering competence in health and safety.

The *MHSWR 1999 (SI 1999 No 3242), Reg 7(1)* states that:

> Every employer shall, appoint one or more competent persons to assist him in undertaking the measures he needs to take to comply with the requirements and prohibitions imposed upon him by or under the relevant statutory provisions.

The *MHSWR 1999 (SI 1999 No 3242), Reg 7(5)* states that:

> A person shall be regarded as competent for the purposes of paragraph (1) where he has sufficient training and experience or knowledge and other qualities to enable him to assist in undertaking the measures referred to in that paragraph.

[T7005] The Approved Code of Practice that accompanies the *MHSWR 1999 (SI 1999 No 3242)* makes some specific conditions regarding the competent person role in these Regulations. In summary these are:

- Employers are solely responsible for ensuring that the person appointed to fulfil the 'competent person' role is capable of applying the principles of risk assessment and prevention together with a current understanding of legislation and health and safety standards.

- Where there is more than one person appointed to fulfil the role, or parts of it, then these people must co-operate.
- Competence in the Regulations does not necessarily depend on the particular possession of particular skills or qualifications and simple situations may require and understanding of relevant current best practices and an awareness of the limitations of one's own knowledge but have an ability to seek external help where necessary.
- More complex situations will require a person with a higher level of knowledge and experience and in these cases employers are advised to check the appropriateness of health and safety qualifications and /or membership of professional bodies. The competence-based qualifications accredited by the Qualification Curriculum Authority (QCA) can act as a guide.

The *MHSWR 1999 (SI 1999 No 3242), Reg 8(1)(b)* imposes a duty on employers to nominate competent persons who have the ability to carry out evacuations of an establishment in the event of serious or imminent danger. *Regulation 8(3)* goes on to say that such a person will have sufficient training and experience or knowledge and other qualities, defining competence in a similar way to the *MHSWR 1999 (SI 1999 No 3242), Reg 7*. Also, in *Regulation 15* of the 1999 Regulations which relates to temporary workers, an employer is again required to determine if a temporary employee has the competence to carry out the job in a healthy and safe manner.

[T7006] The *MHSWR 1999 (SI 1999 No 3242)* make several statements particularly in relation to the requirement for training. *Regulation 3(5)* relates specifically to young people and *Regulation 13* specifically relates to capabilities and training for all employees. *Regulation 13(2)* particularly states that every employer shall ensure that his employees are provided with adequate health and safety training, and very specifically relates to the various times that this should be given, ie:

- on recruitment;
- on being exposed to new or increased risks because of being transferred or having a change of responsibility;
- on the introduction of new work equipment or a significant change to it;
- on the introduction of new work equipment;
- on the introduction of a new system of work or a change to an existing system.

The *MHSWR 1999 (SI 1999 No 3242), Reg 13* also says that this training should be repeated periodically and be adapted to take account of any changes or new risks and should also take place during working hours.

[T7007] Overall the *MHSWR 1999 (SI 1999 No 3242)* impose a very comprehensive requirement for both training and the recognition of competent advice in the workplace. They have highlighted and made a legal requirement of what has been good practice in workplaces for many years. In addition to these Regulations there are several more hazard-specific pieces of legislation, which also require training to be delivered and competence to be demonstrated.

The policy unit of HSE has published an outline map on competence, training and certification. The map gives an overview for competence and training as described in legislation and can be found on the HSE website. The mapping process identifies that there are four different groups in which training and competence requirements can be identified:

Group 1

- General goal-setting requirements where there is no precise detail of how the requirement should be met – there are 37 pieces of legislation that have this approach;

Group 2

- General goal-setting requirements qualified by some specific requirements for people or bodies with particular responsibilities – there are 8 instances of this in legislation;

Group 3

- General goal-setting requirements qualified by particularly strong specific guidance, sometimes agreed with the industry/sector – there are 6 instances of this in legislation;

Group 4

- Specific requirements for competence or training, certification/qualifications and approval, including statutory requirements to meet a performance standard – there are 40 specific requirements of this nature in legislation.

Within group 1 there are five high-risk areas where an employer is required to have arrangements for training and levels of competence written into plans or reports. These are:

- construction (site supervisor's plan);
- control of major hazards (in accordance with the *Control of Major Accident Hazards Regulations 1999 (SI 1999 No 743)*;
- offshore safety cases;
- railway safety cases (in accordance with the *Railways (Safety Case) Regulations 2000 (SI 2000 No 2688)*;
- nuclear site licences.

There are specific requirements for people or bodies within group 2. This includes medical personnel responsible for medical surveillance under the *Control of Substances Hazardous to Health Regulations 2002 (SI 2002 No 2677 as amended by SI 2003 No 978)*, and the requirement of specific training within the *Provision and Use of Work Equipment Regulations 1998 (SI 1998 No 2306 as amended by SI 2002 No 2174)*.

In group 3 there is strong specific guidance advocating the use of particular programmes of training. Regulations included in this group are: the *Noise at Work Regulations 1989 (SI 1989 No 1790)*, the *Manual Handling Operations Regulations 1992 (SI 1992 No 2793 as amended by SI 2002 No 2174)*, and the *Health and Safety (Display Screen Equipment) Regulations 1992 (SI 1992 No 2792 as amended by SI 2002 No 2174)*.

Group 4 lists those areas where there are specific legal requirements for named training programmes. Activities worth particular noting in this area are radiation protection advisors, gas fitting (CORGI), adventure activity licensing, mining, the carriage of dangerous goods and diving.

Employers should make themselves aware of any specifically cited requirements in their sphere of business in addition to the more general management requirements described in the *MHSWR 1999 (SI 1999 No 3242)*. They should give the necessary training required that will lead their workforce to perform competently with regards to health and safety.

Setting national standards in health and safety

[T7008] During the 1980s it was identified by a Government Working Party, that there was no effective system of vocational qualifications within the UK. What did exist had evolved from individual employment sectors and as such there was no commonality of level and the standards varied. This was a very difficult situation for employers who were unable to determine the validity of qualifications; whilst some sectors had highly respected qualifications others had none. A system was needed that would recognise the skills people needed and was reliable, consistent and well structured. In 1986 the Government established the National Council for Vocational Qualifications ('NCVQ') to set up a comprehensive framework of vocational standards covering all occupations and industries. One notable fact was that all the standards were required to have a unit of health and safety within their structure.

The standards were developed by organisation know as 'Lead Bodies'. These were independent groups of professional representing employers, trade unions, government departments, local authorities and other relevant groups. In the health and safety area the organisation setting the standards was known as the Occupational Health and Safety Lead Body ('OHSLB'). HSE formed the secretariat for the Lead Body and included representatives from the CBI, TUC, and Department for Employment and Local Authorities. They developed the national standards for Occupational Health and Safety Practice, Regulation and Radiation Protection. These standards were launched in 1995 and became available as the National and Scottish Vocational Qualifications ('NVQ/SVQ').

HSE adopted the OHSLB Enforcement Standards, in addition to its established academic training programme, in the Guidance Note attached to the *Management of Health and Safety at Work Regulations 1992* – recognising the Practice Standards, as representing levels of competence for those giving advice on health and safety. The Institution of Occupational Safety and Health ('IOSH'), the chartered professional body for practitioners in health and safety also recognised the standards as underpinning their competence categories of membership; level 4 at their full member category and level 3 at technician level. (See **T7010** below for clarification of levels.)

The ownership of the standards passed from the OHSLB to a new organisation, the Employment National Training Organisation ('ENTO'), in 1998, following a government rethink of the structure of how the national vocational standards should be organised. Recent developments in the standards setting world and the establishment of Sector Skills Councils (SSCs) has lead to many standards changing 'ownership'.

The current 'owners' of the OSH standards is Pro-Skills, the SSC for the manufacturing sector. Current projects are re-defining the existing National Occupational Standards (NOS) into the latest format required by the Qualification and Credit Framework, which replaces the older frameworks operating in the UK.

[T7009] The standards in Occupational Health and Safety Practice ('OHS') were launched on the 13 August 2002 in conjunction with the leading players in the health and safety world: IOSH, the Royal Society for the Prevention of Accidents (RoSPA), the National Examining Board for Occupational Safety and Health ('NEBOSH') and the British Safety Council. The OHS standards themselves had undergone a fundamental revision during this process and now became available at level 4 and 5 – the latter reflecting the increasing requirement for higher standards of competence for those who strategically manage or advise on health and safety issues. HSE also adopted the revised level 5 standards for enforcement. These standards have recently been reviewed by ENTO in line with the normal quality standards timelines for national standards. The changes have allowed for a more user friendly wording of the standards in line with the current guidelines for standing writing and also the qualification structure attached to the standards has also been reviewed in the light of experience of use with the standards. However, the essence of the standards remains essential unchanged.

An earlier initiative by ENTO had identified that the mandatory health and safety units included in the wider range of vocational standards were numerous (well over 400) and variable and in the light of current changes to the standards setting bodies already mentioned it would be undesirable to return to this situation. It was its belief, as the training organisation for people who work with people at work, that a collection of stand-alone units for people who needed the competence to fulfil certain health and safety responsibilities, but who were not health and safety practitioners, was necessary to bring a level of consistency in the competence requirements across the whole range of vocational standards. These were produced and launched in 1999, and as such they have no particular level and can be fitted into other sets of vocational standards as appropriate. The eight separate units known as 'Health and Safety for People at Work' can be used to demonstrate a level of health and safety competence for the whole range of employees from workers to managers and directors. They cover the range of knowledge, skills and practical experience, which a person has to have to be able to do a particular task in a safe manner and are designed to allow people to show that they are competent in the safety issues described.

These standards, now 'owned' by Pro-Skills are currently being adapted to the new format required by the QCF and should be complete by the end of 2010. More information on how the QCF works is available on www.qcda.gov.uk.

In conclusion, there are standards available within all the UK national frameworks to recognise the level of competence required by all people in the workforce. These standards, as well being capable of being used as the standards against which the N/SVQ qualifications are assessed, also form the basis for other types of qualification in the health and safety area.

Education and training towards national standards of competence

[T7010] This section will focus in more detail on the required competence levels in health and safety across the whole spectrum of the workforce. This will include strategic health and safety practitioners or managers, health and safety advisers and practitioners, health and safety technicians, managers and supervisors with line responsibilities for health and safety, and last but by no means least, the workforce itself. Some of the qualifications that have been traditionally associated with health and safety such as those from the British Safety Council or NEBOSH will be equated to the national frameworks to give an indication of the parity of the different routes available.

The frameworks of competence standards and associated vocational-related qualifications ('VRQs') is organised in the constituent bodies by the nationally associated bodies highlighted in **T7008** above. For brevity they will be referred to using the English framework the QCF However, this should be taken as meaning all the bodies, other than SQA, which do in fact have a working association. SQA's framework is essential the same but reflects the slightly different levelling of qualifications within the Scottish Sector. The QCF is an eight point framework.

In addition to these frameworks there is another UK qualification framework specifically for the higher education sector. This is known as the Quality Assurance Agency for Higher Education and is referred to as 'QAA'. For a clearer understanding, the relationship of qualifications on the QAA framework to the QCA framework are shown in the following chart:

	QCF Vocational Qualifications	QAA
Level 5	Level 8 , Level 7	Doctorate, Masters Degree, Post Graduate Diploma
Level 4	Level 6 , Level 5 , Level 4	Bachelors Degree , Ordinary Degree , HE Diploma
Level 3	—	'A' Level (access to higher education)
Level 2	—	—
Level 1	—	—

This chart is a rough guide only for the purposes of this chapter. It has no legal standing as to the equivalence of level.

Competence requirements for occupational health and safety practitioners

[T7011] IOSH is the only chartered professional body in the UK for those involved in occupational health and safety practice. It was founded in 1945 and incorporated by Royal Charter in 2003 which was modified in November 2005 to allow for the title 'Chartered Safety and Health Practitioner' to be used by IOSH members in the Chartered Fellowship and Chartered Membership categories. IOSH now has over 37,000 members which makes it one of the largest bodies in the world for people involved in safety. Of these IOSH, as the chartered professional body is required to set the standards for those in the practice of health and safety. However, since their inception, IOSH has taken an active part in the development of the national vocational standards for OHS, and Health and Safety for People at Work, and now uses these national vocational standards as part of the definition of its membership criteria.

[T7012] The national vocational standards for OSH Practice are recognised by the profession as representing the core competence requirements for those in health and safety practice. There are, however, many ways of achieving these standards through the various routes of qualification together coupled with the development of skills and experience. The standards, which are based on safety management models such as that developed in the HSE publication *Successful health and safety management* (HSG65) or OHAS 18001reflect the current management based approach to safety which has gone a long way to improving health and safety within companies within the UK.

The titles of the revised OHS Standards, which are still available on the ENTO website, are as follows:

HSP2   Promote a positive health and safety culture

HSP3   Develop and implement the health and safety policy

HSP4   Develop and implement effective communication systems for health and safety information

HSP5   Develop and maintain individual and organisational competence in health and safety matters

HSP6   Identify and control health and safety risks

HSP7   Develop and implement proactive monitoring systems for health and safety

HSP8   Develop and implement reactive monitoring systems for health and safety

HSP10   Develop, implement and test health and safety emergency response systems and procedures

HSP11   Develop and implement health and safety review systems

H13   Influence and keep pace with improvements in health and safety practice

These 10 units are mandatory in the S or NVQ qualification which is currently at level 4 although it may be amended during the new developments.

Candidates for an NVQ are assessed against these standards. This type of qualification is an assessment of the ability of a person to perform to the laid down standards and involves an assessor who will monitor and lead the candidate through the process of developing a practice portfolio. For an VQ at this higher level it is also necessary for the assessor backed up by the verification processes required by the awarding bodies, to determine that a candidate has the requisite domain knowledge to carry out the competence based tasks required.

This type of qualification has proved very popular in occupational health and safety practice as it is particularly suited to those who have many years of experience in giving health and safety advice, based on sound experience but who are not happy or comfortable with the traditional exam-based type of qualifications available.

[T7013] The national vocational standards are also used by the QCF to assess the relevance of vocationally-related qualifications to the overall national framework. Organisations applying to the QFA to be placed on this national framework must show that they have robust systems of quality control in place and have an examination process totally divorced from the delivery of their courses, this is known as Awarding Body Accreditation. They also must also show that their syllabi deliver the underpinning knowledge requirements of the national vocational standards, in the process known as accreditation of qualifications. NEBOSH and British Safety Council Awards (BSCA) are organisations which currently have qualifications accredited at level 6 on the QCF framework, although there may be some amendment to this following the implementation.

The QCF accreditation verifies that the qualifications on its framework are robust and have achievement recognised at the specified levels. Universities offering qualifications in occupational health and safety also have to undergo similar processes. Universities are themselves responsible for awarding qualifications within the QAA frameworks and have introduced stringent and searching quality control procedures internally, by a system of internal validation and external examination by appointed examiners. The QAA holds the overarching quality control on the qualifications offered. Also it is normal for universities offering vocationally-related degrees to seek external accreditation from the professional body in that area. IOSH undertakes this accreditation process for occupational health and safety degrees and post-graduate qualifications and originally published its own criteria for the conditions of accreditation and syllabus content in 1997. These were revised during 2009 and to maintain IOSH accreditation universities will be expected to adjust their syllabuses to meet the knowledge requirements from next few year.

There are a range of different levels of qualification available from the higher education ('HE') sector degrees which roughly equate to level 6 on the QCF. The more academically challenging Post-Graduate degree programmes, relate to level 7. There are currently 34 universities and related colleges in the UK offering these types of programme, which are proving very popular. In fact this is now by far the biggest route of entry into the professional membership category of IOSH.

It should be clarified that academic qualifications such as those from the HE sector, NEBOSH, BSCA or other similar bodies, are not assessments of competence in the same way as an NVQ qualification. They are a demonstration that a programme of academic work leading to the development of knowledge and some skills has been undertaken and formally assessed by examination and/or other suitable form of academic assessment. This by itself would not be sufficient to demonstrate competence but with suitable experience and further development of skills in a practical setting, the knowledge gained can form the vital knowledge part of the knowledge, skills and experience equation. However, it should also be recognised that although the achievement of an NVQ actually demonstrates competence, this is only in the workplace setting in which the person undertaking the qualification is employed. This means that the broader knowledge base required of a competent practitioner has not necessarily been covered or assessed to the same extent to a person who has taken the academic qualifications.

[T7014] There are clearly pros and cons for each type of qualification and this is a debate, which continues throughout the safety profession and the deliverers of training and education. This is not helpful to employers and those needing to employ competent health and safety practitioners who need to be able to assess the similarities and differences between the varying types of initial qualification. Health and safety is a diverse profession and attracts people from many different backgrounds, so it is advantageous and necessary to have the variety of different ways to qualification currently available. IOSH as the professional body currently accepts all the different qualification routes but recognises that all have particular strengths and by inference weaknesses. In the revised IOSH membership structure operating from November 2005, further assessment via an Initial Professional Development (IPD) process in addition to initial qualifications is necessary to show overall competence. This is tailored to the initial qualification so that those with an NVQ will need to undertake a further formal assessment of knowledge, in the form of an open assessment whilst those with knowledge based qualifications will need to demonstrate the acquisition of skills by developing a skills based portfolio. Both groups undergo a professional peer review interview to determine the validity of the knowledge and skills and additionally the experience of a potential Chartered Member of IOSH. From November 2005 holders of IOSH accredited qualifications enter the organisation as a Graduate Member (Grad IOSH). Potential employers will be able to use the IOSH membership structure and particularly the requirements for its chartered category membership as a benchmark for competence of safety practitioners irrespective of whether they are a member of the professional body.

Competence in health and safety practice at strategic management level

[T7015] In the current set of national vocational standards, a level 5 set of competencies in OHS is available. This reflects the fact that many more health and safety professional are now moving into more strategic management roles within organisations and also the fact that many, who would not consider themselves to be health and safety professionals also operate at this strategic management level role. Although there are a plethora of health and safety training courses specifically targeted at mangers, very few of them are actually pitched at this higher level of management expertise. This is perhaps not surprising, as most managers would incorporate health and safety within their workplace role on day-to-day basis, implementing the rules and policies that have already been developed, using their personal knowledge of the processes of the organisation for which they work.

It has been a long-standing request from the health and safety community that health and safety should be included in MBA programmes so that potential future leaders of industry should have health and safety included within their academic programmes. As yet this has not borne much fruit but one or two organisations do now do this. Perhaps of more potential is the fact that HSC included in its *Revitalising Health and Safety* document the fact that the education for those in safety-critical professions should include health and safety specifically.

The titles of standards for OHS at level 5 are listed as follows:

HSP1	Develop and review the organisation's health and safety strategy.
HSP2	Promote a positive health and safety culture
HSP3	Develop and implement the health and safety policy
P11	Develop a strategy and plan for people resourcing change management
HSP13	Influence and keep pace with improvements in health and safety practice.I
L12	Identify organisational learning and development needs
B3 (MSC)	Manage and use financial resources
B5 (MSC)	Secure financial resources for your organisation

These standards are reflective of health and safety strategies but are in reality management standards. Some of the standards are in fact taken from other sets of standards, such as those from the management-training organisation (MSc). The NVQ level 5 reflects this fact in that it is now called Management of Health and Safety.

Many of those in higher level strategic positions in health and safety management prefer to take a more traditional route to qualification and select the MSc programmes in health and safety mentioned previously.

For those graduates entering health and safety as a second career this appears to be a preferred route.

Competence in health and safety for managers and supervisors

[T7016] Although those in safety practice can set the standards and form the systems to ensure that health and safety risks are controlled, it is actually those in everyday contact with the people and work interface who can actually make the most difference to the overall health and safety performance. The competence of this group can be achieved by a variety of ways and means and it is this group who have the knowledge of the day-to-day workings of an organisation and therefore who will be able to have the most effect in the workplace.

The ENTO suite of standards Health and Safety for People at Work, now called Standalone Units were designed as competence standards for all people in the workplace and many of them were targeted at the managerial and supervisory groups. A level 3 NVQ in occupational health and safety can be obtained by being assessed against five of these standards HSS1-4 and 6 and two selected from HSS5, 7, 8 or 9. Units can also be taken individually for those people who don't require the whole range of standards.

HSS1   Make sure your own actions reduce risks to health and safety

HSS2   Develop procedures to safely control work operations

HSS3   Monitor procedures to safely control work operations

HSS4   Promote a healthy and safe culture within the workplace

HSS5   Investigate and evaluate health and safety incidents and complaints in the workplace

HSS6   Conduct a health and safety assessment of a workplace

HSS7   Make sure your own actions within the workplace aim to protect the environment.

HSS8   Review health and safety procedures in workplaces

HSS9   Supervise the health, safety and Welfare of a learner in the workplace

These standards can be assessed as competence via the NVQ route but at this level the majority of industries still favour the more traditional training approach to providing knowledge and skills. There are numerous programmes, which take this approach. Some are assessments of both knowledge and also have an element of skills assessment, whilst others programmes take a more traditional knowledge approach assuming that the skills can be gained in the actual workplace. This section highlights a few of these programmes but it is not exhaustive and normal training selection criteria should be applied when selecting a programme.

[T7017] The NEBOSH National General Certificate in Occupational Safety and Health is an examined-basis qualification designed to help non-specialists to discharge their duties or functions in workplace health and safety. Such people include managers, supervisors and employee representatives. This certificate covers much of the underpinning knowledge requirements for the Level 3 Standalone Units for Health and Safety.. It is normally taught over 80–100 hours by all modes of delivery including distance learning. NEBOSH makes it clear in its literature that being awarded a certificate does not imply competence but the underpinning knowledge towards competence. Although not specifically designed as an introduction to safety practice this qualification is often uses as a starter for people moving into the profession, who would then progress to the NEBOSH Diploma. It is accredited by IOSH as meeting the academic requirement for the Technician member of IOSH (Tech IOSH) membership category. (from November 2005)

[T7018] The Chartered Institute of Environmental Health ('CIEH') Advanced Health and Safety Certificate is also classified as a level 3 qualification on the QCA framework. However, it is a much shorter programme than the NEBOSH Certificate. It is at a similar level but does not have the full breadth of syllabus found in the NEBOSH Certificate.

[T7019] The Royal Society for the Promotion of Health ('RSPH') Advanced Diploma in Health and Safety in the Workplace is again a QCA level 3 course. It has 36 guided learning hours and assessments are carried out by coursework assignments. The title is a little misleading, as this is a basic health and safety qualification. The QCA framework does not specify the titles of programmes only the levels.

[T7020] The British Safety Council Level 3 Certificate in Occupational Safety and Health is a single unit qualification with two elements 'Principles of Health and Safety' and 'Applied Health and Safety' and is intended for those with some responsibility for health and safety in the workplace. It is assessed by 2x 3 hour examinations including a risk assessment exercise and is approximately 60 guided learning hours with self-study time required as well. It is accredited by IOSH as meeting the academic requirements for Tech IOSH membership. (from November 2005)

[T7021] The IOSH Managing Safety course has been specifically written for managers and focuses on two to three units of the ENTO Standalone Units forHealth and Safety standards. The course particularly focuses on the risk management process and how to use it in day-to-day workplace activities. This course does not currently appear on the QCF as it is run as a collaboration between IOSH and its training delivering members. The course is flexible and can be tailored to meet the specific requirements of organisations. It is awareness training and as such does not claim to be a qualification as it should be used as an integral part of a line manager or supervisor's training programmes.

[T7022] In addition to these qualifications run by health and safety bodies most of the large awarding bodies also offer health and safety programmes. Both OCR and City and Guilds have qualifications at QCF level 2 as well as offering the level 3 NVQ in Occupational Health and Safety.

IOSH also offer a Safety for Senior Executives briefing programme which is tailored to give executives, directors and the most senior managers in organisations the knowledge and information they require to run their organisations in a safe and healthy manner. This group is vital to the facilitation of good health and safety standards in the workplace, and they must show a commitment towards making sure standards are maintained and where possible improved on a continuing and regular basis. Training at this level needs to specifically recognise the requirements of this group of people, and routine training hammering out day-to-day hazards and specific details of the legislative requirements is not appropriate. More in depth consideration and exploration of the financial and moral aspects of health and safety is required, together with an understanding of the legal responsibilities and possible penalties for failure to ensure health and safety in the organisations for which they are responsible.

Most of the organisations also offer courses of a slightly shorter nature for those in supervisory roles. These are shortened versions of the manager programmes. It can be argued, however, that supervisors are the real front-line for safety and their training should be exactly the same as for managers but the reality is that financial constraints often mean that these shorter programmes are used.

The advent of the QCF will mean that the differing length of courses and content will be recognised by different nomenclature, based on the number of credits allocated to the course. This should clarify the differences between programmes that are offered.

Competence in health and safety for safety representatives

[T7023] Many people who perform the function of safety representatives in the workplace, as defined by the *Safety Representatives and Safety Committees Regulations 1977 (SI 1977 No 500)*, actually train in health and safety in a similar way to their colleagues undertaking similar courses to those described for managers and supervisors. There is, however a specific programme run by the TUC under the auspices of the National Open College Network (NOCN), which has been specifically designed to meet the requirements of this group. This programme, the TUC Certificate in Occupational Health and Safety, is a level 3 programme that builds on the TUC basic stage 1 and stage 2 health and safety courses or other union's equivalent courses. The assessment of the course is by a competence-based type of assessment similar to the NVQ qualifications with the addition of workplace-biased projects of practical value to the candidate's workplace and union.

The course is available for union representatives on either a part-time day release basis or by distance learning. The course additionally gives access to higher education for those who wish to progress further.

The TUC Education Service normally pays for the course, provided that the candidate is nominated by a trade union and is a valuable contribution towards competent safety advice in the workplace. This is recognised by IOSH who currently offer TechIOSH membership for those with this qualification and able to demonstrate at least two years pro-rata full-time equivalent experience.

Competent performance by the workforce

[T7024] Last but by no means least in the range of the competence chain that leads to better health and safety performance in the workplace is the workforce itself. The *HSWA 1974* makes it quite clear that employers have a legal duty to ensure, as far as is reasonably practicable, the health, safety and welfare at work of the people for whom they are responsible and the people who may be affected by the work that they do. The 1974 Act further requires employers to provide information, instruction training and supervision as is necessary to ensure, as far as reasonably practicable, the health and safety at work of employees.

It is the training aspect of these requirements, which has a huge variance across employers, on which this section focuses.

The ENTO recognised the fact there should be a standard available for everyone in the different suites of occupational standards, ie the Health and Safety for People at Work suite of standards. Within this suite there is a unit specifically targeted at everyone in a working environment, regardless of position or the number of hours worked. The unit A in the suite titled 'Make sure your own actions reduce risks to health and safety' is about making sure that risks to health and safety are not created or ignored.

The unit identifies what a person should be able to do to act in a competent manner towards health and safety issues in the workplace. Clearly this is competence which is developed whilst undertaking an employment role but it will be necessary to include training that delivers the knowledge that is required to perform to this level. This is normally carried out at an early stage in a person's employment, normally induction training on starting a job.

These standards are straightforward and form a good series of competence requirements for a workforce. The series of performance criteria for identifying hazards and evaluating risks includes such things as:

• correctly naming and locating the persons responsible for health and safety in the workplace;
• identifying workplace policies which are relevant to working practices;
• identifying working practices in any part of a job role which could harm yourself or other people;
• identifying those aspects of the workplace which could harm yourself or other people;
• evaluating which of the potentially harmful working practices and the potentially harmful aspects of the workplace are those with the highest risk to yourself or to others;
• reporting those hazards which present a high risk to the persons responsible for health and safety in the workplace;
• dealing with hazards with low risks in accordance with workplace policies and legal requirements

The attainment of this level of competence would be assessed for an NVQ across a wide range of workplace policies including methods of working, equipment, hazardous substances, smoking, eating, drinking and drugs and what to do in the event of an emergency.

These standards can be used to plan health and safety training and they can also be used to analyse where training is actually needed by comparing existing skills and competence of the workforce with the standards and then identifying if there are any gaps, which can be covered by training. This can be done on an individual or group basis and the training offered as a result of this analysis can be offered in a specifically targeted way.

[T7025] There are a variety of training programmes at this primary level, which develop the knowledge requirements for health and safety in the workplace. Many are quality controlled by either awarding bodies or bodies with a health and safety remit. Some of the well-known programmes are IOSH Working Safely, CIEH Basic Health and Safety Certificate, RSPH Foundation Certificate in Health and Safety, the Workplace and awards from City and Guilds, OCR and Edexcel. Many organisations seek this external verification to show that they have carried out their training requirements with regard to their duties under the *HSWA 1974*. However, there are many perfectly good induction programmes run by organisations themselves which allow this knowledge part of the competence requirements to be achieved. It is the fact that good training is available that is of vital importance, and the national standards of competence give an indication of what needs to be achieved from any training programme.

Competence for contractors and safety passports

[T7026] Leading on from the requirement of the development of a workforce that is competent to perform its role in a healthy and safe manner, is the requirement to look at all those people who also may be present in a workplace.

Although organisations have the ability to control the training and competence of individuals that they have directly employed, by developing programmes which meet their specific requirements, there is one area that they need to control but do not necessarily have complete control over – that is outside contractors coming into their workplaces. This is particularly relevant in some industrial sectors such as construction and related industries where the use of contracted labour is extensive. However, at some point all employing organisations may have people other than their own employees within their workplaces. Many organisations control this situation by having 'permit to work' systems that ensure that anyone entering a workplace has sufficient knowledge of the health and safety arrangements that are in place.

However within the last few years there has been a move to a more transferable skills type of scheme that can allow those who do move between workplaces to show that they have sufficient underpinning knowledge of health and safety matters to allow briefer induction programmes to be given when moving between different workplaces and sites. These are the safety passport schemes.

These passports available from several industry-related schemes, allow employers to determine who may be given access to their workplaces. The employing organisation makes their workplace a passport-controlled environment. The passports themselves generally take the form of a credit-card sized plastic card, containing the name, photograph and other relevant details of an individual, they also verify what training a person has undertaken towards receiving this passport. The passport usually also has an expiry date.

The advantages of passport schemes are:

- they can save time and money through reduced induction training as they are the starting point for workers training

- they demonstrate an organisation's commitment to a safe and health working environment;
- they can create good relationships with the supply chain between organisations and their suppliers;
- they can be used by employees of companies as well as sub-contractors;
- contractors can move between companies and demonstrate that they have the necessary health and safety awareness;
- they can be used to control access to worksites;
- probably most importantly, they can increase a positive safety culture within organisations, which can lead to a reduction in accidents or ill health;

However, it does need to be borne in mind that a passport cannot identify that a worker is competent or be an effective substitute for workplace management including risk assessment and site-specific information.

[T7027] Many large organisations are now taking up this approach to as part of their safety management systems, this in turn will lead to smaller organisations, who are often suppliers to or sub-contractors for the larger organisations to also adopt the schemes as a business decision.

The largest of the schemes currently available is the Construction Skills Certification Scheme ('CSCS') which to date has issues over a quarter of a million passports. This organisation has a management board that has members from the Construction Confederation; Federation of Master Builders; GMB Trade Union; National Specialist Contractors Council; Transport and General Workers Union and the Union of Construction Allied Trades and Technicians. It also includes observer members from relevant government departments and the Confederation of Construction Clients.

The Construction Industry Training Board (CITB) manages the operation of the CSCS and delivers training through its series of regional offices. To obtain a passport the candidates have to sit a 40-question multiple-choice test through the same centres as used for driving tests, although this is expected to be expanded to other centres using an on-line technique in the near future.

The Client/Contractor National Safety Group ('CCNSG') is one of the oldest schemes and was developed for contractor site personnel. Representatives from major clients are nominated by local safety groups to serve on the CCNSG and trade unions; contractors and safety groups also have places. The training is a basic two-day training programme. The Engineering Construction Industry Training Board (ECITB) manages the scheme. Around 160,000 people currently hold this passport.

Other passport initiatives are: Airport Construction Training Alliance (ACTA); Gas and Water Industry; Register of Energy Sector Engineers, Technologist and Support Staff (RESETS); Engineering Services SKILL Card; Sentinel for Railway Track Workers; Federation of Bakers' Contractors Passport (FBCP);IOSH in partnership with other organisations; National Microelectronics Institute (NMI); Metal Industry Skills and Performance Institute (MetSkill); Safety Certification for Contractors (SCC), Safety Pass Alliance and the Port Skills and Safety Limited (PSSL).

In addition to this list there are some other initiatives within specific industries. Clearly some rationalisation of the schemes would be useful to improve the currency of the passports and this may well be a factor in this market over the next few years. To accommodate such a rationalisation approach the HSE have produced a leaflet INDG381 *Passport Schemes—a good practice guide*. This leaflet covers:

- What are health, safety and environment passports?
- How do passport schemes work?
- Benefits and advantages of passport schemes?
- How passports help employers and workers stay within the law?
- How a worker can get a passport?
- Withdrawal arrangements for training, refresher training, and monitoring

Helpfully the HSE have also produced a core syllabus containing all the items that the schemes should include within their training and assessment programmes.

Maintaining competence

[T7028] As we have seen competence is the development of sufficient knowledge, skills and experience to be able to perform a task to the laid down standards in a workplace. These standards may be prescribed by a legislative requirement, national standards or by best practice in the workplace. However competence is prescribed or developed – there is a cycle that needs to be recognised in the attainment and then in the maintenance of the desired level.

Competence is actually a developed attribute and passes through several stages, these can generally be recognised as distinct stages:

- *unconscious incompetence*—a person is unaware of the requirements to perform in a competent manner;
- *conscious incompetence*—a person begins to undertake training and becomes aware of what they do not know;
- *conscious competence*—a person has undergone sufficient training to be able to complete a task in a competent manner and is aware of this;
- *unconscious competence*—a person now performs tasks in a fully competent manner and has now developed to become unaware of this as it is part of their behavioural patterns.
  At this point in the cycle there can be two pathways, a person can continue to perform in a competent manner or they can unconsciously develop bad habits, which can lead to a diminishment in their performance.

The final bullet point indicates that there is a need for those responsible for the workforce to revisit both their own current knowledge and that of their employees on a regular basis to ensure that performance relating to health and safety is in fact to the level they believe it to be.

There can be a number of triggers that can identify that there may be a need for refresher or further training. This is not necessarily simply because of a time lapse since the initial training took place – other issues such as the introduction of new equipment, changes to methods of use, the re-site of equipment, changes of shift patterns etc. should be considered. Also safety audits may identify that there have been behavioural changes that need addressing.

Many of the external agencies, which play a role in the quality control of training programmes, will specify an expiry date for their accreditation. The safety passport schemes have particular expiry dates usually 3–5 years built into the actual passport, whilst organisations such as IOSH have a three-year expiry term on their Managing and Working series of programmes.

For professionals who practice in health and safety there is a particular need for the maintenance and development of their skills. The area of health and safety is constantly changing both in terms of legislative requirements and in the knowledge of best practice in health and safety issues. All members of IOSH other than Affiliates and Retired Members are now required to maintain CPD records via the on-line system, MyIOSH, that is now available. Affiliate Members are also recommended to undertake CPD as many of them are in a developmental stage of their careers The benchmark for the amount of CPD is 5 days per year but as CPD is very subjective and based on reflective practice this will vary depending on the stage a person is at in their career, in the Chartered categories of IOSH are currently required to. Employers looking to recruit either an employee or consultant are advised that they should ask to see evidence of CPD as an indication of continuing competence. This is an extension of the concept of life-long learning which is now being incorporated into the initial training programmes for those in occupational health and safety practice and in fact features within the national vocational standards. The IOSH CPD on-line allows members to maintain their CPD as efficiently as possible without mounds of paperwork. Also as part of this revision a practitioner also takes overall ownership of their own CPD requirements and allows them to be more flexible in the type of activities that they can use to demonstrate CPD. Although there are a number of well designed CPD taught courses available these form only part of the requirement for CPD and practitioners are being encouraged to think beyond attendance at events.

To sum up it is critical for good performance in health and safety in the workplace, that the initially developed knowledge and skills are constantly maintained. This can only be achieved by regular checking of the competence levels of the entire workforce in organisations. Some industry sectors have formalised this approach within their safety management systems and here competence frameworks are developed and monitored on a formal basis. This is good practice and should be considered by those who have not yet taken this approach.

Useful contacts

[T7029] The following is a list of useful contacts for organisations that have been mentioned in this chapter:

Airport Construction Training Alliance
www.acta.easitrack.com

British Safety Council
70 Chancellors Road
London W6 9RS
Tel: 020 8741 1231
Fax: 020 8741 4555
www.britsafe.org

Chartered Institute of Environmental Health (CIEH)
Chadwick Court
15 Hatfields
London SE1 8DJ
Tel: 020 7928 6006
Fax: 020 7828 5866
www.cieh.org

City and Guilds
1 Giltspur Street
London EC1A 9DD
Tel: 020 7294 2468
Fax: 020 7294 2400
www.city-and-guilds.co.uk

**Client Contractor National Safety Group (CCNSG)/
Engineering Construction Industry Training Board**
Blue Court
Church Lane
Kings Langley
Hertfordshire WD4 8JP
Tel: 01502 712 329
www.ecitb.org.ik/safetypassport/safetypassport.html

Construction Skills Certification Scheme (CSCS)
Tel: 01485 578 777
www.cscs.uk.com

Council for the Curriculum Examinations and Assessment (Northern Ireland)
www.ccea.org.uk

Department for Children, Education, Lifelong Learning and Skills (Wales)
www.cqfw.net

Employment National Training Organisation (ENTO)
Kimberley House
47 Vaughan Way
Leicester LE1 4SG
Tel: 0116 251 7979
Fax: 0116 251 1464
www.empnto.co.uk

Engineering Services Skill Card
Old Mansion House
Eamont Bridge
Penrith
Cumbria CA10 2BX
Tel: 01768 860 406
Fax: 01768 860 401
www.skillcard.org.uk

Federation of Bakers (FBCP)
6 Catherine Street
London
WC2B 5JW
Tel: 020 7420 7190
www.bakersfederation.org.uk

Gas and Water Industries National Training Organisation (GWINTO)
The Business Centre
Edward Street
Redditch
Worcestershire B97 6HA
Tel: 01527 584 848
www.gwinto.co.uk

Health and Safety Commission (HSC) (Contact HSE)
Health and Safety Executive (HSE)
Rose Court
2 Southwark Bridge
London SE1 9HS
Tel: 020 7717 6000
Fax: 020 7717 6996
www.hse.gov.uk

Institution of Occupational Safety and Health (IOSH)
The Grange
Highfield Drive
Wigston
Leicester
LE18 1NN
Tel: 0116 257 3100
Fax: 0116 257 3101
www.iosh.co.uk

Metal Industries Skills and Performance (MetSkill)
5–6 Meadowcroft
Amos Road
Sheffield
S9 1BX
www.sinto.co.uk

National Examination Board in Occupational Safety and Health (NEBOSH)
5 Dominus Way
Meridian Business Park
Leicester LE19 1QW

Tel: 0116 263 4700
Fax: 0116 282 4000
www.nebosh.org.uk

National Microelectronics Institute (NMI)
1 Michaelson Square
Kirkton Campus
Livingston
EH54 7DP
Tel: 0131 449 8507
www.nmi.org.uk

OCR Examinations
Westwood Way
Coventry
CV4 8JQ
Tel: 0247 647 0033
Fax: 0247 642 1944
www.meg.org.uk

Port Skills and Safety Ltd
Africa House
64–78 Kingsway
London WC2B 6AH
Tel: 020 7242 3538
www.bpit.co.uk

Proskills
Centurion Court
85B Milton Park
Abingdon
Oxfordshire OX14 4RY
Tel: 01235 833844
info@proskills.co.uk
www.proskills.co.uk

Qualification Assurance Agency for Higher Education (QAA)
Southgate House
Southgate Street
Gloucester GL1 1UB
Tel: 01452557000
Fax: 01452 557 070
www.qca.ac.uk

Qualification and Curriculum Development Agency
83 Piccadilly
London W1J 8QA
Tel: 020 7509 5555
Fax: 020 7509 6666
www.qcda.gov.uk

Royal Society for the Promotion of Health
38A St George's Drive
London SW1V 4BH

Tel: 020 7630 0121
Fax: 020 7976 6847
www.rsph.org

Register of Energy Sector Engineers, Technologists and Support Staff (RESETS)
Operations Enterprise House
Cherry Orchard Lane
Salisbury SP2 7LD
Tel: 01722 427 226
Fax: 01722 414 165
www.resets.org

Safety Pass Alliance
P.O. Box 7772
Braunstone
Leicester LE3 3XY
Tel: 0116 2221228
Fax: 0116 2220329
www.safetypassalliance.co.uk

Scottish Curriculum Authority (SQA)
The Optima Building
58 Robertson St
Glasgow G2 8DQ
Tel: 0845 279 1000
Fax: 0845 213 5000
www.sqa.org.uk

Trade Union Congress (TUC)
Congress House
Great Russell Street
London WC1B 3LS
Tel: 020 7636 4030
www.tuc.org.uk

Ventilation

Chris Hartley

Introduction

[V3001] Ventilation involves the movement of air with the supply and/or removal of air by natural or mechanical means to or from an enclosed space or workplace. In protecting the health and safety of people at work, ventilation provides sufficient fresh air to workplaces and provides a means of controlling exposure to hazardous airborne contaminants arising from work processes.

Ventilation provides fresh air for:

— human respiration, with the provision of oxygen and the dilution and removal of carbon dioxide;
— the dilution of unpleasant odours;
— the removal of excess heat in the work environment and the maintenance of thermal comfort; and
— the control of harmful airborne contamination arising from work activities.

Industrial processes and activities generate a very wide range of airborne contaminants, ie dusts, fumes and vapours, many of which represent an inhalation health hazard to the workers exposed to them. Management of the risks arising from such exposures is required by health and safety legislation. In general, this requires identification of hazards and assessment of risks, followed by action to prevent, so far as is reasonably practicable, or adequately control those risks. In workplace contaminant control, ventilation may be divided into two categories: local exhaust ventilation (LEV) and dilution or general ventilation. These ventilation approaches represent options in the wider portfolio of control measures that may be chosen in a particular risk situation. A hierarchy of control options is available where prevention is not reasonably practicable. These measures include amongst other things:

— Substitution of a substance for a less toxic one;
— Substitution of a process for a safer method;
— Changing a process (for example, reducing operating temperatures may reduce the emission of fume; remote handling of substances);
— Segregation – total enclosure of the process;
— LEV to control at source;
— Dilution ventilation to reduce the concentration of a contaminant;
— Personal protective equipment (PPE) including respiratory protection.

By controlling airborne contaminant exposure at source, LEV represents a powerful tool to control the work environment. Its application represents the major focus of this chapter. In an LEV system, the collection point (enclosure, hood or slot) is positioned as close to the source of the contaminant as possible. Air is drawn into the system ductwork by means of a suitable fan, passes through an air cleaning device and is finally expelled at a point outside the workplace.

With dilution or general ventilation, fresh air is induced to enter the workplace air and this causes dilution of the airborne contaminants present. This air flow may be induced either by natural means (for example, by opening windows or roof vents) or by mechanical means (for example, by electrical fans in walls or roofs). With dilution ventilation there is no special control at the source of emission.

Legal requirements relating to the provision and use of ventilation systems are also considered where appropriate.

Under the *Health and Safety at Work etc Act 1974* employers have a duty to protect employees and others against risks to their health and safety, so far as is reasonably practicable. This includes the risks arising from the use, handling, storage and transportation of substances together with the provision and maintenance of safe plant and systems of work, which includes ventilation systems.

Under the *Management of Health and Safety at Work Regulations 1999 (SI 1999 No 3242)*, the employer has a duty to carry out a risk assessment where a hazard is identified and to take steps to eliminate the hazard or adequately control exposure.

In workplaces where people are exposed to airborne contaminant hazards, statutory regulations require hazard identification, risk assessment and prevention or, where this is not reasonably practicable, control measures to achieve adequate control of exposure. This requirement is found in the *Control of Substances Hazardous to Health Regulations (as amended) 2002 (SI 2002 No 2677)* (COSHH), the *Control of Asbestos Regulations 2006 (SI 2006 No 2739)* and the *Control of Lead at Work Regulations (as amended) 2002 (SI 2002 No 2676)* with respect to 'substances hazardous to health', asbestos and lead hazards respectively.

LEV may also be applied as a control measure where there is exposure to ionising radiations and dangerous substances and this is reflected in the *Ionising Radiations Regulations (as amended) 1999 (SI 1999 No 3232)* and the *Dangerous Substances and Explosive Atmospheres Regulations 2002 (SI 2002 No 2776)*. These particular applications are not discussed further in this chapter.

Where ventilation is used as an air contaminant control measure, the various regulations require competent maintenance and, for LEV, statutory examination and testing at specified intervals. This is to ensure that ventilation systems continue to work effectively. In addition, the *Workplace (Health, Safety and Welfare) Regulations 1992 (SI 1992 No 3004)* require that workplaces ar esupplied with sufficient quantities of fresh air. These legal aspects are discussed further below.

This chapter will focus on the principles of operation, design features and the application of LEV systems as well as the requirements for maintenance, examination and testing. The statutory requirements for the provision and use of these systems will also be considered. The Health and Safety Executive (HSE) has recently published revised guidance on LEV (*'Controlling Airborne Contaminants at Work – A Guide to Local Exhaust Ventilation'* (2011), HSG 258, HSE Books) replacing previous HSE guidance: HSG37 *An introduction to local exhaust ventilation and HSG54 Maintenance, examination and testing of local exhaust ventilation*. Other aspects of 'ventilation', including dilution ventilation and the provision of fresh air to workplaces, are also discussed.

Sources of contamination

[V3002] Many industrial operations and tasks, including machining metals, welding, paint spraying, charging reactor vessels and heating processes, create sources of airborne contamination. By their physical state, airborne contaminants can be broadly divided into aerosols and gases which in turn can be classified into various groups:

An aerosol may be defined as any disperse system of liquid or solid particles suspended in a gas (usually air). These include the following:

— Dusts – solid particles made airborne by the mechanical disintegration of bulk solid material;
— Spray – large liquid droplets (of a few microns upwards) produced during condensation or atomisation;
— Mist – fine liquid droplets (up to a few microns) produced during condensation or atomisation – a micron (1 µm)is one millionth of a meter;
— Fume – small solid particles (usually the aggregates of much smaller primary particles) produced by condensation of vapours or gaseous combustion products;
— Smoke – solid or liquid particles resulting from incomplete combustion, aggregates of very small particles;
— Bioaerosol – solid or liquid particles containing biologically viable organisms (viruses, bacteria, fungal spores), ranging from submicron (ie less than a micron) to over 100µm.

Aerosol particles above 100µm do not remain airborne and therefore are unlikely to be an inhalation hazard. Particles in the range 3–20µm are known as the 'thoracic fraction' and if inhaled they may be deposited beyond the larynx in the respiratory tract. Particles in the range 0.5–7µm are very fine particles, invisible under normal lighting conditions, but are hazardous in that they can avoid respiratory defences and penetrate to the deepest parts of the lungs. (Particles in this size range are known as the 'respirable fraction'). When particles reach the deep lung they may then be absorbed into the systemic blood circulation and cause toxic harm by accumulating in target organs. For example, exposure to cadmium oxide fume can cause irreversible kidney damage. Alternatively, they may be deposited as insoluble particles in the deep lung and provoke serious tissue reactions, for example, crystalline silica particles and silicosis.

These fine particles move with the air in which they are suspended and only sediment out slowly. It is important to emphasise that it is the aerodynamic diameter of aerosols which determines how they move in the atmosphere and where they will deposit in the respiratory system.

Gases include substances which are normally in the gaseous state at room temperature and 'vapours', the gaseous form of a substance which is normally a liquid at room temperature, for example toluene and benzene vapours. Toxic vapour/air mixtures have a density virtually identical to that of air and disperse with movement of the air in which it is located.

That harmful aerosols and gases disperse with the movement of the air in which they are located allows the beneficial application of control at source by LEV, where the system collection point is positioned close to the source of emission allowing the contaminated air to be drawn into the system and removed.

Local exhaust ventilation (LEV)

[V3003] The objective of an LEV system is to remove contaminant efficiently using the minimum volume of exhaust air possible. Unfortunately, there are many examples of poor design of these systems which do not achieve this objective. There is a wide variation in the size and complexity of LEV systems and the appropriate system will be determined by the particular risk situation and other factors such as building constraints. For example, a small hood can be used to extract fume from a manual soldering iron while a large walk-in booth may be used for paint-spraying large components. With an LEV system there are important energy cost implications since air expelled from the workplace has normally to be replaced with warmed clean air. Therefore, it is important that the system is well designed to achieve satisfactory control with a minimum volume of exhaust air.

The basic components of an LEV system are:

—   A collection point, ie enclosure, hood or slot to collect and remove contaminant close to its source;
—   Ducting to transport the contaminant and exhaust air to the air cleaner and ultimately to the discharge point;
—   An air-cleaning device to remove the contaminant from the airstream (this is normally placed between the hood and the fan to protect the integrity of the fan from corrosive or hot airborne contaminants);
—   A fan which provides the power to move a sufficient volume of air through the system;
—   A discharge point to ensure efficient removal of contaminated air from the system.

It is convenient to divide LEV contaminant collection points into partial enclosures, hoods and slots. The collection point is a key part of the system and requires careful design. It is crucial that it is positioned as close as possible to the source of emission. The type of system collection point that is chosen in a specific situation will depend on the particular industrial process and requirements for access for materials and people to carry out tasks. Where the work process allows this, it is desirable to enclose the contaminant emitting process as far as is possible in a box-like structure. However, it may not be possible to enclose a particular process at all and in this case hoods and slots may be used as the inlet point for the local removal of contaminant.

A simple LEV system is shown in Figure 1 and the next section will consider the various components.

Please note: HSG54 has been replaced with HSG258 (see attached). See diagram in Figure 1 on p6 of this document.

Partial enclosures

[V3004] A partial enclosure is a box-like structure which surrounds a process which is emitting airborne contaminant. This type of collection point is used where operator access or the entry/egress requirements for process materials only permit a partial enclosure rather than a total enclosure of the process. The amount of exhaust air needed for a partial enclosure depends on the size of the openings required in the enclosure and the velocity of the air needed to overcome the tendency of the contaminated air inside to escape. Turbulent air conditions both inside and outside the enclosure can influence the likelihood of contaminant escape.

Face velocities ranging from 0.5–2.0 ms^1 are used and the total flow volume can be calculated from the equation Q = VA. (A is open face area in metres (m); V is air velocity in metres per second (ms^1); Q is total volume flow rate in cubic metres per second (m^{3}s^1). The volume flow rate equals the area of the opening multiplied by the face velocity necessary to prevent contaminant escaping. The face velocity required depends on the toxicity of a contaminant and its momentum on release. For example, a higher face velocity would be needed in paint spraying than in a paint dipping operation. With toxic substances, a face velocity in excess of 1.5 ms^1 will be required. Both the size of the enclosure and the size of the openings should be kept to a practicable minimum. A three-sided booth, such as a laboratory fume cupboard or paint spray booth, is a common type of partial enclosure.

In a fume cupboard, there is a sliding front panel which the operator can adjust. Modern fume cupboards are fitted with a by-pass mechanism or variable fan arrangement which allows the same face velocity to be achieved for all positions of the panel. The contaminant source is located within the fume cupboard and the sliding panel not only provides access for the operator but also provides an opportunity for the contaminant to escape. Therefore, in order to prevent escape, sufficient air velocity is required across the face of the cupboard to induce the contaminant to enter the system ducting. A fume cupboard should be aerodynamically designed to allow a smooth pattern of inward air flow and it should be carefully sited in the workplace to avoid external local turbulent conditions (for example, busy passageways). The minimum face velocity will depend on the nature of the contaminant, the operator's required movement and local air conditions near the partial enclosure.

In large ventilated enclosures (as used in paint spraying, for example), the operator may carry out work entirely inside the booth. Paint spray aerosol particles rebound from sprayed surfaces at high speed and a sufficient air flow rate is required to overcome this in order to prevent escape and to induce these particles to enter the system ducting. In this case, a face velocity of 0.7 ms^1 is recommended with an even and stable face velocity across the whole booth. Often a perforated plate is placed at the back of the booth to ensure an even laminar air flow inside the booth. It is important that the operator is not placed between the contaminant source and the booth exhaust point. Where possible the worker should stand at the side of the work. In some situations a rotating turntable for the work piece is used to prevent this occurring.

An average air velocity of 0.5 ms^1 is normally required for a partial enclosure. Some processes are more toxic and emit particles at high velocity, in which case a higher face velocity of at least 1.5 ms^1 may be required.

Hoods

[V3005] Often it is not possible to enclose a process because of operator and/or process material access needs. In such cases, a hood or slot collection point may be required, but these are not as effective as enclosures.

Hoods vary from small apertures to extensive canopies above, below or at the side of the emission source. They should not be placed above if operators have to lean over and put themselves between the contaminant source and the ducting exhaust point. Hoods should be located as close as practicable to the contaminant source. In a large hood, the face velocity distribution may vary across the hood (ie higher close to the duct entrance and lower at the extremities). In some work processes, contaminant is generated with considerable energy and a receptor hood may be placed in the path of the contaminant to collect and remove it. For example, a receptor hood may be placed above a source that is emitting a hot contaminant, which then rises and enters the hood of its own volition.

On the other hand, a captor hood pulls contaminant into the system collection point and here the air flow needs to be sufficient to overcome any forces for the contaminant to disperse. (Capture velocity is the air velocity required at the source of contaminant emission which is sufficient to move contaminant to the mouth of the extract to be successfully captured by the system.) Small aerosol particles and gases and vapours will move with the air flow although larger particles will be more difficult to divert from their natural pathway. However, captor systems can be designed to intercept such particles. A minimum capture velocity of 0.25–1.0msec^1 is required for low velocity particle release with up to 2.5–10msec^1 needed to capture high speed particles. For example, a capture velocity of 0.25msec^1 would be needed to capture the gentle evaporation of solvent from a degreasing tank whilst a capture velocity of around 10msec^1 would be required to capture high velocity particles from a grinding process.

Other factors that are relevant to hood design are the toxicity of the contaminants, the quantities emitted and the required size of the hood. The objective, as stated earlier, is to achieve sufficient capture velocity utilising the minimum volume of exhaust air. Hood size is based on the size of the source or what is practicable and convenient. Applying flanges (ie adding raised edges) to the entrance, of a hood reduces the amount of air drawn in from behind and improves air entry conditions, giving an even velocity distribution across the hood and a lower entry pressure loss. Captor systems should not be sited in areas of local air turbulence – for example, next to busy passageways. Captor hoods require a greater total volume flow than enclosures for equivalent effectiveness and every effort should be made to convert captor hoods into partial enclosures with the use of flanges and general boxing in.

Low-volume high-velocity (LVHV) systems employ very small hoods, with capture velocities from 50–100msec[1], placed very close to the source of emission. They are used for high speed particles (for example, those generated by grinding or sanding) or for gases and vapours generated during welding and soldering. For successful application, such hoods need to be integrated into the design of the tool or they may be cumbersome and difficult to use.

Slots

[V3006] Where the aspect ratio (length:breadth) of a capture hood is less than 0.2, it is called a captor slot. Hoods may be replaced by slots where there are limitations on space and where there are access or process reasons. For example, slots are used on degreasing and electroplating tanks to remove vapours produced at the surface of these tanks. Where there is a wide surface area of potential contaminant release, two slots are often employed to minimise capture distance. According to FS Gill (*Ventilation* in '*Occupational Hygiene*' eds Gardiner, K and Harrington, JM, Blackwell, 2005), it is difficult to pull air into a slot from more than 750mm.

Ducting

[V3007] Enclosures, hoods and slots are connected to air cleaners, fans and the discharge point by ductwork. This consists of straight pieces, bends and changes in cross-section, dampers and other fittings which connect the inlet with the discharge point. The volume flow rate of the system is determined by the air velocity required at the system inlet and its cross-sectional area. The size of the ducting will be determined by the air velocity required or the building space available.

Energy losses due to friction are expressed as a pressure loss and this is proportional to velocity squared (v^2). The larger the cross-section of the duct the lower will be the air velocity and the pressure loss. Conversely, the smaller the cross-section of the duct the higher will be the air velocity, pressure loss and noise levels.

The pressure loss for each section of ductwork, including any dust collection device, is calculated and the total pressure loss of the system is determined. The duty of the system fan is calculated based on the calculated volume flow rate at the total pressure loss for the system. If particles of powder or dusts are to be transported by an LEV system then a sufficient transport velocity needs to be maintained in the system to keep these particles suspended in the airstream and prevent them from settling out in the ducting.

If only gases and vapours are transported by an LEV system, transport velocity is less important. The values for suitable transport velocities vary from 5–10 ms[1]for gases, vapours, smoke and fumes to up to 25 ms[1] for heavy dusts. If a particular transport velocity is required, once the desired volume flow rate has been determined for the system, the duct diameter can be fixed at a size which will allow that transport velocity to be achieved. Ducting is normally made out of galvanised sheet steel unless corrosive gases or vapours are to be transported, in which case stainless steel or plastics are used. The design of the system should avoid abrupt changes in direction and changes in section should be smooth to minimise energy losses. There needs to be adequate access for internal cleaning of ductwork and inspection ports for monitoring equipment.

Fans

[V3008] These provide the motive power for moving air through the system.

A fan needs to be matched with the volume flow rate required and the total pressure drop of the system. With fans of similar design, relative performance depends on their diameter and speed of rotation. A fan has two fundamental components: the impeller which rotates on a shaft and the casing which guides air to and from the impeller.

Two basic types of fan are used, axial and centrifugal. Axial fans are installed in-line in the contaminant airstream. They have a cylindrical casing, are compact and fit readily into ducting. Axial fans can overcome only low resistance to flow and they are generally used in wall and roof-mounted units. The fan motor is positioned in the airstream and therefore should not be used with high temperature or corrosive contaminant exhausts.

With centrifugal fans, air is drawn into the centre of the impeller and thrown off at high velocity into the fan casing and discharge. Centrifugal fans can deliver the required airflow against high system resistance and are the most common type of fan found in LEV systems. Centrifugal fans can be used with abrasive dusts and corrosive chemicals and, although the blades wear out quickly, they can be readily replaced. Fans should be sited outside the building or as near to the discharge as possible since when a leak occurs in front of the fan the ductwork will be under negative pressure and contaminant is less likely to escape.

Air cleaning devices

[V3009] Before exhausted air is discharged to atmosphere or re-circulated it needs to pass through an appropriate air cleaning device. Usually such devices are located in front of the fan to give the fan some protection from damage by contaminants. A range of devices are available depending on the contaminant in question. Dry dusts may be removed from the airstream by air filters; contaminants with larger particles may be removed by cyclones; and those with smaller particles by bag filters. Dust particles that will easily take an electrical charge can be removed by electrostatic precipitators as long as there is no fire and explosion hazard. Wetted particles can be removed by a venturi scrubber method. Organic vapours can be removed by activated charcoal or a filter with a suitable adsorbent.

Discharge

[V3010] The behaviour of an airstream close to the entry and the exit of the LEV system is very different. At the suction end the influence of the system will be very localised whilst at the discharge end the system will have a much greater influence and cause considerably more disturbance in the air. Therefore, it is important that the discharge point is located away from any air inlets so as to avoid re-circulation of exhaust air. Discharge stacks should be positioned so as to ensure efficient mixing with the atmosphere and this may mean extending them to at least 3m above roof level.

Further information on the fundamentals of local exhaust ventilation systems can be found in the HSE publication *Controlling airborne contaminants at work: A guide to local exhaust ventilation* (LEV) HSG258

The LEV design process

[V3011] The basic components of local exhaust ventilation systems have been outlined above. Unfortunately, there are many examples of inadequate systems – for example: poorly designed hoods which are placed in the wrong position based upon misunderstandings about how aerosols and vapour/air mixtures behave in air; captor hoods placed too far from the emission source; inadequate total volume flow; and lack of provision for ease of maintenance. Designing LEV systems that will efficiently and reliably collect airborne contaminants on a continuing basis is a difficult and complex task. The following is adapted from a sequence suggested by Piney et al ('*British Occupational Hygiene Society Technical Guide 7: Controlling airborne contaminants in the workplace*', 1987).

(1)   From knowledge of the health effects of the contaminant(s) decide on tolerable exposure level. Consider Workplace Exposure Limits and application of the principles of good hygiene practice; consider reducing exposure as low as is reasonably practicable for carcinogens and asthmagens.
(2)   Identify all sources of emission sources and rank in order of magnitude.
(3)   Examine process and operator work methods.
(4)   Perform occupational hygiene measurements (static and personal monitoring) and re-rank emission sources in terms of contribution to operator personal exposure.
(5)   Consider control by methods which do not involve ventilation systems; consider how the number of sources and/or emission rates can be reduced.

Control by local exhaust ventilation

(6)   Decide on the shape and size of the hood, slot or enclosure. (Consider possible changes to the work process itself.)
(7)   Decide on appropriate capture velocities and distances for captor hoods and face velocities for receptor hoods.
(8)   Calculate the required volume flow-rate for the system.

(9)   Design ductwork to transport exhaust air in the most energy efficient manner (to minimise losses due to friction); consider required duct transport velocity where this is relevant.

(10)   Determine what air cleaning device will be required.

(11)   Consider how replacement air will be provided to the work area.

(12)   Choose an appropriate fan taking into account to the total volume flow rate required and the total pressure drop for the system.

(13)   Plan for commissioning and maintenance, examination and testing of the LEV system to comply with the *Control of Substances Hazardous to Health Regulations (as amended) 2002 (SI 2002 No 2677)*.

Maintenance, examination and testing of LEV systems

[V3012] In addition to well designed local exhaust ventilation there is a requirement for a high and continuing level of performance. Accordingly, LEV systems, as with any workplace plant and equipment, must be appropriately maintained with specified tasks carried out at defined frequencies by designated personnel. However, 'maintenance' is more than the tasks carried out by maintenance workers and this should include visual checks, inspection, servicing, reviewing systems of work and any remedial work to ensure system effectiveness. A thorough examination normally includes a visual check, quantitative measurement of performance and an assessment of control.

To evaluate performance, amongst other things, the following tasks need to be carried out: the measurement of air velocity, volume flow rate and pressure at various points in the system together with observation of the characteristics of local air movement. There is a need to answer the following questions:

(1)   Are airborne contaminants successfully captured near to their point of release?

(2)   How do the current measured parameters differ from the design specification values?

(3)   What are air velocities at different points in the system?

(4)   What pressure or suction is the fan developing?

(5)   How much pressure (energy) is lost in the different parts of the system? (This may be particularly important at filters or dust collectors).

(6)   Is the exposure of operators under adequate control?

Ventilation system pressures

[V3013] Monitoring ventilation system pressure and its change is an important measure of a system's performance. A pressure difference is required for air to flow in a ventilation system and air will flow from a region of high pressure to one of low pressure. Such pressure differences can be induced naturally or artificially, for example, by a chimney and a fan. Pressure appears in two forms: static pressure and velocity pressure. Static pressure is the pressure exerted in all directions by a stationary fluid (gas or liquid). In a moving fluid, static pressure is measured at right angles to the direction of flow to eliminate the effects of velocity. Static pressure can be negative or positive with respect to normal atmospheric pressure. Velocity pressure is a measure of the kinetic energy of a fluid in motion. It provides the driving force of 'wind'. Total pressure is the sum of velocity pressure and static pressure.

Comparing static pressure measurements at different points with those normally found in an LEV system can be used for checking performance and fault finding. The duct pressures on both sides of a fan are illustrated in Figure 1. A U-tube is used to measure static pressure (p_s), velocity pressure (p_v) and total pressure (p_t).

Figure 2: Duct pressures

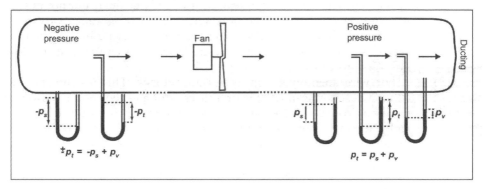

At the suction side of a fan static pressure is negative and on the discharge side it is positive. Velocity pressure gives the push to move air through the system and this is always positive.

Making some assumptions, for air, this can be represented by the simple equation:

$$Pv = 0.6v^2$$

(v is expressed in metres per second (ms^{-1}), velocity pressure (Pv) is in newtons per square metre (Nm^2) or pascals (Pa)).

Total volume flow

When a quantity of air is moving within a system or duct, the volume flow can be calculated by the following formula:

$$Q = vA$$

(v is the average air velocity across the cross-section of the duct (or face velocity) in ms1; A is the cross-sectional area of the ducting or face in square metres (m2); and Q is in cubic metres per second (m3s1)).

Monitoring equipment

Monitoring equipment used in evaluating the performance of a ventilation system includes the following:

(1)   A pitot-static tube measures system pressures (and air velocity indirectly). It measures velocity pressure inside a ventilation system irrespective of the static pressure at the point of measurement. Velocity

pressure can be converted into air velocity using the equation above. Measurements should be carried out in a straight section of ducting with streamlined air flow and minimal turbulence. This instrument is sensitive to duct air velocities above 3ms^1. The instrument should only be used in ducting and it is not suitable for use at the face of hoods or booths. It can also be used to measure static pressure.

(2)   A rotating vane anemometer measures air velocity. The rotating vane is connected electrically or mechanically to a display dial. The axis of the rotating vane should be positioned parallel to the airstream contours to avoid errors. This instrument can be used in large ducts but not in very narrow channels or extract slots since the rotating vanes vary from 25mm to 100mm in diameter. This instrument should not be used for very dusty or corrosive atmospheres. However, there are intrinsically safe versions of this instrument.

(3)   A hot wire anemometer measures air velocity. This instrument uses either a heated thermocouple or a thermistor and the cooling power of the air is converted into an electrical reading with an allowance for ambient temperature taken into account. The probe is narrow and can be fitted into a small hole in ducting to take measurements. Its working range is 0.1–30 ms^1. To avoid errors the axis of the sampling head should be parallel to the airstream contours. Regular calibration of anemometers is required to avoid measurement errors.

(4)   A dust lamp enables visualisation of air flow patterns. This allows study of the movement and distribution of dust aerosols. Many harmful airborne dust particles are too small to be seen by the naked eye. When a parallel beam of light is shone through a dust cloud the particles reflect light by forward light scattering to an observer whose eyes are shielded from the main light source. (This is similar to visualising dust particles floating in a ray of sunlight.) This is not a quantitative method but does allow direct observation of the performance of a ventilation system. Other direct reading dust monitors can be used to make observations.

(5)   Smoke tracers allow visualisation of air flow patterns. These devices generate a plume of smoke and can be used to visualise air movement patterns generated by ventilation systems.

(6)   Air monitoring equipment enables evaluation of environmental levels and personal exposure of operators indicating whether adequate control of exposure is being achieved. Static and personal monitoring using air collection pumps and sampling devices (as appropriate to the specific contaminants) can be carried out. In addition direct monitoring equipment may also be used to assess control, for example, infra-red spectrophotometers may be used for gases and vapour which absorb infra-red radiation.

In checking system performance static pressure is measured behind each enclosure or hood and at various other points in the system whereas air velocity is measured at the face of the enclosure or point of emission and at various points in the ducting. The key question is whether the system is performing as was intended in the design specification.

Dilution ventilation

[V3014] Insufficient fresh air can lead to tiredness, lethargy, headaches, dry or itchy skin or eye irritation in employees. Dilution or 'general' ventilation refers to the movement of air into and out of a work area. A minimum standard of ventilation is required in all enclosed workplaces. Fresh air needs to be supplied and stale, hot or humid air replaced at a reasonable rate. Dilution ventilation may also be used in some situations where it can remove small quantities of low toxicity airborne contaminants generated by work activities.

Careful positioning of inlets is required to ensure that 'fresh' air is contaminant free and sometimes incoming air may need to be filtered to remove particulates. In some cases natural ventilation with doors and windows may be sufficient but often mechanical systems will be required to provide fresh air and dilution and removal of contaminants. Natural ventilation is generated by temperature and pressure changes in buildings where differences can produce an upward movement of air known as the 'stack effect'. This is where cooler air enters at a low level and, as it is warmed and mixes with contaminant, it rises and finally escapes through a roof vent or high level opening.

Dilution ventilation can be used to control exposure to airborne contaminants where:

— local exhaust ventilation is not practicable;
— the contaminant is of low toxicity;
— only small quantities of contaminant are released at a uniform rate;
— contaminants can be sufficiently diluted to control operator exposure adequately.

In re-circulation systems there needs to be adequate filtration to remove impurities and there should also be a significant fresh air component. With air-conditioning, both the temperature and the humidity of re-circulated air need to be adjusted, since the aim is to provide comfort conditions for workplace occupants. With natural ventilation there is limited scope to control air pressure and temperature differences; where mechanical systems are used, however, these parameters are more controllable. As with local exhaust ventilation, mechanical systems including air conditioning should be cleaned, maintained, examined and tested. Where a mechanical system provides dilution ventilation to dilute and remove airborne contaminants, a warning device is required in case of breakdown.

HSE guidance recommends that the fresh air supply rate should be at least 5–8 litres per second per building occupant. Relevant factors to be considered are the amount of floor space for each building occupant, the processes and equipment involved and whether the work is strenuous.

Current Building Regulations (Part F) set out that an air change rate of 8 litres per second must be achieved for new build offices.

To retain comfort conditions in a workplace an air flow velocity at normal temperatures should be around $0.1–0.15ms^1$ and up to $0.25ms^1$ in summer. If gas, coal or oil-fired equipment is used then fresh air requirements will depend on the type of flue arrangement in use. Where dilution ventilation is used to control airborne contaminants from work processes, the correct air supply rate must be determined. This requires calculation of the rate of release of the contaminant into the space to be ventilated.

Legal Requirements

[V3015] There are statutory requirements regarding the application and maintenance, examination and testing of local exhaust ventilation and dilution or general ventilation systems. An outline of the principal duties is given below.

The Health and Safety at Work etc Act 1974 (ch 37)

Under *s* 2 of the Act, an employer must provide and maintain a working environment that is, so far as is reasonably practicable, safe and without risk to health.

The Control of Substances Hazardous to Health Regulations (as amended) 2002 (SI 2002 No 2677)

'Assessment of risk' (reg 6) and 'Adequate control' (reg 7)

The *Control of Substances Hazardous to Health Regulations (as amended) 2002 (SI 2002 No 2677)* require an assessment of the risks to health created by work involving 'substances hazardous to health' (as defined in *reg 2*). Such risks must be prevented (ie the hazard eliminated) or, where this is not reasonably practicable, adequately controlled.

Where appropriate to the activity and the risk assessment, control of exposure at source is required as part of a hierarchy of control and this includes the provision of adequate ventilation systems and their associated maintenance (*regs 7(3)(b), 7(4)(b)*). Measures to control the working environment including appropriate general ventilation are also required under *Control of Substances Hazardous to Health Regulations (as amended) 2002 (SI 2002 No 2677), reg 7(4)(d)*.

'Adequate control' is achieved only if:

— the principles of good practice for control as specified in *Sch 2A reg 7(7)(a)* are applied;
— Workplace Exposure Limits (WELs) are not exceeded (*reg 7(7)(b)*); and
— for specified groups of carcinogens and asthmagens, exposure levels are reduced as low as is reasonably practicable (*reg 7(7)(c)*).

WELs are intended to prevent exposure exceeding a set level and represent the personal maximum exposure concentration averaged over a reference time period (the sampling device is placed in the 'breathing zone' of the person), either 8 hours or 15 minutes. WELs can be used to evaluate the application and effectiveness of ventilation control. They are not to be regarded as the boundary between 'safe' and 'unsafe' conditions. Current WEL values are listed in guidance document '*HSE Workplace Exposure Limits*', EH40/2005,

2005 and the accompanying supplement (2007). In view of the uncertainties relating to the health effects of many substances hazardous to health the general philosophy should always be to reduce exposures as low as is reasonably practicable utilising the principles of good practice specified in *Sch 2A* to the Regulations.

Use of control measures (reg 8)

The employer has a duty to ensure that LEV and other ventilation systems are properly used and applied and employees have a duty to ensure that any defects are immediately reported to the employer. The use of portable LEV systems may be problematic and devices provided for use with welding operations, for example, are often found abandoned in the work area and not in use. This may be because they are cumbersome to use and to move around the workplace.

Maintenance, examination and testing of control measures (reg 9)

Any engineering control including local exhaust ventilation and other ventilation systems must be: "maintained in an efficient state, in efficient working order, in good repair and in a clean condition" (*reg 9(1)(a)*)

Engineering controls must also be subjected to thorough examination and testing and in the case of local exhaust ventilation this must be at least once every 14 months (*reg 9(2)*).

Similarly, the provision of associated systems of work and supervision must be reviewed at suitable intervals (*reg 9(1)(b)*).

Maintenance

The aim of maintenance is to ensure that an LEV system continues to perform as originally intended. Maintenance includes visual checks, inspection, servicing, observations of systems of work and any remedial work to ensure system effectiveness. Deterioration in function must be detected and remedied and the frequency of checks will depend on the nature of the risks involved and the importance of the particular system. For LEV and work enclosures, visual checks should be carried out at least once per week (see para 164 '*L5 Control of Substances Hazardous to Health Regulations 2002 (as amended): Approved Code of Practice and Guidance*', 2005). The employer should ensure that the person carrying out the maintenance, examination and testing is competent to do so in accordance with *Reg 12(4)*). Further guidance* is recommended for the selection of competent persons to undertake maintenance, examination and testing (see para 167 of the COSHH ACOP noted above).

* UK Accreditation Service, '*Accreditation for the inspection of local exhaust ventilating (LEV) plant*', RG4, 2000 (available at www.ukas.com/Library/downloads/publications/RG4.pdf).

With biological agents it is particularly important at all times to maintain ventilation controls to prevent infective organisms escaping (paras 161, 163 COSHH ACOP). *Schedule 3* contains additional provisions relating to work with biological agents with containment measures specified in *Parts II* and *III*.

Examination and testing

Thorough examination and testing is specifically required for LEV systems every 14 months, unless the process is specified in *Sch 4 and reg 9(2)(a)*. A suitable record of the examination and test of the local exhaust ventilation system must be kept available for at least five years (*reg 9(4)*). For all LEV systems, whether portable or fixed and including biological safety cabinets, the examination and test is to ensure that the system is still functioning at the level as originally intended. Guidance is given on the maintenance, examination and testing of LEV systems in the 2011 HSE publication (HSG 258 *'Controlling Airborne Contaminants at Work – A Guide to Local Exhaust Ventilation'* referenced at **V3001**.

A thorough examination will normally have a visual check, a measurement of LEV performance and an assessment of control. Where air is re-circulated there must be an assessment of the air cleaning device. The minimal content of a suitable examination and test record is given in para 176 of the COSHH ACOP and the type of information required is outlined in para 177. This includes information on:

— Enclosures/hoods: maximum number in use; position; static pressure behind each hood or extraction point; face velocity;
— Ducting: dimensions; transport velocity; volume flow;
— Filter/collection point: specification; volume flow; static pressure at inlet; outlet and across filter;
— Fan/air mover specification: volume flow; static pressure at inlet; direction of rotation;
— Systems which re-circulate air to the workplace: filter efficiency; concentration of contaminant in returned air.

Control of Lead at Work Regulations (as amended) 2002 (SI 2002 No 2676)

Similar duties to those found in *Control of Substances Hazardous to Health Regulations (as amended) 2002 (SI 2002 No 2677)* are found in these regulations and in the *Control of Asbestos Regulations 2006 (SI 2006 No 2739)* discussed below.

'Assessment of risk' (reg 5) and 'Adequate control' (reg 6)

The *Control of Lead at Work Regulations (as amended) 2002 (SI 2002 No 2676)* require an assessment of risks to health arising from work involving exposure to lead (*reg 5*) with a requirement for the prevention and control of those risks (*reg 6*). Protective measures must be applied as appropriate to the situation and the risk assessment and, for control at source amongst other things, local exhaust ventilation must be considered (*reg 6(3)(b)*). Suitable maintenance procedures must be adopted for local exhaust ventilation systems (*reg 6(4)(b)*) and control of the environment by 'appropriate general ventilation' is also required (*reg 6(4)(d)*).

For adequate control of exposure to lead, the occupational exposure limit for lead should not be exceeded or, where it is, the employer should identify the reasons and take immediate action to remedy the situation. The occupational exposure limit for inorganic lead is an airborne concentration of 0.15 mg m^3 (averaged over an eight hour work period) (*reg 2(1)*). The limit for organic lead is 0.10 mg m^3.

Maintenance, examination and test of control measures (reg 8)

Local exhaust ventilation systems must be: 'maintained in an efficient state, in efficient working order, in good repair and in a clean condition' (*reg 8(1)*)

In addition, thorough examination and testing of control measures is required (*reg 8(2)*) and for local exhaust ventilation systems this must be carried out at least every 14 months (*reg 8(2)(a)*) with a suitable record kept for at least five years (*reg (8)(4)*). Thorough examination and testing should include the following:

— A thorough examination both internally and externally, where appropriate, of the condition of all parts of the system, ie exhaust openings, collection hoods or suction points, ductwork, dust collection and filtration units, fans or air movers;

— Measurements of static pressure at a point immediately behind each exhaust opening, collection hood or suction when the equipment is simultaneously extracting from all points;

— Measurements of air velocity at the plane of openings to enclosures, collection hoods or suction points for which the standard velocities have been specified;

— An assessment of whether lead dust, fume or vapour is being effectively controlled at each opening, collection hood or suction point. This would be complemented by lead in air monitoring.

Paragraph 185 '*L132 Control of Lead at Work Regulations 2002: Approved Code of Practice*', HSE, 2002.

The Control of Asbestos Regulations 2006 (SI 2006 No 2739)

Where work is liable to result in exposure to asbestos the employer must carry out a suitable and sufficient assessment of risks to health (*reg 6(1)(a)*) that considers the effects of control measures which have been or will be taken (*reg 6(2)(c)*). Where prevention of exposure is not reasonably practicable, adequate control is required. Measures to avoid or minimise the release of asbestos fibres must be applied together with control at source, which includes 'adequate ventilation systems' (*reg 11(2)(b)*). Where maintenance of air extraction plant contaminated with asbestos is carried out, appropriate precautions should be taken and it may be necessary to use LEV (para 188, '*L143 Work with Materials Containing Asbestos: Approved Code of Practice*', HSE, 2006). There are similar requirements to the *Control of Substances Hazardous to Health Regulations (as amended) 2002 (SI 2002 No 2677)* regarding the use of controls, ie where controls are provided the employer has a duty to ensure they are used (*reg 12(1)*) and employees have a duty to make use of controls provided and report defects (*reg 12(2)*).

In addition, controls must be 'maintained in an efficient state, in efficient working order, in good repair and in a clean condition' (*reg 13(1)(a)*).

Systems of work and supervision must be reviewed at suitable intervals and revised where necessary *(reg 13(1)(b))*. For local exhaust ventilation systems, thorough examination and testing must be carried out at 'suitable intervals' by a competent person *(reg 13(2))*. Portable equipment and enclosures may require more examination and testing than static systems (for example, during asbestos removal operations). Suitable records of examination, test and repair of LEV systems must be kept available for five years *(reg 13(3))*.

For asbestos activities where a licence is mandatory, enclosures are normally required, whilst for work not subject to a licence, enclosures may be needed but their requirements may vary. For example, there is no requirement for extraction in an enclosure for the removal of textured decorative coatings (para 206, '*L143 Work with Materials Containing Asbestos: Approved Code of Practice*', HSE, 2006) The employer must ensure that a thorough visual inspection of the enclosure, airlocks and ducting from the air extraction equipment is carried out at least at the beginning of each shift. Air extraction equipment should be operated in an enclosure continually throughout work and for 60 minutes after the end of work activities (see para 207 of the Asbestos ACOP noted above). Air monitoring should be carried out at discharge points if it is not reasonably practicable to discharge outside a building. All air extraction equipment which is necessary (air movers or negative pressure units, including hygiene facilities and those in laboratories handling asbestos) should have a daily visual inspection when in use and this equipment should be thoroughly examined and tested every 6 months by a competent person (see para 211 of the Asbestos ACOP noted above). With reference to the design, operation and airflow testing for negative pressure units reference is made to BS 8520-2:2009 Equipment used in the controlled removal of asbestos-containing materials: negative pressure units — specification.

Workplace (Health, Safety and Welfare) Regulations (as amended) 1992 (SI 1992 No 3004)

Every enclosed workplace must be ventilated by a sufficient quantity of fresh or purified air *(reg 6(1))*. The aim is that stale, hot or humid air should be replaced at a reasonable rate. Careful positioning of air inlet points is required and sometimes filtration may be needed to remove particulates. In some cases natural ventilation may be sufficient but often mechanical ventilation systems will be required to achieve sufficient quantities of fresh air. Where air is re-circulated adequate filtration to remove impurities will be required. Mechanical systems including air conditioning need to be cleaned, maintained and tested at regular intervals (para 33, L24 '*Workplace (Health, Safety and Welfare) Regulations: Approved Code of Practice*', HSE, 1996). A warning device in case of system breakdown is required where this is necessary for health and safety reasons, for example, where a mechanical system is employed to dilute airborne contaminant concentrations (reg 6(2)). It is recommended that the fresh air supply rate is not below 5–8 litres per minute per occupant (para 38, L24 '*Workplace (Health, Safety and Welfare) Regulations: Approved Code of Practice*', HSE, 1996).

Summary

[V3016] This chapter has addressed ventilation in the workplace in general and has focused in particular on local exhaust ventilation as a very important workplace environmental control. Where prevention of exposure is not reasonably practicable, control at source may be possible, by using LEV to limit inhalation exposure to hazardous substances.

The constituent components of local exhaust ventilation systems have been discussed illustrating how careful design is crucial. This applies particularly to the design of the system collection point, ie the enclosure, hood or slot. It is very important that such systems continue to perform as intended in their original design specification and this was seen to require regular checks and maintenance by competent personnel with thorough examination and testing at appropriate intervals. In addition, the role of dilution or general ventilation in providing sufficient fresh air to workplaces and in contaminant control is important in maintaining healthy and comfortable enclosed working environments.

Statutory obligations regarding the application of LEV systems and the provision of maintenance, examination and testing have also been reviewed. It can be seen that there are broadly similar statutory requirements in the *Control of Substances Hazardous to Health Regulations (as amended) 2002 (SI 2002 No 2677)* the *Control of Lead at Work Regulations (as amended) 2002 (SI 2002 No 2676)* and the *Control of Asbestos Regulations 2006*. The statutory requirement to provide a healthy general environment in the workplace has also been discussed with reference to the ventilation requirements of the *Workplace (Health, Safety and Welfare) Regulations (as amended) 1992 (SI 1992 No 3004)*.

Vibration

Roger Tompsett

The scope of this chapter

[V5001] This chapter deals with the effects of vibration on people and the obligations of employers to assess, monitor and reduce the risks of adverse health effects of hand-arm and whole-body vibration on their workforce. This area emerged from relative regulatory obscurity following an EC Directive (2002/44/EC), the Physical Agents (Vibration) Directive, which was implemented in Great Britain by the *Control of Vibration at Work Regulations 2005*.

The regulations came into operation in July 2005 but included transitional provisions concerning *regulation 6(4)*, which deals with exceeding exposure limits set out in the Regulations. For work equipment first provided before 6 July 2007, this regulation came into effect on 6 July 2010. For the agriculture and forestry sectors, *regulation 6(4)* will not apply to whole-body vibration until 6 July 2014 in respect of work equipment which was first provided to employees before 6 July 2007 and which does not permit compliance with the exposure limit value for whole-body vibration.

However, in using such equipment the employer must still take into account the latest technical advances and comply with *regulation 6(2)* of the Regulations. This sets out that where it is not reasonably practicable to eliminate risk at source and an exposure action value is likely to be reached or exceeded, the employer must reduce exposure to as low a level as is reasonably practicable by establishing and implementing a programme of organisational and technical measures which is appropriate to the activity.

The Regulations are covered in detail in a later section of this chapter, which also provides some basic information on vibration and general advice on reducing vibration exposure at work.

An introduction to the effects of vibration on people

[V5002] The effects of vibration on people can be divided into three broad classes:

- building vibration perceptible to occupiers;
- the exposure of people in vehicles and industrial situations to whole-body vibration; and
- the exposure of people operating certain tools or machines to hand-arm vibration

It is convenient to consider these classes separately, as the cause, effect and method of control is different in each case.

Hand-arm vibration and vibration-induced white finger (VWF)

[V5003] The effects of industrial vibration on people received relatively little attention until an award of damages to seven ex-British Coal miners in July 1998, in a landmark Court of Appeal decision (*Armstrong v British Coal Corporation* (1998), unreported. The miners received compensation ranging from £5,000 to £50,000 for the effects of vibration white finger ('VWF'), a form of hand-arm vibration syndrome characterised by the fingers becoming numb and turning white. In its early stages, the disease is reversible, but continued exposure leads to permanent damage and even gangrene, resulting in the loss of fingers or even a complete hand. The decision was important not only for the size of the awards, but also in setting out the terms in which British Coal were negligent and the standards of exposure that would be reasonable.

The miners received compensation ranging from £5,000 to £50,000 for the effects of vibration white finger ('VWF'), a form of hand-arm vibration (HAV) characterised by the fingers becoming numb and turning white. In its early stages, the disease is reversible, but continued exposure leads to permanent damage and even gangrene, resulting in the loss of fingers or even a complete hand. The decision was important not only because of the size of the awards, but also in setting out the terms in which British Coal were negligent and the standards of exposure that would be reasonable.

The then Labour government set up a compensation scheme for miners to avoid the need for individual litigation. The deadline for claims to be submitted under the scheme was 31 October 2002 for live claimants and 30 January 2003 in respect of deceased claimants. The scheme was finally closed by the Court on 1 May 2009 and the Department for Energy and Climate Change (DECC) reported that by June 2011, 169,609 of a total of 169,611 claims had been settled, with £1.7 billion being paid out in compensation. This was far higher in terms of the number of claims and the amount of compensation than had originally been expected.

The landmark award in *Armstrong V British Coal Corporation* was inevitably followed by many others. For example, in August 2000, eight workers at North-West Water received a £1.2 million settlement. They contracted VWF whilst using jackhammers and whackers when breaking and re-instating concrete surfaces.

The Health and Safety Executive (HSE) has also prosecuted several organisations. For example vehicle manufacturer Land Rover was prosecuted for failing to take into account the risks associated with workers at its Solihull plant using vibrating hand tools. In April 2011 it was fined £20,000 and ordered to pay £60,606 costs after pleading guilty to breaching *section 2(1)* of the *Health and Safety at Work Act 1974*. An HSE investigation carried out after two cases of Hand Arm Vibration Syndrome (HAVS) were reported in December 2006 found that vibrating hand tools were being used across the plant with a lack of assessment and management of risk. When a health surveillance regime was then made effective, other cases came to light.

A Medical Research Council survey in 1997–98 gave an estimate of 288,000 sufferers from VWF in Great Britain – more than eight times the number of self-reported sufferers previously identified in the HSE's self-reported work-related illness surveys in 1995. Approximately 40 per cent of those reporting VWF in the HSE survey also reported work-related deafness or other ear problems, indicating that work which exposes people to hand-arm vibration is often noisy. In recent years, VWF has become the most common prescribed disease under the Industrial Injuries Scheme. The number of new cases of Vibration White Finger (VWF) assessed for Industrial Injuries Disability Benefit in 2008/09 was 850.

Carpal Tunnel Syndrome

[V5004] Use, either frequent or intermittent, of hand-held vibratory tools, can result in injury to the wrist: Carpal tunnel syndrome ('CTS') is one example of this. It is thought to arise in part from the trapping or compression of nerves in the wrist. It can arise from repetitive twisting or gripping movements of the hand as well as from the use of vibrating tools. CTS arising from the use of vibrating tools was prescribed as occupational disease A12 in April 1993. [*Social Security (Industrial Injuries) (Prescribed Diseases) Amendment Regulations 1993 (SI 1993 No 862), Reg 6(2)*].

The number of new cases of CTS recognised by the Department for Work and Pensions (DWP) rose steadily from 1993 until 2002/03 when the number peaked at 1030. While as set out above, Carpal Tunnel Syndrome may have other occupational causes, such as repetitive twisting or gripping movements of the hand, such cases do not qualify for compensation by the DWP.

Whole-body vibration

[V5005] The four principal effects of whole-body vibration ('WBV') are considered to be:

- health risks;
- impaired ability to perform activities;
- impaired comfort; and
- motion sickness;

Exposure to WBV causes a complex distribution of oscillatory motions and forces within the body. These may cause unpleasant sensations giving rise to discomfort or annoyance, resulting in impaired performance (e.g. loss of balance, degraded vision) or present a health risk. The most widely reported WBV injury is back pain.

Buildings and structures

[V5006] Vibration can affect buildings and structures, but a detailed treatment of this is beyond the scope of the present chapter. People are more sensitive to vibration than are buildings or structures, so vibration within a building is likely to become unacceptable to the occupants at values well below those which pose a threat to a structurally sound building. For detailed technical guidance on the measurement and evaluation of the effects of vibration on buildings, British Standard BS ISO 4866:2010 *Mechanical vibration and shock. Vibration of fixed structures. Guidelines for the measurement of vibrations and evaluation of their effects on structures* and BS 7385: Part 2: 1993 may be consulted.

Vibration measurement

Measurement units

[V5007] Vibration is the oscillatory motion of an object about a given position. The rate at which the object vibrates (i.e. the number of complete oscillations per second) is called the frequency of the vibration and is measured in Hertz (Hz). The frequency range of principal interest in vibration is from about 0.5 Hz to 100 Hz, i.e. below the range of principal interest in noise control.

The magnitude of the vibration is now generally measured in terms of the acceleration of the object, in metres per second squared (i.e. metres per second, per second) denoted $m.s^{-2}$ or m/s^2. (In mathematical notation, the minus sign in front of the index, such as in s^{-2}, means that s^{-2} is on the bottom line. Thus in $m.s^{-2}$, m is 'divided by' s^2. In vibration work, other values of index may be encountered, such as 1 and 1.75.)

Vibration magnitude can also be measured in terms of peak particle velocity, in metres per second (denoted $m.s^{-1}$ or m/s), or in terms of maximum displacement (in metres or millimetres).

For simple vibratory motion, it is possible to convert a measurement made in any one of these terms to either of the other terms, provided the frequency of the vibration is known, so the choice of measurement term is to some extent arbitrary. However, *acceleration* is now the preferred measurement term because modern electronic instruments generally employ an *accelerometer* to detect the vibration. As the name suggests, this responds to the acceleration of the vibrating object and hence this characteristic can be measured directly.

Direction and frequency

[V5008] The human body has different sensitivity to vibration in the head to foot direction, the side to side direction, and in the front to back direction, and this sensitivity varies according to whether the person is standing, sitting or lying down. In order to assess the effects of vibration, it is necessary to measure its characteristics in each of the three directions and to take account of the recipient's posture. The sensitivity to vibration is also highly frequency-dependent. BS 6841: 1987 *Measurement and evaluation of human exposure to whole-body mechanical vibration and repeated shock* provides a set of six frequency-weighting curves for use in a variety of situations. The curves may be considered analogous to the 'A'-weighting used for noise measurement. However, few instruments have these weightings built into them.

Vibration dose value

[V5009] In most situations, vibration magnitudes do not remain constant, and the concept of vibration dose value ('VDV') has been developed to deal with such cases. This is analogous to the concept of noise dose used in the *Noise at Work Regulations 1989 (SI 1989 No 1790)*. Again, few instruments are capable of measuring VDV. By applying the frequency-weighting system, it is possible to introduce a single-number vibration rating.

The Control of Vibration at Work Regulations 2005 set a daily exposure limit value of 1.15 m/s^2 A(8) and the daily exposure action value is 0.5 m/s^2 A(8) for whole body vibration (see **V50015** below).

Structural vibration of buildings can be felt by the occupants at low vibration magnitudes. It can affect their comfort, quality of life and working efficiency. Low levels of vibration may provoke adverse comments, but certain types of highly sensitive equipment (e.g. electron microscopes) or delicate tasks may require even more stringent criteria.

Adverse comment regarding vibration is likely when the vibration magnitude is only slightly in excess of the threshold of perception and, in general, criteria for the acceptability of vibration in buildings are dependent on the degree of adverse comment rather than other considerations such as short-term health hazard or working efficiency.

Vibration levels greater than the usual threshold may be tolerable for temporary or infrequent events of short duration, especially when the risk of a startle effect is reduced by a warning signal and a proper programme of public information.

Detailed guidance on this subject may be found in BS 6472-1:2008 *Guide to evaluation of human exposure to vibration in buildings. Vibration sources other than blasting* and BS 6472-2:2008 *Guide to evaluation of human exposure to vibration in buildings. Blast-induced vibration.*

Causes and effects of hand-arm vibration exposure

[V5010] Intense vibration can be transmitted to the hands and arms of operators from vibrating tools, machinery or work materials. Examples include the use of pneumatic, electric, hydraulic or engine-driven chain-saws, percussive tools, grinders or sanders. The vibration may affect one or both arms, and may be transmitted through the hand and arm to the shoulder.

The vibration may be a source of discomfort, and possibly reduced proficiency. Habitual exposure to hand-arm vibration has been found to be linked with various diseases affecting the blood vessels, nerves, bones, joints, muscles and connective tissues of the hand and forearm, most commonly VWF.

VWF arises from progressive loss of blood circulation in the hand and fingers, sometimes resulting in necrosis (death of tissue) and gangrene for which the only solution may be amputation of the affected areas or complete hand. Initial signs are mild tingling and numbness of the fingers. Further exposure results in blanching of the fingers, particularly in cold weather and early in the morning. The condition is progressive to the base of the fingers, sensitivity to attacks is reduced and the fingers take on a blue-black appearance. The development of the condition may take up to five years according to the degree of exposure to vibration and the duration of such exposure.

Vibration-induced white finger as an occupational disease

[V5011] VVibration white finger ('VWF') isa reportable disease under the Reporting of *Reporting of Injuries, Diseases and Dangerous Occurrences Regulations 1995 (RIDDOR) (SI 1995 No 3163)*.

VWF is also prescribed occupational disease A11. It is described as: 'episodic blanching, occurring throughout the year, affecting the middle or proximate phalanges or in the case of a thumb the proximal phalanx, of:

(a)   in the case of a person with five fingers (including thumb) on one hand, any three of those fingers;

(b)   in the case of a person with only four such fingers, any two of those fingers; or

(c)   in the case of a person with less than four such fingers, any one of those fingers or . . . the remaining one finger.'

[*Social Security (Industrial Injuries) (Prescribed Diseases) Regulations 1985 (SI 1985 No 967)*].

(Carpal tunnel syndrome in relation to work involving the use of hand-held vibrating tools is also reportable under RIDDOR and CTS arising from the use of vibrating tools was prescribed as occupational disease A12 in April 1993 (see **V5004** above.)

Action at common law

[V5012] In July 1998, the Court of Appeal upheld an award of damages to seven employees of British Coal claiming damages for VWF. The court determined that after January 1976, British Coal should have implemented a range of precautions, including training, warnings, surveillance and job rotation where exposure to vibration was significant.

The court went on to consider what degree of exposure would be reasonable. They supported the standards set out in the HSE Guidance booklet *Hand-arm vibration* published in 1994, which established limits for vibration dose measured as an eight-hour average or A8. The court considered that an exposure of 2.8 m/sec^2 A8 would be an appropriate level for prudent employers to use. This level of exposure would lead to a 10 per cent risk of developing finger blanching (the first reversible stage of VWF) within eight years.

The court also suggested a form of wording to warn employees: 'If you are working with vibrating tools and you notice that you are getting some whitening or discolouration of any of your fingers then, in your own interests, you should report this as quickly as possible. If you do nothing, you could end up with some very nasty problems in both hands.'

Stages of vibration white finger

[V5013] Damages are awarded according to the stage of the disease at the time of the action from the date the employer should have known of the risk. The Taylor-Pelmear Scale is usually used to describe these stages as follows.

Taylor-Pelmear Scale System		
Stage	*Grade*	*Description*
0	—	No attacks
1	Mild	Occasional attacks affecting the tips of one or more fingers
2	Moderate	Occasional attacks affecting the tips and the middle of the fingers (rarely the base of the fingers) on one or more fingers
3	Severe	Frequent attacks affecting the entire length of most fingers
4	Very severe	As in stage 3, with damaged skin and possible gangrene in finger tips

Prescription of hand-arm vibration syndrome

[V5014] Foreseeably, prescription may extend to hand-arm vibration syndrome ('HAVS') instead of VWF. The former would cover recognised neurological effects as well as the currently recognised vascular effects of vibration. Neurological effects will include numbness, tingling in the fingers and reduced sensibility. Moreover, the current list of occupations, for which VWF is prescribed, may be replaced by a comprehensive list of tools/rigid materials against which such tools are held, including:

- percussive metal-working tools (e.g. fettling tools, riveting tools, drilling tools, pneumatic hammers, impact screwdrivers);
- grinders/rotary tools;
- stone working, mining, road construction and road repair tools;
- forest, garden and wood-working machinery (e.g. chain saws, electrical screwdrivers, mowers/shears, hedge trimmers, circular saws); and
- miscellaneous process tools (e.g. drain suction machines, jigsaws, pounding-up machines, vibratory rollers, concrete levelling vibratibles).

The Control of Vibration at Work Regulations, 2005 (SI 2005 No 1093)

[V5015] The Control of Vibration at Work Regulations 2005 implement EC directive 2002/44/EC, the Physical Agents (Vibration) Directive on the protection of workers from exposure to vibration, and came into force on 6 July 2005. However, they included transitional provisions concerning *regulation 6(4)*, which deals with exceeding exposure limits set out in the Regulations. For work equipment first provided before 6 July 2007, this regulation came into effect on 6 July 2010. For the agriculture and forestry sectors, *regulation 6(4)* will not apply to whole-body vibration until 6 July 2014 in respect of work equipment which was first provided to employees before 6 July 2007 and does not permit compliance with the exposure limit value for whole-body vibration.

The Regulations apply to work activities on mainland Britain, on offshore workplaces and in aircraft in flight over Britain, although exemption certificates may be granted to emergency services, air transport and the Ministry of Defence. Separate regulations were introduced in Northern Ireland, The *Control of Vibration at Work Regulations (Northern Ireland) 2005*, and for sea transport, the *Merchant Shipping and Fishing Vessels (Control of Vibration at Work) Regulations 2007*.

The *Control of Vibration at Work Regulations* require that an employer shall make a suitable and sufficient assessment of the risk to health and safety of employees arising from exposure to vibration at work, and if an exposure action value is likely to be exceeded, the employer shall reduce the exposure to the lowest level that is reasonably practicable by appropriate technical and organisational means.

If exposure exceeds an exposure limit value, the exposure must be reduced below the limit value forthwith, the reason for exceeding the limit must be identified and appropriate technical and organisational means used to ensure that the limit is not exceeded again.

For *hand-arm vibration*, the **daily exposure action value is 2.5 m/s²A(8)**, and the **daily exposure limit value is 5 m/s²A(8)** which must not be exceeded except in certain pre-permitted circumstances.

For *whole-body vibration*, the **daily exposure action value is 0.5 m/s² A(8)**, and the **daily exposure limit value is 1.15 m/s² A(8)**, which must not be exceeded except in certain pre-permitted circumstances.

The daily personal exposure values are to be adjusted (or normalised) to an 8-hour reference period. This can be averaged over a seven-day period where vibration exposure varies considerably from day to day.

The risk assessment shall assess whether any employees are likely to be exposed to vibration at or above an exposure action value or limit value by means of:

(a)   observation of specific working practices;
(b)   reference to information on the probable magnitude of vibration from the equipment used in the particular working conditions;

(c)   if necessary, measurement of the magnitude of vibration exposure.

The risk assessment should consider the magnitude, type and duration of the exposure, including the effect of intermittent vibration or repeated shocks. Any health or working conditions that might worsen the risk must also be considered, such as working in low temperatures. In the case of whole body vibration, consideration should be given to exposure beyond normal working hours, including vibration in rest facilities.

The effect of vibration on the ability to perform the task is also relevant, including the effect on handling of controls, reading of indicators and the stability of structures and joints.

Risk assessments must be recorded and updated as necessary. Details should include significant findings and measures taken to control or eliminate vibration exposure and the provision of information, instruction and training to employees, including entitlement to health surveillance.

Health surveillance is required for all employees where there is a risk to health, which includes all employees likely to be exposed to vibration at or above an exposure action value. The health surveillance shall prevent or diagnose any health effect linked with vibration exposure and if such an effect is detected, then it will be necessary to prevent further exposure of the person concerned, to review the vibration protection measures in force and to check whether any other employee is affected.

Other relevant legislation

[V5016] that their processes entail, even though In addition to the requirements imposed by *Control of Vibration at Work Regulations 2005* there are specific obligations on employers contained in other health and safety regulations , in addition to the general duty of care under the *Health and Safety at Work etc. Act 1974*.

Such obligations include, amongst others, those under:

*   the *Reporting of Injuries, Diseases and Dangerous Occurrences Regulations 1995 (SI 1995 No 3163)*;
*   the *Provision and Use of Work Equipment Regulations 1998 (SI 1998 No 2306)*; and
*   the *Supply of Machinery (Safety) Regulations 1992 (SI 1992 No 3073)*, as amended by *SI 2004 No 2063*.

In addition, the *Management of Health and Safety at Work Regulations 1999 (SI 1999 No 3242)* set out in *regulation 16* that where there are women of child-bearing age in the workforce, risk assessments must cover the risks that are specific to new and expectant mothers.

The HSE guidance *New and expectant mothers at work – A guide for employers* contains an appendix setting out hazards, risks and ways of avoiding them including physical, chemical and biological agents and working conditions listed in Annexes 1 and 2 to the EC Directive on Pregnant Workers (92/85/EEC). These cover physical risks where they are considered as agents causing foetal lesions and/or likely to disrupt placental attachment, and include particular reference to shocks and vibration.

In addition, *regulation 19* requires that the employer ensure that young workers are protected from any risks to their health or safety which arise from their lack of experience, or awareness of existing or potential risks, or their lack of maturity. Work that young people must not be employed to carry out includes that where there is a risk to health from extreme cold or heat, noise or vibration.

Advice and obligations in respect of hand-arm vibration

Vibration magnitudes

[V5017] The energy level of the hand tool is significant. Percussive action tools, such as compressed air pneumatic hammers, operate within a frequency range of 33-50 Hz. These cause considerable damage whereas rotary hand tools, which operate within the frequency range 40-125 Hz, are less dangerous.

The Health and Safety Executive (HSE) publishes guidance on the vibration magnitudes of a variety of tools in a selection of typical working conditions. Vibration can vary not only with the type and design of the tool, but also with other things such as the task, the operator's technique and the material being worked (see http://www.hse.gov.uk/vibration/hav/advicetoemployers/assessris ks.htm).

For example, angle grinders create vibration of between 4 and 8 m/s^2, giving between three hours' and 45 minutes' use before the exposure action value is reached, and between 12 hours' and three hours' use before the exposure limit value is reached. Road breakers create between 5 and 20 m/s^2, giving between two hours' and ten minutes' use, with 12 m/s^2 being typical, giving 20 minutes' use before the exposure action value is reached, and between eight hours and 30 minutes before the exposure limit value is reached.

It is clear from this data that most of these tools cannot be used for a whole working day without exceeding the vibration limit value, and few can be used without exceeding the exposure action value. Indeed, it appears that most currently-available vibrating hand tools will exceed the vibration exposure action value, and are very likely to exceed the exposure limit value if used for a whole working day.

Managing vibration exposure

[V5018] The HSE has produced general guidance aimed at the construction, steel fabrication and foundry industries. The guidance focuses on finding alternatives to the use of hand-held vibratory tools. It has been found that methods of working that produce less vibration also produce more accurate work and less user fatigue, so multiple benefits can flow from a change of working methods.

In the construction industry, work can be planned better, for example, casting-in ducts, and detailing box-outs, so that there is less need to break through new masonry. Where breaking-out is necessary, hand-held breakers can be replaced by machine-mounted breakers, floor saws, diamond drilling, concrete crushers and bursters.

The HSE has issued definitive advice that the removal of pile caps using hand-held breakers is not acceptable. They advise on various alternatives of hydraulic bursting and hydraulic cropping. They also advise that concrete scabbling for purely aesthetic effect is not acceptable and alternatives should be used.

Where the use of a vibratory tool cannot be avoided, then it is necessary to assess and reduce the risk that it causes. Where a tool is used for a well-defined process, it might be helpful to control the vibration exposure by limiting the number of at-risk tasks in a day, rather than the number of hours use. For example, it may be found that an operator can drill 50 to 60 holes 100 mm deep in concrete before reaching the exposure action value, and 200 to 230 such before reaching the exposure limit (this is a theoretical example).

Many work processes that involve hand-held power tools will require employees to undertake a variety of tasks, possibly using different tools, each with different vibration exposures. In this case, it could be more difficult to set a limit on the number of operations that are allowable. An alternative is to use a 'points' system.

For example, an employer might allocate points to vibration exposure such that 100 points per day is equal to the exposure action value. The employer assesses the vibration 'dose' of an angle grinder as 30 points per hour, and of an impact drill as 120 points per hour. This means that the grinder gives an exposure of five points every 10 minutes and the impact drill gives 20 points for every 10 minutes use. Thus 20 minutes use of the drill plus 120 minutes use of the grinder would cause the exposure action value to be reached.

The allocation of points to each tool will have to be calculated using a formula which depends on the vibration magnitude of the tool and process. Note that the system described here is only intended to be indicative, as vibration dose does not strictly obey the rules of simple arithmetic.

The HSE guidance L140, *Hand-arm vibration*, (see http://www.hse.gov.uk/p ubns/priced/l140.pdf) provides advice on using an exposure points system or rea dy-reckoner to calculate daily vibration exposures.

Personal protection against hand-arm vibration

[V5019] The risk of injury to workers from hand-arm vibration can be reduced by ensuring that they can keep the blood flowing while working, for example by keeping warm, and exercising the hands and fingers. Gloves and other warm clothing should be provided when working in cold conditions. Tools should be designed for the job, both to lessen vibration and to reduce the strength of grip and amount of force needed. The equipment should be used in short bursts rather than long sessions. Workers should be encouraged to report any symptoms to the employer.

There are a number of 'anti-vibration' gloves on the market but, these may seriously limit dexterity and can in some circumstances actually increase the risk of vibration injury.

The HSE advises: "Gloves marketed as 'anti-vibration', which aim to isolate the wearer's hands from the effects of vibration, are available commercially. There are several different types, but many are only suitable for certain tasks, they are not particularly effective at reducing the frequency-weighted vibration associated with risk of HAVS and they can increase the vibration at some frequencies. It is not usually possible to assess the vibration reduction provided in use by anti-vibration gloves, so you should not generally rely on them to provide protection from vibration. However, gloves and other warm clothing can be useful to protect vibration-exposed workers from cold, helping to maintain circulation."

Manufacturers' warning of risk

[V5020]–[V5021] The *Supply of Machinery (Safety) Regulations 2008 (SI 2008 No 1597)*, came into force in December 2009 and replaced the *Supply of Machinery (Safety) Regulations 1992 (SI 1992 No 3073)*, The regulations establish essential health and safety requirements for machinery supplied in the European Economic Area and set out that "machinery must be designed and constructed in such a way that risks resulting from vibrations produced by the machinery are reduced to the lowest level, taking account of technical progress and the availability of means of reducing vibration, in particular at source."

Machinery manufacturers and suppliers have a duty to warn whenever a machine carries a vibration risk, and to provide information for hand-held or hand-guided machines where vibration emissions exceed 2.5 m/s^2. This is the vibration emission of the machine, not the daily exposure of the user, which will depend on the amount of use of this and other machines. It should be noted that 2.5 m/s^2 is not to be regarded as an acceptable target for the amount of vibration a machine may emit – there is an overriding duty to reduce vibration emission as far as possible within the current state of the art.

Manufacturers are required to test vibration emissions in accordance with test procedures aimed at obtaining a reproducible value, and they must declare both the typical measured value and the amount of uncertainty (variation) of measurements from the typical value. Thus, the typical value plus the uncertainty value should give the 'worst case', but it may still not represent the total amount of vibration in real use. This can complicate the choice of tool, to the extent that sometimes a tool that gives slightly more vibration than another tool may still be better if it is more comfortable, appropriate or quicker for the task.

Moreover, tools that are incorrectly assembled, used wrongly, with worn or loose parts can give many times the amount of vibration than normal.

Formal assessments

[V5022] It is quite difficult to measure the vibration emission of equipment in actual use, as the measuring device must be attached to the equipment in such a place and in such a manner as to represent the vibration being transmitted to the operator. The measuring system must be capable of determining the amount of vibration over a range of frequencies and in different directions. Hence, it is usually necessary to rely on general advice.

The first priority should therefore be to try to find ways to eliminate the use of vibrating tools, to use the most appropriate equipment to get the job done quickly and safely, and to make sure that the tool is correctly assembled and used, with the correct cutting bit kept sharp.

Advice and obligations in respect of whole-body vibration

Incidence of whole-body vibration

[V5023] Although millions of people are exposed to whole-body vibration ('WBV') at work, most of them are users of road vehicles and are not likely to be at risk of injury from WBV. Nevertheless, more than 1.3 million workers are thought to be exposed to vibration above the exposure action value and more than 20,000 are thought to be exposed to more than the WBV limit value. WBV can cause or worsen back pain, and the effect is aggravated when the vibration contains severe shocks and jolts. These conditions occur particularly in off-road driving, such as in farming, construction and quarrying, but can also occur in small, fast boats at sea and in some helicopters. Old railway vehicles running on trackwork in poor condition can also give rise to high levels of WBV.

Back pain can arise from many causes other than WBV, and can be aggravated by poor driving posture, sitting for long periods without being able to change position, awkwardly placed controls that require the operator to stretch or twist to reach them, manual lifting, and repeatedly climbing into or jumping out of a high cab.

Jobs which combine two or more of these factors can increase the risk further, for example, driving over rough ground whilst twisting round to check on the operation of equipment being operated behind the vehicle.

Assessing the risk of WBV problems

[V5024] Road transport drivers are not usually at risk unless there are other aggravating factors such as long hours of driving, or use of unmade or poor road surfaces. However, if there is a history of back pain in the job, this could be indicative of a problem, and it may be wise to consider the regulations and guidance on manual handling of materials in addition to the vibration regulations. Operators of off-road vehicles are at greater risk. Indicators of risk include the operator being thrown around in the cab, or receiving shocks and jolts through the seat, or if the manufacturer warns of a risk. A history of back pain is again an indicative factor.

Measurements of whole-body vibration in construction machines have shown that smaller loaders and excavators can produce WBV levels that exceed the exposure action value over a typical working day, although it is unlikely that the exposure limit value would be exceeded. However, it is possible to exceed the limit in smaller machines that are intensively used for long periods of the day.

Keeping records

[V5025] Risk assessments should be recorded. These should show those jobs or processes identified as being at risk from WBV; an estimate of the amount of exposure; comparison of exposure with the action and limit values; the available risk controls; the plans to control and monitor risks; and the effectiveness of the control measures.

Reducing the risk of WBV problems

[V5026] When a risk is identified the obvious initial actions are to ensure that the machine is suitable for the job, is used correctly and driven at safe speeds for the ground conditions, and that seats are correctly adjusted for the driver's size and weight. Access routes on work sites should be maintained in good condition, for example by using a scalping blade and roller.

Exposure assessments

[V5027] The HSE advises: "In most cases it is simpler to make a broad assessment of the risk rather than try to assess exposure in detail, concentrating your main effort on introducing controls."

It advises employers of drivers and operators to observe work tasks and talk to managers, employees and others in order to collect basic information to allow them to make a broad assessment of the risk and introduce simple control measures to reduce the risk.

It advises that exposures may be high where:

- machine or vehicle manufacturers' handbook warns of risks from whole-body vibration;
- machines or vehicles are unsuitable for the tasks for which they are being used;
- operators and drivers are using poor techniques;
- employees are operating or driving, for several hours a day, machines or vehicles likely to cause high vibration exposures;
- employees are being jolted, continuously shaken or, when going over bumps, rising visibly in the seat;
- vehicle roadways or work areas are potholed, cracked or covered in rubble;
- road-going vehicles are regularly driven off-road or over poorly-paved surfaces for which they are not suitable; or
- operators or drivers report back problems.

It advises: "This kind of broad risk assessment can be done without needing to estimate or measure vibration exposure. Most employers of drivers or operators will not need to do any measurements or employ vibration specialists to help with the risk assessment."

Health surveillance

[V5028] The HSE currently considers that because there is no way in which specific occurrences of back pain can be attributed to WBV, it is not appropriate to undertake health surveillance. It advises employers to use other methods to monitor the health of employees exposed to the risks, such as encouraging symptom reporting and checking sickness records.

HSE guidance on hand-arm and whole-body vibration

[V5029] The HSE publication *Hand-arm vibration* (L140), ISBN 0 7176 6125 3 (2005), superseded the previous HSG88. It also publishes a book of 51 case studies dealing with vibration, *'Vibration solutions: Practical ways to reduce the risk of hand-arm vibration injury'*, (HSG 170). It also provides an online vibration calculator to assist in calculating exposures for hand-arm vibration at: www.hse.gov.uk/vibration/hav/vibrationcalc.htm; and advice on vibration exposure monitoring is available at: www.hse.gov.uk/vibration/hav/a dvicetoemployers/vibration-exposure-monitoring-qa.pdf#?eban=rss-vibrat ion.

In addition to the above, HSE publishes a number of guidance leaflets on vibration, including:

- *Control the risks from hand-arm vibration-* Advice for employers on the Control of Vibration at Work Regulations 2005 INDG175(rev2) www.hse.gov.uk/pubns/indg175.pdf
- *Control back-pain risks from whole-body vibration* – Advice for emp loyers on the Control of Vibration at Work Regulations INDG242(rev1) www.hse.gov.uk/pubns/indg242.pdf
- *Hand-arm vibration* – advice for employees INDG296(rev1) www.hse. gov.uk/pubns/indg296.pdf
- *Drive away bad backs* INDG404 www.hse.gov.uk/pubns/indg404.pdf

There are also a number of leaflets focussed on specific industries which can all be downloaded at: www.hse.gov.uk/pubns/vibindex.htm:

- Health hazards from whole-body vibration caused by mobile agricultura l machinery (Agriculture Information Sheet No 20);
- Hazards associated with foundry processes: Hand-arm vibration - the current picture (Foundries Sheet No 8);
- Hazards associated with foundry processes; hand-arm vibration - sympt oms and solutions (Foundries Sheet No 9);
- Whole-body vibration in ports; and
- Hand-arm vibration in the cast stone industry: reducing the risk.

Violence in the Workplace

Bill Fox and Andrea Oates

Introduction

[V8001] The Health and Safety Executive (HSE) has identified work-related violence, which it defines as "Any incident in which a person is abused, threatened or assaulted in circumstances related to their work", to be a problem of significant magnitude and wide-ranging consequences. Whereas violence was once seen exclusively as a problem for occupations such as police officers and prison staff, it is now widely acknowledged that almost anyone can be exposed to it. At particular risk of exposure are those who come into contact with the public, those who handle money, those who, on occasions must become a service denier rather a service provider, and those who deal with people in pain and distress or under the influence of alcohol and other drugs. There are many jobs to which some, or all, of these dimensions apply, e.g. teachers, nurses, taxi drivers, transport staff, emergency service workers, shop workers, public house staff, social workers and 'helpline' staff.

The HSE emphasises that the process of managing the risk of violence is the same as for any other health and safety risk. They acknowledge, however, that violence is a problem within society and the risk of violence in the workplace depends on the interaction of a range of factors some of which lie outside of the employer's ability to control. There are no easy solutions or short cuts and there are no universal blueprints for dealing with the risk of violence. What works for one organisation may not work for another. When developing a workplace violence prevention programme it is imperative that organisations make discernible judgements based on the peculiarities of their particular working environments.

HSE guidelines suggest the adoption of a risk management framework with a recursive cycle of activities to ensure continuous improvement in the assessment and management of risk, beginning with an appraisal of the nature and extent of the problem, and followed by the design, implementation and evaluation of appropriate preventive measures.

The nature and extent of the problem

[V8002] The HSE uses figures from the British Crime Survey (BCS) as well as its own sources in order to estimate the extent of workplace violence.

The British Crime Survey (BCS) has been collecting data about workplace violence for several years and provides a count of crime across in England and Wales, including crimes not reported to or recorded by the police. It is conducted on a yearly basis.

The definition of workplace violence used in the survey is:

'All assaults or threats which occurred while the victim was working and were perpetrated by a member of the public'.

The HSE says that trends in violence at work are difficult to interpret, with survey estimates tending to fluctuate from year to year, but recent figures have been fairly stable.

Estimates from the 2006/07 British Crime Survey (BCS) indicated that there were some 397,000 threats of violence and 288,000 physical assaults by members of the public on British workers during the 12 months prior to the interviews carried out for the survey.

According to figures on violence at work gathered through the Reporting of Injuries, Diseases and Dangerous Occurrences Regulations 1995 (RIDDOR), there were 6,404 reported injuries caused by violence at work during the financial year 2006/07. These included four deaths, 932 major injuries and 5,468 non-major injuries that resulted in absence from work for at least three days. This was down on the figures for 2005/6, when 6,624 injuries caused by violence were reported under RIDDOR.

The HSE report, *Violence at work: Findings from the 2005/06 and 2006/07 British Crime Survey* provides an overview of the extent of violence at work in England and Wales.

The 2006/07 BCS indicated that 1.7% of working adults were the victim of one or more violent incidents at work. Around 355,000 workers had experienced at least one incident of violence at work, representing a fall of 40% from the peak of 592,000 in 1997. (However the HSE reports that the 2006/07 figure is marginally higher than in 2005/06, when the number of victims was 333,000.)

There were an estimated 684,000 incidents of violence at work according to the 2006/07 figures, including 288,000 assaults and 397,000 threats. The number has fallen by just over 50% from the peak of more than 1.4 million in 1995.

With regard to the nature of violence, in the majority of incidents (62%) the victim did not know the offender. In 15% of cases the offender was a client or a member of public known to the victim through work, and in 6% of cases the offender was a workmate or colleague. Out of work incidents are more likely to involve young people from the local area or other known people, including friends, neighbours and tradespeople.

In more than a third (35%) of incidents, the victims reported that the offender was under the influence of alcohol and in 15% of cases, they said they were under the influence of drugs.

The 2005/6 BCS survey found that while overall 2% of all adults at work were very worried about being assaulted by a member of public while at work, this varied considerably across different occupations. Nearly two thirds (32%) of workers in protective service occupations, such as police officers, who had contact with the public were very or fairly worried about assaults at work, compared with 8% of business and public service associate professionals. Twenty two percent of workers who had contact with the public thought it very or fairly likely that they would be threatened at work over the following year; and 9% of workers with face-to-face contact with the public thought it very or fairly likely that they would be assaulted.

Who is at risk?

[V8003] According to the HSE *Violence at work: Findings from the 2005/06 and 2006/07 British Crime Survey* those in the protective service occupations, including police officers, were most at risk of violence at work. The following table (which is set out in Appendix A of the report) shows the occupations most at risk from threats and assaults:

Table A2.4 Risk of violence at work, by occupation, 2005/06 and 2006/07 BCS interviews

Percentage victims once or more	Assaults	Threats	All vio-lence at work	Un-weighted N
Managers and Senior Officials	0.4	0.8	1.2	7,420
Corporate managers	0.5	0.6	0.9	5,509
Managers and proprietors in agri-culture and services	0.4	1.6	2.0	1,911
Professional Occupations	0.2	0.4	0.6	5,435
Science and technology profes-sionals	0.0	0.0	0.0	1,322
Health professionals	0.0	1.4	1.4	475
Teaching and research profession-als	0.4	0.4	0.8	2,444
Business and public service pro-fessionals	0.3	0.5	0.6	1,194
Associate Professionals and Tech-nical Occupations	0.7	0.2	0.9	7,070
Science and technology associate professionals	0.0	0.0	0.1	828
Health and social welfare associ-ate professionals	0.5	0.3	0.7	1,973
Protective service occupations	5.5	0.9	6.4	605
Culture, media and sports occu-pations	0.8	0.0	0.8	1,027
Business and public service asso-ciate professionals	0.0	0.0	0.1	2,637
Administrative and Secretarial Occupations	0.0	0.3	0.4	5,916
Administrative occupations	0.0	0.3	0.4	4,628

Secretarial and related occupations	0.0	0.3	0.3	1,288
Skilled Trades Occupations	**0.0**	**0.1**	**0.1**	**5,631**
Skilled agricultural trades	0.0	0.1	0.1	527
Skilled metal and electrical trades	0.0	0.2	0.2	2,124
Skilled construction and building trades	0.1	0.1	0.1	2,015
Textiles, printing and other skilled trades	0.0	0.0	0.0	965
Personal Service Occupations	**0.6**	**0.1**	**0.7**	**3,963**
Caring personal service occupations	0.4	0.1	0.4	976
Leisure and other personal service occupations	1.1	0.2	1.3	987
Sales and Customer Service Occupations	**0.2**	**0.9**	**0.1**	**3,375**
Sales occupations	0.2	0.9	0.1	2,755
Customer service occupations	0.2	1.2	1.4	620
Process, Plant and Machine Operatives	**0.2**	**0.6**	**0.8**	**3,824**
Process, plant and machine operatives	0.0	0.9	0.9	1,824
Transport and mobile machine drivers and operatives	0.3	0.3	0.6	2,000
Elementary Occupations	**0.3**	**0.2**	**0.5**	**5,245**
Elementary trades, plant and storage-related occupations	0.1	0.1	0.2	1,713
Elementary administration and service occupations	0.4	0.2	0.6	3,532
All	**0.3**	**0.4**	**0.7**	**47,879**

Notes

1. Source 2005/06 and 2006/07 BCS.

2. Based on adults of working age, in employment.

The BCS shows a large variation in the risk of violence at work across occupational groups. People in protective service occupations, for example police officers, fire service officers and prison service officers, had the highest estimated risk, with 6.4% having experienced one or more incidents of actual or threatened violence while working during the year prior to their interview.

Work-related violence – the legislation

[V8004] In common with most countries, in the UK there is little legislation or regulation designed to deal specifically with the issues of work-related violence. Instead, we need to look to several existing laws that provide for regulation, enforcement and remedy in such cases.

In simple terms, the law provides:

- Regulation to ensure that employers provide a working environment in which the risks of being subjected to violence are minimised and the fears of employees are taken into account. Employees also owe a duty of care to themselves and to others who may be affected by their acts or omissions;
- Enforcement to provide sanctions if employers fail to comply with the regulation and punishment of individuals who act violently towards another person in a workplace context; and
- Remedy for employees who become victims of workplace violence. This will usually be in the form of compensation through an employment tribunal, civil court, or criminal injuries compensation board;

The legal responsibilities and rights relating to workplace violence are contained in a range of provisions that include:

- Health and safety legislation.
- Employment law.
- Criminal law.
- Civil law.

Statutory health and safety requirements

Health and Safety at Work etc Act 1974

[V8005] Employers have a legal duty under the *Health and Safety at Work etc Act 1974 (HSWA 1974)* to ensure, so far as is reasonably practicable, the health, safety and welfare at work of their employees. [*HSWA 1974, s 2(1)*].

Besides a common law 'duty of care' to others, employers have a statutory duty to do 'everything reasonable and practicable' to 'eradicate or minimise' the risk of harm from all hazards to health – including violence.

Employers are also required to conduct their undertaking in such a way as to ensure, so far as reasonably practicable, the safety of other people who are not their employees, and to whom the premises have been made available.

The 1974 Act is considered as the most important piece of legislation dealing with health and safety at work. Although it was implemented to cater for the risks associated with industries such as chemicals, mining and construction work, the basic principles are translated relatively easily within the context of the protection required against the risks of workplace violence. Lord Skelmersdale, who chaired the DHSS Advisory Committee on Violence to Staff wrote in the 1988 report:

> Where violent incidents are foreseeable employers have a duty under section 2 [of the Health and Safety at Work etc Act 1974] to identify the nature and the extent of the risk and to devise measures which provide a safe workplace and a safe system of work.

The Committee was set up in response to the death of a social worker in 1986.

The Health and Safety Executive publication, *Successful Health and Safety Management*, (HSG65), says that accidents and ill health are seldom random events. They generally arise from the failure of control and involve multiple contributory causes.

The duty of care also embraces employees. Employees are required under *section 7* of the *HSWA 1974 to* 'take reasonable care for the health and safety of himself and of other persons who may be affected by his acts or omissions at work'. The duties placed on the employee do not reduce the responsibility of the employer to comply with his health and safety duties under the 1974 Act.

Management of Health and Safety at Work Regulations 1999

[V8006] The *Management of Health and Safety at Work Regulations 1999 (MHSWR) (SI 1999 No 3242)* requires employers to undertake a 'suitable and sufficient' assessment of the risks to which employees are exposed while they are at work and put in place measures to eliminate, minimise or control the risks. In many workplaces the potential for violence is a significant hazard and thus the duty on employers extends to the risk of violence.

Employers must:

- Establish how significant these risks are;
- Identify what can be done to prevent or control the risks; and
- Produce a clear management plan to achieve this.

To satisfy the requirements employers must put into place a comprehensive policy for dealing with work-related violence. This will include effective measures for identifying and assessing risk, clear guidelines and procedures for dealing with violent incidents and their aftermath, and the provision of appropriate training and equipment.

Reporting of Injuries, Diseases and Dangerous Occurrences Regulations 1995

[V8007] Employers must notify their enforcing authority in the event of an accident at work to any employee resulting in death, major injury or incapacity to work for three or more days. This includes any act of non-consensual physical violence inflicted on a person at work. [*Reporting of Injuries, Diseases and Dangerous Occurrences Regulations 1995 (RIDDOR) (SI 1995 No 3163)*].

Safety Representatives and Safety Committees Regulations 1977 and the Health and Safety (Consultation with Employees) Regulations 1996

[V8008] Employers must inform and consult with employees in good time on matters relating to their health and safety. Employee representatives either:

- appointed by recognised trade unions under the *Safety Representatives and Safety Committees Regulations 1977 (SI 1977 No 500)*; or
- elected under the *Health and Safety (Consultation with Employees) Regulations 1996 (SI 1996 No 1513)*,

may make representations to their employer on matters affecting the health and safety of those they represent.

This places a responsibility on the employer to consult and inform on issues of violence at work.

Emergency Workers (Scotland) Act 2005 and Emergency Workers (Obstruction) Act 2006

These pieces of legislation were introduced to give additional legal protection from assault for emergency workers. The *Emergency Workers (Scotland) Act 2005* makes it a specific offence to assault, obstruct, or hinder someone providing an emergency service or someone assisting an emergency worker in an emergency situation. The maximum penalty is a 12-month prison sentence and/or a £10,000 fine.

The *Emergency Workers (Obstruction) Act 2006* came into force in February 2007 and applies to England, Wales and Northern Ireland. Under the Act, it is an offence to "obstruct or hinder" particular groups of emergency workers responding to "blue light" situations. The Act defines emergency workers as firefighters, ambulance workers and those transporting blood, organs or equipment on behalf of the NHS, coastguards and lifeboat crews. (The police already have their own obstruction offence in the Police Act 1996.) The maximum penalty for an offence is £5,000.

Employment law

The Employment Rights Act 1996

[V8009] *Section 44* of the *Employment Rights Act 1996 (ERA 1996)* relates to 'Health and safety cases' and reminds us that responsibility for safety at work is something shared between employers and employees. The section covers:

- Employees statutory entitlement to a safe way of working, and to be able to fulfil their responsibility under the *HSWA 1974, s 7* to take care of themselves and others, without fear of recriminations ('any detriment') from their employer for doing so.
- The circumstances in which an employee should withdraw from danger (see **V8014** below).
- Employees duty and right to be able to draw attention to safety deficiencies that may exist and to take 'appropriate action' to withdraw/remove themselves from 'serious and imminent' danger that they would be unable to avert.
- Warning to employees that if they continue to knowingly and recklessly undertake work that is unsafe and get injured they may not be able to make a claim against their employer for liability.

Contract law

[V8010] A contract of employment imposes two obligations on the part of the employer:

- The employer must provide a workplace in which employees are subjected to minimal exposure to risk.
- The employer must provide 'trust and support' to the employee in carrying out their role.

It is well established that workplace violence should be regarded as a risk in this context. The employer is expected to provide 'trust and support' to ensure that the employee is working under minimum stress and that the employer responds to issues of perceived risk as well as real risk.

Failure to provide appropriate support might reach a point where an individual can no longer tolerate the working conditions and decides to leave. This could be construed as constructive dismissal where the situation is treated as if the employee had been dismissed. The individual will usually seek a remedy through an industrial tribunal or a county court.

The obligations of an employer are unambiguous. There is a clear requirement to assess the possible risk of an employee being subjected to violence and to provide appropriate measures which will minimise that risk. There is also a requirement to recognise and respond to the perceived fears of employees by providing appropriate support in relation to those fears.

Criminal law

[V8011] The laws in relation to violence, disorder, assault and threats are well established and cover almost every incident of work-related violence. Recent legislation offers greater protection against harassment and more serious penalties for racist incidents. Thankfully, the incidence of serious assault is comparatively rare but this brings with it a difficulty in relation to prosecuting the offender. In general, it requires the police to pursue the matter to prosecute the offender and many different issues have to be addressed when making the decision.

Quite often the nature of the assault is not serious enough for a criminal charge to be pursued or, for a variety of reasons, the circumstances do not warrant it. It is unlikely that an assault that results in only minor bruising, for example, would find its way into a court. Whilst these decisions often make economic, procedural or legal sense, they leave the victims feeling frustrated and unsupported. A common cause of dissatisfaction felt by employees who have been assaulted is related to the apparent lack of action that follows the incident. Organisations need to be clear as to their stance on prosecution and their preparedness to support a private/civil action.

Civil Law

Negligence

[V8012] The tort of negligence imposes a duty to take reasonable care to avoid acts or omissions that may cause reasonably foreseeable harm to others. This common law duty of care runs in parallel to statutory health and safety requirements.

In cases where the victim has suffered serious injury, either mental or physical, they will undoubtedly look for substantial compensation for things like their loss of earnings, pain, suffering, and inconvenience. Although it is clear that the person who actually committed the act of violence is directly at fault, it is unlikely that the individual will have the funds to provide for any substantial compensation award. Consequently, the victim will be more likely to seek redress through his or her employer by claiming that the employer was negligent in providing appropriate measures to prevent the incident from happening, and/or in failing to provide adequate support post-incident.

In such cases, key questions will be asked:

*   Was violence in the workplace something that ought to have been assessed?
*   If so, what were the risks of violence that would have been identified in the risk assessment?
*   What reasonably practicable control measures would the risk assessment have identified should it have been put in place?

An employer who has failed to assess the risk of violence will have great difficulty in defending a legal action for injury caused through employer negligence.

Vicarious liability

[V8013] In recent years the boundaries have been extended even further to cover a company's responsibility for its employees' actions. 'Vicarious liability' is the legal concept whereby a company can be liable for the negligent conduct of its employees. A claim for compensation for personal injury will be successful if the conduct occurred during the course of employment.

In *Fennelly v Connex South Eastern Ltd* [2001] IRLR 390, a passenger was assaulted by a ticket inspector, and the Court of Appeal held that the rail company was responsible for the conduct of the inspector and was therefore liable.

Withdrawing services

[V8014] At what point is it appropriate for an employee to say that they are not prepared to continue because they feel they are in danger? This is a question which is faced from time to time where an employee feels that they are in danger of being physically hurt and they are not prepared to continue to deal with a situation. Legislation makes clear the duties upon employers to protect staff, and *s 7* of the *HSWA 1974*, *Reg 8* of the *MHSWR 1999 (SI 1999 No 3242)* and *s 44(1)(d)* of the *ERA 1996* also place clear requirements upon employees to take action if their safety is compromised.

The law relating to the use of force and self defence

[V8015] The use of force is considered under both common law and statute law and it is important that those employees likely to need to use force are aware of the law relevant to their circumstances. Although the powers are drawn from various sources, most specify that force can only be used where it is absolutely necessary, and that the amount of force used must be reasonable in the circumstances. The *Human Rights Act 1998* adds the term 'proportionate'.

Compensation cases and prosecutions of employers arising from work-related violence

A number of employers have found themselves in court following violent incidents in the workplace in addition to Connex (see **V8013** above). These include the following:

- Merseycare NHS Trust was found to be guilty of breaching the Health and Safety at Work Act when it failed to carry out sufficient procedural checks, resulting in a care worker being beaten unconscious by a schizophrenic who threatened to kill her. One of her colleagues was also injured. The Trust was ordered to pay a fine of £12,000 (*R v Merseycare NHS Trust* (Ormskirk MC)(5 September 2002, unreported));

- In May 2005 South West London and St George's Mental Health Trust was prosecuted by the HSE for breaching the Health and Safety at Work Act and fined £28,000 with £14,000 costs after a healthcare assistant was attacked and killed by a psychiatric patient. The junior member of staff was working alone without clear procedures, and with inadequate measures in place to check on his safety.

- In *Cook v Bradford Community Health NHS Trust* [2002] EWCA Civ 1616, [2002] All ER (D) 329 (Oct) a psychiatric hospital worker was attacked by a patient in the "seclusion suite" of a unit for violent patients she had entered in order to take cups of coffee to colleagues. The Court of Appeal said that her employer had a duty not to place her unnecessarily in a position where there was a risk of foreseeable danger.

- In *Patterson v Tees and NE Yorkshire NHS Trust* (2007), unreported a senior psychiatric nurse was assaulted by a patient with a history of violence. In this case, although *Patterson* was aware that a patient had absconded and then returned to the hospital following a broadcast about him warning the public to take care, he had not been warned that the patient had made abusive comments about him. The Trust was found to be negligent because of this failure, and it had also failed to continuously risk assess patients on an event-by-event basis.

- In *Collins v First Quench Retailing Ltd* [2003] SLT 1220, an employer was ordered to pay £179,000 in compensation to an off licence manager who was attacked in an armed robbery. The woman had been working alone in the off licence, which had a history of incidents including previous armed robberies. In addition, she had requested that the employer provide better security by fitting screens and ensuring that there were always two members of staff present. Although the employer fitted panic buttons and CCTV and arranged for two staff to work on the evening shifts, the Judge in the case said that lone workers were generally easier to attack, and that having two employees in the shop at all times would have materially reduced the risk.

- In *Smith v Welsh Ambulance Service* (Chester County Court) (22 March 2007, unreported), a paramedic went into a derelict building alone to treat an unconscious drug addict and was threatened by other drug addicts. The trial judge accepted that members of the emergency services can face risky situations, but said that control room staff should have been trained to do risk assessments to help them make "an intelligent decision" about whether to send in lone workers into a particular situation. In addition,the paramedic had not been given adequate training as to whether he had a choice about giving assistance in risky circumstances. The employer was found to have failed in their duty of care.

(The cases of *Patterson*, *Collins* and *Smith* were highlighted by the trade union solicitors Thompsons in *Personal Injury Law Review Autumn 2007 Issue 004* www.thompsons.law.co.uk).

Developing and implementing policy

[V8016] Effective policies translate priorities into action and set a clear direction for the organisation to follow with defined accountabilities. In any organisation that has workers at risk from violence, a comprehensive and workable policy is the cornerstone for building an effective response. Effective policy development requires the active commitment of senior management and the involvement of workers affected by violent incidents. Unless an implementation plan is put in place supported by adequate resources to drive and embed change the policy is likely to remain as words on a page. Once implemented, the processes, procedures, equipment, training and incidence of violence need to be monitored, analysed and evaluated to ensure a proactive and problem-solving approach to the ever-changing risks associated with work-related violence and aggression.

Problem recognition

[V8017] The most important first step developing policy is to define what 'workplace violence' means within the organisation. Most organisations use the definition offered by the HSE:

Any incident in which a person is abused, threatened, or assaulted in circumstances relating to their work.

This definition encompasses a range of unacceptable behaviour likely to cause physical or psychological harm. Importantly, it includes incidents where someone uses abusive language or behaviour that does not amount to a threat of or actual physical assault. It also covers incidents that might occur when people are not actually at work, for example, being followed home or confronted when arriving at or leaving work.

Conducting a review

[V8018] Having defined violence the next step in developing an effective policy is to understand the nature and extent of the problem facing the organisation and the effectiveness, or otherwise, of existing policy, procedures and guidance. This is best achieved through a systematic review of existing arrangements. The review should include 'top down' and 'bottom up' elements of assessment to ensure that problems facing front line workers are understood and a 'reality check' made on the effectiveness of existing policy and procedures.

The purpose of such a review is to establish:

- the nature and extent of the problem;
- the specific roles that are at risk;
- the nature of the risks;
- the solutions to minimise the risks identified; and
- the effectiveness or otherwise of existing policy and procedures.

Collaboration

[V8019] An important element in the eventual success of such a review will be the approach adopted from the outset. It is relatively easy to secure employees' collaboration because the intention of the review is primarily to ensure they operate in an environment where risks from work-related violence are minimised.

Sources of information

[V8020] To carry out an effective review, there are a variety of sources from which data and information can be gathered. These include:

- Generic violence risk assessment of specific roles;
- Interviews, observation, focus groups and questionnaires; and
- Analysis of data.

Generic violence risk assessment of specific roles

[V8021] The people who face the customers, clients, patients and public every day are best placed to identify and understand the risks they face from violence and aggression, and risk assessment of roles is a statutory requirement under health and safety legislation.

Interviews, observation, focus groups and questionnaires

[V8022] It is important to gather views from front line staff about the effects that the risks of workplace violence have upon them.

The information can be gathered through informal interviews with affected staff, focus groups of staff from a particular role, or through a questionnaire which is designed to extract the appropriate information. Often, a combination of these techniques is used.

There are issues around confidentiality and individuals often fear their views will 'get back to management'. Confidentiality must be guaranteed, be evident in the procedure and be completely respected in practice.

Analysis of data

[V8023] The data available will vary according to the nature of the organisation and the sophistication of the reporting and recording procedures available. In the best case, a dedicated reporting system will inform on not only the number of incidents of violence and aggression, but also provide information about times of day, locations, nature of injuries, frequency and types of injury, roles most at risk, and so on.

Designing the policy

Consultation and collaboration

[V8024] Once the nature and extent of the problem of violence is fully understood consideration can be given to designing policy that will provide direction and focus for action.

It is important that a policy development group is formed and made up of people who understand the issues and practical implications of the decisions that are made. In practice, such a group will probably exist for the general health and safety issues in the form of a health and safety committee or group (see figure 1 below).

Figure 1: Model for effective policy development group

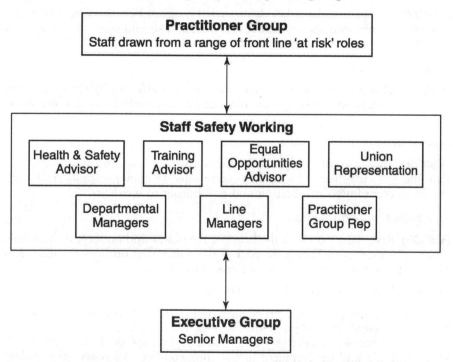

In this model, the practitioner group drives the policy formation and therefore the front line staff have ownership of the issues from the beginning and have been fully involved in the creation of the solution.

Policy content

[V8025] These are the basic elements of an effective policy:

- Purpose.
- Scope and definitions.
- Who the policy affects and their specific responsibilities.
- Risk assessment, reporting and review process.
- Risk reduction and training.

- Response to incidents:
  - – Incident management.
  - – Post-incident management.
- Reporting.
- Involvement of other agencies and sanctions.
- Maintenance.
- Communication strategy.
- Policy review and revision.

Purpose

[V8026] This part of the policy sets out the context and aims, and expresses the corporate values that support the existence of the policy.

Scope and definitions

[V8027] This part of the policy defines what is meant by 'workplace violence'. The HSE definition is recommended. This section should state clearly what is included within the policy and what is not.

Who the policy affects

[V8028] This part of the policy considers who is affected by the policy and identifies these individuals' specific roles and responsibilities in relation to the policy.

Risk assessment, reporting and review process

[V8029] This part of the policy outlines the risk assessment process and requirements, together with clear expectations of managers and staff with regard to recording, classification and monitoring of incidents.

Risk reduction and training

[V8030] Risk reduction is at the heart of the policy and underpins the whole point of its existence. This is the part of the policy that outlines the measures that are to be taken to minimise the risks identified.

The key elements of risk reduction will be:

- advice, guidance, and procedures aimed at safer working practices;
- equipment and design of the working environment;
- improved service delivery and reduction in triggers of violence;
- training and development in handling conflict and potentially violent situations.

Response to incidents

[V8031] The response will typically cover two key areas:

(1) Incident management
This part of the policy outlines the procedures that managers and staff are required to follow should a violent incident occur, and provides clarity on roles, responsibilities and communications. It also includes communications with other agencies such as police, security and media. It should state the position with regard to use of physical intervention and the provision of appropriate guidance and training.

(2)   Post-incident management

This part of the policy clearly defines post-incident management procedures include reporting requirements, operational support, and immediate and longer-term staff support. It should also state the responsibilities of line managers and their health and safety and occupational health functions.

Involvement of other agencies and sanctions

[V8032] This part of the organisation policy will outline guidance about the involvement of other agencies such as the police and local authorities. It should also outline the stance on prosecution and the options available including legal/financial support for employees and, where appropriate, the policy and guidance concerning exclusion and withdrawal of services.

Maintenance

[V8033] Performance indicators need to be established and managers held accountable for ensuring that breaches of policy and guidance are addressed.

Communication strategy

[V8034] This part of the policy states the corporate position with regard to internal and external publicity, and identifies key messages/statements to be promoted. It should also include a strategy for managing media attention following an incident.

Policy review and revision

[V8035] This part of the policy establishes a monitoring and review process that allows periodic re-evaluation of the strategy and policy.

Implementation plan

[V8036] Implementation is about making the policy happen. It includes educating people about new processes and procedures, ensuring managers understand and monitor new practices, introducing and using new equipment, and setting up and running a programme of training and development. It is important that the implementation plan is adequately resourced to ensure that change is driven forward and embedded in the organisation.

Communication

[V8037] There are two aspects to this – internal communication to the people who work within the organisation – and external communication to the people who make up the clients, service users and the general public.

Internal communication

[V8038] As always this will depend upon the structure, size and nature of the particular organisation. If it has a department that deals with communication and the media, their help will be invaluable. There are some factors that will help to make the communication process successful.

•   Plan

Thought needs to be given to the target audience, the communication structures in the organisation and how to make the information accessible.

- Policy launch
  It is much more positive if people recognise the start of a new process.
- Workplace briefings
  Front line managers should be given a full understanding of the policy first, and then should communicate this to their staff.
- Easy access 'help desk'
  A system is needed where people can quickly have their queries, worries and doubts dealt with.
- Reinforcement
  Reinforcement of the messages with a second wave of communication needs to be made some weeks later by checking out how well things are going.

External communication

[V8039] The media will naturally look for a story or a slant and care must be taken that the original good intentions are not skewed to provide an interesting angle – at the expense of the launch.

New systems, procedures, practices and equipment

[V8040] The implementation of a policy will involve the introduction of new systems of working, new procedures, practices, and perhaps equipment, to learn about and to follow or use.

Implementing the policy

[V8041] If the organisation has developed an effective strategy and plan, the actual implementation should be relatively straightforward.

To safeguard the organisation it is necessary to show a training and development plan that clearly outlines the phases of training being undertaken and the thinking that has gone into the development of those phases and the sequence of training for different roles.

Evaluating the policy

[V8042] It is very likely that polices and practices which made perfect sense in theory will run into trouble when applied in the workplace. It is important therefore to monitor and evaluate the implementation of the policy (see figure 2 opposite).

Figure 2: A model for monitoring and evaluating policy

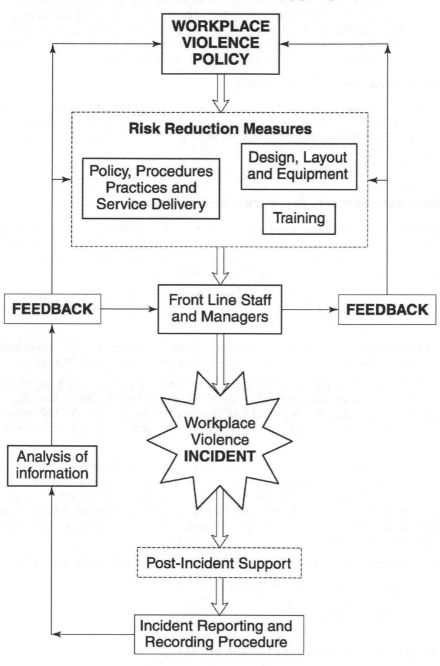

In this model there are two feedback loops.

- An 'immediate loop' where the quality of the measure is evaluated when first received. This will include information about the initial perceptions of the staff about quality of the training, or the usefulness of the equipment, or the workability of a particular policy or procedure.
- The second feedback loop is through the reporting and recording procedure and the quality of this will be defined by the quality of the information and the information gathering techniques.

Co-ordination and revision of policy

[V8043] It is advisable to have a single point of reference for monitoring the effectiveness of the policy and its measures. All the immediate feedback and information from the reporting and recording process should go through this point of reference.

Risk assessment for workplace violence

[V8044] Risk assessment lies at the heart of an effective organisational response to workplace violence. Without a deep understanding of the nature and extent of the risk it is impossible to develop a targeted preventative strategy that will successfully address the hazards.

The legal requirement for employers to carry out risk assessments is contained in the *Management of Health and Safety at Work Regulations 1999 (SI 1999 No 3242), Reg 3* which states:

> Every employer shall make a suitable and sufficient' assessment of . . . the risks to health and safety of his employees to which they are exposed whilst at work . . . .

'Suitable and sufficient' is not defined but detailed guidance is set out in the Management of Health and Safety at Work: Approved Code of Practice & Guidance L 21 (ACOP). National Occupational Standards (see **V8142** below) also re-enforce the importance of the risk assessments in managing work-related violence.

The above Regulations, Code of Practice and standards also require employers to review, and if necessary, modify their risk assessments following any adverse event. Assessment is not to be regarded as a once-and-for-all activity. This idea links together the elements, risk identification, risk assessment, risk control, incident report, incident investigation and risk review into a continuous improvement cycle.

There are three basic ways in which the risk of violence can be assessed:

- Generic violence risk assessment of role.
- Risk assessment of pre-planned event.
- Dynamic Risk Assessment.

Generic violence risk assessment of role

[V8045] This is perhaps the most recognised form of risk assessment. It involves an analysis of all the activities commonly associated with each job role that carry a level of risk from work-related violence. Each activity is examined, the possible risks are identified, and a judgement is made about the likelihood of the perceived risk and the severity of the consequences.

A good example to illustrate the point is any role that has an element of working alone and visiting clients or service users in their homes.

Some of the risks inherent in such a role are:

- Being targeted by criminals when travelling and in car parks, isolated locations, entrances to flats and apartments.
- Violence from a service user or other individual inside the person's home.
- Being held against will in premises.
- Being injured in an isolated location without access to help.

Measures to eliminate, minimise or control could include:

- Procedures in relation to callout, providing details of location, client visits, times of appointments to a central source and call in when complete.
- Personal safety guidance to reduce vulnerability when travelling and visiting.
- Training in Dynamic Risk Assessment (see **V8047** below), conflict resolution, communication skills etc.
- Provision of equipment where appropriate such as a mobile phone or attack alarm.

Each role should be examined in this way to determine all the risks and measures appropriate for that role.

Once this assessment and the resulting control measures have been established for a role, then everyone performing that role should be required to have the appropriate equipment, conform to the procedures and undertake the training required.

Management should ensure that the generic violence risk assessment is carried out on all roles where there is a potential danger and that the control measures are clearly stated and understood by everyone in each role.

Violence risk assessment – pre-planned event

[V8046] The generic violence risk assessment caters well for the main activities of a role. However, in every job there are events and activities that are out of the ordinary or, perhaps, done only rarely. It is quite possible that the generic risk assessment will not cover every circumstance and, in such cases, it is important to assess the risk involved in a special event or activity as part of the planning process. The risk assessment will identify if there is an increased risk of potential for violence and a need for extra measures in these circumstances.

The risk assessment should be recorded and included in the planning documentation for the event itself.

Dynamic Risk Assessment

[V8047] There are many occasions when front line staff are confronted with situations which are unique and not catered for in a generic risk assessment or an assessment of a planned event. These are often the situations where people get hurt because they do not have a way of assessing the risks in the situation they are confronted with, and do not respond appropriately.

'Dynamic Risk Assessment' is a process which helps an individual to effectively assess a situation from a personal safety perspective, as it is unfolding. The person can continuously assess the circumstances and adjust his or her response to meet the risk presented moment by moment. Training which develops understanding and skills in Dynamic Risk Assessment can help an employee to assess situations more effectively and respond in the most appropriate way.

Risk assessment – who should do it?

[V8048] There is little doubt that the individual who is working at the customer interface is best placed to be able to identify the risks faced. For generic violence risk assessment of roles and pre-planned events, the process needs to include someone who is trained and experienced in identifying risks and who can spot the hidden dangers in workplace environments, situations and incidents. The most effective solution is to combine the two – a person who is trained and competent in risk assessment and has experience in the management of violence, should examine the roles alongside the people who carry out those roles on a day-to-day basis.

Risk assessment – how should it be approached?

[V8049] It is recommended that risk assessment be approached on a role-specific basis rather than purely on a geographical area or departmental basis. By definition, the risk of workplace violence will always involve some sort of interaction with another person. This unique feature means the risks of work-related violence are much more difficult to predict and to control because it is hard to predict the range of responses that someone might use in a situation involving conflict.

For each role the following main areas need to be assessed:

- What contact is made with the clients, patients, service users or members of the public? This will usually be described in the job description. For example, enforcing role, caring profession, advising or dealing with complaints.
- What sort of people generally make up this contact? For example, travelling public, offenders, patients, clients or relatives.
- Where does the contact take place? For example, interview rooms, reception, platform, home address, classroom or on the streets.
- What sort of 'state' will the client/patient etc be in? For example, will they be influenced by drink or drugs, anxious, frustrated or hostile?

- What specific tasks are performed? For example, handling complaints, delivering 'bad news', denying benefits, evicting or arresting, or even treating an individual.

Content of the risk assessment

[V8050] The first question that should be asked is 'Is there any likelihood that someone doing this role will be abused, threatened or assaulted in circumstances relating to their work?' If the answer is 'yes', then a risk assessment for workplace violence needs to be carried out.

Figure 3 below outlines five steps that provide a basic risk assessment model for both the generic risk assessment and for the risk assessment of a pre-planned event.

Figure 3: Risk assessment model

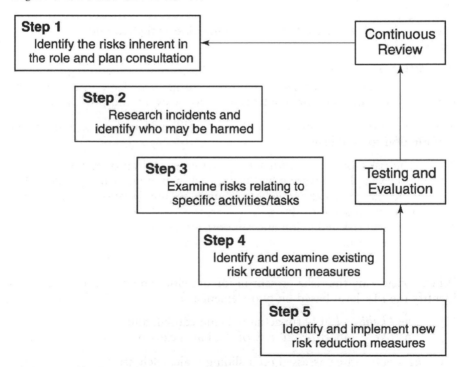

Step 1 – Identify the risks inherent in the role and plan consultation

[V8051] An examination of the job description should show the potential situations where the jobholder will be at risk. However, few job descriptions will capture all the possible scenarios that might arise and other sources of information should be used.

It is important to research other potentially useful sources of information such as HSE and sector-specific guidance and research.

Consultation with workers needs to be planned to ensure this covers a cross section of staff performing the roles being risk assessed. The consultation can take the form of a combination of the following:

- Structured and informal interviews.
- Small focus groups (typically 3–6 staff).
- Questionnaires.
- Workplace observation.

Step 2 – Research incidents and identify who may be harmed

[V8052] An examination and analysis of past-incident reports, local log books etc associated with workplace violence will provide invaluable data in a risk assessment. Unfortunately violence is under-reported in most organisations so it is important to try to establish the true extent of the problem by consulting with managers and staff, and gathering anecdotal evidence of occurrences. Interviews, workplace observations, discussions and surveys should be carried out.

Step 3 – Examine risks relating to the specific activities and tasks performed

[V8053] Assessing the level of risk has its difficulties. By reviewing the actual activities and tasks performed it is possible to get a realistic measure of the risks involved and to ensure controls are both relevant and proportionate.

This stage involves an analysis if the data obtained in Step 2 and further consultation to establish:

- The risks associated with specific activities and tasks performed
  For example, delivering bad news, cash handling, or fare evaders. The risks may be compounded at certain times and locations, or by factors relating to drink or drugs.
- Practical risk reduction and support measures
  Those performing the job will often have the best ideas for reducing risk.

When considering the risks present in job activities and tasks, it is helpful to identify the risk level based on two elements:

(1) the *likelihood* of harm actually being caused, and
(2) the likely impact or *severity* of the harm caused.

The *likelihood* can be assessed on a sliding scale, such as:

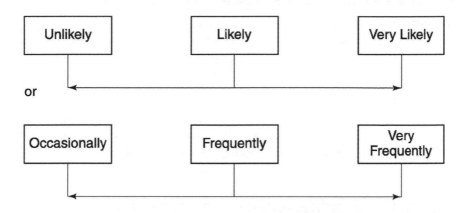

or

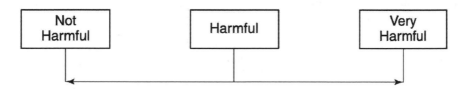

The *severity* of harm can be assessed on a similar scale such as:

From this, a 'risk rating' can be established which identifies the risk as being low, medium or high. When considering the severity of an incident there are number of considerations including:

- Physical harm to those involved.
- Emotional impact on staff.
- Impact on the perpetrator and other service users.
- Numbers of people involved.
- Duration of the incident.
- Effect on operations and service delivery.
- Financial impact on both individuals and organisation.

The likelihood/severity approach to risk assessment is long established and is helpful in providing a broad assessment of risk. However, it does not take into account the fear of violence experienced by the person doing the job. An employer has a duty to include this in the risk assessment and respond to this fear, for example through rationalising fears in training and introducing guidance to further minimise the risk.

Step 4 – Identify and examine existing risk reduction measures

[V8054] There will be risk reduction measures already in place. These existing measures need to be compared with the risks identified in the assessment and an evaluation made about their effectiveness and how relative they are to the current risks.

Step 5 – Identify and implement new risk reduction measures

[V8055] Effective measures for reducing the assessed risks should be identified and implemented. These may be a combination of safer working practices and procedures, policy, training and equipment.

Continuous review

[V8056] The risk assessment for each role should be a 'live' document. Staff should have easy access to it and should be encouraged to involve themselves in the process of keeping the assessment live and up-to-date.

Reporting, recording and monitoring system

[V8057] Central to an effective strategy for combating workplace violence is an understanding of the nature and extent of the problem. Integral to this is a system that monitors and provides data and information about the trends surrounding violence in the workplace for a particular organisation.

Figure 4 opposite shows a typical model of an effective reporting and monitoring system for workplace violence.

Figure 4: A model for reporting and monitoring incidents of workplace violence

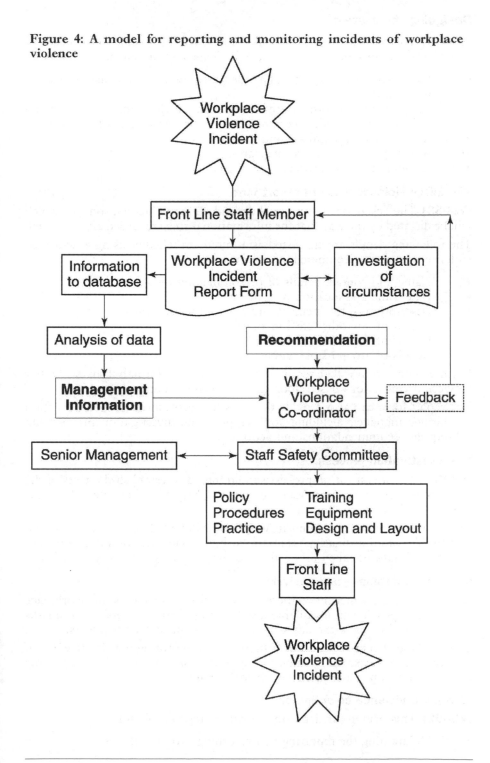

Designing the system

[V8058] The outcomes required of the system will be specific to the needs of a particular organisation but the essential elements of a system are:

• Information gathered through a dedicated workplace violence incident report form.
An investigation process that will examine the circumstances of an incident and provide recommendations for improving risk reduction.
• An information database and analysis process.
• A workplace violence co-ordinator.
• A communication process.

Workplace violence incident report form

[V8059] The design of the form will depend on many factors, some of which will be dictated by the way that the information is to be recorded and analysed.

The following details are suggested as a minimum whether using a dedicated violence report, or a combined use report form:

• Name and contact details of person making the report.
• Date, time and location of incident.
• Description of the incident and any injuries sustained.
• Details of any other staff involved.
• Details of the assailant(s).
• Details of any possible witnesses.

Further information will be required from the incident form, but this will differ between organisations based upon the management data required.

The report form can also provide the basis for recording post-incident action by the line manager, including staff support, the investigation process and subsequent recommendations and actions.

The investigation process

[V8060] The investigation process should intend to establish the facts of the incident and make recommendations that will help to reduce the risk of this type of incident happening again.

The recommendations from this investigation will be fed directly to the person responsible for co-ordinating issues of workplace violence for a decision as to how they should be taken forward.

Information database and analysis

[V8061] The information gathered from all the incidents of workplace violence can provide an invaluable insight into the trends, hotspots and risks that are not obvious in the day-to-day management of such incidents.

The database also provides a powerful tool for monitoring the effectiveness of risk reduction measures by monitoring over a similar period to see if the measures have impacted upon the identified risks.

Workplace violence co-ordinator

[V8062] This person (or department) will be responsible for:

• Maintaining the reporting and recording procedure.

- Monitoring the quality of the content and the investigation process.
- Monitoring the recommendations from the reporting and investigation process.
- Searching and monitoring the database for useful management information, trends, and hotspots that need further investigation.
- Informing the Staff Safety Committee of recommendations and work-related violence issues for discussion and further action.
- Monitoring the effectiveness of risk reduction measures.
- Liaising with the police and other relevant agencies.
- Administering or assisting in the process of litigation or action against the attacker.

A communication process

[V8063] A clear communication process should be built into the system so that the front line staff are kept fully aware of the most up-to-date policy, procedure and practice for their role. A common complaint is that individuals complete incident forms and then hear nothing more about the outcomes.

Implementation

[V8064] The following areas are essential for the successful implementation of a reporting process:

- Management involvement and training.
- Staff training.
- Monitoring the completion of reports.
- Provision of feedback to those affected and information to the wider organisation.

Manager involvement and training

[V8065] Managers at every level need to understand and appreciate the complete reporting, recording and monitoring process.

Staff training

[V8066] Training helps to establish best practice and a sound consistent approach. It ensures that individuals understand the importance of the process and that completion is part of their duty of care.

Monitoring the completion of reports

[V8067] Four fundamental questions need to be asked on a routine basis:

- Are all relevant incidents being recorded?
- Are reports correctly entered?
- Are accounts of incidents adequate and professional?
- Is anything actually being done as a consequence?

Feedback and regular provision of information across the organisation

[V8068] Following the introduction of a reporting process, particular effort must be made to provide immediate and ongoing feedback in relation to reported incidents. This applies not only to the affected individual, but also across the whole organisation.

A *formal review* of the reporting system should be undertaken some months after initial implementation to establish whether the system is being used, whether it can be improved and what is being done with the information generated.

Measurable outcomes

[V8069] Measurable outcomes using information from the reporting process may include:

- The number of injuries to staff.
- The number of injuries to service users.
- The extent and type of injuries suffered.
- The number of working days lost due to assaults on staff.
- The number of prosecutions taken against those assaulting staff.
- Extent to which safety procedures have been followed.
- Which techniques taught in training have been used and their effectiveness.
- Incident and assault patterns.
- The groups at most risk.
- The highest risk tasks and activities.
- Profiles of perpetrators of violence.

The more sophisticated the database, the more comparisons and measures available.

Under-reporting

[V8070] It has been mentioned several times that most organisations suffer from under-reporting of workplace violence-related incidents. There are many reasons for this, for example, over complicated and lengthy reporting forms and processes, fear of criticism after having brought an incident to light.

The organisation should work to eliminate as many barriers to reporting as possible.

Disclosure rules

[V8071] Written records of any incident that results in a court case may be called upon as evidence. This would include any internal incident reports, notes, debriefing notes, statements or correspondence.

The purpose of such 'disclosure' is to ensure that all potential sources of evidence are made available to defendants and the courts. In criminal cases the rules are drawn from the *Police and Criminal Evidence Act 1984* and for civil cases they are drawn from the *Civil Disclosure Rules* and *Civil Procedure Rules*.

From a practical point of view the purpose of the reporting process is to ensure that reports are essentially factual and as free as possible from conjecture and opinion.

Risk reduction measures

[V8072] The *Management of Health and Safety at Work Regulations 1999 (SI 1999 No 3242), Reg 4* and *Sch 1* of require employers to design and implement protective and preventive measures on the basis of hierarchy of general preventative principles. The associated Approved Code of Practice & Guidance L 21 (ACOP) contains further guidance. The principles may be summarised as follows:

- Eliminate risk by avoiding or removing the hazard
- Substitute the process etc by one that carries less risk
- Control the risk at source
- Devise safe systems of work
- Provide adequate training and instructions
- Ensure adequate supervision is provided
- Provide personal protective equipment as a last resort

Most risk reduction measures fall under one of the following broad categories:

- Training and Information
- Work environment
- Job design

It may not be possible to completely eradicate violence in a workplace but it is possible to minimise the risk of violence and its impact by introducing appropriate control measures based upon an understanding of the particular problems. Effective risk reduction is a blend of measures that embrace all the areas of policy, effective risk assessment, practice and procedures, training, equipment, design and layout.

Risk reduction – achieving a balance

[V8073] Appropriate risk reduction measures are a balance between understanding the identified risk and providing a balanced response in terms of acceptability, cost and relative effectiveness. This is not an easy equation to balance and it requires some thought before a reduction measure is introduced. The appropriate balance is, in the end, a matter of judgement, which can only be made by the organisation itself. The cost of a glass screen may be relatively modest and the simple balance may suggest it should be installed. However, other aspects can be added to the equation; is a glass screen acceptable in terms of the organisation's ethos and approach? Does it, in fact, reduce the level of violence – or does it provide a trigger for the level of aggression to increase? On the other hand, if there is a high risk of serious violence involving weapons this may outweigh other considerations and it may be imperative that bullet-proof glass screen is in place.

Service delivery

[V8074] Workplace conflict can occur for many different reasons. Sometimes organisations and staff can create or exacerbate the environment within which a conflict develops and increase the risk of violence by the way they deliver services or approach their work.

Working practice

Staffing levels

[V8075] Queuing and apparent 'queue jumping', long waiting times, cancellations, missed connections and flights, poorly informed staff and rude or abrupt service are all the common sparks which can eventually ignite violence. Organisations should ensure that the ratios of staff to customers, clients or service users are realistic and allow for the cost effective provision of the best service.

Safe practice guidance

[V8076] Some roles have inherent risks associated with them and require specific guidance. For example, safe practice and communication systems need to be put in place for workers working alone, and for those working out in the community who can be isolated and vulnerable to crime.

Staff are more likely to adopt guidance that they have actually been involved in developing.

Systems and procedures

[V8077] Effective risk assessment will anticipate and deal proactively with situations likely to cause conflict. High demands on service, running out of stock, maintenance work likely to cause delays and particularly unpalatable news are some of the situations that can be anticipated and contingencies for minimising and resolving conflict put in place.

Those working in certain urban areas may be more likely to be exposed to theft, robbery, aggressive begging, prostitution, public disorder, drug dealing and drug abuse. Rural areas present different challenges, as staff can be a long way from help.

Some roles, particularly in health care, necessarily involve close contact with a person to carry out checks, tests and medical procedures. Some workers deal with aggressive people in confined spaces where there are risks to other people and little prospect of immediate help from the police or security staff.

Each risk should be examined and appropriate guidance and practice designed which will minimise the risks involved in the working practice or situation.

Communications

[V8078] Poor information and communication often causes or contributes to conflict. Effective communication can quickly calm and resolve potential conflict. Many people will accept the inevitability of a delay or cancellation – as long as they know how long it will be or what alternatives are available.

Design and layout

[V8079] Designers are becoming more aware of the issue of work-related violence and significant developments are being made in this area as it becomes clear that personal safety and good service delivery go hand in hand. Some pub chains for example use clever layouts to reduce potential frustrations and conflict flash points, thereby 'enhancing the customer experience', selling more drinks and reducing the risk of violence and aggression.

'Queue jumping' is a very common cause of conflict and effective signage and queue management can reduce this.

Risk can be further reduced with the introduction of wider desks, raised floors, and areas offering access to those with special needs or requiring greater privacy. Staff need to have access to a secure place should a serious incident occur, and private areas need to be clearly identified and secured.

Proactive service delivery

[V8080] The quality of service delivery has a major bearing on the incidence of violence in the workplace. A climate for trouble is created when people are frustrated and their expectations are not met and staff respond unprofessionally to their concerns.

Sometimes, staff cannot meet the customers' expectations because of circumstances beyond their control. The response should be to influence and adjust the customer expectation. The development of basic interpersonal skills and a service-oriented culture lays the foundations for a safer environment. This area is probably the most under-rated as a control measure.

Simple inexpensive gestures like providing refreshments, upgrades, free use of a telephone or concessions can help to reduce tension, and it is important that staff are empowered to offer these, with appropriate support guidance.

Warning 'flags'

[V8081] A contentious issue for many organisations is the placing of warning 'flags' against an individual service user or address. Many organisations cite client/patient confidentiality, data protection and human rights as reasons for not placing warnings. Some also argue that if staff know someone could be a problem then this will affect the way they deal with them and risk a self fulfilling prophecy. Although each of these issues is valid and needs to be considered, employees also have rights and a balance must be sought in order to ensure staff safety.

Data protection agencies provide guidance on the use of such warning systems and their management.

Security responses

[V8082] Security responses available to organisations break down into three areas:

- Security personnel.
- Security procedures.
- Physical measures.

Security personnel

[V8083] Some organisations use external security contractors, others choose to employ an in-house team or mix the two. There are benefits and drawbacks to each.

If considering an external provider it is important to establish the level of investment that the provider makes in staff training and support. In a traditionally low paid area of work with a high turnover of staff, contractors are reluctant to invest in more than the bare minimum required for training.

More control over selection, training and performance is achieved through developing an in-house security team.

Whichever option is chosen it is important to remember that security staff need to be actively involved in the initiatives, so that they can win the trust and support of other staff and made to feel part of the team.

Selection

[V8084] Security staff, by the very nature of their role, will deal with conflict on a regular basis. It is important to select staff who demonstrate the attitudes and behaviours that help to deal positively with situations where conflict is inevitable.

Training

[V8085] Security staff play a key role in service delivery and are often the first and last point of contact for a customer or service user. It is therefore a worthwhile investment to develop their communication and interpersonal skills.

Security staff are often expected to confront potential criminals, respond to violent situations, and to protect staff and service users. They need to be equipped with the necessary knowledge and skills to manage conflict and violent behaviour.

Security procedures

[V8086] Security staff face very difficult and sometimes dangerous situations where their actions may be subjected to detailed scrutiny. It is important therefore that staff are clear as to the expectations upon them for preventing and responding to violent incidents. Clear procedures need to be put in place that outline the action to be taken in areas such as incident management, refusing entry, searching and evicting people and making an arrest.

Other employees will also need guidance as to what they can expect of their security colleagues, and how they can help security staff when the need arises.

Physical measures

[V8087] There is a vast range of physical measures that can be introduced to reduce the risk of physical injury through a workplace violence incident.

Some of these measures can be extremely expensive and it is worth emphasising here that a thorough risk assessment needs to be undertaken before committing to any expenditure on specialised equipment.

Access control

[V8088] There are many variations of access control ranging from barriers to open plan receptions. However, funding and existing building design are major influences on this. A well-designed and managed reception can contribute greatly to access control, yet be seen as offering a friendly and helpful service rather than looking intimidating and overtly tight on security.

Closed circuit television (CCTV)

[V8089] Although the outlay on CCTV can be high, it is one of the most popular security responses available for deterring and managing crime and violence. However, organisations need to be realistic about its purpose and use. A Home Office commissioned report suggested that CCTV is more effective as a detection tool than as a deterrent. CCTV needs to be monitored and used proactively to provide effective prevention.

Alarms

[V8090] There are many types of alarm system, each designed for a different purpose. The key alarm systems involved with enhancing personal safety are:

- Intruder alarms – commonly used to protect buildings after hours, or to protect restricted areas.
- Panic alarms – increasingly used in reception areas, interview rooms and isolated areas.
- Personal alarms – carried on the person as a means of attracting attention and temporarily distracting an assailant.

Before purchasing an alarm, be sure of the organisation's needs and consider them in the overall strategy for tackling violence.

Radios, paging and public address systems

[V8091] These communication systems have advanced greatly in recent years and can be an asset to staff safety. Some personal radios and pagers have built in panic alarms and sophisticated tracking facilities that pinpoint the location of the unit.

Mobile phones

[V8092] Mobile phones can provide a means of communication and comfort to lone workers. They are useful for keeping colleagues informed of movements and of unforeseen travel/schedule difficulties.

Protective vests

[V8093] Often referred to as 'body armour' these vests can be designed to protect against the threats presented by firearms and edged weapons.

Any organisation can purchase protective vests as no specific authority or licence is required, and they can be a necessary risk reduction measure in some areas of work. It is however important that the decision to adopt protective vests is not taken lightly, as it can be a sensitive issue for employees, service providers, and the public generally. Great care needs to be taken when considering purchasing this expensive equipment and effective risk assessment, incident reporting and monitoring processes should be carefully scrutinised to help make any decision.

Restraints

[V8094] Restraints are common in policing, custodial services, airlines and some areas of healthcare. Although the concept of using restraints will understandably concern most employers, they may be a necessary measure to prevent a disruptive passenger putting an aircraft at risk, or to prevent a patient harming themselves or others. Use of such equipment needs to be justified by the role requirements and risk assessments, and staff need comprehensive training and guidance in its use.

Advice should be sought before adopting handcuffs and other restraints to ensure an appropriate and lawful choice is made. Restraints such as the emergency response belt are popular within some areas of work, as they allow a great deal of control without pain.

Batons

[V8095] Batons are deemed offensive weapons and can only be used by the police.

Training

[V8096] Providing effective training is an essential element of an employer's duty of care to employees. Through training employees can identify ways in which they can reduce risk and develop the skills they need to deal with potentially violent situations. Appropriate training for workplace violence can take many forms and can be sourced internally or externally. For training to be effective it is essential that it is not viewed as a 'stand alone' solution but integrated into a wider learning programme that is actively supported by the organisation.

A training development model

[V8097] A well designed training or development activity is developed through a three-stage process (see figure 5 below):

(1)   Training Needs Analysis.
(2)   Training design, development and testing.
(3)   Implementing, monitoring and evaluation.

Each stage of this model is essential to the development of an effective training solution and should be evident in any training programme being offered whether that be internal or external.

Figure 5: General training and development model

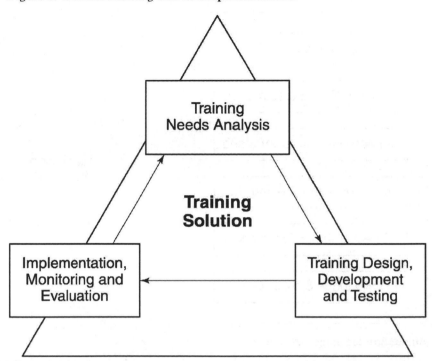

Training Needs Analysis

[V8098] The first step in a Training Needs Analysis is to identify the gap between how things are now and how things ought to be.

Any training or development solution should begin by asking two questions:

(1)   How has the need for the proposed training been identified?
(2)   What are the intended learning outcomes of the proposed training?

Identifying the need

[V8099] The variety of ways through which a training need can be identified are shown in figure 6 below.

A complete review of the risk of conflict and workplace violence across an organisation will identify a range of needs. Amongst these will be specific training and development requirements for staff performing specific roles and for line managers who are responsible for reducing risk and managing the issues that result from workplace incidents.

Figure 6: Identifying the need

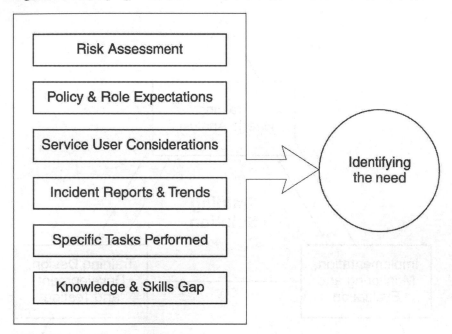

Defining the learning outcomes

[V8100] The outcomes should link directly to the identified needs. A good way to start is by saying: 'At the end of the training the delegates will . . . ' – and then describe the outcome. The outcome will describe what individuals will be able to do after they have attended the programme.

Here is a simple example. A review has shown that the workplace violence reporting form is not being filled in correctly in 80% of cases. A training need has been identified. The training outcome will be: 'At the end of the training the delegates will be able to correctly complete the reporting form.'

The above example also highlights the importance of ensuring that the need is correctly identified – in this case it may well be the form that is ambiguous, unclear or difficult to follow.

Checklist

[V8101] *Training Needs Analysis*

- How has the need for the training been identified?
  For example, through risk assessment; staff consultation; statistical data showing an increase in assaults.
- Who needs the training and why do they need it?
  For example, specific individuals, specific roles, specific departments, lone workers.
- What are the specific roles – when, where and how are they exposed to the risk of conflict or violence?
- What is the level of risk in terms of frequency and seriousness faced in each role and what are the expectations on staff in dealing with this?

- What challenges do service users and the environment present?
- What statistical data are available to help in the analysis of the risk?
  For example, data from reporting system on types of assaults, times of day, locations.
- What research is required to investigate and pinpoint the specific issues relating to each role?
  For example, interviews, focus groups.
- Learning outcomes
  For example, what should each role be able to understand, appreciate, do, or do differently, at the end of the proposed training?

Training design, development and testing

[V8102] The aims, objectives and outcomes of the training should be clearly identified from the desired learning outcomes. Decisions then need to be made about the learning methodologies to be used and the models, tutorials, activities and assessment criteria required to match the specification.

The design of a learning activity, course or programme is a specialist task and should be undertaken by the chosen training provider. It is essential to work closely with them to ensure that the design meets all the requirements.

Generic and 'off-the-shelf' courses are unlikely to meet the needs of staff working in different roles and who have different problems and concerns.

Modern technology provides a great deal of choice in methods of delivering training including distance learning, video, audio and computer-based learning, and virtual classrooms as well as traditional classrooms, workshop and workplace-based learning such as coaching and mentoring. This choice can be somewhat bewildering and it must be recognised that not all of these methods suit all occasions. Distance learning, in particular, can be attractive as a cheap alternative to other methods but it will fail to deliver some important areas. Generally, distance learning is effective as a support element to a training programme, particularly for the 'knowledge' aspects. The classroom is good for practising skills, discussing issues and experiences and problem solving. The workplace is good for coaching and applying the knowledge and skills.

Soft skills and physical intervention training

[V8103] Managing conflict demands a unique set of skills which fall into two distinct areas:

(1)   *Soft skills* – communication and interpersonal skills that can be used to calm and control situations.
(2)   *Physical intervention skills* – used when it is necessary to engage in physical contact with another person.

Core content – soft skills

[V8104] The Training Needs Analysis should identify the learning needs of people who perform a particular role. In general the following constitute the core ingredients of any training solution which is intended to meet soft skills needs.

- *Organisational policy and values* – core values which underpin the organisation's approach and the policy on workplace violence.

- *Definitions of workplace violence* – what the organisation defines as 'violence' and what type of incidents should be considered for reporting.
- *Risk assessment* – dynamic assessment of a situation to assess the risk and the consequent appropriate action.
- *Risk reduction and safety systems* – safe working practice, security procedures, the location, testing and use of alarms, panic buttons and CCTV equipment.
- *Theoretical models of aggression, violence and conflict management* – understanding the physiological and psychological processes that occur when people become angry, frightened or aggressive.
- *Triggers and escalation* – recognising the triggers and signs of increasing aggression so that employees can anticipate and defuse situations before they become more serious.
- *Verbal and non-verbal communication skills* – understanding and practising the specific skills of communication and controlling personal space.
- *Special groups of people* – eg drugs, alcohol, mental illness, learning disabilities, cultural differences and the elderly.
- *De-escalation and calming skills and exit strategies* – de-escalation and calming through empathy, problem solving, win-win thinking and strategies for getting out of difficult situations.
- *Legal issues, self defence and use of force* – employees' understanding of their rights and responsibilities when confronted with aggressive and violent behaviour.
- *Support for staff* – post-incident – what support is available through the organisation from line manger support to specialist counselling.
- *Post-incident reporting and debriefing* – understanding how to properly report an incident and be aware of the reasons why it this is important.

Core content – physical intervention skills

[V8105] It is more difficult to identify the core content for physical skills, as it will vary a great deal across different sectors. The Training Needs Analysis and risk assessment will identify whether there is a specific need for physical intervention in a particular role. Some self-protection systems include strikes and locks to joints in order to gain compliance through pain. Others are based on methods that do not employ aggressive techniques or use pain as a way of gaining compliance. Most organisations requiring physical intervention training are well advised to opt for an effective non-aggressive system, as this will reduce the likelihood of an escalation and the risk of injury to both staff and the assailant.

The following are the key pre-requisites to physical skills training:

- *Core soft skills* – core content described for soft skills should be regarded as an essential pre-requisite of any course involving physical intervention skills.
- *Skills for protecting oneself or another from unlawful assault* – commonly referred to as 'breakaway' or 'disengagement' techniques; skills to protect against strikes and skills to release holds and remove oneself or another from danger.

- *Interventions that are used to hold and restrain another person* – often referred to as 'control and restraint' or 'holding skills'; used to prevent someone from escaping lawful arrest or detention, or preventing them harming themselves or others.
- *Legal issues* – employees must be critically aware of their powers in relation to detaining someone.

Additional physical skills can be added to these core areas, including safer approaches to day-to-day tasks such as escorting or guiding.

Specialist skills can also be added to the highest-level training such as the use of restraint equipment.

Reviewing the training content

[V8106] It is important at this stage to ensure that the solution will satisfy three basic criteria:

(1)   Relevance
      Is the training content based upon a thorough Training Needs Analysis that considers the role and tasks performed and the risks associated with these?
(2)   Legal review
      Is the content of the course legally correct? Will it stand up to examination in legal proceedings if tested?
(3)   Medical review
      Are the learning methods and content safe? Will this stand up to examination in legal proceedings if tested?

The legal and medical basis for the training provided may be challenged in any legal proceedings and it is important that the organisation is confident that the training being provided will stand up to such scrutiny.

It is the responsibility of the training provider to show the appropriate legal and medical basis that underpins the learning solution being designed.

Testing the solution

[V8107] Before implementing the course or programme, it is a good idea to test it out using a pilot course or courses. The pilot course should be delivered under the same conditions that the fully implemented course will be delivered and feedback about the programme should be sought from delegates.

Implementation

[V8108] The rollout of the programme needs consideration, with some thought about who should receive training first. There are no hard and fast rules but it is important to develop a cohesive, prioritised strategy for implementation.

Checklist

Design and Implementation

- What are the aims and objectives of the training programme?
- How does the content of the programme achieve the aims and objectives?
- What areas of knowledge and understanding are required?
  For example, theory, models, legislation, policies and procedures.
- What skills are needed by staff?
  For example, communication, problem solving, disengagement, escorting, restraint, detaining.
- What are the appropriate attitudes and behaviours required?
  For example, avoiding physical contact, assertiveness.
- What learning methodologies are being used to achieve the development and how appropriate are they to these circumstances?
  For example, distance learning, group discussion, practical role play, practice physical skills.
- Has the training been tailored to the role and tasks performed?
- Have staff been consulted in the design process?
- Can any of the training be delivered in the workplace for realism and problem-solving opportunities?
- How is the content reviewed to ensure it is tactically effective, legally correct and medically safe?
- How will the programme be piloted and who should take part in the pilot?
- How will the pilot be evaluated?
- Who needs to know about the programme and how will it be communicated?
- What backing and support (and from whom) is available at senior level?
- How will the programme be rolled out? Who will receive it first – or will it be a mixed rollout? Can some groups, roles, individuals be trained together?

Monitoring and evaluating the solution

[V8110] The monitoring and evaluation of the programme needs to fit with the existing infrastructure of the organisation.

Figure 7 below shows a typical process that combines the individual development needs of the delegate with the feedback required to monitor and evaluate the programme. From an individual point of view, the delegate should define his or her personal learning objectives before the course, preferably involving his or her line manager. A short time after the course they should review the objectives and establish the learning outcomes achieved.

From an organisational point of view, delegates can complete an immediate post-course evaluation, which will establish reactions to, and satisfaction with, the programme. About four to six weeks later, delegates should provide further feedback about the impact that the programme has had in they way they perform in the workplace. This is then fed into the process of Training Needs Analysis and design for continuous improvements to the programme.

Figure 7: A typical delegate feedback process combining individual and organisational needs

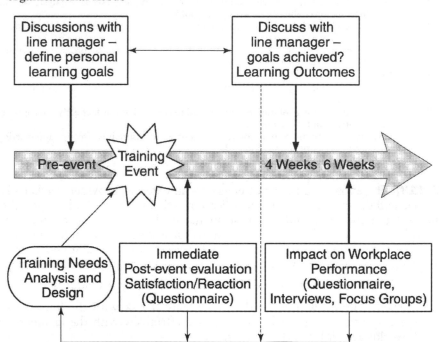

Internal or external training providers

[V8111] The decision about whether to use an internal training department or an external training provider is quite a difficult one and needs some careful consideration.

Internal provider

[V8112] The most compelling reason for choosing an internal training department is usually that it costs less. However, there are hidden costs involved which include specialist research into the subject of violence in the workplace, design of the programme, and specific training for trainers necessary to deliver an effective conflict management training programme.

If an organisation's workplace violence issues require physical intervention training, then it must be remembered that the risks to the organisation increase with the level of physical intervention required. The internal provider must be able to satisfy managers that the interventions are both appropriate and medically and legally defensible.

Most internal training departments attract 'can do' people who are willing and eager to help and enthusiastic about new challenges. Care is needed to ensure that they can provide training specific to requirements – not what they may think is needed. The following checklist will help in this regard.

Internal provider

- How well is the department equipped to do a Training Needs Analysis on the requirements for dealing with workplace violence?
- Can they design a properly researched solution that will provide effective development for the employees?
- Are the trainers suitably experienced, qualified and credible to provide specialised training in managing conflict and workplace violence?
- Does the trainer development programme include coaching and ongoing support?
- Does the Training Needs Analysis indicate a requirement for skills in physical intervention?
- How has the training solution been reviewed to ensure that it is tactically appropriate, legally correct and medically safe?
- How will the training be monitored and evaluated to ensure it is effective and meeting the current needs of the workplace?

External provider

[V8113] The extra investment of engaging an external provider needs to be justified and the most compelling justification is that the provider has specialist knowledge, experience and expertise in the field of managing conflict and workplace violence. Of course, most external providers will claim to have such expertise in abundance and the difficulty lies in making the right choice from the plethora of providers who claim to have just the solution being sought.

In simple terms the client will need to find out from the provider if:

- Their solution will equip the individuals at risk with the knowledge, skills, attitudes and behaviour to deal effectively with the incidents of workplace violence they may face.
- Their organisation has processes which ensure that the training need is correctly identified, the solution is properly designed, and that there is an evaluation process that provides feedback to validate and improve the product.
- Their training is properly designed to meet specific needs and is sufficiently robust to be medically and legally defensible if required.
- The people used to deliver the training are properly trained and experienced in delivering this type of training and understand the particular problems of the client.

The following checklist provides the basic questions that need to be asked.

External providers

- What evidence can they provide of their expertise in the field of managing conflict and workplace violence?
- What experience do they have in the organisation's specific sector and work?
- What do the other organisations they have worked with say about them?
- What evidence can they provide as to the effectiveness of their training in the workplace?
- What methodology will they use to identify the training needs of the organisation in the area of conflict management and workplace violence?
- How will they develop their understanding of the business and organisation?
- What learning methods are they proposing to use? Are they practically-based and in line with the training needs identified?
- How do they ensure that the content of their programmes is appropriate and legally correct?

- How do they ensure that the physical interventions used are medically safe?
- How do they conduct their training to ensure the delegates are trained in a safe environment?
- How do they develop their trainers?
- What level of resources will they commit to the programme and how will they deal with short-term issues such as a trainer becoming ill?
- What level of programme evaluation do they offer and how will it be fed back to the organisation?
- Are they likely to provide credible expert witness support if required?

Combining internal and external provision

[V8114] Combining the two options can provide a compromise and solution to this problem. It can be achieved by engaging an external provider to undertake a Training Needs Analysis and design a solution. The provider could be asked to deliver training directly to the higher risk and more demanding staff groups, and they could train internal trainers to deliver the bulk of the remaining programmes.

This solution is attractive because the visible costs are kept relatively low and the benefits of using an external provider are achieved to some degree.

Combine internal and external providers

- What experience can the external provider evidence to show they are capable of trainer training?
- Will the training enable internal trainers to provide the programme in a way that is professional and credible?
- What level of ongoing trainer coaching and support is offered?
- Does the training require physical interventions? If so, review the decision and be sure the trainers can teach these safely and effectively.

Training the managers

[V8115] Managers are often forgotten when it comes to violence management training. The needs of managers, particularly line managers, are two fold.

Firstly, they are often called into disputes involving customers or members of the public which have escalated and become too difficult for the member of staff to deal with. They may have already made the situation worse. Additionally, they are often dealing with situations when things have gone wrong and the potential for conflict is high. Consequently, there is a clear need for line managers in these situations to be personally skilled in conflict resolution as part of their role. They also need to be aware of the techniques and skills in this area provided to the people they manage.

Secondly, they have a vital role to play in the management of incidents of workplace violence and the provision of front line support both during and after the process. They are key players in ensuring that such incidents are properly recorded and reported and that staff carry out their roles in a way which achieves the organisation's goals in a safe and customer-friendly way. Staff can be affected by incidents in many different ways and need an individual response from their line manager.

The second point above is particularly important. Line managers have a great deal of influence in relation to the approach to dealing with conflict and violence in the workplace. Their individual attitude towards the whole issue will greatly influence how seriously the people who work for them approach the subject. This influence will affect attitudes towards risk assessment, approach to conflict situations, reporting and aftercare. Some people are badly affected by incidents that other members of staff might take in their stride. Sometimes the after effects of an incident do not set in until several hours or several days later and managers need to be aware of and look for the signs which indicate that the incident remains unresolved. The way in which the manager deals with the individual will have an influence upon the speed that he or she will recover from the incident.

Incident management

[V8116] Until this point the focus has been on preventing violence or at least reducing it. This section focuses on the effective management of an incident that does occur, and how its impact can be reduced. It will then consider the steps that should be taken following the incident to support staff and learn from what has happened.

Although every eventuality cannot be planned for, it is important to plan and practise the response to both the most common and most serious scenarios that are likely to be experienced. Even though events will not always 'go to plan', the rigour of the planning process will help ensure that the systems are in place and that staff and managers are better prepared to deal with the situations that arise.

It is particularly important to prepare and practise a response to violent incidents, as during these highly emotive times employees will find it difficult to think clearly and objectively.

Figure 8: Incident management – planning and practice

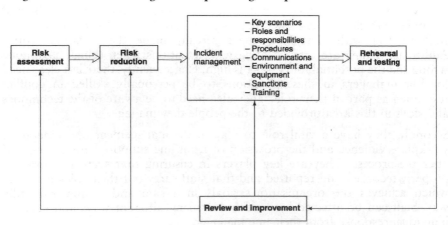

Figure 8 reinforces the importance of undertaking a violence risk assessment to establish key risk areas, and putting in place measures to reduce these risks.

This process will help with the identification of key incident scenarios that the organisation can then plan and train for.

Key scenarios

[V8117] The risk assessment process will have highlighted key areas of risk and control measures will be put in place to tackle these. In preparing for the management of incidents it is important to use the risk assessment findings as the basis for the scenarios to be planned for and practised.

Roles and responsibilities

[V8118] It is important to be clear as to roles and responsibilities prior to an incident occurring to prevent a breakdown in communication when under pressure. This extends further than a specific job role but to the wider team working in an area, for example, what part will the reception, security, domestic and office staff play, when an incident occurs in their area? Who takes control, how are they supported, who calls for help, who takes down descriptions – are all questions that need to be addressed. Everyone needs to be involved and has a part to play – even if it is in the background. In lone working environments communications and expectations need to be clear between lone workers, office colleagues, control centre or switchboard colleagues.

Response teams

[V8119] Some organisations such as hospitals operate incident response teams. These multi-disciplinary teams follow a similar concept to 'crash teams' that respond to cardiac arrest. There are pro's and con's to this approach. Some organisations feel the response team is an intrusion and can damage relationships between local staff and patients/service users. If approached well staff can however benefit from the knowledge that they will get competent help quickly. Careful selection of response teams and a high level of training are vital if they are to operate safely and earn the respect of staff.

Leadership

[V8120] The most critical role in incident management as with any crisis, is leadership. Supervisors and managers need to have the confidence and skills to take control of an incident and its immediate aftermath, and staff will look to them for direction and example. Managers often miss out on staff conflict management training, yet they need these valuable skills and the additional knowledge in how to manage an incident and aftermath. Some managers will find difficulty in adjusting from their preferred leadership style to a more directive style required in a crisis. The concept of situational leadership is useful in training to help managers understand that they need to adapt their style to the situation and that staff will want clear direction and decisions. Senior management needs to provide clarity as to the level of authority given to managers and staff during an incident and as to how and when decisions should be escalated to a senior level.

Procedures

[V8121] The importance of clear role expectations and communications has already been discussed, and these need to be written down in the form of clear guidance and procedures. Flow charts will provide a useful means of simplifying and communicating responsibilities. Most organisations will have clear fire procedures in place and the same approach should be taken with violent incidents.

One key area requiring guidance and procedure is that of *preserving evidence*.

Violent incidents often result in the prosecution of the perpetrator. The success of the prosecution will depend on the quality of the evidence that supports it, and unfortunately many cases collapse at court due to simple errors made at the time of the incident in gathering and protecting evidence. Things that may seem trivial here can have a dramatic impact in court as the defence seeks to discredit prosecution evidence.

The rules of disclosure highlight the duty of the prosecution to disclose all evidence, whether it will stand for or against the defendant. Organisations must therefore realise that anything that could be deemed as evidence relating to the case, such as any reports or records, will form part of the legal process. There are limited exceptions such as medical records.

Some basic tips and common sense will help to strengthen a prosecution and to ensure justice is done. Guidance should cover:

- Importance of securing witnesses.
- Identification evidence.
- Use of Closed Circuit Television (CCTV) as evidence.
- Preserving a scene.

Communications

[V8122] Another important consideration in incident management is communications with other agencies such as the police, specialist consultants and the media. It is important to be proactive in responding to media interest following a violent incident. Considerable harm can be done to both the individuals involved and the reputation of the organisation if this is not handled well.

Environment and equipment

[V8123] Design, layout, security and equipment will be put to the test should an incident occur, and it is wise to ensure that these are considered at the earliest opportunity and tested regularly. Panic alarms, for example, are all too often found to be disconnected or out of order.

Sanctions

[V8124] Many airlines and hospital trusts operate warning systems, for example they issue a letter to a disruptive individual or a yellow or red warning card. Clear sanctions play an important part in preventing and responding to violent behaviour. However, it must be recognised that the confronting of individual/s could be the trigger of an assault. It is essential that staff are taught how to accurately assess people and situations and are provided with training in when, where and how to confront a violent person. Some organisations are quick to launch these schemes but do not consider the position of the manager or member of staff who has to carry out these difficult and high risk tasks.

Training

[V8125] Training will focus firstly on raising safety awareness and reducing risk. This is largely *knowledge* based and staff also need to develop the *skills* for dealing with the incident that does occur. If staff are to develop confidence in managing violence the training will need to be dynamic and realistic, and cover the key scenarios identified.

Teamwork will be tested during stressful incidents and this should also be a key consideration in staff development to ensure staff communicate effectively, understand their responsibilities and respect each other. This should also extend to the wider team operating in a certain area, as communication often breaks down across roles or functions.

Physical fitness is particularly important and vital in some roles where staff are expected to respond to violence. Looking and feeling fit and behaving professionally will help to earn respect and deter assault. Fitness will also play a vital part when restraining a violent person or even running away from one.

Rehearsal and testing

[V8126] Without doubt rehearsal and testing is one of the most under-rated and least performed aspects of incident management. With the exception of some areas of the emergency services, psychiatric care and the armed forces, organisations rarely practise their response to violent incidents. This is somewhat strange as fire drills to test communications and procedures are regularly conducted within organisations.

In sport coaches are often heard telling athletes 'you play as you train' and the same applies to incident management. Realistic practise and scenario-based training will make a big difference to the effectiveness of the incident response. Unforeseen problems with communications, equipment and procedures will be highlighted through a rehearsal and staff can be actively involved in solving these.

In some areas of work it will be beneficial to set up multi-agency exercises involving police and other agencies. This helps to clarify expectations and iron out any problems.

Review and improvement

[V8127] Training, rehearsal and real incidents provide valuable feedback, and it is important that procedures are continuously reviewed to respond to this.

CHECKLIST

- Assess the risk – does the organisation face a foreseeable risk of serious incidents of violence towards staff?
- Are roles, responsibilities and communication lines clearly identified for the management of an incident of workplace violence?
- Are the individuals and departments concerned aware of the roles, responsibilities and communication lines?
- Is the risk to staff and other service users great enough to warrant the development of specially trained response teams?
- Are the leadership roles and decision-making levels clear and have the appropriate people received training for taking a lead role in such incidents?
- Are there clear procedures outlined for managing the different aspects of an incident? In particular, for serious incidents are there clear procedures for securing and preserving evidence?
- Is there a clear strategy for managing communications and the media in the event of an incident?
- Has specialist equipment been identified and have staff been trained in its use in the event of an incident?
- Have the all the people who might be called upon to respond to an incident received appropriate training?
- Is there a robust and regular method in place for rehearsing and testing the complete response to an incident?
- Is there a review and evaluation process in place to provide feedback about the effectiveness of the incident management either after a test or a real incident?

Post-incident management

Investigation

[V8129] The post incident management process includes investigating the circumstances of the incident, attending to the needs of the victim and, where appropriate, dealing with the perpetrator.

Investigating incidents is an integral element of the risk management process. By establishing the factual circumstances surrounding an incident of workplace violence and gaining an understanding of why and how the incident took place it may be possible to take remedial action to prevent a similar happening in the future. As with all other phases of the risk management process it is important that employees are consulted and their suggestions taken into consideration.

An internal investigation should:

- collect facts on who, what, when, where and how the incident occurred;
- record information;
- identify contributing causes;
- encourage appropriate follow-up; and
- consider changes in controls, procedures and policy.

Incident investigation forms part of the wider legal requirement under the *Management of Health and Safety at Work Regulations (SI 1999 No 3242), Reg 5* for an employer to monitor and review preventive and protective measures. The accompanying Approved Codes of Practice (ACOP L21), HSE management guidance (HSG 65) and National Occupational Standards (see V8142) for managing work-related violence provide further guidance. The guidance acknowledges that the amount of time and effort put into the investigation will vary dependent upon the nature and severity of the event.

HSE research clearly demonstrates the link between workplace violence and abuse and stress. Insensitive questioning of the victim during the early stages of an investigation may compound the traumatic effects of the incident itself leading to longer-term psychological damage. Those responsible for conducting the investigation should be aware of this dynamic and carefully balance the priority of providing support to the victim against the requirement to investigate, question and learn from what happened to prevent recurrence.

How people are affected by workplace violence

[V8130] Perhaps the most important thing to recognise is that everyone has a different way of responding to and dealing with the aftermath of a violent or aggressive incident. There is no 'right' or 'wrong' way to react and people must be allowed to deal with it in their own way.

Being the victim of violence is particularly traumatic because it involves an interaction with another person at a very personal level and this can produces some difficult and complex emotional reactions. These reactions will vary over the short, mid and longer term and it is important that support is provided at each stage, where problems persist, professional support should be provided (see figure 9 below). It is important that those involved recognise that these reactions are quite 'normal' following an abnormal event, also that no one says they have to experience them.

Figure 9: Timescale of reactions to workplace violence

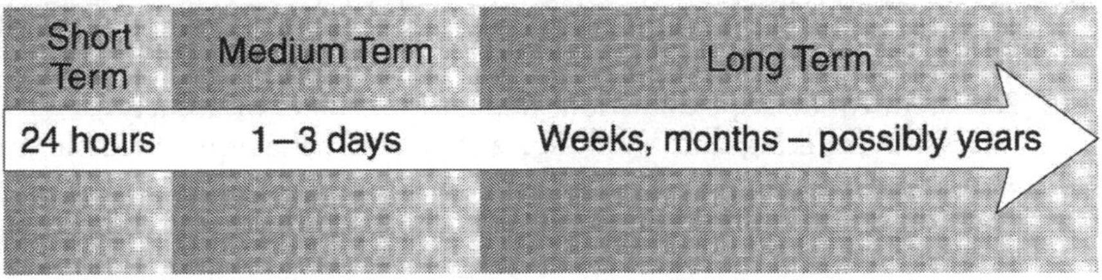

Short Term	Medium Term	Long Term
24 hours	1–3 days	Weeks, months – possibly years

Common reactions to an abnormal event such as an assault

Short-term reactions

[V8131] In the first few hours following the incident, the victim will have some initial reactions to the aggression and violence inflicted upon them. These reactions are predominantly emotional and are a direct response to the incident. Many factors will influence the severity of the reaction, not least of which is the individual's level of resilience towards traumatic situations.

The level of aggression, suddenness of the confrontation and physical injury sustained are also some of the factors which will influence how the victim will react. The following are the most likely reactions:

- Shock, confusion, disbelief, fear, helplessness.
- Anger, embarrassment, feeling of violation.

Many of these initial reactions will begin to lessen as the victim moves into the next phase.

Medium-term reactions

[V8132] The short-term reactions are characterised by their 'immediate' nature, formed before the victim has had any time to think about, and begin to rationalise, what has happened. The medium-term reactions begin to appear when the victim has had a chance to consider the incident, to work though what happened and to think about the consequences, near misses and alternatives. This will be around 24 hours after the incident. Reactions can include:

- Feelings of loss, guilt, shame, embarrassment, humiliation.
- Exhaustion and tiredness, lack of sleep.
- Denial of effects, ready to get back to work.
- Anger, frustration and resentment.
- Lack of confidence, anxiety about similar situations or meeting the aggressor.

Moving successfully through this medium-term phase is often the key to recovery. Once the victim has acknowledged what has happened and come to terms with it then he or she can then move back towards a normal life. Line managers can provide vital support in this phase and are pivotal in the successful recovery of most of the victims.

Long-term reactions

[V8133] Generally, reactions that persist beyond a couple of weeks after the incident are indicative that the victim is finding difficulty in coming to terms with the incident and that he or she probably needs professional specialist help. Examples include:

- Persistent tiredness, exhaustion, depression, bouts of anxiety.
- Excessive drinking and smoking, anti-social behaviour, irritable and aggressive behaviour.
- Nightmares, flashbacks, headaches, nausea, difficulty in eating and sleeping.

A victim who displays these long-term reactions clearly needs specialised help, and an organisation which wants to provide a complete response to the range of issues that result from workplace violence, will need to set up a procedure to facilitate this.

Returning to normal

[V8134] Many factors will influence the speed of recovery including the victim's life circumstances and their emotional resilience at the time of the incident. The turning point for most victims is the acceptance of what has happened to them. Once they accept the incident as a reality, they stop going through the 'if only . . . ' scenarios and stop blaming themselves for what took place.

Most people reach a point where they can move on from the event and get back to their normal daily lives. They achieve this when they regain confidence and self esteem and recognise that, although life has changed in some aspects as a result of what occurred, they can become positive about being back in their working environment.

Although these various reactions have been described in discrete stages it will be rare for anyone to pass smoothly through them all. In reality, many things will cause a victim to progress and regress in the move towards normality. The return to work, for example, can be quite difficult and bring back feelings of insecurity and fear. Having to appear in court as a witness or learning of a colleague who has been involved in a similar incident may well trigger a reoccurrence of one or more of the reactions previously experienced.

For a few people, a return to 'normal' is virtually impossible, particularly if they have been permanently disabled by physical or mental injury. In such cases, the victim will need the most specialised care and support in trying to come to terms with their circumstances.

Thankfully, for most people the support of family, friends, colleagues and managers will be enough to help them recover from the trauma of the incident and return to a normal working life.

Providing post-incident support

[V8135] An appropriate post-incident support system should be an integral part of an organisation's overall response to workplace violence (see figure 10 below). The sophistication of the support system will depend upon the level of risk to which the staff are being exposed.

Staff working in the emergency services can often experience violent and other traumatic events, and therefore require a high level of post-incident support. This will range from line manager support, formal debriefing processes, to occupational health services, which will include access to specialist help for the most serious consequences of work-related violence. Some organisations adopt formal psychological debriefing processes, sometimes referred to a 'critical incident stress debriefing'. Although these approaches have considerable support in some areas, there is ongoing debate about their value, and concern at the potential risks associated with victims 're-living' the traumatic experience in the debriefing process. Whether or not the organisation adopts a formal debriefing process, it is essential that support mechanisms and procedures are put in place for managers and staff.

Even where the frequency of serious incidents is low, the organisation should have a post-incident support procedure and line mangers should be trained in the skills appropriate to helping a victim through the first stages of coping with an incident of work-related violence. It should also be possible to access specialist help if necessary.

Figure 10: Supporting the victim

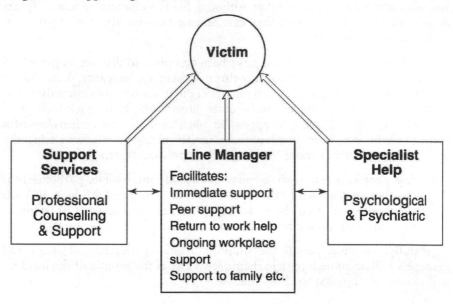

The role of the line manager

[V8136] In the vast majority of cases the support for the victim will be provided through his or her line manager. The short and medium-term reactions following an incident have been described earlier in this chapter and the focus of the line manager's support will be concentrated on helping the victim work though those reactions and in facilitating further support where necessary. Successfully dealing with these phases is vital in providing the optimum conditions for the individual to recover and return to a normal working life.

There are three points at which this support is crucial:

• Immediately after the incident has happened.
• During any absence from work.
• Preparing for and returning to work.

Organisations should provide guidance and training for managers to help them to perform their key role sensitively and effectively.

Other immediate post-incident considerations

[V8137] There will be a need to complete the workplace violence report form, which will require a formal narrative of the incident. For evidential reasons, this should wherever possible be completed personally by the victim as soon as practicable after the incident. Every effort should be made to get the victim to complete at least the basic requirements whilst it is fresh in his or her memory.

It is tempting to leave this task but it can have far reaching consequences if the incident becomes the subject of criminal or civil proceedings. An accurate and early account of the incident can make all the difference to the outcome of a case.

Support services

[V8138] There is a point where some victims need more help than the line manager can be expected to offer. If an individual is suffering from the sort of persistent reactions already discussed in the medium to long term following the incident then they should be referred to more professional help. In this context it will probably consist of professional counselling.

Occasionally, a victim may find it impossible to move on from the incident. Some may develop a condition called post-traumatic stress disorder or exhibit some of the symptoms of it. In all these cases, the victim will need specialist psychological and psychiatric help. Remember the reactions outlined earlier that may indicate that an individual is in need of specialist help. Sometimes colleagues will bring attention to the fact that the individual is simply not 'his normal self'.

Preparing for the worst

[V8139] Organisations need to be in a position to quickly access expert management advice following incidents that are likely to have a severe impact on individuals and/or the organisation. It is important to plan this in advance as poor quality decisions can easily be made when emotions are high and everyone is under pressure. Plans should include access to advice and support concerning:

- Professional help for employees, service users, families and partners.
- Advice on legal issues.
- Advice on managing media attention.

The police investigation and court case

[V8140] Most victims want to see their aggressor brought to justice. However, it may prove to be quite a daunting process and it can help to be aware of how the process works and what might be expected. There are several prosecution routes including:

- Prosecution through police.
- Private prosecution (a similar process but instigated through a solicitor, not police).
- Civil prosecution (instigated through a solicitor in a civil court).

Prosecution through police is the most common route. However employers should have clear policy and guidance on other avenues should the police or prosecution service decide not to progress the matter. This should include the degree of support they can provide to the victim, such as funding and providing time off for the legal process in preparing and giving evidence.

Criminal and civil courts provide further protection from harassment and violence in a number of ways, including:

- Injunctions.

- Restraining orders (e.g. preventing convicted persons contacting the victim/s).
- Anti-social behaviour orders.

When a court order is in place, the police are in a strong position to take action should it be breached.

Note: Offences can carry extra penalties if proven to be racially aggravated, and such offences will be recorded as crimes by the police and can constitute a more serious offence under the *Crime and Disorder Act 1998*.

Giving evidence in court will understandably cause anxiety and it is especially difficult for the victim of violence 're-living' the trauma of what happened. It is vital that support is provided through this process by the organisation.

Victim Support (www.victimsupport.org.uk) is a national organisation with a great deal of experience in helping people to cope with being the victim of a crime, regardless of whether it has been reported to the police. Victim Support also runs the Witness Support Service.

It provides confidential, practical and emotional support through a network of local schemes across England, Wales, Northern Ireland and the Republic of Ireland (www.crimevictimshelpline.ie). For the website for Victim Support Scot land is www.victimsupport.org.uk. It also provides a range of information lea flets details of which are available on the website.

Figure 11: The process of prosecution

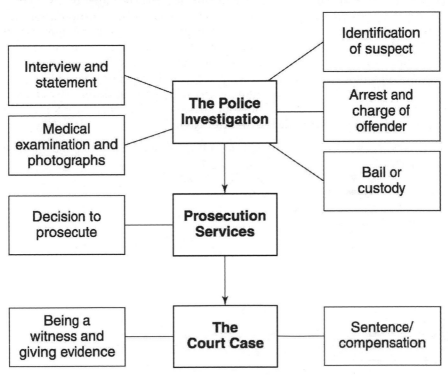

Support during the process of the investigation and court case

[V8141] For anyone who has to undergo the process of a court case it can prove to be a very difficult time.

As mentioned earlier, Victim Support is an organisation with many years of experience in helping individuals to cope with all the different aspects of surviving the trauma of being the victim of a crime. They have developed a Witness Support service, which provides practical help and support for anyone who has to attend a Crown Court as a witness.

Sources of further information

[V8142]–[V8143] The HSE has a work-related violence area on its website at: www.hse.gov.uk/violence/index.htm. This includes information about the National Occupational Standards on the prevention and management of work-related violence which HSE has worked on with the Employment Nationa l Training Organisation (ENTO) and other stakeholders. These provide a fra mework for employers to develop their policies and procedures for tackling violence, as well as guidance for individuals and managers for dealing with violent situations and their aftermath. They can be downloaded from the ENTO webs ite at www.ento.co.uk/standards/wrv/index.php.

The website www.workplaceviolence.co.uk is a one-stop access point to other useful sites.

Figure 12: Maybo Risk Management Model

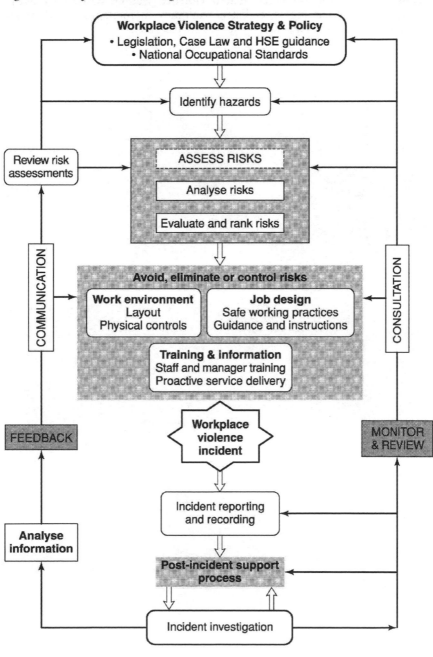

Acknowledgement

[V8144] This section on work-related violence was co-written by Maybo Limited, the UK's leading specialist consultancy in managing work related conflict and violence (www.workplaceviolence.co.uk).

Vulnerable Persons

Mike Bateman

Introduction

[V12001] This chapter deals with groups of people who, for one reason or another, are more vulnerable to health and safety risks. Such people include:

- children and young persons;
- new or expectant mothers;
- lone workers;
- disabled persons; and
- inexperienced workers.

Where these persons are at work they are subject to the protection of the full range of health and safety legislation – the *Health and Safety at Work etc. Act 1974 ('HSWA 1974')* and the various regulations made under it. Of particular relevance are the *Management of Health and Safety at Work Regulations 1999 (SI 1999 No 3242) ('the Management Regulations')* and their requirements for risk assessments. The potential vulnerability of some members of the work-force must be taken into account during the risk assessment process. In addition, the *Management Regulations* contain specific requirements in respect of children and young persons and also new or expectant mothers. These will be explained later in the chapter.

However, *HSWA 1974, s 3* places duties on employers and the self-employed in relation to persons not in their employment such as visitors, occupants of premises, service users, neighbours or passers-by. Risks to such persons must also be considered during risk assessments with appropriate attention given to those who are particularly vulnerable. The general principles of risk assessment are described in the chapter dealing with that subject whilst some specific aspects are covered in the chapter entitled COMMUNITY HEALTH AND SAFETY.

This chapter concentrates on legislation of particular relevance to vulnerable persons and many of the practical considerations to be taken into account when carrying out risk assessments relating to such people.

Relevant legislation

[V12002] Whilst the *Management Regulations 1999 (SI 1999 No 3242)* impose specific requirements in relation to children and young persons, together with new or expectant mothers, there are other regulations where the vulnerability of some people is either recognised directly or must be taken into account indirectly through some form of risk assessment. One example is that the capabilities of the individual must be considered when carrying out assessments of manual handling operations.

Management Regulations requirements

[V12003]–[V12005] The terms 'child' and 'young person' are defined in *Regulation 1(2)* of the *Management Regulations 1999 (SI 1999 No 3242)*, whilst *subsections (4)* and *(5)* of *Regulation 3* contain requirements relating to risk assessments and young people. *Regulation 19* of the Regulations deals with precautions, which must be taken to ensure the 'Protection of Young Persons'.

'New or expectant mother' is also defined in *Regulation 1(2)* whilst there are three regulations relating directly to them:

*   *Regulation 16* – deals with risk assessments;
*   *Regulation 17* – covers possible suspension from night work;
*   *Regulation 18* – contains requirements in respect of notification to the employer of pregnancy or having given birth.

All these definitions and requirements are explained in full later in the chapter.

Other regulations directly affecting vulnerable persons

Children and young persons
*   Several sets of regulations contain age-related prohibitions or restrictions on specified activities. These are set out and explained in TABLE **A** later in the chapter (see **V12015** below).
    Also of relevance are the *Health and Safety (Training for Employment) Regulations 1990 (SI 1990 No 1380)* which give students on work experience programmes and trainees on training for employment programmes the status of 'employees'. The immediate provider of the training is treated as their 'employer'. Whilst this particularly affects children and young persons, it also applies to adult participants in training programmes. (Courses at schools, colleges and universities etc. are not covered by these requirements.)

New or expectant mothers
*   Some regulations contain requirements for employees engaged in certain activities to be free from medical or physical conditions that make them unfit for that activity. Typical of these are the *Work in Compressed Air Regulations 1996 (SI 1996 No 1656)* and the *Diving at Work Regulations 1997 (SI 1997 No 2776)*.
    Both the *Ionising Radiations Regulations 1999 (SI 1999 No 3232)* and the *Control of Lead at Work Regulations 2002 (SI 2002 No 2676)* impose tighter exposure standards on *all* women of reproductive capacity whilst the latter Regulations would almost certainly result in the suspension from lead work of a woman once her pregnancy was notified. (These requirements are explained in more detail later in the chapter – see **V12018** below.)

Regulations with indirect implications relating to vulnerable persons

Construction (Design and Management) Regulations (CDM)

• When construction work is to be carried out in or close to premises occupied by children (or those with special educational needs) the health and safety planning required by the CDM regulations must take particular account of their lack of awareness of risks and general curiosity. If normal access routes in premises are affected this may create risks or problems for pregnant women, disabled persons or the elderly. (From 6 April 2007, the original CDM Regulations passed in 1994 are replaced by new regulations.)

Manual Handling Operations Regulations 1992 (SI 1992 No 2793)

• The limited manual handling capabilities of, and potential risks to, both children and young persons, and new or expectant mothers must be taken into account during risk assessments of manual handling operations. Similarly, risks to lone workers (with no source of assistance), persons with temporary or permanent physical disabilities and inexperienced workers (who may be unaware of relevant handling techniques) must also be considered.

Control of Substances Hazardous to Health Regulations 2002 (SI 2002 No 2677) (COSHH)

• Children, persons with special educational needs or those with fading mental faculties may all be at extra risk from hazardous substances as they may not be able to read or understand warnings or instructions. COSHH assessments should take account of the potential presence of such vulnerable persons and ensure the secure storage of hazardous substances where this is appropriate, e.g. by locking cleaners' cupboards.

Personal Protective Equipment at Work Regulations 1992 (SI 1992 No 2966)

• Use of certain types of personal protective equipment ('PPE') may present difficulties for pregnant women or workers with physical difficulties. Young and inexperienced workers are likely to need more detailed explanations of the purposes of different types of PPE and how and when they are to be used, whilst those with limited intellectual capabilities will need even closer attention.

Health and Safety (Display Screen Equipment) Regulations 1992 (SI 1992 No 2792)

• The suitability of a display screen equipment ('DSE') workstation for a pregnant woman will need to be assessed, particularly as her size increases. Disabled workers or other staff with physical problems (e.g. past injuries) are likely to require special attention during workstation assessments, especially in relation to the suitability of chairs and other ergonomic aspects.

Children and young persons

[V12006] For many years work done by children and young persons was subject to an often bewildering array of prohibitions – equipment that children and young persons were not allowed to use or processes or activities that they must not be involved in. However, since 1997 the emphasis has switched from such prohibitions towards restrictions, based upon a process of risk assessment.

Many of the previous age-related prohibitions were revoked but some still remain. The more important of these are listed in TABLE **A** at **V12015** below. However, the ongoing programme of modernisation of health and safety legislation is continually sweeping away many of these residual prohibitions. (Any cases of doubt should be referred to the HSE.)

Following the publication of the Better Regulation Task Force report 'The regulation of child employment', the government have announced that it is to adopt the report's recommendations and to simplify the legal framework that regulates the employment of children aged from 13 to 16.

Definitions

[V12007] The terms 'child' and 'young person' are defined in *Regulation 1(2)* of the *Management Regulations 1999 (SI 1999 No 3242).*

Child is defined as under the minimum school leaving age ('MSLA') in accordance with:

* *section 8* of the *Education Act 1996* (for England and Wales);
* *section 31* of the *Education (Scotland) Act 1980* (for Scotland).

(The MSLA is just before or just after the age of sixteen)

Young Person is defined as 'any person who has not attained the age of eighteen'.

Some of the prohibitions remaining from older health and safety regulations use different cut off ages – where relevant these are referred to in TABLE **A** at **V12015** below. Further detail is also provided in the HSE guidance booklet HSG 165 'Young people at work'.

Risk assessment requirements

[V12008] *Regulation 3* of the *Management Regulations 1999 (SI 1999 No 3242)* contains important requirements (first introduced in 1997) in relation to risk assessments and young persons. These are:

(4) An employer shall not employ a young person unless he has, in relation to risks to the health and safety of young persons, made or reviewed an assessment in accordance with paragraphs (1) and (5).

(5) In making or reviewing the assessment, an employer who employs or is to employ a young person shall take particular account of—

(a) the inexperience, lack of awareness of risks and immaturity of young persons;

(b)   the fitting-out and layout of the workplace and the workstation;
(c)   the nature, degree and duration of exposure to physical, biological and chemical agents;
(d)   the form, range and use of work equipment and the way in which it is handled;
(e)   the organisation of processes and activities;
(f)   the extent of the health and safety training provided or to be provided to young persons; and
(g)   risks from agents, processes and work listed in the Annex to Council Directive 94/33/EC(b) on the protection of young persons at work.

(Note: *Paragraph (1)* of the regulation contains the basic requirement for employers to carry out risk assessments.)

Protection of young persons

[V12009] In assessing the risks to young persons, the employer must take particular note of *Regulation 19* of the *Management Regulations 1999 (SI 1999 No 3242)* which states:

(1)   Every employer shall ensure that young persons employed by him are protected at work from any risks to their health or safety which are a consequence of their lack of experience, of absence of awareness of existing or potential risks or the fact that young persons have not yet fully matured.
(2)   Subject to paragraph (3), no employer shall employ a young person for work—
(a)   which is beyond his physical or psychological capacity;
(b)   involving harmful exposure to agents which are toxic or carcinogenic, cause heritable genetic damage or harm to the unborn child or which in any way chronically affect human health;
(c)   involving harmful exposure to radiation;
(d)   involving the risk of accidents which it may reasonably be assumed cannot be recognised or avoided by young persons owing to their insufficient attention to safety or lack of experience or training; or
(e)   in which there is a risk to health from—
(i)   extreme cold or heat;
(ii)   noise; or
(iii)   vibration,
and in determining whether work will involve harm or risk for the purposes of this paragraph, regard shall be had to the results of the assessment.
(3)   Nothing in paragraph (2) shall prevent the employment of a young person who is no longer a child for work—
(a)   where it is necessary for his training;
(b)   where the young person will be supervised by a competent person; and
(c)   where any risk will be reduced to the lowest level that is reasonably practicable.

Some of the practical implications of these provisions are considered later in the chapter.

Information on risks to children

[V12010] *Regulation 10* of the *Management Regulations 1999 (SI 1999 No 3242)* deals with 'information for employees' and contains two paragraphs relating to the employment of children. These state:

(2)   Every employer shall, before employing a child, provide a parent of the child with comprehensible and relevant information on—

  (a)   the risks to his health and safety identified by the assessment;
  (b)   the preventive and protective measures; and
  (c)   the risks notified to him in accordance with Regulation 11(1)(c).

(3)   The reference in paragraph (2) to a parent of the child includes—

  (a)   in England and Wales, a person who has parental responsibility, within the meaning of section 3 of the Children Act 1989, for him; and

  (b)   in Scotland, a person who has parental responsibility, within the meaning of section 8 of the Law Reform (Parent and Child) (Scotland) Act 1986, for him.

This requirement to provide information to parents includes situations where children are on work experience programmes (where they have the status of employees by virtue of the *Health and Safety (Training for Employment) Regulations 1990 (SI 1990 No 1380)*) and also includes part time or temporary work carried out by children who have not reached the minimum school leaving age.

Some of the practical implications of this requirement are considered later in this chapter.

Disapplication of requirements

[V12011] The wide-ranging impact of these provisions is lessened to a limited extent by *paragraph (2)* in R*egulation 2* of the *Management Regulations 1999 (SI 1999 No 3242)* – 'Disapplication of these Regulations'.

(2)   Regulations 3(4), (5), and 10(2) and 19 shall not apply to occasional work or short-term work involving—

  (a)   domestic service in a private household; or
  (b)   work regarded as not being harmful, damaging or dangerous to young people in a family undertaking.

However, the term 'family undertaking' is not defined in the Regulations. HSE guidance indicates that this should be interpreted as meaning a firm, owned by, and employing members of the same family, i.e. husbands, wives, fathers, mothers, grandfathers, grandmothers, stepfathers, stepmothers, sons, daughters, grandsons, granddaughters, stepsons, stepdaughters, brothers, sisters, half-brothers and half-sisters.

As far as is known this narrow interpretation of 'family undertaking' has not been tested in the courts. Common usage of the term would suggest small and medium-sized businesses controlled and managed by members of the same family but not necessarily only employing family members, as implied by the words used by the HSE.

Capabilities and training

[V12012] *Regulation 13* of the *Management Regulations 1999 (SI 1999 No 3242)* states in *paragraph (1)* that 'Every employer shall, in entrusting tasks to his employees, take into account their capabilities as regards health and safety'. Quite clearly the capabilities of children and young persons will be somewhat different from those of more experienced and mature employees. (This requirement is also of relevance in considering other vulnerable persons e.g. disabled or inexperienced workers.)

Risk assessment in practice

[V12013] The requirements of *Regulations 3(4)*, *(5)* and *19* of the *Management Regulations 1999 (SI 1999 No 3242)* are based upon the contents of a European Directive (94/33/EC) and it is not particularly easy for the employer to identify exactly what he or she must (or must not) do.

A good starting point is to consider the three characteristics associated with young people which are mentioned in both *Regulations 3(5)* and *19(1)*:

- lack of experience;
- lack of awareness of existing or potential risks; and
- immaturity (in both the physical and psychological sense).

All young people share these characteristics but to differing extents – for example, one would have different expectations of a school leaver who had already been playing a prominent role in a family business, e.g. a farm, as opposed to a work experience student with no previous exposure to the world of work. Employers must also be aware that the degrees of physical and psychological maturity of young people vary hugely.

These three characteristics must then be considered in respect of the risks involved in the employer's work activities and particularly those identified in *Regulation 3(5)(b)–(g)* and *Regulation 19(2)*.

The contents of those Regulations (including those risks listed in the Annex to Council Directive 94/33/EC) have been consolidated into a single checklist for employers which is contained in **TABLE B**.

The purpose of all risk assessments is to identify what measures the employer needs to take in order to comply with the law. Additional measures which the employer must consider in order to provide adequate protection for young persons are:

- not exposing the young person to the risk at all;
- providing additional training;
- providing close supervision by a competent person;
- carrying out additional health surveillance (as required by *Regulation 6* of the *Management Regulations 1999 (SI 1999 No 3242)* or other regulations e.g. *COSHH (SI 2002 No 2677)*);
- taking other additional precautions.

Regulation 19(3) allows more latitude in respect of young persons who are no longer children in relation to the requirement to completely exclude them from the types of risks specified in *Regulation 19(2)*.

In deciding what precautions are required, the employer must consider both young people generally and the characteristics of individual young persons. Additional training and/or supervision may be necessary in respect of young people with 'special needs'.

Young people should gradually acquire more experience, awareness of risks and maturity, particularly as they pass through formal training programmes within the NVQ system. As this occurs, restrictions on their activities may be progressively removed.

Provision of information

[V12014] *Regulation 10(1)* of the *Management Regulations 1999 (SI 1999 No 3242)* requires employers to provide all employees with comprehensible and relevant information on:

- risks to their health and safety identified by risk assessments (or notified by other employers);
- preventive and protective measures (i.e. appropriate precautions);
- emergency procedures and arrangements.

This requirement is of particular importance in relation to young persons. *Regulation 10(2)* requires employers *also* to provide information to parents of children (under the MSLA) on the risks the children will be exposed to and the precautions, which are in place.

The information may be provided either orally or in writing or possibly both. Key items (e.g. critical restrictions or prohibitions) should be recorded. Means of providing information might include:

- induction training programmes;
- employee handbooks or rulebooks;
- job descriptions;
- formal operating procedures;
- trainee agreement forms (increasingly common for young persons on formal training programmes e.g. modern apprenticeships);
- information forms for parents of work experience students.*

The information must be comprehensible – special arrangements may be necessary for young people whose command of English is poor or for those with special needs. The type of information required to be provided will obviously relate to the work activities and the risks involved. The content must be relevant – both to the workplace and the young person. It might include the following types of information:

- general risks present in the workplace
  e.g. fork lift trucks are widely used in the warehouse.
- general precautions taken in respect of those risks
  e.g. all fork lift drivers are trained to the standard required by the Approved Code of Practice ('ACoP').
- specific precautions in respect of the young person
  e.g. the induction tour includes identification of areas where fork lift trucks operate and indication of warning signs.

- restrictions or prohibitions on the young person
  e.g. X will not be allowed to drive fork lift trucks or any other vehicles (he will be considered for fork lift truck training after attaining the age of 17).
- supervision arrangements
  e.g. X will be supervised by the warehouse foreman (or other persons designated by him).
- PPE requirements
  e.g. safety footwear must be worn by all employees working in the warehouse (this is supplied by the company).

Where restrictions or prohibitions are removed (e.g. after successful completion of training programmes) an appropriate record should be made, either on the original restriction/prohibition or within the individual's training record.

---

* Further guidance on work experience is provided in the HSE booklet HSG 199 'Managing health and safety on work experience: A guide for organisers'.

[V12015]

Table A
Prohibitions on children and young persons
Outright prohibitions on work by children and young persons continue to be revoked, and to be replaced by a risk assessment approach. Prohibitions known still to be in place at the time of publication in 2000 of the HSE booklet 'Young people at work: A guide for employers' (HSG 165) were:
Carriage of explosives and dangerous goods
Under-18s may not be employed as a driver or attendant of an explosives vehicle, be responsible for the security of the explosives and may only enter the vehicle under the direct supervision of someone over the age of 18. (There are some exceptions where the risks are low). [*Carriage of Explosives by Road Regulations 1996 (SI 1996 No 2093]*. Under-18s may not supervise road tankers or vehicles carrying dangerous goods nor supervise the unloading of petrol from a road tanker at a petrol filling station. [*Carriage of Dangerous Goods by Road Regulations 1996 (SI 1996 No 2095* as amended].
Agriculture
Under-13s may not drive or ride on vehicles and machines including tractors, unprotected trailers etc. [*Prevention of Accidents to Children in Agriculture Regulations 1998 (SI 1998 No 3262]*. HSE guidance also states that children (under the minimum school leaving age) should not operate certain machines and tractors carrying out certain operations.
Ionising radiation
Under-18s may not be designated as 'classified persons'. Dose exposure limits are lower for under-18s. [*Ionising Radiations Regulations 1999 (SI 1999 No 3232]*.
Lead

Under-18s may not be employed in certain lead processes – lead smelting and refining, lead-acid battery manufacturing – or to clean places where such processes are carried out.
Mines and quarries
Various restrictions exist for under-18s and under-16s relating to use of winding and rope haulage equipment, conveyors at work faces, locomotives, shunting and shot firing. Some restrictions also apply to under-21s and under-22s.
Shipbuilding and shiprepairing
Under-18s, until they have been employed in a shipyard for six months, may not be employed on staging or in any part of a ship where they are liable to fall more than two metres or into water where there is a risk of drowning. [*Shipbuilding and Shiprepairing Regulations 1960 (SI 1960 No 1932)*].
Docks
Under-18s may not operate powered lifting appliances in dock operations unless undergoing a suitable course of training under proper supervision of a competent person (serving members of HM Forces are exempt). [*Docks Regulations 1988 (SI 1988 No 1655)*].

Work presenting increased risks for children and young persons

[V12016] The checklist below (**TABLE B**) is based upon *Regulations 3(5)* and *19(2)* of the *Management Regulations 1999 (SI 1999 No 3242)* and the Annex to the European Council Directive 94/33/EC. It is intended to assist employers conducting risk assessments in respect of work by children and young persons.

These types of work or situations of exposure to risk are not necessarily prohibited, although the requirements of *Regulation 19(2)* must be taken into account. However, restrictions could be required for young persons (particularly children) and additional precautions may be required to provide them with adequate protection from risk. Many situations present no greater risk to young persons than adults and restrictions (e.g. close supervision) may only be necessary until the employer is sure that the young person is fully aware of the risks and is capable of taking the necessary precautions.

Table B
Excessively physically demanding work
• Manual handling operations where the force required or the repetitive nature could injure someone whose body is still developing (including production line work);
• Certain types of piece work (particularly if peer pressure may result in them tackling tasks or working at speeds that are too much for them).
Excessively psychologically demanding work
• Work with difficult clients or situations where there is a possibility of violence or aggression;
• Difficult emotional situations e.g. dealing with death, serious illness or injury;

Table B
• Decision making under stress.
Harmful exposure to physical agents
• Ionising radiation (separate exposure limits apply to young persons);
• Non-ionising electromagnetic radiation, e.g. lasers, UV from electric arc welding or lengthy exposure to sunlight, infra red from furnaces or burning/welding;
• Risks to health from extreme cold or heat;
• Excessive noise;
• Hand-arm vibration, e.g. from portable tools;
• Whole-body vibration, e.g. from off-road vehicles;
• Work in pressurised atmospheres and diving work.
Harmful exposure to biological or chemical agents
• Very toxic, toxic, harmful, corrosive and irritant substances;
• Substances causing heritable genetic damage or harming the unborn child;
• Carcinogenic substances;
• Asbestos (including asbestos-containing materials);
• Lead and lead compounds;
• Biological risks, e.g. legionella, leptospirosis (Weil's disease), zoonoses.
Work equipment
Where there is an increased risk of injury due to the complexity of precautions required or the level of skill required for safe operation e.g:
• Woodworking machines (particularly saws, surface planing and vertical spindle moulding machines);
• Food slicers and other food processing machinery;
• Certain types of portable tools such as chainsaws;
• Power presses;
• Vehicles such as fork lift trucks, mobile cranes, construction vehicles;
• Other cranes and lifting hoists;
• Mobile elevating work platforms;
• Firearms.
Some young people may not have the physical size or strength necessary to operate equipment which has been designed for adults.
Dangerous materials or activities
• Work with explosives, including fireworks;
• Work with fierce or poisonous animals, e.g. on farms, in zoos or veterinary work;
• Certain types of electrical work, e.g. exposure to high voltage or live electrical equipment;
• Handling of flammable liquids or gases, e.g. petrol, acetylene, butane, propane;
• Work with pressurised gases;

Table B
• Work in large slaughterhouses (risks from animal handling, use of stunning equipment); • Holding large quantities of cash or valuables.
Dangerous workplaces or workstations • Work at heights, e.g. on high ladders or other unprotected forms of access; • Work close to deep or fast-flowing water; • Work in confined spaces, particularly where the risks specified in the *Confined Spaces Regulations 1997 (SI 1997 No 1713)* are present; • Work where there is a risk of structural collapse, e.g. in construction or demolition activities or inside old buildings.

Further guidance on risks to young people and appropriate precautions is contained in HSE booklet HSG 165 '*Young people at work: a guide for employers*'. HSE leaflet AS 10 deals specifically with '*Preventing accidents to children on farms*'.

New or expectant mothers

[V12017] Amendments made in 1994 to the previous *Management Regulations (SI 1992 No 2051 now revoked)* implemented the European Directive on Pregnant Workers (92/85/EEC), requiring employers in their risk assessments to consider risks to new or expectant mothers. These amendments were subsequently incorporated into the *Management Regulations 1999 (SI 1999 No 3242)*. *Regulation 1* of the *Management Regulations* contains two relevant definitions:

'New or expectant mother' means an employee who is pregnant; who has given birth within the previous six months; or who is breastfeeding.

'Given birth' means 'delivered a living child or, after twenty-four weeks of pregnancy, a stillborn child'.

Requirements of the Management Regulations

[V12018] The requirements for 'Risk assessment in respect of new and expectant mothers' are contained in *Regulation 16* of the *Management Regulations 1999 (SI 1999 No 3242)* which states in *paragraph (1)*:

Where—

(a)   the persons working in an undertaking include women of child-bearing age; and

(b)   the work is of a kind which could involve risk, by reason of her condition, to the health and safety of a new or expectant mothers, or to that of her baby, from any processes or working conditions, or physical, biological or chemical agents, including those specified in Annexes I and II of Council Directive 92/85/EEC on the introduction of measures to encourage improvements in the safety and health at work of pregnant workers and workers who have recently given birth or are breastfeeding, the assessment required by Regulation 3(1) shall also include an assessment of such risk.

Paragraph (4) of the Regulation states that in relation to risks from infectious or contagious diseases an assessment must only be made if the level of risk is in addition to the level of exposure outside the workplace.

The types of risk that are more likely to affect new or expectant mothers are described later in the chapter. *Paragraphs (2)* and *(3)* of *Regulation 16* set out the actions employers are required to take if these risks cannot be avoided. *Paragraph (2)* states:

> Where, in the case of an individual employee, the taking of any other action the employer is required to take under the relevant statutory provisions would not avoid the risk referred to in paragraph (1) the employer shall, if it is reasonable to do so, and would avoid such risks, alter her working conditions or hours of work.

Consequently where the risk assessment required under *Regulation 16(1)* shows that control measures would not sufficiently avoid the risks to new or expectant mothers or their babies, the employer must make reasonable alterations to their working conditions or hours of work.

In some cases restrictions may still allow the employee to substantially continue with her normal work but in others it may be more appropriate to offer her suitable alternative work.

Any alternative work must be:

- suitable and appropriate for the employee to do in the circumstances;
- on terms and conditions which are no less favourable.

Paragraph (3) of *Regulation 16* states:

> If it is not reasonable to alter the working conditions or hours of work, or if it would not avoid such risk, the employer shall, subject to section 67 of the 1996 Act, suspend the employee from work for so long as is necessary to avoid such risk.

The 1996 Act referred to here is the *Employment Rights Act 1996* which provides that any such suspension from work on the above grounds is on full pay. However, payment might not be made if the employee has unreasonably refused an offer of suitable alternative work. Employment continues during such a suspension, counting as continuous employment in respect of seniority, pension rights etc. Contractual benefits other than pay do not necessarily continue during the suspension. These are a matter for negotiation and agreement between the employer and employee, although employers should not act unlawfully under the *Equal Pay Act 1970* and the *Sex Discrimination Act 1975*. Enforcement of employment rights is through employment tribunals.

Regulation 17 of the *Management Regulations 1999 (SI 1999 No 3242)* deals specifically with night work by new or expectant mothers and states:

> Where—
>
> (a)   a new or expectant mother works at night; and
> (b)   a certificate from a registered medical practitioner or a registered midwife shows that it is necessary for her health or safety that she should not be at work for any period of such work identified in the certificate,

the employer shall, subject to section 67 of the 1996 Act, suspend her from work for so long as is necessary for her health or safety.

Such suspension (on the same basis as described above) is only necessary if there are risks arising from work. The HSE does not consider there are any risks to pregnant or breastfeeding workers or their children from working at night per se. It suggests that any claim from an employee that she cannot work nights should be referred to an occupational health specialist. The HSE's own Employment Medical Advisory Service are likely to have a role to play in such cases.

The requirements placed on employers in respect of altered working conditions or hours of work and suspensions from work only take effect when the employee has formally notified the employer of her condition. *Regulation 18* of the *Management Regulations 1999 (SI 1999 No 3242)* states:

(1) Nothing in paragraph (2) or (3) of regulation 16 shall require the employer to take any action in relation to an employee until she has notified the employer in writing that she is pregnant, has given birth within the previous six months, or is breastfeeding.

(2) Nothing in paragraph (2) or (3) of regulation 16 or in regulation 17 shall require the employer to maintain action taken in relation to an employee—

    (a) in a case—

        (i) to which regulation 16(2) or (3) relates; and

        (ii) where the employee has notified her employer that she is pregnant, where she has failed, within a reasonable time of being requested to do so in writing by her employer, to produce for the employer's inspection a certificate from a registered medical practitioner or a registered midwife showing that she is pregnant;

    (b) once the employer knows that she is no longer a new or expectant mother; or

    (c) if the employer cannot establish whether she remains a new or expectant mother.

Risks to new or expectant mothers

[V12019]–[V12020] The HSE booklet HSG 122 'New and expectant mothers at work: A guide for employers' provides considerable guidance on those risks which may be of particular relevance to new or expectant mothers, including those listed in the EC Directive on Pregnant Workers (92/85/EEC). This guidance is both summarised and augmented below. HSE leaflet CAIS 19 provides guidance on '*Health and safety of new and expectant mothers in the catering industry*'.

Physical agents

Manual handling

    Pregnant women are particularly susceptible to risk from manual handling activities as also are those who have recently given birth, especially after a caesarean section. Manual handling assessments (as required by the *Manual Handling Operations Regulations 1992 (SI 1992 No 2793)* are dealt with in the chapter MANUAL HANDLING in this publication.

Noise

The HSE does not consider that there are any specific risks from noise for new or expectant mothers. Compliance with the requirements of the *Control of Noise at Work Regulations 2005 (SI 2005 No 1643)* should provide them with sufficient protection. See the chapter NOISE AT WORK for more details of the regulations.

Ionising radiation

The foetus may be harmed by exposure to ionising radiation, including that from radioactive materials inhaled or ingested by the mother. The *Ionising Radiations Regulations 1999 (SI 1999 No 3232)* set an external radiation dose limit for the abdomen of any woman of reproductive capacity and also contain a specific requirement to provide information to female employees who may become pregnant or start breast-feeding. Systems of work should be such as to keep exposure of pregnant women to radiation from all sources as low as reasonably practicable. Contamination of a nursing mother's skin with radioactive substances can create risks for the child and special precautions may be necessary to avoid such a possibility.

Several HSE publications provide detailed guidance on work involving ionising radiation. (See L121 'Work with ionising radiation: Ionising Radiations Regulations 1999: Approved Code of Practice and Guidance' (2000); and INDG334 'Working safely with radiation: Guidelines for expectant and breast-feeding mothers' (2001), in particular.)

Other electromagnetic radiation

The HSE does not consider that new or expectant mothers are at any greater risk from other types of radiation, with the possible exception of over exposure to radio-frequency radiation that could raise the body temperature to harmful levels. Compliance with the exposure standards for electric and magnetic fields published by the National Radiological Protection Board should provide adequate protection.

Work in compressed air

Although pregnant women may not be at greater risk of developing the 'bends', potentially gas bubbles in the circulation could seriously harm the foetus should this condition arise. There is also evidence that women who have recently given birth have an increased risk of the bends. The *Work in Compressed Air Regulations 1996 (SI 1996 No 1656), Reg 16(2)* states:

> . . . the compressed air contractor shall ensure that no person works in compressed air where the compressed air contractor has reason to believe that person to be subject to any medical of physical condition which is likely to render that person unfit or unsuitable for such work.

This would appear to prohibit such work by pregnant women or those who have recently given birth. In its booklet HS(G)122 the HSE states that there is no physiological reason why a breastfeeding mother should not work in compressed air although they point out that practical difficulties would exist. HSE booklet L96 'A guide to the Work in Compressed Air Regulations 1996' (1996) provides detailed guidance on the requirements of the *Work in Compressed Air Regulations*.

Diving work

The HSE draws attention to the possible effects of pressure on the foetus during underwater diving by pregnant women and states that they should not dive at all.

Under the *Diving at Work Regulations 1997 (SI 1997 No 2776)*, Reg 15 divers must have a certificate of medical fitness to dive and the HSE guidance to doctors issuing such certificates advises that pregnant workers should not dive.

Regulation 13(1) of the *Diving at Work Regulations 1997 (SI 1997 No 2776)* states that 'No person shall dive in a diving project . . . if he knows of anything (including any illness or medical condition) which makes him unfit to dive.'

A series of HSE booklets provide general guidance on different types of diving projects.

Shock, vibration etc.

Major physical shocks or regular exposure to lesser shocks or low frequency vibration may increase the risk of a miscarriage. Activities involving such risks (e.g. the use of vehicles off road) should be avoided by pregnant women. The *Control of Vibration at Work Regulations 2005 (SI 2005 No 1093)* introduced specific requirements for the assessment of vibration risks — see the chapter entitled VIBRATION.

Movement and posture

Fatigue from standing and other physical work has been associated with miscarriage, premature birth and low birth weight. Ergonomic considerations will increase as the pregnancy advances and these could affect display screen equipment workstations, work in restricted spaces (e.g. for some maintenance or cleaning activities) or work on ladders or platforms. Underground mining work is likely to involve movement and posture problems and will also be subject to some of the other 'physical agents' described in this section. Driving for extended periods or travel by air may also present postural problems.

Pregnant women should be allowed to pace their work appropriately, taking longer and more frequent breaks. They may need to be restricted from carrying out certain tasks. Seating may need to be provided for work that is normally done standing and adjustments to display screen equipment and other workstations may be necessary.

Physical and mental pressure

Excessive physical or mental pressure could cause stress and lead to anxiety and raised blood pressure. Workplace stress is a complex issue which has recently been receiving increased attention. Fatigue issues are referred to above but other possible causes of stress may also need to be considered. These might be associated with the workload of individual pregnant employees (e.g. for those in management or administrative roles), the pressure of decision-making (e.g. in the health care or financial sectors) or the trauma of potential work situations (e.g. serious accidents dealt with by the emergency services).

Extreme cold or heat
Pregnant women are less tolerant of heat and may be more prone to fainting or heat stress. Although the risk is likely to reduce after birth, dehydration may impair breastfeeding. Exposure to prolonged heat at work, e.g. at furnaces or ovens, should be avoided.

Maintenance or cleaning work in hot situations should also be avoided, particularly if this involves use of less secure forms of access such as ladders, where fainting could result in a serious fall.

The HSE does not consider that there are any specific problems from working in extreme cold although obviously appropriate precautions should be taken as for other workers e.g. the provision of warm clothing.

Biological agents

[V12021]–[V12022] Many biological agents in hazard groups 2, 3 and 4 (as categorised by the Advisory Committee on Dangerous Pathogens) can affect the unborn child should the mother be infected during pregnancy. Some agents can cause abortion of the foetus and others can cause physical or neurological damage. Infections may also be passed on to the child during or after birth, e.g. while breastfeeding.

Agents presenting risks to children include hepatitis B, HIV, herpes, TB, syphilis, chickenpox, typhoid, rubella (German measles), cytomegalovirus and chlamydia in sheep. Most women will be at no more risk from these agents at work than living within the community but the risks are likely to be higher in some work sectors, e.g. laboratories, health care, the emergency services and those working with animals or animal products.

Details of appropriate control measures for biological agents are contained in *Schedule 3* and *Appendix 2* of the ACoP booklet to the *COSHH Regulations 2002 (SI 2002 No 2677)*. There is a separate HSE publication on '*Infection risks to new and expectant mothers in the workplace*' (1997). When carrying out a risk assessment in respect of new or expectant mothers, normal containment or hygiene measures may be considered sufficient, but there may be the need for special precautions such as use of vaccines. Where there is a high risk of exposure to a highly infectious biological agent it may be necessary to remove the worker entirely from the high-risk environment.

Chemical agents

Substances labelled with certain risk phrases
The *Chemicals (Hazard Information and Packaging for Supply) Regulations 2002 (SI 2002 No 1689)* ('*CHIP*') require many types of hazardous substances to be labelled with specified risk phrases. Several of those are of relevance to new or expectant mothers:
R40: limited evidence of a carcinogenic effect;
R45: may cause cancer;
R46: may cause heritable genetic damage;
R49: may cause cancer by inhalation;
R61: may cause harm to the unborn child;
R63: possible risk of harm to the unborn child;
R64: may cause harm to breastfed babies;
R68: possible risk of irreversible effects.

The *CHIP Regulations 2002* are regularly subject to changes and employers should be alert for other risk phrases which indicate risks in respect of new or expectant mothers.

Control of such substances at work is already required by the *COSHH Regulations 2002 (SI 2002 No 2677)* or the separate regulations governing lead and asbestos. Risk assessments in relation to pregnant women or those who have recently given birth may indicate that normal control measures are adequate to protect them also. However, additional precautions may be necessary, e.g. improved hygiene procedures, additional PPE or even restriction from work involving certain substances.

The ACoPs relating to the *COSHH Regulations 2002 (SI 2002 No 2677)* (contained in HSE booklet L5) together with HSG 193 '*COSHH Essentials: Easy steps to control chemicals*' provide further details of the types of precautions which may be appropriate.

Mercury and mercury derivatives

The HSE states that exposure to organic mercury compounds can slow the growth of the unborn baby, disrupt the nervous system and cause the mother to be poisoned. It considers that there is no clear evidence of adverse effects on the foetus from mercury itself and inorganic mercury compounds. Mercury and its derivatives are subject to the *COSHH Regulations 2002 (SI 2002 No 2677)* and normal control measures (see above) may be adequate to protect new or expectant mothers.

(See HSE publications L5 and HSG193 as mentioned above; and also MS12 '*Mercury: Medical guidance notes*' (1996).)

Antimitotic (cytotoxic) drugs

These drugs (which may be inhaled or absorbed through the skin) can cause genetic damage to sperm and eggs and some can cause cancer. Those workers involved in preparation or administration of such drugs and disposal of chemical or human waste are at greatest risk, e.g. pharmacists, nurses and other health care workers. Antimitotic drugs can present a significant risk to those of either gender who are trying to conceive a child as well as to new or expectant mothers, and all those working with them should be made aware of the hazards.

The *COSHH Regulations 2002 (SI 2002 No 2677)* again apply to the control of these substances (see above). Since there is no known threshold limit for them, exposure must be reduced to as low a level as is reasonably practicable.

Agents absorbed through the skin

Various chemicals including some pesticides may be absorbed through the skin causing adverse effects. These substances are identified in the tables of occupational exposure limits in HSE booklet EH40 '*Workplace exposure limits*' (an updated version is published regularly), in which some substances are accompanied by an annotation 'Sk'. Many such agents (particularly pesticides which are subject to the *Control of Pesticides Regulations 1986 (SI 1986 No 1510)*) will also be identified by product labels.

Effective control of these chemicals is obviously important in respect of all employees (the *COSHH Regulations 2002 (SI 2002 No 2677)* applying once again) although risk assessments may reveal the need for additional precautions in respect of new or expectant mothers, e.g. modified handling methods, additional PPE, or even their restriction from activities involving exposure to such substances.

Carbon monoxide

Exposure of pregnant women to carbon monoxide can cause the foetus to be starved of oxygen with the level and duration of exposure both being important factors. There are not felt to be any additional risks from carbon monoxide to mothers who have recently given birth or to breast-fed babies. Once again it is important to protect all members of the workforce from high levels of carbon monoxide (the *COSHH Regulations 2002 (SI 2002 No 2677)* require it) but it may also be necessary to ensure that pregnant women are not regularly exposed to carbon monoxide at lower levels, e.g. from use of gas-fired equipment or other processes or activities. HSE Guidance Note EH43 provides general guidance on carbon monoxide.

Lead and lead derivatives

High occupational exposure to lead has historically been linked with high incidence of spontaneous abortion, stillbirth and infertility. Decreases in the intellectual performance of children have more recently been attributed to exposure of their mothers to lead. Since lead can enter breast milk there are potential risks to the child if breastfeeding mothers are exposed to lead.

All women of reproductive capacity are prohibited from working in many lead processing activities. All work involving exposure to lead is subject to the *Control of Lead at Work Regulations 2002 (SI 2002 No 2676)*. Even where women of reproductive capacity are allowed to work with lead or its compounds the blood-lead concentrations contained within the Regulations as an 'action level' and a 'suspension level' are set at half the figures for adult males. This is intended to ensure that women who may become pregnant already have low blood lead levels.

Once pregnancy is confirmed, the doctor carrying out medical surveillance (as required by *Regulation 10* of *SI 2002 No 2676*) would normally be expected to suspend the woman from work involving significant exposure to lead. There is, however, no specific requirement in the Regulations themselves that this must happen. Detailed guidance on the Regulations is contained in HSE booklet L132 'Control of lead at work' (2002).

Working conditions

[V12023] HSE guidance in HS(G)122 in respect of the 'working conditions' referred to in the *Management Regulations (SI 1999 No 3242), Reg 16(1)* repeats the HSE's stated position on work with DSE contained in Appendix 2 of the HSE booklet L26 'Work with display screen equipment'. Its view is that radiation from DSE is well below the levels set out in international recommendations and that scientific studies taken as a whole do not demonstrate any link between this work and miscarriages or birth defects.

There is no need for pregnant women to cease working with DSE. However, the HSE recommends that, to avoid problems from stress or anxiety, women are given the opportunity to discuss any concerns with someone who is well informed on the subject.

Other aspects of pregnancy

[V12024] Although the HSE booklet on 'New and expectant mothers at work' (HSG 122) draws attention to other features of pregnancy which employers may wish to take into account, it suggests that employers have no legal obligation to do so. However, this view has not been tested by the courts and an argument could be advanced that some of these represented 'working conditions' which involve risk to the mother or her baby within the terms of the *Management Regulations 1999 (SI 1999 No 3242), Reg 16(1)(b)*.

The impact of these aspects will vary during the course of the pregnancy and employers are likely to need to keep the situation under review. A modified version of the appendix from the HSE booklet dealing with these 'other aspects' is provided in TABLE C below.

ASPECTS OF PREGNANCY THAT MAY AFFECT WORK

Table C	
Aspects of pregnancy	*Factors in work*
Morning sickness	Early shift work Exposure to nauseating smells Availability for early meetings Difficulty in leaving job
Frequent visits to toilet	Difficulty in leaving job/site of work
Breastfeeding	Access to private area Use of secure, clean refrigerators to store milk
Tiredness	Overtime/long working hours Evening work More frequent breaks Private area to sit or lie down
Increasing size	Difficulty in using protective clothing Work in confined areas Manual handling difficulties Posture at DSE workstations (Dexterity, agility, co-ordination, reach and speed of movement may also be impaired)
Comfort	Problems working in confined or congested workspaces
Backache	Standing for extended periods Posture for some activities Manual handling
Travel i.e. fatigue, stress, static posture	Reduce need for lengthy journeys Change travel methods
Varicose veins	Standing/sitting for long periods
Haemorrhoids	Working in hot conditions/sitting

Table C	
Aspects of pregnancy	*Factors in work*
Vulnerability to passive smoking	An effective smoking policy and ensuring compliance with legislative requirements on smoking
Vulnerability to stress	Modified duties Adjustments to working conditions or hours
Vulnerability to falls	Restrictions on work at heights or on potentially slippery surfaces
Vulnerability to violence (e.g. from contact with the public)	Changes to workplace layout or staffing Modified duties
Lone working	Effective communication and supervision Possible need for improved emergency arrangements

A practical approach to assessing risks

[V12025] Essentially risk assessment in respect of new or expectant mothers must be carried out by an employer if:

- there are women of childbearing age; and
- their work could involve risks to new or expectant mothers or their babies.

Most organisations employ women of childbearing age and, particularly in the case of large businesses, there may be a number of work activities that could create relevant risks. In some cases (e.g. those involving exposure to hazardous substances or radiation) it may be appropriate to stipulate that women are not allowed to work in certain activities, processes or departments once their pregnancy has been notified and/or for a finite period after they return to work after giving birth.

However, in many workplaces the issues will be far from clear cut – it would be neither necessary nor practical to prohibit pregnant workers from carrying out *any* manual handling or from standing up, sitting down or climbing up ladders. Also it is difficult for any employer to identify in advance exactly what steps need to be taken to alter working conditions or hours of work so that risks can be avoided for every female employee who may become pregnant.

A much more practical approach is for the employer to develop a checklist similar to the sample provided at the end of this section. Such a checklist can be prepared by carrying out a review of the organisation's activities in order to identify risks which are present which *may* be of relevance in relation to new or expectant mothers. The risks described in the previous part of this chapter provide a good starting point from which a workplace-specific checklist can be developed. In some workplaces it may be appropriate to develop more than one such checklist (e.g. for production, maintenance and administration) because the profiles of risks are different.

Such a checklist should be completed by a suitable employer's representative (a personnel or health and safety specialist, or perhaps the pregnant woman's manager) together with the pregnant woman herself. The checklist is intended to provoke discussion about the employee's possible exposure to the risks that the employer has identified, so that any necessary additional precautions (or changes to work practices) can be agreed. There is also the opportunity to identify whether any further review of the situation is necessary. The process would be repeated once the mother returned to work after giving birth.

The sample checklist provided below is for a bakery. Female staff are likely to be involved in production work where there is significant potential for risk to new or expectant mothers, and they may also be employed in laboratory or maintenance work. Risks for staff working in administrative activities must also be considered, e.g. those associated with DSE workstations. However, the greatest problems could be for female staff delivering to retail outlets. As well as risks from manual handling, falls and possibly excessive driving, the difficulties for such staff in accessing suitable rest and toilet facilities must also be taken into account.

BUNN THE BAKERS	NEW OR EXPECTANT MOTHERS	RISK ASSESSMENT CHECKLIST
Name of Employee:		Work location:
	Possible risks to consider	Precautions/changes agreed
Bakery and delivery work	Manual handling (production, maintenance, deliveries) Excessive standing (production lines) Awkward sitting (production lines) Difficult access (cleaning, maintenance) Excessive heat (near to ovens) Hazardous substances (e.g. cleaning materials, maintenance, laboratory) Significant risk of falls (e.g. slippy floors, maintenance work, trips on delivery work) Excessive driving (deliveries) Other	
Administrative activities	DSE workstation layout Unsuitable meeting times Manual handling (e.g. stationery, records) Significant travel (especially by car or air) Other	
All activities	Ability to leave workplace temporarily Availability of rest area/toilets Major pressure or stress situations Possible effects of fatigue Other	
Signature (for Bunn the Bakers):		Signature (new/expectant mother):
Date:		Date for further review (if any):
This form should be completed when the pregnancy is first notified *and* when the new mother returns to work.		

Risks to non-employees

[V12026] As part of their general risk assessments, employers must also consider risks to new or expectant mothers who are not part of their own workforce. Such persons may be employed by another organisation or be self-employed but they may also be service users or members of the public. Even though other employers have duties to assess risks to their staff, 'host' employers still have a statutory duty (and a 'duty of care') to other people's employees, particularly in respect of risks of which they may be aware, but which might not be immediately apparent to others. In this context, risks to pregnant women especially could include:

- slips, trips and falls, e.g. due to slippy or uneven surfaces;
- accidental impact from persons, e.g. running children or accidental contacts related to sports activities;
- accidental collision with mobile equipment;
- use of off-road transport;
- assault, e.g. on police or prison premises;
- ionising radiation;
- hazardous substances (see the earlier reference to risk phrases at V12022 above).

Lone workers

[V12027] There are no general legal restrictions on working alone – indeed many work activities would be extremely impractical if there were. However, the activities of lone workers must be subject to the same sort of risk assessment process as any other type of work. Even those regulations imposing the greatest restrictions on lone working (i.e. *Electricity at Work Regulations 1989 (SI 1989 No 635)* and the *Confined Spaces Regulations 1997 (SI 1997 No 1713)*) still require the application of risk assessment techniques in identifying suitable safe systems of work.

The various types of lone workers

[V12028] There are many different types of lone working. These can be divided into three categories:

People working alone in fixed workplaces
  This category includes:
- workers in small shops and kiosks;
- people working in more remote parts of larger premises, e.g. isolated reception or enquiry desks, private interview rooms, security gatehouses;
- laboratories, stores etc.;
- staff working late, at nights or at weekends;
- staff responding to alarm calls etc, e.g. 'keyholders';
- staff working at home.

Staff working on a variety of premises belonging to other employers
Such people will be working independently of colleagues from their own employer, but may have others working around them for at least part of the time. They include:

- persons carrying out construction or plant installation work;
- maintenance and repair workers;
- cleaners, painters, decorators etc;
- other types of contracting staff.

Persons working within the wider community
These staff may be working in urban, rural or remote locations, sometimes in places which are accessible to the public but also on private property (often in domestic premises). This includes:

- mobile security staff;
- domestic installation, maintenance and repair staff;
- delivery staff (including postal workers);
- sales and technical representatives;
- architects, property surveyors and estate agents;
- rent collectors and meter readers;
- social workers, home carers and district nurses;
- police, traffic wardens and probation staff;
- agricultural and forestry workers;
- utility workers (electricity, gas, water, sewage, telecommunications etc.);
- vehicle recovery staff;

and many other similar types of work.

Risks for lone workers

[V12029] In some situations lone workers may be at risk because they are not physically capable of carrying out certain activities safely on their own. These may relate to:

Manual handling activities
- – where team handling is identified as being necessary.

Use of certain types of equipment
- – particularly heavy and bulky equipment, or where another person needs to be available to carry out adjustments.

Operation of controls etc.
- – controls may be physically separate from each other or from related instrumentation, requiring a second person to simultaneously activate controls or relay information (e.g. where a banksman may be required for lifting operations).

Achieving safe access
- – a second person may be necessary to foot a ladder or ensure safe use of other forms of temporary access equipment.

Some work activities may involve such a high degree of risk that another person needs to be present to assist in the identification of risks or act as a deterrent to such risks. These situations include:

Some types of maintenance and repair work
– ensuring that equipment has been correctly isolated and other precautions taken (two heads are better than one!).

Live electrical work
– ensuring that the person carrying out the work is fully aware of what is live and what is not.

Work involving potential risks of aggression or even violence
– handling complaints, passing on unwelcome news, dealing with persons with aggressive or violent histories, working in high risk areas of the community, handling cash or valuables.

Activities where staff feel highly vulnerable
– one to one encounters (possibly with unknown persons), particularly in remote locations, empty premises or domestic property (female staff often feel more vulnerable but males should also be made aware of risks of unwarranted accusations against them)

There may also be risks to lone workers where a second person may be needed to take appropriate emergency actions such as:

Isolating electrical supplies
• – particularly where live work is being carried out.

Rescuing a worker from inside a confined space
• – e.g. by use of a rescue line or wearing suitable respiratory protective equipment ('RPE') and PPE.

Giving immediate first aid attention
• – particularly important for electrical accidents or persons seriously affected by hazardous substances, or asphyxiated.

Safely recovering a worker who has fallen whilst wearing a safety harness

The general potential for accidents and illness to lone workers from less foreseeable causes must also be considered, even though these may not justify the immediate presence of another person.

The above examples are of the more common types of risks that lone workers may encounter. Employers will no doubt be able to add to the list, particularly after consulting those who carry out lone work and their representatives.

Controlling risks to lone workers

[V12030] The previous section identified the types of risks that lone workers may be subject to. Set out below is a variety of measures which may be necessary to control risks. Whilst some measures will be relevant to most, if not all risk situations, others may only have limited application. This section is in effect a menu from which to choose the appropriate precautions.

General precautions

- Lone workers must be made aware of the types of risks that they may encounter and the precautions which must be taken to control those risks. This will require information, instruction and training, supported by effective ongoing supervision.

  Some types of lone working may justify the creation of formal written procedures or rules whilst in others the lone workers themselves may be expected to carry out a degree of dynamic risk assessment (see **R3017** in RISK ASSESSMENT). Lone workers must be familiar with relevant procedures and rules and have received appropriate training if they need to make dynamic risk assessments.

  As a result of the employer's risk assessment some activities may be identified as being ones which workers must never carry out alone. These must be communicated effectively to the lone worker, who must also be given guidance on where the necessary assistance can be obtained. This may be achieved by the temporary presence of work colleagues or assistance may be forthcoming from the customer's personnel or from others working in the vicinity.

Communication

- Good communications are essential to ensuring the safety of lone workers, including those who may work from a home base. This is likely to involve:
    - provision of suitable communication equipment, e.g. mobile phones, portable radios, direct radio links, monitored CCTV surveillance;
    - availability of emergency alarms, e.g. fixed alarm buttons, portable attack alarms (some types of alarms detect an absence of movement by the alarm user or require a regular response from the user);
    - availability of information on lone workers' whereabouts to others, e.g. work colleagues, control centres, security staff, partners, friends etc. (this may involve the use of location boards, computerised or desk diaries, itinerary sheets or just word of mouth);
    - ongoing communication from the lone worker, e.g. about delays, problems, changes in plans;
    - ongoing checks on the lone worker's wellbeing;
    - pre-emptive contacts from lone workers encountering unexpected risks (explaining where they are, who they're with or what they're doing and requesting a return call in an agreed period of time);
    - checks to ensure that the lone worker has safely returned to base (or home);
    - arrangements in place for the repair, inspection, maintenance or replacement of equipment that they use (including communication and alarm equipment).

Specific precautions
- There are many specific precautions which can be taken to reduce the risks to lone workers. The nature of these will depend upon the location of the lone worker and the type of activity involved. Examples of these include:
  - physical separation of the lone worker from people they must deal with, e.g. by use of screens or wide desks;
  - ensuring that assistance is readily available to the lone worker when required, e.g. for certain manual handling operations or live electrical work;
  - placing lone workers where assistance from members of the public is at hand, e.g. arranging encounters, particularly with higher risk contacts, in public buildings or areas, including hotels, cafes etc.

Emergency arrangements
- Here too, the nature of the arrangements will vary according to the location and activity of the lone worker. Arrangements are likely to include:
  - response to alarms – who is expected to respond and what actions should they take;
  - follow-up of lone workers who do not return when expected or do not respond to calls;
  - provision of basic first aid equipment for lone workers, so that they can carry out simple first aid for themselves or receive treatment from others coming to their aid.

Homeworkers
- There are several risks likely to affect homeworkers which need to be properly controlled:
  - suitable choice of equipment for use in the home environment;
  - training of staff in the correct use of equipment;
  - arrangements for inspection and maintenance of equipment (particularly in respect of electrical risks);
  - assessments of DSE workstations (the type of self-assessment checklist contained in the display screen equipment chapter is useful for homeworkers;
  - ensuring that loads can be manually handled safely (e.g. by using suitably sized containers);
  - suitable arrangements for safe storage and use of any hazardous or dangerous substances kept by the homeworker;
  - effective communication arrangements when working away from the home base (see above).

HSE leaflet INDG 226 provides further guidance on 'Homeworking'.

Monitoring of lone workers

[V12031] Those responsible for managing and supervising lone workers must ensure that their activities are monitored effectively. Since only a limited amount of observation of their work in the field is likely to be practicable, regular consultation with the staff involved will be important. Aspects to monitor for are:

- new risks which have become apparent;
- existing risks which may have increased;
- the effectiveness of existing precautions;
- workers' views on new precautions which may be appropriate.

It will also be necessary to actively monitor the effectiveness of precautions through actions such as:

- checking the effectiveness of communication equipment;
- testing alarms;
- enquiring as to the current whereabouts of lone workers;
- testing the awareness of staff on how they should react to emergency situations.

Disabled persons

Types of disability

[V12032] Disabilities come in many different forms. HSE guidance to local authorities on the *Disability Discrimination Act 1995* refers to impairments affecting the ability to perform normal day to day activities affecting at least one of the following:

- mobility;
- manual dexterity;
- physical co-ordination;
- continence;
- ability to lift, carry or otherwise move everyday objects;
- speech, hearing or eyesight;
- memory or ability to concentrate, learn or understand; or
- perception of the risk of physical danger.

The last two on this list have a close relationship with the consideration of risks to young persons referred to earlier in the chapter.

Although not included in the list above, account must also be taken of those with relevant **health conditions** and **allergies,** particularly when assessing risks relating to hazardous substances.

Disabled persons at risk

[V12033] Employers in their risk assessments must take account of all disabled persons who may be affected by their activities, including:

- permanent and temporary employees;
- persons on training or work experience placements (who have the status of employees);

- visitors;
- customers or service users;
- neighbours and passers-by.

Access for disabled persons

[V12034] The *Health and Safety (Miscellaneous Amendments) Regulations 2002 (SI 2002 No 2174)* introduced an additional requirement into *Regulation 25* of the *Workplace (Health, Safety and Welfare) Regulations 1992 (SI 1992 No 3004)*:

> Where necessary, those parts of a workplace (including in particular doors, passageways, stairs, showers, washbasins, lavatories and workstations) used or occupied directly by disabled persons at work shall be organised to take account of such persons.

Factors which must be taken into account in complying with the above requirement and in preventing accidents to disabled persons are:

- the locations where disabled persons work or must make access to;
- the width of access routes;
- maintaining access routes free from obstructions and slipping or tripping hazards;
- standards of guardrails and handrails;
- heights of door handles, light switches, alarms, controls etc.

(It is not the intention of this chapter to consider the general accessibility of work premises to disabled members of the public.)

Other issues of relevance

[V12035] There are a number of other issues which must be considered when carrying out risk assessments in respect of disabled persons, whether members of the workforce or not, including:

- hazardous substances:
  - their potential effects on those with relevant health conditions or allergies;
  - the abilities of exposed persons to understand the risks or implement precautions.
- work equipment:
  - physical ability to handle the equipment;
  - intellectual capability to understand risks and implement necessary precautions.
- hot (or very cold) items or surfaces:
  - the ability to identify such items or surfaces;
  - a perception of the risks involved;
  - the ability to keep clear of such risks.
- workstation layout:
  - ergonomically suited to the needs of the disabled worker;
  - provision of suitable seating is of particular importance.
- fire and other emergency arrangements:

- the ability to hear or see alarm signals;
- the physical ability to react to the alarm with the necessary speed;
- the potential for panic amongst those with some types of disability.

Controlling risks to disabled persons

[V12036] As a result of the risk assessment process employers must control risks to the standards required by legislation – either the general standards required by the *HSWA 1974* or the more specific requirements of individual sets of regulations. In respect of employees they must also comply with the requirements of *Regulation 13(1)* of the *Management Regulations 1999 (SI 1999 No 3242)* which states:

> Every employer shall, in entrusting tasks to his employees, take into account their capabilities as regards health and safety.

This will require employers to:

- Identify the disabilities of individual workers and others via:
  - application forms and questionnaires;
  - interviews and discussions;
  - medical examinations;
  - registration forms e.g. at hotel receptions.
- Identify situations where persons with such disabilities may be at risk:
  - the types of issues referred to above.

In order to control risks to the standards required, it may be necessary for the employer to:

- modify existing equipment;
- provide alternative equipment;
- make modifications to the workplace;
- provide a work location suited to the individual's disability;
- provide additional or modified training (making special arrangements for those with learning difficulties);
- ensure emergency arrangements take account of disabled persons;
- restrict the activities and/or locations of disabled persons.

Employers have a duty under the *Disability Discrimination Act 1995* ('DDA') to make 'reasonable adjustment' to the workplace or working arrangements of disabled persons.

Avoiding discrimination

[V12037] Employers must find a balance between avoiding exposing disabled persons to risk and discriminating against them unfairly. HSE guidance (Local Authority Circular 63/3) states that:

> The DDA does not prevent employers from continuing to stipulate essential health requirements for particular types of employment but they must be able to justify these and show that it would not be reasonable to waive them in any individual case.

Employers may enquire about disability or ask a disabled person to undergo a medical examination but this must be justified by the relevance of the disability to the job in question.

There are concerns that employers may use health and safety reasons as an excuse for discriminating against disabled persons. Some of these concerns were found to be justified by research carried out on behalf of the HSE entitled 'Extent of use of health and safety requirements as a false excuse for not employing sick or disabled persons'. Although this showed that health and safety was frequently being used as the rationale for non-recruitment or dismissal, it found considerably more evidence of employers overcoming health and safety difficulties to retain workers with a disability, health condition or injury.

There are many practical actions that can be taken to accommodate disabled workers safely in the workplace, as described in the section above. Even where restrictions must be placed on the disabled, these can often relate to a very limited range of activities or locations, rather than preventing their employment altogether.

Inexperienced workers

[V12038] Workers of all ages may (like young persons) be lacking in experience of some types of work or lack awareness of the risks associated with particular work activities. *Regulation 13* of the *Management Regulations 1999 (SI 1999 No 3242)* as well as requiring employers to take into account the capabilities of employees in entrusting tasks to them (see above), also contains several requirements about training.

Paragraph (2) of the *Regulation 13* states:

> Every employer shall ensure that his employees are provided with adequate health and safety training – on their being recruited into the employer's undertaking; and on their being exposed to new or increased risks . . .

Such new or increased risks are specified as:

- transfer or change of responsibilities;
- due to new or changed work equipment;
- because of new or changed systems of work.

Training and supervision of inexperienced workers

[V12039] General health and safety induction training and job specific training is necessary in all workplaces but particular account of how to cope with these training needs must be taken in:

- work activities with rapid staff turnover;
- employment utilising large numbers of temporary workers, e.g. to meet seasonal needs or special orders (special arrangements will be necessary for workers without a good command of English);
- major events utilising temporary staff (possibly including volunteers);

- workplaces providing placements for those on government-funded training and re-training schemes;
- work activities involving workers with special educational needs.

Even when inexperienced workers have received appropriate training they will still require ongoing supervision, in many respects similar to that required for young persons. The imposition of restrictions as described earlier in this section (in relation to children and young persons) may also be appropriate.

Work at Height

George Ventris and Gordon Prosser

Introduction

[W9001] Carbon dating has indicated that nearly four thousand years ago people were faced with the risks associated with level differences, the need for working above floor level, on split levels or near shafts, in the copper mines of Alderley Edge in North Cheshire.

In spring 1962, work was being carried out on a flat roofed one storey building. The work had been commissioned by Pontins at their holiday camp in Blackpool. No provision had been made in the contract for any fence or barrier and a fall occurred. The Court of Appeal held that 'it would have been apparent to them, *if they had contemplated the position*, that danger would arise'. This case, *McArdle v Andmac Roofing Co. and ors* [1967] 1 All ER 583, had the effect of extending the range of persons to whom an organisation may be under a duty to take care. This duty relating to safe systems of work was subsequently built into the criminal law in the *Health and Safety at Work etc. Act 1974 (HSWA 1974), s 3* and more explicitly in the *Work at Height Regulations 2005 (SI 2005 No 735)*. Though a Civil action, the case has been cited in the Court of Appeal Criminal division.

The European Parliament and the Council, in their Directive 2001/45/EC (27 June 2001), identified that:

> Work at a height may expose workers to particularly severe risks to their health and safety, notably to the risks of falls from a height and other serious occupational accidents, which account for a large proportion of all accidents, especially of fatal accidents.

> Self-employed persons and employers, where they themselves pursue an occupational activity and personally use work equipment intended for carrying out temporary work at height, may affect *other persons*' health and safety.

In the Consultative document issued by the Health and Safety Commission (now the Health and Safety Executive (HSE)) in compliance with its duty to consult, under *HSWA 1974, ss 16(2) and 50(3)* before submitting proposals for making new Regulations, the following issues were identified:

- 'Falls from height are the biggest single cause of fatal injuries, and the second biggest cause of major injuries, caused by accidents at work – each year around 50-60 fatalities and 4,000 major injuries are caused by falls at work.'
- 'Reducing this toll is one of the HSC's Priority Programmes, which aims to reduce the incidence of fatal and major injuries by 10 per cent over 10 years from 1999/2000.'
- 'Implementing the Directive 2001/45/EC gives us an opportunity to make a substantial impact on the problem. This is an opportunity for us to:

(i) bring together for the first time in one place all the legal requirements for work at height;

(ii) this will be in a single set of goal-setting Regulations, applying to all industries;

(iii) the regulation will establish the key principles whilst allowing:

(a) flexibility for those who work at height in the very wide range of jobs where this is done; and

(b) allow for technical innovation in the development of equipment for safe work at height.'

- 'In implementing the Directive we are determined to ensure that the Regulations make sense, tackling the problems in high-risk areas whilst avoiding over-elaboration, inconsistencies or unworkable requirements.'

- 'Research carried out in support of the Falls from Height Priority Programme has shown that around *60 per cent of all major injuries* are caused by falls from *heights below 2 metres*. We propose, therefore, to cover all work at height where there is a risk of personal injury.'

Arguably the key factor of these proposals was the application of the regulations to all industries. Previous legislation and regulation had tended to be focused on only on the construction industry. It was clear that other industries, including the agricultural sector, transport and goods delivery, warehousing, education, manufacturing, the broadcasting and entertainment industries, telecommunications, retail and service industries, not to mention the simple day-to-day tasks associated with the occupation of premises (such as access to a light fitting in a ceiling for a lamp change) were, in comparison, neglected.

The HSE commissioned research, acknowledging as has the HSE website, that the problem was not *work* at height but *falls* from height. The research, using existing data from the *Reporting of Injuries, Diseases and Dangerous Occurrences Regulations 1995 (RIDDOR) (SI 1995 No 3163)*, reports, trade associations and other sources, as well as direct contact with representatives across industry, was able to look at causation influence factors, strategies for risk control, and assess the effectiveness of prevention methods by using cost-benefit analysis. Information was gathered across a variety of different industries, enabling factors common to all industry to be identified. The causation factors identified point to the importance of planning and organisation reflected in the *Work at Height Regulations 2005*.

Sir Chris Bonington, when comparing his own expeditions with those of others is quoted as saying: 'The mountaineer is exposed to some level of risk at almost all times he is on the mountain, but it is fairly rare for a good climber to be killed because the climb is too hard, or even when caught out by bad weather or some other kind of emergency. Then his concentration is complete, with every nerve stretched towards survival. It is on easy ground that accidents occur; that momentary lack of concentration, a slip where there happens to be a long drop, a hidden crevasse or risk of avalanche.'

The Work at Height Regulations 2005 (SI 2005 No 735)

[W9002] The HSE made it clear that the *Work at Height Regulations 2005 (SI 2005 No 735)* would not be accompanied by an approved code of practice

(ACoP) or interlinear guidance. In that respect they differ from the format of earlier Regulations such as the *Provision and Use of Work Equipment Regulations 1998 (PUWER) (SI 1998 No 2306)*, the *Lifting Operations and Lifting Equipment Regulations 1998 (LOLER) (SI 1998 No 2307)*, and the *Management of Health and Safety at Work Regulations 1999 (SI 1999 No 3242)*.

However, the HSE has developed a range of practical general guidance, generic advice and sector specific guidance which can be found on its website at http://www.hse.gov.uk/falls/index.htm.

This includes information relating to goods storage and warehousing, agriculture, forestry, the education sector, road tankers and car transporters, with cross industry information on tasks such as gutter cleaning, window cleaning, installation of roof fans and the like. A number of case studies highlighting how falls have occurred in a variety of industries are highlighted. The HSE also uses press releases to draw attention to particular problems, such as incidents when operatives were trapped between the basket of a boom type mobile elevating work platform and part of an adjacent structure. This risk is covered by specific wording in *SI 2005 No 735, Reg 5(c)(ii)*. The Regulations state that basic safety, for example slips and trips, and ergonomic comfort must be considered as part of the planning, organisation and risk assessment process.

The most comprehensive guidance to date on aspects of the new regulations is contained in British Standard BS 8437: 2005 (ISBN 0 580 45817 2).

With an eye on the *Construction (Design and Management) Regulations 2007 (SI 2007 No 320)* and the duties for designers under *SI 2005 No 735, Reg 13*, the Standard should not only help designers eliminate unnecessary work at height, but where it will be required in future (for maintenance, cleaning or other routine tasks carried out by future occupants), ensure that the work can be carried out without unnecessary risk or cost.

The following notes follow the order of the regulations and the incorporated schedules which expand on specific areas covered in the regulations. For example, ladders and step ladders, the subject of much misunderstanding and misinformation, are covered by *SI 2005 No 735, Sch 6*. There is now an increasing volume of guidance and information on the correct use of ladders, with pre-prepared tool box talks, equipment check lists and the like.

Commencement and scope

[W9003] The Regulations came into force on 6 April 2005. They apply in Great Britain, and outside Great Britain only as *HSWA 1974 ss 1–59* and *80–82* apply by virtue of the *Health and Safety at Work etc. Act 1974 (Application outside Great Britain) Order 2001 (SI 2001 No 2127)*.

Equivalent Regulations, the *Work at Height Regulations (Northern Ireland) 2005, (SR 2005 No 279)*, apply in Northern Ireland and came into operation on 11 July 2005.

Application

[W9004] The Regulations apply in relation to work. They apply irrespective of employment status. Requirements are imposed on employers, employees, the self-employed, and any persons under the control of any of the above. In a case involving the injury of an employee of a tenant in a multi-unit shopping development, both the tenant and the landlord managing the development were successfully prosecuted. The judge ruled that both parties had failed to control access to the area of fragile material through which the employee fell.

Height or level difference can be anywhere, below, above or at ground level. Sitting on a chair is not work at height. Standing on that chair, or indeed using it to climb onto a table or desk is. (Specific guidance on this issue for teachers and class room assistants is contained in the leaflet 'Keeping safe when working at height' because of the number of incidents in the education sector as a result of standing on chairs and desks.) It can also be downloaded at http ://www.hse.gov.uk/pubns/schoolsfall.pdf.

Access and egress, the route for getting to and from the place of work, has to be considered, unless 'by a staircase in a permanent workplace'. Such staircases are covered by the *Workplace (Health, Safety and Welfare) Regulations 1992 (SI 1992 No 3004), Reg 12* and the duty to assess as a route under the *Manual Handling Operations Regulations 1992 (SI 1992 No 2793), Reg 4 and Sch 1.* (See also **A1007 – ACCESS, TRAFFIC ROUTES AND VEHICLES**)

The duty to cover or fence every tank, pit or structure where there is a risk of a person falling into a dangerous substance is covered by *SI 1999 No 3004, Reg 13(5)–(7).*

Regulation 35 of the *Construction (Design and Management) Regulations 2007* is concerned with the prevention of drowning and includes requirements concerning the provision of suitable rescue equipment and minimising the risk of drowning in the event of a fall into water.

The provision of instruction or leadership of persons in connection with caving or climbing by way of sport, recreation, team building or similar activities, indoor or outdoor, are covered by separate regulations. The *Work at Height (Amendment) Regulations 2007 (SI 2007 No 114)* came into effect on 6 April 2007 to bring workers paid to lead or train others in climbing and caving activities in the adventure activity sector within the scope of the *Work at Height Regulations 2005.*

Organisation and planning, competence and avoidance of risks

[W9005] The order of the first three Regulations of substance is significant. The first, *SI 2005 No 735, Reg 4*, covers 'Organisation and Planning', the second, *SI 2005 No 735, Reg 5*, covers 'Competence'. Only then, in *SI 2005 No 735, Reg 6*, is the subject of Risk Assessment and the Hierarchy of Measures (or Controls), as outlined in the Directive, addressed.

Organisation and planning

[W9006] *SI 2005 No 735, Reg 4* states that work at height must be:

(a)   properly planned (including the selection of work equipment, as expanded on in *SI 2005 No 735, Reg 7*);

(b)   appropriately supervised (as the duty in *HSWA 1974, s 2(2)(c)*); and,

(c)   as so far as is reasonably practicable carried out in a safe manner.

[*Note*: (a) and (b) above are 'must do' absolute duties.]

Emergencies and rescue must be planned for. Other than the emergency services acting in an emergency, work at height must not be carried out if weather conditions jeopardise the health and safety of persons involved.

Competence

[W9007] *SI 2005 No 735, Reg 5* provides for persons to gain competence to carry out their work by being trained under the supervision of a competent person. It requires employers to ensure that no person engages in any activity involving work at height or work equipment for use in work at height, including organisation, planning and supervision, unless they are competent to do so or are being supervised by a competent person if they are being trained.

No specific interpretation of competence is given in *SI 2005 No 735*, but *Reg 2* of the *Management of Health and Safety at Work Regulations 1999 (SI 1999 No 3242), Reg 13* addresses the requirement to assess capabilities and training.

Only having addressed the issues of organisation, planning, supervision and competence does *SI 2005 No 735, Reg 6(1)* require that the measures for the work to be carried out in safe manner must be identified with reference to the Risk Assessment carried out under *Management of Health and Safety at Work Regulations (SI 1999 No 3242), Reg 3*.

Avoidance of risks from work at height

[W9008] *SI 2005 No 735, Reg 6(2)* introduces the hierarchy, avoid work at height where it is reasonably practicable to do so. A simple example is to have lighting units in premises that can be lowered to ground level remotely, for routine maintenance and lamp changes.

Where work at height cannot be avoided, *SI 2005 No 735, Reg 6(3)* stipulates that suitable and sufficient measures to prevent any person falling 'a distance liable to cause personal injury' be taken, so far as is reasonably practicable. Collective protection measures must be given priority over personal protection measures.

The so-called 'two metre' rule, from the revoked *Construction (Health Safety and Welfare) Regulations 1996 (SI 1996 No 1592), Reg 6(3)*, is no longer part of the wording. There was a second round of consultation on this specific issue, and despite some strong lobbying for a two metre threshold to be retained, the argument for goal setting Regulations prevailed, supported by data showing that the majority of serious injuries have been caused by falls from a height of less than two metres.

SI 2005 No 735, Reg 6(4)(a) introduces *SI 2005 No 735, Sch 1* which stipulates the requirements for work areas from which persons could fall and which are built or incorporated into premises, including the access to and egress from them. Examples of work areas covered by this Regulation include plant rooms and roof areas with roof mounted equipment, and platforms and gantries provided for cleaning and maintenance of glazing and atrium roofs.

They must:

(a)   be stable and of sufficient strength and rigidity for the purpose for which they are intended to be or are being used;

(b)   where applicable, rest on (or be suspended from) a stable, sufficiently strong surface;

(c)   be of sufficient dimensions to permit the safe passage of persons and the safe use of any plant or materials required to be used;

(d)   provide a safe working area, having regard to the work to be carried out there;

Note: no absolute figures or dimensions are given here and in *(c)* above, as it is recognised that there may be some localised obstruction that does not preclude an otherwise safe route or platform being used effectively. The ergonomics of how work is carried out have to be considered. This is emphasised in the Directive;

(e)   possess suitable and sufficient means for preventing a fall, including for example barriers, guardrails, and toe boards;

(f)   possess a surface which has no unguarded gap:
        (i)   through which a person could fall;
        (ii)   through which any material or object could fall and injure a person; or
        (iii)   which would give rise to other risk of injury to any person;

(g)   be so constructed and used, and maintained in such condition, as to prevent, so far as is reasonably practicable:
        (i)   the risk of slipping or tripping; or
        (ii)   any person being caught between it and any adjacent structure;

(h)   be prevented from moving inadvertently during work at height, by 'appropriate devices' such as a parking brake or brakes, if applicable.

SI 2005 No 735, Reg 6(4)(b) covers the requirement for temporary work platforms and access, where they are not built into premises. Guardrails, toe boards, barriers and scaffolds are typical examples, covered in more detail in *SI 2005 No 735, Reg 8* and *SI 2005 No 735, Schs 2 and 3*.

SI 2005 No 735, Sch 2, which sets out the requirements for guard rails, toe-boards, barriers and similar collective means of protection, is linked to *SI 2005 No 735, Reg 8(a)*. The schedule specifies minimum requirements, including height, [increased to minimum of 950 mm from the former figure of 910 mm in *Construction (Health Safety and Welfare) Regulations 1996 (SI 1996 No 1592), Sch 1*. The figure of 950 mm equates to the European figure of 1 metre/1000 mm, which is quoted with an allowable tolerance of +/– 50mm]. There are situations, for example adjacent to a sloping roof, where the risk assessment shows that a greater height is required, or where close boarding or full height brick-guard mesh is required. No size for height of toe board is quoted, (previously 150 mm minimum was specified in *SI 1996 No 1592, Sch 1*, but more commonly a 200/225 mm scaffold board was used). A toe board must be 'suitable and sufficient to prevent the fall of any person, or any material or object'. Investigation of incidents by the HSE has indicated the importance of toe boards in preventing persons from falling, (not just materials or tools). The allowable gap between guard rails or toe-boards is unchanged at 470 mm. *Paragraph 5(1)* states that a lateral opening may be made at a point of access to a ladder 'where an opening is necessary'. However as proprietary self closing/push to open scaffold gates are readily available and in common usage, it is hard to envisage a situation where an un-gated opening is 'necessary'. *Paragraph 5(2)* and *(3)* covers the requirements for removing or opening the outer barriers of loading bays, but specify that 'compensatory safety measures' must be in place to protect operatives on or adjacent to the loading bay. Typically this could be achieved by a gate mechanism that protects the back of the loading bay if the front is open, or by the use of harness and tether (a work restraint system as per *SI 2005 No 735, Sch 5, Pt 5*, so designed that it prevents the user from getting into a position from which a fall can occur) should a banksman need to operate on the platform when the outer gate is open.

SI 2005 No 735, Sch 3, Pt 1 specifies the requirements for all working platforms, and covers plant and machinery, (mobile elevating work platforms including scissor lifts and boom–type cherry pickers) as well as mobile aluminium towers, scaffolds and similar fabricated platforms.

The floor, ground or surface on which the equipment relays for support must be adequate for any load that may be imposed. Beware of inspection covers and the like that a wheel or outrigger, or a scaffold standard may rest on.

SI 2005 No 735, Sch 3, Pt 2 specifies the additional requirements for scaffolds. No mention is made of the former British standards BS5973/BS 5974 or the European limit state design standard BS EN 1808. Nor is mention made of following a generally recognised standards set out National Access and Scaffolding Confederation (NASC) guides and codes. However the schedule calls for:

- calculations, erection and dismantling plans (depending on the complexity of the scaffold) to be available and held on site; at its simplest, this means instructions showing the correct method of assembly and dismantling be delivered with hired mobile alloy towers, and for more significant scaffolds, the number of ties required, including the specification for proof load testing of any drilled anchors;
- the assembly, modification and dismantling by trained persons under competent supervision; and
- the marking by signs and 'delineation by physical means preventing access' when not available for use, typically during erection, modification or dismantling.

Descent down the hierarchy continues. *SI 2005 No 735, Reg 6(5)(a)(i)* stipulates the requirement to reduce the distance and consequences of a fall, if it is not reasonably practicable to provide a guarded work platform.

This is followed by *paragraph (ii)*, which states that the consequences of a fall must be minimised where is not reasonably practicable to minimise the distance fallen. Fall arrest nets, air bags and crash decks are examples of this type of measure.

Reliance on individual fall arrest harnesses, which provide only individual not collective protection, would have to have good justification. It is for example, legitimate to consider the risk to persons erecting and dismantling collective measures, in reaching a decision to use individual fall protection only. Use of harnesses may be complimentary to a safe system of work. Scaffolders follow a sequence of work that provides, for example, a three-board working platform and single hand rails as work progresses, but they clip on when there is a suitable anchor point at four metres height or above. Clipping on provides added protection for example, when raising and lowering tubes and fittings. Users of boom-type mobile elevating platforms (cherry pickers), with designated anchor points in the basket/working platform, clip on as their work often requires them to lean out or reach over the guardrails fitted to the basket.

Finally in *SI 2005 No 735, Reg 6(5)(b)* the requirement for *additional* training and instruction is emphasised, so that the user fully understands the limitations and correct method of use of chosen work equipment. Choosing inherently safer equipment alone is not enough. The number of incidents involving mobile elevating platforms overturning, or causing injury, illustrates this. The operator may have been shown how the controls work, but not trained to assess the factors that affect stability or the safety of others.

Work equipment

[W9009] *SI 2005 No 735, Reg 7* outlines the issues to be considered when selecting work equipment for work at height:

- measures that will provide protection for all are to be given priority over measures that protect only one person (similar to the concept in *Control of Substances Hazardous to Health Regulations 2002 (COSHH) (SI 2002 No 2677)* that dust suppression at source is better than a dust mask for an individual);
- duration and frequency of use;
- the length and/or height of means of access;
- consequences of a fall;
- how suitable for emergency evacuation or for rescue?

Note: As stated above it is valid to consider the risks associated with the erection and removal of the work equipment – is the cumulative risk of erecting and dismantling a scaffold higher than a lower number of persons working for a shorter period on an item of equipment that may not provide as high a level of protection? The regulations fully recognise that the use of ladders and step ladders has its place.

SI 2005 No 735, Sch 6 contains the requirements for ladders. The Regulatory Impact Assessment that accompanied the consultative document Proposals for Work at Height Regulations CD192 RIA assumed only a *5 per cent switch* from the existing levels of use of ladders to other equipment, based on industry sources and the knowledge that there is already a trend towards alternative means of access. Thus the importance of *SI 2005 No 735, Sch 6* (and the wide range of guidance available) is for it to lead to users understanding *the correct methods of use.*

Quoting the Directive, a ladder (or step ladder) is to be used only if it is for (*a*) low-risk work of short duration; or (*b*) where there are existing features on site which cannot be altered. This must be demonstrated by the risk assessment showing that more suitable work equipment is not justified. Specific require-ments relating to how the chosen ladder is used are listed in *paragraph 10* of *Schedule 6.* A secure handhold and secure support must always be available to the user. The work must be such that the user can maintain a safe handhold when carrying a load unless, in the case of a step ladder, the maintenance of a handhold is not practicable when a load is carried.

Having decided on the use of a ladder a pre-use check as required by *SI 2005 No 735, Reg 13* should be carried out. Examples of pre-use check lists are available on the HSE website.

The suitability of the ground must be assessed, with particular attention to the stability of the ladder. *Paragraph 5* requires that a portable ladder be prevented from slipping. It is suggested that this is achieved by securing, strapping or tying both stiles to the wall or fixed object, or by the use of an effective anti-slip or stability device. Research is ongoing into the relative effectiveness of available anti-slip and stability devices. The traditional footing of a ladder by a work colleague can be effective for short ladders and in circumstances where there is not a better solution available, but it is not a solution encouraged by the regulations. Training and supervision is required. Assessment of the ground should also check for contaminants including mud, oil and grease that would reduce the grip of footwear on the ladder rungs during use.

The limit of 9 metres vertical distance previously contained in *SI 1996 No 1592, Sch 5* is retained. However, in the *Workplace (Health, Safety and Welfare) Regulations (SI 1992 No 3004)*, the ACoP to *SI 2005 No 735, Reg 13* quoted a maximum height of 6 metres, with access limited to use by 'specially trained and proficient people' if the height of 6 metres is exceeded. The ACoP also implies that hoops fitted to fixed ladders are a safety feature, though investigation since 1992 of incidents, and recent research has shown the limitations of such hoops (HSE Research Report RR258 http://www.hse. gov.uk/research/rrpdf/rr258.pdf); this should be considered in any risk assess ment where access or egress is via a fixed hooped ladder.

According to the HSE, every year some 12 people die at work falling from ladders and over 1,200 suffer major injuries. Ladders remain the most common factor involved and account for more than a quarter of all falls from height.

The HSE believes that misuse of ladders at work can be partly explained by the way they are used in the home. As with all work equipment, users need adequate information and training to be able to use ladders and stepladders safely. Adequate supervision is needed so that safe practices continue to be used.

Safety in window cleaning using portable ladders (HSE Information Sheet MISC613), the HSE guide on the Safe Use Ladders and Step Ladders (INDG402), *A toolbox talk on leaning ladder and stepladder safety* (INDG 403), *Top tips for ladder and stepladder safety* (INDG 405) are examples of the practical guidance available on the HSE website as downloadable PDF files at http://www.hse.gov.uk/falls/publications.htm. The HSE states that this guidance is to 'help employers know when to use a ladder, decide how to go about selecting the right sort of ladder for the particular job, understand how to use it, know how to look after it, and how to take sensible safety precautions'.

Fragile surfaces

[W9010] Fragile surfaces are given a broader definition; a 'fragile surface' means a surface which would be liable to fail if any reasonably foreseeable loading were to be applied to it. *SI 2005 No 735, Reg 9* requires the same level of consideration as a fall into an open void. In addition, prominent warning signs are called for, or equivalent or better means of warning people, which may for example, include site inductions for visitors. With the proliferation of buildings with near flat roofs, deterioration of fixings or poor workmanship can result in a roof failing to support a person even if the material itself is not inherently fragile. Asbestos cement roofs and sky lights subject to ultraviolet degradation, 'traditional' fragile roof materials, will continue to require control of access for many years to come.

Falling objects and danger areas

[W9011] *SI 2005 No 735, Reg 10* relates to protecting people from falling, thrown or tipped material. This includes the need for consideration of how and where materials and objects are stored.

SI 2005 No 735, Reg 11 places a duty to exclude persons who may be harmed by falling objects by the use of:

(i)   barriers – ' . . . devices preventing unauthorised persons entering', and,

(ii)   signs – 'ensure . . . such area is clearly indicated'.

Inspection

[W9012] *SI 2005 No 735, Reg 12* describes the requirements for the inspection of work equipment. With regards to equipment to facilitate safe work at height, it supersedes the requirement of the *Provision and Use of Work Equipment Regulations 1998 (PUWER) (SI 1998 No 2306), Reg 6 SI 2005 No 735, Sch 7* lists the particulars to be recorded in a report of inspection. The list is unchanged from the list contained in the previous *Construction (Health, Safety and Welfare) Regulations 1996 (SI 1998 No 1592), Sch 8*, other than the description of 'the place of work' (or part of that place) must identify and include any plant and equipment and materials, if any, that have been inspected.

SI 2005 No 735, Reg 12(5) includes the requirement for 'physical evidence' of inspection to accompany equipment sent from one undertaking to another. Equipment delivered from a hire company is a simple example. This mirrors the requirement of the *Lifting operations and Lifting Equipment Regulations 1998 (LOLER) (SI 1998 No 2307), Reg 9(4)*, which still applies even for equipment used for lifting and lowering people, including cranes, as well as conventional mobile elevating work platforms.

The requirements for recording reports of inspection are similar to those contained previously in *SI 1996 No 1592*. They must be completed in the same shift as the inspection, and delivered to the client on whose behalf the inspection was carried out within 24 hours. Records must be kept for at least three months after the end of the project during which the work equipment was used.

SI 2005 No 735, Reg 13 contains the sensible requirement for inspection on each occasion before use.

Duties of persons at work

[W9013] *SI 2005 No 735, Reg 14* repeats the duties of individuals to be responsible for their own safety and the safety of others as contained in *HSWA 1974, s 7*.

SI 2005 No 735, Reg 14 (1) imposes the duty to report an activity or defect that may endanger that person or others.

SI 2005 No 735, Reg 14(2) imposes the duty to use equipment in accordance with the training and instructions given.

Protection from falls

[W9014] *SI 2005 No 735, Sch 4* relates to collective safeguards for arresting falls, such as nets, airbags landing mats and the like. The risk assessment must demonstrate that they can work effectively for the work activity taking place. For example if the fall of a person is likely to occur with or following the fall of materials such as sheet materials, steel or pre-cast concrete components, a fall protection system linked to an anchor above the work area may prove to be more effective. Is there enough clearance from objects beneath that could cause injury? If the system requires support from newly built walls, do they have enough strength, allowing for temperature and weather conditions?

SI 2005 No 735, Sch 5, Pt 2 outlines the requirements for personal fall protection systems. These are divided into four categories:

(i)   work positioning systems. *SI 2005 No 735, Sch 5, Pt 2* calls for the inclusion and use of a backup to prevent or arrest falls. Where this is not *reasonably* practicable, all practicable measures (*note*: not *reasonably* practicable) must be taken 'to ensure that the system does not fail'.

(ii)   Rope access and positioning techniques. *SI 2005 No 735, Sch 5, Pt 3* describes the use of two separately anchored lines, a working line and a safety line. This has been standard practice for many years, and competency has been overseen by the Industrial Rope Access Trade Association (IRATA) (www.irata.org), who certify a three level training scheme for operatives, supervisors and trainers. Single rope systems are used, but the onus is on the user or person in control to demonstrate that the use of a second line would entail higher risks and that appropriate measures have been taken to ensure safety without their being a seconda ry or backup safety line.

(iii)   Fall arrest systems. Personal fall arrest shall be used only if the risk ass essment demonstrates that the use of other safer work equipment is not reasonably practicable. The anchor point must be suitable and sufficient and have sufficient strength and stability to support any foreseeable impa ct and static load. A shock absorber, usually incorporated at the harness connection end of any lanyard or attachment line, is required to limit the forces applied to the user's body through the harness in the event of a fall. Assessment of the site needs to ensure that sharp edges that could cut a lanyard are avoided and that there is sufficient clearance to swing without striking objects that could cause injury in the event of a fall. Susp ension trauma, similar to the effect of fainting on the parade ground when insufficient blood reaches the brain, is now well documented. If the blood supply to the brain of a person suspended in a near vertical posit ion is not restored, death is the inevitable consequence. Therefore provis ion for rescue of any person suspended in a harness after a fall has to be planned beforehand, and carried out within minutes.

(iv)   Work restraint. The benefits of a well designed work restraint system is that a fall should be prevented, eliminating shock loads on both the a nchor point and the human body wearing the harness, the need for rescue and the like. Frequently lanyards used for work restraint do incorporate shock absorbers, so that there is no possibility of them causing injury if used incorrectly as part of a fall arrest system.

BSI published British Standard BS 8437: 2005 (ISBN 0 580 45817 2). It was written with input from HSE representatives and gives recommendations and guidance, with illustrative examples, on the following key areas of working at height:

•   Selection, use and maintenance of personal fall protection systems;
•   Systems and equipment suitable for use in rescue;
•   Rescue of persons in the event of an accident;
•   Risk assessment; and
•   The training of users of the equipment.

Conclusion

[W9015] It is known that cost has to be an integral part of business culture in this country. Any message about health and safety needs to recognise this and the economic benefits of good health and safety management need to be demonstrated. The cost benefit analysis in the form of the Regulatory Impact Assessment that accompanied the consultative document *Proposals for Work at Height Regulations, CD192*, demonstrated that a reduction of accidents of only 10 per cent would more than cover any costs associated with the regulations.

In July 2011, two companies were fined £200,000 with £36,186 costs and £100,000 with £180,093 costs respectively after pleading guilty to safety breaches in relation to an incident in which a roadworker died of his injuries after falling 12 metres while working over the M5 motorway.

Designers, and the clients who commission them, have the ability to eliminate the hazards, or significantly reduce the risks associated with situations at work where falls could occur.

When looking at any work at heights it is important not to view the *Work At Height Regulations* in isolation. Other legislation needs to be considered in conjunction with it, examples being the *Construction (Design and Management) Regulations 2007*, and the *Workplace (Health, Safety and Welfare) Regulations 1992* (see also ACCESS, TRAFFIC ROUTES AND VEHICLES). Not only the actual work activity at height needs to be managed, but also the access to and egress from the area where the work is to be undertaken.

There is no shortage of available guidance, or of equipment that can provide safe solutions for work at height. There is however a lack of awareness of the potential risks and the potential solutions amongst those commissioning the work, those managing or controlling the work, and amongst those carrying out the work. This is illustrated by reference to the number of prosecutions being taken and improvement and prohibition notices being issued, as recorded on the HSE enforcement database.

Working Time

Dorothy Henderson and Adam Rice

Introduction

[W10001] The *Working Time Regulations 1998 (SI 1998 No 1833)* ('the Regulations') came into force on 1 October 1998 and were the first piece of English law to set out specific rules governing working hours, rest breaks and holiday entitlement for the majority of workers (as opposed to those in specialised industries). The Regulations were introduced in order to implement the provisions of the Working Time Directive (93/104/EC) and the Young Workers Directive (94/33/EC). This chapter deals only with the Regulations as they govern adult workers, ie those aged 18 or over.

The Working Time Directive was adopted on 23 November 1993 as a health and safety measure under what was then Article 118a of the Treaty of Rome 1957 (now Article 154 of the Treaty on the Functioning of the European Union), thereby requiring only a qualified majority, rather than unanimous, vote. As such, the UK Government, which was opposed to the Directive, was unable to avoid its impact. The Government challenged the Directive on the basis that it was not truly a health and safety measure and should, in fact, have been introduced under then Article 100 (now Article 115) as a social measure which would have required a unanimous vote. The European Court of Justice (ECJ) has nevertheless upheld the status of the Directive as a health and safety measure.

The Working Time Directive (93/104/EC) was amended in 2000 to cover certain excluded sectors and activities. It was ultimately consolidated by the Working Time Directive (2003/88/EC), which came into force on 2 August 2004.

The stated purpose of the Working Time Directive ('the Directive'), acknowledged by the ECJ, is to lay down a minimum requirement for health and safety as regards the organisation of working time. This is, of course, significant to the extent that the English courts and tribunals will interpret the Regulations in accordance with the Directive by adopting the 'purposive' approach, which is now accepted under English law.

When they were first introduced, the Regulations (and accompanying guidance produced by the former Department of Trade and Industry (DTI), now the Department for Business Innovation and Skills (BIS)) were met with vociferous criticism from the business world. Just a year later the Government sought to address these initial concerns. Following a limited consultation in July of 1999, amendments were tabled to two key areas of the Regulations (see **W10013** and **W10033** below) and, in March 2000, the then DTI issued a revised version of its guidance. Government guidance on the Regulations can now be found on the Directgov website (for workers) and the Business Link website (for employers), although it is not yet known how much weight tribunals will attach to these publications which do not have the force of law.

Despite all the controversy regarding the Regulations and the implementation of the Directive it would appear that the Regulations have had little impact on the long-hours culture within the UK. This may of course change if the European Commission restricts the use of the 48-hour limit 'opt out' as recommended following a review conducted in 2003. However, there has been little consensus amongst member states about this proposal (see **W10013**).

The structure of the Regulations is to prescribe various limits and entitlements for workers and then to set out a sequence of exceptions and derogations from these provisions. The structure of this chapter will roughly follow that format.

Definitions

[W10002] The *Working Time Regulations 1998 (SI 1998 No 1833), reg 2* sets out a number of basic concepts (some more familiar than others), which recur consistently and which underpin the legislation. These are as follows.

Working time

[W10003] Working time is defined, in relation to a worker, as:

(a)   any period during which he is working, *at his employer's disposal* and carrying out his activity or duties;

(b)   any period during which he is receiving relevant training; and

(c)   any *additional* period which is to be treated as working time for the purposes of the Regulations under a relevant agreement (see **W10006** below).

This definition initially raised numerous uncertainties, in particular, in relation to workers who are 'on call' or who operate under flexible working arrangements.

Despite this initial confusion, the ECJ decision in the cases of *SIMAP v Conselleria de Sandidad y Consumo de la Generalitat Valencia*: C303/98 [2001] IRLR 845 and *Landeshauptstadt Kiel v Jaeger*: C-151/02 [2003] IRLR 804 confirmed that time spent on call by doctors at health centres is 'working time' provided the doctors are available and required to be physically present at the workplace, even if they are allowed to sleep while on call. By contrast, time spent on call away from the health centre does not count as 'working time' unless the doctor is actually working. This reasoning was followed by the ECJ in *Vorel v Nemocnice Český Krumlov*: C-437/05. While these cases concerned doctors, a similar approach has been taken in other areas. In *MacCartney v Oversley House Management* [2006] ICR 210, the EAT ruled that the manager of sheltered accommodation who was required to remain on call 24 hours to respond to residents was 'working' while on call, even though she was permitted to sleep and rest while on call at a flat provided by the employer at the workplace. A similar approach was taken in relation to a residential home care assistant in *Hughes v Jones and anor t/a Graylyns Residential Home* (UKEAT/0159/08) and in *Anderson v Jarvis Hotels plc* (EAT/0062/05), where a hotel manager who was required to sleep overnight at the hotel and remain on call to cover emergencies was also held to be 'working' although he was free to sleep. In both cases, the EAT ruled that the likelihood of being called out is irrelevant. The cases show that time spent on call at work or at another place specified by the employer where the worker must be available to work on short notice will amount to working time, but time spent on call away from the workplace will usually not.

Because of the difficulty this has caused, in 2003, the European Commission proposed amending the Directive so that 'inactive time' on call, that is, time where a worker is available for work at his place of work but does not carry out any duties, would not count as working time. The proposals lapsed in 2009 without agreement being reached, but this issue is likely to be revisited by member states in future.

In terms of whether work-related travel counts as working time, the Business Link guidance states that working time 'includes travelling if part of the job' but does not 'include travelling between home and work'. In addition, the Directgov guidance states that 'travelling outside of normal working hours' is not working time. This seems to leave open the difficult question of non-routine travel. For example, it has never been clear whether someone who is required to travel from London to Edinburgh on a Sunday, to be at a meeting in Edinburgh at 9am on Monday morning, can count that time as working time. The Directgov guidance suggests that it would not be working time. In practice, however, it seems that it would depend on the facts.

It should be noted that the definition of working time can be clarified in a relevant agreement (see **W10009**) between the parties, but only by specifying any *additional* period which can be treated as working time. The relevant agreement cannot alter, and in particular cannot narrow, the statutory definition of working time.

Worker

[W10004] A worker is any individual who has entered into or works under (or where the employment has ceased), worked under:

(a)   a contract of employment; or
(b)   any other contract, whether express or implied and (if express) whether oral or in writing, whereby the individual undertakes to do or perform personally any work or services for another party to the contract whose status is not by virtue of the contract that of a client or customer of any profession or business undertaking carried on by the individual.

This definition is wider than just employees and covers any individuals who are carrying out work for an employer, unless they are genuinely self-employed, in that the work amounts to a business activity carried out on their own account (see also the reference to agency workers at **W10005**). When the Regulations were first introduced, the definition of 'worker' was a novel concept. It has now become a familiar one and appears in several other contexts such as the *Part Time Workers (Prevention of Less Favourable Treatment) Regulations 2000 (SI 2000 No 1551)* and the *National Minimum Wage Act 1998.*

A requirement to perform services "personally" is essential to qualify as a worker. Whether or not a contract includes an obligation to do work personally is determined by construing the contract in the light of the circumstances in which it was made and the common intention of the parties at that time: see *Wright v Redrow Homes (Yorkshire) Ltd* [2004] IRLR 720. Where the contract allows the individual to provide a substitute, this can cause difficulties. In *Byrne Bros (Formwork) Ltd v Baird & Others* [2002] ICR 667 the EAT concluded, in relation to carpenters and labourers who had been offered work on a building site, that a limited power to appoint substitutes was not inconsistent with an obligation to provide work on a personal basis. Subsequent cases suggest that the circumstances in which substitution is permitted by the contract will be the relevant consideration. For example, in *James v Redcats (Brands) Ltd* [2007] IRLR 296 the EAT ruled that there was an obligation to perform work personally where a courier's contract allowed her to provide a substitute when she was "unable" to work, but not when she was simply unwilling. By contrast, in *Premier Groundworks Ltd v Jozsa* (EAT/0494/08) the EAT concluded that a groundworker was not a 'worker' in circumstances where he had an unfettered right to delegate work to someone who was "at least as capable experienced and qualified". The groundworker had no obligation to provide the services personally.

In *Bacica v Muir* [2006] IRLR 35 the EAT confirmed that the mere rendering of a service personally does not make an individual a worker. If an individual performs work for a person who is a customer of his profession, business or undertaking, then he is not to be regarded as a worker. In that case, a painter who provided personal services was held not to be a worker, as he was taxed at a special rate under the Construction Industry Scheme whereby he paid his own national insurance contributions, had business accounts prepared by an accountant, was free to work for others and had in fact done so, and was paid a rate that included an allowance for overheads. In *Cotswold Developments Construction Ltd v Williams* [2006] IRLR 181, the EAT suggested that the focus is on whether the individual markets his services to the public generally (in which case he is more likely to be self-employed) or whether he was recruited to work as an integral part of the principal's business (in which case he is more likely a worker). In *James v Redcats (Brands) Ltd* [2007] IRLR 296 the EAT stressed that no one factor is determinative and each case will depend on all the elements of the particular relationship. Elias P suggested that it may be helpful to consider whether the obligation to perform work personally is the dominant feature of the contract, as opposed to being merely incidental or secondary. If the dominant purpose is not personal service, but a particular outcome or objective, the individual will not be a worker.

Agency workers

[W10005] The *Working Time Regulations 1998 (SI 1998 No 1833), Regulation 36* makes specific provision in relation to agency workers who do not otherwise fall into the general definition of workers. This provides that where:

(i)   any individual is engaged to work for a principal under an arrangement made between an agent and that principal, and

(ii)   the individual is not a worker because of the absence of a contract between the individual and the agent or the individual and the principal, then the Regulations will apply as if that individual were a worker employed by whichever of the agent or principal is responsible for paying or actually pays the worker in respect of the work.

Again, individuals who are genuinely self-employed are excluded from this definition.

Collective, workforce and relevant agreements

[W10006] These play a significant role in the Regulations as employers and employees can, by entering into such agreements (where the Regulations so allow), effectively supplement or derogate from the strict application of the Regulations. Employers and employees who need a certain amount of flexibility in their working arrangements may well find one or other of these agreements will facilitate compliance with the Regulations.

Collective agreement

[W10007] This is an agreement with an independent trade union within the meaning of the *Trade Union and Labour Relations (Consolidation) Act 1992, s 178*. The definition in *s 178* is wide enough to cover agreements reached between trade unions and employers under the statutory recognition process.

Workforce agreement

[W10008] This is a new concept under English law and since its introduction by the Regulations has not been widely utilised. It is an agreement between an employer and the duly elected representatives of its employees or, in the case of small employers (ie those with 20 or fewer employees), potentially the employees themselves. In order for a workforce agreement to be valid, it must comply with the conditions set out in *Schedule 1* to the Regulations, which provides that a workforce agreement must:

(a)   be in writing;

(b)   have effect for a specified period not exceeding five years;

(c)   apply either to

    (i)   all of the relevant members of the workforce, or

    (ii)   all of the relevant members of the workforce who belong to a particular group;

(d)   be signed by

    (i)   the representatives of the workforce or of the particular group of workers, or

    (ii)   where an employer employs 20 or fewer workers on the date on which the agreement is first made available for signature, either appropriate representatives or by a majority of the workers employed by him;

(e)   before the agreement was made available for signature, the employer must have provided all the workers to whom it was intended to apply with copies of the text of the agreement and such guidance as they might reasonably require in order to understand it fully.

'Relevant members of the workforce' are defined as all of the workers employed by a particular employer (excluding any worker whose terms and conditions of employment are provided for wholly or in part in a collective agreement). Therefore, as soon as a collective agreement is in force in respect of any worker, the provisions of any workforce agreement in respect of that worker would cease to apply.

Paragraph 3 of *Schedule 1* to the Regulations sets out the requirements relating to the election of workforce representatives. These are as follows:

(a)   the number of representatives to be elected shall be determined by the employer;

(b)   candidates for election as representatives for the workforce must be relevant members of the workforce, and the candidates for election as representatives of a particular group must be members of that group;

(c)   no worker who is eligible to be a candidate can be unreasonably excluded from standing for election;

(d)   all the relevant members of the workforce must be entitled to vote for representatives of the workforce and all the members of a particular group must be entitled to vote for representatives of that group;

(e)   the workers must be entitled to vote for as many candidates as there are representatives to be elected;

(f)   the election must be conducted so as to secure that (i) so far as is reasonably practicable those voting do so in secret, and (ii) the votes given at the election are fairly and accurately counted.

Relevant agreement

[W10009] This is an 'umbrella provision' and means a workforce agreement which applies to a worker, any provision of a collective agreement which forms part of a contract between a worker and his employer, or any other agreement in writing which is legally enforceable as between the worker and his employer. A relevant agreement could, of course, include the written terms of a contract of employment. It would only include the provisions of staff handbooks, policies etc where it could be shown that these were 'legally enforceable' as between the parties.

Maximum weekly working time

48-hour working week

[W10010] It is, of course, the 48-hour working week which initially caused so much controversy. The *Working Time Regulations 1998 (SI 1998 No 1833), reg 4* provides that an employer shall take all reasonable steps, in keeping with the need to protect the health and safety of workers, to ensure that a worker's average working time (including overtime) shall not exceed 48 hours for each 7-day period.

Reg 4(6) provides a specific formula for calculating a worker's average working time over a reference period, which the Regulations have determined as 17 weeks (for exceptions, see **W10012**). This is as follows:

$$\frac{a+b}{c}$$

where

a is the total number of hours worked during the reference period;
b is the total number of hours worked during the period which
    (i)   begins immediately after the reference period, and
    (ii)   consists of the number of working days equivalent to the number of 'excluded days' during the reference period; and
c is the number of weeks in the reference period.

'Excluded days' are days comprised of annual leave, sick leave or maternity leave and any period in respect of which an individual has opted out of the 48-hour limit (see **W10013** below).

When the case of *Barber v RJB Mining (UK) Ltd* [1999] IRLR 308 was first reported it was initially thought that this decision would add considerably to the importance of this provision. The facts of this case were as follows. For the 17 weeks from 1 October 1998 (the date the Regulations came into force), the plaintiffs worked more than the average of 48 hours per week. Letters seeking the workers' agreement to opt out of the 48-hour limit were sent out on 7 December 1998. Each of the plaintiffs refused to sign the opt-out. On 25 January 1999, each plaintiff refused to carry out further work until his average working hours fell to within the specified limit. Each was required to continue working and did so 'under protest' and without prejudice to their rights in the proceedings.

In his judgement, Mr Justice Gage said:

> 'It seems to me clear that Parliament intended that all contracts of employment should be read so as to provide that an employee should work no more than an average of 48 hours in any week during the reference period. In my judgement, this is a mandatory requirement which must apply to all contracts of employment.'

He stated that although *reg 4(1)* does not prohibit an employer from requiring his employees to work longer hours, it does not preclude this interpretation as the obligation is in keeping with the stated objective of the Directive, of providing for the health and safety of employees.

The plaintiffs had clearly worked more than their 48-hour average. The judge continued:

> 'Having held that para(1) of *reg 4* provides free-standing legal rights and obligations under their contract of employment, it must follow that to require the plaintiffs to continue to work before sufficient time has elapsed to bring the weekly average below 48 hours is a breach of *reg 4(1)*.'

In theory, the ruling could add an extra string to the worker's bow in the form of a claim for breach of contract if their employer fails to comply with their obligations under *reg 4(1)*. Workers may be able to claim damages, a declaration or possibly an injunction in the event of a breach. However, there have been no reported decisions where a worker has succeeded in bringing a claim for breach of contract for non-compliance with *reg 4(1)*. It has also been held that a worker cannot claim tortious damages for breach of statutory duty based on the employer's failure to enforce the 48-hour limit, as the provision was not intended to confer private law rights of action: *Sayers v Cambridgeshire County Council* [2006] EWHC 2029 (QB), [2007] IRLR 29.

An employee who claims to have suffered psychiatric illness as a result of excessive working hours may seek to rely on a breach of *reg 4(1)* to bolster a claim in negligence. In *Hone v Six Continents Retail Ltd* [2005] EWCA Civ 922, [2006] IRLR 49, an employee succeeded in a negligence claim for psychiatric injury caused by work and the fact that he had been working well in excess of the 48-hour limit was one factor pointing to the conclusion that his injury was reasonably foreseeable. However, the Court of Appeal confirmed in *Pakenham-Walsh v Connell Residential* [2006] EWCA Civ 90, [2006] All ER (D) 275 (Feb) that, while the employer's failure to comply with the 48-hour limit will be relevant, it can not automatically lead to a finding that the employer has been negligent.

The issue of an employer's obligations under the Regulations where they are not the only employer of the worker is problematic. *Reg 4(2)* states 'an employer shall take all reasonable steps, in keeping with the need to protect the health and safety of workers, to ensure that the limit specified in paragraph (1) is complied with in the case of each worker employed by him in relation to whom it applies'. The Directgov guidance suggests that workers with two jobs can either sign an opt-out, if the total time worked exceeds 48 hours a week, or reduce their hours to meet the 48-hour limit. In terms of how far the employer's obligation extends, the old DTI guidance suggested that employers may wish to make an enquiry of their workforce about any additional employment. If a worker does not tell an employer about other employment and the employer has no reason to suspect that the worker has another job, it is extremely unlikely, the guidance suggested, that the employer would be found not to have complied. However, the Business Link guidance, which replaced the old DTI guidance, does not address this.

In *Brown v Controlled Packaging Services Ltd* (1999), unreported, a worker who had another job in a bar on Tuesday and Thursday evenings was told at his interview that his basic working week would be 40 hours but that on occasion he would be required to work overtime. He soon discovered that the amount of overtime expected of him conflicted with the bar job, as on a number of occasions he was required to start work at 6 am instead of 8 am and he did not want to start work early on Wednesdays or Fridays. After the Regulations came into force, the worker was asked to sign an opt-out agreement which was viewed by his manager as a formality to ensure that existing overtime arrangements complied with the Regulations. Brown refused to sign the agreement because long hours would be incompatible with his evening job. Brown left and claimed constructive dismissal. The Employment Tribunal found that the dismissal was automatically unfair under the *Employ-*

ment Rights Act 1996, s 101A – ie for a reason connected with the employee exercising his rights under the Regulations.

Young workers

[W10010A] The Regulations also provide for a maximum working time for young workers of 8 hours a day or 40 hours per week [*Working Time Regulations 1998 (SI 1998 No 1833), reg 5A*]. A young worker is defined as a worker between the ages of 15 and 18 and who is over compulsory school age as defined in the relevant legislation.

Reference period

[W10011] The crucial issue in calculating the average number of hours worked is the question of when the reference period starts. The reference period is any period of 17 weeks in the course of a worker's employment, unless a relevant agreement provides for the application of successive 17-week periods. [*Working Time Regulations 1998 (SI 1998 No 1833), reg 4(3)*]. This means that unless the parties specify that the reference period is a defined period of 17 weeks, followed by a successive period of 17 weeks, then the reference period will become a 'rolling' 17-week period. This could be significant where there is a marked variation in the hours that an individual works in a particular period, from week to week. In such a situation an employer would be recommended to include provision in a relevant agreement for successive 17-week reference periods to ensure that it can comply with the maximum weekly working requirements. Such a provision may be desirable in any event, for ease of administration.

For the first 17 weeks of employment, the average is calculated by reference to the number of weeks actually worked. [*Working Time Regulations 1998 (SI 1998 No 1833), reg 4(4)*].

Exceptions and derogations to Regulation 4

[W10012] There are currently various exceptions to the maximum working week.

(a)   *reg 5* provides that an individual can agree with his employer to opt out of the 48-hour week (see **W10013** below).

(b)   The 48-hour limit does not apply at all in the case of certain workers whose working time is unmeasured; in respect of other workers, part of whose working time is unmeasured, it only applies to that part of their working time which is measured. [*Working Time Regulations 1998 (SI 1998 No 1833), reg 20* – see **W10033** below].

(c)   In special cases (as described in *reg 21* – see **W10034** below), the 17-week reference period over which the 48 hours are averaged will be automatically extended to 26 weeks. [*Working Time Regulations 1998 (SI 1998 No 1833), reg 4(5)*].

(d)   A collective or workforce agreement can extend the 17-week reference period to a period not exceeding 52 weeks, if there are objective or technical reasons concerning the organisation of work. [*Working Time Regulations 1998 (SI 1998 No 1833), reg 23*].

Special rules apply to doctors in training, for whom the working time limit was phased in gradually. The maximum working week for doctors in training was initially 58 hours per week from 1 August 2004, and this reduced to 56 hours per week on 1 August 2007. For most doctors in training, the 48-hour limit now applies as of 1 August 2009. However, a further limited derogation applies from 1 August 2009 until 31 July 2011 to some doctors in training employed in specific areas at named hospitals, whose maximum working week is 52-hours per week [*Working Time Regulations 1998 (SI 1998 No 1833), reg 25A* and *Schedule 2A*]. The reference period for calculating average weekly working time for all doctors in training is any 26-week period in the course of employment, unless a relevant agreement provides for the application of successive 26-week periods.

Agreement to exclude 48-hour working week

[W10013] The 48-hour limit on weekly working time will not apply where a worker agrees with his employer in writing that the maximum 48-hour working week should not apply in the individual's case, provided that certain requirements are satisfied. These are:

(a)   the agreement must be in writing;
(b)   the agreement may specify its duration or be of an indefinite length – however, it is always open to the worker to terminate the agreement by giving notice, which will be the length specified in the agreement (subject to a maximum of 3 months), or, if not specified, 7 days; and
(c)   the employer must keep up-to-date records of all workers who have signed such an agreement.

Until December 1999, employers were under controversial and onerous record keeping requirements in relation to workers who had signed opt out agreements. Since December 1999, employers only need to keep records of who has opted out.

A worker cannot be forced to sign an opt-out agreement. Moreover, the Regulations provide protection where an employee has suffered any detriment on the grounds that he/she has refused to waive any benefit conferred on him/her by the Regulations. [*Employment Rights Act 1996, s 45A(1)(b)*, inserted by the *Working Time Regulations 1998 (SI 1998 No 1833), reg 31(1)(b)* – see **W10041** below]. Furthermore, a worker will always have the right to terminate the opt-out agreement so that the 48-hour limit applies, on giving, at most, three months' notice.

On a practical note, the agreement by a worker to exclude the 48-hour working week must be a written agreement between the employer and the individual worker. It cannot take the form of or be incorporated in a collective or workforce agreement. It could, of course, be part of the contract of employment, (although it would be advisable for an employer to clearly delineate the agreement to work more than 48 hours from the rest of the contract of employment). It seems that employers' record keeping obligations in this regard would be satisfied by keeping an up-to-date list of workers who have signed opt-out agreements.

The ability of employers to seek the agreement of workers to 'opt out' of the 48-hour week (which is specifically permitted by the Directive) remains controversial as it arguably runs a 'coach and horses' through the spirit of the legislation. In 2003, the use of the opt-out was reviewed by the European Council, which recommended that further restrictions be placed on its use. Following years of negotiations, a Common Position was adopted by the European Council of Ministers that would have seen further safeguards for workers – for example, opt-outs could not be signed at the same time as an employment contract or within 4 weeks of starting work, they would have to be renewed in writing after a year, workers could withdraw with immediate effect in the first six months of employment or with two months' notice thereafter and any worker who opted-out would be subject to a maximum average working week of 60 hours per week or 65 hours per week if inactive on-call time was included (see **W10003** above). However, the European Parliament wanted to see the opt-out phased out altogether. Following negotiations, in which the UK Government in particular lobbied to retain the opt-out, no agreement was reached. As a result, the Directive and the Regulations remain unchanged and workers are still able to opt-out of the weekly working time limit. It remains to be seen if fresh proposals will be tabled by the European Commission.

Rest periods and breaks

Daily rest period

[W10014] Adult workers are entitled to an uninterrupted rest period of not less than 11 consecutive hours in each 24-hour period. [*Working Time Regulations 1998 (SI 1998 No 1833), reg 10*].

Employers must ensure that workers are allowed to take their rest breaks and that work arrangements provide a proper opportunity for workers to take their breaks. This was confirmed by the ECJ in a successful challenge brought by the European Commission to the Government's original guidance on rest breaks: see *Commission of the European Communities v United Kingdom*: C-484/04 [2006] ECR I-7471. The guidance originally stated that *"employers must make sure that workers can take their rest, but are not required to make sure they do take their rest"*. The ECJ ruled that this statement failed to properly implement the Directive. While employers do not need to force workers to take their breaks, according to the ECJ, the UK is bound to ensure that the minimum entitlements under the Directive are observed. In the ECJ's view, the words *"but are not required to make sure they do take their rest"* failed to do this, by suggesting that employers were not required to ensure workers actually exercised their rights. The offending words in the guidance have been removed – the Business Link guidance now states, *"It is your duty to ensure that your workers can take their breaks"*. In practice, while employers do not have to force workers to take their breaks, the employer's duty is to ensure there is an opportunity for workers to take their breaks, for example, by arranging work patterns in a way that enables them to do so.

Exceptions and derogations

[W10015] The provisions regarding daily rest periods do not apply where a worker's working time is unmeasured. [*Working Time Regulations 1998 (SI 1998 No 1833), reg 20(1)*].

Derogations from the rule may be made with regard to shift work [*reg 22*], or by means of collective or workforce agreements [*reg 23*] or where there are special categories of workers [*reg 21*]. However, in all cases (except for workers with unmeasured working time under *reg 20(1)* – see **W10033** below) compensatory rest must be provided [*reg 24*].

Weekly rest period

[W10016] Adult workers are entitled to an uninterrupted rest period of not less than 24 hours in each seven-day period. [*Working Time Regulations 1998 (SI 1998 No 1833), reg 11*]. However, if his employer so determines, an adult worker will be entitled to either two uninterrupted rest periods (each of not less than 24 hours) in each 14-day period or one uninterrupted rest period of not less than 48 hours in each 14-day period.

The entitlement to weekly rest is in addition to the 11-hour daily rest entitlement which must be provided by virtue of *reg 10*, except where objective or technical reasons concerning the organisation of work would justify incorporating all or part of that daily rest into the weekly rest period.

For the purposes of calculating the entitlement, the 7 or 14-day period will start immediately after midnight between Sunday and Monday, unless a relevant agreement provides otherwise. The Regulations do not require Sunday to be included in the minimum weekly rest period.

Derogations from the entitlement to weekly rest periods are the same as those for daily rest (see **W10015** above).

Rest breaks

[W10017] By virtue of *reg 12* of the Regulations, adult workers are entitled to a rest break where their daily working time is more than six hours. Details of this rest break, including duration and the terms on which it is granted, can be regulated by a collective or workforce agreement. If no such agreement is in place, the rest break will be for an uninterrupted period of not less than 20 minutes and the worker will be entitled to spend that break away from his/her workstation, if he/she has one. The Business Link guidance states that the employer can determine the timing of the break, but it must be offered during the shift and not at the start or end of it.

The essence of a rest break is that a worker knows at the start of it that he or she has 20 minutes free from work to do with as he or she pleases. A period of 'downtime' during working time therefore cannot constitute a rest break, nor could, for example, requiring workers to eat their lunch at their desks. In *Gallagher v Alpha Catering Services Ltd* [2005] IRLR 102 the Court of Appeal ruled that drivers and loaders of a catering business had not been given adequate breaks, despite the fact that they had several periods of uninterrupted downtime during the day, as they were required to be on duty and await instructions during that time.

Derogations from the entitlement to rest breaks include cases specified in *reg 21* of the Regulations (see above), where working time is unmeasured and cannot be predetermined [*reg 20(1)*], and also by means of a collective or workforce agreement [*reg 23*]. As before, compensatory rest must be provided other than where *reg 20* applies [*reg 24*].

Monotonous work

[W10018] The *Working Time Regulations 1998 (SI 1998 No 1833), reg 8* provides that where the pattern according to which an employer organises work is such as to put the health and safety of a worker employed by him at risk, in particular because the work is monotonous or the work rate is predetermined, the employer shall ensure that the employee is given adequate rest breaks. This Regulation is phrased in virtually identical terms to Article 13 of the Working Time Directive and unfortunately, its incorporation into the Regulations has not clarified its meaning in any way!

It is not clear how rest breaks in *reg 8* would differ from those under *reg 12*. It may be that in the case of monotonous work, an employer might have to consider giving employees shorter breaks more frequently as opposed to one longer continuous break. This was the suggestion in the Government's consultative document.

No derogations from these provisions apply except in the case of domestic workers.

Night work

Definitions

[W10019] Before considering the detailed provisions in relation to night work contained in *reg 6*, an understanding of the definitions of 'night time' and 'night worker' contained in *reg 2* of the Regulations is necessary. These are as follows.

Night time in relation to a worker means a period:

(a)   the duration of which is not less than 7 hours; and
(b)   which includes the period between midnight and 5 a.m.,

which is determined for the purposes of the Regulations by a relevant agreement or, in the absence of such an agreement, the period between 11pm and 6am.

Night worker means a worker:

(i)   who *as a normal course* works at least three hours of his daily working time during night time (for the purpose of this definition, it is stated in the Regulations that (without prejudice to the generality of that expression) a person works hours 'as a normal course' if he works such hours on a majority of days on which he works – a person who performs night work as part of a rotating shift pattern may also be covered); or

(ii)   who is likely during night time to work at least such proportion of his annual working time as may be specified for the purposes of the Regulations in a collective or workforce agreement.

The question of when someone works at least 3 hours of his daily working time 'as a normal course' was addressed in the case of *R v A-G for Northern Ireland, ex p Burns* [1999] IRLR 315. In this case the Northern Ireland High Court held that a worker who spent one week in each 3-week cycle working at least 3 hours during the night was a 'night worker'. The High Court said that 'as a normal course' meant nothing more than as a regular feature. The Directgov guidance states that someone will be a night worker if they 'regularly' work at least three hours during the night time on most of the days they work or often enough to be able to say they work those hours on a regular basis. It seems that occasional and or ad hoc work at night does not make someone a 'night worker'.

The question is certainly open, following the *Burns* case, of how frequent something has to be in order to become 'regular'.

Length of night work

[W10020] If a worker falls within these definitions, then he or she is a night worker for the purposes of the Regulations. The *Working Time Regulations 1998 (SI 1998 No 1833), reg 6* then goes on to provide that an employer shall take all reasonable steps to ensure that the normal working hours of a night worker do not exceed an average of 8 hours in any 24-hour period. This is averaged over a 17-week reference period which is calculated in the same way as in *reg 4* (see **W10011** above).

As with the maximum working week, there is a formula for calculating a night worker's average normal hours for each 24-hour period as follows:

$$\frac{a}{b - c}$$

where:

    a is the normal (not actual) working hours during the reference period;
    b is the number of 24-hour periods during the applicable reference period; and
    c is the number of hours during that period which comprise or are included in weekly rest periods under *reg 11*, which is then divided by 24.

Although the Regulations originally excluded night shift overtime hours from those which count towards 'normal' hours, this provision has been repealed. It is therefore possible for overtime to count towards 'normal hours' where it is regularly worked and forms part of a night worker's normal hours of work.

If the night work involves 'special hazards or heavy physical or mental strain', then a strict eight-hour time limit is imposed on working time in each 24-hour period and no averaging is allowed over a reference period. The identification of night work with such characteristics is by means of either a collective or workforce agreement which takes account of the specific effects and hazards of night work, or by the risk assessment which all employers are required to carry out under the *Management of Health and Safety at Work Regulations 1999 (SI 1999 No 3242)*. (see RISK ASSESSMENT).

The derogations in *reg 21* of the Regulations (special categories of workers) apply to the provisions on length of night work. Workers whose working time is unmeasured or cannot be predetermined (and where it is only partly unmeasured or predetermined, to the extent that it is unmeasured) are also excluded. [*Working Time Regulations 1998 (SI 1998/1833), reg 20*]. Other exemptions may be made by means of collective or workforce agreement. [*Working Time Regulations 1998 (SI 1998/1833), reg 23*].

Health assessment and transfer of night workers to day work

[W10021] An employer must, before assigning a worker to night work, provide him with the opportunity to have a free health assessment. [*Working Time Regulations 1998 (SI 1998 No 1833), reg 7*]. The purpose of the assessment is to determine whether the worker is fit to undertake the night work. While there is no reliable evidence as to any specific health factor which rules out night work, a number of medical conditions could arise or could be made worse by working at night, such as diabetes, cardiovascular conditions or gastric intestinal disorders.

Employers are under a further duty to ensure that each night worker has the opportunity to have such health assessments 'at regular intervals of whatever duration may be appropriate in his case'. [*Working Time Regulations 1998 (SI 1998 No 1833), reg 7(1)(b)*].

The Regulations do not specify the way in which the health assessment must be carried out, nor is there specific reference to medical assessments, so that strictly speaking, such assessments could be carried out by qualified health professionals rather than by a medical practitioner. The Business Link guidance contains a sample health questionnaire.

Contrast this with the position as regards the transfer from night to day work. If a night worker is found to be suffering from health problems that are recognised as being connected with night work, he or she is entitled to be transferred to suitable day work 'where it is possible'. In such a situation, the Regulations provide that a 'registered medical practitioner' must have advised the employer that the worker is suffering from health problems which the practitioner considers to be connected with the performance by that night worker of night work.

Annual leave

Entitlement to annual leave

[W10022] The right under the Regulations to paid holiday was the first time under English law that workers were given a statutory right to paid holiday. The Regulations originally provided for an entitlement to 4 weeks' paid holiday each year, which included leave taken on bank and public holidays. Although many employers developed a practice of giving workers paid leave on bank and public holidays in addition to the 4-week minimum, this disadvantaged those workers who received only the basic statutory entitlement.

The entitlement has now been increased to give all workers (subject to some exceptions) the right to 5.6 weeks' paid holiday each year, which equates to 28 days for individuals working five days a week. This comprises a 4-week basic entitlement [*Working Time Regulations 1998 (SI 1998 No 1833), reg 13*] and an additional entitlement of 1.6 weeks [*Working Time Regulations 1998 (SI 1998 No 1833), reg 13A*]. The statutory entitlement is subject to a statutory cap of 28 days for all workers [*Working Time Regulations 1998 (SI 1998 No 1833), reg 13A(3)*].

The purpose behind the increase was to give all workers the equivalent of eight bank and public holidays in addition to the existing entitlement of 4 weeks' paid holiday each year. However, the additional entitlement does not have to be given or taken on bank or public holidays and, as such, there is no statutory right to time off on bank or public holidays.

To ease the cost burden on employers, the increased entitlement was introduced in phases. On 1 October 2007, the statutory minimum entitlement increased from 4 to 4.8 weeks each year and, from 1 April 2009, this further increased to 5.6 weeks per year.

The right to paid annual leave begins on the first day of employment. Although the Regulations originally provided for a 13-week qualifying period for annual leave, this was ruled unlawful by the ECJ in *R (on the application of the Broadcasting, Entertainment, Cinematographic and Theatre Union) v Secretary of State for Trade and Industry*: C-173/99 [2001] ECR I-4881 (ECJ) and was therefore removed.

Where a worker began work after 25 October 2001, the employer has the option to use an accrual system during the first year of employment. The amount of leave which the worker may take builds up monthly in advance at the rate of one twelfth of the annual entitlement each month. If the calculation does not produce a whole number the accrued leave entitlement is rounded up either to a half day (if the calculation is less than a half day) or a whole day (if the calculation exceeds a half day). Any rounded up element of leave taken would, of course, be deducted from the worker's annual entitlement to paid leave.

For the purposes of *regs 13* and *13A* of the Regulations, a worker's leave year begins on any day provided for in a relevant agreement, or, where there is no such provision in a relevant agreement, 1 October 1998 or the anniversary of the date on which the worker began employment (whichever is the later). In the majority of cases, written statements of terms and conditions contain a reference to the holiday or leave year. If they do not, employers are recommended to ensure that they do so. Failure in this regard could result in administrative confusion as each employee could have a different holiday year for the purposes of calculations under the Regulations.

The statutory leave entitlement may be taken in instalments, but the 4-week entitlement under *reg 13* can only be taken in the leave year to which it relates and a payment in lieu cannot be made except where the worker's employment is terminated (but see comments at **W10024** below). In relation to the additional entitlement under *reg 13A*, a relevant agreement may provide for this portion of the leave entitlement to be carried forward but only to the next leave year. There was a limited right to make payment in lieu of the additional entitlement between 1 October 2007 and 1 April 2009, but this is now gone and no payment in lieu is permitted except where the worker's employment is terminated.

These rules relate only to the statutory entitlement under the Regulations, so that if an employer provides for annual leave over and above the statutory entitlement, the enhanced element of the holiday can be carried forward or paid for in lieu as agreed between the parties. In *Miah v La Gondola Ltd* (1999), unreported, the Employment Tribunal found that a worker who was paid an additional week's pay instead of taking a week's holiday (due under the Regulations) during his employment, was entitled, on the termination of that contract, to payment for that proportion of annual leave which was due to him in accordance with *reg 14(3)(b)* and the payment paid to him during his employment in lieu of the week's statutory holiday had to be disregarded.

For many workers (and employers), the introduction of rights to paid annual leave was significant. It is understood that before the Regulations came into force some 2.5 million workers had no right to a holiday at all. Employment Tribunals have had to resolve many disputes arising from this right.

Compensation related to entitlement to annual leave

[W10023] Where an employee has outstanding leave due to him when the employment relationship ends, *reg 14* of the Regulations specifically provides that an allowance is payable in lieu. It states that in the absence of any relevant agreement to the contrary, the amount of such allowance will be determined by the formula

$$(a \times b) - c$$

where:

(a) is the period of leave to which the worker is entitled under the Regulations;
(b) is the proportion of the worker's leave year which expired before the effective date of termination;
(c) is the period of leave taken by the worker between the start of the leave year and the effective date of termination [*reg 14(3)*].

Regulation 14(4) provides that where there is a relevant agreement which so provides, an employee shall compensate his employer – whether by way of payment, additional work or otherwise – in relation to any holiday entitlement taken in excess of the statutory entitlement.

Interestingly in the case of *Witley & District Mens Club v Mackay* [2001] IRLR 595 the EAT held that where a relevant agreement (in this case a collective agreement) purported to allow an employer to dismiss an employee for gross misconduct without payment for any accrued holiday pay this provision would be void as *reg 14* specifically provides for a sum to be paid. It is open to debate whether a provision of a relevant agreement which permitted the payment of a notional sum (say, £1) on termination in the event of gross misconduct would be valid.

Payment, and notice requirements, for annual leave

[W10024] The *Working Time Regulations 1998 (SI 1998/1833), reg 16* specifies the way in which a worker is paid in respect of any period of annual leave: this is at the rate of a week's pay in respect of each week of leave.

In order to calculate the amount of a 'week's pay', employers are referred to *ss 221–224* of the *Employment Rights Act 1996* (although the calculation date is to be treated as the first day of the period of leave in question, and the relevant references to *ss 227* and *228* do not apply). The application of these provisions can lead to a disproportionately high level of holiday pay for workers paid on a commission only basis. This is because *s 223(2)* stipulates that if, during any of the 12 weeks preceding the worker's holiday, no remuneration was payable by the employer to the worker concerned, account should be taken of remuneration in earlier weeks so as to bring the number of weeks of which account is taken up to 12. In other words, it seems that the calculation of a 'commission only' worker's holiday entitlement will be based solely on the worker's earnings for any weeks (up to a maximum of 12) in which they received commission payments-any slow, unproductive weeks will be disregarded. Where commission is paid in addition to a base salary, the commission payments are normally excluded from the calculation of holiday pay: see *Evans v Malley Organisation Ltd (t/a First Business Support)* [2003] IRLR 156. However, in *Herring v Co-operative Insurance Society Ltd* [2006] All ER (D) 171 (Oct), the EAT held that it was at least arguable that insurance brokers who received a small base salary plus commission should have been paid projected commission during annual leave.

Whether bonus payments should be included as part of the holiday pay will depend on the nature of the bonus and whether it can properly be said that pay varies with the amount of work done: see *Adshead v May Gurney Ltd* [2006] All ER (D) 388 (Jul), EAT (UKEAT/0150/06). Overtime payments will not usually be counted towards a week's pay, unless the overtime is fixed under the employment contract: *Bamsey v Albon Engineering and Manufacturing plc* [2004] EWCA Civ 359, [2004] IRLR 457. In the case of *Smith v Chubb Security Personnel Ltd* (1999), unreported, employees who were obliged to work such hours as were detailed in their duty roster were paid holiday pay on the basis of a 40-hour week when in fact they worked a 3-week cycle of fourteen 12 hour days which averaged out at 56 hours per week. The Tribunal decided that their pay should be calculated on the basis of a 56-hour week.

The *Working Time Regulations 1998 (SI 1998 No 1833), reg 15* deals with the dates on which the leave entitlement can be taken, and sets out detailed requirements as to notice. This Regulation is extremely complicated and before going into the details, it should be noted that alternative provisions concerning the notice requirements can be contained in a relevant agreement. Bearing in mind the detailed nature of the provisions, it is recommended that employers consider specifying such notice requirements in their contracts of employment, thereby avoiding the need for confusion at a later stage.

Regulation 15 provides that a worker must give his or her employer notice equivalent to twice the amount of leave he or she is proposing to take. An employer can then prevent the worker from taking the leave on a particular date by giving notice equivalent to the number of days' leave which the employer wishes to prohibit. An employer can also, by giving notice equivalent to twice the number of days' leave in question, require an employee to take all or part of his or her leave on certain dates. This would, of course, be useful with regard to seasonal shutdowns over summer and Christmas holidays etc. An employee has no right to serve a counter notice and so, provided the employer has given proper notice, the employee must take the holiday as required. The Regulations do not require any notice to be given in writing. However, given the level of detail that must be included in any notice, employers would be advised to give the notice in writing and require their employees to do the same. This would also ensure that employers have accurate records of required holiday periods.

It is open for an employer to require a worker to take leave in single days rather than in a block. In addition, any day upon which a worker could be required to work can be designated by the employer as leave, even if the worker would not necessarily be required to work that day. In *Sumsion v BBC Scotland* [2007] IRLR 678, the EAT held that an employer could require a worker to take annual leave every second Saturday where the worker's contract required him to be available for work for up to 6 days per week. This was notwithstanding that work was rarely done on Saturdays, as the worker was required to be available. See also *Russell and Others v Transocean International Resources Ltd and Others* [2010] CSIH 82 where the Scottish Court of Session ruled that time off spent onshore by offshore workers as part of a rotating shift could constitute annual leave.

The employer's right to require a worker to take leave at a particular time is restricted during periods of maternity leave. In *Gomez v Continental Industrias Del Caucho S*: C-342/01 [2004] IRLR 407, the ECJ ruled that an employer was required to allow an employee to take her statutory leave entitlement at the end of her maternity leave, even though this was outside the agreed time for taking annual leave contained in a collective agreement.

Employers who give more than the statutory minimum entitlement and who are contractually obliged to pay their workers more for the statutory entitlement than they pay for the remaining leave will be interested in *Barton v SCC Ltd* (1999), unreported. In this case, an employer sought to argue that monies paid by way of holiday pay for leave in excess of the statutory minimum leave under the Regulations can go towards the employer's liability to pay a week's pay for each week of leave taken under the Regulations (calculated under the *Working Time Regulations 1998 (SI 1998 No 1833), reg 16*). In addition, the employer argued that they had the right to choose which days off would be paid at the statutory rate and which at the contractual rate. The Employment Tribunal found in favour of the workers on both issues. *Regulations 16(1)* and *(5)* make it clear that the Employment Tribunal has to focus on the individual holiday week in question and the payment made for that individual week. With regard to which holiday is paid at which rate, the Employment Tribunal said that the right to holiday pay is an entitlement and it is a matter for the worker to choose which holiday weeks he or she seeks to take pursuant to his or her statutory entitlement.

Rolled-up holiday pay

Historically, there were conflicting decisions in the UK as to whether it is permissible to roll-up holiday pay into the worker's hourly rate (see for example, *MPB Structures Ltd v Munro* [2003] IRLR 350 and *Caulfield v Marshalls Clay Products Ltd* [2004] EWCA Civ 422, [2004] IRLR 564). However, the ECJ in *Robinson-Steele v R D Retail Services Ltd*: C-131/104 [2006] IRLR 386 ruled that rolled-up holiday pay is incompatible with the Directive. The ECJ reasoned that the purpose of holiday pay is to put the worker in the same position, as regards pay, as he or she would have been if he or she were at work, so the worker should receive his/her pay at the time he/she takes the leave. Accordingly, it is now clear that the practice of rolling up holiday pay is unlawful and employers must pay statutory holiday pay at the time leave is actually taken. This is confirmed in the Business Link guidance.

The ECJ in *Robinson-Steele* did go on to state that payments already made to a worker under genuine rolled up holiday arrangements could be offset against the worker's entitlement to pay during leave, provided the payments had been paid transparently and comprehensibly as holiday pay. Workers who historically benefited from rolled-up holiday pay could not, therefore, gain a windfall by claiming additional holiday pay when they took their leave. For the amounts to be off-set, there would have to have been a clear agreement that part of the hourly rate is attributable to holiday pay which is set out in the employment contract or the worker's payslips: see for example *Lyddon v Englefield Brickwork Limited* [2008] IRLR 198 and *Smith v J Morrisroes and Sons Limited* [2005] IRLR 72.

However, there is some uncertainty as to whether employers who continue to operate rolled-up holiday pay arrangements can benefit from this protection. Whilst the ECJ allowed genuine, historical rolled-up holiday payments to be offset against the employer's obligation to pay holiday pay, it also ruled that member states must take appropriate measures to ensure that practices such as rolled-up holiday pay are not continued. Following the ruling, the UK Government indicated that what was intended by the ECJ was to allow a transitional period for employers who had previously operated genuine rolled-up holiday pay arrangements to make alternative arrangements. Unfortunately no timeframe was provided, but given the time that has elapsed since the ECJ's judgement, operating a rolled-up holiday pay arrangement is fraught with risk and employers should therefore avoid it. From a practical point of view, this is likely to be very difficult for employers who engage casual workers or operate in industries with irregular or unpredictable working patterns, where rolled up holiday pay has become commonplace.

Holidays and sickness absence

An area which has caused great difficulty under the Regulations is the question of a worker's entitlement to annual leave while on sick leave. Does a worker continue to accrue statutory holiday entitlement whilst on sick leave? Can the worker take annual leave during a period that he or she is already on sick leave? What happens to annual leave which has not been taken due to periods of sickness absence on termination of employment?

In *Kigass Aero Components Ltd v Brown* ([2002] ICR 697 (EAT/481/00)) the EAT held that workers accrue, and are entitled to take, their paid annual leave under the Regulations even if they are on sick leave during that period provided that they comply with the notification requirements under the Regulations. The EAT also ruled that where employment has been terminated while the worker is on sick leave, the worker is entitled to claim a payment in lieu of unused statutory leave. The *Kigass* decision was initially overruled by the Court of Appeal in *Ainsworth v IRC* [2005] EWCA Civ 441, [2005] IRLR 465 but later reinstated by the then House of Lords in the same case which became known as *Revenue and Customs v Stringer* [2009] UKHL 31, [2009] 4 All ER 1205, [2009] ICR 985.

The claimants in *Ainsworth/Stringer* were all on long term sickness absence and were no longer receiving pay as they had exhausted their contractual and statutory sick pay entitlements. One of the claimants sought to designate part of her sickness absence as holiday and claimed holiday pay whilst still employed. Four other claimants had been dismissed whilst absent on unpaid sick leave and sought payment in lieu of their accrued annual leave entitlements. The employment tribunals and the EAT allowed the claims. However, on appeal, the Court of Appeal overturned the EAT, ruling that a worker on long-term sick leave is not entitled under *reg 13* to 4 weeks' paid annual leave in a year when he has not been able to attend for work. According to the Court, it followed from this that a worker whose employment was terminated during a year when he is absent from work on long-term sick leave was not entitled to compensation under *reg 14* for unused leave. The case was further appealed to the House of Lords, which referred various questions to the ECJ for determination.

The ECJ confirmed in *Stringer* and the combined case of *Schultz-Hoff v Deutsche Rentenversicherung Bund*: C-350/06 [2009] All ER (EC) 906, [2009] IRLR 214 that, under the Directive, a worker continues to accrue annual leave during a period of sickness absence. A requirement that the worker must actually have performed some work to qualify for annual leave would be contrary to the Directive. The ECJ went on to say that workers on long-term sick leave should not lose the opportunity to take their annual leave under the Directive. Accordingly, member states must either allow them to take it during their sick leave or at a later time when they are well enough, even if this means carrying it forward to a subsequent holiday year. When Stringer returned to the House of Lords, the parties agreed that the decision of the EAT should be reinstated. As a result, their Lordships did not address the issues relating to sickness absence and annual leave in any detail.

The case makes it clear that workers accrue their statutory annual leave entitlement during any period of sickness absence, even if this lasts for a full holiday year or more. Workers must also be allowed to take holiday, at full pay, even if they are already on sick leave and any entitlement to sick pay has run out. In addition, if a worker's employment ends during a period of sickness absence, they should be paid for any unused statutory annual leave for that leave year, even if they were absent for all or part of the leave year.

In practice, the ruling creates complications for workers in receipt of long-term sickness benefits, such as permanent health insurance (PHI), which typically pay part of the worker's salary. It is not clear, for example, whether workers who take annual leave while in receipt of PHI are entitled to full holiday pay on top of the PHI payments or whether it is possible to "top up" the difference between full salary and benefits received. Depending on the terms of the PHI scheme, taking holiday whilst in receipt of PHI payments may also cause the benefits to cease if the worker is no longer deemed under the scheme to be incapacitated.

It is also remains unclear whether workers in the UK who are ill during all or part of a holiday year are entitled to carry forward statutory annual leave. The *Working Time Regulations 1998 (SI 1998 No 1833), reg 13(9)* does not permit statutory annual leave to be carried forward to a subsequent leave year (apart from the additional 1.6 weeks' leave under *reg 13A*). However, in *Stringer* and *Schultz-Hoff* the ECJ said that it would be contrary to the Directive for a worker's annual leave entitlement to be extinguished in circumstances where they have not had the opportunity to take it due to illness. It is arguable that the Regulations may not comply with the Directive in this regard and it remains to be seen whether *reg 13(9)* can be interpreted purposively to allow for carry over of annual leave by workers who are prevented from taking it due to illness (as has happened in the case of *Shah v First West Yorkshire Limited* (ET/1809311/2009) – see below). In the meantime, the safest course for employers would appear to be to allow workers to take annual leave during a period of sickness absence.

The ECJ has also considered the question of whether workers who fall sick during their annual leave are entitled to reschedule their annual leave for another time. In the Spanish case of *Pereda v Madrid Movilidad SA*: C-227/08 [2010] 1 CMLR 103, [2009] IRLR 959, the claimant suffered an injury at work which mean he was unable to take prearranged holiday. The ECJ ruled that a worker in this situation who falls ill during a period of previously scheduled annual leave has the right to take his or her annual leave at a later period when he or she has recovered. This is so even if it means allowing the rescheduled leave to be taken in a subsequent leave year. Accordingly, an employee who falls ill during annual leave is entitled to ask for that period to be treated instead as sick leave and to have "replacement" annual leave at another time. The *Pereda* ruling was applied in the UK in *Shah v First West Yorkshire Limited* (ET/1809311/2009). In that case, an Employment Tribunal implied into the Regulations an exception to the "use it or lose it" principle in *reg 13(9)* for workers who are prevented by illness from taking their holiday within the relevant leave year; in which case, they must be given the opportunity of taking that holiday in the following leave year.

Following *Pereda*, it is not clear whether the notice provisions in *reg 15(2)* can be used to require workers on sick leave to take their annual leave before the end of the holiday year. Arguably an employer can seek to rely on *reg 15(2)* but this will be subject to the worker's right to request to take the annual leave at a later time.

In terms of remedies, the House of Lords ruled in *Stringer* that non-payment of holiday pay gives a worker a right to claim unlawful deduction of wages under *section 13* of the *Employment Rights Act 1996*. In this regard, their Lordships overruled the Court of Appeal which had held previously that such a claim was not possible. A worker who is denied their holiday pay can therefore choose between a claim under the *Working Time Regulations 1998 (SI 1998 No 1833), reg 30* or *section 13* of the *Employment Rights Act 1996*.

One consequence of this is that a worker on long-term sick leave, who has been denied their annual leave over several years, can potentially claim back-pay for previous years' annual leave. The failure to pay holiday pay is likely to be seen as a series of deductions and any claim would only need to be brought within 3 months of the last in the series of deductions (*section 23* of the *Employment Rights Act 1996*). By contrast, a claim for holiday pay under the Regulations must be brought within 3 months of the end of the holiday year and can only relate to the most recent holiday year (see **W10040** below).

There is conflicting authority on the issue of whether the worker must have either taken or requested annual leave before a claim for unlawful deductions can be made. In *Kigass Aero Components Ltd v Brown* [2002] IRLR 312 the EAT said that a worker must ask to take their leave before the entitlement to holiday pay arises. However, in *List Design Group Ltd v Douglas* [2003] IRLR 14 the EAT took a different view, holding that a worker can bring an unlawful deductions claim for unpaid statutory holiday regardless of whether they have actually taken or requested leave. The EAT followed *List Design* in *Canada Life Ltd v Gray* [2004] ICR 673 (UKEAT/0657/03/SM) and drew a distinction between leave taken during employment (where leave must be requested before the entitlement arises) and payment in lieu of unused leave on termination of employment (which does not depend on the worker having requested leave). Given these decisions pre-dated *Stringer*, they are unlikely to be the final word on the issue.

These issues were considered by the Employment Tribunal in *Khan v Martin McColl* (ET/1702926/09), which took the view that, following the ECJ decision Stringer, a worker must have requested holiday and been denied it by the employer before he or she can claim back-pay for previous years' statutory holiday. In addition, the decision suggests that an employer can defeat a claim for several years' unpaid holiday by making a payment in lieu of unused holiday for the most recent holiday year on termination of employment. The employer in Khan successfully argued that payment for the final year's holiday broke the series of deductions, so that claims for previous years were out of time. Whilst the ruling is not binding, it will be of interest to employers faced with historical holiday pay claims.

Records

[W10025] The publicity and controversy surrounding the introduction of the Regulations related primarily to the maximum working week and the annual leave entitlement. However, in practice the provisions relating to record keeping proved very significant for employers, in particular those with workers who had signed opt-outs, where the rigorous record keeping requirements arguably undermined the benefit to an employer of the freedom given by the opt-out. Accordingly, the UK Government significantly watered down the requirements for employers *vis a vis* opted out workers, and the current position is outlined below.

The *Working Time Regulations 1998 (SI 1998 No 1833), reg 9* introduces an obligation on employers to keep specific records of working hours. All employers (except in relation to workers serving in the armed forces – *reg 25(1)*) are under a duty to keep records which are adequate to show that certain specified limits are being complied with in the case of each entitled worker employed by him. These limits are:

(a)   the maximum working week (see **W10010** above);
(b)   the length of night work (including night work which is hazardous or subject to heavy mental strain) (see **W10020** above); and
(c)   the requirement to provide health assessments for night workers (see **W10021** above).

These records are required to be maintained for two years from the date on which they were made.

In addition, where a worker has agreed to exclude the maximum working week under *reg 5* of the Regulations (see **W10013** above), an employer is required to keep up-to-date records of all workers who have opted out, which must also be kept for two years from the date on which they were made.

Excluded sectors

[W10026] Prior to 1 August 2003, the following sectors of activities (listed below) were excluded from the terms of the Regulations by *reg 18*:

(a)   air, rail, road, sea, inland waterway and lake transport;
(b)   sea fishing;
(c)   other work at sea;
(d)   the activities of doctors in training.

Following the adoption of five Directives aimed at extending the working time provisions to previously excluded sectors, this is no longer the case and most of the above sectors have gained some protection under the Regulations. The Directives are:

- Horizontal Amending Directive.
- Road Transport Directive.
- Seafarers' Directive (outside the scope of this chapter).
- Seafarers' Enforcement Directive (outside the scope of this chapter).
- Aviation Directive (outside the scope of this chapter).

The Horizontal Amending Directive was implemented by the UK with effect from 1 August 2003 *Working Time (Amendment) Regulations 2003 (SI 2003 No 1684)*.

The Road Transport Directive was implemented in the UK with effect from 4 April 2005 (see the *Road Transport (Working Time) Regulations 2005 (SI 2005 No 639)*.

Application of the new directives to previously excluded sectors

Mobile and non-mobile workers

[W10027] A mobile worker is defined in the *Working Time Regulations 1998 (SI 1998 No 1833), reg 2* as: 'any worker employed as a member of travelling or flying personnel by an undertaking which operates transport services for passengers or goods by road or air'. Non-mobile workers are undefined and are therefore any worker who is not a mobile worker, for example, all office based staff, warehouse staff etc. Unfortunately this does not address the situation where a worker carries out a combination of mobile and non-mobile activities (for example, a worker who spends part of his time based in a depot loading goods onto lorries and the remainder of his time driving an HGV delivering the goods.)

Road transport sector

Non-mobile workers

[W10028] Following the implementation of the *Working Time (Amendment) Regulations 2003 (SI 2003 No 1684)* on 1 August 2003, the Regulations were extended in full to non-mobile workers in the road transport sector.

Mobile workers

[W10029] Mobile workers who are not covered by the Road Transport Directive:

From 1 August 2003, mobile workers who are *not* covered by the Road Transport Directive (such as drivers of vehicles of less than 3.5 tonnes) are entitled to the following provisions of the Regulations:

(a)   the average 48-hour working week (under *reg 4* of the Regulations – see **W10010** above);

(b)   paid holiday (under *regs 13* and *13A* of the Regulations – see – see **W10022** above);

(c)   appropriate health checks if they are night workers (under *reg 7* of the Regulations – see **W10021** above); and

(d)   provision for adequate rest (unless there are exceptional circumstances
         as provided for in *reg 21(e)* of the Regulations *Working Time Regula-
         tions 1998) (SI 1998 No 1833), Regulation 24A(2)*].

Regulation 24A of the Regulations provides that they are not entitled to the
daily rest requirements set out in *reg 10(1)* or the weekly rest periods set out
in *reg 11(1)* and *(2)*. In addition, they are not entitled to the rest breaks set out
in *reg 12(1)* or the limit on hours of work for night workers' in *reg 6(1)*.
Further, the exemptions for special categories of workers may apply (see
W10034 below).

Mobile workers who are covered by the Road Transport Directive:

From 1 August 2003, mobile workers who are covered by the Road Transport
Directive (mainly drivers of vehicles of over 3.5 tonnes) are entitled to the
following provisions under the Regulations:

(a)   paid holiday (under *regs 13* and *13A* of the Regulations – see **W10022**
         above); and
(b)   appropriate health checks if they are night workers (under *reg 7* of the
         Regulations – see **W10021** above).

In addition, mobile workers who are covered by the Road Transport Directive
are entitled under the *Road Transport (Working Time) Regulations 2005 (SI
2005 No 639)* to a maximum working week of 60 hours (provided always that
the average working week over a 17-week period does not exceed 48 hours).
Mobile worker's should not work for more than 6 hours without a break and
are entitled to a 30 minute break where the worker's working time exceeds 6
hours but less than 9 hours and a 45 minute break where the worker's working
time exceeds 9 hours. Night work is also restricted to 10 hours in any 24 hour
period. Self-employed drivers are excluded from the ambit of the Road
Transport Directive.

Rail sector

[W10030] Following the implementation of the *Working Time (Amendment)
Regulations 2003 (SI 2003 No 1684)* on 1 August 2003, the Regulations are,
in principle, extended in full to all workers in the rail transport sector.

It is, however, possible to derogate from the Regulations in respect of railway
workers whose activities are intermittent, whose working time is spent on
board trains or whose activities 'are linked to transport timetables and to
ensuring the continuity and regularity of traffic'. For more information on
these partial exemptions see **W10034** below.

Special provisions are made for workers on cross-border rail services by the
*Cross-Border Railway Services (Working Time) Regulations 2008 (SI 2008 No
1660)*, which came into force on 27 July 2008. In practice, these regulations
are only relevant for workers on rail services operating through the Channel
Tunnel. The regulations provide more favourable rest breaks, daily and weekly
rest periods and limits on drivers' hours, and can be enforced by individual
claims to an Employment Tribunal. The *Working Time Regulations 1998 (SI
1998 No 1833)* still apply subject to these more favourable provisions, apart
from *reg 24* of the Regulations which deals with compensatory rest.

Other sectors

[W10031] It should be noted that the Horizontal Amending Directive together with the other sector-specific directives detailed above provide different entitlements for mobile workers in the sea-fishing sector, aviation sector, inland waterway and lake transport sectors and also in relation to offshore working. In relation to offshore workers, *reg 25B* of the Regulations provides specific exclusions for workers employed in offshore work.

In addition, in relation to doctors in training the Horizontal Amending Directive provides for a gradual phasing in of the Regulations. From 1 August 2004, the Regulations have applied in full to doctors in training except that the limit on average weekly hours of work was implemented in stages (see **W10012** above).

Partial exemptions

Domestic service

[W10032] The following provisions of the Regulations do not apply in relation to workers employed as domestic servants in private households:

(a)   the 48-hour working week (see **W10010** above);
(b)   length of night work (see **W10020** above);
(c)   health assessments for night workers (see **W10021** above);
(d)   monotonous work (see **W10018** above).

The provisions regarding minimum daily and weekly rest periods and rest breaks, the requirement to keep records and annual leave do apply.

Unmeasured working time

[W10033] The requirements of the Regulations listed below are, by virtue of *reg 20(1)* of the Regulations, not applicable to workers where, on account of the specific characteristics of the activity in which they are engaged, the duration of their working time is not measured or pre-determined or can be determined by the workers themselves. The provisions excluded are:

(a)   the 48-hour working week (see **W10010** above);
(b)   minimum daily and weekly rest periods and rest breaks (see **W10014** above);
(c)   length of night work (see **W10020** above).

The provisions regarding monotonous work, the requirement for health assessments for night workers, the requirement to keep records and annual leave do apply. It should also be noted that there is no requirement to provide such workers with compensatory rest where the requirements of the Regulations are not complied with.

This provision is often thought to exclude the Regulations in relation to 'managing executives or other persons with autonomous decision-making powers'. However, it should be noted that this is simply one of the three examples given in *reg 20(1)*, as regards the type of worker with unmeasured working time. The application of the regulation could be much wider than this, and will depend on whether the worker in question is genuinely able to control his work, to the extent of being able to determine how many hours he works.

In December 1999, the UK Government introduced *reg 20(2)* to give a partial exemption from the 48-hour maximum working week for those workers who had some part of their working time that was unmeasured. *Regulation 20(2)* provided that the working time of workers did not effectively count towards their weekly working time to the extent that it was unmeasured, even if their time was not wholly unmeasured. However, *reg 20(2)* was later repealed with effect from 6 April 2006, following a complaint by the European Commission that it went beyond the permitted derogation in the Directive. Workers are therefore either within the *reg 20* 'unmeasured' exemption or they are outside it.

Special categories of workers

[W10034] The *Working Time Regulations 1998 (SI 1998 No 1833), Regulation 21* provides that in the case of certain categories of employees, the following Regulations do not apply:

(a)   length of night work (see **W10020** above);
(b)   minimum daily rest and weekly rest periods and rest breaks (see **W10014** above).

This leaves the 48-hour working week, requirements relating to monotonous work, the requirement for health assessments for night workers, the requirement to keep records and annual leave. However, it should be noted that where any of the allowable derogations are utilised, an employer will have to provide compensatory rest periods (see **W10037** below).

Further, for these categories of workers, the reference period over which the 48-hour working week is averaged is 26 weeks and not 17 [*Working Time Regulations 1998 (SI 1998 No 1833), Regulation 4(5)*].

The special categories of worker are as follows:

(i)   where the worker's activities mean that his place of work and place of residence are distant from one another, or the worker has different places of work which are distant from one another (including cases where the worker carries out offshore work);
(ii)   workers engaged in security or surveillance activities, which require a permanent presence in order to protect property and persons – examples given are security guards and caretakers;
(iii)   where the worker's activities involve the need for continuity of service or production – specific examples are:
    (A)   services relating to the reception, treatment or care provided by hospitals or similar establishments (including the activities of doctors in training), residential institutions and prisons;
    (B)   workers at docks or airports;
    (C)   press, radio, television, cinematographic production, postal and telecommunications services, and civil protection services;
    (D)   gas, water and electricity production, transmission and distribution, household refuse collection and incineration;
    (E)   industries in which work cannot be interrupted on technical grounds;

(F)   research and development activities;
(G)   agriculture;
(H)   the carriage of passengers on regular urban transport services;
(iv)   any industry where there is a foreseeable surge of activity – specific cases suggested are:
(A)   agriculture;
(B)   tourism;
(C)   postal services;
(v)   where the worker's activities are affected by:
(A)   unusual and unforeseeable circumstances beyond the control of the employer;
(B)   exceptional events which could not be avoided even with the exercise of all due care by the employer;
(C)   an accident or the imminent risk of an accident;
(vi)   where the worker works in railway transport and:
(A)   his activities are intermittent; or
(B)   he works on board trains; or
(C)   his activities are linked to transport timetables and to ensuring the continuity and regularity of traffic.

Shift workers

[W10035] Shift workers are defined by the *Working Time Regulations 1998 (SI 1998/1833), reg 22* as workers who work in a system whereby they succeed each other at the same work station according to a certain pattern, including a rotating pattern which may be continuous or discontinuous, entailing the need for workers to work at different times over a given period of days or weeks.

In the case of shift workers the provisions for daily rest periods (see **W10014** above) and weekly rest periods (see **W10016** above) can be excluded in order to facilitate the changing of shifts, on the understanding that compensatory rest must be provided (see **W10037** below). In addition, those provisions also do not apply to workers engaged in activities involving periods of work split up over the day – the example given is that of cleaning staff.

Collective and workforce agreements

[W10036] By virtue of the *Working Time Regulations 1998 (SI 1998 No 1833), reg 23(a)*, employers and employees have the power to exclude or modify the following provisions:

(a)   length of night work (see **W10020** above);
(b)   minimum daily rest and weekly rest periods and rest breaks (see **W10014** above),

by way of collective or workforce agreements. Compensatory rest must be provided (see **W10037** below).

In addition, *reg 23(b)* of the Regulations allows the reference period for calculating the maximum working week to be extended to up to 52 weeks, if there are objective or technical reasons concerning the organisation of work to justify this.

This Regulation gives the parties a good deal of flexibility (on the understanding that they are prepared to consent with each other) to opt out of significant provisions contained in the Regulations.

Compensatory rest

[W10037] The *Working Time Regulations 1998 (SI 1998 No 1833), reg 24* provides that where a worker is not strictly governed by the working time rules because of:

(a)   a derogation under *Regulation 21* (see **W10034** above); or
(b)   the application of a collective or workforce agreement under *Regulation 23(a)* (see **W10036** above);or
(c)   the special shift work rules under *Regulation 22* (see **W10035** above);

and as a result, the worker is required by his employer to work during what would otherwise be a rest period or rest break, then the employer must wherever possible allow him to take an equivalent period of compensatory rest. In exceptional cases, in which it is not possible for objective reasons to grant such a rest period, the employer must afford the employee such protection as may be appropriate in order to safeguard his health and safety.

Enforcement

[W10038] The Regulations divide the enforcement responsibilities between the Health and Safety Executive and the Employment Tribunals. In essence, the working time limits are enforced by the Health and Safety Executive and local authority environmental health departments. The entitlements to rest and holiday are enforced through Employment Tribunals.

Interestingly, despite the assertions that the Directive is a health and safety measure, there appear to be only two successful prosecutions by the Health and Safety Executive since the Regulations were introduced. In contrast, there have been numerous cases lodged at the Employment Tribunal or High Court since the implementation of the Regulations.

Health and safety offences

[W10039] The *Working Time Regulations 1998 (SI 1998/1833), reg 28* provides that certain provisions of the Regulations (referred to as 'the relevant requirements') will be enforced by the Health and Safety Executive (except to the extent that a local authority may be responsible for their enforcement by virtue of *reg 28(3)*). The relevant requirements are:

(a)   the 48-hour working week;
(b)   length of night work;
(c)   health assessment and transfers from night work;
(d)   monotonous work;
(e)   record keeping;
(f)   failure to provide compensatory rest, where the provision concerning the length of night work is modified or excluded; and

(g)   failure to provide adequate rest to mobile workers, where the provision concerning the length of night work is excluded.

Any employer who fails to comply with any of the relevant requirements will be guilty of an offence and shall be liable on summary conviction (in the magistrates' court) to a fine not exceeding the statutory maximum and on conviction on indictment (in the Crown Court) to a fine.

In addition, the Health and Safety Executive can take enforcement proceedings utilising certain provisions of the *Health and Safety at Work etc Act 1974*, as set out in *regs 28* and *29*. Employers may face criminal liability under these provisions, the sanctions for which range, according to the offence, from a fine to two years' imprisonment. (For enforcement of health and safety legislation generally, see ENFORCEMENT.)

Employment tribunals

Enforcement of the Regulations

[W10040] By virtue of the *Working Time Regulations 1998 (SI 1998/1833)*, *reg 30*, certain provisions of the Regulations may be enforced by a worker presenting a claim to an Employment Tribunal where an employer has refused to permit him to exercise such rights. These are:

(a)   daily rest period;
(b)   weekly rest period;
(c)   rest break;
(d)   annual leave;
(e)   failure to provide compensatory rest, insofar as it relates to situations where daily or weekly rest periods or rest breaks are modified or excluded;
(f)   failure to provide adequate rest, insofar as it relates to situations where daily or weekly rest breaks are excluded;
(g)   the failure to pay the whole or any part of the amount relating to paid annual leave, or payment on termination in lieu of accrued but untaken holiday.

A complaint must be made within (i) three months (other than in the case of members of the armed forces – see below) of the act or omission complained of (or in the case of a rest period or leave extending over more than one day, of the date on which it should have been permitted to begin) or (ii) such further period as the tribunal considers reasonable, where it is satisfied that it was not reasonably practicable for the complaint to be presented within that time. (The time limit in respect of members of the armed forces is six months.)

Where an Employment Tribunal decides that a complaint is well-founded, it must make a declaration to that effect and can award compensation to be paid by the employer to the worker. This shall be such amount as the Employment Tribunal considers just and equitable in all the circumstances, having regard to (i) the employer's default in refusing to permit the worker to exercise his right and (ii) any loss sustained by the worker which is attributable to the matters complained of. With regard to complaints relating to holiday pay or payment in lieu of accrued holiday on termination, the tribunal can also order the employer to pay the worker the amount which it finds properly due.

There is no qualifying service period with regard to such complaints being presented to a tribunal.

In addition to a claim under the *Working Time Regulations 1998 (SI 1998/1833), reg 30*, failure to pay holiday pay also gives the worker the right to claim unlawful deduction of wages under *Section 13* of the *Employment Rights Act 1996* (see **W10024** above). Claims must be brought within 3 months of the deduction or the last in a series of deductions, which may allow a worker to claim back-pay in respect of previous holiday years.

Protection against detriment, and against unfair dismissal

[W10041] The *Working Time Regulations 1998 (SI 1998/1833), reg 31* inserts a new *section 45A* into the *Employment Rights Act 1996*, to protect a worker who is subjected to any detriment, where the worker has:

(a)   refused (or proposed to refuse) to comply with a requirement which the employer imposed (or proposed to impose) in contravention of the Regulations;

(b)   refused (or proposed to refuse) to forgo a right conferred on him by the Regulations;

(c)   failed to sign a workforce agreement or make any other agreement provided for under the Regulations, such as an individual opt-out from the maximum weekly working time limit;

(d)   been a candidate in an election of work place representatives or, having been elected, carries out any activities as such a representative or candidate; or

(e)   in good faith, (i) made an allegation that the employer has contravened a right under the Regulations, or (ii) brought proceedings under the Regulations.

The right to bring these claims would (as with discrimination claims) allow an individual to pursue a claim relating to a breach of the Regulations whilst continuing in employment.

By virtue of the inserted *section 101A* of the *Employment Rights Act 1996*, the dismissal of an employee on all but ground (*e*) above is automatically unfair, although this will only apply to employees and not to the wider definition of worker (for the position of workers who are not employees, see below). The compensation available would be subject to any cap on unfair dismissal compensation (from 1 February 2010 the cap is £65,300 but it is revised in February each year). Employees whose contracts were terminated on ground (*e*) above would be protected from dismissal for assertion of a statutory right, by virtue of *section 104* of the 1996 Act.

An employee would also be protected if selected for redundancy on any of the grounds listed above. This would be automatically unfair selection, by virtue of *section 105* of the 1996 Act.

Workers (ie those who are not employees) whose contracts are terminated for any of the grounds set out above can claim that they have suffered a detriment, and they may claim compensation, which would be capped in the same way as an award for unfair dismissal.

Breach of contract claims

[W10042] Following the *RJB Mining case* (see **W10010**), there is an argument that workers may also be able to claim for breach of contract, if their employer breaches the obligations in *reg 4(1)* of the Regulations. (In practice, however, this type of claim has not arisen since the determination of this case.) It has also been held that a worker cannot claim tortious damages for breach of statutory duty based on the employer's failure to enforce the 48-hour limit, as the provision was not intended to confer private law rights of action: *Sayers v Cambridgeshire County Council* [2006] EWHC 2029 (QB), [2007] IRLR 29.

If the claim is for damages and is outstanding at the termination of the employment, then the worker could in principle take their case to the Employment Tribunal (within 3 months of the effective date of termination of the contract giving rise to the claim). If the claim is for a declaration or an injunction, then the worker will have to apply to the High Court, where special rules apply.

Contracting out of the Regulations

[W10043] Under the *Working Time Regulations 1998 (SI 1998/1833), reg 35*, an agreement to contract out of the provisions of the Regulations can be made via a conciliation officer or by means of a compromise agreement, and the provisions are similar and consistent with the current provisions in the *Employment Rights Act 1996, s 203* (as amended by the *Employment Rights (Dispute Resolution) Act 1998*).

Conclusion

[W10044] The *Working Time Regulations 1998 (SI 1998/1833)* broke new ground in English law. Despite the adverse publicity surrounding them, their introduction does not appear to have unduly changed current industrial practice and does not appear to have led to a reduction in the long hours culture which is prevalent within the UK.

The impact of the Regulations might be increased if the European Commission restricted the use of the 48-hour limit 'opt out' as has been proposed in the past. However, there has been little consensus amongst member states on this issue and there are currently no proposals before the Commission. It remains to be seen if fresh proposals can be agreed between member states in the future.

Workplaces – Health, Safety and Welfare

Andrea Oates

Introduction

[W11001] The *Workplace (Health, Safety and Welfare) Regulations 1992 (SI 1992 No 3004)*, as amended by *SI 2002 No 2174*, set out a wide range of basic health, safety and welfare standards applying to most places of work. The Workplace Regulations cover not only offices, shops and factories covered by earlier legislation in this area, but apply to a much wider range of workplaces, including schools, hospitals, theatres, cinemas and hotels, for example.

The *Health and Safety (Miscellaneous) Amendments Regulations 2002 (SI 2002 No 2174)* mean that the 1992 Regulations now apply to previously excluded workplaces such as factories and mines. In addition *Regulation 25A* states that employers must consider the needs of disabled workers.

The Health and Safety Executive (HSE) reviewed the Workplace Regulations for the European Commission in 2003 and concluded that they were working well.

The Workplace Regulations set down detailed standards for premises used as places of work in the following areas:

(a)   maintenance [*SI 1992 No 3004, Reg 5*] (see **W11003**);
(b)   ventilation [*SI 1992 No 3004, Reg 6*] (see VENTILATION);
(c)   temperature [*SI 1992 No 3004, Reg 7*] (see **W11005**);
(d)   lighting [*SI 1992 No 3004, Reg 8*] (see LIGHTING);
(e)   cleanliness and waste storage [*SI 1992 No 3004, Reg 9*] (see **W11007**);
(f)   room dimensions and space [*SI 1992 No 3004, Reg 10*] (see **W11008**);
(g)   workstations and seating [*SI 1992 No 3004, Reg 11*] (see **W11009**);
(h)   conditions of floors and traffic routes [*SI 1992 No 3004, Reg 12*] (see **W11010**);
(i)   freedom from falls and falling objects [*SI 1992 No 3004, Reg 13*] (see WORK AT HEIGHTS);
(j)   windows and transparent or translucent doors, gates and walls (see **W11012**);
(k)   windows, skylights and ventilators [*SI 1992 No 3004, Reg 18*] (see **W11013**);
(l)   window cleaning [*SI 1992 No 3004, Regs 14—16*] (see WORK AT HEIGHTS);
(m)   organisation of traffic routes [*SI 1992 No 3004, Reg 17*] (see **W11015**);
(n)   doors and gates [*SI 1992 No 3004, Reg 18*] (see **W11016**);
(o)   escalators and travelators [*SI 1992 No 3004, Reg 19*] (see **W11017**);
(p)   sanitary conveniences/washing facilities [*SI 1992 No 3004, Regs 20, 21*] (see **W11019–W11023**);

(q)   drinking water [*SI 1992 No 3004, Reg 22*] (see **W11024**);

(r)   clothing accommodation and facilities for changing clothing [*SI 1992 No 3004, Reg 23*] (see **W11026–W11027**); and

(s)   rest/meal facilities [*SI 1992 No 3004, Reg 25*] (see **W11027**).

In addition to these Regulations, there is other legislation which contain health, safety and welfare standards for particularly high risk industries and workplaces, including construction sites, mines, quarries and the railway industry. This legislation is dealt with in other chapters.

This chapter looks at the provisions of the *Workplace (Health, Safety and Welfare) Regulations 1992*, together with guidance provided in the associated approved code of practice (ACoP). It also sets out the law banning smoking in enclosed workplaces and public places.

In addition, it outlines the requirements of the *Building Regulations 1991 (SI 1991 No 2768)*, which apply where workplaces are being built, extended or modified, and briefly looks at the *Equality Act 2010* which replaced a raft of equalities legislation including the *Disability Discrimination Act 1995* and requires that reasonable adjustments be made to workplaces where necessary to ensure that disabled workers are not put at a substantial disadvantage.

It also sets out the provisions of the *Health and Safety (Safety Signs and Signals) Regulations 1996 (SI 1996 No 341)*.

Workplace (Health, Safety and Welfare) Regulations 1992 (SI 1992 No 3004)

Definitions

[W11002] A 'workplace' is any non-domestic premises available to any person as a place of work, including:

(a)   canteens, toilets;

(b)   parts of a workroom or workplace (eg corridor, staircase, or other means of access/egress other than a public road);

(c)   a completed modification, extension, or conversion of an original workplace;

but excluding

(a)   boats, ships, hovercraft, trains and road vehicles (although the requirements in *regulation 13*, which deal with falls and falling objects, apply when aircraft, trains and road vehicles are stationary inside a workplace);

(b)   building operations/works of engineering construction;

(c)   mining activities.

[*Workplace (Health, Safety and Welfare) Regulations 1992 (SI 1992 No 3004), Regs 2, 3 and 4*].

The definition of 'work' includes work carried out by employees and self-employed people and the definition of 'premises' includes outdoor places. The

Regulations do not apply to private dwellings, but they do apply to hotels, nursing homes and to parts of premises where domestic staff are employed, for example in the kitchens of hostels.

Regulations 20 to *25* (which deal with toilets, washing, changing facilities, clothing accommodation, drinking water and eating and rest facilities) apply to temporary work sites, but only so far as is reasonably practicable.

General maintenance of the workplace

[W11003] All workplaces, equipment and devices should be maintained

(a)   in an efficient state,
(b)   in an efficient working order, and
(c)   in a good state of repair.

[*Workplace (Health, Safety and Welfare) Regulations 1992 (SI 1992 No 3004), Reg 5*].

Dangerous defects should be reported and acted on as a matter of good housekeeping (and to avoid possible subsequent civil liability). Defects resulting in equipment/plant becoming unsuitable for use, though not necessarily dangerous, should lead to decommissioning of plant until repaired – or, if this might lead to the number of facilities being less than required by statute, repaired forthwith (eg a defective toilet).

To this end, a suitable maintenance programme must be instituted, including:

(a)   regular maintenance (inspection, testing, adjustment, lubrication, cleaning);
(b)   rectification of potentially dangerous defects and the prevention of access to defective equipment;
(c)   record of maintenance/servicing.

There are more detailed regulations dealing with plant and equipment used at work, including the *Provision and Use of Work Equipment Regulations 1998 (SI 1998 No 2306 as amended by SI 2002 No 2174)*, which are examined in detail in MACHINERY SAFETY.

Ventilation

[W11004] *Workplace (Health, Safety and Welfare) Regulations 1992 (SI 1992 No 3004), Reg 6* deals with the provision of sufficient ventilation in enclosed workplaces. There must be effective and suitable ventilation in order to supply a sufficient quantity of fresh or purified air. If ventilation plant is necessary for health and safety reasons, it must give warning of failure.

The ACoP to the Regulations sets out that ventilation should not cause uncomfortable draughts, and that it should be sufficient to provide fresh air for the occupants to breathe, to dilute any contaminants and to reduce odour.

The ACoP also sets out that the fresh air supply rate to workplaces should not normally fall below 5 to 8 litres per second, per occupant, and that in deciding the appropriate rate, employers should consider factors including the floor area per person, the processes and equipment involved, and whether the work is strenuous.

Ventilation is dealt with in detail in VENTILATION.

Temperature

[W11005] The temperature in all workplaces inside buildings should be reasonable during working hours. [*Workplace (Health, Safety and Welfare) Regulations 1992 (SI 1992 No 3004), Reg 7*]. Workroom temperatures should enable people to work (and visit sanitary conveniences) in reasonable comfort, without the need for extra or special clothing. Although the Regulations themselves do not specify a maximum or minimum indoor workplace temperature, the approved code of practice (ACoP) sets out that the minimum acceptable temperature is 16°C at the workstation, except where work involves considerable physical effort, when it reduces to 13°C (dry bulb thermometer reading). Space heating of the average workplace should be 16°C, and this should be maintained throughout the remainder of the working day. However, this is a minimum temperature. The method of heating or cooling should not result in dangerous or offensive gases or fumes entering the workplace.

The following temperatures for different types of work are recommended by the Chartered Institute of Building Services Engineers (CIBSE):

(a)   heavy work in factories 13°C;
(b)   light work in factories 16°C;
(c)   hospital wards and shops 18°C; and
(d)   office and dining rooms 20°C.

Maintenance of such temperatures may not always be feasible, as, for instance, where hot/cold production/storage processes are involved, or where food has to be stored. In such cases, an approximate temperature should be maintained. With cold storage, this may be achievable by keeping a small chilling area separate or by product insulation; whereas, in the case of hot processes, insulation of hot plant or pipes, provision of cooling plant, window shading and positioning of workstations away from radiant heat should be considered in order to achieve a reasonably comfortable temperature. Moreover, where it is necessary from time to time to work in rooms normally unoccupied (eg storerooms), temporary heating should be installed. Thermometers must be provided so that workers can periodically check the temperatures.

Where, despite the provision of local heating or cooling, temperatures are still not reasonably comfortable, suitable protective clothing or rest facilities should be provided, or there should be systems of work in place, such as job rotation, to minimise the length of time workers are exposed to uncomfortable temperatures.

HSE guidance on the Regulations, *Workplace health, safety and welfare: A short guide for managers*, sets out how to carry out an assessment of the risk to workers' health from working in either a hot or cold environment. This advises that employers should look at personal factors, such as body activity, the amount and type of clothing and duration of exposure, together with environmental factors, including the ambient temperature and radiant heat, and if the work is outdoor, sunlight, wind velocity and the presence of rain or snow.

It sets out that any assessment needs to consider:

- Measures to control the workplace environment, particularly heat sources;
- Restriction of exposure by, for example, reorganising tasks to build in rest periods or other breaks from work;
- Medical pre-selection of employees to ensure that they are fit to work in these environments;
- Use of suitable clothing;
- Acclimatisation of workers to the working environment;
- Training in the precautions to be taken; and
- Supervision to ensure that the precautions the assessment identifies are taken.

HSE produces further guidance on temperature at work on its websitewww.hse.gov.uk.

Lighting

[W11006] Every workplace must be provided with suitable and sufficient lighting. [*Workplace (Health, Safety and Welfare) Regulations 1992 (SI 1992 No 3004), Reg 8*]. This should be natural lighting, so far as is reasonably practicable. There should also be suitable and sufficient emergency lighting where necessary. The ACoP sets out that in order to be suitable and sufficient, lighting must enable people to work and move about safely. HSE guidance on the Regulations advises that where necessary, local lighting should be provided at individual workstations, and at places of particular risk such as crossing points on traffic routes.

The HSE publication, Lighting at work, gives detailed guidance in this area and is available to download free from the HSE website.

The chapter on LIGHTING also deals with this area in greater detail.

General cleanliness

[W11007] All furniture and fittings of every workplace must be kept sufficiently clean. Surfaces of floors, walls and ceilings must be capable of being kept sufficiently clean and waste materials must not accumulate other than in waste receptacles. [*Workplace (Health, Safety and Welfare) Regulations 1992 (SI 1992 No 3004), Reg 9*].

The level and frequency of cleanliness will vary according to workplace use and purpose. Obviously a factory canteen should be cleaner than a factory floor. Floors and indoor traffic routes should be cleaned at least once a week, though dirt and refuse not in suitable receptacles should be removed at least daily, particularly in hot atmospheres or hot weather. Interior walls, ceilings and work surfaces should be cleaned at suitable intervals and ceilings and interior walls painted and/or tiled so that they can be kept clean. Surface treatment should be renewed when it can no longer be cleaned properly. In addition, cleaning will be necessary to remove spillages and waste matter from drains or sanitary conveniences. Methods of cleaning, however, should not expose anyone to substantial amounts of dust, and absorbent floors likely to be contaminated by oil or other substances difficult to remove, should be sealed or coated, say, with non-slip floor paint (not covered with carpet!).

Workroom dimensions/space

[W11008] Every room in which people work should have sufficient

(a)   floor area,
(b)   height, and
(c)   unoccupied space

for health, safety and welfare purposes. [*Workplace (Health, Safety and Welfare) Regulations 1992 (SI 1992 No 3004), Reg 10*].

Workrooms should have enough uncluttered space to allow people to go to and from workstations with relative ease. The number of people who may work in any particular room at any time will depend not only on the size of the room but also on the space given over to furniture, fittings, equipment and general room layout. Workrooms should be of sufficient height to afford staff safe access to workstations. If, however, the workroom is in an old building, say, with low beams or other possible obstructions, this should be clearly marked, eg 'Low beams, mind your head'.

The total volume of the room (when empty), divided by the number of people normally working there, should be 11 cubic metres (minimum) per person, although this does not apply to:

(i)   retail sales kiosks, attendants' shelters etc,
(ii)   lecture/meeting rooms etc.

Workplace (Health, Safety and Welfare) Regulations 1992 (SI 1992 No 3004), Sch 1

In making this calculation, any part of a room which is higher than 3 metres is counted as being 3 metres high.

Where furniture occupies a considerable part of the room, 11 metres may not be sufficient space per person. Here more careful planning and general room layout is required. Similarly, rooms may need to be larger or have fewer people working in them depending on the contents and layout of the room and the nature of the work.

Workstations and seating

[W11009] The Regulations stipulate the following.

Workstations

(a)   Every workstation must be so arranged that:
       (i)   it is suitable for:
              (a)   any person at work who is likely to work at the work-
                     station, and
              (b)   any work likely to be done there;
       (ii)   so far as reasonably practicable, it provides protection from
              adverse weather;
       (iii) it enables a person to leave it swiftly or to be assisted in an
              emergency;
       (iv)   it ensures any person is not likely to slip or fall.
       [*Workplace (Health, Safety and Welfare) Regulations 1992 (SI 1992
       No 3004), Reg 11(1), (2)*].

       Workstation seating
(b)   A suitable seat must be provided for each person at work whose work
       (or a substantial part of it) can or must be done seated. The seat should
       be suitable for:
       (i)   the person doing the work, and
       (ii)   the work to be done.
       Where necessary, a suitable footrest should be provided.
       [*Workplace (Health, Safety and Welfare) Regulations 1992 (SI 1992
       No 3004), Reg 11(3), (4)*].

It should be possible to carry out work safely and comfortably. Work materials
and equipment in frequent use (or controls) should always be within easy
reach, so that people do not have to bend or stretch unduly, and the worker
should be at a suitable height in relation to the work surface. Workstations,
including seating and access, should be suitable for special needs, for instance,
disabled workers. The workstation should allow people likely to have to do
work there adequate freedom of movement and ability to stand upright,
thereby avoiding the need to work in cramped conditions. More particularly,
seating should be suitable, providing adequate support for the lower back and
a footrest provided, if feet cannot be put comfortably flat on the floor.

Workstations with visual display units (VDUs) are subject to the *Health and
Safety (Display Screen Equipment) Regulations 1992 (SI 1992 No 2792)*, as
amended by *SI 2002 No 2174*.

The Health and Safety Executive (HSE) revised its guidance on seating in 1998,
and advises employers to use risk assessments in order to ensure that safe
seating is provided. *Seating at Work*, HS(G)57 is available, to download free
from the HSE website.

Condition of floors and traffic routes

[W11010] The principal dangers connected with industrial and commercial
floors are slipping, tripping and falling. Slip, trip and fall resistance are a
combination of the right floor surface and the appropriate type of footwear.
Employers should ensure that level changes, multiple changes of floor surfaces,
steps and ramps etc are clearly indicated. Safety underfoot is at bottom a
trade-off between slip resistance and ease of cleaning. Floors with rough

surfaces tend to be more slip-resistant than floors with smooth surfaces, especially when wet; by contrast, smooth surfaces are much easier to clean but less slip-resistant. Use of vinyl flooring in public areas – basically slip-resistant – is on the increase. Vinyl floors should be periodically stripped, degreased and resealed with slip-resistant finish; linoleum floors similarly.

Apart from being safe, floors must also be hygienically clean. In this connection, quarry tiles have long been 'firm favourites' in commercial kitchens, hospital kitchens etc but can be hygienically deceptive. In particular, grouted joints can trap bacteria as well as presenting endless practical cleaning problems. Hence the gradual transition to seamless floors in hygiene-critical areas. Whichever floor surface is appropriate and whichever treatment is suitable, underfoot safety depends on workplace activity (office or factory), variety of spillages (food, water, oil, chemicals), nature of traffic (pedestrian, cars, trucks).

Thus, floors in workplaces must:

(a)   be constructed so as to be suitable for use. [*Workplace (Health, Safety and Welfare) Regulations 1992 (SI 1992 No 3004), Reg 12(1)*].
They should always be of sound construction and adequate strength and stability to sustain loads and passing internal traffic; they should never be overloaded (see *Greaves v Baynham Meikle* [1975] 3 All ER 99 for possible consequences in civil law).
(i)   not have holes or slopes, or
(ii)   not be uneven or slippery

(b)   so as to expose a person to risk of injury. [*Workplace (Health, Safety and Welfare) Regulations 1992 (SI 1992 No 3004), Reg 12(2)(a)*].
The surfaces of floors and traffic routes should be even and free from holes, bumps and slipping hazards that could cause a person to slip, trip or fall, or drop or lose control of something being lifted or carried; or cause instability or loss of control of a vehicle.
Holes, bumps or uneven surfaces or areas resulting from damage or wear and tear should be made good and, pending this, barriers should be erected or locations conspicuously marked. Temporary holes, following, say, removal of floorboards, should be adequately guarded. Special needs should be catered for, for instance, disabled walkers or those with impaired sight. (Deep holes are governed by *Workplace (Health, Safety and Welfare) Regulations 1992 (SI 1992 No 3004), Reg 13* (see **W9013** WORK AT HEIGHTS).) Where possible, steep slopes should be avoided, and otherwise provided with a secure handrail. Ramps used by disabled persons should also have handrails.

(c)   be kept free from
(i)   obstructions, and
(ii)   articles/substances likely to cause persons to slip, trip or fall
so far as reasonably practicable. [*Workplace (Health, Safety and Welfare) Regulations 1992 (SI 1992 No 3004), Reg 12(3)*].
Floors should be kept free of obstructions impeding access or presenting hazards, particularly near or on steps, stairs, escalators and moving walkways, on emergency routes or outlets, in or near doorways or

gangways or by corners or junctions. Where temporary obstructions are unavoidable, access should be prevented and people warned of the possible hazard. Furniture being moved should not be left in a place where it can cause a hazard.

In *Lowles v Home Office* [2004] EWCA Civ 985, [2004] All ER (D) 538 (Jul) the judge ruled that a two-inch step on a ramp leading into Armley prison, on which Ms Lowles slipped and injured herself, could be considered an obstruction.

(d)   have effective drainage. [*Workplace (Health, Safety and Welfare) Regulations 1992 (SI 1992 No 3004), Reg 12(2)(b)*].

Where floors are likely to get wet, effective drainage (without drains becoming contaminated with toxic, corrosive substances) should drain it away, eg in laundries, potteries and food processing plants. Drains and channels should be situated so as to reduce the area of wet floor and the floor should slope slightly towards the drain and ideally have covers flush with the floor surface. Processes and plant which cause discharges or leaks of liquids should be enclosed and leaks from taps caught and drained away. In food processing and preparation plants, work surfaces should be arranged so as to minimise the likelihood of spillage. Where a leak or spillage occurs, it should be fenced off or mopped up immediately.

Staircases should be provided with a handrail. Any open side of a staircase should have minimum fencing of an upper rail at 900mm or higher, and a lower rail.

It is important also to consider the dangers posed by snow and ice upon, for example, external fire escapes.

Lewis v Avidan [2005] EWCA Civ 670, [2005] All ER (D) 136 (Apr) involved a claim from a care assistant who was injured slipping on a patch of water which had resulted from a burst water pipe. In this case, the Court of Appeal dismissed the claim as the judge ruled that an unexpected flood does not mean that the floor was inadequately maintained.

Falls and falling objects

[W11011] *Workplace (Health, Safety and Welfare) Regulations 1992 (SI 1992 No 3004), Reg 13* deals with falls and falling objects, which are covered in detail in WORK AT HEIGHTS.

It also requires that tanks, pits and other structures containing dangerous substances are securely covered or fenced where there is a risk of a person falling. Traffic routes over such open structures should also be securely fenced.

Windows and transparent or translucent doors, gates and walls

[W11012] Transparent or translucent surfaces in windows, doors, gates, walls and partitions should be constructed of safety material or be adequately protected against breakage, where necessary for health and safety reasons, where:

(a)   any part is at shoulder level or below in doors and gates; or

(b)   any part is at waist level or below in windows, walls and partitions, with the exception of glass houses.

Screens or barriers can be used as an alternative to the use of safety materials. Narrow panels of up to 250mm width are excluded from the requirement.

Transparent or translucent surfaces should be marked to make them apparent where this is necessary for health and safety reasons.

Workplace (Health, Safety and Welfare) Regulations 1992 (SI 1992 No 3004), Reg 14

Windows, skylights and ventilators

[W11013] Openable windows, skylights and ventilators must be capable of being opened, closed and adjusted safely. They must not be positioned so as to pose a risk when open.

They should be capable of being reached and operated safely, with window poles or similar equipment, or stable platforms, made available where necessary. Where there is the danger of falling from a height, devices should be provided to prevent this by ensuring the window cannot open too far. They should not cause a hazard by projecting into an area where people are likely to collide with them when open. The bottom edge of opening windows should normally be at least 800mm above floor level, unless there is a barrier to prevent falls.

[Workplace (Health, Safety and Welfare) Regulations 1992 (SI 1992 No 3004), Reg 15].

Ability to clean windows etc safely

[W11014] *Workplace (Health, Safety and Welfare) Regulations 1992 (SI 1992 No 3004), Reg 16* deals with the safe cleaning of windows and skylights where these cannot be cleaned from the ground or other suitable surface. This is dealt with in detail in WORK AT HEIGHTS.

Organisation of traffic routes

[W11015] Traffic routes in workplaces should allow pedestrians and vehicles to circulate safely, be safely constructed, be suitably indicated where necessary for health and safety reasons, and be kept clear of obstructions.

They should be planned to give the safest route, wide enough for the safe movement of the largest vehicle permitted to use them, and they should avoid vulnerable items like fuel or chemical plants or pipes, and open and unprotected edges.

There should be safe areas for loading and unloading. Sharp or blind bends should be avoided where possible, and if they cannot be avoided, one-way systems or mirrors to improve visibility should be used. Sensible speed limits should be set and enforced. There should be prominent warning of any limited headroom or potentially dangerous obstructions such as overhead electric cables. Routes should be marked where necessary and there should be suitable and sufficient parking areas in safe locations.

Traffic routes should keep vehicles and pedestrians apart and there should be pedestrian crossing points on vehicle routes. Traffic routes and parking and loading areas should be soundly constructed on level ground. Health and Safety Executive (HSE) guidance in this area can be found in the publication, *Workplace transport safety – guidance for employers*, HS(G)136, price £7.50, available to download free from the HSE website.

[*Workplace (Health, Safety and Welfare) Regulations 1992 (SI 1992 No 3004), Reg 17*].

This area is dealt with in detail in ACCESS, TRAFFIC ROUTES AND VEHICLES.

Doors and gates

[W11016] Doors and gates must be suitably constructed and fitted with safety devices. In particular,

(i)   a sliding door/gate must have a device to prevent it coming off its track during use;

(ii)   an upward opening door/gate must have a device to prevent its falling back;

(iii)   a powered door/gate must
    (a)   have features preventing it causing injury by trapping a person (eg accessible emergency stop controls),
    (b)   be able to be operated manually unless it opens automatically if the power fails;

(iv)   a door/gate capable of opening, by being pushed from either side, must provide a clear view of the space close to both sides.

[*Workplace (Health, Safety and Welfare) Regulations 1992 (SI 1992 No 3004), Reg 18*].

Doors and gates that swing in both directions should have a transparent panel, unless they are low enough to see over.

Escalators and travelators

[W11017] Escalators and travelators must:

(a)   function safely;
(b)   be equipped with safety devices;
(c)   be fitted with emergency stop controls.

[*Workplace (Health, Safety and Welfare) Regulations 1992 (SI 1992 No 3004), Reg 19*].

Welfare facilities

[W11018] 'Welfare facilities' is a wide term, embracing both sanitary and washing accommodation at workplaces, provision of drinking water, clothing accommodation (including facilities for changing clothes) and facilities for rest and eating meals (see **W11027** below). The need for sufficient suitable hygienic lavatory and washing facilities in all workplaces is obvious. Sufficient facilities must be provided to enable everyone at work to use them without undue delay. They do not have to be in the actual workplace but ideally should be situated in the building(s) containing them and they should provide protection from the weather, be well-ventilated, well-lit and enjoy a reasonable temperature. Where disabled workers are employed, special provision should be made for their sanitary and washing requirements. Wash basins should allow washing of hands, face and forearms and, where work is particularly strenuous, dirty, or results in skin contamination (eg molten metal work), showers or baths should be provided. In the case of showers, they should be fed by hot and cold water and fitted with a thermostatic mixer valve. Washing facilities should ensure privacy for the user and be separate from the water closet, with a door that can be secured from the inside. It should not be possible to see urinals or the communal shower from outside the facilities when the entrance/exit door opens. Entrance/exit doors should be fitted to both washing and sanitary facilities (unless there are other means of ensuring privacy). Windows to sanitary accommodation, showers/bathrooms should be obscured either by being frosted, or by blinds or curtains (unless it is impossible to see into them from outside).

This section examines current statutory requirements in all workplaces. For requirements relating to sanitary conveniences and washing facilities on construction sites see **C8056** CONSTRUCTION AND BUILDING OPERATIONS.

Sanitary conveniences in all workplaces

[W11019] Suitable and sufficient sanitary conveniences must be provided at readily accessible places. In particular,

(a)   the rooms containing them must be adequately ventilated and lit;

(b)   they (and the rooms in which they are situated) must be kept clean and in an orderly condition;

(c)   separate rooms containing conveniences must be provided for men and women except where the convenience is in a separate room which can be locked from the inside.

[*Workplace (Health, Safety and Welfare) Regulations 1992 (SI 1992 No 3004), Reg 20*].

Washing facilities in all workplaces

[W11020] Suitable and sufficient washing facilities (including showers where necessary (see **W11018** above)), must be provided at readily accessible places or points. In particular, facilities must:

(a)   be provided in the immediate vicinity of every sanitary convenience (whether or not provided elsewhere);

(b)   be provided in the vicinity of any changing rooms – whether or not provided elsewhere;

(c)   include a supply of clean hot and cold or warm water (if possible, running water);

(d)   include soap (or something similar);

(e)   include towels (or the equivalent);
(f)   be in rooms sufficiently well-ventilated and well-lit;
(g)   be kept clean and in an orderly condition (including rooms in which they are situate);
(h)   be separate for men and women, except where they are provided in a lockable room intended to be used by one person at a time, or where they are provided for the purposes of washing hands, forearms and face only, where separate provision is not necessary.

[*Workplace (Health, Safety and Welfare) Regulations 1992 (SI 1992 No 3004), Reg 21*].

Minimum number of facilities – sanitary conveniences and washing facilities

(a)   *People at work*

[W11021]

Number of people at work	Number of WCs	Number of wash stations
1 to 5	1	1
6 to 25	2	2
26 to 50	3	3
51 to 75	4	4
76 to 100	5	5

(b)   *Men at work*

Number of men at work	Number of WCs	Number of urinals
1 to 15	1	1
16 to 30	2	1
31 to 45	2	2
46 to 60	3	2
61 to 75	3	3
76 to 90	4	3
91 to 100	4	4

For every 25 people above 100 an additional WC and wash station should be provided; in the case of WCs used only by *men*, an additional WC per every 50 men above 100 is sufficient (provided that at least an equal number of additional urinals is provided). [*Workplace (Health, Safety and Welfare) Regulations 1992 (SI 1992 No 3004), Sch 1, Part II*].

Particularly dirty work etc.

[W11022] Where work results in heavy soiling of hands, arms and forearms, there should be one wash station for every 10 people at work up to 50 people; and one extra for every additional 20 people. And where sanitary and wash facilities are also used by members of the public, the number of conveniences and facilities should be increased so that workers can use them without undue delay.

Temporary work sites

[W11023] At temporary work sites suitable and sufficient sanitary conveniences and washing facilities should be provided so far as is reasonably practicable (see **E15039** ENFORCEMENT for meaning). If possible, these should incorporate flushing sanitary conveniences and washing facilities with running water.

Drinking water

[W11024] An adequate supply of wholesome drinking water must be provided for all persons at work in the workplace. It must be readily accessible at suitable places and conspicuously marked, unless non-drinkable cold water supplies are clearly marked. In addition, there must be provided a sufficient number of suitable cups (or other drinking vessels), unless the water supply is in a jet. [*Workplace (Health, Safety and Welfare) Regulations 1992 (SI 1992 No 3004), Reg 22*].

Where water cannot be obtained from the mains supply, it should only be provided in refillable containers. The containers should be enclosed to prevent contamination and refilled at least daily. So far as reasonably practicable, drinking water taps should not be installed in sanitary accommodation, or in places where contamination is likely, for instance, in a workshop containing lead processes.

Clothing accommodation

[W11025] Suitable and sufficient accommodation must be provided for:

(a)   any person at work's own clothing which is not worn during working hours; and

(b)   special clothing which is worn by any person at work but which is not taken home, for example, overalls, uniforms and thermal clothing.

[*Workplace (Health, Safety and Welfare) Regulations 1992 (SI 1992 No 3004), Reg 23(1)*].

Accommodation is not suitable unless it:

(i)   provides suitable security for the person's own clothing where changing facilities are required;

(ii)   includes separate accommodation for clothing worn at work and for other clothing, where necessary to avoid risks to health or damage to clothing; and

(iii)   is in a suitable location.

[*Workplace (Health, Safety and Welfare) Regulations 1992 (SI 1992 No 3004), Reg 23(2)*].

Work clothing is overalls, uniforms, thermal clothing and hats worn for hygiene purposes. Workers' own clothing should be able to hang in a clean, warm, dry, well-ventilated place. If this is not possible in the workroom, then it should be put elsewhere. Accommodation should take the form of a separate hook or peg. Clothing which is dirty, damp or contaminated owing to work should be accommodated separately from the worker's own clothes.

Facilities for changing clothing

[W11026] Suitable and sufficient facilities must be provided for any person at work in the workplace to change clothing where:

(a)   the person has to wear special clothing for work, and

(b)   the person cannot be expected to change in another room.

Facilities are not suitable unless they include:

(i)   separate facilities for men and women, or

(ii)   separate use of facilities by men and women.

[*Workplace (Health, Safety and Welfare) Regulations 1992 (SI 1992 No 3004), Reg 24*].

Changing rooms (or room) should be provided for workers who change into special work clothing and where they remove more than outer clothing; also where it is necessary to prevent workers' own clothes being contaminated by a harmful substance. Changing facilities should be easily accessible from workrooms and eating places. They should contain adequate seating and clothing accommodation, and showers or baths if these are provided (see **W11018** above). Privacy of user should be ensured. The facilities should be large enough to cater for the maximum number of persons at work expected to use them at any one time without overcrowding or undue delay.

Post Office v Footitt [2000] IRLR 243, involved an employers' appeal against an improvement notice requiring the construction of a separate changing room for women postal workers to change into and out of their uniforms, and looked at the definition of 'special clothing' and at the concept of propriety.

An environmental health officer had served an improvement notice under the *Workplace (Health, Safety and Welfare) Regulations 1992 (SI 1992 No 3004), Reg 24* as she had found that any female employees wishing to change their clothing could only do so in the general area of the women's toilet facilities.

In the High Court, the judge held that the uniform worn by postal workers was 'special clothing' for the purposes of *Regulation 24*. It was held that 'special clothing' is not merely limited to clothing that is worn only at work. Therefore the fact that postal workers wear their uniform to and from work does not prevent it from being 'special clothing'. The changing facilities for women provided by the Post Office were therefore not 'suitable and sufficient' within the meaning of the Regulation.

The court also held that the fact that the changing facilities for men and women were separated was not in itself enough to satisfy the concept of propriety referred to in *Regulation 24(2)*. There is no reason why requiring one female to undress in the presence of another cannot be said to offend against the principles of propriety. The fact that many people would have no objection to changing in the company of others of the same sex does not absolve the employer from providing facilities for those who may prefer privacy.

Rest and eating facilities

[W11027] Suitable and sufficient rest facilities must be provided at readily accessible places. [*Workplace (Health, Safety and Welfare) Regulations 1992 (SI 1992 No 3004), Reg 25(1)*].

(a)   Rest facilities

A rest facility is:

(i)   in the case of a new workplace, extension or conversion – a rest room (or rooms);
(ii)   in other cases, a rest room (or rooms) or rest area; including
(iii)   (in both cases):

- appropriate facilities for eating meals where food eaten in the workplace would otherwise be likely to become contaminated;
- suitable arrangements for protecting non-smokers from tobacco smoke (The regulations came into force before legislation banning smoking in enclosed workplaces and public places. See **W11028** below);
- a facility for a pregnant or nursing mother to rest in.

Canteens or restaurants may be used as rest rooms provided that there is no obligation to buy food there (ACoP). [*Workplace (Health, Safety and Welfare) Regulations 1992 (SI 1992 No 3004), Regs 25(2)–(4)*].

(b)   Eating facilities

Where workers regularly eat meals at work, facilities must be provided for them to do so. [*Workplace (Health, Safety and Welfare) Regulations 1992 (SI 1992 No 3004), Reg 25(5)*].

In offices and other workplaces where there is no risk of contamination, seats in the work area are sufficient, although workers should not be interrupted excessively during breaks, for example, by the public. In other cases, rest areas or rooms should be provided and in the case of new workplaces, this should be a separate rest room. Rest facilities should be large enough, and have enough seats with backrests and tables, for the number of workers likely to use them at one time.

Where workers regularly eat meals at work, there should be suitable and sufficient facilities. These should be provided where food would otherwise be contaminated, by dust or water for example. Seats in work areas can be suitable eating facilities, provided the work area is clean. There should be a means to prepare or obtain a hot drink, and where persons work during hours or at places where hot food cannot be readily obtained, there should be the means for heating their own food. Eating facilities should be kept clean.

Smoking

[W11028] Legislation banning smoking in enclosed workplaces and public places in England came into effect on 1 July 2007, following legislation introducing a ban in Scotland in March 2006, Northern Ireland in April 2006, and Wales in April 2007.

The *Health Act 2006* defines enclosed public places and workplaces as being offices, factories, shops, pubs, bars, restaurants, membership clubs, public transport and work vehicles that are used by more than one person. The legislation introducing smoke-free workplaces is set out in *Part 1* of the *Health Act 2006*.

The Act sets out a number of broad provisions for smoke-free legislation, and provides several legal powers to enable the more detailed aspects of smoke-free legislation to be dealt with by regulations. Five sets of smokefree regulations set out the detail of the smokefree legislation:

- The Smoke-free (Premises and Enforcement) Regulations 2006 (SI 2006 No 3368);
- The Smoke-free (Signs) Regulations 2007 (SI 2007 No 923);
- The Smoke-free (Exemptions and Vehicles) Regulations 2007 (SI 2007 No 765);
- The Smoke-free (Penalties and Discounted Amounts) Regulations 2007 (SI 2007 No 764); and
- The Smoke-free (Vehicle Operators and Penalty Notices) Regulations 2007 (SI 2007 No 760).

The Smokefree England campaign is advising the country's 3.7 million businesses, including nearly 200,000 pubs, bars, restaurants and other leisure outlets on the legislation. Its website is at www.smokefreeengland.co.uk and the telephone number is 0800 169 1697.

It has provided a factsheet which sets out the requirements of the five sets of regulations, which is summarised below:

The Smoke-free (Premises and Enforcement) Regulations 2006 (SI 2006 No 3368)

The Smoke-free (Premises and Enforcement) Regulations apply only in England and set out the definition of 'enclosed' and substantially enclosed' premises, and outline the enforcement provisions.

Section 2 of the *Health Act 2006* sets out that premises that are:

- open to the public, or are used as a place of work by more than one person; or
- where members of the public might attend to receive or provide goods or services, are to be smoke-free in areas that are enclosed or substantially enclosed.

The Regulations set out that premises are enclosed if they have a ceiling or roof and, except for doors, windows or passageways, are wholly enclosed, whether permanently or on a temporary basis.

Premises are substantially enclosed if they have a ceiling or roof, but have permanent openings in the walls which are less than half of the total areas of walls, including other structures which serve the purpose of walls and constitute the perimeter of premises. No account can be taken of openings in which doors, windows or other fittings that can be open or shut. This is known as the 50% rule.

A roof includes any fixed or movable structures, such as canvas awnings. Tents, marquees or similar will also be classified as enclosed premises if they fall within the definition.

The Regulations are enforced by lower-tier local authorities and port health authorities within the areas for which they have responsibilities.

The Smoke-free (Signs) Regulations 2007 (SI 2007 No 923)

These Regulations set out the requirements for no-smoking signs, together with the legal responsibilities concerning the display of no-smoking signs in smoke-free vehicles. They apply only in England.

Under the *Health Act 2006*, it is against the law to not display required no-smoking signs meeting the requirements of the regulations.

The Regulations require all smoke-free premises to display a no-smoking sign in a prominent position at each entrance that:

- is the equivalent of A5 in area;
- displays the international no-smoking symbol in colour, a minimum of 70mm in diameter; and
- carries the words in characters that can be easily read: 'No smoking. It is against the law to smoke in these premises'. The words 'these premises' may be changed to refer to the particular premises where the sign is displayed, for example 'this hotel' or 'this NHS clinic'.

The Regulations also set out that a no-smoking sign that displays the international no smoking symbol in colour, and is at least 70mm in diameter is the minimum requirement at entrances to smoke-free premises which:

- are for staff only (on the basis that the premises displays at least one A5-sized sign with words, as set out above): or
- are located within other smoke-free premises (for example, a shop within an indoor shopping centre).

The Regulations require any person with management responsibilities for a smoke-free vehicle to display a no-smoking sign in each enclosed compartment that can accommodate people. These no-smoking signs must display the international no smoking symbol in colour, and be at least 70mm in diameter.

Signs (together with guidelines) meeting the requirements of these Regulations are available free of charge (in reasonable quantities) from Smokefree England.

The Smoke-free (Exemptions and Vehicles) Regulations 2007 (SI 2007 No 765)

These Regulations, which apply only in England:

- set out the limited exemptions from the smoke-free requirements of section 2 of the Health Act 2006; and
- specify that most public and work vehicles are to be smoke-free under section 5 of the Health Act 2006.

The Regulations also set out a number of exemptions from smoke-free legislation for premises that would otherwise be required to be entirely smoke-free in enclosed parts under smoke-free legislation. These concern:

- private dwellings;
- accommodation for guests and members;
- other residential accommodation;
- offshore installations;
- research and testing facilities;
- temporary exemption for residential mental health units; and
- specialist tobacconists.

They also set out provisions regarding performers: Where the artistic integrity of a performance makes it appropriate for a person who is taking part in that performance to smoke, the regulations allow for parts of premises in which a person performs to be not smoke-free in relation to that person only during the time of the performance.

The Regulations require enclosed vehicles to be smoke-free at all times, if they are used:

- by members of the public or a section of the public (whether or not for reward or hire); or
- in the course of paid or voluntary work by more than one person, even if those people use the vehicle at different times, or only intermittently.

Vehicles required to be smoke-free will not need to be smoke-free when they are conveying persons if they have a removable or stowable roof during the time the roof is completely removed or stowed.

Vehicles will not be required to be smoke-free if they are used primarily for the private purposes of a person who owns it, or has a right to use it, which is not restricted to a particular journey.

The Regulations apply to all vehicles, except aircraft, and there are exceptions regarding ships and hovercrafts.

The Smoke-free (Penalties and Discounted Amounts) Regulations 2007 (SI 2007 No 764)

These Regulations apply to England and Wales.

The Regulations set out that the maximum fine on conviction is £1,000, although a fixed-penalty procedure can be used, where the fine is £200, discounted to £150 if the penalty is paid within 15 days from when the notice is issued.

With regard to smoking in a smoke-free place, the fine is a maximum of £200 on conviction, with a £50 fixed penalty, discounted to £30 if the penalty is paid within 15 days from when the notice is issued.

Section 8 of the *Health Act 2006* places a duty on any person who controls or is concerned in the management of smoke-free premises to cause a person there to stop smoking. The maximum fine on conviction for failing to prevent smoking in a smoke-free place is £2500 and there is no fixed penalty provision.

The Smoke-free (Vehicle Operators and Penalty Notices) Regulations 2007 (SI 2007 No 760)

These Regulations apply only in England and:

- set out who has legal duties corresponding to that in *section 8(1)* of the *Health Act 2006* to cause any person who is smoking in a smoke-free vehicle to stop smoking; and
- specifies the form of the fixed penalty notice for use by enforcement authorities.

The following have a legal duty to cause any person who is smoking in a smoke-free vehicle to stop smoking:

- the driver;
- any person with management responsibilities for the vehicle; and
- any person in a vehicle who is responsible for order or safety on it.

These Regulations specify the form of fixed penalty notices to be used by enforcement authorities and how these can be adapted by enforcement authorities.

The Government has produced an official guide to the smokefree law. The official *Everything you need* guide to the new smokefree law is available to download on the Smoke Free England website. This booklet was included in the full guidance pack sent to businesses in April 2007: employing businesses that were actively trading and registered with Companies House were sent a full guidance pack by post. This pack contains examples of compliant signage for smokefree premises and vehicles, a sample smokefree policy and suggested steps to take if someone smokes in a smokefree place.

Civil liability

[W11029] There is no specific reference to civil liability in the Regulations. However, safety regulations are actionable, even if silent (as here), and, if a person suffered injury/damage as a result of breach by an employer, he could sue. Certainly, there is civil liability for breach of the *Building Regulations 1991 (SI 1991 No 2768)* (see **W11044** below).

Safety signs at work – Health and Safety (Safety Signs and Signals) Regulations 1996 (SI 1996 No 341)

[W11030] Traditionally, safety signs, communications and warnings have played a residual role in reducing the risk of injury or damage at work, the need for them generally having been engineered out or accommodated in the system of work – a situation unaffected by these Regulations.

Types of signs

[W11031] Safety signs and signals can be of the following types:

(a)   permanent (eg signboards);
(b)   occasional (eg acoustic signals or verbal communications – acoustic signals should be avoided where there is considerable ambient noise).

Interchanging and combining signs

[W11032] Examples of interchanging and combining signs are:

(a)   a safety colour (see **W11033** below) or signboard to mark places where there is an obstacle;
(b)   illuminated signs, acoustic signals or verbal communication; and
(c)   hand signals or verbal communication.

[*Health and Safety (Safety Signs and Signals) Regulations 1996 (SI 1996 No 341), Sch 1, Part I, para 3*].

Safety colours

[W11033]

Colour	Meaning or purpose	Instructions and information
Red	Prohibition sign	Dangerous behaviour
	Danger	Stop, shutdown, emergency cut-out services
		Evacuate
	Fire-fighting equipment	Identification and location
Yellow or Amber	Warning sign	Be careful, take precautions
		Examine
Blue	Mandatory sign	Specific behaviour or action
		Wear personal protective equipment
Green	Emergency escape, first-aid sign	Doors, exits, routes, equipment and facilities
	No danger	Return to normal

[*Health and Safety (Safety Signs and Signals) Regulations 1996 (SI 1996 No 341), Sch 1, Part I, para 4.*]

Varieties of safety signs and signals

[W11034] Safety signs and signals include, comprehensively:

(a)   safety signs – providing information about health and safety at work by means of a signboard, safety colour, illuminated sign, acoustic signal, hand signal or verbal communication;
(b)   signboards – signs giving information by way of a simple pictogram, lighting intensity providing visibility (these should be weather-resistant and easily seen);
(c)   mandatory signs – signs prescribing behaviour (eg safety boots must be worn);
(d)   prohibition signs – signs prohibiting behaviour likely to cause a health and safety risk (eg no smoking);
(e)   hand signals – movement or position of arms/hands for guiding persons carrying out operations that could endanger employees;

(f)   verbal communications – predetermined spoken messages communicated by human or artificial voice, preferably short, simple and as clear
as possible.
[*Health and Safety (Safety Signs and Signals) Regulations 1996 (SI
1996 No 341), Reg 2*].

Duty of employer

[W11035] It is only where a risk assessment carried out under the *Management of Health and Safety at Work Regulations 1999 (SI 1999 No 3242)*
indicates that a risk cannot be avoided, engineered out or reduced significantly
by way of a system of work that resort to signs and signals becomes necessary.
In these circumstances, all employers (including offshore employers) must:

(a)   provide   and   maintain   any   appropriate   safety   sign(s)   (see
W11037–W11041 below) (including fire safety signals) but not a hand
signal or verbal communication;

(b)   so far as is reasonably practicable, ensure that correct hand signals or
verbal communications are used;

(c)   provide and maintain any necessary road traffic sign (where there is a
risk to employees in connection with traffic); and

(d)   provide employees with comprehensible and relevant information,
training and instruction and measures to be taken in connection with
safety signs.
[*Health and Safety (Safety Signs and Signals) Regulations 1996 (SI
1996 No 341), Regs 4, 5*]

Schedule 1 to the Regulations sets out the minimum requirements concerning
safety signs and signals with regard to the type of signs to be used in particular
circumstances, interchanging and combining signs, signboards, signs on
containers and pipes, the identification and location of fire-fighting equipment,
signs for obstacles and dangerous locations, for marking traffic routes,
illuminated signs, acoustic signals, verbal communication and hand signals.

Exclusions

[W11036] Excluded from the operation of these Regulations are:

(a)   the supply of dangerous substances or products;

(b)   the transportation of dangerous goods;

(c)   road traffic signs (except where there is a particular risk to employees.
Where there is a risk arising from the movement of traffic and the risk
is addressed by a sign stipulated in the *Road Traffic Regulations Act
1984* (eg speed restriction sign), these signs must be used, whether or
not the Act applies to that place of work. In effect this means that where
road speed and other signs are needed on a company's road, these must
replicate the signs used for the purpose on public roads; and

(d)   activities on board ship.
[*Health and Safety (Safety Signs and Signals) Regulations 1996 (SI
1996 No 341), Reg 3(1)*].

Examples of safety signs

Prohibitory signs

[W11037] Intrinsic features:

- round shape
- black pictogram on white background, red edging and diagonal line (the red part to take up at least 35% of the sign area) .

fig. 1 Safety signs (prohibitory)

No smoking	Smoking and naked flames forbidden	No access for pedestrians
Do not extinguish with water	Not drinkable	No access for unauthorised persons
No access for industrial vehicles		Do not touch

Warning signs

[W11038] Intrinsic features are a triangular shape; and a black pictogram on a yellow background with black edging (the yellow part to take up at least 50% of the area of the sign).

fig. 2 Safety signs (warning)

Flammable material
or high temperature

Explosive material

Toxic material

Corrosive material

Radioactive material

Overhead load

Industrial vehicles

Danger: electricity

General danger

Laser beam   Oxidant material   Non-ionizing radiation

Strong magnetic field

Obstacles

Drop

Biological risk

Low temperature

Harmful or irritant
material

Mandatory signs

[W11039] Intrinsic features:

- round shape
- white pictogram on a blue background (the blue part to take up at least 50% of the area of the sign).

fig. 3 Safety signs (mandatory)

Eye protection
must be worn

Safety helmet
must be worn

Ear protection
must be worn

Respiratory equipment
must be worn

Safety boots
must be worn

Safety gloves
must be worn

Safety overalls
must be worn

Face protection
must be worn

Safety harness
must be worn

Pedestrians must
use this route

General mandatory sign
(to be accompanied where
necessary by another sign)

Emergency escape or first-aid signs

[W11040] Intrinsic features:

- rectangular or square shape
- white pictogram on a green background (the green part to take up at least 50% of the area of the sign).

fig. 4 Safety signs (emergency escape or first-aid)

Emergency exit/escape route

This way
(supplementary information sign)

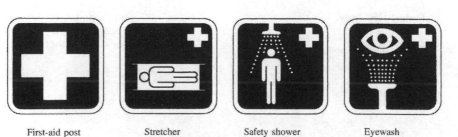

First-aid post   Stretcher   Safety shower   Eyewash

Emergency telephone for first-aid or escape

Fire-fighting signs

[W11041] Intrinsic features:

- rectangular or square shape
- white pictogram on a red background (the red part to take up at least 50% of the area of the sign).

fig. 5 Safety signs (fire-fighting)

Fire hose	Ladder	Fire extinguisher	Emergency fire telephone

This way
(supplementary information sign)

Examples of hand signals

[W11042]

Meaning	*Description*	*Illustration*

A. General signals

fig. 6 Hand signals

START Attention Start of Command	both arms are extended horizontally with the palms facing forwards.	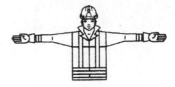

STOP
Interruption
End of movement

the right arm points
upwards with the palm
facing forwards.

END
of the operation

both hands are clasped at
chest height.

B. Vertical movements

RAISE

the right arm points
upwards with the palm
facing forward and slowly
makes a circle.

LOWER

the right arm points downwards
with the palm facing inwards
and slowly makes a circle.

VERTICAL DISTANCE

the hands indicate the relevant
distance.

C. Horizontal movements

MOVE FORWARDS

both arms are bent with the palms facing upwards, and the forearms make slow movements towards the body.

MOVE BACKWARDS

both arms are bent with the palms facing downwards, and the forearms make slow movements away from the body.

RIGHT
to the signalman's

the right arm is extended more or less horizontally with the palm facing downwards and slowly makes small movements to the right.

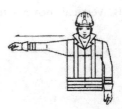

LEFT
to the signalman's

the left arm is extended more or less horizontally with the palm facing downwards and slowly makes small movements to the left.

HORIZONTAL
DISTANCE

the hands indicate the relevant distance.

D. Danger

DANGER
Emergency stop

both arms point upwards with
the palms facing forwards.

QUICK

all movements faster.

SLOW

all movements slower.

Equality Act 2010

[W11043] From 1 October 2010, the Equality Act replaced most of the *Disability Discrimination Act (DDA) 1995*. The new Act aims to protect disabled people and prevent disability discrimination. It provides legal rights for disabled people in areas including employment, education, and access to goods, services and facilities.

It sets out rights for people not to be directly discriminated against or harassed because they have an association with a disabled person. This can apply to a carer or parent of a disabled person. In addition, it means that people must not be directly discriminated against or harassed because they are wrongly perceived to be disabled.

In the Act, a person has a disability if they have a physical or mental impairment and the impairment has a substantial and long-term adverse effect on their ability to perform normal day-to-day activities.

According to the ACAS "The Act has made it easier for a person to show that they are disabled and protected from disability discrimination. Under the Act, a person is disabled if they have a physical or mental impairment which has a substantial and long term adverse effect on their ability to carry out normal day-to-day activities, which would include things like using a telephone, reading a book or using public transport."

The *DDA 1995* required employers to make reasonable adjustments to working conditions or to the workplace to avoid putting disabled workers at a substantial disadvantage by any physical feature of the premises, and could therefore involve making adjustments to the physical features of workplace premises.

As before, the Equality Act puts a duty on employers to make reasonable adjustments for staff to help them overcome disadvantage resulting from an impairment (for example by providing assistive technologies to help visually impaired staff use computers effectively).

It also includes a new protection from discrimination arising from disability. It is discrimination to treat a disabled person unfavourably because of something connected with their disability. ACAS gives the example of a tendency to make spelling mistakes arising from dyslexia.

This type of discrimination is unlawful where the employer or other person acting for the employer knows, or could reasonably be expected to know, that the person has a disability and is only justifiable if an employer can show that it is a proportionate means of achieving a legitimate aim.

Indirect discrimination now covers disabled people. A job applicant or employee could claim that a particular rule or requirement in place disadvantages people with the same disability. Unless employers can justify this, it would be unlawful.

Finally, the Act also includes a new provision which makes it unlawful, except in certain circumstances, for employers to ask about a candidate's health before offering them work.

More detailed guidance for employers on complying with disability discrimination legislation is provided by the Government Equalities Office at www.equalities.gov.uk.

Building Regulations 2010 (SI 2010 No 2214)

[W11044] The *Building Regulations 2000* and the *Building (Approved Inspectors etc.) Regulations 2000*, and all amendments to both were revoked from 1 October 2010. The *Building Regulations 2010 (SI 2010 No 2214)* and the *Building (Approved Inspectors etc.) Regulations 2010 (SI 2010 2215)* consolidate the *Building Regulations 2000* and the *Building (Approved Inspectors etc.) Regulations 2000*.

There are also a number of minor amendments. These include removing the requirement where building work involves inserting insulation into a cavity wall, to submit along with a building notice a statement about the proposed insulating material. The Regulations also insert into Schedule 3 three new types of work that may be carried out under a self-certification scheme, and revise the list of bodies which are able to register people for the purposes of self-certification – installation of cavity wall insulating material, replacement of roof coverings and installation of systems to produce electricity, heat or cooling by microgeneration or from renewable sources.

The explanatory note to the regulations explains: "The Regulations impose requirements on people carrying out "building work" which is defined in regulation 3 as the erection or extension of a building; the provision or extension of a controlled service or fitting; the material alteration of a building or controlled service or fitting; work required in relation to a material change of use; insertion of insulating material into a cavity wall; work involving underpinning of a building; work required to replace or renovate thermal elements; work relating to a change in a building's energy status and work relating to improvement of the energy status of certain large existing buildings."

Building work must be carried out so that it complies with requirements set out in Parts A to P of Schedule 1 of the Regulations (regulation 4). These relate to structure (Part A), fire safety (Part B), site preparation and resistance to contaminants and moisture (Part C), toxic substances (Part D), resistance to the passage of sound (Part E), ventilation (Part F), sanitation, hot water safety and water efficiency (Part G), drainage and waste disposal (Part H), combustion appliances and fuel storage systems (Part J), protection from falling, collision and impact (Part K), conservation of fuel and power (Part L), access to and use of buildings (Part M), glazing – safety in relation to impact, opening and cleaning (Part N) and electrical safety (Part P). (Not all provisions of Schedule 1 apply to all building work.)

Further guidance on the regulations is available on the Department for Communities and Local Government (DCLG) at www.communities.gov.uk/planningandbuilding/buildingregulations/brlegislation.

Factories Act 1961

[W11045] Although large parts of the *Factories Act 1961* have now been repealed and replaced by more recent legislation, some provisions are still in force. Additionally, civil actions for injury, relating to breach of health and safety provisions of the *Factories Act* (though not welfare) may well continue for some time, since actions for personal injury can be initiated for up to three years after injury/disease has occurred. [*Limitation Act 1980, s 11*].

Residual application of the Factories Act 1961

[W11046] The *Factories Act 1961* applies to factories, as defined in *s 175*, including 'factories belonging to or in the occupation of the Crown, to building operations and works of engineering construction undertaken by or on behalf of the Crown, and to employment by or under the Crown of persons in painting buildings', eg hospital painters. [*Factories Act 1961, s 173(1)*].

Enforcement of the Factories Act 1961 and regulations

[W11047] Offences under the Act are normally committed by occupiers rather than owners of factories. Unless they happen to occupy a factory as well, the owners of a factory would not normally be charged. Offences therefore relate to physical occupation or control of a factory (for an extended meaning of 'occupier', see OCCUPIERS' LIABILITY). Hence the person or persons or body corporate having managerial responsibility in respect of a factory are those who commit an offence under *s 155(1)*. This will generally be the managing director and board of directors and/or individual executive directors. Moreover, if a company is in liquidation and the receiver is in control, he is the person who will be prosecuted and this has in fact happened (*Meigh v Wickenden* [1942] 2 KB 160; *Lord Advocate v Aero Technologies* 1991 SLT 134 where the receiver was 'in occupation' and so under a duty to prevent 'accidents by fire or explosion', for the purposes of the *Explosives Act 1875, s 23*).

Defence of factory occupier

[W11048] The main defence open to a factory occupier charged with breach of the *Factories Act 1961* is that the Act itself, or more likely regulations made under it, placed the statutory duty on some person other than the occupier. Thus, where there is a contravention by any person of any regulation or order under the *Factories Act 1961*, 'that person shall be guilty of an offence and the occupier or owner . . . shall not be guilty of an offence, by reason only of the contravention of the provision . . . unless it is proved that he failed to take all reasonable steps to prevent the contravention . . . '. [*Factories Act 1961, s 155(2)*].

Before this defence can be invoked by a factory occupier or company, it is necessary to show that:

(a)   a statutory duty had been laid on someone other than the factory occupier by a regulation or order passed under the Act;

(b)   the factory occupier took all reasonable steps to prevent the contravention (a difficult test to satisfy).

NB. This statutory defence is not open to a building contractor (in his capacity as a notional factory occupier).

Effect on possible civil liability

[W11049] Whether conviction of an employee under the *Factories Act 1961, s 155(2)* would prejudice a subsequent claim for damages by him against a factory occupier, must be regarded as an open question. Thus, in *Potts v Reid* [1942] 2 All ER 161 the court said 'Criminal and civil liability are two separate things . . . The legislation (the *Factories Act 1937*) might well be unwilling to convict an owner who failed to carry out a statutory duty of a crime with which he was not himself directly concerned, but still be ready to leave the civil liability untouched'. Similarly in *Boyle v Kodak Ltd* [1969] 2 All ER 439 it was said, 'When considering the civil liability engrafted by judicial decision upon the criminal liability which has been imposed by statute, it is no good looking to the statute and seeing from it where the criminal liability would lie, for we are concerned only with civil liability. We must look to the cases' (per Lord Diplock). Moreover, a breach of general duties of *HSWA 1974* gives rise only to civil liability at common law and not under statute. (Though this is not the position where there is a breach of a specific regulation under *HSWA 1974*.) On the other hand, there is at least one isolated instance of an employee being denied damages where he was in breach of specific regulations (*ICI Ltd v Shatwell* [1964] 2 All ER 999). It is thought, however, that this decision would not apply in the case of breach of a *general* statutory duty, such as *s 155(2)*.

Table of Cases

A

A v National Blood Authority [2001] 3 All ER 289, [2001] Lloyd's Rep Med 187, 60 BMLR 1, [2001] All ER (D) 298 (Mar) P9032

AM v WC [1999] ICR 1218, [1999] IRLR 410, EAT H1714

AMF International Ltd v Magnet Bowling Ltd [1968] 2 All ER 789, [1968] 1 WLR 1028, 66 LGR 706, 112 Sol Jo 522 .... O3015

Adams v Southern Electricity Board (1993) Times, 21 October, CA O3017

Addie (Robert) & Sons (Collieries) Ltd v Dumbreck [1929] AC 358, 98 LJPC 119, 34 Com Cas 214, [1929] All ER Rep 1, 140 LT 650, 45 TLR 267, HL O3008

Adsett v K and L Steelfounders and Engineers Ltd [1953] 2 All ER 320, [1953] 1 WLR 773, 51 LGR 418, 97 Sol Jo 419, CA R3004

Adshead v May Gurney Ltd [2006] All ER (D) 388 (Jul), EAT W10024

Advocate (Lord) v Aero Technologies Ltd (in receivership) 1991 SLT 134, Ct of Sess
.... W11047

Ainsworth v IRC [2005] EWCA Civ 441, [2005] IRLR 465, [2005] NLJR 744, (2005) Times, 16 May , [2005] All ER (D) 328 (Apr), sub nom IRC v Ainsworth [2005] ICR 1149; revsd sub nom Revenue and Customs v Stringer (sub nom Ainsworth v IRC) [2009] UKHL 31, [2009] ICR 985, [2009] IRLR 677, (2009) Times, 15 June, [2009] All ER (D) 168 (Jun) W10024

Alcock v Chief Constable of South Yorkshire Police [1992] 1 AC 310, [1991] 4 All ER 907, [1991] 3 WLR 1057, 8 BMLR 37, [1992] 3 LS Gaz R 34, 136 Sol Jo LB 9, HL C6033

Allison v London Underground Ltd [2008] EWCA Civ 71, [2008] ICR 719, [2008] IRLR 440, [2008] PIQR P185, 152 Sol Jo (no 8) 34, [2008] All ER (D) 185 (Feb)
.... M1007

Aluminium Wire and Cable Co Ltd v Allstate Insurance Co Ltd [1985] 2 Lloyd's Rep 280 .... E13020

Anderson v Jarvis Hotels plc (EAT/0062/05) (30 May 2006, unreported) W10003

Anderson v Newham College of Further Education [2002] EWCA Civ 505, [2003] ICR 212, [2002] All ER (D) 381 (Mar) O3011

Anyanwu v South Bank Student Union (Commission for Racial Equality, interveners) [2001] UKHL 14, [2001] 2 All ER 353, [2001] 1 WLR 638, [2001] ICR 391, [2001] IRLR 305, [2001] ELR 511, [2001] 21 LS Gaz R 39, 151 NLJ 501, [2001] All ER (D) 272 (Mar) H1714

Aparau v Iceland Frozen Foods plc [1996] IRLR 119, EAT; revsd [2000] 1 All ER 228, [2000] IRLR 196, [1999] 45 LS Gaz R 31, CA E14004

Arafa v Potter. See Potter v Arafa

Archibald v Fife Council [2004] UKHL 32, [2004] 4 All ER 303, [2004] ICR 954, [2004] IRLR 651, 82 BMLR 185, [2004] 31 LS Gaz R 25, (2004) Times, 5 July, 2004 SC 942, 2004 SCLR 971, 148 Sol Jo LB 826, [2004] All ER (D) 32 (Jul) E16511

Armour v Skeen [1977] IRLR 310, 1977 JC 15, 1977 SLT 71, HC of Justiciary (Sc)
.... E15040, S7002

Armstrong v British Coal Corpn (1998), unreported V5003

Arnold v Central Electricity Generating Board [1988] AC 228, [1986] 3 WLR 171, 130 Sol Jo 484, [1986] LS Gaz R 2090; revsd [1988] AC 228, [1987] 2 WLR 245, 131 Sol Jo 167, [1987] LS Gaz R 743, CA; affd [1988] AC 228, [1987] 3 All ER 694, [1987] 3 WLR 1009, 131 Sol Jo 1487, [1987] LS Gaz R 3416, [1987] NLJ Rep 1014, HL
.... N3022

Arthur v Anker [1997] QB 564, [1996] 3 All ER 783, [1996] 2 WLR 602, 72 P & CR 309, [1996] RTR 308, [1996] NLJR 86, CA O3017

Aspden v Webbs Poultry & Meat Group (Holdings) Ltd [1996] IRLR 521 .... E14004

Associated Dairies Ltd v Hartley [1979] IRLR 171, Ind Trib E15024

Attia v British Gas plc [1988] QB 304, [1987] 3 All ER 455, [1987] 3 WLR 1101, [1987] BTLC 394, 131 Sol Jo 1248, [1987] LS Gaz R 2360, [1987] NLJ Rep 661, CA C6033
Austin Rover Group Ltd v HM Inspector of Factories [1990] 1 AC 619, [1989] 3 WLR 520, [1990] ICR 133, [1989] IRLR 404, sub nom Mailer v Austin Rover Group plc [1989] 2 All ER 1087, HL O3019

B

Bacica v Muir [2006] IRLR 35, EAT W10004
Baker v T E Hopkins & Son Ltd [1959] 3 All ER 225, [1959] 1 WLR 966, 103 Sol Jo 812, CA A1022
Balgobin v Tower Hamlets London Borough Council [1987] ICR 829, [1987] IRLR 401, [1987] LS Gaz R 2530, EAT H1709, H1713, H1720, H1738
Bamsey v Albon Engineering & Manufacturing plc [2004] EWCA Civ 359, [2004] 2 CMLR 1353, [2004] IRLR 457, [2004] 17 LS Gaz R 31, (2004) Times, 15 April, 148 Sol Jo LB 389, [2004] All ER (D) 482 (Mar) W10024
Barber v RJB Mining (UK) Ltd [1999] 2 CMLR 833, [1999] ICR 679, [1999] IRLR 308, 143 Sol Jo LB 141, [1999] All ER (D) 244 .... E14002, W10010, W10042
Barber v Somerset County Council [2002] EWCA Civ 76, [2002] 2 All ER 1, [2002] ICR 613, [2002] IRLR 263, 68 BMLR 115, (2002) Times, 11 February, [2002] All ER (D) 53 (Feb); revsd sub nom Barber v Somerset County Council [2004] UKHL 13, [2004] 2 All ER 385, [2004] 1 WLR 1089, [2004] ICR 457, [2004] IRLR 475, 77 BMLR 219, (2004) Times, 5 April, 148 Sol Jo LB 419, [2004] All ER (D) 07 (Apr) R2006, S11009
Barclays Bank plc v Kapur (No 2) [1995] IRLR 87, CA H1708
Barker v Corus UK Ltd [2006] UKHL 20, [2006] 2 AC 572, [2006] 3 All ER 785, [2006] 2 WLR 1027, [2006] ICR 809, 89 BMLR 1, [2006] NLJR 796, [2006] PIQR P390, (2006) Times, 4 May, [2006] 5 LRC 271, [2006] All ER (D) 23 (May) P9039
Barton v SCC Ltd (1999), unreported W10024
Barton v Wandsworth Council (1995) IDS Brief 549 .... E14009
Baxter v Harland & Wolff plc [1990] IRLR 516, NI CA N3022
Beattie v Secretary of State for Social Security [2001] EWCA Civ 498, [2001] 1 WLR 1404, 145 Sol Jo LB 120, [2001] All ER (D) 93 (Apr) C6041
Belhaven Brewery Co v McLean [1975] IRLR 370, Ind Trib E15024
Berry v Stone Manganese & Marine Ltd [1972] 1 Lloyd's Rep 182, 12 KIR 13, 115 Sol Jo 966 .... N3022
Berryman v Hounslow London Borough Council (1996) 30 HLR 567, [1997] PIQR P 83, CA O3004
Bett v Dalmey Oil Co (1905) 7F (Ct of Sess) 787 .... S3003
Betts v Tokley (t/a PDQ Cars Minibuses Limousines & Couriers) [2002] All ER (D) 99 (Jan), CA O3011
Bews v Scottish Hydro Electric plc 1992 SLT 749, Ct of Sess C6053
Biesheuval v Birrell [1999] PIQR Q 40 .... C6023
Bishop v Baker Refractories Ltd [2002] EWCA Civ 76, [2002] 2 All ER 1, [2002] ICR 613, [2002] IRLR 263, 68 BMLR 115, (2002) Times, 11 February, [2002] All ER (D) 53 (Feb); revsd sub nom Barber v Somerset County Council [2004] UKHL 13, [2004] 2 All ER 385, [2004] 1 WLR 1089, [2004] ICR 457, [2004] IRLR 475, 77 BMLR 219, (2004) Times, 5 April, 148 Sol Jo LB 419, [2004] All ER (D) 07 (Apr) R2006
Bixby, Case, Fry and Elliott v Ford Motor Group (1990), unreported N3022
Bottomley v Todmorden Cricket Club [2003] EWCA Civ 1575, [2003] 48 LS Gaz R 18, (2003) Times, 13 November, 147 Sol Jo LB 1309, [2003] All ER (D) 102 (Nov) O3014
Boyle v Kodak Ltd [1969] 2 All ER 439, [1969] 1 WLR 661, 113 Sol Jo 382, HL W11049
Bracebridge Engineering Ltd v Darby [1990] IRLR 3, EAT H1706, H1720, H1723
Bradburn v Great Western Rly Co (1874) LR 10 Exch 1, 44 LJ Ex 9, 23 WR 48, [1874–80] All ER Rep 195, 31 LT 464 .... C6053, E13007
Bradley v Eagle Star Insurance Co Ltd [1989] AC 957, [1989] 1 All ER 961, [1989] 2 WLR 568, [1989] 1 Lloyd's Rep 465, [1989] ICR 301, [1989] BCLC 469, 133 Sol Jo 359, [1989] 17 LS Gaz R 38, [1989] NLJR 330, HL E13024

Brice v Brown [1984] 1 All ER 997, 134 NLJ 204 .... C6033
Brioland Ltd v Searson [2005] EWCA Civ 55, sub nom Searson v Brioland Ltd [2005]
   05 EG 202 (CS), [2005] All ER (D) 197 (Jan) O3002
British Airways Ltd v Moore [2000] IRLR 296 .... E14016
British Home Stores Ltd v Burchell [1980] ICR 303n, [1978] IRLR 379, 13 ITR
   560, EAT E14013
British Railways Board v Herrington [1972] AC 877, [1972] 1 All ER 749, [1972] 2
   WLR 537, 116 Sol Jo 178, 223 Estates Gazette 939, HL F5073, O3008
British Transport Commission v Gourley [1956] AC 185, [1955] 3 All ER 796, [1956] 2
   WLR 41, [1955] 2 Lloyd's Rep 475, 34 ATC 305, 49 R & IT 11, [1955] TR 303, 100
   Sol Jo 12, HL C6022, C6026
Brown v Controlled Packaging Services Ltd (1999), unreported W10010
Brumfitt v Ministry of Defence [2005] IRLR 4, 148 Sol Jo LB 1028, [2004] All ER (D)
   479 (Jul), EAT H1707
Bunker v Charles Brand & Son Ltd [1969] 2 QB 480, [1969] 2 All ER 59, [1969] 2
   WLR 1392, 113 Sol Jo 487 .... O3005
Burnett v British Waterways Board [1973] 2 All ER 631, [1973] 1 WLR 700, [1973] 2
   Lloyd's Rep 137, 117 Sol Jo 203, CA O3017
Burton v De Vere Hotels Ltd [1997] ICR 1, [1996] IRLR 596, EAT H1706
Bux v Slough Metals Ltd [1974] 1 All ER 262, [1973] 1 WLR 1358, [1974] 1
   Lloyd's Rep 155, 117 Sol Jo 615, CA P3001, P3026
Byrne Brothers (Formwork) Ltd v Baird [2002] ICR 667, [2002] IRLR 96, [2001] All ER
   (D) 321 (Nov), EAT W10004

C

Calder v Secretary of State for Work and Pensions (UKEAT/0512/08/LA) [2009] All ER
   (D) 106 (Aug) J3011
Campbell v Mylchreest [1999] PIQR Q 17, CA C6041
Campion v Hughes [1975] IRLR 291, Ind Trib E15024
Canterbury City Council v Howletts & Port Lympne Estates Ltd. See Langridge v
   Howletts & Port Lympne Estates Ltd
Capitano v Leeds Eastern Health Authority (1989) Current Law, October O3006
Cappoci v Bloomsbury Health Authority (21 January 2000, unreported) C6023
Carragher v Singer Manufacturing Ltd 1974 SLT (Notes) 28, Ct of Sess N3022
Castellain v Preston (1883) 11 QBD 380, 52 LJQB 366, 31 WR 557, [1881–5] All ER
   Rep 493, 49 LT 29, CA E13007
Caulfield v Marshalls Clay Products Ltd [2004] EWCA Civ 422, [2004] 2 CMLR 1040,
   [2004] ICR 1502, [2004] IRLR 564, 148 Sol Jo LB 539, [2004] All ER (D) 292
   (Apr) W10024
Century Insurance Co Ltd v Northern Ireland Road Tranport Board [1942] AC 509,
   [1942] 1 All ER 491, 111 LJPC 138, 167 LT 404, HL A1020
Chadwick v British Transport Commission (or British Railways Board) [1967] 2 All ER
   945, [1967] 1 WLR 912, 111 Sol Jo 562 .... C6033
Chessington World of Adventures Ltd v Reed. See A v B, ex p News Group
   Newspapers Ltd
Chief Adjudication Officer v Faulds [2000] 2 All ER 961, [2000] 1 WLR 1035, [2000]
   ICR 1297, (2000) Times, 16 May, 2000 SC (HL) 116, 2000 SLT 712, HL C6005
Chief Adjudication Officer v Rhodes [1999] ICR 178, [1999] IRLR 103, [1998]
   36 LS Gaz R 32, 142 Sol Jo LB 228, CA C6006
Chief Constable of Lincolnshire Police v Stubbs [1999] ICR 547, [1999] IRLR
   81, EAT H1706, H1712
Chrysler (UK) Ltd v McCarthy [1978] ICR 939 .... E15016
Clare v Perry (t/a Widemouth Manor Hotel) [2005] EWCA Civ 39, [2005] All ER (D) 67
   (Jan) O3002
Clark v Hosier & Dickinson Ltd (in liq) [2003] EWCA Civ 1467, [2003] All ER (D) 225
   (Oct) O3015
Clark v Nomura International plc [2000] IRLR 766 .... M1513
Coates v Jaguar Cars Ltd [2004] EWCA Civ 337, [2004] All ER (D) 87 (Mar) O3018

Coleman v Attridge Law: C-303/06 [2008] All ER (EC) 1105, [2008] 3 CMLR 777, [2008] ICR 1128, [2008] IRLR 722, (2008) Times, 29 July, [2008] All ER (D) 245 (Jul), ECJ M1508

Coleman v EBR Attridge Law LLP [2010] 1 CMLR 846, [2010] ICR 242, [2010] IRLR 10, (2009) Times, 5 November, 153 Sol Jo (no 42) 28, [2009] All ER (D) 14 (Nov), EAT H1703

College of Ripon and York St John v Hobbs [2002] IRLR 185, [2001] All ER (D) 259 (Nov), EAT E16510

Collins v First Quench Retailing Ltd [2003] SLT 1220, 2003 SCLR 205, OH V8015

Collins v Wilcock [1984] 3 All ER 374, [1984] 1 WLR 1172, 79 Cr App Rep 229, 148 JP 692, [1984] Crim LR 481, 128 Sol Jo 660, [1984] LS Gaz R 2140, DC . H1723, H1730

Condogianis v Guardian Assurance Co [1921] 2 AC 125, 3 Ll L Rep 40, 90 LJPC 168, 125 LT 610, 37 TLR 685, PC E13005

Conn v Sunderland City Council [2007] EWCA Civ 1492, [2008] IRLR 324, [2007] All ER (D) 99 (Nov) H1726

Cook v Bradford Community Health NHS Trust [2002] EWCA Civ 1616, [2002] All ER (D) 329 (Oct), CA V8015

Cotswold Developments Construction Ltd v Williams [2006] IRLR 181, [2005] All ER (D) 355 (Dec), EAT W10004

Cotton v Derbyshire Dales District Council (1994) Times, 20 June, CA O3017

Courtaulds Northern Spinning Ltd v Sibson [1988] ICR 451, [1988] IRLR 305, 132 Sol Jo 1033, CA E14004

Coventry City Council v Ackerman Group plc [1995] Crim LR 140, DC P9008

Cox v H C B Angus Ltd [1981] ICR 683 .... A1021

Cox v Hockenhull [1999] 3 All ER 577, [2000] 1 WLR 750, [1999] RTR 399, CA C6037

Coxall v Goodyear Great Britain Ltd [2002] EWCA Civ 1010, [2003] 1 WLR 536, [2003] ICR 152, [2002] IRLR 742, (2002) Times, 5 August, [2002] All ER (D) 303 (Jul) R2006

Crawley v Mercer (1984) Times, 9 March C6052

Cresswell v Eaton [1991] 1 All ER 484, [1991] 1 WLR 1113 .... C6039

Crouch v British Rail Engineering Ltd [1988] IRLR 404, CA P3001

Cruickshank v VAW Motorcast [2002] ICR 729, [2002] IRLR 24, [2001] All ER (D) 372 (Oct), EAT E16510

Cunningham v Harrison [1973] QB 942, [1973] 3 All ER 463, [1973] 3 WLR 97, 117 Sol Jo 547, CA C6053

D

Da'Bell v National Society for Prevention of Cruelty to Children [2010] IRLR 19, [2009] All ER (D) 219 (Nov), EAT H1715

Dacas v Brook Street Bureau (UK) Ltd [2004] EWCA Civ 217, [2004] ICR 1437, [2004] IRLR 358, (2004) Times, 19 March, [2004] All ER (D) 125 (Mar) E13016

Darby v National Trust [2001] EWCA Civ 189, [2001] All ER (D) 216 (Jan) O3017

Dashiell v Luttitt [2000] 3 QR 4 .... C6023

Davie v New Merton Board Mills Ltd [1959] AC 604, [1959] 1 All ER 346, [1959] 2 WLR 331, 103 Sol Jo 177, HL P9025

Daw v Intel Corpn (UK) Ltd [2007] EWCA Civ 70, [2007] 2 All ER 126, [2007] ICR 1318, [2007] NLJR 259, [2007] All ER (D) 96 (Feb), sub nom Intel Corpn (UK) Ltd v Daw [2007] IRLR 355 .... M1524

Day v T Pickles Farms Ltd [1999] IRLR 217, EAT E14019, E14025

Deary v Mansion Hide Upholstery Ltd [1983] ICR 610, [1983] IRLR 195, 147 JP 311 .... E15021

Deeley v Effer (1977) COIT No 1/72, Case No 25354/77, unreported E15023

Dhaliwal v Richmond Pharmacology [2009] ICR 724, [2009] IRLR 336, [2009] All ER (D) 158 (Feb), EAT H1702, H1706, H1708

Dhanjal, SS v British Steel plc (Case 50740/91) (1994), unreported C8018

Dickins v O2 plc [2008] EWCA Civ 1144, [2009] IRLR 58, [2008] All ER (D) 154 (Oct) M1524

Dietrich v Westdeutscher Rundfunk: C-11/99 [2000] ECR I-5589, ECJ D8202

Doleman v Deakin (1990) Times, 30 January, CA C6035
Donaldson v Hays Distribution Services Ltd [2005] CSIH 48, 2005 SLT 733, 2005 SCLR 717 .... O3018
Donoghue (or McAlister) v Stevenson. See M'Alister (or Donoghue) v Stevenson
Dornan v Department of the Environment for Northern Ireland [1993] NI 1, NI CA O3010
Driskel v Peninsula Business Services Ltd [2000] IRLR 151, EAT H1707
Dunham v Ashford Windows [2005] IRLR 608, [2005] ICR 1584, [2005] All ER (D) 104 (Jun), EAT E16510
Dunn v British Coal Corpn [1993] ICR 591, [1993] IRLR 396, [1993] 15 LS Gaz R 37, [1993] PIQR P 275, 137 Sol Jo LB 81, CA C6023
Dunn v Ovalcode Ltd (t/a UKR) [2003] All ER (D) 241 (Apr), EAT J3030

E

EC Commission v United Kingdom: C-484/04 [2006] ECR I-7471, [2006] 3 CMLR 1322, [2007] ICR 592, [2006] IRLR 888, (2006) Times, 21 September, [2006] All ER (D) 32 (Sep), ECJ W10014
Edwards v National Coal Board [1949] 1 KB 704, [1949] 1 All ER 743, 93 Sol Jo 337, 65 TLR 430, CA E15039, R3003, R3004
English v Thomas Sanderson Blinds Ltd [2008] ICR 607, [2008] IRLR 342, [2008] All ER (D) 282 (Feb), EAT; revsd [2008] EWCA Civ 1421, [2009] 2 All ER 468, [2009] 2 CMLR 437, [2009] ICR 543, [2009] IRLR 206, (2009) Times, 5 January, 153 Sol Jo (no 1) 31, [2008] All ER (D) 219 (Dec) H1707
Equal Opportunities Commission v Secretary of State for Trade and Industry [2007] EWHC 483 (Admin), [2007] 2 CMLR 1351, [2007] ICR 1234, [2007] IRLR 327, [2007] All ER (D) 183 (Mar) H1702, H1706
Evans v Kosmar Villa Holidays plc [2007] EWCA Civ 1003, [2008] 1 All ER 530, [2008] 1 All ER (Comm) 721, [2008] 1 WLR 297, [2008] PIQR P126, [2007] All ER (D) 330 (Oct) O3017
Evans v Malley Organisation Ltd (t/a First Business Support) [2002] EWCA Civ 1834, [2003] ICR 432, [2003] IRLR 156, (2003) Times, 23 January, [2002] All ER (D) 397 (Nov) W10024

F

Fairchild v Glenhaven Funeral Services Ltd (1978) Ltd [2002] UKHL 22, [2003] 1 AC 32, [2002] 3 All ER 305, [2002] ICR 798, [2002] IRLR 533, 67 BMLR 90, [2002] NLJR 998, (2002) Times, 21 June, [2003] 1 LRC 674, [2002] All ER (D) 139 (Jun) P9039
Fairhurst v St Helens and Knowsley Health Authority [1994] 5 Med LR 422, [1995] PIQR Q 1 .... C6028
Farrant v Thanet Disctrict Council (11 June 1996, unreported) O3006
Faulkner v Chief Adjudication Officer [1994] PIQR P 244, CA C6006
Fearon v Chief Constable of Derbyshire, UKEAT/0445/02/RN, [2004] All ER (D) 101 (Jan), EAT H1705
Fennelly v Connex South Eastern Ltd [2001] IRLR 390, [2000] All ER (D) 2233, CA V8013
Fildes v International Computers (1984), unreported A1001
Fish v British Tissues (1994), unreported C6058
Fisher v CHT Ltd (No 2) [1966] 2 QB 475, [1966] 1 All ER 88, [1966] 2 WLR 391, 109 Sol Jo 933, CA O3004
Flanagan v Watts Bearne & Co plc [1992] PIQR P 144, CA C6052
Fletcher v Ministry of Defence [2010] IRLR 25, [2009] All ER (D) 187 (Nov), EAT H1706

Fletcher v Rylands and Horrocks (1865) 3 H & C 774, 34 LJ Ex 177; revsd (1866) LR
    1 Exch 265, 30 JP 436, 4 H & C 263, 35 LJ Ex 154, 12 Jur NS 603, 14 LT 523; affd
    sub nom Rylands v Fletcher (1868) LR 3 HL 330, 33 JP 70, 37 LJ Ex 161, 14 WR
    799, [1861–73] All ER Rep 1, 19 LT 220, HL G1001
Fraser v State Hospitals Board for Scotland (2000) Times, 12 September, 2001 SLT
    1051, 2001 SCLR 357, Ct of Sess E14003
Frost v Aylesbury Dairy Co [1905] 1 KB 608, 74 LJKB 386, 53 WR 354,
    [1904–7] All ER Rep 132, 49 Sol Jo 312, 92 LT 527, 21 TLR 300, CA P9047, P9054
Frost v Chief Constable of South Yorkshire Police. See White v Chief Constable of South
    Yorkshire Police

G

Gallagher v Alpha Catering Services Ltd (t/a Alpha Flight Services) [2004] EWCA Civ
    1559, [2005] ICR 673, [2005] IRLR 102, [2004] All ER (D) 121 (Nov) W10017
Gallagher v The Drum Engineering Co Ltd, COIT 1330/89, (1989), unreported J3011
Geddling v Marsh [1920] 1 KB 668, 89 LJKB 526, [1920] All ER Rep 631, 122 LT 775,
    36 TLR 337 .... P9053
General Cleaning Contractors Ltd v Christmas [1953] AC 180, [1952] 2 All ER 1110,
    [1953] 2 WLR 6, 51 LGR 109, 97 Sol Jo 7, HL O3014
Gentle v Perkins Engines Peterborough Ltd (formerly Perkins Group Ltd) [2001] All ER
    (D) 360 (Jul), EAT E14022
Gerrard v Staffordshire Potteries Ltd [1995] ICR 502, [1995] PIQR P 169, CA P3002
Gibson v Skibs AS Marina (or Marena) and Orkla Grobe (or Grube) A/B and
    Smith Coggins Ltd [1966] 2 All ER 476, [1966] 2 Lloyd's Rep 39 .... T7002
Gitsham v C H Pearce & Sons plc [1992] PIQR P57, CA O3018
Gómez v Continental Industrias del Caucho SA: C-342/01 [2004] ECR I-2605,
    [2004] 2 CMLR 38, [2005] ICR 1040, [2004] IRLR 407, [2004] All ER (D) 350
    (Mar), ECJ W10024
Goodchild v Vaclight [1965] CLY 2669 .... P9038
Goodwin v Bennetts UK Ltd [2008] EWCA Civ 1374, [2008] All ER (D) 220 (Dec)
    .... D8209
Goodwin v Patent Office [1999] ICR 302, [1999] IRLR 4, EAT E16510
Goold (W A) (Pearmak) Ltd v McConnell [1995] IRLR 516, EAT E14008
Gough v Thorne [1966] 3 All ER 398, [1966] 1 WLR 1387, 110 Sol Jo 529, CA ... O3011
Graney v Liverpool County Council (1995), unreported O3002
Grant v Australian Knitting Mills Ltd [1936] AC 85, 105 LJPC 6, [1935] All ER Rep
    209, 79 Sol Jo 815, 154 LT 18, 52 TLR 38, PC P9053
Greaves & Co (Contractors) Ltd v Baynham, Meikle & Partners [1975] 3 All ER 99,
    [1975] 1 WLR 1095, [1975] 2 Lloyd's Rep 325, 119 Sol Jo 372, 4 BLR 56, CA .. W11010
Greene v Chelsea Borough Council [1954] 2 QB 127, [1954] 2 All ER 318, [1954] 3
    WLR 12, 52 LGR 352, 118 JP 346, 98 Sol Jo 389, CA O3004
Griffiths v Peter Conway Ltd [1939] 1 All ER 685, CA P9053
Grovehurst Energy Ltd v Strawson, HM Inspector, COIT No 5035/90, (1990),
    unreported E15024
Gunion v Roche Products Ltd (1994) Times, 4 November A1001
Gwilliam v West Hertfordshire Hospital NHS Trust [2002] EWCA Civ 1041, [2003] QB
    443, [2003] 3 WLR 1425, [2003] PIQR P99, (2002) Times, 7 August, [2002] All ER
    (D) 345 (Jul) O3014

H

Hadley v Hancox (1986) 85 LGR 402, 151 JP 227, DC E15011
Hairsine v Kingston upon Hull City Council [1992] ICR 212, [1992] IRLR 211, EAT
    .... J3011

Hampstead Heath Winter Swimming Club v Corpn of London [2005] EWHC 713 (Admin), [2005] LGR 481, [2005] NLJ 827, (2005) Times, 19 May, [2005] All ER (D) 353 (Apr), sub nom R (on the application of Hampstead Heath Winter Swimming Club) v Corpn of London [2005] 1 WLR 2930 .... O3019

Harris v Evans [1998] 3 All ER 522, [1998] 1 WLR 1285, [1998] NLJR 745, [1998] All ER (D) 148, CA E15028

Harrison (TC) (Newcastle-under-Lyme) Ltd v Ramsey [1976] IRLR 135, Ind Trib .. E15024

Hartley v British Railways Board (1981) 125 Sol Jo 169, (1981) Times, 2 February, CA O3014

Hartley v Sandholme Iron Co Ltd [1975] QB 600, [1974] 3 All ER 475, [1974] 3 WLR 445, [1974] STC 434, 17 KIR 205, 118 Sol Jo 702 .... C6052

Hartley (R S) Ltd v Provincial Insurance Co Ltd [1957] 1 Lloyd's Rep 121 .... E13020

Hartman v South Essex Mental Health and Community Care NHS Trust [2005] EWCA Civ 6, [2005] ICR 782, [2005] IRLR 293, 85 BMLR 136, (2005) Times, 21 January, [2005] All ER (D) 141 (Jan) S11009

Harvest Press Ltd v McCaffrey [1999] IRLR 778, EAT E14009

Hatton v Sutherland [2002] EWCA Civ 76, [2002] 2 All ER 1, [2002] ICR 613, [2002] IRLR 263, 68 BMLR 115, (2002) Times, 11 February, [2002] All ER (D) 53 (Feb); revsd sub nom Barber v Somerset County Council [2004] UKHL 13, [2004] 2 All ER 385, [2004] 1 WLR 1089, [2004] ICR 457, [2004] IRLR 475, 77 BMLR 219, (2004) Times, 5 April, 148 Sol Jo LB 419, [2004] All ER (D) 07 (Apr) R2006, S11009

Hay (or Bourhill) v Young [1943] AC 92, [1942] 2 All ER 396, 111 LJPC 97, 86 Sol Jo 349, 167 LT 261, HL C6033

Health and Safety Executive v George Tancocks Garage (Exeter) Ltd [1993] Crim LR 605, DC E15014

Healy v Corpn of London (24 June 1999, unreported) E14004

Heil v Rankin [2001] QB 272, [2000] 3 All ER 138, [2000] 2 WLR 1173, [2000] Lloyd's Rep Med 203, [2000] IRLR 334, [2000] NLJR 464, [2000] PIQR Q 187, 144 Sol Jo LB 157, CA C6032

Herring v Co-operative Insurance Society Ltd [2006] All ER (D) 171 (Oct), EAT .. W10024

Heslop v Metalock (Great Britain) Ltd (1981) Observer, 29 November N3022

Hewett v Alf Brown's Transport Ltd [1992] ICR 530, [1992] 15 LS Gaz R 33, CA P3001

Hewson v Downs [1970] 1 QB 73, [1969] 3 All ER 193, [1969] 2 WLR 1169, 6 KIR 343, 113 Sol Jo 309 .... C6053

Hinz v Berry [1970] 2 QB 40, [1970] 1 All ER 1074, [1970] 2 WLR 684, 114 Sol Jo 111, CA C6033

Hone v Six Continents Retail Ltd [2005] EWCA Civ 922, [2006] IRLR 49 .... S11009, W10010

Hood v Mitie Property Services (Midlands) Ltd [2005] All ER (D) 11 (Jul) O3014

Horkulak v Cantor Fitzgerald International [2003] EWHC 1918 (QB), [2004] ICR 697, [2003] IRLR 756, [2003] All ER (D) 542 (Jul); revsd in part [2004] EWCA Civ 1287, [2005] ICR 402, [2004] IRLR 942, 148 Sol Jo LB 1218, [2004] All ER (D) 170 (Oct) H1707, H1720

Horne-Roberts v SmithKline Beecham plc [2001] EWCA Civ 2006, [2002] 1 WLR 1662, 65 BMLR 79, (2002) Times, 10 January, 146 Sol Jo LB 19, [2001] All ER (D) 269 (Dec) P9044

Houweling v Wesseler [1963] 40 DLR (2d) 956 .... O3005

Howard and Peet v Volex plc (HSIB 181) J3011

Hunt v Severs [1994] 2 AC 350, [1994] 2 All ER 385, [1994] 2 WLR 602, [1994] 2 Lloyd's Rep 129, [1994] 32 LS Gaz R 41, [1994] NLJR 603, 138 Sol Jo LB 104, HL C6028

Hunter v British Coal Corpn [1999] QB 140, [1998] 2 All ER 97, [1998] 3 WLR 685, [1999] ICR 72, [1998] 12 LS Gaz R 27, 142 Sol Jo LB 85, CA C6033

Hussain v New Taplow Paper Mills Ltd [1988] AC 514, [1988] 1 All ER 541, [1988] 2 WLR 266, [1988] ICR 259, [1988] IRLR 167, 132 Sol Jo 226, [1988] 10 LS Gaz R 45, [1988] NLJR 45, HL C6052

I

ICI Ltd v Shatwell [1965] AC 656, [1964] 2 All ER 999, [1964] 3 WLR 329, 108 Sol Jo
578, HL W11049
IPC Magazines Ltd v Clements (EAT/456/99) E14022
Indermaur v Dames (1866) LR 1 CP 274, 35 LJCP 184, Har & Ruth 243, 12 Jur NS
432, 14 WR 586, [1861–73] All ER Rep 15, 14 LT 484; affd (1867) LR 2 CP 311, 31
JP 390, 36 LJCP 181, 15 WR 434, 16 LT 293, Ex Ch O3014
IRC v Ainsworth. See Ainsworth v IRC
Insitu Cleaning Co Ltd v Heads [1995] IRLR 4, EAT H1706
Intel Corpn (UK) Ltd v Daw. See Daw v Intel Corpn (UK) Ltd
Interfoto Picture Library Ltd v Stiletto Visual Programmes Ltd [1989] QB 433, [1988]
1 All ER 348, [1988] 2 WLR 615, [1988] BTLC 39, 132 Sol Jo 460, [1988] 9 LS Gaz
R 45, [1987] NLJ Rep 1159, CA O3016
Ionic Bay, The. See Mills v Hassall
Irvine v Metropolitan Police Comr [2004] EWHC 1536 (QB), [2004] All ER (D) 79
(May) O3018

J

James v London Borough of Greenwich [2008] EWCA Civ 35, [2008] ICR 545, [2008]
IRLR 302, [2008] All ER (D) 54 (Feb) E13016
James v Redcats (Brands) Ltd [2007] ICR 1006, [2007] IRLR 296, [2007] All ER (D)
270 (Feb), EAT W10004
Jayes v IMI (Kynoch) Ltd [1985] ICR 155, [1984] LS Gaz R 3180, CA O3011
Joel v Law Union and Crown Insurance Co [1908] 2 KB 863, 77 LJKB 1108, 52 Sol Jo
740, 99 LT 712, 24 TLR 898, CA E13004
Johnson v Coventry Churchill International Ltd [1992] 3 All ER 14 .... E13010
Johnson v Unisys Ltd [2001] UKHL 13, [2003] 1 AC 518, [2001] 2 All ER 801, [2001]
2 WLR 1076, [2001] ICR 480, [2001] IRLR 279, [2001] All ER (D) 274 (Mar) H1728
Johnstone v Bloomsbury Health Authority [1992] QB 333, [1991] 2 All ER 293, [1991]
2 WLR 1362, [1991] ICR 269, [1991] IRLR 118, [1991] 2 Med LR 38, CA E14002,
E14003
Jolley v Sutton London BC [2000] 3 All ER 409, [2000] 1 WLR 1082, [2000] LGR 399,
[2000] 2 Lloyd's Rep 65, [2000] 2 FCR 392, [2000] All ER (D) 689, HL O3010
Jones v Sainsbury's Supermarkets Ltd (2000), unreported E14013
Jones v Sandwell Metropolitan Borough Council [2002] EWCA Civ 76, [2002] 2 All ER
1, [2002] ICR 613, [2002] IRLR 263, 68 BMLR 115, (2002) Times, 11 February,
[2002] All ER (D) 53 (Feb); revsd sub nom Barber v Somerset County Council [2004]
UKHL 13, [2004] 2 All ER 385, [2004] 1 WLR 1089, [2004] ICR 457, [2004] IRLR
475, 77 BMLR 219, (2004) Times, 5 April, 148 Sol Jo LB 419, [2004] All ER (D) 07
(Apr) R2006
Jones v Tower Boot Co Ltd [1995] IRLR 529, EAT; revsd [1997] 2 All ER 406, [1997]
ICR 254, [1997] IRLR 168, [1997] NLJR 60, CA H1706, H1711, H1712, H1723
Jordan v Achara (1988) 20 HLR 607, CA O3004

K

Kendall (Henry) & Sons (a firm) v William Lillico & Sons Ltd [1969] 2 AC 31, [1968]
2 All ER 444, [1968] 3 WLR 110, [1968] 1 Lloyd's Rep 547, 112 Sol Jo 562, HL
.... P9054
Keown v Coventry Healthcare NHS Trust [2006] EWCA Civ 39, [2006] 1 WLR 953,
[2006] 08 LS Gaz R 23, 150 Sol Jo LB 164, [2006] All ER (D) 27 (Feb) O3009
Kerr v Nathan's Wastesavers Ltd (1995) IDS Brief 548 .... E14009
Kigass Aero Components Ltd v Brown [2002] ICR 697, [2002] IRLR 312,
[2002] All ER (D) 341 (Feb), EAT W10024
King v Smith [1995] ICR 339, CA O3014
King v Sussex Ambulance NHS Trust [2002] EWCA Civ 953, [2002] ICR
1413, 68 BMLR 177, (2002) Times, 25 July, [2002] All ER (D) 95 (Jul) M3002

Knott v Newham Healthcare NHS Trust [2002] EWHC 2091 (QB), [2002] All ER (D)
216 (Oct); affd [2003] All ER (D) 164 (May), CA M3002
Knowles v Liverpool City Council [1993] 4 All ER 321, [1993] 1 WLR 1428, 91 LGR
629, [1994] 1 Lloyd's Rep 11, [1994] ICR 243, [1993] IRLR 588, [1993] NLJR
1479n, [1994] PIQR P 8, HL M1005

L

Lambert v Lewis. See Lexmead v Lewis
Landeshauptstadt Kiel v Jaeger: C-151/02 [2003] ECR I-8389, [2004] All ER (EC) 604,
[2003] 3 CMLR 493, [2004] ICR 1528, [2003] IRLR 804, 75 BMLR 201, (2003)
Times, 26 September, [2003] All ER (D) 72 (Sep), ECJ W10003
Langridge v Howletts & Port Lympne Estates Ltd (1996) 95 LGR 798, sub nom
Canterbury City Council v Howletts & Port Lympne Estates Ltd [1997] ICR 925,
[1997] JPIL 51 .... E15022
Larby v Thurgood [1993] ICR 66, [1992] 39 LS Gaz R 35, [1993] PIQR P 218, 136 Sol
Jo LB 275, HC C6026
Lask v Gloucester Area Health Authority (1995) Times, 13 December, CA J3015
Latham v R Johnson and Nephew Ltd [1913] 1 KB 398, 77 JP 137, 82 LJKB 258,
[1911–13] All ER Rep 117, 57 Sol Jo 127, 108 LT 4, 29 TLR 124, CA O3010
Lau v DPP [2000] All ER (D) 224 , DC H1726
Law Hospital NHS Trust v Rush [2001] IRLR 611, Ct of Sess E16510
Lea v British Aerospace plc [1991] 1 AC 362, [1991] 1 All ER 193, [1991] 2 WLR 27,
135 Sol Jo 16, HL C6043
Leonard v Southern Derbyshire Chamber of Commerce [2001] IRLR 19, [2000] All ER
(D) 1327, EAT E16510
Lewis v Avidan Ltd [2005] EWCA Civ 670, 149 Sol Jo LB 479, [2005] All ER (D) 136
(Apr) O3018, W11010
Lexmead (Basingstoke) Ltd v Lewis [1982] AC 225, sub nom Lambert v Lewis [1980]
1 All ER 978, [1980] 2 WLR 299, [1980] RTR 152, [1980] 1 Lloyd's Rep 311, 124
Sol Jo 50, CA; revsd sub nom Lexmead (Basingstoke) Ltd v Lewis [1982] AC 225,
268, [1981] 2 WLR 713, [1981] RTR 346, [1981] 2 Lloyd's Rep 17, 125 Sol Jo 310,
sub nom Lambert v Lewis [1981] 1 All ER 1185, HL P9055
Lidl Italia Srl v Comune di Arcole (VR): C-315/05 [2006] ECR I-11181, [2006] All ER
(D) 331 (Nov), ECJ F9023
Liffen v Watson [1940] 1 KB 556, [1940] 2 All ER 213, 109 LJKB 367, 84 Sol Jo 368,
162 LT 398, 56 TLR 442, CA C6053
Lim Poh Choo v Camden and Islington Area Health Authority [1980] AC 174, [1979]
2 All ER 910, [1979] 3 WLR 44, 123 Sol Jo 457, HL C6027, C6052
Lindop v Goodwin Steel Castings Ltd (1990) Times, 19 June C6043
List Design Group Ltd v Catley [2002] ICR 686, [2003] IRLR 14, [2002] All ER (D) 215
(Mar), EAT W10024
Lister v Romford Ice and Cold Storage Co Ltd [1957] AC 555, [1957] 1 All ER 125,
[1957] 2 WLR 158, [1956] 2 Lloyd's Rep 505, 121 JP 98, 101 Sol Jo 106, HL E13027
Livingstone v Rawyards Coal Co (1880) 5 App Cas 25, 44 JP 392, 28 WR 357, 42 LT
334, HL H1725
Longden v British Coal Corpn [1998] AC 653, [1998] 1 All ER 289, [1997] 3 WLR
1336, [1998] ICR 26, [1998] IRLR 29, [1998] 01 LS Gaz R 25, [1997] NLJR 1774,
[1998] PIQR Q 11, 142 Sol Jo LB 28, HL C6053
Lowles v Home Office [2004] EWCA Civ 985, [2004] All ER (D) 538 (Jul) W11010
Lyddon v Englefield Brickwork Ltd [2008] IRLR 198, [2007] All ER (D) 198
(Nov), EAT W10024

M

M'Alister (or Donoghue) v Stevenson [1932] AC 562, 101 LJPC 119, 37 Com Cas 350, 48 TLR 494, 1932 SC (HL) 31, sub nom Donoghue (or McAlister) v Stevenson [1932] All ER Rep 1, 1932 SLT 317, sub nom McAlister (or Donoghue) v Stevenson 76 Sol Jo 396, 147 LT 281, HL H1724, O3014, P9024, P9037

McArdle v Andmac Roofing Co [1967] 1 All ER 583, [1967] 1 WLR 356, 111 Sol Jo 37, CA W9001

McCafferty v Metropolitan Police District Receiver [1977] 2 All ER 756, [1977] 1 WLR 1073, [1977] ICR 799, 121 Sol Jo 678, CA N3022

McCamley v Cammell Laird Shipbuilders Ltd [1990] 1 All ER 854, [1990] 1 WLR 963, CA C6053

MacCartney v Oversley House Management [2006] ICR 510, [2006] IRLR 514, [2006] All ER (D) 246 (Jan), EAT W10003

McDermid v Nash Dredging and Reclamation Co Ltd [1987] AC 906, [1987] 2 All ER 878, [1987] 3 WLR 212, [1987] 2 Lloyd's Rep 201, [1987] ICR 917, [1987] IRLR 334, 131 Sol Jo 973, [1987] LS Gaz R 2458, HL O7029

Macdonald v Advocate General for Scotland [2003] UKHL 34, [2004] 1 All ER 339, [2003] ICR 937, [2003] IRLR 512, [2003] ELR 655, [2003] 29 LS Gaz R 36, (2003) Times, 20 June, 2003 SLT 1158, 2003 SCLR 814, 147 Sol Jo LB 782, [2004] 2 LRC 111, [2003] All ER (D) 259 (Jun) H1702, H1706

McEwan v Lothian Buses plc [2006] CSOH 56, 2006 SCLR 592, OH O3011

McFarlane v EE Caledonia Ltd [1994] 2 All ER 1, [1994] 1 Lloyd's Rep 16, [1993] NLJR 1367, [1994] PIQR P 154, CA C6033

McGinlay (or Titchener) v British Railways Board. See Titchener v British Railways Board

McGivney v Golderslea Ltd (1997), unreported O3003

McGuinness v Kirkstall Forge Engineering Ltd (1979), unreported N3022

McIlgrew v Devon County Council [1995] PIQR Q 66, CA C6027

McKay v Unwin Pyrotechnics Ltd [1991] Crim LR 547, DC P9012

McLoughlin v O'Brian [1983] 1 AC 410, [1982] 2 All ER 298, [1982] 2 WLR 982, [1982] RTR 209, 126 Sol Jo 347, [1982] LS Gaz R 922, HL C6033

McNulty v Marshalls Food Group Ltd 1999 SC 195, OH C6027

Mailer v Austin Rover Group plc. See Austin Rover Group Ltd v HM Inspector of Factories

Majrowski v Guy's and St Thomas's NHS Trust [2005] EWCA Civ 251, [2005] QB 848, [2005] 2 WLR 1503, [2005] ICR 977, [2005] IRLR 340, (2005) Times, March 21, 149 Sol Jo LB 358, [2005] All ER (D) 273 (Mar); affd sub nom Majrowski v Guy's and St Thomas' NHS Trust [2006] UKHL 34, [2007] 1 AC 224, [2006] 4 All ER 395, [2006] 3 WLR 125, [2006] ICR 1199, [2006] IRLR 695, 91 BMLR 85, [2006] NLJR 1173, (2006) Times, 13 July, 150 Sol Jo LB 986, [2006] All ER (D) 146 (Jul) H1726

Makepeace v Evans Brothers (Reading) (a firm) [2001] ICR 241, [2000] BLR 737, (2000) Times, 13 June, [2000] All ER (D) 720, CA O3019

Malcolm v Broadhurst [1970] 3 All ER 508 .... H1724

Malik v BCCI SA (in liq) [1998] AC 20, [1997] 3 All ER 1, [1997] 3 WLR 95, [1997] ICR 606, [1997] IRLR 462, [1997] 94 LS Gaz R 33, [1997] NLJR 917, HL H1728

Marinello v City of Edinburgh Council [2010] CSOH 17, 2010 SLT 349, OH H1726

Masiak v City Restaurants (UK) Ltd [1999] IRLR 780, EAT E14009

Meigh v Wickenden [1942] 2 KB 160, [1942] 2 All ER 68, 40 LGR 191, 106 JP 207, 112 LJKB 76, 86 Sol Jo 218, 167 LT 135, 58 TLR 260, DC W11047

Miah v La Gondola Ltd (1999), unreported W10022

Mills v Hassall [1983] ICR 330 .... C6053

Moloney v Lambeth London Borough Council (1966) 64 LGR 440, 110 Sol Jo 406, 198 Estates Gazette 895 .... O3003, O3004

Moonsar v Fiveways Express Transport Ltd [2005] IRLR 9, [2004] All ER (D) 110 (Nov), EAT H1706

Moorcock, The (1889) 14 PD 64, 58 LJP 73, 6 Asp MLC 373, 37 WR 439, [1886–90] All ER Rep 530, 60 LT 654, 5 TLR 316, CA E14004

Morris v Ford Motor Co Ltd [1973] QB 792, [1973] 2 All ER 1084, [1973] 2 WLR 843, [1973] 2 Lloyd's Rep 27, 117 Sol Jo 393, CA E13027

Moualem v Carlisle City Council (1994) 158 JP 1110, [1995] ELR 22, DC O3019
Muir (William) (Bond 9) Ltd v Lamb [1985] IRLR 95, EAT E14013
Munro v MPB Structures Ltd [2004] 2 CMLR 1032, [2004] ICR 430, [2003] IRLR 350,
(2003) Times, 24 April, 2003 SLT 551, 2003 SCLR 542 .... W10024
Murphy v Bradford Metropolitan Borough Council [1992] 2 All ER 908, [1992] PIQR
P 68, CA O3002

N

Nagarajan v London Regional Transport [2000] 1 AC 501, [1999] 4 All ER 65, [1999]
3 WLR 425, [1999] ICR 877, [1999] IRLR 572, [1999] 31 LS Gaz R 36, 143 Sol Jo
LB 219, HL H1736
National Trust for Places of Historic Interest or Natural Beauty v Haden Young Ltd
(1993) 66 BLR 88, [1993] BLM (October) 1; affd (1994) 72 BLR 1, [1994] BLM
(September) 1, CA F5029
Naylor (t/a Mainstreet) v Payling [2004] EWCA Civ 560, (2004) Times, 2 June, 148 Sol
Jo LB 573, [2004] All ER (D) 83 (May), sub nom Payling v Naylor (t/a Mainstreet)
[2004] 23 LS Gaz R 33 .... O3014
New Southern Railway Ltd v Quinn [2006] ICR 761, [2006] IRLR 266, [2005] All ER
(D) 367 (Nov), EAT E14025
Newsholme Bros v Road Transport and General Insurance Co Ltd [1929] 2 KB 356, 24
Ll L Rep 247, 98 LJKB 751, 34 Com Cas 330, [1929] All ER Rep 442, 73 Sol Jo 465,
141 LT 570, 45 TLR 573, CA E13003
Nico Manufacturing Co Ltd v Hendry [1975] IRLR 225, Ind Trib E15024
Nixon v Ross Coates Solicitors, UKEAT/0108/10ZT (6 August 2010,unreported) ... H1703
Norouzi v Sheffield City Council (2011) UKEAT/0497/10/RN, [2011] All ER (D) 77
(Aug), EAT H1706

O

O'Brien v MGN Ltd [2001] EWCA Civ 1279, (2001) Times, 8 August, [2001] All ER
(D) 01 (Aug) O3016
O'Byrne v Sanofi Pasteur MSD Ltd: C-127/04 [2006] ECR I-1313, [2006] All ER (EC)
674, [2006] 1 WLR 1606, [2006] 2 CMLR 656, 91 BMLR 175, (2006) Times,
15 February, [2006] All ER (D) 117 (Feb), ECJ P9044
Ogwo v Taylor [1988] AC 431, [1987] 3 All ER 961, [1987] 3 WLR 1145, 131 Sol Jo
1628, [1988] 4 LS Gaz R 35, [1987] NLJ Rep 1110, HL O3014
O'Hanlon v Revenue and Customs Comrs [2006] ICR 1579, [2006] IRLR 840,
[2006] All ER (D) 53 (Aug), EAT; affd [2007] EWCA Civ 283, [2007] ICR 1359,
[2007] IRLR 404, (2007) Times, 20 April, [2007] All ER (D) 516 (Mar) M1511
Okerago v May & Baker Ltd (trading as Sanofi-Aventis Pharma) (2010)
UKEAT/0278/09/ZT, [2010] IRLR 394, [2010] All ER (D) 79 (Mar), EAT . H1706, H1714
O'Neill v DSG Retail Ltd [2002] EWCA Civ 1139, [2003] ICR 222, [2002] 40 LS Gaz
R 32, (2002) Times, 9 September, [2002] All ER (D) 500 (Jul) M3002
Osborne v Bill Taylor of Huyton Ltd [1982] ICR 168, [1982] IRLR 17, DC M2031,
S7002
O'Shea v Kimberley-Clark Ltd (1982) Guardian, 8 October N3022

P

PCW Syndicates v PCW Reinsurers [1996] 1 All ER 774, [1996] 1 WLR 1136, [1996] 1
Lloyd's Rep 241, CA E13004
Page v Freight Hire (Tank Haulage) Ltd [1981] 1 All ER 394, [1981] ICR 299, [1981]
IRLR 13, EAT E14023
Page v Sheerness Steel plc [1995] PIQR Q 26, HC C6052

Page v Smith [1996] AC 155, [1995] 2 All ER 736, [1995] 2 WLR 644, [1995] RTR 210, [1995] 2 Lloyd's Rep 95, 28 BMLR 133, [1995] 23 LS Gaz R 33, [1995] NLJR 723, [1995] PIQR P 329, HL C6033

Pakenham-Walsh v Connell Residential [2006] EWCA Civ 90, [2006] 11 LS Gaz R 25, [2006] All ER (D) 275 (Feb) W10010

Palfrey v Greater London Council [1985] ICR 437 .... C6052

Palmer v R [1971] AC 814, [1971] 1 All ER 1077, [1971] 2 WLR 831, 55 Cr App Rep 223, 115 Sol Jo 264, 16 WIR 499, 511, PC O3009

Pape v Cumbria County Council [1992] 3 All ER 211, [1992] ICR 132, [1991] IRLR 463 .... P3026

Parry v Cleaver [1970] AC 1, [1969] 1 All ER 555, [1969] 2 WLR 821, [1969] 1 Lloyd's Rep 183, 113 Sol Jo 147, HL C6053

Patterson v Tees and NE Yorkshire NHS Trust (2007), unreported V8015

Paul Sayers v Smithkline Beecham plc (MMR/MR litigation) [1999] MLC O117 .... P9031

Payling v Naylor (t/a Mainstreet). See Naylor (t/a Mainstreet) v Payling

Pearce v Governing Body of Mayfield School [2003] UKHL 34, [2004] 1 All ER 339, [2003] ICR 937, [2003] IRLR 512, [2003] ELR 655, [2003] 29 LS Gaz R 36, (2003) Times, 20 June, 2003 SLT 1158, 2003 SCLR 814, 147 Sol Jo LB 782, [2004] 2 LRC 111, [2003] All ER (D) 259 (Jun) H1702

Phipps v Rochester Corpn [1955] 1 QB 450, [1955] 1 All ER 129, [1955] 2 WLR 23, 53 LGR 80, 119 JP 92, 99 Sol Jo 45 .... O3010

Pickford v Imperial Chemical Industries plc [1998] 3 All ER 462, [1998] 1 WLR 1189, [1998] ICR 673, [1998] IRLR 435, [1998] 31 LS Gaz R 36, [1998] NLJR 978, 142 Sol Jo LB 198, [1998] All ER (D) 302, HL C6058

Ping v Esselte-Letraset [1992] PIQR P 74 .... C6058

Polkey v A E Dauton (or Dayton) Services Ltd [1988] AC 344, [1987] 3 All ER 974, [1987] 3 WLR 1153, [1988] ICR 142, [1987] IRLR 503, 131 Sol Jo 1624, [1988] 1 LS Gaz R 36, [1987] NLJ Rep 1109, HL E14013, M1517

Pollard v Tesco Stores Ltd [2006] EWCA Civ 393, [2006] All ER (D) 186 (Apr) P9032

Porcelli v Strathclyde Regional Council [1986] ICR 564, sub nom Strathclyde Regional Council v Porcelli [1986] IRLR 134, Ct of Sess .. H1702, H1706, H1707, H1723

Post Office v Footitt [2000] IRLR 243 .... W11026

Potter v Arafa [1995] IRLR 316, sub nom Arafa v Potter [1994] PIQR Q 73, CA ... C6027

Potts (or Riddell) v Reid [1943] AC 1, [1942] 2 All ER 161, 111 LJPC 65, 167 LT 301, 58 TLR 335, HL W11049

Pratley v Surrey County Council [2003] EWCA Civ 1067, [2004] ICR 159, [2003] IRLR 794, [2003] All ER (D) 438 (Jul) S11009

Preist v Last [1903] 2 KB 148, 72 LJKB 657, 51 WR 678, 47 Sol Jo 566, 89 LT 33, 19 TLR 527, [1900–3] All ER Rep Ext 1033, CA P9053

Q

Quinn v McGinty 1999 SLT 27, Sh Ct E13008

R

R v Adomako [1995] 1 AC 171, [1994] 3 All ER 79, [1994] 3 WLR 288, 158 JP 653, [1994] Crim LR 757, [1994] 5 Med LR 277, 19 BMLR 56, [1994] NLJR 936, [1994] 2 LRC 800, HL C9021

R v Associated Octel Co Ltd [1996] 4 All ER 846, [1996] 1 WLR 1543, [1996] ICR 972, [1997] IRLR 123, [1997] Crim LR 355, [1996] NLJR 1685, HL Intro9

R v A-G for Northern Ireland, ex p Burns [1999] IRLR 315 .... W10019

R v Balfour Beatty Rail Infrastructure Services Ltd [2006] EWCA Crim 1586, [2007] 1 Cr App Rep (S) 370, [2007] ICR 354, (2006) Times, 18 July, [2006] All ER (D) 47 (Jul) E15036

R v Barker (1996), unreported E15017

R v Boal [1992] QB 591, [1992] 3 All ER 177, [1992] 2 WLR 890, 95 Cr App Rep 272, [1992] ICR 495, [1992] IRLR 420, 156 JP 617, [1992] BCLC 872, [1992] 21 LS Gaz R 26, 136 Sol Jo LB 100, CA E15040

R (on the application of Aineto) v Brighton and Hove District Coroner [2003] EWHC 1896 (Admin), [2003] All ER (D) 353 (Jul) A3019

R (on the application of Junttan Oy) v Bristol Magistrates' Court [2003] UKHL 55, [2004] 2 All ER 555, [2003] ICR 1475, [2003] 44 LS Gaz R 30, (2003) Times, 24 October, 147 Sol Jo LB 1243, [2003] All ER (D) 386 (Oct) P9009

R v British Steel plc [1995] 1 WLR 1356, [1995] ICR 586, [1995] IRLR 310, [1995] Crim LR 654, CA E15039

R v Chapman (1992), unreported E15040

R v Chargot Ltd (t/a Contract Services) [2007] EWCA Crim 3032, [2008] 2 All ER 1077, [2008] ICR 517, [2007] All ER (D) 198 (Dec); affd [2008] UKHL 73 , [2009] 2 All ER 645, [2009] 1 WLR 1, (2008) Times, 16 December, 153 Sol Jo (no 1) 32, [2008] All ER (D) 106 (Dec) E15031

R (on the application of Hampstead Heath Winter Swimming Club) v Corpn of London. See Hampstead Heath Winter Swimming Club v Corpn of London

R v D'Albuquerque, ex p Bresnahan [1966] 1 Lloyd's Rep 69 .... C6007

R v F Howe & Son (Engineers) Ltd [1999] 2 All ER 249, [1999] 2 Cr App Rep (S) 37, [1999] IRLR 434, 163 JP 359, [1999] Crim LR 238, [1998] 46 LS Gaz R 34, [1998] All ER (D) 552, CA E15043

R v Gateway Foodmarkets Ltd [1997] 3 All ER 78, [1997] 2 Cr App Rep 40, [1997] ICR 382, [1997] IRLR 189, [1997] Crim LR 512, [1997] 03 LS Gaz R 28, 141 Sol Jo LB 28, CA E15039

R v HTM Ltd [2006] EWCA Crim 1156, [2007] 2 All ER 665, [2006] ICR 1383, [2006] All ER (D) 299 (May) E15039, O3019

R v Jarvis Facilities Ltd [2005] EWCA Crim 1409, [2006] 1 Cr App Rep (S) 247, (2005) Times, 6 June, [2005] All ER (D) 429 (May) E15036

R v Kerr (1996), unreported E15017

R v Merseycare NHS Trust (Ormskirk MC) (5 September 2002, unreported) V8015

R v National Insurance Comr, ex p Michael [1977] 2 All ER 420, [1977] 1 WLR 109, [1977] ICR 121, 120 Sol Jo 856, CA C6006

R v Nelson Group Services (Maintenance) Ltd [1999] IRLR 646 .... E15039

R v Owino [1996] 2 Cr App Rep 128, [1995] Crim LR 743, CA O3009

R v Rollco Screw and Rivet Co Ltd [1999] 2 Cr App Rep (S) 436, [1999] IRLR 439, CA E15043

R (on the application of the Broadcasting, Entertainment, Cinematographic and Theatre Union) v Secretary of State for Trade and Industry: C-173/99 [2001] ECR I-4881, [2001] All ER (EC) 647, [2001] 1 WLR 2313, [2001] 3 CMLR 109, [2001] ICR 1152, [2001] IRLR 559, [2001] All ER (D) 272 (Jun), ECJ W10022

R v Swan Hunter Shipbuilders Ltd [1982] 1 All ER 264, [1981] ICR 831, [1981] IRLR 403, [1981] Crim LR 833, sub nom R v Swan Hunter Shipbuilders Ltd and Telemeter Installations Ltd, HMS Glasgow [1981] 2 Lloyd's Rep 605, CA A1022

Rae v Mars (UK) Ltd [1990] 1 EGLR 161, [1990] 03 EG 80 .... O3006

Railtrack plc v Smallwood [2001] EWHC Admin 78, [2001] ICR 714, [2001] 09 LS Gaz R 40, 145 Sol Jo LB 52, [2001] All ER (D) 103 (Jan) E15015

Railways Comr (NSW) v Cardy (1961) 104 CLR 274, Aus HC O3007

Read v J Lyons & Co Ltd [1947] AC 156, [1946] 2 All ER 471, [1947] LJR 39, 91 Sol Jo 54, 175 LT 413, 62 TLR 646, HL G1001

Readmans v Leeds City Council [1993] COD 419 .... E15015

Redfearn v Serco Ltd (t/a West Yorkshire Transport Service) [2006] EWCA Civ 659, [2006] ICR 1367, [2006] IRLR 623, (2006) Times, 27 June, 150 Sol Jo LB 703, [2006] All ER (D) 366 (May) H1703

Reed v Stedman [1999] IRLR 299, EAT E14019, H1702, H1706, H1707, H1708

Reid v Rush & Tompkins Group plc [1989] 3 All ER 228, [1990] 1 WLR 212, [1990] RTR 144, [1989] 2 Lloyd's Rep 167, [1990] ICR 61, [1989] IRLR 265, CA E13008

Reilly v Merseyside Health Authority [1995] 6 Med LR 246, 23 BMLR 26, CA C6032

Revenue and Customs v Stringer (sub nom Ainsworth v IRC). See Ainsworth v IRC

Revill v Newbery [1996] QB 567, [1996] 1 All ER 291, [1996] 2 WLR 239, [1996] NLJR 50, 139 Sol Jo LB 244, CA O3009

Rhind v Astbury Water Park Ltd [2003] EWHC 1029 (QB), [2003] All ER (D) 217 (May); affd [2004] EWCA Civ 756, 148 Sol Jo LB 759, [2004] All ER (D) 129 (Jun) O3004

Ribee v Norrie (2000) 81 P & CR D37, 33 HLR 777, [2000] NPC 116, [2001] PIQR P 8, (2000) Times, 22 November, [2000] All ER (D) 1644, CA O3004

Richardson v Pitt-Stanley [1995] QB 123, [1995] 1 All ER 460, [1995] 2 WLR 26, [1995] ICR 303, [1994] PIQR P 496, [1994] JPIL 315, CA E13008

Ricketts v Torbay [2003] EWCA Civ 613 .... O3018

Roberts v Day (1977) COIT No 1/133, Case No 3053/77, unreported E15022

Roberts v Dorothea Slate Quarries Co Ltd [1948] 2 All ER 201, [1948] LJR 1409, 92 Sol Jo 513, 41 BWCC 154, HL C6005

Robertson and Rough v Forth Road Bridge Joint Board [1995] IRLR 251, Ct of Sess C6033

Robinson v Harman (1848) 18 LJ Ex 202, 1 Exch 850, 154 ER 363, [1843–60] All ER Rep 383, 13 LTOS 141, Exch Ct H1728

Robinson-Steele v RD Retail Services Ltd: C-257/04 [2006] ECR I-2531, [2006] All ER (EC) 749, [2006] ICR 932, [2006] IRLR 386, (2006) Times, 22 March, [2006] All ER (D) 238 (Mar), ECJ W10024

Rose v Plenty [1976] 1 All ER 97, [1976] 1 WLR 141, [1976] 1 Lloyd's Rep 263, [1975] ICR 430, [1976] IRLR 60, 119 Sol Jo 592, CA A1020

Rylands v Fletcher. See Fletcher v Rylands and Horrocks

S

Sabloff v Yamaha Motor Co 113 NJ Super 279, 273 A 2d 606 (1971) P9036

Saini v All Saints Haque Centre [2009] 1 CMLR 1060, [2009] IRLR 74, [2008] All ER (D) 250 (Oct), EAT H1703

Sarkar West London Mental Health NHS Trust [2009] IRLR 512, [2009] All ER (D) 298 (Mar), EAT E14012

Sayers v Cambridgeshire County Council [2006] EWHC 2029 (QB), [2007] IRLR 29 .... W10010, W10042

Scarth v East Hertfordshire District Council (HSIB 181) J3013

Scott and Swainger v Associated British Ports and British Railways Board [2000] All ER (D) 1937 .... O3017

Searson v Brioland Ltd. See Brioland Ltd v Searson

Secretary of State for Employment v Associated Society of Locomotive Engineers and Firemen (No 2) [1972] 2 QB 455, [1972] 2 All ER 949, [1972] 2 WLR 1370, [1972] ICR 19, 13 KIR 1, 116 Sol Jo 467, CA E14007

Select Managements Ltd v Westminster City Council. See Westminster City Council v Select Management Ltd

Sheldrake v DPP [2004] UKHL 43, [2005] 1 AC 264, [2005] 1 All ER 237, [2004] 3 WLR 976, [2005] RTR 13, [2005] 1 Cr App Rep 450, 168 JP 669, [2004] 43 LS Gaz R 33, (2004) Times, 15 October, 148 Sol Jo LB 1216, 17 BHRC 339, [2005] 3 LRC 463, [2004] All ER (D) 169 (Oct) E15031

Shipham & Co Ltd v Skinner [2001] All ER (D) 201 (Dec), EAT J3030

Shirlaw v Southern Foundries (1926) Ltd [1939] 2 KB 206, [1939] 2 All ER 113, CA; affd sub nom Southern Foundries (1926) Ltd v Shirlaw [1940] AC 701, [1940] 2 All ER 445, 109 LJKB 461, 84 Sol Jo 464, 164 LT 251, 56 TLR 637, HL E14004

Showboat Entertainment Centre Ltd v Owens [1984] 1 All ER 836, [1984] 1 WLR 384, [1984] ICR 65, [1984] IRLR 7, 128 Sol Jo 152, [1983] LS Gaz R 3002, 134 NLJ 37, EAT H1703

Sibbald v Sher Bros (1980) Times, 1 February O3014

Siddorn v Patel [2007] EWHC 1248 (QB), [2007] All ER (D) 453 (Mar) O3009

Simpson v AI Dairies Farms Ltd [2001] EWCA Civ 13, [2001] All ER (D) 31 (Jan) O3014

Sindicato de Médicos de Asistencia Pública (Simap) v Conselleria de Sanidad y Consumo de la Generalidad Valenciana: C-303/98 [2000] ECR I-7963, [2001] All ER (EC) 609, [2001] 3 CMLR 932, [2001] ICR 1116, [2000] IRLR 845, [2000] All ER (D) 1236, ECJ W10003

Smith v A J Morrisroes & Sons Ltd [2005] ICR 596, [2005] IRLR 72, [2004] All ER (D)
291 (Dec), EAT W10024
Smith v Chubb Security Personnel Ltd (1999), unreported W10024
Smith v Churchills Stairlifts plc [2005] EWCA Civ 1220, [2006] ICR 524, [2006] IRLR
41, [2005] All ER (D) 318 (Oct) E16511
Smith v Leech Brain & Co Ltd [1962] 2 QB 405, [1961] 3 All ER 1159, [1962] 2 WLR
148, 106 Sol Jo 77 .... C6022
Smith v Manchester City Council (or Manchester Corpn) (1974) 17 KIR 1, 118 Sol Jo
597, CA C6026
Smith v Northamptonshire County Council [2008] EWCA Civ 181, [2008] 3 All ER
1054, [2008] ICR 826, [2008] All ER (D) 132 (Mar); affd [2009] UKHL 27, [2009]
4 All ER 557, [2009] ICR 734, 110 BMLR 15, [2009] PIQR P292, (2009) Times,
21 May, 153 Sol Jo (no 20) 39, [2009] All ER (D) 170 (May) M1005
Smith v Welsh Ambulance Service (Chester County Court) (22 March 2007,
unreported) V8015
Smoker v London Fire and Civil Defence Authority [1991] 2 AC 502; [1991] 2 All ER
449 .... C6053
Snowball v Gardner Merchant Ltd [1987] ICR 719, [1987] IRLR 397, [1987] LS Gaz R
1572, EAT H1715
South Surbiton Co-operative Society v Wilcox [1975] IRLR 292, Ind Trib E15025
Southern Foundries (1926) Ltd v Shirlaw. See Shirlaw v Southern Foundries (1926) Ltd
Southern Portland Cement Ltd v Cooper [1974] AC 623, [1974] 1 All ER 87, [1974] 2
WLR 152, 118 Sol Jo 99, PC O3010
Speed v Thomas Swift & Co Ltd [1943] KB 557, [1943] 1 All ER 539, 112 LJKB 487,
87 Sol Jo 336, 169 LT 67, CA S3003
Spencer-Franks v Kellogg Brown and Root Ltd [2008] UKHL 46, [2009] 1 All ER 269,
[2008] ICR 863, [2008] NLJR 1004, [2008] PIQR P389, (2008) Times, 3 July, 2008
SLT 675, 2008 SCLR 484, 152 Sol Jo (no 27) 30, [2008] All ER (D) 26 (Jul) M1005
Staples v West Dorset District Council (1995) 93 LGR 536, [1995] 18 LS Gaz R 36,
[1995] PIQR P 439, 139 Sol Jo LB 117, CA O3005, O3017
Stark v Post Office [2000] All ER (D) 276, CA P9010
Stevens v Blaenau Gwent County Borough [2004] EWCA Civ 715, [2004] HLR 1039,
(2004) Times, 29 June, [2004] All ER (D) 116 (Jun) O3003
Strathclyde Regional Council v Porcelli. See Porcelli v Strathclyde Regional Council
Stringer v HM Revenue and Customs: C-520/06 [2009] All ER (EC) 906,
[2009] 2 CMLR 657, [2009] ICR 932, [2009] IRLR 214, (2009) Times, 28 January,
[2009] All ER (D) 147 (Jan), ECJ M1514
Sumsion v BBC (Scotland) [2007] IRLR 678, EAT W10024
Swain v Denso Martin Ltd (2000) Times, 24 April M3002

T

TSB Bank plc v Harris [2000] IRLR 157, EAT E14019
Tawling v Wisdom Toothbrushes Ltd (1997), unreported W11043
Taylor v Furness, Withy & Co Ltd [1969] 1 Lloyd's Rep 324, 6 KIR 488 .... E14001
Tesco Stores Ltd v Kippax COIT No 7605-6/90 (1990), unreported E15015
Tesco Supermarkets Ltd v Nattrass [1972] AC 153, [1971] 2 All ER 127, [1971] 2 WLR
1166, 69 LGR 403, 135 JP 289, 115 Sol Jo 285, HL E15039
Thomas v Brighton Health Authority [1999] 1 AC 345, [1998] 3 All ER 481, [1998] 3
WLR 329, [1998] IRLR 536, [1998] 2 FLR 507, [1998] Fam Law 593, 43 BMLR 99,
[1998] NLJR 1087, [1998] PIQR Q 56, 142 Sol Jo LB 245, [1998] All ER (D)
352, HL C6027
Thomas v Bunn [1991] 1 AC 362, [1991] 1 All ER 193, [1991] 2 WLR 27, 135 Sol Jo
16, [1990] NLJR 1789, HL C6043
Thomas Sanderson Blinds Ltd v English, UKEAT/0316/10/JOJ (21 February 2011,
unreported) H1705
Thompson v Smith's Shiprepairers (North Shields) Ltd [1984] QB 405, [1984] 1 All ER
881, [1984] 2 WLR 522, [1984] ICR 236, [1984] IRLR 93, 128 Sol Jo 225,
[1984] LS Gaz R 741 .... N3022

Thompson v T Lohan (Plant Hire) Ltd (JW Hurdiss Ltd, third party) [1987] 2 All ER 631, [1987] 1 WLR 649, [1987] IRLR 148, [1987] BTLC 221, 131 Sol Jo 358, [1987] LS Gaz R 979, CA P9057

Thornett and Fehr v Beers & Son [1919] 1 KB 486, 88 LJKB 684, 24 Com Cas 133, 120 LT 570 .... P9052

Titchener v British Railways Board 1981 SLT 208, OH; on appeal 1983 SLT 269, Ct of Sess; affd [1983] 3 All ER 770, 127 Sol Jo 825, 1984 SLT 192, sub nom McGinlay (or Titchener) v British Railways Board [1983] 1 WLR 1427, HL O3010, O3017

Tomlinson v Congleton Borough Council [2003] UKHL 47, [2004] 1 AC 46, [2003] 3 All ER 1122, [2003] 3 WLR 705, [2003] 34 LS Gaz R 33, [2003] NLJR 1238, [2003] 32 EGCS 68, (2003) Times, 1 August, 147 Sol Jo LB 937, [2003] All ER (D) 554 (Jul) O3017, O3019

Trent Strategic Health Authority v Jain [2009] UKHL 4, [2009] 1 AC 853, [2009] 1 All ER 957, [2009] 2 WLR 248, [2009] PTSR 382, 106 BMLR 88, (2009) Times, 22 January, 153 Sol Jo (no 4) 27, [2009] All ER (D) 148 (Jan) E15028

Trim Joint District School Board of Management v Kelly [1914] AC 667, 83 LJPC 220, 58 Sol Jo 493, 111 LT 305, 30 TLR 452, [1914–15] All ER Rep Ext 1453, 7 BWCC 274, HL C6005

Tripp v Ministry of Defence [1982] CLY 1017 .... N3022

Twine v Bean's Express Ltd [1946] 1 All ER 202, 62 TLR 155; affd (1946) 202 LT 9, 62 TLR 458, CA A1020

V

Vacwell Engineering Co Ltd v BDH Chemicals Ltd [1971] 1 QB 88, [1969] 3 All ER 1681, [1969] 3 WLR 927, 7 KIR 286, 113 Sol Jo 639; revsd [1971] 1 QB 111n, [1970] 3 All ER 553n, [1970] 3 WLR 67n, 114 Sol Jo 472, CA P9035

Veakins v Kier Islington Ltd [2009] EWCA Civ 1288, [2010] IRLR 132, (2010) Times, 13 January, [2009] All ER (D) 34 (Dec) H1726

Vento v Chief Constable of West Yorkshire Police [2002] EWCA Civ 1871, [2003] ICR 318, [2003] IRLR 102, [2003] 10 LS Gaz R 28, (2002) Times, 27 December, 147 Sol Jo LB 181, [2002] All ER (D) 363 (Dec) H1715

Viasystems (Tyneside) Ltd v Thermal Transfer (Northern) Ltd [2005] EWCA Civ 1151, [2006] QB 510, [2005] 4 All ER 1181, [2006] 2 WLR 428, [2006] ICR 327, [2005] IRLR 983, [2005] 44 LS Gaz R 31, [2005] All ER (D) 93 (Oct) E13027

W

Walker v BHS Ltd [2005] All ER (D) 146 (May), EAT H1705

Walker v Northumberland County Council [1995] 1 All ER 737, [1995] ICR 702, [1995] IRLR 35, [1995] ELR 231, [1994] NLJR 1659 ... E14003, H1724, H1728, R2004, S11009

Waltons & Morse v Dorrington [1997] IRLR 488, EAT E14003, E14019

Waters v Metropolitan Police Comr [1997] ICR 1073, [1997] IRLR 589, CA; revsd [2000] 4 All ER 934, [2000] 1 WLR 1607, [2000] ICR 1064, HL H1712

Watson (administrators of) v Willmott [1991] 1 QB 140, [1991] 1 All ER 473, [1990] 3 WLR 1103 .... C6039

Wattleworth v Goodwood Road Racing Co Ltd [2004] EWHC 140 (QB), [2004] All ER (D) 51 (Feb) O3011, O3017

Waugh v British Railways Board [1980] AC 521, [1979] 2 All ER 1169, [1979] 3 WLR 150, [1979] IRLR 364, 123 Sol Jo 506, HL A3029, J3015

Waugh v London Borough of Sutton (1983) HSIB 86 .... J3011

Weathersfield Ltd (t/a Van & Truck Rentals) v Sargent [1999] ICR 425, [1999] IRLR 94, 143 Sol Jo LB 39, CA H1703

Webster v Blue Ship Tea Room Inc 347 Mass 421, 198 NE 2d 309 (1964) P9031

Wells v Wells [1999] 1 AC 345, [1998] 3 All ER 481, [1998] 3 WLR 329, [1998] IRLR 536, [1998] 2 FLR 507, [1998] Fam Law 593, 43 BMLR 99, [1998] NLJR 1087, [1998] PIQR Q 56, 142 Sol Jo LB 245, [1998] All ER (D) 352, HL C6027

West (H) & Son Ltd v Shephard [1964] AC 326, [1963] 2 All ER 625, [1963] 2 WLR 1359, 107 Sol Jo 454, HL C6030

West Cumberland By-Products Ltd v DPP [1988] RTR 391, DC E15042

West Midland Co-operative Society Ltd v Tipton [1986] AC 536, [1986] 1 All ER 513, [1986] 2 WLR 306, [1986] ICR 192, [1986] IRLR 112, 130 Sol Jo 143, [1986] LS Gaz R 780, [1986] NLJ Rep 163, HL E14013

Western Excavating (ECC) Ltd v Sharp [1978] QB 761, [1978] 1 All ER 713, [1978] 2 WLR 344, [1978] ICR 221, [1978] IRLR 27, 121 Sol Jo 814, CA H1720

Westminster City Council v Select Management Ltd [1985] 1 All ER 897, [1985] 1 WLR 576, 83 LGR 409, [1985] ICR 353, 129 Sol Jo 221, [1985] 1 EGLR 245, [1985] LS Gaz R 1091, sub nom Select Managements Ltd v Westminster City Council [1985] IRLR 344, CA O3001, O3019

Westwood v Post Office [1974] AC 1, [1973] 3 All ER 184, [1973] 3 WLR 287, 117 Sol Jo 600, HL A1001

Wheat v E Lacon & Co Ltd [1966] 1 QB 335, [1965] 2 All ER 700, [1965] 3 WLR 142, 109 Sol Jo 334, CA; affd [1966] AC 552, [1966] 1 All ER 582, [1966] 2 WLR 581, [1966] RA 193, 110 Sol Jo 149, [1966] RVR 223, HL O3004

Whirlpool (UK) Ltd and Magnet Ltd v Gloucestershire County Council (1993) 159 JP 123 ...... P9008

White v Chief Constable of South Yorkshire Police [1999] 2 AC 455, [1999] 1 All ER 1, [1998] 3 WLR 1509, [1999] ICR 216, 45 BMLR 1, [1999] 02 LS Gaz R 28, [1998] NLJR 1844, [1999] 3 LRC 644, sub nom Frost v Chief Constable of South Yorkshire Police 143 Sol Jo LB 51, HL C6033

White v Pressed Steel Fisher [1980] IRLR 176, EAT J3011

White v St Albans City and District Council (1990) Times, 12 March, CA O3009

Wieland v Cyril Lord Carpets Ltd [1969] 3 All ER 1006 ...... C6022

Wileman v Minilec Engineering Ltd [1988] ICR 318, [1988] IRLR 144, EAT H1715

Williams v Gloucestershire County Council (10 September 1999, unreported) C6023

Willson v Ministry of Defence [1991] 1 All ER 638, [1991] ICR 595 ...... C6040

Wilson v Graham [1991] 1 AC 362, [1991] 1 All ER 193, [1991] 2 WLR 27, 135 Sol Jo 16, HL C6043

Wilson v National Coal Board 1978 SLT 129, OH; affd 1981 SLT 67, HL C6053

Wilsons and Clyde Coal Co Ltd v English [1938] AC 57, [1937] 3 All ER 628, 106 LJPC 117, 81 Sol Jo 700, 157 LT 406, 53 TLR 944, HL S3003

Winterbottom v Wright (1842) 11 LJ Ex 415, 10 M & W 109, 152 ER 402 ...... P9024

Withers v Perry Chain Co Ltd [1961] 3 All ER 676, [1961] 1 WLR 1314, 59 LGR 496, 105 Sol Jo 648, CA R2006

Witley and District Men's Club v Mackay [2001] IRLR 595, [2001] All ER (D) 31 (Jun), EAT W10023

Woking Borough Council v BHS plc (1994) 93 LGR 396, 159 JP 427, 159 JPN 387 ...... A3008.1

Wood v British Coal Corpn [1991] 2 AC 502, [1991] 2 All ER 449, [1991] 2 WLR 1052, [1991] ICR 449, [1991] IRLR 271, HL C6053

Woolfall and Rimmer Ltd v Moyle [1942] 1 KB 66, [1941] 3 All ER 304, 71 Ll L Rep 15, 111 LJKB 122, 86 Sol Jo 63, 166 LT 49, 58 TLR 28, CA E13020

Woolmington v DPP [1935] AC 462, 25 Cr App Rep 72, 104 LJKB 433, 30 Cox CC 234, [1935] All ER Rep 1, 79 Sol Jo 401, 153 LT 232, 51 TLR 446, HL E15031

Workvale Ltd (No 2), Re [1992] 2 All ER 627, [1992] 1 WLR 416, [1992] BCLC 544, [1992] BCC 349, CA E13024

Worsley v Tambrands Ltd (No 2) [2000] MCR 0280 ...... P9035

Wren v Holt [1903] 1 KB 610, 67 JP 191, 72 LJKB 340, 51 WR 435, 88 LT 282, 19 TLR 292, [1900–3] All ER Rep Ext 1152, CA P9053

Wright v British Railways Board [1983] 2 AC 773, [1983] 2 All ER 698, [1983] 3 WLR 211, 127 Sol Jo 478, HL C6043

Wright v Redrow Homes (Yorkshire) Ltd [2004] EWCA Civ 469, [2004] 3 All ER 98, [2004] ICR 1126, [2004] IRLR 720, 148 Sol Jo LB 666, [2004] All ER (D) 221 (Apr) W10004

Y

Young v Kent County Council [2005] EWHC 1342 (QB), [2005] All ER (D) 217
(Mar) O3010, O3011
Young v Post Office [2002] EWCA Civ 661, [2002] IRLR 660, [2002] All ER (D) 311
(Apr) R2004, R2008

Table of Statutes

A

Access to Justice Act 1999
.... C6021
Access to Health Records Act 1990
.... O1017
Access to Health Records Act 1999
.... O1017
Access to Medical Reports Act 1988
.... M1505, M1512, O1017
Activity Centres (Young Persons' Safety)
Act 1995..................... C6508
Administration of Justice Act 1982
s 1(1) C6023, C6034
(2) C6037
3(1) C6037
5 .... C6052
6(1), (2) C6040
15 .... C6043
Sch 1 .... C6043
Aviation and Maritime Security Act 1990
.... O7001

B

Building Act 1984 .... F5031, G1026
Boiler Explosions Act 1882
.... P7001

C

Celluloid and Cinematograph Film Act
1922................. D0938, D0939
Charities Act 2006
.... C9002
Children Act 1989
Pt 4 (ss 31–42) C9008
Pt 5 (ss 43–52) C9008
Children (Scotland) Act 1995
Pt 2 (ss 16–93) C9008
Civil Aviation Act 1982
.... O7001
Civil Contingencies Act 2004
.... B8046, D6001, D6005
Pt 1 (ss 1–18) .. B8043, B8044, B8045,
D6003, D6005, D6006
s 1(1)–(3) D6006
2 .... D6006
(1)(a)–(d) D6007
(1)(g) D6007
(3) D6007
4 .... D6006
(1) D6007
13 .... D6006
17 .... D6006
Pt 2 (ss 19–33) B8044, D6003,
D6005, D6006

Civil Contingencies Act 2004 – *cont.*
s 19(1)–(6) D6006
20 .... D6006
23 .... D6006
24(3) D6006
26 .... D6006
28 .... D6006
Pt 3 .... B8044
Sch 1
Pt 1
para 1...................... D6007
Pt 3 .... D6005
Civil Defence Act 1948
.... D6005
Civil Jurisdiction and Judgments Act 1982
.... P9046
Civil Liability (Contribution) Act 1978
.... E13027
Civil Protection in Peacetime Act 1986
.... D6005
Clean Air Act 1956
.... E5061
Clean Air Act 1968
.... E5061
Clean Air Act 1993
Pt I (ss 1–3) E5061
Pt II (ss 4–17) E5061
Pt III (ss 18–29) E5061
Clean Neighbourhoods and Environment
Act 2005
s 54 .... C8020
Communications Act 2003
.... F9012
Companies Act 1985
s 141 .... E13024
310 .... E15041
651(5), (6) E13024
722, 723 .... E13013
Companies Act 2006
.... C9017, E16005
s 232 .... E15041
Company Directors' Disqualification Act
1986
s 2(1) E15040
Compensation Act 2006
.... E16024
Pt 1 (ss 1–3) E13033
s 1–3 .... E13033
Pt 2 (ss 4–15) E13033
Pt 3 (ss 16–18) E13033
Consumer Protection Act 1987
... Intro 19, A5030, F9004, F9009,
F9010, G1026, M1002, P9027, P9038
Pt I (ss 1–9) P9044
s 1(2)(a)–(c) P9029
2(1) P9028
(2) P9028

Consumer Protection Act 1987 – *cont.*
s 2(2)(b), (c) P9030
  (3) P9030
  (5) P9039
  3(1) P9031
  (2) P9032
  4(1)(a)–(f) P9041
  5(1)–(4) P9040
  6(4) P9034
  7 .... P9001, P9043
  41(1) P9026
  46(1)–(3) P9042
  Sch 1 .... P9044
  Sch 3 .... P9009, P9021, P9022, P9023
Consumer Safety Act 1978
  .... E3029
Continental Shelf Act 1964
  .... O7001, O7007
  s 1(7) O7009
Control of Pollution Act 1974
  .... E5061
  Pt 1 (ss 1–30) N3017
Control of Pollution (Amendment) Act
  1989......................... E5061
Corporate Manslaughter and Corporate
  Homicide Act 2007
  .. Intro 4, Intro 8, Intro 9, Intro 15,
    C9001, C9010, D6029, D6030,
    E13001, E15001, E15038, M2103
  s 1 .... C9023
  (1)–(5) C9002
  (6), (7) C9011
  2 .... C9002
  (1), (2) C9003, E15038.1
  (4)–(6) C9003
  3(1)–(4) C9003
  4 .... C9005, E15038.1
  5 .... E15038.1
  (1)–(3) C9006
  6 .... E15038.1
  (1) C9007
  (3), (4) C9007
  (7), (8) C9007
  8 .... C8520, E15038.1
  (1), (2) C9009
  (4) C9009
  9(1) C9008, C9012
  (2), (3) C9012
  (4), (5) C9012, C9013
  10 .... E15038.1
  (1) C9013
  11 .... C9002
  (1)–(5) C9014
  12 .... C9015
  13 .... C9016
  14(1), (2) C9017
  15(4) C9018
  16(1)–(3) C9019
  17 .... C9020
  18 .... C9021

Corporate Manslaughter and Corporate Homicide Act 2007 – *cont.*
  s 19 .... C9022
  20 .... C9023
  21 .... C9002, C9017
  22 .... C9024
  25 .... C9002
  27(4) C9023
  Sch 1 ... C9002, C9014, C9018, C9024
Countryside and Rights of Way Act 2000
  .... O3007
Crime and Disorder Act 1998
  .... H1729, V8140
  s 28 .... H1733
  32(1)(a), (b) H1733
Criminal Justice Act 1982
  s 37(2) E13014, E13015
Criminal Justice and Court Services Act
  2000
  Pt 1 Ch 1 (ss 1–10)
  .... C9008
Criminal Justice and Public Order Act 1994
  s 34 .... E15026

D

Damages Act 1996
  .... C6001, C6025
  s 1(1) C6027
  2 .... C6021
Data Protection Act 1998
  ... A3025, F7020, M1528, O1017,
    R2007
  s 1(1)(a)–(d) D6007
Defective Premises Act 1972
  .... C9003
Disability Discrimination Act 1995
  . A1001, E14002, E14020, E14021,
    E14022, E17070, H1701, V12032,
    V12036, W11043
  s 6 .... W11001
  18B E16511
Disability Discrimination Act 2005
  .. E17070, O1003, O1050, O1052,
    R2005, R2006

E

Education Act 1996
  s 8 .... R3024, V12007
Education (Scotland) Act 1980
  s 31 .... R3024, V12007
Emergency Powers Act 1920
  .... D6005, D6006
Emergency Workers (Obstruction) Act
  2006 .... V8008
Emergency Workers (Scotland) Act 2005
  .... V8008

Employers' Liability (Compulsory Insurance) Act 1969
.... Intro 19, C6001, E13011, E13017, E13020, E13021, O7030, P9025, P9037
s 1(1) E13008
(3) E13008
2(2)(a) E13010
3 .... E13012
4(1) E13013
5 .... E13008
Employers' Liability (Defective Equipment) Act 1969............. C6001, E13001
Employment Act 2002
Pt 3 (ss 29–40) M1505
Employment Act 2008
.... M1505, M1517
Employment Act 1989
s 11, 12 .... C8018
Employment Relations Act 1999
.... E14013, E14019, J3005
s 10 .... E14008
Employment Rights Act 1996
.... E14001, E14002, E14017, M1530, O7001, O7055, R2007, R3026
s 1 .... E14006, M1505, M1513
13 .... E14014, W10024
43B(1)(d) J3030
44 .... E14011, V8009
(1) J3030
(d) V8014
(e) E14010
(2), (3) E14009
45A W10041
(1)(b) W10013
47B J3030
48 .... E14009, J3030
49 .... J3030
(2) E14009
64 .... E14016
67 .... E14015, E14025, V12018
68 .... E14016
70(1) E14016
(4) E14015
72(1) E14028
86 .... E14002, E14006, E14011, E14018
95 .... E14019
(1)(c) H1719, H1720
98 .... E14010, E14020, M1516
(4) E14021
100 .... E14009, E14011, E14022, J3030
(1)(e) E14010
101A W10010, W10041
103A J3030
104 .... E14011, W10041
105 .... W10041
108 .... J3030

Employment Rights Act 1996 – cont.
s 108(3)(c) E14010
109 .... J3030
113–118 .... H1721
124(1A) J3030
203 .... W10043
221, 222 .... W10024
223(2) W10024
224 .... W10024
227, 228 .... W10024
Energy Act 1976 .... O7001
Environment Act 1995
..... A5030, D6001, D6035, E5068
s 14–18 .... D6025
95(2)–(4) E15040
108, 109 .... E15027
Environment and Safety Information Act 1988
s 1 .... E15029
2(2), (3) E15029
3 .... E15029
Sch E15029
Environmental Protection Act 1990
.... A5030
Pt 1 (ss 1–28) E5044
s 20 .... E15029
Pt 2 (ss 29–78) E5027
s 33–46 .... C8020
Pt 3 (ss 79–85) E5062
s 79 .... C8019
157 .... E15040
Equal Pay Act 1970
.... E14002, V12018
Equality Act 2010 ..... Intro 18, A1001, E16501, E16502, E16509, E16512, E16515, M1505, M1506, M1507, M1508, M1509, M1511, M1523, R2005, R2006, R2008
s 4 .... H1701
6 .... E16510
5(2) E16513
9 .... H1703
(1) E16517
(2)(b) E16517
(3) E16517
Pt 2, Ch 2 (ss 13–27)
.... H1709
s 13 .... E16503
(2) E16513
(6) E16520
14 .... E16505
15(1)(b) E16506
(2) E16506
19 .... E16504
20(3)–(5) E16511
(9) E16511
(11) E16511
21, 22 .... E16511
24 .... H1705

Equality Act 2010 – *cont.*
s 26 .... E16507, H1701, H1702
  (1) H1703
    (b) H1707
  (2), (3) H1703
  (5) H1703
  27(3) E16508
  40 .... H1703
    (3) H1706
  109(1) H1711, H1714
    (4) H1713, H1714
  110 .... H1714
  112 .... H1714
  120 .... E16522
  136 .... H1707
  Sch 1
    para 3 .... E16510
  Sch 8 .... E16511
    Pt 2 .... E16511
    Pt 3
    para 2Ω E16511
Explosives Acts 1875
  .... E15012
s 23 .... W11047
  25 .... C8002
Explosives Acts 1923
  .... E15012

F

Factories Act 1937
  .... W11049
s 14(3) E13020
Factories Act 1961
  . Intro 10, C8518, D0939, E15012,
    E15015, E15045, F5031, P7001,
    P7020, P9009, W11045
s 14(1) E13008
  29 .... O3019
    (1) A1001
  30 .... A1023
  31(3), (4) G1034
  123–126 .... G1010
  155(1) W11047
    (2) W110487, W11049
  173(1) W11046
  175 .... G1010, W11046
Factory and Workshop Act 1901
  .... P7001
Fatal Accidents Act 1976
  .... C6040, E15001
s 1A C6035
  4 .... C6036
  5 .... C6037
Fatal Accidents and Sudden Deaths Inquiry
  (Scotland) Act 1976
  .... O7001
Financial Services and Markets Act 2000
  .... E13024
Fire and Rescue Services Act 2004
  .... E15009

Fire Precautions Act 1971
  .... E15029, F5030, F5031, F5032,
    F5033, R3012
s 23 .... E15040
Fires Prevention (Metropolis) Act 1774
  .... F5031
Fire Safety and Safety of Places of Sport Act
  1987 .... F5031, F5033, F5056
Fire (Scotland) Act 2005
  .... C8002, C8518, F5033
s 6 .... E15009
Fire Services Act 1947
  .... F5031
s 3(1)(dd) O7028
Food and Environment Protection Act
  1985 .... F9007, O7001
s 1, 2 .... F9020
  19 .... E15029
Food Safety Act 1990
  ..... F9001, F9002, F9003, F9004,
    F9022, F9072, F9078, O7001
s 1–6 .... F9007
  7 ...... F9006, F9008, F9009, F9010,
    F9023
  8 .... F9006, F9008, F9011
    (2)(a) F9010
  9 ...... F9008, F9010, F9013, F9014
  10 ..... F9008, F9010, F9013, F9015,
    F9015, F9016
  11 ..... F9008, F9010, F9013, F9016,
    F9017, F9018
  12, ... F9008, F9010, F9013, F9018
  13 ..... F9008, F9010, F9013, F9019,
    F9020
  14 ..... F9008, F9011, F9012, F9024
  15 .... F9008, F9012
  16–18 .... F9008, F9020
  19 .... F9008, F9021
  20 .... F9008
  21 .... F9008, F9070
    (3), (4) F9023
  22 .... F9008
  26 .... F9020
  29–33 .... F9012
  35 .... F9024
  37 .... F9015
  39 .... F9015
  40 .... F9007, F9025
  54(2) E15045
Food Standards Act 1999
  .... F9005, F9007
Freedom of Information Act 2000
  .... E16009, E5041

G

Gas Act 1986 .... G1001, O7001
  Pt I (ss 1–48) G1002
s 4A G1002
  7 .... G1002, G1005

Gas Act 1986 – *cont.*
s 8–11 .... G1002
18, 18A G1002
21–22A G1002
23 .... G1002
33 .... G1002
Gas Act 1995 .... G1001

H

Health Act 2006 .... C8002, E13031
Pt 1 (ss 1–13) W11028
s 2 .... W11028
5 .... W11028
8(1) W11028
Health and Safety at Work etc Act 1974
.... Intro 5, Intro 9, Intro 10, Intro
18, A1001, A5030, C6501, C6505,
C6527, C7004, C7005, C7009,
C8508, C8509, C8516, C9011,
D0401, D0948, D8203, E3003,
E13015, E13020, E14003, E14027,
E15001, E15002, E15013, E17045,
E17061, F3034, F5031, G1026,
J3023, L5002, M1004, M1139,
M2005, M2006, M2015, M2103,
M2125, N3002A, O1056, O3003,
O7001, O7003, O7009, O7010,
O7033, O7046, O7055, P3027,
P7031, P9010, P9011, R3001, S7004,
T7024, T7025, V3001, V5016,
V12036, W10039, W11049
Pt 1 (ss 1–53) E15012, O7004
s 1 .... E15005, W9003
(1)(a)–(d) D6009
2 . Intro 15, A1009, E14001, E14006,
E14012, E14023, E15022, E15032,
E15036, E15037, E15039, E16509,
M1002, M2002, M2014, M2016,
O3019, P3004, P9009, R1010, R3005,
V3015, W9003
(1) .... Intro 11, D0951, E15031,
L3017, M2003, P3005, P7022,
R3002, S3001, S3004, S11006,
V5003, V8004
(2) ...... Intro 11, E15031, M2003,
R3002, T7003
(a) C7008, L3001, R3017,
S3004
(b) L3001
(c) S11040, W9006
(3) ...... Intro 11, Intro 23, M2003,
M2031, S3004, S7001, S7002,
S7003
(4) J3004
(6) J3002
(7) J3017
3 .. C6504, C6512, E15032, E15037,
E15039, M2002, M2014, P9009,
R3005, V12001, W9001, W9003

Health and Safety at Work etc Act 1974 –
cont.
s 3(1) .... Intro 8, Intro 11, C6502,
C6503, C9001, E15031, E15036,
F3027, G1001, G1029, L3003,
M2004, O3019, P7022, R3002,
S3004
(a) S3001
(2) . G1028, G1033, M2004, R3002
4 .... Intro 11, E15032, E15036,
E15039, M2014, O3001, R3002,
R3005, W9003
(1)–(3) O3019
5 . E15032, E15036, E15039, W9003
6 .... Intro 12, E15032, E15036,
E15039, P3002, P3012, P9001, P1013,
P9026, R3002, R3005, W9003
(1) M1002
(a) . P1012, P9014, P9020, P9022
(b)–(d) P9014
(2) P9016
(3) P9017
(4) P9015
(a) P9022
(5) P9018
(6) P9019
(8) P9020, P9021
(8A) P9009, P9021
(10) P9022
7 . Intro 12, A1020, E15032, E15036,
J3009, M2017, S3004, V8004, V8009,
V8014, W9003, W9013
8 . Intro 12, E15032, E15036, W9003
9 . Intro 12, E15032, E15036, P3004,
W9003
11 .... E15005, W9003
12, 13 .... W9003
14 .... E15005, E15032, W9003
(1) D6009
15 .... T7003, W9003
16 .... W9003
(1) C8501
(2) W9001
17 .... W9003
18 .... W9003
(1) E15005, E15008
(7)(a) E15008
19 .... O7041, O7042, W9003
(1) E15007, T7003
20 .... G1033, O7041, W9003
(1)(j) E15026
(7) E15026
21 ... E15014, G1033, J3003, O7041,
W9003
22 .... O7041, W9003
(1)–(3) E15015
(4) E15016
23 .... O7041, W9003
(5) E15014
24 .... E15024, O7041, W9003
(2) E15019

Health and Safety at Work etc Act 1974 – cont.
s 24(3)(a) E15019
25 .... E15032, O7041, W9003
(1)–(3) E15027
25A . E15032, O7041, P9023, W9003
26 .... E15028, O7041, W9003
27 .... O7041, W9003
27A O7041, P9023, W9003
28 .... O7041, W9003
(2) D6007
(7) D6007
(8) J3016, O7050
29–32 .... W9003
33 .... Intro 15, M1115, W9003
(1) E15032
(a) L3036
(b) E15036
(c)–(f) E15036
(g) ... E15017, E15031, E15036
(h)–(k) E15036
(n), (o) E15036
(2A)(b) E15017
34 .... M1115, W9003
(1) E15032
35 .... M1115.W9003
36 .... M1115, W9003
(1) E15042
37 .... Intro 15, C9022, E13032,
      E15031, M1115, O7002, S3004,
      W9003
(1) E15040
38, 39 .... M1115, W9003
40 . E15031, E15031, M1115, R3003,
      W9003
41 .... M1115, W9003
42 .... C9012, E15032, E15036,
      M1115,W9003
43–46 .... W9003
47 .... P9026, W9003
(2) O3018
48 .... W9003
(1), (2) E15045
49 .... W9003
50 .... W9003
(3) W9001
51, 52 .... W9003
53 .... E15015, W9003
(1) E15012, E15027, P9012
54–59 .... W9003
80, 81 .... W9003
82 .... W9003
(1)(c) E15019
Sch 1 .... E15012
Sch 3A E15032
Health and Safety (Offences) Act 2008
    . Intro 4, Intro 8, Intro 10, E15001,
      E15032, E15036
Highways Act 1980
s 1 .... D6006
168, 169 .... C8019

Highways Act 1980 – cont.
s 174 .... C8019
Housing Act 2004
s 163 .... G1025
Human Rights Act 1998
      .... D6006, V8015

I

Industrial Diseases (Notification) Act 1981
      .... A3010
International Headquarters and Defence
      Organisations Act 1964
      .... E15009, F5056
Interpretation Act 1978
Sch 1 .... E15032

J

Judgments Act 1838
s 17 .... C6043

L

Law Reform (Contributory Negligence) Act
      1945
s 1(1) O3011
Law Reform (Personal Injuries) Act 1948
      .... C6001
Law Reform (Miscellaneous Provisions)
      Act 1934 .... C6036, C6038
Legislative and Regulatory Reform Act
      2006 .... E15002
Licensing Act 2003
      .... F5046
Limitation Act 1980
      .... E13025
s 11 .... W11045
11A(3)–(5) P9044
33 .... E13024
Limited Partnerships Act 1907
      .... C9002
Local Government Act 1972
s 138 .... D6005
Local   Government   (Miscellaneous
      Provisions) Act 1982
      .... F5046

M

Magistrates' Courts Act 1980
s 127(1) E15032
Medicines Act 1968
      .... P9001
Merchandise Marks Act 1887
      .... F9023
Merchant Shipping Act 1988
      .... C9001, O7067

Merchant Shipping Act 1995
.... O7008
s 100A O7008
Mineral Workings (Offshore Installations)
Act 1971 ... E15012, O7001, O7003,
O7004, O7004, O7039, O7046
s 5 .... O7011
Mines and Quarries Act 1954
.... E15012, G1010, O7009
s 180 .... F5034

N

National Health Service and Community
Care Act 1990
s 60 .... E15045
National Minimum Wage Act 1998
.... W10004
New Roads and Street Works Act 1991
s 50 .... C8019
Nuclear Installations Act 1965
s 1, 2 .... F5056

O

Occupiers' Liability Act 1957
.. Intro 19, C6508, C8019, C9003,
O3001, O3011, O3016, O3019
s 1(2) O3004, O3006
(3) O3003
(a) O3006
(4) O3007
2(1) O3002
(2) O3002, O3003, O3017
(3)(a) O3010
(b) O3014
(4)(b) O3015
(5) O3017
(6) O3006
Occupiers' Liability Act 1984
.. Intro 19, C6508, C8019, C9003,
F5073, O3001, O3008, O3010,
O3016
s 1 .... O3007, O3013
(1)(a) O3009
(3)–(5) O3009
(6) O3017
(9) O3009
Occupiers' Liability (Scotland) Act 1960
.... O3001
Offences Against the Person Act 1861
s 5 .... C9021
Offices, Shops and Railway Premises Act
1963 ...... Intro 10, E15012, E15025
s 16(1) E15023
Offshore Petroleum Development
(Scotland) Act 1975
.... O7001

Offshore Safety Act 1992
.... E15012, O7001, O7033
s 1(1) O7004
Offshore Safety (Protection against
Victimisation) Act 1992
.... O7055
Oil and Pipelines Act 1985
.... O7001

P

Partnership Act 1890
.... C9002, F5031
Petroleum Act 1879
.... D0402, F5031
Petroleum Act 1987
.... O7004
s 21 .... O7007
23 .... O7001
Petroleum Act 1998
.... O7001, O7002, O7014
s 16 .... O7034
Pt IV (ss 29–45) O7015
Petroleum and Submarine Pipe-lines Act
1975 .... O7004
Petroleum (Consolidation) Act 1928
... D0402, D0938, D0948, E15005,
F5031, H2141, M1103, R3014
Petroleum (Production) Act 1934
.... O7001
Petroleum (Transfer of Licences) Act 1936
.... E15005, F5031
Pipe-lines Act 1962
.... O7004
Planning (Hazardous Substances) Act 1990
.... M1136
Pneumoconiosis   etc.
(Workers' Compensation) Act 1979
.... C6059, C6061
Police Act 1996 .... V8008
Police and Criminal Evidence Act 1984
.... E15026, V8071
Pollution Prevention and Control Act 1999
.... E5027, O7067
Prevention of Oil Pollution Act 1971
.... O7001, O7067
Prices Act 1974 .... F9004
Protection from Harassment Act 1997
.... H1702, H1707, H1729
s 1(2) H1731
3 .... H1726
7(4) H1726
Public Health Act 1936
.... F5031
s 205 .... E14028
Public Health Act 1961
.... E15012, F5031
Public Health (Control of Disease) Act
1984
s 2(4) D6006

Public Interest Disclosure Act 1998
.... E14001, E14009
Public Order Act 1986
.... H1729
s 4A H1731

R

Race Relations Act 1976
... E14002, H1701, H1702, H1706
s 32(1) H1711
Radioactive Substances Act 1993
.... E15029, R1018
Road Traffic Act 1960
.... F7012
Road Traffic Act 1988
s 145 .... E13001
185 .... P7013
Road Traffic (NHS Charges) Act 1999
.... E13031
Road Traffic Regulation Act 1984
.... A1005, C8019

S

Safety of Sports Grounds Act 1975
.... F5033, F5056
s 10 .... E15029
Sale and Supply of Goods Act 1994
.... P9047, P9048, P9056
Sale of Goods Act 1893
.... P9048
Sale of Goods Act 1979
..... P9001, P9047, P9048, P9049,
P9057
s 14 .... P9054, P9056
(2) ... P9050, P9052, P9053, P9055
(3) P9053, P9055
15(2) P9051
Supply of Goods (Implied Terms) Act 1973
.... P9048, P9049
Serious Crimes Act 2007
.... E13029
Sex Discrimination Act 1975
.. E14002, H1701, H1707, O7001,
V12018
s 1(1)(a) H1702
4A H1702
6(2)(b) H1706
41(1) H1714
(3) H1714
42 .... H1714
51(1) E14023
Sex Discrimination Act 1986
.... E14002
Social Security Act 1989
s 22(4)(c) C6036
Social Security Act 1998
s 29(4) C6012

Social Security Administration Act 1992
.... C6001
Social Security (Contributions and Benefits)
Act 1992 ..... C6001, C6059, O7001
s 2 .... C6008
94 .... C6004
95 .... C6008
97–101 .... C6007
103 .... C6011
110 .... C6060
Pt XI (ss 151–163)
.... M1505
s 171A–171G C6002
Sch 6 .... C6011
Social Security (Incapacity for Work) Act
1994................. C6001, C6002
Social Security (Recoupment of Benefits)
Act 1997...................... C6001
Social Security (Recovery of Benefits) Act
1997
s 4 .... C6045
6 .... C6045
8 .... C6044
9 .... C6045
23 .... C6049
Social Work (Scotland) Act 1968
s 27 .... C9008
Statutory Sick Pay Act 1994
.... C6001
Supply of Goods and Services Act 1982
.... P9047
s 2 .... P9049
Supply of Goods (Implied Terms) Act 1973
.... P9047
Supreme Court Act 1981
s 35A C6043

T

Telecommunications Act 1984
.... O7001
Territorial Sea Act 1987
.... O7001
s 1 .... O7009
Third Parties (Rights Against Insurers) Act
1930
s 1(1)(b) E13024
Trade Descriptions Act 1968
.... F9003, F9023
s 1(1) F9012
Trade Disputes Act 1906
.... J3009
Trade Union and Labour Relations
(Consolidation) Act 1992
.... E14011, O7001
s 5, 6 .... J3005
8 .... J3005
20 .... J3009
168 .... J3029

Trade Union and Labour Relations
  (Consolidation) Act 1992 – *cont.*
  s 178 .... J3032, W1007
      (3) J3005
     179(1) E14005
     199 .... E14008
  Sch A1 .... J3005

U

Unfair Contract Terms Act 1977
      .... P9001, P9047, P9048
  s 2 .... E14002, P9057
     (2) O3016
     (3) O3017
    3 .... P9057

Unfair Contract Terms Act 1977 – *cont.*
  s 6(2) P9057
    11 .... O3016

V

Vehicle Excise and Registration Act 1994
  .... F5034
Visiting Forces Act 1952
  s 12 .... E15009

W

Weights and Measures Act 1985
  .... F9004

Table of Statutory Instruments

A

Abstract of Special Regulations (Highly Flammable Liquids and Liquefied Petroleum Gases) Order 1974 (SI 1974 No 1587) D0939

Adventure Activities Licensing Regulations 2004 (SI 2004 No 1309) C6508

Agriculture (Tractor Cabs) Regulations 1974 (SI 1974 No 2034) E15005
reg 3(3) N3015

Air Navigation (Dangerous Goods) Regulations 1994 (SI 1994 No 3187) O7068

Air Quality Limit Values Regulations 2003 (SI 2003 No 2121) E5069

Air Quality Standards Regulations 2007 (SI 2007 No 64) E5037, E5066

Ammonium Nitrate Materials (High Nitrogen Content) Safety Regulations 2003 (SI 2003 No 1082)
Sch 2 .... M1104

Animal By-Products Regulations 2005 (SI 2005 No 2347) F9014, F9044

Asbestos (Licensing) Regulations 1983 (SI 1983 No 1649) A5005, A5006, A5007, H2138, O1046, O7009

Asbestos (Prohibitions) Regulations 1992 (SI 1992 No 3067) A5004, A5005, A5007

B

Borehole Sites and Operations Regulations 1995 (SI 1995 No 2038) F5034

Building Regulations 1991 (SI 1991 No 2768) G1026, W11001, W11029

Building Regulations 2000 (SI 2000 No 2531) F5035, G1019
Sch 1 .... W11044
Pt P E3033
Pt L
L2 .... F3029, F3034

Building Regulations 2010 (SI 2010 No 2214) A1001

Building Standards (Scotland) Regulations 1990 (SI 1990 No 2179) G1019

Business Protection from Misleading Marketing Regulations 2008 (SI 2008 No 1276) F9003

C

Carriage of Dangerous Goods (Amendment) Regulations 1998 (SI 1998 No 2885) D0445, O7011

Carriage of Dangerous Goods (Amendment) Regulations 1999 (SI 1999 No 303) D0445

Carriage of Dangerous Goods and Transportable Pressure Vessels (Amendment) Regulations 2003 (SI 2003 No 1431) D0445

Carriage of Dangerous Goods and Use of Transportable Pressure Equipment Regulations 2004 (SI 2004 No 568) D0402, D0444, D6025, D6027, O7036, P7001, P7012, P7013, P7031

Carriage of Dangerous Goods and Use of Transportable Pressure Equipment Regulations 2007 (SI 2007 No 1573) D0402, D0415, D0424, D0438, D0439, D0448, D6025, D6027
reg 9 .... D0444
  12(4) D0443
  13, 14 .... D0443
  19 .... D0443
  24(6) D0435
  26 .... D0443
  30 .... D0443
  31 .... D0446
  39 .... D0426
  42 .... D0446
  43(1) D0443
  47 .... D0406
  51 .... D0413
  57 .... D0418, D0446
  62 .... D0426
  63 .... D0431
    (7) D0435
  64, 65 .... D0431
Pt 4 (regs 71–80) D0401
reg 91 .... D0418, D0421, D0422
  93 .... D0413

Carriage of Dangerous Goods and Use of Transportable Pressure Equipment Regulations 2009 (SI 2009 No 1348) .... Intro 13, A5030

Carriage of Dangerous Goods (Classification, Packaging and Labelling) and Use of Transportable Pressure Receptacles Regulations 1996 (SI 1996 No 2092) D0445, D0936, D0939, P7001

Carriage of Dangerous Goods by Rail Regulations 1996 (SI 1996 No 2089) A5030, D0936, P7013

Carriage of Dangerous Goods by Road (Driver Training) Regulations 1996 (SI 1996 No 2094) D0445

Carriage of Dangerous Goods by Road Regulations 1996 (SI 1996 No 2095) D0402, D0936, D0938, D0939, D6001, P7013, V12015

Carriage of Explosives by Road Regulations 1996 (SI 1996 No 2093) D0445, D0936, D6027, V12015

Celluloid etc Factories and Workshops Regulations 1921 (SR&O 1921 No 1825) D0939, F5062

Chemicals (Hazard Information and Packaging for Supply) (Amendment) Regulations 2008 (SI 2008 No 2337) .... Intro 9

Chemicals (Hazard Information and Packaging for Supply) Regulations 2002 (SI 2002 No 1689) D0909, D0936, H2101, H2117, H2137, H2142, P9001, P9035, R2508, V12021
reg 4 .... M1104

Children (Northern Ireland) Order 1995 (SI 1995 No 755 (NI 2))
Pt 5 (arts 49–61) C9008
Pt 6 (arts 62–71) C9008

Cinematograph Film Stripping Regulations 1939 (SR&O 1939 No 571) D0939, F5062

Civil Contingencies Act 2004 (Contingency Planning) Regulations 2005 (SI 2005 No 2042) D6001, D6006
reg 3(1) D6007
Pt 2 (regs 4–12) D6007
reg 4–12 .... D6007
Pt 3 (regs 13–18) D6007
reg 13–18 .... D6007
Pt 4 (regs 19–26) D6007
reg 21–26 .... D6007
Pt 5 (reg 27) D6007
Pt 6 (regs 28–35) D6007

Civil Contingencies Act 2004 (Contingency Planning) Regulations 2005 (SI 2005 No 2042) – cont.
reg 29 .... D6007
31–35 .... D6007
Pt 7 (regs 36–44) D6007
reg 36(a) D6007
39–44 .... D6007
Pt 8 (regs 45–54) D6007
reg 45–54 .... D6007
Pt 9 (regs 55, 56) D6007
Pt 10 (regs 57, 58) D6007

Civil Defence (Grant) Regulations 1953 (SI 1953 No 1777) D6005

Civil Defence (General Local Authority Functions) Regulations 1993 (SI 1993 No 1812) D6008

Civil Defence (Scotland) Regulations 2001 (SI 2001 No 139) D6008

Civil Jurisdiction and Judgments Order 2001 (SI 2001 No 3929) P9046

Civil Jurisdiction (Offshore Activities) Order 1987 (SI 1987 No 2197) O7009

Civil Procedure Rules 1998 (SI 1998 No 3132) .... Intro 19, E13029, V8071
Pt 25 .... C6041
Pt 36
r 11 .... C6042
Pt 41
r 2(2) C6040

Classification and Labelling of Explosives Regulations 1983 (SI 1983 No 1140) D0936

Confined Spaces Regulations 1997 (SI 1997 No 1713) C7001, C7002, C7008, C7009, C7030, C8018, M2027, P7016, R3017, S3018, V12016, V12027
reg 1 .... C7003
(2) A1023, O7065
3 .... A1023, C7004
4 .... A1023
(1) C7005
(2) C7006
5 .... C7007
(1)–(3) D6024
6 .... A1023

Construction (Design and Management) Regulations 1994 (CDM) (SI 1994 No 3140) .... Intro 9, Intro 13, C8501, C8510, C8514, G1010, M2022, V12003
reg 15 .... S3004

Construction (Design and Management)
Regulations 2007 (SI 2007 No 320)
.... Intro 9, Intro 13, A5030, C7002,
C7005, C8001, C8019, C8520, D6023,
F3021, F3028, F3032, F3033, F3034,
F5062, M1007, M2002, M2015,
M2037, R3017, V12003, W9002,
W9015
Pt 1 (regs 1–3) C8502, M2022
reg 2 .... C8501
    (1) F5056
    (3) E15009
Pt 2 (regs 4–13) C8002, C8502,
C8504, C8513, M2022,
M2023
reg 4 .... C8002, C8505
    (1) M2023
    5 .... C8506, C8515, M2023
    6 .... C8507, C8515, M2023
    7 .... C8508, M2023
    8 .... M2023
    9 .. C6506, C8002, C8503, C8509,
C8518, M2024
    (2) C8513
    10 .... C8509, M2024
    11 .... C8510, C8515, M2025
    12 .... C8510, M2025
    (4) C8002
    13 . C8002, C8511, C8518, M2026
Pt 3 (regs 14–24) C8502, C8504,
C8511, C8512, C8516,
M2022, M2027
reg 14 .... M2027
    (d) C8513
    15–17 .... C8514, M2027
    18, 19 .... C8515, M2027
    20 .... C8513, M2027
    21 . C8513, C8515, C8518, M2027
    22 . C8002, C8516, C8518, M2027
    23, 24 .... C8516, M2027
Pt 4 (regs 25–44) C8002, C8502,
M2022
reg 25, 26 .... C8002
    27 .... C8002, C8018
    28 .... C8002
    29 .... C8002, C8502
    30 .... C8002
    31(3), (4) C8002
    32 .... C8002
    33 .... C8518
    (1)(b) C8002
    34 .... C8002
    35 .... C8002, W9004
    36 .... C8002, C8018
    37 .... C8002
    38 .... C8002
    39 .... D6022
    40(4) C8002
    41, 42 .... C8002
    44 .... C8002, L5005

Construction (Design and Management)
Regulations 2007 (SI 2007 No 320) –
cont.
Pt 5 (regs 45–48) C8502, C8517,
M2022
reg 48 .... C8518
Sch 1 .. C8509, C8513, C8518, M2023
Sch 2 ... C8002, C8509, C8514, C8518
Sch 3 .... C8516, C8518
Sch 4 .... C8502, C8518
Sch 5 .... C8502, C8518
App 2 .... C8513, C8514
App 3 .... C8514, C8515
Construction (Head Protection) Regulations
1989 (SI 1989 No 2209)
.... C8018
reg 3 .... P3008
Construction (Health, Safety and Welfare)
Regulations 1996 (SI 1996 No
1592)Intro 13, ... C8001, C8502
reg 6(3) W9008
    15 .... C8018
    20 .... D6023
    21 .... C8002
Sch 1 .... W9008
Sch 5 .... W9009
Sch 6 .... C8002
Sch 8 .... W9012
Consumer Protection Act 1987
(Commencement No 1) Order 1987 (SI
1987 No 1680) A5030
Consumer Protection from Unfair Trading
Regulations 2008 (SI 2008 No 1277)
.... F9003, F9012
Consumer Transactions (Restrictions
on Statements) Order 1976 (SI 1976
No 1813) P9057
Control of Asbestos at Work Regulations
1987 (SI 1987 No 2115)
.... O7009, R3001, R3013
Control of Asbestos at Work Regulations
2002 (SI 2002 No 2675)
...... A5007, C8018, F3024, H2101,
H2138, R3013, S3018
reg 4 .... A5006
Control of Asbestos Regulations 2006 (SI
2006 No 2739) A5003, A5005,
A5006, A5009, A5020,
A5022, A5024, A5030,
E15005, F3024, F3025,
O1004, O1018, O1046,
R3023, S3018, V3001,
V3015, V3016
reg 3(1) A5007
    4 .... Intro 13, A5017, R3013
    6 .... R3013
    (1)(a) V3015
    (2)(c) V3015
    8 .... A5012

Control of Asbestos Regulations 2006 (SI 2006 No 2739) – *cont.*
reg 10 .... A5007
  11(2)(b) V3015
  12(1), (2) V3015
  13(1)(a), (b) V3015
    (2), (3) V3015
  14 .... P3008
  15 .... D6016
  21 .... A5011
  22 .... A5007

Control of Asbestos in the Air Regulations 1990 (SI 1990 No 556)
  .... A5030

Control of Explosives Regulations 1991 (SI 1991 No 1531) C8002, O7009

Control of Industrial Major Accident Hazards Regulations 1984 (SI 1984 No 1902) D6001, M1101, M1103, O7003

Control of Lead at Work Regulations 1980 (SI 1980 No 1248)
  .... O7009, R3001

Control of Lead at Work Regulations (SI 2002 No 2676) C8018, E14002, H2101, H2140, O1004, R3016, V3001, V3016, V12003
reg 2(1) V3015
  5 .... V3015
  6 .... P3008, V3015
    (3)(b) V3015
    (4)(b) V3015
    (d) V3015
  8 .... P3008
    (1) V3015
    (2)(a) V3015
    (4) V3015
  10 .... V12021

Control of Major Accident Hazards Regulations 1999 (SI 1999 No 743)
  ...... C6506, D6001, D6003, D6007, D6011, D6035, D6039, D6045, D6054, D6066, D6071, H2101, H2139, M1102, M1109, M1115, M1121, M1125, M1131, M1133, M1137, M1139, M1141, P7031, R3016, T7007
reg 2(1) .... Intro 14
  3(1), (2) .... Intro 14
  4 .... M1103, M1116
  5 .... M1103
    (1) M1117
    (4), (5) M1117
  6 .... M1103, M1110, M1111, M1129
  7 .... Intro 14, M1103, M1108, M1110
    (1) M1118
    (5) M1118

Control of Major Accident Hazards Regulations 1999 (SI 1999 No 743) – *cont.*
reg 7(12), (13) M1120
  8 .... Intro 14, M1103, M1108
  9 .... Intro 14, M1103, M1108
    (1) D6012, M1122
    (3)(a) M1122
  10 ...... Intro 14, M1103, M1108, M1127
    (3) M1122
    (6), (7) D6013, M1124
  11 .... Intro 14, D6067, M1103, M1108, M1127
    (1) D6014, M1126
  12 .... Intro 14, D6014, M1103, M1108, M1126
  13 .... Intro 14, M1103, M1108
  14 .... Intro 14, M1103, M1108
    (1) M1128
  15(1), (2) M1103
    (3) M1103, M1112
    (4) A3002, M1103
  16 .... M1103
  17 .... M1103
    (1)(a) M1118
  18 .... M1103
    (1) M1113
  19 .... M1103, M1112
  20 .... M1103
    (2) M1114
  21–23 .... M1103
Sch 1 .... M1103, M1138
Pt 2 .... M1104, M1108
  column 2 .... Intro 14
        3 .... Intro 14
Pt 3 .... M1108, M1129
  column 2 .... Intro 14
        3 .... Intro 14
Sch 2 .... M1103, M1120
  para 4 .... M1117
Sch 3 .... M1103, M1110
Sch 4 .... M1103
Pt 1 .... M1119
  para 1 .... M1121
      3(a) M1118
Pt 2 .... M1118, M1120
Sch 5 .... M1103
Pt 1 .. D6012, D6013, M1122, M1124
Pt 2 .... D6012, M1122
Sch 6 .... M1103, M1128, M1129
Sch 7 .... M1103
Sch 8 .... M1103, M1111

Control of Major Accident Hazards Regulations (Amendment) Regulations 2005 (SI 2005 No 1088)
  ..... D6007, D6011, D6071, M1106, M1141
reg 3 .... M1110
  6(3A) M1110
  6(4)(ba) M1110
  7 .... M1118

Control of Major Accident Hazards Regulations (Amendment) Regulations 2005 (SI 2005 No 1088) – *cont.*
reg 8(1), (2) M1121
(4) M1121
9(3)(a) D6012
11 .... D6013, M1124, M1126
12 .... D6014
13 .... M1128
Sch 1
Pt 3 .... M1104
note 3 .... M1105
Sch 3
para 5 .... M1110

Control of Noise at Work Regulations 2005 (SI 2005 No 1643)
.. Intro 10, Intro 14, E17032, N3002, N3009, N3012, N3013, N3023, N3029, N3035, O1025, P3017, V12019
reg 5 .... R3008
7 .... P3008
9 .... O1026
(1) C8018
10 .... P9014
Schedule N3007

Control of Pesticides Regulations 1986 (SI 1986 No 1510) V12021

Control of Substances Hazardous to Health Regulations 1988 (SI 1988 No 1657)
.... Intro 8, R3001

Control of Substances Hazardous to Health Regulations 1999 (SI 1999 No 437)
.... D0401, H2140

Control of Substances Hazardous to Health Regulations 2002 (SI 2002 No 2677)
.... Intro 2, Intro 14, C7008, C8002, C8018, D0902, D0904, D0909, D0912, D6016, D6017, D6071, H2101, H2117, H2118, H2120, H2122, H2123, H2126, H2137, H2138, M1024, M2009, M3002, O1005, O1018, O1041, O1045, O1046, O7009, R3023, R3028, S3018, T7007, V3001, V3011, V3016, V12003, V12013, V12021, W9009
reg 2 .... V3015
(1) H2104
3(1) C6505, H2105, H2106
4, 5 .... H2105
6 .... H2105, R3007
(1)(a) H2107
(2) H2106
(3) H2125
7 ... H2105, H2114, H2135, P3008
(1), (2) H2107
(3) H2107, H2108, H2113
(b) V3015
(4) H2107, H2108
(b) V3015
(d) V3015

Control of Substances Hazardous to Health Regulations 2002 (SI 2002 No 2677) – *cont.*
reg 7(5), (6) H2108, H2113
(7) H2111
(a)–(c) V3015
(9) H2110
(10) H2113
(11) H2111
8 .. H2105, H2114, H2127, P3008, V3015
9 ... H2105, H2114, H2125, P3008
(1) H2128, H2131
(a), (b) V3015
(2) H2128, H2129
(a) V3015
(3) H2128, H2131
(4) H2128, V3015
(5)–(7) H2128, H2131
10 .... H2105, H2114, H2125
(1)–(7) H2132
11 .... H2105, H2114, H2125
(1)–(11) H2133
12 .... H2105, H2114
(1)–(3) H2134
(4) H2115, H2129, H2134, V3015
(5) H2134
13 .... H2105, H2114
(3)(b) D6015
(4), (5) H2135
21 .... E15037
Sch 2 .... H2105
Sch 2A H2111, V3015
Sch 3 .... H2113
Sch 4 .... H2130, V3015
Sch 5 .... H2132
Sch 6 .... H2133

Control of Substances Hazardous to Health (Amendment) Regulations 2004 (SI 2004 No 3386) C8018, O1041, O1046

Control of Vibration at Work Regulations 2005 (SI 2005 No 1093)
... Intro 14, E17033, O1031, R3016, V5009, V5015, V5016, V12019
reg 6(1) C8018
(4) V5001, V5015
7 .... C8018

Control of Vibration at Work Regulations (Northern Ireland) 2005 (SI 2005 No 397) V5015

Controlled Waste Regulations 1992 (SI 1992 No 588) A5030

Controlled Waste (Registration of Carriers and Seizure of Vehicles) Regulations 1991 (SI 1991 No 1624)
.... C8020

Controlled Waste (Registration of Carriers and Seizure of Vehicles) (Amendment) Regulations 1998 (SI 1998 No 605) A5030

Criminal Jurisdiction (Offshore Activities) Order 1987 (SI 1987 No 2198) O7009

D

Dangerous Substances and Explosive Atmospheres Regulations 2002 (SI 2002 No 2776) ..... Intro 15, D0401, D0901, D0902, D0903, D0904, D0905, D0906, D0916, D0921, D0928, D0931, D0940, D0941, D0942, D0949, D0950, D0951, D0952, E3027, F5001, F5031, F5045, F5062, G1001, H2101, H2141, H2142, S3024, V3001
reg 1 .... D0911
2 .... D0908, D0909
3 .... D0908
4(1)(a), (b) D0907
5 .. D0925, D0943, D0944, D0945, D0946, D0948, H2143, R3014
(1), (2) D0912
(3) D0914
(4) D0913
(5) D0913, D0914
6 .. D0943, D0944, D0945, D0946, D0948, H2143, R3014
(1) D0915
(3) D0917, D0918
(4) D0917
(5) D0918
(6) D0919
(8) D0919, G1016
7 .. D0943, G1026, H2143, R3014
(2) D0921, D0924
(3) D0922
(4) D0923
(5) D0929
8 .. D0943, D6017, H2143, R3014
(1) D0934
(2)(a), (b) D0934
(3)(a) D0934
(4) D0934
9 .. D0935, D0943, H2143, R3014
10 .... D0919, D0936, D0943
11 .... D0907, D0943, G1026
12 .... D0908
13, 14 .... D0937
15 .... D0938
16 .... D0939
17 .... D0911, D0943

Dangerous Substances and Explosive Atmospheres Regulations 2002 (SI 2002 No 2776) – cont.
Sch 1 .... D0919, D0943
Sch 2
para 1 .... D0920
2 .... D0921
Sch 3 .... D0924
Sch 4 .... D0922
Sch 5 .... D0936
Sch 6 .... D0943
Pt 1, Pt 2 .... D0938
Sch 7 .... D0943
Pt 1, Pt 2 .... D0939

Dangerous Substances in Harbour Areas Regulations 1987 (SI 1987 No 37) D0404, D0938, D0939, F5062

Dangerous Substances (Notification and Marking of Sites) Regulations 1990 (SI 1990 No 304) H2101
Sch 3 .... H2141

Detention of Food (Prescribed Forms) Regulations 1990 (SI 1990 No 2614) F9014

Disability Discrimination (Employment Relations) Regulations 1996 (SI 1996 No 1456) W11043

Diving at Work Regulations 1997 (SI 1997 No 2776) ... O7004, O7057, V12003
reg 3(1), (2) O7036
4 .... O7036
5(1) O7037
(2) O7038
6 .... O7037
(3)(b) P7013
7 .... O7037
8 .... O7037, O7038
9 .... O7037
10, 11 .... O7037.1
12(1) O7037.2
(3) O7037.2
13(1) O7037.2, V12019
14(1) O7036
15 .... V12019
(6) O7036
Annex 2 .... O7037.2

Diving Operations at Work Regulations 1981 (SI 1981 No 399) O7036

Docks Regulations 1988 (SI 1988 No 1655) V12015

Dry Cleaning (Metrication) Regulations 1983 (SI 1983 No 977) D0939

Dry Cleaning Special Regulations 1949 (SI 1949 No 2224) D0939

E

Electrical Equipment (Safety) Regulations
1994 (SI 1994 No 3260)
.... E3029, E3031, L3002, M1017
Electricity at Work Regulations 1989 (SI
1989 No 635) .... Intro 15, C7008,
C8018, E3009, E3032,
E3033, O7063, V12027
Pt I (regs 1–3) E3003
reg 2 .... E3003, E3004, E3021
Pt II (regs 4–16) E3003, E3010
reg 4 .... E3010
(2) E3021
(3) E3011, E3022, E3028
(4) E3024
5 .... E3010, E3012
6 .... E3010, E3013, E3025, E3026
(d) F5062
7 .... E3005, E3010, E3014, E3024
8 .... E3005, E3010, E3015
9 .... E3010, E3016
10 .... E3010, E3017
11 .... E3010, E3018
12 .... E3010, E3019, E3023
13 .... E3010, E3019, E3022
14 .... E3010, E3022, E3024
15 .... E3010, E3020, L5006
16 .... E3010, E3011
Pt III (regs 17–28) E3003
Pt IV (regs 29–33) E3003
Electricity at Work Regulations (Northern
Ireland) 1991 (SI 1991 No 13)
.... E3003
Electricity (Factories Act) Special
Regulations 1944 (SI 1944 No 739)
.... E3020
Electricity Safety, Quality and Continuity
Regulations 2002 (SI 2002 No 2665)
.... E3032
Electromagnetic Compatibility Regulations
1992 (SI 1992 No 2372)
.... L3002, M1017
Electromagnetic Compatibility Regulations
2006 (SI 2006 No 3418)
.... M1017
Employed Earners' Employments for
Industrial Injuries Purposes Regulations
1975 (SI 1975 No 467)
reg 2–7 .... C6008
Sch 1, Sch 2 .... C6008
Sch 3 .... C6009
Employers' Health and Safety
Policy Statements (Exception)
Regulations 1975 (SI 1975 No 1584)
.... M2031, S7002
Employer's Liability (Compulsory
Insurance) Regulations 1998 (SI 1998
No 2573) .... Intro 15

Employer's Liability (Compulsory Insurance)
Regulations 1998 (SI 1998 No 2573) –
cont.
reg 1(2) E13009, E13018
2 .... E13017, E13021
(1)(a)–(d) E13020
(2) E13020
3(1) E13011
4(3)–(5) E13013
5(4) E13013
7, 8 .... E13013
9 .... E13001
Sch 2 .... E13012
para 14 .... E13001
Employers' Liability (Compulsory
Insurance) (Amendment) Regulations
2008 (SI 2008 No 1765)
.... Intro 15, E13013
Employers' Liability (Compulsory
Insurance) (Amendment) Regulations
2004 (SI 2004 No 2882)
.... E13012
Employment Act 2002 (Dispute Resolution)
Regulations 2004 (SI 2004 No 752)
.... E14008
Employment Equality (Age) Regulations
2006 (SI 2006 No 1031)
.... E14002, H1701
reg 6 .... H1702
Employment Equality (Religion and Belief)
Regulations 2003 (SI 2003 No 1660)
.... E14002, H1701
reg 5 .... H1702
Employment Equality (Repeal of Retirement
Age Provisions) Regulations 2011 (SI
2011 No 1069) M1505, M1511
Employment Equality (Sex Discrimination)
Regulations 2005 (SI 2005 No 2467)
.... H1702
Employment Equality (Sexual Orientation)
Regulations 2003 (SI 2003 No 1661)
.... E14002, H1701, H1707
reg 5 .... H1702
Employment Tribunals (Constitution and
Rules of Procedure) Regulations 2004
(SI 2004 No 1861)
Sch 4 .... E15019, E15021
Environmental Information Regulations
2004 (SI 2004 No 3391)
.... E5041
Environmental Permitting (England and
Wales) Regulations 2007 (SI 2007 No
3538) A5030, E5027, E5040
reg 7 .... E5041
17 .... E5041
22 .... E5041
34(2) E5041
37 .... E5041

Environmental Permitting (England and
    Wales) Regulations 2007 (SI 2007 No
    3538) – *cont.*
reg 57 .... E5041
Sch 1 .... E5041
Sch 5 .... E5041
Sch 7–Sch 18B E5041
Sch 19
para 1 .... E5041
Environmental Protection (Controls on
    Ozone-Depleting Substances)
    Regulations 2002 (SI 2002 No 528)
    .... E5026.1
Environmental Protection (Duty of Care)
    Regulations 1991 (SI 1991 No 2839)
    .... A5027, A5030
Equality Act 2010 (Commencement No. 4,
    Savings, Consequential, Transitional,
    Transitory and Incidental Provisions
    and Revocation) Order 2010
    (Amendment) Order 2010 (SI 2010 No
    2337)
art 7 .... H1702
Equipment and Protective Systems Intended
    for Use in Potentially Explosive
    Atmospheres Regulations 1996 (SI
    1996 No 192) D0902, D0924,
                        D0928, E3027

F

Factories (Testing of Aircraft Engines and
    Accessories) (Metrication) Regulations
    1983 (SI 1983 No 979) (Now revoked)
    .... D0939
Factories (Testing of Aircraft Engines and
    Accessories) Special Regulations 1952
    (SI 1952 No 1689)
    .... D0939, F5062
Fire and Rescue Services (Northern Ireland)
    Order 2006 (SI 2006 No 1254)
    .... F5033
Fire Certificates (Special Premises)
    Regulations 1976 (SI 1976 No 2003)
    .... D0938, F5031
Fire Precautions (Workplace) Regulations
    1997 (SI 1997 No 1840)
    .... F5031, F5033, R3012
Sch 1 .... D0938
Fire Safety (Scotland) Regulations 2006 (SSI
    2006 No 456) F5033
Fixed-term Employees (Prevention of Less
    Favourable Treatment) Regulations
    2002 (SI 2002 No 2034)
    .... E14002

Food Hygiene (England) Regulations 2006
    (SI 2006 No 14) F9002, F9004,
                        F9021, F9023, F9026,
                        F9035, F9036, F9046,
                        F9060
reg 6 .... F9015
    7 .... F9016
    8 .... F9018
    9 .... F9019
    27 .... F9014
    30 .... F9034, F9052, F9058
Sch 4 .... F9034, F9050, F9052
para 2 .... F9053
    (c) F9066
    4 .... F9058, F9066
Food Hygiene (Scotland) Regulations 2006
    (SI 2006 No 3) F9063
reg 30 .... F9064
Sch 4 .... F9064
Food Labelling Regulations 1996 (SI 1996
    No 1499) .... Intro 15, F9003, F9012,
                        F9051
Food Labelling (Amendment) Regulations
    1998 (SI 1998 No 1398)
    .... F9003
Food Labelling (Amendment) (England) (No
    2) Regulations (SI 2004 No 2824)
    .... F9003
Food Labelling (Amendment) (England) (No
    2) Regulations (SI 2005 No 2057)
    .... F9003
Food Labelling (Declaration of Allergens)
    (England) Regulations (SI 2007 No
    3256) F9003
Food Labelling (Declaration of Allergens)
    (England) Regulations 2008 (SI 2008
    No 1188) .... Intro 15, F9003
Food Premises (Registration) Regulations
    1991 (SI 1991 No 2825)
    .... F9021, F9026, F9034
Food Safety Act 1990 (Amendment)
    Regulations 2004 (SI 2004 No 2990)
    .... F9002, F9006
Food Safety (General Food Hygiene)
    Regulations 1995 (SI 1995 No 1763)
    .... F9034
Food Safety (Improvement and
    Prohibition—Prescribed Forms)
    Regulations 1991 (SI 1991 No 100)
    .... F9015
Food Safety (Temperature Control)
    Regulations 1995 (SI 1995 No 2200)
    .... F9034, F9050
Freight Containers (Safety Convention)
    Regulations 1984 (SI 1984 No 1890)
    .... E15005, O7009

G

Gas Appliances (Safety) Regulations 1992
(SI 1992 No 711) G1015

Gas Appliances (Safety) Regulations 1995
(SI 1995 No 1629)
.... G1026

Gas Cylinders (Pattern Approval)
Regulations 1987 (SI 1987 No 116)
.... D0445

Gas Safety (Installation and Use)
Regulations 1984 (SI 1984 No 1358)
.... G1028

Gas Safety (Installation and Use)
Regulations 1994 (SI 1994 No 1886)
.... G1028, G1029, G1031
reg 35A G1033

Gas Safety (Installation and Use)
Regulations 1998 (SI 1998 No 2451)
.... G1001, G1026
reg 2 .... G1010, G1020
3 .... G1011, G1014
(1) G1013
(3) G1033
4 .... G1013
5(1) G1013
(2), (3) G1015
6(1)–(10) G1016
9 .... G1020
12(6) G1022
13 .... G1022
14(1)(a)–(c) G1022
(2)–(4) G1022
15–17 .... G1022
18–23 .... G1023
26(1) G1033
(2) G1017
(6)–(9) G1017
27 .... G1017
(1) G1019, G1033
(2)–(5) G1019
28, 29 .... G1017
30(1) G1018
(2), (3) G1018, G1024
(4) G1018
31 .... G1017
33(1), (2) G1024
34 .... G1017
35 .... G1024
36(2) G1024
(3) G1024, G1025
(6)(a), (b) G1025
(7), (8) G1025
(10) G1024
37 .... G1010
(1)–(3) G1012
(8) G1012
38 .... G1035

Gas Safety (Management) Regulations 1996
(SI 1996 No 551) G1001, G1009
reg 3 .... G1003
(1) G1004
5 .... G1005
6, 7 .... G1004, G1005
8 .... G1004
16 .... E15005
Sch 1 .... G1004

Gas Safety Regulations 1972 (SI 1972 No
1178) G1030

Gas Safety (Rights of Entry) Regulations
1996 (SI 1996 No 2535)
.... G1001
reg 4 .... G1006
5–8 .... G1007
9 .... G1008

General Food Regulations 2004 (SI 2004
No 3279) F9002, F9006, F9008

General Product Safety Regulations 1994
(SI 1994 No 2328)
.... P9002

General Product Safety Regulations 2005
(SI 2005 No 1803)
.... P9001, P9026
reg 2 .... P9003
3(1) P9002
5 .... P9006
(1) P9002
6(1), (2) P9003
7(1) P9006
(a), (b) P9002
(3) P9002, P9006
(4) P9002
8(1)(a) P9002, P9006
9(1) P9002, P9006
(2)(b) P9002
(4) P9002
11–15 .... P9002
20(2) P9006
29(1)–(4) P9008
21 .... P9007

Genetically Modified Food (England)
Regulations 2004 (SI 2004 No 2335)
.... F9020

Genetically Modified Organisms (Deliberate
Release) Regulations 2002 (SI 2002 No
2443) F9020

Genetically Modified Organisms
(Traceability and Labelling) (England)
Regulations 2004 (SI 2004 No 2412)
.... F9020

Good Laboratory Practice Regulations 1997
(SI 1997 No 654) D0936

Greenhouse Gas Emissions Trading Scheme
Regulations 2005 (SI 2005 No 925)
.... E5069

H

Hazardous Waste (England and Wales)
Regulations 2005 (SI 2005 No 894)
.... A5027, A5030

Health and Safety at Work etc. Act 1974
(Application Outside Great Britain)
Order 1977 (SI 1977 No 1232)
.... O7003, O7009

Health and Safety at Work etc. Act 1974
(Application Outside Great Britain)
Order 1989 (SI 1989 No 840)
.... O7009

Health and Safety at Work etc. Act 1974
(Application Outside Great Britain)
Order 2001 (SI 2001 No 2127)
...... C8501, D0908, O7005, O7009,
O7029, W9003

Health and Safety at Work etc. Act 1974
(Application Outside Great Britain)
(Variation) Order 1995 (SI 1995 No
263)
reg 2(1) O7009

Health and Safety (Consultation with
Employees) Regulations 1996 (SI 1996
No 1513) .. Intro 15, D0948, E14009,
J3002, J3019, J3022, J3023,
J3030, J3031, V8008
reg 3 .... J3005, J3021
5(1) A3028
(3) J3027
7(1) J3028
(b) J3024
(2) J3025
Sch 1 .... J3026
Sch 2 .... J3029

Health and Safety (Display Screen
Equipment) Regulations 1992 (SI 1992
No 2792) ... Intro 15, D8201, D8203,
D8210, D8213, D8220,
E17044, E17052, E17064,
L5002, O1005, O1036,
O1051, T7007, V12003,
W11009
reg 1 .... D8202, D8211
2 .... D8204, R3010
3 .... D8205
4 .... D8206
5 .... D8207
6 .... D8208
7 .... D8209
Schedule D8205, S11015

Health and Safety (Enforcing Authority)
Regulations 1998 (SI 1998 No 494)
.... D0940, E15008
reg 2 .... E15009
4(4)(b) E15009
3(1) E15009, E15010
(3) E15009

Health and Safety (Enforcing Authority)
Regulations 1998 (SI 1998 No 494) –
cont.
reg 3(4)(b) E15009
(5) E15009
5 .... E15011
6(1) E15011
Sch 1 .... E15009, E15010
Sch 2 .... E15009

Health and Safety (Fees) Regulations 2005
(SI 2004 No 676) O7004, O7022

Health and Safety (Fees) Regulations 2008
(SI 2008 No 736)
reg 2–5 .... E15005
8, 9 .... E15005
11 .... E15005
13 .... E15005
15 .... E15005

Health and Safety (Fees) Regulations 2009
(SI 2009 No 515) A5030

Health and Safety (First-Aid) Regulations
1981 (SI 1981 No 917)
..... Intro 16, D6021, D6071, F7001

Health and Safety Information for
Employees (Amendment) Regulations
2009 (SI 2009 No 606)
.... Intro 10

Health and Safety (Miscellaneous
Amendments) Regulations 2002 (SI
2002 No 2174) ..... Intro 10, D8201,
E17064, M3002, P3002,
R3028, V12034,
W11001
reg 4(1) P3004

Health and Safety (Safety Signs and Signals)
Regulations 1996 (SI 1996 No 341)
..... Intro 16, C8002, C8019, D0936,
N3013, W11001, W11030
reg 2 .... W11034
3(1) W11036
4, 5 .... W11035
Sch 1 .... W11035
Pt I
para 3 .... W11032
4 .... W11033

Health and Safety (Training for
Employment) Regulations 1990 (SI
1990 No 1380) ..... Intro 16, C6518,
R3025, V12003, V12010

Health and Safety (Young Persons)
Regulations 1997 (SI 1997 No 135)
.... R3025

High-activity Sealed Radioactive Sources
and Orphan Sources Regulations 2005
(SI 2005 No 2686)
.... R1018

Highly Flammable Liquids and Liquefied
  Petroleum Gases Regulations 1972 (SI
  1972 No 917) D0901, D0939,
                F5062, H2141, R3014

I

Income Support (General) Regulations 1987
  (SI 1987 No 1967)
  .... C6041
Information and Consultation of Employees
  Regulations 2004 (SI 2004 No 3426)
  .... J3001, J3002, J3033
Ionising Radiations Regulations 1985 (SI
  1985 No 1333) O7009, O7066
Ionising Radiations Regulations 1999 (SI
  1999 No 3232) E13020, E15005,
                E15012, H2101, O1004,
                O1056, R1018, R3016,
                R3023, V3001, V12003,
                V12015, V12019
  reg 7 .... H2145, R1019
    8(5) R1019
    9 .... P3008, R1019
    12 .... R1021
    13 .... R1019
    15–20 .... R1019
    21(5) R1019
    24–26 .... R1019
  Sch 4 .... R1019
Ionising Radiation (Medical Exposure)
  Regulations 2000 (SI 2000 No 1059)
  .... R1018
Ionising Radiation (Medical Exposures)
  (Amendment) Regulations 2006 (SI
  2006 No 2523) R1018

J

Justification of Practice Involving Ionising
  Radiation Regulations 2004 (SI 2004
  No 1769) R1018

L

Lifting Operations and Lifting Equipment
  Regulations 1998 (SI 1998 No 2307)
  ..... Intro 16, C8018, F3034, L3001,
  L3032, L3037, L3039, M1001, M1004,
  O7070, R3017, R3051, W9002
  reg 4 .... L3015
    5(1)(b) L3016
    6 .... L3017
    7 .... L3014, L3018
    8 .... L3019, L3020

Lifting Operations and Lifting Equipment
  Regulations 1998 (SI 1998 No 2307) –
  cont.
  reg 9 .... L3023
    (1), (2) L3021, L3027
    (3) . L3021. L3026, L3027, S3004
      (b) L3028
    (4) L3021, W9012
    10 .... L3023, L3026
    (1), (2) L3027
    11(1) L3027
    (2)(a), (b) L3027
  Sch 1 .... L3026
Lifts Regulations 1997 (SI 1997 No 831)
  .... L3002, L3021
  reg 3–7 .... L3004
    8 .... L3007, L3010, L3012
    (1) L3004
    (2) L3008, L3011
    (3) L3004
    9 .... L3007, L3010, L3012
    (1) L3006
    (2) . L3004, L3006, L3008, L3011
    (3) L3006
    10 .... L3007, L3008, L3010
    11 .... L3008, L3012
    12 ... L3004, L3006, L3007, L3010
    13 .... L3011, L3012, L3013
    (1) L3004, L3006
    (2)(a)–(c) L3008
    14 .... L3012
    15 .... L3013
    18(1) L3004, L3006
    22 .... L3008, L3008
  Sch 3 .... L3004, L3006
  Sch 5–Sch 13 .... L3011
  Sch 14 .... L3004
Low Voltage Electrical Equipment (Safety)
  Regulations 1989 (SI 1989 No 728)
  .... E3031, P9008

M

Magnesium (Grinding of Castings and
  Other Articles) Special Regulations
  1946 (SR&O 1946 No 2017)
  .... D0939, F5062
Management of Health and Safety at Work
  and Fire Precautions (Workplace)
  (Amendment) Regulations 2003 (SI
  2003 No 2457) R2004
Management of Health and Safety at Work
  Regulations 1992 (SI 1992 No 2051)
  ...... C8018, G1028, J3002, M2005,
                T7008, V12017
  reg 6 .... T7004

Management of Health and Safety at Work
    Regulations 1999 (SI 1999 No 3242)
    ... Intro 16, Intro 18, A1010, A1013,
    A5030, C6501, C7002, C7008, C8508,
    D0904, D0948, D6015, D6067, E3003,
    E13006, E14023, E14024, E14025,
    E15012, E17044, E17062, F7004,
    G1026, H2127, H2135, J3002, J3019,
    L3019, M2001, M2002, M2103,
    M2125, O1005, O1036, O1041, O1045,
    O1051, O1055, P7031, R1018, R3019,
    R3020, R3021, R3043, R3047, T7007,
    V3001, V5016, V8006,V12001,
    V12002, W9002, W10020, W11035
reg 1 .... D6010, R3026, V12017
    (2) R3025, V12003, V12007
2(2) V12011
3   . Intro 8, A1023, M2005, M2008,
    M2020, R1010, R3005, S3004,
    S3016, S7006, S11007, V8044,
    W9007
    (1) C6505, M1006, M1007,
    M2019, R3018, V12008, V12018
    (a) S3001
    (2) R3018
    (3) A3030, R3018, R3022
    (4) M2006, R3018, V12003,
    V12008, V12013
    (5) M2006, T7006, V12003,
    V12008, V12016
    (b)–(g) V12013
    (6) R3018, R3048
4   ... M2007, R3023, S3004, V8072
5   . C6505, D6039, M1131, M2005,
    M2008, R1010, R3005, R3023,
    R3024, R3048, S3004, S7006,
    S7007, V8129
6   . M2009, M3007, O1009, R3023,
    V12013
7   .... Intro 24, H2115, M1007,
    M2005, R1010, S3016, S7003,
    T7005
    (1) T7004
    (5) T7004
    (8) M2010, R3029
8   . D0934, D6039, M1130, M1131,
    M2005, M2011, M2039, R3023,
    V8014
    (1) D6010
    (b) T7005
    (2) D6010
    (3) D6010, T7005
9   . A1015, D6010, D6039, M1130,
    M1131, M2012
10(1) V12014
    (2) J3015, M2013, V12010,
    V12014
    (3) V12010
11   .... A3005, M2014, M2015,
    R1019
12   .... M2015

Management of Health and Safety at Work
    Regulations 1999 (SI 1999 No 3242) –
    cont.
reg 13 .... M2016, W9007
    (1) V12012, V12036
    (2) T7006, V12038
14 .... M2017
15 .... M2018, T7005
16 .... M2006, M2019, R3018
    (1) R3026, V12018, V12023
    (b) V12024
    (2), (3) E14025, E14027,
    R3026, V12018
    (4) R3026, V12018
17 .... E14026, M2019, R3026,
    V12018
18 .... E14027, M2019, V12018
    (2), (3) R3026
19 .... R3018, V12003
    (1) V12009, V12013
    (2) ... V12009, V12013, V12016
    (3) V12009, V12013
20 .... M2021
21 .... E15039
22 .... E15039, R2004
23–30 .... E15039
Sch 1 .. C8519, M2007, R3023, V8072

Management of Health and Safety at Work
    (Amendment) Regulations 2006 (SI
    2006 No 438)
reg 6 .... R2004

Manual Handling Operations Regulations
    1992 (SI 1992 No 2793)
    .. Intro 10, Intro 17, C8018, E17044,
    E17052, E17063, M3001, M3008,
    M3017, O1005, O1051, R3051, T7007,
    V12003, V12019
reg 2(2) M3002
4 .... R3009, W9004
    (1) M3002
    (2) M3002, M3026
    (3) M3002
    (c) M3015
    (e) M3007
    5–8 .... M3002
Sch 1 .... M3002, M3003, W9004

Manufacture and Storage of Explosives
    Regulations 2005 (SI 2005 No 1082)
    .... E15005

Manufacture of Cinematograph Film
    Regulations 1928 (SR&O 1928 No 82)
    .... D0939, F5062

Maternity and Parental Leave etc
    Regulations 1999 (SI 1999 No 3312)
    .... E14002

Medical Devices Regulations 1994 (SI 1994
    No 3017) P7013

Medicines (Administration of Radioactive Substances) (Amendment) Regulations 1995 (SI 1995 No 2147) R1018

Merchant Shipping and Fishing Vessels (Control of Vibration at Work) Regulations 2007 (SI 2007 No 3077) V5015

Merchant Shipping (Dangerous Goods and Marine Pollutants) Regulations 1990 (SI 1990 No 2605) D0404

Merchant Shipping (Dangerous Goods and Marine Pollutants) Regulations 1997 (SI 1997 No 2367) O7068

Merchant Shipping (Prevention of Pollution) (Law of the Sea Convention) Order 1996 (SI 1996 No 282) O7067

Merchant Shipping (Prevention of Pollution) Order 1983 (SI 1983 No 1106) O7067

Merchant Shipping (Prevention of Pollution) Regulations 1996 (SI 1996 No 2154) reg 1(2) O7067

N

Noise at Work Regulations 1989 (SI 1989 No 1790) N3002, N3012, N3023, O7063, T7007, V5009

Noise Emission in the Environment by Equipment for Use Outdoors Regulations 2001 (SI 2001 No 1701) N3016

Notification of Installations Handling Hazardous Substances Regulations 1982 (SI 1982 No 1357) .... M1103 Sch 2 .... D0909

Notification of Installations Handling Hazardous Substances (Amendment) Regulations 2002 (SI 2002 No 2979) D0909

O

Offshore Chemicals Regulations 2002 (SI 2002 No 1355) O7067

Offshore Combustion Installations (Prevention and Control of Pollution) Regulations 2001 (SI 2001 No 1091) O7067

Offshore Electricity and Noise Regulations 1997 (SI 1997 No 1993) E3003, O7039, O7063

Offshore Installations and Pipeline Works (First-Aid) Regulations 1989 (SI 1989 No 1671) .... F7003, O7039 reg 5(1) .... O7057 (a), (b) .... O7058 (c) O7062

Offshore Installations and Pipeline Works (Management and Administration) Regulations 1995 (SI 1995 No 738) ...... O7004, O7005, O7011, O7029 reg 2(1) .... O7009 3 .... F5034 (3) .... O7006 5, 6 .... O7030 7 .... O7031 8 .... O7032 9–19 .... O7030

Offshore Installations and Wells (Design and Construction, etc.) Regulations 1996 (SI 1996 No 913) ..... O7004, O7011, O7012, O7016, O7022, O7024, O7063 reg 5(1)(a)–(e) .... O7025 (2) .... O7025 6–11 .... O7025 13–19 .... O7026 20 .... O7027 21 .... O7026 Sch 1 .... O7025

Offshore Installations (Construction and Survey) Regulations 1974 (SI 1974 No 289) .... O7012 reg 26 .... O7024 Sch 2 .... O7024

Offshore Installations (Inspections and Casualties) Regulations 1973 (SI 1973 No 1842) .... O7004, O7039, O7041 reg 2 .... O7042 4 .... O7043 5 .... O7044

Offshore Installations (Logbooks and Registration of Death) Regulations 1972 (SI 1972 No 1542) .... O7004, O7039 reg 8–11 .... O7040 Schedule O7040

Offshore Installations (Prevention of Fire and Explosion, and Emergency Response) Regulations 1995 (SI 1995 No 743) ...... F5062, O7004, O7011 reg 4–10 .... O7028 11(1) .... O7028 12, 13 .... O7028 14(1)–(4) .... O7028 15–22 .... O7028

Offshore Installations (Safety Case) Regulations 1992 (SI 1992 No 2885) ...... O7004, O7010, O7012, O7033 reg 4(1) O7022

Offshore Installations (Safety Case) Regula-
    tions 1992 (SI 1992 No 2885) – *cont.*
    reg 4(2) O7013
        6 .... O7022
        (1), (2) O7013
        7 .... O7013, O7022
        8 .... O7013, O7064
        (1)(b) O7023
        9 .... O7013
        (2) O7022
        10 .... O7019
        11 .... O7014
        12 .... O7016
        17 .... O7020, O7022
        19 .... O7017
        20 .... O7018
    Sch 1, Sch 2 .... O7013
    Sch 4 .... O7019
    Sch 5 .... O7014
    Sch 6 .... O7020
    Sch 7 .... O7017, O7018
Offshore Installations (Safety Case)
    Regulations 2005 (2005 No 3117)
    .... O7004, O7010, O7012, O7022,
                O7028, O7029, O7046
    reg 19–21 .... O7024
    Sch 7 .... O7024
Offshore Installations (Safety
    Representatives and
    Safety Committees) Regulations 1989
    (SI 1989 No 971) J3002, O7004,
                        O7010, O7039,
                        O7046, O7053
    reg 5–10 .... O7047
        16 .... O7048
        17 .... O7049, O7050
        (4)(a) O7051
        18, 18A O7050
        19, 20 .... O7052
        22–24 .... O7054
        26, 27 .... O7055
        28(1), (2) O7055
Offshore Installations (Safety Zones) Order
    2004 (SI 2004 No 343)
    .... O7007
Offshore Installations (Safety Zones)
    Regulations 1987 (SI 1987 No 1331).
    .... O7007
Offshore Petroleum Activities (Oil Pollution
    Prevention and Control) Regulations
    2005 (SI 2005 No 2055)
    .... O7067
Offshore Safety (Miscellaneous
    Amendments) Regulations 2002 (SI
    2002 No 2175) O7029
Offshore Safety (Repeals and Modifications)
    Regulations 1993 (SI 1993 No 1823)
    .... O7004

P

Packaging, Labelling and Carriage of
    Radioactive Material by Rail
    Regulations 1996 (SI 1996 No 2090)
    .... D0936
Packaging of Explosives for Carriage
    Regulations 1991 (SI 1991 No 2097)
    .... D0445
Part-time Workers (Prevention of Less
    Favourable Treatment) Regulations
    2000 (SI 2000 No 1551)
    .... E14002, W10004
Personal Protective Equipment at Work
    Regulations 1992 (SI 1992 No
    2966)Intro 17, A5030,
    .... C7008, E3024, E17066, O1026,
    O1031, O1036, O1041, O1046, P3001,
    P3002, P3004, P3008, P3017, P3027,
                            V12003
    reg 4(1) P3006
        (3) P3006
        (e) P3013
        5 .... P3006
        6 .... R3011
        (1)–(3) P3006
        7–9 .... P3006
        10 .... P3006
        (2) P3007
        (4) P3007
        11 .... P3007
Personal Protective Equipment Regulations
    2002 (SI 2002 No 1144)
    .... P3012, R1018, R1019
Personal Protective Equipment (EC
    Directive) (Amendment) Regulations
    1994 (SI 1994 No 2326)
    .... P3002, P3004, P3025
Personal Protective Equipment (EC
    Directive) Regulations 1992 (SI 1992
    No 3139) P3001, P3002, P3004,
            P3009, P3012, P3013, P3014
Petroleum (Consolidation) Act 1928
    (Enforcement) Regulations 1979 (SI
    1979 No 427) D0938
Petroleum (Carbide of Calcium) Order
    1929 (SI 1929 No 992)
    .... D0939
Petroleum (Carbide of Calcium) Order
    1947 (SI 1947 No 1442)
    .... D0939
Petroleum (Compressed Gases) Order 1930
    (SI 1930 No 34) D0939
Petroleum (Liquid Methane) Order 1957 (SI
    1957 No 859) D0938
Petroleum Spirit (Motor Vehicles etc)
    Regulations 1929 (SI 1929 No 952)
    .... D0938, M1103

Petroleum Spirit (Plastic Containers)
  Regulations 1982 (SI 1982 No 630)
    .... D0938, M1103
Pipelines Safety Regulations 1996 (SI 1996
  No 825) G1001, O7024, P7031
  reg 3 .... E15009
    (4) G1009
    5–10 .... G1009, O7035
    11 .... G1009, O7016, O7035
    12–14 .... G1009, O7035
    16 .... G1009, O7035
    17, 18 .... G1009
    19 .... O7035
    20, 21 .... G1009, O7035
    22 .... G1009
      (1), (2) O7035
    23–26 .... G1009, O7035
Pipelines Safety (Amendment) Regulations
  2003 (SI 2003 No 2563)
    .... G1009
Planning (Control of Major Accident
  Hazards) Regulations 1999 (SI 1999
  No 981) M1102, M1141
Planning (Hazardous Substances)
  Regulations 1992 (SI 1992 No 656)
    .... M1136, M1138, M1141
Plugs and Sockets etc. (Safety) Regulations
  1994 (SI 1994 No 1768)
    .... E3029, P9003
Pneumoconiosis etc.
  (Workers' Compensation)
  (Determination of Claims) Regulations
  1985 (SI 1985 No 1645)
  reg 4(1), (2) C6061
    (3), (4) C6062
Pneumoconiosis etc.
  (Workers' Compensation) (Specified
  Diseases) Order 1985 (SI 1985 No
  2034) C6062
Pollution Prevention and Control (England
  and Wales) Regulations 2000 (SI 2000
  No 1973) E5027
Pressure Equipment Regulations 1999 (SI
  1999 No 2001) P7001, P7006,
                        P7007, P7010, P7013,
                        P7016, P7017
  Sch IV P7008
Pressure Equipment (Amendment)
  Regulations 2002 (SI 2002 No 1267)
  reg 25(a) P7010
Pressure Systems and Transportable
  Gas Containers Regulations 1989 (SI
  1989 No 2169) P7001, P7022
  reg 9(1)(a) P7022
Pressure Systems Safety Regulations 2000
  (SI 2000 No 128) D0401, G1026,
                        P7001, P7014, P7018,
                        P7019, P7020, P7031

Pressure Systems Safety Regulations 2000 (SI
  2000 No 128) – cont.
  reg 1 .... P7011
    2 .... P7012, P7015
    3(5) P7011
    4 .... P7016
    5 .... P7017
    9 .... P7022, P7024
    10 .... P7025
    11, 12 .... P7028
    13 .... P7029
    14 .... P7030
    15 .... P7027
    16 .... P7011
  Sch1
  Pt II P7013
  Pt II P7013
  Sch 2 .... P7011
  Sch 3 .... P7017
Pressure Vessels (Verification) Regulations
  1988 (SI 1988 No 896)
    .... D0445
Prevention of Accidents to Children in
  Agriculture Regulations 1998 (SI 1998
  No 3262) V12015
Probation Board (Northern Ireland) Order
  1982 (SI 1982 No 713 (NI 10))
  art 4 .... C9008
Provision and Use of Work Equipment
  Regulations 1992 (SI 1992 No 2932)
    .... E17065, M1003, M1004, O7041
  reg 4 .... L3001
    6(1) E15022
Provision and Use of Work Equipment
  Regulations 1998 (SI 1998 No 2306)
    ..... Intro 17, A1009, A1010, A5030,
    C6510, C7008, C8002, E3003, F3032,
    G1026, L3051, M1001, M1004,
  M1006, M3002, P7031, R3051, S11015,
    T7007, V5016, W9002, W11003
  Pt I (regs 1–3) M1005
  Pt II (regs 4–24) M1005, M1007
  reg 4 .... L3109, M1007
    (4) M1003
    5 .. A1011, A1015, A1020, M1007
    (1) P9010
    6 .... A1012, W9012
    (5) M1003, M1007
    7 .... M1007, O1031
    8 .... M1007
    9 .... M1007, P7026
    (1) S3001
    10 .... D0928, M1005
    (1), (2) M1003, M1007
    11(1) M1007
    (2) M1003
    (d) M1007
    (3) M1007
    12 .... E3026, M1007, M1008

Provision and Use of Work Equipment Regulations 1998 (SI 1998 No 2306) – *cont.*
  reg 13–17 .... M1007
      18 .... M1007
      (1) M1003
      19 .... E3019, E3023, M1007
      20 .... M1007
      21 .... L5004, M1007
      22–24 .... M1007
  Pt III (regs 25–30) M1005, M1008
  reg 25 .... M1008
      26 .... C8018, M1008
      27–30 .... M1008
  Pt IV (regs 31–35) M1005
  Pt V (regs 36–39) M1005
  Sch 1 .... M1007
Public Information for Radiation
  Emergencies Regulations 1992 (SI 1992
  No 2997) D6018, D6071
  reg 3 .... D6019
  Sch 2 .... D6019

R

Race Relations Act 1976 (Amendment)
  Regulations 2003 (SI 2003 No 1626)
  .... H1702
Radiation (Emergency Preparedness and
  Public Information) Regulations 2001
  (SI 2001 No 2975)
  ...... D6071, E15005, R1018, R1021
  reg 4–14 .... D6018
      16, 17 .... D6018
  Sch 5–Sch 10 .... D6018
Radioactive Material (Road Transport)
  Regulations 2002 (SI 2002 No 1093)
  .... D0445, D0446, R1018
Radioactive Material (Road Transport)
  (Amendment) Regulations 2003 (SI
  2003 No 1867) D0445, D0446,
                              R1018
Radioactive Material (Road Transport)
  (Great Britain) Regulations 1996 (SI
  1996 No 1350) D0936
Railways (Safety Case) Regulations 2000 (SI
  2000 No 2688) T7007
Registration of Births and Deaths
  Regulations 1987 (SI 1987 No 2088)
  Sch 2
  Form 14 .... A3010
Regulatory Reform (Fire Safety) Order
  2005 (SI 2005 No 1541)
  .... Intro 17, C8002, F5006, F5009,
  F5020, F5022, F5030, F5031, F5032,
  F5033, F5034, F5035, F5036, F5037,
                   F5045, F5061A, R3012
  Pt 2 .... F5046
  Pt 5 .... F5046
  art 8 .... C6507

Regulatory Reform (Fire Safety) Order 2005
  (SI 2005 No 1541) – *cont.*
  art 10 .... F5038
      12 .... F5044
      16 .... F5043
      17 .... F5047
      18 .... F5048
      (5) F5049
      19 .... F5050
      (2) F5052
      20 .... F5051
      21 .... F5053
      22 .... F5054
      23 .... F5055, F5060
      25, 26 .... F5056
      29 .... E15029, F5057
      30 .... E15029, F5058
      31 .... E15029, F5059
      37 .... F5046, F5060
      38 .... F5046
      40 .... F5060
      49 .... F5061
Reporting of Injuries, Diseases and
  Dangerous Occurrences Regulations
  1985 (SI 1985 No 2023)
  .... N3012
Reporting of Injuries, Diseases and
  Dangerous Occurrences Regulations
  1995 (SI 1995 No 3163)
  .... Intro 17, A3002, A3003, A3004,
  A3022, A3029, A3030, A3031, A5030,
  C6004, C6506, C8516, D6003, D6020,
  D6039, D6071, E3003, E17067, F7020,
  H2103, J3027, M1112, M1131, M2036,
  M2041, O1056, O7004, O7009, O7040,
  O7050, S7005, V5011, V5016, V8007,
                                W9001
  reg 2(1) A3001, A3007
      (a)–(c) A3005
      (2)(c) A3008.1
      3 .... A3019
      (1) A3001, A3011
      (2) A3008, A3012
      4 .... A3012, A3019
      5(1) A3010, A3012
      (2) A3012
      6(1) A3011, A3020
      (2) A3021
      7 .... A3025
      (3) A3014
      10(1) A3013
      (2) A3011, A3019
      (3) A3013
      11 .... A3023
  Sch 1 .... A3007
  Sch 2 .... A3009, Appendix A
  Sch 3 .... A3010, Appendix B
  Sch 4
  Pt I A3015
  Pt II A3016

Road Transport (Working Time)
Regulations 2005 (SI 2005 No 639)
.... M2103, W10026, W10029
Road Vehicles (Construction and Use)
Regulations 1986 (SI 1986 No 1078)
.... M2103

S

Sale and Supply of Goods to Consumers
Regulations 2002 (SI 2002 No 3045)
.... P9001
reg 3 .... P9050
5 .... P9048
14 .... P9057
Safety Representatives and
Safety Committees Regulations 1977
(SI 1977 No 500) .. Intro 15, Intro 17,
D0948, J3002, J3006,
J3010, J3011, J3017,
J3021, J3030, T7003,
T7023, V8008
reg 2 .... J3003
3 .... J3003
(2) J3005
4(1) J3007, J3009
(2) J3011, J3013
4A(1) J3007
(2) J3004
5 .... J3008
(3) J3014
6 .... J3007, J3008
7 .... J3007, J3027
(1) J3015
(2) A3028, J3015
8 .... J3005
9(1), (2) J3018
11 .... J3013
Sch 2 .... J3012, J3013
Sex Discrimination Act 1975 (Amendment)
Regulations 2008 (SI 2008 No 656)
.... H1702
Shipbuilding and Ship-repairing Regulations
1960 (SI 1960 No 1932)
.... D0939, V12015
reg 50, 51 .... P3008
Simple Pressure Vessel Regulations 1991 (SI
1991 No 2749) P7001, P7003
Simple Pressure Vessels (Safety) Regulations
1991 (SI 1991 No 2749)
.... P7003
Sch 1 .... P7005
Smoke-free (Exemptions and Vehicles)
Regulations 2007 (SI 2007 No 765)
.... W11028
Smoke-free (Penalties and Discounted
Amounts) Regulations 2007 (SI 2007

Smoke-free (Penalties and Discounted
Amounts) Regulations 2007 (SI 2007 –
cont.
No 764) W11028
Smoke-free (Premises and Enforcement)
Regulations 2006 (SI 2006 No 3368)
.... F5018, W11028
Smoke-free (Signs) Regulations 2007 (SI
2007 No 923) F5018, W11028
Smoke-free (Vehicle Operators and Penalty
Notices) Regulations 2007 (SI 2007 No
760) W11028
Social Security Benefits Up-rating Order
2003 (SI 2003 No 526)
.... C6017
Social Security (Claims and Payments)
Regulations 1979 (SI 1979 No 628)
reg 24(1) A3025
(3) A3025
25 .... A3025
31 .... A3025
Social Security (Claims and Payments)
Regulations 1987 (SI 1987 No 1968)
reg 19 .... C6012
Sch 4
para 3 .... C6012
Social Security (General Benefit) Regulations
1982 (SI 1982 No 1408)
reg 11 .... C6011
Social Security (Incapacity for Work)
(General) Regulations 1995 (SI 1995
No 311)
reg 13A C6002
17 .... C6002
Social Security (Industrial Injuries)
(Prescribed Diseases) Regulations 1985
(SI 1985 No 967) C6061, N3012,
N3018, N3019, V5011
reg 4(5) N3020
25(2) N3020
33 .... N3020
34 .... N3021
Sch 1 .... C6058
Pt 1 .... N3020
Sch 3
Pt 2 .... N3021
Social Security (Industrial Injuries)
(Prescribed Diseases) Amendment
Regulations 1989 (SI 1989 No 1207)
reg 4(2) N3021
(5) N3018
Social Security (Industrial Injuries)
(Prescribed Diseases) Amendment
Regulations 1990 (SI 1990 No 2269)
.... N3021

Social Security (Industrial Injuries)
(Prescribed Diseases) Amendment
Regulations 1993 (SI 1993 No 862)
reg 6(2) V5004

Social Security (Industrial Injuries)
(Prescribed Diseases) (Amendment No
2) Regulations 1987 (SI 1987 No
2112) N3019

Social Security (Industrial Injuries)
(Prescribed Diseases) Amendment
Regulations 1994 (SI 1994 No 2343)
.... N3018, N3019

Social Security (Recovery of Benefits)
(General) Regulations 1997 (SI 1997
No 2205)
reg 3 .... C6049
5 .... C6045, C6049
6 .... C6049
9 .... C6048
10 .... C6047

Special Waste Regulations 1996 (SI 1996
No 972) C8020

Submarine Pipe-lines (Inspectors etc.)
Regulations 1977 (SI 1977 No 835)
.... O7004

Supply of Machinery (Safety) (Amendment)
Regulations 1994 (SI 1994 No 2063)
reg 4 .... M1009
Sch 2
para 10 .... M1009

Supply of Machinery (Safety) Regulations
1992 (SI 1992 No 3073)
..... E3003, L3002, M1001, M1018,
N3014, N3016, R3016, V5016, V5020
reg 2(2) M1009
13–15 .... M1013
29(a) P9009
Sch 4 .... M1013

Supply of Machinery (Safety) Regulations
2008 (SI 2008 No 1597)
.... V5020

T

Town and Country Planning (General
Development Procedure) Order 1995
(SI 1995 No 419)
art 10(d) M1139
(zb) M1139

Transfer of Undertakings (Protection of
Employment) Regulations 2006 (SI
2006 No 246) E14011

Transnational Information and Consultation
of Employees Regulations 1999 (SI
1999 No 3323) .. J3001, J3002, J3031
Pt III J3033
Pt IV J3033
reg 17(1) J3033

Transnational Information and Consultation
of Employees Regulations 1999 (SI 1999
No 3323) – cont.
reg 17(3) J3033
18(1) J3033
Schedule
para 6 .... J3033
para 7(3) J3033

Transport of Dangerous Goods (Safety
Advisers) Regulations 1999 (SI 1999
No 257) D0402, D0445

Transportable Pressure Vessels Regulations
2001 (SI 2001 No 1426)
.... D0445, P7001

U

Unfair Terms in Consumer Contracts
Regulations 1999 (SI 1999 No 2083)
.... P9001, P9048
reg 4(1) P9047
5(1), (2) E13004, P9058
(4) E13004, P9058
6(2) P9058
7 .... E13004, P9058
8(1), (2) P9058
10, 11 .... P9058
12 .... E13004, P9058
Sch 1 .... E13004, P9058
Sch 2 .... E13004, P9059

W

Waste Management Licensing Regulations
1994 (SI 1994 No 1056)
.... A5030, C8020

Work at Height (Amendment) Regulations
2007 (SI 2007 No 114)
.... W9004

Work at Height Regulations 2005 (SI 2005
No 735) .... Intro 18, C6506, C7008,
C8018, C8502, F3025, O7072,
W9001, W9004, W9015
reg 2 .... W9007
4 .... W9005, W9006
5 .... W9005, W9007
(c)(ii) W9002
6 .... R3015, W9005
(1) W9007
(2), (3) W9008
(4)(a), (b) W9008
(5)(a)(i) W9008
(b) W9008
7 .... W9006, W9009
8(a) W9008
9 .... W9010
10, 11 .... W9011
12(5) W9012
13 .... W9002, W9009, W9012

Work at Height Regulations 2005 (SI 2005 No 735) – *cont.*
reg 14(1), (2) W9013
Sch 1 .... W9008
Sch 2
para 5(1)–(3) W9008
Sch 3
Pt 1 .... W9008
Pt 2 .... W9008
Sch 4 .... W9014
Sch 5
Pt 2 .... W9014
Pt 3 .... W9014
Pt 5 .... W9008
Sch 6 .... W9002
para 5 .... W9009
10 .... W9009
Sch 7 .... W9012
Work at Height Regulations (Northern Ireland) 2005 (SR 2005 No 279)
.... W9003
Work in Compressed Air Regulations 1996 (SI 1996 No 1656)
.... C8018, P7013, V12003
reg 14 .... F5062
16(2) V12019
Work in Compressed Air Special Regulations 1958 (SI 1958 No 61)
.... C8018
Working Time Regulations 1998 (SI 1998 No 1833) .. E14001, E14002, E14011, J3001, J3031, M1505, S11008, S11014, S11019, W10001, W10006, W10032, W10044
reg 2 .... J3032, W10002, W10019, W10027
4 .... J3002, W10012, W10020, W10029
(1) W10010, W10042
(2) W10010
(3), (4) W10011
(5) W10011, W10034
(6) W10010
5 .... W10012
6 . J3002, J3032, W10019, W10020
(1) W10029
7 .... W10029
(1)(b) W10021
8 .... W10018
9 .... W10025
10 .... J3002, J3032, W10014, W10016
(1) W10029
11 .... J3002, J3032, W10016
(1), (2) W10029
12 .... J3002, J3032, W10017, W10018
(1) W10029
13 .... W10022, W10024, W10029

Working Time Regulations 1998 (SI 1998 No 1833) – *cont.*
reg 13A W10029
(3) W10022
14 .... W10024
(3) W10023
(b) W10022
(4) W10023
15 .... W10024
16(1) W10024
(5) W10024
18 .... W10026
(2) W10031
20 .... W10012, W10020
(1) . W10015, W10017, W10033
(2) W10033
21 ... W10015, W10020, W10034, W10037
(e) W10029
22 .... W10015, W10035, W10037
23 ... W10012, W10015, W10017, W10020
(a) J3032, W10036, W10037
(b) J3032, W10036
24 ...... J3032, W10015, W10017, W10037
24A(2) W10029
25(1) W10025
25A W10012, W10031
28(3) W10039
29 .... W10039
30 .... W10040
31 .... W10041
35 .... W10043
36 .... W10005
Sch 1 .... J3002, J3032
para 3 .... W10008
Working Time (Amendment) Regulations 2003 (SI 2003 No 1684)
.... W10026, W10028, W10030
Working Time (Amendment) Regulations 2007 (SI 2007 No 2079)
.... W10022
Workplace (Health, Safety and Welfare) Regulations 1992 (SI 1992 No 3004)
..... Intro 18, A1001, A1002, A5030, C7008, C8503, D0939, D8203, E15012, E15023, E17029, E17034, M2103, P7031, R3028, R3050, S11014, S11015, V3001, V3016, W9015, W9009
reg 2–4 .... W11002
5 .... W11001, W11003
(1) O3018
(3) O3018
6 .... W11001, W11004
(1), (2) V3015
7 .... W11001, W11005
8 .... W11001, W11006
(3) L5002
9 .... W11001, W11007

Workplace (Health, Safety and Welfare)
    Regulations 1992 (SI 1992 No 3004) –
    cont.
  reg 10 .... W11001, W11008
    11 .... W11001
    (1)–(4) W11009
    12 .... W9004, W11001
    (1) O3018, W11010
    (2)(a), (b) W11010
    (3) O3018, W11010
    13 ..... A1015, W11001, W11002,
                W11010, W11011
    (1)–(3) D0948
    (5)–(7) W9004
    14 .... W11001, W11012
    15 .... W11001, W11013
    16 .... W11001, W11014
    17 .... W11001, W11015
    (1) A1003

Workplace (Health, Safety and Welfare)
    Regulations 1992 (SI 1992 No 3004) –
    cont.
  reg 17(2) A1004
    (3), (4) A1004, A1005
    18 .... W11001, W11016
    19 .... W11001, W11017
    20 .... W11001, W11002, W11019
    21 .... W11001, W11002, W11020
    22 .... W11001, W11002, W11024
    23 .... W11001, W11002
    (1), (2) W11025
    24 .... W11002
    (2) W11026
    25 ..... V12034, W11001, W11002
    (1)–(5) W11027
    25A W11001
  Sch 1 .... W11008
  Pt 2 .... W11021

Index

A

ABATEMENT OF STATUTORY
  NUISANCE, E5062
ABATTOIR,
  controls as to, F9005, F9007
ABRASION, M1025
ABSENCES, see Managing Absence, Sickness
  Absence
ACAS,
  code on disciplinary procedures in
    industry E14008, E14014, E14021
  dismissal,
    Code of Practice on Disciplinary and
      Grievance Procedures, M1517
  warnings, E14014
ACCESS,
  access/egress points,
    confined spaces, C7023
    employers to provide and
      maintain, A1001
    hazards, A1016–A1019
    safe workplace transport
      systems, A1002, A1003
  business continuity planning, B8008
  common law duty of care, A1001, A1021
  community health and safety, C6516,
    C6517
  confined spaces, see Confined Spaces
  definition, A1001
  disabled persons, facilities for, A1001,
    V12034, W11043
  disaster and emergency planning, D6010
  electricity, E3020
  employer's statutory duty, A1001
  factories, O3018
  hazards, A1016–A1019
  Health and Safety Executive
    inspections, Intro 29
  heights, working at, W9004, W9008
  occupiers' liability, O3018
  parking spaces, A1004
  pedestrians, A1002, A1004, A1007
  safe workplace transport systems, A1003
  speed limits regarding, A1005
  traffic routes, see Traffic Routes

ACCESS, – cont.
  trespassers, C65013
  vehicles, Intro 29, A1003–A1019
    access to, A1015
    contractors, A1014
    drivers, training of, A1013, A1014
    hazardous operations with, A1008,
      A1016–A1019
    maintenance, A1012
    pedestrians and, A1002, A1004, A1007
    provision of, A1011
    statutory requirements
      concerning, A1008, A1009
    subcontractors, A1014
    unauthorised lifts in, A1020
    work vehicles, A1010–A1015
  violence in the workplace, V8088
  workplace transport systems,
    organisation of safe, A1002, A1003
    parking areas, A1004
    pedestrians, A1004, A1007
    safe traffic routes, A1006, A1007
    suitability of traffic routes, A1004
    vehicles, A1005
ACCIDENT LINE, C6021
ACCIDENT
  INVESTIGATIONS, A3030–A3034
  allocation of blame and liability, A3030
  causes, analysis of, A3033
  disciplinary procedures, A3030
  elements of, main, A3031–A3034
  evidence, collating, A3032, A3033
  guidance, A3030
  Health and Safety Executive, A3031
  information, A3030, A3033
  interviewing witnesses, A3032
  lessons, learning, A3030
  monitoring, A3030
  performance, A3030
  rationale for, A3030
  reassurance and explanation, A3030
  reports, A3031, A3033–A3034
  risk assessment, A3030
  witnesses, A3032

ACCIDENT REPORTING, Intro 7
accident book, A3002
action to be taken by employers, A3017
basic rules, A3001
biological agents, A3035
breathing apparatus, A3035
building work, A3035
causes of accidents, A3001
coverage, A3006
dangerous occurrences, A3009
    list of reportable, A3035, A3036
death of employee, reporting, A3018
defence for breach of Regulations, A3023
disaster and emergency planning, D6035
disclosure of accident data,
    generally, A3027
    legal representatives, A3029
    safety representatives, A3027
disease,
    duty to keep records of, A3016
    prescribed form, A3038
    reportable, list of, A3002, A3016,
        A3035
diving operations, A3005, A3035
electricity, A3035
employees, duties on, A3024–A3026
enforcing authority, A3003
ergonomics and, E17067
exceptions to notification and
    reporting, A3013
explosion, A3035
explosives, A3035
fairgrounds, A3035
fatal accidents, A3004, A3007
fines, A3022
fire, A3035
flammable substance, escape of, A3035
freight containers, A3035
gas fitters, A3021
gas incidents, A3002, A3020, A3021
generally, A3001
Health and Safety Executive Regional
    Offices, A3039
Health and Safety Commission priority
    areas, Intro 29
incapacity for more than three
    days, A3008
incident reporting centre, A3003, A3004
injuries and dangerous occurrences,
    duty to keep records of, A3015, A3031
    major, A3006
Law Society Accident Line, C6021
legal representatives, A3029
level crossings, A3035
lifting machinery, A3035

ACCIDENT REPORTING, – *cont.*
major accidents, Intro 29, R3016, R3035
major injuries or conditions, A3007,
    A3011
managing health and safety, M2041
manual handling, M3019
members of public, A3001
mines, A3005, A3035
notify, duty to, A3011
objectives, A3026
offshore operations, A3005, A3035
out of hours, A3034
overhead electric lines, A3035
penalties, A3022
persons responsible for notification and
    reporting, A3005
pipelines, A3005, A3035
prescribed form,
    dangerous occurrences, A3037
    diseases, A3038
    injuries, A3036
pressure systems, A3035
priority areas, Intro 29
quarries, A3005, A3035
radiation, A3035
records, duty to keep, A3014–A3017
    action to be taken by employers, A3017
    disease, A3016
    duration, A3014
    general rule, A3014
    injuries and dangerous
        occurrences, A3015, A3036
relevant enforcing authority, A3003
report, duty to, A3012, A3008.1
reportable dangerous occurrences
    listed, A3035
requirements, A3001
responsible persons, A3002, A3005,
    A3011
RIDDOR, A3002–A3023
    accident book, A3002
    dangerous occurrences, A3009, A3029
    death of employee, reporting, A3018
    defences, A3023
    duties under, A3002
    enforcing authority, A3003
    exceptions to notification and
        reporting, A3013
    gas incidents, A3020, A3021
    generally, C6004
    incapacity for more than three days,
        injuries, A3008
    major injuries or conditions, A3007
    notify, duty to, A3011
    penalties, A3022

ACCIDENT REPORTING, – *cont.*
  RIDDOR, – *cont.*
    persons responsible for notification and reporting, A3005
    records, duty to keep, *see* records, duty to keep *above*
    relevant enforcing authority, A3003
    report, duty to, A3012
    responsible persons, A3002, A3005
    road accidents, A3002, A3019
    scope of regulations, A3002, A3006
    specified dangerous occurrences, A3009
    three days, injuries or incapacity for more than, A3008
  road accidents, A3019, A3035
  safety representatives, A3027, A3028
  scaffolding, A3035
  scope of duty, A3001, A3002
  scope of provisions, A3002, A3006
  Seveso Directive, Intro 29
  specified dangerous occurrences, A3009
  statistics, A3001
  substances, escape of, A3035
  three days, injuries or incapacity for more than, A3009
  trains, A3035
  transport systems, A3035
  transportation of dangerous substances, A3035
  violence, acts of, A3008.1
  visitors, A3001
  wells, A3005, A3035
ACCIDENTS,
  causation, M2028
  community safety, C6516
  costs of, Intro 26
  fatal injuries, *see* Fatal Accidents
  hazardous substances in workplace, where, H2105, H2114, H2135, H2139
  industrial injuries benefit, C6003, C6006
  major accident hazards, *see* Major Accident Hazards Regulations
  manual handling, M3019
  reporting, *see* Accident Reporting
  Royal Society for the Prevention of Accidents, Intro 28
ACCOUNTING STANDARDS BOARD, E16005
ACCOUNTS MODERNISATION DIRECTIVE,
  Business Review, E16005, E16024
  environment management, E16005, E16024, E16025

ACID RAIN, E5001, E5014, E5015, E5063
ACOPs, *see* Approved codes of practice
ACOUSTIC BOOTHS, N3037
ACOUSTIC SIGNALS,
  safety signs regulations, W11030
ACTORS,
  safety representatives, J3005
ACTUAL DISMISSAL,
  case law illustrating, E7025
  test of dismissal, E7017
ADJUSTMENTS,
  disability discrimination, E17070, M1511, R2005
  examples of, R2005
  health and safety, R2006
  justification, R2006
  making, R2004
  reasonable, E17070, R2005–R2006
  rehabilitation, R2004–R2006
  sickness absence, R2008
ADVENTURE ACTIVITIES
  regulations, C6508, T7007, W9004
ADVERSE ENVIRONMENT,
  harassment, H1702
ADVERTISEMENT,
  food, as to, F9020
ADVERTISING STANDARDS AUTHORITY (ASA), F9012
ADVICE ON HEALTH AND SAFETY LAW,
  small businesses, Intro 4
  sources of, Intro 28
  training and competence, T7001, T7002, T7007
ADVISORY, CONCILIATION & ARBITRATION SERVICE (ACAS),
  Code of Practice, disciplinary procedures, E14008, E14021
AGE,
  Employment Quality (Repeal of Retirement Age Provisions) Regulations 2011, M1505
  Equality Act 2010, E16513
  harassment, H1702
  insurance exemption, M1511
  manual handling, M3007, M3015
  protected characteristic, H1703
  risk assessment, E16515
  violence in the workplace, V8032
  workforce, E16514
  young workers, E16516
AGENCY WORKERS,
  employers' liability insurance, E13016
  visual display units, D8202
  working time, W10005

AGENTS,
liability of principal, E16501
AGRICULTURE,
air pollution and, E5015
children and young persons, V12015
climate change, effect of, E5023
enforcing authority, E15009
risk factors, E16521
tractor cabs, E15005
vibration, V5001, V5015
AIR,
cleaning, types of, V3010
compressed air, see Compressed Air, Work in
general duty on employers, provision of pure and uncontaminated air, V3004
pollution, see Air Pollution
quality, see Air Quality
specific processes, control of dust and fumes, V3006
ventilation of excavations and pits, V3009
AIR CONDITIONING, L5007
AIR PASSENGER DUTY (APD), E16006
AIR POLLUTION,
generally, see Emissions into the Atmosphere
polluting agents, V3003
AIR QUALITY,
assessment, E5033–E5039
emissions, see Emissions into the Atmosphere
local authority duty, E5002
Local Air Quality Management (LAQM), E5069
pollutants, priority E5018-E5019
strategy (AQS), E5068
AIR TRANSPORT,
transport of dangerous goods, D0404
vibration, V5015
working time, W10026, W10027, W10031
AIRBORNE PARTICLES,
control of, V3010
AIRCRAFT,
food on board, F9032
occupiers' liability, O3005
vibration, V5015
ALARMS,
fire, F5017
violence in the workplace, V8090
ALCOHOL,
work in18sed air, and, C8018
ALLERGIC REACTION,
allergic rhinitis, O1042
extrinsic allergic alveolitis, O1042

ALLERGIC REACTION, – cont.
food, to, F9009, F9011, F9075
respiratory diseases, O1042–O1043
AMENITY, LOSS OF,
kinds of, C6031, C6034
AMMONIUM NITRATE, D0704, D0908, D6013
ANIMAL PRODUCTS, F9002, F9004, F9014, F9028, F9034
ANIMAL WORKERS,
respiratory diseases, O1043
ANNUAL LEAVE,
accrual,
sickness absence, during, M1514
system, W10022
claim for unlawful deduction of wages, W10040
compensation related to entitlement, W10023
enforcement of provisions, W10040
entitlement, W10001, W10022
instalments, W10022
notice requirements, W10024
payment, W10024
rolling up, W10024
road transport, W10029
sick leave and, W10024
ANTHROPOMETRIC DATA, E17007, E17010
ANTIMITOTIC DRUGS, V12022
ANTI-SOCIAL BEHAVIOUR ORDERS, V8140
ANXIETY, O1052–O1055
APC, see Emissions into the Atmosphere
APPEAL,
compensation for work injuries, C6018, C6050–C6058
enforcement, E14014
improvement notice,
against, E15019–E15025
ionising radiation, R1019
prohibition notice,
against, E15019–E15025
APPROVED CODES OF PRACTICE, Intro 10
asbestos, A5005, A5012, A5022, A5030
confined spaces, C7008
control of substances hazardous to health, H2122
dangerous substances, D0902, D0933–D0933.6, D0934
explosions, D0933, D0933.2
factories, O3018
machinery safety, M1007
management of health and safety, M2005

APPROVED CODES OF PRACTICE, – *cont.*
  non-ionising radiation, R1019
  occupational health and safety, O1009
  occupiers' liability, O3018
  record assessments, H2122
  training and competence, T7005
  ventilation, D0933.2
ARMED FORCES,
  corporate manslaughter, C9001
  manual handling, M3002
  personal protective equipment, P3018
ARM,
  vibration syndrome, *see* Hand Arm
    Vibration Syndrome
  work-related Upper Limb Disorders, *see*
    Work-Related Upper Limb Disorders
ARSON, F5074
ASBESTOS, Intro 29, A5001–A5030
  action levels, A5003
  Approved Code of Practice, A5005,
    A5012, A5022, A5030
  asbestos containing
    material, A5011–A5012, H2101,
      H2138–H2138, R3013
    knowledge of, A5013
    management, A5013
  asbestos zones, A5003
  asbestosis, A5002
  buildings,
    maintenance workers, exposure
      of, A5006
    removal from, A5005
    surveys, A5024–A5026
  clearance indicators, A5005
  compliance strategy, development
    of, A5016
  contractors, A5026
  contractual agreements, A5021
  control of exposure,
    control limits, V3010
    generally, Intro 18
    historical, A5003
    seven step programme, A5013–A5019
    statutory requirements
      governing, H2101, H2138
  Control of Asbestos at Work
    Regulations, A5007, D6016, H2138
  decorative textured coatings, A5028
  definition, A5001
  diseases, A5002
  duty holders, A5007–A5019, R3013
    identification of, A5009
    third parties, A5020
  employers' duties, A5007–A5019

ASBESTOS, – *cont.*
  employers' liability insurance, E13001,
    E13026
  exposure of building maintenance
    workers, A5006
  facilities management, F3024
  fines for offences, Intro 15
  historical exposures, A5003
  HSE publications, A5030, R3013
  identification of exposure, A5006
  inspection, A5010
  laboratories, A5012, A5025
  leases, A5021
  legislation, A5030–A5031
  licensing of works, A5029
  licensing regulations, E15005, R3016,
    R3035
  long development period of, E13001
  lung cancer, A5002, A5003
  maintenance, A5006, A5012
    building maintenance workers, exposure
      of, A5006
    records, A5011
  management,
    asbestos-containing materials, A5013
    leases, A5021
    long-term plans, A5018
    plans, A5012, A5018–A5019, A5022
    third parties, A5020
  mesothelioma, A5002, A5003
  minimum requirements, A5022
  mixtures, control limit, A5002, A5003,
    A5005
  personal protective equipment, P3004,
    P3008
  plans,
    long-term, A5018
    monitoring, A5019
    review of, A5019
    risk management, A5012, A5018,
      A5019, A5022
  premises,
    affected, A5008
    control of work on, A5014
    guidance, A5005
  priorities, A5017
  prohibitions, A5004
  records, A5011
  register, A5011, A5022
  repairs,
    leases, A5021
  respiratory diseases, O1042, O1046
  risk assessment, Intro 16, A5003, A5010,
    A5017, R3013, R3016, R3035
  samples, A5011, A5025

ASBESTOS, – *cont.*
self-employment, period of, A5007
sporadic or minor intensity, work of, A5028
surveys, A5024–A5026
removal specifications, A5026
risk assessments, A5025
type 1, identification and condition, A5024
type 2, identification, sampling and condition, A5025
type 3, quantitative for refurbishment and demolition, A5026
tenancy agreements, A5021
third parties, A5020
training requirements, A5028
types of, A5001
UK Accreditation Service laboratories, A5012
ventilation regulations, V3015
waste, A5027
removal, A5027
special, A5027
works, licensing of, A5029, A5030
ASPHYXIATION,
Confined spaces, C7001
ASSAULT,
harassment, H1723, H1729-H1730
violence in the workplace, V8002–V8003, V8011, V8131–V8133
ASSEMBLY OR PRODUCTION LINES, M3004
ASSOCIATION,
discrimination by, M1508
ASTHMA,
employers' liability insurance, E13026
hazardous substances in workplace, where, H2103
occupational disease, C6057
respiratory diseases, O1042, O1044
ATEX DIRECTIVE, D0933.6, E3027
ATMOSPHERES,
explosive,
classification, D0920-D0921
sources of, D0921
ATMOSPHERIC HAZARDS,
confined spaces, A1020
ATTENDANCE MANAGEMENT, R2007
AUDIOMETERS, O1023
AUDIT REPORT,
environment management system, E16025
risk assessment, use for, R3031
AUDITING,
business continuity planning, B8008, B8034

AUDITING, – *cont.*
disaster and emergency planning, D6018, D6067, D6068
stress, S11037
AUTOMATION, M3010
AUXILLIARY AID,
reasonable adjustments, M1511
AVIATION,
managing health and safety, M2042, M2045

B

BACK,
manual handling, M3004
pain, 01047, V5005, V5023–V5024, V5028
vibration, V5005, V5023–V5024, V5028
whole-body vibration, V5023–V5024
BAGS, M3013
BANKS,
environmental management, E16007
Equator Principles, E16007
BARRIERS,
heights, working at, W9008
mobile work equipment, M1008
BATH HOISTS, L3036
BATON,
violence in the workplace, V8095
BATTERY,
electricity, E3027
rooms, E3027
BATTERY AND ASSAULT,
common law, H1730
criminal action, H1729
harassment, H1730
trespass to the person, H1723
BCF FIRE EXTINGUISHERS, F5014
BEAT HAND AND BEAT ELBOW, 01047
BEHAVIOURAL MEASURES,
machinery safety, M1026
BELIEF,
harassment, protected characteristic, H1703
BENCHMARKING, M2039
consumer groups, E16012
BENEFIT RATES, *see* Industrial Injuries
BEREAVEMENT,
non-pecuniary losses, recoverability, C6031, C6035
BIOLOGICAL AGENTS,
accident reporting, A3035
children and young persons, V12016

BIOLOGICAL AGENTS, – *cont.*
new or expectant mothers, V12021
risk assessment/control of
exposure, R3028
unborn child, hazardous to, R3028
BLASTING OPERATIONS,
personal protective equipment, P3003,
P3019
'BODY ARMOUR', USE OF, V8093
BOILERS
gas safety, G1029
maintenance, G1029
BP DEEPWATER HORIZON,
oil rig explosion, E16010
BREACH OF CONTRACT,
harassment, H1727
remedies, H1728
working time, and, W10042
BREACH OF STATUTORY DUTY, Intro
24, E13002, E15021–E15035
BREAKDOWN MAINTENANCE, F3015
BREAKS, REST, W10017
stress, S11008, S11014, S11019–S11020
visual display screens, D8206
BREASTFEEDING MOTHERS,
ionising radiation, R1019
risk assessment, E14024, E14025, R3006
BREATHING APPARATUS,
accident reporting, A3035
BRITISH SAFETY COUNCIL
advice from, Intro 28
training and competence, T7009, T7010,
T7013, T7020
BRITISH STANDARDS,
fire extinguishers, F5016
fire prevention and control, F5017
heights, working at, W9002, W9014
managing health and safety, M2029
vibration, V5009
BROADCASTING
enforcing authority, E15009
BROKERS
employers' liability insurance, E13003
professional indemnity insurance, E13003
BSE, F9020
BUILDING
see also Facilities Management
electricity, E3033
head protection, *see* Head Protection
health and safety regulations, W11042
lifts, L3009
vibration, *see* Vibration

**BUILDING AND CONSTRUCTION
OPERATIONS**
see also Fire; Fire Precautions; Ladders;
Work at Heights; Working Platforms
asbestos
introduction, A5006
surveys, A5023–A5026
accidents, prevention, C8018
CDM Regulations 2007. *see* Construction,
Design and Management Regulations
changing rooms, C8017
children and young persons, C8019,
V12005
civil liability, case law, C8040–C8041
cofferdams and caissons, C8010
compressed air, work in, C8018
confined spaces
and see Confined Spaces
generally, C8018
Construction Skills Certification
Scheme, T7027
construction work,
definition, Intro 23
demolition or dismantling, C8007
drinking water, C8017
drowning, prevention of, C8013
electricity supply, safety
precautions, C8018
emergency procedures, F5026
energy distribution systems, C8012
excavations, C8009, C8018
explosives, safety provisions C8008
Engineering Construction Industry
Training, T7027
facilities management, F3033
fire prevention and control, C8016
building and construction
operations, C8016
code of practice, F5024, F5029
designing out fire, F5027
emergency procedures, F5026
emergency routes and exits, C8016
fire extinguishers, F5028
insurance, F5029
plant, F5028
precautions, F5029
risk assessment, F5024
site fire safety co-
ordinator, F5024–F5025
first-aid areas, F7025
fresh air, C8017
good order, C8005
head protection, C8018

BUILDING AND CONSTRUCTION OPERATIONS – *cont.*
health and safety file, Intro 23
health and safety plan, Intro 23
hazardous substances, C8018
induction, C8031
inspection, reports, C8011
insurance cover, C8033, F5029
introduction, C8001
ladders, *see* Ladders,
lifting operations, A1024
lighting, C8013, L5004
lockers, C8017
maintenance projects, Intro 23
new or expectant mothers, V12005
noise, C8018
occupational diseases, *see* Occupational Diseases
personal protective equipment, P3003, P3004, P3008
practical issues
   changing rooms, C8017
   drinking water, C8017
   fresh air, C8017
   good order, C8005
   lighting, C8017
   lockers, C8017
   rest facilities, C8017
   safe place of work, C8004
   sanitary conveniences, C8017
   temperature, C8017
   washing facilities, C8017
   weather, protection from, C8017
   welfare facilities, C8017
Regulations. *See* Building Regulations
rest facilities, C8017
risk assessment, F5024, R3017
safe place of work, C8004
sanitary conveniences, C8017, W11018
signing, C8018
site fire safety co-ordinator, F5024–F5026
site inductions and security, C8019
temperature, C8017
traffic routes, C8014, C8018
training and competence, T7027
typical projetc issues, C8002
vehicles, C8015, C8018, A1014
vibration, risks relating to, C8018
visitor protection, C8019
vulnerable persons, V12005
washing facilities, C8017
weather, protection from, C8017
welfare facilities, C8017

BUILDING AND CONSTRUCTION OPERATIONS – *cont.*
working platforms, *see* Work at Heights; Working Platforms
BUILDING REGULATIONS, Intro 23
fire prevention and control, and, F5035
gas safety, G1019
requirements, W11044
BUILDING WORK,
see also Construction and Building Operations
accident reporting, A3035
controls over,
   access for disabled persons, W11043
definition, W11044
Factories Act, relating to, W11045–W11049
violence in the workplace, *see* Workplace violence
BULK CONTAINER,
intermediate, D0416
BUNCEFIELD OIL DEPOT, D0951
BURDEN OF PROOF,
criminal cases, E15031
employers' liability insurance, E13002
presumption of innocence, E15031
reverse, E15031
BURGOYNE REPORT, O7003
BURNS,
electricity, E3006
BUSINESS CONTINUITY PLANNING, B8001–B8042
access, effects of denial of, B8008
after the crisis, communication, B8042
audits, B8008, B8034
British Standard, B8006
business case for, B8010
Business Continuity Institute, B8001
business impact analysis, B8011, B8016, B8018
buy-in, securing, B8027
Cadbury guidance, B8006
call deluge, B8040
commercial drivers, B8005
command and control team, B8024
communication, B8038–B8042
component approach, B8014, B8022
component testing, B8036
context of, B8002
creditworthiness, loss of, B8017
crisis communication, B8038–B8042
critical business approach, B8015, B8018, B8022, B8037
customer base, erosion of, B8017
damage assessment team, B8024

BUSINESS CONTINUITY PLANNING, –
cont.
departmental teams, B8024
desk check/unit, B8034
development of, B8011, B8020, B8025
dilemma, the planning, B8021
disasters,
imported, B8008
natural, effects of, B8008
public relations, B8009, B8038
recovery, B8001–B8003, B8023
drivers, B8004–B8009
during crisis, communications, B8041
emergency response, B8023
fallback procedures, B8023
financial drivers, B8007
full simulation, B8037
hierarchical structure, B8024
implementation, B8011, B8026–B8032
importance of, B8003
individual responsibilities, B8030
insurance, B8007
key staff, loss of, B8017
maintenance, B8011, B8033–B8037
media training, B8040
model, B8011, B8015
objectives, B8001
outputs, B8018
phased roll-out, B8031
policy, B8028
pre-crisis measures, B8040
press statements, B8040
project authorisation, B8029
public relations, B8038–B8042
disasters, B8009, B8038
quality drivers, B8006
quantification of business impact, B8017
regulatory and contracted impact, B8017
regulatory drivers, B8006
reputation, loss of, B8017
resumption of business, B8023
risk assessment, B8012–B8018
risk management, B8007
simulation, B8037
specialist services, B8032
standards, B8006
strategy, B8011, B8019
testing, B8011, B8033–B8037
Turnbull Committee, B8001, B8006
understanding your business, B8011,
B8012
walkthrough test, B8035

BUSINESS IN THE COMMUNITY
(BITC), E16002, E16008
CR Academy, E16002
BUSINESS IN THE
ENVIRONMENT, E16007
BUSINESS REVIEW
Accounts Modernisation
Directive, E16005, E16024
Companies Act 2006, E16005., E16024
disclosure, E16005, E16024
environmental reporting, E16005, E16024
BUTCHER'S SHOPS, F9021
BYSSINOSIS,
occupational disease, C6062

C

CABLES,
electricity, E3028
tripping over, O3015
CADBURY REPORT, Intro 5, B8006
CAISSON,
Building and construction sites, C8010
CALL CENTRES,
noise and, N3028
risk factors, E16521
CALL DELUGE, B8040
CANCER,
carcinogens, O1056
employers' liability insurance, E13026
mobile phones, R1008
non-ionising radiation, R1008
respiratory, H2103
skin, H2103, O1038
CANNED FOOD, F9056
CAPABILITY,
definition, M1516
managing health and safety, M2016
manual handling, M3007, M3015
CAPITAL,
legal aid and, C6021
CARBON ACTION SCHEME, E16007
CARBON DIOXIDE FIRE
EXTINGUISHERS, F5015
CARBON DISCLOSURE PROJECT
(CDP) E16007-E16008
CARBON MONOXIDE,
boilers, maintenance of, G1029
gas safety, G1017, G1029, G1033
new or expectant mothers, V12022
poisoning, G1017, G1029, G1033

CARBON REDUCTION COMMITMENT,
sustainability, F3018A
CARBON TRUST, E16008
CARCINOGENS, D0704, O1043, O1056
**CARPAL TUNNEL SYNDROME
(CTS),** 01047, V5004
reporting, V5011
CARRIAGE OF DANGEROUS GOODS. See
Transportation of Dangerous Substances
CARRYING, M3003
CASH AND CARRY WAREHOUSE,
food from, F9060
CATARACTS, O1033
CATERING AND HOSPITALITY,
See also Food
gas safety, G1017
CAVERS, W9004
CDM Regulations 1994
generally, C80305
CE MARK,
equipment, M1007
lifts, L3005–L3006, L3010
machinery safety, M1009, M1016, M1026
personal protective equipment, P3012,
P3013
**CENTRE FOR RADIATION, CHEMICAL
AND ENVIRONMENTAL
HAZARDS,** R1017
CERTIFICATES,
certificate in safety management, T7020
emergencies, V5015
gas safety, G1025
vibration, emergency certificates
for, V5015
CHAINS, L3043–L3044
failure, L3044
types of, L3043
CHANGES IN WORK, Intro 6
CHANGING FACILITIES
construction and building operations,
and, C8017
generally, W11026
CHARITY,
corporate manslaughter, C9001
CHARITY EVENTS,
food at, F9007
**CHARTERED INSTITUTE OF BUILDING
SERVICES ENGINEERS
(CIBSE),** L5002
**CHARTERED INSTITUTE OF
ENVIRONMENTAL HEALTH,**
advice from, Intro 28
training and competence, T7013, T7018

CHEMICAL HAZARDS,
Building and construction
operations, C8018
confined spaces, A1022
major accident hazards, see Major
Accident Hazards Regulations
new or expectant mothers, V12022
supply chain, E16008
CHEMICAL PLANT,
first-aid areas, F7025
CHEMICAL SUBSTANCES,
children and young persons, V12016
dangerous goods, carriage of, D0402
EU law, D0901, D0908
European Chemicals Agency
(ECHA), R2502, R2511
European Classification, Labelling and
Packaging (CLP) Regulation, H2137
hazards, see Chemical Hazards; Major
Accident Hazards Regulations
information, H2101, H2137
packaging requirements, H2101, H2137
REACH See Registration, valuation and
Authorisation of Chemicals in Europe
skin diseases, O1038, O1041
CHEWING GUM, F9007
**CHILDREN AND YOUNG
PERSONS,** V12001–V120026
adventure activities, C6508
age, V12006
agriculture, V12015
biological agents, exposure to, V12016
characteristics of, V12013
checklist, V12016
chemical agents, exposure to, V12016
child, definition of, V12007
construction work, C8019 ,V12005
contributory negligence, O3011
control of substances hazardous to
health, V12005
dangerous materials or activities, V12016
dangerous workplaces or
workstations, V12016
definition, V12003, V12007
docks, V12015
employers' liability insurance, E13016
equipment, V12005, V12016
explosives, V12015
family undertakings, V12011
increased risks, works presenting, V12016
information, V12010, V12014
ionising radiation, V12015
lead processes, E14002, V12015
legislation, V12001–V12005

CHILDREN AND YOUNG PERSONS, – *cont.*
Management of Health and Safety at Work Regulations, V12001–V12003, V12009–V12016
managing health and safety, M2006, M2013, M2020
manual handling, V12005, V12016
maturity, V12013
mines and quarries, V12015
occupiers' liability, O3010, O3011, O3017
parents, information to, V12010, V12014
personal protective equipment, V12005
physical agents, harmful exposure to, V12016
physically demanding work, excessively, V12016
precautions, V12013–V12014, V12016
psychologically demanding work, excessively, V12016
risk assessment, M2013, V12002, V12006, V12008, V12013
shipbuilding and repairing, V12015
train surfing, 03017
trainees in relevant training, R3025
training and competence, T7006, V12004, V12012
trespass, C65012
volenti non fit injuria, 03017
warnings, O3010
work experience, V12004, V12010
young person, definition of, V12006
CHLORACNE, O1038
CHRONIC UPPER LIMB PAIN, 01047
CHUTES, M3010
CIVIL LIABILITY,
see also Negligence
breach of statutory duty, Intro 24, W11049
commencement of actions, Intro 24
construction and building operations, and
   case law, C8040–C8041
   generally, C8039
contributory negligence, Intro 24
damages, assessment of, Intro 24
defences,
   contributory negligence, Intro 24
   factory occupier, W11048
   foreseeability of injuries, Intro 24
   generally, Intro 24
   voluntary assumption of risk, Intro 24
director, E15041
duty of care, Intro 24
employers' liability insurance, Intro 24
evidence, Intro 24

CIVIL LIABILITY, – *cont.*
expert evidence, Intro 24
Factories Act,
   effect of conviction under, W11049
fast track system, Intro 24
foreseeability of injuries, Intro 24
generally, Intro 24
initiating action, Intro 24
manual handling, F5073
occupiers' liability, Intro 24
pre-action protocol, Intro 24
procedure, Intro 24
product safety, Intro 24, P9028–P9048
proof, Intro 24
settlement of cases, Intro 24
stress, S11009
time limits, Intro 24
training and competence, T7002
vicarious liability, Intro 24
violence in the workplace, V8012, V8140
volenti non fit injuria, Intro 24
voluntary assumption of risk, Intro 24
Woolf Report, Intro 24
CIVIL PARTNERSHIP,
Equality Act 1910, E16501
harassment, H1703
CIVIL PROCEDURE RULES,
employers' liability insurance, E13029
pre-action protocols, E13029
CLAIMS FARMERS, CORPORATE KILLING AND, Intro 14
CLEANER,
fire prevention and control, F5018
risk assessment, R3032
risk factors, E16521
CLEANING,
control of airborne particles, V3010
dangerous substances, D0933.5
facilities management, F3030
windows, *see* Window Cleaning
CLEANLINESS,
food premises, in, F9025, F9026
workplaces, of, W11001, W11007
CLIENT/CONTRACTOR NATIONAL SAFETY GROUP, T7027
CLIMATE CHANGE,
acidifying substances, E5015
air passenger duty, E16006
Carbon Disclosure Project (CDP) E16007
consequences of, E5023
disclosure, Global Framework for climate risk, E16007
financial consequences of, E16007
Global Reporting Initiative, E16007, E16024

CLIMATE CHANGE, – *cont.*
greenhouse effect, E5022
international protocols, E5026
levy, E16006
Treaty, E5026-E5027
CLIMBERS, W9001, W9004
CLOSED CIRCUIT TELEVISION,
violence in the workplace, V8089
CLOTHING,
accommodation for,
workplaces, S11014, W11001,
W11025, W11026
dangerous substances, D0923
explosions, D0923
manual handling, M3004, M3007
workplaces, S11014, W11001, W11025,
W11026
CLUBS,
noise, and, N3027
COAL,
see also Coal Miners
burning, air pollution and, E5003, E5021,
E5061
COAL MINERS,
compensation, vibration white finger
for, V5003, V5012
vibration, V5003, V5012
CODES OF PRACTICE,
See also Approved Codes of Practice
disability discrimination, A1001
disaster and emergency systems, D6056
explosions, D0919
food, as to, F9006, F9009, F9025, F9035,
F9038
hazardous substances in the
workplace, H2105
noise, N3017
pressure systems, and, P7014
Safety Case Regulations, O7016
tracing employers, E13026
vibration, N3017
violence in the workplace, V8044, V8072,
V8129
COFFERDAM,
building and construction sites, C8010
COLLECTIVE AGREEMENTS, E14005
COLOUR BLINDNESS, O1033, O1034,
O1035
COLOUR DISCRIMINATION,
Equality Act 2010, H1703
**COMBINED CODE ON CORPORATE
GOVERNANCE,**
see also UK Corporate
Governance Codereplacement, E16005

COMMERCIAL PREMISES,
health and safety law, enforcement
of, E15007
COMMON LAW,
gross negligence manslaughter, C9021
**COMMUNICATION EQUIPMENT
ROOMS, FIRE AND,** F5006
**COMMUNITY HEALTH AND
SAFETY** C6502–C6528
access, C6516, C6517
accidents and emergencies, C6516
awareness, importance of worker, C6511
capabilities of customers, C6516
communication, C6523
community groups, C6518
construction projects, C6506
contractors, use of, C6527
control of substances hazardous to
health, C6506
customers and users of facilities and
services, C6515– C6517
dangerous activities, C6510
planning of, C6511
dangerous substances, C6510
electrical equipment, C6510
emergencies, C6511, C6517
customers and users, arrangements
for, C6517
equipment, C6514
volunteers, C6522
equipment, C6510
emergency, C6514
personal protective, C6520
selection and maintenance of, C6517
volunteers, C6522
Event Safety Guide, C6524
facilities and services, customers and users
of, C6515– C6517
fencing, C6514
fines, C6503–C6504
fire safety, C6507, C6522
groups, taking into account, C6518–
C6520
guidance, S7010
hazardous substances, C6510
health surveillance, C6506
height, working at, C6506, C6510
home, entry into clients' C6522
individuals, taking into account, C6518–
C6520
informal visits, C6518
information, C6523, C6527
legal requirements, C6502–C6508
lifting, C6510
planning of operations, C6511

COMMUNITY HEALTH AND SAFETY –
cont.
major accident hazards, see Major
Accident Hazards Regulations
major public events C6524–C6526, C6527
Management Regulations, C6504
manual handling, C6522
monitoring of customers, C6517
neighbours, duties to, C6509–C6511
notices, C6511, C6514
occupiers' liability, C6508
passers-by, duties to, C6509–C6511
personal protective equipment, C6520
precautions,
    contractors, C6527
    customers and users of facilities and
        services, C6517
    neighbours, C6511
    passers-by, C6511
    trespassers, C6514
    visiting individuals and groups, C6520
products, risks from, C6516
prosecutions, C6503–C6504
public events, major, C6524–C6526
restricted areas, activities and
    groups, C6520, C6523
risk assessment, C6504, C6509–C6511,
    C6526, C6527
Safety Advisory Groups, C6524
schoolchildren, C6518
screening, C6517, C6523
security of the workplace, C6511
services, customers and users of, C6515–
    C6517
signs, C6511, C6514
standards, C6527
storage locations and facilities, C6511,
    C6514
students, C6518
supervision and control, C6517, C6520,
    C6523
traffic management, C6511
training, C6523
travel risks, C6522
trespassers on work premises, C6512–C
    C6514
users of facilities and services, C6515–
    C6517
vehicle traffic, C6510
violence, C6522
visiting individuals and groups, C6518–
    C6520
volunteers, C6521– C6523
worker awareness, C6511

COMPANIES ACT 2006,
Business Review, E16005, E16024
corporate social responsibility, E16005,
    E16024
environmental reporting, E16005, E16024
COMPANY,
Company Law Review, E16005
director, see Director
health and safety
    offences, E15039–E15042
junior staff, delegation to, E15039
mergers and acquisitions, E16007
occupiers' liability, O3001
reform of company law, E16005
COMPENSATION,
acting in the course of
    employment, C6006, C6007
all work test, C6002
appeals, C6018, C6050–C6058
asthma, C6057
capitalisation of future losses, C6027
coal miners, vibration white finger
    and, V5003
Compensation Bill, E13001
compensation culture, Intro 14, E13001
compensator, duties of, C6045
complications, C6046–C6049
constant attendance allowance, C6014
contributory negligence, C6049
corporate killing, compensation culture
    and, Intro 14
damages, see Damages
deductible social security benefits, see
    Damages
discrimination claims, E16522
employers' liability insurance, C6001,
    E13001
    database, proposal, E13026
employers,
    persons treated as, C6009
    work-related violence, V8015
exceptionally severe disablement
    allowance, C6014
fatal injuries, see Fatal Injuries,
Financial Services Compensation
    Scheme, E13024
future earnings, loss of, C6026
home care, expenses of, C6028
incapacity benefit, C6002
incapacity for work test, C6002
industrial injuries benefits,
    benefits overlap, C6020
    scope of, C6003
    social security benefits, C6002
    weekly benefit rates payable, C6017

COMPENSATION, – *cont.*
industrial injuries disablement benefit,
claims in accident cases, C6011
entitlement and assessment, C6011
two rates of benefit, C6013
information provisions, C6049
institutional care, expenses of medical
treatment, C6028
interest on damages, C6043
interim awards, C6042
lost years, damages for, C6029
main types of injury to qualify, C6023
medical examination procedures, C6001
non-pecuniary losses,
bereavement, C6022, C6031, C6035
loss of amenity, C6030, C6031, C6034
nervous shock, C6031, C6033
pain and suffering, C6031, C6032
quantification of, C6030
recoverable, C6031
occupational deafness, C6055
own occupation test, C6002
pain and suffering, C6032
payments into court, C6042
pecuniary losses, assessment of, C6022,
C6042
personal injuries provisions,
acting in course of employment
duties, C6006, C6007
basis of claim for damages, C6022
caused by accident, C6005
non-pecuniary losses, C6023
pecuniary losses, C6023
relevant accident, C6005
pneumoconiosis, *see* Pneumoconiosis
pre-existing disabilities, C6011
prescribed diseases and, C6055, C6062
provisional awards, C6040
receipt of benefit, financial effects
of, C6019
recovery of state benefit, *see* Damages
relevant employment, C6008
repetitive strain injury, C6059
rise in, Intro 14
social security benefits, C6002,
C6015–C6020, C6044, C6045
specified occupational diseases, claims
for, C6043
statutory sick pay, C6015
structured settlements, C6025, C6047
test criterion for, C6002
unfair dismissal, for,
additional award, E14022
basic award, E14022
compensatory award, E14022

COMPENSATION, – *cont.*
unfair dismissal, for, – *cont.*
formula for, E14017
special award, E14022
unlawful discrimination, E16522
vibration white finger, V5003
Woolf reforms, Intro 29
workplace violence, V8004, V8012,
V8013, V8015
COMPETENCE. *See also* **Training and
competence**
definition, T7002
COMPETENT PERSONS,
appointment of, M2010
lifting equipment, L3019–L3022
noise, N3029
permit to work systems, S3016
pressure systems, and, P7018
training requirement, T7001
COMPETITION,
EC law, M1009
machinery safety, M1009
COMPLIANCE,
hazardous substances, Intro 29
large companies, Intro 5
COMPRESSED AIR, WORK IN,
generally, C8018
new or expectant mothers, V12004,
V12020
protective devices, P7020
COMPRESSED NATURAL GAS,
UNLOADING, D09033.6
COMPUTER,
ergonomic design, *see* Ergonomic Design
laptop, E17017, E17064
CONCERT HALL,
noise exposure levels, N3027.1
CONDITIONAL SALE AGREEMENT,
product liability, P9055
CONDUCT, UNWANTED,
purpose or effect, H1703
CONFIDENTIALITY,
harassment in the workplace, H1736
CONSTRUCTION INDUSTRY,
community health and safety, C6506
CONSTRUCTIVE DISMISSAL,
harassment, and, H1720
CONFINED SPACES, C7001–C7030
access and egress, A1001, A1022, C7001,
C7024
asphyxiation C7001
atmosphere, testing and monitoring
the, C7016
building and construction
operations, C8018

CONFINED SPACES, – *cont.*
Code of Practice, C7009
communications, A1023, C7015
competence, A1023, C7012
Confined Spaces Regulations 1997, C7003
  persons with duties under, C7004
contents and residues, removal of, C7019
contractors, C7004
control measures, C7010
definition of, A1023, C7001, C7002
designing safe system of work, A1023
duty holders, C7004
electrical equipment, C7021
electricity, A1022, A1023, C7026
emergencies, C7007, C7029
employers' duties, C7004
entry, C7004–C7005, C7010, C7028
exemption certificates, A1023
fire precautions, A1023, C7025
formalised permit to work system, A1022
free flowing solid, definition, C7003
gas, A1022, A1023
  equipment, C7022
  flammable, C7001
  purging, C7017
  toxic, C7001
hazards, C7001
  atmospheric, A1022
  chemical, A1022
  physical, A1022
hot work, C7027
identification of risks, C7010
independent contractors, C7004
internal combustion engines, C7022
introduction, C8014A
isolations, C7020
Job Safe Procedure, C7011
Legal requirements, relevant, C7002
lighting, C7026
offshore operations, O7065
permit-to-work, C7010–C7012
personal protective equipment, C7023
precautions, C7014–C7028
prohibition, A1023
publications, C7030
regulations as to, C7001–C7007
respiratory protective equipment, C7023
risk assessment, A1023, C7005–C7006,
  C7008, R3017
roof space, C7003
safe system of work, A1023,
  C7007–C7013
self-employed, C7004
smoking, C7027

CONFINED SPACES, – *cont.*
specified risk, A1023, C7003
Standard Practice Instruction, C7011
static electricity, C7027
statutory requirements relating to work
  in, A1001, A1023
supervision and training, A1023, C7013
system of work, C7003
task procedures, C7011
training, C7013
ventilation, A1023, C7018
working time, A1023, C7028
CONFORMITY ASSESSMENT,
generally, M1013–M1016
pressure systems, and, P7008
CONSERVATION,
Building Regulations, W11044
energy, W11044
**CONSTANT ATTENDANCE
  ALLOWANCE,**
industrial injuries disablement
  benefit, C6014
CONSTITUTION OF THE EU, Intro 9
**CONSTRUCTION AND BUILDING
  OPERATIONS**
see also Fire; Fire Precautions; Ladders;
  Work at Heights; Working Platforms
accidents, prevention, C8018
asbestos
  introduction, A5006
  surveys, A5023–A5026
CDM Regulations 2007. *see* Construction,
  Design and Management Regulations
changing rooms, C8017
children and young persons, C8019,
  V12005
civil liability
  case law, C8040–C8041
  generally, C8039
compressed air, work in, C8018
confined spaces, C8018
Construction Skills Certification
  Scheme, T7027
construction work,
  definition, Intro 23, M2002
criminal liability
  case law, C8035–C8038
  generally, C8034
drinking water, C8017
emergency procedures, F5026
Engineering Construction Industry
  Training, T7027
facilities management, F3033
fire prevention and control
  code of practice, F5024, F5029

CONSTRUCTION AND BUILDING
    OPERATIONS – cont.
fire prevention and control – cont.
    designing out fire, F5027
    emergency procedures, F5026
    fire extinguishers, F5028
    insurance, F5029
    plant, F5028
    precautions, F5029
    risk assessment, F5024
    site fire safety co-
        ordinator, F5024–F5025
    first-aid areas, F7025
fresh air, C8017
good order, C8005
head protection, C8018
health and safety file, Intro 23
health and safety plan, Intro 23
induction, C8031
insurance cover, C8033, F5029
introduction, C8001
ladders, see Ladders,
lifting operations, A1024
lighting
    generally, C8013
    statutory requirements, L5004
lockers, C8017
maintenance projects, Intro 23
new or expectant mothers, V12005
occupational diseases, see Occupational
    Diseases
personal protective equipment, P3003,
    P3004, P3008
practical issues
    changing rooms, C8017
    drinking water, C8017
    fresh air, C8017
    good order, C8005
    lighting, C8017
    lockers, C8017
    rest facilities, C8017
    safe place of work, C8004
    sanitary conveniences, C8017
    temperature, C8017
    washing facilities, C8017
    weather, protection from, C8017
    welfare facilities, C8017
Regulations, Intro 23
responsible person C8002
rest facilities, C8017
risk assessment, F5024, R3017
safe place of work, C8004
sanitary conveniences, C8017
signing, C8019

CONSTRUCTION AND BUILDING
    OPERATIONS – cont.
site fire safety co-ordinator, F5024–F5026
site inductions and security, C8019
temperature, C8017
training and competence, T7027
typical project issues, C8002
vehicles, C8015, C8018, A1014
visitor protection, C8019
vulnerable persons, V12005
washing facilities, C8017
weather, protection from, C8017
welfare facilities, C8017
working platforms, see Work at Heights;
    Working Platforms
CONSTRUCTION, DESIGN
    AND MANAGEMENT
    REGULATIONS
aims, C8503
building costs, C8503
client's responsibilities, C8509, C8514
bureaucracy, unnecessary, C8503
contractor's responsibilities, C8511
demolition or dismantling, C8502
designer's responsibilities, C8510
generally, C8501, C8502, M2022
hazards, identification, C8503
management duties, C8504
    competence, C8505
    co-operation, C8506
    co-ordination, C8507
    prevention, principles of, C8508
meaning of construction, C8501
notifiable projects, C8501
    appendices, information in, C8519
    CDM co-ordinator, C8502, C8503
    demolition or dismantling work, C8502
    duty holders, C8502
    employers, application to, C8501
    health and safety file, C8502
    information in Schedules, C8518
    notifiable threshold, C8502, C8518
    pre-construction information, C8502
    self-employed, application to, C8501
planning and management, C8503
public, protection of, C8511
speculative developments, C8509
sub-contractors, C8511
welfare facilities, C8511
CONSTRUCTION SITES, see Construction
    and Building Operations
CONSTRUCTIVE DISMISSAL, E14014,
    E14019, V8010
CONSULTANTS,
    health and safety, Intro 28

CONSULTATION, Intro 29
food safety, F9006, F9020
generally, J3001
joint, Intro 29, J3002
regulatory framework, J3002
safety committees, see Safety Committees
safety representatives, see Safety
Representatives
violence in the workplace, V8002
CONSUMER AWARENESS, E16012
benchmarking, E16012
CONSUMER CONTRACTS,
unfair terms, P9061–P9063, P9065
CONSUMER PROTECTION,
food legislation and, F9008, F9077
product safety, see Product Liability;
Product Safety
unfair contract terms
legislation, P9061–P9063, P9065
CONTAMINATED PREMISES,
cleaning of, F9074
CONTAMINATION,
food, of, F9010, F9036, F9038, F9047,
F9074
ionising radiation, R1016
CONTINGENCY PLANNING, M2037
CONTINUING PROFESSIONAL
DEVELOPMENT, T7028
CONTRACT,
employment, of, see Contract of
Employment
facilities management, F3022
sale of goods legislation, see Sale of Goods
unfair contract terms, P9061–P9063
CONTRACT MEDICAL ADVISER,
CONTRACT OF EMPLOYMENT,
disciplinary procedures, see Disciplinary
Procedures
additional duties, as to, E14001
collective agreements, E14005
disciplinary procedures, E14006, E14008
express terms, E14002
grievance procedures, E14006, E14008
implied duties, E14003, E14006, E14008
implied terms, E14004
other contracts contrasted, E14001
sickness absence, M1513, M1518
sources of, E14002–E14008
statement of terms and conditions, E14006
statutory minimum notice, E14002,
E14018
violence in the workplace and, V8010
working hours, E14002, E14006
works rules, E14007, E14008

CONTRACTORS,
asbestos removal, A5026
Client/Contractor National Safety
Group, T7027
community health and safety, C6527
enforcing authority, E15009
generally, see Construction and Building
Operations
health and safety policy and, S7004
health and safety standards, Intro 17
licensing, A5026
major public events, C6527
occupiers' liability, O3015
permit to work systems, S3015
risk assessment, R3017
training and competence, T7026–T7027
vicarious liability, E13027
CONTRIBUTORY NEGLIGENCE, Intro
24, O3011, P9032
children, O3011
employers' liability insurance, E13017
occupier's liability, O3011
product liability, in cases of P9038
CONTROL DEVICES, M1036
CONTROL OF LEAD AT WORK, see Lead
at Work
CONTROL OF MAJOR ACCIDENT
HAZARDS REGULATIONS,
see Major Accident Hazards Regulations
CONTROL OF SUBSTANCES
HAZARDOUS TO HEALTH,
see also Hazardous Substances
biological agents, see Biological Agents
children and young persons, V12005
community health and safety, C6506
dangerous substances, D0901, D0903,
D0910
lead at work, see Lead at Work
respiratory diseases, O1046
vulnerable persons, V12005
CONVICTION,
health and safety offence, for publication
of, E15044
COOLING, F5005, N3035
CORGI, G1017, T7007
CORPORATE GOVERNANCE, Intro 1,
Intro 5
Cadbury Committee, Intro 5, Intro 14
claims farmers, Intro 14
Combined Code, E16005
compensation culture, Intro 14
directing mind, Intro 14
Greenbury Report, Intro 5
Hampel Report, Intro 5
large companies, Intro 5

CORPORATE GOVERNANCE, – *cont.*
risk management, Intro 5
Turnbull Report, Intro 5
CORPORATE HOMICIDE. *see also*
  Corporate Manslaughter
Scotland, E15038.1
CORPORATE KILLING, E15001, E15038,
  E15038.1, E13032, M2001
see also Corporate Manslaughter,
Act, effect and requirements, C9001
proposal for offence, Intro 7, Intro 12,
  Intro 20
prosecutions, E13001
small firms, E13001
CORPORATE MANSLAUGHTER
Act, effect and requirements, C9001,
  M2001
aggravating factors, C9011
armed forces, C1001, C9005, C9013
assessing seriousness, C9011
causation, C9002
charities, C9002
charge, E13032
child protection, functions relating
  to, C9008
common law,
  gross negligence prosecutions, C9021
corporations, C9002, E15038.1
Crown bodies, C9001, C9012, C9023
directing mind, C9001
disaster and emergency management
  systems, D6029
DPP consent for proceedings, C1009,
  C9018, C9023
duty of care, C9001, C9002, E15038.1
  relevant, meaning, C9003
emergency services, C9007
employers' organisations, C9002
evidence, C9016
exclusively public functions, C9004
exemptions, C9001
fines C9011
government departments, C9006, C9023,
  E15038.1
gross breach of duty, C9001, C9002,
  C9009
guidance,
  Health and Safety Executive, C9023
  Ministry of Justice, C9023
health and safety legislation, conviction
  under, C9021
identification principle, C9001
individuals, liability, C9019
investigation, C9023
management, senior, C9002

CORPORATE MANSLAUGHTER – *cont.*
management systems and practices, C9002
mitigating factors, C9011
negligence, C9001
no fault clause, where, C9003
law enforcement bodies, C9006
offence, generally, C9002, E15038,
  E15038.1
organisation, C9002
partnerships, C9002, C9015, E15038.1
penalties, C9001, C9002, C9009, E13032,
  E15043
  fines, C9001, C9002, C9009, E13032,
    E15043
  publicity orders, C9011
  remedial orders, C9010
police forces, C1001, C9002, C9014,
  E15038.1
probationary service, C9008
procedure, C9016
prosecution, procedure, C9019
public policy decisions, C9001, C9004
publicity orders, C9013
remedial orders, C9012
sentencing, C9010, C9016, E15043
Sentencing Guidelines Council, C9010,
  E15001, E15043
seriousness of offence, C9011
statutory inspections, C9004
subcontractors, C9002
territorial application, C9022
trade unions, C9002, E15038.1
transfer of functions, C9017
voluntary organizations, C9002
**CORPORATE SOCIAL RESPONSIBILITY
  (CSR),** Intro 1, Intro 5, E16002
Advisory Group E16027
Companies Act 2006, E16005
CR Academy, E16002
CR index E16007
international standard on, E16002
International Strategic Framework
  for, E16002
ISO international standard, E16002
Memorandum of Understanding, E16002
training and qualifications E16002
UN Global Compact Office, E16002
CORROSIVE SUBSTANCES,
manual handling, M3002
COSTS,
director's liability, E15041
employers' liability insurance, E13017
environmental management, E16002

COSTS OF ACCIDENTS AND ILL-HEALTH, Intro 26
large companies, Intro 26
small companies, Intro 26
COUNSELLING, D6061, S11009
COURT OF APPEAL,
criminal jurisdiction, Intro 20
CR ACADEMY,
services, E16002
CR INDEX, E16007
CRAMP, 01047
CRANES, L3018, L3029–L3031
appointed persons, L3030
collapse, L3029
drivers, L3030
failure of, L3031
mobile,
offshore operations, O7070
safe lifting by, L3030
overturning, L3029
risk assessment, L3018
signallers, L3030
slingers, L3030
supervisors, L3030
types of, L3029
tower crane, L3029
uses of, L3029
CREDIT SALE AGREEMENT,
product liability, P9055
breach of statutory duty,
for, E15021–E15035
CRIMINAL OFFENCES,
breach of statutory duty,
for, E15021–E15035
burden of proof, E15031
construction and building operations, and
case law, C8035–C8038
generally, C8034
corporate manslaughter, Intro 7, Intro 12,
E13032
See also Corporate Manslaughter
custodial sentences, Intro 7
employers' liability insurance, E13008
generally, Intro 20
fire prevention and control, F5060
harassment, H1729
battery, H1730
Crime and Disorder Act 1998, H1733
Protection from Harassment Act
1997, H1731
Public Order Act 1986, H173
human rights legislation, E15031
innocence, presumption of, E15031
lifting, L3017
lifts, L3008–L3009

CRIMINAL OFFENCES, – *cont.*
machinery safety, M1018
penalties, Intro 7, Intro 12, Intro 15, Intro
20
personal protective equipment, P3027
product liability, as to, P9015–P9027
public service cases, E15036
training and competence, T7002
workplace violence, V8011
CRISIS,
communication, B8038–B8042
CROWN, THE,
enforcement, E15009, E15013
fire prevention and control, F5061
food premises regulations, F9033
health and safety law and, E15045
immunity, E15045
CROWN COURT,
jurisdiction, Intro 20
CRUSHING, M1025
CSR, *see* Corporate Social Responsibility
CUTTING AND SEVERING, M1025
CULLEN REPORT, Intro 14
see also Offshore Installations, O7003,
O7046
CUSTODIAL SENTENCES, Intro 7
CUSTOM BUILT ARTICLES, P9024
CUSTOMERS,
benchmarking, E16012
community health and safety, C6515–
C6517
consumer awareness, E16012
customer base, erosion of, B8017
facilities management, F3002
precautions, C65016
small businesses, Intro 4

D

DAIRY,
hygiene, F9025
registration, F9028
DAMAGES,
assessment, Intro 24, C6039, R2004
awards, C6024, C6044
bereavement, for, C6031
categories of injuries, C6043
coal miners, V5003, V5012
contributory negligence, C6046
cost of care, C6054
deductible payments, C6051, C6052
director's liability, E15041
employers' liability insurance, E13002

DAMAGES, – *cont.*
 fatal injuries, C6036–C6039
 general, C6021, C6024
 interest on, C6043
 interim awards, C6041
 liquidated damages, C6021
 loss of amenity, C6034
 loss of future earnings, C6026
 lost years, C6029
 lump sum, C6021
 noise, N3012
 non-pecuniary losses, *see* Non-pecuniary
  losses
 occupational health and safety, O1002,
  O1005
 occupational injuries and diseases,
  basis of an award, C6022
  capitalisation of future losses, C6027
  conditional fee system, C6021
  hazardous substances, as to, H2103
  liquidated, C6021
  loss of future earnings, C6026
  lump sum, C6021
  non-pecuniary losses, C6023, C6030
  pecuniary losses, C6023, C6030
  prescribed diseases, for, C6055–C6062
  structured settlements, C6021, C6047
  unliquidated, C6021
 pain and suffering, C6031–C6033
 payments into court, C6042
 pecuniary losses, assessment of, C6026
 personal injuries at work,
  accident, caused by, C6005
  basis of award, C6022
  fireman, C6005
  non-pecuniary losses, C6023
  pecuniary losses, C6023
 personal protective equipment, P3027
 prescribed diseases, C6055–C6062
 product liability, *see* Product Liability
 psychiatric injury, S11009
 recovery of state benefits,
  appeals against certificates of
   recoverable benefits, C6050
  compensator's duties, C6045
  complex cases, C6046
  contributory negligence, C6046
  exempt payments, C6044
  information provisions, C6045, C6046
  new regulations, C6044
  rationale, C6044
  retrospective effect of provisions, C6044
  structured settlements, C6025, C6046
 rehabilitation, R2004, 2006

DAMAGES, – *cont.*
 set-off,
  deductible items, C6042
  non-deductible items, C6042
 special, C6021, C6024
 stress, S11009
 structured settlements, C6025, C6047
 unliquidated damages, C6021
 vibration white finger, V5003, V5012,
  V5013
 wages, loss of, C6053
 wrongful dismissal, for, E14017
DANGER,
 areas, M2011
 serious and imminent danger, procedure
  for, M2011
DANGEROUS ACTIVITIES,
 community health and safety, C6510,
  C6511
 Health and Safety Executive, C6508
 neighbours and passers-by, C65011
 planning, C6511
 young people, C6508
DANGEROUS GOODS. See DANGEROUS
  SUBSTANCE
DANGEROUS MACHINERY, *see* Machinery
  Safety
DANGEROUS OCCURRENCES,
 list of reportable, A3035, A3037
DANGEROUS PRODUCTS,
 liability for, P9054
 withdrawal of, P90055, P9009, P9059
DANGEROUS SUBSTANCES,
 See also Hazardous Substances; Major
  Accident Hazards Regulations;
  Transportation of Dangerous
  Substances
 aggregation, D6015
 Approved Codes of Practice, D0902,
  D0933–D0933.6, D0934
 case histories, D0901.1
 chemical agents, D0901, D0908
 children and young persons, V12016
 CHIP Regulations, Intro 11
 cleaning, D0933.5
 clothing, D0923
 communication systems, D0933.1
 community health and safety, C6510
 compressed natural gas,
  unloading, D0933.6
 containers, D0933.1
  identification of hazardous contents
   of, D0927
 control measures, D0915, D0933.4

DANGEROUS SUBSTANCES, – cont.
control of substances hazardous to
health, D0901, D0903, D0910
Dangerous Substances and Explosive
Atmospheres Regulations
2002, D0901–D0917
application, general conditions
for, D0905
commencement, D0909, D0924
exemptions, D0928
modernisation of
legislation, amendments to, D0929
repeals and revocations of existing
legislation, D0930
scope of, D0903
definition, D0908, H2142
design of plant equipment and
workplaces, D0933.2
dusts, safe handling of
combustible, D0933.6
duty holders, D0906
elimination of risk, D0914
emergencies, dealing with, D0925,
D0933.1
employees,
information, D0933.1
training, D0933.1
enforcement of legislation, D0931
equipment, D0920–D0924
EU laws, D0901
explosive atmosphere D0920-D0921,
H2141, H2143
fire, F5031
risk of, D0910
safety legislation, H2101, H2141,
H2142
garages, D0901.1
general safety measures, D0917
guidance, D0933.1–D0933.6, D0934
Health and Safety Executive
guidance, D0914, D0933.6
heat, sparks and flame, generation
of, D0933.5
hot work, D0933.5
ignition sources, D0933.3, D0933.4,
D0933.6
incidents, dealing with, D0925
information provision, D0926, D0933.1
inspection, D0933.5
instructions, D0926
labelling, D0933.1
leaflet, D0933.6
liquefied petroleum gas, unloading
of, D0933.6
maintenance, D0933.5
mitigation measures, D0916, D0933.4

DANGEROUS SUBSTANCES, – cont.
motor vehicle repair garages, D0901.1
national security, D0928
Permit-to-Work system, D0933.5
petrol, D0901.1
road tankers, unloading, D0933.6
personal protective equipment, D0906,
P3017
petroleum, D0929
pipes, identification of hazardous contents
of, D0927, D0933.1
plant equipment, design of, D0933.2
precautions, D0920
protection systems, D0920–D0924
radiation emergencies, see Radiation
Emergencies
radioactive substances, see Radioactive
Substances
recording findings, D0911
reduction of risk, D0914
regulatory impact assessment, D0932
repairs, D0933.5
risk assessment, D0902, D0910–D0911,
D0933.1, D0933.6, R3007, R3014,
R3035
equipment, D0920
guidance, D0933.1
reviews of, D0912
risk management, D0902, D0910,
D0913–D0917
road tankers, unloading, D0933.6
ships, D0904
sites, notification and marking of, H2144
sources, D0934
storage, D0933.3
training, D0926, D0933.1
trespassers, C65013
typical activities, D0904
ventilation,
Approved Code of Practice, D0933.4
verification of safety prior to first
use, D0922
warning signs, D0921, D0933.1
waste, D0933.3
workplaces,
definition of, D0907
design of, D0933.2
DATA,
ergonomic design and, E17057, E17058
workplace violence, V8023
DATA PROTECTION,
occupational health and safety, O1017
rehabilitation, R2007
sickness absences, R2007
violence in the workplace, V8081

DATABASE,
Employers' Liability Insurance,
proposal, E13026
DE QUERVAIN'S DISEASE, 01047
DEAFNESS,
occupational,
compensation for, C6055
generally, *see* Noise-induced Hearing
Loss
DEALER,
product safety, P9056
DEATH,
see also Fatal Injuries
corporate manslaughter, Intro 12, E13032
See also Corporate Manslaughter
employers' liability insurance, E13025
offshore installations, O7040
reporting, A3018
work-related,
single greatest cause, A5001
DECLARATIONS,
Declarations of Conformity,
European Union, M1015–M1016
example of, M1015
incorporation into other
machinery, M1015
machinery safety, M1015–M1016
DECOMPRESSION PROCEDURES,
work in compressed air, and, C8019
DEDUCTIONS FROM WAGES, E14014
DEFECTIVE PRODUCTS,
liability for, *see* Product Liability
DEFENCES,
factory occupier, W11049
food legislation, under, F9022, F9023,
F9070, F9082
HSWA, under, E15029, E15031
occupier's liability, O3017
offshore operations, O7056
DELIVERY VEHICLES,
traffic routes for, A1001, A1005, A1006
DEMOLITION,
building and construction sites, C8007,
C8502
DEMOTION, E14014
DEMS, *see* Disaster and Emergency
Management Systems
**DEPARTMENT FOR ENERGY AND
CLIMATE CHANGE (DECC),**
environmental management, E16024
**DEPARTMENT FOR ENVIRONMENT,
FOOD AND RURAL AFFAIRS
(DEFRA),**
environmental management, E16024

**DEPARTMENT FOR WORK AND
PENSIONS,**
health, work and recovery
programme, Intro 7
DEPARTMENT OF HEALTH,
food, role as to, F9005
DEPRESSION, O1052–O1055
DEREGULATION, Intro 12
DERMATITIS, H2103, O1037–O1038,
O1040–O1041, P3026
DESIGN,
dangerous substances, D0933.2
designers, M1002, M1009
electricity, E3012–E3021, E3025, E3031
ergonomic, *see* Ergonomic Design
facilities management, F3028,
F3031–F3032
fire prevention and control, F5027
floor, M3014
heights, working at, W9002
lifts, L3009, L3011
lighting, L5014, E17031
machinery safety, M1002, M1003,
M1009, M1026
manual handling, M3009, M3012, M3014
plant equipment, D0933.2
pressure systems, and, P7016
processes or activities, M3009
role of, E17001, E17002
stress, S11013
task layout, M3012
violence in the workplace, V8058–V8064,
V8079, V8109
workplaces, D0933.2
DESIGNER,
product safety, as to, *see* Product Safety
**DETERRING/IMPEDING
DEVICES,** M1036
DEVELOPMENT RISK,
defence, as P9049
DIFFUSE MESOTHELIOMA
compensation for, C6062
DIGNITY
harassment, violation H1701, H1708
purpose or effect, H1707
DILUTION VENTILATION, V3011
**DIPLOMA IN
SAFETY MANAGEMENT,** T7013,
T7020
DIRECT SALES, E13003
DIRECTIVES, EU, *see* European Union (EU
Laws),
DIRECTORS,
civil liability, E13008, E15041
criminal liability, E15040

DIRECTORS, – *cont.*
disclosure, E16005
employers' liability insurance, E13008
environmental management, E16005
Gross Negligence Manslaughter, E13032
guidance, Intro 7
Health and Safety Executive, Intro 5
health and safety responsibilities, E15040,
    M2030
indemnities, E15041
insurance, E15041
large companies, Intro 5
liability insurance, E13023
offences committed by, E15032
Operating and Financial Review, E16005
training and competence, T7001
DISABILITY DISCRIMINATION,
adjustments, reasonable, A1001, A1005,
    R2005
association, discrimination by, M1508
auxiliary aid, M1511
avoiding, V12037
before offer of work, W11043
codes of practice, A1001
disabled person, meaning, M1506,
    W11043
duty to make reasonable
    adjustments, E16511, R2005
Equality Act 2010, A1001, E16509-
    E16512
  disability, definition, E16510, M1506
  discrimination by association, M1508
  rehabilitation, R2005
  sickness absence, M1505
ergonomics, E17070
harassment, H1703
 health and safety excuse,R2006
indirect discrimination, M1508, M1510,
    W11043
legal rights, W11043
long-term, R2005
musculoskeletal disorders, 01050
protected characteristic, H1703, R2005
reasonable adjustments, A1001, E17070,
    M1511, R2005
sickness absence, M1506–M1511
stress, O1052
unfair dismissal, and, E14020–E14022
workplace, in, W11043
DISABILITY TAX CREDIT, C6003, C6016
**DISABILITY WORKING
    ALLOWANCE,** C6003, C6016
DISABLED PERSONS, V12032–V12037
access, facilities for, A1001, R3028,
    W11043, V12034

DISABLED PERSONS, – *cont.*
'associated with', M1508
conditions, M1509
  sight impairment exception, M1509
definition of, A1001, E16510, M1506,
    W11043
disability tax credit, C6003, C6016
disability working allowance, C6003,
    C6016
disablement benefit, *See* Disablement
    benefit
discrimination, *See* Disability
    discrimination
disfigurement, M1506
duty to make reasonable
    adjustments, E16511, R2005
Equality Act 2010, E16501, R2005-
    R2006
HSE guidance, E16511, V12032, V12037
impairment, E16510
long term, E16510
normal day to day
    activities, R2005–R2006, V12032
rehabilitation, R2005
risk assessment, E16512, R3028, R3032,
    V12033, V12035–V12036
sickness absence, M1506–M1511
tax credits, C6003, C6016
types of disability, V12032
DISABLEMENT BENEFIT,
occupational deafness, for
    assessment of benefit, N3021
    conditions under which payable, N3020
    entitlement to, N3002, N3018
vibration white finger, N3012
**DISASTER AND
    EMERGENCY MANAGEMENT
    SYSTEMS,** Intro 29, D6001–D6090
access, D6010
active policy and pre-planning, D6056
aggregation, D6015
ammonium nitrate, D6013
audits, D6018, D6067, D6068
building contractors, D6061
business continuity, D6055, D6060
business management, D6002
carcinogens, D6013
CIMAH, pre D6001
civil contingencies, D6001
Civil Contingencies Act
    2004, D6005–D6007
civil defence, D6008
codes, D6056
command and control chart of
    arrangements, D6041

DISASTER AND
    EMERGENCY MANAGEMENT
    SYSTEMS, – *cont.*
communication, establishing
    effective, D6044
competence, D6046
confined spaces, D6024
construction industry, D6022–D6023
contingency planning, D6006
control, D6047
co-operation, D6007, D6045
corporate culture and practice, D6037
corporate manslaughter, D6029, E13032
    See also Corporate Manslaughter
corporate reasons, D6031
counselling support, D6061
criminal offences, D6031, D6029
culture, D6051
dangerous substances, D6009, D6017
definitions, D6003, D6013
design and architecture, D6037
effective DEM, need for, D6004–D6032
    legal reasons, D6005–D6011
emergency, meaning, D6006
employers' duties, D6010
enforcement, D6032
environment, D6033, D6036
    Environment Act 1995, D6028
    Environment Agency, D6011
establishing policy on, D6038–D6042
European Union, D6001, D6011
    radiation, D6018
exclusions, D6016
exercises, D6054
explosives, D6009, D6017, D6027
external factors, D6035–D6036
Federal Emergency Management Agency
    (FEMA), D6001
fertiliser, D6013
first aid, D6021, H6016
hazardous substances, D6013
Health and Safety Executive, D6011
historical development, D6001–D6003
holistic phase, D6001
human error, D6037
humanitarian reasons, D6032
information,
    public, D6018–D6019
    radiation, D6019
    sources of, D6036, D6071
inquiries, D6062
insurance, D6002, D6030, D6061
internal factors, D6035, D6037
investigations, D6059, D6062

DISASTER AND
    EMERGENCY MANAGEMENT
    SYSTEMS, – *cont.*
lead categories, D6007
liberalisation, D6001
local government, D6008
loss adjusters, D6061
lower tier duties, D6017–D6020
Major Accident Prevention Policy
    (MAPP), D0717, D6018
major disasters, E17002
    arrangements for, D6040–D6042
management failure, D6002
Management of Health and Safety at Work
    Regulations, D6010
manslaughter, D6029
maritime and coastal emergencies, D6006
monitoring, D6018, D6066
national security, D6007
Northern Ireland, D6007
notifications, D6019
occupational health and safety, D6002
organisation, D6043–D6047
origins of, D6002
petroleum products, D6013
phases, D6001
Piper Alpha, D6029
plans, D6007, D6012–D6014, D6048-
    –D6065
    after the event, D6057–D6065
    before the event, D6049–D6055
    decisional planning, D6051
    during the event, D6056
    long-term, D6064–D6065
    medium-term, D6063
    monitoring, D6066
    off-site, D6013
    on-site, D6012
    reviewing, testing and
        implementing, D6014
    short-term, D6062
    testing, D6052
police, D6061
policy,
    establishing, D6038–D6042
    statement of, D6039
    summary, D6042
potassium nitrate, D6013
procedures, D6010
protected emergency centres, D6008
radiation, D6018
    public information, D6019
    regulations, D6018
reporting duties, D6020, D6035
responders, D6007

**DISASTER AND
EMERGENCY MANAGEMENT
SYSTEMS, – cont.**
responsiveness, D6006
reviews, D6067, D6069
risk assessment, D6007, D6010, D6066
road transport, D6025–D6027
safety reports, D6021–D6024
  contents, D6021, D6023
  preparation, D6021
  purpose of, D6022
  review and revision of, D6024
Scottish Environment Protection
  Agency, D6011
security management, D6002
sensitive information, D6007
September 11, 2001, terrorist
  attacks, D6001
Seveso II Directive, D6011
testing, D6052
top-tier duties, D6021–D6028
training, D6010, D6053
transport,
  explosives, D6027
  road, D6025–6027
United States, D6001
use, powers to prohibit, D6031
visitors, D6061
warnings, D6007
DISASTERS,
see also Disaster and Emergency
  Management Systems
business continuity
  planning, B8001–B8003, B8008,
  B8023
natural, B8008
public relations, B8009
recovery, B8001
DISCIPLINARY PROCEDURES,
ACAS procedures, E14013, E14014
accident investigations, A3030
appeal, E14013
contract of employment, in, E14006,
  E14008
essential elements, E14013
hearing, E14013
investigation, E14013
rules, scope of, E14012
sanctions,
  demotion, E14014
  dismissal, *see* Dismissal
  fines or deductions, E14014
  suspension, E14014, E14015
  warning, E14014
sickness absence, M1529

DISCOS, N3027
DISCRIMINATION,
see also Harassment
age, H1702-H1703
agents, E16501
associative, H1703, H1707
colour, H1703
combined, E16501, E16505
compensation, E16522
direct, E17501
  definition, E16503
  test, E16503
disability, E16501, E16506, H1703
employees E16501
employers, liability, E16501
Equality Act 2010, E16501-E16522
gender reassignment, H1703
general claim, H1703
history, H1701-H1702
indirect, E16501, E16504
marriage, H1703
maternity, H1703
nationality, H1703
perceived, H1703, H1707
pregnancy, H1703
principals, liability, E16501
racial, H1703
remedies, E16522
sexual, H1703
training, H1739
unlawful, H1717-H1718
DISEASES,
accident reporting, A3002, A3016, A3035
  prescribed form, A3038
asbestos-related, *see* Asbestos-related
  Diseases
cancer, *see* Cancer
compensation for workplace diseases, *see*
  Compensation for Work
  Injuries/Diseases
food hygiene, and, F9023
list of reportable, A3035
livestock, H2103
new or expectant mothers, V12018
non-ionising radiation, R1008
pneumoconiosis, *see* Pneumoconiosis
prescribed industrial, benefits for, C6010
record keeping duty, A3016
specified, claims for, C6048
DISFIGUREMENT,
disability, M1506
DISMISSAL,
Code of Practice on Disciplinary and
  Grievance Procedures, M1517
constructive, E14014, E14019

DISMISSAL, – *cont.*
health and safety grounds, E14010
sickness absence, M1516–M1517, M1530
unfair, *see* Unfair Dismissal
wrongful, E14017, E14018
DISPLAY SCREEN EQUIPMENT. *See* **Visual display units**
DIVING OPERATIONS,
accident reporting, A3035
equipment, P3017
manual handling, M3002
new or expectant mothers, V12020
offshore operations,
diving contractor, O7037
general requirements, O7036
planning, dive, O7038
risk assessment, O7038
safety devices for, P3017, P3018
training and competence, T7007
DOCKS,
children and young persons, V12015
DOCTOR,
training, W10026, W10031, W10034
working hours, W10003, W10012, W10031
DOMESTIC SERVICE,
working time, W10032
DOORS AND GATES,
workplaces, safety regulations, W11001, W11016
DOW JONES SUSTAINABILITY INDEX, E16008
DRAINS,
fire fighters, O3014
occupiers' liability, O3014
DRILLING MACHINES, M1030
DRINKING WATER,
construction and building operations, and, C8017
general provision, W11024
DRIVE SHAFTS, M1008
DRIVERS,
access, when using vehicles, *see* Access
training, generally, A1013
whole-body vibration, V5023–V5024, V5026
DROWNING,
prevention of, C8013, P3017, P3024
DRUGS,
DRY POWDER FIRE EXTINGUISHERS, F5013
DUCTS, F5007
DUST,
dangers of exposure to, D0933.6, V3002
explosions, D0919, D0933.6

DUST, – *cont.*
fire, D0919
nuisance, E5016
polluting agent, as, V3003
safe handling, D0933.6
ventilation, *see* Ventilation,
workplace, origins of, V3002
DUTY OF CARE, Intro 24
occupational health and safety, O1002, O1005
rehabilitation, R2004
right to roam, O3007
trespassers, O3007
violence in the workplace, V8004–V8005

E

E COLI OUTBREAK, F9021
EAR,
noise-induced hearing loss, *see* Noise-induced Hearing Loss
EARTHING, E3015
EATING FACILITIES,
provision of, W11027
ECO-LABEL SCHEME, E16008
ECO-MANAGEMENT AND AUDIT SCHEME (EMAS) E16014, E16026
differences with ISO 14001, E16016
ECZEMA, P3026
EDUCATION,
See also Training and competence
higher, T7010, T7013
safety representatives, role, J3002
EFAW,
emergency first aid at work training, F7001, F7018
EGGS,
production of, F9030
ELECTRICAL HAZARDS,
confined work space, in, A1022, A1023
fire, F5004, F5015
extinguishers, F5004, F5015
temporary installations, C6516, C6517
ELECTRICITY,
access, E3020
accident reporting, A3035
adverse or hazardous environments, E3013
alternating current, E3002
arcing, E3008
ATEX Directives, E3027
battery rooms, E3027
Building Regulations, W11044
buildings, E3033

ELECTRICITY, – *cont.*
burns, E3006
cables, E3028
community health and safety, C6510
competence, E3011, E3024
conductors, E3016
confined spaces, C7020, C7026
connections, E3017
construction of system, Intro 23
current, measurement of, E3002
dangers, E3004–E3009
design and construction, E3012–E3021,
    E3025, E3031
direct current, E3002
earthing, E3015
electric shocks, E3005, E3014–E3015,
    E3018
electromagnetic compatibility, E3029
equipment, E3003, E3031
    confined spaces, C7020, C7022
    dangerous substances, D0920
    explosions, D0920
    isolated equipment, work on, E3023
    live work, E3024
    potentially explosive
        atmospheres, E3027
    strength and capability, E3012
excess current protection, E3018
explosions, D0920, E3009, E3027
fatal accidents, E3001
fencing, E3014
fire, E3007
flashovers, E3008, E3009
guidance, E3003, E3021, E3022, E3028
injuries, E3004
insulation, E3014, E3015, E3024
isolation, E3019, E3023
legal background, E3003
lighting, E3020
lightning protection systems, E3025
live parts and work, E3004, E3005,
    E3011, E3024
maintenance, Intro 23, E3021
national vocational qualifications
    (NVQs), E3011
offshore operations, O7063
Ohm's Law, E3002
overhead lines, E3028
permit-to-work, E3023
plugs, E3030
potentially explosive atmospheres, E3027
precautions, E3014–E3015
qualifications, E3011
quality, E3032
quarries, E3003

ELECTRICITY, – *cont.*
rectification process, E3002
referenced conductors, E3016
residential dwellings, E3033
residual current drive, E3016
resistance, E3002
risk assessment, E3009, R3016 (App B)
risks, controlling, E3010
safe systems of work, E3022–E3024
shocks, E3005, E3014–E3015, E3018
sockets, E3030
standards, E3003, E3024, E3027, E3033
static electricity, C7026, E3026
trade associations, E3011
training, E3011
understanding, E3002
voltage, E3004, E3014, E3031
watts, E3002
wiring regulations, E3003
working space, E3020
ELECTROMAGNETIC
    RADIATION, R1001–R1002,
    R1004–R1010, V12020
EMAS (ECO-MANAGEMENT AND AUDIT
    SCHEME), E16014, E16026
differences with ISO
    14001, E16016EMERGENCIES,
    vibration, emergency certificates
        and, V5015
EMERGENCY ARRANGEMENTS,
    business continuity planning, B8023
    community health and safety, C6511,
        C6517
    customers and users, arrangements
        for, C6517
    equipment, C6514
    volunteers, C6522
    confined spaces, C7006, C7028
    customers and users, arrangements
        for, C6517
    dangerous substances, D0925, D0933.1
    carriage of *see* Transportation of
        Dangerous Substances
    equipment, C6514
    escapes, D0925
    explosions, D0925
    heights, working at, W9006
    managing health and safety, M2039
    noise, N3041
    sound systems, N3041
    violence in the workplace, V8094,V8135
    volunteers, C6522
    warnings, D0925
EMERGENCY EXITS,
    fire prevention and control, F5040

EMERGENCY PLANS,
see also Disaster and Emergency
Management Systems; Emergency
Procedures
ionising radiation, R1021
EMERGENCY PROCEDURES,
construction sites, F5026
disaster and emergency management
systems, see Disasters and Emergency
Management Systems
food safety, F9015, F9018, F9019
hazardous substances, where, H2139
fire prevention and control, F5026
lone workers, V12029, V12030
offshore installations, O7028
radiation emergencies, see Radiation
Emergencies
**EMERGENCY PROHIBITION NOTICES
AND ORDERS,** F9018–F9019
EMERGENCY SERVICES,
employer's duties as to, R3002, R3005,
R3018, R3029
managing health and safety, M2012
EMERGENCY WORKERS,
legal protection from assault, V8008
EMISSIONS INTO THE ATMOSPHERE,
acidification, E5001, E5014, E5015,
E5063
air quality assessment,
daughter directives, E5034–E5039
framework directive, E5033
UK regulations, E5040
Air Quality Strategy,
Local air quality management
(LAQM), E5069
preparation of, E5002
priority pollutants, E5018, E5020
climate change, see Climate change
CLRTAP, role of, E5021, E5026
coal burning, E5003, E5021, E5061
dark smoke, E5062
distance scales, range of, E5001
ecosystems, damage to, E5014, E5015
Environment Agency,
IPPC, implementation of, E5002
local authorities and, E5046
PPC
role of, E5002, E5043
Europe,
air quality assessment, see air quality
assessment, above
BREF Notes, E5028, E5029
IPPC directive, see integrated pollution
prevention control, below
national emissions ceilings, E5037

EMISSIONS INTO THE ATMOSPHERE, –
cont.
Europe, – cont.
policy makers, European E5026
protocols, E5026
solvent emissions, E5026
UNECE, role of, E5021, E5026
waste incineration directive, E5026
furnaces E5061
global warming, E5002
human health,
effect on, E5003–E5006
protection of, E5020
integrated pollution control ('IPC'),
predecessor of IPPC, see intergrated
pollution prevention control E5061
integrated pollution prevention control
('IPPC'),
application of, E5027
BAT concept, E5027, E5028
BREF Notes, E5028, E5029
European Bureau, E5028
information exchange forum, E5028
LAPC system, replacement of, E5040
introduction of E5040
role of, E5027
technical working groups, E5028
UK law, introduction into, E5041
local air pollution control ('LAPC'),
replacement of, E5040
transfer to IPPC, E5040
local authorities,
Environment Agency and, E5046
IPPC, implementation of, E5002
role of, E5002
nuisance, E5016, E5062
permits,
BAT concept, based on, E5028
pollutants,
ammonia, E5014
arsenic, E5033
benzene, E5003, E5019, E5033, E5036
cadium, E5033
carbon monoxide, E5007, E5019,
E5034, E5036
dioxins, E5026
examples of, E5001
industrial processes, from, E5020
lead, E5007, E5019, E5033
medical effects of, E5003–E5006
nickel, E5033
nitrogen dioxide, E5004, E5019, E5026,
E5033
1-3 butadiene, E5003, E5019
ozone, E5003, E5019, E5033, E5037

EMISSIONS INTO THE ATMOSPHERE, –
cont.
pollutants, – *cont.*
particulate matter, E5003, E5033,
E5034
persistent organic, E5003
polycyclic aromatic
hydrocarbons, E5003, E5019,
E5033
sulphur dioxide, E5003, E5019, E5021,
E5033
UK priority pollutants, E5019
pollution prevention control ('PPC'),
Act of 1999, role of, E5027
regulation,
regulatory drivers, E5003–E5016
sources of, E5002
smog, E5003, E5019
smoke, E5061, E5062
stratospheric ozone depletion, E5002,
E5024, E5025
trans-boundary pollution, E5014, E5021,
E5064
UNECE, role of, E5021, E5026
United Kingdom,
acronyms, definition of, E5041
air quality standards, E5063–E5067
air quality strategy, E5068
clean air legislation, E5061
effects-based policies, E5018, E5063
Environment Agency, *see* Environment
Agency, *above*
Integrated Pollution Prevention Control
(IPPC), E5041
local authorities, *see* local authority,
above
Northern Ireland, E5045, E5069
nuisance, E5016, E5062
priority pollutants, E5018–E5020
regulators, E5044–E5047
Scotland, E5044
terms, definition of, E5041
useful websites, E5034
World Health Organisation
guidelines, E5063–E5065
EMISSIONS TRADING E16006
EU Emissions Trading System, E16006
EMPLOYEES,
accident reporting, A3024–A2036
consultation with, *see* Consultation, Joint;
Safety Committees; Safety
Representatives
contract, *see* Contract of Employment
dangerous substances, D0933.1
defective equipment supplied to, P9024

EMPLOYEES, – *cont.*
disciplinary procedures, *see* Disciplinary
Procedures
duties of,
manual handling, M3002
control of hazardous
substances, H2105, H2127, S7005
personal protective equipment, as
to, P3007
statutory duties, Intro 21
employment protection rights, E14009
enforcement of health and safety laws
against, E15040–E15042
Equality Act 2010, E16501
heights, working at, W9013
information, D09033.1
managing health and safety, M2017
manual handling, M3002
negligence, *see* Negligence
safety rights,
not to be dismissed on certain
grounds, E14010, E14011, E14022
not to suffer a detriment, *see also*
Employment Protection, E14009
small businesses, focus on, Intro 4
suspension from work, *see* Suspension
from Work
training and competence, D0933.1,
T7001, T7024–T7025
vibration, V5016
women workers, *see* Women Workers
workplace violence, *see* Workplace
Violence
**EMPLOYERS, *see also* Employers' Liability
Insurance**
asbestos, A5007–A5019
breach of contract, H1720
compensation *see* Compensation
constructive dismissal, H1720
confined spaces, C7003
delegation to junior staff, E15039
discriminatory acts of employees, E16501
disaster and emergency planning, D6010
duty of care, Intro 24, M1524
employees' safety, E14009
employees who drive road
vehicles, M2103
enforcement
safety rules, E14012, E14021
unlawful discrimination, E16522
Equality Act 2010, E16501
harassment, liability,
causal link with employment, H1712
course of employment, H1711
defence, , H1713

EMPLOYERS, *see also* **Employers' Liability Insurance** – *cont.*
  harassment, liability, – *cont.*
    legal action, H1716-H1733
    narrowing extent, H1715
    remedies, H1718, H1721, H1725, H1728
    third party, protection, H1703
    tort, H1717, H1722-H1725
    unfair dismissal, H1719
    unlawful discrimination, H1717-H1718
    vicarious liability, H1710
  harassment, policy, H1734-H1740
    complaints procedure, H1735
    confidentiality, H1736
    formal policy, H1738
    informal action, H1737
    monitoring and review, H1740
    training, H1739
  hazardous substances, where, H2105
  health and safety policy statement, *see* Health and Safety Policy Statement
  health and safety legislation, under, R3001–R3006
  insurance, as to, E13002
  lighting, as to, L5002
  managing health and safety, M2004
  manual handling, M3002
    duty of care, Intro 24
    vicarious liability, principles of, Intro 24
  negligence,
    workplace stress, M1524
  occupational diseases, as to, *see* Occupation Diseases
  personal protective equipment, as to, P3006, P3007
  risk assessment, *see* Risk Assessment
  safe system of work, S3003
  statutory duties,
    civil liability for breach, Intro 24, W11045
    corporate manslaughter, Intro 20, E13032
    criminal liability for breach, Intro 20
    health and safety legislation, under, Intro 21, R3001–R3006
    risk assessments, Intro 16
    written statement of health and safety policy, *see* Health and Safety Policy Statement
  stress, S11006
  third parties and, R3002, R3005, R3018
  unlawful discrimination, E16522
  violence in the workplace, *see* also Workplace Violence

EMPLOYERS, *see also* **Employers' Liability Insurance** – *cont.*
  violence in the workplace, *see* also Workplace Violence – *cont.*
    prosecutions, V8015
  volenti non fit injuria, Intro 24
  vulnerable persons V12005
EMPLOYERS' LIABILITY INSURANCE, Intro 29
  admission of liability, E13029
  agency workers, E13016
  apologies, E13033
  apportionment, E13026
  asbestos, E13001, E13026
  assessment of premium, E13022
  Association of British Insurers, E13028
  asthma, E13026
  book rate, E13022
  breach of statutory duty, E13002
  brokers, E13003
    Financial Services Authority, E13003
    proposal forms, completion of, E13003
    regulation of, E13003
  burden of proof, reversal of, E13002
  cancer, E13026
  caveat emptor, E13004
  certificates of insurance,
    display, E13013
    issue, E13013
    retention, E13013
  children, E13016
  Civil Procedure Rules, E13029
  claims handling, E13017, E13033
  Code of Practice, E13023
  compensation, E13002, E13017
    Compensation Act, E13033
    compensation culture, E13001
  conditions of cover, E13019
    prohibition of certain conditions, E13020
  contributory negligence, E13017
  corporate killing, E13001, E13032
  costs, E13017
  cover provided by policy, E13016–E13019
    business cover, E13016
    compensation, E13017
    conditions, E13019
    costs, E13017
    FSCS scheme. E13024
    geographical limits, E13018
    interpretation of, E130165
    legal expenses, E13023
    necessary, E13011
    persons, E13016

EMPLOYERS' LIABILITY INSURANCE, – *cont.*
cover provided by policy, – *cont.*
  related covers, E13023
  scope of cover, E13017
criminal offences,
  corporate killing, E13001
  failure to insure, E13009
damages, E13017
database, proposal, E13027
death of claimant, E13025
defence, E13002
degree of cover necessary, E13011
direct sales, E13003
directors' personal liability, E13008, E13023
disclosure, duty of, E13004
display of certificate, E13013
  failure, E13015
duty to take out, Intro 24, E13008
employees,
  children and young people, E13016
  contributions, from, E13017
  covered by Act, E13009
  definition, E13016
  duty of employers to, E13002
  indemnities, from, E13007
  not covered by Act, E13010
failure,
  criminal offences, E13009
  display, to, E13015
  insure, to, E13001, E13014
  long development period, diseases with, E13001
  maintain insurance, to E13014
Financial Services Compensation Scheme, E1302
fire,
  mitigation, E13006
  sprinklers, installation of, E13006
first-aid, F7012
generally, Intro 24, E13001
geographical limits, E13018
good faith, E13004
hazardous work, conditions imposed for, E13021
identification of insurer, E13026
indemnities, E13007, E13024
industrial diseases,
  death of claimant, E13025
  limitation periods, E13025
insolvent employers, E13024
  indemnities, E13024
  revival of dissolved companies, E13024

EMPLOYERS' LIABILITY INSURANCE, – *cont.*
insurance contracts, law relating to, E13003–E13006
  disclosure, duty of, E13004
  loss mitigation, E13006
  mitigation of loss, E13006
  proposal form, filling in, E13005
issue of certificate, E13013
knowledge,
  date of, E13025
legal expenses insurance, E13023
legal liability policy, as, E13002
limitation periods, E13025
  accrual, date of, E13025
  death of claimant, E13025
  industrial diseases, E13025
  knowledge, date of, E13025
  mesothelioma, E13025
long tail diseases, E13026
loss mitigation, E13006
lung cancer, E13026
maintain, duty to, E13008, E13014
Making the Market Work Scheme, E13028
measure of risk, E13022
medical evidence, E13029
mesothelioma, E13025, E13026, E13033
NHS recoveries, E13031
noise, E13026
notification of claims, E13017
offshore operations, E13001
penalties, E13014, E13015
  display, failure to, E13015
  failure to display, E13015
  failure to insure, E13014
  failure to maintain, E13014
  insure, failure to, E13014
  maintain, failure to, E13014
persons covered by policy, E13016
Policyholders Protection Board, E13024
potential liability, E13033
pre-action protocol, E13029
  voluntary nature, E13029
premiums, E13022
proposal form, filling in, E13005
public liability policy, E13002, E13016
purpose, E13002
retention of certificate, E13013
risk, measure of, E13022
scope of cover, E13017
settlements, E13017, E13029
Small Business Performance Indicator, E13030

EMPLOYERS' LIABILITY INSURANCE, – *cont.*

sole-employee incorporated companies (SEICs), E13012

statutory duty to take out, E13001

subrogation, E13007

tracing employers, code of practice on, E13026

tracing insurers, E13001

trade associations, E13028

trade endorsements for certain types of work, E13021

treatment, offer of, E13033

uberrima fides, E13004

vibration, E13026

vicarious liability, E13002, E13027

voluntary workers, E13016

Woolf reforms, E13017, E13029

work, making the market, E13028

young persons, E13016

EMPLOYMENT EQUALITY (REPEAL OF RETIREMENT AGE PROVISIONS) REGULATIONS 2011,

insurance benefits, M1511

sickness absence, M1505

EMPLOYMENT NATIONAL TRAINING ORGANISATION,

role, T7008, T7009, T7016–T7017, T7021, T7024

EMPLOYMENT PROTECTION, Intro 29, E14001–E14028

civil action, right of, Intro 24

constructive dismissal, E14014, E14019

contractual terms of employment, *see* Contract of Employment

employees' rights,

disclosures, as to, E14009

dismissal, in respect of, *see* Dismissal

health and safety cases, E14009, E14010

safety representatives, for J3030

scope of, E14009–E14011

medical adviser, E15005

suspension from work, E14015, E14016

women workers, *see* Women Workers

EMPLOYMENT TRIBUNALS,

complaints to, E14009

discrimination claims, E16522

health and safety law, E15014, E15016–E15025

new or expectant mothers, V12018

working time, enforcement of provisions, W10040–W10043

ENABLING DEVICES, M1036

ENERGY CONSERVATION, W11044

ENERGY EFFICIENCY,

Lighting, L5001, L5002

ENERGY MANAGEMENT, F3018, F3029, F3034

Display Energy Certificate (DEC), F3034

ENERGY PERFORMANCE CERTIFICATE (EPC), F3034

ENERGY SOURCES, ISOLATION FROM, M1007

ENFORCEMENT,

appeals, E14013

corporate manslaughter, E15001, E13032

See also Corporate Manslaughter

Crown, position of, E15045

dangerous substances legislation, D0931

deductions, E14014

demotion, E14014–E14016

disaster and emergency planning, D6032

employee, against, E15040–E15042

Enforcement Policy Statement, E15001

enforcing authority, E15001

appropriate, E15008

commercial premises, for, E15007

indemnification by, E15028

industrial premises, for, E15007

responsibilities of, E15007

environmental health officers, role of, F9005

explosions, D0931

fines, E14014

fire prevention and control, F5056

food legislation, *see* Food

gross negligence manslaughter, E15001

Health and Safety Commission, E15001

Health and Safety Executive, E15001, E15005

constitution of, E15002

enforcing authority, as, Intro 20, E15007, E15008, E15009

fees, E15005

publications, E15006

role of, E15002, E15005

transfer of responsibility by, E15011

health and safety policy statement, E15040, S7002

hearings, E14013

Improvement Notice, *see* Improvement Notice

inspectors' powers,

improvement notice, as to, *see* Improvement Notice

ENFORCEMENT, – *cont.*
  inspectors' powers, – *cont.*
    indemnification, E15028
    investigation, E15026
    prohibition notice, as to, *see* Prohibition
      Notice
    search and seizure, E15027
  investigations, E14013
  local authority,
    enforcing authority, as, E15007,
      E15008, E15010
    role of, E15007
    transfer of responsibility by, E15011
  manslaughter charges, E15001, E13038
  notices, F5056
  offences,
    company officers, E15040
    corporate offences, E15039
    defences, E15037, E15039
    directors, E15040–E15041
    due to act of another, E15042
    main offences, E15032–E15035
    manslaughter, E15038
    method of trial, E15032
    penalties for, E15036
    prosecution, *see* prosecution, *below*
    publication of convictions for, E15044
    sentencing guidelines, E15043
    statistics, E15030
  penalties, E15036
  procedures, E14012
  product safety directive, P9011, P9012
  Prohibition Notice, *see* Prohibition Notice
  prosecution, E15001
    breach of statutory provisions, E15030
    burden of proof, E15031
    HSE name and shame website, E15044
    offences, *see* offences, *above*
    publication of convictions, E15044
    service of notice, coupled with, E15018
  public register of notices, E15029
  'relevant statutory provisions',
    breach of, prosecution for, E15030
    execution of, E15028
    scope of, E15012
  risk assessment, E15001
  rules, E14012
  sanctions other than dismissal, E14014
  sentencing guidelines, E15043
  suspension without pay, E14014
  training and competence, T7008
  violence in the workplace, V8004
  warnings, E14014

ENGINEERING CONSTRUCTION
    INDUSTRY TRAINING, T7027
ENTANGLEMENT, M1025
ENVIRONMENT,
  dangerous goods, carriage of, D0448
  disaster and emergency planning, D6028,
    D6033
  ergonomic approach, *see* Ergonomics
  ergonomic design and, *see* Ergonomic
    design
  management of, *see* Environmental
    Management
  thermal, E17034
ENVIRONMENT AGENCY,
  disaster and emergency planning, D6011
  environmental management and, E16005
  local authority associations and, E5046
  pollution control, *see also* Emissions into
    the Atmosphere E5043
  regulation, E5002
  Scottish Environment Protection Agency
    compared, E5044
  website, E5042
ENVIRONMENT, WORK,
  offensive, H1707, H1708
  violence, V8123
ENVIRONMENTAL MANAGEMENT, Intro
    29
  Accounting Standards Board, E16005
  Accounts Modernisation Review, E16005,
    E16024, E16025
  aggregates levy, E16006
  air passenger duty (APD), E16006
  auditing of systems, E16025
  banks, E16007
  benefits of, E16026
  BP Deepwater Horizon, E16010
  Business in the Community, E16002,
    E16008
  Business in the Environment, E16007
  Business Review, E16005, E16024
  Carbon Action Scheme, E16007
  Carbon Disclosure Project (CDP) E16007
  Carbon Trust, E16008
  chemicals, E16008
  climate change levy, E16006
  communications, E16022
  Companies Act 2006, E16005
  Company Law Review, E16005
  consumers, E16012
  corporate social responsibility, E16002,
    E16005, E16027
  costs, E16002
  CR index, E16007
  customers, E16012

ENVIRONMENTAL MANAGEMENT, –
cont.
directors, E16005
disclosure, E16005, E16024
Department for Energy and Climate
    Change (DECC), E16024
Department for Environment, Food and
    Rural Affairs (DEFRA), E16024
Dow Jones Sustainability Index, E16008
Eco-label Scheme, E16008
effective management, E16026
efficiency in business, E16027
EMAS (Eco-Management and Audit
    Scheme) E16014
    regulatory control, lessening of,
        E16026
    revision, E16014
emissions trading, E16006
Environment Agency, role of, E16005
Environment Index, E16007
Equator Principles (EPs), E16007
Europe and, E16002, E16005
external drivers,
    business community, E16001, E16011
    competitors' interests, E16001
    community pressure groups, E16009
    customers, E16001, E16012
    environmental crises, E16010
    environmental pressure groups, E16001,
        E16009
    financial community, E16001, E16007
    legislation, E16001, E16005
    market mechanisms, E16001, E16006
    role of, E16001
facilities management, F3018, F3029,
    F3031, F3034
financial performance, disclosure
    of, E16005, E16024
future of, E16027
guidelines,
    Eco-Management and Audit Scheme
        (EMAS), E16013, E16014,
        E16016, E16017
    GRI Sustainability Reporting
        Guidelines, E16024
    International Standards, E16013,
        E16015, E16016
    standards for environmental
        management, E16013
    WBCSD guidelines, E16024
hazardous chemicals, E16008
Hermes principle, E16007
implementation, E16016–E16024
information management, E16023
insurers' interest, E16007

ENVIRONMENTAL MANAGEMENT, –
cont.
internal controls, E16005
internal drivers,
    corporate social responsibility, E16001,
        E16002, E16005
    employees, E16001, E16004
    management information
        needs, E16001, E16003
international corporate social
    responsibility, E16002
investors, and, E16001, E16007, E16024
Landfill Communities Fund, E16006
landfill tax, E16006
loans for large projects, E16007
meaning, E16001
mergers and acquisitions, E16007
Minister for Corporate Social
    Responsibility E16002
objectives, E16020
Operator and Pollution Risk Appraisal
    (OPRA) regulation scheme, E16026
Operating and Financial Review
    (OFR), E16005
operational controls, E16023
policy as to, E16019
practical requirements, E16017
pressure groups, E16001, E16009
Principles for Responsible Investment
    (PRI), E16007
procurement, E16008
project finance, E16007
public procurement, E16008
public reporting, E16024
regulatory control, lessening of, E16026
review, E16018, E16025
socially responsible investment
    (SRI), E16007
shareholders, and, E16001, E16007
small and medium-sized enterprises
    (SMEs), E16005
Stakeholder
    Engagement Standard, E16009
standards, E16015
Strategy for Sustainable
    Development, E16027
supply chain, E16008
Supply Chain Leadership Collaboration
    (SCLC), E16007
Sustainable Consumption and
    Production, E16027
sustainable development, E16011, E16027
Sustainable Workplace website, E16011
training, E16022
UK Corporate Governance Code, E16005

ENVIRONMENTAL MANAGEMENT, – *cont.*

UN Environment Programme, E16007
vehicle and fuel duty, E16006
EPICONDYLITIS, 01047
EQUAL PAY LEGISLATION,
employment contract and, E14002
EQUALITY ACT 2010,age, E16513-
    E16516
    ageing workers, E16514
    risk assessment, E16515
    young workers, E16516
belief, E16501
civil partnership, E16501
combined discrimination, E16505
direct discrimination, E16503
disability, E16502
    discrimination, E16506
    duty to make reasonable
        adjustments, E16511, R2005
    example of protection, E16506
    impairment, E16510
    long term, E16510
    meaning, E16510, M1506
    risk assessment, E16512
Disability Discrimination Act, A1001
    replacement, R2005
discrimination, E16503-E16506
    association, by, M1508
    combined, E16505
    direct, E16503
    enforcement, E16522
    remedies, E16522
duty to make reasonable
    adjustments, E16511
enforcement, E16522
gender, E16520-E16521
    risk assessment, E16521
gender reassignment, E16501
harassment, E16502
    definition, H1703
    types of, E16507
indirect discrimination, E16502, E15504
marriage, E16501
maternity, E16501
migrant workers, E16518
    risk assessment, E16519
pregnancy, E16501
prohibited conduct, E16502
protected act, E16508
protected characteristics, E16501, H1703,
    R2005
race E16517-E16519
    migrant workers, E16518
    risk assessment, E16519

EQUAL PAY LEGISLATION, – *cont.*
reasonable adjustments, E16511, M1511,
    R2005
religion, E16501
remedies, E16522
risk assessment, E16512, E16515, E16519
    gender issues, E16521
sexual orientation, E16501
sickness absence, M1505
unlawful discrimination, E16522
    compensation, E16522
victimisation, E16502, E16508
young workers, E16516
EQUATOR PRINCIPLES, E16007
EQUIPMENT,
see also Personal protective equipment,
    Visual display units
Approved Code of Practice, M1007
buying, N3035, M1007
cartridge-operated tools, noise
    from, N3003
CE markings, M1007
checklist, M1007
children and young persons, V12016
community health and safety, C6510,
    C6520
    emergency, C6514
    selection and maintenance of, C6517
    volunteers, C6522
confined spaces, C7018, C7022
controls and control systems, M1007
dangerous parts of machinery, M1007
dangerous substances, D0920, D0933.2
design, D0933.2
directive on, M1003–M1004
electricity, D0920, E3003, E3031
    dangerous substances, D0920
    explosions, D0920
    isolated equipment, work on, E3023
    lighting requirements, L5006, L5008
    live work, E3024
    made dead, E3019
    potentially explosive
        atmospheres, E3027
    strength and capability, E3012
emergencies, C6514
energy sources, isolation from, M1007
ergonomic design, see Ergonomic Design
European Union, M1003–M1004, M1007
explosions, D0920
fire, F5001, F5019–F5027
food, F9038, F9039, F9043
guards, M1007
guidance, M1006, M1007
hardware, M1005

EQUIPMENT, – *cont.*
HSE publications, M1006, M1007
information, M1007
inspection, M1007
instructions, M1007
lighting requirements, L5006, L5008,
    M1007
maintenance, C6517, M1007
manufacturers, M1007, N3013
markings, M1007
mobile work, M1005, M1007–M1008
noise, N3003, N3013, N3016, N3035
outdoors, noise and equipment
    used, N3016
permit to work systems, S3017
plant equipment, D0933.2
power presses, M1005
precautions, D0920
Provision and Use of Work Equipment
    Regulations, M1003, M1004–M1008
structure if, M1005
summary of main requirements
    of, M1007
respiratory protective equipment, C7022
risk assessment, D0920, M1006, M1007
risk, definition of, M1006
second-hand, M1007
self-propelled, C6510, M1008
'six-pack' M1004
software, M1005
stability, M1007
standards, M1007
stress, S11021
suitability, M1007
suppliers, M1007
temperature, M1007
training, M1007
Use of Work Equipment
    Directive, M1003–M1004
vibration, V5015
violence in the workplace, V8123
volunteers, C6522-C6523
warnings, M1007
work, definition of, Intro 23
ERGONOMIC DESIGN, Intro 29
anthropometric data, E17007, E17010
capabilities,
    machinery, of, E17009
    person, of, E17009
disability discrimination, E17070
equipment,
    controls, design of, E17013, E17015
    design of, generally, E17011
    display screen equipment, E17044,
        E17064

ERGONOMIC DESIGN, – *cont.*
equipment, – *cont.*
    displays, design of, E17013–E17015
    hand tool design, E17012
    information instruction, E17016
    labelling, E17016
    legislation as to, E17060, E17064
    personal protective equipment, E17066
    provision of, E17042
    software design, E17017
    suitability of, E17065
    user trial, E17056
guidelines as to, E17003, E17005–E17043,
    E17064-E17069
job design,
    enlargement of job, E17036, E17038
    enrichment of job, E17036, E17039
    rotation of work, E17036, E17037
    organisational support and, E17043
    scheduling of work, E17036, E17040
    shift-work, E17036, E17041
legislation promoting, E17003,
    E17060–E17067
machinery, *see* equipment *above*
musculoskeletal disorders and, *see*
    Musculoskeletal disorders
people, for, E17005
physical work environment,
    lighting and, E17031
    noise and, E17032
    physical hazards, E17035
    role of, E17030
    thermal environment, E17034
    vibration and, E17032, E17033
population stereotypes, E17008
posture, role of, E17006, E17010,
    E17021, E17031
risk assessment, E17059, E17062, E17064,
    E17068
stress, S11013
task,
    allocation of, E17009
    designing, E17010
tools assisting, E17003, E17051–E17058
workstation,
    factors to be
        considered, E17018–E17029
    arrangement of items on, E17024
    characteristics of, E17020
    height of, E17019
    layout of, E17021–E17024
    lighting, E17031, E17035
    normal working area, E17023
    sitting versus standing, E17029
    user trial, E17056

ERGONOMIC DESIGN, – *cont.*
  workstation, – *cont.*
    visual considerations, E17025–E17028
    zone of convenient reach, E17022
ERGONOMICS,
  aim of, E17001
  application of, E17002
  core disciplines, E17003
  data,
    collection of, E17057
    existing, use of, E17058
  design, *see* Ergonomic design
  display screen equipment, E17052,
    E17064
  employee involvement, E17058
  ergonomic approach,
    illustration of, E17004
    overview of, E17003, E17004
  frequency analysis, E17055
  laptops, E17064
  legislation as to, E17003, E17060–E17067
  manual handling, E17044, E17052,
    E17063
  occupational health and safety, O1014
  origin of term, E17001
  risk assessment, E17052, E17059, E17062,
    E17064, 17068
  role of, E17001
  task analysis, E17053
  tools of, E17051–E17059
  upper limb disorders E17052
  use of term, E17001
  workflow analysis E17054
ESCALATORS,
  workplaces, safety regulations, W11001,
    W11017
EU Emissions Trading System, E16006
EUROPEAN AGENCY FOR SAFTEY AND
  HEALTH AT WORK (EASHW),
  risk assessment, E16521
EUROPEAN COMMISSION,
  air pollution, and, E5026, E5028
  CSR strategy, E16002
  Directorate General Environment, E5026
  website, E5042
EUROPEAN STANDARDS, D0919
EUROPEAN CHEMICALS AGENCY
  (ECHA),
  establishment, R2511
  registration with, R2502
  role, R2511
EUROPEAN UNION (EU),
  Accounts Modernisation
    Directive, E16005

EUROPEAN UNION (EU), – *cont.*
  air pollution, control of, *see* Emissions into
    the Atmosphere
  ATEX Directive, D0933.6
  Business Review, E16005, E16024
  CE marking, M1007, M1009, M1016,
    M1026, P3012, P3013
  chemical agents, D0901, D0908
  competition, M1009
  Constitution of the EU, Intro 9
  corporate social responsibility, E16005
    strategy, E16002
  dangerous substances, D0901, D0908
    carriage of, D0404
  Declarations
    of Conformity, M1015–M1016
  directives, Intro 9, Intro 10
    air quality and emissions, *see* Emissions
      into the Atmosphere
    ATEX Directives, E3027
    chemical agents, D0901
    dangerous substances, D0901
    electricity, E3027
    equipment, M1003–M1004
    explosion, D0901
    food, as to, F9001, F9002, F9025,
      F9050, F9052, F9063
    Framework Directive, Intro 9
    health and safety, E15004, J3002,
      R3005
    implementation, Intro 10
    machinery safety, M1001–M1003,
      M1009–M1016
    manual handling, M3001
    new or expectant
      mothers, V12017–V12018,
      V12019
    physical agents, on, Intro 11, Intro 29
    Physical Agents (Vibration)
      Directive, V5001, V5015–V5016
    training and competence, T7004
    vibration, V5001, V5015–V5016
    working time, M1514, W10001,
      W10026
  disaster and emergency planning, D6018
  Eco-label Scheme, E16008
  electricity, E3027
  Emissions Trading System, E16006
  environmental management, E16002,
    E16005
  equipment, M1003–M1004, M1007
  European Commission, *see*
    European Commission
  explosion, D0901, D0908, D0933.6
  fire extinguishers, F5004, F5014

EUROPEAN UNION (EU), – *cont.*
Framework Directive, Intro 9
halon fire extinguishers, F5004
heights, working at, W9001, W9009
Information and Consultation Procedures
(ICPs), J3032
judgments, transportability of, P9045
ladders, W9009
machinery safety, M1001–M1003,
M1009–M1016
CE marking, M1009, M1016, M1026
competition, M1009
Declarations
of Conformity, M1015–M1016
Transposed
Machinery Standards, M1001,
M1003, M1010, M1013–M1015
manual handling, M3001
new or expectant
mothers, V12017–V12018, V12019
noise, Intro 29, N3002, N3016, O1026
non-ionising radiation, R1010
outdoors, noise and equipment for
use, N3016
Physical Agents (Vibration)
Directive, N3002, O1026, O1031
product safety, *see* Product Liability;
Product Safety,
public procurement, E16008
radiation, D6018
road transport, W10029
role of, Intro 9
Seveso Directive, Intro 29
Social Charter declaration on health and
safety, Intro 9
standards, M1001, M1003, M1010,
M1013–M1015
training and competence, T7004
Transposed Machinery Standards, M1001,
M1003, M1010, M1013–M1015
vibration, Intro 29, V5001, V5015–V5016
working time, W10001, W10026,
W10027–W10031
Works Councils, J3002, J3033
EVACUATION SIGNALS & SIGNS,
safety sign regulations, W11033, W11040
EVENT SAFETY GUIDE, C6524
EVIDENCE,
accident investigations, A3032, A3033
violence in the workplace, V8121
EXAMINATION,
lifting equipment, L3021, L3023–L3026,
L3028
lifts, L3011
pressure systems, and,
fitness for service, P7023

EXAMINATION, – *cont.*
pressure systems, and, – *cont.*
frequency, P7020
general, P7022
imminent danger, P7025
nature, P7021
report, P7024
scope, P7019
EXCAVATION,
Building and construction
regulations, C8008
EXECUTIVES,
training and competence, T7022
EXPECTANT MOTHERS,
see also New or Expectant Mothers,
musculoskeletal disorders, E17049
risk assessment and, E14024–E14026,
R3018, R3026, R3027, R3030
EXPERIENCE,
manual handling, M3007
EXPLOSIONS,
approved codes of practice, D0933,
D0933.2
ATEX Directive, D0933.6
Buncefield Oil Depot, D0901.1
case histories, D0901.1
classification of places with explosive
atmospheres, D0918, D0919
clothing, D0923
codes, D0919
containers, D0918, D0927
Dangerous Substances and Explosive
Atmospheres Regulations
2002, D0901, D0909
commencement, D0924
enforcement, D0931
exemptions, D0928
modernisation of legislation, D0929
repeals and revocations, D0930
drawings and plans, D0919
dust, D0919
safe handling of combustible, D0933.6
duty holders, D0906
electricity, D0920, E3009, E3027
emergencies, dealing with, D0925
enforcement, D0931
equipment, D0920–D0924
EU law, D0901, D0908
European standards, D0919
explosive atmospheres, D0918–D0919
explosives, *see* EXPLOSIONS
fire, D0919
fuel, D0918
gas explosion, G1001, G1030
guidance, D0933.1, D0933.2

EXPLOSIONS, – *cont.*
ignition sources, D0933.2, D0933.4
incidents, dealing with, D0925
information provision, D0926
instructions, D0926
national security, D0928
noise and hearing loss, N3003
offshore operations, O7028
petroleum, D0929
pipes, D0918, D0927
risk assessment and management, D0902,
    D0919
precautions, D0918
regulatory impact assessment, D0932
sources, D0934
spillages, D0918
storage, D0918
training, D0926
verification of safety prior to first
    use, D0922
warning signs on entry, D0921
workplace, definition of, D0907
zoning, D0918, D0919
EXPLOSIVES,
accident reporting, A3035
building and construction
    regulations, C8008
carriage by road, D0407
children and young persons, V12015
disaster and emergency planning, D6009,
    D6014, D6038
fire prevention and control, F5031
hazardous substances, where, D0704,
    D0706, H2141, H2143
noise and, E17032
offences, Intro 20
risk assessment, R3014, R3016 (App B)
EXPORTS,
product safety, P9002, P9009
EXTRINSIC ALLERGIC
    ALVEOLITIS, O1042
EYEBOLTS, L3046–L3047
EYES,
binocular vision, O1034
cataracts, O1033
causes of diseases and disorders, O1033
colour blindness, O1033, O1034, O1035
colour vision, O1033, O1034, O1035
detection, O1034
diseases and disorders, O1032–O1036
display screen equipment, D8201, D8207,
    O1033, O1036
dryness and low humidity, O1033
eye protection, O1035, O1036
irritation, O1033

EYES, – *cont.*
legal requirements, O1036
management, O1035
offshore installations, injured on, O7073
photokeratitis and
    photoconjunctivitis, O1033
protection of, P3021, P3026
risk assessment, O1035, O1036
tests, D8201, D8207, O1034
visual acuity, O1033, O1035

F

FACILITIES,
see also Facilities management, Welfare
    Facilities
disabled person, for, W11043
gas safety, Intro 29
managers, Intro 29
FACILITIES AND SERVICES, CUSTOMERS
    AND USERS OF, C6515–C6517
FACILITIES MANAGEMENT, F3001–F3035
acquisition and disposal of
    buildings, F3019
advantages, F3009–F3010, F3028
asbestos, F3024
breakdown maintenance, F3015
change management, F3023
cleaning, F3030
communications, F3010
company approach to, F3006
Computer Aided Facilities Management
    (CAFM) software, F3010
construction, F3033
contract management, F3022
contractors, management of, F3027
cost certainty, F3010
coverage, F3003
customer service, F3017
definition, F3002
design of buildings, F3028, F3031–F3032
energy management, F3018, F3029, F3034
    Display Energy Certificate
        (DEC), F3034
    Energy Performance Certificate
        (EPC), F3034
environmental management, F3018,
    F3029, F3031, F3034
floor cleaning, F3030
government departments, recognition
    from, F3029
hard, F3004
Health and Safety Executive, F3029

FACILITIES MANAGEMENT, – *cont.*
height, working at, F3025
help desks, F3030
information technology, F3030
in-house, F3008
innovations, F3030, F3035
integrated approach, F3009
intelligent buildings, F3030
key performance indicator (KPI), F3017
legislation, F3031–F3034
life cycle costing, F3016
lifts, F3032, F3034
maintenance,
    breakdown, F3015
    planned corrective, F3014
    planned preventative, F3013
    policy, F3012
management, F3005
operation, F3034
outsourcing, F3008
planned corrective maintenance, F3014
planned preventative maintenance, F3013
project management, F3021
qualification, F3029
reduced cost, F3010
recognition of, F3029, F3035
Reliability Centered maintenance
    (RCM), F3015A
responsibilities of, F3011–F3023
security industry, F3030
service delivery, F3010
Six Sigma, F3010
soft, F3004
software, F3030
solutions, development and implementation
    of, F3007
space planning, F3020
strategy, F3007
support services, F3010
telecommunications, F3030
university degrees in, F3029
water quality, F3026
work equipment, F3032
FACTORIES,
absence of overcrowding, W11008
access, O3018
    disabled persons, W11043
changing clothes facilities, W11026
clothing accommodation, W11025
defence of factory occupier, W11049
doors and gates, W11016
drinking water, supply of, W11024
eating facilities, W11027
enforcement of regulations, W11047

FACTORIES, – *cont.*
floors, condition of, safety
    underfoot, W11010
gas in, G1034
general cleanliness, W11007
occupiers, actions against, O3018
occupiers, defences of, W11048
occupiers' liability, O3018, W11047
owner, liability of, W11047–W11049
pedestrian traffic routes, W11015
rest facilities, W11027
sanitary conveniences, W11019–W11021
sedentary comfort, W11009
slips, trips and falls, O3018
temperature control, W11005
traffic routes, O3018
washing facilities, W11020
work clothing, W11025
workroom dimensions/space, W11008
workstations, W11009
FAIRGROUNDS,
accident reporting, A3035, R3002
enforcing authority, E15009
FALLING,
arrest systems, W9008, W9014
fatal accidents, W9001
harnesses, W9008
heights, working at, W9001, W9004,
    W9008
    arrest systems, W9008, W9014
    fatal, W9001
    harnesses, W9008
    objects, W9011
    protection from, W9014
factories, O3018
mobile work equipment, M1008
objects, W9011, W11011
FAMILY BUSINESSES,
children and young persons, V12011
FARM,
livestock on, F9030
FAST TRACK CLAIMS, E13029
FAW
first aid at work training, F7001, F7018
FATAL INJURIES,
assessment of damages, C6039
corporate manslaughter, Intro 20, E13012
damages under fatal accident
    legislation, C6036, C6037
electricity, E3001
falls, W9001
heights, working at, W9001
interim awards, C6041
provisional awards, C6040

FATAL INJURIES, – *cont.*
  reporting, A3005, A3007
  survival of actions, C6036, C6038
  types of action for damages, C6036
FEDERAL EMERGENCY MANAGEMENT
    AGENCY (FEMA), D6001
FENCING,
  community health and safety, C6514
  electricity, E3014
FERTILISER, D6013, D0908
FINANCIAL SERVICES COMPENSATION
    SCHEME,
  employers' liability insurance and, E13024
FINES,
  see also Enforcement
  accident reporting, A3022
  aggravating features, Intro 15
  asbestos-related offences, Intro 15
  community health and
      safety, C6503–C6504
  corporate manslaughter, E15043
  disasters, major public, Intro 15
  financial means, assessing, Intro 15
  Heathrow rail tunnel collapse, Intro 15
  level of, Intro 15
  machinery safety, M1018
  Ramsgate Port walkway collapse, Intro 15
  smoking, W11028
FIREARMS,
  noise and hearing loss, N3003
FIRE EXTINGUISHERS,
  BCF extinguishers, F5014
  British Standard, F5016
  carbon dioxide extinguishers, F5015
  colour coding, F5016
  construction sites, F5028
  dangerous goods, carriage of, D0445
  distribution of portable
      extinguishers, F5016
  dry powder extinguishers, F5013
  electrical fires, F5004, F5015
  European Union, F5004, F5014
  fitting, F5010
  foam extinguishers, F5012
  halon, EU regulations on, F5004, F5014
  portable, distribution of, F5016, F5028
  self-contained systems, F5007
  siting of, F5010, F5028
  suitability of, F5003
  training, F5015
  types of, F5010–F5015
  water extinguishers, F5011
FIRE FIGHTERS,
  calling, F5008
  damages for psychological injury, C6005

FIRE FIGHTERS, – *cont.*
  drains, O3014
FIRE PRECAUTIONS,
FIRE PREVENTION AND CONTROL
  active fire protection measures, F5005
  alarms in certificated premises,
    British Standards, F5017
  alterations notice, F5057
  appeals, F5060
  arson, F5074
  automatic detection systems,
    British Standards, F5017
    generally, F5007
  building and construction
      operations, C8016
  Building Regulations, F5035, W11044
  Buncefield Oil Depot, D0901.1
  breaching of fire resistant walls, F5007
  certificated premises, alarms in, F5017
  child employees, and, F5052
  civil liability, F5073
  classification, F5003
  cleaners, F5018
  common causes of, F5002, F5018
  common law, civil liability at, F5073
  communication equipment rooms, damage
      to, F5006
  community safety, C6507, C6522
  competent persons, F5049
  confined spaces, C7024
  construction sites, on
    code of practice, F5024, F5029
    designing out fire, F5027
    emergency procedures, F5026
    insurance, F5029
    plant, F5028
    precautions, F5029
    risk assessment, F5024
    site fire safety co-
        ordinator, F5024–F5025
  cooling, F5005
  co-operation and co-ordination, F5054
  costs of improvements, F5001
  criminal offences, F5060
  Crown,
    RR(FS)O, and, F5061
  dampers, F5007
  dangerous processes, F5062
  dangerous substances, F5043–F5045
  detection, C8016
  emergency routes, F5040
  designated premises, F5030, F5032, F5041
  designing out fire, F5027
  detection, F5007, F5017, F5042

FIRE PREVENTION AND CONTROL –
cont.
ducts, F5007
electrical, E3007, F5004, F5015
elements of, F5001
emergency procedures, F5026
emergency routes and exits, F5040
employees' duties. F5055
employers' liability insurance, E13006
enforcement,
    alterations notice, F5057
    enforcement notice, F5058
    generally, F5056
    prohibition notice, F5059
enforcement notices,
    appeals, F5036
    generally, F5058
    introduction, F5056
equipment, F5001, F5009–F5017
evacuation, M2011
explosive atmospheres, F5031
external fire spread, F5035
extinguishers, *see* Fire extinguishers
fire brigade,
    calling the, F5008
    facilities for, F5035
fire certificates
    change of conditions affecting
        exemptions, F5042
    introduction, F5032
fire detection, F5042
fire doors, F5007
fire drills, F5022–F5023
    notice, F5023
fire escapes, F5007, F5020
    emergency routes and exits, F5040
    improvement notices, F5035
    means of, F5035
    unsatisfactory means of, F5021
fire fighters,
    and see Fire fighters
    switches, F5046
fire fighting, F5042
fire procedures, F5008
Fires Prevention (Metropolis) Act
    1774, F5073
fire risk assessment,
    carrying out, F5038
    controlling residual risk, F5039
    dangerous substances, F5043–F5045
    emergency routes, F5040
    fire detection provision, F5042
    fire fighting provision, F5042
    introduction, F5037

FIRE PREVENTION AND CONTROL –
cont.
fire risk assessment, – *cont.*
    policy and procedures, F5041
fuel,
    generally, F5001
    tanks, F5028
gases, F5003
good housekeeping, F5018, F5022
guidance documents, F5061
hazardous materials, F5031
HSE guidance, F5001
improvement notices
    access, F5035
    appeals, F5036
    escape, means of, F5035
    external fire spread, F5035
    fire service, facilities for the, F5035
    internal fire spread, F5035
    warning, means of, F5035
information provision, F5051
insurance, F5006, F5029, F5074
internal fire spread, F5035
introduction, INTRO 29
legislation,
    Building Regulations, F5035
    current position, F5031
    fire certification, F5032
    introduction, F5030
    Regulatory Reform (Fire Safety)
        Order, F5033–F5034
liquids and liquefiable solids, F5003
maintenance, F5047
managing health and safety,
    evacuation procedure, M2011
    risks, assessment of, M2006
means of escape, F5020–F5021
metals, F5003
new buildings, F5030
occupiers' liability of, F5073
offences, F5060
offshore operations, O7028
oils and fats, cooking, F5003
passive fire protection, F5007
personal protective equipment, P3017
petroleum, D0929, F5031
plant, F5028
precautions, F5030–F5032
    construction sites, F5029
    criminal offences, F5060
    enforcement, F5056
    legislation, F5030–F5031
    rationalisation of legislation on, F5031
    regulations, specific, F5032

FIRE PREVENTION AND CONTROL – cont.
pre-planning of fire prevention
  generally, F5019
  means of escape, F5020–F5021
procedures, F5008
prohibition notice, F5059
property risk, F5006–F5007
Regulatory Reform (Fire Safety) Order,
  application, F5034
  generally, F5033, F5035
risk assessment, F5001, F5007, F5032,
  F5037, M2006, R3012, R3018,
  R3033, R3035
risk management, F5006
rubbish, F5018
safety assistance, F5048
separation, F5006
signs, F5032, W11041, W11042
site fire safety co-ordinators,
  emergency procedures, F5026
  role of, F5025
smoking, F5018
smothering, F5005
solid organic materials, F5003
special risks, F5001
sports, safety certificates and, F5032
sprinkler systems, E13006, F5007, F5035
standards, F5030
starvation, F5005
training, F5053
triangle, F5001
visitors, liability to, F5073
volunteers, C6523
walls, breach of fire resistant, F5007
warnings, means of, F5035
waste, F5018
water, F5005
FIRST AID, Intro 29
absences of first-aiders or appointed
  persons, F7002, F7011
annual leave, F7011
appointed persons,
  absence, F7002, F7011
  appointment of, F7003
  role of, F7019
arrangements,
  employers, F7001
assessment of first-aid needs,
  absences of first-aiders or appointed
    persons, F7011, F7002
  aim, F7003
  annual leave, F7011
  basic rule, F7003
  checklist, F7025 (App B)

FIRST AID, – cont.
assessment of first-aid needs, – cont.
  distance to emergency medical
    services, F7008
  distant workers, F7009
  doctors, qualified medical, F7003
  factors affecting, F7003
  hazards and risk in the
    workplace, F7004
  history of accidents in
    organisation, F7006
  HSE approval, E15005
  insurance, employers' liability, F7012
  lone workers, F7009
  multi-occupied sites, F7010
  nature and distribution of the
    workforce, F7007
  no need to appoint first-aiders,
    where, F7003
  nurses, registered, F7003
  occupational health service
    advice, F7003
  public, members of the, F7012
  reassessment of needs, F7013
  requirements, F7003
  shared sites, employees on, F7010
  size of organisation, F7005
  trainees, F7012
  travelling workers, F7009
  workforce, nature and distribution of
    the, F7007
certificates, F7018
checklist, first aid needs, CH 3 App B
containers, F7022
courses, F7018
disaster and emergency planning, D6021,
  H6016
distance to emergency medical
  services, F7008
distant workers, F7009
doctors, qualified medical, F7003
emergency,
  services, calling, F7019
emergency first aid at work
  (EFAW), F7001, F7018
employer's duty,
  arrangements, F7001
  informing employees of
    arrangements, F7014
  assessment of first-aid needs. *See*
    assessment of first-aid needs *above*
  first-aiders, provision of, F7002
  requirements, F7002
  scope of duty, F7002
equipment, provision of, Intro 23

FIRST AID, – cont.
eye irrigation, F7023
first aid at work (FAW), F7001, F7018
first-aiders,
abilities, F7017
absence of first-aider, F7002, F7011
certificates, F7018
competencies, F 7018, F7025 (App D)
courses, F7018
generally, F7016
guide to provision of, CH 3 app E
no need to appoint first-aiders,
where, F7003
numbers required, F7016, F7017
offshore operations, O7058
position in company, F7017
provision of, F7002, F7016–F7018
qualifications, F7018
refresher training, F7018
selection, F7017
special training, F7018
standard, F7017
training, F7018
unusual risks/hazards, F7018
updating skills, F7014, F7018
generally, F7001
hazards,
common, listing, CH 3 App A
risk in the workplace, assessment
of, F7004
history of accidents in organisation, F7006
insurance, employers' liability, F7012
internal memos as to, F7014
key employees, nomination of, F7014
lone workers, F7009
meaning, F7001
multi-occupied sites, F7010
new employees, provision for, F7014
nature and distribution of the
workforce, F7007
nurses, registered, F7003
offshore operations, O7057–O7062
installation manager's duty, O7062
medical personnel, O7058, O7057,
O7058, O7057
offshore first-aider, O7058
offshore medic, O7058
sick bays, O7059–O7061
suitable persons, O7057, O7058
public, members of the, F7012
qualifications of first-aid personnel, F7018
reassessment of needs, F7013
record keeping, F7020
records, F7020
first aid provision, CH 3 App C

FIRST AID, – cont.
refresher training, F7018
resources, first-aid, F7021–F7024
additional resources, F7023
containers, F7022
eye irrigation, F7023
materials, F7022
protective equipment, F7023
provision, F7021
stock of first-aid materials, F7022
travelling, kits for, F7024
rooms, first-aid, F7025
selection of first-aiders, F7017
self-employed, F7015
shared sites, employees on, F7010
signs, *see also* Safety Signs W11040,
W11041
size of organisation, assessment of, F7005
specially trained staff, action by, F7001
summary of requirements, F7001
trainees, F7012
training of first-aid personnel, F7001,
F7018, F7019
travelling, kits for, F7024
travelling workers, F7009
updating of skills and
arrangements, F7014, F7018
workforce, nature and distribution of
the, F7007
FIT NOTE, M1504
options for return to work, M1504
rehabilitation, M1515, R2004
FITNESS FOR SERVICE,
pressure systems, and, P7023
FITNESS FOR WORK, M1504
FITNESS PROGRAMMES, M3015
FIVE CAPITALS MODEL,
sustainability, F3018A
FIXED PENALTY NOTICES,
smoking, W11028
FLAMMABLE SUBSTANCE,
accident reporting, A3035
explosive atmosphere, D0921
major accident hazards
regulations, D0704, D0705
FLEXIBLE WORK PATTERNS, Intro 6
FLOORS,
condition of, in workplaces, M3006,
M3014, W11010
design, M3014
facilities management, F3030
free from obstructions, W11010
hygienically clean, W11010
manual handling, M3006, M3014
variation in levels of, M3006

FLUORESCENT TUBES,
  stroboscopic effect, L5008, L5020
FOAM FIRE EXTINGUISHERS, F5012
FOCUS GROUPS, V8022
FOOD,
  adulteration of, F9001
  advertising, F9012, F9020
  Advertising Standards Authority
    (ASA), F9012
  aircraft, on, F9032
  allergens, pre-packed foods, in, F9003,
    F9075
  allergic reaction to, F9009–F9011, F9075
  analysis of, F9013
  animal products, F9002, F9004, F9014,
    F9028, F9034
  business,
    licensing, F9021
    occasional F9041
    registration, F9021
    scope of, F9007
    temporary, F9041
  butcher's shops, F9021
  canned foods, F9056
  change in activities, F9027
  charity events, at, F9007
  chewing gum, F9007
  chilled storage, F9052–F9064
  choking, risk of, F9010
  codes of practice, F9007
    Codex Alimentarius, F9035, F9038
    range of, F9006, F9025
  complaints procedure, F9080
  composite products, establishments
    producing, F9028
  composition of, F9020
  consultation, F9006, F9020
  consumer concerns, F9077
  consumer protection legislation, F9008
  contamination, F9010, F9036, F9038,
    F9047, F9074
  cooling, F9036, F9062, F9066
  criminal offences,
    defences, F9022, F9023, F9070, F9082
    enforcement officers, by, F9013
    information requirements, F9027
    prosecution of, F9013, F9015
    scope of, F9008–F9012
    strict liability of, F9023
  dairy hygiene, F9025
  definitions, F9006, F9077
  Department of Health's role, F9005
  descriptions, F9012
  dietary preferences, F9076
  display, F9059, F9061

FOOD, – cont.
  domestic premises, exemption of, F9031
  due diligence, F9049
  emergency control orders, F9019–F9020
  emergency prohibition notices and
    orders F9018
  enforcement procedures,
    detention of food, F9013, F9014
    emergency control order, F9019
    emergency prohibition, F9015, F9018
    enforcement officers'
      powers, F9013–F9019
    entry to premises, powers of, F9013
    government policy, F9005
    improvement notice, F9015
    licensing, F9021
    offences, prosecution, F9013, F9015
    personal prohibition, F9017
    prohibition, order, scope of, F9016
    registration of businesses, F9021
    regulations as to, F9004
    scope of, F9013
    seizure of food, F9013–F9019
  equipment, F9038, F9039, F9043, F9074
  EU,
    directives, F9001, F9002, F9025, F9050
    medicinal products, licensed, F9007
    regulations, F9006, F9034
  exempt premises, F9029–F9033
  fair trading, F9078
  falsely describing, F9008, F9012
  falsely presenting, F9008, F9012
  food business, definition, F9026
  food rooms, F9038, F9040
  Food Safety Manual,
    contents of, F9070–F9081
    value of F9082
  Food Standards Agency,
    emergency control orders, issuing
      of, F9019–F9020
    establishment of, F9005
    role of, F9005
    website, F9019
  frozen storage, F9047, F9050
  gelatine, F9069
  genetically modified, F9020, F9077
  genetically modified crops, F9020
  handler,
    personal hygiene, F9038, F9046, F9074
    policy as to, F9074
    risk factors, E16521
    supervision, F9048, F9081
    training, F9048, F9081
  hazard analysis, F9034–F9037,
    F9073–F9074

FOOD, – *cont.*
hazard warnings, use of, F9019
heat processing, F9047.2
herbal remedies, F9007
hot food, F9061, F9065
hotels, in, F9007
human consumption, fitness for, F9007, F9010, F9035
hygiene,
definition, F9035
emergency prohibition notice, F9018
hazard analysis, F9035–F9037
improvement notices, F9015
industry guides, F9049
Meat Hygiene Service, F9005, F9028
'pre-requisites', F9038
regulations, scope of, F9034–F9049
temperature control, *see* temperature control, *below*
improvement notices, F9015–F9016
industry guides, F9025, F9049
information requirements, F9027
injurious to health, F9008–F9010
inspection, F9027
labelling requirements, F9003, F9012, F9020, F9023, F9075, F9078
licensing, F9027–F9028
local authorities
meat, F9028
monitoring of, F9005
role of, F9005, F9007
mail order foods, F9054
markets,
industry guides, F9049
registration of, F9027
stalls, F9032
meat,
export cutting premises etc. for, F9028
licensing sale of, F9021
Meat Hygiene Service, F9005, F9028
product plants, F9028
medicinal products, unlicensed, F9007
Medicines and Healthcare Products Regulatory Agency (MHRA), F9007
microbiological standards, F9020, F9036
Ministry of Agriculture Fisheries and Food, and F9005
motor vehicles, F9027, F9032
moveable premises, F9027
national disaster, for, F9033
nature demanded, not of, F9011
new premises, notification of, F9027
nutrition labelling, F9078
packaging, F9047.1
parasites, F9047

FOOD, – *cont.*
pathogenic organisms, F9009, F9047
pests, F9047, F9074
poisoning,
microbiological, F9036
risk of, F9050
premises,
cleaning, F9074
enforcement officers' powers, F9013
exemption from registration, F9029–F9033
facilities, F9039
mobile food premises, F9026, F9038
occasionally used only, F9029
pest control, F9074
registration, F9004, F9021
regulations, scope of, F9026–F9033
temporary, F9038
ventilation, F9039
prices, display of, F9004
processed food, F9057
product recall, F9079
prohibition orders and notices, F9016–F9018
quality demanded, not of, F9011
raw products, F9028, F9057
registration, F9026–F9033
reheating, F9068
remedial action notices, F9019
safety, Intro 29
definition of terms, F9007
enforcement procedures, *see* enforcement procedures, *above*
legislation, F9001, F9006, F9007
offences under, *see* offences, *above*
principal sections of, F9006
scope of, F9001, F9007
sale of,
ban on, F9020
meaning of, F9007
offences as to, F9010, F9011
'use by' date, F9007
vending machine, through, F9033, F9041
samples, taking of, F9013
Scottish legislation, *see* Scotland
sanitary conveniences, F9039
short 'shelf life', F9055, F9064
standards, Intro 29
Agency, *see* Food Standards Agency, *above*
control of, F9001, F9003
substance demanded, not of, F9011
supervision of staff, F9048, F9081

FOOD, – *cont.*
  temperature control,
    England and Wales, in, F9052–F9062
    four-hour rule, F9059
    regulations, scope of, F9034, F9039,
      F9047, F9050–F9069
    Scotland, in, F9050, F9063–F9069
    time and temperature, relationship
      of, F9036, F9051
  thawing frozen food, F9047
  trade descriptions legislation, F9012
  training of staff, F9058, F9081
  transport, F9038, F9042
  unfit for human consumption, F9010
  vegetarian, F9076
  vehicles, F9027, F9032
  vending machines, use of, F9033, F9041
  waste, F9038, F9044
  water, F9007, F9038, F9045
  weights and measures, F9004
  wrapping, F9047.1
FOOTWEAR,
  manual handling, M3004, M3007
FOREIGN DESIGNER,
  product safety, P9025
FOREIGN MANUFACTURER,
  product safety, P9025
FORESTRY,
  air pollution and, E5015, E5021
  vibration, V5001, V5015
FORK-LIFT TRUCKS, L3034–L3035
  counterbalance trucks, L3034
  failure, L3035
  narrow aisle trucks, L3034
  order pickers, L3034
  overturning, M1008
  pedestrian-operated stackers, L3034
  reach trucks, L3034
FOSSIL FUELS, E16006
FRAGILE SURFACES, W9010
FREIGHT CONTAINERS,
  accident reporting, A3035
FRESH AIR
  construction and building operations,
    and, C8017
FRICTION, M1025
FUEL,
  see also Petroleum
  Building Regulations, W11044
  Clean Air Act 1993, E5061
  emission controls, and, E5035
  explosions, D0918
  fire prevention and control, F5001, F5028

FUMES,
  polluting agent, as, V3003
  ventilation, *see* Ventilation,
  welding, V3002

G

GANGWAYS,
  manual handling, M3014
GARAGES, D0901.1
GAS SAFETY,
  advice and assistance, E15005
  appliances,
    installation and use, *see* installation and
      use, *below*
    maintenance, G1024
    temporary installations, C6516, C6517
  assessment of domestic operatives, G1014
  boilers, maintenance of, G1029
  Building Regulations, G1019
  Bulk LPG installations, protective
    devices, P7020
  carbon monoxide poisoning, G1017,
    G1029, G1033
  catering and hospitality, G1017
  certificates, G1025
  chimney flues, G1019, G1032
  competent person,
    qualifications, G1014
    quality control, G1013
    supervision, G1014
  compressed natural gas, D09033.6
  confined spaces, C7016
  CORGI, G1017, T7007
  court cases, summary of, G1027–G1032
  dampers, G1019
  dangers of gas, G1001
  definition of gas, G1001
  Director General of Gas Supply, G1002
  disconnection, G1007
  domestic customers, G1002, G1014
  duty holder, compliance etc by, G1005
  entry, *see* rights of entry *below*
  equipment, C7021
  escape incidents,
    liability for, G1001
    responsible persons' duties, G1005
    suspecting leak, G1005
  explosion, G1001, G1030
  explosive atmosphere, D0921
  facilities managers, Intro 29
  factories, G1033
  fire, F5003

GAS SAFETY, – *cont.*
flues, G1019, G1032
'gas', G1001, G1019
Gas Consumer's Council, G1002
gas fittings, G1020, G1028
gas incidents, reporting, G1005
generally, G1001
Health and Safety Commission
    Fundamental Review, G1014
Home Information Packs, inclusion of gas
    certificates in, G1025
Individual Gas Fitting Operatives, scheme
    for, G1014
inspection, G1007, G1025
installation and use,
    appliances, G1010, G1012, G1017,
        G1018, G1033
    case law, G1026–G1032
    checks, safety, G1024
    competent persons, G1013, G1014
    dampers, G1019
    duties, G1024
    emergency controls, G1021
    escape of gas, G1010, G1012
    excluded supplies, G1010
    flues, G1019
    gas fittings, G1020, G1028, G1032
    general safety precautions, G1016
    landlords, G1024, G1028, G1029,
        G1031
    liability, G1026–G1032
    maintenance requirements, G1025
    materials, G1015, G1022
    meters, G1022
    pipework, installation, G1023
    pressure fluctuations, G1034
    pressure systems, *see* Pressure Systems
    prosecutions, G1026–G1032
    quality control, G1013
    records, G1025
    regulations,
        interface with other
            legislation, G1026
        scope of, G1010
    regulators, G1022
    responsibilities, G1011
    room-sealed appliances, G1018
    student deaths, G1029, G1030
    supervision, G1014
    temporary installations, C6516
    tenants, G1011
    testing gas appliances, G1024
    unsafe gas appliances, G1017
    use requirement, G1010
    venting of waste gas, G1010

GAS SAFETY, – *cont.*
installation and use, – *cont.*
    work, G1010
    workmanship, G1015
landlords, Intro 29, G1017
leaflets, G1017
leak, suspected, G1007, G1008
legislation,
    gas supply, as to, G1002
    interface of, G1026
    scope of, G1001
liability for escape, G1001
maintenance, G1024, G1025
management of supply,
    bad gas management, G1029
    compliance and co-operation, G1005
    regulations as to, G1003
    safe case document, G1003
materials, fittings, G1015
medical advice, G1017
meters, G1022
National Grid, G1001
Nationally Accredited Certification Scheme
    (NACS), G1014
network emergency co-ordinator, G1003
owners' responsibility, G1002
pipelines, G1001, G1009
pipework installation, G1023
precautions, G1016, G1020
pressure fluctuations, G1035
pressure systems, *see* Pressure Systems
protective device, P7020
purging, C7016
re-assessments of domestic
    operatives, G1014
reconnection prohibition, G1008
regulations, G1001, G1010, G1026
regulators, G1018
relevant authority, G1002
reporting gas incidents, A3002, A3020,
    A3021, G1005
responsible persons, G1005
review of, G1014
rights of entry,
    disconnection, G1007
    generally, G1006
    inspection, G1007
    reconnection prohibition, G1008
    testing, G1007
room-sealed appliances, G1018
safety case, G1003, G1004
Scotland, G1001
student death, G1031

GAS SAFETY, – *cont.*
  supply,
    domestic customers, G1002
    management, G1003
    owners, G1002
    relevant authority, G1002
    responsibilities, G1002
    safe, G1002
  suspecting leak, G1005
  testing, G1007, G1024
  training and competence, T7007
  unloading compressed natural
    gas, D0933.6
  ventilation, G1017
  website, G1017
  workmanship of fitters, G1015
GELATINE, F9069
GENDER,
  Equality Act 2010, E16520
  hearing loss, N3002
  manual handling, M3007
  risk assessment, E16521
    table of relevant hazards, E16521
  sexual harassment, H1702
GENDER REASSIGNMENT,
  harassment, H1702
  protected characteristic, H1703
GENETICALLY MODIFIED ORGANISMS,
  applications as to, E15005
  food regulations and, F9020, F9077
  notifications as to, E15005
  risk assessment, R3016, R3035
GLASS,
  Building Regulations, W11044
GLOBAL REPORTING INITIATIVE
  (GRI) E16007, E16024
GLOBAL LEGAL FRAMEWORK, Intro 8
GLOVES, V5019
GOOD ORDER
  construction and building operations,
    and, C8005
GREEN FORM LEGAL ADVICE,
  renaming of, C6021
GREEN PUBLIC PROCUREMENT
  (GPP), E16008
GREENBURY REPORT, Intro 5
GRIEVANCE PROCEDURES,
  contract of employment, in, E14006,
    E14008
GRIPS, M3013
GROSS NEGLIGENCE MANSLAUGHTER,
  director, E13032
GUARDS,
  adjustable, M1030
  automatic, M1031

GUARDS, – *cont.*
  drilling machines, M1030
  equipment, M1007
  fixed, M1029–M1030
  heights, working at, W9008
  interlocking, M1032
    mechanical, M1035
    types of, M1033
  mobile work equipment, M1008
  position sensors, M1032
  power isolation, M1034
  risk assessment, M1032
  self-adjusting, M1030
  standards, M1029, M1032
  sweep away, M1031
  switches, M1032, M1034–M1035
  telescopic, M1030
  trapped key interlocking, M1034

H

HALON FIRE EXTINGUISHERS, F5004,
  F5014
HAMPEL REPORT, Intro 5
HAND ARM VIBRATION SYNDROME
  (HAVS),
  advice, V5017–V5018
  carpal tunnel syndrome (CTS), V5004,
    V5011
  cases, V5003
  causes, O1027,V5010
  daily exposure action value, V5015
  detection of, O1029
  diseases caused by, V5010
  effects, V5010, V5014
  emissions, testing, V5020
  ergonomic approach to, E17012, E17033
  E.U. law, Intro 11, O1031, V5015
  examples of causes of, V5010
  exposure action value, V5015
  frequency, V5017
  gloves, V5019
  guidance, V5017-V5018
  Health and Safety Executive (HSE), V5003
  health surveillance, O1029–O1030, V5015
  legal requirements, O1031
  magnitude, V5017
  maintenance, O1031
  managing exposure, O1030, V5018
  noise, V5003
  occupational diseases, O1027–O1031
  occupations, examples of, V5014
  personal protection against, V5019–V5022

HAND ARM VIBRATION SYNDROME (HAVS), – cont.
Physical Agents (Vibration) Directive, Intro 11, O1031, V5015
prescription of, V5014
risk assessment, O1030–O1031, V5015, V5022
symptoms, O1027
tools, vibration magnitudes of, V5017, V5020
vibration, meaning of, O1027
warnings, V5020
HAND RAILS, W9008
HAND SIGNALS, W11032, W11042
HANDCUFFS, V8094
HANDLES, M3013
HARASSMENT,
adverse environment, H1702
age, grounds of, H1702-H1703
assault, H1723, H1730
associative discrimination, H1707
battery,
    criminal action, H1729-H1730
    tort action, H1723
breach of contract, H1727
    remedies, H1728
civil partnership, in respect of, H1703
claim
    elements of, H1702
commission of, H1704
compensation, H1718, H1721
concept H1702
constructive dismissal, H1720
course of conduct cases, H1706
criminal action,
    battery, H1730
    Crime and Disorder Act 1998, H1733
    Protection from Harassment Act 1997, H1731
    Public Order Act 1986, H1732
definition, H1702-H1703
    limitations, H1709
dignity, H1701
disability, H1703
discrimination claim, general, H1703
effect, purpose or, H1702-H1703, H1707
elements of claim, H1703
employers' liability, H1710-H1715
    causal link with employment, H1712
    course of employment, H1711
    defence, H1713
    narrowing extent, H1715
    preventing claims, H1734, H1739-H1740
    vicarious liability, H1710

HARASSMENT, – cont.
employers' policy, H1734-H1740
    complaints procedure, H1735
    confidentiality, H1736
    formal policy, H1738
    informal action, H1737
    monitoring and review, H1740
    training, H1739
environment, work, H1707-H1708
Equality Act 2010, E16507, H1701
    definition of harassment, H1703
    employees' protection, E16501
    protected characteristics, H1703
failures to investigate or protect, H1706
free standing claim, H1701
gender reassignment, H1702
    protected characteristic, H1703
general discrimination claim, H1703
generally, Intro 29
heads of claim, H1702
history, H1702
intimidating work environment, H1707-H1708
legal action, H1716-H1733
    breach of contract, H1720
    constructive dismissal, H1720
    criminal action, H1729
    remedies, H1718, H1721, H1725, H1728
    tort, H1717, H1722-H1725
    unfair dismissal, H1719
    unlawful discrimination, H1717-H1718
marriage, in respect of, H1703
maternity, in respect of, H1703
monitoring, H1740
nature of conduct, H1706
negligence, H1724
perceived discrimination, H1707
policy, H1734-H1740
pregnancy, in respect of, H1703
protected characteristics, H1703
protected statutory grounds, H1702
purpose or effect, H1702-H1703, H1707
racial, H1702, H1703 H1733
religion, H1702-H1703
remedies, H1718, H1721
    breach of contract, H1728
    tort, H1725
sexual H1701-H1703
    definition, H1702
sexual orientation, H1702
single incident cases, H1706
third party, protection, H1703

HARASSMENT, – *cont.*
 torts,
  assault and battery, H1723
  negligence, H1724
  remedies, H1725
  unlawful discrimination, H1717
 training, H1739
 trespass to the person, H1723
 types of, E16507, H1701, H1703
 unfair dismissal,
  claim, H1719
  constructive dismissal, H1720
  proof, H1720
  remedies, H1721
 unlawful discrimination,
  elements of the tort, H1717
  remedies, H1718
 unwanted conduct, H1702, H1704
 vicarious liability, H1710-H1715
  causal link with employment, H1712
  course of employment, H1711
  defence, H1713
  liability of employees, H1714
  narrowing extent, H1715
 violation of dignity, H1701
 violence in the workplace, V8011
 work environment, H1707-H1708
HARNESSES, M3013, W9008
HATFIELD RAIL CRASH, Intro 14
HAZARDOUS SUBSTANCES,
 accidents where, H2105, H2114, H2135,
  H2139
 ammonium nitrate, D0908
 Approved Code of Practice, H2122
 assessment of risk,
  discussion with staff, H2119
  further investigations, H2121
  guidance as to, H2120
  information gathering for, H2117
  ongoing review of, H2125
  organisation of, H2116
  records as to, H2122, H2124
  requirements as to, H2101,
   H2104–H2106, H2114–H2126
  who should carry out, H2115
  workplace observations, H2118
 biological problems, H2103
 chemical substances, *see* Chemical hazards,
  Chemical Substances
 Codes of Practice, H2105
 community health and safety, C6510
 compliance, Intro 29
 control of exposure,
  adequacy of, H2111, H2112

HAZARDOUS SUBSTANCES, – *cont.*
 control of exposure, – *cont.*
  biological agents, where, H2113
  carcinogens, where, H2113,
  examination of, H2128
  maintenance etc of, H2114, H2126,
   H2128, H2129
  monitoring, H2114
  other than PPE, H2109
  PPE, using, H2107, H2110, H2118,
   H2131
  steps required, H2107
  suitability and sufficiency of, H2114
  testing of, H2128
  use of, H2107, H2110, H2114, H2127
 disaster and emergency planning, D6013
 emergencies, H2105, H2135
 employees' duties, H2105
 employers' duties, H2105
 employers' liability insurance, conditions
  on E13021
 fire prevention and control, F5031
 first aid needs, F7004
 guidance as to, H2105
 harm to the body, H2102
 health surveillance, H2105, H2114,
  H2133
 incidents, H2105, H2135
 instruction as to, H2105, H2134
 legislation,
  allied regulations, H2101,
   H2138–H2145
  definitions, H2101, H2104
  duties under. H2105
  main regulations, H2101
  summary of, H2101, H2104, H2105
 major accident hazards, *see* Major
  Accident Hazards Regulations
 maximum exposure limits, replacement
  with workplace exposure limits
  of H2112
 monitoring exposure, H2111, H21012,
  H2132
 mutagens H2113
 notification requirements, D0908
 personal protective equipment, P3008,
  P3017
 poisoning, H2103
 PPE, use of H2109, H2110, H2118
 prevention of exposure,
  methods of achieving, H2108
  steps requiring, H2105, H2107
 REACH regulation, H2137
 recommendations,
  implementation of, H2123

HAZARDOUS SUBSTANCES, – cont.
recommendations, – cont.
  review of, H2123
records, preparation of assessment, H2122
risk assessments, R3007, R3018, R3035
self-employed, duties of, H2105
skin conditions, H2103
training as to, H2105, H2114, H2134
transportation of chemicals, see
    Transportation of Dangerous
    Substances
trespassers, C65013
ventilation, H2130
workplace exposure limits, H2112
HAZARDS,
See also Hazardous Substances
analysis, M1024
checklist, M1024–M1025
continuing, M1024
definition of, M1023
examples of, M1025
food, as to, F9019
identification of, M1024–M1025
machinery safety,
  analysis, M1024
  checklist, M1024–M1025
  continuing, M1024
  definition of, M1023
  examples of, M1025
  identification of, M1024–M1025
warnings, F9019
workplace, CH 3 App A
HEAD PROTECTION,
British Standard, P3025
construction and building operations,
    and, C8018
HEALTH,
air pollution and, E5003–E5013, E5020
definition, Intro 8
Health and Safety Commission, see Health
    and Safety Commission
Health and Safety Executive, see Health
    and Safety Executive
health and safety law, see Health and
    Safety Law
manual handling, M3001, M3002, M3007
occupational health and safety, O1007
World Health Organisation, Intro 8,
    E5063–E5065
HEALTH AND SAFETY COMMISSION,
accident reporting, Intro 29
agricultural equipment, E15005
codes of practice, Intro 10
consultative documents, Intro 10
EU Directives, implementation of, Intro 10

**HEALTH AND SAFETY COMMISSION, –
cont.**
funding, Intro 7
guidance,
  generally, Intro 10, Intro 28
  offshore operations, O7028
heights, working at, W9001
merger with Health and Safety
    Executive, E15002
lobby groups and, Intro 14
public opinion, Intro 14
Revitalising Health and Safety, Intro 7,
    T7015
Securing Health Together, Intro 7
training and competence, aims of strategy
    on, T7001
**HEALTH AND SAFETY EXECUTIVE
    (HSE),**
access traffic routes and vehicles, Intro 29
accident investigations, A3031
Area Offices, A3039 (App E)
composition of, E15002
dangerous substances, D0933.6
directors' duties, Intro 5
disabled persons, V12032, V12037
  reasonable adjustments, E16511
disaster and emergency planning, D6011
enforcement, E15001
enforcing authorities, Intro 20,
    E15007–E15009
equipment, M1006, M1007
facilities management, F3029
fees payable to, E15005
functions, E15005
funding, Intro 7
heights, working at, Intro 29, W9002,
    W9009
infoline, Intro 28
inspectors, E15005
Internet, on, Intro 28
ladders, W9009
lead body, as, T7008
lobby groups and, Intro 14
migrant workers, E16518
mines and quarries, E15005, E15009
naming and shaming, Intro 7
new or expectant mothers, V12019,
    V12023–V12024
occupational health, O1056, R2009
policy unit, T7007
priority areas, Intro 29
prosecutions by, Intro 15
publications, Intro 28, E15005, E15006
regional offices, A3039
rehabilitation, R2003

HEALTH AND SAFETY EXECUTIVE (HSE), – *cont.*
risk assessment, relating to,
  age, E16515
  disability, E16512
  gender, E16521
  migrant workers, E16519
  young people, E16516
training and competence, T7003
  lead body, as, T7008
  policy unit, T7007
transfer of responsibilities, E15011
vibration, V5003, V5017-V5018, V5029
violence in the workplace, V8001–V8002, V8005
website E15044
working at heights, Intro 29, W9001
HEALTH AND SAFETY LAW,
advice, sources of, Intro 28
civil law, *see* Civil Liability
consultants, Intro 28
criminal law, Intro 20
enforcement, *see* Enforcement
European Union,
  directives, Intro 10
  role of, Intro 9
Health and Safety at Work Act 1974,
  application, Intro 21
  duties under, Intro 21
  employers, Intro 21
  levels of duty, Intro 21
  premises, as to, Intro 21
  reasonably practicable, Intro 21
enforcement of, *see* Enforcement
general public, protection for, Intro 21
Health and Safety Commission, *see* Health and Safety Commission
Health and Safety Executive, *see* Health and Safety Executive
lobby groups, Intro 14
management of health and safety,
  arrangements, Intro 25
  organisational responsibilities, Intro 25
  policies, Intro 25
  risk assessment, *see* Risk Assessment
  sources of advice, Intro 28
outside advice, Intro 28, S7005
penalties and prosecutions, Intro 15, Intro 20
public opinion, role of, Intro 14
regulations, Intro 23
risk assessments, *see* Risk Assessment
self-regulation, Intro 15
structure, Intro 19–Intro 24
supply-chain pressure, Intro 17

HEALTH AND SAFETY LAW, – *cont.*
training, Intro 28
HEALTH AND SAFETY OFFENCES,
penalties, E15043
sentencing guidelines, E15043
Sentencing Guidelines Council, E15001, E15043
HEALTH AND SAFETY OFFICER,
responsibilities of, S7003, S7005, S7009
HEALTH AND SAFETY POLICY STATEMENTS,
arrangements for implementation of, S7003, S7006
arrangements to be considered, S7003, S7006
communication of, S7003, S7008
consultation mechanisms, S7003
contents of, S7003–S7006
contractors and, S7004
effectively, ensuring arrangements work, S7007
emergency procedures, S7003
failure to implement, E15040, S7002
format, S7003
guidance, S7011
health and safety inspections, S7003, S7006
Intranet, S7008
legal requirements, S7001, S7002
management cycle, S7007
managing health and safety, M2032
organisation, S7003, S7005
organisation for implementation of, S7003, S7005
outside advice, S7005
policy, S7009
review of, S7003, S7009
risk assessments, recording of, S7006
role of,
  health and safety officer, S7003, S7005, S7009
  employees, S7005
  maintenance engineer, S7005
  managers, S7005
  managing director, S7005, S7009
  occupational hygienist, S7005
  personnel staff, S7005
  supervisors, S7005
statement of intent, Intro 25, E14012, S7003, S7004
visitors and, S7004, S7006
young people and, S7008
HEALTH RECORDS, O1016–O1018
HEALTH SCREENING,
risk assessment, R3006, R3016 (App B)

HEALTH SURVEILLANCE,
  community health and safety, C6506
  hand-arm vibration
    syndrome, O1029–O1030, V5015
  hazardous substances in workplace,
    where, H2105, H2114, H2133
  ionising radiation, R1019
  managing health and safety, M2009
  manual handling, M3007, M3015, M3019
  night workers, W10033, W10034
  noise and hearing loss, N3013, O1025
  non-ionising radiation, R1010
  occupational health and safety, O1004,
    O1009
  respiratory diseases, O1044, O1045
  risk assessment, R3006, R3016 (App B)
  skin diseases, O1041
  vibration, V5015
    whole-body vibration, O1029–O1030,
      V5028
  working time, W10033, W10034
HEALTH WORKERS,
  risk factors, E16521
  violence in the workplace, V8077
HEARING,
  disciplinary, E14013
HEARING LOSS, *see* **Noise and occupational
  hearing loss**
**HEAT PROCESSING, FOOD SAFETY
  AND,** F9047.2
**HEATHROW RAIL TUNNEL
  COLLAPSE,** Intro 15
HEAVY METAL,
  protocol as to, E5026
HEIGHTS, WORKING AT, *see* **Working at
  Heights**
HERALD OF FREE ENTERPRISE, D6029
HERBAL REMEDIES,
  food safety regulations, F9007
HERMES PRINCIPLE, E16007
**HIGH PRESSURE FLUID
  INJECTION,** M1025
HIGHLY FLAMMABLE LIQUIDS,
  ventilation regulations, V3006
HIGHER EDUCATION, T7010, T7013
HIRE CONTRACTS,
  product liability, P9053, P9060
HIRE PURCHASE,
  product liability, P9053, P9060
HOISTS, L3032–L3033, L3036
  failure, L3033
  patient/bath hoist, L3036
  regulations, L3032

HOLD-TO-RUN DEVICES, M1036
HOLIDAYS, *see* **Annual Leave**
HOME,
  care workers, C6028
  community health and safety, C6522
  compensation for work injuries, C6028
  entry into client's, C6522
  volunteers, C6522
HOME INFORMATION PACKS,
  gas certificates, G1025
HOME WORKERS,
  display screen equipment, D8202, D8213
  risk assessment, D8213
HOMICIDE, *see also* **Manslaughter**
  corporate, E15038
  Scotland, E15038.1
HONEY,
  food premises regulations, F9030
HOSPITAL,
  occupiers' liability, O3018
HOT DESKING, D8218
HOT WORKS, C7026, D0933.5
HOTEL,
  food premises regulations, F9007, F9030
HOUSE OF LORDS,
  criminal jurisdiction, Intro 20
HSE, *see* **Health and Safety Executive**
HUMAN ERROR,
  disaster and emergency planning, D6051
HUMAN RIGHTS,
  criminal cases and, E15031
HUMIDITY,
  manual handling, M3006
HYGIENE,
  food legislation and, *see* Food
  skin diseases, O1040
HYPERTHERMIA, E17034

I

ICE AND SNOW, SLIPPING ON, M3014
IGNITION SOURCES, D0933.2, D0933.3,
    D0933.4, D0933.6
ILL-HEALTH,
  costs of, Intro 26
  manual handling, M3019
ILLUMINANCE,
  average, L5010 (Table 26)
  light measuring instruments, L5010
  lighting, *see* Lighting,
  maximum ratios, adjacent areas, L5011
    (Table 27)
  minimum measured, L5011

ILLUMINANCE, – *cont.*
ratios, L5012
technical measurement of, L5009
ILLUMINATED SIGNS,
safety signs regulations, W11033
ILLUMINATING ENGINEERING
SOCIETY,
limited glare index, L5014
IMPORTER,
machinery safety, M1013
product safety, duties as to, *see* Product
Safety
IMPROVEMENT NOTICE, E15001
appeal against, E15019–E15025
contents, E15014
contravention, Intro 20, E15015, E15017,
E15030
Crown, position of, E15045
fire prevention and control, F5036
food safety, F9009, F9010, F9015–F9016
fundamentally flawed, E15021
Prohibition Notice compared, E15016
prosecution, coupled with, E15018
public register, E15029, E15044
service, E15001, E15014
use of, E15013
withdrawal, E15014
INCAPACITY FOR WORK,
all work test, C6002
own occupation test, C6002
'welfare to work beneficiaries', C6002
INCIDENT CONTACT CENTRE, A3004
INDEMNITY,
directors, for, E15041
employers' liability insurance, E13007,
E13017, E13024
inspectors, for, E15028
vicarious liability, E13027
INDEPENDENT CONTRACTORS,
cables, O3015
occupiers' liability, O3015, 03019
safe system of work, 03019
supervision, O3015, 03019
INDEX-LINKED GOVERNMENT STOCK,
inflation proof compensation, C6028
INDIRECT DISCRIMINATION,
claim by job applicant, employee, W11043
Equality Act 2010, E16502, E16504
INDUCTION PROGRAMMES,
construction and building operations,
and, C8031
generally, T7011–T7012

INDUSTRIAL ATMOSPHERIC
POLLUTION, *see* Emissions into the
Atmosphere
INDUSTRIAL DISEASES,
compensation for, C6010
employers' liability insurance, E13001,
E13025
limitation periods, E13025
long tail diseases, E13026
INDUSTRIAL INJURIES,
benefit, *see* Industrial Injuries Benefit
disablement benefit, *see* Industrial Injuries
Disablement Benefit
disablement pension, C6003, C6011
social security benefits, C6002
INDUSTRIAL INJURIES BENEFIT,
accidents, C6003, C6006
appeals relating to, C6016
benefit overlaps, C6020
constant attendance allowance, C6003,
C6014
disabled person's tax benefit, C6003,
C6016
disablement benefit, *see* Industrial Injuries
Disablement Benefit
disablement pension, C6003, C6011
entitlement to, C6002
exceptionally severe disablement
allowance, C6003, C6014
payment of, C6017
personal injuries provisions,
acting in course of employment
duties, C6006, C6007
caused by accident, C6005
receipt of benefit, financial benefits
of, C6019
statutory sick pay, C6015
weekly benefit rates payable, C6017
INDUSTRIAL INJURIES DISABLEMENT
BENEFIT,
accident cases, claims, C6012
assessment, C6011
entitlement, C6011
rate of benefit, C6013
INDUSTRIAL PREMISES,
health and safety law,
enforcement of, E15007
INDUSTRIAL PRODUCTS,
product safety, P9013, P9023
INDUSTRIAL RELATIONS,
policy and procedure, E17043
INDUSTRY GUIDE,
food, as to, F9025, F9049

INEXPERIENCED WORKERS, R3028,
V12038–V12039
supervision, V12040
training, V12039, V12040
INFECTIOUS DISEASES, O1043, O1045
INFORMATION PROVISION,
accident investigations, A3030, A3033
children and young persons, V12010,
V12014
community health and safety, C6523,
C6527
dangerous substances, D0926, D0933.1
display screen equipment, D8209
environmental management, E16024
equipment, M1007
explosions, D0926
food safety, F9027
heights, working at, W9009
ladders, W9009
lifting equipment, L3026–L3027
managing health and safety, M2013,
M20306
manual handling, M3007, M3019
permit to work systems, S3014
radiation emergencies, see Radiation
Emergencies
safety reports, D6023
sensitive information, D6007
small businesses, Intro 4
stress, S11040
violence in the workplace, V8008, V8020,
V8061, V8068, V8141
INFORMATION TECHNOLOGY,
facilities management, F3030
INJUNCTION,
violence in the workplace, V8140
INJURED EMPLOYEES,
accident reporting, see Accident Reporting
INLAND REVENUE,
disclosures to, E14009
INLAND WATERWAYS,
staff working hours, W10026, W10031
INNOCENCE, PRESUMPTION OF, E15031
INQUIRIES,
disaster and emergency planning, D6062
INSECTS,
plagues of, E5023
INSOLVENCY,
employers' liability insurance, E13024
INSPECTIONS,
access traffic routes and vehicles, Intro 29
arrangements for, S7003, S7006
asbestos, A5015

INSPECTIONS, – cont.
building and construction
operations, C8011
dangerous substances, D0933.5
enforcement of law, for, see Enforcement
equipment, M1007
food safety, F9027
heights, working at, W9012
lifting equipment, L3021
managing health and safety, M2041
INSTITUTION OF OCCUPATIONAL
SAFETY AND HEALTH, Intro 25,
Intro 28
continuing professional development, Intro
28
Register of Safety Practitioners, T7028
Safety Pass Alliance Scheme, T7027
training and competence, T7011–T7014,
T7021–T7023, T7027, T7028
INSTITUTIONAL CARE,
compensation for work injuries, C6028
INSTRUCTIONS,
equipment, M1007
INSULATION, E3014, E3015, E3024
INSURANCE,
age discrimination, exemption, M1511
business continuity planning, B8007
construction and building operations,
and, C8033, F5029
construction sites, F5029
directors, E15041
disaster and emergency planning, D6002,
D6030, D6061
employers' liability, see Employers'
Liability Insurance,
expenses, liability policies, E13023
fire prevention and control, F5006, F5029,
F5074
managing health and safety, M2001,
M1511
occupiers' liability, O3014
professional indemnity insurance, E13003
public liability insurance, O3003, O3014,
E13016
stress, S11009
INTEREST,
damages, on, C6041
INTERIM AWARDS,
damages, on, C6041
INTERLOCKING
GUARDS, M1032–M1033, M1035
INTERMEDIARIES,
product liability and, P9041

INTERNAL COMBUSTION
  ENGINE, C7021
INTERNATIONAL COMMISSION ON
  NON IONISING RADIATION
  PROTECTION (ICNIRP) R1009,
  R1010
INTERNATIONAL LABOUR
  ORGANISATION, Intro 8
INTERNATIONAL STANDARDS,
  environmental management, E16015
  ISO 14000 series, E16015
  EMAS differences, E16016
INTRANET,
  health and safety policy statements, S7008
INTRUDER ALARMS, V8090
INVESTIGATIONS,
  See also Accident Investigations
  disaster and emergency planning, D6059,
    D6062
  enforcement, E14013
  violence in the workplace, V8060, V8129,
    V8140–V8141
INVESTMENT,
  environmental management, E16024
  Principles for Responsible Investment
    (UN), E16007
IONISING RADIATION, R1011–R1021
  appeals, R1019
  Approved Code of Practice, R1019
  approved dosimetry service, R1019
  breast-feeding, R1019
  Centre for Radiation, Chemical and
    Environmental Hazards, R1017
  children and young persons, V12015
  classified persons, designation of, R1019
  consumer products, R1014
  contamination, R1016
  co-operation between employers, R1019
  cosmic radiation, R1013
  definition, R1011
  designated areas, R1019
  dose assessment and recording, R1019
  emergency planning, R1021
  enforcing authority, E15009
  female employees, R1019
  guidance, R1019
  harmful events, R1015
  health surveillance, R1019
  irradiation, R1016
  management, R1019
  man-made radiation, R1014

IONISING RADIATION, – cont.
  medicine, R1014, R1018
  National Arrangements for Incidents
    involving Radioactivity
    (NAIR), R1019
  new or expectant mothers, R1019,
    V12004, V12020
  nuclear power, R1014
  occupational diseases, O1056
  outside workers, R1019
  overexposure, R1019
  passbooks, R1019
  personal protective equipment, P3008,
    P3014, P3017, R1019
  pregnant workers, R1019, V12004,
    V12020
  protection against radiation, R1017
  Radiation Protection Advisor, R1019
  Radiation Protection Division, R1017
  radioactivity and radiation
    dose, R1012–R1013, R1015, R1019
  radiography, R1014
  radon, R1013
  regulation of, see Radioactive Substances
  regulations, R1018–R1019
  risk assessment, H2101, H2145, R1019,
    R3016, R3027, R3035
  types of, R1011
  units of measurement, R1012–R1013
IOSH, See Institution of Occupational Safety
  and Health
IRRADIATION, R1016
ISO,
  social responsibility E16002
  ISO 14000 series, E16015
  EMAS differences, E16016
ISOLATION PROCEDURES, S3023

J

JOB SAFETY ANALYSIS, S3009
JOB SAFETY INSTRUCTIONS, S3010
JOINT AND SEVERAL LIABILITY,
  product liability, P9043
JUDGMENT,
  transportability of, P9050

K

KEY PERFORMANCE INDICATOR
(KPI), F3017
KITEMARK, P3009

L

LABELLING,
dangerous goods, carriage of, D0420
dangerous substances, D0933.1
food, requirements as to, F9003, F9012,
F9020, F9023, F9075, F9078
small businesses, Intro 4
LADBROKE GROVE RAIL
DISASTER, Intro 14, D6029, E17002
LADDERS,
guidance, W9009
portable ladders, W9009
pre-check lists, W9009
step-ladders, W9009
LAKE TRANSPORT,
working time, W10026
LANDFILL COMMUNITIES
FUND, E16006
LANDFILL TAX, E16006LANDLORD,
gas safety, Intro 29, G1017
health and safety duties, Intro 21
LAPTOP COMPUTERS, D8219, E17064
LARGE BUSINESSES,
compliance, Intro 5
corporate governance, Intro 5
corporate social responsibility, Intro 5
costs of accidents and ill-health, Intro 26
directors, duties of, Intro 5
performance, improving, Intro 5
plan-do-check management systems, Intro
5
public sector, Intro 5
risk management, Intro 5
scandals, Intro 5
voluntary sector, Intro 5
LEAD,
children and young persons, V12002,
V12016
employers' duties, V3008
exhaust ventilation system, V3008
fumes, control of, V3007
hazardous substances regulations
and, H2101, H2140
new or expectant mothers, V12004,
V12022
regulations, H2140
risk assessment, R3016, R3027, R3035

LEAD, – cont.
women, employment of, E14002
LEASE,
asbestos, A5021
LEAVE, see Annual Leave
LEGAL ADVISER,
disclosures to, E14009
LEGAL AID,
availability of, C6021
LEGAL EXPENSES INSURANCE, E13023
LEGAL FRAMEWORK FOR HEALTH
AND SAFETY, Intro 7–Intro 11
LEGAL REPRESENTATIVES,
accident reporting, A3029
LEGIONNAIRES' DISEASE, C6510, F3026,
H2103, O1043
LEPTOSPIROSIS, H2103
LEUCODERMA, O1038
LEVEL CROSSINGS,
accident reporting, A3035
LIABILITY INSURANCE. See Employers'
Liability Insurance, Insurance
LICENSING,
adventure activities, C6508, T7007
asbestos, A5026,
contractors, asbestos removal
and, A5026
regulations as to, E15005, R3016,
R3034
butcher's shops, F9021
contractors, asbestos removal and, A5026
training and competence, T7007
LIFELONG LEARNING, T7028
LIFTING EQUIPMENT,
accessories, L3038–L3047
bath hoist, L3036
chains, L3043–L3044
community health and safety, C6510,
C6511
competent person, L3019–L3022
cranes, L3018, L3029–L3031
criminal offences, L3017
declarations of conformity, L3027
defects, L3025
duty holders, L3014
electrical equipment, controlled by, L3002
examination, L3021, L3023–L3026,
L3028
eyebolts, L3046–L3047
failure of equipment, L3028–L3031
fork lift trucks, L3034–L3035
general requirements, L3015–L3027
hoists, L3032–L3033, L3036
individual operations, planning of, L3020
information, prescribed, L3026–L3027

LIFTING EQUIPMENT, – *cont.*
inspection, L3021
installation, L3017, L3021
lifts, *see* Lifts
manufacturers and owners/users,
relationship between, L3002
markings, L3018
mobile elevating work platforms, L3016
mobile equipment, L3016, L3051
once only use accessories, L3050
organisation, L3019
passers-by, C65011
patient/bath hoist, L3036
persons, lifting, L3016
planning, C6511, L3019–L3020
positioning, L3017
publications, L3019
records, L3027
regulations, L3014–L3027
reports, L3025
risk assessment, L3017
ropes, L3039–L3042
shackles, L3048–L3049
slings, L3045
so far as is reasonably practicable
(SFAIRP) L3016
statutory requirements, L3002
strength and stability, L3015
terminology, L3001
vehicle lifting table, L3037
LIFTING OPERATIONS, Intro 29,
L3001–L3051
accident reporting, A3035
duty holders, L3014
equipment, *see* Lifting equipment
lifts, *see* Lifts
LOLER, L3001, L3014–L3021
manual handling, M3010
offshore operations, O7070
principals, L3001
regulations, L3014–L3027
risk assessment, R3017
terminology, L3001
LIFTS,
building or construction, L3009
CE markings, L3005–L3006, L3010
conformity assessment procedure, L3005,
L3011
criminal offences, L3008–L3009
declarations of conformity, L3005–L3006,
L3009
design, L3009, L3011
due diligence, L3008–L3009
examination, L3011
exhibitions, L3010

LIFTS, – *cont.*
facilities management, F3032, F3034
failure, L3033
general requirements, L3005–L3006
goods, L3003
information, supply of, L3009
installations, L3011–L3012
maintenance, L3032
meaning, L3003
notified bodies, L3013
obstruction of lift shafts, L3009
penalties, L3008
persons, L3003
placing on the market, L3006, L3010,
L3012
powered working platforms, L3032
quality assurance, L3011
regulations on, L3004
responsible persons, L3005
safety components, L3006–L3007, L3010,
L3012
service safety components, putting
into, L3006, L3010
standards, L3005–L3006
supply, L3007
technical documentation, retention
of, L3005–L3006, L3009
testing, L3011
LIGHTING, Intro 29, L5001–L5020
brightness, L5014, L5017
Chartered Institute of Building Services
Engineers (CIBSE), L5002
colour rendition, L5014, L5019
confined spaces, C7025
construction and building operations, and
generally, C8017
statutory requirements, L5004
design, L5014
diffusion, L5014, L5018
disability glare, L5015
discomfort glare, L5015
distribution, British Zonal
Method, L5014, L5016
economic lamp replacement policy, L5013
electricity, E3020, L5009
energy
efficiency, L5001, L5002
saving bulbs, L5009
equipment, M1007
ergonomic design and, E17004, E17031,
E17035
fitments, maintenance of, L5013
fluorescent tubes, L5008, L5020
glare, L5015
heights, working at, W9008

LIGHTING, – *cont.*
  illuminance,
      average, L5011 (Table 26)
      light measuring instruments, L5010
      minimum measured, L5011
      ratios, L5012
      technical measurement of, L5009
  lumen, meaning, L5009
  maintenance of fitments, L5013
  manual handling, M3006, M3014
  natural, L5007
  qualitative aspects, L5014
  reflected glare, L5015
  Society of Light and Lighting, L5002
  solar, L5008
  sources of light, L5007, L5008
  specific processes, L5003–L5006
  standards of, L5009–L5013
  statutory requirements, general, L5002
  specific processes, L5003–L5006
  stroboscopic effect, L5020
  traffic routes, L5002
  VDUs, L5002, L5003
  work equipment, L5004
  workplace, W11006
  workstations, W11006
LIGHTNING PROTECTION
  SYSTEMS, E3025
LIMITATION OF ACTIONS, P9048,
  W11045
  accrual, date of, E13025
  death of claimant, E13025
  employers' liability insurance, E13001,
      E13025
      accrual, date of, E13025
      death of claimant, E13025
  industrial diseases, E13025
  knowledge, date of, E13025
  mesothelioma, E13025
LIMITING DEVICES, M1036
LIQUEFIED PETROLEUM GASES,
  dangerous substances, D0933.6
  unloading, D0933.6
  ventilation regulations, V3006
LIQUIDATED DAMAGES,
  breach of contract, C6021
LIQUIDS AND LIQUIFIABLE SOLIDS,
  FIRE AND, F5003
LOADING BAYS,
  workplace access, generally, A1001
LOADS, D0441
  see also Manual Handling,

LOADS, – *cont.*
  compressed natural gas, D0933.6
  dangerous substances,
      carriage of, D0439, D0440, D0443
      compressed natural gas, D0933.6
  regulated, D0439
  vehicles, A1006
  segregation, D0430
LOCAL AUTHORITY,
  air pollution, role as to, *see* Emissions into
      the Atmosphere
  disaster and emergency planning, D6008,
      D6028
  Environment Agency and, E5046
  food, role as to, F9005, F9007. F9028
  land-use planning, E5069
  occupiers' liability, O3001
  pollution control, role as to, E5046
  statutory nuisance, abatement of, E5062
  traffic management, E5069
LOCKERS
  construction and building operations,
      and, C8017
LONE WORKERS, V12027–V12031
  communication, V12030
  community, persons working in
      the, V12028
  emergencies, V12029, V12030
  fixed workplaces, V12028
  precautions, V12030
  premises belonging to other employers,
      working at, V12028
  risk assessment, R3028, V12027
  risks for, V12029
      control of, V12030
  safe systems of work, V12027
  types of, V12028
  violence in the workplace, V8092
  vulnerable workers, R3028
LONG TAIL DISEASES, E13026
LONG-TERM SICKNESS
  ABSENCE, M1502–M1505, O1002,
      O1007, R2002
  definition, M1502
  employment laws, M1505
  statement of fitness for work, M1504
LOSS ADJUSTERS, D6061
LOSS OF AMENITY,
  compensation for, C6030, C6031, C6034
LUNG CANCER, A5002–A5003, E13026,
      O1042

M

MACHINERY SAFETY, Intro 29,
    M1000–M1035
abrasion, M1025
behavioural measures, M1026
CE marking, M1009, M1016, M1026
checklist, M1024
competition, EC law and, M1009
conformity assessment, M1013–M1016
control devices, M1036
criminal offences, M1018
crushing, M1025
cutting and severing, M1025
Declarations of Conformity,
    European Union, M1015–M1016
    example of, M1015
    incorporation into other
        machinery, M1015
definition of a machine, M1011
design and, *see* Ergonomic Design M1002,
    M1003
    technical measures to be taken
        at, M1026
designers, M1002, M1009
deterring/impeding devices, M1036
directives on, M1001, M1003,
    M1009–M1017
drawing in, M1025
enabling devices, M1036
entanglement, M1025
equipment, *see* Equipment
essential health and safety
    requirements, M1012–M1016,
    M1026
European Union, M1001–M1003,
    M1009–M1016
    CE marking, M1009, M1016, M1026
    competition, M1009
    Declarations
        of Conformity, M1015–M1016
    Transposed
        Machinery Standards, M1001,
        M1003, M1010, M1013–M1015
fines, M1018
foreseeable use, M1002
friction, M1025
goal setting, M1001
guards, M1029–M1035
harm, definition of, M1023
hazards,
    analysis, M1024
    checklist, M1024–M1025
    continuing, M1024
    definition of, M1023
    examples of, M1025

MACHINERY SAFETY, – *cont.*
hazards, – *cont.*
    identification of, M1024–M1025
Health and Safety at Work etc. Act
    1974, M1002
high pressure fluid injection, M1025
hold-to-run devices, M1036
impact, M1025
importers, M1013
incorporation into other
    machinery, M1015
legal requirements, M1002
limiting devices, M1036
limited movement control devices, M1036
manufacturers, duties on M1001, M1002,
    M1009–M1010, M1013
markings, M1009, M1016, M1026
noise and hearing loss, O1024
procedural measures, M1026
product liability, M1002, M1003
    directive on, M1009
puncture, M1025
records, M1014
reduction in risk, options for, M1026
risk,
    analysis, M1024
    definition of, M1023
    evaluation, M1024
    reduction, options for, M1026
risk assessment, M1001, M1003, M1014,
    M1021–M1022
    definition, M1022
    elements of, M1024
    framework, M1024
safe, demonstration that the machinery
    is, M1018
safeguards and safety devices,
    characteristics of, M1028
    types of, M1027–M1035
shearing, M1025
stabbing, M1025
standards,
    American National Institute, M1019
    B type, M1021
    British, M1019, M1021, M1024
    C type, M1021
    objectives, M1020
    risk assessment, M1021
    status and scope of, M1020
    structure of, M1021
    Transposed Harmonised European
        Machinery, M1001, M1003,
        M1010, M1013–M1015

MACHINERY SAFETY, – *cont.*
suppliers, M1002, M1003,
M1009–M1010, M1013,
M1017–M1018
Supply of Machinery (Safety) Regulations
1992, M1009–M1018
technical file, M1014
trapping, M1025
trip devices, M1036
two-hand control devices, M1036
MAGISTRATES' COURT,
jurisdiction, Intro 20
MAINTENANCE,
asbestos,
building maintenance workers, exposure
of, A5006
plans, A5006, A5012
records, A5011
business continuity planning, B8011,
B8033–B8037
dangerous substances, D0933.5
electricity, E3021
equipment, M1007
facilities management, F3012–F3015
fire prevention and control, F5047
hand arm vibration, O1031
permit to work systems, S3011
pressure systems, and, P7028
records, A5011
training and competence, T7028
vehicles, A1011
whole body vibration, O1031
workplace, of, W11001, W11003
MAINTENANCE ENGINEER,
responsibilities of, S7005
**MAJOR ACCIDENT HAZARDS
REGULATIONS,**
aggregation of substances, D0707– D0708
application, D0704– D0708
arrangements for major incident
management, D0731
as low as is reasonably practicable
(ALARP), D0716
changes in risk, D0721
command and control chart of
arrangements, D0732
community health and safety, C6506,
H2101, H2139
consultation distances, D0740
dangerous substance, meaning, D0704
duties of operator, D0703, D0712, D0713,
D0716– D0718
enforcement, D0702, D0714– D0715
environment, substances dangerous
for, D0704, D0738

**MAJOR ACCIDENT HAZARDS
REGULATIONS, – *cont.***
excluded sites and activities, D0709
exercises, D0733–D0734
explosive substances, D0704, D0706,
D0738
fees chargeable, D0703, D0727
flammable substances, D0704, D0705,
D0738
hazardous installation, meaning, D0738
history, D0701– D0703
information, sources of, D0741
land-use planning, D0735–D0740
list of substances covered by, D0704
loss of control, D0704
lower-tier, D0703, D0704, D0716– D0717
major accident, what constitutes, D0703
notice of prohibition, D0713
notification requirements, D0710
off-site emergency plan, requirement for,
D0724–D0726
on-site emergency plan, requirement for,
D0722–D0723, D0726
operators, D0703
oxidising substances, D0704, D0738
places where applicable, D0704, D0709
processing of chemicals, D0703
prohibition of use, D0713
public information, requirement for,
D0728–D0729
public register, maintenance, D0711
pyrotechnics, D0706
quantity of substance or
substances, D0703
reporting requirements, Intro 29, D0712
risk contours, D0740
risk control mechanisms, D0739
safety reports, D0718–D0727
statement of policy for management,
D0730
Seveso Directives, Intro 29, D0701–
D0702
storage of chemicals, D0703
testing and validation of
plans, D0733–D0734
thresholds for named substances, D0738
top-tier, D0703, D0704, D0718
toxic substances, D0704, D0738
training, D0734
**MAJOR ACCIDENT PREVENTION
POLICY (MAPP),** D0717, D6018
MAJOR PUBLIC EVENTS,
community health and
safety, C6524–C6526, C6527

MANAGEMENT,
See also Disaster and Emergency
    Management Systems, Environmental
    Management, Facilities management,
    Managing health and safety, Risk
    management
agents, Intro 21
arrangements, Intro 25
attendance, R2007
Certificate in Safety Management, T7020
community safety, C6504
Diploma in Safety Management, T7013,
    T7020
directors, S7005, S7009
disaster and emergency planning, D6002
eye diseases and disorders, O1035
facilities management, F3005
generally, Intro 23
guidance, Intro 25
hand arm vibration, O1030
health and safety policy statements, S7005
ionising radiation, R1019
musculoskeletal disorder, 01049
occupational health and safety
    management systems, Intro 25,
    R2009
Occupational Safety and Health guidance
    on, Intro 25
offshore workers, O7029–O7032
organisation, Intro 25
plan-do-check management systems, Intro
    5, Intro 29
policies, Intro 25, S7005
premises, duties at to, Intro 21
respiratory diseases, O1045
sickness absences, see Managing absence
small businesses, beliefs and attitudes of
    managers in, Intro 4
stress, Intro 29, O1054, S11009,
    S11011–S11033, S11036–S11040
supply, Intro 4, G1003, G1005, G1029
third parties
    asbestos, A5020
traffic, E5069
training and competence, T7001,
    T7004–T7007, T7015–T7022
violence, V8117–V8129
whole body vibration, O1030
working time, W10033
workplaces, Intro 29
MANAGING ABSENCE, M1501–M1531
causes, M1523
certification, M1521
disability,
    conditions that may amount to, M1509

MANAGING ABSENCE, – cont.
disability, – cont.
    definition, M1507
    discrimination, M1507–M1511
    reasonable adjustments, M1511, R2005
    where condition amounts
        to, M1506–M1511
discrimination, M1507–M1511
dismissal,
    Code of Practice on Disciplinary and
        Grievance Procedures, M1517
    fair, M1516–M1517, M1530
    short-term absence, M1530
    time limits, M1517
employment contracts, M1513, M1518
fit for work, M1504, R2007
GP medical statements, R2007
HSE guidance, R2010
improvement, setting time limits
    for, M1527
insurance benefits, M1511
keeping in touch with absent
    employees, M1504
long-term absence, M1502
    employment laws, M1505
medical reports, access to, M1512
    General
        Medical Council's Guidance, M1512
musculoskeletal injuries, M1501
normal day to day activities, M1507
notify, requirement to, M1520
occupational medicine, R2009
pandemics, R2007
patterns, looking for, M1525
popular sporting events, R2007
progressive conditions, M1509
reasonable adjustments, M1511, R2005
records, M1528
rehabilitation, M1515, R2005
return to work interviews, M1522
reviews, M1503, M1524
short-term absences, M1518–M1531
sick pay, M1513–M1514
snow, R2007
statement of fitness for work, M1504
statutory sick pay, M1514
stress, M1501, M1518
    negligence, M1524
substantial, effect, M1507
targets, setting, M1527
terms and conditions of
    employment, M1513-M1517
trigger points, defining, M1526
warnings, giving formal, M1529
workplace factors, reviewing, M1524

MANAGING AGENT,
premises, duties at to, Intro 21
MANAGING DIRECTOR,
responsibilities of, S7005, S7009
MANAGING HEALTH AND
SAFETY, M2001–M2046
accident reporting and investigation
systems, M2041
Approved Code of Practice, M2005
asbestos,
asbestos-containing materials, A5013
leases, A5021
long-term plans, A5018
plans, A5012, A5018–A5019, A5022
third parties, A5020
assistance with health and safety, M2010
audits, M2042, M2045
bench-marking, M2039
British Standards, M2029
capabilities, M2016
causes of accidents, M2028
checklist, M2030
children, risk assessment and, M2013
client, duties of, M2024, M2027
communication, M2036, M2045
competence, M2037, M2045
competent persons, appointment
of, M2010
contingency planning, M2037
contractor,
duties of, M2024, M2027
notifiable projects, M2027
principal, M2027
control, M2034, M2045
co-operation and co-ordination, M2014,
M2035, M20345
Corporate Manslaughter and Corporate
Homicide Act 2007, M2001
danger,
areas, M2011
serious and imminent danger, procedure
for, M2011
designer, duties of, M2024, M2027
emergencies,
plans and procedures, M2039
services, M2012
employees' duties, M2017
employers' duties, M2003–M2004
employees, to, M2004
third parties, M2004
European Agency for Health and Safety at
Work, M2001
external sources, contacts with, M2012
financial issues, M2001

MANAGING HEALTH AND SAFETY, –
cont.
fire,
evacuation procedure, M2011
risks, assessment of, M2006
health surveillance, M2009
host employers, persons working
with, M2015
humanitarian issues, M2001
implementation, M2040, M2045
improvement, continual, M2029, M2044
information for employees, M2013,
M2036
inspections, M2041
insurance, M2001
investigation systems, M2041
leases, A5021
legal issues, M2001
legal requirements, M2002, M2039
measurement performance, M2045
monitoring, M2041
new and expectant mothers and their
babies, risk assessment and, M2006,
M2019
objectives, setting, M2039
occupational health and safety
management systems, M2029,
M2038, M2040, M2044
organisation, M2033
performance,
measurement, M2045
reviews, M2045
planning, M2038–M2039, M2045
policy, M2031–M2032
POPIMAR mnemonic, M2029, M2032,
M2042, M2045
prevention principles, M2007
reasonably practicable, meaning
of, M2003
records, M2008
recruitment and appointment, M2037
regulations, M2005–M2021
reviews, M2043
checklist, M2030
performance, M2045
risk assessment, M2002, M2005–M2009,
M2013, M2019
risk control systems, M2039
self-employed, M2015
small firms, M2010
standards, M2029, M2032, M2034
statements, contents of policy, M2032
succession and contingency
planning, M2037
systems for, M2029

MANAGING HEALTH AND SAFETY, –
cont.
temporary workers, M2018
training, M2016, M2037
visitors, M2015
young persons, M2006, M2013, M2020
MANDATORY SIGNS, W11038
see also Safety Signs
MANSLAUGHTER,
corporate killing, Intro 12, Intro 20
    D6029, E13001, E13032, E15001,
    E15038, M2001
corporate. *See* Corporate Manslaughter
director, E13032
disaster and emergency planning, D6042
duty of care, E15038.1
enforcement, E15001
fines, E15038.1
gross negligence, C9001, E13032, E15038
Involuntary Homicide Bill, D6042
reckless killing, D6042
Southall rail crash, Intro 15
unlawful act manslaughter, D6042
workplace, in, E15038
MANUAL HANDLING
    OPERATIONS, M3001–M3027
accident investigation reports, M3019
accident treatment records, M3019
age, M3007, M3015
armed forces, M3002
assembly or production lines, M3004
automation, M3010
avoidance of risks, M3008–M3015
back, lower, M3004
bags, M3013
body movements, M3004
capability of the individual, M3007,
    M3015
carrying, M3003
charts, M3003
children and young persons, V12005,
    V12016
chutes, M3010
clothing, M3004, M3007
community health and safety, C6522
conveyors, M3010
definitions, M3002
deliveries, requiring direct, M3009
design,
    floor, M3014
    processes or activities, of, M3009
    task layout and, M3012
directive on, M3001
discussions, M3022
diving, M3002

MANUAL HANDLING OPERATIONS, –
cont.
elimination of handling, M3009
employees' duties, M3002
employers' duties, M3002
ergonomic approach, E17044, E17052,
    E17063
excessive movement of loads, M3004
exemptions, M3002
experience, M3007
fire procedures, M3008
fitness programmes, M3015
floors,
    condition, M3006, M3014
    design, M3014
    uneven, slippery or unstable, M3006
    variation in levels of, M3006
footwear, M3004, M3007
frequent or prolonged physical
    effort, M3004
gangways, M3014
gender, M3007
grasp, making loads easier to, M3013
gravity feed transfer systems, M3010
grips, M3013
handles, M3013
harnesses, M3013
health sector, M3001, M3002, M3007
health surveillance, M3007, M3015,
    M3019
holding or manipulating loads at a distance
    from the trunk, M3004
HSE guidance, M3003, M3012, M3015,
    M3017, M3019, M3024, M3027
humidity, M3006
ice and snow, slipping on, M3014
ill-health records, M3019
improving the work environment, M3014
indents, M3013
information and guidance, M3007,
    M3019, M3028
    work sector, M3019
levels, differing work, M3014
lifting and lowering, M3003
lifting tools, M3010
lighting, M3006, M3014
loads,
    bulkiness or unwieldiness, M3005,
        M3008
    grasp, difficult to, M3005
    load related measures, M3013
    meaning, M3002
    related factors, M3005
    sharp objects, M3005
    size, M3005, M3008, M3011

MANUAL HANDLING OPERATIONS, – *cont.*
loads, – *cont.*
   stability, M3005, M3013
   temperature, M3005
   unwieldiness, M3005, M3008
   weight, M3005, M3008, M3011
long-term restrictions, M3015
manually-operated lifting devices, M3010
mechanical handling aids, M3001,
   M3010, M3012, M3013, M3025
mechanisation, M3010
medical screening, M3015
musculoskeletal problems, people
   with, M3007
new or expectant mothers, V12005,
   V12020
non-powered conveyors, M3010
notes, M3023
observations, M3021
offshore activities, M3002
operating procedures, M3019
physique and stature, M3007
planning and preparation, M3016–M3019
pneumatic transfer systems, M3010
posture, M3004
powered conveyors, M3010
pregnancy, M3007, M3015
prevention of risks, M3008–M3015
previous injury or ill-health, M3007
prolonged physical effort, M3004
pushing and pulling, M3003, M3004
   excessive, M3004
rate of work imposed by a
   process, M3004
'reasonably practicable', M3002, M3008,
   M3011
recommendations, review and
   implementation of, M3025
records, M3019, M3024
reduction of loads, M3011
reduction of risks, M3008, M3013
regulations on, M3001–M3003
rest and recovery, insufficient, M3004
risk assessment, Intro 16, Intro 23, M3002
   after, M3024–M3027
   arranging, M3018
   checklist, M3020, M3024
   guidelines, M3003
   information gathering, M3019
   making, M3020–M3023
   planning and
      preparation, M3016–M3019
   recommendations, review of, M3025
   reviews, M3026

MANUAL HANDLING OPERATIONS, – *cont.*
risk assessment, – *cont.*
   team, M3017
risk of injury, M3002–M3015
   avoiding or preventing, M3007–M3015
routines, M3012
seated, handling when, M3003
ship's crew, M3002
short-term restrictions, M3015
sickness absence, Intro 29
size reduction, M3013
slings, M3013
slips, trips or falls, M3006, M3014
social work sector, M3001, M3002
space,
   clear, providing, M3014
   constraints, M3006
stability, M3005, M3013
stooping, M3004
sudden movement, risk of, M3004
surfaces, variation in levels of, M3006
task related factors, M3004
task related measures, M3012
team handling, M3012
temperature, extremes of, M3006, M3013,
   M3014
toxic or corrosive substances, leaks
   of, M3002
training, M3001, M3007, M3015, M3027
transport, M3009
trolleys and trucks, M3010
twisting, M3004
ventilation, M3006, M3014
volunteers, C6523
vulnerable persons, V12005
weight,
   limits, M3003
   reduction, M3013
whole-body vibration, V5024
wind, gusts of, M3006, M3014
work sector information, M3019
working environment
   factors relating to the, M3006
   improving the, M3014
MANUFACTURER,
equipment, M1007, N3013
hand-arm vibration syndrome, V5020
health and safety duties, Intro 21
machinery safety, M1001, M1002,
   M1009–M1010, M1013
noise and hearing loss, N3013
product liability, *see* Product Liability;
   Product Safety

MARITIME AND COASTAL
    EMERGENCIES, D6006
MARKET,
    food premises regulations, *see* Food
MARKINGS,
    CE Mark, *see* CE Mark
    lifting equipment, L3018
    pressure systems, and, P7017
MARQUEE,
    food premises regulations, F9032
MARRIAGE,
    harassment, in respect of, H1703
MATERNITY. *See* New or expectant
    mothers
MATERNITY LEAVE,
    entitlement, E14002, E14028
MEALS,
    facilities for, W11027
MEANS OF ESCAPE,
    fire prevention and control,
        and, F5020–F5021
MEAT,
    food safety legislation, *see* Food
MECHANICAL HANDLING AIDS, M3001,
    M3010, M3012, M3013, M3025
MECHANICAL WORK,
    risk assessment, R3016 (App B)
MECHANISATION, M3010
MEDIA,
    business continuity planning, B8009,
        B8038–B8042
    crisis communication, B8038–B8042
    disaster and emergency planning, D6062
MEDICAL ADVISER,
    carbon monoxide poisoning, G1017
    employment cases, E15005
    gas safety, G1017
MEDICAL CERTIFICATES, M1504
    statement of fitness for work, M1504,
        M1515
MEDICAL DEVICES, R1007
MEDICAL GROUNDS,
    suspension from work, E14016
MEDICAL REPORTS,
    access to, M1512
MEDICAL SERVICES,
    work injuries, for, C6028
MEDICAL STATEMENTS, M1504, M1515
MEDICAL SURVEILLANCE,
    manual handling, M3015
    training and competence, T7007

MEDICINAL PRODUCTS, F9007, R1014,
    R1018
MEDICINES AND HEALTHCARE
    PRODUCTS REGULATORY AGENCY
    (MHRA), F9007
MENTAL HEALTH,
    stress, M1518
MERCURY POISONING,
    new or expectant mothers, V12022
MESOTHELIOMA, A5002–A5003, C6010,
    C6062, E13025, E13026
METAL FUME FEVER, H2103
METALS, FIRE AND, F5003
METER,
    gas, G1022
MIGRANT WORKERS,
    race, E16518
    risk assessment, E16519
MILK,
    distribution, of, F9028
MINES,
    accident reporting, A3005, A3035
    children and young persons, V12015
    heights, working at, W9001
    training and competence, T7007
MINIMUM SPACE PER PERSON,
    factories, offices and shops, W11008
MOBILE PHONE,
    cancer risks, R1008
    display screen equipment, D8219
    provision for travelling or otherwise absent
        employees, F7009
    non-ionising radiation, R1008
    work place violence and, V8092
MOBILE WORK EQUIPMENT,
    definition, L3051, M1008
    drive shafts, M1008
    employees carried on, M1008
    falling, M1008
    falling object protective structures, M1008
    forklift trucks, *see* Fork-lift trucks
    guards and barriers, M1008
    lifting, L3016, L3051
    mobile elevating work platforms, L3016
    mobile workers, W10027–W10030
    overturning, L3051
    Provision and Use of Work Equipment
        Regulations M1005, M1007–M1008
    remote controlled, M1008
    restraining systems, M1008
    rolling over, M1008

MOBILE WORK EQUIPMENT, – *cont.*
self-propelled, C6510, M1008
speed adjustment, M1008
MONOTONOUS WORK,
working time, W10019, W10033,
W10034
MOTHERS, *see* New or Expectant Mothers
MOTOR INSURERS BUREAU
(MIB), E13026
MOTOR VEHICLES. *See* VEHICLES
MULTI-TRACK CLAIMS, E13029
MUSCULOSKELETAL DISORDERS, Intro
29, O1047–O1054
accidental reporting, E17067
back disorders, E17045, O1047
causes, O1048
clinical effects, E17044
detection, O1049
disability discrimination, O1050
guidance as to, E17067, E17068
heavy manual work, O1050
HSE priority area, M3001
individual differences, E17049
legislation as to, E17044, O1051
litigation, E17044
lower limb discomfort, E17047
management, O1050
managing absence, M1501
manual handling, M3007
occupational health professional, reporting
to, O1050
osteo-arthritis, O1047
prevention of, E17050
psychosocial risk factors, E17048
repetitive strain injury, O1047
reporting, O1050
risk assessment, O1050–O1051
scope of, E17003, E17044–E17050
sickness absence, O1049
statistics as to, E17002
symptoms, O1049, O1050
task design and, E17010
tendons, O1047
upper limb disorders, E17046, E17052,
E17067, O1047
use of term, E17044
vibration, exposure to, E17033, V5003
work-related upper limb disorders, O1047
MUSIC VENUE, OUTDOOR, N3027.2
MUSICIANS,
noise exposure regulations, N3002,
N3026–N3027.2
safety representatives, J3005
MUSTER AREAS,
offshore operations, O7028

MUTAGENS, H2113

N

NAMING AND SHAMING, Intro 7
NATIONAL ARRANGEMENTS FOR
INCIDENTS INVOLVING
RADIOACTIVITY (NAIR), R1019
NATIONAL DISASTER,
food in event of, F9033
NATIONAL EXAMINING BOARD IN
OCCUPATIONAL SAFETY AND
HEALTH, Intro 28, T7009, T7010,
T7013, T7017
NATIONAL GENERAL CERTIFICATE IN
OCCUPATIONAL SAFETY AND
HEALTH, T7017
NATIONAL GRID, G1001
NATIONAL OCCUPATIONAL
STANDARDS,
setting, T7008
work-related violence, V8142
NATIONAL OPEN COLLEGE
NETWORK, T7023
NATIONAL RADIOLOGICAL
PROTECTION BOARD, R1009
NATIONAL SECURITY, D0928
disaster and emergency planning, D6097
NATIONAL VOCATIONAL
STANDARDS, E3011, T7002, T7008,
T7009, T7012–T7016, T7022–T7025
NATIONALITY
discrimination, H1703
NATURAL GAS, D0933.6
NATURAL LIGHTING,
source of lighting, L5007
NEBOSH DIPLOMA, T7013, T7017
NEGLIGENCE,
contributory negligence,
damages and, C6046
employee's, Intro 24
product liability, as to, P9033
duty of care, Intro 24
gross, E15038
harassment in the workplace, H1724
legal aid, availability of, C6021
manslaughter, E15038
noise, N3022
occupiers' liability and, O3003
product liability,
comparison with, P9042
contributory negligence, P9038
vicarious liability, principles of, Intro 24
workplace violence and, V8012

NEIGHBOURS,
community safety, C6509–C6511
dangerous activities, planning of, C65011
employer's duty to, R3002, R3032
precautions, C6511
signs and notices, C65011
traffic management and
   precautions, C65011
NERVOUS SHOCK,
recoverable non-pecuniary losses, C6033
NETTLE RASH, O1038
NEW OR EXPECTANT
   MOTHERS, V12017–V12026
antimitotic drugs, V12022
biological agents, V12021
carbon monoxide, V12022
checklist, V12025
chemical agents, V12022
compressed air, work in, V12004, V12020
construction work, V12005
definition, V12003, V12017
directive on, V12017–V12018, V12019
discrimination, H1703
diseases, infectious or contagious, V12018
diving, V12020
electromagnetic radiation, V12020
employment tribunal, V12018
Equality Act 2010, E16520
European Union, V12017–V12018,
   V12019
HSE guidance, V12019, V12023–V12024
ionising radiation, R1019, V12004,
   V12020
lead, V12004, V12022
legislation, V12001–V12005
Management of Health and Safety at Work
   Regulations, V12001–V12003,
   V12017–V12018
managing health and safety, M2006,
   M2019
manual handling, M3007, M3015,
   V12005, V12020
maternity leave E14002, E14028
mercury and mercury derivatives, V12022
movement, V12020
musculoskeletal disorders, E17049
night work, V12018
noise, V12020
non-employees, risks to, V12026
personal protective equipment, V12005
physical agents, V12020
posture, V12020
pressure, physical and mental, V12020

NEW OR EXPECTANT MOTHERS, – *cont.*
risk assessment, M2006, M2019, R3018,
   R3026, R3027, R3032, R3035,
   V12017–V12018, V12021,
   V12025–V12026
risks to, V12019–V12024
shock, V12020
skin, agents absorbed through
   the, V12022
stress, S11014
suitable alternative work, V12018
suspension from work, E14015, V12018
table of hazards, E16521
temperature, V12020
vibration V5016, V12020
visual display equipment, V12005,
   V12023
working conditions, V12018, V12023
working time, V12018
NIGHT CLUBS, NOISE AND, N3027
NIGHT WORK,
definitions, W10019
health assessment, W10021, W10033,
   W10034
length of, W10020
limits on, E14002
new or expectant mothers, E14026,
   V12018
night time, W10019
night worker, W10019
road transport, W10029
stress, E17041
transfer to day work, W10021
violence in the workplace, V8003
working time, W10029, W10033,
   W10034
NOISE AND OCCUPATIONAL HEARING
   LOSS, N3003, O1021–O1026
acoustic booths, N3037
acoustic signals, W11031
action against employer at common
   law, N3022
agriculture, specific legal
   requirements, N3015
assessments, formal, N3013,
   N3029–N3034
audiometers, O1023
background noise level, N3010
barriers to noise, N3039
building and construction
   operations, C8018
British Standard, N3003, N3036, N3041
buying policy for equipment, N3035
C-weighting, N3003
call centres, N3028

NOISE AND OCCUPATIONAL HEARING LOSS, – *cont.*

cartridge-operated tools, N3003
causes, O1022
clubs, N3027
code of practice, N3017
competent person, N3029
concert halls, N3027.1
conditions when deafness
    prescribed, N3019
construction and building
    operations, N3016–N3017
construction sites,
    specific legal requirements, N3016,
        N3017
continuous, N3005–N3006
contribution, N3022
cooling, N3035
daily personal noise exposure, N3007,
    N3013, N3031
damage occurs, how, N3003
damping material, N3035
deafness,
    noise-induced hearing loss, N3003
    statistics, N3002
defensive mechanisms of the ear, N3003
definition, O1021
designers of machinery, duties of, N3014
disablement benefit, payment of, N3002
    assessment of benefit, N3021
    claimants, number of, N3002
    conditions under which payable, N3020
    entitlement to, N3002, N3018
    vibration white finger, N3012
discos, N3027
dosimeters, N3003, N3032
ear muffs, N3036
ear plugs, N3027, N3036
ear protection, N3002, N3013, N3022,
    N3027, N3036–N3041, P3018
ear protection zones, N3012, N3014
emergency purposes, sound systems
    for, N3041
employers' duties, N3013
employers' liability insurance, E13026
equipment,
    buying policy, N3035
    manufacturers, information provided
        by, N3013
ergonomic approach to, E17004, E17032
E.U. law, N3002, N3016
evidence, N3022
explosives, N3003
exposure limit values, N3013
firearms, N3003

NOISE AND OCCUPATIONAL HEARING LOSS, – *cont.*

frequencies, O1022
gender differences, N3002
general legal requirements, N3012–N3017,
    N3023–N3024
hair cells, O1022
hazardous environments, N3033
health surveillance, N3002, N3013,
    O1025
hearing disability through ageing, N3003
hearing loss, N3003
    disablement benefit, N3002
hearing protection zones, Intro 23, N3013,
    N3031, N3034
HSE guidance, N3023–N3025
human ear, effect of sound on, N3004
human responses to, N3005
importers of machinery, duties of, N3014
indices, N3005–N3011
legal requirements relating
    to, N3012–N3017, O1025
liability of employer at common
    law, N3022
machinery, use of, O1024
manufacturers' duties, N3014
maximum sound pressure level, N3008
measurement of sound, N3002,
    N3005–N3011, N3029–N3030,
    O1023
meters, N3029–N3030, N3033
music systems, N3037, N3040
musicians, N3002, N3026–N3027.2
negligence, N3022
new or expectant mothers, V12020
night clubs, N3027
noise-induced hearing loss, *see* Noise-
    Induced Hearing Loss
Northern Ireland, N3022
nuisance, E5062
occupational deafness,
    assessment of disablement
        benefit, N3021
    compensation for, N3018–N3021
    conditions for benefit to be
        payable, N3019
    damages for, N3022
    hearing loss, degree of
        disablement, N3021 (Table 29)
organisational and technical measures,
    reduction by, N3013
outdoor music venues, N3027.2
outdoors, equipment used, N3016
PA and music systems, N3040
peak sound pressure, N3009

NOISE AND OCCUPATIONAL HEARING LOSS, – *cont.*
personal exposure, N3007
personal protective equipment, P3004, P3008
Physical Agents Directive, Intro 29, N3002, O1026
precautions, N3002
reducing noise N3035
specific environments, in, N3026–N3028
reduction measures, Intro 23
regulations, Intro 23, N3001, N3013, N3023, P3018
reports, N3034
risk assessment, Intro 16, N3002, N3013, R3008, R3018, R3032, R3035
competent persons, N3002, N3029
sufficient information, instruction and training, assessors having, N3002, N3029
rotation of work, N3038
scale of the problem, N3002
ships, N3002
sound level meters, N3029–N3030, N3033, O1023
sound, perception of N3004, O1021
sound power level, N3011
sound systems, N3027, N3040–N3041
statistical evidence, N3022
statutory controls over, N3012–N3013
successive employers, N3022
survey of, N3002
theatres, N3027.1
tinnitus, O1022
tractor cabs, N3015
upper exposure action levels, N3013
ventilation, N3035
vibration, E17033, N3012, N3013, N3017
vibration white finger, N3012, V5003
weekly personal noise exposure level, N3007, N3013
weighted sound pressure level, N3006, N3008
NON-IONISING RADIATION, R1001–R1010
body, interaction with the, R1003
cancer, R1008
definition, R1001
direct effects, R1005
diseases, R1008
EC law, R1010
electromagnetic spectrum, R1001–R1002, R1004–R1010

NON-IONISING RADIATION, – *cont.*
frequencies, R1001–R1010
guidelines on exposure, R1009
health surveillance, R1010
indirect effects, R1006–R1008
International Commission on Non Ionising Radiation Protection (ICNIRP), R1009, R1010
known risks, R1003
legal requirements, R1010
medical devices, implanted, R1007
mobile phone cancer risks, R1008
National Radiological Protection Board, R1009
precautions, R1010
risk assessment, R1010
uncertain health risks, R1003–R1004, R1009
NON-PECUNIARY LOSSES,
bereavement, C6031
loss of amenity, C6031, C6034
pain and suffering,
illness following post-accident surgery, C6031, C6032
nervous shock, C6033
principal types, C6023, C6031
recoverable, C6033
NORTHERN IRELAND,
air pollution, regulation of, E5046, E5069
disaster and emergency planning, D6007
Food Standards Agency, F9005
heights, working at, W9003
Industrial Pollution and Radiochemical Inspectorate, E5069
vibration regulation, V5015
NOTICES,
community health and safety, C6511, C6514
neighbours and passers-by, C65011
NOTIFICATION,
see also Accident Reporting
disaster and emergency planning, D6019
hazardous substances, D0908
violence in the workplace, V807
work in compressed air, and, C8016
NUCLEAR SAFETY, R1014
NUISANCE,
air pollution, E5016, E5062
statutory nuisance, E5062
NURSERY WORKERS,
risk factors, E16521
NURSES,
occupational health nurses, qualifications of, R2009
rehabilitation, R2009

NURSES, – *cont.*
risk factors, E16521
NVQs (NATIONAL VOCATIONAL
STANDARDS), T7002, T7008-T7009,
T7012–T7016, T7022–T7025

O

OCCUPATIONAL DEAFNESS,
awards of damages for, C6055
OCCUPATIONAL
DISEASES, O1021–O1056
cancer, O1056
carginogens, O1056
eye, diseases and disorders of
the, O1032–O1036
hand arm vibration and whole body
vibration, O1027–O1031
ionising radiation, O1056
musculoskeletal disorders, O1047–O1051
noise and hearing loss, O1021–O1026
psychological disorders, O1052–O1055
radiation, O1056
reportable, A3002, A3016, A3035
respiratory diseases, O1042–O1046
skin diseases, O1037–O1041
stress, anxiety and
depression, O1052–O1055
vibration white finger, V5011–V5014
whole body vibration, O1027–O1031
work-related upper limb disorders
(WRULDs), *see* Work-Related Upper
Limb Disorders (WRULDs)
OCCUPATIONAL HEALTH. *See*
Occupational safety and health
OCCUPATIONAL HYGIENIST,
responsibilities, of, O1013, S7005
OCCUPATIONAL MEDICINE,
rehabilitation, R2009
OCCUPATIONAL SAFETY AND
HEALTH, O1001–O1020, O1056
Approved Code of Practice, O1009
buying services, O1019
confidentiality, O1016, O1018
damages, O1002, O1005
data protection, O1017
definition, O1001
disaster and emergency planning, D6002
duty of care, O1002, O1005
ergonomics and, *see* Ergonomic Design;
Ergonomics
ergonomists, O1014
guidance, O1009

OCCUPATIONAL SAFETY AND HEALTH,
– *cont.*
Health and Safety Executive, reporting
to, O1056
health education and promotion, O1007
health records, O1016–O1018
confidentiality, O1016, O1018
data protection, O1017
keeping and making use of, O1018
obtaining, O1017
subject access, O1017
health surveillance, O1004, O1009
importance of, O1001–O1005
initiatives and targets, O1020
legal duties, O1004
long-term sickness, O1002, O1007
medical reports, obtaining, O1017
musculoskeletal disorders, *see*
Musculoskeletal Disorders
national support programme, Intro 7
management systems, M2029, M2038,
M2040, M2044
occupational health nurses, O1010–O1011
duties of, O1011
qualifications of, O1011, R2009
occupational hygienists, O1013
occupational physicians, O1012, R2009
duties of, O1012
qualifications of, O1012, R2009
occupational health team, O1010–O1015
Occupational Safety Consultants Register
(OSCHR), R3029
personnel support, O1003
pilots, O1020
prevention rather than
cure, O1007–O1009
priorities, O1019
proactive, being, O1007
reactive, being, O1007
records, O1016–O1018
rehabilitation, R2003–2004, 2009
relationship between occupational health
and safety, O1006
reporting, O1056
risk assessment, O1005, O1008
Robens Centre for Occupational Health
and Safety, R2009
self-surveillance, O1009
sickness absence, O1002, O1007
short-term sickness, O1002
support, access to, O1020
targets, O1020
temperature, O1056
training and competence, *see* Safety
Training and Competence

OCCUPATIONAL SAFETY AND HEALTH, – *cont.*
treatment services, O1015
vulnerable persons, O1005, O1008
OCCUPIER,
dangers to guard against, O3012
definition, O3004
duty to provide,
    sanitary conveniences, W11019
    washing facilities, W11020
factory, actions against, O3018
fire prevention and control, F5073
liability, *see* Occupiers' Liability
OCCUPIERS' LIABILITY,
access, O3018
causation, O3017
children, O3010, O3011, O3017
    trespass, O3010
    warning signs, O3010
common law duty of care, O3002
community health and safety, C6508
contributory negligence, O3011
dangers to guard against, O3012
drains, O3014
duties under,
    HSWA, O3019
    OLA 1957, Intro 24, O3002
factories, O3018, W11047
    access, O3018
    Approved Code of Conduct, O3018
    slips, trips and falls, O3018
    traffic routes, O3018
fire, as to, O3014
foolhardy and negligent conduct, O3002, O3014
foreseeability, O3010
free will, O3017
heights, working at, O3014
independent contractors, O3015, O3019
intruders, O3009
latent dangers, O3010
maintenance and repair of
    premises, O3004
nature of duty, O3003
obvious risks, O3017
occupier, definition, O3004
premises,
    definition, O3005
    maintenance and repair, O3004
    visitors to, O3006
    work being done, dangers of, O3015
public liability insurance, O3014
public right of way, O3006
recklessness, O3014

OCCUPIERS' LIABILITY, – *cont.*
risks willingly accepted, O3017
safe system of work, O3019
scope of duty, O3004
Scotland, in, O3001
slips, trips and falls, O3018
standard of care, C6508
structural defects, O3013
swimming, O3017
supervision, O3006
traffic routes, O3018
train surfing, O3017
trespassers,
    children, O3010
    common law definition, O3007
    common law duty to, O3008
    duty to, under OLA, O3009
    latent dangers, O3010
unfair contract terms, O3016
visitors,
    community health and safety, C6508
    invitees, O3003, O3006
    licensees, O3003, O3006
volenti non fit injuria, O3017
waiver of duty, O3016
warning notices, O3006, O3009, O3016, O3017
    children, O3010
window cleaners, to, O3014
workmen on premises, O3014, O3015
ODOUR NUISANCE, E5016, E5062
OFFENSIVE WORK ENVIRONMENT,
harassment H1707, H1708
OFFICES,
cleanliness, level and frequency
    of, W11006
window cleaning, *see* Window Cleaning
OFFSHORE OIL INSTALLATIONS,
IPPC Directive, E5027
UK regulations, E5040
OFFSHORE OPERATIONS, O7001–O7073, O7019
abandonment of offshore installations,
    provisions, O7001
access hazards, O7072
accident reporting, A3035
administration. *See* management and
    administration *below*
applicable law, O7009
application of criminal and civil
    law, O7002
application outside Great Britain
    clause, O7009
    activities within territorial
        waters, O7009

OFFSHORE OPERATIONS, – *cont.*
application outside Great Britain clause, – *cont.*
associated structure, O7009
fixed installation, O7009
jurisdiction, O7009
mobile installation, O7009
relevant waters, O7009
territorial sea, O7009
territorial waters, O7009
vessel, O7009
within territorial waters, O7009
Approved Code of Practice, O7028
Burgoyne Report, O7003
burns, O7071
carriage of dangerous substances, O7068
civil law, application of, O7002
combined operations,
safety case requirements, O7013
confined spaces, working in, O7065
cranes, O7070
criminal law, application of, O7002
Cullen Report, O7003, O7046
death, registration of, O7040
defences, O7056
design and construction, O7024–O7027
duties of dutyholders, O7025
generally, O7024
integrity of installations, integrity of, O7025
personnel, duties of, O7027
wells, integrity of, O7026
diseases, reportable, A3035
diving,
diving contractor, O7037
general requirements, O7036
planning, dive, O7038
risk assessment, O7038
duty holders,
management and administration, O7030
electricity, O7063
emergency response, O7028
employers' liability insurance, E13001
escapes, O7028
evacuation, O7028
exclusion zones, temporary, O7008
exploration/exploitation licences, O7001
explosions, O7028
eye injuries, O7071
fire, O7028
first-aid, O7057–O7062
installation manager's duty, O7062
medical personnel, O7057, O7058
offshore first-aider, O7058

OFFSHORE OPERATIONS, – *cont.*
first-aid, – *cont.*
offshore medic, O7058
sick bays, O7059–O7061
suitable persons, O7058, O7057
fixed installation, O7009
abandonment, O7015
decommissioning, O7014, O7015
management and administration, O7031, O7030
removal of person from, O7031
safety case requirements, O7013
generally, O7001
hazards,
access hazards, O7072
burns, O7071
cranes, O7070
eye injuries, O7071
gassing incidents, O7073
lifting equipment, O7070
health and safety law applicable to, O7009
HSC Guidance, O7028
information provision, O7028
inspection, O7041–O7045
assist inspector, duties to, O7044
background, O7041
casualty, O7041
disclosure of information, O7045
equipment, O7041, O7043
functions of inspectors, O7042
general requirements, O7041
powers of inspectors, O7042
legislative background, O7001
lifting equipment, O7070
logbooks, O7040
major accident, hazardous operations with potential to cause, O7033
management and administration, O7029–O7032
delegation of duties, O7029
duty holders, O7030
general requirements, O7029
installation manager's duties, O7031
operators of fixed installation, O7030, O7031
owners of mobile installations, O7030
persons on installation, duties of, O7032
removal of person from fixed installation, O7031
manual handling, M3002
meaning, O7001
mobile installation, O7009
duties on owners, O7030

OFFSHORE OPERATIONS, – *cont.*
mobile installation, – *cont.*
management and administration, O7030
safety case requirements, O7013
model clauses, O7001
muster areas, O7028
noise at work, O7063
notification of hazardous
activities, O7019–O7021
construction activities, O7021
generally, O7019
well operations, O7020
offences, O7056
offshore activities,
provisions, O7001
offshore installation,
abandonment, O7001
applicable law, O7009
application outside Great Britain clause.
See application outside Great
Britain clause *above*
definition, O7005
design and construction. *See* design and
construction *above*
exclusion zones, temporary, O7008
health and safety law applicable
to, O7009
management and administration. *See*
management and administration
above
parts of, O7006
Safety Case Regulations. *See* Safety Case
Regulations
safety zones, O7007
temporary exclusion zones, O7008
oil pollution, O7067
petroleum,
definition, O7001
provisions, O7001
Pipelines Safety Regulations, O7035
Piper Alpha disaster, O7046
protective equipment, emergency, O7028
radiation, O7066
removal of person from fixed
installation, O7031
repeal and revocation of outdated
legislation, O7004
Robens Report, O7003
safety committees, O7052–O7054
co-opted members, O7052
establishment of committee, O7052
functions, O7054
meetings, O7053
qualifications, O7052
training, O7055

OFFSHORE OPERATIONS, – *cont.*
safety representatives,
constituencies, O7047
establishment of constituencies, O7047
functions, O7048
generally, O7046
HSE Guidance, O7046
imminent risk of serious personal
injury, O7051
inspection of equipment, O7049
risk of serious personal injury, O7051
safety case, inspection of, O7050
safety documents, inspection of, O7050
training, O7055
safety zones, O7007
scope of provisions, O7001
Sea Gem incident, O7001
sick bays, O7059–O7061
specific legislation, O7064, O7039
stages, O7001
submarine pipelines,
construction, O7034
provisions, O7001
use, O7034
temporary exclusion zones, O7008
training, O7069
training of safety representatives, O7055
verification of safety-critical
elements, O7017, O7018, O7012
vibration, V5015
warnings, O7028
wells, integrity of, O7026
working time, W10031
written procedures for dealing with
hazards, O7069
OIL FOLLICULITIS, O1038
OIL POLLUTION,
offshore operations, O7067
OIL RIG EXPLOSION,
BP Deepwater Horizon, E16010
OILS AND FATS, FIRE AND, F5003
**OPERATING AND FINANCIAL REVIEW
(OFR),** E16005
**OPERATOR AND POLLUTION RISK
APPRAISAL (OPRA) REGULATION
SCHEME,** E16026
ORANGE BOOK, D0404, D0407
OSTEOARTHRITIS, 01047
OUTDOORS, EQUIPMENT USED, N3016
OUTSOURCING,
facilities management, F3008
OVERCROWDING, W11008
OVERHEAD ELECTRIC LINES,
accident reporting, A3035
electricity, E3028

OVERTURNING,
cranes, L3029
mobile equipment, L3051
platform lifts, L3032
OZONE,
stratospheric ozone depletion, E5002,
E5024, E5025

P

PA SYSTEMS, N3040
PACKAGING, F9047.1
Dangerous goods, transport
of, D0413–D0416
PADDINGTON RAIL CRASH, Intro 14,
E17002
PAGING, V8091
PAIN AND SUFFERING,
damages for, C6031–C6033
PANIC ALARMS, V8090
PARENTAL LEAVE,
right to, E14002
PARKING AREAS,
workplace access, statutory duties, A1009
PART-TIME WORKERS,
treatment of, E14002
PASSERS-BY,
community health and
safety, C6509–C6511
dangerous activities, planning of, C65011
duties to, C6509–C6511
employer's duties to, R3002, R3032
lifting operations, planning of, C65011
signs and notices, C65011
PASSPORTS,
safety, T7026–T7027, T7028
training and competence, T7026–T7027,
T7028
PATIENT,
detained, corporate
manslaughter, E15038.1
PATIENT/BATH HOIST, L3036
PAY,
definition, J3026
representatives of employee safety, J3026
PAYMENT INTO COURT,
damages and, C6042
PECUNIARY LOSSES,
principal types, C6023
PEDESTRIAN TRAFFIC ROUTES,
workplaces, safety regulations, W11015
PEDESTRIANS,
workplace access, statutory duties, A1007

PENALTIES,
corporate manslaughter, C9010, E15043
fines, C9011
health and safety, E15043
PENSION,
employee, for, E14006
state second pension, E14006
PERFORMANCE,
accident investigations, A3030
drivers for, Intro 5
improvements for, Intro 3–Intro 5
large businesses, Intro 5
managing health and safety,
measurement, M2045
reviews, M2045
public sector, Intro 5
small businesses, Intro 3–Intro 4
voluntary sector, Intro 5
PERITENDINITIS, O1047
PERMIT-TO-WORK
SYSTEM, S3011–S3022
application, general, S3020
competent persons, S3016
confined spaces, C7009–C7011
contractors, S3015
dangerous substances, D0933.5
definition, S3011
description of work, S3017
documentation, S3017
electricity, E3023
equipment, S3017
essential features of, S3013–S3019
hazards, S3018
HSE guidance, S3013
information, provision of, S3014
isolation procedures, S3023
maintenance, S3011
operating principles, S3022
outline of, S3021
plant and equipment, S3017
precautions, S3018
procedures, S3019
risk assessments, S3016
selection, S3015
time for, S3012
training, S3015
written, S3007
PERSONAL ALARMS, V8090
PERSONAL HYGIENE,
food handlers, and, F9038, F9046, F9074
PERSONAL INJURIES,
limitation of action, W11045
pre-action protocols, E13029

PERSONAL ORGANISERS, D8219
PERSONAL PROTECTIVE
EQUIPMENT, Intro 29
accommodation for, P3002, P3006
armed forces, for, P3018
asbestos work, P3004, P3008
assessment of, Intro 23
breach of common law
requirements, P3027
British Standards, P3002, P3009, P3010,
P3025 (Table 1)
CE mark of conformity, P3012, P3013
CEN standard, P3009, P3011, P3013
certification procedures, P3009–P3014
children and young persons, V12005
clothing, P3001, P3025 (Table 1)
common law requirements, P3001, P3026,
P3027
community health and safety, C6520
confined spaces, C7022
construction and building work, P3003,
P3004, P3008
criminal offences, P3027
damages, P3027
dangerous substances, for, D0906, P3017
employees' duty to wear, P3007
employers' duties, P3006
ergonomic requirements, E17035, E17066
exclusions, P3018
eye protection, P3002, P3020, P3025,
P3026
eczema, P3026
face protection, P3020, P3025
floor protection, P3022
foot protection, P3022, P3025
general purpose, P3016
general requirements, P3015
hand protection, P3021, P3025, P3026
hazardous substances, H2109, H2110,
H2118, P3008
head protection, P3008, P3019, P3025
health and safety
requirements, P3015–P3018, P3025
hearing protection, N3013, N3022,
P3025
ionising radiation, R1019
kitemark, P3009
lead, P3008
maintenance of, H2131, P3002, P3005,
P3006
mechanical impact risks, P3017
mortal/serious dangers, protection
against, P3014
new or expectant mothers, V12005
noise, for, P3004, P3008, P3017

PERSONAL PROTECTIVE EQUIPMENT, –
cont.
particular risks, additional
requirements, P3017
provision of, Intro 23, P3002, P3006
quality control system, P3015
radiation protection, P3008, P3014,
P3017, P3025
refusal to wear, P3027
respiratory protection, P3008, P3014,
P3023, P3025
risk assessment, Intro 16, P3002, P3006,
R3011, R3018, R3033, R3035
safe system of work, P3026
safety training, P3002, P3006
selection of, P3002
self-defence, for, P3018
self-employed persons, P3006
shipbuilding operations, P3008
statutory requirements, P3004, P3008
testing of, P3010, P3011
uniform European standards, increased
importance of, P3009–P3013
use of, P3002, P3006, P3007
visitors, C6520
vulnerable persons, V12005
whole body protection, P3024
work activities requiring, P3003
work regulations, P3001–P3008
PERSONNEL,
occupational health and safety, O1003,
R2009
PESTS, F9047
PETROLEUM,
dangerous substances, D0704, D0933.6
disaster and emergency planning, D6013
fire, F5031
legislation, D0938
road tankers, unloading, D0933.6
PHOTOKERATITIS AND
PHOTOCONJUNCTIVITIS, O1033
PHYSICAL AGENTS,
children and young persons, V12016
directive, Intro 11, Intro 29, O1026,
O1031
hand arm vibration, O1031
new or expectant mothers, V12020
noise and hearing loss, O1026
skin diseases, O1038
whole body vibration, O1031
PHYSICAL HAZARDS,
confined spaces, A1020
PHYSIQUE AND STATURE, M3007
PIPELINES,
accident reporting, A3005, A3035

PIPELINES, – *cont.*
 dangerous substances, D0927, D0933.1
 explosives, D0918, D0927
 gas safety, G1001, G1009
 iron pipelines, G1001
 offshore operations safety
  regulations, O7035
PIPER ALPHA DISASTER, D6029
PITS,
 heights, working at, W9004
  PLACARDING, D0421, D0422
**PLAN-DO-CHECK MANAGEMENT
 SYSTEMS,** Intro 5, Intro 29
PLANS, EMERGENCY, *see* **Emergency Plans**
PLANT,
 dangerous substances, D0933.2
 design, D0933.2
 fire prevention and control, F5028
 permit to work systems, S3017
 work in compressed air, and, C8018
PLANT DISEASES, E5023
PLATFORM LIFTS,
 disabled persons, W11043
 location, L3032
 overturning, L3032
 powered working platforms, L3032
PLUGS, E3030
PNEUMOCONIOSIS, O1042, O1043
 compensation claims for, C6059–C6062
 disablement benefit, C6059, C6060
 hazardous substances and, H2103
 specified disease, as a, C6062
 workers compensation, C6059, C6061
POINTING DEVICES, D8220
POISONING,
 carbon monoxide, G1017
 hazardous substances and, H2103
POLICE,
 corporate manslaughter, C1001, C9002,
  C9014, E15038.1
 death of detainee, C9001
 disaster and emergency planning, D6061
 violence, S11001, V8140–V8141
POLICE AUTHORITY,
 Enforcing authority, E15009
POLICIES, HEALTH AND SAFETY, *see*
 Health and Safety Policy Statement
**POLICYHOLDERS PROTECTION
 BOARD,**
 claims handled by, E13024
 priority, E5018
POLLUTION CONTROL,
 air, *see* Emissions into the Atmosphere
 Environment Agency, *see* Environment
  Agency

POLLUTION CONTROL, – *cont.*
 environmental management, *see*
  Environmental Management
 industrial, *see* Emissions into the
  Atmosphere
 integrated, *see* Emissions into the
  Atmosphere
POPIMAR MNEMONIC, M2029, M2032,
 M2042, M2045
POSITION SENSORS, M1038
**POST TRAUMATIC STRESS
 DISORDER,** V8139
POSTURE, D8214, D8210
POTASSIUM NITRATE, D0704, 6013
POWER ISOLATION, M1034
POWER PRESSES, M1007
PRE-ACTION PROTOCOLS,
 Civil Procedure Rules, E13029
 employers' liability insurance, E13029
 fast track claims, E13029
 multi-track claims, E13029
 personal injury, E13029
 settlement, E13029
 small claims, E13029
PRECAUTIONS,
 community health and safety,
  contractors, C6527
  customers and users of facilities and
   services, C6517
  neighbours, C6511
  passers-by, C6511
  trespassers, C6514
  visiting individuals and groups, C6520
 confined spaces, C7009, C7013–C7027
 contractors, C6527
 customers and users of facilities and
  services, C6517
 electricity, E3014–E3015
 neighbours, C6511
 noise and hearing loss, N3002
 non-ionising radiation, R1010
 passers by, C6511
 permit to work systems, S3018
 trespassers, C6514
 volunteers, C6520
PREGNANT WORKERS, *see* **New or
 Expectant Mothers**
PREMISES,
 occupiers' liability, O3005
PRESENTEEISM, R2007
PRESSURE GROUPS,
 environmental management,
  and, E16001–E16009
 tactics, E16009

PRESSURE SYSTEMS,
accident reporting, A3035
certificate of adequacy, P7004
code of practice, P7014
competent persons, P7018
conformity assessment procedure, P7008
construction, P7016
design, P7016
documentation, P7017
duty holders, P7011
EC Directive, P7006
examination scheme,
    fitness for service, P7023
    frequency, P7020
    general, P7022
    imminent danger, P7025
    nature, P7021
    report, P7024
    scope, P7019
excepted systems and equipment, P7013
fitness for service, P7023
guidance, P7014
health and safety legislation, and, P7031
imminent danger, P7025
installation, P7016
introduction, P7001
maintenance, P7028
marking, P7017
modification, P7029
notified bodies, P7009
operation, P7026
overpressure prevention
    precautions, P7027
protective device, meaning, P6020
Pressure Equipment Regulations,
    amendment, P7010
    conformity assessment
        procedure, P7008
    generally, P7006
    notified bodies, P7009
    safety requirements, P7007
Pressure Systems Safety Regulations,
    code of practice, P7014
    coverage, P7012
    definitions, P7015
    duty holders, P7011
        excepted systems and equipment, P7013
        guidance, P7014
    record keeping, P7030
    relevant systems and equipment, P7012
    repair, P7029
    safe operation, P7026
    safety requirements, P7007

PRESSURE SYSTEMS, – *cont.*
simple pressure vessels,
    generally, P7002
    Regulations, P7003
Sound Engineering Practice (SEP), P7008
type-examination certificate, P7005
PRESUMPTION OF INNOCENCE, E15031
PRINCIPALS,
liability for discriminatory acts, E16501
PRINCIPLE C,
UK Corporate Governance Code, E16005
**PRINCIPLES FOR RESPONSIBLE
    INVESTMENT (PRI),** E16007
PRISONER,
corporate manslaughter, E15038.1
PROCUREMENT,
environmental considerations, E16008
European Union, E16008
sustainable, E16008
PRODUCT LIABILITY,
see also Product Safety
action, time limit for bringing, P9048
civil liability, P9028–P9030
community health and safety, C6516
contractual liability, P9051–P9060
criminal liability, P9015–P9027
damages,
    contributory negligence, P9038
    persons liable for, P9032
dangerous products, for, P9009, P9059
defect in product,
    definition of, P9035
    key to liability, as P9035
    material time for existence of, P9040
development risk as a defence, P9049
EC Directive, *see* Product Safety
hire contract, P9053, P9060
hire purchase, where, P9053, P9060
intermediaries, role of, P9041
joint and several liability, P9043
judgments, transportability of, P9050
machinery safety, M1002, M1003, M1009
negligence,
    comparison with, P9042
    contributory negligence, P9038
parameters of, P9044
producers,
    damages, liability for, P9032
    defences open to, P9045, P9049
    definition of, P9033
sale of goods legislation, *see* Sale of Goods
    Acts
strict liability,
    introduction of, P9032

PRODUCT LIABILITY, – *cont.*
strict liability, – *cont.*
no contracting out of, P9047
substandard products, contractual liability
for, P9051
suppliers' liability, P9034
supply,
attachment to, P9037
definition of, P9046
time of, P9037
unfair contract terms, P9061–P9063
warnings, absence or inadequacy
of, P9039
PRODUCT SAFETY,
see also, Product Liability, Product Safety
articles for use at work,
definition of, P9016
designers' duties, P9017, P9018, P9020
importers' duties, P9017, P9018
installers' duties, P9021
manufacturers' duties, P9017, P9019,
P9020
suppliers' duties, P9017–P9020
civil liability,
breach of statutory duty, P9030
generally, Intro 24
historical background, P9029
consumer expectation of, P9036
consumer products, P9002–P9007
criminal liability for breach of statutory
duties, P9015
custom built articles, P9024
dangerous products, P9059
defective equipment supplied to
employees, P9029
designers,
duties of, P9017, P9018, P9020
foreign, offences by, P9025
distributors, duties of, P9002
EC Directives,
Commission guide, P9009
enforcement, P9011, P9012
exports and, P9002, P9009
implementation, P9002, P9008, P9013,
P9014
machinery safety, M1009
market surveillance, P9011
safety requirements, conformity
with, P9010
scope of, P9001, P9009
general regulations,
conformity with, presumption of, P9003
defences under, P9004, P9007
distributors' duties, P9002
producers' duties, P9002

PRODUCT SAFETY, – *cont.*
general regulations, – *cont.*
offences and penalties
under, P9004–P9006
scope of, P9002
importers,
duties of, P9017–P9019
liability of, for
foreigner's offences, P9025
industrial products, P9013, P9023
information, availability of, P9012
liability of suppliers, P9034
machinery safety, 1009
manufacturers,
duties of, P9017–P9020, P9022
foreign, offences by, P9025
negligence, comparison with, P9042
producers,
duties of, P9002, P9009
liability of, *see* Product Liability
regulations,
general regulations, *see* general
regulations, *above*
HSWA 1974, under, P9014
substance for use at work,
definition of, P9014
designers' duties, P9017
importers' duties, P9017, P9019
manufacturers' duties, P9017, P9019
suppliers' duties, P9017, P9019
duties of, P9017–P9019
industrial products, of, P9023
sale of goods legislation, *see* Sale of
Goods Acts
unsafe imported products, powers to deal
with, P9027
unsafe products, civil liability
for, P9028–P9030
voluntary national standards, P9003
work equipment, P9014
PRODUCTION LINES, M3004
PRODUCTION METHODS,
risk assessment, R3016 (App B)
PROFESSIONAL ADVICE,
rehabilitation, R2009
**PROFESSIONAL INDEMNITY
INSURANCE,** E13003
PROHIBITED CONDUCT,
Equality Act 2010, E16502
PROHIBITION NOTICE, E15001
appeals against, E15019–E15025
contents of, E15015
contravention of, E15015, E15017,
E15030
Crown, position of, E15045

PROHIBITION NOTICE, – *cont.*
food safety, F9010
emergency, F9011
improvement notice compared, E15016
prosecution, coupled with, E15018
public register, E15029, E15044
service, E15001
temporary cessation of an activity, E15015
uses of, E15013, E15015
PROHIBITORY SIGNS, W11037
see also Safety Signs
PROSECUTION,
community health and
safety, C6503–C6504
corporate killing, E13001
enforcement of legislation by, *see*
Enforcement
small firms, E13001
work-related violence, V8015
PROTECTED ACTS,
victimisation, E16508
PROTECTED CHARACTERISTICS,
Equality Act 2010, E16501, H1703
rehabilitation, R2005
PROTECTIVE DEVICES
meaning, P7020
PROTECTIVE VESTS,
violence in the workplace, V8002
PSYCHOLOGICAL INJURY,
damages for, C6005, S11009
stress, S11009
PUBLIC ADDRESS SYSTEMS, V8091
PUBLIC EVENTS, MAJOR
contractors, C6527
Event Safety Guide, C6525
health and safety, C6524–C6526
risk assessment, C6526
Safety Advisory Groups (SAGS), C0024
PUBLIC LIABILITY INSURANCE, O3003
PUBLIC PROCUREMENT,
environmental considerations, E16008
European Union, E16008
sustainable, E16008
PUBLIC SECTOR,
large companies, Intro 5
performance, Intro 5
PUNCTURE, M1025

Q

**QUALIFICATION AND CREDIT
FRAMEWORK,**
standards, T7008

QUALIFICATIONS, T7005, T7008, T7010,
T7012–T7025
occupational health nurses, R2009
occupational physicians, R2009
rehabilitation, R2009
stress, S11034
**QUALIFICATIONS AND CURRICULUM
AUTHORITY,** T7013, T7019
**QUALITY ASSURANCE AGENCY FOR
HIGHER EDUCATION,** T7010,
T7013, T7022
QUARRIES,
accident reporting, A3005, A3035
children and young persons, V12015
electrical safety, E3003
risk assessment, C6510
**QUEUING AND QUEUE
JUMPING,** V8075, V8079

R

RACE
Equality Act 2010, E16517-E16519
harassment H1702, H1733
migrant workers, E16518
protected characteristic, E16517, H1703
risk assessment, E16519
RACIST INCIDENTS, V8011, V8140
RADIATION,
accident reporting, A3035
disaster and emergency
planning, D6018–D6019
emergencies, *see* Radiation Emergencies
European Union, D6018
exposure to, *see* Radiation Exposure
guidance, D6018
ionising radiation, *see* Ionising radiation
non-ionising, *see* Non-ionising radiation
occupational diseases, O1056
offshore operations, O7066
radioactive substances, *see* Radioactive
Substances
risk assessment, R3016, R3027, R3035
training and competence, T7007
RADIATION EXPOSURE,
new or expectant mothers, V12020
protection from, P3008, P3015, P3017,
P3025
RADIO,
workplace violence, use in case of, V8091
RADIOACTIVE SUBSTANCES,
carriage of, D0446–D0447
exposure to, *see* Radiation Exposure

RADIOACTIVE SUBSTANCES, – *cont.*
ionising radiation's, R1012–R1013,
R1015, R1019
assessment of risk, R3016, R3027,
R3035
recommendations, D04012
regulations on, D04013
RADON, R1013
RAILWAYS,
accident reporting, A3035
accidents, Intro 14
cross-border services, W10030
dangerous goods, carriage of, D0404,
D0427
Ladbroke Grove rail disaster, Intro 14,
D6029, E17002
staff working hours, W10026, W10030,
W10034
trespassers, C65013
RAMPS, W11010
RAMSGATE PROSECUTION, Intro 15,
Intro 17
RATE OF WORK IMPOSED BY A
PROCESS, M3004
REGISTRATION, EVALUATION AND
AUTHORISATION OF CHEMICALS
IN EUROPE (REACH),
adequate control, R2513
candidate substances, R2513
chemicals covered by, R2504
customs supervision, substances
under, R2504
dangerous substances, H2137
distributors, R2505
downstream users, R2505, R2510, R2516
duty holders, R2505
enforcement of provisions, R2516
evaluation of registration packages, R2512
exempt chemicals, R2504
free movement within EU, R2502
food and foodstuff additives, R2504
helpdesk and further information, R2515,
R2518
import limits, R2504
information, failure to supply, R2516
introduction, R2501, R2517
medicines, human and veterinary, R2504
preparation for, R2506
prioritised substances, annual list, R2512
purpose, R2502, R2503, R2517
radioactive substances, R2504
registration,
enforcement, R2516
information required, R2509
process, R2507

REGISTRATION, EVALUATION AND
AUTHORISATION OF CHEMICALS
IN EUROPE (REACH), – *cont.*
registration, – *cont.*
timescale for, R2508
registrants, R2505
restrictions, power to impose, R2514
Safety Data Sheets, R2510, R2517
transport, substances undergoing, R2504
UK competent authority, R2515
very high concern, substances of, R2513
waste, R2504
RECKLESS KILLING, D6042
RECORDS,
accidents, as to, *see* Accident Reporting
asbestos, A5011
dangerous substances, D0911
display screen equipment, D8216
first-aid, F7020
lifting, L3027
machinery safety, M1014
managing health and safety, M2008
manual handling, M3019, M3024
occupational health and
safety, O1016–O1018
pressure systems, and, P7030
risk assessments, D8216
Safety Case Regulations, O7023
sickness absence, M1528
violence in the workplace, V8071
whole-body vibration, V5025
working time, W10033, W10034
REDUNDANCY,
selection for, E14010
REGISTER OF SAFETY
PRACTITIONERS, T7028
REGULATORY IMPACT ASSESSMENTS,
dangerous substances, D0932
explosions, D0932
heights, working at, W9009, W9015
REHABILITATION, R2001–2009
adjustments,
disability discrimination, R2005
examples of, R2005
health and safety, R2006
justification, R2006
making, R2004
reasonable adjustments, R2005–R2006
sickness absence, R2008
advice, using professional, R2009
apportionment of damages, R2004
attendance management, R2007
breach of duty, R2004
business argument, R2002

REHABILITATION, – cont.
causation, R2004
common law, R2004
damages, R2004, R2006
data protection, R2007
definition, R2001
disability discrimination, R2005
definition, R2005
long-term disability, R2005
normal day-to-day
activities, R2005–R2006
reasonable adjustments, R2005
duty of care, R2004
exacerbation of illness, R2004
foreseeability, R2004
guidance, R2003
Health and Safety Executive (HSE), R2003
importance of, R2002–R2006
legal argument, R2003
long-term absence, R2002, R2008
occupational health, R2003–R2004,
R2009
occupational health nurses, R2009
occupational physicians, R2009
presenteeism, R2007
conditions resulting from, R2007
professional advice, using, R2009
quantification of damages, R2004
reading, R2010
reasonable adjustments, R2005–R2006
return to work,
adjustments, R2008
alteration of hours, R2008
arrangements for, R2004
changing jobs, R2008
interviews, R2007
phased return, R2008
plans, R2008
preventing, R2006
risk assessments, R2004
Robens Centre for Occupational Health
and Safety, R2009
sickness absences, R2002
attendance management, R2007
data protection, R2007
long-term, R2002, R2007, R2008
lost time rate, R2007
managing, M1515, R2007
policy, contents of, R2007
recording, R2007
short-term, R2002, R2007
size of undertakings, R2004
stress, R2004, S11009
vocational, R2001

REHABILITATION, – cont.
working hours, alteration of, R2008
RELIGION,
harassment, H1702
protected characteristic, H1703
REMOTE-CONTROLLED MOBILE WORK
EQUIPMENT, M1008
REPAIR
dangerous substances, D0933.5
pressure systems, and, P7029
REPETITIVE STRAIN INJURY (RSI),
damages for, C6058
visual display units, see Visual Display
Units
REPORTABLE DANGEROUS
OCCURRENCES, see Accident
Reporting
REPORTABLE OCCUPATIONAL
DISEASES,
see also Accident Reporting
accident reporting, A3015, A3029
carpal tunnel syndrome (CTS), V5011
vibration white finger (VWF), V5011,
V5012
REPUTATION, LOSS OF, B8017
RESCUE,
heights, working at, W9006
RESEARCH,
heights, working at, W9001
product safety, P9020, P9022, P9023,
P9039
violence in the workplace, V8112
RESIDENTIAL HOMES, FIRE
PREVENTION AND, F5035
RESIDENTIAL PROPERTY, ELECTRICITY
AND, E3033
RESPIRATORY DISEASES, O1042–O1046
allergic responses, O1043
allergic rhinitis, O1042
animal workers, O1043
asbestos, O1042, O1046
asthma, O1042, O1044
carcinogens, O1043
causes, O1043
control of substances hazardous to
health, O1046
detection of, O1044
engineering solutions, O1045
extrinsic allergic alveolitis, O1042
health surveillance, O1044, O1045
industries prone to causing, O1042
infectious diseases, O1043, O1045
legal implications, O1046
Legionnaire's disease, O1043
lung cancer, O1042

RESPIRATORY DISEASES, – *cont.*
management, O1045
meaning, O1042
pneumoconiosis, O1042, O1043
respiratory protection, O1045
risk assessment, O1045, O1046
sensitisation, O1043
silicosis, O1043
spirometers, O1044
symptoms, O1044, O1045
tuberculosis, O1042, O1043
x-rays, O1044
zoonoses, O1043
RESPIRATORY PROTECTION,
confined spaces, C7022
dust respirators, V3011
generally, V3011
personal protective
equipment/clothing, P3023, P3024
RESPONSE TEAMS, V8119
REST FACILITIES,
construction and building operations,
and, C8017
general provision, W11027
REST PERIODS,
breaks, W10017
daily, W10014, W10015, W10034
manual handling, M3004
monotonous work, W10018
stress, S11008, S11014, S11019–S11020
weekly, W10016, W10034
RESTRAINING ORDERS, V8140
RESTRAINTS, M1008, V8094W9014
RETURN TO WORK,
adjustments, R2008
alteration of hours, R2008
arrangements for, R2004
changing jobs, R2008
interviews, R2007
phased return, R2008
plans, R2008
preventing, R2006
REVERSING, A1017
REVITALISING HEALTH AND SAFETY.
HEALTH AND SAFETY
COMMISSION, Intro 7
RHINITIS, O1043
RIDDOR, *see* **Accident Reporting**
RIGHT TO ROAM, O3007
RISK,
See also Risk Assessment
analysis, M1024
definition of, M1006, M1023
evaluation, M1024
lone workers, V12029, V12030

RISK, – *cont.*
machinery safety, M1023–M1024, M1026
reduction, options for, M1026
RISK ASSESSMENT Intro 29
absolute requirements, R3004
accident investigations, A3030
age, E16515
asbestos, Intro 16, A5003, A5010, A5017,
A5025, R3013, R3016, R3035
assessment unit, R3030
breastfeeding mothers, E14024
business continuity
planning, B8012–B8018
CDM health and safety plans, R3017
check list, M3020, M3024, R3035
children and young persons, Intro 29,
M2013, R3025, R3032, V12002,
V12006, V12008, V12013
cleaners, R3002, R3032
community health and safety, C6504,
C6509–C6511, C6526, C6527
confined spaces, C7004–C7005
construction operations, F5024
contractors, R3002, R3032
control of, R3024
cranes, L3018
customers, R3002
danger areas, R3023
dangerous substances, D0902,
D0910–D0911, D0933.1, D0933.6,
R3007, R3014, R3035
disability discrimination, reasonable
adjustments and, E16511, W11043
disabled persons, E16512, R3028,
V12033, V12035–V12036
disaster and emergency planning, D6007,
D6010, D6066
discussions, R3038
display screen equipment, Intro 16
diving at work, O7038
dynamic risk assessment, R3017
emergency services, R3002
employees, R3002, R3032
enforcement, E15001
equipment, M1006, M1007
ergonomic tool, as, E17052, E17062,
E17064, E17068
expectant mothers, E14024, R3026,
R3027
explosives, D0902, D0919, R3014, R3016
(App B)
eye diseases and disorders, O1035, O1036
fire precautions, Intro 16, F5001, F5007,
F5031–F5032, F5037, R3014,
R3018, R3033, R3035
gender, E16521

RISK ASSESSMENT – *cont.*

genetically modified organisms, R3016
guards, M1032
guidance, D0933.1
hand-arm vibration
    syndrome, O1030–O1031, V5015,
    V5022
hazardous substances, R3007
health and safety at work
    requirements, R3002–R3004
health and safety policy statement, S7006
health surveillance, R3023
heights, working at, R3015, W9004,
    W9007, W9008
HSE publications and website, R3031,
    R3053
identification of issues, R3033
inexperienced workers, R3028
information required, M3019, R3031
ionising radiation, H2101, H2145, R1019,
    R3016, R3027, R3035, R3016
lead, R3016
lifting equipment, L3017
lone workers, R3028, V12027
machinery safety, M1001, M1003,
    M1014, M1021–M1022, M1024
major incident potential, R3016
major public events, C6526
making the assessment, R3036–R3041
management regulations,
    ACOP to, R3019, R3020, R3042
    application of management
        cycle, R3024
    children and young persons, R3025
    evaluation of precautions, R3020
    hazards and risks, R3019
    new or expectant mothers, R3026,
        R3027
    scope of, R3005, R3018, R3023
    'suitable and sufficient' R3021
managing health and safety, M2002,
    M2005–M2009, M2013, M2019
manual handling operations, Intro 16,
    Intro 23, M3002, R3009
    after, M3024–M3027
    arranging, M3018
    checklist, M3020, M3024
    guidelines, M3003
    information gathering, M3019
    making, M3020–M3023
    planning and
        preparation, M3016–M3019
    recommendations, review of, M3025
    reviews, M3026
    team, M3017

RISK ASSESSMENT – *cont.*

method statements, R3017
migrant workers, E16519
monitoring, R3024
musculoskeletal disorders, O1050, O1051
new mothers, E14024, M2006, M2019,
    R3026-R3027, V12017–V12018,
    V12021, V12025–V12026
newspaper publisher, R3051
noise, Intro 16, N3013, R3008, R3018,
    R3032, R3035
    competent persons, N3002, N3029
race, E16519
requirements as to, R3003
sufficient information, instruction and
    training, assessors having, N3002,
    N3029
non-ionising radiation, R1010
notes, R3041
observation, R3037
occupational health and safety, O1005,
    O1008
Occupational Safety Consultants Register
    (OSCHR), R3029
organisation, R3024
permit to work, R3017
personal protective equipment, Intro 16,
    R3011, R3018, R3033, R3035
personnel, as to, R3002, R3032
planning, M3016–M3019, R3024, R3036
practical application of requirements, Intro
    16
pregnant workers, E14025
preventative measures, R3023
public at large, R3002
reasonable practicality, R3003
recommendations, M3025, R3045, R3046
regulations,
    common regulations, R3006–R3012
    specialist regulations, R3016
rehabilitation, R2004
respiratory diseases, O1045, O1046
review, M3026, R3022, R3024, R3045,
    R3047
safe systems of work, Intro 29, R3017,
    S3001
sales representatives, R3032
self-employment, R3002
serious and imminent danger,
    where, R3023
skin diseases, O1039–O1041
source of risk, R3035
staff absences, R3034
stress, O1054, S11007, S11009
substances hazardous to health, R3007

RISK ASSESSMENT – *cont.*
supermarket, R3050
team, M3017
temporary workers, R3032
test, R3039
training and competence, R3025, R3031,
    R3040, T7003, T7005
variations of work practices, R3034
vibration, V5015, V5027
violence in the workplace, V8001, V8006,
    V8021, V8044–V8055, V8057,
    V8077, V8094
visual display units, Intro 16, D8201,
    D8204, D8210–D8216, R3010,
    R3018, R3027, R3036
    after the, D8221
    approach to, D8212
    checklist, D8211–D8212, D8216
    discussions with users and
        operators, D8215
    homeworkers, D8213
    observations at the, D8214
    persons carrying out the, D8210
    re-assessments, D8222
    records, D8216
    reviews, D8222
    teleworkers, D8213
    users, identification of, D8211
visitors, R3002, R3032
volunteer workers, R3032
vulnerable persons, R3028, V12001
who might be at risk, R3032
who should carry out, R3032
whole-body vibration, O1030–O1031,
    V5015, V5024–V5025
work activities examined under, R3032
workplace violence, *see* Workplace
    Violence
young persons, R3025, R3032
RISK MANAGEMENT,
asbestos, A5012, A5018–A5019, A5022
business continuity planning, B8007
corporate governance, Intro 5
dangerous substances, D0902,
    D0913–D0917
fire precautions, F5001, F5006
generally, R3052
HSE publications and website, R3031,
    R3053
large companies, Intro 5
reports, Intro 5
risk aversion, R3052
violence in the workplace, V8001, V8129,
    V8141

ROAD ACCIDENTS,
reporting, A3010, A3035
ROAD TRANSPORT,
annual leave, W10029
dangerous substances, transportation, *see*
    Transportation of Dangerous
    Substances
directive on, W10029
disaster and emergency
    planning, D6025–D6027
night workers, W10029
staff working hours, W10026, W10027,
    W10028–W10029
ROAM, RIGHT TO, O3007
**ROBENS CENTRE FOR OCCUPATIONAL
    HEALTH AND SAFETY,** R2009
ROBENS COMMITTEE,
enforcement powers of inspectors, E15013
offshore installations, O7003
report, O7003
ROLLING OVER, M1008
ROOFS,
heights, working at, W9008
ROPES, L3039–L3042
breakage, L3039
failure, L3040, L3042
fibre rope failure, L3040
heights, working at, W9014
wire rope, L304–L3042
**ROYAL SOCIETY FOR THE
    PREVENTION OF ACCIDENTS,** Intro
    28, T7009
**ROYAL SOCIETY FOR THE
    PROMOTION OF HEALTH,** T7013,
    T7019
RUBBISH, FIRE AND, F5018

S

SAFE OPERATING PROCEDURES, S3010
SAFE PLACE OF WORK
construction and building operations,
    and, C8004
SAFE SYSTEMS OF WORK, S3001–S3024
common law, S3003, S3005–S3006
communication, S3010
components of, S3002–S3006
confined spaces, C7008
development of, S3008–S3010
electricity, E3022–E2024
employers' duty, S3003
hardware, S3006
heights, working at, W9001

SAFE SYSTEMS OF WORK, – *cont.*
independent contractors, 03019
isolation procedures, S3023
job safety analysis, S3009
job safety instructions, preparation
of, S3010
legal components, S3003–S3006
lone workers, V12027
matrix, S3007
occupational health and safety
practitioners, S3001
occupiers' liability, 03019
permit to work systems, *see* Permit to
work systems
pitfalls of not utilising, S3001
policy statements, S3004
publications, S3024
review, S3007
risk assessment, S3001
safe operating procedures, preparation
of, S3010
software, S3006
statute law, S3004–S3006
written arrangements, S3004, S3007
SAFEGUARDS AND SAFETY DEVICES,
characteristics of, M1028
types of, M1027–M1035
SAFETY ADVISERS, R3008
SAFETY ADVISORY GROUPS
(SAGS), C6524
SAFETY CASE REGULATIONS,
abandonment programme, O7015
audit, O7011
background, O7010
before commencement of operations,
combined operations, O7013
fixed installations, O7013
mobile installations, O7013
co-operation, O7016
codes of practice, O7016
compliance with health and safety
requirements, O7011
copies, O7023
Cullen Report, O7010
decommissioning fixed
installations, O7015, O7014
documents required to be kept on
installation, O7023
drafting of Regulations, O7010
duty holder, O7011
fixed installations,
abandonment programme, O7015
before commencement of
operations, O7013
decommissioning, O7015, O7014

SAFETY CASE REGULATIONS, – *cont.*
generally, O7010
living document, as, O7011
major accident, O7011
methodologies, O7016
organisation and arrangements for detailed
design, O7016
previous legislation, O7010
prior notice of hazardous
operations, O7011
quality control, O7016
record keeping, O7023
recording of verification schemes, O7018
regulations to underpin, O7011
reports, O7023
review of verification schemes, O7018
revision of verification schemes, O7018
safety representative inspection, O7050
safety-critical elements, verification
of, O7012, O7018, O7017
specifications, O7016
co-operation, O7016
codes of practice, O7016
methodologies, O7016
organisation and arrangements for
detailed design, O7016
quality control, O7016
standards, O7016
standards, O7016
summary of contents, O7022
verification of safety-critical elements of
installations, O7012, O7018, O7017
SAFETY COLOURS, W11032
see also Safety Signs
SAFETY COMMITTEES,
'Brown Book', J3019
consultation with, Intro 23, J3018
duty to establish, J3017
establishment, Intro 23, J3017, J3018
function, Intro 23, J3019
offshore operations, O7052–O7054
co-opted members, O7052
establishment of committee, O7052
functions, O7054
meetings, O7053
qualifications, O7052
training, O7055
risk assessment, involvement in, R3007
role, J3019
SAFETY FOR SENIOR
EXECUTIVES, T7022
SAFETY HELMETS,
personal protective equipment, P3020,
P3025

SAFETY OF FOOD, *see* Food
SAFETY PASS ALLIANCE SCHEME, T7027
SAFETY PASSPORTS, T7026–T7027,
T7028
SAFETY POLICY, *see* Health and Safety
Policy Statement
SAFETY REPORTS, D6021–D6024
SAFETY REPRESENTATIVES,
see also Safety Committees
accident reporting, A3027, A3028
actors, J3005
appointment,
generally, Intro 23
unionised workforce, J3005
armed forces, J3002
'Brown Book', J3019
Charter for, J3002
disclosure of information, J3005, J3015,
J3030
duties of, J3007
education sector, in, J3002
employment protection legislation, overlap
with, J3002
European developments, J3001, J3002,
J3033
European Works Councils, J3002, J3033
extension of duty to consult, J3002
facilities to be provided by
employer, J3014
general duty,
non-unionised workforce, J3021
unionised workforce, J3004
hazardous substances, discussions as
to, H2119
independent, J3005
information disclosure,
non-unionised workforce, J3027
unionised workforce, J3015
musicians, J3005
non-unionised workforce,
general duty, J3021
generally, J3020
information provision, J3002, J3027
number of representatives for
workforce, J3022
recourse for representative, J3030
rights and duties of
representatives, J3024–J3029
role of safety representatives, J3023
time off with pay, J3024–J3026
training, relevant, J3028
number of representatives for workforce,
non-unionised workforce, J3022
unionised workforce, J3006
obligation to consult, J3001, J3002

SAFETY REPRESENTATIVES, – *cont.*
offshore operations,
functions of representative, O7048
generally, O7046
HSE Guidance, O7046
imminent risk of serious personal
injury, O7051
inspection of equipment, O7049
risk of serious personal injury, O7051
Safety Case, inspection of, O7050
safety documents, inspection of, O7050
training, O7055
prosecutions for non-compliance, J3002
protected disclosures, J3030
regulatory framework, J3002
rights and duties of safety representatives,
generally, Intro 23
non-unionised workforce, J3024–J3030
unionised workforce, J3010–J3013
risk assessment, involvement in, R3032
role,
non-unionised workforce, J3023
unionised workforce, J3007
statutory protection for, J3030
technical information, J3016
time off with pay, J3010–J3013, J3029
non-unionised workforce, J3024–J3026
unionised workforce, J3010–J3013
training and competence, E14001, J3028,
T7003, T7023
unionised workforce,
appointment of safety
representatives, J3005
disclosure of information, J3015
duty to consult, J3003, J3004
facilities to be provided by
employer, J3014
functions of safety
representatives, J3007
general duty, J3004
generally, J3003
immunity, legal, J3009
'in good time', consultation, J3007
information disclosure, J3015
number of representatives for
workforce, J3006
rights and duties of safety
representatives, J3001,
J3010–J3013
role of safety representatives, J3007
technical information, J3016
terms of reference, J3007
time off with pay, J3010–J3013
workplace inspections, J3008
workplace inspections, J3008

SAFETY RULES,
enforcement by employer, *see* Disciplinary
Procedures
SAFETY SIGNS,
see also Labelling of Hazardous Substances
colour, Intro 23
community health and safety, C6511,
C6514
design, Intro 23
employers' duties, W11035
emergency escape, W11040
exclusions, W11036
fire, F5032, W11041, W11042
hand signals, W11032, W11042
mandatory signs, W11039
neighbours and passers-by, C65011
prohibitory signs, W11037
provision of, Intro 23
warning signs W11038
SAFETY TRAINING AND COMPETENCE.
See **Training and competence**
SALE OF GOODS,
consolidation of legislation, P9053
dangerous products, P9059
dealer, sales by, P9056
fitness for purpose, P9052–P9054, P9057
food, *see* Food
history of, P9052
implied terms, P9052–P5057
quality of product, P9052–P9054
sample, sale by, P9055
strict liability under s14, P9055, P9058
SALES REPRESENTATIVE,
risk assessment, R3032
SANITARY CONVENIENCES,
Building Regulations, W11044
construction and building operations,
and, C8017
disabled persons, for, W11043
factories, W11018–W11021
food premises, for, F9039
minimum number of facilities, W11021
occupier's duty to provide, W11019
temporary work sites, W11023
SCAFFOLDING,
see also Working at Heights
accident reporting, A3029
guidance, W9008
heights, working at, W9008, W9009
SCHOOLS,
community health and safety, C6518
occupiers' liability, O3018
visiting schoolchildren, C6518
SCOTLAND,
corporate homicide, E15038.1

SCOTLAND, – *cont.*
disaster and emergency planning, D6011
enforcement agencies, F9005
penalties for offences, F9024
temperature control, F9050,
F9063–F9069
gas safety, G1001
local air quality management, E5069
occupiers' liability, O3001
Scottish Environment Protection
Agency, D6011, E5044
Scottish Executive, E5042
Scottish Vocational Qualifications, T7008,
T7009
smoking, prohibition of, W11028
SCREENS, COMPUTER, *see* **Visual Display
Units**
SEA FISHING,
working time, W10026, W10031
SEA TRANSPORT,
manual handling, M3002
radiation emergencies, *see* Radiation
emergencies
staff working hours, W10026
vibration, V5015
SEATING, WW11001, 11008
see also Workstation
SECOND-HAND EQUIPMENT, M1007
SECTOR SKILLS COUNCILS (SSCs),
standard setting, T7008
SECURING HEALTH TOGETHER.
**HEALTH AND SAFETY
COMMISSION,** Intro 7
SECURITY,
community health and safety, C6511
risk assessment, R3016 (App B)
violence in the workplace, V8082–V8095
working time, W10034
workplace, C6511
SECURITY PERSONNEL, F3030,
V8083–V8086
SELF-ADJUSTING GUARDS, M1030
SELF-DEFENCE,
personal protective equipment for, P3018
use of force, V8015
violence in the workplace, V8015
SELF-EMPLOYMENT,
asbestos, A5007
confined spaces, C7003
first-aid, F7015
hazardous substances, duties as to, H2105
managing health and safety, M2015
personal protective equipment, P3006
risk assessment, R3002
vulnerable persons, V12001

SELF-PROPELLED MOBILE
    EQUIPMENT, M1008
SENTENCING,
    health and safety offences, E15043
    mitigating factors, C9011
    seriousness of offence, C9011
Sentencing Guidelines Council
    (SGC), E15001, E15043
    corporate manslaughter, C9001, C9010-
        C9011
SERVICE INDUSTRIES, Intro 6
SERVICES, CUSTOMERS AND USERS
    OF, C6515–C6517
SETTLEMENT,
    employers' liability insurance, E13029
    pre-action protocols, E13029
SEVESO DISASTER, Intro 29, D6011
SEX DISCRIMINATION,
    health and safety legislation, E14023
SEXUAL HARASSMENT,
    see also Harassment
    concept, H1702
    definition, H1702
    detriment, H1702
    nature of, H1701
    protected characteristic, H1703
    third party, by,employer's liability, H1710
SEXUAL ORIENTATION,
    harassment H1702
SHACKLES, L3048–L3049
SHARED WORKSTATIONS, D8218
SHAREHOLDERS,
    environmental management and, E16001,
        E16007
SHEARING, M1025
SHIFT WORKERS,
    ergonomics and, E17036, E17041
    stress, S11019
    working time, W10035
SHIPBUILDING,
    first aid areas, F7025
    personal protective equipment, P3008
SHIPS, D0907
    building and repair, V12015
    children and young persons, V12015
    enforcing authority, E15009
    noise and hearing loss, N3002
    transportation of dangerous goods, D0404
        UN Numbers and Proper Shipping
            Names, D0408
SHOCK, V12020
SHOPS,
    access for disabled persons, W11043
    doors and gates, W11016
    drinking water, supply of, W11024

SHOPS, – cont.
    eating facilities, W11027
    escalators, W11017
    first-aid, F7017
    floors, safety regulations, W11010
    pedestrian traffic routes, W11015
    rest facilities, W11027
    sanitary conveniences, W11018–W11021
    supermarket, R3050
    temperature control, W11005
    washing facilities, W11020
SHOREHAM DOCKS, Intro 14
SICK BUILDING SYNDROME, S11015
SICK PAY, E14006, M1513–M1514
SICKNESS ABSENCES, R2002
    see also Managing absence
    adjustments, R2008
    attendance management, R2007
    data protection, R2007
    fit notes, R2007
    GP medical statements, R2007
    holidays and, W10024
    HSE guidance, R2010
    insurance benefits, M1511
    long-term, O1002, O1007, R2002
    lost time rate, R2007
    managing, see Managing absence
    manual handling, Intro 29
    musculoskeletal disorders, 01049
    occupational health and safety, O1002,
        O1007
    occupational medicine, R2009
    pandemics, R2007
    policy, contents of, R2007
    recording, R2007
    short-term, O1020, R2002
    stress, O1052, S11017
SIGNALLERS,
    cranes and, L3030
SIGNALS, see Safety Signs
SIGNS, see Safety Signs
SIKH,
    head protection, P3008
SILICOSIS, O1043
SITE INDUCTION'S
    construction and building operations,
        and, C8031
SIX SIGMA,
    facilities management, F3010
SKI ACTIVITY,
    Enforcing authority, E15009
SKIN CONDITIONS AND
    DISEASES, H2103, O1037–O1041
    causes, O1038

SKIN CONDITIONS AND DISEASES, – *cont.*
chemical substances, O1038, O1041
chloracne, O1038
dermatitis, O1037–O1038, O1040–O1041
   allergic contact, O1037–O1038, O1040
   irritant contact, O1037–O1038
detection, O1039
engineering, O1040
experts, O1039
health surveillance, O1041
hygiene, O1040
legal requirements, O1041
leucoderma, O1038
management, O1040
oil folliculitis, O1038
personal protection, O1041
physical agents, O1038
risk assessment, O1039, O1040–O1041
skin cancers, O1038
skin ulcers, O1038
urticaria, O1038
SLAUGHTERHOUSE,
controls as to, F9028
SLINGERS,
cranes and, L3030
SLINGS, L3045, M3013
failure, L3045
safe working loads, L3045
SMALL AND MEDIUM-SIZED ENTERPRISES, (SMEs),
see also Small businesses
environmental legislation, E16005
SMALL BUSINESSES,
beliefs and attitudes of owners/managers in, Intro 4
control of substances hazardous to health, Intro 4
corporate killing, E13001
costs of accidents and ill-health, Intro 26
customers, focus on, Intro 4
Eco-Management Audit Scheme (EMAS), E16014
employees, focus on, Intro 4
environmental legislation, E16005
influences over, Intro 3
labels, Intro 4
managing health and safety, M2010
operating conditions, Intro 4
performance, Intro 3–Intro 4
prosecutions, E13001
rise in, Intro 6

SMALL BUSINESSES, – *cont.*
Small Business Performance Indicator, E13030
sole-employee incorporated companies (SEICs), E13012
suppliers, Intro 4
supply chain management, Intro 4
SMALL CLAIMS, E13029
SMEs (SMALL AND MEDIUM-SIZED ENTERPRISES,
see also Small businesses
environmental legislation, E16005
SMOG, E5003, E5019
SMOKE,
statutory nuisance, E5062
SMOKING,
confined spaces, C7026
fire, F5018
public places, prohibition in, C8017, F5018, H2109, W11028
smoking shelters or areas, provision C8017
stress, S11014
workplaces, prohibition in, C8017, F5018, W11028
SOCIAL CHARTER DECLARATION ON HEALTH AND SAFETY, Intro 9
SOCIAL RESPONSIBILITY, *see* Corporate Social Responsibility
SOCIAL SECURITY BENEFITS,
compensation for work injuries, C6002, C6044–C6062
incapacity for work, C6002
SOCIAL WORK SECTOR,
manual handling, M3001, M3002
SOCIALLY RESPONSIBLE INVESTMENT (SRI), E16007
SOCIETY OF LIGHT AND LIGHTING, L5002
SOCKETS, E3014
SOFTWARE, D8206, F3030
SOLAR LIGHTING, L5008
government grants, L5008
SOLE-EMPLOYEE INCORPORATED COMPANIES (SEICs), E13012
SOLICITOR,
personal injuries cases, C6021
SOLID ORGANIC MATERIALS, FIRE AND, F5003
SOLVENTS,
control of, E5042
SOUND,
Building Regulations, W11044

SOUND ENGINEERING PRACTICE
(SEP), P7008
SOUND SYSTEMS, N3027, N3040–N3041
SOUTHALL RAIL CRASH, Intro 14, Intro
15
SPECIFIED DISEASES,
claims for, C6045–C6050
SPEED,
mobile work equipment, M1008
SPILLAGES, D0918
SPIROMETERS, O1044
SPORT, FIRE SAFETY AND, F5032
SRI (SOCIALLY RESPONSIBLE
INVESTMENT), E16007
STABBING, M1025
STABILITY, M1007
Structural, building and construction
operations C8006
STAFF FACILITIES, S11014
STAIRS,
workplace, safety regulations on, W11010
STAKEHOLDER ENGAGEMENT
STANDARD, E16009
STANDARD FORM CONTRACT,
unfair contract terms, P9056–P9058
STANDARDS,
American National Institute, M1019
British standards, see British Standards
business continuity planning, B8006
community health and safety, C6527
corporate social responsibility. E16002
dangerous substances, carriage of, D0421,
D0422
electricity, E3003, E3024, E3027, E3033
Employment National Training
Organisation, T7008, T7009, T7024
enforcement, T7008
environmental management, E16015
equipment, M1007
European, D0919
machinery safety, M1001, M1003,
M1010, M1013–M1015
explosions, D0919
fire extinguishers, F5016
fire prevention and control, F5030
guards, M1029, M1032
Institution of Occupational Safety and
Health (IOSH), T7011–T7012
lead bodies, T7008
lifts, L3005–L3006
machinery safety,
American National Institute, M1019
B type, M1021

STANDARDS, – cont.
machinery safety, – cont.
British, M1019, M1021, M1024
C type, M1021
objectives, M1020
risk assessment, M1021
status and scope of, M1020
structure of, M1021
Transposed Harmonised European
Machinery M1001, M1003,
M1010, M1013–M1015
managing health and safety, M2029,
M2032, M2034
National Occupational Standards
(NOS), T7008
National Vocational Standards, T7002,
T7008, T7009, T7012–T7016,
T7022–T7025
noise, N3003, N3036
Occupational Health and Safety
Practice, T7009
packaging, D0421
practice, T7008, T7009
Qualification and Credit
Framework, T7008
revision of, T7008, T7012
setting national, T7008–T7009
stakeholder environment standard, E16009
stress, Intro 29
title of, T7012, T7015
training and competence, T7002,
T7008–T7016, T7022–T7025
Transposed Harmonised European
Machinery M1001, M1003, M1010,
M1013–M1015
violence in the workplace, V8044
STATEMENT OF FITNESS TO
WORK, M1504
options for return to work, M1504
rehabilitation, M1515
STATEMENT OF TERMS AND
CONDITIONS OF
EMPLOYMENT, E14006
STATEMENTS OF HEALTH AND SAFETY
POLICY, see Health and Safety
Policy Statements
STATUTORY NUISANCE, E5062
STATUTORY SICK PAY, M1514
STEAM,
pressure systems, see Pressure systems
STORAGE,
community health and safety, C6511,
C6514
dangerous substances, D0933.3

STORAGE, – cont.
explosions, D0918
STRATOSPHERIC OZONE
DEPLETION, E5002, E5024, E5025
STRESS, ANXIETY AND
DEPRESSION, Intro 13, O1052–O1055
causes, O1053
definition, O1052
disability discrimination, O1052
ergonomics, role of, E17002
guidance, O1054
legal requirements, O1055
lost working days, O1052
management, Intro 29, O1054
managing absence, M1501
modifying factors, O1052
policies, O1054
pressure, adverse reaction to, O1052
rehabilitation, R2004
rise in incidence of, Intro 6
risk assessment, O1054
support systems, O1054
symptoms, O1052
work, *see* STRESS AT WORK
STRESS AT WORK.
administrative support and, S11023
ambiguity, role of, S11022
auditing of stress, S11037
Beacons of Excellence report, S11009
breaks S11008, S11014, S11019–S11020
British Occupational Health Research
Foundation (BOHRF) report, S11009
changes and arrangements,
organisational, S11027
civil liability, S11009
clerical support and, S11023, S11026
communication, S11027
company structure, relevance
of, S11025–S11031
competencies, S11034
conditions of work, S11011–S11015
conflict, role of, S11022
conflicting roles and
functions, S11017–S11018
control
hierarchy of, S11026
work, over, S11033
counselling, offers of, S11009
cover during staff absence, S11024
criminal liability, S11009
damages, S11009
deadlines, role of, S11002, S11017
decoration, S11012
definition, S11002
design, S11013

STRESS AT WORK. – *cont.*
employers' duties, S11006
ergonomics, role of, S11013
excessive pressure, S11002
extent of problem, S11001
factors to be taken into account, guidance
on, S11009
foreseeability, S11009
guidance, S11009
hierarchy of control, S11026
individual perception of company, S11029
information and training, S11040
insurance, S11009
interventions,
primary, S11009, S11040
secondary, S11009, S11040
tertiary, S11009, S11040
job organisation, relevance
of, S11025–S11031
layout of workplace, S11013
maintenance, S11012
management standards, S11009,
S11011–S11033
management techniques S11036–S11040
auditing of stress, S11037
information and training, S11040
management policy, S11038
on-going monitoring S11039
role of, S11004–S11009
surveys and, S11037
on-going monitoring S11039
organisation of job, S11025–S11031
perceptions of role and
company, S11029–S11030
personal factors, S11032–S11035
personal issues, S11035
physical environment and, S11011
pregnant workers, S11014
previous experience, S11031
psychiatric injury, damages for, S11009
qualifications, S11034
rehabilitation, S11009
research, S11009
response to, understanding, S11003
responsibility, S11028, S11033
rest times and facilities, S11008, S11014,
S11019–S11020
retention strategies, S11009
risk assessment, S11007, S11009
risk factors, S11010–S11015
self-esteem, perception of one's, S11030
shift patterns, S11019
sick building syndrome, S11015
sickness absences, S11017
smoking, S11014

STRESS AT WORK. – *cont.*
staff absences, cover during, S11024
staff facilities, S11014
standards, S11009, S11011–S11033
surveys and, S11037
targets, role of, S11002, S11017
technology, role of, S11021
time demands, S11018
training, S11040
travel and, S11020
work equipment, role of, S11021
work environment, O1052, S11009, S11011–S11015
work itself, S11016–S11024
working time, S11009, S11014, S11019
STRICT LIABILITY,
product safety, as to, P9032, P9037
STRUCTURED SETTLEMENTS,
method of paying damages, C6025
STUDENTS,
visitors, C6518
SUBCONTRACTOR,
corporate manslaughter, C9001
SUBROGATION,
employer's liability insurance, E13007
SUBSTANCES, ESCAPE OF,
accident reporting, A3029
SUBSTANDARD PRODUCTS,
contractual liability for, P9051
SUITABLE ALTERNATIVE WORK, E14015
SUPERMARKET,
risk assessment, R3050
SUPERVISION,
community health and safety, C6517, C6520, C6523
SUPPLIER,
equipment, M1007
health and safety duties, Intro 21
machinery safety, M1002, M1003, M1009–M1010, M1013, M1017–M1018
product liability, *see* Product Liability; Product Safety
small businesses, Intro 4
SUPPLY CHAIN,
Carbon Trust, Carbon Footprints E16008
hazardous chemicals, E16008
regulation, Intro 17
chain management, Intro 4
SUPPORT SERVICES,
facilities management, F3010

SURVEILLANCE,
night workers, W10034
SURVEYS, A5023–A5026
SURVIVAL OF ACTIONS, C6038
SUSPENSION FROM WORK,
maternity grounds, E14015, E14016, E14025–E14027, V12018
medical grounds, E14016
remuneration, E14016
suitable alternative work, E14015
without pay, E14014
SUSTAINABLE DEVELOPMENT, E16011, E16027, F3018A
Carbon Reduction Commitment, F3018A
definition, F3018A
Five Capitals model, F3018A
Sustainable Procurement Taskforce, E16008
SUZY LAMPLUGH TRUST, R3016 (App C)
SWIMMING,
occupiers' liability, O3017
warnings, O3017
SWITCHES, M1032, M1034–M1035

T

TANKS,
dangerous goods, carriage of, D0416, D0421, D0422
heights, working at, W9001
TAX CREDIT,
disability tax credit, C6016
TELECOMMUNICATIONS, F3030
TELEWORKERS,
visual display screens, D8202, D8213
risk assessment, D8213
TEMPERATURE CONTROL,
construction and building operations, and, C8017
equipment, M1007
food, as to, *see* Food
manual handling, M3006, M3013, M3014
new or expectant mothers, V12020
occupational health, O1056
sun protection, W11005
vibration, V5015
workplaces, W11001, W11005
TEMPORARY WORK AND WORKERS,
managing health and safety, M2018

TENANCY AGREEMENTS,
asbestos, A5021
TENDINITIS, 01047
TENDONS, 01047
TENOSYNOVITIS, 01047
TEN-POINT GOVERNMENT STRATEGY, Intro 7
TERRORISM, D6001
TESTING,
disaster and emergency planning, D6066
TEXTILES,
workers, risk factors, E16521
THALIDOMIDE,
product safety and product liability, P9001
THEATRES,
noise exposure levels, N3027.1
THIRD PARTY,
harassment, H1703
THREATS,
violence in the workplace, as, V8002–V8003
THREE MILE ISLAND DISASTER, E17002
TINNITUS, O1022
TOE-BOARDS, W9008
TORTS,
harassment, and, H1722
assault and battery, H1723
negligence, H1724
remedies, H1725
TOXIC SUBSTANCES,
Building Regulations, W11044
major accident hazards, D0704
manual handling, M3002
TOXICOLOGY,
air pollution, E5003
TRACING,
code on, E13026
employers' liability insurance, E13026
TRACTORS,
noise, N3015
TRADE UNIONS,
collective agreements, E14005
recognition of, E14001
TUC Certificate in Occupational Health and Safety, T7023
TRAFFIC MANAGEMENT, C6511
TRAFFIC ROUTES,
above ground level, access routes to, A1006
access, Intro 29
building and construction sites, C8014
checklist,
pedestrians, A1007
vehicles, A1006
deliveries, for, A1001, A1005, A1006

TRAFFIC ROUTES, – *cont.*
Health and Safety Executive inspections, Intro 29
heights, working at, W9004
internal traffic, A1001, A1005
lighting, L5002
markings, A1005
number of, A1004
obstructions, A1005
pedestrians, A1002, A1004, A1007
positions, A1004
safe access, A1004
sheeting operations, access to load tops, A1008
speed limits, A1005
speed ramps, A1005
statutory duties, A1001, A1004
sufficiency, A1004, A1005
suitability, A1002, A1004
vehicles, for, A1005, A1006
TRAFFIC SYSTEMS,
see also Access,
common law duty of care, A1001, A1021
factories, O3018
occupiers' liability, O3018
organisation of safe, A1002, A1003
safe traffic routes, A1006, A1007
traffic routes, suitability of, A1004
vehicles allowed to circulate safely, A1005
TRAINING AND COMPETENCE, Intro 28, Intro 29, T7001–T7029
adventure activity licensing, T7007
advice, T7001, T7002, T7007
approved codes of practice, T7005
British Safety Council, T7009, T7010, T7013, T7020
business continuity planning, B8040
Certificate in Safety Management, T7020
Chartered Institute of Environmental Health, T7013, T7018
children and young persons, V12004, V12012
civil liability, T7002
Client/Contractor National Safety Group, T7027
community health and safety, C6523
competent person, definition of, T7002, T7004–T7005
confined spaces, C7011
Construction Skills Certification Schemes, T7027
construction workers, T7027
contacts, useful, T7029
continuing professional development, T7028

TRAINING AND COMPETENCE, – cont.
contractors, T7026–T7027
criminal offences, T7002
dangerous substances, D0926, D0933.1
transport of, D0433, D0449, T7007
definition of competence, T7001, T7002
Diploma in Safety Management, T7013,
T7020
directives, T7004
directors, T7001
disaster and emergency planning, D6010,
D6046, D6053
diving, T7007
doctors, W10031
electricity, E3011, E3024
employees, of, T7001, T7024–T7025
Employment National Training
Organisation, T7008, T7009,
T7016–T7017, T7021, T7024
enforcement standards, T7008
Engineering Construction Industry
Training, T7027
environmental management, E16022
equipment, M1007
executives, T7022
expiry of accreditation schemes, T7028
explosions, D0926
fire extinguishers, F5015
food handlers, of, F9035
food legislation, as to, F9058, F9081
gas fitters, CORGI, T7007
goal-setting, T7007
guidance, T7007–T7008
harassment, and, H1739
hazardous substances, as to, H2105,
H2114, H2134
Health and Safety at Work etc. Act
1974, T7003
Health and Safety Commission,
Revitalising Health and Safety, T7015
strategy on, aims of, T7001
Health and Safety Executive, T7003
lead body, as, T7008
policy unit, T7007
health and safety law,
food legislation, as to, F9058, F9081
relevant training, Intro 23
heights, working at, W9007, W9008,
W9009
higher education, T7010, T7013
inexperienced workers, V12039, V12040
Institution of Occupational Safety and
Health (IOSH), T7011–T7014,
T7021–T7023, T7028
Chartered Members and Fellows T7028

TRAINING AND COMPETENCE, – cont.
Institution of Occupational Safety and
Health (IOSH), – cont.
Register of Safety Practitioners, T7028
Safety Pass Alliance Scheme, T7027
Tech IOSH, T7017
knowledge, experience and skills, T7002,
T7005, T7013, T7016, T7028
lack of, E17002
ladders, W9009
legal obligations for, T7003–T7007
lifelong learning, T7028
maintenance of, T7028
management, M2016, M2037, M2045,
T7001, T7004–T7007, T7015–T7022
regulations, T7004–T7007
strategic level, at, T7015
manual handling, M3001, M3007,
M3015, M3033
medical personnel, T7007
mining, T7007
National Examining Board for
Occupational Health and
Safety, T7009, T7010, T7013, T7017
National General Certificate in
Occupational Safety and
Health, T7017
National Occupational Standards
(NOS), T7008
National Open College Network, T7023
National Vocational Standards, T7002,
T7008, T7009, T7012–T7016,
T7022–T7025
NEBOSH Diploma, T7013, T7017
need for, assessment of, E17010
occupational health and safety
practitioners, competence
requirements for, T7011–T7014
passports, safety, T7026–T7027, T7028
permit to work systems, S3015, T7026
practice standards, T7008, T7009
programmes, guidance on
particular, T7007
Qualification and Credit
Framework, T7008
qualifications, T7005, T7008, T7010,
T7012–T7025
Qualifications and Curriculum
Authority, T7013, T7019
Quality Assurance Agency for Higher
Education, T7010, T7013, T7022
radiation protection advisers, T7007
Register of Safety Practitioners, T7028
regulations, power to make, T7003–T7007
repeated, T7006, T7028
resources, E17042

TRAINING AND COMPETENCE, – *cont.*
risk assessment, R3025, R3031, R3040, T7003, T7005
Royal Society for the Prevention of Accidents, T7009
Royal Society for the Promotion of Health, T7013, T7019
Safety for Senior Executives, T7022
Safety Pass Alliance Scheme, T7027
safety passports, T7026–T7027, T7028
safety representatives, J3028, T7003, T7023
Scottish Vocational Qualifications, T7008, T7009
Sector Skills Councils (SSCs), T7008
stages of, T7028
standards, T7002, T7010
  Employment National Training Organisation, T7008, T7009, T7024
  enforcement, T7008
  Institution of Occupational Safety and Health (IOSH), T7011–T7012
  lead bodies, T7008
  National Vocational Standards, T7002, T7008, T7009, T7012–T7016, T7022–T7025
  Occupational Health and Safety Practice, T7009
  practice, T7008, T7009
  revision of, T7008, T7012
  setting national, T7008–T7009
  title of, T7012, T7015
stress, S11040
students, R3025
supervisors, T7016–T7022
Tech IOSH, T7017
TUC Certificate in Occupational Health and Safety, T7023
universities, T7013
visual display units, D8208
work experience, V12004, V12010
work in compressed air, and, C8018
workplace violence, as to, *see* Workplace Violence
young people, T7006
TRAINS,
accident reporting, A3035
children and young persons, 03017
occupiers' liability, 03017
surfing, 03017
TRANSPORT,
accident reporting, A3035
disaster and emergency planning, D6025–D6027
enforcing authority, E15009

TRANSPORT, – *cont.*
Herald of Free Enterprise, D6029
manual handling, M3009
systems, A3035
TRANSPORTATION OF DANGEROUS SUBSTANCES,
ADR Agreement, D0402, D0403
  consignment procedures, D0419
  documentation, D0425
  exemptions, D0440
  handling, D0427
  loading and unloading, D0427
  marking and labelling, D0420
  packaging and containment, D0414–D0416
  vehicle requirements, D0431–D0438
air, carriage by, D0404
Approved List, D0410, D0411
bulk container and tank systems, intermediate, D0416
  placarding and marking, D0421, D0422
carriage, D0426–D0430
categories, D0402, D0407
classification system, D0406–D04012
consignment procedures, D0417–D0422
container, marking, D0421, D0422
crew and vehicle, D0431–D0438
Dangerous Goods Safety Advisor, D0405
documentation, D0423–D0425
driver, D0431–D0438
European Union, D0404
emergency procedures and information, D0438
exemptions, D0439–D0444
generally, D0401, D0402
handling, D0426–D0430
identification, D0406–D0412
labelling, D0420
legal framework, A1009–A1014, D0403–D0404, W11001
limited quantity packages, D0441
loading and unloading, D0426–D0430
marking and labelling,
  packages, D0420
  vehicles, tanks and containers, D0421, D0422
operational procedures, D0444–D0449
Orange Book, D0404, D0407
package marking and labelling, D0420
package selection, D0408
packing groups, D0412
packaging and containment, D0413–D0416
placarding, D0421, D0422
Proper Shipping Names, D0418

TRANSPORTATION OF DANGEROUS SUBSTANCES, – *cont.*
quantity exemptions,
limited, D0431–D0432
radioactive materials, D0446–D0447
rail, carriage by, D0402, D0404, D0419
revoked Regulations, D0445
road, carriage by, D0402, D0403, D0409
training, D0433, D0449
vehicle marking, D0421, D0422
vehicle requirements, D0434–D0438
sea, carriage by, D0404
segregation, D0430
small loads, D0442
stowage, D0429
tank systems, D0416
training and competence, D0435, T7007
transport documents, D0409,
D0428–D0430
United Nations,
classification/identification
system, D0407
Orange Book, D0404, D0407
Proper Shipping Names (PSNs), D0408
packaging and containment, D0413
UN Numbers, D0408
vehicle requirements, D0431–D0438
emergency information and
documentation, D0438
equipment, D0437
filling and loading, D0441
placarding and marking, D0421, D0422
stowage, D0429
TRAPPING, M1025
TRAVEL,
community health and safety, C6522
stress, S11020
travelling time, W10003
volunteers, C6522
TRAVELATORS,
workplaces, safety regulations, W11017
TREE FELLING,
personal protective equipment, P3003,
P3024
TRESPASSERS,
access, C65013
children, C65012
dangerous equipment or
materials, C65013
duty to care, O3007
employer's duties to, C6512–C C6514,
R3025, R3032
hazardous or dangerous
substances, C65013
occupier's liability, *see* Occupiers' Liability

TRESPASSERS, – *cont.*
precautions, C65013
railway lines, C65013
right to roam, O3007
visitors, O3007
water, C65013
workplaces, C6512–C C6514
TRESPASS TO THE PERSON,
harassment, H1723
TRIBUNAL, *see* Employment Tribunal
TRIP DEVICES, M1036
TROLLEYS AND TRUCKS, M3010
TUBERCULOSIS, O1042, O1043
TURNBULL REPORT, Intro 5, B8001,
B8006, E16005
TWO-HAND CONTROL DEVICES, M1036
TYPE-EXAMINATION,
pressure systems, and, P7005

U

UK ATOMIC ENERGY AUTHORITY,
premises of, E15005
UK CORPORATE GOVERNANCE CODE,
principles and standards, E16005
UN ENVIRONMENT PROGRAMME, E16007
UNFAIR CONTRACT TERMS,
occupiers' liability, O3016
product liability and, P9061–P9063
UNFAIR DISMISSAL,
see also Enforcement
assertion of statutory right, for, E14011
compensation for,
additional award, E14022
basic award, E14022
compensatory award, E14022
formula for, E14017
health and safety cases, E14010
special award, E14022
constructive dismissal, E14014, E14019
harassment, and, H1719
compensation, E14022
constructive dismissal, H1720
proof, H1720
remedies, H1721
health and safety grounds, on, E14010,
E14011, E14020
procedural fairness, E14012, E14013
protection against, E14017, E14019,
J3030
qualifying period for, E14019

UNFAIR DISMISSAL, – *cont.*
  reasonableness of
      employer's action, E14021
  reasons for dismissal, E14020
  remedies for,
    compensation, E14022
    re-engagement, E14022
    reinstatement, E14022
  wrongful dismissal and, E14017
UNITED NATIONS,
  Economic Commission for Europe, E5021,
    E5026
  Committee of Experts, D0415
  dangerous goods, carriage of, D0404,
    D0407, D0415
    documentation, D0429
    Proper Shipping Names (PSNs), D0408
    UN Numbers, D0408
  Orange Book, D0403–D0405, D04011
  training, D0433
UNLAWFUL ACT MANSLAUGHTER, D6042
UNLIQUIDATED DAMAGES,
  assessed by judges, C6021
  basis for, C6022
UNSAFE PRODUCTS,
  liability for, P9027, P9028
UNWANTED CONDUCT,
  element of claim, H1704
  harassment, H1702
  purpose or effect, H1703
  sexual nature, H1703
UPPER LIMB DISORDERS, E17052, 01047
URTICARIA, O1038
UTMOST GOOD FAITH, E13004

V

VAPOUR,
  explosive atmospheres, D0921
VDU,
  see Visual Display Units
VEGETARIAN FOOD, F9076
VEHICLE EXCISE DUTY (VED), E16006
VEHICLES,
  access to, A1015
  building and construction
    operations, C8015
  community health and safety, C6510
  co-operation requirements, A1015
  contractors, A1014
  dangerous products,
    carriage of, D0434–D0438
    filling and loading, D0441

VEHICLES, – *cont.*
  dangerous products, – *cont.*
    petrol, D0901.1
    placarding and marking, D0421, D0422
    stowage, D0429
  deaths and injuries caused by, A1008
  fall prevention, A1008, A1015
  food safety. F9027, F9032
  garages, dangerous substances in, D0901.1
  giving unauthorised lifts, A1020
  hazardous operations, *see* potentially
    hazardous operations *below*
  loading, A1008, A1018
  maintenance, A1012
  Motor Insurers Bureau (MIB), E13026
  off-road, whole-body vibration
    and, V5024
  parking areas, A1009
  petrol, D0901.1
  potentially hazardous
    operations, A1016–A1019
    generally, A1016
    loading, A1008, A1018
    reversing, A1008, A1017
    tipping, A1008, A1019
    unloading, A1008, A1018
  private vehicles, A1009
  provision, A1011
  repairs, D0901.1
  reversing, A1008, A1017
  safety of, A1008–A1010
  speed limits, A1009
  statutory requirements, A1009
  subcontractors, A1014
  tipping, A1008, A1019
  traffic routes, *see* Traffic Routes
  training of drivers, A1013
  unauthorised lifts, A1020
  unloading, A1008, A1018
  vehicle and fuel duty, E16006
  vehicle lifting table, L3037
  work equipment, A1009
  work vehicles, A1010–A1015
    access to, A1015
    contractors, A1014
    co-operation requirements, A1015
    fall prevention, A1015
    generally, A1010
    maintenance, A1012
    provision, A1011
    subcontractors, A1014
    training of drivers, A1013
VENDING MACHINE,
  food, sale of, F9033, F9041

VENTILATION, Intro 29
adequate control, V3015
aerosol contamination, V3002
air cleaning devices, V3009
Approved Code of Practice, D0922.4
asbestos, V3025
Building Regulations, W11044
confined work spaces, A1023, C7017
contamination,
  aerosols, V3002
  gases, V3002
  generally, V3001
  sources of, V3002
  toxicity, V3004, V3005, V3014
control options, V3001
dangerous substances, D0933.4
dilution (general), V3001, V3014
  fresh air, V3014
dilution vent, V3014
discharge, V3010
ducting, V3003, V3007
  pressure, V3013
electrostatic precipitator, V3009
explosive atmospheres, D0921
fans, V3008
fresh air,
  incoming, V3014
  supply rate, V3014
fume cupboard, V3004
gases, V3002
hoods, V3003, V3005, V3006
incoming air, V3014
lead processes, see Lead Processes
legal requirements, V3001, V3015
local exhaust ventilation (LEV), H2127,
  H2130, V3001, V3003
  alternative to, V3014
  components, V3003–V3009
  design process, V3002
  examination and testing, V3012, V3015
  maintenance, V3012, V3015
low-volume high-velocity (LVHV), V3005
maintenance, V3012, V3015
manual handling, M3006, M3014
noise, N3035
natural, V3014
objectives, V3001
partial enclosures, V3004
personal protective equipment, V3001
re-circulation systems, V3014
risk assessment, V3015
slots, V3003, V3006
system pressure, V3013
  monitoring equipment, V3013

VENTILATION, – *cont.*
system pressure, – *cont.*
  total volume flow, V3013
temperature control, W11005
testing, V3012, V3015
venturi scrubber, V3009
workplace, W11005, W11013
VIBRATION,
agriculture sector, V5001, V5015
aircraft in flight. V5015
back pain, V5005
body, E17033
buildings and structures, V5006, V5009
code of practice, noise and, N3017
compensation scheme for miners,
  government, V5003
construction work, C8018, N3017
controls, V5027
daily personal exposure values, V5015
damages, V5003
direction and frequency, V5008
disablement benefit, vibration white finger
  and, N3012
dose value, V5009
effects on people of, V5002–V5006
emergency certificates, V5015
employers, impact on, V5016
employers' liability insurance, E13026
equipment, transitional period for, V5015
EU law, Intro 11, V5001, V5015–V5016
exposure limit values, V5009, V5015
fishing vessels, V5015
forestry sector, V5001, V5015
frequency of, V5008
Hand Arm Vibration Syndrome (HAVS),
  see Hand arm vibration syndrome
  (HAVS)
health surveillance, V5015
measurement, V5007–V5009
  direction and frequency, V5008
  dose value, V5009
  units, V5007
new or expectant mothers, V12020
noise and, N3012, N3013, E17032,
  V5003
offshore workplaces, V5015
organisation and technical means,
  reduction of exposure by, V5015
Physical Agents (Vibration) Directive, Intro
  11, Intro 29, V5001, V5015–V5016
regulations relating to, N3012, V5001,
  V5015
risk assessment, V5015
sea transport, V5015
shipping vessels, V5015

VIBRATION, – cont.
temperature, V5015
Vibration White Finger (VWF). *See* VIBRATION WHITE FINGER
Whole-body vibration, *see* Whole-body vibration
young workers, V5016

VIBRATION WHITE FINGER (VRF),
coal miners, V5003, V5012
common law, action at, V5012
damages, V5003, V5012, V5013
definition, V5011
dosage, V5012
guidance, V5012
number of sufferers of, V5003
occupational disease, as, V5011–V5014
reporting, V5011–V5012
stages of, V5013
Taylor-Pelmear Scale, V5013
warnings, wording of, V5012
water supply workers, V5003
whole-body vibration, V5003

VICARIOUS LIABILITY, Intro 24
contractors, E13027
definition, E13027
harassment, H1710-H1715
course of employment, H1711
employer's defence, H1713
employer's liability, narrowing, H1715
liability of employees, H1714
indemnities, E13027
insurance, E13002
workplace violence, for, V8013

VICTIM SUPPORT SERVICES, V8140–V8141

VICTIMISATION,
employees' protection, E16501
Equality Act 2010, E16508
protected acts, E16508

VIOLATION OF DIGNITY,
harassment, H1701-H1702, H1708
purpose or effect, H1707

VIOLENCE IN THE WORKPLACE, V8001–8.143
access control, V8088
age groups, V8003
agencies, involvement of other, V8032
alarms, V8090
anti-social behaviour orders, V8140
assaults, V8002–V8003, V8011, V8131–V8133
batons, V8095
body armour, V8093
British Crime Survey, V8002–V8003
CCTV, V8089

VIOLENCE IN THE WORKPLACE, – *cont.*
checklist, V8109, V8128
civil law proceedings, V8012, V8140
codes of practice, V8044, V8072, V8129
collaboration, V8019, V8024
communication, V8078, V8121, V8122
external, V8039
internal, V8038
process of, V8063
strategy, V8034, V8037
community health and safety, C6522
compensation, V8004, V8012, V8013, V8015
constructive dismissal, V8010
consultation, V8008, V8024, V8051
contract law, V8010
co-ordinator, role of workplace violence, V8062
court case, V8140–V8141
criminal law, V8011
data, analysis of, V8023
data protection, warning flags and, V8081
definition of, V8001, V8017
delegation V8110
designing the system, V8058–V8064, V8079, V8109
disclosure, V8071, V8121
distance learning, V8102
duty of care, V8004–V8005
emergencies,
post-incident support, V8135
response belts, V8094
employee representatives, V8008
enforcement, V8004
environment, V8123
equipment, V8123
evening and night, working in the, V8003
evidence, preserving, V8121
experts, V8139
external providers of training, V8113, V8114
feedback, V8068
focus groups, V8022
forms,
design of report, V8059
post-incident considerations, V8137
handcuffs, V8094
harassment, V8011
health and safety, V8001–V8002, V8005–V8008
Health and Safety Executive, V8001–V8002, V8005
health care workers, V8077
identification of persons who may be harmed, V8052

VIOLENCE IN THE WORKPLACE, – *cont.*
identification of risks, V8051
implementation, V8036–V8041,
    V8064–V8068, V8108–V8109
improvements, V8127
incidents,
    key scenarios, V8117
    leadership, V8120
    management, V8116–V8128
    response teams, V8119
    response to, V8031
    roles and responsibilities, V8118–V8120
information, V8008, V8020, V8061,
    V8068, V8141
informing and consulting
    employees, V8008
injunctions, V8140
internal training providers, V8111–V8112,
    V8114
interviews, V8022
intruder alarms, V8090
investigation process, V8060, V8129,
    V8140–V8141
job description, V8051
layout, V8079
legislation, V8004–V8016
lone workers, V8092
long-term reactions, V8133
maintenance, V8033
managers,
    involvement, V8067
    post-incident support, V8135
    training, V8115
Maybo Risk Management Model, V8144
measurable outcomes, V8069
medium-term reactions, V8132
mobile phones, V8092
monitoring and evaluating the
    solution, V8110–V8114
nature and extent of problem, V8002
negligence, V8012
notification, V8007
observation, V8022
paging, V8091
panic alarms, V8090
personal alarms, V8090
persons at risk, V8003
physical measures, V8087–V8095
police investigation process V8140–V8141
policy,
    affected by, persons, V8028
    content of, V8025
    co-ordination of, V8043
    designing the, V8024–V8035
    development of, V8016–V8043

VIOLENCE IN THE WORKPLACE, – *cont.*
policy, – *cont.*
    evaluation of, V8042
    group, policy development, V8024
    implementation of, V8016–V8043
    monitoring and evaluating, V8042
    new systems, procedures, practices and
        equipment, V8040
    purpose, V8026
    review and revision, V8035, V8043
    scope and definitions, V8027
post-incident management, V8129–V8141
    common reactions, V8131–V8133
    effect on people, V8130
    managers, role of, V8136
    normal, return to, V8134
    report forms, V8137
    support, providing, V8135–V8139
    support services, V8138
    worst, preparing for the, V8139
prevention programmes, V8001
proactive service delivery, V8080
problem recognition, V8017
procedures, V8121
prosecution of employers, V8015
protective vests, V8093
public address systems, V8091
public, contact with the, V8001
questionnaires, V8022
queuing and queue jumping, V8075,
    V8079
racist incidents, V8011, V8140
radios, paging and public address
    systems, V8091
receptions, V8088
records, disclosure of, V8071
rehearsals, V8126
remedies, V8004
reporting, A3008.1, V8029, V8059,
    V8137
research, V8112
response teams, V8119
restraining orders, V8140
restraints, V8094
reviews, V8127
    conducting, V8018–V8023
    process, V8029
risk assessment, V8001, V8006, V8021,
    V8029, V8044–V8056
    continuous review, V8057
    contents, V8050–V8055
    dynamic, V8047
    generic, V8045
    methods of, V8049
    persons who should carry out, V8048

VIOLENCE IN THE WORKPLACE, – *cont.*
risk assessment, – *cont.*
pre-planned events, V8046
rating of risks, V8053
reduction measures, V8054–V8055
restraints, V8094
specific activities and tasks
performed, V8053
suitable and sufficient, V8044
systems and procedures, V8077
risk management, V8001, V8129, V8141
risk reduction, V8030, V8072–V8073
safe practice guidance, V8076
sanctions, V8004, V8032, V8124
security personnel, V8083–V8086
procedures, V8086
selection, V8084
training, V8085
security procedures, V8086
security responses, V8082–V8095
self-defence, V8015
service delivery, V8074–V8081
short-term reactions, V8131
soft skills, V8103
staff, violence by or to, A3008.1
staffing levels, V8075
standards, V8044
statistics, V8002
support,
court case, during, V8140–V8141
providing, V8135
services, V8138, V8140
sources of, V8142
systems and procedures, V8077
testing, V8102–V8107, V8107, V8126
threats, V8002–V8003
training, V8030, V8096–V8115, V8125
checklist, V8101
core content, V8104–V8105
design, V8102–V8107
development model, V8097
external providers, V8113, V8114
identification of needs, V8099
internal providers, V8111–V8112,
V8114
learning outcomes, defining, V8100
managers, V8115
physical intervention training, V8103,
V8105
reviewing the content, V8106
soft skills, V8103–V8104
technology. V8102
testing, V8102–V8107, V8107
Training Needs Analysis, V8098–V8101

VIOLENCE IN THE WORKPLACE, – *cont.*
under-reporting, V8070
use of force, self-defence and, V8015
vests, V8093
vicarious liability, V8013
victim support services, V8140–V8141
warnings, V8009, V8081, V8124
withdrawing services, V8014
working practice, V8075
VIRAL HEPATITIS, H2103
VISITORS,
accident reporting, A3001
activities to be carried out, C6519
areas to be visited, C6519
capabilities of visitors, C6519
community health and safety, C6518–
C6520
construction and building operations,
and, C8018
disaster and emergency planning, D6061
employers' duties to, C0016–C0018,
R3002, R3032
fire prevention and control, F5073
health and safety policy statement, S7004,
S7006
informal individual visits, C6519
information and instruction, C6520
managing health and safety, M2015
occupiers' liability, C6508
personal protective equipment, C6520
precautions, C6520
restricted activities, C6520
restricted areas and routes, C6520
schoolchildren and students, C6519
VISUAL DISPLAY UNITS, D8201–D8222
agency workers, D8202
breaks, D8206
causes of problems, D8215
daily work routine of users, D8206
definitions, D8202
discomfort glare, L5014
discussions with users and
operators, D8215
ergonomic approach, E17044, E17052,
E17064
exclusions, D8203
eye diseases and disorders, O1033, O1036
eyesight tests, Intro 23, D8201, D8207
homeworkers, D8202, D8213
hot desking, D8218
information, provision of, D8209
laptops, D8219, E17064
lighting, L5002, L5003
mobile phones, D8219

VISUAL DISPLAY UNITS, – *cont.*
new or expectant mothers, V12005,
    V12023
personal organisers, D8219
pointing devices, D8220
portable, D8219
posture, D8214, D8210
radiation exposure from, R3028
recommendations, review and
    implementation of, D8221
records, D8216
reflected glare, L5014
regulations on, D8201–D8213
risk assessment, Intro 16, D8201, D8204,
    D8210–D8216, R3010, R3018,
    R3027, R3036
  after the, D8221
  approach to, D8212
  checklist, D8211–D8212, D8216
  discussions with users and
      operators, D8215
  homeworkers, D8213
  observations at the, D8214
  persons carrying out the, D8210
  re-assessments, D8219
  records, D8216
  reviews, D8222
  teleworkers, D8213
  users, identification of, D8211
self-assessment checklists, D8216
shared workstations, D8218
software tools, D8206
special situations, D8217–D8220
teleworkers, D8202, D8213
training, D8208
users, definition of, D8208
vision, D8214
vulnerable persons, V12005
workstations, requirements for, D8205,
    D8210–D8216
VOLATILE ORGANIC COMPOUNDS,
Protocol, as to, E5026
VOLENTI NON FIT INJURIA,
children, 03017
defence to negligence, employer's Intro 24
occupiers' liability claims, O3017
VOLUNTARY SECTOR,
activities that volunteers can be involved
    in, C6521
communication, C6523
community health and safety, C6521–
    C6523
corporate manslaughter, C9002
emergencies, C6522
employers' liability insurance, E13016

VOLUNTARY SECTOR, – *cont.*
entry into client's homes, C6522
equipment, C6522
fire, C6522
information, C6523
large companies, Intro 5
manual handling, C6522
monitoring, C6523
one-off volunteers, C6521
performance, Intro 5
precautions, C6522
regular volunteers, C6521
restrictions, C6523
risk assessment, C6522
screening, C6523
supervision and support, C6523
training, C6523
travel, C6522
violence, C6522
work environment, C6522
VULNERABLE WORKERS,
see also Children and young persons,
    Disabled persons, Lone workers,
new or expectant
    mothers V12001–V12039
construction, V12005
control of substances hazardous to
    health, V12005
display screen equipment, V12005
employers' duties, V12001
Health and Safety at Work etc. Act
    1974, V12001
inexperienced workers, R3028,
    V12038–V12039
legislation, V12002–V12005
Management of Health and Safety at Work
    Regulations 1999, V12001, V12003
manual handling, V12005
occupational health and safety, O1005,
    O1008
personal protective equipment, V12005
risk assessment, R3028, V12001
self-employed, V12001

W

WAGES,
deductions from, E14014
suspension from work, E14016
WALES,
fire prevention and control, F5031
Food Standards Agency, F9005

WAREHOUSES,
fire protection, F5035
WARNING SIGNS, W11038
see also Safety Signs
dangerous substances, D0921, D0933.1
entry, on, D0921
explosions, D0921
occupiers' liability and, O3006, O3009,
O3010, O3016
WARNINGS,
see also Warning Signs
ACAS Code, E14014
children and young persons, O3010
disaster and emergency planning, D6007
emergencies, D0925
fire prevention and control, F5035
formal, M1529
hand-arm vibration syndrome, V5020
occupiers' liability, O3010, O3017
offshore operations, O7028
product liability and, P9039
sickness absence, M1529
swimming, O3017
vibration white finger, V5012
violence in the workplace, V8009, V8081,
V8124
WASHING FACILITIES,
construction and building operations,
and, C8017
generally, W11021
WASTE DISPOSAL,
asbestos, A5027
dangerous substances, D0933.3
food, in respect of, F9038, F9045
incineration, EC Directive, E5026
WATER,
see also Legionnaires' disease
atmospheric pollution and, E5014–E5015
facilities management, F3026
fire extinguishers, F5011
fire prevention and control, F5005
food safety, F9007
quality, F3026
trespassers, C65013
WATER EFFICIENCY,
Building Regulations, W11044
WATER SUPPLY,
food hygiene, F9038, F9045
vibration white finger, V5003
WATER SYSTEM,
cleaning, C6510
WEATHER,
construction and building operations,
and, C8017
risk assessment and, R3034

WEBSITE,
Department for Environment Food and
Rural Affairs (DEFRA) E5042
Environment Agency, E5042
European Commission, E5042
European Integrated Pollution Prevention
and Control Bureau, E5042
gas safety, G1017
Scottish Executive, E5042
Sustainable Workplace, E16011
WEIGHTS,
food, for, F9004
manual handling, M3003, M3013
measures and, F9004
WELDING FUMES,
ventilation requirements, V3002
WELFARE FACILITIES,
clothing accommodation, W11025
construction and building operations,
and, C8005
current specific statutory
requirements, W11018–W11029
drinking water, provision of adequate
supply, W11024
facilities for changing clothes, W11026
minimum number of facilities, W11021
occupier's duties,
sanitary conveniences, W11019
washing facilities, W11020
particularly dirty work, wash stations
required, W11022
rest and eating facilities, W11027
temporary work sites, W11023
WELLS,
accident reporting, A3005, A3035
WHEELCHAIR STAIRLIFTS,
disabled persons, provided for, W11043
WHITE FINGER,
vibration induced, Intro 24
WHOLE-BODY VIBRATION,
advice, V5023–V5028
back pain, V5023–V5024, V5028
causes, O1028
daily exposure action value, V5015
detection of, O1029
driving, V5023–V5026
effects of, V5005
exposure assessments, V5015
guidance, V5029
health surveillance, O1029–O1030, V5028
HSE guidance, V5029
incidence of, V5023
maintenance, O1031
management, O1030
manual handling, V5024

WHOLE-BODY VIBRATION, – *cont.*
occupational diseases, O1027–O1031
off-road vehicles, V5024
Physical Agents (Vibration) Directive, Intro 11, O1031, V5015
record-keeping, V5025
risk assessment, O1030–O1031, V5015, V5024–V5025, V5027
risk reduction V5026
symptoms, O1027
WIND, GUSTS OF, M3006, M3014
WINDOWS AND WINDOW CLEANING
general safety requirements, W11001, W11014
occupiers' liability, O3014
position and use of, W11012–W11013
WIRING REGULATIONS, E3003
WITNESSES,
accident investigations, A3032
interviewing, A3032
WOMEN WORKERS,
breast feeding mothers, E14024, E14025, R3017, R3026, R3027, R3032
dismissal on medical grounds, Intro 24
general statutory requirements, E14023
generally, E14023
HGV driving, E14023
hours of work, E14002
ionising radiation, R1019
lead processes, E14002
maternity leave, E14028
new and breast feeding mothers, safety provisions, E14024, E14026
night work, E14002, E14026
notification of pregnancy, birth or breastfeeding, E14027
pregnant workers, safety provisions, E14024, E14027
risk assessment, E14025, R3018
sex discrimination legislation, effect of, E14002, E14023
statutory requirements, E14002
suspension from work, health and safety reasons, E14015
unfair dismissal, qualifying period for, E14009
unlawful sex discrimination, E14023
WOOLF REFORMS, Intro 24, Intro 29, E13018, E13029
WORK ENVIRONMENT,
see also Harassment
intimidating, H1708
violence, V8123
WORK EQUIPMENT,
definition, M1005

WORK EQUIPMENT, – *cont.*
ergonomic design, *see* Ergonomic Design
facilities management, F3032
heights, working at, W9009
machine safety, *see* Machinery Safety
risk assessment, M1006
self-propelled, M1007
training, M1007
WORK EXPERIENCE, V12004, V12010
WORLD HEALTH ORGANISATION, Intro 8
WORK PATTERNS, Intro 6
WORK PLATFORMS, W9008
WORK RELATED ROAD SAFETY,
assessing risks, M2111
business case for, M2104
evaluation, of M2117
getting help M2124
health and safety management systems, M2105
measuring performance, M2109
organising, M2107
performance, reviewing and auditing, M2110
planning and implementing, M2108
policy, M2106
legal responsibilities, M2103
managing, M2101
monitoring and evaluation, M2117
active monitoring, M2118
reactive monitoring, M2119
evaluation M2122
gathering data, M2120
investigation, M2121
problem with, M2102
road risk, control measures, M2112
control measures,
driver, M2115
implementing measures, M2116
journey task, M2113
useful publications, M2125
vehicle, M2114
WORK-RELATED UPPER LIMB DISORDERS, O1047
WORK VEHICLES, *see also* Vehicles
access to, A1015
drivers of, A1013
loading, A1018
maintenance, A1012
provision of, A1011
statutory requirements relating to, A1009
tipping, A1019
training of drivers, A1013
reversing, A1017
unauthorised lifts, giving of, A1020

WORKERS,
definition of, E14001, W11001
part-time, E14002
women, *see* Women Workers
WORKING AT HEIGHTS, W9001–W9015
access and egress, W9004, W9008
Adventure Activities Licensing Authority
(AALA), C6508
barriers, W9008
British Standards, W9002, W9008,
W9014
causation, W9001
caving, W9004
climbers, W9001, W9004
community health and safety, C6506,
C6510
competence, W9005–W9008
danger areas, W9011
design, W9002
duties of persons at work, W9013
EC law, W9001, W9009
emergencies, W9006
employees' duties, W9013
equipment, W9009
facilities management, F3025
falls, W9001, W9008
arrest systems, W9008, W9014
fatal, W9001
harnesses, W9008
objects, W9011
protection from, W9014
fatal accidents, W9001
fragile surfaces, W9010
guard rails, W9008
guidance, W9002, W9004, W9014
Health and Safety Commission, W9001
Priority Programme, W9001
Health and Safety Executive,
advice, W9002
ladders, W9009
priority area, Intro 29
inspection, W9012
instruction, W9008
ladders, W9009
guidance, W9009
portable, W9009
pre-check lists, W9009
step-ladders, W9009
level differences W9004, W9009
lighting, W9008
mines, W9001
National Access and
Scaffolding Confederation
(NASC), W9008
Northern Ireland, W9003

WORKING AT HEIGHTS, – *cont.*
occupiers' liability, O3014
organisation, W9005–W9008
planning, W9005–W9008
regulations on, W9002–W9015
application, W9004
commencement and scope, W9003
Regulatory Impact Assessment, W9009,
W9015
rescue, W9006
research, W9001
restraints, W9014
risk assessment, R3015, W9004,
W9007–W9008
risks, avoidance of, W9005–W9008
roofs, W9008
ropes, W9002, W9014
safe systems of work, W9001
scaffolding, W9008, W9009
sectors, list of relevant, W9001
standards, W9002, W9008, W9014
toe boards, W9008
training, W9007, W9008
two metre rule, W9008
work equipment, W9009
work platforms, W9008
work positioning systems, W9014
WORKING TIME, Intro 29
48 hour working week, W10010–W10013
agency workers, W10005
air transport, W10027, W10031
annual leave, *see* Annual Leave
breach of contract claims, W10042
breaks, W10017
case law, W10001
collective agreements,
definition, W10007
excluded sectors, W10036
role of, W10006
commencement of regulations, W10001
compensatory rest, W10029, W10033,
W10034, W10037, W10039
continuity of service or production, need
for, W10034
contracting out of regulations, W10043
daily rest periods, W10014, W10015,
W10029, W10034
decision-making powers, persons with
autonomous, W10033
definitions, W10001–W10009
agency workers, W10005
collective agreements, W10007
relevant agreement, W10009
worker, W10004
workforce agreement, W10008

WORKING TIME, – *cont.*
  definitions, – *cont.*
    working time, W10003
  detriment, protection against, W10041
  directives, M1514, W10001, W10026,
    W10027–W10031
  distant from each other, work and
    home, W10034
  domestic workers, W10032
  doctors, W10001, W10026
  DTI booklet, W10001, W10017
  EC Directives, W10001, W10026,
    W10027–W10031
  employment tribunals enforcement of
    provisions, W10040–W10044
  enforcement, W10038–W10043
    detriment, protection against, W10041
    employment
      tribunals, W10040–W10044
    generally, W10038
    health and safety offences, W10039
    unfair dismissal, W10041
  exclusions,
    collective agreements, by, W10036
    directives to previously excluded sectors,
      application of W10027–W10031
    partial, W10032–W10036
    shift workers, W10035
    special categories of workers, W10029,
      W10034
    workforce agreements, W10036
  generally, W10001
  health assessments, W10021, W10033,
    W10034
  holidays, *see* Annual Leave
  impact of regulations, W10044
  inland waterways, workers on, W10031
  lake transport, workers on, W10031
  leave, *see* Annual Leave
  48 hour working
    week, W10010–W10013
    agreement to exclude 48 hour working
      week, W10001, W10013, W10038
    derogations, W10012
    exceptions, W10012
    excluded days, W10010
    reference period, W10011
  management, W10033
  mobile workers, W10027–W10030
    non-mobile activities and, combination
      of, W10027
  monotonous work, W10018, W10033,
    W10034, W10039
  new or expectant mothers, V12018
  night work, W10034

WORKING TIME, – *cont.*
  night work, – *cont.*
    definitions, W10019
    health assessment, W10021, W10033,
      W10034
    length of, W10020, W10029, W10039
    night time, W10019
    night worker, W10019
    offences, W10039
    overtime hours, W10020
    road transport, W10029
    transfer to day work, W10021
  non-mobile workers, W10027
  notice, W10024
  offshore workers, W10031
  on-call workers, W10001
  purpose of provisions, W10001
  rail sector, W10030, W10034
  records, W10025, W10033, W10034,
    W10039
  rehabilitation, R2008
  relevant agreements,
    definition, W10009
    role of, W10006
  rest periods,
    breaks, J3002, J3032, W10017,
      W10034
    compensatory rest, W10029, W10037,
      W10039
    daily, J3002, J3032, W10014, W10015,
      W10029
    monotonous work, W10018
    weekly, J3002, J3032, W10016,
      W10029, W10034
  return to work, R2008
  road transport workers, W10026,
    W10027–W10029, W10038
    daily rest, W10029
    directive on, W10029
    maximum working week, W10029
    night workers, W10029
    non-mobile workers and, W10028
  scope of provisions, W10001
  sea fishing, W10031
  seafarers, W10026
  security workers, W10034
  shift workers, W10035, W10037
  special categories of workers, W10034
  stress and, S11014, S11019
  structure of regulations, W10001
  surges of activity, foreseeable, W10034
  surveillance, W10034
  transport workers, W10026–W10031,
    W10034, W10038
  travelling time, W10003

WORKING TIME, – *cont.*
unfair dismissal, protection against, W10041
unmeasured time, W10033
weekly rest periods, W10016
worker, W10004
workforce agreements,
definition, W10008
excluded sectors, W10036
role of, W10006
'working time', W10003

WORKPLACE VIOLENCE, *see* Violence in the workplace

WORKPLACES, Intro 23
buildings, *see* Building Work,
civil liability, W11029, W11051
cleanliness, W11001, W11007
clothing accommodation, S11014, W11001, W11025, W11026
condition of floor, S11012, W11010
dangerous substances, D0907
definition of, D0907, W11002
dimensions/space, W11008
doors and gates, W11001, W11016
drinking water, provision of, W11001, W11024
eating facilities, W11001, W11027
enforcement of legislation as to, *see* Enforcement
escalators and travelators, W11001, W11017
explosions, D0907
factories, *see* Factories,
falls from a height, safety measures, W11001, W11011
fire precautions, *see* Fire Precautions
floor coverings, W11001, W11010
general cleanliness, W11001, W11007
general maintenance of, W11003
general requirements, W11001
hazardous substances in, see Hazardous Substances
hazards, CH 3 App A
inspections by safety representatives, J3007
layout, S11013
lighting, *see also* Lighting W11001, W11006
loading bays, statutory duties, A1009
maintenance requirements, W11001, W11003
management, Intro 29
minimal standards, W11001
organisation of safe, A1004
overcrowding, W11008

WORKPLACES, – *cont.*
parking areas, statutory duties, A1009
pedestrian access, A1007
pedestrian traffic routes, W11001, W11015
ramps W11010
rest facilities, W11026
safety signs, *see* Safety Signs
sanitary conveniences, S11014, W11001, W11019, W11021, W11023
security, C6511
smoking, W11028
staff facilities, S11014, W11026
steep slopes provided with handrail, W11010
stress, negligence claim, M1524
temperature controls, W11001, W11005
traffic routes, *see* Traffic Routes
vehicle traffic routes, *see also* Traffic Routes, W11001
ventilation of, *see also* Ventilation, W11001
washing facilities, W11020
welfare facilities, W11018
window cleaning, *see* Window Cleaning
workroom dimensions, W11001, W11008
workstations, W11001, W11009

WORKS COUNCILS, J3002, J3033

WORKS MANAGER,
responsibilities of, S7005

WORKS RULES,
incorporation into contract of employment, E14007, E14008

WORKSTATIONS,
adequate freedom of movement, W11008
children and young persons, V12016
design of, *see* Ergonomic Design
musculoskeletal disorders, *see* Musculoskeletal Disorders
rest breaks, E17040
sedentary comfort, W11008
suitable seating, W11008
user trial, E17056

WORLD HEALTH ORGANISATION,
air quality guidelines, E5064–E5065

WRITTEN POLICY STATEMENT, *see* Health and Safety Policy Statement

WRONGFUL DISMISSAL, E14017, E14018

Y

YOUNG PERSON,
adventure activities, C6508

YOUNG PERSON, – *cont.*
 children and, *see* Children and Young
  Persons
 health and safety policy and, S7008
 risk assessment, E16516
 vibration, V5016

Z

ZONING, D0918, D0921, D0919
ZOONOSES, O1043